CHEMISTRY

THE SCIENCE IN CONTEXT

THIRD EDITION

Thomas R. Gilbert
NORTHEASTERN UNIVERSITY

Rein V. Kirss
NORTHEASTERN UNIVERSITY

Natalie Foster
LEHIGH UNIVERSITY

Geoffrey Davies
NORTHEASTERN UNIVERSITY

W. W. NORTON & COMPANY
NEW YORK · LONDON

About the cover: Chemistry offers us a way to describe the material world from the molecular point of view, and the pictures in the DNA helix on the cover illustrate how widely applicable that point of view is. The molecule in the upper left is made from glucose and is used in medicine to assess metabolism and image tumors. Glucose itself is one of the main products of photosynthesis and is a source of energy for most organisms, from bacteria to ants to humans. In addition to requiring energy, all life on Earth also depends on water. Scientists have found water throughout the universe, which leads to the question: If there is water out there, is life out there, too? Throughout the book we use examples spanning the universe, from living cells to galaxies, to show how chemistry and the molecular point of view enable us not only to understand our world but also to improve the lives of all who live in it.

W. W. Norton & Company has been independent since its founding in 1923, when William Warder Norton and Mary D. Herter Norton first published lectures delivered at the People's Institute, the adult education division of New York City's Cooper Union. The Nortons soon expanded their program beyond the Institute, publishing books by celebrated academics from America and abroad. By mid-century, the two major pillars of Norton's publishing program—trade books and college texts—were firmly established. In the 1950s, the Norton family transferred control of the company to its employees, and today—with a staff of four hundred and a comparable number of trade, college, and professional titles published each year—W. W. Norton & Company stands as the largest and oldest publishing house owned wholly by its employees.

Editor: Erik Fahlgren
Project Editor: Carla L. Talmadge
Editorial Assistant: Mary Lynch
Developmental Editors: Irene Nunes, Andrew Sobel
Production Manager: Chris Granville
Marketing Manager: Kelsey Volker
Managing Editor, College: Marian Johnson
Book Designer: Lissi Sigillo
Design Director: Rubina Yeh
Photo Editor: Stephanie Romeo
Photo Researcher: Donna Ranieri
Associate Media Editor, Sciences: Matthew A. Freeman
Science Media Editor: Robert Bellinger
Composition: Prepare, Inc.
Illustrations: Precision Graphics; Prepare, Inc.
Manufacturing: Transcontinental Interglobe

Library of Congress Cataloging-in-Publication Data
Chemistry : the science in context / Thomas R. Gilbert . . . [et al.]. — 3rd ed.
   p. cm.
Includes index.
ISBN 978-0-393-93431-1 (hardcover)
1. Chemistry. I. Gilbert, Thomas R.
QD33.2.G55 2011
540—dc22
      2010016096

W. W. Norton & Company, Inc., 500 Fifth Avenue, New York, NY 10110
www.wwnorton.com

W. W. Norton & Company Ltd., Castle House, 75/76 Wells Street, London W1T 3QT

2 3 4 5 6 7 8 9 0

Brief Contents

Contents

1 Matter, Energy, and the Origins of the Universe 2

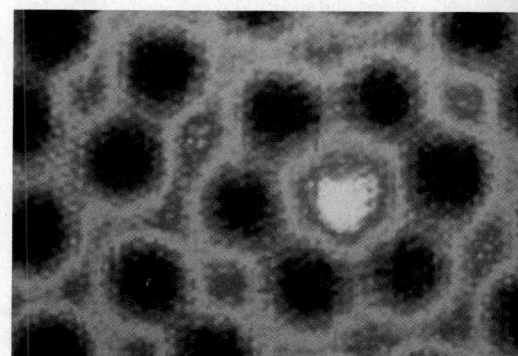

Just how small are these atoms?
(Chapter 1)

2 Atoms, Ions, and Compounds 42

What are atoms made of? *(Chapter 2)*

What unit could be considered the "chemist's dozen"? *(Chapter 3)*

Why is water necessary for all known forms of life? *(Chapter 4)*

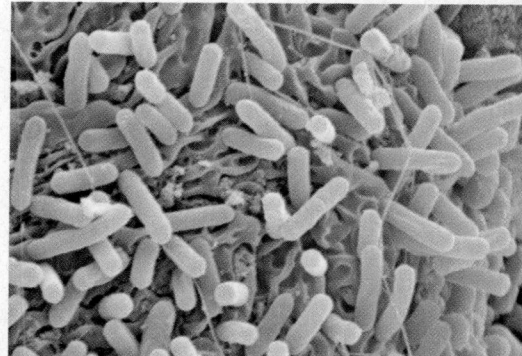

Have bacteria ever lived on Mars?
(Chapter 4)

What molecule's shape resembles a
geodesic dome? *(Chapter 5)*

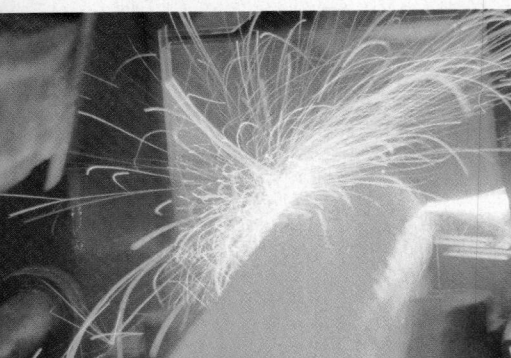

How does the glow of hot metal relate
to its temperature? *(Chapter 7)*

How can you make tomatoes ripen
faster? *(Chapter 9)*

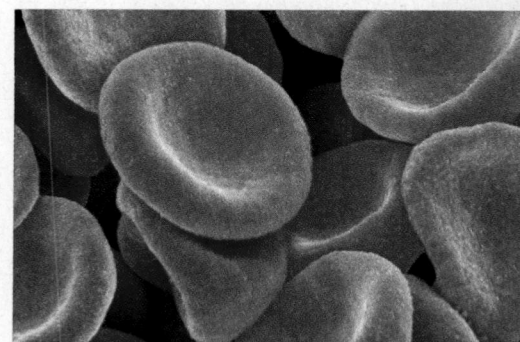

How do red blood cells transport
oxygen throughout the body?
(Chapter 10)

Why do fish have trouble breathing in
very warm water? *(Chapter 10)*

What molecule accounts for the aroma
of both almonds and cherries?
(Chapter 11)

What kinds of materials make the
strongest bridge cables? *(Chapter 12)*

What do the soccer ball and the seats have in common? *(Chapter 13)*

How do living systems convert food into energy? *(Chapter 14)*

Why does warm soda fizz when its
container is opened? *(Chapter 16)*

How can color tell us the pH of a
solution? *(Chapter 17)*

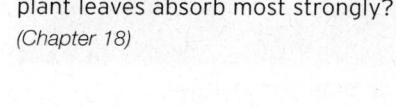

What wavelengths of light do green plant leaves absorb most strongly?
(Chapter 18)

What metal ion causes the color of aquamarine crystals? *(Chapter 18)*

19 Electrochemistry and the Quest for Clean Energy 916

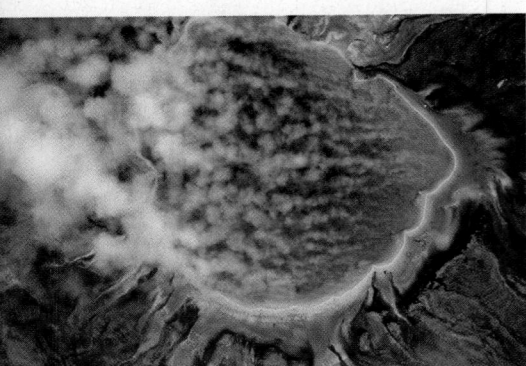

How do organisms harness geothermal energy? *(Chapter 20)*

20 Biochemistry: The Compounds of Life 956

How do ancient living trees help us achieve accurate radiocarbon dating? *(Chapter 21)*

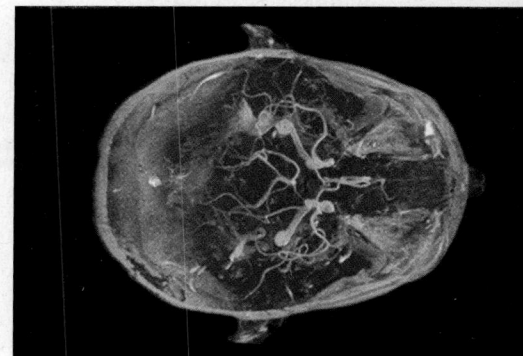

What elements are used in diagnostic imaging, therapy, and medical devices?
(Chapter 22)

Descriptive Chemistry Boxes

Applications

▶❚❚ ChemTours

About the Authors

Thomas R. Gilbert has a BS in chemistry from Clarkson University and a PhD in analytical chemistry from MIT. After 10 years with the Research Department of the New England Aquarium in Boston, he joined the faculty of Northeastern University, where he is currently an associate professor of chemistry and chemical biology and an academic director of biotechnology programs. His research interests are in chemical and science education. He teaches general chemistry and science education courses, and conducts professional development workshops for K–12 teachers. He has won Northeastern's Excellence in Teaching Award and Outstanding Teacher of First-Year Engineering Students Award.

Rein V. Kirss received both a BS in chemistry and a BA in history as well as an MA in chemistry from SUNY Buffalo. He received his PhD in inorganic chemistry from the University of Wisconsin, Madison, where the seeds for this textbook were undoubtedly planted. After two years of postdoctoral study at the University of Rochester, he spent a year at Advanced Technology Materials, Inc., before returning to academics at Northeastern University in 1989. He is an associate professor of chemistry with an active research interest in organometallic chemistry.

Natalie Foster is an associate professor of chemistry at Lehigh University in Bethlehem, Pennsylvania. She received a BS in chemistry from Muhlenberg College and MS, DA, and PhD degrees from Lehigh University. Her research interests include studying poly(vinyl alcohol) gels by NMR as part of a larger interest in porphyrins and phthalocyanines as candidate contrast enhancement agents for MRI. She teaches the introductory chemistry class every fall to engineering, biology, and other nonchemistry majors. Natalie also regularly teaches a spectral analysis course at the graduate level. Natalie was the 2008 recipient of the Christian & Mary Lindback Award for Distinguished Teaching at Lehigh University.

Geoffrey Davies holds BSc, PhD, and DSc degrees in chemistry from Birmingham University, England. He joined the faculty at Northeastern University in 1971 after postdoctoral research on the kinetics of very rapid reactions at Brandeis University, Brookhaven National Laboratory, and the University of Kent at Canterbury. He is now a Matthews Distinguished University Professor at Northeastern University. His research group has explored experimental and theoretical redox chemistry, alternative fuels, transmetalation reactions, tunable metal–zeolite catalysts and, most recently, the chemistry of humic substances, the essential brown animal and plant metabolites in sediments, soils, and water. He edits a column on experiential and study-abroad education in the *Journal of Chemical Education* and a book series on humic substances. He is a Fellow of the Royal Society of Chemistry and was awarded Northeastern's Excellence in Teaching Award in 1981, 1993, and 1999, and its first Lifetime Achievement in Teaching Award in 2004.

Preface

Dear Student:

Chemistry: The Science in Context, Third Edition, introduces chemical principles using contexts drawn from many disciplines including biology, environmental science, materials science, astronomy, geology, and medicine. We believe that these contexts make chemistry more interesting, relevant, understandable, and memorable. We hope this book helps you to understand the principles of chemistry and how you can apply them to solving global problems such as treating and curing disease, or making more efficient use of Earth's natural resources. Chemists' unique perspective on natural processes and our insights into the properties of substances from high-performance alloys to the products of biotechnology are based on understanding these processes and substances at the atomic and molecular level. A major goal of this book is to help you develop this microscale perspective.

Another major goal of the book is to help you improve your problem-solving skills. To solve problems, you first need to apply your understanding of chemical principles to recognize the connections between the information provided in a problem and the answer you are asked to find. Sometimes the hardest part of solving a problem is distinguishing between information that is relevant and information that is not. Once you are clear on where you are starting and where you are going, planning for and carrying out a solution become much easier.

To help you hone your problem-solving skills, we have developed a framework for doing so that is introduced in Chapter 1. It is a four-step approach we call **COAST**, which is our acronym for (1) **C**ollect and **O**rganize, (2) **A**nalyze, (3) **S**olve, and (4) **T**hink about it. We use these four steps in *every* Sample Exercise and in the solutions to *odd* problems in the Student Solutions Manual. They are also used in the hints and feedback embedded in the SmartWork online homework program. To summarize the four steps:

Collect and Organize helps you understand where to begin to solve the problem. In this step we often rephrase the problem and the answer that is sought, and we identify the relevant information that is provided in the problem statement or available elsewhere in the book.

Analyze is where we map out a strategy for solving the problem. As part of that strategy we often estimate what a reasonable answer might be.

Solve applies our analysis of the problem from the second step to the information and relations from the first step to actually solve the problem. We walk you through each step in the solution so that you can follow the logic and the math.

Think about It reminds us that an answer is not the last step in solving a problem. We should check the accuracy of the solution and think about the value of a quantitative answer. Is it realistic? Are the units correct? Is the number of significant figures appropriate? Does it agree with our estimate from the Analyze step?

Glucose ($C_6H_{12}O_6$) is a simple sugar formed by photosynthesis in plants. The complete combustion of 0.5763 g of glucose in a calorimeter ($C_{calorimeter}$ = 6.20 kJ/°C) raises the temperature of the calorimeter by 1.45°C. What is the food value of glucose in Calories per gram?

Collect and Organize We are asked to determine the food value of glucose, which means the energy given off when 1 g is burned. We have the mass of glucose burned and the calorimeter constant. We can relate the energy given off by the glucose to the energy gained by the calorimeter (Equation 5.15) and the energy given off by the glucose to the temperature change and the calorimeter constant (Equation 5.12).

Analyze We use the data from the calorimetry experiment to determine how much heat in kilojoules is given off when the stated amount of glucose is burned. We can convert that quantity into Calories by using the conversion factor 1 Cal = 4.184 kJ.

Solve

$$q_{calorimeter} = C_{calorimeter}\,\Delta T = (6.20 \text{ kJ/°C})(1.45°C) = 8.99 \text{ kJ}$$

To convert this quantity of energy to a food value, we divide by the sample mass:

$$\frac{8.99 \text{ kJ}}{0.5763 \text{ g}} = 15.6 \text{ kJ/g}$$

and then convert into Calories:

$$(15.6 \text{ kJ/g})\left(\frac{1 \text{ Cal}}{4.184 \text{ kJ}}\right) = 3.73 \text{ Cal/g}$$

Think about It One gram of a doughnut, which is mostly carbohydrates, and 1 g of glucose have about the same food value (19 kJ/g and 15.6 kJ/g, respectively), so the answer to this Sample Exercise seems reasonable.

CONNECTION We introduced enthalpy change (ΔH) in Chapter 5 as a thermodynamic quantity that describes heat flow into or out of a system.

 **CHEMTOUR**

The energy lost by the beverages inside the 72 cans in the preceding discussion was more than 100 times the heat lost by the cans. What factors contributed to this large difference between the energy lost by the cans and the energy lost by their contents?

(Answers to Concept Tests are in the back of the book.)

Many students use the **Sample Exercises** more than any other part of the book. Sample Exercises take the concept being discussed and illustrate how to apply it to solve a problem. We hope that repeated application of the COAST framework will help you refine your problem-solving skills and that the approach becomes habit-forming for you. When you finish a Sample Exercise, you'll find a Practice Exercise to try on your own. If you have the ebook, the Practice Exercises are "live," meaning that you can solve them interactively using the tutorial features in SmartWork to guide you with hints and answer-specific feedback. The next few pages describe how to use the tools built into each chapter to gain a conceptual understanding of chemistry.

Each chapter begins with **A Look Ahead**. These sections provide glimpses of how the chemistry in the chapter that follows connects to the world. We have used topics that should be familiar to you, but we place them in chemical contexts that may surprise you.

If you are trying to decide what is most important in a chapter, check the **Learning Outcomes** listed on the first page. Whether you are reading the chapter from first page to last or reviewing it for an exam, the Learning Outcomes help you focus on the key information you need to know and the skills you should acquire.

As you study each chapter, you will find **key terms** in boldface in the text and in a running glossary in the margin. We have deliberately duplicated these definitions so that you can continue reading without interruption but quickly find them when doing homework or reviewing for a test. All key terms are also defined in the Glossary in the back of the book.

Many concepts are related to others described earlier in the book. We point out these relationships with **Connection** icons in the margins. We hope they help you see the big picture and draw your own connections between the major themes covered in the book.

To help you develop your own microscale view of matter, we use molecular art to enhance photos and figures, and to illustrate what is happening at the atomic and molecular levels. The third edition has more molecular art than previous editions of this book.

If you're looking for additional help visualizing a concept, we have more than 100 **ChemTours**, denoted by the ChemTour icon, available on StudySpace (see p. 18). ChemTours demonstrate dynamic processes and help you visualize events at the molecular level. For the third edition, we have added audio to the ChemTours so you can focus on an animation while listening to an explanation of it. Many of the ChemTours allow you to manipulate variables and observe the resulting changes. Questions at the end of the ChemTour tutorials offer step-by-step assistance in solving problems and provide useful feedback.

Concept Tests are short, conceptual questions that serve as a self-check by asking you to stop and answer a question relating to what you just read. We designed them to help you see for yourself whether you have grasped a key concept and can apply it. For the third edition, we have an average of one Concept Test per section and many have a visual component. You may find some Concept Tests challenging. Don't be discouraged if you can't answer all of them, and remember that all of the answers are in the back of the book.

Near the end of many chapters is a **Descriptive Chemistry** box. While the biochemical properties of the elements and their roles in medicine are the focus of Chapter 22, Descriptive Chemistry boxes throughout the textbook summarize the properties and uses of individual elements or groups of elements that were highlighted in a particular chapter. We discuss where the substances occur in nature and how they are used in ways that touch our lives and shape our world

At the end of each chapter is a thematic **Summary** and a **Problem-Solving Summary**. The first is a brief synopsis of the chapter, organized by section. Key figures have been added to this Summary to provide visual cues as you review. The Problem-Solving Summary is unique to this general chemistry book—it outlines the different types of problems you should be able to solve, where to find examples of them in the Sample Exercises, and reiterates relevant concepts and equations.

Following the summaries are groups of questions and problems. The first group consists of **Visual Problems**. In many of them, you are asked to interpret a molecular view of a sample or a graph of experimental data.

Concept Review Questions and Problems come next, arranged by topic in the same order as they appear in the chapter. Concept Reviews are qualitative and often ask you to explain why or how something happens. Problems are paired and can be quantitative, conceptual, or a combination of both. **Contextual problems** have a title that describes the context in which the problem is placed. **Additional Problems** can come from any section or combination of sections in the chapter. Some of them incorporate concepts from previous chapters. Problems marked with an asterisk (*) are more challenging and often take multiple steps to solve.

We want you to have confidence in using the answers in the back of the book as well as the Student Solutions Manual, so we used a rigorous triple-check accuracy program for the third edition. Each end-of-chapter question and problem was solved independently by the Solutions Manual author, Karen Brewer, and by two additional chemical educators. Karen compared her solutions to those from the two reviewers and resolved any discrepancies. This process is designed to ensure clearly written problems and accurate answers in the appendices and Solutions Manual.

Changes to the Third Edition

Dear Instructor:

Whether you used the second edition of this book or not, you might find it useful to know how this edition compares to its predecessor. Here are some of the general changes we made throughout this edition:

➤ Nearly every **Sample Exercise** has been revised to uniformly implement the **COAST** framework for problem solving. In most quantitative exercises we estimate what a reasonable answer might be in the **Analyze** step and then check the solution against this estimate in the **Think about It** step.

PROBLEM-SOLVING SUMMARY

TYPE OF PROBLEM	CONCEPTS AND EQUATIONS	SAMPLE EXERCISES
Identifying endothermic and exothermic processes, and calculating internal energy change (ΔE) and P–V work	For the system: $$\Delta E = q + w \quad (5.5)$$ where $w = -P\Delta V$.	5.1–5.3
Determining the flow of energy (q) associated with a change of state or with changing the temperature of a substance	Melting a solid at its melting point: $$q = n\Delta H_{fus} \quad (5.9)$$ vaporizing a liquid at its boiling point: $$q = n\Delta H_{vap} \quad (5.10)$$ or heating a substance: $$q = nc_P \, \Delta T \quad (5.8)$$	5.4–5.6
Measuring the heat capacity (calorimeter constant) of a calorimeter	$$C_{calorimeter} = q/\Delta T \quad (5.13)$$ where $C_{calorimeter}$ is the heat capacity of calorimeter, q is the heat released by a standard combustion reaction, and ΔT is the temperature change of calorimeter.	5.7
Recognizing and writing formation reactions	In a formation reaction, the reactants are elements in their standard states and the product is 1 mole of a single compound.	5.8
Calculating standard enthalpies of reaction from heats of formation	$$\Delta H^\circ_{rxn} = \sum n_{products} \, \Delta H^\circ_{f,products} - \sum n_{reactants} \, \Delta H^\circ_{f,reactants} \quad (5.17)$$	5.9
Calculating fuel values and food value	The fuel value or food value of a substance is the energy released by the complete combustion of 1 g of the substance.	5.10, 5.11
Calculating standard enthalpies of reaction using Hess's law	Reorganize the information so that the reactions add together as desired. Reversing a reaction changes the sign of the reaction's ΔH°_{rxn} value. Multiplying the coefficients in a reaction by a factor means the reaction's ΔH°_{rxn} value has to be multiplied by the same factor.	5.12

Enthalpy and Enthalpy Changes

CONCEPT REVIEW

5.31. What is meant by an *enthalpy change*?

5.32. Describe the difference between an internal energy change (ΔE) and an enthalpy change (ΔH).

5.33. Why is the sign of ΔH negative for an exothermic process?

5.34. What happens to the magnitude and sign of the enthalpy change when a process is reversed?

PROBLEMS

5.35. A Clogged Sink Adding Drano to a clogged sink causes the drainpipe to get warm. What is the sign of ΔH for this process?

5.36. Cold Pack for Injuries Breaking a small pouch of water inside a larger bag containing ammonium nitrate activates chemical cold packs, used by sports trainers for injured athletes. What is the sign of ΔH for the process taking place in the cold pack?

5.37. Break a Bond The stable form of oxygen at room temperature and pressure is the diatomic molecule O_2. What is the sign of ΔH for the following process?

$$O_2(g) \rightarrow 2\,O(g)$$

doubled? Is the same true for the specific heat?

5.43. Are the heats of fusion and vaporization of a given substance usually the same?

5.44. An equal amount of heat is added to pieces of metal A and metal B having the same mass. Does the metal with the larger heat capacity reach the higher temperature?

*5.45. Cooling an Automobile Engine Most automobile engines are cooled by water circulating through them and a radiator. However, the original Volkswagen Beetle had an air-cooled engine. Why might car designers choose water cooling over air cooling?

*5.46. Nuclear Reactor Coolants The reactor-core cooling systems in some nuclear power plants use liquid sodium as the coolant. Sodium has a thermal conductivity of 1.42 J/(cm · s · K), which is quite high compared with that of water [6.1 × 10⁻³ J/(cm · s · K)]. The respective molar heat capacities are 28.28 J/(mol · K) and 75.31 J/(mol · K). What is the advantage of using liquid sodium over water in this application?

PROBLEMS

5.47. How much heat is needed to raise the temperature of 100.0 g of water from 30.0°C to 100.0°C?

5.48. At an elevation where the boiling point of water is 93°C, 100.0 g of water at 30°C absorbs 290.0 kJ of heat from a mountain climber's stove. Is this amount of energy sufficient to heat the water to its boiling point?

➤ There are more molecular views in the figures and more **Concept Tests** throughout the book so that nearly every section has at least one.

➤ Changes to the media package include an audio component for the ChemTours animations, and the ebook includes "live" Practice Exercises with hints, answer-specific feedback, and fully worked solutions.

➤ More than 700 new problems have been added to SmartWork to support the Third Edition, including Practice Exercises and 45 new Tutorial Problems.

This edition has one more chapter than the second because we decided to devote a single chapter (10) to intermolecular forces: their types, relative strengths, and impacts on the properties of substances. Chapter 11 is now focused on the properties of solutions, including colligative properties and the behavior of homogeneous mixtures of volatile substances (formerly in the organic chemistry chapter), which also frames our discussion of the dependence of vapor pressure on temperature.

Other changes include reorganizing our approach to drawing Lewis structures (Section 8.2), and rewriting Section 12.3 as a stand-alone introduction to crystal structures that can be used separately from the rest of Chapter 12 on the chemistry of solids. Our coverage of chirality now falls much earlier in the book than before, with a brief introduction to the subject in Chapter 9 and a full treatment in Chapter 13. The concept of the solubility product is now introduced in our chapter on equilibrium in the aqueous phase, rather than the transition metals chapter. In Chapter 19, the equation for calculating standard cell potentials has been revised to conform to guidelines established by the International Union of Pure and Applied Chemistry (IUPAC).

Ancillaries for Students

wwnorton.com/smartwork

SMARTWORK: AN ONLINE TUTORIAL AND HOMEWORK PROGRAM FOR GENERAL CHEMISTRY

Created by chemistry educators, SmartWork is the most intuitive online tutorial and homework-management system available for general chemistry. Powerful engines support an unparalleled range of question types, which include graded molecule drawing, math and chemical equations, and graphs. Answer-specific feedback, hints, and stepwise tutorials coach students through solving problems. Integration of ebook and multimedia content completes this chemistry learning system. Assigning, editing, and administering homework within SmartWork is easy. WYSIWYG (What You See Is What You Get) authoring tools allow instructors to modify existing problems or develop new content.

Problems in SmartWork use the same language and notation as the Third Edition of *Chemistry*. Every problem in the system has hints and answer-specific feedback using the same COAST steps from the text. Multistep Tutorial Problems are provided for the more challenging topics. If a student answers a Tutorial Problem incorrectly, SmartWork coaches the student through a series of steps using COAST that map a path to the correct answer. Each step starts with a question. Hints on how to answer it are provided if needed.

After completing the tutorial, the student returns to the original problem ready to apply the problem-solving insights gained from the tutorial. Students who can answer challenging questions without help will do their homework quickly and efficiently, because supplementary instruction is provided only to students who have asked for it. Students also have the option to complete only a part of a tutorial sequence and return to the original problem when they feel they are ready.

Wherever possible, SmartWork makes use of algorithmic variables so that students see slightly different versions of the same problem. Assignments are graded automatically and SmartWork includes sophisticated yet flexible tools for managing class data. Instructors can use the Item Analysis feature to assess how students have done on specific problems within an assignment and go deeper to review an individual student's work on a problem.

SmartWork also allows students to quickly and easily access the ebook version of *Chemistry*, Third Edition. Reference links are available from each question. Combined access to both ebook and SmartWork is available with a new text at no extra cost.

EBOOK

An affordable and convenient alternative, the ebook retains the content and design of the print book and allows students to highlight and take notes with ease, print chapters as needed, and search the text. The online version of *Chemistry*, Third Edition provides students with stand-alone Interactive Practice Exercises. These are self-grading SmartWork problems that allow students to practice solving problems, and receive hints and feedback, with no penalty. The online ebook also allows students one-click access to the 100+ ChemTours.

www.nortonebooks.com

The online ebook is available with the print text at no extra cost, or it may be purchased separately with SmartWork for one-third the cost of the printed textbook.

Norton also offers a downloadable PDF version of the ebook at one-third the cost of the printed book.

STUDYSPACE: THE STUDENTS' PLACE FOR A BETTER GRADE

Students use online resources to help them succeed in their courses—StudySpace is unmatched in providing a one-stop solution that's closely aligned with their textbook. This free and easy-to-navigate website offers students an impressive range of exercises, interactive learning tools, and assessment and review materials, including:

wwnorton.com/studyspace

➤ **ChemTours:** More than 100 ChemTour tutorial animations, which now include audio.

➤ **Study Plans** featuring the Problem-Solving Summary for each chapter.

➤ **Diagnostic Quizzes** and Flashcards.

➤ **Links** to premium content in the ebook and SmartWork, and Chemistry in the News features.

STUDENT'S SOLUTIONS MANUAL by Karen Brewer, Hamilton College

The Student's Solutions Manual provides students with fully worked solutions to select end-of-chapter problems using the **COAST** four-step method (**C**ollect and **O**rganize, **A**nalyze, **S**olve, and **T**hink about It).

Ancillaries for Instructors

CLICKERS IN ACTION: INCREASING STUDENT PARTICIPATION IN GENERAL CHEMISTRY by Margaret Asirvatham, University of Colorado-Boulder

An instructor-oriented resource providing information on implementing clickers in general chemistry courses. *Clickers in Action* contains more than 250 class-tested, lecture-ready questions with histograms showing student responses, as well as insights and suggestions for implementation. Question types include macroscopic observation, symbolic representation, and atomic/molecular views of processes.

INSTRUCTOR'S SOLUTIONS MANUAL by Karen Brewer, Hamilton College

Revised for the Third Edition, the Instructor's Solutions Manual provides instructors with fully worked solutions to every end-of-chapter Concept Review and Problem. Each solution uses the **COAST** four-step method (**C**ollect and **O**rganize, **A**nalyze, **S**olve, and **T**hink about It).

INSTRUCTOR'S RESOURCE MANUAL by Rein Kirss, Northeastern University

Revised for the Third Edition, and written by one of the textbook authors, each chapter of the Instructor's Resource Manual begins with a brief overview of the text chapter, followed by suggestions for integrating the contexts featured in the book into a lecture, sample lecture outlines, and alternate contexts to use with each chapter. New to this edition are suggestions on how to use *Clickers in Action* clicker questions in your lecture. Instructor notes for ChemConnections Activities supplement the ChemConnections worksheets, which are also included. Summaries of the ChemTours available on the Norton Media Library and Instructor's Resource Disc round out each chapter.

INSTRUCTOR'S RESOURCE DISC

This helpful classroom presentation tool features:

➤ Lecture PowerPoint slides that include integrated figures from the text, ChemTours, and stick-or-switch clicker questions. These are particularly helpful to first-time teachers of the introductory course.

➤ All ChemTours, ready for off-line use.

➤ *Clickers in Action* clicker questions for each chapter provide instructors with class-tested questions they can integrate into their course.

➤ Photographs, drawn figures, and tables from the text available in PowerPoint and JPEG.

wwnorton.com/instructors

DOWNLOADABLE INSTRUCTOR'S RESOURCES

➤ Lecture Power Points with stick-or-switch clicker questions.

➤ Test bank in PDF, Word RTF, *ExamView* Assessment Suite formats.

➤ Solutions Manual in PDF and Word, so that instructors may edit solutions.

➤ All of the end-of-chapter questions and problems are available in Word along with the key equations.

➤ Photographs, drawn figures, and tables from the text available in PowerPoint and JPEG.

➤ *Clickers in Action* clicker questions.

➤ BlackBoard and WebCT materials.

BLACKBOARD AND WEBCT COURSE CARTRIDGES

Course cartridges for BlackBoard and WebCT include access to the ChemTours, a Study Plan for each chapter, multiple-choice tests, and links to premium content in the ebook and SmartWork.

TEST BANK by David M. Hanson and Troy Wolfskill, Stony Brook University

Norton uses an innovative, evidence-based model to deliver high-quality and pedagogically effective quizzes and testing materials. Thoroughly revised for the Third Edition using an evidence-centered approach, the Test Bank contains more than 2,200 questions. Each chapter of the Test Bank is structured around a Concept Map and evaluates student knowledge on three distinct levels:

➤ **Factual** questions that test students' basic understanding of facts and concepts.

➤ **Applied** questions that require students to apply knowledge in the solution of a problem.

➤ **Conceptual** questions that require students to engage in a qualitative reasoning and to explain why things are as they are.

Questions are further classified by section and difficulty, making it easy to construct tests and quizzes that are meaningful and diagnostic according to instructor need. Questions are multiple-choice and short answer.

The Test Bank is available in *ExamView* Assessment Suite, Word RTF, and PDF formats.

Acknowledgments

The decision to revise a textbook unleashes a storm of activity. Our first order of thanks must go to W. W. Norton for having enough confidence in the idea behind the first two editions to commit to the massive labor of the third. The people at W. W. Norton with whom we work most closely deserve much more than the feeble thanks and first billing we give them here. First, we appreciate the work of our editor and navigator through the storm, Erik Fahlgren. He has been an indefatigable source of guidance, help, inspiration, and the occasional prodding with carrots or sticks that was crucial to keep four academicians moving toward a common goal in a timely fashion. Erik's involvement in the project is the single greatest reason for its completion, and our greatest thanks are too small an offering for his unwavering focus.

We are pleased to acknowledge the contributions of two developmental editors in creating this book. Irene Nunes went through the second edition with a very fine-toothed comb, suggesting ways it could be improved. Our second developmental editor, Andrew Sobel, blended Irene's reviews with those of several other educators, as well as his own keen insights, as he contributed abundant and unfailingly perceptive suggestions on issues of content, tone, level, and clarity of text and illustrations. Our project editor and doyenne of all things artistic in the book, Carla Talmadge, wrangled illustrations, figures, and tables with skill, finesse, and unfailing good humor. Editorial assistant Mary Lynch kept the paper flowing and all of us on the same page. Thanks as well to Lissi Sigillo for a design that is appealing and easy to navigate; Debra Morton Hoyt for a spectacular cover; Stephanie Romeo and Donna Ranieri for finding great photos; production manager Chris Granville for his work behind the scenes; Matthew Freeman for managing the print ancillaries; Rob Bellinger for his diligence on the media; and Kelsey Volker for her words of encouragement, unfailing sense of humor, and marketing prowess. The entire Norton team was so highly competent and professional that they actually made the daunting task of bookmaking fun.

This book has benefited greatly from the care and thought that many reviewers, listed here, gave to their readings of earlier drafts. We owe an extra special thanks to Karen Brewer for her dedicated and precise work on the Solutions Manual. She, along with Steven S. Trail, Thomas J. Anderson, Jordan L. Fantini, C. Alton Hassell, and David E. Phippen are the triple-check accuracy team who solved each problem and reviewed each solution for accuracy. We are deeply grateful to Resa Kelly for her thoughtful review of the art program and Stephen R. Parker, Mauro Di Renzo, Matthew G. K. Thompson, and Laurie Tyler for participating in the diary reviews and providing detailed comments on nearly every chapter. Their insights into teaching, learning, and chemistry have had a profound and positive impact on the content. Finally, we greatly appreciate Karen Frindell, Stephen Goldberg, Jason Kautz, Albert Martin, Trilisa Perrine, Prasad Polavarapu, Karla Radke, and Edwin Sibert for checking the accuracy of the myriad facts that form the framework of the science.

Thomas R. Gilbert
Rein V. Kirss
Natalie Foster
Geoffrey Davies

Third Edition Reviewers

Thomas J. Anderson, Francis Marion University
Mikhail V. Barybin, University of Kansas
Simon Bott, University of Houston
David Cedeno, Illinois State University
Andrew L. Cooksy, San Diego State University
Charles Cornett, University of Wisconsin, Platteville
Mitchel Cottenoir, South Plains College
John Davison, Irvine Valley College
Mauro Di Renzo, Vanier College
Bill Durham, University of Arkansas
Jordan L. Fantini, Denison University
Amy Flanagan-Johnson, Eastern Michigan University
George Flowers, Darton College
Karen Frindell, Santa Rosa Junior College
Stephen Z. Goldberg, Adelphi University
Margie Haak, Oregon State University
C. Alton Hassell, Baylor University
Paul Higgs, University of Tennessee, Martin
Kimberly Hill Edwards, Oakland University
Matthew Horn, Utah Valley University
Tim Jackson, University of Kansas
Lori Jones, University of Guelph
Jason Kautz, University of Nebraska, Lincoln
Resa Kelly, San Jose State University
Robert Kerber, Stony Brook University
Elizabeth Kershisnik, Oakton Community College
Larry Kolopajlo, Eastern Michigan University
Jailson de Lima, Vanier College
Willem Leenstra, University of Vermont
Laura MacManus-Spencer, Union College
Albert Martin, Moravian College
Thomas McGrath, Baylor University
Dan Moriarty, Siena College

Richard Nafshun, Oregon State University
Mya Norman, University of Arkansas
Ken O'Connor, Marshall University
Gregory Oswald, North Dakota State University
Stephen R. Parker, Montana Tech
Jessica Parr, University of Southern California
Trilisa Perrine, Ohio Northern University
David E. Phippen, Shoreline Community College
Prasad Polavarapu, Vanderbilt University
John Pollard, University of Arizona
Lisa Ponton, Elon University
Karla Radke, North Dakota State University
Alan Richardson, Oregon State University
Mark Rockley, Oklahoma State University
Joel W. Russell, Oakland University
Nancy Savage, University of New Haven
Mark Schraf, West Virginia University
Fatma Selampinar, University of Connecticut
Edwin Sibert, University of Wisconsin, Madison
Virginia Smith, United States Naval Academy
Xianzhi Song, University of Pittsburgh, Johnstown
William Steel, York College of Pennsylvania
Matthew G. K. Thompson, Trent University
Craig Thulin, Utah Valley University
Steve S. Trail, Elgin Community College
Laurie Tyler, Union College
Kris Varazo, Francis Marion University
Andrew Vreugdenhil, Trent University
Wayne Wesolowski, University of Arizona
Cynthia Woodbridge
Mingming Xu, West Virginia University
Tim Zauche, University of Wisconsin, Platteville

Previous Editions' Reviewers

William Acree, Jr., University of North Texas
R. Allendoefer, State University of New York, Buffalo
Sharon Anthony, The Evergreen State College
Jeffrey Appling, Clemson University
Marsi Archer, Missouri Southern State University
Margaret Asirvatham, University of Colorado, Boulder
Robert Balahura, University of Guelph
Anil Banerjee, Texas A&M University, Commerce
Sandra Banks, Mills College
Mufeed Basti, North Carolina Agricultural & Technical State University
Robert Bateman, University of Southern Mississippi
Kevin Bennett, Hood College
H. Laine Berghout, Weber State University
Eric Bittner, University of Houston
David Blauch, Davidson College
Robert Boggess, Radford University
Simon Bott, University of Houston
Michael Bradley, Valparaiso University
Karen Brewer, Hamilton College
Timothy Brewer, Eastern Michigan University
Julia Burdge, Florida Atlantic University
Robert Burk, Carleton University
Andrew Burns, Kent State University
Sharmaine Cady, East Stroudsburg University
Chris Cahill, George Washington University
Kevin Cantrell, University of Portland
Nancy Carpenter, University of Minnesota, Morris
Patrick Caruana, State University of New York, Cortland
David Cedeno, Illinois State University
Tim Champion, Johnson C. Smith University
William Cleaver, University of Vermont
Penelope Codding, University of Victoria
Jeffery Coffer, Texas Christian University
Renee Cole, Central Missouri State University
Brian Coppola, University of Michigan
Richard Cordell, Heidelberg College
Robert Cozzens, George Mason University
Margaret Czrew, Raritan Valley Community College
Laura Deakin, University of Alberta
Anthony Diaz, Central Washington University
Klaus Dichmann, Vanier College
Mauro Di Renzo, Vanier College
Kelley J. Donaghy, State University of New York, College of Environmental Science and Forestry
Michelle Driessen, University of Minnesota
Dan Durfey, Naval Academy Preparatory School
Stefka Eddins, Gardner-Webb University
Dwaine Eubanks, Clemson University
Lucy Eubanks, Clemson University
Nancy Faulk, Blinn College, Bryan
Tricia Ferrett, Carleton College
Matt Fisher, St. Vincent College

Richard Foust, Northern Arizona University
David Frank, California State University, Fresno
Cynthia Friend, Harvard University
Barbara Gage, Prince George's Community College
Brian Gilbert, Linfield College
Jack Gill, Texas Woman's University
Arthur Glasfeld, Reed College
Frank Gomez, California State University, Los Angeles
John Goodwin, Coastal Carolina University
Steve Gravelle, Saint Vincent College
Tom Greenbowe, Iowa State University
Stan Grenda, University of Nevada
Margaret Haak, Oregon State University
Todd Hamilton, Adrian College
Robert Hanson, St. Olaf College
David Harris, University of California, Santa Barbara
Holly Ann Harris, Creighton University
Donald Harriss, University of Minnesota, Duluth
C. Alton Hassell, Baylor University
Dale Hawley, Kansas State University
Brad Herrick, Colorado School of Mines
Vicki Hess, Indiana Wesleyan University
Donna Hobbs, Augusta State University
Angela Hoffman, University of Portland
Tamera Jahnke, Southern Missouri State University
Shahid Jalil, John Abbot College
Kevin Johnson, Pacific University
Martha Joseph, Westminster College
David Katz, Pima Community College
Phillip Keller, University of Arizona
Angela King, Wake Forest University
John Krenos, Rutgers University
C. Krishnan, State University of New York, Stony Brook
Richard Langley, Stephen F. Austin State University
Sandra Laursen, University of Colorado, Boulder
Richard Lavrich, College of Charleston
David Laws, The Lawrenceville School
George Lisensky, Beloit College
Jerry Lokensgard, Lawrence University
Boon Loo, Towson University
Roderick M. Macrae, Marian College
John Maguire, Southern Methodist University
Susan Marine, Miami University, Ohio
Diana Mason, University of North Texas
Garrett McGowan, Alfred University
Craig McLauchlan, Illinois State University
Heather Mernitz, Tufts University
Stephen Mezyk, California State University, Long Beach
Rebecca Miller, Lehigh University
John Milligan, Los Angeles Valley College
Timothy Minger, Mesa Community College
Ellen Mitchell, Bridgewater College
Stephanie Myers, Augusta State College

Melanie Nilsson, McDaniel College
Sue Nurrenbern, Purdue University
Gerard Nyssen, University of Tennessee, Knoxville
Jodi O'Donnell, Siena College
Jung Oh, Kansas State University, Salinas
MaryKay Orgill, University of Nevada, Las Vegas
Robert Orwoll, College of William and Mary
Jason Overby, College of Charleston
Greg Owens, University of Utah
Giuseppe Petrucci, University of Vermont
Julie Peyton, Portland State University
Alexander Pines, University of California, Berkeley
Gretchen Potts, University of Tennessee, Chattanooga
Robert Pribush, Butler University
Gordon Purser, University of Tulsa
Robert Quandt, Illinois State University
Casey Raymond, State University of New York, Oswego
Beatriz Ruiz Silva, University of California, Los Angeles
Pam Runnels, Germanna Community College
Jerry Sarquis, Miami University, Ohio
Barbara Sawrey, University of California, Santa Barbara
Truman Schwartz, Macalester College
Shawn Sendlinger, North Carolina Central University

Susan Shadle, Boise State University
Peter Sheridan, Colgate University
Ernest Siew, Hudson Valley Community College
Roberta Silerova, John Abbot College
Sally Solomon, Drexel University
Estel Sprague, University of Cincinnati
Steven Strauss, Colorado State University
Mark Sulkes, Tulane University
Duane Swank, Pacific Lutheran University
Keith Symcox, University of Tulsa
Agnes Tenney, University of Portland
Edmund Tisko, University of Nebraska, Omaha
Brian Tissue, Virginia Polytechnic Institute and State University
Mike van Stipdonk, Wichita State University
William Vining, University of Massachusetts
Andrew Vruegdenhill, Trent University
Ed Walton, California State Polytechnic University, Pomona
Charles Wilkie, Marquette University
Ed Witten, Northeastern University
Stephen Wood, Brigham Young University
Noel Zaugg, Brigham Young University, Idaho
James Zimmerman, Missouri State University
Martin Zysmilich, George Washington University

CHEMISTRY

1

Matter, Energy, and the Origins of the Universe

Learning Outcomes

- Describe forms of matter and their structures at the atomic level
- Relate chemical formulas to molecular structures and vice versa
- Distinguish between physical processes and chemical reactions and between physical and chemical properties
- Use a systematic approach (COAST) to solve problems
- Describe the three states of matter and the transitions between them at the macroscopic and atomic levels
- Describe the scientific method
- Distinguish between exact and uncertain values
- Express values with the appropriate number of significant figures

A LOOK AHEAD

Blinded by the Sun

Today we know things about the universe that our grandparents could not even imagine; someday the same will be true of our grandchildren and their grandchildren. We know about the processes that take place in stars and how those processes release energy and produce elements, including those elements that make up everything on our planet Earth. All of this knowledge, however, has not diminished the number of questions we still ask about the universe and our place in it. Our tendency to ask *how* and *why* is part of human nature, and one truth about scientific explorations is that every answer spawns a host of new questions.

Even though people throughout history have asked how and why natural events occur, only in the last thousand years have they moved away from mythological speculation and toward a rational approach based on science and the scientific method. This approach involves careful observation followed by developing testable explanations called hypotheses. The process of testing hypotheses by running experiments is fundamental to scientific inquiry. An early example of this process ended an ancient debate about how vision works. Mathematicians, including Euclid, used geometry to reason that light traveled from the eye to any object being observed. Aristotle and other philosophers proposed that light traveled from the object to the eye. Both theories were logical, but which one was right? Around the year 1000, a Persian philosopher named Ibn al-Haytham suggested an experiment to resolve the dilemma: stare at the sun. Everyone knew you could not do that; if you stare at the sun, your eyes will be burned; you might even be

Planetary Nebula NGC 2818 The Hubble Space Telescope recorded this ▶ image of planetary nebula NGC 2818 in November 2008. Despite its classification, this is not a planet, but rather an expanding cloud of hot gases produced by a dying giant star that has exploded, sending the elements it synthesized during its lifetime into interstellar space. These elements include nitrogen (the source of the orange color in the photograph) and oxygen (shown in blue).

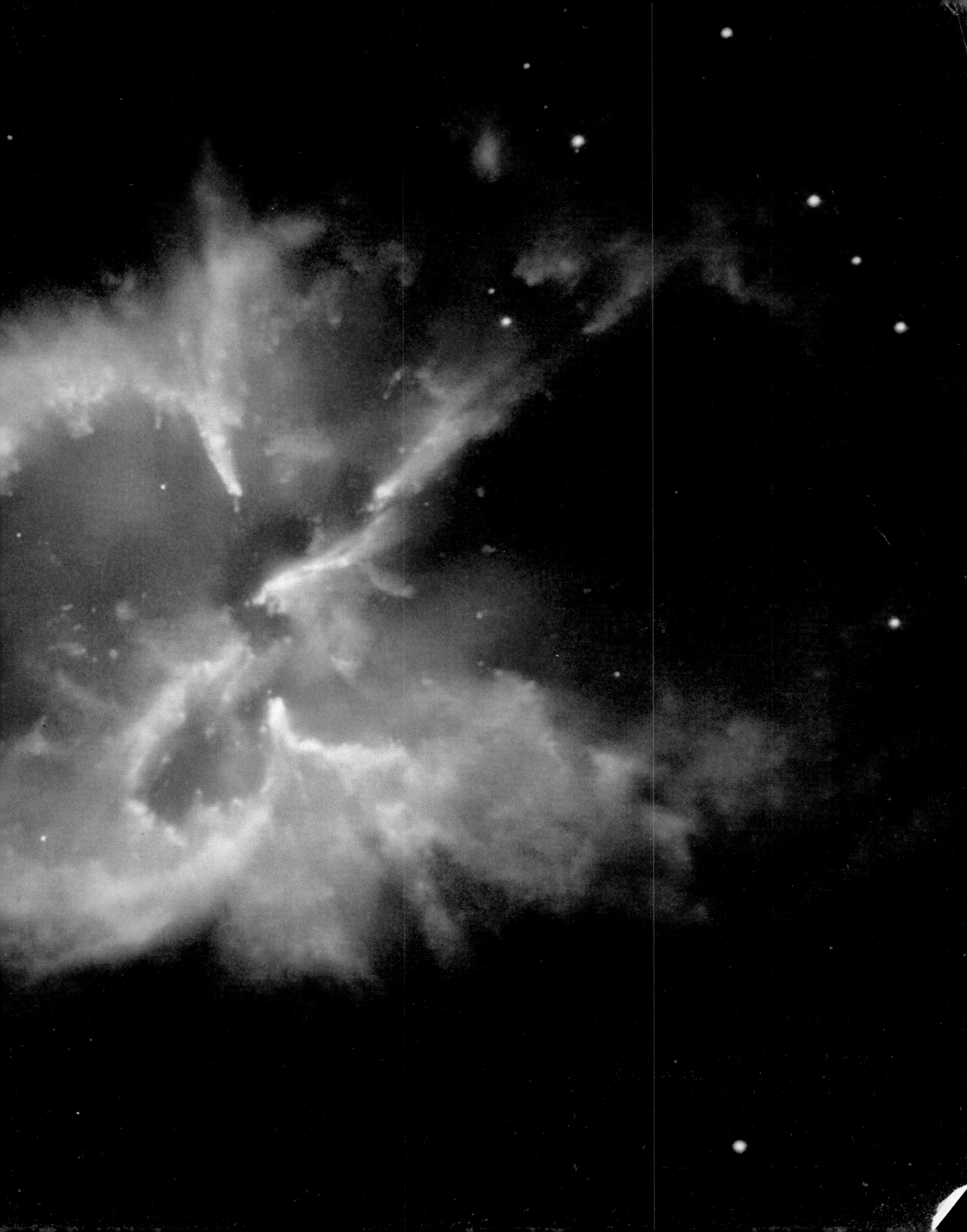

blinded. This can happen only if light from the sun enters the eye and damages it. This simple experiment resolved a centuries-old debate. Since the time of Ibn al-Haytham, testing hypotheses has led to scientific theories about how the natural world works. No process has been more influential in shaping our understanding of nature.[1]

Scientific investigation requires experimentation, and most experiments require measurements. For example, evidence for how stars synthesize elements comes from analysis of the energy that reaches us from the stars and from the cosmic explosions that mark their deaths. Although we have learned a great deal about the origin of the elements, many of the processes that take place in stars are still not understood, so our current explanations for how elements form will doubtless be revised and refined as more observations lead to more hypotheses, which will be tested by more experiments (Figure 1.1). Science is a dynamic process. ■

1.1 Classes of Matter

Consider the world around us. All things in it that are physically real—from the air we breathe to the ground we walk on to the sun's rays that warm us—are forms of either matter or energy. Scientists define **matter** as everything in the universe that has **mass** and occupies space. **Energy** is defined as the capacity to transfer heat or do work. **Chemistry** is the study of the composition, structure, and properties of matter. Chemists observe the changes that matter undergoes and measure the amount of energy produced or consumed during those changes. The science of chemistry has led to the synthesis of many new forms of matter that affect the way we live and the planet on which we live. Modern life is difficult to imagine without plastics, computers, cell phones, aspirin, and the countless other synthetic materials and technological innovations made possible through the study of chemistry.

(a)

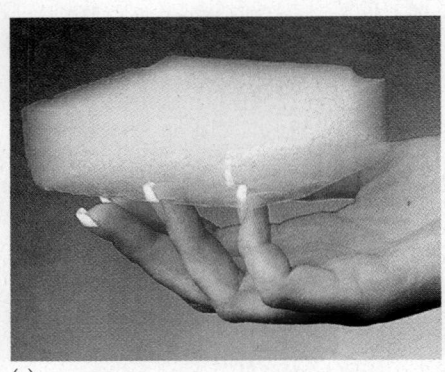

(b)

(c)

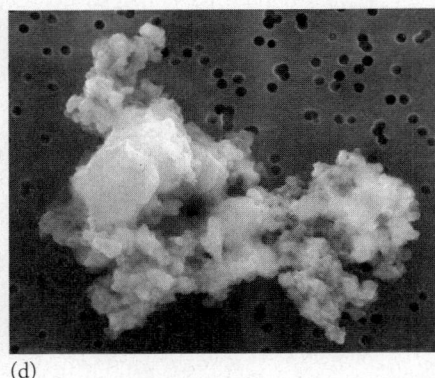

(d)

FIGURE 1.1 Scientists have learned about the composition and chemical history of the universe by analyzing particles of interstellar dust. NASA's *Stardust* probe returned to Earth in 2006 after a 7-year voyage in which it collected interstellar dust and sampled the tail of a comet. Comets are believed to contain material left over when the solar system formed over 4.5 billion years ago. This sequence shows (a) an artist's rendition of *Stardust* approaching the tail of the comet; (b) Jet Propulsion Laboratory scientist Dr. Peter Tsou holding a *Stardust* sample tray; (c) a sample of the aerogel (a material somewhat like dried Jell-O with very tiny pores) used to trap the dust particles; (d) a photomicrograph of a cosmic dust fragment on the aerogel.

[1] Richard Powers, "Best Ideas: Eyes Wide Open," *New York Times,* April 18, 1999.

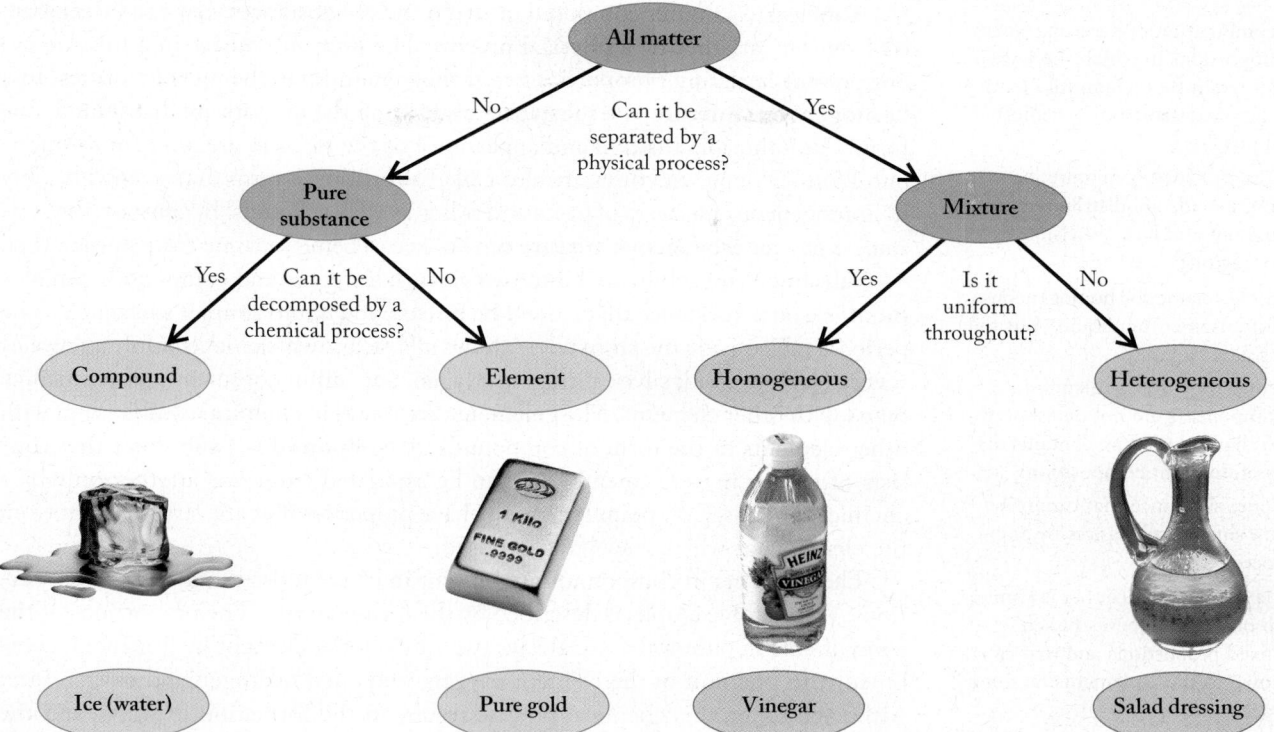

FIGURE 1.2 Matter is classified as shown in this diagram. The two principal categories are pure substances and mixtures. A substance may be a compound (such as water) or an element (such as gold). When the substances making up a mixture are distributed uniformly, as they are in vinegar (a mixture of acetic acid and water), the mixture is homogeneous. When the substances making up a mixture are not distributed uniformly, as when solids are suspended in a liquid and then settle to the bottom of the container as they do in Italian salad dressing, the mixture is heterogeneous.

The different forms of matter can be organized according to the classification scheme shown in Figure 1.2. The principal classes are pure substances and mixtures (Figure 1.3). A pure **substance** has a constant composition that does not vary from one sample to another. For example, the composition of pure water does not vary, no matter what its source. It is matter that cannot be separated into simpler substances by any physical process. By **physical process**, we mean a transformation of a sample of matter, such as a change in its physical state, that does not alter the chemical identities of any of the substances in the sample.

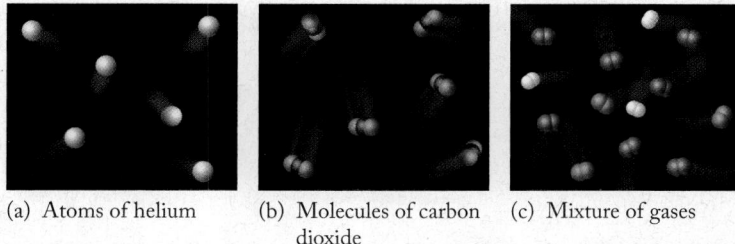

(a) Atoms of helium   (b) Molecules of carbon dioxide   (c) Mixture of gases

FIGURE 1.3 All matter is made up of either pure substances (of which there are few in nature) or mixtures. (a) The element helium (He), the second most abundant element in the universe, is one example of a pure substance. (b) The compound carbon dioxide (CO_2), the gas used in many fire extinguishers, is also a substance. (c) This homogeneous mixture contains three substances: nitrogen (N_2, blue), hydrogen (H_2, white), and oxygen (O_2, red).

matter anything that has mass and occupies space.

mass the property that defines the quantity of matter in an object.

energy the capacity to transfer heat or do work.

chemistry the study of the composition, structure, and properties of matter and of the energy consumed or given off when matter undergoes a change.

substance matter that cannot be broken down to simpler matter by any physical process; also known as *pure substance*.

physical process a transformation of a sample of matter, such as a change in its physical state, that does not alter the chemical identity of any substance in the sample.

mixture a combination of pure substances in variable proportions in which the individual substances retain their chemical identities and can be separated from one another by a physical process.

homogeneous mixture a mixture in which the components are distributed uniformly throughout and have no visible boundaries or regions.

solution another name for homogeneous mixture. Solutions are often liquids, but they may also be solids or gases.

heterogeneous mixture a mixture in which the components are not distributed uniformly, so that the mixture contains distinct regions of different compositions.

element a pure substance that cannot be separated into simpler substances by any chemical process.

compound a pure substance that is composed of two or more elements linked together in fixed proportions and that can be broken down to those elements by some chemical process.

law of constant composition all samples of a particular compound contain the same elements combined in the same proportions.

A **mixture** is matter composed of two or more substances that can be separated from one another by a physical process. The pure substances in a mixture are not present in definite proportions, and they retain their chemical identities. In a **homogeneous mixture**, the substances making up the mixture are distributed uniformly, and the composition and appearance of the mixture are uniform throughout. Homogeneous mixtures are also called **solutions**, a term that scientists apply to homogeneous mixtures of gases and solids as well as liquids. In contrast, the substances in a **heterogeneous mixture** can be seen as being separate from one another.

Substances are subdivided into two groups: elements and compounds. An **element** is a pure substance that cannot be broken down into simpler substances. The periodic table inside the front cover shows all the known elements. Only a few elements (such as gold, silver, nitrogen, oxygen, and sulfur) occur in nature uncombined with other elements. Most elements are found in chemical combination with other elements in the form of compounds. A **compound** is a substance that consists of two or more elements that can be separated from one another only by a chemical process. Compounds typically have properties that are very different from the elements of which they are composed.

The elements in compounds are present in characteristic and definite proportions. Water, for example, is described by the formula H_2O. The proportions of the components in pure water are always two units of the element hydrogen, H, combined with one unit of the element oxygen, O. When hydrogen and oxygen react with each other, this chemical process results in the formation of water, and the combining ratio is *always* two volumes of hydrogen gas for every one volume of oxygen gas. If we reverse the process and decompose water into hydrogen and oxygen (Figure 1.4), which is another chemical process, we always obtain two volumes of hydrogen gas for every one volume of oxygen. This consistency illustrates the **law of constant composition**: every sample of a particular compound always contains the same elements combined in the same proportions.

CONCEPT TEST

A compound with the formula NO is present in the exhaust gases leaving a car's engine. As NO travels through the car's exhaust system, some of it decomposes into nitrogen and oxygen gas. What is the volume ratio of nitrogen to oxygen formed from NO?

(Answers to Concept Tests are in the back of the book.)

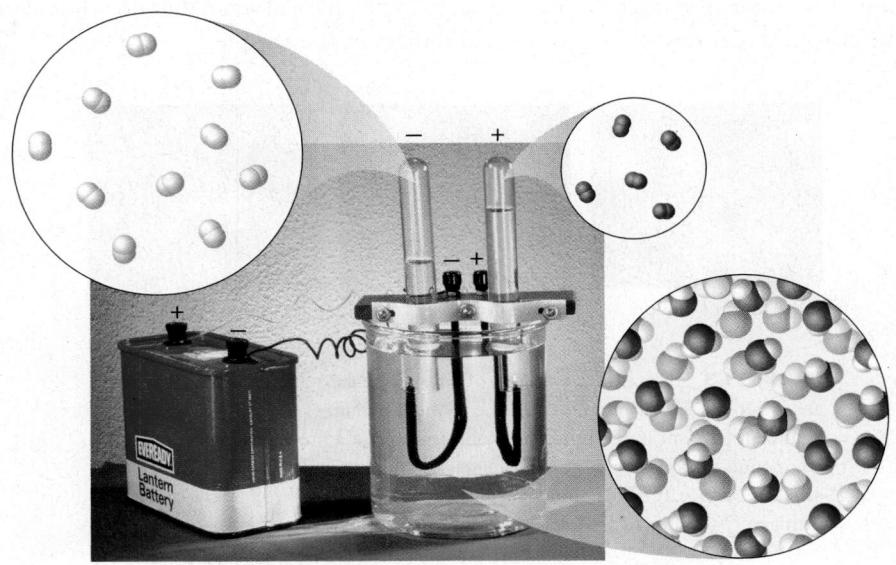

FIGURE 1.4 An electric current passed through water decomposes the water into oxygen gas and hydrogen gas. The volume ratio of the gases produced is always two volumes of hydrogen for every one volume of oxygen. A quantitative observation like this illustrates the law of constant composition.

1.2 Matter: An Atomic View

An **atom** is the smallest representative particle of an element. If, for example, you were to grind a sample made up solely of the element silicon into the finest dust imaginable, there would be a limit to how tiny a particle of the dust could be and still be silicon. That limit is an atom of silicon.

Chemists view matter and its properties on the atomic level, which is also sometimes called the microscopic level. Scientists have even developed instruments to produce images of matter at that level. For example, if the surface of a silicon wafer (Figure 1.5a), like those used to make computer chips or photovoltaic (solar) cells, is magnified over 100 million times using a device called a scanning tunneling microscope (STM), the result is an image of individual silicon atoms (Figure 1.5b).

atom the smallest particle of an element that retains the chemical characteristics of the element.

molecule a collection of atoms chemically bonded together in characteristic proportions.

chemical formula a notation for representing elements and compounds; consists of the symbols of the constituent elements and subscripts identifying the number of atoms of each element in one molecule.

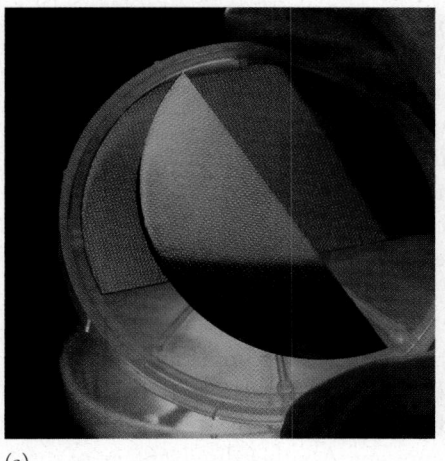
(a)

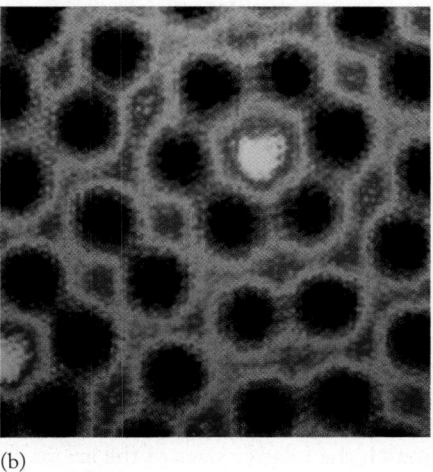
(b)

FIGURE 1.5 (a) Silicon wafers are widely used to make computer chips and photovoltaic (solar) cells. (b) Since the 1980s, scientists have been able to image individual atoms using an instrument called a scanning tunneling microscope (STM). In this STM image, the fuzzy spheres are individual silicon atoms. The radius of each atom is 117 picometers (pm), or 117 trillionths of a meter. These atoms are the tiniest particles of silicon that still retain the chemical characteristics of silicon.

Just as elements are made of particles called atoms, many compounds are made of multiatom particles called molecules. A **molecule** is a collection of atoms chemically bonded together in a characteristic pattern and proportion.

Let's take an atomic view of the combination of hydrogen and oxygen to form water. In Figure 1.6, we represent the reaction using models depicting atoms of hydrogen (the white spheres) and oxygen (the red spheres). Hydrogen and oxygen, like many other elements that are gases at room temperature, exist as *diatomic* (two-atom) molecules and therefore are represented by the chemical formulas H_2 and O_2. A **chemical formula** consists of the symbols of elements and subscripts that indicate the proportions of the elements. For example, the chemical formula H_2O tells us that every molecule of water contains two hydrogen atoms and one oxygen atom.

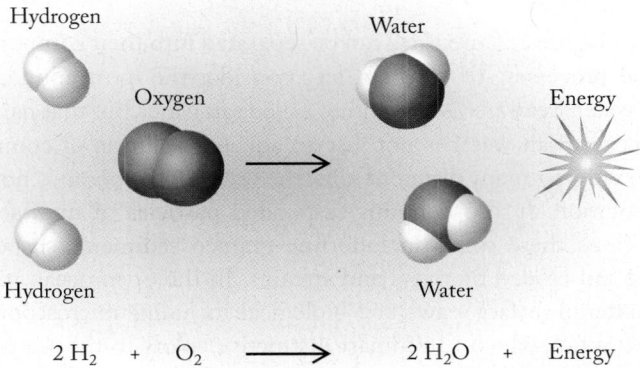

$$2\,H_2 \quad + \quad O_2 \quad \longrightarrow \quad 2\,H_2O \quad + \quad Energy$$

FIGURE 1.6 The reaction between hydrogen and oxygen is depicted here with space-filling molecular models (white and red spheres) and in the form of a chemical equation. Note that energy is also a product of the reaction.

(a) Chemical formulas: H_2O CH_3COOH

(b) Structural formulas:

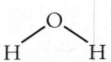 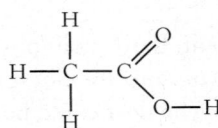

(c) Ball-and-stick models:

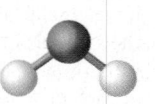 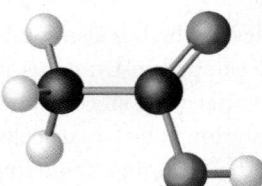

(d) Space-filling models:

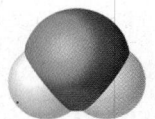

Water

Acetic acid

FIGURE 1.7 Four ways to represent the arrangement of atoms in molecules of water and acetic acid (a principal ingredient in vinegar): (a) chemical formulas; (b) structural formulas; (c) ball-and-stick models, where white spheres represent hydrogen atoms, black spheres represent carbon atoms, and red spheres represent oxygen atoms; (d) space-filling models.

FIGURE 1.8

We noted in the previous section that whenever hydrogen and oxygen combine to form water, two molecules of H_2 combine with one molecule of O_2 to form two molecules of H_2O. This relation is represented in Figure 1.6 by the molecular models and the chemical equation beneath them. In a **chemical equation**, chemical formulas represent the identities and their coefficients express the quantities of substances involved in a **chemical reaction**.

Chemical formulas, such as those shown in Figure 1.7(a), provide information about the proportions of the elements in a compound, but they do not tell us how the atoms of those elements are connected in a molecule of the compound, nor do they tell us anything about the shape of the molecules. We use structural formulas and molecular models to show atom-to-atom connections and molecular shapes. The atoms in a molecule are linked together by **chemical bonds** in a fixed arrangement, and a *structural formula* (Figure 1.7b) shows the atoms and the bonds between them; it shows how the atoms are connected but does not necessarily indicate the correct angles between bonds and the three-dimensional shape of the molecule.

Ball-and-stick models (Figure 1.7c) use spheres to represent atoms and sticks to represent chemical bonds. The advantage of ball-and-stick models is that they show the correct angles between the bonds. However, the relative sizes of the spheres don't always match the relative sizes of the atoms they represent. Another disadvantage of these models is that the atoms must be spaced far apart to accommodate the stick bonds. In real molecules, however, the atoms touch each other. Both of these disadvantages are overcome with *space-filling models* (Figure 1.7d), in which the spheres are drawn to scale and abut one another as atoms do in real molecules. The disadvantage of space-filling models is that the bond angles between atoms may be hard to see.

CONCEPT TEST ••••••••••••••••••••

Figure 1.8 is the space-filling model of formaldehyde. What is its chemical formula?

(Answers to Concept Tests are in the back of the book.)

1.3 Mixtures and How to Separate Them

As noted in Figure 1.2, mixtures can be separated into their component substances by physical processes. To see how, let's consider the most abundant mixture on Earth: seawater. Seawater is sometimes called salt water, but that name may be misleading because salt water is not just an aqueous solution of common table salt. Besides containing many different salts, the water in the ocean is not even a homogeneous solution. It also contains suspended particles of undissolved solids. In coastal regions these solids include fine-grained sediments suspended by wave action and soil eroded by rivers and streams. In the open ocean most of the suspended matter in surface seawater is biological, including microscopic plants known as phytoplankton, which can impart distinctive colors to the sea when their con-

chemical equation notation in which chemical formulas express the identities and their coefficients express the quantities of substances involved in a chemical reaction.

chemical reaction the transformation of one or more substances into different substances.

chemical bond the energy that holds two atoms in a molecule together.

(a)

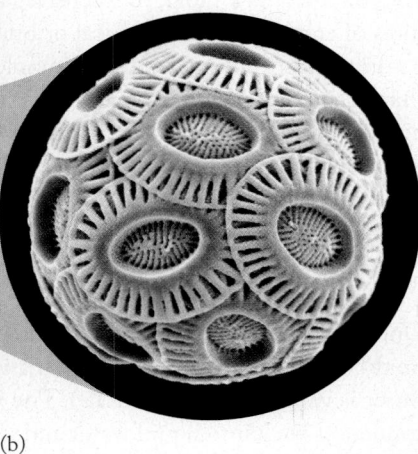

(b)

FIGURE 1.9 (a) Very high concentrations (called "blooms") of phyto-plankton known as *coccolithophores* in the Bering Sea were photographed by a NASA satellite in April 1998. During intense blooms, cocco-lithophores turn the color of the ocean a milky aquamarine. The color comes from the chlorophyll in the phytoplankton, the milkiness from sunlight scattering off the organisms. (b) Intricately designed plates (or *coccoliths*) surround each cell of the organism. The coccoliths are so tiny that 500 of them placed end to end would fit in a length of 1 mm.

centrations are unusually high (Figure 1.9). Marine scientists who study phyto-plankton can separate them from seawater by **filtration**, which involves passing the heterogeneous seawater sample through a filter that traps the phytoplankton cells but allows water and the salts dissolved in it through (Figure 1.10).

The particles caught on the filter can be further treated by a technique called sol-vent extraction. This process involves soaking the filter in a vial of solvent that dis-solves many of the compounds present inside the phytoplankton cells (Figure 1.11a).

FIGURE 1.10 Particles in suspension, such as a culture of phytoplankton in seawater, can be separated by filtration.

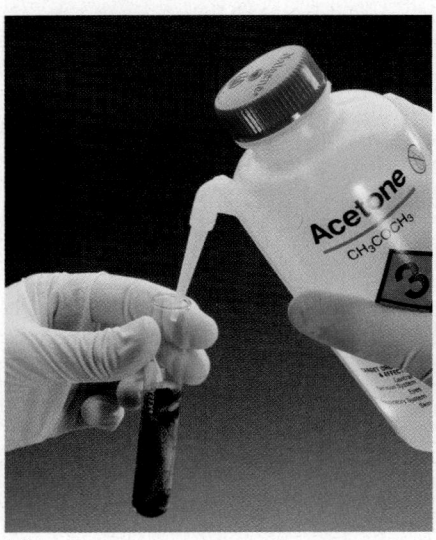

(a)

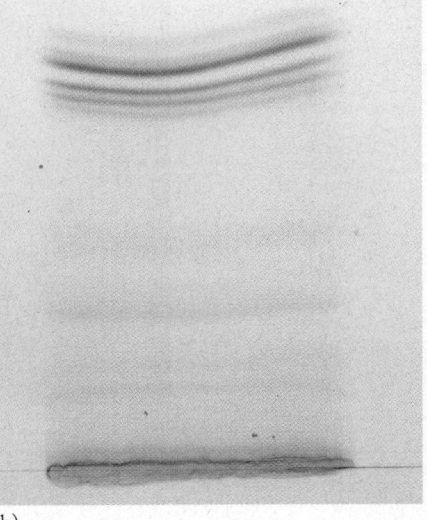

(b)

FIGURE 1.11 (a) Scientists may extract the material trapped on a filter with a solvent such as acetone that dissolves the chlorophyll and other pigments. (b) Dissolved pigments can be separated from one another using thin layer chromatography, producing a distinctive pattern.

filtration a process for separating particles suspended in a liquid or a gas from that liq-uid or gas by passing the mixture through a medium that retains the particles.

distillation a separation technique in which the more *volatile* (more easily vaporized) components of a mixture are vaporized and then condensed, thereby separating them from the less volatile components.

These compounds include chlorophyll and other pigments that have distinctive colors. Once dissolved in such a solvent, the pigments can be separated from each other using a technique called thin layer chromatography (TLC). In TLC a small volume of the pigment extract is applied to the porous surface coating a thin plate. Then the edge of the plate nearest the sample is immersed in a shallow bath of solvent. The solvent is drawn up the face of the plate by capillary action, the same way liquids wick up a paper towel. Different pigments migrate upward with the rising flow of solvent at different rates, producing a distinctive pattern as shown in Figure 1.11(b). TLC is a rapid, inexpensive, and widely used method for separating mixtures of compounds of chemical or biological interest.

Filtration is based on the principle that an object cannot pass through a pore that is smaller than the object. Air may also be filtered to remove particles suspended in it. Air filters range in size from the screens used to prevent insects from entering buildings, to HEPA (high-efficiency particulate air) filters used in clean rooms to prevent dust, bacteria, or viruses from contacting samples such as high-purity silicon chips or cells in culture that must be protected from contamination.

Seawater is unfit to drink because the concentrations of salts in it are too high. One way to render seawater drinkable is through another physical separation method—**distillation**—whereby seawater is warmed to a temperature at which water is vaporized (Figure 1.12). The water vapor contacts a cool surface where it condenses back into liquid water and is collected as purified *distillate*. Any dissolved salts and suspended particles remain behind because they are much less volatile than water.

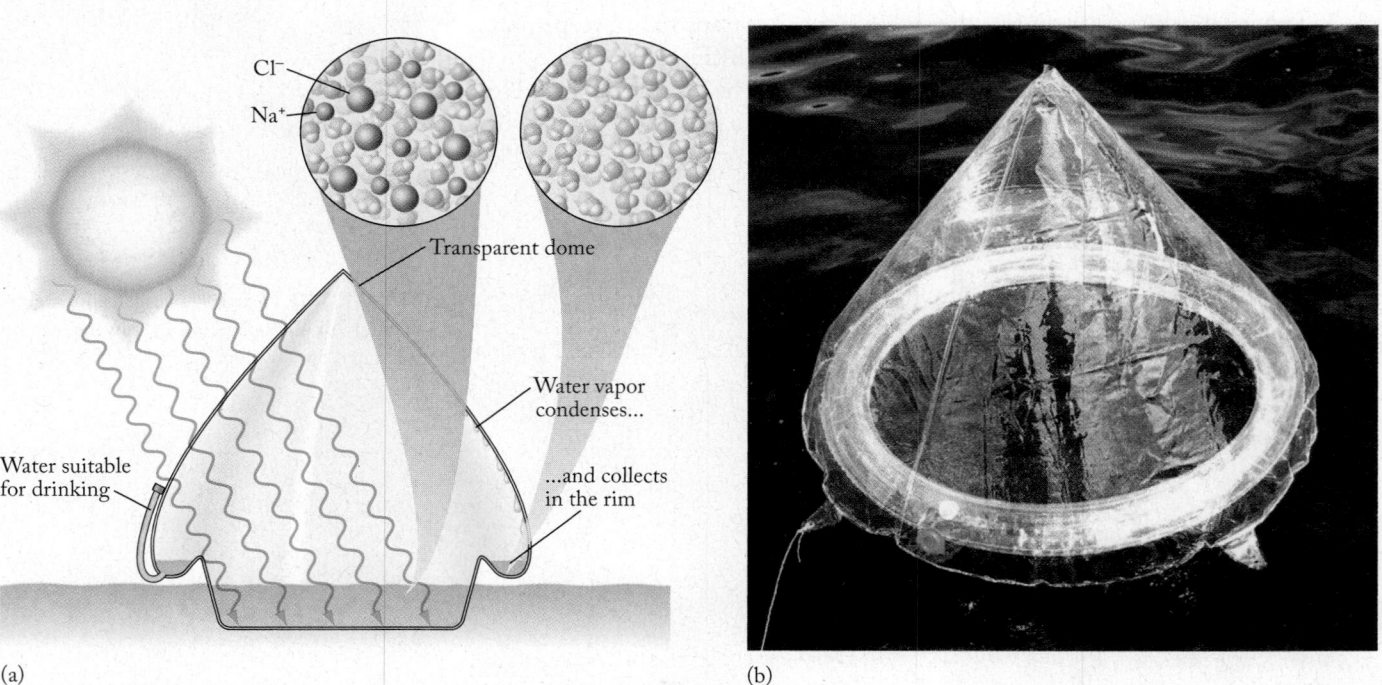

(a)

(b)

FIGURE 1.12 (a) In a solar still used in survival gear to provide freshwater from seawater, sunlight passes through the transparent dome and heats a pool of seawater. Water vapor rises from the pool, contacts the inside of the transparent dome, which is relatively cool, and condenses. The distilled water collects in the depression around the rim and then passes into the attached tube, from which one may drink it. (b) A solar still in use.

CONCEPT TEST

The whole milk that is sold in most grocery stores is labeled "homogenized." Is homogenized milk a solution? Explain why you think it is, or why you think it is not.

(Answers to Concept Tests are in the back of the book.)

1.4 Properties of Matter

Every day we use information from our senses to describe the world around us, and in doing so we categorize matter in a variety of ways. We breathe a gaseous mixture of matter called air to bring oxygen into our bodies to metabolize food. Oxygen is odorless, as is nitrogen, the main component of air on Earth, so when we smell something, we know we are inhaling some other gas along with the oxygen and nitrogen.

Pure substances have distinctive properties. Some substances burn; others put out fires. Some substances are shiny; others are dull. Some are brittle; others are malleable. These and other properties do not vary from one sample of a given pure substance to the next. Consider gold, for example: its color (very distinctive), its hardness (soft for a metal), its malleability (the fact that it can be hammered into very thin sheets called gold leaf), and its melting temperature (1064°C) all apply to any sample of pure gold. These are examples of **intensive properties**, properties that characterize matter independent of the quantity of the material present. On the other hand, an object made of gold, such as an ingot, has a particular length, width, mass, and volume. These properties of a particular sample of a pure substance, which depend on how much of the substance is present, are **extensive properties**.

CONCEPT TEST

Which of these properties of a sample of pure iron are intensive? (a) its mass; (b) its density; (c) its volume; (d) its hardness

(Answers to Concept Tests are in the back of the book.)

Properties fall into two other general categories: physical and chemical. **Physical properties** are the properties of a pure substance that can be observed or measured without changing the substance into another substance. Pure gold, for example, has a distinctive yellow color and metallic luster that we can observe without even touching it. Gold is relatively soft compared with other metals and its **density (d)**, the ratio of the mass (*m*) of a gold nugget to its volume (*V*),

$$d = \frac{m}{V} \tag{1.1}$$

is higher than the densities of other pure substances found in rocks and minerals.

As noted in Section 1.1, gold is one of the few elements found in nature uncombined with other elements. Such *free elements* may exist as single atoms, like He, or as molecules containing atoms of only the element, like O_2 and S_8. Most elements are not found free in nature but rather are combined with other elements in compounds, like hydrogen in H_2O or sodium in NaCl. As we have already

intensive property a property that is independent of the amount of substance present.

extensive property a property that varies with the quantity of the substance present.

physical property a property of a substance that can be observed without changing it into another substance.

density (d) the ratio of the mass (*m*) of an object to its volume (*V*).

chemical property a property of a substance that can be observed only by reacting it to form another substance.

discussed, hydrogen combines with oxygen in a chemical reaction that produces water and energy. This reaction is an example of combustion, and H_2 is the fuel. High flammability is a **chemical property** of hydrogen. Like any other chemical property, flammability can only be determined by reacting one substance with another substance and determining that a different material is thereby produced. A substance's chemical reactivity, including the rates of the reactions, the identities of the other reacting substances, and the identities of the products formed, all define the chemical properties of the substance.

CONCEPT TEST

Is the solubility of sugar in water a chemical or physical property of sugar? Explain your answer.

(Answers to Concept Tests are in the back of the book.)

The physical and chemical properties of a compound can be very different from those of the elements that combine to form it. Water is a liquid at room temperature, for example, whereas hydrogen and oxygen are gases at room temperature. Water expands when it freezes at 0°C; hydrogen (freezing point, −259°C) and oxygen (−219°C) do not. Oxygen supports combustion reactions and hydrogen is a highly flammable fuel, but water neither supports combustion nor is flammable. Because of its physical and chemical properties, water is widely used to put out fires.

CONCEPT TEST

Provide another example of a compound formed by the combination of two elements that has chemical and physical properties very different from the properties of either of the two elements.

(Answers to Concept Tests are in the back of the book.)

1.5 A Framework for Solving Problems

Throughout this book we include questions and problems that test your comprehension of the material. Your success in this course depends, in part, on your ability to solve these problems. Let's look at some suggested guidelines for how to do this. We follow these guidelines throughout the text whenever we develop solutions to problems.

Solving chemistry problems is like playing a musical instrument: the more you practice, the better you become. In this section we present a framework for solving problems that we follow in the Sample Exercises throughout this book. Each Sample Exercise is followed by a Practice Exercise involving the same type of problem, which can be solved using a similar approach. We strongly encourage you to hone your problem-solving skills by working all the Practice Exercises as you read the chapters. You should find the approach described here useful in solving the exercises in this book as well as problems you encounter in other courses and other contexts.

We use the acronym COAST (**C**ollect and **O**rganize, **A**nalyze, **S**olve, and **T**hink about the answer) to represent the four steps in this approach. As you read about it here and use it later, keep in mind that COAST is merely a *framework* for solving problems, not a recipe. You can use it as a guide to developing your own approach to solving problems.

Collect and Organize The first step in solving a problem is to decide how to begin. In COAST you start by collecting both the given information and your own ideas. This step is based on your understanding of the problem, including the fundamental chemical principles on which it is based. In collecting and organizing relevant information, you complete these tasks:

➤ Identify the skill(s) required to solve the problem.

➤ Identify the key concept of the problem. Identify and define the key terms used to express that concept. You may find it useful to restate the problem in your own words.

➤ Sort through the information given in the problem, separating what is pertinent from what is not.

➤ Assemble any supplemental information that may be needed, including equations, definitions, and constants.

Analyze The next step is to analyze the information you have to determine how to connect it to the answer you seek. Sometimes it is easier to work backward to create these links: Consider the nature of the answer first and think about how you might get to it from the information provided in the problem and other sources. If the problem is quantitative and requires a numerical answer, frequently the units of initial values and the final answer help you identify how they are connected and which equation(s) may be useful. This step may include rearranging equations to solve for an unknown or setting up conversion factors. For some problems, drawing a sketch based on molecular models or an experimental setup may help you visualize how the starting points and final answer are connected. You should also look at the numbers involved and estimate your answer.

Solve The solutions to most qualitative problems that test your understanding of a concept flow directly from your analysis of the problem. To solve quantitative problems, you need to insert the starting values and appropriate constants into the relevant equations or conversion factors and calculate the answer. In this step, make sure that units are consistent and cancel out as needed, and that the certainty of the quantitative information is reflected in the number of significant figures in your final answer.

Think about It Finally, you need to think about your result and answer such questions as: Does this answer make sense based on my own experience and based on what I have just learned? Is the value of a quantitative answer reasonable? Are the units correct and the number of significant figures appropriate? Then ask yourself how confident you are that you could solve another problem, perhaps drawn from another context but based on the same chemical concept. Sometimes you may also think about how this problem relates to other observations you may have made about matter in your daily life.

The COAST approach should help you solve problems in a logical way and avoid certain pitfalls, such as grabbing an equation that seems to have the right variables and plugging in numbers or resorting to trial and error. As you study the steps in each Sample Exercise, try to answer these questions about each step:

➤ **What** is done in this step?

➤ **How** is it done?

➤ **Why** is it done?

After answering these questions, you will be ready to solve the Practice Exercises and end-of-chapter Questions and Problems in a systematic way.

SAMPLE EXERCISE 1.1 **Distinguishing Physical and Chemical Properties**

Which of the following properties of gold are chemical and which are physical?
a. Gold metal, which is insoluble in water, can be made soluble by reacting it with a mixture of nitric and hydrochloric acids known as aqua regia.
b. Gold melts at 1064°C.
c. Gold can be hammered into sheets so thin that light passes through them.
d. Gold metal can be recovered from gold ore by treating the ore with a solution containing cyanide, which reacts with and dissolves gold.

Collect and Organize We are asked to determine whether the properties listed are chemical properties or physical properties. This is an exercise in classification. We need to collect the relevant definitions from the text: chemical properties describe how a substance reacts with other substances; physical properties can be observed or measured without changing one substance into another.

Analyze Properties a and d involve reactions that chemically change gold metal, which does not dissolve in water, into compounds of gold that do dissolve. Properties b and c describe processes in which elemental gold remains elemental gold. Gold, when it melts, changes its physical state, but not its chemical identity. Gold leaf is still solid, elemental gold.

Solve Properties a and d are chemical properties, and b and c are physical properties.

Think about It When possible, fall back on your experiences and observations. You know that jewelry made from gold does not dissolve in water. Therefore, dissolving gold metal requires a change in its chemical identity: it can no longer be elemental gold, but rather a soluble compound of gold. On the other hand, physical changes, such as melting, do not alter the chemical identity of the gold. Gold can be melted, poured into a mold, and then cooled to produce solid gold again.

Practice Exercise Which of the following properties of water are chemical and which are physical?
a. It normally freezes at 0.0°C.
b. It normally is useful for putting out most fires.
c. A cork floats in it, but a piece of copper sinks.
d. During digestion, starch reacts with water to form sugar.

(Answers to Practice Exercises are in the back of the book.)

solid a form of matter that has a definite shape and volume.

liquid a form of matter that occupies a definite volume but flows to assume the shape of its containers.

gas a form of matter that has neither definite volume nor shape, and that expands to fill its containers; also known as *vapor*.

sublimation transformation of a solid directly into a vapor (gas).

deposition transformation of a vapor (gas) directly into a solid.

1.6 States of Matter

Matter exists in one of three phases or physical states: solid, liquid, or gas. You are probably familiar with the characteristic properties of these states:

➤ A **solid** has a definite volume and shape.

➤ A **liquid** has a definite volume but not a definite shape.

➤ A **gas** (or *vapor*) has neither a definite volume nor a definite shape. Rather, it expands to occupy the entire volume and shape of its container.

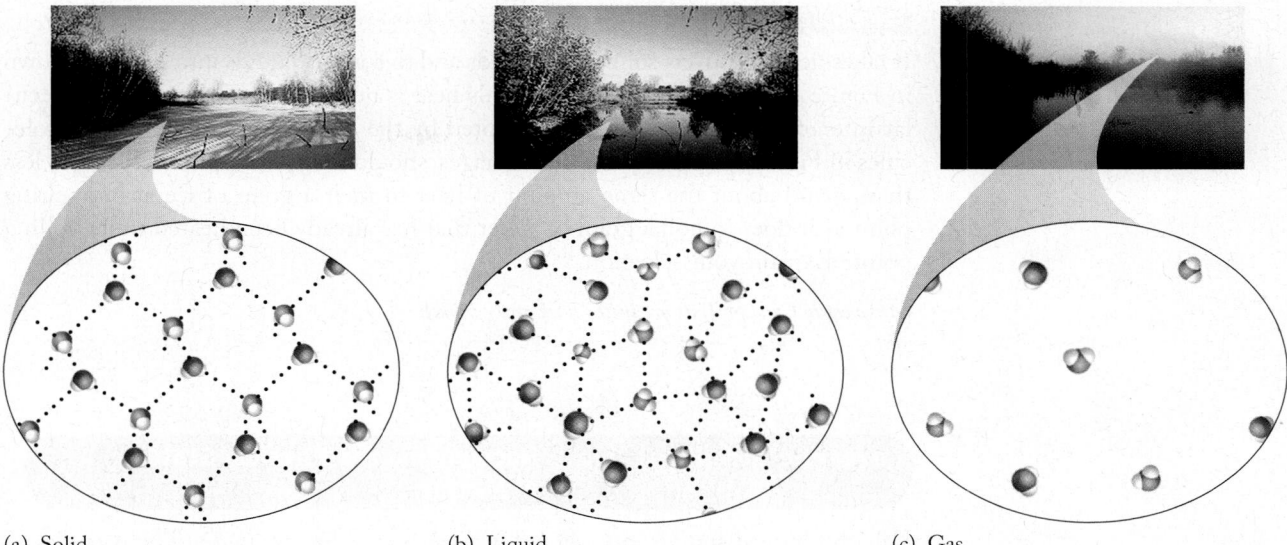

(a) Solid   (b) Liquid   (c) Gas

FIGURE 1.13 Water can exist in three states: (a) solid (ice), (b) liquid (water), or (c) gas (water vapor). (a) In the solid state, each H_2O molecule in ice is held in place by strong interactions (represented by dotted lines) in a rigid, three-dimensional array. (b) In the liquid state, the H_2O molecules are close together and still interact with each other, but they are free to flow over one another. (c) In the gas state, the H_2O molecules are far apart, largely independent of one another, and move freely. Water vapor is invisible, but we can see clouds and fog that form when atmospheric water vapor condenses to liquid H_2O.

A gas, unlike a solid or a liquid, is compressible, which means it can be squeezed into a smaller volume if its container is not rigid and if pressure is applied to it.

The differences between solids, liquids, and gases can be understood if we view the three states on the atomic level. As shown in Figure 1.13, each H_2O molecule in ice is mostly surrounded by other H_2O molecules and is locked in place in a three-dimensional array. Molecules in this structure may vibrate, but they are not free to move past the molecules that surround them; they have the same nearest neighbors over time. The H_2O molecules in liquid water are free to flow past one another, but they are still in close proximity to one another; their nearest neighbors change over time. In water vapor the H_2O molecules are widely separated, and the volume they occupy is negligible relative to the volume occupied by the vapor. This separation accounts for the compressibility of gases. The molecules in the gas phase are relatively independent and move rapidly throughout the space occupied by the vapor.

We can transform water from one physical state to another by raising or lowering its temperature. Ice forms on a pond when the temperature drops in the winter and the water freezes. This process is reversed when warmer temperatures return in the spring and the ice melts. The heat of the sun may vaporize liquid water during the daytime, but colder temperatures at night may cause the water vapor to condense as dew.

Some solids change directly into gases with no intervening liquid phase. For example, snow may be converted to water vapor on a very cold, sunny winter day even though the air temperature remains well below freezing. This transformation of solid directly to vapor is called **sublimation**. The reverse process—in which water vapor forms a layer of frost on a cold night—is an example of a gas being transformed directly into a solid without ever being a liquid, a process called **deposition**.

The physical changes that matter undergoes when it is transformed from one physical state to another are illustrated in Figure 1.14. The transformations represented by upward-pointing arrows require the addition of heat energy; those represented by downward-pointing arrows release heat.

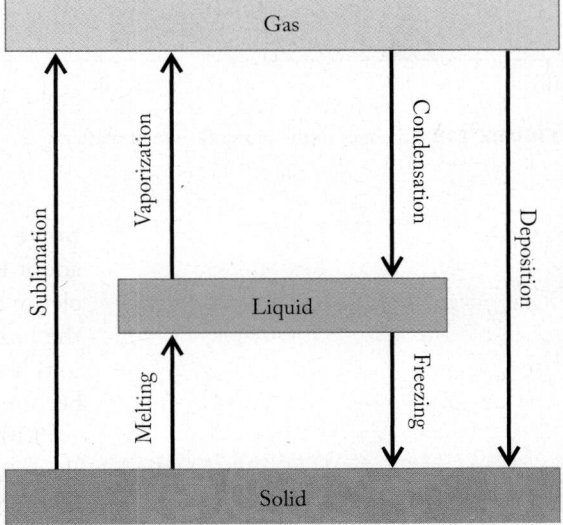

FIGURE 1.14 Matter changes from one state to another when heat is either added or removed. Arrows pointing upward represent transformations that require the addition of heat; arrows pointing downward represent transformations that release heat.

CONCEPT TEST

It takes heat to convert solids into liquids and to convert liquids into gases, as shown in Figure 1.14. In the case of water, this heat is needed to overcome intermolecular interactions such as those represented by the dotted lines between the molecules in Figure 1.13. Based on these images, should it require (a) more than, (b) less than, or (c) about the same amount of heat to melt a gram of ice at its melting point as it does to boil a gram of water that has already been heated to its boiling point? Explain your selection.

(Answers to Concept Tests are in the back of the book.)

SAMPLE EXERCISE 1.2   Recognizing the Physical States of Matter

Which physical state is represented in each box of Figure 1.15? (The particles could be atoms or molecules.) What changes of state are indicated by the two arrows? What would the changes of state be if both arrows pointed in the opposite direction?

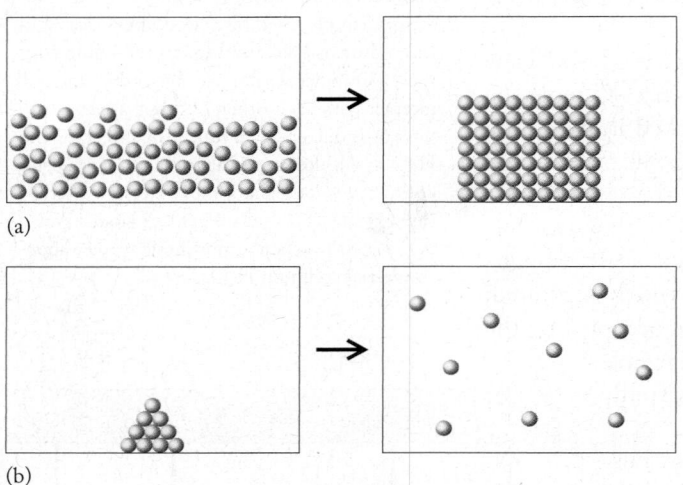

(a)

(b)

FIGURE 1.15 Changes in the physical state of matter.

Collect and Organize The particles in each box represent atoms or molecules in a solid, liquid, or gas. The first stage of this exercise involves recognizing the patterns of atomic-level particles in these different states. Once we have defined the initial and final states represented by each pair of boxes, we can name the transition using the information in Figure 1.14.

Analyze We look for the following patterns among the particles:

➤ An ordered arrangement of particles that does not fill a box or match the shape of the box represents a solid.

➤ A less-ordered arrangement of particles that partially fills a box and conforms to its shape represents a liquid.

➤ A dispersed array of particles distributed throughout a box represents a gas.

Solve The particles in the left box in Figure 1.15(a) partially fill the box and adopt the shape of the container; therefore they represent a liquid. The particles in the box on the right are ordered and form a shape different from that of the box, so they must represent a solid. The arrow represents a liquid turning into a solid, and thus the transition is freezing. The reverse process—a solid becoming a liquid—is melting. The particles in the box on the left in Figure 1.15(b) represent a solid because they are ordered and do not adopt the shape of the box. Those in the right box are dispersed throughout the box and represent a gas. The arrow represents a solid turning into a vapor, and the state change is sublimation. The reverse process—a vapor becoming a solid—is deposition.

Think about It In this exercise we first recognized patterns in the distribution of particles in different states of matter and then assigned names to the phase changes associated with transitions between different states. All three states are represented in this exercise. The following Practice Exercise further tests your ability to identify transitions between phases.

Practice Exercise (a) What physical state is represented by the particles in each box of Figure 1.16 and which change of state is represented? (b) Which change of state would be represented if the arrow pointed in the opposite direction?

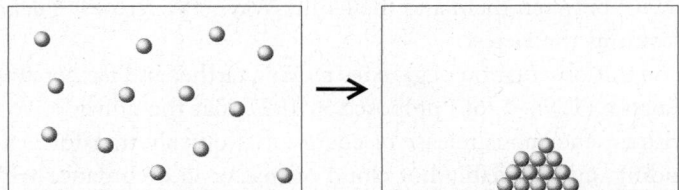

FIGURE 1.16

(Answers to Practice Exercises are in the back of the book.)

1.7 The Scientific Method: Starting Off with a Bang

In Sections 1.4 and 1.6 we examined the properties and states of matter. In this section and in Section 1.10 we examine the origins of matter and a theory of how the universe came into being.

The ancient Greeks believed that at the beginning of time there was no matter, only a vast emptiness they called Chaos, and that from that emptiness emerged the first supreme being, Gaia (also known as Mother Earth), who gave birth to Uranus (Father Sky). Other cultures and religions have described creation in similar terms of supernatural beings creating ordered worlds out of vast emptiness. The opening verses of the book of Genesis, for example, describe darkness "without form and void" from which God created the heavens and Earth. Similar stories are part of Asian, African, and Native American cultures (Figure 1.17).

Today we have the technological capacity to take a different approach to explaining how the universe was formed and what physical forces control it. Our approach is based on the **scientific method** of inquiry (Figure 1.18).

This approach evolved in the late Renaissance, during a time when economic and social stability gave people an opportunity to study nature and question old beliefs. By the early 17th century the English philosopher Francis Bacon (1561–1626) had published his *Novum Organum* (*New Organ* or *New Instrument*) in which he described how humans can acquire knowledge and understanding of the natural world through observation, experimentation, and reflection. In the process he described, observations of a natural phenomenon lead to a tentative explanation, or **hypothesis**, of what causes the phenomenon. The validity of the hypothesis is then tested through additional experimentation. One measure of the validity of a hypothesis is that it enables scientists to predict accurately the results of future experiments and observations. Further testing and observation might support a hypothesis or disprove it, or perhaps require that it be modified so that it adequately explains all of the experimental results. A hypothesis that withstands the tests of many experiments over time and explains the results of further

FIGURE 1.17 For centuries, primitive cultures have passed on creation stories in which a supreme being is responsible for creating the sun, moon, Earth, and its inhabitants. According to a creation myth of Native Americans of the Pacific Northwest, a deity in the form of a raven was responsible for releasing the sun into the sky.

scientific method an approach to acquiring knowledge based on observation of phenomena, development of a testable hypothesis, and additional experiments that test the validity of the hypothesis.

hypothesis a tentative and testable explanation for an observation or a series of observations.

scientific theory (model) a general explanation of a widely observed phenomenon that has been extensively tested and validated.

▶❙❙ **CHEMTOUR** Big Bang

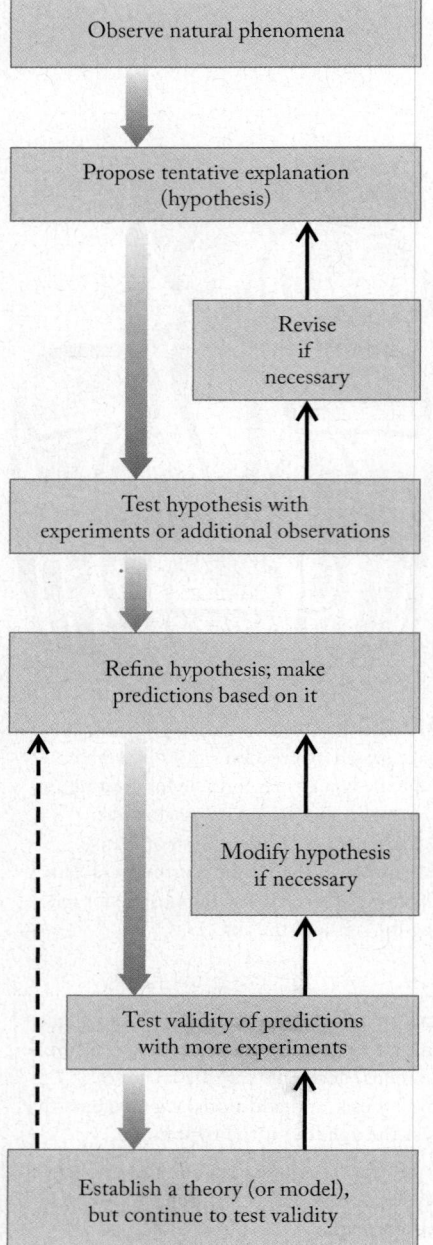

observation and experimentation may be elevated to the rank of a **scientific theory**, or **model**.

Let's examine how the scientific method works by addressing a profound question: how did the universe form? In the early 20th century, astronomers using increasingly powerful telescopes discovered that the universe contains billions of galaxies, each containing billions of stars. They also discovered that (1) the other galaxies in the universe are moving away from our own Milky Way and from one another and (2) the speeds with which the galaxies are receding are proportional to the distance between them and the Milky Way—the farthest galaxies are moving away from us the fastest.

Based on this observation of galaxies moving farther and farther away, Georges-Henri Lemaître (1894–1966) proposed in 1927 that the universe we know today formed with an enormous release of energy that quickly transformed into a rapidly expanding, unimaginably hot cloud of matter in accordance with Einstein's famous formula:

$$E = mc^2$$

where E is energy, m is the equivalent mass of matter, and c is the speed of light. To understand Lemaître's explanation of our expanding universe, consider what we would see if the motion of the matter in the universe had somehow been recorded since the beginning of time and we were able to play the recording in reverse. The other galaxies would be moving toward ours (and toward one another), with the ones farthest away closing in the fastest, catching up with those galaxies that were closer to start with but moving more slowly. Eventually, near the beginning, all the matter in the universe would approach the same point at the same time. During this compression, the density of the universe would become enormous and it would get very hot. At the very beginning, the universe would be squeezed into an infinitesimally small space of unimaginably high temperature. At such a temperature the distinction between matter and energy becomes blurred. Indeed, at the very beginning there is no matter, only energy. Now, if we play our recording in the forward direction, we see an instantaneous release of an enormous quantity of energy in an event that has come to be known as the Big Bang.

Lemaître's explanation of how the universe began has been the subject of extensive testing—and controversy—ever since it was first proposed. Indeed, the term Big Bang was first used by British astronomer Sir Fred Hoyle (1915–2001) to poke fun at the idea. However, the results of many experiments and observations conducted since the 1920s support what is now described as the Big Bang theory. We use the term *theory* to describe an explanation of natural phenomena that is succinct, comprehensive, validated by the results of extensive experimentation and observation, and that can be used to predict accurately the results of other experiments and observations. In the next section we examine a few of the experiments that have supported the validity of the Big Bang theory.

CONCEPT TEST

If the volume of the universe is expanding and if the mass of the universe is not changing, is the density of the universe (a) increasing, (b) decreasing, or (c) constant?

(Answers to Concept Tests are in the back of the book.)

FIGURE 1.18 In the scientific method, observations lead to a tentative explanation, or hypothesis, which leads to more observations and testing, which may lead to the formulation of a succinct, comprehensive explanation called a theory. The process, which repeats in a self-correcting fashion, represents a highly reasoned way to understand nature.

1.8 Making Measurements and Expressing the Results

Let's examine some of the experiments designed to test the Big Bang theory. Before we do, we need to comment on the critical importance of accurate measurements in science, whether we are exploring the nature of the cosmos or determining whether a medication contains the right quantity of a particular drug. Accurate measurements are essential to our ability to characterize many of the physical and chemical properties of matter. The rise of scientific inquiry in the 17th and 18th centuries brought about a heightened awareness of the need for accurate measurements and of the need for expressing those measurements in ways that were understandable to others. Standardization of the units of measurement was essential.

SI Units

In 1791 French scientists proposed a standard unit of length, which they called the **meter** (m), after the Greek *metron,* which means "measure." They based the length of the meter on 1/10,000,000 of the distance along an imaginary line running from the North Pole to the equator. By 1794 hard work by teams of surveyors had established the length of the meter that is still in use today.

These French scientists also settled on a decimal-based system for designating lengths that are multiples or fractions of a meter (Table 1.1). They chose Greek prefixes for lengths much greater than 1 m, such as *deka-* and *kilo-* for lengths of 10 meters (1 dekameter) and 1000 m (1 kilometer), and Latin prefixes for lengths

TABLE 1.1   Commonly Used Prefixes for SI Units

PREFIX		VALUE	
Name	Symbol	Numerical	Exponential
zetta	Z	1,000,000,000,000,000,000,000	10^{21}
exa	E	1,000,000,000,000,000,000	10^{18}
peta	P	1,000,000,000,000,000	10^{15}
tera	T	1,000,000,000,000	10^{12}
giga	G	1,000,000,000	10^{9}
mega	M	1,000,000	10^{6}
kilo	k	1,000	10^{3}
hecto	h	100	10^{2}
deka	da	10	10^{1}
deci	d	0.1	10^{-1}
centi	c	0.01	10^{-2}
milli	m	0.001	10^{-3}
micro	μ	0.000001	10^{-6}
nano	n	0.000000001	10^{-9}
pico	p	0.000000000001	10^{-12}
femto	f	0.000000000000001	10^{-15}
atto	a	0.000000000000000001	10^{-18}
zepto	z	0.000000000000000000001	10^{-21}

meter the standard unit of length, named after the Greek *metron,* which means "measure," and equivalent to 39.37 inches.

TABLE 1.2   SI Base Units

Quantity or Dimension	Unit Name	Unit Abbreviation
Mass	kilogram	kg
Length	meter	m
Temperature	kelvin	K
Time	second	s
Electric current	ampere	A
Amount of a substance	mole	mol
Luminosity	candela	cd

much smaller than a meter, such as *centi-* and *milli-* for lengths of 1/100 of a meter (1 centimeter) and 1/1000 of a meter (1 millimeter).

Eventually these decimal prefixes carried over to the names of standard units for other dimensions. In July 1799 a platinum (Pt) rod 1 meter long and a platinum block having a mass of 1 kilogram (1000 grams) were placed in the French National Archives to serve as the legal standards for length and mass. These objects served as references for the metric system for expressing measured quantities.

Since 1960 scientists have, by international agreement, used a modern version of the French metric system: the *Système International d'Unités,* commonly abbreviated SI. There are seven *base SI units* (Table 1.2), and all other SI units are derived from them. For example, a common SI unit for volume, the cubic meter (m^3), is derived from the length base unit, the meter, and a common SI unit for speed, meters per second (m/s), is derived from the length and time base units. Table 1.3 contains some of these derived units and their equivalents in the U.S. Customary System of units. They include the volume corresponding to 1 cubic decimeter (a cube 1/10 meter on a side), which we call a liter (L).

Modern science requires that the length of the meter, as well as the dimensions of other SI units, be known or defined by quantities that are much more constant than the length of a platinum rod in Paris. Two such quantities are the speed of light (*c*) and time. In 1983 1 m was redefined as the distance traveled in 1/299,792,458 of a second by the light emitted from a helium–neon laser. This modern definition of the meter is consistent with the one adopted in France in 1794.

TABLE 1.3   Conversion Factors for SI and Other Commonly Used Units

Quantity or Dimension	Equivalent Units
Mass	1 kg = 2.205 pounds (lb); 1 lb = 0.4536 kg = 453.6 g 1 g = 0.03527 ounce (oz); 1 oz = 28.35 g
Length (distance)	1 m = 1.094 yards (yd); 1 yd = 0.9144 m (exactly) 1 m = 39.37 inches (in); 1 foot (ft) = 0.3048 m (exactly) 1 in = 2.54 cm (exactly) 1 km = 0.6214 miles (mi); 1 mi = 1.609 km
Volume	$1\ m^3 = 35.31\ ft^3$; $1\ ft^3 = 0.02832\ m^3$ $1\ m^3 = 1000$ liters (L) (exactly) 1 L = 0.2642 gallon (gal); 1 gal = 3.785 L 1 L = 1.057 quarts (qt); 1 qt = 0.9464 L

Significant Figures

All scientific measurements have one thing in common: there is a limit to how accurate they can be. Nobody is perfect, and no analytical method is perfect either. Every method has an inherent limit in its capacity to produce accurate results, and we need ways to express experimental results that reflect these limits. We do so by expressing numerical data from experiments to the appropriate number of **significant figures**.

The number of significant figures reported in a measured value or reported in a calculation based on one or more measured values indicates how certain we are of the measured value or values. For example, suppose we determine the mass of a penny on the two balances shown in Figure 1.19, with both balances working properly. The balance on the right determines the mass of objects to the nearest 0.0001 gram (g); the balance on the left determines the mass of objects only to the nearest 0.01 g. According to the balance on the left, our penny has a mass of 2.53 g; according to the balance on the right, it has a mass of 2.5271 g. The mass obtained with the left balance has three significant figures: the 2, 5, and 3 are considered *significant*, which means we are confident in their values. The mass obtained using the right balance has five significant figures (2, 5, 2, 7, and 1). We may conclude that the mass of the penny can be determined with greater certainty with the balance on the right.

Now suppose a puff of air blows across the tops of both balances as someone walks by. The value displayed on the left balance will probably not be affected, but the balance on the right is so sensitive that there will likely be a change in the last digit of the displayed value. This change illustrates an important point: for many measured values there is some uncertainty in the rightmost digit. However, the last measurable digit is considered significant even though we are less certain of its value than we are of the value of the other digits. The significant figures in a number include all the digits we know with certainty plus one digit that is uncertain.

Now consider this experimental result: an aspirin tablet is placed on the balance on the right in Figure 1.19, and the display reads 0.0810 g. How many significant figures are there in this value? You might be tempted to say five because that is the number of digits displayed. However, the first two zeros are not considered significant because they serve only to determine the location of the decimal point. These two zeros function the way exponents do when we express values using scientific notation. Expressing 0.0180 g using scientific notation gives us 1.80×10^{-2} g. Only the three digits in the decimal part indicate how precisely we know the value; the "−2" in the exponent does not. (See Appendix 1 for a review of how to express values using scientific notation.)

Why is the rightmost zero in 0.0810 g significant? The answer is related to the ability of the balance to measure masses to the nearest 0.0001 g. If the balance is operating correctly, we may assume that the mass of the tablet is 0.0810 g, not 0.0811 g or 0.0809 g. We may assume that the last digit really is zero, and we need a way to express that. If we dropped the zero and recorded a value of only 0.081 g, we would be implying that we knew the value to only the nearest 0.001 g, which is not the case. The following guidelines will help you handle zeros (highlighted in green) in deciding the number of significant figures in a value:

1. Zeros at the beginning of a value, as in **0.0**592, are never significant. In this example, they just set the decimal place.
2. Zeros at the end of a value and after a decimal point, as in $3.\mathbf{00} \times 10^8$, are always significant.

significant figures all the certain digits in a measured value plus one estimated digit. The greater the number of significant figures, the greater the certainty with which the value is known.

FIGURE 1.19 The mass of a penny can be measured to the nearest 0.01 gram with the balance on the left and to the nearest 0.0001 gram with the balance on the right.

▶II CHEMTOUR Significant Figures

▶II CHEMTOUR Scientific Notation

3. Zeros at the end of a value that contains no decimal point, as in 96,500, may or may not be significant. They may be there only to set the decimal place. We should use scientific notation to avoid this ambiguity. Here, 9.65×10^4 (three significant figures) and 9.6500×10^4 (five significant figures) are two of the possible interpretations of 96,500.
4. Zeros between nonzero digits, as in 101.3, are always significant.

CONCEPT TEST ....

How many significant figures are there in the values used as examples in guidelines 1, 2, and 4 above?

(Answers to Concept Tests are in the back of the book.)
....

Significant Figures in Calculations

Now let's consider how significant figures are used to express the results of calculations involving measured quantities. An important rule to remember is that significant-figure rules should be used only at *the end of a calculation*, never on intermediate results. A reasonable guideline to follow is that at least one digit to the right of the last significant digit should be carried forward in all intermediate steps.

Suppose we believe that a small nugget of yellow metal is pure gold. We could test our belief by determining the mass and volume of the nugget and then calculating its density. If its density matches that of gold (19.3 g/mL), chances are good that the nugget is pure gold because few minerals are that dense. We find that the mass of the nugget is 4.72 g and its volume is 0.25 mL. What is the density of the nugget, expressed in the appropriate number of significant figures?

Using Equation 1.1 to calculate the density (d) from the mass (m) and volume (V) produces the following result:

$$d = \frac{m}{V} = \frac{4.72 \text{ g}}{0.25 \text{ mL}} = 18.88 \text{ g/mL}$$

This density value appears to be slightly less than that of pure gold. However, we need to answer the question, "How well do we know the result?" The mass value is known to three significant figures, but the volume value is known only to two. At this point we need to invoke the *weak-link principle*, which is based on the idea that a chain is only as strong as its weakest link. In calculations involving measured values, this principle means that we can know the answer of a calculation only as well as we know the least well-known value used in the calculation. In calculations involving multiplication or division the weak link is the value with the fewest significant figures. In this example the weak link is the value of the volume, because it has only two significant figures. Our final answer cannot have more than two significant figures, so we must convert 18.88 to a number with two significant figures. We do this by a process called rounding off.

Rounding off a value means dropping the *insignificant digits* (all digits to the right of the first uncertain digit) and then rounding that first uncertain digit either up or down. If the first digit in the string of insignificant digits dropped is greater than 5, we round up; if it is less than 5, we round down. The question remains of how to handle cases in which the first dropped digit is 5. If there are nonzero digits to the right of the 5, then round up. If there are no nonzero digits to the right of the 5, then a good rule to follow is to round to the nearest even number. For example, rounding 45.450001 to three significant figures makes it 45.5 (we rounded up because there was a nonzero digit to the right of the second 5). Rounding 45.45 or 45.450 to three significant figures makes either value 45.4 because the 4 in the tenths place is the nearest even number. However, we round off 45.55 or 45.550 to 45.6 because the 6 in the tenths place is the nearest even digit.

In the case of our gold nugget, the density value resulting from our calculation, 18.88, must be rounded to two significant figures because the weakest-link value in our calculation, 0.25 mL, has two significant figures. Because the first insignificant digit to be dropped (shown in red) is greater than 5, we round up the blue 8 to 9:

$$18.88 \text{ g/mL} = 19 \text{ g/mL}$$

The density of pure gold expressed to two significant figures is also 19 g/mL, so based on how well we know the measured values, we conclude that the nugget could indeed be pure gold.

CONCEPT TEST ....

Does our measured density value prove that the nugget is pure gold, or does it simply fail to prove that the nugget is not pure gold? Explain the difference between these two conclusions, and explain why you prefer one over the other.

(Answers to Concept Tests are in the back of the book.)
....

The weak-link principle for significant figures also applies to calculations requiring addition and subtraction. To illustrate, consider how the volume of a gold nugget might be determined. One approach is to measure the volume of water the nugget displaces. Suppose we add 50.0 mL of water to a 100 mL graduated cylinder as shown in Figure 1.20. Note that the cylinder is graduated in milliliters. However, the space between the graduations allows us to estimate the volumes of samples to the nearest tenth of a milliliter. For example, the meniscus of the water in Figure 1.20 is aligned exactly with the 50 mL graduation. We can correctly record the volume of water as 50.0 mL because we can read the "50" part of this value directly from the cylinder, and we can estimate that the next digit is ".0".

Then we place the nugget in the graduated cylinder, being careful not to splash any water out. The level of the water is now about halfway between the 58 and 59 mL graduations and so we estimate the volume of the combined sample is 58.5 mL. Taking the difference of the two estimated values we have the volume of our nugget:

$$\begin{array}{r} 58.5 \text{ mL} \\ -50.0 \text{ mL} \\ \hline 8.5 \text{ mL} \end{array}$$

The result has only two significant figures for this reason: because the initial and final volumes are known to the nearest tenth of a milliliter, we can know the difference between them to only the nearest tenth of a milliliter. When measured numbers are added or subtracted, the result has the same number of digits to the right of the decimal as the measured number with the fewest digits to the right of the decimal.

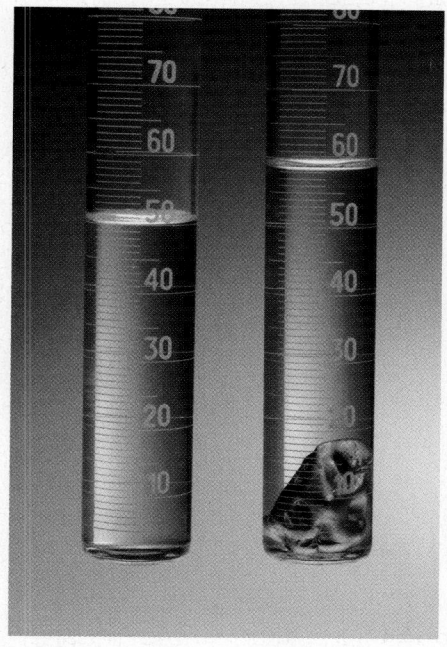

FIGURE 1.20 A nugget believed to be pure gold is placed in a graduated cylinder containing 50.0 mL of water. The volume rises to 58.5 mL, which means that the volume of the nugget is 8.5 mL.

Consider one more example. Suppose the U.S. Mint defines the average mass of a penny to be 2.53 g and stipulates that a roll of pennies must contain 50 pennies. We use the balance on the left in Figure 1.19 to determine the mass of a roll of pennies. Using the tare feature, which enables us to correct for the mass of the wrapper, we find that the mass of just the pennies in the roll is 124.01 g. Dividing this value by the average mass of a penny and canceling out the grams units in the numerator and denominator, we get

$$\frac{124.01 \text{ g}}{2.53 \text{ g/penny}} = 49.0 \text{ pennies}$$

If we were measuring uncountable substances, rounding off to 49.0 would be appropriate (because under these circumstances, the measured value of 2.53 g/penny would be the weak link in the calculation). However, because we use only whole numbers to count items, we round to 49 and conclude that the roll contains 49 pennies. That means we have to add one penny to make the roll complete.

Suppose the penny we add is one we know has a mass of 2.5271 g because we determined its mass on the balance on the right in Figure 1.19, which determines masses to the nearest 0.0001 g. What is the mass of all 50 pennies to the appropriate number of significant figures? Adding the mass of the 50th penny to the mass of the other 49, we have

$$\begin{array}{r} 124.01 \text{ g} \\ + \quad 2.5271 \\ \hline 126.5371 \text{ g} \end{array}$$

The mass of the first 49 pennies is known to two decimal places, and the mass of the 50th is known to four decimal places. Therefore, by the weak-link principle we can know the sum of the two numbers to only two decimal places. This makes the last two digits in the sum, shown in red, not significant, so we must round off the mass to 126.54 g.

SAMPLE EXERCISE 1.3   **Using Significant Figures in Calculations**

A nugget of a shiny yellow mineral has a mass of 30.01 g. Its volume is determined by placing it in a 100 mL graduated cylinder containing 56.3 mL of water. The volume after the nugget is added is 62.6 mL. Is the nugget made of gold?

Collect and Organize We have seen in this section how knowing the mass and calculating the volume of an irregularly shaped object by water displacement can be used to calculate the object's density and to help identify what it's made of. In this exercise we are given the mass of a shiny yellow nugget, and we are given water displacement data that can be used to calculate its volume. We know from Equation 1.1 that the density of an object is its mass divided by its volume. We know from data given in this section that the density of gold is 19.3 g/mL.

Analyze We can use Equation 1.1 to calculate the density of the nugget from its mass (30.01 g) and its volume: the amount of water displaced, which is the difference between 62.6 and 56.3 mL. We need to express the result of this calculation using the weak-link rule for determining the number of significant figures in a value calculated from experimental data. Then we compare our result with the known density of gold to determine whether the nugget could be gold.

Solve

$$d = \frac{m}{V}$$

$$= \frac{30.01 \text{ g}}{(62.6 - 56.3) \text{ mL}} = \frac{30.01 \text{ g}}{6.3 \text{ mL}} = \frac{4.7635 \text{ g}}{\text{mL}}$$

Because we know the volume of the nugget (6.3 mL) to two significant figures, we can know its density to only two significant figures, so we must round off the result to 4.8 g/mL. The density of gold is 19.3 g/mL, so the nugget cannot be pure gold.

Think about It As in the problem involving gold in the text, the weak link in this exercise is the volume of the sample calculated from water displacement. Balances such as the one on the right in Figure 1.19 are capable of determining masses of objects weighing tens of grams to the nearest tenth of a milligram. This means that the results of these mass measurements can have as many as six significant figures, which is many more than can be obtained by the common methods for determining the volumes of irregularly shaped samples.

Practice Exercise Express the result of the following calculation to the appropriate number of significant figures:

$$\frac{(0.391)(0.0821)(273 + 25)}{8.401}$$

(Answers to Practice Exercises are in the back of the book.)

CONCEPT TEST

Lap speed in stock car racing is calculated by dividing the distance around a track (Figure 1.21) by the time it takes a car to go around the track. If lap time is measured to the nearest hundredth of a second, which dimension, distance, or time is probably the weak link in the calculation? Explain your answer.

(Answers to Concept Tests are in the back of the book.)

Measurements always have some degree of uncertainty, which limits the number of significant figures we can use to report any measurement. On the other hand, some values are known exactly, such as 12 eggs in a dozen and 60 seconds in a minute. Because these values are definitions, there is no uncertainty in them and they are not considered when determining significant figures in the answer to a mathematical calculation in which they appear.

FIGURE 1.21 Dirt track for stock car racing.

SAMPLE EXERCISE 1.4   Distinguishing Exact from Uncertain Values

Which of the following numerical values associated with the Washington Monument in Washington, DC, are exact numbers and which are not exact? (a) the monument is made of 36,491 white marble blocks; (b) the monument is 169 m tall; (c) there are 893 steps to the top; (d) the mass of the aluminum capstone is 2.8 kg; (e) the area of the foundation is 1487 m^2

Collect and Organize This exercise involves distinguishing between values based on an exact number, such as 12 eggs in a dozen, from those that are not exact numbers, such as the mass of an egg.

Analyze One way to distinguish exact from inexact values is to answer the question, "Which values represent quantities that can be counted?" Values that can be counted are exact.

Solve (a) The number of marble blocks and (c) the number of stairs are quantities we can count, so they are exact numbers. The other three quantities are based on measurements of (b) length, (d) mass, and (e) area and therefore are not exact.

Think about It The "you can count them" property of exact numbers worked in this exercise and will work in others even when doing so may take a long time (the number of hairs on a dog) or if help is needed in visualizing the objects (the number of cilia on a paramecium). Actually, these examples bring up a point we have not discussed: might there be uncertainty in the counting process itself? Can you think of a famous example in American politics of an uncertain counting process?

Practice Exercise Which of the following statistics associated with the Golden Gate Bridge in San Francisco, CA, are exact numbers and which have some inherent uncertainty? (a) the roadway is six lanes wide; (b) the width of the bridge is 27.4 m; (c) the bridge has a mass of 3.808×10^8 kg; (d) the length of the bridge is 2740 m; (e) the toll for a car traveling south increased to $6.00 on September 1, 2008

(Answers to Practice Exercises are in the back of the book.)

■ ....

Precision and Accuracy

Two terms—precision and accuracy—are used to describe how well a measured quantity or a value calculated from a measured quantity is known. **Precision** indicates how repeatable a measurement is. Suppose we used the balance on the left in Figure 1.19 to determine the mass of a penny over and over again, and suppose the reading on the balance was always 2.53 g. These results tell us that our determinations of the mass of the penny were precisely 2.53 g. Said another way, the results were precise to the nearest 0.01 g.

Now suppose we use the balance on the right in Figure 1.19 to determine the mass of the same penny five more times and obtain these values:

Measurement	Mass (g)
1	2.5270
2	2.5271
3	2.5272
4	2.5271
5	2.5271

precision the extent to which repeated measurements of the same variable agree.

Note that there is a small variability in the last decimal place. Such variability is not unusual when using a balance that can report masses to the nearest 0.0001 g.

These results are quite consistent with one another, so we can say that the balance is precise. In addition to air moving over the balance, many other factors—particles of dust landing on the balance, vibration of the laboratory bench, or the transfer of moisture from our fingers to the penny as we handle it—could all produce a change in mass of 0.0001 g or more.

One way to express the precision of these results is to cite the range between the highest and lowest values, in this example, 2.5270 to 2.5272. Range can also be expressed using the average value (2.5271 g) and the range above (0.0001) and below (0.0001) the average that includes all the observed results. A convenient way to express the observed range in this case is 2.5271 ± 0.0001, where the symbol ± means "plus or minus" the value that follows it.

While precision relates to the agreement among repeated measurements, **accuracy** reflects how close the measured value is to the true value. Suppose the true mass of our penny is 2.5267 g. That means the average result obtained with the balance on the right in Figure 1.19—2.5271 g—is 0.0004 g too high. Thus the measurements made on this balance may be precise to within 0.0001 g, but they are not accurate to within 0.0001 g of the true value. A way to visualize the difference between accuracy and precision is presented in Figure 1.22.

How can we be sure that the results of a measurement are accurate? The accuracy of a balance can be checked by determining the mass of objects of known mass. A thermometer can be calibrated by measuring the temperature at which a substance changes state. Ice normally melts to liquid water at 0.0°C, for instance, and an accurate thermometer dipped into a mixture of ice and liquid water reads 0.0°C. At sea level, liquid water boils and changes to water vapor at 100.0°C, which means an accurate thermometer dipped into the boiling water reads that temperature. A measurement that is validated by calibration with an accepted standard material is considered accurate.

1.9 Unit Conversions and Dimensional Analysis

Throughout this book we frequently must convert quantities from one unit to another. Sometimes it is simply a matter of using the appropriate prefix, such as expressing the distance between Toronto and Montreal in kilometers instead of meters. Other times it is a matter of converting a value from one unit system to another, such as converting a gasoline price per liter to an equivalent price per U.S. gallon.

To do these calculations and many others, we use an approach sometimes called the *unit factor method* but more often called *dimensional analysis*. The approach makes use of **conversion factors**, which are fractions in which the numerators and denominators have different units but represent equivalent quantities. This equivalency means that multiplying a quantity by a conversion factor is like multiplying by 1: the intrinsic value that the quantity represents does not change; it is simply expressed in different units.

The key to using conversion factors correctly is to set them up correctly, with the appropriate units in the numerator and denominator. For example, the odometer of a Canadian car traveling between the airports serving Vancouver, British Columbia and Seattle, Washington records the distance as 241 km. What is that distance in miles? To find out we need to translate the following equivalency from Table 1.3 (page 20):

$$1 \text{ km} = 0.6214 \text{ mi}$$

(a)

(b)

(c)

FIGURE 1.22 (a) Three dart throws meant to hit the center of the target are both accurate and precise. (b) The three throws meant to hit the center of the target are precise but not accurate. (c) This set of three throws is neither precise nor accurate.

accuracy agreement between an experimental value and the true value.

conversion factor a fraction in which the numerator is equivalent to the denominator but is expressed in different units, making the value of the fraction one.

into this conversion factor:

$$\frac{0.6214 \text{ mi}}{1 \text{ km}}$$

When we multiply the distance in kilometers by this conversion factor, the km in the initial value and in the denominator of the conversion factor cancel each other out, and the answer has the desired units of miles:

$$241 \, \cancel{\text{km}} \times \frac{0.6214 \text{ mi}}{1 \, \cancel{\text{km}}} = 150 \text{ mi} = 1.50 \times 10^2 \text{ mi}$$

The general form of the equation for converting a value from one unit to another is

$$\cancel{\text{initial units}} \times \frac{\text{desired units}}{\cancel{\text{initial units}}} = \text{desired units}$$

Most conversion factors such as the one for converting kilometers to miles also have a "1" in the numerator or denominator. All of these ones are *exactly* one; there is no uncertainty in their value.

▶❙❙ **CHEMTOUR** Dimensional Analysis

SAMPLE EXERCISE 1.5   Converting One Unit to Another

The Star of Africa (Figure 1.23) is one of the world's largest diamonds, having a mass of 106.04 g. What is its mass in milligrams and in kilograms?

Collect and Organize We must convert a mass expressed in grams (g) to milligrams (mg) and to kilograms (kg). From Table 1.1, we know that the prefixes *milli* and *kilo-* represent units that are 1/1000 and 1000 times the value of the given unit "grams."

Analyze We need two conversion factors, one for converting grams to milligrams and one for converting grams to kilograms. Table 1.1 tells us that 1 mg = 0.001 g and 1 kg = 1000 g. Therefore our possibilities for conversion factors are

$$\frac{1 \text{ mg}}{0.001 \text{ g}} \qquad \frac{0.001 \text{ g}}{1 \text{ mg}} \qquad \frac{1 \text{ kg}}{1000 \text{ g}} \qquad \frac{1000 \text{ g}}{1 \text{ kg}}$$

Because we want factors showing our desired units in the numerator, we use the first and third possibilities.

Solve We multiply the given mass by each conversion factor to obtain the desired results:

$$106.04 \, \cancel{\text{g}} \times \frac{1 \text{ mg}}{0.001 \, \cancel{\text{g}}} = 106,040 \text{ mg} \quad 106.04 \, \cancel{\text{g}} \times \frac{1 \text{ kg}}{1000 \, \cancel{\text{g}}} = 0.10604 \text{ kg}$$

Think about It The results make sense because there should be many more of the smaller units (106,040 mg versus 106.04 g) in a given mass and many fewer of the larger units (0.10604 kg versus 106.04 g).

Practice Exercise The Eiffel Tower in Paris is 324 m tall, including a 24 m television antenna that was not there when the tower was built in 1889. What is the height of the Eiffel Tower in kilometers and in centimeters?

(Answers to Practice Exercises are in the back of the book.)

FIGURE 1.23 The Star of Africa is one of the largest cut diamonds in the world. It is mounted in the handle of the Royal Sceptre in the British crown jewels displayed in the Tower of London.

SAMPLE EXERCISE 1.6   Converting Customary U.S. Units to SI Units

The summit of Mount Washington in New Hampshire is famous for its bad weather: on average it experiences hurricane-force winds 110 days per year (Figure 1.24). Remarkably, the summit is only 6288 feet above sea level. What is this altitude in meters?

Collect and Organize We are asked to convert the height of a mountain from units of feet to meters. The fact that the summit is a very windy place may be interesting, but it is irrelevant to the altitude calculation. Table 1.3 contains conversion factors for converting lengths from customary U.S. units to SI units.

Analyze The conversion factor from Table 1.3 with the initial unit in the denominator is

$$\frac{0.3048 \text{ m}}{1 \text{ ft}}$$

Because 1 meter is larger than 1 foot, we expect our final answer to be smaller than the number of feet. Because 1 foot is about one-third of a meter, we can estimate that our answer should be about one-third of 6000, or 2000 m.

Solve Multiplying the summit height in feet by this conversion factor, we get

$$6288 \text{ ft} \times \frac{0.3048 \text{ m}}{1 \text{ ft}} = 1916.5824 \text{ m}$$

Rounding off the answer to four significant figures, we have 1917 m.

Think about It Our answer is reasonable based on our estimate. The answer is rounded to four significant figures because the initial value has four significant figures.

Practice Exercise Perhaps the most famous horse race ever run in the United States was the 1938 race between two thoroughbreds, War Admiral and Seabiscuit, both of whom stood about 15 hands high. Seabiscuit won the race. The length unit *hand*, used to measure horses, is exactly 4 inches. How tall were both horses in centimeters?

(Answers to Practice Exercises are in the back of the book.)

FIGURE 1.24 A warning to Mount Washington hikers from the U.S. Forest Service.

SAMPLE EXERCISE 1.7   U.S. Customary-to-SI Unit Conversions with Multiple Steps

On April 12, 1934, a wind gust of 231 miles per hour was recorded at the summit of Mount Washington in New Hampshire. What is this wind speed in meters per second?

Collect and Organize Speed is expressed as a distance unit divided by a time unit. In this conversion, both the given distance unit and the given time unit must change. Table 1.3 contains the equivalence 1 km = 0.6214 mi. The desired unit of distance is meters, so we also need to convert kilometers to meters (1 km = 1000 m, Table 1.1). Finally, we need to convert hours to seconds.

Analyze The conversion factors we need are

$$\frac{1\ km}{0.6214\ mi} \qquad \frac{1000\ m}{1\ km} \qquad \frac{1\ hr}{60\ min} \qquad \frac{1\ min}{60\ s}$$

The first conversion factor for distance is written with the given unit in the denominator; in the second conversion factor, the unit in the denominator is the result of the first conversion, and the desired unit is (as usual) in the numerator. This same relation between units is also used in the conversion factors for time. Hour is the unit in the denominator of the initial value. This means that for hour to cancel it must appear in the numerator of a conversion factor for time. To estimate our answer, note that 1000 m in our numerator is divided by about $2/3 \times 60 \times 60$ or about 2400, so we expect our answer in m/s to be a little less than half of the value of the speed in mi/hr, or about 100 m/s.

Solve

$$231\ \frac{mi}{hr} \times \frac{1\ km}{0.6214\ mi} \times \frac{1000\ m}{1\ km} \times \frac{1\ hr}{60\ min} \times \frac{1\ min}{60\ s} = 103\ \frac{m}{s}$$

Think about It The answer seems very reasonable based on our estimate. We rounded the result to three significant figures because that is the number of significant figures in 231 mi/hr.

Practice Exercise If light travels exactly 1 meter in 1/299,792,458 of a second, how many kilometers does it travel in one year?

(Answers to Practice Exercises are in the back of the book.)

1.10 Testing a Theory: The Big Bang Revisited

Let's return to our discussion of experiments that have tested the validity of the Big Bang theory. If the matter in the universe started as a dense cloud of very hot gas that formed following a enormous release of energy and if the universe has been expanding ever since, then the universe must have been cooling throughout time. Gases cool as they expand; this is the principle on which refrigerators and air conditioners operate.

Temperature Scales

▶❚❚ **CHEMTOUR** Temperature Conversion

If the universe is still expanding and still cooling, some residual warmth must be left over from the Big Bang. By the 1950s some scientists not only thought so but also predicted how much leftover warmth there should be: enough to give interstellar space a temperature of 2.73 K, where K represents a temperature value on the Kelvin scale.

Several temperature scales are in general use today. In the United States the Fahrenheit scale is still the most popular in terms of public use. In the rest of the world and in science, temperatures are most often expressed in degrees Celsius or on the Kelvin scale. The Fahrenheit and Celsius scales differ from each other in

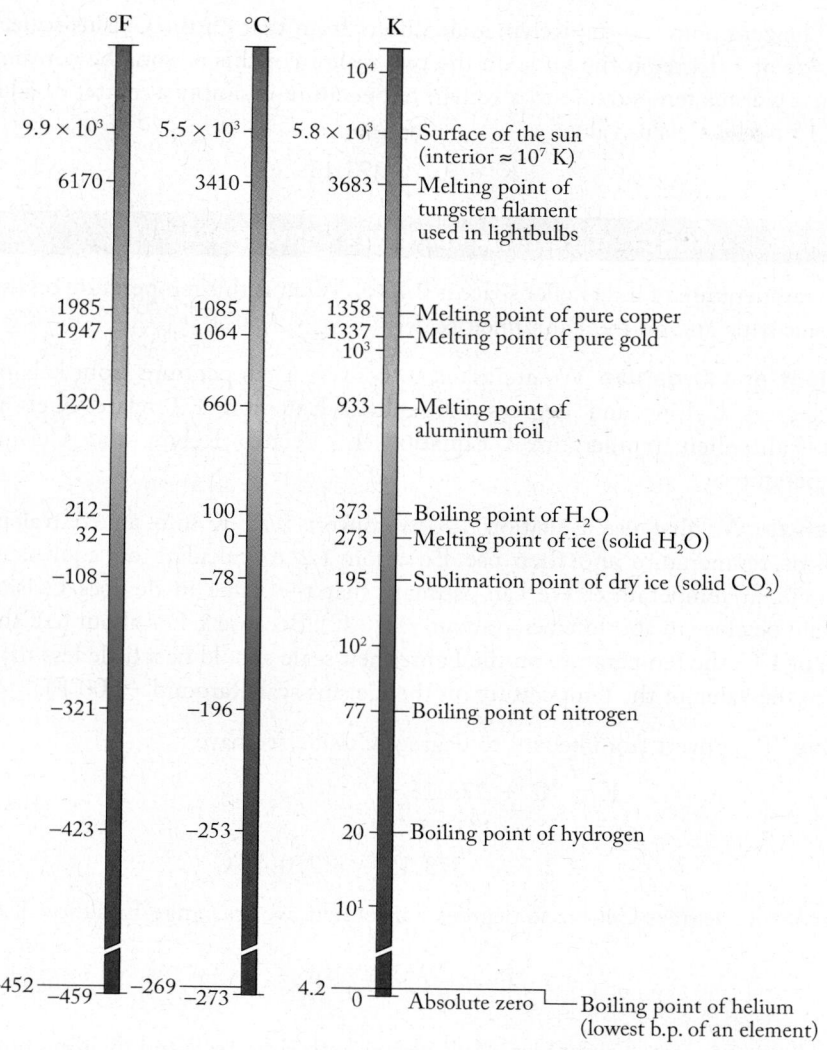

FIGURE 1.25 Three temperature scales are commonly used today, although the Fahrenheit scale is rarely used in scientific work.

two ways, as shown in Figure 1.25. First, their zero points are different. Zero degrees Celsius (0°C) is the temperature at which water freezes under normal conditions, but that temperature is 32 degrees on the Fahrenheit scale (32°F). The other difference is in the size of the temperature change corresponding to 1 degree. The difference between the freezing and boiling points of water is $212 - 32 = 180$ degrees on the Fahrenheit scale but only $100 - 0 = 100$ degrees on the Celsius scale. This difference means that a Fahrenheit degree is 100/180, or $\frac{5}{9}$, as large as a Celsius degree.

To convert temperatures from Fahrenheit into Celsius, we need to account for the differences in zero point and in degree size. The following equation does both:

$$^\circ C = \frac{5}{9}(^\circ F - 32) \qquad (1.2)$$

The SI unit of temperature (Table 1.2) is the **kelvin (K)**. The zero point on the Kelvin scale is not related to the freezing of a particular substance; rather, it is the coldest temperature—called **absolute zero (0 K)**—that can theoretically exist. It is equivalent to $-273.15°C$. No one has ever been able to chill matter to absolute zero, nor is it theoretically possible to do so, but scientists have come very close, cooling samples to less than 10^{-9} K.

kelvin (K) the SI unit of temperature.

absolute zero (0 K) the zero point on the Kelvin temperature scale; theoretically the lowest temperature possible.

The zero point on the Kelvin scale differs from that on the Celsius scale, but the size of 1 degree is the same on the two scales. For this reason, the conversion from a Celsius temperature to a Kelvin temperature is simply a matter of adding 273.15 to the Celsius value:

$$K = °C + 273.15 \qquad\qquad (1.3)$$

SAMPLE EXERCISE 1.8   Temperature Conversions

The temperature of interstellar space is 2.73 K. What is this temperature on the Celsius scale and on the Fahrenheit scale?

Collect and Organize We are asked to convert a temperature from kelvins to degrees Celsius and degrees Fahrenheit. Equation 1.2 relates Celsius and Fahrenheit temperatures; Equation 1.3 relates Kelvin and Celsius temperatures.

Analyze We first use Equation 1.3 to convert 2.73 K into an equivalent Celsius temperature and then use Equation 1.2 to calculate an equivalent Fahrenheit temperature. We can estimate that the value in degrees Celsius should be close to absolute zero (about $-273°C$). Because $1°F$ is about half the size of $1°C$, the temperature on the Fahrenheit scale should be a little less than twice the value of the temperature on the Celsius scale (around $-500°F$).

Solve To convert from kelvins to degrees Celsius, we have

$$K = °C + 273.15$$
$$°C = K - 273.15$$
$$= 2.73 - 273.15 = -270.42°C$$

To convert degrees Celsius to degrees Fahrenheit, we rearrange Equation 1.2

$$°C = \frac{5}{9}(°F - 32)$$

to solve for degrees Fahrenheit. Multiplying both sides by 9 and dividing both by 5 gives us

$$\frac{9}{5}°C = °F - 32$$

$$°F = \frac{9}{5}°C + 32 = \frac{9}{5}(-270.42) + 32 = -454.76°F$$

The value $32°F$ is considered a definition and so is not used in determining the number of significant figures in the answer. The number that determines the accuracy to which we can know this value is $-270.42°C$.

Think about It The calculated Celsius value of $-270.42°C$ makes sense because it represents a temperature only a few degrees above absolute zero, just as we estimated. The Fahrenheit value is within 10% of our estimate and so is reasonable too.

Practice Exercise The temperature of the moon's surface varies from $-233°C$ at night to $123°C$ during the day. What are these temperatures on the Kelvin and Fahrenheit scales?

(Answers to Practice Exercises are in the back of the book.)

An Echo of the Big Bang

In the early 1960s, Princeton University physicist Robert Dicke (Figure 1.26), who had hypothesized the presence of a residual energy left over from the Big Bang, was anxious to test this hypothesis. He proposed building an antenna that could detect microwave energy reaching Earth from outer space. Why did he pick microwaves? Even matter as cold as 2.73 K emits a "glow" (an energy signature), but not a glow you can see, like the visible light emitted by the sun, or feel, like the infrared rays emitted by any warm object (Figure 1.27). Instead, the glow from a 2.73 K object is in the form of microwave energy.

Dicke's microwave detector was never built because of events that unfolded just a short distance from Princeton. By the early 1960s the United States had launched the first communication satellites, *Echo* and *Telstar*. These early satellites were reflective spheres designed to bounce microwave signals to receivers on Earth. An antenna designed to receive the microwave signals had been built at a Bell Laboratories facility in Holmdel, NJ (Figure 1.28). Two Bell Labs scientists, Robert W. Wilson and Arno A. Penzias, were working to improve the antenna's reception when they encountered a problem. They observed that no matter where they directed their antenna, it picked up a background microwave signal much like the hissing sound radios make when tuned between stations. They hypothesized that the signal was due to some flaw in their antenna or in one of the instruments connected to it. (At one point they came up with an alternative hypothesis: that the source of the background signal was a pair of pigeons roosting on the antenna and coating parts of it with their droppings, but when the droppings were cleaned up, the problem persisted.) Additional testing led them to discard the flawed-instrument hypothesis, but it left unanswered the question of where the signal was coming from.

The nuisance signal picked up by the Wilson–Penzias antenna matched the microwave echo of the Big Bang Dicke had predicted. When the scientists at Bell Labs learned about Dicke's prediction, they realized the significance of the signal they had discovered; others did too. Wilson and Penzias shared the Nobel Prize in Physics in 1978 for discovering the cosmic microwave background radiation of the universe. Dicke did not share in the prize, even though he had accurately predicted what Wilson and Penzias discovered by accident.

FIGURE 1.26 Robert Dicke (1916–1997) predicted the existence of cosmic microwave background radiation. His prediction was confirmed by the serendipitous discovery of this radiation by Robert Wilson and Arno Penzias.

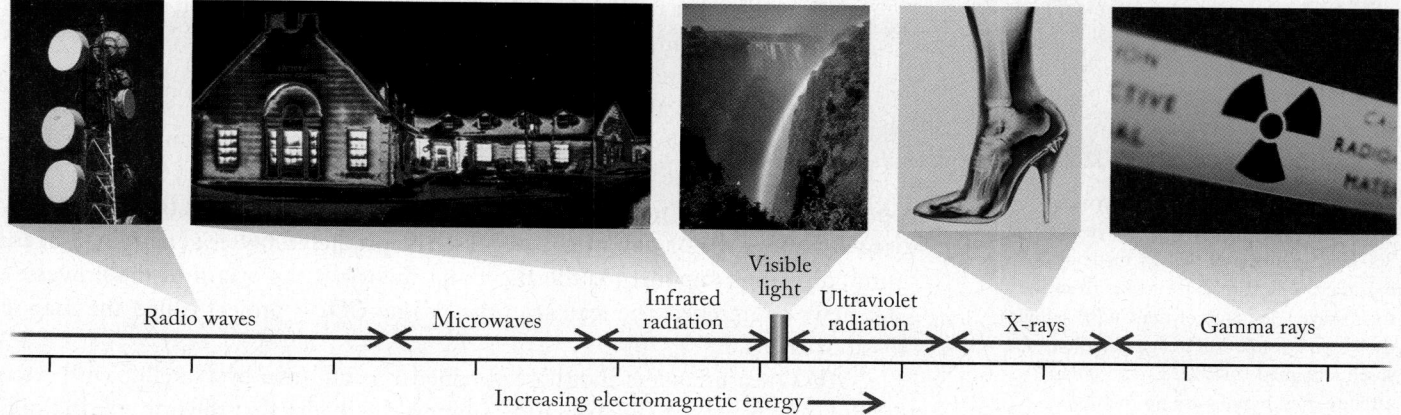

Radio waves   Microwaves   Infrared radiation   Visible light   Ultraviolet radiation   X-rays   Gamma rays

Increasing electromagnetic energy ⟶

FIGURE 1.27 The electromagnetic spectrum includes (but is not limited to) visible light, infrared radiation emitted by warm objects (and thus used to evaluate how much heat is lost from buildings), and the microwaves used for cell phone communications.

FIGURE 1.28 In 1965 Robert Wilson (left) and Arno Penzias discovered the microwave echo of the Big Bang while tuning this highly sensitive "horn" antenna in Holmdel, NJ.

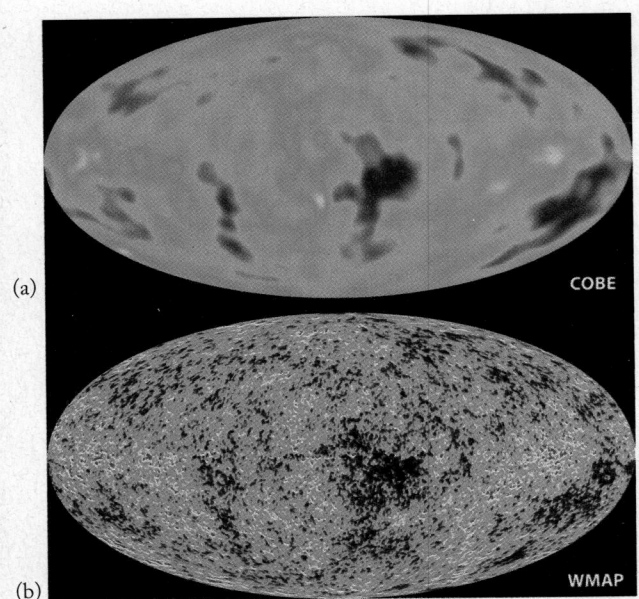

(a)

(b)

FIGURE 1.29 (a) Map of the cosmic microwave background released in 1992. It is a 360-degree image of the sky made by collecting microwave signals for a year from the microwave telescopes of the *COBE* satellite. (b) Higher-resolution image based on measurements made in 2002 by the *WMAP* satellite. Red regions are up to 200 μK warmer than the average interstellar temperature of 2.73 K, and blue regions are up to 200 μK colder than 2.73 K.

As happens often in science, a major discovery leads to new questions. The discovery of cosmic microwaves simultaneously reinforced the Big Bang theory and raised questions about it. The Bell Labs antenna picked up the same microwave signal no matter where in the sky it was pointed. In other words, the cosmic microwaves appeared to be uniformly distributed throughout the universe. If the afterglow from the Big Bang really was uniform, how could galaxies have formed? Some heterogeneity had to arise in the expanding universe—some clustering of the matter in it—to allow galaxies to form.

Scientists doing work related to Dicke's proposed that, if that clustering had occurred, a record of it should exist as subtle heterogeneities in the cosmic background radiation. Unfortunately, a microwave antenna in New Jersey, or any place on Earth, cannot detect such subtle differences in signals because too many other sources of microwaves interfere with the measurements. One way for scientists to find these heterogeneities, if they did exist, would be from a radio antenna in space.

In late 1989 the United States launched such an antenna in the form of the *Cosmic Background Explorer* (*COBE*) satellite. After many months of collecting data and many more months of analyzing it, the results were released in 1992. The image in Figure 1.29(a) appeared on the front pages of newspapers and magazines around the world. This discovery was a major news story because the predicted heterogeneity had been found, providing support for the Big Bang theory of the origin of the universe. At a news conference, the lead scientist on the *COBE* project called the map a "fossil of creation."

COBE's measurements and those obtained a decade later by a satellite with even higher resolving power (Figure 1.29b) support the theory that the universe did not expand and cool uniformly. The blobs and ripples in the images in Figure 1.29 indicate that galaxy "seed clusters" formed early in the history of the universe. Cosmologists believe that these ripples are a record of the next stage after the Big

Bang in the creation of matter. We examine this stage in Chapter 2 as we discuss how some elements may have formed just after the Big Bang, and how others have been, and continue to be, formed by the nuclear reactions that fuel our sun and all the other stars in the universe.

CONCEPT TEST

When Lemaître first proposed how the universe may have formed following a cosmic release of energy, would it have been more appropriate to call his explanation a *hypothesis* or a *theory*? Why?

(Answers to Concept Tests are in the back of the book.)

SUMMARY

Section 1.1 **Matter** is everything in the universe that has **mass** and occupies space. **Chemistry** is the science of matter. The principal categories of matter are pure **substances**, which may be either **elements** or **compounds** (elements chemically combined together according to the **law of constant composition**), and **mixtures**. Mixtures may be **homogeneous** (these mixtures are also called **solutions**) or **heterogeneous**.

Section 1.2 An **atom** is the smallest representative particle of an element; a **molecule** is a collection of atoms held together by **chemical bonds** in a characteristic pattern and proportion. A **chemical formula** expresses the elemental composition of a compound or a polyatomic form of an element. A **chemical equation** is used to describe the proportions of the substances involved in a **chemical reaction**. Space-filling and ball-and-stick models are used to show molecular structure, the three-dimensional arrangement of atoms in a molecule.

Section 1.3 The components of mixtures are not present in fixed proportions. They can be separated by physical processes, such as **filtration** and **distillation**.

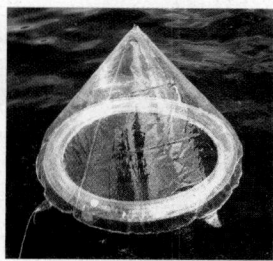

Section 1.4 The properties of a substance are either **intensive properties**, which are independent of quantity, or **extensive properties**, which are related to the quantity of the substance. The **physical properties** of a substance can be observed without changing the substance into another one; the **chemical properties** of a substance (such as flammability) can be observed only through chemical reactions involving the substance. The **density** of an object is the ratio of its mass to its volume.

Section 1.5 The COAST framework used in this book to solve problems has four components: **C**ollect and **O**rganize information and ideas, **A**nalyze the information to determine how it can be used to obtain the answer, **S**olve the problem (often the math-intensive step), and **T**hink about the answer.

Section 1.6 The states (or phases) of matter include **solid**, in which the particles are locked in place; **liquid**, in which the particles are free to move past each other; and **gas** (or vapor), in which the particles have the most freedom and completely fill their container. The transformation of a solid directly into a gas is **sublimation**; the reverse process is **deposition**.

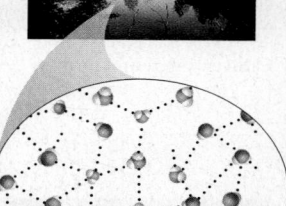

Section 1.7 The **scientific method** is an inquiry method based on observing either a natural phenomenon or a laboratory experiment; developing a tentative explanation, or **hypothesis**, for the observed results; testing the hypothesis through further experimentation; and then formulating a **scientific theory**.

Section 1.8 The International System of units (SI), in which the kilogram is the standard unit of mass and the meter is the standard unit of length, evolved from the metric system and is widely used in science to express the results of measurements. Prefixes naming powers of 10 are used with SI base units to express quantities much larger or much smaller than the base units. The appropriate number of **significant figures** is used to express the certainty in the result of a measurement or calculation. The **precision** of any set of measurements indicates how repeatable the measurement; the **accuracy** of a measurement indicates how close to the true value the measured value is.

Section 1.9 Dimensional analysis uses **conversion factors** (fractions in which the numerators and denominators have different units but represent the same quantity) to convert a value from one unit into another unit.

Section 1.10 Temperatures are expressed using the Fahrenheit, Celsius, and Kelvin temperature scales. Zero on the Kelvin scale is **absolute zero**, the coldest possible temperature.

PROBLEM-SOLVING SUMMARY

TYPE OF PROBLEM	CONCEPTS AND EQUATIONS		SAMPLE EXERCISES
Distinguishing physical properties from chemical properties	The chemical properties of a substance can be determined only by reacting it with another substance; physical properties can be determined without altering the substance's composition.		1.1
Using particles to represent states of matter	Particles in a solid are ordered; particles in a liquid are randomly arranged but close together; particles in a gas are separated by space and entirely fill the volume of their container.		1.2
Using significant figures in calculations	Apply the weak-link rule: the number of significant figures allowed in a calculated quantity involving multiplication or division can be no greater than the numberof significant figures in the least-certain value used to calculate it.		1.3
Calculating density from mass and volume	$$d = \frac{m}{V}$$	(1.1)	1.3
Distinguishing exact from uncertain values	Quantities that can be counted are exact. Measured quantities or conversion factors that are not exact values are inherently uncertain.		1.4
Doing dimensional analysis and converting units	Convert values from one set of units to another by multiplying by conversion factors set up so that the original units cancel.		1.5–1.7
Converting temperatures	$$°C = \frac{5}{9}(°F - 32)$$	(1.2)	1.8
	$$K = °C + 273.15$$	(1.3)	

VISUAL PROBLEMS

(Answers to boldface end-of-chapter questions and problems are in the back of the book.)

1.1. For each image in Figure P1.1, identify what class of pure substance is depicted (element or compound) and identify the physical state(s).

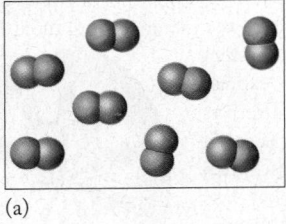
(a)

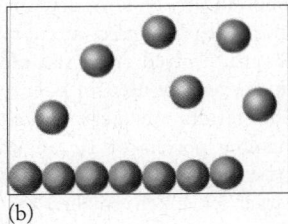
(b)

FIGURE P1.1

1.2. For each image in Figure P1.2, identify what class of matter is depicted (an element, a compound, a mixture of elements, or a mixture of compounds) and identify the physical state.

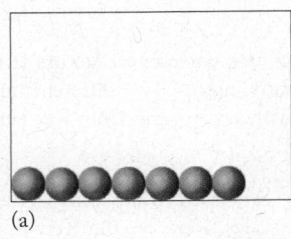
(a)

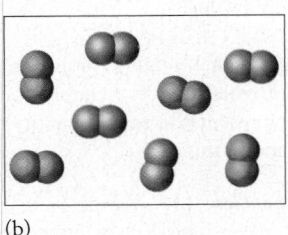
(b)

FIGURE P1.2

1.3. How would you describe the change depicted in Figure P1.3?

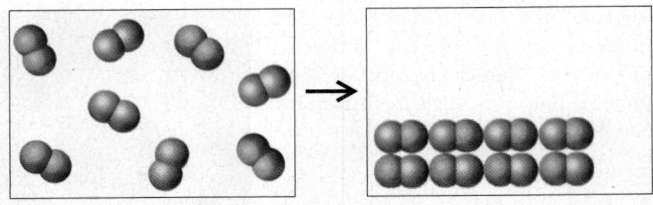

FIGURE P1.3

a. A mixture of two gaseous elements undergoes a chemical reaction, forming a gaseous compound.
b. A mixture of two gaseous elements undergoes a chemical reaction, forming a solid compound.
c. A mixture of two gaseous elements undergoes deposition.
d. A mixture of two gaseous elements condenses.

1.4. How would you describe the change depicted in Figure P1.4?

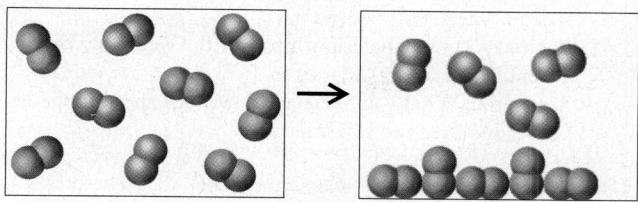

FIGURE P1.4

a. A mixture of two gaseous elements is cooled to a temperature at which one of them condenses.
b. A mixture of two gaseous compounds is heated to a temperature at which one of them decomposes.
c. A mixture of two gaseous elements undergoes deposition.
d. A mixture of two gaseous elements reacts together to form two compounds, one of which is a liquid.

1.5. A space-filling model of methanol is shown in Figure P1.5. What is the chemical formula of methanol?

FIGURE P1.5

1.6. A ball-and-stick model of acetone is shown in Figure P1.6. What is the chemical formula of acetone?

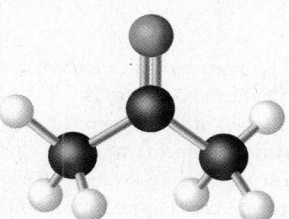

FIGURE P1.6

QUESTIONS AND PROBLEMS

Matter

CONCEPT REVIEW

1.7. Two students get into an argument over whether the sun is an example of matter or energy. Which point of view is correct? Why?

1.8. List three differences and three similarities between a compound and an element.

1.9. List one chemical and four physical properties of gold.

1.10. Describe three physical properties that gold and silver have in common, and three physical properties that distinguish them.

1.11. How might you use filtration to separate a mixture of salt and sand?

1.12. How can distillation be used to desalinate seawater?

1.13. Which of the following processes is a chemical reaction? (a) distillation; (b) combustion; (c) filtration; (d) condensation

1.14. Gasohol is a fuel that contains ethanol dissolved in gasoline. Is gasohol a heterogeneous mixture or a homogeneous one?

1.15. Which of the following foods is a heterogeneous mixture? (a) solid butter; (b) a Snickers bar; (c) grape juice; (d) an uncooked hamburger

1.16. Which of the following foods is a homogeneous mixture? (a) freshly brewed coffee; (b) vinegar; (c) a slice of white bread; (d) a slice of ham

1.17. Which of the following foods is a heterogeneous mixture? (a) apple juice; (b) cooking oil; (c) solid butter; (d) orange juice; (e) tomato juice

1.18. Indicate which of the following is a homogeneous mixture: (a) a wedding ring; (b) sweat; (c) Nile River water; (d) human blood; (e) compressed air in a scuba tank.

1.19. Give three properties that enable a person to distinguish between table sugar, water, and oxygen.

1.20. Give three properties that enable a person to distinguish between table salt, sand, and copper.

1.21. Indicate whether each of the following properties is a physical or a chemical property of sodium (Na):
a. Its density is greater than that of kerosene and less than that of water.
b. It has a lower melting point than most metals.
c. It is an excellent conductor of heat and electricity.
d. It is soft and can be easily cut with a knife.
e. Freshly cut sodium is shiny, but it rapidly tarnishes in contact with air.
f. It reacts very vigorously with water to form hydrogen gas (H_2) and sodium hydroxide (NaOH).

1.22. Indicate whether each of the following is a physical or chemical property of hydrogen gas (H_2):
a. At room temperature, its density is less than that of any other gas.
b. It reacts vigorously with oxygen (O_2) to form water.
c. Liquefied H_2 boils at a very low temperature ($-253°C$).
d. H_2 gas does not conduct electricity.

1.23. Enzymes are proteins. Proteins are constituents of egg whites. Assume we have a sample of an enzyme dissolved in water. Would filtration or distillation be a suitable way of separating the enzyme from the water?

1.24. Suggest a way of separating ice and water.

1.25. What is the state of each of these elements at ordinary temperature and pressure? Fe, O_2, Hg

1.26. Which of these mixtures is a solid at room temperature and pressure? sea salt, ketchup, ready-to-eat Jell-O

1.27. Can an extensive property be used to identify a substance? Explain why or why not.

1.28. Which of these are intensive properties of a sample of a substance? freezing point, heat content, temperature

The Scientific Method: Starting Off with a Bang

CONCEPT REVIEW

1.29. What kinds of information are needed to formulate a hypothesis?

1.30. How does a hypothesis become a theory?

1.31. Is it possible to disprove a scientific hypothesis?

1.32. Why is the theory that matter consists of atoms universally accepted?

1.33. How do people use the word *theory* in normal conversation?

1.34. Can a theory be proven?

Making Measurements and Expressing the Results; Unit Conversions and Dimensional Analysis

CONCEPT REVIEW

1.35. Describe in general terms how the SI and U.S. customary systems of units differ.

1.36. Suggest two reasons why SI units are not more widely used in the United States.

PROBLEMS

NOTE: The physical properties of the elements are in Appendix 3.

1.37. The speed of light in a vacuum is 2.9979×10^8 m/s. Calculate the speed of light in km/hr.

1.38. **Boston Marathon** To qualify to run in the 2009 Boston Marathon, a distance of 26.2 miles, an 18-year-old woman had to have completed another marathon in 3 hours and 40 minutes or less. To qualify, what must a woman's average speed have been (a) in miles per hour and (b) in meters per second?

1.39. **Olympic Mile** An Olympic "mile" is actually 1500 m. What percentage is an Olympic mile of a U.S. mile (5280 feet)?

*1.40. The price of a popular soft drink is $1.00 for 24 fluid ounces (fl oz) or $0.75 for 0.50 L. Which is a better buy? 1 qt = 32 fl oz

1.41. **Nearest Star** At a distance of 4.3 light-years, Proxima Centauri is the nearest star to our solar system. What is the distance to Proxima Centauri in kilometers?

*1.42. The level of water in an Olympic size swimming pool (50.0 meters long, 25.0 meters wide, and about 2 meters deep) needs to be lowered 3.0 cm. If water is pumped out at a rate of 5.2 liters per second, how long will it take to lower the water level 3.0 cm?

*1.43. If a wheelchair-marathon racer moving at 13.1 miles per hour expends energy at a rate of 665 Calories per hour, how much energy in Calories would be required to complete a marathon race (26.2 miles) at this pace?

1.44. An American sport-utility vehicle has an average mileage rating of 18 miles per gallon. How many gallons of gasoline are needed for a 389-mile trip?

1.45. A single strand of natural silk may be as long as 4.0×10^3 m. Convert this length into miles.

*1.46. **Automotive Engineering** The original (1955) Ford Thunderbird (Figure P1.46, left) was powered by a 292-cubic-inch (displacement) V-8 engine. The 2005 Thunderbird (Figure P1.46, right) was powered by a 3.9-liter V-8 engine. Which engine was bigger?

FIGURE P1.46

1.47. Suppose a runner completes a 10K (10.0 km) road race in 41 minutes and 23 seconds. What is the runner's average speed in meters per second?

1.48. **Kentucky Derby** The fastest time for the Kentucky Derby is 1 minute and 59 seconds, set in 1973 by a horse named Secretariat. What was Secretariat's average speed in meters per second over the 1.25-mile race?

1.49. What is the mass of a magnesium block that measures 2.5 cm × 3.5 cm × 1.5 cm?

1.50. What is the mass of an osmium block that measures 6.5 cm × 9.0 cm × 3.25 cm? Do you think you could lift it with one hand?

1.51. A chemist needs 35.0 g of concentrated sulfuric acid for an experiment. The density of concentrated sulfuric acid at room temperature is 1.84 g/mL. What volume of the acid is required?

1.52. What is the mass of 65.0 mL of ethanol? (Its density at room temperature is 0.789 g/mL.)

1.53. A brand new silver U.S. dollar weighs 0.934 ounces. Express this mass in grams and kilograms. 1 oz = 28.35 g

1.54. A U.S. dime weighs 2.5 g. What is the dollar value of 1.0 kg of dimes?

1.55. What volume of gold would be equal in mass to a piece of copper with a volume of 125 cm³?

*1.56. A small hot-air balloon is filled with 1.00×10^6 L of air ($d = 1.20$ g/L). As the air in the balloon is heated, it expands to 1.09×10^6 L. What is the density of the heated air in the balloon?

1.57. What is the volume of 1.00 kg of mercury?

1.58. A student wonders whether a piece of jewelry is made of pure silver. She determines that its mass is 3.17 g. Then she drops it into a 10 mL graduated cylinder partially filled with water and determines that its volume is 0.3 mL. Could the jewelry be made of pure silver?

*1.59. The average density of Earth is 5.5 g/cm³. The mass of Venus is 81.5% of Earth's mass, and the volume of Venus is 88% of Earth's volume. What is the density of Venus?

1.60. Earth has a mass of 6.0×10^{27} g and an average density of 5.5 g/cm³.
  a. What is the volume of Earth in cubic kilometers?
  * b. Geologists sometimes express the "natural" density of Earth after doing a calculation that corrects for gravitational squeezing (compression of the core because of high pressure). Should the natural density be more or less than 5.5 g/cm³?

*1.61. **Utility Boats for the Navy** A plastic material called HDPE or high-density polyethylene was once evaluated for use in impact-resistant hulls of small utility boats for the Navy. A cube of this material measures 1.20×10^{-2} m on a side and has a mass of 1.70×10^{-3} kg. Seawater at the surface of the ocean has a density of 1.03 g/cm³. Will this cube float on water?

1.62. **The Sun** The sun is a sphere with an estimated mass of 2×10^{30} kg. If the radius of the sun is 7.0×10^5 km, what is the average density of the sun in units of grams per cubic centimeter? The volume of a sphere is $\frac{4}{3}\pi r^3$.

1.63. Diamonds are measured in carats, where 1 carat = 0.200 g. The density of diamond is 3.51 g/cm³. What is the volume of a 5.0-carat diamond?

*1.64. If the concentration of mercury in the water of a polluted lake is 0.33 μg (micrograms) per liter of water, what is the total mass of mercury in the lake, in kilograms, if the lake has a surface area of 10.0 km² and an average depth of 15 m?

1.65. The cartoon in Figure P1.65 applies accuracy and precision to the measurement of body mass.
   a. Give definitions of accuracy and precision.
   b. Is the lawyer using the two terms correctly?
   c. Is it possible to be "precisely accurate"?
   d. What does the sign "Precise Weight" say about the uncertainty in the measurements?

FIGURE P1.65

1.66. **Healthy Snack?** Three different analytical techniques were used to determine the quantity of sodium in a Mars Milky Way candy bar. Each technique was used to analyze five portions of the same candy bar, with the following results (expressed in milligrams of sodium per candy bar):

Technique 1	Technique 2	Technique 3
109	110	114
111	115	115
110	120	116
109	116	115
110	113	115

The actual quantity of sodium in the candy bar was 115 mg. Which techniques would you describe as precise, which as accurate, and which as both? What is the range of the values for each technique?

*1.67. The widths of copper lines in printed circuit boards must be close to a specified value. Three manufacturers were asked to prepare circuit boards with copper lines that are 0.500 μm (micrometers) wide (1 μm = 1 × 10⁻⁶ m). Each manufacturer's quality control department reported the following line widths on five sample circuit boards (given in micrometers):

Manufacturer #1	Manufacturer #2	Manufacturer #3
0.512	0.514	0.500
0.508	0.513	0.501
0.516	0.514	0.502
0.504	0.514	0.502
0.513	0.512	0.501

   a. What is the range of the data provided by each manufacturer?
   b. Can any of the manufacturers justifiably advertise that they produce circuit boards with "high precision"?
   c. Is there a data set for which this claim is misleading?

*1.68. **Patient Data** Measurements of a patient's temperature are routinely done several times a day in hospitals. Digital thermometers are routinely used, and it is important to evaluate new thermometers and select the best ones. The accuracy of these thermometers is checked by immersing them in liquids of known temperature. Such liquids include an ice–water mixture at 0.0°C and boiling water at 100.0°C at exactly 1 atmosphere pressure (boiling point varies with atmospheric pressure). Suppose the data shown in the following table were obtained on three available thermometers and you were asked to select the "best" one of the three.

Thermometer	Measured Temperature of Ice Water, °C	Measured Temperature of Boiling Water, °C
A	−0.8	99.9
B	0.3	99.8
C	0.3	100.3

Explain your choice of the "best" thermometer for use in the hospital.

1.69. Which of the following quantities have four significant figures?
   a. 0.0592   d. 5420
   b. 0.08206   e. 5.4 × 10³
   c. 8.314   f. 3.752 × 10⁻⁵

1.70. Which of the following numbers have just three significant figures?
   a. 7.02   d. 6.02 × 10²³
   b. 6.452   e. 12.77
   c. 302   f. 3.43

1.71. Perform each of the following calculations and express the answer with the correct number of significant figures:
a. $0.6274 \times 1.00 \times 10^3/[2.205 \times (2.54)^3] =$
b. $6 \times 10^{-18} \times (1.00 \times 10^3) \times 17.4 =$
c. $(4.00 \times 58.69)/(6.02 \times 10^{23} \times 6.84) =$
d. $[(26.0 \times 60.0)/43.53]/(1.000 \times 10^4) =$

1.72. Perform each of the following calculations, and express the answer with the correct number of significant figures:
a. $[(12 \times 60.0) + 55.3]/(5.000 \times 10^3) =$
b. $(2.00 \times 183.9)/[6.02 \times 10^{23} \times (1.61 \times 10^{-8})^3] =$
c. $0.8161/[2.205 \times (2.54)^3] =$
d. $(9.00 \times 60.0) + (50.0 \times 60.0) + (3.00 \times 10^1) =$

Testing a Theory: The Big Bang Revisited

CONCEPT REVIEW
1.73. Can a temperature in °C ever have the same value in °F?
1.74. What is meant by an *absolute* temperature scale?

PROBLEMS
1.75. Liquid helium boils at 4.2 K. What is the boiling point of helium in degrees Celsius?
1.76. Liquid hydrogen boils at −253°C. What is the boiling point of H_2 on the Kelvin scale?

1.77. **Topical Anesthetic** Ethyl chloride is supplied to physicians and athletic trainers as a liquid in a spray bottle propelled by its own vapor pressure. It acts as a mild topical anesthetic because it chills the skin when sprayed on it. It dulls the pain of injury and may also be applied when splinters need to be removed. The boiling point of ethyl chloride is 12.3°C. What are its boiling points on the Fahrenheit and Kelvin scales?

1.78. The temperature of the dry ice (solid carbon dioxide) in ice cream vending carts is −78°C. What is this temperature on the Fahrenheit and Kelvin scales?

1.79. A person has a fever of 102.5°F. What is this temperature in degrees Celsius?

1.80. Physiological temperature, or body temperature, is considered to be 37.0°C. What is this temperature in °F?

1.81. **Record Low** The lowest temperature measured on Earth is −128.6°F, recorded at Vostok, Antarctica, in July 1983. What is this temperature on the Celsius and Kelvin scales?

1.82. **Record High** The highest temperature ever recorded in the United States is 134°F at Greenland Ranch, Death Valley, CA, on July 13, 1913. What is this temperature on the Celsius and Kelvin scales?

1.83. The coolant in an automobile radiator freezes at −39°C and boils at 11°C. What are these temperatures on the Fahrenheit scale?

1.84. Silver and gold melt at 962°C and 1064°C, respectively. Convert these two temperatures to the Kelvin scale.

1.85. **Critical Temperature** The discovery of new "high temperature" superconducting materials in the mid-1980s spurred a race to prepare the material with the highest superconducting temperature. The *critical temperatures* (T_c)—the temperatures at which the material becomes superconducting—of $YBa_2Cu_3O_7$, Nb_3Ge, and $HgBa_2CaCu_2O_6$ are 93.0 K, −250.0°C, and −231.1°F, respectively. Convert these temperatures into a single temperature scale, and determine which superconductor has the highest T_c value.

1.86. As air is cooled, which gas condenses first: N_2, O_2, or Ar?

Additional Problems

****1.87.** **Agricultural Runoff** A farmer applies 1500 kg of a fertilizer that contains 10% nitrogen to his fields each year. Fifteen percent of the fertilizer washes into a stream that runs through the farm. If the stream flows at an average rate of 1.4 cubic meters per minute, what is the additional concentration of nitrogen (expressed in milligrams of nitrogen per liter) in the stream water due to the farmer's yearly application of fertilizer?

1.88. Your laboratory instructor has given you two shiny, light gray metal cylinders. Your assignment is to determine which one is made of aluminum ($d = 2.699$ g/mL) and which one is made of titanium ($d = 4.54$ g/mL). The mass of each cylinder was determined on a balance to five significant figures. The volume was determined by immersing the cylinders in a graduated cylinder as shown in Figure P1.88. The initial volume of water was 25.0 mL in each graduated cylinder. The following data were collected:

	Mass (g)	Height (cm)	Diameter (cm)
Cylinder A:	15.560	5.1	1.2
Cylinder B:	35.536	5.9	1.3

a. Calculate the volume of each cylinder using the dimensions of the cylinder only.
b. Calculate the volume from the water displacement method.
c. Which volume measurement allows for the greater number of significant figures in the calculated densities?
d. Express the density of each cylinder to the appropriate number of significant figures.

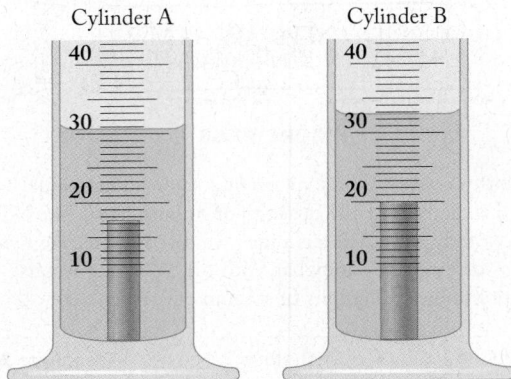

FIGURE P1.88

****1.89.** Sodium chloride (NaCl) contains 1.54 g Cl for every 1.00 g Na. Which of the following mixtures would react to produce sodium chloride with no Na or Cl left over?
a. 11.0 g Na and 17.0 g Cl   c. 6.5 g Na and 12.0 g Cl
b. 6.5 g Na and 10.0 g Cl   d. 6.5 g Na and 8.0 g Cl

****1.90.** **Toothpaste Chemistry** Most of the toothpaste sold in the United States contains about 1.00 milligram of fluoride per gram of toothpaste. The fluoride compound that is most often used in toothpaste is sodium fluoride, NaF, which is 45% fluoride by mass. How many milligrams of NaF are in a typical 8.2 ounce tube of toothpaste?

*1.91. **Test for HIV** Tests called ELISAs (enzyme-linked immunosorbent assays) detect and quantify substances such as HIV antibodies in biological samples. A "sandwich" assay traps the HIV antibody between two other molecules. The trapping event causes a detector molecule to change color. To make a sandwich assay for HIV, you need the following components: one plate to which the molecules are attached; a 0.550 mg sample of the recognition molecule that "recognizes" the HIV antibody; 1.200 mg of the capture molecule that "captures" the HIV antibody in a sandwich; and 0.450 mg of the detector molecule that produces a visible color when the HIV antibody is captured. You need to make 96 plates for an assay. You are given the following quantities of material: 100.00 mg of the recognition molecule; 100.00 mg of the capture molecule; 50.00 mg of the detector molecule.
   a. Do you have sufficient material to make 96 plates?
   b. If you do, how much of each material is left after 96 sandwich assays are assembled? If you don't have sufficient material to make 96 assays, how many assays can you assemble?

1.92. Some people believe that large doses of vitamin C can cure the common cold. One commercial over-the-counter product consists of 500.0 mg tablets that are 20% by mass vitamin C. How many tablets are needed for a 1.00 g dose of vitamin C?

1.93. We are building bicycles from separate parts. Each bicycle needs a frame, a front wheel, a rear wheel, two pedals, a set of handlebars, a bike chain, and a set each of front and rear brakes. How many complete bicycles can we make from 111 frames, 81 front wheels, 95 rear wheels, 112 pedals, 47 sets of handlebars, 38 bike chains, 17 front brakes, and 35 rear brakes?

1.94. Each Thursday the 11 kindergarten students in Miss Goodson's class are each allowed one slice of pie, one cup of orange juice, and two "doughnut holes." The leftovers will be given to the custodian on the night shift. This Thursday the caterer has left two pies that each can be cut into 8 slices, 18 cups of orange juice, and 24 doughnut holes. How many slices of pie, cups of orange juice, and doughnut holes are left for the custodian?

1.95. Manufacturers of trail mix have to control the distribution of items in their products. Deviations of more than 2% outside specifications cause supply problems and downtime in the factory. A favorite trail mix is designed to contain 67% peanuts and 33% raisins. Bags of trail mix were sampled from the assembly line on different days. The bags were opened and the contents counted, with the following results:

Day	Peanuts	Raisins	Day	Peanuts	Raisins
1	50	32	21	48	34
11	56	26	31	52	30

On which day(s) did the product meet the specification of 65% to 69% peanuts in the bag?

*1.96. Gasoline and water do not mix. Regular grade (87 octane) gasoline has a lower density (0.73 g/mL) than water (1.00 g/mL). A 100 mL graduated cylinder with an inside diameter of 3.2 cm contains 34.0 g of gasoline and 34.0 g of water. What is the combined height of the two liquid layers in the cylinder? The volume of a cylinder is $\pi r^2 h$, where r is the radius and h is the height.

Additional study materials including ChemTours and Diagnostic Quizzes are available at StudySpace at www.wwnorton.com/studyspace.

2

Atoms, Ions, and Compounds

Learning Outcomes

- Describe subatomic particles and how they are distributed inside atoms

- Explain how the experiments of Thomson, Millikan, and Rutherford contributed to our understanding of atomic structure

- Identify isotopes and use natural abundance data to calculate average atomic mass

- Use the periodic table to predict the chemical properties of elements

- Describe the general differences in the properties of metals, nonmetals, and metalloids

- Name common ionic and molecular compounds and write their formulas

- Describe how elements are synthesized in the cores of giant stars

A LOOK AHEAD

When Artillery Shells Bounced off Tissue Paper

The Greek philosopher Democritus (ca. 400 BC) proposed that matter is composed of indestructible particles called atoms. For over 2000 years most of the philosophers and scientists who believed in Democritus' atomic model also assumed that *indestructible* meant *indivisible*. However, a series of discoveries made during the last years of the 19th century through the 1930s showed that this assumption was wrong, and that atoms contain particular combinations of three subatomic particles: positively charged protons, negatively charged electrons, and particles with no electrical charge called neutrons.

By the beginning of the 20th century scientists knew the mass and charge of an electron, but they were unclear on how electrons and positive charges were distributed inside atoms. Then, in 1909, students of Ernest Rutherford at the University of Manchester in England aimed a beam of positively charged *alpha particles* at a piece of thin gold foil. Rutherford and his students were shocked by what they observed: while most of the particles went straight through the foil, as they had expected, a few were deflected well away from the incident beam, and a very few bounced right back at the source. Rutherford described his amazement at the result: "It is about as incredible as if you had fired a 15-inch shell[1] at a piece of tissue paper and it came back and hit you."

In light of these experimental results, Rutherford proposed the model of the atom we still use today. Its features include a dense, heavy nucleus at the center that contains the atom's positive charges and that is surrounded by a cloud of electrons. Additional insights in the 1930s and 1940s into the structure of atoms and into the enormous energies

The Large Hadron Collider This particle accelerator located near Geneva, Switzerland, tests predictions about the subatomic particles that make up all atoms. ▶

[1] A shell with a diameter of 15 inches was the largest projectile that could be fired by a British battleship in 1909.

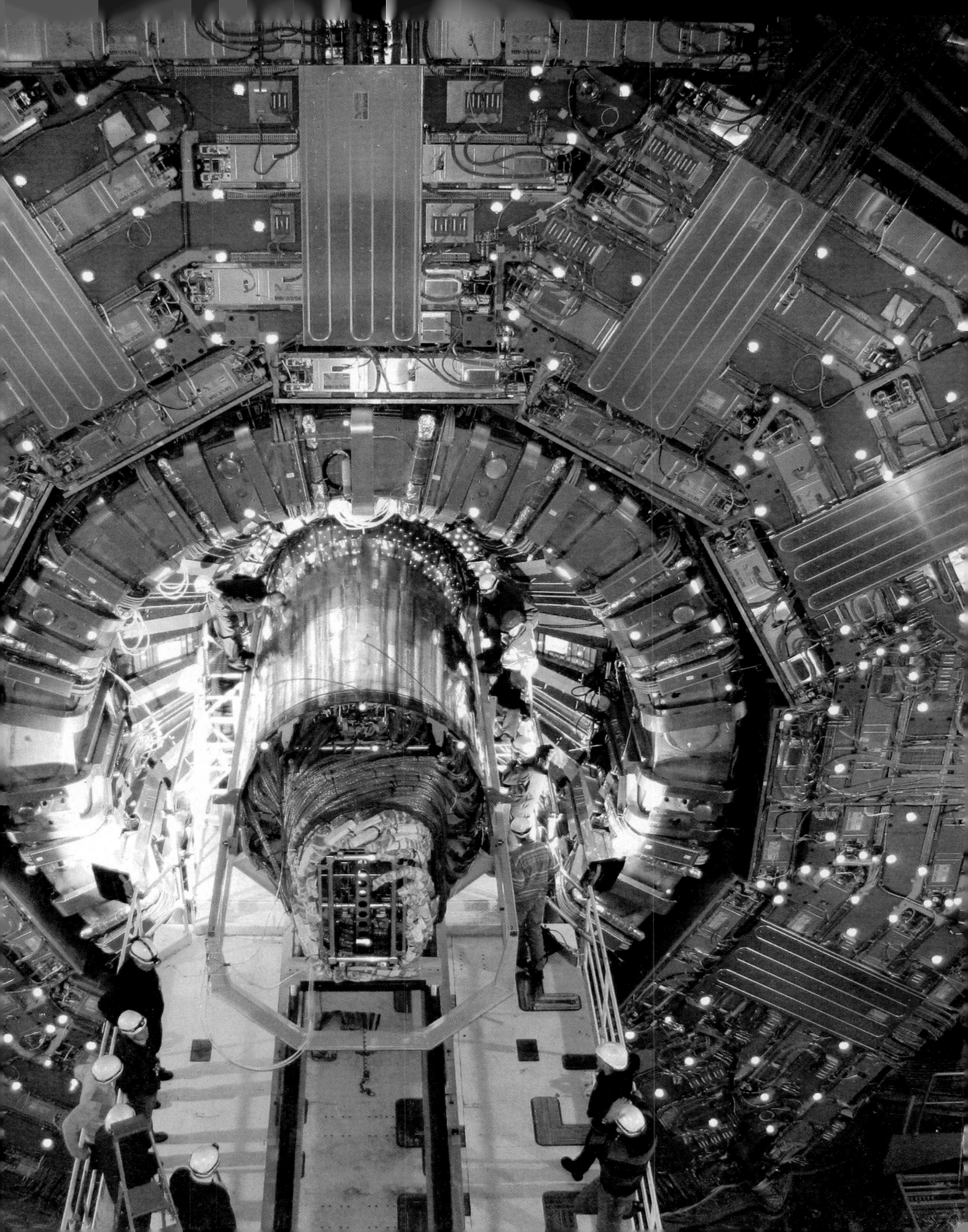

subatomic particles the neutrons, protons, and electrons in an atom.

cathode rays streams of electrons emitted by the cathode in a partially evacuated tube.

FIGURE 2.1 J. J. Thomson (1856–1940) discovered electrons in 1897 using a cathode-ray tube, but he was not sure where electrons fit into the structure of atoms.

▶❙❙ **CHEMTOUR** Cathode-Ray Tube

that are released by changes in the composition of nuclei led to amazing discoveries, technological advances, and to the dawn of the nuclear age. These discoveries also had an impact on the evolution of the Big Bang theory we examined in Chapter 1 because much of that theory involves nuclear transformations and the concept of nucleosynthesis: formation of the nuclei of elements in the early moments of the universe and later in the cores of giant stars. We explore both processes in Chapter 2 as well as the structures of the atomic, ionic, and molecular building blocks of matter. ■

2.1 The Rutherford Model of Atomic Structure

By the end of the 19th century, many scientists realized that atoms are not the smallest particles of matter, but instead are made up of even smaller **subatomic particles**. This realization came in part from the seminal research of British scientist Joseph John (J. J.) Thomson (Figure 2.1).

Electrons

Figure 2.2 depicts the apparatus Thomson used in his experiments. It is called a cathode-ray tube (CRT) and consists of a glass tube from which most of the air has been removed. Electrodes within the tube are attached to the poles of a high-voltage power supply. The electrode called the cathode is connected to the negative terminal of the power supply, and the anode is connected to the positive terminal. When these connections are made, electricity passes through the glass tube in the form of a beam of **cathode rays** emitted by the cathode. Cathode rays are invisible to the naked eye, but when the end of the CRT opposite the cathode is coated with a phosphorescent material, a glowing spot appears where the beam hits it.

Thomson observed that cathode-ray beams are deflected by magnetic fields (Figure 2.2a) and by electric fields (Figure 2.2b). The directions of these deflections established that cathode rays are negatively charged particles. By adjusting

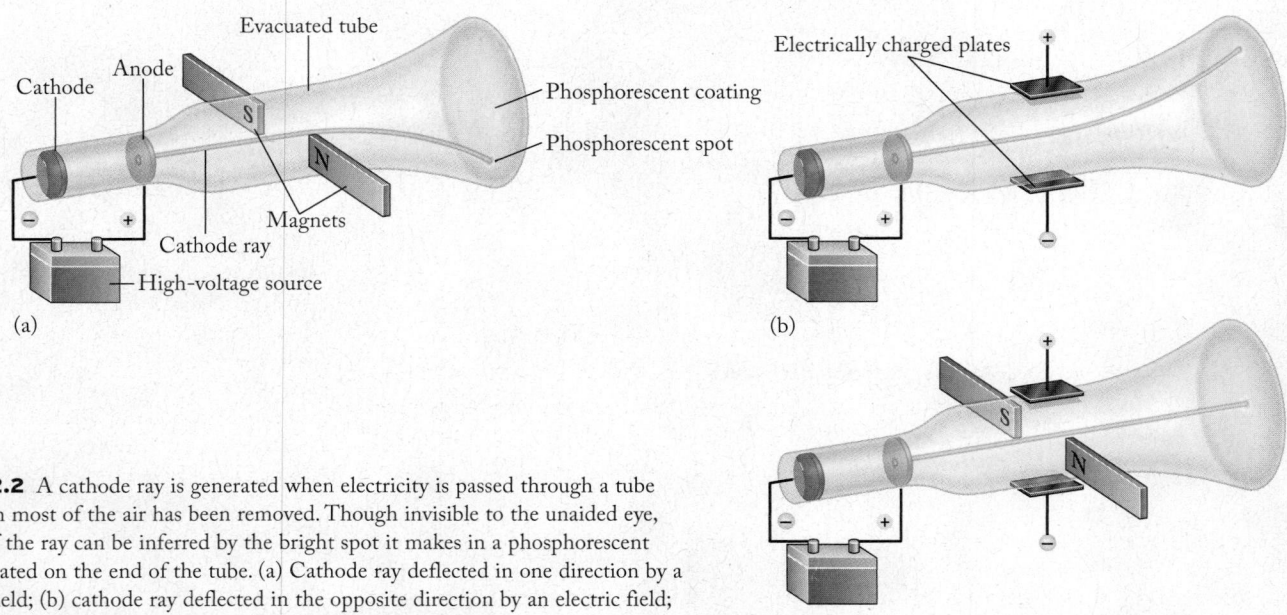

FIGURE 2.2 A cathode ray is generated when electricity is passed through a tube from which most of the air has been removed. Though invisible to the unaided eye, the path of the ray can be inferred by the bright spot it makes in a phosphorescent material coated on the end of the tube. (a) Cathode ray deflected in one direction by a magnetic field; (b) cathode ray deflected in the opposite direction by an electric field; (c) electric and magnetic fields tuned to balance out the deflections.

the strengths of the electric and magnetic fields, Thomson balanced out the deflections (Figure 2.2c). From the strengths of the two opposing fields he calculated the mass-to-charge (m/e) ratio of the particles. Thomson and others observed that these particles always behave the same way and always have the same mass-to-charge ratio, no matter what cathode material is used. This observation established that the particles in the cathode rays, which are now known as **electrons**, are fundamental particles present in all forms of matter.

In 1909 American physicist Robert Millikan (1868–1953) advanced Thomson's work by determining the charge on the electron. Thomson had calculated the mass-to-charge ratio of the electron, and if Millikan could determine the charge, then the mass would also be known. Figure 2.3 illustrates Millikan's experiment: Highly energetic X-rays remove electrons from molecules of air in the lower of two connected chambers. Oil drops falling from the upper chamber into the lower one pick up these electrons and develop a charge. Millikan measured the mass of the drops in the absence of an electric field, when their rate of fall was governed by gravity. He then turned on and adjusted the electric field to make the drops fall at different rates, and even suspended some of them in midair. From the strength of the electric field and the rate of fall, he calculated the charge on a drop. By measuring the charges on hundreds of drops, Millikan determined that the charge on each one was a whole-number multiple of a minimum charge. He concluded that this minimum charge had to be the charge on one electron. Millikan's value was within 1% of the modern value: -1.602×10^{-19} C. (The coulomb, abbreviated C, is the SI unit for electric charge.) Knowing Thomson's value of m/e for the electron, Millikan calculated the mass of the electron: 9.109×10^{-28} g.

The discovery of the electron raised the possibility of there being other subatomic particles. Scientists in the 1890s knew that matter was electrically neutral, but they did not know how the electrons and the positive charges were arranged at the atomic level. Thomson proposed a *plum-pudding model* in which the atom was a diffuse sphere of positive charge with negatively charged electrons embedded in the sphere, like raisins in a plum pudding (Figure 2.4). Thomson's plum-pudding model did not last long. Its demise was linked to another scientific discovery in the 1890s: radioactivity.

Radioactivity and the Nuclear Atom

In 1896 French physicist Henri Becquerel (1852–1908) discovered that *pitchblende*, a brownish-black mineral that is the principal source of uranium, produces radiation that can be detected on photographic plates. Becquerel and his contemporaries initially thought that this radiation consisted of the X-rays that had just been discovered by German scientist Wilhelm Conrad Röntgen (1845–1923).[2]

electron a subatomic particle that has a negative charge and essentially zero mass.

▶❚❚ CHEMTOUR Millikan Oil-Drop Experiment

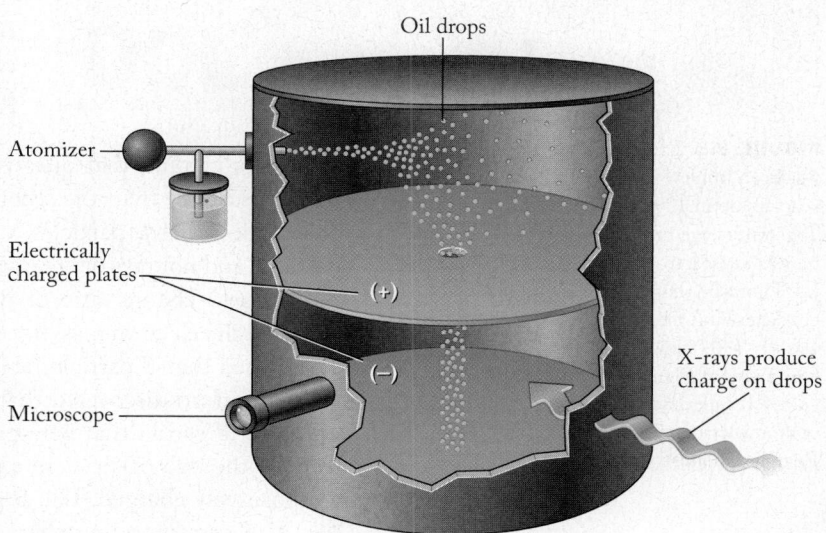

FIGURE 2.3 Millikan's oil-drop experiment.

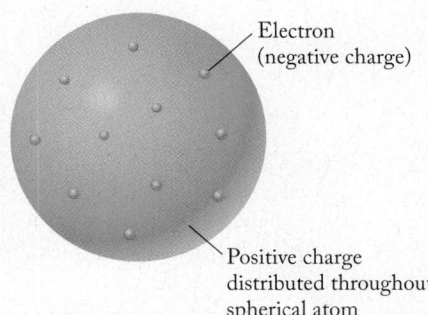

FIGURE 2.4 In J. J. Thomson's plum-pudding model, atoms consist of electrons distributed throughout a massive, positively charged, but very diffuse sphere. The plum-pudding model lasted only a few years before it was replaced by a model based on experiments carried out under the direction of Thomson's former student Ernest Rutherford.

[2] Röntgen discovered X-rays in experiments with a cathode-ray tube much like the apparatus used by J. J. Thomson. After completely encasing the tube in a black carton, Röntgen discovered that invisible rays escaped the carton and were detected by a photographic plate. Because he knew so little about these rays, he called them X-rays.

FIGURE 2.5 Ernest Rutherford (1871–1937) was born in New Zealand and was awarded a scholarship in 1894 that enabled him to go to Trinity College in Cambridge, England. There he was a research assistant in the laboratory of J. J. Thomson. His contributions included characterizing the properties of α and β particles. By 1907 he was a professor at the University of Manchester where his famous gold-foil experiments led to our modern view of atomic structure. He received the Nobel Prize in Chemistry in 1908.

radioactivity the spontaneous emission of high-energy radiation and particles by materials.

beta (β) particle a radioactive emission that is a high-energy electron.

alpha (α) particle a radioactive emission with a charge of 2+ and a mass equivalent to that of a helium nucleus.

nucleus (of an atom) the positively charged center of an atom that contains nearly all the atom's mass.

Additional experiments by Becquerel, by the French wife-and-husband team of Marie (1867–1934) and Pierre (1859–1906) Curie, and by British scientist Ernest Rutherford (Figure 2.5) showed that this radiation is actually several types of rays. The paths of two types of rays were deflected by electrical and magnetic fields, indicating that they were actually composed of charged particles. Today we use the term **radioactivity** for the spontaneous emission of high-energy radiation and particles by materials such as pitchblende.

In studying the particles emitted by pitchblende, Rutherford found that one type, which he named **beta (β) particles**, penetrate materials better than the second type, which he named **alpha (α) particles**. The observation that β particles can be deflected by a magnetic field proved they are particles and not rays of energy. The degree of deflection in a magnetic field allowed the mass-to-charge ratio of β particles to be calculated, and the results exactly matched the electron mass-to-charge ratio determined by J. J. Thomson. These data established that β particles are simply high-energy electrons.

Rutherford discovered that α particles are deflected by an electric field in the opposite direction that beta particles are deflected in the same field; the same is true for the two particles in a magnetic field. Therefore, he concluded α particles are positively charged. The β particle (electron) was assigned a relative charge of $1-$. The corresponding charge of an α particle was $2+$. In addition, α particles have nearly the same mass as an atom of helium, making them over 10^3 times more massive than β particles.

The plum-pudding model was discarded following an experiment directed by Rutherford and carried out by two of his students at Manchester University: Hans Geiger (1882–1945; for whom the Geiger counter was named) and Ernest Marsden (1889–1970). Geiger and Marsden bombarded a thin foil of gold with a beam of α particles emitted from a radioactive source and measured how many particles were deflected and to what extent they were deflected. Rutherford's hypothesis was that, if Thomson's model were correct, most of the α particles would pass straight through the diffuse positive spheres of gold atoms (each atom a "pudding" like the one shown in Figure 2.4) but a few of the particles would interact with the electrons (the "raisins") embedded in these spheres and be deflected slightly (Figure 2.6a).

Geiger and Marsden observed something completely unexpected. For the most part, the α particles did indeed pass directly through the gold. However, about 1 in every 8000 particles was deflected from the foil through an average angle of 90 degrees (Figure 2.6b) and a very few (perhaps 1 out of 100,000) bounced back in the direction from which the particles came.

The results of the gold-foil experiments ended the short life of the plum-pudding model because it could not account for the large angles of deflection. Rutherford concluded that those deflections occurred because the α particles occasionally encountered small regions of high positive charge and large mass. Rutherford determined that the region of positive charge is only about 1/10,000 of the overall size of a gold atom. His model of the atom became the basis for our current understanding of atomic structure. It assumes that an atom consists of a massive but tiny positively charged **nucleus** surrounded by a diffuse cloud of negatively charged electrons.

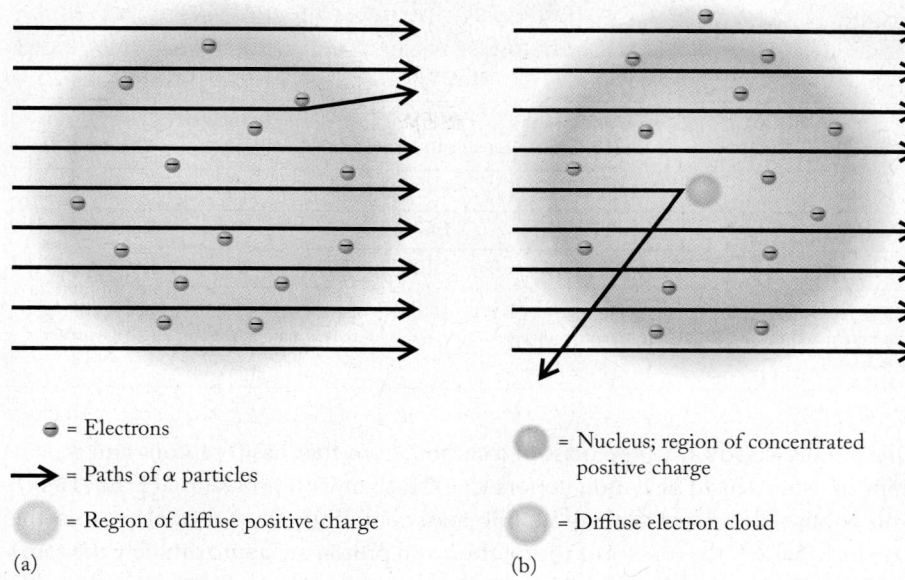

FIGURE 2.6 (a) If Thomson's plum-pudding model had been correct, most of the α particles would have passed through the gold foil and a few would have been deflected slightly in Rutherford's experiment. (b) In fact, most did pass straight through, but a few were scattered widely. This unexpected result led to the theory that an atom has a small, positively charged nucleus that contains most of the mass of the atom.

⊖ = Electrons

⟶ = Paths of α particles

⬤ = Region of diffuse positive charge

⬤ = Nucleus; region of concentrated positive charge

⬤ = Diffuse electron cloud

(a)   (b)

Protons and Neutrons

In the decade following the gold-foil experiments, Rutherford and others observed that bombarding elements with α particles could change, or *transmute*, these elements into other elements. They also discovered that hydrogen nuclei were frequently produced during transmutation reactions. By 1920 a consensus was growing that hydrogen nuclei, which Rutherford called **protons** (from the Greek *protos*, meaning "first"), were part of all nuclei. For example, to account for the mass and charge of an α particle, Rutherford assumed that it was made of four protons, two of which had combined with two electrons to form two electrically neutral particles, which he called **neutrons**. Repeated attempts to produce neutrons by neutralizing protons with electrons were unsuccessful. However in 1932 one of Rutherford's students, James Chadwick (1891–1974), successfully isolated and characterized free neutrons. With the discovery of neutrons the current model of atomic structure was complete, as illustrated with the gold atom in Figure 2.7.

Table 2.1 summarizes the properties of neutrons, protons, and electrons. For convenience, the masses of these tiny particles are expressed in **atomic mass units (amu)**. These units are also called **daltons (Da)** or *unified atomic mass units* (u).

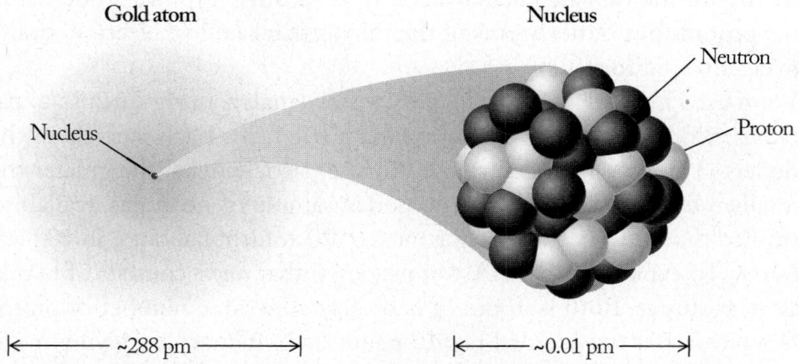

▶❚❚ **CHEMTOUR** Rutherford Experiment

Gold atom   Nucleus

                                         Neutron

Nucleus   Proton

|← ~288 pm →| |← ~0.01 pm →|

FIGURE 2.7 This modern view of Rutherford's model of the gold atom includes a nucleus that is about 1/10,000 the overall size of the atom. Note the scales are given in picometers (pm = 10^{-12} m); the nucleus would be too small to see if drawn to scale in the left drawing.

proton a positively charged subatomic particle present in the nucleus of an atom.

neutron an electrically neutral (uncharged) subatomic particle found in the nucleus of an atom.

atomic mass unit (amu) unit used to express the relative masses of atoms and subatomic particles that is exactly 1/12 the mass of one atom of carbon with 6 protons and 6 neutrons in its nucleus.

dalton (Da) a unit of mass identical to 1 atomic mass unit.

TABLE 2.1	Properties of Subatomic Particles				
		MASS		CHARGE	
Particle	Symbol	In Atomic Mass Units (amu)	In Grams (g)	Relative Value	Charge (C[a])
Neutron	^1_0n	$1.00867 \approx 1$	1.67493×10^{-24}	0	0
Proton	^1_1p	$1.00728 \approx 1$	1.67262×10^{-24}	1+	$+1.602 \times 10^{-19}$
Electron	$^0_{-1}e$	$5.485799 \times 10^{-4} \approx 0$	9.10939×10^{-28}	1−	-1.602×10^{-19}

[a] The *coulomb* (C) is the SI unit of electric charge. When a current of 1 ampere (see Table 1.2) passes through a conductor for 1 second, the quantity of electric charge that moves past any point in the conductor is 1 C.

One amu is exactly 1/12 the mass of a carbon atom that has 6 protons and 6 neutrons in its nucleus. The dalton honors English chemist John Dalton (1766–1844), who published the first table of atomic masses in 1803. As you can see from the data in Table 2.1, the masses of the neutron and proton are approximately the same and both are assigned a mass of 1 amu. If you compare the amu values in the table with the masses of the particles in grams, you will see that these particles are tiny indeed: 1 amu is only 1.66054×10^{-24} g.

2.2 Isotopes

Even as Thomson investigated the properties of cathode rays in 1897, other scientists designed and built devices to produce beams of positively charged particles. One of Thomson's former students, Francis W. Aston (1877–1945), built modified cathode-ray tubes that were evacuated except for small quantities of *fill gases*, such as neon. With these tubes he detected conventional beams of cathode rays, but he also detected secondary beams of positively charged particles. Their charges were not the only thing different about the particles in these secondary beams. Whereas conventional cathode rays consist of electrons that all have the same mass and charge no matter what the cathode material or the fill gas is, the masses of Aston's positive rays varied depending on the identity of the fill gas. The particles in the positive rays were not just individual protons but rather atoms of the fill gas that had lost electrons and formed positively charged **ions**.

Aston used his device, dubbed a positive-ray analyzer (Figure 2.8), to pass positively charged beams of gas through a magnetic field. Each particle in the beam was deflected along a path determined by the particle mass: the greater the mass, the smaller the deflection. Using the purest sample of neon gas available, Aston determined that most particles had a mass of 20 amu, but about 1 in 10 had a mass of 22 amu. To explain his data, Aston proposed that neon consists of two kinds of atoms, or **isotopes**. Both isotopes of neon have the same number of protons (10) in the nucleus, but one isotope has 10 neutrons in its nucleus, giving it a mass of 20 amu, while the other isotope has 12 neutrons in its nucleus, giving it a mass of 22 amu. The individual isotopes of neon or any element with particular combinations of neutrons or protons are called **nuclides**.

Since the time of John Dalton, scientists had defined an element as *matter composed of identical atoms, all of which have the same mass.* Aston's work required a mod-

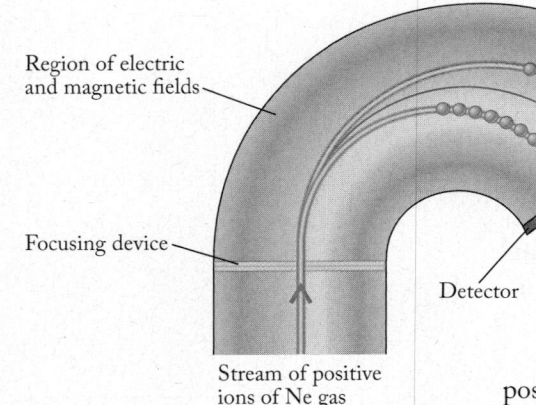

Region of electric and magnetic fields

Focusing device

22 amu

20 amu

Detector

Stream of positive ions of Ne gas

FIGURE 2.8 Aston's positive-ray analyzer. A beam of positively charged ions of neon gas is passed through a focusing slit into a region of an electric and magnetic field. The ions are separated according to mass: those with a mass of 20 amu—90% of the sample—hit the detector at one spot, and those with a mass of 22 amu—the remaining 10% percent—hit the detector at a different spot. Aston's positive-ray analyzer was the forerunner of the modern mass spectrometer.

ification of this definition: henceforth an element was defined as *matter composed of atoms all having the same number of protons in their nuclei.* This number of protons is called the **atomic number (Z)** of the element. The total number of **nucleons** (neutrons and protons) in the nucleus of an atom defines its **mass number (A)**. The isotopes of a given element thus all have the same atomic number Z but different mass numbers A. The modern **periodic table of the elements** (inside the front cover of this book) displays the elements in order of atomic number.

The general format for identifying a particular nuclide is

$$^A_Z X$$

where X represents the one- or two-letter symbol for the element. For example, two isotopes of oxygen (O) and lead (Pb) are written

$$^{16}_8O \quad ^{18}_8O \quad ^{206}_{82}Pb \quad ^{208}_{82}Pb$$

Because Z and X provide the same information—each by itself identifies the element—the subscript Z is frequently omitted, so that the isotope symbol is often written $^A X$. This same information—mass number and element name—is frequently spelled out. For example, the names of the two isotopes of neon that Aston discovered may be written neon-20 and neon-22.

ion an atom or group of atoms that has a positive or negative charge.

isotopes atoms of an element containing the same number of protons but different numbers of neutrons.

nuclide a specific isotope of an element.

atomic number (Z) the number of protons in the nucleus of an atom.

nucleon either a proton or a neutron in a nucleus.

mass number (A) the number of nucleons in an atom.

periodic table of the elements a chart of the elements in order of their atomic numbers and in a pattern based on their physical and chemical properties.

SAMPLE EXERCISE 2.1   **Writing Symbols of Nuclides**

Write symbols in the form $^A_Z X$ for the nuclides that have (a) 6 protons and 6 neutrons, (b) 11 protons and 12 neutrons, and (c) 92 protons and 143 neutrons.

Collect and Organize We know the number of protons and neutrons in the nuclei of three nuclides and are to write symbols of the form $^A_Z X$ where Z is the atomic number, A is the mass number, and X is the symbol of the element.

Analyze The number of protons in the nucleus of an atom defines its atomic number (Z), which in turn defines which element it is (X). The sum of the nucleons (protons plus neutrons) is the mass number (A).

Solve

a. This nuclide has 6 protons and therefore $Z = 6$; it must be an isotope of carbon. Six protons plus six neutrons gives the isotope a mass number of 12. This isotope of carbon is carbon-12, $^{12}_6C$.

b. This nuclide has 11 protons and therefore $Z = 11$; it must be an isotope of sodium. Eleven protons and 12 neutrons means that the isotope has a mass number of 23. This isotope of sodium is sodium-23, $^{23}_{11}Na$.

c. This nuclide has 92 protons and therefore $Z = 92$; it must be an isotope of uranium. The mass number is $92 + 143 = 235$. This isotope of uranium is uranium-235, $^{235}_{92}U$.

Think about It In working through this exercise, did you use the periodic table of the elements to identify the symbol of the element once you used the number of protons in a nucleus to determine the atomic number? Finding a symbol and identifying the element it represents is easy because the elements in the periodic table are arranged in order of increasing atomic number.

Practice Exercise Use the format $^A X$ to write the symbols of the nuclides having (a) 26 protons and 30 neutrons, (b) 7 protons and 8 neutrons, (c) 17 protons and 20 neutrons, and (d) 19 protons and 20 neutrons.

(Answers to Practice Exercises are in the back of the book.)

2.3 Average Atomic Mass

At the center of each of the cells in the periodic table on the inside front cover is the symbol of an element. The number above the symbol is the element's atomic number Z and the number below the symbol is the element's **average atomic mass**. More precisely, the number below the symbol is the *weighted average* of the masses of all the isotopes of the element.

To understand the meaning of a weighted average, consider the masses and **natural abundances** of the three isotopes of neon in the table to the left. Natural abundances are usually expressed in percentages. Thus, 90.4838% of all neon atoms are neon-20, 9.2465% are neon-22, and only 0.2696% are neon-21. This abundance of neon-21 is so small Aston could not detect it with his positive-ray analyzer. Modern mass spectrometers, which are the source of natural abundance data such as these, are vastly more sensitive and more precise than Aston's prototype.

To determine the average atomic mass, we multiply the mass of each isotope by its natural abundance (in the language of mathematics, we *weight* the isotope's mass using natural abundance as the *weighting factor*) and then sum the three weighted masses. To simplify the calculation we first convert the percent abundance values into their decimal equivalents:

$$\begin{aligned}
\text{Average atomic mass of neon} = \quad & (19.9924 \text{ amu} \times 0.904838) \\
+ \ & (20.9940 \text{ amu} \times 0.002696) \\
+ \ & \underline{(21.9914 \text{ amu} \times 0.092465)} \\
& 20.1799 \text{ amu}
\end{aligned}$$

It is important to note that no single atom of neon has this average atomic mass; every atom of neon in the universe must have a mass equal to one of the three neon isotopes. The value we just calculated is the weighted average of these three isotopic masses.

This method for calculating average atomic mass works for all elements. The general equation for doing the calculation is

$$m_X = a_1 m_1 + a_2 m_2 + a_3 m_3 + \cdots \tag{2.1}$$

where m_X is the average atomic mass of an element X, which has isotopes with masses $m_1, m_2, m_3, \ldots$, the natural abundances of which expressed in decimal form are $a_1, a_2, a_3, \ldots$.

Isotope	Mass (amu)	Natural Abundance (%)
Neon-20	19.9924	90.4838
Neon-21	20.9940	0.2696
Neon-22	21.9914	9.2465

average atomic mass a weighted average of masses of all isotopes of an element, calculated by multiplying the natural abundance of each isotope by its mass in atomic mass units and then summing these products.

natural abundance the proportion of a particular isotope, usually expressed as a percentage, relative to all the isotopes of that element in a natural sample.

SAMPLE EXERCISE 2.2 Calculating an Average Atomic Mass

The precious metal platinum ($Z = 78$) has six isotopes with these natural abundances:

Symbol	Mass (amu)	Natural Abundance (%)
^{190}Pt	189.96	0.014
^{192}Pt	191.96	0.782
^{194}Pt	193.96	32.967
^{195}Pt	194.97	33.832
^{196}Pt	195.97	25.242
^{198}Pt	197.97	7.163

Use these data to calculate the average atomic mass of platinum.

Collect and Organize We know the masses and natural abundances of each of the six isotopes of platinum and are asked to calculate the average atomic mass. Equation 2.1 may be used to do such a calculation.

Analyze Using Equation 2.1, we multiply the mass of each isotope by its natural abundance expressed as a decimal, and then add the products together.

Solve

$$
\begin{aligned}
\text{Average atomic mass} = \quad & (189.96 \text{ amu})(0.00014) \\
+ \; & (191.96 \text{ amu})(0.00782) \\
+ \; & (193.96 \text{ amu})(0.32967) \\
+ \; & (194.97 \text{ amu})(0.33832) \\
+ \; & (195.97 \text{ amu})(0.25242) \\
\underline{+ \; & (197.97 \text{ amu})(0.07163)} \\
& 195.08 \text{ amu}
\end{aligned}
$$

Think about It Note that the six values of natural abundances expressed as decimals should add up to 1.00000, and they do. Sometimes this is not the case (check the neon abundances above). Uncertainties in the last decimal place may be due to uncertainties in measured values or in rounding them off. The calculated average atomic mass of platinum is consistent with the value given inside the front cover.

Practice Exercise Silver (Ag) has two stable isotopes: ^{107}Ag, 106.90 amu, and ^{109}Ag, 108.90 amu. If the average atomic mass of silver is 107.87 amu, what is the natural abundance of each isotope? *Hint*: Let x be the natural abundance of one of the isotopes. Then $1 - x$ is the natural abundance of the other.

(Answers to Practice Exercises are in the back of the book.)

2.4 The Periodic Table of the Elements

Long before chemists knew about subatomic particles and the concept of atomic numbers, they knew that groups of elements, such as Li, Na, and K, or F, Cl, and Br, had similar chemical (and sometimes physical) properties, and that when the elements were arranged by increasing atomic mass, there were repeating patterns of similar properties among the elements. This *periodicity* in the chemical properties of the elements inspired several 19th-century scientists to create tables of the elements in which the elements were arranged in patterns based on similarities in their chemical properties.

By far the most successful of these scientists was Russian chemist Dmitri Mendeleev (1834–1907). In 1872 he published a table (Figure 2.9) that is widely considered the forerunner of the modern periodic table (Figure 2.10). In addition to organizing all the elements that were known at the time, Mendeleev realized that there might be elements in nature that were yet to be discovered. This insight meant that he could leave empty cells in his table for unknown elements. Doing so allowed him to align the known elements so that those in each column had similar chemical properties. Based on the locations of these empty cells, Mendeleev was able to predict the chemical properties of the missing elements. These predictions greatly facilitated the subsequent discovery of these elements by other scientists. Mendeleev

FIGURE 2.9 Mendeleev organized his periodic table on the basis of chemical and physical properties and atomic masses. He assigned three elements with similar properties to group VIII in rows 4, 6, and 10. Because he did this, the elements in the rows that followed lined up in appropriate groups. In this way, rows 4 and 5 together contain spaces for 18 elements, corresponding to the 18 groups in the modern periodic table.

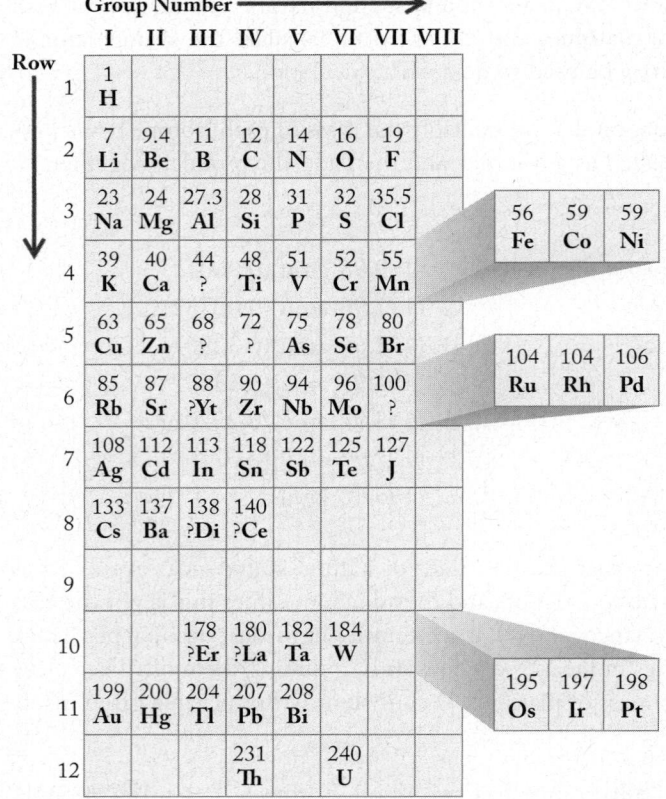

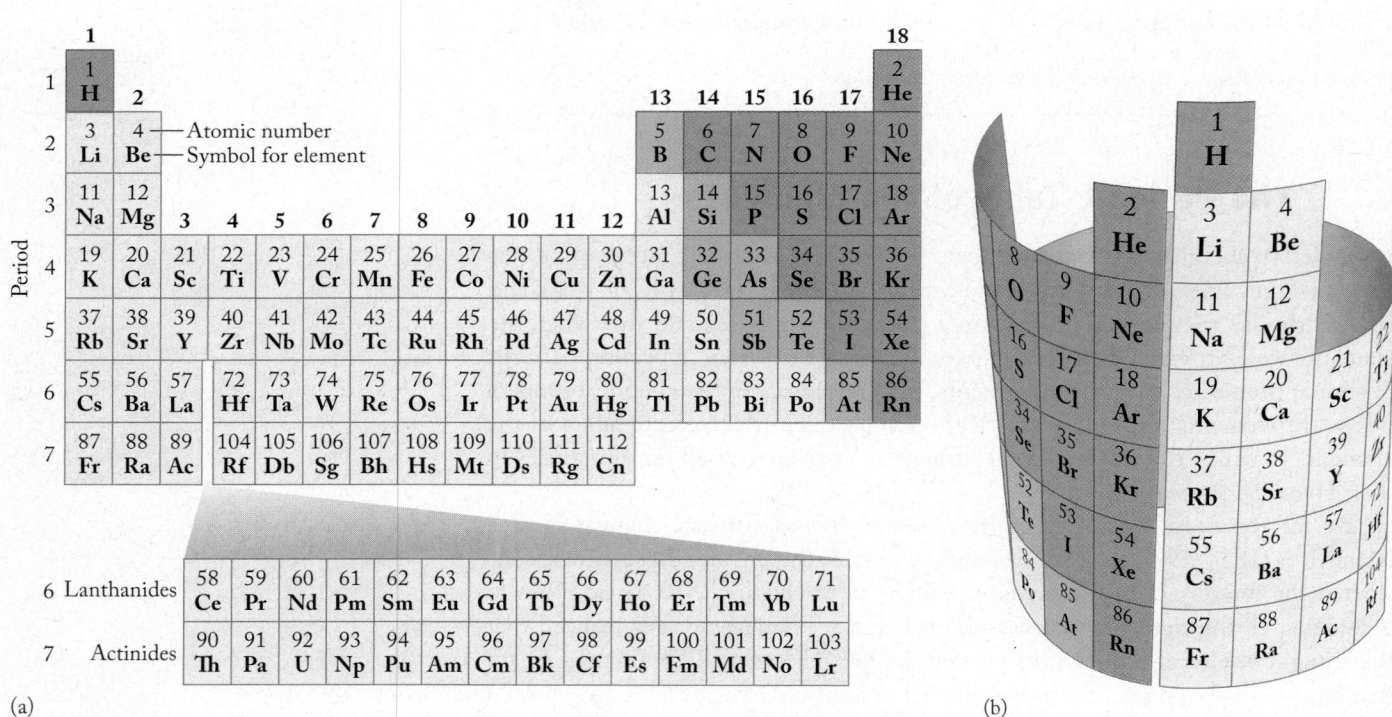

FIGURE 2.10 (a) In the modern periodic table, the elements are arranged in order of atomic number (Z) and in a pattern related to their physical and chemical properties. The elements shown in tan are classified as metals, those shown in blue, nonmetals, and those shown in green, metalloids (also called semimetals). (b) In this view of the periodic table, the sides have been joined together to illustrate the sequence of the elements from one row to the next.

arranged the elements in his periodic table in order of increasing atomic mass. In modern periodic tables the elements appear in order of their atomic numbers.

CONCEPT TEST ···

Suggest a reason why the elements in Mendeleev's version of the periodic table are in order of atomic mass and not atomic number.

(Answers to Concept Tests are in the back of the book.)

periods the horizontal rows in the periodic table.

group or **family** all elements in the same column of the periodic table.

halogen an element in group 17 of the periodic table.

alkali metal an element in group 1 of the periodic table.

Navigating the Modern Periodic Table

The modern periodic table (Figure 2.10) contains seven horizontal rows (called **periods**) and 18 columns (called **groups** or **families**) (Figure 2.11a). The periods are numbered at the far left of each row. The group numbers appear at the top of each column. The periodic table inside the front cover shows a second set of column headings containing numbers followed by the letter A or B. These secondary headings were widely used in earlier versions of the table, and many scientists still find them useful.

The first row contains only two elements—hydrogen and helium—and the second and third rows contain only eight. Starting with the fourth row, all 18 columns are full. Actually, the sixth and seventh rows contain more elements than there is space for in an 18-column array. These additional elements appear in the two separate rows at the bottom of the main table. Elements in the row with atomic numbers from 58 to 71 are called the lanthanides (after element 57, lanthanum) and those with atomic numbers between 90 and 103 are called actinides. Most of the nuclides of the actinide elements are radioactive and none of those with atomic numbers above 94 occur in nature. Therefore, they have no *natural* abundance.

Several of the groups have names in addition to a number. Their names are typically based on properties common to the elements in that group (Figure 2.11b). For example, the elements in group 17 are called **halogens**. The word *halogen* is derived from the Greek for "salt former." Chlorine is a typical halogen: it forms a 1:1 compound, for example, NaCl (table salt), with the elements in group 1, and 2:1 compounds, for example, $CaCl_2$, with the elements in group 2. For their part, group 1 elements, which are called **alkali metals**, form 1:1 compounds with all the group 17 elements and 2:1 compounds, such as Na_2S, with the group 16 elements.

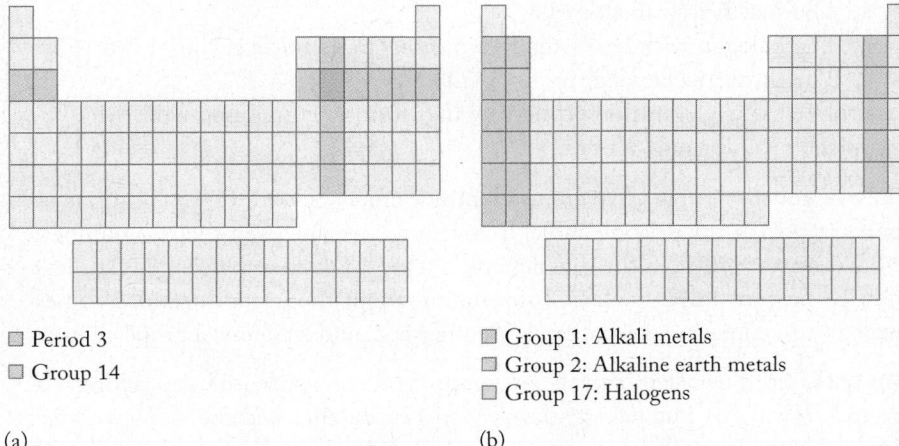

☐ Period 3
☐ Group 14

(a)

☐ Group 1: Alkali metals
☐ Group 2: Alkaline earth metals
☐ Group 17: Halogens

(b)

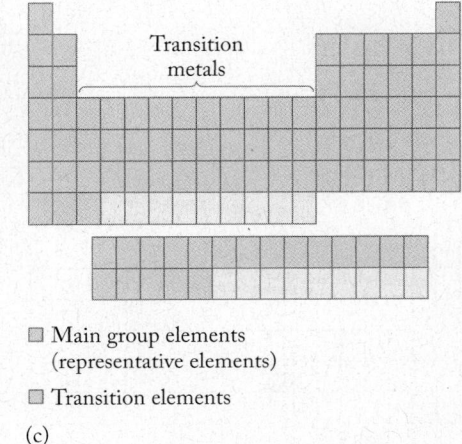

Transition metals

☐ Main group elements (representative elements)

☐ Transition elements

(c)

FIGURE 2.11 (a) Periodic tables consist of elements in rows, called periods, and columns, called groups or families. (b) The commonly used names of groups 1, 2, and 17. (c) The *main group* (or *representative*) elements are in groups 1, 2, and 13–18. They are separated by the *transition metals* in groups 3–12.

alkaline earth metal an element in group 2 of the periodic table.

metals the elements on the left side of the periodic table that are typically shiny solids that conduct heat and electricity well and are malleable and ductile.

nonmetals elements with properties opposite those of metals including poor conductivity of heat and electricity.

metalloids or **semimetals** elements along the border of the periodic table between metals and nonmetals; they have some metallic and some nonmetallic properties.

main group elements or **representative elements** the elements in groups 1, 2, and 13 through 18 of the periodic table.

transition metals the elements in groups 3 through 12 of the periodic table.

noble gases the elements in group 18 of the periodic table.

The group 2 elements, which are called **alkaline earth metals**, form 1:1 compounds, such as MgO, with the group 16 elements and 1:2 compounds with all the group 17 elements. These and other reactivity patterns were the basis for Mendeleev's arrangement of the elements in his early periodic table, and they illustrate what we mean by "similar chemical properties."

The elements in the periodic table are also divided into three broad categories highlighted by the three different cell colors in Figure 2.10. Elements in the tan cells are **metals**. They tend to conduct heat and electricity well; they tend to be malleable (capable of being shaped by hammering) or ductile (capable of being drawn out in a wire), and all but mercury (Hg) are shiny solids at room temperature. Elements in the blue cells are **nonmetals**. They are poor conductors of heat and electricity; most are gases at room temperature, the solids among them tend to be brittle, and bromine is a low-boiling liquid. The elements in the green cells are called **metalloids** or **semimetals**, so named because they tend to have the physical properties of metals but the chemical properties of nonmetals.

In the modern periodic table, groups 1, 2, and 13 through 18 are referred to collectively as **main group elements** or **representative elements** (Figure 2.11c). They include the most abundant elements in the solar system and many of the most abundant on Earth. Note that these are the "A" elements in the older group labeling system shown on the table inside the front cover. The elements in groups 3 through 12 are called **transition metals**; these are the "B" elements. They are nearly all classic metals: hard, shiny, ductile, malleable, and excellent conductors of heat and electricity.

Take a moment to compare Figures 2.9 and 2.10. First note the similarity in the arrangements of the lighter (smaller atomic number) elements through calcium ($Z = 20$): All of the elements in groups 1, 2, and 13 through 17 of the modern table appear in the same order in Mendeleev's table, but group 18 (the **noble gases**) is missing from Mendeleev's table. There is a good reason for this. Helium was the first noble gas to be discovered, and that was not until 1895, many years after Mendeleev published his table. Noble gases are chemically unreactive, and so were elusive substances for early chemists to isolate and identify. Because Mendeleev arranged his table largely on the basis of reactivity, he had no reason to predict the existence of the noble gases.

> **SAMPLE EXERCISE 2.3 Navigating the Periodic Table**

Using Figure 2.10, give the symbol and name of each element:
a. The fourth row alkali metal
b. The halogen with fewer than 16 protons in its nucleus
c. The third row element in group 14
d. The metal (X) in the second row that forms a compound with the chemical formula XBr_2

Collect and Organize We are to identify elements based on the locations of their symbols in the periodic table. In a, c, and d we are given the row number; in b we are not provided the row directly, but we know that the element has less than 16 protons in its nucleus. Information about the group locations comes from group names in a and b, group number in c, and a chemical property in d.

Analyze (a) The alkali metals are group 1 elements, so the cell address is group 1, row 4. (b) The halogens are group 17 elements, and the only group 17 element with fewer than 16 protons in its nucleus ($Z < 16$) is fluorine (F). (c) The cell address is group 14, row 3. (d) The group 2 elements form 1:2 compounds with Br and the other halogens, so the cell address is group 2, row 2.

Solve (a) K, potassium; (b) F, fluorine; (c) Si, silicon; (d) Be, beryllium (Figure 2.12).

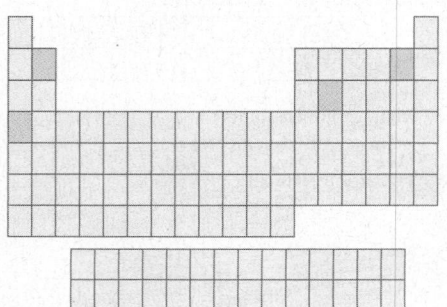

FIGURE 2.12

Think about It Each element has a unique location in the periodic table determined by its atomic number, which defines the row it is in, and its reactivity with other elements, which defines the group it is in. We assumed that beryllium was the only metal in the second row that could form a 1:2 compound with bromine. This is a valid assumption because the only other metal in the second row is Li, which is a group 1 element whose compound with Br has the formula LiBr.

Practice Exercise Write the symbol and name of each element (or elements):

a. The metalloid in family 15 closest in mass to the noble gas krypton
b. A representative element in the fourth row that is an alkaline earth metal
c. A transition metal in the sixth row you expect to have chemical properties similar to zinc ($Z = 30$)
d. A nonmetal in the third row that should have chemical properties similar to oxygen

(Answers to Practice Exercises are in the back of the book.)

> **law of multiple proportions** the ratio of the two masses of one element that react with a given mass of another element to form two different compounds is the ratio of two small whole numbers.

2.5 Trends in Compound Formation

In Section 2.1 we noted John Dalton's pioneering work in determining the atomic masses of the elements. Dalton also played a key role in documenting and predicting patterns in how elements combine to form compounds. We have already explored this topic in our discussion in the previous section of how Mendeleev placed elements in different groups in his early periodic table. Both Dalton and Mendeleev knew that when elements combine to form compounds, they do so in characteristic proportions. These proportions are reflected in the chemical formulas of compounds. For example, the formula of carbon dioxide, CO_2, tells us that in every molecule of CO_2 there is 1 atom of carbon combined with 2 atoms of oxygen.

Dalton's atomic view of compounds also explains another property of some elements: if the same two elements (for example, S and O) can form more than one compound (as in SO_2 and SO_3), the ratio of the different masses of O that react with a given mass of S to form the two compounds can be expressed as the ratio of two small whole numbers. This principle was observed experimentally and is known as Dalton's **law of multiple proportions**.

To see what this principle means, consider specifically SO_2 and SO_3. We determine in an experiment that, under one set of conditions, 10 g of oxygen reacts with 10 g of sulfur to form SO_2 but under different conditions, 15 g of oxygen reacts with 10 g of sulfur to form SO_3. The ratio of the two masses of oxygen is 10:15, or 2:3, which is a ratio of two small whole numbers. This example illustrates the law of multiple proportions.

Similarly, we can confirm experimentally that the mass of oxygen that reacts with a given mass of nitrogen to form NO_2 (22.8 g O for every 10.0 g N) is twice as much as the mass of oxygen that reacts with the same mass of nitrogen to form NO (11.4 g O for every 10.0 g N). The ratio of the two oxygen masses is 22.8 : 11.4, or 2:1, again a ratio of small whole numbers.

> **SAMPLE EXERCISE 2.4** **Using Chemical Formulas and the Law of Multiple Proportions**

Carbon combines with oxygen to form either CO or CO_2 (Figure 2.13), depending on reaction conditions. If 26.6 g of oxygen reacts with 10.0 g of carbon to make CO_2, how many grams of oxygen reacts with 10.0 g of carbon to make CO?

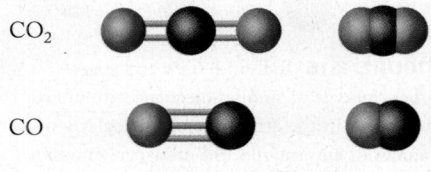

CO_2

CO

FIGURE 2.13

molecular compound a compound composed of atoms held together in molecules by covalent bonds.

covalent bond a bond between two atoms created by sharing one or more pairs of electrons.

molecular formula a notation showing the number and type of atoms present in one molecule of a molecular compound.

ionic compound a compound composed of positively and negatively charged ions held together by electrostatic attraction.

cation positively charged particle created when an atom or molecule loses one or more electrons.

anion negatively charged particle created when an atom or molecule gains one or more electrons.

empirical formula a formula showing the smallest whole-number ratio of elements in a compound.

formula unit the smallest electrically neutral unit of an ionic compound.

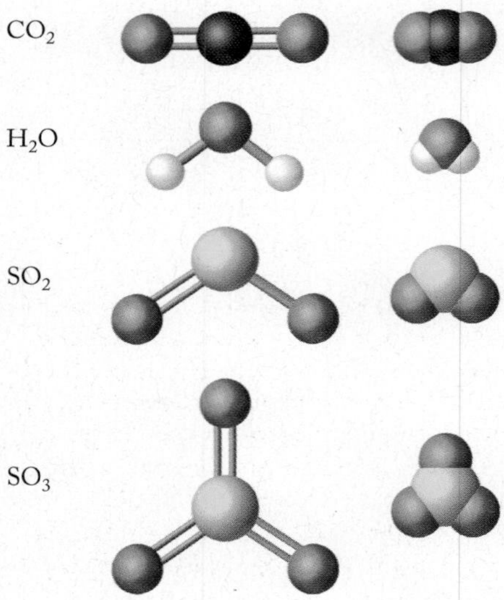

Collect and Organize The two compounds contain the same two elements but in different proportions, so Dalton's law of multiple proportions applies. We have formulas for both compounds, CO and CO_2, and are told that both reactions involve 10.0 g of carbon.

Analyze The ratio of the O atoms to C atoms in CO is 1:1. The ratio of O atoms to C atoms in CO_2 is 2:1. Therefore, half as much oxygen will react with 10.0 g of carbon to make CO as reacts with 10.0 g of carbon to make CO_2.

Solve (26.6 g of oxygen) $\times \frac{1}{2} = 13.3$ g of oxygen

Think about It Dalton's atomic view of these compounds is expressed by their chemical formulas. We used these chemical formulas in this exercise to calculate the different masses of oxygen required to react completely with a given mass of carbon to form the two compounds. In actual practice, the reverse is done: chemists analyze the masses of the elements in a compound and use that information to determine its molecular formula.

Practice Exercise Predict the mass of oxygen required to react with 14.0 g of nitrogen to make N_2O_5 if 16.0 g of oxygen reacts with 14.0 g of nitrogen to make N_2O_2 (Figure 2.14).

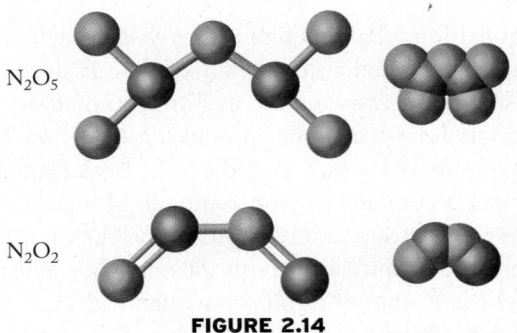

N_2O_5

N_2O_2

FIGURE 2.14

(Answers to Practice Exercises are in the back of the book.)

Molecular Compounds

All of the compounds we have examined so far in this section have been binary (two-element) **molecular compounds** (Figure 2.15). The building blocks of these compounds are molecules that contain atoms of two nonmetals (the elements with symbols in the blue cells of the periodic table inside the front cover). The atoms are connected together by shared pairs of electrons called **covalent bonds** (the sticks in the ball-and-stick models in Figure 2.15).

All of the compounds in Figure 2.15 may be present in the air we breathe, especially if we happen to live in a heavily populated area where air quality is impacted by exhaust from vehicles, power plants, and factories. All of the compounds contain oxygen and another nonmetal, and so are examples of *nonmetal oxides*. Each of their chemical formulas specifies the number of atoms of each element in one molecule of the compound. Therefore, these chemical formulas are **molecular formulas**. The fact that the same two elements can form compounds with different molecular formulas means that there are different ways to form covalent bonds between atoms of the same two elements. We explore the impact of different bonding patterns on the chemical properties of molecular compounds in several of the later chapters of this book.

CO_2

H_2O

SO_2

SO_3

NO

NO_2

FIGURE 2.15 Ball-and-stick and space-filling models of some molecular compounds that occur in the atmosphere, particularly near sources of automobile and industrial emissions.

Ionic Compounds

Now we shift our focus to binary **ionic compounds**. They each contain a metallic element (shown in tan in the periodic table in Figure 2.10) combined with a nonmetal (shown in blue). In an ionic compound each atom of the metal has lost one or more electrons, forming a positively charged ion called a **cation**, and each atom of the nonmetal has gained one or more electrons, forming a negatively charged **anion**. Because these ions start out as single atoms, they are called *monatomic* ions. The cations and anions in an ionic compound are held together by the electrostatic attraction that ions of opposite charge have for each other.

Consider the formation of a familiar ionic compound, sodium chloride, also known as table salt. We can use single atoms of sodium and chlorine to visualize how these elements combine. Each sodium atom loses an electron and forms a sodium ion:

$$Na \rightarrow Na^+ + e^-$$

Correspondingly, each chlorine atom gains an electron and forms a chloride ion:

$$Cl + e^- \rightarrow Cl^-$$

Figure 2.16(a) illustrates the loss and gain of electrons by these atoms. Note that the size of a sodium atom shrinks when it loses an electron and a chlorine atom expands when it gains one.

Figure 2.16(b) shows several crystals of sodium chloride and an atomic view of part of a crystal revealing the three-dimensional array of sodium ions and chloride ions. Within the crystal each sodium ion is surrounded by six chloride ions, and each chloride ion is surrounded by six sodium ions. However, the smallest whole-number ratio of sodium ions to chloride ions in the crystal is simply 1:1. Formulas based on the lowest whole-number ratio of the component elements in a compound are called **empirical formulas**. The chemical formulas of ionic compounds, such as NaCl for sodium chloride, are examples of empirical formulas. The empirical formula of an ionic compound describes a **formula unit**, the smallest electrically neutral unit within the crystal.

The periodic table helps us predict the charges on the monatomic ions that elements form and thereby helps us predict the chemical formulas of ionic compounds. For example, atoms of group 1 elements each lose one electron and form 1+ ions; atoms of group 2 elements each lose two electrons and form 2+ ions (Figure 2.17).

▶II **CHEMTOUR** NaCl Reaction

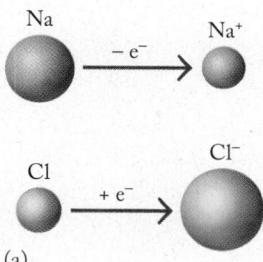

(a)

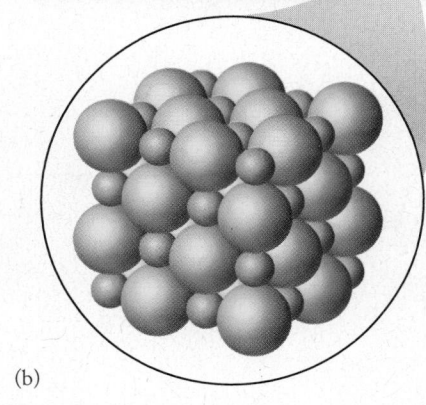

(b)

FIGURE 2.16 (a) A sodium atom forms a Na^+ cation by losing one electron. A chlorine atom forms a Cl^- anion by gaining one electron. (b) Crystals of sodium chloride. The cubic shape of the crystals mirrors the cubic array of Na^+ and Cl^- ions that make up its structure. The empirical formula NaCl describes the smallest whole-number ratio of cations to anions in the structure, which is electrically neutral.

FIGURE 2.17 The most common charges on the ions of some common elements. With the representative elements (groups 1, 2, and 13–18), all the ions in the group typically have the same charge. All the ions shown are made of only one atom except for the mercury ion Hg_2^{2+}.

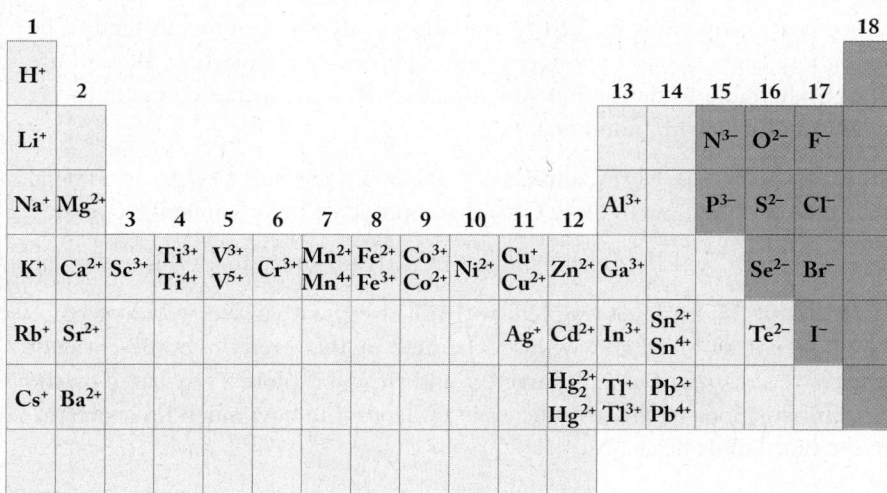

Note that the charges on these monatomic cations match the group numbers. Unfortunately, no strong correlation exists between group number and cation charge among the transition metals and the metallic elements on the right side of the periodic table, as you can see in Figure 2.17. Still, similarities are apparent within groups. For example, the most common charge of the group 13 monatomic ions is 3+.

As metallic elements lose electrons in forming ionic compounds, their nonmetallic partners gain them so that the overall charge on the resultant compounds is zero. As Figure 2.17 shows, the charge on the monatomic anions formed by the group 17 elements is 1−; the charge on the monatomic anions formed by the group 16 nonmetals is 2−, and the charge is 3− for the nonmetals in group 15. Note that the charge of each of these anions is the group number minus 18.

CONCEPT TEST ....

One of the following empirical formulas does not represent an electrically neutral compound. Which one? *Hint*: Base your selection on the charges of the common ions in Figure 2.17. (a) KBr; (b) MgF_2; (c) CsN; (d) TiO_2; (e) AgCl

....

CONCEPT TEST ....

Figure 2.18 shows a space-filling model of a compound that may be familiar to you, hydrogen peroxide. What are the molecular formula and empirical formula of hydrogen peroxide?

(Answers to Concept Tests are in the back of the book.)

....

FIGURE 2.18

SAMPLE EXERCISE 2.5   Classifying Compounds as Molecular or Ionic

Identify each of the following compounds as ionic or molecular: (a) sodium bromide (NaBr); (b) carbon dioxide (CO_2); (c) lithium iodide (LiI); (d) magnesium fluoride (MgF_2); (e) calcium chloride ($CaCl_2$).

Collect and Organize We are to distinguish between ionic and molecular compounds based on their names and chemical formulas. In this section we learned that compounds formed by reacting metals with nonmetals tend to be ionic; those that contain only nonmetallic elements are molecular. We can use the periodic table to determine which of the elements in the compounds are metallic and which are nonmetallic.

Analyze NaBr, LiI, MgF_2, and $CaCl_2$ all contain a group 1 or group 2 metal and a group 17 nonmetal. Only CO_2 is composed of two nonmetals.

Solve (a) NaBr, (c) LiI, (d) MgF_2, and (e) $CaCl_2$ are ionic; (b) CO_2 is molecular.

Think about It In later chapters we will discover that the world of compounds is not so black and white as painted in this exercise. Some covalent bonds have a degree of ionic "character," and we will explore a way based on the elements' positions in the periodic table to determine how much ionic character covalent bonds have.

Practice Exercise Which of these binary compounds are molecular and which are ionic? (a) carbon disulfide (CS_2); (b) carbon monoxide (CO); (c) ammonia (NH_3); (d) water (H_2O); (e) sodium iodide (NaI)

(Answers to Practice Exercises are in the back of the book.)

2.6 Naming Compounds and Writing Formulas

At this point we need to establish some rules for naming compounds and writing their chemical formulas. These names and formulas are a foundation of the language of chemistry. The periodic table is a valuable resource for naming simple compounds, for translating those names into chemical formulas, and for translating chemical formulas into names.

Binary Molecular Compounds

Translating the molecular formula of a binary molecular compound into the compound name is straightforward:

TABLE 2.2	Naming Prefixes for Molecular Compounds
one	*mono-*
two	*di-*
three	*tri-*
four	*tetra-*
five	*penta-*
six	*hexa-*
seven	*hepta-*
eight	*octa-*
nine	*nona-*
ten	*deca-*

1. Start with the name of the first element in the formula.
2. Change the ending of the name of the second element to *-ide*.
3. Add prefixes (Table 2.2) to the first and second names to indicate the number of atoms of each type in the molecule. (However, do not use the prefix *mono-* with the first element in a name.)

For example, NO is nitrogen monoxide (not *mono*nitrogen monoxide), NO_2 is nitrogen dioxide, SO_2 is sulfur dioxide, and SO_3 is sulfur trioxide. When prefixes ending in *o-* or *a-* (like *mono-* and *tetra-*) precede a name that begins with a vowel (such as *oxide*), the *o* or *a* at the end of the prefix is deleted to make the combination of prefix and name easier to pronounce: monoxide, not *mono*oxide.

The order in which the elements are named and given in formulas corresponds to their relative positions in the periodic table: the element with the lower group number appears first. When the elements are in the same group—for example, sulfur and oxygen—the element with the higher atomic number is named first.

SAMPLE EXERCISE 2.6 **Naming Binary Molecular Compounds**

What are the names of the compounds with these chemical formulas: (a) N_2O; (b) N_2O_4; (c) N_2O_5?

Collect and Organize All three compounds are binary nonmetal oxides and hence molecular compounds. Therefore, we use prefixes from Table 2.2 in the names to indicate the number of atoms of each element present in one molecule.

Analyze The first element in all three compounds is nitrogen, so the first word in each name is *nitrogen* with the appropriate prefix. There are two nitrogen atoms in the formulas in a, b, and c, so add the prefix *di-* to the name *nitrogen*. The second element in all three compounds is oxygen, so the second word in each name is *oxide* with the appropriate prefixes: *mono-* to indicate one O atom in a; *tetra-* to indicate four O atoms in b; *penta-* to indicate five O atoms in c.

Solve
  a. dinitrogen monoxide
  b. dinitrogen tetroxide
  c. dinitrogen pentoxide

Think about It To avoid back-to-back vowels in the middle of the second terms in all three names, we deleted the last letter of the three prefixes before *oxide*.

Practice Exercise Name these compounds: (a) P_4O_{10}; (b) CO; (c) NCl_3.

(Answers to Practice Exercises are in the back of the book.)

Binary Ionic Compounds

To name a binary ionic compound:

  1. Start with the name of the cation, which is simply the name of the parent element.
  2. Add the name of the anion, which is the name of the element except that the ending is changed to -*ide*.

Prefixes are not used in naming binary ionic compounds of representative elements because group 1 and 2 metals and aluminum in group 13 all have characteristic positive charges, as do the monatomic anions formed by the group 16 and 17 elements. Ionic compounds are electrically neutral, so the negative and positive charges in an ionic compound must balance, which dictates the number of each of the ions in the chemical formula. Therefore, the name *magnesium fluoride*, for example, is unambiguous: it can mean only MgF_2.

SAMPLE EXERCISE 2.7   Writing Formulas of Binary Ionic Compounds

Write the chemical formula of (a) potassium bromide, (b) calcium oxide, (c) sodium sulfide, (d) magnesium chloride, and (e) aluminum oxide.

Collect and Organize The name of each compound consists of the name of one main group metal and one main group nonmetal, which tells us that these are binary ionic compounds. To write formulas of ionic compounds, we assign the charges on the ions based on the group numbers of the parent elements.

Analyze Locate each element in the periodic table and predict the charge of its most common ion based on location and group number: K^+, Br^-, Ca^{2+}, O^{2-}, Na^+, S^{2-}, Mg^{2+}, Cl^-, and Al^{3+}. If you have difficulty predicting ionic charge, refer to Figure 2.17. Writing chemical formulas of the compounds is an exercise in balancing positive and negative charges.

Solve We must balance the positive and negative charges in each compound:
  a. In potassium bromide, the ionic charges are 1+ and 1− (K^+ and Br^-). A 1:1 ratio of the ions is required for electrical neutrality, making the formula KBr.
  b. In calcium oxide, the ionic charges are 2+ and 2−. A 1:1 ratio of Ca^{2+} to O^{2-} ions balances their charges, making the formula CaO.

   c. In sodium sulfide, the ionic charges are $1+$ and $2-$. A 2:1 ratio of Na^+ to S^{2-} ions is needed: Na_2S.

   d. In magnesium chloride, the ionic charges are $2+$ and $1-$. A 1:2 ratio of Mg^{2+} to Cl^- ions is needed: $MgCl_2$.

   e. In aluminum oxide, the ionic charges are $3+$ and $2-$. If we use two Al^{3+} ions for every three O^{2-} ions, the charges will balance. The formula is Al_2O_3.

Think about It Different approaches may be used to work out the formulas of ionic compounds. The basic principle is that the sum of the total positive and negative charges must balance to give a net charge of zero. If you had difficulty writing the formula of aluminum oxide, try this shortcut: use the charge on each ion as the subscript for the other ion. Thus the $3+$ charge on Al^{3+} becomes a subscript $_3$ after O, and the $2-$ charge on the oxide ion becomes a subscript $_2$ after Al. The result is Al_2O_3:

$$Al^{3+}O^{2-} \rightarrow Al_2O_3$$

Practice Exercise Write the chemical formulas of (a) strontium chloride, (b) magnesium oxide, (c) sodium fluoride, and (d) calcium bromide.

(Answers to Practice Exercises are in the back of the book.)

Binary Compounds of Transition Metals

Some metallic elements, including many of the transition metals, form several cations carrying different charges. For example, most of the copper found in nature is present as Cu^{2+}; however, some copper compounds contain Cu^+ ions. Because the name *copper chloride* could apply to either $CuCl_2$ or $CuCl$, systematic names are needed to distinguish between the two compounds. One system uses a Roman numeral after the word *copper* in the name of the compound, which defines the charge on the copper ion. Thus copper(II) chloride is the chloride of Cu^{2+} ($CuCl_2$), and copper(I) chloride (CuCl) is the chloride of Cu^+.

   Chemists for many years used different names to identify different cations of the same element. For instance, Cu^+ is called the *cuprous* ion and Cu^{2+} is called *cupric*. Similarly, the ions Fe^{2+} and Fe^{3+} are called *ferrous* and *ferric*, respectively. Note that, in both these pairs of ions, the name of the ion with the lower charge ends in *-ous* and the name of the ion with the higher charge ends in *-ic*.

> **SAMPLE EXERCISE 2.8**   **Writing Formulas of Transition Metal Compounds**

(a) Write the chemical formulas of iron(II) sulfide and iron(III) oxide. (b) Write alternative names for these compounds that do not use Roman numerals to indicate the charge on the iron ions.

Collect and Organize We are to write chemical formulas for two ionic compounds. Because iron is a transition metal, Roman numerals are used to indicate the charges on the iron ions.

Analyze The Roman numerals (II) and (III) indicate that the charges on the iron cations are $2+$ and $3+$, respectively. Oxygen and sulfur are both in group 16.

polyatomic ions charged groups of two or more atoms joined together by covalent bonds.

oxoanions polyatomic ions that contain oxygen in combination with one or more other elements.

Therefore the charge on both the sulfide ion and oxide ion is 2−. In the alternate naming system, Fe^{2+} is the *ferrous* ion and Fe^{3+} is the *ferric* ion.

Solve

a. A charge balance in iron(II) sulfide is achieved with equal numbers of Fe^{2+} and S^{2-} ions, so the chemical formula is FeS. To balance the different charges on the Fe^{3+} and O^{2-} ions in iron(III) oxide, we need three O^{2-} ions for every two Fe^{3+} ions. Thus the formula of iron(III) oxide is Fe_2O_3.

b. The alternate names of FeS and Fe_2O_3 are ferrous sulfide and ferric oxide, respectively.

Think about It We use the Roman numeral system for designating the charges on transition metal ions, but you may encounter *-ous/-ic* nomenclature in older books and articles.

Practice Exercise Write the formulas of manganese(II) chloride and manganese(IV) oxide.

(Answers to Practice Exercises are in the back of the book.)

Polyatomic Ions

Table 2.3 lists commonly encountered ions, several of which are **polyatomic ions** that consist of more than one kind of atom joined by covalent bonds. The ammonium ion (NH_4^+) is the only common polyatomic cation; all the others are anions.

Polyatomic ions containing oxygen and one or more other elements are called **oxoanions**. Most oxoanions have a name based on the name of the element that

TABLE 2.3	**Names and Charges of Some Common Ions**		
Name	**Chemical Formula**	**Name**	**Chemical Formula**
Acetate	CH_3COO^-	Hydrogen phosphate	HPO_4^{2-}
Ammonium	NH_4^+	Hydrogen sulfite or bisulfite	HSO_3^-
Azide	N_3^-	Hydroxide	OH^-
Bromide	Br^-	Nitrate	NO_3^-
Carbonate	CO_3^{2-}	Nitride	N^{3-}
Chlorate	ClO_3^-	Nitrite	NO_2^-
Chloride	Cl^-	Oxide	O^{2-}
Chromate	CrO_4^{2-}	Perchlorate	ClO_4^-
Cyanide	CN^-	Permanganate	MnO_4^-
Dichromate	$Cr_2O_7^{2-}$	Peroxide	O_2^{2-}
Dihydrogen phosphate	$H_2PO_4^-$	Phosphate	PO_4^{3-}
Disulfide	S_2^{2-}	Sulfate	SO_4^{2-}
Fluoride	F^-	Sulfide	S^{2-}
Hydride	H^-	Sulfite	SO_3^{2-}
Hydrogen carbonate or bicarbonate	HCO_3^-	Thiocyanate	SCN^-

appears first in the formula, but the ending is changed to either *-ite* or *-ate*, depending on the number of oxygen atoms in the formula. Thus, for example, SO_4^{2-} is the sulfate ion and SO_3^{2-} is the sulfite ion. Note that the *-ate* oxoanions have more oxygen atoms than the corresponding *-ite* oxoanions.

If an element forms more than two kinds of oxoanions, as chlorine and the other group 17 elements do, prefixes are used to distinguish among them (Table 2.4). The oxoanion with the largest number of oxygen atoms has the prefix *per-*, and the one with the smallest number of oxygen atoms may have the prefix *hypo-* in its name. Because these rules do not enable you to predict the chemical formula of an oxoanion either from the name or from the charge on the anion, you need to memorize the formulas, charges, and names of the common oxoanions in Table 2.4.

TABLE 2.4   Oxoanions of Chlorine and Their Corresponding Acids

Ions		Acids	
ClO^-	hypochlorite	$HClO$	hypochlorous acid
ClO_2^-	chlorite	$HClO_2$	chlorous acid
ClO_3^-	chlorate	$HClO_3$	chloric acid
ClO_4^-	perchlorate	$HClO_4$	perchloric acid

SAMPLE EXERCISE 2.9   Writing the Formulas of Compounds Containing Oxoanions

Write the chemical formulas of (a) sodium sulfate and (b) magnesium phosphate.

Collect and Organize We are given the names of two compounds containing oxoanions and are to write their chemical formulas. The cations in these compounds are those formed by Na and Mg atoms.

Analyze To write the formulas of these ionic compounds, we need to know the formulas and charges of the ions. Sodium is in group 1, and magnesium is in group 2. The charges on their ions are 1+ and 2+, respectively. The sulfate ion is SO_4^{2-}, and phosphate is PO_4^{3-}.

Solve
   a. To balance the charges on Na^+ and SO_4^{2-}, we need twice as many Na^+ ions as SO_4^{2-} ions. Therefore the formula is Na_2SO_4.
   b. To balance the charges on Mg^{2+} and PO_4^{3-}, we need three Mg^{2+} ions for every two PO_4^{3-} ions, which gives us $Mg_3(PO_4)_2$.

Think about It To complete this exercise we had to know the formulas and charges of the sulfate and phosphate oxoanions. The charges on the cations could be inferred from the positions of the elements in the periodic table. In writing the formula, we used parentheses around the phosphate ion in magnesium phosphate to make it clear that the subscript 2 applies to the entire oxoanion.

Practice Exercise Write the chemical formulas of (a) strontium nitrate and (b) potassium sulfite.

(Answers to Practice Exercises are in the back of the book.)

SAMPLE EXERCISE 2.10 **Naming Compounds Containing Oxoanions**

Name the following compounds: (a) $CaCO_3$, (b) $LiNO_3$, (c) $MgSO_3$, (d) $RbNO_2$, (e) $KClO_3$, and (f) $NaHCO_3$.

Collect and Organize We are to name six compounds each containing an oxoanion. The names of ionic compounds begin with the names of the parent elements of the cations followed by the names of the oxoanions.

Analyze The cations in these compounds are those formed by atoms of the elements (a) calcium, (b) lithium, (c) magnesium, (d) rubidium, (e) potassium, and (f) sodium. The names of the oxoanions, as listed in Table 2.3, are (a) carbonate, (b) nitrate, (c) sulfite, (d) nitrite, (e) chlorate, and (f) hydrogen carbonate.

Solve Combining the names of these cations and oxoanions, we get (a) calcium carbonate, (b) lithium nitrate, (c) magnesium sulfite, (d) rubidium nitrite, (e) potassium chlorate, and (f) sodium hydrogen carbonate.

Think about It Sodium hydrogen carbonate is often called sodium bicarbonate. The prefix *bi-* is sometimes used to indicate that there is a hydrogen ion (H^+) attached to an oxoanion.

Practice Exercise Name these compounds: (a) $Ca_3(PO_4)_2$; (b) $Mg(ClO_4)_2$; (c) $LiNO_2$; (d) $NaClO$; (e) $KMnO_4$.

(Answers to Practice Exercises are in the back of the book.)

Acids

Certain compounds have special names that highlight particular chemical properties. Among these are acids. We discuss acids in greater detail in later chapters, but for now it is sufficient to say that acids are compounds that release hydrogen ions (H^+) when they dissolve in water. Their chemical formulas begin with H. For example, the binary compound HCl is hydrogen chloride. When HCl dissolves in water, it produces the acidic solution we call hydrochloric acid. To name this and other binary acids:

1. Affix the prefix *hydro-* to the name of the element other than hydrogen.
2. Replace the last syllable in the name from step 1 with the suffix *-ic* and add *acid*.

The most common binary acids are compounds of hydrogen and the halogens. Their aqueous solutions are hydrofluoric, hydrochloric, hydrobromic, and hydroiodic acid.

The scheme for naming the acids of oxoanions, called *oxoacids*, is illustrated in Table 2.4. If the oxoanion name ends in *-ate*, the name of the corresponding oxoacid ends in *-ic*; if the oxoanion name ends in *-ite*, the name of the oxoacid ends in *-ous*. Thus, the acid of the sulf*ate* (SO_4^{2-}) ion is sulfur*ic* acid (H_2SO_4) and the acid of the nitr*ite* (NO_2^-) ion is nitr*ous* acid (HNO_2).

SAMPLE EXERCISE 2.11   **Naming Oxoacids**

Name the oxoacids formed by the following oxoanions: (a) SO_3^{2-}; (b) ClO_4^-; (c) NO_3^-.

Collect and Organize We are given the formulas of three oxoanions and are to name the oxoacids formed when they combine with H^+ ions.

Analyze According to Table 2.3, the names of the oxoanions are (a) sulfite, (b) perchlorate, and (c) nitrate. When the oxoanion name ends in *-ite*, the corresponding oxoacid name ends in *-ous*. When the anion name ends in *-ate*, the oxoacid name ends in *-ic*.

Solve Making the appropriate changes to the endings of the oxoanion names and adding the word *acid*, we get (a) sulfurous acid, (b) perchloric acid, and (c) nitric acid.

Think about It Once we know the names of the common oxoanions, naming the corresponding oxoacids is simply a matter of changing the ending of the oxoanion name from *-ate* to *-ic*, or from *-ite* to *-ous*, and then adding the word *acid*.

Practice Exercise Name these acids: (a) $HClO$; (b) $HClO_2$; (c) H_2CO_3.

(Answers to Practice Exercises are in the back of the book.)

Living organisms produce another class of acids called **carboxylic acids**. For example, fermentation reactions may produce acetic acid, which has the molecular structure shown in Figure 2.19(a). Vinegar is a solution of acetic acid. The molecular structures of all carboxylic acids include this combination of atoms:

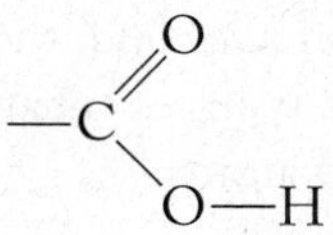

$$
\begin{array}{c}
\quad\quad O \\
\quad\ \parallel \\
-C \\
\quad\ \backslash \\
\quad\ O-H
\end{array}
$$

which is called the carboxylic acid **functional group**. Ionization of the hydrogen atom in the group (Figure 2.19b) is the source of the acidity of these compounds.

Other types of functional groups are responsible for the characteristic chemical and physical properties of other classes of **organic compounds**. All organic

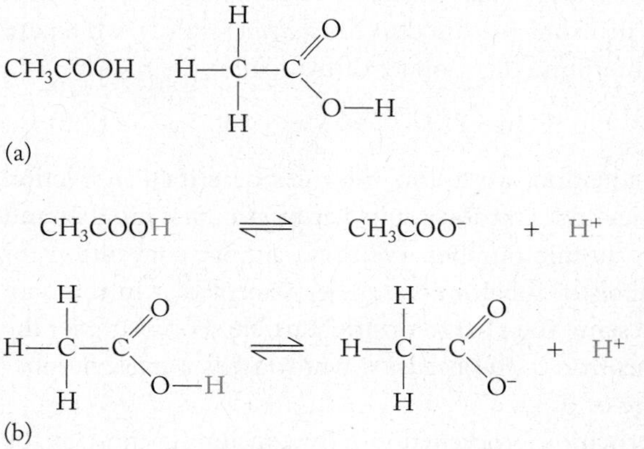

(a) CH_3COOH

(b) $CH_3COOH \rightleftharpoons CH_3COO^- + H^+$

FIGURE 2.19 (a) The chemical and structural formulas of acetic acid. (b) Only the hydrogen atom on the carboxylic acid group (–COOH) ionizes.

carboxylic acid an organic compound containing the –COOH functional group.

functional group a structural subunit in organic molecules that imparts characteristic chemical and physical properties.

organic compounds compounds containing carbon, and commonly including certain other elements such as hydrogen, oxygen, and nitrogen.

quarks elementary particles that combine to form neutrons and protons.

nucleosynthesis the natural formation of nuclei as a result of fusion and other nuclear processes.

compounds are carbon-containing substances that were originally considered the products of living systems only. The definition has broadened and now includes most compounds of carbon.

2.7 Nucleosynthesis

We began this chapter by discussing some remarkable advances that took place in the early 20th century that provided science with a clearer view of the structure of atoms and with new insights into why elements react the way they do. At the same time these discoveries were taking place, scientists were making remarkable strides in advancing our understanding of how the elements may have formed in the first place. We began to explore these concepts and the evolution of the Big Bang theory in Chapter 1. Now it is time to discuss how the two most abundant elements in the universe, hydrogen and helium, formed shortly after the Big Bang and how the other elements formed, and continue to be synthesized, in the cores of giant stars.

Theoretical physicists believe that within a few microseconds after the Big Bang much of the energy released at the instant of creation had transformed into matter, and that this matter consisted of the smallest of subatomic particles: electrons and **quarks** (Figure 2.20). Less than a millisecond later, the universe had expanded and "cooled" to a mere 10^{12} K and quarks combined with one another to form neutrons and protons. Thus in less than a second, the matter in the universe consisted of the three types of subatomic particles that would eventually make up atoms.

Primordial Nucleosynthesis

By about 4 minutes after the Big Bang, the universe had expanded and cooled to 10^9 K. In this hot, dense, subatomic "soup," neutrons and protons that collided with one another began to fuse together in a process called primordial **nucleosynthesis**. In one step, protons (p) and neutrons (n) fused to form *deuterons* (d), which are nuclei of the deuterium ($_1^2$H) isotope of hydrogen:

$$_1^1\text{p} + _0^1\text{n} \rightarrow _1^2\text{d} \qquad (2.2)$$

In writing this equation, we follow the rules described in Section 2.2 for writing nuclide symbols: a superscript is a mass number and a subscript is an atomic number. We use a similar convention for writing the symbols of subatomic particles, except that in this case a subscript represents the charge on the particle. For example, the symbol of the neutron is $_0^1$n because a neutron has a mass number of 1 and a charge of 0.

Deuteron formation proceeded rapidly, consuming most of the neutrons in the universe in a matter of seconds. No sooner did

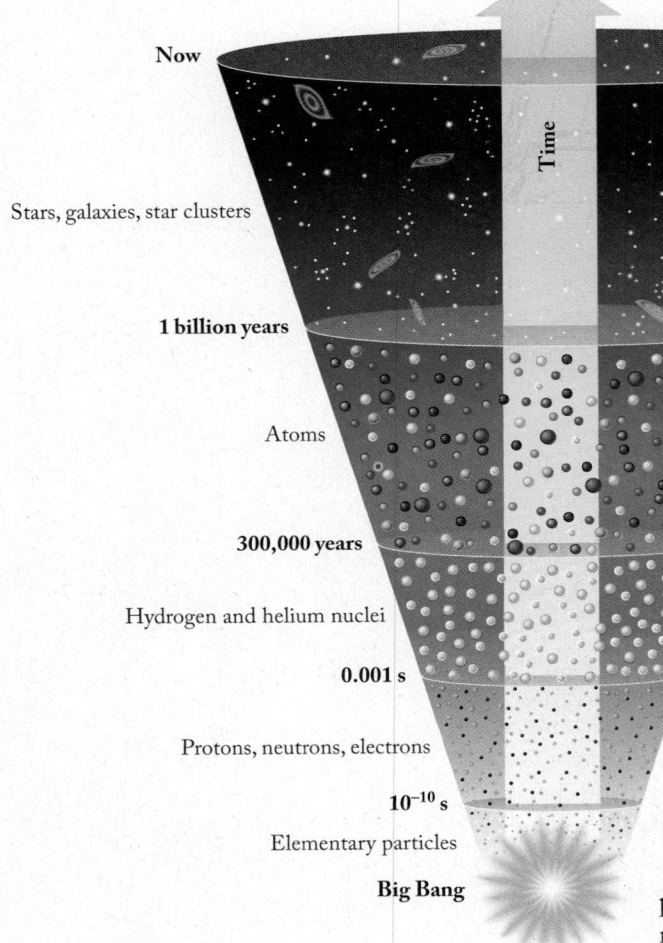

FIGURE 2.20 Timeline for energy and matter transformations believed to have occurred since the universe began. In this model, protons, neutrons, and electrons were formed from quarks in the first millisecond after the Big Bang, followed by hydrogen and helium nuclei. Whole atoms of H and He did not form until after 300,000 years of expansion and cooling, and other elements did not form until the first galaxies appeared, around 1 billion years after the Big Bang. According to this model, our solar system, our planet, and all life forms on it are composed of elements synthesized in stars that were born, burned brightly, and then disappeared millions to billions of years after the Big Bang.

deuterons form than they too were rapidly consumed by colliding with one another and fusing to make $_2^4$He nuclei, which are the same as α particles:

$$2\,_1^2\text{d} \rightarrow\,_2^4\alpha \qquad\qquad (2.3)$$

Equations 2.2 and 2.3 are called *nuclear equations*, and they are related to the chemical equations we will begin writing in Chapter 3 in that they are *balanced*. This means that in each equation the sum of the masses (superscripts) of the particles to the left of the arrow is equal to the sum of the masses of the particles to the right of the arrow. Similarly the sum of the charges (subscripts) of the particles on the left side is equal to the sum of the charges of the particles on the right.

About 5 minutes after the Big Bang and as a result of the fusion reactions shown in Equations 2.2 and 2.3, the matter of the universe was about 75% (by mass) protons and 25% α particles. Primordial nucleosynthesis then came to a halt. Why did it halt? Why didn't α particles and protons fuse together to make ^{5}Li:

$$_2^4\alpha +\,_1^1\text{p} \xrightarrow{?}\,_3^5\text{Li}$$

or why didn't two α particles fuse together to make a nucleus of ^{8}Be?

$$2\,_2^4\alpha \xrightarrow{?}\,_4^8\text{Be}$$

Neither process took place because neither ^{5}Li nor ^{8}Be is stable under the conditions present in the early universe. In fact, there are no stable nuclei with mass numbers of 5 or 8. As the universe continued to expand and cool, its temperature dropped below 10^8 K, which is too low to sustain nuclear fusion. Cooler temperatures allowed electrons to combine with protons and alpha particles to form neutral atoms of hydrogen and helium. As a result, the elemental composition of the universe remained 75% hydrogen and 25% helium for millions of years.

Stellar Nucleosynthesis

That matter in the universe today still consists mostly of atoms of ^{1}H and ^{4}He is strong evidence supporting the Big Bang theory. But how did the other elements in the periodic table form, including those that make up most of our planet? Scientists theorize that the synthesis of elements more massive than helium had to wait until nuclear fusion resumed in the first generation of stars. Inside the coalescing masses of hydrogen and helium that would turn into the first stars, these gases underwent enormous compression heating and the nuclear furnaces that are the source of the energy in all stars were ignited as protons began to fuse, making more α particles.[3]

Some of these stars, known as *red giants*, had cores so extraordinarily hot and dense that sometimes groups of three α particles collided with each other simultaneously and fused together. This *triple-alpha process* produces a nucleus with 6 protons and 6 neutrons, which is the stable isotope carbon-12:

$$3\,_2^4\alpha \rightarrow\,_6^{12}\text{C}$$

With the formation of ^{12}C, the barrier to the nucleosynthesis of larger nuclei, which had stopped the process only minutes after the Big Bang, had been overcome. At the even higher temperatures in the cores of stars called *supergiants*, ^{12}C nuclei fuse with α particles to form ^{16}O (Figure 2.21). Then ^{16}O fuses with an α particle to

[3] The hydrogen–helium fusion process going on today in stars such as our sun is not the same as in early nucleosynthesis because today there are few free neutrons in these stars. Details of stellar hydrogen fusion are described in Chapter 21.

▶❚❚ **CHEMTOUR** Synthesis of Elements

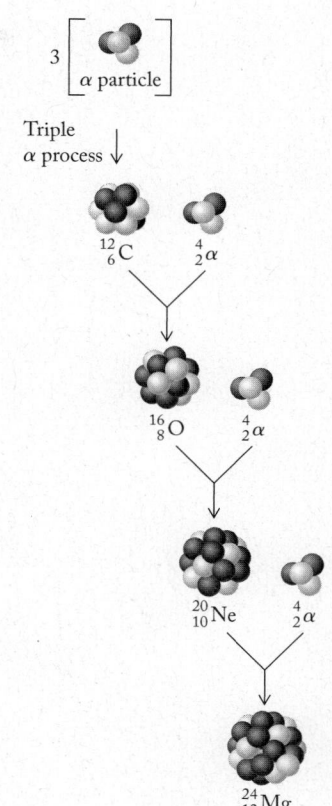

FIGURE 2.21 Fusion of three α particles, which forms carbon-12 in the triple-alpha process, followed by fusion of successively more massive nuclei to form oxygen-16, neon-20, magnesium-24, and so on. These fusion processes release the energy that fuels the nuclear furnaces of stars today.

FIGURE 2.22 The star η-Carinae is believed to be evolving toward an explosion. The outer regions are still fueled by energy released as hydrogen isotopes fuse, but the star is increasingly hotter and denser closer to the center. This central heating allows the fusion of larger nuclei and results in the production of ^{56}Fe in the core.

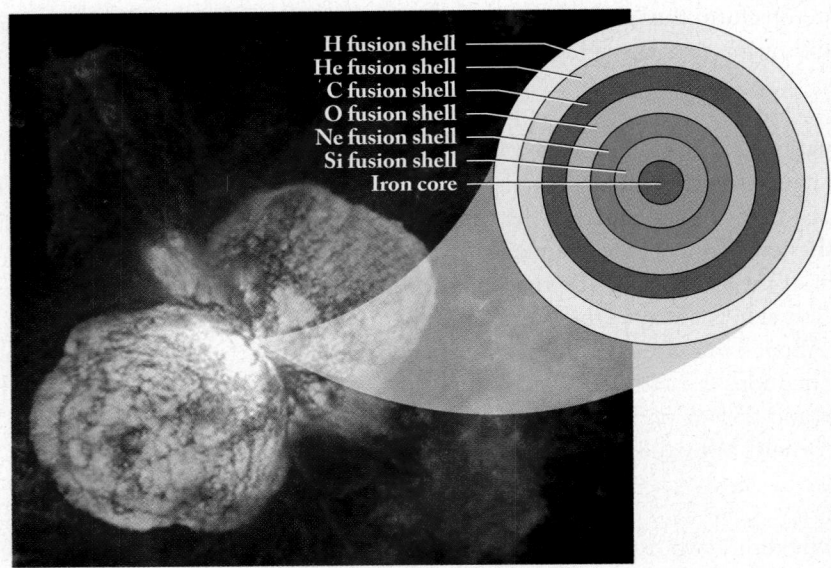

form ^{20}Ne, and so on. Additional fusion reactions involving larger nuclei with increasingly greater positive charges are possible in these intensely hot (10^9 K) stars because nuclei move fast enough to overcome the electrostatic repulsion experienced by particles with large positive charges.

For the past 13 billion years, fusion reactions of the sort described above have simultaneously fueled the nuclear furnaces of stars and produced isotopes as heavy as ^{56}Fe (Figure 2.22). However, once the core of a star turns into iron, the star is in trouble because fusion reactions involving iron nuclei do not release energy; instead they *consume* it. For example, fusing an alpha particle with ^{56}Fe to make ^{60}Ni,

$$^{56}_{26}\text{Fe} + {}^{4}_{2}\alpha \rightarrow {}^{60}_{28}\text{Ni}$$

requires the *addition* of energy. A star with an iron core has essentially run out of fuel. Its nuclear furnace goes out, and the star begins to cool and collapse into itself.

As the star collapses, compression reheats its core to above 10^9 K. At such temperatures, nuclei begin to disintegrate into free protons and neutrons. Free neutrons readily fuse with atomic nuclei in a process called **neutron capture**. When a stable nucleus captures enough neutrons, it becomes unstable. For example, if a nucleus of ^{56}Fe captures 3 neutrons it forms the unstable nuclide ^{59}Fe:

$$^{56}_{26}\text{Fe} + 3\,{}^{1}_{0}\text{n} \rightarrow {}^{59}_{26}\text{Fe}$$

Iron-59 spontaneously undergoes **beta (β) decay** (it emits a β particle). We can write a nuclear equation for the decay process by starting with what we know:

$$^{56}_{26}\text{Fe} \rightarrow ? + {}^{0}_{-1}\beta \qquad (2.4)$$

We use the symbol ${}^{0}_{-1}\beta$ to represent a beta particle, with the subscript indicating a charge of $1-$. We assign the particle a mass number of 0 because an electron has less than 1/1000 the mass of a neutron or proton (Table 2.1). We can identify the unknown product by balancing the superscripts and subscripts on both sides of the reaction arrow. To do that requires that the unknown nuclide have a superscript (mass number) of 59 and a subscript (atomic number) of 27. A check of the periodic table reveals that the element with atomic number 27 is cobalt, so the product is cobalt-59, which is a stable isotope. The overall equation describing the neutron capture and β decay is

$$^{56}_{26}\text{Fe} + 3\,{}^{1}_{0}\text{n} \rightarrow {}^{59}_{27}\text{Co} + {}^{0}_{-1}\beta \qquad (2.5)$$

neutron capture the absorption of a neutron by a nucleus.

beta (β) decay the process by which a neutron decays into a proton and a β particle.

In figure (labels): H fusion shell / He fusion shell / C fusion shell / O fusion shell / Ne fusion shell / Si fusion shell / Iron core

Note that the product ^{59}Co has 27 protons but the starting material ^{56}Fe has only 26 protons. The additional proton was produced during the β decay shown in Equation 2.4 as a neutron disintegrated into a proton, which stayed in the nucleus and thus transformed ^{59}Fe to ^{59}Co, and a β particle, which was emitted from the nucleus:

$$\,^1_0n \rightarrow \,^1_1p \,+\, \,^{\,0}_{-1}\beta \qquad\qquad (2.6)$$

SAMPLE EXERCISE 2.12   Predicting the Product of β Decay

An isotope of krypton, Kr ($Z = 36$), undergoes β decay. Which element is produced?

Collect and Organize We know that an isotope of krypton undergoes β decay. When it does, one neutron in each nucleus decays into a β particle and a proton as described in Equation 2.6.

Analyze Beta decay increases the number of protons in a nucleus by one. Krypton has 36 protons, so the new isotope will have 37 protons and an atomic number of 37.

Solve The element with $Z = 37$ is rubidium (Rb).

Think about It Neutron capture followed by β decay is a key process in the stellar nucleosynthesis of elements with atomic numbers greater than 26. Rubidium is one such element.

Practice Exercise A nuclide of arsenic ($Z = 33$) undergoes neutron capture followed by β decay. Which element is produced?

(Answers to Practice Exercises are in the back of the book.)

Repeated neutron capture and β decay events in the cores of collapsing stars produce nuclei of the most massive elements in the periodic table. However, one more chapter in the element-manufacturing story needs to be told: distribution of the products. The enormous heating that occurs when a dying star collapses produces a gigantic explosion. Cosmologists call such an event a *supernova*. In addition to finishing the job of synthesizing the elemental building blocks found in the universe, a supernova serves as its own element-distribution system, blasting its inventory of elements throughout its galaxy (Figure 2.23). The legacies of supernovas are found in the elemental composition of later-generation stars like our sun and in the planets that orbit these stars. Indeed, everything that exists in our solar system, including all life, is made from the residues of exploding stars.

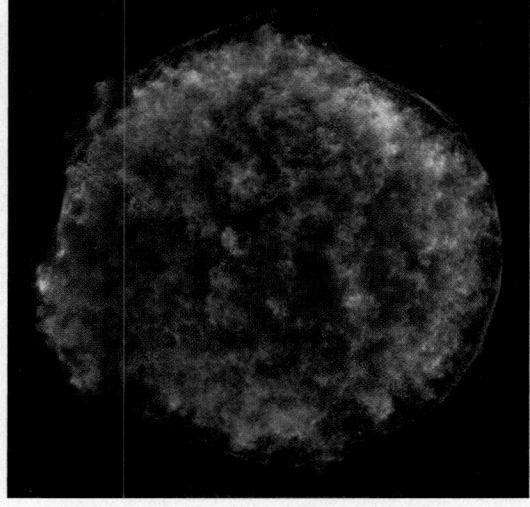

FIGURE 2.23 This colorized picture from the Chandra X-ray Observatory shows the remains of the supernova of the star Tycho in the constellation Cassiopeia. The expanding bubble of debris colored red and green is a cloud of hot ionized gas inside a more rapidly moving shell of extremely high-energy electrons in blue.

Section 2.1 Atoms are composed of negatively charged **electrons** surrounding a **nucleus**, which contains positively charged **protons** and electrically neutral **neutrons**. The values of the charge and mass of the electron were determined by J. J. Thomson's studies using cathode-ray tubes and by Robert Millikan's oil-drop experiments. The research of Ernest Rutherford's group, who bombarded thin gold foil with α **particles** and determined how the particles were deflected, showed that all the positive charge and nearly all the mass of an atom are contained in its nucleus.

Section 2.2 The number of protons in the nucleus of an element defines its **atomic number (Z)**; the number of **nucleons** (protons and neutrons) in the nucleus defines the element's **mass number (A)**. The different **isotopes** of an element consist of atoms with the same number of protons per nucleus but different numbers of neutrons.

Section 2.3 The **average atomic mass** of an element is found by multiplying the mass of each of its stable isotopes by the **natural abundance** of that isotope, and then summing these products.

Section 2.4 Elements are arranged in the periodic table of the elements in order of increasing atomic number and in a pattern based on their physical and chemical properties. Elements in the same vertical column are said to be in the same **group**, or **family**. Among the **main group** (or **representative**) **elements** are the **alkali metals** (group 1), the **alkaline earth metals** (group 2), and the **halogens** (group 17). Most of the elements are **metals**, which means that they are malleable, ductile solids (except mercury) and are good conductors of heat and electricity. **Nonmetals** include elements in all three physical states that are poor conductors of heat and electricity. **Metalloids**, or **semimetals**, have many of the physical properties of metals but the chemical properties of nonmetals; in the periodic table they are located between the metals to the left and the nonmetals to the right.

Section 2.5 John Dalton's atomic theory of matter explains the **law of multiple proportions**, which states that the ratio between two different masses of element Y, reacting with a given mass of element X to form two different compounds, is the ratio of two small whole numbers. When nonmetals react with other nonmetals or with metalloids, they

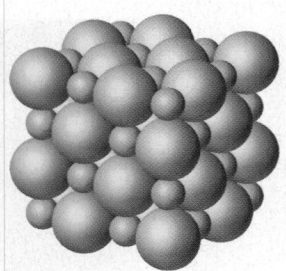

form **molecular compounds**, in which atoms are held together by **covalent bonds**. A **molecular formula** describes the exact number and type of atoms present in one molecule of a compound. When metals react with nonmetals, the metals form positively charged ions, called **cations**, and the nonmetals form negatively charged ions, called **anions**. These ions are held together by electrostatic attraction in solid **ionic compounds**. The charge of the ions formed by a main group element can be predicted on the basis of which group the element is in. The **empirical formula** of a molecular or ionic compound gives the smallest whole-number ratio of the atoms (or ions) in it.

Section 2.6 To write the name of binary compounds, first write the name of the element that is to the left of, or, if the elements are in the same group, below the other one in the periodic table. In the names of molecular compounds, prefixes indicate the number of atoms of the second element in each molecule and the number of atoms of the first one when there is more than one atom per molecule. The ending of the name of the second element is changed to -*ide*. For binary ionic compounds, the name or symbol of the metallic element is followed by the name or symbol of the nonmetal, but the ending of the name of the nonmetal is changed to -*ide*. Roman numerals in parentheses indicate the charges of cations formed by transition metals. The names of **oxoanions** (**polyatomic ions** containing oxygen atoms) end in -*ate* or -*ite* and may have a prefix of *per-* or *hypo-* or no prefix, depending on the number of oxygen atoms per ion. In the formulas of most acids hydrogen is written first. The names of binary acids begin with the prefix *hydro-* and end with -*ic acid*. The names of acids based on -*ate* and -*ite* oxoanions end in -*ic* and -*ous*, respectively.

Section 2.7 Neutrons, protons, and electrons formed within seconds of the Big Bang. During primordial **nucleosynthesis** protons and neutrons fused to produce nuclei of helium. After galaxies formed, the nuclei of atoms with $Z \leq 26$ formed when the nuclei of lighter elements fused together in the cores of giant stars (stellar nucleosynthesis). The nuclei of elements with $Z > 26$ formed by a combination of **neutron capture**, β **decay**,

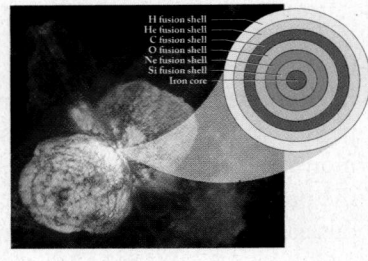

and other nuclear reactions that occur during supernovas (explosions of giant stars). As a result of these explosions, the elements produced were distributed throughout galaxies for possible inclusion in later-generation stars and in planets such as our own. Stellar nucleosynthesis continues today.

PROBLEM-SOLVING SUMMARY

TYPE OF PROBLEM	CONCEPTS AND EQUATIONS	SAMPLE EXERCISES
Writing symbols of isotopes	To the left of the element symbol, place a superscript for the mass number (A) and a subscript for the atomic number (Z).	2.1
Calculating the average atomic mass of an element	Multiply the mass (m) of each stable isotope of the element times the natural abundance (a) of that isotope; then sum these products: $$m_X = a_1 m_1 + a_2 m_2 + a_3 m_3 + \cdots \qquad (2.1)$$	2.2
Calculating the quantity of an element in one compound based on the quantity of the same element in another compound	Use the chemical formulas of the two compounds and the law of multiple proportions.	2.4
Classifying compounds as ionic or molecular	Ionic compounds contain metallic and nonmetallic elements; molecular compounds contain nonmetals and/or metalloids.	2.5
Naming binary inorganic compounds and writing their formulas	Apply the naming rules in Section 2.6 summary on page 70.	2.6, 2.7
Naming transition metal compounds and writing their formulas	Use a Roman numeral to indicate the charge on the transition metal cation.	2.8
Naming compounds containing oxoanions and writing their formulas	Apply the naming rules in Section 2.6 summary on page 70.	2.9–2.11
Predicting the product of neutron capture followed by β decay	Neutron capture adds to the mass number of a nucleus; emission of a β particle increases its atomic number by one.	2.12

VISUAL PROBLEMS

(Answers to boldface end–of–chapter questions and problems are in the back of the book.)

2.1. In Figure P2.1 the blue spheres represent nitrogen atoms and the red spheres represent oxygen atoms. The figure as a whole represents which of the following gases? (a) N_2O_3; (b) N_7O_{11}; (c) a mixture of NO_2 and NO; (d) a mixture of N_2 and O_3

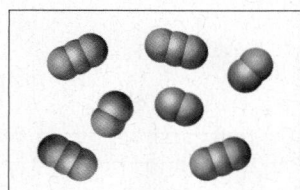

FIGURE P2.1

2.2. In Figure P2.2 the black spheres represent carbon atoms and the red spheres represent oxygen atoms. Which of the following statements about the two equal-volume compartments is or are true?

a. The compartment on the left contains CO_2; the one on the right contains CO.

b. The compartments contain the same mass of carbon.

c. The ratio of the oxygen to carbon in the gas in the left compartment is twice that of the gas in the right compartment.

d. The pressures inside the two compartments are equal. (Assume that the pressure of a gas is proportional to the number of particles in a given volume.)

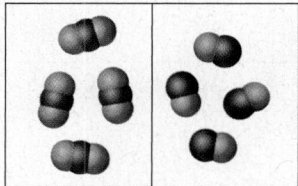

FIGURE P2.2

2.3. Which of the highlighted elements in Figure P2.3 formed, according to the theory of the Big Bang, before the first generation of galaxies formed?

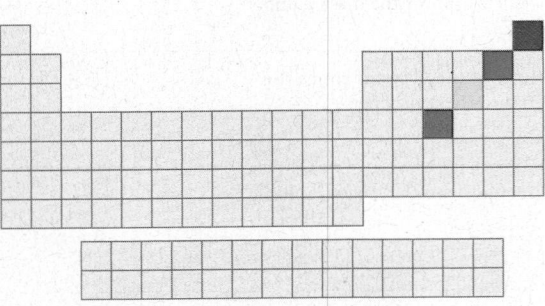

FIGURE P2.3

2.4. Which of the highlighted elements in Figure P2.4 *is not* formed by fusion of lighter elements in the cores of giant stars?

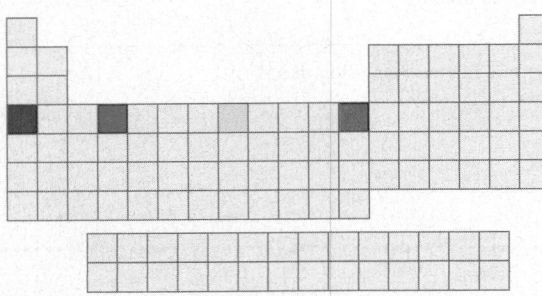

FIGURE P2.4

2.5. Which of the highlighted elements in Figure P2.5 is (a) a reactive nonmetal; (b) a chemically inert gas; (c) a reactive metal?

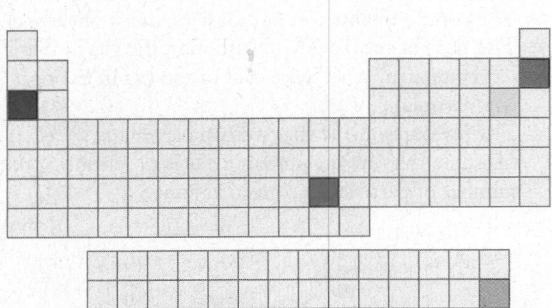

FIGURE P2.5

2.6. Which of the highlighted elements in Figure P2.6 forms monatomic ions with a charge of (a) 1+; (b) 2+; (c) 3+; (d) 1−; (e) 2−?

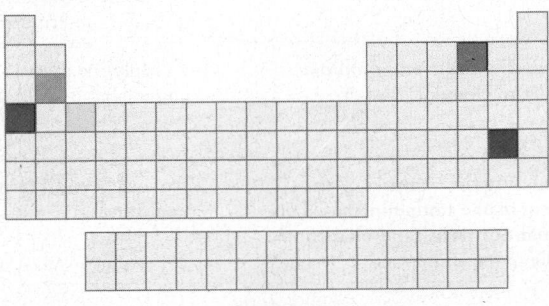

FIGURE P2.6

2.7. Which of the highlighted elements in Figure P2.7 forms an oxide with the following formula: (a) XO; (b) X_2O; (c) XO_2; (d) X_2O_3?

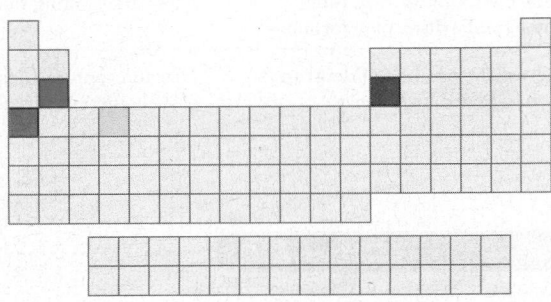

FIGURE P2.7

2.8. Which of the highlighted elements in Figure P2.8 forms an oxoanion with the following generic formula: (a) XO_4^-; (b) XO_4^{2-}; (c) XO_4^{3-}; (d) XO_3^-?

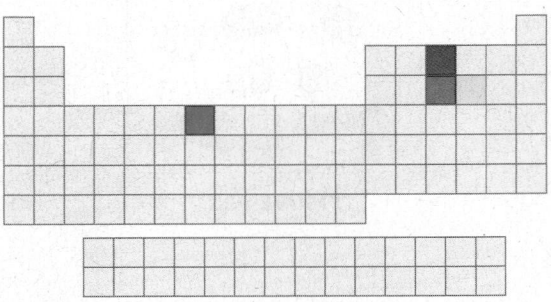

FIGURE P2.8

QUESTIONS AND PROBLEMS

The Rutherford Model of Atomic Structure

CONCEPT REVIEW

2.9. Explain how the results of the gold-foil experiment led Rutherford to dismiss the plum-pudding model of the atom and create his own model based on a nucleus surrounded by electrons.

2.10. Had the plum-pudding model been valid, how would the results of the gold-foil experiment have differed from what Geiger and Marsden actually observed?

2.11. What properties of cathode rays led Thomson to conclude that they were not pure energy, but rather particles with an electric charge?

*2.12. Alpha particles, with a charge of 2+ and a mass of 4 amu, are actually helium nuclei. The element helium was first discovered in a sample of pitchblende, an ore of radioactive uranium oxide. How did helium get in the ore?

Isotopes; Average Atomic Mass

CONCEPT REVIEW

2.13. What is meant by a *weighted average*?

2.14. Explain how percent natural abundances are related to average atomic masses.

PROBLEMS

2.15. If the mass number of an isotope is more than twice the atomic number, is the neutron-to-proton ratio less than, greater than, or equal to 1?

2.16. In each of the following pairs of isotopes, which isotope has more protons and which has more neutrons? (a) ^{127}I or ^{131}I; (b) ^{188}Re or ^{188}W; (c) ^{14}N or ^{14}C

2.17. Boron, lithium, nitrogen, and neon each have two stable isotopes. In which of the following pairs of isotopes is the heavier isotope more abundant?
  a. ^{10}B or ^{11}B (average atomic mass, 10.81 amu)
  b. ^6Li or ^7Li (average atomic mass, 6.941 amu)
  c. ^{14}N or ^{15}N (average atomic mass, 14.01 amu)
  d. ^{20}Ne or ^{22}Ne (average atomic mass, 20.18 amu)

2.18. Naturally occurring copper contains a mixture of 69.17% copper-63 (62.9296 amu) and 30.83% copper-65 (64.9278 amu). What is the average atomic mass of copper?

2.19. Naturally occurring chlorine consists of two isotopes: 75.78% ^{35}Cl, and 24.22% ^{37}Cl. Calculate the average atomic mass of chlorine.

2.20. Naturally occurring sulfur consists of four isotopes: ^{32}S (31.97207 amu, 95.04%); ^{33}S (32.97146 amu, 0.75%); ^{34}S (33.96787 amu, 4.20%); and ^{36}S (35.96708 amu, 0.01%). Calculate the average atomic mass of sulfur in atomic mass units.

2.21. Chemistry of Mars The 1997 mission to Mars included a small robot, the Sojourner, which analyzed the composition of Martian rocks. Magnesium oxide from a boulder dubbed "Barnacle Bill" was analyzed and found to have the following isotopic composition:

Mass (amu)	Natural Abundance (%)
39.9872	78.70
40.9886	10.13
41.9846	11.17

If essentially all of the oxygen in the Martian MgO sample is oxygen-16 (which has an exact mass of 15.9948 amu), is the average atomic mass of magnesium on Mars the same as on Earth (24.31 amu)?

2.22. Using the following table of abundances and masses of the three naturally occurring argon isotopes, calculate the mass of ^{40}Ar.

Symbol	Mass (amu)	Natural Abundance (%)
^{36}Ar	35.96755	0.337
^{38}Ar	37.96272	0.063
^{40}Ar	?	99.60
Average	39.948	

2.23. From the following table of abundances and masses of five naturally occurring titanium isotopes, calculate the mass of ^{48}Ti.

Symbol	Mass (amu)	Natural Abundance (%)
^{46}Ti	45.95263	8.25
^{47}Ti	46.9518	7.44
^{48}Ti	?	73.72
^{49}Ti	48.94787	5.41
^{50}Ti	49.9448	5.18
Average	47.87	

2.24. Strontium has four isotopes: ^{84}Sr, ^{86}Sr, ^{87}Sr, and ^{88}Sr.
  a. How many neutrons are there in each isotope?
  b. The natural abundances of the four isotopes are 0.56% ^{84}Sr (83.9134 amu); 9.86% ^{86}Sr (85.9094 amu); 7.00% ^{87}Sr (86.9089 amu); and 82.58% ^{88}Sr (87.9056 amu). Calculate the average atomic mass of strontium and compare it to the value in the periodic table on the inside front cover.

The Periodic Table of the Elements

CONCEPT REVIEW

2.25. Mendeleev ordered the elements in his version of the periodic table on the basis of their atomic masses instead of their atomic numbers. Why?

2.26. Why did Mendeleev not include the noble gases in his version of the periodic table?

PROBLEMS

2.27. How many protons, neutrons, and electrons are there in the following atoms? (a) ^{14}C; (b) ^{59}Fe; (c) ^{90}Sr; (d) ^{210}Pb

2.28. How many protons, neutrons, and electrons are there in the following atoms? (a) ^{11}B; (b) ^{19}F; (c) ^{131}I; (d) ^{222}Rn

2.29. Fill in the missing information in the following table of four neutral atoms:

Symbol:	^{23}Na	?	?	?
Number of protons:	?	39	?	79
Number of neutrons:	?	50	?	?
Number of electrons:	?	?	50	?
Mass number:	?	?	118	197

2.30. Fill in the missing information in the following table of four neutral atoms:

Symbol:	^{27}Al	?	?	?
Number of protons:	?	42	?	92
Number of neutrons:	?	56	?	?
Number of electrons:	?	?	60	?
Mass number:	?	?	143	238

2.31. Fill in the missing information in the following table of ions:

Symbol:	^{37}Cl$^-$	?	?	?
Number of protons:	?	11	?	88
Number of neutrons:	?	12	46	?
Number of electrons:	?	10	36	86
Mass number:	?	?	81	226

2.32. Fill in the missing information in the following table of ions:

Symbol:	^{137}Ba^{2+}	?	?	?
Number of protons:	?	30	?	40
Number of neutrons:	?	34	16	?
Number of electrons:	?	28	18	36
Mass number:	?	?	32	90

2.33. Which element is most likely to form a cation with a 2+ charge? (a) S; (b) P; (c) Be; (d) Al

2.34. Which element is most likely to form an anion with a 2− charge? (a) S; (b) P; (c) Be; (d) Al

2.35. Which species contains the greatest number of electrons? (a) F; (b) O^{2-}; (c) S^{2-}; (d) Cl

2.36. Which species contains the smallest number of electrons? (a) F; (b) O^{2-}; (c) S^{2-}; (d) Cl

2.37. Which ion has the same number of electrons as an atom of argon? (a) S^{2-}; (b) P^{3-}; (c) Be^{2+}; (d) Ca^{2+}

2.38. Which ion has the same number of electrons as an atom of krypton? (a) Se^{2-}; (b) As^{3-}; (c) Ca^{2+}; (d) K$^+$

2.39. Which element is a nonmetal? (a) Si; (b) Br; (c) Ca; (d) Ru

2.40. Which element is a metalloid? (a) Si; (b) Br; (c) Ca; (d) Ru

Trends in Compound Formation

CONCEPT REVIEW

2.41. **Cations in Blood and Urine** Reports from standard blood and urine tests indicate the amounts of sodium, potassium, calcium, and magnesium cations present. Chloride ion is the most abundant anion in both blood and urine; urine also contains some sulfate ion. Write formulas for the chlorides and sulfates of the four cations.

2.42. Explain why the law of constant composition is classified a scientific *law*, whereas Dalton's view of the atomic structure of matter is classified a scientific *theory*.

2.43. How does Dalton's atomic theory of matter explain the fact that when water is decomposed into hydrogen and oxygen gas, the volume of hydrogen is always twice that of oxygen?

2.44. **Pollutants in Automobile Exhaust** In the internal combustion engines that power most automobiles, nitrogen and oxygen may combine to form NO. When NO in automobile exhaust is released into the atmosphere, it reacts with more oxygen, forming NO$_2$, a key ingredient in smog. How do these reactions illustrate Dalton's law of multiple proportions?

PROBLEMS

2.45. Cobalt forms two sulfides: CoS and Co$_2$S$_3$. Predict the ratio of the two masses of sulfur that combine with a fixed mass of cobalt to form CoS and Co$_2$S$_3$.

2.46. Lead forms two oxides: PbO and PbO$_2$. Predict the ratio of the two masses of oxygen that combine with a fixed mass of lead to form PbO and PbO$_2$.

2.47. When 5.0 grams of sulfur is combined with 5.0 grams of oxygen, 10.0 grams of sulfur dioxide is formed. What mass of oxygen would be required to convert 5.0 grams of sulfur into sulfur trioxide?

*2.48. Nitrogen monoxide (NO) is 46.7% nitrogen by mass. Use the law of multiple proportions to calculate the mass percentage of nitrogen in nitrogen dioxide (NO$_2$).

2.49. **Seawater** The most abundant anion in seawater is the chloride ion. Write the formulas for the chlorides and sulfates of the most abundant cations in seawater: sodium, magnesium, calcium, potassium, and strontium.

2.50. The most abundant cation in seawater is the sodium ion. The evaporation of seawater gives a mixture of ionic compounds containing sodium combined with chloride, sulfate, carbonate, bicarbonate, bromide, fluoride, and tetrahydroxyborate, B(OH)$_4$$^-$. Write the chemical formulas of all these compounds.

2.51. Which of these compounds consist of molecules and which consist of ions? (a) CH$_3$COOH; (b) SrCl$_2$; (c) MgCO$_3$; (d) H$_2$SO$_4$

2.52. Which of these compounds consist of molecules, and which consist of ions? (a) LiOH; (b) Ba(NO$_3$)$_2$; (c) HNO$_3$; (d) CH$_3$(CH$_2$)$_3$OH

Naming Compounds and Writing Formulas

CONCEPT REVIEW

2.53. Consider a mythical element X, which forms only two oxoanions: XO_2^{2-} and XO_3^{2-}. Which of the two has a name that ends in *-ite*?

2.54. Concerning the oxoanions in Problem 2.53, would the name of either of them require a prefix such as *hypo-* or *per-*? Explain why or why not.

2.55. What is the role of Roman numerals in the names of the compounds formed by transition metals?

2.56. Why do the names of the ionic compounds formed by the alkali metals and by the alkaline earth metals not include Roman numerals?

PROBLEMS

2.57. **Toxicity of Nitrogen Oxides** Nitrogen oxides form naturally during the combustion of nitrogen-containing compounds such as coal, diesel fuel, and green plants. Small quantities of all except N_2O (laughing gas) are irritating to eyes, skin, and the respiratory tract. Name the binary compounds of nitrogen and oxygen: (a) NO_3; (b) N_2O_5; (c) N_2O_4; (d) NO_2; (e) N_2O_3; (f) NO; (g) N_2O; (h) N_4O.

2.58. More than a dozen binary compounds containing sulfur and oxygen have been identified. Give the chemical formulas for the following six:
a. sulfur monoxide
b. sulfur dioxide
c. sulfur trioxide
d. disulfur monoxide
e. hexasulfur monoxide
f. heptasulfur dioxide

2.59. Predict the formula and give the name of the binary ionic compound containing the following:
a. sodium and sulfur
b. strontium and chlorine
c. aluminum and oxygen
d. lithium and hydrogen

2.60. Predict the formula and give the name of the binary ionic compound containing the following:
a. potassium and bromine
b. calcium and hydrogen
c. lithium and nitrogen
d. aluminum and chlorine

2.61. Give the chemical names of the cobalt oxides that have the following formulas: (a) CoO; (b) Co_2O_3; (c) CoO_2.

2.62. Give the formula of each of the following copper minerals:
a. cuprite, copper(I) oxide
b. chalcocite, copper(I) sulfide
c. covellite, copper(II) sulfide

2.63. Give the formula and charge of the oxoanion in each of the following compounds:
a. sodium hypobromite
b. potassium sulfate
c. lithium iodate
d. magnesium nitrite

*2.64. Give the formula and charge of the oxoanion in each of the following compounds:
a. potassium tellurite
b. sodium arsenate
c. calcium selenite
d. potassium chlorate

2.65. Give chemical names of the following ionic compounds: (a) $NiCO_3$; (b) $NaCN$; (c) $LiHCO_3$; (d) $Ca(ClO)_2$.

2.66. Give chemical names of the following ionic compounds: (a) $Mg(ClO_4)_2$; (b) NH_4NO_3; (c) $Cu(CH_3COO)_2$; (d) K_2SO_4.

2.67. Give the name or chemical formula of each of the following acids: (a) HF; (b) $HBrO_3$; (c) phosphoric acid; (d) nitrous acid.

2.68. Give the name or chemical formula of each of the following acids: (a) HBr; (b) HIO_4; (c) selenous acid; (d) hydrocyanic acid.

2.69. Name these compounds: (a) Na_2O; (b) Na_2S; (c) Na_2SO_4; (d) $NaNO_3$; (e) $NaNO_2$.

2.70. Name these compounds: (a) K_3PO_4; (b) K_2O; (c) K_2SO_3; (d) KNO_3; (e) KNO_2.

2.71. Write the chemical formulas of these compounds:
a. potassium sulfide
b. potassium selenide
c. rubidium sulfate
d. rubidium nitrite
e. magnesium sulfate

2.72. Write the chemical formulas of these compounds:
a. rubidium nitride
b. potassium selenite
c. rubidium sulfite
d. rubidium nitrate
e. magnesium sulfite

2.73. Name these compounds: (a) MnS; (b) V_3N_2; (c) $Cr_2(SO_4)_3$; (d) $Co(NO_3)_2$; (e) Fe_2O_3.

2.74. Name these compounds: (a) RuS; (b) $PdCl_2$; (c) Ag_2O; (d) WO_3; (e) PtO_2.

2.75. Which compound is sodium sulfite? (a) Na_2S; (b) Na_2SO_3; (c) Na_2SO_4; (d) $NaHS$

2.76. Which compound is calcium nitrate? (a) Ca_3N_2; (b) Ca_2NO_3; (c) $Ca_2(NO_3)_2$; (d) $Ca(NO_3)_2$

2.77. Which element is a halogen? (a) N_2; (b) Cl_2; (c) Xe; (d) H_2

2.78. Which element is a noble gas? (a) N_2; (b) Cl_2; (c) Xe; (d) I_2

2.79. Which element is an alkali metal? (a) Na; (b) Cl_2; (c) Xe; (d) Br_2

2.80. Which element is an alkaline earth metal? (a) Na; (b) Ca; (c) Xe; (d) H_2

Nucleosynthesis

CONCEPT REVIEW

2.81. Write brief (one-sentence) definitions of *chemistry* and *cosmology*, and then give as many examples as you can of how the two sciences are related.

2.82. In the history of the universe, which of these particles formed first and which formed last? (a) deuteron; (b) neutron; (c) proton; (d) quark

2.83. Chemists don't include quarks in the category of subatomic particles—can you think of a reason why?

2.84. Why did early nucleosynthesis last such a short time?

2.85. In the current cosmological model, the volume of the universe is increasing with time. How might this expansion affect the density of the universe?

2.86. **Components of Solar Wind** Most of the ions that flow out from the sun in the solar wind are hydrogen ions. The ions of which element should be next most abundant?

2.87. Nucleosynthesis in Giant Stars A star needs a core temperature of about 10^7 K for hydrogen fusion to occur. Core temperatures above 10^8 K are needed for helium fusion. Why does helium fusion require much higher temperatures?

2.88. Why was the triple-alpha process unlikely to happen in a rapidly cooling universe soon after the Big Bang?

*__2.89.__ It takes nearly twice the energy to remove an electron from a helium atom as it does to remove an electron from a hydrogen atom. Propose an explanation for this.

2.90. Origins of the Elements Our sun contains carbon even though its core is not hot or dense enough to sustain carbon synthesis through the triple-alpha process. Where could the carbon have come from?

2.91. Early nucleosynthesis produced a universe that was more than 99% hydrogen and helium with less than 1% lithium. Why were the other elements not formed?

2.92. What is the effect of β decay on the ratio of neutrons to protons in a nucleus?

PROBLEMS

2.93. What nuclide is produced in the core of a giant star by each of the following fusion reactions?
a. $^{12}_{6}C + ^{4}_{2}\alpha \rightarrow$
b. $^{20}_{10}Ne + ^{4}_{2}\alpha \rightarrow$
c. $^{32}_{16}S + ^{4}_{2}\alpha \rightarrow$

2.94. What nuclide is produced in the core of a giant star by each of the following fusion reactions?
a. $^{28}_{14}Si + ^{4}_{2}\alpha \rightarrow$
b. $^{40}_{20}Ca + ^{4}_{2}\alpha \rightarrow$
c. $^{24}_{12}Mg + ^{4}_{2}\alpha \rightarrow$

2.95. What nuclide is produced in the core of a collapsing giant star by each of the following reactions?
a. $^{56}_{26}Fe + 3\,^{1}_{0}n \rightarrow \underline{\qquad} + ^{0}_{-1}\beta$
b. $^{118}_{50}Sn + 3\,^{1}_{0}n \rightarrow \underline{\qquad} + ^{0}_{-1}\beta$
c. $^{108}_{47}Ag + ^{1}_{0}n \rightarrow \underline{\qquad} + ^{0}_{-1}\beta$

2.96. What nuclide is produced in the core of a collapsing giant star by each of the following reactions?
a. $^{65}_{29}Cu + 3\,^{1}_{0}n \rightarrow \underline{\qquad} + ^{0}_{-1}\beta$
b. $^{68}_{30}Zn + 2\,^{1}_{0}n \rightarrow \underline{\qquad} + ^{0}_{-1}\beta$
c. $^{88}_{38}Sr + ^{1}_{0}n \rightarrow \underline{\qquad} + ^{0}_{-1}\beta$

2.97. Radioactive ^{137}I decays to ^{137}Xe, which is also radioactive and decays to ^{137}Cs. Do either, or both, of these decay processes involve emission of a β particle?

2.98. Isotopes in Geochemistry The relative abundances of the stable isotopes of the elements are not entirely constant. For example, in some geological samples (soils and rocks) the ratio of ^{87}Sr to ^{86}Sr is affected by the presence of a radioactive isotope of another element, which slowly undergoes β decay to produce more ^{87}Sr. What is this other isotope?

Additional Problems

2.99. In April 1897, J. J. Thomson presented the results of his experiment with cathode-ray tubes (Figure P2.99) in which he proposed that the rays were actually beams of negatively charged particles, which he called "corpuscles."
a. What is the name we use for these particles today?
b. Why did the beam deflect when passed between electrically charged plates, as shown in Figure P2.99?

c. If the polarity of the plates were switched, how would the position of the light spot on the phosphorescent screen change?
d. If the voltage on the plates were reduced by half, how would the position of the light spot change?

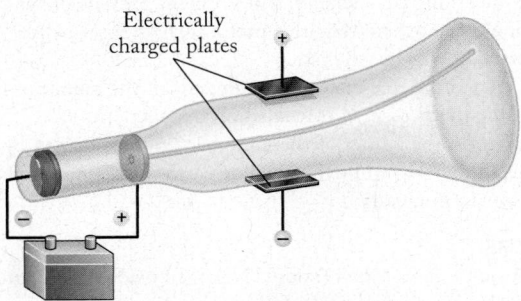

Electrically charged plates

FIGURE P2.99

*__2.100.__ Suppose the electrically charged discs at the end of the cathode-ray tube were replaced with a radioactive source, as shown in Figure P2.100. Also suppose the radioactive material inside the source emits α and β particles, plus rays of energy with no charge. The only way for any of the three kinds of particles or rays to escape the source is through a narrow channel drilled through a block of lead.
a. How many light spots do you expect to see on the phosphorescent screen?
b. What are their positions relative to the electrical plates, and which particle produces which spot?

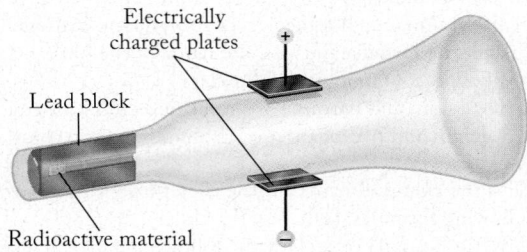

Electrically charged plates

Lead block

Radioactive material

FIGURE P2.100

*__2.101.__ Suppose the radioactive material inside the source in the apparatus shown in Figure P2.100 emits protons and α particles, and suppose both kinds of particles have the same velocities.
a. How many light spots do you expect to see on the phosphorescent screen?
b. What are their positions on the screen (above, at, or below the center)? Which particle produces which spot?

2.102. Cosmologists estimate that the matter in the early universe was 75% by mass hydrogen-1 and 25% helium-4 when atoms first formed.
a. Assuming these proportions are correct, what was the ratio of hydrogen to helium *atoms* in the early universe?
b. The ratio of hydrogen to helium atoms in our solar system is slightly less than 10:1. Compare this value with the value you calculated in part a.

c. Propose a hypothesis that accounts for the difference in composition between the solar system and the early universe.

d. Describe an experiment that would test your hypothesis.

2.103. Sources of Breathable Air Potassium forms three compounds with oxygen: K_2O (potassium oxide), K_2O_2 (potassium peroxide), and KO_2 (potassium superoxide). Potassium is rarely encountered; it reacts violently with water and is very corrosive to human tissue. Potassium superoxide is used in self-contained breathing apparatus as a source of oxygen in mines, submarines, and spacecraft. Potassium peroxide binds carbon dioxide and is used to scrub (remove) toxic CO_2 from the air in submarines. Predict the ratio of the masses of oxygen that combine with a fixed mass of potassium in K_2O, K_2O_2, and KO_2.

2.104. Stainless Steel The gleaming metallic appearance of the Gateway Arch (Figure P2.104) in St. Louis, Missouri, comes from the stainless steel used in its construction. This steel is made mostly of iron but it also contains 19% by mass chromium and 9% by mass nickel.

a. Stainless steel maintains its metallic sheen because the chromium and nickel in it combine with oxygen from the atmosphere, forming a layer of Cr_2O_3 and NiO that is too thin to detract from the luster of the steel but that protects the metal beneath from further corrosion. What are the names of these two ionic compounds?

b. What are the charges of the cations in Cr_2O_3 and NiO?

FIGURE P2.104

2.105. Bronze Age Historians and archaeologists often apply the term "Bronze Age" to the period in Mediterranean and Middle Eastern history when bronze was the preferred material for making weapons, tools, and other metal objects. Ancient bronze was an alloy prepared by blending molten copper (90%) and tin (10%) by mass. What is the ratio of copper to tin atoms in a piece of bronze with this composition?

*2.106. In his version of the periodic table, Mendeleev arranged elements based on the formulas of the compounds they formed with hydrogen and oxygen. The elements in one of his eight groups formed compounds with these generic formulas: MH_3 and M_2O_5, where M is the symbol of an element in the group. Which Roman numeral did Mendeleev assign to this group?

2.107. In the Mendeleev table in Figure 2.9, there are no symbols for elements with predicted atomic masses of 44, 68, and 72.

a. Which elements are these?

b. Mendeleev anticipated the later discovery of these three elements and gave them tentative names: ekaaluminum, ekaboron, and ekasilicon, reflecting the probability that their properties would resemble those of aluminum, boron, and silicon, respectively. What are the modern names of ekaaluminum, ekaboron, and ekasilicon?

c. When were these elements finally discovered? To answer this question you may wish to consult a reference such as webelements.com.

2.108. Medical and Commercial Compounds Many common compounds have old-fashioned, nonsystematic names or newer commercial names that are widely used. Search for the systematic names and chemical formulas of compounds with these nonsystematic names: (a) magnesia; (b) Epsom salt; (c) K-Dur; (d) lime; (e) baking soda; (f) caustic soda; (g) muriatic acid; (h) zirconia.

*2.109. The ruby shown in Figure P2.109 has a mass of 12.04 carats (1 carat = 200.0 mg). Rubies are made of a crystalline form of Al_2O_3.

a. What percentage of the mass of the ruby is aluminum?

b. The density of rubies is $4.02\ g/cm^3$. What is the volume of the ruby?

FIGURE P2.109

2.110. In chemical nomenclature, the prefix *thio-* is used to indicate that a sulfur atom has replaced an oxygen atom in the structure of a molecule or a polyatomic ion.

a. With this rule in mind, write the formula for the thiosulfate ion.

b. What is the formula of sodium thiosulfate?

2.111. There are two stable isotopes of gallium. Their masses are 68.92558 and 70.9247050 amu. If the average atomic mass of gallium is 69.7231 amu, what is the natural abundance of the lighter isotope?

2.112. There are two stable isotopes of bromine. Their masses are 78.9183 and 80.9163 amu. If the average atomic mass of bromine is 79.9091 amu, what is the natural abundance of the heavier isotope?

*2.113. Start with the information in the previous question, and then do the following:

a. Predict the possible masses of individual molecules of Br_2.

b. Calculate the natural abundance of molecules with each of the masses predicted in part a in a sample of Br_2.

*2.114. There are three stable isotopes of magnesium. Their masses are 23.9850, 24.9858, and 25.9826 amu. If the average atomic mass of magnesium is 24.3050 amu and the natural abundance of the lightest isotope is 78.99%, what are the natural abundances of the other two isotopes?

Additional study materials including ChemTours and Diagnostic Quizzes are available at StudySpace at www.wwnorton.com/studyspace.

3

Chemical Reactions and Earth's Composition

Learning Outcomes

- Use Avogadro's number and the definition of the mole in calculations
- Write balanced chemical equations that describe chemical reactions
- Use balanced chemical equations to relate the mass of a reactant to the mass of a product
- Determine an empirical formula from the percent composition of a substance
- Determine a molecular formula from the empirical formula and molar mass of a substance
- Use data from combustion reactions in determining empirical formulas of substances
- Determine the limiting reactant in a chemical reaction
- Calculate the theoretical and percent yields in a chemical reaction

A LOOK AHEAD

Early Earth

Chapter 2 ended with a description of how elements are synthesized in the cores of giant stars and how these elements are dispersed throughout galaxies when these stars self-destruct during events called supernovas. These dispersed elements become the building blocks of later generations of stars. Our sun is one of these latter-day stars. It and the planets orbiting it are made of elements synthesized in giant stars that existed, and then cataclysmically ceased to exist, many billions of years ago.

Scientists estimate that our solar system formed about 4.6 billion years ago from a gigantic swirling mass of gas called the solar nebula. In cooler regions of the nebula millions of kilometers from its center, some of the swirling matter condensed into structures called planetesimals. As these structures collided and fused with each other, the planets of the solar system formed.

The elements in the nebula were not evenly distributed among the new planets. The lightest and most volatile elements were swept away from the planets closest to the sun by a combination of solar wind (high-velocity charged particles emitted by the sun) and solar heat. This separation left the inner planets—Mercury, Venus, Earth, and Mars—rich in nonvolatile elements such as Fe, Si, Mg, and Al and the outermost planets, called the gas giants, rich in hydrogen and helium.

This distribution of elements also left Earth rich in oxygen. Why wasn't oxygen carried away with the other gases? The reason is that oxygen forms stable, nonvolatile compounds with elements such as Fe, Si, Mg, and Al. In this chapter we explore the chemical properties of these and other elements as well as their tendencies to combine to form compounds. We begin our exploration with the newly formed molten planet Earth and follow some of the physical processes and chemical changes that have altered the composition of its surface and the atmosphere above it, producing the world we live in today. ■

Fire, Water, and Salt Nate Smith demonstrates a fire vortex on the Great ▶
Salt Lake in Utah. Fire is an example of a chemical change, a combustion reaction.

chemical equation a description of the identities and quantities of **reactants** (substances consumed during a chemical reaction) and **products** (substances formed).

combination reaction a reaction in which two (or more) substances combine to form one product.

mole (mol) an amount of material (atoms, ions, or molecules) that contains Avogadro's number (N_A = 6.022 × 10²³) of particles.

Avogadro's number (N_A) the number of carbon atoms in exactly 12 grams of the carbon-12 isotope; N_A = 6.022 × 10²³. It is the number of particles in one mole.

3.1 Chemical Reactions and Earth's Early Atmosphere

The Earth that formed 4.6 billion years ago was a hot, molten sphere that gradually separated into distinct regions based on differences in density and melting point. The densest elements, notably iron and nickel, sank to the center of the planet. A less dense mantle rich in compounds containing aluminum, magnesium, silicon, and oxygen formed around the core. As time passed and Earth cooled, the mantle fractionated further, allowing a solid crust to form from the components of the mantle that were the least dense and had the highest melting points. The core also separated into a solid inner core and a molten outer core. The elemental compositions of these layers are shown in Figure 3.1.

Earth's early crust was torn by the impact of asteroids and widespread volcanic activity. The gases released by these impacts and eruptions generated a primitive atmosphere with a chemical composition very different from the air that we breathe. Earth's early atmosphere was nearly devoid of oxygen but was rich in oxygen-containing compounds, including carbon dioxide (CO_2), water vapor (H_2O), and other volatile oxides including sulfur dioxide (SO_2), sulfur trioxide (SO_3), nitrogen monoxide (NO), and nitrogen dioxide (NO_2). The most abundant gases released today by volcanic systems like Mount St. Helens are water vapor, CO_2, and SO_2 (Figure 3.2).

Sometimes these compounds combined together to make substances with more elaborate molecular structures. For example, sulfur trioxide gas combined with water vapor, producing liquid sulfuric acid, H_2SO_4. We use the formulas of these substances in an expression called a **chemical equation** to describe the reaction:

$$SO_3(g) + H_2O(g) \rightarrow H_2SO_4(\ell) \qquad (3.1)$$

Another view is provided by the space-filling molecular models in Figure 3.3.

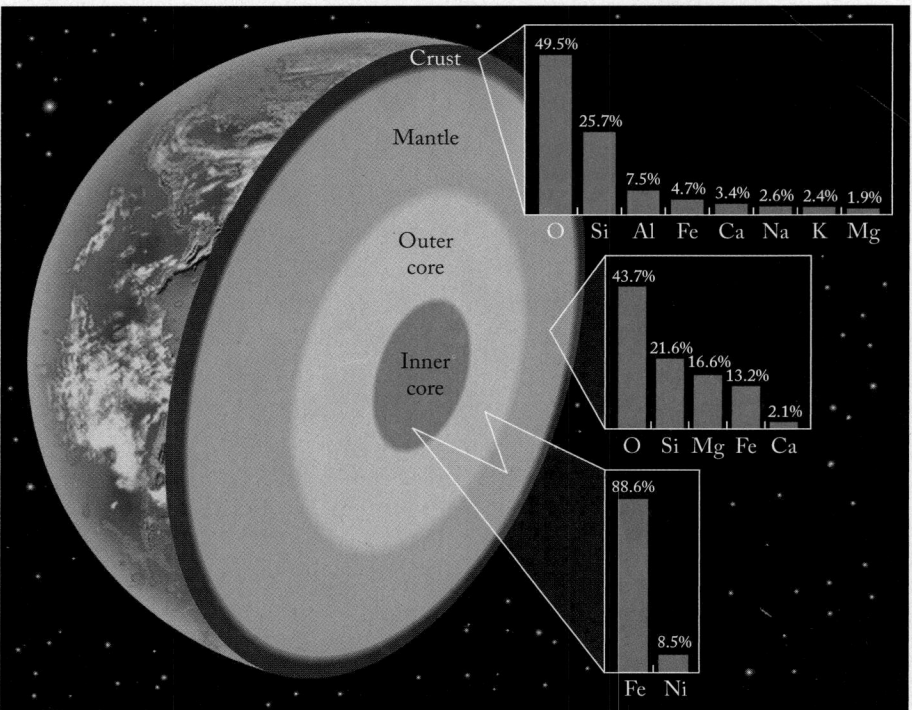

FIGURE 3.1 Earth is composed of a solid inner core consisting mostly of nickel and iron surrounded by a molten outer core of similar composition. A liquid mantle composed mostly of oxygen, silicon, magnesium, and iron lies between the outer core and a relatively thin solid crust.

The reaction between SO_3 and H_2O is an example of a **combination reaction** in which two (or more) substances combine to form one **product**. In this reaction SO_3 and H_2O are called **reactants** and H_2SO_4 is the product. We use a reaction arrow to link the reactants and product and show the direction of the reaction. The symbols in parentheses after the chemical formulas indicate the physical states of the reactants and product: g for gas and ℓ for liquid. An important feature of this or any chemical equation is that it is balanced: every atom that is present in the reactants is also present in the products. This conservation of atoms means that there is also a conservation of mass: the sum of the masses of the reactants always equals the sum of the masses of the products.

The sulfuric acid that formed in Earth's early atmosphere eventually fell to the planet's surface in the form of highly acidic rain. This rain would have landed on a crust made up mostly of metal and metalloid oxides, including the mineral hematite, Fe_2O_3. When that sulfuric acid encountered hematite another chemical reaction took place—one that produced water-soluble iron sulfate, $Fe_2(SO_4)_3$, and liquid water. This reaction is described by the following chemical equation:

$$Fe_2O_3(s) + 3\,H_2SO_4(aq) \rightarrow 3\,H_2O(\ell) + Fe_2(SO_4)_3(aq) \qquad (3.2)$$

In Equation 3.2 the symbol s means solid and aq stands for aqueous solution, meaning that the substance is dissolved in water. The "3"s in front of both H_2SO_4 and H_2O are coefficients indicating that three molecules of each compound are needed to complete the reaction. We need these coefficients to balance the number of atoms of each element on both sides of the reaction arrow. The absence of a coefficient in front of Fe_2O_3 and $Fe_2(SO_4)_3$ is the same as having a coefficient of "1" in front of them. The chemical formula itself stands for one molecule (or one formula unit) of the substance.

3.2 The Mole

Equation 3.1 describes a chemical reaction between a molecule of sulfur trioxide and a molecule of water. In our macroscopic (visible) world, chemists rarely work with individual atoms or molecules: they are too small to manipulate easily. Instead, they usually deal with quantities of reactants and products large enough to see and work with comfortably. Such measurable quantities contain enormous numbers of particles—either atoms, molecules, or ions—and consequently chemists need a unit that can relate measurable quantities of substances to the number of particles they contain. That unit is the **mole (mol)**, the SI base unit for expressing quantities of substances (see Table 1.2).

Moles allow us to relate the mass of a pure substance to the number of particles it contains. One mole of any substance is defined as the quantity of the substance that contains the same number of particles as the number of carbon atoms in exactly 12 g of the isotope carbon-12. This number is 6.022×10^{23} carbon-12 atoms. This very large value is called **Avogadro's number (N_A)** after the Italian scientist Amedeo Avogadro (1776–1856), whose research enabled other scientists to accurately determine the atomic masses of the elements. One mole of water also contains 6.022×10^{23} particles—in this case molecules of water. To put a number of this magnitude in perspective, it would take 9.5 trillion computer flash drives, each capable of storing 64 gigabytes (6.4×10^{10} bytes) of data, to store a mole of bytes.

Dividing the number of particles in a sample by Avogadro's number yields the number of moles of those particles. For example, a jet airplane flying at an altitude

FIGURE 3.2 Volcanic eruption of Mount St. Helens in Washington State on May 18, 1980. The most abundant gas released in this eruption was water vapor.

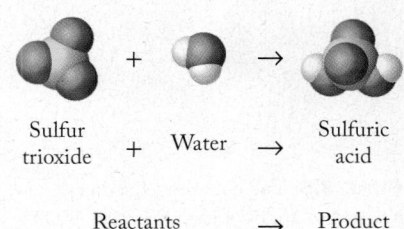

Sulfur trioxide + Water → Sulfuric acid

Reactants → Product

FIGURE 3.3 A molecule of SO_3 and a molecule of H_2O react to produce a molecule of H_2SO_4.

▶❙❙ **CHEMTOUR** Avogadro's Number

of 11,300 m (37,000 ft) is flying through air that contains about 7.0×10^{21} particles per liter. This number is equivalent to 1.2×10^{-2} moles of particles:

$$7.0 \times 10^{21} \; \text{particles} \times \frac{1 \; \text{mol}}{6.022 \times 10^{23} \; \text{particles}} = 1.2 \times 10^{-2} \; \text{mol}$$

On the other hand, multiplying a number of moles by Avogadro's number gives us the number of particles in that many moles. For example, a liter bottle of seltzer water contains 55 moles of H_2O. The equivalent number of molecules of water in that liter bottle is a very large number:

$$55 \; \text{mol } H_2O \times \frac{6.022 \times 10^{23} \; \text{molecules } H_2O}{1 \; \text{mol } H_2O} = 3.3 \times 10^{25} \; \text{molecules } H_2O$$

These conversions are illustrated in Figure 3.4, and the quantities of some common elements equivalent to 1 mol are illustrated in Figure 3.5. While it is sometimes useful to know the number of molecules of a substance in a sample, we'll mostly be concerned about knowing how many *moles* are in a particular quantity of a substance.

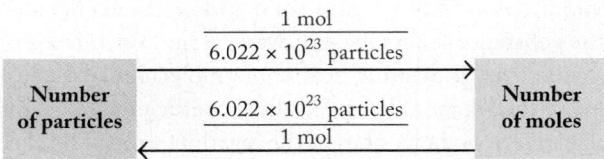

FIGURE 3.4 Converting between a number of particles and an equivalent number of moles (or vice versa) is a matter of dividing (or multiplying) by Avogadro's number.

FIGURE 3.5 The quantities shown are equivalent to 1 mol of each material: 4.003 g of helium gas in the balloon and, left to right in front of the balloon, 32.06 g of solid sulfur, 63.55 g of copper metal, and 200.59 g of liquid mercury.

SAMPLE EXERCISE 3.1   Converting Number of Moles into Number of Particles

It's not unusual for the polluted air above a large metropolitan area to contain as much as 5×10^{-10} moles of SO_2 per liter of air. What is this concentration of SO_2 in molecules per liter?

Collect and Organize The problem gives the number of moles of SO_2 per liter of air. Avogadro's number defines the number of particles in 1 mole of anything. Given 5×10^{-10} mol of SO_2 in 1 L of air and Avogadro's number (6.022×10^{23} molecules/mol), we can determine number of molecules of SO_2 in 1 L.

Analyze We can convert the number of moles into the number of molecules by multiplying by a conversion factor that has *molecules* in the numerator and moles in the denominator; that is, 6.022×10^{23} molecules/mol. We can estimate the answer by considering the approximate value of Avogadro's number (about 6×10^{23}) and the number of moles of SO_2 (5×10^{-10}). The product of the two is 3×10^{14}.

Solve

$$\frac{5 \times 10^{-10} \; \text{mol } SO_2}{1 \; \text{L air}} \times \frac{6.022 \times 10^{23} \; \text{molecules } SO_2}{1 \; \text{mol } SO_2}$$

$$= \frac{3 \times 10^{14} \; \text{molecules } SO_2}{1 \; \text{L air}}$$

Think about It The result tells us that a tiny fraction of 1 mole of a molecular substance (10^{-10}) is equivalent to a very large number of molecules (10^{14}), which makes sense given the immensity of Avogadro's number (10^{23}).

Practice Exercise If 1.0 mL of seawater contains about 2.5×10^{-14} moles of dissolved gold, how many atoms of gold are in that volume of seawater?

(Answers to Practice Exercises are in the back of the book.)

molar mass ($\mathcal{M}$) the mass of 1 mole of a substance. The molar mass of an element in grams per mole is numerically equal to that element's average atomic mass in atomic mass units.

CONCEPT TEST

How does a unit of measure such as a gross of marbles (144) relate to the concept of the mole?

(Answers to Concept Tests are in the back of the book.)

Molar Mass

The mole provides an important link from the values of atomic mass in the periodic table to masses of macroscopic quantities of elements and compounds. Recall from Section 2.3 that the mass value shown in the cell of any element in the periodic table is the average atomic mass of the atoms of the element, expressed in atomic mass units (amu) or daltons (Da). That same value is also the mass of 1 mole of atoms of that element, expressed in grams. Thus the average mass of one atom of helium is 4.003 amu, and the mass of 1 mole of helium is 4.003 g. The mass in grams of 1 mole of any substance is called the **molar mass ($\mathcal{M}$)** of the substance, and we say that the molar mass of helium is 4.003 g/mol (Figure 3.6).

As mentioned earlier, one of the reasons we use moles in chemistry is to allow us to compare the number of particles in differing masses of two different substances. Molar mass can be used to convert between the mass of a sample of any substance and the number of moles in that sample (Sample Exercises 3.2 and 3.3).

The mole enables us to count the number of particles in any sample of a given substance simply by knowing the mass of the sample. We can do this because the mole represents both a fixed number of particles (Avogadro's number) and a specific mass (the molar mass) of the substance. It may be useful to think about moles and mass using the following analogy. A dozen golf balls and a dozen baseballs have very different masses, but each contains 12 balls (Figure 3.7). Indeed the mole is sometimes referred to as "the chemist's dozen."

2
He
4.003

Atomic mass of He	4.003 amu/atom
Mass of 1 mol of He	4.003 g
Molar mass of He	4.003 g/mol

FIGURE 3.6 The atomic mass (in amu/atom) and the molar mass (in g/mol) of helium have the same numerical value.

FIGURE 3.7 A dozen golf balls weighs less than a dozen baseballs, but both contain the same number of balls.

SAMPLE EXERCISE 3.2   **Converting Mass into Number of Moles**

Some antacid tablets contain 425 mg of calcium (as Ca^{2+} ions). How many moles of calcium are in each tablet?

Collect and Organize The number of moles of a substance (for example, Ca^{2+}) is related to the mass of the substance by its molar mass. Because the mass of electrons is insignificant compared with that of protons and neutrons (Table 2.1), the mass of a calcium ion is essentially the same as the mass of a calcium atom, even though the ion has two fewer electrons than the atom.

Analyze The mass of the calcium is given in milligrams (mg). Converting to grams requires an additional mathematical step using the conversion factor $1 \text{ g} = 10^3 \text{ mg}$. To convert grams of Ca^{2+} into moles of Ca^{2+} we divide the mass by the molar mass of Ca^{2+}. The average atomic mass of an atom of calcium is 40.078 amu, which means the molar mass of calcium is 40.08 g/mol when rounded to four significant figures.[1] We can estimate the answer to be about 0.01 mole because we are dividing about 4×10^{-1} g by about 40 g/mol $(4.0 \times 10^1$ g/mol).

Solve

$$425 \text{ mg Ca}^{2+} \times \frac{1 \text{ g}}{10^3 \text{ mg}} \times \frac{1 \text{ mol Ca}^{2+}}{40.08 \text{ g Ca}^{2+}} = 0.0106 \text{ mol Ca}^{2+}$$

Think about It People seem most comfortable dealing with numbers between 1 and 1000, and that is why the mass is given in milligrams in the antacid tablet. The answer here seems appropriate based on the small number of grams of calcium in the tablet.

Practice Exercise The mass of the diamond in Figure 3.8 is 3.25 carats (1 carat = 0.200 g). Diamonds are nearly pure carbon. How many moles of carbon are in the diamond?

(Answers to Practice Exercises are in the back of the book.)

FIGURE 3.8 A 3.25-carat diamond.

SAMPLE EXERCISE 3.3 **Converting Number of Moles into Mass in Grams**

A helium balloon sold at an amusement park contains 0.462 mole of He. How many grams of He are in the balloon?

Collect and Organize The number of grams of He in 0.462 mole represents the mass of helium particles present in this quantity of He. The molar mass of helium is 4.003 g/mol.

Analyze We need to convert moles of He into grams of He. The conversion factor that relates moles of He to grams of helium may be written two ways: 1 mol He/4.003 g He and 4.003 g He/1 mol He. Because our answer must be in grams, we choose the conversion factor in which grams is in the numerator: 4.003 g He/1 mol He. Because the balloon contains approximately $\frac{1}{2}$ mole of helium, we estimate the answer should be around 2 grams.

Solve

$$0.462 \text{ mol He} \times \frac{4.003 \text{ g He}}{1 \text{ mol He}} = 1.85 \text{ g He}$$

Think About It As we predicted, multiplying a number just a little more than 4 by about one-half results in an answer (1.85 g He) that is close to 2 grams.

Practice Exercise How many grams of gold are there in 0.250 moles of gold?

(Answers to Practice Exercises are in the back of the book.)

---

[1] In Sample Exercises involving molar masses, we will usually round off the values given in the periodic table to four significant figures except for elements with molar masses over 100.

Which contains more atoms—1 g of gold (Au) or 1 g of silver (Ag)?

(Answers to Concept Tests are in the back of the book.)

molecular mass the mass of one molecule of a molecular compound.

formula mass the mass in atomic mass units of one formula unit of an ionic compound.

Molecular Masses and Formula Masses

The **molecular mass** of a molecular compound is the sum of the atomic masses of the atoms in that molecule. Because atomic masses are given in atomic mass units, molecular masses are also reported in amu. Thus the molecular mass of sulfur trioxide, SO_3, is the sum of the masses of 1 sulfur atom and 3 oxygen atoms:

$$32.06 \text{ amu} + (3 \times 16.00 \text{ amu}) = 80.06 \text{ amu}$$

Just as the molar mass of an element in grams per mole is numerically the same as the atomic mass of one atom of the element in atomic mass units (Figure 3.6), the molar mass ($\mathcal{M}$) of a molecular compound in grams per mole is numerically the same as its molecular mass in atomic mass units. Thus the molar mass of SO_3 is 80.06 g/mol (Figure 3.9).

As with atoms and atomic mass, we can use the concept of moles to scale up molecular masses to molar masses, which are quantities we can measure and manipulate. The result is that atomic mass units become grams, atoms and molecules become moles, and we finally have the molar mass of the compound. For example, the molar mass of SO_3 in grams per mole is the sum of the masses of 1 mole of sulfur atoms and 3 moles of oxygen atoms:

$$\mathcal{M}_{SO_3} = 32.06 \text{ g/mol} + 3(16.00 \text{ g/mol}) = 80.06 \text{ g/mol } SO_3$$

The concept of molar mass applies to ionic as well as molecular compounds. For example, 1 mole of BaS contains 1 mole (137.33 g) of Ba^{2+} ions and 1 mole (32.06 g) of S^{2-} ions. Therefore, the molar mass of BaS is

$$\mathcal{M}_{BaS} = 137.33 \text{ g/mol} + 32.06 \text{ g/mol} = 169.39 \text{ g/mol}$$

Keep in mind that there are no discrete molecules of BaS. This compound is ionic and as we saw in Chapter 2, its crystals consist of ordered arrays of cations and anions (Figure 3.10). The formula unit (Section 2.5) defines the smallest integer ratio of positive and negative ions, 1:1 in the case of BaS, that describes the composition of an ionic compound. The mass of one formula unit of an ionic compound is called its **formula mass**.

Molecular mass of SO_3	80.06 amu/molecule
Mass of 1 mol of SO_3	80.06 g
Molar mass of SO_3	80.06 g/mol

FIGURE 3.9 The molecular mass (in amu/molecule) and the molar mass (in g/mol) of sulfur trioxide have the same numerical value.

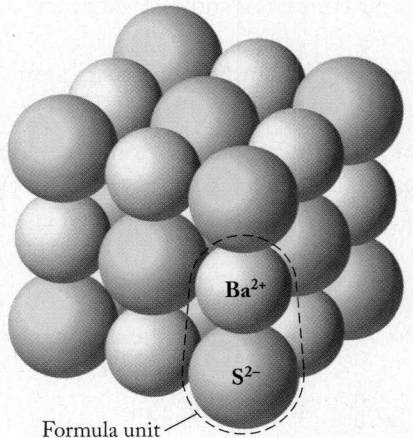

Formula unit

FIGURE 3.10 A crystal of barium sulfide consists of a three-dimensional array of Ba^{2+} ions and S^{2-} ions. The 1:1 ratio of Ba^{2+} ions to S^{2-} ions is represented in the formula unit of the crystal (enclosed in a dashed oval) and in the empirical formula of the compound (BaS).

FIGURE 3.11 The mass of a pure substance can be converted into the equivalent number of moles or number of particles (atoms, ions, or molecules) and vice versa.

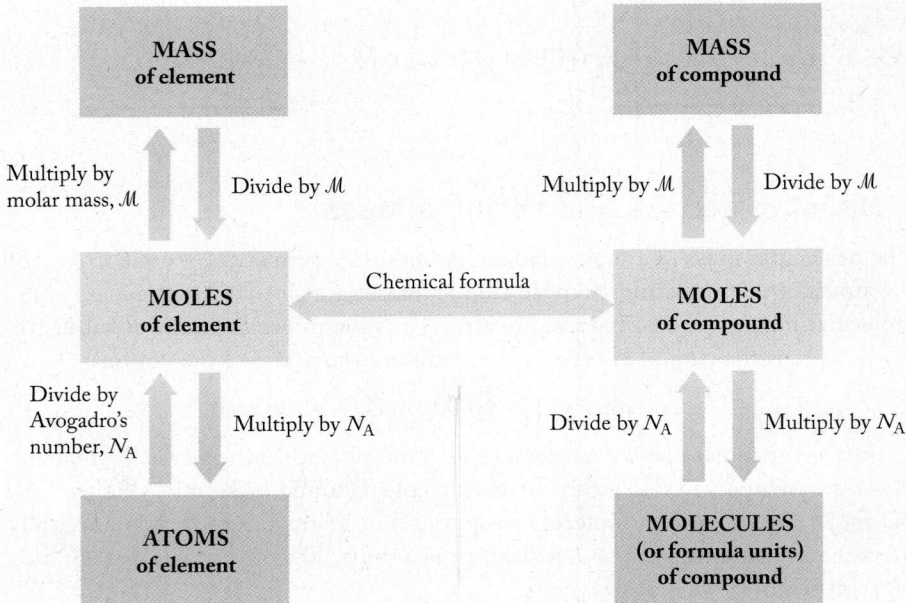

Figure 3.11 summarizes how to use Avogadro's number, chemical formulas, and molar masses to convert between the mass, the number of moles, and the number of particles in a given quantity of an element or compound. The calculation of molar mass is further illustrated in Sample Exercise 3.4.

SAMPLE EXERCISE 3.4   Calculating the Molar Mass of a Compound

Calculate the molar masses of (a) H_2O and (b) H_2SO_4.

Collect and Organize We are asked to calculate the molar mass of two compounds given their chemical formulas. The molar mass of a molecular compound is the sum of the molar masses of the elements in the molecular formula of the compound, each multiplied by the number of atoms of that element in the molecular formula. The molar mass of an element in grams per mole is numerically equivalent to the atomic mass of the element expressed in atomic mass units.

Analyze The chemical formulas of H_2O and H_2SO_4 tell us that:
   a. One mole of H_2O contains 2 moles of H atoms and 1 mole of O atoms.
   b. One mole of H_2SO_4 contains 2 moles of H atoms, 1 mole of S atoms, and 4 moles of O atoms.

Because H_2SO_4 has a total of 7 moles of atoms including 4 moles of oxygen atoms, we predict that the molar mass of H_2SO_4 will be greater than that of H_2O, which contains only 3 moles of atoms in total and only 1 mole of oxygen atoms.

Solve
   a. The molar mass of H_2O is 2 times the molar mass of H plus the molar mass of O:

$$2(1.008 \text{ g/mol}) + 16.00 \text{ g/mol} = 18.02 \text{ g/mol } H_2O$$

b. The molar mass of H_2SO_4 is 2 times the molar mass of H plus the molar mass of S plus 4 times the molar mass of O:

$$2(1.008 \text{ g/mol}) + 32.06 \text{ g/mol} + 4(16.00 \text{ g/mol}) = 98.08 \text{ g/mol } H_2SO_4$$

Think about It The molar mass of a compound reflects the masses of the elements in that compound. Because H_2SO_4 contains more atoms and heavier atoms than H_2O, we expect the molar mass of sulfuric acid to be larger than that for water.

Practice Exercise Green plants take in water (H_2O) and carbon dioxide (CO_2) and produce glucose ($C_6H_{12}O_6$) and oxygen gas (O_2). Calculate the molar masses of carbon dioxide, oxygen gas, and glucose.

(Answers to Practice Exercises are in the back of the book.)

SAMPLE EXERCISE 3.5 | **Interconverting Grams, Moles, and Molecules of a Compound**

Calculate the number of moles and the number of formula units of calcium carbonate contained in 1.28 g of $CaCO_3$.

Collect and Organize In this problem we are asked to express a given mass of a compound in terms of the number of moles of the compound and the number of formula units that mass represents. Figure 3.11 illustrates the general method we can apply for the conversion between grams of $CaCO_3$, moles of $CaCO_3$, and formula units of $CaCO_3$. From Chapter 2 we recall that calcium carbonate is an ionic compound. We need to determine the formula mass of $CaCO_3$ and to use Avogadro's number.

Analyze To convert grams of $CaCO_3$ to moles of $CaCO_3$, we need to divide by the formula mass of $CaCO_3$. Multiplying the moles of $CaCO_3$ by Avogadro's number yields the number of formula units of $CaCO_3$ contained in 1.28 g of $CaCO_3$. Because we have a relatively small mass of $CaCO_3$, we expect the number of moles of $CaCO_3$ to be less than one. Avogadro's number, however, is very large, and we expect the number of formula units of $CaCO_3$ to be very large.

Solve First we need to determine the formula mass of $CaCO_3$ by adding the molar mass of Ca plus the molar mass of C plus 3 times the molar mass of O:

$$40.08 \text{ g/mol} + 12.01 \text{ g/mol} + 3(16.00 \text{ g/mol}) = 100.09 \text{ g/mol } CaCO_3$$

Converting from grams $CaCO_3$ to moles $CaCO_3$ gives

$$1.28 \; \cancel{\text{g } CaCO_3} \times \frac{1 \text{ mol } CaCO_3}{100.09 \; \cancel{\text{g } CaCO_3}} = 0.01279 \text{ mol } CaCO_3$$

reported with three significant figures as 0.0128 mol.

Carrying on the calculation with the intermediate value and multiplying by Avogadro's number gives

$$0.01279 \; \cancel{\text{mol } CaCO_3} \times \frac{6.022 \times 10^{23} \text{ formula units } CaCO_3}{1 \; \cancel{\text{mol } CaCO_3}}$$

$$= 7.702 \times 10^{21} \text{ formula units } CaCO_3$$

The final answer must have three significant figures, so the number of formula units of $CaCO_3$ is 7.70×10^{21}.

Think about It Because atoms and ions are tiny, we expect a large number of formula units in a relatively small mass of $CaCO_3$. We can't "count" individual formula units of an ionic compound, but we can convert the number of formula units to mass, a quantity we can readily measure.

Practice Exercise How many sodium ions are there in 1.28 g of $NaHCO_3$?

(Answers to Practice Exercises are in the back of the book.)

■ ···

Moles and Chemical Equations

The concepts of mole and molar mass provide three more interpretations of Equation 3.1:

$$SO_3(g) + H_2O(g) \rightarrow H_2SO_4(\ell)$$

1. The coefficients in the chemical equation tell us that 1 *mole* of SO_3 reacts with 1 *mole* of H_2O, producing 1 *mole* of H_2SO_4.
2. The molar masses of the reactants and products allow us to say that 80.06 *grams* of SO_3 reacts with 18.02 *grams* of H_2O, producing 98.08 *grams* of H_2SO_4.
3. Avogadro's number tells us that 6.022×10^{23} *molecules* of SO_3 react with 6.022×10^{23} *molecules* of H_2O, forming 6.022×10^{23} *molecules* of H_2SO_4; this means that, in terms of lowest whole-number ratios, 1 *molecule* of SO_3 reacts with 1 *molecule* of H_2O, forming 1 *molecule* of H_2SO_4.

The several interpretations of this equation are not limited to 1 mole of SO_3 reacting with 1 mole of H_2O. In general, the equation tells us that *any* number of moles (x mol) of SO_3 reacts with an equal number of moles (x mol) of water to produce a specific amount (x mol) of sulfuric acid. This interpretation is valid because the mole ratio (the ratio of the coefficients) of SO_3 to H_2O to H_2SO_4 in the equation is 1:1:1. The mole ratio is specific for a given equation and is different when different substances combine. For example, for the reaction in Equation 3.2 between sulfuric acid and Fe_2O_3 (hematite):

$$Fe_2O_3(s) + 3\,H_2SO_4(aq) \rightarrow 3\,H_2O(\ell) + Fe_2(SO_4)_3(aq)$$

the mole ratio of Fe_2O_3 to H_2SO_4 is 1:3.

From Equation 3.1, we can determine either how much of one reactant is needed to completely react with any quantity of the other reactant or how much product can be made from any quantity of reactants by multiplying x by the appropriate molar mass:

	$SO_3(g)$	+	$H_2O(g)$	$\rightarrow$	$H_2SO_4(\ell)$
	1 molecule	+	1 molecule	$\rightarrow$	1 molecule
Mole ratios:	1 mol	+	1 mol	$\rightarrow$	1 mol
Mass ratios:	80.06 g	+	18.02 g	$\rightarrow$	98.08 g
General case (moles):	x mol	+	x mol	$\rightarrow$	x mol
General case (masses):	x (80.06 g)	+	x (18.02 g)	$\rightarrow$	x(98.08 g)

stoichiometry the quantitative relation between the reactants and products in a chemical reaction.

The quantitative relation between the reactants and products involved in a chemical reaction is called the **stoichiometry** of the reaction. In a reaction where all reactants are completely converted into products, the sum of the masses of the reactants equals the sum of the masses of the products (Figure 3.12).

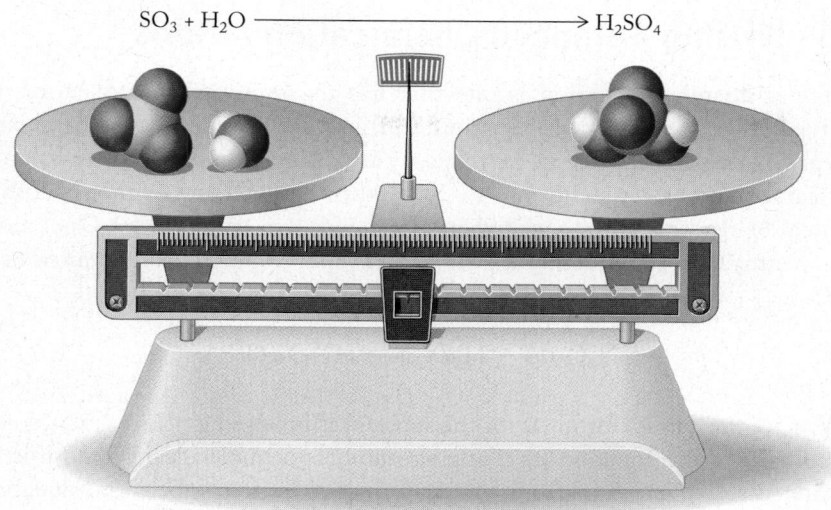

$$SO_3 + H_2O \longrightarrow H_2SO_4$$

FIGURE 3.12 The law of conservation of mass states that the total mass of reactants consumed in a chemical reaction equals the total mass of products formed in the reaction. In the reaction shown here, the combined mass of SO_3 and H_2O consumed equals the mass of the H_2SO_4 that forms. The number of each kind of atom is the same on the left and right sides of the balanced equation.

This fact illustrates a fundamental relation known as the **law of conservation of mass**, which applies to all chemical reactions. This law works for two reasons: (1) the total number of atoms (and hence moles) of each element to the left of the reaction arrow in the equation representing any chemical reaction must match the total number of atoms (and moles) of that element to the right of the reaction arrow, and (2) the identity of atoms does not change in a chemical reaction.

CONCEPT TEST

Until the last decades of the 20th century, flashbulbs were used for indoor photography (Figure 3.13). Flashbulbs are based on the reaction between magnesium and oxygen described by the chemical equation:

$$2\,Mg(s) + O_2(g) \rightarrow 2\,MgO(s)$$

Electricity is used to initiate the reaction, which produces a bright white flash as the magnesium burns. The mass of magnesium used in the reaction does not equal the mass of oxygen used in the reaction, but the total mass of the flashbulb remains constant before and after it is used. Why?

(a)　　　　　　　　(b)

$$2\,Mg(s)\ +\ O_2(g)\ \rightarrow\ 2\,MgO(s)$$

FIGURE 3.13 A flashbulb contains a fixed amount of magnesium and air. Electricity is used to start the reaction that converts $Mg(s)$ to $MgO(s)$. Because the law of conservation of mass applies, the mass of the flashbulb before it goes off (a) is the same as after (b).

(Answers to Concept Tests are in the back of the book.)

law of conservation of mass the sum of the masses of the reactants in a chemical reaction is equal to the sum of the masses of the products.

3.3 Writing Balanced Chemical Equations

A *balanced* chemical equation is one that has the same number of atoms of each type on both sides of the equation. In this section we look at some chemical reactions that took place in Earth's early atmosphere to learn how to balance chemical equations. In addition to SO_3, this atmosphere contained other nonmetal oxides that form acids when combined with water vapor. One, dinitrogen pentoxide (N_2O_5), combines with water vapor to produce liquid nitric acid (HNO_3):

$$N_2O_5(g) + H_2O(g) \rightarrow HNO_3(\ell) \qquad (3.3)$$

We have the correct formulas for the substances involved in the reaction, so we have to balance the equation by changing numbers of molecules. We do this by adjusting the *coefficients* in front of the formulas. In the unbalanced equation (Equation 3.3), the coefficients are understood to be 1, as the formula itself stands for one molecule (or 1 mole) of the substance. Never change the subscripts in the formulas in chemical equations when balancing equations because doing so changes the identity of the substance.

There are many ways to balance chemical equations. Here is a three-step approach you may find useful.

1. *Write an equation using the correct chemical formulas for reactants and products. Include symbols indicating physical states.* Here you are given the reaction of gaseous dinitrogen pentoxide (N_2O_5) with water vapor (H_2O) to make liquid nitric acid (HNO_3):

$$N_2O_5(g) + H_2O(g) \rightarrow HNO_3(\ell)$$

Check whether the equation is balanced by adding up the different types of atoms on each side of the reaction arrow. This equation as written is unbalanced; the number of N, O, and H atoms in the reactants does not equal the number of N, O, and H on the product side.

Element	Reactant Side	Product Side
N	2	1
O	5 + 1 = 6	3
H	2	1

2. *Choose an element that appears in only one reactant and product to balance first.* In this case, let's choose nitrogen, which appears only in N_2O_5 and HNO_3. To make the number of N atoms the same on both sides of the equation, we need to add a coefficient of 2 in front of HNO_3:

$$\underline{}\ N_2O_5(g) + \underline{}\ H_2O(g) \rightarrow \underline{2}\ HNO_3(\ell)$$

3. *Choose coefficients for the other substances so that the numbers of atoms of each element are the same on the two sides of the equation.* Remember, balancing is

done *only* with coefficients—you *never* change subscripts in a chemical formula:

$$N_2O_5(g) \; + \; H_2O(g) \rightarrow \; 2\,HNO_3(\ell)$$

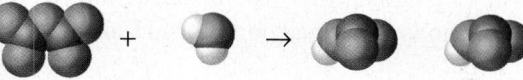

This equation now has 2 H atoms on the left and on the right. There are a total of 6 O atoms on the left (5 from N_2O_5 and 1 from H_2O) and 6 O atoms on the right, which means the equation is balanced.

Element	Reactant Side	Product Side
N	2	$2 \times 1 = 2$
O	$5 + 1 = 6$	$2 \times 3 = 6$
H	2	$2 \times 1 = 2$

The chemical reaction of N_2O_5 or any substance with water is called **hydrolysis**. Hydrolysis reactions that involve nonmetal oxides such as N_2O_5 result in acidic solutions, and such reactions were the source of the acidic environment on early Earth described in Section 3.1.

Let's practice balancing equations by looking at the reaction in which dinitrogen pentoxide is formed from nitrogen gas and oxygen gas.

1. *Write the correct formulas for reactants and products.* For this reaction we need to know that nitrogen and oxygen gases exist in nature as diatomic molecules, so their molecular formulas are N_2 and O_2. Indeed, all of the gaseous nonmetallic elements except for the noble gases exist in nature as diatomic molecules: H_2, N_2, O_2, F_2, and Cl_2; Br_2, a liquid at room temperature, and I_2, a solid, are also diatomic (Figure 3.14). When these substances are involved in chemical reactions, they must be written in the form of diatomic molecules.

$$N_2(g) + O_2(g) \rightarrow N_2O_5(g)$$

Element	Reactant Side	Product Side
N	2	2
O	2	5

The equation as written is not balanced.

2. *Choose an element that appears in only one reactant and product to balance first.* In this case we could choose either nitrogen or oxygen. If we choose nitrogen, we see that there are already two N atoms on each side of the equation so no adjustment of the coefficients appears to be necessary.

3. *Choose coefficients for the other substances to equalize the numbers of atoms on the two sides.* Oxygen is the only remaining element to balance in the equation. With 2 O atoms on the reactant side and 5 O atoms on the product side, there is no whole-number coefficient that we can place in front of the O_2 to make the number of O atoms equal on both sides of the equation; five is 2.5 (5/2) times 2, but we need to use whole numbers.

hydrolysis the reaction of water with another material. The hydrolysis of nonmetal oxides produces acids.

▶❚❚ **CHEMTOUR** Balancing Equations

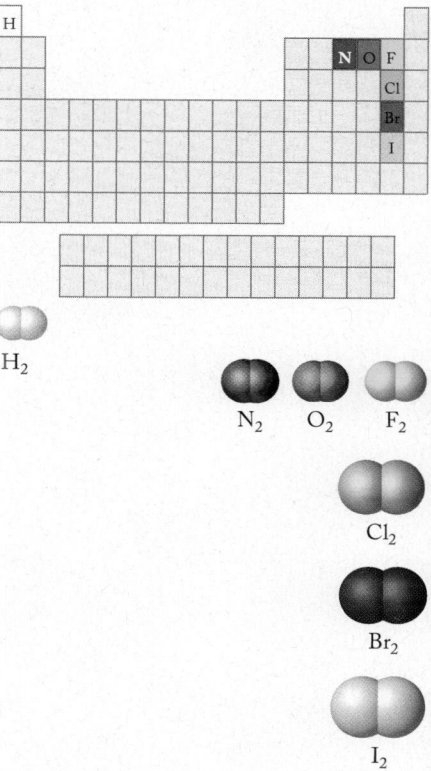

FIGURE 3.14 The diatomic elements are hydrogen, nitrogen, oxygen, and the group 17 halogens. The halogen astatine (At) at the bottom of group 17 is the rarest terrestrial element, and virtually none of its bulk physical properties are known.

Recognizing that 5/2 times 2 is the whole number 5, we can place a coefficient of 5 in front of O_2 and a coefficient of 2 in front of N_2O_5:

$$\underline{}\, N_2(g) + \underline{5}\, O_2(g) \rightarrow \underline{2}\, N_2O_5(g)$$

As a result of assigning these coefficients, however, the number of nitrogen atoms is no longer equal on both sides of the equation:

Element	Reactant Side	Product Side
N	2	$2 \times 2 = 4$
O	$5 \times 2 = 10$	$2 \times 5 = 10$

We can correct this by placing a coefficient of 2 in front of N_2 to finish the balancing of the equation:

$$\underline{2}\, N_2(g) + \underline{5}\, O_2(g) \rightarrow \underline{2}\, N_2O_5(g)$$

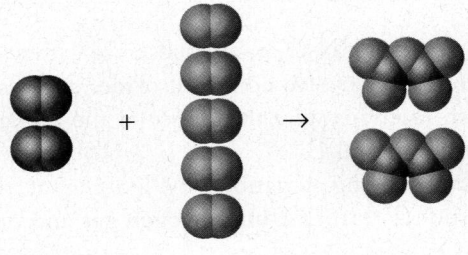

Element	Reactant Side	Product Side
N	$2 \times 2 = 4$	$2 \times 2 = 4$
O	$5 \times 2 = 10$	$2 \times 5 = 10$

The equation is now balanced. These steps do involve some trial and error; however, with practice you will find that this approach works for most chemical reactions.

SAMPLE EXERCISE 3.6   Balancing a Chemical Equation

Write a balanced chemical equation for the gas-phase reaction between SO_2 and O_2 that forms SO_3. This reaction, which describes one of the origins of acid rain, occurs when fuels containing sulfur are burned. Acid rain is responsible for the degradation of limestone statues. It is harmful to aquatic life and is also damaging to terrestrial plants.

Collect and Organize We are given the chemical formulas and physical states of the reactants and products. In a balanced chemical equation, we may not change the formulas of the compounds but we must have the same number of atoms of each type on both sides of the equation.

Analyze To write a balanced chemical equation, we must assign coefficients as needed to make the numbers of atoms of each element on the reactant (left) and product (right) sides of the reaction arrow equal.

Solve

1. Write an equation using correct formulas.

$$SO_2(g) + O_2(g) \rightarrow SO_3(g)$$

Element	Reactant Side	Product Side
S	1	1
O	2 + 2 = 4	3

The number of atoms in the reactants does not equal the number of atoms in the products; the equation is not balanced.

2. Sulfur appears in only one compound on each side of the equation. Choosing sulfur to balance first, we see that the number of S atoms on the left equals the number of S atoms on the right.

3. The number of O atoms on the left (4) does not equal the number of O atoms on the right (3). Changing the coefficient of SO_2 means that the number of S atoms on the left-hand side also changes. No single whole-number coefficient can be assigned for O_2 that makes the number of O atoms equal on both sides. We need to use two atoms of O (because oxygen exists in nature as molecules of O_2), so we also need two molecules of SO_2. We can accomplish this by placing a coefficient of 2 in front of both SO_2 and SO_3 to keep the number of S atoms equal.

$$\underline{2}\ SO_2(g) + O_2(g) \rightarrow \underline{2}\ SO_3(g)$$

Element	Reactant Side	Product Side
S	2 × 1 = 2	2 × 1 = 2
O	(2 × 2) + 2 = 6	2 × 3 = 6

The equation is balanced.

Think about It Balancing this equation required some trial and error as we had two reactants containing oxygen. Choosing which compound to start with when balancing equations may require you to try several different combinations of coefficients before arriving at the balanced equation.

Practice Exercise Balance the chemical equations for the reaction between elemental phosphorus $P_4(s)$ and oxygen to make $P_4O_{10}(s)$ and the hydrolysis of $P_4O_{10}(s)$ to produce phosphoric acid, $H_3PO_4(\ell)$.

(Answers to Practice Exercises are in the back of the book.)

CONCEPT TEST

Why can't we balance the equation for the reaction between SO_2 and O_2 in Sample Exercise 3.6 in the following fashion?

$$SO_2(g) + O_2(g) \rightarrow SO_3(g) \quad \text{unbalanced equation}$$
$$SO_2(g) + O(g) \rightarrow SO_3(g) \quad \text{``balanced'' equation}$$

(Answers to Concept Tests are in the back of the book.)

As a final note, a balanced equation is analogous to a mathematical equation. The equation in Sample Exercise 3.6 is still balanced if we multiply all the coefficients by the same number. For example, if we multiply all the coefficients by 2, the equation

$$4\,SO_2(g) + 2\,O_2(g) \rightarrow 4\,SO_3(g)$$

is still balanced. However, it is conventional to report a balanced equation with the smallest whole-number coefficients.

3.4 Combustion Reactions

In addition to carbon dioxide, Earth's early atmosphere contained significant concentrations of carbon monoxide (CO) and methane (CH_4). Both methane and carbon monoxide are still present in our atmosphere, although at much lower concentrations. Atmospheric methane can be traced to a number of diverse sources including wetlands, cattle ranching, rice production, and oil drilling operations. Carbon monoxide in the atmosphere arises from burning carbon-containing compounds. Natural sources of CO include volcanic activity and forest fires. In our industrialized world, the greatest use of methane (natural gas) and petroleum-based fuels is energy production. Fuels like methane are members of a class of organic compounds known as **hydrocarbons** because they are composed of only hydrogen and carbon.

Natural gas and other hydrocarbon fuels are burned to heat buildings and to warm water. Burning CH_4 is an example of an important class of chemical reactions known as **combustion reactions**, with the process usually referred to simply as *combustion*. Combustion refers to the reaction of a fuel with oxygen, such as the burning of a substance in air, which contains 21% oxygen.

When the combustion of a hydrocarbon is complete, the only products are carbon dioxide and water. In the presence of O_2, the most stable form of carbon is CO_2, not CO, so carbon monoxide also reacts with oxygen to form carbon dioxide. Let's balance chemical equations describing the combustion reactions between O_2 and CO and between O_2 and CH_4.

1. Starting with CO, the first step is to write the correct formulas of the reactants and products and indicate states:

$$CO(g) + O_2(g) \rightarrow CO_2(g)$$

Element	Reactant Side	Product Side
C	1	1
O	1 + 2 = 3	2

Carbon appears in only one reactant and product and the number of C atoms is already the same on both sides, so both CO and CO_2 have coefficients of 1.

2. Carbon dioxide has one more O atom than CO, so by analogy to the conversion of SO_2 to SO_3 in Sample Exercise 3.6, we need to put a coefficient of 2 in front of both CO and CO_2 to balance the equation:

$$\underline{2}\,CO(g) + O_2(g) \rightarrow \underline{2}\,CO_2(g)$$

Element	Reactant Side	Product Side
C	2 × 1 = 2	2 × 1 = 2
O	(2 × 1) + 2 = 4	2 × 2 = 4

hydrocarbons a class of organic compounds containing molecular compounds composed of only hydrogen and carbon.

combustion reaction a reaction between oxygen and another element or compound that produces heat.

We do not need to continue with step 3 because the equation is now balanced.

In the combustion reaction between methane and oxygen, the carbon becomes carbon dioxide when a sufficient supply of oxygen is available. The hydrogen in the fuel combines with oxygen to form water vapor.

1. Our first step is to write the reaction in the form of an equation with reactants on the left and products on the right. Write the equation:

$$CH_4(g) + O_2(g) \rightarrow H_2O(g) + CO_2(g)$$

Element	Reactant Side	Product Side
C	1	1
H	4	2
O	2	$1 + 2 = 3$

The carbon atoms are balanced but the hydrogen and oxygen atoms are not.

2. Like carbon, H appears in only one reactant and one product. Adding a coefficient of 2 in front of the H_2O will make the number of H atoms on both sides of the equation equal but increases the number of O atoms on the product side:

$$CH_4(g) + \underline{\ \ } O_2(g) \rightarrow \underline{2}\ H_2O(g) + CO_2(g)$$

Element	Reactant Side	Product Side
C	1	1
H	4	$2 \times 2 = 4$
O	2	$(2 \times 1) + 2 = 4$

3. This leaves only the O atoms to balance. A coefficient of 2 in front of O_2 makes the atoms of O equal on both sides of the equation:

$$CH_4(g) + \underline{2}\,O_2(g) \rightarrow \underline{2}\,H_2O(g) + CO_2(g)$$

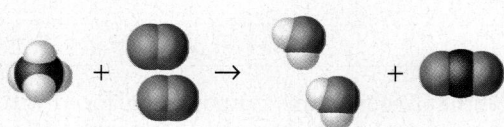

Element	Reactant Side	Product Side
C	1	1
H	4	$2 \times 2 = 4$
O	$2 \times 2 = 4$	$(2 \times 1) + 2 = 4$

Balancing first C, then H, then O is useful in writing the equations for many reactions in which oxygen reacts with compounds containing only carbon and hydrogen.

SAMPLE EXERCISE 3.7 **Writing and Balancing the Equation for a Combustion Reaction**

Methane (CH_4) is the principal ingredient in natural gas, but significant concentrations of the gases ethane (C_2H_6) and propane (C_3H_8) are also present in most natural gas samples. Write and balance the equation describing the complete combustion of C_2H_6.

Collect and Organize Ethane is a hydrocarbon, and combustion means that oxygen is the other reactant. The products of complete combustion of a hydrocarbon with oxygen are carbon dioxide and water. The balanced chemical equation will have the same number of atoms on both sides of the equation.

Analyze

1. The chemical equation describing the combustion of ethane is

$$C_2H_6(g) + O_2(g) \rightarrow CO_2(g) + H_2O(g)$$

Element	Reactant Side	Product Side
C	2	1
H	6	2
O	2	2 + 1 = 3

Solve

2. Because this reaction is between oxygen and a hydrocarbon, we balance first C, then H, then O. Balance the carbon atoms first by giving CO_2 in the product a coefficient of 2:

$$\_\_ C_2H_6(g) + \_\_ O_2(g) \rightarrow \underline{2}\ CO_2(g) + \_\_ H_2O(g)$$

Element	Reactant Side	Product Side
C	2	2 × 1 = 2
H	6	2
O	2	(2 × 2) + 1 = 5

3a. Then balance the hydrogen atoms by giving H_2O a coefficient of 3:

$$\_\_ C_2H_6(g) + \_\_ O_2(g) \rightarrow \underline{2}\ CO_2(g) + \underline{3}\ H_2O(g)$$

Element	Reactant Side	Product Side
C	2	2 × 1 = 2
H	6	3 × 2 = 6
O	2	(2 × 2) + (3 × 1) = 7

3b. At this stage, the oxygen atoms cannot be balanced with a simple whole-number coefficient for O_2; we need 7/2 O_2. However, if we give O_2 a coefficient of 7 and double the coefficients for ethane, carbon dioxide, and steam, we can write

$$\underline{2}\ C_2H_6(g) + \underline{7}\ O_2(g) \rightarrow \underline{4}\ CO_2(g) + \underline{6}\ H_2O(g)$$

Element	Reactant Side	Product Side
C	2 × 2 = 4	4 × 1 = 4
H	2 × 6 = 12	6 × 2 = 12
O	7 × 2 = 14	(4 × 2) + (6 × 1) = 14

This equation is balanced.

Think about It Balancing the equations for combustion reactions of hydro-carbons often requires several iterations of the coefficients before all of the elements are balanced.

Practice Exercise Balance the chemical equation describing the complete combustion of propane (C_3H_8).

(Answers to Practice Exercises are in the back of the book.)

3.5 Stoichiometric Calculations and the Carbon Cycle

Earth's atmosphere underwent a major change beginning about 2.5 billion years ago with the evolution of green plants and the onset of *photosynthesis*. This process, driven by the energy in visible light, involves several steps, but the overall reaction is

$$6\,CO_2(g) + 6\,H_2O(\ell) \rightarrow C_6H_{12}O_6(aq) + 6\,O_2(g)$$
$$\text{Glucose}$$

and the reverse reaction, *respiration*,

$$C_6H_{12}O_6(aq) + 6\,O_2(g) \rightarrow 6\,CO_2(g) + 6\,H_2O(\ell)$$
$$\text{Glucose}$$

is the major source of energy for all living things on Earth. The appearance of bacteria capable of photosynthesis is believed to be the initial source of oxygen in our atmosphere, but today it comes mostly from green plants.

Photosynthesis and respiration are key reactions in the *carbon cycle* (Figure 3.15). The two processes are nearly but not exactly in balance in Earth's biosphere. If they were exactly in balance, no net change would have taken place in the concentrations of atmospheric carbon dioxide or oxygen in the past 2.5 billion years. However, about 0.01% of the decaying mass of plants and animals (called *detritus*) is incorporated into sediments and soil when organisms die. Shielded in this way from exposure to oxygen, the carbon in this mass is not converted back into CO_2. Although 0.01% may not seem like much, over hundreds of millions of years it has added up to the removal of about 10^{20} kg of carbon dioxide from the atmosphere. About 10^{15} kg of this buried carbon is in the form of fossil fuels: coal, petroleum, and natural gas.

As a result of human activity and the combustion of fossil fuels, the natural balance that limited the concentration of CO_2 in the atmosphere is being altered. Annually about 6.8 trillion (6.8×10^{12}) kg of carbon is reintroduced to the atmosphere as CO_2 as a result of the combustion of fossil fuels, and deforestation adds another 2×10^{12} kg each year. The effects of these additions on global climate have been the subject of considerable debate, and we will examine them again in Chapter 8. Here, though, we use a typical reaction involving CO_2 to learn how to use a balanced chemical equation to calculate the mass of products in a chemical reaction.

If combustion of fossil fuels adds 6.8×10^{12} kg of carbon to the atmosphere each year as CO_2, what is the mass of the carbon dioxide added? The mass must be more than 6.8×10^{12} kg, because this amount is only the mass due to carbon.

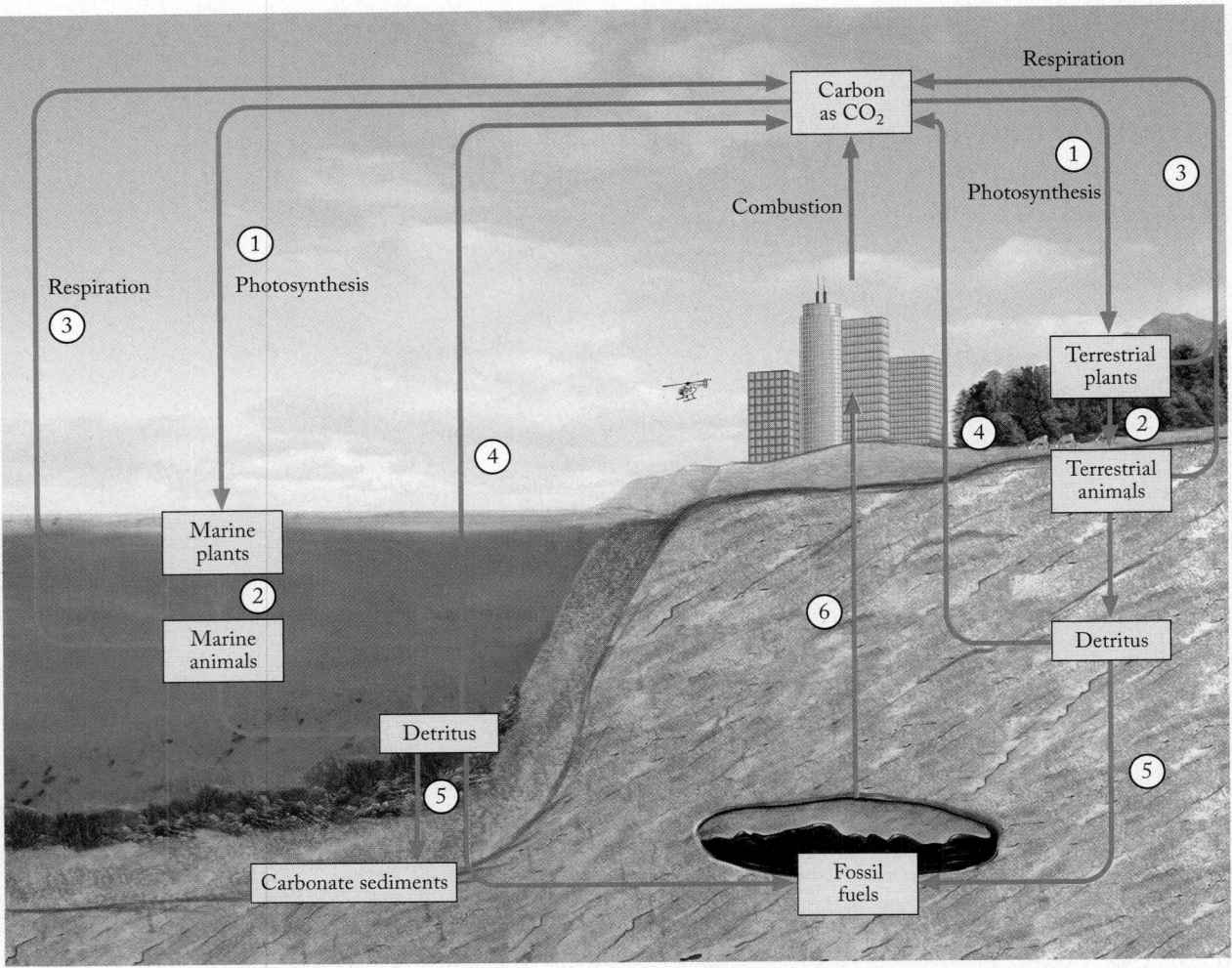

FIGURE 3.15 The carbon cycle. ① Green plants and marine plants incorporate CO_2 into their biomass. ② Some of the plant biomass becomes the biomass of animals. ③ As plants and animals respire, they release CO_2 back into the environment. ④ When they die, the decay of their tissues releases most of their carbon content as CO_2, but about 0.01% is incorporated into carbonate minerals and deposits of coal, petroleum, and natural gas (fossil fuels, ⑤). ⑥ Mining and the combustion of fossil fuels for human use are shifting the natural equilibrium that has controlled the concentration of CO_2 in the atmosphere.

The balanced chemical equation for the combustion of carbon to carbon dioxide in excess O_2 is

$$C(s) + O_2(g) \rightarrow CO_2(g)$$

To use this equation to determine the mass of CO_2 released, we first use the molar mass of carbon, 12.01 g/mol, to convert mass into moles:

$$6.8 \times 10^{12} \, \cancel{\text{kg C}} \times \frac{10^3 \, \cancel{\text{g}}}{1 \, \cancel{\text{kg}}} \times \frac{1 \text{ mol C}}{12.01 \cancel{\text{g C}}} = 5.7 \times 10^{14} \text{ mol C}$$

We know from the coefficients for CO_2 and C in the balanced equation that the mole ratio of CO_2 to C is 1:1, and so the amount of CO_2 produced is also 5.7×10^{14} moles:

$$5.7 \times 10^{14} \, \cancel{\text{mol C}} \times \frac{1 \text{ mol } CO_2}{1 \, \cancel{\text{mol C}}} = 5.7 \times 10^{14} \text{ mol } CO_2$$

To convert moles of CO_2 to mass, we first calculate the molar mass of CO_2:

$$12.01 \text{ g/mol} + 2(16.00 \text{ g/mol}) = 44.01 \text{ g/mol } CO_2$$

Multiplying the moles of CO_2, by the molar mass of CO_2, and converting grams to kilograms, gives us the mass of CO_2:

$$5.7 \times 10^{14} \text{ mol } CO_2 \times \frac{44.01 \text{ g } CO_2}{1 \text{ mol } CO_2} \times \frac{1 \text{ kg}}{10^3 \text{ g}} = 2.5 \times 10^{13} \text{ kg } CO_2$$

▶❙❙ **CHEMTOUR** Carbon Cycle

Notice that the answer in each of these steps is the starting point for the next step. Therefore we can combine the three separate calculations into a single calculation:

$$6.8 \times 10^{12} \text{ kg C} \times \frac{10^3 \text{ g}}{1 \text{ kg}} \times \frac{1 \text{ mol C}}{12.01 \text{ g C}} \times \frac{1 \text{ mol } CO_2}{1 \text{ mol C}} \times \frac{44.01 \text{ g } CO_2}{1 \text{ mol } CO_2} \times \frac{1 \text{ kg}}{10^3 \text{ g}}$$

$$= 2.5 \times 10^{13} \text{ kg } CO_2$$

This procedure can be applied to determining the mass of any substance (reactant or product) involved in any chemical reaction if we know (1) the mass of another substance in the reaction and (2) the *stoichiometric relation* between the two substances, that is, their mole ratio in the balanced chemical equation.

SAMPLE EXERCISE 3.8   Calculating a Product Mass from a Reactant Mass

In 2008, electric power plants in the United States consumed about 1.14×10^{11} kg of natural gas. Natural gas is mostly methane, CH_4, so we can approximate the combustion reaction generating the energy by the equation:

$$CH_4(g) + 2\,O_2(g) \rightarrow 2\,H_2O(g) + CO_2(g)$$

How many kilograms of CO_2 were released into the atmosphere from these power plants in 2008?

Collect and Organize We are to calculate the mass of CO_2 released by the combustion of methane. The balanced chemical equation relates the moles of product (CO_2) to the moles of reactant (CH_4). The mass of CO_2 is calculated by converting the moles of CO_2 to grams of CO_2 using the molar mass, then converting to kilograms.

Analyze This is a stoichiometry problem. The coefficients for CH_4 and CO_2 in the balanced equation tell us that 1 mole of carbon dioxide is produced for every 1 mole of methane consumed. We can convert the mass of CH_4 into moles of CH_4 by dividing the mass of CH_4 by the molar mass of CH_4. In this reaction the moles of CH_4 consumed is equal to the number of moles of CO_2 produced; the mole ratio is 1:1. Finally we convert moles of CO_2 into kilograms of CO_2.

Solve We check the chemical equation to confirm that it is balanced as written—and it is. To convert between grams and moles of CH_4 and CO_2, we need to calculate the molar masses of these compounds. The molar mass of CH_4 is

$$12.01 \text{ g/mol} + 4(1.008 \text{ g/mol}) = 16.04 \text{ g/mol } CH_4$$

and the molar mass of CO_2 is

$$12.01 \text{ g/mol} + 2(16.00 \text{ g/mol}) = 44.01 \text{ g/mol } CO_2$$

1. Convert the mass of CH_4 into moles:

$$1.14 \times 10^{11} \text{ kg } CH_4 \times \frac{10^3 \text{ g}}{1 \text{ kg}} \times \frac{1 \text{ mol } CH_4}{16.04 \text{ g } CH_4} = 7.107 \times 10^{12} \text{ mol } CH_4$$

2. Convert moles of CH_4 into moles of CO_2 using the coefficients from the balanced chemical equation:

$$7.107 \times 10^{12} \text{ mol } CH_4 \times \frac{1 \text{ mol } CO_2}{1 \text{ mol } CH_4} = 7.107 \times 10^{12} \text{ mol } CO_2$$

3. Convert moles of CO_2 into mass of CO_2:

$$7.107 \times 10^{12} \text{ mol } CO_2 \times \frac{44.01 \text{ g } CO_2}{1 \text{ mol } CO_2} \times \frac{1 \text{ kg}}{10^3 \text{ g}} = 3.13 \times 10^{11} \text{ kg } CO_2$$

We can combine the three separate calculations into a single calculation, and in subsequent problems we may not show the individual steps:

$$1.14 \times 10^{11} \text{ kg } CH_4 \times \frac{10^3 \text{ g}}{1 \text{ kg}} \times \frac{1 \text{ mol } CH_4}{16.04 \text{ g } CH_4} \times \frac{1 \text{ mol } CO_2}{1 \text{ mol } CH_4}$$

$$\times \frac{44.01 \text{ g } CO_2}{1 \text{ mol } CO_2} \times \frac{1 \text{ kg}}{10^3 \text{ g}} = 3.13 \times 10^{11} \text{ kg } CO_2$$

Think about It The mole ratio of CO_2 to CH_4 is 1:1 but the molar mass of CO_2 (44.01 g/mol) is almost three times the molar mass of CH_4 (16.01 g/mol), so we expect the mass of CO_2 produced to be greater than the mass of CH_4 consumed.

Practice Exercise Disposable lighters burn butane (C_4H_{10}) and produce CO_2 and H_2O. Balance the chemical equation for this combustion reaction, and determine how many grams of CO_2 are produced by burning 1.00 g of C_4H_{10}.

(Answers to Practice Exercises are in the back of the book.)

■ •

3.6 Determining Empirical Formulas from Percent Composition

During the Industrial Revolution, access to geological deposits of iron oxides was a key to economic development. The more accessible an iron-containing mineral is and the higher its iron content, the more valuable it is. We typically express contents such as this in terms of **percent composition**: the composition of a compound expressed in terms of the percentages of the masses of the constituent elements with respect to the total mass of the compound. Let's consider two iron-containing minerals—wustite, FeO, and hematite, Fe_2O_3. Which has the higher iron content, and what is the content of the richer element, expressed as the percentage of the mass of iron with respect to the total mass of the compound?

percent composition the composition of a compound expressed in terms of the percentage by mass of each element in the compound.

One way to answer the first question is to examine the mole ratios of Fe to O, which are 1:1 in FeO and 2:3 (or $1:1.5$) in Fe_2O_3. The fact that in hematite (Fe_2O_3) the oxygen has a higher value (1.5) in the mole ratio tells us there is a higher proportion of O and hence a lower proportion of Fe in Fe_2O_3; therefore FeO must have the higher iron content.

To define the percent composition, we must calculate the iron content of FeO. Suppose we have 1 mole of FeO, which has a molar mass of

$$55.85 \text{ g Fe/mol} + 16.00 \text{ g O/mol} = 71.85 \text{ g/mol}$$

Of this 71.85 g, Fe accounts for 55.85 g. Therefore, the Fe content of FeO is

$$\frac{\text{mass of Fe}}{\text{total mass}} = \frac{55.85 \text{ g Fe}}{71.85 \text{ g FeO}} = 0.7773$$

$$\text{or} \quad 0.7773 \times 100\% = 77.73\% \text{ Fe}$$

It follows that the oxygen content is

$$\frac{\text{mass of O}}{\text{total mass}} = \frac{16.00 \text{ g O}}{71.85 \text{ g FeO}} = 0.2227 \quad \text{or} \quad 22.27\%$$

Because FeO contains only Fe and O, we could also determine the percent O by subtracting the percent Fe from 100%:

$$100.00\% - 77.73\% = 22.27\%$$

Thus the percent composition of FeO is 77.73% Fe and 22.27% O.

▶❚❚ **CHEMTOUR** Percent Composition

SAMPLE EXERCISE 3.9 **Calculating Percent Composition from a Chemical Formula**

What is the percent composition of the mineral forsterite, Mg_2SiO_4?

Collect and Organize We are given the chemical formula of forsterite and asked to calculate its percent composition, which is its composition expressed in terms of the percentages by mass of its component elements.

Analyze We need the molar masses of each of the elements to calculate the molar mass of Mg_2SiO_4 and the contribution of each element to it. To calculate the percent composition of forsterite, we must determine the percentage of each element in one mole of forsterite. These percentages can be calculated by dividing the mass of each element in 1 mole of forsterite by the molar mass of forsterite.

Solve The molar mass of Mg_2SiO_4 is

$$2(24.31 \text{ g/mol}) + 28.09 \text{ g/mol} + 4(16.00 \text{ g/mol}) = 140.71 \text{ g/mol}$$

The percent composition of this compound is therefore

$$\%\text{Mg} = \frac{48.62 \text{ g Mg}}{140.71 \text{ g}} \times 100\% = 34.55\% \text{ Mg}$$

$$\%\text{Si} = \frac{28.09 \text{ g Si}}{140.71 \text{ g}} \times 100\% = 19.96\% \text{ Si}$$

$$\%\text{O} = \frac{64.00 \text{ g O}}{140.71 \text{ g}} \times 100\% = 45.48\% \text{ O}$$

Think about It The percentage of the mass of forsterite due to Mg is nearly twice the percentage due to Si, which makes sense because the molar masses of Mg and Si are not that different and there are 2 moles of Mg for every 1 mole of Si. It also makes sense that oxygen accounts for nearly half the mass of forsterite because, even though O has the smallest molar mass of the three elements, there are 4 moles of O for every 3 moles of the other elements combined.

Practice Exercise Determine the percent composition of enstatite, $MgSiO_3$, a mineral found in Earth's crust.

(Answers to Practice Exercises are in the back of the book.)

Note in Sample Exercise 3.9 that the three percentages sum to 99.99%. The percent composition values should always sum to 100% or very close to 100% if we have accounted for all the elements that make up the total mass of the compound. The total may deviate slightly from 100% because of rounding.

We typically determine the percent composition of a substance in the laboratory by measuring the amount of each element in a given mass of the substance. We can use these data to determine the empirical formula of the substance. There are four steps to deriving an empirical formula from percent composition data: (1) Assume you have 100.00 g of the substance so that the percent composition values are equivalent to the values of the masses of the elements expressed in grams. As an initial check, we add the mass percentages given in the problem to make sure they total 100%. (2) Convert the mass of each element into moles. (3) Compute the mole ratio by reducing one of the mole values to 1. (4) If necessary, convert the ratio from step 3 into a ratio of whole numbers.

Suppose, for example, we run an elemental analysis on a sample of an iron oxide and find that the percent composition is 77.73% Fe and 22.27% O. In a 100.00 g sample there are

$$77.73\% \text{ Fe in } 100.00 \text{ g of sample} = 77.73 \text{ g Fe}$$

$$22.27\% \text{ O in } 100.00 \text{ g of sample} = 22.27 \text{ g O}$$

Next we determine the number of moles of Fe and O in our 100.00 g sample by dividing the mass of each element by its molar mass:

$$\frac{77.73 \text{ g Fe}}{55.85 \text{ g Fe/mol Fe}} = 1.392 \text{ mol Fe} \qquad \frac{22.27 \text{ g O}}{16.00 \text{ g O/mol O}} = 1.392 \text{ mol O}$$

The mole ratio of iron to oxygen is 1.392 : 1.392, or 1:1. This means that in 1 mole of our substance, there is 1 mole of iron atoms (or 1 Fe atom) for every 1 mole of oxygen atoms. Thus the chemical formula of this iron oxide is FeO, or iron(II) oxide.

Let's see how this procedure works with another iron compound. Elemental analysis reveals that the percent composition of a sample is 69.94% Fe and 30.06% O. Our calculations are then

$$69.94\% \text{ Fe in } 100.00 \text{ g of sample} = 69.94 \text{ g Fe}$$

$$30.06\% \text{ O in } 100.00 \text{ g of sample} = 30.06 \text{ g O}$$

and the numbers of moles are

$$\frac{69.94 \; \text{g Fe}}{55.85 \; \text{g Fe/mol Fe}} = 1.252 \; \text{mol Fe} \qquad \frac{30.06 \; \text{g O}}{16.00 \; \text{g O/mol O}} = 1.879 \; \text{mol O}$$

The Fe:O mole ratio is $1.252 : 1.879$. To convert this ratio to small whole numbers, we divide both values by the smaller value. This guarantees that one of the numbers in the ratio is 1 and the other number is greater than 1:

$$\frac{1.252 \; \text{mol Fe}}{1.252} = 1.000 \; \text{mol Fe} \qquad \frac{1.879 \; \text{mol Fe}}{1.252} = 1.501 \; \text{mol O} \approx 1.5 \; \text{mol O}$$

To change the mole ratio $1 : 1.5$ to a ratio of whole numbers, we multiply both numbers in the ratio by a factor that results in two whole numbers. Multiplying 1.5 (or 3/2) by 2 yields 3, so we multiply both terms in the ratio $1 : 1.5$ by 2 to obtain an Fe:O ratio of 2:3. Thus the chemical formula is Fe_2O_3, and the substance is hematite.

The formulas FeO and Fe_2O_3 are empirical formulas, which we discussed in Chapter 2 as the typical formulas of ionic compounds. *Empirical,* which is related to the word *experiment,* means "derived from experimental data," and empirical formulas result from the analysis of data from experiments run to determine percent composition. Both FeO and Fe_2O_3 are ionic compounds made up of three-dimensional arrays of Fe^{2+} and O^{2-} ions in FeO and Fe^{3+} and O^{2-} ions in Fe_2O_3. Empirical formulas represent the relative proportions of the ions in the formula unit of an ionic compound. In FeO, the Fe^{2+} and O^{2-} ions are arranged as shown in Figure 3.16, in what is often called the *halite structure* because it is the structure seen in the mineral halite (NaCl). In an ionic compound the chemical formula, the empirical formula, and the formula unit are all the same.

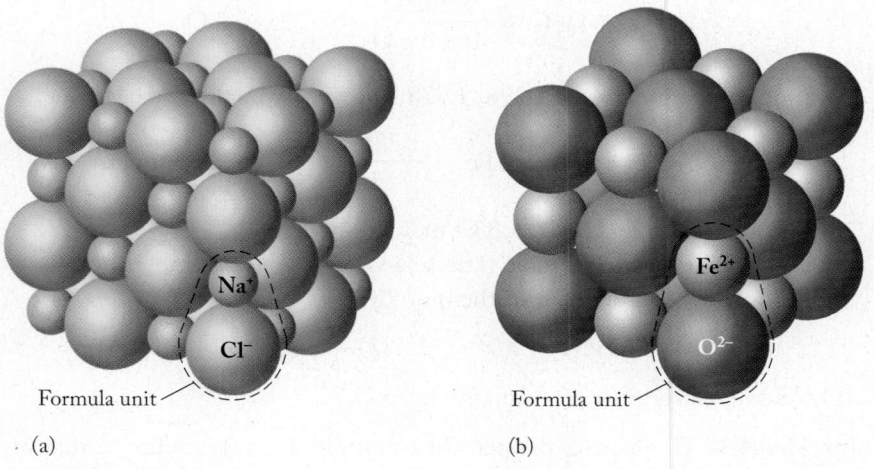

Formula unit

(a)

Formula unit

(b)

FIGURE 3.16 The halite structure is the three-dimensional pattern of the Na^+ and Cl^- ions in (a) NaCl; the same pattern is found in the Fe^{2+} and O^{2-} ions in (b) FeO.

CONCEPT TEST

Which of the following compounds have the same empirical formula and same percent composition?
   a. Ethylene (C_2H_4), a gas used to ripen bananas.
   b. Eicosene ($C_{20}H_{40}$), a compound used to attract Japanese beetles to traps.
   c. Acetylene (C_2H_2), a gas used in welding.
   d. Benzene (C_6H_6), a known carcinogen found in gasoline.

(Answers to Concept Tests are in the back of the book.)

FIGURE 3.17 A team of geologists and students from the University of British Columbia encountered this 8-metric ton (8000 kg) magnetite boulder on Vancouver Island during a 2005 field trip to an abandoned mine site.

SAMPLE EXERCISE 3.10 **Deriving an Empirical Formula from Percent Composition**

The mineral magnetite (Figure 3.17) is composed only of iron and oxygen and is 72.36% iron by mass. What is its empirical formula?

Collect and Organize We are given the percent composition by mass of Fe in magnetite and asked to determine the simplest whole-number ratio of iron and oxygen in the chemical formula of magnetite.

Analyze Because the only elements in the mineral are iron and oxygen, the percent composition by mass of the oxygen may be obtained by subtracting the percent Fe from 100%. We can use the four steps described in the preceding discussion to determine the empirical formula of magnetite.

Solve We are given the percent Fe but we also need the percent O in the sample: %O = 100.00% − %Fe = 100.00% − 72.36% = 27.64%. Now we have the data we need to determine the empirical formula of magnetite:

1. Convert percent composition values into masses by assuming a sample size of 100.00 g:

$$72.36\% \text{ Fe in } 100.00 \text{ g of sample} = 72.36 \text{ g Fe}$$
$$27.64\% \text{ O in } 100.00 \text{ g of sample} = 27.64 \text{ g O}$$

2. Convert masses into moles:

$$72.36 \text{ g Fe} \times \frac{1 \text{ mol Fe}}{55.85 \text{ g Fe}} = 1.296 \text{ mol Fe}$$

$$27.64 \text{ g O} \times \frac{1 \text{ mol O}}{16.00 \text{ g O}} = 1.728 \text{ mol O}$$

3. Simplify the mole ratio 1.296 : 1.728 by dividing by the smaller value:

$$\frac{1.296 \text{ mol Fe}}{1.296} = 1.000 \text{ mol Fe} \qquad \frac{1.728 \text{ mol O}}{1.296} = 1.333 \text{ mol O}$$

The mole ratio Fe:O is 1 : 1.333, or 1 : $\frac{4}{3}$.

4. Convert the fractional mole ratio to a ratio of whole numbers by multiplying each number in the ratio by 3:

$$\text{Fe:O} = 3 \times (1 : 1.333) = 3 : 4$$

The empirical formula of magnetite is Fe_3O_4.

Think about It Comparing the percent composition of magnetite to that of the two other ores of iron we have evaluated, magnetite (Fe_3O_4) is 72.36% Fe, a value that is in between the value for wustite (FeO), 77.73%, and hematite (Fe_2O_3), 69.94% Fe. This makes sense in terms of the empirical formulas, because the ratio of Fe to O in magnetite (1 : 1.333) is smaller than the 1:1 ratio in FeO and greater than the 1 : 1.5 ratio in Fe_2O_3.

Practice Exercise For thousands of years the mineral chalcocite (pronounced KAL-kuh-site; Figure 3.18) has been a highly prized source of copper. Its chemical composition is 79.85% Cu and 20.15% S. What is its empirical formula?

(Answers to Practice Exercises are in the back of the book.)

FIGURE 3.18 Chalcocite, a copper-containing ore.

SAMPLE EXERCISE 3.11	**Deriving an Empirical Formula of a Compound Containing More than Two Elements**

A sample of the carbonate mineral dolomite is determined in the laboratory to be 21.73% Ca, 13.18% Mg, 13.03% C, and 52.06% O. What is the empirical formula of dolomite?

Collect and Organize We are given the percent composition by mass of dolomite and asked to determine the simplest whole-number ratio of the four elements (Ca, Mg, C, and O) that will enable us to define the chemical formula of dolomite.

Analyze We follow the same steps in deriving this empirical formula as we did in Sample Exercise 3.10, except that here we must calculate more than one mole ratio to determine the empirical formula. The sum of the mass percentages should be very close to 100%; in this case, 21.73% + 13.18% + 13.03% + 52.06% = 100.00%. For this exercise we also need the molar masses of the four component elements: C, 12.01 g/mol; O, 16.00 g/mol; Mg, 24.31 g/mol; and Ca, 40.08 g/mol.

Solve

1. Assuming we have a sample of dolomite that is exactly 100.00 g, the sample contains 21.73 g of Ca, 13.18 g of Mg, 13.03 g of C, and 52.06 g of O.

2. Converting these masses into moles, we have

$$21.73 \text{ g Ca} \times \frac{1 \text{ mol Ca}}{40.08 \text{ g Ca}} = 0.5422 \text{ mol Ca}$$

$$13.18 \text{ g Mg} \times \frac{1 \text{ mol Mg}}{24.31 \text{ g Mg}} = 0.5422 \text{ mol Mg}$$

$$13.03 \text{ g C} \times \frac{1 \text{ mol C}}{12.01 \text{ g C}} = 1.085 \text{ mol C}$$

$$52.06 \text{ g O} \times \frac{1 \text{ mol O}}{16.00 \text{ g O}} = 3.254 \text{ mol O}$$

3. We divide each mole value by the smallest value to obtain a simple ratio of the four elements where at least one of the numbers is 1:

$$\frac{0.5422 \text{ mol Ca}}{0.5422} = 1.000 \text{ mol Ca} \qquad \frac{0.5422 \text{ mol Mg}}{0.5422} = 1.000 \text{ mol Mg}$$

$$\frac{1.085 \text{ mol C}}{0.5422} = 2.001 \text{ mol C} \qquad \frac{3.254 \text{ mol O}}{0.5422} = 6.001 \text{ mol O}$$

4. All these results are either whole numbers or very close to whole numbers, so no further multiplication of terms is needed. Our mole ratio is Ca:Mg:C:O = 1:1:2:6, which means the empirical formula is $CaMgC_2O_6$.

Think about It In this Sample Exercise we see how a greater percentage by mass does not necessarily correspond to a greater number of moles of an element in the empirical formula. The sample contains 21.73% Ca and 13.03% C but there are 2 moles of C and only 1 mole of Ca in a mole of the compound.

In addition, we are told the mineral is a carbonate; the carbonate ion is CO_3^{2-}, so a better formula of the compound is $CaMg(CO_3)_2$.

Practice Exercise Determine the empirical formula of a mineral having the percent composition 28.59% O, 24.95% Fe, and 46.46% Cr.

(Answers to Practice Exercises are in the back of the book.)

■ ·

3.7 Empirical and Molecular Formulas Compared

An empirical formula tells us the simplest whole-number ratio of the elements contained in a compound but does not necessarily indicate the molecular formula of the compound. To illustrate the difference between empirical and molecular formulas, let's look at the organic molecule glycolaldehyde (Figure 3.19). About one in 20 of the meteorites that fall to Earth today contains a variety of organic compounds, including glycolaldehyde. Some scientists view the presence of these compounds as evidence that the molecular building blocks of life on Earth may have come from space. Glycolaldehyde is also present in our bodies as a product of the metabolism of sugars and proteins.

Elemental analysis of glycolaldehyde gives the following percent composition data: 40.00% C, 6.71% H, and 53.28% O. We can use this information to calculate the empirical formula of the compound and then compare our result with the molecular structure in Figure 3.19. Using the method developed in Section 3.6, we obtain these results:

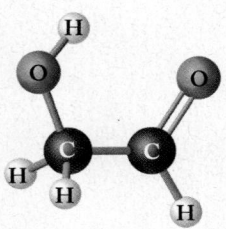

FIGURE 3.19 Glycolaldehyde is one of over 100 organic compounds detected in interstellar gases. It is also the smallest molecule among those identified as sugars.

$$40.00 \text{ g C} \times \frac{1 \text{ mol C}}{12.01 \text{ g C}} = 3.331 \text{ mol C}$$

$$6.71 \text{ g H} \times \frac{1 \text{ mol H}}{1.008 \text{ g H}} = 6.66 \text{ mol H}$$

$$53.28 \text{ g O} \times \frac{1 \text{ mol O}}{16.00 \text{ g O}} = 3.330 \text{ mol O}$$

The mole ratio of carbon to hydrogen to oxygen is 3.331 : 6.66 : 3.330 = 1 : 2 : 1, which defines an empirical formula of CH_2O.

Comparing this empirical formula with the molecule in Figure 3.19 reveals that the two are related but not equivalent. Counting up the atoms in the model, we arrive at the molecular formula $C_2H_4O_2$, which represents the actual numbers of C, H, and O atoms in one molecule of glycolaldehyde. In this case, we must multiply each subscript in the empirical formula by 2 to obtain the molecular formula.

─────── CONCEPT TEST ·······································

Which of the following chemical formulas must represent empirical formulas and which ones must represent molecular formulas: $C_2H_6O_2$, ethylene glycol, used in antifreeze; C_3H_8O, isopropanol, also known as rubbing alcohol; and $C_6H_{12}O_6$, glucose, also known as "blood sugar" and the primary source of energy for our body's cells.

(Answers to Concept Tests are in the back of the book.)

· ·

As the example of glycolaldehyde demonstrates, the results of elemental analysis do not necessarily reveal the molecular formula of a compound: they simply provide the simplest mole ratios of the elements in it and hence the empirical formula. An empirical formula represents the simplest mole ratio of the elements in a compound, whereas a molecular formula expresses the actual number of atoms of each element in a molecule.

Occasionally, empirical and molecular formulas are identical. For example, formaldehyde—which like glycolaldehyde has the percent composition 40.00% C, 6.71% H, 53.28% O—has the molecular formula CH_2O (Figure 3.20). Many other compounds also have the empirical formula CH_2O, as illustrated in Sample Exercise 3.12.

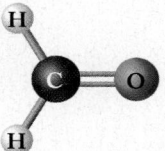

FIGURE 3.20 The empirical formula of the molecular compound formaldehyde is identical to its molecular formula.

SAMPLE EXERCISE 3.12   Deriving an Empirical Formula from a Molecular Formula

Molecular structures of four common sugars are shown in Figure 3.21. Which one(s) has the empirical formula CH_2O? (The wedges and heavy lines in Figure 3.21 denote bonds that stick out of the plane of the paper.) An empirical formula represents the simplest mole ratio of the elements in a compound, whereas a molecular formula expresses the actual number of atoms of each element in a molecule of a compound.

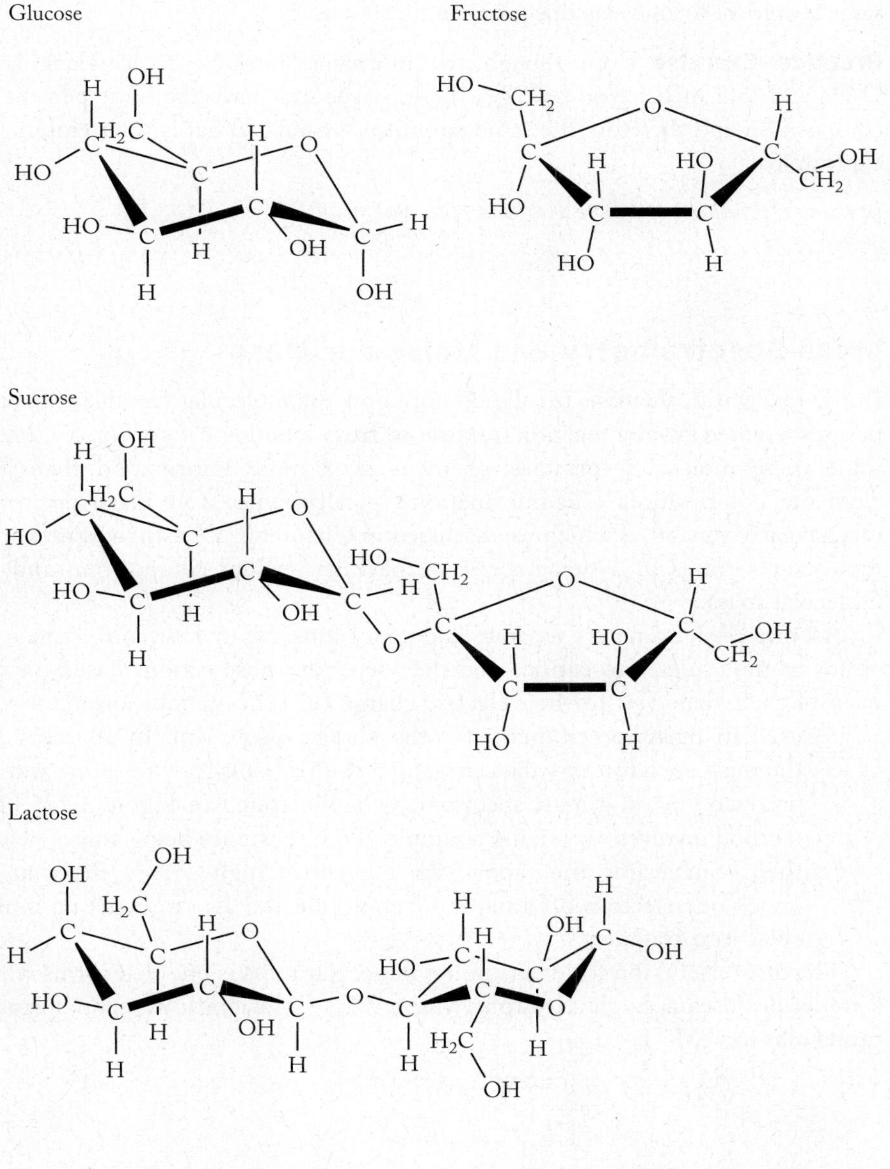

Glucose

Fructose

Sucrose

Lactose

FIGURE 3.21 The molecular structures of glucose, fructose, sucrose, and lactose.

Collect and Organize The problem gives the molecular structures of the sugars.

Analyze We determine the molecular formulas of the sugars from the molecular structures and then evaluate whether the ratios of carbon to hydrogen to oxygen in the molecules simplify to 1:2:1, the ratio in the empirical formula CH_2O in the problem statement.

Solve Counting up the atoms in the structures reveals the molecular formulas $C_6H_{12}O_6$ for glucose, $C_6H_{12}O_6$ for fructose, $C_{12}H_{22}O_{11}$ for sucrose, and $C_{12}H_{22}O_{11}$ for lactose. The C:H:O ratios in glucose and fructose are 6:12:6, which simplify to 1:2:1, resulting in an empirical formula of CH_2O for these two sugars. The ratios for lactose and sucrose, both 12:22:11, clearly do not simplify to 1:2:1, and indeed cannot be simplified further. The formulas for lactose and sucrose are both empirical formulas and molecular formulas. Fructose and glucose have the empirical formula CH_2O.

Think about It Sugars are all members of a major class of biological compounds called *carbohydrates. All contain carbon, hydrogen, and oxygen*, and have the generic molecular formula $C_x(H_2O)_y$. However, they do not all have the same empirical formula, as this problem illustrates.

Practice Exercise Even though the molecular formula of acetylene is C_2H_2 and that of benzene is C_6H_6, these compounds have the same percent composition and therefore the same empirical formula. What is their empirical formula?

(Answers to Practice Exercises are in the back of the book.)

Mass Spectrometry and Molecular Mass

In the real world, chemists usually do not know the molecular formulas of compounds isolated from a reaction mixture or from a biological system. To determine these molecular formulas, chemists need more information than just elemental composition. This information typically comes from mass spectrometry, an early version of which we discussed in Chapter 2. We can determine the molecular formula of a substance if we have its percent composition and its molecular mass.

All mass spectrometers separate and count ions. Many first convert neutral atoms or molecules into cations and then separate those cations based on the ratio of their mass (m) to their electric charge (z). The symbol lowercase z is used in mass spectrometry for the charge on an ion. In all cases for the mass spectrometry data presented in this book, $z = 1+$. One way to produce ions in a mass spectrometer is illustrated in Figure 3.22. This method involves vaporizing a sample of the substance being analyzed and then bombarding the vapor with a beam of high-energy electrons to knock one electron off an atom or molecule and also to break up molecules into fragments.

The most useful ion for determining molecular mass is one that forms when a molecule loses a single electron (whose mass is insignificant) and forms a **molecular ion ($M^{+\cdot}$)**:

$$M \rightarrow M^{+\cdot} + e^-$$

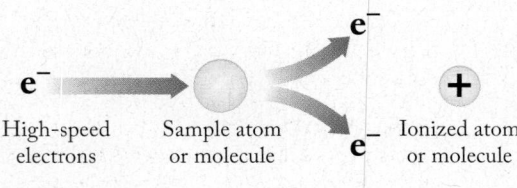

FIGURE 3.22 In some mass spectrometers, the atoms or molecules of substances are bombarded with a beam of high-energy electrons to make atomic or molecular ions.

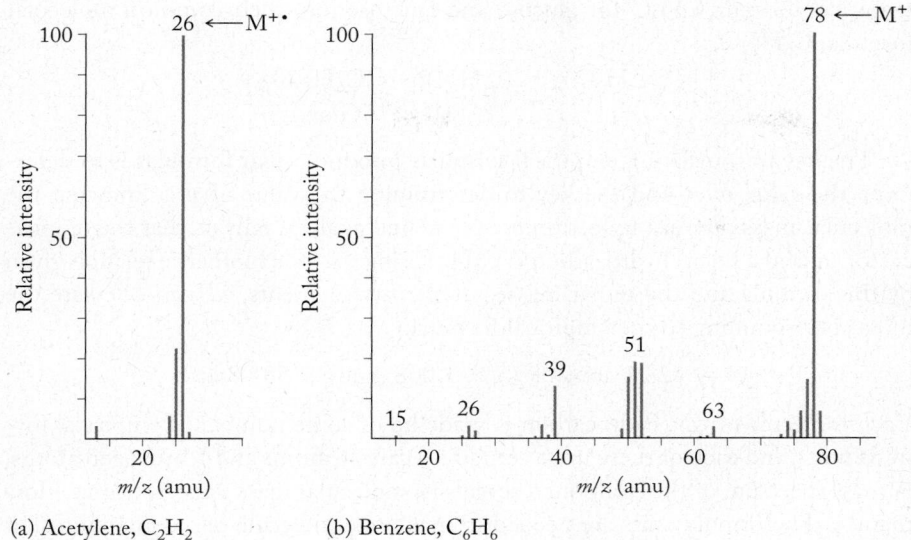

(a) Acetylene, C_2H_2

(b) Benzene, C_6H_6

FIGURE 3.23 Mass spectra of (a) acetylene and (b) benzene. The molecular-ion peak in both spectra is labeled $M^{+\cdot}$. The dot indicates that the molecular ion has an odd number of electrons. This is the case for most molecular ions. The peaks at much smaller m/z values are produced when the ionizing beam also breaks apart the molecules into fragments. The values for m are in amu and z is unitless, so m/z has units of amu.

Data produced by a mass spectrometer are displayed in a graph, called a **mass spectrum** (plural *spectra*), in which the m/z values of the ions reaching the detector are plotted on the horizontal axis against the intensity (number of particles) on the vertical axis. Because the charge on every ion is $1+$, the m/z ratio is $m/1$, so we can read the mass of the particle directly from the position of the peak on the horizontal axis.

Figure 3.23 shows the mass spectra for acetylene and benzene. Mass spectra such as these allow scientists to know with high precision and accuracy the molecular mass of compounds. For now, we concentrate on the molecular-ion peak, labeled $M^{+\cdot}$, as it is the peak that provides the information on the molecular mass of a substance. The molecular-ion peak is not necessarily the tallest peak in the spectrum, but it always is the peak of highest mass (not counting the minor isotope peaks due to the presence of ^{13}C at $m/z = 27$ and $m/z = 79$ in Figure 3.23a and b). The other peaks in a mass spectrum are fragments of the molecule that have masses smaller than the molecular ion, and an analyst uses them to determine the structure of the molecule.

CONCEPT TEST ●●●●●●●●●●●●●●●●●●●●●●●●●●●●●●●●●●●

Does the mass of a molecular-ion peak represent the mass of the empirical formula unit or the mass associated with the molecular formula of a compound?

(Answers to Concept Tests are in the back of the book.)

●●●

Using Percent Composition and a Mass Spectrum to Determine a Molecular Formula

Mass spectrometry gives the molecular mass of a compound, and percent composition gives its empirical formula. With these two pieces of information, we can determine the molecular formula of the compound.

Each subscript in a molecular formula is a multiple of the corresponding subscript in the empirical formula. For example, a multiplier n of 6 converts the

molecular ion ($M^{+\cdot}$) an ion formed in a mass spectrometer when a molecule loses an electron after being bombarded with high-energy electrons. The molecular ion has a charge of $1+$ and has essentially the same molecular mass as the molecule from which it came.

mass spectrum a graph of the data from a mass spectrometer, where m/z ratios of the deflected particles are plotted against the number of particles with a particular mass. Because the charge on the ions typically is $1+$, $m/z = m/1 = m$, and the mass of the particle may be read directly from the m/z axis.

empirical formula CH_2O for glucose and fructose into their common molecular formula, $C_6H_{12}O_6$:

$$(CH_2O)_n = (CH_2O)_6 = C_6H_{12}O_6$$
$$\text{Glucose}$$

The key to translating empirical formulas into molecular formulas is to determine the value of n, and the key to determining the value of n is knowing the molecular mass. For example, suppose elemental analysis tells us that the empirical formula of a liquid hydrocarbon is C_4H_5. Using the stoichiometric relation given by this formula and the molar masses of the two elements, we can calculate the mass corresponding to the empirical formula:

$$(4 \times 12.01 \text{ amu}) + (5 \times 1.008 \text{ amu}) = 53.08 \text{ amu}$$

Each molecule of this hydrocarbon is made up of some number of empirical formula units, and each of these units contains 4 carbon atoms and 5 hydrogen atoms. A mass spectrum of the compound reveals its molecular mass to be 106 amu. How many C_4H_5 formula units are needed to make one molecule of a compound that has a molar mass of 106 g/mol?

$$\frac{\text{molar mass}}{\text{mass of formula unit}} = \frac{106 \text{ g/mol}}{53 \text{ g/mol}} = 2$$

This 2 is the multiplier n we must apply to the subscripts of the empirical formula to obtain the molecular formula of the compound:

$$(C_4H_5)_n = (C_4H_5)_2 = C_8H_{10}$$

SAMPLE EXERCISE 3.13   **Determining a Molecular Formula**

Pheromones are chemical substances secreted by members of a species to stimulate a response in other individuals of the same species. For example, certain pheromones are secreted by females of a species to attract males for mating. The percent composition of eicosene, a compound similar to the Japanese beetle mating pheromone, is 85.63% C and 14.37% H. Its mass spectrum shows a peak for the molecular ion at 280 amu. Determine the molecular formula of eicosene.

Collect and Organize We are given the percent composition of eicosene from which we can determine the empirical formula of the compound. Empirical formulas are connected to molecular formulas by the molar mass of the compound. In addition to the information provided, we need the molar masses of carbon and hydrogen, 12.01 g/mol and 1.008 g/mol, respectively.

Analyze We can follow the procedure used in Section 3.6 to determine the empirical formula of eicosene and then use the molar masses of the elements in the molecule to determine the value of the multiplier needed to convert the empirical formula into a molecular formula.

Solve Assuming a 100.00 g sample, we have 85.63 g of carbon and 14.37 g of hydrogen, which we convert to moles:

$$85.63 \text{ g C} \times \frac{1 \text{ mol C}}{12.01 \text{ g C}} = 7.130 \text{ mol C}$$

$$14.37 \text{ g H} \times \frac{1 \text{ mol H}}{1.008 \text{ g H}} = 14.26 \text{ mol H}$$

The mole ratio of C to H is 7.130 : 14.26, or 1:2, which means the empirical formula is CH_2.

To determine the molecular formula, we first determine the formula mass of CH_2:

$$1 \times 12.01 \text{ amu} + 2(1.008 \text{ amu}) = 14.03 \text{ amu}$$

Next we divide the molecular mass known from the mass spectrum, 280 amu, by the formula mass to determine the multiplier n:

$$n = \frac{\text{molecular mass}}{\text{formula mass}} = \frac{280 \text{ amu}}{14.03 \text{ amu}} = 20$$

The molecular formula is

$$(CH_2)_n = (CH_2)_{20} = C_{20}H_{40}$$

Think about It The molecular formula of eicosene represents the number of C and H atoms in a molecule of eicosene. The compound has the same empirical formula as a large number of compounds, all of which have the empirical formula CH_2.

Practice Exercise Determine the empirical and molecular formulas of a compound that contains 56.36% O and 43.64% P and has a molar mass of 284 g/mol.

(Answers to Practice Exercises are in the back of the book.)

> **combustion analysis** a laboratory procedure for determining the composition of a substance by burning it completely in oxygen to produce known compounds whose masses are used to determine the composition of the original material.

3.8 Combustion Analysis

As noted in Section 3.4, combustion reactions involve burning substances in oxygen. In **combustion analysis**, the complete combustion of a compound followed by an analysis of the products enables chemists to determine the chemical composition of that compound. To ensure that combustion is complete, the process is carried out in excess oxygen, which means more oxygen is present than the stoichiometric amount. This type of analysis relies on complete combustion. For an organic compound, for example, this means converting all of the carbon in the compound to CO_2 and all of the hydrogen to H_2O:

$$C_aH_b + \text{excess } O_2(g) \rightarrow a\ CO_2(g) + b/2\ H_2O(g)$$

Consider burning a hydrocarbon of unknown composition in a chamber through which a stream of pure oxygen flows (Figure 3.24). The $CO_2(g)$ and $H_2O(g)$ produced flow first through a tube packed with $Mg(ClO_4)_2(s)$, which selectively

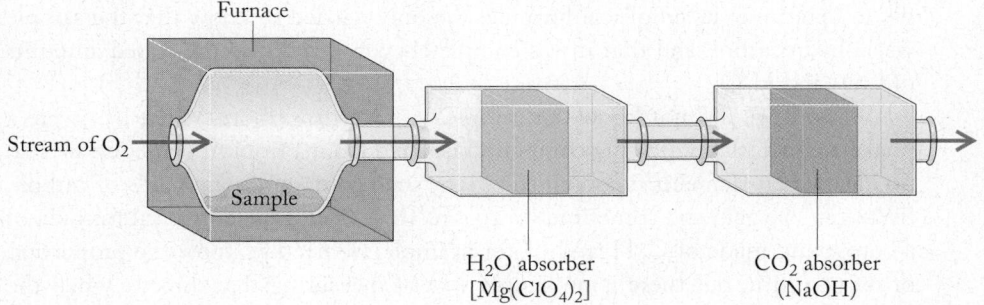

Furnace

Stream of O_2

Sample

H_2O absorber
$[Mg(ClO_4)_2]$

CO_2 absorber
(NaOH)

FIGURE 3.24 A carbon/hydrogen elemental analyzer relies on the complete combustion of organic compounds in excess oxygen. The products are CO_2 and H_2O vapor. Water vapor is absorbed by a $Mg(ClO_4)_2$ filter, and carbon dioxide is absorbed by a NaOH filter. The empirical formula of the compound is calculated from the masses of H_2O and CO_2 absorbed.

absorbs the $H_2O(g)$, and then through a tube containing $NaOH(s)$, which absorbs the $CO_2(g)$. The masses of these tubes are measured before and after combustion. Suppose the mass of the tube that traps $CO_2(g)$ increases by 1.320 g, and the mass of the tube that traps $H_2O(g)$ increases by 0.540 g. How can we use these results to determine the empirical formula of the hydrocarbon?

First, let's establish what we know about the hydrocarbon in this example:

1. Being a hydrocarbon, it contains only carbon and hydrogen.
2. Complete conversion of its carbon into CO_2 produces 1.320 g of CO_2.
3. Complete conversion of its hydrogen into H_2O produces 0.540 g of H_2O.

To derive an empirical formula for the hydrocarbon, we must determine the number of moles of carbon in 1.320 g of carbon dioxide and the number of moles of hydrogen in 0.540 g of water vapor. These quantities are directly related to the number of moles of carbon and hydrogen in the sample we burned. Converting the mass of carbon dioxide to moles of carbon and the mass of water to moles of hydrogen:

$$1.320 \text{ g } CO_2 \times \frac{1 \text{ mol } CO_2}{44.01 \text{ g } CO_2} \times \frac{1 \text{ mol C}}{1 \text{ mol } CO_2}$$

$$= 0.02999 \text{ mol C} \approx 0.0300 \text{ mol C}$$

$$0.540 \text{ g } H_2O \times \frac{1 \text{ mol } H_2O}{18.02 \text{ g } H_2O} \times \frac{2 \text{ mol H}}{1 \text{ mol } H_2O}$$

$$= 0.05993 \text{ mol H} \approx 0.0600 \text{ mol H}$$

The empirical formula is based on these proportions of C and H. If we divide both molar amounts by the smaller one:

$$\frac{0.0300 \text{ mol C}}{0.0300 \text{ mol C}} = 1 \qquad \frac{0.0600 \text{ mol H}}{0.0300 \text{ mol C}} = 2$$

we find that the empirical formula of the hydrocarbon is CH_2.

If we want to extend this analysis to determine a molecular formula, we need to know the molecular mass of the hydrocarbon. Suppose its mass spectrum has a molecular ion with a mass of 84 amu. The formula mass of CH_2 is 14 amu. Dividing the molecular mass by the formula mass, we get the multiplier, n:

$$n = \frac{84 \text{ amu}}{14 \text{ amu}} = 6$$

The molecular formula is therefore

$$(CH_2)_6 = C_6H_{12}$$

Note that in this problem we did not need to know the initial mass of the sample to determine its empirical formula. We only needed to know that the sample was a hydrocarbon and that it was completely converted into the stated amounts of CO_2 and H_2O.

What if we did not know our sample was a hydrocarbon? What if it were a pharmacologically promising compound isolated from a tropical plant and we had no idea which elements it contained? Many such compounds are made of carbon, hydrogen, oxygen, and sometimes nitrogen. To calculate the empirical formula of a compound made of C, H, and O, for example, we need to know the proportion of oxygen in it, but there is no simple way of measuring that directly when the compound is burned in excess oxygen. Combustion analyses yield the percentages by mass of all atoms in a sample *except* oxygen. When given data from a combustion analysis, always check to see whether the percentages add up to 100%. If they do, all the elements in the sample are accounted for in the results. If they do not, the missing mass is probably due to oxygen.

SAMPLE EXERCISE 3.14   Determining an Empirical Formula from Combustion Analysis

Combustion of 1.000 g of an organic compound known to contain carbon, hydrogen, and oxygen produces 2.360 g of CO_2 and 0.640 g of H_2O. What is the empirical formula of the compound?

Collect and Organize We are given the initial mass of the sample and the masses of CO_2 and H_2O produced by its combustion. This information enables us to determine the masses of C and H in the sample and the empirical formula of the compound.

Analyze First we determine the numbers of moles and the masses of C and H in the CO_2 and H_2O. The only source of these elements is the burned sample, so these numbers are equal to the numbers of moles and the masses of C and H in the sample. Once we know the masses of C and H, we can apply the law of conservation of mass to determine the mass of O in the sample and then calculate the number of moles of O in the sample. We then determine the C:H:O mole ratio and convert that ratio into a ratio of small whole numbers.

Solve The moles of C and H in the CO_2 and H_2O collected during combustion are

$$2.360 \text{ g CO}_2 \times \frac{1 \text{ mol CO}_2}{44.01 \text{ g CO}_2} \times \frac{1 \text{ mol C}}{1 \text{ mol CO}_2} = 0.05362 \text{ mol C}$$

$$0.640 \text{ g H}_2\text{O} \times \frac{1 \text{ mol H}_2\text{O}}{18.02 \text{ g H}_2\text{O}} \times \frac{2 \text{ mol H}}{1 \text{ mol H}_2\text{O}} = 0.07103 \text{ mol H}$$

and the masses are

$$0.05362 \text{ mol C} \times \frac{12.01 \text{ g C}}{1 \text{ mol C}} = 0.6440 \text{ g C}$$

$$0.07103 \text{ mol H} \times \frac{1.008 \text{ g H}}{1 \text{ mol H}} = 0.07160 \text{ g H}$$

The sum of these two masses (0.6440 g + 0.07160 g = 0.7156 g) is less than the mass of the sample (1.000 g). The difference must be the mass of oxygen in the sample:

$$\text{Mass of oxygen} = 1.000 \text{ g} - 0.7156 \text{ g} = 0.2844 \text{ g O}$$

The number of moles of O atoms in the sample is thus

$$0.2844 \text{ g O} \times \frac{1 \text{ mol O}}{16.00 \text{ g O}} = 0.01778 \text{ mol O}$$

The mole ratio of the three elements in the sample is

$$0.05362 \text{ mol C} : 0.07103 \text{ mol H} : 0.01778 \text{ mol O}$$

limiting reactant a reactant that is consumed completely in a chemical reaction. The amount of product formed depends on the amount of the limiting reactant available.

Dividing through by the smallest value (0.01778 mol) gives a mole ratio of 3:4:1, making the empirical formula of the sample C_3H_4O.

Think about It The subscripts in our final answer are relatively small whole numbers, which should give us confidence that our answer is a plausible one.

Practice Exercise Vanillin is the compound containing carbon, hydrogen, and oxygen that gives vanilla beans their distinctive flavor. The combustion of 30.4 mg of vanillin produces 70.4 mg of CO_2 and 14.4 mg of H_2O. The mass spectrum of vanillin shows a molecular ion peak at 152 amu. Use this information to determine the molecular formula of vanillin.

(Answers to Practice Exercises are in the back of the book.)

3.9 Limiting Reactants and Percent Yield

Let's return to photosynthesis, the process responsible for the O_2 in our present-day atmosphere and for the energy that sustains life on Earth's surface. The stoichiometry of the reaction calls for equal parts on a mole basis, referred to as *equimolar amounts*, of CO_2 and H_2O:

$$6\,CO_2(g) + 6\,H_2O(\ell) \rightarrow C_6H_{12}O_6(aq) + 6\,O_2(g)$$
$$\text{Glucose}$$

Because in nature there is little likelihood of having exactly equal proportions of reactants at a reaction site, let's consider what happens when more than six molecules of water are available for every six molecules of CO_2; having water in excess is a common occurrence in biological systems. The photosynthetic production of glucose continues only until all the CO_2 is consumed, leaving the extra molecules of water unreacted. In this example, carbon dioxide is the **limiting reactant**, meaning that the extent to which the reaction proceeds is determined by the quantity of CO_2 available and not by the quantity of H_2O, which is in excess. Figure 3.25 illustrates the concept of a limiting reactant.

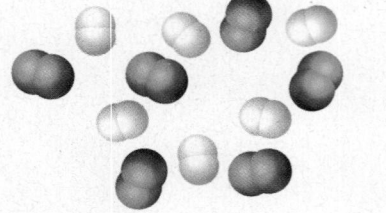
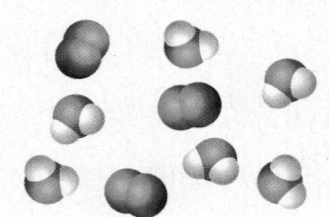

FIGURE 3.25 The reaction of hydrogen (H_2, white molecules) with oxygen (O_2, red molecules) produces water (H_2O). A mixture containing equal numbers of hydrogen and oxygen molecules produces only as many water molecules as the number of H_2 molecules available. In this case, H_2 is the limiting reactant and O_2 is in excess.

CONCEPT TEST

Suppose you need 50 nuts and 50 bolts for a household project. At the hardware store, why is it better to count the nuts and bolts rather than buy them by the pound?

(Answers to Concept Tests are in the back of the book.)

Calculations Involving Limiting Reactants

How do we know if one reactant is limiting when we are asked to calculate the mass of a product from given masses of reactants? It is tempting to select the reactant present in the smaller amount by mass. Avoid this temptation and instead take a systematic approach based on the stoichiometry of the reaction. Several approaches are possible, two of which we present here—one that uses stoichiometric calculations and one that uses mole ratios.

In our first method, we calculate how much product would be formed if reactant A was completely consumed. Then we repeat the calculation for reactant B and any other reactants. Let's try this approach with the reaction between sulfur trioxide gas and water vapor to form sulfuric acid,

$$SO_3(g) + H_2O(g) \rightarrow H_2SO_4(\ell) \qquad (3.1)$$

described in Section 3.1 as contributing to the acidity of Earth's early atmosphere. Suppose we carry out this reaction in the laboratory using 20.00 g of $SO_3(g)$ and 10.00 g of $H_2O(g)$. Which is the limiting reactant, and how many grams of H_2SO_4 are produced from these masses of reactants?

We first calculate the molar masses and then the numbers of moles of SO_3 and H_2O from the two masses given:

▶❚❚ **CHEMTOUR** Limiting Reactant

$$\mathcal{M}_{SO_3} = 32.06 \text{ g/mol} + 3(16.00 \text{ g/mol}) = 80.06 \text{ g/mol}$$
$$\mathcal{M}_{H_2O} = 2(1.008 \text{ g/mol}) + 16.00 \text{ g/mol} = 18.02 \text{ g/mol}$$

$$20.00 \text{ g SO}_3 \times \frac{1 \text{ mol SO}_3}{80.06 \text{ g SO}_3} = 0.2498 \text{ mol SO}_3$$

$$10.00 \text{ g H}_2\text{O} \times \frac{1 \text{ mol H}_2\text{O}}{18.02 \text{ g H}_2\text{O}} = 0.5549 \text{ mol H}_2\text{O}$$

Then we calculate the mass of sulfuric acid produced if all of the SO_3 is consumed:

$$0.2498 \text{ mol SO}_3 \times \frac{1 \text{ mol H}_2\text{SO}_4}{1 \text{ mol SO}_3} \times \frac{98.08 \text{ g H}_2\text{SO}_4}{1 \text{ mol H}_2\text{SO}_4} = 24.50 \text{ g H}_2\text{SO}_4$$

Next we carry out the same calculation to determine how much sulfuric acid is produced assuming 10.00 g of water reacts completely:

$$0.5549 \text{ mol H}_2\text{O} \times \frac{1 \text{ mol H}_2\text{SO}_4}{1 \text{ mol H}_2\text{O}} \times \frac{98.08 \text{ g H}_2\text{SO}_4}{1 \text{ mol H}_2\text{SO}_4} = 54.42 \text{ g H}_2\text{SO}_4$$

The SO_3 is the limiting reactant because, when it is completely consumed, the smaller amount of product is formed. Thus when 20.00 g of SO_3 and 10.00 g of H_2O are combined, the maximum amount of product that can form is 24.50 g of H_2SO_4. Making that amount of sulfuric acid consumes all the available SO_3 but not all the available H_2O.

The second method compares the mole ratio of the reactants to the mole ratio required by the balanced chemical equation. Using this approach, we

1. Convert masses of reactants A and B into moles.
2. Calculate the mole ratio of A to B.
3. Compare this mole ratio with the stoichiometric mole ratio from the balanced chemical equation. If

$$\left(\frac{\text{mol A}}{\text{mol B}}\right)_{given} > \left(\frac{\text{mol A}}{\text{mol B}}\right)_{stoichiometric} \qquad (3.4)$$

B is the limiting reactant. If

$$\left(\frac{\text{mol A}}{\text{mol B}}\right)_{given} < \left(\frac{\text{mol A}}{\text{mol B}}\right)_{stoichiometric} \qquad (3.5)$$

theoretical yield the maximum amount of product possible in a chemical reaction for given quantities of reactants; also known as *stoichiometric yield*.

then A is the limiting reactant. If the two ratios are equal, the masses are the stoichiometric amounts and both reactants are consumed completely.

For the same example using 20.00 g of SO_3 and 10.00 g of H_2O, we calculate the mole ratio of SO_3 to H_2O after calculating the moles of SO_3 and H_2O available:

$$\frac{\text{mol } SO_3}{\text{mol } H_2O} = \frac{20.00 \text{ g } SO_3 \times \dfrac{1 \text{ mol } SO_3}{80.06 \text{ g } SO_3}}{10.00 \text{ g } H_2O \times \dfrac{1 \text{ mol } H_2O}{18.02 \text{ g } H_2O}} = \frac{0.2498 \text{ mol } SO_3}{0.5549 \text{ mol } H_2O} = 0.4502$$

Equation 3.1 tells us that the stoichiometric ratio of SO_3 to H_2O is 1/1 = 1. The SO_3/H_2O mole ratio for our reaction conditions is 0.2498/0.5549 = 0.4502. Expressing this outcome as a mathematical relationship, we get

$$\left(\frac{\text{mol } SO_3}{\text{mol } H_2O}\right)_{\text{given}} < \left(\frac{\text{mol } SO_3}{\text{mol } H_2O}\right)_{\text{stoichiometric}}$$

This inequality matches Equation 3.5, which tells us that SO_3 (reactant A in Equation 3.5) must be the limiting reactant.

These two approaches to determining a limiting reactant lead to the same conclusion and are of similar complexity. In both cases, because the masses of the reactants are given in grams, we must convert those masses into moles. The advantage of the first method is that it not only determines the identity of the limiting reactant; it also determines the **theoretical yield**, the maximum amount of product that can be produced by the given mixture. The theoretical yield is sometimes called the *stoichiometric yield* or *100% yield*.

SAMPLE EXERCISE 3.15   Identifying the Limiting Reactant

During the launch of a U.S. space shuttle (Figure 3.26), high-pressure pumps deliver 4,400 kg of H_2 fuel and 31,000 kg of O_2 to each main engine every minute. Is one of these reactants a limiting reactant, or is the mixture stoichiometric? If one reactant is limiting, which one is it?

Collect and Organize We are given quantities of two reactants (H_2 and O_2) and asked to determine whether one of them is limiting and, if so, which one. We can make that determination by calculating the ratio of H_2 to O_2 in the balanced chemical equation for the reaction. That means we'll use the second method described in the preceding discussion.

Analyze An unbalanced equation for this combustion reaction is

$$H_2(g) + O_2(g) \rightarrow H_2O(g)$$

After balancing this equation, we compare the ratio of the amounts of reactants described in the problem statement with the stoichiometric ratio in the balanced equation to determine whether or not one of the reactants is limiting.

Solve First we need to balance the combustion equation:

$$\underline{\quad} H_2(g) + \underline{\quad} O_2(g) \rightarrow \underline{\quad} H_2O(g)$$
$$2\,H_2(g) + O_2(g) \rightarrow 2\,H_2O(g)$$

FIGURE 3.26 The power that thrusts a space shuttle into orbit comes from two solid-fuel booster rockets and three main engines that burn hydrogen. Both hydrogen and oxygen are stored in liquid form in the large brown tank under the shuttle. The reaction is $2\,H_2(g) + O_2(g) \rightarrow 2\,H_2O(g)$.

This balanced equation tells us that the stoichiometric ratio of H_2 to O_2 is 2:1. Next we convert the masses of reactants given to moles:

$$31,000 \text{ kg O}_2 \times \frac{10^3 \text{ g O}_2}{1 \text{ kg O}_2} \times \frac{1 \text{ mol O}_2}{32.00 \text{ g O}_2}$$

$$= 9.69 \times 10^5 \text{ mol O}_2 \approx 9.7 \times 10^5 \text{ mol O}_2$$

$$4,400 \text{ kg H}_2 \times \frac{10^3 \text{ g H}_2}{1 \text{ kg H}_2} \times \frac{1 \text{ mol H}_2}{2.016 \text{ g H}_2}$$

$$= 2.18 \times 10^6 \text{ mol H}_2 \approx 2.2 \times 10^6 \text{ mol H}_2$$

and calculate the mole ratio:

$$\frac{2.2 \times 10^6 \text{ mol H}_2}{9.7 \times 10^5 \text{ mol O}_2} = 2.3$$

This ratio is greater than the stoichiometric $H_2{:}O_2$ ratio of 2:1, so hydrogen is in slight excess and oxygen is the limiting reactant.

Think about It The calculation indicates that the ratio of hydrogen to oxygen used in the space shuttle is nearly equal to the stoichiometric ratio. This makes sense because there is little advantage to carrying excess fuel (hydrogen).

Practice Exercise Any fuel–oxygen mixture that contains more oxygen than is needed to burn the fuel completely is called a *lean mixture*, and a mixture containing too little oxygen to allow complete combustion of the fuel is called a *rich mixture*. A high-performance heater that burns propane, $C_3H_8(g)$, is adjusted so that 100.0 g of $O_2(g)$ enters the system for every 100.0 g of propane. Is this mixture rich or lean?

(Answers to Practice Exercises are in the back of the book.) ∎

Actual Yields versus Theoretical Yields

When we calculated the mass of sulfuric acid formed by 10.00 g of water and 20.00 g of sulfur trioxide, we defined the value as a theoretical yield: the maximum amount of product possible from the given quantities of reactants. In nature, industry, or the laboratory, the **actual yield** is often less than the theoretical yield for several reasons. In addition to the reaction we want, the reactants may also undergo other reactions, yielding products in addition to the ones we expect. Sometimes the rate of a reaction is so slow that some reactants remain unreacted even after an extended time. Other reactions do not go to completion no matter how long they are allowed to run, yielding a mixture of reactants and products, the composition of which does not change with time. For these and other reasons, it is useful to distinguish between the theoretical and the actual yields of a chemical reaction and to calculate the **percent yield**:

$$\text{Percent yield} = \frac{\text{actual yield}}{\text{theoretical yield}} \times 100\% \qquad (3.6)$$

actual yield the amount of product obtained from a chemical reaction, which is often less than the theoretical yield.

percent yield the ratio, expressed as a percentage, of the actual yield of a chemical reaction to the theoretical yield.

Hydrogen and Helium: The Bulk of the Universe

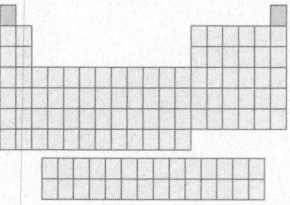

Hydrogen and helium, the only two elements in the first row of the periodic table, are the least dense of all the elements and are the most abundant in the universe, accounting for over 99% of all atoms. They are much less abundant on Earth as free elements: Hydrogen gas (H_2) constitutes only about 0.00005% of the atmosphere, by volume. The atmospheric concentration of helium is a factor of 10 higher—0.0005% by volume—which still makes He a very minor component.

These two elements have similar physical properties, but the chemical properties of hydrogen are remarkably different from those of helium. Helium is chemically inert, while hydrogen is extremely reactive and occurs in more compounds than any other element. Compounds containing carbon and hydrogen are the major components of all living matter, and water (H_2O) is ubiquitous and essential for life on Earth.

Hydrogen is currently being hailed as the fuel of the future, with the potential for large-scale use in internal combustion engines and fuel cells (Figure 3.27). It is already produced and stored on a large scale for the U.S. space program. The principal advantage of hydrogen over fossil fuels is its high fuel value—the amount of energy it releases per unit of mass. Hydrogen is also a clean fuel, producing only water and

energy and not oxides of carbon, nitrogen, or sulfur, which are products of fossil-fuel combustion and contribute to atmospheric pollution and climate change.

Because hydrogen does not occur as the free element in significant amounts in nature, it must be extracted from compounds. Most of the H_2 produced in the United States and about half that produced worldwide comes from methane (CH_4), the principal component of natural gas. Hydrogen is generated from methane in a process called *steam–methane reforming*:

$$CH_4(g) + H_2O(g) \xrightarrow{1000°C} CO(g) + 3\,H_2(g)$$

Additional H_2 is generated when the $CO(g)$ is reacted with more steam in the *water–gas shift reaction*:

$$CO(g) + H_2O(g) \xrightarrow{200°C} CO_2(g) + H_2(g)$$

Another way to generate hydrogen is to split H_2O molecules into H_2 and O_2 by *electrolysis*:

$$2\,H_2O(\ell) \xrightarrow{\text{electric current}} 2\,H_2(g) + O_2(g)$$

Currently, developing hydrogen as a widely used fuel is hampered by the fact that it cannot be generated in bulk quantities by any efficient means that do not also consume fossil fuels.

Probably the most important use of hydrogen is in producing ammonia in the *Haber–Bosch process*:

$$N_2(g) + 3\,H_2(g) \rightarrow 2\,NH_3(g)$$

The hydrogen used to make ammonia comes from steam–methane reforming, which means the production of hydrogen and ammonia are tightly linked commercially. Ammonia is used as a fertilizer and as a source of nitrogen for manufacturing other nitrogen-containing compounds, including explosives.

Another major use of hydrogen is as a reactant with liquid vegetable oils to make solid edible fats such as margarine. This reaction is called *hydrogenation*, and the products are called *hydrogenated fats*. Some of the oils from partial hydrogenation are *trans fats*, which differ in structure from natural vegetable oils. Because trans fats have been used in many processed foods, they are currently the object of much scrutiny because of their negative impact on human health.

The *deuterium* and *tritium* isotopes of hydrogen are also useful. Deuterium (^2H or D) accounts for about 0.015% of the hydrogen in nature. Heavy water, D_2O, is an essential component in nuclear reactors where plutonium is

FIGURE 3.27 A hydrogen fueling station for vehicles.

generated from uranium and is also crucial to the production of nuclear weapons. Heavy water is so important to this technology that its production is monitored and the material kept under strict export controls. Heavy water has a slightly higher boiling point (101.4°C) than H_2O (100.0°C), and in principle D_2O can be separated from H_2O by distillation. In practice, D_2 is separated from liquefied H_2 and then used to make deuterium compounds that can be purified by subsequent processing. For example, D_2 is reacted to form D_2S, which is treated with H_2O and converted to D_2O. The production of 1 L of D_2O requires about 340,000 L of H_2O.

Because tritium (^3H) is radioactive, it can be detected easily, which gives rise to its utility. It is used in devices where dim light is needed but no source of electricity is available, such as emergency exit signs in buildings that must function in times of power failure, dials in aircraft, rifle sights, and even in the luminous dials of some watches. In these applications, the radiation from tritium interacts with a thin coating of a luminescent material, causing it to emit light. The level of radioactivity in these devices is not harmful.

On early Earth, hydrogen was retained because it formed compounds with heavier elements, but *helium* forms no compounds and was too light to be held by Earth's gravitational field. Therefore any helium now on Earth is the product of radioactive decay. Most helium in the world is obtained from natural gas deposits in Texas, Oklahoma, Kansas, the Middle East, and Russia. These locations have significant deposits of uranium ore, and the decay of uranium produces helium, which mixes with underground deposits of natural gas. Helium is separated from the natural gas, in which its concentration may be as high as 0.3% by mass.

Helium has a number of uses beyond keeping blimps, weather balloons, and decorative balloons aloft. It is mixed with oxygen in air tanks used in deep-sea diving to create nitrogen-free gas mixtures. "Regular" air is about four-fifths nitrogen. Because nitrogen is soluble in blood, divers breathing regular air, with its high nitrogen content, run the risk of developing nitrogen narcosis, a disorienting condition in which the nervous system becomes saturated with nitrogen. Helium does not have the same narcotic properties as nitrogen. Helium mixtures also help divers avoid decompression sickness (the bends), a painful and life-threatening condition caused by nitrogen bubbles forming in the blood and joints as divers ascend to the surface. Helium protects divers from these dangers because it is less soluble in blood than nitrogen.

Helium is increasingly used as a *cryogen*, which is a gas that has been liquefied by cooling to a very low temperature. Helium has the lowest boiling point of any element (−268.9°C), and it is used to create and maintain superconducting magnets in instruments ranging from equipment in chemistry and physics laboratories to magnetic-resonance-imaging (MRI) units in hospitals and clinics (Figure 3.28).

In its liquid phase, helium has some extraordinary physical properties. It exhibits superfluidity, which means it flows so freely that it can flow up the walls of containers. Liquid helium is also a nearly perfect conductor of heat and electricity.

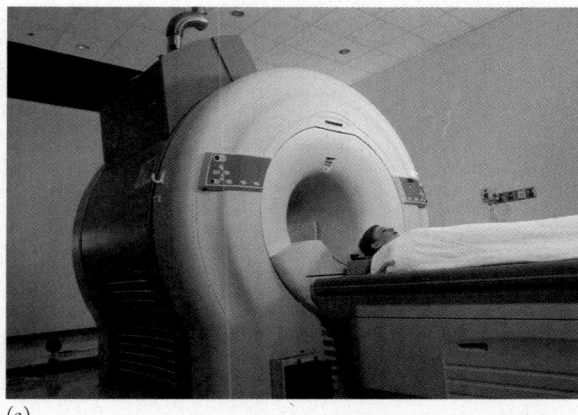

(a)

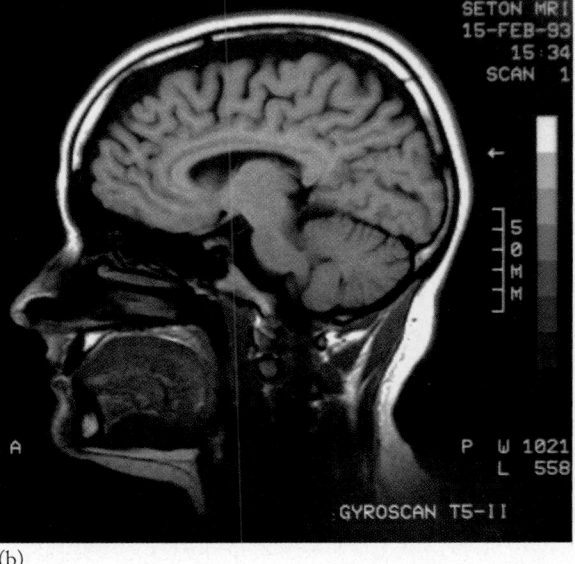

(b)

FIGURE 3.28 (a) The magnet in a magnetic-resonance-imaging (MRI) unit maintains its superconductivity by being submerged in liquid helium inside the large white cylinder shown here. (b) An MRI image.

SAMPLE EXERCISE 3.16   Calculating Percent Yield

The industrial process for making the ammonia used in fertilizer, explosives, and many other products is based on the reaction between nitrogen and hydrogen at high temperature and pressure:

$$N_2(g) + 3\,H_2(g) \rightarrow 2\,NH_3(g)$$

If 18.20 kg of NH_3 is produced by a reaction mixture that initially contains 6.00 kg of H_2 and an excess of N_2, what is the percent yield of the reaction?

Collect and Organize We know that the actual yield of NH_3 is 18.20 kg. We also know that H_2 must be the limiting reactant because the problem specifies the presence of excess N_2 and that the reaction mixture initially contained 6.00 kg of H_2.

Analyze We need to use the mass of H_2 to calculate how much NH_3 could have been produced—the theoretical yield. We then use the theoretical yield and the actual yield to calculate percent yield. We get the theoretical yield by (1) converting the mass of H_2 to moles of H_2; (2) converting moles of H_2 to moles of NH_3; and then (3) converting moles of NH_3 to mass of NH_3. We need to work with the molar masses of H_2 and NH_3 and the stoichiometry of the reaction, which tells us that 2 moles of NH_3 are produced for every 3 moles of H_2 consumed. Finally we calculate the percent yield by comparing actual yield to theoretical yield using Equation 3.6.

Solve The molar masses we need are

$$\mathcal{M}_{H_2} = 2(1.008 \text{ g/mol}) = 2.016 \text{ g/mol}$$
$$\mathcal{M}_{NH_3} = 14.01 \text{ g/mol} + 3(1.008 \text{ g/mol}) = 17.03 \text{ g/mol}$$

We calculate the theoretical yield of NH_3:

$$6.00 \text{ kg H}_2 \times \frac{10^3 \text{ g H}_2}{1 \text{ kg H}_2} \times \frac{1 \text{ mol H}_2}{2.016 \text{ g H}_2} \times \frac{2 \text{ mol NH}_3}{3 \text{ mol H}_2} \times \frac{17.03 \text{ g NH}_3}{1 \text{ mol NH}_3}$$
$$\times \frac{1 \text{ kg NH}_3}{10^3 \text{ g NH}_3} = 33.8 \text{ kg NH}_3$$

and divide that by the actual yield to determine the percent yield:

$$\frac{18.20 \text{ kg NH}_3}{33.8 \text{ kg NH}_3} \times 100\% = 53.8\%$$

Think about It A yield of about 54% may seem low, but it may be the best that can be achieved for a particular process.

Practice Exercise The combustion of 58.0 g of butane (C_4H_{10}) produces 158 g of CO_2. What is the percent yield of the reaction?

(Answers to Practice Exercises are in the back of the book.)

At the beginning of this chapter we asked several questions including: how do we describe chemical changes involving different compounds using chemical reactions, how do we determine the composition of a compound, and how do we determine the chemical formula of a compound? Chemical reactions are described by balanced chemical equations that contain the information needed to calculate the amounts of reactants needed to prepare certain amounts of products. Data from chemical reactions can also be used to determine the percentage composition of compounds as well as empirical and molecular formulas. While the examples we used in this chapter relate to processes that have changed Earth, the concepts introduced have broad applicability in all areas of chemistry. The ideas of the mole and stoichiometry are fundamental to understanding chemistry on the microscopic level. The calculations you have mastered are part of your growing "toolbox" of skills that we will apply in subsequent chapters.

SUMMARY

Section 3.1 In a **chemical equation**, the chemical formulas of **reactants** appear first, followed by a reaction arrow and the chemical formulas of the **products**. The proportions of the reactants and products are expressed by coefficients preceding their formulas, and symbols in parentheses are used to indicate physical states: (g) for gas, (ℓ) for liquid, (s) for solid, and (aq) for substances dissolved in water.

Section 3.2 The **molecular mass** of a molecular compound in atomic mass units per molecule is the sum of the atomic masses of the atoms in one molecule of the compound. The **mole (mol)** is the SI base unit for quantity of substance. One mole of any substance contains **Avogadro's number** ($N_A = 6.022 \times 10^{23}$) of particles. The mass of 1 mole of a substance is its **molar mass** ($\mathcal{M}$); the molar mass of a compound is the

sum of the molar masses of the elements in the compound, each elemental molar mass multiplied by the number of moles of that element in 1 mole of the compound. The mass of one formula unit of an ionic compound is called its **formula mass**. The mole ratios of reactants and products in a chemical reaction are called the **stoichiometry** of the reaction. The **law of conservation of mass** states that the sum of the masses of the reactants in a chemical reaction is equal to the sum of the masses of the products.

Section 3.3 In a balanced chemical equation, the number of atoms of each element is the same on the two sides of the reaction arrow. In **hydrolysis**, water reacts with another substance.

Section 3.4 Combustion reactions occur between oxygen and another element or compound. In the complete combustion of compounds made of carbon and hydrogen, called **hydrocarbons**, the

hydrogen combines with oxygen to form H_2O, and the carbon combines with oxygen to form CO_2.

Section 3.5 In the carbon cycle, CO_2 from the atmosphere is converted to glucose during photosynthesis in green plants. Most, but not all, of the carbon in green plants returns to the atmosphere as CO_2 during respiration.

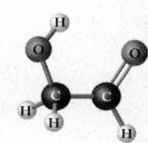

Section 3.6 The **percent composition** of a compound is the percentage by mass of each element in the compound.

Section 3.7 The empirical formula of a compound may or may not be the same as its molecular formula, which indicates the number of each type of atom in one molecule. Converting the empirical formula of a compound into a molecular formula requires knowing the molecular mass of the compound, which can be determined by identifying the **molecular ion ($M^{+\cdot}$)** in its **mass spectrum**.

Section 3.8 In **combustion analysis**, a known mass of an organic compound is burned in a stream of oxygen gas, the carbon in the sample is converted into CO_2, and the hydrogen is converted into H_2O. The masses of CO_2 and H_2O are measured and used to determine the masses of C and H in the organic compound and then the compound's empirical formula.

Section 3.9 The **limiting reactant** in a reaction mixture is the reactant that controls how much product can be made. The maximum amount of product that can form in a chemical reaction is the **theoretical yield**. The measured amount of product formed in a reaction is the **actual yield**. The ratio of actual yield to theoretical yield expressed as a percentage is the **percent yield** for the reaction.

PROBLEM-SOLVING SUMMARY

TYPE OF PROBLEM	CONCEPTS AND EQUATIONS	SAMPLE EXERCISES
Converting number of particles into number of moles (or vice versa)	Convert number of particles to moles by dividing by Avogadro's number ($N_A = 6.022 \times 10^{23}$ particles/mol).	3.1
	Convert number of moles to particles by multiplying by Avogadro's number ($N_A = 6.022 \times 10^{23}$ particles/mol).	
Converting mass of a substance into number of moles (or vice versa)	Convert mass of substance to moles by dividing by the molar mass ($\mathcal{M}$) of the substance.	3.2, 3.3, 3.5
	Convert moles of substance to mass by multiplying by the molar mass ($\mathcal{M}$) of the substance.	
Calculating molar mass of a compound	Multiply molar mass of each element by its subscript in the compound's formula; then add the products.	3.4
Balancing a chemical reaction	Change coefficients in the equation so that the number of atoms of each element are the same on the two sides of the reaction arrow.	3.6
Writing and balancing a chemical equation for a combustion reaction	The C and H in organic compounds react with O_2 to form CO_2 and H_2O, for example: $$CH_4(g) + 2\,O_2(g) \rightarrow 2\,H_2O(g) + CO_2(g)$$ Balance moles of C first, then H, then O.	3.7
Calculating mass of a product from mass of a reactant	Use $\mathcal{M}_{reactant}$ to convert mass of reactant into moles of reactant; use reaction stoichiometry to calculate moles of product and then use $\mathcal{M}_{product}$ to calculate mass of product.	3.8
Calculating percent composition from a chemical formula	Calculate molar mass of the compound represented by the chemical formula. Determine mass of each element in grams from the chemical formula. Divide each element mass by the formula mass. Express each result as a percentage; percentages should add up to 100%.	3.9
Determining empirical formula from percent composition	Assuming a 100.00 g sample, assign the mass (g) of each element to equal its percentage. Divide the mass of each element by its molar mass to get moles. Simplify mole ratios to lowest whole numbers, and use those numbers as subscripts in empirical formula of the compound.	3.10, 3.11
Relating empirical and molecular formulas	Calculate conversion factor n by dividing the compound's molar mass by its formula mass. Multiply subscripts in the empirical formula by this conversion factor.	3.12, 3.13
Deriving empirical formula from combustion analysis	For hydrocarbons, convert given masses of CO_2 and H_2O into moles of CO_2 and H_2O, and then to moles of C and H. For compounds containing O, convert moles of C and H into masses of C and H and subtract the sum of these values from the sample mass to calculate mass of O. Convert mass of O to moles of O. Simplify the mole ratio of C to H to O.	3.14
Identifying limiting reactant	Method 1: Calculate how much product each reactant could make; reactant making the least amount of product is limiting reactant.	3.15
	Method 2: Convert given masses of reactants into moles. Compare mole ratio of reactants to corresponding mole ratio in stoichiometric equation (Equations 3.4 and 3.5). If $$\left(\frac{mol\ A}{mol\ B}\right)_{given} > \left(\frac{mol\ A}{mol\ B}\right)_{stoichiometric} \qquad (3.4)$$ then B is the limiting reactant.	

TYPE OF PROBLEM	CONCEPTS AND EQUATIONS	SAMPLE EXERCISES

If

$$\left(\frac{\text{mol A}}{\text{mol B}}\right)_{\text{given}} < \left(\frac{\text{mol A}}{\text{mol B}}\right)_{\text{stoichiometric}} \qquad (3.5)$$

then A is the limiting reactant.

Calculating percent yield	Calculate theoretical yield of product using mass of limiting reactant. Divide actual yield (given) by theoretical yield (Equation 3.6):	3.16

$$\text{Percent yield} = \frac{\text{actual yield}}{\text{theoretical yield}} \times 100\% \qquad (3.6)$$

VISUAL PROBLEMS ●● ■

(Answers to boldface end-of-chapter questions and problems are in the back of the book.)

3.1. Each of the pairs of containers pictured in Figure P3.1 contains substances composed of elements X (red balls) and Y (blue balls). For each pair, write a balanced chemical equation describing the reaction that takes place. Be sure to indicate the physical states of the reactants and products using the appropriate symbols in parentheses.

3.2. Identify the limiting reactant in each of the pairs of containers pictured in Figure P3.2. The red balls represent atoms of element X, the blue balls are atoms of element Y. Each question mark means that there is unreacted reactant left over.

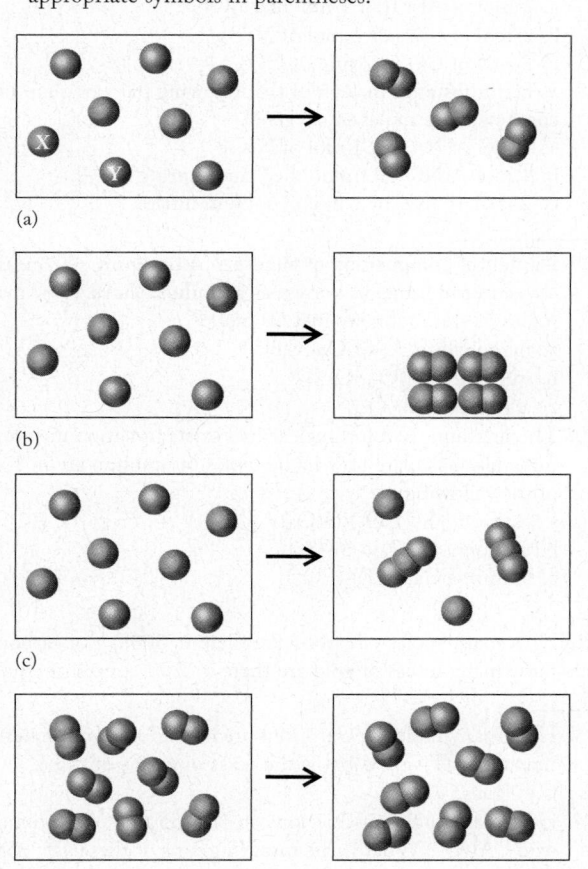

FIGURE P3.1

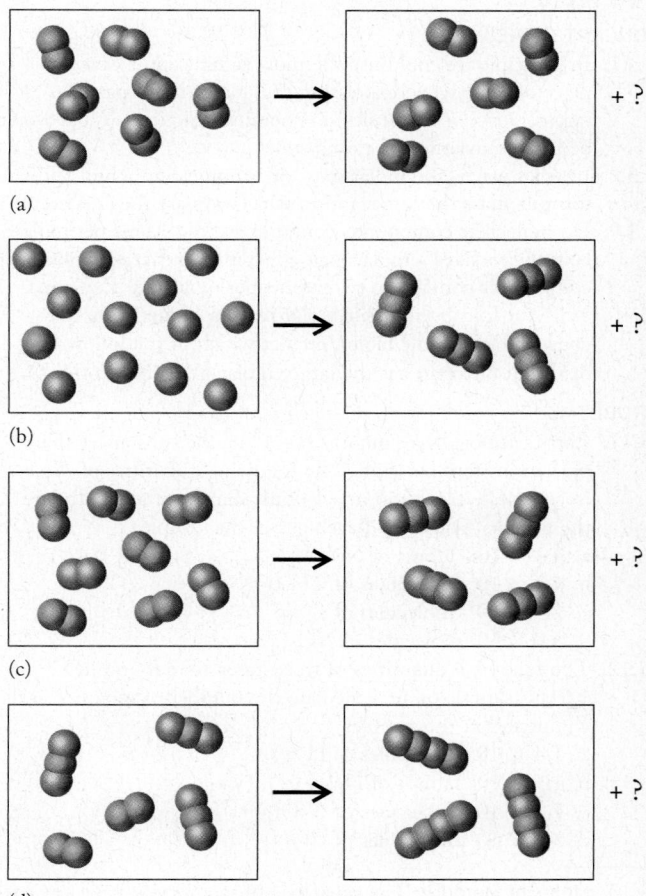

FIGURE P3.2

Chemical Reactions and Earth's Early Atmosphere

CONCEPT REVIEW

3.3. On the basis of the distribution of the elements in Earth's layers (see Figure 3.1), which of the following substances should be the most dense? $SiO_2(s)$; $Al_2O_3(s)$; $Fe(\ell)$

3.4. On the basis of the compositions and physical states of Earth's various layers (Figure 3.1), which of the following substances has the highest melting point? Al_2O_3; Fe; Ni; S

3.5. In a combination reaction, is the number of different products equal to, less than, or greater than the number of different reactants?

*3.6. The proportions of the elements that make up the asteroid 433 Eros are similar to those that make up Earth. Scientists believe that this similarity means that the asteroid and Earth formed around the same time. If the asteroid formed after Earth formed a solid crust, and if the asteroid was the product of a collision between Earth and an even larger asteroid, how would the composition of 433 Eros be different from what it actually is?

The Mole

CONCEPT REVIEW

3.7. In principle we could use the more familiar unit *dozen* in place of mole when expressing the quantities of particles (atoms, ions, or molecules) in chemical reactions. What would be the disadvantage in doing so?

3.8. In what way is the molar mass of an ionic compound and its formula mass the same, and in what ways are they different?

3.9. Do molecular compounds containing three atoms per molecule always have a molar mass greater than that of molecular compounds containing two atoms per molecule? Explain.

3.10. Without calculating their molar masses (though you may consult the periodic table), predict which of the following oxides of nitrogen has the larger molar mass: NO_2 or N_2O.

PROBLEMS

3.11. Earth's atmosphere contains many volatile substances that are present in trace amounts. The following quantities of these trace gases were found in a 1.0 mL sample of air. Calculate the number of moles of each gas in the sample:
a. 4.4×10^{14} atoms of Ne(*g*)
b. 4.2×10^{13} molecules of CH_4(*g*)
c. 2.5×10^{12} molecules of O_3(*g*)
d. 4.9×10^9 molecules of NO_2(*g*)

3.12. The following quantities of trace gases were found in a 1.0 mL sample of air. Calculate the number of moles of each compound in the sample:
a. 1.4×10^{13} molecules of H_2(*g*)
b. 1.5×10^{14} atoms of He(*g*)
c. 7.7×10^{12} molecules of N_2O(*g*)
d. 3.0×10^{12} molecules of CO(*g*)

3.13. **Moles of Memory** The capacities of memory-storage devices in computers are represented by the number of bytes of information that can be stored. How many moles of bytes are there in each of the following?
a. a 1.5-terabyte hard drive (where 1 terabyte = 10^{12} bytes)
b. a 2.0-gigabyte flash drive (where 1 gigabyte = 10^9 bytes)

3.14. Express the following population estimates for the year 2010 in nanomoles of people:
a. United States and Canada, 348 million people
b. Europe, 719 million people
c. Asia, 4.15 billion people
d. the world, 6.83 billion people

3.15. How many atoms of titanium are there in 0.125 moles of each of the following?
a. ilmenite, $FeTiO_3$
b. titanium(IV) chloride
c. Ti_2O_3
d. Ti_3O_5

3.16. How many atoms of iron are there in 2.5 moles of each of the following?
a. wolframite, $FeWO_4$
b. pyrite, FeS_2
c. magnetite, Fe_3O_4
d. hematite, Fe_2O_3

3.17. Which substance in each of the following pairs of quantities contains more moles of oxygen?
a. 1 mol of Al_2O_3 or 1 mol of Fe_2O_3
b. 1 mol of SiO_2 or 1 mol of N_2O_4
c. 3 mol of CO or 2 mol of CO_2

3.18. Which substance in each of the following pairs of quantities contains more moles of oxygen?
a. 2 mol of N_2O or 1 mol of N_2O_5
b. 1 mol of NO or 1 mol of calcium nitrate
c. 2 mol of NO_2 or 1 mol of sodium nitrite

3.19. **Elemental Composition of Minerals** Aluminum, silicon, and oxygen form minerals known as aluminosilicates. How many moles of aluminum are in 1.50 mol of
a. pyrophyllite, $Al_2Si_4O_{10}(OH)_2$?
b. mica, $KAl_3Si_3O_{10}(OH)_2$?
c. albite, $NaAlSi_3O_8$?

3.20. The uranium used for nuclear fuel exists in nature in several minerals. Calculate how many moles of uranium are in 1 mol of the following:
a. carnotite, $K_2(UO_2)_2(VO_4)_2$
b. uranophane, $CaU_2Si_2O_{11}$
c. autunite, $Ca(UO_2)_2(PO_4)_2$

3.21. How many moles of carbon are there in 500.0 g of carbon?

3.22. How many moles of gold are there in 2.00 ounces of gold?

3.23. How many moles of Ca^{2+} ions are in 0.25 mol of calcium titanate, $CaTiO_3$? What is the mass in grams of these Ca^{2+} ions?

3.24. How many moles of O^{2-} ions are in 0.55 mol of aluminum oxide, Al_2O_3? What is the mass in grams of these O^{2-} ions?

3.25. How many moles of iron are there in 1 mole of the following compounds? (a) FeO; (b) Fe_2O_3; (c) $Fe(OH)_3$; (d) Fe_3O_4

3.26. How many moles of sodium are there in 1 mol of the following compounds? (a) NaCl; (b) Na_2SO_4; (c) Na_3PO_4; (d) $NaNO_3$

3.27. Calculate the molar masses of the following atmospheric molecules: (a) SO_2; (b) O_3; (c) CO_2; (d) N_2O_5.

3.28. Determine the molar masses of the following minerals:
   a. rhodonite, $MnSiO_3$
   b. scheelite, $CaWO_4$
   c. ilmenite, $FeTiO_3$
   d. magnesite, $MgCO_3$

3.29. Flavoring Additives Calculate the molar masses of the following common flavors in food:
   a. vanillin, $C_8H_8O_3$
   b. oil of cloves, $C_{10}H_{12}O_2$
   c. anise oil, $C_{10}H_{12}O$
   d. oil of cinnamon, C_9H_8O

3.30. Sweeteners Calculate the molar masses of the following common sweeteners:
   a. sucrose, $C_{12}H_{22}O_{11}$
   b. saccharin, $C_7H_5O_3NS$
   c. aspartame, $C_{14}H_{18}N_2O_5$
   d. fructose, $C_6H_{12}O_6$

3.31. Suppose pairs of balloons are filled with 10.0 g of the following pairs of gases. Which balloon in each pair has the greater number of particles? (a) CO_2 or NO; (b) CO_2 or SO_2; (c) O_2 or Ar

3.32. If you had equal masses of the substances in the following pairs of compounds, which of the two would contain the greater number of ions? (a) NaBr or KCl; (b) NaCl or $MgCl_2$; (c) $BaCl_2$ or Li_2CO_3

3.33. How many moles of SiO_2 are there in a quartz crystal (SiO_2) that has a mass of 45.2 g?

3.34. How many moles of NaCl are there in a crystal of halite that has a mass of 6.82 g?

3.35. What is the mass of 0.122 mol of $MgCO_3$?

3.36. What is the volume of 1.00 mol of benzene (C_6H_6) at 20°C? The density of benzene is 0.879 g/mL.

*3.37. The density of uranium (U; 19.05 g/cm³) is more than five times as great as that of diamond (C; 3.514 g/cm³). If you have a cube (1 cm on a side) of each element, which cube contains more atoms?

*3.38. Aluminum ($d = 2.70$ g/mL) and strontium ($d = 2.64$ g/mL) have nearly the same density. If we manufacture two cubes, each containing 1 mole of one element or the other, which cube will be smaller? What are the dimensions of this cube?

Writing Balanced Chemical Equations

CONCEPT REVIEW

3.39. In a balanced chemical equation, does the number of moles of reactants always equal the number of moles of products?

3.40. In a balanced chemical equation, does the sum of the coefficients for the reactants always equal the sum of the coefficients for the products?

3.41. In a balanced chemical equation, must the sum of the masses of all the gaseous reactants always equal the sum of the masses of the gaseous products?

3.42. In a balanced chemical equation, must the sum of the volumes occupied by the gaseous reactants always equal the sum of the volumes occupied by the gaseous products?

PROBLEMS

3.43. Balance the following reactions, which are believed to take place in the environment of a protostar, like that of our early sun:
   a. $CH_4(g) + H_2O(g) \rightarrow CO(g) + H_2(g)$
   b. $NH_3(g) \rightarrow N_2(g) + H_2(g)$
   c. $CO(g) + H_2O(g) \rightarrow CO_2(g) + H_2(g)$

3.44. Chemistry of Volcanic Gases Balance the following reactions that occur among volcanic gases:
   a. $SO_2(g) + O_2(g) \rightarrow SO_3(g)$
   b. $H_2S(g) + O_2(g) \rightarrow SO_2(g) + H_2O(g)$
   *c. $H_2S(g) + SO_2(g) \rightarrow S_8(s) + H_2O(g)$

*3.45. Chemical Weathering of Rocks and Minerals Balance the following chemical reactions, which contribute to weathering of the iron–silicate minerals ferrosilite ($FeSiO_3$), fayalite (Fe_2SiO_4), and greenalite [$Fe_3Si_2O_5(OH)_4$]:
   a. $FeSiO_3(s) + H_2O(\ell) \rightarrow Fe_3Si_2O_5(OH)_4(s) + H_4SiO_4(aq)$
   b. $Fe_2SiO_4(s) + CO_2(g) + H_2O(\ell) \rightarrow$
      $FeCO_3(s) + H_4SiO_4(aq)$
   c. $Fe_3Si_2O_5(OH)_4(s) + CO_2(g) + H_2O(\ell) \rightarrow$
      $FeCO_3(s) + H_4SiO_4(aq)$

*3.46. Copper was one of the first metals used by humans because it can be refined from a wide variety of copper minerals, such as cuprite (Cu_2O), chalcocite (Cu_2S), and malachite [$Cu_2CO_3(OH)_2$]. Balance the following reactions for converting these minerals into copper metal:
   a. $Cu_2O(s) + C(s) \rightarrow Cu(s) + CO_2(g)$
   b. $Cu_2O(s) + Cu_2S(s) \rightarrow Cu(s) + SO_2(g)$
   c. $Cu_2CO_3(OH)_2(s) + C(s) \rightarrow Cu(s) + CO_2(g) + H_2O(g)$

3.47. Physiologically Active Nitrogen Oxides The oxides of nitrogen are biologically reactive substances now known to be formed endogenously in the human lung: NO is a powerful agent for dilating blood vessels; N_2O is the anesthetic known as laughing gas; NO_2 has an acrid odor and is corrosive to lung tissue. Balance the following reactions for the formation of nitrogen oxides:
   a. $N_2(g) + O_2(g) \rightarrow NO(g)$
   b. $NO(g) + O_2(g) \rightarrow NO_2(g)$
   c. $NO(g) + NO_3(g) \rightarrow NO_2(g)$
   d. $N_2(g) + O_2(g) \rightarrow N_2O(g)$

3.48. Chemistry of Geothermal Vents Some scientists believe that life on Earth may have originated near deep-ocean vents. Balance the following reactions, which are among those taking place near such vents:
   a. $CH_3SH(aq) + CO(aq) \rightarrow CH_3COSCH_3(aq) + H_2S(aq)$
   b. $H_2S(aq) + CO(aq) \rightarrow CH_3CO_2H(aq) + S_8(s)$

*3.49. Write a balanced chemical equation for each of the following reactions:
   a. Dinitrogen pentoxide reacts with sodium metal to produce sodium nitrate and nitrogen dioxide.
   b. A mixture of nitric acid and nitrous acid is formed when water reacts with dinitrogen tetroxide.
   c. At high pressure, nitrogen monoxide decomposes to dinitrogen monoxide and nitrogen dioxide.

3.50. Write a balanced chemical equation for each of the following reactions:
   a. Carbon dioxide reacts with carbon to form carbon monoxide.
   b. Potassium reacts with water to give potassium hydroxide and the element hydrogen.
   c. Phosphorus (P_4) burns in air to give diphosphorus pentoxide.

3.51. Write a balanced chemical equation for the combustion of acetylene (C_2H_2).

3.52. Write a balanced chemical equation for the combustion of octane (C_8H_{18}).

Combustion Reactions; Stoichiometric Calculations and the Carbon Cycle

CONCEPT REVIEW

3.53. If the sum of the masses of the reactants in a chemical equation equals the sum of the masses of the products, must the equation be balanced?

***3.54.** There are two ways to write the equation for the combustion of ethane:

$$C_2H_6(g) + 7/2\,O_2(g) \rightarrow 3\,H_2O(g) + 2\,CO_2(g)$$
$$2\,C_2H_6(g) + 7\,O_2(g) \rightarrow 6\,H_2O(g) + 4\,CO_2(g)$$

Do these two different ways of writing the equation affect the calculation of how much CO_2 is produced from a known quantity of C_2H_6?

PROBLEMS

3.55. Land Management The United Nations Intergovernmental Panel on Climate Change reported in June 2000 that better management of cropland, grazing land, and forests would reduce the amount of carbon dioxide in the atmosphere by 5.4×10^9 kg of carbon per year.
   a. How many moles of carbon are present in 5.4×10^9 kg of carbon?
   b. How many kilograms of carbon dioxide does this quantity of carbon represent?

3.56. Energy generation results in the addition of an estimated 27 billion metric tons of CO_2 to the atmosphere each year.
   a. How many moles of CO_2 does 27 billion tons represent?
   b. How many grams of carbon are in 27 billion tons of CO_2?

3.57. When $NaHCO_3$ is heated above 270°C, it decomposes to $Na_2CO_3(s)$, $H_2O(g)$, and $CO_2(g)$.
   a. Write a balanced chemical equation for the decomposition reaction.
   b. Calculate the mass of CO_2 produced from the decomposition of 25.0 g of $NaHCO_3$.

3.58. Egyptian Cosmetics Pb(OH)Cl, one of the lead compounds used in ancient Egyptian cosmetics (see Problem 3.74), was prepared from PbO according to the following recipe:

$$PbO(s) + NaCl(aq) + H_2O(\ell) \rightarrow Pb(OH)Cl(s) + NaOH(aq)$$

How many grams of PbO and how many grams of NaCl would be required to produce 10.0 g of Pb(OH)Cl?

3.59. The manufacture of aluminum includes the production of cryolite (Na_3AlF_6) from the following reaction:

$$6\,HF(g) + 3\,NaAlO_2(s) \rightarrow Na_3AlF_6(s) + 3\,H_2O(\ell) + Al_2O_3(s)$$

How much $NaAlO_2$ (sodium aluminate) is required to produce 1.00 kg of Na_3AlF_6?

3.60. Chromium metal can be produced from the high-temperature reaction of Cr_2O_3 [chromium(III) oxide] with silicon or aluminum by each of the following reactions:

$$Cr_2O_3(s) + 2\,Al(\ell) \rightarrow 2\,Cr(\ell) + Al_2O_3(s)$$
$$2\,Cr_2O_3(s) + 3\,Si(\ell) \rightarrow 4\,Cr(\ell) + 3\,SiO_2(s)$$

   a. Calculate the number of grams of aluminum required to prepare 400.0 g of chromium metal by the first reaction.
   b. Calculate the number of grams of silicon required to prepare 400.0 g of chromium metal by the second reaction.

***3.61.** Suppose 25 metric tons of coal that is 3.0% sulfur by mass is burned at an electric power plant (1 metric ton = 10^3 kg). During combustion, the sulfur is converted into sulfur dioxide. How many tons of sulfur dioxide are produced?

3.62. Charcoal (C) and propane (C_3H_8) are used as fuel in backyard grills.
   a. Write balanced chemical equations for the complete combustion reactions of C and C_3H_8.
   b. How many grams of carbon dioxide are produced from burning 500.0 g of each of the two fuels?

***3.63.** The uranium minerals found in nature must be refined and enriched in ^{235}U before the uranium can be used as a fuel in nuclear reactors. One procedure for enriching uranium relies on the reaction of UO_2 with HF to form UF_4, which is then converted into UF_6 by reaction with fluorine:

$$(1)\ UO_2(g) + 4\,HF(aq) \rightarrow UF_4(g) + 2\,H_2O(\ell)$$
$$(2)\ UF_4(g) + F_2(g) \rightarrow UF_6(g)$$

   a. How many kilograms of HF are needed to completely react with 5.00 kg of UO_2?
   b. How much UF_6 can be produced from 850.0 g of UO_2?

3.64. The mineral bauxite, which is mostly Al_2O_3, is the principal industrial source of aluminum metal. How much aluminum can be produced from 1.00 metric ton of Al_2O_3?

***3.65.** Chalcopyrite ($CuFeS_2$) is an abundant copper mineral that can be converted into elemental copper. How much Cu could be produced from 1.00 kg of $CuFeS_2$?

***3.66. Mining for Gold** Unlike most metals, gold is found in nature as the pure element. Miners in California in 1849 searched for gold nuggets and gold dust in streambeds, where the denser gold could be easily separated from sand and gravel. However, larger deposits of gold are found in veins of rock and can be separated chemically in a two-step process:

$$(1)\ 4\,Au(s) + 8\,NaCN(aq) + O_2(g) + 2\,H_2O(\ell) \rightarrow$$
$$4\,NaAu(CN)_2(aq) + 4\,NaOH(aq)$$
$$(2)\ 2\,NaAu(CN)_2(aq) + Zn(s) \rightarrow$$
$$2\,Au(s) + Na_2[Zn(CN)_4](aq)$$

If a 1.0×10^3 kg sample of rock is 0.019% gold by mass, how much Zn is needed to react with the gold extracted from the rock? Assume that reactions (1) and (2) are 100% efficient.

Determining Empirical Formulas from Percent Composition; Empirical and Molecular Formulas Compared

CONCEPT REVIEW

3.67. What is the difference between an empirical formula and a molecular formula?

3.68. Do the empirical and molecular formulas of a compound have the same percent composition values?

3.69. Is the element with the largest atomic mass always the element present in the highest percentage by mass in a compound?

3.70. Sometimes the composition of a compound is expressed as a mole percent or atom percent. Are the values of these parameters likely to be the same for a given compound, or different?

PROBLEMS

3.71. Calculate the percent composition of (a) Na_2O, (b) $NaOH$, (c) $NaHCO_3$, and (d) Na_2CO_3.

3.72. Calculate the percent composition of (a) sodium sulfate, (b) dinitrogen tetroxide, (c) strontium nitrate, and (d) aluminum sulfide.

3.73. Organic Compounds in Space The following compounds have been detected in space. Which of them contains the greatest percentage of carbon by mass?
a. naphthalene, $C_{10}H_8$
b. chrysene, $C_{18}H_{12}$
c. pentacene, $C_{22}H_{14}$
d. pyrene, $C_{16}H_{10}$

3.74. Toxicity of Lead Compounds Ancient Egyptians used the following lead compounds as white pigments in cosmetics (see Problem 3.58) and their use in pigments for paints continued into modern times: PbS, $PbCO_3$, $PbCl(OH)$, and $Pb_2Cl_2CO_3$. Calculate the percentage of lead in each compound and order them from lowest to highest with respect to the percent lead content.

3.75. Of the nitrogen oxides—N_2O, NO, N_2O_3, and NO_2—which is more than 50% oxygen by mass?

3.76. Of the sulfur oxides—S_2O, SO, SO_2, and SO_3—which is more than 50% oxygen by mass?

3.77. Do any two of the following compounds, which have been detected in outer space, have the same empirical formula?
a. naphthalene, $C_{10}H_8$
b. chrysene, $C_{18}H_{12}$
c. anthracene, $C_{14}H_{10}$
d. pyrene, $C_{16}H_{10}$
e. benzoperylene, $C_{22}H_{12}$
f. coronene, $C_{24}H_{12}$

3.78. Which, if any, of the following nitrogen oxides have the same empirical formula? (a) N_2O; (b) NO; (c) NO_2; (d) N_2O_2; (e) N_2O_4

3.79. Surgical Grade Titanium Medical implants and high-quality jewelry items for body piercings are frequently made of a material known as G23Ti or surgical-grade titanium. The percent composition of the material is 64.39% titanium, 24.19% aluminum, and 11.42% vanadium. What is the empirical formula for surgical-grade titanium?

3.80. A sample of an iron-containing compound is 22.0% iron, 50.2% oxygen, and 27.8% chlorine by mass. What is the empirical formula of this compound?

3.81. In an experiment, 2.43 g of magnesium reacts with 1.60 g of oxygen, forming 4.03 g of magnesium oxide.
a. Use these data to calculate the empirical formula of magnesium oxide.
b. Write a balanced chemical equation for this reaction.

3.82. Ferrophosphorus (Fe_2P) reacts with pyrite (FeS_2), producing iron(II) sulfide and a compound that is 27.87% P and 72.13% S by mass and has a molar mass of 444.56 g/mol.

a. Determine the empirical and molecular formulas of this compound.
b. Write a balanced chemical equation for this reaction.

3.83. Asbestosis Asbestosis is a lung disease caused by inhaling asbestos fibers. In addition, fiber from one form of asbestos called chrysotile is considered to be a human carcinogen by the U.S. Department of Health and Human Services. Chrysotile has the composition 26.31% magnesium, 20.20% silicon, an 1.45% hydrogen with the remainder of the mass as oxygen. Determine the empirical formula of chrysotile.

3.84. Chemistry of Soot A candle flame produces easily seen specks of soot near the edges of the flame, especially when the candle is moved. A piece of glass held over a candle flame will become coated with soot, which is the result of the incomplete combustion of candle wax. Elemental analysis of a compound extracted from a sample of this soot gave these results: 7.74% H and 92.26% C by mass. Calculate the empirical formula of the compound.

3.85. What is the empirical formula of the compound that is 24.2% Cu, 27.0% Cl, and 48.8% O by mass?

3.86. The compound made of chlorine and oxygen that has been used to kill anthrax spores in contaminated buildings is 52.6% Cl by mass. What is its empirical formula?

Combustion Analysis

CONCEPT REVIEW

3.87. Explain why it is important for combustion analysis to be carried out in an excess of oxygen.

3.88. Why is the quantity of CO_2 obtained in a combustion analysis not a direct measure of the oxygen content of the starting compound?

3.89. Can the results of a combustion analysis ever give the true molecular formula of a compound?

3.90. What additional information is needed to determine a molecular formula from the results of an elemental analysis of an organic compound?

PROBLEMS

3.91. The combustion of 135.0 mg of a hydrocarbon produces 440.0 mg of CO_2 and 135.0 mg H_2O. The molar mass of the hydrocarbon is 270 g/mol. Determine the empirical and molecular formulas of this compound.

3.92. A 0.100 g sample of a compound containing C, H, and O is burned in oxygen, producing 0.1783 g of CO_2 and 0.0734 g of H_2O. Determine the empirical formula of the compound.

3.93. GRAS List for Food Additives The compound geraniol is on the GRAS (generally recognized as safe) list and can be used in foods and personal care products. By itself, geraniol has a rose-like odor but it is frequently blended with other fragrances on the GRAS list and then added to products to produce a pleasant peach- or lemon-like aroma. In an analysis, the complete combustion of 175 mg of geraniol produced 499 mg CO_2 and 184 mg H_2O. What is the empirical formula for geraniol?

*3.94. The combustion of 40.5 mg of a compound containing C, H, and O, and extracted from the bark of the sassafras tree, produces 110.0 mg of CO_2 and 22.5 mg of H_2O. The molar mass of the compound is 162 g/mol. Determine its empirical and molecular formulas.

Limiting Reactants and Percent Yield

CONCEPT REVIEW

3.95. If a reaction vessel contains equal masses of Fe and S, a mass of FeS corresponding to which of the following could theoretically be produced?
  a. the sum of the masses of Fe and S
  b. more than the sum of the masses of Fe and S
  c. less than the sum of the masses of Fe and S

3.96. A reaction vessel contains equal masses of magnesium metal and oxygen gas. The mixture is ignited, forming MgO. After the reaction has gone to completion, the mass of the MgO is less than the mass of the reactants. Is this result a violation of the law of conservation of mass? Explain your answer.

3.97. Explain how the parameters of theoretical yield and percent yield differ.

3.98. Can the percent yield of a chemical reaction ever exceed 100%?

3.99. Give two reasons why the actual yield from a chemical reaction is usually less than the theoretical yield.

3.100. A chemical reaction produces less than the expected amount of product. Is this result a violation of the law of conservation of mass?

PROBLEMS

3.101. Making Hollandaise Sauce A recipe for 1 cup of hollandaise sauce calls for $\frac{1}{2}$ cup of butter, $\frac{1}{4}$ cup of hot water, 4 egg yolks, and the juice of a medium-sized lemon. How many cups of this sauce can be made from a pound (2 cups) of butter, a dozen eggs, 4 medium lemons, and an unlimited supply of hot water?

3.102. A factory making toy wagons has 13,466 wheels, 3360 handles, and 2400 wagon beds in stock. What maximum number of wagons can the factory make?

3.103. Potassium superoxide, KO_2, reacts with carbon dioxide to form potassium carbonate and oxygen:

$$4\,KO_2(s) + 2\,CO_2(g) \rightarrow 2\,K_2CO_3(s) + 3\,O_2(g)$$

This reaction makes potassium superoxide useful in a self-contained breathing apparatus. How much O_2 could be produced from 2.50 g of KO_2 and 4.50 g of CO_2?

3.104. A reaction vessel contains 10.0 g of CO and 10.0 g of O_2. How many grams of CO_2 could be produced according to the following reaction?

$$2\,CO(g) + O_2(g) \rightarrow 2\,CO_2(g)$$

3.105. Ammonia rapidly reacts with hydrogen chloride, making ammonium chloride. Write a balanced chemical equation for the reaction, and calculate the number of grams of excess reactant when 3.0 g of NH_3 reacts with 5.0 g of HCl.

3.106. Sulfur trioxide dissolves in water, producing H_2SO_4. How much sulfuric acid can be produced from 10.0 mL of water ($d = 1.00$ g/mL) and 25.6 g of SO_3?

3.107. The reaction of 3.0 g of carbon with excess O_2 yields 6.5 g of CO_2. What is the percent yield of this reaction?

3.108. Baking soda ($NaHCO_3$) can be made in large quantities by the following reaction:

$$NaCl(aq) + NH_3(aq) + CO_2(aq) + H_2O(\ell) \rightarrow$$
$$NaHCO_3(s) + NH_4Cl(aq)$$

If 10.0 g of NaCl reacts with excesses of the other reactants and 4.2 g of $NaHCO_3$ is isolated, what is the percent yield of the reaction?

3.109. Chemistry of Fermentation Yeast converts glucose ($C_6H_{12}O_6$) into ethanol ($d = 0.789$ g/mL) in a process called fermentation. An equation for the reaction can be written as follows:

$$C_6H_{12}O_6(aq) \rightarrow C_2H_5OH(\ell) + CO_2(g)$$

  a. Write a balanced chemical equation for this fermentation reaction.
  b. If 100.0 g of glucose yields 50.0 mL of ethanol, what is the percent yield for the reaction?

*3.110. Composition of Seawater A 1-liter sample of seawater contains 19.4 g of Cl^-, 10.8 g of Na^+, and 1.29 g of Mg^{2+}.
  a. How many moles of each ion are present?
  b. If we evaporated the seawater, would there be enough Cl^- present to form the chloride salts of all the sodium and magnesium present?

Additional Problems

*3.111. Artificial Bones for Medical Implants The material often used to make artificial bones is the same material that gives natural bones their structure. Its common name is hydroxyapatite, and its formula is $Ca_5(PO_4)_3OH$.
  a. Propose a systematic name for this compound.
  b. What is the mass percentage of calcium in it?
  c. When treated with hydrogen fluoride, hydroxyapatite becomes fluorapatite [$Ca_5(PO_4)_3F$], a stronger structure. The strengthening is due to the altered mass percentage of calcium in the substance. Does the percent mass of Ca increase or decrease as a result of this substitution?

*3.112. As a solution of copper sulfate slowly evaporates, beautiful blue crystals made of Cu(II) and sulfate ions form such that water molecules are trapped inside the crystals. The overall formula of the compound is $CuSO_4 \cdot 5H_2O$.
  a. What is the percent water in this compound?
  b. At high temperatures the water in the compound is driven off as steam. What mass percentage of the original sample of the blue solid is lost as a result?

3.113. Aluminum is mined as the mineral bauxite, which consists primarily of Al_2O_3 (alumina).
  a. How much aluminum is produced from 1 metric ton (1 metric ton = 10^3 kg) of Al_2O_3?

$$2\,Al_2O_3(s) \rightarrow 4\,Al(s) + 3\,O_2(g)$$

  b. The oxygen produced in part a is allowed to react with carbon to produce carbon monoxide:

$$O_2(g) + 2\,C(s) \rightarrow 2\,CO(g)$$

Balance the following equation describing the reaction of alumina with carbon:

$$Al_2O_3(s) + C(s) \rightarrow Al(s) + CO(g)$$

  c. How much CO is produced from the O_2 made in part a?

*3.114. Chemistry of Copper Production "Native," or elemental, copper can be found in nature, but most copper is mined as oxide or sulfide minerals. Chalcopyrite ($CuFeS_2$) is one copper mineral that can be converted to elemental copper in a series of chemical steps. Reacting chalcopyrite with oxygen at high temperature produces a mixture of copper sulfide and iron oxide. The iron oxide is separated from CuS by reaction with sand. CuS is converted to Cu_2S in the process and the Cu_2S is burned in air to produce Cu and SO_2:

(1) $2\,CuFeS_2(s) + 3\,O_2(g) \rightarrow$
$$2\,CuS(s) + 2\,FeO(s) + 2\,SO_2(g)$$

(2) $FeO(s) + SiO_2(s) \rightarrow FeSiO_3(s)$

(3) $2\,CuS(s) \rightarrow Cu_2S(s) + \frac{1}{8}\,S_8(s)$

(4) $Cu_2S(s) + O_2(g) \rightarrow 2\,Cu(s) + SO_2(g)$

An average copper penny minted in the 1960s has a mass of about 3.0 g.

a. How much chalcopyrite had to be mined to produce one dollar's worth of pennies?

b. How much chalcopyrite had to be mined to produce one dollar's worth of pennies if reaction 1 above had a percent yield of 85% and reactions 2, 3, and 4 had percent yields of essentially 100%?

c. How much chalcopyrite had to be mined to produce one dollar's worth of pennies if each reaction involving copper proceeded with an 85% yield?

*3.115. **Mining for Gold** Gold can be extracted from the surrounding rock using a solution of sodium cyanide. While effective for isolating gold, toxic cyanide finds its way into watersheds, causing environmental damage and harming human health.

$$4\,Au(s) + 8\,NaCN(aq) + O_2(g) + 2\,H_2O(\ell) \rightarrow$$
$$4\,NaAu(CN)_2(aq) + 4\,NaOH(aq)$$

$$2\,NaAu(CN)_2(aq) + Zn(s) \rightarrow 2\,Au(s) + Na_2[Zn(CN)_4](aq)$$

a. If a sample of rock contains 0.009% gold by mass, how much NaCN is needed to extract the gold from 1 metric ton (1 metric ton = 10^3 kg) of rock as $NaAu(CN)_2$?

b. How much zinc is needed to convert the $NaAu(CN)_2$ from part a to metallic gold?

c. The gold recovered in part b is manufactured into a gold ingot in the shape of a cube. The density of gold is 19.3 g/cm^3. How big is the block of gold?

*3.116. Phosgenite, a lead compound with the formula $Pb_2Cl_2CO_3$, is found in Egyptian cosmetics. Phosgenite was prepared by the reaction of PbO, NaCl, and CO_2. An unbalanced equation of the reactant mixture is

$$PbO(s) + NaCl(aq) + H_2O(\ell) + CO_2(g) \rightarrow$$
$$Pb_2Cl_2CO_3(s) + NaOH(aq)$$

a. Balance this equation.

b. How many grams of phosgenite can be obtained from 10.0 g of PbO and 10.0 g NaCl in the presence of excess water and CO_2?

c. Phosgenite can be considered a mixture of two lead compounds. Which compounds appear to be combined to make phosgenite?

*3.117. Uranium oxides used in the preparation of fuel for nuclear reactors are separated from other metals in minerals by converting the uranium to $UO_x(NO_3)_y(H_2O)_z$, where uranium has a positive charge ranging from 3+ to 6+.

a. Roasting $UO_x(NO_3)_y(H_2O)_z$ at 400°C leads to loss of water and decomposition of the nitrate ion to nitrogen oxides, leaving behind a product with the formula U_aO_b that is 83.22% U by mass. What are the values of a and b? What is the charge on U in U_aO_b?

b. Higher temperatures produce a different uranium oxide, U_cO_d, with a higher uranium content, 84.8% U. What are the values of c and d? What is the charge on U in U_aO_b?

c. The values of x, y, and z in $UO_x(NO_3)_y(H_2O)_z$ are found by gently heating the compound to remove all of the water. In a laboratory experiment, 1.328 g of $UO_x(NO_3)_y(H_2O)_z$ produced 1.042 g of $UO_x(NO_3)_y$. Continued heating generated 0.742 g of U_nO_m. Using the information in parts a and b, calculate x, y, and z.

*3.118. Large quantities of fertilizer are washed into the Mississippi River from agricultural land in the Midwest. The excess nutrients collect in the Gulf of Mexico, promoting the growth of algae and endangering other aquatic life.

a. One commonly used fertilizer is ammonium nitrate. What is the chemical formula of ammonium nitrate?

b. Corn farmers typically use 5.0×10^3 kg of ammonium nitrate per square kilometer of cornfield per year. Ammonium nitrate can be prepared by the following reaction:

$$NH_3(aq) + HNO_3(aq) \rightarrow NH_4NO_3(aq)$$

How much nitric acid would be required to make the fertilizer needed for 1 km^2 of cornfield per year?

c. The ammonium ions can be converted into NO_3^- by bacterial action.

$$NH_4^+(aq) + 2\,O_2(g) \rightarrow NO_3^-(aq) + H_2O(\ell) + 2\,H^+(aq)$$

If 10% of the ammonium component of 5.0×10^2 kg of fertilizer ends up as nitrate, how much oxygen would be consumed?

3.119. **Composition of Over-the-Counter Medicines** Calculate the number of molecules or formula units of compound in each of the following common, over-the-counter medications:

a. ibuprofen, a pain reliever and fever reducer that contains 200.0 mg of the active ingredient, $C_{13}H_{18}O_2$

b. an antacid containing 500.0 mg of calcium carbonate

c. an allergy tablet containing 4 mg Chlor-Trimeton $(C_{16}H_{19}N_2Cl)$

3.120. **Chemistry of Pain Relievers** The common pain relievers aspirin $(C_9H_8O_4)$, acetaminophen $(C_8H_9NO_2)$, and naproxen sodium $(C_{14}H_{13}O_3Na)$ are all available in tablets containing 200.0 mg of the active ingredient. Which compound contains the greatest number of molecules per tablet? How many molecules of the active ingredient are present in each tablet?

3.121. **Fiber in the Diet** Dietary fiber is a mixture of many compounds including xylose $(C_5H_{10}O_5)$ and methyl galacturonate $(C_7H_{12}O_7)$.

a. Do these compounds have the same empirical formula?

b. Write balanced chemical equations for the combustion of xylose and methyl galacturonate.

3.122. Some catalytic converters in automobiles contain the manganese oxides Mn_2O_3 and MnO_2.

a. Give the names of Mn_2O_3 and MnO_2.

b. Calculate the percent manganese by mass in Mn_2O_3 and MnO_2.

c. Explain how Mn_2O_3 and MnO_2 are consistent with the law of multiple proportions.

*3.123. A number of chemical reactions have been proposed for the formation of organic compounds from inorganic precursors. Here is one of them:

$$H_2S(g) + FeS(s) + CO_2(g) \rightarrow FeS_2(s) + HCO_2H(\ell)$$

a. Identify the ions in FeS and FeS_2. Give correct names for each compound.

*b. How much HCO_2H is obtained by reacting 1.00 g of FeS, 0.50 g of H_2S, and 0.50 g of CO_2 if the reaction results in a 50.0% yield?

*3.124. The formation of organic compounds by the reaction of iron(II) sulfide with carbonic acid is described by the following chemical equation:

$$2\,FeS + H_2CO_3 \rightarrow 2\,FeO + \tfrac{1}{n}(CH_2O)_n + 2\,S$$

a. How much FeO is produced starting with 1.50 g FeS and 0.525 mol of H_2CO_3 if the reaction results in a 78.5% yield?

b. If the carbon-containing product has a molar mass of 3.00×10^2 g/mol, what is the chemical formula of the product?

*3.125. **Marine Chemistry of Iron** On the seafloor, iron(II) oxide reacts with water to form Fe_3O_4 and hydrogen in a process called serpentization.

a. Balance the following equation for serpentization:

$$FeO(s) + H_2O(\ell) \rightarrow Fe_3O_4(s) + H_2(g)$$

b. When CO_2 is present, the product is methane, not hydrogen. Balance the following chemical equation:

$$FeO(s) + H_2O(\ell) + CO_2(g) \rightarrow Fe_3O_4(s) + CH_4(g)$$

3.126. Titanium dioxide and zinc oxide are two of the 16 active ingredients approved by the FDA for use in sunscreens.

a. What are the chemical formulas of these compounds?

b. How would you modify the names for the two compounds based on the rules for naming given in Chapter 2?

c. Which compound contains the higher percentage of oxygen by mass?

3.127. The solar wind is made up of ions, mostly protons, flowing out from the sun at about 400 km/s. Near Earth, each cubic kilometer of interplanetary space contains on average 6×10^{15} solar-wind ions. How many moles of ions are in a cubic kilometer of near-Earth space?

3.128. The famous Hope Diamond at the Smithsonian National Museum of Natural History has a mass of 45.52 carats (see Figure P3.128). Diamond is a crystalline form of carbon.

a. How many moles of carbon are in the Hope Diamond (1 carat = 200.0 mg)?

b. How many carbon atoms are in the diamond?

FIGURE P3.128

*3.129. E-85 is an alternative fuel for automobiles and light trucks that consists of 85% (by volume) ethanol, C_2H_5OH, and 15% gasoline. The density of ethanol is 0.79 g/mL. How many moles of ethanol are in a gallon of E-85?

3.130. With reference to the previous question, how many moles of carbon dioxide are produced by the complete combustion of the quantity of ethanol in a gallon of E-85 fuel?

*3.131. A 100.00 g sample of white powder (substance A) is heated to 550°C. At that temperature the powder decomposes, giving off colorless gas B, which is denser than air and which is neither flammable nor does it support combustion. The products also include 56 g of a second white powder C. When gas B is bubbled through a solution of calcium hydroxide, substance A re-forms. What are the identities of substances A, B, and C?

3.132. A sealed chamber contains 1.604 g of CH_4 and 6.800 g of O_2. The mixture is ignited. How many grams of CO_2 are produced?

*3.133. You are given a 0.6240 g sample of a substance with the generic formula $MCl_2(H_2O)_2$. After completely drying the sample (which means removing the 2 mol of H_2O per mole of MCl_2), the sample has a mass of 0.5471 g. What is the identity of element M?

3.134. A compound found in crude oil consists of 93.71% C and 6.29% H by mass. The molar mass of the compound is 128 g/mol. What is its molecular formula?

3.135. A reaction vessel for synthesizing ammonia by reacting nitrogen and hydrogen is charged with 6.04 kg of H_2 and excess N_2. A total of 28.0 kg of NH_3 is produced. What is the percent yield of the reaction?

3.136. If a cube of table sugar, which is made of sucrose, $C_{12}H_{22}O_{11}$, is added to concentrated sulfuric acid, the acid "dehydrates" the sugar: removing the hydrogen and oxygen from it and leaving behind a lump of carbon. What percentage of the initial mass of sugar is carbon?

*3.137. A power plant burns 1.0×10^2 metric tons of coal that contains 3.0% (by mass) sulfur (1 metric ton = $.10^3$ kg). The sulfur is converted to SO_2 during combustion.

a. How many metric tons of SO_2 are produced?

b. When SO_2 escapes into the atmosphere it may combine with O_2 and H_2O, forming sulfuric acid, H_2SO_4. Write a balanced chemical equation describing this reaction.

c. How many metric tons of sulfuric acid, a component of acid rain, could be produced from the quantity of SO_2 calculated in part a?

3.138. **Reducing SO₂ Emissions** With respect to the previous question, one way to reduce the formation of acid rain involves trapping the SO_2 by passing smokestack gases through a spray of calcium oxide and O_2. The product of this reaction is calcium sulfate.

a. Write a balanced chemical equation describing this reaction.

b. How many metric tons of calcium sulfate would be produced from each ton of SO_2 that is trapped?

3.139. In the early 20th century, Londoners suffered from severe air pollution caused by burning high-sulfur coal. The sulfur dioxide that was emitted into the air mixed with London fog, forming sulfuric acid. For every gram of sulfur that was burned, how many grams of sulfuric acid could have formed?

*3.140. **Gas Grill Reaction** The burner in a gas grill mixes 24 volumes of air for every one volume of propane (C_3H_8) fuel. Like all gases, the volume that propane occupies is directly proportional to the number of moles of it at a given temperature and pressure. Air is 21% (by volume) O_2. Is the flame produced by the burner fuel rich (excess propane in the reaction mixture), fuel lean (not enough propane), or stoichiometric (just right)?

3.141. A common mineral in Earth's crust has the chemical composition 34.55% Mg, 19.96% Si, and 45.49% O. What is its empirical formula?

*3.142. **Ozone Generators** Some indoor air-purification systems work by converting a little of the oxygen in the air to ozone, which oxidizes mold and mildew spores and other biological air pollutants. The chemical equation for the ozone generation reaction is

$$3\,O_2(g) \rightarrow 2\,O_3(g)$$

It is claimed that one such system generates 4.0 g of O_3 per hour from dry air passing through the purifier at a flow of 5.0 L/min. If 1 liter of indoor air contains 0.28 g of O_2,

a. what fraction of the molecules of O_2 is converted to O_3 by the air purifier?

b. what is the percent yield of the ozone generation reaction?

Additional study materials including ChemTours and Diagnostic Quizzes are available at StudySpace at www.wwnorton.com/studyspace.

4

Solution Chemistry and the Hydrosphere

Learning Outcomes

- Express the concentrations of solutions in different units and convert from one set of units to another
- Calculate the molar concentration (molarity) of a solution, the mass of a solute, or the volume of a stock solution required to make a solution of specified molarity
- Calculate the concentration of a solute from stoichiometry and titration data
- Identify strong electrolytes, weak electrolytes, and nonelectrolytes from conductivity experiments
- Write molecular, overall ionic, and net ionic equations for reactions
- Predict precipitation reactions using solubility rules and quantify results from precipitation titrations
- Identify redox reactions, oxidizing agents, and reducing agents, and balance redox reactions

A LOOK AHEAD

Solutions

Since their original formation, the oceans of Earth have contained essentially the same volume of seawater, and their chemical content has changed very little over time. How can this be, considering the continuous influx of soluble and suspended material from rivers? The answer is that an elaborate system of physical and chemical processes operates on a grand scale to maintain the composition of the oceans.

Let's begin our look at this system with water-soluble compounds dissolving from the land into freshwater streams that flow to the sea. These compounds contain the ions that make ocean water salty, but they also contain soluble nutrients that ultimately may become trapped in the mud on the ocean floor. These nutrients eventually return to Earth's surface—either as oceans recede and the mud is exposed along new coastlines, or as the seafloor is uplifted to make mountain ranges composed of the sedimentary rocks formed when the mud is compressed. This monumental cycle repeats: rain dissolves compounds when it falls on land, rivers carry dissolved compounds and suspended matter to the sea, and this mixture finds its way into the mud on the ocean floor, only to be recycled again in the form of sedimentary rock with the next shift in Earth's crust.

Life on Earth may have begun in the sea. According to one theory, life originated deep beneath the surface of the oceans in hydrothermal vents, where hot gases mixed with water, forming mineral-rich deposits in the mud. Whether this particular theory is correct or not, water is considered essential for the chemistry of all biological systems because it provides the medium in which the transport of molecules and ions can occur during chemical reactions. Because water is necessary for all life, scientists look for evidence of liquid water wherever they search for life, whether it is somewhere on Earth or even somewhere else in our solar system or beyond.

Earth, Water, and Life During their lives these salmon live in freshwater and in seawater: aquatic environments with very different concentrations of dissolved substances. ▶

The fluids within cells are saline solutions, although they are not quite as salty as seawater. The concentration of dissolved matter within the cells in our bodies also stays remarkably constant in healthy individuals. Cells contain sodium, potassium, chloride, and hydrogen carbonate ions, and when blood tests are done to assess a patient's well-being, the concentrations of these ions are a crucial part of the picture. A shift in their concentrations indicates problems ranging from kidney disease to drug abuse.

Many other solutions are part of daily life. Vinegar in salad dressing is a solution of acetic acid in water. People with contact lenses probably wash them in a solution containing sodium chloride and cleaning agents. The liquid in an automobile battery is a solution of sulfuric acid in water. Most beverages are solutions—even bottled water contains dissolved ions that give it a refreshing flavor.

Whether dealing with seawater, freshwater, the fluids within cells, or the myriad solutions we come in contact with in daily life, issues of determining the identity and amount of materials dissolved in aqueous solutions and knowledge of the major types of reactions that take place when solutions are mixed are important in countless situations. How we describe the content of solutions and reactions taking place in solution, and how reactions in solution produce substances important environmentally, medically, and commercially are the subjects of this chapter. ■

4.1 Solutions on Earth and Other Places

Earth is widely referred to as the "water planet." Life exists here because liquid water is abundant: depressions in Earth's crust contain about 1.5×10^{21} L of liquid H_2O, nature's solvent. Debate over the existence of life elsewhere in the universe hinges on the prospect of liquid water existing on the planets. The belief that Martian meteorites collected in Antarctica contain fossilized life-forms (Figure 4.1) is consistent with evidence that there may have been water on the surface of Mars in the distant past. The Grand Canyon in the southwestern United States was formed by the action of water on rocks, and similar features have been observed on Mars (Figure 4.2). In addition, a series of images from the *Mars Orbital Camera* indicate that water may have flowed on the surface of Mars during the past several years (Figure 4.3). These observations support the idea that liquid water may now exist just under the surface of Mars, and where there is water there could be life. The biochemical reactions in all living cells, from single-celled organisms to human beings, require the presence of liquid water, and understanding the chemistry of the biosphere—just like the hydrosphere—requires understanding the principles of chemical reactions between substances dissolved in water.

All natural waters, whether saltwater or fresh, contain ionic and molecular compounds dissolved in them. When one element or compound dissolves in another, a solution forms. As defined in Section 1.1, a *solution* is a homogeneous mixture of two or more substances (Figure 4.4). The substance present in the greatest proportion is called the **solvent**, and all the other substances in the solution are called **solutes**. When the solvent is water, the solution is an *aqueous solution*, and in this chapter you may assume all of the solutions we discuss are aqueous. In the most general case, the solvent does not even have to be a liquid. Many materials in Earth's crust, for example, are *solid solutions*, which are uniform mixtures of solid substances with variable composition.

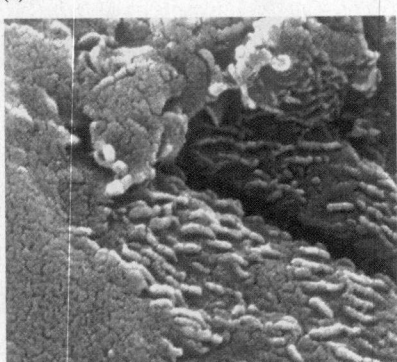

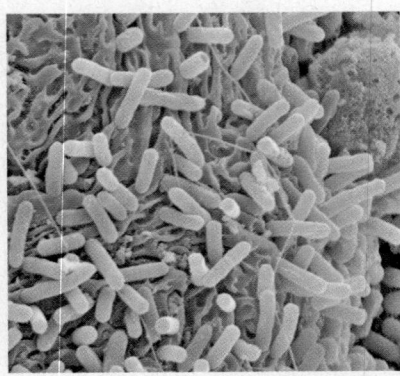

FIGURE 4.1 (a) Meteorite ALH84001, thought to have originated from Mars, contains (b) microscopic features believed to be fossilized bacteria. (c) An Earth-grown colony of the bacterium *Escherichia coli*, for comparison.

(a)   (b)

solvent the component of a solution that is present in the largest amount.

solute any component in a solution other than the solvent. A solution may contain one or more solutes.

FIGURE 4.2 (a) Arizona's Grand Canyon is a dramatic example of how Earth's surface is continually modified by flowing water. (b) Similar topographic features on Mars suggest that water once flowed on the Martian surface and may occasionally flow there even now.

CONCEPT TEST

Which, if any, of these are solutions? (a) muddy river water; (b) helium gas; (c) clear cough syrup; (d) filtered dry air

(Answers to Concept Tests are in the back of the book.)

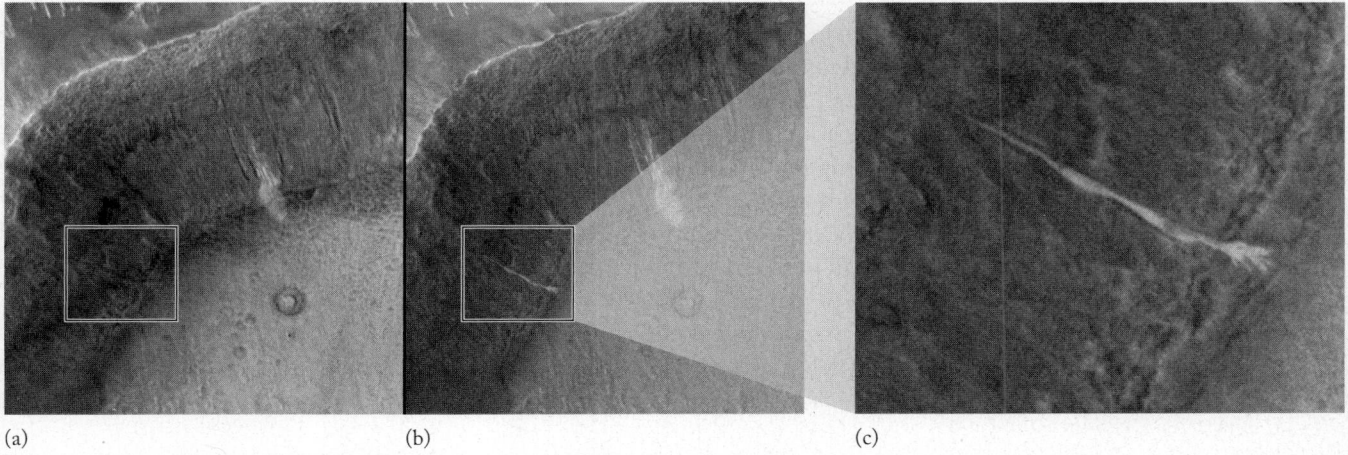

(a)   (b)   (c)

FIGURE 4.3 Photographs from *Mars Orbital Camera*. (a) Gully site in 2001. (b) Mosaic of two 2005 images of the same gully shows light-colored material that appears to have flowed into the gully. (c) Enlargement of the 2005 image.

Table sugar

Water

Homogeneous solution

FIGURE 4.4 Adding table sugar (the solute), represented by red spheres, to water (the solvent) produces a solution of sugar molecules evenly distributed among water molecules.

4.2 Concentration Units

The concentration of any solution can be expressed in different ways. Many are based on mass-to-mass ratios, such as milligrams of solute per kilogram of solution. Others are based on mass-to-volume ratios, such as milligrams of solute per liter of solution. For example, when environmental regulatory agencies establish limits[1] on the concentrations of contaminants permitted in drinking water, the limits may be expressed as milligrams of contaminant (solute) per liter of drinking water (solution), and when these agencies express the maximum recommended human exposure to these contaminants, a common unit is milligrams of contaminant per kilogram of body mass.

Two convenient units for expressing very small concentrations are *parts per million* (ppm) and *parts per billion* (ppb). A solution with a concentration of 1 ppm contains 1 part solute for every million parts of solution, which means 1 g of solute

[1] The U.S. Environmental Protection Agency (EPA) sets contaminant concentration limits called *maximum contaminant levels* (MCLs) for air and water, including drinking water. Similar limits have been established by Environment Canada, the World Health Organization, and other environmental agencies.

for every million (10^6) grams of solution. Put another way, each gram of a 1 ppm solution contains one-millionth of a gram (10^{-6} g = 1 μg) of solute. A solution of 1 ppm is also the same as 1 mg of solute per kilogram of solution:

$$1 \text{ ppm} = \frac{1 \text{ } \cancel{\mu\text{g solute}}}{1 \text{ } \cancel{\text{g solution}}} \times \frac{1 \text{ mg solute}}{10^3 \text{ } \cancel{\mu\text{g solute}}} \times \frac{10^3 \text{ } \cancel{\text{g solution}}}{1 \text{ kg solution}} = \frac{1 \text{ mg solute}}{1 \text{ kg solution}}$$

A concentration of 1 ppb is 1/1000 as concentrated as a concentration of 1 ppm. Since 1 ppm is the same as 1 mg of solute per kilogram of solution, 1 ppb is equivalent to 1 μg of solute per kilogram of solution:

$$1 \text{ ppb} = \frac{1 \text{ ppm}}{1000} = \frac{0.001 \text{ } \mu\text{g solute}}{1 \text{ } \cancel{\text{g solution}}} \times \frac{1000 \text{ } \cancel{\text{g solution}}}{1 \text{ kg solution}} = \frac{1 \text{ } \mu\text{g solute}}{1 \text{ kg solution}}$$

Table 4.1 lists the major ions in seawater and their concentrations using three different units. Let's express the smallest concentration in the table (0.00130 g F^-/kg seawater) in parts per million. To do so, we convert grams of F^- into milligrams of F^- because 1 mg of solute per kilogram of solution is the same as 1 ppm:

$$\frac{0.00130 \text{ } \cancel{\text{g F}^-}}{1 \text{ kg seawater}} \times \frac{1 \text{ mg F}^-}{0.001 \text{ } \cancel{\text{g F}^-}} = \frac{1.30 \text{ mg F}^-}{1 \text{ kg seawater}} = 1.30 \text{ ppm F}^-$$

As you can see, parts per million is a particularly convenient unit for expressing the concentration of F^- ion in seawater because it avoids the use of exponents or lots of zeroes to set the decimal place.

TABLE 4.1	Average Concentrations of the 11 Major Constituents of Seawater		
Constituent	g/kg[a]	mmol/kg[b]	mmol/L[c]
Na^+	10.781	468.96	480.57
K^+	0.399	10.21	10.46
Mg^{2+}	1.284	52.83	54.14
Ca^{2+}	0.4119	10.28	10.53
Sr^{2+}	0.00794	0.0906	0.0928
Cl^-	19.353	545.88	559.40
SO_4^{2-}	2.712	28.23	28.93
HCO_3^-	0.126	2.06	2.11
Br^-	0.0673	0.844	0.865
$B(OH)_3$	0.0257	0.416	0.426
F^-	0.00130	0.068	0.070
Total	35.169	1119.87	1147.59

[a] g/kg = grams of solute per kilogram of solution.
[b] mmol/kg = millimoles of solute per kilogram of solution.
[c] mmol/L = millimoles of solute per liter of solution.

SAMPLE EXERCISE 4.1 **Comparing Ion Concentrations in Aqueous Solutions**

The average concentration of chloride ion in seawater is 19.353 g Cl^-/kg solution. The World Health Organization recommends that the concentration of Cl^- ions in drinking water not exceed 250 ppm. How many times as much chloride ion is there in seawater than in the maximum concentrations allowed in drinking water?

Collect and Organize The concentrations of Cl^- ion are given in different units: g Cl^-/kg and ppm. We know how to create conversion factors from equalities. Our task is to determine the value of the ratio of the chloride ion concentration in seawater to its concentration in drinking water.

Analyze We can convert the seawater concentration to milligrams of Cl^- per kilogram of seawater using the conversion factor 10^3 mg/1 g and then use the fact that 1 mg solute/kg solution = 1 ppm. We expect the value in mg/kg to be much larger than the value given in g/kg.

Solve

$$19.353 \ \frac{\text{g } Cl^-}{\text{kg}} \times \frac{10^3 \text{ mg}}{\text{g}} = 19,353 \ \frac{\text{mg } Cl^-}{\text{kg}} = 19,353 \text{ ppm } Cl^-$$

Next we take the ratio of the two concentrations:

$$\frac{19,353 \text{ ppm } Cl^- \text{ in seawater}}{250 \text{ ppm } Cl^- \text{ in drinking water}} = 77.4$$

There is 77.4 times as much chloride ion in seawater as in acceptable drinking water.

Think about It The question posed could also be answered by converting the drinking water concentration to grams of Cl^- per kilogram of drinking water and comparing that number with 19.353 g Cl^-/kg seawater. It doesn't matter which units you choose to make the comparison as long as the units of the two numbers are the same. It makes sense that seawater is much saltier than drinking water. Drinking seawater induces nausea and vomiting, and may even cause death. To meet the drinking water needs of people in arid countries, seawater is distilled or desalinated by other means.

Practice Exercise The World Health Organization (WHO) drinking water standard for arsenic is 10.0 μg/L. Water from some wells in Bangladesh was found to contain as much as 1.2 mg of arsenic per liter of water. How many times above the WHO standard is this level? Assume that all the samples have a density of 1 g/mL.

(Answers to Practice Exercises are in the back of the book.)

∞ **CONNECTION** In Chapter 1 we saw an example of a solar still used in survival gear to generate freshwater from saltwater.

▶|| **CHEMTOUR** Molarity

molarity (M) the number of moles of solute divided by solution volume in liters: $M = n/V$. A 1.0 M solution contains 1.0 mol of solute per liter of solution; also known as *molar concentration*.

Most scientists interested in studying chemical processes in natural waters or laboratory solutions prefer to work with concentrations based on moles of solute rather than mass. For these scientists, the preferred concentration unit is moles of solute per liter of solution. This ratio is called **molarity (M)**.[2] A 1.0 molar (1.0 M) solution contains 1.0 mole of solute for every liter of solution:

---

[2] Note that molarity is symbolized by M while molar mass is $\mathcal{M}$ throughout this text.

$$\text{Molarity} = \frac{\text{moles of solute}}{\text{liters of solution}}$$

$$M = \frac{n}{V} \qquad\qquad (4.1)$$

where M is molarity, n is the number of moles of solute, and V is the volume of the solution in liters.

If we know V and M for any solution, we can readily calculate the mass of solute in the solution. First we rearrange Equation 4.1 to

$$n = V \times M$$

Next, we convert the number of moles of solute n into mass in grams by multiplying by the molar mass ($\mathcal{M}$):

$$\text{Mass of solute (g)} = n \times \mathcal{M}$$

Substituting $V \times M$ for n in the above equation yields

$$\text{Mass of solute (g)} = (V \times M) \times \mathcal{M} \qquad\qquad (4.2)$$

Equation 4.2 is useful when we need to calculate the mass of a solute needed to prepare a solution of a desired volume and molarity or when we have a solution of known molarity and want to know the mass of solute in a given volume of the solution.

In many environmental and biological systems, solute concentrations are often much less than 1.0 M. In Table 4.1, for instance, the concentration units in column 4 are not moles per liter but *millimoles* per liter, which means we are describing concentration in terms of *millimolarity* (mM, 1 m$M = 10^{-3}$ M) rather than molarity. On an even smaller scale, the concentrations of minor and trace elements in seawater are often expressed in terms of *micromolarity* (μM, 1 $\mu M = 10^{-6}$ M), *nanomolarity* (nM, 1 n$M = 10^{-9}$ M), and even *picomolarity* (pM, 1 p$M = 10^{-12}$ M). The concentration ranges of many biologically active substances in blood, urine, and other biological liquids are also so small that they are often expressed in units such as these.

Column 3 in Table 4.1 shows concentrations as millimoles of solute per kilogram of seawater. Oceanographers prefer this unit to one based on solution volume because the volume of a given mass of water varies with changing temperature and pressure, whereas its mass remains constant.

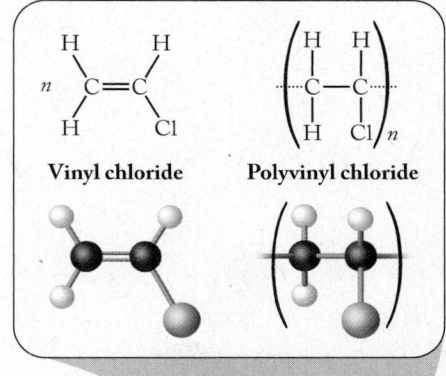

Vinyl chloride   **Polyvinyl chloride**

CONCEPT TEST

Identify which of these aqueous sodium chloride solutions is the least concentrated: (a) 0.0053 M NaCl; (b) 54 mM NaCl; (c) 550 μM NaCl; (d) 56,000 nM NaCl.

(Answers to Concept Tests are in the back of the book.)

SAMPLE EXERCISE 4.2   Calculating Molarity from Mass and Volume

Pipes made of polyvinyl chloride (PVC, Figure 4.5) are widely used in homes and office buildings. However, because vinyl chloride may leach from them, PVC pipes are only used for drain pipes, never for pipes bringing water into a building. The maximum concentration of vinyl chloride (CH_2CHCl) allowed in drinking water in the United States is 0.002 mg CH_2CHCl/L solution. What molarity is this?

FIGURE 4.5 Vinyl chloride, a suspected carcinogen, enters drinking water by leaching from pipes made of polyvinyl chloride (PVC). Vinyl chloride is the common name of a small molecule, called a monomer (Greek for "one unit"), from which the very large molecule called a polymer ("many units") is formed. The polymer PVC usually consists of hundreds of molecules of vinyl chloride bonded together.

Collect and Organize We are asked to convert a concentration from milligrams of solute per liter of solution to molarity, which means moles of solute per liter of solution. The molar mass of a substance relates the mass and number of moles of a given quantity of the substance.

Analyze Because the solute mass is given in milligrams, we need to convert this mass first to grams and then to moles. Because the number of grams is very small, we can expect that the molarity will be smaller still.

Solve The conversion factors we need are the molar mass of CH_2CHCl and the equality $1 \text{ g} = 10^3$ mg. The molar mass of vinyl chloride is

$$\mathcal{M} = 2(12.01 \text{ g C/mol}) + 3(1.008 \text{ g H/mol}) + 35.45 \text{ g Cl/mol} = 62.49 \text{ g/mol}$$

and the molarity is

$$\frac{0.002 \text{ mg}}{\text{L}} \times \frac{1 \text{ g}}{10^3 \text{ mg}} \times \frac{1 \text{ mol}}{62.49 \text{ g}} = \frac{3.2 \times 10^{-8} \text{mol}}{\text{L}} = 3 \times 10^{-8} M \text{ or } 0.03 \,\mu M$$

Think about It This concentration may seem very low, but remember that the amount of vinyl chloride (0.002 mg, or 2×10^{-6} g) in 1 L of water is very small. The extremely low limit of only 0.002 mg/L or 3×10^{-8} M reflects recognition of the danger vinyl chloride poses to human health.

Practice Exercise When 1.00 L of water from the surface of the Dead Sea is evaporated, 179 g of $MgCl_2$ is recovered. What is the molarity of $MgCl_2$ in the original sample?

(Answers to Practice Exercises are in the back of the book.)

SAMPLE EXERCISE 4.3   Calculating Molarity from Density

A water sample from the Great Salt Lake in Utah contains 83.6 mg of Na^+ per 1.000 g of water. What is the molarity of Na^+ if the density of the water is 1.160 g/mL?

Collect and Organize Our task is to convert concentration units from mg of Na^+ per gram of solution to moles of Na^+ per liter of solution.

Analyze To convert this concentration into molarity, we need to convert the solute mass into moles and the solution mass into a volume in liters. To convert milligrams of Na^+ to moles, we need the molar mass of sodium ion: 22.99 g/mol. For the solution, we need to convert its mass in grams into a volume in milliliters by dividing by the density given in the problem and then convert milliliters into liters.

Solve For moles of solute, we have

$$83.6 \text{ mg Na}^+ \times \frac{1 \text{ g}}{1000 \text{ mg}} \times \frac{1 \text{ mol Na}^+}{22.99 \text{ g Na}^+} = 3.64 \times 10^{-3} \text{ mol Na}^+$$

The volume of solution is

$$1.000 \text{ g} \times \frac{1 \text{ mL}}{1.160 \text{ g}} \times \frac{1 \text{ L}}{1000 \text{ mL}} = 8.621 \times 10^{-4} \text{ L}$$

CONNECTION In Chapter 1 we introduced density, the ratio of mass of a quantity of material to its volume or $d = m/V$, as an intensive physical property.

The molarity of sodium ion in the water from the Great Salt Lake is

$$\frac{mol}{L} = \frac{3.64 \times 10^{-3} \text{ mol Na}^+}{8.621 \times 10^{-4} \text{ L}} = 4.22 \ M$$

Think about It According to Table 4.1, second column, the Na^+ concentration in seawater is about 10 g of Na^+ per kilogram of seawater, or about 10 mg of Na^+ per gram of seawater, which is approximately 10 mg/mL assuming the density of the water is about 1.0 g/mL. The concentration given for Na^+ in the Great Salt Lake, 83.6 mg/g lake water, is more than 8 times the seawater concentration, so the molarity should be more than 8 times the millimolar concentration given in column 4 of Table 4.1 (480.57 mM ≈ 0.48 M). Our answer, 4.22 M, is reasonable in light of that comparison. The Great Salt Lake is the result of evaporation of ocean water over thousands of years. Evaporation of the solvent from any solution reduces the volume of solution and increases its concentration because the amount of solute (n) is unchanged.

Practice Exercise If the density of ocean water at a depth of 10,000 m is 1.071 g/mL and if 25.0 g of water at that depth contains 190 mg of potassium chloride, what is the molarity of potassium chloride in the sample?

(Answers to Practice Exercises are in the back of the book.)

SAMPLE EXERCISE 4.4 **Calculating the Quantity of Solute Needed to Prepare a Solution of Known Concentration**

In setting up a saltwater aquarium, you need an aqueous solution called Kalkwasser, which has a calcium hydroxide concentration of 0.0225 M. How many grams of $Ca(OH)_2$ do you need to make 500.0 mL of Kalkwasser?

Collect and Organize We know the volume and molarity of the solution we must prepare. We also know the identity of the solute, which enables us to calculate its molar mass.

Analyze We use Equation 4.2 to calculate the mass of $Ca(OH)_2$ needed. We also must convert the volume in milliliters to liters to obtain the mass in grams of solute required. The molar mass of $Ca(OH)_2$ is

$$40.08 \text{ g/mol} + 2(16.00 \text{ g/mol}) + 2(1.008 \text{ g/mol}) = 74.10 \text{ g/mol}$$

and 500.0 mL is 0.5000 L.

Solve Equation 4.2 gives us the mass of $Ca(OH)_2$ needed:

$$\text{Grams } Ca(OH)_2 = \mathcal{M} \times V \times M$$

$$= \frac{74.10 \text{ g}}{\text{mol}} \times 0.5000 \text{ L} \times \frac{0.0225 \text{ mol}}{\text{L}} = 0.834 \text{ g}$$

Figure 4.6 shows the technique for making this solution. A key feature is that we dissolve 0.834 g of solute in 200–300 mL of water and then add additional water until the final solution has the volume specified.

Think about It To solve a problem like this, all we need to remember is the definition of molarity: moles of solute per liter of solution. The solution we want has a concentration of 0.0225 M, or 0.0225 mol/L. Because we only need

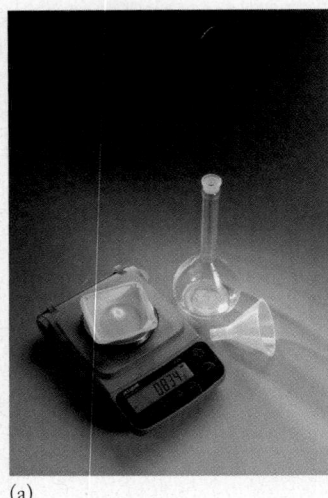
(a)

(b)

(c)
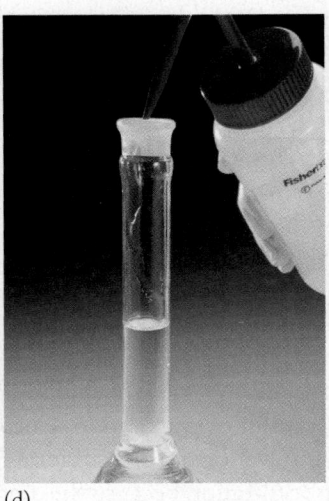
(d)

FIGURE 4.6 Preparing 500.0 mL of 0.0225 M Ca(OH)$_2$. (a) Weigh out the desired quantity of solid Ca(OH)$_2$; (b) transfer the Ca(OH)$_2$ to a 500 mL volumetric flask and add 200 to 300 mL of water; (c) swirl to dissolve all the solute; (d) dilute to exactly 500 mL. Stopper the flask and invert it several times to thoroughly mix the solution.

0.5000 L of this solution, we do not need 0.0225 moles of Ca(OH)$_2$ but only half that amount (0.0112 mol). Finally, we calculate the number of grams of calcium hydroxide in 0.0112 moles of calcium hydroxide (0.834 g). The amount of Ca(OH)$_2$ needed is less than a teaspoon.

Practice Exercise An aqueous solution known as Ringer's lactate is administered intravenously to trauma victims suffering from blood loss or severe burns. The solution contains the chloride salts of sodium, potassium, and calcium and is 4.00 mM in sodium lactate (NaC$_3$H$_5$O$_3$). How many grams of sodium lactate are needed to prepare 10.0 L of Ringer's lactate?

(Answers to Practice Exercises are in the back of the book.)

4.3 Dilutions

In laboratories such as those that test drinking water quality, **stock solutions** of substances, such as pesticides and toxic metals, are available commercially in concentrations too high to be used directly in analytical procedures. In the laboratory, a stock solution is diluted to prepare a solution with a solute concentration of the substance close to its concentration in the sample being tested. **Dilution** is the process of lowering the concentration of a solution by adding solvent to a known volume of the initial solution. The stock solution is described as being *concentrated*, because it contains a much larger solute to solvent ratio than does the *dilute* solution. A dilute solution contains a very small amount of solute compared to the amount of solvent. These are relative terms—what constitutes a large amount of solute versus a small amount?—but they are used constantly to compare solutions containing different quantities of solutes.

We can use the definition of molarity given in Equation 4.1, $M = n/V$, to determine how to dilute a stock solution to make a solution with a desired concentration.

▶❚❚ **CHEMTOUR** Dilution

stock solution a concentrated solution of a substance used to prepare solutions of lower concentration.

dilution the process of lowering the concentration of a solution by adding more solvent.

Suppose we need to prepare 250.0 mL of an aqueous solution that is 0.0100 M in Cu^{2+} by starting with a stock solution that is 0.1000 M in Cu^{2+}. What volume of the stock solution do we need?

Rearranging Equation 4.1 to solve for n gives the moles of Cu^{2+} needed in the dilute solution we want to make:

$$n_{diluted} = V_{diluted} \times M_{diluted}$$
$$= (0.2500 \text{ L})(0.0100 \text{ mol/L}) = 2.50 \times 10^{-3} \text{ mol } Cu^{2+}$$

The subscript "diluted" refers to this solution. Using the same equation, we can calculate the *initial* quantities, that is, what volume of the stock solution we need to deliver 2.50×10^{-3} moles of Cu^{2+}:

$$n_{initial} = V_{initial} \times M_{initial}$$

$$V_{initial} = \frac{n_{initial}}{M_{initial}} = \frac{2.50 \times 10^{-3} \text{ mol}}{0.1000 \text{ mol/L}} = 2.50 \times 10^{-2} \text{ L} = 25.0 \text{ mL}$$

To make the diluted solution, we add enough water to 25.0 mL of the stock solution to make 250.0 mL of the solution we want (Figure 4.7).

Because the number of moles of solute Cu^{2+} ions is the same in the diluted solution and in the portion of stock solution required ($n_{diluted} = n_{initial}$), we can combine these two equations into one that applies to all situations in which we know any three of these four variables: the initial volume $V_{initial}$ of stock solution, the initial solute concentration $M_{initial}$, the volume $V_{diluted}$ of the diluted solution, and the solute concentration $M_{diluted}$ in the diluted solution:

$$V_{initial} \times M_{initial} = n_{initial} = n_{diluted} = V_{diluted} \times M_{diluted}$$

or simply

$$V_{initial} \times M_{initial} = V_{diluted} \times M_{diluted} \qquad (4.3)$$

Equation 4.3 works because each side of the equation represents a quantity of solute that does not change because of dilution. Furthermore, this equation can be used for *any* units of volume and concentration, as long as the units used to express the initial and diluted volumes are the same and the units for the initial and diluted concentrations are the same.

Equation 4.3 applies to systems much larger than a laboratory stock solution to be diluted. For instance, in estuaries, where rivers flow into the sea, salinity is reduced because the seawater is diluted by the freshwater that enters it. Suppose a coastal bay contains 3.8×10^{12} L of ocean water mixed with 1.2×10^{12} L of river water for a total volume of 5.0×10^{12} L. We can calculate the Na^+ concentration in the bay if we know that the Na^+ concentration in the open ocean is 0.48 M and if we assume that the river water does not contribute significantly to the Na^+ content in the bay. Solving Equation 4.3 for $M_{diluted}$, the Na^+ concentration in the bay, and inserting the given values, we have

$$M_{diluted} = \frac{V_{initial} \times M_{initial}}{V_{diluted}} = \frac{(3.8 \times 10^{12} \text{ L}) \times 0.48 \text{ } M}{5.0 \times 10^{12} \text{ L}} = 0.36 \text{ } M$$

The bay is less salty than the ocean because it is diluted by the freshwater from the river.

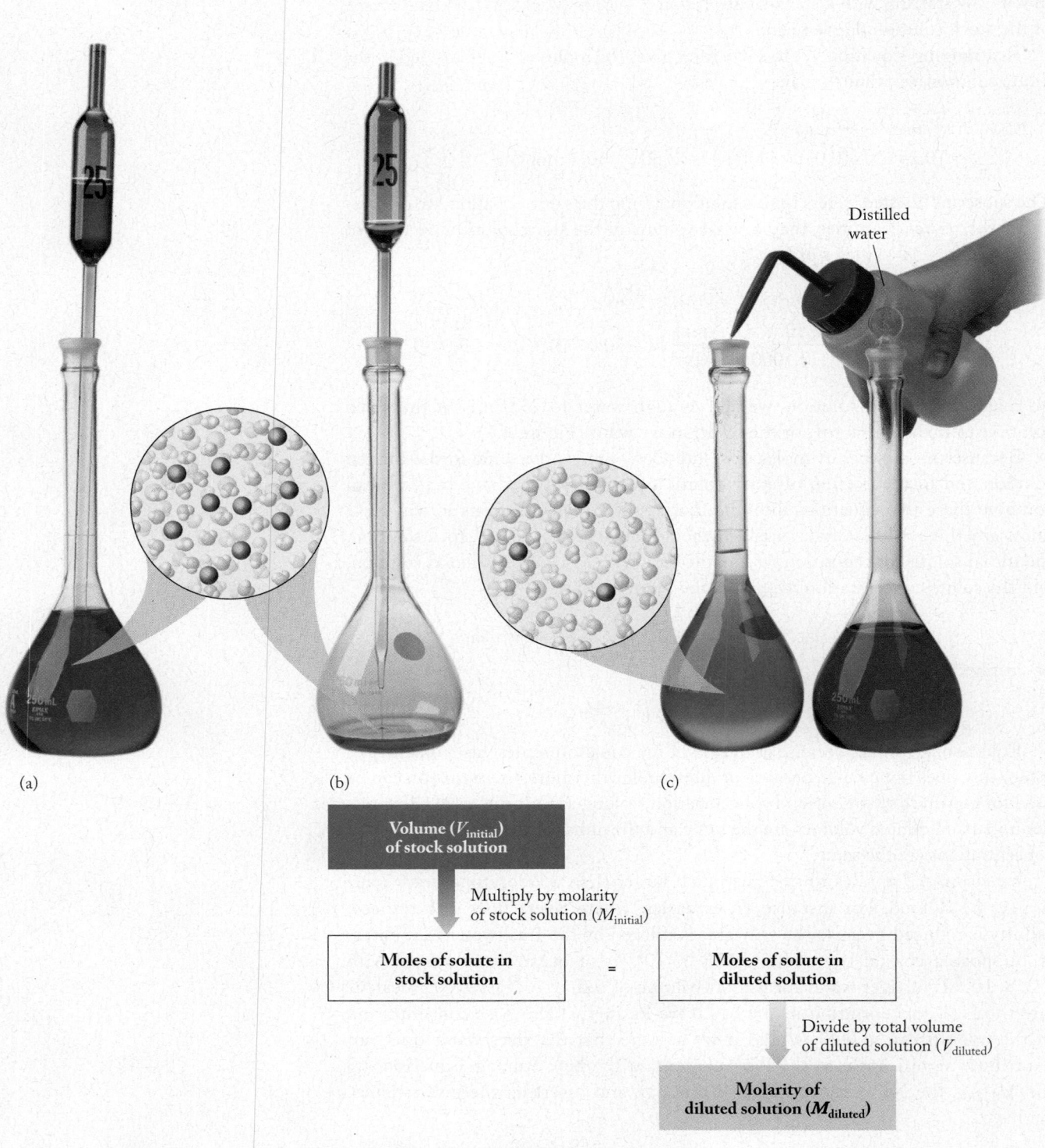

(a) (b) (c)

Distilled water

Volume ($V_{initial}$) of stock solution

Multiply by molarity of stock solution ($M_{initial}$)

| Moles of solute in stock solution | = | Moles of solute in diluted solution |

Divide by total volume of diluted solution ($V_{diluted}$)

Molarity of diluted solution ($M_{diluted}$)

FIGURE 4.7 To prepare 250.0 mL of a solution that is 0.0100 M in Cu^{2+}, (a) a pipet is used to withdraw 25.0 mL of a 0.1000 M stock solution. (b) This volume of stock solution is transferred to a 250.0 mL volumetric flask. (c) Distilled water is added to bring the volume of the diluted solution to 250.0 mL.

SAMPLE EXERCISE 4.5 Calculating Dilutions

The solution used in hospitals for intravenous infusion—called *physiological saline* or *saline solution*—is 0.155 M in NaCl. It is typically prepared by diluting a stock solution, the concentration of which is 1.76 M, with water. What volume of stock solution is required to prepare 60.0 L of physiological saline?

Collect and Organize We know the volume ($V_{diluted}$ = 60.0 L) and concentration ($M_{diluted}$ = 0.155 M) of the diluted solution and the concentration of stock solution available ($M_{initial}$ = 1.76 M).

Analyze We must first find $V_{initial}$, the volume of the stock solution required to make 60.0 L of physiological saline.

Solve Solving Equation 4.3 for $V_{initial}$ and using the values given:

$$V_{initial} = \frac{V_{diluted} \times M_{diluted}}{M_{initial}} = \frac{60.0 \text{ L} \times 0.155 \ \cancel{M}}{1.76 \ \cancel{M}} = 5.28 \text{ L}$$

Think about It Because the concentration of the stock solution is about ten times the concentration of the diluted solution, the volume of stock solution required should be about one-tenth the final volume. Our result of 5.28 L is reasonable.

Practice Exercise The concentration of Pb^{2+} in a stock solution is 1.000 mg/mL. What volume of this solution should be diluted to 500.0 mL to produce a solution in which the Pb^{2+} concentration is 0.0575 mg/L?

(Answers to Practice Exercises are in the back of the book.)

CONCEPT TEST

Why is a stock solution always more concentrated than the solutions made from it?

CONCEPT TEST

Figure 4.8 shows several solutions of red cough syrup dissolved in water. Order these solutions, from the most dilute to the most concentrated.

(a)   (b)   (c)   (d)   (e)

FIGURE 4.8 Aqueous solutions of cough syrup.

(Answers to Concept Tests are in the back of the book.)

▶❙❙ **CHEMTOUR** Migration of Ions in Solution

4.4 Electrolytes and Nonelectrolytes

The high concentrations of NaCl and other salts in seawater (Table 4.1) make it a good conductor of electricity. Suppose we immerse two *electrodes* (solid conductors) in distilled water (Figure 4.9a) and connect them to a battery and a lightbulb. For electricity to flow and the bulb to light up, the circuit must be completed by mobile charge carriers (ions) in the solution. The lightbulb does not light when the electrodes are placed in distilled water because there are very few ions in distilled water. When the electrodes are pressed into solid NaCl (Figure 4.9b), no electric current flows and the bulb does not light because the Na^+ and Cl^- ions in solid NaCl are fixed in position. However, when the electrodes are immersed in aqueous 0.50 M NaCl (Figure 4.9c), the bulb lights because the many mobile ions in the solution act as charge carriers between the two electrodes. The saline solution conducts electricity because Na^+ ions are attracted to and migrate toward the electrode connected to the negative terminal of the battery and Cl^- ions are attracted to and migrate toward the positive electrode, completing an electric circuit.

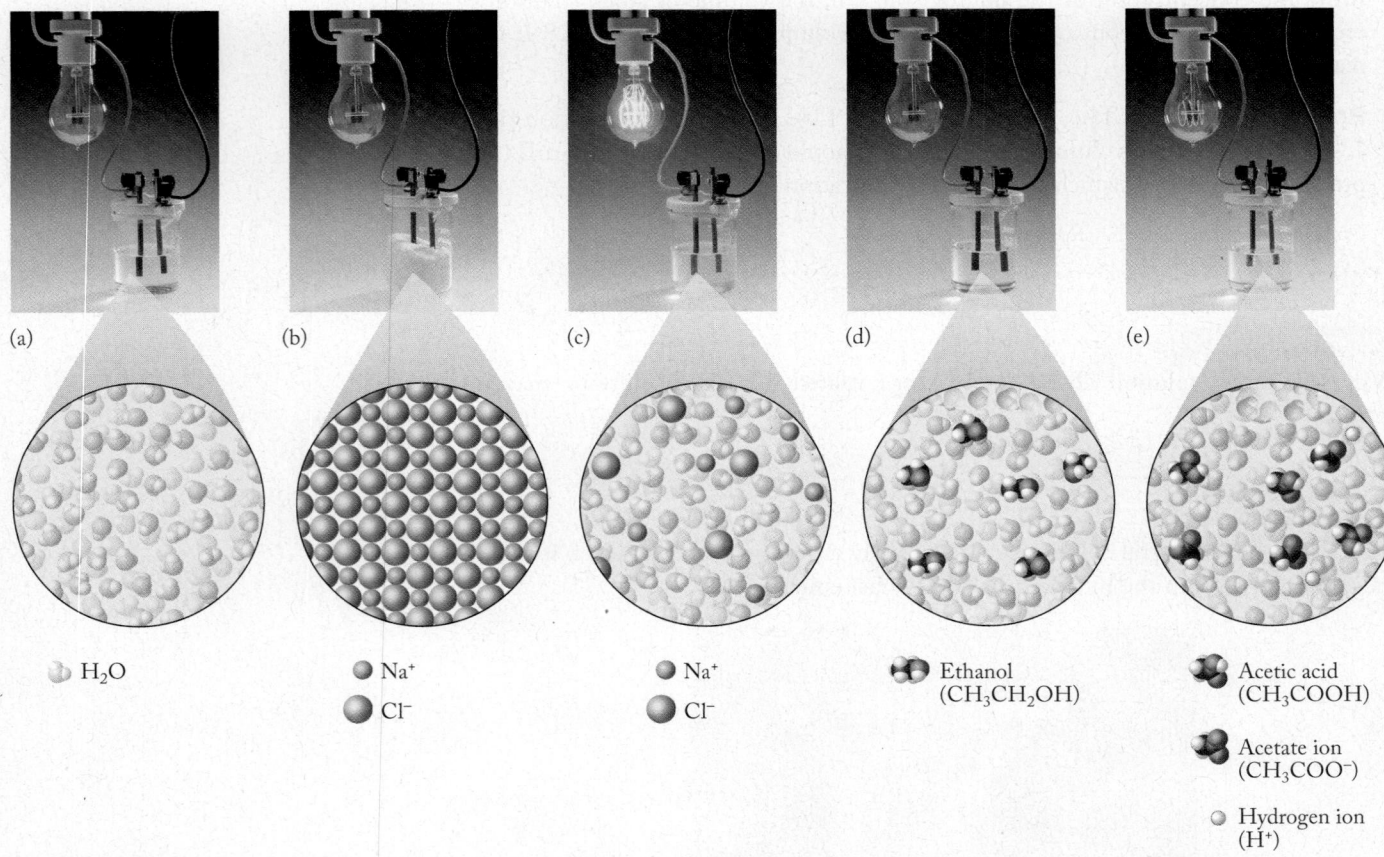

(a)   (b)   (c)   (d)   (e)

🝔 H_2O

● Na^+
● Cl^-

● Na^+
● Cl^-

🝔 Ethanol (CH_3CH_2OH)

🝔 Acetic acid (CH_3COOH)

🝔 Acetate ion (CH_3COO^-)

○ Hydrogen ion (H^+)

FIGURE 4.9 (a) The unlit lightbulb indicates that pure water (H_2O) conducts electricity very poorly; because water is a molecular material, it contains very few ions. (b) The unlit lightbulb indicates that solid sodium chloride (NaCl) does not conduct electricity. (c) The brightly lit bulb indicates that a 0.50 M solution of NaCl conducts electricity very well. (d) The unlit bulb indicates that a 0.50 M solution of ethanol (CH_3CH_2OH) does not improve the conductivity of water. (e) The dimly lit bulb indicates that a 0.50 M solution of acetic acid (CH_3COOH) conducts electricity better than pure water or the ethanol solution but to a much smaller extent than the NaCl solution.

Any solute that imparts electrical conductivity to an aqueous solution is called an **electrolyte**. Sodium chloride is considered a **strong electrolyte** because it *dissociates* completely in water, which means it completely breaks up into its component ions when it dissolves, releasing charge-carrying Na^+ cations and Cl^- anions.

Many substances do not dissociate when dissolved in water. For instance, a 0.50 *M* solution of ethanol (CH_3CH_2OH) in water conducts electricity no better than pure water because each CH_3CH_2OH molecule stays intact in the solution and no ions are formed (Figure 4.9d). Solutes that do not form ions when dissolved in water are called **nonelectrolytes**.

Besides strong electrolytes (complete dissociation into ions) and nonelectrolytes (no dissociation), **weak electrolytes** also exist. Weak electrolytes dissociate partially when dissolved in water. One example of a weak electrolyte is acetic acid, CH_3COOH. In Figure 4.9e, the beaker contains a 0.50 *M* aqueous solution of acetic acid; the acetic acid molarity is equivalent to that of the solutes in parts c and d of the figure. Note that the bulb connected to the acetic acid solution lights but not very brightly. The acetic acid molecules dissociate to some extent, forming hydrogen ions and acetate ions, but the process does not go to completion:

$$CH_3COOH(aq) \rightleftharpoons H^+(aq) + CH_3COO^-(aq)$$

The arrow with the top half pointing right and the bottom half pointing left indicates that all species are present in solution: undissociated CH_3COOH, hydrogen ion, and acetate ion. This type of arrow is used to indicate that the dissociation process is not complete.

CONCEPT TEST

Why were equivalent molar concentrations used in the solutions in Figure 4.9?

(Answers to Concept Tests are in the back of the book.)

electrolyte a substance that dissociates into ions when it dissolves, enhancing the conductivity of the solvent.

strong electrolyte a substance that dissociates completely into ions when it dissolves in water.

nonelectrolyte a substance that does not dissociate into ions and therefore does not enhance the conductivity of water when it dissolves.

weak electrolyte a substance that only partly dissociates into ions when it dissolves in water.

hydronium ion (H_3O^+) an H^+ ion plus a water molecule, H_2O; the form in which the hydrogen ion is found in an aqueous solution.

4.5 Acid-Base Reactions: Proton Transfer

When oxides of sulfur, nitrogen, and other nonmetals dissolve in water, they produce hydrogen ions in solution. Examples of these processes are

$$N_2O_5(g) + H_2O(\ell) \rightarrow 2H^+(aq) + 2NO_3^-(aq)$$

$$SO_3(g) + H_2O(\ell) \rightarrow H^+(aq) + HSO_4^-(aq)$$

We focus in this section on the processes that occur in solution when hydrogen ions are produced.

Let's look at the behavior of a common binary acid: hydrochloric acid (HCl). Pure HCl is a molecular compound and a gas. However, when it dissolves in water it ionizes completely, producing $H^+(aq)$ and $Cl^-(aq)$. Generally we use $H^+(aq)$ to describe the cation produced in an aqueous medium by an acid. However, hydrogen ions (H^+, which actually are protons, the extremely small nuclei of hydrogen atoms) do not have an independent existence in water because each proton combines with a water molecule to form a **hydronium ion (H_3O^+)**:

$$H^+(aq) + H_2O(\ell) \rightarrow H_3O^+(aq)$$

CONNECTION In Chapter 3, we described the hydrolysis reactions of nonmetal oxides that produce acidic solutions.

CONNECTION We discussed the naming of binary acids in Section 2.6.

acid (Brønsted-Lowry acid) a proton donor.

base (Brønsted-Lowry base) a proton acceptor.

neutralization reaction a reaction that takes place when an acid reacts with a base and produces a solution of a salt in water.

salt the product of a neutralization reaction; it is made up of the cation of the base in the reaction plus the anion of the acid.

In aqueous acid–base chemistry, hydrogen ion, proton, and hydronium ion all refer to the same species.

When a molecule of HCl dissolves, it donates a proton to a water molecule:

$$HCl(g) \quad + \quad H_2O(\ell) \quad \rightarrow \quad H_3O^+(aq) + Cl^-(aq) \quad (4.4)$$
$$\underset{\text{(acid)}}{\text{proton donor}} \quad \underset{\text{(base)}}{\text{proton acceptor}}$$

This behavior fits our definition of an **acid** as a *proton donor*. Because the water molecule accepts the proton, water in this case is a *proton acceptor*, which is the corresponding definition of a **base**. These definitions originally were proposed by chemists Johannes Brønsted (1879–1947) and Thomas Lowry (1874–1936), and the acids and bases so identified are called **Brønsted–Lowry acids** and **Brønsted–Lowry bases**. We will develop this concept more completely later, but for now it provides a convenient way to define acids and bases in aqueous solutions.

A base may also be defined as a substance that produces hydroxide ions in an aqueous solution. With the Brønsted–Lowry definition of a base as a proton acceptor, the hydroxide ion is still classified as a base, but other substances can be categorized as bases as well.

In the reaction between aqueous solutions of HCl and NaOH,

$$HCl(aq) + NaOH(aq) \rightarrow NaCl(aq) + H_2O(\ell) \quad (4.5)$$

the HCl functions as an acid by donating a proton to the hydroxide ion, and the hydroxide ion functions as a base by accepting the proton; the H^+ and OH^- ions combine to form a molecule of water. The remaining two ions—the acid anion Cl^- and the base cation Na^+—form a salt, sodium chloride, which remains dissolved in the aqueous solution. This reaction is called a **neutralization reaction** because the acid and the base have been eliminated as a result of the reaction, leaving a chemically *neutral* (neither acidic nor basic) solution. The products of the neutralization of an acid like HCl with a base like NaOH are always water and a salt. This provides us with the definition of a **salt** as a substance formed along with water as a product of a neutralization reaction.

CONCEPT TEST

Which drawing in Figure 4.10 describes the particles in an aqueous solution of an acid that is a weak electrolyte and which describes the particles in an aqueous solution of an acid that is a strong electrolyte?

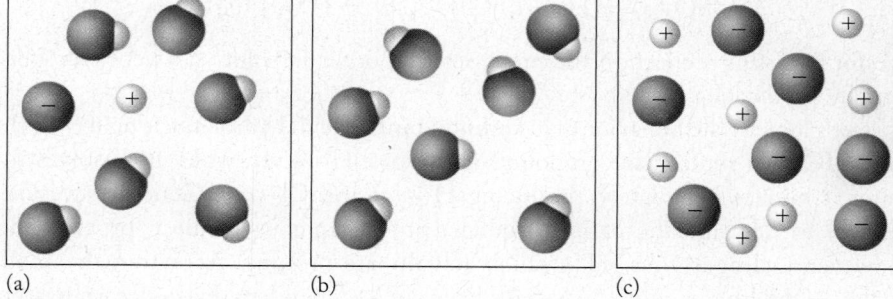

+ Hydrogen ion

− Anion

(a)   (b)   (c)

FIGURE 4.10 Aqueous solutions of binary compounds.

(Answers to Concept Tests are in the back of the book.)

Match each of the substances NaCl, HCl, NaOH, and CH₃COOH (acetic acid) with as many of the words describing their behavior in aqueous solution as you can: (a) strong electrolyte; (b) weak electrolyte; (c) nonelectrolyte; (d) acid; (e) neutral; (f) base.

(Answers to Concept Tests are in the back of the book.)

Equation 4.5 is written as a **molecular equation**, meaning each reactant and product is written as a neutral compound. Molecular equations are sometimes the easiest equations to balance, so reactions involving substances that dissociate in solution are often written first as molecular equations to indicate each reactant and product and to simplify balancing the equation.

Another equation representing the reaction in Equation 4.5 is one that shows each reactant and product the way it is present in solution, which means any species that dissociates completely is written as a cation/anion pair:

$$\underbrace{HCl(aq)}_{H^+(aq) + Cl^-(aq)} + \underbrace{NaOH(aq)}_{Na^+(aq) + OH^-(aq)} \rightarrow \underbrace{NaCl(aq)}_{Na^+(aq) + Cl^-(aq)} + H_2O(\ell) \quad (4.6)$$

This form, called an **overall ionic equation**, distinguishes ionic substances from molecular substances in a chemical reaction taking place in solution.

Notice in Equation 4.6 that chloride ions and sodium ions appear on both sides of the reaction arrow. If this were an algebraic equation, we would remove the terms that are the same on both sides. When we do the same thing in chemistry, the result is a **net ionic equation**, as written in Equation 4.7. The ions that are removed (in this case, sodium ion and chloride ion) are **spectator ions**, and the resulting equation shows only the species taking part in the reaction:

$$H^+(aq) + OH^-(aq) \rightarrow H_2O(\ell) \quad (4.7)$$

The spectator ions are unchanged by the reaction and remain in solution. The chemical change in a neutralization reaction occurs when a hydrogen ion reacts with a hydroxide ion to produce a molecule of water. The net ionic equation enables us to focus on those species that actually participate in the reaction.

SAMPLE EXERCISE 4.6 Writing Neutralization Reaction Equations

Write the balanced (a) molecular, (b) overall ionic, and (c) net ionic equations that describe the reaction that takes place when an aqueous solution of sulfuric acid is neutralized by an aqueous solution of potassium hydroxide.

Collect and Organize Solutions of sulfuric acid and potassium hydroxide take part in a neutralization reaction. We know that potassium hydroxide (KOH) is the base and sulfuric acid (H_2SO_4) is the acid. The products of a neutralization reaction are water and a salt. Our task is to write three different equations for the neutralization reaction.

molecular equation a balanced equation that describes a reaction in solution in which the reactants are written as undissociated molecules.

overall ionic equation a balanced equation that shows all the species, both ionic and molecular, present in a reaction occurring in aqueous solution.

net ionic equation a balanced equation that describes the actual reaction taking place in aqueous solution; it is obtained by eliminating the spectator ions from the overall ionic equation.

spectator ion an ion that is present in a reaction vessel when a chemical reaction takes place but is unchanged by the reaction; spectator ions appear in an overall ionic equation but not in a net ionic equation.

Analyze (a) All species are written as neutral compounds in the molecular equation. The products are water and the salt potassium sulfate, made from the cation of the base and the anion of the acid. (b) We can write each substance in the balanced molecular equation in ionic form to generate the overall ionic equation. (c) Removing spectator ions from the overall ionic equation gives us the net ionic equation.

Solve

a. The molecular equation is

$$H_2SO_4(aq) + KOH(aq) \rightarrow K_2SO_4(aq) + H_2O(\ell) \qquad \text{(unbalanced)}$$
$$H_2SO_4(aq) + 2\,KOH(aq) \rightarrow K_2SO_4(aq) + 2\,H_2O(\ell) \qquad \text{(balanced)}$$

b. The overall ionic equation is

$$2\,H^+(aq) + SO_4^{2-}(aq) + 2\,K^+(aq) + 2\,OH^-(aq) \rightarrow$$
$$2\,K^+(aq) + SO_4^{2-}(aq) + 2\,H_2O(\ell)$$

c. The spectator ions are K^+ and SO_4^{2-}. Removing them gives us the net ionic equation:

$$2\,H^+(aq) + 2\,OH^-(aq) \rightarrow 2\,H_2O(\ell)$$

or

$$H^+(aq) + OH^-(aq) \rightarrow H_2O(\ell)$$

Think about It We treat chemical equations just like algebraic equations and cancel out spectator ions that are unchanged in the reaction and appear on both sides of the reaction arrow in the *overall* ionic equation. After canceling the spectator ions, we are left with the net ionic equation, which focuses on the species that were changed as a result of the reaction. It is also important to remember that sulfate is a polyatomic ion (see Table 2.3) that retains its identity in solution and does not separate into atoms and ions. The net ionic equation is the same as in the reaction of an aqueous solution of HCl with aqueous NaOH.

Practice Exercise Write balanced (a) molecular, (b) overall ionic, and (c) net ionic equations for the reaction between an aqueous solution of phosphoric acid, $H_3PO_4(aq)$, and an aqueous solution of sodium hydroxide. The products are sodium phosphate and water.

(Answers to Practice Exercises are in the back of the book.)

FIGURE 4.11 In Carlsbad Caverns in New Mexico, stalagmites of limestone grow up from the cavern floor and stalactites grow downward from the ceiling.

For billions of years, neutralization reactions have played key roles in the chemical transformations of Earth's crust that geologists call *chemical weathering*. One type of chemical weathering occurs when carbon dioxide dissolves in rainwater to create a weakly acidic solution of carbonic acid, $H_2CO_3(aq)$. This solution dissolves calcium carbonate, which occurs as chalk, limestone, and marble and is insoluble in pure water. The reaction between carbonic acid and calcium carbonate is responsible for the formation of stalactites and stalagmites (Figure 4.11). This reaction and others caused by even more acidic rain are also responsible for the degradation of statues and the exteriors of buildings made of marble (Figure 4.12). Table 4.2 lists some common nonmetal oxides that are classified as atmospheric pollutants because they produce acid rain.

TABLE 4.2	Volatile Nonmetal Oxides and Their Acids	
Oxide	**Acid**
SO_2	H_2SO_3
SO_3	H_2SO_4
NO_2	HNO_2, HNO_3
N_2O_5	HNO_3
CO_2	H_2CO_3

1935

1994

FIGURE 4.12 Atmospheric sulfuric acid, made when $SO_3(g)$ dissolves in rainwater, attacks marble statues and converts the calcium carbonate to calcium sulfate, which is more soluble and is slowly washed away by rain and melting snow.

Nonmetal oxides other than carbon dioxide form solutions that are much more acidic than a solution of carbonic acid. Dinitrogen pentoxide, for example, forms nitric acid in water:

$$N_2O_5(g) + H_2O(\ell) \rightarrow 2\,HNO_3(aq) \qquad HNO_3(aq) \rightarrow H^+(aq) + NO_3^-(aq)$$

and sulfur trioxide forms sulfuric acid:

$$SO_3(g) + H_2O(\ell) \rightarrow H_2SO_4(aq) \qquad H_2SO_4(aq) \rightarrow H^+(aq) + HSO_4^-(aq)$$

Nitric acid and sulfuric acid both are identified as **strong acids**, because they dissociate completely in water. They and other common strong acids are listed in Table 4.3. Because one molecule of nitric acid donates one proton, nitric acid is called a *monoprotic acid*. Hydrochloric acid is another strong monoprotic acid. Sulfuric acid can potentially donate two protons and so is a *diprotic acid*. However, in some H_2SO_4 solutions only one of the two H atoms in each molecule ionizes. This behavior means that sulfuric acid is a strong acid in terms of donation of the first proton. This complete dissociation results in the formation of the $HSO_4^-(aq)$ ion (hydrogen sulfate), which is also an acid because in solution some HSO_4^- ions dissociate to H^+ and SO_4^{2-}. However, because not all HSO_4^- ions dissociate, the hydrogen sulfate anion is a **weak acid**, defined as one that dissociates only partially in water.

The two equations for the dissociation of the two protons of sulfuric acid are written to reflect this behavior. First, the dissociation of the strong acid H_2SO_4:

$$H_2SO_4(aq) \rightarrow H^+(aq) + HSO_4^-(aq) \qquad (\rightarrow \text{means complete ionization})$$

Second, the dissociation of the weak acid HSO_4^-:

$$HSO_4^-(aq) \rightleftharpoons H^+(aq) + SO_4^{2-}(aq) \qquad (\rightleftharpoons \text{means incomplete ionization})$$

For any diprotic acid, the HX^- anion formed when the first proton dissociates is a weaker acid than the original acid H_2X. The pattern continues for triprotic acids (H_3X) and other *polyprotic acids* as well: in terms of their strength as acids, the proton donors in a triprotic acid are in the order $H_3X > H_2X^- > HX^{2-}$.

TABLE 4.3	Strong Acids
Acid	**Molecular Formula**
Hydrochloric acid	HCl
Hydrobromic acid	HBr
Hydroiodic acid	HI
Nitric acid	HNO_3
Sulfuric acid	H_2SO_4
Perchloric acid	$HClO_4$

strong acid an acid that completely dissociates into ions in aqueous solution.

weak acid an acid that only partially dissociates in aqueous solution and so has a limited capacity to donate protons to the medium.

A molecule called EDTA has four acidic protons. We can symbolize it as H_4E to highlight this property. Pick the more acidic member of each pair listed: (a) H_2E^{2-} or HE^{3-}; (b) H_4E or H_3E^-.

(Answers to Concept Tests are in the back of the book.)

Strong acids are strong electrolytes, and weak acids are weak electrolytes. A balance, called a *dynamic equilibrium,* is achieved in solutions of weak electrolytes, at which point the concentrations of the reactants and the products in the dissociation reaction do not change. Solutions of weak electrolytes, such as weak acids, are characterized by dissociated and undissociated species existing together in dynamic equilibrium.

In a classification scheme similar to that used for acids, bases are classified as **strong bases** or **weak bases** depending on the extent to which they dissociate in aqueous solution. Strong bases include the hydroxides of groups 1 and 2 metals, which dissociate completely when dissolved in water. Table 4.4 lists some common basic minerals that are weak bases. Another example of a weak base is ammonia (NH_3). Ammonia in its molecular form is a gas at room temperature. When NH_3 dissolves in water it produces a solution that conducts electricity weakly, which means that ammonia is a weak electrolyte:

$$NH_3(g) \quad + \quad H_2O(\ell) \; \rightleftharpoons \; NH_4^+(aq) + OH^-(aq) \qquad (4.8)$$

$$\underset{\text{(proton acceptor)}}{\text{base}} \qquad \underset{\text{(proton donor)}}{\text{acid}}$$

Note that water behaves as a base in Equation 4.4 and as an acid in Equation 4.8. Is water an acid or a base? The answer depends on what is dissolved in it. Water is called an **amphiprotic** substance because it can function as either a proton acceptor (Brønsted–Lowry base) in solutions of acidic solutes, or as a proton donor (Brønsted–Lowry acid) in solutions of basic solutes. In the hydrolysis of a nonmetal oxide like SO_3, water functions as an acid; in the hydrolysis of ammonia, it functions as a base.

Which, if any, of these species is amphiprotic? (a) H_2SO_4; (b) HSO_4^-; (c) SO_4^{2-}

(Answers to Concept Tests are in the back of the book.)

4.6 Titrations

Our understanding of the reactions of acids and bases can be used to analyze solutions and determine the concentration of dissolved substances. An approach called **titration** is a common analytical method based on measured volumes of reactants, and acid–base neutralization reactions are frequently the basis of analyses done by titration.

One use of titrations is in the analysis of water draining from abandoned coal mines. Sulfide-containing minerals called pyrites may be exposed when coal is removed from a site. The ensuing reaction between the minerals and microorgan-

TABLE 4.4	Names and Formulas of Some Basic Minerals
Name[a]	Formula
Calcite	$CaCO_3$
Gibbsite	$Al(OH)_3$
Dolomite	$MgCa(CO_3)_2$

[a] All these minerals are considered weak bases. Strong bases include all the hydroxides of the group 1 and group 2 elements except Be and Mg.

strong base a base that completely dissociates into ions in aqueous solution.

weak base a base that only partially dissociates in aqueous solution and so has a limited capacity to accept protons in the medium.

amphiprotic a substance that can behave as either a proton acceptor or a proton donor.

titration an analytical method for determining the concentration of a solute in a sample by reacting the solute with a standard solution of known concentration.

titrant the standard solution added to the sample in a titration.

standard solution a solution of known concentration used in titrations.

equivalence point the point in a titration where the number of moles of titrant added is stoichiometrically equal to the number of moles of the substance being analyzed.

end point the point in a titration that is reached when just enough standard solution has been added to cause the indicator to change color.

isms in the presence of oxygen produces very acidic solutions of sulfuric acid. To manage these hazardous wastes appropriately, we must know the concentrations of acid present in them. Titrations give us this information.

Suppose we have 100.0 mL of drainage water containing an unknown concentration of sulfuric acid. We can use an acid–base titration in which the sulfuric acid in the water sample is neutralized by reaction with a standard solution of 0.00100 M NaOH. The NaOH solution in this case is called the **titrant**; it is a **standard solution**, meaning its concentration is known accurately. The neutralization reaction is

$$H_2SO_4(aq) + 2\,NaOH(aq) \rightarrow Na_2SO_4(aq) + 2\,H_2O(\ell)$$

To determine the concentration of sulfuric acid in the sample, we need to determine the volume of titrant needed to neutralize a known volume of the sample.

Suppose 22.40 mL of the NaOH solution is required to react completely with the H_2SO_4 in the sample. Because we know the volume and molarity of the NaOH solution, we can calculate the number of moles of NaOH consumed:

$$n = V \times M$$

$$= 0.02240\ \cancel{L} \times 1.00 \times 10^{-3}\ \frac{mol\ NaOH}{\cancel{L}} = 2.24 \times 10^{-5}\ mol\ NaOH$$

We know from the stoichiometry of the reaction that 2 moles of NaOH are required to neutralize 1 mole of H_2SO_4, so the number of moles of H_2SO_4 in the 100.0 mL sample must be

$$= 2.24 \times 10^{-5}\ \cancel{mol\ NaOH} \times \frac{1\ mol\ H_2SO_4}{2\ \cancel{mol\ NaOH}} = 1.12 \times 10^{-5}\ mol\ H_2SO_4$$

making the concentration of H_2SO_4

$$\frac{1.12 \times 10^{-5}\ mol\ H_2SO_4}{0.1000\ L} = 1.12 \times 10^{-4}\ \frac{mol\ H_2SO_4}{L} = 1.12 \times 10^{-4}M\ H_2SO_4$$

How did we find out that exactly 22.40 mL of the NaOH solution was needed to react with the sulfuric acid in the sample? A setup for doing an acid–base titration is illustrated in Figure 4.13. The standard solution of NaOH is poured into a *buret*, a narrow glass cylinder with volume markings. The solution is gradually added to the sample until an indicator that changes color in response to $H^+(aq)$ concentration signals that the reaction is complete. The point in the titration when just enough standard solution has been added to completely react with all the solute in the sample is called the **equivalence point** of the titration. If the correct indicator has been chosen, the equivalence point is very close to the **end point**, the point at which the indicator changes color.

To detect the end point in our sulfuric acid–sodium hydroxide titration, we might use the indicator *phenolphthalein*, which is colorless in acidic solutions but pink in basic solutions. The part of the titration that requires the most skill is adding just enough NaOH solution to reach the end point, the point at which a pink color first persists in the solution being titrated. To catch this end point, the standard solution must be added no faster than one drop at a time with thorough mixing between drops.

(a)

(b)

FIGURE 4.13 Determining a sulfuric acid concentration. (a) A known volume of the H_2SO_4 solution is placed in the flask. The buret is filled with an aqueous NaOH solution of known concentration. A few drops of phenolphthalein indicator solution are added to the flask. (b) Sodium hydroxide is carefully added to the flask until the indicator changes from colorless to pink, signaling that the acid has been neutralized.

SAMPLE EXERCISE 4.7 **Calculating Molarity from Titration Data**

Vinegar is an aqueous solution of acetic acid (CH_3COOH) that can be made from any source containing starch or sugar. Apple cider vinegar is made from apple juice that is fermented to produce alcohol which then reacts with oxygen in the air in the presence of certain bacteria to produce vinegar. Commercial vinegar must contain no less than 4 grams of acetic acid per 100 mL of vinegar. Suppose the titration of a 25.00 mL sample of vinegar requires 11.20 mL of a 5.95 M solution of NaOH. What is the molarity of the vinegar? Could this be a commercial sample of vinegar?

Collect and Organize We are given the results of a titration of an acid with a standard solution of base. We have the volume and concentration of the titrant and the volume of the sample.

Analyze We need a balanced chemical equation for the neutralization reaction. We can calculate the molarity of the vinegar, and we can determine the number of grams of acetic acid in the sample and thereby determine if the sample is commercial grade.

Solve The balanced chemical equation for the neutralization reaction is

$$CH_3COOH(aq) + NaOH(aq) \rightarrow H_2O(\ell) + CH_3COONa(aq)$$

The stoichiometry of the reaction shows that 1 mole of base reacts with 1 mole of acid. The number of moles of base used in the titration is

$$0.01120 \text{ L NaOH} \times \frac{5.95 \text{ mol NaOH}}{1 \text{ L NaOH}} = 0.0666 \text{ mol NaOH}$$

which is equivalent to 0.0666 mol CH_3COOH in the sample because of the 1:1 stoichiometry. The molarity of the vinegar is

$$\frac{0.0666 \text{ mol } CH_3COOH}{0.02500 \text{ L vinegar}} = 2.66 \ M$$

The number of grams of acetic acid in the sample is

$$0.0666 \text{ mol } CH_3COOH \times \frac{60.05 \text{ g } CH_3COOH}{1 \text{ mol } CH_3COOH} = 4.00 \text{ g } CH_3COOH$$

The original sample had a volume of 25.00 mL, so the number of grams of acetic acid per 100 mL of sample is

$$\frac{4.00 \text{ g } CH_3COOH}{25.00 \text{ mL vinegar}} \times 100 \text{ mL vinegar} = 16.0 \text{ g}$$

The sample has 16.0 g in 100 mL of solution, so it could certainly be a commercial vinegar.

Think about It About 11 mL of titrant was needed to neutralize the sample; that's about half the volume of the sample, so the molarity of the vinegar should be about half that of the titrant. The answer for the concentration of the vinegar seems reasonable. Vinegar this strong is typically used for pickling. Vinegar with 4 to 8% acetic acid is table vinegar used in salad dressings.

Practice Exercise Citric acid ($C_6H_8O_7$) is a triprotic acid found naturally in lemon juice, and it is widely used as a flavoring in beverages. What is the molarity of $C_6H_8O_7$ in commercially available lemon juice if 14.26 mL of 1.751 M NaOH is required in a titration to neutralize 20.00 mL of the juice? How many grams of citric acid are in 100 mL of the juice?

(Answers to Practice Exercises are in the back of the book.)

4.7 Precipitation Reactions

Some of the most abundant elements in Earth's crust, including silicon (Si), aluminum (Al), and iron (Fe), are not abundant in seawater for the simple reason that most compounds containing these elements are insoluble in water. Earth's crust must be made of compounds with limited water solubility, or else they would never survive as solids in the presence of all the water on our planet. The reasons why some compounds are insoluble in water while others are soluble involve many factors. For now, Table 4.5 summarizes solubility rules for common ionic compounds. Chemists make use of the relative solubilities of ionic compounds to synthesize materials and to analyze solutions.

TABLE 4.5 Solubility Rules for Common Ionic Compounds in Water
All compounds containing the following ions are soluble: • Cations: Group 1 ions (alkali metals) and NH_4^+ • Anions: NO_3^- and CH_3COO^- (acetate)
Compounds containing the following anions are soluble except as noted: • Group 17 ions (halides), except the halides of Ag^+, Cu^+, Hg_2^{2+}, and Pb^{2+} • SO_4^{2-}, except the sulfates of Ba^{2+}, Ca^{2+}, Hg_2^{2+}, Pb^{2+}, and Sr^{2+}
Insoluble compounds include the following: • All hydroxides except those of group 1 cations and $Ca(OH)_2$, $Sr(OH)_2$, and $Ba(OH)_2$ • All sulfides except those of group 1 cations and NH_4^+, CaS, SrS, and BaS • All carbonates except those of group 1 cations and NH_4^+ • All phosphates except those of group 1 cations and NH_4^+

Making Insoluble Salts

When two solutions containing ions are mixed, a solid product called a **precipitate** may form. This reaction is called a *precipitation reaction*. We can use Table 4.5 to predict when a precipitate will form on mixing two solutions. For example, does a precipitate form when aqueous solutions of potassium nitrate (KNO_3) and sodium iodide (NaI) are mixed? To answer this question we need to:

1. *Recognize that both salts are in solution, which indicates that they are both soluble in water.* Table 4.5 specifies that all compounds containing cations from group 1 in the periodic table are soluble in water. Potassium and sodium are in group 1, so salts containing them are soluble:

$$\text{Solution 1:} \quad KNO_3(aq) \rightarrow K^+(aq) + NO_3^-(aq)$$
$$\text{Solution 2:} \quad NaI(aq) \rightarrow Na^+(aq) + I^-(aq)$$

precipitate a solid product formed from a reaction in solution.

From this information we can set up the reactant side of the equation:

$$K^+(aq) + NO_3^-(aq) + Na^+(aq) + I^-(aq) \rightarrow ?$$

2. *Determine whether any of the possible combinations of ions produces an insoluble product.* KNO_3, KI, NaI, and $NaNO_3$ are all soluble according to Table 4.5, so no precipitate forms when the two solutions are mixed:

$$K^+(aq) + NO_3^-(aq) + Na^+(aq) + I^-(aq) \rightarrow \text{no reaction}$$

What happens when aqueous solutions of lead(II) nitrate $[Pb(NO_3)_2]$ and sodium iodide are mixed? All nitrate salts are soluble, so $Pb(NO_3)_2$ dissolves in water and forms ions. The ions present in the mixed solutions are

$$Pb^{2+}(aq) + 2\,NO_3^-(aq) + Na^+(aq) + I^-(aq) \rightarrow ?$$

The solubility rules in Table 4.5 indicate that halide ions such as I^- of group 17 of the periodic table form insoluble compounds with $Pb^{2+}(aq)$, so we predict that $PbI_2(s)$ will precipitate (Figure 4.14).

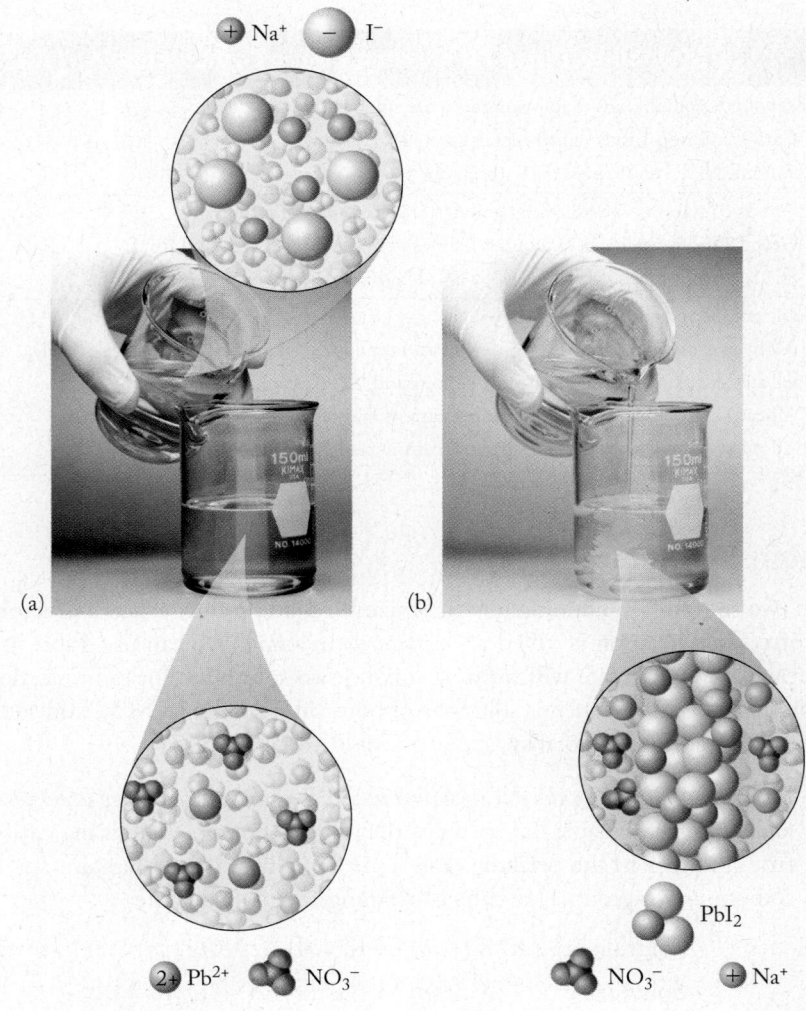

FIGURE 4.14 (a) One beaker contains a 0.1 M solution of $Pb(NO_3)_2$, and the other contains a 0.1 M solution of NaI. Both solutions are colorless. (b) As the NaI solution is poured into the $Pb(NO_3)_2$ solution, a yellow precipitate of PbI_2 forms.

Because a reaction takes place, let's write a balanced equation that describes the process, starting with the molecular equation:

Unbalanced: $Pb(NO_3)_2(aq) + NaI(aq) \rightarrow PbI_2(s) + NaNO_3(aq)$

Balanced: $Pb(NO_3)_2(aq) + 2\,NaI(aq) \rightarrow PbI_2(s) + 2\,NaNO_3(aq)$

From the balanced molecular equation we get the overall ionic equation:

$$Pb^{2+}(aq) + 2\,NO_3^-(aq) + 2\,Na^+(aq) + 2\,I^-(aq) \rightarrow$$
$$PbI_2(s) + 2\,Na^+(aq) + 2\,NO_3^-(aq)$$

Removing the spectator ions NO_3^- and Na^+ gives us the net ionic equation:

$$Pb^{2+}(aq) + 2\,I^-(aq) \rightarrow PbI_2(s)$$

Soluble and *insoluble* are qualitative terms. In principle, all ionic compounds dissolve in water to some extent. In practice, we consider a compound to be insoluble in water if the maximum amount that dissolves gives a concentration of less than 0.01 M. At this level, a solid appears to be insoluble to the naked eye; the tiny amount that dissolves is negligible compared with the amount that remains in the solid state and in contact with the solvent.

SAMPLE EXERCISE 4.8   **Writing Equations for Precipitation Reactions**

A precipitate forms when aqueous solutions of ammonium sulfate and barium chloride are mixed. Write the net ionic equation for the reaction.

Collect and Organize We know that mixing two solutions causes an insoluble compound to form. We need to identify it.

Analyze The two salts are in solution, so they both must be soluble in water. This is consistent with the solubility rules in Table 4.5. We need to determine which combinations of the ions produce insoluble solids. First, we need to write correct formulas of the salts. Then we can check Table 4.5 to see which of the possible combinations of ions are insoluble in water.

Solve

Solution 1: $(NH_4)_2SO_4(aq) \rightarrow 2\,NH_4^+(aq) + SO_4^{2-}(aq)$

Solution 2: $BaCl_2(aq) \rightarrow Ba^{2+}(aq) + 2\,Cl^-(aq)$

The new combinations are $BaSO_4$ and NH_4Cl. Table 4.5 indicates that all ammonium salts are soluble, and so the ammonium and chloride ions remain dissolved. Barium sulfate is insoluble, however, and that salt precipitates. The net ionic equation describes the formation of the precipitate:

$$Ba^{2+}(aq) + SO_4^{2-}(aq) \rightarrow BaSO_4(s)$$

Think about It To identify whether or not a precipitation reaction occurs when two solutions containing ions are mixed, we check to see whether an insoluble material is formed when the ions change partners. In this exercise, the sulfate ion, originally in the salt ammonium sulfate, partners with barium ions originally in the salt barium chloride, forming insoluble $BaSO_4$.

The other new combination of ions, NH_4^+ and Cl^-, remains in solution. Solubility rules (Table 4.5) enable us to predict if a precipitate is formed when solutions containing particular anions and cations are mixed. For example, Table 4.5 indicates that AgCl has very low solubility in water, which is another way of saying that $Ag^+(aq)$ and $Cl^-(aq)$ cannot coexist at significant concentrations in solution.

Practice Exercise Does a precipitate form when you mix aqueous solutions (a) of sodium acetate and ammonium sulfate; (b) of calcium chloride and mercury(I) nitrate? (c) If you answered yes in either case, write the net ionic equation for the reaction.

(Answers to Practice Exercises are in the back of the book.)

Precipitation reactions can be used to synthesize water-insoluble salts. For example, barium sulfate, used in medical procedures to image the gastrointestinal tract, can be made by mixing an aqueous solution of any water-soluble barium salt with a solution of a soluble sulfate salt. The precipitate from such a reaction can be collected by filtration, dried, and used for whatever purpose we may have. The soluble compound remaining in solution as hydrated ions can also be collected. The ammonium chloride left in solution in Sample Exercise 4.8, for instance, can be isolated by boiling off the water to leave a residue of solid NH_4Cl.

CONCEPT TEST

Lead(II) dichromate ($PbCr_2O_7$; the dichromate ion is $Cr_2O_7^{2-}$) is a water-insoluble pigment called school bus yellow that is used to paint lines on highways. Design a synthesis of school bus yellow that uses a precipitation reaction.

(Answers to Concept Tests are in the back of the book.)

Using Precipitation in Analysis

Chemists use the insolubility of ionic compounds to determine concentrations of ions in solution. For example, NaCl (in the form of *rock salt*) is used to melt ice and snow on roads during the winter. As the resulting NaCl solutions run off into nearby water supplies, that water becomes contaminated with high levels of sodium and chloride ions. To determine if sources of drinking water have become contaminated with such runoff, analytical chemists can determine the concentration of chloride ion by reacting a sample of the water with a solution of silver nitrate, $AgNO_3(aq)$. Any chloride ion in the water combines with Ag^+ to form a precipitate of AgCl:

$$NaCl(aq) + AgNO_3(aq) \rightarrow AgCl(s) + NaNO_3(aq) \qquad (4.9)$$

This precipitate can be filtered and dried, and its mass determined. From its mass, the concentration of chloride in the water sample can be calculated.

Equation 4.9 is the molecular equation describing the formation of AgCl(s). Because both reactants are soluble ionic compounds and completely dissociate in solution, the overall ionic equation is

$$Na^+(aq) + Cl^-(aq) + Ag^+(aq) + NO_3^-(aq) \rightarrow$$
$$AgCl(s) + Na^+(aq) + NO_3^-(aq) \qquad (4.10)$$

Eliminating the spectator ions $Na^+(aq)$ and $NO_3^-(aq)$ yields the net ionic equation:

$$Ag^+(aq) + Cl^-(aq) \rightarrow AgCl(s) \qquad (4.11)$$

When performing this analysis, we must add enough $Ag^+(aq)$ to ensure that all the $Cl^-(aq)$ is precipitated. Specifically, the molar ratio $Ag^+(aq)/Cl^-(aq)$ must be greater than one. In this situation, the $Ag^+(aq)$ is *in excess*. Precipitation reactions such as this may also include indicators that produce a visual signal (like a color change) that defines when the reaction of interest is over and the reagent in excess accumulates.

CONNECTION In Chapter 3 we introduced the idea of limiting reactants and reactants in excess when calculating the yield of reactions.

SAMPLE EXERCISE 4.9   Calculating the Mass of a Precipitate

The clinical management of burns on the skin from exposure to hydrofluoric acid, $HF(aq)$, may include the injection of a soluble calcium salt in the skin near the site of contact. As F^- ions from the acid move through the skin, they combine with Ca^{2+} ions and form deposits of insoluble CaF_2. If 1.00 mL of a 2.24 M aqueous solution of Ca^{2+} is injected, what is the mass of CaF_2 produced if all the calcium ions react with fluoride ions?

Collect and Organize We are given the volume and concentration of a solution of Ca^{2+} ions and asked to determine the maximum amount of CaF_2 precipitate that could be formed on reaction with F^- ions.

Analyze The net ionic reaction for the process involved is

$$Ca^{2+}(aq) + 2F^-(aq) \rightarrow CaF_2(s)$$

The problem states that all the calcium ions react, so we calculate the mass of product assuming Ca^{2+} ions are the limiting reactant.

Solve We can calculate the number of moles of $Ca^{2+}(aq)$ and use the stoichiometric relationships in the net ionic equation to determine the mass of product:

$$1.00 \times 10^{-3} \text{ L} \times \frac{2.24 \text{ mol Ca}^{2+}}{1 \text{ L}} \times \frac{1 \text{ mol CaF}_2}{1 \text{ mol Ca}^{2+}} \times \frac{78.08 \text{ g CaF}_2}{1 \text{ mol CaF}_2}$$
$$= 0.175 \text{ g CaF}_2$$

Think about It This problem asks for the mass of CaF_2 that could be formed, assuming Ca^{2+} ions were the limiting reactant. However, when this treatment is actually used, it is more likely that the Ca^{2+} ions will be in excess to make sure all the fluoride ions are consumed.

Practice Exercise Vermillion is a very rare and expensive solid natural pigment. Known as Chinese red, it is the pigment used to print the author's signature on works of art. It is mercury(II) sulfide (HgS), and it is also very insoluble in water. What is the maximum number of grams of vermillion you can make from the reaction of 50.00 mL of a 0.0150 M aqueous solution of mercury(II) nitrate [$Hg(NO_3)_2$] with an excess of a concentrated aqueous solution of sodium sulfide (Na_2S)?

(Answers to Practice Exercises are in the back of the book.)

SAMPLE EXERCISE 4.10   **Calculating a Solute Concentration from Mass of a Precipitate**

To determine the concentration of chloride ion in a 100.0 mL sample of groundwater, a chemist adds a large enough volume of a solution of $AgNO_3$ to the sample to precipitate all the Cl^- as AgCl. The mass of the resulting AgCl precipitate is 71.7 mg. What is the chloride concentration in milligrams of Cl^- per liter of groundwater?

Collect and Organize We are given the sample volume, 100.0 mL, and the mass of AgCl formed, 71.7 mg. Our task is to determine the chloride concentration in milligrams of Cl^- per liter of groundwater. We need a balanced chemical equation for the reaction. Although we can use the molecular, the overall ionic, or the net ionic equation, we choose the net ionic equation because it contains only those species involved in the precipitation reaction.

Analyze The net ionic equation is

$$Ag^+(aq) + Cl^-(aq) \rightarrow AgCl(s)$$

It shows that 1 mole of $Cl^-(aq)$ reacts with 1 mole of $Ag^+(aq)$ to produce 1 mole of AgCl(s). The molar mass of Cl^- is 35.45 g/mol, and that of AgCl is 143.32 g/mol. We use the methods developed in Chapter 3 to determine the mass of chloride ion in 100.0 mL of groundwater and then use this mass to calculate the concentration we need.

Solve Because our molar masses are in grams per mole, a logical first step is to convert the AgCl(s) mass to grams:

$$71.7 \text{ mg} \times \frac{10^{-3} \text{ g}}{1 \text{ mg}} = 0.0717 \text{ g}$$

The mass of Cl^- ions in the 100.0 mL sample of groundwater is therefore

$$0.0717 \text{ g AgCl(s)} \times \frac{1 \text{ mol AgCl(s)}}{143.32 \text{ g AgCl(s)}} \times \frac{1 \text{ mol Cl}^-(aq)}{1 \text{ mol AgCl(s)}}$$

$$\times \frac{35.45 \text{ g Cl}^-(aq)}{1 \text{ mol Cl}^-(aq)} = 0.0177 \text{ g Cl}^-(aq) = 17.7 \text{ mg Cl}^-(aq)$$

The Cl^- ion concentration in milligrams per liter of water is

$$\frac{17.7 \text{ mg Cl}^-}{100.00 \text{ mL}} \times \frac{1000 \text{ mL}}{L} = 177 \text{ mg Cl}^- / L \text{ groundwater}$$

An alternative way to solve this problem is to determine the mass of Cl^- ions in 1 g of AgCl(s):

$$\frac{1 \text{ mol Cl}^-}{1 \text{ mol AgCl}} \times \frac{35.45 \text{ g Cl}^-/\text{mol Cl}^-}{143.32 \text{ g AgCl/mol AgCl}} = 0.2473 \text{ g Cl}^-/\text{g AgCl}$$

Converting this factor to milligrams and applying it to the mass of AgCl(s) precipitated, we get

$$71.7 \text{ mg AgCl} \times \frac{247.3 \text{ mg Cl}^-}{1000 \text{ mg AgCl}} = 17.7 \text{ mg Cl}^- \text{ precipitated as AgCl(s)}$$

From here, the problem proceeds as before.

Think about It In Sample Exercise 4.1 we noted that the chloride concentration of drinking water should not exceed 250 ppm. Since mg/L is equivalent to ppm for dilute solutions, the concentration of Cl^- in the sample for this exercise is 177 ppm, which meets the guidelines for drinking water.

Practice Exercise The concentration of $SO_4^{2-}(aq)$ in a 50.0 mL sample of river water is determined using a precipitation titration in which the titrant cation is $Ba^{2+}(aq)$ and $BaSO_4(s)$ is the precipitate. What is the $SO_4^{2-}(aq)$ concentration in molarity if 6.55 mL of a 0.00100 M $Ba^{2+}(aq)$ solution is consumed?

(Answers to Practice Exercises are in the back of the book.)

Saturated Solutions and Supersaturation

A solution containing the maximum amount of solute that can dissolve is a **saturated solution** (Figure 4.15). The maximum amount of solute that can dissolve in a given quantity of solvent at a given temperature is called the **solubility** of that solute in the solvent. Precipitates form when the solubility of a solute is exceeded.

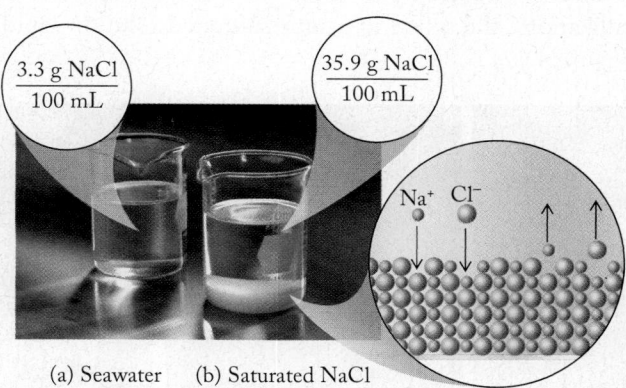

$$\frac{3.3 \text{ g NaCl}}{100 \text{ mL}}$$

$$\frac{35.9 \text{ g NaCl}}{100 \text{ mL}}$$

Na^+ Cl^-

(a) Seawater   (b) Saturated NaCl

FIGURE 4.15 (a) The liquid, although a concentrated aqueous solution of NaCl, is far from saturated. (b) This solution is in equilibrium with solid NaCl at 20°C (the white mass at the bottom of the beaker), so the liquid is a saturated solution of NaCl. Crystals in the solid are constantly dissolving, while ions in the solution are constantly precipitating. Because of this dynamic equilibrium, the concentration of solute [$Na^+(aq)$ and $Cl^-(aq)$ ions] in the solvent remains constant.

Solubility depends on temperature; often, the higher the temperature, the greater the solubility of solids in water. For example, more table sugar dissolves in hot water than in cold, a fact applied in making rock candy (Figure 4.16). After a large amount of sugar is dissolved in hot water, the solution is slowly cooled. An object with a slightly rough surface suspended in the solution, like a string or a wooden stirrer, serves as a site for crystallization. As the solution cools below the temperature at which the solubility of sugar is exceeded, crystals of sugar grow on the rough surface. The formation of crystals as the temperature of a saturated solution is lowered is a form of precipitation.

Sometimes, more solute dissolves in a volume of liquid than the amount predicted by the solute's solubility in that liquid, creating a **supersaturated solution**.

▶‖ **CHEMTOUR** Saturated Solutions

saturated solution a solution that contains the maximum concentration of a solute possible at a given temperature.

solubility the maximum amount of a substance that dissolves in a given quantity of solvent at a given temperature.

supersaturated solution a solution that contains more than the maximum quantity of solute predicted to be soluble in a given volume of solution at a given temperature.

(a) *T* = close to boiling point

(b) Put in a wooden stirrer

(c) Cool to room temperature

(d) Rock candy

FIGURE 4.16 Making rock candy. (a) A large quantity of table sugar is dissolved in hot water. The solution is not saturated at this high temperature, however. (b) A wooden stirrer is added and the solution is allowed to cool. (c) As the solution cools, it reaches the temperature at which the concentration of sugar exceeds the maximum that can be in solution and crystals ("rocks") of solid sugar ("candy") precipitate and attach to the stirrer. (d) Rock candy.

This can happen when the temperature of a saturated solution drops slowly or when the volume of an unsaturated solution is reduced through slow evaporation. Slow evaporation of the solvent is in part responsible for the formation of stalactites and stalagmites as well as of the salt flats in the American Southwest. Sooner or later, usually in response to a disruption such as a change in temperature, a mechanical shock, or the addition of a *seed crystal* (a small crystal of the solute that provides a site for crystallization), the solute in a supersaturated solution rapidly comes out of solution (Figure 4.17).

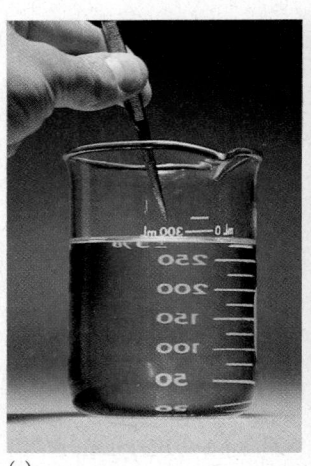

(a)   (b)   (c)

FIGURE 4.17 (a) A seed crystal is added to a supersaturated solution of sodium acetate. (b) The seed crystal becomes a site for rapid growth of sodium acetate crystals. (c) Crystal growth continues until the solution is no longer supersaturated but merely saturated with sodium acetate.

4.8 Ion Exchange

We just saw how dissolved ions can exchange partners, forming insoluble precipitates. This idea of **ion exchange** is important in water purification. Water containing certain metal ions—principally Ca^{2+} and Mg^{2+}—is called *hard water*, and it causes problems in industrial water supplies and in homes. Because hard water combines with soap to form a gray scum, clothes washed in hard water appear gray and dull. Hard water forms scale (an incrustation) in boilers, pipes, and kettles, diminishing their ability to conduct heat and carry water; and hard water sometimes has an unpleasant taste. In one common method of *water softening* (removing the ions responsible for hard water), hard water is passed through a system that exchanges calcium and magnesium ions for innocuous ones (Figure 4.18).

ion exchange a process by which one ion is displaced by another.

zeolites natural crystalline minerals or synthetic materials consisting of three-dimensional networks of channels that contain sodium or other 1+ cations.

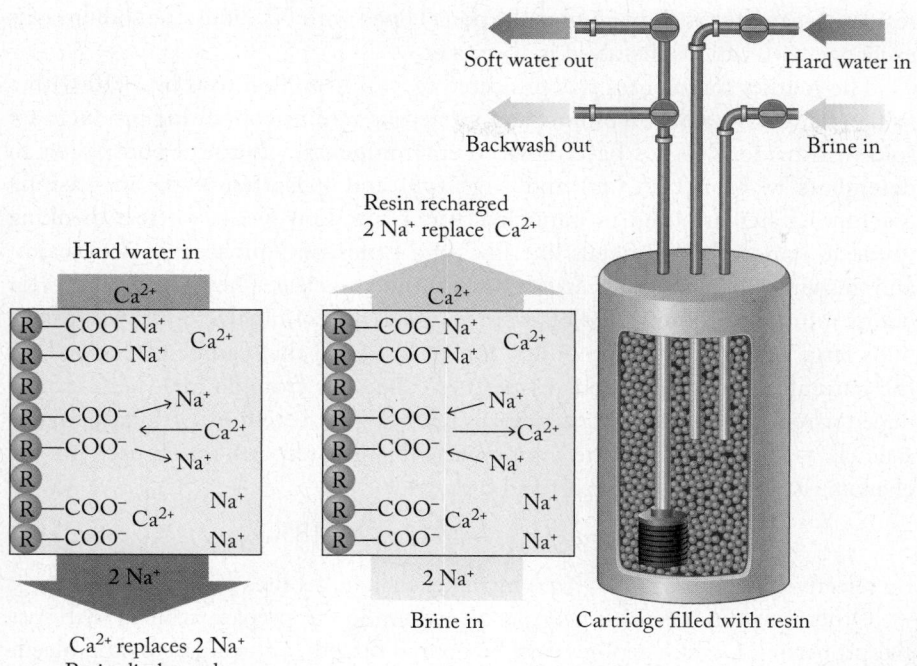

FIGURE 4.18 Residential water softeners use ion exchange to remove 2+ ions (such as Ca^{2+}) that make water hard. The ion-exchange resin contains cation-exchange sites that are initially occupied by Na^+ ions. These ions are replaced by 2+ "hardness" ions as water flows through the resin. Eventually most of the ion-exchange sites are occupied by 2+ ions, and the system loses its water-softening ability. The resin is then backwashed with a saturated solution of NaCl (*brine*), displacing the hardness ions (which wash down the drain), thus restoring the resin to its Na^+ form.

The system consists of a cartridge packed with beads of a porous plastic resin (R) bonded to anions capable of binding with hard-water cations—mainly Ca^{2+}, Mg^{2+}, and Fe^{2+}. The carboxylate anion (COO^-) is often used, and water in the ion-exchange cartridge contains Na^+ to balance the negative charges on the anions. We can therefore think of the resin as beads containing $(R\!-\!COO^-)Na^+$ units. As hard water flows through the cartridge, 2+ ions in the water exchange places with sodium ions on the resin. This is because cations with 2+ and 3+ charges bind more strongly to the $R\!-\!COO^-$ groups on the resin than sodium cations do. The ion-exchange reaction, with calcium ion as an example, is

$$2(R\!-\!COO^-)Na^+(s) + Ca^{2+}(aq) \rightarrow (R\!-\!COO^-)_2Ca^{2+}(s) + 2\,Na^+(aq)$$

Hard water that has been softened in this way contains increased concentrations of sodium ion. Although this may not be a problem for healthy children and adults, people suffering from high blood pressure often must limit their intake of Na^+ and so should not drink water softened by this kind of ion-exchange reaction.

Not only man-made materials perform ion exchange. Naturally occurring minerals called **zeolites** are extensively mined in many parts of the world and have huge commercial and technological utility as water softeners and purifiers, livestock feed additives, and odor suppressants. Zeolites, formed in nature as a result of the chemical reaction between molten lava from volcanic eruptions and salt water, are crystalline solids with tiny pores. Synthetic zeolites have also been manufactured that have structures similar to the natural materials. All zeolites have a rigid three-dimensional structure like a honeycomb (Figure 4.19) consisting of a network of interconnecting tunnels and cages. They work just like the plastic

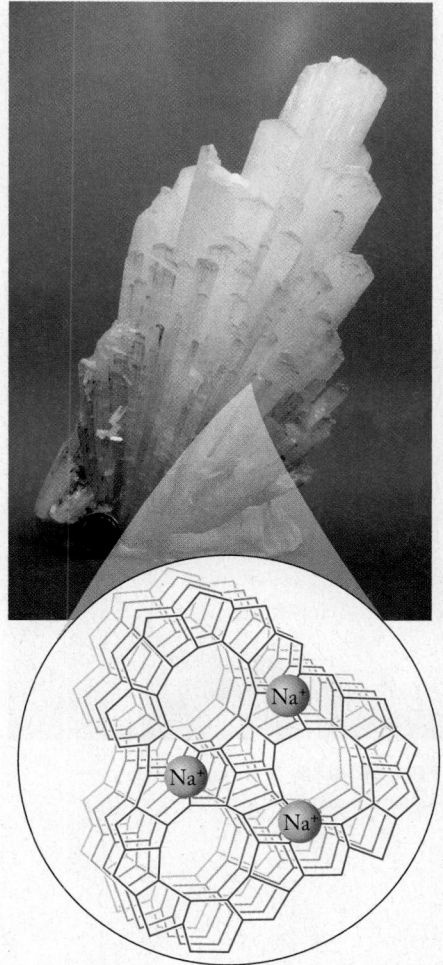

FIGURE 4.19 A sample of zeolite with a drawing showing the regular pattern of pores containing sodium ions that exchange for other cations dissolved in water that flows through the material.

resins: as water flows through tunnels (pores) lined with Na^+ ions, the sodium ions exchange with cations dissolved in the water.

The market for zeolites was projected to be 7.4 million tons by 2010 with a value of approximately $3 billion, and numerous zeolite-containing products are sold worldwide. Zeolites have replaced environmentally harmful phosphates in detergents to bind $Ca^{2+}(aq)$ and $Mg^{2+}(aq)$ and to soften water in washing machines. They are used in municipal water filtration plants to treat drinking water, to remove heavy metals like $Pb^{2+}(aq)$ from contaminated waste streams, and in swimming pools to keep the water clean and clear. They are even used for odor control in barns and feedlots where animals are confined. Animal waste contains large quantities of ammonium ion $[NH_4^+(aq)]$ that can be exchanged for the naturally occurring cations in zeolites. The odor from animal waste is produced when $NH_4^+(aq)$ ions participate in acid–base reactions with basic materials (B = base) and liberate ammonia $[NH_3(g)]$, responsible in part for the characteristic smell of cat boxes and barnyards:

$$NH_4^+(aq) + B \rightarrow NH_3(g) + HB^+(aq)$$

By selectively removing $NH_4^+(aq)$ by ion exchange, zeolites reduce odors.

Other ions can be exchanged for the sodium ion to prepare zeolites with special properties. Certain zeolites may be poured directly on wounds to stop bleeding by absorbing water from the blood, thereby concentrating clotting factors to promote coagulation. If the cations in these zeolites are exchanged for silver ion (Ag^+), the resulting product also has an antimicrobial effect and not only stems bleeding but reduces the risk of infection.

4.9 Oxidation-Reduction Reactions: Electron Transfer

One of the earliest descriptions of the properties of oxygen came from Leonardo da Vinci in the 15th century: "Where flame cannot live, no animal that draws breath can live." Although oxygen was not isolated and recognized as an element until over 250 years later, Leonardo captured an important property of the substance in his observation. Oxygen makes up about 50% by mass of Earth's crust, it is almost 89% by mass of the water that covers 70% of Earth's surface, and it is about 20% of the volume of Earth's atmosphere. It is essential to the metabolism of all creatures that "draw breath," just as it is essential for combustion reactions (Figure 4.20). Oxygen's name is applied to perhaps the most important type of chemical processes implied in Leonardo's statement: oxidation–reduction reactions.

Combination reactions of oxygen with nonmetals such as carbon, sulfur, and nitrogen produce volatile oxides. Combination reactions of O_2 with metals and semimetals produce solid oxides. All of these are examples of oxidation–reduction reactions. *Oxidation* was first defined as a reaction that increased the oxygen content of a substance. Hence two reactions involving oxygen—with methane to produce $CO_2(g)$ and $H_2O(\ell)$, and with elemental iron to produce $Fe_2O_3(s)$—are both oxidations from the point of view of the CH_4 and the Fe because the products contain more oxygen than the starting materials:

$$CH_4(g) + 2O_2(g) \rightarrow CO_2(g) + 2H_2O(\ell)$$
Methane
$$4Fe(s) + 3O_2(g) \rightarrow 2Fe_2O_3(s)$$

FIGURE 4.20 All fires are oxidation–reduction reactions.

Reduction reactions were originally defined in a similar fashion—reactions in which the oxygen content of a substance is reduced, a classic example being the reduction of iron ore to metallic iron:

$$Fe_2O_3(s) + 3\,CO(g) \rightarrow 2\,Fe(s) + 3\,CO_2(g)$$
Iron ore

The oxygen content of the iron ore is reduced—Fe_2O_3 is converted into Fe—as a result of its reaction with $CO(g)$. Note that the carbon atom in $CO(g)$ is oxidized because the oxygen content of the carbon species has been increased—CO is converted into CO_2.

> **CONCEPT TEST** ···

The following equation describes the conversion of one iron mineral called magnetite (Fe_3O_4) into another, called hematite (Fe_2O_3):

$$4\,Fe_3O_4(s) + O_2(g) \rightarrow 6\,Fe_2O_3(s)$$

Identify which element is oxidized and which is reduced and explain your choices.

(Answers to Concept Tests are in the back of the book.)

···

Oxidation Numbers

Ultimately, the meanings of oxidation and reduction were expanded, so that **oxidation** now refers to any chemical reaction in which a substance *loses electrons* and **reduction** refers to any reaction in which a substance *gains electrons*.[3] The processes of oxidation and reduction always occur together. If one substance loses electrons, another substance must gain them; if one species is oxidized, another must be reduced. The defining event in oxidation–reduction is the *transfer* of electrons from one substance to another. The recognition that oxidation and reduction must occur together gives rise to the common way of referring to them as *redox reactions*.

To keep track of electron loss and gain and to distinguish between oxidation and reduction in redox reactions, chemists use a system of assigning oxidation numbers to atoms in reactants and products. If the oxidation number of an atom changes as a result of a reaction—if the oxidation number of an atom is different on the left-hand side of an equation than on the right-hand side—then the reaction is redox. The change in oxidation number reflects the change in the number of electrons associated with an atom. This change in the number of electrons happens because electrons are transferred from one atom to another during the course of the reaction.

The **oxidation number (O.N.)**, or **oxidation state**, of an atom in a molecule or ion may be a positive or negative number or zero. The oxidation number represents the charge an atom has as determined by the following rules:

1. The oxidation numbers of the atoms in a neutral molecule sum to zero; those of the atoms in an ion sum to the charge on the ion.

oxidation a chemical change in which a species loses electrons; the oxidation number of the species increases.

reduction a chemical change in which a species gains electrons; the oxidation number of the species decreases.

oxidation number (O.N.) (also called **oxidation state**) a positive or negative number based on the number of electrons the atom gains or loses when it forms an ion, or that it shares when it forms a covalent bond with another element; pure elements have an oxidation number of zero.

---

[3] The mnemonic OIL RIG may be helpful for remembering these expanded definitions: "Oxidation Is Loss; Reduction Is Gain."

2. Each atom in a pure element has an oxidation number of zero:

F_2 O.N. = 0 for each F O_2 O.N. = 0 for each O

Fe   O.N. = 0   Na   O.N. = 0

3. In monatomic ions, the oxidation number is the charge on the ion:

F^- O.N. = −1 O^{2-} O.N. = −2

Fe^{3+} O.N. = +3 Na^+ O.N. = +1

Note that the number comes first when we write the symbol of an ion, Fe^{3+}, but the sign comes first when we write an oxidation number, +3.

4. In compounds containing fluorine and one or more other elements, the oxidation number of the fluorine is *always* −1:

KF   O.N. = −1 for F, which means O.N. = +1 for K (rule 1)

OF_2 O.N. = −1 for each F, which means O.N. = +2 for O (rule 1)

CF_4 O.N. = −1 for each F, which means O.N. = +4 for C (rule 1)

5. In most compounds, the oxidation number of hydrogen is +1 and that of oxygen is −2. Exceptions include hydrogen in metal hydrides (for example, LiH), where O.N. = −1 for H, and oxygen in the peroxide ion, O_2^{2-} (for example, H_2O_2), where O.N. = −1 for each O.

6. Unless combined with oxygen or fluorine, chlorine and bromine have an oxidation number of −1:

$CaCl_2$ O.N. = −1 for Cl, which means O.N. = +2 for Ca (rule 1)

$AlBr_3$ O.N. = −1 for Br, which means O.N. = +3 for Al (rule 1)

but

ClO_4^- O.N. = −2 for O (rule 5), which means O.N. = +7 for Cl (rule 1)

SAMPLE EXERCISE 4.11   **Determining Oxidation Numbers**

What is the oxidation number of sulfur in (a) SO_2, (b) Na_2S, and (c) $CaSO_4$?

Collect and Organize We are to assign the oxidation number of sulfur in three of its compounds.

Analyze To do so we apply the rules for determining the oxidation numbers of the elements in these compounds. SO_2 is the molecule sulfur dioxide. Na_2S is a binary ionic compound, and the oxidation number of an ion in an ionic compound equals its charge. Sulfur in $CaSO_4$ is part of the sulfate ion, which means its oxidation number added to those of the four O atoms must add up to the charge of the ion; we determine the oxidation number for calcium from rule 1.

Solve

a. From rule 5, O.N. = −2 for the O in SO_2. Rule 1 says that the sum of the oxidation numbers for S and the two O in the neutral molecule must be zero. Letting O.N. for S be x:

$$x + 2(-2) = 0$$
$$x = +4 \qquad \text{O.N. for S in } SO_2 = +4$$

b. Sodium forms only one ion, Na^+, which means that, according to rule 3, O.N. = +1. To balance the O.N. values in Na_2S (rule 1), we let y stand for the O.N. of sulfur:

$$2(+1) + y = 0$$
$$y = -2 \qquad \text{O.N. for S in } Na_2S = -2$$

c. The charge on the calcium ion is always 2+. This means that the charge on the sulfate ion must be 2−. Assigning O.N. = −2 for oxygen (rule 5) and z for sulfur:

$$z + 4(-2) = -2$$
$$z = +6 \qquad \text{O.N. for S in } CaSO_4 = +6$$

Think about It We found oxidation numbers for sulfur ranging from −2 to +6 in Na_2S, SO_2, and $CaSO_4$. Many other elements have a range of oxidation numbers in their compounds. For example, the oxidation numbers of iodine range from −1 in KI to +7 in KIO_4.

Practice Exercise Determine the oxidation number of nitrogen in (a) NO_2; (b) N_2O; (c) HNO_3.

(Answers to Practice Exercises are in the back of the book.)

Considering Electron Transfer in Redox Reactions

Because redox reactions involve the transfer of electrons among atoms, we must learn how to track changes in oxidation numbers when we consider these reactions. To see how this works, let's look at a reaction that we balanced in Chapter 3: the combustion of methane (CH_4) to produce CO_2 and H_2O.

$$CH_4(g) + 2O_2(g) \rightarrow CO_2(g) + 2H_2O(g) \qquad (4.12)$$

First we assign oxidation numbers to all the atoms in the reactants and products:

OXIDATION NUMBERS		
Atoms in Reactants		**Atoms in Products**
C in CH_4:	−4	C in CO_2: +4
H in CH_4:	+1	H in H_2O: +1
O in O_2:	0	O in CO_2: −2
		O in H_2O: −2

The oxidation numbers of the carbon and oxygen atoms change; such a change for any element defines a reaction as redox. One carbon atom on the left goes from O.N. = −4 in CH_4 to +4 in CO_2 for a net loss of 8 electrons (e^-). Loss of electrons is oxidation, so carbon is oxidized in this reaction. Oxygen atoms go from O.N. = 0 in O_2 to −2 in CO_2 and H_2O; each oxygen atom gains 2 e^-. In the balanced equation, there are 4 O atoms on the left and 4 O atoms on the right, so the total number of electrons gained by all the oxygen atoms is $(4 \times 2e^-) = 8e^-$. Gain of electrons is reduction, so oxygen is reduced in this reaction. The electrons gained by oxygen (8 e^-) offset the electrons lost by carbon (8 e^-). In redox

oxidizing agent a substance in a redox reaction that accepts electrons from another species, thereby oxidizing that species; the oxidizing agent is reduced in the reaction.

reducing agent a substance in a redox reaction that gives up electrons to another species, thereby reducing that species; the reducing agent is oxidized in the reaction.

reactions, we can keep track of the oxidation numbers and the electrons transferred by annotating the equation in the following way:

Carbon: Δ O.N. = (+4) − (−4) = +8

8 electrons lost

−4   oxidation   +4

$$CH_4(g) + 2\,O_2(g) \rightarrow CO_2(g) + 2\,H_2O(\ell)$$

0 reduction −2   −2

Oxygen: Δ O.N. = [2(−2) + 2(−2)] − 4(0) = −8

8 electrons gained

Note that when we assigned oxidation numbers to the atoms in this reaction, the value for hydrogen did not change. This means that hydrogen is neither oxidized nor reduced in this reaction: it is not involved in our consideration of electron transfer, and we do not need to incorporate it into our annotation above.

Because oxygen in Equation 4.12 takes electrons from the methane carbon atom, oxygen is called the **oxidizing agent** in the combustion reaction. We say that "oxygen oxidizes methane." In the process of functioning as an oxidizing agent, the oxygen is reduced. This is always the case: the *oxidizing* agent in any redox reaction is always *reduced* in the process. Any species that is reduced experiences a *reduction in oxidation number*, as, for instance, the oxidizing agent in Equation 4.12, where oxygen goes from O.N. = 0 to O.N. = −2.

The methane carbon atom is the species being oxidized in Equation 4.12, which means this C gives electrons to another substance and reduces that substance. Methane in this reaction is therefore called the **reducing agent**. The *reducing* agent is the source of the electrons that are transferred and it is always *oxidized*. The material that is oxidized experiences an *increase in oxidation number*; in this reaction, the carbon atom in methane goes from O.N. = −4 to O.N. = +4.

Every redox reaction has an oxidizing agent and a reducing agent. The reducing agent is the electron donor and the oxidizing agent is the electron acceptor. Electrons are transferred from the donor to the acceptor.

CONCEPT TEST

Determine whether the following are redox reactions. For the redox reactions, identify the oxidizing agent and the reducing agent.
 a. $Sn^{2+}(aq) + Br_2(aq) \rightarrow Sn^{4+}(aq) + 2\,Br^-(aq)$
 b. $2\,F_2(g) + 2\,H_2O(\ell) \rightarrow 4\,HF(aq) + O_2(g)$
 c. $NaHCO_3(aq) + HCl(aq) \rightarrow NaCl(aq) + CO_2(g) + H_2O(\ell)$

(Answers to Concept Tests are in the back of the book.)

SAMPLE EXERCISE 4.12 **Identifying Oxidizing Agents and Reducing Agents and Number of Electrons Transferred**

The reaction of oxygen with hydrazine:

$$O_2(aq) + N_2H_4(aq) \rightarrow 2\,H_2O(\ell) + N_2(g)$$

is used to remove dissolved oxygen gas from aqueous solutions, and the combustion of hydrazine produces enough energy that the substance is used as a rocket fuel. Identify the species oxidized, the species reduced, the oxidizing agent, the reducing agent, and the number of electrons transferred in the balanced chemical equation.

Collect and Organize The reactants are the element oxygen and the compound hydrazine. The products are water and nitrogen. The O.N. = +1 of hydrogen is unchanged.

Analyze The O.N. of the oxygen atoms in O_2 changes from zero to -2 in H_2O. The O.N. of nitrogen changes from -2 in N_2H_4 to zero in N_2.

Solve

Nitrogen: $\qquad\qquad$ Δ O.N. = 2(0) − 2(−2) = +4

$$\text{4 electrons lost}$$

$$\begin{array}{ccc} -2 & \text{oxidation} & 0 \end{array}$$

$$O_2(aq) + N_2H_4(aq) \rightarrow 2\,H_2O(\ell) + N_2(g)$$

$$\begin{array}{ccc} 0 & \text{reduction} & -2 \end{array}$$

Oxygen: $\quad$ Δ O.N. = 2(−2) − 2(0) = −4

$$\text{4 electrons gained}$$

We see that oxygen is the oxidizing agent because the nitrogen in N_2H_4 loses electrons and is oxidized. That means that O_2 is reduced and N_2H_4 is the reducing agent. Four electrons are transferred in the process.

Think about It We tend to identify the specific *atoms* that are oxidized or reduced, while whole *molecules or ions* are identified as the oxidizing agents or reducing agents. In the equation in this exercise, nitrogen is oxidized but N_2H_4 is the reducing agent.

Practice Exercise In the reaction between $O_2(g)$ and $SO_2(g)$ to make $SO_3(g)$, identify the species oxidized, the species reduced, the oxidizing agent, the reducing agent, and the number of electrons transferred.

(Answers to Practice Exercises are in the back of the book.)

Balancing Redox Reactions Using Half-Reactions

When a piece of copper wire [elemental copper, $Cu(s)$] is placed in a colorless solution of silver nitrate [$Ag^+(aq) + NO_3^-(aq)$], the solution gradually turns blue, the color of a solution of $Cu^{2+}(aq)$, and branchlike structures of $Ag(s)$ form in the medium (Figure 4.21). The process can be summarized as

Silver: $\qquad$ +1 $\quad$ reduction $\quad$ 0

$$Ag^+(aq) + Cu(s) \rightarrow Ag(s) + Cu^{2+}(aq)$$

Copper: $\qquad\qquad$ 0 $\quad$ oxidation $\quad$ +2

At first glance, this equation may seem balanced—and in a material sense it is: the number of silver and copper atoms or ions is the same on both sides.

(a) $\qquad\qquad$ (b)

FIGURE 4.21 (a) When a Cu wire is immersed in a solution of $AgNO_3$, Cu metal oxidizes to Cu^{2+} ions as Ag^+ ions are reduced to Ag metal. (b) A day later, the solution has the blue color of a solution of $Cu(NO_3)_2$, and the wire is coated with Ag metal.

half-reactions one of the two halves of an oxidation–reduction reaction; one half-reaction is the oxidation component, and the other is the reduction component.

However, the charges are not balanced: the sum of the charges is 1+ for the reactants but 2+ for the products. The reason for this imbalance is that one electron is involved in the reduction reaction but two are involved in the oxidation reaction. Equal numbers of electrons must be lost and gained, and so to balance redox reactions like this one we must develop a method that accounts for electron transfer.

Think of this reaction as consisting of one oxidation and one reduction. Each of these reactions represents *half* of the overall reaction; hence, each is called a **half-reaction**. We balance the equation by following these five steps.

1. *Write one equation for the oxidation half-reaction and a separate equation for the reduction half-reaction*:

   Oxidation: $\qquad\qquad\qquad\qquad Cu(s) \rightarrow Cu^{2+}(aq)$

   Reduction: $\qquad\qquad\qquad\qquad Ag^+(aq) \rightarrow Ag(s)$

2. *Balance the number of particles in each half-reaction*. This example requires no changes at this point because both equations are balanced in terms of mass.

3. *Balance the charge in each half-reaction* by adding electrons to the appropriate side. Always *add* electrons; never subtract them. Adding 2 electrons to the product side of the Cu half-reaction:

   Oxidation: $\qquad\qquad\qquad\qquad Cu(s) \rightarrow Cu^{2+}(aq) + 2\,e^-$

   and 1 electron to the reactant side of the Ag half-reaction:

   Reduction: $\qquad\qquad 1\,e^- + Ag^+(aq) \rightarrow Ag(s)$

   gives us a total charge of 0 on both sides of both half-reactions. Note that electrons are added on the reactant side of the reduction equation and on the product side of the oxidation equation. This is always the case.

4. *Multiply each half-reaction by the appropriate whole number* to make the number of electrons in the oxidation equation equal the number in the reduction equation:

   Oxidation: $\qquad\qquad\qquad\qquad Cu(s) \rightarrow Cu^{2+}(aq) + 2\,e^-$

   Reduction: $\qquad 2 \times [1\,e^- + Ag^+(aq) \rightarrow Ag(s)]$

   As a result we have the following:

   Oxidation: $\qquad\qquad\qquad\qquad Cu(s) \rightarrow Cu^{2+}(aq) + 2\,e^-$

   Reduction: $\qquad 2\,e^- + 2\,Ag^+(aq) \rightarrow 2\,Ag(s)$

5. *Add the two half-reactions to generate the equation representing the redox reaction*:

   Oxidation: $\qquad\qquad\qquad\qquad Cu(s) \rightarrow Cu^{2+}(aq) + \cancel{2\,e^-}$

   Reduction: $\qquad \cancel{2\,e^-} + 2\,Ag^+(aq) \rightarrow 2\,Ag(s)$

   ───

   Overall equation: $2\,Ag^+(aq) + Cu(s) \rightarrow 2\,Ag(s) + Cu^{2+}(aq)$

If we carried out step 4 correctly, the number of electrons in the oxidation half-reaction is the same as the number of electrons in the reduction half-reaction and they cancel out when we write the net ionic equation. The reaction is now balanced in terms of mass and charge. The overall equation is a net ionic equation, which means that only species involved in the redox reaction are shown. The nitrate ions from the silver nitrate solution (Figure 4.21) are spectator ions and are not shown in the final net ionic equation or in any of the intermediate equations we used to balance it.

> **SAMPLE EXERCISE 4.13** **Balancing Redox Reactions with Half-Reactions**

Iodine is slightly soluble in water and dissolves to make a yellow-brown solution of $I_2(aq)$. When colorless $Sn^{2+}(aq)$ is dissolved in it (Figure 4.22), the solution turns colorless as $I^-(aq)$ forms and $Sn^{2+}(aq)$ is converted into $Sn^{4+}(aq)$. (a) Is this a redox reaction? (b) Balance the equation that describes the reaction.

Collect and Organize The reactants in this process are $I_2(aq)$ and $Sn^{2+}(aq)$; the products are $I^-(aq)$ and $Sn^{4+}(aq)$. If this is redox, we write tin and iodine in separate equations when we write the respective half-reactions.

Analyze The oxidation number of tin changes from +2 in $Sn^{2+}(aq)$ to +4 in $Sn^{4+}(aq)$. The oxidation number of iodine changes from 0 in $I_2(aq)$ to −1 in $I^-(aq)$.

Solve

a. Because the oxidation numbers change, this is a redox reaction.
b. The unbalanced equation is

$$Sn^{2+}(aq) + I_2(aq) \rightarrow I^-(aq) + Sn^{4+}(aq)$$

We balance it via our five-step procedure.

1. Separate the half-reactions.

   Oxidation: $Sn^{2+}(aq) \rightarrow Sn^{4+}(aq)$
   Reduction: $I_2(aq) \rightarrow I^-(aq)$

2. Balance mass.

   Oxidation: $Sn^{2+}(aq) \rightarrow Sn^{4+}(aq)$
   Reduction: $I_2(aq) \rightarrow 2\,I^-(aq)$

3. Balance the charge by adding electrons.

   Oxidation: $Sn^{2+}(aq) \rightarrow Sn^{4+}(aq) + 2\,e^-$
   Reduction: $2\,e^- + I_2(aq) \rightarrow 2\,I^-(aq)$

4. Balance the numbers of electrons. This is not needed here because the numbers of electrons are the same in the two half-reactions.

5. Add the two half-reactions.

   Oxidation: $Sn^{2+}(aq) \rightarrow Sn^{4+}(aq) + \cancel{2\,e^-}$
   Reduction: $\cancel{2\,e^-} + I_2(aq) \rightarrow 2\,I^-(aq)$
   ──────────────────────────────
   Overall equation: $Sn^{2+}(aq) + I_2(aq) \rightarrow 2\,I^-(aq) + Sn^{4+}(aq)$

Check the overall equation for mass balance: $1\,Sn + 2\,I = 2\,I + 1\,Sn$. Check the charge balance: reactant side 2+, product side $2(1-) + (4+) = 2+$.

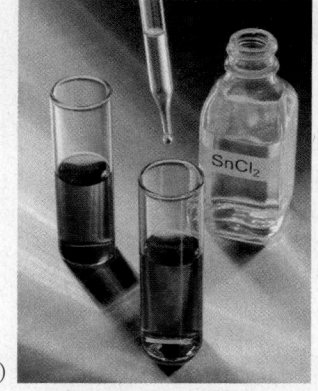

(a)

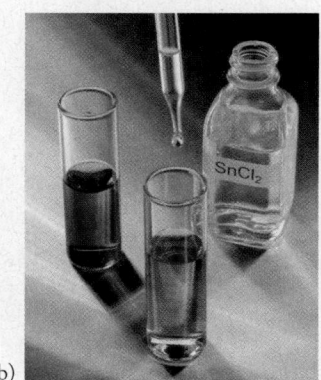

(b)

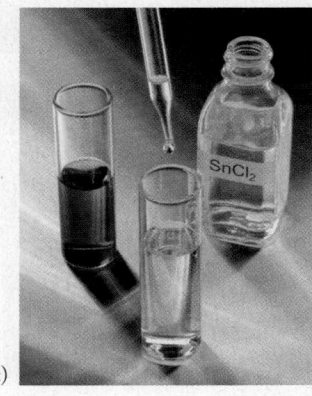

(c)

FIGURE 4.22 The test tubes in these photos contain an aqueous solution of iodine. (a) When drops of a solution of $SnCl_2$ are added to the middle tube, the yellow-brown color of the iodine solution begins to fade (b) as Sn^{2+} ions reduce I_2 to colorless I^- ions. (c) The color disappears completely when enough Sn^{2+} has been added to reduce all the I_2.

FIGURE 4.23 The orange-red rock in Bryce Canyon contains oxides of the iron(III) ion that result from weathering of the rock.

(a)

(b)

FIGURE 4.24 (a) The soil in this wetland has a blue-gray color because of the presence of iron(II) compounds. (b) Orange-red mottling indicates the presence of iron(III) oxides formed as a result of O_2 permeation through channels made by plant roots.

Think about It We can balance redox reactions using half-reactions. Adding the half-reactions gives us the net ionic reaction and tells us the number of electrons transferred from the reducing agent to the oxidizing agent. Two moles of electrons are transferred to iodine for each mole of $Sn^{2+}(aq)$ that is oxidized.

Practice Exercise A nail made of $Fe(s)$ that is placed in an aqueous solution of a soluble palladium(II) salt [containing $Pd^{2+}(aq)$] gradually disappears as the iron enters the solution as $Fe^{3+}(aq)$ and palladium metal [$Pd(s)$] forms. (a) Is this a redox reaction? (b) Balance the equation that describes this reaction.

(Answers to Practice Exercises are in the back of the book.)

Redox in Nature

Redox processes play a major role in determining the character of Earth's rocks and soils. For example, the orange-red rocks of Bryce Canyon National Park in Utah contain an iron(III) oxide mineral called hematite (Figure 4.23). Rocks containing hematite are relatively weak and easily broken, causing the fascinating shapes of such deposits. Iron(III) oxide is also the form of iron known as rust, the crumbly, red-brown solid that forms on iron objects exposed to water and oxygen.

Soil color is influenced by mineral content. Wetlands, areas that are either saturated or flooded with water during much of the year, are protected environments in many areas of the United States. Defining a given area as a wetland relies in part on the characteristics of the soils in the area, and soil color is an important diagnostic feature (Figure 4.24).

The redox reactions of iron(II) and iron(III) compounds can be modeled in the laboratory. Figure 4.25a shows a solution of iron(II) ammonium sulfate, $(NH_4)_2Fe(SO_4)_2(aq)$, upon addition of $NaOH(aq)$. The pale, greenish color of the liquid is characteristic of $Fe^{2+}(aq)$. Addition of $NaOH(aq)$ causes $Fe(OH)_2(s)$ to precipitate as a blue-gray solid, similar in color to the wetland soils shown in Figure 4.24. When this mixture is filtered, the iron(II) hydroxide residue immediately starts to darken. After about 20 minutes of exposure to oxygen in the air (Figure 4.25c), the precipitate turns the orange-red color of iron(III) hydroxide. The iron(II) has been oxidized—exactly the way the iron(II) compounds in a wetland soil are oxidized when exposed to oxygen.

Many redox reactions take place either in acidic solutions or in basic solutions, and the $H^+(aq)$, $OH^-(aq)$, or even the water may play a role in the reaction. Let's look at what we must do to balance the equation for such reactions.

A method for determining the concentration of $Fe^{2+}(aq)$ in an acidic solution involves its oxidation to $Fe^{3+}(aq)$ by the intensely purple permanganate ion, $MnO_4^-(aq)$ (Figure 4.26). In the reaction, the permanganate ion is reduced to $Mn^{2+}(aq)$:

$$Fe^{2+}(aq) + MnO_4^-(aq) \rightarrow Fe^{3+}(aq) + Mn^{2+}(aq)$$

| Appears colorless | Deep purple | Yellow-orange | Pale pink |

 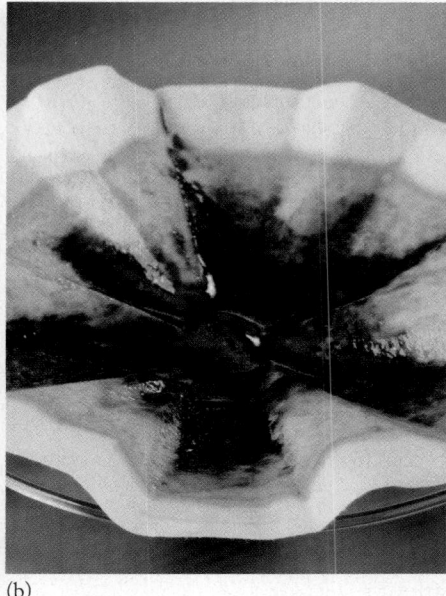 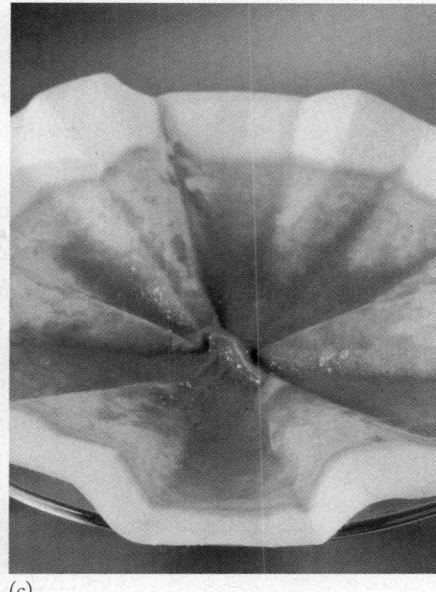

(a)   (b)   (c)

FIGURE 4.25 (a) A solution of iron(II) ammonium sulfate upon addition of NaOH(aq). The precipitate is water-insoluble iron(II) hydroxide. (b) The iron(II) hydroxide precipitate immediately after filtration. (c) The same precipitate after about 20 minutes of exposure to air. The color change indicates oxidation of Fe(II) to Fe(III).

This reaction occurs in an acidic solution, which suggests that it involves protons, and it is clearly redox, because $Fe^{2+}(aq)$ is oxidized to $Fe^{3+}(aq)$. The permanganate ion must be the oxidizing agent, so it must be reduced. Indeed, the O.N. of the Mn atom in MnO_4^- is +7 and it is reduced to +2 in Mn^{2+}.

The reaction will be balanced when (a) the number of atoms of each element on both sides of the reaction is the same and (b) the total charges on each side of the reaction arrow are the same. We proceed in the same way as for the previous examples in neutral medium, but we must add a few new steps to manage the role of the protons in the reaction.

1. Separate iron and manganese in their respective half-reactions.

   Oxidation: $Fe^{2+}(aq) \rightarrow Fe^{3+}(aq)$

   Reduction: $MnO_4^-(aq) \rightarrow Mn^{2+}(aq)$

2. Step 2, balancing mass, needs to be broken down into substeps to account for any role played by the aqueous acid. First we balance all the elements *except hydrogen and oxygen:*

   2a.   Oxidation: $Fe^{2+}(aq) \rightarrow Fe^{3+}(aq)$ (mass balanced)

       Reduction: $MnO_4^-(aq) \rightarrow Mn^{2+}(aq)$ (O mass not balanced)

   Next we balance oxygen by adding water as a product.

   2b.   Oxidation: $Fe^{2+}(aq) \rightarrow Fe^{3+}(aq)$ (mass balanced)

       Reduction: $MnO_4^-(aq) \rightarrow Mn^{2+}(aq) + 4H_2O(\ell)$

                      (O mass balanced, H mass not)

FIGURE 4.26 The test tubes on the left and right initially contain the same solution of Fe^{2+} ions, which appear colorless. When drops of the purple solution of $KMnO_4$ from the middle test tube are added to the solution on the right, the mixture turns yellow-orange, which is the color of Fe^{3+} ions in solution. The purple color disappears as MnO_4^- ions are reduced to Mn^{2+} ions.

In acidic solutions, the only source of H to make $H_2O(\ell)$ is H^+ from a strong acid, so we balance the hydrogen by adding $H^+(aq)$:

2c.   Oxidation: $Fe^{2+}(aq) \rightarrow Fe^{3+}(aq)$ (mass balanced)

        Reduction: $8\,H^+(aq) + MnO_4^-(aq) \rightarrow Mn^{2+}(aq) + 4\,H_2O(\ell)$

                                                      (mass balanced)

3. Balance charge by adding electrons.

    Oxidation: $Fe^{2+}(aq) \rightarrow Fe^{3+}(aq) + 1\,e^-$

    Reduction: $5\,e^- + 8\,H^+(aq) + MnO_4^-(aq) \rightarrow Mn^{2+}(aq) + 4\,H_2O(\ell)$

4. Balance numbers of electrons.

    Oxidation: $5 \times [Fe^{2+}(aq) \rightarrow Fe^{3+}(aq) + 1\,e^-]$

    Reduction: $1 \times [5\,e^- + 8\,H^+(aq) + MnO_4^-(aq) \rightarrow Mn^{2+}(aq) + 4\,H_2O(\ell)]$

5. Add the two equations.

    Oxidation: $5\,Fe^{2+}(aq) \rightarrow 5\,Fe^{3+}(aq) + \cancel{5\,e^-}$

    Reduction: $\cancel{5\,e^-} + 8\,H^+(aq) + MnO_4^-(aq) \rightarrow Mn^{2+}(aq) + 4\,H_2O(\ell)$

Overall equation:
$$8\,H^+(aq) + MnO_4^-(aq) + 5\,Fe^{2+}(aq) \rightarrow 5\,Fe^{3+}(aq) + Mn^{2+}(aq) + 4\,H_2O(\ell)$$

Always check the mass and charge balance of the final net ionic reaction. The reactant side contains 8 H, 1 Mn, 4 O, and 5 Fe atoms and the product side contains 8 H, 1 Mn, 4 O, and 5 Fe, so mass is balanced. On the reactant side, the charge is $(8+) + (1-) + (10+) = 17+$, and on the product side it is $(15+) + (2+) = 17+$, so charge is balanced.

This method works for balancing the equation for any redox reaction taking place in an aqueous acid. The oxidation in Figure 4.25, however, was carried out in a basic medium, and so now let's look at how to balance the equation for such a reaction. We balance such reactions in almost the same way we do in an acidic medium: in fact, we carry out exactly the same steps outlined in the example we just worked as though the reaction were run in acidic medium. We just add a final step in which sufficient hydroxide ion is added to both sides of the equation to convert any $H^+(aq)$ into $H_2O(\ell)$. This step converts the medium into an aqueous basic solution, as specified by the conditions.

SAMPLE EXERCISE 4.14   **Balancing Redox Reactions That Involve Hydroxide Ions**

Wetland soil is blue-gray due to $Fe(OH)_2(s)$, while well-aerated soils are often orange-red due to the presence of $Fe(OH)_3(s)$ (Figure 4.24). Give the balanced equation for the reaction of $O_2(g)$ with $Fe(OH)_2(s)$ in soil that produces $Fe(OH)_3(s)$ in basic solution.

Collect and Organize We know the reactants and product and can write the unbalanced equation:

$$Fe(OH)_2(s) + O_2(g) \rightarrow Fe(OH)_3(s)$$

Our task is to balance the equation with respect to both mass and charge.

Analyze This reaction takes place with both iron compounds in the solid phase in the presence of hydroxide ions in water. The oxidation half-reaction is clear: iron(II) hydroxide is oxidized to iron(III) hydroxide. For the reduction half-reaction, O_2 must be converted into the additional OH^- ion in the iron(III) hydroxide product.

Solve

1. Separate the equations.

   Oxidation: $\qquad Fe(OH)_2(s) \rightarrow Fe(OH)_3(s)$
   Reduction: $\qquad\quad O_2(g) \rightarrow OH^-(aq)$

2a. Balance all masses except H and O. This step is not needed because the only mass not attributed to H and O is Fe, which is balanced.

2b. Balance O by adding water as needed.

   Oxidation: $\qquad H_2O(\ell) + Fe(OH)_2(s) \rightarrow Fe(OH)_3(s)$
   Reduction: $\qquad\qquad\quad O_2(g) \rightarrow OH^-(aq) + H_2O(\ell)$

2c. Balance H by adding $H^+(aq)$ as needed.

   $$H_2O(\ell) + Fe(OH)_2(s) \rightarrow Fe(OH)_3(s) + H^+(aq)$$
   $$3\,H^+(aq) + O_2(g) \rightarrow OH^-(aq) + H_2O(\ell)$$

3. Balance charge.

   $$H_2O(\ell) + Fe(OH)_2(s) \rightarrow Fe(OH)_3(s) + H^+(aq) + 1\,e^-$$
   $$4\,e^- + 3\,H^+(aq) + O_2(g) \rightarrow OH^-(aq) + H_2O(\ell)$$

4. Balance numbers of electrons.

   $$4 \times [H_2O(\ell) + Fe(OH)_2(s) \rightarrow Fe(OH)_3(s) + H^+(aq) + 1\,e^-]$$
   $$4\,e^- + 3\,H^+(aq) + O_2(g) \rightarrow OH^-(aq) + H_2O(\ell)$$

5. Add the two equations.

   $$3\ 4\,H_2O(\ell) + 4\,Fe(OH)_2(s) \rightarrow 4\,Fe(OH)_3(s) + 4\,H^+(aq) + \cancel{4\,e^-}$$
   $$\cancel{4\,e^-} + \cancel{3\,H^+(aq)} + O_2(g) \rightarrow OH^-(aq) + \cancel{H_2O(\ell)}$$

This gives us the balanced equation:

$$3\,H_2O(\ell) + 4\,Fe(OH)_2(s) + O_2(g) \rightarrow 4\,Fe(OH)_3(s) + OH^-(aq) + H^+(aq)$$

Now we must switch to basic solution by adding the same number of $OH^-(aq)$ ions as we have $H^+(aq)$ ions to both sides of the equation:

$$OH^-(aq) + 3\,H_2O(\ell) + 4\,Fe(OH)_2(s) + O_2(g) \rightarrow$$
$$4\,Fe(OH)_3(s) + OH^-(aq) + H^+(aq) + OH^-(aq)$$

The $H^+(aq)$ is now removed from the equation and the species that are the same on the reactant and product sides of the reaction are cancelled:

$$\cancel{OH^-(aq)} + \overset{2}{\cancel{3}}\,H_2O(\ell) + 4\,Fe(OH)_2(s) + O_2(g) \rightarrow$$
$$4\,Fe(OH)_3(s) + \cancel{OH^-(aq)} + \underbrace{H^+(aq) + OH^-(aq)}_{\cancel{H_2O(\ell)}}$$

The final equation is

$$2\,H_2O(\ell) + 4\,Fe(OH)_2(s) + O_2(g) \rightarrow 4\,Fe(OH)_3(s)$$

Think about It This reaction is favored in neutral and basic soils because H_2O and O_2 combine to form the OH^- ions needed in the conversion of $Fe(OH)_2(s)$ to $Fe(OH)_3(s)$.

Practice Exercise The hydroperoxide ion, $HO_2^-(aq)$, reacts with permanganate ion, $MnO_4^-(aq)$ to produce $MnO_2(s)$ and oxygen gas. Balance the equation for the oxidation of hydroperoxide ion to $O_2(g)$ by permanganate ion in a basic solution.

(Answers to Practice Exercises are in the back of the book.)

Calcium: In the Limelight

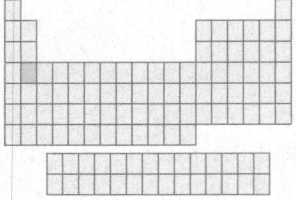

Calcium is the fifth most abundant element in Earth's crust. Vast deposits of calcium carbonate, $CaCO_3$, occur over large areas of the planet as limestone, marble, and chalk. These minerals are the fossilized remains of life in Earth's early oceans. Calcium carbonate is called a biomineral because it is an inorganic substance synthesized as an integral part of a living organism. Corals and seashells are mainly calcium carbonate, and islands like the Florida Keys and the Bahamas are made of large beds of $CaCO_3$.

(a)

(b)

FIGURE 4.27 Prized examples of calcium minerals are (a) pearls, which are heterogeneous mixtures of calcium carbonate and water, and (b) alabaster, fine-grained calcium sulfate.

Calcium sulfate, $CaSO_4$, commonly called gypsum, is another important calcium-containing mineral. If you are reading this in a room whose walls are finished with drywall, you are surrounded by gypsum because it is the primary component of drywall.

Although mostly known for their use as bulk chemicals in heavy industry and construction, special forms of calcium carbonate and calcium sulfate are valued for their beauty. Pearls (Figure 4.27a) are gemstones consisting mainly of calcium carbonate, specifically a heterogeneous mixture of $CaCO_3(s)$ and a small amount of water suspended within the solid. Light travels one way through the solid material and a different way through the suspended water. The same phenomenon occurs when a thin layer of oil or gasoline floats on the surface of water: we see an array of colors. Diffraction of the light gives a high-quality pearl its special iridescence. The substance known as alabaster, prized for its translucence and subtle patterns, is a fine-grained form of calcium sulfate (Figure 4.27b).

Both calcium carbonate and calcium sulfate are used in *desiccants*, those small packets of granules included in boxes of electronic devices, shoes, pharmaceuticals, and foods. Desiccants (from the Latin *siccus*, "dry") remove water from the air, thereby keeping dry the items with which they are packed.

Another calcium compound used as a desiccant is calcium silicate, $CaSiO_3$. About 0.5% by mass is added to table salt to keep it flowing freely in damp weather. It is estimated that calcium silicate can absorb almost 600 times its mass in water and still remain a free-flowing powder.

Calcium hypochlorite, $Ca(OCl)_2$, is one of the sources of chlorine used to disinfect the water in swimming pools.

Calcium carbonate is an industrial chemical of great utility. High-quality paper contains $CaCO_3(s)$, which improves brightness as well as the paper's ability to bind ink. The pharmaceutical industry sells calcium carbonate as an antacid and in dietary supplements, but $CaCO_3$ has an even greater industrial use as the precursor of two other important materials: quicklime (CaO) and slaked lime [$Ca(OH)_2$]. To produce quicklime, calcium carbonate (derived usually from mining it or from oyster shells) is roasted:

$$CaCO_3(s) \xrightarrow{\text{heat}} CaO(s) + CO_2(g)$$

Slaked lime is produced from the reaction of quicklime with water:

$$CaO(s) + H_2O(\ell) \rightarrow Ca(OH)_2(s)$$

To be "in the limelight" is now often used to describe anyone who is the center of attention. The expression derives from the calcium oxide lighting used in 19th-century theaters. When heated in an oxygen–hydrogen flame, calcium oxide emits a brilliant but soft white light. By the mid-1860s, limelights fitted with reflectors to direct and spread illumination were in wide use in theaters.

In modern times, lime has moved out of the entertainment business and into heavy industry. About 75 kg of lime is used to produce 1 ton of steel. Lime added to molten iron removes phosphorus, sulfur, and silicon to purify the metal. It is used in large amounts in municipal water supplies to coagulate suspended solids. Lime is used in smokestack scrubbers to remove oxides of sulfur from the gases emitted by coal-burning power plants (Figure 4.28). During combustion, the sulfur in coal forms $SO_2(g)$ and $SO_3(g)$. In the atmosphere, these nonmetal oxides mix with atmospheric water to make acid rain. To "scrub" these gases from stack emissions, the emission gases are passed through a suspension of calcium oxide in water. Here is the set of scrubber reactions for trapping SO_2:

$$CaO(s) + H_2O(\ell) \rightarrow Ca(OH)_2(aq)$$
(hydrolysis of a metal oxide to make a base)

$$SO_2(g) + CaOH_2(aq) \rightarrow CaSO_3(s) + H_2O(\ell)$$
(neutralization makes water and a salt)

The SO_2 gas is trapped as solid calcium sulfite, which is then oxidized to $CaSO_4$ (gypsum):

$$CaSO_3(s) + \tfrac{1}{2}O_2(g) \rightarrow CaSO_4(s)$$

Calcium is essential for life, and most animals—including humans—must absorb calcium every day from their diet. Over 95% of the calcium in the human system is in bones and teeth, and the average blood concentration of calcium is 100 mg per liter of blood. The body needs calcium to clot blood, transmit nerve impulses, and regulate the heartbeat. If there is insufficient calcium in the blood for these uses, your body breaks down bone to extract it. Hence, blood calcium concentration is not an accurate indicator of your calcium status; bone density measurements are required to determine that status.

In addition, bones are not static: even in the absence of demand from other parts of your body, your bones are constantly being remodeled. This means that a small amount of old bone is removed each day and new bone formed in its place. After about age 35, however, bone is removed faster than it is deposited, resulting in net bone loss as we grow older.

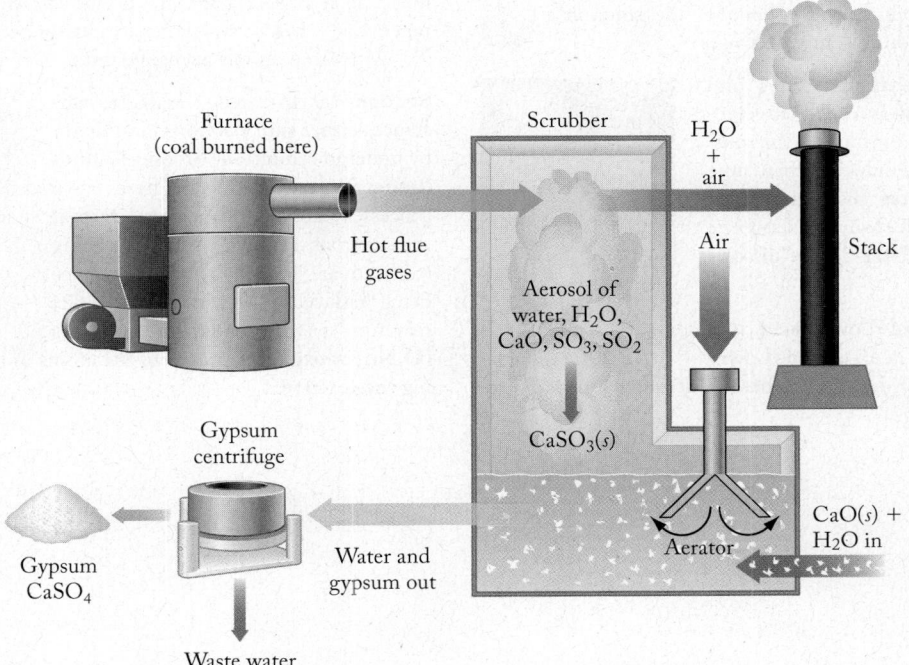

FIGURE 4.28 A scrubber capitalizes on the acid–base chemistry of aqueous solutions of calcium oxide and the oxides of sulfur to trap emission gases that would produce acid rain if released into the atmosphere.

Reactions in aqueous solutions are an integral part of our daily lives. On a large scale in oceans, rivers, and rain they shape our physical world. Many reactions to make the substances that are part of modern life—from paint pigments to drugs—are run in water, and many analytical procedures rely on reactions in water to determine the content of aqueous solutions that we drink, that we swim in, and that we use in countless consumer products like car batteries and shampoos. On a small scale in the cells of our bodies and in all living organisms, reactions in water make possible the chemical processes that are essential to life. On Earth, there may be water without life, but there is no life without water.

SUMMARY

Section 4.1 A *solution* is a homogeneous mixture of two or more substances. The substance in greatest molar proportion is the **solvent**; all other substances in the solution are **solutes**.

Section 4.2 The concentration of solute in a solution can be expressed as mass of solute per mass of solution, such as grams of solute per kilogram of solution, parts per million (1 ppm = 1 μg solute/g solution = 1 mg solute/kg solution), and parts per billion (1 ppb = 1 μg solute/kg solution). Solute concentration can also be expressed as mass of solute per volume of solvent and as moles of solute per liter of solution. The latter concentration scale is called **molarity (*M*)**.

Section 4.3 During **dilution** the quantity of solute in solution does not change, but as the volume of the solution increases, the concentration of solute decreases.

Section 4.4 A solute that dissociates into ions in aqueous solution is called an **electrolyte**. Mobility of these ions makes the solution a better electrical conductor than pure water. **Strong electrolytes** dissociate completely in water, **weak electrolytes** dissociate partially, and **nonelectrolytes** do not dissociate at all.

Section 4.5 A **Brønsted–Lowry acid** is a proton (H^+) donor, and a **Brønsted–Lowry base** is a proton acceptor. In an acid–base **neutralization reaction**,

H^+ ions from the acid combine with OH^- ions from the base, forming H_2O and a **salt**. The **net ionic equation** of a reaction includes only the species that change during the reaction and omits the **spectator ions**.

Section 4.6 **Titrations** are often based on acid–base neutralization or precipitation reactions.

Section 4.7 In a precipitation reaction, soluble reactants in solution form an insoluble **precipitate**. A **saturated solution** contains the maximum amount of solute possible at a given temperature. A dynamic equilibrium exists in a saturated solution. A **supersaturated solution** contains more than the maximum concentration of a solute.

Section 4.8 In an **ion-exchange** reaction, ions in solution displace ions held at ion-exchange sites on the surface of a solid. In a water softener, Ca^{2+}, Mg^{2+}, and other cations present in hard water displace Na^+ ions from an ion-exchange resin.

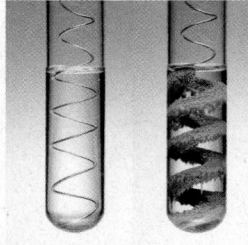

Section 4.9 In a redox reaction, substances either gain electrons (and thereby undergo **reduction**) or lose electrons (undergo **oxidation**). These two processes are complementary so that the electrons lost by the substance being oxidized are gained by the substance being reduced. A reaction is a redox reaction if the **oxidation numbers (O.N.)**, or **oxidation states**, of the atoms in the reactants change during the reaction.

PROBLEM-SOLVING SUMMARY ●● ■

TYPE OF PROBLEM	CONCEPTS AND EQUATIONS	SAMPLE EXERCISES
Comparing ion concentrations in aqueous solutions	Use conversion factors to express concentrations in the same units.	4.1
Calculating molarity from mass and volume or from density	Convert the solute mass first to grams and then to moles. Convert the solution mass into volume by dividing by the density of the solvent. Divide moles of solute by liters of solution: $$\text{Molarity} = \frac{\text{moles of solute}}{\text{liters of solution}} \quad \text{or} \quad M = \frac{n}{V} \qquad (4.1)$$	4.2, 4.3
Calculating quantities of solute as a (a) pure solid or as a (b) stock solution to prepare solutions	a. Multiply the known concentration (in mol/L) by the target volume (in L) to obtain the moles of solute needed. Then multiply moles of solute by solute molar mass $\mathcal{M}$ to get mass of solute needed: $$\text{Mass of solute (g)} = \mathcal{M} \times V \times M \qquad (4.2)$$ b. Given three of the four variables, use $$V_{\text{initial}} \times M_{\text{initial}} = V_{\text{diluted}} \times M_{\text{diluted}} \qquad (4.3)$$ to solve for the fourth.	4.4, 4.5
Writing neutralization reaction equations	Balance the molecular equation by balancing the moles of H^+ ions donated by acid and accepted by the base. Next create the overall ionic equation by writing strong electrolytes in their ionic form. Finally, create the net ionic equation by eliminating spectator ions from the overall ionic equation.	4.6
Calculating molarity from titration data	Use the volume and concentration of titrant used to neutralize a sample along with a balanced chemical equation to calculate the number of moles in a sample of known volume and determine its molarity.	4.7
Predicting precipitation reactions	Write all the ions present in the solutions being mixed. If any cation/anion pair forms an insoluble compound, that compound will precipitate from the solution.	4.8
Calculating the mass of a precipitate	Find the limiting reactant. Use the stoichiometry of the net ionic reaction to find the moles of precipitate, then convert moles to mass of precipitate with the precipitate's molar mass.	4.9
Calculating a solute concentration from a precipitate mass	Convert precipitate mass into moles by dividing by its molar mass. Convert moles of precipitate into moles of solute. Calculate molarity of the solute in the sample by dividing the moles of solute by the volume of sample in liters.	4.10
Determining oxidation numbers (O.N.)	O.N. for a monatomic ion is equal to the ion's charge. O.N. for a pure element is 0. To assign O.N. in a molecule containing more than one type of atom, assign O.N. +1 to H, −2 to O, and then calculate O.N. for any remaining atoms such that all the O.N. sum to 0. For polyatomic ions, the sum of the O.N. must equal the charge on the ion.	4.11
Identifying oxidizing and reducing agents and number of electrons transferred	The oxidizing agent contains an atom whose O.N. decreases during the reaction; the reducing agent contains an atom whose O.N. increases; the change in O.N. determines the number of electrons transferred.	4.12
Balancing redox reactions with half-reactions	Multiply one or both half-reactions by the appropriate coefficient(s) to balance the loss and gain of electrons. Combine the two half-reactions and simplify.	4.13
Balancing redox reactions that involve acidic or basic conditions	Balance half-reactions for elements except H and O. In acid, balance H by adding H^+, O by adding H_2O. In base, add OH^- to each side to neutralize H^+ and simplify.	4.14

VISUAL PROBLEMS

(Answers to boldface end-of-chapter questions and problems are in the back of the book.)

4.1. In Figure P4.1, which shows a solution containing three binary acids, one of the three is a weak acid and the other two are strong acids. Which color sphere represents the dissociated weak acid?

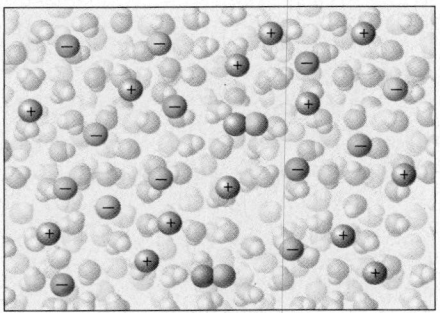

FIGURE P4.1

4.2. Solutions of sodium chloride and silver nitrate are mixed together and vigorously shaken. Which colored spheres in Figure P4.2 represent the following species? NOTE: The silver spheres represent Ag^+ ions. (a) Na^+; (b) Cl^-; (c) NO_3^- ions

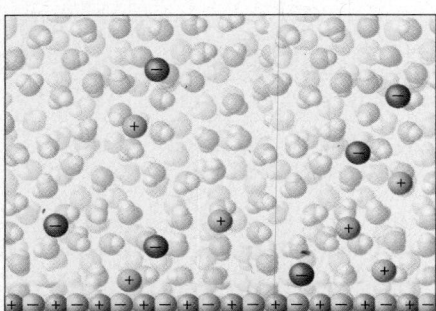

FIGURE P4.2

4.3. Which of the highlighted elements in Figure P4.3 forms an acid with the following generic formula? (a) HX; (b) H_2XO_4; (c) HXO_3; (d) H_3XO_4

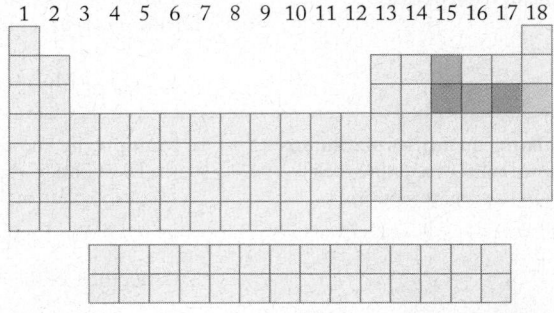

FIGURE P4.3

4.4. In which of the highlighted groups of elements in Figure P4.4 will you find an element that forms the following? (a) insoluble halides; (b) insoluble hydroxides; (c) hydroxides that are soluble; (d) binary compounds with hydrogen that are strong acids

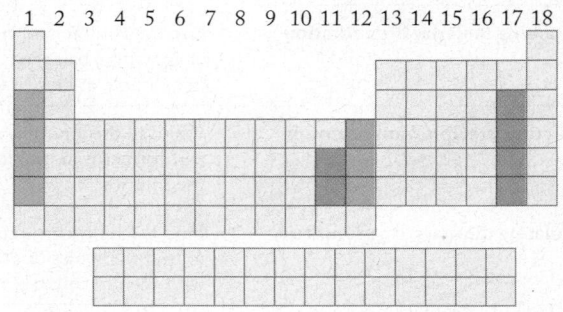

FIGURE P4.4

QUESTIONS AND PROBLEMS

Concentration Units

CONCEPT REVIEW

4.5. How do we decide which component in a solution is the solvent?

4.6. Can a solid ever be a solvent?

4.7. What is the molarity of a solution that contains 1.00 mmol of solute per milliliter of solution?

*4.8. A beaker contains 100 g of 1.00 *M* NaCl. If you transfer 50 g of the solution to another beaker, what is the molarity of the solution remaining in the first beaker?

PROBLEMS

4.9. Calculate the molarity of each of the following solutions:
a. 0.56 mol of $BaCl_2$ in 100.0 mL of solution
b. 0.200 mol of Na_2CO_3 in 200.0 mL of solution
c. 0.325 mol of $C_6H_{12}O_6$ in 250.0 mL of solution
d. 1.48 mol of KNO_3 in 250.0 mL of solution

4.10. Calculate the molarity of each of the following solutions:
a. 0.150 mol of urea (CH_4N_2O) in 250.0 mL of solution
b. 1.46 mol of $NH_4C_2H_3O_2$ in 1.000 L of solution
c. 1.94 mol of methanol (CH_3OH) in 5.000 L of solution
d. 0.045 mol of sucrose ($C_{12}H_{22}O_{11}$) in 50.0 mL of solution

_____

4.11. Calculate the molarity of each of the following ions:
a. 0.33 g Na^+ in 100.0 mL of solution
b. 0.38 g Cl^- in 100.0 mL of solution
c. 0.46 g SO_4^{2-} in 50.0 mL of solution
d. 0.40 g Ca^{2+} in 50.0 mL of solution

4.12. Calculate the molarity of each of the following:
a. 64.7 g LiCl in 250.0 mL of solution
b. 29.3 g $NiSO_4$ in 200.0 mL of solution
c. 50.0 g KCN in 500.0 mL of solution
d. 0.155 g $AgNO_3$ in 100.0 mL of solution

_____

4.13. How many grams of solute are needed to prepare each of the following solutions?
a. 1.000 L of 0.200 M NaCl
b. 250.0 mL of 0.125 M CuSO$_4$
c. 500.0 mL of 0.400 M CH$_3$OH

4.14. How many grams of solute are needed to prepare each of the following solutions?
a. 500.0 mL of 0.250 M KBr
b. 25.0 mL of 0.200 M NaNO$_3$
c. 100.0 mL of 0.375 M CH$_3$OH

4.15. River Water The Mackenzie River in northern Canada contains, on average, 0.820 mM Ca^{2+}, 0.430 mM Mg^{2+}, 0.300 mM Na$^+$, 0.0200 M K$^+$, 0.250 mM Cl$^-$, 0.380 mM SO$_4^{2-}$, and 1.82 mM HCO$_3^-$. What, on average, is the total mass of these ions in 2.75 L of Mackenzie River water?

4.16. Zinc, copper, lead, and mercury ions are toxic to Atlantic salmon at concentrations of 6.42×10^{-2} mM, 7.16×10^{-3} mM, 0.965 mM, and 5.00×10^{-2} mM, respectively. What are the corresponding concentrations in milligrams per liter?

4.17. Calculate the number of moles of solute contained in the following volumes of aqueous solutions of four pesticides:
a. 0.400 L of 0.024 M lindane
b. 1.65 L of 0.473 mM dieldrin
c. 25.8 L of 3.4 mM DDT
d. 154 L of 27.4 mM aldrin

4.18. A sample of crude oil contains 3.13 mM naphthalene, 12.0 mM methylnaphthalene, 23.8 mM dimethylnaphthalene, and 14.1 mM trimethylnaphthalene. What is the total number of moles of all the naphthalene compounds combined in 100.0 mL of the oil?

4.19. DDT Affects Neurons The pesticide DDT (C$_{14}$H$_9$Cl$_5$) kills insects such as malaria-carrying mosquitoes by opening sodium ion channels in neurons, causing them to fire spontaneously and leading to spasms and eventual death. However, its toxicity in wildlife and humans led to the banning of its use in the United States in 1972. Determination of DDT concentrations in groundwater samples between 1969 and 1971 in Pennsylvania yielded the following results:

Location	Sample Size	Mass of DDT
Orchard	250.0 mL	0.030 mg
Residential	1.750 L	0.035 mg
Residential after a storm	50.0 mL	0.57 mg

Express these concentrations in millimoles per liter.

4.20. Pesticide concentrations in the Rhine River between Germany and France between 1969 and 1975 averaged 0.55 mg/L of hexachlorobenzene (C$_6$Cl$_6$), 0.06 mg/L of dieldrin (C$_{12}$H$_8$Cl$_6$O), and 1.02 mg/L of hexachlorocyclohexane (C$_6$H$_6$Cl$_6$). Express these concentrations in millimoles per liter.

4.21. Effluent from municipal sewers often contains high concentrations of zinc. A sewer pipe discharges effluent that contains 10 mg Zn^{2+}/L. What is the molarity of Zn^{2+} in the effluent?

*4.22. The concentration of copper(II) sulfate in one brand of soluble plant fertilizer is 0.07% by mass. If a 20 g sample of this fertilizer is dissolved in 2.0 L of solution, what is the molarity of Cu^{2+}?

*4.23. For which of the following compounds is it possible to make a 1.0 M solution at 0°C?
a. CuSO$_4 \cdot 5$H$_2$O, solubility = 23.1 g/100 mL
b. AgNO$_3$, solubility = 122 g/100 mL
c. Fe(NO$_3$)$_2 \cdot 6$H$_2$O, solubility = 113 g/100 mL
d. Ca(OH)$_2$, solubility = 0.185 g/100 mL

4.24. Gold in the Ocean About 6×10^9 g of gold is thought to be dissolved in the oceans of the world. If the total volume of the oceans is 1.5×10^{21} L, what is the average molarity of gold in seawater?

4.25. The concentration of Mg^{2+} in a sample of coastal seawater is 1.09 g/kg. What is the molarity of Mg^{2+} in this seawater with a density of 1.02 g/mL?

4.26. Hemoglobin in Blood A typical adult body contains 6.0 liters of blood. The hemoglobin content of blood is about 15.5 g/100.0 mL of blood. The approximate molar mass of hemoglobin is 64,500 g/mol. How many moles of hemoglobin are present in a typical adult?

Dilutions

PROBLEMS

4.27. Calculate the final concentrations of the following aqueous solutions after each has been diluted to a final volume of 25.0 mL:
a. 1.00 mL of 0.452 M Na$^+$
b. 2.00 mL of 3.4 mM LiCl
c. 5.00 mL of 6.42×10^{-2} mM Zn^{2+}

4.28. Chemists who analyze samples for trace metals may buy standard solutions that contain 1000.0 mg/L concentrations of the metals. If a chemist wishes to dilute such a stock solution to prepare 500.0 mL of a working standard that has a concentration of 5.00 mg/L, what volume of stock solution is needed?

*4.29. A puddle of coastal seawater, caught in a depression formed by some coastal rocks at high tide, begins to evaporate on a hot summer day as the tide goes out. If the volume of the puddle decreases to 23% of its initial volume, what is the concentration of Na$^+$ after evaporation if initially it was 0.449 M?

4.30. What volume of 2.5 M SrCl$_2$ is needed to prepare 500.0 mL of 5.0 mM solution?

4.31. Dilution of Adult-Strength Cough Syrup A standard dose of an over-the-counter cough suppressant for adults is 20.0 mL. A portion this size contains 35 mg of the active pharmaceutical ingredient (API). Your pediatrician suggests you may give this medication to your 6-year-old child, but the child may only take 10.0 mL at a time and receive a maximum of 4.00 mg of the API. What is the concentration in mg/mL of the adult-strength medication, and how many millimeters of it would you need to dilute to make 100.0 mL of child-strength cough syrup?

*4.32. Mixing Fertilizer The label on a bottle of "organic" liquid fertilizer concentrate states that it contains 8 grams of phosphate per 100.0 mL and that 16 fluid ounces should be diluted with water to make 32 gallons of fertilizer to be applied to growing plants. What is the phosphate concentration in grams per liter in the diluted fertilizer? (1 gallon = 128 fluid ounces.)

Electrolytes and Nonelectrolytes

CONCEPT REVIEW

4.33. A solution of table salt is a good conductor of electricity, but a solution containing an equal molar concentration of table sugar is not. Why?

4.34. Metallic fixtures on the bottom of a ship corrode more quickly in seawater than in freshwater. Why?

4.35. Explain why liquid methanol, CH_3OH, cannot conduct electricity whereas molten NaOH can.

4.36. **Fuel Cells** The electrolyte in an electricity-generating device called a *fuel cell* consists of a mixture of Li_2CO_3 and K_2CO_3 heated to 650°C. At this temperature the ionic solids melt. Explain how this mixture of molten carbonates can conduct electricity.

4.37. Rank the following solutions on the basis of their ability to conduct electricity, starting with the most conductive: (a) 1.0 *M* NaCl; (b) 1.2 *M* KCl; (c) 1.0 *M* Na_2SO_4; (d) 0.75 *M* LiCl.

4.38. Rank the conductivities of 1 *M* aqueous solutions of each of the following solutes, starting with the most conductive: (a) acetic acid; (b) methanol; (c) sucrose (table sugar); (d) hydrochloric acid.

PROBLEMS

4.39. What is the molarity of Na^+ ions in a 0.025 *M* aqueous solution of (a) NaBr; (b) Na_2SO_4; (c) Na_3PO_4?

4.40. What is the molarity of each ion in a 0.025 *M* aqueous solution of (a) KCl; (b) $CuSO_4$; (c) $CaCl_2$?

4.41. Which of the following solutions has the greatest number of particles (atoms or ions) of solute per liter? (a) 1 *M* NaCl; (b) 1 *M* $CaCl_2$; (c) 1 *M* ethanol; (d) 1 *M* acetic acid

4.42. Which of the following solutions contains the most solute particles per liter? (a) 1 *M* KBr; (b) 1 *M* $Mg(NO_3)_2$; (c) 4 *M* ethanol; (d) 4 *M* acetic acid

Acid-Base Reactions: Proton Transfer

CONCEPT REVIEW

4.43. What name is given to a proton donor?

4.44. What is the difference between a strong acid and a weak acid?

4.45. Give the formulas of two strong acids and two weak acids.

4.46. Why is $HSO_4^-(aq)$ a weaker acid than $H_2SO_4(aq)$?

4.47. What name is given to a proton acceptor?

4.48. What is the difference between a strong base and a weak base?

4.49. Give the formulas of two strong bases and two weak bases.

4.50. Write the net ionic equation for the neutralization of a strong acid by a strong base.

PROBLEMS

4.51. For each of the following acid–base reactions, identify the acid and the base and then write the net ionic equation:
a. $H_2SO_4(aq) + Ca(OH)_2(aq) \rightarrow CaSO_4(s) + 2H_2O(\ell)$
b. $PbCO_3(s) + H_2SO_4(aq) \rightarrow PbSO_4(s) + CO_2(g) + H_2O(\ell)$
c. $Ca(OH)_2(s) + 2CH_3COOH(aq) \rightarrow$
$\qquad Ca(CH_3COO)_2(aq) + 2H_2O(aq)$

4.52. Complete and balance each of the following neutralization reactions, name the products, and write the net ionic equations.
a. $HBr(aq) + KOH(aq) \rightarrow$
b. $H_3PO_4(aq) + Ba(OH)_2(aq) \rightarrow$
c. $Al(OH)_3(s) + HCl(aq) \rightarrow$
d. $CH_3COOH(aq) + Sr(OH)_2(aq) \rightarrow$

4.53. Write a balanced molecular equation and a net ionic equation for the following reactions:
a. Solid magnesium hydroxide reacts with a solution of sulfuric acid.
b. Solid magnesium carbonate reacts with a solution of hydrochloric acid.
c. Ammonia gas reacts with hydrogen chloride gas.

4.54. Write a balanced molecular equation and a net ionic equation for the following reactions:
a. Solid aluminum hydroxide reacts with a solution of hydrobromic acid.
b. A solution of sulfuric acid reacts with solid sodium carbonate.
c. A solution of calcium hydroxide reacts with a solution of nitric acid.

4.55. **Toxicity of Lead Pigments** The use of lead(II) carbonate and lead(II) hydroxide as white pigments in paint has been discontinued, because children have been known to eat paint chips. The pigments dissolve in stomach acid, and lead ions enter the nervous system and interfere with neurotransmissions in the brain, causing neurological disorders. Using net ionic equations, show why lead(II) carbonate and lead(II) hydroxide dissolve in acidic solutions.

4.56. **Lawn Care** Many homeowners treat their lawns with $CaCO_3(s)$ to reduce the acidity of the soil. Write a net ionic equation for the reaction of $CaCO_3(s)$ with a strong acid.

Titrations

PROBLEMS

4.57. How many milliliters of 0.100 *M* NaOH are required to neutralize the following solutions?
a. 10.0 mL of 0.0500 *M* HCl
b. 25.0 mL of 0.126 *M* HNO_3
c. 50.0 mL of 0.215 *M* H_2SO_4

4.58. How many milliliters of 0.100 *M* HNO_3 are needed to neutralize the following solutions?
a. 45.0 mL of 0.667 *M* KOH
b. 58.5 mL of 0.0100 *M* $Al(OH)_3$
c. 34.7 mL of 0.775 *M* NaOH

*4.59. The solubility of slaked lime, $Ca(OH)_2$, in water at 20°C is 0.185 g/100.0 mL. What volume of 0.00100 *M* HCl is needed to neutralize 10.0 mL of a saturated $Ca(OH)_2$ solution?

4.60. The solubility of magnesium hydroxide, $Mg(OH)_2$, in water is 9.0×10^{-4} g/100.0 mL. What volume of 0.00100 *M* HNO_3 is required to neutralize 1.00 L of saturated $Mg(OH)_2$ solution?

4.61. A 10.0 mL dose of the antacid in Figure P4.61 contains 830 mg of magnesium hydroxide. What volume of 0.10 *M* stomach acid (HCl) could one dose neutralize?

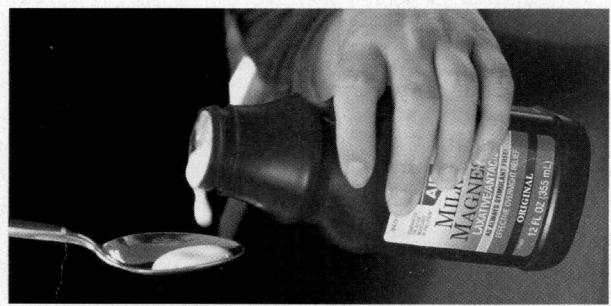

FIGURE P4.61

*4.62. **Exercise Physiology** The ache, or "burn," you feel in your muscles during strenuous exercise is caused by the accumulation of lactic acid, which has the structure shown in Figure P4.62. Only the hydrogen atom in the –COOH group is acidic, that is, can release a H^+ ion in aqueous solutions. To determine the concentration of a solution of lactic acid, a chemist titrates a 20.00 mL sample of it with 0.1010 M NaOH, and finds that 12.77 mL of titrant is required to reach the equivalence point. What is the concentration of the lactic acid solution in moles per liter?

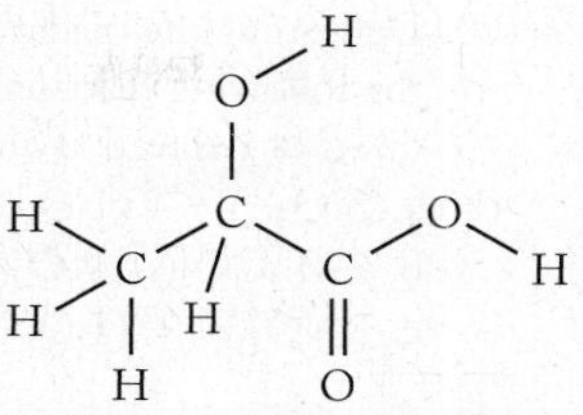

CH_3—$CHOH$—$COOH$

FIGURE P4.62

Precipitation Reactions

CONCEPT REVIEW

4.63. What is the difference between a saturated solution and a supersaturated solution?

4.64. What are common solubility units?

4.65. What is a precipitation reaction?

4.66. A precipitate may appear when two completely clear aqueous solutions are mixed. What circumstances are responsible for this event?

4.67. Is a saturated solution always a concentrated solution? Explain.

4.68. Honey is a concentrated solution of sugar molecules in water. Clear, viscous honey becomes cloudy after being stored for long periods. Explain how this transition illustrates supersaturation.

PROBLEMS

4.69. According to the solubility rules in Table 4.5, which of the following compounds have limited solubility in water? (a) barium sulfate; (b) barium hydroxide; (c) lanthanum nitrate; (d) sodium acetate; (e) lead hydroxide; (f) calcium phosphate

4.70. **Ocean Vents** The black "smoke" that flows out of deep ocean hydrothermal vents (Figure P4.70) is made of insoluble metal sulfides suspended in seawater. Of the following cations that are present in the water flowing up through these vents, which ones could contribute to the formation of the black smoke? Na^+, Li^+, Mn^{2+}, Fe^{2+}, Ca^{2+}, Mg^{2+}, Zn^{2+}, Pb^{2+}, Cu^{2+}

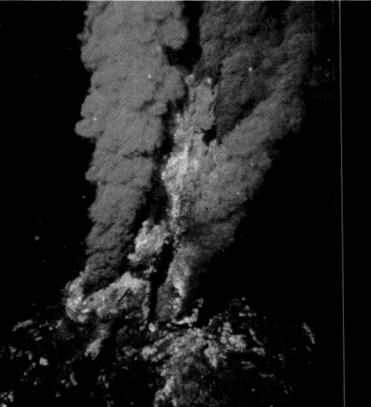

FIGURE P4.70

4.71. Complete and balance the chemical equations for the precipitation reactions, if any, between the following pairs of reactants, and write the net ionic equations:

a. $Pb(NO_3)_2(aq) + Na_2SO_4(aq) \rightarrow$

b. $NiCl_2(aq) + NH_4NO_3(aq) \rightarrow$

c. $FeCl_2(aq) + Na_2S(aq) \rightarrow$

d. $MgSO_4(aq) + BaCl_2(aq) \rightarrow$

*4.72. **Wastewater Treatment** Show with appropriate net ionic equations how Cr^{3+} and Cd^{2+} can be removed from wastewater by treatment with solutions of sodium hydroxide.

*4.73. An aqueous solution containing Ca^{2+}, Cl^-, CO_3^{2-}, and NO_3^- is allowed to evaporate. Which compound will precipitate first?

4.74. Ten milliliters of a 5×10^{-3} M solution of Cl^- ions is reacted with a 0.500 M solution of $AgNO_3$. What is the maximum mass of AgCl that precipitates?

4.75. Calculate the mass of $MgCO_3$ precipitated by mixing 10.0 mL of a 0.200 M Na_2CO_3 solution with 5.00 mL of 0.0500 M $Mg(NO_3)_2$ solution.

4.76. Toxic chromate can be precipitated from an aqueous solution by bubbling SO_2 through the solution. How many grams of SO_2 are required to treat 3.0×10^8 L of 0.050 mM Cr(VI)?

$$2CrO_4^{2-}(aq) + 3SO_2(g) + 4H^+(aq) \rightarrow Cr_2(SO_4)_3(s) + 2H_2O(\ell)$$

4.77. Fe(II) can be precipitated from a slightly basic aqueous solution by bubbling oxygen through the solution, which converts Fe(II) to insoluble Fe(III):

$$4Fe(OH)^+(aq) + 4OH^-(aq) + O_2(g) + 2H_2O(\ell) \rightarrow$$
$$4Fe(OH)_3(s)$$

How many grams of O_2 are consumed to precipitate all of the iron in 75 mL of 0.090 M Fe(II)?

4.78. Given the following equation, how many grams of $PbCO_3$ will dissolve when 1.00 L of 1.00 M H^+ is added to 5.00 g of $PbCO_3$?

$$PbCO_3(s) + 2H^+(aq) \rightarrow Pb^{2+}(aq) + H_2O(\ell) + CO_2(g)$$

*4.79. **Treating Drinking Water** Phosphate can be removed from drinking-water supplies by treating the water with $Ca(OH)_2$:

$$5Ca(OH)_2(aq) + 3PO_4^{3-}(aq) \rightarrow Ca_5OH(PO_4)_3(s) + 9OH^-(aq)$$

How much $Ca(OH)_2$ is required to remove 90% of the PO_4^{3-} from 4.5×10^6 L of drinking water containing 25 mg/L of PO_4^{3-}?

4.80. Toxic cyanide ions can be removed from wastewater by adding hypochlorite:

$$2CN^-(aq) + 5OCl^-(aq) + H_2O(\ell) \rightarrow$$
$$N_2(g) + 2HCO_3^-(aq) + 5Cl^-(aq)$$

How many liters of 0.125 M OCl^- are required to remove the CN^- in 3.4×10^6 L of wastewater in which the CN^- concentration is 0.58 mg/L?

Ion Exchange

CONCEPT REVIEW

4.81. Explain how a mixture of anion and cation exchangers can be used to deionize water.

4.82. Describe the process by which the ion exchanger in a home water softener is regenerated for further use.

4.83. If an ion-exchange resin is to be used to deionize water, what ions must be at the cation and anion exchange sites on the resin?

*4.84. A piece of Zn metal is placed in a solution containing Cu^{2+} ions. At the surface of the Zn metal, Cu^{2+} ions react with Zn atoms, forming Cu atoms and Zn^{2+} ions. Is this reaction an example of ion exchange? Explain why or why not.

Oxidation-Reduction Reactions: Electron Transfer

CONCEPT REVIEW

4.85. How are the gains or losses of electrons related to changes in oxidation numbers?

4.86. What is the sum of the oxidation numbers of the atoms in a molecule?

4.87. What is the sum of the oxidation numbers of all the atoms in each of the following polyatomic ions? (a) OH^-; (b) NH_4^+; (c) SO_4^{2-}; (d) PO_4^{3-}

4.88. Gold does not dissolve in concentrated H_2SO_4 but readily dissolves in H_2SeO_4 (selenic acid). Which acid is the stronger oxidizing agent?

4.89. Silver dissolves in sulfuric acid to form silver sulfate and H_2, but gold does not dissolve in sulfuric acid to form gold sulfate. Which of the two metals is the better reducing agent?

4.90. What is meant by a half-reaction?

4.91. What are the half-reactions that take place in the electrolysis of molten NaCl?

4.92. Electron gain is associated with _____ half-reactions and electron loss is associated with _____ half-reactions.

PROBLEMS

4.93. Give the oxidation number of chlorine in each of the following molecules and ion: (a) hypochlorous acid (HClO); (b) chloric acid ($HClO_3$); (c) perchlorate ion (ClO_4^-).

4.94. Give the oxidation number of nitrogen in each of the following molecules and ion: (a) elemental nitrogen (N_2); (b) hydrazine (N_2H_4); (c) ammonium ion (NH_4^+).

4.95. Balance the following half-reactions by adding the appropriate number of electrons. Identify the oxidation half-reactions and the reduction half-reactions.
a. $Br_2(\ell) \rightarrow 2\,Br^-(aq)$
b. $Pb(s) + 2\,Cl^-(aq) \rightarrow PbCl_2(s)$
c. $O_3(g) + 2\,H^+(aq) \rightarrow O_2(g) + H_2O(\ell)$
d. $8\,H_2S(g) \rightarrow S_8(s) + 16\,H^+(aq)$

4.96. Balance the following half-reactions by adding the appropriate number of electrons. Which are oxidation half-reactions and which are reduction half-reactions?
a. $Fe^{2+}(aq) \rightarrow Fe^{3+}(aq)$
b. $AgI(s) \rightarrow Ag(s) + I^-(aq)$
c. $VO_2^+(aq) + 2\,H^+(aq) \rightarrow VO^{2+}(aq) + H_2O(\ell)$
d. $I_2(s) + 6\,H_2O(\ell) \rightarrow 2\,IO_3^-(aq) + 12\,H^+(aq)$

*4.97. Natural Weathering of Ores Iron is oxidized in a number of chemical weathering processes. Write a half-reaction for the oxidation of magnetite (Fe_3O_4) to hematite (Fe_2O_3) in acidic groundwater. Add H_2O, H^+, and electrons as needed to balance the half-reaction.

*4.98. The mineral rhodochrosite [manganese(II) carbonate, $MnCO_3$] is a commercially important source of manganese. Write a half-reaction for the oxidation of the manganese in $MnCO_3$ to MnO_2 in neutral groundwater where the principal carbonate species is HCO_3^-. Add H_2O, H^+, and electrons as needed to balance the half-reaction.

4.99. Earth's Crust The following chemical reactions have helped to shape Earth's crust. Determine the oxidation numbers of all the elements in the reactants and products, and identify which elements are oxidized and which are reduced:
a. $3\,SiO_2(s) + 2\,Fe_3O_4(s) \rightarrow 3\,Fe_2SiO_4(s) + O_2(g)$
b. $SiO_2(s) + 2\,Fe(s) + O_2(g) \rightarrow Fe_2SiO_4(s)$
c. $4\,FeO(s) + O_2(g) + 6\,H_2O(\ell) \rightarrow 4\,Fe(OH)_3(s)$

4.100. Determine the oxidation numbers of each of the elements in the following reactions, and identify which of them are oxidized or reduced, if any:
a. $SiO_2(s) + 2\,H_2O(\ell) \rightarrow H_4SiO_4(aq)$
b. $2\,MnCO_3(s) + O_2(g) \rightarrow 2\,MnO_2(s) + 2\,CO_2(g)$
c. $3\,NO_2(g) + H_2O(\ell) \rightarrow 2\,NO_3^-(aq) + NO(g) + 2\,H^+(aq)$

4.101. Combine the following oxidation half-reactions (which are based on common iron minerals) with the half-reaction for the reduction of O_2
$$O_2(aq) + 4\,H^+(aq) + 4\,e^- \rightarrow 2\,H_2O(\ell)$$
to develop complete redox reactions:
a. $2\,FeCO_3(s) + H_2O(\ell) \rightarrow$
$\qquad Fe_2O_3(s) + 2\,CO_2(g) + 2\,H^+(aq) + 2\,e^-$
b. $3\,FeCO_3(s) + H_2O(\ell) \rightarrow$
$\qquad Fe_3O_4(s) + 3\,CO_2(g) + 2\,H^+(aq) + 2\,e^-$
c. $2\,Fe_3O_4(s) + H_2O(\ell) \rightarrow 3\,Fe_2O_3(s) + 2\,H^+(aq) + 2\,e^-$

4.102. Uranium is found in Earth's crust as UO_2 and an assortment of compounds containing UO_2^{n+} cations. Add the following pairs of reduction and oxidation equations to develop overall equations for converting soluble uranium polyatomic ions into insoluble UO_2:
a. $6\,H^+(aq) + UO_2(CO_3)_3^{4-}(aq) + 2\,e^- \rightarrow$
$\qquad UO_2(s) + 3\,CO_2(g) + 3\,H_2O(\ell)$
$Fe^{2+}(aq) + 3\,H_2O(\ell) \rightarrow Fe(OH)_3(s) + 3\,H^+(aq) + e^-$
b. $6\,H^+(aq) + UO_2(CO_3)_3^{4-}(aq) + 2\,e^- \rightarrow$
$\qquad UO_2(s) + 3\,CO_2(g) + 3\,H_2O(\ell)$
$HS^-(aq) + 4\,H_2O(\ell) \rightarrow SO_4^{2-}(aq) + 9\,H^+(aq) + 8\,e^-$
c. $2\,e^- + UO_2(HPO_4)_2^{2-}(aq) \rightarrow UO_2(s) + 2\,HPO_4^{2-}(aq)$
$3\,OH^-(aq) \rightarrow H_2O(\ell) + HO_2^-(aq) + 2\,e^-$

4.103. Nitrogen in the hydrosphere is found primarily as ammonium ions and nitrate ions. Complete and balance the following chemical equation describing the oxidation of ammonium ions to nitrate ions in acid solution:
$$NH_4^+(aq) + O_2(g) \rightarrow NO_3^-(aq)$$

4.104. In sediments and waterlogged soil, dissolved O_2 concentrations are so low that the microorganisms living there must rely on other sources of oxygen for respiration. Some bacteria can extract the oxygen from sulfate ions, reducing the sulfur in them to hydrogen sulfide gas and giving the sediments or soil a distinctive rotten-egg odor.
a. What is the change in oxidation state of sulfur as a result of this reaction?
b. Write the balanced net ionic equation for the reaction under acidic conditions that releases O_2 from sulfate and forms hydrogen sulfide gas.

4.105. The solubilities of Fe and Mn in freshwater streams are affected by changes in their oxidation states. Complete and balance the following redox reaction in which soluble Mn^{2+} becomes solid MnO_2:
$$Fe(OH)_2^+(aq) + Mn^{2+}(aq) \rightarrow MnO_2(s) + Fe^{2+}(aq)$$

4.106. A method for determining the quantity of dissolved oxygen in natural waters requires a series of redox reactions. Balance the following chemical equations in that series under the conditions indicated:
a. $Mn^{2+}(aq) + O_2(g) \rightarrow MnO_2(s)$ (basic solution)
b. $MnO_2(s) + I^-(aq) \rightarrow Mn^{2+}(aq) + I_2(s)$ (acidic solution)
c. $I_2(s) + S_2O_3^{2-}(aq) \rightarrow I^-(aq) + S_4O_6^{2-}(aq)$ (neutral solution)

4.107. Silver can be extracted from rocks by using cyanide ion. Complete and balance the following reaction for this process:

$$Ag(s) + CN^-(aq) + O_2(g) \rightarrow Ag(CN)_2^-(aq) \quad \text{(basic solution)}$$

4.108. Permanganate ion (MnO_4^-) is used in water purification to remove oxidizable substances. Complete and balance the following reactions for the removal of sulfide, cyanide, and sulfite. Assume that reaction conditions are basic.
a. $MnO_4^-(aq) + S^{2-}(aq) \rightarrow MnS(s) + S_8(s)$
b. $MnO_4^-(aq) + CN^-(aq) \rightarrow CNO^-(aq) + MnO_2(s)$
c. $MnO_4^-(aq) + SO_3^{2-}(aq) \rightarrow MnO_2(s) + SO_4^{2-}(aq)$

4.109. Bacteriocide and Viruscide The water-soluble gas ClO_2 is known as an oxidative biocide. It destroys bacteria by oxidizing their cell walls and viruses by attacking their viral envelopes. ClO_2 may be prepared for use as a decontaminating agent from several different starting materials in slightly acidic solutions. Complete and balance the following chemical reactions for the synthesis of ClO_2.
a. $ClO_3^-(aq) + SO_2(g) \rightarrow ClO_2(g) + SO_4^{2-}(aq)$
b. $ClO_3^-(aq) + Cl^-(aq) \rightarrow ClO_2(g) + Cl_2(g)$
c. $ClO_3^-(aq) + Cl_2(g) \rightarrow ClO_2(g) + O_2(g)$

4.110. Biochemical Oxidation Nitrification is a multistep process in which the nitrogen in organic and inorganic compounds is biochemically oxidized. Bacteria and fungi are responsible for a part of the *nitrification process* described by this half-reaction:

$$NH_4^+(aq) + 2H_2O(\ell) \rightarrow 8H^+(aq) + NO_2^-(aq) + 6e^-$$

What are the oxidation numbers of nitrogen in the reactant and product of this reaction?

Additional Problems

4.111. To determine the concentration of SO_4^{2-} ion in a sample of groundwater, 100.0 mL of the sample is titrated with 0.0250 M $Ba(NO_3)_2$, forming insoluble $BaSO_4$. If 3.19 mL of the $Ba(NO_3)_2$ solution is required to reach the end point of the titration, what is the molarity of the SO_4^{2-}?

4.112. Antifreeze Ethylene glycol is the common name for the liquid used to keep the coolant in automobile cooling systems from freezing. It is 38.7% carbon, 9.7% hydrogen, and 51.6% oxygen by mass. Its molar mass is 62.07 g/mol, and its density is 1.106 g/mL at 20°C.
a. What is the empirical formula of ethylene glycol?
b. What is the molecular formula of ethylene glycol?
c. In a solution prepared by mixing equal volumes of water and ethylene glycol, which ingredient is the solute and which is the solvent?

4.113. According to the label on a bottle of concentrated hydrochloric acid, the contents are 36.0% HCl by mass and have a density of 1.18 g/mL.
a. What is the molarity of concentrated HCl?
b. What volume of it would you need to prepare 0.250 L of 2.00 M HCl?
c. What mass of sodium hydrogen carbonate would be needed to neutralize the spill if a bottle containing 1.75 L of concentrated HCl dropped on a lab floor and broke open?

4.114. Synthesis and Toxicity of Chlorine Chlorine was first prepared in 1774 by heating a mixture of NaCl and MnO_2 in sulfuric acid:

$$NaCl(aq) + H_2SO_4(aq) + MnO_2(s) \rightarrow$$
$$Na_2SO_4(aq) + MnCl_2(aq) + H_2O(\ell) + Cl_2(g)$$

a. Assign oxidation numbers to the elements in each compound, and balance the redox reaction in acid solution.
b. Write a net ionic equation describing the reaction for formation of chlorine.
c. If chlorine gas is inhaled, it causes pulmonary edema (fluid in the lungs) because it reacts with water in the alveolar sacs of the lungs to produce the strong acid HCl and the weaker acid HOCl. Balance the equation for the conversion of Cl_2 to HCl and HOCl.

*4.115. When a solution of dithionate ions ($S_2O_4^{2-}$) is added to a solution of chromate ions (CrO_4^{2-}), the products of the ensuing chemical reaction that occurs under basic conditions include soluble sulfite ions and solid chromium(III) hydroxide. This reaction is used to remove Cr(VI) from wastewater generated by factories that make chrome-plated metals.
a. Write the net ionic equation for this redox reaction.
b. Which element is oxidized and which is reduced?
c. Identify the oxidizing and reducing agents in this reaction.
d. How many grams of sodium dithionate would be needed to remove the Cr(VI) in 100.0 L of wastewater that contains 0.00148 M chromate ion?

4.116. A prototype battery based on iron compounds with large, positive oxidation numbers was developed in 1999. In the following reactions, assign oxidation numbers to the elements in each compound and balance the redox reactions in basic solution:
a. $FeO_4^{2-}(aq) + H_2O(\ell) \rightarrow FeOOH(s) + O_2(g) + OH^-(aq)$
b. $FeO_4^{2-}(aq) + H_2O(\ell) \rightarrow Fe_2O_3(s) + O_2(g) + OH^-(aq)$

4.117. Polishing Silver Silver tarnish is the result of silver metal reacting with sulfur compounds, such as H_2S, in the air. The tarnish on silverware (Ag_2S) can be removed by soaking in a solution of $NaHCO_3$ (baking soda) in a basin lined with aluminum foil.
a. Write a balanced equation for the tarnishing of Ag to Ag_2S, and assign oxidation numbers to the reactants and products. How many electrons are transferred per mole of silver?
b. Write a balanced reaction for the reaction of Ag_2S with Al metal and water to produce $Al(OH)_3$, H_2S, H_2, and Ag metal.

4.118. Give the formula and name of the acids formed in the following chemical reactions of chlorine oxides.
a. $ClO + H_2O \rightarrow$ ____ + ____
b. $Cl_2O + H_2O \rightarrow HCl +$ ____
c. $Cl_2O_6 + H_2O \rightarrow$ ____ + ____

4.119. Many nonmetal oxides react with water to form acidic solutions. Give the formula and name for the acids produced from the following reactions:
a. $P_4O_{10} + 6H_2O \rightarrow$?
b. $SeO_2 + H_2O \rightarrow$?
c. $B_2O_3 + 3H_2O \rightarrow$?

4.120. Write overall and net ionic equations for the reactions that occur when
a. a sample of acetic acid is titrated with a solution of KOH.
b. a solution of sodium carbonate is mixed with a solution of calcium chloride.
c. calcium oxide dissolves in water.

4.121. One way to determine the concentration of hypochlorite ions (ClO^-) in solution is by first reacting them with I^- ions. Under acidic conditions the products of this reaction are I_2 and Cl^- ions. Then the I_2 produced in the first reaction is titrated with a solution of thiosulfate ions ($S_2O_3^{2-}$). The products of the titration reaction are $S_4O_6^{2-}$ and I^- ions. Write net ionic equations for these two reactions.

*4.122. **Fluoride Ion in Drinking Water** Sodium fluoride is added to drinking water in many municipalities to protect teeth against cavities. The target of the fluoridation is hydroxyapatite, $Ca_{10}(PO_4)_6(OH)_2$, a compound in tooth enamel. There is concern, however, that fluoride ions in water may contribute to skeletal fluorosis, an arthritis-like disease.

a. Write a net ionic equation for the reaction between hydroxyapatite and sodium fluoride that produces fluorapatite, $Ca_{10}(PO_4)_6F_2$.

b. The EPA currently restricts the concentration of F^- in drinking water to 4 mg/L. Express this concentration of F^- in molarity.

c. One study of skeletal fluorosis suggests that drinking water with a fluoride concentration of 4 mg/L for 20 years raises the fluoride content in bone to 6 mg/g, a level at which a patient may experience stiff joints and other symptoms. How much fluoride (in milligrams) is present in a 100 mg sample of bone with this fluoride concentration?

*4.123. **Rocket Fuel in Drinking Water** Near Las Vegas, NV, improper disposal of perchlorates used to manufacture rocket fuel has contaminated a stream that flows into Lake Mead, the largest artificial lake in the United States and a major supply of drinking and irrigation water for the American Southwest. The EPA has proposed an advisory range for perchlorate concentrations in drinking water of 4 to 18 μg/L. The perchlorate concentration in the stream averages 700.0 μg/L, and the stream flows at an average rate of 161 million gallons per day (1 gal = 3.785 L).

a. What are the formulas of sodium perchlorate and ammonium perchlorate?

b. How many kilograms of perchlorate flow from the Las Vegas stream into Lake Mead each day?

c. What volume of perchlorate-free lake water would have to mix with the stream water each day to dilute the stream's perchlorate concentration from 700.0 to 4 μg/L?

d. Since 2003, Maryland, Massachusetts, and New Mexico have limited perchlorate concentrations in drinking water to 0.1 μg/L. Five replicate samples were analyzed for perchlorates by laboratories in each state, and the following data (μg/L) were collected:

MD	MA	NM
1.1	0.90	1.2
1.1	0.95	1.2
1.4	0.92	1.3
1.3	0.90	1.4
0.9	0.93	1.1

Which of the labs produced the most precise analytical results?

*4.124. **Water from Mines** Water draining from abandoned mines on Iron Mountain in California is extremely acidic, and leaches iron, zinc, and other metals from the underlying rock (Figure P4.124). One liter of drainage contains as much as 80.0 g of dissolved iron and 6 g of zinc.

a. Calculate the molarity of iron and of zinc in the drainage.

b. One source of the dissolved iron is the reaction between water containing H_2SO_4 and solid $Fe(OH)_3$. Complete the following chemical equation, and write a net ionic equation for the process.

$$2\,Fe(OH)_3(s) + 3\,H_2SO_4(aq) \rightarrow ?$$

c. Sources of zinc include the mineral smithsonite, $ZnCO_3$. Write a balanced net ionic equation for the reaction between smithsonite and H_2SO_4 that produces $Zn^{2+}(aq)$.

d. One member of a class of minerals called ferrites is found to contain a mixture of Zn(II), Fe(II), and Fe(III) oxides. The generic formula for the mineral is $Zn_xFe_{1-x}O \cdot Fe_2O_3$. If acidic mine waste flowing through a deposit of this mineral contains 80 g of Fe and 6 g of Zn as a result of dissolution of the mineral, what is the value of x in the formula of the mineral in the deposit?

FIGURE P4.124

*4.125. **Making Apple Cider Vinegar** Some people who prefer natural foods make their own apple cider vinegar. They start with freshly squeezed apple juice that contains about 6% natural sugars. These sugars, which all have nearly the same empirical formula, CH_2O, are fermented with yeast in a chemical reaction that produces equal numbers of moles of ethanol (Figure P4.125a) and carbon dioxide. The product of this fermentation, called hard cider, undergoes an acid fermentation step in which ethanol and dissolved oxygen gas react together to form acetic acid (Figure P4.125b) and water. This acetic acid is the principal solute in vinegar.

Ethanol

$CH_3 — CH_2 — OH$

(a)

Acetic acid

$CH_3 — COOH$

(b)

FIGURE P4.125

a. Write a balanced chemical equation describing the fermentation of natural sugars to ethanol and carbon dioxide. You may use in the equation the empirical formula given in the above paragraph.

b. Write a balanced chemical equation describing the acid fermentation of ethanol to acetic acid.

c. What are the oxidation states of carbon in the reactants and products of the two fermentation reactions?

d. If a sample of apple juice contains 1.00×10^2 g of natural sugar, what is the maximum quantity of acetic acid that could be produced by the two fermentation reactions?

*4.126. A food chemist determines the concentration of acetic acid in a sample of apple cider vinegar (see Problem 4.125) by acid–base titration. The density of the sample is 1.01 g/mL. The titrant is 1.002 M NaOH. The average volume of titrant required to titrate 25.00 mL subsamples of the vinegar is 20.78 mL. What is the concentration of acetic acid in the vinegar? Express your answer the way a food chemist probably would: as percent by mass.

*4.127. One way to follow the progress of a titration and detect its equivalence point is by monitoring the conductivity of the titration reaction mixture. For example, consider the way the conductivity of a sample of sulfuric acid changes as it is titrated with a standard solution of barium hydroxide before and then after the equivalence point.

a. Write the overall ionic equation for the titration reaction.

b. Which of the four graphs in Figure P4.127 comes closest to representing the changes in conductivity during the titration? (The zero point on the y-axis of these graphs represents the conductivity of pure water; the break points on the x-axis represent the equivalence point.)

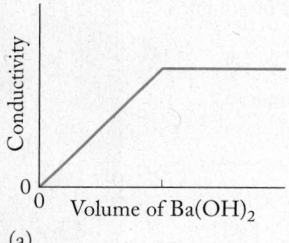
(a)

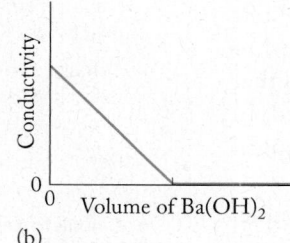
(b)

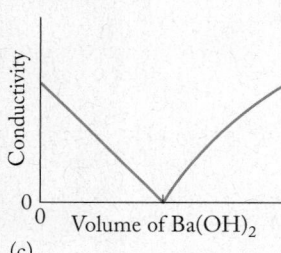
(c)

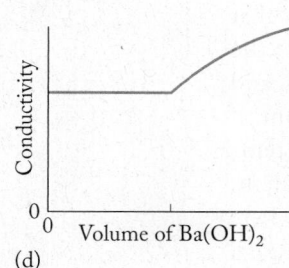
(d)

FIGURE P4.127

*4.128. Which of the graphs in Figure P4.128 best represents the changes in conductivity that occur before and after the equivalence point in each of these following titrations:

a. sample, $AgNO_3(aq)$; titrant, $KCl(aq)$

b. sample, $HCl(aq)$; titrant, $LiOH(aq)$

c. sample, $CH_3COOH(aq)$; titrant, $NaOH(aq)$

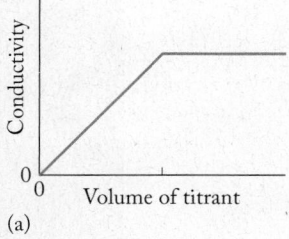
(a)

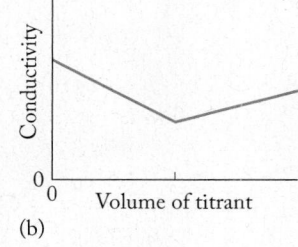
(b)

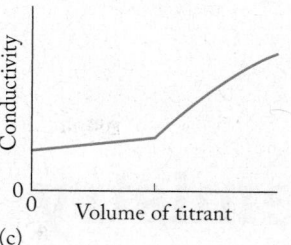
(c)

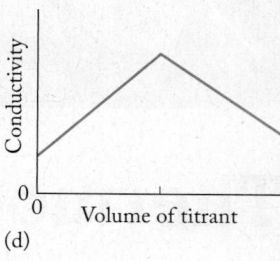
(d)

FIGURE P4.128

Calcium: In the Limelight

4.129. **Rocks in Caves** The stalactites and stalagmites in most caves are made of limestone (calcium carbonate) (see Figure 4.11). However, in the Lower Kane Cave in Wyoming they are made of gypsum (calcium sulfate). The presence of $CaSO_4$ is explained by the following sequence of reactions:

$$H_2S(aq) + 2O_2(g) \rightarrow H_2SO_4(aq)$$
$$H_2SO_4(aq) + CaCO_3(s) \rightarrow CaSO_4(s) + H_2O(\ell) + CO_2(g)$$

a. Which (if either) of these reactions is a redox reaction? How many electrons are transferred?

b. Write a net ionic equation for the reaction of H_2SO_4 with $CaCO_3$.

c. How would the net ionic equation be different if the reaction were written as follows?

$$H_2SO_4(aq) + CaCO_3(s) \rightarrow CaSO_4(s) + H_2CO_3(aq)$$

4.130. The alkaline earth elements react with nitrogen to form nitrides (with the general formula M_3N_2) for M = Be, Mg, Ca, Sr, and Ba. Like the alkaline earth oxides (general formula MO), the nitrides react with water to form alkaline earth hydroxides, $M(OH)_2$. Predict the other product for the reaction, and balance the equation:

$$M_3N_2(s) + H_2O(\ell) \rightarrow M(OH)_2(s) + \underline{\quad}$$

4.131. Which of the following reactions of calcium compounds is or are redox reactions?

a. $CaCO_3(s) \rightarrow CaO(s) + CO_2(g)$

b. $CaO(s) + SO_2(g) \rightarrow CaSO_3(s)$

c. $CaCl_2(s) \rightarrow Ca(s) + Cl_2(g)$

d. $3Ca(s) + N_2(g) \rightarrow Ca_3N_2(s)$

4.132. HF is prepared by reacting CaF_2 with H_2SO_4:

$$CaF_2(s) + H_2SO_4(\ell) \rightarrow 2HF(g) + CaSO_4(s)$$

HF can in turn be electrolyzed when dissolved in molten KF to produce fluorine gas:

$$2HF(\ell) \rightarrow F_2(g) + H_2(g)$$

Fluorine is extremely reactive, so it is typically sold as a 5% mixture by volume in an inert gas such as helium. How much CaF_2 is required to produce 500.0 L of 5% F_2 in helium? Assume the density of F_2 gas is 1.70 g/L.

5

Thermochemistry

Learning Outcomes

- Identify familiar endothermic and exothermic processes
- Calculate changes in the internal energy of a system
- Calculate the amount of heat transferred in physical or chemical processes
- Calculate thermochemical values using data from calorimetry experiments
- Recognize and write equations for formation reactions
- Calculate enthalpies of reaction
- Calculate and compare fuel and food values and fuel densities

The Sunlight Unwinding

Energy—to power an automobile, heat a home, or support life—may seem an abstract idea. Energy has no mass and it has no volume. However, we see energy changing matter from one state to another—sunlight melts snow, a gas flame boils water and converts it into steam—and we see energy being transformed from one form to another, as when the chemical energy of gasoline moves a car. Part of the energy in the gasoline contributes nothing to moving the vehicle and is lost to the surroundings as heat. By adding the energy used to move the vehicle and the energy lost to the surroundings, we find that the total energy equals the energy in the gasoline that was burned. In other words, energy is neither created nor destroyed during chemical reactions.

Chemical reactions produce nearly all of the energy we consume. For example, a pot of water on a gas stove gets hot because energy is given off when natural gas (mostly methane, CH_4) burns. We know that energy flows from the gas flame to the pot, and from the pot to the water inside it because heat flows from hot objects to cooler ones, and never the other way around. We can't directly measure the amount of energy in the water, or any object, but by measuring changes in the temperature of the water, we can calculate *the change* in its energy content.

We can roast marshmallows using energy from a campfire, but where does that energy come from? R. Buckminster Fuller (1895–1983), a 20th-century architect, inventor, and futurist, described a burning log like this: Trees gather the energy in sunlight and combine it with water and carbon dioxide to make the molecules that compose wood. When the wood is burned, the chemical products are carbon dioxide and water, and the fire is, as Fuller said, "all that sunlight unwinding." The sunlight unwinding is the release of chemical energy stored in the molecules of the wood. Through the transforming power of green plants, sunlight is the source of the chemical energy stored in all the substances we consume as food and fuel.

Forest in Tennessee The sun is the ultimate source of energy for most forms ▶ of life on Earth.

thermochemical equation the chemical equation of a reaction that includes heat as a reactant or a product.

energy the capacity to transfer heat or do work.

heat the energy transferred between objects because of a difference in their temperatures.

work a form of energy: the energy required to move an object through a given distance.

thermodynamics the study of energy and its transformations.

thermochemistry the study of the relation between chemical reactions and changes in energy.

heat transfer the process of heat energy flowing from one object into another.

Nearly all chemical reactions, such as combustion of fuels and reactions in solution, involve energy as either a product or a reactant, and all physical changes of matter involve changes in energy. By following these changes in energy, we can explain events such as ponds freezing in winter and thawing in spring, and we can answer questions such as whether methane is a better fuel than propane. Studying energy changes also gives important insights into the way nature works and helps us to address the impact of human activities on our world.

How do we measure the amount of energy involved in physical and chemical processes? When we keep track of energy, we can relate its changes to the identities and amounts of reactants and products involved in changes of state and in chemical reactions. Observing and quantifying changes in matter with respect to the flow of energy enables us to predict the physical and chemical behavior of substances and to rank fuels and foods in terms of their energy content. ■

5.1 Energy: Basic Concepts and Definitions

As in many chemical reactions, the one in which hydrogen combines with oxygen to form water releases energy: this energy may be converted into motion, as when spacecraft lift off, or into electrical energy, as happens in fuel cells. The equation representing a reaction in which energy is either a reactant or product is called a **thermochemical equation**, because it describes whether energy is absorbed or released when the reaction occurs. A thermochemical equation can be written for any reaction. For instance, we represent the reaction of hydrogen and oxygen to form water with the thermochemical equation

$$2H_2(g) + O_2(g) \rightarrow 2H_2O(\ell) + \text{energy}$$

The traditional definition of **energy** is the capacity either to transfer **heat** or to do **work**. What does this mean in terms of chemistry and the material world? In the most general sense, it means that when energy is transferred from one object to another, that energy does work, heats the object, or both. Energy that does work includes electrical, mechanical, light, and sound energy. Whatever the form, energy used to do work causes motion: the location of an object is changed when an energy source does work on the object. A transfer of energy changes the temperature of an object, making the object warmer or colder. Changes in energy can cause changes in the state of a material, as when a solid melts or a liquid freezes. The study of energy and its transformation from one form to another is called **thermodynamics**. The part of thermodynamics in which changes in energy that accompany chemical reactions are studied is known as **thermochemistry**.

When we put an ice cube initially at −18°C (0°F), the typical temperature of a freezer, into room-temperature water (25°C), the ice cube melts and the water cools because energy moves from the room-temperature water into the colder ice cube. The process by which energy moves from a warmer object to a cooler object is called **heat transfer**. The difference in temperature defines the direction of heat transfer when two objects come into contact: heat always flows from a hotter object into a colder object (Figure 5.1). Heat transfer changes the temperature of matter; it can also cause a change in physical state. The ice cube, for example, changes state from solid to liquid as heat is transferred to it from the water. The water remains in the liquid

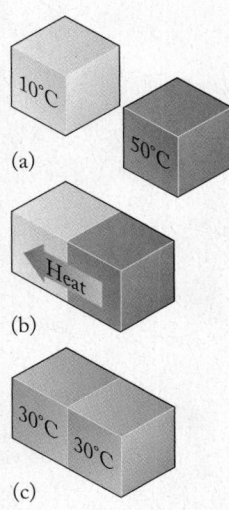

FIGURE 5.1 (a) Two identical blocks at different temperatures are brought into contact (b). Heat is transferred from the one at higher temperature to the one at lower temperature until (c) thermal equilibrium (same temperature) is reached.

state, but its temperature drops as heat from it is transferred to the ice cube. Ultimately, these two portions of matter achieve the same temperature, higher than the initial temperature of the ice cube but lower than the initial temperature of the water. At this point, **thermal equilibrium** has been reached, which means the temperature is the same throughout the combined material and no further heat transfer occurs.

thermal equilibrium a condition in which temperature is constant throughout a material and no heat flows from one point to another.

potential energy (PE) the energy stored in an object because of its position.

Work, Potential Energy, and Kinetic Energy

In the physical sciences, work (w) is done whenever a force (F) moves an object through a distance (d). The amount of mechanical work done is

$$w = F \times d \tag{5.1}$$

Consider Equation 5.1 as it relates to skiers ascending a mountain (Figure 5.2). The work (w) done by the lift on a skier equals the length of the ride (d) times the force (F) needed to overcome gravity and transport the skier up the mountain. Some of the work done is stored in the skier as **potential energy (PE)**, which is the energy an object has because of its position. The farther up the mountain the skier is carried, the greater his potential energy. The mathematical expression for the skier's potential energy is

$$PE = m \times g \times h \tag{5.2}$$

where m is the skier's mass, g is the acceleration due to the force of gravity, and h is the vertical distance between the skier's location on the mountain and his starting point. How he gets to that position is not important. Because the potential energy of any object does not depend on how the object gets to a

FIGURE 5.2 Work is done as skiers ascend to the top of a mountain. The amount of work may differ, depending on, for example, whether the skiers (1) ride a gondola on a direct route to the top or (2) hike to the top along a winding path.

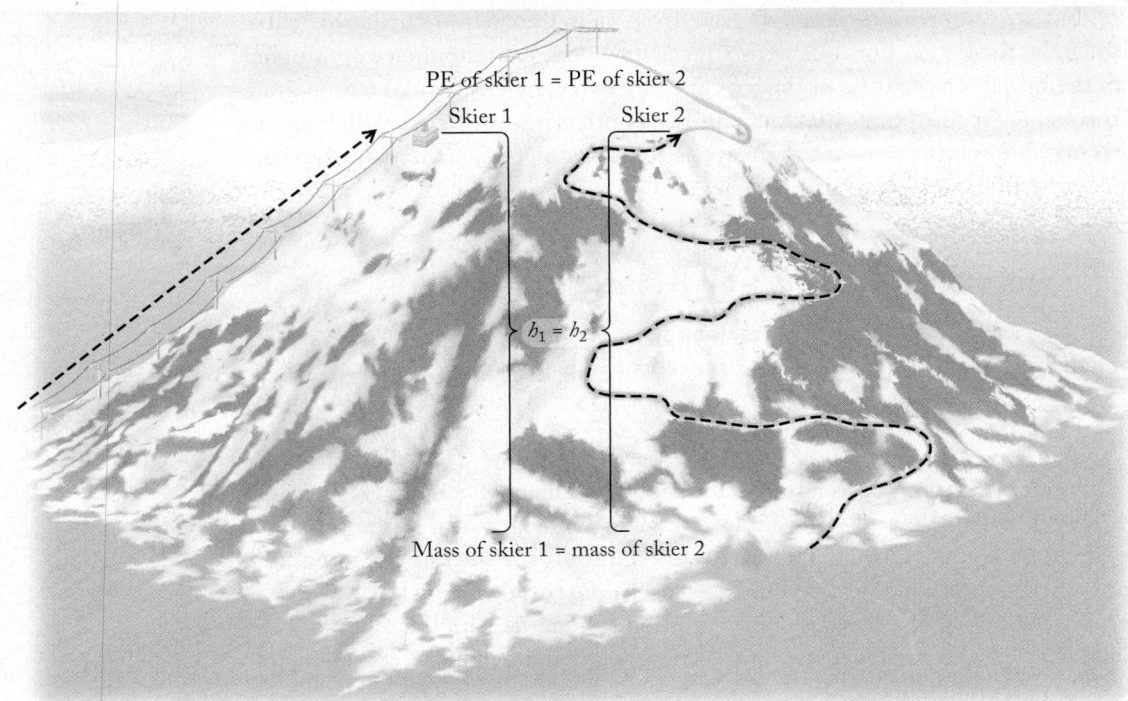

PE of skier 1 = PE of skier 2

Skier 1   Skier 2

$b_1 = b_2$

Mass of skier 1 = mass of skier 2

FIGURE 5.3 The potential energy of a skier depends only on the skier's mass and height above the base. If two skiers are at the same height ($b_1 = b_2$) and both skiers have the same mass, then they have the same potential energy, no matter how each skier got to that height.

particular point, potential energy is a **state function**, which means it is independent of the path followed to acquire the potential energy. Only position is important in considering potential energy (Figure 5.3). The term *state function* refers to a property of a system that is determined by the position or condition of the system; don't confuse it with the term *change of state*, which means a phase change.

Now consider the potential energy of a skier standing still at the top of a ski jump (Figure 5.4a). At this position, her energy is all potential energy, but as she moves down the slope, her potential energy is converted into **kinetic energy (KE)**, the energy of motion (Figure 5.4b). At any moment between the start of the run and coming to a stop at the bottom of the hill, the jumper's kinetic energy is proportional to the product of her mass (m) times the square of her speed (u):

$$KE = \tfrac{1}{2}mu^2$$

This equation tells us that a heavier skier (larger m) moving at the same speed as a lighter skier (smaller m) has more kinetic energy. Our intuition and experience tell us this as well. If you were standing at the bottom of the ski jump, how would your fate differ if a 136 kg (300 lb) skier ran into you rather than a 45 kg (100 lb) skier going the same speed? The difference lies in their relative kinetic energies.

According to the **law of conservation of energy**, energy cannot be created or destroyed. However, it can be converted from one form to another, as this example illustrates. Potential energy at the top of the slope becomes kinetic energy

▶❙❙ **CHEMTOUR** State Functions and Path Functions

state function a property of an entity based solely on its chemical or physical state or both, but not on how it achieved that state.

kinetic energy (KE) the energy of an object in motion due to its mass (m) and its speed (u):

$$KE = \tfrac{1}{2}mu^2$$

law of conservation of energy energy cannot be created or destroyed.

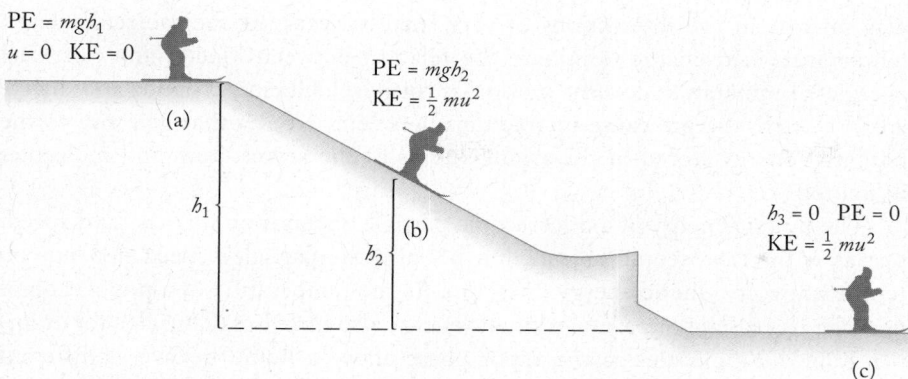

$PE = mgh_1$
$u = 0 \quad KE = 0$

(a)

$PE = mgh_2$
$KE = \frac{1}{2} mu^2$

h_1

(b)

h_2

$h_3 = 0 \quad PE = 0$
$KE = \frac{1}{2} mu^2$

(c)

FIGURE 5.4 (a) A skier at the starting gate of a ski jump has potential energy (PE) due to her position (h_1) above the bottom of the slope, her mass (m), and the force of gravity (g): $PE = mgh_1$. (b) During her run, the skier's potential energy is converted into kinetic energy: $KE = \frac{1}{2} mu^2$. While she moves down the slope, she has both KE and PE. (c) At the end of the run, the skier's potential energy is 0. Her KE decreases from its maximum value to 0 as she slows to a stop at some point along the flat region.

during the run. The total energy at any position on the hill is the sum of the skier's potential and kinetic energies.

CONCEPT TEST

Two skiers with masses m_1 and m_2, where $m_1 > m_2$, are poised at the starting gate of a downhill course (Figure 5.5a). Is the potential energy of skier 1 the same as that of skier 2? If the energies are different, which skier has more potential energy?

CONCEPT TEST

Two skiers with masses m_1 and m_2, where $m_1 > m_2$, go past the same elevation on parallel race courses at the same time (Figure 5.5b). At that moment, is the potential energy of skier 2 more than, less than, or equal to the PE of skier 1? If the two are moving at the same speed, which has the greater kinetic energy?

(Answers to Concept Tests are in the back of the book.)

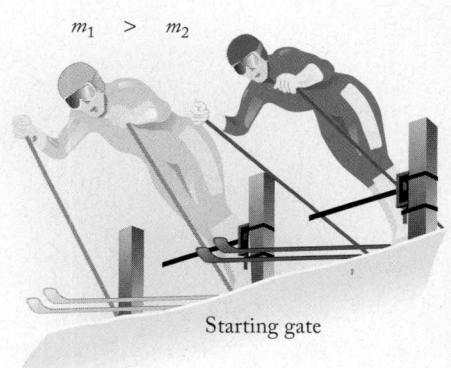

$m_1 > m_2$

Starting gate

(a)

Kinetic Energy and Potential Energy at the Molecular Level

The relation just described between kinetic and potential energies holds for atoms and molecules as well. There is no direct analogy with the kinetic and potential energies of a skier, however, because gravitational forces that control the skier

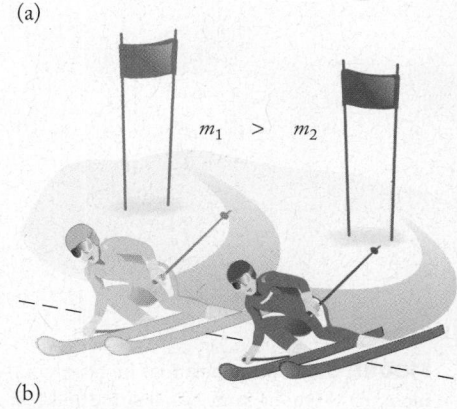

$m_1 > m_2$

(b)

FIGURE 5.5 (a) Two skiers of different mass are at the same position at the start of a race with respect to the bottom of the hill; (b) the same two skiers during the course of a race pass the same elevation at the same time.

play no role in the interactions of very small objects. At the molecular level, temperature and charge dominate the relation between kinetic and potential energies. Temperature governs motion at this level. Chemical bonds and differential electric charges cause interactions between particles that give rise to the potential energy stored in the arrangements of the atoms, ions, and molecules in matter.

The kinetic energy of a microscopic particle depends on its mass and speed, just as with macroscopic objects, but because the particle's speed depends on temperature, its kinetic energy does, too. As the temperature of a population of particles increases, their average kinetic energy also increases. Consider, for example, how the molecules in the vapor phase above a liquid behave at different temperatures. We pick the gas phase because the molecules in a gas at normal pressures are widely separated and behave independently; they do not interact with each other, and they all behave the same way regardless of their identities. If we have two samples of water vapor (molecular mass 18.02 amu) at room temperature, the two populations of H_2O molecules have the same average kinetic energy. The average speeds of the molecules are the same because their masses are identical. If the temperature of one sample is increased, that population of H_2O molecules acquires a higher average kinetic energy and the average speed of the molecules increases. An equivalent population of ethanol molecules (molecular mass 46.07 amu) in the gas phase at room temperature has the same average kinetic energy as the water molecules at room temperature, but the average speed of the ethanol molecules is lower because their mass is higher (Figure 5.6).

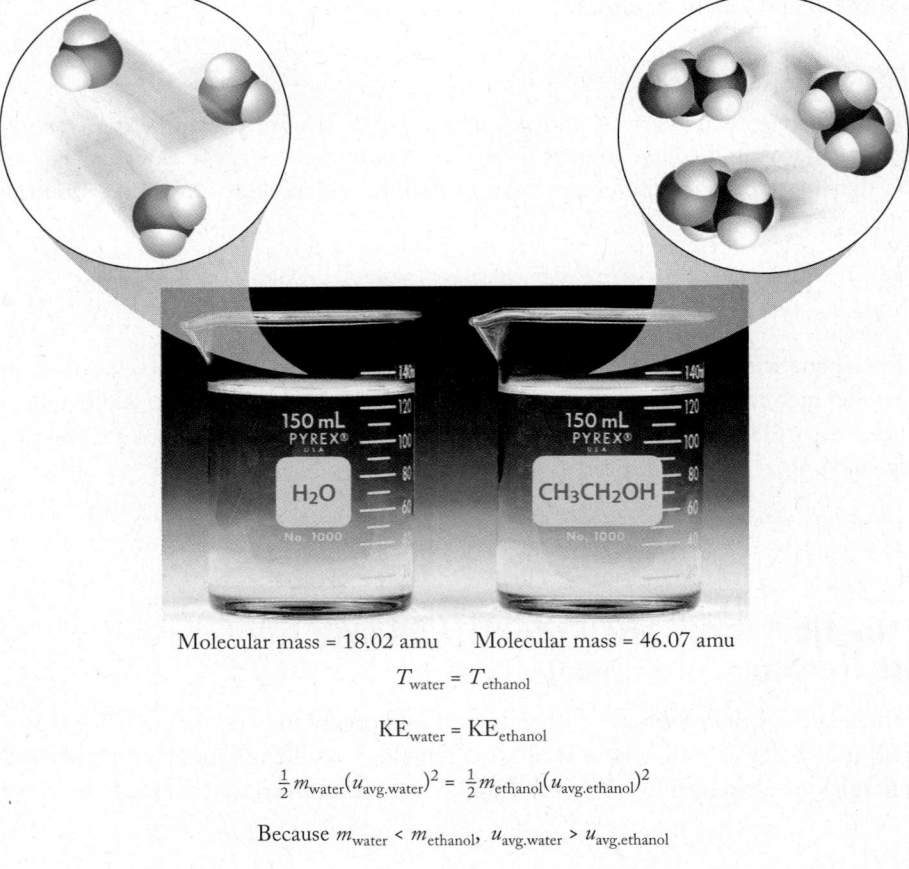

Molecular mass = 18.02 amu   Molecular mass = 46.07 amu

$$T_{\text{water}} = T_{\text{ethanol}}$$

$$\text{KE}_{\text{water}} = \text{KE}_{\text{ethanol}}$$

$$\tfrac{1}{2} m_{\text{water}}(u_{\text{avg.water}})^2 = \tfrac{1}{2} m_{\text{ethanol}}(u_{\text{avg.ethanol}})^2$$

Because $m_{\text{water}} < m_{\text{ethanol}},\ u_{\text{avg.water}} > u_{\text{avg.ethanol}}$

FIGURE 5.6 Two populations of gas-phase molecules have the same temperature and therefore the same average kinetic energy. Because ethanol molecules have a greater mass than water molecules, the average speed of the water molecules in the water vapor above the liquid water is greater than the average speed of the gas-phase ethanol molecules above the liquid ethanol.

The kinetic energy associated with the random motion of molecules is called **thermal energy**, and the thermal energy of a given sample of matter is proportional to the temperature of the sample. However, thermal energy also depends on the number of particles in a sample. The water in a swimming pool and in a cup of water taken from the pool have the same temperature, so their molecules have the same average kinetic energy. The water in the pool has much more thermal energy than the water in the cup, however, simply because there is a larger number of molecules in the pool. A large number of particles at a given temperature has a higher total thermal energy than a small number of particles at the same temperature.

thermal energy the kinetic energy of atoms, ions, and molecules.

electrostatic potential energy (E_{el}) the energy a particle has because of its position relative to another particle; it is directly proportional to the product of the charges of the particles and inversely proportional to the distance between them.

> **CONCEPT TEST**
>
> If we heat a cup of water from a swimming pool almost to the boiling point, will its thermal energy be more than, less than, or the same as the thermal energy of all the water in the pool?
>
> *(Answers to Concept Tests are in the back of the book.)*

An important form of potential energy at the atomic level arises from electrostatic interactions between charged particles. Just as the potential energy of skiers is determined by their positions above the bottom of the slope, the **electrostatic potential energy (E_{el})** of charged particles is determined by the distance between them. The magnitude of this electrostatic potential energy, also known as *coulombic interaction*, is directly proportional to the product of the charges (Q_1 and Q_2) on the particles and is inversely proportional to the distance (d) between them:

$$E_{el} \propto \frac{Q_1 \times Q_2}{d} \tag{5.3}$$

where the symbol $\propto$ means "is proportional to." Coulombic interactions determine the potential energy of matter at the atomic level because they determine the relative positions of particles.

If the two charges in Equation 5.3 are either both positive or both negative, their product is positive, the particles repel each other, and E_{el} is positive. If one particle is positive and the other negative, their product is negative, the particles attract each other, and E_{el} is negative (Figure 5.7b–c). A lower

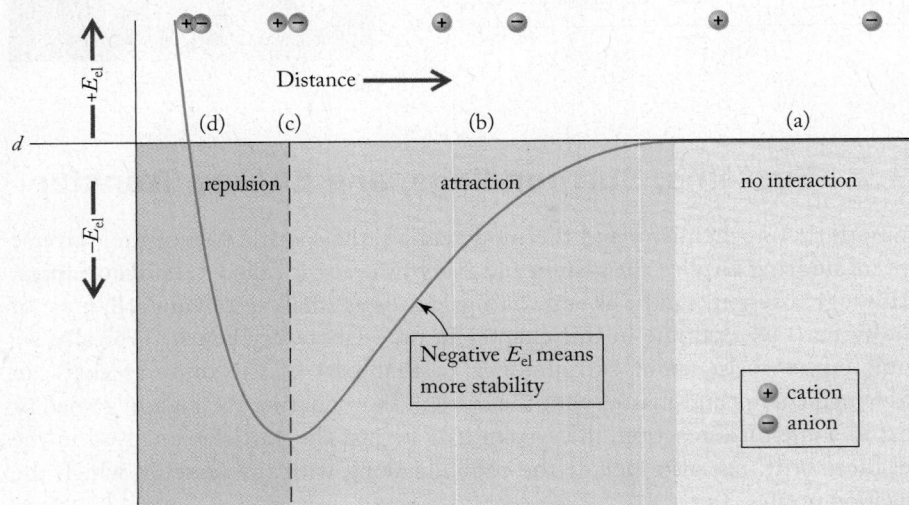

FIGURE 5.7 Electrostatic potential energy. (a) A positive ion and a negative ion are so far apart (d is large) that they do not interact at all. (b) As the ions move closer together (d decreases), the electrostatic potential energy between them becomes more negative. (c) At this distance the attraction between them produces an arrangement that is the most favorable energetically because the ions have the lowest electrostatic potential energy. (d) If the ions are forced even closer together, they repel each other as their nuclei begin to interact.

electrostatic potential energy (a more negative value of E_{el}) corresponds to greater stability, so particles that attract each other because of their charge form an arrangement with a lower electrostatic potential energy than particles that repel each other. However, there is a limit to how close two particles can be. Remember that ions, whatever their charge, have positively charged nuclei, and these nuclei repel each other if they are pushed together too closely (Figure 5.7d).

Ions are not the only particles that experience coulombic interactions. Neutral species such as water molecules attract each other as well due to distortions in the electron distribution about the nuclei of their atoms. The same ideas we used to describe the behavior of oppositely charged ions apply to molecules as well. Whether dealing with matter composed of atoms, molecules, or ions, the total energy at the microscopic level is the sum of the kinetic energy due to the random motion of particles and the potential energy due to their arrangement.

The energy given off or absorbed during a chemical reaction is due to the difference in the potential energy of the reactants and products. For example, when hydrogen molecules burn in oxygen, the products are water and a considerable amount of energy (Figure 5.8). The energy given off by this reaction powers rockets used to launch satellites and is now being used to run some buses and automobiles. The fact that energy is given off in the reaction tells us that the product molecules must be at a lower potential energy than the reactant molecules. The difference between the energy of the products and the energy of the reactants is the energy released. In the case of hydrogen combustion (Figure 5.9a), this energy is sufficient to run vehicles now powered by fossil fuels (Figure 5.9b).

FIGURE 5.8 Energy from the combustion of hydrogen can be used to launch rockets.

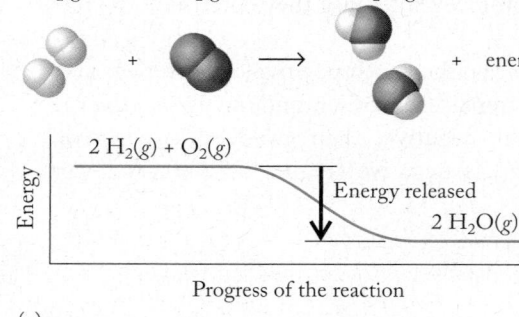

FIGURE 5.9 (a) Hydrogen reacts with oxygen to produce water. Because this reaction releases energy as it runs, the molecules in the product are at a lower energy than those in the reactants. (b) Fueling a hydrogen-powered vehicle.

$2 H_2(g)$ + $O_2(g)$ ⟶ $2 H_2O(g)$ + energy

+ ⟶ + energy

Energy

$2 H_2(g) + O_2(g)$

Energy released

$2 H_2O(g)$

Progress of the reaction

(a)

(b)

5.2 Systems, Surroundings, and Energy Transfer

In both thermochemistry and thermodynamics, the specific part of the universe we are studying is called the **system** and everything else is called the **surroundings**. Although a system can be as large as a galaxy or as small as a living cell, most of the systems we examine in this chapter fit on a laboratory bench. Typically, we limit our concern about surroundings to that part of the universe that can exchange energy and matter with the system. In evaluating the energy gained or lost in a chemical reaction, the system may be just the particles involved in the reaction, or it may also include the contents along with the vessel in which the reaction occurs.

system the part of the universe that is the focus of a thermochemical study.

surroundings everything that is not part of the system.

(a) **Isolated system:**
A thermos bottle
containing hot soup with
the lid screwed on tightly

(b) **Closed system:** A cup
of hot soup with a lid

(c) **Open system:** An open cup of
hot soup

Steam

Pepper
mill

Grated cheese   Crackers

FIGURE 5.10 Transfer of energy and matter in isolated, closed, and open systems: (a) Hot soup in a tightly sealed thermos bottle approximates an isolated system: no vapor escapes, no matter is added or removed, and no energy escapes to the surroundings. (b) Hot soup in a cup with a lid is a closed system; the soup transfers heat to the surroundings as it cools; however, no matter escapes and none is added. (c) Hot soup in a cup with no lid is an open system; it transfers both matter (steam) and energy (heat) to the surroundings as it cools. Matter in the form of pepper, grated cheese, crackers, or other matter from the surroundings may also be added to the soup.

Isolated, Closed, and Open Systems

In discussions involving energy transferred, or work, or both, three types of systems are common: isolated, closed, and open (Figure 5.10). These designations are important because they define the system we are dealing with and the part of the universe the system interacts with. In the three cases in the following discussion, the system is hot soup.

Consider hot soup in an ideal closed thermos bottle. An ideal thermos bottle takes no energy away from the soup, thereby completely insulating the soup from the rest of the universe, and it loses no energy. This is, of course, impossible, but sometimes in thermochemistry we must discuss systems that do not exist in the real world to define the total range of possibilities. The soup in the ideal thermos bottle is an example of an **isolated system**, which is a system that exchanges no energy or matter with its surroundings. The ideal thermos bottle prevents matter from being exchanged with the surroundings, and the thermal insulation provided by the ideal thermos also prevents heat from being transferred from the system to the surroundings, including the bottle itself. From the soup's point of view, the soup *is* the entire universe; it neither picks up nor donates any matter to the rest of the world and it loses no energy. It is *isolated*, and in an isolated system, the system has no surroundings; that's why it is called isolated. The mass of the system does not change, and its energy content is constant; the soup stays at one constant temperature. Of course, even the best real thermos bottle can't maintain the soup over time as an isolated system because energy will leak out and the contents will cool, but for short time periods the soup in a good thermos bottle approximates an isolated system.

Hot soup in a cup with a lid is an example of a **closed system**, which is a system that exchanges energy but not matter with its surroundings. Because the cup has a lid, no vapor (which is matter) escapes from the soup and no matter can be added. Only energy is exchanged between the soup and its surroundings. Heat from the soup is transferred—first into the cup walls, then into the air and the tabletop—and the soup gradually cools. Many real systems are closed systems.

isolated system a system that exchanges neither energy nor matter with the surroundings.

closed system a system that exchanges energy but not matter with the surroundings.

open system a system that exchanges both energy and matter with the surroundings.

exothermic process one in which energy flows from a system into its surroundings.

endothermic process one in which energy flows from the surroundings into the system.

CONNECTION In Chapter 3 we defined combustion as the reaction of oxygen with another element.

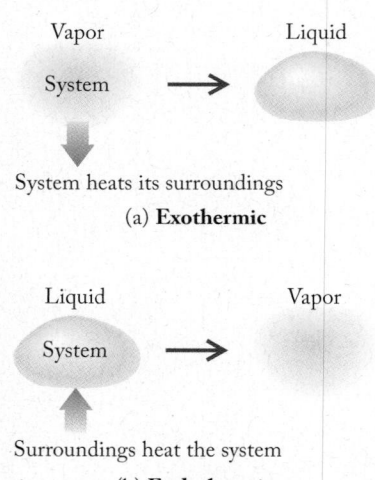

(a) **Exothermic**

System heats its surroundings

(b) **Endothermic**

Surroundings heat the system

FIGURE 5.11 A process that is exothermic in one direction, such as (a) the condensation of a vapor, is endothermic in the reverse direction, (b) the evaporation of a liquid. Reversing a process changes the direction in which energy is transferred but not the quantity of energy transferred.

Soup in an open cup is an **open system**, one that can exchange both energy and matter with its surroundings. Energy from the soup is transferred to the surroundings (cup, air, tabletop), and matter in the form of water vapor leaves the system and enters the air or vice versa. We may also add matter to the system from the surroundings by sprinkling on a little grated cheese, some ground pepper, or a few crumbled crackers.

Most of the real systems we deal with are closed, and we may treat some of them as isolated, at which point we deal with behavior that is ideal as opposed to real. We do this with real systems because isolated and closed systems are easier to deal with quantitatively than open systems. However, many important systems are open—including cells, organisms, and Earth itself.

CONCEPT TEST

Identify the following systems as isolated, closed, or open: (a) the water in a pond; (b) a carbonated beverage in a sealed bottle; (c) a sandwich wrapped in thermally conducting plastic wrap; (d) a live chicken.

(Answers to Concept Tests are in the back of the book.)

Exothermic and Endothermic Processes

Chemists classify thermochemical processes based on whether they give off or absorb energy. A chemical reaction or a change of state that results in the transfer of heat from a system to its surroundings is **exothermic** from the point of view of the system (Figure 5.11a). This energy can be detected because it causes an increase in the temperature of the surroundings. Combustion reactions (the reactants are the system) release energy and are examples of exothermic reactions. In contrast, a chemical reaction or change of state that absorbs energy from the surroundings is **endothermic** from the point of view of the system (Figure 5.11b). For example, ice cubes (the system) in a glass of warm water (the surroundings) absorb energy from the water, which causes the cubes to melt. The process is endothermic because energy enters the system.

In another phase change, water vapor (the system) from humid air condensing into drops of liquid water on the outside of a glass containing an ice-cold drink (the surroundings) requires that energy flow from the system to the surroundings. From the point of view of the system, this process is exothermic because heat leaves the system (the water vapor). If we pour out the cold drink and pour hot coffee into the glass, the water drops (the system) on the outside surface of the glass absorb energy from the coffee and vaporize in a process that is endothermic from the point of view of the system. This illustrates an important concept: a process that is exothermic in one direction (vapor → liquid; releases energy) is endothermic in the reverse direction (liquid → vapor; absorbs energy).

We use the symbol q to represent the *quantity* of energy produced or consumed by a chemical reaction or a change of state. If the reaction or process is exothermic, q is negative, meaning that the system *loses* energy to its surroundings. If the reaction or process is endothermic, q is positive, indicating that energy is *gained* by the system. In Figure 5.12, endothermic changes of state are represented by arrows pointing upward: solid → liquid, liquid → gas, and solid → gas. The opposite changes: liquid → solid, gas → liquid, and gas → solid, represented by arrows pointing downward, are exothermic. To summarize:

$$\text{Exothermic:} \quad q < 0 \qquad \text{Endothermic:} \quad q > 0$$

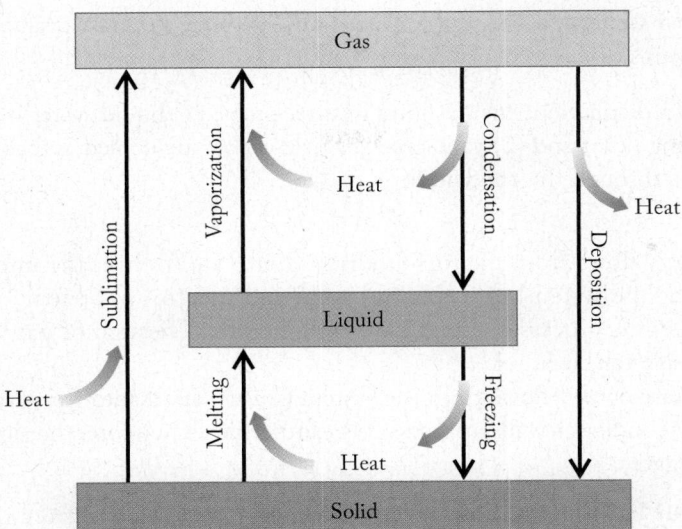

FIGURE 5.12 Matter can be transformed from one physical state to another by adding or removing heat. Upward-pointing black arrows represent endothermic processes (heat enters the system from the surroundings; $q > 0$). Downward-pointing arrows represent exothermic processes (heat leaves the system and enters the surroundings; $q < 0$).

SAMPLE EXERCISE 5.1   **Identifying Exothermic and Endothermic Processes**

Describe the flow of heat during the purification of water by distillation (Figure 5.13), identify the steps in the process as either endothermic or exothermic, and give the sign of q associated with each step. Consider the water being purified to be the system.

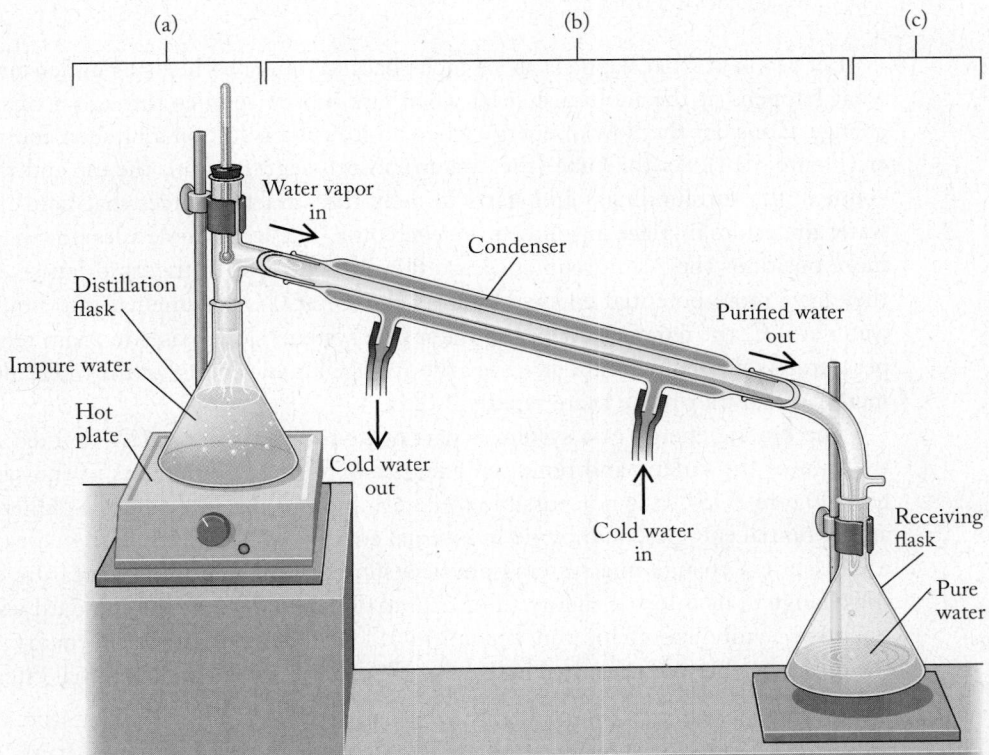

FIGURE 5.13 A laboratory setup for distilling water. (a) Impure water is heated to the boiling point in the distillation flask. (b) Water vapor rises and enters the condenser, where it is liquefied. (c) The purified liquid is collected in the receiving flask.

Collect and Organize Since the water is the system, we must evaluate how the water gains or loses energy during distillation.

Analyze In distillation, energy flows in three steps: (1) liquid water is heated to the boiling point and (2) vaporizes. (3) The vapors are cooled and condense as they pass through the condenser.

Solve

a. Energy flows from the surroundings (hot plate) to heat the impure water (the system) to its boiling point and then to vaporize it. Therefore, processes 1 and 2 are endothermic. The sign of q is positive for both.

b. Because energy flows from the system (water vapor) into the surroundings (condenser walls), process 3 is exothermic. Therefore, the sign of q is negative.

Think about It "Endothermic" means that energy is transferred from the surroundings into the system, the water in the distillation flask. When the water vapor is cooled in the condenser, energy flows from the vapor as it is converted from a gas to a liquid; the process is exothermic.

Practice Exercise What is the sign of q as (a) a match burns, (b) drops of molten candle wax solidify, and (c) perspiration evaporates from skin? In each case, define the system and indicate whether the process is endothermic or exothermic.

(Answers to Practice Exercises are in the back of the book.)

∞ CONNECTION In Chapter 1 we discussed the arrangement of molecules in ice, water, and water vapor.

▶❚❚ CHEMTOUR Internal Energy

internal energy (E) the sum of all the kinetic and potential energies of all of the components of a system.

first law of thermodynamics the energy gained or lost by a system must equal the energy lost or gained by the surroundings.

calorie (cal) the amount of energy necessary to raise the temperature of 1 g of water by 1°C.

joule (J) the SI unit of energy; 4.184 J = 1 cal.

Let's look at what happens as ice melts because doing so helps us understand what happens at the molecular level when any substance goes through a phase change. Consider the flow of energy when an ice cube is left on a kitchen counter (Figure 5.14). As the cube (the system) absorbs energy from the air and the counter (the surroundings) and starts to melt, the attractive forces that hold the water molecules in place in solid ice are overcome. The water molecules now have more positions they can occupy because they are not held in the rigid lattice, so they have more potential energy. After all the ice at 0°C has melted into liquid water at 0°C, the temperature of the water (the system) slowly rises to room temperature. As the temperature increases, the average kinetic energy of the molecules increases and they move more rapidly.

The kinetic energy of a system is part of its **internal energy (E)**, defined as the sum of the kinetic and potential energies of all the components of the system (Figure 5.15). It is not possible to determine the absolute values of kinetic and potential energies, but *changes* in internal energy (ΔE) are fairly easy to measure because a change in a system's physical state or temperature is a measure of the change in its internal energy. (The capital Greek delta, Δ, is the standard way scientists symbolize change in a quantity.) The change in internal energy is the difference between the final internal energy of the system and its initial internal energy:

$$\Delta E = E_{\text{final}} - E_{\text{initial}} \qquad (5.4)$$

Internal energy is a state function because ΔE depends only on the initial and final states. How the change occurs in the system does not matter.

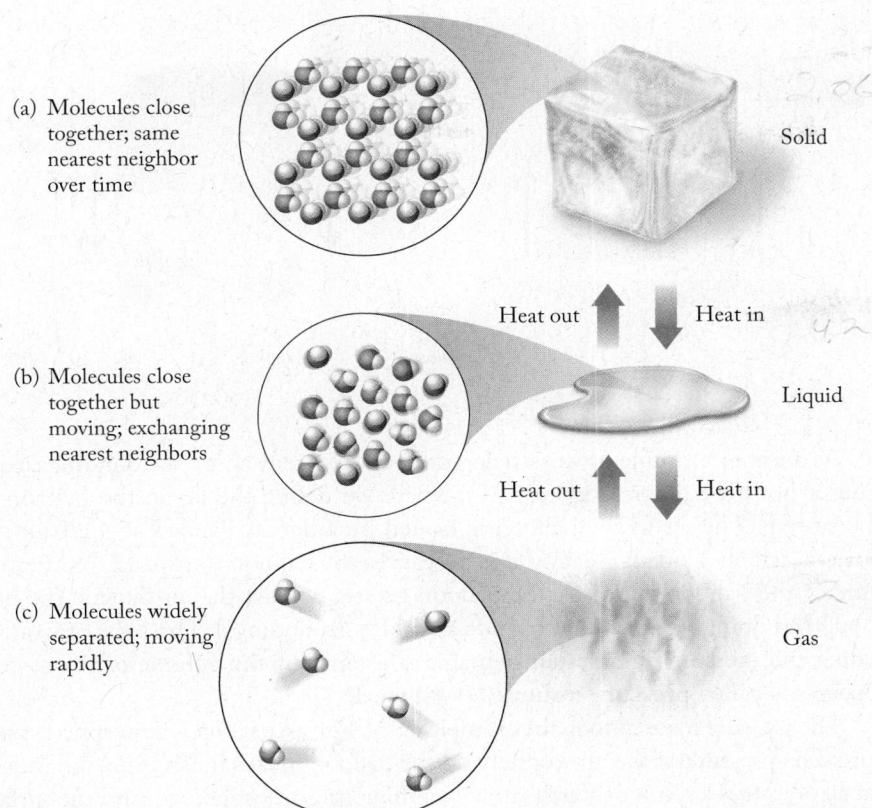

(a) Molecules close together; same nearest neighbor over time

Solid

Heat out / Heat in

(b) Molecules close together but moving; exchanging nearest neighbors

Liquid

Heat out / Heat in

(c) Molecules widely separated; moving rapidly

Gas

FIGURE 5.14 Changes of state. (a) Solid ice absorbs energy and is converted to liquid water. The molecules in the solid are held together in a rigid three-dimensional arrangement; they have the same nearest neighbors over time. (b) When the solid melts, the molecules in the liquid state exchange nearest neighbors and occupy many more positions relative to one another than were possible in the solid. (c) The liquid absorbs energy and is converted to a gas. The molecules in a gas are widely separated from one another and move rapidly. The reverse of these processes occurs when water in the gas state loses energy and condenses to a liquid. The liquid also loses energy when it is converted to a solid.

The law of conservation of energy (Section 5.1) applies to the transfer of energy in materials. The total energy change experienced by a system must be balanced by the total energy change experienced by its surroundings. The law of conservation of energy is a statement of the **first law of thermodynamics**. The energy changes of the system and the surroundings are equal in magnitude but opposite in sign, so their sum is zero. Energy is neither created nor destroyed; it is conserved.

Energy Units and *P-V* Work

Energy changes that accompany chemical reactions and changes in physical state are sometimes expressed in calories. A **calorie (cal)** is the quantity of energy required to raise the temperature of 1 g of water from 14.5°C to 15.5°C. The SI unit of energy, used throughout this text, is the **joule (J)**; 1 cal = 4.184 J. The *Calorie* (Cal; note the capital C) in nutrition is actually one kilocalorie (kcal): 1 Cal = 1 kcal = 1000 cal.

Doing work on a system is a way to add to its internal energy. For example, compressing a quantity of gas (the system) into a smaller volume does work on the gas, and that work causes the temperature of the gas to rise, meaning its internal energy increases. The total increase in the internal energy of a system is the sum of the work done on it (w) and any energy (q) gained by heating:

$$\Delta E = q + w \tag{5.5}$$

When work is done *by* a system on its surroundings, the internal energy of the system decreases. For example, when fuel in the cylinder of a diesel engine ignites and produces hot gases, the gases (the system) expand and do work on the surroundings by pushing on the piston (Figure 5.16).

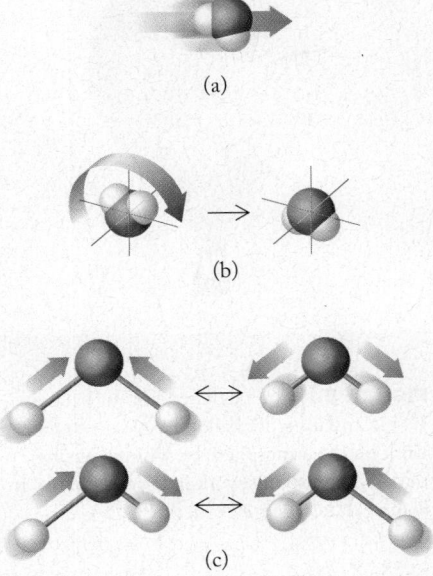

(a)

(b)

(c)

FIGURE 5.15 Some of the types of molecular motion that contribute to the overall internal energy of a system: (a) translational motion, motion from place to place along a path; (b) rotational motion, motion about a fixed axis; and (c) vibrational motion, movement back and forth from some central position.

FIGURE 5.16 Work performed by changing the volume of a gas. (a) Highest position of the piston after it compresses the gas in the cylinder, causing the fuel to ignite in a diesel engine. The piston does work on the gas in decreasing the gas volume. (b) The exploding fuel releases energy, causing the gas in the cylinder to expand and push the piston down. The gas has done work on the piston.

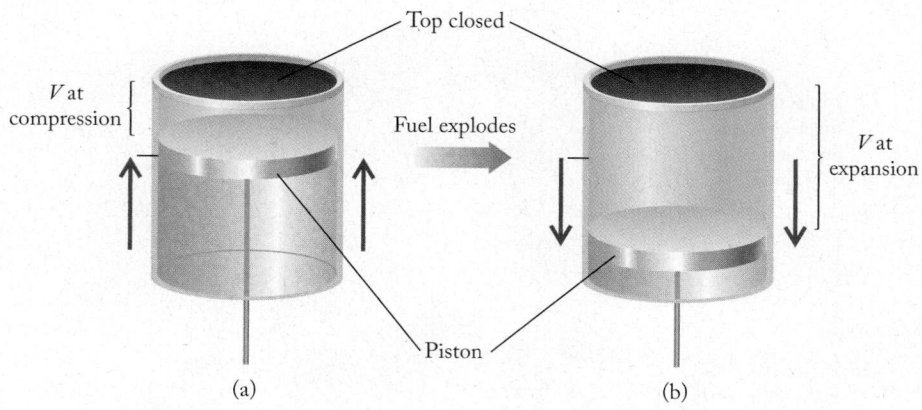

(a)   (b)

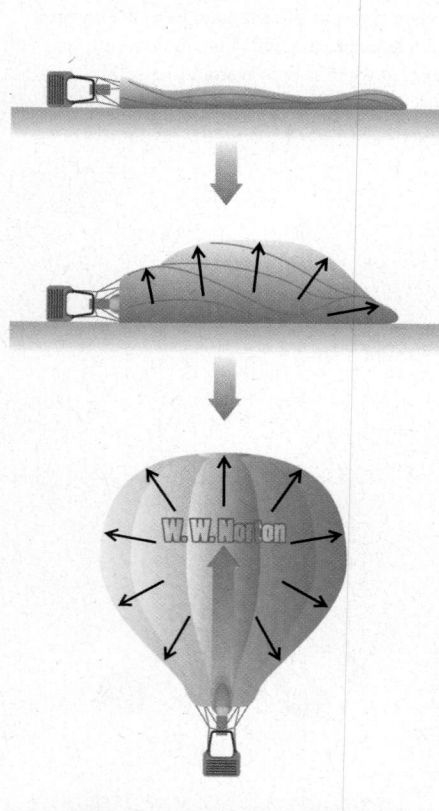

FIGURE 5.17 In a deflated hot-air balloon, $V = 0$. Inflating the balloon causes it to do work on the atmosphere by pushing against the air. Because this work involves a change in volume, it is called $P–V$ work.

▶‖ **CHEMTOUR** Pressure-Volume Work

....

pressure-volume ($P–V$) work the work associated with the expansion or compression of a gas.

....

As another example of work being done by a system on its surroundings, consider a hot-air balloon (Figure 5.17), where we define the air in the balloon as the system. The air in the balloon is heated by a burner located at the balloon's bottom, which is open. Heating this air causes the balloon to expand. As the volume of the balloon increases, the balloon presses against the air outside the balloon, thus doing work on that outside air (the surroundings). This type of work in which the pressure on a system remains constant but the volume of the system changes is called **pressure–volume ($P–V$) work**.

The pressure referred to in the example of the hot-air balloon is atmospheric pressure. The pressure of the atmosphere on the balloon in this illustration and indeed on all objects is a result of Earth's gravity pulling the atmosphere toward the surface of the planet, where it exerts a force on all things because of its mass. The pressure at sea level on a dry day is about 1.00 atmosphere (atm), which equals a force of 14.7 pounds pressing on each square inch of surface area. Atmospheric pressure varies slightly with the weather, but in an example involving $P–V$ work like this one with the balloon, the important feature is that the pressure is constant.

Let's examine what happens to the balloon in terms of heat transferred and work done under constant pressure. First, adding hot air to the balloon increases its internal energy and causes the balloon to expand. Second, the expansion of the balloon against the pressure of its surroundings is $P–V$ work done by the system on its surroundings. The internal energy of the system decreases as it performs this $P–V$ work. We can relate this change in internal energy to the energy gained by the system by heating (q) and the work done by the system ($P\Delta V$) by writing Equation 5.5 in the form

$$\Delta E = q + (-P\Delta V) = q - P\Delta V \qquad (5.6)$$

The negative sign in front of $P\Delta V$ is appropriate in this case because, when the system expands (positive change in volume ΔV), it loses energy (negative change in internal energy ΔE) as it does work on its surroundings. Correspondingly, when the surroundings do work on the system (for example, when a gas is compressed), the quantity $(-P\Delta V)$ has a positive value.

The sign of q may also be either positive or negative (Figure 5.18). If the system is heated by its surroundings, then q is positive ($q > 0$). Energy is added to the balloon, for instance, when it is being inflated, because heat flows from the surroundings (the burner) into the system (the air in the balloon). When the balloon expands, energy flows out of the system into the surroundings, and the sign of the work done by the system is negative. If heat is transferred from the system into the surroundings, its sign is also negative ($q < 0$). From Equation 5.6, we see that the change in internal energy of a system is positive when more energy enters the system than leaves and negative when more energy leaves the system than enters.

Surroundings

System

Heat in ➤ $+q$

Heat out ➤ $-q$

Work done on system ➤ $+w$

Work done by system ➤ $-w$

$\Delta E = q + w$

FIGURE 5.18 Energy entering a system by heating or by work done on the system by the surroundings are both positive quantities because both increase the internal energy of the system. Energy released by a system to the surroundings and work done by a system on the surroundings are both negative quantities because both decrease the internal energy of a system.

SAMPLE EXERCISE 5.2 Calculating Changes in Internal Energy

Figure 5.19 shows a simplified version of a piston and cylinder in an engine. Suppose combustion of fuel injected into the cylinder produces 155 J of energy. The hot gases in the cylinder expand, pushing the piston down. In doing so, the gases do 93 J of *P–V* work on the piston. If the system is the gases in the cylinder, what is the change in internal energy of the system?

Collect and Organize The change in internal energy is related to the work done by a system or on a system and the heat gained or lost by the system (Equation 5.5). First we have to decide whether q and w are positive or negative according to the sign convention shown in Figure 5.18.

Analyze The system (gases in the cylinder) absorbs energy, so $q > 0$, and the system does work on the surroundings (the piston), so $w < 0$.

Solve

$$\Delta E = q + w = (155\ \text{J}) + (-93\ \text{J}) = 62\ \text{J}$$

Think about It More energy enters the system (155 J) than leaves it (93 J), so a positive value of ΔE is reasonable.

Practice Exercise In another event, the piston in Figure 5.19 compresses the air in the cylinder by doing 64 J of work on the gas. As a result, the air gives off 32 J of energy to the surroundings. If the system is the air in the cylinder, what is the change in its internal energy?

(Answers to Practice Exercises are in the back of the book.)

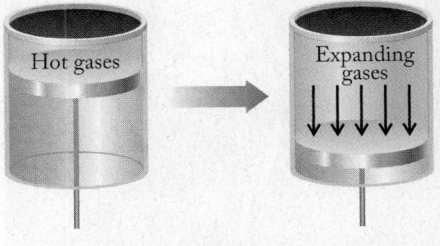

Hot gases

Expanding gases

FIGURE 5.19

SAMPLE EXERCISE 5.3 Calculating *P–V* Work

A tank of compressed helium is used to inflate balloons for sale at a carnival on a day when the atmospheric pressure is 1.01 atm. If each balloon is inflated from an initial volume of 0.0 L to a final volume of 4.8 L, how much *P–V* work is done by 100 balloons on the surrounding atmosphere when they are inflated? The atmospheric pressure remains constant during the filling process.

Collect and Organize Each of 100 balloons goes from empty ($V = 0.0$ L) to 4.8 L, which means $\Delta V = 4.8$ L, and the atmospheric pressure P is constant at

1.01 atm. The identity of the gas used to fill the balloons doesn't matter, because under normal conditions all gases behave the same way, regardless of their identity. Our task is to determine how much $P–V$ work is done by the 100 balloons on the air that surrounds them.

Analyze We focus on the work done on the atmosphere (the surroundings) by the system; the balloons and the helium they contain are the system.

Solve The volume change (ΔV) as all the balloons are inflated is

$$100 \text{ balloons} \times 4.8 \text{ L/balloon} = 480 \text{ L}$$

The work (w) done by our system (the balloons) as they inflate against an external pressure of 1.01 atm is

$$w = -P\Delta V = 1.01 \text{ atm} \times 480 \text{ L}$$
$$= -480 \text{ L} \cdot \text{atm}$$

Because work is done *by* the system on its surroundings, the work is negative from the point of view of the system: $w = -480 \text{ L} \cdot \text{atm}$.

Think about It The internal energy of the balloons (the system) decreases when they are inflated. Air injected into a balloon cools when the balloon expands and does work on the surroundings. The work is done by the system.

Practice Exercise The balloon *Spirit of Freedom* (Figure 5.20), flown around the world by American Steve Fossett in 2002, contained 550,000 cubic feet of helium. How much $P–V$ work was done by the balloon on the surrounding atmosphere while the balloon was being inflated, assuming atmospheric pressure was 1.00 atm? ($1 \text{ m}^3 = 1000 \text{ L} = 35.3 \text{ ft}^3$.)

(Answers to Practice Exercises are in the back of the book.)

FIGURE 5.20 The balloon *Spirit of Freedom* was flown around the world in 2002.

The units "L · atm" (liters times atmospheres) may seem strange for expressing work. We have a conversion factor, 101.32 J/(L · atm), that relates joules, the SI unit for energy and work, to liters times atmospheres. Using this factor for the work done in filling the balloons in Sample Exercise 5.3, we get

$$w = -480 \text{ L} \cdot \text{atm}[101.32 \text{ J}/(\text{L} \cdot \text{atm})] = -4.9 \times 10^4 \text{ J} = -49 \text{ kJ}$$

5.3 Enthalpy and Enthalpy Changes

Many physical and chemical changes take place at constant atmospheric pressure (P). The thermodynamic parameter that relates the flow of energy into or out of a system during chemical reactions or physical changes at constant pressure is called the **enthalpy change (ΔH)**. We symbolize this enthalpy change at constant pressure as q_P, where the subscript P specifies a process taking place at constant pressure. We then rearrange Equation 5.6 to define ΔH as

$$\Delta H = q_P = \Delta E + P\Delta V \qquad (5.7)$$

enthalpy change (ΔH) the energy absorbed by the reactants (endothermic reaction) or the energy given off by the products (exothermic reaction) for a reaction carried out at constant pressure.

Thus the change in enthalpy is the energy transferred at constant pressure when $P–V$ work is done.

The **enthalpy (H)** of a thermodynamic system is the sum of the internal energy and the pressure–volume product ($H = E + PV$), but as we saw with internal energy, determining the absolute values of these parameters is difficult whereas determining *changes* is fairly easy. We therefore concentrate on the *change* in enthalpy (ΔH) of a system or the surroundings. Equation 5.7 tells us that for a reaction run at constant pressure, the enthalpy change is equal to q_P, the heat gained or lost by the system during the reaction. This statement also means that the units for ΔH are the same as those for q. Heat has units of joules (J), and if it is reported with respect to the quantity of a substance, J/g or J/mol.

When heat flows out of a system, q is negative according to our sign convention (Figure 5.18), so the enthalpy change is negative: $\Delta H < 0$. When heat flows into a system, q is positive, so $\Delta H > 0$. As we saw in Section 5.2, positive q values indicate endothermic processes and negative q values indicate exothermic processes. Knowing that, we can relate the terms *exothermic* and *endothermic* to enthalpy changes as well. For example, heat flows into a melting ice cube (the system) from its surroundings; this process is endothermic, meaning that q—and therefore ΔH—is positive. However, to make ice cubes in a freezer, the water (the system) must lose heat, which means the process is exothermic and both q and ΔH are negative. The enthalpy changes for the two processes—melting and freezing—have different signs but, for a given amount of water, they have the same absolute value. As Figure 5.12 illustrates, the heat required to melt a given quantity of a substance has the same magnitude but is opposite in sign to the heat given off when that same quantity of material freezes. For example, $\Delta H_{fus} = 6.01$ kJ if 1 mol of ice melts, but $\Delta H_{solid} = -6.01$ kJ if 1 mole of water freezes. We add the subscript *fus* or *solid* to ΔH to identify which process is occurring: melting (*fusion*) or freezing (*solidification*). The magnitudes and signs of enthalpies associated with other paired phase changes—vaporization and condensation, and sublimation and deposition—also work this way.

In terms of symbols, we put subscripts on ΔH to clarify not only the specific process but also the part of the universe to which the value applies. If we wish to indicate the enthalpy change associated with the system or the surroundings, we may write ΔH_{sys} or ΔH_{surr}, respectively.

enthalpy (H) the sum of the internal energy and the pressure–volume product of a system; $H = E + PV$.

SAMPLE EXERCISE 5.4   Determining the Value and Sign of ΔH

Between periods of a hockey game, an ice-refinishing machine spreads 855 L of water across the surface of a hockey rink.

   a. If the system is the water, what is the sign of ΔH_{sys} as the water freezes?
   b. To freeze this volume of water at 0°C, what is the value of ΔH_{sys}? The density of water is 1.00 g/mL.

Collect and Organize The water from the ice-refinishing machine is identified as the system, so we can determine the sign of ΔH by determining whether heat transfer is into or out of the water. We are told the volume of the water and its density, so we can calculate the mass of water involved in the change of state from liquid water to ice.

Analyze (a) Because the freezing takes place at constant pressure, $\Delta H_{sys} = q_P$. (b) To calculate the amount of energy lost from the water as it freezes, we must convert 855 L of water into moles of water, because the conversion factor between the quantity of energy removed and the quantity of water that freezes is the enthalpy of solidification of water, $\Delta H_{solid} = -6.01$ kJ/mol.

Solve

a. For the water (the system) to freeze, energy must be removed from it. Therefore, ΔH_{sys} must be a negative value.

b. We convert the volume of water into moles:

$$855 \text{ L} \times \frac{1000 \text{ mL}}{1 \text{ L}} \times \frac{1.00 \text{ g}}{1 \text{ mL}} \times \frac{1 \text{ mol}}{18.02 \text{ g}} = 4.745 \times 10^4 \text{ mol}$$

and then calculate ΔH_{sys}:

$$\Delta H_{sys} = 4.745 \times 10^4 \text{ mol} \times \frac{-6.01 \text{ kJ}}{1 \text{ mol}} = -2.85 \times 10^5 \text{ kJ}$$

Think about It ΔH changes with pressure but the magnitude of the change is negligible for pressures near normal atmospheric pressure. The magnitude of the answer to part b is reasonable given the large amount of water that is frozen to refinish the rink's ice.

Practice Exercise The flame in a torch used to cut metal is produced by burning acetylene (C_2H_2) in pure oxygen. Assuming the combustion of 1 mole of acetylene releases 1251 kJ of energy, what mass of acetylene is needed to cut through a piece of steel if the process requires 5.42×10^4 kJ of energy?

(Answers to Practice Exercises are in the back of the book.)

5.4 Heating Curves and Heat Capacity

Winter hikers and high-altitude mountain climbers use portable stoves fueled by propane or butane to prepare hot meals. Their only source of water may be ice or snow. In this section, we use this scenario to examine the transfer of energy into water that begins as snow and ends up as vapor.

Hot Soup on a Cold Day

Let's consider the changes of temperature and state that water undergoes as some hikers melt snow as a first step in preparing soup from a dry soup mix. Suppose they start with a saucepan filled with snow at $-18°C$ at constant pressure. They place the pan above the flame of a portable stove, and energy begins to flow into the snow. The temperature of the snow immediately begins to rise. If the flame is steady so that energy flow is constant, the temperature of the snow changes as described on the graph shown in Figure 5.21. First, heating increases the temperature

molar heat capacity (c_P) the energy required at constant pressure to raise the temperature of 1 mole of a substance by 1°C.

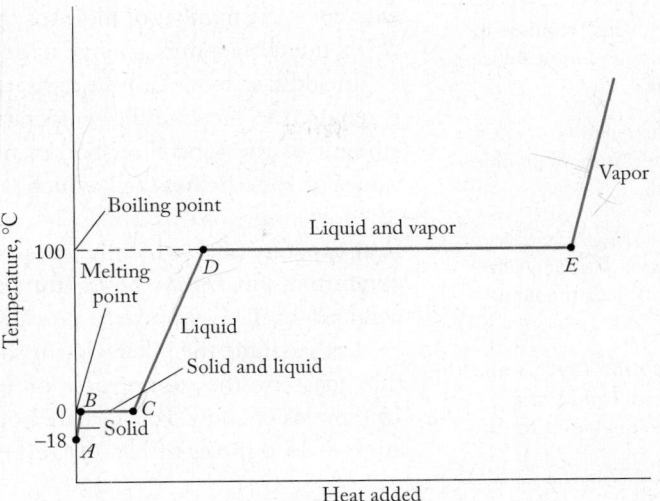

FIGURE 5.21 The energy required to melt snow and boil the resultant water is illustrated by the four line segments on the heating curve of water: heating snow to its melting point ($\overline{AB}$); melting the snow to form liquid water ($\overline{BC}$); heating the water to its boiling point ($\overline{CD}$); and boiling the water to convert it to vapor ($\overline{DE}$).

of the snow to its melting point, 0°C. This temperature rise and the heat transfer into the snow are represented by line segment $\overline{AB}$ in Figure 5.21. The temperature of the snow then remains steady at 0°C while the snow continues to absorb energy and melt. This state change that produces liquid water from snow takes place at constant temperature and is represented by the constant-temperature (horizontal) line segment $\overline{BC}$. Phase changes of pure materials take place at constant temperature and pressure.

When all the snow has melted, the temperature of the liquid water rises (line segment $\overline{CD}$) until it reaches 100°C. At 100°C, the temperature of the water remains steady while another state change takes place: liquid water becomes water vapor. This constant-temperature process is represented by line segment $\overline{DE}$ in Figure 5.21. If all of the liquid water in the pan were converted into vapor, the temperature of the vapor would then rise as long as heat was added, as indicated by the final (slanted) line segment in Figure 5.21.

The difference in the *x*-axis coordinates for the beginning and end of each line segment in Figure 5.21 indicates how much energy is required in each step in this process. In the first step (line segment $\overline{AB}$), the energy required to raise the temperature of the snow from −18°C to 0°C can be calculated if we know how many moles of snow we have and how much heat is required to change the temperature of 1 mole of snow.

Molar heat capacity is the quantity of energy required to raise the temperature of 1 mole of a substance by 1°C. The symbol we use for molar heat capacity is c_p, where the subscript P again indicates the value for a process taking place at *constant pressure*. Molar heat capacities for several substances are given in Table 5.1, from which we see that ice has a molar heat capacity of 37.1 J/(mol · °C) at constant pressure. From these units, we see that if we know the number of moles of snow we have and the temperature change it experiences as we bring it to its melting point, we can calculate q for the process:

$$q = n c_p \Delta T \qquad (5.8)$$

▶II **CHEMTOUR** Heating Curves

TABLE 5.1	Molar Heat Capacities at 25°C of Selected Substances
Substance	**c_p [J/(mol · °C)]**
$H_2O(s)$	37.1
$H_2O(\ell)$	75.3
$H_2O(g)$	33.6
$CH_3CH_2OH(\ell)$, ethanol	113.1
$C(s)$, graphite	8.54
$Al(s)$	24.4
$Cu(s)$	24.5
$Fe(s)$	25.1

specific heat (c_s) the energy required to raise the temperature of 1 g of a substance 1°C at constant pressure.

heat capacity (C_p) the quantity of energy needed to raise the temperature of an object 1°C at constant pressure.

molar heat of fusion (ΔH_{fus}) the energy required to convert 1 mole of a solid substance at its melting point into the liquid state.

molar heat of vaporization (ΔH_{vap}) the energy required to convert 1 mole of a liquid substance at its boiling point to the vapor state.

where n is the number of moles of the substance absorbing or releasing energy and ΔT is the temperature change in degrees Celsius.

In addition to molar heat capacity, there are other measures of how much energy is required to increase the temperature of a substance, each referring to a specific amount of the substance. For example, some tables of thermodynamic data list values of **specific heat (c_s)**, which is the energy required to raise the temperature of *1 g of a substance* by 1°C; c_s has units of J/(g · °C). Later in this chapter we use **heat capacity (C_p)**, which is the quantity of energy required to increase the temperature of *a particular object* (for example, a calorimeter, used to measure energy changes) by 1°C at constant pressure.

Let's assume the hikers decide to cook their meal with 270 g of snow. Dividing that mass by the molar mass of water (18.02 g/mol), we find that they melt 15.0 moles of snow. To calculate how much energy is needed to raise the temperature of 15.0 moles of $H_2O(s)$ from −18°C to 0°C, we use Equation 5.8:

$$q = nc_p \, \Delta T$$

$$= 15.0 \; \cancel{mol} \times \frac{37.1 \, J}{\cancel{mol \cdot °C}} \times (+18 \; \cancel{°C}) = 1.0 \times 10^4 \, J = 10 \, kJ$$

Notice that this value is positive, which means that the system (the snow) gains energy as it warms up.

The energy absorbed as the snow melts, or *fuses* (line segment $\overline{BC}$ in Figure 5.21), can be calculated using the enthalpy change that takes place as 1 mole of snow melts. This enthalpy change is called the **molar heat of fusion (ΔH_{fus})**, and the energy absorbed as n moles of a substance melts is given by

$$q = n\Delta H_{fus} \tag{5.9}$$

The molar heat of fusion for water is 6.01 kJ/mol, and using this value and the known number of moles of snow, we get

$$q = 15.0 \; \cancel{mol} \times \frac{6.01 \, kJ}{\cancel{mol}} = 90.0 \, kJ$$

This value is positive because heat enters the system. This energy overcomes the attractive forces between the water molecules in the solid, and the snow becomes a liquid. Note that no factor for temperature appears in this calculation, because state changes of pure substances take place at constant temperature, as the two horizontal line segments in Figure 5.21 indicate. Snow at 0°C becomes liquid water at 0°C.

While the water temperature increases from 0°C to 100°C, the relation between temperature and energy absorbed is again defined by Equation 5.8, but this time c_p represents not the molar heat capacity of $H_2O(s)$ but the molar heat capacity of water(ℓ), 75.3 J/(mol · °C) (Table 5.1):

$$q = nc_p \, \Delta T$$

$$= 15.0 \; \cancel{mol} \times \frac{75.3 \, J}{\cancel{mol \cdot °C}} \times 100 \; \cancel{°C} = 1.13 \times 10^5 \, J = 113 \, kJ$$

This value is positive because the system takes in energy as its temperature rises.

At this point in our story, let's assume our hiker–chefs accidentally leave the boiling water unattended and it vaporizes completely. As line segment $\overline{DE}$ in Figure 5.21 shows, the temperature of the water remains at 100°C until all of it is vaporized. The enthalpy change associated with changing 1 mole of a liquid to a gas is the **molar heat of vaporization (ΔH_{vap})**, and the quantity of energy absorbed as n moles of a substance vaporizes is given by

$$q = n\Delta H_{vap} \qquad\qquad (5.10)$$

The molar heat of vaporization for water is 40.67 kJ/mol, so for our hikers' vaporizing water we have

$$q = (15.0 \text{ mol})(40.67 \text{ kJ/mol}) = +610 \text{ kJ}$$

This value is positive because the system takes in energy as it converts liquid into vapor. Again no factor for temperature appears because the phase change takes place at constant temperature. Only after all the water has vaporized does its temperature increase above 100°C, along the line above point E in Figure 5.21. The molar heat capacity of steam, 33.6 J/(mol · °C), is used to calculate the energy required to heat the vapor to any temperature above 100°C.

The c_p values for $H_2O(s)$, $H_2O(\ell)$, and $H_2O(g)$ are different. Nearly all substances have different molar heat capacities in their different physical states.

Notice in Figure 5.21 that line segment $\overline{DE}$, the phase change from liquid water to water vapor, is much longer than the line segment $\overline{BC}$ representing the change from solid snow to liquid water. The relative lengths of these lines indicate that the molar heat of vaporization of water (40.67 kJ/mol) is much larger than the molar heat of fusion of snow (6.01 kJ/mol). Why does it take more energy to boil 1 mole of water than to melt 1 mole of snow? The answer is related to the extent to which attractive forces between molecules must be overcome in each process and to how the internal energy of the system changes as energy is added (Figure 5.22).

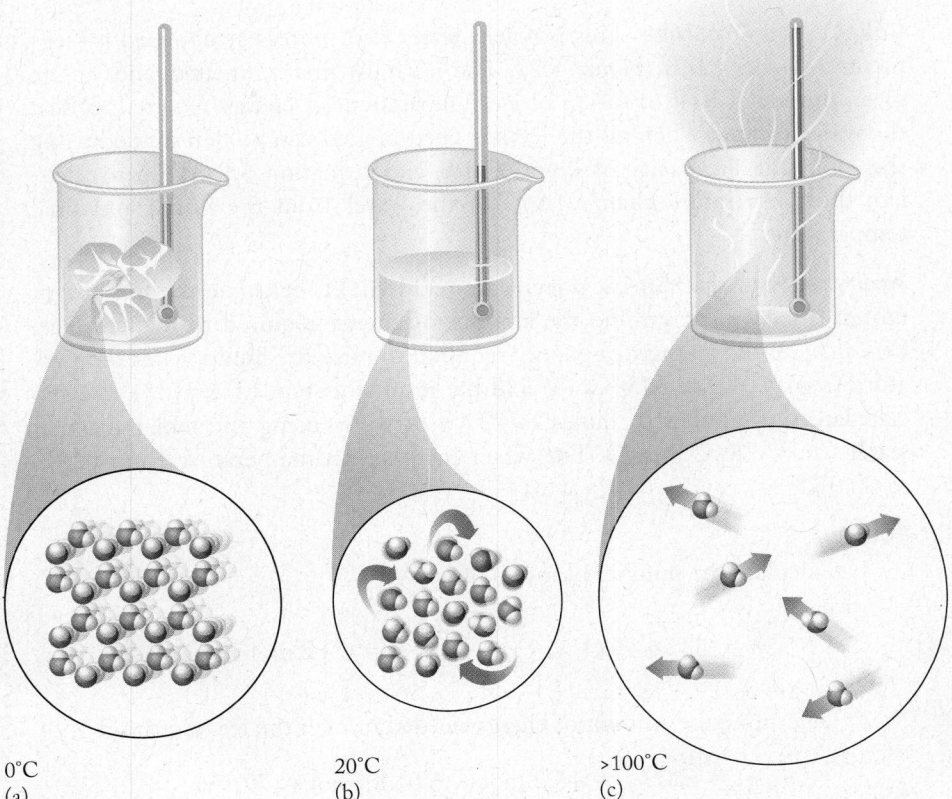

0°C
(a)

20°C
(b)

>100°C
(c)

FIGURE 5.22 Macroscopic and molecular-level views of (a) ice, (b) water, and (c) water vapor.

Attractive forces determine both the organization of the molecules in any substance and their relation to their nearest neighbors. The kinetic and potential energies of the molecules in snow are low, and attractive forces are strong enough to hold the molecules in place relative to one another. Melting the snow requires that these attractive forces be overcome. The energy added to the system at the melting point is sufficient to overcome the attractive forces and change the arrangement (and hence the potential energy) of the molecules.

Intermolecular attractive forces still exist in liquid water, but the energy added at the melting point increases the energy of the molecules and enables them to move with respect to their nearest neighbors. The molecules are still almost as close together in the liquid as they are in the solid, but their relative positions constantly change; they have different nearest neighbors over time.

Once the snow has completely melted, added energy causes the temperature of the liquid water to rise, increasing its internal energy. When water vaporizes at the boiling point, the attractive forces between water molecules must be overcome to separate the molecules from one another as they enter the gas state. Separating the molecules widely in space requires work (that is, energy), and the amount required is much larger than that for the solid-to-liquid state change. This difference is consistent with the relative lengths of $\overline{BC}$ and $\overline{DE}$ in Figure 5.21.

SAMPLE EXERCISE 5.5 **Calculating the Energy Required to Raise the Temperature of Water**

Calculate the amount of energy required to raise the temperature of 237 g of solid ice from 0.0°C to 80.0°C. The molar heat of fusion (ΔH_{fus}) of ice is 6.01 kJ/mol. The molar heat capacity of liquid water is 75.3 J/(mol · °C).

Collect and Organize This problem refers to a process symbolized by segments $\overline{BC}$ and $\overline{CD}$ in Figure 5.21. The ice must first be melted, and we are given the molar heat of fusion of ice. The amount of energy required to heat the water formed when all the ice has been melted can be determined using the molar heat capacity of liquid water, and Equation 5.8. We can calculate the temperature change (ΔT) of the water from the initial and final temperatures.

Analyze Four mathematical steps are required: (1) calculate the number of moles of ice; (2) determine the amount of energy required to melt the ice; (3) calculate the amount of energy required to raise the liquid water temperature from 0.0°C to 80.0°C; (4) add the results of steps (2) and (3). We can calculate the number of moles in 237 g of water using the molar mass of water ($\mathcal{M} = 18.02$ g/mol). The water increases in temperature from 0.0°C to 80.0°C.

Solve

1. Calculate the number of moles of water:

$$n = 237 \text{ g H}_2\text{O} \times \frac{1 \text{ mol H}_2\text{O}}{18.02 \text{ g H}_2\text{O}} = 13.2 \text{ mol H}_2\text{O}$$

2. Determine the amount of energy needed to melt the ice (Equation 5.9):

$$q_1 = n\Delta H_{fus} = 13.2 \text{ mol} \times 6.01 \text{ kJ/mol} = 79.3 \text{ kJ}$$

3. Determine the amount of energy needed to warm the water (Equation 5.8):

$$q_2 = nc_p \, \Delta T$$

$$= 13.2 \; \text{mol} \times \frac{75.3 \text{ J}}{\text{mol} \cdot {}^\circ\text{C}} \times (80.0 - 0.00){}^\circ\text{C} = 79{,}517 \text{ J} = 79.5 \text{ kJ}$$

4. Add the results of parts 2 and 3:

$$q_1 + q_2 = 79.3 \text{ kJ} + 79.5 \text{ kJ} = 158.8 \text{ kJ}$$

Think about It Using the definitions of molar heat of fusion and molar heat capacity, we can solve this exercise without referring back to Equations 5.8 and 5.9. Molar heat of fusion defines the amount of energy needed to melt 1 mole of ice. Multiplying that value (6.01 kJ/mol) by the number of moles (13.2 mol) gives us the energy required to melt the given amount of ice. By the same token, the molar heat capacity of water [75.3 J/(mol · °C)] defines the amount of energy needed to raise the temperature of 1 mole of liquid water by 1°C. We know the number of moles (13.2 mol), and we know the number of degrees by which we want to raise the temperature (80.0°C − 0.00°C = 80.0°C); multiplying those factors together gives us the heat needed to raise the temperature of the water.

Practice Exercise Calculate the change in energy when 125 g of water vapor at 100.0°C condenses to liquid water and then cools to 25.0°C.

(Answers to Practice Exercises are in the back of the book.)

Water is an extraordinary substance for many reasons, but its high molar heat capacity is one of the more important. The ability of water to absorb large quantities of energy is one reason it is used as a *heat sink,* both in automobile radiators and in our bodies. The term "heat sink" is often used to identify matter that can absorb energy without changing phase or significantly changing its temperature. Weather and climate changes are largely driven and regulated by cycles involving retention of energy by our planet's oceans, which serve as giant heat sinks for solar energy.

Cold Drinks on a Hot Day

Let's consider another useful situation involving heat transfer. Suppose we throw a party and plan to chill three cases (72 aluminum cans, each containing 355 mL) of beverages by placing the cans in an insulated cooler and covering them with ice cubes. If the temperature of the ice (sold in 10-pound bags) is −8.0°C and the temperature of the beverages is initially 25.0°C, how many bags of ice do we need to chill the cans and their contents to 0.0°C (as in "ice cold")?

You may already have an idea that more than 1 bag, but probably fewer than 10, will be needed. We can use the heat transfer relationships we have defined to predict more accurately how much ice is required. In doing so, we assume that whatever energy is absorbed by the ice is lost by the cans and the beverages in them. As the ice absorbs heat from the cans and the beverages, the temperature of the ice increases from −8.0°C to 0.0°C, and, as we saw in analyzing Figure 5.21, the resulting liquid water remains at 0.0°C until all the ice has melted. We need enough ice so that the last of it melts just as the temperature of the beverages and the cans reaches 0.0°C. Our analysis of this cooling process is a little simpler than our snow/liquid water/water vapor analysis because here there is no change of state. All of the heat transferred goes only into cooling the cans and beverages to 0.0°C. The cans stay in the solid state, and the beverages stay in the liquid state.

First let's consider the energy lost in the cooling process. Two materials are to be chilled: 72 aluminum cans and 72 × 355 mL = 25,600 mL of beverages. The beverages are mostly water. The other ingredients are present in such small concentrations that they will not affect our calculation, so we can assume that we need to reduce the temperature of 25,600 mL of water by 25.0°C. We can calculate the amount of energy lost with Equation 5.8 if we first calculate the number of moles of water in 25,600 mL, assuming a density of 1.000 g/mL:

$$25{,}600\ \cancel{\text{mL H}_2\text{O}} \times 1.000\ \cancel{\text{g}}/\cancel{\text{mL}} \times \frac{1\ \text{mol H}_2\text{O}}{18.02\ \cancel{\text{g H}_2\text{O}}} = 1420\ \text{mol H}_2\text{O}$$

We can use the molar heat capacity of water and Equation 5.8 to calculate the energy lost by 1420 moles of water as its temperature decreases from 25.0°C to 0.0°C:

$$\begin{aligned} q &= nc_\text{P}\,\Delta T \\ &= 1420\ \cancel{\text{mol}} \times \frac{75.3\ \text{J}}{\cancel{\text{mol}}\cdot\cancel{^\circ\text{C}}} \times (-25.0\cancel{^\circ\text{C}}) \\ &= -2.67 \times 10^6\ \text{J} \end{aligned}$$

We must also consider the energy released in lowering the temperature of 72 aluminum cans by 25.0°C. The typical mass of a soda can is 12.5 g. The molar heat capacity of solid aluminum (Table 5.1) is 24.4 J/(mol · °C), and the molar mass of aluminum is 26.98 g/mol. Thus

$$\begin{aligned} q &= nc_\text{P}\,\Delta T \\ &= 72\ \cancel{\text{cans}} \times \frac{12.5\ \cancel{\text{g Al}}}{\cancel{\text{can}}} \times \frac{1\ \cancel{\text{mol}}}{26.98\ \cancel{\text{g Al}}} \times \frac{24.4\ \text{J}}{\cancel{\text{mol}}\cdot\cancel{^\circ\text{C}}} \times (-25.0\cancel{^\circ\text{C}}) \\ &= -2.03 \times 10^4\ \text{J} \end{aligned}$$

The total quantity of energy that must be removed from the cans and beverages is

$$\begin{aligned} q_\text{total} = q_\text{beverage} + q_\text{cans} &= [(-2.67 \times 10^6) + (-2.03 \times 10^4)]\ \text{J} \\ &= (-2.67 - 0.0203) \times 10^6\ \text{J} = -2.69 \times 10^6\ \text{J} \\ &= -2.69 \times 10^3\ \text{kJ} \end{aligned}$$

This quantity of energy must be absorbed by the ice as it warms to its melting point and then melts. The calculation of the amount of ice needed is an algebra problem. Let n be the number of moles of ice needed. The energy absorbed is the sum of (1) the energy needed to raise the temperature of n moles of ice from −8.0° to 0.0°C and (2) the energy needed to melt n moles of ice. These quantities can be calculated with Equation 5.8 for step 1 and Equation 5.9 for step 2. Note that the c_P values in Table 5.1 have units of joules, while ΔH_fus values given earlier have units of kilojoules. We need to have both terms in the same units, so we use 0.0371 kJ/mol · °C for c_P:

$$\begin{aligned} q_\text{total gained} &= q_1 + q_2 \\ &= nc_\text{P}\,\Delta T + n\Delta H_\text{fus} \\ &= n\left(\frac{0.0371\ \text{kJ}}{\text{mol}\cdot{}^\circ\text{C}}\right)(8.0^\circ\text{C}) + n(6.01\ \text{kJ/mol}) \\ &= n(6.31\ \text{kJ/mol}) \end{aligned}$$

The energy lost by the cans and the beverages balances the energy gained by the ice:

$$-q_{\text{total lost}} = +q_{\text{total gained}}$$
$$2.69 \times 10^3 \text{ kJ} = n(6.31 \text{ kJ/mol})$$
$$n = 4.26 \times 10^2 \text{ mol ice}$$

Converting 4.26×10^2 mol of ice into pounds and adjusting the significant figures gives us

$$(4.26 \times 10^2 \text{ mol}) \times \frac{18.02 \text{ g}}{\text{mol}} \times \frac{1 \text{ lb}}{453.6 \text{ g}} = 17 \text{ lb of ice}$$

Thus we need at least two 10-pound bags of ice to chill three cases of our favorite beverages.

CONCEPT TEST ·

The energy lost by the beverages inside the 72 cans in the preceding discussion was more than 100 times the heat lost by the cans. What factors contributed to this large difference between the energy lost by the cans and the energy lost by their contents?

(Answers to Concept Tests are in the back of the book.)

· ·

SAMPLE EXERCISE 5.6 **Calculating the Temperature of Iced Tea**

If you add 250.0 g of ice initially at $-18.0°C$ to 237 g (1 cup) of freshly brewed tea initially at $100.0°C$ and the ice melts, what is the final temperature of the tea? Assume that the mixture is an isolated system (in an ideal insulated container) and that tea has the same molar heat capacity, density, and molar mass as water.

Collect and Organize We know the mass of tea, its initial temperature, and the molar heat capacity of tea, for which we just use the c_p value for water. The amount of energy released when the tea is cooled will be the same as the amount of energy gained by the ice and, once it is melted, the amount gained by the water that is formed as it warms to the final temperature. We know the amount of ice, its initial temperature, and the heat of fusion of ice. Our task is to find the final temperature of the tea ice–water mixture.

Analyze Before solving this problem, we need to think about the changes that take place when the tea and ice come into contact. Assuming the ice melts completely, three heat transfers account for the energy lost by the tea:
1. q_1: raising the temperature of the ice to $0.0°C$.
2. q_2: melting the ice.
3. q_3: bringing the mixture to the final temperature; the temperature of the water from the melted ice rises and the temperature of the tea ($T_{\text{initial}} = 100.0°C$) falls to the final temperature of the mixture ($T_{\text{final}} = ?$).

The energy gained by the ice equals the energy lost by the tea:

$$q_{\text{ice}} = -q_{\text{tea}}$$

Based on our analysis, we know that

$$q_{\text{ice}} = q_1 + q_2 + q_3$$

Solve The energy lost by the tea as it cools from 100.0°C to T_{final} is, from Equation 5.8,

$$\begin{aligned} q_{\text{tea}} &= nc_{\text{P}}\,\Delta T_{\text{tea}} \\ &= 237\ \cancel{\text{g}} \times \frac{1\ \cancel{\text{mol}}}{18.02\ \cancel{\text{g}}} \times \frac{75.3\ \text{J}}{\cancel{\text{mol}} \cdot {}^\circ\text{C}} \times (T_{\text{final}} - 100.0^\circ\text{C}) \\ &= (990\ \text{J/}^\circ\text{C})(T_{\text{final}} - 100.0^\circ\text{C}) \end{aligned}$$

The transfer of this heat is responsible for the changes in the ice. We can treat the heat transfer from the hot tea to the ice in terms of the three processes we identified in Analyze. In step 1, the ice is warmed from −18.0°C (T_{initial}) to 0.0°C (T_{final} in this step):

$$\begin{aligned} q_1 &= n_{\text{ice}}\, c_{\text{ice}}\, \Delta T_{\text{ice}} \\ &= 250.0\ \cancel{\text{g}} \times \frac{1\ \cancel{\text{mol}}}{18.02\ \cancel{\text{g}}} \times \frac{37.1\ \text{J}}{\cancel{\text{mol}} \cdot \cancel{{}^\circ\text{C}}} \times [0.0\cancel{{}^\circ\text{C}} - (-18.0\cancel{{}^\circ\text{C}})] \\ &= 9.26 \times 10^3\ \text{J} \end{aligned}$$

In step 2, the ice melts, requiring the absorption of energy:

$$\begin{aligned} q_2 &= n_{\text{ice}}\, \Delta H_{\text{fus, ice}} \\ &= 250.0\ \cancel{\text{g}} \times \frac{1\ \cancel{\text{mol}}}{18.02\ \cancel{\text{g}}} \times \frac{6.01\ \text{kJ}}{\cancel{\text{mol}}} \\ &= 83.4\ \text{kJ} \end{aligned}$$

In step 3, the water from the ice, initially at 0.0°C, warms to the final temperature (where $\Delta T = T_{\text{final}} - T_{\text{initial}} = T_{\text{final}} - 0.0^\circ\text{C}$):

$$\begin{aligned} q_3 &= n_{\text{water}}\, c_{\text{water}}\, \Delta T_{\text{water}} \\ &= 250.0\ \cancel{\text{g}} \times \frac{1\ \cancel{\text{mol}}}{18.02\ \cancel{\text{g}}} \times \frac{75.3\ \text{J}}{\cancel{\text{mol}} \cdot {}^\circ\text{C}} \times (T_{\text{final}} - 0.0^\circ\text{C}) \\ &= (1045\ \text{J/}^\circ\text{C})\,(T_{\text{final}}) \end{aligned}$$

The sum of the quantities of energy absorbed by the ice and the water from it during steps 1 through 3 must balance the energy lost by the tea:

$$q_{\text{ice}} = q_1 + q_2 + q_3 = -q_{\text{tea}}$$

$$9260\ \text{J} + 83.4\ \text{kJ} + (1045\ \text{J/}^\circ\text{C})(T_{\text{final}}) = -[(990\ \text{J/}^\circ\text{C})(T_{\text{final}} - 100.0^\circ\text{C})]$$

Expressing all values in kilojoules:

$$9.26\ \text{kJ} + 83.4\ \text{kJ} + (1.045\ \text{kJ/}^\circ\text{C})(T_{\text{final}}) = -[(0.990\ \text{kJ/}^\circ\text{C})\,(T_{\text{final}} - 100.0^\circ\text{C})]$$

and rearranging the terms to solve for T_{final}, we have

$$\begin{aligned} (2.04\ \text{kJ/}^\circ\text{C})T_{\text{final}} &= -9.26\ \text{kJ} - 83.4\ \text{kJ} + 99.0\ \text{kJ} = 6.34\ \text{kJ} \\ T_{\text{final}} &= 3.1^\circ\text{C} \end{aligned}$$

Think about It This calculation was carried out assuming the system (the tea plus the ice) to be isolated and the vessel to be a perfect insulator. Our answer, therefore, is an "ideal" answer and reflects the coldest temperature we can expect the tea to reach. In the real world, the ice would absorb some energy from the surroundings (the container and the air) and the tea would lose some energy to the surroundings, so the final temperature of the beverage could be different from the ideal value we calculated.

Practice Exercise Calculate the final temperature of a mixture of 350 g of ice initially at −18°C and 237 g of water initially at 100.0°C.

(Answers to Practice Exercises are in the back of the book.)

5.5 Calorimetry: Measuring Heat Capacity and Calorimeter Constants

Up to this point we have discussed molar heat capacities c_P and the enthalpy changes ΔH_{fus} and ΔH_{vap} associated with phase changes, but we have not broached the issue of how we know the values of these parameters. The experimental method of measuring the quantities of energy associated with chemical reactions and physical changes is called **calorimetry**. The device used to measure the heat released or absorbed during a process is a **calorimeter**.

Determining Molar Heat Capacity and Specific Heat

When we determined the amount of ice needed to cool 72 aluminum beverage cans in Section 5.4, we used the molar heat capacity of aluminum to determine the quantity of energy lost by the cans. How are heat capacities measured?

We can apply the first law of thermodynamics and design an experiment to determine the specific heat of aluminum, from which we can calculate its molar heat capacity. Recall from Section 5.4 that specific heat c_s is defined as the quantity of energy required to raise the temperature of 1 g of a substance by 1°C. The units of specific heat, J/(g · °C), tell us exactly what we have to do. "Determine the specific heat of aluminum" means determine the quantity of energy (in joules) required to change the temperature of 1 g of aluminum by 1°C. If we determine how much energy is required to change the temperature of any known mass of aluminum by any measured number of degrees, we can calculate how much energy is required to change the temperature of 1 g of aluminum by 1°C—in other words, we'll have the value of the specific heat of aluminum.

Suppose we have beads of pure aluminum with a total mass of 23.5 g and want to transfer a known amount of energy to or from the aluminum. Refer to Figure 5.23 for this discussion.

First, we must heat the known quantity of aluminum to a known temperature. We can do this by boiling some water in a beaker containing a test tube holding the aluminum beads (the system). We wait a few minutes while the beads and the boiling water bath all come to thermal equilibrium at 100.0°C (Figure 5.23a).

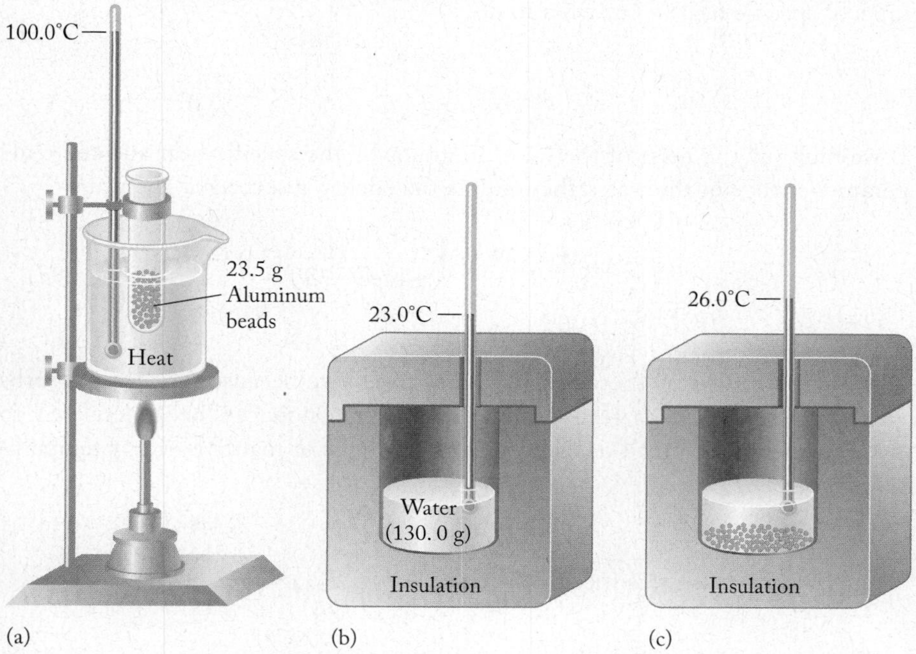

calorimetry the measurement of the quantity of heat transferred during a physical change or chemical process.

calorimeter a device used to measure the absorption or release of energy by a physical change or chemical process.

FIGURE 5.23 Experimental setup to determine the molar heat capacity of a metal. (a) Pure aluminum beads having a combined mass of 23.5 g are heated to 100.0°C in boiling water; (b) 130.0 g of water at 23.0°C is in a Styrofoam box; (c) the hot Al beads are dropped into the water, and the temperature at thermal equilibrium is 26.0°C.

Because water (or any pure liquid) boils at a constant temperature, we can use this step to bring the beads to a known temperature. No matter how rapidly we boil the water, it will maintain a temperature of 100.0°C.

While the metal is heating, we place a measured mass of water (in this case, 130.0 g) in a beaker in a Styrofoam box (Figure 5.23b). We consider this box a perfect insulator that effectively isolates its contents from the rest of the universe. Through a small hole in the box lid (no energy escapes through the hole because the box is a perfect insulator), we insert a thermometer into the water, read the temperature (23.0°C), and leave the thermometer in place. When we have waited long enough for the beads to heat up to 100.0°C, we remove the test tube from the boiling water, remove the lid from the insulated box, pour the beads out of the test tube into the water in the beaker, and quickly close the lid (Figure 5.23c). The temperature of the water rises because it is now in contact with the hot aluminum. Because we treat the box as a perfect insulator, we assume that all the heat coming from the beads goes into the water in the beaker. Suppose the temperature of the aluminum–water mixture in the box rises to 26.0°C and stays at that temperature.

Energy flows from the beads to the water. From the first law of thermodynamics we know

$$-q_{aluminum} = q_{water}$$

where we show a minus sign on $q_{aluminum}$ because energy leaves the aluminum and a plus sign on q_{water} because energy enters the water. To determine the amount of energy transferred from the aluminum, we need to know how much energy enters the water. We know the mass of water (130.0 g) and the temperature change the water experiences ($\Delta T_{water} = 26.0°C - 23.0°C = 3.0°C$); we need the specific heat of water to calculate q_{water}.

When we defined units of energy in Section 5.2, we defined 4.184 joules (J) as the amount of energy needed to raise the temperature of 1 g of water by 1°C; that value is the specific heat of water. We could also calculate $c_{s,water}$ by dividing the molar heat capacity of water [75.3 J/(mol · °C), Table 5.1] by its molar mass ($\mathcal{M} = 18.02$ g/mol). With that value we now have everything we need to calculate the specific heat c_s of aluminum.

First, we determine the amount of energy gained by the water in the box. The units of specific heat tell us what to do:

$$c_{s,water} = \frac{4.184 \text{ J}}{\text{g} \cdot \text{°C}}$$

If we multiply the mass of the water in grams by the specific heat and the temperature change of the water, the result is the energy absorbed by the water:

$$q_{water} = (130.0 \, \cancel{\text{g H}_2\text{O}}) \left(\frac{4.184 \text{ J}}{\cancel{\text{g H}_2\text{O}} \cdot \cancel{°C}} \right) (3.0 \, \cancel{°C})$$

$$= 1600 \text{ J}$$

The energy gained by the water has a positive value because the water absorbs energy from the beads (the system). This expression can be generalized to calculate energy absorbed or released from any mass of matter over any temperature change:

$$q = mc_s \Delta T \tag{5.11}$$

where m is the mass, c_s is specific heat, and $\Delta T = T_{final} - T_{initial}$.

To find the specific heat of aluminum, we recognize that the 1600 J of energy that increased the temperature of the water came from the aluminum beads. Therefore,

$$q_{aluminum} = -q_{water} = -1600 \text{ J} = mc_s\Delta T_{aluminum}$$

where m is the mass of aluminum, c_s is the specific heat of aluminum, and $\Delta T_{aluminum} = T_{final} - T_{initial} = (26.0 - 100.0)°C = -74.0°C$. Note that ΔT for the aluminum is different than ΔT of the water. The aluminum is initially at 100°C and drops to a final temperature of 26.0°C. Thus

$$-1600 \text{ J} = (23.5 \text{ g})(c_s)(-74.0°C)$$

where c_s is the unknown value we are after.

Solve for c_s:

$$c_s = \frac{-1600 \text{ J}}{(23.5 \text{ g})(-74.0°C)} = \frac{0.92 \text{ J}}{\text{g} \cdot °C}$$

From the specific heat capacity, we can also calculate the molar heat capacity of aluminum by multiplying c_s by the mass of 1 mole of aluminum:

$$c_P = c_s \times \mathcal{M} = \frac{0.92 \text{ J}}{\text{g} \cdot °C} \times 26.98 \frac{\text{g}}{\text{mol}} = 25 \frac{\text{J}}{\text{mol} \cdot °C}$$

> **enthalpy of reaction (ΔH_{rxn}) or heat of reaction** the energy absorbed or given off by a chemical reaction.
>
> **bomb calorimeter** a constant-volume device used to measure the energy released during a combustion reaction.

▶❙❙ **CHEMTOUR** Calorimetry

CONCEPT TEST

Thermometers capable of measuring to the nearest 0.001°C are used in experiments to determine very precise values of specific heats. What impact would this change in equipment have on the numerical values reported for specific heats?

(Answers to Concept Tests are in the back of the book.)

Measuring Calorimeter Constants

The heat transfer accompanying any chemical reaction is defined by a quantity known as the **enthalpy of reaction (ΔH_{rxn})**, also called the **heat of reaction**. The subscript may be changed to reflect a specific type of reaction being studied: for example, ΔH_{comb} for the enthalpy of a combustion reaction.

Heats of reaction are measured with a device called a **bomb calorimeter** (Figure 5.24). To measure the heat of a combustion reaction, the sample is placed in a sealed vessel (called a *bomb*) capable of withstanding high pressures and submerged in a large volume of water in a heavily insulated container. Oxygen is introduced into the bomb, and the mixture is ignited with an electric spark. As combustion occurs, energy generated by the reaction flows into the walls of the bomb and then into the water surrounding the bomb. A good bomb calorimeter keeps the system (the chemical reaction) contained within the bomb and ensures that all energy generated by the reaction stays in the calorimeter. The system consists of the bomb, the water, the insulated container, and minor components (stirrer, thermometer, and any other materials).

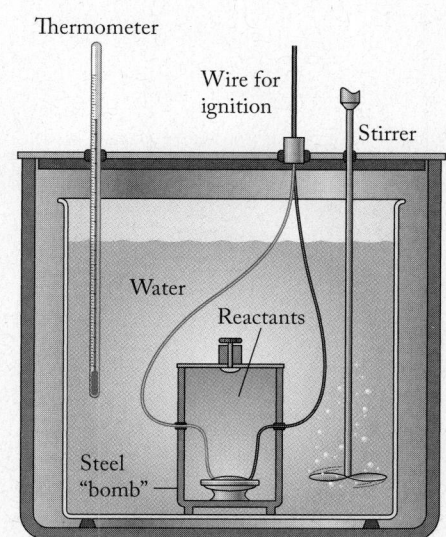

FIGURE 5.24 A bomb calorimeter.

The energy produced by the reaction is determined by measuring the temperature of the water before and after the reaction. The water is at the same temperature as the parts of the calorimeter it contacts—the walls of the bomb, the thermometer, and the stirrer—so the temperature change of the water takes the entire calorimeter into account.

Measuring the change in temperature of the water is not the whole story, however. We also need to know the heat capacity of the calorimeter. As mentioned previously, heat capacity (C_p) is the quantity of energy required to raise the temperature of a particular object by 1°C. Because this value is unique to every calorimeter, it is frequently referred to as that **calorimeter's constant ($C_{\text{calorimeter}}$)**. If we know the value of $C_{\text{calorimeter}}$ and if we can measure the change in water temperature, we can calculate the quantity of energy that flowed from the reactants into the calorimeter ($q_{\text{calorimeter}}$) to cause the temperature change of the water:

$$q_{\text{calorimeter}} = C_{\text{calorimeter}} \, \Delta T \tag{5.12}$$

Rearranging terms:

$$C_{\text{calorimeter}} = \frac{q_{\text{calorimeter}}}{\Delta T} \tag{5.13}$$

This equation indicates that heat capacity is expressed in units of energy divided by temperature, usually kilojoules per degree Celsius (kJ/°C).

Equation 5.13 can be used to determine $C_{\text{calorimeter}}$ for a bomb calorimeter. To do this, we burn a quantity of material in the calorimeter that produces a known quantity of energy when it burns—in other words, a material whose ΔH_{comb} value is known. Benzoic acid ($C_7H_6O_2$) is often used for this purpose because it can be obtained in a form that is very pure. Once $C_{\text{calorimeter}}$ has been determined, the calorimeter can be used to determine ΔH_{comb} for other substances, and the observed increases in water temperature can be used to calculate the quantities of energy produced by combustion reactions on a per-gram or per-mole basis.

Because there is no change in the volume of the reaction mixture in a bomb calorimeter, this technique is referred to as *constant-volume calorimetry*. No *P–V* work is done, so according to Equation 5.12 the energy gained by the calorimeter during the combustion equals the internal energy lost by the reaction system during the combustion:

$$q_{\text{calorimeter}} = -\Delta E_{\text{comb}} \tag{5.14}$$

The pressure inside a bomb calorimeter may change as a result of a combustion reaction, and in such cases ΔE_{comb} is not *exactly* the same as ΔH_{comb}. However, the pressure effects are usually so small that ΔE_{comb} is *nearly* the same as ΔH_{comb}, and we do not worry about the very small differences. Hence we discuss enthalpies of reactions (ΔH_{rxn}) throughout and apply the approximate relation

$$q_{\text{calorimeter}} = -\Delta H_{\text{comb}}$$

or even more generally

$$q_{\text{calorimeter}} = -\Delta H_{\text{rxn}} \tag{5.15}$$

Other types of calorimeters allow the volume to change while the pressure remains constant, so $q_{\text{calorimeter}}$ in those cases is exactly the same as the enthalpy of reaction (ΔH_{rxn}).

SAMPLE EXERCISE 5.7   Determining a Calorimeter Constant

Before we can determine the enthalpy change of a reaction run in a calorimeter, we must determine the heat capacity (the calorimeter constant) of the calorimeter. What is the calorimeter constant of a bomb calorimeter if burning 1.000 g of benzoic acid in it causes the temperature of the calorimeter to rise by 7.248°C? The heat of combustion of benzoic acid is $\Delta H_{comb} = -26.38$ kJ/g.

Collect and Organize We are asked to find the calorimeter constant of a calorimeter. We are given data describing how much the temperature of the calorimeter rises when a known amount of benzoic acid is burned in it, and we are given the heat of combustion of benzoic acid.

Analyze We need to determine the amount of energy required to raise the temperature of the calorimeter by 1°C. The heat capacity of a calorimeter can be calculated using Equation 5.13 and the knowledge that the combustion of 1.000 g of benzoic acid produces 26.38 kJ of heat.

Solve

$$C_{calorimeter} = \frac{q_{calorimeter}}{\Delta T}$$
$$= \frac{26.38 \text{ kJ}}{7.248°C}$$
$$= 3.640 \text{ kJ/°C}$$

This calorimeter can now be used to determine the heat of combustion of any combustible material.

Think about It The calorimeter constant is determined for a specific calorimeter. If anything changes—if the thermometer breaks and has to be replaced, or if the calorimeter loses any of the water it contains—a new constant must be determined.

Practice Exercise Assume that when 0.500 g of a mixture of hydrocarbons is burned in the bomb calorimeter from Sample Exercise 5.7 its temperature rises by 6.76°C. How much energy (in kilojoules) is released during combustion? How much energy is released with the combustion of 1.000 g of the same mixture?

(Answers to Practice Exercises are in the back of the book.)

5.6 Enthalpies of Formation and Enthalpies of Reaction

As noted in Section 5.2, it is impossible to measure the *absolute* value of the internal energy of a substance. The same is true for the enthalpy of a substance. However, we can establish *relative* enthalpy values that are referenced to a convenient standard. This approach is similar to using the freezing point of water as the zero point on the Celsius temperature scale or sea level as the zero point for expressing altitude. Enthalpy values referenced to this zero point are the **standard enthalpies of formation ($\Delta H_f°$)**, defined as the enthalpy change that takes place at constant pressure when 1 mole of a substance is formed from its constituent elements in

standard enthalpy of formation ($\Delta H_f°$) the enthalpy change of a formation reaction; also known as *standard heat of formation* or *heat of formation*.

TABLE 5.2	Standard Enthalpies at 25°C for Selected Substances

A. STANDARD ENTHALPIES OF FORMATION

Substance	ΔH_f° (kJ/mol)
$O_2(g)$	0
$H_2(g)$	0
$H_2O(g)$	−241.8
$H_2O(\ell)$	−285.8
$C_{graphite}(s)$	0
$CH_4(g)$, methane	−74.8
$C_2H_2(g)$, acetylene	226.7
$C_2H_4(g)$, ethylene	52.26
$C_2H_6(g)$, ethane	−84.68
$C_3H_8(g)$, propane	−103.8
$C_4H_{10}(g)$, butane	−125.6
$CO_2(g)$	−393.5
$CO(g)$	−110.5
$N_2(g)$	0
$NH_3(g)$, ammonia	−46.1
$N_2H_4(g)$, hydrazine	95.4
$N_2H_4(\ell)$	50.63
$NO(g)$	90.3
$Br_2(\ell)$	0
$CH_3OH(\ell)$, methanol	−238.7
$CH_3CH_2OH(\ell)$, ethanol	−277.7
$CH_3COOH(\ell)$, acetic acid	−484.5

B. STANDARD ENTHALPIES OF COMBUSTION

Substance	ΔH_{comb}° (kJ/mol)
$CO(g)$	−283.0
$C_5H_{12}(\ell)$, pentane	−3535
$C_9H_{20}(\ell)$, avg. gasoline compound	−6160
$C_{14}H_{30}(\ell)$, avg. diesel compound	−7940

their standard states. The superscript degree in ΔH_f° always signals that the symbol refers to some *standard* parameter. A reaction that fits this description is known as a **formation reaction**.

The **standard state** of an element is its most stable physical form under **standard conditions** of pressure (1 bar) and some specified temperature: in their standard states, oxygen is a gas, water is a liquid, and carbon is solid graphite. The value of pressure of 1 bar is very close to 1 atm; for the level of precision required in the data used in this book, a standard pressure of 1 bar will be considered equivalent to 1 atm.

By definition, for a pure element in its most stable form under standard conditions, $\Delta H_f^\circ = 0$. This is the zero point of enthalpy values. Standard heats of formation for several substances are given in Table 5.2. The symbol of the enthalpy change associated with a reaction that takes place under standard conditions is ΔH_{rxn}°, and the value is called either a **standard enthalpy of reaction** or a *standard heat of reaction*. Implied in our notion of standard states and standard conditions is the assumption that parameters such as ΔH change with temperature and pressure. That assumption is correct, although the changes are so small that we ignore them in the calculations in this chapter and those that follow.

Because the definition of a formation reaction specifies 1 mole of product, writing balanced equations for formation reactions may require the use of something we avoided in Chapter 3: fractional coefficients in the final form of our balanced equations. For example, the reaction for the production of ammonia from nitrogen and hydrogen is usually written

$$N_2(g) + 3H_2(g) \rightarrow 2NH_3(g) \qquad \Delta H_{rxn}^\circ = -92.2 \text{ kJ}$$

Although all reactants and products in this equation are in their standard states, it is not a formation reaction because 2 moles of product are formed, which is why we denote the enthalpy change of this reaction by ΔH_{rxn}° rather than ΔH_f°. The formation reaction for ammonia must show 1 mole of product being formed, so we divide each coefficient in the above equation by 2. Because energy is a stoichiometric quantity, the heat of reaction is divided by 2 as well. Thus, the equation representing the formation reaction of ammonia is

$$\tfrac{1}{2}N_2(g) + \tfrac{3}{2}H_2(g) \rightarrow NH_3(g) \qquad \Delta H_f^\circ = -46.1 \text{ kJ}$$

SAMPLE EXERCISE 5.8   Recognizing Formation Reactions

Which of the following reactions are formation reactions at 25°C? For those that are not, explain why not.

a. $H_2(g) + \tfrac{1}{2}O_2(g) \rightarrow H_2O(g)$

b. $C_{graphite}(s) + 2H_2(g) + \tfrac{1}{2}O_2(g) \rightarrow CH_3OH(\ell)$
   (CH_3OH is methanol, a liquid in its standard state.)

c. $CH_4(g) + 2O_2(g) \rightarrow CO_2(g) + 2H_2O(\ell)$

d. $P_4(s) + 2O_2(g) + 6Cl_2(g) \rightarrow 4POCl_3(\ell)$
   (P_4 is a solid in its standard state; Cl_2 is a gas, and $POCl_3$ is a liquid.)

Collect and Organize We are given four balanced chemical equations and information about the standard states of specific reactants and products.

Analyze For a reaction to be a formation reaction, it must meet the criteria that it produces 1 mole of a substance from its component elements in their standard states. Each reaction must therefore be evaluated for the quantity of product and for the state of each reactant and product.

Solve

a. The reaction shows 1 mole of water vapor formed from its constituent elements in their standard states. This is the formation reaction for $H_2O(g)$, and its enthalpy of reaction is correctly symbolized ΔH_f°.

b. The reaction shows 1 mole of liquid methanol formed from its constituent elements in their standard states. This reaction therefore is a formation reaction, and its change in enthalpy is correctly symbolized ΔH_f°.

c. This is not a formation reaction because the reactants are not elements in their standard states and because more than one product is formed. The change in enthalpy of this reaction would be symbolized ΔH_{comb}.

d. This is not a formation reaction because the product is 4 moles of $POCl_3$. Note, however, that all of the constituent elements are in their standard states and that only one compound is formed. We could therefore call the heat of reaction (ΔH_{rxn}) for this reaction a *standard* heat of reaction, which we indicate by adding a superscript degree to the symbol: ΔH_{rxn}°. We could easily convert this into a formation reaction by dividing all the coefficients and the standard heat of reaction (ΔH_{rxn}) by 4. Once we do this, $\Delta H_{rxn}^\circ = \Delta H_f^\circ$.

Think about It Just because we can write formation reactions for substances like methanol does not mean that anyone would ever use that reaction to make methanol. Remember that formation reactions are defined to provide a standard against which other reactions can be compared when their thermochemistry is evaluated.

Practice Exercise Write formation reactions for (a) $CaCO_3(s)$; (b) $CH_3COOH(\ell)$ (acetic acid); (c) $KMnO_4(s)$.

(Answers to Practice Exercises are in the back of the book.)

formation reaction a reaction in which 1 mole of a substance is formed from its component elements in their standard states.

standard state the most stable form of a substance under 1 bar pressure and some specified temperature (25°C unless otherwise stated).

standard conditions in thermodynamics: a pressure of 1 bar (~1 atm) and some specified temperature, assumed to be 25°C unless otherwise stated; for solutions, a concentration of 1 *M* is specified.

standard enthalpy of reaction (ΔH_{rxn}°) the energy associated with a reaction that takes place under standard conditions; also known as *standard heat of reaction*.

Table 5.2 lists standard enthalpies of formation for some substances, including hydrocarbons. A more complete list can be found in Appendix 4.[1] Note that the standard enthalpies of formation of acetylene and ethylene are positive, which means that the formation reactions for these compounds are endothermic. The other hydrocarbon fuels in Table 5.2 have negative enthalpies of formation, but the values for all of them are less negative than the values for water and carbon dioxide. Recall that the products of the complete combustion of hydrocarbons are carbon dioxide and water. The more negative the heat of formation ΔH_f° for a reaction, the more stable the substance. The implication of this is something stated earlier: the reaction of fuels with oxygen to produce CO_2 and H_2O is exothermic. The reactions produce energy, which is why hydrocarbons are useful as fuels. Note also that the values for the C_2 to C_4 hydrocarbons become more negative with increasing molar mass.

---

[1] The NIST Web site contains much additional data: http://webbook.nist.gov.

FIGURE 5.25 Methane (CH_4) is a renewable source of energy because it can be produced by degradation of organic matter by methanogenic bacteria. Methane is produced in swamps; hence it is commonly called swamp gas. Experiments in the bulk production of methane from natural sources, like the one shown here of a plastic tarp covering a lagoon of animal waste, are being carried out worldwide to augment the fuel supply.

Standard enthalpies of formation ΔH_f° are used to predict standard heats of reaction ΔH_{rxn}°. We can calculate the standard heat of reaction for any reaction by determining the difference between the ΔH_f° values of the products and the ΔH_f° values of the reactants. To see how this approach works, consider the reaction

$$CH_4(g) + 2\,O_2(g) \rightarrow CO_2(g) + 2\,H_2O(g) \qquad (5.16)$$

which represents the combustion of methane, an energy source of increasing importance for the future because of its possible bulk production from renewable sources (Figure 5.25). In this calculation, we use data from Table 5.2 in Equation 5.17:

$$\Delta H_{rxn}^\circ = \sum n_{products}\,\Delta H_{f,products}^\circ - \sum n_{reactants}\,\Delta H_{f,reactants}^\circ \qquad (5.17)$$

where $n_{products}$ is the number of moles of each product in the balanced equation and $n_{reactants}$ is the number of moles of each reactant. Equation 5.17 states that the value of ΔH_{rxn}° for any chemical reaction equals the sum (Σ) of the ΔH_f° value for each product times the number of moles of that product in the balanced equation minus the sum of the ΔH_f° value for each reactant times the number of moles of that reactant in the balanced chemical equation. Thus we multiply the value of ΔH_f° for $O_2(g)$ in Equation 5.17 by 2 before summing with ΔH_f° for $CH_4(g)$ because the O_2 coefficient is 2 in the balanced equation. We do the same with ΔH_f° of water in the product.

Once we insert ΔH_f° values from Table 5.2 and remember that for O_2 $\Delta H_f^\circ = 0$ because it is a pure element in its most stable form, Equation 5.17 becomes

$$\Delta H_{rxn}^\circ = [(1\ \text{mol } CO_2)(-393.5\ \text{kJ/mol}) + (2\ \text{mol } H_2O)(-241.8\ \text{kJ/mol})]$$
$$- [(1\ \text{mol } CH_4)(-74.8\ \text{kJ/mol}) + (2\ \text{mol } O_2)(0.0\ \text{kJ/mol})]$$
$$= [(-393.5\ \text{kJ}) + (-483.6\ \text{kJ})] - [(-74.8\ \text{kJ}) + (0.0\ \text{kJ})]$$
$$= -802.3\ \text{kJ}$$

Calculations of standard enthalpies of reaction ΔH_{rxn}° from standard enthalpies of formation ΔH_f° can be carried out for all kinds of chemical reactions, including reactions that occur in solution. Consider the reaction between methane and steam, which yields a mixture of hydrogen and carbon monoxide known as *water gas*:

$$CH_4(g) + H_2O(g) \rightarrow CO(g) + 3\,H_2(g) \qquad (5.18)$$

This reaction is used to synthesize the hydrogen used in the steel industry to remove impurities from molten iron and in the chemical industry to make hundreds of compounds including ammonia, NH_3, for use in fertilizers and as a refrigerant. The reaction is also important in the manufacture of hydrogen fuel for fuel cells, which are used to generate electricity directly from a chemical reaction.

Inserting the appropriate values for the compounds from Table 5.2 into Equation 5.17 along with the coefficient 3 for H_2, we calculate

$$\Delta H_{rxn}^\circ = [(1\ \text{mol } CO)(-110.5\ \text{kJ/mol}) + (3\ \text{mol } H_2)(0.0\ \text{kJ/mol})]$$
$$- [(1\ \text{mol } CH_4)(-74.8\ \text{kJ/mol}) + (1\ \text{mol } H_2O)(-241.8\ \text{kJ/mol})]$$
$$= +206.1\ \text{kJ}$$

The positive standard enthalpy of reaction tells us that this reaction between water vapor and methane (called steam-methane reforming, or simply *steam reforming*) is endothermic. Therefore, energy must be added to make the reaction take place. (It is typically conducted at temperatures near 1000°C.) Thus, although hydrogen is attractive as a fuel because it burns vigorously and produces only water as a product,

$$2\,H_2(g) + O_2(g) \rightarrow 2\,H_2O(\ell) + \text{energy}$$

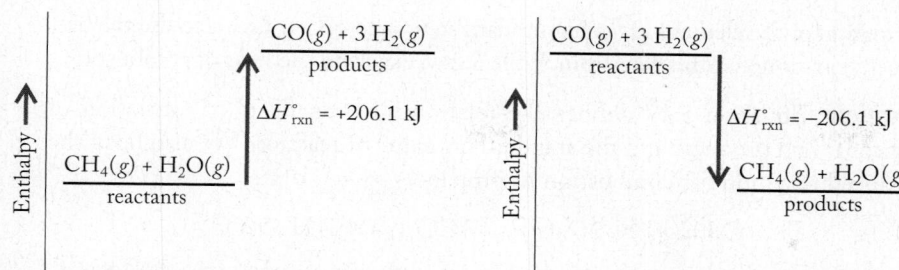

FIGURE 5.26 (a) The reaction of methane with water to produce carbon monoxide and hydrogen is endothermic. (b) The reverse reaction, carbon monoxide plus hydrogen producing methane and water, is exothermic. The value of the enthalpy change of the two reactions has the same magnitude but is positive for the endothermic reaction and negative for the exothermic reaction.

its production via the reaction in Equation 5.18 does not reduce the use of fossil fuels because that reaction consumes methane, most of which currently comes from fossil fuel. (This could of course change if production of methane from renewable sources becomes more feasible.) Fossil fuels are also burned to generate the heat that must be added to the reaction in Equation 5.18 to make it run. Using hydrogen-powered vehicles merely shifts consumption of fossil fuels from the consumption of gasoline at the filling station to an earlier step in the process.

Hydrocarbons other than methane are also used as the starting materials in the production of $H_2(g)$; the reactions are still usually endothermic and require the input of energy, which means fossil fuels are consumed to generate that energy. Also, CO_2 is released into the environment as a result of the production of hydrogen in this way.

Enthalpy is a state function, and, as noted in Section 5.1, the value of a state function is independent of the path taken to achieve that value. Thus values for enthalpies of reaction ΔH_{rxn} are independent of pathway. It does not matter what path we take to get from the reactants to the products; the enthalpy difference between them will always be the same, as Figure 5.26 illustrates. Just as we saw in Section 5.2 with state changes, once we know the value of the energy associated with running a reaction in one direction, we also know the value of the energy associated with running the reaction in the reverse direction because it has the same magnitude but the opposite sign. This means that, having calculated a value of $\Delta H_{rxn}° = +206.1$ kJ for the water gas reaction (Equation 5.18), we can write

$$CO(g) + 3\,H_2(g) \rightarrow CH_4(g) + H_2O(g) \qquad \Delta H_{rxn}° = -206.1 \text{ kJ}$$

for the exothermic reverse reaction.

SAMPLE EXERCISE 5.9 **Calculating Enthalpy of Reaction**

Using the appropriate values from Table 5.2, calculate $\Delta H_{rxn}°$ for the combustion of the fuel propane (C_3H_8) in air.

Collect and Organize The reactants are propane and elemental oxygen, O_2, and the products are carbon dioxide and water. Even though combustion is an exothermic reaction, we assume that water is produced as liquid, i.e., $H_2O(\ell)$, so that all products and reactants are in their standard states. The heats of formation of propane, carbon dioxide, and $H_2O(\ell)$ are given in Table 5.2. The heat of formation of the element O_2 in its standard state is zero. Our task is to use the balanced equation and the data from Table 5.2 to calculate the heat of combustion.

Analyze Equation 5.17 defines the relation between heats of formation of reactants and products and the standard enthalpy of reaction. We also need the balanced equation for combustion of propane:

$$C_3H_8(g) + 5\,O_2(g) \rightarrow 3\,CO_2(g) + 4\,H_2O(\ell)$$

Solve Inserting ΔH_f° values for the products $[CO_2(g)$ and $H_2O(\ell)]$ and reactants $[C_3H_8(g)$ and $O_2(g)]$ from Table 5.2 and the coefficients in the balanced chemical equation into Equation 5.17, we get

$$\Delta H_{rxn}^\circ = [(3 \text{ mol } CO_2)(-393.5 \text{ kJ/mol}) + (4 \text{ mol } H_2O)(-285.8 \text{ kJ/mol})]$$
$$- [(1 \text{ mol } C_3H_8)(-103.8 \text{ kJ/mol}) + (5 \text{ mol } O_2)(0.0 \text{ kJ/mol})]$$
$$= -2219.9 \text{ kJ}$$

Think about It The result of the calculation has a large negative value, which means that the combustion reaction is highly exothermic, as expected for a hydrocarbon fuel.

Practice Exercise Calculate ΔH_{rxn}° for the *water-gas shift reaction*, in which the carbon monoxide formed in the reaction shown in Equation 5.18 is reacted with more steam, producing CO_2 and H_2:

$$CO(g) + H_2O(g) \rightarrow CO_2(g) + H_2(g)$$

(Answers to Practice Exercises are in the back of the book.)

Hydrocarbons such as methane, ethane, and propane are components of natural gas and are excellent fuels. However, they are currently classified as nonrenewable fuels and their combustion produces carbon dioxide, contributing to climate change.

> **CONCEPT TEST**
>
> What is the enthalpy of reaction ΔH_{rxn}° for the production of one mole of C_3H_8 and oxygen from $CO_2(g)$ and $H_2O(\ell)$? Explain the reasoning you used to arrive at your answer. (See information in Sample Exercise 5.9, and do not do any mathematical calculations in answering this question.)

(Answers to Concept Tests are in the back of the book.)

A Practical Application of Thermochemistry

Analyses of the energy required to carry out industrial procedures are frequently done to assess costs of operations and support new approaches to producing materials. The following example of a simple evaluation of the energy requirements associated with the production and recycling of aluminum illustrates a practical application of thermochemical concepts.

Over the last century, aluminum both alone and in combination with other metals has replaced steel for use where high strength-to-weight ratios and corrosion resistance are paramount. Airplanes, motor vehicles, and the facades of buildings may now all be made from aluminum. The industrial process for converting Al_2O_3 (alumina), the form of aluminum in the ore bauxite, into aluminum was developed by two 23-year-old chemists, Charles Hall and Paul Louis-Toussaint Héroult (both 1863–1914), working independently in the United States (Hall) and France (Héroult) (Figure 5.27). The process is based on passing an electric current through a solution of alumina dissolved in molten cryolite (Na_3AlF_6). As electricity passes through the solution, aluminum ions are reduced to aluminum metal while the positively charged carbon electrode is oxidized to carbon dioxide. The process is described by the following reaction:

$$2\,Al_2O_3(\text{in molten } Na_3AlF_6) + 3\,C(s) \rightarrow 4\,Al(\ell) + 3\,CO_2(g)$$

The principal energy cost of the Hall–Héroult process is the electricity needed to reduce Al_2O_3; the major cost in recycling is the energy required to melt aluminum metal. We can use thermochemistry principles to estimate the energy requirements for the Hall–Héroult process and compare it to the energy needed for recycling.

We can calculate the standard energy change for the reduction reaction of alumina from the standard enthalpies of formation using Equation 5.17:

$$\Delta H^\circ_{rxn} = [3(\Delta H^\circ_{f,CO_2} + 4(\Delta H^\circ_{f,Al})] - [2(\Delta H^\circ_{f,Al_2O_3}) + 3(\Delta H^\circ_{f,C})]$$

$$= \left[(3\ \text{mol CO}_2)\left(\frac{-393.5\text{ kJ}}{1\ \text{mol CO}_2}\right) + (4\ \text{mol Al})\left(\frac{10.79\text{ kJ}}{1\ \text{mol Al}}\right)\right]$$

$$- \left[(2\ \text{mol Al}_2O_3)\left(\frac{-1675.7\text{ kJ}}{1\ \text{mol Al}_2O_3}\right) + (3\ \text{mol C})\left(\frac{0.0\text{ kJ}}{1\ \text{mol C}}\right)\right]$$

$$= +2214.1\text{ kJ}$$

Dividing this value by the 4 moles of aluminum produced in the reaction as written, we get

$$\frac{2214.1\text{ kJ}}{4\text{ mol Al}} = 553.52\text{ kJ/mol}$$

or an energy cost of 553.52 kJ for every mole of aluminum produced.

We can estimate the energy required to recycle 1.00 mole of aluminum by heating it from 25°C to its melting point (660°C) until all the aluminum melts. The energy needed to heat 1 mole of aluminum from 25°C to 660°C can be calculated from its molar heat capacity [24.4 J/(mol · °C)] and Equation 5.8:

$$q = nc_p\Delta T$$

$$= 1.00\ \text{mol} \times 24.4\ \frac{\text{J}}{\text{mol}\cdot{}^\circ\text{C}} \times (660 - 25)^\circ\text{C} \times \frac{1\text{ kJ}}{1000\text{ J}}$$

$$= 15.5\text{ kJ}$$

Once the aluminum is at its melting point, the energy required to melt 1 mole of Al is its heat of fusion ($\Delta H_{fus} = 10.79$ kJ/mol):

$$10.79\text{ kJ/mol} \times 1.00\ \text{mol} = 10.8\text{ kJ}$$

The estimated total energy to heat and melt 1.00 mole of aluminum is

$$15.5\text{ kJ} + 10.8\text{ kJ} = 26.3\text{ kJ}$$

This value represents

$$\frac{26.3\text{ kJ}}{553.52\text{ kJ}} \times 100\% = 4.75\%$$

of the energy needed electrically to produce 1 mole of aluminum from its ore. The high cost of electricity makes recycling aluminum economically attractive in addition to environmentally sound.

FIGURE 5.27 Charles M. Hall and Paul Louis-Toussiant Héroult independently developed the same electrolytic process for producing aluminum metal from alumina. Hall's sister Julia, also a chemistry major at Oberlin College, assisted her brother in the lab, and her business skills made their aluminum production company a financial success. The company became the Aluminum Company of America, shortened to Alcoa, Inc.

Of course other energy costs arise in both the production and recycling of aluminum and the numbers calculated here should be viewed as estimates based on ideal situations; overall, however, recycling saves aluminum manufacturers about 95% of the energy required to produce the metal from the ore. This energy saving has inspired the rapid growth of a global aluminum recycling industry, and in the United States alone, aluminum recycling is a $1 billion per year business. Junkyards in the United States currently recycle 85% of the aluminum in cars and over 50% of the aluminum in food and beverage containers.

5.7 Fuel Values and Food Values

From Sample Exercise 5.9 and the text preceding it, we learned that the enthalpy of reaction for one mole of propane (-2219.9 kJ) is much greater than for one mole of methane—that is, much more energy is released in the combustion of propane. Does this make propane an inherently better (higher-energy) fuel? Not necessarily. Expressing ΔH_{rxn}° values on a per-mole basis is the only way to ensure that we are talking about the same number of molecules. However, we do not purchase fuels, or anything else for that matter, in units of moles. Depending on the fuel, we buy it either by mass (coal by the ton) or by volume (gasoline by the gallon or the liter).

Fuel Value

To calculate the enthalpy change that takes place when 1 g of methane or 1 g of propane burns in air producing CO_2 and liquid water, we divide the absolute value of ΔH_{rxn} (in kilojoules per mole) for each reaction by the molar mass of the hydrocarbon to determine the number of kilojoules of energy released per gram of substance:

$$CH_4: \quad \frac{802.3 \text{ kJ}}{\text{mol}} \times \frac{1 \text{ mol}}{16.04 \text{ g}} = 50.02 \text{ kJ/g}$$

$$C_3H_8: \quad \frac{2219.9 \text{ kJ}}{\text{mol}} \times \frac{1 \text{ mol}}{44.10 \text{ g}} = 50.34 \text{ kJ/g}$$

These quantities of energy per gram of fuel are called **fuel values**. If we carry out similar calculations for the other hydrocarbons in Table 5.2, we can determine their fuel values. What we find when we do is that the values decrease with increasing molar mass. Why is that the case?

The answer lies in the hydrogen-to-carbon ratios in these compounds. As the number of carbon atoms per molecule increases, the hydrogen-to-carbon ratio decreases. Consider that there are 12 times as many H atoms in 1 g of hydrogen as there are C atoms in 1 g of carbon; this means that, *gram for gram,* 1 g of hydrogen can form six times as many moles of H_2O as 1 g of carbon can form moles of CO_2. More energy is released in a combustion reaction by the formation of 1 mole of CO_2 (393.5 kJ) than by 1 mole of $H_2O(g)$ (241.8 kJ), but even if we take this difference into account, *gram for gram* hydrogen has many times the fuel value of carbon.

Another term is frequently used to compare the energy content of liquid fuels: **fuel density**. Fuel density describes the amount of energy available per unit volume of a liquid fuel and is typically reported as energy released when 1 liter of liquid is completely burned. Both fuel value and fuel density are reported as positive numbers, and it is simply understood that these values refer to the energy released from the fuel when it is burned.

fuel value the energy released during the complete combustion of 1 g of a substance.

fuel density the amount of energy released during the complete combustion of 1 liter of a liquid fuel.

CONCEPT TEST ···

Without doing any calculations, predict which compound in each pair releases more energy during combustion in air: (a) 1 mole of CH_4 or 1 mole of H_2; (b) 1 g of CH_4 or 1 g of H_2.

(Answers to Concept Tests are in the back of the book.)

···

SAMPLE EXERCISE 5.10 **Comparing Fuel Values and Fuel Densities**

Most automobiles run on either gasoline or diesel fuel. Although both fuels are mixtures, the energy content in gasoline can be approximated by considering it to have the formula C_9H_{20} ($d = 0.718$ g/mL), while diesel fuel may be considered to be $C_{14}H_{30}$ ($d = 0.763$ g/mL). Using these two formulas, compare (a) fuel value per gram of each fuel and (b) fuel density per liter of each fuel.

Collect and Organize We are given representative chemical formulas for gasoline (C_9H_{20}) and diesel fuel ($C_{14}H_{30}$) and can calculate the molar mass for each fuel. Heat of combustion data for C_9H_{20} and $C_{14}H_{30}$ are given in Table 5.2. Our task is to use these data and the respective molar masses to obtain the fuel values. We can then use the given density of each fuel to convert the fuel value from kJ/g to kJ/L.

Analyze The enthalpies of combustion in Table 5.2 are given in kilojoules per mole, so to answer this question in terms of grams and liters, we need molar masses to convert moles to grams in part a and for part b we need density ($d = m/V$, grams per milliliter) to convert grams to liters. Taking C_9H_{20} and $C_{14}H_{30}$ as the average molecules in regular gasoline and diesel fuel, respectively, we can determine their molar masses. Then the fuel value can be calculated from the heats of combustion. Using the densities given in the problem, we can calculate fuel densities in kJ/L.

Solve

a. Fuel values

$$\text{Gasoline as } C_9H_{20}\text{:}\quad 6160\ \frac{\text{kJ}}{\text{mol}} \times \frac{1\ \text{mol}}{128.25\ \text{g}} = 48.0\ \text{kJ/g}$$

$$\text{Diesel as } C_{14}H_{30}\text{:}\quad 7940\ \frac{\text{kJ}}{\text{mol}} \times \frac{1\ \text{mol}}{198.38\ \text{g}} = 40.0\ \text{kJ/g}$$

b. Fuel densities

$$\text{Gasoline as } C_9H_{20}\text{:}\quad 48.0\ \frac{\text{kJ}}{\text{g}} \times 0.718\ \frac{\text{g}}{\text{mL}} \times 1000\ \frac{\text{mL}}{1\ \text{L}} = 34{,}500\ \text{kJ/L}$$

$$\text{Diesel as } C_{14}H_{30}\text{:}\quad 40.0\ \frac{\text{kJ}}{\text{g}} \times 0.763\ \frac{\text{g}}{\text{mL}} \times \frac{1000\ \text{mL}}{1\ \text{L}} = 30{,}500\ \text{kJ/L}$$

Think about It We wouldn't expect the fuel values of gasoline and diesel fuel to be very different, or otherwise there would be a strong preference for gasoline-fueled cars over diesel or vice versa. The values are similar to the fuel values of methane and propane calculated in the text, so these answers seem reasonable. In the United States, fuel for cars and trucks is bought by the gallon and its consumption is rated in miles per gallon, whereas in most countries it is bought by the liter (1 U.S. gal = 3.785 L). Because of the way we buy automobile fuels, it makes sense to compare fuel densities rather than fuel values. On either basis, diesel is a slightly inferior fuel compared to gasoline.

Practice Exercise Kerosene, used as a fuel in high-performance aircraft and in space heaters, is a hydrocarbon intermediate in composition between gasoline and diesel fuel and may be approximated as $C_{12}H_{26}$ ($d = 0.750$ g/mL; $\Delta H^\circ_{comb} = -7050$ kJ/mol). Estimate the fuel value and the fuel density of kerosene.

(Answers to Practice Exercises are in the back of the book.)

Food Value

Food serves the same purpose in living systems as fuel does in mechanical systems. The chemical reactions that convert food into energy resemble combustion but consist of many more steps that are much more highly controlled. Carbon dioxide and water are the ultimate end products, however, and in a fundamental way metabolism of food by a living system and combustion of fuel in an engine are the same process. The **food value** of the material we eat—the amount of energy produced when food is burned completely—can be determined using the same equipment and applying the same concepts of thermochemistry we have developed to evaluate fuels for vehicles. We can analyze the relative food value of the items we consume in the same way we analyzed fuel value: by burning material in a bomb calorimeter and measuring the quantity of energy released.

As an illustration, let's consider the food value of a jelly doughnut. To determine its heat energy content per unit of mass, we burn the doughnut in a calorimeter. A fresh jelly doughnut may be hard to burn because of its high water content, so we first prepare the doughnut by drying it. This is a necessary step to get an accurate value for the enthalpy of reaction ΔH_{rxn}, because we need to make sure that the great majority of the sample mass is due to its carbon-containing components that will burn completely to produce CO_2 and H_2O.

Once it is dry, suppose our jelly doughnut has a mass of 55 g. We put it in a calorimeter for which $C_{calorimeter} = 41.8$ kJ/°C and burn it completely in excess oxygen. If the temperature of the calorimeter rises by 25.0°C, what is the food value of the jelly doughnut?

To answer this question, we use Equation 5.12 to determine the quantity of energy that flowed from the doughnut to the calorimeter:

$$q_{calorimeter} = C_{calorimeter}\, \Delta T = (41.8 \text{ kJ/°C})(25.0°C) = +1050 \text{ kJ}$$

where the plus sign reminds us that energy flowed into the calorimeter. The energy lost by the doughnut as it burned has the same value but opposite sign as the energy absorbed by the calorimeter:

$$-q_{doughnut} = q_{calorimeter}$$
$$= 1050 \text{ kJ}$$

food value the quantity of energy produced when a material consumed by an organism for sustenance is burned completely; it is typically reported in Calories (kilocalories) per gram of food.

Just as with fuel values, however, food value is reported as a positive number, and it is understood that the energy contained in the food is released when the food is metabolized. The doughnut's food value is therefore

$$\frac{1050 \text{ kJ}}{1 \text{ \sout{doughnut}}} \times \frac{1 \text{ \sout{doughnut}}}{55 \text{ g}} = 19 \text{ kJ/g}$$

Most of us still think in terms of nutritional Calories (kilocalories), and we can use the definition of Calorie from Section 5.2 to convert energy in kilojoules into the familiar unit used by nutritionists:

$$\frac{1050 \text{ \sout{kJ}}}{\text{jelly doughnut}} \times \frac{1 \text{ Cal}}{4.184 \text{ \sout{kJ}}} = \frac{251 \text{ Cal}}{\text{jelly doughnut}}$$

Between 8 and 10 jelly doughnuts would provide all the Calories needed by an adult human with normal activity in one day. A bicyclist riding in the Tour de France would need about 32 jelly doughnuts a day.

SAMPLE EXERCISE 5.11   Calculating Food Value

Glucose ($C_6H_{12}O_6$) is a simple sugar formed by photosynthesis in plants. The complete combustion of 0.5763 g of glucose in a calorimeter ($C_{\text{calorimeter}}$ = 6.20 kJ/°C) raises the temperature of the calorimeter by 1.45°C. What is the food value of glucose in Calories per gram?

Collect and Organize We are asked to determine the food value of glucose, which means the energy given off when 1 g is burned. We have the mass of glucose burned and the calorimeter constant. We can relate the energy given off by the glucose to the energy gained by the calorimeter (Equation 5.15) and the energy given off by the glucose to the temperature change and the calorimeter constant (Equation 5.12).

Analyze We use the data from the calorimetry experiment to determine how much heat in kilojoules is given off when the stated amount of glucose is burned. We can convert that quantity into Calories by using the conversion factor 1 Cal = 4.184 kJ.

Solve

$$q_{\text{calorimeter}} = C_{\text{calorimeter}} \, \Delta T = (6.20 \text{ kJ/°C})(1.45°C) = 8.99 \text{ kJ}$$

To convert this quantity of energy to a food value, we divide by the sample mass:

$$\frac{8.99 \text{ kJ}}{0.5763 \text{ g}} = 15.6 \text{ kJ/g}$$

and then convert into Calories:

$$(15.6 \text{ \sout{kJ}/g})\left(\frac{1 \text{ Cal}}{4.184 \text{ \sout{kJ}}}\right) = 3.73 \text{ Cal/g}$$

Think about It One gram of a doughnut, which is mostly carbohydrates, and 1 g of glucose have about the same food value (19 kJ/g and 15.6 kJ/g, respectively), so the answer to this Sample Exercise seems reasonable.

CONNECTION In Chapter 3, we introduced the sugar glucose as a carbohydrate: a compound composed of carbon, hydrogen, and oxygen.

▶❚❚ CHEMTOUR Hess's Law

Practice Exercise Sucrose (table sugar) has the formula $C_{12}H_{22}O_{11}$ ($\mathcal{M}$ = 342.30 g/mol) and a food value of 16.4 kJ/g. Determine the calorimeter constant of the calorimeter in which the combustion of 1.337 g of sucrose raises the temperature by 1.96°C.

(Answers to Practice Exercises are in the back of the book.)

■ ....

5.8 Hess's Law

In Section 5.6 we described a process for synthesizing hydrogen gas from methane and steam at high temperatures. The reaction is carried out on an industrial scale in two steps. The first step is an endothermic reaction between methane and a limited supply of high-temperature steam, producing carbon monoxide and hydrogen gas in a reaction that has an enthalpy of reaction of +206 kJ:

$$(1) \quad CH_4(g) + H_2O(g) \rightarrow CO(g) + 3\,H_2(g) \qquad \Delta H_1 = +206 \text{ kJ}$$

In the second step, the carbon monoxide from the first reaction is allowed to react with more steam, producing carbon dioxide and more hydrogen gas in a reaction that has an enthalpy of reaction of −41 kJ:

$$(2) \quad CO(g) + H_2O(g) \rightarrow CO_2(g) + H_2(g) \qquad \Delta H_2 = -41 \text{ kJ}$$

We can write an overall reaction equation that results from adding reactions 1 and 2 and simplifying:

$$(1) \quad CH_4(g) + H_2O(g) \rightarrow \cancel{CO(g)} + 3\,H_2(g)$$
$$(2) \quad \cancel{CO(g)} + H_2O(g) \rightarrow CO_2(g) + H_2(g)$$
$$\overline{(3) \quad CH_4(g) + 2\,H_2O(g) \rightarrow CO_2(g) + 4\,H_2(g)}$$

Just as we obtain the overall chemical equation 3 by adding equations 1 and 2, we obtain the enthalpy of reaction for the overall reaction by adding the ΔH values for reactions 1 and 2. The thermochemical equation for the overall reaction is

$$\Delta H_1 + \Delta H_2 = \Delta H_3$$
$$+206 \text{ kJ} + (-41 \text{ kJ}) = +165 \text{ kJ}$$

This calculation for ΔH_{rxn} is an application of **Hess's law**. Also known as *Hess's law of constant heat of summation*, it states that the enthalpy of reaction ΔH_{rxn} for a process that is the sum of two or more other reactions is equal to the sum of the ΔH_{rxn} values of the constituent reactions. This relation is illustrated for the reaction to synthesize hydrogen shown in Figure 5.28.

FIGURE 5.28 Hess's law predicts that the enthalpy change for the production of 4 moles of $H_2(g)$ and 1 mole of $CO_2(g)$ from 1 mole of $CH_4(g)$ and 2 moles of $H_2O(g)$ (ΔH_3) is the sum of the enthalpies of two reactions: the formation of 1 mole of $CO(g)$ and 3 moles of $H_2(g)$ from 1 mole of $CH_4(g)$ and 1 mole of $H_2O(g)$ (ΔH_1), and the formation of 4 moles of $H_2(g)$ and 1 mole of $CO_2(g)$, from 1 mole of $CO(g)$, 1 mole of $H_2O(g)$, and 3 moles of $H_2(g)$ (ΔH_2).

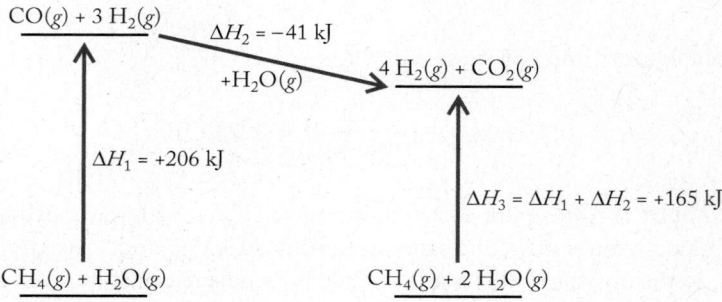

Hess's law is especially useful for calculating enthalpy changes that are difficult to measure directly. For example, CO_2 is the principal product of the combustion of carbon in the form of charcoal:

Reaction A: $\quad C(s) + O_2(g) \rightarrow CO_2(g) \quad \Delta H°_{comb} = -393.5 \text{ kJ}$

When the oxygen supply is limited, however, the products include carbon monoxide:

Reaction B: $\quad C(s) + \frac{1}{2}O_2(g) \rightarrow CO(g)$

It is difficult to measure the standard enthalpy of combustion of this reaction directly because, as long as any oxygen is present, some of the $CO(g)$ formed reacts with the O_2 to form $CO_2(g)$, yielding a mixture of CO and CO_2 as the product. However, we can use Hess's law to obtain this value *indirectly* by working with $\Delta H°_{comb}$ values we can measure.

Because we can run reaction A with excess oxygen and thereby force it to completion, we can measure the standard enthalpy of combustion, which is -393.5 kJ. We can also react a sample of pure $CO(g)$ with oxygen and measure $\Delta H°_{comb}$ for that reaction:

Reaction C: $\quad CO(g) + \frac{1}{2}O_2(g) \rightarrow CO_2(g) \quad \Delta H°_{comb} = -283.0 \text{ kJ}$

Hess's law gives us a way to calculate $\Delta H°_{comb}$ for reaction B from the measured values for reactions A and C. To do this, we must find a way to combine the equations for reactions A and C so that the sum equals reaction B. Once we have that combination, because we know two of the $\Delta H°_{comb}$ values, we can calculate the one we do not know.

One approach to this analysis is to focus on the reactants and products in the reaction whose $\Delta H°_{comb}$ value is unknown—reaction B in this example. This reaction has carbon and oxygen as reactants and carbon monoxide as a product. Note that reaction C has carbon monoxide as a reactant. If we add reaction C to reaction B, the carbon monoxide cancels out, and we end up with reaction A as the sum of C and B:

Reaction B:	$C(s) + \frac{1}{2}O_2(g) \rightarrow \cancel{CO(g)}$	$\Delta H°_{comb} = ?$
Reaction C:	$\cancel{CO(g)} + \frac{1}{2}O_2(g) \rightarrow CO_2(g)$	$\Delta H°_{comb} = -283.0 \text{ kJ}$
Reaction A:	$C(s) + O_2(g) \rightarrow CO_2(g)$	$\Delta H°_{comb} = -393.5 \text{ kJ}$

We now use algebra to find $\Delta H°_{comb}$ for reaction B:

$$\Delta H°(B) + \Delta H°(C) = \Delta H°(A)$$
$$\Delta H°(B) = \Delta H°(A) - \Delta H°(C)$$
$$= -393.5 \text{ kJ} - (-283.0 \text{ kJ}) = -110.5 \text{ kJ}$$

Always remember that ΔH is a state function. This means we can manipulate equations in two important ways, should the need arise when applying Hess's law. (1) We can multiply the coefficients in a balanced equation and the ΔH for the reaction by a factor to change the quantity of material we are dealing with. (2) We can reverse a reaction (make the reactants the products and the products the reactants) if we also change the sign of ΔH.

Hess's law the standard enthalpy of reaction $\Delta H°_{rxn}$ for a reaction that is the sum of two or more reactions is equal to the sum of the $\Delta H°_{rxn}$ values of the constituent reactions; also known as *Hess's law of constant heat of summation.*

Carbon: Diamonds, Graphite, and the Molecules of Life

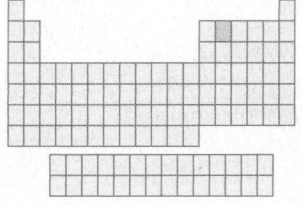

Carbon is an element central to our existence, both because we are a carbon-based life-form and because carbon is the central element in hydrocarbons—currently our most significant sources of energy and causes of environmental changes. We also value carbon in its crystalline forms (Figure 5.29). One of these is *diamond;* we prize diamonds for their beauty and take advantage of their hardness by using them in drill bits and as abrasives. Another crystalline form of carbon, *graphite,* is black and opaque, and prized for its strength and flexibility (graphite fibers are used in golf clubs and tennis rackets) and for its softness and smoothness (flakes are used as lubricants and baked with clay to make the "lead" in pencils). A large variety of specialized noncrystalline materials known as *activated carbon* are synthesized for wastewater treatment, gas purification, and sugar refining (activated carbon decolorizes solutions of raw sugar by selectively binding colored impurities).

The 1996 Nobel Prize in Chemistry was awarded to Robert Curl, Harold Kroto, and Richard Smalley for their 1985 discovery of a third crystalline form of carbon, buckminsterfullerene. Each molecule consists of a cluster of carbon atoms with the formula C_{60}; this molecule is also known as a *buckyball* because it looks like a soccer ball and is reminiscent of the architectural designs of Buckminster Fuller (Figure 5.30). Buckminsterfullerene is the most abundant form of a class of carbon-clustering molecules called *fullerenes.* Their discovery has spawned research into applications ranging from rocket fuels to drug-delivery systems for treating cancer and AIDS.

Carbon is found in minor amounts in Earth's crust as the free element. Diamonds are found in ancient volcanic pipes (openings in the crust through which lava flows); the exact way they were formed is still debated. Graphite

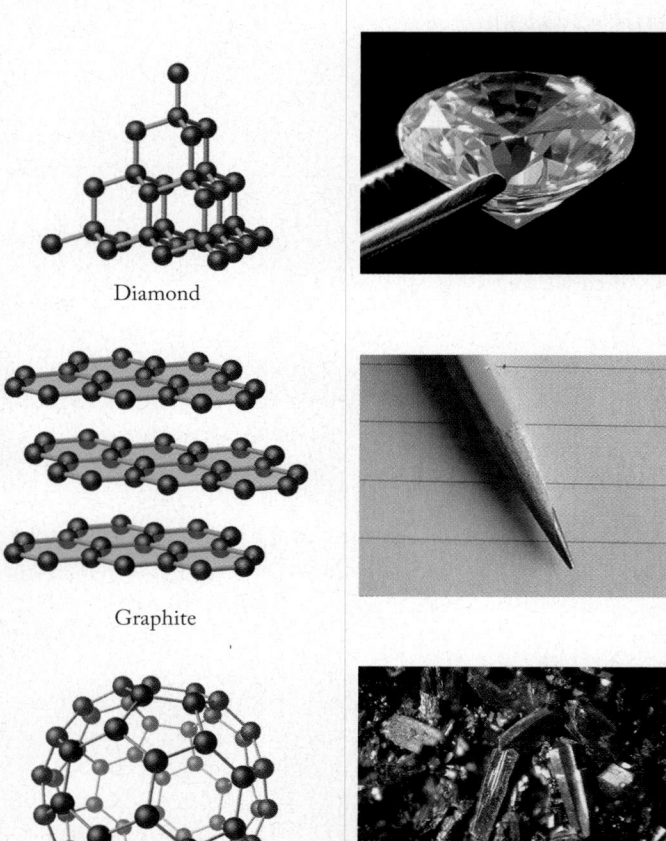

Diamond

Graphite

Buckyball C_{60}

FIGURE 5.29 Three crystalline forms of carbon: diamond, graphite, and buckminsterfullerene (a "buckyball").

FIGURE 5.30 One of Buckminster Fuller's architectural designs was the geodesic dome. Compare this structure to the buckyball in Figure 5.29.

deposits occur worldwide. Over half of the carbon on Earth is contained in carbonate minerals, such as limestone, dolomite, and chalk, in which carbon is in its highest oxidation state as the CO_3^{2-} ion (O.N. = +4), and in fossil fuels (petroleum, coal, natural gas), in which carbon is in reduced states (O.N. as low as −4). These two carbon reservoirs are interconnected by a dynamic system described in Section 3.5, the *carbon cycle*.

Atmospheric CO_2 is the source of all carbon-containing compounds produced in the leaves of green plants by photosynthesis:

$$\text{Sunlight} + 6\,CO_2(g) + 6\,H_2O(\ell) \rightarrow$$
$$C_6H_{12}O_6(aq) + 6\,O_2(g)$$

The most fundamental carbon-containing product of photosynthesis, the sugar glucose ($C_6H_{12}O_6$), is the main food for biological systems. Glucose is also the primary structural subunit of cellulose, which is the chief component of plant fibers such as wood and cotton. A single molecule of cellulose may contain over 1500 glucose units. Plants produce over 100 billion tons of cellulose each year. The oxidation number of carbon in glucose is 0, so green plants may be thought of as carbon-reducing machines. Other life-forms eat plants and derive large amounts of energy from metabolism through a series of oxidation reactions that are the reverse of photosynthesis, which takes carbon at O.N. = 0 (glucose) back to O.N. +4 (carbon dioxide). Life-forms are carbon-oxidizing machines.

When plants and animals die, their carbon-containing molecules are oxidized to CO_2 and water if decomposition is complete. However, plant material deprived of oxygen may be further reduced by bacteria and end up as petroleum and coal. When these fuels react with oxygen in a combustion reaction, energy is released because reduced carbon is oxidized when the hydrocarbons are converted into CO_2.

Some of the CO_2 removed from the atmosphere in photosynthesis is returned to the atmosphere as a result of respiration by animals and plants, but our burning of fossil fuels inserts additional carbon into the atmosphere as CO_2 that was formerly sequestered underground. Atmospheric CO_2 dissolves in seawater to form carbonic acid, which increases the acidity of the ocean. Changing the acidity affects the solubility of $CaCO_3$ in seawater, which in turn has potentially

FIGURE 5.31 A healthy coral reef.

disastrous consequences for marine life, including the many oceanic life-forms that build shells, reefs, and other structures from $CaCO_3$ and other carbonates (Figure 5.31).

Carbon forms millions of compounds, many of which play critical roles in biological processes. Carbon atoms attach themselves to other carbon atoms to an extent not possible for atoms of any other element. The result is the formation of chains and rings that combine to create an almost incomprehensible variety of structures that can link together to form gigantic molecules containing hundreds and even thousands of carbon atoms. Over 16 million compounds of carbon are known, and over 90% of the thousands of new materials synthesized each year contain carbon. The number of compounds that contain carbon is far greater than the number of compounds that do not. An entire branch of chemistry, called *organic chemistry*, is devoted to the study of their structures and properties.

The chemistry of carbon is the chemistry of pharmaceuticals, soft contact lenses, paints, synthetic fibers, plastics, and gasoline. It is the chemistry of the food we eat and the fuel that powers our cars and airplanes. Fossil fuels are the primary source of molecules that industry modifies to make the items we regard as part of modern life, so they can be regarded as fuel for the manufacturing industry as well as for transportation. Combined in precise ways with a small number of other elements—such as sulfur, phosphorus, oxygen, nitrogen, and hydrogen—carbon serves as the foundation of life.

SAMPLE EXERCISE 5.12 **Calculating Enthalpies of Reaction Using Hess's Law**

Hydrocarbons burned in a limited supply of air may not burn completely, and $CO(g)$ may be generated. One reason furnaces and hot-water heaters fueled by natural gas need to be vented is that incomplete combustion can produce toxic carbon monoxide:

Reaction A: $2\,CH_4(g) + 3\,O_2(g) \rightarrow 2\,CO(g) + 4\,H_2O(g)$ $\Delta H^{\circ}_{comb} = ?$

Use the reactions

Reaction B: $CH_4(g) + 2\,O_2(g) \rightarrow CO_2(g) + 2\,H_2O(g)$ $\Delta H^{\circ}_{comb} = -802\,kJ$

Reaction C: $2\,CO(g) + O_2(g) \rightarrow 2\,CO_2(g)$ $\Delta H^{\circ}_{comb} = -566\,kJ$

to calculate the ΔH°_{comb} for reaction A.

Collect and Organize We are given two reactions (B and C) with thermochemical data and a third (A) for which we are asked to find ΔH°_{comb}. All the reactants and products of reaction A are present in reactions B and/or C.

Analyze We can manipulate the equations for reactions B and C so that they sum to give the equation for which ΔH°_{comb} is unknown. Then we can calculate this unknown value by applying Hess's law.

The reaction of interest (A) has methane on the reactant side. Because reaction B also has methane as a reactant, we can use B as written. Reaction A has CO as a product. Reaction C involves CO as a reactant, so we have to reverse C in order to get CO on the product side. Once we reverse C, we must change the sign of ΔH°_{comb}. If the coefficients as given do not allow us to sum the two reactions to yield reaction A, we can multiply one or both reactions by other factors.

Solve We start with reaction B as written and add the reverse of reaction C, remembering to change the sign of $\Delta H^{\circ}_{comb}(C)$:

B: $CH_4(g) + 2\,O_2(g) \rightarrow CO_2(g) + 2\,H_2O(g)$ $\Delta H^{\circ}_{comb} = -802\,kJ$

C (reversed): $2\,CO_2(g) \rightarrow 2\,CO(g) + O_2(g)$ $-\Delta H^{\circ}_{comb} = +566\,kJ$

Because methane has a coefficient of 2 in reaction A, we multiply all the terms in reaction B, including $\Delta H^{\circ}_{comb}(B)$, by 2:

$2 \times [CH_4(g) + 2\,O_2(g) \rightarrow CO_2(g) + 2\,H_2O(g)]$ $2[\Delta H^{\circ}_{comb}(B) = -802\,kJ]$

Because the carbon monoxide in reaction A has a coefficient of 2, we do not need to multiply reaction C by any factor. Now we add $(2 \times B)$ to the reverse of C and cancel out common terms:

$2 \times B$: $2\,CH_4(g) + \overset{3}{4}O_2(g) \rightarrow \cancel{2\,CO_2(g)} + 4\,H_2O(g)$ $\Delta H^{\circ}_{comb} = -1604\,kJ$

C (reversed): $\cancel{2\,CO_2(g)} \rightarrow 2\,CO(g) + \cancel{O_2(g)}$ $\Delta H^{\circ}_{comb} = +566\,kJ$

A: $2\,CH_4(g) + 3\,O_2(g) \rightarrow 2\,CO(g) + 4\,H_2O(g)$ $\Delta H^{\circ}_{comb} = -1038\,kJ$

Think about It We used Hess's law to calculate the heat of combustion of methane to make CO, which is impossible to achieve in an experiment because any CO produced can react with O_2 to give CO_2, which results in a mixture of CO and CO_2 as a product. The answer of -1038 kJ is a little less negative than the heat of combustion for the production of CO_2 from methane, so the answer is reasonable.

Practice Exercise It does not matter how you assemble the equations in a Hess's law problem. Show that reactions A and C can be summed to give reaction B and result in the same value for $\Delta H^{\circ}_{comb}(B)$.

(Answers to Practice Exercises are in the back of the book.)

When we used standard enthalpies of formation (ΔH°_f) to determine standard enthalpies of reaction (ΔH°_{rxn}) in Section 5.6, we were applying Hess's law. To see that this is the case, consider the reaction of ammonia with oxygen to make $NO(g)$ and liquid water in the first step of the industrial synthesis of nitric acid:

$$4\,NH_3(g) + 5\,O_2(g) \rightarrow 4\,NO(g) + 6\,H_2O(\ell)$$

First we write an equation for the formation reaction for each reactant and product that is not an element in its standard state (remember $\Delta H^{\circ}_f = 0$ for elements in their standard states), using the ΔH°_f values in Table 5.2:

$$\tfrac{1}{2}N_2(g) + \tfrac{3}{2}H_2(g) \rightarrow NH_3(g) \qquad \Delta H^{\circ}_f = -46.1 \text{ kJ/mol}$$

$$\tfrac{1}{2}N_2(g) + \tfrac{1}{2}O_2(g) \rightarrow NO(g) \qquad \Delta H^{\circ}_f = +90.3 \text{ kJ/mol}$$

$$H_2(g) + \tfrac{1}{2}O_2(g) \rightarrow H_2O(\ell) \qquad \Delta H^{\circ}_f = -285.8 \text{ kJ/mol}$$

We reverse the NH_3 equation so that the NH_3 is a reactant and multiply each equation by a factor that matches the product's or reactant's coefficient in the balanced equation.

$$4[NH_3(g) \rightarrow \tfrac{1}{2}N_2(g) + \tfrac{3}{2}H_2(g)] \qquad 4 \text{ mol } (\Delta H^{\circ}_f = +46.1 \text{ kJ/mol})$$

$$4[\tfrac{1}{2}N_2(g) + \tfrac{1}{2}O_2(g) \rightarrow NO(g)] \qquad 4 \text{ mol } (\Delta H^{\circ}_f = +90.3 \text{ kJ/mol})$$

$$6[H_2(g) + \tfrac{1}{2}O_2(g) \rightarrow H_2O(\ell)] \qquad 6 \text{ mol } (\Delta H^{\circ}_f = -285.8 \text{ kJ/mol})$$

Their sum yields the reaction of interest:

$$4\,NH_3(g) \rightarrow 2\,\cancel{N_2(g)} + 6\,\cancel{H_2(g)} \qquad \Delta H^{\circ}_{rxn} = +184.4 \text{ kJ}$$

$$2\,\cancel{N_2(g)} + 2\,O_2(g) \rightarrow 4\,NO(g) \qquad \Delta H^{\circ}_{rxn} = +361.2 \text{ kJ}$$

$$\underline{6\,\cancel{H_2(g)} + 3\,O_2(g) \rightarrow 6\,H_2O(\ell) \qquad \Delta H^{\circ}_{rxn} = -1714.8 \text{ kJ}}$$

$$4\,NH_3(g) + 5\,O_2(g) \rightarrow 4\,NO(g) + 6\,H_2O(\ell) \qquad \Delta H^{\circ}_{rxn} = -1169.2 \text{ kJ}$$

This is exactly the same mathematical operation that results when we use Equation 5.17:

$$\Delta H^{\circ}_{rxn} = \sum n_{products} \, \Delta H^{\circ}_{f,products} - \sum n_{reactants} \, \Delta H^{\circ}_{f,reactants}$$

$$= [4 \text{ mol } NO(+90.3 \text{ kJ/mol}) + 6 \text{ mol } H_2O(-285.8 \text{ kJ/mol})]$$

$$- [4 \text{ mol } NH_3(-46.1 \text{ kJ/mol}) + 5 \text{ mol } O_2(0 \text{ kJ/mol})] = -1169.2 \text{ kJ}$$

Hess's law is a consequence of the fact that enthalpy is a state function. This statement means that for a particular set of reactants and products, the enthalpy change of the reaction is the same whether the reaction takes place in one step or in a series of steps. The concept of enthalpy and its expression in Hess's law are very useful because the heats associated with a large number of reactions can be calculated from a few that have been measured.

In this chapter we have considered the flow of energy and its role in defining the behavior of physical, chemical, and biological processes. The interaction of energy with matter—the transfer of energy into and out of materials—causes phase changes and alters the temperature of matter. The magnitude of these changes and the temperatures and temperature ranges over which they occur are characteristic of the quantity and identity of the matter involved. Energy is also a product or a reactant in virtually all chemical reactions and as such behaves stoichiometrically, which means the chemical changes that a specific quantity of material undergoes are characterized by the release or consumption of a specific amount of energy. Understanding the energy contained in fuels is particularly important, because fuels provide the bulk of the power needed by the equipment and devices of modern life—from cars and airplanes to computers, air conditioners, and lightbulbs. Just as significant is the energy contained in foods that sustain life. How we deal with the needs for energy in the near future—in terms of both fuel and food—will determine the quality of all our lives and the health of our planet.

SUMMARY ∎

Section 5.1 Energy is the capacity to transfer **heat** (q) or to do **work** (w). A **thermochemical equation** includes the energy absorbed or released in a chemical reaction. **Thermodynamics** is the study of energy and its transformations. Differences in temperature result in **heat transfer** from one object to another. No heat is transferred when two objects in thermal contact with each other have the same temperature. **Potential energy (PE)** is the energy of position or composition and is a **state function**. **Kinetic energy (KE)** is the energy of motion. The **law of conservation of energy** states that energy cannot be created or destroyed. Heating a sample increases the average kinetic energy of the atoms in the sample. Energy is stored in compounds, and energy is absorbed or released when they are transformed into different compounds or when a change of state occurs.

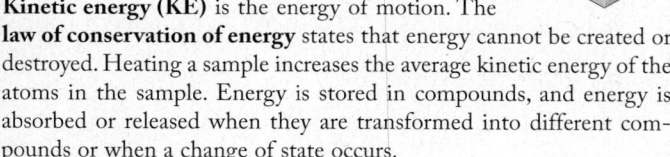

Section 5.2 A **system** is the part of the universe under study. Everything not part of the system is the **surroundings**. **Isolated systems** exchange neither energy nor matter with their surroundings, **closed systems** exchange only energy, and **open systems** exchange energy and matter. In an **exothermic process** the system loses energy by heating

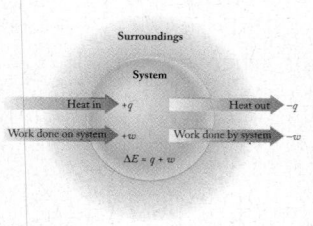

its surroundings ($q < 0$); in an **endothermic process** the system absorbs energy ($q > 0$) from its surroundings. The sum of the kinetic and potential energies of a system is called its **internal energy (E)**. The **first law of thermodynamics** states that the energy gained or lost by a system equals the energy lost or gained by the surroundings. Common energy units are **calories (cal)** and **joules (J)**. The internal energy of a system is increased ($\Delta E = E_{final} - E_{initial}$ is positive) when it is heated ($q > 0$) or if work is done on it ($w > 0$).

Section 5.3 The **enthalpy (H)** of a system is given by $H = E + PV$. The **enthalpy change (ΔH)** of a system is equal to the energy (q_p) added to or removed from the system at constant pressure: $\Delta H > 0$ for endothermic reactions and $\Delta H < 0$ for exothermic reactions.

Section 5.4 The **heat capacity (C_p)** of an object at constant pressure is the amount of energy required to increase the temperature of the object by 1°C. The **molar heat capacity (c_p)** of a substance at constant pressure is the amount of energy required to increase the temperature of 1 mole of the substance by 1°C. The **specific heat (c_s)** of a substance is the amount of energy required to increase the temperature of 1 g of the substance by 1°C. The **molar heat of**

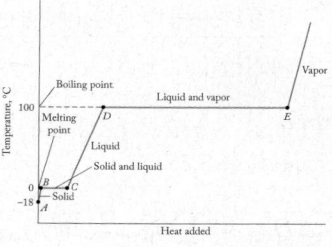

fusion (ΔH_{fus}) of a substance is the amount of energy required to convert 1 mole of the solid substance to a liquid at the melting point of the substance. The **molar heat of vaporization (ΔH_{vap})** of a substance is the amount of energy required to convert 1 mole of the liquid substance to a vapor at the boiling point of the substance.

Section 5.5 A **calorimeter** is a device used to measure the amounts of energy involved in physical and chemical processes. A **bomb calorimeter** is a device used to measure the enthalpy change of a combustion reaction (ΔH_{comb}). The enthalpy change associated with a reaction is defined by the **enthalpy of reaction (ΔH_{rxn})** or the **heat of reaction**.

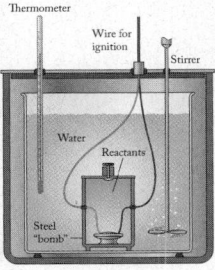

Section 5.6 The **standard enthalpy of formation** of a substance, ΔH_f°, is the amount of energy involved in a **formation reaction**, in which 1 mole of the substance is made from its constituent elements in their **standard states** (under **standard conditions**). The standard enthalpy of formation of an element in its standard state is zero. Enthalpy changes for physical changes and chemical reactions can be calculated from the enthalpies of formation of the reactants and products.

Section 5.7 **Fuel value** is the amount of energy released on complete combustion of 1 g of a fuel. **Food value** is the amount of energy released when a material consumed by an organism for sustenance is burned completely; nutritionists often express food values in Calories (kilocalories) rather than in the SI unit kilojoules.

Section 5.8 **Hess's law** states that the standard enthalpy of a reaction (ΔH_{rxn}°) that is the sum of two or more other reactions is equal to the sum of the ΔH_{rxn}° values of the constituent reactions. It can be used to calculate enthalpy changes in reactions that are hard or impossible to measure directly.

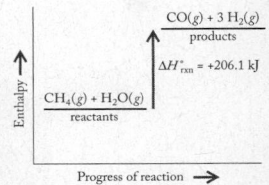

PROBLEM-SOLVING SUMMARY

TYPE OF PROBLEM	CONCEPTS AND EQUATIONS	SAMPLE EXERCISES
Identifying endothermic and exothermic processes, and calculating internal energy change (ΔE) and P–V work	For the system: $$\Delta E = q + w \qquad (5.5)$$ where $w = -P\Delta V$.	5.1–5.3
Determining the flow of energy (q) associated with a change of state or with changing the temperature of a substance	Melting a solid at its melting point: $$q = n\Delta H_{fus} \qquad (5.9)$$ vaporizing a liquid at its boiling point: $$q = n\Delta H_{vap} \qquad (5.10)$$ or heating a substance: $$q = nc_p \Delta T \qquad (5.8)$$	5.4–5.6
Measuring the heat capacity (calorimeter constant) of a calorimeter	$$C_{calorimeter} = q/\Delta T \qquad (5.13)$$ where $C_{calorimeter}$ is the heat capacity of calorimeter, q is the heat released by a standard combustion reaction, and ΔT is the temperature change of calorimeter.	5.7
Recognizing and writing formation reactions	In a formation reaction, the reactants are elements in their standard states and the product is 1 mole of a single compound.	5.8
Calculating standard enthalpies of reaction from heats of formation	$$\Delta H_{rxn}^\circ = \sum n_{products} \Delta H_{f,products}^\circ - \sum n_{reactants} \Delta H_{f,reactants}^\circ \qquad (5.17)$$	5.9
Calculating fuel values and food value	The fuel value or food value of a substance is the energy released by the complete combustion of 1 g of the substance.	5.10, 5.11
Calculating standard enthalpies of reaction using Hess's law	Reorganize the information so that the reactions add together as desired. Reversing a reaction changes the sign of the reaction's ΔH_{rxn}° value. Multiplying the coefficients in a reaction by a factor means the reaction's ΔH_{rxn}° value has to be multiplied by the same factor.	5.12

VISUAL PROBLEMS

(Answers to boldface end-of-chapter questions and problems are in the back of the book.)

5.1. A brick lies perilously close to the edge of the flat roof of a building (Figure P5.1). The roof edge is 50 ft above street level, and the brick has 500 J of potential energy with respect to street level. Someone (we don't know who) edges the brick off the roof, and it begins to fall. What is the brick's kinetic energy when it is 35 ft above street level? What is its kinetic energy the instant before it hits the street surface?

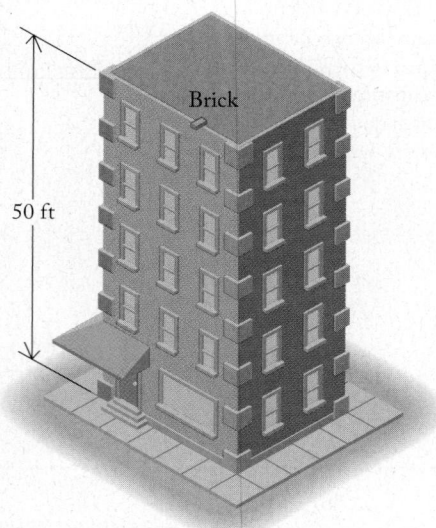

FIGURE P5.1

5.2. Figure P5.2 diagrams pairs of cations and anions in contact. The energy of each interaction is proportional to Q_1Q_2/d, where Q_1 and Q_2 are the charges on the cation and anion, and d is the distance between their nuclei. Which pair has the greatest interaction energy and which has the smallest?

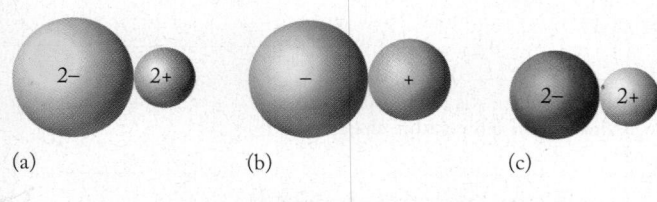

FIGURE P5.2

5.3. Figure P5.3 shows a sectional view of an assembly consisting of a gas trapped inside a stainless steel cylinder with a stainless steel piston and a block of iron on top of the piston to keep it in place.
   a. Sketch the situation after the assembly has been heated for a few seconds with a blowtorch.
   b. Is the piston higher or lower in the cylinder?
   c. Has heat (q) been added to the system?
   d. Has the system done work (w) on its surroundings, or have the surroundings done work on the system?

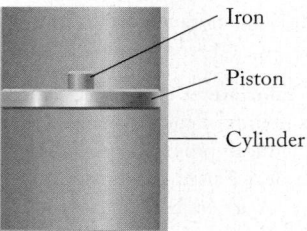

FIGURE P5.3

5.4. Figure P5.4 represents the energy change in a chemical reaction. The reaction results in a very small decrease in volume.
   a. Has the internal energy of the system decreased or increased as a result of the reaction?
   b. What sign is given to ΔE for the reaction system?
   c. Is the reaction endothermic or exothermic?

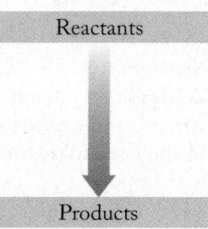

FIGURE P5.4

5.5. The closed rigid metal box lying on the wooden kitchen table in Figure P5.5 is about to be involved in an accident.
   a. Are the contents of the box an isolated system, a closed system, or an open system?
   b. What will happen to the internal energy of the system if the table catches fire and burns?
   c. Will the system do any work on the surroundings while the fire is burning or after the table has burned away and collapsed?

FIGURE P5.5

5.6. The diagram in Figure P5.6 shows how a chemical reaction in a cylinder with a piston affects the volume of the system.
   a. In this reaction, does the system do work on the surroundings?
   b. If the reaction is endothermic, does the internal energy of the system increase or decrease when the reaction is proceeding?

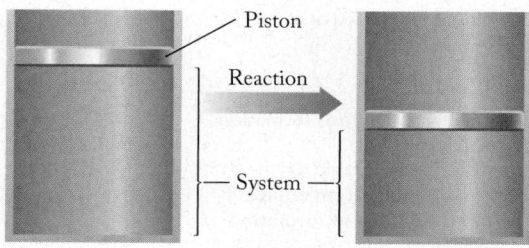

FIGURE P5.6

5.7. The enthalpy diagram in Figure P5.7 indicates the enthalpies of formation of four compounds made from the elements listed on the "zero" line of the vertical axis.
   a. Why are the elements all put on the same horizontal line?
   b. Why is $C_2H_2(g)$ sometimes called an "endothermic" compound?

c. How could the data be used to calculate the heat of the reaction that converts a stoichiometric mixture of $C_2H_2(g)$ and $O_2(g)$ to $CO(g)$ and $H_2O(g)$?

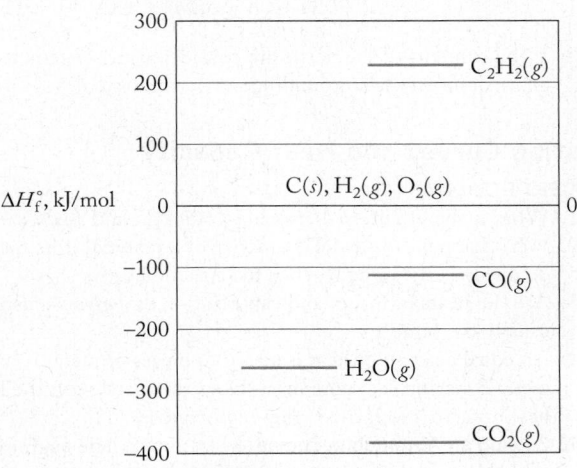

FIGURE P5.7

5.8. The process illustrated in Figure P5.8 takes place at constant pressure. (The figure is not to scale; the molecules are actually much, much smaller than shown here.)
a. Write a balanced equation for the process.
b. Is w positive, negative, or zero for this reaction?
c. Using data from Appendix 4, calculate ΔH_{rxn}° for the formation of 1 mole of the product.

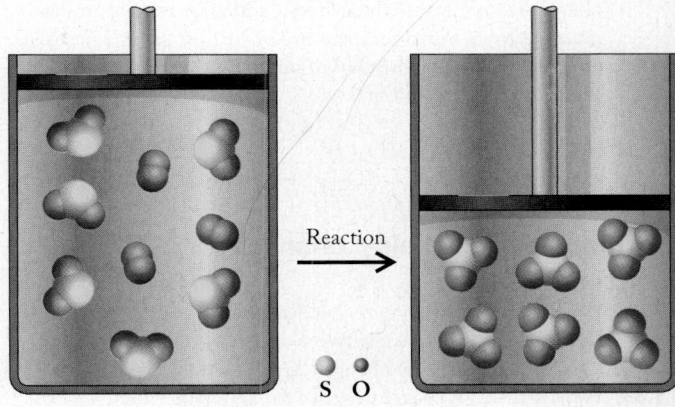

FIGURE P5.8

◼

QUESTIONS AND PROBLEMS

Energy: Basic Concepts and Definitions

CONCEPT REVIEW

5.9. How are energy and work related?

5.10. Explain the difference between potential energy and kinetic energy.

5.11. Explain what is meant by a state function.

5.12. Are kinetic energy and potential energy both state functions?

5.13. Explain the nature of the potential energy in the following: (a) the new battery for your remote control; (b) a gallon of gasoline; (c) the crest of a wave before it crashes onto shore.

5.14. Explain the kinetic energy in a stationary ice cube.

Systems, Surroundings, and Energy Transfer

CONCEPT REVIEW

5.15. What is meant by the terms *system* and *surroundings*?

5.16. What is the difference between an exothermic process and an endothermic one?

5.17. Give two ways of increasing the internal energy of a gas sample.

5.18. In each of the following processes describe the system and give the sign of q_{system}: (a) driving an automobile; (b) applying ice to a sprained ankle; (c) cooking a hot dog.

PROBLEMS

5.19. Which of the following processes are exothermic, and which are endothermic? (a) a match burns; (b) a molten metal solidifies; (c) rubbing alcohol feels cold on the skin

5.20. Which of the following processes are exothermic, and which are endothermic? (a) ice cubes solidify in the freezer; (b) water evaporates from a glass left on a windowsill; (c) dew forms on grass overnight

5.21. What happens to the internal energy of a liquid at its boiling point when it vaporizes?

5.22. What happens to the internal energy of a gas when it expands (with no heat flow)?

5.23. How much P–V work does a gas system do on its surroundings at a constant pressure of 1.00 atm if the volume of gas triples from 250.0 mL to 750.0 mL? Express your answer in $L \cdot atm$ and joules (J).

5.24. An expanding gas does 150.0 J of work on its surroundings at a constant pressure of 1.01 atm. If the gas initially occupied 68 mL, what is the final volume of the gas?

5.25. Calculate ΔE for the following situations:
a. $q = 100.0$ J; $w = -50.0$ J
b. $q = 6.2$ kJ; $w = 0.70$ J
c. $q = -615$ J; $w = -325$ J

5.26. Calculate ΔE for a system that absorbs 726 kJ of heat from its surroundings and does 526 kJ of work on its surroundings.

5.27. Calculate ΔE for the combustion of a gas that releases 210.0 kJ of heat to its surroundings and does 65.5 kJ of work on its surroundings.

5.28. Calculate ΔE for a chemical reaction that produces 90.7 kJ of heat but does no work on its surroundings.

*5.29. The following reactions take place in a cylinder equipped with a movable piston at atmospheric pressure (Figure P5.29). Which reactions will result in work being done on the surroundings? Assume the system returns to an initial temperature

FIGURE P5.29

of 110°C. *Hint*: The volume of a gas is proportional to number of moles (n) at constant temperature and pressure.

a. $CH_4(g) + 2O_2(g) \rightarrow CO_2(g) + 2H_2O(g)$
b. $C_3H_8(g) + 5O_2(g) \rightarrow 3CO_2(g) + 4H_2O(g)$
c. $N_2(g) + 2O_2(g) \rightarrow 2NO_2(g)$

*5.30. In which direction will the piston shown in Figure P5.29 move when the following reactions are carried out at atmospheric pressure inside the cylinder and after the system has returned to its initial temperature of 110°C? *Hint*: The volume of a gas is proportional to number of moles (n) at constant temperature and pressure.

a. $N_2(g) + 3H_2(g) \rightarrow 2NH_3(g)$
b. $C(s) + O_2(g) \rightarrow CO_2(g)$
c. $CH_3CH_2OH(g) + 3O_2(g) \rightarrow 2CO_2(g) + 3H_2O(g)$

Enthalpy and Enthalpy Changes

CONCEPT REVIEW

5.31. What is meant by an *enthalpy change*?

5.32. Describe the difference between an internal energy change (ΔE) and an enthalpy change (ΔH).

5.33. Why is the sign of ΔH negative for an exothermic process?

5.34. What happens to the magnitude and sign of the enthalpy change when a process is reversed?

PROBLEMS

5.35. A Clogged Sink Adding Drano to a clogged sink causes the drainpipe to get warm. What is the sign of ΔH for this process?

5.36. Cold Pack for Injuries Breaking a small pouch of water inside a larger bag containing ammonium nitrate activates chemical cold packs, used by sports trainers for injured athletes. What is the sign of ΔH for the process taking place in the cold pack?

5.37. Break a Bond The stable form of oxygen at room temperature and pressure is the diatomic molecule O_2. What is the sign of ΔH for the following process?

$$O_2(g) \rightarrow 2O(g)$$

5.38. Plaster of Paris Gypsum is the common name of calcium sulfate dihydrate ($CaSO_4 \cdot 2H_2O$). When gypsum is heated to 150°C, it loses most of the water in its formula and forms plaster of Paris ($CaSO_4 \cdot 0.5H_2O$):

$$2CaSO_4 \cdot 2H_2O(s) \rightarrow 2CaSO_4 \cdot 0.5H_2O(s) + 3H_2O(g)$$

What is the sign of ΔH for making plaster of Paris from gypsum?

5.39 Metallic Hydrogen A solid with metallic properties is formed when hydrogen gas is compressed under extremely high pressures. Predict the sign of the enthalpy change for the following reaction:

$$H_2(g) \rightarrow H_2(s)$$

5.40. Kitchen Chemistry A simple "kitchen chemistry" experiment requires placing some vinegar in a soda bottle. A deflated balloon containing baking soda is stretched over the mouth of the bottle. Adding the baking soda to the vinegar starts the following reaction and inflates the balloon:

$$NaHCO_3(aq) + CH_3COOH(aq) \rightarrow$$
$$CH_3COONa(aq) + CO_2(g) + H_2O(\ell)$$

If the contents of the bottle are considered the system, is work being done on the surroundings or on the system?

Heating Curves and Heat Capacity

CONCEPT REVIEW

5.41. What is the difference between *specific heat* and *heat capacity*?

5.42. What happens to the heat capacity of a material if its mass is doubled? Is the same true for the specific heat?

5.43. Are the heats of fusion and vaporization of a given substance usually the same?

5.44. An equal amount of heat is added to pieces of metal A and metal B having the same mass. Does the metal with the larger heat capacity reach the higher temperature?

***5.45. Cooling an Automobile Engine** Most automobile engines are cooled by water circulating through them and a radiator. However, the original Volkswagen Beetle had an air-cooled engine. Why might car designers choose water cooling over air cooling?

***5.46. Nuclear Reactor Coolants** The reactor-core cooling systems in some nuclear power plants use liquid sodium as the coolant. Sodium has a thermal conductivity of 1.42 J/(cm · s · K), which is quite high compared with that of water [6.1×10^{-3} J/(cm · s · K)]. The respective molar heat capacities are 28.28 J/(mol · K) and 75.31 J/(mol · K). What is the advantage of using liquid sodium over water in this application?

PROBLEMS

5.47. How much heat is needed to raise the temperature of 100.0 g of water from 30.0°C to 100.0°C?

5.48. At an elevation where the boiling point of water is 93°C, 100.0 g of water at 30°C absorbs 290.0 kJ of heat from a mountain climber's stove. Is this amount of energy sufficient to heat the water to its boiling point?

5.49. Use the following data to sketch a heating curve for 1 mole of methanol. Start the curve at −100°C and end it at 100°C.

Boiling point	65°C
Melting point	−94°C
ΔH_{vap}	37 kJ/mol
ΔH_{fus}	3.18 kJ/mol
Molar heat capacity (ℓ)	81.1 J/(mol · °C)
(g)	43.9 J/(mol · °C)
(s)	48.7 J/(mol · °C)

5.50. Use the following data to sketch a heating curve for 1 mole of octane. Start the curve at −57°C and end it at 150°C.

Boiling point	125.7°C
Melting point	−56.8°C
ΔH_{vap}	41.5 kJ/mol
ΔH_{fus}	20.7 kJ/mol
Molar heat capacity (ℓ)	254.6 J/(mol · °C)
(g)	316.9 J/(mol · °C)

5.51. Keeping an Athlete Cool During a strenuous workout, an athlete generates 2000.0 kJ of heat energy. What mass of water would have to evaporate from the athlete's skin to dissipate this much heat?

5.52. The same quantity of energy is added to 10.00 g pieces of gold, magnesium, and platinum, all initially at 25°C. The molar heat capacities of these three metals are 25.41 J/(mol · °C), 24.79 J/(mol · °C), and 25.95 J/(mol · °C), respectively. Which piece of metal has the highest final temperature?

*5.53. Exactly 10.0 mL of water at 25.0°C is added to a hot iron skillet. All of the water is converted into steam at 100.0°C. The mass of the pan is 1.20 kg and the molar heat capacity of iron is 25.19 J/(mol · °C). What is the temperature change of the skillet?

*5.54. A 20.0 g piece of iron and a 20.0 g piece of gold at 100.0°C were dropped into 1.00 L of water at 20.0°C. The molar heat capacities of iron and gold are 25.19 J/(mol · °C) and 25.41 J/(mol · °C), respectively. What is the final temperature of the water and pieces of metal?

Calorimetry: Measuring Heat Capacity and Calorimeter Constants

CONCEPT REVIEW

5.55. Why is it necessary to know the heat capacity of a calorimeter?

5.56. Could an endothermic reaction be used to measure the heat capacity of a calorimeter?

5.57. If we replace the water in a bomb calorimeter with another liquid, do we need to redetermine the heat capacity of the calorimeter?

5.58. When measuring the heat of combustion of a very small amount of material, would you prefer to use a calorimeter having a heat capacity that is small or large?

PROBLEMS

5.59. Calculate the heat capacity of a calorimeter if the combustion of 5.000 g of benzoic acid led to a temperature increase of 16.397°C.

5.60. Calculate the heat capacity of a calorimeter if the combustion of 4.663 g of benzoic acid led to an increase in temperature of 7.149°C.

5.61. The complete combustion of 1.200 g of cinnamaldehyde (C_9H_8O, one of the compounds in cinnamon) in a bomb calorimeter ($C_{calorimeter} = 3.640$ kJ/°C) produced an increase in temperature of 12.79°C. Calculate the molar enthalpy of combustion of cinnamaldehyde (ΔH_{comb}) in kilojoules per mole of cinnamaldehyde.

5.62. Aromatic Spice The aromatic hydrocarbon cymene ($C_{10}H_{14}$) is found in nearly 100 spices and fragrances including coriander, anise, and thyme. The complete combustion of 1.608 g of cymene in a bomb calorimeter ($C_{calorimeter} = 3.640$ kJ/°C) produced an increase in temperature of 19.35°C. Calculate the molar enthalpy of combustion of cymene (ΔH_{comb}) in kilojoules per mole of cymene.

5.63. Hormone Mimics Phthalates used to make plastics flexible are among the most abundant industrial contaminants in the environment. Several have been shown to act as hormone mimics in humans by activating the receptors for estrogen, a female sex hormone. In characterizing the compounds completely, the value of ΔH_{comb} for dimethyl phthalate ($C_{10}H_{10}O_4$) was determined to be −4685 kJ/mol. Assume that 1.00 g of dimethyl phthalate is combusted in a calorimeter whose heat capacity ($C_{calorimeter}$) is 7.854 kJ/°C at 20.215°C. What is the final temperature of the calorimeter?

5.64. Flavorings The flavor of anise is due to anethole, a compound with the molecular formula $C_{10}H_{12}O$. The ΔH_{comb} value for anethole is −5541 kJ/mol. Assume 0.950 g of anethole is combusted in a calorimeter whose heat capacity ($C_{calorimeter}$) is 7.854 kJ/°C at 20.611°C. What is the final temperature of the calorimeter?

Enthalpies of Formation and Enthalpies of Reaction

CONCEPT REVIEW

5.65. Oxygen and ozone are both forms of elemental oxygen. Are the standard enthalpies of formation of oxygen and ozone the same?

5.66. Why are the standard enthalpies of formation of elements in their standard states assigned a value of zero?

PROBLEMS

5.67. For which of the following reactions does ΔH°_{rxn} represent an enthalpy of formation?
   a. $C(s) + O_2(g) \rightarrow CO_2(g)$
   b. $CO_2(g) + C(s) \rightarrow 2CO(g)$
   c. $CO_2(g) + H_2(g) \rightarrow H_2O(g) + CO(g)$
   d. $2H_2(g) + C(s) \rightarrow CH_4(g)$

5.68. For which of the following reactions does ΔH°_{rxn} also represent an enthalpy of formation?
   a. $2N_2(g) + 3O_2(g) \rightarrow 2NO_2(g) + 2NO(g)$
   b. $N_2(g) + O_2(g) \rightarrow 2NO(g)$
   c. $2NO_2(g) \rightarrow N_2O_4(g)$
   d. $N_2(g) + 2O_2(g) \rightarrow 2NO_2(g)$

5.69. Methanogenesis Use standard enthalpies of formation from Appendix 4 to calculate the standard enthalpy of reaction for the following methane-generating reaction of methanogenic bacteria:

$$4H_2(g) + CO_2(g) \rightarrow CH_4(g) + 2H_2O(\ell)$$

5.70. Use standard enthalpies of formation from Appendix 4 to calculate the standard enthalpy of reaction for the following methane-generating reaction of methanogenic bacteria, given ΔH°_f of $CH_3NH_2(g) = -22.97$ kJ/mol:

$$4CH_3NH_2(g) + 2H_2O(\ell) \rightarrow 3CH_4(g) + CO_2(g) + 4NH_3(g)$$

5.71. Ammonium nitrate decomposes to N_2O and water vapor at temperatures between 250°C and 300°C. Write a balanced chemical reaction describing the decomposition of ammonium nitrate, and calculate the heat of reaction using the appropriate enthalpies of formation from Appendix 4.

5.72. Military Explosives Explosives called amatols are mixtures of ammonium nitrate and TNT introduced during World War I when TNT was in short supply. The mixtures can provide 30% more explosive power than TNT alone. Above 300°C, ammonium nitrate decomposes to N_2, O_2, and H_2O. Write a balanced chemical reaction describing the decomposition of ammonium nitrate, and determine the standard heat of reaction by using the appropriate standard enthalpies of formation from Appendix 4.

5.73. Explosives Mixtures of fertilizer (ammonium nitrate) and fuel oil (a mixture of long-chain hydrocarbons similar to decane, $C_{10}H_{22}$) are the basis for a powerful explosion. Determine the enthalpy change of the following explosive reaction by using the appropriate enthalpies of formation ($\Delta H^\circ_{f,C_{10}H_{22}} = 249.7$ kJ/mol):

$$3\,NH_4NO_3(s) + C_{10}H_{22}(\ell) + 14\,O_2(g) \rightarrow$$
$$3\,N_2(g) + 17\,H_2O(g) + 10\,CO_2(g)$$

*5.74. A Little TNT Trinitrotoluene (TNT) is a highly explosive compound. The thermal decomposition of TNT is described by the following chemical equation:

$$2\,C_7H_5N_3O_6(s) \rightarrow 12\,CO(g) + 5\,H_2(g) + 3\,N_2(g) + 2\,C(s)$$

If ΔH_{rxn} for this reaction is $-10{,}153$ kJ/mol, how much TNT is needed to equal the explosive power of 1 mol of ammonium nitrate in Problem 5.73?

Fuel Values and Food Values

CONCEPT REVIEW

5.75. What is meant by *fuel value*?

5.76. What are the units of fuel values?

5.77. How are fuel values calculated from molar heats of combustion?

5.78. Is the fuel value of liquid propane the same as that of propane gas?

PROBLEMS

5.79. If all the energy obtained from burning 1.00 pound of propane is used to heat water, how many kilograms of water can be heated from 20.0°C to 45.0°C?

5.80. A 1995 article in *Discover* magazine on world-class sprinters contained the following statement: "In one race, a field of eight runners releases enough energy to boil a gallon jug of ice at 0.0°C in ten seconds!" How much "energy" do the runners release in 10 seconds? Assume that the ice has a mass of 128 ounces.

5.81. The Joys of Camping Lightweight camping stoves typically use *white gas*, a mixture of C_5 and C_6 hydrocarbons.
a. Calculate the fuel value of C_5H_{12}, given that $\Delta H^\circ_{comb} = -3535$ kJ/mol.
b. How much heat is released during the combustion of 1.00 kg of C_5H_{12}?
c. How many grams of C_5H_{12} must be burned to heat 1.00 kg of water from 20.0°C to 90.0°C? Assume that all the heat released during combustion is used to heat the water.

5.82. The heavier hydrocarbons in white gas are hexanes (C_6H_{14}).
a. Calculate the fuel value of C_6H_{14}, given that $\Delta H^\circ_{comb} = -4163$ kJ/mol.
b. How much heat is released during the combustion of 1.00 kg of C_6H_{14}?
c. How many grams of C_6H_{14} are needed to heat 1.00 kg of water from 25.0°C to 85.0°C? Assume that all of the heat released during combustion is used to heat the water.
d. Assume white gas is 25% C_5 hydrocarbons and 75% C_6 hydrocarbons; how many grams of white gas are needed to heat 1.00 kg of water from 25.0°C to 85.0°C?

Hess's Law

CONCEPT REVIEW

5.83. How is Hess's law consistent with the law of conservation of energy?

5.84. Why is the standard enthalpy of formation of $CO(g)$ difficult to measure experimentally?

5.85. Explain how the use of ΔH°_f to calculate ΔH°_{rxn} is an example of Hess's law.

5.86. Why is it important for Hess's law that enthalpy is a state function?

PROBLEMS

5.87. How can the first two of the following reactions be combined to obtain the third reaction?
a. $CO(g) + \frac{1}{2}O_2(g) \rightarrow CO_2(g)$
b. $C(s) + O_2(g) \rightarrow CO_2(g)$
c. $C(s) + \frac{1}{2}O_2(g) \rightarrow CO(g)$

5.88. How can the standard enthalpy of formation of $CO(g)$ be calculated from the standard enthalpy of formation ΔH°_f of $CO_2(g)$ and the standard heat of combustion ΔH°_{comb} of $CO(g)$?

5.89. Calculate the standard enthalpy of formation of $SO_2(g)$ from the standard enthalpy changes of the following reactions:

$$2\,SO_2(g) + O_2(g) \rightarrow 2\,SO_3(g) \quad \Delta H^\circ_{rxn} = -196\,kJ$$
$$\tfrac{1}{4}S_8(s) + 3\,O_2(g) \rightarrow 2\,SO_3(g) \quad \Delta H^\circ_{rxn} = -790\,kJ$$
$$\tfrac{1}{8}S_8(s) + O_2(g) \rightarrow SO_2(g) \quad \Delta H^\circ_f = ?$$

5.90. Ozone Layer The destruction of the ozone layer by chlorofluorocarbons (CFCs) can be described by the following reactions:

$$ClO(g) + O_3(g) \rightarrow Cl(g) + 2\,O_2(g) \quad \Delta H^\circ_{rxn} = -29.90\text{ kJ}$$
$$2\,O_3(g) \rightarrow 3\,O_2(g) \quad \Delta H^\circ_{rxn} = 24.18\text{ kJ}$$

Determine the value of heat of reaction for the following:

$$Cl(g) + O_3(g) \rightarrow ClO(g) + O_2(g) \quad \Delta H^\circ_{rxn} = ?$$

5.91. The mineral spodumene ($LiAlSi_2O_6$) exists in two crystalline forms called α and β. Use Hess's law and the following information to calculate ΔH°_{rxn} for the conversion of α-spodumene into β-spodumene:

$$Li_2O(s) + 2\,Al(s) + 4\,SiO_2(s) + \tfrac{3}{2}O_2(g) \rightarrow$$
$$2\,\alpha\text{-LiAlSi}_2O_6(s) \quad \Delta H^\circ = -1870.6\text{ kJ}$$
$$Li_2O(s) + 2\,Al(s) + 4\,SiO_2(s) + \tfrac{3}{2}O_2(g) \rightarrow$$
$$2\,\beta\text{-LiAlSi}_2O_6(s) \quad \Delta H^\circ = -1814.6\text{ kJ}$$

5.92. Use the following data to determine whether the conversion of diamond into graphite is exothermic or endothermic:

$$C_{diamond}(s) + O_2(g) \rightarrow CO_2(g) \qquad \Delta H° = -395.4 \text{ kJ}$$
$$2\,CO_2(g) \rightarrow 2\,CO(g) + O_2(g) \qquad \Delta H° = 566.0 \text{ kJ}$$
$$2\,CO(g) \rightarrow C_{graphite}(s) + CO_2(g) \qquad \Delta H° = -172.5 \text{ kJ}$$
$$C_{diamond}(s) \rightarrow C_{graphite}(s) \qquad \Delta H° = ?$$

5.93. You are given the following data:

$$\tfrac{1}{2}N_2(g) + \tfrac{1}{2}O_2(g) \rightarrow NO(g) \qquad \Delta H°_{rxn} = +90.3 \text{ kJ}$$
$$NO(g) + \tfrac{1}{2}Cl_2(g) \rightarrow NOCl(g) \qquad \Delta H°_{rxn} = -38.6 \text{ kJ}$$
$$2\,NOCl(g) \rightarrow N_2(g) + O_2(g) + Cl_2(g) \quad \Delta H°_{rxn} = ?$$

a. Which of the $\Delta H°_{rxn}$ values represent enthalpies of formation?
b. Determine $\Delta H°_{rxn}$ for the decomposition of NOCl.

5.94. The enthalpy of decomposition of NO_2Cl is -114 kJ. Use the following data to calculate the heat of formation of NO_2Cl from N_2, O_2, and Cl_2:

$$NO_2Cl(g) \rightarrow NO_2(g) + \tfrac{1}{2}Cl_2(g) \quad \Delta H°_{rxn} = -114 \text{ kJ}$$
$$\tfrac{1}{2}N_2(g) + O_2(g) \rightarrow NO_2(g) \qquad \Delta H°_f = +33.2 \text{ kJ}$$
$$\tfrac{1}{2}N_2(g) + O_2(g) + \tfrac{1}{2}Cl_2(g) \rightarrow NO_2Cl(g) \qquad \Delta H°_f = ?$$

Additional Problems

5.95. Polychlorinated biphenyls (PCBs) are efficient coolants whose use in transformers and other electrical devices has been banned because of their toxicity. What is the specific heat (c_s) of the PCB with the molecular formula $C_{12}Cl_{10}$ if its molar heat capacity (c_P) is 345.7 J/mol · K?

5.96. Carbon tetrachloride (CCl_4) was at one time used as a fire-extinguishing agent. It has a molar heat capacity c_P of 131.3 J/mol · °C. How much heat is required to raise the temperature of 275 g of CCl_4 from room temperature (22°C) to its boiling point (77°C)?

5.97. Ethylene glycol ($HOCH_2CH_2OH$) is mixed with the water in radiators to cool car engines. How much heat will 725 g of pure ethylene glycol remove from an engine as it is warmed from 0°C to its boiling point of 196°C? The c_P of ethylene glycol is 149.5 J/mol · °C.

5.98. Sodium may be used as a heat-storage material in some devices. The specific heat (c_s) of sodium metal is 1.23 J/g · K. How many moles of sodium metal are required to absorb 1.00×10^3 kJ of heat?

5.99. The water in a calorimeter was replaced with the organic compound methylene chloride (CH_2Cl_2). Burning 2.23 g of glucose ($C_6H_{12}O_6$; $\Delta H_{comb} = -2801$ kJ/mol) in the calorimeter causes its temperature to rise 9.64°C. What is the heat capacity of the calorimeter?

5.100. The standard enthalpy of formation of NH_3 is -46.1 kJ/mol. What is $\Delta H°$ for the following reactions?
a. $N_2(g) + 3\,H_2(g) \rightarrow 2\,NH_3(g)$
b. $NH_3(g) \rightarrow \tfrac{1}{2}N_2(g) + \tfrac{3}{2}H_2(g)$

*5.101. Hung Out to Dry Laundry forgotten and left outside to dry on a clothesline in the winter slowly dries by "ice vaporization" (sublimation). The increase in internal energy of water vapor produced by sublimation is less than the amount of heat absorbed. Explain.

5.102. Chlorofluorocarbons (CFCs) such as CF_2Cl_2 are refrigerants whose use has been phased out because of their destructive effect on Earth's ozone layer. The standard enthalpy of evaporation of CF_2Cl_2 is 17.4 kJ/mol, compared with $\Delta H_{vap} = 41$ kJ/mol for liquid water. How many grams of liquid CF_2Cl_2 are needed to cool 200.0 g of water from 50.0°C to 40.0°C? The specific heat of water is 4.184 J/(g · °C).

5.103. A 100.0 mL sample of 1.0 M NaOH is mixed with 50.0 mL of 1.0 M H_2SO_4 in a large Styrofoam coffee cup; the cup is fitted with a lid through which passes a calibrated thermometer. The temperature of each solution before mixing is 22.3°C. After adding the NaOH solution to the coffee cup and stirring the mixed solutions with the thermometer, the maximum temperature measured is 31.4°C. Assume that the density of the mixed solutions is 1.00 g/mL, that the specific heat of the mixed solutions is 4.18 J/(g · °C), and that no heat is lost to the surroundings.
a. Write a balanced chemical equation for the reaction that takes place in the Styrofoam cup.
b. Is any NaOH or H_2SO_4 left in the Styrofoam cup when the reaction is over?
c. Calculate the enthalpy change per mole of H_2SO_4 in the reaction.

5.104. Varying the scenario in Problem 5.97 assumes this time that 65.0 mL of 1.0 M H_2SO_4 is mixed with 100.0 mL of 1.0 M NaOH and that both solutions are initially at 25.0°C. Assume that the mixed solutions in the Styrofoam cup have the same density and specific heat as in Problem 5.97 and that no heat is lost to the surroundings. What is the maximum measured temperature in the Styrofoam cup?

*5.105. An insulated container is used to hold 50.0 g of water at 25.0°C. A 7.25 g sample of copper is placed in a dry test tube and heated for 30 minutes in a boiling water bath at 100.1°C. The heated test tube is carefully removed from the water bath with laboratory tongs and inclined so that the copper slides into the water in the insulated container. Given that the specific heat of solid copper is 0.385 J/(g · °C), calculate the maximum temperature of the water in the insulated container after the copper metal is added.

5.106. The mineral magnetite (Fe_3O_4) is magnetic, whereas iron(II) oxide is not.
a. Write and balance the chemical equation for the formation of magnetite from iron(II) oxide and oxygen.
b. Given that 318 kJ of heat is released for each mole of Fe_3O_4 formed, what is the enthalpy change of the balanced reaction of formation of Fe_3O_4 from iron(II) oxide and oxygen?

5.107. Which of the following substances has a standard heat of formation $\Delta H°_f$ of zero? (a) Pb at 1000°C; (b) $C_3H_8(g)$ at 25.0°C and 1 atm pressure; (c) solid glucose at room temperature; (d) $N_2(g)$ at 25.0°C and 1 atm pressure

*5.108. The standard heat of formation of liquid water is -285.8 kJ/mol.
a. What is the significance of the negative sign associated with this value?
b. Why is the magnitude of this value so much larger than the heat of vaporization of water ($\Delta H_{vap} = 41$ kJ/mol)?
c. Calculate the amount of heat produced in making 50.0 mL of water from its elements under standard conditions.

5.109. Endothermic compounds are unusual: they have positive heats of formation. An example is acetylene, C_2H_2 ($\Delta H_f^\circ = 226.7$ kJ/mol). The combustion of acetylene is used to melt and weld steel. Use Appendix 4 to answer the following questions.

a. Give the chemical formulas and names of three other endothermic compounds.

b. Calculate the standard molar heat of combustion of acetylene.

*5.110. Balance the following chemical equation, name the reactants and products, and calculate the standard enthalpy change using the data in Appendix 4.

$$FeO(s) + O_2(g) \rightarrow Fe_2O_3(s)$$

*5.111. Add reactions 1, 2, and 3, and label the resulting reaction 4. Consult Appendix 4 to find the standard enthalpy change for balanced reaction 4.

$$(1) \quad Zn(s) + \tfrac{1}{8}S_8(s) \rightarrow ZnS(s)$$
$$(2) \quad ZnS(s) + 2O_2(g) \rightarrow ZnSO_4(s)$$
$$(3) \quad \tfrac{1}{8}S_8(s) + O_2(g) \rightarrow SO_2(g)$$

5.112. Conversion of 0.90 g of liquid water to steam at 100.0°C requires 2.0 kJ of heat. Calculate the molar enthalpy of evaporation of water at 100.0°C.

5.113. The specific heat of solid copper is 0.385 J/(g · °C). What thermal energy change occurs when a 35.3 g sample of copper is cooled from 35.0°C to 15.0°C? Be sure to give your answer the proper sign. This amount of heat is used to melt solid ice at 0.0°C. The molar heat of fusion of ice is 6.01 kJ/mol. How many moles of ice are melted?

*5.114. **Metabolism of Methanol** Methanol is toxic because it is metabolized in a two-step process *in vivo* to formic acid (HCOOH). Consider the following overall reaction under standard conditions:

$$O_2(g) + 2CH_3OH(\ell) \rightarrow$$
$$2HCOOH(\ell) + 2H_2O(\ell) + 1019.6 \text{ kJ}$$

a. Is this reaction endothermic or exothermic?

b. What is the value of ΔH_{rxn}° for this reaction?

c. How much heat would be absorbed or released if 60.0 g of methanol were metabolized in this reaction?

d. In the first step of metabolism, methanol is converted into formaldehyde (CH_2O), which is then converted into formic acid. Would you expect ΔH_{rxn}° for the metabolism of 1 mole of $CH_3OH(\ell)$ to give 1 mole of formaldehyde to be larger or smaller than 509.8 kJ?

5.115. Using the information given below, write an equation for the enthalpy change in reaction 3:

$$(1) \quad B(g) + A(s) \rightarrow C(g) \quad \Delta H_1$$
$$(2) \quad C(g) \rightarrow C(s) \quad \Delta H_2$$
$$(3) \quad C(s) \rightarrow A(s) + B(g) \quad \Delta H_3$$

5.116. From the information given below, write an equation for the enthalpy change in reaction 3:

$$(1) \quad F(g) \rightarrow D(s) + E(g) \quad \Delta H_1$$
$$(2) \quad F(s) \rightarrow F(g) \quad \Delta H_2$$
$$(3) \quad D(s) + E(g) \rightarrow F(s) \quad \Delta H_3$$

5.117. The reaction of $CH_3OH(g)$ with $N_2(g)$ to give $HCN(g)$ and $NH_3(g)$ requires 164 kJ/mol of heat.

a. Write a balanced chemical equation for this reaction.

b. Should the thermal energy involved be written as a reactant or as a product?

c. How much heat is involved in the reaction of 60.0 g of $CH_3OH(g)$ with excess $N_2(g)$ to give $HCN(g)$ and $NH_3(g)$ in this reaction?

5.118. Calculate ΔH_{rxn}° for the reaction

$$2Ni(s) + \tfrac{1}{4}S_8(s) + 3O_2(g) \rightarrow 2NiSO_3(s)$$

from the following information:

$$(1) \quad NiSO_3(s) \rightarrow NiO(s) + SO_2(g) \qquad \Delta H_{rxn}^\circ = 156 \text{ kJ}$$
$$(2) \quad \tfrac{1}{8}S_8(s) + O_2(g) \rightarrow SO_2(g) \qquad \Delta H_{rxn}^\circ = -297 \text{ kJ}$$
$$(3) \quad Ni(s) + \tfrac{1}{2}O_2(g) \rightarrow NiO(s) \qquad \Delta H_{rxn}^\circ = -241 \text{ kJ}$$

5.119. Use the following information to calculate the amount of heat involved in the complete reaction of 3.0 g of carbon to form $PbCO_3(s)$ in reaction 4. Be sure to give the proper sign (positive or negative) with your answer.

$$(1) \quad Pb(s) + \tfrac{1}{2}O_2(g) \rightarrow PbO(s) \qquad \Delta H_{rxn}^\circ = -219 \text{ kJ}$$
$$(2) \quad C(s) + O_2(g) \rightarrow CO_2(g) \qquad \Delta H_{rxn}^\circ = -394 \text{ kJ}$$
$$(3) \quad PbCO_3(s) \rightarrow PbO(s) + CO_2(g) \qquad \Delta H_{rxn}^\circ = 86 \text{ kJ}$$
$$(4) \quad Pb(s) + C(s) + \tfrac{3}{2}O_2(g) \rightarrow PbCO_3(s)$$

*5.120. **Ethanol as Automobile Fuel** Brazilians are quite familiar with fueling their automobiles with ethanol, a fermentation product from sugarcane. Calculate the standard molar enthalpy for the complete combustion of liquid ethanol using the standard enthalpies of formation of the reactants and products as given in Appendix 4.

5.121. **Formation of CO₂** Baking soda decomposes on heating as follows, creating the holes in baked bread:

$$2NaHCO_3(s) \rightarrow Na_2CO_3(s) + CO_2(g) + H_2O(\ell)$$

Calculate the standard enthalpy of formation of $NaHCO_3(s)$ from the following information:

$$\Delta H_{rxn}^\circ = -129.3 \text{ kJ} \qquad \Delta H_f^\circ[Na_2CO_3(s)] = -1131 \text{ kJ/mol}$$
$$\Delta H_f^\circ[CO_2(g)] = -394 \text{ kJ/mol} \qquad \Delta H_f^\circ[H_2O(\ell)] = -286 \text{ kJ/mol}$$

*5.122. **Specific Heats of Metals** In 1819, Pierre Dulong and Alexis Petit reported that the product of the atomic mass of a metal times its specific heat is approximately constant, an observation called the *law of Dulong and Petit*. Use the following data to answer the following questions.

Element	$\mathcal{M}$ (g/mol)	c_s [J/(g · °C)]	$\mathcal{M} \times c_s$
Bismuth		0.120	
Lead	207.2	0.123	25.5
Gold	197.0	0.125	
Platinum	195.1		
Tin	118.7	0.215	
		0.233	
Zinc	65.4	0.388	
Copper	63.5	0.397	
		0.433	
Iron	55.8	0.460	
Sulfur	32.1		
		Average value:	

a. Complete each row in the table by multiplying each given molar mass and specific heat pair (one result has been entered in the table). What are the units of the resulting values in column 4?

b. Next, calculate the average of the values in column 4.

c. Use the mean value from part b to calculate the missing atomic masses in the table. Do you feel confident in identifying the elements from the calculated atomic masses?

d. Use the average value from part b to predict the missing specific heat values in the table.

*5.123. Odor of Urine Urine odor gets worse with time because urine contains the metabolic product urea $[CO(NH_2)_2]$, a compound that is slowly converted to ammonia, which has a sharp, unpleasant odor, and carbon dioxide:

$$CO(NH_2)_2(aq) + H_2O(\ell) \rightarrow CO_2(aq) + 2\,NH_3(aq)$$

This reaction is much too slow for the enthalpy change to be measured directly using a temperature change. Instead, the enthalpy change for the reaction may be calculated from the following data:

Compound	ΔH_f° (kJ/mol)
Urea(aq)	−319.2
CO_2(aq)	−412.9
H_2O(ℓ)	−285.8
NH_3(aq)	−80.3

Calculate the standard molar enthalpy change for the reaction.

5.124. Experiment with a Metal Explain how the specific heat of a metal sample could be measured in the lab. *Hint*: You'll need a test tube, a boiling water bath, a Styrofoam cup calorimeter containing a known mass of water, a calibrated thermometer, and a known mass of the metal.

*5.125. Rocket Fuels The payload of a rocket includes a fuel and oxygen for combustion of the fuel. Reactions 1 and 2 describe the combustion of dimethylhydrazine and hydrogen, respectively. Pound for pound, which is the better rocket fuel, dimethylhydrazine or hydrogen?

(1) $(CH_3)_2NNH_2(\ell) + 4\,O_2(g) \rightarrow$
$\quad N_2(g) + 4\,H_2O(g) + 2\,CO_2(g) \quad \Delta H_{rxn}^\circ = -1694\ kJ$

(2) $H_2(g) + \frac{1}{2}O_2(g) \rightarrow H_2O(g) \quad\quad \Delta H_{rxn}^\circ = -286\ kJ$

5.126. At high temperatures, such as those in the combustion chambers of automobile engines, nitrogen and oxygen form nitrogen monoxide:

$$N_2(g) + O_2(g) \rightarrow 2\,NO(g) \quad\quad \Delta H_{comb}^\circ = +180\ kJ$$

Any NO released into the environment is oxidized to NO_2:

$$2\,NO(g) + O_2(g) \rightarrow 2\,NO_2(g) \quad\quad \Delta H_{comb}^\circ = -112\ kJ$$

Is the overall reaction

$$N_2(g) + 2\,O_2(g) \rightarrow 2\,NO_2(g)$$

exothermic or endothermic? What is ΔH_{comb}° for this reaction?

Carbon: Diamonds, Graphite, and the Molecules of Life

*5.127. Industrial Use of Cellulose Research is being carried out on cellulose as a source of chemicals for the production of fibers, coatings, and plastics. Cellulose consists of long chains of glucose molecules ($C_6H_{12}O_6$), so for the purposes of modeling the reaction we can consider the conversion of glucose to formaldehyde.

a. Is the reaction to convert glucose into formaldehyde an oxidation or a reduction?

b. Calculate the heat of reaction for the conversion of 1 mole of glucose into formaldehyde, given the following thermochemical data:

$\quad \Delta H_{comb}^\circ$ of formaldehyde gas $\quad$ −572.9 kJ/mol
$\quad \Delta H_f^\circ$ of solid glucose $\quad\quad\quad$ −1274.4 kJ/mol

$\quad C_6H_{12}O_6(s) \rightarrow 6\,CH_2O(g)$
$\quad\quad$ Glucose $\quad\quad$ Formaldehyde

5.128. Converting Diamond to Graphite The standard state of carbon is graphite. $\Delta H_{f,diamond}^\circ$ is +1.896 kJ/mol. Diamond masses are normally given in carats, where 1 carat = 0.20 g. Determine the standard enthalpy of the reaction for the conversion of a 4-carat diamond into an equivalent mass of graphite. Is this reaction endothermic or exothermic?

Additional study materials including ChemTours and Diagnostic Quizzes are available at StudySpace at www.wwnorton.com/studyspace.

6

Properties of Gases: The Air We Breathe

Learning Outcomes

- Distinguish gases from liquids and solids
- Calculate changes in the volume, temperature, pressure, and number of moles of a gas using the combined gas law and the ideal gas law
- Use balanced chemical equations to relate the volume of a gas-phase reactant to the amount of a product using the stoichiometry of the reaction and the ideal gas law
- Calculate the density of any gas, and determine the mole fraction and the partial pressure of a gas in a mixture
- Use kinetic molecular theory to explain the behavior of gases
- Calculate the root-mean-square speed of a gas and relative rates of effusion and diffusion
- Use the van der Waals equation to correct for nonideal behavior

A LOOK AHEAD

An Invisible Necessity

How often do we think about air? We are reminded how important it is when we dive into a pool of water or hike a tall mountain. Otherwise we usually do not think about breathing or about the mixture of gases that make up what we call "air." We may not think much about the oxygen we must inhale to stay alive because it is colorless, tasteless, and odorless, in addition to being free—we do not have to pay for it. Despite our lack of attention to the process, the exchange of gases between our lungs and the surrounding atmosphere is crucial to our lives; air is an invisible necessity.

In a hospital operating room, having the right mixture of gases in our bodies is so important during surgery that anesthesiologists constantly monitor blood levels of oxygen and carbon dioxide to ensure that they are in acceptable ranges. Managing the delicate balance of gases entering and leaving a patient can mean the difference between normal recovery and irreversible coma.

Gases are intimately involved in chemical reactions in living systems as well as in the material world. Most life in our biosphere requires oxygen. Fish, insects, birds, mammals, and even plants must bring $O_2(g)$ into their systems to metabolize nutrients. All the fuels we burn in vehicles, conventional power plants, airplanes, fireplaces, and most other devices that provide heat and light for our buildings and power for our transportation systems rely on the presence of O_2 in the air to support combustion. The products of the complete metabolism of food and the complete combustion of most fuels are CO_2, H_2O, and all the energy we need for life and for the inanimate objects that make life easier.

Gases are also important and useful because of their physical properties. We rely on their expansion to fill the space available or their ability to be compressed to save space in myriad processes. Our ability to understand how the chemical and physical properties of gases fit into

Scuba Diver and Sea Goldies The gas exhaled by the diver forms bubbles ▶ that increase in size as they rise to the surface.

our lives—whether in living rooms, hospitals, scuba tanks, tires, welding torches, or balloons—is based on our knowledge of how gases behave in response to changing volume, temperature, and pressure. Why do hot air balloons rise? Why does an undersea explorer carry a particular mixture of gases in a scuba tank? Why does the pressure in automobile tires change as we drive? In this chapter we answer these questions and others as we explore the properties of gases on both the macroscopic and molecular levels. ■

6.1 The Gas Phase

Gases have neither definite volumes nor definite shapes; they expand to occupy the entire volume of their container and assume the container's shape. Other properties also distinguish gases from liquids and solids under everyday conditions.

1. Unlike the volume occupied by a liquid or solid, the volume occupied by a gas changes significantly with pressure. If we carry an inflated balloon from sea level (0 m) to the top of a 1600 m mountain, the balloon volume increases by about 20%. The volume of a liquid or solid is unchanged under these conditions.

2. The volume of a gas changes with temperature. For example, the volume of a balloon filled with room-temperature air decreases when the balloon is taken outside on a cold winter's day. A temperature decrease from 20°C to 0°C leads to a volume decrease of about 7%, whereas the volume of a liquid or solid remains practically unchanged by this modest temperature change.

3. Gases are **miscible**, which means they can be mixed together in any proportion (unless they chemically react with one another). A hospital patient experiencing respiratory difficulties may be given a mixture of nitrogen and oxygen in which the proportion of oxygen is much higher than its proportion in air, and a scuba diver may leave the ocean surface with a tank of air containing a homogeneous mixture of 17% oxygen, 34% nitrogen, and 49% helium. In contrast, many liquids are immiscible, such as oil and water.

4. Gases are typically much less dense than liquids or solids. One indicator of this large difference is that gas densities are expressed in grams per *liter* but liquid densities are expressed in grams per *milliliter*. The density of dry air at 20°C at typical atmospheric pressure is 1.20 g/L, for example, whereas the density of liquid water under the same conditions is 999 g/L, over 800 times greater than the density of dry air.

These four observations about gases are consistent with the idea that the molecules or atoms in a gas are farther apart than in solids and liquids. The larger spaces between the gas molecules in air, for example, account for the ready compressibility of air into scuba tanks. Greater distances between molecules account for both the lower densities and the miscibility of gases. The molecules in gases are also in constant, random motion. Because of this motion, gas molecules have measurable kinetic energies. The kinetic energy and speed at which gas molecules travel depend on their temperature. Moving molecules also experience collisions.

miscible capable of being mixed in any proportion (without reacting chemically).

pressure (*P*) the ratio of force to surface area over which the force is applied.

atmospheric pressure (*P*~atm~) the force exerted by the gases surrounding Earth on Earth's surface and on all surfaces of all objects.

barometer an instrument that measures atmospheric pressure.

Nitrogen dioxide, NO_2, is a red-brown gas that contributes to smog. Which of the drawings in Figure 6.1 best depicts a sealed container of NO_2? Which drawing best shows a homogeneous liquid, such as a solution of bromine in carbon tetrachloride?

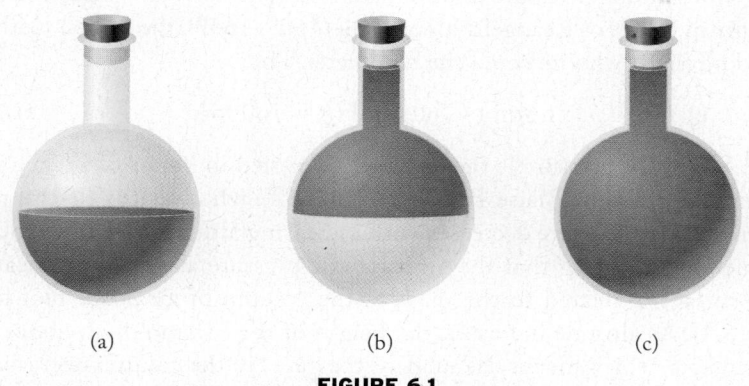

(a)   (b)   (c)

FIGURE 6.1

(Answers to Concept Tests are in the back of the book.)

6.2 Atmospheric Pressure

Earth is surrounded by a layer of gases 50 km thick. We call this mixture of gases either *air* or *the atmosphere*. By volume it is composed primarily of nitrogen (78%), oxygen (21%), and lesser amounts of other gases (Table 6.1). To put the thickness of the atmosphere in perspective, if Earth were the size of an apple, the atmosphere would be about as thick as the apple's skin.

Earth's atmosphere is pulled toward Earth by gravity and exerts a force that is spread across the entire surface of the planet (Figure 6.2). The ratio of force (F) to surface area (A) is called **pressure (P)**,

$$P = \frac{F}{A} \qquad (6.1)$$

and the force exerted by the atmosphere on Earth's surface is called **atmospheric pressure (P_{atm})**. Atmospheric pressure is measured with an instrument called a **barometer**. A simple but effective barometer design consists of a tube nearly 1 m long, filled with mercury, and closed at one end (Figure 6.3). The tube is inverted and its open end is placed into a pool of mercury that is open to the atmosphere. Gravity pulls the mercury in the tube downward, creating a vacuum at the top of the tube, while atmospheric pressure pushes the mercury in the pool up into the tube. The net effect of these opposing forces is indicated by the height of the mercury in the tube, which provides a measure of atmospheric pressure.

TABLE 6.1	Composition of Earth's Atmosphere
Component	**% (by volume)**
Nitrogen	78.08
Oxygen	20.95
Argon	0.934
Carbon dioxide	0.0386
Methane	2×10^{-4}
Hydrogen	5×10^{-5}

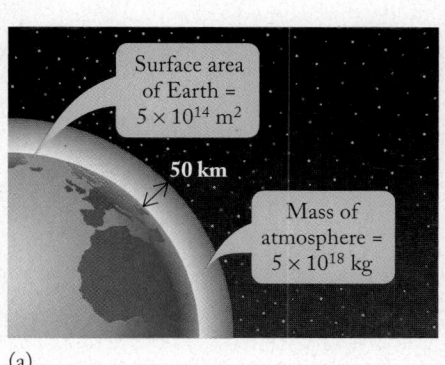

Surface area of Earth = 5×10^{14} m^2

50 km

Mass of atmosphere = 5×10^{18} kg

(a)

FIGURE 6.2 (a) Atmospheric pressure results from the force exerted by the atmosphere on Earth's surface. (b) If you stretch out your hand palm upward, the mass of the column of air above your palm is about 100 kg. This textbook has a mass of about 2.5 kg, so the mass of the atmosphere on your palm is equivalent to the mass of about 40 textbooks. (b)

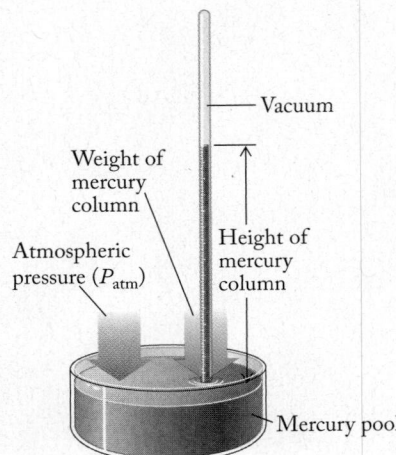

FIGURE 6.3 The height of the mercury column in this simple barometer designed by Evangelista Torricelli is proportional to atmospheric pressure.

Atmospheric pressure varies from place to place and with changing weather conditions. Several units are used to express pressure. The **standard atmosphere (1 atm)** of pressure is the pressure capable of supporting a column of mercury 760 mm high in a barometer. This column height was chosen because it equals the average height of mercury in a barometer at sea level. Another pressure unit, **millimeters of mercury (mmHg)**, is more explicitly related to column height, where 1 atm = 760 mmHg. Pressure in millimeters of mercury is also expressed in units called **torr** in honor of Evangelista Torricelli (1608–1647), the Italian mathematician and physicist who invented the barometer. Thus

$$1 \text{ atm} = 760 \text{ mmHg} = 760 \text{ torr}$$

The SI unit of pressure is the *pascal (Pa)*, named in honor of French mathematician and physicist Blaise Pascal (1623–1662), who was the first to propose that atmospheric pressure decreases with increasing altitude. We can explain this phenomenon by noting that the atmospheric pressure at any given location on Earth's surface is related to the mass of the column of air *above* that location (Figure 6.4). As altitude increases, the height of the column of air above a location decreases, which means the mass of the gases in the column decreases. Less mass means a smaller force exerted by the air, and, according to Equation 6.1, a smaller force means less pressure.

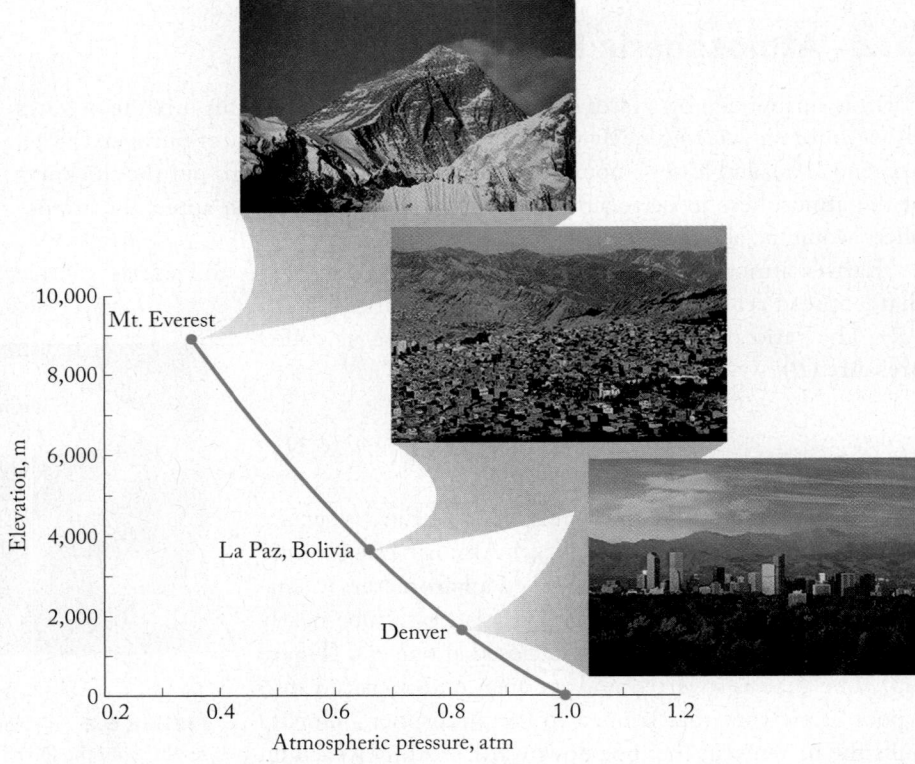

FIGURE 6.4 Atmospheric pressure decreases with increasing altitude because the mass of the column of air above a given area decreases with increasing altitude. If the weather were the same nice, clear day in all three of these locations, identical barometers would read 1.00 atm for the atmospheric pressure at a city at sea level 0.83 atm in Denver, 0.62 atm in La Paz, and a mere 0.35 atm on the summit of Mt. Everest.

standard atmosphere (1 atm) the pressure capable of supporting a column of mercury 760 mm high in a barometer.

millimeters of mercury (mmHg) (also called **torr**) a unit of pressure where 1 atm = 760 mmHg = 760 torr.

The pascal is a derived SI unit; it is defined using the SI base units (Table 1.2) kilogram, meter, and second:

$$1 \text{ Pa} = \frac{1 \text{ kg}}{\text{m} \cdot \text{s}^2}$$

To understand the logic of this combination of units, consider the relationship in physics that states that pushing an object of mass m with a force F causes the object to accelerate at a rate a:

$$F = ma \tag{6.2}$$

Combining Equations 6.1 and 6.2 we have

$$\frac{F}{A} = P = \frac{ma}{A} \tag{6.3}$$

If m is in kilograms, a is in meters per second-squared (m/s^2), and A is in square meters (m^2), the units for P are

$$\frac{\text{kg}\,\dfrac{\text{m}}{\text{s}^2}}{\text{m}^2} = \text{kg}\,\frac{\cancel{\text{m}}}{\text{s}^2} \times \frac{1}{\text{m}^{\cancel{2}}} = \frac{\text{kg}}{\text{m} \cdot \text{s}^2}$$

The relation between atmospheres and pascals is

$$1 \text{ atm} = 101{,}325 \text{ Pa}$$

Clearly, 1 Pa is a tiny quantity of pressure. In many applications it is more convenient to express pressure in kilopascals.

For many years, meteorologists have expressed atmospheric pressure in *millibars* (*mbar*). The weather maps prepared by the U.S. National Weather Service show changes in atmospheric pressure by constant-pressure contour lines, called *isobars*, spaced 4 mbar apart (Figure 6.5). There are exactly 10 mbar in 1 kPa; thus

$$1 \text{ atm} = (101.325 \cancel{\text{ kPa}})(10 \text{ mbar}/\cancel{\text{kPa}})$$
$$= 1013.25 \text{ mbar}$$

Other units of pressure are derived from masses and areas in the U.S. Customary System, such as pounds per square inch (lb/in^2, psi) for tire pressures and inches of mercury for atmospheric pressure in weather reports. The relationships between different units for pressure are summarized in Table 6.2.

FIGURE 6.5 Small differences in atmospheric pressure are associated with major changes in weather. Adjacent isobars on this map of Hurricane Katrina differ by 4 mbar of pressure.

TABLE 6.2	Units for Expressing Pressure
Unit	**Value**
Atmosphere (atm)	1 atm
Pascal (Pa)	1 atm = 1.01325 × 10^5 Pa
Kilopascal (kPa)	1 atm = 101.325 kPa
Millimeter of mercury (mmHg)	1 atm = 760 mmHg
Torr	1 atm = 760 torr
Bar	1 atm = 1.01325 bar
Millibar (mbar)	1 atm = 1013.25 mbar
Pounds per square inch (psi)	1 atm = 14.7 psi
Inches of mercury	1 atm = 29.92 inches of Hg

The accelerations due to gravity on Jupiter (25.95 m/s²), Mars (3.77 m/s²), and Saturn (11.08 m/s²) are different from the value on Earth (9.80 m/s²). On which planet would you exert the least pressure on the surface?

SAMPLE EXERCISE 6.1 Calculating Atmospheric Pressure

The mass of Earth's atmosphere is estimated to be 5.3×10^{18} kg. The surface area of Earth is 5.1×10^{14} m². The force of gravity pulling the atmosphere toward Earth gives the air an acceleration of 9.80 m/s². From these values calculate an average atmospheric pressure in kilopascals.

Collect and Organize We are given the mass of the atmosphere in kilograms, the surface area of Earth in square meters, and the acceleration due to gravity in meters per second-squared. These units can be combined to calculate atmospheric pressure in pascals.

Analyze Equation 6.3 describes the relationship between pressure, force, mass, and the acceleration due to gravity. We can estimate the magnitude of the force by considering that the mass is about 10^{18} kg and a is approximately 10, so the force is about 10^{19}. If the force is ~10^{19} and the area is ~10^{14}, then the pressure will be about 10^5 Pa or 10^2 kPa.

Solve Substituting in Equation 6.3 gives us

$$P = \frac{ma}{A}$$

$$= \frac{(5.3 \times 10^{18} \text{ kg})(9.80 \text{ m/s}^2)}{5.1 \times 10^{14} \text{ m}^2}$$

$$= \frac{1.0 \times 10^5 \text{ kg}}{\text{m} \cdot \text{s}^2} = 1.0 \times 10^5 \text{ Pa}$$

$$= 1.0 \times 10^2 \text{ kPa}$$

Think about It The calculated atmospheric pressure, 1.0×10^2 kPa, is the same as our estimate. This value is consistent with how we defined the average value of atmospheric pressure at sea level: 1 atm = 101.325 kPa or about 10^2 kPa.

Practice Exercise Calculate the pressure in pascals exerted on a tabletop by a cube of iron that is 1.00 cm on each side and has a mass of 7.87 g.

(Answers to Practice Exercises are in the back of the book.)

Scientists conducting experiments with gases must often measure the pressures exerted by the gases. A **manometer** is an instrument used to measure gas pressures in closed systems. Two types of manometers are illustrated in Figure 6.6. In each case, a U-shaped tube filled with mercury (or another dense liquid) is connected to an evacuated flask that is attached to the container holding the gas whose pressure is being measured. When valve 1 is opened, gas flows into the evacuated flask and fills it. Then valve 2 is opened, allowing the gas to exert pressure on the mercury.

The difference between the two manometers is whether the end of the tube not connected to the evacuated flask is closed or open to the atmosphere. In a closed-end manometer, the difference in the height of the mercury columns in the two arms

manometer an instrument for measuring the pressure exerted by a gas.

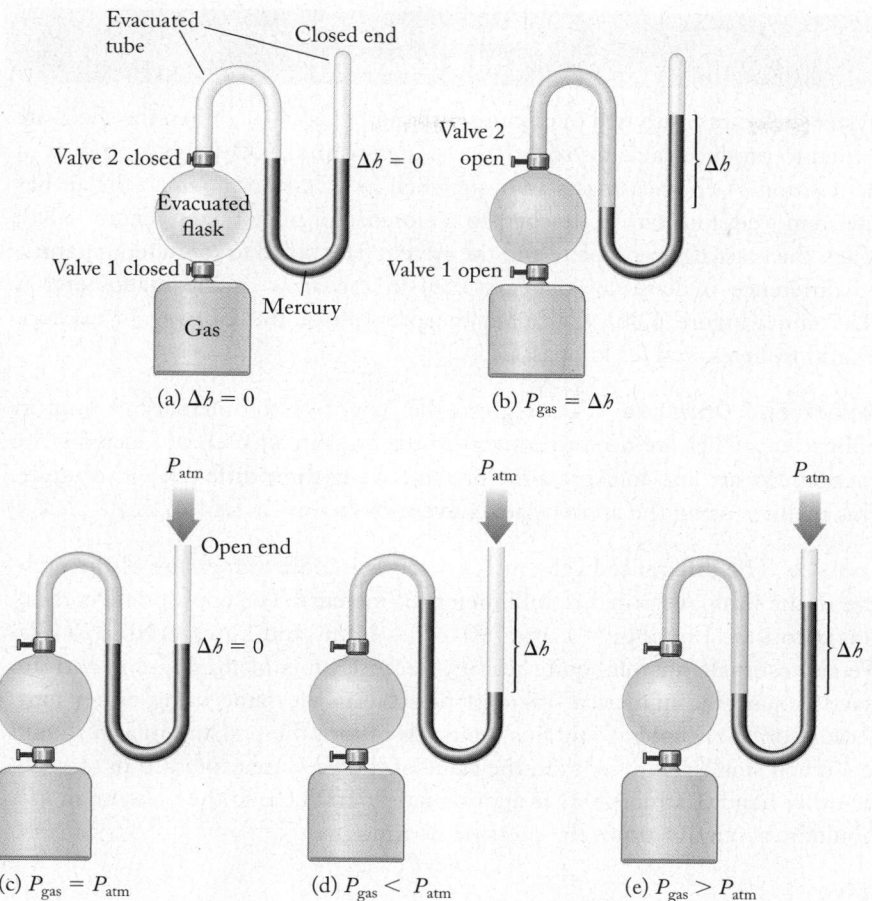

(a) $\Delta h = 0$

(b) $P_{gas} = \Delta h$

(c) $P_{gas} = P_{atm}$

(d) $P_{gas} < P_{atm}$

(e) $P_{gas} > P_{atm}$

FIGURE 6.6 (a and b) A closed-end manometer: (a) The air has been removed from the flask attached to the container of gas, and the mercury levels in the two arms of the U-tube are equal because both ends of the tube are exposed to a vacuum. (b) When valve 1 is opened, gas enters the flask; when valve 2 is opened, the gas exerts pressure on the mercury, causing the mercury level to drop in the left arm and rise in the right arm. The difference in the height of the mercury columns (Δh) is the pressure exerted by the gas. (c–e) An open-end manometer; note that both valves are open in all setups: (c) When the gas in the flask and the gas in the container are at atmospheric pressure, the mercury level is the same in both arms. (d) When the pressure of the gas in the flask and container is less than atmospheric pressure, the mercury level is higher in the left arm than in the right arm: $P_{gas} = P_{atm} - \Delta h$. (e) When the pressure of the gas in the flask and container is greater than atmospheric pressure, the mercury level is lower in the left arm than in the right arm: $P_{gas} = P_{atm} + \Delta h$.

of the U-tube is a direct measure of the gas pressure (Figure 6.6b). In an open-end manometer, the difference in height represents the difference between the pressure in the flask and atmospheric pressure. When the pressure in the flask is lower than atmospheric pressure (Figure 6.6d), the level in the arm attached to the flask is higher than the level in the other arm; when the pressure in the flask is greater than atmospheric pressure (Figure 6.6e), the level in the arm attached to the flask is lower than the level in the other arm.

Manometers have been largely displaced by pressure sensors based on flexible metallic or ceramic diaphragms. As the pressure on one side of the diaphragm increases, it distorts away from that side. This is the same mechanism used to sense changes in atmospheric pressure in most barometers, including the recording barometer, or *barograph*, shown in Figure 6.7.

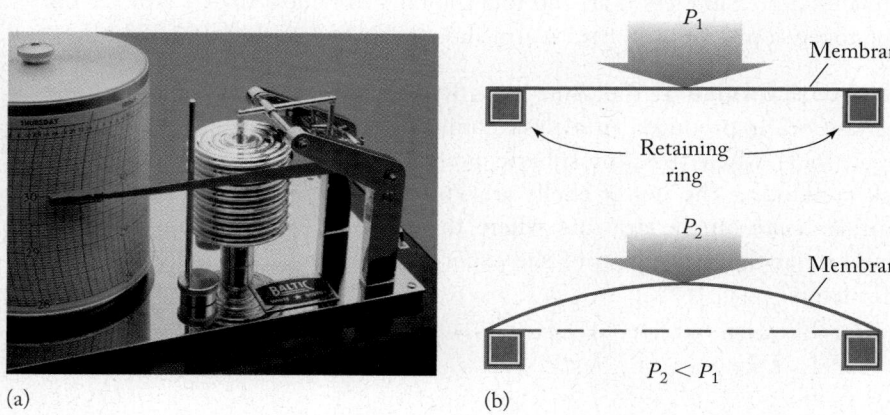

(a)

(b)

FIGURE 6.7 (a) The pressure sensor in this barograph is a partially evacuated corrugated metal can. (b) As atmospheric pressure decreases, the lid of the can distorts outward. This motion is amplified by a series of levers and transmitted via the horizontal arm to a pen tip that records the pressure on the graph paper as the drum slowly turns. A week's worth of barometric data can be recorded in this way.

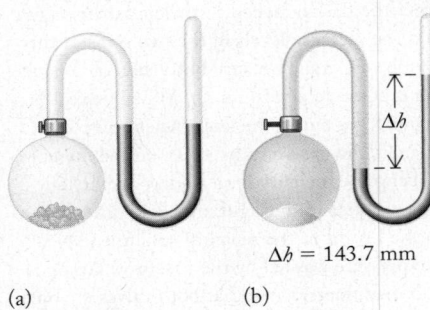

$\Delta h = 143.7$ mm

(a)   (b)

FIGURE 6.8 Roasting CaCO$_3$ in a closed-end manometer. (a) An evacuated flask containing oyster shells. (b) The same setup after the shells have been roasted, a decomposition reaction that releases CO$_2(g)$ in the flask.

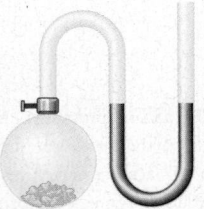

FIGURE 6.9 Before roasting in an open-end manometer.

SAMPLE EXERCISE 6.2   Measuring Gas Pressure with a Manometer

Oyster shells are composed of calcium carbonate (CaCO$_3$). When the shells are roasted to produce calcium oxide [CaO(s), quicklime], CO$_2(g)$ is a product of the reaction. A chemist roasts an oyster shell in a flask from which the air has been removed and that is attached to a closed-end manometer (Figure 6.8a). When the roasting is complete and the system has cooled to room temperature, the difference in levels of mercury (Δh) in the arms of the manometer is 143.7 mm (Figure 6.8b). Calculate the pressure of the CO$_2(g)$ in (a) torr, (b) atmospheres, and (c) kilopascals.

Collect and Organize We are given the height of the mercury column in millimeters, which we need to convert to the pressure of CO$_2$ produced in the reaction. We are also to express the pressure using three different sets of units. This requires using the appropriate conversion factors in Table 6.2.

Analyze The difference in the mercury levels is a direct measure of the pressure in the flask, measured in millimeters of mercury. The appropriate conversion factors are 1 mmHg = 1 torr, 760 torr = 1 atm, and 1 atm = 101.325 kPa. We can estimate answers quite readily. Because units of mmHg and torr are exactly equal, the numerical value of pressure is the same using either unit. Because one atmosphere contains hundreds of torr, the pressure in atm should be a much smaller number than the value of the pressure expressed in torr. On the other hand, 1 atmosphere is approximately 100 kPa, so the pressure in kPa should be about 10^2 times the pressure in atm.

Solve

a. First we convert millimeters of mercury to torr:

$$143.7 \; \cancel{\text{mmHg}} \times \frac{1 \; \text{torr}}{1 \; \cancel{\text{mmHg}}} = 143.7 \; \text{torr}$$

b. Then we convert torr to atmospheres:

$$143.7 \; \cancel{\text{torr}} \times \frac{1 \; \text{atm}}{760 \; \cancel{\text{torr}}} = 0.1891 \; \text{atm}$$

c. Finally we convert atmospheres to kilopascals:

$$0.1891 \; \cancel{\text{atm}} \times \frac{101.325 \; \text{kPa}}{1 \; \cancel{\text{atm}}} = 19.16 \; \text{kPa}$$

Think about It The calculated values for pressure make sense based on our estimates: a pressure less than 760 torr should also be less than 1 atm; by the same token, a pressure less than 1 atm should also be less than 101.325 kPa.

Practice Exercise If the same quantity of CO$_2(g)$ produced in Sample Exercise 6.2 is produced in a flask connected to an open-end manometer (Figure 6.9) when the atmospheric pressure and the pressure inside the flask containing the oyster shells are 760 torr at the start of the experiment, indicate on the drawing where the mercury levels in the two arms should be at the conclusion of the experiment. What is the value of Δh in millimeters?

6.3 The Gas Laws

In Section 6.1, we summarized some of the properties of gases and described the effect of pressure and temperature on volume, mostly in qualitative terms. Our knowledge of the quantitative relationships among P, T, and V goes back more than three centuries, to a time before the field of chemistry as we know it even existed. Some of the experiments that led to our understanding of how gases behave were driven by human interest in hot-air balloons, and today balloons are still used to study weather and atmospheric phenomena (Figure 6.10). Their successful and safe use requires applying an understanding of gas properties that was first gained in the 17th and 18th centuries.

Boyle's Law: Relating Pressure and Volume

Gases are compressible, a property that allows us to store large amounts of gases in relatively small metal cylinders. The ability to compress gases is the cornerstone of the compressed-gas industry because it makes the packaging and transport of large quantities of gas in a small space economical.

The relation between the pressure and volume of a fixed quantity of gas (constant value of n, where n is the number of moles) at constant temperature was investigated by British chemist Robert Boyle (1627–1691) and is known as **Boyle's law**. Boyle observed that the volume of a given amount of a gas is inversely proportional to its pressure when kept at a constant temperature (Figure 6.11):

$$P \propto \frac{1}{V} \qquad (T \text{ and } n \text{ fixed}) \qquad (6.4)$$

$P = 1\ \text{atm} = 760\ \text{mmHg}$

(a)

$P = 266\ \text{mmHg}$

(b)

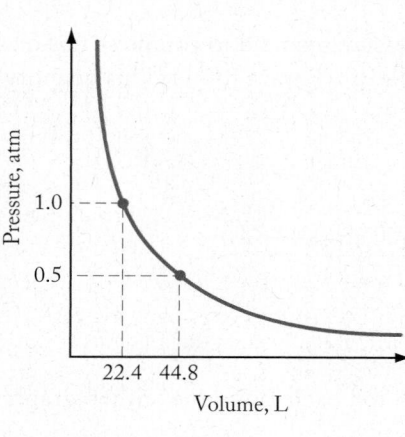

(c)

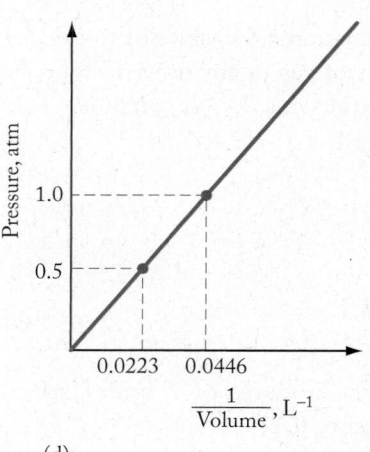

(d)

Boyle's law the volume of a given amount of gas at constant temperature is inversely proportional to its pressure.

FIGURE 6.10 Weather balloons are used to carry instruments aloft in the atmosphere.

FIGURE 6.11 The change in the size of the balloon demonstrates the relationship between pressure and volume. (a) The balloon inside the bell jar is at a pressure of 760 mmHg (1 atm). (b) A slight vacuum applied to the bell jar drops the interior pressure to 266 mmHg. (c) At constant temperature, the pressure of a given quantity of gas is inversely proportional to the volume occupied by the gas; the graph of an inverse proportion is a hyperbola. (d) The inverse proportion between P and V means that a plot of P versus $1/V$ is a straight line.

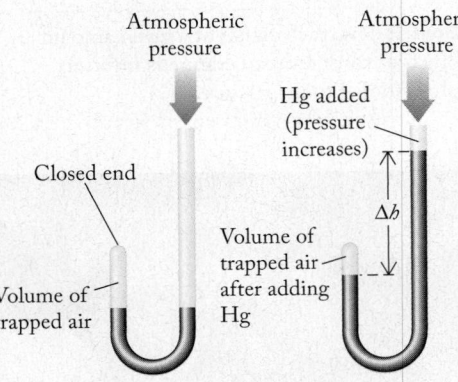

Atmospheric pressure

Atmospheric pressure

Hg added (pressure increases)

Closed end

Δh

Volume of trapped air

Volume of trapped air after adding Hg

FIGURE 6.12 Boyle used a J-shaped tube for his experiments on the relation between P and V. The volume of the column of air trapped in the closed end of the tube changes in proportion to the difference in height of mercury in the tube, which changes as mercury is added to the open arm. The final pressure on the gas is equal to the pressure exerted by the mass of the mercury plus the atmospheric pressure.

In other words, as the pressure on a constant amount of gas at a constant temperature increases, the volume of the gas decreases, and conversely, as the pressure on the gas decreases, its volume increases.

How can we use Boyle's law to predict exactly how much a balloon expands when the pressure decreases? In his experiments, Boyle used a J-shaped tube open at one end and closed at the other end (Figure 6.12). When a small amount of mercury was poured into the tube, a column of air was trapped at the closed end. The pressure exerted on this trapped air depended on the difference in the height of the mercury in the two sides of the tube and the atmospheric pressure. The *amount* (number of moles) of trapped air remained constant, and the temperature was also held constant. The pressure on the trapped air was changed by varying the amount of mercury poured into the open arm. Remember that the more mercury added, the greater the force exerted by the mercury ($F = ma$), and hence the greater the force per unit area ($P = F/A$) on the column of trapped air in the closed arm. As Boyle added mercury to the open arm, the volume occupied by the trapped air decreased in a way described by Equation 6.4. Mathematically we can replace the proportionality symbol with an equals sign and a constant,

$$P = (\text{constant})\frac{1}{V}$$

$$PV = \text{constant} \qquad (6.5)$$

where the value of the constant depends on the mass of trapped air in the sample and the temperature.

Because the value of the product PV in Equation 6.5 does not change for a given mass of trapped air at constant temperature, any two combinations of pressure and volume are related as follows:

$$P_1V_1 = P_2V_2 \qquad (6.6)$$

This relationship, which applies to all gases, is illustrated by the dashed lines in Figure 6.11. When, for example, 44.8 L (V_1) of gas is held at a pressure of 0.500 atm (P_1), we have

$$P_1V_1 = (0.500 \text{ atm})(44.8 \text{ L}) = 22.4 \text{ L} \cdot \text{atm}$$

Equation 6.6 enables us to calculate the pressure required to compress this quantity of gas to any other volume. For example, if we want to hold this quantity of the gas in a 22.4 L container (V_2):

$$P_1V_1 = P_2V_2$$

$$P_2 = \frac{P_1V_1}{V_2} = \frac{(0.500 \text{ atm})(44.8 \text{ L})}{22.4 \text{ L}}$$

$$= 1.00 \text{ atm}$$

The products of P and V are the same for each point on either graph in Figure 6.11.

CONCEPT TEST

Which of the graphs in Figure 6.13 correctly describes the relationship between the product of pressure and volume (PV) as a function of pressure (P) for a given quantity of gas at constant temperature?

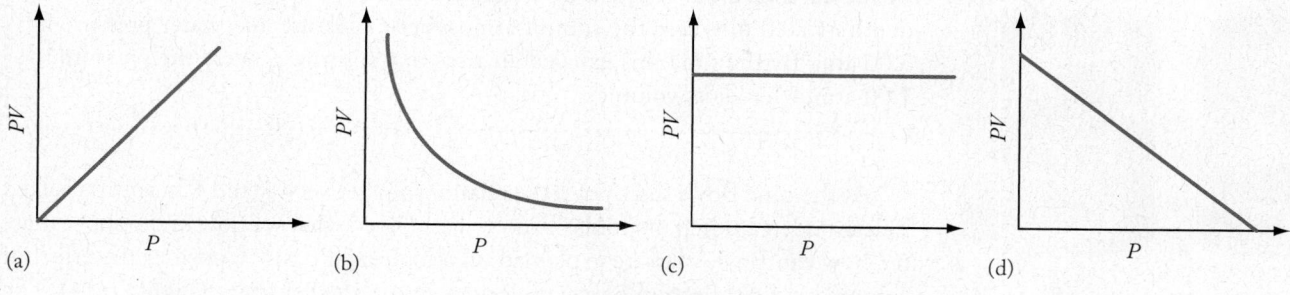

(a)   (b)   (c)   (d)

FIGURE 6.13

SAMPLE EXERCISE 6.3   Using Boyle's Law

A balloon is partly inflated with 5.00 L of helium at sea level, where the atmospheric pressure is 1.00 atm. The balloon ascends to an altitude of 1600 m, where the pressure is 0.83 atm.

  a. What is the volume of the balloon at the higher altitude?
  b. What is the percent increase in volume? Assume the temperature of the helium does not change during the ascent.

Collect and Organize We are given the volume ($V_1 = 5.00$ L) of a gas at a given pressure ($P_1 = 1.00$ atm) and asked to find the volume V_2 of the gas when the pressure changes ($P_2 = 0.83$ atm). The problem states that the temperature of the gas does not change.

Analyze The balloon contains a fixed amount of gas and its temperature is constant, so pressure and volume are related by Boyle's law (Equation 6.6). We predict that the volume should increase because pressure decreases.

Solve

  a. Rearranging Equation 6.6 to solve for V_2 and inserting the given values for P_1, P_2, and V_1 give us

$$V_2 = \frac{P_1 V_1}{P_2}$$

$$V_2 = \frac{(1.00 \text{ atm})(5.00 \text{ L})}{0.83 \text{ atm}} = 6.0 \text{ L}$$

  b. To calculate the percent increase in the volume, we must determine the difference between V_1 and V_2 and compare the result with V_1:

$$\text{\% increase in volume} = \frac{V_2 - V_1}{V_1} \times 100\%$$

$$= \frac{6.0 \text{ L} - 5.0 \text{ L}}{5.0 \text{ L}} \times 100\% = 20\%$$

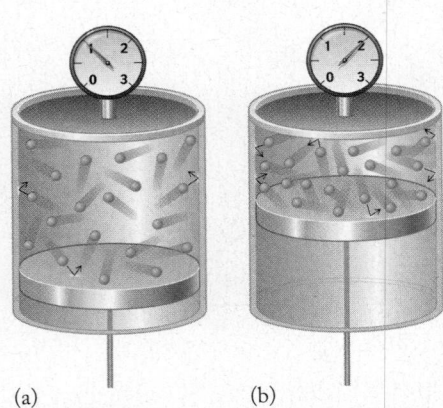

(a)   (b)

FIGURE 6.14 (a) Gas molecules are in constant random motion, exerting pressure through collisions with the interior surface of their container. (b) When a quantity of gas is squeezed into half its original volume, the frequency of collisions per unit of interior surface area increases by a factor of 2, and so does the pressure.

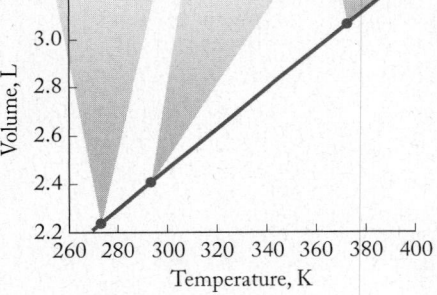

FIGURE 6.15 A balloon attached to a flask inflates as the temperature of the gas inside the flask increases from 273 K to 373 K at constant atmospheric pressure. This behavior is described by Charles's law.

Think about It The prediction we made about the volume increasing is confirmed. Notice how Equation 6.6 can also be used to calculate the change in pressure that takes place at constant temperature when the volume of a quantity of gas changes.

Practice Exercise A scuba diver exhales 3.50 L of air while swimming at a depth of 20.0 m where the sum of atmospheric pressure and water pressure is 3.00 atm. By the time this exhaled air rises to the surface where the pressure is 1.00 atm, what is its volume?

■

At the time Boyle discovered the relationship between P and V, scientists lacked a clear understanding of atoms or molecules. Given what we now know about matter, how can Boyle's law be explained on a molecular basis? Consider the effect of compressing a collection of gas molecules into a smaller space (Figure 6.14). The molecules are in constant random motion, which means that they collide with one another and with the interior surface of their container. The collisions with the interior surface are responsible for the pressure exerted by the gas. If the molecules are squeezed into a smaller space—if the volume of the container is decreased—more collisions occur per unit time. The more frequent the collisions, the greater is the force exerted by the molecules against the interior surface and thus the greater the pressure. Dividing the force of the collisions by the area over which they occur yields the pressure (Equation 6.1). The interior surface area decreases when the size of the container is decreased, which in turn causes F/A to increase. Therefore the value of P increases as the size of the container decreases: as volume goes down, pressure goes up because the frequency of collisions increases. The converse is also true: as the volume of a container is increased, A increases, the collision frequency decreases, and the pressure drops.

Charles's Law: Relating Volume and Temperature

Nearly a century after Boyle's discovery of the inverse relation between the pressure exerted by a gas and the volume of the gas, French scientist Jacques Charles (1746–1823) documented the linear relation between the volume and temperature of a fixed quantity of gas at constant pressure. Now known as **Charles's law**, the relation states that, when the pressure exerted on a gas is held constant, the volume of a fixed quantity of gas is directly proportional to the **absolute temperature**, the temperature expressed in kelvins, of the gas:

$$V \propto T \qquad (P \text{ and } n \text{ fixed}) \qquad (6.7)$$

The effects of Charles's law can be seen in Figure 6.15, in which a balloon has been attached to a flask, trapping a fixed amount of gas in the apparatus. Heating the flask causes the gas to expand, inflating the balloon. The higher the temperature, the greater is the volume occupied by the gas and the bigger the balloon.

As with Boyle's law, we can replace the proportionality symbol in Equation 6.7 by an equals sign if we include a proportionality constant:

$$V = (\text{constant})T$$

$$\frac{V}{T} = \text{constant} \qquad (6.8)$$

The value of the constant depends on the number of moles of gas in the sample (n) and on the pressure of the gas (P). When those two parameters are held constant, the ratio V/T does not change, and any two combinations of volume and temperature are related as follows:

$$\frac{V_1}{T_1} = \frac{V_2}{T_2} \qquad (6.9)$$

Applying Charles's law allows us to predict (see Section 6.1) that a decrease in temperature from 20°C to 0°C reduces volume by about 7%. Let's see how we arrive at those numbers. Consider what happens when a balloon containing 2.00 L of air at 20°C is taken outside on a day when the temperature is 0°C. The amount of gas in the balloon is fixed, and the atmospheric pressure is constant. Experience tells us that the volume of the balloon decreases; Charles's law enables us to quantify that change. We can calculate the final volume with Equation 6.9, provided we express T_1 and T_2 in kelvins, not degrees Celsius. We must *always* use absolute temperatures in gas equations. Because the lowest temperature on the absolute temperature scale is 0 K, temperatures expressed in kelvins are always positive numbers. Thus Equation 6.9 gives us

$$\frac{V_1}{T_1} = \frac{V_2}{T_2}$$

$$V_2 = \frac{V_1 T_2}{T_1} = \frac{2.00\ \text{L} \times 273\ \cancel{\text{K}}}{293\ \cancel{\text{K}}} = 1.86\ \text{L}$$

As predicted, the volume of the balloon decreases when the temperature decreases. The percent change in the volume of the balloon is

$$\% \text{ decrease} = \frac{V_1 - V_2}{V_1} \times 100\% = \frac{2.00\ \cancel{\text{L}} - 1.86\ \cancel{\text{L}}}{2.00\ \cancel{\text{L}}} \times 100\% = 6.8\%$$

CONCEPT TEST

Graph a is a plot of volume (V) as a function of temperature (T) for 1 mole of a gas at a pressure of 1.00 atm; graph b is a graph of volume (V) as a function of (T) for 1 mole of a gas at 2.00 atm pressure. How do the slopes of the two graphs differ?

SAMPLE EXERCISE 6.4   Using Charles's Law

Charles and his compatriot Joseph Gay-Lussac (1778–1850) were drawn to the study of gases because of the interest in hot-air balloons in the late 18th century. We can imagine an early experiment these two scientists might have performed: what Celsius temperature is required to increase the volume of a sealed balloon from 2.00 L to 3.00 L if the initial temperature is 15°C and the atmospheric pressure is constant?

Collect and Organize We are given the volume ($V_1 = 2.00$ L) of a gas at an initial Celsius temperature and asked to calculate the Celsius temperature needed to make the gas reach a different volume ($V_2 = 3.00$ L). The container (a balloon) is sealed, so n is constant, as is atmospheric pressure.

CONNECTION In Chapter 1, we defined a temperature on the Kelvin scale as equal to the temperature in degrees Celsius plus 273.15: K = °C + 273.15.

Charles's law the volume of a fixed quantity of gas at constant pressure is directly proportional to its absolute temperature.

absolute temperature temperature expressed in kelvins on the absolute temperature scale, on which 0 K is the lowest possible temperature.

Analyze The relationship between volume and temperature is given by Charles's law in the form of Equation 6.9. Charles's law tells us that for volume to increase, temperature must also increase. If the volume increases by 50%, we predict that the absolute temperature must also increase by 50%.

Solve First we rearrange Equation 6.9 to solve for T_2:

$$T_2 = \frac{V_2 T_1}{V_1}$$

We then substitute for V_1, V_2, and T_1, remembering to convert temperature to kelvins:

$$T_2 = \frac{V_2 T_1}{V_1} = \frac{(3.00 \; \cancel{L})(288 \; K)}{2.00 \; \cancel{L}} = 432 \; K$$

The final step is to convert the final temperature to degrees Celsius:

$$T_2 = 432 \; K - 273 = 159°C$$

Think about It To increase the volume by 1.5 times [(1.5)(2.00 L) = 3.00 L], the temperature must also increase 1.5 times [432 K = (1.5)(288 K)], as expected from the statement that volume is directly proportional to the absolute temperature. The volume of the gas is *not* directly proportional to temperature in degrees Celsius: 159°C ≠ (1.5)(15°C).

Practice Exercise Hot expanding gases can be used to perform useful work in a cylinder fitted with a movable piston, as in Figure 6.16. If the temperature of a gas confined to such a cylinder is raised from 245°C to 560°C, what is the ratio of the initial volume to the final volume if the pressure exerted on the gas remains constant?

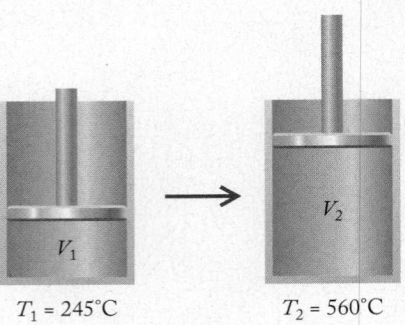

$T_1 = 245°C$ $T_2 = 560°C$

FIGURE 6.16 The volume of a gas in a cylinder changes as the temperature of the gas increases from 245°C to 560°C.

Charles's law also allowed the original determination that absolute zero (0 K) is equal to −273.15°C. Consider what happens when we plot the volume of a fixed quantity of gas at constant pressure as a function of temperature. Some typical results are graphed in Figure 6.17, showing that the decrease in

FIGURE 6.17 The volume of a gas for which *n* and *P* are held constant plotted against temperature on the Celsius and Kelvin scales. As predicted by Charles's law, volume decreases as the temperature decreases; the relationship is linear on both scales. The dashed line shows the extrapolation from the experimental data to a point corresponding to zero volume. This temperature is known as absolute zero: −273.15°C on the Celsius scale and 0 K on the Kelvin scale.

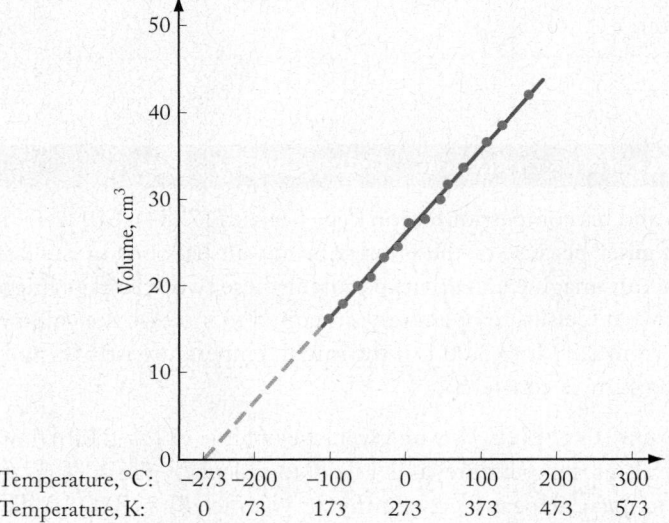

volume with decreasing temperature is linear. In the laboratory we are limited by how low we can make the temperature of a gas because at some temperature the gas liquefies; at this point the gas laws no longer describe its behavior because it is no longer a gas. However, we can *extrapolate* from our measured data to the point where the volume of a gas would reach zero if liquefaction did not occur. The dashed line in Figure 6.17 crosses the temperature axis at a volume of 0 at −273.15°C, and that temperature is defined as 0 K or absolute zero.

Avogadro's Law: Relating Volume and Quantity of Gas

Boyle's and Charles's experiments involved a constant quantity of gas in a confined sample. Now consider what happens when the quantity of gas is not constant, as when a balloon is inflated. The larger the quantity of gas blown into the balloon, the larger the balloon gets until it finally bursts. Clearly, the volume depends on the quantity (number of grams or moles) of the gas. If a hole is poked in the balloon, the gas escapes, decreasing both the quantity of gas and the volume. The relationship between the pressure and the quantity of the gas can be illustrated in several ways.

Consider a bicycle tire that is inflated enough to hold its shape but is nevertheless too soft to ride on. We can increase the pressure inside the tire by pumping more air into it. In the process, the tire does not expand much because the heavy rubber is much more rigid than the material in a balloon. As another example, consider a carnival vendor filling balloons from a helium tank fitted with a pressure gauge (Figure 6.18). When sales are brisk, it is possible to observe the pressure decreasing as the vendor fills more and more balloons. The pressure in the tank decreases as the amount of gas in the cylinder decreases.

A few decades after Charles discovered the relation between *V* and *T*, Amedeo Avogadro (1776–1856), whom we know from Avogadro's number (N_A), recognized the relationship between *V* and *n*. This relation is described by **Avogadro's law**,

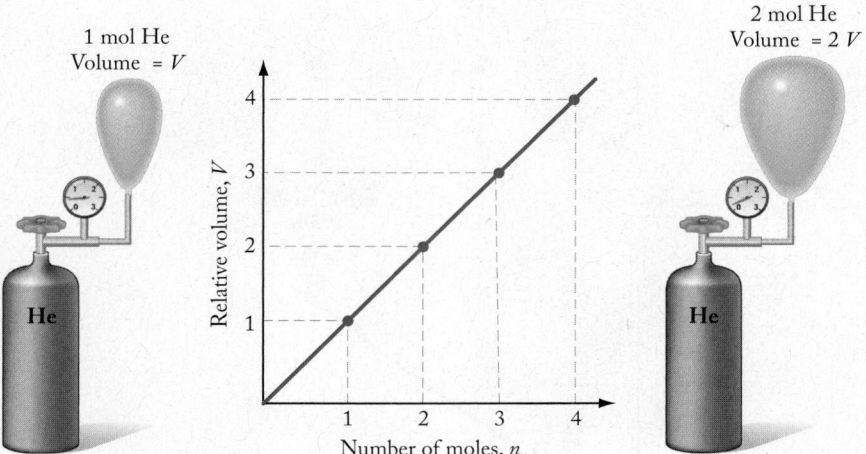

FIGURE 6.18 Because it is filled with twice the quantity of helium, the balloon on the right has twice the volume of the balloon on the left. This direct relation between volume and number of moles of a gas at constant temperature and pressure is Avogadro's law.

which states that the volume of a gas at a given temperature and pressure is proportional to the quantity of the gas in moles:

$$V \propto n \quad \text{or} \quad \frac{V}{n} = \text{constant} \qquad (P \text{ and } T \text{ fixed}) \qquad (6.10)$$

⊙⊙ CONNECTION In Chapter 3 the number of particles in a mole was defined as Avogadro's number, in honor of his early work with gases that lead to determining atomic masses.

CONCEPT TEST ·

Which graph in Figure 6.19 correctly describes the relationship between the value of V/n as n is increased (at constant P and T)?

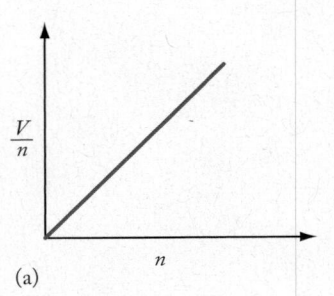 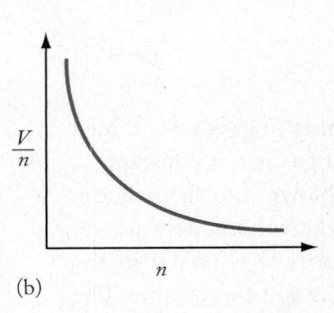 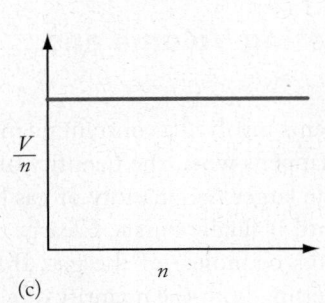 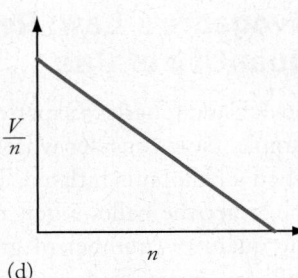

(a) (b) (c) (d)

FIGURE 6.19

· ·

Amontons's Law: Relating Pressure and Temperature

Both Boyle's law ($PV = \text{constant}$) and Charles's law ($V/T = \text{constant}$) describe effects on volume, but is there a relation between pressure and temperature? Experiments show that pressure is directly proportional to temperature when n and V are constant:

$$P \propto T \quad \text{or} \quad \frac{P}{T} = \text{constant} \qquad (V \text{ and } n \text{ fixed}) \qquad (6.11)$$

This relationship means that as the absolute temperature of a fixed amount of gas held at a constant volume increases, the pressure of the gas increases (Figure 6.20).

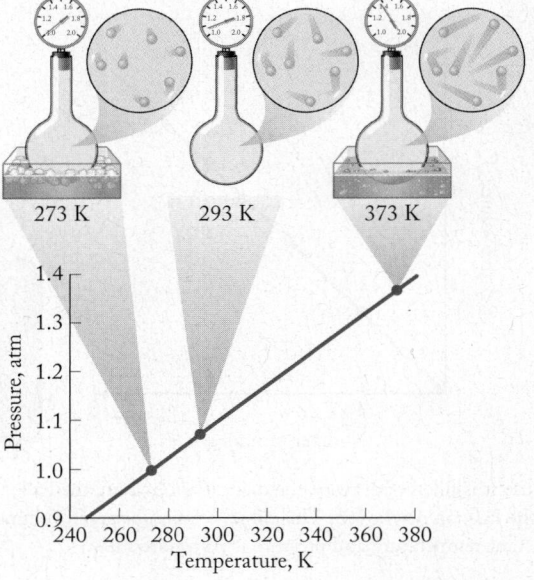

FIGURE 6.20 The pressure of a given quantity of gas is directly proportional to its absolute temperature when the gas volume is held constant. Each flask has the same volume and contains the same number of molecules. The relative speeds of the molecules (represented by the lengths of their "tails") increase with increasing temperature, causing more frequent and more forceful collisions with the walls of the flasks and hence higher pressures.

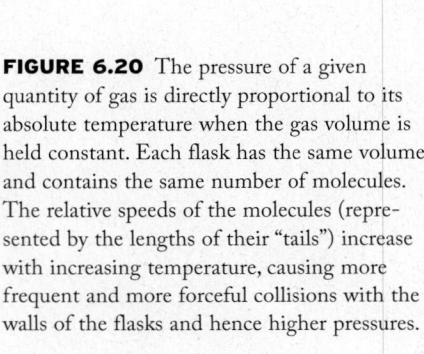

This statement is referred to as **Amontons's law** in honor of French physicist Guillaume Amontons (1663–1705), a contemporary of Robert Boyle's who constructed a thermometer based on the observation that the pressure of a gas is directly proportional to its temperature.

Where do we see evidence of Amontons's law? Consider a bicycle tire inflated to a prescribed pressure of 100 psi (6.8 atm) on an afternoon in late autumn when the temperature is 25°C. The next morning, following an early freeze in which the temperature drops to 0°C, the pressure in the tire drops to about 91 psi (6.2 atm). Did the tire leak? Perhaps, but the decrease in pressure can also be explained by Amontons's law.

We can confirm that the decrease in tire pressure is consistent with the decrease in temperature by applying the logic we used to develop Equations 6.6 and 6.9. Relating initial pressure and temperature to final pressure and temperature, we can write

$$\frac{P_1}{T_1} = \frac{P_2}{T_2} \tag{6.12}$$

Solving for P_2, we get

$$P_2 = \frac{P_1 T_2}{T_1} = \frac{(6.8 \text{ atm})(273 \text{ K})}{298 \text{ K}} = 6.2 \text{ atm}$$

Why is pressure directly proportional to temperature? Like pressure, temperature is directly related to molecular motion. The average speed at which a population of molecules moves increases with increasing temperature (Figure 6.20). For a given number of gas molecules, increasing the temperature increases the average speed and therefore increases the frequency and force with which the molecules collide with the walls of their container. Because pressure is related to the frequency and force of the collisions, it follows that higher temperatures produce higher pressures, as long as other factors such as volume and quantity of gas are constant.

> **Amontons's law** as the absolute temperature of a fixed amount of gas increases, the pressure increases as long as the volume and quantity of gas remain constant.

SAMPLE EXERCISE 6.5   Using Amontons's Law

Labels on aerosol cans caution against incineration because the cans may explode when the pressure inside them exceeds 3.00 atm. At what temperature in degrees Celsius will an aerosol can burst if initially the pressure inside the can is 2.20 atm at 25°C?

Collect and Organize We are given the temperature ($T_1 = 25°C$) and pressure ($P_1 = 2.20$ atm) of a gas and asked to determine the temperature (T_2) at which the pressure ($P_2 = 3.00$ atm) causes the can to explode.

Analyze Because the gas is enclosed in an aerosol can, we know that the quantity of gas and volume are constant. We can use Amontons's law (Equation 6.12) because it describes what happens to pressure as the temperature of a gas is changed. We can estimate the answer by considering that the pressure in the can must increase by about 50% in order for the pressure to exceed 3 atm. Pressure is directly proportional to temperature, so the absolute temperature must also increase by about 50%. Before proceeding, we should recognize that the initial temperature is in degrees Celsius, and we must report our answer in degrees Celsius. Because Equation 6.12 works only with absolute temperatures, we need to convert the initial Celsius temperature to kelvins.

Solve We rearrange Equation 6.12 to solve for T_2:

$$T_2 = \frac{T_1 P_2}{P_1}$$

After converting T_1 from degrees Celsius to kelvins, we have

$$T_2 = \frac{(25 + 273)\ \text{K} \times 3.00\ \cancel{\text{atm}}}{2.20\ \cancel{\text{atm}}} = 406\ \text{K}$$

Converting T_2 to degrees Celsius gives us

$$T_2 = 406\ \text{K} - 273 = 133°\text{C}$$

Think about It This temperature is certainly higher than the original temperature. In fact 406 K is about 1.5 times (or 50% higher than) the initial temperature, 298 K, so the answer makes sense.

Practice Exercise The air pressure in the tires of an automobile is adjusted to 28 psi at a gas station in San Diego, where the air temperature is 68°F. The air in the tires is at the same temperature as the atmosphere. The automobile is then driven east along a hot desert highway, and the temperature inside the tires reaches 140°F. What is the pressure in the tires?

Gases that behave in accordance with the linear relations discovered by Boyle, Charles, Avogadro, and Amontons are called **ideal gases**. Most gases exhibit ideal behavior at the pressures and temperatures typically encountered in the atmosphere. Under these conditions, the volumes occupied by the gas molecules or atoms are insignificant compared with the overall volume occupied by the gas. All the open space between molecules makes gases compressible. In an ideal gas, the atoms or molecules are assumed not to interact with one another; rather, they move independently with speeds that are related to their masses and to the temperature of the gas.

6.4 The Ideal Gas Law

ideal gas a gas whose behavior is predicted by the linear relations defined by Boyle's, Charles's, Avogadro's, and Amontons's laws.

ideal gas equation (also called **ideal gas law**) relates the pressure, volume, number of moles, and temperature of an ideal gas; expressed as $PV = nRT$, where R is the universal gas constant.

universal gas constant the constant R in the ideal gas equation; its value and units depend on the units used for the variables in the equation.

What if we launched a weather balloon from the surface of Earth and allowed it to drift to an elevation of 10,000 m? The volume of the balloon would expand as the atmospheric pressure decreased, but the air temperature would decrease as the balloon ascended. How do we determine the final volume of the balloon when we are changing three variables (P, V, and T) while keeping just one variable (n) fixed? Taken individually, none of the four gas laws and their accompanying mathematical relations fit this situation. Nevertheless, we can derive a relation, known as the **ideal gas equation**, that combines all four variables:

$$PV = nRT \tag{6.13}$$

where R is the **universal gas constant**. This equation is also called the **ideal gas law**.

As Table 6.3 shows, R has different values depending on the units used for pressure, volume, and temperature. For many calculations in chemistry, it is convenient to use $R = 0.08206$ L · atm/(mol · K). As the units tell us, however, we can use this value only when the quantity of gas is expressed in moles, the volume in liters, the pressure in atmospheres, and the temperature in kelvins.

Where does the ideal gas equation come from? Boyle's law expresses that volume and pressure are inversely proportional. Charles's law tells us that volume is directly proportional to temperature. Putting Boyle's and Charles's laws together (Equations 6.5 and 6.8) allows us to write an equation relating P, V, and T and combining the constants:

$$PV = \text{constant}$$

$$\frac{V}{T} = \text{constant}$$

$$\frac{PV}{T} = \text{combined constant} \qquad (6.14)$$

Avogadro's law tells us that volume is directly proportional to the number of moles of gas when T and P are constant, so we can include n in Equation 6.14 and combine its constant with the "combined constant" in that equation:

$$\frac{PV}{nT} = \text{combined constant} \quad \text{or} \quad PV = (\text{combined constant}) \times nT \quad (6.15)$$

We turn Equation 6.15 into a meaningful equality by inserting the appropriate constant. By convention, this constant is the universal gas constant R, giving us the ideal gas law (Equation 6.13).

We can derive another relation involving these four variables that is useful when a system starts in an initial state (P_1, V_1, T_1, n_1) and moves to a final state (P_2, V_2, T_2, n_2):

$$P_1V_1 = n_1RT_1 \quad \text{so} \quad \frac{P_1V_1}{n_1T_1} = R \quad \text{and}$$

$$\qquad (6.16)$$

$$P_2V_2 = n_2RT_2 \quad \text{so} \quad \frac{P_2V_2}{n_2T_2} = R$$

Because both the initial-state and final-state expressions are equal to R, they are also equal to each other, allowing us to simplify Equation 6.16 to

$$\frac{P_1V_1}{n_1T_1} = \frac{P_2V_2}{n_2T_2} \qquad (6.17)$$

Furthermore, because we frequently work with systems like weather balloons and gas canisters in which the amount of gas is constant ($n_1 = n_2$), but pressure, temperature, and volume vary, we define an especially useful form of this equation:

$$\frac{P_1V_1}{T_1} = \frac{P_2V_2}{T_2} \qquad \text{for constant } n \qquad (6.18)$$

Equation 6.18 is known as the **general gas equation**, or **combined gas law**. Sample Exercise 6.6 gives an example of how this equation is used in determining the effect of changes in P and T on the volume (V) of a weather balloon. Different reduced versions of Equation 6.17 are certainly possible when other

TABLE 6.3	Values for the Universal Gas Constant (R)
Value of R	**Units**
0.08206	L · atm/(mol · K)
8.314	kg · m²/(s² · mol · K)
8.314	J/(mol · K)
8.314	m³ · Pa/(mol · K)
62.37	L · torr/(mol · K)

▶❚❚ **CHEMTOUR** The Ideal Gas Law

general gas equation (also called **combined gas law**) based on the ideal gas law and used when one or more of the four gas variables are held constant while the remaining variables change.

standard temperature and pressure (STP) 0°C and 1 bar as defined by IUPAC; in the United States, 0°C and 1 atm.

molar volume volume occupied by 1 mole of an ideal gas at STP; 22.4 L.

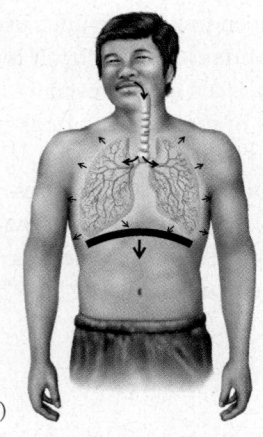 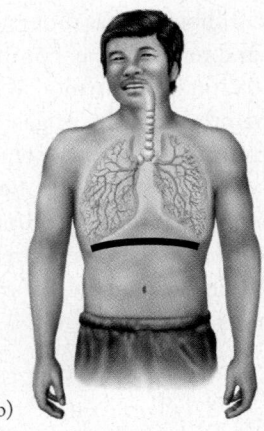

(a)   (b)

FIGURE 6.21 Breathing illustrates the relationship between P, V, and n. (a) When you inhale, your rib cage expands and your diaphragm moves down, (b) increasing the volume of your lungs. Increased volume decreases the pressure inside your lungs, in accord with Boyle's law: PV = constant. Decreased pressure allows more air to enter until the pressure inside your lungs matches atmospheric pressure.

variables besides n are fixed (Figure 6.21), but only Equation 6.18 is known as the combined gas law. Figure 6.22 summarizes the four laws expressed within the ideal gas law.

CONCEPT TEST ....

Which of the following variations of Equation 6.17 are incorrect?

a. $\dfrac{n_2 T_2}{P_2} = \dfrac{n_1 T_1}{P_1}$ at constant V b. $\dfrac{n_2 V_2}{P_2} = \dfrac{n_1 V_1}{P_1}$ at constant T

c. $\dfrac{n_2 T_2}{V_2} = \dfrac{n_1 T_1}{V_1}$ at constant P d. $\dfrac{T_1}{n_1 V_1} = \dfrac{T_2}{n_2 V_2}$ at constant P

A useful reference point defined by the International Union of Pure and Applied Chemistry (IUPAC) used in studying the properties of gases is **standard temperature and pressure (STP)**, defined as 0°C and 1 bar. The more familiar unit of 1 atm is very close to 1 bar, and at the level of accuracy in calculations in this text this substitution makes little difference, so we consider STP to be 0°C and 1 atm.

Another reference point is **molar volume**, which is the volume that 1 mole of an ideal gas occupies at STP (Figure 6.23). We can calculate the molar volume from the ideal gas equation by solving the equation for V and inserting the values of n, P, and T at standard conditions:

$$V = \frac{(1\ \text{mol})\left(0.08206\ \dfrac{\text{L} \cdot \text{atm}}{\text{mol} \cdot \text{K}}\right)(273\ \text{K})}{1\ \text{atm}} = 22.4\ \text{L}$$

Many chemical and biochemical processes take place at pressures near 1 atm and at temperatures between 0°C and 40°C. Within this range, the volume that 1 mole of gaseous reactant or product occupies is no more than about 15% greater than the molar volume. Therefore volumes can be estimated easily if molar amounts are known. An important feature of molar volume is that it applies to any ideal gas, independent of its chemical composition. In other words, at STP 1 mole of helium occupies the same volume—22.4 L—as 1 mole of methane (CH_4), of carbon dioxide (CO_2), or even of a compound like uranium hexafluoride (UF_6), which has a molar mass almost 90 times that of helium.

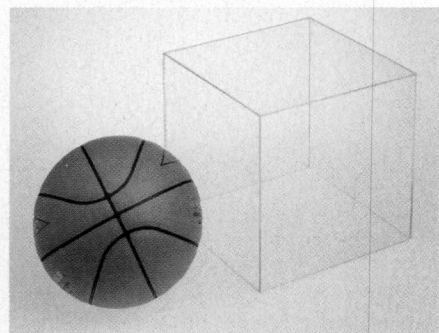

FIGURE 6.23 The box contains 1 mole of gas; the molar volume of an ideal gas is 22.4 L at 0°C and 1 atm of pressure. A basketball fits loosely into a box having this volume.

(a) Boyle's law: volume inversely proportional to pressure; n and T fixed

$$PV = \text{constant}$$

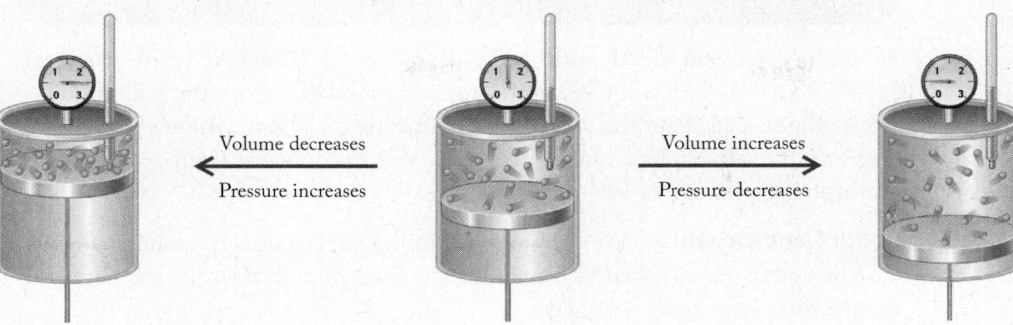

Volume decreases

Pressure increases

Volume increases

Pressure decreases

(b) Charles's law: volume directly proportional to temperature; n and P fixed

$$\frac{V}{T} = \text{constant}$$

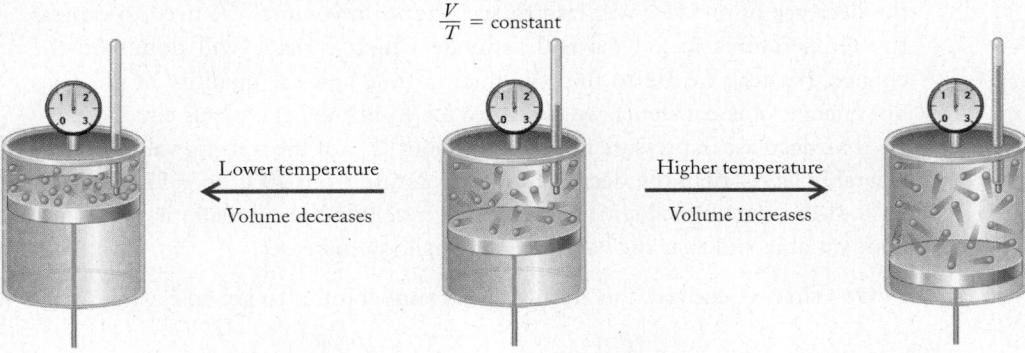

Lower temperature

Volume decreases

Higher temperature

Volume increases

(c) Avogadro's law: volume directly proportional to number of moles; T and P fixed

$$\frac{V}{n} = \text{constant}$$

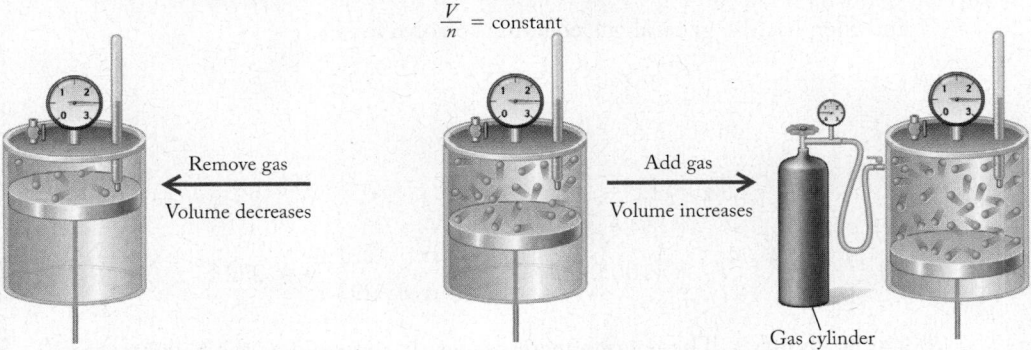

Remove gas

Volume decreases

Add gas

Volume increases

Gas cylinder

(d) Amontons's law: pressure directly proportional to temperature; n and V fixed

$$\frac{P}{T} = \text{constant}$$

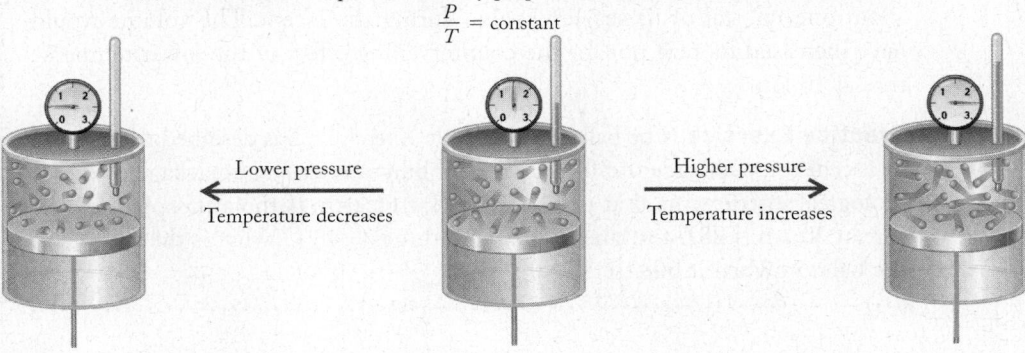

Lower pressure

Temperature decreases

Higher pressure

Temperature increases

FIGURE 6.22 Relationships between pressure, volume, temperature, and/or moles of gas.
(a) Increasing or decreasing V at constant n and T: Boyle's law. (b) Increasing or decreasing T at
constant n and P: Charles's law. (c) Increasing or decreasing n at constant T and P: Avogadro's law.
(d) Increasing or decreasing P at constant n and V: Amontons's law.

SAMPLE EXERCISE 6.6   **Calculations Involving Changes in *P*, *V*, and *T***

A weather balloon filled with 100.0 L of He is launched from sea level ($T = 20°C$, $P = 755$ torr). No gas is added or removed from the balloon during its flight. Calculate the volume at an altitude of 10 km, where the temperature of the atmosphere and the gas in the balloon are both $-52°C$ and atmospheric pressure is 195 torr.

Collect and Organize We are given the initial temperature, pressure, and volume of a gas, and are asked to determine the final volume after the pressure and temperature have both changed.

Analyze The decrease in temperature will lead to a decrease in volume, but the decrease in pressure will lead to an increase in volume. We need to express the temperatures in kelvins and estimate which variable will dominate the change. Because we are to find the final volume and the quantity of gas does not change (n is constant), we can solve for V_2 in the general gas equation.

The decrease in pressure to 195 torr (about 25% of the starting value) is considerably larger than the decrease in temperature from 20°C to $-52°C$ (about 75% of the starting value). Therefore, we conclude that pressure is the dominant variable and that the volume of the balloon increases.

Solve First we convert the given Celsius temperatures to kelvins:

$$T_1 = (20°C + 273) = 293 \text{ K}$$
$$T_2 = -52°C + 273 = 221 \text{ K}$$

and then use the general gas equation to solve for V_2:

$$\frac{P_1 V_1}{T_1} = \frac{P_2 V_2}{T_2}$$

$$V_2 = V_1 \times \frac{P_1}{P_2} \times \frac{T_2}{T_1}$$

$$V_2 = (100.0 \text{ L}) \times \frac{755 \text{ torr}}{195 \text{ torr}} \times \frac{221 \text{ K}}{293 \text{ K}} = 292 \text{ L}$$

Think about It The volume increases nearly threefold as the balloon ascends to 10 km. This result makes sense because atmospheric pressure decreases to about one-quarter of its sea-level value during the ascent. The volume would have increased more if not for the countervailing effect of the lower temperature at 10 km.

Practice Exercise The balloon in Sample Exercise 6.6 is designed to continue its ascent to an altitude of 30 km, where it bursts, releasing a package of meteorological instruments that parachute back to Earth. If the atmospheric pressure at 30 km is 28.0 torr and the temperature is $-45°C$, what is the volume of the balloon when it bursts?

The ideal gas law describes the relation among number of moles, pressure, volume, and temperature for any gas provided it behaves as an ideal gas. Because many gases behave ideally at typical atmospheric pressures, we can apply the ideal gas equation to many situations outside the laboratory. In Sample Exercise 6.7, we calculate the mass of oxygen in an alpine climber's compressed-oxygen cylinder.

SAMPLE EXERCISE 6.7   **Applying the Ideal Gas Law**

Bottles of compressed O_2 carried by climbers ascending Mt. Everest have an internal volume of 5.90 L. Assume that such a bottle has been filled with O_2 to a pressure of 2025 psi at 25°C.

a. How many moles of O_2 are in the bottle?
b. What is the mass in grams of O_2 in the bottle? Assume that O_2 behaves as an ideal gas.

Collect and Organize We are given the pressure, volume, and temperature of a gas and asked to determine its quantity in moles and its mass in grams.

Analyze The ideal gas equation enables us to use P, V, and T to calculate n, the number of moles of O_2. Then we can use its molar mass (32.00 g/mol) to calculate the mass of O_2 in the bottle. We can estimate the answer by considering that 1 mole of gas at STP occupies 22.4 L. Our oxygen bottle has a volume of about 6 L, which is about one-quarter of the molar volume at STP but the pressure (2025 psi) is more than 100 times greater than 1 atm (see Table 6.2); thus, we predict that the bottle contains more than 25 times more O_2.

Solve

a. Let's start with the ideal gas equation rearranged to solve for n:

$$PV = nRT \qquad\qquad P = \frac{nRT}{V} \qquad V = \frac{nRT}{P}$$

$$n = \frac{PV}{RT} \qquad\qquad R = \frac{PV}{nT} \qquad T = \frac{PV}{nR}$$

Before using this expression for n, we need to convert pressure into atmospheres and temperature into kelvins:

$$P = (2025 \ \cancel{psi})\left(\frac{1\ atm}{14.7\ \cancel{psi}}\right) = 138\ atm \qquad T = 25°C + 273 = 298\ K$$

$$n = \frac{(138\ \cancel{atm})(5.90\ \cancel{L})}{\left(0.08206\ \dfrac{\cancel{L} \cdot \cancel{atm}}{mol \cdot \cancel{K}}\right)(298\ \cancel{K})} = 33.3\ mol$$

b. Converting moles into grams is a matter of multiplying by the molar mass:

$$(33.3\ \cancel{mol})\left(\frac{32.00\ g}{1\ \cancel{mol}}\right) = 1066\ g = 1.07 \times 10^3\ g$$

Think about It Our answer in part a is certainly reasonable based on our estimate. Most climbers require many bottles to climb Mt. Everest and return.

Practice Exercise Starting with the moles of O_2 calculated in Sample Exercise 6.7, calculate the volume of O_2 the bottle could deliver to a climber at an altitude where the temperature is $-38°C$ and the atmospheric pressure is 0.35 atm.

■

6.5 Gases in Chemical Reactions

Chemical reactions that use gases as reactants or produce gaseous products are quite common in our daily lives. When a commercial airliner flying at high altitudes loses cabin pressure, oxygen masks are deployed automatically for the passengers to use until the aircraft can descend or the cabin can be repressurized. The oxygen that flows in the masks is generated by the decomposition of sodium chlorate. Similarly, when an air bag deploys during a automobile accident, the nitrogen gas that rapidly inflates the protective air bag is the product of the decomposition of sodium azide. Gases are reactants in, for example, the combustion of charcoal in a backyard grill (Equation 6.19). Solid carbon reacts with oxygen gas in the air to produce carbon dioxide and the heat used to cook our food.

$$C(s) + O_2(g) \rightarrow CO_2(g) + \text{heat} \qquad (6.19)$$

In any chemical reaction that involves a gas as either reactant or product, the volume of the gas indirectly defines the amount of it in the reaction. If T and P are known, we can use the ideal gas equation to relate volume to the number of moles of gas in the system. Once we know that, we can use stoichiometric calculations to relate quantities of gas to quantities of other reactants and products, including heat. For example, consider the volume of oxygen needed to completely burn 500.0 g (about 1 lb) of charcoal at 1.00 atm of pressure on an average summer day (25°C). Before starting the calculation we must first write a balanced chemical equation for the reaction. Equation 6.19 is indeed balanced, and 1 mole of C reacting with 1 mole of oxygen should yield 1 mole of CO_2. If we start with 500.0 g of C we have

$$500.0 \text{ g C} \times \frac{1 \text{ mol C}}{12.01 \text{ g C}} = 41.63 \text{ mol C}$$

The stoichiometry of the reaction tells us that we need 41.6 moles of O_2 to completely react with the given amount of carbon. The volume of O_2 that corresponds to 41.6 moles of O_2 is calculated using the ideal gas equation by first rearranging the equation to solve for V and then substituting the values of n, R, T (in kelvins), and P:

$$V = \frac{nRT}{P} = \frac{(41.63 \text{ mol})\left(0.08206 \frac{\text{L} \cdot \text{atm}}{\text{mol} \cdot \text{K}}\right)(298 \text{ K})}{1.00 \text{ atm}} = 1.02 \times 10^3 \text{ L}$$

> SAMPLE EXERCISE 6.8   **Combining Stoichiometry and the Ideal Gas Law**

Oxygen generators in some airplanes are based on the chemical reaction between sodium chlorate and iron:

$$NaClO_3(s) + Fe(s) \rightarrow O_2(g) + NaCl(s) + FeO(s)$$

The resultant O_2 is blended with cabin air to provide 10–15 minutes of breathable air for passengers. How many grams of $NaClO_3$ are needed in a typical generator to produce 125 L of O_2 gas at 1.00 atm and 20.0°C?

Collect and Organize We are given the volume of gas we need to prepare at a particular pressure and temperature. With this information we can determine the mass of $NaClO_3$ needed based on the stoichiometric relations in the balanced chemical equation.

Analyze The solution requires two calculations. We start by recognizing that the balanced chemical equation tells us that 1 mole of $NaClO_3$ is needed to produce 1 mole of oxygen. If we can determine how many moles of O_2 occupy a volume of 125 L at 1.00 atm pressure and 20.0°C, we can determine the number of moles of $NaClO_3$ we need. To determine the moles of oxygen (n_{O_2}), we can use the ideal gas law (Equation 6.13) to calculate the amount of oxygen when V = 125 L, P = 1.00 atm, and T = 20.0°C. Then we use our calculated value of n_{O_2} and the balanced chemical equation to determine the number of moles and number of grams of $NaClO_3(s)$ required.

We can estimate the answer by comparing the volume of gas we wish to make (125 L) with the molar volume of an ideal gas at STP (22.4 L). Considering that the difference in temperature at STP (0°C = 273 K) and the temperature in this problem (20°C = 293 K) is relatively small, we estimate that we will need about 5 moles of O_2, requiring 5 moles of $NaClO_3$.

Solve We use the rearranged ideal gas equation to solve for the moles of O_2:

$$n_{O_2} = \frac{P_{O_2}V_{O_2}}{RT_{O_2}}$$

$$n_{O_2} = \frac{(1.00 \ \text{atm})(125 \ \text{L})}{\left(0.08206 \ \dfrac{\text{L} \cdot \text{atm}}{\text{mol} \cdot \text{K}}\right)(273 + 20.0) \ \text{K}} = 5.20 \ \text{mol} \ O_2$$

To convert moles of O_2 into an equivalent mass of $NaClO_3$, we use the stoichiometry of the reaction to calculate the equivalent number of moles of $NaClO_3$, and use the conversion factor of molar mass to determine the number of grams of $NaClO_3$ required:

$$5.20 \ \text{mol} \ O_2 \times \frac{1 \ \text{mol} \ NaClO_3}{1 \ \text{mol} \ O_2} \times \frac{106.44 \ \text{g} \ NaClO_3}{1 \ \text{mol} \ NaClO_3}$$

$$= 5.53 \times 10^2 \ \text{g} \ NaClO_3$$

FIGURE 6.24 An automobile air bag inflates when solid NaN_3 rapidly decomposes, producing N_2 gas.

FIGURE 6.25 The release of CO_2 from volcanic centers on the Dieng Plateau in 1979 killed many people and animals as the dense gas flowed over the valley floor, effectively displacing the air (and oxygen) that the inhabitants and their livestock needed to survive.

Think about It We predicted that about 5 moles of $NaClO_3$ would be needed to produce 125 L of oxygen, which is quite close to the calculated value of 5.20 moles of $NaClO_3$. Multiplying by the molar mass of $NaClO_3$ (106 g/mol or about 10^2 g/mol) makes the final answer of 5.53×10^2 grams of sodium chlorate a reasonable value.

Practice Exercise Automobile air bags (Figure 6.24) inflate during a crash or sudden stop by the rapid generation of nitrogen gas from sodium azide, according to the reaction

$$2\,NaN_3(s) \rightarrow 2\,Na(s) + 3\,N_2(g)$$

How many grams of sodium azide are needed to provide sufficient nitrogen gas to fill a $45.0 \times 45.0 \times 25.0$ cm bag to a pressure of 1.20 atm at 15°C?

6.6 Gas Density

Carbon dioxide is a relatively minor component of our atmosphere. Because CO_2 is produced by the combustion of fuels, however, the amount of it in the atmosphere has increased since the Industrial Revolution. This increase is of great concern because of the impact on climate change.

Another source of atmospheric carbon dioxide is volcanoes. Because $CO_2(g)$ is denser than air, sudden releases of large quantities from areas of volcanic activity are life-threatening events. On the Dieng Plateau in Indonesia in 1979, 149 people died from asphyxiation in a valley after a massive, sudden release of carbon dioxide from one of the volcanoes around the valley (Figure 6.25). Being denser than air, the carbon dioxide settled in the valley, displacing the air containing the oxygen necessary for life.

The density of a gas at STP can be calculated by dividing its molar mass by the molar volume. Carbon dioxide, for example, has a molar mass of 44.01 g/mol. Therefore the density of CO_2 at STP is

$$\frac{44.01 \text{ g/mol}}{22.4 \text{ L/mol}} = 1.96 \text{ g/L}$$

The density of air at STP is about 1.3 g/L, so it is not surprising that the CO_2 in the Dieng Plateau disaster was concentrated close to the surface of the land. That the density of CO_2 is greater than that of air also gives rise to an important use of the gas in fighting fires. CO_2 effectively blankets a fuel, thereby separating it from the O_2 it needs to burn and thus extinguishing the fire.

CONCEPT TEST

Which gas has the highest density at STP: CH_4, Cl_2, Kr, or C_3H_8?

Using the ideal gas equation, we can calculate the density of an ideal gas at any temperature and pressure. Because density is the mass of a sample divided by its volume, $d = m/V$, we need to identify the factors in the ideal gas equation that represent the density. Sample mass can be determined from the number of moles, and of course the V in the ideal gas equation is the volume. We can rearrange Equation 6.13 so that these two factors are isolated:

$$\frac{P}{RT} = \frac{n}{V} \tag{6.20}$$

For a given sample of gas, n equals the mass of the sample in grams divided by the molar mass of the gas: $m/\mathcal{M}$. We can substitute this ratio for n in Equation 6.20 and solve the resulting expression for m/V (the density):

$$\frac{P}{RT} = \frac{m/\mathcal{M}}{V} = \frac{m}{\mathcal{M}V}$$

$$d = \frac{m}{V} = \frac{P\mathcal{M}}{RT}$$

If the term on the right in this last equation has P in atmospheres, $\mathcal{M}$ in grams per mole, R in L·atm/(mol·K), and T in kelvins, then the units of density are g/L:

$$\frac{(\cancel{\text{atm}})\left(\dfrac{\text{g}}{\cancel{\text{mol}}}\right)}{\left(\dfrac{\text{L} \cdot \cancel{\text{atm}}}{\cancel{\text{mol}} \cdot \cancel{\text{K}}}\right)(\cancel{\text{K}})} = \text{g/L} = \text{density}$$

Thus we have an expression for the density d of any gas in terms of its pressure, temperature, and molar mass:

$$d = \frac{P\mathcal{M}}{RT} \tag{6.21}$$

Figure 6.26 illustrates the relationship between density and molar mass for three pure gases and air. Note how the balloon containing $CO_2(g)$ does not float at all but rather sinks below the benchtop because the density of this pure gas

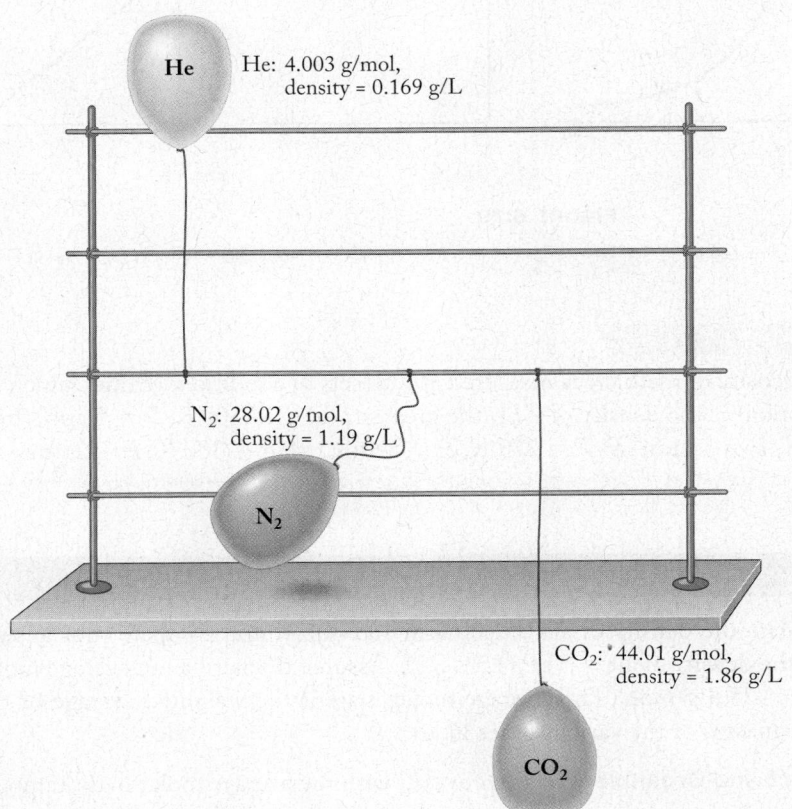

He: 4.003 g/mol, density = 0.169 g/L

N_2: 28.02 g/mol, density = 1.19 g/L

CO_2: 44.01 g/mol, density = 1.86 g/L

FIGURE 6.26 At 15°C and 1 atm, the density of the gas inside each balloon determines whether the balloon floats (helium), hovers just slightly above the benchtop (nitrogen), or sinks (carbon dioxide). Note the correlation between molar mass and density: the higher the molar mass of a gas, the higher the density.

FIGURE 6.27 Hot air in a balloon displaces more air than cooler air, making the balloon buoyant.

is greater than the density of air. In the absence of mixing, carbon dioxide moves to the lowest level accessible to it, just as it filled the valley in the Dieng Plateau disaster and cut the inhabitants off from their supply of oxygen, and just as it separates fuel from oxygen and extinguishes fires.

Equation 6.21 also explains why a balloon inflated with hot air rises (Figure 6.27). Put another way, the density of a given quantity of gas decreases with increasing temperature, giving the balloon buoyancy.

CONCEPT TEST

Which graph in Figure 6.28 best approximates (a) the relationship between density and pressure (n and T constant) and (b) the relationship between density and temperature (n and P constant) for an ideal gas?

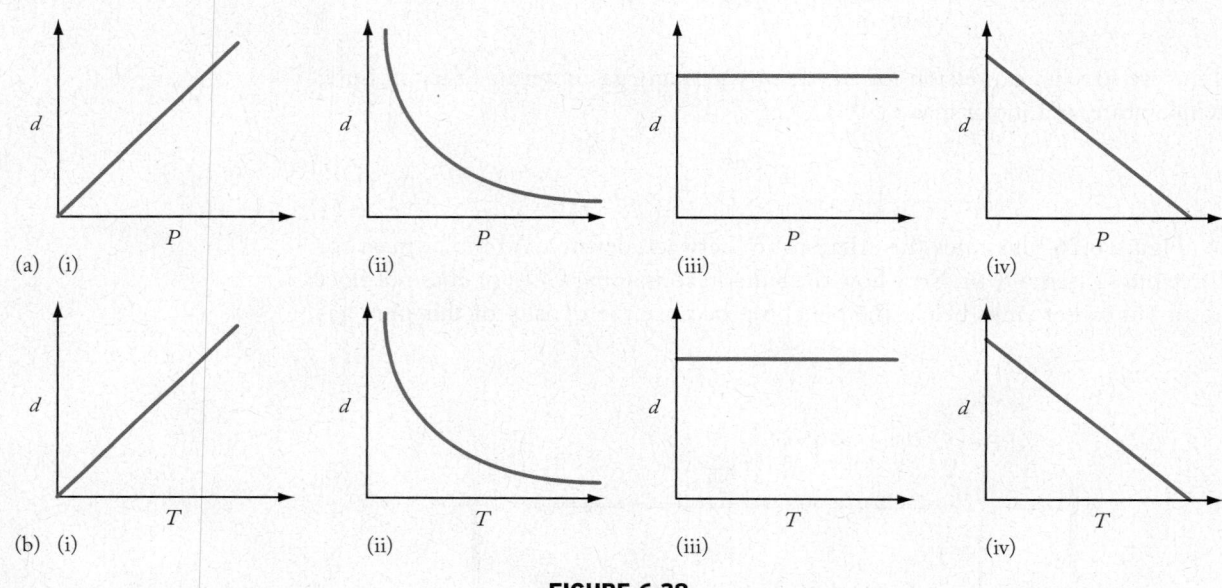

(a) (i)   (ii)   (iii)   (iv)

(b) (i)   (ii)   (iii)   (iv)

FIGURE 6.28

CONCEPT TEST

The density of methane is measured at four sets of conditions. Under which set of conditions is the density of CH_4 the greatest? (a) $T = 300$ K, $P = 1$ atm; (b) $T = 325$ K, $P = 1$ atm; (c) $T = 275$ K, $P = 2$ atm; (d) $T = 300$ K, $P = 2$ atm

SAMPLE EXERCISE 6.9   Calculating the Density of a Gas

Calculate the density of air at 1.00 atm and 302 K and compare your answer with the density of air at STP (1.29 g/L). Assume that air has an average molar mass of 28.8 g/mol. (This average molar mass is the weighted average of the molar masses of the various gases in air.)

Collect and Organize We are provided with the average molar mass, temperature, and atmospheric pressure of air and asked to calculate the density of air.

Analyze The given quantities are related by Equation 6.21. The density of a gas is inversely proportional to temperature. The temperature given in this problem (302 K) is higher than the temperature at STP, so we predict that the density of the air will be less than 1.29 g/L.

Solve Inserting the values of P, T, and $\mathcal{M}$ into Equation 6.21, we have

$$d_{air} = \frac{P_{air}\,\mathcal{M}_{air}}{RT_{air}} = \frac{\left(28.8\,\dfrac{g}{mol}\right)(1.00\,\cancel{atm})}{[0.08206\;L \cdot \cancel{atm}/(\cancel{mol} \cdot \cancel{K})](302\,\cancel{K})} = 1.16\;g/L$$

Think about It The solution confirms that the density of air calculated at 302 K (1.16 g/L) is less than the value at 273 K (1.29 g/L).

Practice Exercise Air is a mixture of mostly nitrogen and oxygen. A balloon is filled with oxygen and released in a room full of air. Will it sink to the floor or float to the ceiling?

Equation 6.21 tells us that density is inversely proportional to temperature, which means that density decreases as temperature increases (Figure 6.29). Thus we can use this equation to calculate the molar mass of a gas from its density at any temperature and pressure. To do so, we solve Equation 6.21 for molar mass:

$$\mathcal{M} = \frac{dRT}{P} \tag{6.22}$$

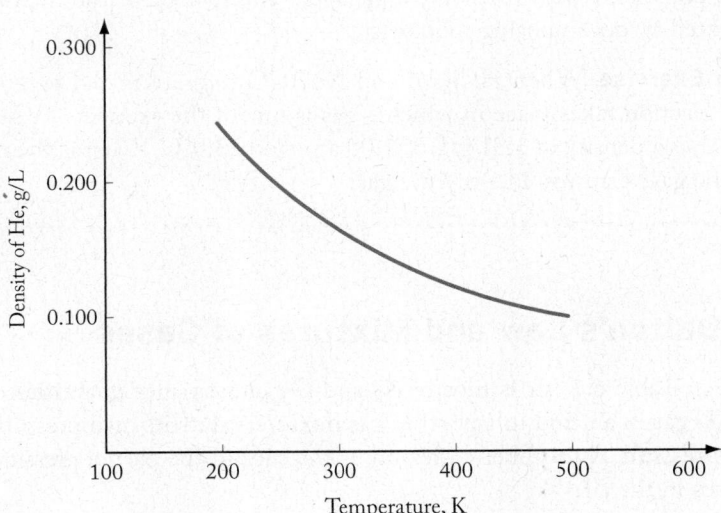

FIGURE 6.29 At 1 atm, the density of helium decreases with increasing temperature. The density of any ideal gas at constant pressure is inversely proportional to its absolute temperature. This graph shows a portion of the hyperbola that describes the behavior of the gas from approximately 200 to 500 K.

Gas density can be measured with a glass tube of known volume attached to a vacuum pump (Figure 6.30). The mass of the tube is determined when it has essentially no gas in it and again when it is filled with the test gas. The difference in the masses divided by the volume of the tube is the density of the gas.

(a)

(b)

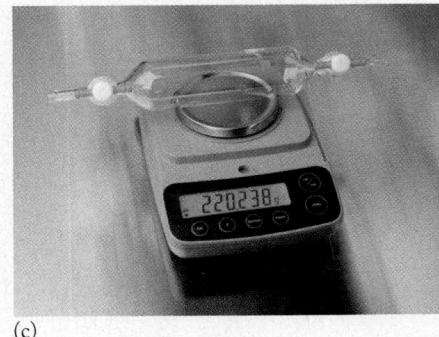

(c)

FIGURE 6.30 (a) A gas collection tube with an internal volume of 235 mL is evacuated by connecting it to a vacuum pump. (b) The mass of the evacuated tube is measured. (c) Then the tube is opened to the atmosphere and refills with air. The difference in mass between the filled and evacuated tubes (0.273 g or 273 mg) is the mass of the air inside it. The density of the air sample is 273 mg/235 mL or 1.16 mg/mL.

> **SAMPLE EXERCISE 6.10** **Calculating Molar Mass from Density**
>
> Vent pipes at solid-waste landfills often emit foul-smelling gases that may be either relatively pure substances or mixtures of several gases. A sample of such an emission has a density of 0.650 g/L at 25.0°C and 757 mmHg. What is the molar mass of the gas emitted? (Note that if the sample is a mixture, the answer will be the weighted average of molar masses of the individual gases.)
>
> **Collect and Organize** We are given the density, temperature, and pressure of a gaseous sample and asked to calculate the molar mass of the gas.
>
> **Analyze** Equation 6.22 relates molar mass to the density, pressure, and temperature of a gas, so we can use it to calculate the molar mass, remembering to convert the temperature to kelvins and the pressure to atmospheres before using these values in the equation.
>
> $$P = 757 \text{ mmHg} \times \frac{1 \text{ atm}}{760 \text{ mmHg}} = 0.996 \text{ atm}$$
>
> Many gases have relatively low molar masses, so we expect that the molar mass should be similar to that of typical gases we have seen thus far.
>
> **Solve**
>
> $$\mathcal{M} = \frac{dRT}{P} = \left(\frac{0.650 \text{ g}}{\text{L}}\right)\left(\frac{0.08206 \text{ L} \cdot \text{atm}}{\text{mol} \cdot \text{K}}\right)\left(\frac{273 + 25.0 \text{ K}}{0.996 \text{ atm}}\right)$$
> $$= 16.0 \text{ g/mol}$$
>
> **Think about It** The molar mass of the gas is 16.0 g/mol, a fairly small molar mass and consistent with methane, a principal component in the mixture of gases emitted by decomposing solid waste.
>
> **Practice Exercise** When $HCl(aq)$ and $NaHCO_3(aq)$ are mixed together, a chemical reaction takes place in which a gas is one of the products. A sample of the gas has a density of 1.81 g/L at 1.00 atm and 23.0°C. What is the molar mass of the gas? Can you identify the gas?

6.7 Dalton's Law and Mixtures of Gases

▶❙❙ **CHEMTOUR** Dalton's Law

As noted in Table 6.1, air is mostly N_2 and O_2 plus smaller quantities of other gases. Each gas in air, and in any other gas mixture, exerts its own pressure, called a **partial pressure**. Atmospheric pressure is the sum of the partial pressures of all of the gases in the air:

$$P_{atm} = P_{N_2} + P_{O_2} + P_{Ar} + P_{CO_2} + \dots$$

partial pressure the contribution to the total pressure made by a component in a mixture of gases.

Dalton's law of partial pressures the total pressure of any mixture of gases equals the sum of the partial pressures of all the gases in the mixture.

A similar expression can be written for any mixture of gases:

$$P_{total} = P_1 + P_2 + P_3 + P_4 + \dots \qquad (6.23)$$

which is a mathematical expression of **Dalton's law of partial pressures**: the total pressure of any mixture of gases equals the sum of the partial pressures of the gases in the mixture.

You might guess that the most abundant gases in a mixture have the greatest partial pressures and contribute the most to the total pressure of the mixture. The mathematical term used to express the abundance of each component x is its **mole fraction (χ_x)**,

$$\chi_x = \frac{n_x}{n_{total}} \tag{6.24}$$

where n_x is the number of moles of the component and n_{total} is the sum of the number of moles of all the components of the mixture.

The concentration scale using mole fractions has three characteristics worth noting: (a) unlike molarity, mole fractions have no units; (b) unlike molarity, mole fractions are based on number of moles and can be used for any kind of mixture or solution (solid, liquid, or gas); and (c) the mole fractions of all the components in a mixture add up to exactly 1.

To see how mole fractions work, consider a sample of the atmosphere (air) that contains 100.0 moles of atmospheric gases. Making up this 100.0 mol total are 21.0 mol of O_2 and 78.1 mol of N_2. The mole fraction of O_2 in the air sample is thus

$$\chi_{O_2} = \frac{n_{O_2}}{n_{total}} = \frac{21.0}{100.0} = 0.210$$

Similarly, the mole fraction of N_2 is

$$\chi_{N_2} = \frac{n_{N_2}}{n_{total}} = \frac{78.1}{100.0} = 0.781$$

Note that mole fraction is a concentration scale based on number of moles of material present. In the example, the numbers of moles indicate there are more nitrogen molecules in air than molecules of oxygen:

N_2 78.1 m̶o̶l̶ $\times$ (6.022 $\times$ 10^{23} molecules/m̶o̶l̶) = 4.70 $\times$ 10^{25} molecules

O_2 21.0 m̶o̶l̶ $\times$ (6.022 $\times$ 10^{23} molecules/m̶o̶l̶) = 1.26 $\times$ 10^{25} molecules

> ▰ CONCEPT TEST ······································
>
> The partial pressure of one component of a gas mixture cannot be directly measured. Why?
>
> ···

The partial pressure of any gas in air is the product of the mole fraction of the gas times total atmospheric pressure. If the total pressure is 1.00 atm, for instance, as it can be at sea level, the partial pressures of O_2 and N_2 are

$$P_{O_2} = \chi_{O_2} P_{total} = (0.210)(1.00 \text{ atm}) = 0.210 \text{ atm}$$
$$P_{N_2} = \chi_{N_2} P_{total} = (0.781)(1.00 \text{ atm}) = 0.781 \text{ atm}$$

These two equations are specific applications of the general equation for the partial pressure of a gas in a mixture of gases:

$$P_x = \chi_x P_{total} \tag{6.25}$$

The pressure of a gas is proportional only to the quantity of the gas in a given volume and does not depend on the identity of the gas or on whether the gas is pure or a mixture (Figure 6.31).

mole fraction (χ_x) the ratio of the number of moles of a component in a mixture to the total number of moles in the mixture.

8 mol N_2

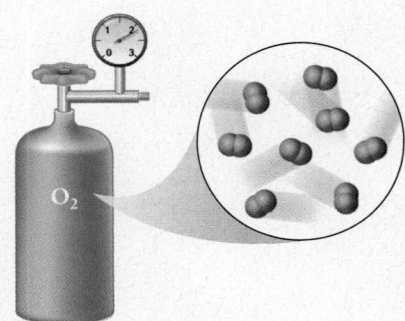

8 mol O_2

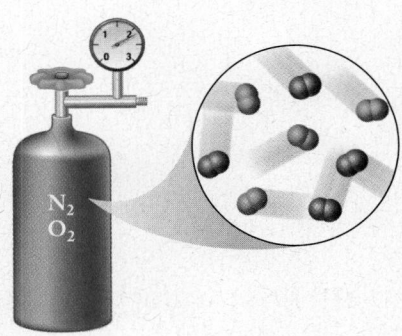

8 mol gas

FIGURE 6.31 Pressure is directly proportional to number of moles of an ideal gas, independent of the identity of the gas and independent of whether the sample is a pure gas or a mixture. In three containers that have the same volume and are at the same temperature, 8 moles of nitrogen, 8 moles of oxygen, and 8 moles of a 50:50 mixture of nitrogen and oxygen all exert the same pressure, as indicated by the gauges.

SAMPLE EXERCISE 6.11   **Calculating Mole Fraction**

Scuba divers who descend more than 45 m below the surface may breathe a gas mixture that is 11.7% He, 56.2% N_2, and 32.1% O_2 by mass. Calculate the mole fraction of each gas in this mixture.

Collect and Organize We are given the composition of a gas mixture in terms of mass percentages and asked to determine mole fractions.

Analyze If we consider a 100.0 g sample of Trimix, we can calculate the mass of each gas from the mass percentages. Then we can convert each mass to moles, from which we can determine the total number of moles in the sample and the mole fraction of each gas. Mole fraction is defined in Equation 6.24, and to calculate it we need the total number of moles of gas in the mixture and the number of moles of each component. We can estimate the answer by considering the relative molar masses of the three gases. The molar masses of nitrogen and oxygen are similar and 7 to 8 times larger, respectively, than the molar mass of helium. Because N_2 weighs less than O_2 and also is a higher percentage of the mass of the mixture than is O_2, we predict that the mole fraction of nitrogen will be larger than that of oxygen. Furthermore, we predict that the mole fractions of He, N_2, and O_2 will be quite different from the mass percentages of the three gases.

Solve In 100.0 g of this mixture we have 11.7 g of He, 56.2 g of N_2, and 32.1 g of O_2. Using molar masses to convert these masses into moles yields

$$(11.7 \ \cancel{g \ He})\left(\frac{1 \ mol \ He}{4.003 \ \cancel{g \ He}}\right) = 2.92 \ mol \ He$$

$$(56.2 \ \cancel{g \ N_2})\left(\frac{1 \ mol \ N_2}{28.02 \ \cancel{g \ N_2}}\right) = 2.01 \ mol \ N_2$$

$$(32.1 \ \cancel{g \ O_2})\left(\frac{1 \ mol \ O_2}{32.00 \ \cancel{g \ O_2}}\right) = 1.00 \ mol \ O_2$$

Mole fractions are calculated with Equation 6.24. The total number of moles in 100.0 g of Trimix is the sum of the number of moles of the constituent gases: He, N_2, and O_2.

$$n_{total} = 2.92 + 2.01 + 1.00 = 5.93 \ mol$$

Thus, X_{He}, X_{N_2}, and X_{O_2} are

$$X_{He} = \frac{2.92 \ \cancel{mol} \ He}{5.93 \ \cancel{mol}} = 0.492$$

$$X_{N_2} = \frac{2.01 \ \cancel{mol} \ N_2}{5.93 \ \cancel{mol}} = 0.339$$

$$X_{O_2} = \frac{1.00 \ \cancel{mol} \ O_2}{5.93 \ \cancel{mol}} = 0.169$$

Alternatively, we could use Equation 6.24 for all but one of the gases in the sample and then calculate the mole fraction of the final component by subtracting the sum of the calculated mole fractions from 1.00. For example, we could use the equation for He and N_2 and then determine the O_2 mole fraction by difference:

$$X_{O_2} = 1 - (X_{He} + X_{N_2}) = 1 - (0.492 + 0.339) = 0.169$$

Think about It We can check the answer by summing the mole fractions; they should add up to 1.

$$\chi_{He} + \chi_{N_2} + \chi_{O_2} = 1$$
$$0.492 + 0.339 + 0.169 = 1.00$$

Finally, as we predicted, the mole fractions of the three gases do not reflect their mass percentages. Although nitrogen is present in the greatest amount by mass in Trimix, helium has the largest mole fraction.

Practice Exercise A gas mixture called Heliox, 52.17% O_2 and 47.83% He by mass, is used in scuba tanks for descents more than 65 m below the surface. Calculate the mole fractions of He and O_2 in this mixture.

CONCEPT TEST

Do each of the gases in an equimolar mixture of four gases have the same mole fraction?

Let's consider the implications of Equation 6.25 in the context of the "thin" air at high altitudes. The mole fraction of oxygen in air is 0.210, which does not change significantly with increasing altitude. However, the total pressure of the atmosphere, and therefore the partial pressure of oxygen, decreases with increasing altitude, as illustrated in the following Sample Exercise.

SAMPLE EXERCISE 6.12   Calculating Partial Pressure

Calculate the partial pressure in atmospheres of O_2 in the air outside an airplane cruising at an altitude of 10 km, where the atmospheric pressure is 190.0 mmHg. The mole fraction of O_2 in the air is 0.210.

Collect and Organize We are given the mole fraction of oxygen and atmospheric pressure and are asked to calculate the partial pressure of oxygen.

Analyze Equation 6.25 relates the partial pressure of a gas in a gas mixture to its mole fraction in the mixture and to the total gas pressure.

Solve

$$P_{O_2} = \chi_{O_2} P_{total} = (0.210)(190.0 \text{ mmHg})\left(\frac{1 \text{ atm}}{760 \text{ mmHg}}\right)$$

$$= 0.0525 \text{ atm}$$

Think about It As noted in the text, the partial pressure of oxygen in air at sea level is 0.210 atm. As we move upward, the atmosphere becomes thinner (less dense). Our answer of 0.0525 atm makes sense because, at an altitude of 10 km, the air is much less dense than at sea level.

Practice Exercise Assume a scuba diver is working at a depth where the total pressure is 5.0 atm (about 50 m below the surface). What mole fraction of oxygen is necessary in order for the partial pressure of oxygen in the gas mixture the diver breathes to be 0.21 atm?

FIGURE 6.32 At high altitudes, atmospheric pressure is low and the partial pressure of oxygen falls below 0.21 atm, the optimum value for humans. Therefore mountaineers carry supplemental oxygen to compensate for the "thinner" air.

Sample Exercises 6.11 and 6.12 illustrate how the properties of gases make our dependence on our atmosphere so great that if we venture very far from Earth's surface, we have to take the right blend of gases with us to sustain life. The scuba diver can't breathe the same mixture of gases as we do on the surface because, at the increased partial pressure of O_2 below 30 m, O_2 can be toxic. Similarly, the higher partial pressure of N_2 can lead to nitrogen narcosis, a dangerous condition caused by a high concentration of nitrogen in the blood that leads to hallucinations. Alpine climbers on the tallest peaks on Earth experience the opposite problem. For them, the lower atmospheric pressure leads to a lower partial pressure of oxygen and makes it hard to get enough oxygen into the blood. Most mountaineers in these locations carry bottled oxygen (Figure 6.32). The lower partial pressure of oxygen at high elevation is also the reason why aircraft cabins are pressurized.

Dalton's law of partial pressures is useful in the laboratory when we need to know the pressure exerted by a gaseous product of a chemical reaction. For example, heating potassium chlorate ($KClO_3$) in the presence of MnO_2 causes it to decompose into $KCl(s)$ and $O_2(g)$. The oxygen gas produced by this reaction can be collected by bubbling the gas into an inverted bottle that is initially filled with water (Figure 6.33). As the reaction proceeds, $O_2(g)$ displaces the water in the bottle. When the reaction is complete, the volume of water displaced provides a measure of the volume of O_2 produced. If the temperature of the water and the barometric pressure are known, the ideal gas law can be used to determine the number of moles of O_2 produced.

The procedure works for any gas that neither reacts with nor dissolves appreciably in water. However, one additional step is needed to apply the ideal gas law in calculating the number of moles of gas produced. At room temperature (nominally 20°C), any enclosed space above a pool of liquid water contains some $H_2O(g)$, meaning some water vapor is in the collection flask in addition to the gas produced by the reaction. Thus in the $KClO_3$ reaction in Figure 6.33, the gas collected is a mixture of $O_2(g)$ and $H_2O(g)$. A mixture of a gas with water vapor is said to be *wet*, in contrast to a gas that contains no water vapor, which is said to be *dry*. Dalton's law of partial pressures gives us the total pressure of the mixture at the point where the water level inside the collection vessel matches the water level outside the vessel:

$$P_{total} = P_{atm} = P_{O_2} + P_{H_2O} \qquad (6.26)$$

FIGURE 6.33 (a) During the thermal decomposition of $KClO_3$, the product $O_2(g)$ is collected by displacing water from an inverted bottle initially filled with water. (b) When the reaction is over, the collection jar is lowered until the surface of the water in the jar is at the same level as the water in the large container. At that point the pressure in the jar equals atmospheric pressure. The temperature of the gas is the same as room temperature.

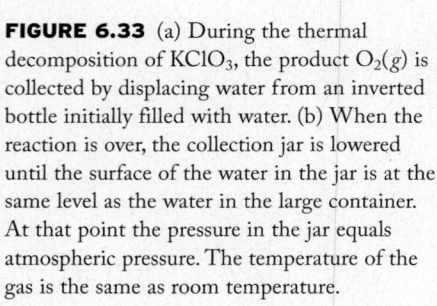

(a)

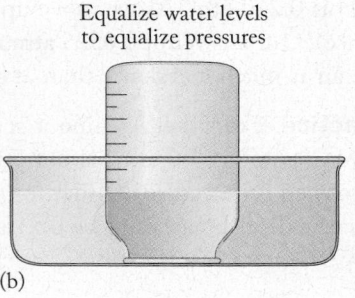

Equalize water levels to equalize pressures

(b)

To calculate the quantity of oxygen produced using the ideal gas law, we must know P_{O_2}, which we get by subtracting P_{H_2O} (Table 6.4) from P_{atm}. If we know the values of P_{O_2}, T, and V, we can calculate the number of moles or number of grams of oxygen produced.

TABLE 6.4	Partial Pressure of $H_2O(g)$ at Selected Temperatures
Temperature (°C)	**Pressure (mmHg)**
5	6.5
10	9.2
15	12.8
20	17.5
25	23.8
30	31.8
35	42.2
40	55.3
45	71.9
50	92.5

SAMPLE EXERCISE 6.13 Determining the Partial Pressure of a Gas Collected over Water

During the decomposition of $KClO_3$, 92.0 mL of gas is collected by the displacement of water at 25.0°C. If atmospheric pressure is 756 mmHg, what mass of O_2 is collected?

Collect and Organize We are given the atmospheric pressure, the volume of oxygen, and the temperature, and asked to calculate the mass of oxygen collected by water displacement.

Analyze A quick estimation can give us an idea about the value of our final answer. The molar volume of oxygen is about 20 L at 25.0°C; if the oxygen were dry, about 100 mL would be collected, or $(0.100 \text{ L})/(20 \text{ L/mol}) = 0.005$ mol of oxygen. Considering that oxygen has a molar mass of 32.00 g, 0.005 mol is about 0.16 g.

When a gas is collected over water, however, the collection vessel contains both water vapor and the gas of interest. According to Equation 6.26, we need the partial pressure of the $H_2O(g)$ at the temperature of the apparatus to determine the partial pressure due to O_2 in the collection flask. We are not given this pressure for the $H_2O(g)$ but can get it from Table 6.4. The presence of water vapor in the gas that we collect means that the partial pressure of the gas being collected is less than the total pressure in the vessel. This also means that our estimated value for the mass of oxygen based on the entire volume of gas being O_2 is a high estimate.

Solve Table 6.4 indicates that P_{H_2O} at 25°C is 23.8 mmHg. To calculate P_{O_2} in the collected gas, we subtract this value from P_{total}:

$$P_{O_2} = P_{total} - P_{H_2O} = (756 - 23.8) \text{ mmHg}$$
$$= 732 \text{ mmHg}$$

We can use the ideal gas equation to calculate the moles of O_2 produced; recall that we need to convert P_{O_2} to atmospheres, V to liters, and T to kelvins, and that we may express n as grams of O_2 per 32.00 g/mol:

$$732 \text{ mmHg} \times \frac{1 \text{ atm}}{760 \text{ mmHg}} = 0.963 \text{ atm}$$

$$92.0 \text{ mL} \times \frac{10^{-3} \text{ L}}{1 \text{ mL}} = 0.0920 \text{ L}$$

$$25.0°C = 298 \text{ K}$$

$$n = \frac{PV}{RT} = \frac{(0.963 \text{ atm})(0.0920 \text{ L})}{[0.08206 \text{ L} \cdot \text{atm/mol} \cdot \text{K}](298 \text{ K})} = 0.00362 \text{ mol}$$

$$m_{O_2} = 0.00362 \text{ mol} \times \frac{32.00 \text{ g}}{1 \text{ mol}} = 0.116 \text{ g}$$

Think about It First, the pressure of the $O_2(g)$ is slightly less than P_{total}, as we predicted. Our answer (0.116 g) is close enough to our estimate that we can be confident our answer is reasonable.

Practice Exercise Electrical energy can be used to separate water into $O_2(g)$ and $H_2(g)$. In one demonstration of this reaction, 27 mL of H_2 is collected over water at 25°C. Atmospheric pressure is 761 mmHg. How many grams of H_2 are collected?

6.8 The Kinetic Molecular Theory of Gases and Graham's Law

The fundamental discoveries of Boyle, Charles, and Amontons all took place before scientists recognized the existence of molecules. Only Avogadro, because he lived after Dalton, had the advantage of knowing Dalton's proposal that matter was composed of tiny particles called atoms. Thus Avogadro recognized that the volume of a gas was directly proportional to the number of particles (moles) of gas. Still, Avogadro's law, $V/n =$ constant (Equation 6.10) does not explain why 1 mole of relatively small $He(g)$ atoms at STP has the same volume as 1 mole of, for example, much larger $SF_6(g)$ molecules. Dalton's law of partial pressures (Equation 6.23) does not explain *why* each gas in a mixture contributes a partial pressure based on its mole fraction. Similarly, Boyle's, Charles's, and Amontons's laws demand explanations for why pressure and volume are inversely proportional to each other but both directly proportional to temperature.

In this section we examine the **kinetic molecular theory** of gases, a unifying theory developed in the late 19th century that explains the relationship described by the ideal gas law and its predecessors, the laws of Boyle, Charles, Dalton, Amontons, and Avogadro.

The main assumptions of the kinetic molecular theory are:

1. Gas molecules have tiny volumes compared with the collective volume they occupy. Their individual volumes are so small as to be considered negligible, allowing particles in a gas to be treated as *point masses*— masses with essentially no volume. Gas molecules are separated by large distances; hence a gas is mostly empty space.
2. Gas molecules move constantly and randomly throughout the volume they collectively occupy.
3. The motion of these molecules is associated with an average kinetic energy that is proportional to the absolute temperature of the gas. All populations of gas molecules at the same temperature have the same average kinetic energy.
4. Gas molecules continually collide with one another and with their container walls. These collisions are *elastic;* that is, they result in no net transfer of energy to the walls. Therefore, the average kinetic energy of gas molecules is not affected by these collisions and remains constant as long as there is no change in temperature.
5. Each gas molecule acts independently of all other molecules in a sample. We assume there are no forces of attraction or repulsion between the molecules.

To illustrate and test the kinetic molecular theory, let's consider the examples provided by air and by the $N_2/O_2/He$ mixture in a scuba tank.

Explaining Boyle's, Dalton's, and Avogadro's Laws

We already have a picture in our minds that describes the origin of gas pressure. Every collision between a gas molecule and a wall of its container generates a force. The more frequent the collisions, the greater the force and thus the greater the pressure. Compressing molecules into a smaller space by reducing the volume of the container means more collisions take place per unit time, and the pressure increases (Boyle's law: $P \propto 1/V$). Because $P \propto n$, we can also increase the num-

kinetic molecular theory a model that describes the behavior of gases; all equations defining relationships between pressure, volume, temperature, and number of moles of gases can be derived from the theory.

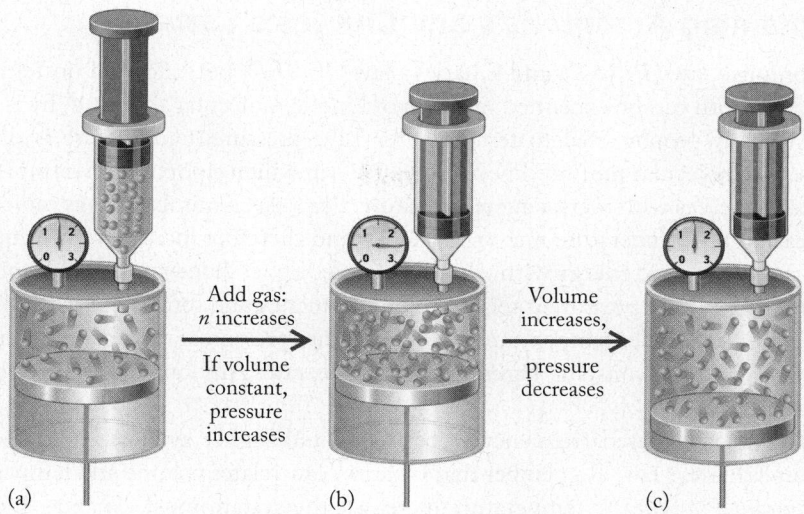

FIGURE 6.34 (a) Gas contained in a cylinder fitted with a movable piston exerts a pressure equal to atmospheric pressure ($P_{gas} = P_{atm}$). (b) When more gas is added to the cylinder (n increases), P_{gas} increases because the number of collisions with the walls of the container increases. (c) In order for the pressure to remain the same as in part a, the volume must increase.

ber of collisions by increasing the number of gas molecules (number of moles, n) in a container of fixed volume. We can increase n either by adding more of the same gas to a container of fixed volume or by adding some quantity of a different gas to the container. If we add two different gases, such as N_2 and O_2, to a container in a 4:1 mole ratio, then the number of collisions involving nitrogen molecules and a container wall will be greater than the number involving oxygen molecules. Therefore, the pressure due to nitrogen (P_{N_2}) will be greater than that due to oxygen (P_{O_2}), and the total pressure will be $P_{total} = P_{O_2} + P_{N_2}$. This statement is precisely what Dalton proposed in his law of partial pressures.

Avogadro's law ($V \propto n$) is also explained by the kinetic molecular theory. Avogadro observed that the volume of a gas *at constant pressure* is directly proportional to the number of moles of the gas. Because additional gas molecules in a given volume lead to increased numbers of collisions and an increase in pressure, the only way to reduce the pressure is to allow the gas to expand (Figure 6.34).

CONCEPT TEST ....

If the collisions between molecules and the walls of the container were not elastic and energy was lost to the walls, would the pressure of the gas be higher or lower than that predicted by the ideal gas law?

....

CONCEPT TEST ....

Which of the graphs in Figure 6.35 best represents the relationship between number of collisions and pressure?

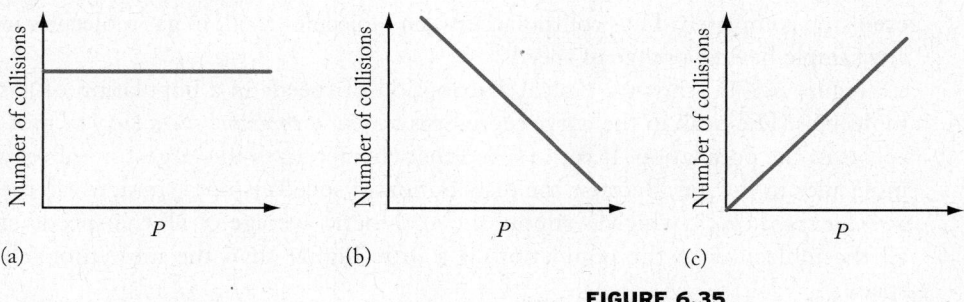
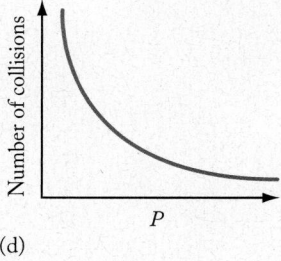

FIGURE 6.35

Explaining Amontons's and Charles's Laws

Amontons's law ($P \propto T$) and Charles's law ($V \propto T$) both depend on temperature, and both can be explained in terms of kinetic molecular theory. Why is pressure directly proportional to temperature? Like pressure, temperature is directly related to molecular motion. The average speed at which a population of molecules moves increases with increasing temperature. For a given number of gas molecules, increasing the temperature increases velocity and therefore increases the frequency and average kinetic energy with which the molecules collide with the walls of their container. Because pressure is related to the frequency and force of these collisions, it follows that higher temperatures produce higher pressures, as long as volume and quantity of gas are constant. Figure 6.36(a, b) illustrates Amontons's law on a molecular level.

How do increased frequency of collisions and higher average kinetic energy explain Charles's law? Remember that Charles's law relates volume and temperature *at constant P and n*. As temperature increases, the system must expand—increase its volume—in order to maintain a constant pressure (Figure 6.36a, c).

FIGURE 6.36 (a) As the temperature of a given amount of gas in a cylinder increases while the volume is held constant, the number and forcefulness of molecular collisions with the walls of the container increase, causing (b) the pressure to increase, as stated by Amontons's law. For the pressure to remain constant as the temperature of the gas increases, (c) the volume must increase until $P_{gas} = P_{atm}$. Hence, as temperature increases at constant pressure, volume increases, as stated by Charles's law.

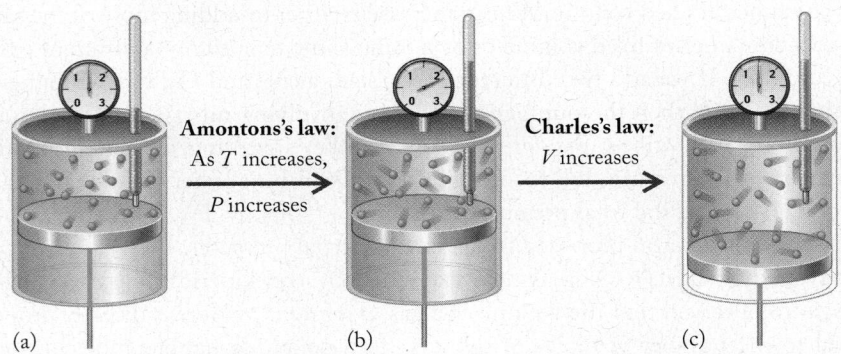

Amontons's law:
As T increases,
P increases

Charles's law:
V increases

(a)   (b)   (c)

Molecular Speeds and Kinetic Energy

Kinetic molecular theory tells us that all populations of gas particles at a given temperature have the same *average* kinetic energy. The kinetic energy of a single molecule or atom of a gas can be calculated by using the equation

$$\text{KE} = \tfrac{1}{2}\,mu^2$$

where m is the mass of a molecule of the gas and u is its speed. At any given moment, however, not all gas molecules in a population are traveling at exactly the same speed. Even elastic collisions between two molecules may result in one of them moving with a greater speed than the other after the collision. One molecule might even stop completely. Thus collisions between molecules result in gas molecules in any sample having a range of speeds.

Figure 6.37(a) shows a typical distribution of speeds in a population of gas molecules. The peak in the curve represents the *most probable speed* (u_m) of molecules in the population. It is the speed that characterizes the largest number of molecules in the gas. Because the distribution of speeds is not symmetrical, the *average speed* (u_{avg}), which is simply the arithmetic average of all the speeds of all the molecules in the population, is a little higher than the most probable speed.

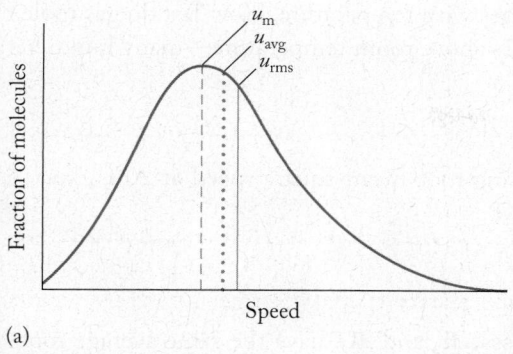

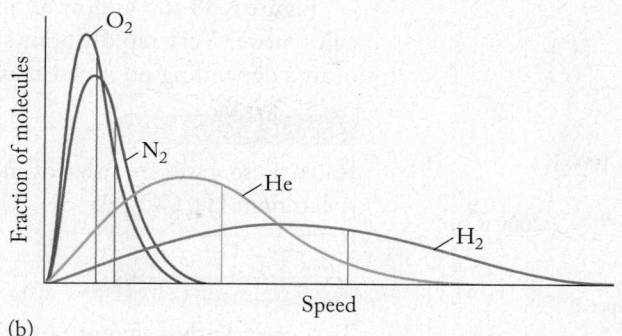

FIGURE 6.37 At any given temperature, the speeds of gas molecules cover a range of values. (a) The most probable speed (u_m, dashed line), at the highest point on the curve, is the speed of the largest fraction of molecules in the population; in other words, more molecules have the most probable speed than any other speed. The average speed (u_{avg}, dotted line), a little higher than the most probable speed, is the arithmetic average of all the speeds. The root-mean-square speed (u_{rms}, solid line), a little higher than the average speed, is directly proportional to the square root of the absolute temperature of the gas and inversely proportional to the square root of its molar mass. (b) Molecular speed distributions for samples of oxygen, nitrogen, helium, and hydrogen gas at the same temperature. On each curve, the vertical solid line is the u_{rms}. The lower the molar mass of the molecules, the higher is their root-mean-square speed (u_{rms}) and the broader the distribution of speeds. Having the lowest molar mass of the four gases, the H_2 molecules have the highest u_{rms} and the broadest curve. Having the highest molar mass in the group, the O_2 molecules have the lowest u_{rms} and the narrowest curve.

A very important value is the **root-mean-square speed (u_{rms})**; this is the speed of a molecule possessing the average kinetic energy. The absolute temperature of a gas is a measure of the average kinetic energy of the population of gas molecules. At a given temperature, the population of molecules in a gas has the same *average* kinetic energy as every other population of gas molecules at that same temperature, and this *average* kinetic energy is defined as

$$KE_{avg} = \tfrac{1}{2}\, m(u_{rms})^2 \qquad (6.27)$$

The root-mean-square speed (u_{rms}) of a gas with molar mass $\mathcal{M}$ at temperature T is defined as

$$u_{rms} = \sqrt{\frac{3RT}{\mathcal{M}}} \qquad (6.28)$$

Because different gases have different molar masses, this equation indicates that more massive molecules move more slowly at a given temperature than lighter molecules. Figure 6.37(b) shows the different distributions of speeds for several gases at a constant temperature.

Because u_{rms} is typically expressed in meters per second, care must be taken in choosing the units for R in Equation 6.28. One value of R (Table 6.3) that has meters in its units is

$$R = 8.314 \ kg \cdot m^2/(s^2 \cdot mol \cdot K)$$

and this is the most convenient value to use when working with Equation 6.28. Using this value for R requires that we express molar mass in kilograms per mole rather than in the more common grams per mole.

▶❚❚ **CHEMTOUR** Molecular Speed

root-mean-square speed (u_{rms}) the square root of the average of the squared speeds of all the molecules in a population of gas molecules; a molecule possessing the average kinetic energy moves at this speed.

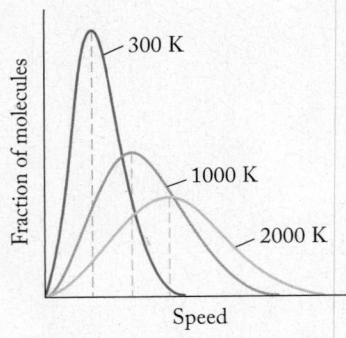

FIGURE 6.38 The most probable speed (u_m, dashed lines) increases with increasing temperature. Notice that the distributions broaden as the temperature increases: a smaller fraction of the molecules moves at any given speed, and more speeds are represented by a significant fraction of the population.

Figure 6.38 shows how u_m increases with temperature. How fast do gas molecules move? Very rapidly at or slightly above room temperature—many hundreds of m/s depending on molar mass.

CONCEPT TEST

Rank these gases in order of increasing root-mean-square speed at 20°C, lowest speed first: H_2, CO_2, Ar, SF_6, UF_6, Kr.

CONCEPT TEST

Two gases with different molar masses, $\mathcal{M}_1$ and $\mathcal{M}_2$, have the same average root-mean-square speed. If $\mathcal{M}_1 > \mathcal{M}_2$, which gas is at the higher temperature?

SAMPLE EXERCISE 6.14   Calculating Root-Mean-Square Speeds

Calculate the root-mean-square speed of nitrogen molecules at 300.0 K in meters per second and miles per hour.

Collect and Organize Given only the temperature of a sample of nitrogen gas, we are asked to calculate the root-mean-square speed of the molecules in the sample. According to Equation 6.28, we need the molar mass of the gas in addition to the temperature. We also need $R = 8.314$ kg·m²/(s²·mol·K) because it gives us a speed in meters per second.

Analyze Nitrogen gas (N_2) has a molar mass of 28.02 g/mol. Having chosen the value of R, we need to express the molar mass in kilograms. We can estimate the magnitude of the answer by considering the relative magnitudes of R, T, and molar mass. When T is expressed in kelvins, it is on the order of 10^2 while the correct value of R is approximately 10. The molar mass of nitrogen expressed as kilograms per mole is a small number, on the order of 10^{-2} kg/mol. When these values are substituted into equation 6.28, the answer will be a large number, even after taking the square root: $(10 \times 10^2/10^{-2})^{1/2} \approx 10^2$ to 10^3 m/s.

Solve

$$u_{\text{rms},N_2} = \sqrt{\frac{3RT}{\mathcal{M}}} = \sqrt{\frac{3\left(\dfrac{8.314 \, \cancel{\text{kg}} \cdot \text{m}^2}{\text{s}^2 \cdot \cancel{\text{mol}} \cdot \cancel{\text{K}}}\right)(300.0 \, \cancel{\text{K}})}{0.02802 \, \dfrac{\cancel{\text{kg}}}{1 \, \cancel{\text{mol}}}}}$$

$$= 516.8 \text{ m/s} = 5.168 \times 10^2 \text{ m/s}$$

$$u_{\text{rms},N_2} = \left(516.8 \, \frac{\cancel{\text{m}}}{\cancel{\text{s}}}\right)\left(\frac{1 \text{ mi}}{1.6093 \times 10^3 \, \cancel{\text{m}}}\right)\left(3600 \, \frac{\cancel{\text{s}}}{\text{hr}}\right)$$

$$= 1156 \text{ mi/hr} = 1.156 \times 10^3 \text{ mi/hr}$$

Think about It The average speed of a nitrogen molecule does indeed fall between 10^2 and 10^3 m/s as we predicted. The relatively small molar mass contributes to the large root-mean-square speed.

Practice Exercise Calculate the root-mean-square speed of helium at 300 K in meters per second, and compare your result with the root-mean-square speed of nitrogen calculated in Sample Exercise 6.14.

It is not immediately obvious why the pressure exerted by two different gases is the same at a given temperature when their root-mean-square speeds are quite different. The answer lies in the assumption in the kinetic molecular theory that all populations of gas molecules have the same average kinetic energy at a given temperature, independent of their molar mass. To examine this issue, let's calculate the average kinetic energy of 1 mole of N_2 and 1 mole of He at 300 K. The solution to Sample Exercise 6.14 reveals that the u_{rms} of N_2 molecules at 300 K is 517 m/s. The root-mean-square speed of He atoms is 1370 m/s (the result for the Practice Exercise of Sample Exercise 6.14). Using these values of speed and the mass of 1 mole of each gas ($m = \mathcal{M}$) in Equation 6.27 gives

$$KE_{N_2} = \tfrac{1}{2}\mathcal{M}_{N_2}(u_{rms,N_2})^2 = \tfrac{1}{2}(2.802 \times 10^{-2}\text{ kg})(517\text{ m/s})^2$$

$$= 3.74 \times 10^3 \text{ kg} \cdot \text{m}^2/\text{s}^2$$

$$KE_{He} = \tfrac{1}{2}\mathcal{M}_{He}(u_{rms,He})^2 = \tfrac{1}{2}(4.003 \times 10^{-3}\text{ kg})(1370\text{ m/s})^2$$

$$= 3.76 \times 10^3 \text{ kg} \cdot \text{m}^2/\text{s}^2$$

This calculation demonstrates that the same quantities of two *different* gases have essentially the same average kinetic energy (when the correct number of significant figures are used). Therefore at the same temperature, they exert the same pressure. The slower N_2 molecules collide with the container walls less often, but because they are more massive than He atoms, the N_2 molecules exert a greater force during each collision.

Equation 6.28 enables us to compare the relative root-mean-square speeds of two gases at the same temperature. Consider an equimolar mixture of $N_2(g)$ and $He(g)$. We have just demonstrated that at any given temperature their average kinetic energies are the same:

$$KE_{N_2} = \tfrac{1}{2}m_{N_2}(u_{rms,N_2})^2 = \tfrac{1}{2}m_{He}(u_{rms,He})^2 = KE_{He}$$

or

$$m_{N_2}(u_{rms,N_2})^2 = m_{He}(u_{rms,He})^2$$

Rearranging this equation to express the ratio of the root-mean-square speeds in terms of the ratio of the molar masses and then taking the square root of each side, we get

$$\frac{(u_{rms,He})^2}{(u_{rms,N_2})^2} = \frac{\mathcal{M}_{N_2}}{\mathcal{M}_{He}}$$

$$\frac{u_{rms,He}}{u_{rms,N_2}} = \sqrt{\frac{\mathcal{M}_{N_2}}{\mathcal{M}_{He}}} = \sqrt{\frac{28.02\text{ g}}{4.003\text{ g}}} = 2.650$$

The root-mean-square speed of helium atoms at 300 K is 2.65 times higher than that of nitrogen molecules. We can test this using the values for u_{rms} we used above to calculate KE_{N_2} and KE_{He}:

$$\frac{u_{rms,He}}{u_{rms,N_2}} = \frac{1370\text{ m/s}}{517\text{ m/s}} = 2.65$$

The relation between root-mean-square speeds and molar masses applies to any pair of gases x and y:

$$\frac{u_{rms,x}}{u_{rms,y}} = \sqrt{\frac{\mathcal{M}_y}{\mathcal{M}_x}} \qquad (6.29)$$

This equation leads to the conclusion that the root-mean-square speed of more massive particles is lower than the root-mean-square speed of lighter particles.

CONCEPT TEST ··

Since root-mean-square speed depends on temperature, why doesn't the ratio of the u_{rms} of two gases change with temperature?

Graham's Law: Effusion and Diffusion

Let's consider what happens to the two balloons in Figure 6.39, one filled with nitrogen and the other with helium. The volume, temperature, and pressure of the gas are identical in the two balloons, and the pressure inside each balloon is greater than atmospheric pressure. Over time, the volume of the helium balloon decreases significantly but the volume of the nitrogen balloon does not.

The skin of any balloon is slightly permeable; that is, it has microscopic holes that allow gas to escape, reducing the pressure inside the balloon. Why does the helium leak out faster than the nitrogen? We now know the relation between root-mean-square speeds and molar masses of two gases, and we have calculated that helium atoms move about 2.65 times faster than nitrogen atoms at the same temperature. The gas with the greater root-mean-square speed (He) leaks out of the balloon at a higher rate. This process of moving through a small opening from a higher-pressure region into a lower-pressure region is called **effusion**.

In the 19th century, Scottish chemist Thomas Graham (1805–1869) recognized that the effusion rate of a gas is related to its molar mass. Today this relation is known as **Graham's law of effusion**, and it states that the effusion rate of any gas is inversely proportional to the square root of its molar mass. We can derive a mathematical representation of Graham's law starting from Equation 6.28 for two gases x and y:

$$u_{rms,x} = \sqrt{\frac{3RT}{\mathcal{M}_x}}$$

$$u_{rms,y} = \sqrt{\frac{3RT}{\mathcal{M}_y}}$$

by substituting rate (r) for root-mean-square speed in Equation 6.30:

$$\frac{r_x}{r_y} = \frac{u_{rms,x}}{u_{rms,y}} = \frac{\sqrt{\dfrac{3RT}{\mathcal{M}_x}}}{\sqrt{\dfrac{3RT}{\mathcal{M}_y}}} \tag{6.30}$$

which simplifies to the usual form of Graham's law:

$$\frac{r_x}{r_y} = \sqrt{\frac{\mathcal{M}_y}{\mathcal{M}_x}} \tag{6.31}$$

Using Equation 6.31 for the helium and nitrogen balloons, we see that the effusion rate of helium is 2.65 times that of nitrogen:

$$\frac{r_{He}}{r_{N_2}} = \sqrt{\frac{\mathcal{M}_{N_2}}{\mathcal{M}_{He}}} = \sqrt{\frac{28.02 \ \cancel{g/mol}}{4.003 \ \cancel{g/mol}}} = 2.650$$

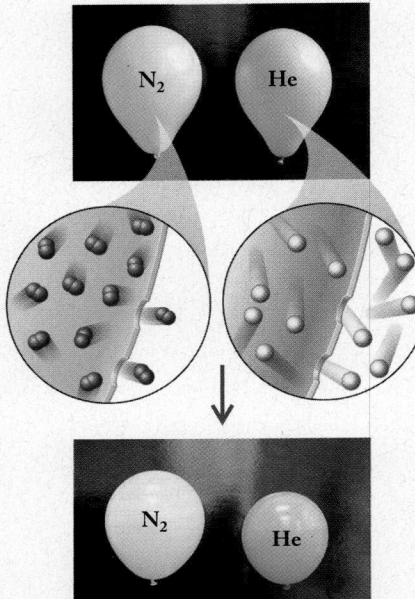

FIGURE 6.39 Two balloons at the same temperature and pressure, one filled with nitrogen gas and the other filled with an equal volume of helium gas. Over time, the volume of the helium balloon decreases much more than the nitrogen balloon. The helium atoms are lighter than the nitrogen molecules and therefore have higher root-mean-square speeds. As a consequence, the helium atoms escape from the balloon faster, causing the helium-containing balloon to shrink more quickly than the nitrogen balloon.

Notice that this expression has the same form—and hence the ratio has the same value—as the ratio of the root-mean-square speeds of the N_2 and He molecules.

Two things about Equation 6.31 are worth noting: (1) like the root-mean-square speed in Equation 6.29, Graham's law involves a square root. (2) With labels x and y for two gases in a mixture, the left side of Equation 6.31 has the x term in the numerator and the y term in the denominator, while the right side of the equation has the opposite: the y term is in the numerator and x term in the denominator. Keeping the labels straight is the key to solving problems involving Graham's law.

How does the kinetic molecular theory explain Graham's law of effusion? The escape of a gas molecule from a balloon requires that the molecules encounter one of the microscopic holes in the balloon. The faster a gas molecule moves, the more likely it is to find one of these holes.

Effusion of gas molecules is related to **diffusion**, which is the spread of one substance through another. If we focus our attention on gases, then diffusion contributes to the passage of odors such as a perfume throughout a room and the smell of baking bread throughout a house. The rates of diffusion of gases depend on the average speeds of the gas molecules and on their molar masses. Like effusion, diffusion of gases is described by Graham's law.

effusion the process by which a gas escapes from its container through a tiny hole into a region of lower pressure.

Graham's law of effusion the rate of effusion of a gas is inversely proportional to the square root of its molar mass.

diffusion the spread of one substance (usually a gas or liquid) through another.

SAMPLE EXERCISE 6.15 **Applying Graham's Law to Diffusion of a Gas**

An odorous gas emitted by a hot spring was found to diffuse 2.92 times slower than helium. What is the molar mass of the emitted gas?

Collect and Organize We are asked to determine the molar mass of an unknown gas based on its rate of diffusion relative to the rate for helium.

Analyze Graham's law (Equation 6.31) relates the relative rate of diffusion of two gases to their molar masses. We know that helium ($\mathcal{M}_{He} = 4.003$ g/mol) diffuses 2.92 times faster than the unidentified gas (y). Because helium diffuses faster than the unknown gas, we predict that the unidentified gas has a larger molar mass. In fact, it should be about 9 (or 3^2) times the mass of He based on Graham's law, or about 36 g/mol, because the ratio of the molar masses will equal the square of the ratio of the diffusion rates.

Solve Rearranging Equation 6.31 we obtain an expression for $\mathcal{M}_y$:

$$\mathcal{M}_{He}\left(\frac{r_{He}}{r_y}\right)^2 = \mathcal{M}_y$$

Substituting for the ratio of the diffusion rates and the mass of one mole of helium gives

$$\mathcal{M}_y = \mathcal{M}_{He}\left(\frac{r_{He}}{r_y}\right)^2 = (4.003 \text{ g/mol})(2.92)^2 = 34.1 \text{ g/mol}$$

Think about It The molar mass of the unidentified gas is 34.1 g/mol, which is consistent with our prediction. One possibility for the identity of this gas is $H_2S(g)$, $\mathcal{M} = 34.08$ g/mol, a foul-smelling and toxic gas frequently emitted from volcanoes and also responsible for the odor of rotten eggs.

Practice Exercise Helium effuses 3.16 times as fast as which other noble gas?

The rate of effusion of a gas increases with temperature. Which graph in Figure 6.40 best describes the ratio of the rates of effusion of two gases (x and y) as a function of temperature?

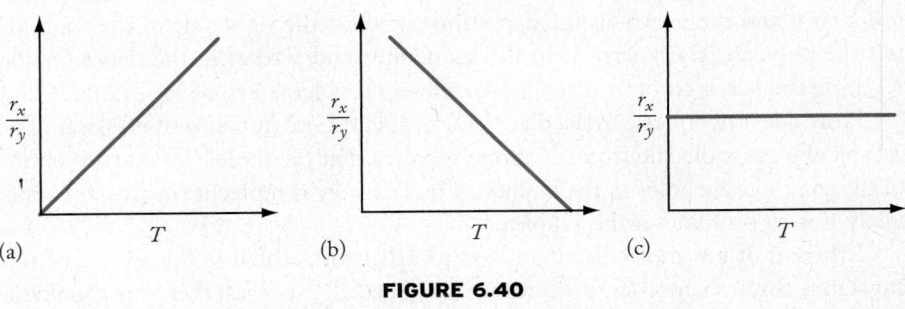

FIGURE 6.40

6.9 Real Gases

Up to now we have treated all gases as ideal, and this is acceptable because, under typical atmospheric pressures and temperatures, most gases *do* behave ideally. We have also assumed, according to kinetic molecular theory, that the volume occupied by individual gas molecules is negligible compared with the total volume occupied by the gas. In addition, we have assumed that no interactions occur between gas molecules other than random elastic collisions. These assumptions are not valid, however, when we begin to compress gases into increasingly smaller volumes.

Deviations from Ideality

Let's start by considering the behavior of 1.0 mole of a gas as we increase the pressure on the gas. From the ideal gas law, we know that $PV/RT = n$, so for 1 mole of gas, PV/RT should remain equal to 1.0 regardless of how we change the pressure. This relationship between PV/RT and P for an ideal gas is shown by the purple line in Figure 6.41. However, the curves for PV/RT versus P for CH_4, H_2, and CO_2 at pressures above 10 atm are not horizontal straight lines like this ideal curve. Not only do the curves diverge from the ideal, but the shapes of the curves also differ for each gas, indicating that when we are dealing with real gases rather than ideal ones, the identity of the gas does matter.

Why don't real gases behave like ideal gases at high pressure? One reason is that the ideal gas law considers gas molecules to have so little volume compared

FIGURE 6.41 The effect of pressure on the behavior of real and ideal gases. The curves diverge from ideal behavior in a manner unique to each gas. When gases deviate from ideal behavior, their identity matters.

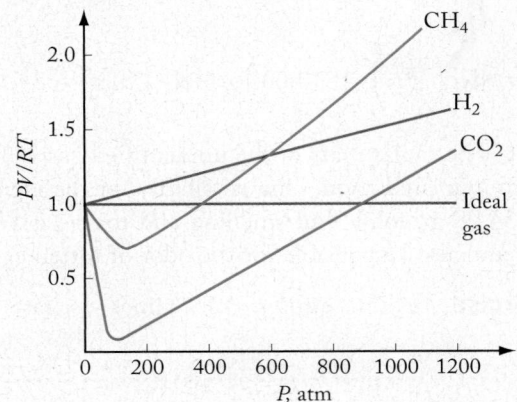

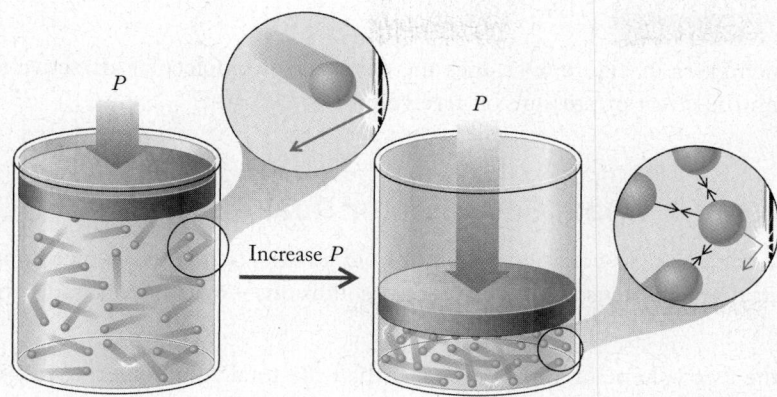

FIGURE 6.42 At high pressures, a greater fraction of the volume of a gas is occupied by the gas molecules. The greater density of molecules also results in more interactions between them (black arrows in expanded view). These interactions reduce the frequency and force of collisions with the walls of the container (red arrow), thereby reducing the pressure.

with the volume of their container that they are assumed to have no volume at all. However, at high pressures, more molecules are squeezed into a given volume (Figure 6.42). Under these conditions the volume occupied by the molecules can become significant. What is the impact of this on the graph of PV/RT versus P?

In PV/RT, the V actually refers to the *free volume* ($V_{\text{free volume}}$), the empty space not occupied by gas molecules. Because $V_{\text{free volume}}$ is difficult to measure, we instead measure V_{total}, the total volume of the container holding the gas:

$$V_{\text{total}} = V_{\text{free volume}} + V_{\text{molecules}}$$

In an ideal gas and in a real gas at low pressure, $V_{\text{free volume}} \gg V_{\text{molecules}}$ so $V_{\text{molecules}}$ is essentially 0, and we can use the ideal gas equation with confidence. Consider, however, what happens to a real gas in a closed but flexible container as we increase the external pressure, causing the container—and hence the volume occupied by the gas—to shrink. As the external pressure increases, the assumption that $V_{\text{molecules}} = 0$ becomes less and less valid because the proportion of the container's volume taken up by the molecules increases. The relation $V_{\text{total}} = V_{\text{free volume}} + V_{\text{molecules}}$ still applies, but now $V_{\text{molecules}} > 0$, which means $V_{\text{free volume}}$ can no longer be approximated by V_{total}. Since $V_{\text{total}} > V_{\text{free volume}}$, the ratio PV/RT is also larger for a real gas than for an ideal gas:

$$\frac{PV_{\text{total}}}{RT} > \frac{PV_{\text{free volume}}}{RT}$$

As a consequence, the curve for a real gas in Figure 6.41 diverges upward from the line for an ideal gas.

A second situation also causes the ratio PV/RT to diverge from 1. Kinetic molecular theory assumes that molecules do not interact, but real molecules do attract one another. The attractive forces arise because of imbalances in the distribution of electrons. These attractive forces function over short distances, so the assumption that molecules behave independently is a good one as long as the molecules are far apart. As pressure increases on a population of gas molecules and they are pushed closer and closer together, intermolecular attractive forces can become significant. This causes the molecules to associate with one another, which decreases the force of their collisions with the walls of their container, thereby decreasing the pressure exerted by the gas. If the value of P in PV/RT is smaller in the real gas than the ideal value, the value of the ratio decreases

$$\frac{P_{\text{real}}V}{RT} < \frac{P_{\text{ideal}}V}{RT}$$

and the curve for a real gas diverges from the ideal gas line by moving below it on the graph.

TABLE 6.5 Van der Waals Constants of Selected Gases

Substance	a (L$^2 \cdot$ atm/mol^2)	b (L/mol)
He	0.0341	0.02370
Ar	1.34	0.0322
H$_2$	0.244	0.0266
N$_2$	1.39	0.0391
O$_2$	1.36	0.0318
CH$_4$	2.25	0.0428
CO$_2$	3.59	0.0427
CO	1.45	0.0395
H$_2$O	5.46	0.0305
NO	1.34	0.02789
NO$_2$	5.28	0.04424
HCl	3.67	0.04081
SO$_2$	6.71	0.05636

CONCEPT TEST

For which gases in Figure 6.41 does the effect of intermolecular attractive forces outweigh the effect of pressure on free volume at 200 atm?

The van der Waals Equation for Real Gases

Because the ideal gas equation does not hold at high pressures, we need another equation that can be used under nonideal conditions—one that accounts for the fact that:

1. the free volume of a real gas is less than the total volume because its molecules occupy significant space; and
2. the observed pressure is less than the pressure of an ideal gas because of intermolecular attractions.

The **van der Waals equation**

$$\left(P + \frac{n^2 a}{V^2}\right)(V - nb) = nRT \qquad (6.32)$$

includes terms to correct for pressure ($n^2 a / V^2$) and volume (nb). The values of a and b, called *van der Waals constants*, have been determined experimentally for many gases (Table 6.5). Both a and b increase with increasing molar mass and with the number of atoms in each molecule of a gas.

SAMPLE EXERCISE 6.16   Calculating Pressure with the van der Waals Equation

Calculate the pressure of 1.00 mol of N$_2$ in a 1.00 L container at 300.0 K using first the van der Waals equation and then the ideal gas equation.

Collect and Organize We are given the amount of nitrogen, its volume, and its temperature, and are to calculate its pressure using the van der Waals equation and the ideal gas equation. Using the van der Waals equation means we need to know the values of the van der Waals constants for nitrogen.

Analyze Table 6.5 lists the van der Waals constants for nitrogen gas as $a = 1.39$ L$^2 \cdot$ atm/mol^2 and $b = 0.0391$ L/mol. The values of a and b represent experimentally determined corrections for the interactions between molecules and the portion of the total volume occupied by the gas molecules. One mole of ideal gas at STP occupies 22.4 L. Compressing the gas to a volume of about 1/20 of the initial volume will require a pressure of about 20 atm. Heating it to 300 K at constant volume will cause a relatively minor change in the pressure. We can estimate the effect of a pressure of about 20 atm on the ideality of the nitrogen from the curves in Figure 6.41. While we don't know which curve best describes the behavior of N$_2$, the scale of the x-axis is rather large. At a pressure of 20 atm, none of the curves deviate substantially from ideality, so we expect a relatively small difference in the pressure calculated for N$_2$ as an ideal gas or using the van der Waals equation.

van der Waals equation an equation that includes experimentally determined factors a and b that quantify the contributions of nonnegligible molecular volume and nonnegligible intermolecular interactions to the behavior of real gases with respect to changes in P, V, and T.

Solve We solve Equation 6.32 for pressure:

$$P = \frac{nRT}{V - nb} - \frac{n^2 a}{V^2}$$

$$P_{N_2} = \frac{(1.00 \ \text{mol})\left(0.08206 \ \frac{\text{L} \cdot \text{atm}}{\text{mol} \cdot \text{K}}\right)(300.0 \ \text{K})}{1.00 \ \text{L} - (1.00 \ \text{mol})\left(0.0391 \ \frac{\text{L}}{\text{mol}}\right)} - \frac{(1.00 \ \text{mol})^2\left(1.39 \ \frac{\text{L}^2 \cdot \text{atm}}{\text{mol}^2}\right)}{(1.00 \ \text{L})^2}$$

$$= 24.2 \ \text{atm}$$

If nitrogen behaved as an ideal gas, we would have

$$P = \frac{nRT}{V}$$

$$P_{N_2} = \frac{(1.00 \ \text{mol})\left(0.08206 \ \frac{\text{L} \cdot \text{atm}}{\text{mol} \cdot \text{K}}\right)(300.0 \ \text{K})}{1.00 \ \text{L}}$$

$$= 24.6 \ \text{atm}$$

Think about It This pressure is about 25 times greater than normal atmospheric pressure, so we expect some difference in the calculated pressures. However, the small deviation from ideality, $0.4/24.6 = 2\%$, supports our prediction. Furthermore, because of the likely deviation from ideal behavior, we expect the pressure calculated using the ideal gas equation to be too high. In this case, we observe that P_{ideal} is greater than P_{real}.

Practice Exercise Assuming the conditions stated in Sample Exercise 6.16, use the van der Waals equation and the ideal gas equation to calculate the pressure for 1.00 mole of He gas. Which gas behaves more ideally at 300 K, He or N_2?

We began this chapter with a description of our atmosphere as an invisible necessity. Looking back, we can now answer the questions we raised and many more about gases and our atmosphere. We now understand how the effect of elevation on the quantity of gas available results in requirements for pressurized cabins on commercial airliners for safe, comfortable travel. We have seen the chemical reactions used to supply oxygen in aircraft under emergency conditions and applied principles of stoichiometry to calculate the amount of material required to produce sufficient breathable air. Undersea explorers carry a different mixture of gases in a scuba tank than we breathe under normal conditions because high pressures under water increase the partial pressure of oxygen and nitrogen to unhealthy levels. Hot-air balloons rise because volume is directly proportional to temperature and because the density of air decreases as temperature rises. Understanding the physical properties of gases, especially their compressibility and their responses to changing conditions, makes their use in many everyday items ranging from automobile tires to weather balloons and fire extinguishers possible because their behavior is reliable and predictable. Our dependence on air for the maintenance of the chemical processes essential to life is so great that if we venture very far from Earth's surface, we have to take the right blend of gases with us to sustain that life.

Nitrogen: Feeding Plants and Inflating Air Bags

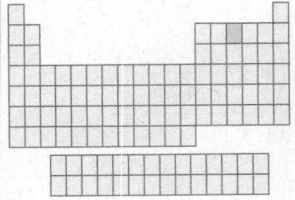

Nitrogen is the most abundant element in the atmosphere, 78% by volume. It is the most abundant free element on Earth and the sixth most abundant element in the universe. However, it is a relatively minor constituent of Earth's crust (only about 0.003% by mass), existing in the crust mostly in deposits of potassium nitrate (KNO_3) or sodium nitrate ($NaNO_3$). Common names for KNO_3 include saltpeter and niter. The latter name is the source of the name of the element: in Greek *nitro-* and *-gen* mean "niter-forming."

Nitrogen in combined forms is essential to all life; it is the nutrient needed in greatest quantity by food crops, it is needed by animals to make proteins, and all genetic material (DNA and RNA) contains nitrogen. The continuous exchange of nitrogen between the atmosphere and the biosphere (Figure 6.43) is mediated by bacteria called diazotrophs ("nitrogen eaters"), which live in the roots of leguminous plants such as peas, beans, and peanuts (Figure 6.44). These bacteria convert atmospheric nitrogen (N_2) into ammonia (NH_3) in a process called *nitrogen fixation*. Other microorganisms then oxidize NH_3 to oxides of nitrogen. Nitrogen

FIGURE 6.44 The root system of a soybean plant contains nodules caused by the presence of nitrogen-fixing bacteria that live in a symbiotic relation with the plants. This plant is a member of the legume family, and all other members of the family have the same type of root system.

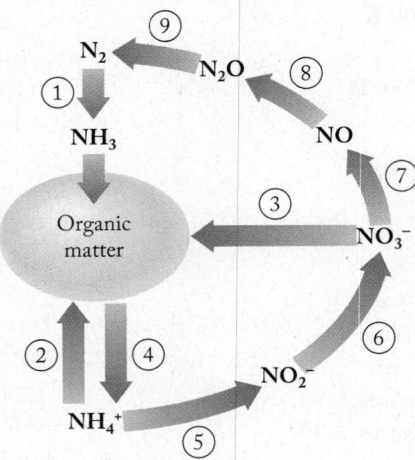

FIGURE 6.43 The nitrogen cycle describes the relationship between atmospheric nitrogen and the compounds of nitrogen necessary to support life on Earth. At any time, a large portion of nitrogen is found in the biosphere in living organisms and their dead remains. When organic matter decomposes, ammonium ion and other simple nitrogen compounds are released. (1) Nitrogen fixation provides ammonia to organisms. (2, 3) Microorganisms take up NH_4^+ and NO_3^-. (4) NH_4^+ is released during decomposition of organisms upon their death. (5, 6) Microorganisms oxidize NH_4^+ to various oxides of nitrogen. (7–9) Microorganisms use NO_3^- as an oxidizing agent and convert nitrate ion back into N_2.

fixation is necessary to sustain life on Earth because ammonia is required for the formation of biologically essential nitrogen-containing molecules, such as proteins and DNA.

A simple laboratory preparation of nitrogen is based on the thermal decomposition of ammonium nitrite:

$$NH_4NO_2(s) \rightarrow N_2(g) + 2\,H_2O(g)$$

On an industrial scale, nitrogen is produced by distilling liquid air. Nitrogen boils at $-196°C$, so it distills before oxygen, which boils at $-183°C$.

Liquid nitrogen is a cryogen, which means a very cold fluid. It is used in commercial refrigeration. Nitrogen's relative lack of reactivity at ordinary temperatures and pressures coupled with its availability makes it the agent of choice as a protective gas in the semiconductor industry to keep oxygen away from sensitive materials and for several industrial processes involving welding or soldering. Oil companies pump nitrogen under high pressure into oil deposits to force crude oil to the surface (Figure 6.45).

In the early 20th century, mined nitrates were widely used in the manufacture of gunpowder and other explosives. For example, a mixture of potassium nitrate, sulfur, and carbon known as *black powder* is still the best fuse material ever discovered. It reacts exothermically and explosively, producing nitrogen and carbon dioxide. Rapid expansion of these hot gases adds to the explosive character of the reaction:

$$16\,KNO_3(s) + S_8(s) + 24\,C(s) \rightarrow$$
$$8\,K_2S(s) + 8\,N_2(g) + 24\,CO_2(g)$$

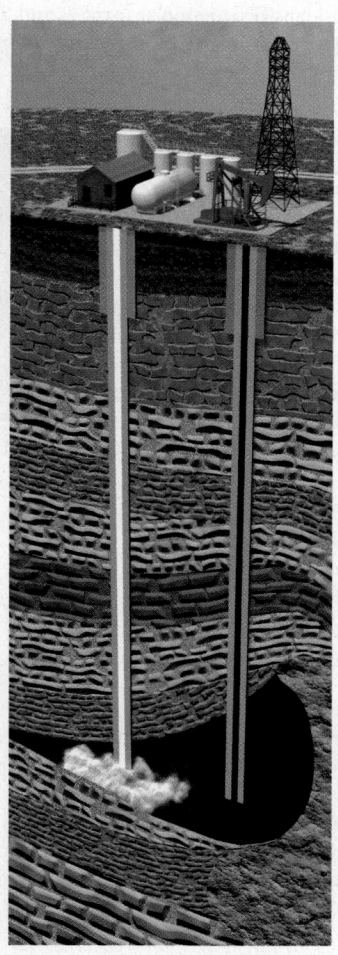

FIGURE 6.45 Nitrogen under high pressure is pumped into oil deposits to bring crude oil to the surface.

When naval blockades cut off Germany's supplies of KNO_3 during World War I, German chemist Fritz Haber (1868–1934) developed a process for making nitrates that starts with the synthesis of ammonia from hydrogen gas and nitrogen from the atmosphere. The *Haber–Bosch process* relies on high temperature and pressure to promote the reaction:

$$N_2(g) + 3H_2(g) \rightarrow 2NH_3(g)$$

In this reaction the very strong nitrogen–nitrogen bond in N_2 must be broken, which requires an energy investment. Today, the *Haber–Bosch process* is the principal source of ammonia and the nitrogen compounds derived from ammonia that are widely used in industry and agriculture.

The first step in converting ammonia into other important nitrogen compounds entails its oxidation to NO_2 and H_2O:

$$4NH_3(g) + 5O_2(g) \rightarrow 4NO(g) + 6H_2O(g)$$
$$2NO(g) + O_2(g) \rightarrow 2NO_2(g)$$

Nitrogen dioxide dissolves in water, producing a mixture of nitrous and nitric acids:

$$2NO_2(g) + H_2O(\ell) \rightarrow HNO_2(aq) + HNO_3(aq)$$

Heating the mixture converts nitrous acid into nitric acid, as steam and NO are given off:

$$3HNO_2(aq) \rightarrow HNO_3(aq) + H_2O(g) + 2NO(g)$$

About 75% of nitric acid produced in this way is combined with more ammonia to produce ammonium nitrate:

$$NH_3(g) + HNO_3(\ell) \rightarrow NH_4NO_3(s)$$

Ammonium nitrate is an important industrial chemical and the source of water-soluble nitrogen in many formulations of fertilizers.

All of this chemistry described for industrial use—some of which takes place at high temperature and pressure and requires energy derived from fossil fuels—also takes place in bacteria but at normal temperatures and pressures. Studying the mechanism of these reactions in bacteria remains an active area of research. To date, efforts at finding alternative methods of producing the large quantities of nitrogen-containing substances needed by commerce and agriculture in ways that minimize the high costs associated with industrial production have not been successful.

Nitrogen has a wide and varied chemistry beyond the synthesis and uses of ammonia. Hydrazine (N_2H_4) is used as a rocket fuel. The reaction of metals with nitrogen produces metal nitrides containing the N^{3-} anion. Iron(III) nitride (FeN) is used to harden the surface of steel. Sodium azide is a bactericide used in many biology laboratories to control bacterial growth. Metal salts containing the azide ion (N_3^-) are often explosive and are used as a ready source of nitrogen gas in some automobile air bags. The overall reaction in an air bag is the sum of several reactions, summarized as:

$$20NaN_3(s) + 6SiO_2(s) + 4KNO_3(s) \rightarrow$$
$$32N_2(g) + 5Na_4SiO_4(s) + K_4SiO_4(s)$$

Azides are contact explosives, which means mechanical contact causes them to detonate. The trigger mechanism in a steering column that causes the azide to explode and produce nitrogen to inflate the bag involves the transmission of the force of an accident to the azide contained in the air bags. Highly corrosive and reactive sodium metal is produced in the azide reaction along with the nitrogen, and other materials—the SiO_2 and KNO_3 included in the overall reaction shown—react with the sodium to convert it into relatively harmless sodium silicates.

Section 6.1 Gases occupy the entire volume of their container. The volume occupied by a gas changes significantly with pressure and temperature. Gases are **miscible**, mixing in any proportion. They are much less dense than liquids or solids.

Section 6.2 The mass of the gases in Earth's atmosphere combined with the force of gravity results in an **atmospheric pressure** on the surface of the planet. **Pressure** is defined as the ratio of force to surface area and is measured with **barometers** and **manometers**. Pressure can be expressed in **atmospheres**, **torr**, **mmHg**, millibars, or pascals.

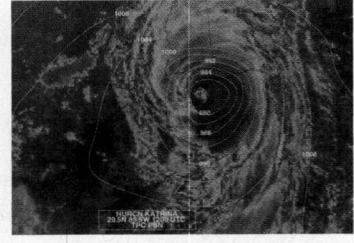

Section 6.3 At constant temperature, the volume of a gas increases with decreasing pressure (**Boyle's law**). At constant pressure, the volume increases with increasing temperature (**Charles's law**). At constant temperature and constant pressure, increasing the number of moles of gas produces a proportionate increase in volume (**Avogadro's law**). At constant volume, the pressure of a gas increases with increasing temperature (**Amontons's law**).

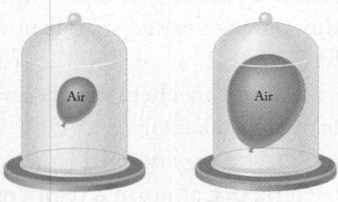

Section 6.4 Boyle's, Charles's, Avogadro's, and Amontons's laws can be combined into the **ideal gas law**, also called the **ideal gas equation**: $PV = nRT$, where R is the **universal gas constant**. The ideal gas equation accurately describes the behavior of most gases under normal conditions. The **molar volume** of any gas is 22.4 L at **standard temperature and pressure (STP)**, which in the United States is defined as 0°C and 1 atm.

Section 6.5 The ideal gas equation and the stoichiometry of a chemical reaction can be used to calculate the volumes of gases required or produced in the reaction.

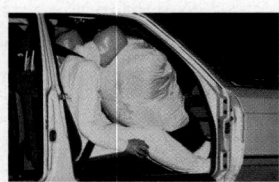

Section 6.6 Gas densities increase with increasing molar mass and with increasing pressure, and they decrease with increasing temperature. The ideal gas equation can be used to calculate gas density from molar mass, P, and T. Calculating the molar mass of a gas is possible from the mass, volume, pressure, and temperature of a sample of the gas.

Section 6.7 In a gas mixture, the contribution each component gas makes to the total gas pressure is called the **partial pressure** of that gas. **Dalton's law of partial pressures** states that the total pressure of a gas mixture is the sum of the partial pressures of its components. Dalton's law allows us to calculate the partial pressure (P_x) of any constituent gas x in a gas mixture if we know its **mole fraction (X_x)** and the total pressure. The total pressure of a gas sample collected over water is the sum of the $H_2O(g)$ partial pressure and the partial pressure of the gas.

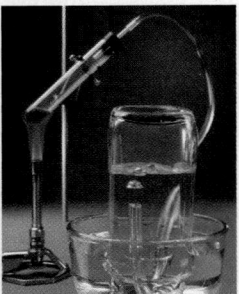

Section 6.8 The behavior of gases is explained by the **kinetic molecular theory**. Gases are composed of particles in constant random motion, moving at **root-mean-square speeds (u_{rms})** that are inversely proportional to the square root of their molar masses and directly proportional to the square root of their temperature. Pressure arises from elastic collisions between gases and the walls of their container. The kinetic molecular theory explains **Graham's law of effusion**, which states that the rate of **effusion** (escape through a pinhole) or **diffusion** (spreading) of a gas at a fixed temperature is inversely proportional to the square root of its molar mass.

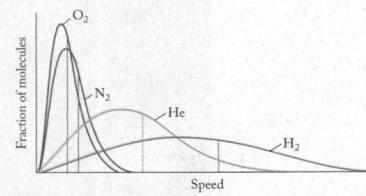

Section 6.9 Ideal gas behavior is observed at moderate temperatures and low pressures, where $PV = nRT$ holds. Ideal gas behavior is characterized by elastic collisions and an absence of attractive forces between gas molecules. At high pressures, the behavior of real gases deviates from the predictions of the ideal gas law. The **van der Waals equation**, a modified form of the ideal gas equation, accounts for real gas properties.

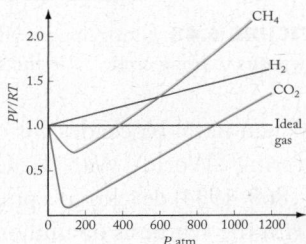

PROBLEM-SOLVING SUMMARY

TYPE OF PROBLEM	CONCEPTS AND EQUATIONS	SAMPLE EXERCISES
Calculating pressure of any gas, calculating atmospheric pressure	Divide force by area over which the force is applied, using the equation $$P = \frac{F}{A} \qquad (6.1)$$	6.1, 6.2
Calculating changes in P, V, and/or T in response to changing conditions	Rearrange $$\frac{P_1 V_1}{T_1} = \frac{P_2 V_2}{T_2} \qquad (6.18)$$ for whichever variable is sought and then substitute given values. (All T must be in kelvins and n is constant.)	6.3–6.6
Determining n from P, V, and T	Rearrange $$PV = nRT \qquad (6.13)$$ for n and then substitute given values of P, T, and V. (T must be in kelvins.)	6.7, 6.8
Calculating the density of a gas and calculating molar mass from density	Substitute values for pressure, absolute temperature, and molar mass into the equation $$d = \frac{P\mathcal{M}}{RT} \qquad (6.21)$$ Substitute values for pressure, absolute temperature, and density into the equation $$\mathcal{M} = \frac{dRT}{P} \qquad (6.22)$$	6.9, 6.10
Calculating mole fraction for one component gas in a mixture	Divide the number of moles of the component gas by the total number of moles in the mixture: $$\chi_x = \frac{n_x}{n_{\text{total}}} \qquad (6.24)$$	6.11
Calculating partial pressure of one component gas in a mixture and total pressure in the mixture	Substitute the mole fraction of the component gas and the total pressure in the equation $$P_1 = \chi_1 P_{\text{total}} \qquad (6.25)$$ Solve the equation $$P_{\text{total}} = P_1 + P_2 + P_3 + P_4 + \ldots \qquad (6.23)$$ for the partial pressure of the component gas and then substitute given values for other partial pressures and total pressure.	6.12, 6.13
Calculating root-mean-square speeds	Substitute absolute temperature, molar mass, and the value 8.314 kg · m²/ (s² · mol · K) for R in the equation $$u_{\text{rms}} = \sqrt{\frac{3RT}{\mathcal{M}}} \qquad (6.28)$$	6.14
Calculating relative rate of effusion or diffusion from molar masses	Substitute values for molar masses into the equation $$\frac{r_x}{r_y} = \sqrt{\frac{\mathcal{M}_y}{\mathcal{M}_x}} \qquad (6.31)$$	6.15
Calculating P for a real gas	Solve the van der Waals equation $$\left(P + \frac{n^2 a}{V^2}\right)(V - nb) = nRT \qquad (6.32)$$ for P and substitute given values of n, V, and T (in kelvins), plus values of a and b from Table 6.5.	6.16

VISUAL PROBLEMS

(Answers to boldface end-of-chapter questions and problems are in the back of the book.)

6.1. Shown in Figure P6.1 are three barometers. The one in the center is located at sea level. Which barometer is most likely to reflect the atmospheric pressure in Denver, CO, where the elevation is approximately 1500 m? Explain your answer.

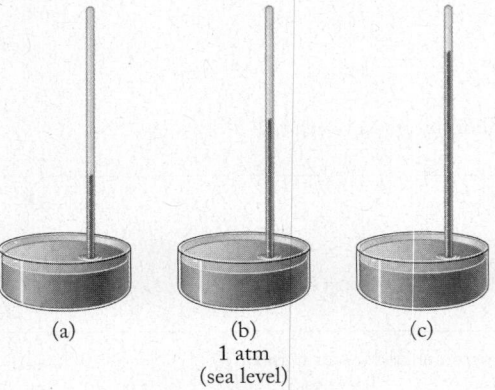

(a) (b) (c)

1 atm
(sea level)

FIGURE P6.1

6.2. A rubber balloon is filled with helium gas. Which of the drawings in Figure P6.2 most accurately reflects the gas in the balloon on a molecular level? The blue-gray spheres represent helium atoms. Explain your answer.

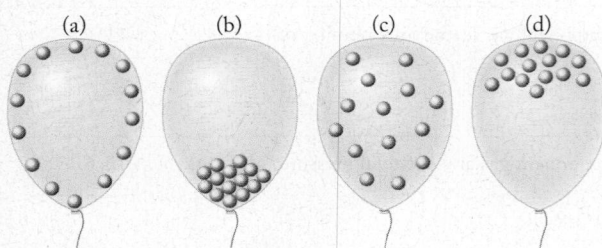

(a) (b) (c) (d)

FIGURE P6.2

6.3. Which of the three changes shown in Figure P6.3 best illustrates what happens when the atmospheric pressure on a helium-filled rubber balloon is increased at constant temperature?

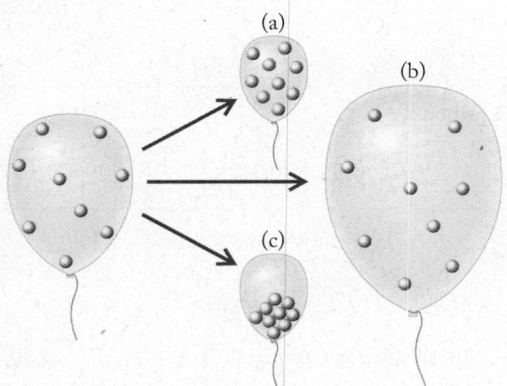

(a)

(b)

(c)

FIGURE P6.3

6.4. Which of the drawings in Figure P6.3 best illustrates what happens when the temperature of a helium-filled rubber balloon is increased at constant pressure?

6.5. Which of the three changes shown in Figure P6.5 best illustrates what happens when the amount of gas in a helium-filled rubber balloon is increased at constant temperature and pressure?

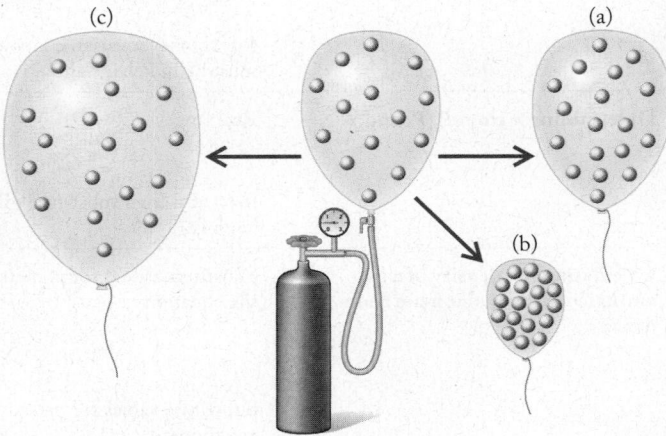

(c) (a)

(b)

FIGURE P6.5

6.6. Which line plotting volume versus reciprocal pressure in Figure P6.6 corresponds to the higher temperature?

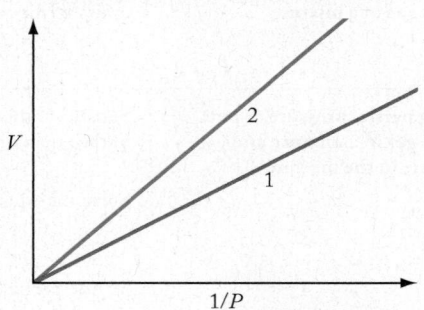

FIGURE P6.6

6.7. In Figure P6.7, which line of volume versus temperature represents a gas at higher pressure? Is the x-axis an absolute temperature scale?

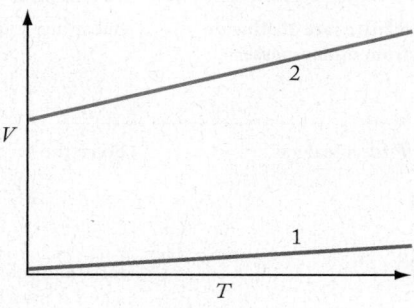

FIGURE P6.7

6.8. In Figure P6.8, which of the two plots of volume versus pressure at constant temperature is not consistent with the ideal gas law?

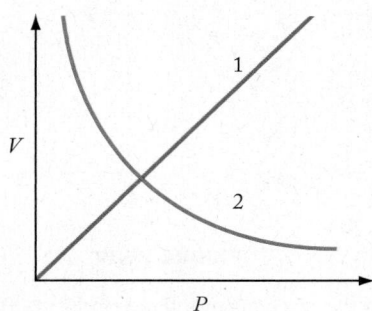

FIGURE P6.8

6.9. In Figure P6.9, which of the two plots of volume versus temperature at constant pressure is not consistent with the ideal gas law?

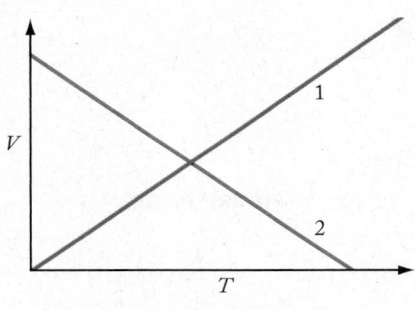

FIGURE P6.9

6.10. In Figure P6.10, which line in the graph of density versus pressure at constant temperature for methane (CH_4) and nitrogen (N_2) should be labeled *methane*?

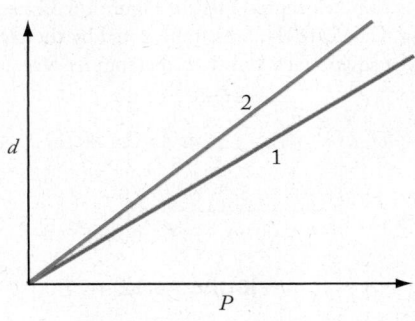

FIGURE P6.10

6.11. Add lines showing the densities of He and NO as a function of pressure to a copy of the graph in Figure P6.10.
6.12. Which of the drawings in Figure P6.12 best depicts the arrangement of molecules in a mixture of gases? The red and blue-gray spheres represent atoms of two different gases such as helium and neon.

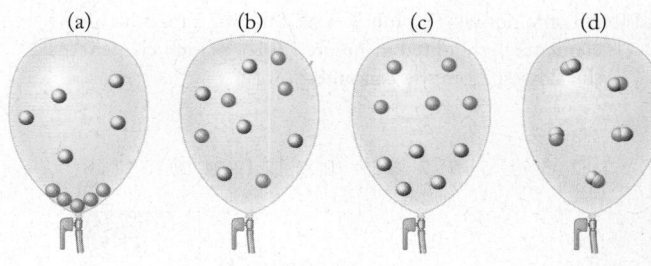

FIGURE P6.12

6.13. The drawings in Figure P6.13 illustrate four mixtures of two diatomic gases (such as N_2 and O_2) in flasks of identical volume and at the same temperature. Is the total pressure in each flask the same? Which flask has the highest partial pressure of nitrogen, depicted by the blue molecules?

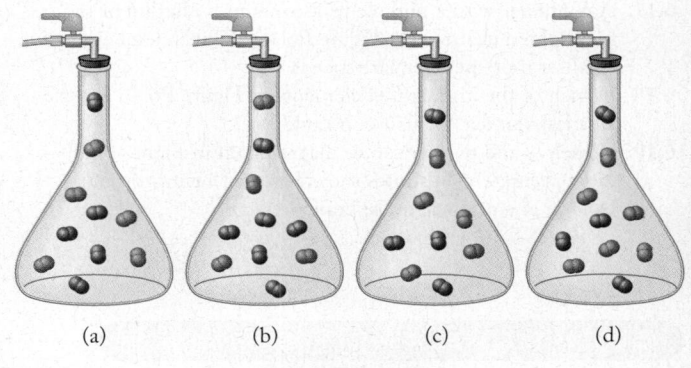

FIGURE P6.13

6.14. Figure P6.14 shows the distribution of molecular speeds of CO_2 and SO_2 molecules at 25°C. Which curve is the profile for SO_2? Which of these profiles should match that of propane (C_3H_8), a common fuel in portable grills?

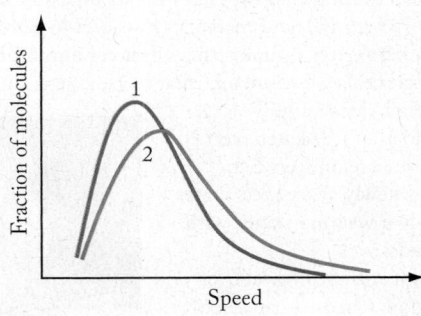

FIGURE P6.14

6.15. How would a graph showing the distribution of molecular speeds of CO_2 at −100°C differ from the curve for CO_2 shown in Figure P6.14?

6.16. A container with a pinhole leak contains a mixture of the elements highlighted in Figure P6.16. Which element leaks the slowest from the container?

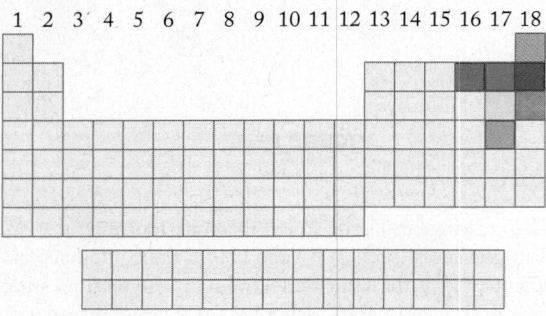

1 2 3 4 5 6 7 8 9 10 11 12 13 14 15 16 17 18

FIGURE P6.16

6.17. A container with a pinhole leak contains a mixture of the highlighted elements in Figure P6.16. Which element has the smallest root-mean-square speed?

6.18. Which of the highlighted elements in Figure P6.16 has the smallest van der Waals *b* constant?

6.19. Which of the two outcomes diagrammed in Figure P6.19 more accurately illustrates the effusion of helium from a balloon at constant atmospheric pressure?

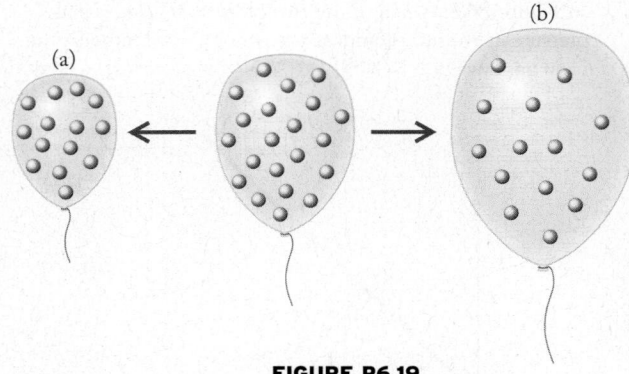

(a) (b)

FIGURE P6.19

6.20. Which of the two outcomes shown in Figure P6.20 more accurately illustrates the effusion of gases from a balloon at constant atmospheric pressure if the red spheres have a greater root-mean-square speed than the blue-gray spheres?

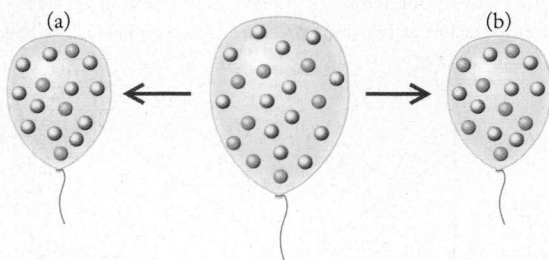

(a) (b)

FIGURE P6.20

QUESTIONS AND PROBLEMS

The Gas Phase; Atmospheric Pressure

CONCEPT REVIEW

6.21. Describe the difference between force and pressure.

6.22. How does Torricelli's barometer measure atmospheric pressure?

6.23. What is the relation between *torr* and *atmospheres* of pressure?

6.24. What is the relation between *millibars* and *pascals* of pressure?

6.25. Three barometers based on Torricelli's design are constructed using water (density $d = 1.00$ g/mL), ethanol ($d = 0.789$ g/mL), and mercury ($d = 13.546$ g/mL). Which barometer contains the tallest column of liquid?

6.26. In constructing a barometer, what advantage is there in choosing a dense liquid?

6.27. Why does an ice skater exert more pressure on ice when wearing newly sharpened skates than when wearing skates with dull blades?

6.28. Why is it easier to travel over deep snow when wearing boots and snowshoes (Figure P6.28) rather than just boots?

FIGURE P6.28

6.29. Why does atmospheric pressure decrease with increasing elevation?

*6.30. Pieces of different metals have exactly the same mass but different densities. Could these objects ever exert the same pressure?

PROBLEMS

6.31. Calculate the downward pressure due to gravity exerted by the bottom face of a 1.00 kg cube of iron that is 5.00 cm on a side.

6.32. The gold block represented in Figure P6.32 has a mass of 38.6 g. Calculate the pressure exerted by the block when it is on (a) a square face and (b) a rectangular face.

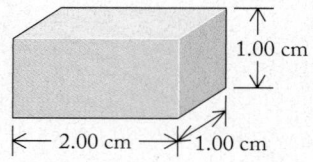

1.00 cm

2.00 cm — 1.00 cm

FIGURE P6.32

6.33. Convert the following pressures into atmospheres: (a) 2.0 kPa; (b) 562 mmHg.

6.34. Convert the following pressures into millimeters of mercury: (a) 0.541 atm; (b) 2.8 kPa.

6.35. **Record High Atmospheric Pressure** The highest atmospheric pressure recorded on Earth was measured at Tosontsengel, Mongolia, on December 19, 2001, when the barometer read 108.6 kPa. Express this pressure in (a) millimeters of mercury, (b) atmospheres, and (c) millibars.

6.36. Record Low Atmospheric Pressure Despite the destruction from Hurricane Katrina in September 2005, the lowest pressure for a hurricane in the Atlantic Ocean was measured several weeks after Katrina. Hurricane Wilma registered an atmospheric pressure of 88.2 kPa on October 19, 2005, about 2 kPa lower than Hurricane Katrina. What was the *difference* in pressure between the two hurricanes in (a) millimeters of mercury, (b) atmospheres, and (c) millibars?

The Gas Laws

CONCEPT REVIEW

6.37. From the molecular perspective, why is pressure directly proportional to temperature at fixed volume (Amontons's law)?

6.38. How do we explain Boyle's law on a molecular basis?

6.39. A balloonist is rising too fast for her taste. Should she increase the temperature of the gas in the balloon, or decrease it?

6.40. Could the pilot of the balloon in Problem 6.39 reduce her rate of ascent by allowing some gas to leak out of the balloon? Explain your answer.

6.41. A quantity of gas is compressed into half its initial volume as it cools from 20°C to 10°C. Does the pressure of the gas increase, decrease, or remain the same?

6.42. If the volume of gasoline vapor and air in an automobile engine cylinder is reduced to 1/10 of its original volume before ignition, by what factor does the pressure in the cylinder increase? (Assume there is no change in temperature.)

PROBLEMS

6.43. The volume of 1.00 mol of ammonia gas at 1.00 atm of pressure is gradually decreased from 78.0 mL to 39.0 mL. What is the final pressure of ammonia if there is no change in temperature?

6.44. The pressure on a sample of an ideal gas is increased from 715 mmHg to 3.55 atm at constant temperature. If the initial volume of the gas is 485 mL, what is the final volume of the gas?

6.45. A scuba diver releases a balloon containing 153 L of helium attached to a tray of artifacts at an underwater archaeological site (Figure P6.45). When the balloon reaches the surface, it has expanded to a volume of 352 L. The pressure at the surface is 1.00 atm; what is the pressure at the underwater site? Pressure increases by 1.0 atm for every 10 m of depth; at what depth was the diver working? Assume the temperature remains constant.

FIGURE P6.45

6.46. Breath-Hold Diving The world record for diving without supplemental air tanks ("breath-hold diving") is about 125 m, a depth at which the pressure is about 12.5 atm. If a diver's lungs have a volume of 6 L at the surface of the water, what is their volume at a depth of 125 m?

6.47. Use the following data to draw a graph of the volume of 1 mole of H_2 as a function of the reciprocal of pressure at 298 K:

P (mmHg)	V (L)
100	186
120	155
240	77.5
380	48.9
500	37.2

Would the graph be the same for the same number of moles of argon?

6.48. The following data for P and V were collected for 1 mole of argon at 300 K. Draw a graph of the volume of 1 mole of Ar as a function of the reciprocal of pressure. Does this graph look like the graph you drew for Problem 6.47?

P (atm)	V (L)
0.10	246.3
0.25	98.5
0.50	49.3
0.75	32.8
1.0	24.6

6.49. Use the following data to draw a graph of the volume of He as a function of temperature for 1.0 mole of He gas at a constant pressure of 1.00 atm:

V (L)	T (K)
7.88	96
3.94	48
1.97	24
0.79	9.6
0.39	4.8

How would the graph change if the amount of gas were halved?

6.50. Use the following data to draw a graph of the volume of He as a function of temperature for 0.50 mole of He gas at a constant pressure of 1.00 atm:

V (L)	T (K)
3.94	96
1.97	48
0.79	24
0.39	9.6
0.20	4.8

Does this graph match your prediction from Problem 6.49?

6.51. A cylinder with a piston (Figure P6.51) contains a sample of gas at 25°C. The piston moves in response to changing pressure inside the cylinder. At what gas temperature would the piston move so that the volume inside the cylinder doubled?

6.52. The temperature of the gas in Problem 6.51 is reduced to a temperature at which the volume inside the cylinder has decreased by 25% from its initial volume at 25.0°C. What is the new temperature?

FIGURE P6.51

6.53. A 2.68 L sample of gas is warmed from 250 K to a final temperature of 398 K. Assuming no change in pressure, what is the final volume of the gas?

6.54. A 5.6 L sample of gas is cooled from 78°C to a temperature at which its volume is 4.3 L. What is this new temperature? Assume no change in pressure of the gas.

6.55. Balloons for a New Year's Eve party in Fargo, ND, are filled to a volume of 2.0 L at a temperature of 22°C and then hung outside where the temperature is −22°C. What is the volume of the balloons after they have cooled to the outside temperature? Assume that atmospheric pressure inside and outside the house is the same.

6.56. The air inside a hot-air balloon is heated to 45°C and then cools to 25°C. By what percentage does the volume of the balloon change?

6.57. Which of the following actions would produce the greatest increase in the volume of a gas sample: (a) lowering the pressure from 760 mmHg to 720 mmHg at constant temperature or (b) raising the temperature from 10°C to 40°C at constant pressure?

6.58. Which of the following actions would produce the greatest increase in the volume of a gas sample: (a) doubling the amount of gas in the sample at constant temperature and pressure or (b) raising the temperature from 244°C to 1100°C?

*6.59. What happens to the volume of gas in a cylinder with a movable piston under the following conditions?
   a. Both the absolute temperature and the external pressure on the piston double.
   b. The absolute temperature is halved, and the external pressure on the piston doubles.
   c. The absolute temperature increases by 75%, and the external pressure on the piston increases by 50%.

*6.60. What happens to the pressure of a gas under the following conditions?
   a. The absolute temperature is halved and the volume doubles.
   b. Both the absolute temperature and the volume double.
   c. The absolute temperature increases by 75%, and the volume decreases by 50%.

6.61. A 150.0 L weather balloon filled with 6.1 mole of helium has a small leak. If the helium leaks at a rate of 10 mmol/hr, what is the volume of the balloon after 24 hr?

6.62. Which has the greater effect on the volume of a gas at constant temperature: doubling the number of moles or doubling the pressure?

6.63. Temperature Effects on Bicycle Tires A bicycle racer inflates his tires to 7.1 atm on a warm autumn afternoon when temperatures reached 27°C. By morning the temperature has dropped to 5.0°C. What is the pressure in the tires if we assume that the volume of the tire does not change significantly?

*6.64. A balloon vendor at a street fair is using a tank of helium to fill her balloons. The tank has a volume of 150.0 L and a pressure of 120.0 atm at 25°C. After a while she notices that the valve has not been closed properly. The pressure had dropped to 110.0 atm. How many moles of gas have been lost?

The Ideal Gas Law; Gases in Chemical Reactions

CONCEPT REVIEW

6.65. What is meant by standard temperature and pressure (STP)? What is the volume of 1 mole of an ideal gas at STP?

6.66. Which of the following are not characteristics of an ideal gas?
   a. The molecules of gas have little volume compared with the volume that they occupy.
   b. Its volume is independent of temperature.
   c. The density of all ideal gases is the same.
   d. Gas atoms or molecules do not interact with one another.

*6.67. What does the slope represent in a graph of pressure as a function of $1/V$ at constant temperature for an ideal gas?

*6.68. How would the graph in Problem 6.67 change if we increased the temperature to a larger but constant value?

PROBLEMS

6.69. How many moles of air must there be in a bicycle tire with a volume of 2.36 L if it has an internal pressure of 6.8 atm at 17.0°C?

6.70. At what temperature will 1.00 mole of an ideal gas in a 1.00 L container exert a pressure of 1.00 atm?

6.71. Hyperbaric Oxygen Therapy Hyperbaric oxygen chambers are used to treat divers suffering from decompression sickness (the "bends") with pure oxygen at greater than atmospheric pressure. Other clinical uses include treatment of patients with thermal burns, necrotizing fasciitis, and CO poisoning. What is the pressure in a chamber with a volume of 2.36×10^3 L that contains 4635 g of $O_2(g)$ at a temperature of 298 K?

6.72. What is the volume of 100 g of H_2O vapor at 120°C and 1.00 atm?

6.73. A weather balloon with a volume of 200.0 L is launched at 20°C at sea level, where the atmospheric pressure is 1.00 atm. The balloon rises to an altitude of 20,000 m, where atmospheric pressure is 63 mmHg and the temperature is 210 K. What is the volume of the balloon at 20,000 m?

6.74. For some reason, a skier decides to ski from the summit of a mountain near Park City, UT (elevation = 9970 ft, $T = -10°C$, and $P_{atm} = 623$ mmHg), to the base of the mountain (elevation = 6920 ft, $T = 25°C$, and $P_{atm} = 688$ mmHg) with a balloon tied to each of her ski poles. If each balloon is filled to a volume of 2.00 L at the summit, what is the volume of each balloon at the base?

6.75. Hydrogen holds promise as an "environment friendly" fuel. How many grams of H_2 gas are present in a 50.0 L fuel tank at a pressure of 2850 lb/in² (psi) at 20°C? Assume that 1 atm = 14.7 psi.

6.76. Liquid Nitrogen–Powered Car Students at the University of North Texas and the University of Washington built a car propelled by compressed nitrogen gas. The gas was obtained by boiling liquid nitrogen stored in a 182 L tank. What volume of N_2 is released at 0.927 atm of pressure and 25°C from a tank full of liquid N_2 (d = 0.808 g/mL)?

6.77. Miners' Lamps Before the development of reliable batteries, miners' lamps burned acetylene produced by the reaction of calcium carbide with water:

$$CaC_2(s) + H_2O(\ell) \rightarrow C_2H_2(g) + CaO(s)$$

A lamp uses 1.00 L of acetylene per hour at 1.00 atm pressure and 18°C.
a. How many moles of C_2H_2 are used per hour?
b. How many grams of calcium carbide must be in the lamp for a 4 hr shift?

6.78. Acid precipitation dripping on limestone produces carbon dioxide by the following reaction:

$$CaCO_3(s) + 2H^+(aq) \rightarrow Ca^{2+}(aq) + CO_2(g) + H_2O(\ell)$$

If 15.0 mL of CO_2 were produced at 25°C and 760 mmHg, then
a. how many moles of CO_2 were produced?
b. how many milligrams of $CaCO_3$ were consumed?

6.79. Oxygen is generated by the thermal decomposition of potassium chlorate:

$$2KClO_3(s) \rightarrow 2KCl(s) + 3O_2(g)$$

How much $KClO_3$ is needed to generate 200.0 L of oxygen at 0.85 atm and 273 K?

6.80. Calculate the volume of carbon dioxide at 20°C and 1.00 atm produced from the complete combustion of 1.00 kg of methane. Compare your result with the volume of CO_2 produced from the complete combustion of 1.00 kg of propane (C_3H_8).

6.81. Healthy Air for Sailors The CO_2 that builds up in the air of a submerged submarine can be removed by reacting it with sodium peroxide:

$$2Na_2O_2(s) + 2CO_2(g) \rightarrow 2Na_2CO_3(s) + O_2(g)$$

If a sailor exhales 150.0 mL of CO_2 per minute at 20°C and 1.02 atm, how much sodium peroxide is needed per sailor in a 24 hr period?

6.82. Rescue Breathing Devices Self-contained self-rescue breathing devices, like the one shown in Figure P6.82, convert CO_2 into O_2 according to the following reaction:

$$4KO_2(s) + 2CO_2(g) \rightarrow 2K_2CO_3(s) + 3O_2(g)$$

How many grams of KO_2 are needed to produce 100.0 L of O_2 at 20°C and 1.00 atm?

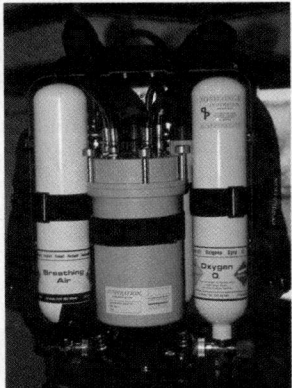

FIGURE P6.82

Gas Density

CONCEPT REVIEW

6.83. Do all gases at the same pressure and temperature have the same density? Explain your answer.

6.84. Birds and sailplanes take advantage of thermals (rising columns of warm air) to gain altitude with less effort than usual. Why does warm air rise?

6.85. How does the density of a gas sample change when (a) its pressure is increased and (b) its temperature is decreased?

6.86. How would you measure the density of a gas sample of known molar mass?

PROBLEMS

6.87. Biological Effects of Radon Exposure Radon is a naturally occurring radioactive gas found in the ground and in building materials. It is easily inhaled and emits α particles when it decays. Cumulative radon exposure is a significant risk factor for lung cancer.
a. Calculate the density of radon at 298 K and 1.00 atm of pressure.
b. Are radon concentrations likely to be greater in the basement or on the top floor of a building?

*__6.88.__ Four empty balloons, each with a mass of 10.0 g, are inflated to a volume of 20.0 L. The first balloon contains He, the second Ne, the third CO_2, and the fourth CO. If the density of air at 25°C and 1.00 atm is 0.00117 g/mL, will any of the balloons float in this air?

6.89. A 150.0 mL flask contains 0.391 g of a volatile oxide of sulfur. The pressure in the flask is 750 mmHg, and the temperature is 22°C. Is the gas SO_2 or SO_3?

6.90. A 100.0 mL flask contains 0.193 g of a volatile oxide of nitrogen. The pressure in the flask is 760 mmHg at 17°C. Is the gas NO, NO_2, or N_2O_5?

6.91. The density of an unknown gas is 1.107 g/L at 300 K and 740 mmHg. Could this gas be CO or CO_2?

6.92. A gas containing chlorine and oxygen has a density of 2.875 g/L at 756 mmHg and 11°C. What is the most likely molecular formula of the gas?

Dalton's Law and Mixtures of Gases

CONCEPT REVIEW

6.93. What is meant by the *partial pressure* of a gas?

6.94. Can a barometer be used to measure just the partial pressure of oxygen in the atmosphere? Why or why not?

6.95. Which gas sample has the largest volume at 25°C and 1 atm pressure? (a) 0.500 mol dry H_2; (b) 0.500 mol dry N_2; (c) 0.500 mol wet H_2 (H_2 collected over water)

6.96. Two identical balloons are filled to the same volume at the same pressure and temperature. One balloon is filled with air and the other with helium. Which balloon contains more particles (atoms and molecules)?

PROBLEMS

6.97. A gas mixture contains 0.70 moles of N_2, 0.20 moles of H_2, and 0.10 moles of CH_4. What is the mole fraction of H_2 in the mixture?

6.98. A gas mixture contains 7.0 g of N_2, 2.0 g of H_2, and 16.0 g of CH_4. What is the mole fraction of H_2 in the mixture?

6.99. Calculate the pressure of the gas mixture and the partial pressure of each constituent gas in Problem 6.97 if the mixture is in a 10.0 L vessel at 27°C.

6.100. Calculate the pressure of the gas mixture and the partial pressure of each constituent gas in Problem 6.98 if the mixture is in a 1.00 L vessel at 0°C.

6.101. A sample of oxygen was collected over water at 25°C and 1.00 atm. If the total sample volume was 0.480 L, how many moles of O_2 were collected?

6.102. Water was removed from the O_2 sample in Problem 6.101. What is the volume of the dry O_2 gas sample at 25°C and 1.00 atm?

6.103. The following reactions were carried out in sealed containers. Will the total pressure after each reaction is complete be greater than, less than, or equal to the total pressure before the reaction? Assume all reactants and products are gases at the same temperature.
   a. $N_2O_5(g) + NO_2(g) \rightarrow 3NO(g) + 2O_2(g)$
   b. $2SO_2(g) + O_2(g) \rightarrow 2SO_3(g)$
   c. $C_3H_8(g) + 5O_2(g) \rightarrow 3CO_2(g) + 4H_2O(g)$

6.104. In each of the following gas-phase reactions, determine whether the total pressure at the end of the reaction (carried out in a sealed, rigid vessel) will be greater than, less than, or equal to the total pressure at the beginning. Assume all reactants and products are gases at the same temperature.
   a. $H_2(g) + Cl_2(g) \rightarrow 2HCl(g)$
   b. $4NH_3(g) + 5O_2(g) \rightarrow 4NO(g) + 6H_2O(g)$
   c. $2NO(g) + O_2(g) \rightarrow 2NO_2(g)$

6.105. **High-Altitude Mountaineering** Alpine climbers use pure oxygen near the summits of 8000 m peaks, where $P_{atm} = 0.35$ atm (Figure P6.105). How much more O_2 is there in a lung full of pure O_2 at this elevation than in a lung full of air at sea level?

FIGURE P6.105

6.106. **Scuba Diving** A scuba diver is at a depth of 50 m, where the pressure is 5.0 atm. What should be the mole fraction of O_2 in the gas mixture the diver breathes to replicate the same gas mixture at sea level?

6.107. Carbon monoxide at a pressure of 680 mmHg reacts completely with O_2 at a pressure of 340 mmHg in a sealed vessel to produce CO_2. What is the final pressure in the flask?

6.108. Ozone reacts completely with NO, producing NO_2 and O_2. A 10.0 L vessel is filled with 0.280 moles of NO and 0.280 moles of O_3 at 350 K. Find the partial pressure of each product and the total pressure in the flask at the end of the reaction.

*6.109. Ammonia is produced industrially from the reaction of hydrogen with nitrogen under pressure in a sealed reactor.

What is the percent decrease in pressure of a sealed reaction vessel during the reaction between 3.60×10^3 mole of H_2 and 1.20×10^3 mole of N_2 if half of the N_2 is consumed?

*6.110. A mixture of 0.156 moles of C is reacted with 0.117 moles of O_2 in a sealed, 10.0 L vessel at 500 K, producing a mixture of CO and CO_2. The total pressure is 0.640 atm. What is the partial pressure of CO?

The Kinetic Molecular Theory of Gases and Graham's Law

CONCEPT REVIEW

6.111. What is meant by the *root-mean-square speed* of gas molecules?

6.112. Why don't all molecules in a sample of air move at exactly the same speed?

6.113. How does the root-mean-square speed of the molecules in a gas vary with (a) molar mass and (b) temperature?

6.114. Does pressure affect the root-mean-square speed of the molecules in a gas? Explain your answer.

6.115. How can Graham's law of effusion be used to determine the molar mass of an unknown gas?

6.116. Is the ratio of the rates of effusion of two gases the same as the ratio of their root-mean-square speeds?

6.117. What is the difference between *diffusion* and *effusion*?

6.118. If gas X diffuses faster in air than gas Y, is gas X also likely to effuse faster than gas Y?

PROBLEMS

6.119. Rank the gases SO_2, CO_2, and NO_2 in order of increasing root-mean-square speed at 0°C.

6.120. In a mixture of CH_4, NH_3, and N_2, which gas molecules are, on average, moving fastest?

6.121. At 286 K, three gases, A, B, and C, have root-mean-square speeds of 360 m/s, 441 m/s, and 472 m/s, respectively. Which gas is O_2?

6.122. Air is approximately 21% O_2 and 78% N_2 by mass. Calculate the root-mean-square speed of each gas at 273 K.

6.123. Calculate the root-mean-square speed of Ne atoms at the temperature at which their average kinetic energy is 5.18 kJ/mol.

6.124. Determine the root-mean-square speed of CO_2 molecules that have an average kinetic energy of 4.2×10^{-21} J per molecule.

6.125. What is the ratio of the root-mean-square speed of D_2 to that of H_2 at constant temperature?

6.126. **Enriching Uranium** The two isotopes of uranium, ^{238}U and ^{235}U, can be separated by diffusion of the corresponding UF_6 gases. What is the ratio of the root-mean-square speed of $^{238}UF_6$ to that of $^{235}UF_6$ at constant temperature?

6.127. Molecular hydrogen effuses 4 times as fast as gas X at the same temperature. What is the molar mass of gas X?

6.128. Gas Y effuses half as fast as O_2 at the same temperature. What is the molar mass of gas Y?

6.129. If an unknown gas has one-third the root-mean-square speed of H_2 at 300 K, what is its molar mass?

6.130. A flask of ammonia is connected to a flask of an unknown acid HX by a 1.00 m glass tube. As the two gases diffuse down the tube, a white ring of NH_4X forms 68.5 cm from the ammonia flask. Identify element X.

6.131. Isotope Use by Plants During photosynthesis, green plants preferentially use $^{12}CO_2$ over $^{13}CO_2$ in making sugars, and food scientists can frequently determine the source of sugars used in foods on the basis of the ratio of ^{12}C to ^{13}C in a sample.
a. Calculate the relative rates of diffusion of $^{13}CO_2$ and $^{12}CO_2$.
b. Specify which gas diffuses faster.

6.132. At fixed temperature, how much faster does NO effuse than NO_2?

6.133. Two balloons were filled with H_2 and He, respectively. The person responsible for filling them neglected to label them. After 24 hr the volumes of both balloons had decreased but by different amounts. Which balloon contained hydrogen?

6.134. Compounds sensitive to oxygen are often manipulated in *glove boxes* that contain a pure nitrogen or pure argon atmosphere. A balloon filled with carbon monoxide was placed in a glove box. After 24 hr, the volume of the balloon was unchanged. Did the glove box contain N_2 or Ar?

Real Gases

CONCEPT REVIEW

6.135. Explain why real gases behave nonideally at low temperatures and high pressures.

6.136. Under what conditions is the pressure exerted by a real gas *less* than that predicted for an ideal gas?

6.137. Explain why the values of the van der Waals constant b of the noble gas elements increase with atomic number.

6.138. Explain why the constant a in the van der Waals equation generally increases with the molar mass of the gas.

PROBLEMS

6.139. The graphs of PV/RT versus P (see Figure 6.41) for 1 mole of CH_4 and 1 mole of H_2 differ in how they deviate from ideal behavior. For which gas is the effect of the volume occupied by the gas molecules more important than the attractive forces between molecules?

6.140. Which noble gas is expected to deviate the most from ideal behavior in a graph of PV/RT versus P?

6.141. At high pressures, real gases do not behave ideally. (a) Use the van der Waals equation and data in the text to calculate the pressure exerted by 40.0 g H_2 at 20°C in a 1.00 L container. (b) Repeat the calculation assuming that the gas behaves like an ideal gas.

6.142. (a) Calculate the pressure exerted by 1.00 mole of CO_2 in a 1.00 L vessel at 300 K, assuming that the gas behaves ideally. (b) Repeat the calculation using the van der Waals equation.

Additional Problems

6.143. The volume of a sample of propane gas at 12.5 atm is 10.6 L. What volume does the gas occupy if the pressure is reduced to 1.05 atm and the temperature remains constant?

6.144. The pressure on a sample of carbon dioxide gas collected during a laboratory experiment is increased from 10.5 mmHg to 765 mmHg. If the initial volume of the gas is 185 mL, what volume does the gas occupy at the higher pressure if the temperature remains constant?

6.145. A 22.4 L sample of hydrogen chloride gas is heated from 15°C to 78°C. What volume does it occupy at the higher temperature if the pressure remains constant?

6.146. What is the new temperature in °C of 6.95 L of helium initially at 25°C if its volume is increased to 15.00 L? Assume there is no change in the pressure of the gas.

6.147. A gas cylinder in the back of an open truck experiences a temperature change from −8.6°C to 110°C as it is driven from high in the Rocky Mountains of Colorado to Death Valley, CA. If the pressure gauge on the tank reads 14.6 atm in Colorado, what do you predict the gauge will read in Death Valley?

6.148. The temperature of a quantity of methane gas at a pressure of 761 mmHg is 18.6°C. Predict the temperature of the gas if the pressure is reduced to 355 mmHg while the volume remains constant.

6.149. A souvenir soccer ball is partially deflated and then put into a suitcase. At a pressure of 0.947 atm and a temperature of 27°C, the ball has a volume of 1.034 L. What volume does it occupy during an airplane flight if the pressure in the baggage compartment is 0.235 atm and the temperature −35°C?

6.150. Asthma Therapy A gas mixture used experimentally for asthma treatments contains 17.5 moles of helium for every 0.938 moles of oxygen. What is the mole fraction of oxygen in the mixture?

6.151. Scientists have used laser light to slow atoms to speeds corresponding to temperatures below 0.00010 K. At this temperature, what is the root-mean-square speed of argon atoms?

6.152. Blood Pressure A typical blood pressure in a resting adult is "120 over 80," meaning 120 mmHg with each beat of the heart and 80 mmHg of pressure between heartbeats. Express these pressures in the following units: (a) torr; (b) atm; (c) bar; (d) kPa.

*6.153. A popular scuba tank is the "aluminum 80," so named because it can deliver 80 cubic feet of air at "normal" temperature (72°F) and pressure (1.00 atm, 14.7 psi) when filled with air at a pressure of 3000 psi. A particular aluminum 80 tank has a mass of 15 kg empty. What is its mass when filled with air at 3000 psi?

6.154. The flame produced by the burner of a gas (propane) grill is a pale blue color when enough air mixes with the propane (C_3H_8) to burn it completely. For every gram of propane that flows through the burner, what volume of air is needed to burn it completely? Assume that the temperature of the burner is 200°C, the pressure is 1.00 atm, and the mole fraction of O_2 in air is 0.21.

6.155. Which noble gas effuses at about half the effusion rate of O_2?

6.156. Anesthesia A common anesthesia gas is halothane, with the formula $C_2HBrClF_3$ and the structure shown in Figure P6.156. Liquid halothane boils at 50.2°C and 1.00 atm. If halothane behaved as an ideal gas, what volume would 10.0 mL of liquid halothane ($d = 1.87$ g/mL) occupy at 60°C and 1.00 atm of pressure? What is the density of halothane vapor at 55°C and 1.00 atm of pressure?

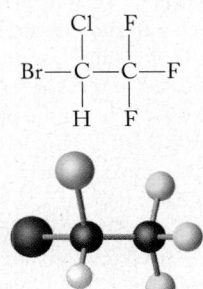

FIGURE P6.156

6.157. A cotton ball soaked in ammonia and another soaked in hydrochloric acid were placed at opposite ends of a 1.00 m glass tube (Figure P6.157). The vapors diffused toward the middle of the tube and formed a white ring of ammonium chloride where they met.

a. Write the chemical equation for this reaction.

b. Should the ammonium chloride ring be closer to the end of the tube with ammonia or the end with hydrochloric acid? Explain your answer.

c. Calculate the distance from the ammonia end to the position of the ammonium chloride ring.

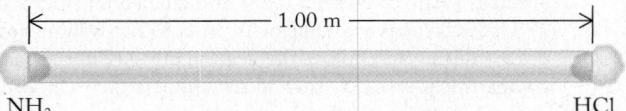

NH₃   HCl

FIGURE P6.157

*6.158.** The same apparatus described in Problem 6.157 was used in another series of experiments. A cotton ball soaked in either hydrochloric acid (HCl) or acetic acid (CH₃COOH) was placed at one end. Another cotton ball soaked in one of three amines (a class of organic compounds)—CH₃NH₂, (CH₃)₂NH, or (CH₃)₃N—was placed in the other end (Figure P6.158).

a. In one combination of acid and amine, a white ring was observed almost exactly halfway between the two ends. Which acid and which amine were used?

b. Which combination of acid and amine would produce a ring closest to the amine end of the tube?

c. Do any two of the six combinations result in the formation of product at the same position in the ring? Assume measurements can be made to the nearest centimeter.

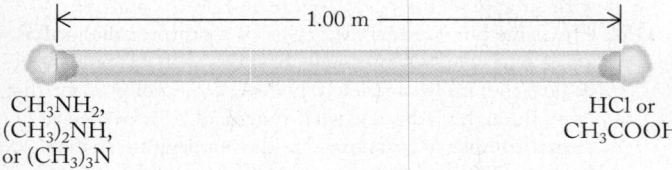

CH₃NH₂,   HCl or
(CH₃)₂NH,   CH₃COOH
or (CH₃)₃N

FIGURE P6.158

6.159. A flashbulb of volume 2.6 mL contains $O_2(g)$ at a pressure of 2.3 atm and a temperature of 26°C. How many grams of $O_2(g)$ does the flashbulb contain?

6.160. The pressure in an aerosol can is 1.5 atm at 27°C. The can will withstand a pressure of about 2 atm. Will it burst if heated in a campfire to 450°C?

6.161. An expandable container at 50.0°C and 1.00 atm pressure contains 30.0 g of $CO(g)$. How many times larger is the volume of the gas when 51.0 g of the gas is in the container at the same temperature and pressure?

6.162. A sample of 11.4 L of an ideal gas at 25.0°C and 735 torr is compressed and heated so that the volume is 7.9 L and the temperature is 72.0°C. What is the pressure in the container?

6.163. A sample of argon gas at STP occupies 15.0 L. What mass of argon is present in the container?

6.164. A sample of a gas has a mass of 2.889 g and a volume of 940 mL at 735 torr and 31°C. What is its molar mass?

6.165. What pressure is exerted by a mixture of 2.00 g of H_2 and 7.00 g of N_2 at 273°C in a 10.0 L container?

6.166. Uranus has a total atmospheric pressure of 130 kPa and consists of the following gases: 83% H_2, 15% He, and 2% CH_4 by volume. Calculate the partial pressure of each gas in Uranus's atmosphere.

6.167. A sample of $N_2(g)$ requires 240 s to diffuse through a porous plug. It takes 530 s for an equal number of moles of an unknown gas X to diffuse through the plug under the same conditions of temperature and pressure. What is the molar mass of gas X?

6.168. The rate of effusion of an unknown gas is 0.10 m/s and the rate of effusion of $SO_3(g)$ is 0.052 m/s under identical experimental conditions. What is the molar mass of the unknown gas?

6.169. Why do gas bubbles exhaled by a scuba diver get larger as they rise to the surface?

6.170. Derive an equation that expresses the ratio of the densities (d_1 and d_2) of a gas under two different combinations of temperature and pressure, (T_1, P_1) and (T_2, P_2).

6.171. Denitrification in the Environment In some aquatic ecosystems, nitrate (NO_3^-) is converted to nitrite (NO_2^-), which then decomposes to nitrogen and water. As an example of this second reaction, consider the decomposition of ammonium nitrite:

$$NH_4NO_2(aq) \rightarrow N_2(g) + 2H_2O(\ell)$$

What would be the change in pressure in a sealed 10.0 L vessel due to the formation of N_2 gas when the ammonium nitrite in 1.00 L of 1.0 M NH_4NO_2 decomposes at 25°C?

*6.172.** When sulfur dioxide bubbles through a solution containing nitrite, chemical reactions that produce gaseous N_2O and NO may occur.

a. How much faster on average would NO molecules be moving than N_2O molecules in such a reaction mixture?

b. If these two nitrogen oxides were to be separated based on differences in their rates of effusion, would unreacted SO_2 interfere with the separation? Explain your answer.

*6.173.** **Using Wetlands to Treat Agricultural Waste** Wetlands can play a significant role in removing fertilizer residues from rain runoff and groundwater; one way they do this is through denitrification, which converts nitrate ions to nitrogen gas:

$$2NO_3^-(aq) + 5CO(g) + 2H^+(aq) \rightarrow$$
$$N_2(g) + H_2O(\ell) + 5CO_2(g)$$

Suppose 200.0 g of NO_3^- flows into a swamp each day. What volume of N_2 would be produced at 17°C and 1.00 atm if the denitrification process were complete? What volume of CO_2 would be produced? Suppose the gas mixture produced by the decomposition reaction is trapped in a container at 17°C; what is the density of the mixture assuming $P_{total} = 1.00$ atm?

6.174. Ammonium nitrate decomposes on heating. The products depend on the reaction temperature:

$$NH_4NO_3(s) \xrightarrow{>300°C} N_2(g) + \tfrac{1}{2}O_2(g) + H_2O(g)$$
$$\xrightarrow{200-260°C} N_2O(g) + 2H_2O(g)$$

A sample of NH_4NO_3 decomposes at an unspecified temperature, and the resulting gases are collected over water at 20°C.

a. Without completing a calculation, predict whether the volume of gases collected can be used to distinguish between the two reaction pathways. Explain your answer.

b. The gas produced during the thermal decomposition of 0.256 g of NH_4NO_3 displaces 79 mL of water at 20°C and 760 mmHg of atmospheric pressure. Is the gas N_2O or a mixture of N_2 and O_2?

6.175. Calculate the pressure in pascals exerted by the atmosphere on a 1.0 m^2 area in La Paz, Bolivia, given a mass of gases of 6.6×10^3 kg.

6.176. A balloon is partly inflated with 5.00 L of helium at sea level where the atmospheric pressure is 1008 mbar. The balloon ascends to an altitude of 3000 meters, where the pressure is 855 mbar. What is the volume of the helium in the balloon at the higher altitude? Assume that the temperature of the gas in the balloon does not change in the ascent.

6.177. Tropical Storms The severity of a tropical storm is related to the depressed atmospheric pressure at its center. Figure P6.177 is a photograph of Typhoon Odessa taken from the space shuttle *Discovery* in August 1985, when the maximum winds of the storm were about 90 mi/hr and the pressure was 40 mbar lower at the center than normal atmospheric pressure. In contrast, the central pressure of Hurricane Andrew was 90 mbar lower than its surroundings when it hit southern Florida with winds as high as 165 mi/hr. If a small weather balloon with a volume of 50.0 L at a pressure of 1.0 atmosphere was deployed at the center of Andrew, what was the volume of the balloon when it reached the center?

FIGURE P6.177

Nitrogen: Feeding Plants and Inflating Air Bags

6.178. Write balanced chemical equations for the following reactions: (a) nitrogen reacts with hydrogen; (b) nitrogen dioxide dissolves in water.

6.179. Write balanced chemical equations for reactions of ammonia with (a) oxygen and (b) nitric acid.

6.180. Military Signal Flares Flares are used for signaling or as defensive countermeasures in both civilian and military applications. Both the reaction of black powder and a possible replacement in handheld signal flares with materials of more consistent ignition properties have been studied by the U.S. Army, among many others. The chemical reaction that occurs when black powder is ignited is

$$2\,KNO_3(s) + \tfrac{1}{8}\,S_8(s) + 3\,C(s) \rightarrow K_2S(s) + N_2(g) + 3\,CO_2(g)$$

a. Identify the oxidizing and reducing agents.
b. How many electrons are transferred per mole of KNO_3?

6.181. A simple laboratory-scale preparation of nitrogen involves the thermal decomposition of ammonium nitrite (NH_4NO_2).

a. Write a balanced chemical equation for the decomposition of NH_4NO_2 to give nitrogen and water.
b. Is the thermal decomposition of NH_4NO_2 a redox reaction?

6.182. Oklahoma City Bombing The 1995 explosion that destroyed the Murrah Federal Office Building in Oklahoma City was believed to be caused by the reaction of ammonium nitrate with fuel oil:

$$3\,NH_4NO_3(s) + C_{10}H_{22}(\ell) + 14\,O_2(g) \rightarrow$$
$$3\,N_2(g) + 17\,H_2O(g) + 10\,CO_2(g)$$

a. What common use for ammonium nitrate made it easy for the conspirators to acquire this compound in large quantities without suspicion?
b. Which elements are oxidized and which are reduced in the reaction?
*c. Write a chemical equation for the thermal decomposition of ammonium nitrate in the absence of fuel and oxygen.

*6.183. Automobile Air Bags Sodium azide is used in automobile air bags as a source of nitrogen gas in the reaction

$$2\,NaN_3(s) \rightarrow 2\,Na(s) + 3\,N_2(g)$$

The reaction is rapid at 350°C but produces elemental sodium. Potassium nitrate and silica (SiO_2) are added to remove the liquid sodium metal formed in the reaction. Why is it necessary from a safety perspective to avoid molten alkali metals in an air bag?

6.184. Manufacturing a Proper Air Bag Here is the overall reaction in an automobile air bag:

$$20\,NaN_3(s) + 6\,SiO_2(s) + 4\,KNO_3(s) \rightarrow$$
$$32\,N_2(g) + 5\,Na_4SiO_4(s) + K_4SiO_4(s)$$

Calculate how many grams of sodium azide (NaN_3) are needed to inflate a $40 \times 40 \times 20$ cm bag to a pressure of 1.25 atm at a temperature of 20°C. How much more sodium azide is needed if the air bag must produce the same pressure at 10°C?

6.185. The first rocket-propelled aircraft used a mixture of hydrogen peroxide and hydrazine as the propellant. Using the thermochemical data in the appendices and ΔH_f° of $N_2H_4(\ell) = 50.63$ kJ/mol, calculate the enthalpy change for the following reaction:

$$2\,H_2O_2(\ell) + N_2H_4(\ell) \rightarrow N_2(g) + 4\,H_2O(g)$$

*6.186. In Problem 6.76, nitrogen gas obtained by boiling liquid nitrogen ($d = 0.808$ g/mL) from a 182 L tank was used to power a car. How much hydrazine (in grams) is needed to produce an equivalent amount of N_2 gas by reaction with hydrogen peroxide in Problem 6.185?

Additional study materials including ChemTours and Diagnostic Quizzes are available at StudySpace at www.wwnorton.com/studyspace.

7

Electrons in Atoms and Periodic Properties

A LOOK AHEAD

Can Nature Be as Absurd as It Seems?

Those who study the history of chemistry sometimes call the 18th century the century of mass. Most of the ideas of stoichiometry that are fundamental to chemistry as a quantitative science come from 18th-century measurements of the masses of reacting substances. Building on this basis for the atomic view of matter, chemistry in the 19th and early 20th centuries focused on the interaction between matter and energy. Such inquiries led to the discovery of electrons, protons, and neutrons as the fundamental particles from which all matter is made.

At the end of the 19th century, some scientists believed that all the major scientific discoveries had already been made and all that was left to do was to solve a few lingering problems. One of these problems, called the photoelectric effect, involved the flow of electrons caused by light shining on a metal surface. Another involved the range of energies given off by hot objects. A third dealt with the light emitted from gases such as neon when an electric current was passed through them.

In this chapter we see that 20th-century solutions to these and related problems resulted in a new vision of matter at the atomic level and a new theory called quantum theory. In addition to giving us powerful insights into the way the world works on the atomic level, these developments forever altered our sense of what we can really know about the world around us.

The world at the atomic level is governed by a set of rules entirely different from those describing behavior in the macroscopic world. Quantum theory explains these rules. Its development was a towering intellectual achievement, and the picture it paints of how electrons and light behave is like nothing any of us have ever seen. In fact, the rules

Learning Outcomes

- Describe the wavelike and particle-like properties of electromagnetic radiation

- Explain the complementary nature of the absorption and emission lines of atomic spectra and describe the relationship of those lines to electron transitions between energy levels in atoms

- Explain the photoelectric effect using quantum theory

- Assign quantum numbers to orbitals and use their values to describe the size, energy, and orientation of orbitals

- Use the aufbau principle and Hund's rule to write electron configurations and draw orbital diagrams of atoms and monatomic ions

- Explain the energies of orbitals based on the concept of effective nuclear charge

- Relate the ionization energies and electron affinities of the elements to their positions in the periodic table

Forging Molten Metal Molten metals emit broad ranges of electromagnetic radiation that depend on the metals' temperature. Scientists' efforts to explain the properties of this radiation led to the development of quantum theory and a whole new way of viewing matter at the atomic level. ▶

electromagnetic spectrum a continuous range of radiant energy that includes radio waves, infrared radiation, visible light, ultraviolet radiation, X-rays, and gamma rays.

electromagnetic radiation any form of radiant energy in the electromagnetic spectrum.

wavelength (λ) the distance from crest to crest or trough to trough on a wave.

frequency (ν) the number of crests of a wave that pass a stationary point of reference per second.

hertz (Hz) the SI unit of frequency with units of reciprocal seconds: 1 Hz = 1 s^{-1} = 1 cycle per second (cps).

amplitude the height of the crest or depth of the trough of a wave with respect to the center line of the wave.

CONNECTION In Chapter 1 we discussed the background microwave radiation of the universe, discovered by Penzias and Wilson, that provided evidence for the Big Bang.

▶‖ **CHEMTOUR** Electromagnetic Radiation

that apply at the atomic level often seem to contradict our understanding of how nature works at the macroscopic level.

Two of the giants of quantum theory, Niels Bohr and Werner Heisenberg, wrestled with these contradictions. They briefly worked together at the University of Copenhagen, and after a particularly heated discussion about a feature of quantum theory, Heisenberg posed this question to himself: "Can nature possibly be as absurd as it seems?" The answer is probably yes.

In this chapter we explore the evolution of quantum theory. How do matter and energy interact at the atomic level? In addressing this question, we link the spectra of atoms to the structure of atoms, and in particular to patterns in the distribution of electrons around the nuclei of atoms. These patterns help explain the properties of the elements, including the charges on the ions they form, and how elements interact with other elements as they form compounds. ■

7.1 Waves of Light

In Chapter 1 we discussed the Big Bang theory. That theory evolved from the discovery by American astronomer Edwin Hubble that the universe has been expanding for billions of years. To understand how Hubble came to this conclusion, we need to examine the nature of the light emitted by our sun and all the other stars in the universe.

Properties of Waves

Visible light is a small part of the **electromagnetic spectrum** (Figure 7.1), which consists of a continuous range of radiant energy that extends from low-energy radio waves to ultrahigh-energy gamma rays. All forms of radiant energy are examples of **electromagnetic radiation**.

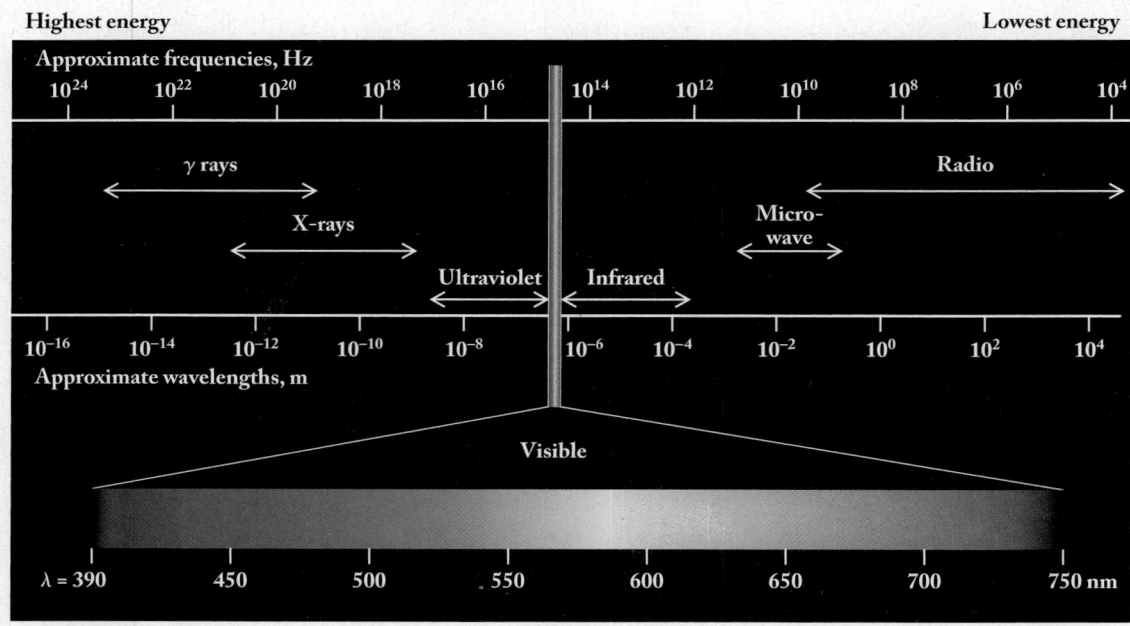

FIGURE 7.1 The visible light region between about 390 and 750 nm is a tiny fraction of the electromagnetic spectrum, which ranges from very-short-wavelength, high-frequency gamma (γ) rays to long-wavelength, low-frequency radio waves. Note the inverse relation between frequency and wavelength: frequencies increase from right to left here, and wavelengths increase from left to right.

The term *electromagnetic* comes from the theory proposed by Scottish scientist James Clerk Maxwell (1831–1879) that radiant energy moving through space (or the air) does so in a way that resembles waves flowing across a body of water. Unlike water waves, which only oscillate up and down, Maxwell's waves of radiant energy have two components: an oscillating electric field and an oscillating magnetic field (Figure 7.2) that are perpendicular to each other and travel together through space. Maxwell derived a set of equations based on his oscillating-wave model that accurately describes nearly all the observed properties of light.

A wave of electromagnetic radiation, like any wave traveling through any medium, has a characteristic **wavelength** (the distance from crest to crest, λ) and **frequency** (the number of crests that pass a stationary point of reference per second, ν), as shown in Figure 7.3(a). Frequencies have units of **hertz (Hz)**, also called *cycles per second* (cps): 1 Hz = 1 cps = 1 s^{-1}. The product of the wavelength and frequency of any electromagnetic radiation is the universal constant c, which is the symbol for the *speed of light* in a vacuum:

$$\lambda\nu = c = 2.99792458 \times 10^8 \text{ m/s} \approx 3.00 \times 10^8 \text{ m/s} \qquad (7.1)$$

Thus wavelength and frequency have a reciprocal relationship: as the wavelength decreases, the frequency increases. Another characteristic of a wave is its **amplitude**, the height of the crest or the depth of the trough with respect to the center line of the wave (Figure 7.3b).

SAMPLE EXERCISE 7.1   **Calculating Frequency from Wavelength**

What is the frequency of the yellow-orange light ($\lambda = 589$ nm) produced by sodium-vapor streetlights?

Collect and Organize We are given light of a specific wavelength in nanometers and asked to find the frequency. Frequency and wavelength are related by Equation 7.1 ($\lambda\nu = c$).

Analyze The value for c given in Equation 7.1 is in meters per second, but the wavelength is given in nanometers. Therefore we need to convert nanometers to meters: 1 nm = 10^{-9} m. The unit labels in Figure 7.1 indicate that frequencies of visible light are in the 10^{14} Hz range, so a correct answer should have a value in this range.

Solve Let's rearrange Equation 7.1 to solve for frequency

$$\nu = \frac{c}{\lambda}$$

and then convert the wavelength units from nanometers to meters:

$$589 \text{ nm} \times \frac{10^{-9} \text{ m}}{1 \text{ nm}} = 589 \times 10^{-9} \text{ m} = 5.89 \times 10^{-7} \text{ m}$$

The frequency is thus

$$\nu = \frac{3.00 \times 10^8 \text{ m/s}}{5.89 \times 10^{-7} \text{ m}}$$
$$= 5.09 \times 10^{14} \text{ s}^{-1} = 5.09 \times 10^{14} \text{ Hz}$$

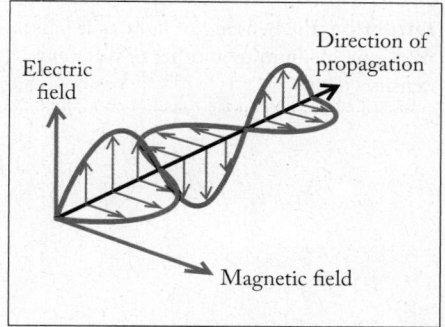

FIGURE 7.2 Electromagnetic waves consist of electric and magnetic fields that oscillate in planes oriented at right angles to each other.

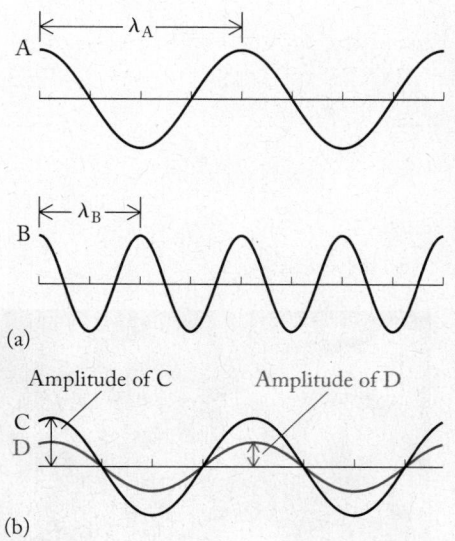

FIGURE 7.3 Every wave has a characteristic wavelength (λ), frequency (ν), and amplitude (intensity). (a) Wave A has a longer wavelength (and lower frequency) than wave B. (b) Waves C and D have the same wavelength and frequency but the amplitude of C is greater than that of D.

refraction the bending of light as it passes from one medium to another of different density.

Think about It This large value is in the range we expected. Remember that λ and ν are reciprocal: as one increases, the other decreases.

Practice Exercise If the radio waves transmitted by a radio station have a frequency of 90.9 MHz, what is the wavelength of the waves, in meters?

(Answers to Practice Exercises are in the back of the book.)

CONCEPT TEST

The ultraviolet (UV) region of the electromagnetic spectrum contains waves with wavelengths from about 10^{-7} m to 10^{-9} m; the infrared (IR) region contains waves with wavelengths from about 10^{-4} m to 10^{-6} m. Are waves in the UV region higher in frequency or lower in frequency than waves in the IR region?

(Answers to Concept Tests are in the back of the book.)

The Behavior of Waves

In the 17th century, British physicist Isaac Newton (1642–1727) used glass prisms to separate sunlight into its component wavelengths, forming the continuum of the colors that we see in rainbows (Figure 7.4). This separation process is known as **refraction** because the path of a beam of light entering a prism (or a raindrop) is bent, or *refracted*, as it moves from air into the denser medium. Different wavelengths of light are bent through different angles: violet light (the

(a)

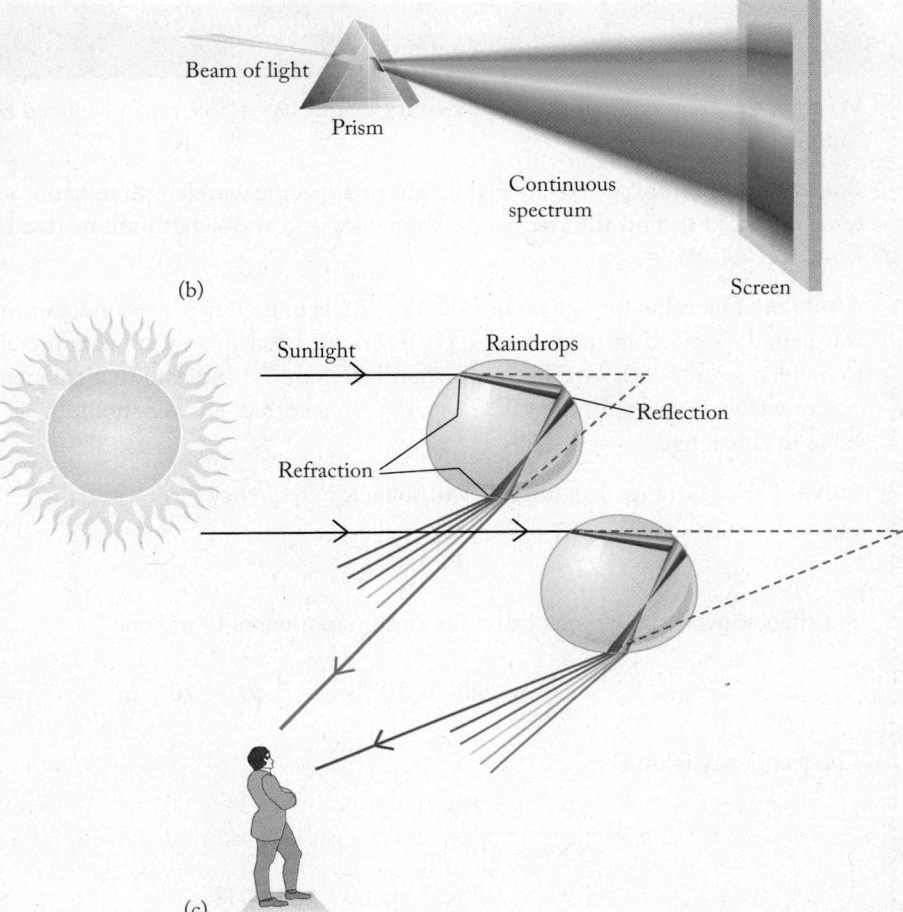

FIGURE 7.4 The components of visible light. (a) Sir Isaac Newton used a prism to separate visible light into a continuous spectrum containing all the colors of the rainbow. (b) As a beam of visible light passes through a prism, the different components of the light are refracted by different amounts. Red light is refracted the least, violet light the most. (c) A combination of refraction and reflection of sunlight inside raindrops produces rainbows. We see the red end of the visible spectrum at the top of the rainbow because of the large angle between the incident and reflected/refracted rays of red light. The corresponding angle for violet rays of light is smaller, as shown by the pairs of dashed lines.

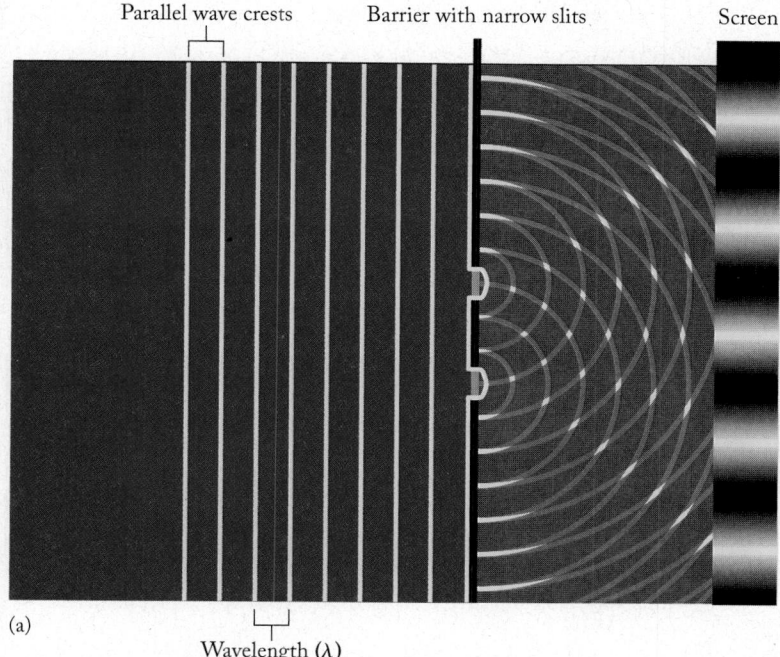

Parallel wave crests   Barrier with narrow slits   Screen

(a)

Wavelength (λ)

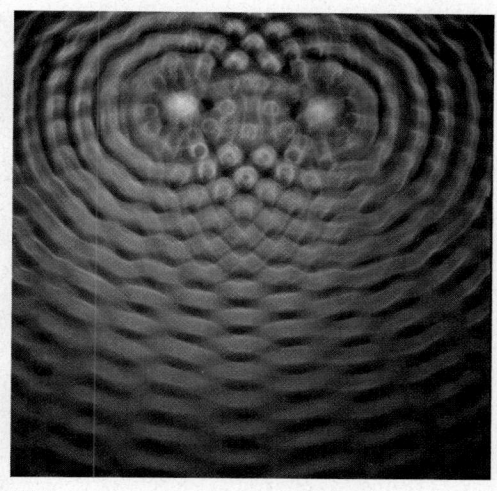

(b)

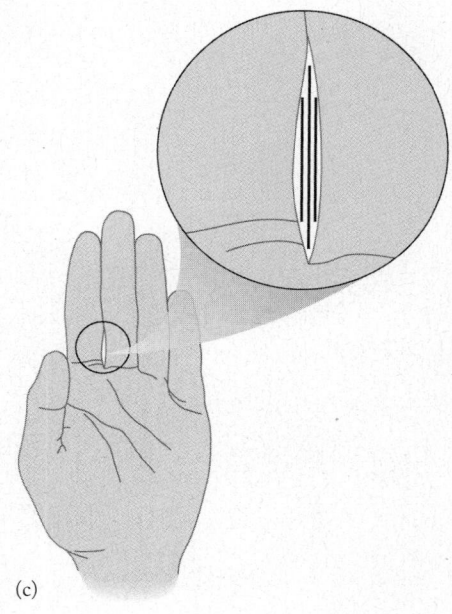

(c)

FIGURE 7.5 Diffraction patterns and interference: (a) When waves approach a barrier that has gaps about the size of the wavelength, the waves are bent around the edges of the gaps and radiate out in circular wave patterns. This is diffraction. When these circular diffracted waves overlap, they produce a pattern of bright and dark bands. (b) A similar pattern is produced when two pebbles are dropped into a pool. (c) Hold your hand in front of a light with your palm facing you; slowly bring your fingers together until they form a narrow slit. The dark lines in the gap between your fingers are an interference pattern generated by diffraction of the light.

shortest wavelengths of visible light) is bent the most, and red light is bent the least. When Newton placed two prisms back to back, the colors produced by refraction in the first prism were recombined into *white light* (light that contains all the colors in the visible spectrum) by the second prism. These results mean that refraction separates light into its component wavelengths but does not otherwise change the light, and that white light is a mixture of all colors of light. Refraction is not unique to visible light; any electromagnetic radiation is refracted when it passes from a substance with one density into a substance with a different density.

Another property of electromagnetic radiation is that it undergoes **diffraction**. Diffraction happens when a beam of light passes through narrow slits as shown in Figure 7.5(a). As the waves emerge from the slits, they spread out in circles. This pattern means that the waves passing through the slits bend around, or are *diffracted* by, the slits.

When two sets of these circular waves spread out from a pair of slits, as in Figure 7.5, they run into one another and in so doing they **interfere** with one another, much like the spreading waves produced when two pebbles are simultaneously dropped in a pool (Figure 7.5b). In both cases, the overlapping waves produce diffraction patterns. The waves in a pool produce series of concentric arcs as shown in Figure 7.5(b). When interfering waves of light diffracted by the pair of slits in Figure 7.5(a) are projected on a screen, we see a pattern of light and dark bands. Figure 7.5(c) shows how you can see an interference pattern generated by diffraction through a slit formed with your fingers.

▶❚❚ **CHEMTOUR** Light Diffraction

∞ **CONNECTION** Diffraction of light caused by water suspended in layers of calcium carbonate was originally mentioned in Chapter 4 as the phenomenon responsible for the iridescence of pearls.

diffraction bending of electromagnetic radiation as it passes around an edge of an object or through a narrow opening.

interference the interaction of waves that results in either reinforcing their amplitudes (constructive interference) or canceling them out (destructive interference).

FIGURE 7.6 Constructive and destructive interference. When the crests of two identical waves overlap, the waves are in phase and experience constructive interference; they enhance each other. When a crest from one pattern overlaps a trough of the other, the waves are out of phase and experience destructive interference; they cancel each other. Constructive interference results in bright bands on the screen on the left, and destructive interference results in dark bands. The bright and dark bands form a diffraction pattern.

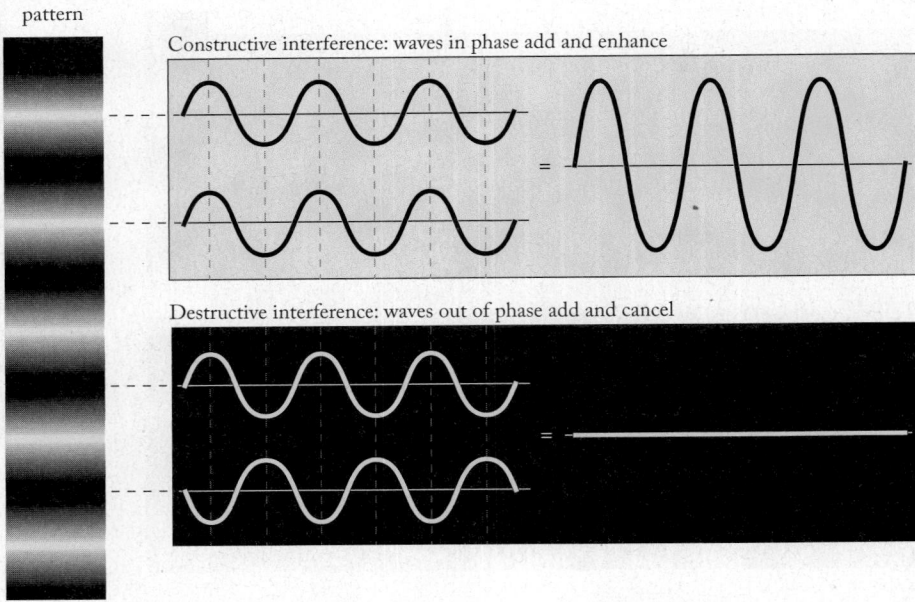

Diffraction pattern

Constructive interference: waves in phase add and enhance

Destructive interference: waves out of phase add and cancel

To understand how these bands form, consider two waves with the same wavelength traveling in the same direction as shown in Figure 7.6. When the crests of the two waves overlap, the waves are said to be *in phase*. They combine *constructively*, and the amplitude of the resulting wave is twice the amplitude of the two original waves. However, when the crest of one wave overlaps the trough of the other, the waves are said to be *out of phase*. They combine *destructively*, the amplitude of the resultant wave is zero, and the wave disappears. We can use this notion of constructive and destructive interference to explain the pattern on the screen in Figure 7.5(a): the light bands on the screen are the result of constructive interference, and the dark bands are caused by destructive interference.

CONCEPT TEST

When two identical waves of red light ($\lambda = 660$ nm) interfere constructively, is there any change in the wavelength and the frequency of the light?

The Expanding Universe

In 1800, English scientist William Hyde Wollaston (1766–1828) made a startling discovery about sunlight. Using carefully ground prisms to separate a beam of sunlight into its component colors, Wollaston discovered that the spectrum was not continuous. Instead, the spectrum contained dark, narrow lines (Figure 7.7). Using even better-quality prisms, German physicist Joseph von Fraunhofer (1787–1826) resolved and mapped the wavelengths of over 500 of these dark lines, now called **Fraunhofer lines** (Figure 7.7).

FIGURE 7.7 The spectrum of sunlight is not continuous but contains numerous narrow gaps, which appear as dark lines called Fraunhofer lines.

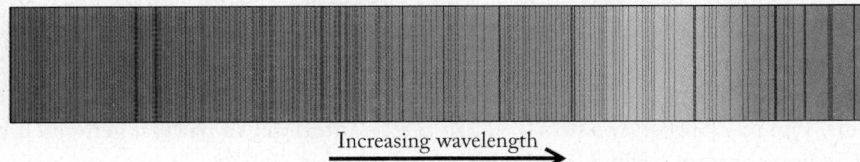

Increasing wavelength

Fraunhofer lines were crucial to the discovery that we live in an expanding universe, because similar patterns of dark lines are seen in the light produced by stars throughout the universe. Though the pattern is much the same, the wavelengths of the lines in galactic spectra are not exactly the same as those in sunlight: they tend to be shifted to longer wavelengths. Because the color of the longest wavelength of visible light is red, scientists call this shift a *redshift*.

To understand why redshifts occur, let's consider a model from closer to home: the different sounds of the whistle of a moving train. Sound is a wave phenomenon, and the *pitch* of any sound is directly related to the frequency of the sound wave. As a train approaches, we perceive a pitch that is higher than if the train were not moving. In other words, the sound waves reaching our ears are slightly compressed and have shorter wavelengths and higher frequencies. After the train has passed and moves away from us, the sound waves are stretched out and we hear a lower pitched, lower frequency sound.

This motion-induced shifting of frequency is called the *Doppler effect*. The Doppler effect also explains the redshift of galactic light: Distant galaxies are moving away from Earth so rapidly that when their radiation reaches us, it appears to have been stretched out to longer wavelengths and lower frequencies. In addition, light from the most distant galaxies has the greatest redshifts, meaning that these galaxies are moving away the most rapidly. We discussed the implications of these observations in Chapter 1 as we considered what would happen if time ran backwards to a moment 13.7 billion years ago when all the matter in the universe was compressed into an infinitesimal volume of infinite density and temperature: the moment we call the Big Bang.

Fraunhofer lines a set of dark lines in the otherwise continuous solar spectrum.

atomic emission spectrum (also called **bright-line spectrum**) a characteristic series of bright lines produced by excited-state atoms.

▶❚❚ **CHEMTOUR** Doppler Effect

7.2 Atomic Spectra

Fraunhofer and his contemporaries knew that there were narrow regions of light missing from the sun's spectrum, but they did not know why. Nearly a half century later, German chemist Robert Wilhelm Bunsen (1811–1899) and physicist Gustav Robert Kirchhoff (1824–1887) collaborated on extensive studies of the light emitted (given off) by elements vaporized in the transparent flame of a burner designed by Bunsen. Unlike the spectrum of sunlight, which displays gaps in a nearly continuous spectrum of all colors, the spectra produced by the elements in these flames consist of only a few bright lines on a dark background. Bunsen and Kirchhoff discovered that these **atomic emission spectra** (or **bright-line spectra**) consist of bright (colored) lines at *exactly the same wavelengths* as some of the dark (black) Fraunhofer lines in the spectrum of sunlight. For example, the Fraunhofer D line (actually a pair of closely spaced lines at 589.0 and 589.6 nm) exactly matched the two bright lines corresponding to the yellow-orange light produced by hot sodium vapor (Figure 7.8).

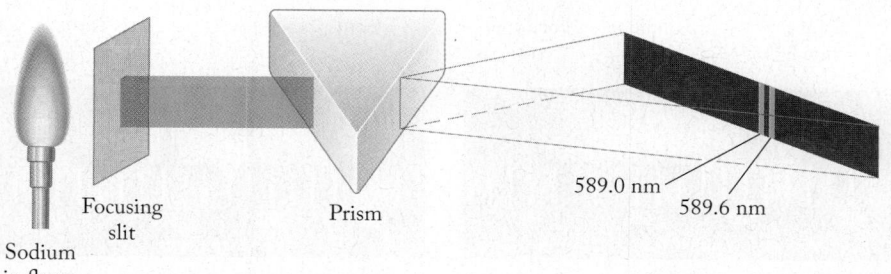

589.0 nm
589.6 nm
Focusing slit
Prism
Sodium in flame

FIGURE 7.8 Sodium atoms heated in a flame emit a bright yellow-orange light. When this light is passed through a prism, the result is an atomic emission (bright-line) spectrum consisting of two bright lines that exactly match the wavelength and frequency of a prominent Fraunhofer line.

atomic absorption spectrum (also called **dark-line spectrum**) a characteristic series of dark lines produced when free, gaseous atoms are illuminated by an external source of radiation.

quantum (plural *quanta*) the smallest discrete quantity of a particular form of energy.

quantum theory a model based on the idea that energy is absorbed and emitted in discrete quantities of energy called quanta.

▶❚❚ **CHEMTOUR** Light Emission and Absorption

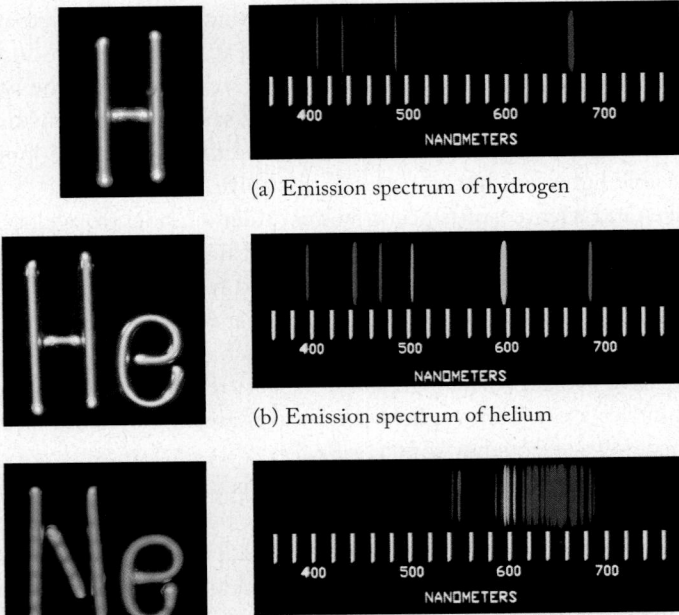

(a) Emission spectrum of hydrogen

(b) Emission spectrum of helium

(c) Emission spectrum of neon

FIGURE 7.9 The colored light emitted from gas discharge tubes filled with various gaseous elements produces atomic emission spectra that are characteristic of the element: (a) hydrogen, (b) helium, (c) neon.

These experiments and others employing light sources called gas discharge tubes showed that each element *emits* characteristic electromagnetic radiation when its atoms are heated to a sufficiently high temperature. This emission is captured in the element's atomic emission spectrum (Figure 7.9). Elements *absorb* electromagnetic radiation when illuminated by an external source of radiation. This absorption of radiation by atoms produces an **atomic absorption spectrum** (or **dark-line spectrum**), which consists of dark lines in an otherwise colored spectrum (Figure 7.10). For any given element, the dark lines in its absorption spectrum are at the same wavelengths as the bright lines in its emission spectrum. Thus the two processes we know as *emission* and *absorption* involve electromagnetic radiation of the same wavelength (and energy). The phenomenon of absorption explains the origins of the Fraunhofer lines: gaseous atoms in the outer regions of the sun absorb characteristic wavelengths of the sunlight passing through them on its way to Earth.

FIGURE 7.10 (a) When gaseous atoms of hydrogen, helium, and neon are illuminated by an external source of white light (containing all colors of the visible spectrum), the resultant atomic absorption spectrum contains dark lines that are characteristic of the element. The dark lines in the atomic absorption spectra of these elements (b) have the same wavelengths as the bright lines in their atomic emission spectra, shown in Figure 7.9.

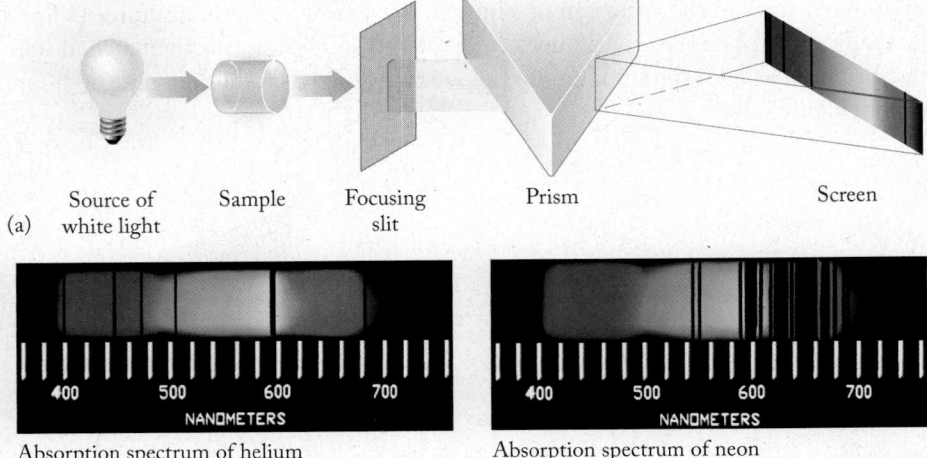

(a) Source of white light   Sample   Focusing slit   Prism   Screen

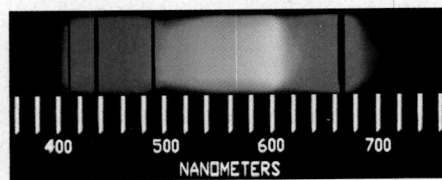

Absorption spectrum of hydrogen

(b)

Absorption spectrum of helium

Absorption spectrum of neon

CONCEPT TEST

The top spectrum in Figure 7.11 is the emission spectrum of mercury vapor. Select the absorption spectrum of mercury vapor from the other four spectra.

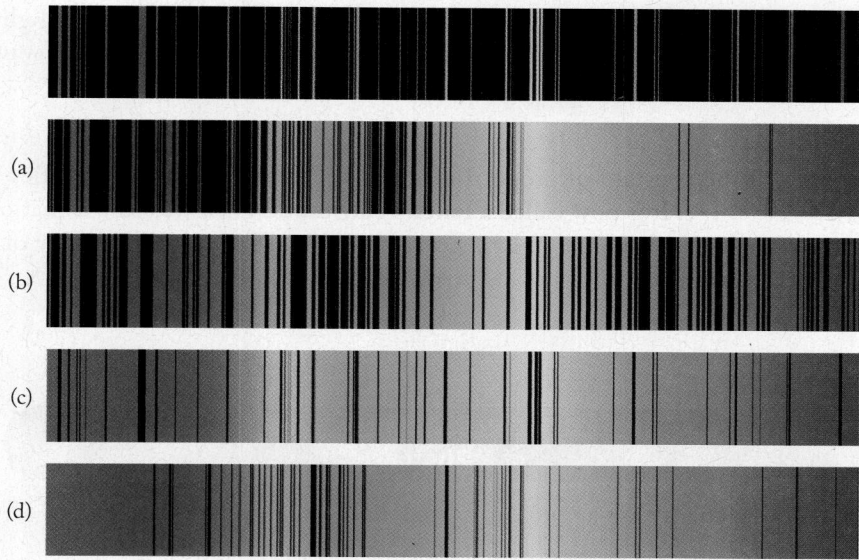

(a)

(b)

(c)

(d)

FIGURE 7.11 Atomic emission spectrum of mercury vapor (top) and four atomic absorption spectra (a–d).

7.3 Particles of Light and Quantum Theory

As studies of electromagnetic radiation progressed in the 19th century, scientists discovered limits to the wave model describing the radiation. One limit was encountered when they tried to account for radiation given off by very hot objects. Consider, for example, what happens when a metal rod is heated in a flame. At first the rod gives off only heat in the form of infrared radiation, which we can feel but not see. With more heating the metal begins to glow a dull red, shown at the top in Figure 7.12. With more heating the intensity of the red glow increases; the color begins to shift to red-orange, orange, yellow, and eventually to white as the metal emits all wavelengths of visible light. Even white-hot metal emits little radiation in the ultraviolet region and none at even shorter wavelengths.

This phenomenon was well known at the end of the 19th century. However, none of Maxwell's equations for electromagnetic radiation could account for the emission spectrum of a heated metal filament. Another explanation—indeed, another model of radiation behavior—was required.

Quantum Theory

In 1900, German scientist Max Planck (1858–1947) proposed such a model. It was based on the view that light and all other forms of electromagnetic radiation have both wavelike properties, as Maxwell had proposed, *and*, at the atomic level, particle-like properties. Planck (Figure 7.13) called a particle of radiant energy a **quantum** (plural *quanta*). In his model, which we call **quantum theory**, quanta are the smallest amounts of radiant energy in nature, the sort that a single atom or molecule might absorb or emit. An object made of

FIGURE 7.12 When a metal rod is heated, it glows first red, then orange, and finally becomes white-hot as light of all colors is emitted.

FIGURE 7.13 German scientist Max Karl Ernst Ludwig Planck is considered the father of quantum physics. He won the 1918 Nobel Prize in Physics for his pioneering work on the quantized nature of electromagnetic radiation. Planck was revered by his colleagues for his personal qualities as well as his scientific accomplishments.

a large but discrete number of atoms or molecules could therefore emit only a discrete number of quanta. In other words, light from such an object must be **quantized**.

Quantum theory also connects the wave and particle natures of electromagnetic radiation. Einstein proposed that the energy E of a quantum of light, which he called a **photon**, was directly proportional to the frequency of the wave of its corresponding radiation:

$$E = h\nu \tag{7.2}$$

The value of the constant of proportionality (*h*) relating these two quantities is 6.626×10^{-34} J·s. It is now called **Planck's constant**. When we solve Equation 7.1, $\lambda\nu = c$, for ν and substitute into Equation 7.2, we find that the energy of a photon is inversely proportional to its wavelength:

$$E = \frac{hc}{\lambda} \tag{7.3}$$

SAMPLE EXERCISE 7.2   Calculating the Energy of a Photon

What is the energy of a photon of red light that has a wavelength of 656 nm?

Collect and Organize We need to calculate the energy of a photon of electromagnetic radiation starting with its wavelength. Equation 7.3:

$$E = \frac{hc}{\lambda}$$

relates the energy of a photon to its wavelength. The value of Planck's constant (*h*) is 6.626×10^{-34} J·s, and the speed of light is 3.00×10^{8} m/s.

Analyze The wavelength is given in nanometers, but the value of the speed of light has units of meters per second. Therefore, we need to convert nanometers to meters so the distance units cancel. The value of *h* is extremely small, so it is likely the results of our calculation, even factoring in the speed of light, will be very small, too.

Solve

$$E = \frac{hc}{\lambda} = \frac{(6.626 \times 10^{-34} \text{J} \cdot \cancel{s})\left(3.00 \times 10^{8} \frac{\cancel{m}}{\cancel{s}}\right)}{656 \cancel{nm} \times \dfrac{10^{-9} \cancel{m}}{1 \cancel{nm}}}$$

$$= 3.03 \times 10^{-19} \text{J}$$

Think about It This quantity of energy is extremely small, as it should be, because a photon is an atomic-level particle of radiant energy.

Practice Exercise Some instruments differentiate individual quanta of electromagnetic radiation based on their energies. Assume such an instrument has been adjusted to detect quanta that have 1.00×10^{-16} J of energy. What is the wavelength of the detected radiation? Give your answer in nanometers and in meters.

---

quantized having values restricted to whole-number multiples of a specific base value.

photon a quantum of electromagnetic radiation.

Planck's constant (*h*) the proportionality constant between the energy and frequency of electromagnetic radiation expressed in $E = h\nu$; $h = 6.626 \times 10^{-34}$ J·s.

To visualize the meaning of Planck's quanta, consider two ways you might get from the sidewalk to the entrance of a building (Figure 7.14). If you walked up the steps, you would be able to stand at only discrete heights above the sidewalk—each height equal to the rise of a single step. You could not stand at a height between two adjacent steps because there would be nothing to stand on at that height. If you walked up the ramp, however, you could stop at any height between the sidewalk and the entrance. The discrete height changes represented by the steps are also a model of Planck's hypothesis that energy is released (analogous to walking down the steps) or absorbed (walking up the steps) in discrete packets, or quanta, of energy.

FIGURE 7.14 Quantized and unquantized heights. A flight of stairs exemplifies quantization: each step rises by a discrete height to the next step. In contrast, the heights on a ramp are not quantized.

> **CONCEPT TEST** ·······························
>
> Which of these quantities vary continuously (are not quantized) and which vary by discrete values (are quantized)?
> a. The volume of water that evaporates from a lake each day during a summer heat wave
> b. The number of eggs remaining in a carton
> c. The time it takes you to get ready for class in the morning
> d. The number of red lights encountered when driving the length of Fifth Avenue in New York City

Although Planck's quantum theory fit the emission spectra of hot objects, Planck had no experimental evidence to support the existence of quanta of energy. That evidence was supplied by Albert Einstein just 5 years after Planck published his quantum theory.

The Photoelectric Effect

When light shines on a metal surface, the surface may give off electrons. Because light produces these electrons, this phenomenon is called the **photoelectric effect**, and the electrons emitted are called *photoelectrons*. (*Photo-* comes from the Greek word meaning "light.")

Photoelectrons flow only when the frequency of the light is above some minimum **threshold frequency (ν_0)** (Figure 7.15). Light of frequencies less than the threshold value produces no photoelectrons, no matter how intense the light. On the other hand, even a dim source of light produces at least a few photoelectrons when the frequencies it emits are equal to or greater than the threshold frequency. Einstein used Planck's quantum theory to explain this behavior. He proposed that the threshold frequency ν_0 is the frequency of the minimum quantum of absorbed energy needed to remove a single electron from the metal surface. This minimum quantity of energy is called the metal's **work function (Φ)**:

$$\Phi = h\nu_0 \qquad (7.4)$$

The value of Φ is related to the strength of the attraction between the nuclei of the metal atoms and the electrons surrounding those nuclei.

When a metal surface is illuminated by a beam of photons of sufficient energy, each photon is absorbed by an individual electron, giving that electron enough energy to break free of the surface. If the beam includes photons with frequencies above

photoelectric effect when light striking a metal surface produces an electric current (a flow of electrons).

threshold frequency (ν_0) the minimum frequency of light required to produce the photoelectric effect.

work function (Φ) the amount of energy needed to dislodge an electron from the surface of a metal.

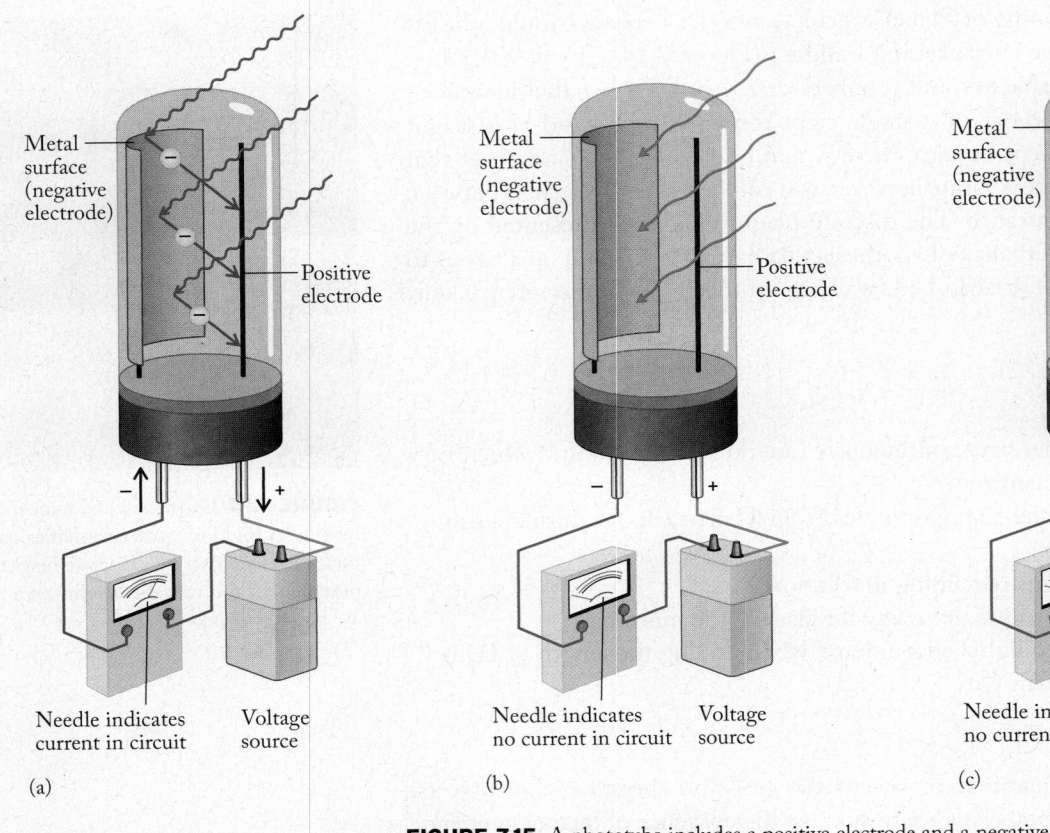

Metal surface (negative electrode)

Positive electrode

Needle indicates current in circuit

Voltage source

(a)

Metal surface (negative electrode)

Positive electrode

Needle indicates no current in circuit

Voltage source

(b)

Metal surface (negative electrode)

Positive electrode

Needle indicates no current in circuit

Voltage source

(c)

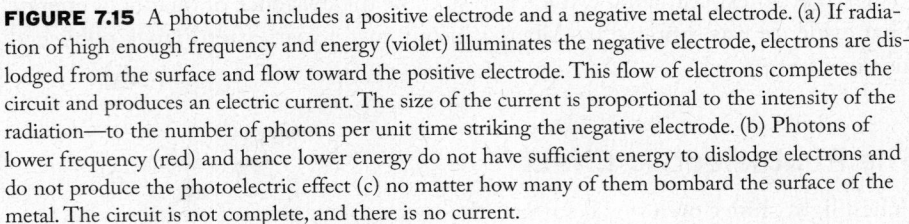

High-frequency light

Low-frequency light

Electron

FIGURE 7.15 A phototube includes a positive electrode and a negative metal electrode. (a) If radiation of high enough frequency and energy (violet) illuminates the negative electrode, electrons are dislodged from the surface and flow toward the positive electrode. This flow of electrons completes the circuit and produces an electric current. The size of the current is proportional to the intensity of the radiation—to the number of photons per unit time striking the negative electrode. (b) Photons of lower frequency (red) and hence lower energy do not have sufficient energy to dislodge electrons and do not produce the photoelectric effect (c) no matter how many of them bombard the surface of the metal. The circuit is not complete, and there is no current.

the threshold frequency ($\nu > \nu_0$), any energy in excess of Φ is imparted to each ejected electron as kinetic energy:

$$KE_{electron} = h\nu - h\nu_0 = h\nu - \Phi \tag{7.5}$$

The higher the frequency above the threshold frequency, the higher the kinetic energy and the higher the velocity of the ejected electrons.

SAMPLE EXERCISE 7.3   Using the Work Function

The work function of mercury is 7.22×10^{-19} J.
 a. What is the minimum frequency of radiation required to eject photoelectrons from a mercury surface?
 b. Could visible light produce the photoelectric effect in mercury?

Collect and Organize We need to convert the value of the work function (Φ) of a metal into the threshold (minimum) frequency (ν_0). Equation 7.4 ($\Phi = h\nu_0$) relates these parameters. Figure 7.1 contains information about the frequencies and wavelengths of the different regions of the electromagnetic spectrum.

Analyze We need to rearrange the terms in Equation 7.4 to solve for ν_0. The result of the calculation must have a ν_0 value in the range of 10^{14} to 10^{15} Hz to be visible light.

Solve

a. Rearranging Equation 7.4, we solve for the threshold frequency for mercury:

$$\Phi = h\nu_0$$

$$\nu_0 = \frac{\Phi}{h} = \frac{7.22 \times 10^{-19} \cancel{J}}{6.626 \times 10^{-34} \cancel{J} \cdot s} = 1.09 \times 10^{15}\ s^{-1}$$

b. According to Figure 7.1, the highest frequency of visible light is at the violet end of the spectrum, with a wavelength of 390 nm or 3.90×10^{-7} m. Equation 7.1 tells us that radiation of this wavelength has a frequency of

$$\nu = \frac{3.00 \times 10^8\ \frac{\cancel{m}}{s}}{3.90 \times 10^{-7}\ \cancel{m}} = 7.69 \times 10^{14}\ s^{-1}$$

This value is less than the calculated threshold frequency and represents less energy than that required to liberate a photoelectron from mercury. Therefore, visible light cannot produce the photoelectric effect with mercury.

Think about It The calculated threshold frequency is close to, but still greater than, that of the highest frequency (violet) visible light.

Practice Exercise The work function of silver is 7.59×10^{-19} J. What is the longest wavelength of electromagnetic radiation that can eject an electron from the surface of a piece of silver?

◼

CONCEPT TEST

Consult Figure 7.1 to answer this question without making a calculation. If a photon of orange light has sufficient energy to eject a photoelectron from the surface of a metal, does a photon of green light have enough energy to do so?

7.4 The Hydrogen Spectrum and the Bohr Model

In formulating his quantum theory, Planck was influenced by the results of investigations of the emission spectra produced by free (gas-phase) atoms, results that led him to question whether any spectrum, even that of an incandescent lightbulb, was truly continuous. Among these earlier results was a discovery made in 1885 by a Swiss mathematician and schoolteacher named Johann Balmer (1825–1898).

The Hydrogen Emission Spectrum

Balmer determined that the frequencies of the four brightest lines in the visible region of the emission spectrum of hydrogen (Figure 7.9a) fit the simple equation

$$\nu = \left(3.2881 \times 10^{15}\ s^{-1}\right)\left(\frac{1}{2^2} - \frac{1}{n^2}\right) \tag{7.6}$$

where *n* is a whole number greater than 2—specifically, 3 for the red line, 4 for the green, 5 for the blue, and 6 for the violet. Without having seen any other lines in the hydrogen emission spectrum, Balmer predicted there should be at least one more (*n* = 7) at the edge of the violet region, and indeed such a line was later discovered.

Balmer also predicted that hydrogen emission lines should exist in regions outside the visible range, lines corresponding to frequencies calculated by replacing $1/2^2$ in Equation 7.6 with $1/1^2$, $1/3^2$, $1/4^2$, and so forth. He was right. In 1908 German physicist Friedrich Paschen (1865–1947) discovered hydrogen emission lines in the infrared region, corresponding to $1/3^2$ instead of $1/2^2$ in Balmer's equation. A few years later, Theodore Lyman (1874–1954) at Harvard University discovered hydrogen emission lines in the UV region corresponding to $1/1^2$. By the 1920s the $1/4^2$ and $1/5^2$ series of emission lines had been discovered. Like the $1/3^2$ lines, they are in the infrared region.

Later, Swedish physicist Johannes Robert Rydberg (1854–1919) revised Balmer's equation by changing frequency to *wave number* $(1/\lambda)$, which is the number of wavelengths per unit of distance. Rydberg's equation is

$$\frac{1}{\lambda} = \left[1.097 \times 10^{-2}\,(\text{nm})^{-1}\right]\left(\frac{1}{n_1^{\,2}} - \frac{1}{n_2^{\,2}}\right) \qquad (7.7)$$

where n_1 is a whole number that remains fixed for a series of emission lines and where n_2 is a whole number equal to $n_1 + 1$, $n_1 + 2$, . . . , for successive bright lines in the spectrum.

When Balmer and Rydberg derived their equations describing the hydrogen spectrum, they didn't know why the equations worked. The discrete frequencies of hydrogen's emission lines indicated that only certain levels of internal energy were available in hydrogen atoms. However, classical (macroscale) physics could not explain the existence of these internal energy levels. A new model was needed that works at the atomic level.

SAMPLE EXERCISE 7.4   Calculating the Wavelength of a Line in the Hydrogen Emission Spectrum

In the visible portion of the hydrogen emission spectrum (Figure 7.9a), what is the wavelength of the bright line corresponding to *n* = 3 in Equation 7.6?

Collect and Organize We are to calculate the wavelength of a line in hydrogen emission spectrum given the *n* values of its initial and final energy levels. Equation 7.7 relates what we know (*n* values) to what we seek (wavelength of light).

Analyze We know that the emission line is in the visible region, so we should obtain a wavelength between 400 and 750 nanometeres. In Equation 7.7, n_2 must be greater than n_1. Fitting this requirement to the given *n* values, we let $n_2 = 3$ and $n_1 = 2$.

Solve

$$\frac{1}{\lambda} = \left[1.097 \times 10^{-2}\,(\text{nm})^{-1}\right]\left(\frac{1}{2^2} - \frac{1}{3^2}\right) = \left[1.097 \times 10^{-2}\,(\text{nm})^{-1}\right]\left(\frac{1}{4} - \frac{1}{9}\right)$$

$$= \left[1.097 \times 10^{-2}\,(\text{nm})^{-1}\right](0.1389) = 1.524 \times 10^{-3}\,(\text{nm})^{-1}$$

$$\lambda = 656 \text{ nm}$$

Think about It The calculated wavelength is in the visible region of the electromagnetic spectrum, so the answer is reasonable.

Practice Exercise What is the wavelength, in nanometers, of the line in the hydrogen spectrum corresponding to $n = 4$ in Equation 7.6?

The Bohr Model of Hydrogen

Scientists in the early 20th century faced yet another dilemma. Ernest Rutherford had established a model of the atom that was mostly empty space occupied by negatively charged electrons with a tiny nucleus containing virtually all the mass and all the positive charge. What kept the electrons from falling into the nucleus? Rutherford had suggested that the electrons had to be in motion and hypothesized that they might orbit the nucleus the way planets orbit the sun. However, classical physics predicted that negative electrons orbiting a positive nucleus would emit energy in the form of electromagnetic radiation and eventually spiral into the nucleus. This does not happen!

The problem of explaining why a hydrogen atom's electron is not pulled into its nucleus was addressed by Danish physicist Niels Bohr (1885–1962), who was well acquainted with the issue because he had studied with Rutherford. Bohr designed a theoretical model based on the electron in a hydrogen atom traveling around the nucleus in one of an array of concentric orbits. Each orbit represents an allowed energy level and is designated by the value of n as shown in Equation 7.8:

$$E = -2.178 \times 10^{-18}\,\text{J}\left(\frac{1}{n^2}\right) \qquad (7.8)$$

where $n = 1, 2, 3, \ldots, \infty$. In the Bohr model an electron in the orbit closest to the nucleus ($n = 1$) has the lowest energy:

$$E = -2.178 \times 10^{-18}\,\text{J}\left(\frac{1}{1^2}\right) = -2.178 \times 10^{-18}\,\text{J}$$

The next closest orbit has an n value of 2 and an electron in it has an energy of

$$E = -2.178 \times 10^{-18}\,\text{J}\left(\frac{1}{2^2}\right) = -5.445 \times 10^{-19}\,\text{J}$$

Note that this value is less negative than the value for the electron in the $n = 1$ orbit. As the value of n increases, the radius of the orbit increases and so, too, does the energy of an electron in the orbit; its value becomes less negative. As n approaches ∞, E approaches zero:

$$E = -2.178 \times 10^{-18}\,\text{J}\left(\frac{1}{\infty^2}\right) = 0$$

Zero energy means that the electron is no longer part of the atom. In other words, the H atom has become a H^+ ion and a free electron.

An important feature of the Bohr model is that it provides a theoretical framework for explaining the experimental observations of Balmer, Rydberg, and others. To see the connection, consider what happens when an electron moves between two allowed energy levels in Bohr's model. If we label the initial energy level (the level where the electron starts) n_{initial}, and we label the second

◉◉ CONNECTION We discussed Rutherford's gold-foil experiment and the development of the idea of the nuclear atom in Section 2.1.

▶ǁ CHEMTOUR Bohr Model of the Atom

ground state the most stable, lowest energy state available to an atom or ion.

excited state any energy state in an atom or ion above the ground state.

electron transition movement of an electron between energy levels.

level (the level where the electron ends up) n_{final}, then the change in energy of the electron is

$$\Delta E = -2.178 \times 10^{-18} \, \text{J} \left(\frac{1}{n_{final}^2} - \frac{1}{n_{initial}^2} \right) \qquad (7.9)$$

If the electron moves to an orbit farther from the nucleus, then $n_{final} > n_{initial}$, and the value of the terms inside the parentheses in Equation 7.9 is negative because $\frac{1}{n_{final}^2} < \frac{1}{n_{initial}^2}$. This negative value multiplied by the negative coefficient gives us a positive ΔE and represents an increase in electron energy. On the other hand, if an electron moves from an outer orbit to one closer to the nucleus, then $n_{final} < n_{initial}$, and the sign of ΔE is negative. This means the electron loses energy.

When the electron in a hydrogen atom is in the lowest ($n = 1$) energy level, the atom is said to be in its **ground state**. If the electron in a hydrogen atom is in an energy level above $n = 1$, then the atom is said to be in an **excited state**. According to the Bohr model the hydrogen electron can move from the $n = 1$ (ground state) energy level to a higher level (for example, $n = 3$) by absorbing a quantity of energy (ΔE) that exactly matches the energy difference between the two states. Similarly, an electron in an excited state can move to an even higher energy level by absorbing a quantity of energy that exactly matches the energy difference between the two excited states. An electron in an excited state can also move to a lower-energy excited state, or to the ground state, by emitting a quantity of energy that exactly matches the energy difference between those two states. This type of electron movement is called an **electron transition**.

*Energy-level diagram*s show the transitions from one energy level to another that electrons in atoms can make. Figure 7.16 is such a diagram for the hydrogen atom. The black arrow pointing upward represents absorption of sufficient energy to completely remove the electron from a hydrogen atom (ionization). The downward-pointing colored arrows represent decreases in the internal energy of the hydrogen atom that occur when photons are emitted as the electron moves from a higher-energy level to a lower-energy level. If the colored arrows pointed up, they would represent *absorption* of photons leading to *increases* in the internal energy of the atom. In every case the energy of the photon matches the absolute value of ΔE.

If you compare Equation 7.9 with Equation 7.7, you will see that they are much alike. The coefficients differ only because of the different units used to express wave number and energy. The key point is that the equation developed to fit the absorption and emission spectra of hydrogen has the same form as the theoretical equation developed by Bohr to explain the internal structure of the hydrogen atom. Thus, atomic emission and absorption spectra reveal the energies of electrons inside atoms.

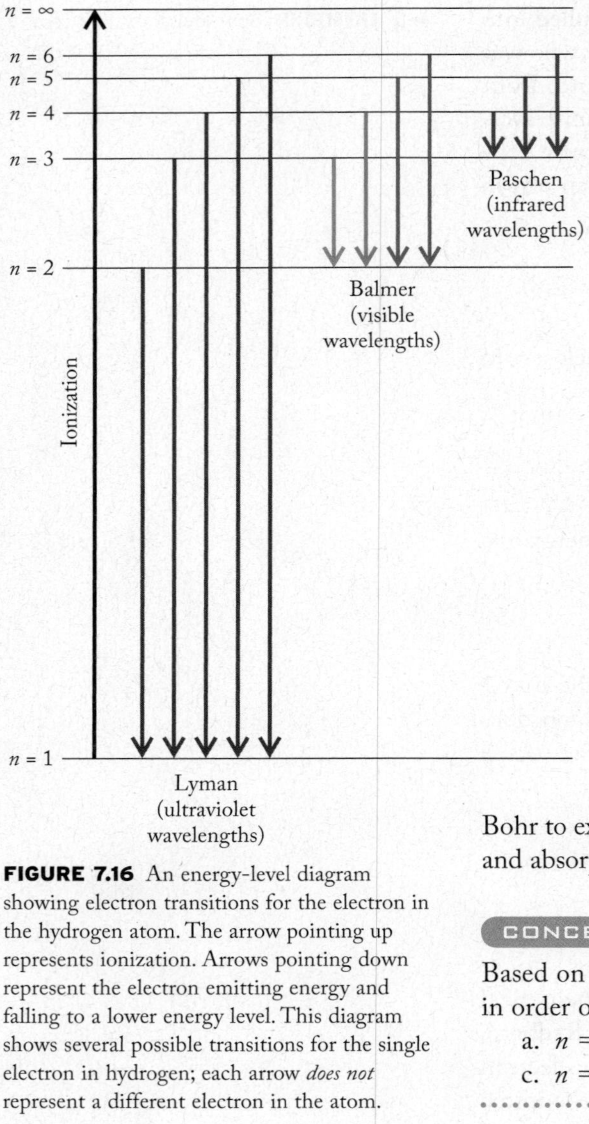

FIGURE 7.16 An energy-level diagram showing electron transitions for the electron in the hydrogen atom. The arrow pointing up represents ionization. Arrows pointing down represent the electron emitting energy and falling to a lower energy level. This diagram shows several possible transitions for the single electron in hydrogen; each arrow *does not* represent a different electron in the atom.

CONCEPT TEST

Based on the lengths of the arrows in Figure 7.16, rank the following transitions in order of greatest change in electron energy to the smallest change:

a. $n = 4 \rightarrow n = 2$ b. $n = 3 \rightarrow n = 2$
c. $n = 2 \rightarrow n = 1$ d. $n = 4 \rightarrow n = 3$

SAMPLE EXERCISE 7.5   Calculating the Energy Needed for an Electron Transition

How much energy is required to ionize a ground-state hydrogen atom?

Collect and Organize We are asked to determine the energy required to remove the electron from a hydrogen atom in its ground state. Equation 7.9 enables us to calculate the energy change associated with any electron transition.

Analyze To use Equation 7.9, we need to identify the initial ($n_{initial}$) and final (n_{final}) energy levels of the electron. The ground state of a H atom corresponds to the $n = 1$ energy level. If the atom is ionized, $n = \infty$ and the electron is no longer associated with the nucleus.

Solve

$$\Delta E = -2.178 \times 10^{-18} \, J \left(\frac{1}{n_{final}^2} - \frac{1}{n_{initial}^2} \right)$$

$$= -2.178 \times 10^{-18} \, J \left(\frac{1}{\infty^2} - \frac{1}{1^2} \right)$$

Dividing by ∞^2 yields zero, so the difference in parentheses simplifies to -1, which gives

$$\Delta E = 2.178 \times 10^{-18} \, J$$

Think about It This is a small amount of energy, but it is comparable to the work function values (see Sample Exercise 7.3) which involved removing an electron from a metal surface. The sign of ΔE is positive because energy must be added to remove an electron from the atom and away from its positive nucleus.

Practice Exercise Calculate the energy, in joules, required to ionize a hydrogen atom when its electron is initially in the $n = 3$ energy level. Before doing the calculation, predict whether this energy is greater than or less than the 2.178×10^{-18} J needed to ionize a ground-state hydrogen atom.

One of the strengths of the Bohr model of the hydrogen atom is that it accurately predicts the energy needed to remove the electron. This energy is called the *ionization energy* of the hydrogen atom. We examine the ionization energies of other elements in Section 7.11. However, the Bohr model applies only to hydrogen atoms and to ions that have only a single electron. The model does not account for the observed spectra of multielectron elements and ions because it does not account for the way electrons interact with each other. Thus, the picture of the atom provided by Bohr's model is limited, but it enabled other scientists to begin using quantum theory to explain the behavior of matter at the atomic level.

CONCEPT TEST

Can the Bohr model be used to explain the emission spectrum of He$^+$ ions? Why, or why not?

7.5 Electrons as Waves

A decade after Bohr published his model of the hydrogen atom, a French graduate student named Louis de Broglie (1892–1987) provided a theoretical basis for the stability of electron orbits and greatly affected our view of the structure of matter. His approach incorporated yet another significant advance in the way early-20th-century scientists viewed atoms and subatomic particles—namely, to think of them not only as particles of matter but also as waves.

De Broglie Wavelengths

De Broglie proposed that if light, which we normally think of as a wave, has particle properties, perhaps the electron, which we normally think of as a particle, has wave properties. If that were true, an electron moving around in an atom should have a characteristic wavelength. De Broglie calculated electron wavelengths using an equation he derived from Einstein's equations relating energy and mass, $E = mc^2$, and the energy of a photon, $E = h\nu = hc/\lambda$:

$$\lambda = \frac{hc}{E} = \frac{hc}{mc^2} = \frac{h}{mc} \qquad (7.10)$$

To apply Equation 7.10 to electrons, de Broglie replaced c (the speed of light) with v, the velocity of an orbiting electron in an atom:

$$\lambda = \frac{h}{mv} \qquad (7.11)$$

▶❚❚ **CHEMTOUR** De Broglie Wavelength

where h is Planck's constant, m is the mass of the electron in kilograms, and v is its velocity in meters per seconds. The wavelength of an electron calculated in this way is often called the *de Broglie wavelength* of the electron.

De Broglie's equation is not restricted to electrons. It tells us that *any moving particle* has wavelike properties. In other words, the particle behaves as a **matter wave**. De Broglie predicted that moving particles much bigger than electrons, such as atomic nuclei, molecules, and even tennis balls and airplanes, have characteristic wavelengths that can be calculated using Equation 7.11. The wavelengths of such large objects are extremely small, of course, given the tiny size of Planck's constant in the numerator of Equation 7.11, and for this reason we never notice the wave nature of large objects in motion.

SAMPLE EXERCISE 7.6 **Calculating the Wavelength of a Particle in Motion**

(a) Compare the wavelength of a 142 g baseball thrown at 44 m/s (98 mi/hr) with the size of the ball, which has a diameter of 7.5 cm. (b) Compare the wavelength of an electron ($m_e = 9.109 \times 10^{-31}$ kg) moving at one-tenth the speed of light in a hydrogen atom with the size of the atom (diameter = 1.06×10^{-10} m).

Collect and Organize We know the masses and velocities of two moving objects and are to calculate the wavelengths of their matter waves. Equation 7.11 may be used to calculate these wavelengths.

Analyze Given the small value of h, it is likely that the wavelength of a pitched baseball is only a tiny fraction of the size of the baseball. The wave-

matter wave the wave associated with any particle.

length of a tiny electron moving at one-tenth the speed of light may be a much greater fraction of the size of the atom it is in. The fraction on the right side of Equation 7.11 has units of joule-seconds in the numerator and mass and velocity in the denominator. To combine these units in a way that gives us a unit of length, we need to use the following conversion factor:

$$1\ \text{J} = 1\ \text{kg}\ (\text{m/s})^2$$

To use this equality, we must express the mass of the baseball in kilograms: 142 g = 0.142 kg.

Solve

a. For the baseball:

$$\lambda = \frac{h}{mv} = \frac{6.626 \times 10^{-34}\ \text{J} \cdot \text{s}}{(0.142\ \text{kg})(44\ \text{m/s})} \times \frac{1\ \text{kg} \cdot \text{m}^2/\text{s}^2}{1\ \text{J}}$$

$$= 1.06 \times 10^{-34}\ \text{m}$$

The wavelength of the baseball is

$$\frac{1.06 \times 10^{-34}\ \text{m}}{0.075\ \text{m}} \times 100\% = 1.4 \times 10^{-31}\%$$

of the ball's diameter.

b. The wavelength of an electron moving at one-tenth the speed of light is

$$\lambda = \frac{h}{mv} = \frac{6.626 \times 10^{-34}\ \text{J} \cdot \text{s}}{(9.109 \times 10^{-31}\ \text{kg})(3.00 \times 10^7\ \text{m/s})} \times \frac{1\ \text{kg} \cdot \text{m}^2/\text{s}^2}{1\ \text{J}}$$

$$= 2.42 \times 10^{-11}\ \text{m}$$

or

$$\frac{2.42 \times 10^{-11}\ \text{m}}{1.06 \times 10^{-10}\ \text{m}} \times 100\% = 23\%$$

of the diameter of a hydrogen atom.

Think about It The matter wave of the baseball is much too small to be observed, so its character contributes nothing to the behavior of the baseball. We expected that. Our experience with objects in the world is that they behave "like matter," not like waves. For the electron, however, the wavelength is a significant percentage of the size of the hydrogen atom.

Practice Exercise The velocity of the electron in the ground state of the hydrogen atom is 2.2×10^6 m/s. What is the wavelength of this electron in meters?

De Broglie applied the following reasoning to explain the stability of the electron levels in Bohr's model of the hydrogen atom. He proposed that the electron in a hydrogen atom behaves like a circular wave oscillating around the nucleus. To understand the implications of this statement, we need to examine what is required to make a stable, circular wave. Consider the motion of a vibrating violin string of length L (Figure 7.17a). Because the string is fixed at both ends, there is no vibration at the ends and the vibration is a maximum in the middle. The wave created by this combination of fixed ends and maximum vibration in the middle is a **standing wave**, that is, a wave that oscillates back and forth within a fixed space rather than moving through space the way waves of light travel

standing wave a wave confined to a given space with a wavelength λ related to the length L of the space by $L = n(\lambda/2)$, where n is a whole number.

node a location in a standing wave that experiences no displacement.

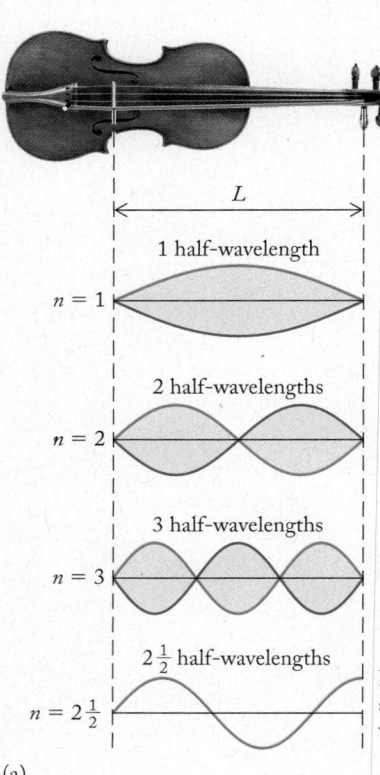

FIGURE 7.17 Linear and circular standing waves. (a) The wavelength of a standing wave in a violin string fixed at both ends is related to the distance L between the ends of the string by the equation $L = n(\lambda/2)$. In the standing waves shown, $n = 1, 2,$ and 3. An n value of $2\frac{1}{2}$ does not produce a standing wave because there can be no string motion at either end. (b) The circular standing waves proposed by de Broglie account for the stability of the energy levels in Bohr's model of the hydrogen atom. Each stable wave must have a circumference equal to $n\lambda$, with n restricted to being an integer, such as the $n = 3$ circular standing wave shown here. If the circumference is not an exact multiple of λ, as shown in the $2\frac{1}{4}$ image, there is no standing wave.

through space. On a standing wave, any points that have zero displacement, such as the two ends of the string, are called **nodes**. The sound wave produced in this way on a violin string is called the *fundamental* of the string. The wavelength of the fundamental is $2L$. It has the lowest frequency and longest wavelength possible for that string. The length of the string is one-half the wavelength of the fundamental ($L = \lambda/2$).

If the string is held down in the middle (creating a third node) and plucked halfway between the middle and one end, a new, higher-frequency wave called the *first harmonic* is produced. The wavelength of this harmonic is equal to L, and $L = 2(\lambda/2) = \lambda$. We can continue generating higher frequencies with wavelengths that are related to the length of the string by the equation

$$L = \frac{n\lambda}{2}$$

where n is a whole number: $3(\lambda/2), 4(\lambda/2), 5(\lambda/2),$ and so on.

CONCEPT TEST

Why are the various waves possible in a violin string examples of quantization?

The standing-wave pattern for a circular wave generated by an electron differs slightly from that of a vibrating violin string in that in the circular wave there are no defined stationary ends. Instead, the electron vibrates in an endless series of waves, but only if, as shown in Figure 7.17(b), the circumference of the circle equals a whole-number multiple of the electron's wavelength:

$$\text{Circumference} = n\lambda \qquad (7.12)$$

Equation 7.12 gives a new meaning to Bohr's quantum number n: it represents the number of matter-waves in a given energy level.

De Broglie's research created a quandary for the graduate faculty at the University of Paris, where he studied. Bohr's model of electrons moving between allowed energy levels had been widely criticized as an arbitrary suspension of well-tested physical laws. De Broglie's rationalization of Bohr's model seemed even more outrageous to many scientists. Before the faculty would accept his thesis, they wanted another opinion, so they sent it to Albert Einstein for review. Einstein wrote back that he found the young man's work "quite interesting." That endorsement was good enough for the faculty: de Broglie's thesis was accepted in 1924 and immediately submitted for publication.

The Heisenberg Uncertainty Principle

After de Broglie proposed that electrons exhibited both particle behavior and wave behavior, questions arose about the impact of wave behavior on our ability to locate the electron. A wave by its very nature is spread out in space. The question "Where is the electron?" has one answer if we treat the electron as a particle and a different answer if we treat it as a wave. This issue was addressed by German physicist Werner Heisenberg (1901–1976), who proposed the following thought experiment: watch an electron with a hypothetical gamma-ray microscope to "see" the electron's path around an atom. The microscope (if it existed) would need to use gamma rays for illumination because they are the only part of the electromagnetic spectrum with wavelengths short enough to match the diminutive size of electrons.

However, Equations 7.2 and 7.3 tell us that the short wavelengths and high frequencies of gamma rays mean that they have enormous energies—so large that any gamma ray striking an electron would knock the electron off course. The only way not to affect the electron's motion would be to use a much lower-energy, longer-wavelength source of radiation to illuminate it, but then we would not be able to see the tiny electron clearly.

This situation presents a quantum mechanical dilemma. The only means for clearly observing an electron make it impossible to know the electron's motion or, more precisely, its momentum, which is defined as an object's velocity times its mass. Therefore we can never know exactly both the position and the momentum of the electron simultaneously. This conclusion is known as the **Heisenberg uncertainty principle** and is mathematically expressed by

$$\Delta x \cdot m \Delta v \geq \frac{h}{4\pi} \qquad (7.13)$$

where Δx is the uncertainty in the position of the electron, m is its mass, Δv is the uncertainty in its velocity, and h is Planck's constant. To Heisenberg, this uncertainty was the essence of quantum mechanics. Its message for us is that there are limits to what we can observe, measure, and therefore know.

Heisenberg uncertainty principle one cannot determine both the position and the momentum of an electron in an atom at the same time.

SAMPLE EXERCISE 7.7   Calculating Uncertainty

Use the data in Sample Exercise 7.6 and compare the uncertainty in the velocity of the baseball with the uncertainty in the velocity of the electron. Assume that the position of the baseball is known to within one wavelength of red light ($\Delta x_{\text{baseball}} = 680$ nm) and that the position of the electron is known to within the radius of the hydrogen atom ($\Delta x_{\text{e}} = 5.3 \times 10^{-11}$ m).

Collect and Organize We are asked to calculate the uncertainty in the velocities of two particles and are given the uncertainties in their positions. From Sample Exercise 7.6 we know that the mass of a baseball is 0.142 kg and the mass of an electron is 9.109×10^{-31} kg. Equation 7.13 provides a mathematical connection between these variables.

Analyze According to Equation 7.13, the uncertainty in the velocity and position of a particle is inversely proportional to its mass. Therefore, we can expect little uncertainty in the velocity of the baseball but much greater uncertainty in the velocity of the electron. We need to rearrange the terms in the equation to solve for the uncertainty in velocity (Δv):

$$\Delta v \geq \frac{h}{4\pi \Delta x m}$$

Solve For the baseball:

$$\Delta v \geq \frac{6.626 \times 10^{-34} \, \text{J} \cdot \text{s}}{4\pi (6.80 \times 10^{-7} \, \text{m})(0.142 \, \text{kg})} \times \frac{1 \, \text{kg} \cdot \text{m}^2/\text{s}^2}{1 \, \text{J}}$$

$$\geq 5.46 \times 10^{-28} \, \text{m/s}$$

For the electron:

$$\Delta v \geq \frac{6.626 \times 10^{-34} \, \text{J} \cdot \text{s}}{4\pi (5.3 \times 10^{-11} \, \text{m})(9.109 \times 10^{-31} \, \text{kg})} \times \frac{1 \, \text{kg} \cdot \text{m}^2/\text{s}^2}{1 \, \text{J}}$$

$$\geq 1.09 \times 10^6 \, \text{m/s}$$

Comparing the two values, we see that the uncertainty in the velocity of the baseball is extremely small (about 10^{-28} m/s), whereas that of the electron is huge (over 1 million m/s).

Think about It The uncertainty in the measurement of the velocity of the baseball is so minuscule it's insignificant. This result is expected for objects in the macroscopic world. The uncertainty in the measurement of the velocity of the electron, however, is huge, which is what one must expect at the atomic level, where "particles" such as the electron also behave like waves.

Practice Exercise What is the uncertainty, in meters, in the position of an electron moving near a nucleus at a speed of 8×10^7 m/s? Assume the uncertainty in the velocity of the electron is 1% of its value—that is, $\Delta v = (0.01)(8 \times 10^7$ m/s).

When Heisenberg proposed his uncertainty principle, he was working with Bohr at the University of Copenhagen. The two scientists had widely different views about the significance of the uncertainty principle and the idea that particles could behave like waves. To Heisenberg, uncertainty was a fundamental characteristic of nature. To Bohr, it was merely a mathematical consequence of the wave–particle duality of electrons; there was no physical meaning to an electron's position and path. The debate between these two gifted scientists was heated at times. Heisenberg later wrote about one particularly emotional debate:

> [A]t the end of the discussion I went alone for a walk in the neighboring park [and] repeated to myself again and again the question: "Can nature possibly be as absurd as it seems?"[1]

7.6 Quantum Numbers and Electron Spin

Many of the leading scientists of the 1920s were unwilling to accept the dual wave–particle nature of electrons proposed by de Broglie until the model could be used to predict the features of the hydrogen emission spectrum. Such application required the development of equations describing the behavior of electron waves. Over his Christmas vacation in 1925, Austrian physicist Erwin Schrödinger (1887–1961) did just that, developing in a few weeks the mathematical foundation for what came to be called **wave mechanics** or **quantum mechanics**.

Schrödinger's mathematical description of electron waves is called the **Schrödinger wave equation**. Although it is not discussed in detail in this book, you should know that solutions to the wave equation are called **wave functions**: mathematical expressions represented by the Greek letter psi (ψ) that describe how the matter wave of an electron in an atom varies both with time and with the location of the electron in the atom. Wave functions define the energy levels in the hydrogen atom. They can be simple trigonometric functions, such as sine or cosine waves, or they can be very complex.

What is the physical significance of a wave function? Actually there is none. However, the *square of a wave function* (ψ^2) does have physical meaning. Initially

wave mechanics (also called **quantum mechanics**) mathematical description of the wavelike behavior of particles on the atomic level.

Schrödinger wave equation a description of how the electron matter wave varies with location and time around the nucleus of a hydrogen atom.

wave function (ψ) a solution to the Schrödinger wave equation.

[1] Werner Heisenberg, *Physics and Philosophy: The Revolution in Modern Science* (Harper & Row, 1958), p. 42.

Schrödinger believed that a wave function depicted the "smearing" of an electron through three-dimensional space. This notion of subdividing a discrete particle was later rejected in favor of the model developed by German physicist Max Born (1882–1970), who proposed that ψ^2 defines an **orbital**: the space around the nucleus in an atom where the probability of finding an electron is high. Born later showed that his interpretation could be used to calculate the probability of a transition between two orbitals, as happens when an atom absorbs or emits a photon.

To help visualize the probabilistic meaning of ψ^2, consider what happens when we spray ink onto a flat surface (Figure 7.18). If we then draw a circle encompassing most of the ink spots, we are identifying the region of maximum probability for finding the spots.

It is important to understand that quantum mechanical orbitals in an atom are not two-dimensional concentric orbits as in Bohr's model of the hydrogen atom or even two-dimensional circles as in the pattern of ink drops in Figure 7.18. Instead, they are three-dimensional regions of space with distinctive shapes, orientations, and average distances from the nucleus. Each orbital is a solution to Schrödinger's wave equation and is identified by a unique combination of three integers called **quantum numbers**, whose values flow directly from the mathematical solutions to the wave equation. The quantum numbers are as follows:

- ➤ The **principal quantum number** n is like Bohr's quantum number n for the hydrogen atom in that it is a positive integer that indicates the relative size and energy of an orbital or a group of orbitals in an atom. Orbitals with the same value of n are in the same *shell*. Orbitals with larger values of n are farther from the nucleus and, in the hydrogen atom, represent higher energy levels, consistent with Bohr's model of the hydrogen atom. In multielectron atoms, the relationship between energy levels and orbitals is more complex, but increasing values of n generally represent higher energy levels.

- ➤ The **angular momentum quantum number** ℓ is an integer with a value ranging from zero to $n - 1$ that defines the shape of an orbital. Orbitals with the same values of n and ℓ are in the same *subshell* and represent equal energy levels. Orbitals with a given value of ℓ are identified with a letter according to the following scheme:

Value of ℓ:	0	1	2	3
Letter identifier:	s	p	d	f

- ➤ The **magnetic quantum number** m_ℓ is an integer with a value from $-\ell$ to $+\ell$. It defines the orientation of an orbital in the space around the nucleus of an atom.

Each subshell in an atom has a two-part designation containing the appropriate value of n and a letter designation for ℓ. For example, orbitals with $n = 3$ and $\ell = 1$ are called $3p$ orbitals, and electrons in $3p$ orbitals are called $3p$ electrons. How many $3p$ orbitals are there? We can answer this question by finding all possible values of m_ℓ. Because p orbitals are those for which $\ell = 1$, they have m_ℓ values of -1, 0, and $+1$. These three values mean that there are three $3p$ orbitals, each with a unique combination of n, ℓ, and m_ℓ values. All the possible

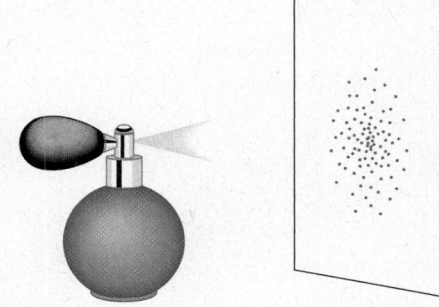

FIGURE 7.18 The probability of finding an ink spot in the pattern produced by a source of ink spray decreases with increasing distance from the center of the pattern, in much the way that electron density in the 1s orbital decreases with increasing distance from the nucleus.

▶❚❚ **CHEMTOUR** Quantum Numbers

orbitals defined by the square of the wave function (ψ^2); regions around the nucleus of an atom where the probability of finding an electron is high and identified by a unique combination of three quantum numbers.

quantum number one of four related numbers that specify the shape and energy of orbitals in an atom; the "address" of an electron in an atom or ion.

principal quantum number (n) a positive integer describing the relative size and energy of an atomic orbital or group of orbitals in an atom.

angular momentum quantum number (ℓ) an integer having any value from 0 to $n - 1$ that defines the shape of an orbital.

magnetic quantum number (m_ℓ) defines the orientation of an orbital in space; an integer that may have any value from $-\ell$ to $+\ell$, where ℓ is the angular momentum quantum number.

combinations of these three quantum numbers for the orbitals of the first four shells are given in Table 7.1.

TABLE 7.1	Quantum Numbers of the Orbitals in the First Four Shells				
Value of n	Allowed Value of ℓ	Subshell Label	Allowed Values of m_ℓ	Number of Orbitals in: Subshell	Shell
1	0	s	0	1	1
2	0	s	0	1	
	1	p	$-1, 0, +1$	3	4
3	0	s	0	1	
	1	p	$-1, 0, +1$	3	
	2	d	$-2, -1, 0, +1, +2$	5	9
4	0	s	0	1	
	1	p	$-1, 0, +1$	3	
	2	d	$-2, -1, 0, +1, +2$	5	
	3	f	$-3, -2, -1, 0, +1, +2, +3$	7	16

SAMPLE EXERCISE 7.8 Identifying the Subshells and Orbitals in an Energy Level

(a) What are the designations of all the subshells in the $n = 4$ shell? (b) How many orbitals are in these subshells?

Collect and Organize We are asked to describe the subshells in the fourth shell and to determine how many orbitals are in all these subshells. Table 7.1 contains an inventory of all the subshells in the first four shells.

Analyze The designations of subshells are based on the possible values of quantum numbers n and ℓ. The allowed values of ℓ depend on the value of n, in that ℓ is an integer between 0 and $n - 1$. The number of orbitals in a subshell depends on the number of possible values of m_ℓ, from $-\ell$ to $+\ell$.

Solve

a. The allowed values of ℓ for $n = 4$ range from 0 to 3 ($n - 1$) and so are 0, 1, 2, and 3. These ℓ values correspond to the subshell designations s, p, d, and f. The appropriate subshell names are thus $4s$, $4p$, $4d$, and $4f$.

b. The possible values of m_ℓ from $-\ell$ to $+\ell$ are as follows:

$\ell = 0$; $m_\ell = 0$: This combination of ℓ and m_ℓ values for the $n = 4$ shell represents a single $4s$ orbital.

$\ell = 1$; $m_\ell = -1, 0,$ or $+1$: These three combinations of ℓ and m_ℓ values for the $n = 4$ shell represent the three $4p$ orbitals.

$\ell = 2$; $m_\ell = -2, -1, 0, +1,$ or $+2$: These five combinations of ℓ and m_ℓ values represent the five $4d$ orbitals.

$\ell = 3$; $m_\ell = -3, -2, -1, 0, +1, +2,$ or $+3$: These seven combinations of ℓ and m_ℓ values represent the seven $4f$ orbitals.

Thus there are $1 + 3 + 5 + 7 = 16$ orbitals in the $n = 4$ shell.

Think about It We determined that there are 16 orbitals in the fourth shell. The number of orbitals in each shell is equal to the square of the principal quantum number of the shell.

Practice Exercise How many orbitals are there in the $n = 2$ shell?

Name, Symbol (Property)	Allowed Values	Quantum Numbers
Principal, n (size, energy)	Positive integers $(1, 2, 3, \ldots)$	
Angular momentum, ℓ (shape)	From 0 to $n-1$	
Magnetic, m_ℓ (orientation)	$-\ell, \ldots, 0, \ldots, +\ell$	

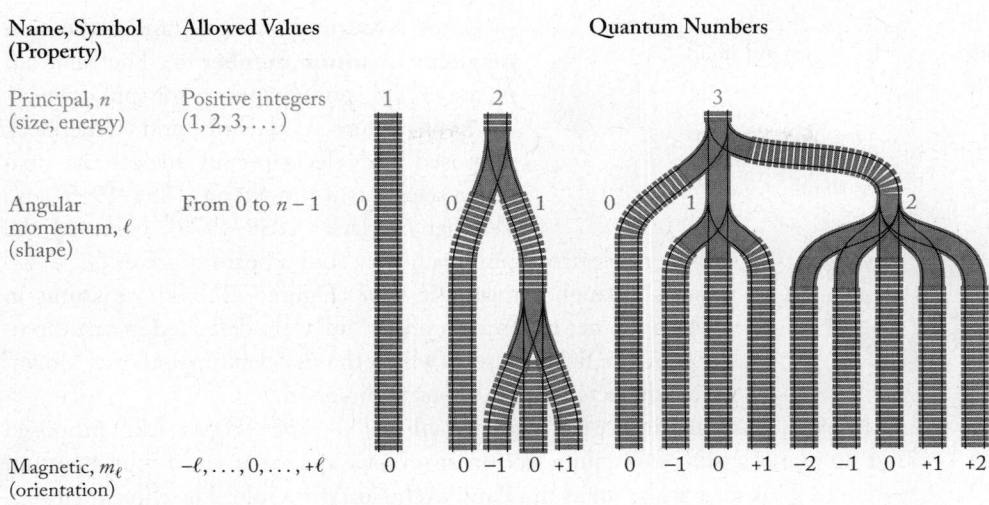

FIGURE 7.19 These three sets of train tracks provide a visual metaphor of the orbitals in the first three energy shells in an atom. The single track labeled **1** represents the first shell and its 1s orbital, which has ℓ and m_ℓ values of 0. The set labeled **2** has two ℓ branches, labeled **0** and **1**, representing the 2s and 2p orbitals. The **1** branch further divides into branches with m_ℓ values of -1, 0, and $+1$ representing the three 2p orbitals. The set labeled **3** divides into three main branches with m_ℓ values of 0, 1, 2 representing the 3p, 3p, and 3d subshells, respectively.

Several relationships are worth noting in the quantum numbering system (Figure 7.19):

➤ There are n subshells in the nth shell: one subshell (1s) in the $n = 1$ shell, two subshells (2s and 2p) in the $n = 2$ shell, and so on.

➤ There are n^2 orbitals in the nth shell: $1^2 = 1$ in the $n = 1$ shell, $2^2 = 4$ in the $n = 2$ shell, and so on.

➤ There are $(2\ell + 1)$ orbitals in each subshell: one s orbital $(2 \times 0 + 1 = 1)$ in each s subshell, three p orbitals $(2 \times 1 + 1 = 3)$ in each p subshell, five d orbitals $(2 \times 2 + 1 = 5)$ in each d subshell, and so on.

The Schrödinger wave equation accounts for most, but not all, aspects of atomic spectra. The emission spectrum of hydrogen, for example, has a pair of red lines at 656 nm where Balmer thought there was only one line (Figure 7.20). There are also pairs of lines in the spectra of multielectron atoms that have a single electron in their outermost shells.

In 1925, two students at the University of Leiden in the Netherlands, Samuel Goudsmit (1902–1978) and George Uhlenbeck (1900–1988), proposed that the pairs of lines, called *doublets,* were caused by a property they called electron spin. In their model, electrons spin in one of two directions designated spin "up" and spin "down." A moving electron (or any charged particle) creates a magnetic field by virtue of its motion. The spinning motion produces a second magnetic field oriented up or down. To account for these two spin orientations, Goudsmit and Uhlenbeck

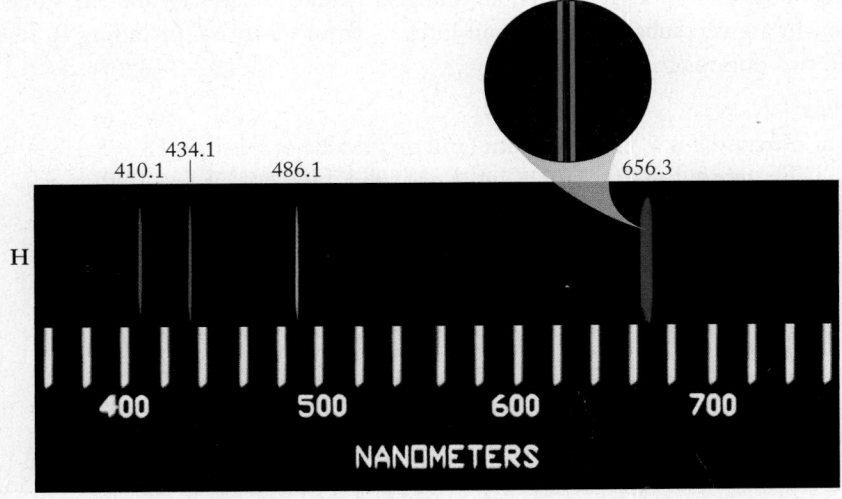

FIGURE 7.20 The Schrödinger equation does not account for the appearance of closely spaced pairs of bright lines in the emission spectra of atoms, such as the red lines at 656.272 and 656.285 nm in the spectrum of hydrogen.

FIGURE 7.21 A narrow beam of silver atoms passed through a magnetic field is split into two beams because of the interactions between the field and the spinning electrons in the atoms. This observation led to proposing the fourth quantum number, m_s.

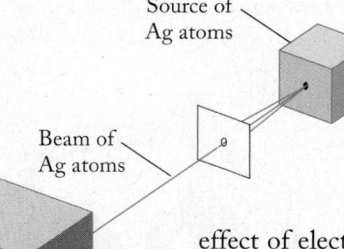

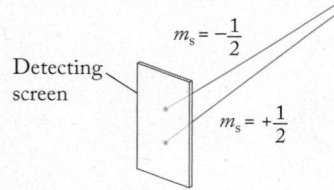

proposed a fourth quantum number, the **spin magnetic quantum number m_s**. The values of m_s are $+\frac{1}{2}$ for spin up and $-\frac{1}{2}$ for spin down.

Even before Goudsmit and Uhlenbeck proposed the electron-spin hypothesis, two other scientists, Otto Stern (1888–1969) and Walther Gerlach (1889–1979), observed the effect of electron spin when they shot a beam of silver ($Z = 47$) atoms through a magnetic field (Figure 7.21). Those atoms in which the net electron spin was "up" were deflected in one direction by the field; those in which the net electron spin was "down" were deflected in the opposite direction.

In 1925, Austrian physicist Wolfgang Pauli (1900–1958) (Figure 7.22) proposed that no two electrons in a multielectron atom have the same set of four quantum numbers. This idea is known as the **Pauli exclusion principle**. The three quantum numbers from Schrödinger's wave equation define the orbitals where an atom's electrons are likely to be. The two allowed values of the spin magnetic quantum number indicate that each orbital can hold two electrons, one with $m_s = +\frac{1}{2}$ and the other with $m_s = -\frac{1}{2}$. Thus, each electron in an atom has a unique "quantum address" defined by a particular combination of n, ℓ, m_ℓ, and m_s values.

SAMPLE EXERCISE 7.9 **Identifying Valid Quantum Number Sets**

Which of these five combinations of quantum numbers are valid?

	n	ℓ	m_ℓ	m_s
(a)	1	0	−1	$+\frac{1}{2}$
(b)	3	2	−2	$+\frac{1}{2}$
(c)	2	2	0	0
(d)	2	0	0	$-\frac{1}{2}$
(e)	−3	−2	−1	$-\frac{1}{2}$

Collect and Organize Table 7.1 contains valid combinations of quantum numbers for the first four shells. The rules for which values of n, ℓ, and m_ℓ are possible are given at the beginning of this section.

Analyze The principal quantum number (n) can be any positive integer. The valid values of ℓ in a given shell are integers from 0 to ($n − 1$), and the values of m_ℓ in a given subshell include all integers from $-\ell$ to $+\ell$ including 0. The only two options for m_s are $+\frac{1}{2}$ or $-\frac{1}{2}$.

Solve
a. Because n is 1, the maximum (and only) value of ℓ is ($n − 1$) = 1 − 1 = 0. Therefore the values of n and ℓ are valid. However, if $\ell = 0$, then m_ℓ must be 0; it cannot be −1. Therefore, this set is not valid. The spin quantum number is a possible one.
b. Because n is 3, ℓ can be 2 and m_ℓ can be −2. Also, $m_s = +\frac{1}{2}$ is a valid choice for the spin magnetic quantum number. This set is valid.
c. Because n is 2, ℓ cannot be 2, making this set invalid. In addition m_s has an invalid value (0).
d. Because n is 2, ℓ can be 0, and for that value of ℓ, m_ℓ must be 0. The value of m_s is also valid, and so is the set.
e. This set contains two impossible values, $n = −3$ and $\ell = −2$, so it is invalid.

FIGURE 7.22 Wolfgang Pauli (left) and Niels Bohr are apparently amused by the behavior of a toy called a tippe-top, which, when spun on its base, tips itself over and spins on its stem. The flip is caused by a combination of friction and the top's angular momentum, and provides a visual metaphor of the two spin orientations of an electron in an orbital.

Think about It The values of n, ℓ, and m_ℓ are related mathematically and m_s can be either $+\frac{1}{2}$ or $-\frac{1}{2}$. The quantum numbers are the address of the electron in an atom, and every electron has its own unique address—its own unique set of four quantum numbers.

Practice Exercise Write all the possible sets of quantum numbers for an electron in the $n = 3$ shell that has an angular momentum quantum number $\ell = 1$ and a spin quantum number $m_s = +\frac{1}{2}$.

spin magnetic quantum number (m_s) either $+\frac{1}{2}$ or $-\frac{1}{2}$, indicating that the spin orientation of an electron is either up or down.

Pauli exclusion principle no two electrons in an atom can have the same set of four quantum numbers.

7.7 The Sizes and Shapes of Atomic Orbitals

As noted earlier, the orbitals in atoms have three-dimensional shapes that are graphical representations of ψ^2. In this section we examine the shapes of atomic orbitals and how those shapes impact the energies of the electrons in them.

s Orbitals

Figure 7.23 provides several representations of the $1s$ orbital of the hydrogen atom. In Figure 7.23(a), electron density is plotted against distance from the nucleus and shows that density decreases with increasing distance. However, Figure 7.23(b) provides a more useful profile of electron distribution. To understand why, think of the hydrogen atom as being like an onion, made of many concentric spherical layers all of the same thickness. A cross section of this image of the atom is shown in Figure 7.23(c). What is the probability of finding the electron in one of these spherical layers? A layer very close to the nucleus has a very small radius, so it accounts for only a small fraction of the total volume of the atom. A layer with a larger radius makes up a much larger fraction of the volume of the atom because the volume of the layers increases as a function of r^2. Even though electron densities are higher closer to the nucleus (as Figure 7.23a shows), the volumes of the spherical layers closest to the nucleus are so small that the chances of the electron being near the center of an atom are extremely low; this is shown in Figure 7.23(b), where the curve starts off at essentially zero for electron distribution values at distances very close to the nucleus. Farther from the nucleus, electron densities are lower but the volumes of the layers are much larger, so the probability of the electron being in one of these layers is relatively high, represented by the peak in the curve of Figure 7.23(b). At greater distances, volumes of the layers are very large but ψ^2 drops to nearly zero (see Figure 7.23a); therefore, the chances of finding the electron layers far from the nucleus are very small.

Thus Figure 7.23(b) represents a combination of two competing factors: increasing layer volume and decreasing probability of finding an electron in a given layer. This combination produces a *radial distribution profile* for the electron.

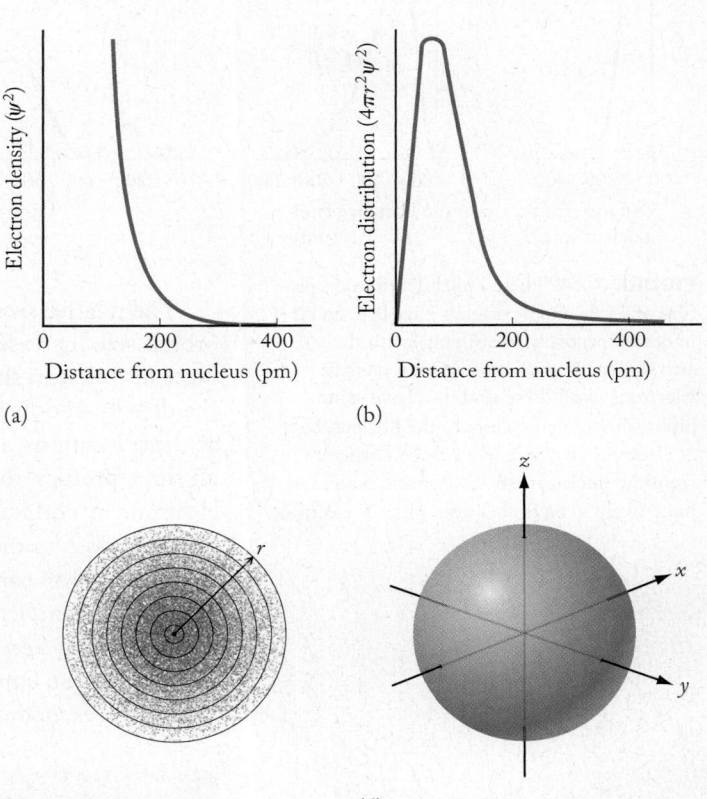

(a)

(b)

(c)

(d)

FIGURE 7.23 (a) Probable electron density in the $1s$ orbital of the hydrogen atom represented by a plot of electron density (ψ^2) versus distance from nucleus. (b) Electron distribution in the $1s$ orbital versus distance from nucleus. The distribution is essentially zero both for very short distances from nucleus and for very long distances from nucleus. (c) Cross section through the hydrogen atom, with the space surrounding the nucleus divided into an arbitrary number of thin, concentric hollow layers. Each layer has a unique value for radius r. The probability of finding an electron in a particul radius r depends on the volume of the layer and the density of elec layer. (d) Boundary–surface representation of a sphere within whi ability of finding a $1s$ electron is 90%.

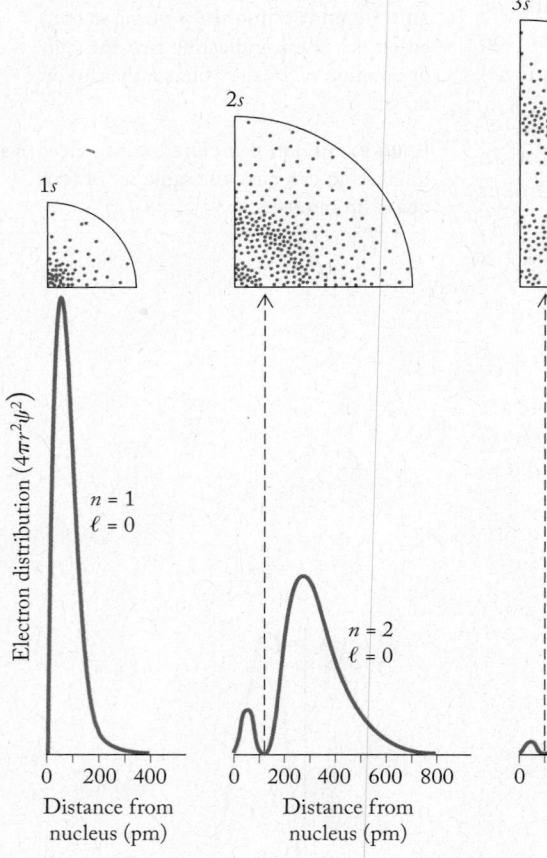

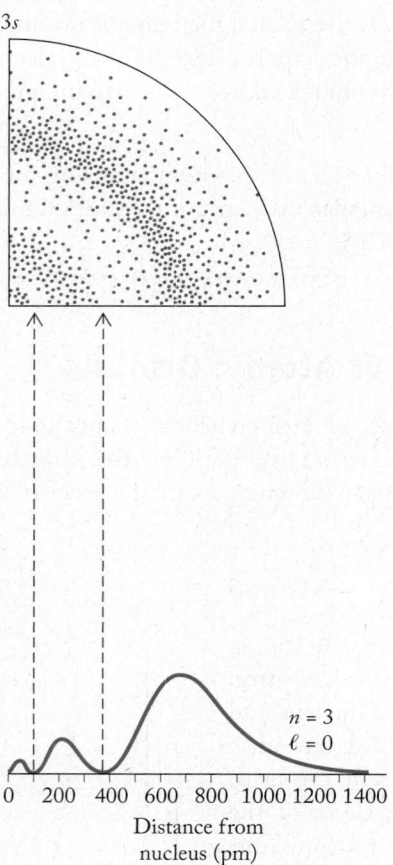

Figure 7.23(b) is a plot, not of ψ^2 versus distance from the nucleus as in Figure 7.23(a), but rather of $4\pi r^2\psi^2$ versus distance from the nucleus. In geometry, $4\pi r^2$ is the formula for the area of a sphere, but here it represents the volume of one of the thin spherical layers in Figure 7.23(c).

A significant feature of the curve in Figure 7.23(b) is that its maximum value corresponds to the most likely radial distance of the electron from the nucleus. The value of r corresponding to this maximum for the 1s orbital of hydrogen is 53 pm.

Figure 7.23(d) provides a view of the spherical shape of this (or any other) s orbital. The surface of the sphere encloses the volume within which the probability of finding a 1s electron is 90%. This type of depiction, called a *boundary–surface representation*, is one of the most useful ways to view the relative sizes, shapes, and orientations of orbitals. All s orbitals are spheres, which have only one orientation and in which electron density depends only on distance from the nucleus. Boundary surfaces are a useful way to depict the shape and relative size of an orbital.

FIGURE 7.24 These radial distribution profiles of 1s, 2s, and 3s orbitals have 0, 1, and 2 nodes, respectively, identifying (with dashed arrows) locations of zero electron density. Electrons in all these s orbitals have some probability of being close to the nucleus, but 3s electrons are more likely to be farther away from the nucleus than 2s electrons, which are more likely to be farther away than 1s electrons.

The relative sizes of 1s, 2s, and 3s orbitals are shown in Figure 7.24. Note that orbital size increases with increasing values of the principal quantum number n. Note also that in the quadrants above the profile curves there are bands in which the density of dots is high. The dots represent the probability of an electron being at these locations, and each band is called a *local maximum* of electron density. In all three profiles, there is a local maximum close to the nucleus. This means that electrons in s orbitals, even s orbitals with high values of n, have some probability of being close to the nucleus.

The local maxima in any s orbital are separated from other local maxima by nodes. The number of nodes in any s orbital is equal to $n - 1$. Nodes have the same meaning here as they do in one-dimensional standing waves: places where the wave has an amplitude of zero. In the context of electrons as three-dimensional matter waves, nodes are locations at which electron density goes to zero.

> **CONCEPT TEST** ....
>
> How many nodes are there in the electron distribution profile of the 6s orbital?
> ....

p and d Orbitals

We have noted that s orbitals are spherical, so in these orbitals electron density depends only on distance from the nucleus, not on angle of orientation. Let's now consider the shapes of p and d orbitals.

All shells with $n \geq 2$ have a subshell containing three p orbitals ($\ell = 1$; $m_\ell = -1, 0, +1$). Each of these orbitals has two teardrop-shaped lobes oriented

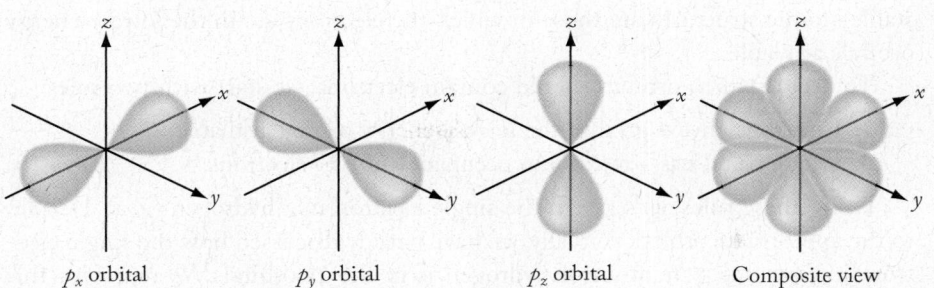

p_x orbital p_y orbital p_z orbital   Composite view

FIGURE 7.25 Boundary–surface views of the three p orbitals, showing their orientation along the x-, y-, and z-axes. The nucleus of the atom in which these orbitals appear is located at the origin.

one on either side of the nucleus along one of the three perpendicular Cartesian axes x, y, z (Figure 7.25). These orbitals are designated $p_x, p_y,$ and p_z, depending on the axis along which the lobes are aligned. The two lobes of a p orbital are sometimes labeled with plus and minus signs, indicating that the sign of the wave function defining them is either $+\psi$ or $-\psi$. An electron in a p orbital occupies both lobes. Because a node of zero probability separates the two lobes, you may wonder how an electron gets from one lobe to the other. One way to think about how this happens is to remember that an electron behaves as a three-dimensional standing wave, and waves have no difficulty passing through nodes. After all, in Figure 7.17 there is a node right in the middle of the $n = 2$ standing wave separating the positive and negative displacement regions in the vibrating string.

The five d orbitals ($\ell = 2, m_\ell = -2, -1, 0, +1, +2$) found in shells of $n \geq 3$ all have different orientations (Figure 7.26). Four of them consist of four teardrop-shaped lobes oriented like the leaves in a four-leaf clover. In three of these orbitals, the lobes lie between, not on, the x-, y-, and z-axes. These orbitals are designated $d_{xy}, d_{xz},$ and d_{yz}. In the fourth orbital in this set, designated $d_{x^2-y^2}$, the four lobes lie along the x- and y-axes. The fifth d orbital, designated d_{z^2}, is mathematically equivalent to the other four but has a much different shape, with two teardrop-shaped lobes oriented along the z-axis and a doughnut shape called a *torus* in the x–y plane that surrounds the middle of the two lobes.

We will not address the shapes and geometries of other types of orbitals in this text, although we will still refer to them. The s, p, and d orbitals are most crucial to the discussions of chemical bonding in the chapters to come.

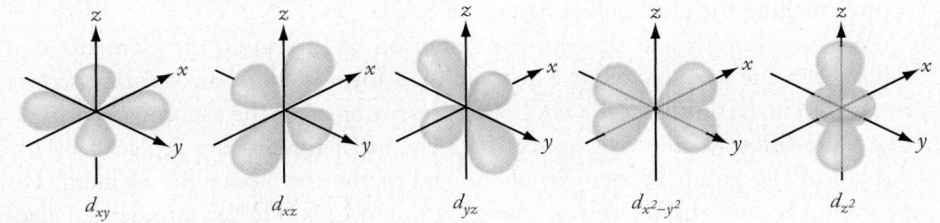

d_{xy} d_{xz} d_{yz} $d_{x^2-y^2}$ d_{z^2}

FIGURE 7.26 Boundary–surface views of the five d orbitals, showing their orientation relative to the x-, y-, and z-axes. The $d_{xy}, d_{xz},$ and d_{yz} orbitals are not aligned along any axis; the $d_{x^2-y^2}$ orbital lies along the x and y axes; the d_{z^2} orbital consists of two teardrop-shaped lobes along the z axis with a donut-shaped torus ringing the point where the two lobes meet.

7.8 The Periodic Table and Filling the Orbitals of Multielectron Atoms

Now that we have developed a model describing the atomic orbitals, we can use that model to fill them in atoms containing more than one electron. To explore the orbital filling sequence, let's start at the beginning of the periodic table with hydrogen and put each successive electron into the lowest-energy orbital available as we move through the table element by element. This method is based on the **aufbau principle** (German *aufbauen,* "to build up"), which states that the most

aufbau principle the method of building electron configurations of atoms by adding one electron at a time as atomic number increases across the rows of the periodic table.

CONNECTION We described in Section 2.4 how Mendeleev developed the first useful periodic table of the elements, not only decades before de Broglie gave us electron waves and Schrödinger pioneered quantum mechanics but even before atoms were known to consist of electrons, protons, and neutrons.

CHEMTOUR Electron Configuration

stable atomic structures are those in which the electrons are in the lowest-energy orbitals available.

To decide which orbitals should contain electrons, we start with two rules:

1. Electrons always go into the lowest-energy orbital available.
2. Each orbital has a maximum occupancy of two electrons.

Using these rules, let's assign the single electron in a hydrogen ($Z = 1$) atom to the appropriate orbital. Actually, we have already discussed how the single electron in a ground-state atom of hydrogen is in the $1s$ orbital. We represent this arrangement with the **electron configuration** $1s^1$, where the first "1" indicates the principal quantum number (n) of the orbital, "s" indicates the type of orbital, and the superscript "1" indicates that there is *one* electron in this $1s$ orbital. Because the hydrogen atom has only one electron, $1s^1$ is the complete electron configuration for a hydrogen atom in its ground state.

The atomic number of helium is 2, which tells us there are two protons in the nucleus and two electrons in the neutral atom. Using the aufbau principle, we simply add another electron to the $1s$ orbital. This orbital already contains one electron, so it has space for one more. The spin quantum numbers for the two electrons cannot be the same. One must be $+\frac{1}{2}$ and the other must be $-\frac{1}{2}$. These two electrons are said to be *spin-paired*. Their presence gives helium a ground-state electron configuration of $1s^2$. With two electrons, the $1s$ orbital is filled to capacity and so is the $n = 1$ shell.

The concept of a *filled shell* is key to understanding chemical properties: elements composed of atoms that have filled s and p subshells in their outermost shells are chemically stable and generally unreactive. Helium is such an element, as are all the other elements in group 18.

The location of lithium ($Z = 3$) in the periodic table—the first element in the second period—is a signal that an atom of lithium has one electron in its $n = 2$ shell. The row numbers in the periodic table correspond to the n values of the outermost shells of the atoms in the rows. The second shell has four orbitals (one $2s$ and three $2p$), so it can hold up to eight electrons. Which of the four orbitals in the second shell contains the third electron in a lithium atom? The answer is linked to the fact that an electron in a $2s$ orbital has a lower average energy than an electron in a $2p$ orbital. Therefore, the third electron in a lithium atom occupies the $2s$ orbital, making the electron configuration $1s^2 2s^1$.

We can simplify the electron configuration of Li and all the elements that follow it in the periodic table. The simplified form is called a *condensed electron configuration*. In this form the symbols representing all of the electrons in orbitals that were filled in the rows above the element of interest are replaced by the symbol of the group 18 element at the end of the row above the element. For example, the condensed electron configuration of Li is [He]$2s^1$. Condensed electron configurations are useful because they eliminate the symbols of the **core electrons** in filled shells and subshells that are not involved in the chemistry of an element. They include only the symbols of the electrons in the outermost shells and subshells that are involved in bond formation. These electrons are called **valence electrons**.

In any atom, the shell containing the valence electrons is referred to as the *valence shell*. Notice that lithium has a single electron in its valence-shell s orbital, as does hydrogen, the element directly above it in the periodic table; therefore both atoms have the valence-shell configuration ns^1, where n represents both the number of the row in which the atom is located on the table and the principal quantum number of the valence shell in the atom.

electron configuration the distribution of electrons among the orbitals of an atom or ion.

core electrons electrons in the filled, inner shells in an atom or ion that are not involved in chemical reactions.

valence electrons electrons in the outermost occupied shell of an atom having the most influence on the atom's chemical behavior.

Beryllium ($Z = 4$) is the fourth element in the periodic table and the first in group 2. The configuration for its four electrons is $1s^2 2s^2$ or [He]$2s^2$. The other elements in group 2 also have two spin-paired electrons in the s orbital of their outermost occupied shell. The second shell is not full at this point because it also has three p orbitals, which are all empty and fill as we move to the next elements in the periodic table.

Boron ($Z = 5$) is the first element in group 13. Its fifth electron is in one of its three $2p$ orbitals, resulting in the condensed electron configuration [He]$2s^2 2p^1$. Which of the three $2p$ orbitals contains the fifth electron is not important because these three orbitals all have the same energy; we say they are **degenerate**.

The next element is carbon ($Z = 6$). It has four electrons in its valence shell (the $n = 2$ shell), so its condensed electron configuration is [He]$2s^2 2p^2$. Note that there are two electrons in the $2p$ orbitals. Are they both in the same orbital? Remember that all electrons have a negative charge and repel one another. Thus they tend to occupy orbitals that allow them to be as far away from each other as possible, which means the two $2p$ electrons in carbon occupy separate $2p$ orbitals. This distribution pattern is an application of **Hund's rule**, which states that, for degenerate orbitals, like the three $2p$ orbitals in the second shell, the lowest-energy electron configuration is the one with the maximum number of unpaired valence electrons, all of which have the same spin.

Hund's rule is the third rule we apply when determining electron configurations. By convention, the first electron placed in an orbital has a positive spin. When electrons fill a degenerate set of orbitals, Hund's rule requires that an electron with a positive spin occupies each valence orbital before any electron with a negative spin enters any orbital. Electrons do not pair in degenerate orbitals until they have to.

To be unambiguous about the electron configuration, we use **orbital diagrams** to show how electrons, represented by single-headed arrows, are distributed among orbitals, which are represented by boxes. A single-headed arrow pointing upward represents an electron with spin up ($m_s = +\frac{1}{2}$), and a downward-pointing single-headed arrow represents an electron with spin down ($m_s = -\frac{1}{2}$). To obey Hund's rule, the orbital diagram for carbon must be

Carbon: $\boxed{\uparrow\downarrow}$ $\boxed{\uparrow\downarrow}$ $\boxed{\uparrow\,|\,\uparrow\,|\,}$
$\quad\quad\quad\;\; 1s \quad\; 2s \quad\quad 2p$

showing the two $2p$ electrons as unpaired and spinning in the same direction.

The next element is nitrogen ($Z = 7$), represented by either $1s^2 2s^2 2p^3$ or [He]$2s^2 2p^3$. According to Hund's rule, the third $2p$ electron resides alone in the third $2p$ orbital, so that the electron distribution is

Nitrogen: $\boxed{\uparrow\downarrow}$ $\boxed{\uparrow\downarrow}$ $\boxed{\uparrow\,|\,\uparrow\,|\,\uparrow}$
$\quad\quad\quad\quad 1s \quad\; 2s \quad\quad 2p$

As we proceed across the second row to neon ($Z = 10$), we fill the $2p$ orbitals as shown in Figure 7.27. The last three $2p$ electrons added (in oxygen, fluorine, and neon) pair up with the first three, so that in neon, the three $2p$ orbitals are all filled to capacity. At this point the $n = 2$ shell is completely filled. Note that neon is directly below helium in group 18. Helium, neon, and all the other noble gases in group 18 have filled s and p orbitals in their valence shells. This same trend is seen throughout the periodic table: *elements in the same column of the periodic table have the same valence-shell configuration.* Argon, krypton, xenon, and radon also have filled s and p orbitals in their outermost occupied shells, and they also are chemically inert gases at room temperature. Thus chemical inertness is associated with filled s and p orbitals in the valence shell.

degenerate describes orbitals of the same energy.

Hund's rule the lowest-energy electron configuration of an atom has the maximum number of unpaired electrons, all of which have the same spin, in degenerate orbitals.

orbital diagram depiction of the arrangement of electrons in an atom or ion using boxes to represent orbitals.

FIGURE 7.27 Orbital diagrams and condensed electron configurations for the first 10 elements show that each orbital (indicated by a square in the orbital diagrams) holds a maximum of two electrons and the two electrons must be of opposite spin. The orbitals are filled in order of increasing quantum numbers n and ℓ. Condensed electron configurations for all elements are given in Appendix 3.

	Orbital Diagram			Electron Configuration	Condensed Configuration
	$1s$	$2s$	$2p$		
H	↑			$1s^1$	
He	↑↓			$1s^2$	
Li	↑↓	↑		$1s^2 2s^1$	[He]$2s^1$
Be	↑↓	↑↓		$1s^2 2s^2$	[He]$2s^2$
B	↑↓	↑↓	↑	$1s^2 2s^2 2p^1$	[He]$2s^2 2p^1$
C	↑↓	↑↓	↑ ↑	$1s^2 2s^2 2p^2$	[He]$2s^2 2p^2$
N	↑↓	↑↓	↑ ↑ ↑	$1s^2 2s^2 2p^3$	[He]$2s^2 2p^3$
O	↑↓	↑↓	↑↓ ↑ ↑	$1s^2 2s^2 2p^4$	[He]$2s^2 2p^4$
F	↑↓	↑↓	↑↓ ↑↓ ↑	$1s^2 2s^2 2p^5$	[He]$2s^2 2p^5$
Ne	↑↓	↑↓	↑↓ ↑↓ ↑↓	$1s^2 2s^2 2p^6$	[He]$2s^2 2p^6$ = [Ne]

Sodium ($Z = 11$) follows neon in the periodic table. It is the third element in group 1 and the first element in the third period. Ten of its electrons are distributed as in neon. The 11th electron is in the lowest-energy orbital available after $2p$ has been filled, which is $3s$. The condensed electron configuration of Na is [Ne]$3s^1$. Just as we write condensed electron configurations that provide detailed information for only the outermost occupied shell, we can condense orbital diagrams in this same way, so that the condensed orbital diagram for sodium is

$$\text{Sodium:} \qquad [\text{Ne}] \;\boxed{\uparrow}\atop{3s}$$

This diagram reinforces the message that the electron configuration of a sodium atom consists of a neon core plus a single electron in the $3s$ orbital in the outermost occupied shell. The sodium atom has the same generic valence-shell configuration as lithium and hydrogen, namely ns^1, where n is the period number. This pattern for elements in the same group continues throughout the periodic table. The electron configuration of magnesium ($Z = 12$), for instance, is [Ne]$3s^2$, and the electron configuration of every other element in group 2 consists of the immediately preceding noble gas core followed by ns^2.

The next six elements in the periodic table—aluminum, [Ne]$3s^2 3p^1$, to argon, [Ne]$3s^2 3p^6$—show a pattern of increasing numbers of $3p$ electrons, a trend that continues until all three $3p$ orbitals are filled (six electrons), which means the s and p orbitals of the $n = 3$ shell are filled (eight electrons). Thus, argon is chemically inert, as predicted by its position in group 18.

After argon comes potassium ($Z = 19$) in group 1 of row 4 ([Ar]$4s^1$), followed by calcium ($Z = 20$) in group 2 ([Ar]$4s^2$). At this point, the $4s$ orbital is filled but the $3d$ orbitals are still empty. Why were the $3d$ orbitals not filled before $4s$? (Review Table 7.1 if you need help in recalling that the $n = 3$ shell contains a d subshell.)

Applying the first aufbau rule—in building atoms, each electron goes in the lowest-energy orbital available—would be straightforward were it not for the fact that the differences in energy between shells get smaller as n gets larger (see Figure 7.16). These smaller differences result in orbitals with large ℓ values in one shell having energies similar to orbitals with small ℓ values in the next higher shell. Note in Figure 7.28 that the energy of the $4s$ orbital is slightly lower than that of the $3d$ orbitals. The $4s$ orbitals in potassium and calcium are the lowest-energy orbitals available and are filled before any electrons go into a $3d$ orbital.

The element after calcium is scandium ($Z = 21$). It is the first element in the central region of the periodic table, the region populated by transition metals. Scandium has the condensed electron configuration $[Ar]3d^1 4s^2$. Note that the orbitals are arranged in order of increasing principal quantum number, not necessarily in the order in which they were filled. The $3d$ orbitals are filled in the transition metals from scandium to zinc ($Z = 30$). This pattern of filling the d orbitals of the shell whose principal quantum number is 1 less than the period number, $(n - 1)d$, is followed throughout the periodic table: the $4d$ orbitals are filled in the transition metals of the fifth period, and so on (Figure 7.29).

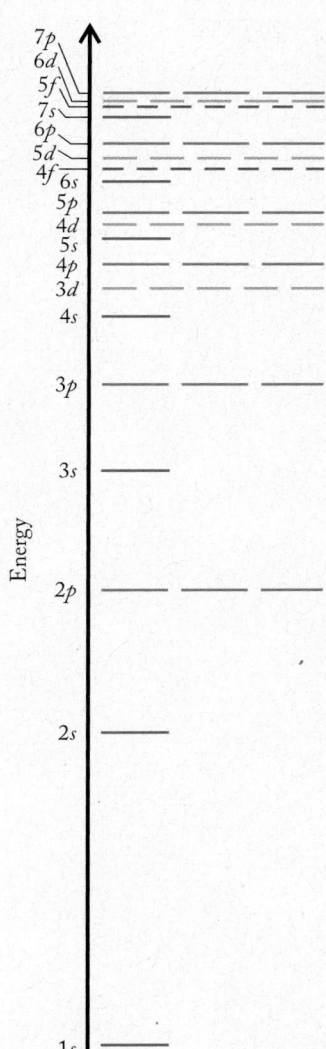

FIGURE 7.28 The energy levels in multi-electron atoms increase with increasing values of n and with increasing values of ℓ within a shell. The difference in energy between adjacent shells decreases with increasing values of n, which may cause the energies of subshells in two adjacent shells to overlap. For example, electrons in $3d$ orbitals have slightly higher energy than those in the $4s$ orbital, resulting in the order of subshell filling $4s \rightarrow 3d \rightarrow 4p$.

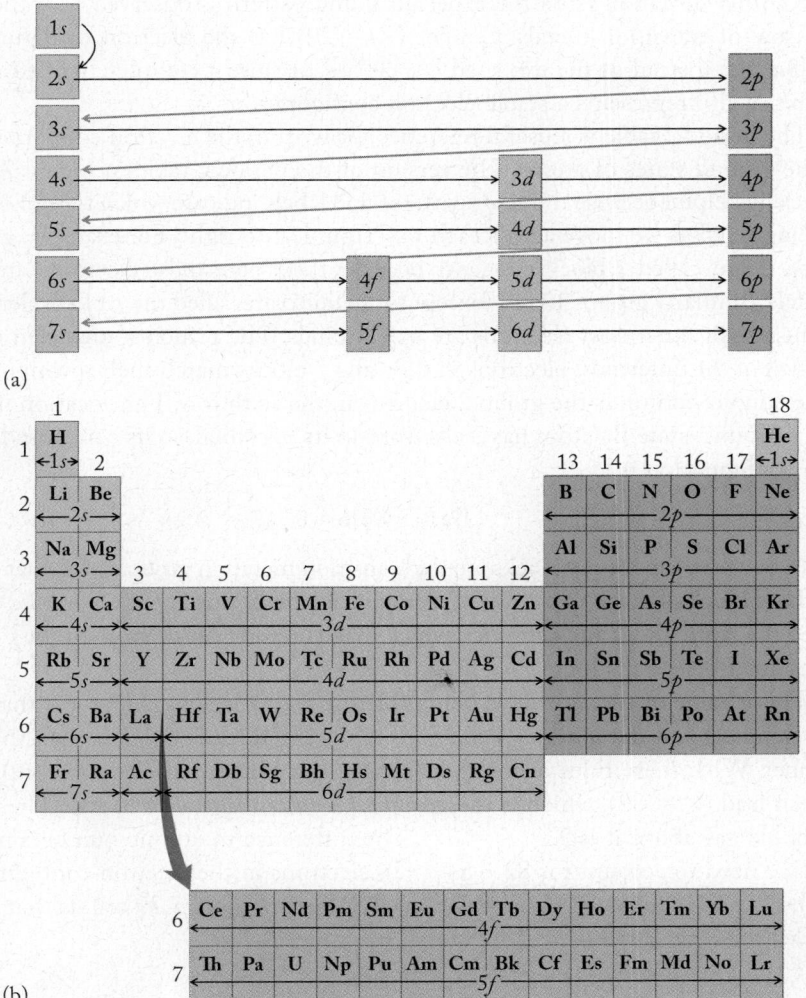

FIGURE 7.29 (a) This diagram shows the sequence in which atomic orbitals fill. (b) The same color coding is used in this version of the periodic table to highlight the four "blocks" of elements in which valence shell s (green), p (red), d (orange), f (purple) orbitals are filled with increasing atomic number across a row of the table.

The element after scandium is titanium ($Z = 22$), which has one more d electron than scandium, so its condensed electron configuration is $[\text{Ar}]3d^24s^2$. At this point, you may feel you can accurately predict the electron configurations of the remaining transition metals in the fourth period. However, because the energies of the $3d$ and $4s$ orbitals are similar, the sequence of d-orbital filling deviates in two spots from the pattern you might expect. Vanadium ($Z = 23$) has the expected configuration $[\text{Ar}]3d^34s^2$, but the next element, chromium ($Z = 24$), has the configuration $[\text{Ar}]3d^54s^1$. The reason for this difference is that this chromium configuration puts one electron in each of the five d orbitals:

Chromium: $[\text{Ar}]$ ⬜⬆ ⬆ ⬆ ⬆ ⬆ $\quad$ ⬆
$$\underbrace{\qquad\qquad}_{3d^5} \quad \underbrace{\quad}_{4s^1}$$

This half-filled set of d orbitals is an energetically favored configuration. Apparently, the stability of having five half-filled $3d$ orbitals compensates for the energy needed to raise a $4s$ electron to a $3d$ orbital. As a result, $[\text{Ar}]3d^54s^1$ is a more stable electron configuration than $[\text{Ar}]3d^44s^2$.

Another deviation from the expected filling pattern is observed near the end of a row of transition metals. Copper ($Z = 29$) has the electron configuration $[\text{Ar}]3d^{10}4s^1$ instead of the expected $[\text{Ar}]3d^94s^2$ because a completely filled set of d orbitals also represents a stable electron configuration.

The periodic table is a useful reference for writing the electron configurations of the ground states of atoms. The version of the periodic table in Figure 7.29 is especially helpful because the color patterns and labels indicate which type of orbital is being filled as we move across each row from left to right. For example, groups 1 and 2 are called s block elements because their outermost electrons are in s orbitals. Similarly groups 13–18 (except for helium) are called the p block elements because their outermost electrons are in p orbitals. The principal quantum numbers (n) of the outermost electrons in the s and p blocks match their row numbers. For example, barium is the group 2 element in the sixth row. This location means that a ground-state Ba atom has 2 electrons in its $6s$ orbital, so its condensed electron configuration is

$$\text{Ba:} \quad [\text{Xe}]6s^2$$

In between the s and p blocks are the transition metals in groups 3–12 that make up the d block and the two rows of elements at the bottom of the periodic table, called the lanthanides and actinides, which make up the f block. Both of the f block series are 14 elements long.

The n value of the d orbitals being filled in a row is always one less than the row numbers and the n value of the f orbitals being filled is two less than the row number. With these rules in mind, let's write the condensed electron configuration of lead ($Z = 82$), which is the group 14 element in the sixth row. The nearest noble gas above it is Xe ($Z = 54$). The difference in atomic numbers means that we need to account for $82 - 54 = 28$ electrons in the electron configuration symbols. The location of Pb and the block labels in Figure 7.29 tell us that these 28 electrons are distributed as follows:

2 electrons in $6s$
14 electrons in $4f$
10 electrons in $5d$
2 electrons in $6p$

The electron configuration of ground-state lead atoms reflects this distribution:

$$Pb: \quad [Xe]4f^{14}5d^{10}6s^26p^2$$

SAMPLE EXERCISE 7.10   Writing Electron Configurations of Atoms

Write the condensed electron configuration of an atom of silver ($Z = 47$).

Collect and Organize In a condensed electron configuration the filled sets of orbitals in the inner shells of the atom are represented by the atomic symbol of the noble gas immediately preceding the element of interest in the periodic table.

Analyze Silver ($Z = 47$) is the group 11 element in the fifth row of the periodic table; krypton ($Z = 36$) is the immediately preceding noble gas at the end of the fourth row. The difference between these atomic numbers means that we need to account for $47 - 36 = 11$ electrons.

Solve We would initially predict the first 2 of the 11 electrons would be in the $5s$ orbital and the next nine would be in $4d$ orbitals, resulting in a condensed electron configuration of $[Kr]4d^95s^2$. However, a completely filled set of d orbitals is more stable than a partially filled set, so silver, like copper just above it in the periodic table, has 10 electrons in its occupied d orbitals and only one electron in its outermost occupied shell, which is the $n = 5$ shell: $[Kr]4d^{10}5s^1$.

Think about It We can generate a tentative electron configuration by simply moving across a row in Figure 7.29 until we come to the element of interest. However, in the transition metals, we have to remember the special stability of half-filled and filled d orbitals and make the appropriate adjustments in our configuration.

Practice Exercise Write the condensed electron configuration of a ground-state atom of cobalt ($Z = 27$).

7.9 Electron Configurations of Ions

To write the electron configuration of an ion, we begin with the electron configuration of the atom from which the ion was formed. If the ion has a positive charge, we remove the appropriate number of electrons from the orbital(s) with the highest principal quantum number. If the ion has a negative charge, we add the appropriate number of electrons to one or more partially filled outer-shell orbitals.

Ions of the Main Group Elements

The s block elements (see Figure 7.29) form monatomic cations by losing all their outer-shell electrons, leaving their ions with the electron configuration of the noble gas immediately preceding them in the periodic table. For example, an atom of sodium forms a Na^+ ion by losing its single $3s$ electron:

$$Na: \quad [Ne]3s^1$$
$$Na^+: \quad [Ne]$$

◯◯ **CONNECTION** In the previous chapters, we learned the charges on the ions of common elements. Electron configurations help us understand why these ions have the charges they do.

Nonmetallic elements of the p block that form monatomic anions do so by gaining enough electrons to completely fill their outer-shell p orbitals, forming ions with the electron configurations of the noble gases at the right ends of their rows in the periodic table. For example, an atom of fluorine forms a F^- ion by gaining one electron, which completely fills its set of three $2p$ orbitals and gives it the electron configuration of neon:

$$F: \quad [He]2s^2 2p^5$$

$$F^-: \quad [He]2s^2 2p^6 = [Ne]$$

Thus, a Na^+ ion and a F^- ion have the same electron configuration as an atom of Ne. We say that these three species, Na^+, F^-, and Ne, are **isoelectronic**, meaning that they have the same electron configuration.

SAMPLE EXERCISE 7.11 **Determining Isoelectronic Species in Main Group Ions**

(a) Write the electron configurations of the ions in NaF, $MgCl_2$, CaO, and KBr. (b) Which ions in part a are isoelectronic with neon atoms?

Collect and Organize In part a, we must determine the electron configuration of each ion in four binary ionic compounds. In part b, our task is to compare the configurations from part a with the configuration for a neon atom and determine which have the same electron configuration.

Analyze The elements in the compounds include

two from group 1, Na and K, which form 1+ cations;

two from group 2, Mg and Ca, which form 2+ cations;

one from group 16, O, which forms a 2− anion;

three from group 17, F, Cl, and Br, which form 1− anions.

Let's arrange these atoms and ions in a table, remembering that an atom loses valence electrons in becoming a cation and gains valence electrons in becoming an anion:

Element	Electron Configuration of Atom	Atomic Number (Z)	Charge on Ion	Electrons per Ion
Na	$[Ne]3s^1$	11	1+	10
K	$[Ar]4s^1$	19	1+	18
Mg	$[Ne]3s^2$	12	2+	10
Ca	$[Ar]4s^2$	20	2+	18
O	$[He]2s^2 2p^4$	8	2−	10
F	$[He]2s^2 2p^5$	9	1−	10
Cl	$[Ne]3s^2 3p^5$	17	1−	18
Br	$[Ar]3d^{10}4s^2 4p^5$	35	1−	36

isoelectronic describes atoms or ions that have identical electron configurations.

Solve

a. The electron configurations for the eight ions are

$$Na^+: \quad [Ne]$$
$$K^+: \quad [Ar]$$
$$Mg^{2+}: \quad [Ne]$$
$$Ca^{2+}: \quad [Ar]$$
$$O^{2-}: \quad [He]2s^22p^6 = [Ne]$$
$$F^-: \quad [He]2s^22p^6 = [Ne]$$
$$Cl^-: \quad [Ne]3s^23p^6 = [Ar]$$
$$Br^-: \quad [Ar]3d^{10}4s^24p^6 = [Kr]$$

b. Four of the ions formed—Na^+, Mg^{2+}, O^{2-}, and F^-—have the same number of electrons as an atom of neon (10) and are isoelectronic with Ne and with one another.

Think about It Each of the ions has an electron configuration of a noble gas atom. This configuration is associated with stable ions of many main group elements.

Practice Exercise Write the electron configurations of the ions in KI, BaO, Rb_2O, and $AlCl_3$. Which of these ions are isoelectronic with Ar?

Transition Metal Cations

As with the main group elements, writing the electron configurations of transition metal cations begins with the atoms from which the cations form. Nickel atoms, like those of many transition metals, form ions with 2+ charges by losing both electrons from the shell of highest n, in this case the outer-shell s electrons:

$$Ni: \quad [Ar]3d^84s^2$$
$$Ni^{2+}: \quad [Ar]3d^8$$

A few transition metals, including silver, have only one outer-shell s electron in their atoms and so form singly charged ions:

$$Ag: \quad [Kr]4d^{10}5s^1$$
$$Ag^+: \quad [Kr]4d^{10}$$

We might have expected Ni and Ag atoms to lose their slightly higher-energy $3d$ and $4d$ electrons, respectively, reasoning that the last orbitals to be filled should be the first to be emptied when an atom forms a positive ion. However, the rule that the electrons in orbitals with the highest n value ionize first applies to these and the other transition metals. Preferential loss of outer-shell s electrons explains why the most frequently encountered charge on transition metal ions is 2+.

Many transition metal atoms also lose one or more d electrons as they form ions with charges $\geq 2+$. Atoms of scandium, $[Ar]3d^14s^2$, for example, lose both $4s$ electrons and the $3d$ electron as they form Sc^{3+} ions. The chemistry of titanium, $[Ar]3d^24s^2$, is dominated by its tendency to lose its $4s$ and $3d$ electrons to form Ti^{4+} ions. In general, when we are determining electron configurations for transition metal ions, we first remove electrons from the outermost occupied s orbital, then from the outermost occupied d subshell, until the charge on the ion is achieved.

SAMPLE EXERCISE 7.12 **Writing Electron Configurations of Transition Metal Ions**

What are the electron configurations of Fe^{2+} and Fe^{3+}?

Collect and Organize We are asked to write the electron configurations of two ions formed by iron ($Z = 26$). Iron is the group 8 element of the fourth row of the periodic table. Figure 7.29 provides information on the order in which orbitals are filled. Transition metal atoms preferentially lose their outermost s electrons when they form ions.

Analyze The location of iron in the periodic table tells us that the atom has two $4s$ and six $3d$ electrons built on an argon core: the electron configuration of Fe is

$$Fe: \quad [Ar]3d^64s^2$$

Solve We remove the two $4s$ electrons to form Fe^{2+} and the two $4s$ and one of the $3d$ electrons to form Fe^{3+}:

$$Fe^{2+}: \quad [Ar]3d^6 \qquad Fe^{3+}: \quad [Ar]3d^5$$

Think about It In terms of the stability we have seen associated with half-filled shells, it makes sense that Fe atoms form ions with a charge of $3+$. The loss of the $4s$ electrons and one of the $3d$ electrons results in an electron configuration with a stable half-filled set of $3d$ orbitals.

Practice Exercise Write the electron configurations for the manganese atom and the ions Mn^{3+} and Mn^{4+}.

We have yet to consider the lanthanides (elements 58 through 71) and actinides (elements 90 through 103), represented by the two purple rows in Figure 7.29. The lanthanides have partly filled $4f$ orbitals, and the actinides have partly filled $5f$ orbitals. There are 14 elements in each group, reflecting the capacity of the seven orbitals in each f subshell ($\ell = 3$, $m_\ell = -3, -2, -1, 0, +1, +2,$ and $+3$). As you can see in Figure 7.29, the $4f$ orbitals are not filled until after the $6s$ orbital has been filled. This order of filling is due to the similar energies of $6s$ and $4f$ orbitals (Figure 7.28). Similarly, the $5f$ orbitals are filled after the $7s$ orbital is filled.

The periodic table is a useful reference for predicting the physical and chemical properties of elements. Our quantum mechanical perspectives on atomic structure provide us with a theoretical basis for explaining why particular families of elements behave similarly. The original table was based on periodic trends in observable chemical properties. Now we know that the chemical properties of an element are closely linked to the electron configurations of the atoms of the element.

The electron configuration of Eu ($Z = 63$) is $[Xe]4f^76s^2$, but the electron configuration of Gd ($Z = 64$) is $[Xe]4f^75d^16s^2$. Suggest a reason why the additional electron in Gd is not in a $4f$ orbital, which would result in the electron configuration $[Xe]4f^86s^2$.

atomic radius (also called **covalent radius**) half the distance between identical nuclear centers in a molecule.

metallic radius half the distance between nuclear centers in the crystal of a metal.

ionic radius radius derived from the distance between nuclear centers in ionic crystals.

7.10 The Sizes of Atoms and Ions

The size of the atoms or ions making up any substance influences both physical and chemical properties of the substance. Because the Heisenberg uncertainty principle prevents us from knowing the exact location of the electrons in atoms and ions, the outer boundary of an atom or ion cannot really be defined. Therefore we must define some convention for how atomic and ionic sizes are measured.

We calculate the sizes of atoms from the distances between nuclei bonded together in molecules. For elements such as N_2, O_2, and the halogens that exist as diatomic molecules, the **atomic radius**, also referred to as a **covalent radius** because it is determined for two atoms in a covalent bond, is simply half the distance between the nuclear centers in a molecule (Figure 7.30a). For metals, the **metallic radius** is defined as half the distance between the nuclear centers in a crystal of the metal (Figure 7.30b). Based on these definitions, the relative sizes of atoms along with their locations in the periodic table are shown in Figure 7.31. The values of **ionic radii** are derived from the distances between nuclear centers in ionic crystals (Figure 7.30c).

Atomic size varies from top to bottom in a given group and from left to right across a given period. Two opposing factors contribute to these variations. The first is the principal quantum number n, which determines the most probable distance of the electrons from the nucleus. As n increases, the probability increases that the electrons are farther from the nucleus. The second factor is the nuclear charge, the positive charge of the protons in the nucleus, which determines the attractive force holding the electrons in the atom. As nuclear charge increases, the positive

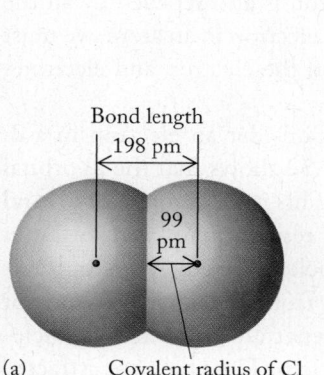

(a) Covalent radius of Cl

Bond length
198 pm
99 pm

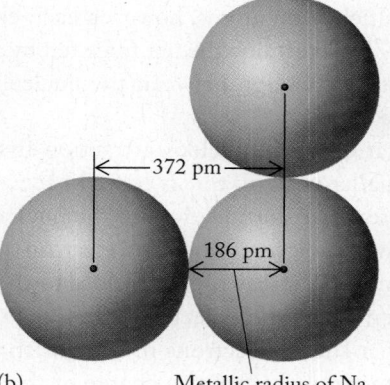

(b) Metallic radius of Na

372 pm
186 pm

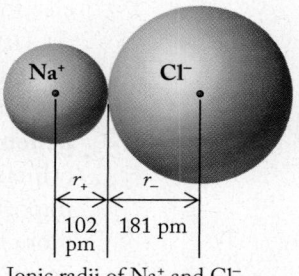

(c) Ionic radii of Na$^+$ and Cl$^-$

Na$^+$ Cl$^-$

r_+ r_-
102 pm | 181 pm

FIGURE 7.30 A comparison of covalent, metallic, and ionic radii. (a) A covalent radius is half the distance between identical nuclei in a covalent bond, such as the bond in the diatomic molecule Cl_2. (b) A metallic radius is based on the distance of closest approach of adjacent atoms in a solid metal. (c) An ionic radius is determined by a series of comparisons among ionic compounds containing the ions of interest.

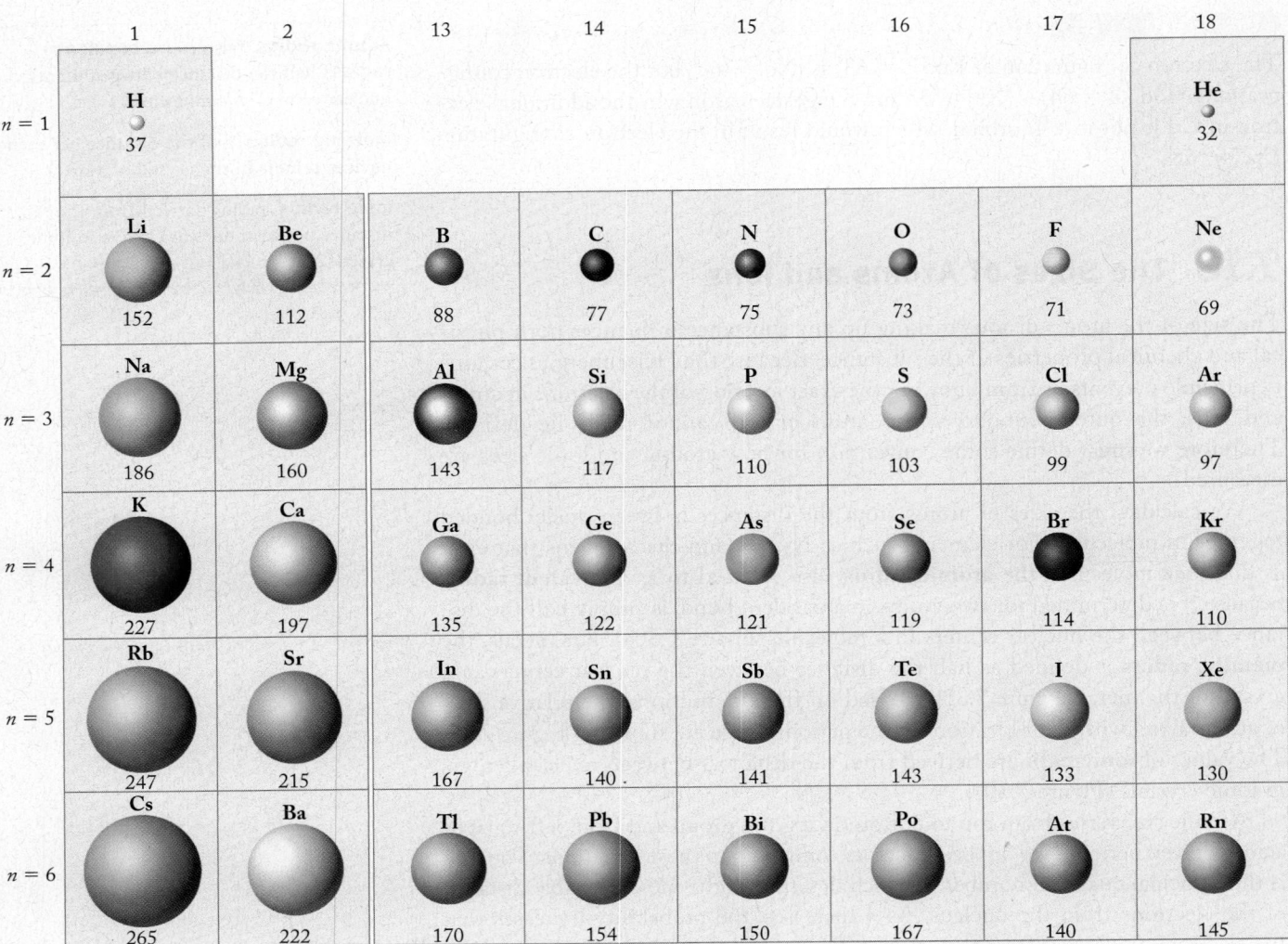

FIGURE 7.31 Atomic radii in picometers of the main group elements. Size generally increases from top to bottom in any group and generally decreases from left to right across any period.

charge felt by the electrons increases, and the electrons are pulled closer to the nucleus. In multielectron atoms, however, each electron is also repelled by all the other electrons. To determine the net force felt by any electron in an atom, we must consider both the attraction between the nucleus and the electron and electron–electron repulsions.

Let's look at electron–nucleus attraction first. Consider an electron in a 2*s* orbital. The smaller peak on the 2*s* curve in Figure 7.32 shows that the 2*s* orbital has some electron distribution close to the nucleus. This is an example of **orbital penetration**, which occurs when an electron that resides mainly in an outer orbital has some probability of being close to the nucleus. Orbital penetration is important when outer-shell electrons are separated from the nucleus by one or more filled inner shells. Electrons in orbitals that penetrate closer to the nucleus experience more of the positive charge of the nucleus. Thus they are attracted more strongly to the nucleus than are electrons in orbitals that do not penetrate as much.

The absence of any secondary peak in the 2*p* curve of Figure 7.32 shows that electrons in a 2*p* orbital penetrate less effectively than those in a 2*s* orbital.

Penetration by electrons in shells farther from the nucleus decreases in the orbitals in a given shell according to the order $s > p > d > f$. Since greater penetration means lower energy, the energies of the orbitals in a given shell are $s < p < d < f$. Because the electrons in the ground state of a multielectron atom occupy the lowest-energy orbitals available, the order of filling orbitals in a given shell, as we see in Figure 7.29, is s first, then (if available) p, then d, then f.

We must also consider electron–electron repulsions in a multielectron atom. For example, an electron in a $3d$ orbital is repelled by every electron that lies between it and the nucleus. Because this repulsive effect cancels some of the electron–nuclear attraction, we can picture the electrons lying between a $3d$ electron and the nucleus as **shielding**, or **screening**, an electron from experiencing the full charge on the nucleus. From the point of view of a $3d$ electron, the nuclear charge is reduced from its actual value to that of an **effective nuclear charge (Z_{eff})**.

In general, the Z_{eff} experienced by outer-shell electrons increases if the orbitals occupied by the electrons penetrate the lower-lying orbitals. For example, a $3s$ electron that penetrates close to the nucleus experiences a higher Z_{eff} than does a $3d$ electron that has little penetration of inner-shell orbitals. Trends in properties that depend on the strength of attraction between electrons and nuclei are determined by how effectively electrons are shielded from nuclear charge.

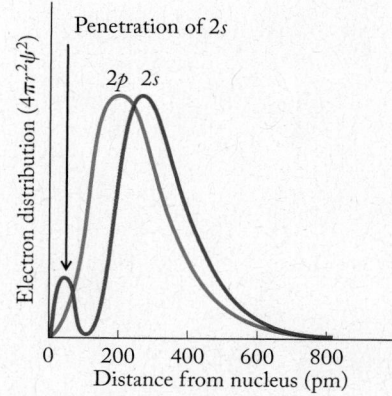

FIGURE 7.32 Because the main peak of the $2s$ curve (blue) is farther from the nucleus than the $2p$ peak (red), the $2s$ orbital appears to be farther from the nucleus than the $2p$ orbital. However, the $2s$ orbital is of lower energy because electrons in it penetrate more closely to the nucleus, indicated by the smaller $2s$ peak at about 50 pm from the nucleus. The result is that $2s$ electrons experience a greater effective nuclear charge and have lower energy than $2p$ electrons.

CONCEPT TEST

Rank the following orbitals in an atom of silver in order of decreasing nuclear charge experienced by the electrons in them: (a) $1s$, (b) $2s$, (c) $3s$, (d) $4s$, (e) $2p$, (f) $3p$, (g) $4p$.

Trends in Atom and Ion Sizes

As one moves down a group in the periodic table, the principal quantum number n increases, and as n increases, the probability that the outermost electrons are farther from the nucleus increases. As a result, atomic radii usually increase. However, as one moves left to right across a row of main group elements or transition metals, the charge on the nucleus increases, but n stays the same. As a result Z_{eff} increases. Increasing Z_{eff} means that atom size generally decreases with increasing atomic number across a row. The major deviation from this trend is seen in the transition metals, where size decreases for the first two elements but then decreases so gradually that it is virtually constant until the last element in the series. The d electrons that fill in the transition series enter an inner shell ($n - 1$) and so do not cause the size to change significantly.

The cations of the main group elements are much smaller than their parent atoms, but the anions are much larger (Figure 7.33). To understand these opposite trends, consider what happens when a Na atom forms a Na^+ ion: it loses its entire valence shell, leaving behind a much smaller neon core of electrons. On the other hand, when a Cl atom acquires an electron and forms a Cl^- ion, it contains more electrons than protons; hence, the attractive force per electron decreases while the electron–electron repulsion increases. Anions are thus always larger than the atoms from which they form.

orbital penetration the probability that an electron in an outer orbital will be as close to the nucleus as an electron in an inner shell.

shielding (also called **screening**) the effect when inner-shell electrons protect outer-shell electrons from experiencing the total nuclear charge.

effective nuclear charge (Z_{eff}) the attractive force toward the nucleus experienced by an electron in an atom; the positive charge on the nucleus reduced by the extent to which other electrons in the atom shield the electron from the nucleus.

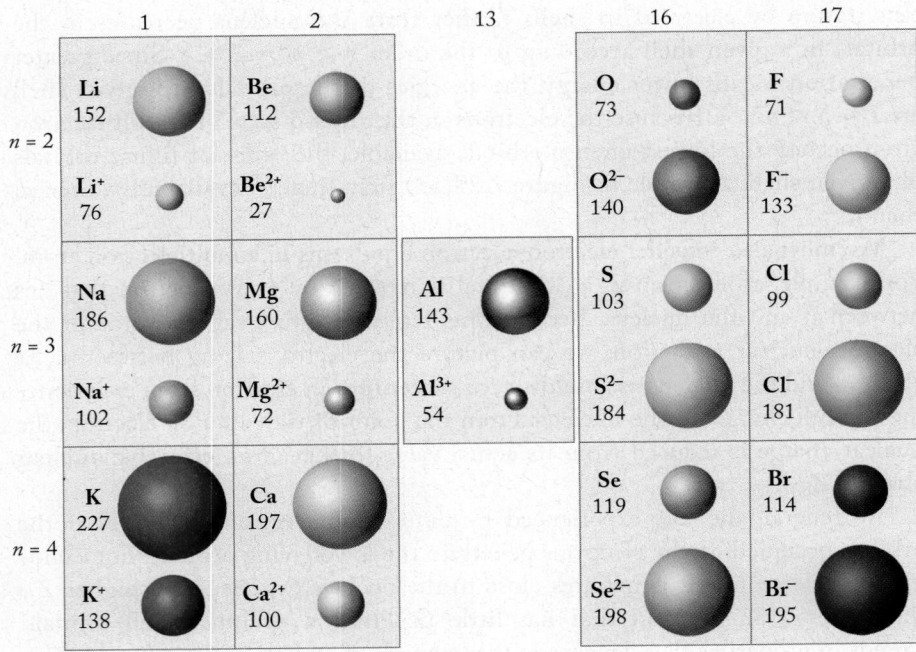

FIGURE 7.33 Comparison of atomic and ionic radii, in picometers.

SAMPLE EXERCISE 7.13   Ordering Atoms and Ions by Size

Arrange each set by size, largest to smallest: (a) O, P, S; (b) Na^+, Na, K.

Collect and Organize We are to rank a set of three atoms based on their size and a set of two atoms and a cation of one of the atoms based on their size. The location of elements in the periodic table can be used to determine the relative sizes of their atoms.

Analyze Sizes decrease as we move left to right across a period and increase as we move down a column. In addition, a cation is always smaller than the atom from which it is made.

Solve

    a. S is below O in group 16, so in terms of atomic size, S > O. S is to the right of P, so P > S. The size order is thus P > S > O.

    b. Cations are smaller than their atoms, so Na > Na^+. Size increases down a group; K is below Na in the alkali metals group, so K > Na. Therefore, the size order is K > Na > Na^+.

Think about It The trend in the sizes of the atoms in set a reflects decreasing atomic size with increasing nuclear charge within a row of elements in the periodic table and increasing atomic size with increasing atomic number down a group. The relative sizes of the particles in set b are linked (1) to increasing atomic size as one goes down a column of elements in the periodic table and (2) to the smaller size of a cation relative its parent atom.

Practice Exercise Arrange each set in order of increasing size (smallest to largest): (a) Cl^-, F^-, Li^+; (b) P^{3-}, Al^{3+}, Mg^{2+}.

7.11 Ionization Energies

In developing electron configurations, we followed a theoretical framework for the arrangement of electrons in orbitals that was developed in the early years of the 20th century. Is there actual experimental evidence for the existence of orbitals representing different energy levels inside atoms? Yes, there is. The evidence includes the measurement of the energies needed to remove electrons from atoms and ions.

Ionization energy (IE) is the energy needed to remove 1 mole of electrons from 1 mole of gas-phase atoms or ions in their ground state. Removing these electrons always costs energy because a negatively charged electron is attracted to a positively charged nucleus, and energy is required to overcome that attractive force. The amount of energy needed to remove 1 mole of electrons from 1 mole of atoms to make 1 mole of 1+ cations is called the *first ionization energy* (IE_1); the energy needed to remove 1 mole of electrons from 1 mole of 1+ cations to make 1 mole of 2+ cations is the *second ionization energy* (IE_2), and so forth. For example,

$$Mg(g) \rightarrow Mg^+(g) + 1\,e^- \qquad (IE_1 = 738 \text{ kJ/mol})$$
$$Mg^+(g) \rightarrow Mg^{2+}(g) + 1\,e^- \qquad (IE_2 = 1451 \text{ kJ/mol})$$

The total energy required to make 1 mole of $Mg^{2+}(g)$ cations from 1 mole of $Mg(g)$ atoms is the sum of these two ionization energies:

$$Mg(g) \rightarrow Mg^{2+}(g) + 2\,e^-$$
$$\text{Total IE} = (738 + 1451) \text{ kJ/mol} = 2189 \text{ kJ/mol}$$

Figure 7.34 shows how the first ionization energies vary in the main group elements. The IE_1 of hydrogen is 1312 kJ/mol. The IE_1 of helium is nearly twice as big: 2372 kJ/mol. This difference seems reasonable because He atoms have two protons per nucleus, whereas H atoms have only one. Twice the nuclear charge leads

ionization energy (IE) the amount of energy needed to remove 1 mole of electrons from 1 mole of ground-state atoms or ions in the gas phase.

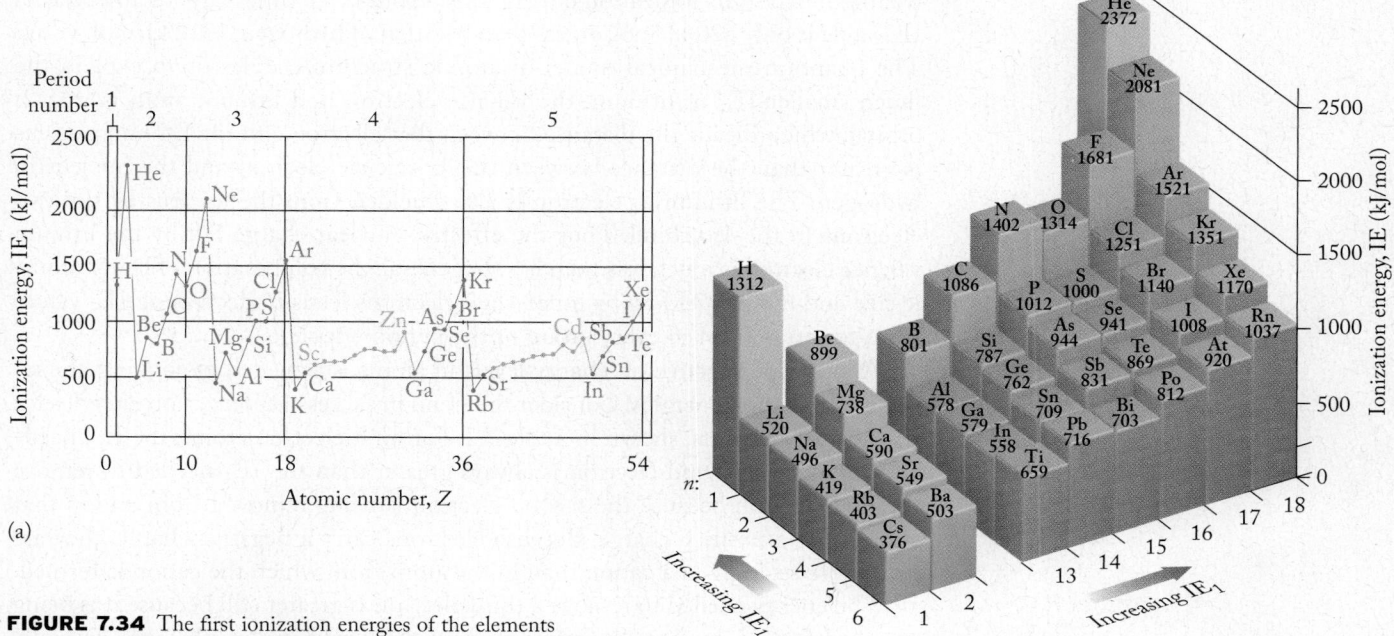

FIGURE 7.34 The first ionization energies of the elements generally increase from left to right in a period and decrease from top to bottom in a group: (a) IE_1 values for the elements in the first five periods; (b) IE_1 values for all the representative group elements.

to twice the nuclear attraction, so twice the ionization energy is needed to pull an electron away from the nucleus. In general, first ionization energies increase from left to right across a period. Thus the easiest element to ionize in a period is the group 1 element in that period and the hardest is the group 18 element. This pattern makes sense because the charge of the nucleus increases across a row and so does the attraction between the nucleus and the surrounding electrons.

Two anomalies occur in the general trend of increasing IE_1 with increasing Z across a row. One shows up between the group 2 and 13 elements that are next to each other in the second and third rows. There is a decrease in IE_1 between Be and B and between Mg and Al. The reason is that B and Al lose a p electron when they ionize, whereas Be and Mg must lose an s electron. It is easier to remove a p electron because it experiences less effective nuclear charge than an s electron in the same shell.

Another anomaly occurs in the values of IE_1 between the group 15 and 16 elements. The decrease here is associated with the enhanced stability that comes from having a half-filled set of p orbitals. We observed in Section 7.9 a similar stability in a half-filled set of d orbitals. For example, when an oxygen atom loses a $2p$ electron, it achieves the slightly more stable configuration of a half-filled set of $2p$ orbitals:

$$O^+: \quad \boxed{\uparrow\downarrow}_{1s} \quad \boxed{\uparrow\downarrow}_{2s} \quad \boxed{\uparrow\,|\,\uparrow\,|\,\uparrow}_{2p}$$

The enhanced stability of the O^+ ion means that less energy is required than would be predicted by applying the general trend of increasing IE_1 with increasing nuclear charge to produce the ion from a neutral O atom.

Now let's see how ionization energy changes as we go down a group of the periodic table. We begin with group 1 and the IE_1 values of hydrogen and lithium. A lithium atom has three times the nuclear charge of a hydrogen atom, so we might expect its ionization energy to be about three times larger. However, its IE_1 value is only 520 kJ/mol, or less than half that of hydrogen, 1312 kJ/mol. Why? The quantum mechanical model of atomic structure enables us to explain the much smaller IE_1 of lithium: the valence electron in a lithium atom is in a $2s$ orbital, which means the distance between that electron and the lithium nucleus is greater than the distance between the $1s$ valence electron and the nucleus in hydrogen. The lithium $2s$ electron is also shielded from the nucleus by the two electrons in the $1s$ orbital. Thus the effective nuclear charge felt by the lithium valence electron is much less than $3+$. In general, the combination of larger atomic size and more shielding by inner-shell electrons leads to decreasing IE_1 values from top to bottom in every group of the periodic table.

Another perspective on energy levels in atoms is provided by looking at successive ionization energies. Consider the trend in successive ionization energies for the first 10 elements, shown in Table 7.2. For multielectron atoms, the IE_2 needed to remove a second electron is always greater than the IE_1 needed to remove the first electron because the second electron is being removed from an ion that already has a positive charge. Because electrons carry a negative charge, they are held more strongly in a cation than in the atom from which the cation is formed.

The energy needed to remove a third electron is greater still because it is being removed from a $2+$ ion. Superimposed on this trend is a much more dramatic increase in ionization energy (defined by the red line in Table 7.2) when all the valence electrons in an ionizing atom have been removed and the next electron must

TABLE 7.2		Successive Ionization Energies[a] of the First 10 Elements									
Element	Z	IE_1	IE_2	IE_3	IE_4	IE_5	IE_6	IE_7	IE_8	IE_9	IE_{10}
H	1	1312									
He	2	2372	5249								
Li	3	520	7296	12040							
Be	4	897	1758	15050	21070						
B	5	801	2426	3660	24682	32508					
C	6	1087	2348	4617	6201	37926	46956				
N	7	1402	2860	4581	7465	9391	52976	64414			
O	8	1314	3383	5298	7465	10956	13304	71036	84280		
F	9	1681	3371	6020	8428	11017	15170	17879	92106	106554	
Ne	10	2081	3949	6140	9391	12160	15231	19986	23057	115584	131236

[a]All values in kilojoules per mole of atoms.

come from an inner shell. A core electron experiences much less shielding and much more effective nuclear charge, and therefore requires much more energy to be removed from an atom than does an outer-shell electron.

SAMPLE EXERCISE 7.14   Ranking Ionization Energies

Arrange argon, magnesium, and phosphorus in order of increasing first ionization energy, from lowest to highest value.

Collect and Organize We are to order three elements on the basis of their IE_1 values. All three elements are in the third row of the periodic table.

Analyze First ionization energies generally increase from left to right across a row.

Solve Assuming that increasing ionization energy with increasing atomic number is the dominant factor, the elements in order of increasing first ionization energies should be:

$$Mg < P < Ar$$

Think about It Magnesium forms stable 2+ cations, so we expect its ionization energy to be smaller than the values for phosphorus and argon, which do not form cations. Argon is a noble gas with a stable valence-shell electron configuration, so its first ionization energy would logically be the highest in the set. We can check our prediction against Figure 7.34(b): Mg 738 kJ/mol, P 1012 kJ/mol, Ar 1521 kJ/mol.

Practice Exercise Arrange cesium, calcium, and neon in order of decreasing first ionization energy, from largest value to smallest value.

CONCEPT TEST

Why does ionization energy decrease as one moves down the noble gas family from helium to neon to argon?

electron affinity (EA) the energy change that occurs when 1 mole of electrons combines with 1 mole of atoms or ions in the gas phase.

7.12 Electron Affinities

In the preceding section we examined the periodic nature of the energy required to ionize atoms. Now we look at a complementary process and examine the change in energy when electrons are added to atoms to form monatomic anions. The energies involved are called **electron affinities (EA)**. They are the energy changes that occur when 1 mole of electrons is added to 1 mole of atoms or ions in the gas phase.

For example, the energy associated with adding 1 mole of electrons to 1 mole of chlorine atoms in the gas phase is

$$Cl(g) + e^- \rightarrow Cl^-(g) \qquad EA_1 = -349 \text{ kJ/mol}$$

The electron affinities of many elements are negative, meaning the formation of 1− anions from free atoms is an exothermic process (see Figure 7.35). This fact seems reasonable because the association of a negative electron with a positively charged nucleus should be energetically favorable.

An examination of the EA values in Figure 7.35 reveals that the trends in this property are not as regular as the trends in size and ionization energy. Electron affinity increases down a column only for the group 1 metals; other groups do not display a clear trend. In general, electron affinity becomes more negative with increasing atomic number across a row, but there are exceptions to that trend, too. The halogens of group 17 have the most negative EA values, which seems logical because each halogen atom needs one more electron to achieve a noble gas electron configuration.

On the other hand, adding an electron to an atom of a noble gas is an endothermic process, which makes sense because their atoms have stable electron configurations already. Beryllium and magnesium have positive electron affinities because the added electrons have to occupy outer-shell p orbitals that are significantly higher in energy than the outer-shell s orbitals. We can rationalize the positive electron affinity of nitrogen by noting that adding an electron to an N atom means that the atom loses the stability associated with a half-filled set of $2p$ orbitals:

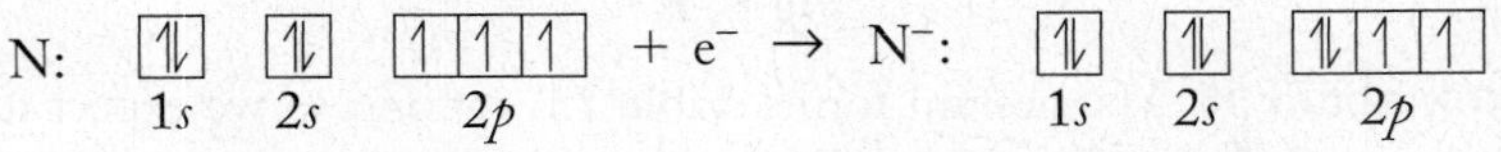

1A (1)							8A (18)
H −72.6	2A (2)	3A (13)	4A (14)	5A (15)	6A (16)	7A (17)	He (0.0)*
Li −59.6	Be >0	B −26.7	C −122	N +7	O −141	F −328	Ne (+29)*
Na −52.9	Mg >0	Al −42.5	Si −134	P −72.0	S −200	Cl −349	Ar (+35)*
K −48.4	Ca −2.4	Ga −28.9	Ge −119	As −78.2	Se −195	Br −325	Kr (+39)*
Rb −46.9	Sr −5.0	In −28.9	Sn −107	Sb −103	Te −190	I −295	Xe (+41)*
Cs −45.5	Ba −14	Tl −19.2	Pb −35.2	Bi −91.3	Po −183.3	At −270*	Rn (+41)*

*Calculated values.

FIGURE 7.35 Electron affinity (EA) values of main group elements are expressed in kilojoules per mole. The more negative the value, the more energy is released when 1 mole of atoms combines with 1 mole of electrons to form 1 mole of anions with a 1− charge. A greater release of energy reflects more attraction between the atoms of the elements and free electrons.

N: [↑↓] [↑↓] [↑ ↑ ↑] + e⁻ → N⁻: [↑↓] [↑↓] [↑↓ ↑ ↑]
   1s   2s   2p   1s   2s   2p

CONCEPT TEST

Describe at least one similarity and one difference in the periodic trends in first ionization energies and electron affinities among the main group elements.

Momentous advances in chemistry and physics were made in the first three decades of the 20th century due to the brilliant discoveries of Einstein, Planck, Rutherford, Bohr, de Broglie, Schrödinger, and others, which have forever changed scientists' view of the fundamental structure of matter and the universe. Figure 7.36 summarizes and connects some of these advances.

Consensus in the scientific community on the ideas of quantum theory did not come easily. We have seen how hot sodium atoms in the excited state emit photons of yellow-orange light as they fall to the ground state. Einstein puzzled over

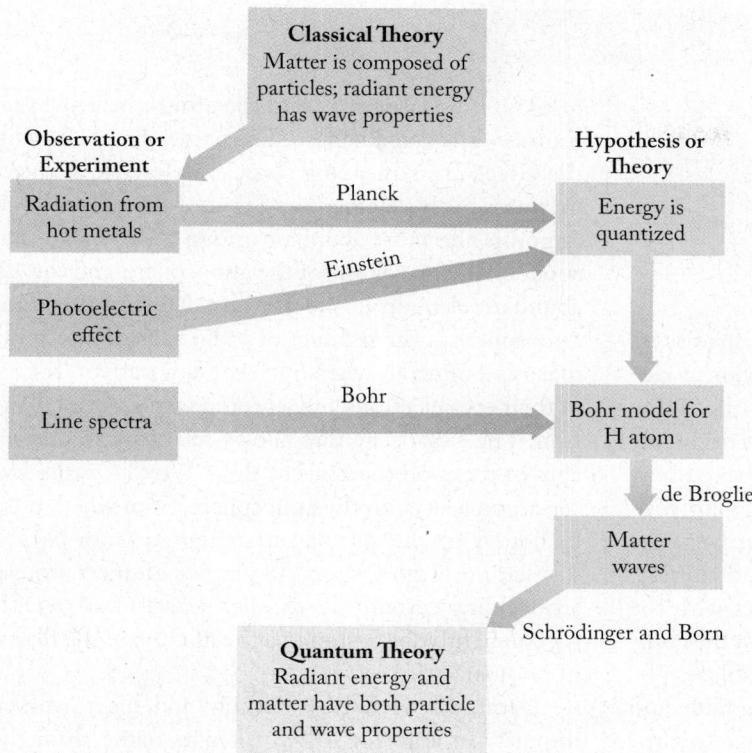

FIGURE 7.36 During the first three decades of the 20th century, quantum theory evolved from classical 19th-century theories of the nature of matter and energy. The arrows trace the development of modern quantum theory, which assumes that radiant energy has both wave properties and particle properties and that mass (matter) also has both properties and particle properties.

this phenomenon for several years before deciding that the moment when emission occurs and the direction of the photon emitted could not be predicted exactly. He concluded that quantum theory allows us to calculate only the probability of a spontaneous electron transition; the details of the event are left to chance. In other words, no force of nature causes a hot sodium atom to fall to a lower energy level at a particular instant.

How different is the view of the interaction of matter and energy provided by quantum mechanics from the laws governing the behavior of large objects? A pebble picked up and dropped immediately falls to the ground. However, electrons remain in excited states for indeterminate (though usually short) times before falling to their ground states. This lack of determinacy bothered Einstein and many of his colleagues. Had they discovered an underlying theme of nature—that some processes cannot be described or known with certainty? Are there fundamental limits to how well we can know and understand our world and the events that change it?

Many scientists in the early decades of the 20th century did not care to make such admissions. They preferred the Newtonian view of the world as a place where events occur for a reason and where there are causes and effects. They believed that the more they studied nature with ever more sophisticated tools, the more they would understand why things happen as they do. Soon after Max Born published a probabilistic interpretation of Schrödinger's wave functions in 1926, Einstein wrote Born a letter in which he contrasted the new theories with the certainties many people find in religious beliefs:

> Quantum mechanics is very impressive. But an inner voice tells me that it is not yet the real thing. The theory produces a great deal but hardly brings us closer to the secret of the Old One. I am at all events convinced that He does not play dice.[2]

---

[2] Letter to Max Born, 12 December 1926; quoted in R. W. Clark, *Einstein: The Life and Times* (New York: HarperCollins, 1984), p. 880.

A Noble Family: Special Status for Special Behavior

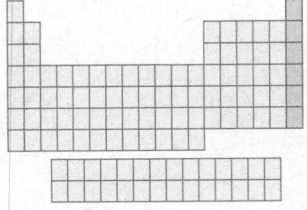

The descriptive chemistry boxes in previous chapters contain information about one or two elements. Now that we have developed electron configurations and highlighted their role in determining the periodic behavior of elements, we can discuss those features of the chemical and physical behavior of elements that vary periodically. The six elements in the rightmost column of the periodic table (group 18) were originally called inert gases, but in the 1960s chemists discovered how to make compounds of several members of the group with very reactive substances like fluorine. After this discovery, the group name was changed to *noble gases*. Just as in human relations the families of "nobility" tend not to interact with the "common people," the noble gases do not react with the common elements of other families on the periodic table.

Helium is the second element in the periodic table and the second most abundant element in the universe (hydrogen is the most abundant). Its presence in the absorption spectra of gases around the sun was detected in 1868 as a set of dark lines that did not correspond with those of any of the known elements on Earth. These dark lines were assigned to helium, whose name derives from *helios*, the Greek word for sun. The element was isolated on Earth by William Ramsay (1852–1916) in 1895. After the hydrogen-filled dirigible Hindenberg exploded and burned in 1937, helium, which is chemically inert and therefore not a fire hazard, became the gas of choice for such craft (Figure 7.37).

Argon was isolated from the atmosphere the year before Ramsay discovered helium. The name of this element is from the Greek *argos*, meaning "lazy," in reference to the chemical inertness of this element and that of the others in group 18. Argon is the most abundant group 18 element, making up about 0.94% by volume of the atmosphere, and the 12th most abundant element in the universe. Most of the argon in the atmosphere is the product of radioactive decay, produced in rocks and minerals when nuclei of radioactive ^{40}K capture one of their own electrons and a proton is transformed into a neutron. The low decay rate allows scientists to determine the ages of rocks on the basis of their $^{40}Ar/^{40}K$ ratio. Eventually the argon leaks into the atmosphere, from which it is isolated by liquefying the air and separating Ar from N_2 and O_2 by distillation. Argon is used to provide an inert atmosphere in arc-welding certain metals, such as stainless steel. It is also used in lightbulbs to improve the lifetime of the filaments and in photoelectric devices.

Neon, discovered by Ramsay and his coworkers as an impurity in argon in 1898, takes its name from the Greek *neos*, "new." Although neon is found in some minerals as a result of radioactive decay, its principal industrial source is the atmosphere. However, its tiny concentration in the atmosphere (0.0018% by volume) means that nearly 100 kg of air must be liquefied to obtain 1 g of neon. When an electric current is passed through a tube containing neon at low pressure, the gas emits the characteristic red-orange light we associate with neon signs. Electrical discharge tubes filled with the other noble gases emit their own unique combinations of emission lines and characteristic colors (Figure 7.38).

(a) (b)

FIGURE 7.37 Both hydrogen gas and helium gas have been used to fill lighter-than-air craft, but (a) the hydrogen-filled Hindenburg ended its last trans-Atlantic flight in flames in Lakehurst, New Jersey, in 1937. (b) Modern blimps get their buoyancy from helium, which is chemically inert and hence not a fire or explosion hazard.

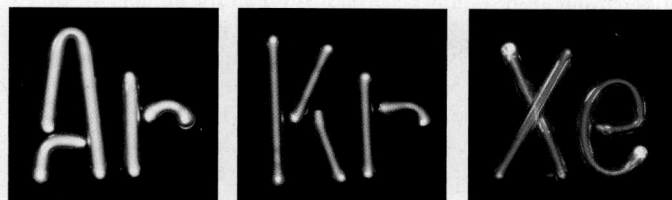

FIGURE 7.38 Gas discharge tubes filled with different noble gases emit characteristic colors.

Neon was the gas analyzed by Francis Aston in his positive-ray analyzer that led to the discovery of stable isotopes and to the development of mass spectrometry.

Krypton (from the Greek *kryptos*, "hidden") and xenon (from the Greek *xenos*, "foreign") are even rarer components of the atmosphere. The concentration of krypton is only 1.1 ppm by volume, which means, on average, 1.1 atoms of krypton for every *million* atoms and molecules of the other components of air. The atmospheric concentration of xenon is even smaller: only 86 ppb by volume. Because the boiling points of krypton and xenon are higher than the boiling points of the principal components of the atmosphere, these two noble gases can be separated by fractional distillation of liquefied air, being recovered in the least-volatile fraction.

All the noble gases have a filled outermost occupied shell. The lack of reactivity of these elements is a direct result of this stable electron configuration. Helium has a filled 1*s* orbital, which corresponds to a full first shell, and only two valence electrons; the other noble gases have filled *s* and *p* orbitals and eight valence electrons.

Each noble gas has the highest ionization energy of all the elements in its period. For example, the energy required to remove an electron from an atom of neon (period 2) is about 20% more than that required to remove an electron from an atom of the element immediately preceding it (fluorine) and 400% more than that required to remove an electron from an atom of the first element in period 2 (lithium). The ionization energies of the noble gases decrease as we move down the group (Figure 7.39) because the increasing numbers of core electrons shield the outermost electron from the positive charge on the nucleus. Atom size increases from top to bottom in the group (Figure 7.40) because the valence electrons, which are a main determinant of atom size, are farther and farther from the nucleus. Because covalent radii are determined by measuring internuclear distances in diatomic molecules, the radii of helium, neon, and argon, for which no stable compounds are known, are estimations.

The noble gases do not easily share electrons to form compounds with other atoms. However, xenon and krypton do react with fluorine to form, for example, XeF_2, XeF_4, XeF_6, and KrF_2. The three xenon compounds react with water, yielding compounds containing xenon, fluorine, and oxygen.

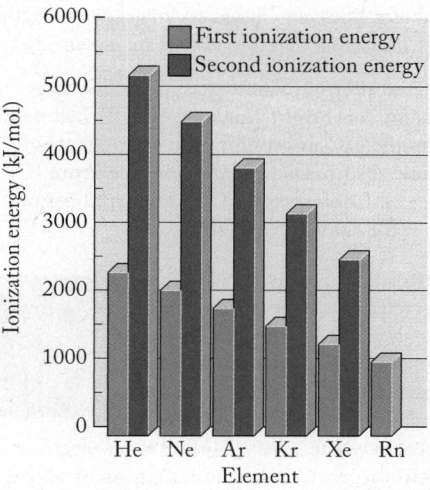

FIGURE 7.39 The first ionization energies of the group 18 elements decrease down the group. The second ionization energies, which exhibit this same trend, are larger than the first ionization energies because of the greater amount of energy required to remove an electron from the cation formed in the first ionization.

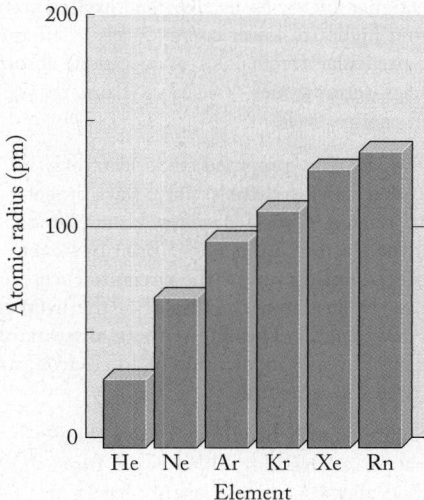

FIGURE 7.40 The atomic radii of the group 18 elements increase down the group because the valence electrons occupy orbitals that are, on average, farther and farther from the nucleus.

SUMMARY ■

Section 7.1 Light, one form of **electromagnetic radiation**, has wave properties described by characteristic visible **wavelengths** and **frequencies**. A beam of sunlight passing  through a prism is dispersed into its colors because of **refraction**. Waves also experience **diffraction** and **interference**. The sun's spectrum contains narrow gaps, called **Fraunhofer lines**, that are also present in the light from distant galaxies. In galactic spectra the Fraunhofer lines are shifted to longer wavelengths due to the Doppler effect, indicating that the galaxies are moving away from Earth.

Section 7.2 Free atoms in flames and in gas discharge tubes produce **atomic emission** (or **bright-line**) **spectra**. When radiation passes through an atomic gas, absorption produces a spectrum of dark lines, called an **atomic absorption** (or **dark-line**) **spectrum**. The bright lines of an element's emission spectrum and the dark lines of its absorption spectrum are at the same wavelengths.

Section 7.3 According to **quantum theory** there are discrete energy levels in atoms, which means they absorb or emit discrete amounts of energy called **quanta**. Einstein called these particles radiant energy **photons** in his explanation of the **photoelectric effect**: a process in which a metal surface emits electrons when illuminated by electromagnetic radiation that is at or above a **threshold frequency**. The radiant energy required to dislodge an electron from a metal surface is called the **work function** of the metal.

Section 7.4 Balmer derived an equation that accounted for the bright lines in the visible emission spectrum of hydrogen and that predicted the existence of bright lines in the UV and IR regions of hydrogen's emission spectrum. Bohr proposed that the lines predicted by Balmer are related to electron energy levels inside the hydrogen atom. A **ground state** atom or ion has all its electrons in the lowest possible energy levels. Higher energy levels are called **excited states**. **Electron transitions** from higher to lower energy levels in an atom cause the atom to emit particular frequencies of radiation; absorption of the same frequencies accompanies transitions from the same lower to higher electron energy levels.

Section 7.5 De Broglie proposed that electrons in atoms, and all other moving particles, have wave properties and can be treated as **matter waves**. He explained  the stability of the electron orbits in the Bohr hydrogen atom in terms of **standing waves**: the circumferences of the allowed orbits had to be whole-number multiples of the hydrogen electron's characteristic wavelength. The **Heisenberg uncertainty principle** states that the position and momentum of an electron cannot both be precisely known at the same time.

Section 7.6 Solutions to **Schrödinger's wave equation**, mathematical expressions called **wave functions** (ψ), define allowed electron energy levels in atoms. Though wave functions have no physical meaning, ψ^2 defines **orbitals**, the three-dimensional 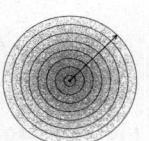 regions inside an atom that describe the probability of finding an electron at a given distance from the nucleus. Each orbital has a unique set

of three **quantum numbers**: **principal quantum number** n, which defines orbital size and energy level; **angular momentum quantum number** ℓ, which defines orbital shape; and **magnetic quantum number** m_ℓ, which defines orbital orientation in space. A fourth quantum number, **spin magnetic quantum number** m_s, is necessary to explain certain characteristics of emission spectra. The two electrons that occupy one orbital have opposite spins ($m_s = +\frac{1}{2}$ and $-\frac{1}{2}$). The **Pauli exclusion principle** states that no two electrons in an atom can have the same four values of n, ℓ, m_ℓ, and m_s.

Section 7.7 Orbitals have characteristic three-dimensional sizes, shapes, and orientations that are depicted by boundary–surface representations. All s orbitals are spheres that increase in size with increasing values of n. The three p orbitals in any $n \geq 2$ shell have two teardrop-shaped lobes on one of the x-, y-, or z-axes. The five d orbitals 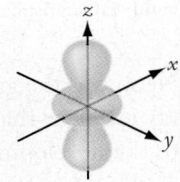 found in $n \geq 3$ shells come in two forms: four are shaped like a four-leaf clover, and the fifth has two lobes oriented along the z-axis and a torus surrounding the middle of the two lobes.

Section 7.8 According to the **aufbau principle**, electrons fill the lowest-energy atomic orbitals of a ground-state atom first. An **electron configuration** is a set of numbers and letters expressing the number of electrons in each occupied orbital in an atom. The electrons in the outermost occupied shell of an atom are **valence electrons**. They are the electrons lost, gained, or shared in chemical reactions. All the orbitals of a given p, d, or f subshell are **degenerate**; they all have the same energy. **Hund's rule** states that in any set of degenerate orbitals, each orbital must contain one electron before any orbital in the set can accept a second electron.

Section 7.9 Atoms of group 1 and group 2 elements tend to lose electrons and form 1+ and 2+ cations, respectively. In so doing they become **isoelectronic** with the noble gas in the preceding period. Atoms in groups 16 and 17 tend to gain electrons to form 2– and 1– anions, respectively, thereby becoming isoelectronic with the noble gas at the end of their period. When transition metals form ions, the electrons are removed from the shell of highest n until the charge on the ion is achieved.

Section 7.10 The **effective nuclear charge (Z_{eff})** is the net nuclear charge felt by an electron when electrons closer to the nucleus **shield** (**screen**) the electron from experiencing the full nuclear charge. Trends in Z_{eff} cause atom size to decrease across a period and increase down a group. **Orbital penetration** by outer electrons also plays a role in atom size: greater pene- 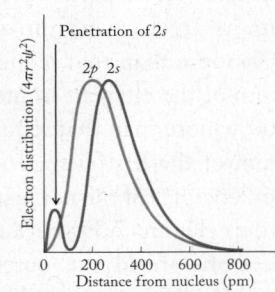 tration means the electrons experience a greater effective nuclear charge, leading to smaller atom size. Anions are larger than their parent atoms due to additional electron–electron repulsion, but cations are smaller than their parent atoms—sometimes much smaller when all the electrons in the valence shell are lost.

Section 7.11 **Ionization energies** generally increase with increasing effective nuclear charge across a period. Exceptions can be explained by the fact that, in a given shell, electrons in p orbitals have more energy than electrons in the s orbital and by the stability of half-filled sets of p and d orbitals. The energy differences between atoms in different shells and subshells are also apparent in the values of successive ionization energies (IE_1, IE_2, IE_3, . . .) for a given element.

Section 7.12 Electron affinity is the energy change that occurs when 1 mole of electrons combines with 1 mole of atoms or ions in the gas phase. This process is most exothermic for the halogens, moderately exothermic for many main group elements, but endothermic for all the noble gases.

PROBLEM-SOLVING SUMMARY

TYPE OF PROBLEM	CONCEPTS AND EQUATIONS		SAMPLE EXERCISES
Calculating frequency from wavelength	$\lambda\nu = c$ where $c \approx 3.00 \times 10^8$ m/s	(7.1)	7.1
Calculating the energy of a quantum of light	$E = \dfrac{hc}{\lambda}$	(7.3)	7.2
Using the work function	$\Phi = h\nu_0$	(7.4)	7.3
Calculating the wavelength of a line in the hydrogen spectrum	$\dfrac{1}{\lambda} = \left[1.097 \times 10^{-2}/(nm)^{-1}\right]\left(\dfrac{1}{n_1{}^2} - \dfrac{1}{n_2{}^2}\right)$	(7.7)	7.4
Calculating the energy of an electron transition in a hydrogen atom	$\Delta E = -2.178 \times 10^{-18}\,J\left(\dfrac{1}{n_{final}{}^2} - \dfrac{1}{n_{initial}{}^2}\right)$	(7.9)	7.5
Calculating the wavelength of particles in motion	$\lambda = \dfrac{h}{mv}$	(7.11)	7.6
Calculating uncertainty	$\Delta x \cdot m\Delta v \geq \dfrac{h}{4\pi}$	(7.13)	7.7
Identifying the subshells and orbitals in an energy level and valid quantum number sets	n is the shell number, ℓ defines both the subshell and the type of orbital; an orbital has a unique combination of allowed n, ℓ, and m_ℓ values. ℓ is any integer from 0 to $n - 1$; m_ℓ is any integer from $-\ell$ to $+\ell$, including zero.		7.8, 7.9
Writing electron configurations of atoms and ions; determining isoelectronic species	Orbitals fill up in the following sequence: $1s^2, 2s^2, 2p^6, 3s^2, 3p^6, 4s^2, 3d^{10}, 4p^6, 5s^2, 4d^{10}, 5p^6, 6s^2, 4f^{14}, 5d^{10}, 6p^6$ (superscripts represent maximum numbers of electrons). Arrange orbitals in electron configuration based on increasing value of "n." In forming transition metal ions, electrons are removed to maximize the number of d electrons; there is enhanced stability in half-filled and filled d subshells.		7.10–7.12
Ordering atoms and ions by size	In general, sizes decrease left to right across a row and increase down a column of the periodic table: cations are smaller and anions are larger than their parent atoms.		7.13
Ranking ionization energies	First ionization energies increase across a row and decrease down a column.		7.14

VISUAL PROBLEMS

(Answers to boldface end-of-chapter questions and problems are in the back of the book.)

7.1. Which of the elements highlighted in Figure P7.1 consists of ground state atoms with
  a. a single *s* electron in their outermost shells? (More than one answer is possible.)
  b. filled sets of *s* and *p* orbitals in their outermost shells?
  c. filled sets of *d* orbitals?
  d. half-filled sets of *d* orbitals?
  e. two *s* electrons in their outermost shells?

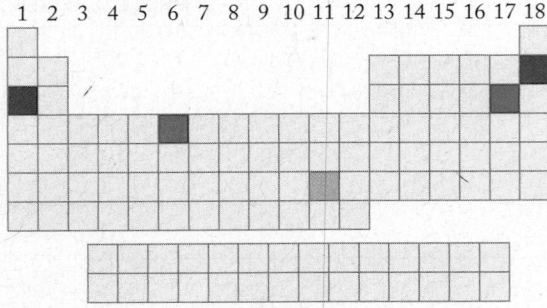

FIGURE P7.1

7.2. Which of the highlighted elements in Figure P7.1 has the greatest number of unpaired electrons per ground state atom?

7.3. Which of the highlighted elements in Figure P7.1 form(s) common monatomic ions that are larger than their parent atoms?

7.4. Which of the elements highlighted in Figure P7.4 forms monatomic ions by
  a. losing an *s* electron?
  b. losing two *s* electrons?
  c. losing two *s* electrons and a *d* electron?
  d. adding an electron to a *p* orbital?
  e. adding electrons to two *p* orbitals?

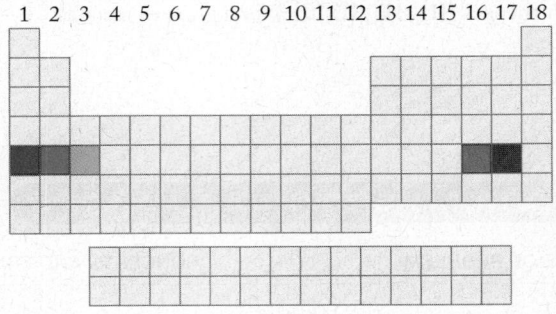

FIGURE P7.4

7.5. Which of the highlighted elements in Figure P7.4 form(s) common monatomic ions that are smaller than their parent atoms?

7.6. Rank the elements highlighted in Figure P7.4 based on increasing size of their atoms.

7.7. Rank the highlighted elements in Figure P7.4 based on increasing size of their most common monatomic ions.

7.8. Which of the elements highlighted in Figure P7.4 has the largest first ionization energy, IE_1?

7.9. Which of the elements highlighted in Figure P7.4 has the largest second ionization energy, IE_2?

7.10. Consider the results of an experiment shown in Figure P7.10. The screen containing two narrow slits is illuminated by a distant source of light off to the left. Instead of just the images of the two slits appearing on the screen on the right, a whole series of slit images (more than the five shown here) appear. Explain how there could be many more than two slit images on the screen on the right.

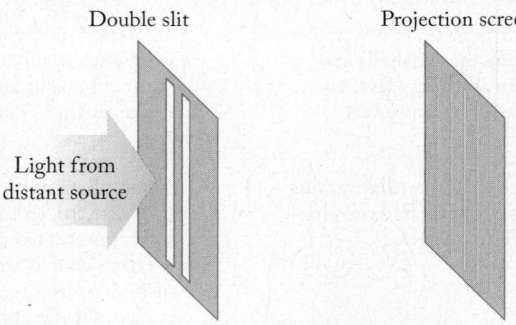

FIGURE P7.10

QUESTIONS AND PROBLEMS

Waves of Light

CONCEPT REVIEW

7.11. Why are the various forms of radiant energy called *electromagnetic* radiation?

7.12. Explain with a sketch why the frequencies of long-wavelength waves of electromagnetic radiation are lower than those of short-wavelength waves.

7.13. **Dental X-rays** When X-ray images are taken of your teeth and gums in the dentist's office, your body is covered with a lead shield. Explain the need for this precaution.

7.14. **UV Radiation and Skin Cancer** Ultraviolet radiation causes skin damage that may lead to cancer, but exposure to infrared radiation does not seem to cause skin cancer. Why do you think this is so?

7.15. Gamma rays are an example of "ionizing" radiation because they have the energy to break apart molecules into molecular ions and free electrons. What other forms of electromagnetic radiation could also be ionizing radiation?

7.16. Explain how redshifts in the spectra of starlight from distant galaxies led to the conclusion that our universe is expanding.

*7.17. The Doppler effect is described by the equation

$$\frac{(\nu - \nu^1)}{\nu} = \frac{u}{c}$$

where ν is the unshifted frequency, ν^1 is the perceived frequency, c is the speed of light, and u is the speed at which the object is moving. If hydrogen in a galaxy that is receding from Earth at half the speed of light emits radiation with a wavelength of 656 nm, will the radiation still be in the visible part of the electromagnetic spectrum when it reaches Earth?

*7.18. If light consists of waves, why don't things look "wavy" to us?

PROBLEMS

7.19. A neon light emits radiation of $\lambda = 616$ nm. What is the frequency of this radiation?

7.20. **Submarine Communications** In the 1990s the Russian and American navies developed extremely low frequency communications networks to send messages to submerged submarines. The frequency of the carrier wave of the Russian network was 82 Hz, while the Americans used 76 Hz.
   a. What was the ratio of the wavelengths of the Russian network to the American network?
   b. To calculate the actual underwater wavelength of the transmissions in either network, what additional information would you need?

7.21. **Broadcast Frequencies** FM radio stations broadcast at different frequencies. Calculate the wavelengths corresponding to the broadcast frequencies of the following radio stations: (a) KKNB (Lincoln, NE), 104.1 MHz; (b) WFNX (Boston, MA), 101.7 MHz; (c) KRTX (Houston, TX), 100.7 MHz.

7.22. Which radiation has the longer wavelength? (a) radio waves from an AM radio station broadcasting at 680 kHz or (b) infrared radiation emitted by the surface of Earth ($\lambda = 15$ mm)

7.23. Which radiation has the lower frequency? (a) radio waves from an AM radio station broadcasting at 1090 kHz or (b) the green light ($\lambda = 550$ nm) from an LED (light-emitting diode) on a stereo system

7.24. Which radiation has the higher frequency? (a) the red light on a bar-code reader at a grocery store or (b) the green light on the battery charger for a laptop computer

7.25. **Speed of Light** How long does it take light to reach Earth from the sun? (The distance from the sun to Earth is 93 million miles.)

7.26. **Exploration of the Solar System** How long would it take an instruction to a Martian rover to travel from a NASA site on Earth to Mars? Assume the signal is sent when Earth and Mars are 75 million kilometers apart.

Atomic Spectra

CONCEPT REVIEW

7.27. Describe the similarities and differences in the atomic emission and absorption spectra of hydrogen.

7.28. Are the Fraunhofer lines in the spectra of stars the result of atomic emission or atomic absorption?

7.29. How did the study of the atomic emission spectra of elements lead to the identification of the Fraunhofer lines in sunlight?

*7.30. What would happen in terms of the appearance of the Fraunhofer lines in the solar spectrum if sunlight were passed through a flame containing high-temperature calcium atoms and then analyzed?

Particles of Light and Quantum Theory

CONCEPT REVIEW

7.31. What is a quantum?

7.32. What is a photon?

7.33. A variable power supply is connected to an incandescent lightbulb. At the lowest power setting, the bulb feels warm to the touch but produces no light. At medium power, the lightbulb filament emits a red glow. At the highest power, the lightbulb emits white light. Explain this emission pattern.

*7.34. Has a photon of radiation that is red-shifted 10% actually lost 10% of its energy, where $E = h\nu$?

PROBLEMS

7.35. Which of the following have quantized values? Explain your selections.
   a. The elevation of the treads of a moving escalator
   b. The elevations at which the doors of an elevator open
   c. The speed of an automobile

7.36. Which of the following have quantized values? Explain your selections.
   a. The pitch of a note played on a slide trombone
   b. The pitch of a note played on a flute
   c. The wavelengths of light produced by the heating elements in a toaster
   d. The wind speed at the top of Mt. Everest

7.37. When a piece of metal is irradiated with UV radiation ($\lambda = 162$ nm), electrons are ejected with a kinetic energy of 5.34×10^{-19} J. What is the work function of the metal?

7.38. The first ionization energy of a gas-phase atom of a particular element is 6.24×10^{-19} J. What is the maximum wavelength of electromagnetic radiation that could ionize this atom?

*7.39. Thin layers of potassium ($\Phi = 3.68 \times 10^{-19}$ J) and sodium ($\Phi = 4.41 \times 10^{-19}$ J) are exposed to radiation of wavelength 300 nm. Which metal emits electrons with the greater velocity? What is the velocity of these electrons?

7.40. **Solar Power** Photovoltaic cells convert solar energy into electricity. Could tantalum ($\Phi = 6.81 \times 10^{-19}$ J) be used to convert visible light to electricity? Assume that most of the electromagnetic energy from the sun is in the visible region near 500 nm.

7.41. With reference to Problem 7.40, could tungsten ($\Phi = 7.20 \times 10^{-19}$ J) be used to construct solar cells?

7.42. Titanium ($\Phi = 6.94 \times 10^{-19}$ J) and silicon ($\Phi = 7.24 \times 10^{-19}$ J) surfaces are irradiated with UV radiation with a wavelength of 250 nm. Which surface emits electrons with the longer wavelength? What is the wavelength of the electrons emitted by the titanium surface?

7.43. The power of a red laser ($\lambda = 630$ nm) is 1.00 watt (abbreviated W, where 1 W = 1 J/s). How many photons per second does the laser emit?

*7.44. The energy density of starlight in interstellar space is 10^{-15} J/m^3. If the average wavelength of starlight is 500 nm, what is the corresponding density of photons per cubic meter of space?

The Hydrogen Spectrum and the Bohr Model

CONCEPT REVIEW

7.45. Why should hydrogen have the simplest atomic spectrum of all the elements?

7.46. For an electron in a hydrogen atom, how is the value of n of its orbit related to its energy?

7.47. Does the energy of electromagnetic energy emitted by an excited-state H atom depend on the individual values of n_1 and n_2, or only on the difference between them ($n_1 - n_2$)?

7.48. Explain the difference between a ground-state H atom and an excited-state H atom.

7.49. Without calculating any wavelength values, predict which of the following electron transitions in the hydrogen atom is associated with radiation having the shortest wavelength.
   a. From $n = 1$ to $n = 2$ c. From $n = 3$ to $n = 4$
   b. From $n = 2$ to $n = 3$ d. From $n = 4$ to $n = 5$

7.50. Without calculating any frequency values, rank the following transitions in the hydrogen atom in order of increasing frequency of the electromagnetic radiation that could produce them.
   a. From $n = 4$ to $n = 6$ c. From $n = 9$ to $n = 11$
   b. From $n = 6$ to $n = 8$ d. From $n = 11$ to $n = 13$

7.51. Electron transitions from $n = 2$ to $n = 3, 4, 5,$ or 6 in hydrogen atoms are responsible for some of the Fraunhofer lines in the sun's spectrum. Are there any Fraunhofer lines due to transitions that start from $n = 3$ in hydrogen atoms?

7.52. In the visible portion of the atomic emission spectrum of hydrogen, are there any bright lines due to electron transitions to the $n = 1$ state?

7.53. Balmer observed a hydrogen emission line for the transition from $n = 6$ to $n = 2$, but not for the transition from $n = 7$ to $n = 2$. Why?

*7.54. In what ways should the emission spectra of H and He$^+$ be alike, and in what ways should they be different?

PROBLEMS

7.55. What is the wavelength of the photons emitted by hydrogen atoms when they undergo transitions from $n = 4$ to $n = 3$? In which region of the electromagnetic spectrum does this radiation occur?

7.56. What is the frequency of the photons emitted by hydrogen atoms when they undergo transitions from $n = 5$ to $n = 3$? In which region of the electromagnetic spectrum does this radiation occur?

*7.57. The energies of the photons emitted by one-electron atoms and ions fit the equation

$$E = (2.178 \times 10^{-18} \text{ J})Z^2\left(\frac{1}{n_1^2} - \frac{1}{n_2^2}\right)$$

where Z is the atomic number, n_2 and n_1 are positive integers, and $n_2 > n_1$.
   a. As the value of Z increases, does the wavelength of the photon associated with the transition from $n = 2$ to $n = 1$ increase or decrease?

   b. Can the wavelength associated with the transition from $n = 2$ to $n = 1$ ever be observed in the visible region of the spectrum?

*7.58. Can transitions from higher energy states to the $n = 2$ state in He$^+$ ever produce visible light? If so, for what values of n_2? (*Hint*: The equation in Problem 7.57 may be useful.)

7.59. The transition from $n = 3$ to $n = 2$ in a hydrogen atom produces a photon with a wavelength of 656 nm. What is the wavelength of the transition from $n = 3$ to $n = 2$ in a Li^{2+} ion? (Use the equation in Problem 7.57.)

*7.60. The hydrogen atomic emission spectrum includes a UV line with a wavelength of 92.3 nm.
   a. Is this line associated with a transition between different excited states, or between an excited state and the ground state?
   b. What is the value of n_1 of this transition?
   c. What is the energy of the longest-wavelength photon that a ground-state hydrogen atom can absorb?

Electrons as Waves

CONCEPT REVIEW

7.61. Identify the symbols in the de Broglie relation $\lambda = h/mv$, and explain how the relation links the properties of a particle to those of a wave.

7.62. Explain how the observation of electron diffraction supports the description of electrons as waves.

7.63. Would the density or shape of an object have an effect on its de Broglie wavelength?

7.64. How does de Broglie's hypothesis that electrons behave like waves explain the stability of the electron orbits in a hydrogen atom?

PROBLEMS

7.65. Calculate the wavelengths of the following objects:
   a. A muon (a subatomic particle with a mass of 1.884×10^{-25} g) traveling at 325 m/s
   b. Electrons ($m_e = 9.10938 \times 10^{-28}$ g) moving at 4.05×10^6 m/s in an electron microscope
   c. An 80 kg athlete running a 4-minute mile
   d. Earth (mass = 6.0×10^{27} g) moving through space at 3.0×10^4 m/s

7.66. Two objects are moving at the same velocity. Which (if any) of the following statements about them are true?
   a. The de Broglie wavelength of the heavier object is longer than that of the lighter one.
   b. If one object has twice as much mass as the other, its wavelength is one-half the wavelength of the other.
   c. Doubling the velocity of one of the objects will have the same effect on its wavelength as doubling its mass.

7.67. Which (if any) of the following statements about the frequency of a particle is true?
   a. Heavy, fast-moving objects have lower frequencies than those of lighter, faster-moving objects.
   b. Only very light particles can have high frequencies.
   c. Doubling the mass of an object and halving its velocity result in no change in its frequency.

7.68. How rapidly would each of the following particles be moving if they all had the same wavelength as a photon of red light ($\lambda = 750$ nm)?

a. An electron of mass 9.10938×10^{-28} g
b. A proton of mass 1.67262×10^{-24} g
c. A neutron of mass 1.67493×10^{-24} g
d. An α particle of mass 6.64×10^{-24} g

7.69. Particles in a Cyclotron The first cyclotron was built in 1930 at the University of California, Berkeley, and was used to accelerate molecular ions of hydrogen, H_2^+, to a velocity of 4×10^6 m/s. (Modern cyclotrons can accelerate particles to nearly the speed of light.) If the uncertainty in the velocity of the H_2^+ ion was 3%, what was the uncertainty of its position?

7.70. Radiation Therapy An effective treatment for some cancerous tumors involves irradiation with "fast" neutrons. The neutrons from one treatment source have an average velocity of 3.1×10^7 m/s. If the velocities of individual neutrons are known to within 2% of this value, what is the uncertainty in the position of one of them?

Quantum Numbers and Electron Spin; The Sizes and Shapes of Atomic Orbitals

CONCEPT REVIEW

7.71. How does the concept of an orbit in the Bohr model of the hydrogen atom differ from the concept of an orbital in quantum theory?

7.72. What properties of an orbital are defined by each of the three quantum numbers n, ℓ, and m_ℓ?

7.73. How many quantum numbers are needed to identify an orbital?

7.74. How many quantum numbers are needed to identify an electron in an atom?

PROBLEMS

7.75. How many orbitals are there in an atom with each of the following principal quantum numbers? (a) 1; (b) 2; (c) 3; (d) 4; (e) 5

7.76. How many orbitals are there in an atom with the following combinations of quantum numbers?
a. $n = 3, \ell = 2$ c. $n = 4, \ell = 2, m_\ell = 2$
b. $n = 3, \ell = 1$

7.77. What are the possible values of quantum number ℓ when $n = 4$?

7.78. Which are the possible values of m_ℓ when $\ell = 2$?

7.79. What set of orbitals corresponds to each of the following sets of quantum numbers?
a. $n = 2, \ell = 0$ c. $n = 4, \ell = 2$
b. $n = 3, \ell = 1$ d. $n = 1, \ell = 0$

7.80. What set of orbitals corresponds to each of the following sets of quantum numbers?
a. $n = 2, \ell = 1$ c. $n = 3, \ell = 2$
b. $n = 5, \ell = 3$ d. $n = 4, \ell = 3$

7.81. How many electrons could occupy orbitals with the following quantum numbers?
a. $n = 2, \ell = 0$ c. $n = 4, \ell = 2$
b. $n = 3, \ell = 1, m_\ell = 0$ d. $n = 1, \ell = 0, m_\ell = 0$

7.82. How many electrons could occupy an orbital with the following quantum numbers?
a. $n = 3, \ell = 2$ c. $n = 3, \ell = 0$
b. $n = 5, \ell = 4$ d. $n = 4, \ell = 1, m_\ell = 1$

7.83. Which of the following combinations of quantum numbers are allowed?
a. $n = 1, \ell = 1, m_\ell = 0, m_s = +\frac{1}{2}$
b. $n = 3, \ell = 0, m_\ell = 0, m_s = -\frac{1}{2}$
c. $n = 1, \ell = 0, m_\ell = 1, m_s = -\frac{1}{2}$
d. $n = 2, \ell = 1, m_\ell = 2, m_s = +\frac{1}{2}$

7.84. Which of the following combinations of quantum numbers are allowed?
a. $n = 3, \ell = 2, m_\ell = 0, m_s = -\frac{1}{2}$
b. $n = 5, \ell = 4, m_\ell = 4, m_s = +\frac{1}{2}$
c. $n = 3, \ell = 0, m_\ell = 1, m_s = +\frac{1}{2}$
d. $n = 4, \ell = 4, m_\ell = 1, m_s = -\frac{1}{2}$

The Periodic Table and Filling the Orbitals of Multielectron Atoms; Electron Configurations of Ions

CONCEPT REVIEW

7.85. What is meant when two or more orbitals are said to be degenerate?

7.86. Explain how the electron configurations of the group 2 elements are linked to their location in the periodic table developed by Mendeleev (see Figure 2.9).

7.87. How do we know from examining the periodic table's structure that the $4s$ orbital is filled before the $3d$ orbital?

7.88. Explain why so many transition metals form ions with a 2+ charge.

PROBLEMS

7.89. List the following orbitals in order of increasing energy in a multielectron atom:
a. $n = 3, \ell = 2$ c. $n = 3, \ell = 0$
b. $n = 5, \ell = 4$ d. $n = 4, \ell = 1, m_\ell = -1$

7.90. Place the following orbitals in order of increasing energy in a multielectron atom:
a. $n = 2, \ell = 1$ c. $n = 3, \ell = 2$
b. $n = 5, \ell = 3$ d. $n = 4, \ell = 3$

7.91. What are the electron configurations of Li, Li^+, Ca, F^-, Na^+, Mg^{2+}, and Al^{3+}?

7.92. Which species listed in Problem 7.91 are isoelectronic with Ne?

7.93. What are the condensed electron configurations of K, K^+, S^{2-}, N, Ba, Ti^{4+}, and Al?

7.94. In what way are the electron configurations of H, Li, Na, K, Rb, and Cs similar?

7.95. Write the electron configurations of the following species: Na, Cl, Mn, and Mn^{2+}.

7.96. Write the electron configurations of the following species: C, S, Ti, and Ti^{4+}.

7.97. How many unpaired electrons are there in the following ground-state atoms and ions? (a) N; (b) O; (c) P^{3-}; (d) Na^+

7.98. How many unpaired electrons are there in the following ground-state atoms and ions? (a) Sc; (b) Ag^+; (c) Cd^{2+}; (d) Zr^{4+}

7.99. Identify the atom whose electron configuration is $[Ar]3d^24s^2$. How many unpaired electrons are there in the ground state of this atom?

7.100. Identify the atom whose electron configuration is $[Ne]3s^23p^3$. How many unpaired electrons are there in the ground state of this atom?

7.101. Which monatomic ion has a charge of 1− and the electron configuration $[Ne]3s^23p^6$? How many unpaired electrons are there in the ground state of this ion?

7.102. Which monatomic ion has a charge of 1+ and the electron configuration $[Kr]4d^{10}5s^2$? How many unpaired electrons are there in the ground state of this ion?

7.103. Predict the charge of the monatomic ions formed by Al, N, Mg, and Cs.

7.104. Predict the charge of the monatomic ions formed by S, P, Zn, and I.

7.105. Which of the following electron configurations represent an excited state?
a. $[He]2s^12p^5$
b. $[Kr]4d^{10}5s^25p^1$
c. $[Ar]3d^{10}4s^24p^5$
d. $[Ne]3s^23p^24s^1$

7.106. Which of the following electron configurations represent an excited state?
a. $[Ne]3s^23p^1$
b. $[Ar]3d^{10}4s^14p^2$
c. $[Kr]4d^{10}5s^15p^1$
d. $[Ne]3s^23p^64s^1$

7.107. In which subshell are the highest-energy electrons in a ground-state atom of the isotope ^{131}I? Are the electron configurations of ^{131}I and ^{127}I the same?

7.108. Although no currently known elements contain electrons in *g* orbitals, such elements may be synthesized some day. What is the minimum atomic number of an element whose ground-state atoms have an electron in a *g* orbital?

The Sizes of Atoms and Ions

CONCEPT REVIEW

7.109. Sodium atoms are much larger than chlorine atoms, but in NaCl sodium ions are much smaller than chloride ions. Why?

7.110. Why does atomic size tend to decrease with increasing atomic number across a row of the periodic table?

7.111. Which of the following group 1 elements has the largest atoms? Li, Na, K, Rb. Explain your selection.

7.112. Which of the following group 17 elements has the largest monatomic ions? F, Cl, Br, I. Explain your selection.

Ionization Energies

CONCEPT REVIEW

7.113. How do ionization energies change with increasing atomic number (a) down a group of elements in the periodic table and (b) from left to right across a period of elements?

7.114. The ionization energies of the main group elements are given in Figure 7.34. Explain the differences in ionization energy between (a) He and Li; (b) Li and Be; (c) Be and B; (d) N and O.

7.115. Explain why it is more difficult to ionize a fluorine atom than a boron atom.

7.116. Do you expect the ionization energies of anions of group 17 elements to be lower or higher than for neutral atoms of the same group?

7.117. Which of the following elements should have the smallest *second* ionization energy? Br, Kr, Rb, Sr, Y

7.118. Why is the first ionization energy (IE_1) of Al ($Z = 13$) less than the IE_1 of Mg ($Z = 12$) *and* less than the IE_1 of Si ($Z = 14$)?

Additional Problems

7.119. Interstellar Hydrogen Astronomers have detected hydrogen atoms in interstellar space in the $n = 732$ excited state. Suppose an atom in this excited state undergoes a transition from $n = 732$ to $n = 731$.
a. How much energy does the atom lose as a result of this transition?
b. What is the wavelength of radiation corresponding to this transition?
c. What kind of telescope would astronomers need in order to detect radiation of this wavelength? (*Hint*: It would not be one designed to capture visible light.)

*__**7.120.** When an atom absorbs an X-ray of sufficient energy, one of its 2*s* electrons may be emitted, creating a hole that can be spontaneously filled when an electron in a higher-energy orbital—a 2*p*, for example—falls into it. A photon of electromagnetic radiation with an energy that matches the energy lost in the $2p \rightarrow 2s$ transition is emitted. Predict how the wavelengths of $2p \rightarrow 2s$ photons would differ between (a) different elements in the fourth row of the periodic table and (b) different elements in the same column (for example, between the noble gases from Ne to Rn).

*__**7.121.** Two helium ions (He^+) in the $n = 3$ excited state emit photons of radiation as they return to the ground state. One ion does so in a single transition from $n = 3$ to $n = 1$. The other does so in two steps: $n = 3$ to $n = 2$ and then $n = 2$ to $n = 1$. Which of the following statements about these two pathways is true?
a. The sum of the energies lost in the two-step process is the same as the energy lost in the single transition from $n = 3$ to $n = 1$.
b. The sum of the wavelengths of the two photons emitted in the two-step process is equal to the wavelength of the single photon emitted in the transition from $n = 3$ to $n = 1$.
c. The sum of the frequencies of the two photons emitted in the two-step process is equal to the frequency of the single photon emitted in the transition from $n = 3$ to $n = 1$.
d. The wavelength of the photon emitted by the He^+ ion in the $n = 3$ to $n = 1$ transition is shorter than the wavelength of a photon emitted by an H atom in an $n = 3$ to $n = 1$ transition.

*__**7.122.** Use your knowledge of electron configurations to explain the following observations:
a. Silver tends to form ions with a charge of 1+, but the elements to the left and right of silver in the periodic table tend to form ions with 2+ charges.
b. The heavier group 13 elements (Ga, In, Tl) tend to form ions with charges of 1+ or 3+ but not 2+.
c. The heavier elements of group 14 (Sn, Pb) and group 4 (Ti, Zr, Hf) tend to form ions with charges of 2+ or 4+.

7.123. Trends in ionization energies of the elements as a function of the position of the elements in the periodic table are a useful test of our understanding of electronic structure.

a. Should the same trend in the first ionization energies for elements with atomic numbers $Z = 31$ through $Z = 36$ be observed for the second ionization energies of the same elements? Explain why or why not.

b. Which element should have the greater second ionization energy: Rb ($Z = 37$) or Kr ($Z = 36$)? Why?

7.124. Chemistry of Photo-Gray Glasses "Photo-gray" lenses for eyeglasses darken in bright sunshine because the lenses contain tiny, transparent AgCl crystals. Exposure to light removes electrons from Cl^- ions forming a chlorine atom in an excited state (indicated below by the asterisk):

$$Cl^- + h\nu \rightarrow Cl^* + e^-$$

The electrons are transferred to Ag^+ ions, forming silver metal:

$$Ag + e^- \rightarrow Ag$$

Silver metal is reflective, giving rise to the photo-gray color.

a. Write condensed electron configurations of Cl^-, Cl, Ag, and Ag^+.

b. What do we mean by the term *excited state*?

c. Would more energy be needed to remove an electron from a Br^- ion or from a Cl^- ion? Explain your answer.

*d. How might substitution of AgBr for AgCl affect the light sensitivity of photo-gray lenses?

7.125. Tin (in group 14) forms both Sn^{2+} and Sn^{4+} ions, but magnesium (in group 2) forms only Mg^{2+} ions.

a. Write condensed ground-state electron configurations for the ions Sn^{2+}, Sn^{4+}, and Mg^{2+}.

b. Which neutral atoms have ground-state electron configurations identical to Sn^{2+} and Mg^{2+}?

c. Which 2+ ion is isoelectronic with Sn^{4+}?

7.126. Oxygen Ions in Space Since 1999 the *Far Ultraviolet Spectroscopic Explorer (FUSE)* satellite has been analyzing the spectra of emission sources within the Milky Way. Among the satellite's findings are interplanetary clouds containing oxygen atoms that have lost five electrons.

a. Write an electron configuration for these highly ionized oxygen atoms.

b. Which electrons have been removed from the neutral atoms?

c. The ionization energies corresponding to removal of the third, fourth, and fifth electrons are 4581 kJ/mol, 7465 kJ/mol, and 9391 kJ/mol, respectively. Explain why removal of each additional electron requires more energy than removal of the previous one.

d. What is the maximum wavelength of radiation that will remove the fifth electron from O^{4+}?

*7.127. Effective nuclear charge (Z_{eff}) is related to atomic number (Z) by a parameter called the shielding parameter (σ) according to the equation $Z_{eff} = Z - \sigma$.

a. Calculate Z_{eff} for the outermost s electrons of Ne and Ar given $\sigma = 4.24$ (for Ne) and 11.24 (for Ar).

b. Explain why the shielding parameter is much greater for Ar than for Ne.

7.128. Fog Lamp Technology Sodium fog lamps and street lamps contain gas-phase sodium atoms and sodium ions. Sodium atoms emit yellow-orange light at 589 nm. Do sodium ions emit the same yellow-orange light? Explain why or why not.

7.129. How can an electron get from the (+) lobe of a p orbital to the (−) lobe without going through the node between the lobes?

7.130. Einstein did not fully accept the uncertainty principle, remarking that "He [God] does not play dice." What do you think Einstein meant? Niels Bohr allegedly responded by saying, "Albert, stop telling God what to do." What do you think Bohr meant?

A Noble Family: Special Status for Special Behavior

*7.131. Compounds have been made from xenon and krypton with fluorine and oxygen, but not from the other noble gases. Suggest a reason, based on periodic properties, why xenon and krypton can be forced to form compounds but helium, neon, and argon cannot.

7.132. The successive ionization energies of the noble gases increase uniformly, in comparison to the successive ionization energies of the other elements, which at some place in the sequence show a rather large jump between values. Why is this the case?

*7.133. Helium is the only element that was discovered in space before it was found on Earth. Suggest a reason for this.

7.134. The first ionization energy of the noble gas neon is 2081 kJ/mol. The sodium ion (Na^+) is isoelectronic with neon, but the energy required to take one electron from the sodium ion is 4560 kJ/mol. Why is it so much harder to remove an electron from a sodium ion than from a neon atom if the two are isoelectronic?

Additional study materials including ChemTours and Diagnostic Quizzes are available at StudySpace at www.wwnorton.com/studyspace.

8

Chemical Bonding and Climate Change

A LOOK AHEAD

The Greenhouse Effect: Good News and Bad

Climate change is a hotly debated topic. Many scientists and non-scientists believe that recent increases in the average temperature of Earth's surface (about half a Celsius degree in the last half century) are linked to increases in the concentration of carbon dioxide and other *greenhouse gases* in the atmosphere. These gases trap Earth's heat in its atmosphere much like panes of glass trap heat in a greenhouse. Increases in atmospheric concentrations of greenhouse gases have been linked to human activity, particularly to increasing rates of fossil fuel combustion and the destruction of forests that would otherwise consume CO_2 during photosynthesis.

Carbon dioxide allows solar radiation to pass through the atmosphere and warm Earth's surface, but it traps heat that would otherwise radiate from the warm surface back into space. Were it not for the presence of atmospheric CO_2 and other gases, Earth would be much colder than it is. Indeed, it would be too cold to be habitable. The presence of these gases has moderated Earth's average global temperature to within a fairly narrow range, $15°–17°C$, for many thousands of years.

But that was then. During the last 50 years, atmospheric concentrations of CO_2 have increased from about 315 ppm to nearly 390. The atmosphere hasn't contained this much CO_2 in over half a million years, long before our species evolved. We will not discuss here the climatic and social consequences of global warming, but we will address the question of what it is about CO_2 that makes it so effective at trapping heat while other gases like O_2 and N_2 that are much more abundant are not a concern.

We noted in Chapter 7 that heat energy is associated with the infrared region of the electromagnetic spectrum. Carbon dioxide is good at trapping heat because it absorbs infrared radiation. To understand

Learning Outcomes

- Describe ways in which covalent, ionic, and metallic bonds are alike and ways in which they differ

- Draw Lewis structures of molecular compounds and polyatomic ions including resonance structures when appropriate

- Predict the polarity of covalent bonds based on differences in the electronegativities of the bonded elements

- Use formal charges to identify preferred resonance structures

- Describe how bond order, bond energy, and bond length are related

Greenhouse Effect The glass windows of a greenhouse allow sunlight to enter ▶ but trap the warm air inside. Greenhouse gases, such as CO_2, behave in a similar fashion. Recent increases in atmospheric CO_2 concentrations have been linked to rising global temperatures and the threat of severe climate change.

ionic bond results from the electrostatic attraction of a cation for an anion.

bond length the distance between the nuclear centers of two atoms joined together in a bond.

bond energy the energy needed to break 1 mole of a particular covalent bond in a molecule or polyatomic ion in the gas phase.

metallic bond consists of the nuclei of metal atoms surrounded by a "sea" of shared electrons.

how CO_2 does this while O_2 and N_2 do not, we need to consider the characteristics of the chemical bonds that hold atoms together in molecules.

How do chemical bonds form when pairs of atoms share their outermost electrons? Does "sharing" imply *equal* sharing? In this chapter we explore how unequal sharing of electron pairs together with the motion of atoms in bonds allows some compounds, such as CO_2, to absorb infrared radiation, including the heat emitted by Earth's surface, and in so doing contribute to global warming. ■

8.1 Chemical Bonds

There are fewer than 100 stable (nonradioactive) elements that make up all the matter in our world. These elements combine to form over 50 million compounds, and the number grows daily as chemists synthesize new ones. This multitude of compounds exists because they are more stable than free elements. That is, atoms that are linked together by chemical bonds are in lower chemical energy states than the free atoms that form them.

Why are bonded atoms more stable? We introduced the principal reason why in Chapter 5 when we discussed the concept of electrostatic potential energy (E_{el}) between charged particles. We noted that E_{el} is directly proportional to the product of the charges (Q_1 and Q_2) on pairs of ions (or subatomic particles) and is inversely proportional to the distance (d) between them:

$$E_{el} \propto \frac{Q_1 \times Q_2}{d} \tag{5.3}$$

When we apply Equation 5.3 to particles of opposite charge, the product of Q_1 and Q_2 is negative and so is the value of E_{el}. Also, E_{el} becomes more negative as the particles approach each other and the distance (d) between them decreases. This trend is illustrated in Figure 5.7 for a cation–anion pair. As such a pair of ions moves even closer to each other, electrostatic repulsion between their two clouds of negatively charged electrons and between their two positive nuclei stops them from getting closer still. At the energy minimum in Figure 5.7 the forces of attraction and repulsion are in balance, and the cation–anion pair is held together by a stable **ionic bond**.

∞ **CONNECTION** Negative values of E_{el} quantify the attraction between particles of opposite charge; positive values relate to the repulsion of particles with the same charge, as described in Chapter 5.

▶❙❙ **CHEMTOUR** Bonding

We can apply the concept of electrostatic potential energy to explain why two neutral atoms may come together to form a stable molecule. Let's consider the case of two hydrogen atoms (Figure 8.1a) that are sufficiently far apart that they do not interact with each other and have zero potential energy with respect to each other. As they approach each other, the proton in the nucleus of one H atom is attracted to the electron of the other, and vice versa. This mutual attraction produces a negative E_{el} (Figure 8.1b). As the distance between atoms decreases, the value of E_{el} continues to decrease, eventually reaching a minimum when the nuclei are 74 pm apart. If they come any closer together, the repulsion between their two positive nuclei more than offsets the mutual proton–electron attractions, and E_{el} begins to rise rapidly with decreasing d (Figure 8.1c). The distance 74 pm is the **bond length** of a H—H covalent bond. In addition, the value of E_{el} at this distance, -432 kJ/mol, is the energy released when 2 moles of H atoms form 1 mole of H_2 molecules. This much energy would have to be added to break up a mole of H_2 molecules into free H atoms. Thus, 432 kJ/mol is the H—H **bond energy** or *bond strength*. We will explore the significance of bond length and bond strength in more detail in Section 8.8.

∞ **CONNECTION** In Chapter 1 we defined chemical bond as the energy that holds two atoms in a molecule together. A covalent bond is defined in Chapter 2 as a bond between two atoms created by sharing pairs of electrons.

We can also use electrostatic potential energy to explain the interactions between the atoms in a solid piece of metal. As with atoms in molecules, the positive nucleus of each atom in a metallic solid is attracted to the electrons of the atoms that surround it. These attractions result in the formation of **metallic bonds** that are unlike covalent bonds in that they are not pairs of electrons shared by pairs of atoms. Instead,

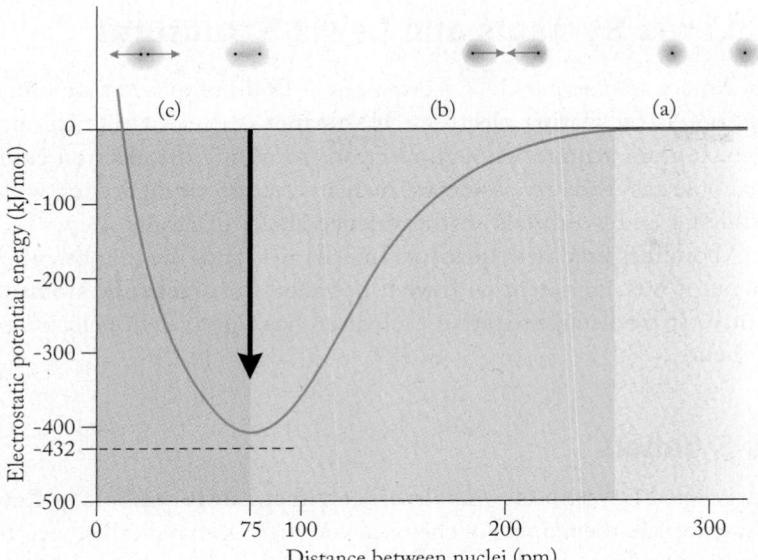

FIGURE 8.1 The electrostatic potential energy (E_{el}) profile of a H—H bond: (a) Two H atoms that are far apart do not interact, and so $E_{el} = 0$. (b) As the atoms approach each other mutual attraction between their positive nuclei and negative electrons causes E_{el} values to decrease. (c) At 74 pm E_{el} reaches a minimum as the two atoms form a H—H bond. (d) If the atoms are even closer together, repulsion between their nuclei causes E_{el} to increase and destabilizes the bond.

the shared electrons in metallic bonds are highly delocalized, forming a "sea" of mobile electrons that move freely among all the atoms in a metallic solid. This electron mobility explains why metals are such good conductors of heat and electricity.

Table 8.1 summarizes some of the similarities and differences between ionic, covalent, and metallic bonds. As you consider these properties keep in mind that many bonds do not fall exclusively into just one category, but rather have some covalent, ionic, and even metallic character.

TABLE 8.1   Types of Chemical Bonds

	Covalent	Ionic	Metallic
Elements involved:	Nonmetals and metalloids	Metals and nonmetals	Metals
Electron distribution:	Shared	Transferred	Pooled
Microview:	Br Br	Na⁺ Cl⁻	Cu
Macroview:			

8.2 Lewis Symbols and Lewis Structures

In 1916, American chemist G. N. Lewis (1875–1946) proposed that atoms form chemical bonds by sharing electrons. He further suggested that through this sharing each atom acquires enough electrons to mimic the electron configuration of a noble gas. Today we associate such an electron configuration with completely filled s and p orbitals in the valence shells of atoms. Lewis's view of chemical bonding predated quantum mechanics and the notion of atomic orbitals, but it was consistent with what he called the **octet rule**: atoms tend to lose, gain, or share electrons so that each atom has eight valence electrons, or an *octet* of them.

Lewis Symbols

Lewis developed a system of symbols, called either **Lewis symbols** or **Lewis dot symbols**, to explain the number of chemical bonds an atom typically forms to complete its octet. In other words, the Lewis symbol of an atom explains its **bonding capacity**. A Lewis symbol consists of the symbol of an element surrounded by dots representing the valence electrons. The dots are placed on the four sides of the symbol (top, bottom, right, and left). The order in which they are placed does not matter *as long as one dot is placed on each side before any dots are paired.* The number of *unpaired* dots in a Lewis symbol indicates the typical bonding capacity of the atom.

Figure 8.2 shows the Lewis symbols of the main group elements. Because all elements in a family have the same number of valence electrons, they all have the same arrangement of dots in their Lewis symbols. For example, the Lewis symbols of carbon and all other group 14 elements have four unpaired electrons. Four unpaired electrons means that a carbon atom tends to form four chemical bonds. In doing so it completes its octet because these four bonds contain the carbon atom's original four valence electrons plus four more from the atoms with which it forms the bonds. Similarly, the Lewis symbols of nitrogen and all other group 15 elements have three unpaired electrons; a nitrogen atom typically forms three bonds to complete its octet.

There are some important exceptions to Lewis's octet rule, several of which involve H, Be, and B. The valence shell of hydrogen has only one (1s) orbital and so can hold only two electrons. The one electron in the Lewis symbol of hydrogen indicates it has a bonding capacity of 1. A hydrogen atom needs only a *duet* (rather than an octet) of electrons to acquire the stable electron configuration of He. The valence shells of beryllium and boron are also underpopulated in some compounds. For example, the Lewis symbol of the beryllium atom has two unpaired dots for a bonding capacity of 2; the beryllium atom in $BeCl_2$ has only four valence electrons instead of eight. The three dots in the Lewis symbol of boron indicate a bonding capacity of 3, and the boron atom in BF_3 has only six valence electrons. Compounds having fewer than eight valence electrons about an atom are referred to as *electron-deficient* compounds in Lewis theory.

In another type of electron deficiency, atoms in some molecules have incomplete octets because there are odd numbers of valence electrons in those molecules. Examples include NO and NO_2. We explore the chemical implications of the unpaired electrons in these molecules in Section 8.7.

CONNECTION The elements in groups 1, 2, and 13 through 18 in the periodic table are called main group elements (see Chapter 2). The monatomic ions of main group elements have filled s and p orbitals in their outermost shell and so are isoelectronic with a noble gas (Chapter 7).

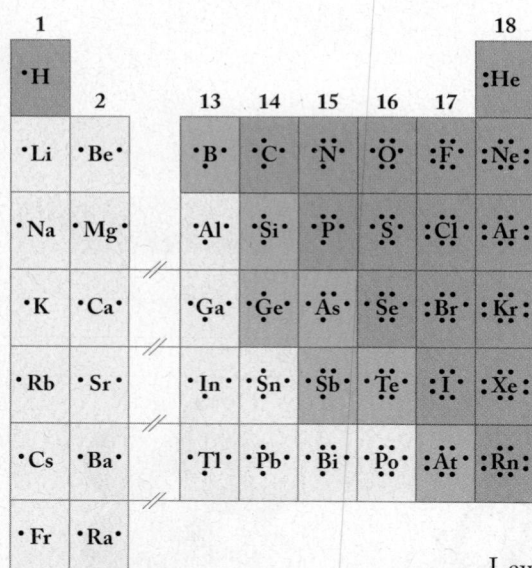

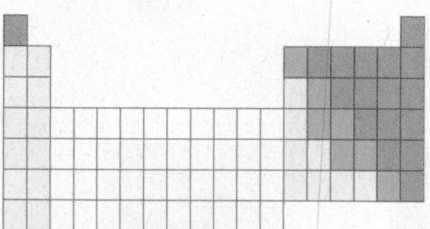

FIGURE 8.2 Lewis symbols of the main group elements of the periodic table. Because elements in a family have similar outer-shell electron configurations, they have the same number of dots in their Lewis symbols, which represent valence electrons.

CONCEPT TEST ....

Devise a formula that relates the bonding capacity of the atoms of the elements in groups 14–17 with their group number.

(Answers to Concept Tests are in the back of the book.)

....

Lewis Structures

The properties of molecular substances depend on how many atoms of which elements are in each of their molecules and how those atoms are bonded together. A **Lewis structure** is a two-dimensional representation of a molecule showing how the atoms in the molecule are connected. Because covalent bonds are *pairs* of shared valence electrons, Lewis structures focus on how these electron pairs, called **bonding pairs**, are distributed among the atoms in a molecule. In a Lewis structure, the pairs of electrons that are shared in chemical bonds are drawn with lines, as in the molecular structures we first saw in Chapter 1. A pair of electrons shared between two atoms is called a **single bond**. Electron pairs that are not involved in bond formation appear as pairs of dots on one atom. These unshared electron pairs are called **lone pairs**. Atoms with one or more pairs of dots in their Lewis symbols frequently have that same number of lone pairs of electrons in the Lewis structures of the molecules they form.

The following guidelines describe a five-step approach to drawing Lewis structures. The guidelines are particularly useful in drawing the structures of molecules and polyatomic ions that have a central atom bonded to atoms that have less bonding capacity. Many inorganic molecules and polyatomic ions and many small organic molecules (fewer than four carbon atoms) fit this description. In addition, to build some larger structures, we may sometimes need to break the structure into subunits and consider more than one "central" atom, as in hydrogen peroxide:

$$\text{H}\!-\!\overset{\text{..}}{\underset{\text{..}}{\text{O}}}\!-\!\overset{\text{..}}{\underset{\text{..}}{\text{O}}}\!-\!\text{H}$$

In these cases, each atom that is bonded to two other atoms—each oxygen atom in H_2O_2—is the "central atom" within a three-atom subunit, as shown by the O atoms highlighted in red:

$$-\!\overset{\text{..}}{\underset{\text{..}}{\text{O}}}\!-\!\overset{\text{..}}{\underset{\text{..}}{\text{O}}}\!-\!\text{H}$$

$$\text{H}\!-\!\overset{\text{..}}{\underset{\text{..}}{\text{O}}}\!-\!\overset{\text{..}}{\underset{\text{..}}{\text{O}}}\!-$$

It is important to note that these initial guidelines are just a starting point in the discussion of molecular structure. We will continue to refine these guidelines as we draw structures for increasingly more challenging molecules and learn more about the distribution of electrons in bonds.

Steps for Drawing Lewis Structures

1. *Determine the number of valence electrons.* For a neutral molecule, count the valence electrons in all the atoms in the molecule. For a polyatomic ion, count the valence electrons and then add or subtract the number of electrons needed to account for the charge on the ion.

octet rule atoms of main group elements make bonds by gaining, losing, or sharing electrons to achieve a valence shell containing 8 electrons, or four electron pairs.

Lewis symbol (also called **Lewis dot symbol**) the chemical symbol for an atom surrounded by one or more dots representing the valence electrons.

bonding capacity the number of covalent bonds an atom forms to have an octet of electrons in its valence shell.

Lewis structure a two-dimensional representation of the bonds and lone pairs of valence electrons in a molecule or polyatomic ion.

bonding pair a pair of electrons shared between two atoms.

single bond results when two atoms share one pair of electrons.

lone pair a pair of electrons that is not shared.

▶II **CHEMTOUR** Lewis Dot Structures

2. *Arrange the Lewis symbols of the atoms in a pattern that shows how they are bonded together and then connect them with single bonds (single pairs of bonding electrons).* Put the atom with the greatest bonding capacity in the center. Place the remaining atoms around the central atom and those that form the fewest bonds (hydrogen) around the periphery.

3. *Complete the octets of the atoms bonded to the central atom by adding lone pairs of electrons.*

4. *Match the number of valence electrons in the Lewis structure to the number determined in step 1.* If the numbers are the same, then all the electrons have been used and none need to be added or taken away. If there are not enough in the structure, add electrons to the central atom; if there are too many in the structure, delete lone pairs from bonded atoms.

5. *Complete the octet on the central atom.* If there is an octet on the central atom, the structure is done. If there is less than an octet on the central atom, create additional bonds to it by converting one or more lone pairs of electrons on atoms surrounding the central atom into bonding pairs.

SAMPLE EXERCISE 8.1 **Drawing the Lewis Structure of Chloroform**

Chloroform is a low-boiling liquid that was once used as an anesthetic in surgery. Draw its Lewis structure.

Collect and Organize Chloroform has the molecular formula $CHCl_3$. We can use the five-step approach described above to generate the Lewis structure.

Analyze The formula $CHCl_3$ tells us that a chloroform molecule contains one carbon atom, one hydrogen atom, and three chlorine atoms. Because it is a group 14 element, the carbon atom has 4 valence electrons and needs 4 more to complete its octet. Hydrogen, in group 1, has 1 valence electron and needs 1 more to complete its duet. Chlorine, in group 17, has 7 valence electrons and needs 1 more to complete its octet.

Solve

Step 1: The number of valence electrons in the $CHCl_3$ molecule is

Element:	C	+	H	+	3 Cl
Valence electrons:	4	+	1	+	$(3 \times 7) = 26$

Step 2: The carbon atom has the most (4) unpaired electrons in its Lewis symbol and hence has the greatest bonding capacity. Therefore C is the central atom in the molecule. Each H and each Cl needs one more electron to achieve the electron configuration of a noble gas, and so each forms one bond to the C atom:

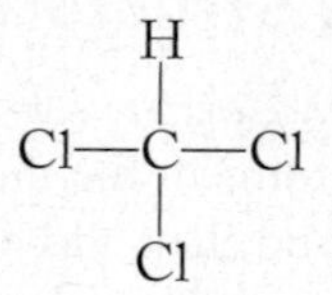

$$\begin{array}{c} H \\ | \\ Cl-C-Cl \\ | \\ Cl \end{array}$$

Step 3: In this structure the hydrogen atom has a duet because it shares a bonding pair of electrons with the central carbon atom. We need to add lone pairs of electrons to complete the octets on the three chlorine atoms:

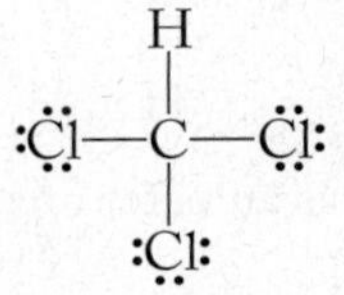

$$
\overset{\displaystyle \text{H}}{\underset{\displaystyle :\overset{..}{\underset{..}{\text{Cl}}}:}{:\overset{..}{\underset{..}{\text{Cl}}}\!-\!\text{C}\!-\!\overset{..}{\underset{..}{\text{Cl}}}:}}
$$

Step 4: This structure contains four pairs of bonding electrons and nine lone pairs, for a total of

$$(4 \times 2) + (9 \times 2) = 26 \text{ electrons}$$

which is the number of electrons determined in step 1.

Step 5: The carbon atom is surrounded by four bonds, which means 8 electrons, so it has a full octet. The structure is done.

Think about It In this example, there was no need to change the structure during steps 4 and 5 because all the electrons are included and all the atoms have an octet (or duet in the case of the hydrogen atom).

Practice Exercise Draw the Lewis structure of methane, CH_4.

(Answers to Practice Exercises are in the back of the book.)

SAMPLE EXERCISE 8.2   **Drawing the Lewis Structure of Ammonia**

Draw the Lewis structure of ammonia, NH_3.

Collect and Organize Ammonia has the molecular formula NH_3.

Analyze The formula NH_3 tells us that each molecule contains one atom of nitrogen and three atoms of hydrogen. Nitrogen is a group 15 element with 5 valence electrons, 3 of which are unpaired for a bonding capacity of 3. Hydrogen atoms have 1 valence electron each and a bonding capacity of 1.

Solve

Step 1: The number of valence electrons in the NH_3 molecule is

Element:	N	+	3 H
Valence electrons:	5	+	$(3 \times 1) = 8$

Step 2: The nitrogen atom has the greater bonding capacity and is the central atom. Connecting each H atom to the nitrogen atom with a covalent bond yields

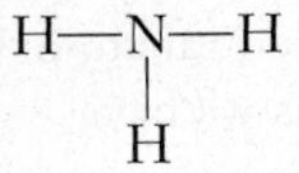

$$
\text{H}\!-\!\overset{\displaystyle }{\underset{\displaystyle \text{H}}{\text{N}}}\!-\!\text{H}
$$

Step 3: Each bonded H atom has a complete duet of electrons.

Step 4: Three lines in the structure of step 3 represent $3 \times 2 = 6$ valence electrons, but we need 8 to match the number determined in step 1. To add the 2 electrons we need, we add a lone pair of electrons to nitrogen:

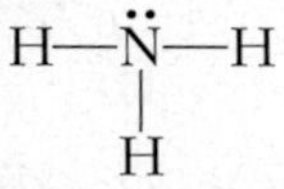

$$H\!-\!\ddot{N}\!-\!H$$
$$|$$
$$H$$

Step 5: The N atom now has an octet of electrons, so the Lewis structure is complete.

Think about It This structure makes sense because the Lewis symbol of the nitrogen atom contains 2 electrons that are paired and 3 that are unpaired. Therefore, it is reasonable that the nitrogen atom in NH_3 has one lone pair and three bonding pairs of electrons.

Practice Exercise Draw the Lewis structure of phosphorus trichloride.

Lewis Structures of Molecules with Double and Triple Bonds

Lewis structures can be used to show the bonding in molecules in which two atoms share more than one pair of bonding electrons. A bond in which two atoms share two pairs of electrons is called a **double bond**. For example, when the two oxygen atoms in a molecule of O_2 share two pairs of electrons, they form an $O\!=\!O$ double bond. When the two nitrogen atoms in a molecule of N_2 share three pairs of electrons, they form a $N\!\equiv\!N$ **triple bond**. In Section 8.8 we discuss the characteristics of these multiple bonds and compare them with those of single bonds.

How do we know when a Lewis structure has a double or triple bond? Typically we find this out when we apply steps 4 and 5 in the guidelines. Suppose we fill the octets of all the atoms attached to the central atom (step 4), and in doing so we use all the valence electrons that we have. We then determine in step 5 that the central atom does not have an octet. We do not have any more electrons available, and the only way we can provide the central atom with an octet is to *convert a lone pair of electrons* from one of the other atoms into a bonding pair. If the central atom still does not have an octet, we may convert a second lone pair into a bonding pair. The following example illustrates this application of the guidelines.

Let's draw the Lewis structure for the organic molecule formaldehyde (H_2CO). We begin with the five-step guidelines.

Step 1: The total number of valence electrons is

Element:	C	+	2 H	+	O
Valence electrons:	4	+	(2×1)	+	$6 = 12$

Step 2: Of the three elements, carbon has the greatest bonding capacity (4). Therefore, C is the central atom. Connecting it with single bonds to the other three atoms we have

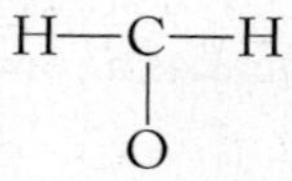

$$H\!-\!C\!-\!H$$
$$|$$
$$O$$

double bond results when two atoms share two pairs of electrons.

triple bond results when two atoms share three pairs of electrons.

Step 3: Each H atom has a single covalent bond (2 electrons) and a complete valence shell. Oxygen needs three lone pairs of electrons to complete its octet:

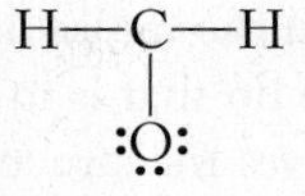

$$H\!-\!\underset{\underset{\displaystyle :\ddot{O}:}{|}}{C}\!-\!H$$

Step 4: There are 12 valence electrons in this structure, which matches the number determined in step 1.

Step 5: The central C atom has only 6 electrons. To provide the carbon atom with 2 more electrons so that it has an octet but without removing any from oxygen, which already has an octet, we convert one of the lone pairs on the oxygen atom into a bonding pair between C and O:

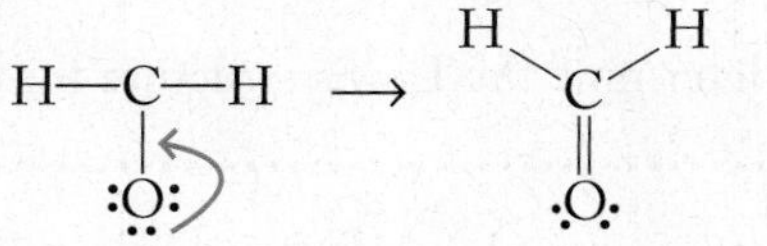

$$H\!-\!\underset{\underset{\displaystyle :\ddot{O}:}{|}}{C}\!-\!H \quad \rightarrow \quad \overset{\displaystyle H \qquad H}{\underset{\displaystyle \cdot\ddot{O}\cdot}{\underset{\|}{C}}}$$

It does not matter which of the three lone pairs we move because they are equivalent. The central carbon atom now has a complete octet, and the oxygen atom still does. This structure makes sense because the four covalent bonds around carbon—two single bonds and one double bond—match its bonding capacity. The double bond to oxygen makes sense because oxygen is a group 16 element with a bonding capacity of 2, and it has two bonds in this structure.

Notice that we have drawn the double bond and the two single bonds on the central carbon atom at an angle of about 120° from each other in the final structure; we have done the same thing with the lone pairs of electrons on oxygen with respect to the double bond. Electrons are negatively charged and repel each other, so we draw them as far apart from each other as possible. This logic will be an important part of drawing three-dimensional structures of molecules in Chapter 9.

SAMPLE EXERCISE 8.3   Drawing the Lewis Structure of Acetylene

Draw the Lewis structure of acetylene, the hydrocarbon fuel used in oxyacetylene torches for welding.

Collect and Organize We are to draw the Lewis structure of acetylene, which has the molecular formula C_2H_2. We follow the steps in the guidelines, using double or triple bonds as needed.

Analyze Each molecule contains two atoms of carbon and two atoms of hydrogen. Carbon is a group 14 element with a bonding capacity of 4.

Solve

Step 1: The total number of valence electrons is

Element:	2 C	+	2 H	
Valence electrons:	(2×4)	+	(2×1)	$= 10$

Step 2: Of the two elements, carbon has the greater bonding capacity. Therefore C is the central atom. Connecting each carbon atom with single bonds to the other atoms we have:

$$H\!-\!C\!-\!C\!-\!H$$

Step 3: Each H atom has a single covalent bond (made up of 2 electrons) and a complete valence shell.

Step 4: There are 6 valence electrons in the structure, but there are 10 valence electrons in the molecule. This means we have to add two pairs. One way to do that is to add two bonds between the carbon atoms. This gives the structure the right number of valence electrons, and it completes the octets of both carbon atoms (step 5):

$$H\!-\!C\!\equiv\!C\!-\!H$$

Think about It This structure makes sense because carbon has a bonding capacity of 4, and there are four covalent bonds around each carbon atom—1 single bond and 1 triple bond—giving both complete octets.

Practice Exercise Determine the Lewis structure for carbon dioxide.

Lewis Structures of Ionic Compounds

We can use the octet rule and Lewis structures to explain the composition of many common ionic compounds, such as sodium chloride (NaCl). Crystals of NaCl are held together by the attractive force between two oppositely charged ions. By giving away its $3s^1$ valence electron, a sodium atom becomes a positively charged Na^+ ion and achieves the same electron configuration as Ne: $1s^2 2s^2 2p^6$. In other words, Na^+ has a vacant valence shell. If a chlorine atom, which has the valence electron configuration $3s^2 3p^5$, gains an electron to form a Cl^- ion, it achieves a filled outermost shell and becomes isoelectronic with the noble gas argon. We can illustrate this behavior using Lewis symbols as

$$Na^{\cdot} + {\cdot}\ddot{\underset{\cdot\cdot}{Cl}}{:} \rightarrow Na^+\left[{:}\ddot{\underset{\cdot\cdot}{Cl}}{:}\right]^-$$

The brackets around the chloride ion are used to emphasize that all 8 valence electrons are associated with it; none of them are associated with the sodium ion.

All alkali metal elements and alkaline earth elements tend to lose, rather than share, their valence electrons, in part because they have low ionization energies. The cations they form by losing these electrons obey the octet rule because they have the electron configurations of the noble gases that precede the parent elements in the periodic table.

∞ CONNECTION We defined ionization energy in Chapter 7 as the amount of energy required to remove 1 mole of electrons from 1 mole of atoms or molecules in the gas phase.

SAMPLE EXERCISE 8.4 **Drawing Lewis Symbols of Monatomic Ions and Lewis Structures of Binary Ionic Compounds**

Draw the Lewis symbols of the monatomic ions formed by calcium and oxygen. Then draw the Lewis structure of calcium oxide.

Collect and Organize We are to draw the Lewis symbols of two ions and the Lewis structure of the ionic compound they form. The charges of common monatomic ions are given in Figure 2.17.

Analyze When a metal combines with a nonmetal, the metal forms a cation and the nonmetal forms an anion. The monatomic ions formed by calcium and oxygen are Ca^{2+} and O^{2-}, respectively.

Solve To form a cation with a 2+ charge, a Ca atom loses both its valence electrons, leaving it with none:

$$\cdot \text{Ca} \cdot \xrightarrow{-2\,e^-} \text{Ca}^{2+}$$

When an O atom accepts 2 electrons it has a complete octet:

$$:\ddot{\text{O}}\cdot \xrightarrow{+2\,e^-} \left[:\ddot{\ddot{\text{O}}}:\right]^{2-}$$

Combining the Lewis symbols of Ca^{2+} and O^{2-} ions, we have the Lewis structure of CaO:

$$\text{Ca}^{2+}\left[:\ddot{\ddot{\text{O}}}:\right]^{2-}$$

Think about It The lack of dots in the symbol of the Ca^{2+} ion reinforces the fact that atoms of Ca, like all main group metal atoms, lose all the electrons in their valence shells when they form monatomic ions. In contrast, atoms of all the nonmetals form monatomic ions with complete octets.

Practice Exercise Draw the Lewis structure of magnesium fluoride.

polar covalent bond results from unequal sharing of bonding pairs of electrons between atoms.

bond polarity a measure of the extent to which bonding electrons are unequally shared due to differences in electronegativity of the bonded atoms.

nonpolar covalent bond a bond characterized by an even distribution of charge; electrons in the bonds are shared equally by the two atoms; pure covalent bonds give rise to nonpolar diatomic molecules.

8.3 Electronegativity, Unequal Sharing, and Polar Bonds

When Lewis proposed that atoms form chemical bonds by sharing electrons, he knew that electron sharing in covalent bonds did not necessarily mean *equal* sharing. For example, he knew, based on the chemical properties of HCl, that the H—Cl bond is a **polar covalent bond**. Lewis explained this **bond polarity** by assuming that the shared pair of electrons in HCl is closer to the chlorine end of the molecule than to the hydrogen end. In Figure 8.3 we show this unequal sharing using an arrow with the plus sign embedded in its tail (↦) to indicate the *direction of polarity:* the arrow points toward the more negative, electron-rich atom in the bond, and the position of the plus sign indicates the more positive, electron-poor atom.

Figure 8.4 shows examples of equal and unequal sharing of bonding pairs of electrons. It includes (a) Cl_2, in which the Cl atoms are connected by a **nonpolar covalent bond** shared equally between the two identical atoms, (b) a polar molecular compound (HCl), and (c) an ionic compound, NaCl, which represents an extreme case of unequal sharing: the bonding pair has been completely transferred from the Na atom to the Cl atom, creating a Na^+ ion and a Cl^- ion. The color bar at the top of Figure 8.4 indicates the values associated with the colors: yellow = no charge separation; blue and red = full 1+ and 1− charges, respectively; intermediate colors = partial charges (δ^+ and δ^-).

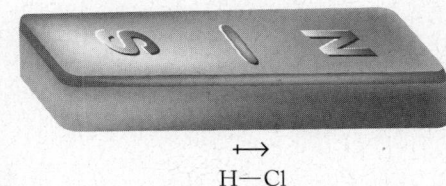

↦
H—Cl

FIGURE 8.3 Just as a bar magnet has a north and a south pole, a polar molecule such as HCl has positive and negative ends, represented here by the arrow above the H—Cl bond. The tail of the arrow is on the hydrogen atom, which has a partial positive charge, and the arrowhead is pointed toward the Cl, which has a partial negative charge.

▶‖ **CHEMTOUR** The Periodic Table

▶‖ **CHEMTOUR** Partial Charges and Bond Dipoles

FIGURE 8.4 Variations in valence electron density are represented using colored surfaces in these molecular models. (a) In the covalent bond in Cl_2, uniform electron density is represented by an evenly yellow surface, indicating that the two atoms share their bonding pair of electrons equally. (b) Unequal sharing of the bonding pair of electrons occurs in HCl, as shown by the orange-red color of the Cl (δ^-) and the green color around the H (δ^+). (c) In ionic NaCl, the blue color on the surface of the sodium ion indicates that it has a full 1+ charge and the red of the chloride ion reflects its charge of 1−.

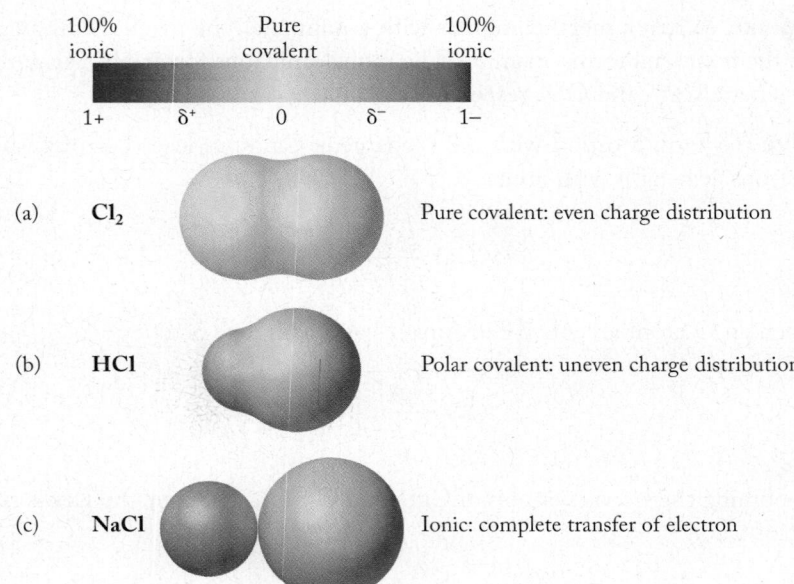

Another American chemist, Linus Pauling (1901–1994), developed the concept of **electronegativity** to explain bond polarity. Pauling assigned electronegativity values to the elements (Figure 8.5) based on the idea that the bonds between atoms of different elements are neither 100% covalent nor 100% ionic, but somewhere in between. The degree of ionic character of a bond depends on the differences in the abilities of the two atoms to attract the electrons they share: the greater the difference, the more ionic is the character of the bond between them.

The data in Figure 8.5 show that electronegativity is a periodic property, with values generally increasing left to right across a row in the periodic table, and decreasing top to bottom down a group. The reasons for these trends are essentially the same ones that produce similar trends in first ionization energies, as we discussed in Chapter 7 (see Figure 7.34). Increasing attraction between the nuclei of atoms and their outer-shell electrons with increasing atomic number across a row produces both higher ionization energies and greater electronegativities (Figure 8.6a). Within a group of elements the weaker attraction between

FIGURE 8.5 The Pauling electronegativity scale. Electronegativity increases from left to right across a period and decreases from top to bottom down a group. The greater the electronegativity value, the greater is that atom's ability to attract electrons in a bond. The Pauling electronegativity values shown below the symbols of the elements are unitless numbers that define the relative ability of an atom in a bond to attract shared electrons to itself.

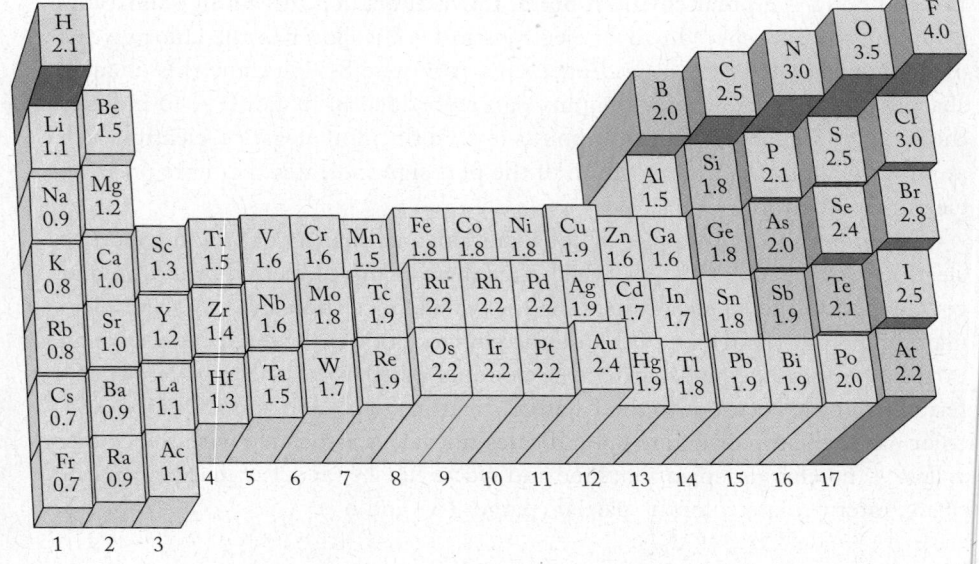

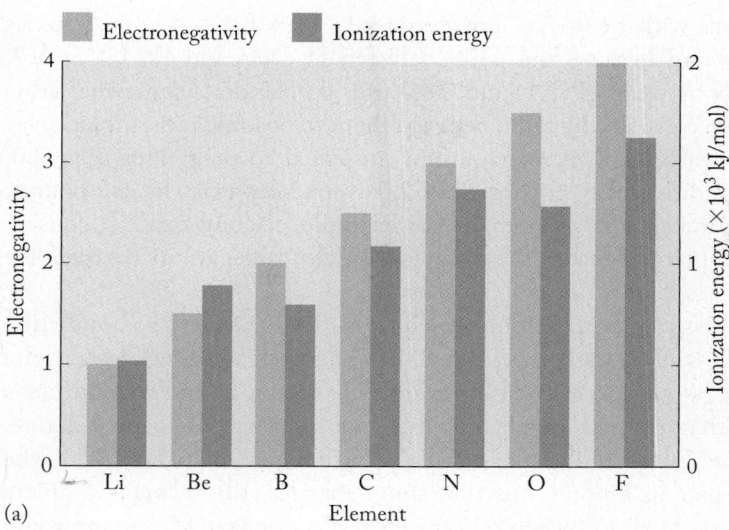

(a)

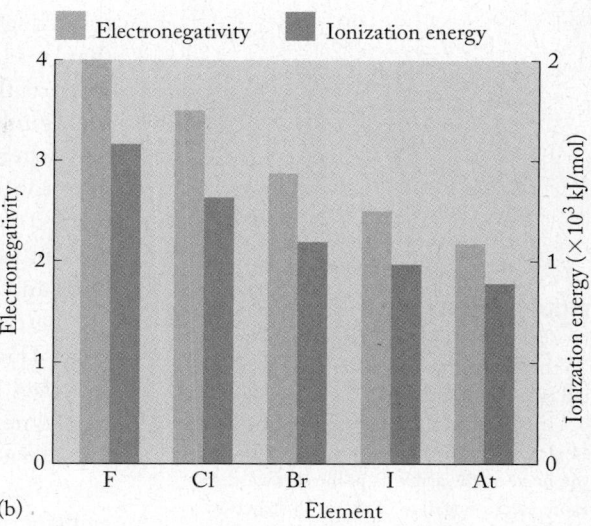

(b)

nuclei and outer-shell electrons with increasing atomic number leads to lower ionization energies and smaller electronegativities (Figure 8.6b). For these two reasons, the most electronegative elements—fluorine, oxygen, and nitrogen—are in the upper right corner of the periodic table, and the least electronegative elements are in the lower left corner.

> **CONCEPT TEST**
>
> Why does Figure 8.5 not include electronegativity values for the noble gases?

> **CONCEPT TEST**
>
> Draw arrows on the Lewis structure of CO_2 to indicate the polarity of the bonds.

Polarity and Type of Bond

Comparing electronegativity values allows us to determine which end of a covalent bond is electron-rich and which end is electron-poor. In addition, the greater the difference in electronegativity (ΔEN), the more uneven is the distribution of electrons and the more polar the bond, as shown in Figure 8.7 for the compounds

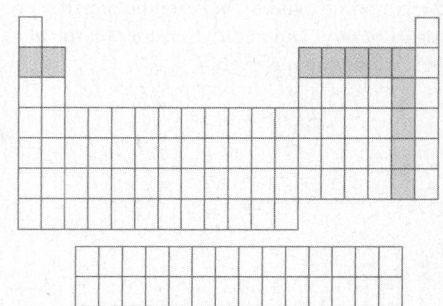

FIGURE 8.6 The trends in the electronegativities of the main group elements follow those of ionization energies: both tend to (a) increase with increasing atomic number across a row and (b) decrease with increasing atomic number within a group.

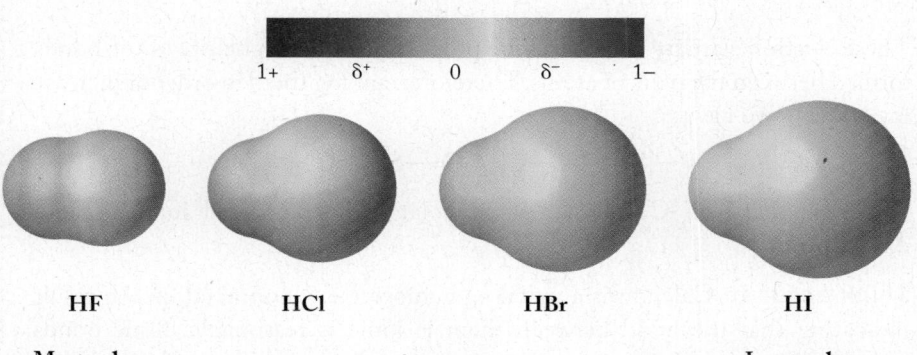

FIGURE 8.7 The greater the electronegativity difference (ΔEN) between two atoms, the more polar is the bond they form. Among the hydrogen halides, HF has the largest ΔEN and HI has the smallest. Therefore the HF bond is the most polar in this group, and the HI bond is the least polar.

electronegativity a relative measure of the ability of an atom in a bond to attract electrons to itself.

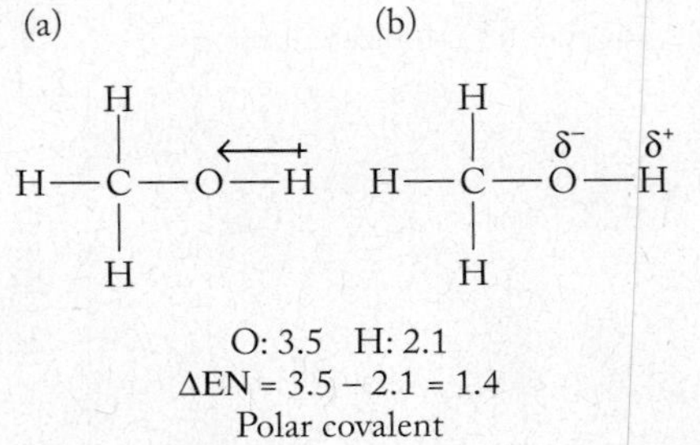

(a)   (b)

O: 3.5   H: 2.1
ΔEN = 3.5 − 2.1 = 1.4
Polar covalent

FIGURE 8.8 Two ways to indicate the polarity of the O—H bond in methanol include (a) the use of an arrow with a plus sign on its tail (see Figure 8.3) and (b) the use of δ^+ and δ^-. In structure a, the tail is placed next to the H atom with the partial positive charge and the head of the arrow is pointed toward the more electronegative O atom. In structure b the symbols δ^+ and δ^- indicate the presence of partial positive and negative charges on the H and O atoms, respectively.

the halogens form with hydrogen. For example, a H—F bond is more polar than a H—Cl bond because the ΔEN between H (2.1) and F (4.0) is 1.9, whereas the ΔEN between H (2.1) and Cl (3.0) is 0.9. Under a somewhat arbitrary guideline, we consider the bond between them to be ionic rather than covalent when ΔEN values between two atoms are equal to or greater than 2.0. Electronegativity differences greater than 2.0 frequently exist in compounds formed between metals and nonmetals. For example, calcium oxide is considered an ionic compound because the difference in electronegativity between Ca (1.0) and O (3.5) is 2.5.

Figure 8.8 illustrates both symbols used to indicate polarity in bonds: the arrow (Figure 8.3) and δ^+ and δ^- (Figure 8.4). The lowercase Greek letter delta followed by a negative sign δ^- is used to indicate which end of a bond has a partial negative charge and δ^+ indicates which end has a partial positive charge. The term *partial charge* indicates that there is unequal distribution of the shared electron pair in a bond, but that some sharing still occurs. Complete transfer of the shared pair to the atom of the more electronegative element would produce a full 1− charge on it and 1+ on the atom of the less electronegative element.

SAMPLE EXERCISE 8.5   Comparing the Polarity of Bonds

Rank, in order of increasing polarity, the bonds formed between O and C, Cl and Ca, N and S, O and Si. Are any of these bonds considered ionic?

Collect and Organize We are given four pairs of atoms and are to rank the pairs according to the polarity of the bond each pair forms and to identify any ionic bonds in the set. We need to refer to the Pauling electronegativities (Figure 8.5) to judge the relative polarity.

Analyze The polarity of a bond is related to the difference in electronegativities of the atoms in the bond. The guideline we apply is as follows: if the electronegativity difference is 2.0 or greater, the bond is considered ionic.

Solve Calculate the electronegativity difference between the atoms:

$$
\begin{array}{lll}
\text{O and C:} & \Delta EN = 3.5 - 2.5 = 1.0 \\
\text{Cl and Ca:} & \Delta EN = 3.0 - 1.0 = 2.0 \\
\text{N and S:} & \Delta EN = 3.0 - 2.5 = 0.5 \\
\text{O and Si:} & \Delta EN = 3.5 - 1.8 = 1.7
\end{array}
$$

These electronegativity differences are proportional to the polarity of the bonds formed between the pairs of atoms. Therefore, ranking them in order of increasing polarity we have

$$ \text{N—S} < \text{O—C} < \text{O—Si} < \text{Cl—Ca} $$

The bond between Cl and Ca is so polar it is considered ionic because ΔEN = 2.0.

Think about It Calcium is a metal and chlorine is a nonmetal, so the result indicating that the bond between them is ionic is reasonable. Ionic bonds tend to be formed between metals and nonmetals. Two of the other bonds, N—S and O—C, connect pairs of nonmetals, and the O—Si bond connects a nonmetal with a metalloid. We expect these three bonds to be covalent.

Practice Exercise Which pair forms the most polar bond: O and S, Be and Cl, N and H, or C and Br? Is the bond between that pair ionic?

8.4 Vibrating Bonds and the Greenhouse Effect

As we noted in Chapter 5 in the discussion of thermal energy, chemical bonds are not rigid. They vibrate a little, stretching and bending like tiny atomic-sized springs (Figure 8.9). There are natural frequencies to these vibrations, frequencies that match the frequencies of infrared electromagnetic radiation. This match, coupled with the unequal sharing of bonding electrons, allows atmospheric molecules with polar bonds, such as CO_2, to absorb infrared radiation emitted by Earth's surface. As a result, heat that might have dissipated into space if these molecules were not present in the atmosphere is trapped in the atmosphere, contributing to what is called the greenhouse effect.

Molecules with polar bonds may absorb and emit radiation in the infrared region because as these bonds vibrate, the tiny electrical fields associated with the separation of partial charges in the molecule fluctuate. These fluctuations can alter the strengths of the fields or even create new ones, depending on the nature of the vibration. When the frequencies of the bond vibrations match the frequencies of photons of infrared radiation, the molecules may absorb those photons. Scientists say these vibrations are *infrared active*. This is the molecular mechanism behind the greenhouse effect.

Not all polar bond vibrations absorb and emit infrared radiation. For example, two kinds of stretching vibrations can occur in a molecule of CO_2. One is a symmetric stretching vibration (Figure 8.9a) in which the two C=O bonds stretch and then compress in opposite directions. In this case the two fluctuating electrical fields produced by the two C=O bonds cancel each other out, and no infrared absorption or emission is possible. This vibration is said to be *infrared inactive*. However, when the bonds stretch in the same direction, that is, one gets shorter as the other gets longer (Figure 8.9b), the changes in charge separation do not cancel. This asymmetric stretch produces a fluctuation in the electrical field surrounding the overall CO_2 molecule and enables it to absorb infrared radiation. Molecules can also bend (Figure 8.9c) in ways that produce fluctuating electrical fields. Because the frequencies of the asymmetric stretching and bending of the bonds in CO_2 are in the same range as the frequencies of infrared radiation emitted from Earth's surface, carbon dioxide is a potent greenhouse gas.

▶II **CHEMTOUR** Greenhouse Effect

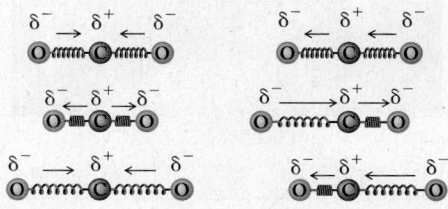

(a) Symmetric stretch (infrared inactive)   (b) Asymmetric stretch (infrared active)

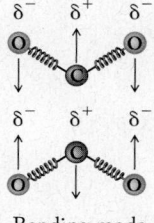

(c) Bending mode (infrared active)

FIGURE 8.9 Three modes of bond vibration in a molecule of CO_2 include (a) symmetric stretching of the C=O bonds, which produces no overall change in the polarity of the molecule. (b) Asymmetric stretching does produce side-to-side fluctuations in polarity that may result in absorption of IR radiation. (c) The bending mode produces up-and-down fluctuations that may also absorb IR radiation.

CONCEPT TEST

Nitrogen and oxygen make up about 99% of the gases in the atmosphere. Could the stretching of the N≡N and O=O bonds in these molecules result in the absorption of infrared radiation? Explain why or why not.

▶II **CHEMTOUR** Vibrational Modes

8.5 Resonance

The atmosphere contains two types of molecular oxygen. Most of it is O_2, but trace concentrations of ozone (O_3) are also present. Ozone in the lower atmosphere is sometimes referred to as "bad ozone" because high levels damage crops, harm trees,

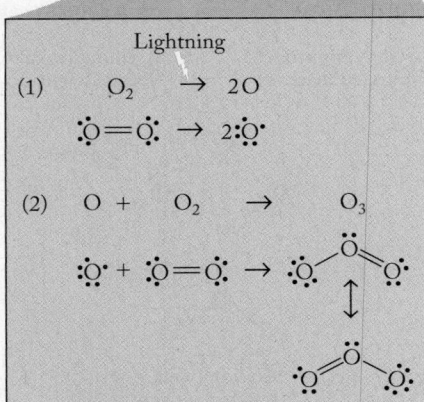

Lightning

(1) O_2 $\rightarrow$ $2O$

$\ddot{O}=\ddot{O}$ $\rightarrow$ $2\,\ddot{\underset{..}{O}}\cdot$

(2) O + O_2 $\rightarrow$ O_3

$\cdot\ddot{\underset{..}{O}}\cdot$ + $\ddot{O}=\ddot{O}$ $\rightarrow$

FIGURE 8.10 Lightning strikes contain sufficient energy to break oxygen–oxygen double bonds. The O atoms formed in this fashion collide with other O_2 molecules, forming ozone (O_3), an allotrope of oxygen.

and lead to human health problems. Ozone is also present in the upper atmosphere where it is considered "good ozone" because it shields life on Earth from potentially harmful UV radiation from the sun.

Ozone is a *triatomic* (three-atom) molecule produced naturally by lightning (Figure 8.10) and accounts for the pungent odor you may have smelled after a severe thunderstorm. Ozone (O_3) and diatomic oxygen (O_2) have the same empirical formula: O. Different molecular forms of the same element, such as O_2 and O_3, are called **allotropes** of the element and have different chemical and physical properties. Ozone, for example, is an acrid, pale blue gas that is toxic even at low concentrations, whereas O_2 is a colorless, odorless gas essential for most life-forms.

The different molecular formulas of allotropes mean that they have different molecular structures. Let's draw the Lewis structure for ozone following our five-step approach. Oxygen is a group 16 element and so has 6 valence electrons. Therefore the total number of valence electrons in an ozone molecule is $3 \times 6 = 18$ (step 1). Connecting the three O atoms with single bonds, we have (step 2)

$$O—O—O$$

Completing the octets of the noncentral atoms gives us (step 3)

$$\ddot{\underset{..}{O}}—O—\ddot{\underset{..}{O}}$$

This structure contains 16 electrons. We determined in step 1 that there are 18 valence electrons in the molecule, so we add 2 to the central oxygen atom (step 4):

$$\ddot{\underset{..}{O}}—\ddot{O}—\ddot{\underset{..}{O}}$$

This structure leaves the central atom 2 electrons short, so we convert one of the lone pairs on the O atom on the left end of the molecule into a bonding pair (step 5):

We could just as well used a lone pair from the O atom on the right, which would have given us

Which structure is correct? Experimental evidence indicates that neither structure is. Scientists have determined that all the bonds in ozone have the same length. As we shall see in Section 8.8, a double bond is always shorter than a single bond between the same two atoms. Figure 8.11 shows that the length of the two bonds in O_3 (128 pm) is about halfway between the length of an O—O single bond (148 pm) and an O=O double bond (121 pm). One way to explain this result is to assume that the bonding pattern of O_3 is halfway between the last two structures:

allotropes different molecular forms of the same element, such as oxygen (O_2) and ozone (O_3).

with the equivalent of 1.5 bonds between each pair of atoms. It is very important to note that this average does *not* mean that the molecule spends half its time as the left-hand structure in Figure 8.11 and half as the right-hand stucture. It always has three bonding pairs spread out evenly on the two sides of the central atom, as shown above.

To better understand how resonance occurs, consider what happens when the bonding electrons and the lone pair electrons in structure a below can be rearranged as shown by the red arrows:

a → b

Note how this rearrangement in a produces b. The process is completely reversible so that the electron pairs in b could just as easily be rearranged into the pattern in structure a:

b → a

We use a double-headed reaction arrow between structures to indicate that they are equivalent and that the actual structure is a blend of the two:

a ↔ b

The ability to draw two equivalent Lewis structures for the ozone molecule illustrates an important concept in Lewis theory called **resonance**: the existence of multiple Lewis structures, called **resonance structures** (or sometimes *resonance hybrids*), for a given arrangement of atoms. To determine whether resonance occurs in a molecule, we need to determine whether there can be alternative bond arrangements inside the molecule: arrangements in which the positions of some bonding pairs of electrons change but the positions of the atoms stay the same. A key indicator of the possibility of resonance is the presence of both single and double bonds from a central atom to two or more atoms of the same element.

Because all resonance structures are by definition Lewis structures, the five-step guidelines apply to drawing resonance structures. Because all the atoms in ozone are oxygen, an oxygen atom is clearly the central atom. Two additional issues, however, must be raised at this point, both of which relate to the concept of bonding capacity. First, note that the oxygen atoms in ozone have different numbers of bonds. Oxygen atoms *typically* form two bonds, but Lewis theory allows both more and fewer bonds than are implied by the Lewis symbol of an atom. Second, in previous examples we have used differences in typical bonding capacity as a criterion to select a central atom. As we deal with more molecules, the situation frequently arises that several atoms in one molecule may have the same bonding capacity. In that case, the selection of the central atom is based on electronegativity, with the least electronegative atom chosen as the central atom in the Lewis structure.

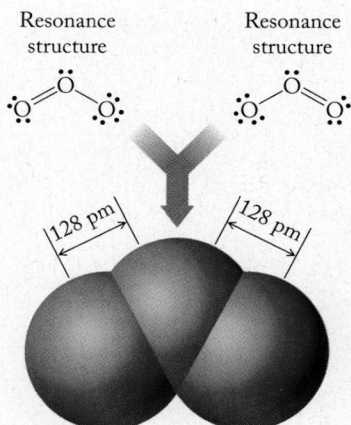

FIGURE 8.11 The molecular structure of ozone is an average of the two resonance structures shown at the top of the figure. Both bonds in ozone are 128 pm long, a value between the average length of an O—O single bond (148 pm) and the average length of an O=O double bond (121 pm). The intermediate value for the ozone bond length indicates that in an ozone molecule the bonds are neither single bonds nor double bonds but something in between.

▶❚❚ **CHEMTOUR** Resonance

resonance characteristic of electron distributions when two or more equivalent Lewis structures can be drawn for one compound.

resonance structure one of two or more Lewis structures with the same arrangement of atoms but different arrangements of bonding pairs of electrons.

SAMPLE EXERCISE 8.6   **Drawing Resonance Structures of a Molecule**

Sulfur trioxide (SO_3) is produced in the atmosphere when the SO_2 from natural and industrial sources combines with O_2. Draw all the possible resonance structures of SO_3.

Collect and Organize Each SO_3 molecule contains one atom of sulfur and three atoms of oxygen. We are to draw all resonance structures, which means at least two different Lewis structures with different bonding patterns but the same atom placement.

Analyze Both sulfur and oxygen are in group 16 and have 6 valence electrons per atom and a typical bonding capacity of 2. Sulfur is less electronegative than oxygen.

Solve

Step 1: The number of valence electrons in the SO_3 molecule is

Element :	S	+	3 O
Valence electrons:	6	+	$(3 \times 6) = 24$

Step 2: Because sulfur is less electronegative, it is selected as the central atom. Connecting it with single bonds to the three O atoms:

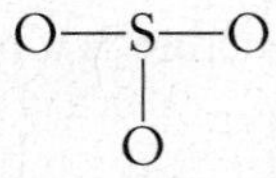

Step 3: Each O atom needs three lone pairs of electrons to complete its octet:

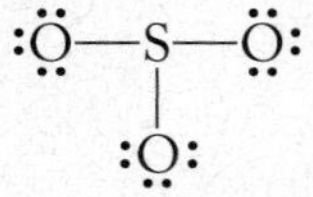

Step 4: There are 24 valence electrons in the structure in step 3, which matches the number determined in step 1.

Step 5: The central S atom has only 6 electrons in the structure of step 3. To complete its octet, we convert a lone pair on one of the oxygen atoms into a bonding pair:

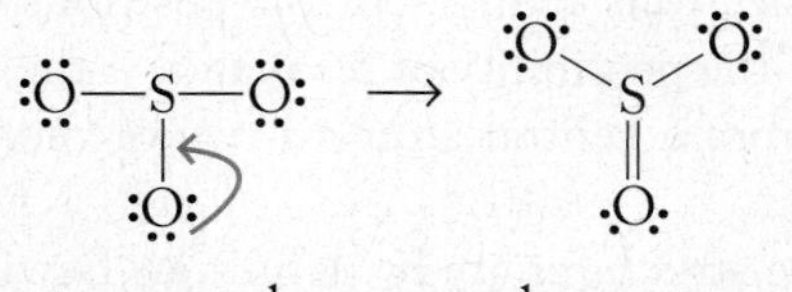

The sulfur atom now has a complete octet, and the Lewis structure is complete.

Because the three oxygen atoms are equivalent in the structure in step 3, we cannot arbitrarily choose one of them and ignore the others in forming a double bond. Therefore, three resonance structures are needed to describe the bonding in SO_3:

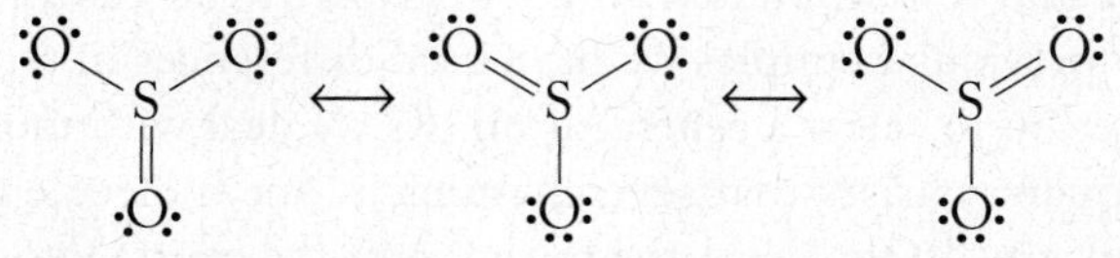

Think about It It makes sense that there are three resonance forms of SO_3 because there are three equivalent O atoms bonded to the central S atom, and

any one of the O atoms could be the one with the double bond. As with O_3, none of the three resonance forms represents the actual bonding pattern in SO_3. Rather, the best representation is the average of all three:

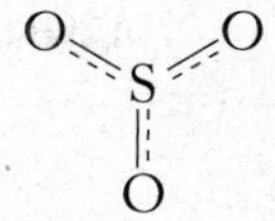

$$O\diagdown_{S}\diagup O$$
$$\underset{O}{\overset{|}{|}}$$

where each combination of a solid and dashed line represents 1 and 1/3 bonds between the central S atom and each O atom.

Practice Exercise Draw all possible resonance structures for sulfur dioxide.

We can also draw resonance structures for polyatomic ions. In doing so, we need to account for the charge on each ion, which means adding the appropriate number of valence electrons to a polyatomic anion and subtracting the appropriate number from a polyatomic cation.

SAMPLE EXERCISE 8.7 Drawing Resonance Structures of a Polyatomic Ion

Draw all the resonance structures for the nitrate ion, NO_3^-.

Collect and Organize We are to draw the resonance structures for the NO_3^- ion, which contains one nitrogen atom and three oxygen atoms. The charge of the ion is $1-$.

Analyze Nitrogen is a group 15 element and has 5 valence electrons per atom and a bonding capacity of 3. Oxygen is in group 16 and so has 6 valence electrons and a bonding capacity of 2. The $1-$ charge means there is an additional valence electron in the ion.

Solve

Step 1: The number of valence electrons is

Element:	N + 3 O
Valence electrons:	$5 + (3 \times 6) = 23$
Additional electron due to the $1-$ charge:	$\underline{1}$
Total valence electrons:	24

Step 2: The nitrogen atom has the higher bonding capacity, so N is the central atom in a NO_3^- ion. Connecting N with single bonds to the three O atoms, we have

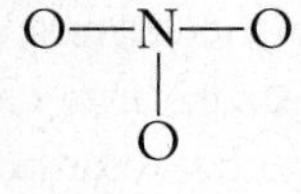

$$O—N—O$$
$$\overset{|}{O}$$

Step 3: Each O atom needs three lone pairs of electrons to complete its octet:

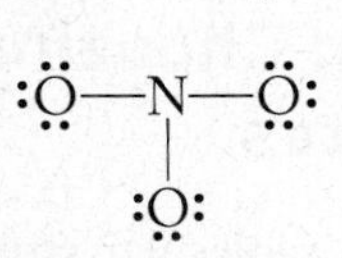

$$:\ddot{O}—N—\ddot{O}:$$
$$\overset{|}{:\underset{..}{O}:}$$

Step 4: There are 24 valence electrons in this structure, which matches the number determined in step 1.

Step 5: The central N atom has only 6 electrons. To provide it with the 2 more it needs to complete its octet, we convert a lone pair on one of the oxygen atoms into a bonding pair:

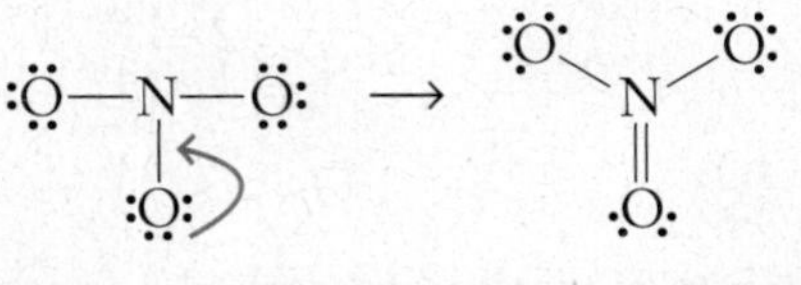

$$:\ddot{O}-N-\ddot{O}: \quad \longrightarrow \quad :\ddot{O}\diagdown N \diagup \ddot{O}:$$
$$:\ddot{O}: \cdot\ddot{O}\cdot$$

The nitrogen atom now has a complete octet. Adding brackets and the ionic charge, we have a complete Lewis structure:

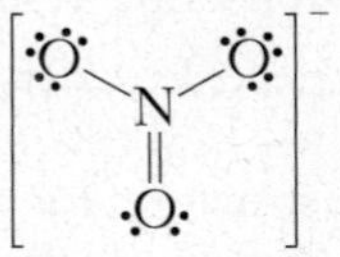

$$\left[\begin{array}{c} :\ddot{O} \diagdown N \diagup \ddot{O}: \\ \cdot\ddot{O}\cdot \end{array} \right]^{-}$$

Using the O atom to the left or right of the central N atom to form the double bond creates two additional resonance forms, or three in all:

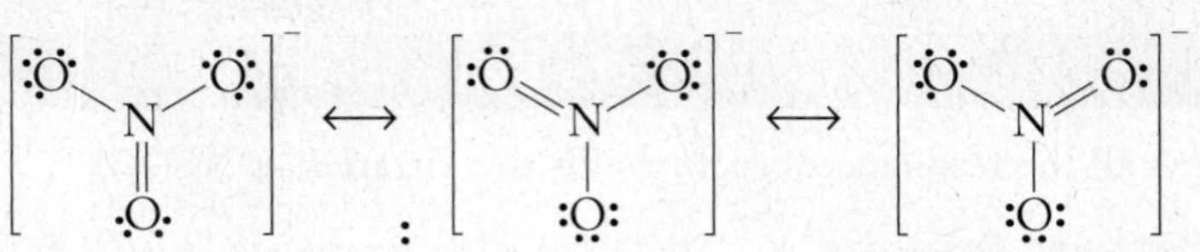

$$\left[\begin{array}{c} :\ddot{O} \diagdown N \diagup \ddot{O}: \\ \cdot\ddot{O}\cdot \end{array} \right]^{-} \leftrightarrow \left[\begin{array}{c} :\ddot{O} \diagdown N \diagup \ddot{O}: \\ :\ddot{O}: \end{array} \right]^{-} \leftrightarrow \left[\begin{array}{c} :\ddot{O} \diagdown N \diagup \ddot{O}: \\ :\ddot{O}: \end{array} \right]^{-}$$

Think about It It makes sense that there are three resonance forms of the NO_3^- ion because there are three equivalent O atoms bonded to the central N atom, and any one of the O atoms could be the one with the double bond. The additional electron from the negative charge on the ion means that it has the same number of valence electrons as SO_3 (Sample Exercise 8.6), which has the same number of resonance structures. Nitrogen in neutral molecules typically has a bonding capacity of 3, but in these resonance structures it makes four bonds.

Practice Exercise Draw all the resonance forms of the azide ion, N_3^-, and the nitronium ion, NO_2^+.

■ ....

FIGURE 8.12 The molecular structure of benzene is an average of the two equivalent structures at the top. The average is frequently represented by a circle inside the hexagonal ring indicating completely uniform distribution of the electrons in the bonds around the ring.

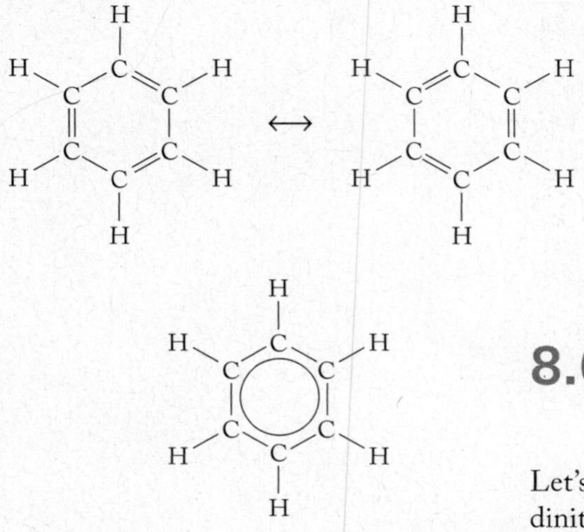

Resonance also occurs in organic molecules that have alternating single and double bonds. Molecules of benzene (C_6H_6), for example, contain six-membered rings of carbon atoms with alternating single and double bonds (Figure 8.12). When we fix the atoms in a molecule of benzene in place, there are two equivalent ways to draw the single and double bonds. To depict their equivalency, chemists frequently draw benzene molecules with circles in the centers, as shown in Figure 8.12. This symbol emphasizes that the six carbon–carbon bonds in the ring are all identical and intermediate in character between single and double bonds.

8.6 Formal Charge: Choosing among Lewis Structures

Let's turn our attention to the molecular structure of another atmospheric gas, dinitrogen monoxide (N_2O), also known as nitrous oxide or, more commonly, laughing gas. Its common name is linked to the fact that people who inhale high concentrations of N_2O usually laugh spontaneously. Nitrous oxide was an early anesthetic used in dentistry because breathing it has a narcotic effect.

Nitrous oxide concentrations in the lower atmosphere range between 0.1 and 1.0 ppm. It is produced naturally by bacterial action in soil and through human activities such as agriculture and sewage treatment. It is occasionally in the news because its concentration in the troposphere (the atmosphere at ground level) has been increasing. Along with CO_2 and other gases, it may be contributing to climate cange.

Let's draw the Lewis structure of nitrous oxide. First we count the number of valence electrons: 5 each from the two nitrogen atoms and 6 from the oxygen atom for a total of $(2 \times 5 + 6) = 16$. The central atom is a nitrogen atom because N has a higher bonding capacity than O. Connecting the atoms with single bonds we have

$$N—N—O$$

Completing the octets of the noncentral atoms gives us a structure with 16 valence electrons, which is the number determined in step 1:

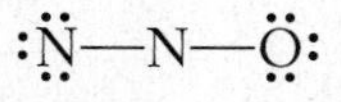

$$:\ddot{N}—N—\ddot{O}:$$

However, there are only two bonds and thus only 4 valence electrons on the central N atom. To give it 4 more electrons, we need to convert lone pairs on the surrounding atoms to bonding pairs. Which lone pairs do we choose? We could use two lone pairs from the N atom on the left to form a $N{\equiv}N$ triple bond:

$$:N{\equiv}N—\ddot{O}:$$

we could use two lone pairs on the O atom to make a $N{\equiv}O$ triple bond:

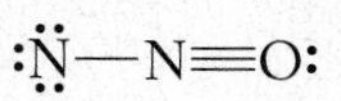

$$:\ddot{N}—N{\equiv}O:$$

or we could use one pair from each terminal atom to make two double bonds:

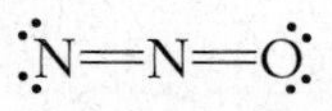

$$\ddot{N}{=}N{=}\ddot{O}$$

Which of these resonance structures is best? We have seen that in some sets of resonance structures, such as those of O_3, SO_3, and $NO_3{}^-$, all the structures are equivalent, and so no one of them is any more important than another in giving us a sense of the actual bonding in the molecule. This is not the case with the nonequivalent resonance forms of N_2O. To help us decide which resonance form in a nonequivalent set is the most important and comes the closest to representing the actual bonding pattern in a molecule, we make use of the concept of formal charge.

A **formal charge (FC)** is not a real charge but rather is a measure of the number of electrons *formally assigned* to an atom in a molecular structure as distinct from the number of electrons in the free atom. We follow a series of steps to calculate the number of electrons formally assigned to each atom in each resonance structure. Once we have determined the formal charges, we then use these two criteria to select the preferred structure:

1. The preferred structure is the one with formal charges of zero; if no such structure can be drawn, the preferred structure is the one with the most formal charges equal to zero or closest to zero.
2. Any negative formal charges should be on the atom(s) of the more/most electronegative element.

formal charge (FC) value calculated for an atom in a molecule or polyatomic ion by determining the difference between the number of valence electrons in the free atom and the sum of lone-pair electrons plus half of the electrons in the atom's bonding pairs.

Calculating Formal Charge of an Atom in a Resonance Structure

Step 1: Determine the number of valence electrons in the free atom.

Step 2: Count the number of lone-pair electrons on the atom in the molecular structure.

Step 3: Count the number of electrons in bonding pairs to the atom and divide that number by 2.

Step 4: Sum the results of steps 2 and 3 and subtract that sum from the number determined in step 1.

Summarizing these steps in the form of an equation we have

$$\text{FC} = \begin{pmatrix} \text{number of} \\ \text{valence e}^- \end{pmatrix} - \left[\begin{matrix} \text{number of} \\ \text{unshared e}^- \end{matrix} + \frac{1}{2} \begin{pmatrix} \text{number of e}^- \\ \text{in bonding pairs} \end{pmatrix} \right] \quad (8.1)$$

The calculation of formal charge assumes that each atom has exclusive "title" to the electrons in its lone pairs and shares its bonding electrons equally with the atom at the other end of the bond. We can confirm that the three resonance forms of N_2O are not equivalent by calculating the formal charges in each structure.

In the table below we have colored the lone pairs of electrons red and the shared pairs green to make it easier to track the quantities of these electrons in the formal charge calculation. The numbers of valence electrons in the free atoms are shown in blue.

Formal Charge Calculations for the Resonance Structures of N_2O										
Step		:N≡N—Ö:			N̈=N=Ö			:N̈—N≡O:		
1	Number of valence electrons	5	5	6	5	5	6	5	5	6
2	Number of lone pair electrons	2	0	6	4	0	4	6	0	2
3	Number of shared electrons	6	8	2	4	8	4	2	8	6
4	FC = [valence − (lone pair + $\frac{1}{2}$ shared)]	0	+1	−1	−1	+1	0	−2	+1	+1

To illustrate one of the formal charge calculations in the table, consider the N atom at the left end of the left resonance structure. The Lewis dot symbol of nitrogen reminds us that free nitrogen atoms have 5 valence electrons. In this structure, this N atom has 2 electrons in a lone pair and 6 electrons in three shared (bonding) pairs. Using Equation 8.1 to calculate the formal charge on this N atom,

$$\text{FC} = 5 - [2 + \tfrac{1}{2}(6)] = 0$$

which is the first value in the bottom row of the table. The results of similar formal carge calculations for all the other atoms in the three resonance structures complete the row. Note that the sum of the formal charges on the three atoms in each structure is zero, as it should be for a neutral molecule. When we do an analysis of the formal charges of atoms in a polyatomic ion, then the formal charges on its atoms must add up to the charge on the ion.

Now we must apply the two criteria for selecting the preferred resonance structure of N_2O. The first thing to note about these sets of formal charges is that in none of them are all three formal charges zero, meaning that the first part of criterion 1 is not met. The next step is to identify which structure has the most formal charge values that are the closest to zero, such as -1 or $+1$. On this count we have a tie between the structure on the left $(0, +1, -1)$ and the one in the middle $(-1, +1, 0)$.

To break the tie, we invoke the second criterion and answer the question, "In which structure is the negative formal charge on the more electronegative atom?" The answer is the structure on the left, which has an oxygen atom with a formal charge of -1. We conclude that this structure is the best of the three in representing the actual bonding in a molecule of N_2O.

In reality, it is known from experimental measurements that the middle structure also contributes to the bonding in N_2O. For example, the length of the bond between the two nitrogen atoms is between the length of a $N{=}N$ bond and the length of a $N{\equiv}N$ bond, and the nitrogen–oxygen bond is a little shorter than a typical $N{-}O$ single bond.

CONCEPT TEST

What is the formal charge on a sulfur atom that has three lone pairs of electrons and one bonding pair?

SAMPLE EXERCISE 8.8   **Selecting Resonance Structures Based on Formal Charges**

Which of these resonance forms best describes the actual bonding in a molecule of CO_2?

$$:\!\ddot{O}\!-\!C\!\equiv\!O\!: \quad \longleftrightarrow \quad :\!\ddot{O}\!=\!C\!=\!\ddot{O}\!: \quad \longleftrightarrow \quad :O\!\equiv\!C\!-\!\ddot{O}\!:$$

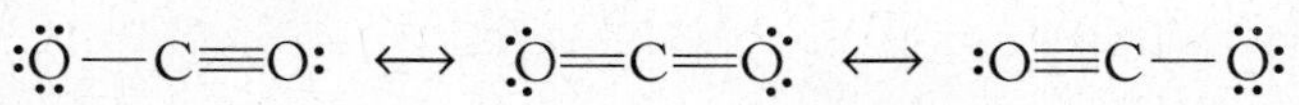

Collect and Organize We are given three resonance forms for CO_2. Formal charges can be used to select the most representative structure.

Analyze The preferred structure is one in which the formal charges are closest to zero and any negative formal charges are on the more electronegative atom. In this case, oxygen is a group 16 element and is more electronegative than carbon, a group 14 element. Each free carbon atom has 4 valence electrons, and free oxygen atoms have 6 valence electrons each.

Solve We use Equation 8.1 to find the formal charge on each atom. We illustrate these results in a table, applying the same color coding scheme used for N_2O structures on page 388:

Formal Charge Calculations for the Resonance Structures of CO_2

Step		$:\ddot{O}-C\equiv O:$			$\ddot{O}=C=\ddot{O}$			$:O\equiv C-\ddot{O}:$		
1	Number of valence electrons	6	4	6	6	4	6	6	4	6
2	Number of lone pair electrons	6	0	2	4	0	4	2	0	6
3	Number of shared electrons	2	8	6	4	8	4	6	8	2
4	FC = [valence − (lone pair + $\frac{1}{2}$ shared)]	−1	0	+1	0	0	0	+1	0	−1

The formal charges are all zero on the atoms in the middle resonance form with the two double bonds. Therefore this structure best represents the actual bonding in a molecule of CO_2.

Think about It Notice that the sum of the formal charges is zero in all three resonance forms. Valid Lewis structures of all neutral molecules have net formal charges of zero.

Practice Exercise Which resonance forms of the azide ion, N_3^-, and the nitronium ion, NO_2^+, contribute the most to the actual bonding in each ion?

Let's take a final look at the resonance structures for N_2O. Our purpose here is to examine the link between the formal charge on an atom in a resonance structure and the bonding capacity of that atom. The Lewis symbol of nitrogen, which has 3 unpaired electrons, tells us that a nitrogen atom can complete its octet by forming three bonds. The 2 unpaired electrons in the Lewis symbol of oxygen tell us that the bonding capacity of an oxygen atom is 2. In the N_2O structures on page 388, the nitrogen atom with three bonds has a formal charge of zero (left N in first structure), and the oxygen atom with 2 bonds has a formal charge of zero (O in the middle structure). As a general rule—and assuming the octet rule is obeyed—atoms have zero formal charges in resonance structures in which the numbers of bonds they form match their bonding capacities. If the number of bonds an atom forms is one more than its bonding capacity, such as an oxygen atom with three bonds (O in third structure in the table), the formal charge is $+1$. If the number of bonds it forms is 1 fewer than the bonding capacity, such as an oxygen atom with one bond (O in first structure in the table), the formal charge is -1.

8.7 Exceptions to the Octet Rule

Earth's atmosphere contains trace concentrations of several compounds that illustrate the limitations of the octet rule. They include two nitrogen oxides, NO and NO_2, that contribute to photochemical smog formation in urban areas (Figure 8.13). Each has an odd number of valence electrons per molecule, which means that at least one atom in each molecule cannot have a complete octet. Another important compound is sulfur hexafluoride (SF_6), which may be the most potent greenhouse gas present in the atmosphere (Figure 8.14). Each of its molecules contains six sulfur–fluorine covalent bonds, which means each S atom is surrounded by 12 valence electrons, not 8.

Odd-Electron Molecules

Nitric oxide (NO) is produced when high temperatures in vehicle engines lead to the reaction

$$N_2(g) + O_2(g) \rightarrow 2\,NO(g) \qquad (8.2)$$
$$\text{Nitric oxide}$$

The nitric oxide then reacts with atmospheric oxygen to produce nitrogen dioxide:

$$2\,NO(g) + O_2(g) \rightarrow 2\,NO_2(g) \qquad (8.3)$$
$$\text{Nitric oxide} \qquad\qquad \text{Nitrogen dioxide}$$

Nitric oxide is highly reactive because it is an odd-electron molecule. To understand the implications of this, let's draw its Lewis structure. Nitric oxide has 11 valence

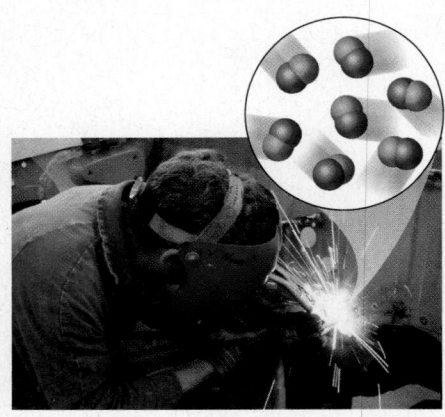

FIGURE 8.13 The high temperatures of arc welding produce significant concentrations of nitric oxide (NO) as a result of the highly endothermic reaction $N_2 + O_2 \rightarrow 2\,NO$. The U.S. Environmental Protection Agency limit on NO concentrations in the air is 25 ppm, or 0.0025% by volume.

electrons: nitrogen contributes 5 and oxygen 6. There is no central atom so we start with a single bond between N and O and then complete the octet around O, which is the more electronegative element:

$$N—\ddot{\ddot{O}}:$$

We then place the remaining 3 electrons around the N atom:

$$\dot{N}—\ddot{\ddot{O}}:$$

This leaves the N atom short of valence electrons. We can increase its number by converting a lone pair on the O atom into a bonding pair:

$$.\dot{N}=\ddot{O}:$$

This change has the added advantage of creating a double-bonded O atom, which gives it a formal charge of zero. The formal charge on the N atom is also zero. The only problem with the structure is that N does not have an octet: it has only 7 electrons. Because nitrogen is less electronegative than oxygen, it is reasonable that we short-change it when there are not enough electrons to complete the octets of both atoms. The fact that NO exists indicates that the requirement of complete octets is not always met. When that happens, the most representative Lewis structures are those that come *as close as possible* to producing zero formal charges and complete octets.

Compounds that have odd numbers of valence electrons are called **free radicals**. They are typically very reactive species because it is often energetically favorable for them to acquire an electron from another molecule or ion. This characteristic makes them excellent oxidizing agents and is responsible for the damage they may cause to materials or living tissue with which they come in contact.

FIGURE 8.14 Electrical transformers use sulfur hexafluoride as an insulator because it is thermally stable and does not react with water. This compound, however, is classified as a possible contributor to global warming because it can leak from transformers and other electrical equipment and enter the atmosphere, where, because it is so unreactive, it may remain for thousands of years. It is also an exceptionally effective absorber of infrared radiation. The Intergovernmental Panel on Climate Change has identified SF_6 as the most potent greenhouse gas it has ever evaluated, with over 10^4 times the global warming potential of CO_2.

> **SAMPLE EXERCISE 8.9 Drawing Lewis Structures of Odd-Electron Molecules**

Draw the resonance structures of nitrogen dioxide (NO_2) and assign formal charges to the atoms.

Collect and Organize We need to draw the resonance forms of NO_2 and use formal charges to determine which structure most nearly reflects the actual bonding in the molecule. Each molecule contains one atom of nitrogen (bonding capacity 3) and two atoms of oxygen (bonding capacity 2).

Analyze Nitrogen dioxide is an odd-electron molecule, so we anticipate that one of the atoms will not have a complete octet. We analyze the resonance structures by assigning formal charges to select the one most representative of the actual bonding in the molecule.

Solve The number of valence electrons is

Element:	N	+	2 O
Valence electrons:	5	+	$(2 \times 6) = 17$

Nitrogen has the greater bonding capacity and so is the central atom:

$$O—N—O$$

Completing the octets on the O atoms gives

$$:\ddot{\ddot{O}}—N—\ddot{\ddot{O}}:$$

---

free radical an odd-electron molecule with an unpaired electron in its Lewis structure.

There are 16 valence electrons in this structure, but we need 17 to match the number available in the molecule. We add 1 more electron to the N atom:

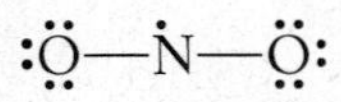

$$:\ddot{O}—\dot{N}—\ddot{O}:$$

There are only 5 valence electrons around the N atom, fewer than the 8 we need. We can increase this number by converting a lone pair on one of the O atoms to a bonding pair, giving the formal charges shown in red:

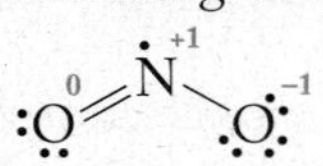

$$\overset{0}{:}O\!\!=\!\!\overset{\overset{\bullet}{+1}}{N}\diagdown\overset{-1}{\ddot{O}:}$$

An equivalent resonance form can be drawn with the double bond on the right side:

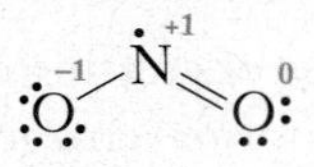

$$\overset{-1}{\ddot{O}}\diagdown\overset{\overset{\bullet}{+1}}{N}\!\!=\!\!\overset{0}{O}:$$

Think about It The two Lewis structures are equivalent because the two O atoms in each structure are equivalent. The structures do not satisfy the octet rule, but the formal charges of the atoms are close to zero, and the negative formal charge is on the atom of the more electronegative element. Both O atoms have complete octets, leaving the less electronegative N atom one electron short in this odd-electron molecule.

Practice Exercise Nitrogen trioxide (NO_3) may form in polluted air when NO_2 reacts with O_3. Draw its Lewis structure(s).

Atoms with More than an Octet

Some atoms can have more than 8 valence electrons in their Lewis structures. Consider, for example, the Lewis structure of SF_6. Its valence electron inventory (step 1) is

Element:	S	+	6 F
Valence electrons:	6	+	$(6 \times 7) = 48$

Sulfur has a greater bonding capacity (2) than fluorine (1), so S is the central atom (step 2). However, connecting six fluorine atoms to the sulfur atom means that sulfur's bonding capacity is significantly exceeded:

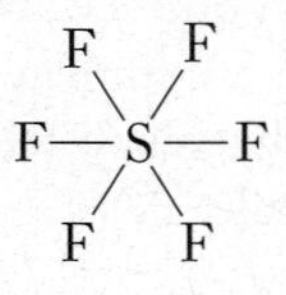

$$\begin{array}{ccc} F & & F \\ & \diagdown & \diagup \\ F—&S&—F \\ & \diagup & \diagdown \\ F & & F \end{array}$$

Completing the octets on the fluorine atoms (step 3),

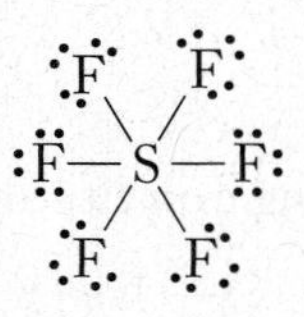

$$\begin{array}{ccc} :\ddot{F} & & \ddot{F}: \\ & \diagdown & \diagup \\ :\ddot{F}—&S&—\ddot{F}: \\ & \diagup & \diagdown \\ :\ddot{F} & & \ddot{F}: \end{array}$$

gives us a structure with 48 valence electrons (step 4), a match with the number determined for the molecule. This structure is correct even though it breaks the octet rule. It tells us that, in order to accommodate six fluorine atoms about a central sulfur atom, the sulfur atom is able to *expand its valence shell* to accommodate more than 8 electrons. Six S—F bonds and 12 valence electrons around S leave the SF_6 molecule with zero formal charge on each atom:

$$\text{Formal charge of S} = 6 - \left[0 + \tfrac{1}{2}(12)\right] = 0$$

$$\text{Formal charge of F} = 7 - \left[6 + \tfrac{1}{2}(2)\right] = 0$$

Based on our criteria for judging Lewis structures, this is a good one, but how can a sulfur atom have more than 8 valence electrons?

The answer has to do with the size of the atom. Atoms of elements with $Z > 12$ have the ability to expand their valence shells by using empty d orbitals. For example, the electron configuration of S is $[Ne]3s^2 3p^4$. From Section 7.6, we know that for principal quantum numbers $n \geq 3$, one possible value for the angular quantum number ℓ is $\ell = 2$, which means d orbitals are available in sulfur's valence shell (see Table 7.1). Thus a S atom (third row; $Z = 16$) can use its $3d$ orbitals to form covalent bonds once its $3p$ orbitals are full.

The fact that some atoms have the ability to expand their valence shell does not mean these atoms *always* do so. Instead, they tend to do so in two situations:

▶❙❙ **CHEMTOUR** Expanded Valence Shells

1. When they form compounds with strongly electronegative elements, particularly F, O, and Cl.
2. When an expanded shell results in smaller formal charges on the atoms in a molecule.

To illustrate a situation in which expanding a valence shell results in smaller formal charges, let's consider the sulfate ion, SO_4^{2-}. When high-sulfur coal is burned, SO_2 is released into the atmosphere and reacts with atmospheric oxygen to form sulfur trioxide,

$$2\,SO_2(g) + O_2(g) \rightarrow 2\,SO_3(g)$$

which reacts with water vapor to form sulfuric acid:

$$SO_3(g) + H_2O(g) \rightarrow H_2SO_4(\ell)$$

Sulfuric acid, a principal component of acidic precipitation in eastern North America and in Europe, dissociates in aqueous solutions:

$$H_2SO_4(aq) \rightarrow 2\,H^+(aq) + SO_4^{2-}(aq)$$

In the Lewis structure of the sulfate ion a central S atom is bonded to four O atoms. Applying the method for drawing Lewis structures and assigning formal charges, we get

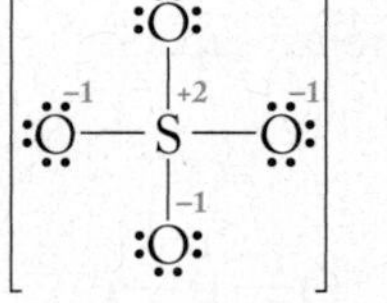

The sum of the formal charges on atoms in the ion is $1(+2) + 4(-1) = -2$. This calculation yields the correct ionic charge, but remember that the goal is to minimize formal charges in Lewis structures, which we can do by expanding the valence shell of sulfur:

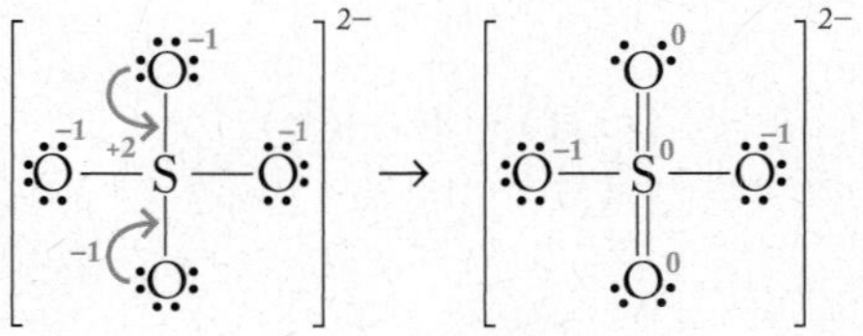

Note that each oxygen atom still has a complete octet, but now the sulfur has expanded its valence shell to accommodate 12 electrons. In this way, the formal charges change from $+2$ on sulfur and -1 on each of the four oxygen atoms to 0 on sulfur, 0 on two of the oxygen atoms, and -1 on the other two. These values sum to -2, which is the value required to give the structure its overall $2-$ charge.

We can draw the two double bonds at any location around the sulfur atom in the Lewis structure for the sulfate ion. Consequently, this structure has several equivalent resonance forms (not shown).

If we now wanted to draw the Lewis structure for sulfuric acid, we could bond two hydrogen ions to the two negative oxygen atoms:

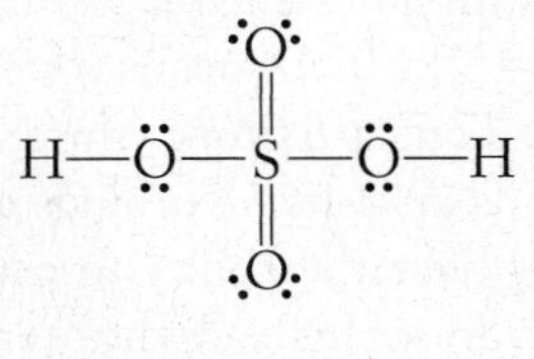

Each hydrogen atom has achieved a duet of electrons, each oxygen atom has an octet, and every atom has a formal charge of zero.

SAMPLE EXERCISE 8.10   Drawing Lewis Structures of Ions with an Expanded Valence Shell

Draw the Lewis structure for the phosphate ion (PO_4^{3-}) that minimizes the formal charges on its atoms.

Collect and Organize Each ion contains one atom of phosphorus and four atoms of oxygen and has an overall charge of $3-$.

Analyze Phosphorus and oxygen are in groups 15 and 16 and have bonding capacities of 3 and 2, respectively. Phosphorus is in row 3 ($Z = 15$), so we can also expand its octet if we need to.

Solve The number of valence electrons is

Element:	P + 4 O
Valence electrons:	$5 + (4 \times 6) = 29$
Additional electrons due to the 3− charge:	3
Total valence electrons:	32

Phosphorus has the greater bonding capacity (3) and so is the central atom:

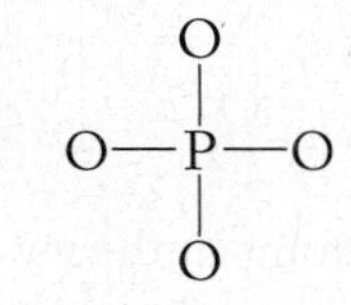

Each O atom needs three lone pairs of electrons to complete its octet:

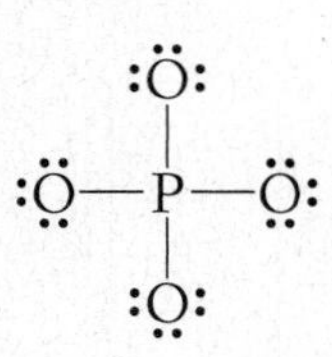

There are 32 valence electrons in this structure, which matches the number determined for the ion. Therefore, it is a complete Lewis structure of a poly-atomic ion once we add the brackets and electrical charge:

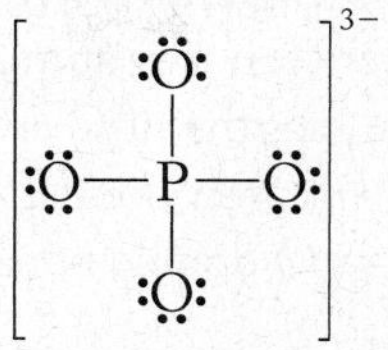

$$\left[\begin{array}{c} :\ddot{O}: \\ | \\ :\ddot{O}-P-\ddot{O}: \\ | \\ :\ddot{O}: \end{array} \right]^{3-}$$

Each O has a single bond and a formal charge of −1; the four bonds around the P atom are one more than its bonding capacity, so its formal charge is +1. The sum of the formal charges, $[+1 + 4(-1)]$ matches the charge on the ion, 3−.

We can reduce the formal charge on P by increasing the number of bonds to it, and we can do that by converting a lone pair on one of the O atoms into a bonding pair:

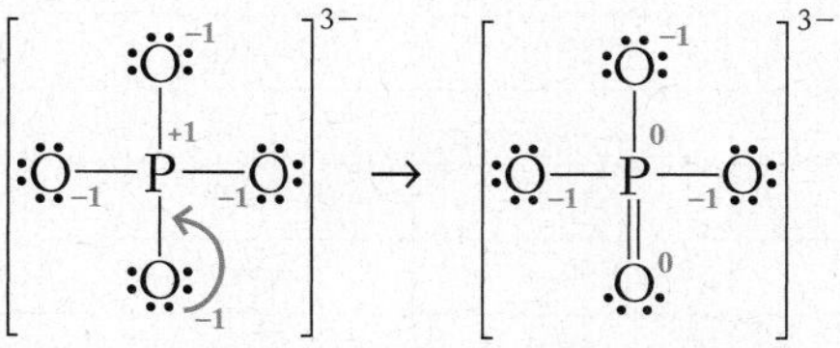

$$\left[\begin{array}{c} :\ddot{O}:^{-1} \\ |^{+1} \\ :\ddot{O}-P-\ddot{O}: \\ _{-1} \quad |_{-1} \\ :\ddot{O}: \\ _{-1} \end{array} \right]^{3-} \rightarrow \left[\begin{array}{c} :\ddot{O}:^{-1} \\ |^{0} \\ :\ddot{O}-P-\ddot{O}: \\ _{-1} \quad ||_{-1} \\ :\ddot{O}: \\ _{0} \end{array} \right]^{3-}$$

At the same time, we change a single-bonded O atom into a double-bonded O atom and thereby make its formal charge zero. Therefore, the structure on the right, in which the P atom has an expanded valence shell, is the best Lewis structure we can draw for the phosphate ion. Other resonance structures can be drawn.

Think about It The phosphorus atom in the final structure has 10 valence electrons. This exception to the octet rule is allowed because phosphorus is in the third row and has empty *d* orbitals available with which to expand its valence shell.

Practice Exercise Draw the resonance structures of the selenite ion (SeO_3^{2-}) that minimize the formal charges on the atoms.

CONCEPT TEST

The noble gases Xe and Kr form some compounds with fluorine, for example, XeF_2, but no compounds of He and Ne have yet been made. Suggest a reason for this difference in chemical reactivity.

8.8 The Lengths and Strengths of Covalent Bonds

In Section 8.5 we discussed the equivalent resonance structures of ozone, and we noted that the true nature of the two oxygen–oxygen bonds in O_3 is reflected in their equal bond length, 128 pm, which is between the length of a typical O=O double bond (121 pm) and an O—O single bond (148 pm) as shown in Figure 8.15. We used these results to conclude that there are effectively 1.5 bonds between the atoms in a molecule of O_3. We explore this use of bond length, and also bond strength, to rationalize and validate molecular structures in this section. We also use bond strengths to esti-mate the enthalpy changes that occur in chemical reactions.

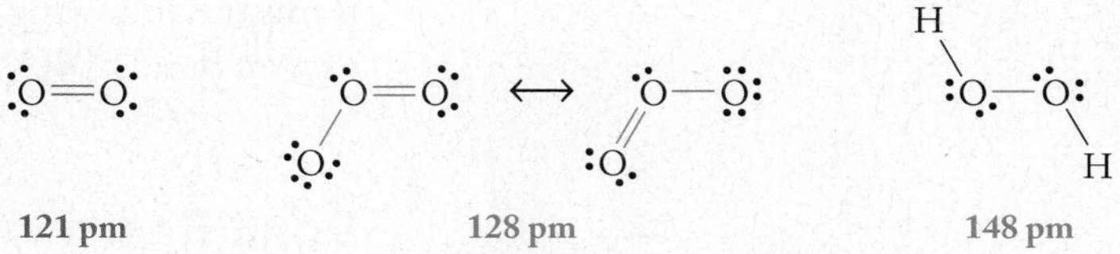

FIGURE 8.15 Resonance influences bond length. Because of resonance, the lengths of the two bonds in ozone are identical and between the length of the O=O double bond in O_2 and the O—O single bond in H_2O_2.

121 pm   128 pm   148 pm

bond order the number of bonds between atoms: 1 for a single bond, 2 for a double bond, and 3 for a triple bond.

Bond Length

The length of the bond between any two atoms depends on the identity of the atoms and on whether the bond is single, double, or triple (Table 8.2). As **bond order**, which is the number of bonds between two atoms, increases, bond length decreases, as we can see by comparing the lengths of C—C, C=C, and C≡C bonds in Table 8.2. Similarly, for carbon–oxygen bonds, the C≡O triple bond in carbon monoxide is shorter than the C=O double bond in carbon dioxide (Figure 8.16a).

TABLE 8.2	Selected Average Covalent Bond Lengths and Bond Energies					
Bond	**Bond Length (pm)**	**Bond Energy (kJ/mol)**		**Bond**	**Bond Length (pm)**	**Bond Energy (kJ/mol)**
C—C	154	348		N≡O	106	678
C=C	134	614		O—O	148	146
C≡C	120	839		O=O	121	495
C—N	143	293		O—H	96	463
C=N	138	615		S—O	151	265
C≡N	116	891		S=O	143	523
C—O	143	358		S—S	204	266
C=O	123	743[a]		S—H	134	347
C≡O	113	1072		H—H	75	436
C—H	110	413		H—F	92	567
C—F	133	485		H—Cl	127	431
C—Cl	177	328		H—Br	141	366
N—H	104	388		H—I	161	299
N—N	147	163		F—F	143	155
N=N	124	418		Cl—Cl	200	243
N≡N	110	941		Br—Br	228	193
N—O	136	201		I—I	266	151
N=O	122	607				

[a] The bond energy of the C=O bond in CO_2 is 799 kJ/mol.

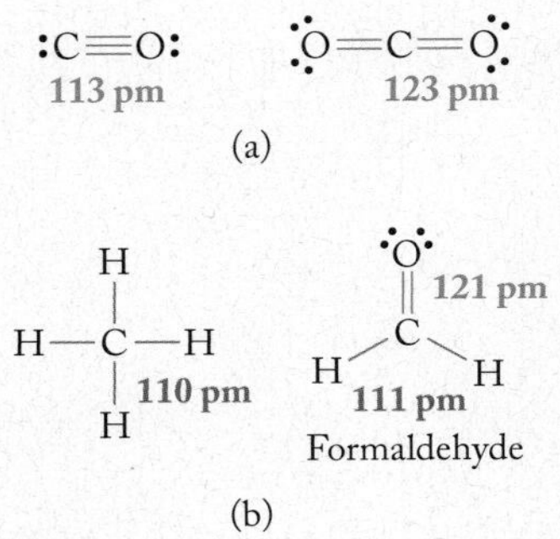

:C≡O: ..Ö=C=Ö..
113 pm   123 pm

(a)

H
|
H—C—H   (formaldehyde structure with O, 121 pm, C, 111 pm, H)
|
H
110 pm

Formaldehyde

(b)

FIGURE 8.16 (a) Bond length depends on the identity of the two atoms forming the bond and on bond order. (b) A bond between the same two atoms can have different lengths in different molecules. Compare the C—H length in the formaldehyde molecule with the C—H length in CH_4. Also compare the C=O length in formaldehyde with the C=O length in CO_2.

Measurements of the bond lengths in many molecules indicate that there are small differences in bond lengths for any given covalent bond. For example, the C—H and C=O bond lengths in formaldehyde (Figure 8.16b) are close to but not exactly the same as the C—H bond length in CH_4 and the C=O bond length in CO_2.

CONCEPT TEST

Rank the following molecules in order of decreasing lengths of their nitrogen–oxygen bonds: NO, NO_2, N_2O.

Bond Energies

The energy changes associated with chemical reactions depend on how much energy is required to break the bonds in the reactants and how much is released as the atoms recombine to form products. For example, in the methane combustion reaction

$$CH_4(g) + 2\,O_2(g) \rightarrow CO_2(g) + 2\,H_2O(g)$$

the C—H bonds in CH_4 and the O=O bonds in O_2 must be broken before the C=O bonds in CO_2 and the O—H bonds in H_2O can form. Breaking bonds is endothermic (the "energy in" arrow in Figure 8.17), and forming bonds exothermic (the "energy out" arrow). If a chemical reaction is exothermic, as methane combustion is, more energy is released in forming the bonds in molecules of products than is consumed in breaking the bonds in molecules of reactants.

Bond energy, or *bond strength*, is usually expressed in terms of the enthalpy change (ΔH) that occurs when 1 mole of bonds in the gas phase are broken. Bond energies for some common covalent bonds are given in Table 8.2. As we noted in Section 8.1, the quantity of energy needed to break a particular bond is equal in magnitude but opposite in sign to the quantity of energy released when that same bond forms.

Like bond lengths, the bond energies in Table 8.2 are average values because bond energies vary depending on the structure of the rest of the molecule. For example, the bond energy of a C=O bond in carbon dioxide is 799 kJ/mol, but C=O bond energy in formaldehyde is only 743 kJ/mol. Bond energies are always positive quantities because breaking bonds is endothermic.

Another view of the variability in bond energy comes from breaking the C—H bonds in CH_4 in a step-by-step fashion, shown in the table to the right. These results tell us that the chemical environment of a bond affects the energy required to break it: breaking the first C—H bond in methane is easier (requires less energy) than breaking the second but more difficult than breaking the third or fourth. The total energy needed to break all four C—H bonds is 1652 kJ/mol, an average of 413 kJ/mol per bond.

The relationship between bond order and bond energy is also apparent in Table 8.2. At 495 kJ/mol, the bond energy of the O=O double bond is more than three times that of the O—O single bond. This correlation between bond order and bond energy is true for other pairs of atoms: the higher the bond order, the greater the bond energy. The bond energy of the N≡N triple bond (941 kJ/mol) is more than twice the bond energy of the N=N double bond and more than five times that of the N—N bond. The large quantity of energy required to break a mole of N≡N bonds is one of the reasons N_2 participates in so few chemical reactions.

In the combustion of 1 mole of CH_4 (Figure 8.17), 4 moles of C—H bonds and 2 moles of O=O bonds must be broken. The formation of 1 mole of CO_2 and 2 moles of H_2O requires the formation of 2 moles of C=O bonds and 4 moles of O—H bonds. The net change in energy resulting from breaking and forming these bonds can be estimated from average bond energies. We start by taking an inventory of the bond energies involved:

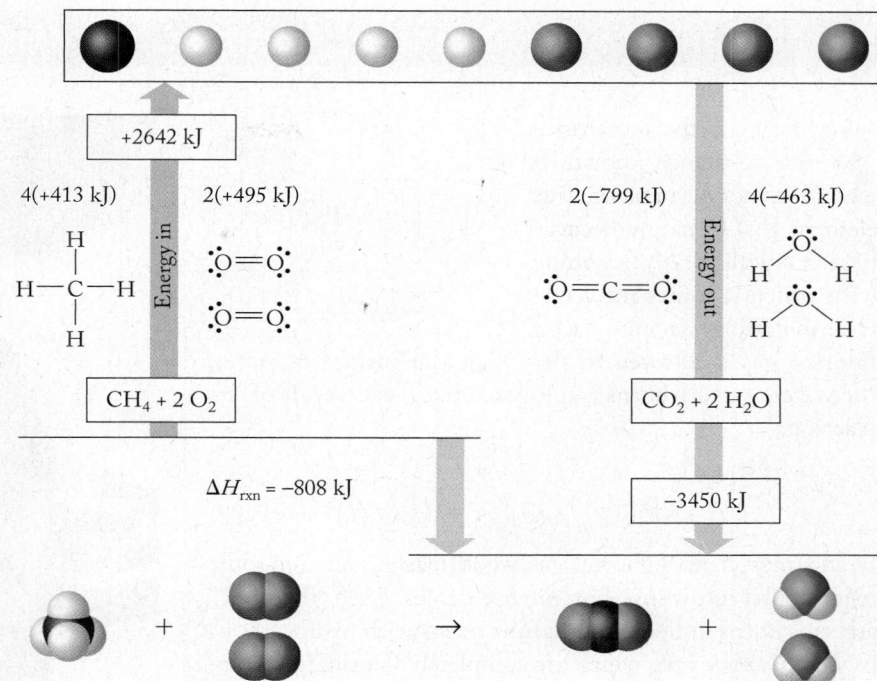

FIGURE 8.17 The combustion of 1 mole of methane requires that 4 moles of C—H bonds and 2 moles of O=O bonds be broken. These processes require an enthalpy change of about +2642 kJ. In the formation of 4 moles of O—H bonds and 2 moles of C=O bonds there is an enthalpy change of about −3450 kJ. The overall reaction is exothermic: 2642 kJ − 3450 kJ = −808 kJ.

Decomposition Step	Energy Needed (kJ/mol)
$CH_4 \rightarrow CH_3 + H$	435
$CH_3 \rightarrow CH_2 + H$	453
$CH_2 \rightarrow CH + H$	425
$CH \rightarrow C + H$	339
Total:	1652
Average:	413

Bond Energies (ΔH) in Methane Combustion			
Bond	Number of Bonds (mol)	Bond Energy (kJ/mol)	Bond Breaking or Forming?
C—H	4	413	Breaking
O=O	2	495	Breaking
O—H	4	463	Forming
C=O	2	799	Forming

Fluorine and Oxygen: Location, Location, Location

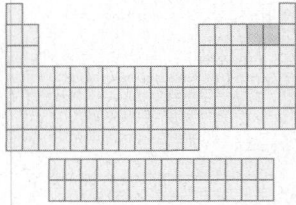

Fluorine is the most reactive substance known. It does not occur as the free element (F_2) in nature because it reacts with nearly anything with which it comes into contact, including Kr and Xe. If fluorine gas is allowed to flow over the surface of water, the water actually burns (rapidly oxidizes) as a result of the reaction:

$$5\,F_2(g) + 5\,H_2O(\ell) \rightarrow$$
$$8\,HF(g) + O_2(g) + H_2O_2(\ell) + OF_2(g)$$

In an atmosphere of fluorine gas, wood, plastic, and even some metals burst into white-hot, intense flames. Even "fireproof" asbestos burns in fluorine. Fluorine reacts with hydrocarbons to yield *fluorocarbons*, which are completely fluorinated molecules, meaning that every hydrogen in the hydrocarbon is replaced by fluorine.

Although fluorine is extremely reactive, fluorocarbons are among the most stable compounds known because of the stability of the carbon–fluorine bond. Fluorocarbons are very rare in nature; most are manufactured. They are generally unreactive toward most chemical reagents, inert to solvents, and nonflammable. The stability of the carbon–fluorine bond has been used to advantage in commercial products. For example, the addition of a single carbon–bound fluorine to certain pharmaceutical compounds improves their potency because it makes the molecules more difficult for the body to degrade.

The anesthetic halothane ($C_2HBrClF_3$) was introduced in 1956 as the first modern inhaled anesthetic. The low toxicity, low flammability, and high stability of halothane enabled it to replace much more hazardous anesthetics, such as ether ($CH_3CH_2OCH_2CH_3$) and chloroform ($CHCl_3$). A synthetic polymer (a very large molecule) made when molecules of tetrafluoroethylene bind together,

$$n \quad \underset{\text{Tetrafluoroethylene}}{\overset{F}{\underset{F}{\underset{\displaystyle|}{C}}} = \overset{F}{\underset{F}{\underset{\displaystyle|}{C}}}} \quad \rightarrow \quad \underset{\text{Teflon}}{\left[\overset{F}{\underset{F}{\overset{|}{\underset{|}{C}}}} - \overset{F}{\underset{F}{\overset{|}{\underset{|}{C}}}} \right]_n} \quad \text{where } n \approx 1000$$

FIGURE 8.18 A frying pan coated with Teflon.

is commonly known as Teflon, a substance so stable to moderate temperatures and so much less reactive than other nonmetallic materials that it is used on no-stick cookware (Figure 8.18), electrical insulation, and shipping containers for fluorine gas.

In contrast to the free element, compounds of the fluoride ion (F^-) are relatively abundant in nature. For decades, small quantities of fluoride salts have been added to municipal water supplies in the United States to prevent cavities in adults and children. Fluorides are also added in small amounts (about 0.15% by mass) to toothpaste for the same reason.

Oxygen is one of the most reactive elements though less reactive than F_2. It is the second most abundant gas in the atmosphere and is essential to many life-forms on Earth. Physical activity for human beings becomes difficult if the partial pressure of oxygen in the air drops from the normal value of 0.21 atm to around 0.12, and breathing air that contains only 0.06 atm oxygen can result in death in 6 to 8 minutes.

Oxygen is also an important commercial bulk chemical, heavily used in the steel industry and in medicine. It is frequently transported and stored as a cryogenic (around $-183°C$) liquid referred to as LOX (*l*iquid *ox*ygen). LOX is extremely reactive and must be handled with great care. Metallic iron burns so vigorously in liquid oxygen that the iron melts from the heat of combustion even though it is surrounded by frigid liquid. Many materials normally considered noncombustible burn at explosive rates in liquid oxygen, and oils, greases, paint, and paper burn immediately on contact. LOX penetrates into porous materials such as wood and asphalt; even when combustion does not occur immediately, the material remains hazardous. Asphalt has been known to burst into flames days after LOX had been spilled on it. Large tanks containing cryogenic oxygen at hospitals or manufacturing sites are always mounted on concrete slabs, not asphalt, as a safety precaution.

The chemistries of oxygen and fluorine intersect in Earth's upper atmosphere. The substances at the intersection are ozone (O_3) and chlorofluorocarbons (CFCs). CFCs were used for years as "clean" fire-extinguishing agents, and as refrigerants and aerosol propellants. CFCs have now been banned because of their role in destroying ozone in the stratosphere.

The quantity of ozone in the upper atmosphere is not large—if compressed by atmospheric pressure at Earth's surface, it would be squashed to a band only a few millimeters thick—but in the upper atmosphere it exists in a layer more than 20 km thick that absorbs UV radiation from the sun and protects life on Earth from these energetic rays. Ozone is produced from O_2 in the upper atmosphere in a process driven by UV radiation:

$$3\,O_2(g) \xrightarrow{\text{UV radiation}} 2\,O_3(g)$$

The conversion of ozone back to O_2 occurs but is slow in the upper atmosphere, so a protective ozone layer remains in place unless disrupted by other agents.

Those disrupting agents are CFCs. When these gases are released into the atmosphere, they persist because of the lack of reactivity that makes them so useful. Wind and air currents carry them to the upper reaches of the atmosphere, where UV radiation provides sufficient energy to break chemical bonds and form chlorine atoms:

$$CF_2Cl_2(g) \xrightarrow{\text{UV radiation}} CF_2Cl(g) + Cl(g)$$

Chlorine atoms are free radicals that react with ozone producing chlorine monoxide (another free radical) and oxygen:

$$Cl(g) + O_3(g) \rightarrow ClO(g) + O_2(g) \qquad (1)$$

Chlorine monoxide reacts with more ozone, generating more oxygen and another chlorine free radical:

$$ClO(g) + O_3(g) \rightarrow Cl(g) + 2\,O_2(g) \qquad (2)$$

If we add Equations 1 and 2, we see that the end result is the conversion of ozone to oxygen:

$$\cancel{Cl(g)} + O_3(g) \rightarrow \cancel{ClO(g)} + O_2(g)$$
$$\cancel{ClO(g)} + O_3(g) \rightarrow \cancel{Cl(g)} + 2\,O_2(g)$$
$$\overline{2\,O_3(g) \rightarrow 3\,O_2(g)}$$

Not only does this process take place much faster than it would without Cl, but chlorine free radicals are not consumed in the reaction. Therefore one chlorine free radical can destroy thousands of ozone molecules before being consumed in a reaction with another free radical in the stratosphere.

The Montreal Protocol, an international agreement now signed by over 190 nations, mandated the phasing out of synthetic substances responsible for depletion of Earth's ozone layer. The search continues for new substances that will be as useful for fire extinguishers and refrigerants as CFCs without damaging the stratospheric ozone layer.

▶II CHEMTOUR Estimating Enthalpy Changes

CONNECTION In Chapter 5 we calculated the difference between the sums of heats of formation of products and of reactants to estimate the heat of reaction.

Next we estimate the enthalpy change of the reaction by calculating the difference between the sum of the bond energies of the reactants and the sum of the bond energies of the products. Equation 8.4 can be used to do this calculation:

$$\Delta H_{rxn} = \sum \Delta H_{\text{bonds breaking}} - \sum \Delta H_{\text{bonds forming}} \qquad (8.4)$$
$$= [(4 \text{ mol} \times 413 \text{ kJ/mol}) + (2 \text{ mol} \times 495 \text{ kJ/mol})]$$
$$- [(4 \text{ mol} \times 463 \text{ kJ/mol}) + (2 \text{ mol} \times 799 \text{ kJ/mol})]$$
$$= -808 \text{ kJ}$$

SAMPLE EXERCISE 8.11 **Estimating Heats of Reaction from Average Bond Energies**

Use the average bond energies in Table 8.2 to estimate ΔH_{rxn} for the reaction in which HCl(g) is formed from $H_2(g)$ and $Cl_2(g)$:

$$\text{H—H} \quad + \quad \ddot{\text{C}}\text{l}—\ddot{\text{C}}\text{l} \rightarrow 2\,\text{H—}\ddot{\text{C}}\text{l}$$

Collect and Organize We are to estimate the value of ΔH_{rxn} of a gas-phase reaction from the bond energies of the reactants and products. Table 8.2 lists the values of these bond energies:

$$\text{H—H} \qquad 436 \text{ kJ/mol}$$
$$\text{Cl—Cl} \qquad 243 \text{ kJ/mol}$$
$$\text{H—Cl} \qquad 431 \text{ kJ/mol}$$

Analyze The reaction between H_2 and Cl_2 requires that 1 mole of H—H bonds and 1 mole of Cl—Cl bonds be broken. Two moles of H—Cl bonds are formed.

Solve Using the above information in Equation 8.4:

$$\Delta H_{rxn} = \sum \Delta H_{\text{bond breaking}} - \sum \Delta H_{\text{bond forming}}$$
$$= [(1 \text{ mol} \times 436 \text{ kJ/mol}) + (1 \text{ mol} \times 243 \text{ kJ/mol})]$$
$$- [(2 \text{ mol} \times 431 \text{ kJ/mol})]$$
$$= -183 \text{ kJ}$$

Think about It An enthalpy change of −183 kJ (note the minus sign) means that the amount of energy consumed as the bonds in the reactants are broken is 183 kJ less than the amount of energy released as the bonds in the products are formed. In other words, the reaction is exothermic. The reaction involves the formation of 2 moles of a gaseous compound from its component elements in their standard states. Therefore the result of this calculation should be close to 2 times the standard heat of formation (ΔH_f°) of HCl. That value (see Appendix 4) is −92.3 kJ/mol. Multiplying by 2 moles we get −184.6 kJ, which is quite close to the estimated value.

Practice Exercise Use average bond energies to calculate ΔH_{rxn} for the reaction of H_2 and N_2 to form ammonia:

$$:\text{N}\equiv\text{N}: \quad + \quad 3\,\text{H—H} \rightarrow 2\,\text{H—}\overset{\displaystyle\cdot\cdot}{\text{N}}\text{—H}$$
$$\underset{\displaystyle\text{H}}{|}$$

CONCEPT TEST ∙∙∙

Suggest a reason, based on bond energies, why oxygen is a much more reactive element than nitrogen.

∙∙

In this chapter we have explored the nature of the covalent bonds that hold together molecules and polyatomic ions, observing that these bonds owe their strength to the presence of pairs of electrons shared between nuclei of atoms. Sharing does not necessarily mean equal sharing, and unequal sharing coupled with bond vibration accounts for the ability of some atmospheric gases to absorb and emit infrared radiation. As a result, these gases function as potent greenhouse gases.

Early in the chapter we noted that moderate concentrations of greenhouse gases are required for climate stability and to make our planet habitable. The escalating concern of many is that Earth's climate is currently being destabilized by too much of a good thing. Policies made by the world's governments in the near future will have a significant impact on the problem of global warming, one way or the other. As an informed member of the world community, you will have the opportunity to influence how those policy decisions are made.

SUMMARY ∙∙∙ ■

Section 8.1 A **chemical bond** results from two atoms sharing electrons (a **covalent bond**) or from two ions being attracted to each other (an **ionic bond**). The atoms in metallic solids pool their electrons in forming **metallic bonds**.

Section 8.2 **Lewis symbols** use dots to represent paired and unpaired electrons in the ground states of atoms. The number of unpaired electrons indicates the number of bonds the element is likely to form, that is, its **bonding capacity**. Chemical stability is achieved when atoms have 8 electrons in their valence s and p orbitals, following the **octet rule**. A **Lewis structure** shows the bonding pattern in molecules; pairs of dots represent **lone pairs** of electrons that do not contribute to bonding. A **single bond** consists of a single pair of electrons shared between two atoms; there are two shared pairs in a **double bond** and three shared pairs in a **triple bond**.

Section 8.3 Unequal electron sharing between atoms of different elements results in **polar covalent bonds**. **Bond polarity** is a measure of how unequally the electrons in covalent bonds are shared. More polarity results from greater differences in the **electronegativities** of the bonded atoms. Electronegativity generally increases with increasing ionization energy.

Section 8.4 Covalent bonds behave more like flexible springs than rigid rods. They can undergo a variety of bond vibrations. The vibrations of polar bonds may create fluctuating electrical fields that allow molecules to absorb infrared electromagnetic radiation. When atmospheric gases absorb IR radiation they contribute to the greenhouse effect.

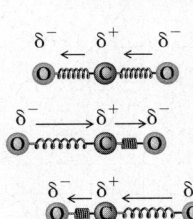

Section 8.5 Two or more equivalent Lewis structures—called **resonance structures**—can be drawn for one molecule or polyatomic ion. The actual bonding pattern in a molecule is an average of equivalent resonance structures.

Section 8.6 The preferred resonance structure of a molecule is one in which the **formal charges** on its atoms are zero or closest to zero and any negative formal charges are on the more electronegative atoms. The formal charge on an atom in a Lewis structure is the difference between the number of valence electrons in the free atom and the sum of the number of electrons in lone pairs and half the number of electrons in bonding pairs on the bonded atom.

Section 8.7 **Free radicals** are reactive molecules that have an odd number of valence electrons. Atoms of elements in the third row of the periodic table with $Z > 12$ and beyond can expand their valence shells by using empty valence-shell d orbitals to accommodate additional electrons.

Section 8.8 **Bond order** is the number of bonding pairs in a covalent bond. **Bond energy** is the enthalpy change, ΔH, required to break 1 mole of a particular covalent bond in the gas phase. As the bond order between two atoms increases, the bond length decreases and the bond energy increases.

PROBLEM-SOLVING SUMMARY

TYPE OF PROBLEM	CONCEPTS AND EQUATIONS	SAMPLE EXERCISES
Drawing Lewis structures for molecules, monatomic ions, and ionic compounds	Connect the atoms with single covalent bonds, distributing the valence electrons to give each noncentral atom 8 valence electrons (except 2 for H); use multiple bonds where necessary to complete the central atom's octet.	8.1–8.4
Comparing bond polarities	Calculate the difference in electronegativity (ΔEN) between the two bonded atoms; if $\Delta EN \geq 2.0$, the bond is considered ionic.	8.5
Drawing resonance structures of molecules and polyatomic ions	Include all possible arrangements of covalent bonds in the molecule if more than one equivalent structure can be drawn.	8.6, 8.7
Selecting resonance structures based on formal charges	Calculate formal charge using $$FC = \begin{pmatrix} \text{number of} \\ \text{valence e}^- \end{pmatrix} - \left[\begin{pmatrix} \text{number of} \\ \text{unshared e}^- \end{pmatrix} + \frac{1}{2} \begin{pmatrix} \text{number of e}^- \\ \text{in bonding pairs} \end{pmatrix} \right] \quad (8.1)$$ Select structures with formal charges closest to zero and with negative formal charges on the most electronegative atoms.	8.8
Drawing Lewis structures of odd-electron molecules	Distribute the valence electrons in the Lewis structure to leave the most electronegative atom(s) with 8 valence electrons and the least electronegative atom with the odd number of electrons.	8.9
Drawing Lewis structures with an expanded valence shell	Distribute the valence electrons in the Lewis structure, allowing atoms of elements in period 3 and beyond to have more than 8 valence electrons if more than four bonds are needed or if the structure with the expanded valence shell results in formal charges closer to zero.	8.10
Estimating heats of reaction from average bond energies	Multiply bond energies by number of bonds and calculate using $$\Delta H_{rxn} = \sum \Delta H_{bonds\ breaking} - \sum \Delta H_{bonds\ forming} \quad (8.4)$$	8.11

VISUAL PROBLEMS

(Answers to boldface end-of-chapter questions and problems are in the back of the book.)

8.1. Which group highlighted in Figure P8.1 contains atoms that have the following? (a) 1 valence electron; (b) 4 valence electrons; (c) 6 valence electrons

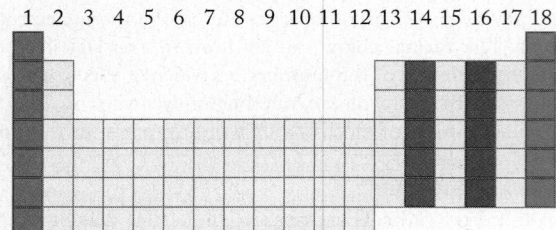

FIGURE P8.1

8.2. Which of the groups highlighted in Figure P8.2 contains atoms with the following? (a) 2 valence electrons; (b) 3 valence electrons; (c) 5 valence electrons

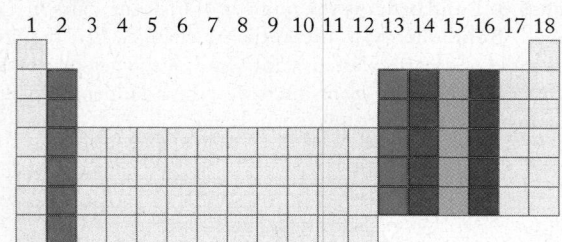

FIGURE P8.2

8.3. Which of the Lewis symbols in Figure P8.3 correctly portrays the most stable ion of magnesium?

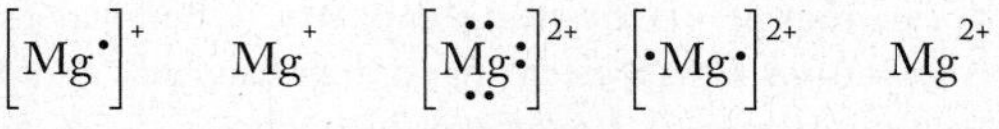

$$\left[\text{Mg}\cdot \right]^+ \qquad \text{Mg}^+ \qquad \left[\ddot{\text{M}}\ddot{\text{g}}\colon \right]^{2+} \qquad \left[\cdot\text{Mg}\cdot \right]^{2+} \qquad \text{Mg}^{2+}$$

FIGURE P8.3

8.4. Which of the Lewis symbols in Figure P8.4 are correct?

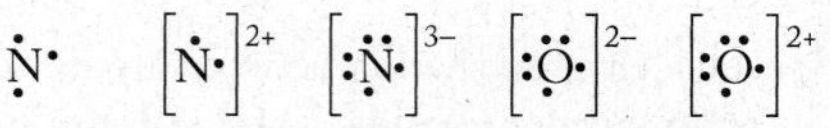

$$\dot{\text{N}}\cdot \qquad \left[\dot{\text{N}}\cdot \right]^{2+} \qquad \left[\colon\!\ddot{\text{N}}\cdot \right]^{3-} \qquad \left[\colon\!\ddot{\text{O}}\cdot \right]^{2-} \qquad \left[\colon\!\ddot{\text{O}}\colon \right]^{2+}$$

FIGURE P8.4

8.5. Which of the highlighted elements in Figure P8.5 has the greatest bonding capacity?

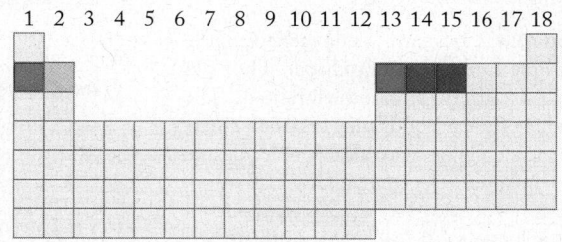

FIGURE P8.5

8.6. Which of the highlighted elements in Figure P8.6 has the greatest electronegativity?

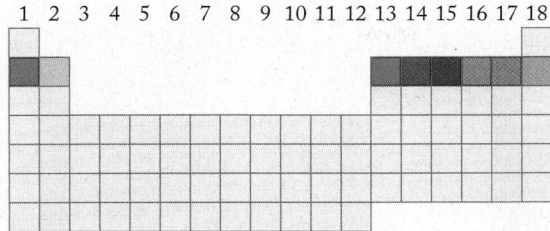

FIGURE P8.6

8.7. Which two of the highlighted elements in Figure P8.6 is the pair that forms the bond with the most ionic character?

NOTE: The color scale used in Problems 8.8, 8.9, 8.12, and 8.13 is the same as in Figure 8.4, where dark blue is a charge of 1+, red is a charge of 1−, and yellow is 0. The larger the size, the greater the electron density.

8.8. Which of the drawings in Figure P8.8 is the best description of the distribution of electrical charge in ClBr?

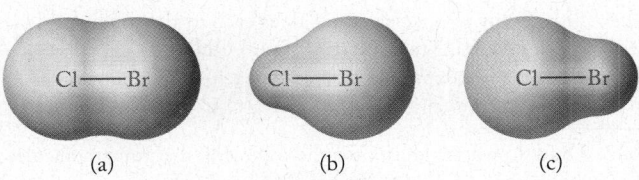

FIGURE P8.8

*8.9. Which of the drawings in Figure P8.9 best describes the distribution of electrical charge in LiF?

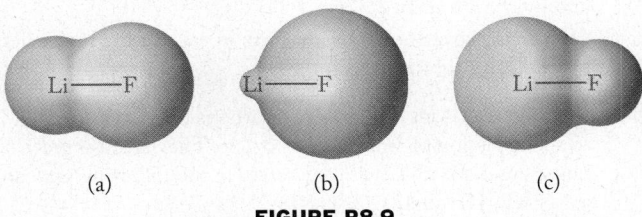

FIGURE P8.9

8.10. Are the three structures in Figure P8.10 resonance forms of the thiocyanate ion (SCN⁻)?

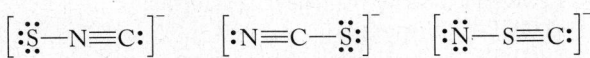

FIGURE P8.10

8.11. Why are the structures in Figure P8.11 not all resonance forms of the molecule S_2O?

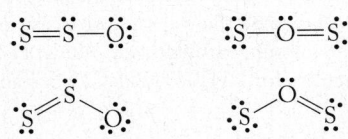

FIGURE P8.11

*8.12. Which of the drawings in Figure P8.12 most accurately describes the distribution of electrical charge in ozone? Explain your answer.

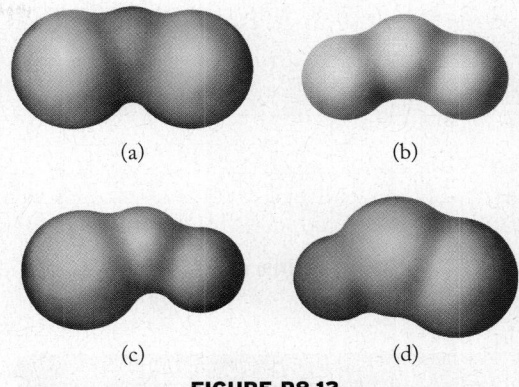

FIGURE P8.12

*8.13. Which of the drawings in Figure P8.13 most accurately describes the distribution of electrical charge in SO_2? Explain your answer.

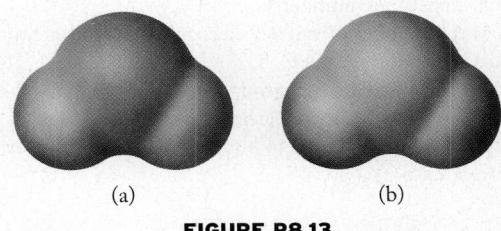

FIGURE P8.13

8.14. Which groups among main group elements in Figure P8.14 have an odd number of valence electrons?

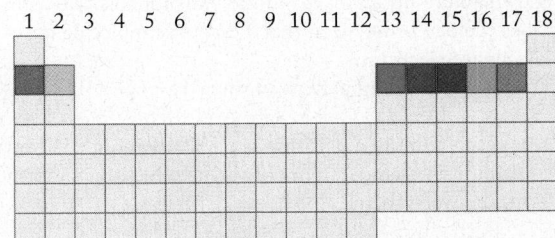

FIGURE P8.14

8.15. Krypton and xenon form compounds with only the most reactive of other elements. Which of the highlighted elements in Figure P8.15 is one of these highly reactive elements?

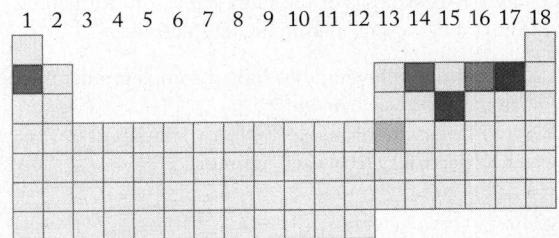

FIGURE P8.15

8.16. Which of the highlighted elements in Figure P8.16 expands its valence shell when bonding to a highly electronegative element?

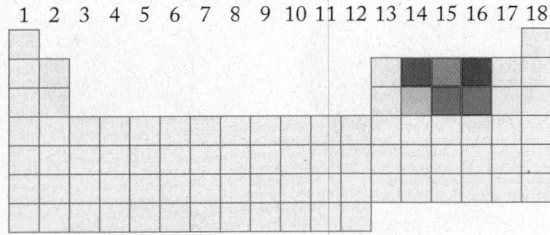

FIGURE P8.16

8.17. Which of the highlighted groups in Figure P8.17 have negative partial charges in diatomic compounds with hydrogen, HX?

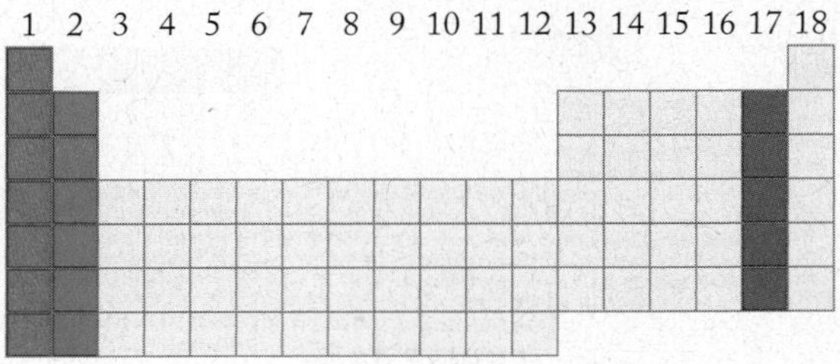

FIGURE P8.17

QUESTIONS AND PROBLEMS

Chemical Bonds

CONCEPT REVIEW

8.18. Which electrons in an atom are considered the *valence* electrons?

8.19. Does the number of valence electrons in a neutral atom ever equal the atomic number?

8.20. Does the number of valence electrons in a neutral atom ever equal the group number?

8.21. Do all the elements in a group in the periodic table have the same number of valence electrons?

8.22. Describe the differences in bonding in *covalent* and *ionic* compounds.

Lewis Symbols and Lewis Structures

CONCEPT REVIEW

8.23. Some of his critics described G. N. Lewis's approach to explaining covalent bonding as an exercise in double counting and therefore invalid. Explain the basis for this criticism.

8.24. Does the octet rule mean that a diatomic molecule must have 16 valence electrons?

8.25. Why is the bonding pattern in water H—O—H and not H—H—O?

8.26. Does each atom in a pair that is covalently bonded always contribute the same number of valence electrons to form the bonds between them?

PROBLEMS

8.27. Draw Lewis symbols of atoms of lithium, magnesium, and aluminum.

8.28. Draw Lewis symbols of atoms of nitrogen, oxygen, fluorine, and chlorine.

8.29. Draw Lewis symbols of Na^+, In^+, Ca^{2+}, and S^{2-}.

8.30. Draw Lewis symbols of the most stable ions formed by lithium, magnesium, aluminum, and fluorine.

8.31. Which of the following ions have a complete valence-shell octet? B^{3+}, I^-, Ca^{2+}, or Pb^{2+}?

8.32. Draw Lewis symbols of Xe, Sr^{2+}, Cl, and Cl^-. How many valence electrons are in each atom or ion?

8.33. Draw the Lewis symbol of an ion that has the following:
a. 1+ charge and 1 valence electron
b. 3+ charge and 0 valence electrons

8.34. Draw the Lewis symbol of an ion that has the following:
a. 1− charge and 8 valence electrons
b. 1+ charge and 5 valence electrons

8.35. How many valence electrons does each of the following species contain? (a) BN; (b) HF; (c) OH^-; (d) CN^-

8.36. How many valence electrons does each of the following species contain? (a) N_2^+; (b) CS^+; (c) CN; (d) CO

8.37. Draw Lewis structures for the following diatomic molecules and ions: (a) CO; (b) O_2; (c) ClO^-; (d) CN^-.

8.38. Draw Lewis structures for the following diatomic molecules and ions: (a) F_2; (b) NO^+; (c) SO; (d) HI.

8.39. How many electron pairs are shared in each of the molecules and ions in Problem 8.37?

8.40. How many covalent bonds are there in each of the molecules and ions in Problem 8.38?

8.41. Greenhouse Gases Chlorofluorocarbons (CFCs) are linked to the depletion of stratospheric ozone. They are also greenhouse gases. Draw Lewis structures for the following CFCs:
a. CF_2Cl_2 (Freon 12)
b. Cl_2FCCF_2Cl (Freon 113, containing a C—C bond)
c. C_2ClF_3 (Freon 1113, containing a C=C bond)

8.42. The replacement of one halogen in a CFC by hydrogen makes the compound more environmentally "friendly." Draw Lewis structures for the following compounds:
a. CHF_2Cl (Freon 22)
b. $CHBr_3$ (bromoform, used as a solvent in geology)
c. CH_2Cl_2 (methylene chloride, a common laboratory solvent)

8.43. Skunks and Rotten Eggs Many sulfur-containing organic compounds have characteristically foul odors: butanethiol $(CH_3CH_2CH_2CH_2SH)$ is responsible for the odor of skunks, and rotten eggs smell the way they do because they produce tiny amounts of pungent hydrogen sulfide, H_2S. Draw the Lewis structures for $CH_3CH_2CH_2CH_2SH$ and H_2S.

8.44. Acid in Ants Formic acid, HCOOH, is the smallest organic acid and was originally isolated by distilling red ants. Draw its Lewis structure given the connectivity of the atoms as shown in Figure P8.44.

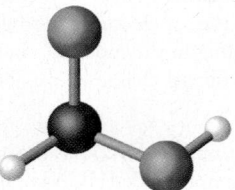

FIGURE P8.44

8.45. Chlorine Bleach Chlorine combines with oxygen in several proportions. Dichlorine monoxide (Cl_2O) is used in the manufacture of bleaching agents. Potassium chlorate ($KClO_3$) is used in oxygen generators aboard aircraft. Draw the Lewis structures for Cl_2O and ClO_3^-. Cl is the central atom in each case.

8.46. Dangers of Mixing Cleansers Labels on household cleansers caution against mixing bleach with ammonia (Figure P8.46) because the reaction produces monochloramine (NH_2Cl) and hydrazine (N_2H_4), both of which are toxic:

$$NH_3(aq) + OCl^-(aq) \rightarrow NH_2Cl(aq) + OH^-(aq)$$

$$NH_2Cl(aq) + NH_3(aq) + OH^-(aq) \rightarrow$$
$$N_2H_4(aq) + Cl^-(aq) + H_2O(\ell)$$

Draw the Lewis structures for monochloramine and hydrazine.

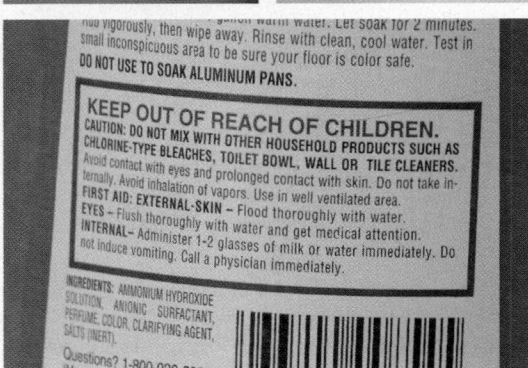

FIGURE P8.46

Electronegativity, Unequal Sharing, and Polar Bonds

CONCEPT REVIEW

8.47. How can we use electronegativity to predict whether a bond between two atoms is likely to be covalent or ionic?

8.48. How do the electronegativities of the elements change across a period and down a group?

8.49. Explain on the basis of atomic structure why trends in electronegativity are related to trends in atomic size.

8.50. Is the element with the most valence electrons in a period also the most electronegative?

8.51. What is meant by the term *polar covalent bond*?

8.52. Why are the electrons in bonds between different elements not shared equally?

PROBLEMS

8.53. Which of the following bonds are polar? C—Se, C—O, Cl—Cl, O=O, N—H, C—H. In the bond or bonds that you selected, which atom has the greater electronegativity?

8.54. Which is the least polar bond: C—Se, C=O, Cl—Br, O=O, N—H, C—H?

8.55. Which of the binary compounds formed by the following pairs of elements contain polar covalent bonds, and which are considered ionic compounds?
a. C and S
c. Al and Cl
b. C and O
d. Ca and O

8.56. Which of the beryllium halides, if any, are considered ionic compounds?

Vibrating Bonds and the Greenhouse Effect

CONCEPT REVIEW

8.57. Describe how atmospheric greenhouse gases act like the panes of glass in a greenhouse.

*__8.58.__ Water vapor in the atmosphere contributes more to the greenhouse effect than carbon dioxide; yet water vapor is not considered an important factor in climate change. Propose a reason why.

8.59. Increasing concentrations of nitrous oxide in the atmosphere may be contributing to climate change. Is the ability of N_2O to absorb IR radiation due to nitrogen–nitrogen bond stretching, nitrogen–oxygen bond stretching, or both? Explain your answer.

8.60. Is the ability of H_2O molecules to absorb photons of IR radiation due to symmetrical stretching or asymmetrical stretching of its O—H bonds, or both? Explain your answer. (*Hint*: The angle between the two O—H bonds in H_2O is 104.5°.)

8.61. Can molecules of carbon monoxide in the atmosphere absorb photons of IR radiation? Explain why or why not.

8.62. One of the activities that may be contributing to climate change is the production of cement. Why? (*Hint*: see p. 176.)

8.63. Why does infrared radiation cause bonds to vibrate but not break (as UV radiation can)?

8.64. Argon is the third most abundant species in the atmosphere. Why isn't it a greenhouse gas?

*8.65. Would the energy required to cause the bond in CO to vibrate be more or less than that required by the carbon–oxygen bond in CO_2?

*8.66. Which compound absorbs IR radiation of a longer wavelength, NO or NO_2?

Resonance

CONCEPT REVIEW

8.67. Explain the concept of resonance.

8.68. How does resonance influence the stability of a molecule or an ion?

8.69. What factors determine whether or not a molecule or ion exhibits resonance?

8.70. What structural features do all the resonance forms of a molecule or ion have in common?

8.71. Explain why NO_2 is more likely to exhibit resonance than CO_2.

8.72. Are these two skeletal structures resonance forms:
X—X—O and X—O—X?

PROBLEMS

8.73. Draw two Lewis structures showing the resonance that occurs in cyclobutadiene (C_4H_4), a cyclic molecule with a structure that includes a ring of four carbon atoms.

*8.74. Pyridine (C_5H_5N) and pyrazine ($C_4H_4N_2$) have structures similar to benzene's. Both compounds have structures with six atoms in a ring. Draw Lewis structures for pyridine and pyrazine showing all resonance forms. The N atoms in pyrazine are across the ring from each other as shown in Figure P8.74.

FIGURE P8.74

*8.75. Oxygen and nitrogen combine to form a variety of nitrogen oxides, including the following two unstable compounds each with two nitrogen atoms per molecule: N_2O_2 and N_2O_3. Draw Lewis structures for these molecules showing all resonance forms.

*8.76. Oxygen and sulfur combine to form a variety of different sulfur oxides. Some are stable molecules and some, including S_2O_2 and S_2O_3, decompose when they are heated. Draw Lewis structures for these two compounds showing all resonance forms.

8.77. Draw Lewis structures for fulminic acid (HCNO) showing all resonance forms.

8.78. Draw Lewis structures for hydrazoic acid (HN_3) showing all resonance forms.

8.79. Draw Lewis structures showing the resonance that occurs in carbonate ions.

8.80. **Bacteria Make Nitrites** Nitrogen-fixing bacteria convert urea [$H_2NC(O)NH_2$] into nitrite ions. Draw Lewis structures for these two species. Include all resonance forms. (*Hint*: There is a C=O bond in urea.)

Formal Charge: Choosing among Lewis Structures

CONCEPT REVIEW

8.81. Describe how formal charges are used to choose between possible molecular structures.

8.82. How do the electronegativities of elements influence the selection of which Lewis structure is favored?

8.83. In a molecule containing S and O atoms, is a structure with a negative formal charge on sulfur more likely to contribute than an alternative structure with a negative formal charge on oxygen?

8.84. In a cation containing N and O, why do Lewis structures with a positive formal charge on nitrogen contribute more to the actual bonding in the molecule than do those structures with a positive formal charge on oxygen?

PROBLEMS

8.85. Hydrogen isocyanide (HNC) has the same elemental composition as hydrogen cyanide (HCN) but the H in HNC is bonded to the nitrogen atom. Draw a Lewis structure for HNC, and assign formal charges to each atom. How do the formal charges on the atoms differ in the Lewis structures for HCN and HNC?

8.86. **Molecules in Interstellar Space** Hydrogen cyanide (HCN) and cyanoacetylene (HC_3N) have been detected in the interstellar regions of space and in comets close to Earth (Figure P8.86). Draw Lewis structures for these molecules, and assign formal charges to each atom. The hydrogen atom is bonded to carbon in both cases.

FIGURE P8.86

8.87. **Origins of Life** The discovery of polyatomic organic molecules such as cyanamide (H_2NCN) in interstellar space has led some scientists to believe that the molecules from which life began on Earth may have come from space. Draw Lewis structures for cyanamide, and select the preferred structure on the basis of formal charges.

8.88. Complete the Lewis structures and assign formal charges to the atoms in five of the resonance forms of thionitrosyl azide (SN_4). Indicate which of your structures should be most stable. The molecule is linear with S at one end.

*8.89. Nitrogen is the central atom in molecules of nitrous oxide (N_2O). Draw Lewis structures for another possible arrangement: N—O—N. Assign formal charges and suggest a reason why this structure is not likely to be stable.

8.90. **More Molecules in Space** Formamide ($HCONH_2$) and methyl formate (HCO_2CH_3) also have been detected in space. Draw Lewis structures for these compounds, based on the skeletal structures in Figure P8.90, and assign formal charges:

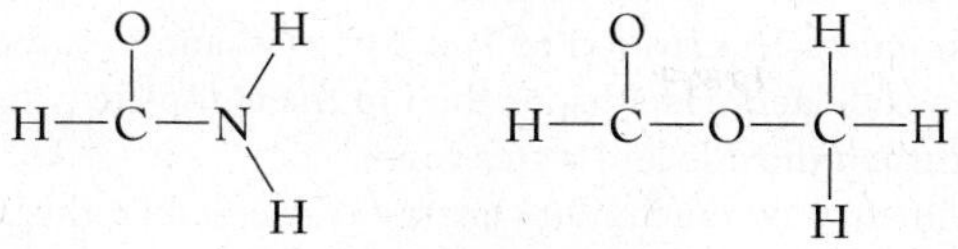

O H
‖ |
H—C—N
      |
      H

O H
‖ |
H—C—O—C—H
            |
            H

FIGURE P8.90

*8.91. Nitromethane (CH_3NO_2) reacts with hydrogen cyanide to produce $CNNO_2$ and CH_4:

$$HCN(g) + CH_3NO_2(g) \rightarrow CNNO_2(g) + CH_4(g)$$

a. Draw Lewis structures for CH_3NO_2, showing all resonance forms.
b. Draw Lewis structures for $CNNO_2$, showing all resonance forms, based on the two possible skeletal structures for it in Figure P8.91. Assign formal charges, and predict which structure is more likely to exist.
c. Are the two structures of $CNNO_2$ resonance forms of each other?

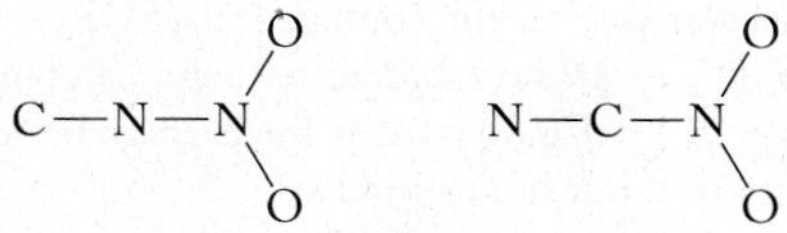

O   O
‖   ‖
C—N—N   N—C—N
      | |
      O   O

FIGURE P8.91

8.92. Use formal charges to determine which resonance form of each of the following ions is preferred: CNO^-, NCO^-, and CON^-.

Exceptions to the Octet Rule

CONCEPT REVIEW

8.93. Are all odd-electron molecules exceptions to the octet rule?
8.94. Describe the factors that contribute to the stability of structures in which the central atoms have more than 8 valence electrons.
8.95. Why do C, N, O, and F atoms in covalently bonded molecules and ions have no more than 8 valence electrons?
8.96. Do atoms in rows 3 and below always expand their valence shell? Explain your answer.

PROBLEMS

8.97. In which of the following molecules does the sulfur atom have an expanded valence shell? (a) SF_6; (b) SF_5; (c) SF_4; (d) SF_2
8.98. In which of the following molecules does the phosphorus atom have an expanded valence shell? (a) $POCl_3$; (b) PF_5; (c) PF_3; (d) P_2F_4 (which has a P—P bond)

8.99. How many electrons are there in the covalent bonds surrounding the sulfur atom in the following species? (a) SF_4O; (b) SOF_2; (c) SO_3; (d) SF_5^-
8.100. How many electrons are there in the covalent bonds surrounding the phosphorus atom in the following species? (a) $POCl_3$; (b) H_3PO_4; (c) H_3PO_3; (d) PF_6^-

*8.101. Draw Lewis structures for NOF_3 and POF_3 in which the group 15 element is the central atom and the other atoms are bonded to it. What differences are there in the types of bonding in these molecules?

*8.102. The phosphate anion is common in minerals. The corresponding nitrogen-containing anion, NO_4^{3-}, is unstable but can be prepared by reacting sodium nitrate with sodium oxide at 300°C. Draw Lewis structures for each anion. What are the differences in bonding between these ions?

8.103. Dissolving NaF in selenium tetrafluoride (SeF_4) produces $NaSeF_5$. Draw Lewis structures for SeF_4 and SeF_5^-. In which structure does Se have more than 8 valence electrons?
8.104. Reaction between NF_3, F_2, and SbF_3 at 200°C and 100 atm pressure gives the ionic compound NF_4SbF_6:

$$NF_3(g) + 2F_2(g) + SbF_3(g) \rightarrow NF_4SbF_6(s)$$

Draw Lewis structures for the ions in this product.

8.105. Ozone Depletion The compound Cl_2O_2 may play a role in ozone depletion in the stratosphere. In the laboratory, reaction of $FClO_2$ with aluminum chloride produces Cl_2O_2 and $AlCl_2F$:

$$FClO_2(g) + AlCl_3(s) \rightarrow Cl_2O_2(g) + AlFCl_2(s)$$

Draw a Lewis structure for Cl_2O_2 based on the arrangement of atoms in Figure P8.105. Does either of the chlorine atoms in the structure have an expanded valence shell?

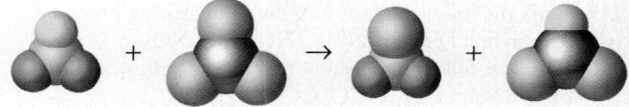

FIGURE P8.105

8.106. Cl_2O_2 decomposes to chlorine and chlorine dioxide as shown in Figure P8.106:

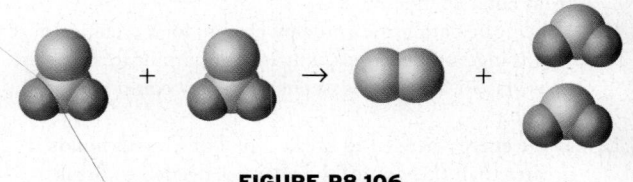

FIGURE P8.106

$$2Cl_2O_2(g) \rightarrow Cl_2(g) + 2ClO_2(g)$$

Draw the Lewis structure for ClO_2. Which atom in the structure does not have 8 valence electrons?

8.107. Which of the following chlorine oxides are odd-electron molecules? (a) Cl_2O_7; (b) Cl_2O_6; (c) ClO_4; (d) ClO_3; (e) ClO_2
8.108. Which of the following nitrogen oxides are odd-electron molecules? (a) NO; (b) NO_2; (c) NO_3; (d) N_2O_4; (e) N_2O_5

8.109. In the following species, which atom is most likely to have an unpaired electron? (a) SO^+; (b) NO; (c) CN; (d) OH
8.110. In the following molecules, which atom is most likely to have an unpaired electron? (a) NO_2; (b) CNO; (c) ClO_2; (d) HO_2

8.111. Which of the Lewis structures in Figure P8.111 contributes most to the bonding in CNO?

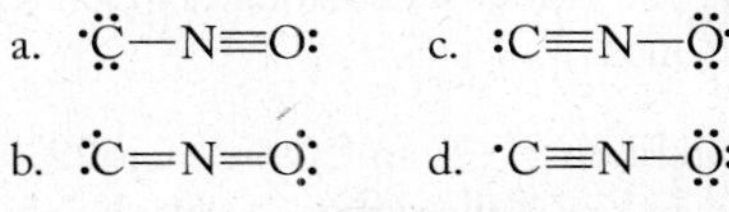

a. :C̈—N≡O: c. :C≡N—Ö:

b. :C̈=N=Ö: d. ·C≡N—Ö:

FIGURE P8.111

8.112. Why is the Lewis structure in Figure P8.112 unlikely to contribute much to the bonding in NCO?

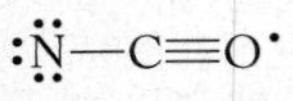

$$:\ddot{N}-C\equiv O\,^{\bullet}$$

FIGURE P8.112

The Lengths and Strengths of Covalent Bonds

CONCEPT REVIEW

8.113. Do you expect the nitrogen–oxygen bond length in the nitrate ion to be the same as in the nitrite ion?

8.114. Why is the oxygen–oxygen bond length in O_3 not the same as in O_2?

8.115. Explain why the nitrogen–oxygen bond lengths in N_2O_4 (which has a nitrogen–nitrogen bond) and N_2O are nearly identical (118 and 119 pm, respectively).

8.116. Do you expect the sulfur–oxygen bond lengths in SO_3^{2-} and SO_4^{2-} ions to be about the same? Why?

8.117. Rank the following ions in order of increasing nitrogen–oxygen bond lengths: NO_2^-, NO^+, and NO_3^-.

8.118. Rank the following ions in order of increasing carbon–oxygen bond lengths: CO, CO_2, and CO_3^{2-}.

8.119. Rank the following ions in order of increasing nitrogen–oxygen bond energy: NO_2^-, NO^+, and NO_3^-.

8.120. Rank the following ions in order of increasing carbon–oxygen bond energy: CO, CO_2, and CO_3^{2-}.

8.121. Why must the stoichiometry of a reaction be known in order to estimate the enthalpy change from bond energies?

8.122. Why must the structures of the reactants and products be known in order to estimate the enthalpy change of a reaction from bond energies?

*8.123. When calculating the enthalpy change for a chemical reaction using bond energies, why is it important to know the phase (solid, liquid, or gaseous) for every compound in the reaction?

*8.124. If the energy needed to break 2 mol of C=O bonds is greater than the sum of the energies needed to break the O=O bonds in 1 mol of O_2 and vaporize 1 mol of carbon, why does the combustion of pure carbon release heat?

PROBLEMS

NOTE: Use the average bond energies in Table 8.2 to answer Problems 8.125 to 8.134.

8.125. Use average bond energies to estimate the enthalpy changes of the following reactions:
   a. $N_2(g) + 3H_2(g) \rightarrow 2NH_3(g)$
   b. $N_2(g) + 2H_2(g) \rightarrow H_2NNH_2(g)$
   c. $2N_2(g) + O_2(g) \rightarrow 2N_2O(g)$

8.126. Use average bond energies to estimate the enthalpy changes of the following reactions:
   a. $CO_2(g) + H_2(g) \rightarrow H_2O(g) + CO(g)$
   b. $N_2(g) + O_2(g) \rightarrow 2NO(g)$
   *c. $C(s) + CO_2(g) \rightarrow 2CO(g)$
   [*Hint*: The heat of sublimation of graphite, C(s), is 719 kJ/mol.]

8.127. The combustion of CO to CO_2 releases 283 kJ/mol. What is the bond energy of the carbon–oxygen bond in carbon monoxide?

8.128. Use average bond energies to estimate the standard enthalpy of formation of HF gas.

8.129. Estimate how much less energy is released during the incomplete combustion of 1 mol of methane to carbon monoxide and water vapor than in the complete combustion to carbon dioxide and water vapor.

8.130. Estimate how much more energy is released by the reaction

$$C(s) + O_2(g) \rightarrow CO_2(g)$$

than by the reaction

$$C(s) + \tfrac{1}{2}O_2(g) \rightarrow CO(g)$$

*8.131. Estimate ΔH_{rxn} for the following reaction:

$$4NH_3(g) + 7O_2(g) \rightarrow 4NO_2(g) + 6H_2O(g)$$

*8.132. The value of ΔH_{rxn} for the reaction

$$2H_2S(g) + 3O_2(g) \rightarrow 2SO_2(g) + 2H_2O(g)$$

is −1036 kJ. Estimate the energy of the bonds in SO_2.

8.133. A molecular view of the combustion of CS_2 is shown in Figure P8.133. If the standard enthalpy of combustion of CS_2 is −1102 kJ/mol, what is the average bond energy for the carbon–sulfur bonds in CS_2?

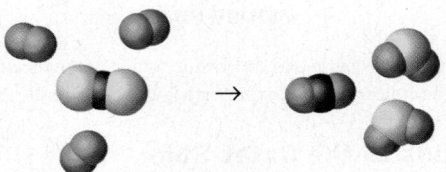

FIGURE P8.133

*8.134. The standard enthalpy of reaction for the decomposition of carbon oxysulfide (COS) to CO_2 and CS_2, as shown in Figure P8.134, is −1.9 kJ/mol COS. Are the apparent bond energies of the carbon–sulfur and carbon–oxygen bonds in COS stronger or weaker than in CS_2 and CO_2, respectively?

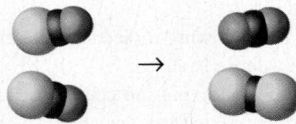

FIGURE P8.134

*8.135. Carbon and oxygen form three oxides: CO, CO_2, and carbon suboxide (C_3O_2). Draw a Lewis structure for C_3O_2 in which the three carbon atoms are bonded to each other, and predict whether the carbon–oxygen bond lengths in the carbon suboxide molecule are equal.

*8.136. Spectroscopic analysis of the linear molecule N_4O reveals that the nitrogen–oxygen bond length is 135 pm and that there are three nitrogen–nitrogen bond lengths: 148, 127, and 115 pm. Draw the Lewis structure for N_4O consistent with these observations.

Additional Problems

8.137. The unpaired dots in Lewis symbols of the elements represent valence electrons available for covalent bond formation. In Figure P8.137, which of the options for placing dots around the symbol for each element is preferred?

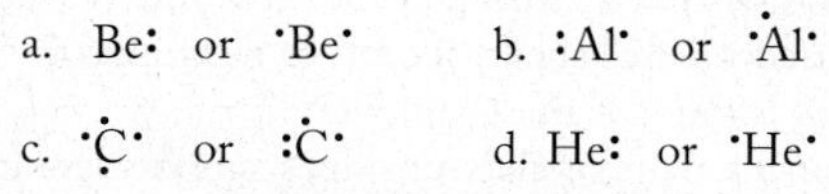

    a. Be: or ·Be·   b. :Al· or ·Äl·

    c. ·C̈· or :C̈·   d. He: or ·He·

FIGURE P8.137

8.138. Based on the Lewis symbols in Figure P8.138, predict to which group in the periodic table element X belongs.

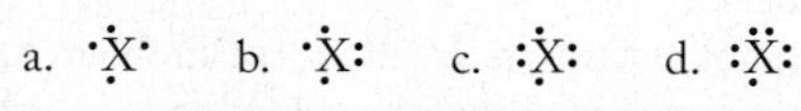

    a. ·Ẍ·   b. ·Ẍ: c. :Ẍ: d. :Ẍ:

FIGURE P8.138

8.139. Use formal charges to predict whether the atoms in carbon disulfide are arranged CSS or SCS.

8.140. Use formal charges to predict whether the atoms in hypochlorous acid are arranged HOCl or HClO.

*8.141. Chemical Weapons Phosgene is a poisonous gas first used in chemical warfare during World War I. It has the formula $COCl_2$ (C is the central atom).
  a. Draw its Lewis structure.
  b. Phosgene kills because it reacts with water in nasal passages, in the lungs, and on the skin to produce carbon dioxide and hydrogen chloride. Write a balanced chemical equation for this process showing the Lewis structures for reactants and products.

8.142. The dinitramide anion $[N(NO_2)_2{}^-]$ was first isolated in 1996. The arrangement of atoms in $N(NO_2)_2{}^-$ is in Figure P8.142.
  a. Complete the Lewis structure for $N(NO_2)_2{}^-$ including any resonance forms, and assign formal charges.
  b. Explain why the nitrogen–oxygen bond lengths in $N(NO_2)_2{}^-$ and N_2O should (or should not) be similar.
  c. $N(NO_2)_2{}^-$ was isolated as $[NH_4{}^+][N(NO_2)_2{}^-]$. Draw the Lewis structure for $NH_4{}^+$.

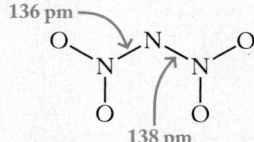

FIGURE P8.142

*8.143. Silver cyanate (AgOCN) is a source of the cyanate ion (OCN⁻), which reacts with a number of small molecules. Under certain conditions the species OCN is an anion with a charge of 1−; under others it is a neutral, odd-electron molecule, OCN.
  a. Two molecules of OCN combine to form OCNNCO. Draw the Lewis structures for this molecule including all resonance forms.
  b. The OCN⁻ ion reacts with BrNO, forming the unstable molecule OCNNO. Draw the Lewis structures for BrNO and OCNNO including all resonance forms.

  c. The OCN⁻ ion reacts with Br_2 and NO_2 to produce N_2O, CO_2, BrNCO, and OCN(CO)NCO. Draw three of the resonance forms of OCN(CO)NCO, which has the arrangement of atoms shown in Figure P8.143.

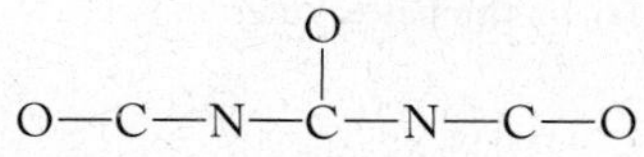

$$
\begin{array}{c}
\text{O} \\
\| \\
\text{O—C—N—C—N—C—O}
\end{array}
$$

FIGURE P8.143

*8.144. During the reaction of the cyanate ion (OCN⁻) with Br_2 and NO_2, a very unstable substance called an *intermediate* forms and then quickly falls apart. Its formula is O_2NNCO.
  a. Draw three of the resonance forms for O_2NNCO, assign formal charges, and predict which of the three contributes the most to the bonding in O_2NNCO. The connectivity of the atoms is shown in Figure P8.144a.

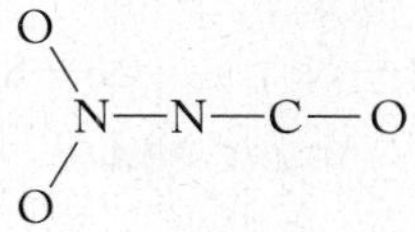

$$
\begin{array}{c}
\text{O} \\
\diagdown \\
\text{N—N—C—O} \\
\diagup \\
\text{O}
\end{array}
$$

FIGURE P8.144a

  b. Which bond in O_2NNCO must break in the reaction with Br_2 to form BrNCO? What other product forms?
  c. Draw Lewis structures for a different arrangement of the N, C, and O atoms in O_2NNCO as shown in Figure P8.144b.

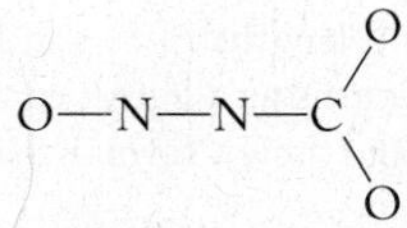

$$
\begin{array}{c}
\text{O} \\
\diagup \\
\text{O—N—N—C} \\
\diagdown \\
\text{O}
\end{array}
$$

FIGURE P8.144b

8.145. A compound with the formula Cl_2O_6 decomposes to a mixture of ClO_2 and ClO_4. Draw two Lewis structures for Cl_2O_6: one with a chlorine–chlorine bond and one with a Cl—O—Cl arrangement of atoms. Draw a Lewis structure for ClO_2.

$$Cl_2O_6 \rightarrow ClO_2 + ClO_4$$

*8.146. A compound consisting of chlorine and oxygen, Cl_2O_7, decomposes by the following reaction:

$$Cl_2O_7 \rightarrow ClO_4 + ClO_3$$

  a. Draw two Lewis structures for Cl_2O_7: one with a chlorine–chlorine bond and one with a Cl—O—Cl arrangement of atoms.
  b. Draw a Lewis structure for ClO_3.

*8.147. The odd-electron molecule CN dimerizes to give cyanogen (C_2N_2).
  a. Draw a Lewis structure for CN, and predict which arrangement for cyanogen is more likely: NCCN or CNNC.
  b. Cyanogen reacts slowly with water to produce oxalic acid $(H_2C_2O_4)$ and ammonia; the Lewis structure for oxalic

acid is shown in Figure P8.147. Compare this structure to your answer in part a. When the actual structures of molecules have been defined experimentally, the structures have been used to refine Lewis structures. Does this structure increase your confidence that the structure you selected in part a may be the better one?

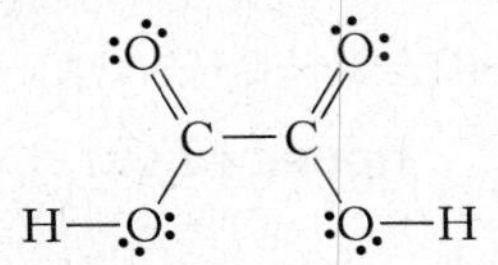

FIGURE P8.147

8.148. The odd-electron molecule SN forms S_2N_2, which has a cyclic structure (the atoms form a ring).
   a. Draw a Lewis structure for SN and complete the possible Lewis structures for S_2N_2 in Figure P8.148.
   b. Which of the two is the preferred structure for S_2N_2?

FIGURE P8.148

*8.149. In electron diffraction, an interference pattern is produced based on the location of atoms when a beam of electrons is fired at a sample. The electrons are diffracted by the arrangement of the atoms in a crystal just the way waves are diffracted by passing through gaps in a barrier, as we saw in Chapter 7. The positions of the atoms can be determined from the interference pattern. The molecular structure of sulfur cyanide trifluoride (SF_3CN) has been shown by electron diffraction studies to have the arrangement of atoms with the indicated bond lengths in Figure P8.149. Using the observed bond lengths as a guide, complete the Lewis structure for SF_3CN and assign formal charges.

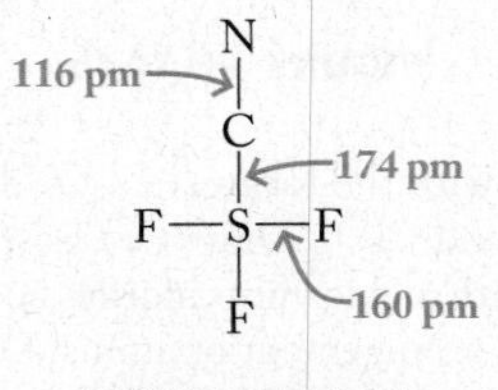

FIGURE P8.149

8.150. **Strike-Anywhere Matches** Heating phosphorus with sulfur gives P_4S_3, a solid used in the heads of strike-anywhere matches (Figure P8.150). P_4S_3 has the Lewis structure framework shown. Complete the Lewis structure so that each atom has the optimum formal charge.

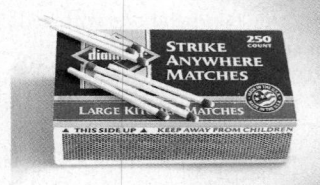

FIGURE P8.150

*8.151. The heavier group 16 elements can expand their valence shell. The $TeOF_6^{2-}$ anion was first prepared in 1993. Draw the Lewis structure for $TeOF_6^{2-}$.

*8.152. **Sulfur in the Environment** Sulfur is cycled in the environment through compounds such as dimethyl sulfide (CH_3SCH_3), hydrogen sulfide (H_2S), and sulfite and sulfate ions. Draw Lewis structures for these four species. Are expanded valence shells needed to minimize the formal charges for any of these species?

8.153. How many pairs of electrons does xenon share in the following molecules and ions? (a) XeF_2; (b) $XeOF_2$; (c) XeF^+; (d) XeF_5^+; (e) XeO_4

8.154. Consider a hypothetical structure of ozone that is cyclic (the atoms form a ring) such that its three O atoms are at the corners of a triangle. Draw the Lewis structure for this molecule.

*8.155. How might electron diffraction (see Problem 8.149) help characterize the bonding in a compound with the formula A_2X in terms of the following?
   a. Distinguishing between these two bonding patterns: X—A—A and A—X—A.
   b. Distinguishing between these resonance forms: A—X≡A, A=X=A, and A≡X—A.

8.156. The highly explosive N_5^+ cation was first isolated in 1999 by reaction of N_2F^+ with HN_3:

$$N_2F^+ + HN_3 \rightarrow [N_5^+] + HF$$

Draw the Lewis structures for the reactants and products, including all resonance forms.

*8.157. **Jupiter's Atmosphere** The ionic compound NH_4SH was detected in the atmosphere of Jupiter (Figure P8.157) by the Galileo space probe in 1995. Draw the Lewis structure for NH_4SH. Why couldn't there be a covalent bond between the nitrogen and sulfur atoms, making NH_4SH a molecular compound?

FIGURE P8.157

8.158. **Antacid Tablets** Antacids commonly contain calcium carbonate and/or magnesium hydroxide. Draw the Lewis structures for calcium carbonate and magnesium hydroxide.

*8.159. An allotrope of nitrogen, N_4, was reported in 2002. The compound has a lifetime of 1 μs at 298 K and was prepared by adding an electron to N_4^+. Since the compound cannot be isolated, its structure is unconfirmed experimentally.

a. Draw the Lewis structures for all the resonance forms of linear N_4. (Linear means that all four nitrogens are in a straight line.)

b. Assign formal charges, and determine which structure is the best description of N_4.

c. Draw a Lewis structure for a ring (cyclic) form of N_4, and assign formal charges.

*8.160. Scientists have predicted the existence of O_4 even though this compound has never been observed. However, O_4^{2-} has been detected. Draw the Lewis structures for O_4 and O_4^{2-}.

8.161. Draw a Lewis structure for $AlFCl_2$, the second product in the synthesis of Cl_2O_2 in the following reaction:

$$FClO_2(g) + AlCl_3(s) \rightarrow Cl_2O_2(g) + AlFCl_2(s)$$

*8.162. Draw Lewis structures for BF_3 and $(CH_3)_2BF$. The B—F distance in both molecules is the same (130 pm). Does this observation support the argument that all the boron–fluorine bonds in BF_3 are single bonds?

8.163. Which of the following molecules and ions contains an atom with an expanded valence shell? (a) Cl_2; (b) ClF_3; (c) ClI_3; (d) ClO^-

8.164. Which of the following molecules contains an atom with an expanded valence shell? (a) XeF_2; (b) $GaCl_3$; (c) ONF_3; (d) SeO_2F_2

*8.165. A linear nitrogen anion, N_5^-, was isolated for the first time in 1999.

a. Draw the Lewis structures for four resonance forms of linear N_5^-.

b. Assign formal charges to the atoms in the structures in part a, and identify the structures that contribute the most to the bonding in N_5^-.

c. Compare the Lewis structures for N_5^- and N_3^-. In which ion do the nitrogen–nitrogen bonds have the higher average bond order?

*8.166. Carbon tetraoxide (CO_4) was discovered in 2003.

a. The four atoms in CO_4 are predicted to be arranged as shown in Figure P8.166. Complete the Lewis structure for CO_4.

b. Are there any resonance forms for the structure you drew that have zero formal charges on all atoms?

c. Can you draw a structure in which all four oxygen atoms in CO_4 are bonded to carbon?

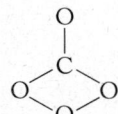

FIGURE P8.166

8.167. Plot the electronegativities of elements with $Z = 3$ to 9 (y-axis) versus their first ionization energy (x-axis). Is the plot linear? Use your graph to predict the electronegativity of neon, whose first ionization energy is 2080 kJ/mol.

8.168. Use the data in Figures 7.34 and 8.5 to plot electronegativity as a function of first ionization energy for the main group elements of the fifth row of the periodic table. From this plot estimate the electronegativity of xenon, whose first ionization energy is 1170 kJ/mol.

8.169. The cation N_2F^+ is isoelectronic with N_2O.

a. What does it mean to be isoelectronic?

b. Draw the Lewis structure for N_2F^+. (*Hint*: The molecule contains a nitrogen–nitrogen bond.)

c. Which atom has the +1 formal charge in the structure you drew in part b?

d. Does N_2F^+ have resonance forms?

e. Could the middle atom in the N_2F^+ ion be a fluorine atom? Explain your answer.

8.170. **Ozone Depletion** Methyl bromide (CH_3Br) is produced naturally by fungi. Methyl bromide has also been used in agriculture as a fumigant, but this use is being phased out because the compound has been linked to ozone depletion in the upper atmosphere.

a. Draw the Lewis structure for CH_3Br.

b. Which bond in CH_3Br is more polar, carbon–hydrogen or carbon–bromine?

Fluorine and Oxygen: Location, Location, Location

8.171. Write a balanced equation for the reaction of fluorine gas with water.

a. Identify the oxidizing and reducing agents.

b. Draw Lewis structures for all the compounds involved in the reaction, and place arrows on each polar bond to indicate the direction of the polarity.

*8.172. Atoms of Xe have complete octets, yet the compound XeO_2 exists. How is this possible?

*8.173. All the carbon–fluorine bonds in tetrafluoroethylene (Figure P8.173) are polar, but experimental results show that the molecule as a whole is nonpolar. Draw arrows on each bond to show the direction of polarity, and suggest a reason why the molecule is nonpolar.

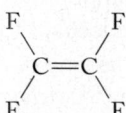

Tetrafluoroethylene

FIGURE P8.173

*8.174. Free radicals increase in stability, and hence decrease in reactivity, if they have more than one atom in their structure that can carry the unpaired electron.

a. Draw Lewis structures for the three free radicals formed from CF_2Cl_2 in the stratosphere—ClO, Cl, and CF_2Cl—and use this principle to rank them in order of reactivity.

b. Free radicals are "neutralized" when they react with each other, forming an electron pair (a single covalent bond) between two atoms. This is called a *termination reaction* because it shuts down any reaction that was powered by the free radical. Predict the formula of the molecules that are produced when the chlorine free radical reacts with each of the free radicals in part a.

Additional study materials including ChemTours and Diagnostic Quizzes are available at StudySpace at www.wwnorton.com/studyspace.

9

Molecular Geometry and Bonding Theories

Learning Outcomes

- Explain the theory of valence-shell electron-pair repulsion (VSEPR)

- Use VSEPR and the concept of steric number to predict the bond angles in molecules and the shapes of molecules with one central atom

- Predict whether a substance is polar or nonpolar based on its molecular structure

- Use valence bond theory to explain bond angles and molecular shapes

- Recognize molecular shapes that are stabilized by delocalization of π electrons

- Recognize chiral molecules

- Draw molecular orbital (MO) diagrams of diatomic molecules and use MO diagrams to predict magnetic properties and explain spectra

A LOOK AHEAD

Biological Activity and Molecular Shape

Hold your hands out in front of you, palms up, fingers extended. Now rotate your wrists inward so that your thumbs point straight up. Your right hand looks the same as the image your left hand makes in a mirror. Does that mean your two hands have the same shape? If you ever tried to put your right hand in a glove made for your left, you know that they do not have the same shape—very similar, but not the same. Many other objects in our world have a "handedness" about them, from headphones to scissors to golf clubs.

This chapter focuses on the importance of shape at the molecular level. For example, the compound that produces the refreshing aroma of spearmint has the molecular formula $C_{10}H_{14}O$. The compound responsible for the musty aroma of caraway seeds has the same molecular formula *and* the same Lewis structure. To understand how two compounds could be so much alike and still have different properties, we have to consider their structures in three dimensions.

We perceive a difference in aromas in part because each molecule has a unique site where it attaches to our nasal membranes. Just as a left hand only fits a left glove, the spearmint molecule fits only the spearmint-shaped site, and the caraway molecule, the caraway site. This phenomenon is called molecular recognition, and it enables biomolecular structures, such as nasal membranes, to recognize and react when a particular molecule with a particular shape fits into a part of the structure known as an *active site*. Many substances in the foods we eat and in the pharmaceuticals we take exert physiological effects because they are recognized by and bind to active sites in biomolecules.

What are the three-dimensional shapes of molecules and what determines those shapes? Can we predict shapes if we know how atoms are bonded together in molecules? In this chapter, we examine

Molecules Accelerate Ripening Ripening tomatoes give off the gas ethylene ▶ that speeds up the ripening process. Ethylene molecules have a unique shape that enables them to fit into a biomolecular site that controls ripening.

theories of bonding that explain molecular shapes, and we begin to explore the impact of molecular shape on the physical, chemical, and biological properties of compounds. ■

9.1 Molecular Shape

The shape of its molecules can affect many of a compound's properties: its physical state at room temperature, its solubility in water and other liquids, its aroma, its biological activity, and its distribution in the environment. In Chapter 8 we drew Lewis structures to account for bonding in molecules, but Lewis structures are only two-dimensional representations of how atoms and the electron pairs that surround them are arranged in molecules. Lewis structures show how atoms are *connected* in molecules, but they don't show how the atoms are *oriented* in three dimensions, nor do they necessarily show the overall shape of the molecule.

To illustrate this point, let's consider the Lewis structures and ball-and-stick models (Figure 9.1) of two compounds: carbon dioxide and methane. The linear array of atoms and bonding electrons in the Lewis structure for CO_2 corresponds to the actual linear shape of the molecule as represented by the ball-and-stick model. The angle between the two $C{=}O$ bonds is 180°, just as in the Lewis structure. On the other hand, the Lewis structure for methane does not convey the true orientation of the four $C{-}H$ bonds in each molecule. The 90° between bonds in the Lewis structure are not close to the actual $H{-}C{-}H$ **bond angles** of 109.5°.

In this chapter, we explore theories of covalent bonding that account for and predict the shapes of molecules. We focus on small molecules with single central atoms, though both theories can be applied to molecules of any size. Both start with the shared pairs and lone pairs of electrons predicted by Lewis theory and then predict how those pairs should be oriented about a central atom to minimize their interactions and produce the most stable molecular structure. As we shall see, both theories accurately predict electron-pair orientations, molecular shapes, and the properties of molecular compounds.

Compound	Carbon Dioxide	Methane
Molecular Formula	CO_2	CH_4
Lewis Structure	:O=C=O:	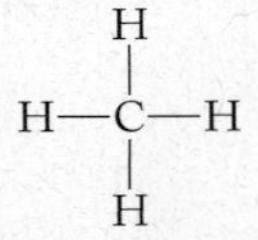 H—C—H (with H above and below)
Ball-and-Stick Model and Bond Angles	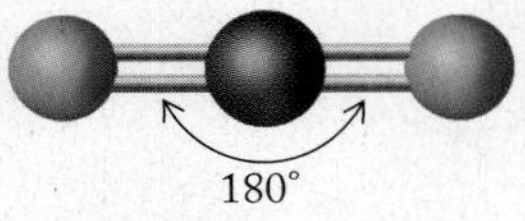 180°	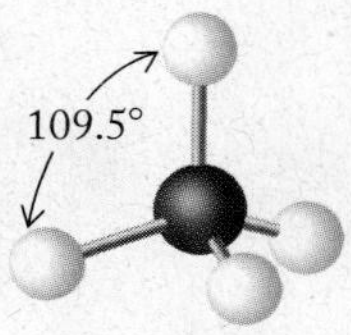 109.5°

FIGURE 9.1 The Lewis structure of CO_2 matches its true molecular structure, but the Lewis structure of CH_4 does not because the $C{-}H$ bonds in CH_4 extend in all three dimensions.

CONCEPT TEST

If the key to molecular shape is minimizing interactions between pairs of electrons, what is it about the Lewis resonance structures for O_3:

$$\overset{\cdot\cdot}{\underset{\cdot\cdot}{O}}=\overset{\cdot\cdot}{\underset{\cdot\cdot}{O}}-\overset{\cdot\cdot}{\underset{\cdot\cdot}{O}}\cdot \quad \longleftrightarrow \quad \cdot\overset{\cdot\cdot}{\underset{\cdot\cdot}{O}}-\overset{\cdot\cdot}{\underset{\cdot\cdot}{O}}=\overset{\cdot\cdot}{\underset{\cdot\cdot}{O}}$$

that might explain why O_3 molecules are bent (bond angle 117°) whereas CO_2 molecules are linear?

(Answers to Concept Tests are in the back of the book.)

9.2 Valence-Shell Electron-Pair Repulsion Theory (VSEPR)

Let's start with a theory based on a fundamental chemical principle: electrons have negative charges and repel each other. **Valence-shell electron-pair repulsion theory (VSEPR)** applies this principle by assuming that pairs of valence electrons are arranged about central atoms in ways that minimize repulsions between the pairs. To predict molecular shape using VSEPR we must consider two things: **electron-pair geometry**, which defines the relative positions in three-dimensional space of all the bonding pairs and lone pairs of valence electrons on the central atom, and **molecular geometry**, which defines the relative positions of the atoms in a molecule. To accurately predict molecular geometry we first need to know electron-pair geometry. If there are no lone pairs of electrons, then the process is simplified because the electron-pair geometry *is* the molecular geometry. Let's begin with this simpler case and consider the shapes of molecules that have various numbers of bonds around a central atom but no lone pairs.

CONCEPT TEST

In your own words explain how electron-pair geometry and molecular geometry are related and how they are different.

Central Atoms with No Lone Pairs

To determine the geometry of a molecule, we start by drawing its Lewis structure. From the Lewis structure, we determine the **steric number (SN)** of the central atom, which is the sum of the number of atoms bonded to that atom and the number of lone pairs on it:

$$SN = \begin{pmatrix} \text{number of atoms} \\ \text{bonded to central atom} \end{pmatrix} + \begin{pmatrix} \text{number of lone pairs} \\ \text{on central atom} \end{pmatrix} \quad (9.1)$$

Because we are focused on molecules in which the central atom has no lone pairs, the *steric number equals the number of atoms bonded to the central atom*. In evaluating the shapes of these molecules, we generate five common shapes that describe both the electron-pair geometries and the molecular geometries of many covalent compounds.

▶❚❚ **CHEMTOUR** VSEPR Model

bond angle the angle (in degrees) defined by lines joining the centers of two atoms to a third atom to which they are chemically bonded.

valence-shell electron-pair repulsion theory (VSEPR) a model predicting the arrangement of valence electron pairs around a central atom that minimizes their mutual repulsion to produce the lowest-energy orientations.

electron-pair geometry the three-dimensional arrangement of bonding pairs and lone pairs of electrons about a central atom.

molecular geometry the three-dimensional arrangement of the atoms in a molecule.

steric number (SN) the sum of the number of atoms bonded to a central atom plus the number of lone pairs of electrons on the atom.

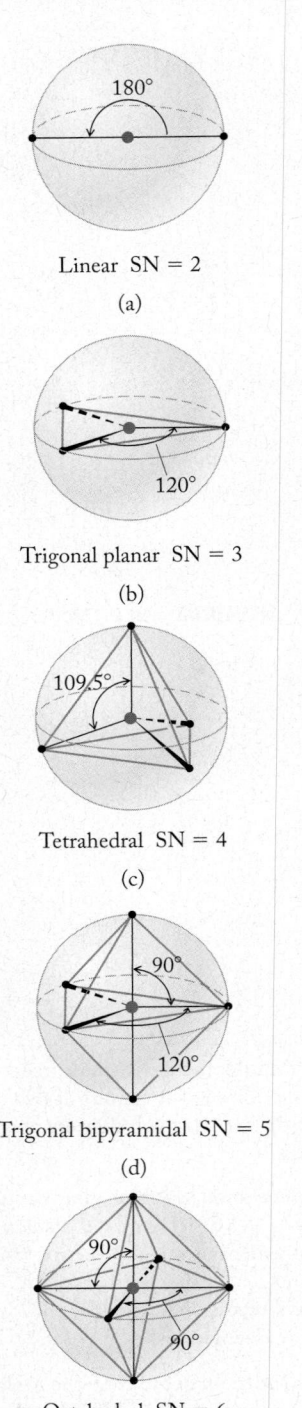

Linear SN = 2
(a)

Trigonal planar SN = 3
(b)

Tetrahedral SN = 4
(c)

Trigonal bipyramidal SN = 5
(d)

Octahedral SN = 6
(e)

FIGURE 9.2 Electron-pair geometries depend on the steric number (SN) of the central atom. In these images there are no lone pairs of electrons on the central atoms (red dots), so these *electron-pair* geometries are also *molecular* geometries. The images show bond angles for different numbers of atoms located on the surface of a sphere and bonded to an atom at the center of the sphere. The blue lines define the geometric forms that give the molecular shapes their names.

Let's start by thinking of the central atom as the center of a sphere with all the other atoms in the molecule placed on the surface of the sphere and linked to the central atom by covalent bonds. If the central atom is bonded covalently to only two other atoms, then SN = 2. How do the electron pairs in the two bonds arrange themselves to minimize their mutual repulsion? The answer is that they are as far from each other as possible on opposite sides of the sphere (Figure 9.2a). This gives a *linear* electron-pair geometry and a *linear* molecular geometry. The three atoms in the molecule are arranged in a straight line, and the bond angle is 180°.

If there are three atoms bonded to a central atom and no lone pairs, then SN = 3. The three bonding pairs are as far apart as possible when they are located at the three corners of an equilateral triangle. The angle between each pair of bonds is the same: 120° (Figure 9.2b). The name of this electron-pair and molecular geometry is **trigonal planar**.

With four atoms around the central atom and no lone pairs (SN = 4), we have the first case in which the atoms are not all in the same plane. Instead, the atoms bonded to the central atom occupy the four vertices of a *tetrahedron*, which is a four-sided pyramid (*tetra* is Greek, meaning "four"). The bonding pairs form bond angles of 109.5° with each other as shown in Figure 9.2(c). The electron-pair geometry and molecular geometry are both **tetrahedral**.

When five atoms are bonded to a central atom with no lone pairs, SN = 5 and the atoms occupy the five corners of two triangular pyramids that share the same base. The central atom of the molecule is in the center of the sphere at the common center of the two bases as shown in Figure 9.2(d). One bonding pair points to the tip of the top pyramid, one points to the tip of the bottom pyramid, and the other three point to the three vertices of the shared triangular base. These three vertices lie along the equator of the sphere, so the atoms that occupy these sites and the bonds that connect them to the center atom are called *equatorial* atoms and bonds. The bond angles between the three equatorial bonds are 120° (just as in the triangle of the trigonal planar geometry in Figure 9.2b). The bond angle between an equatorial bond and either vertical, or *axial*, bond is 90°, and the angle between two axial bonds is 180°. A molecule in which the atoms are arranged this way is said to have a **trigonal bipyramidal** electron-pair and molecular geometry.

For SN = 6, picture two pyramids that have a square base (Figure 9.2e). Put them together, base to base, and you form a shape in which all six positions around the sphere are equivalent. We can think of the six bonding pairs of electrons as three sets of two electron pairs each. The two pairs in each set are oriented at 180° to each other and at 90° to the other two pairs, just like the axes of an *xyz* coordinate system. Four equatorial atoms lie at the four vertices of the common square base and are 90° apart, another atom lies at the tip of the top pyramid, and a sixth lies at the tip of the bottom pyramid. This arrangement defines **octahedral** electron-pair and molecular geometries.

CONCEPT TEST ··································

Assume that all of the bonds in molecules with the five shapes described above are single bonds and that none of the central atoms have lone pairs of electrons. Which SN values correspond to central atoms with less than an octet of valence electrons, which values correspond to central atoms with an expanded octet, and which value has a central atom with exactly 8 valence electrons?

··································

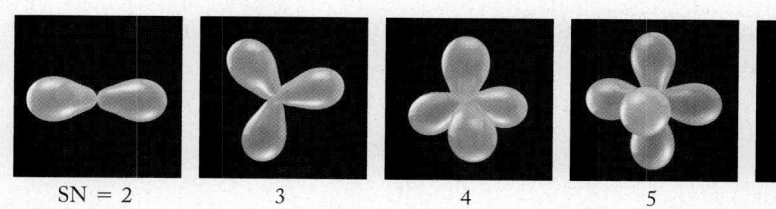

SN = 2   3   4   5   6

FIGURE 9.3 Balloons charged with static electricity repel each other and align themselves as far apart as possible when tied together. In doing so they mimic the locations of different numbers of electron pairs about a central atom. Each balloon represents one electron pair. Note the similarities between the patterns of the balloons and the diagrams in Figure 9.2.

Now let's consider a real-world analogy to the above bond orientations. We start with a batch of fully inflated yellow balloons. We tie them together in clusters of two, three, four, five, and six balloons (Figure 9.3). Let us assume that all the balloons have acquired a static electrical charge so that they repel each other. If the tie points of the clusters represent the central atom in our balloon model, then the opposite ends represent the atoms that are bonded to the central atom. Note how our clusters of two, three, four, five, and six balloons produce the same orientations that resulted from selecting points on a sphere that were as far apart as possible. The long axes of the balloons in Figure 9.3 provide an accurate representation of the bond directions in Figure 9.2.

Now let's look at five simple molecules that have no lone pairs about the central atom and apply VSEPR and the concept of steric number to predict their electron-pair and molecular geometries. To do so, we follow these three steps:

1. Draw the Lewis structure for the molecule.
2. Determine the steric number of the central atom.
3. Use the steric number to predict the electron-pair and molecular geometries using the images in Figure 9.2.

EXAMPLE I: CARBON DIOXIDE, CO_2

1. Lewis structure:

$$\ddot{O}{=}C{=}\ddot{O}$$

2. The central carbon atom has two atoms bonded to it and no lone pairs, so the steric number is 2.
3. Because SN = 2, the O—C—O bond angle is 180° (see Figure 9.2a) and the electron-pair and molecular geometries are both linear.

EXAMPLE II: BORON TRIFLUORIDE, BF_3

1. Lewis structure:

$$\ddot{F}\diagdown_{\displaystyle B}\diagup\ddot{F}$$
$$|$$
$$\ddot{F}$$

2. There are three fluorine atoms bonded to the central boron atom, and there are no lone pairs on boron. Therefore, SN = 3.
3. If SN = 3 then all four atoms lie in the same plane and form a triangle with F—B—F bond angles of 120°. The electron-pair and molecular geometries are trigonal planar (Figure 9.4).

trigonal planar molecular geometry about a central atom with a steric number of 3 and no lone pairs of electrons.

tetrahedral molecular geometry about a central atom with a steric number of 4 and no lone pairs of electrons.

trigonal bipyramidal molecular geometry about a central atom with a steric number of 5 and no lone pairs of electrons in which three atoms occupy equatorial sites and two other atoms occupy axial sites above and below the equatorial plane.

octahedral molecular geometry about a central atom with a steric number of 6 and no lone pairs of electrons in which all six sites are equivalent.

CONNECTION We learned in Chapter 8 that some compounds have central atoms that do not have an octet of electrons. The boron atom in BF_3 has only three pairs of electrons, but because this distribution results in formal charges of 0 on all the atoms, it is the preferred structure.

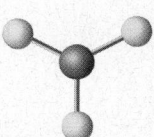

FIGURE 9.4 The ball-and-stick model of BF_3 shows the orientation of the atoms. All F—B—F bond angles are 120° in this trigonal planar molecular geometry.

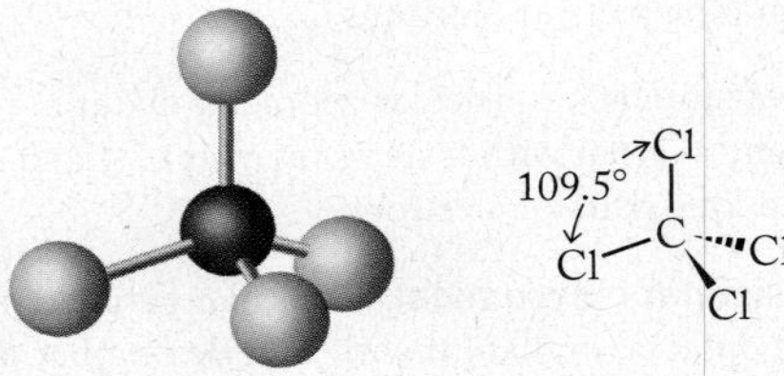

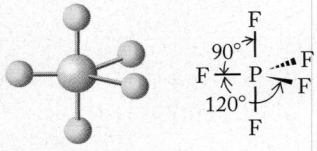

FIGURE 9.5 The ball-and-stick model shows the actual orientation of the atoms in carbon tetrachloride in three dimensions. All Cl—C—Cl bond angles are 109.5° in this tetrahedral molecular geometry.

EXAMPLE III: CARBON TETRACHLORIDE, CCl_4

1. Lewis structure:

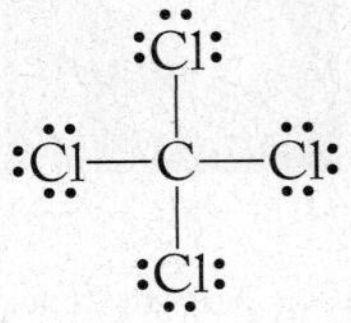

2. There are four chlorine atoms bonded to the central carbon atom, which gives carbon a full octet. Therefore, SN = 4.
3. If SN = 4 then the chlorine atoms are located at the vertices of a tetrahedron and all four Cl—C—Cl bond angles are 109.5°, producing tetrahedral electron-pair and molecular geometries (Figure 9.5).

Before we explore any more molecular geometries, some comments about the conventions used in drawing three-dimensional structures on two-dimensional paper are in order. To convey the structure of a molecule in three dimensions, we use a solid wedge (—▸) to indicate a bond that comes out of the paper toward the viewer. The solid wedge in the CCl_4 structure in Figure 9.5, for example, means that the chlorine in that position points toward the viewer at a downward angle. A dashed wedge (⋯⋯) indicates a bond that goes into the paper away from the viewer. Solid lines are used for bonds that lie in the plane of the paper. We typically orient a structure so that the maximum number of bonds lie in the plane of the paper.

EXAMPLE IV: PHOSPHORUS PENTAFLUORIDE, PF_5

1. Lewis structure:

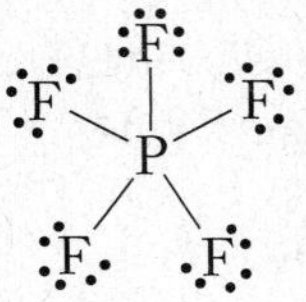

2. There are five fluorine atoms bonded to the central phosphorus atom, which has no lone pairs. Therefore, SN = 5.
3. If SN = 5 then the fluorine atoms are located at the five vertices of a trigonal bipyramid and the electron-pair geometry and molecular geometry are trigonal bipyramidal (Figure 9.6).

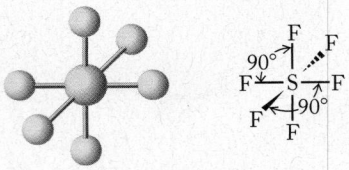

FIGURE 9.6 The ball-and-stick model shows the actual orientation of the atoms in phosphorus pentafluoride. The P—F bonds located around the equator of the sphere are 120° apart. Each of them is 90° from the two bonds connecting F atoms in the axial positions (at the north and south poles). The resulting molecular geometry is called trigonal bipyramidal.

EXAMPLE V: SULFUR HEXAFLUORIDE, SF_6

1. Lewis structure:

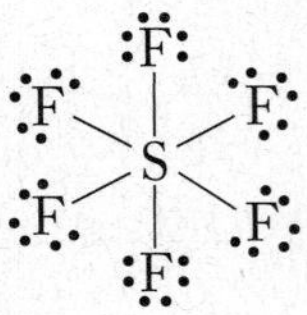

2. There are six fluorine atoms bonded to the central sulfur atom, which has no lone pairs. Therefore, SN = 6.
3. If SN = 6 then the fluorine atoms are located at the vertices of an octahedron and the electron-pair and molecular geometries are octahedral (Figure 9.7).

FIGURE 9.7 The ball-and-stick model shows how the six fluorine atoms are oriented in three dimensions about the sulfur atom. All the F—S—F bond angles are 90° in this octahedral molecular geometry.

Formaldehyde, CH_2O, is a gas at room temperature (boiling point $-21°C$). Aqueous solutions of formaldehyde are used to preserve biological samples. It is also a product of the incomplete combustion of hydrocarbons. Use VSEPR to predict the molecular geometry of formaldehyde.

Collect and Organize We are given the molecular formula of formaldehyde. Assuming there are no lone pairs of valence electrons on the central atom, the solution requires (1) drawing the Lewis structure, (2) determining the steric number, and (3) identifying the molecular geometry using Figure 9.2.

Analyze Carbon is the likely central atom of the molecule because it has a bonding capacity of 4 and is less electronegative than oxygen, whose bonding capacity is 2. If the three atoms bonded to C are as far from each other as possible, the probable molecular geometry is trigonal planar.

Solve Following the procedure developed in Chapter 8 for drawing Lewis structures, we obtain this for formaldehyde:

$$\overset{\cdot\cdot}{\underset{}{O}}$$
$$\parallel$$
$$\underset{}{C}$$
$$H \qquad H$$

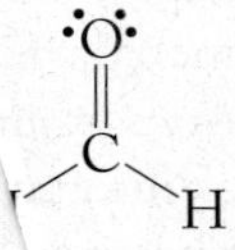

As we predicted, SN = 3, and according to Figure 9.2, the molecular geometry is trigonal planar.

Think about It The key to predicting the correct molecular geometry of a molecule with no lone pairs on its central atom is to determine the SN, which is simply a matter of counting the number of atoms bonded to the central atom.

Practice Exercise Use VSEPR to determine the molecular geometry of the chloroform molecule, $CHCl_3$, and draw the molecule using the solid-wedge, dashed-wedge convention.

(Answers to Practice Exercises are in the back of the book.)

Measurements of the bond angles in formaldehyde show that the H—C—H bond angle is slightly smaller than the $120°$ predicted for a trigonal planar geometry (Figure 9.2b). The $C=O$ double bond consists of two pairs of bonding electrons (4 electrons) and exerts greater repulsion than a single bond would. This greater repulsion decreases the H—C—H bond angle. VSEPR does not enable us to predict the actual values of the bond angles in molecules containing double bonds to the central atom, but it does allow us to correctly predict how a bond angle deviates from the ideal value due to the presence of a double bond.

CONCEPT TEST

Rank the following bond angles from largest to smallest.
  a. the F—B—F bond angles in BF_3
  b. the O—C—H bond angles in CH_2O
  c. the H—C—H bond angle in CH_2O

Central Atoms with Lone Pairs

We have established the electron-pair and molecular geometries of molecules whose central atoms have no lone pairs of electrons. Now let's explore what happens when a central atom has one or more lone pairs. For SN = 2, the only bonding pattern possible is two atoms bound to a central atom. If we replace one bonding pair with a lone pair, we have a molecule with only two atoms, which means there is no central atom and no bond angle; remember that it takes three atoms to define a bond angle. Such diatomic molecules must have a linear geometry. Therefore, we begin the discussion of molecules containing lone pairs with SN = 3.

Consider sulfur dioxide, one of the gases produced when high-sulfur coal is burned. The resonance structures showing the distribution of electrons are

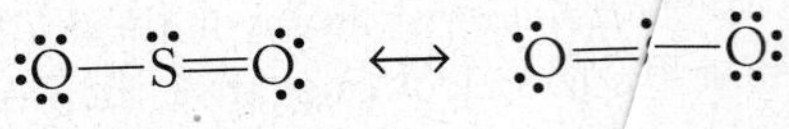

$$:\ddot{O}\!-\!\ddot{S}\!=\!\ddot{O}: \quad \longleftrightarrow \quad :\ddot{O}\!=\!\ddot{S}\!-\!\ddot{O}: $$

To calculate the SN of the central S atom we need to add up the number of atoms and lone pairs of electrons that surround it. In both resonance structures, S is bonded to two atoms and has one lone pair of electrons. Therefore its SN value in either resonance structure is 2 + 1 = 3. When SN = 3, the arrangement of atoms and lone pairs about the sulfur atom is trigonal planar (Figure 9.2b). This means that the electron-pair geometry, which takes into account both the atoms and the nonbonding pairs of electrons around the central atom, is trigonal planar (as shown in Figure 9.8a). However, the gray lines in the structures in Figure 9.8(a) do not indicate a bond but rather point to a lone pair of electrons. The lone pair is located on the S atom, and the attachment of two oxygen atoms gives the molecular geometry shown in Figure 9(b). This molecular geometry is called *bent* or *angular*.

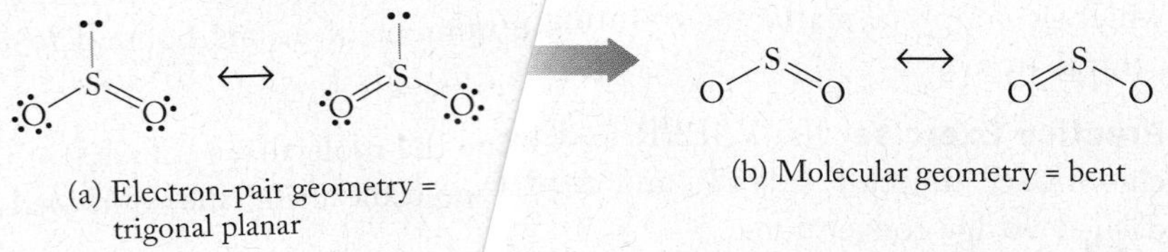

FIGURE 9.8 (a) The electron-pair geometry of SO_2 is trigonal planar because the steric number of the S atom is 3 (it is bonded to two atoms and has one lone pair of electrons). (b) The molecular geometry is bent because there is no bonded atom, only a lone pair of electrons, above the central sulfur atom in the structure.

(a) Electron-pair geometry = trigonal planar

(b) Molecular geometry = bent

Experimental measurements establish that SO_2 is indeed a bent molecule, but the O—S—O bond angle is actually a little smaller than 120°. We explain the smaller angle using VSEPR by comparing the space occupied by bonding electrons with the space occupied by electrons in a lone pair. Because the bonding electrons are attracted to two nuclei, they have a high probability of being located between the two atomic centers that share them. In contrast, the lone pair is not shared with a second atom and is spread out around the sulfur atom as shown in Figure 9.9. This puts the lone pair of electrons closer to the bonding pairs and produces greater repulsion. As a result, the lone pair pushes the bonding pairs closer together, thereby reducing the bond angle. In general,

> repulsion between lone pairs and bonding pairs is greater than the repulsion between bonding pairs;

> repulsion caused by a lone pair is greater than the repulsion caused by a double bond;

> repulsion caused by a double bond is greater than that caused by a single bond;

> two lone pairs of electrons on a central atom exert a greater repulsive force on the atom's bonding pairs than does one lone pair.

O—S—O bond angle < 120°

FIGURE 9.9 The lone pair of electrons on the central sulfur atom in SO_2 occupies more space (larger blue region) than the electron pairs in the S—O bonds (smaller blue regions). The increased repulsion (represented by the double-headed arrows) resulting from the larger volume occupied by the lone pair forces the oxygen atoms closer together, making the O—S—O bond angle slightly less than the ideal value of 120°.

SAMPLE EXERCISE 9.2   **Predicting Relative Sizes**
                       of Bond Angles

Rank NH_3, CH_4, and H_2O in order of decreasing bond angles in their molecular structures.

Collect and Organize We are asked to predict the relative size of the bond angles in three molecules. We are given their molecular formulas. The information in Figure 9.2 links steric numbers to electron-pair geometries and bond angles. Bond angles are also influenced by the presence of double bonds and lone pairs of electrons on the central atom.

Analyze We need to determine the steric numbers of the central atoms in these three molecules. To do that we first need to translate the molecular formulas into Lewis structures to determine how many lone pairs or double bonds they have.

Solve Using the method for drawing Lewis structures from Chapter 8 we obtain these results:

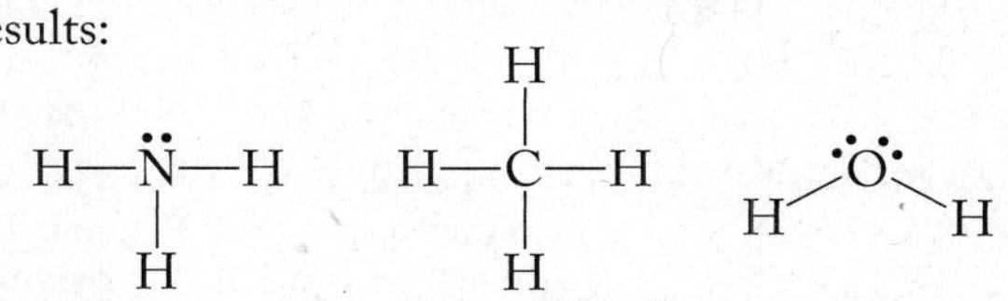

In each of these structures the central atom is surrounded by a total of four atoms or lone pairs of electrons, which means a steric number of 4. There are no double bonds.

According to Figure 9.2, the electron-pair geometry for SN = 4 is tetrahedral. In a molecule such as CH_4 in which all four tetrahedral electron pairs are equivalent bonding pairs, all bond angles are 109.5°. However, repulsion from the lone pairs of electrons in NH_3 and H_2O squeezes their bonds together, reducing the bond angles. The two lone pairs on the O atom in H_2O exert a greater repulsive force on its bonding pairs than the single lone pair on the N atom exerts on the bonding pairs in NH_3. Therefore, the bond angle in H_2O should be less than the bond angles in NH_3. Ranking the three molecules in order of decreasing bond angle we have:

$$CH_4 > NH_3 > H_2O$$

Think about It The logic used in answering this problem is supported by experimental evidence: the bond angles in molecules of CH_4, NH_3, and H_2O are 109.5°, 107.0°, and 104.5°, respectively.

Practice Exercise Are the O—S—O bond angles greater in SO_2 or SO_3?

As we saw in the preceding Sample Exercise, three possible combinations of atoms and lone pairs are possible about a central atom with a SN of 4: four atoms and no lone pairs, three atoms and one lone pair, or two atoms and two lone pairs. The first case is illustrated by the molecular structure of methane, CH_4, in which a tetrahedral electron-pair geometry translates into a tetrahedral molecular geometry. In a molecule of ammonia, NH_3, there are only three bonding pairs and one lone pair. This means that the Lewis structure in Figure 9.10(a) translates into a tetrahedral electron-pair geometry in which one of the vertices is the lone pair on the N atom (Figure 9.10b).

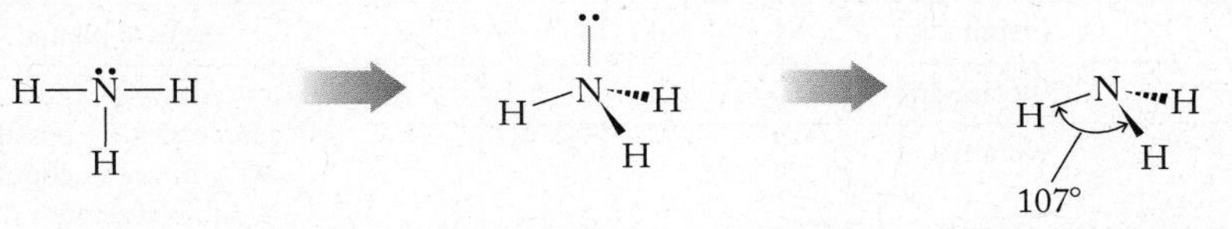

(a) Lewis structure

(b) Tetrahedral electron-pair geometry

(c) Trigonal pyramidal molecular geometry

FIGURE 9.10 The steric number of the N atom in NH_3 is 4, so its electron-pair geometry is tetrahedral. However, one of the vertices of the tetrahedron is occupied by a lone pair of electrons, not an atom, so the molecular geometry is trigonal pyramidal.

The resulting molecular geometry, which is essentially a flattened tetrahedron, is *trigonal pyramidal* (Figure 9.10c). As we discussed in Sample Exercise 9.2, the strong repulsion produced by the diffuse lone pair of electrons on the N atom pushes the N—H bonds closer together in NH_3 and reduces the angles between them as compared to the H—C—H bond angles in CH_4. Thus, the H—N—H bond angles in ammonia are only 107.0°.

The central O atom in a molecule of H_2O has a SN of 4 because it is bonded to two atoms and has two lone pairs of electrons. The result is a tetrahedral electron-pair geometry (Figure 9.11). The presence of the two lone pairs means that two of the four tetrahedral vertices are missing, which leaves us with a bent (or angular) molecular geometry. As we also discussed in Sample Exercise 9.2, the H—O—H bond angle in water is smaller than 109.5° due to repulsion between the two lone pairs and each bonding pair. The actual bond angle is 104.5°.

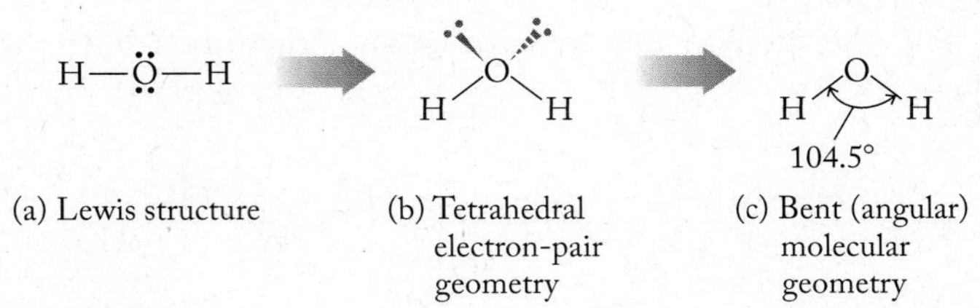

FIGURE 9.11 The steric number of the O atom in H_2O is 4 (it is bonded to two atoms and has two lone pairs of electrons). Therefore, its electron-pair geometry is tetrahedral. However, two of the vertices of the tetrahedron are occupied by lone pairs of electrons, meaning only three atoms define the molecular geometry, which is bent.

(a) Lewis structure

(b) Tetrahedral electron-pair geometry

(c) Bent (angular) molecular geometry

CONCEPT TEST

The bond angles in silane (SiH_4), phosphine (PH_3), and hydrogen sulfide (H_2S) are 109.5°, 93.6°, and 92.1°, respectively. Use VSEPR to explain this trend.

Molecules with trigonal bipyramidal electron-pair geometry have four possible molecular geometries depending on the number of lone pairs per molecule (see Table 9.1). These options are linked to the fact that there are axial and equatorial

seesaw molecular geometry about a central atom with a steric number of 5 and one lone pair of electrons in an equatorial position.

T-shaped molecular geometry about a central atom with a steric number of 5 and two lone pairs of electrons that occupy equatorial positions; the atoms occupy two axial sites and one equatorial site.

TABLE 9.1	**Electron-Pair Geometries and Molecular Geometries**			
SN = 4	**Electron-Pair Geometry**	**No. of Bonded Atoms**	**No. of Lone Pairs**	**Molecular Geometry**
	Tetrahedral	4	0	Tetrahedral
	Tetrahedral	3	1	Trigonal pyramidal
	Tetrahedral	2	2	Bent (angular)
SN = 5				
	Trigonal bipyramidal	5	0	Trigonal bipyramidal
	Trigonal bipyramidal	4	1	Seesaw
	Trigonal bipyramidal	3	2	T-shaped
	Trigonal bipyramidal	2	3	Linear
SN = 6				
	Octahedral	6	0	Octahedral
	Octahedral	5	1	Square pyramidal
	Octahedral	4	2	Square planar
	Octahedral	3	3	Although these geometries are possible, we will not encounter any molecules with them.
	Octahedral	2	4	

vertices in a trigonal bipyramid (Figure 9.2d). VSEPR enables us to predict which vertices are occupied by atoms and which are occupied by lone pairs. The key to these predictions is the fact that the repulsions between pairs of electrons decrease as the angle between them increases: two electron pairs at 90° experience a greater mutual repulsion than two at 120°, which have a greater repulsion than two at 180°. To minimize repulsions involving lone pairs, VSEPR predicts that they preferentially occupy equatorial rather than axial vertices. Why? Because an equatorial lone pair has *two* 90° repulsions with the two axial electron pairs as shown in Figure 9.12(a), but an axial lone pair has *three* 90° repulsions (with three equatorial electron pairs) as shown in Figure 9.12(b).

When we assign one, two, or three lone pairs of valence electrons to equatorial vertices, we get three of the molecular geometries for SN = 5 in Table 9.1. When a single lone pair occupies an equatorial site we get a molecular geometry called **seesaw** (Figure 9.13) because its shape, when rotated 90° clockwise, resembles a playground seesaw. (The formal name for this shape is *disphenoidal*.)

(a) Equatorial lone pair   (b) Axial lone pair

FIGURE 9.12 A lone pair of electrons in an equatorial position of a molecule with trigonal bipyramidal electron-pair geometry interacts through 90° with two other electron pairs. A lone pair in an axial position interacts through 90° with *three* other electron pairs. Fewer 90° interactions reduce internal electron-pair repulsion and lead to greater stability.

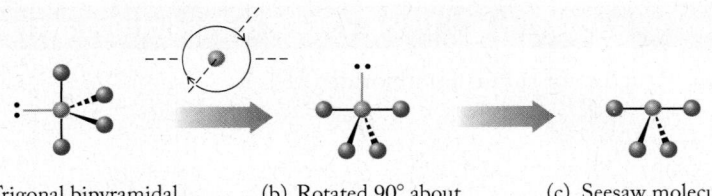

(a) Trigonal bipyramidal electron-pair geometry

(b) Rotated 90° about horizontal axis

(c) Seesaw molecular geometry

FIGURE 9.13 A single lone pair of electrons in an equatorial position of a trigonal bipyramidal electron-pair geometry produces a seesaw molecular geometry.

When two lone pairs occupy equatorial sites (Figure 9.14a), the molecular geometry that results is called **T-shaped**. This designation becomes more apparent when the structure in Figure 9.14(a) is rotated 90° counterclockwise. Lone-pair repulsions result in bond angles in T-shaped molecules that are slightly less than the 90° and 180° angles we would expect from a perfectly shaped "T" geometry.

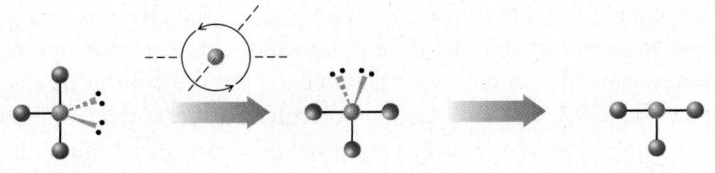

(a) Trigonal bipyramidal electron-pair geometry

(b) Rotated 90° about horizontal axis

(c) T-shaped molecular geometry

FIGURE 9.14 Two lone pairs of electrons in equatorial positions of a trigonal bipyramidal electron-pair geometry produce a T-shaped molecular geometry.

Finally, a SN = 5 molecule with three lone pairs and two bonded atoms has a linear geometry because all three lone pairs occupy equatorial sites and the bonding pairs are in the two axial positions (Figure 9.15).

(a) Trigonal bipyramidal electron-pair geometry

(b) Linear molecular geometry

FIGURE 9.15 Three lone pairs of electrons in equatorial positions of a trigonal bipyramidal electron-pair geometry produce a linear molecular geometry.

CONCEPT TEST

The bond angles in a trigonal bipyramidal molecular structure are either 90°, 120°, or 180°. Are the corresponding bond angles in a seesaw structure likely to be larger than, the same as, or smaller than these values?

(a) Octahedral
electron-pair
geometry

(b) Square pyramidal
molecular geometry

FIGURE 9.16 A lone pair of electrons in an octahedral electron-pair geometry produces a square pyramidal molecular geometry.

(a) Octahedral
electron-pair
geometry

(b) Square planar
molecular geometry

FIGURE 9.17 Two lone pairs of electrons on opposite sides of an octahedral electron-pair geometry produce a square planar molecular geometry.

Molecules that have a central atom with a steric number of 6 have an octahedral electron-pair geometry (Figure 9.2e) and the molecular geometries listed in Table 9.1. With one lone pair of electrons, there is only one possible molecular geometry because all the sites in an octahedron are equivalent (Figure 9.16). That one molecular geometry is called **square pyramidal**: a pyramid with a square base, four triangular sides, and the central atom "embedded" in the base.

Because of stronger repulsion between lone pairs and bonding pairs, we predict the bond angles in a square pyramidal molecule to be slightly less than the ideal angle of 90°. The molecule BrF_5, for example, has a square pyramidal molecular geometry, and the angles between its equatorial and axial bonds are 85°.

When two lone pairs are present at the central atom in a molecule with octahedral electron-pair geometry, they occupy vertices on opposite sides of the octahedron to minimize the interactions between them. The resultant molecular geometry is called **square planar** because the molecule is shaped like a square and all five atoms reside in the same plane (Figure 9.17). Because the two lone pairs are on opposite sides of the bonding pairs, the presence of the lone pairs does not distort the bond angles: they are all 90°.

SAMPLE EXERCISE 9.3   **Using VSEPR to Predict Geometry II**

The Lewis structure of sulfur tetrafluoride (SF_4) is

What is its molecular geometry and what are the angles between the S—F bonds?

Collect and Organize We are given the Lewis structure of SF_4 and can determine from it the steric number of the central atom in the molecule and its electron-pair geometry.

Analyze Four atoms are bonded to the central S atom, which also has one lone pair of electrons. This means that SN = 5.

Solve With a steric number of 5 for its central atom, the electron-pair geometry of SF_4 is trigonal bipyramidal. The presence of one lone pair of electrons on the S atom means that its molecular geometry is not the same as its electron-pair geometry. Table 9.1 tells us that the molecular geometry that results when a lone pair occupies one of the three equatorial positions in a trigonal bipyramidal electron-pair geometry is seesaw:

Electron-pair geometry =
trigonal bipyramidal

Molecular geometry =
seesaw

Frequently drawn from
this perspective as well

The equatorial lone pair reduces slightly the bond angles from their normal values of 90° between the axial and equatorial bonds, 120° between the two equatorial bonds, and 180° between the two axial bonds.

Think about It Any molecule with SN = 5 and one lone pair should have the same geometry as SF_4.

Practice Exercise Determine the molecular geometry and the bond angles of SO_2Cl_2.

square pyramidal molecular geometry about a central atom with a steric number of 6 and one lone pair of electrons; as typically drawn, the atoms occupy four equatorial and one axial site.

square planar molecular geometry about a central atom with a steric number of 6 and two lone pairs of electrons that occupy axial sites; the atoms occupy four equatorial positions.

9.3 Polar Bonds and Polar Molecules

Water is the only atmospheric gas that is a liquid at standard temperature and pressure. Why is H_2O a liquid when all other compounds of comparable molar mass, such as N_2, O_2, CO_2, and CH_4, are gases? The answer in part is that water is a polar substance. As we saw in Chapter 8, polar molecules contain polar bonds. However, some nonpolar molecules also contain polar bonds. For example, the C=O double bonds in CO_2 are polar because carbon and oxygen have different electronegativities. The carbon atom in the molecule has a partial positive charge and the oxygen atoms have a partial negative charge:

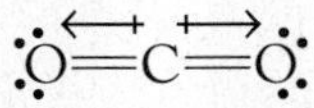

See Figure 8.3 to review the use of these symbols to indicate bond polarity.

As we noted in Chapter 8, an unequal distribution of bonding electrons between two atoms produces a partial negative charge on one end of the bond and a partial positive charge on the other. This charge separation makes a **bond dipole**. The overall polarity of a molecule can be determined by summing the polarities of all the bond dipoles in the molecule. This summing must take into account both the strengths of the individual bond dipoles and their orientations with respect to one another. In CO_2, for example, the two dipoles are equivalent in strength because they involve the same atoms and the same kind of bond (C=O). The linear shape of the molecule means that the direction of one C=O bond dipole is opposite the direction of the other. Thus, one exactly offsets the other so that, overall, CO_2 is a nonpolar substance.

Similarly, CH_4 is a nonpolar substance even though its molecules contain polar C—H bonds. The four C—H bond dipoles are all the same because the same atoms are involved. The tetrahedral molecular geometry means that the bond dipoles offset one another so that, overall, the CH_4 molecule is nonpolar:

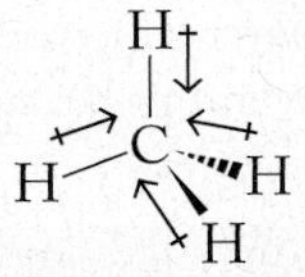

Water molecules are bent, not linear like molecules of CO_2. Therefore, the dipoles of the two O—H bonds in a molecule of H_2O do not offset each other. Instead, the molecule has an overall dipole with the negative end directed toward the O atom:

The presence of this overall dipole means that water is a polar molecule. This polarity leads to interactions between H and O atoms on adjacent H_2O molecules in liquid water or solid ice. Other polar molecules exhibit similar interactions. We discuss these interactions in Chapter 10.

CONCEPT TEST

Each molecule of ethane contains six polar C—H bonds, yet ethane:

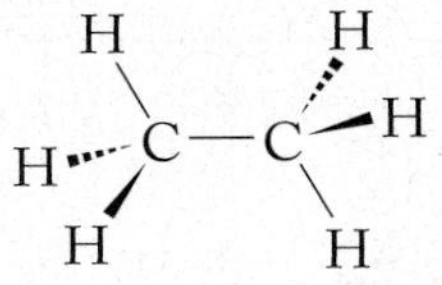

is a nonpolar substance. Explain how this is possible.

bond dipole separation of electrical charge created when atoms with different electronegativities form a covalent bond.

CONNECTION In Chapter 8, we introduced electronegativity and the unequal distribution of electrons in polar covalent bonds.

CONNECTION In Chapter 8 we discussed the impact of bond polarity on the physical properties of compounds like the hydrogen halides.

FIGURE 9.18 Gaseous HF molecules are oriented randomly in the absence of an electric field but align themselves when an electric field is applied to two metal plates. The negative (fluorine) end of each molecule is directed toward the positively charged plate; the positive (hydrogen) end, toward the negative plate.

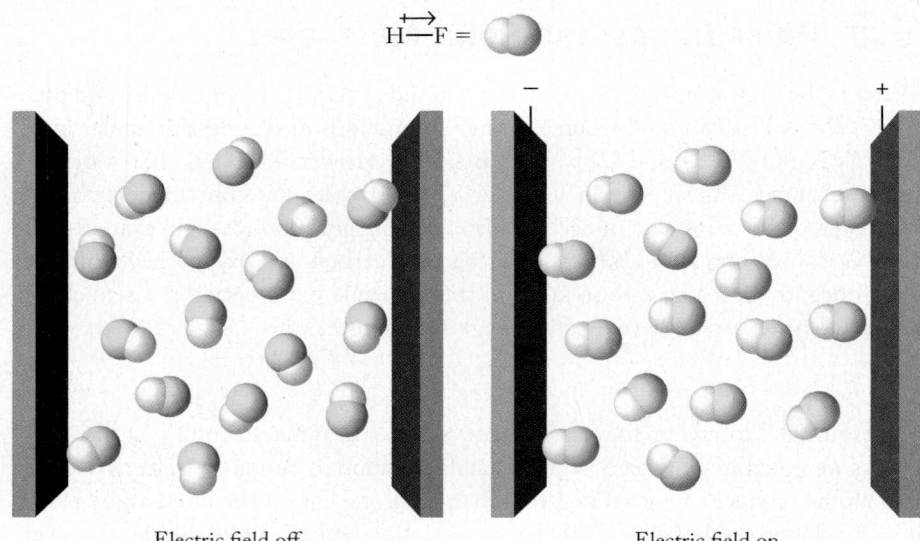

Electric field off   Electric field on

We have seen how the overall polarity of a molecule depends on the differences in the electronegativities of the bonded pairs of atoms in its molecular structure and on the arrangement of those bonded atoms. Actually, the polarity of a molecule can be determined experimentally by measuring its permanent **dipole moment (μ)**. The value of μ expresses the extent of the overall separation of positive and negative charge in the molecules of a substance and is determined by measuring the degree to which the molecules are aligned when they are placed in a strong electric field (Figure 9.18). Polar molecules align with the field so that their negative ends are oriented toward the positive plate and their positive ends are oriented toward the negative plate. The more polar the molecules are, the more strongly they align with the field. Dipole moments are usually expressed in units of *debyes* (D), where $1\ D = 3.34 \times 10^{-30}$ coulomb-meter. The dipole moments of several polar substances are shown in Table 9.2.

TABLE 9.2	Permanent Dipole Moments of Several Polar Molecules	
Formula	**Structure with Bond Dipole(s)**	**Dipole Moment (debyes)**
HF	H—F	1.91
H_2O	H—O—H	1.85
NH_3	H—N—H (H)	1.47
$CHCl_3$	H—C(Cl)(Cl)Cl	1.04
CCl_3F	F—C(Cl)(Cl)Cl	0.45

dipole moment (μ) a measure of the degree to which a molecule aligns itself in an applied electric field; a quantitative expression of the polarity of a molecule.

Note the structures in the last two rows of the table. Chloroform ($CHCl_3$) has a relatively strong dipole moment because its bond dipoles differ both in the *degree* of polarity and in the *direction* of the polarity (Figure 9.19a). Chlorine is more electronegative than carbon, and the two electrons in each C—Cl bond are pulled away from C and toward Cl. Because hydrogen is less electronegative than carbon, the two electrons in the H—C bond are pulled away from H and toward C. Consequently the electron distribution is away from the top part of the molecule as drawn and toward the bottom.

In trichlorofluoromethane (CCl_3F) all four bond dipoles point away from the central carbon atom because fluorine and chlorine are more electronegative than carbon (Figure 9.19b). However, a C—F bond dipole is greater than a C—Cl bond dipole because fluorine is more electronegative than chlorine. As a result, bonding electrons are pulled more toward the C—F (top) side of the molecule than toward the C—Cl (bottom) side. The net direction of the dipole is upward.

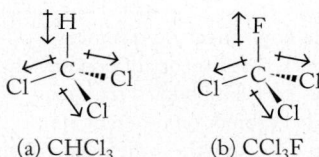

(a) $CHCl_3$ (b) CCl_3F

FIGURE 9.19 (a) The four bonds in chloroform ($CHCl_3$) are not equivalent in terms of direction or degree of polarity. The molecule has a dipole moment. (b) The C—F bond dipole in CCl_3F is stronger than the C—Cl bond dipoles and is not offset by them. This molecule also has a dipole moment.

SAMPLE EXERCISE 9.4   Predicting the Polarity of a Substance

Does formaldehyde gas (CH_2O) have a permanent dipole moment?

Collect and Organize To predict whether a molecular compound has a permanent dipole moment, we need to determine whether or not its molecules contain polar bonds and whether the bond dipoles offset each other. A bond's polarity depends on the difference in the electronegativity of the bonded pair of atoms. The extent to which bond dipoles offset each other depends on the molecular geometry of the molecule. In Sample Exercise 9.1 we determined that formaldehyde has a trigonal planar structure:

Analyze The electronegativities of the elements in the compound are H = 2.1, C = 2.5, and O = 3.5. Given the differences in the electronegativities of the atoms bonded to the central atom, it is likely that formaldehyde will have a permanent dipole moment.

Solve The hydrogen atoms are the least electronegative of the atoms in the molecule and oxygen the most, so each of the bonds in the molecule has a bond dipole directed toward the oxygen atom:

Thus, the formaldehyde molecule has a permanent dipole moment.

Think about It The presence of different atoms bonded to a central atom makes it highly likely that the molecule will have a permanent dipole moment.

Practice Exercise Does carbon disulfide (CS_2), a gas present in small amounts in crude petroleum, have a dipole moment?

valence bond theory a quantum mechanics–based theory of bonding that assumes covalent bonds form when half-filled orbitals on different atoms overlap or occupy the same region in space.

overlap a term in valence bond theory describing bonds arising from two orbitals on different atoms that occupy the same region of space.

sigma (σ) bond a covalent bond in which the highest electron density lies between the two atoms along the bond axis.

CONCEPT TEST

Water and hydrogen sulfide both have a bent molecular geometry with dipole moments of 1.85 D and 0.98 D, respectively. Why is the dipole moment of H_2S less than that of H_2O?

9.4 Valence Bond Theory

Drawing Lewis structures and applying VSEPR enable us to predict the geometry of many molecules reliably. However, we have avoided making a connection between molecular geometry and the picture of electrons in atoms developed in Chapter 7, namely, that these electrons exist in atomic orbitals. The absence of a connection between the electronic structure of atoms and the electronic structure of molecules led to the development of bonding theories that use atomic orbitals to account for the molecular geometries predicted by VSEPR. We explore one of these theories next.

Orbital Overlap and Hybridization

Valence bond theory arose in the late 1920s, largely as a result of the genius and efforts of Linus Pauling, who developed this theory of molecular bonding based on quantum mechanics. Valence bond theory assumes (a) that a chemical bond between two atoms results from **overlap** of the atoms' atomic orbitals and (b) that the greater the overlap, the stronger and more stable the bond. Shared electrons in a chemical bond are located between the nuclei of two atoms and are attracted to both. This attraction leads to lower potential energy and greater chemical stability than if the atoms were completely free (see Figure 8.1). This view of chemical bonding is especially useful for analyzing the physical and chemical properties of covalent substances because it provides a picture of where the electrons are in a molecule. Just as the locations of electrons in atoms define atomic properties, the locations of electrons in molecules define the properties of those molecules.

Let's begin by applying Pauling's valence bond theory to the simplest of diatomic molecules: H_2. A free H atom has a single electron in a 1s atomic orbital. According to valence bond theory, the overlap between two of these half-filled 1s orbitals produces a single H—H bond (Figure 9.20). Overlapping two orbitals from two atoms in this way increases electron density along the axis connecting the two nuclei. Whenever the region of high electron density lies along the bond axis, the resulting covalent bond is called a **sigma (σ) bond**.

In molecules other than simple ones like H_2, an inconsistency arises between the molecular geometries we predict based on VSEPR and the orbital shapes and orientations described in Chapter 7. Methane, for example, has a carbon atom at its center. We established in Chapter 7 that carbon atoms have the electron configuration $[He]2s^2 2p^2$ and that the 2s orbital is spherical and the three 2p orbitals, p_x, p_y, and p_z, are teardrop-shaped orbitals oriented at 90° to one another (Figure 9.21). Further, we learned in Chapter 8 that carbon atoms form four covalent bonds to complete their octets. How can four equivalent bonds oriented at 109.5° to one another in a tetrahedral arrangement form from a filled 2s orbital and two partly filled 2p orbitals? Furthermore, how can two double bonds 180° apart form around the central carbon atom in CO_2, or one

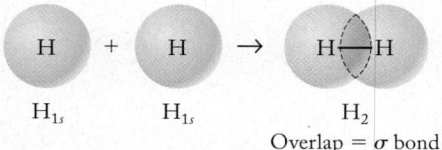

FIGURE 9.20 The overlap of the 1s orbitals on two hydrogen atoms produces a single σ bond that holds the two hydrogen atoms together in a H_2 molecule.

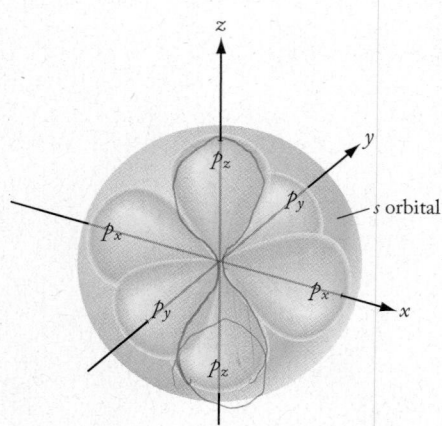

FIGURE 9.21 Relative orientation in space of one 2s orbital (spherical) and three 2p orbitals (teardrop-shaped) oriented at 90° to one another along the x-, y-, and z-axes of a coordinate system.

double bond and two single bonds 120° apart form around the central carbon in CH_2O?

To account for the geometry of methane, carbon dioxide, formaldehyde, and many other molecules, valence bond theory describes atomic orbitals of different shapes and energies as being mixed together in a process called **hybridization**, to form **hybrid atomic orbitals**. Covalent bonds then result either from overlap of a hybrid orbital on one atom with an unhybridized orbital on another or from overlap of two hybrid orbitals on two atoms. Let's begin by looking at the types of hybrid orbitals we can form on carbon and how these orbitals account for observed molecular geometries.

Tetrahedral Geometry: sp^3 Hybrid Orbitals

We have noted that methane has a tetrahedral molecular geometry with H—C—H bond angles of 109.5°. To account for this geometry using orbital hybridization, we start with a free carbon atom with its $2s^2 2p^2$ valence-shell electron configuration (Figure 9.22a). Then we promote one electron from the $2s$ orbital to the empty $2p$ orbital. We now have four orbitals containing one electron each on the carbon atom.

Next the four orbitals are *hybridized* (mixed and averaged) to make a set of four equivalent hybrid orbitals, each containing one electron. An atom's steric number always indicates how many of its orbitals must be mixed to generate the hybrid set, and the number of hybrid orbitals in a set is always equal to the number of atomic orbitals mixed. The carbon atom in methane has SN = 4, so four atomic orbitals on carbon are mixed to form four hybrid orbitals.

The four orbitals are called sp^3 **hybrid orbitals** because they result from the mixing of one s and three p orbitals. Their energy level is the weighted average of the energies of the s and p orbitals used to form them. They are oriented 109.5° from one another and point toward the vertices of a tetrahedron (Figure 9.22b). Now four hydrogen atoms can form σ bonds with the sp^3 hybridized orbitals on carbon and form a methane molecule that has the required tetrahedral geometry.

hybridization in valence bond theory the mixing of atomic orbitals to generate new sets of orbitals that then are available to form covalent bonds with other atoms.

hybrid atomic orbital in valence bond theory one of a set of equivalent orbitals about an atom created when specific atomic orbitals are mixed.

sp^3 **hybrid orbitals** a set of four hybrid orbitals with a tetrahedral orientation produced by mixing one s and three p atomic orbitals.

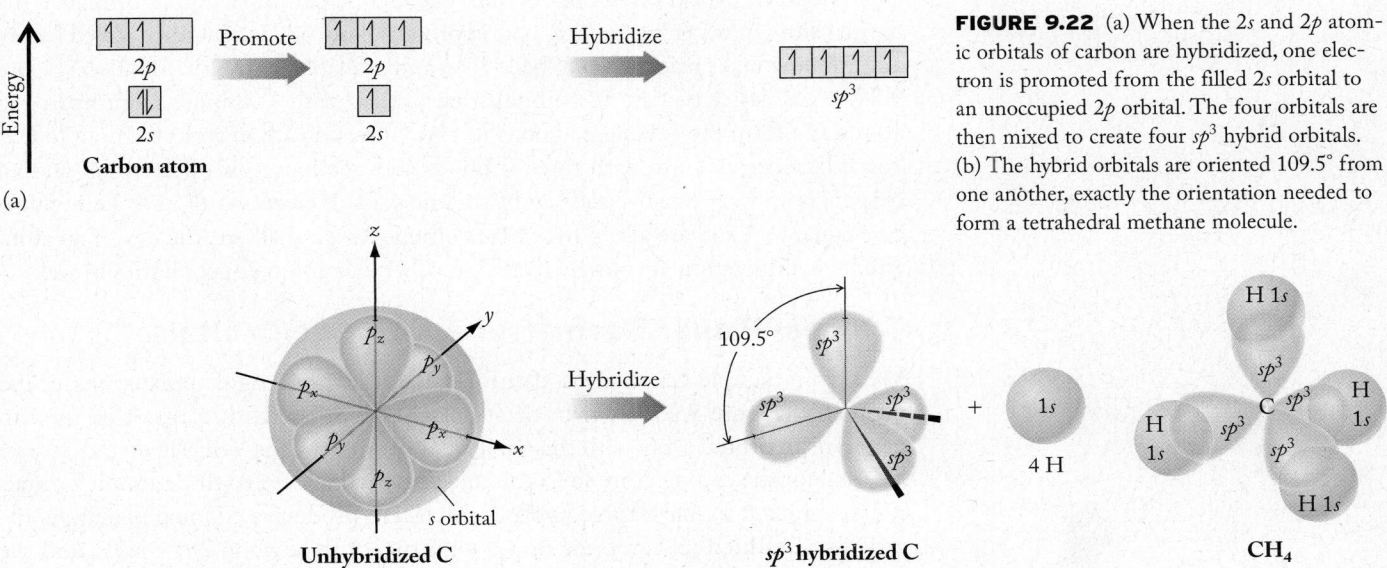

(a)

(b)

FIGURE 9.22 (a) When the $2s$ and $2p$ atomic orbitals of carbon are hybridized, one electron is promoted from the filled $2s$ orbital to an unoccupied $2p$ orbital. The four orbitals are then mixed to create four sp^3 hybrid orbitals. (b) The hybrid orbitals are oriented 109.5° from one another, exactly the orientation needed to form a tetrahedral methane molecule.

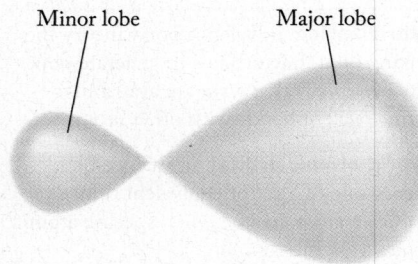

Minor lobe Major lobe

FIGURE 9.23 The sp^3 hybrid orbitals each consist of a major and minor lobe. Because the minor lobes are not involved in orbital overlap leading to bond formation, they have been omitted from the orbital models in this book.

FIGURE 9.24 A set of four sp^3 hybrid orbitals point toward the vertices of a tetrahedron. (a) The nonbonding electron pair of the N in NH_3 occupies one of the four hybrid orbitals, giving the molecule a trigonal pyramidal molecular geometry. (b) The O atom in H_2O is also sp^3 hybridized. Only two of the four orbitals contain bonding pairs of electrons, giving H_2O a bent molecular geometry.

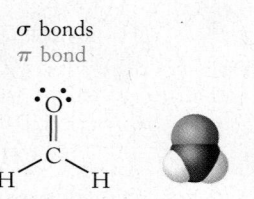

σ bonds
π bond

FIGURE 9.25 Formaldehyde has 3 σ bonds and 1 π bond.

The sp^3 hybrid orbitals in Figure 9.22(b) are shown as having one lobe each, but in fact every sp^3 hybrid orbital consists of a major lobe and a minor lobe (Figure 9.23). Because the minor lobe is not involved in bonding, we ignore it in this chapter. It becomes important in the discussion of chemical reactions that are beyond the scope of this book.

According to valence bond theory, any atom with a set of four equivalent sp^3 hybrid orbitals has a tetrahedral orientation of its valence electrons. This includes atoms in which one or more hybrid orbitals are filled before any bonding takes place. For example, sp^3 hybridization of the valence electrons on the nitrogen atom in ammonia produces three hybrid orbitals that are half-filled and one hybrid orbital that is completely filled (Figure 9.24a). The three half-filled orbitals form the three σ bonds in ammonia by overlapping with the $1s$ orbitals of three hydrogen atoms.

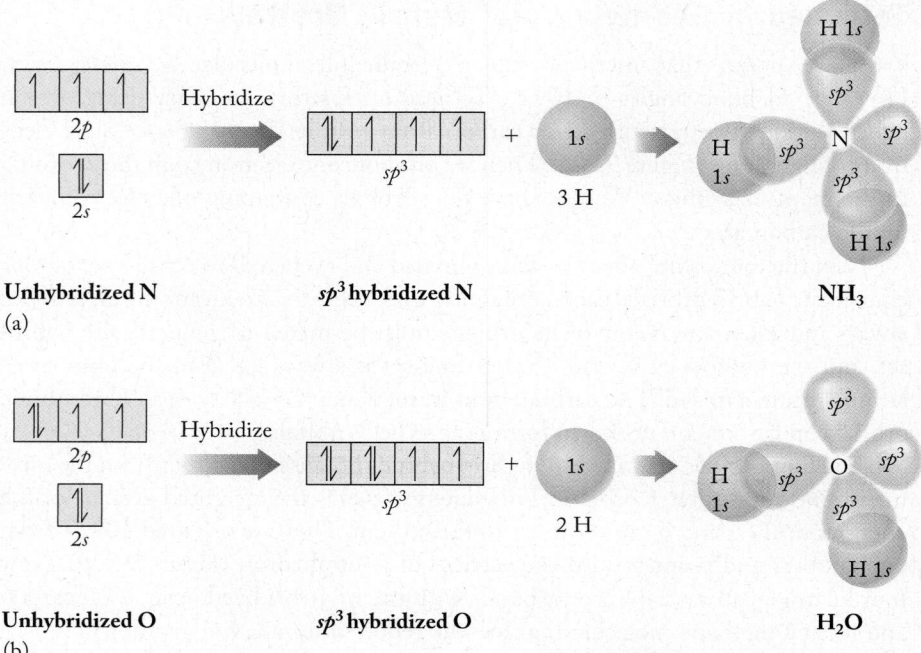

The one filled hybrid orbital of N contains the lone pair of electrons. Similarly, the oxygen atom in water has four sp^3 hybrid orbitals in its valence shell, two filled before any bonding takes place and two half-filled and available for bond formation (Figure 9.24b). The latter two are the orbitals that overlap with $1s$ orbitals from hydrogen atoms and form the two sigma bonds in H_2O. Thus, a carbon atom forming four σ bonds, a nitrogen atom with three σ bonds and one lone pair of electrons, and an oxygen atom with two σ bonds and two lone pairs of electrons all have steric numbers equal to 4 and are all sp^3 hybridized atoms. As we shall see, the SN of an atom and its hybridization are closely related in other electron pair geometries as well.

Trigonal Planar Geometry: sp^2 Hybrid Orbitals

We saw in Sample Exercise 9.1 that formaldehyde has a trigonal planar molecular geometry (Figure 9.25). A hybridization scheme other than sp^3 must be used to generate an orbital array with this molecular geometry. The VSEPR model defines SN = 3 for the carbon atom in a molecule of formaldehyde, so three atomic orbitals must be mixed to make three hybrid orbitals. To produce a trigonal planar geometry, the $2s$ orbital on carbon is mixed with two of the carbon $2p$ orbitals, and the third $2p$ orbital is left unhybridized (Figure 9.26a).

Mixing and averaging one s and two p orbitals generates three hybrid orbitals called **sp^2 hybrid orbitals**. The energy level of sp^2 orbitals is slightly lower than that

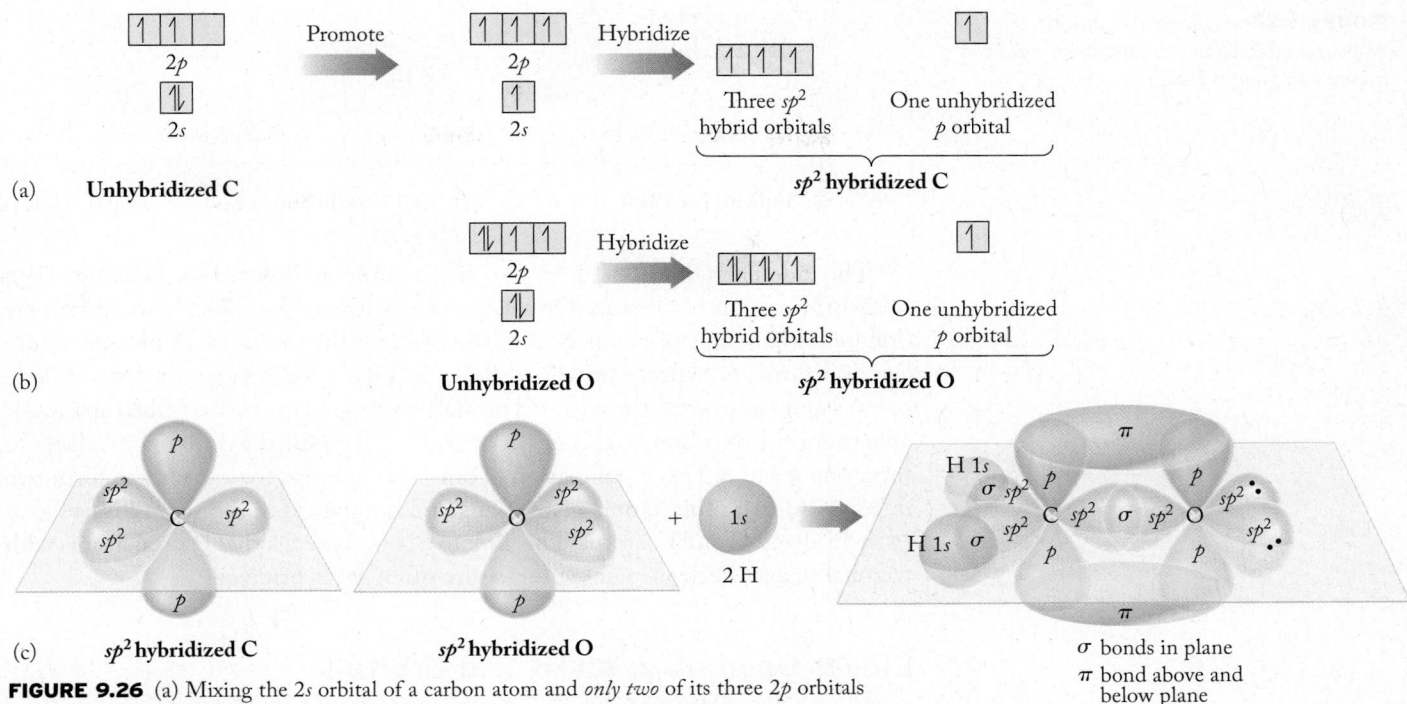

FIGURE 9.26 (a) Mixing the 2*s* orbital of a carbon atom and *only two* of its three 2*p* orbitals produces three *sp*² hybrid orbitals, leaving one unhybridized *p* orbital available for π bond formation. (b) The 2*s* and 2*p* orbitals of oxygen can hybridize in the same way. (c) An *sp*² hybridized carbon forms three σ bonds in the plane with its hybrid orbitals and one π bond with the unhybridized *p* orbital above and below the plane defined by the three hybrid orbitals. An *sp*² hybridized oxygen atom forms one σ bond in the plane with one of its *sp*² hybrid orbitals, forms one π bond above and below the plane with an unhybridized *p* orbital, and has two lone pairs of electrons in the remaining two *sp*² hybrid orbitals. This hybridization scheme for the central C atom in a molecule of CH_2O accounts for its double bond to an O atom and two single bonds to H atoms.

of *sp*³ orbitals because only two *p* orbitals are mixed with one *s* orbital in a set of *sp*² orbitals. The orbitals in an *sp*² hybridized atom all lie in the same plane and are 120° apart. The two lobes of the unhybridized *p* orbital lie above and below the plane of the triangle defined by the *sp*² hybrid orbitals. An *sp*² hybridized carbon atom can form three σ bonds with its three hybridized orbitals.

Valence electrons in *s* and *p* orbitals in other atoms can also be *sp*² hybridized. To complete the valence bond picture of formaldehyde, the oxygen atom must be *sp*² hybridized as well (Figure 9.26b). That both the carbon and the oxygen atoms in formaldehyde are hybridized points out another difference between VSEPR, which considers only the central atom in a molecule, and valence bond theory, which considers all the atoms in the molecule.

The valence bond view of the bonding in formaldehyde shows the 1*s* orbitals of two hydrogen atoms overlapping with two carbon *sp*² hybrid orbitals and the third carbon *sp*² hybrid orbital overlapping with one oxygen *sp*² hybrid orbital (Figure 9.26c). These overlapping orbitals constitute the sigma-bonding framework of the molecule. The unhybridized carbon 2*p* orbital is parallel to the unhybridized oxygen 2*p* orbital, and the overlap of these two orbitals above and below the plane of the σ-bonding network forms a **pi (π) bond**. The widths of the lobes on the *p* orbitals and the distances between atoms are not drawn to scale in this and similar figures. The lobes are actually much closer together than they appear and they *do* overlap.

Pi bonds have their greatest electron density above and below the internuclear axis (or in front of and in back of the internuclear axis). They can be formed only by the overlap of partially filled *p* orbitals that are parallel to each other and perpendicular to the sigma bond joining the atoms.

sp² hybrid orbitals three hybrid orbitals in a trigonal planar orientation formed by mixing one *s* and two *p* orbitals.

pi (π) bond a covalent bond in which electron density is greatest on opposite sides of the bonding axis.

FIGURE 9.27 A nitrogen atom with sp^2 hybrid orbitals has the capacity to form two σ bonds and one π bond.

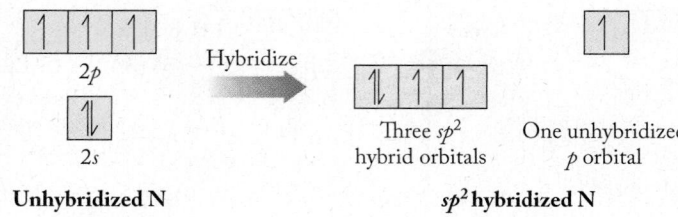

The lone pairs of electrons on the oxygen atom are located in the two oxygen sp^2 hybrid orbitals not involved in the bonding with carbon. These two orbitals are oriented at an angle of about 120° in the plane of the molecule. A nitrogen atom can also form sp^2 hybrid orbitals as shown in Figure 9.27. When it does, its lone pair occupies one of the three hybrid orbitals, the other two are half-filled and available to form two σ bonds, and the unhybridized half-filled p orbital is available to form one π bond. This combination of one lone pair plus the capacity to form two σ bonds to two other atoms gives N a steric number of 3, the same SN value as sp^2 hybridized C and O atoms. This link to SN = 3 means that central atoms with trigonal planar electron-pair geometry are often sp^2 hybridized.

Linear Geometry: *sp* Hybrid Orbitals

We have seen hybrid orbitals generated by mixing one s orbital with three p orbitals and by mixing one s orbital with two p orbitals. The remaining mixture, one s orbital with one p orbital, forms the final set of important hybrid orbitals made from s and p atomic orbitals.

In the linear molecule acetylene, C_2H_2, both carbon atoms have SN = 2 according to VSEPR theory. Because steric number always indicates the number of hybrid orbitals that must be created, we must mix two atomic orbitals to make two hybrid orbitals on each carbon atom.

The results of mixing one $2s$ and one $2p$ orbital on carbon and leaving the other two $2p$ orbitals unhybridized are shown in Figure 9.28(a). The two **sp hybrid orbitals** have major lobes that are on opposite sides of the carbon atom. One set of the unhybridized p orbitals form lobes above and below the axis of the sp hybrid orbitals; the second set is in front and in back of the plane of the hybrid orbitals.

FIGURE 9.28 (a) Mixing one $2s$ orbital and one $2p$ orbital on a carbon atom creates two sp hybrid orbitals and leaves the carbon atom with two unhybridized $2p$ orbitals. (b) Two sp hybridized carbon atoms each bond to one hydrogen atom and to the other carbon atom via σ bonds to form the linear molecule C_2H_2. The two unhybridized p orbitals overlap above and below the plane to form one π bond and in front of and in back of the plane to form a second π bond.

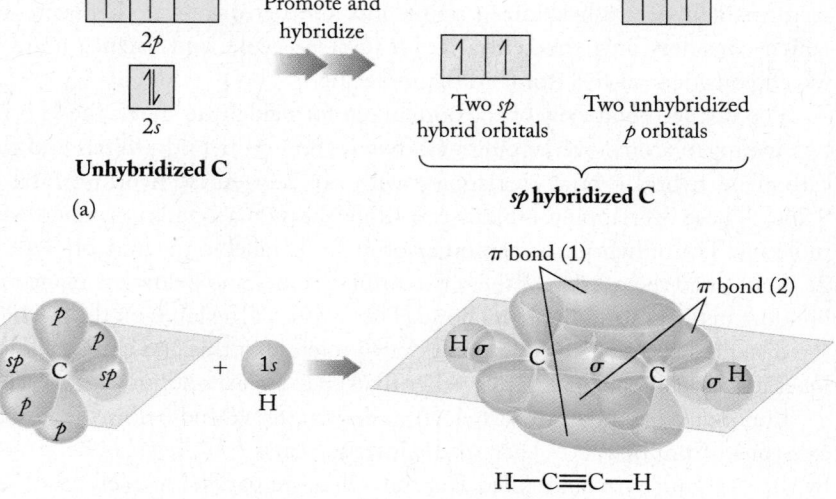

The valence bond view of C_2H_2 shows a σ-bonding framework in which a hydrogen $1s$ orbital overlaps with one sp hybrid orbital on each carbon atom. The atom's other sp hybrid orbital overlaps with one sp hybrid orbital on the other carbon atom (Figure 9.28b). This arrangement brings the two sets of unhybridized p orbitals on the two carbon atoms into parallel alignment with each other. They then overlap to form two π bonds between the two carbon atoms, so that the one σ bond and the two π bonds form the triple bond of $HC\equiv CH$. The steric number of the sp hybridized carbon atom in acetylene is 2 (no lone pairs and bonds to two atoms). The link between $SN = 2$ and sp hybridization also applies to nitrogen atoms in N_2. A lone pair occupies one of the two sp orbitals on each nitrogen atom and the other sp hybrid orbital forms the σ bond (Figure 9.29).

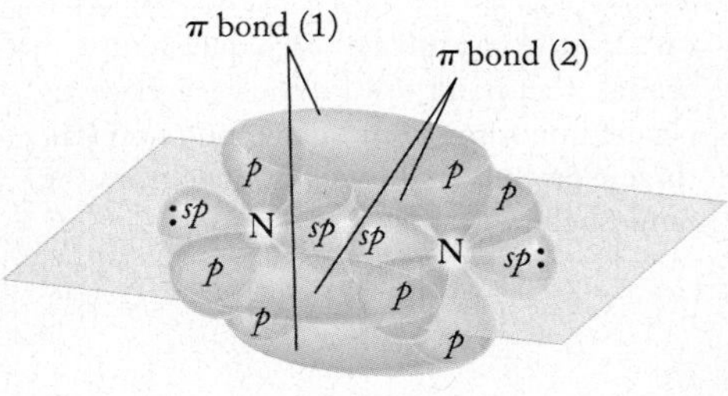

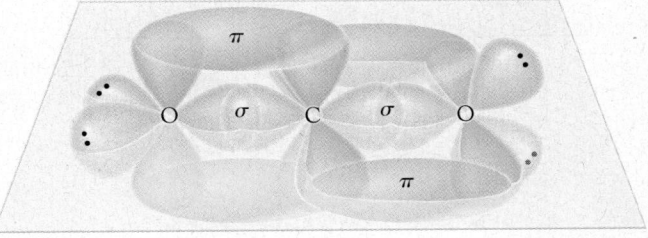

FIGURE 9.29 Valence bond view of a molecule of N_2 and its sp hybrid N atoms.

SAMPLE EXERCISE 9.5   **Describing Bonding in a Molecule**

Use valence bond theory to account for the linear molecular geometry of CO_2, determine the hybridization of carbon and oxygen in this molecule, and describe the orbitals that overlap to form the bonds. Draw the molecule, showing the orbitals that overlap to form the bonds.

Collect and Organize We know that CO_2 is linear and that the Lewis structure of the molecule is

$$\ddot{O}=C=\ddot{O}$$

Analyze Each double bond is composed of one σ bond and one π bond. Therefore, the carbon atom forms two σ bonds and two π bonds. It has no lone pairs and so $SN = 2$. Each oxygen atom forms one σ bond and one π bond.

Solve The carbon atom must be sp hybridized because half-filled sp hybrid orbitals have the capacity to form two σ bonds and the two half-filled unhybridized p orbitals are available to form the two π bonds (Figure 9.30). The lobes of sp hybrid orbitals are at an angle of 180°, which is consistent with the linear molecular geometry of CO_2. The oxygen atoms must be sp^2 hybridized because that hybridization leaves them with one unhybridized p orbital to form a π bond. The σ bonds form when each sp orbital on the carbon overlaps with one sp^2 orbital on an oxygen atom.

Notice that the two unhybridized p orbitals of carbon are oriented 90° to each other and that the plane containing the three sp^2 orbitals of one oxygen atom is rotated 90° with respect to the plane containing the three sp^2 orbitals of the other oxygen atom. This orientation is necessary for the formation of the two π bonds, because the two unhybridized p orbitals that overlap to form a π bond must be parallel. The unshared pairs of electrons on each oxygen atom and the σ bond to carbon lie in a trigonal plane 120° apart. The sp^2 hybridization of the oxygen accounts for all these features.

Think about It The sp hybridization of the carbon atom is consistent with: (1) its steric number ($SN = 2$), (2) its capacity to form two σ and two π bonds, and (3) the linear geometry of the molecule.

Practice Exercise In which of these molecules does the central atom have sp^3 hybrid orbitals? (a) CCl_4; (b) HCN; (c) SO_2; (d) PH_3

FIGURE 9.30 The bonding pattern in CO_2: the π bond to the left in the drawing is above and below the plane of the molecule; the π bond to the right is in front of and in back of the plane.

sp **hybrid orbitals** two hybrid orbitals on opposite sides of the hybridized atom formed by mixing one s and one p orbital.

sp^3d^2 hybrid orbitals six equivalent orbitals that point toward the vertices of an octahedron form from mixing one s orbital, three p orbitals, and two d orbitals from the same shell.

CONCEPT TEST

Why can a carbon atom with sp^3 hybrid orbitals not form π bonds?

Octahedral and Trigonal Bipyramidal Geometry: sp^3d^2 and sp^3d Hybrid Orbitals

Valence bond theory can also account for the molecular geometries of molecules with central atoms that have more than eight valence electrons. For example, the central sulfur atom in a molecule of SF_6 must expand its octet to bind six fluorine atoms. As we discussed in Chapter 8, only atoms with unoccupied d orbitals in their valence shells can expand their octets. Valence bond theory provides a way to describe the expansion of an octet by including these d orbitals in hybridization schemes. To form six σ bonds to six fluorine atoms, six atomic orbitals on sulfur—one 3s orbital, three 3p orbitals, and two 3d orbitals—hybridize producing six equivalent **sp^3d^2 hybrid orbitals** (Figure 9.31a).

Other molecules, such as phosphorus pentachloride (PCl_5), have a trigonal bipyramidal geometry, and the central phosphorus atom shares 10 valence electrons. Five atomic orbitals—one 3s orbital, three 3p orbitals, and one 3d orbital—can be

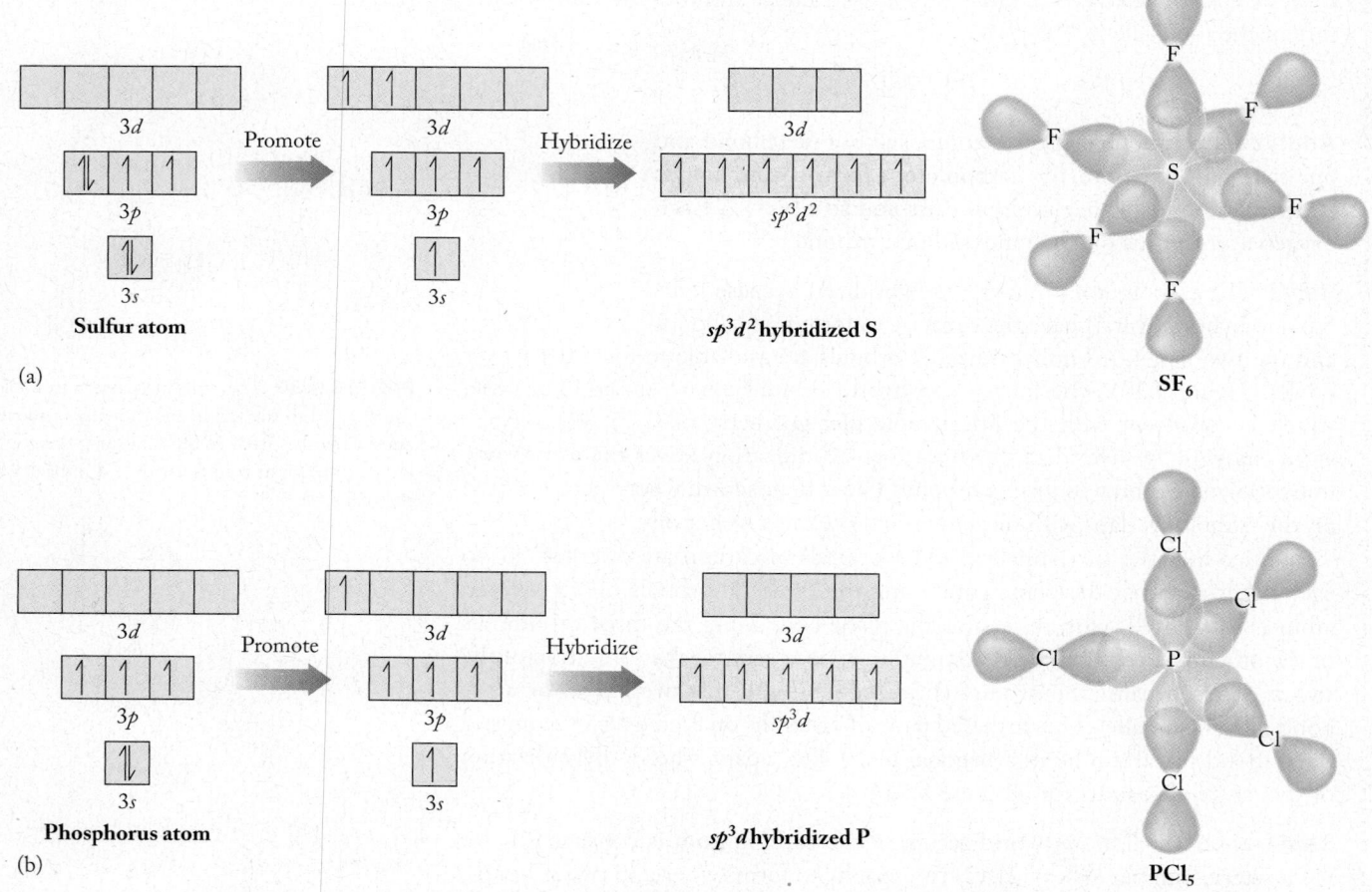

FIGURE 9.31 Hybrid orbitals can be generated by combining d orbitals with s and p orbitals in atoms that expand their octet. (a) One 3s, three 3p, and two 3d orbitals mix on the central sulfur atom in SF_6 to form six sp^3d^2 hybrid orbitals. (b) One 3s orbital, three 3p orbitals, and one 3d orbital mix on the central phosphorus atom in PCl_5 to form five sp^3d hybrid orbitals. For simplicity, only the chlorine and fluorine orbitals that overlap with the hybrid orbitals are shown.

hybridized to produce five equivalent *sp³d* **hybrid orbitals** with lobes that point toward the vertices of a trigonal bipyramid (Figure 9.31b). A summary of the shapes associated with all of the hybridization schemes we have discussed so far is given in Figure 9.32.

Hybridization	Orientation of Hybrid Orbitals	Number of σ Bonds	Molecular Geometries	Angles between Hybrid Orbitals
sp		2	Linear	180°
sp²		3 2	Trigonal planar Bent	120°
sp³		4 3 2	Tetrahedral Trigonal pyramidal Bent	109.5°
sp³d		5 4 3 2	Trigonal bipyramidal Seesaw T-shaped Linear	90°, 120°, 180°
sp³d²		6 5 4	Octahedral Square pyramidal Square planar	90°, 180°

FIGURE 9.32 Summary of hybridization schemes and the orientations of orbitals derived from them. Yellow orbitals are hybrids; blue are unhybridized atomic orbitals.

▶❚❚ **CHEMTOUR** Hybridization

SAMPLE EXERCISE 9.6 **Recognizing Hybridized Atoms in Molecules**

Here is the Lewis structure of SeF₄.

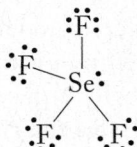

What is the shape of the molecule and the hybridization of the selenium atom in SeF₄?

Collect and Organize We are provided with the Lewis structure of SeF₄ and are to determine the hybridization of the central atom. Hybridization schemes are used to explain molecular shapes, so we need to determine the shape of the molecule first.

Analyze The central atom has four single (σ) bonds and one lone pair, giving it SN = 5.

---

sp³d **hybrid orbitals** five equivalent hybrid orbitals with lobes pointing toward the vertices of a trigonal bipyramid that form by mixing one *s* orbital, three *p* orbitals, and one *d* orbital from the same shell.

molecular recognition the process by which molecules interact with other molecules to produce a biological effect.

Solve The electron-pair geometry associated with SN = 5 is trigonal bipyramidal (Figure 9.2). The presence of the lone pair means that the molecular geometry is not the same as the electron-pair geometry. Instead, according to Table 9.1, the molecular shape is seesaw.

A steric number of 5 also means that the Se atom has an expanded octet of 10 valence electrons. To accommodate this many electrons, one d orbital must be involved in the hybridization scheme along with one s and three p orbitals, which means sp^3d hybridization.

Think about It The information in Figure 9.32 confirms that a seesaw molecular geometry is consistent with sp^3d hybridization.

Practice Exercise What is the hybridization of the iodine atom in IF_5 that is consistent with this Lewis structure:

$$
\begin{array}{c}
\ddot{\ddot{F}} \\
\ddot{\ddot{F}} \diagdown \overset{|}{I} \diagup \ddot{\ddot{F}} \\
\ddot{\ddot{F}} \diagup \overset{|}{|} \diagdown \ddot{\ddot{F}} \\
\ddot{\ddot{F}}
\end{array}
$$

9.5 Shape and Interactions with Large Molecules

Up to now, most of the molecules and ions we have considered have small molar masses and typically have a single central atom bonded to two or more other atoms. It is important to learn to apply the skills we have developed to larger molecules, however, because molecular shape is an important factor in determining the physical, chemical, and biological properties of all substances.

Living things respond to molecules that interact with regions in their tissues called *receptors* or *active sites*. The process by which these molecules and sites interact is known as **molecular recognition**. This recognition does not usually involve covalent bond formation. These noncovalent interactions require that the biologically active molecules and the receptors that respond to them fit tightly together, which means that they must have complementary three-dimensional shapes. An example of a biological effect caused by molecular recognition is the process by which produce such as green tomatoes ripen. Tomatoes ripen faster when stored in a paper or plastic bag instead of sitting on a kitchen counter. The reason why is that the tomatoes give off ethylene gas as they ripen, and this gas accelerates the ripening process when it is trapped in a bag.

The molecular structure of ethylene is shown in Figure 9.33. Both carbon atoms are bonded to three atoms and have no lone pairs of electrons. Therefore SN = 3. This means that the geometry around each carbon atom is trigonal planar and that the carbon atoms are both sp^2 hybridized (see Figure 9.32). Two of these orbitals form σ bonds with hydrogen $1s$ orbitals. The third forms the C=C σ bond. The C=C π bond is formed by overlap of the unhybridized $2p$ orbitals on the carbon atoms. Taken together, the two trigonal planar carbon atoms produce an overall planar geometry for ethylene, which means that all six atoms lie in the same plane.

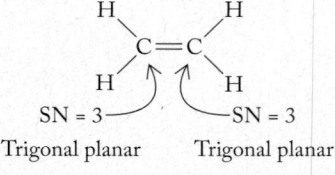

SN = 3 — Trigonal planar   — SN = 3 — Trigonal planar

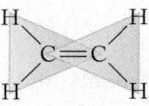

FIGURE 9.33 Ethylene molecules contain two carbon atoms that are at the centers of two overlapping triangular planes that are also coplanar. This combination means that all of the atoms are in the same plane.

SAMPLE EXERCISE 9.7   **Comparing Structures Using Valence Bond Theory**

Use valence bond theory to describe the bonding and the molecular geometry of ethane, CH_3—CH_3, and compare them to the geometry and bonding in ethylene.

Collect and Organize We are given the molecular formula of ethane and asked to describe its geometry and bonding and then compare them with the bonding and shape of ethylene.

Analyze We can use the Lewis structure for ethane and VSEPR to determine the molecular geometry and then determine the hybridization of its carbon atoms. We can then compare atom locations and electron distributions in the two molecules.

Solve There are four single bonds and no lone pairs around each carbon atom in ethane, which means (1) SN = 4 for both carbon atoms, (2) tetrahedral geometry around both, and (3) sp^3 hybridization of both. Comparing the overall molecular structures of ethane and ethylene (Figure 9.34), we see the distinctive three-dimensionality of tetrahedral environments: only two of the six C—H bonds in ethane are in the plane of the page, whereas all the bonds in ethylene are coplanar.

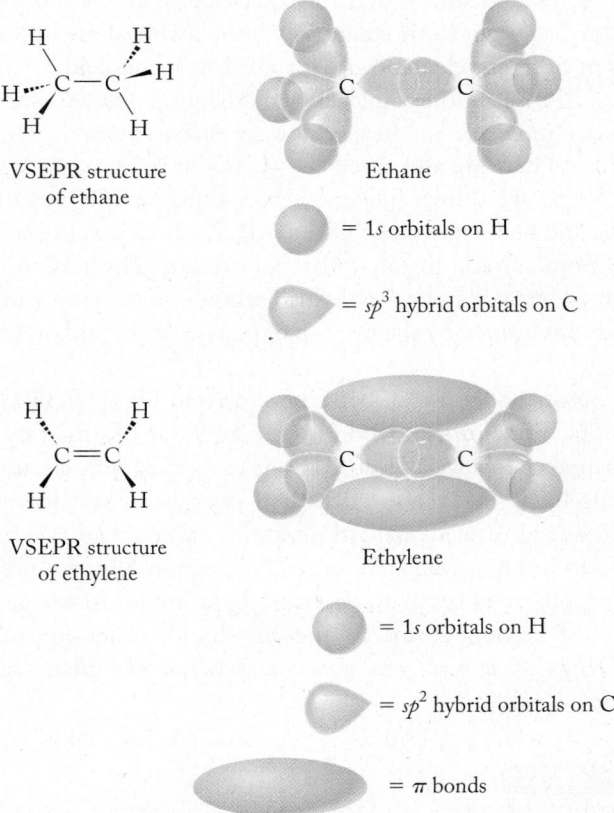

VSEPR structure of ethane

Ethane

   = 1s orbitals on H

   = sp^3 hybrid orbitals on C

VSEPR structure of ethylene

Ethylene

   = 1s orbitals on H

   = sp^2 hybrid orbitals on C

   = π bonds

FIGURE 9.34 Molecular structures of ethane and ethylene.

Think about It Remember the analogy at the beginning of the chapter about how molecules fit receptors like hands fit gloves. It's clear from the structures that ethylene and ethane would fit into very different gloves.

Practice Exercise Diazene (N_2H_2) and hydrazine (NH_2NH_2) are reactive nitrogen compounds. Use valence bond theory to compare the bonding in these two molecules, and describe the differences in their molecular structures.

delocalization (*adjective*: **delocalized**) when electrons in alternating single and double bonds are spread over three or more atoms in a molecule.

As Sample Exercise 9.7 illustrates, ethane and ethylene have similar formulas and molar masses, but very different molecular geometries. Ethane does not trigger the ripening process because its shape does not allow it to fit the receptor in plant tissue that binds the planar molecule ethylene.

Now let's consider a biologically active molecule with three "central" atoms. Acrolein is one of the components of barbeque smoke that contributes to the distinctive odor of a cookout. It is also a possible cancer-causing compound. The Lewis structure of acrolein:

shows that each molecule contains a C=O double bond and a C=C double bond. Both are formed by trigonal planar, sp^2 hybridized carbon atoms.

The pattern of alternating single and double bonds means that the electrons in the π bonds are **delocalized** over the three carbon atoms and the oxygen atom. Delocalization can occur both when the atoms involved are all carbon atoms or when atoms of different elements are involved, as in acrolein.

Benzene, C_6H_6, is another molecule in barbeque smoke. As we saw in Chapter 8, each benzene molecule is a hexagon of six carbon atoms, each bonded to one hydrogen atom. There are also three C=C bonds in benzene. Resonance structures for benzene are shown in Figure 9.35 along with a view of the π bonds located above and below the plane of the ring. Each carbon in benzene has a trigonal planar geometry and forms sp^2 hybrid orbitals. The carbon ring is made of σ bonds formed by overlapping sp^2 hybrid orbitals on adjacent carbon atoms. The C—H bonds are formed by the overlap of carbon sp^2 hybrid orbitals with hydrogen 1*s* orbitals.

The two resonance forms of benzene shown in Figure 9.35(a) correspond to shifts in the locations of the π bonds (Figure 9.35b) formed by the overlap of carbon 2*p* orbitals. However, because all the carbon 2*p* orbitals are identical, they are all equally likely to overlap with their neighbors, and so the π bonds are actually delocalized over all six carbon atoms rather than fixed between alternating pairs of carbon atoms. As noted in Section 8.5, the presence of these delocalized π bonds is often represented by a circle drawn in the middle of the hexagon of carbon atoms in the line-bond structure corresponding to continuous rings of π electrons above and below the plane of the molecule (Figure 9.35c).

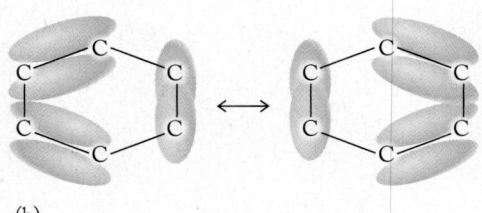

(a)

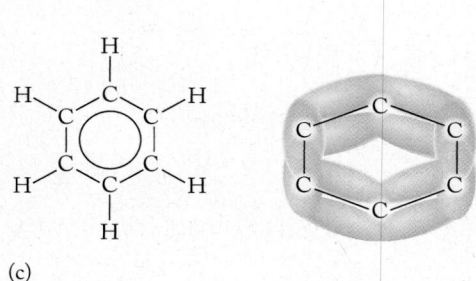

(b)

(c)

FIGURE 9.35 (a) The Lewis resonance structures of benzene can be explained using valence bond theory by showing localized π bonds (b) between adjacent carbon atoms. (c) Resonance leads to complete delocalization of the π bonds around the benzene ring. Hydrogen atoms have been excluded in the valence bond images to simplify the graphics.

CONCEPT TEST

In which of the molecules and polyatomic ions in Figure 9.36 are the π bonds delocalized?

FIGURE 9.36 Organic compounds with double bonds: (a) butadiene, used to make polymers; (b) 2,4-pentanedione, a reactive organic compound used in synthesis; and (c) the oxalate ion, present in spinach.

(a)   (b)   (c)

The molecular structures of many other compounds in addition to benzene contain carbon rings with delocalized π electrons above and below the plane of the ring. They are called **aromatic compounds**. An important class of these compounds known as *polycyclic aromatic hydrocarbons* (PAHs) consists of molecules containing several benzene rings joined together (Figure 9.37). PAHs are formed any time coal, oil, gas, and most hydrocarbon fuels are burned, and they are found in cigarette smoke and vehicle exhaust. In 2004 they were discovered in interstellar space.

The shape of PAH molecules gives rise to a particular health hazard. After we inhale or ingest them, some PAHs associate with DNA in a process called *intercalation*. Because PAHs are flat, they can intercalate in DNA—they slide into the two strands of DNA as shown in Figure 9.37. Once there, they sometimes form covalent bonds, altering or preventing DNA replication and thereby damaging or killing cells. Intercalation in DNA is one step in the process by which PAHs induce cancer.

The planarity of ethylene and of aromatic compounds as described by VSEPR and valence bond theories plays a key role in determining the behavior of such molecules in biological systems. The shape of these molecules matters because it influences their interaction with other molecules, which in turn determines their biological activity.

9.6 Chirality and Molecular Recognition

Before ending our exploration of molecular shape, we need to address the subject of handedness introduced in A Look Ahead. There we described how two molecules with the same molecular formula and Lewis structure interact differently with receptors in our nasal membranes. As a result, one produces the smell of caraway seeds, and the other spearmint leaves. You may find these different odors surprising when you consider the molecular structures of these two compounds (Figure 9.38). At first glance they may seem identical, but look closely. Notice in particular the bonding pattern around the carbon atom at the bottom of the ring. (The ring structures have been simplified by not including the carbon atoms in the ring and most of the H atoms bonded to them.) The H atom bonded directly to the bottom carbon atom is on the front side of the ring in the spearmint compound, but it is on the back side in the caraway compound. This minor difference in bond orientation creates a difference in molecular shape that is easily recognized by receptors in our noses.

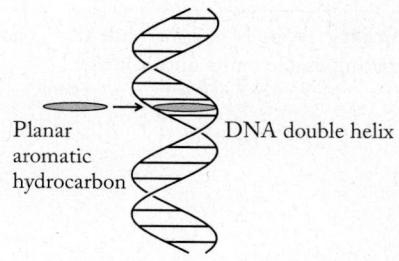

Planar aromatic hydrocarbon — DNA double helix

Intercalation of PAH in DNA

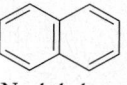

Naphthalene

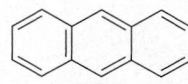

Anthracene

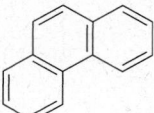

Phenanthrene

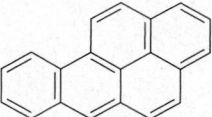

Benzo[*a*]pyrene

FIGURE 9.37 The molecules of these polycyclic aromatic hydrocarbons consist of fused benzene rings whose π bonds are delocalized over all the rings in each molecule. Molecules with this shape can slip in between the strands of DNA and disrupt cell replication, which can lead to cell death or induce malignancy.

(a) Spearmint   (b) Caraway

FIGURE 9.38 The distinctive aromas of (a) spearmint and (b) caraway are primarily due to two compounds with nearly identical molecular structures. The only difference between the two is the orientation of the two groups attached to the carbon atoms highlighted with the red circles. Note that the hydrogen atom is in front of the plane of the paper and down in structure a but is in back of the plane of the paper and down in b.

aromatic compound a cyclic, planar compound with delocalized π (pi) electrons above and below the plane of the molecule.

chirality property of a molecule that is not superimposable on its mirror image.

The structures in Figure 9.38 are called *optical isomers*. The term "optical" refers to the ways these compounds interact with a special kind of light called *plane-polarized* light. We explore this topic in more detail in Chapter 13, but it is sufficient now for you to know that when plane-polarized light passes through a solution of the caraway compound, the light twists in one direction, but when the light passes through a solution of the spearmint compound, it twists in the opposite direction. The term "isomer" is derived from *iso* (Greek for same) and *mer* (Greek for unit or part). Isomers are compounds formed from the same atoms, but within their molecules the atoms are arranged differently in three-dimensional space.

Optical isomerism has another name: **chirality**. Many molecules of biological importance, including the proteins and carbohydrates that we consume each day, are composed of chiral compounds. Chirality comes from the Greek word *chier* ("hand") and is quite correctly called "handedness." Although several features within molecules can lead to chirality, the most common one is the presence of a carbon atom that has four different atoms or groups of atoms attached to it.

To see how chirality works, let's look at a compound that contains a central carbon atom bonded to four different atoms. The compound is bromochlorofluoromethane (CHBrClF) (Figure 9.39a). It is used in fire extinguishers on airplanes. We will compare its molecular structure to that of dibromochloromethane (CHBr$_2$Cl), a compound that may form during the purification of municipal water supplies with chlorine, and that has only three different atoms bonded to its central carbon atom (Figure 9.39b). In the first step in our comparison we generate mirror images of both molecules. Then we rotate the mirror images 180° in an attempt to superimpose each mirror image on its original image. If the reflected, rotated image is superimposable on the original image, then the

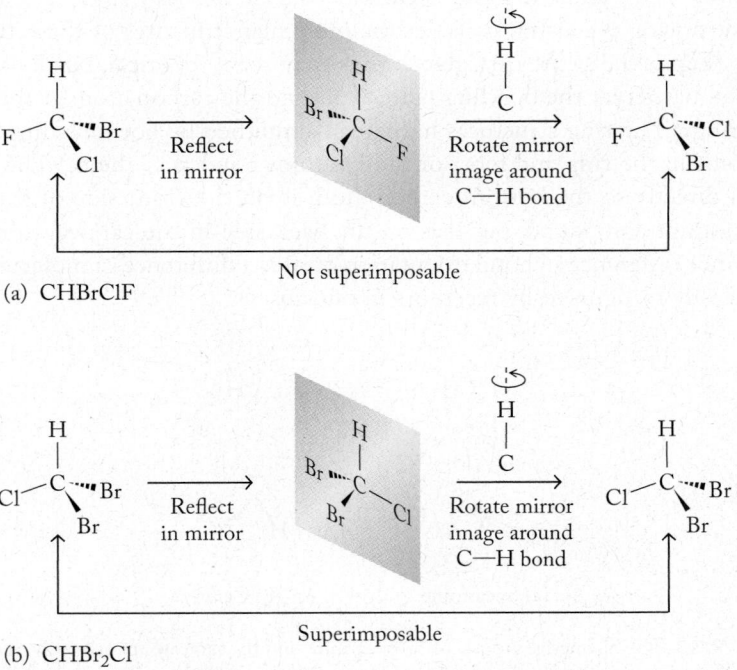

FIGURE 9.39 (a) A molecule of a chiral compound, such as CHBrClF, is not superimposable on its mirror image. (b) A molecule of a compound that is not chiral, such as CHBr$_2$Cl, is superimposable on its mirror image.

substance is not chiral. Scientists call it *achiral*. The molecular images of $CHBr_2Cl$ can be superimposed in this way, so it is not a chiral compound. However, the images of CHBrClF cannot be superimposed. For example, when we superimpose the F, C, and H atoms of the two images in Figure 9.39(a), the Br and Cl atoms are not aligned. This means that CHBrClF *is* a chiral compound.

Any structure like CHBrClF that has four different groups attached to an sp^3 hybridized carbon atom is chiral. It has two optical isomers that are mirror images of each other and distinctly different compounds. You may be wondering where the chiral carbon atom is in the molecules of the caraway and spearmint compounds. If you guessed the circled carbon atoms, you were right. Two of the four different groups are easy to see: a hydrogen atom and the

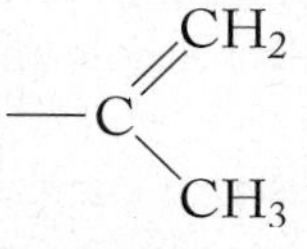

$$-C\diagup\diagdown\begin{matrix}CH_2\\CH_3\end{matrix}$$

group. The other two "groups" on the bottom carbon atom are really the two sides of the ring. The left side contains a C=C double bond, and the right side contains a C=O double bond. These differences mean that the two sides are not equivalent, and so the carbon atom is attached to four different groups, which makes it a chiral carbon atom, and the two compounds are optical isomers of each other.

> **CONCEPT TEST**
>
> Identify the molecules in Figure 9.40 that are chiral.

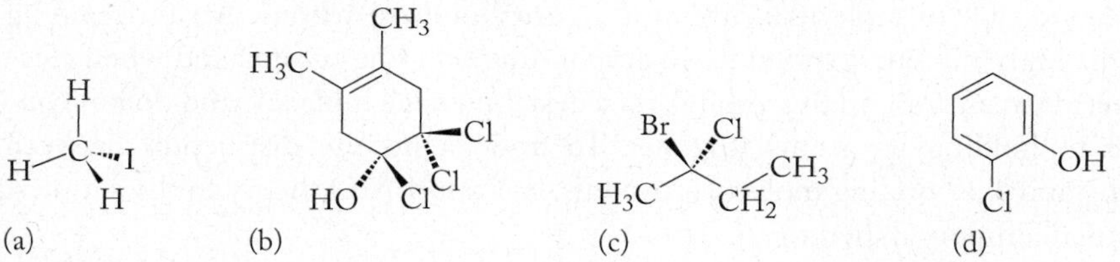

(a)   **(b)**   **(c)**   **(d)**

FIGURE 9.40

9.7 Molecular Orbital Theory

Lewis structures and valence bond theory help us understand the bonding capacities of individual atoms and the bonding patterns in molecules and molecular ions, while VSEPR and valence bond theories account for their molecular shapes. However, none of these models explain why O_2 is attracted to a magnetic field while N_2 is slightly repelled by one. For that explanation, we turn to **molecular orbital (MO) theory**.

In addition to magnetic behavior, MO theory also explains the absorption and emission of light by molecules and molecular ions including those in Earth's upper atmosphere that are produced in the shimmering colorful displays known as an *aurora* or northern lights (Figure 9.41). The colors of an aurora are caused by collisions between atoms, molecules, and molecular ions in the atmosphere with electrons and positive particles in the solar wind that are attracted to Earth's magnetic poles. These collisions produce excited-state species. As they return to their ground states, these species emit characteristic colors of light. We discussed in Chapter 7 how

molecular orbital (MO) theory a bonding theory based on the mixing of atomic orbitals of similar shapes and energies to form molecular orbitals that belong to the molecule as a whole.

FIGURE 9.41 Auroras are spectacular displays of color produced when the solar wind collides with Earth's upper atmosphere.

▶️II **CHEMTOUR** Chemistry of the Upper Atmosphere

CONNECTION In Chapter 7 we discussed the role of atomic emission and absorption spectra in the development of quantum mechanics.

CONNECTION We first met Friedrich Hund in Chapter 7 in the discussion of Hund's rule, which describes how electrons fill atomic orbitals.

molecular orbital a region of characteristic shape and energy where electrons in a molecule are located.

bonding orbital term in MO theory describing regions of increased electron density between nuclear centers that serve to hold atoms together in molecules.

antibonding orbital term in MO theory describing regions of electron density in a molecule that destabilize the molecule because they do not increase the electron density between nuclear centers.

sigma (σ) molecular orbital in MO theory, the lowest-energy orbital that forms when atomic orbitals mix; electrons in σ molecular orbitals form sigma (σ) bonds.

molecular orbital diagram in MO theory, an energy-level diagram showing the relative energies and electron occupancy of the molecular orbitals for a molecule.

excited-state atoms emit characteristic atomic spectra. In this section we explore how excited-state molecules and molecular ions, such as N_2 and N_2^+, do the same thing. Our exploration requires a different view of the bonding in these species; MO theory provides us that view.

The historical roots of MO theory can be traced to the same decades as Pauling's pioneering work in valence bond theory, but MO theory did not appeal to a wide number of chemists until the late 1950s and early 1960s. German physicist Friedrich Hund (1896–1997) and American chemist Robert S. Mulliken (1896–1986) were instrumental in developing MO theory.

Like valence bond theory, molecular orbital theory invokes the mixing of atomic orbitals. In valence bond theory, the mixing results in hybrid atomic orbitals, while molecular orbital theory is based on the formation of **molecular orbitals**. A key difference between the two types of orbitals is that hybrid atomic orbitals are associated with a particular atom in the molecule and molecular orbitals are spread out over all the atoms in a molecule.

Molecular orbitals represent discrete energy states in molecules just as atomic orbitals represent allowed energy states in free atoms. As with atomic orbitals, electrons enter and fill the lowest-energy MOs first; higher-energy MOs are filled as more electrons are added. Electrons in molecules can be raised to higher-energy MOs by absorbing quanta of electromagnetic radiation. When they return to lower-energy MOs, distinctive wavelengths of UV and visible radiation are emitted, including some of the shimmering colors in an aurora.

According to MO theory, when two atomic orbitals combine, they form two molecular orbitals. One of the two molecular orbitals is a **bonding orbital**. When electrons occupy a bonding orbital, they hold the molecule together by increasing electron density between the atoms; in other words, they form a covalent bond. The second type of molecular orbital is an **antibonding orbital**. An antibonding orbital is higher in energy than its corresponding bonding orbital, and when electrons reside in an antibonding orbital, they destabilize the molecule and do not contribute to holding its atoms together. To understand the distinction between bonding and antibonding molecular orbitals, let's apply MO theory to the simplest molecular compound, hydrogen.

CONCEPT TEST

In your own words describe one way in which hybrid orbitals and molecular orbitals are similar and one way in which they are different.

Molecular Orbitals of H₂

According to MO theory, a hydrogen molecule is formed when the $1s$ atomic orbitals on two hydrogen atoms combine to form two molecular orbitals. Molecular orbital theory stipulates that mixing two atomic orbitals creates two molecular orbitals and that these two orbitals represent two different energy states (Figure 9.42). As a general rule, *the number of molecular orbitals formed in a molecule equals the number of atomic orbitals combined*.

When two $1s$ atomic orbitals combine, the lower-energy, bonding molecular orbital formed is oval and spans the two atomic centers. Its shape corresponds to enhanced electron density between the two atoms that donated their atomic orbitals. This enhanced electron density is a covalent bond. When two electrons occupy a bonding MO, a single bond is formed. When the region of highest density lies along the bond axis, as it does in the bonding MO in H_2, the MO is designated a **sigma (σ) molecular orbital** and the covalent bond is a sigma (σ) bond.

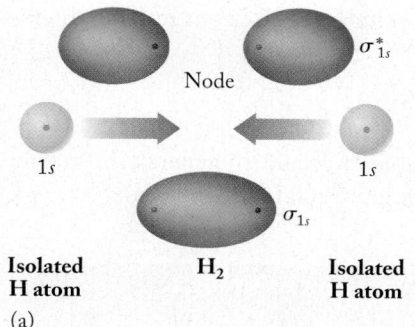

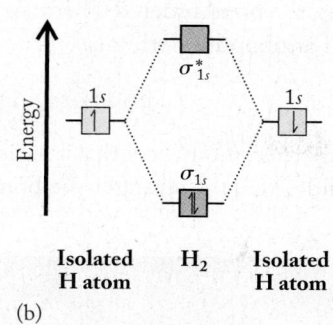

Isolated H atom **H₂** **Isolated H atom**
(a)

Isolated H atom **H₂** **Isolated H atom**
(b)

FIGURE 9.42 Mixing the 1s orbitals of two hydrogen atoms creates two molecular orbitals: a filled bonding σ_{1s} orbital containing two electrons and an empty antibonding σ_{1s}^* orbital. (a) The lower red oval is the bonding orbital. The two red ovals at the top together make up the antibonding orbital. *Note*: The two top ovals represent only *one* molecular orbital with a node of zero electron density in between. Dots show the locations of hydrogen nuclei. (b) A molecular orbital diagram shows the relative energies of bonding and antibonding molecular orbitals and of the atomic orbitals that formed them.

The σ bonding molecular orbital in H_2 is labeled σ_{1s} in Figure 9.42(a) because it is formed by mixing two 1s atomic orbitals.

The higher-energy (less stable), antibonding molecular orbital formed from two hydrogen atomic orbitals is designated σ_{1s}^* (pronounced "sigma star"). This antibonding orbital has two separate lobes of electron density and a region of zero electron density (a node) between the two hydrogen atoms, as shown in Figure 9.42(a).

Figure 9.42(b) is a **molecular orbital diagram**, analogous to an energy-level diagram for atomic orbitals. It shows that the σ_{1s} MO is lower in energy and therefore more stable than the 1s atomic orbitals by nearly the same amount that the σ_{1s}^* MO is higher in energy than the 1s atomic orbitals. Therefore, the formation of the two MOs does not significantly change the total energy of the system. A hydrogen molecule has two valence electrons, one from each H atom, both residing in the lower-energy σ_{1s} orbital because that is the lowest-energy orbital available. As in atomic orbitals, these two σ_{1s} electrons must have opposite spins. The electron configuration that corresponds to the molecular orbital diagram in Figure 9.42 is written $(\sigma_{1s})^2$ where the superscript indicates that there are two electrons in the σ_{1s} molecular orbital. Because the energy of the electrons in a $(\sigma_{1s})^2$ configuration is lower than the energy of the electrons in two isolated hydrogen atoms, MO theory explains why hydrogen is a diatomic gas: H_2 molecules are lower in energy and so are more stable than H atoms.

Hydrogen is a diatomic gas, but helium exists as free atoms and not as molecular He_2. MO theory explains why. One helium atom has two valence electrons in a 1s atomic orbital. Mixing two He 1s orbitals yields the same set of molecular orbitals we generated for H_2, as shown in Figure 9.43. Unlike H_2, the two helium atoms have a total of four valence electrons. Each molecular orbital in Figure 9.43—the bonding orbital and the antibonding orbital—has a maximum capacity of two electrons. Adding four valence electrons to the orbitals in Figure 9.43 means filling both orbitals. The presence of two electrons in the σ_{1s}^* orbital cancels the stability gained from having two electrons in the σ_{1s} orbital. Because there is no net gain in stability, He_2 does not form.

Another way of comparing the bonding in H_2 and He_2 is to look at the *bond order* in the two molecules. We have previously defined bond order as the number of bonds between two atoms: a bond order of 1 for X—X, 2 for X=X, and 3 for X≡X. In MO theory we define bond order as follows:

$$\text{Bond order} = \frac{1}{2}\left(\begin{array}{c}\text{number of}\\\text{bonding electrons}\end{array}\right) - \left(\begin{array}{c}\text{number of}\\\text{antibonding electrons}\end{array}\right) \quad (9.2)$$

In a molecule of H_2, there are two electrons in the bonding MO and none in the antibonding MO, so

$$\text{Bond order in } H_2 = \frac{1}{2}(2 - 0) = 1$$

CONNECTION We used energy-level diagrams in Chapter 7 to show the transitions that electrons in atoms can make from one energy level to another.

CONNECTION We first defined bond order in Chapter 8 when we related the length and strength of bonds to the number of electron pairs shared by two atoms.

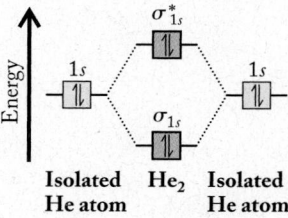

Isolated He atom **He₂** **Isolated He atom**

FIGURE 9.43 The molecular orbital diagram for the fictitious molecule He_2 indicates that the same number of electrons occupy the antibonding orbital and the bonding orbital. Therefore the bond order is 0; the molecule is not stable.

For He_2, the bond order is 0 because an equal number of electrons reside in bonding and antibonding orbitals:

$$\text{Bond order in } He_2 = \tfrac{1}{2}(2 - 2) = 0$$

A bond order of 0 means that He_2 is not a stable molecule. In general, the greater the bond order, the stronger the bond and the more stable the molecule.

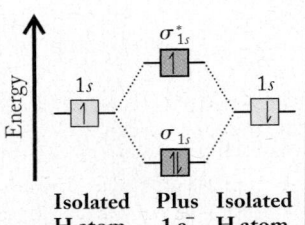

FIGURE 9.44 The molecular orbital diagram of H_2^-.

SAMPLE EXERCISE 9.8 **Using MO Diagrams to Predict Bond Order I**

Draw the MO diagram for the molecular ion H_2^-, determine the bond order of the ion, and predict whether or not the ion is stable.

Collect and Organize We apply MO theory to draw the MO diagram for the molecular ion H_2^-, and then determine bond order using Equation 9.2. If the value of the bond order is greater than zero, the ion may be stable.

Analyze We should be able to base the MO diagram for H_2^- on the MO diagram for H_2 (Figure 9.42) because the H_2^- ion has only one more electron than H_2 and the empty σ_{1s}^* orbital in H_2 can accommodate up to two more electrons.

Solve The σ_{1s} orbital is filled in H_2. The third electron goes into the σ_{1s}^* orbital, so the MO diagram is as shown in Figure 9.44. The notation for this electron configuration is $(\sigma_{1s})^2(\sigma_{1s}^*)^1$ (listing the molecular orbitals in order of increasing energy). The bond order is

$$\text{Bond order} = \tfrac{1}{2}(2 - 1) = \tfrac{1}{2}$$

The H_2^- ion is not as stable as H_2, but it is more stable than He_2.

Think about It We encountered the idea of fractional bonds in Chapter 8 in the discussion of resonance, and we encounter it again here in the MO treatment of H_2^-. A bond order of $\tfrac{1}{2}$ in MO theory means that the bond between the two atoms in H_2^- is weaker than the single bond in H_2, making H_2^- a less stable species.

Practice Exercise Use MO theory to predict whether the H_2^+ ion can exist.

Molecular Orbitals of Homonuclear Diatomic Molecules

Molecular orbital diagrams for homonuclear (same atom) diatomic molecules like N_2 and O_2 are more complex than that of H_2 because of the greater number and variety of atomic orbitals in N_2 and O_2. Not all combinations of atomic orbitals result in effective bonding, but there are some general guidelines for constructing the molecular orbital diagram for any molecule:

1. The number of molecular orbitals equals the number of atomic orbitals used to create them.
2. Atomic orbitals with similar energy and shape mix more effectively than do those that have different energies and shapes. For example, an s atomic orbital mixes more effectively with another s atomic orbital than with a p orbital.

3. Atomic orbitals of different principal quantum numbers (for example, $1s$ and $2s$) have different sizes and energies, resulting in less effective mixing than two $1s$ or two $2s$ orbitals. Better mixing leads to a larger energy difference between bonding and antibonding orbitals and thus greater stabilization of the bonding MOs.

4. A molecular orbital can accommodate a maximum of two electrons; two electrons in the same MO have opposite spins.

5. Electrons occupy the lowest-energy molecular orbitals available, following Hund's rule.

In mixing atomic orbitals to create molecular orbitals, we consider *only the valence electrons* on the atoms because core electrons do not participate in bonding. Focusing on N_2 and O_2 as examples, we first mix their $2s$ orbitals. The mixing process is analogous to the one we used for H_2, except that the resulting MOs are designated σ_{2s} and σ_{2s}^*.

Next we mix the three pairs of $2p$ orbitals, producing a total of six MOs. The different spatial orientations of the $2p_x$, $2p_y$, and $2p_z$ atomic orbitals result in different kinds of MOs (Figure 9.45). The $2p_z$ atomic orbitals point toward each other.

CONNECTION In Chapter 7 we discussed Hund's rule, which states that the lowest-energy electron configuration of an atom is the one with the maximum number of unpaired electrons, all having the same spin.

FIGURE 9.45 Two atoms come together and their p atomic orbitals mix to form six molecular orbitals. (a) The $2p_z$ atomic orbitals create a σ_{2p} bonding orbital and a σ_{2p}^* antibonding orbital. (b) The $2p_x$ and $2p_y$ atomic orbitals mix to form two π_{2p} bonding molecular orbitals and two π_{2p}^* antibonding molecular orbitals.

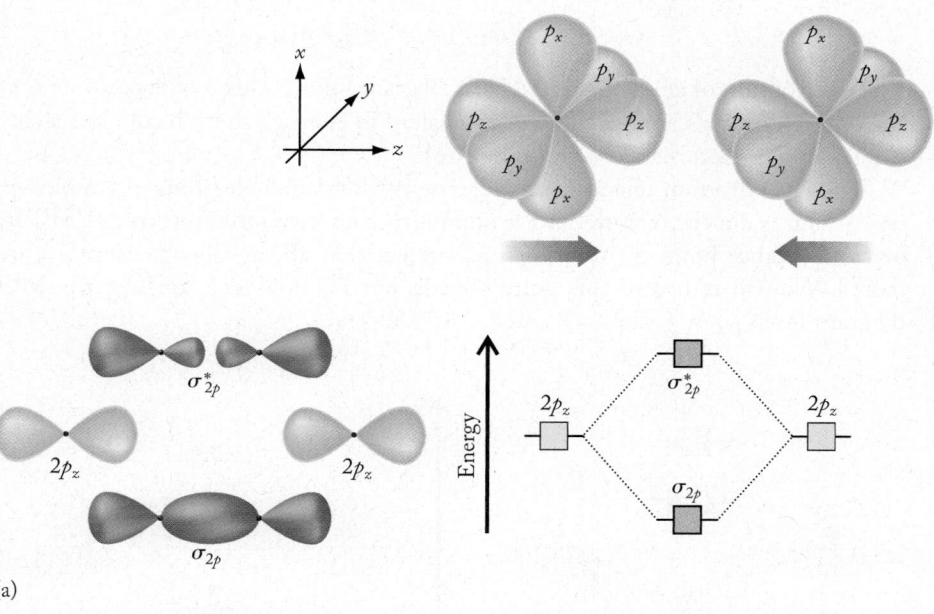

(a)

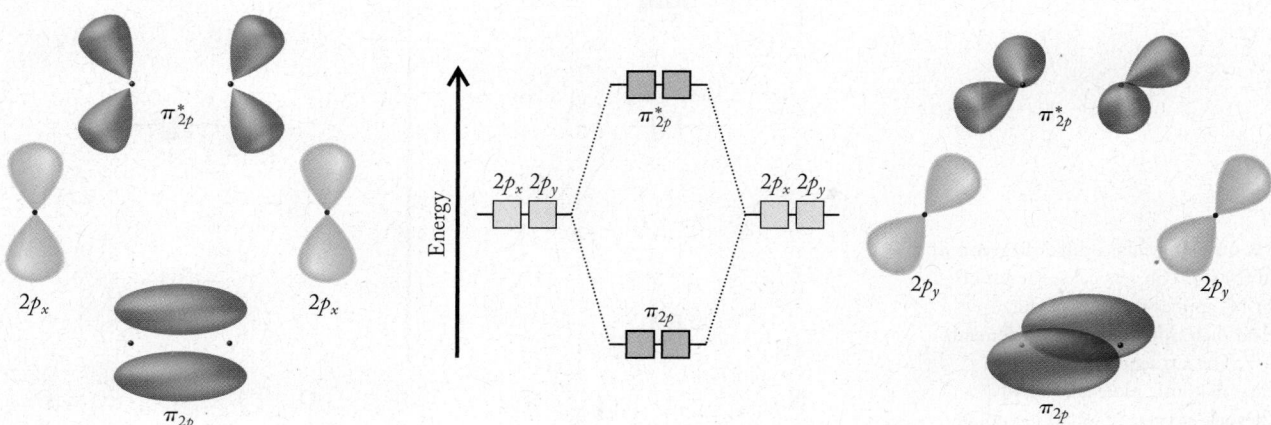

(b)

pi (π) molecular orbitals in MO theory, formed by the mixing of atomic orbitals oriented above and below, or in front of and behind the bonding axis.

▶‖ **CHEMTOUR** Molecular Orbitals

When they mix, two molecular orbitals form, a σ_{2p} bonding orbital and a σ_{2p}^* antibonding orbital. The lobes of the $2p_y$ and $2p_x$ atomic orbitals are oriented at 90° to the bonding axis and also at 90° to each other. When the $2p_x$ orbitals mix together, and when the $2p_y$ orbitals mix together, they do so around the bonding axis instead of along it. This mixing produces two **pi (π) molecular orbitals** and two π^* molecular orbitals. When electrons occupy a π orbital, they form a π bond.

The relative energies of σ and π molecular orbitals for N_2 and O_2 are shown in Figure 9.46. In each molecule, the energies of the MOs derived from two $2s$ atomic orbitals (σ_{2s} and σ_{2s}^*) are lower than the energy of the σ_{2p} MO for the same reason that a $2s$ atomic orbital is lower in energy than a $2p$ atomic orbital.

Now let's consider the relative energies of the MOs formed by mixing the $2p$ orbitals in N_2 and O_2. We begin with O_2 (Figure 9.46b) because it is representative of most homonuclear diatomic molecules including all the halogens. In order of increasing energy the MOs are σ_{2p}, π_{2p}, π_{2p}^*, and σ_{2p}^*. Keep in mind that there are groups of two π_{2p} and two π_{2p}^* orbitals. This means that each group of two can hold 4 electrons. Adding 12 valence electrons (6 from each O atom) in the O_2 molecule into these MOs, starting with the lowest-energy MO first and working our way up, we get the following electron configuration:

$$O_2: \quad (\sigma_{2s})^2(\sigma_{2s}^*)^2(\sigma_{2p})^2(\pi_{2p})^4(\pi_{2p}^*)^2$$

The distribution of electrons in the MO diagram follows this sequence. Note that the two π_{2p}^* orbitals are degenerate (equivalent in energy), so each contains a single electron, in accordance with Hund's rule.

The MO diagram tells us that there are two unpaired electrons in a molecule of O_2. This is not the picture that we obtain from a Lewis structure, from VSEPR, or from valence bond theory, which all predict that all the valence electrons are paired. We will return to this point shortly, but for now let's consider the MO diagram for N_2.

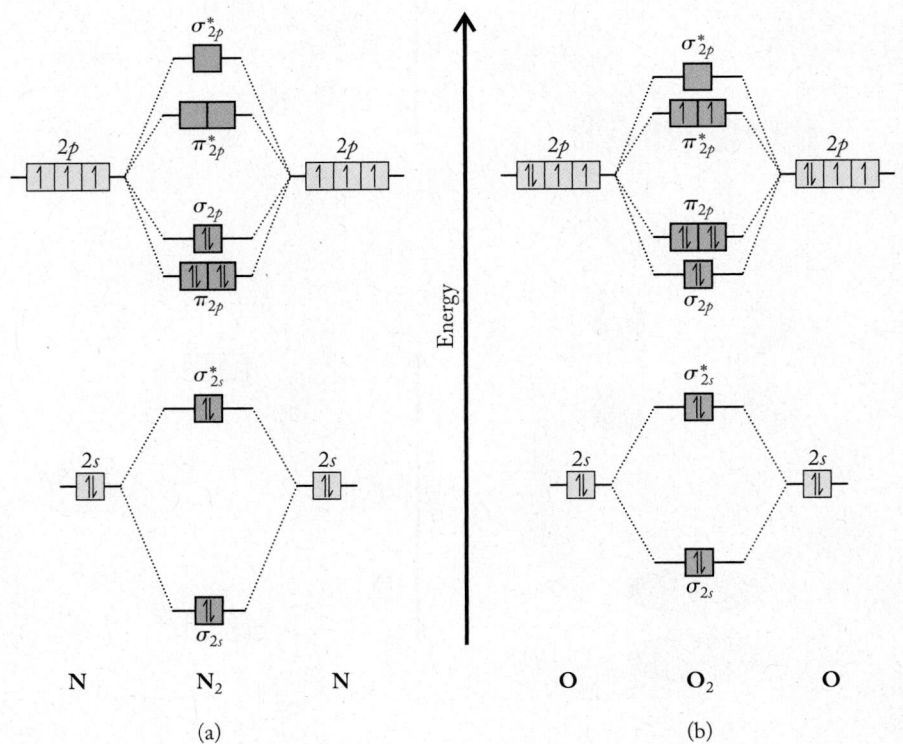

FIGURE 9.46 Molecular orbital diagrams of (a) N_2 and (b) O_2. The vertical sequence of orbitals for N_2 applies to MOs of the homonuclear diatomic molecules of elements with $Z \leq 7$. The O_2 sequence applies to homonuclear diatomic molecules of all elements beyond oxygen ($Z \geq 8$) including the halogens.

When we compare the MO diagrams of N_2 and O_2 in Figure 9.46, we see a difference in the relative energies of two of their MOs. The π_{2p} molecular orbital is lower in energy than the σ_{2p} orbital in N_2. This switch of energy levels from their relative positions in O_2 and many other diatomic molecules is thought to be due to the stability (and lower energy) of the three half-filled $2p$ orbitals in N atoms ($2s^2 2p^3$), which brings their energy closer to that of the $2s$ orbital. This proximity of orbitals has the effect of lowering the energy of the σ_{2s} molecular orbital and raising the energy of the σ_{2p} orbital—enough to put it above π_{2p}. Adding 10 valence electrons to this stack of MOs in N_2, lowest energy first, produces the following electron configuration:

$$N_2: \quad (\sigma_{2s})^2(\sigma_{2s}^*)^2(\pi_{2p})^4(\sigma_{2p})^2$$

The distribution of electrons in the MO diagram in Figure 9.46(a) reflects this electron configuration. There are a total of 8 electrons in bonding MOs and 2 in antibonding MOs. Using Equation 9.2 to calculate the bond order for N_2, we get

$$\text{Bond order} = \tfrac{1}{2}(8 - 2) = 3$$

In O_2, 8 electrons occupy bonding orbitals and 4 electrons occupy antibonding orbitals, so

$$\text{Bond order} = \tfrac{1}{2}(8 - 4) \doteq 2$$

On the basis of their Lewis structures, we predicted in Section 8.2 a triple bond in N_2 and a double bond in O_2, and molecular orbital theory leads us to the same predictions.

As we indicated previously, one of the strengths of MO theory is that it enables us to explain properties of molecular compounds that can't be explained by other bonding theories. The magnetic behavior of homonuclear diatomic molecules is a case in point. Electrons in atoms have two possible spin orientations depending on the value of their spin magnetic quantum number m_s. Figure 9.47 shows the MO-based electron configurations in diatomic molecules of the period 2 elements. Note that in most of these molecules all of the electrons are paired. The molecules that make up most substances contain only paired electrons.

CONNECTION In Chapter 7 we introduced the spin magnetic quantum number (m_s), which has a value of either $+\tfrac{1}{2}$ or $-\tfrac{1}{2}$ and defines the orientation of an electron in a magnetic field.

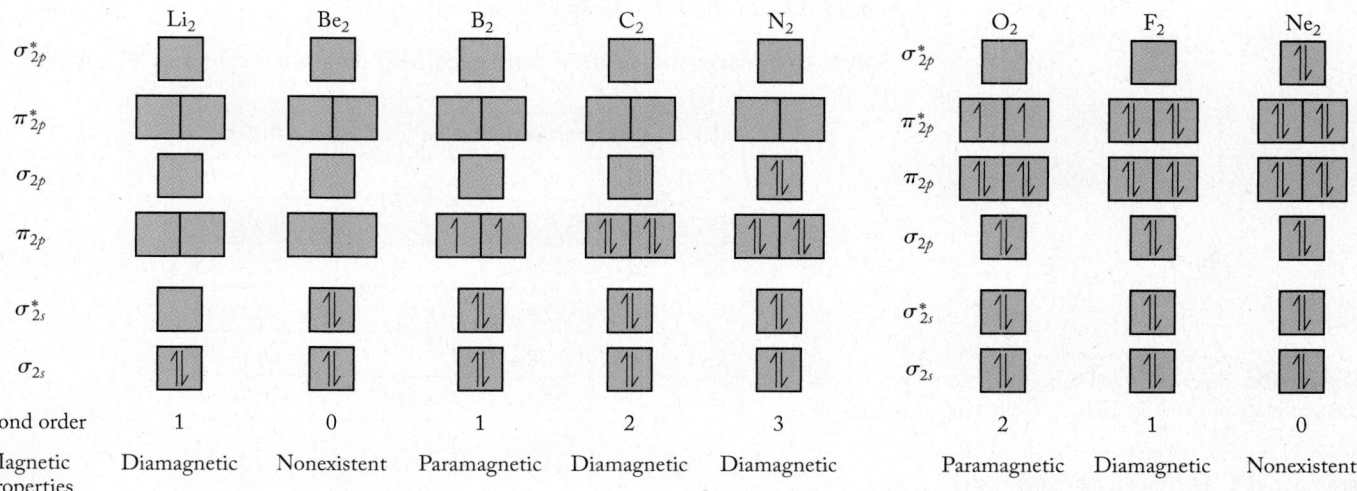

FIGURE 9.47 Valence-shell molecular orbital diagrams and magnetic properties of the homonuclear diatomic molecules of the second row elements.

FIGURE 9.48 Liquid O_2 poured from a Styrofoam cup is suspended in the space between the poles of this magnet because the unpaired electrons in its molecules make O_2 paramagnetic. Paramagnetic substances are attracted to magnetic fields.

This complete electron pairing means that these substances are repelled slightly by a magnetic field. These substances are said to be **diamagnetic**. If a substance's molecules contain unpaired electrons, as do those of O_2, then it is attracted by a magnetic field, as shown in Figure 9.48, and is **paramagnetic**. The more unpaired electrons in a molecule, the greater the paramagnetism. Only MO theory accounts for the magnetic behavior of oxygen and of many other substances as well.

CONCEPT TEST

Can liquid N_2 and O_2 be separated from each other with a magnet?

Figure 9.47 enables us to make several predictions about diatomic molecules of the second row elements. First, we predict that Be_2 and Ne_2 do not exist for the same reason that He_2 does not exist: both Be_2 and Ne_2 have as many antibonding electrons as they have bonding electrons and therefore have a net bond order of 0. Second, Li_2, B_2, and F_2 have a bond order of 1, whereas C_2 has a bond order of 2. Like O_2, B_2 is paramagnetic, whereas Li_2, C_2, N_2, and F_2 are diamagnetic.

SAMPLE EXERCISE 9.9   **Using MO Diagrams to Predict Bond Order II**

In which molecules in Figure 9.47 is there an increase in bond order when one electron is removed from the molecule?

Collect and Organize We are to determine which molecules in Figure 9.47 acquire a higher bond order when one electron is removed. Equation 9.2 relates bond order to the difference in the numbers of electrons in bonding and antibonding orbitals.

Analyze Removing an electron from Li_2, Be_2, B_2, C_2, N_2, O_2, and F_2 will result in the molecular ions Li_2^+, Be_2^+, B_2^+, C_2^+, N_2^+, O_2^+, and F_2^+. We may assume that the molecular ions have MO diagrams with orbital energies in the same order as their parent molecules. Thus, the MO diagrams for the molecular ions are the same as those in Figure 9.47, but with one electron removed from the highest-energy orbital.

Solve Removing one electron from each MO diagram in Figure 9.47 gives us

Ion	Electron Configuration	Bond Order
Li_2^+	$(\sigma_{2s})^1$	$\frac{1}{2}(1-0) = 0.5$
Be_2^+	$(\sigma_{2s})^2(\sigma_{2s}^*)^1$	$\frac{1}{2}(2-1) = 0.5$
B_2^+	$(\sigma_{2s})^2(\sigma_{2s}^*)^2(\pi_{2p})^1$	$\frac{1}{2}(3-2) = 0.5$
C_2^+	$(\sigma_{2s})^2(\sigma_{2s}^*)^2(\pi_{2p})^3$	$\frac{1}{2}(5-2) = 1.5$
N_2^+	$(\sigma_{2s})^2(\sigma_{2s}^*)^2(\pi_{2p})^4(\sigma_{2p})^1$	$\frac{1}{2}(7-2) = 2.5$
O_2^+	$(\sigma_{2s})^2(\sigma_{2s}^*)^2(\sigma_{2p})^2(\pi_{2p})^4(\pi_{2p}^*)^1$	$\frac{1}{2}(8-3) = 2.5$
F_2^+	$(\sigma_{2s})^2(\sigma_{2s}^*)^2(\sigma_{2p})^2(\pi_{2p})^4(\pi_{2p}^*)^3$	$\frac{1}{2}(8-5) = 1.5$

Comparing these values with the bond orders listed in Figure 9.47, we see that bond order increases for only Be_2^+, O_2^+, and F_2^+.

diamagnetic describes a substance with no unpaired electrons that is weakly repelled by a magnetic field.

paramagnetic describes a substance with unpaired electrons that is attracted to a magnetic field.

Think about It Removing an electron from Be_2, O_2, or F_2 reduces the number of electrons in antibonding molecular orbitals while leaving the number of electrons in bonding molecular orbitals unchanged. The result is an increase in bond order. In the other four homonuclear diatomic molecules, removing an electron reduces the number of electrons in bonding molecular orbitals while leaving the number of electrons in antibonding orbitals unchanged. The result is a reduction in bond order for Li_2^+, B_2^+, C_2^+, and N_2^+.

Practice Exercise Which molecules in Figure 9.47 show an increase in bond order when one electron is added to the molecule?

Molecular Orbitals of NO and Other Heteronuclear Diatomic Molecules

Molecular orbital theory also enables us to account for the bonding in *heteronuclear* diatomic molecules, which are molecules containing two different atoms. The bonding in some of these molecules is difficult to explain using other bonding theories. For example, it is often difficult to draw a single Lewis structure for an odd-electron molecule like nitric oxide (NO). In Chapter 8, we considered several arrangements of the valence electrons in nitrogen monoxide, such as

$$:\ddot{N}=\ddot{O}: \qquad :\ddot{N}=\ddot{O}·$$

We predicted that oxygen was more likely to have a complete octet of valence electrons because it is the more electronegative element. In addition, experimental evidence allowed us to rule out structures with unpaired electrons on the oxygen atom. Our preferred structure was therefore the one shown in red. However, the bond length in NO (115 pm) is considerably shorter than the value in Table 8.2 for an average $N=O$ double bond (122 pm). Molecular orbital theory is useful for explaining both the bonding in NO and the deviation from the expected bond length.

Let's look at the bonding first. The MO diagram for nitric oxide is different from the diagrams of homonuclear diatomic gases. Nitrogen and oxygen atoms have different numbers of protons and electrons, and the difference in effective nuclear charge in N and O atoms means that their atomic orbitals have different energies, as Figure 9.49 shows for the 2s and 2p orbitals.

In constructing the MO diagram for NO, the guidelines described previously still apply. The number of MOs formed must equal the number of atomic orbitals combined, and the energy and orientation of the atomic orbitals being mixed must be considered. One additional factor influences the energies of the MOs in heteronuclear diatomic molecules: *bonding* MOs tend to be closer in energy to the atomic orbitals of the more electronegative atom and *antibonding* MOs tend to be closer in energy to the atomic orbitals of the less electronegative atom. The MO diagram for NO in Figure 9.49 illustrates this phenomenon. Note how the energy of the bonding σ_{2s} orbital is closer to that of the 2s orbital of the O atom, and the energy of the antibonding

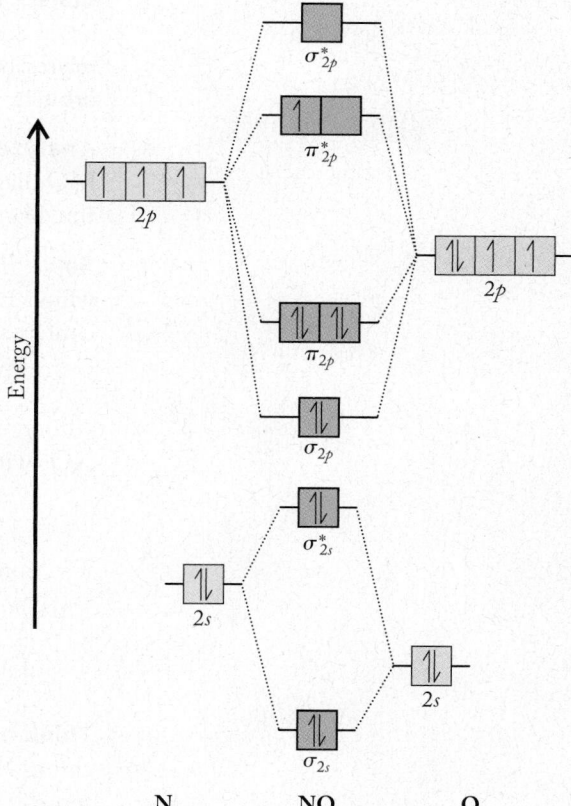

FIGURE 9.49 The molecular orbital diagram for NO shows that the unpaired electron occupies a π_{2p}^* antibonding orbital, which is closer in energy to the 2p atomic orbitals of nitrogen than to the 2p atomic orbitals of oxygen. As a result of this proximity, the electron has more nitrogen character and is more likely to be located on the nitrogen atom than on the oxygen atom.

σ_{2s}^* orbital is closer to that of the $2s$ orbital of the N atom. Similarly, the π_{2p} MOs in NO are closer in energy to the $2p$ orbitals of oxygen, and the π_{2p}^* MOs are closer in energy to the $2p$ orbitals of nitrogen. The proximity of the nitrogen $2p$ atomic orbitals to the π_{2p}^* MOs means that the single electron in the π_{2p}^* MOs is more likely to be on nitrogen than on oxygen. This prediction is consistent with our Lewis structure in which the odd electron in NO is on the nitrogen atom.

Molecular orbital theory also enables us to rationalize the relatively short bond length in NO. Equation 9.2 tells us that the bond order is $\frac{1}{2}(8 - 3) = 2.5$, halfway between the bond orders for N$=$O and N$\equiv$O and consistent with a bond length of 115 pm, halfway between the lengths of the N$=$O bond (122 pm) and the N$\equiv$O bond (106 pm).

SAMPLE EXERCISE 9.10 Using MO Diagrams for Heteronuclear Diatomic Molecules

Nitrogen monoxide reacts with many transition metals, including the iron in our blood. In these compounds, NO is sometimes considered to be NO^+ and at other times NO^-. Use Figure 9.49 to predict the bond order of NO^+ and NO^-.

Collect and Organize We are to predict the bond order of two diatomic ions based on the MO diagram of their parent molecule (Figure 9.49). Equation 9.2 relates bond order to the numbers of electrons in bonding and antibonding orbitals.

Analyze NO has 11 valence electrons. We can remove one electron from the MO diagram for NO to get the diagram for NO^+ and add one electron to get the diagram for NO^-.

Solve To generate NO^+, we remove the highest-energy electron in NO, which is the one in the π_{2p}^* molecular orbital. This gives NO^+ the electron configuration

$$NO^+: \quad (\sigma_{2s})^2(\sigma_{2s}^*)^2(\sigma_{2p})^2(\pi_{2p})^4$$

Adding an electron to the lowest-energy MO available in NO (also π_{2p}^*) yields NO^- with the electron configuration

$$NO^-: \quad (\sigma_{2s})^2(\sigma_{2s}^*)^2(\sigma_{2p})^2(\pi_{2p})^4(\pi_{2p}^*)^2$$

The bond orders of the two ions are

$$NO^+: \quad \text{bond order} = \tfrac{1}{2}(8 - 2) = 3$$

$$NO^-: \quad \text{bond order} = \tfrac{1}{2}(8 - 4) = 2$$

Think about It The bond orders in N_2 and O_2 are 3 and 2, respectively. The cation NO^+ is isoelectronic with N_2, so our calculated bond order for it makes sense. The anion NO^- is isoelectronic with O_2, and so our calculated bond order for this ion is also reasonable.

Practice Exercise Using Figure 9.49 as a guide, draw the MO diagram for carbon monoxide, and determine the bond order for the carbon–oxygen bond.

Molecular Orbitals of N_2^+ and Spectra of Auroras

In addition to predicting the magnetic properties of molecules, MO theory is particularly useful for predicting their spectroscopic properties—and the colors of auroras. In Section 7.4, we learned that the light emitted by excited free atoms is quantized and can be related to the movement of electrons between atomic orbitals. Broadly speaking, the same is true in molecules: electrons can move from one molecular orbital to another by absorbing or emitting light.

We can use this information to look again at the phenomenon described at the opening of this section: how the colors of the aurora are produced. The principal chemical species involved are listed in Table 9.3. An asterisk indicates a molecule or molecular ion in an excited state. Excited N_2^+ ions produce blue-violet (391–470 nm) light, and excited N_2 molecules produce deep crimson red (650–680 nm) light. The MO diagrams for these species are shown in Figure 9.50. Comparing the MO diagrams of N_2^* and N_2 in Figure 9.50(a) we find that one of the two electrons originally in the σ_{2p} MO in N_2 has been raised to a π_{2p}^* orbital in N_2^*, leaving an unpaired σ_{2p} electron behind. Figure 9.50(b) shows us that N_2^{+*} also has one electron in a π_{2p}^* orbital, but its σ_{2p} orbital is empty because the other σ_{2p} electron originally in the N_2 molecule was lost when the molecule was ionized. As π_{2p}^* electrons return from their antibonding, excited-state orbitals to the bonding σ_{2p} orbital in the ground state, the distinctive blue-violet and crimson emissions of N_2^+ and N_2 appear as shown in the photograph in Figure 9.41.

TABLE 9.3	Origins of Colors in the Aurora	
Wavelength (nm)	Color	Chemical Species
650–680	Deep red	N_2^*
630	Red	O^*
558	Green	O^*
391–470	Blue violet	N_2^{+*}

CONCEPT TEST

Are the bond orders of the excited-state species in Figure 9.50 the same as the ground-state species?

In Chapter 8 and in this chapter we have presented several theories of chemical bonding. Each theory has its strengths and weaknesses. The best one to apply in a given situation depends on the question being asked and on the level of sophistication required in the answer. Molecular orbital theory may provide the most complete picture of covalent bonding, but it is also the most difficult to apply to large molecules.

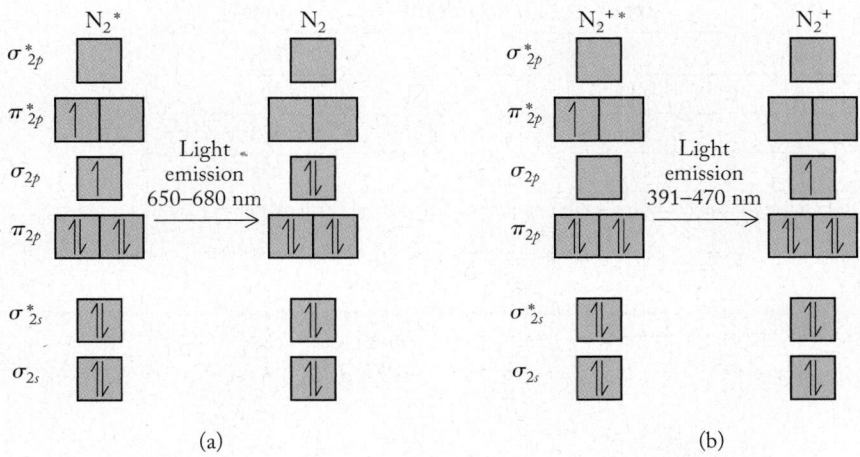

(a)   (b)

FIGURE 9.50 Molecular orbital diagrams for (a) N_2 and (b) N_2^+ show electronic transitions that result in the emission of visible light. Collisions with ions in the solar wind result in the promotion of electrons from the σ_{2p} orbitals in N_2 molecules to π_{2p}^* orbitals. Higher-energy collisions create N_2^{+*} molecular ions, which also have electrons in π_{2p}^* orbitals. When these electrons return to the ground state, red and blue-violet light is emitted.

From Alcohol to Asparagus: The Nose Knows

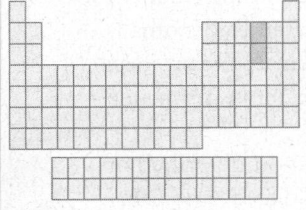

Oxygen, sulfur, and the other group 16 elements of the periodic table are called *chalcogens*. Their atoms have valence-shell electron configurations ns^2np^4. Consequently, the formation of structures in which there are two bonds on each atom determines the common chemical properties of these elements, since forming two bonds fills their valence octets.

We have already discussed oxygen in comparison to its neighbor fluorine to the right in the periodic table. In this section, we compare oxygen to the element directly below it in the periodic table, sulfur. A significant difference between the two elements is illustrated in some of the compounds discussed in this chapter. In molecular compounds, oxygen tends to form two covalent bonds. Sulfur does too, but it also can expand its octet to accommodate five or even six electron pairs in its valence shell. Hence the bonding patterns and molecular shapes of sulfur compounds show greater variability. This in turn affects the properties of the compounds.

Compounds of hydrogen with oxygen, sulfur, and the other group 16 elements provide interesting contrasts with respect to molecular shape and properties. The data in the table below show how different water is from the other compounds. It is a liquid at ordinary temperatures and pressures; it has a large negative heat of formation, and it has a much larger bond angle than the other three hydrides. We also know that water is odorless and is absolutely essential for life. The hydrides of sulfur (S), selenium (Se), and tellurium (Te) are all gases under ordinary conditions, have bond angles close

to 90°, and are foul-smelling and poisonous. Hydrogen sulfide is responsible for the smell of rotten eggs. It is especially dangerous because it tends to very quickly fatigue the nasal sensory sites responsible for detecting it. This means that the intensity of the odor is a very poor guide to the concentration of H_2S in the air. Headache and nausea set in at air concentrations of H_2S as low as 5 ppm, and at 100 ppm paralysis and death result.

Organic compounds of sulfur similarly have properties that differ from those of their oxygen-containing counterparts, and many of them have characteristic odors. Methanol is an organic alcohol with the formula CH_3—OH. It is a liquid at room temperature and has an odor usually described as slightly alcoholic. Methanethiol, CH_3—SH, is a gas at room temperature and has the pungent odor of rotten cabbage. It is produced in the intestinal tract of animals by the action of bacteria on proteins and is one of the sulfur compounds responsible for the characteristic aroma of a feedlot or a barnyard.

If we go up one more carbon unit in size, the oxygen-containing compound is a liquid at room temperature called ethanol (beverage grade alcohol), CH_3—CH_2—OH. The corresponding sulfur compound is a very low boiling liquid at room temperature called ethanethiol, which has a penetrating and unpleasant odor like very powerful green onions. The human nose can detect the presence of ethanethiol at levels as low as 1 ppb (part per billion) in the air. This gives rise to its use as an odorant in natural gas. Natural gas has no odor, and leaks of natural gas are such enormous fire hazards that ethanethiol is added to natural gas streams to make leaks immediately detectable.

Hydride[a]	Melting Point (°C)	Boiling Point (°C)	Bond Length (pm)	Bond Angle (degrees)	Heat of Formation (kJ/mol)
H_2O	0	100	96	104.5	−285.8
H_2S	−86	−60	134	92	−20.17
H_2Se	−66	−41	146	91	73.0
H_2Te	−51	−4	169	90	99.6

[a] H_2Po is excluded; too little is known of its chemistry. Polonium has no stable isotopes and is present on Earth only in very small quantities.

If we rearrange the atoms in ethanol and ethanethiol, we produce two new compounds. In the case of ethanol we get CH_3-O-CH_3, dimethyl ether, a colorless gas used in refrigeration systems. Its counterpart, dimethyl sulfide, CH_3-S-CH_3, is one of the compounds responsible for the "low-tide" smell of ocean shorelines.

Three of the sulfur compounds described—hydrogen sulfide, methanethiol, and dimethyl sulfide—are referred to as volatile sulfur compounds (VSCs) by dentists. They are produced by bacteria in the mouth and are the principal compounds responsible for bad breath. One of the reasons the odors of these compounds differ from those of their oxygen counterparts is that their molecular sizes and shapes are slightly different. Also their polarities differ because of the electronegativity difference between oxygen and sulfur. In A Look Ahead, we discussed the importance of molecular shape in determining the extent of interaction of a compound with receptors in nasal membranes. In part, the vast differences in odor and sensory detectability of these compounds are due to their shapes and electron distributions.

Not all sulfur compounds have an odor, but many odiferous compounds do contain sulfur. The characteristic and unpleasant smell of urine produced by some people after eating asparagus results from the inability of their bodies to convert odiferous sulfur compounds (Figure 9.51) into odor-free sulfate ions. Not all people are able to convert the sulfur compounds in asparagus to sulfate, and not all people are able to smell the odiferous sulfur compounds. Apparently, genetic differences determine how we metabolize these compounds and how well we can sense their odors.

The odor of skunk is due mostly to butanethiol, $CH_3-CH_2-CH_2-CH_2-SH$ (Figure 9.52), and the odor of well-used athletic shoes is primarily due to the presence of sulfur compounds produced by bacteria. Not all sulfur compounds have aromas as unpleasant as these, however. A compound with the formula $C_{10}H_{18}S$ is responsible for the aroma of grapefruit. If the orientation of two atoms on one of the carbon atoms in the molecule is switched, the resulting molecule has the same Lewis structure but now has no aroma at all.

H₃C—S
     \
      H
Methanethiol

Dimethyl sulfide

Dimethyl disulfide

Bis(methylthio)methane

Dimethyl sulfoxide

FIGURE 9.51 Structures of some of the volatile sulfur compounds responsible for the smell of "asparagus" urine. Compounds shown here toward the top have stronger (and more unpleasant) odors.

FIGURE 9.52 The pungent smell of skunk spray is due to butanethiol, $CH_3(CH_2)_3SH$.

Although we have focused on the small gas-phase molecules found in the atmosphere—nitrogen, oxygen, water, carbon dioxide, methane, ozone, and others—it is important to realize that the principles described in this chapter apply to larger and more complex molecules. We return to the importance of molecular shape, particularly in defining the biological activity of both large and small molecules, in later chapters of this book.

SUMMARY

Section 9.1 The shape of a molecule reflects the arrangement of the atoms in three-dimensional space and is determined largely by characteristic **bond angles**.

Section 9.2 Minimizing repulsion between pairs of valence electrons (the **VSEPR** model) results in the lowest-energy orientations of bonding and nonbonding electron pairs and accounts for the observed **molecular geometries** of molecules. The shape of a molecule can be determined by its **steric number** (the sum of the number of  bonded atoms and lone pairs around a central atom) and the **electron-pair geometry**, or arrangement of atoms and lone pairs. Molecules with SN = 2 and no lone pairs on the central atom have a linear electron-pair geometry and linear molecular geometry, while the electron-pair geometries of molecules with steric numbers 3 to 6 are **trigonal planar**, **tetrahedral**, **trigonal bipyramidal**, and **octahedral**, respectively. Molecular geometries called bent (or angular), **trigonal pyramidal**, **seesaw**, **T-shaped**, **square pyramidal**, and **square planar** result from different combinations of bonded atoms and lone pairs about a central atom. The observed bond angles in molecules deviate from the ideal values as a result of unequal repulsions between lone pairs and bonding pairs of electrons.

Section 9.3 Two covalently bonded atoms with different electronegativities have partial electrical charges of opposite sign, creating a **bond dipole**. If the individual bond dipoles in a molecule do not offset each other, the molecule is polar. If they do offset each other, the molecule is nonpolar. A polar molecule has a **dipole moment (μ)**, which is a quantitative measure of its polarity.

Section 9.4 According to **valence bond theory**, the **overlap** of half-filled orbitals results in covalent bonds between pairs of atoms in molecules. Molecular geometry is explained by the mixing, or **hybridizing**, of atomic orbitals to create **hybrid atomic orbitals**. Mixing one s and three p orbitals forms four sp^3 hybrid orbitals. Overlap between sp^3 orbitals and other atomic or hybrid orbitals results in up to four **sigma (σ) bonds** and a tetrahedral orientation of valence electrons. Mixing one s and two p orbitals forms three sp^2 **hybrid orbitals**. Overlap between sp^2 orbitals and other atomic or hybrid orbitals results in up to three σ bonds and a trigonal planar orientation of valence electrons. Mixing one s and one p orbital forms two sp **hybrid**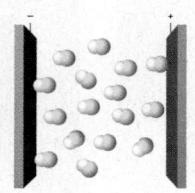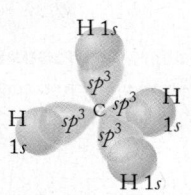

orbitals. Overlap between two sp hybrid orbitals results in up to two σ bonds 180° apart. Covalent bonds in which the electron density is greatest either above and below or in front of and behind the bonding axis are **pi (π) bonds**. Mixing an s orbital, three p orbitals, and two d orbitals gives six equivalent sp^3d^2 **hybrid orbitals** that point toward the vertices of an octahedron. Overlap between sp^3d^2 orbitals and other atomic or hybrid orbitals results in up to six σ bonds. Mixing an s orbital, three p orbitals, and one d orbital yields five equivalent sp^3d **hybrid orbitals** with lobes that point toward the vertices of a trigonal bipyramid. Overlap between sp^3d orbitals and other atomic or hybrid orbitals results in up to five σ bonds.

Section 9.5 The shape of a molecule with more than one central atom is a result of overlapping geometries around these atoms. Molecules with only sp^2 hybridized central atoms have planar geometries. The 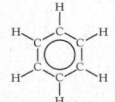  molecules of **aromatic compounds** contain six-carbon rings with alternating single and double bonds whose electrons are **delocalized** over the entire ring system.

Section 9.6 **Chiral** molecules exist in left- and right-handed forms that have different properties. Many contain an sp^3 hybridized carbon atom with four different groups attached.

Section 9.7 **Molecular orbital theory** is based on the formation of **molecular orbitals**, which are orbitals belonging to an entire molecule. MO theory explains the magnetic and spectroscopic properties of molecules but does not explain their shapes. Mixing two atomic orbitals creates one **bonding orbital** and one **antibonding orbital**. The region of highest electron density lies along the bond axis in a **sigma (σ) molecular orbital**. Electrons in sigma molecular orbitals form sigma bonds. The regions of highest electron density are on opposite sides of the bonding axis in a **pi (π) molecular orbital**. Electrons occupying π orbitals form π bonds. A **molecular orbital diagram** shows relative energies of the molecular orbitals in a molecule. MO electron configurations use the designations σ, σ^*, π, and π^* to describe the type of molecular orbitals occupied by electrons, subscripts to identify the atomic orbitals that led to the MOs, and superscripts to indicate the number of electrons in each molecular orbital. Atoms, ions, and molecules with no unpaired electrons are **diamagnetic** and are slightly repelled by an applied magnetic field. Atoms, ions, and molecules containing at least one unpaired electron are **paramagnetic** and are attracted by an external magnetic field.

PROBLEM-SOLVING SUMMARY

TYPE OF PROBLEM	CONCEPTS AND EQUATIONS	SAMPLE EXERCISES
Predicting molecular geometry	Draw a Lewis structure for the molecule. Determine the steric number (SN) of the central atom, where $$SN = \left(\begin{array}{c}\text{number of atoms}\\\text{bonded to central atom}\end{array}\right) + \left(\begin{array}{c}\text{number of lone pairs}\\\text{on central atom}\end{array}\right) \quad (9.1)$$ Choose a geometry that minimizes repulsion between electron pairs.	9.1, 9.3
Predicting relative sizes of bond angles	Lone pairs on a central atom push bonded atoms closer together and decrease bond angles.	9.2
Predicting polarity of a substance	Assign the direction of polarity to each bond dipole and use molecular geometry to determine whether the dipoles offset each other.	9.4
Describing bonding in molecules and the shape of molecules using hybrid orbitals	Identify the hybrid orbitals in molecules that result from mixing different numbers of s, p, and d orbitals that result in the observed molecular geometry: $s + p =$ two sp hybrid orbitals $s +$ two $p =$ three sp^2 $s +$ three $p =$ four sp^3 $s +$ three $p + d =$ five sp^3d $s +$ three $p +$ two $d =$ six sp^3d^2 hybrid orbitals	9.5–9.7
Using MO diagrams to predict bond order	$$\text{Bond order} = \tfrac{1}{2}\left(\begin{array}{c}\text{number of}\\\text{bonding electrons}\end{array}\right) - \left(\begin{array}{c}\text{number of}\\\text{antibonding electrons}\end{array}\right) \quad (9.2)$$	9.8–9.10

VISUAL PROBLEMS

(Answers to boldface end-of-chapter questions and problems are in the back of the book.)

9.1. Two compounds with the same formula, S_2F_2, have been isolated. The structures in Figure P9.1 show the arrangements of the atoms in these different compounds. Can these two compounds be distinguished by their dipole moments?

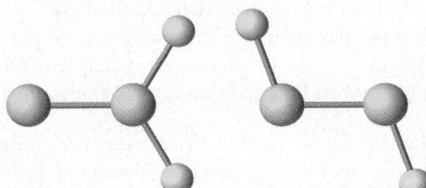

FIGURE P9.1

9.2. Could you distinguish between the two structures of N_2H_2 shown in Figure P9.2 by the magnitude of their dipole moments?

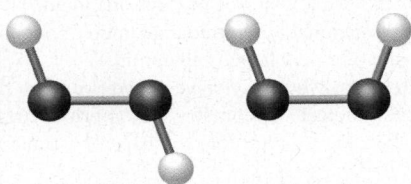

FIGURE P9.2

9.3. Which of the molecules shown in Figure P9.3 are planar, that is, have all atoms in a single plane? Are there delocalized π electrons in any of these molecules?

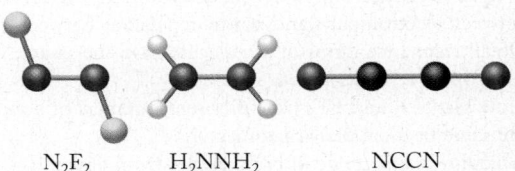

N_2F_2 H_2NNH_2 NCCN

FIGURE P9.3

9.4. Which of the molecules shown in Figure P9.4 is *not* planar? Are there delocalized π electrons in any of these molecules?

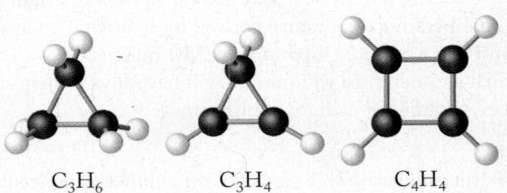

C_3H_6 C_3H_4 C_4H_4

FIGURE P9.4

9.5. Use the MO diagram in Figure P9.5 to predict whether O_2^+ has more or fewer electrons in antibonding molecular orbitals than O_2^{2+}.

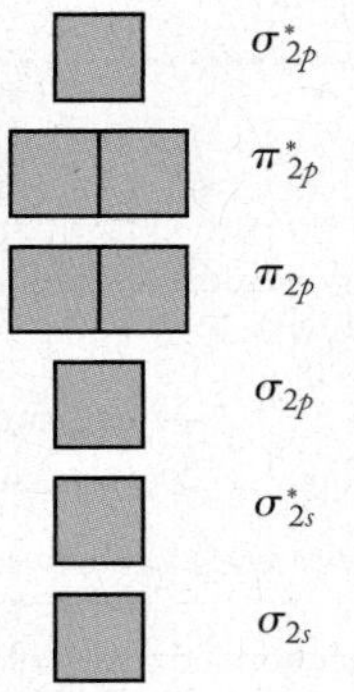

σ_{2p}^*

π_{2p}^*

π_{2p}

σ_{2p}

σ_{2s}^*

σ_{2s}

FIGURE P9.5

9.6. Under appropriate conditions, I_2 can be oxidized to I_2^+, which is bright blue. The corresponding anion, I_2^-, is not known. Use the molecular orbital diagram in Figure P9.6 to explain why I_2^+ is more stable than I_2^-.

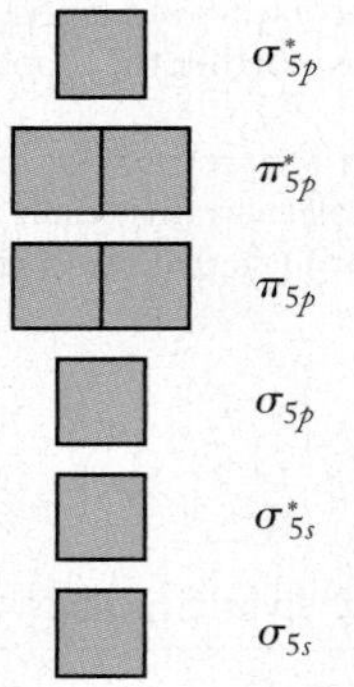

σ_{5p}^*

π_{5p}^*

π_{5p}

σ_{5p}

σ_{5s}^*

σ_{5s}

FIGURE P9.6

9.7. The molecular geometry of ReF_7 is an uncommon structure called a pentagonal bipyramid, which is shown in Figure P9.7. What are the bond angles in a pentagonal bipyramid?

FIGURE P9.7

*9.8. The molecular geometry of the transition metal–containing anions MF_8^{2-} (M = Mo and W) is not known. What would be the F—M—F bond angles in MF_8^{2-} if the anion has a cubic geometry as shown in Figure P9.8?

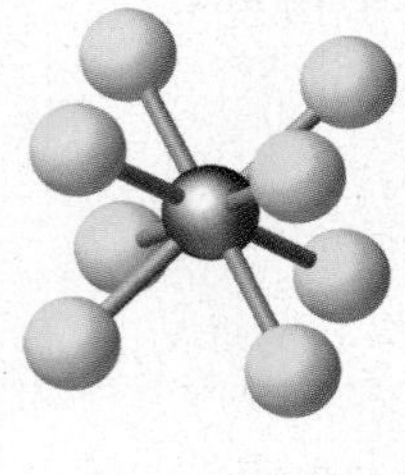

FIGURE P9.8

QUESTIONS AND PROBLEMS

Molecular Shape; Valence-Shell Electron-Pair Repulsion Theory (VSEPR)

CONCEPT REVIEW

9.9. Why is the shape of a molecule determined by repulsions between electron pairs and not by repulsions between nuclei?

9.10. Do all resonance forms of a molecule have the same molecular geometry? Explain your answer.

9.11. How can SO_3 and BF_3 have different numbers of bonds but the same trigonal planar geometry?

9.12. Account for the range of bond angles from about 104° to 180° in triatomic molecules.

9.13. In a molecule of ammonia, why is the repulsion between the lone pair and a bonding pair of electrons on nitrogen greater than the repulsion between two N—H bonding pairs?

9.14. Why is it important to draw a correct Lewis structure for a molecule before predicting its geometry?

9.15. Why does the seesaw structure have lower energy than a trigonal pyramidal structure derived by removing an apical atom from a trigonal bipyramidal AB_5 molecule?

*9.16. Which geometry do you predict will have lower energy: a square pyramid or a trigonal bipyramid?

PROBLEMS

9.17. Rank the smallest bond angle for the following molecular geometries in order of increasing bond angle: (a) trigonal planar; (b) octahedral; (c) tetrahedral.

9.18. Rank the smallest bond angle for the following molecular geometries in order of increasing bond angle: (a) seesaw; (b) tetrahedral; (c) square pyramidal.

9.19. Which of the molecular geometries discussed in this chapter have more than one characteristic bond angle?

*9.20. Which molecular geometries for molecules of the general formula AB_x (x = 2–6) discussed in this chapter have the same bond angles when lone pairs replace one or more atoms?

9.21. Which of the following molecular geometries does not lead to linear triatomic molecules after removing one or more atoms? (a) tetrahedral; (b) octahedral; (c) T-shaped

9.22. Which of the following molecular geometries does not lead to linear triatomic molecules after removing one or more atoms? (a) trigonal bipyramidal; (b) seesaw; (c) trigonal planar

*9.23. Describe the molecular geometries that result from replacing one atom with a lone pair of electrons in an AB_7 molecule with a pentagonal bipyramidal geometry. (See Figure P9.7 for the shape of a pentagonal bipyramid.)

*9.24. Which atoms would you have to remove from the cubic AB_8 molecule to create a geometry that approximates an octahedron? (See Figure P9.8 for the shape of a cubic molecule.)

9.25. Determine the molecular geometries of the following molecules: (a) GeH_4; (b) PH_3; (c) H_2S; (d) $CHCl_3$.

9.26. Determine the molecular geometries of the following molecules and ions: (a) NO_3^-; (b) NO_4^{3-}; (c) S_2O; (d) NF_3.

9.27. Determine the molecular geometries of the following ions: (a) NH_4^+; (b) CO_3^{2-}; (c) NO_2^-; (d) XeF_5^+.

9.28. Determine the geometries of the following ions: (a) SCN^-; (b) $CH_3PCl_3^+$ (P is the central atom and this cation contains a C—P bond); (c) ICl_2^-; (d) PO_3^{3-}.

9.29. Determine the geometries of the following ions and molecules: (a) $S_2O_3^{2-}$; (b) PO_4^{3-}; (c) NO_3; (d) NCO.

9.30. Determine the geometries of the following molecules: (a) ClO_2; (b) ClO_3; (c) IF_3; (d) SF_4.

9.31. Which of the following triatomic molecules, O_3, SO_2, and CO_2, have the same molecular geometry?

9.32. Which of the following species, N_3^-, O_3, and CO_2, have the same molecular geometry?

9.33. Which of the following ions, SCN^-, CNO^-, and NO_2^-, have the same geometry?

9.34. Which of the following molecules, N_2O, S_2O, and CO_2, have the same molecular geometry?

9.35. **The Venusian Atmosphere** A number of sulfur oxides not found in Earth's atmosphere have been detected in the atmosphere of Venus (Figure P9.35), including S_2O and S_2O_2. Draw Lewis structures for S_2O and S_2O_2, and determine their molecular geometries.

FIGURE P9.35

9.36. The structures of NOCl, NO_2Cl, and NO_3Cl were determined in 1995. They have the skeletal structures shown in Figure P9.36. Draw Lewis structures for these three compounds and predict the bonding geometry at each nitrogen atom.

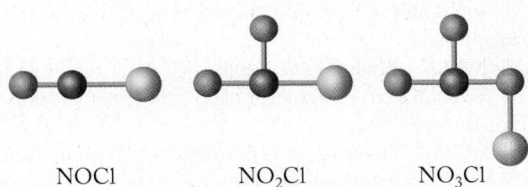

NOCl NO_2Cl NO_3Cl

FIGURE P9.36

*9.37. For many years, it was believed that the noble gases could not form covalently bonded compounds. However, xenon reacts with fluorine and oxygen. Reaction between xenon tetrafluoride and fluoride ions produces the pentafluoroxenate anion:

$$XeF_4 + F^- \rightarrow XeF_5^-$$

Draw Lewis structures for XeF_4 and XeF_5^-, and predict the geometry around xenon in XeF_4. The crystal structure of XeF_5^- compounds indicates a pentagonal bipyramidal orientation of valence pairs around Xe. Sketch the structure for XeF_5^-.

*9.38. The first compound containing a xenon–sulfur bond was isolated in 1998. Draw a Lewis structure for HXeSH and determine its molecular geometry.

*9.39. **Chemical Terrorism** In 1995 a gang attacked the Tokyo subway system with the nerve gas Sarin and focused world attention on the dangers of chemical warfare agents. The structure in Figure P9.39 shows the connectivity of the atoms in the Sarin molecule. Complete the Lewis structure by adding bonds and lone pairs as necessary. Assign formal charges to the P and O atoms, and determine the molecular geometry around P.

$$
\begin{array}{ccccc}
 & & & & H \\
 & & & & | \\
H & O & & H—C—H \\
| & \| & & | \\
H—C—P—O—C—H \\
| & | & & | \\
H & F & & H—C—H \\
 & & & & | \\
 & & & & H
\end{array}
$$

Sarin

FIGURE P9.39

9.40. Determine the bonding geometry around the nitrogen atom in the following unstable nitrogen oxides: (a) N_2O_2; (b) N_2O_5; (c) N_2O_3. (N_2O_2 and N_2O_3 have N–N bonds; N_2O_5 does not.)

Polar Bonds and Polar Molecules

CONCEPT REVIEW

9.41. Explain the difference between a polar bond and a polar molecule.

9.42. Must a polar molecule contain polar covalent bonds? Why?

9.43. Can a nonpolar molecule contain polar covalent bonds?

9.44. What does a dipole moment measure?

PROBLEMS

9.45. The following molecules contain polar covalent bonds. Which of them are polar molecules and which are nonpolar? (a) CCl_4; (b) $CHCl_3$; (c) CO_2; (d) H_2S; (e) SO_2

9.46. Photolysis of Cl_2O_2 is thought to produce compounds with the skeletal structures shown in Figure P9.46. Do the two compounds have the same dipole moment?

FIGURE P9.46

9.47. **Freon Bar** Compounds containing carbon, chlorine, and fluorine are known as Freons or chlorofluorocarbons (CFCs). Widespread use of these substances was banned because of their effect on the ozone layer in the upper atmosphere. Which of the following CFCs are polar and which are nonpolar? (a) Freon 11 ($CFCl_3$); (b) Freon 12 (CF_2Cl_2); (c) Freon 113 (Cl_2FCCF_2Cl)

9.48. Which of the following chlorofluorocarbons (CFCs) are polar and which are nonpolar? (a) Freon C318 (C_4F_8, cyclic structure); (b) Freon 1113 (C_2ClF_3); (c) $Cl_2HCCClF_2$

9.49. Predict which molecule in each of the following pairs is the more polar: (a) Freon 13 ($CClF_3$) or Freon 13B1 ($CBrF_3$); (b) Freon 12 (CF_2Cl_2) or Freon 22 (CHF_2Cl); (c) Freon 113 (Cl_2FCCF_2Cl) or Freon 114 (ClF_2CCF_2Cl).

9.50. Which molecule in each of the following pairs is more polar? (a) NH_3 or PH_3; (b) CCl_2F_2 or CBr_2F_2

9.51. **Chemical Warfare Gas** A series of carbonyl dihalide compounds of formula COX_2 (where X = I, Cl, or Br) has been prepared, and $COCl_2$ has been used as a chemical warfare agent. All these compounds are irritants and cause blistering of tissue. Their reactivity with tissue is influenced by their polarity. Place these compounds—COI_2, $COCl_2$, and $COBr_2$—in order of increasing polarity. Explain your reasoning.

9.52. Simple diatomic molecules detected in interstellar space include CO, CS, SiO, SiS, SO, and NO. Arrange these molecules in order of increasing dipole moment based on the location of the constituent elements in the periodic table, and then calculate the electronegativity differences from the data in Figure 8.5.

Valence Bond Theory

CONCEPT REVIEW

9.53. Are hybrid orbitals ever constructed from atomic orbitals with different principal quantum numbers?

9.54. Why aren't the orbitals on free atoms hybridized?

9.55. Do all resonance forms of N_2O have the same hybridization at the central N atom?

PROBLEMS

9.56. What is the hybridization of nitrogen in each of the following ions and molecules? (a) NO_2^+; (b) NO_2^-; (c) N_2O; (d) N_2O_5; (e) N_2O_3

9.57. What is the hybridization of sulfur in each of the following molecules? (a) SO; (b) SO_2; (c) S_2O; (d) SO_3

9.58. **Airbags** Azides such as sodium azide, NaN_3, are used in automobile airbags as a source of nitrogen gas. Another compound with three nitrogen atoms bonded together is N_3F. What differences are there in the arrangement of the electrons around the nitrogen atoms in the azide ion (N_3^-) and N_3F? Is there a difference in the hybridization of the central nitrogen atom?

9.59. N_3F decomposes to nitrogen and N_2F_2 by the following reaction:

$$2\,N_3F \rightarrow 2\,N_2 + N_2F_2$$

N_2F_2 has two possible structures as shown in Figure P9.59. Are the differences between these structures related to differences in the hybridization of nitrogen in N_2F_2? Identify the hybrid orbitals that account for the bonding in N_2F_2. Are they the same as those in acetylene, C_2H_2?

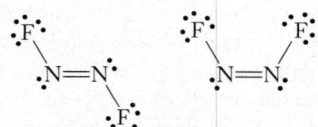

FIGURE P9.59

9.60. How does the hybridization of the sulfur atom change in the series SF_2, SF_4, SF_6?

9.61. How does the hybridization of the central atom change in the series CO_2, NO_2, O_3, and ClO_2?

*9.62. Draw the Lewis structure of the chlorite ion, ClO_2^-, which is used as a bleaching agent. Include all resonance structures in which formal charges are closest to zero. What is the shape of the ion? Suggest a hybridization scheme for the central chlorine atom that accounts for the structures you have drawn.

*9.63. **Perchlorate Ion and Human Health** Perchlorate ion adversely affects human health by interfering with the uptake of iodine in the thyroid gland, but because of this behavior it also provides a useful medical treatment for hyperthyroidism, or overactive thyroid. Draw the Lewis structure of the perchlorate ion, ClO_4^-. Include all resonance structures in which formal charges are closest to zero. What is the shape of the ion? Suggest a hybridization scheme for the central chlorine atom that accounts for this shape.

9.64. Draw a Lewis structure for Cl_3^+. Determine its molecular geometry and the hybridization of the central Cl atom.

9.65. Synthesis of the first compound of argon was reported in 2000. HArF was made by reacting Ar with HF. Draw a Lewis structure for HArF, and determine the hybridization of Ar in this molecule.

9.66. The Lewis structure of N_4O, with the skeletal structure O–N–N–N–N, contains one N—N single bond, one N=N double bond, and a N≡N triple bond. Is the hybridization of all the nitrogen atoms the same?

9.67. The trifluorosulfate anion (Figure P9.67) was isolated in 1999 as the tetramethylammonium salt $(CH_3)_4NSOF_3$. Determine the geometry around sulfur in the anion and describe the bonding according to valence bond theory.

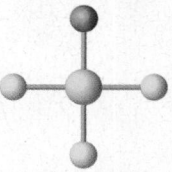

FIGURE P9.67

Shape and Interactions with Large Molecules; Chirality and Molecular Recognition

CONCEPT REVIEW

9.68. Why is it difficult to assign a single geometry to a molecule with more than one central atom?

9.69. Can molecules with more than one central atom have resonance forms?

*9.70. Can hybrid orbitals be associated with more than one atom?

*9.71. Are resonance structures examples of electron delocalization? Explain your answer.

9.72. Can sp^2 and sp hybridized carbon atoms be chiral centers?

9.73. Which of the following objects are chiral? (a) a baseball bat with no lettering on it; (b) a pair of scissors; (c) a boot; (d) a fork

PROBLEMS

9.74. Cyclic structures exist for many compounds of carbon and hydrogen. Describe the molecular geometry and hybridization around each carbon atom in benzene (C_6H_6), cyclobutane (C_4H_8), and cyclobutene (C_4H_6) (Figure P9.74).

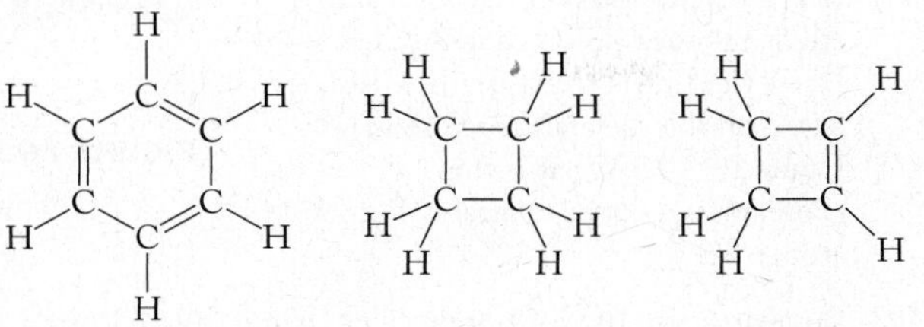

Benzene   Cyclobutane   Cyclobutene

FIGURE P9.74

9.75. The two nitrogen atoms in nitramide are connected, with two oxygen atoms on one terminal nitrogen and two hydrogen atoms on the other (Figure P9.75). What is the molecular geometry of each nitrogen atom in nitramide? Is the hybridization of both nitrogens the same?

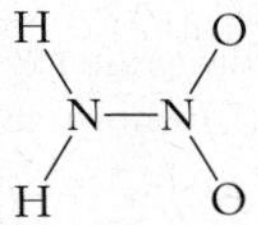

FIGURE P9.75

9.76. What is the geometry around each sulfur atom in the disulfate anion shown in Figure P9.76? What is the hybridization of the central oxygen atom?

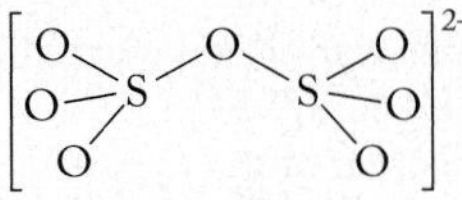

FIGURE P9.76

9.77. What is the molecular geometry around sulfur and nitrogen in the sulfamate anion shown in Figure P9.77? Which atomic or hybrid orbitals overlap to form the S–O and S–N bonds in the sulfamate anion?

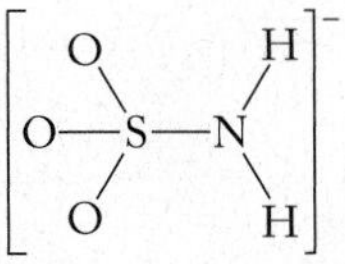

FIGURE P9.77

9.78. Which molecules in Figure P9.78 are chiral?

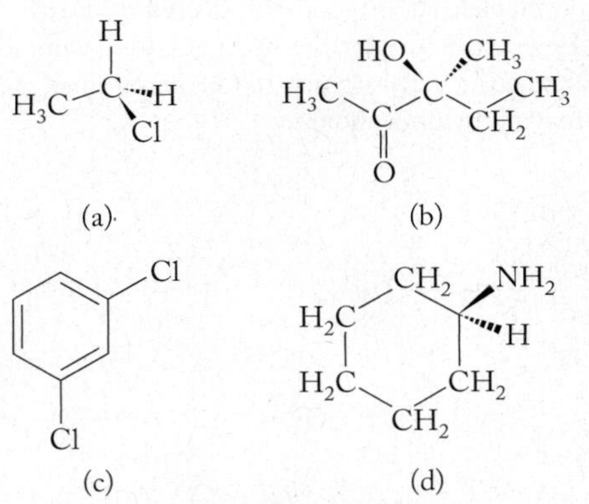

FIGURE P9.78

9.79. Which molecules in Figure P9.79 are chiral?

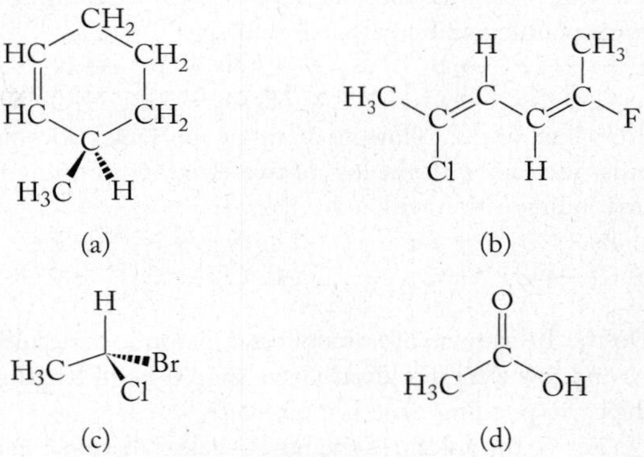

FIGURE P9.79

Molecular Orbital Theory

CONCEPT REVIEW
9.80. Which atomic orbitals are more likely to mix to form a set of molecular orbitals—a $2s$ and a $3p$ orbital or a $4s$ and a $5p$ orbital?

9.81. Which better explains molecular geometry: valence bond theory or molecular orbital theory?

9.82. Which better explains the magnetic properties of a diatomic molecule: valence bond theory or molecular orbital theory?

9.83. Do all σ molecular orbitals result from the overlap of s atomic orbitals?

9.84. Do all π molecular orbitals result from the overlap of p atomic orbitals?

9.85. Are s atomic orbitals with different principal quantum numbers (n) as likely to overlap and form MOs as s atomic orbitals with the same value of n?

9.86. Are atomic orbitals with the same principal quantum number (n) but different angular momentum quantum numbers (ℓ) as likely to overlap and form MOs as orbitals with the same values of n and ℓ?

PROBLEMS
9.87. Make a sketch showing how two $1s$ orbitals overlap to form a σ_{1s} bonding molecular orbital and a σ_{1s}^* antibonding molecular orbital.

9.88. Make a sketch showing how two $2p_y$ orbitals overlap "sideways" to form a π_{2p} bonding molecular orbital and a π_{2p}^* antibonding molecular orbital.

9.89. Use MO theory to predict the bond orders of the following molecular ions: N_2^+, O_2^+, C_2^+, and Br_2^{2-}. Do you expect any of these species to exist?

9.90. Diatomic noble gas molecules, such as He_2 and Ne_2, do not exist. Would removing an electron create molecular ions, such as He_2^+ and Ne_2^+, that are more stable than He_2 and Ne_2?

9.91. Which of the following molecular ions is expected to have one or more unpaired electrons? (a) N_2^+; (b) O_2^+; (c) C_2^{2+}; (d) Br_2^{2-}

9.92. Which of the following molecular ions is expected to have one or more unpaired electrons? (a) O_2^-; (b) O_2^{2-}; (c) N_2^{2-}; (d) F_2^+

9.93. Which of the following anions have electrons in π antibonding orbitals? (a) C_2^{2-}; (b) N_2^{2-}; (c) O_2^{2-}; (d) Br_2^{2-}

9.94. Which of the following molecular cations have electrons in π antibonding orbitals? (a) N_2^+; (b) O_2^+; (c) C_2^{2+}; (d) Br_2^{2+}

9.95. For which of the following diatomic molecules does the bond order increase with the gain of two electrons, forming the corresponding anion with a 2− charge?
a. $B_2 + 2e^- \rightarrow B_2^{2-}$
c. $N_2 + 2e^- \rightarrow N_2^{2-}$
b. $C_2 + 2e^- \rightarrow C_2^{2-}$
d. $O_2 + 2e^- \rightarrow O_2^{2-}$

9.96. For which of the following diatomic molecules does the bond order increase with the loss of two electrons, forming the corresponding cation with a 2+ charge?
a. $B_2 \rightarrow B_2^{2+} + 2e^-$
c. $N_2 \rightarrow N_2^{2+} + 2e^-$
b. $C_2 \rightarrow C_2^{2+} + 2e^-$
d. $O_2 \rightarrow O_2^{2+} + 2e^-$

9.97. Do the 1+ cations of homonuclear diatomic molecules of the second row elements always have shorter bond lengths than the corresponding neutral molecules?

9.98. Do any of the anions of the homonuclear diatomic molecules formed by B, C, N, O, and F have shorter bond lengths than those of the corresponding neutral molecules? Consider only the anions with 1− or 2− charge.

Additional Problems

9.99. Draw the Lewis structure for the two ions in ammonium perchlorate (NH_4ClO_4), which is used as a propellant in solid fuel rockets, and determine the molecular geometries of the two polyatomic ions.

9.100. Pressure-Treated Lumber By December 31, 2003, concerns over arsenic contamination had prompted the manufacturers of pressure-treated lumber (Figure P9.100) to voluntarily cease producing lumber treated with CCA (chromated copper arsenate) for residential use. CCA-treated lumber has a light greenish color and was widely used to build decks, sand boxes, and playground structures. Draw the Lewis structure for the arsenate ion (AsO_4^{3-}) that yields the most favorable formal charges. Predict the angles between the arsenic–oxygen bonds in the arsenate anion.

FIGURE P9.100

9.101. Consider the molecular structure of the amino acid glycine in Figure P9.101. What is the angle formed by the N—C—C bonds in this structure? What are the O—C—O and C—O—H bond angles?

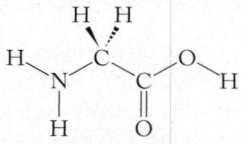

FIGURE P9.101

9.102. Cl_2O_2 may play a role in ozone depletion in the stratosphere. In the laboratory, a reaction between ClO_2F and $AlCl_3$ produces Cl_2O_2 and $AlCl_2F$. Draw the Lewis structure for Cl_2O_2 based on the skeletal structure in Figure P9.102. What is the geometry about the central chlorine atom?

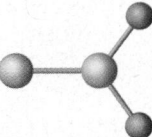

FIGURE P9.102

9.103. Bombardment of Cl_2O_2 molecules (Figure P9.102) with intense radiation is thought to produce the two compounds with the skeletal structures shown in Figure P9.103. Do both of these molecules have linear geometry?

FIGURE P9.103

*9.104. Complete the Lewis structure for the cyclic structure of Cl_2O_2 shown in Figure P9.104. Is the cyclic Cl_2O_2 molecule planar?

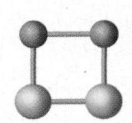

FIGURE P9.104

9.105. In 1999, the ClO^+ ion, a potential contributor to stratospheric ozone depletion, was isolated in the laboratory.
a. Draw the Lewis structure for ClO^+.
b. Using the molecular orbital diagram for ClO^+ in Figure P9.105, determine the order of the Cl—O bond in ClO^+.

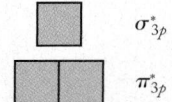
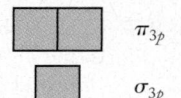
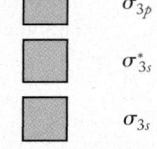

σ_{3p}^*
π_{3p}^*
π_{3p}
σ_{3p}
σ_{3s}^*
σ_{3s}

FIGURE P9.105

9.106. Life on Earth The photochemical reaction of sodium hydrogen phosphite with formaldehyde is illustrated in Figure P9.106. This reaction may have played a role in the formation of nucleic acids before life existed on Earth. Complete the Lewis structure for the hydrogen phosphite ion. Complete the Lewis structure for the product, hydroxymethylphosphonate.

$$Na_2HPO_3(aq) + \underset{H}{\overset{O}{\underset{\displaystyle}{\|}}}\!\!\!C\!\!-\!H(aq) \xrightarrow{h\nu}$$

$$H-O-\underset{H}{\overset{H}{\underset{\displaystyle|}{\overset{\displaystyle|}{C}}}}-\underset{O^-}{\overset{O}{\underset{\displaystyle|}{\overset{\displaystyle\|}{P}}}}-O^-(aq) + 2\,Na^+(aq)$$

FIGURE P9.106

9.107. Cola Beverages Phosphoric acid imparts a tart flavor to cola beverages. The skeletal structure of phosphoric acid is shown in Figure P9.107. Complete the Lewis structure for phosphoric acid in which formal charges are closest to zero. What is the molecular geometry around the phosphorus atom in your structure?

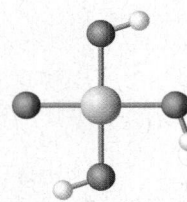

FIGURE P9.107

9.108. Fluoroaluminate anions AlF_4^- and AlF_6^{3-} have been known for over a century, but the structure of the pentafluoroaluminate ion, AlF_5^{2-}, was not determined until 2003. Draw the Lewis structures for AlF_3, AlF_4^-, AlF_5^{2-}, and AlF_6^{3-}. Determine the molecular geometry of each molecule or ion. Describe the bonding in AlF_3, AlF_4^-, AlF_5^{2-}, and AlF_6^{3-} using valence bond theory.

*9.109. Thermally unstable compounds can sometimes be synthesized using matrix isolation methods in which the compounds are isolated in a nonreactive medium such as frozen argon. The reaction of boron with carbon monoxide produces compounds with these skeletal structures: B–B–C–O and O–C–B–B–C–O. For each of these compounds, draw the Lewis structure that minimizes formal charges. Do any of your structures contain atoms with incomplete octets? Predict the molecular geometries of BBCO and OCBBCO.

*9.110. The products of the reaction between boron (B) and NO can be trapped in solid Ar matrices. Among the products is BNO. Draw the Lewis structure for BNO including any resonance forms. Assign formal charges and predict which structure provides the best description of the bonding in this molecule. Do any of your structures contain atoms without complete octets? Predict the molecular geometry of BNO.

9.111. Compounds May Help Prevent Cancer Broccoli, cabbage, and kale contain compounds that break down in the human body to form isothiocyanates, whose presence may reduce the risk of certain types of cancer. The simplest isothiocyanate is methyl isothiocyanate, CH_3NCS. Draw the Lewis structure for CH_3NCS, including all resonance forms. Assign formal charges and determine which structure is likely to contribute the most to bonding. Predict the molecular geometry of the molecule at both carbon atoms.

9.112. Toxic to Insects and People Methyl thiocyanate (CH_3SCN) is used as an agricultural pesticide and fumigant. It is slightly water soluble and is readily absorbed through the skin, but it is highly toxic if ingested. Its toxicity stems in part from its metabolism to cyanide ion. Draw three resonance structures for methyl thiocyanate. Assign formal charges and predict which structure would be the most stable. Predict the molecular geometry of the molecule at both carbon atoms.

9.113. Borazine, $B_3N_3H_6$ (a cyclic compound with alternating B and N atoms in the ring), is isoelectronic with benzene (C_6H_6). Are there delocalized π electrons in borazine?

9.114. Draw a molecular orbital diagram for F_2. How many electrons are found in antibonding molecular orbitals in F_2?

*9.115. Some chemists think HArF consists of H^+ ions and ArF^- ions. Using an appropriate MO diagram, determine bond order of the Ar–F bond in ArF^-.

*9.116. Assuming HArF is a molecular compound:
  a. Draw its Lewis structure.
  b. What are the formal charges on Ar and F in the structure you drew?

  c. What is the shape of the molecule?
  d. Is HArF polar?

9.117. Which of the following unstable nitrogen oxides, N_2O_2, N_2O_5, and N_2O_3, are polar molecules? (N_2O_2 and N_2O_3 have N–N bonds; N_2O_5 does not.)

9.118. Explain why O_2 is paramagnetic.

9.119. Using an appropriate molecular orbital diagram, show that the bond order in the disulfide anion S_2^{2-} is equal to 1. Is S_2^{2-} diamagnetic or paramagnetic?

9.120. Use molecular orbital diagrams to determine the bond order of the peroxide (O_2^{2-}) and superoxide ions (O_2^-). Are these bond order values consistent with those predicted from Lewis structures?

9.121. Elemental sulfur has several allotropic forms including cyclic S_8 molecules. What is the orbital hybridization of sulfur atoms in this allotrope? The bond angles are about 108°.

From Alcohol to Asparagus: The Nose Knows

9.122. All of the group 16 elements form compounds with the generic formula H_2E (E = O, S, Se, or Te). Which compound is the most polar? Which compound is the least polar?

*9.123. Ozone (O_3) has a dipole moment (0.54 D). How can a molecule with only one kind of atom have a dipole moment?

*9.124. The bond angle in H_2O is 104.5°; the bond angle in H_2S, H_2Se, and H_2Te is very close to 90°. Which theory would you apply to describe the geometry in H_2S, H_2Se, and H_2Te: VSEPR? Valence bond without invoking hybrid orbitals? Valence bond theory using hybrid orbitals? Why?

9.125. Garlic Garlic contains the molecule alliin (Figure P9.125). When garlic is crushed or chopped, a reaction occurs that converts alliin into the molecule allicin, which is primarily responsible for the aroma we associate with garlic.
  a. Describe the molecular geometry about the sulfur atoms in both compounds.
  b. Do any of the sulfur atoms in allicin have the same geometry as the sulfur atoms in the volatile sulfur compounds that cause bad breath (H_2S, CH_3—SH, and CH_3—S—CH_3)?

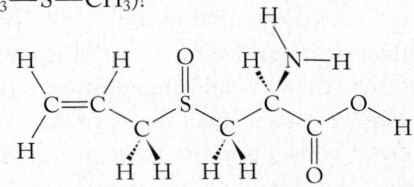

Alliin

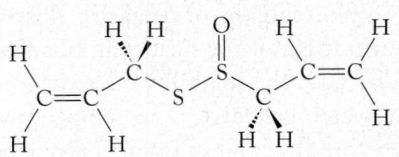

Allicin

FIGURE P9.125

Additional study materials including ChemTours and Diagnostic Quizzes are available at StudySpace at www.wwnorton.com/studyspace.

10

Forces between Ions and Molecules

Learning Outcomes

- Estimate the relative strengths of ion-ion interactions
- Explain the origins of ion-dipole forces, dipole-dipole forces, hydrogen bonds, and dispersion forces
- Explain the effect of intermolecular forces on the boiling points of compounds and the behavior of gases
- Calculate the solubility of gases using Henry's law
- Identify the regions of a phase diagram and explain the effect of temperature and pressure on phase changes
- Describe the role of hydrogen bonding in the unique properties of water

A LOOK AHEAD

Ubiquitous, Essential, and Remarkable

Water is everywhere in our world. It covers 71% of Earth's surface, comprises 75% of lean muscle tissue by weight, and accounts for 95% of the mass of blood. We take water for granted because it is such a familiar substance, but water exhibits remarkable physical properties. We sometimes observe water in all three phases at the same time on an early spring day, as solid ice melts to liquid water in the sun while white clouds of condensed water vapor dot the sky. We know that ice cubes float in a glass of water and that ice floats on a pond or river during the spring thaw. These properties and others result from the strong interactions between water molecules.

Seawater contains both dissolved ionic compounds like sodium chloride and dissolved gases such as oxygen, albeit in greatly different concentrations. The oxygen in seawater, freshwater, and the water in biological systems is essential to life. With every breath we take, oxygen gas enters tiny, moist sacs in our lungs, where it dissolves and is transported into the blood. Fish and other aquatic creatures also need oxygen to survive; they extract oxygen from water, where it is present at much lower concentrations than in air. At a partial pressure of 0.21 atm, the solubility of oxygen in water is about 10 mg per liter, but that low level is sufficient to sustain aquatic life.

Changes in temperature can have a disastrous effect on aquatic life. We all have seen photographs of dead fish floating in water that has become too warm. It is not the heat that kills the fish, but rather the lack of oxygen, because the solubility of oxygen decreases as the temperature of the water increases. The temperature does not have to increase much before the level of oxygen, drops below the concentration necessary for survival.

Life also relies on the solubility of other compounds and nutrients in water and their ability to be transported as solutes through

Healthy Red Blood Cells Red blood cells contain hemoglobin, which binds oxygen and transports it from the lungs to tissue. ▶

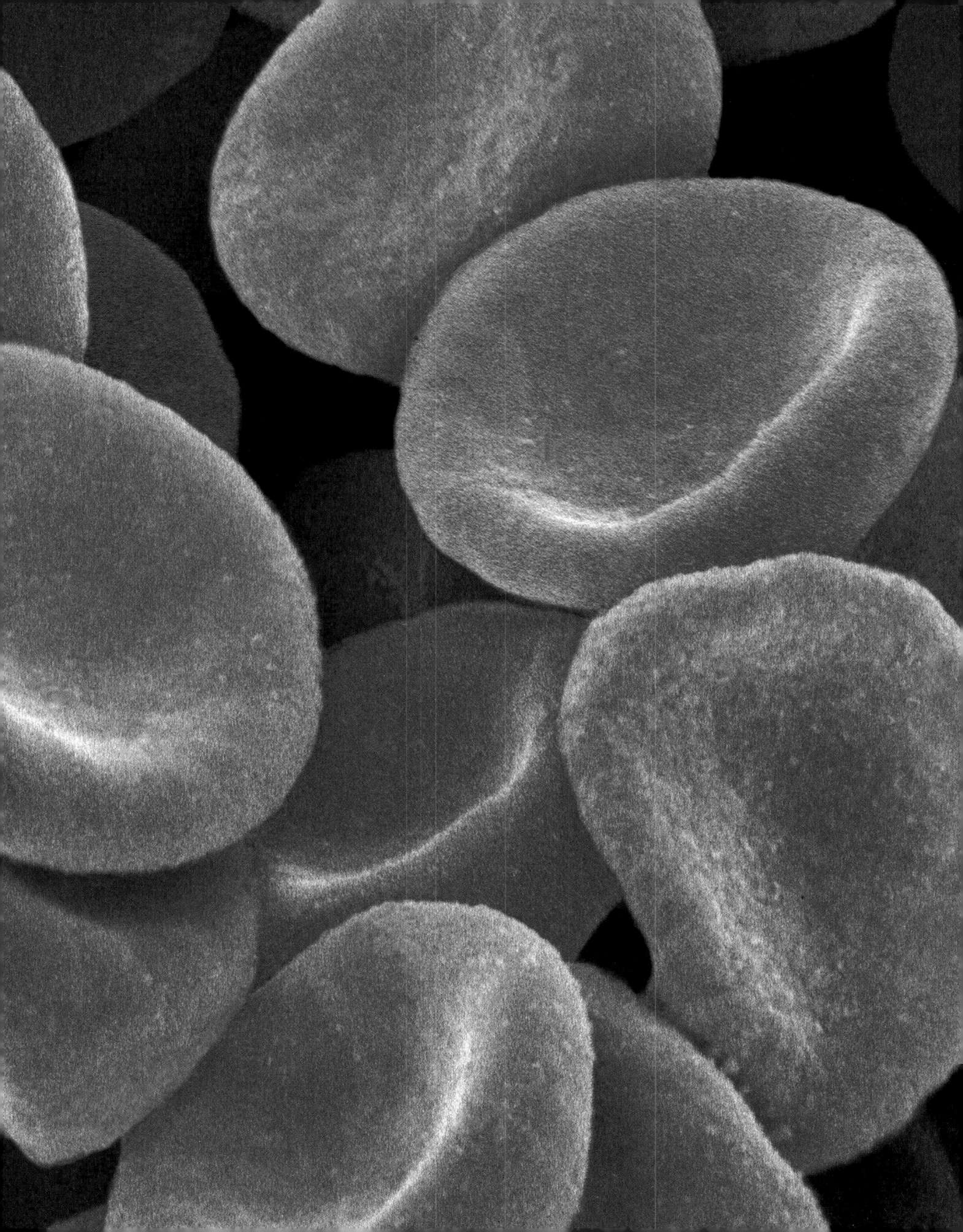

blood and other biological fluids. Although we discussed solubility in Chapter 4, we did not address the structures of solutes at the molecular level or the interactions between solute and solvent that influence solubility. In this chapter we discuss the influence of structure on solubility, interactions between molecules, and the answers to such questions as why does ice float, how does aquatic life survive in ice-covered lakes, why do oil and water not mix, why can small insects walk on water, and why does water boil at a relatively high temperature? ■

10.1 Interactions between Ions in Salts

Ocean waves crashing on a rocky shore create plumes of sea spray carrying small drops of seawater into the atmosphere, where they evaporate. As these drops evaporate, their concentrations of sea salts made of Cl^-, Na^+, Mg^{2+}, Ca^{2+}, SO_4^{2-}, and other ions increase. Eventually, the decreased volume of the drops produces supersaturated solutions of the salts, which begin to precipitate. Among the first solids to form is $CaSO_4$. Among the last is NaCl, which does not precipitate until 90% of the seawater in a drop has evaporated. This sequence takes place even though the concentrations of Ca^{2+} and SO_4^{2-} ions are much lower than the concentrations of Na^+ and Cl^- ions. Why does $CaSO_4$ precipitate before NaCl? Put another way: Why is NaCl more soluble in water than $CaSO_4$? And why are NaCl and $CaSO_4$ solids under ordinary conditions of temperature and pressure while water is a liquid under the same conditions? To answer these questions, we need to examine the attractive forces between the particles that make up these substances.

Think about the three common states of matter (Figure 10.1) and how they differ based on the kinetic energy of the particles in them and their ability to overcome the attractive forces between particles. In a solid, the kinetic energy of the particles is insufficient to overcome these forces of attraction. Consequently, the particles have the same nearest neighbors over time and do not move much. In a liquid, the kinetic energy of the particles is sufficient to overcome some of the attractive forces; particles in a liquid experience more freedom of motion, and can move past each other. In gases, the kinetic energy of the particles is sufficient to overcome essentially all of the attractive forces between them, imparting to the widely separated particles nearly compete freedom of motion.

The stronger the attractive forces among the particles in a substance, the greater the amount of energy needed to overcome those forces and allow the substance to melt or vaporize. Thus a substance made of particles that interact relatively strongly has high melting and boiling points, which means it is likely to be a solid at room temperature and normal pressure. Under the same conditions, a substance with

CONNECTION In Chapter 5, we looked at the flow of energy accompanying phase changes of water (solid ice ⇌ liquid water ⇌ water vapor) in terms of the changes in kinetic and potential energies of the molecules.

FIGURE 10.1 (a) The molecules of H_2O in solid ice are locked in place by the strength of intermolecular attraction. (b) In liquid water, molecules of H_2O have more energy and are free to flow past one another. (c) In water vapor, the molecules have enough energy to overcome nearly all intermolecular attraction and move freely throughout the space they occupy.

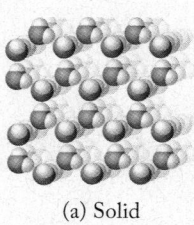  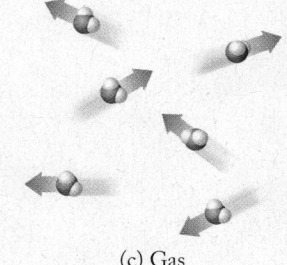

(a) Solid (b) Liquid (c) Gas

somewhat weaker particle–particle interactions has a lower melting point and is more likely to be a liquid, and a substance with very weak particle–particle interactions has even lower melting and boiling points, and is more likely to be a gas.

Ion-Ion Attractions

Ionic compounds are among the substances most likely to be solids at room temperature. They are solids because **ion–ion attraction** between ions of opposite charge is the strongest kind of interactive force between particles; the interaction results in the formation of ionic bonds. The strengths of ion–ion attractions are defined by *coulombic interaction* and the potential energy *E* between charged particles:

$$E \propto \frac{(Q_1 Q_2)}{d} \tag{10.1}$$

The value of *E* in Equation 10.1 is negative when Q_1 and Q_2 have opposite signs, which means *E* is negative for any salt because it is made of cations $(+Q)$ and anions $(-Q)$. Oppositely charged ions attract each other to form an arrangement characterized by a lower potential energy than the potential energy of the separated ions. When the ions are far apart, the *d* term in Equation 10.1 is large and *E* is a small negative number. When the ions are close together in a salt, *d* is small and *E* is a large negative number, which corresponds to lower energy (see Figure 5.7).

Whenever cations and anions in the gas phase combine to make a solid ionic compound, energy is released and the process is exothermic. The same amount of energy must be absorbed to separate the compound into its cations and anions. We can use Equation 10.1 to compare the relative strengths of interactive forces between ions in salts if we know the charges on the ions and their radii; the attractive force between two ions increases as ionic charge increases and as ionic size decreases.

The interion distance *d* for ionic compounds is the sum of the ionic radii. Recall that size of atoms and ions is a periodic property. In Chapter 7 we learned that cations are always smaller than the atoms from which they form and anions are always larger (see Figure 10.2). Ions in a group in the periodic table having the same charge increase in size as one moves down the group.

CONCEPT TEST

Put the following pairs of ions in order of increasing distance (*d*) between their nuclear centers: (a) KF, LiF, NaF; (b) $CaBr_2$, $CaCl_2$, CaF_2.

(Answers to Concept Tests are in the back of the book.)

To address the question of why $CaSO_4$ precipitates from sea spray before NaCl, we can use Equation 10.1 to compare the relative strengths of their ionic interactions. For NaCl the numerator is $(+1)(-1) = -1$, and for $CaSO_4$ the numerator is $(+2)(-2) = -4$, four times as large as for NaCl. For NaCl, the ionic radii are 102 pm for Na^+ and 181 pm for Cl^-, so *d* is $102 + 181 = 283$ pm. For $CaSO_4$ the radii are 100 pm for Ca^{2+} and 230 pm for SO_4^{2-} (Table 10.1), making $d = 330$ pm. The denominator for $CaSO_4$ is about 1.2 times the size of the NaCl denominator, so the factor-of-4 difference in the numerators dominates this comparison. Thus, the attraction between Ca^{2+} ions and SO_4^{2-} ions is stronger than the attraction between Na^+ ions and Cl^- ions. The stronger attraction between its ions is not the only reason $CaSO_4$ precipitates first from sea spray, but it is a major factor.

ion-ion attraction an attractive force between ions of opposite charge that results in an ionic bond.

CONNECTION We introduced Equation 10.1 in Chapter 5 when discussing the electrostatic interaction between particles in an ionic bond.

CONNECTION In Chapter 5 we learned that the energy associated with a process has the same absolute value but the opposite sign of the energy associated with the reverse process.

CONNECTION In Chapter 7 we discussed the use of ionic radii to calculate the distance between the nuclear centers in an ionic bond.

TABLE 10.1	Estimated Radii of Polyatomic Ions
Polyatomic Ion	**Radius (pm)**
CO_3^{2-}	185
NO_3^-	189
SO_4^{2-}	230
PO_4^{3-}	238

Values from R. B. Heslop and K. Jones, *Inorganic Chemistry: A Guide to Advanced Study* (Elsevier, 1976), p. 123.

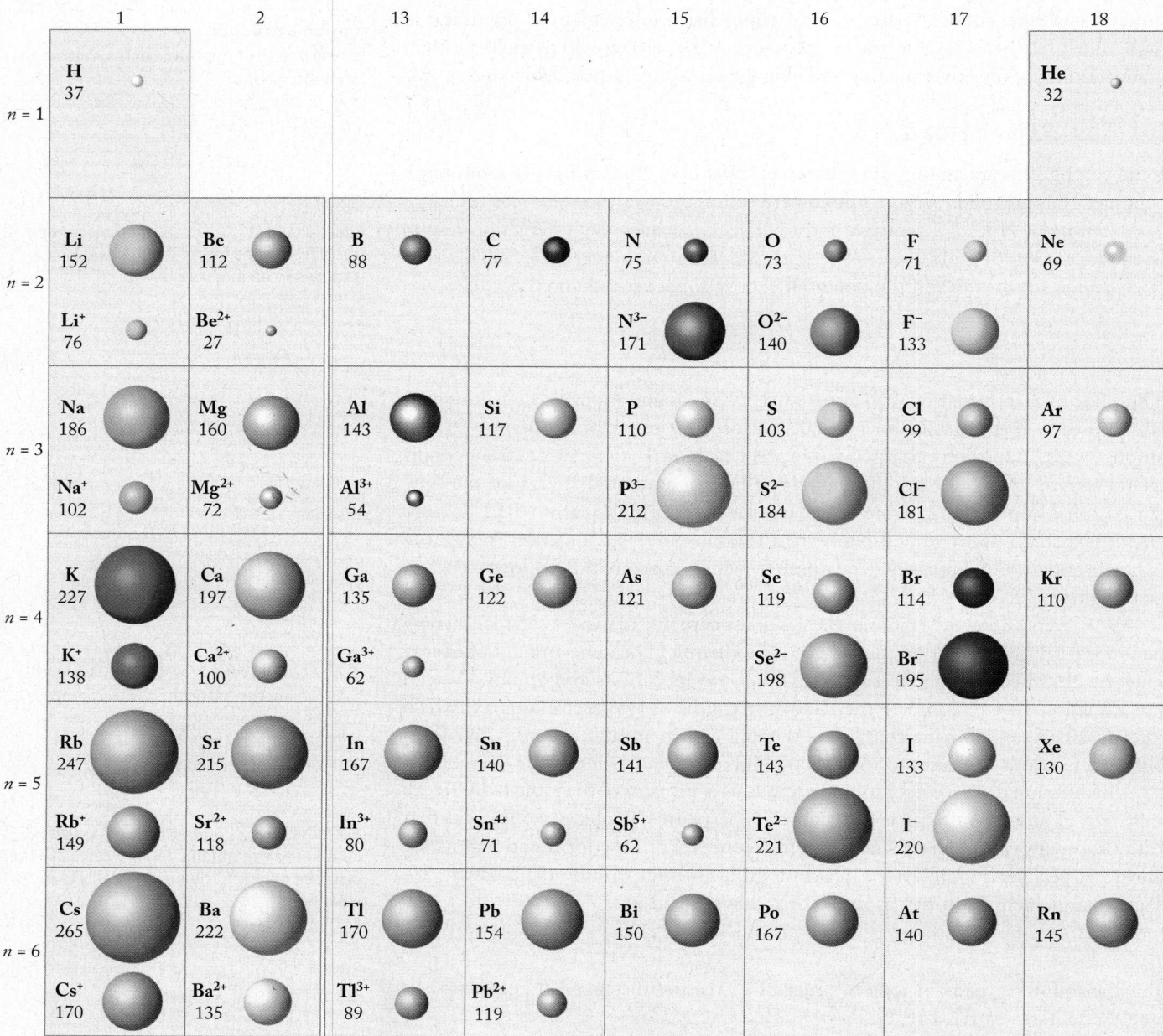

	1	2	13	14	15	16	17	18
n = 1	H 37							He 32
n = 2	Li 152	Be 112	B 88	C 77	N 75	O 73	F 71	Ne 69
	Li⁺ 76	Be²⁺ 27			N³⁻ 171	O²⁻ 140	F⁻ 133	
n = 3	Na 186	Mg 160	Al 143	Si 117	P 110	S 103	Cl 99	Ar 97
	Na⁺ 102	Mg²⁺ 72	Al³⁺ 54		P³⁻ 212	S²⁻ 184	Cl⁻ 181	
n = 4	K 227	Ca 197	Ga 135	Ge 122	As 121	Se 119	Br 114	Kr 110
	K⁺ 138	Ca²⁺ 100	Ga³⁺ 62			Se²⁻ 198	Br⁻ 195	
n = 5	Rb 247	Sr 215	In 167	Sn 140	Sb 141	Te 143	I 133	Xe 130
	Rb⁺ 149	Sr²⁺ 118	In³⁺ 80	Sn⁴⁺ 71	Sb⁵⁺ 62	Te²⁻ 221	I⁻ 220	
n = 6	Cs 265	Ba 222	Tl 170	Pb 154	Bi 150	Po 167	At 140	Rn 145
	Cs⁺ 170	Ba²⁺ 135	Tl³⁺ 89	Pb²⁺ 119				

FIGURE 10.2 The radii of anions formed by the main group elements are larger than the radii of their parent atoms. The radii of cations are smaller than the radii of their parent atoms. All values are in picometers. Values from N. N. Greenwood and A. Earnshaw, *Chemistry of the Elements*, 2nd ed. (Boston: Butterworth-Heinemann, 1997); Catherine E. Housecroft and Alan G. Sharpe, *Inorganic Chemistry*, 3rd ed. (Upper Saddle River, NJ: Pearson Prentice Hall, 2008).

SAMPLE EXERCISE 10.1 **Predicting Relative Strengths of Ion-Ion Attractive Forces**

List the ionic compounds CaO, NaF, and CaF_2 in order of decreasing strength of the attraction between their ions.

Collect and Organize According to Equation 10.1, the strength of an ion–ion attractive force depends on the charges on the ions and the distances between them. We should be familiar with the charges on the ions in binary ionic compounds from Chapter 2. Interion distances are determined from ionic radii in Figure 10.2.

Analyze The charges on the calcium, sodium, fluoride, and oxide ions are $2+$, $1+$, $1-$, and $2-$, respectively. The distances between the ions in CaO, NaF, and CaF_2 are equal to the sum of their ionic radii. Equation 10.1 tells us that the strength of an ion–ion attractive force is directly proportional to the product of the charges on the ions and inversely proportional to the distance between them.

Solve Each of these compounds has a different Q_1Q_2 value:

$$Ca^{2+} \quad O^{2-} \quad Q_1Q_2 = (+2)(-2) = -4$$
$$Na^+ \quad F^- \quad Q_1Q_2 = (+1)(-1) = -1$$
$$Ca^{2+} \quad F^- \quad Q_1Q_2 = (+2)(-1) = -2$$

The values of d in Equation 10.1 are

$$d_{CaO} = Ca^{2+} \quad 100 \text{ pm} + O^{2-} \text{ 140 pm} = 240 \text{ pm}$$
$$d_{NaF} = Na^+ \quad 102 \text{ pm} + F^- \text{ 133 pm} = 235 \text{ pm}$$
$$d_{CaF_2} = Ca^{2+} \quad 100 \text{ pm} + F^- \text{ 133 pm} = 233 \text{ pm}$$

Substituting the values of d and Q_1Q_2 into Equation 10.1 for each compound:

$$E_{CaO} \propto (-4)/240 = -0.017$$
$$E_{NaF} \propto (-1)/235 = -0.0043$$
$$E_{CaF_2} \propto (-2)/233 = -0.0086$$

Thus, the predicted order of decreasing strength of ionic interactions is $CaO > CaF_2 > NaF$.

Think about It The ion–ion distances are nearly the same for all three compounds, which means that the product of the ionic charges determines the relative interaction strengths.

Practice Exercise Arrange the ionic compounds $CaCl_2$, BaO, and NaCl in order of decreasing strength of the attraction between their ions.

(Answers to Practice Exercises are in the back of the book.)

We have seen two instances where the product of the charges Q_1Q_2 was the dominant factor in determining the relative strengths of the ion–ion interactions. This result is general. The ionic radii of cations in Figure 10.2 range from 27 pm (Be^{2+}) to 170 pm (Cs^+), and those of monatomic anions and polyatomic anions (Figure 10.2 and Table 10.1) range from 133 pm (F^-) to 238 pm (PO_4^{3-}). If we put together the smallest cation and anion (BeF_2) and the largest cation and anion (Cs_3PO_4), the range of nucleus-to-nucleus distances is 160 to 408 pm, which is a factor of 2.6. Moreover, the radii of common ions cover an even smaller size range. We have seen the charge product Q_1Q_2 range from -1 to -4 (a factor of 4) in Sample Exercise 10.1, and for common ionic compounds it is indeed the charge product that is nearly always responsible for large differences in the strengths of ionic interactions. Solid ionic compounds have more than one ion–ion interaction in a three-dimensional lattice. We will learn more about the strengths of multiple ion–ion interactions in Chapter 11, and we will examine the structures of the resultant solids in Chapter 12.

ion-dipole interaction an attractive force between an ion and a molecule that has a permanent dipole moment.

sphere of hydration the cluster of water molecules surrounding an ion in aqueous medium; the general term applied to such a cluster forming in any solvent is *sphere of solvation*.

10.2 Interactions Involving Polar Molecules

Ionic and covalent bonds involve the strong interactions that hold ions together in an ionic solid and atoms together in a molecule. Typical bond energies for covalent and ionic compounds range from several hundred to several thousand kilojoules per mole. In contrast, interactions between molecules and ions, such as those responsible for salts dissolving in water, are 10- to 100-fold weaker than bonding interactions. Even though much weaker than bonds, molecule–ion interactions do influence the properties of substances. For example, polar covalent molecules interact with ionic substances. These interactions are responsible for the solubility of salts in water. Polar molecules also attract each other, and these dipole–dipole interactions are responsible for a range of physical properties. For example, boiling a liquid requires that nearly all the intermolecular interactions between molecules in the liquid be overcome so that they can be separate from one another in the gas phase. Thus, polar substances with stronger dipole–dipole interactions tend to have higher boiling points than substances with similar molar mass but weaker dipole–dipole interactions.

Ion-Dipole Interactions

One factor influencing the solubility of ionic compounds in water is the attraction between their ions and the regions of polar water molecules carrying the opposite partial charges. These are **ion–dipole interactions**, and they occur between ions and water molecules in all aqueous solutions.

When an ionic solid dissolves in water, ion–dipole interactions between the ions and water molecules pull ions from the solid into solution (Figure 10.3). As an ion is pulled away from its solid-state neighbors, it is surrounded by water molecules forming a **sphere of hydration**. If the solvent were something other than water, the cluster would be called a *sphere of solvation*. These dissolved ions are said to be *hydrated* or, for other solvents, *solvated*. The strengths of many ion–dipole interactions help to overcome the ion–ion interactions in an ionic solid, allowing these spheres of hydration (or solvation) to form and the compound to dissolve. Within a sphere of hydration, the water molecules closest to the ion are oriented so that their oxygen atoms (negative poles) are directed toward a cation and their hydrogen atoms (positive poles) are directed toward an anion (Figure 10.4).

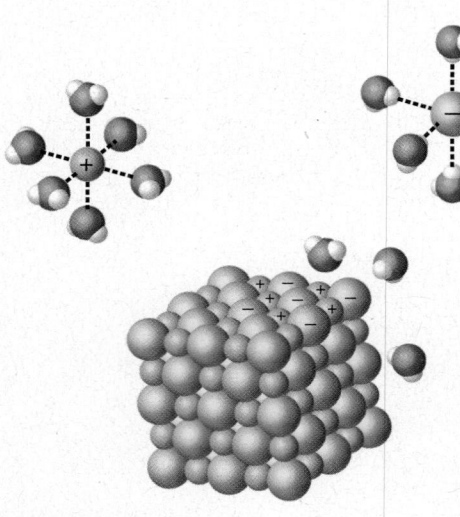

FIGURE 10.3 The hydrogen ends (positive poles) of H_2O molecules are attracted to the Cl^- ions of NaCl and the O atoms (negative poles) are attracted the Na^+ ions. Multiple ion–dipole interactions overcome the attractive forces holding ions at the surface of the solid NaCl, causing the NaCl to dissolve.

FIGURE 10.4 Each hydrated Na^+ ion and Cl^- ion is surrounded by six water molecules in an inner hydration sphere. Water molecules in an outer hydration sphere surround the inner sphere. The outer sphere is the result of dipole–dipole interactions between the rest of the water, known as bulk water, and the water molecules of the inner sphere. Beyond the outer hydration sphere, dipole–dipole interactions also occur between outer-sphere water molecules and molecules in bulk water.

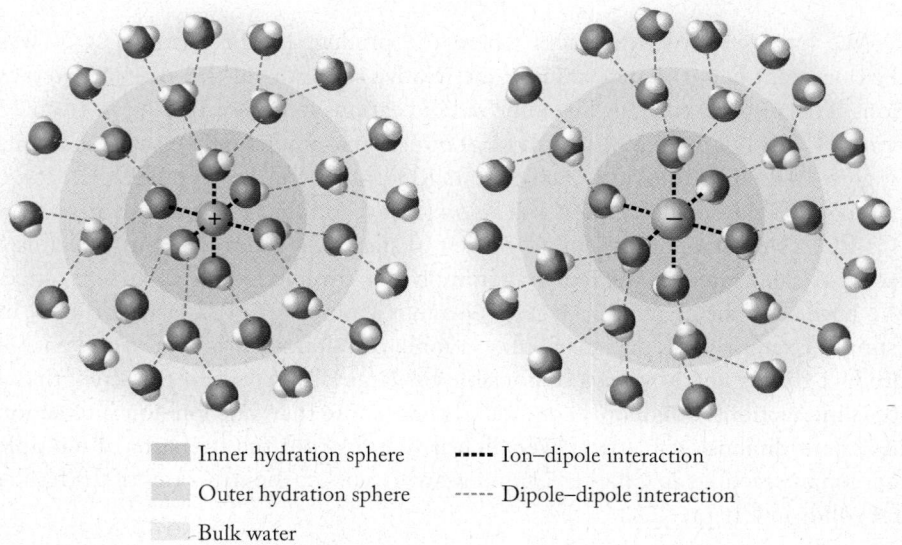

▦ Inner hydration sphere	···· Ion–dipole interaction
▦ Outer hydration sphere	---- Dipole–dipole interaction
▦ Bulk water	

The number of water molecules oriented in this way depends on the size of the ion. Typically six water molecules hydrate an ion, but the number can range from four to nine. As Figure 10.4 shows, six water molecules surround the Na^+ and Cl^- ions in an aqueous solution of NaCl.

Dipole-Dipole Interactions

The water molecules closest to the ions in Figure 10.4 are surrounded by other water molecules that form an outer hydration sphere. The molecules in this outer sphere are more randomly oriented than those in the inner sphere but not completely so. What ordering there is among outer-sphere water molecules is caused by another intermolecular force, **dipole–dipole interactions**, which operate between molecules that have permanent dipole moments—in other words, between polar molecules. In water molecules, the partial charges on oxygen and hydrogen atoms result in attractions between a hydrogen atom of one molecule and an oxygen atom of another. Dipole–dipole interactions are not as strong as ion–dipole interactions because dipole–dipole interactions involve only partial charges caused by unequal sharing of electrons. In contrast, the ion involved in an ion–dipole interaction has completely lost or gained electrons and has a full positive or negative charge. The magnitude of the dipole moment is reflected in the strength of the dipole–dipole interactions between molecules.

> **CONNECTION** In previous chapters, we have indicated the presence of a hydration sphere around an ion in aqueous solution by placing (*aq*) after the symbol.

> **CHEMTOUR** Intermolecular Forces

> **CONNECTION** In Chapter 9 we learned that permanent dipole moments are experimentally measured values expressed in units of debyes that define the polarity of molecules.

> **CONCEPT TEST** ....

Dimethyl ether, CH_3OCH_3, and acetone, $CH_3C(O)CH_3$ (Figure 10.5), have similar formulas and molar masses. However, their dipole moments are quite different: 1.30 D for dimethyl ether and 2.88 D for acetone. Predict which compound has the higher boiling point.

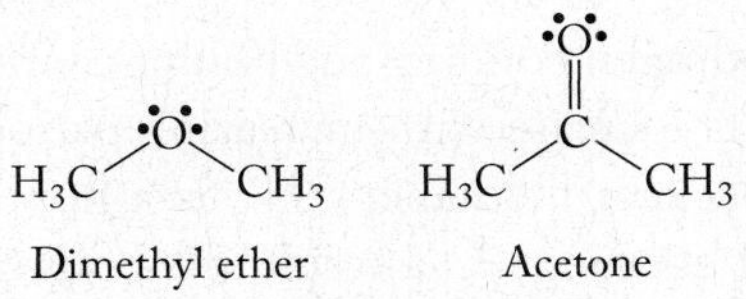

Dimethyl ether   Acetone

FIGURE 10.5

Hydrogen Bonds

In polar molecules containing either an O—H bond or a N—H bond and in the molecule HF, the hydrogen atom is bonded to a small, highly electronegative atom, which leads to relatively large separations of charge, large dipole moments, and stronger-than-average dipole–dipole interactions. Because of its strength, this particular dipole–dipole interaction involving a H atom on one molecule and an O, N, or F atom on an adjacent molecule merits special distinction: it is called a **hydrogen bond** (Figure 10.6). Hydrogen bonds are the strongest dipole–dipole interactions and are about one-tenth the strength of covalent bonds; hydrogen bonds range in strength from about 5 to about 30 kJ/mol. Hydrogen bonds in water play a key role in defining the remarkable behavior of H_2O.

One physical property strongly influenced by hydrogen bonds is boiling point. Water, HF, and ammonia all have boiling points that are much higher than the boiling points of other compounds of similar mass. Another way to look at the

dipole-dipole interaction an attractive force between polar molecules.

hydrogen bond the strongest dipole–dipole interaction. It occurs between a hydrogen atom bonded to a small, highly electronegative element (O, N, F) and an atom of oxygen or nitrogen in another molecule. Molecules of HF also form hydrogen bonds.

FIGURE 10.6 (a) Hydrogen bonds (dashed lines) occur between hydrogen atoms bonded to O, N, or F in one molecule and an O, N, or F in an adjacent molecule. Hydrogen bonds occur whenever –NH or –OH groups are present in molecules. (b) Hydrogen bonding interactions are so strong in a carboxylic acid like acetic acid that two molecules stay together as a unit called a dimer in the liquid phase. (c) Hydrogen bonds between nitrogen and hydrogen in ammonia form extensive three-dimensional networks.

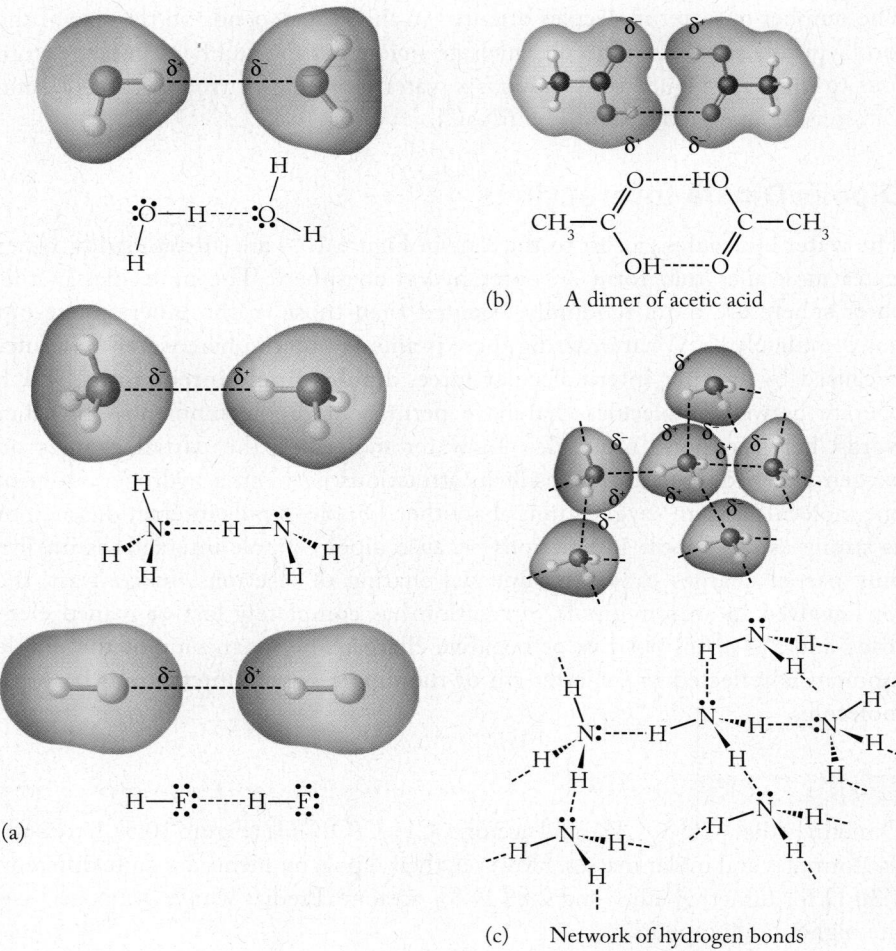

(b) A dimer of acetic acid

(a)

(c) Network of hydrogen bonds in ammonia

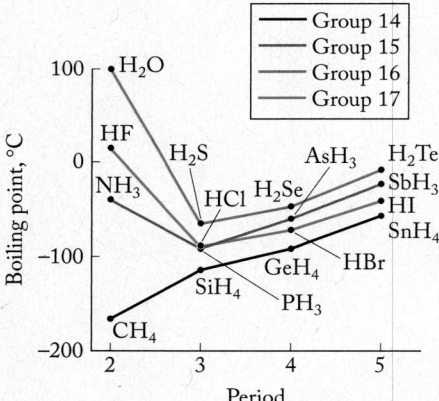

FIGURE 10.7 The boiling points of binary hydrides plotted against the period in the periodic table of the atom bonded to hydrogen. The boiling points of H_2O (18.02 g/mol), NH_3 (17.03 g/mol), and HF (20.01 g/mol) are not only significantly higher than that of CH_4 (16.04 g/mol) but also much higher than one would predict based on periodic trends.

relationship between hydrogen bonding and boiling point is shown in Figure 10.7. Note how boiling points increase with increasing period number for the period 3, 4, and 5 hydrides. Because methane, CH_4, is a nonpolar molecule, it cannot participate in hydrogen bonding. As a result, the boiling-points of the group 14 hydrogen compounds increase with increasing period number (and molar mass), as shown by the black line in Figure 10.7. Each of the three hydrogen-bonding substances, however, has a boiling point much higher than the boiling points of other hydrogen-containing compounds in the same family (blue, green, and red lines in Figure 10.7).

CONCEPT TEST

Predict the compound in each pair that has the higher boiling point:
(a) CH_3Cl, CH_3Br; (b) CH_3CH_2OH, CH_3OH; (c) CH_3NH_2, $(CH_3)_3N$.

Hydrogen bonds have a defining impact on the three-dimensional shape of many polymers and large biological molecules, such as proteins and DNA. As we have seen with small molecules, shape often determines behavior. For example, the ability of the two strands of DNA to stay together in a double helix is due to many hydrogen bonds that hold the molecule together in aqueous solution (Figure 10.8). Because the hydrogen bonds are weaker than covalent bonds, however, the DNA strands can be pulled apart during DNA replication.

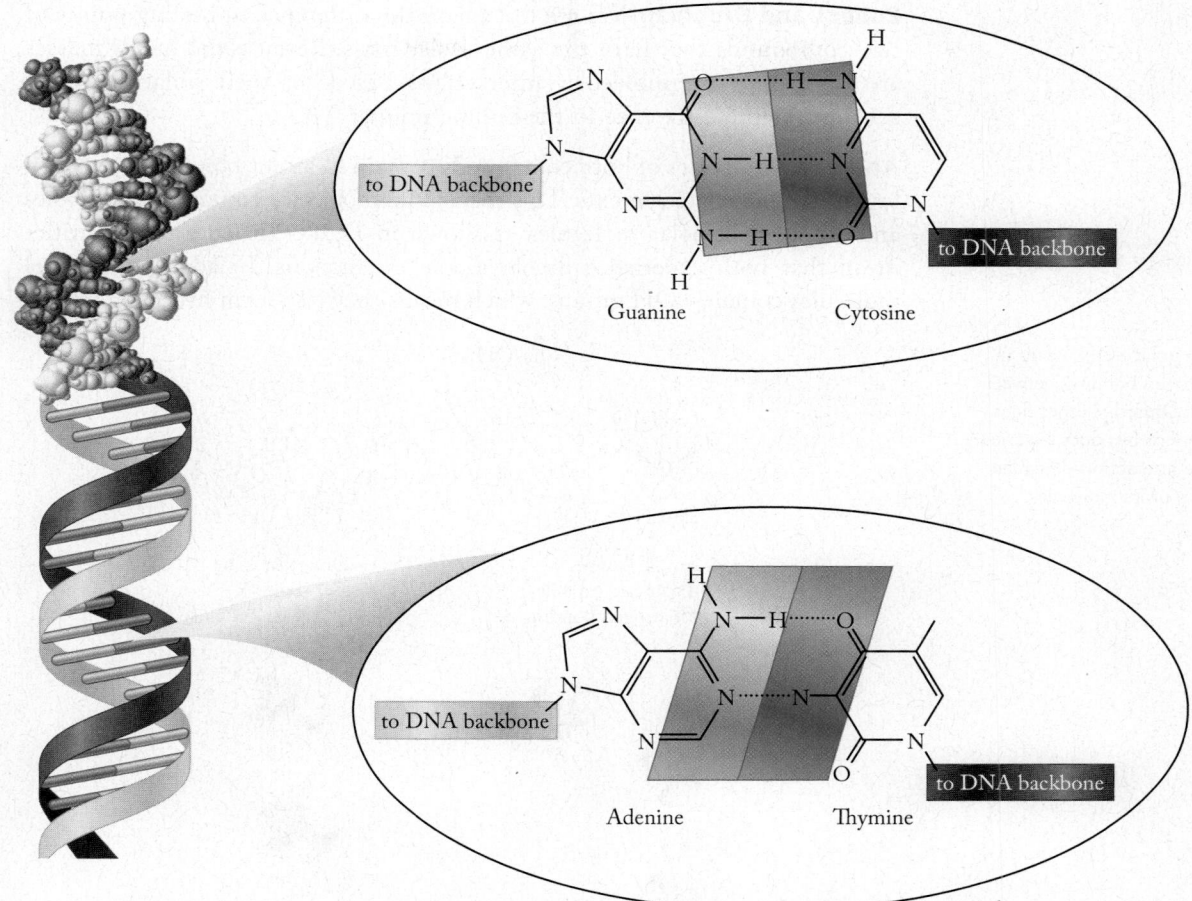

FIGURE 10.8 Hydrogen bonds (dashed lines) occur between hydrogen and nitrogen or oxygen in adjacent strands of DNA contributing to the double-helix structure. The two detailed views of the DNA double helix contain the names of four of the building blocks of DNA: *guanine, cytosine, adenine,* and *thymine.* Note that there are three H-bonding sites between pairs of guanine and cytosine, but only two between adenine and thymine. This difference means that guanine is always paired with cytosine and adenine is always paired with thymine in the twin helices of double-stranded DNA.

SAMPLE EXERCISE 10.2 **Explaining Differences in Boiling Points**

Dimethyl ether (C_2H_6O) has a molar mass of 46.07 g/mol and a boiling point of −24.9°C. Ethanol (C_2H_6O) has the same formula and therefore the same molar mass but a boiling point of 78.5°C. Explain this difference in boiling points. The structures are shown in Figure 10.9.

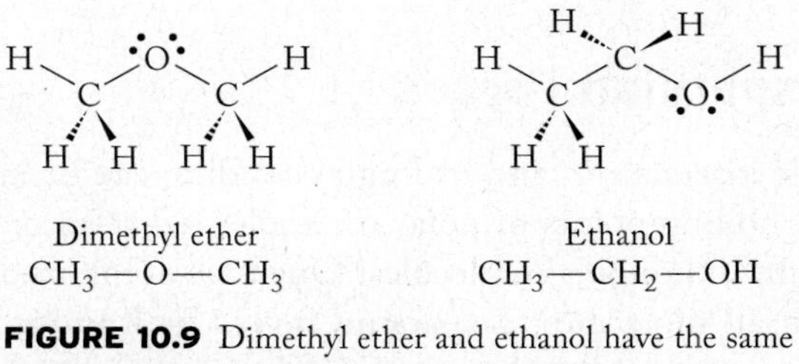

Dimethyl ether
CH_3—O—CH_3

Ethanol
CH_3—CH_2—OH

FIGURE 10.9 Dimethyl ether and ethanol have the same molecular formula but different boiling points.

Collect and Organize We are to explain the difference in boiling points of two compounds that have the same molar mass. Because the molar masses are the same, intermolecular interactions based on their polarities must account for the difference in the boiling points.

Analyze Molecules of both compounds contain O atoms bonded to atoms of less-electronegative elements. This means the molecules contain bond dipoles and are, overall, polar molecules as shown in Figure 10.10. Their polarities mean that both experience dipole–dipole interactions. In addition, ethanol molecules contain –OH groups, which means they also form hydrogen bonds.

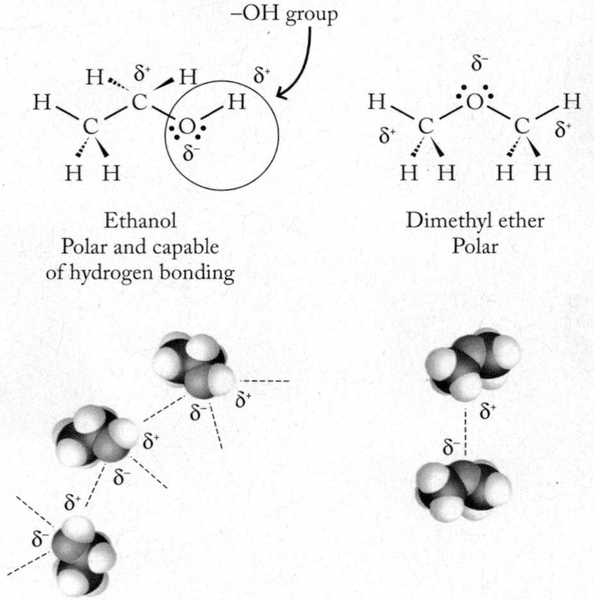

FIGURE 10.10 The polar –OH group in ethanol leads to hydrogen bonding between molecules of ethanol. Dimethyl ether does not form hydrogen bonds but does experience weaker dipole–dipole interactions between the δ^+ and δ^- regions of its molecules.

Ethanol
Polar and capable
of hydrogen bonding

Dimethyl ether
Polar

Solve Hydrogen bonds are stronger than other types of dipole–dipole interactions. Therefore, the intermolecular interactions in ethanol are stronger than those in dimethyl ether; more energy is required to overcome these stronger interactions, and that is why ethanol has a higher boiling point.

Think about It Dimethyl ether and ethanol have the same molecular formula and similar molecular structures, yet their boiling points differ by more than 100°C: a difference directly linked to the –OH group that each molecule of ethanol has, and that a molecule of dimethyl ether does not.

Practice Exercise Isopropanol (molar mass 60.10 g/mol), the familiar rubbing alcohol in your medicine cabinet, boils at 82°C. Ethylene glycol, used as automotive antifreeze, has almost the same molar mass (62.07 g/mol) but boils at 196°C. Why do these substances (Figure 10.11) have such different boiling points?

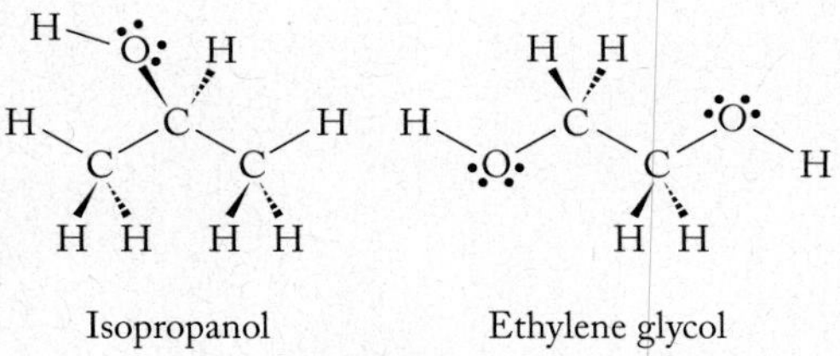

Isopropanol   Ethylene glycol

FIGURE 10.11

dispersion force (also called **London force**) an intermolecular force between nonpolar molecules caused by the presence of temporary dipoles in the molecules.

temporary dipole (also called **induced dipole**) the separation of charge produced in an atom or molecule by a momentary uneven distribution of electrons.

10.3 Dispersion Forces

If dipole–dipole interactions and hydrogen bonding can be used to explain differences in the boiling points of polar molecules, what accounts for the different boiling points of nonpolar molecules? One type of intermolecular interaction occurs between all molecules: **dispersion forces** or **London forces**, named in honor of German-American physicist Fritz London (1900–1954), whose work explained attractions between atoms of the noble gases and between nonpolar

TABLE 10.2 Relative Strengths of Intermolecular Forces and Some Phenomena They Explain

Type of Force	Relative Strength	Phenomenon
Ion–dipole		NaCl dissolves in water 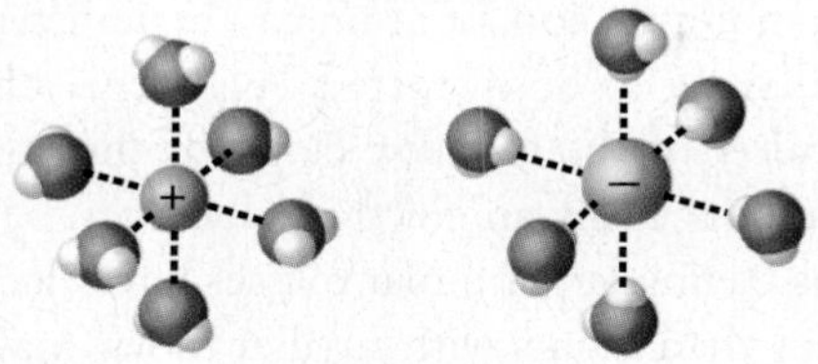
Hydrogen bonding		Water expands when it freezes 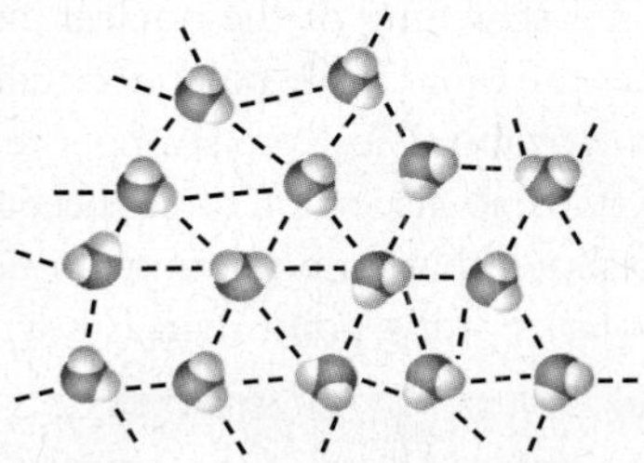
Dipole–dipole		The boiling point of dimethyl ether ($\mu = 1.30$ D, on the left) is 19°C higher than that of non-polar propane 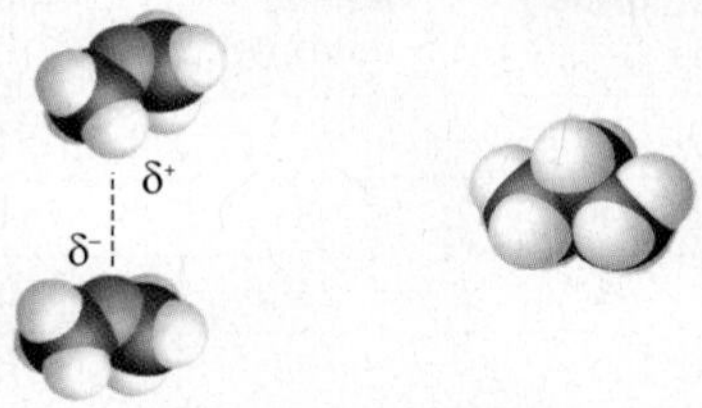
		δ^+
		δ^-
Dipole–induced dipole		O_2 dissolves in water 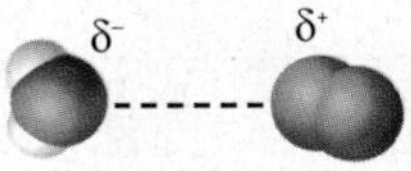
		δ^- δ^+
Dispersion		At 298 K: Cl_2 is a gas Br_2 is a liquid I_2 is a solid 

molecules. Dispersion forces are the only intermolecular force between nonpolar molecules. Among small molecules dispersion forces may by weaker than other intermolecular forces, but in large molecules they can be stronger. The relative strengths of intermolecular forces are shown in Table 10.2.

London determined that the intermolecular attractive forces between atoms of the group 18 elements and between nonpolar molecules are caused by **temporary dipoles** (or **induced dipoles**) produced by momentary changes in the electron distribution in the atoms or molecules. The mutual repulsion between the electrons of the two atoms (or two molecules) causes the electrons to be distributed unevenly, creating two temporary dipoles (Figure 10.12). The presence of temporary dipoles in a noble gas or nonpolar molecular substance causes the atoms or molecules to interact more than they would in the absence of the temporary dipoles.

(a)

(b)

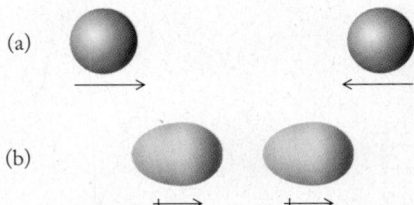

FIGURE 10.12 (a) Two atoms of a noble gas with a spherical distribution of electrons approach each other and (b) create two temporary dipoles when their electron clouds repel one another. As soon as the atoms move away from each other, the electrons in each atom return to their original uniform distribution.

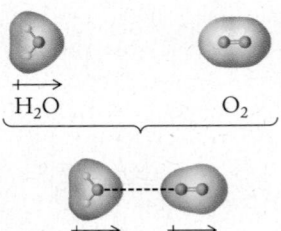

FIGURE 10.13 The approach of a polar water molecule induces a temporary dipole in an initially nonpolar oxygen molecule by distorting the distribution of the electrons.

⊙⊙ **CONNECTION** The concept of screening of outer electrons by inner electrons was presented in Chapter 7 when we discussed trends in the sizes of atoms and ions.

polarizability the relative ease with which the electron cloud in a molecule, ion, or atom can be distorted, inducing a temporary dipole.

A molecule with a permanent dipole can induce a temporary dipole in a nonpolar molecule by perturbing the electron distribution in the nonpolar molecule (Figure 10.13). The intermolecular interaction in this case is weaker than the dipole–dipole force between two polar molecules and is of the same order of magnitude as the dispersion forces between temporary dipoles in nonpolar molecules (see Table 10.2).

The magnitude of an induced dipole depends on the ease with which the electrons in a molecule, ion, or atom can be redistributed. The relative tendency of the electron density to be distorted by another charged particle is a measure of the **polarizability** of the electron cloud of the atom, ion, or molecule. What factors determine how easily an electron cloud can be polarized?

Atoms having larger molar masses have electron density in orbitals farther from the nucleus than atoms with smaller molar masses. Electrons in these larger atoms are held less tightly by the nucleus because of both their greater average distance from the nucleus and the screening of the nuclear charge by electrons in lower orbitals. Consequently, they are more easily polarized than electrons in smaller atoms or molecules. Greater polarizability leads to stronger temporary dipoles and stronger intermolecular interactions, so dispersion forces become stronger as atoms or ions become larger. Table 10.3 shows that for both monatomic noble gases and diatomic halogens (which are nonpolar), boiling point increases with increasing molar mass.

TABLE 10.3 Molar Masses and Boiling Points of the Halogens and Noble Gases

Halogen	$\mathcal{M}$ (g/mol)	Boiling Point (K)	Noble Gas	$\mathcal{M}$ (g/mol)	Boiling Point (K)
			He	4	4
F_2	38	85	Ne	20	27
Cl_2	71	239	Ar	40	87
Br_2	160	332	Kr	84	120
I_2	254	457	Xe	131	165
			Rn	222	211

CONCEPT TEST ••••••••••••••••••••••••••••••••••••

Rank the following gases in order of increasing polarizability: argon, hydrogen, krypton, neon.

•••

Polarizability is one factor in determining the strength of dispersion forces between molecules. Molecular shape also plays a role in the magnitude of dispersion forces. The three hydrocarbons in Figure 10.14, for instance, have the same formula, C_5H_{12}, and therefore the same molar mass, 72.15 g/mol, but different shapes and different boiling points. All these molecules are nonpolar, so the only intermolecular interactions are due to dispersion forces. We can think of molecules of *n*-pentane as being relatively long and straight—think of them as pencils. They can interact with one another over a relatively large surface area and therefore have more possibilities for dispersion forces to hold them together. In contrast, molecules of 2,2-dimethylpropane are almost spherical—think of them as Ping-Pong balls. They have relatively less surface area to interact with adjacent molecules; less interaction means weaker dispersion forces, resulting in the lowest boiling point in this set of compounds. The remaining molecule, 2-methylbutane, is neither as straight as *n*-pentane nor as spherical as 2,2-dimethylpropane—think of it as a short, forked stick. It should come as no surprise that this compound boils at a temperature lower than *n*-pentane but higher than 2,2-dimethylpropane.

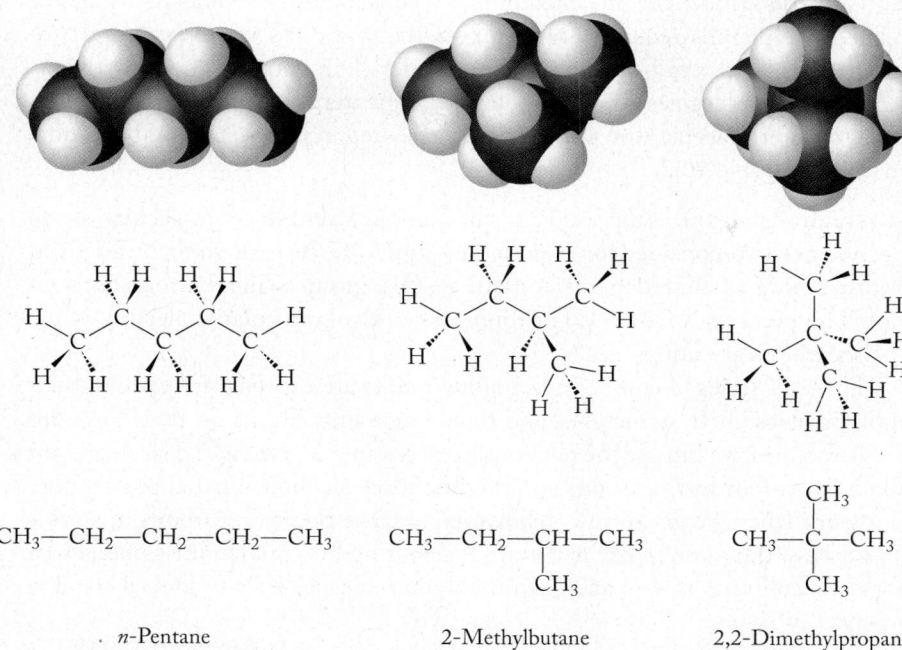

FIGURE 10.14 Three molecular shapes are possible for C_5H_{12}. The more spread out the molecule, the greater its polarizability, the stronger the opportunity for dispersion forces between molecules and the higher the boiling point.

$CH_3-CH_2-CH_2-CH_2-CH_3$

n-Pentane
Boiling point 36°C

$CH_3-CH_2-CH-CH_3$
                   $|$
                  CH_3

2-Methylbutane
Boiling point 28°C

$CH_3-\overset{\displaystyle CH_3}{\underset{\displaystyle CH_3}{\overset{|}{\underset{|}{C}}}}-CH_3$

2,2-Dimethylpropane
Boiling point 9°C

Dispersion forces are weak, but they add up, and for large molecules they are particularly strong. A large number of weak interactions can sometimes dominate a much smaller number of strong interactions in a system, as we will see in the discussion of polarity and solubility in Section 10.4.

CONCEPT TEST

Do you think an attractive force exists between ions and nonpolar molecules? How would you describe such an interaction? Where would you place such an intermolecular force in Table 10.2?

SAMPLE EXERCISE 10.3   Explaining Trends in Boiling Points

Table 10.4 gives the formulas, molar masses, and boiling points of several hydrocarbons and alcohols that have comparable molar masses. Figure 10.15 shows the molecular structures of the compounds in rows 4 and 5 of the table. (a) Explain the trend in boiling points of the five hydrocarbons and the four alcohols. (b) Explain the difference in boiling point for each hydrocarbon and the alcohol of comparable molar mass.

Row 4

$CH_3CH(CH_3)CH_3$
Boiling point −11.7°C

$CH_3CH(OH)CH_3$
Boiling point 82°C

Row 5

$CH_3CH_2CH_2CH_3$
Boiling point −0.5°C

$CH_3CH_2CH_2OH$
Boiling point 97°C

FIGURE 10.15 Structures of isobutane (a hydrocarbon) and isopropanol (an alcohol) from row 4 and *n*-butane (a hydrocarbon) and *n*-propanol (an alcohol) from row 5 in Table 10.4.

TABLE 10.4	Boiling Point Data for Sample Exercise 10.3					
HYDROCARBON			**ALCOHOL**			
Molecular Formula	$\mathcal{M}$ (g/mol)	Boiling Point (°C)	Molecular Formula	$\mathcal{M}$ (g/mol)	Boiling Point (°C)	
1. CH_4	16.04	−161.5				
2. CH_3CH_3	30.07	−88	CH_3OH	32.04	64.5	
3. $CH_3CH_2CH_3$	44.09	−42	CH_3CH_2OH	46.07	78.5	
4. $CH_3CH(CH_3)CH_3$	58.12	−11.7	$CH_3CH(OH)CH_3$	60.09	82	
5. $CH_3CH_2CH_2CH_3$	58.12	−0.5	$CH_3CH_2CH_2OH$	60.09	97	

Collect and Organize Interpretation of trends in boiling points of a series of related compounds requires evaluation of the types and relative strengths of intermolecular forces between their molecules. The stronger the interactions between the molecules of a compound, the greater its boiling point. Several types of intermolecular forces have been presented in this section and are summarized in Table 10.2.

Analyze All of the compounds in this Sample Exercise are molecular, so we do not need to consider ion–ion or ion–dipole forces. Alcohols differ from hydrocarbons in that alcohols contain a –OH group while hydrocarbons do not. The presence of the –OH group makes alcohols polar molecules while hydrocarbons are nonpolar.

For both series of compounds, boiling points are expected to relate to their molar masses, molar structures, and dipole moments. Based on their formulas and structures, we can use the relationship between molar mass and boiling point for the first four hydrocarbons and the first three alcohols. That approach does not work when we get to row 4, however, because the hydrocarbons in rows 4 and 5 have the same molar mass but different boiling points; the same is true for the alcohols in rows 4 and 5. In these molecules we have to look at the differences in shape.

Solve (a) The trend in boiling point for both series of compounds in Table 10.4 reflects an increase in boiling point as molar mass increases. Larger molecules are more easily polarized leading to larger induced dipoles and greater dispersion forces between molecules. Based on the structures shown, the molecules in row 4 (Figure 10.15) are more spherical and have smaller surface areas than the molecules in row 5. The molecules in row 4 have fewer points of contact, and therefore dispersion forces are weaker. Consequently, both compounds in row 4 boil at lower temperatures than their counterparts in row 5.

(b) The molar masses of ethane and methanol (row 2) are very similar, as are the molar masses of each hydrocarbon–alcohol pair; however, the alcohols all have much higher boiling points than their hydrocarbon counterparts. The presence of –OH groups means that the alcohols are capable of hydrogen bonding. Hydrogen bonds are the strongest dipole–dipole interactions, and the occurrence of hydrogen bonding among the alcohol molecules accounts for their much higher boiling points. All the hydrocarbons interact only via dispersion forces.

Think about It Identifying the types of intermolecular interactions between molecules enables us to evaluate the relative strengths of the forces and therefore their relative contributions to determining the boiling point. Shape matters, too, especially in terms of the contribution of dispersion forces to molecular interactions.

Practice Exercise Which of these substances has the largest dipole–dipole interactions, the largest dispersion forces, and the lowest boiling point: (a) H_2NNH_2; (b) $H_2C=CH_2$; (c) Ne; (d) $CH_3CH_2CH_2CH_2CH_3$?

■

CONCEPT TEST

Explain why CF_4 is a gas at room temperature but CCl_4 is a liquid.

Chapter 6 ended with a discussion of how the behavior of real gases cannot be described exactly by the ideal gas equation ($PV = nRT$), especially at high pressures and/or low temperatures. High pressures push molecules closer together, and low temperatures reduce their average speeds, meaning that they take longer to pass by one another and hence have more chance to interact. Thus both high pressures and low temperatures favor intermolecular interactions that are not accounted for in the ideal gas equation. The van der Waals equation, $(P + n^2a/V^2)(V - nb) = nRT$ (Section 6.9), more accurately relates the pressure, volume, and temperature of a real gas because the pressure correction term, n^2a/V^2, accounts for interactions between atoms or molecules in the gas phase. Because these interactions are considered in the van der Waals equation, the term **van der Waals forces** is frequently used collectively to indicate all types of attractive forces possible between molecules: dipole–dipole interactions, induced dipole interactions, and dispersion forces.

CONNECTION One of the assumptions about an ideal gas is that atoms or molecules do not interact.

CONNECTION The second correction term in the van der Waals equation, $-nb$, accounts for the contribution of the volume of the particles in a gas to the overall volume the gas occupies, as was discussed in Chapter 6.

SAMPLE EXERCISE 10.4 **Assessing the van der Waals Constant *a***

The van der Waals constant a for SO_2, 6.71 $L^2 \cdot atm/mol^2$, is nearly twice the value of a for CO_2, 3.59 $L^2 \cdot atm/mol^2$. Suggest a reason for this difference.

Collect and Organize The values of their van der Waals constants a reflect the strengths of the intermolecular forces in the two gases. Table 10.2 summarizes the principal intermolecular forces and their relative strengths.

Analyze For molecules of similar mass and shape, the relative strengths of intermolecular forces are hydrogen bonding > dipole–dipole interactions. Neither molecule has a hydrogen atom attached to an O, N, or F atom, so hydrogen bonding is not a factor. We need to look at Lewis structures and shapes of the molecules (Figure 10.16) to assess whether either has a permanent dipole. The more polar molecules experience stronger intermolecular forces and have larger values for constant a. If neither has a permanent dipole, we then must consider the relative polarizability of the two molecules. The more polarizable molecules will have stronger dispersion forces, which will contribute to a larger value of a.

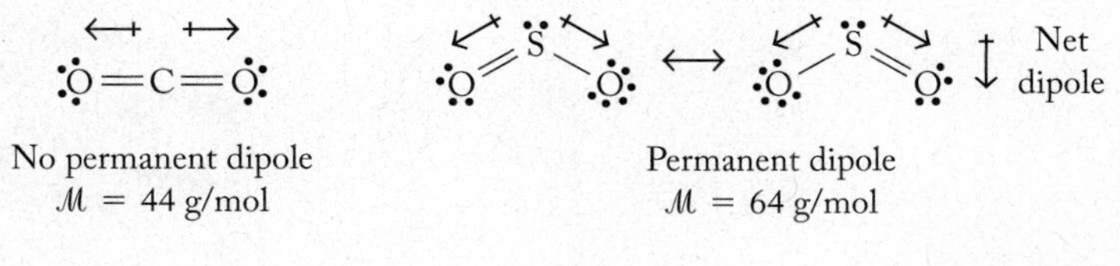

No permanent dipole
$\mathcal{M} = 44$ g/mol

Permanent dipole
$\mathcal{M} = 64$ g/mol

FIGURE 10.16

Solve Carbon dioxide molecules are linear and nonpolar, so they interact only through dispersion forces. Sulfur dioxide molecules are bent and therefore have a permanent dipole. In addition, SO_2 has a molar mass of 64 g/mol, as compared to 44 g/mol for CO_2, which means SO_2 is more polarizable and experiences stronger dispersion forces. The presence of dipole–dipole interactions and stronger dispersion forces translates into a larger value of the a constant for SO_2.

van der Waals forces all types of attractive forces possible between molecules: hydrogen bonds, other dipole–dipole interactions, and dispersion forces. The term applies only to interactions between molecules; ion–ion and ion–dipole interactions are *not* van der Waals forces.

Think about It We predicted that a larger polar molecule would have a larger value of *a* in the van der Waals equation than a smaller nonpolar molecule. In addition to being polar, SO_2 is more polarizable than CO_2, which also contributes to the larger value of *a* for SO_2. The larger value of *a* for the polar molecule, sulfur dioxide, is consistent with our prediction.

Practice Exercise Without looking up the values in Chapter 6, rank H_2O, O_2, and CO in order of increasing value of the van der Waals constant *a*.

■ ⋯⋯⋯⋯⋯⋯⋯⋯⋯⋯⋯⋯⋯⋯⋯⋯⋯⋯⋯

10.4 Polarity and Solubility

The process by which an ionic salt dissolves in water is explained in Section 10.2 in terms of ion–dipole interactions. The dissolving of one liquid in another and the dissolving of a gas in a liquid can also be explained in terms of intermolecular forces. To look at these processes, we consider the situation where the solvent is water or some other liquid and the solute is a covalent molecule. To understand and predict whether a given solute is soluble in a given solvent, we look at the balance between solute–solute interactions (that is, interactions between two or more solute molecules), solvent–solvent interactions, and solvent–solute interactions.

Just as ionic compounds dissolve in polar solvents because of strong ion–dipole interactions, polar solutes tend to dissolve in polar solvents because of dipole–dipole interactions between solute and solvent molecules (Figure 10.17a).

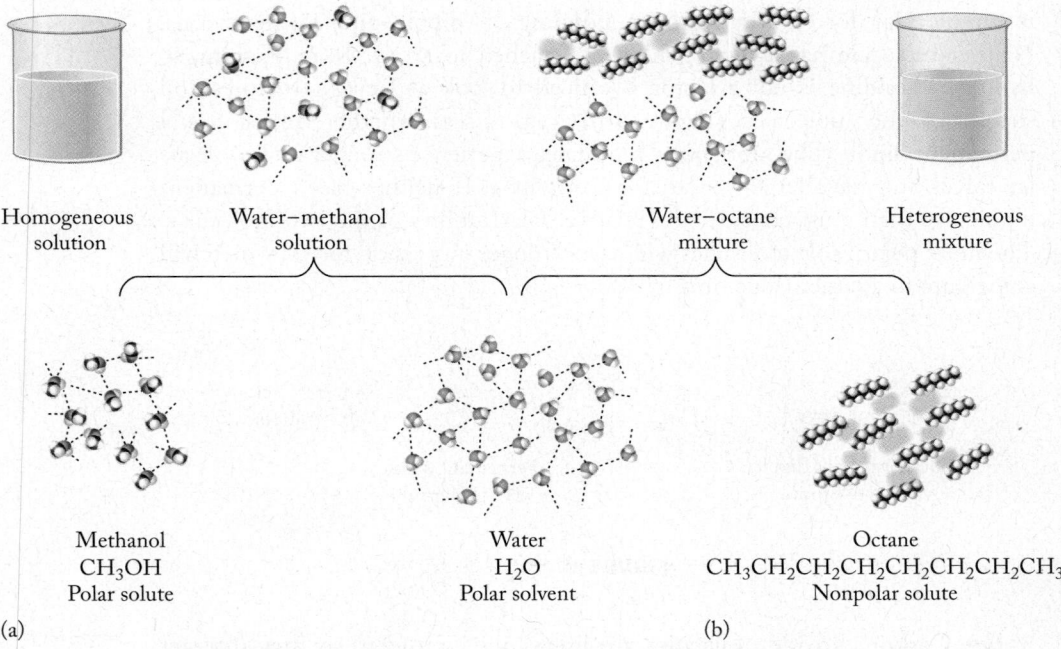

FIGURE 10.17 (a) A polar solvent like water dissolves polar materials like methanol because of favorable dipole–dipole interactions. In this specific case, both types of molecules are capable of hydrogen bonding, and solute–solvent interactions are of the same order of magnitude as solvent–solvent or solute–solute interactions. Methanol is miscible in all proportions in water. (b) Dispersion forces between long hydrocarbon chains provide the attractive force that holds octane molecules together in the liquid phase. The highly polar water molecules interact preferentially with each other, and any solvent–solute interactions are too weak to compete with the hydrogen-bonding network within the solvent. As a result, the hydrophobic octane is virtually insoluble in water.

Nonpolar solutes tend not to dissolve in polar solvents because the solvent–solute interactions that promote dissolution are weaker than those that keep solute molecules together and solvent molecules together (Figure 10.17b). This observation is the source of a common phrase used to describe solubility: like dissolves like.

Factors other than polarity influence solubility: temperature and pressure are two very important ones. Nevertheless, we can use the van der Waals forces present in any mixture of molecules as a guide to predict the relative solubilities of solutes in a given solvent.

SAMPLE EXERCISE 10.5 Predicting Solubility in Water

Which of these compounds should be very soluble in water and which should have limited solubility in water: carbon tetrachloride (CCl_4), ammonia (NH_3), hydrogen fluoride (HF), oxygen (O_2)?

Collect and Organize We are to predict the solubilities of several compounds in water. We can make our predictions based on polarity. Because water is polar, it is likely that compounds soluble in it are also polar.

Analyze To determine whether the compounds are is polar or nonpolar, we must draw their structures, evaluate each bond for its polarity, and assess whether the molecule has a permanent dipole moment. We predict that only polar compounds exhibit significant solubility in the polar solvent, water.

Solve Applying the concepts from Chapter 9, we determine the structures and overall dipole moments for the molecules (Figure 10.18). On the premise that like dissolves like, we predict that the polar molecules (NH_3 and HF) are soluble in water and the nonpolar molecules (CCl_4 and O_2) have only limited solubility in water. The ability of ammonia and hydrogen fluoride to participate in hydrogen bonding is another property they share with water, making them even more likely to be water soluble.

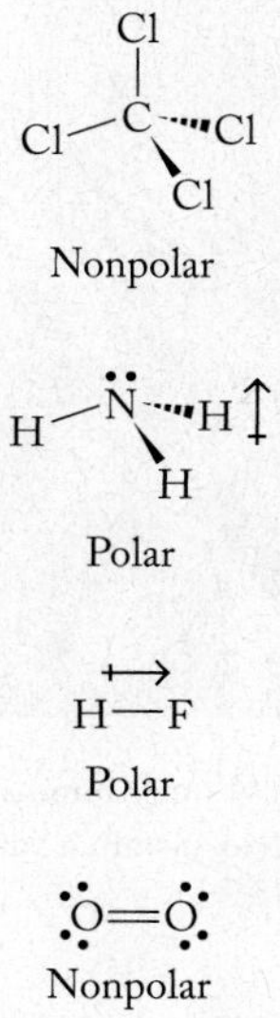

Think about It The three-dimensional structure of a molecule and the electronegativity difference between the atoms in each bond determine the polarity of a molecule. The polarity serves as a guide to solubility.

Practice Exercise In Chapter 6, we learned that some deep-sea divers breathe a mixture of gases rich in helium because the solubility of helium gas in blood is lower than the solubility of nitrogen gas in blood. Assuming that blood behaves like water in terms of dissolving substances, why should helium be less soluble in water than nitrogen?

■

FIGURE 10.18

In Sample Exercise 10.5 we predicted that ammonia and hydrogen fluoride are water soluble, while carbon tetrachloride and oxygen are much less soluble. These predictions are correct. Ammonia and hydrogen fluoride are both very soluble in water. They not only dissolve in water, *they react with it*. The hydrogen bonds between NH_3 and H_2O molecules are so strong they ionize water molecules, producing OH^- ions and H^+ ions that attach to NH_3 molecules, forming ammonium (NH_4^+) ions. The hydrogen bonds between molecules of HF and H_2O are so strong they ionize HF molecules, producing F^- ions and H^+ ions that attach to H_2O molecules, forming hydronium (H_3O^+) ions. We will return to these reactions in our discussion of acids and bases in Chapter 17.

FIGURE 10.19 Water contains sufficient dissolved oxygen to sustain a variety of aquatic life.

∞ **CONNECTION** The partial pressure of a gas in a mixture was defined in Chapter 6 as the contribution to the total pressure made by a component gas in the mixture.

▶❙❙ **CHEMTOUR** Henry's Law

....

Henry's law the concentration of a sparingly soluble, chemically unreactive gas in a liquid is proportional to the partial pressure of the gas.

....

In contrast, the nonpolar molecules carbon tetrachloride and oxygen have very limited solubility in water because solvent–solute interactions are very weak. The large permanent dipole of a water molecule interacts more favorably with other water dipoles than it does with the weaker induced dipoles in the nonpolar materials. Solvent–solute interactions between water and nonpolar molecules cannot compete with the much stronger solvent–solvent interactions. However, the induced dipoles in molecules like CCl_4 and O_2 are of the same order of magnitude as the dispersion forces between molecules of a nonpolar solute; hence the solvent–solute interactions in that case compete favorably with the solvent–solvent interactions in a nonpolar solvent. Nonpolar molecules are very sparingly soluble in polar solvents but can be quite soluble in nonpolar solvents. We explore these effects further in Chapter 14. Even though oxygen is only slightly soluble in water, its limited solubility of about 10 mg/L at normal atmospheric pressure is sufficient to sustain aquatic life.

Calculating the Solubility of Gases in Water: Henry's Law

The O_2 level in natural waters is normally sufficient to support life (Figure 10.19), but you may have noticed fish in rivers gasping for air in very warm weather. The warmed water lacks sufficient dissolved oxygen because the solubility of O_2 (and of most other gases) in water decreases with increasing temperature. Relatively weak dipole–induced dipole interactions between water molecules and the nonpolar molecules of oxygen gas account for the solubility of O_2 in water. As temperature increases, the kinetic energy of the molecules increases and more energy is available to disrupt these intermolecular attractions, reducing solubility. In other words, at higher temperatures more O_2 molecules have the necessary kinetic energy to overcome the dipole–induced dipole forces between O_2 and H_2O molecules, allowing O_2 to escape from the solution.

The solubility of oxygen gas in a liquid such as water also depends on the partial pressure of the O_2 in the air above the surface of the liquid. While the partial pressure of O_2 at sea level remains fairly constant at about 0.21 atm, the low partial pressure of oxygen at high altitudes can result in lower than normal concentrations of oxygen in liquids, including the blood and tissues of humans and animals. For example, climbers in the Himalayas may become weak and unable to think clearly because of lack of oxygen to the brain, a condition known as *anoxia*.

The relationship between gas solubility in a liquid (the concentration of the dissolved gas) and the partial pressure of the gas in the environment surrounding the liquid applies to all sparingly soluble gases. This is known as **Henry's law** in honor of William Henry (1775–1836), a British physician who first proposed the relationship:

$$C_{gas} = k_H P_{gas} \qquad (10.2)$$

where C_{gas} represents the concentration (solubility) of a gas in a particular solvent, k_H is the Henry's law constant for the gas in that solvent, and P_{gas} is the partial pressure of the gas in the environment surrounding the solvent. When C_{gas} is expressed in molarity, the units of the Henry's law constant are moles per liter-atmosphere, mol/(L · atm). Table 10.5 lists k_H values for several common gases in water.

Henry's law states that the concentration of dissolved oxygen in blood is proportional to the partial pressure of oxygen in the air we inhale and thus proportional to atmospheric pressure. This is an accurate statement of Henry's law, but residents of Denver, Colorado, or Kimberley, Canada (average atmospheric pressure 0.85 atm), do not live with less blood oxygen than residents of New York City or Rome, Italy (average atmospheric pressure 1.00 atm), because of the way the oxygen transport system in our bodies responds to local conditions.

The amount of oxygen in the blood is related to the concentration of hemoglobin and to the fraction of the hemoglobin sites that contain oxygen as the blood leaves the lungs. This saturation of binding sites depends on the partial pressure of oxygen as well as proper lung function. For most people, breathing air with $P_{O_2} > 0.11$ atm results in nearly 100% saturation of hemoglobin binding sites. If P_{O_2} decreases to about 0.066 atm (as it does on high mountains), the percent saturation decreases to about 80%. Over several weeks, the body responds to lower oxygen partial pressures by producing more red blood cells and more hemoglobin. This increase in the number of O_2 carriers compensates for the lower partial pressure of O_2. Even though the level of saturation decreases, the actual number of carriers of O_2 increases, so the same amount of O_2 is delivered to tissues.

TABLE 10.5	Henry's Law Constants for Gas Solubility in Water at 20°C
Gas	k_H [mol/(L · atm)]
He	3.5×10^{-4}
O_2	1.3×10^{-3}
N_2	6.7×10^{-4}
CO_2	3.5×10^{-2}

SAMPLE EXERCISE 10.6   **Calculating Gas Solubility Using Henry's Law**

Calculate the solubility of oxygen in water in moles per liter at 1.00 atm pressure and 20°C. The mole fraction of O_2 in air is 0.209. (Remember that the sum of the mole fractions of all the gases in a mixture equals 1.)

Collect and Organize We are to determine the solubility of oxygen gas in water at 1.00 atm and 20°C. We are given the mole fraction of oxygen in air, and the total (atmospheric) pressure. Henry's law (Equation 10.2) relates solubility to partial pressure. The Henry's law constant for oxygen at 20°C is 1.3×10^{-3} mol/L · atm.

Analyze We need the partial pressure of oxygen for Henry's law. Partial pressure is the product of the mole fraction (X_{O_2}) of a gas times the total pressure. We can then substitute the partial pressure into Equation 10.2 and calculate the solubility of oxygen. From Sample Exercise 10.5 we expect the solubility of O_2 in water to be quite low.

Solve The partial pressure of oxygen is calculated using Equation 6.24:

$$P_{O_2} = X_{O_2}P_{total} = (0.209)(1.00 \text{ atm}) = 0.209 \text{ atm}$$

Substituting this value for P_{O_2} and k_H for O_2 in water in Equation 10.2 gives

$$C_{O_2} = k_H P_{O_2} = \left(\frac{1.3 \times 10^{-3} \text{ mol}}{\text{L} \cdot \text{atm}}\right)(0.209 \text{ atm}) = 2.7 \times 10^{-4} \text{ mol/L}$$

Think about It We predicted that oxygen is not very soluble in water because O_2 is a nonpolar solute and water is a polar solvent, and the answer agrees with that prediction.

Practice Exercise Calculate the solubility of oxygen in water at the top of Mt. Everest, where atmospheric pressure is 0.35 atm.

CONNECTION We defined mole fraction in Chapter 6 in the discussion of partial pressures of gases.

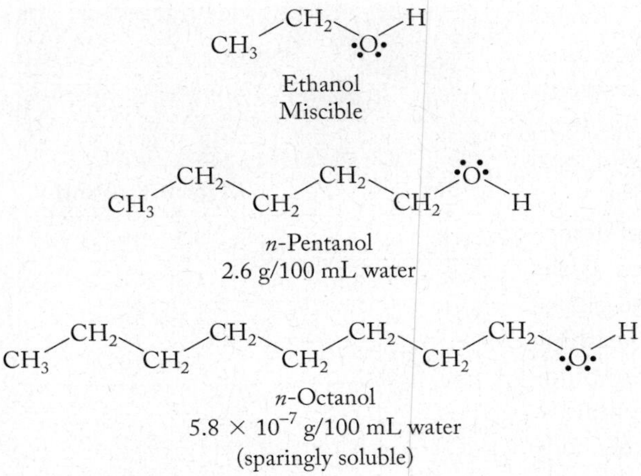

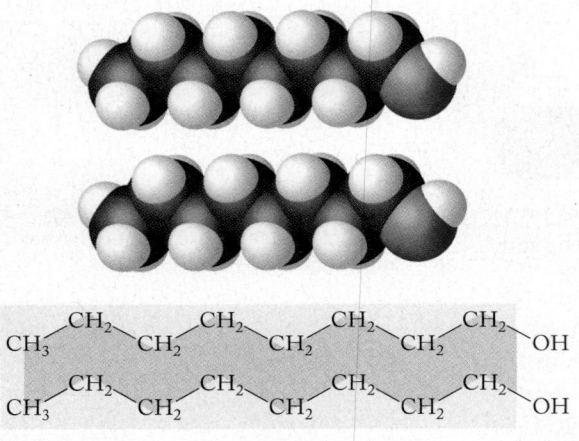

FIGURE 10.20 Ethanol, *n*-pentanol, and *n*-octanol have different solubilities in water.

Ethanol
Miscible

n-Pentanol
2.6 g/100 mL water

n-Octanol
5.8×10^{-7} g/100 mL water
(sparingly soluble)

Dispersion forces

FIGURE 10.21 Even though the interaction between two –CH₂– units is weak, the interactions add up along the long hydrocarbon chains of *n*-octanol molecules and give rise to an attractive force that keeps the octanol molecules together in water despite the presence of the polar –OH group in the alcohol.

hydrophobic a "water-fearing" or repulsive interaction between a solute and water that diminishes water solubility.

hydrophilic a "water-loving" or attractive interaction between a solute and water that promotes water solubility.

phase diagram a graphical representation of the dependence of the stabilities of the physical states of a substance on temperature and pressure.

Combinations of Intermolecular Forces

More than one type of intermolecular force may need to be considered when larger molecules dissolve in a liquid solvent. Consider the solubilities in water of three organic alcohols with the structures shown in Figure 10.20. All three alcohols are liquids at room temperature, so we are looking at the solubility of one liquid in another. All three compounds contain a –OH group, which means they are like water in that they make hydrogen bonds. Considering that they are all polar molecules, why do their solubilities differ so greatly in the polar solvent water?

Ethanol is miscible with water, which means it is soluble in all proportions. Pentanol has a finite solubility, which means that once a solution of pentanol in water is saturated, adding more pentanol results in a heterogeneous mixture in which two layers of liquid are visible. Octanol is sparingly soluble in water, and almost any measurable amount of octanol added to water results in a heterogeneous mixture.

The feature that differentiates these molecules from one another is the number of –CH₂– groups. The long series of –CH₂– groups in octanol makes the non-OH part of the molecule very nonpolar and very much unlike water. The dispersion forces between the long –CH₂– chains of adjacent octanol molecules are quite strong (Figure 10.21), and the sum of these attractive forces keeps the octanol molecules together, even though the polar ends of the molecules are attracted to water molecules. The minuscule solubility of octanol in water illustrates a situation where dispersion forces contribute more to intermolecular interactions than do dipole–dipole forces and hydrogen bonding localized in one small region of a molecule.

In ethanol, the polarity and hydrogen bonding ability of the –OH group dominate the interaction of ethanol with water because the hydrocarbon portion of the molecule is too short (only two carbon atoms long) to have significant dispersion forces. The solubility of pentanol is less than that of ethanol but much greater than that of octanol because of the difference in lengths of the –CH₂– chains. The competing dispersion forces that limit pentanol solubility and the hydrogen-bonding interactions that promote it combine to produce moderate solubility for this alcohol.

Nonpolar interactions like the dispersion forces between hydrocarbon chains are called **hydrophobic** (literally, "water fearing") interactions, whereas interactions that promote solubility in water are called **hydrophilic** ("water loving") interactions. For molecules that contain both polar and nonpolar groups, as these alcohols do, the solubility in water is due to the balance between hydrophilic and hydrophobic interactions. As the hydrophobic portion of the molecule increases in size, the entire molecule becomes more hydrophobic and solubility in water decreases.

10.5 Phase Diagrams: Intermolecular Forces at Work

The strength of forces between the particles making up any substance and the mass of those particles control whether the substance exists as a solid, liquid, or gas at a given temperature. However, temperature is not the only factor

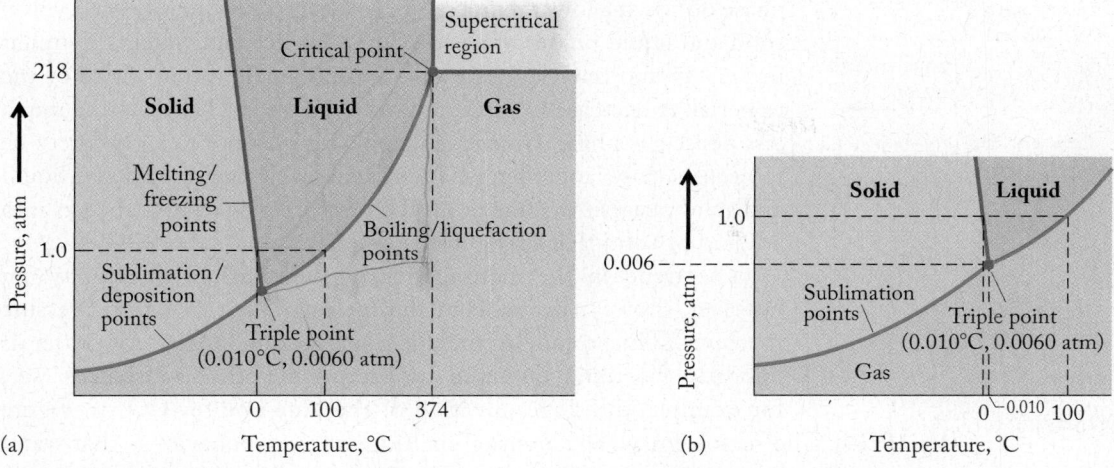

FIGURE 10.22 (a) The phase diagram for water indicates in which phase water exists at various combinations of pressure and temperature. (b) An expanded view of the phase diagram at low pressure and temperature (axes are not linear).

that influences the state of a substance. Pressure also plays a role. To represent how these two variables relate to the phases of substances, scientists use **phase diagrams**.

Phases and Phase Transformations

The phase diagram for water, like that for many other pure substances, has three regions corresponding to the three phases of matter, plus a fourth region called a *supercritical region* (Figure 10.22). The lines separating the regions are called *equilibrium lines* because the two states bordering them are at equilibrium. The blue equilibrium line separating the solid and liquid regions represents a series of freezing (or melting) points; the points on this line are combinations of temperature and pressure at which the solid and liquid states coexist. The red line separating the liquid and gas regions of Figure 10.22 represents a series of boiling points or *liquefaction* points (gas turning into liquid); the points on this line are combinations of temperature and pressure at which the liquid and gaseous states coexist. The green line separating the solid and gaseous states represents a series of sublimation points (solid turning to gas) or deposition points (gas turning to solid); the points on this line are combinations of temperature and pressure at which the solid and gaseous states coexist.

Notice that the red line curves from the lower left to the upper right, separating the liquid and gas regions of the phase diagram. This line represents the changing boiling point of the liquid as a function of pressure. Its shape makes sense because when the pressure above a liquid is increased, as in a pressure cooker (Figure 10.23), the temperature required to overcome that pressure and allow liquid molecules to enter the gas phase (i.e., the boiling point) increases. The shape of the solid–gas curve also makes sense: higher pressures make it more difficult for molecules of ice to sublime to water vapor.

The trends in the boiling/liquefaction equilibrium line and the sublimation/deposition equilibrium line in Figure 10.22 show that both the boiling point (liquid ⇌ gas) and the sublimation point (solid ⇌ gas) of water decrease as pressure decreases. In both cases a phase transition from a dense phase to the vapor

FIGURE 10.23 Steam cannot escape from a pressure cooker and pressure inside the pot increases as water is heated inside it. The pressure increase causes the water to boil at a higher temperature, so food inside cooks faster.

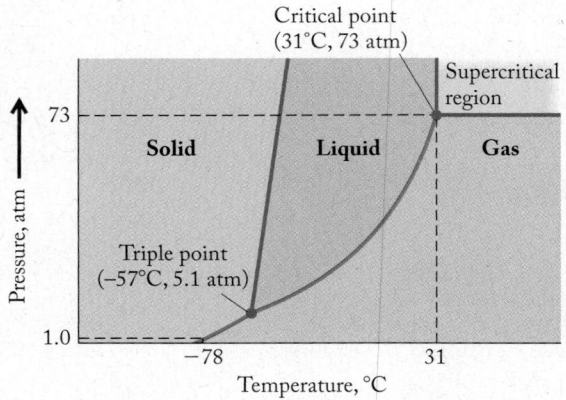

FIGURE 10.24 Phase diagram for carbon dioxide (axes are not linear).

▶❚❚ **CHEMTOUR** Phase Diagrams

triple point the temperature and pressure where all three phases of a substance coexist. Freezing and melting, boiling and liquefaction, and sublimation and deposition all proceed at the same rate, so no net change takes place in the system.

critical point a specific temperature and pressure at which the liquid and gas phases of a substance have the same density and are indistinguishable from each other.

supercritical fluid a substance at conditions above its critical temperature and pressure, where the liquid and vapor phases are indistinguishable and have some characteristics of both a liquid and a gas.

phase occurs at a lower temperature when the pressure is lower. Typical solid and liquid phases of a compound, like ice and water, are similar in density and many times more dense than the vapor phase of the material. A decrease in pressure favors the phase change to the much less dense gas phase. Conversely, applying pressure to a gas forces its molecules closer together, which in turn favors the more dense liquid and solid phases. At some point the pressure may change the gas into a liquid or solid that takes up much less volume.

The trend in the melting/freezing equilibrium line for water, however, shows a decrease in the melting point of ice as pressure increases. This trend in the melting/freezing points for water is opposite the trend observed for almost all other substances (see, for example, the slope direction of the blue line for CO_2 in Figure 10.24). The reason for water's unusual melting/freezing behavior is that water expands when it freezes. Most other substances are denser in the solid state than in the liquid state because they contract when they freeze. Water, however, expands as it freezes because hydrogen bonds between molecules of water in the solid phase create a structure that is more open than the structure in liquid water, making ice less dense than liquid water. Applying enough pressure to ice forces it into a physical state (liquid water) in which it takes up less volume.

CONCEPT TEST ···

A truck with a mass of 2000 kg is parked on an icy driveway where the ice is at $-4°C$. Is it possible that the ice under the tires of this vehicle melts, even though the temperature remains constant?

···

A point of special interest on a phase diagram is the point where all three lines describing the phase transitions meet. Known as the **triple point**, it identifies the temperature and pressure at which all three states (liquid, solid, and vapor) coexist. For water, the triple point is just above the normal melting point, at $0.010°C$, but at a very low pressure of 0.0060 atm.

Another point of interest is the place where the boiling/liquefaction equilibrium line ends. At this **critical point**, the liquid and gaseous states are indistinguishable from each other. This point is reached because thermal expansion at this high temperature causes the liquid to become less dense, while the high pressure compresses the gas into a small volume, increasing its density. At the critical point the densities of the liquid and gaseous states are equal, so one cannot be distinguished from the other.

At temperature–pressure combinations above its critical point, a substance exists as a **supercritical fluid**. A supercritical fluid has a density similar to that of a liquid but a viscosity similar to that of a gas. It can penetrate materials like a gas and dissolve materials like a liquid. Supercritical carbon dioxide is used in the food-processing industry to decaffeinate coffee and remove fat from potato chips. Supercritical carbon dioxide and water are sometimes mixed to generate a range of fluids able to selectively dissolve materials of interest from mixtures while leaving other components untouched.

The phase diagram of CO_2 is shown in Figure 10.24. The dashed line at $P = 1.0$ atm defines the phases that exist at 1 atm pressure, but note that the blue region representing liquid CO_2 does not extend below 5.1 atm. This means that solid CO_2 does not melt into a liquid at normal temperatures and pressures. Rather, it sublimes directly to CO_2 gas. This behavior gives rise to the common name for solid CO_2, *dry ice*; it is a solid that keeps things cool, as ice does, but it forms no "wet" liquid. The critical point of CO_2 is at 31°C and 73 atm, a pressure easily achieved with compressors in laboratories, factories, and food-processing plants, which means CO_2 is readily available for use as a supercritical fluid.

SAMPLE EXERCISE 10.7   Reading a Phase Diagram

Describe the phase changes that take place as the pressure on a sample of water at 0°C is increased from 0.0001 atm to 200 atm.

Collect and Organize We are to describe the phase changes water undergoes at one temperature as the pressure is increased. To solve this problem, we can use Figure 10.22, the phase diagram for water. To read a phase diagram, we need to remember that every point is characterized by a temperature and a pressure, and that every time we cross an equilibrium line, the phase changes.

Analyze The changes we must describe take place along the line that intersects the temperature axis at 0°C and runs parallel to the pressure axis.

Solve Starting at the bottom of the phase diagram and following the dashed line at 0°C upward as pressure increases, we approximate the location of the starting pressure, 0.0001 atm, a little below and to the left of the triple point ($T = 0.010$°C and $P = 0.0060$ atm). Water at 0°C and 0.0001 atm is a gas, as indicated by the phase diagram. As we follow the 0°C dashed line up from the temperature axis, the point where this line intersects the green equilibrium line indicates that the gas solidifies to ice at a pressure below 1 atm. Increasing the pressure above 1 atm, we see the dashed line intersect the blue equilibrium line at that pressure. This intersection means that at this pressure the ice melts to liquid water. Mentally extending the 0°C dashed line farther upward tells us that the water remains a liquid at pressures up to and beyond 200 atm, the upper limit we were asked for.

Think about It The phase changes in water from gas to solid to liquid with increasing pressure at 0°C make sense from a molecular point of view because higher pressures favor the most dense phase. Water is more dense than ice, so the transitions from gas to solid to liquid are transitions from the least dense to the most dense phase.

Practice Exercise Describe the phase changes that occur when the temperature of CO_2 is increased from −100°C to 200°C at a pressure of 25 atm.

CONCEPT TEST

Look at the phase diagram in Figure 10.24. If a phase diagram favors the more dense phase as pressure increases, does solid CO_2 float on liquid CO_2 at any point along the blue line?

surface tension the energy needed to separate the molecules at the surface of the liquid.

meniscus the concave or convex surface of a liquid.

10.6 Some Remarkable Properties of Water

Water has many remarkable properties. Its melting and boiling points are much higher than those of all other molecular substances with similar molar masses (Table 10.6), and its solid form (ice) is less dense than liquid water. These phenomena and many others are related to the strength of the hydrogen bonds that attract water molecules to one another. Let's look now at some other unique properties resulting from the existence of hydrogen bonds in water.

TABLE 10.6	Melting Points and Boiling Points of Four Compounds of Similar Molar Mass		
Substance	$\mathcal{M}$ (g/mol)	Melting Point (°C)	Boiling Point (°C)
H_2O	18.02	0	100
HF	20.01	−83	19.5
NH_3	17.03	−78	−33
CH_4	16.04	−182	−164

Surface Tension and Viscosity

▶❚❚ **CHEMTOUR** Hydrogen Bonding in Water

Surface tension is the resistance of a liquid to any increase in its surface area. Surface tension represents the energy required to move molecules apart so that an object can break through the surface.

Hydrogen bonding in water creates such a high surface tension, 7.29×10^{22} J/m² at 25°C, that a carefully placed steel needle floats on water and insects called water striders can walk on it (Figure 10.25). The same needle and insect would sink in oil or gasoline.

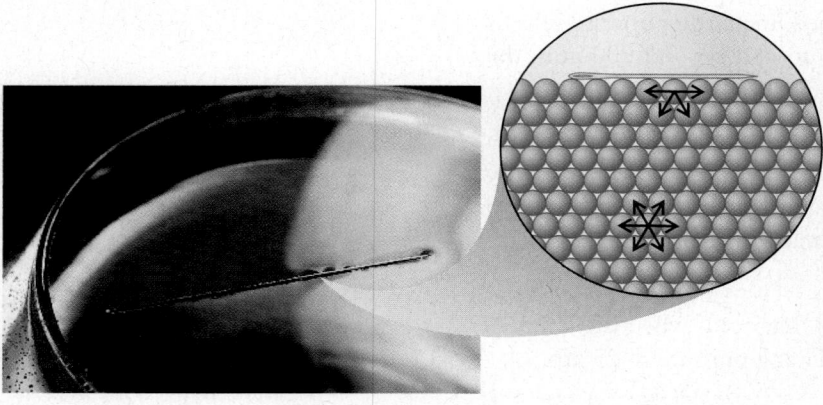

(a)

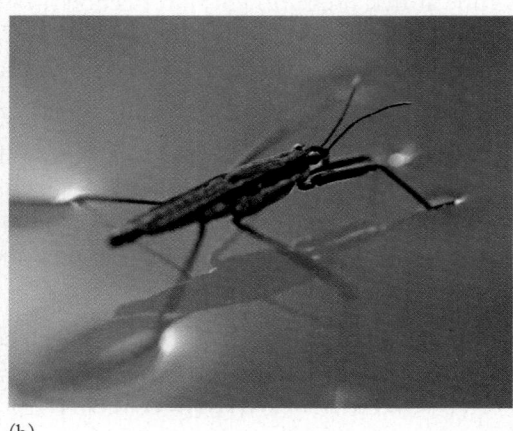

(b)

FIGURE 10.25 (a) Intermolecular forces, including hydrogen bonding, are exerted equally in all directions in the interior of a liquid. Because forces in opposite directions cancel, a molecule in the interior does not experience surface tension. However, there is no liquid water above a surface to exert attractive intermolecular forces. The resulting imbalance causes the surface water molecules to adhere tightly to one another, creating surface tension. When the magnitude of the surface tension exceeds the downward force exerted on the surface water molecules by a needle, the needle cannot break through and floats on the surface. (b) Surface tension also allows a water strider to rest on top of water without sinking.

Another illustration of the intermolecular forces acting in liquids is seen in the shape of the liquid surface when a liquid is in a graduated cylinder or other small-diameter tube (Figure 10.26). The surface of water is concave in such a tube, but the surface of liquid mercury is convex. Either curved surface, concave or convex, is called a **meniscus**. In both liquids, the meniscus is the result of two competing forces: *cohesive forces*, which are interactions between like particles, and *adhesive forces*, which are interactions between unlike particles. In the water sample in Figure 10.26, the cohesive forces are hydrogen bonds between water molecules and the adhesive forces are dipole–dipole interactions between water molecules and polar Si—O—Si groups on the surface of the glass. The adhesive forces are strong enough to cause the water to climb upward on the glass, creating the concave meniscus. The adhesive forces are greater than the cohesive forces in this case because surface water molecules next to the glass have less contact with other water molecules and therefore experience smaller cohesive forces.

In the mercury, the cohesive forces are metallic bonds between mercury atoms and the adhesive forces are interactions between induced dipoles in the mercury atoms and the polar Si—O—Si groups on the glass surface. In this case, the adhesive forces are much weaker than the cohesive forces, and mercury atoms are more attracted to one another than to the glass. As a result they do not adhere to the wall and form a convex meniscus.

The observation that adhesive forces outweigh cohesive forces in water has consequences in other situations. In a very narrow tube, such as a capillary tube, adhesion to the tube wall again draws the outer ring of water molecules upward. At the same time, cohesive forces between the outer ring molecules and those adjacent to them draw the adjacent molecules upward (Figure 10.27).

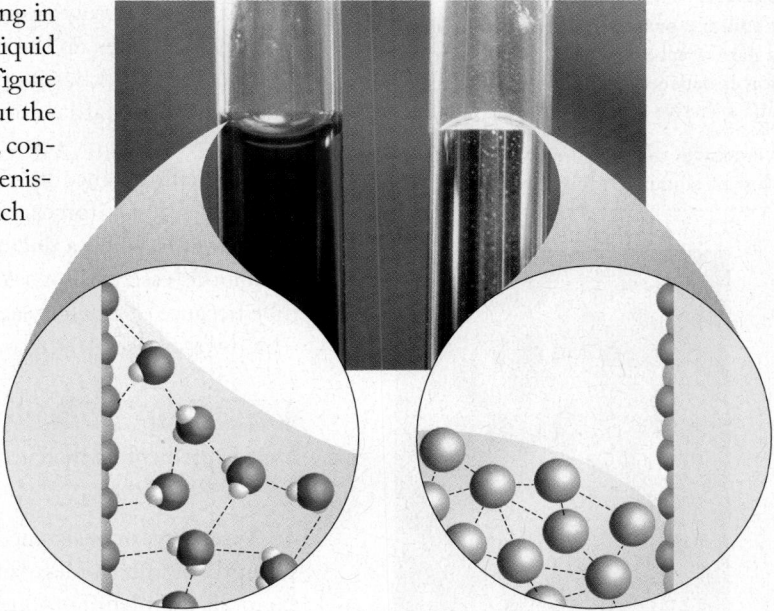

FIGURE 10.26 Hydrogen bonds between the H of water molecules and the O of the silicon dioxide that makes up the glass are an adhesive force that causes water molecules to adhere to the glass surface and form a concave meniscus. Because mercury atoms have no such attraction to the silicon dioxide in the glass, mercury forms a convex meniscus.

▶❚❚ **CHEMTOUR** Capillary Action

(a)

(b)

FIGURE 10.27 Because of capillary action, water rises in (a) a capillary tube and (b) in a stick of celery.

capillary action the rise of a liquid in a narrow tube as a result of adhesive forces between the liquid and the tube and cohesive forces within the liquid.

viscosity the measure of the resistance to flow of a liquid.

FIGURE 10.28 High viscosity results from strong intermolecular forces between large polar molecules. The red liquid is more viscous than the green, which pours freely.

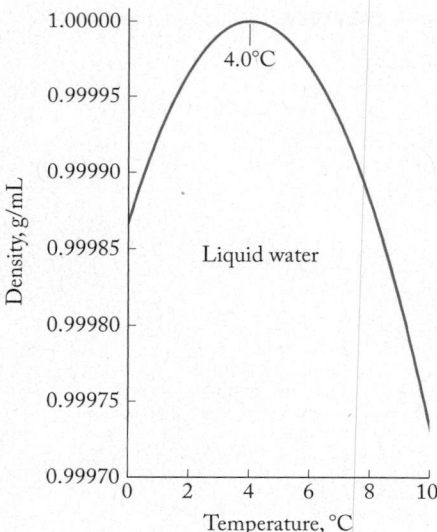

FIGURE 10.29 As water is cooled, its density increases until the temperature has been lowered to 4°C. At this temperature the density has its maximum value, 1.000 g/mL. As the water cools from 4°C to its freezing point at 0°C, the density decreases.

If the tube is narrow enough, this combination of adhesion and cohesion draws a column of water up the tube in a phenomenon known as **capillary action**. The water column reaches its maximum height when the downward force of gravity balances the upward adhesive and cohesive forces.

In a test tube or pipette, water molecules in contact with the glass move only a very small distance up the glass before the force of gravity balances the adhesive and cohesive forces.

However, when a doctor takes a blood sample after a pin prick in your finger, the blood (essentially an aqueous solution) spontaneously moves into a capillary tube because of capillary action. Capillary action also contributes to the process by which water rises 100 m or more up the trunks of tall trees.

> **CONCEPT TEST** ···
>
> Do you predict that mercury would be spontaneously drawn into a glass capillary tube?

Viscosity, or resistance to flow, is another property of liquids related to the strength of intermolecular forces (Figure 10.28). In nonpolar compounds—for example, petroleum products like the fuels pentane and hexane—viscosity increases with increasing molar mass. Lubricating oil, for example, is much more viscous than gasoline because the hydrocarbons in a typical lubricating oil have molar masses that are two to three times those of the hydrocarbons in gasoline. Larger molar masses mean stronger dispersion forces between the molecules in lubricating oil. Stronger interactions mean that molecules of lubricating oil do not slide past one another as easily as the shorter-chain hydrocarbons in gasoline, and bulk quantities of the larger molecules do not flow as easily when poured.

For polar molecules, viscosity also reflects the strengths of intermolecular forces. Comparing two of the polar alcohols from Figure 10.20, we see that octanol has a higher viscosity than ethanol, reflecting the greater intermolecular forces (dipole–dipole and dispersion forces) in octanol. Water is more viscous than gasoline even though water molecules are much smaller than the nonpolar molecules in gasoline. The remarkable viscosity of water is another property directly related to the hydrogen bonds between water molecules.

> **CONCEPT TEST** ···
>
> Is the viscosity of seawater greater than that of pure water? Why or why not?

Another property unique to water is how its density changes with temperature. The density increases as water is cooled to 4°C (Figure 10.29), a pattern observed for most liquids and solids and for all gases. However, as water is cooled from 4°C to 0°C, it expands, and its density decreases as the pattern of its hydrogen bonds changes with temperature. As water freezes at 0°C, its density drops even more, to about 0.92 g/mL for ice, causing ice to float on liquid water (Figure 10.30). This unusual behavior is caused by the formation of a network of hydrogen bonds in ice. With each oxygen atom covalently bonded to two hydrogen atoms and hydrogen-bonded to two other hydrogen atoms, the molecules form an extensive and open hexagonal network in the solid phase. Because of the space between the molecules in the network, the same number of molecules occupies more volume in ice than in liquid water. When ice melts, some of the hydrogen bonds in the rigid array break, allowing the molecules in the liquid to be arranged more compactly, up to 4°C, whereupon the density begins to decrease once more.

 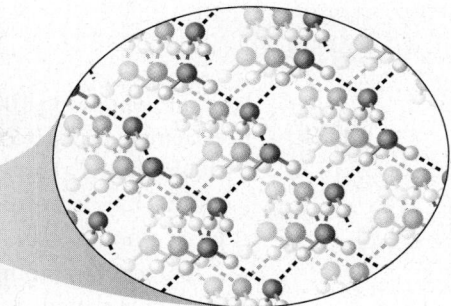

FIGURE 10.30 Because of the changes in the density of water near the freezing point and because ice has a lower density (0.92 g/mL) than water, ice and ice-cold water float on top of warmer water (4°C) in lakes and rivers. In ice, each oxygen atom is linked to four hydrogen atoms: two by covalent bonds and two by hydrogen bonds.

Water and Aquatic Life

The expanded structure of ice plays a crucial ecological role in temperate and polar climates. The lower density of ice means that lakes, rivers, and polar oceans freeze from the top down, allowing fish and other aquatic life to survive in the liquid water below. Each fall, surface water cools first, and its density increases until its temperature reaches 4°C, the peak in the graph of Figure 10.29. This maximum-density water at 4°C sinks to the bottom, bringing warmer water to the surface, which in turn cools to 4°C and sinks, pushing the next layer of warm water to the surface in a continuing cycle (Figure 10.31). This autumnal turnover stirs up dissolved nutrients, making them available for life in the sunlit surface during the next growing season. When all of the water has reached 4°C and the surface water cools further, it becomes less dense and ice may eventually form. The layer of ice insulates the 4°C water beneath it, allowing aquatic life to survive.

In spring, the ice melts and the surface water warms to 4°C. At this temperature, the entire column of water has nearly the same temperature and density; dissolved nutrients for plant growth are evenly distributed, and the stage is set for a burst of photosynthesis and biological activity called the spring bloom.

Further warming of the surface water creates a warm upper layer separated from colder, denser water by a *thermocline*, a sharp change in temperature between the two layers. Biological activity depletes the pool of nutrients above the thermocline as decaying biomass settles to the bottom. Consequently, photosynthetic activity drops from its spring maximum during the summer, even though there is much more energy available from the sun. The thermocline persists until the autumn turnover mixes the water column and the cycle begins anew.

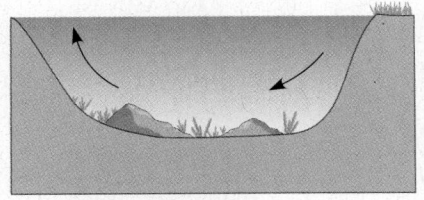 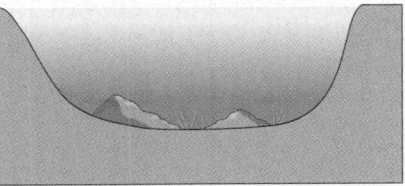

(a) Autumn   (b) Winter

FIGURE 10.31 (a) As the surface water of a pond cools to 4°C, its density increases and it sinks to the bottom, bringing the warmer, less dense water to the surface. (b) Continued cooling of the surface water below 4°C produces a less dense layer that may eventually freeze while the more dense water deeper in the pond remains at 4°C.

The Halogens: The Salt of the Earth

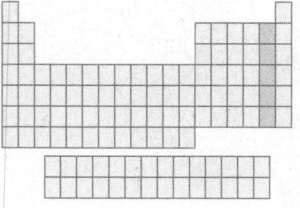

The halogen family, group 17 of the periodic table, consists of four common elements—fluorine, chlorine, bromine, and iodine—and one, astatine, that is rarely encountered because all of its 24 isotopes are radioactive. Nothing is known of the bulk physical properties of astatine, and it may be the rarest naturally occurring terrestrial element. Estimates suggest that the outermost kilometer of Earth's crust contains less than 45 mg of astatine. We discussed fluorine in the descriptive chemistry section in Chapter 8, and we address the remaining halogens here.

The word *halogen* means salt-former. Although the chemistry of the halogens as components of salts is important, these elements have a rich molecular chemistry as well.

Physical properties of the common halogens follow the expected trends:

	Fluorine (F$_2$)	Chlorine (Cl$_2$)	Bromine (Br$_2$)	Iodine (I$_2$)
Color:	Pale yellow	Yellow-green	Red-brown	Violet-black
State under standard conditions:	Gas	Gas	Liquid	Solid
Melting point (°C):	−219	−101	−7	114
Boiling point (°C):	−188	−34	59	185
Atomic radius (pm):	71	99	114	133
Ionic radius (pm):	133	181	195	220
Energy of H—X bond (kJ/mol):	567	431	366	299

Under standard conditions, fluorine and chlorine are both gases, bromine is a liquid, and iodine is a solid (Figure 10.32). Bromine and iodine have low boiling and sublimation points, respectively, so that a dark-red vapor always accompanies liquid bromine at room temperature, and violet vapor is frequently visible above solid iodine. Bromine is the only liquid nonmetallic element.

All the halogens have strong, penetrating odors and are hazardous to humans. A concentration of chlorine in the air greater than 1 ppm is damaging to health, and a few breaths of air containing 1000 ppm of chlorine are fatal. Bromine is extremely corrosive to human tissue, and exposure of the skin causes burns that are painful and slow to heal. Iodine, as iodide ion, is an essential trace element, but ingestion of large amounts of iodine is dangerous, and consumption of 2 to 3 g is fatal to humans.

The halogens may be the most important group in the periodic table in terms of general industrial use. By far the most important use of chlorine is in the manufacture of chemicals used to sterilize water for drinking and for filling swimming pools, bleach for the paper industry, explosives, dyes, insecticides, cleaning agents, and plastics. The synthetic sweetener Splenda is made from sucrose by replacing three −OH groups with chlorine atoms. The molecule retains its sweetness but is not metabolized in our bodies and hence is considered calorie-free. Chloride ions are ubiquitous in living systems and play a vital role in biology. They balance the positive charge on the sodium and potassium ions in body fluids and thereby maintain electrical neutrality.

Bromine is used in making fumigants and insecticides, dyes, compounds for purifying water, and flame-proofing agents.

Tincture of iodine, typically part of emergency survival and first aid kits, is a solution of up to 10% iodine in ethanol that is used to disinfect wounds and purify water. The radioactive isotope ^{131}I is used to treat thyroid cancer and other diseases of the thyroid. Two other radioactive isotopes, ^{123}I and ^{125}I, are used as imaging agents to evaluate thyroid function. Table salt is often enriched with iodide ion (iodized salt) to

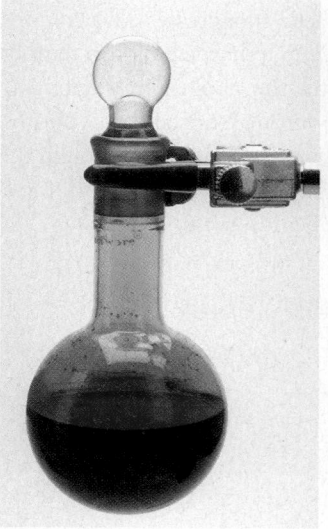

FIGURE 10.32 Liquid bromine and solid iodine both vaporize considerably at room temperature.

ensure proper production of the thyroid hormones that regulate metabolism. Iodine is necessary for human health at the dietary level of 150 μg per day. Iodine deficiency in infants is a leading cause of preventable mental retardation and a serious health problem in developing nations.

None of the halogens exist in nature as free diatomic elements. One source of chloride, bromide, and iodide ions is the ocean, from which these ions are extracted and then oxidized to produce the pure elements. It is estimated that 10^{16} tons of chloride ion are available in Earth's oceans. However, only about one-third of the NaCl used commercially is claimed from the currently existing oceans; the bulk of it is mined from rock salt deposits, which are the residues from the evaporation of ancient seas.

Chlorine is produced industrially by electrolysis of molten NaCl or aqueous solutions of NaCl. In the first reaction, sodium metal is produced; in the second, hydrogen gas and sodium hydroxide are produced, giving rise to the name *chlor-alkali process* for the reaction.

$$2\,NaCl(\ell) \rightarrow 2\,Na(s) + Cl_2(g)$$

$$2\,NaCl(aq) + 2\,H_2O(\ell) \rightarrow H_2(g) + Cl_2(g) + 2\,NaOH(aq)$$

Bromine is obtained from ocean water and from highly concentrated brine sources like the Dead Sea (Figure 10.33). The bromide ion is oxidized by treatment with Cl_2:

$$2\,Br^-(aq) + Cl_2(g) \rightarrow Br_2(\ell) + 2\,Cl^-(aq)$$

Elemental iodine is obtained from natural brines as well, and its collection involves the oxidation of I^- to I_2. Another industrial source is Chilean saltpeter, in which iodine is in the form of iodate salts (IO_3^-). The iodate is reduced to I^-, which is then oxidized to I_2 by treatment with more IO_3^-:

$$IO_3^-(aq) + 3\,HSO_3^-(aq) \rightarrow I^-(aq) + 3\,SO_4^{2-}(aq) + 3\,H^+(aq)$$

$$5\,I^-(aq) + IO_3^-(aq) + 6\,H^+(aq) \rightarrow 3\,I_2(s) + 3\,H_2O(\ell)$$

Elemental chlorine reacts with water to form a mixture of hydrochloric and hypochlorous acids:

$$Cl_2(g) + H_2O(\ell) \rightarrow HCl(aq) + HOCl(aq)$$

All the other halogens react with water in this fashion to make the corresponding acids. Addition of NaOH to solutions of HOCl (a neutralization reaction) produces aqueous NaOCl, the active ingredient in household bleach:

$$HOCl(aq) + NaOH(aq) \rightarrow NaOCl(aq) + H_2O(\ell)$$

Hypochlorites are also used in the paper industry to bleach wood pulp to produce white paper, and to sterilize water in swimming pools. The chlorine odor of pools is due to compounds called chloramines, produced when hypochlorous acid reacts with ammonia and nitrogen-containing compounds in bacteria:

$$HOCl(aq) + NH_3(aq) \rightarrow NH_2Cl(aq) + H_2O(\ell)$$

Chloramine is used as an alternative to direct chlorination to disinfect municipal drinking water. The reason some municipalities prefer chloramine to Cl_2 is related to the hazards associated with chlorine gas. Salts containing chlorate (ClO_3^-) and chlorite (ClO_2^-) ions are also used as bleaching agents. The perchlorate ion (ClO_4^-) reacts with organic matter rapidly, and some perchlorate salts are extremely reactive as contact explosives or as oxidizing agents when mixed with other substances that are easily oxidized. The oxidizers in many fireworks are perchlorate salts. When potassium perchlorate is mixed with charcoal, the material ignites spontaneously:

$$KClO_4(s) + 2\,C(s) \rightarrow KCl(s) + 2\,CO_2(g)$$

Mixtures of $KClO_4$, S, and Al provide the white flash and noise in fireworks. The flash powder used in stage shows is a mixture of Mg metal and $KClO_4$. Ammonium perchlorate decomposes explosively with heat and is also shock sensitive:

$$2\,NH_4ClO_4(s) \rightarrow N_2(g) + Cl_2(g) + 4\,H_2O(g) + 2\,O_2(g)$$

This behavior gives rise to its use as a solid propellant in space-shuttle booster rockets. Oxoanions of the other halogens, including bromate, iodate, perbromate, and periodate, have chemistries that are similar to those of the corresponding chlorine oxoanions.

FIGURE 10.33 Salt formations at the shore of the Dead Sea indicate the high salt content of the water.

SUMMARY

Section 10.1 The strengths of intermolecular attractive forces determine whether a compound is a gas, a liquid, or a solid under given conditions. Strong attractive ion–ion interactions hold ionic solids together.

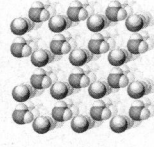

Section 10.2 Ions interact with water through **ion–dipole interactions**. **Dipole–dipole interactions** take place between water molecules and between other polar molecules. The strongest dipole–dipole interactions are **hydrogen bonds**, which form between water molecules and other polar molecules containing O—H, N—H, or F—H covalent bonds.

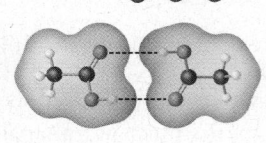

Section 10.3 Dispersion (London) forces are due to the **polarizability** of atoms and molecules and the existence of **temporary (induced) dipoles**. These interactions may be weak compared with ion–ion, ion–dipole, and dipole–dipole interactions. The strongest dispersion forces exist between the largest atoms, ions, and molecules, which are most easily polarized. The van der Waals equation accounts for the behavior of gases at high pressures and low temperatures. The magnitude of its correction constant *a* for any gas reflects the strength of all intermolecular forces (**van der Waals forces**) in the gas.

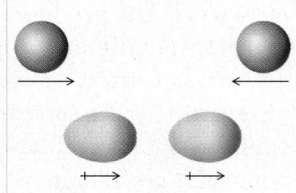

Section 10.4 Polar solutes dissolve in polar solvents when the dipole–dipole interactions between solute and solvent molecules offset the interactions that keep either solute molecules or solvent molecules together. The limited solubility of nonpolar solutes in polar solvents is a result of interactions between dipoles and induced dipoles. **Hydrophilic** substances are more soluble in water than are **hydrophobic** substances, which are more soluble in nonpolar solvents. **Henry's law,** $C_{gas} = k_H P_{gas}$, gives the maximum concentration (the solubility) of a sparingly soluble gas in a liquid solvent.

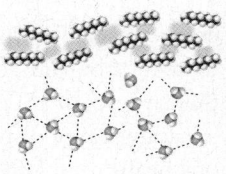

Section 10.5 The **phase diagram** of a substance indicates whether it exists as a solid, liquid, gas, or **supercritical fluid** at a particular pressure and temperature. Two adjoining regions in a phase diagram are separated by an equilibrium line between phases. All three states (solid, liquid, and gas) exist in equilibrium at the **triple point**. Above their critical temperatures and critical pressures, substances exist as supercritical fluids.

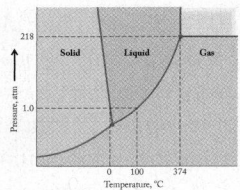

Section 10.6 The remarkable behavior of water, including its high melting and boiling points, its **surface tension**, its **capillary action**, and its **viscosity**, results from the strength of intermolecular hydrogen bonds.

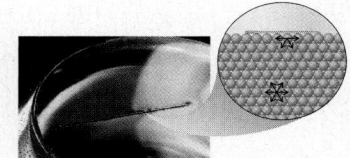

PROBLEM-SOLVING SUMMARY

TYPE OF PROBLEM	CONCEPTS AND EQUATIONS	SAMPLE EXERCISES
Predicting relative strengths of ion–ion attractive forces	To predict relative interaction strengths, use $$E \propto (Q_1 Q_2)/d \qquad (10.1)$$ More negative values of E correspond to stronger ion–ion attractions.	10.1
Explaining differences in boiling points of liquids and trends in boiling points of pure substances	Large molecules usually have higher boiling points than smaller molecules, with notable exceptions. The presence of polar –OH and –NH groups in molecules of a liquid leads to intermolecular hydrogen bonding that markedly increases the boiling point of the liquid.	10.2, 10.3
Assessing van der Waals constant *a*	Molecular size, polarity, and intermolecular hydrogen bonding all tend to increase the value of the constant *a* in the $n^2 a/V^2$ term of the van der Waals equation for real gases.	10.4
Predicting solubility in water	Polar molecules are more soluble in water than nonpolar molecules. Molecules that form hydrogen bonds are more soluble in water than molecules that cannot form these bonds. Like dissolves like.	10.5
Calculating solubility of an unreactive gas using Henry's law	Use Henry's law, $$C_{gas} = k_H P_{gas} \qquad (10.2)$$ where C_{gas} is the solubility of the gas, k_H depends on the gas, the solvent, and the temperature, and P_{gas} is the pressure of the gas (or partial pressure if the gas is part of a gas mixture).	10.6
Reading a phase diagram	Locate the point (T and/or P) specified by the question as the starting point for the exercise. Temperature is on the horizontal axis, pressure is on the vertical axis. When you draw a line either horizontally rightward from the pressure axis or vertically up from the temperature axis, a phase change occurs wherever the line crosses an equilibrium line.	10.7

(Answers to boldface end-of-chapter questions and problems are in the back of the book.)

10.1. Look at the pairs of ions in the structures of KF and KI represented in Figure P10.1. Which substance has the stronger cation–anion attractive forces and the higher melting point?

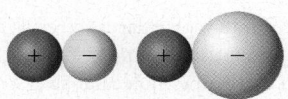

FIGURE P10.1

10.2. In Figure P10.2, identify the physical state (solid, liquid, or gas) of xenon and classify the attractive forces between the xenon atoms.

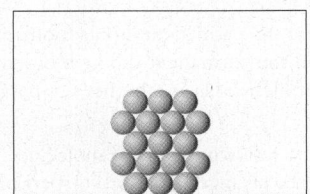

FIGURE P10.2

10.3. Figure P10.3 depicts molecules of XH_3 and YH_3 (not shown to scale) and the boiling points at 1 atm pressure of XH_3 and YH_3, respectively. One substance is phosphine (PH_3) and the other substance is ammonia (NH_3). Which molecule is phosphine? Explain your answer.

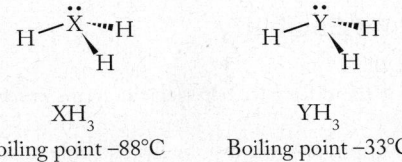

XH_3
Boiling point –88°C

YH_3
Boiling point –33°C

FIGURE P10.3

10.4. Figure P10.4 shows representations of the molecules *n*-pentane, C_5H_{12}, and *n*-decane, $C_{10}H_{22}$. Which substance has the lower freezing point? Explain your answer and use the Web to verify your prediction.

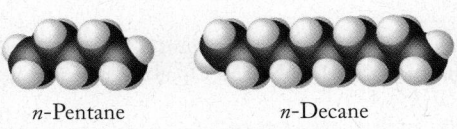

n-Pentane   *n*-Decane

FIGURE P10.4

10.5. Which of the drawings in Figure P10.5 best describes the effect of pressure on the solubility of a gas?

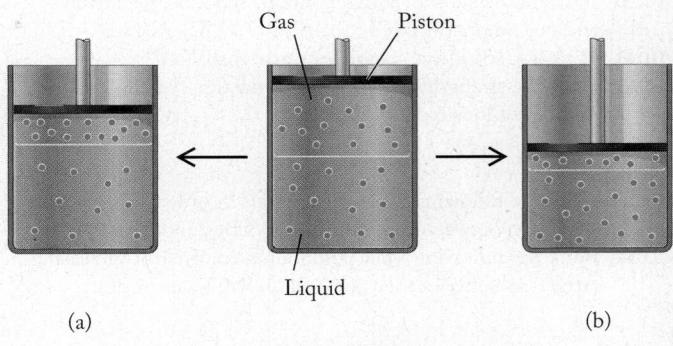

(a)   (b)

FIGURE P10.5

10.6. Which of the drawings in Figure P10.6 represents the gas with the largest Henry's law constant, k_H?

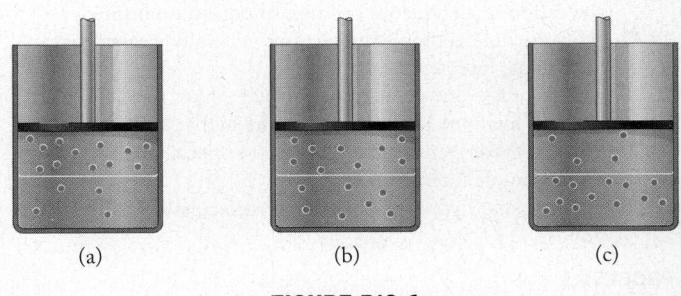

(a)   (b)   (c)

FIGURE P10.6

10.7. Examine the phase diagram of substance Z in Figure P10.7. Does the freezing point of the substance increase or decrease with increasing pressure?

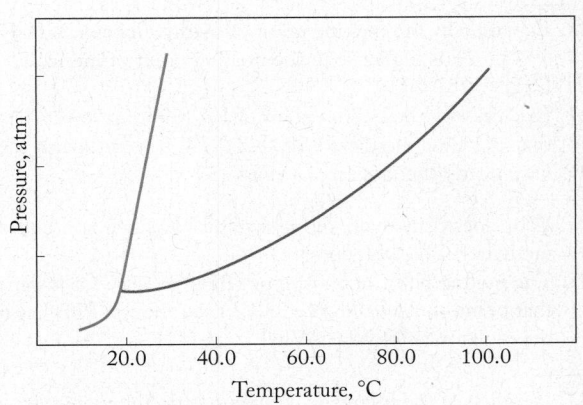

FIGURE P10.7

10.8. Refer to Figure P10.7. Do you predict the solid phase of substance Z would float on the liquid phase? Why?

Interactions between Ions in Salts

CONCEPT REVIEW

10.9. Indicate the substance that contains the largest anion. (a) $BaCl_2$; (b) AlF_3; (c) KI; (d) $SrBr_2$

10.10. Indicate the substance that contains the most negatively charged anion. (a) $BaCl_2$; (b) Al_2O_3; (c) Mg_3N_2; (d) SrS

10.11. Why is $CaSO_4$ less soluble in water than NaCl?

10.12. Does the strength of an ion–ion attraction depend on the number of ions in the compound?

PROBLEMS

10.13. Rank the following ionic compounds in order of increasing attraction between their ions: KBr, $SrBr_2$, and CsBr.

10.14. Rank the following ionic compounds in order of increasing attraction between their ions: BaO, $BaCl_2$, and CaO.

Interactions Involving Polar Molecules

CONCEPT REVIEW

10.15. How are the water molecules preferentially oriented around the anion in an aqueous solution of sodium chloride?

10.16. How are the water molecules preferentially oriented around the cation in an aqueous solution of potassium bromide?

10.17. Why are dipole–dipole interactions generally weaker than ion–dipole interactions?

10.18. Two liquids—one polar, one nonpolar—have the same molar mass. Which one is likely to have the higher boiling point?

10.19. Why are hydrogen bonds considered a special class of dipole–dipole interactions?

10.20. Can all polar hydrogen-containing molecules form hydrogen bonds?

PROBLEMS

10.21. In an aqueous solution containing chloride and iodide salts, which anion would you expect to be the more strongly hydrated?

10.22. In an aqueous solution containing Fe(II) and Fe(III) salts, which cation would you expect to be the more strongly hydrated?

10.23. Explain why the melting point of methyl fluoride, CH_3F ($-142°C$), is higher than the melting point of methane, CH_4 ($-182°C$).

10.24. Explain why the boiling point of Br_2 ($59°C$) is lower than that of iodine monochloride, ICl ($97°C$), even though they have nearly the same molar mass.

10.25. Why doesn't methane (CH_4) exhibit hydrogen bonding while methanol, CH_3OH, does?

10.26. The boiling point of phosphine (PH_3) ($-88°C$) is lower than that of ammonia (NH_3) ($-33°C$) even though PH_3 has twice the molar mass of NH_3. Why?

10.27. In which of the following compounds do the molecules experience the strongest dipole–dipole attractions? (a) CF_4; (b) CF_2Cl_2; (c) CCl_4

10.28. Which of the following compounds, CO_2, NO_2, SO_2, or H_2S, is expected to have the weakest interactions between its molecules?

Dispersion Forces

CONCEPT REVIEW

10.29. Which type of intermolecular force exists in all substances?

10.30. Why do the strengths of London (dispersion) forces generally increase with increasing molecular size?

10.31. Why do gases behave nonideally at high pressures and low temperatures?

10.32. What properties of real gas molecules are associated with parameters a and b of the van der Waals equation?

PROBLEMS

10.33. The permanent dipole moment of CH_2F_2 (1.93 D) is larger than that of CH_2Cl_2 (1.60 D), yet the boiling point of CH_2Cl_2 ($40°C$) is much higher than that of CH_2F_2 ($-52°C$). Why?

10.34. How is it that the permanent dipole moment of HCl (1.08 D) is larger than the permanent dipole moment of HBr (0.82 D), yet HBr boils at a higher temperature?

10.35. In each of the following pairs of molecules, which compound experiences the stronger London (dispersion) forces? (a) CCl_4 or CF_4; (b) CH_4 or C_3H_8

10.36. What kinds of intermolecular forces must be overcome as (a) solid CO_2 sublimes; (b) $CHCl_3$ boils; (c) ice melts?

10.37. Explain why the van der Waals constant a for Ar is greater than a for He.

10.38. The van der Waals constant a for CO_2 is 3.59 $L^2 \cdot atm/mol^2$. Would you expect the value of a for CS_2 to be larger or smaller than 3.59 $L^2 \cdot atm/mol^2$?

Polarity and Solubility

CONCEPT REVIEW

10.39. What is the difference between the terms *miscible* and *insoluble*?

10.40. Which substances are essentially insoluble in water? (a) benzene, $C_6H_6(\ell)$; (b) KBr(s); (c) $Br_2(\ell)$

10.41. Why does the solubility of sparingly soluble gases in most liquids increase with increasing gas pressure?

10.42. Why does the solubility of most gases in most liquids increase with decreasing temperature?

10.43. Which term, k_H or P, in Henry's law is affected by temperature?

10.44. Air is primarily a mixture of nitrogen and oxygen. Is the Henry's law constant for the solubility of air in water the sum of k_H for N_2 and k_H for O_2? Explain why or why not.

10.45. Why is the Henry's law constant for CO_2 so much larger than those for N_2 and O_2 at the same temperature?

10.46. A student observes bubbles while heating a sample of water in a beaker at 60°C. What are the gases in the bubbles and where did they come from?

10.47. In what context do the terms *hydrophobic* and *hydrophilic* relate to the solubilities of substances in water?

10.48. How does the presence of increasingly longer hydrocarbon chains in the structure affect the solubility of a series of structurally related molecules in water?

PROBLEMS

10.49. In each of the following pairs of compounds, which compound is likely to be more soluble in water?
a. CCl_4 or $CHCl_3$
b. CH_3OH or $C_6H_{11}OH$
c. NaF or MgO
d. CaF_2 or BaF_2

10.50. In each of the following pairs of compounds, which compound is likely to be more soluble in CCl_4?
a. Br_2 or $NaBr$
b. CH_3CH_2OH or CH_3OCH_3
c. CS_2 or KOH
d. I_2 or CaF_2

*10.51. Arterial Blood Arterial blood contains about 0.25 g of oxygen per liter at 37°C and standard atmospheric pressure. What is the Henry's law constant (mol/L · atm) for O_2 dissolution in blood? The mole fraction of O_2 in air is 0.209.

*10.52. The solubility of O_2 in water is 6.5 mg/L at an atmospheric pressure of 1 atm and temperature of 40°C. Calculate the Henry's law constant of O_2 at 40°C. The mole fraction of O_2 in air is 0.209.

*10.53. Oxygen for Climbers and Divers Use the Henry's law constant for O_2 dissolved in arterial blood from Problem 10.51 to calculate the solubility of O_2 in the blood of (a) a climber on Mt. Everest ($P_{atm} = 0.35$ atm) and (b) a scuba diver at 100 feet ($P \approx 3$ atm).

*10.54. The solubility of air in water is approximately 7.9×10^{-4} M at 20°C and 1.0 atm. Calculate the Henry's law constant for air. Is the value of k_H for air equal to the sum of k_H for O_2 and for N_2?

*10.55. Use the graph of solubility of O_2 versus temperature in Figure P10.55 to calculate the value of the Henry's law constant k_H for O_2 at 10°C, 20°C, and 30°C. Assume $P_{O_2} = 1$ atm.

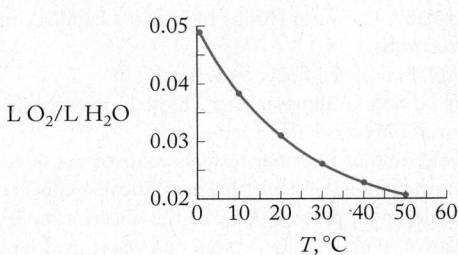

FIGURE P10.55

10.56. Based on the data in Figure P10.55, which has a greater effect on the solubility of oxygen in water: (a) decreasing the temperature from 20°C to 10°C or (b) raising the pressure from 1.00 atm to 1.25 atm?

10.57. Which of the following compounds is likely to be the most soluble in water? (a) $NaCl$; (b) KI; (c) $Ca(OH)_2$; (d) CaO

10.58. Which sulfur oxide would you predict to be more soluble in nonpolar solvents: SO_2 or SO_3?

10.59. Which of these substances is the least soluble in water?
a. $CH_3(CH_2)_2CH_2OH$
b. $CH_3(CH_2)_4CH_2OH$
c. $CH_3(CH_2)_6CH_2OH$
d. $CH_3(CH_2)_8CH_2OH$

10.60. Which of these substances is the most soluble in water?
a. $CH_3(CH_2)_2CH_2NH_2$
b. $CH_3(CH_2)_4CH_2Cl$
c. $CH_3(CH_2)_6CH_2Br$
d. $CH_3(CH_2)_8CH_2I$

Phase Diagrams: Intermolecular Forces at Work

CONCEPT REVIEW

10.61. Explain the difference between sublimation and evaporation.

10.62. Can ice be melted merely by applying pressure? Explain your answer.

10.63. Explain what is meant by the term *equilibrium line*.

10.64. Explain how the solid–liquid line in the phase diagram of water differs in character from the solid–liquid line in the phase diagrams of most other substances, such as CO_2.

10.65. Which phase of a substance (gas, liquid, or solid) is more likely to be the stable phase: (a) at low temperatures and high pressures; (b) at high temperatures and low pressures?

10.66. At what temperatures and pressures does a substance behave as a supercritical fluid?

10.67. Preserving Food Freeze-drying is used to preserve food at low temperature with minimal loss of flavor. Freeze-drying works by freezing the food and then lowering the pressure with a vacuum pump to sublime the ice. Must the pressure be lower than the pressure at the triple point of H_2O?

10.68. Solid helium cannot be converted directly into the vapor phase. Does the phase diagram of He have a triple point?

PROBLEMS

For help in answering Problems 10.69 through 10.72, consult Figures 10.22 and 10.24.

10.69. List the steps you would take to convert a 10.0 g sample of water at 25°C and 1 atm pressure to water at its triple point.

10.70. List the steps you would take to convert a 10.0 g sample of water at 25°C and 2 atm pressure to ice at 1 atm pressure. At what temperature would the water freeze?

10.71. What phase changes, if any, does liquid water at 100°C undergo if the initial pressure of 5.0 atm is reduced to 0.5 atm at constant temperature?

10.72. What phase changes, if any, occur if CO_2 initially at $-80°C$ and 8.0 atm is allowed to warm to $-25°C$ at 5.0 atm?

For help in answering Problems 10.73 and 10.74, consult Figure 10.24.

10.73. Below what temperature can solid CO_2 (dry ice) be converted into CO_2 gas simply by lowering the pressure?

10.74. What is the maximum pressure at which solid CO_2 (dry ice) can be converted into CO_2 gas without melting?

For help in answering Problems 10.75 and 10.76, consult Figure 10.22.

10.75. Predict the phase of water that exists under the following conditions:
   a. 2 atm of pressure and 110°C
   b. 200 atm of pressure and 380°C
   c. 6.0×10^{-3} atm of pressure and 0°C

10.76. Which phase or phases of water exist under the following conditions?
   a. 0.32 atm and 120°C
   b. 300 atm and 400°C
   c. 1 atm and 0°C

Some Remarkable Properties of Water

CONCEPT REVIEW

10.77. Explain why a needle floats on the surface of water but sinks in a container of methanol (CH_3OH).

10.78. Explain why different liquids do not reach the same height in capillary tubes of the same diameter.

10.79. Explain why pipes filled with water are in danger of bursting when the temperature drops below 0°C.

10.80. A hot needle sinks when put on the surface of cold water. Will a cold needle float in hot water?

10.81. The meniscus of mercury in a thermometer (Figure P10.81) is convex, rather than concave. Explain why.

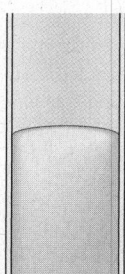

FIGURE P10.81

*10.82. The mercury level in a capillary tube placed in a dish of mercury is below the surface of the mercury in the dish. Explain why.

10.83. Describe the origin of surface tension at the molecular level.

10.84. What is the origin of the high viscosity of molasses?

10.85. Describe how the surface tension and viscosity of a liquid are affected by increasing temperature.

10.86. Explain how strong intermolecular forces are expected to result in a relatively high surface tension and viscosity of a liquid.

PROBLEMS

10.87. One of two glass capillary tubes of the same diameter is placed in a dish of water and the other in a dish of ethanol (CH_3CH_2OH). Which liquid will rise higher in its tube?

10.88. Would you expect water to rise to the same height in a tube made of a polyethylene plastic as it does in a glass capillary tube of the same diameter? The molecular structure of polyethylene is shown in Figure P10.88.

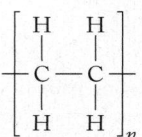

FIGURE P10.88

10.89. The normal boiling points of liquids A and B are 75.0°C and 151°C, respectively. Which of these liquids would you expect to have the higher surface tension and viscosity at 25°C? Explain your answer.

10.90. One beaker contains pure water and the other beaker contains pure methanol at the same temperature. Which liquid has the higher surface tension and viscosity? Explain your answer.

Additional Problems

10.91. Which substance contains the smallest cation? (a) CsF; (b) RbF; (c) KCl; (d) NaBr

10.92. Which substance contains the most positively charged cation? (a) $MgCl_2$; (b) AlF_3; (c) KI; (d) SrO

10.93. Why does methanol (CH_3OH) boil at a lower temperature than water, even though CH_3OH has the greater molar mass?

10.94. Why is methanol (CH_3OH) miscible with water, while CH_4 is practically insoluble in water?

10.95. Does the sublimation point of ice increase or decrease with increasing pressure?

10.96. Sketch a phase diagram for element X, which has a triple point (152 K, 0.371 atm), a boiling point of 166 K at a pressure of 1.00 atm, and a normal melting point of 161 K.

*10.97. The melting point of hydrogen is 14.96 K at 1.00 atm pressure. The temperature of its triple point is 13.81 K. Does H_2 expand or contract when it freezes?

10.98. Explain why water climbs higher in a capillary tube than in a test tube.

10.99. Explain why ice floats on water.

10.100. **Fish Dying in Summer Heat** Explain why fish in a pond die if water becomes too warm.

*10.101. **Evaluation of Pharmaceuticals** One of the tests that was routinely done on new pharmaceutical agents early in the development process required the observation of their relative solubilities in octanol and water. A drug had to be sufficiently soluble in water (hydrophilic) to be carried in the bloodstream but also be sufficiently hydrophobic (octanol soluble) to move across cell membranes. Pick the molecule from Figure P10.101 that you predict might have comparable solubility in both solvents.

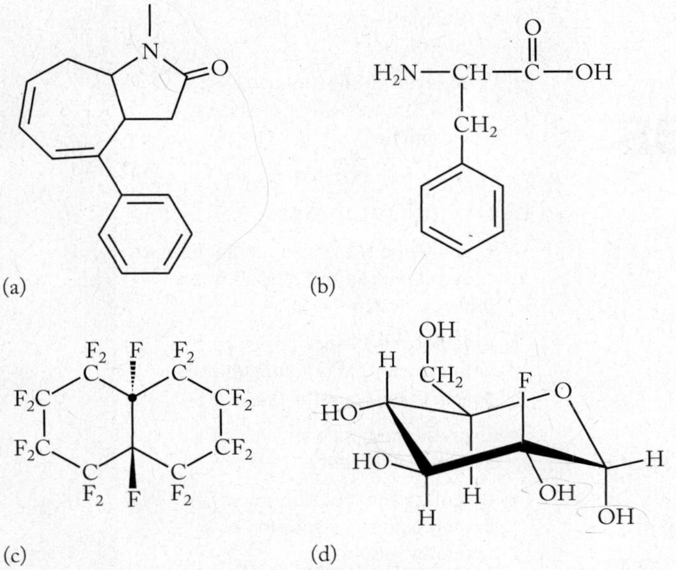

(a)

(b)

(c)

(d)

FIGURE P10.101

10.102. First-Aid for Bruises Compounds with low boiling points may be sprayed on the skin as a topical anesthetic to chill it as they evaporate and provide short-term relief from injuries. Predict which compound among those in Figure P10.102 has the lowest boiling point.

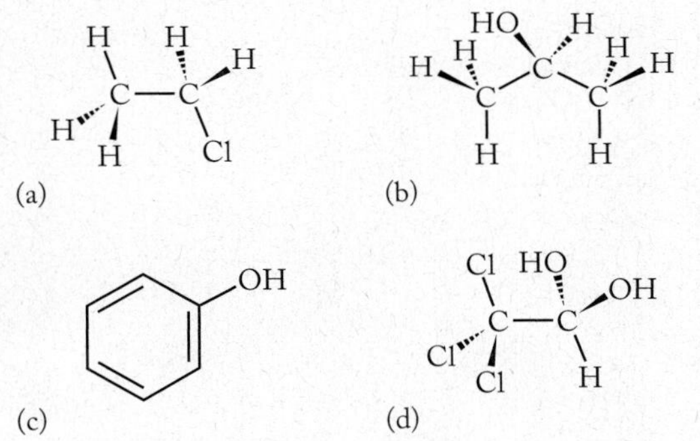

(a)

(b)

(c)

(d)

FIGURE P10.102

The Halogens: The Salt of the Earth

10.103. What are the major natural sources of chlorine and bromine?

10.104. Under normal conditions, bromine is a volatile red liquid and iodine is a violet solid. Why are Br_2 and I_2 so volatile?

10.105. How do the boiling points of the halogens change in the progression F_2, Cl_2, Br_2, I_2?

10.106. Which is by far the least abundant of the halogens?

10.107. Give two industrial uses each of (a) chlorine and (b) bromine.

10.108. Give the chemical formula of chloramine. What shape is the chloramine molecule? Would you expect chloramine to be soluble in water? Why?

10.109. Which of each halide pair is the most easily oxidized? (a) fluoride, iodide; (b) bromide, chloride; (c) chloride, fluoride; (d) iodide, bromide

10.110. Give the formula of each of the following halogen species: (a) hypochlorite; (b) chlorite; (c) chlorate; (d) perchlorate.

10.111. Give the species with the higher oxidation number of the halogen in each of the following pairs: (a) hypochlorous acid, Cl^-; (b) chlorous acid, $HClO_3$; (c) chloric acid, BrO_2^-; (d) perchloric acid, IO^-.

10.112. Balance the equation for the reduction of iodate by hydrogen sulfite ions to give iodide and sulfate in basic aqueous solution.

11

Solutions and Their Colligative Properties

Learning Outcomes

- Understand the energy changes that accompany the dissolution of an ionic compound
- Calculate the vapor pressure of a solution containing a nonvolatile solute using Raoult's law
- Express the concentration of a solution in molality
- Calculate the freezing point and boiling point of a solution of a nonvolatile solute
- Understand the significance of the van't Hoff factor
- Predict the direction of solvent flow in osmosis and calculate osmotic pressure
- Apply colligative properties to the determination of molar mass

A LOOK AHEAD

Water, Water, Everywhere

In his "Rime of the Ancient Mariner," British poet Samuel Taylor Coleridge (1772–1834) wrote, "Water, water, everywhere, and all the boards did shrink. Water, water, everywhere, nor any drop to drink." Coleridge's verse reminds us that most of Earth's surface water is ocean water, an aqueous solution whose primary solute is sodium chloride. Despite the necessity of water for life, exposure to salt water can be damaging—wooden boards exposed to it shrink—and drinking seawater causes dehydration and death.

Why does seawater have these effects? How can we obtain potable water from salt water? Such questions are particularly relevant as the availability of suitable drinking water has become an international concern. Lakes and rivers provide adequate supplies of freshwater in many parts of the world, but arid regions turn to the sea for water. It is essential to remove the ions from salt water to make it drinkable.

We learn in this chapter why differences in solute concentrations between seawater and the interior of their cells causes both plants and animals to lose water, the implication of Coleridge's rhyme. However, the presence of some solutes is essential in certain situations: water dispensed intravenously in hospitals to deliver medications or restore fluids must contain small amounts of solutes to prevent dehydration as well as to avoid retention of excess water by our bodies. How does either too much or too little salt in water cause these problems? Clearly, determining the right amount of solute in a solution for a specific use is crucial.

Other properties of water besides potability are influenced by the presence of solutes. Seawater is less volatile (less easily vaporized) than pure water; a glass of seawater evaporates more slowly than a glass of distilled water. Seawater freezes at a lower temperature than pure water, just as a solution of antifreeze in an automobile radiator freezes at a lower temperature than pure water. Compared to a pure solvent, how

Sea Spray Ocean waves crashing on a shoreline produce a spray that carries the ions dissolved in the ocean into the air. ▶

enthalpy of solution ($\Delta H_{solution}$) the overall heat change when a solute is dissolved in a solvent.

enthalpy of hydration ($\Delta H_{hydration}$) the heat change when gas-phase ions dissolve in a solvent.

lattice energy (U) the energy released when 1 mole of an ionic compound forms from its free ions in the gas phase.

do solutes affect the properties of solutions? In this chapter we explore how the properties of solutions depend on the concentration of solutes as we grapple with the science behind Coleridge's poetry. ■

11.1 Energy Changes during Dissolution

The taste of the sea and the scent of ocean air remind us that seawater contains salt. The fact that we are more buoyant in the ocean than in freshwater further indicates that solutions have different properties than pure solvents. Allowing a container of seawater to evaporate in the sun leaves behind a white mixture of salts, mostly NaCl (see Table 4.1). Adding water to this residue dissolves the solids and restores the "seawater." How do we describe the dissolution of ionic compounds like NaCl at the microscopic level? In Chapter 10, we learned that ionic compounds are held together by ion–ion attractions while water molecules interact through the strong dipole–dipole forces of hydrogen bonds. In an aqueous solution of ionic compounds, the ions interact with water through ion–dipole forces. What interactions between solvent and solute during dissolution overcome these attractive forces in pure substances, and how do we describe the changes in energy associated with these new interactions?

Applying what we learned in Chapter 8 about bond dissociation energies, we can describe the enthalpy change that accompanies dissolution of an ionic compound, or **enthalpy of solution ($\Delta H_{solution}$)**, in a polar solvent as

$$\Delta H_{solution} = \Delta H_{ion-ion} + \Delta H_{dipole-dipole} + \Delta H_{ion-dipole} \quad (11.1)$$

Of the quantities in Equation 11.1, we can readily measure $\Delta H_{solution}$ using calorimetric methods. Measuring the remaining three terms independently is more difficult. The energy associated with ion–ion attraction was introduced in Equation 10.1 and is proportional to the product of the charges on the ions and inversely proportional to the distance between them:

$$E \propto \frac{(Q_1 Q_2)}{d} \quad (10.1)$$

We can combine the energies of dipole–dipole and ion–dipole interactions into a single term called the **enthalpy of hydration ($\Delta H_{hydration}$)** for the compound,

$$\Delta H_{hydration} = \Delta H_{dipole-dipole} + \Delta H_{ion-dipole} \quad (11.2)$$

where *hydration* refers to the formation of solvated ions. Combining Equations 11.1 and 11.2:

$$\Delta H_{solution} = \Delta H_{ion-ion} + \Delta H_{hydration} \quad (11.3)$$

Let's examine each of these terms in more detail.

The strength of ion–ion interactions is described by the **lattice energy (U)** of an ionic compound, which is the energy released when free ions in the gas phase combine to form 1 mole of a solid ionic compound. The lattice energies of some common binary ionic compounds are given in Table 11.1. The formula for lattice energy

$$U = \frac{k(Q_1 Q_2)}{d} \quad (11.4)$$

resembles Equation 10.1 except that it includes a proportionality constant k, the value of which depends on the structure of the ionic solid. We examine the structures of

∞ CONNECTION Calorimetry as a method for measuring enthalpy changes was described in Chapter 5.

▶II CHEMTOUR Lattice Energy

solids in Chapter 12, at which point we will learn about the different arrangements possible for ions in an ionic solid. The same value of k is used for all compounds that have the same or nearly the same arrangement of ions. For now, we need only consider the charges on ions and the distances between them and use Equation 10.1 to predict relative values of lattice energies. It is the lattice energy of an ionic compound, U, that we substitute in Equation 11.3 for $\Delta H_{\text{ion–ion}}$:

$$\Delta H_{\text{solution}} = -U + \Delta H_{\text{hydration}}$$

or

$$\Delta H_{\text{solution}} = \Delta H_{\text{hydration}} - U \qquad (11.5)$$

Equation 11.5 has a minus sign in front of the lattice energy term because U is defined as the energy change when gas-phase ions *combine* to form an ionic solid. In Equation 11.5, we are calculating the energy change associated with *separating* the ions in an ionic compound. Recall from Chapter 5 that the enthalpy change for a process in one direction has the same magnitude but opposite sign of the process in the reverse direction.

The lattice energy of an ionic solid not only affects its solubility in water, lattice energy determines the temperature at which the solid melts. Melting an ionic structure in which the ions are held together tightly should require more thermal energy (a higher temperature) than melting a structure in which the ions are held together less tightly—a trend we observe experimentally. Consider two ionic compounds: LiF ($U = -1047$ kJ/mol) and MgO ($U = -3791$ kJ/mol). The greater lattice energy of MgO, nearly four times that of LiF, is reflected in its higher melting point, 2825°C, versus 848°C for LiF, and its higher boiling point, 3600°C for MgO, versus 1673°C for LiF.

TABLE 11.1	Lattice Energies (U) of Common Binary Ionic Compounds
Compound	**U (kJ/mol)**
LiF	−1047
LiCl	−864
NaCl	−786
KCl	−720
KBr	−691
MgCl$_2$	−2540
MgO	−3791

SAMPLE EXERCISE 11.1   **Ranking Lattice Energies and Melting Points**

Rank these three ionic compounds in order of increasing lattice energy and increasing melting point: NaF, KF, and RbF. Assume that these compounds have the same solid structure, which means they have the same value of k in Equation 11.4.

Collect and Organize We can determine relative lattice energies from Equation 11.4. To do so we need to establish the charges and radii of the ions in each compound using Figure 10.2.

Analyze All the cations are alkali metal cations and have a charge of 1+; all the anions are fluoride ions, with a charge of 1−. We are told that k is the same for all three solids. Therefore, any differences in lattice energy must be related to differences in the nucleus-to-nucleus distance d between ions. Because the fluoride ion is common to all the salts, variations in d depend only on the size of the cation.

Solve Periodic trends in size predict, and Figure 11.1 confirms, that the cation sizes are Na$^+$ < K$^+$ < Rb$^+$. Therefore, the compounds in order of increasing value of d are NaF < KF < RbF (Figure 11.1). As d increases, lattice energy decreases, so RbF has the lowest lattice energy, followed by KF, followed by NaF with the highest. The same trend occurs in melting points: RbF < KF < NaF.

Think about It We predicted the order of lattice energies and melting points for three alkali metal fluorides. This predicted order is confirmed by the experimentally measured melting points: 775°C for RbF, 846°C for KF, and 988°C for NaF.

Na$^+$ = 102 pm
F$^-$ = 133 pm

K$^+$ = 138 pm
F$^-$ = 133 pm

Rb$^+$ = 149 pm
F$^-$ = 133 pm

FIGURE 11.1 The distance d between the ions in NaF, KF, and RbF is in the order $d_{\text{RbF}} > d_{\text{KF}} > d_{\text{NaF}}$.

Born–Haber cycle a series of steps with corresponding enthalpy changes that describes the formation of an ionic solid from its constituent elements.

(a)

∞ **CONNECTION** We used Hess's law in Chapter 5 to determine heats of reactions that are difficult to measure experimentally.

Practice Exercise Predict which compound has the highest melting point: $CaCl_2$, $PbBr_2$, or TiO_2. All three compounds have nearly the same structure and therefore the same value of k in Equation 11.3. The radius of Ti^{4+} is 60.5 pm.

(Answers to Practice Exercises are in the back of the book.)

■ ....

Calculating Lattice Energies Using the Born-Haber Cycle

Lattice energies are difficult to measure directly, but the lattice energy of a binary ionic compound can be calculated from its standard enthalpy of formation (ΔH_f°). Hess's law is used to determine the enthalpies of reaction, ΔH_{rxn}, associated with a series of reactions that take the constituent elements from their standard states to ions in the gas phase and then to ions in the ionic solid.

Consider the formation of NaCl:

$$Na(s) + \tfrac{1}{2}Cl_2(g) \rightarrow NaCl(s) \qquad \Delta H_f^\circ = -411.2 \text{ kJ}$$

The reaction is exothermic, and the heat produced, 411.2 kJ per mole of NaCl formed, has been measured with a calorimeter and is among those tabulated in Appendix A4.3. This enthalpy of formation can be viewed as the algebraic sum of all the enthalpy changes associated with five reactions that together form a **Born–Haber cycle** (Figure 11.2):

1. Sublimation of 1 mole of Na(s) atoms into 1 mole of Na(g) atoms: ΔH_{sub}, the molar heat of sublimation of sodium.
2. Breaking covalent bonds in 0.5 moles of $Cl_2(g)$ molecules to make 1 mole of Cl(g) atoms: $\tfrac{1}{2}\Delta H_{BE}$, where ΔH_{BE} is the enthalpy change needed to break 1 mole of $Cl_2(g)$ bonds.
3. Ionization of 1 mole of Na(g) atoms to 1 mole of $Na^+(g)$ ions and 1 mole of electrons: IE_1, the first ionization energy of sodium.

FIGURE 11.2 (a) The reaction between sodium metal and chlorine gas releases more than 400 kJ of energy per mole of NaCl produced. (b) The Born–Haber cycle shows that the principal reason for this violently exothermic reaction is the energy released in step 5 when free sodium ions $Na^+(g)$ and free chloride ions $Cl^-(g)$ combine to form NaCl(s). (c) Solid sodium chloride.

(b)

(3) IE_1

Na(g)   Cl(g)

$\tfrac{1}{2}$ Na(g) $\tfrac{1}{2}Cl_2(g)$ (2) $\tfrac{1}{2}\Delta H_{BE}$

$\tfrac{1}{2}$ Na(s) $\tfrac{1}{2}Cl_2(g)$ (1) ΔH_{sub}

ΔH_f°

$Na^+(g)$ Cl(g)

(4) EA_1

$Na^+(g)$ $Cl^-(g)$

(5) U

NaCl(s)

(c)

4. Combination of 1 mole of $Cl(g)$ atoms with 1 mole of electrons to form 1 mole of $Cl^-(g)$ ions: EA_1, the first electron affinity of chlorine.
5. Formation of 1 mole of $NaCl(s)$ from 1 mole of $Na^+(g)$ ions and 1 mole of $Cl^-(g)$ ions: $U = \Delta H_{lattice}$, the lattice energy of NaCl.

This reaction sequence is summarized in Table 11.2. To use the Born–Haber cycle to calculate the lattice energy of $NaCl(s)$, we start with an equation relating the value of ΔH_f° for NaCl to the sum of the enthalpy changes of the five reactions in Table 11.2:

$$\Delta H_f^\circ = \Delta H_{step\ 1} + \Delta H_{step\ 2} + \Delta H_{step\ 3} + \Delta H_{step\ 4} + \Delta H_{step\ 5}$$

$$= \Delta H_{sub} + \tfrac{1}{2}\Delta H_{BE} + IE_1 + EA_1 + U$$

Inserting the values from Table 11.2 and solving for U:

$$-411.2\ kJ = (+109\ kJ) + \tfrac{1}{2}(+240\ kJ) + (+495\ kJ) + (-349\ kJ) + U$$

$$U = (-411.2\ kJ) - (+109\ kJ) - (+120\ kJ) - (+495\ kJ) - (-349\ kJ) = -786\ kJ$$

In addition to its usefulness for calculating lattice energies, the Born–Haber cycle can also be used to calculate other values. For example, electron affinities can be very difficult to measure, and if the thermochemical values are known for all other steps in the cycle, the Born–Haber cycle can be used to calculate electron affinities.

TABLE 11.2   Born–Haber Cycle for Formation of NaCl(s)

Step	Process	Enthalpy Change (kJ)
1	$Na(s) \rightarrow Na(g)$	$\Delta H_{sub} = +109$
2	$\tfrac{1}{2}Cl_2(g) \rightarrow Cl(g)$	$\tfrac{1}{2}\Delta H_{BE} = \tfrac{1}{2}(240) = +120$
3	$Na(g) \rightarrow Na^+(g) + e^-$	$IE_1 = +495$
4	$Cl(g) + e^- \rightarrow Cl^-(g)$	$EA_1 = -349$
5	$Na^+(g) + Cl^-(g) \rightarrow NaCl(s)$	$\Delta H_{lattice} = U$

SAMPLE EXERCISE 11.2   Calculating Lattice Energy

In Sample Exercise 10.1, we predicted that the ion–ion attraction in CaF_2 is greater than in NaF. Confirm this prediction by calculating the lattice energies of (a) NaF and (b) CaF_2 given the following information:

$\Delta H_{sub}\ Na(s) = +109\ kJ/mol \qquad \Delta H_{sub}\ Ca(s) = +154\ kJ/mol$

$\Delta H_{BE}\ F_2(g) = +154\ kJ/mol \qquad IE_1\ Ca(g) = +590\ kJ/mol$

$EA_1\ F(g) = -328\ kJ/mol \qquad IE_2\ Ca(g) = +1145\ kJ/mol$

$IE_1\ Na(g) = +495\ kJ/mol$

Collect and Organize We are to calculate the lattice energy of two ionic compounds using enthalpy values for processes that can be summed to describe an overall process that produces a salt from its constituent elements. We can use a Born–Haber cycle to calculate the unknown lattice energy. Table A4.3 in Appendix 4 contains standard enthalpy of formation values for NaF and CaF_2: -575.4 kJ/mol and -1228.0 kJ/mol, respectively. These enthalpy changes apply to the formation of one mole of the two compounds by combining their component elements in their standard states.

Analyze The standard enthalpy of formation of NaF is

$$Na(s) + \tfrac{1}{2}F_2(g) \rightarrow NaF(s) \qquad \Delta H_f^\circ = -575.4 \text{ kJ/mol}$$

This value is the enthalpy change for the entire process, ΔH_{rxn}. The fluorine in the equation has a coefficient of $\tfrac{1}{2}$; this means that the energy needed to break a mole of F—F bonds to make atomic fluorine must be multiplied by $\tfrac{1}{2}$.

Following the same steps for part b:

$$Ca(s) + F_2(g) \rightarrow CaF_2(s) \qquad \Delta H_f^\circ = -1228.0 \text{ kJ/mol}$$

This value is the enthalpy change for the entire process. For calcium, we must include both the first and second ionization energies because Ca loses two electrons when it forms Ca^{2+} ions. Since two fluorine atoms are needed to react with a single calcium atom, we do not need the factor of $\tfrac{1}{2}$ in front of the term for energy to break a mole of F—F bonds. However, we need to multiply the electron affinity of F by 2 because two moles of fluorine atoms gain two moles of electrons to form two moles of fluoride ions.

Figure 11.3 summarizes the Born–Haber cycles for calculating the lattice energies of NaF and CaF_2. Based on the charges on the ions, their ionic radii, and Equation 11.4, the lattice energy of CaF_2 should be greater than the lattice energy of NaF.

Solve a. The overall enthalpy change in the reaction producing NaF is

$$\Delta H_f^\circ = \Delta H_{sub} \, Na(s) + \tfrac{1}{2}\Delta H_{BE} \, F_2(g) + IE_1 \, Na(g) + EA_1 \, F(g) + U \, NaF(s)$$

Substituting the values given:

$$-575.4 \text{ kJ} = (+109 \text{ kJ}) + \tfrac{1}{2}(+154 \text{ kJ}) + (+495 \text{ kJ}) + (-328 \text{ kJ}) + U$$

Solving for U:

$$U = (-575 \text{ kJ}) - (+109 \text{ kJ} + 77 \text{ kJ} + 495 \text{ kJ} - 328 \text{ kJ})$$
$$= -928 \text{ kJ/mol NaF}(s) \text{ formed}$$

FIGURE 11.3 Born–Haber cycles for the formation of (a) NaF and (b) CaF$_2$.

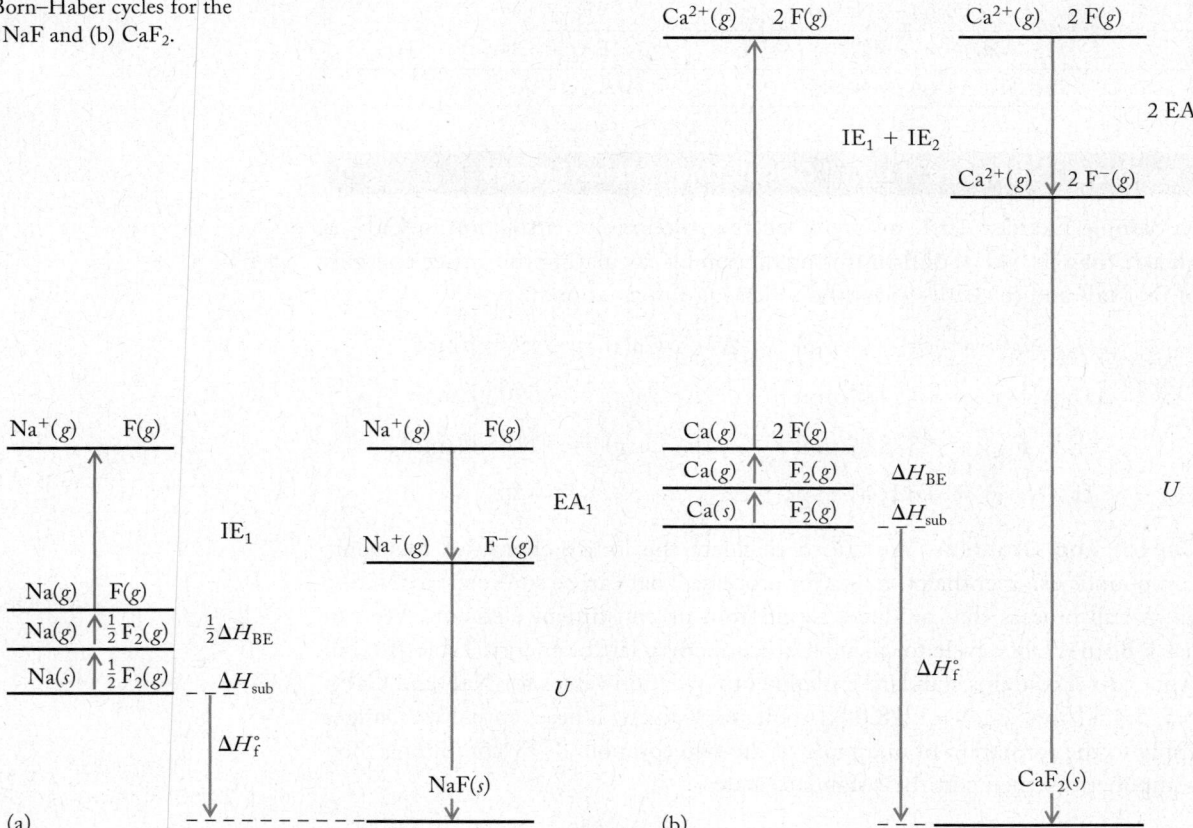

b. The overall enthalpy change in the reaction producing CaF_2 is

$$\Delta H_f^\circ = \Delta H_{sub} Ca(s) + \Delta H_{BE} F_2(g) + [IE_1 Ca(g) + IE_2 Ca(g)]$$
$$+ 2EA_1 F(g) + U(CaF_2)(s)$$

Substituting the values given:

$$-1228.0 \text{ kJ} = (+154 \text{ kJ}) + (+154 \text{ kJ}) + [(+590 \text{ kJ}) + (+1145 \text{ kJ})]$$
$$+ 2(-328 \text{ kJ}) + U$$

Solving for U:

$$U = (-1228.0 \text{ kJ}) - [+154 \text{ kJ} + 154 \text{ kJ} + (590 \text{ kJ} + 1145 \text{ kJ}) - 656 \text{ kJ}]$$
$$= -2615 \text{ kJ/mol CaF}_2(s) \text{ formed}$$

Think about It Based on Equation 11.4, we predicted that the lattice energy of CaF_2 would be larger than that of NaF. The calculated values of U confirm that $U_{CaF_2} > U_{NaF}$.

Practice Exercise Burning magnesium metal in air produces MgO and a very bright white light, making the reaction popular in fireworks and signaling devices:

$$Mg(s) + \tfrac{1}{2}O_2(g) \rightarrow MgO(s) + \text{light}$$

The energy change that accompanies this reaction is -602 kJ/mol MgO. Calculate the lattice energy of MgO from the following energy changes:

Process	Enthalpy Change (kJ/mol)	Process	Enthalpy Change (kJ/mol)
$Mg(s) \rightarrow Mg(g)$	150	$Mg(g) \rightarrow Mg^{2+}(g) + 2e^-$	2188
$O_2(g) \rightarrow 2O(g)$	499	$O(g) + 2e^- \rightarrow O^{2-}(g)$	603

The lattice energy of an ionic compound also affects its solubility in water. As we discussed in Chapter 10, the positive and negative regions of the water molecules interact with the anions and cations in an ionic solid and pull those ions into solution. The greater the lattice energy of the solid, the greater is the force required for water molecules to pull apart the ions and the less soluble the solid is likely to be. We might predict that NaF is more soluble in water than CaF_2 on the basis of the greater lattice energy of CaF_2, and that prediction is correct: at 20°C, the solubility of NaF in water is 4.0 g/100 mL and the solubility of CaF_2 is only 0.0015 g/100 mL.

Enthalpies of Hydration

Once we have calculated the lattice energy of an ionic compound, we can use Equation 11.5 to calculate its enthalpy of hydration, $\Delta H_{hydration}$, as shown in Figure 11.4. Figure 11.4 represents another example of a Born–Haber cycle. If the measured enthalpy

CONNECTION The solubility trends for ionic compounds given in Chapter 4 are related to the strengths of intermolecular attractive forces.

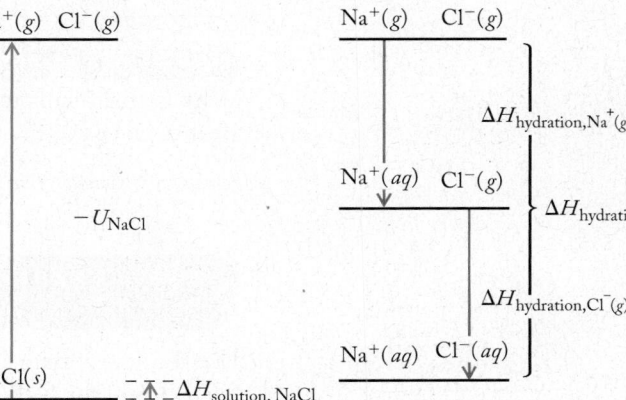

FIGURE 11.4 Born–Haber cycle for the dissolution of an ionic compound in water. The enthalpy of solution ($\Delta H_{solution}$) is the enthalpy of hydration ($\Delta H_{hydration}$) minus the lattice energy (U). The size of $\Delta H_{solution}$ is not drawn to scale; it is very small relative to the other values.

of solution of sodium chloride is 4 kJ/mol and the lattice energy of NaCl is −786 kJ/mol, then:

$$\Delta H_{\text{solution,NaCl}} = \Delta H_{\text{hydration,NaCl}(aq)} - U_{\text{NaCl}}$$

$$\Delta H_{\text{hydration,NaCl}(aq)} = \Delta H_{\text{solution,NaCl}} + U_{\text{NaCl}}$$

$$= 4 \text{ kJ/mol} + (-786 \text{ kJ/mol}) = -782 \text{ kJ/mol}$$

The value for $\Delta H_{\text{hydration,NaCl}(aq)}$ is negative, indicating that when Na$^+$ and Cl$^-$ ions in the gas phase dissolve in water, the reaction is exothermic. The sign of $\Delta H_{\text{hydration,NaCl}(aq)}$ also tells us something about the relative strengths of the dipole–dipole interactions in water and the ion–dipole interactions in aqueous NaCl. If $\Delta H_{\text{hydration,NaCl}(aq)}$ is less than zero, then according to Equation 11.4 the formation of ion–dipole interactions must release more energy than is needed to disrupt the dipole–dipole interactions (hydrogen bonds) in water.

With careful measurements and calculations, we can even separate $\Delta H_{\text{hydration,NaCl}(aq)}$ into enthalpies of hydration for the individual ions:

$$\Delta H_{\text{hydration,NaCl}(aq)} = \Delta H_{\text{hydration,Na}^+(g)} + \Delta H_{\text{hydration,Cl}^-(g)}$$

The values for $\Delta H_{\text{hydration}}$ for selected cations and anions are listed in Table 11.3 and can be used to calculate $\Delta H_{\text{solution}}$ for an ionic compound if we know the lattice energy, or to calculate the lattice energy if we measure the enthalpy of solution. The values in Table 11.3 are the results of several different experiments and include an inherent uncertainty. The sum of the values for Na$^+$ and Cl$^-$ is −786 kJ/mol; that is close enough to the value calculated from the measured enthalpy of solution (−782 kJ/mol) to be acceptable.

TABLE 11.3	Enthalpies of Hydration for Selected Cations and Anions[a]		
Cation	**$\Delta H_{\text{hydration}}$ (kJ/mol)**	**Anion**	**$\Delta H_{\text{hydration}}$ (kJ/mol)**
Li$^+$	−536	F$^-$	−502
Na$^+$	−418	Cl$^-$	−368
K$^+$	−335	Br$^-$	−335
Rb$^+$	−305	I$^-$	−293
Cs$^+$	−289	ClO$_4^-$	−238
Mg^{2+}	−1903	NO$_3^-$	−301
Ca^{2+}	−1591	SO$_4^{2-}$	−1017

[a] Based on enthalpy of hydration of H$^+$ as −1105 kJ/mol.

CONCEPT TEST ....

Why is it difficult to calculate the lattice energy of sodium perchlorate using a Born–Haber cycle?

(Answers to Concept Tests are in the back of the book.)

....

SAMPLE EXERCISE 11.3   **Calculating Lattice Energy Using Enthalpies of Hydration**

Use the appropriate enthalpy of hydration values in Table 11.3 and $\Delta H_{\text{solution}} = 0.91$ kJ/mol for NaF to calculate the lattice energy of NaF.

Collect and Organize The enthalpies of hydration for Na$^+$ and F$^-$ are −418 kJ/mol and −502 kJ/mol, respectively. We can use the enthalpy of solution and Equation 11.5 to calculate the lattice energy.

Analyze Before we can use Equation 11.5 to calculate U_{NaF}, we must add the values of $\Delta H_{hydration}$ for the two ions to obtain a value for $\Delta H_{hydration}$ for NaF. From our calculations in Sample Exercise 11.2, we expect that the lattice energy of NaF should be close to -927 kJ/mol.

Solve Solving for the enthalpy of hydration of NaF:

$$\Delta H_{hydration,NaF} = \Delta H_{hydration,Na^+(g)} + \Delta H_{hydration,F^-(g)}$$
$$= -418 \text{ kJ/mol} + (-502 \text{ kJ/mol}) = -920 \text{ kJ/mol}$$

Substituting $\Delta H_{hydration,NaF(aq)}$ and $\Delta H_{solution,NaF}$ into Equation 11.5:

$$\Delta H_{solution,NaF} = \Delta H_{hydration,NaF(aq)} - U_{NaF}$$
$$+0.91 \text{ kJ/mol} = (-920 \text{ kJ/mol}) - U_{NaF}$$

Solving for U_{NaF}:

$$U_{NaF} = -920 \text{ kJ/mol} - (0.91 \text{ kJ/mol}) = -921 \text{ kJ/mol}$$

Think about It The value of U_{NaF} calculated from $\Delta H_{solution}$ and $\Delta H_{hydration}$ is within 1% of the value calculated in Sample Exercise 11.2 using different kinds of measurements, suggesting that the value calculated this way accurately reflects the lattice energy of NaF.

Practice Exercise Calculate the lattice energy for $NaClO_4$ using the data in Table 11.3 and $\Delta H_{solution,NaClO_4} = 14$ kJ/mol.

11.2 Vapor Pressure

vapor pressure the pressure exerted by a gas at a given temperature in equilibrium with its liquid phase.

Water in a glass left on a countertop slowly disappears. Of course, it does not actually disappear, it just enters the gas phase and we can no longer see it. Molecules on the surface of the liquid vaporize, or evaporate, over time; they escape from the surface of the liquid and enter the gas phase. The rate at which the molecules make this transformation depends on the temperature, on the surface area of the liquid, and on the strength of the intermolecular forces that hold the molecules together in the liquid:

1. The higher the temperature, the larger the number of molecules with sufficient kinetic energy to break the attractive forces that hold them together in the liquid and enable them to enter the gas phase.
2. The greater the surface area of the liquid, the larger the number of molecules on the surface in a position to enter the gas phase.
3. The stronger the intermolecular forces, the greater the kinetic energy needed for a molecule to escape the surface, and the smaller the number of molecules in the population that have this energy.

If that same glass of water is covered (Figure 11.5), a different situation arises. Molecules at the surface still evaporate, but now they are confined to the space above the water. Some of them *condense* at the liquid surface and return to the liquid phase. In a short time, the two processes of evaporation and condensation equalize, and the same number of molecules leave the surface as reenter it. At this point of dynamic equilibrium, no further change takes place in the level of the liquid in the glass, although molecules continue to evaporate and condense constantly. In such a situation at constant temperature, the pressure exerted by the gas in equilibrium with its liquid is called the **vapor pressure** of the liquid.

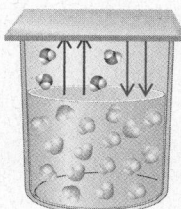

↑ Evaporation   ↓ Condensation

FIGURE 11.5 A covered glass of water achieves a dynamic equilibrium where the rate at which liquid water is lost to evaporation equals the rate at which liquid water is gained by condensation.

FIGURE 11.6 When a beaker containing seawater is placed in a sealed chamber with a beaker of pure water, the slightly higher vapor pressure of the pure water leads to a net transfer of water from the beaker of pure water to the beaker of seawater.

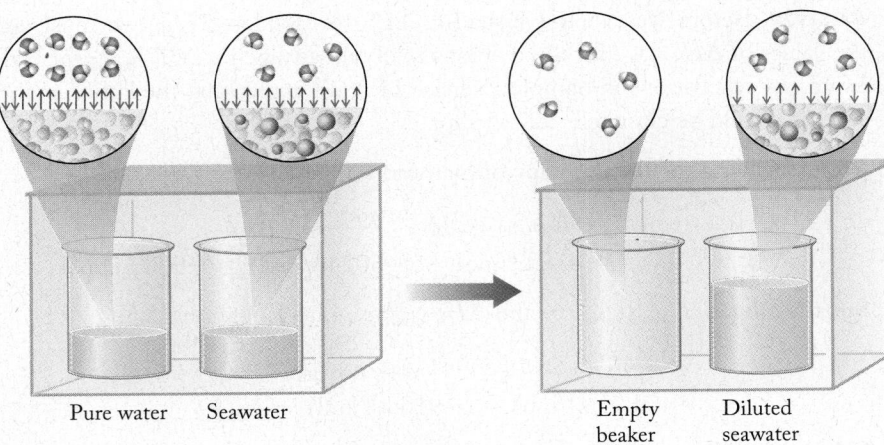

Pure water   Seawater   Empty   Diluted
                                beaker   seawater

▶️|| **CHEMTOUR** Molecular Motion

Figure 11.6 provides evidence that adding a solute to a pure liquid changes the vapor pressure of the liquid. When a beaker of seawater (water containing numerous dissolved solutes) and a beaker of pure water are placed in a sealed chamber, the volume in the seawater beaker increases over time, and the volume in the pure-water beaker decreases at the same rate. Ultimately nearly all the water ends up in the seawater beaker. This transfer of the pure water is due to the dissolved solutes in the seawater.

As the water in both beakers evaporates, the concentration of water vapor in the air space of the sealed chamber increases. As the concentration of water vapor increases, the pressure the water vapor exerts on the two liquid surfaces increases. At constant temperature, this pressure eventually stabilizes at a value equal to the vapor pressure of water at that temperature. At this point, the rate of evaporation from the beakers is equal to the rate of condensation. If the evaporation/condensation rate for the pure water were the same as the evaporation/condensation rate for the seawater, the liquid levels in the beakers would not change over time.

The liquid levels do change, however, until the pure-water beaker is nearly empty. Because the $H_2O(g)$ molecules in the sealed chamber are free to condense into either beaker, the rate of condensation is the same into both beakers. Therefore, in order for most of the pure water to end up in the seawater beaker, the pure water must have a higher rate of evaporation than the seawater. This conclusion leads to another: if the seawater evaporates at a lower rate, it must have a lower vapor pressure than the pure water at the same temperature.

Because the vapor pressure of the pure water is greater than the vapor pressure of the seawater, more water vapor enters the air in the chamber from the pure-water beaker than from the seawater beaker. Because the condensation rates are the same, the pure-water beaker over time loses more water than is restored to it by condensation, while the seawater beaker gains more water than it loses by evaporation. The process in Figure 11.6 is an illustration of a more general observation: at a given temperature, the vapor pressure of a solvent in a solution containing nonvolatile solutes is less than the vapor pressure of the pure solvent.

CONCEPT TEST ....

Which of the following solutes would be considered nonvolatile when dissolved in water? Methanol (bp 64.5°C), ethylene glycol (bp 197°C), isopropanol (rubbing alcohol, bp 82°C), triethylene glycol (bp 285°C)

....

Vapor Pressure of Solutions: Raoult's Law

The connection between the vapor pressure of a solvent and the concentration of nonvolatile solutes dissolved in the solvent was studied extensively by French chemist François Marie Raoult (1830–1901). He discovered that the relation between the vapor pressure of a solution, $P_{solution}$, and that of the pure solvent, $P°_{solvent}$, is

$$P_{solution} = X_{solvent} P°_{solvent} \qquad (11.6)$$

where $X_{solvent}$ is the mole fraction of solvent. This relationship is now known as **Raoult's law**.

Let's take another look at the two beakers in Figure 11.6 and assume that the temperature in the chamber is 20°C. At 20°C the vapor pressure of pure water is 0.0231 atm. What is the vapor pressure produced by evaporation of water from seawater if the mole fraction of water in the sample is 0.980? We can use Raoult's law as expressed by Equation 11.6 to answer this question:

$$P_{solution} = (0.980)(0.0231 \text{ atm}) = 0.0226 \text{ atm}$$

Because the vapor pressure of seawater at 20°C is slightly lower than the vapor pressure of pure water at 20°C, fewer molecules evaporate from the seawater than from the pure water; the evaporation rates differ, as we stated in our earlier discussion of Figure 11.6.

Lowering the vapor pressure of a solution relative to the vapor pressure of pure solvent depends only on the concentration of solute particles, not on their identity. Properties of solutions that depend only on the concentration of particles and not on their identity are called **colligative properties**. We examine them in detail in Section 11.4.

Solutions that obey Raoult's law are called **ideal solutions**. Raoult's law works best for solutions where solute and solvent experience similar intermolecular forces. Deviations from ideal behavior occur when solute–solvent interactions are much stronger than solvent–solvent interactions. Then the rate of evaporation of the solvent is slower and vapor pressure is less than predicted by Raoult's law. Because solvent molecules are held at the surface by solute–solvent attractive interactions, more energy is required to separate them from the surface and fewer vaporize at a given temperature. If solute–solvent interactions are much weaker than solvent–solvent interactions, less energy is required to separate solvent molecules from the surface, and more solvent molecules vaporize. In this case the vapor pressure is greater than the value predicted by Raoult's law.

Raoult's law represents one of several colligative properties of solutions. The definition of colligative properties states that the identity of the solute particles does not matter, but that is not always the case if intermolecular forces intervene. We treat solutions for the most part as ideal systems, but you should be aware that deviations from ideal behavior exist in liquids just as they do in gases.

....

Is the vapor pressure of a pure solvent an intensive or an extensive property? Is the vapor pressure of a solution an intensive or an extensive property?

Raoult's law the vapor pressure of a solution containing nonvolatile solutes is proportional to the mole fraction of the solvent.

colligative properties characteristics of solutions that depend on the concentration and not the identity of particles dissolved in the solvent.

ideal solution one that obeys Raoult's law.

∞ **CONNECTION** We used mole fractions when discussing Henry's law in Chapter 10.

▶❚❚ **CHEMTOUR** Raoult's Law

∞ **CONNECTION** As we discussed in Chapter 1, intensive properties of matter are independent of the amount of material present, while extensive properties vary with the quantity of substance present.

SAMPLE EXERCISE 11.4 **Calculating the Vapor Pressure of a Solution**

The liquid used in automobile cooling systems is prepared by dissolving ethylene glycol ($HOCH_2CH_2OH$, molar mass 62.07 g/mol) in water. What is the vapor pressure of a solution prepared by mixing 1.000 L of ethylene glycol (density 1.114 g/mL) with 1.000 L of water (density 1.000 g/mL) at 100.0°C? Assume that the mixture obeys Raoult's law.

Collect and Organize We are asked to calculate the vapor pressure of a solution of ethylene glycol in water. The vapor pressure of a solution is a colligative property that depends on the number of solute particles, and hence the concentration. Volume and density must be used to determine the concentration of ethylene glycol. Ethylene glycol is a nonvolatile solute and its solution obeys Raoult's law, which means it behaves ideally. The vapor pressure curves in Figure 11.7 show that the vapor pressure of pure water at 100°C (its normal boiling point) is 1.00 atm, whereas that of ethylene glycol is less than 0.05 atm.

Analyze We have a mixture of equal volumes of two liquids. The solvent is the one present in the greater number of moles. We can determine the numbers of moles of each by converting their volumes into masses by multiplying by their densities and then calculate the number of moles of each by dividing by their molar masses. From these values we can decide which liquid is the solvent and calculate its mole fraction in the mixture.

Solve For ethylene glycol:

$$\frac{1.114 \text{ g}}{1 \text{ mL}} \times 1000 \text{ mL} \times \frac{1 \text{ mol}}{62.07 \text{ g}} = 17.95 \text{ mol}$$

For water:

$$\frac{1.000 \text{ g}}{1 \text{ mL}} \times 1000 \text{ mL} \times \frac{1 \text{ mol}}{18.02 \text{ g}} = 55.49 \text{ mol}$$

The mole fraction of water is

$$\chi_{\text{water}} = \frac{55.49 \text{ mol}}{55.49 \text{ mol} + 17.95 \text{ mol}} = 0.7556$$

Ethylene glycol is essentially nonvolatile, so the vapor pressure of the solution is due only to the solvent, and $P_{\text{solvent}} = P_{H_2O} = 1.00$ atm. Using this value and the calculated mole fraction of water in the mixture yields

$$P_{\text{solution}} = \chi_{H_2O} \times P_{H_2O}^\circ = (0.756)(1.00 \text{ atm}) = 0.756 \text{ atm}$$

Think about It The presence of a nonvolatile solute causes the vapor pressure of the solution to be less than 1 atm, giving us confidence in our result. In addition, our selection of water as the solvent is justified because it is the major component in terms of numbers of moles of substances present: 55.49 moles of water with 17.95 moles of ethylene glycol dissolved in it.

Practice Exercise Glycerol [$HOCH_2CH(OH)CH_2OH$] is considered to be a nonvolatile, water-soluble liquid. Its density is 1.25 g/mL. Predict the vapor pressure of a solution of 275 mL of glycerol in 375 mL of water at the normal boiling point of water.

To boil the solution described in Sample Exercise 11.4, we must heat it to above 100°C, to a temperature at which the vapor pressure of the solution is 1 atm. This is how antifreeze works in an automobile engine: it not only lowers the freezing point of water, it raises the temperature at which water boils, making it possible for the coolant to remove more heat from the engine while staying in the liquid state.

11.3 Mixtures of Volatile Solutes

Thus far, we have considered only the effect of essentially nonvolatile solutes on the vapor pressure of a solvent. In our everyday lives we encounter many solutions containing volatile solutes. The natural gas used for heating and the gasoline we use to power vehicles are mixtures of hydrocarbons. Gasoline comes from **crude oil**, a complex mixture of compounds composed mostly of carbon and hydrogen. Crude oil contains hydrocarbons with five or more carbon atoms in their molecular structures. The hydrocarbons with one to four carbon atoms are usually found in deposits of natural gas, although they are also dissolved in crude oil. Gasoline is separated from crude oil by distillation. Most hydrocarbons in gasoline have between five and nine atoms per molecule. Each of these compounds is volatile and has a measurable vapor pressure at 25°C. In this section we explore distillation on a molecular level and consider the effects of volatile solutes on the vapor pressure of a solution.

Volatility and the Clausius-Clapeyron Equation

As we learned in Section 11.2, the vapor pressure of a liquid increases as temperature increases. Before we can understand what happens when a solution of volatile substances is distilled, we need to examine the vapor pressure of a pure substance as a function of temperature.

Figure 11.7 shows that vapor pressure increases with increasing temperature. At some point as temperature increases, the vapor pressure of any substance reaches 1 atm; the temperature at which this occurs is the **normal boiling point** of the substance. For water, to take one example from Figure 11.7, the vapor pressure reaches 1 atm at 100°C, which we recognize as the normal boiling point of water. The reason for calling it the *normal* boiling point arises from the observation that most chemical reactions in nature, in our bodies, and in the laboratory take place at or near an atmospheric pressure of 1 atm.

crude oil a combustible liquid mixture of hydrocarbons and other organic molecules formed under Earth's surface.

normal boiling point the temperature at which the vapor pressure of a liquid equals 1 atm (760 torr).

∞ **CONNECTION** We discussed simple distillation in Chapter 1 as a way to purify seawater.

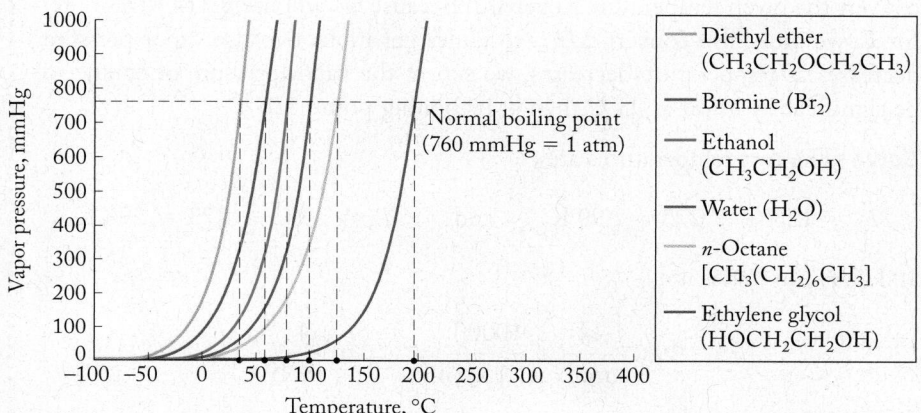

FIGURE 11.7 A graph of vapor pressure versus temperature for six liquids shows that vapor pressure increases with increasing temperature. The temperature at which the vapor pressure equals 1 atm is the normal boiling point of the liquid.

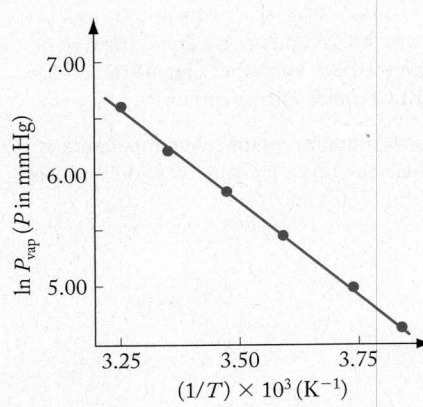

FIGURE 11.8 Plotting the natural logarithm of the vapor pressure versus the reciprocal of the absolute temperature gives a straight line described by the Clausius–Clapeyron equation. The graph shows the plot for *n*-pentane.

The relationship between the vapor pressure of a pure substance and absolute temperature is not linear. However, if we graph the natural logarithm of the vapor pressure versus $1/T$, where T is the absolute temperature, we get a straight line (Figure 11.8) described by the equation

$$\ln(P_{vap}) = -\frac{\Delta H_{vap}}{R}\left(\frac{1}{T}\right) + C$$

where ΔH_{vap} is the enthalpy of vaporization, R is the gas constant, and C is a constant that depends on the identity of the liquid. We can solve this equation for C and write it in terms of two temperatures:

$$\ln(P_{vap,T_1}) + \left(\frac{\Delta H_{vap}}{RT_1}\right) = C = \ln(P_{vap,T_2}) + \left(\frac{\Delta H_{vap}}{RT_2}\right)$$

which can be rearranged to

$$\ln\left(\frac{P_{vap,T_1}}{P_{vap,T_2}}\right) = \frac{\Delta H_{vap}}{R}\left(\frac{1}{T_2} - \frac{1}{T_1}\right) \tag{11.7}$$

This expression, called the **Clausius–Clapeyron equation**, can be used to calculate ΔH_{vap} if the vapor pressures at two temperatures are known, and to calculate either the vapor pressure P_{vap,T_2} at any given temperature T_2, or the temperature T_2 at any given vapor pressure P_{vap,T_2}, if ΔH_{vap} and P_{vap,T_1} and T_1 are known. Because the units of ΔH_{vap} are typically joules (J) or kilojoules (kJ), the value used for R in the Clausius–Clapeyron equation is 8.314 J/(mol · K).

SAMPLE EXERCISE 11.5   Calculating Vapor Pressure

At its normal boiling point of 126°C, *n*-octane, C_8H_{18}, has a vapor pressure of 760 torr. What is its vapor pressure at 25°C? The enthalpy of vaporization of *n*-octane is 39.07 kJ/mol.

Collect and Organize The vapor pressure of octane at any temperature can be calculated using Equation 11.7, the Clausius–Clapeyron equation, given the enthalpy of vaporization and the vapor pressure of octane at its normal boiling point.

Analyze We can use the Clausius–Clapeyron equation, but to do so we must convert the given temperature to kelvins. Because we will use 8.314 J/(mol · K) for R, we must also convert ΔH_{vap} to joules per mole. Because vapor pressure decreases as temperature decreases, we expect the vapor pressure of octane to be significantly lower at 25°C than at its boiling point, 126°C.

Solve The two temperatures are

$$T_1 = 126°C + 273 = 399 \text{ K} \quad \text{and} \quad T_2 = 25°C + 273 = 298 \text{ K}$$

and ΔH_{vap} in joules is

$$39.07 \ \frac{\cancel{kJ}}{mol} \times \frac{1000 \text{ J}}{1 \ \cancel{kJ}} = 39,070 \ \frac{J}{mol}$$

Clausius–Clapeyron equation relates the vapor pressures of a substance at different temperatures to its heat of vaporization.

Inserting these values in Equation 11.7:

$$\ln\!\left(\frac{P_{vap,T_1}}{P_{vap,T_2}}\right) = \frac{\Delta H_{vap}}{R}\left(\frac{1}{T_2} - \frac{1}{T_1}\right)$$

$$\ln\!\left(\frac{760 \text{ torr}}{P_{vap,T_2}}\right) = \frac{39{,}070\ \dfrac{\cancel{J}}{\cancel{mol}}}{8.314\ \dfrac{\cancel{J}}{\cancel{mol}\cdot\cancel{K}}}\left(\frac{1}{298\ \cancel{K}} - \frac{1}{399\ \cancel{K}}\right)$$

$$P_{vap,T_2} = 14.0 \text{ torr}$$

fractional distillation a method of separating a mixture of compounds on the basis of their different boiling points.

Think about It We expect octane to have a low vapor pressure at a temperature much lower than its boiling point, so this number seems reasonable.

Practice Exercise *n*-Pentane, C_5H_{12}, boils at 36°C. What is its molar heat of vaporization in kilojoules per mole if its vapor pressure at 25°C is 505 torr?

CONCEPT TEST

Diesel fuel is made of hydrocarbons with an average of 13 carbon atoms per molecule, and gasoline is made of hydrocarbons with an average of 7 carbon atoms per molecule. Which fuel has the higher vapor pressure at room temperature?

Vapor Pressures of Mixtures of Volatile Solutes

A process called **fractional distillation** (Figure 11.9) is used to separate the volatile components of a mixture. As the mixture is heated, the vapor that rises and fills the space above the liquid has a composition different from the composition of the mixture: the concentration of the component with the lowest boiling point is higher in the vapor than in the liquid. If this enriched vapor is collected, condensed, and redistilled, the vapor this time is even richer in the component with the lowest boiling point. In a fractional distillation apparatus, repeated distillation steps allow components with only slightly different boiling points to be separated from one another.

To see how fractional distillation works, let's examine the heating curves of a pure substance and a solution of two volatile liquids. Figure 11.10(a) is the heating curve for pure *n*-octane (C_8H_{18}, bp 126°C). As in the heating curves of Section 5.4, the phase change from liquid to vapor takes place at a constant temperature of 126°C. Figure 11.10(b) shows the heating curve for a mixture of *n*-heptane (C_7H_{16}, bp 98°C) and *n*-octane. Here the portion of the curve from (1) to (2) again represents a vapor–liquid phase change, but in this case the temperature is not constant during the phase change. Instead, the two components co-distill over a range of temperatures.

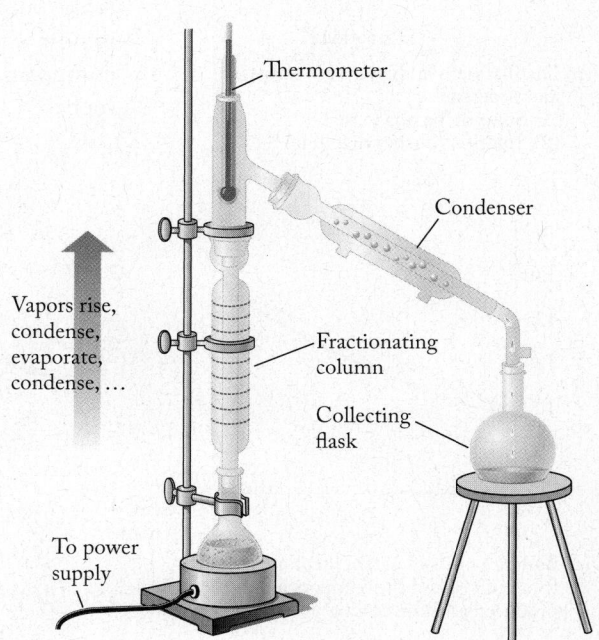

FIGURE 11.9 Fractional distillation separates mixtures. Vapors rise through a fractionating column, where they repeatedly condense and vaporize. The most volatile component distills first and is first to pass through the condenser and into the collecting flask. Increasingly less volatile, higher-boiling components are distilled in turn. The progress of the distillation process is monitored using the thermometer at the top of the fractionating column.

▶‖ **CHEMTOUR** Fractional Distillation

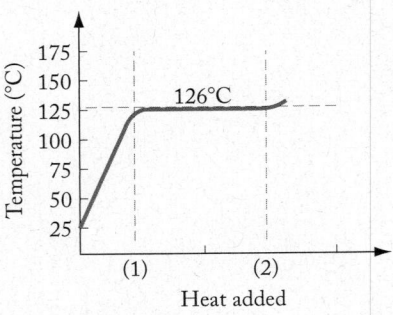

(a) Distillation of *n*-octane
  (1) Octane begins to distill
  (2) Octane finishes distilling

(b) Distillation of a mixture of *n*-heptane and *n*-octane
  (1) Solution begins to distill
  (2) Solution finishes distilling

(c) Boiling points of octane/heptane mixtures (blue curve) and the composition of the vapors produced at those boiling points (red curve).

① Solution boils at this temperature
② Vapor has this composition
③ Vapor condenses, then boils
④ Vapor has this composition
⑤ Vapor condenses, then boils

FIGURE 11.10 Heating curves describe the behavior of boiling liquids. (a) When pure *n*-octane is distilled, the distillation takes place at a constant temperature, indicated by the horizontal line at *T* = 126°C. The liquid in the collecting flask is pure *n*-octane. (b) When a mixture of *n*-heptane (bp 98°C) and *n*-octane is distilled, the distillation takes place over a range of temperatures. (c) The blue line shows the temperatures where solutions of a given composition boil. The red line shows the composition of the vapor arising from those solutions. The stair-step line illustrates what happens in a fractionating column.

Now let's analyze the graphs describing this process to understand how fractional distillation works. Figure 11.10(c) shows the composition of the vapor above a series of boiling solutions, and herein lies the key to how distillation separates mixtures: in the vapor in equilibrium with a solution of two volatile substances, the concentration of the lower-boiling (more volatile) component is always greater in the vapor than in the liquid. Suppose we use the distillation apparatus from Figure 11.9 to heat 100 mL of a solution that is 50% by volume *n*-heptane and 50% *n*-octane. The blue curve in Figure 11.10(c) gives the temperature of the boiling solution, and point 1 on it tells us that the boiling point of the solution (50:50) is initially about 108°C. The red line on the graph gives the composition of the vapor above the solution at a given temperature. To find the composition of the vapor at 108°C, we move horizontally to the left along the dashed line between points 1 and 2. This horizontal line corresponds to a temperature of 108°C on the vertical (temperature) axis. Point 2 corresponds to a different composition (on the *x*-axis) for the vapor than for the solution at point 1. The vapor at point 2 is enriched in the lower-boiling component: it is about 65% *n*-heptane and only 35% *n*-octane.

Suppose this 65:35 vapor rises up in the distillation column, cools, and condenses in the next region of the column, a process represented by the red arrow from point 2 to point 3. The condensed liquid in this plate is 65% *n*-heptane and 35% *n*-octane. Continued heating of the column warms this liquid, and it begins boiling at about 104°C, which the temperature is at point 3. To find the concentration of the vapor above this boiling liquid, we move left from point 3 until we intersect the red curve (point 4). Reading down from point 4 to the concentration axis, we see that the vapor concentration is now about 80% *n*-heptane and only 20% *n*-octane. This vapor with 80:20 composition rises up, where it cools and condenses, and the distillation is repeated.

If we continue this process of redistilling mixtures with increasing concentrations of *n*-heptane and then cooling and condensing the vapors, we eventually obtain a condensate that is pure *n*-heptane. If we monitor the temperature at which vapors condense at the very top of our distillation column, we will see a profile of temperature versus volume of distillate produced like that shown in Figure 11.11. The first liquid to be produced is pure *n*-heptane, which has a boiling point of 98°C. Ideally over time, all 50 mL of *n*-heptane in the original sample is recovered. As we continue to add heat at the bottom of the column, the temperature at the top rises to 126°C, signaling that the second component (pure *n*-octane, bp 126°C) is being collected. In the real world, the separation may not be as complete as described in this idealized presentation.

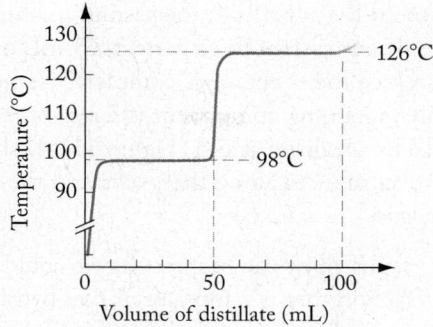

FIGURE 11.11 Fractional distillation of a mixture of 50 mL of *n*-heptane and 50 mL of *n*-octane produces two plateaus at the boiling points of the two components. If fractionation is perfect, the first 50 mL of distillate is pure *n*-heptane and the second 50 mL is pure *n*-octane.

CONCEPT TEST

Dimethyl ether, CH_3OCH_3, and acetone, $CH_3C(O)CH_3$, shown in Figure 11.12, have similar molar masses but different dipole moments: 1.30 D and 2.88 D, respectively. Which compound would you expect to distill first from a mixture of acetone and dimethyl ether?

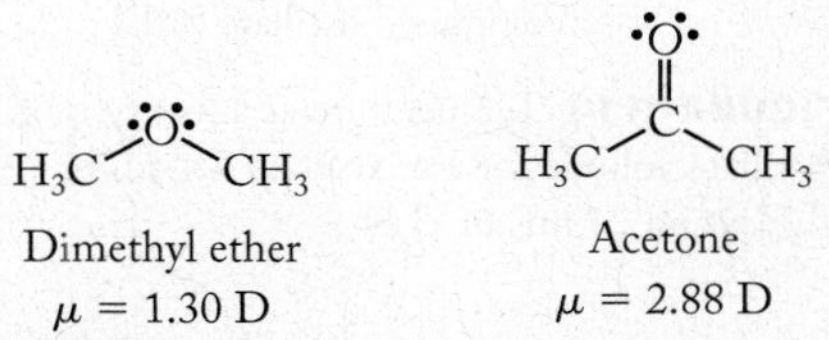

Dimethyl ether
$\mu = 1.30$ D

Acetone
$\mu = 2.88$ D

FIGURE 11.12 Lewis structures for dimethyl ether and acetone.

SAMPLE EXERCISE 11.6   Interpreting Data from Fractional Distillation

How would the graph in Figure 11.11 be different if (a) we started with a mixture of 75 mL of *n*-octane (C_8H_{18}) and 25 mL of *n*-heptane (C_7H_{16}) and (b) we started with a mixture of 75 mL of *n*-octane (C_8H_{18}) and 25 mL of *n*-nonane (C_9H_{20})? The normal boiling points of *n*-heptane, *n*-octane, and *n*-nonane are 98°C, 126°C, and 151°C, respectively.

Collect and Organize The graph in Figure 11.11 is an idealized plot of the temperature of the solution as a function of the volume of distillate for a 50:50 mixture of C_7H_{16}:C_8H_{18}. We are asked to describe the changes in the graph if we change the ratio of the components and if we change the identity of one of the components. We are given the normal boiling points of all compounds.

Analyze Fractional distillation separates solutions of volatile liquids into pure substances on the basis of their different boiling points. The components distill in the order of their boiling points, with the lowest-boiling component distilling first. Ideally the total volume of distillate equals the total volume of the solution, and the volume of each fraction reflects the volume of the component in the original solution.

Solve

a. The first component that distills is the lower-boiling *n*-heptane. If our system were perfect, 25 mL of *n*-heptane would distill at 98°C. When the *n*-heptane was completely removed from the solution, the only remaining component (75 mL of *n*-octane) would distill at 126°C. The idealized graph (Figure 11.13) shows the boiling points of the two substances along the *y*-axis and the volume of each distilled along the *x*-axis. The lengths of the horizontal lines at 98°C and 126°C are in a 1:3 ratio reflecting the composition of the mixture, 25 mL of C_7H_{16} and 75 mL of C_8H_{18}.

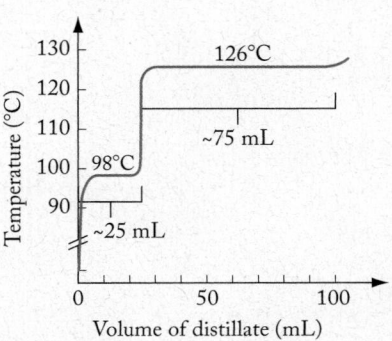

FIGURE 11.13 Temperature as a function of distillate volume for a mixture of 25 mL of C_7H_{16} and 75 mL of C_8H_{18}.

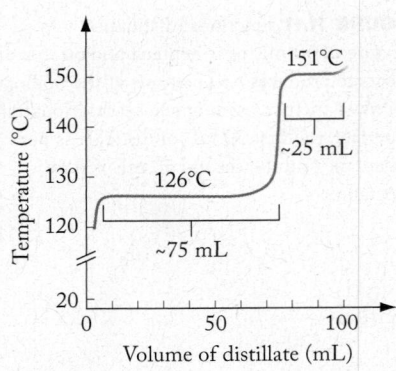

FIGURE 11.14 Temperature as a function of distillate volume for a mixture of 75 mL of C_8H_{18} and 25 mL of C_9H_{20}.

b. The first component that distills in the second mixture is the lower-boiling *n*-octane. If our system were perfect, 75 mL of *n*-octane would distill at 126°C. Once the *n*-octane is completely removed from the solution, the only remaining component (25 mL of *n*-nonane) would distill at 151°C. The idealized graph (Figure 11.14) shows the boiling points of the two substances along the *y*-axis and the volume of distillate along the *x*-axis.

Think about It Our graphs show the best results we could achieve. In a real-world system, a small fraction that is a mixture of two hydrocarbons distills at temperatures between the boiling points of the two pure components.

Practice Exercise Draw an idealized graph of the temperature versus volume of distillate collected when a solution consisting of 30 mL of *n*-hexane (boiling point 69°C), 50 mL of *n*-heptane (boiling point 98°C), and 20 mL of *n*-nonane (boiling point 151°C) is fractionally distilled.

■ ·······································

Now that we know how fractional distillation works, the next step is to understand *why* it works. We return to Raoult's law, which we used in Section 11.2 to describe the influence of nonvolatile solutes on the boiling point of a pure solvent. Raoult's law also applies to homogeneous mixtures of volatile compounds, such as crude oil. Because the solutes in a solution of crude oil are volatile, they contribute to the solution's overall vapor pressure. The total vapor pressure equals the sum of the vapor pressures of each component (P_x° is the equilibrium vapor pressure of the pure component at the temperature of interest) multiplied by the mole fraction of that component in the solution (X_x):

$$P_{total} = X_1 P_1^\circ + X_2 P_2^\circ + X_3 P_3^\circ + \cdots \tag{11.8}$$

SAMPLE EXERCISE 11.7   **Calculating the Vapor Pressure of a Solution of Volatile Substances**

Calculate the vapor pressure of a solution prepared by dissolving 13 g of *n*-heptane (C_7H_{16}) in 87 g of *n*-octane (C_8H_{18}) at 25°C. By what factor does the concentration of the more volatile component in the vapor exceed the concentration of this component in the liquid? The vapor pressures of *n*-octane and *n*-heptane at 25°C are 11 torr and 31 torr, respectively.

Collect and Organize We can use Equation 11.8 to determine the total vapor pressure of the solution from the vapor pressures of the components after we determine the composition of the solution in terms of mole fractions.

Analyze To calculate mole fractions, we need the molar masses of *n*-heptane and *n*-octane. The mole fraction is then equal to the number of moles of each component divided by the total number of moles of material in the solution. The vapor pressure of the solution must lie between the vapor pressures of the pure compounds.

Solve The number of moles of each component is

$$87 \text{ g } C_8H_{18} \times \frac{1 \text{ mol } C_8H_{18}}{114.23 \text{ g } C_8H_{18}} = 0.762 \text{ mol } C_8H_{18}$$

$$13 \text{ g } C_7H_{16} \times \frac{1 \text{ mol } C_7H_{16}}{100.20 \text{ g } C_7H_{16}} = 0.130 \text{ mol } C_7H_{16}$$

The mole fraction of each component in the mixture is

$$X_{octane} = \frac{0.762 \text{ mol}}{(0.762 + 0.130) \text{ mol}} = 0.854$$

$$X_{heptane} = 1 - X_{octane} = 0.146$$

Using these mole fraction values and the vapor pressures of the two hydrocarbons in Equation 11.8, we have

$$P_{total} = X_{heptane}P°_{heptane} + X_{octane}P°_{octane}$$
$$= 0.146(31 \text{ torr}) + 0.854(11 \text{ torr})$$
$$= 4.5 \text{ torr} + 9.4 \text{ torr} = 13.9 \text{ torr}$$

To calculate how enriched the vapor phase is in the more volatile component (the component with the greater vapor pressure, *n*-heptane), we need to recall Dalton's law of partial pressures (Section 6.7) and the concept that the partial pressure of a gas in a mixture of gases is proportional to its mole fraction in the mixture. Therefore, the ratio of the mole fraction of *n*-heptane to that of *n*-octane is the ratio of their two vapor pressures:

$$\frac{4.5 \text{ torr}}{9.4 \text{ torr}} = 0.48$$

The mole ratio of *n*-heptane to *n*-octane in the liquid mixture is

$$\frac{0.13 \text{ mol}}{0.76 \text{ mol}} = 0.17$$

Therefore, the vapor phase is enriched in *n*-heptane by a factor of

$$\frac{0.48}{0.17} = 2.8$$

Think about It As expected, the vapor pressure of the mixture (13.9 torr) is between the vapor pressures of the separate components. This result illustrates how fractional distillation works. The vapor is enriched in the lower-boiling component.

Practice Exercise Benzene (C_6H_6) is a trace component of gasoline. What is the mole ratio of benzene to *n*-octane in the vapor above a solution of 10% benzene and 90% *n*-octane by mass at 25°C? The vapor pressures of *n*-octane and benzene at 25°C are 11 torr and 95 torr, respectively.

Solutions such as the hydrocarbons in crude oil obey Raoult's law when the strengths of solvent–solvent, solute–solute, and solute–solvent interactions are similar. Under these conditions, a solution behaves ideally. However, if the solute–solvent interactions are stronger than solvent–solvent or solute–solute interactions, the solute inhibits the solvent from vaporizing and the solvent inhibits the solute from vaporizing. This situation produces negative deviations from the vapor pressures predicted by Raoult's law, as shown in Figure 11.15(a).

If solute–solvent interactions are weaker than solvent–solvent and solute–solute interactions, it is easier for solvent and solute molecules to vaporize from the solution, which leads to greater-than-predicted vapor pressures and positive deviations from Raoult's law (Figure 11.15b). A mixture of hydrocarbons is expected to behave like an ideal solution because intermolecular interactions between the components are all London forces acting on molecules of similar structure and size.

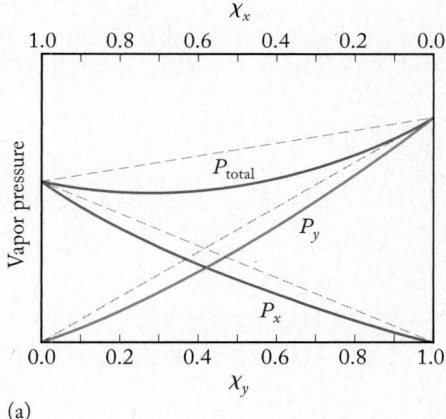

(a)

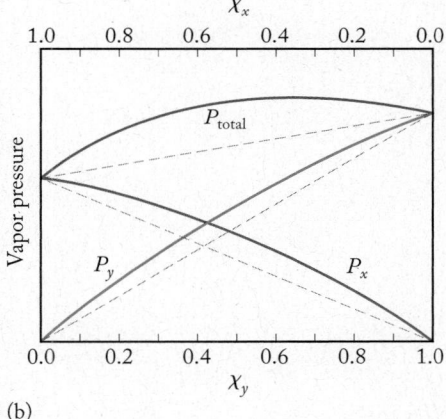

(b)

FIGURE 11.15 In a mixture of two volatile substances *x* and *y*, the vapor pressures P_x and P_y may deviate from the ideal behavior predicted by Raoult's law and described by the dashed lines. (a) If solute–solvent interactions are stronger than solvent–solvent or solute–solute interactions, the deviations from Raoult's law are negative. (b) If solute–solvent interactions are weaker than solvent–solvent or solute–solute interactions, the deviations are positive.

Which of the following solutions is least likely to follow Raoult's law? The structures of the compounds are given in Figure 11.16. (a) acetone/ethanol; (b) *n*-pentane/*n*-hexane; (c) pentanol/water

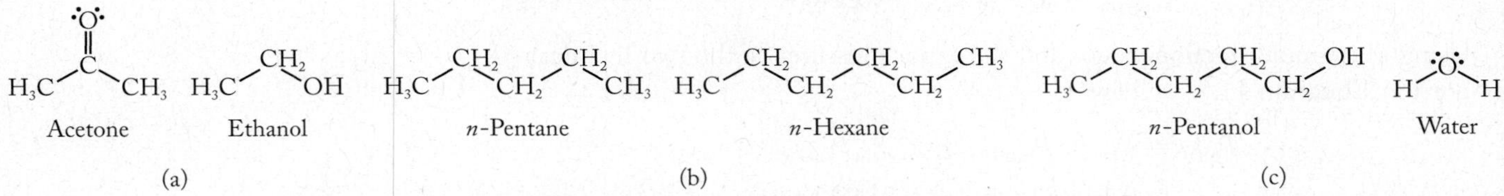

Acetone   Ethanol

(a)

n-Pentane   *n*-Hexane

(b)

n-Pentanol   Water

(c)

FIGURE 11.16 Lewis structures for acetone, ethanol, *n*-pentane, *n*-hexane, *n*-pentanol, and water.

11.4 Colligative Properties of Solutions

We learned in Section 11.2 that the vapor pressure of a solution containing a non-volatile solute is lower than the vapor pressure of the pure solvent. In this section we will see that many other physical properties of solvents are changed when a non-volatile solute is added to the solvent. Solutions generally have greater densities than the solvent alone, and an aqueous solution containing a nonvolatile solute has a higher boiling point and a lower freezing point than pure water. As we saw earlier, antifreeze, an aqueous solution of ethylene glycol used to cool automobile engines, boils at a temperature above 100°C and also has a lower freezing point than pure water.

The combined phase diagram of water and an aqueous solution of a non-volatile solute is shown in Figure 11.17. Note that the red line representing boiling/liquefaction points of the solution lies below the red line for pure water. The blue melting/freezing line for the solution lies to the left of the blue line for pure water. Following the dashed line from left to right at $P = 1$ atm, we see that the solution freezes at a lower temperature than pure water freezes and that the solution boils at a higher temperature than pure water. Furthermore, the higher boiling point for the solution indicates that the solution has a lower vapor pressure than pure water. Boiling point elevation and freezing point depression

FIGURE 11.17 Combined phase diagram for pure water and a solution of a nonvolatile solvent in water. Notice that the solution's boiling point is higher than the boiling point of the water and that the solution's freezing point is lower than the freezing point of the water.

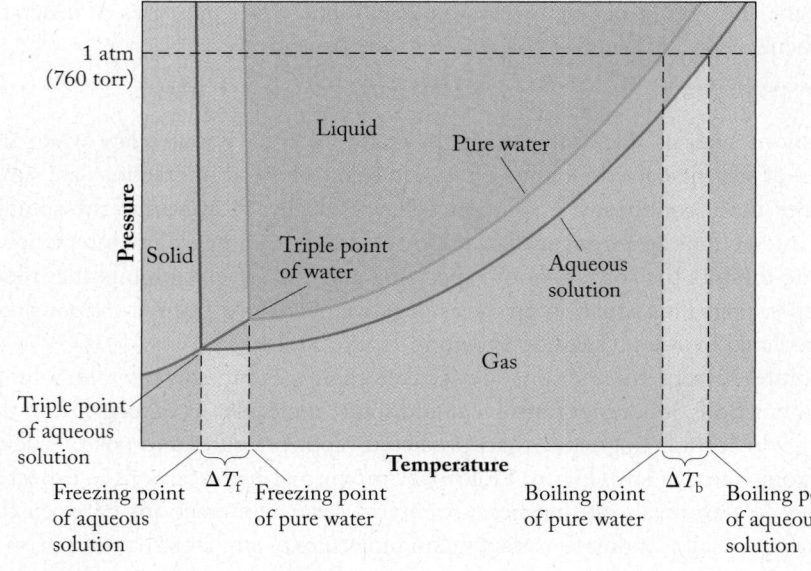

are both colligative properties. As noted in Section 11.2, only the number of particles in solution determines the impact of the solute on the colligative properties of the solvent. Before we can quantify these properties, we must introduce a new concentration unit.

molality (*m*) concentration expressed as the number of moles of solute per kilogram of solvent.

Molality

In Chapter 4, we introduced the concentration unit of molarity. If we want to quantify colligative properties, however, the conventional approach is to use a different concentration unit, **molality (*m*)**, defined as the number of moles of solute (n_{solute}) per kilogram of solvent:

$$m = \frac{n_{solute}}{kg \ solvent} \tag{11.9}$$

The difference between these two similar-sounding concentration units is that molarity is the number of moles of solute *per liter of solution,* whereas molality is the number of moles of solute *per kilogram of solvent.* Molality is used when discussing colligative properties when temperature changes are involved. Because solvent volume changes with temperature but solvent mass does not change, a concentration expressed in molality does not change with temperature.

To illustrate the difference between the molarity and molality of a solution, let's use the procedure outlined in Figure 11.18 to calculate the molality of 1 L of an aqueous solution of sodium chloride that is 0.558 *M* in NaCl (approximately the NaCl concentration in seawater). To calculate molality, we need the density of the solution so that we can convert liters of solution into kilograms of solvent. The density of the solution is 1.022 g/mL at 25°C, so 1 L of solution has a mass of

$$\frac{1.022 \ g}{1 \ mL} \times 1000 \ mL = 1022 \ g$$

Of this 1022 g of solution, the mass of dissolved NaCl is

$$\frac{0.558 \ mol \ NaCl}{1 \ L \ solution} \times \frac{58.44 \ g \ NaCl}{1 \ mol \ NaCl} = 32.6 \ g \ NaCl \ in \ 1 \ L \ of \ solution$$

If the mass of 1 L of solution is 1022 g, and 32.6 g is due to the dissolved NaCl, then the mass of the solvent is (1022 g − 32.6 g) = 989 g = 0.989 kg. The molality of the solution (number of moles of solute per kilogram of solvent) is

$$\frac{0.558 \ mol \ NaCl}{0.989 \ kg \ solvent} = 0.564 \ m$$

The two concentrations are numerically close, but molality is always slightly higher than molarity.

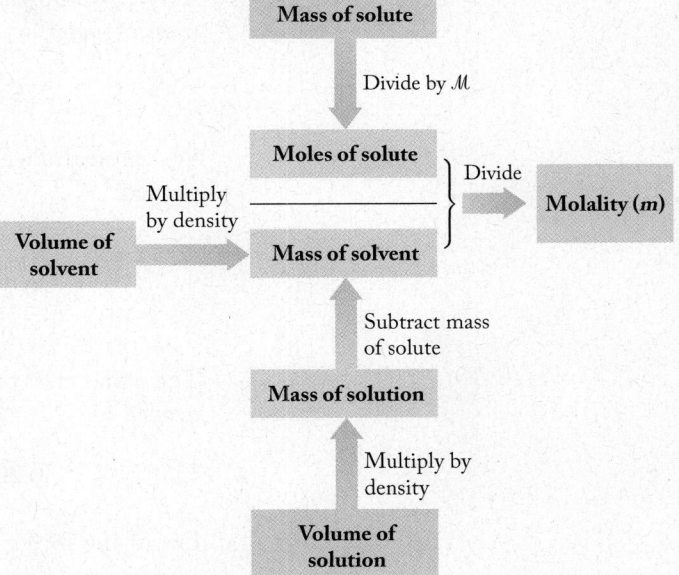

FIGURE 11.18 Flow diagram for calculating molality.

CONCEPT TEST

The difference between the molar concentration and molal concentration of any dilute aqueous solution is small. Why?

SAMPLE EXERCISE 11.8 **Preparing a Solution of Known Molality**

How many grams of Na_2SO_4 should be added to 275 mL of water to prepare a 0.750 *m* solution of Na_2SO_4? Assume the density of water is 1.000 g/mL.

Collect and Organize We are asked to determine the mass of a solute, sodium sulfate, needed to prepare a 0.750 *m* solution. Molality is moles of solute per kilogram of solvent. The volume of water, the solvent, is 275 mL.

Analyze Following the procedure shown in Figure 11.18, we use the density of water to convert 275 mL of water to kilograms. We can convert molality into a number of moles of Na_2SO_4 needed, and then convert that number into grams using the molar mass of Na_2SO_4. Our goal is to prepare a relatively small volume ($\sim\frac{1}{3}$ L) of a dilute solution (< 1 *m*), so we predict that the mass of solute needed will probably be less than $\frac{1}{3}$ mol of solute, which has a molar mass of 142 g/mol, or less than 50 g.

Solve Multiplying 275 mL of water by the density of water, we find that the mass of water is

$$275 \text{ mL water} \times \frac{1.000 \text{ kg water}}{1000 \text{ mL water}} = 0.275 \text{ kg water}$$

We can rearrange Equation 11.9 to determine the number of moles of solute needed:

$$n_{\text{solute}} = (m)(\text{kg solvent})$$

$$n_{Na_2SO_4} = \frac{0.750 \text{ mol } Na_2SO_4}{1 \text{ kg water}} \times 0.275 \text{ kg water} = 0.206 \text{ mol } Na_2SO_4$$

The molar mass of Na_2SO_4 is 142.04 g/mol, so the number of grams of Na_2SO_4 needed is

$$0.206 \text{ mol } Na_2SO_4 \times \frac{142.04 \text{ g}}{1 \text{ mol}} = 29.3 \text{ g } Na_2SO_4$$

Dissolving 29.3 g of Na_2SO_4 in 275 mL of water produces a 0.750 *m* solution.

Think about It The calculated value of 29 g Na_2SO_4 is consistent with our prediction that less than 50 g of solute would be required.

Practice Exercise What is the molality of a solution prepared by dissolving 78.2 g of ethylene glycol, $HOCH_2CH_2OH$, in 1.50 L of water? Assume the density of water is 1.00 g/mL.

Boiling Point Elevation

In Section 11.2 we discussed how the vapor pressure of a solution is reduced and the boiling point is elevated with respect to pure solvent, and at the beginning of this section we examined the phase diagrams of solutions. Now we can look quantitatively at how much the boiling points and freezing points of liquids change when solutes are present.

Boiling point elevation is a colligative property of the solvent. It is described in equation form as

$$\Delta T_b = K_b m \tag{11.10}$$

where ΔT_b is the increase in temperature above the boiling point of the pure solvent, K_b is the *boiling-point-elevation constant* of the solvent, and m is the molality of the solution. The units of K_b are $°C/m$, so the concentration of particles in solution must also have units of molality (m):

$$\Delta T_b = K_b m = \frac{°C}{\cancel{m}} \times \cancel{m}$$

The K_b of water is $0.52°C/m$, which means that for every mole of particles that dissolve in 1 kg of water (1 m) the boiling point of the solution is $0.52°C$ above the normal boiling point of $100.0°C$:

$$\Delta T_b = K_b m = \frac{0.52°C}{\cancel{m}} \times 1 \, \cancel{m} = 0.52°C$$

SAMPLE EXERCISE 11.9 **Calculating the Boiling Point Elevation of an Aqueous Solution**

What is the boiling point of seawater if the concentration of ions in seawater is 1.15 m?

Collect and Organize We are asked to calculate the elevation in the boiling point of an aqueous solution in which the ion concentration is 1.15 m. The boiling-point-elevation constant of water is $K_b = 0.52°C/m$, and the normal boiling point of water is $100.0°C$.

Analyze The boiling point elevation is the product of the boiling-point-elevation constant, $0.52°C/m$, and the solute concentration. Since the concentration is close to 1 m, we predict that the boiling point elevation will be close to $0.5°C$.

Solve

$$\Delta T_b = K_b m = \frac{0.52°C}{\cancel{m}} \times 1.15 \, \cancel{m} = 0.60°C$$

The temperature at which this seawater boils is $0.60°C$ higher than the normal boiling point of pure water: $100.0°C + 0.60°C = 100.6°C$.

Think about It As predicted, the boiling point of the solution is only slightly higher than the boiling point of the pure solvent.

Practice Exercise Crude oil pumped out of the ground may be accompanied by *formation water*, a solution that contains high concentrations of NaCl and other salts. If the boiling point of a sample of formation water is $2.3°C$ above the boiling point of pure water, what is the molality of dissolved particles in the sample?

Freezing Point Depression

As we mentioned when discussing Figure 11.17, *freezing point depression* is a colligative property that is put to good use in car radiators to ensure that they do not freeze in cold weather. The magnitude of the freezing point depression is directly proportional to the molal concentration of dissolved solute:

$$\Delta T_f = K_f m \tag{11.11}$$

where ΔT_f is the change in the freezing temperature of the solvent; K_f is the *freezing-point-depression constant* of the solvent; and m is the molality of the solution. Figure 11.19 summarizes calculations of freezing point depression and boiling point elevation.

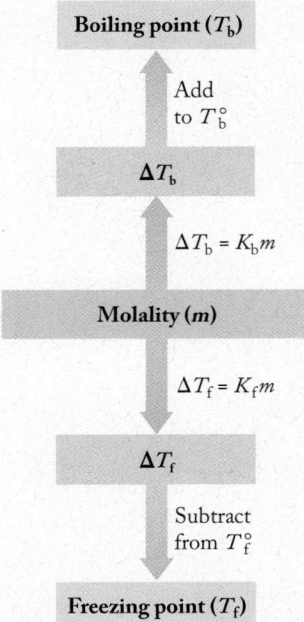

FIGURE 11.19 Flow diagram for calculating freezing points and boiling points of solutions of known molality ($T_b°$ = normal boiling point; $T_f°$ = normal freezing point).

SAMPLE EXERCISE 11.10   **Calculating the Freezing Point of a Solution**

What is the freezing point of radiator fluid prepared by mixing 1.00 L of ethylene glycol ($HOCH_2CH_2OH$, density 1.114 g/mL) with 1.00 L of water (density 1.000 g/mL)? The freezing-point-depression constant of water, K_f, is 1.86°C/m.

Collect and Organize We are asked to determine the freezing point of a solution of ethylene glycol in water. We are given the volumes of the two liquids, their densities, and K_f for water, and we have Equation 11.11.

Analyze To determine the freezing point, we need to calculate ΔT_f for the solution and then subtract that value from water's normal freezing point. Because using Equation 11.11 requires us to know m, the molality of the solution, we need to convert volumes of solute and solvent into moles of solute and kilograms of solvent; knowing their densities allows us to make both conversions, and we use ethylene glycol's formula to calculate its molar mass. Based on Sample Exercise 11.4, we choose water as the solvent. It is a logical choice, because ethylene glycol has a much higher molar mass and a density very similar to water, so we can assume that 1 L of ethylene glycol contains fewer moles of material than 1 L of water. The relatively large volume (10^3 mL) and a density close to 1 g/mL for the solute allow us to predict that the mass of solute will be on the order of 10^3 g. With a solute molar mass of 62 g/mol, the molality of the solution should be between 10 and 20 m. Considering $K_f = 1.86$°C/m, we predict ΔT_f to be in the range of 20° to 40°C.

Solve The solvent mass is

$$1.00 \; \cancel{L} \times \frac{1 \; \cancel{mL}}{0.001 \; \cancel{L}} \times \frac{1.000 \; \cancel{g}}{1 \; \cancel{mL}} \times \frac{0.001 \; kg}{1 \; \cancel{g}} = 1.00 \; kg$$

After calculating the ethylene glycol molar mass to be 62.07 g/mol, we have for the solute

$$1.00 \; \cancel{L} \; \text{solute} \times \frac{1000 \; \cancel{mL}}{1 \; \cancel{L}} \times \frac{1.114 \; \cancel{g}}{1 \; \cancel{mL}} \times \frac{1 \; mol}{62.07 \; \cancel{g}} = 17.9 \; \text{mol solute}$$

Therefore the molal concentration is

$$m = \frac{17.9 \; \text{mol solute}}{1.00 \; \text{kg solvent}} = 17.9 \; m$$

Using Equation 11.11 gives

$$\Delta T_f = K_f m$$
$$= \frac{1.86°C}{\cancel{m}} \times 17.9 \; \cancel{m} = 33.3°C$$

Subtracting this temperature change from the normal freezing point of water, 0.0°C, we have

Freezing point of radiator fluid = 0.0°C − 33.3°C = −33.3°C

Think about It The answer makes sense because the reason we add antifreeze to radiators is to depress the freezing point of water, and the solution we evaluated here certainly has a freezing point lower than that of pure water. To be an effective radiator fluid, the solution must have a freezing point sufficiently below the coldest expected temperatures, so our answer of −33°C is reasonable and matches our prediction.

Practice Exercise What is the boiling point of the automobile radiator fluid in the preceding Sample Exercise? The K_b of water is 0.52°C/m.

The van't Hoff Factor

Recall that some compounds, called *electrolytes*, dissociate into ions when in solution, whereas *nonelectrolytes* remain intact. Think about what this means in terms of how we count the number of dissolved particles that various solutes produce upon dissolution. If a solute is a nonelectrolyte, then every mole of solute yields one mole of particles. If, however, the solute is an electrolyte, then the number of moles of particles depends on the chemical formula of the compound. For example, if we approximate seawater as 0.574 m sodium chloride, then each kilogram of water contains 1.148 moles of particles because NaCl forms Na^+ and Cl^- ions on dissolving.

Because freezing point depression and boiling point elevation are colligative properties, the dissolution of 1 mole of a strong electrolyte such as NaCl in a given quantity of water produces the same changes in freezing point and boiling point as 1 mole of the electrolyte KNO_3, even though the latter has a much higher molar mass. Each of these salts adds 2 moles of particles (1 mole of cations and 1 mole of anions) to the water for every 1 mole of salt that dissolves. However, when a nonelectrolyte such as ethylene glycol is dissolved in water, 1 mole of the solute produces only 1 mole of particles (1 mole of molecules) in the solution. Therefore, the boiling point and freezing point of 2 m ethylene glycol solution are nearly the same as the boiling point and freezing point of 1 m NaCl or 1 m KNO_3.

Dutch chemist Jacobus van't Hoff (1852–1911) studied colligative properties and defined a term *i*, now called the **van't Hoff factor** (or *i* **factor**), that is a ratio of the experimentally measured value of a colligative property to the value expected if the solute were a nonelectrolyte (that is, no dissociation into ions). For example, suppose we make a solution that is 0.010 m in NaCl by dissolving 0.5844 g of NaCl in 1 kg of water. If we treat the NaCl as a nonelectrolyte, then according to Equation 11.11, the freezing point of the solution should be lower than that of pure water by

$$\Delta T_f = K_f m = \frac{1.86°C}{1 \text{ mol NaCl/1 kg water}} \times \frac{0.0100 \text{ mol NaCl}}{1 \text{ kg water}} = 0.0186°C$$

However, the freezing point of the solution measured in the laboratory is −0.0372°C, which is twice as low as the value predicted for a nonelectrolyte. The ratio of the experimentally measured value to the theoretical value, which is the $K_f m$ term in Equation 11.11,

$$\frac{\Delta T_{f,\text{measured}}}{K_f m} = \frac{0.0372°C}{0.0186°C} = 2 = i$$

is the van't Hoff factor for sodium chloride. This makes sense, because we know that sodium chloride is a strong electrolyte and forms two moles of ions from each mole of NaCl.

The significance of the *i* factor is based on the definition of colligative properties: changes due to the *total concentration of dissolved particles* present in solution. When 1 mole of a strong electrolyte such as solid sodium chloride dissolves, it produces 2 moles of particles. Thus, we should expect that the freezing point depression (or boiling point elevation) that results from the dissolution of a given

CONNECTION In Chapter 4, we defined an electrolyte as a substance that dissociates into ions when it dissolves.

▶❙❙ **CHEMTOUR** Boiling and Freezing Points

van't Hoff factor (also called *i* factor) the ratio of the experimentally measured value of a colligative property to the theoretical value expected for that property if the solute were a nonelectrolyte.

number of moles of NaCl should be twice as great as that produced when the same number of moles of a nonelectrolyte dissolves.

Equations 11.10 and 11.11 can be modified to include the i factor:

$$\Delta T_b = iK_b m \qquad (11.12)$$

$$\Delta T_f = iK_f m \qquad (11.13)$$

If the solute is molecular (such as ethylene glycol) and therefore a nonelectrolyte, then $i = 1$ because each mole of solute produces 1 mole of dissolved particles. If the solute is a strong electrolyte, then $i =$ the number of ions in one formula unit. For NaCl, therefore, $i = 2$; for Na_2SO_4, $i = 3$ (two Na^+ ions and one SO_4^{2-} ion). Note that we do *not* break a polyatomic ion such as SO_4^{2-} into its atoms when determining an i factor; polyatomic ions stay intact.

SAMPLE EXERCISE 11.11   Using the van't Hoff Factor

The salt lithium perchlorate ($LiClO_4$) is one of the most water-soluble salts known. At what temperature does a 0.130 m solution of $LiClO_4$ freeze? The K_f of water is 1.86°C/m; assume $i = 2$ for $LiClO_4$ and the freezing point of pure water is 0.00°C.

Collect and Organize We are asked to determine the freezing point of a salt solution. We know the formula of the solute, its molal concentration, and the K_f of the solvent. We know that the freezing point of the pure solvent is 0.00°C. Equation 11.13 relates freezing point depression to solute concentration when we take the van't Hoff factor into consideration. We are given the value of the van't Hoff factor as $i = 2$.

Analyze We need Equation 11.13 to solve for the freezing point depression. The value of K_f is close to 2, the value of i is 2, and the concentration of solute is close to 0.1 m, so we predict that the freezing point of the solution will be about 0.4°C lower than that of pure water.

Solve

$$\Delta T_f = iK_f m = (2)(1.86°C/m)(0.130\ m) = 0.48°C$$

The freezing point of the solution is $(0.00 - 0.48)°C = -0.5°C$.

Think about It Lithium perchlorate dissolves in aqueous solution to form 2 moles of ions for every mole of solute that dissolves: 1 mole of Li^+ cations and 1 mole of ClO_4^- anions, consistent with $i = 2$. The calculated value is consistent with our prediction.

Practice Exercise Determine the value of the van't Hoff factor and calculate the boiling point of a 1.75 m aqueous solution of barium nitrate, $Ba(NO_3)_2$. The K_b of water is 0.52°C/m.

CONCEPT TEST

Which aqueous solution has the lowest freezing point: (a) 3 m glucose ($C_6H_{12}O_6$), (b) 2 m potassium iodide (KI), or (c) 1 m sodium sulfate (Na_2SO_4)?

ion pair a cluster formed when a cation and an anion associate with each other in solution.

Using Equations 11.12 and 11.13 to calculate freezing point depressions and boiling point elevations for concentrated solutions of strong electrolytes often gives values smaller than the experimentally measured values. The reason is that the cations and anions produced when strong electrolytes dissolve may not be

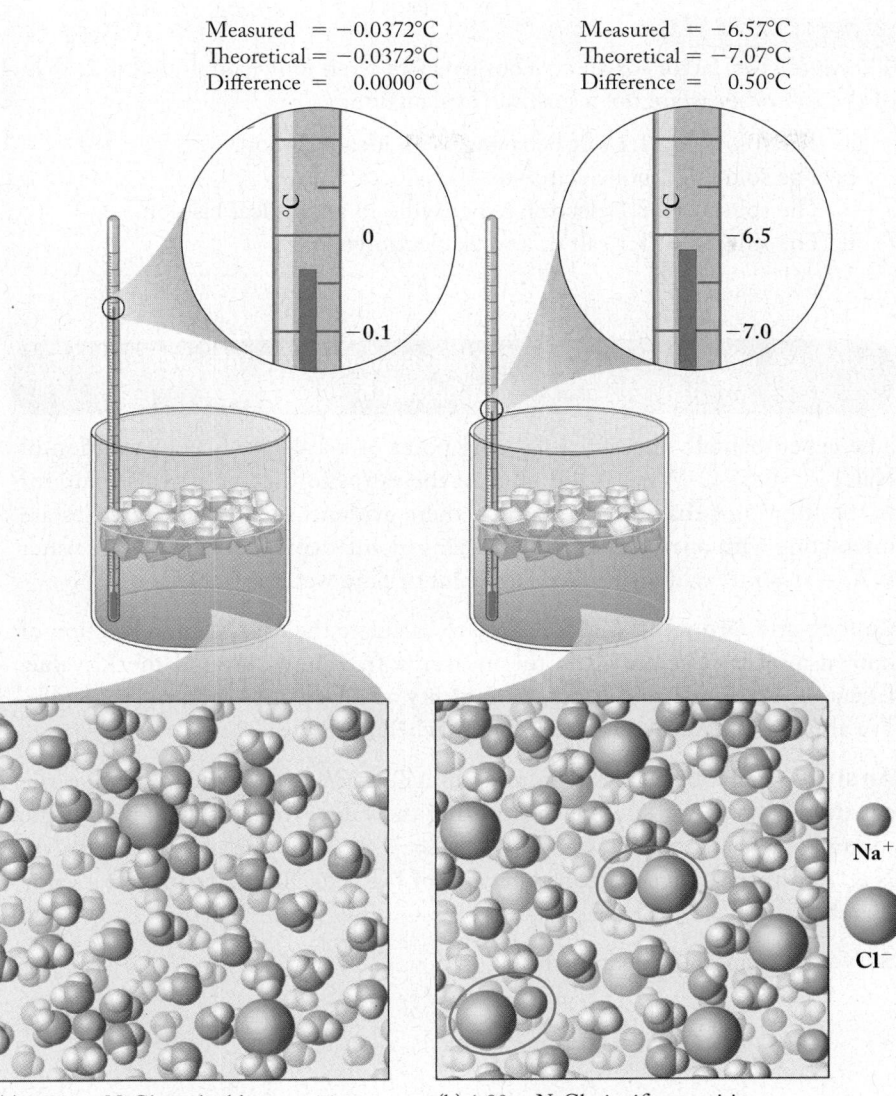

Measured   = −0.0372°C
Theoretical = −0.0372°C
Difference =   0.0000°C

Measured   = −6.57°C
Theoretical = −7.07°C
Difference =   0.50°C

Na⁺

Cl⁻

(a) 0.010 *m* NaCl: negligible ion pairing

(b) 1.90 *m* NaCl: significant pairing

FIGURE 11.20 (a) The experimentally measured freezing point of a 0.010 *m* solution of NaCl is the same as the theoretical value obtained with Equation 11.13. This agreement means that the solution behaves ideally and little or no ion pairing takes place. (b) The experimentally measured freezing point of a 1.90 *m* solution is about 0.5°C higher than the theoretical value because some of the Na^+ and Cl^- ions form ion pair, as shown inside the red ovals. The formation of ion pairs causes the concentration of solute particles to be less than the theoretical number, and as a result the van't Hoff factor for the solution is less than the theoretical value of 2, and the decrease in the freezing point is less than expected.

totally independent of one another. As concentration increases, cations and anions may form ionic clusters. The simplest cluster, an **ion pair**, consists of a cation and an anion that associate in solution, acting as a single particle. Thus, the overall concentration of particles is reduced when ion pairs form, and experimentally measured freezing point depressions and boiling point elevations are smaller than the theoretical values obtained with Equations 11.12 and 11.13 (Figure 11.20).

The extent to which free ions form when a strong electrolyte dissolves is expressed by the van't Hoff factor. The van't Hoff factor for NaCl in water is 2 if the solution behaves ideally, because ideally 2 moles of ions are produced for each mole of NaCl that dissolves. The value of *i* is 2 for 0.0100 *m* NaCl, but only 1.9 for 0.100 *m* NaCl. Whenever a calculation for *i* gives a non-integer value, solute particles are associating in solution and the behavior is nonideal. Figure 11.21 gives some theoretical and experimentally measured values of the van't Hoff factor for several substances.

FIGURE 11.21 Theoretical and experimentally measured values for the van't Hoff factors for 0.1 *m* solutions of several electrolytes and the nonelectrolyte ethanol. The higher the charge on the ions, the greater the difference between theoretical and experimentally measured values.

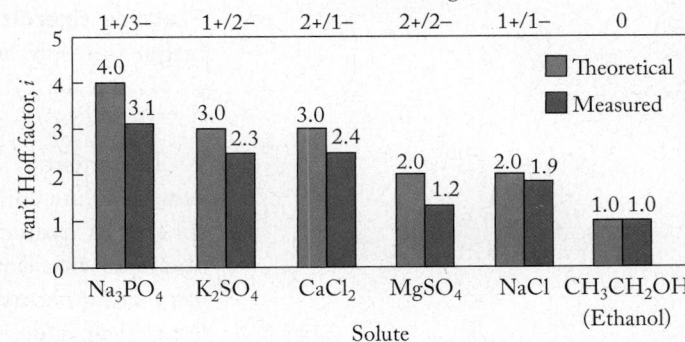

CONCEPT TEST

The van't Hoff factor for an aqueous solution of an ionic compound is 2. Which of the following is/are not a possible explanation?

    a. The solute is a 1:1 salt behaving as an ideal solution.
    b. The solute is a nonelectrolyte.
    c. The solute is a 2:1 electrolyte behaving in a nonideal fashion.
    d. The solute is a 1:1 salt of a weak electrolyte.

SAMPLE EXERCISE 11.12   **Assessing Particle Interactions in Solution**

The experimentally measured freezing point of a 1.90 m aqueous solution of NaCl is −6.57°C. What is the value of the van't Hoff factor for this solution? Is the solution behaving ideally, or is there evidence that solute particles are interacting with one another? The freezing-point-depression constant of water is $K_f = 1.86°C/m$, and the freezing point of pure water is 0.00°C.

Collect and Organize We are asked to calculate the i factor for a solution of known molality. We are given the measured freezing point and the K_f value. Equation 11.13 relates ΔT_f, K_f, the molality of the solution, and the value of i. We are also asked whether the solution is behaving ideally.

Analyze Equation 11.13 tells us that $\Delta T_f = iK_f m$. We can rearrange the equation to solve for i before substituting the values for ΔT_f, K_f, and m. If the solution behaves ideally, we would expect $i = 2$ for the strong electrolyte NaCl; however, given the concentration of NaCl (1.90 m), we predict a value less than 2.

Solve Rearranging Equation 11.13 gives us

$$i = \frac{\Delta T_{f,\text{measured}}}{K_f m}$$

$$i = \frac{6.57°\text{C}}{\left(\dfrac{1.86°\text{C}}{m}\right) 1.90\ m} = 1.86$$

That i is not an integer tells us that the solution is not behaving ideally. Ion pairs must be forming in solution.

Think about It As predicted, the value of i for this solution is less than the theoretical value of 2.

Practice Exercise The van't Hoff factor for a 0.050 m aqueous solution of magnesium sulfate is 1.3. What is the freezing point of the solution?

The extent of ion pairing in a solution of a strong electrolyte generally increases with solute concentration as the solution "runs out of water" needed to form spheres of hydration around the ions. For any salt, the theoretical value of i obtained with Equation 11.12 or 11.13 is an upper limit of possible values. If ion pairing occurs, the experimentally measured value must be smaller than the theoretical value.

Osmosis and Osmotic Pressure

The final colligative property we look at in this chapter is *osmotic pressure*, the result of the process called **osmosis**—the movement of a solvent through a semipermeable membrane from a region of low solute concentration to a region of higher solute concentration. A *semipermeable membrane* is one in which the pores are so small that water molecules can pass through them but most solute particles are too large to pass through. As an example of the importance of osmosis, we start with the observation that we all need water to survive. More than 97% of the water on Earth is seawater, and none of it is fit to drink. Let's look at the reason on a cellular level. The liquid inside each cell in the body is a complex solution of numerous solutes, with the average concentration of these solutes being about one-third the concentration of solutes in seawater. When cells are exposed to seawater, this substantial difference in solute concentration is the driving force for osmosis, with the cell membrane acting as the necessary semipermeable membrane (Figure 11.22). Because the solute concentration is higher outside a cell, water in the cell crosses the cell membrane and enters the seawater, moving from the low-solute-concentration side of the membrane to the high-solute-concentration side. As water leaves the cell, it shrivels and ultimately ceases to function.

> **CONCEPT TEST**
>
> Why do cucumbers shrivel when pickled in brine (salt water)?

Water molecules migrate through a cell membrane or through any other semipermeable membrane because a force makes it happen—a force caused by the different concentrations of solutes on the two sides of the membrane. When we divide the magnitude of this force F by the surface area A of the membrane, we get pressure: $P = F/A$. **Osmotic pressure (π)** is the pressure required to halt the flow of solvent from the more dilute solution across the membrane (Figure 11.23). Osmotic pressure exactly balances the pressure (F/A) driving solvent through the membrane so that no net flow of solvent takes place.

The Greek letter pi (π) is used as the symbol for osmotic pressure to distinguish it from the pressure exerted by gases. In Figure 11.23(a), the solution on the right has a lower concentration and lower osmotic pressure (π_{NaCl}) than the solution on

osmosis the flow of a fluid through a semipermeable membrane to balance the concentration of solutes in solutions on the two sides of the membrane. The solvent molecules' flow proceeds from the more dilute solution into the more concentrated one.

osmotic pressure (π) the pressure applied across a semipermeable membrane to stop the flow of solvent from the compartment containing pure solvent or a less concentrated solution to the compartment containing a more concentrated solution. The osmotic pressure of a solution increases with solute concentration M and with solution temperature T.

▶❚❚ **CHEMTOUR** Osmotic Pressure

FIGURE 11.22 The membrane of a red blood cell is semipermeable, which means that water easily flows by osmosis into and out of the cell to equalize the solute concentrations on the two sides of the membrane. (a) When a cell is bathed in a solution in which the solute concentration equals the solute concentration inside the cell (isotonic conditions), the flow of water into the cell is exactly balanced by the flow of water out of the cell, and the cell size does not change. (b) When the cell is bathed in a solution in which the solute concentration is higher than the solute concentration inside the cell (hypertonic conditions), water flows by osmosis from the region of low solute concentration to the region of high solute concentration—in other words, out of the cell—and the cell shrinks. (c) When the cell is bathed in pure water, the solute concentration is higher inside the cell than outside (hypotonic conditions), water flows by osmosis from the region of zero solute concentration to the region of high solute concentration—into the cell—and the cell expands.

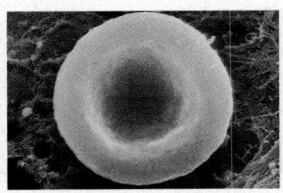
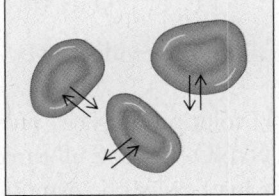

(a) Isotonic: total solute concentration in the solution matches that inside the cell

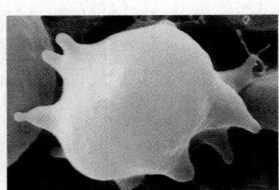
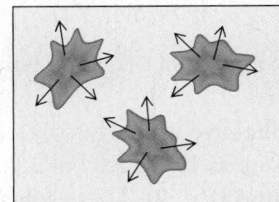

(b) Hypertonic: total solute concentration in the solution is greater than that inside the cell

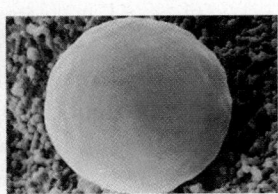
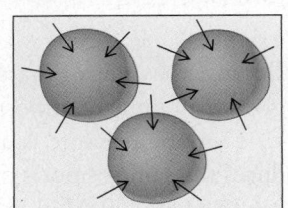

(c) Hypotonic: total solute concentration in the solution is less than that inside the cell

FIGURE 11.23 (a) The solute concentration of seawater is approximately 1.15 M. When equal volumes of seawater and a 0.10 M solution of NaCl are separated by a semipermeable membrane, water moves by osmosis from 0.10 M NaCl (low solute concentration) to the seawater (high solute concentration) side. (b) The volume on the NaCl side decreases until the osmotic pressure of the solution equals the osmotic pressure (and concentration) of the diluted seawater solution, resulting in a difference in the heights of the liquid levels. This difference in volume is proportional to the original difference in osmotic pressures ($\Delta\pi$) between the two solutions.

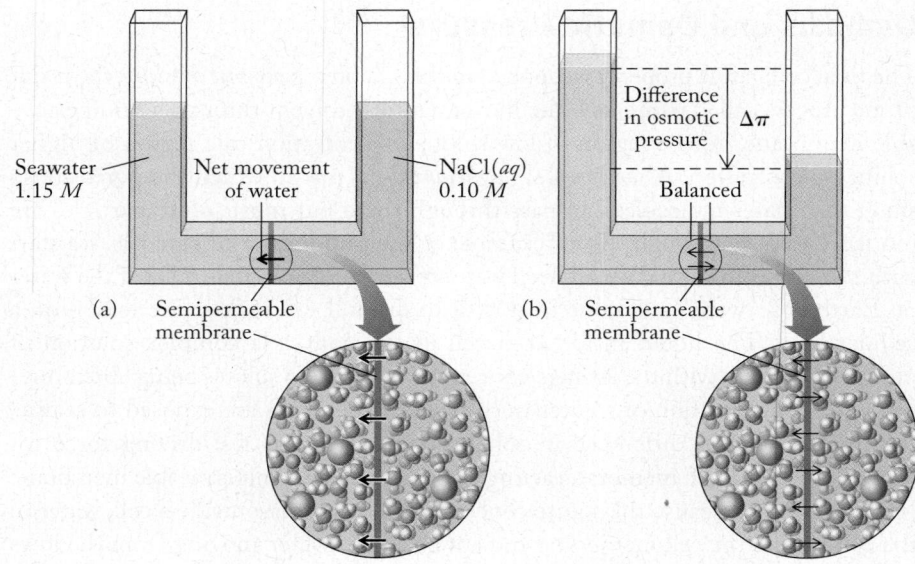

the left (π_{seawater}): $\pi_{\text{NaCl}} < \pi_{\text{seawater}}$. The result is a net flow of solvent (water) from right to left until the pressure on both sides is equal. Figure 11.23(b) shows how the volumes of both solutions change: the difference in volume reflects the difference in osmotic pressures, $\Delta\pi = \pi_{\text{seawater}} - \pi_{\text{NaCl}}$, in the original solutions. This process is profoundly important in living systems, where the solution inside each cell exerts an osmotic pressure on the cell membrane, a pressure pushing toward the outside of the cell. At the same time, blood or other liquid outside the cell exerts an osmotic pressure on the cell membrane, and this pressure pushes toward the cell's interior.

The magnitude of the osmotic pressure π required to stop the net flow of solvent across a semipermeable membrane separating a pure solvent from a solution depends on the solute concentration, the absolute temperature, and the constant R, 0.0821 L · atm/(mol · K):

$$\pi = MRT$$

where M is the molarity of the solute. Molarity is used to express concentration in calculating π because the expression for π can be derived from an equation similar to the ideal gas equation:

$$\pi = P = \left(\frac{n}{V}\right)RT = MRT$$

Note that the term n/V has units of moles per liter, which matches the definition of molarity. Osmotic pressure is a colligative property because π is proportional to the concentration of solute and does not depend on the identity of the solute. Therefore the molarity must be multiplied by the van't Hoff factor i for the solute:

$$\pi = iMRT \qquad (11.14)$$

Because the units of R are L · atm/(mol · K) and i has no units, the product $iMRT$ gives π in units of atmospheres.

Osmotic pressure is a colligative property, so a 1.0 M solution of NaCl produces the same osmotic pressure as 1.0 M KCl or 1.0 M NaNO$_3$ because all three solutions are 2.0 M in total ions ($i = 2$). These solutions have twice the osmotic pressure of a 1.0 M solution of glucose because glucose is a molecular substance ($i = 1$) and produces a solution that is only 1.0 M in dissolved particles (glucose molecules).

If a solution that is 1.0 M in glucose is on one side of a semipermeable membrane and a solution that is 1.0 M in KCl is on the other side, in which direction does the water flow?

SAMPLE EXERCISE 11.13   Calculating Osmotic Pressure I

At the beginning of this section, we mentioned that the concentration of solutes in a red blood cell is about a third of that of seawater—more precisely, about 0.30 M. If red blood cells are bathed in pure water, they swell, as shown in Figure 11.22(c). Calculate the osmotic pressure at 25°C of red blood cells across the cell membrane from pure water.

Collect and Organize We are to calculate an osmotic pressure across a membrane that separates pure water from red blood cells. The osmotic pressure of a solution is related to the total particle concentration (0.30 M) and temperature (25°C) of the solution by Equation 11.14. We must convert the temperature to kelvins.

Analyze Osmotic pressure is related to the total concentration of all particles in solution and the absolute temperature. The total particle concentration, 0.30 M, represents the iM term in Equation 11.14.

Solve First we need to convert the temperature in degrees Celsius to kelvins: $T(°C) + 273 = T(K)$. Inserting the values of iM, R, and T in Equation 11.14 we have

$$\pi = iMRT = \frac{0.30 \text{ mol}}{\text{L}} \times \frac{0.0821 \text{ L} \cdot \text{atm}}{\text{mol} \cdot \text{K}} \times (25 + 273)\text{K} = 7.3 \text{ atm}$$

Think about It The calculated pressure across the membranes of red blood cells in pure water is over 7 atmospheres: enough to rupture the membranes.

Practice Exercise Calculate the osmotic pressure across a semipermeable membrane separating pure water from seawater at 25°C. The total concentration of all the ions in seawater is 1.15 M.

Which has the greater effect on the osmotic pressure of a 1.0 M solution: increasing the temperature from 10°C to 20°C, or adding enough solute to raise the concentration to 2.0 M?

Calculating the osmotic pressure of a solution relative to pure solvent is straightforward. In Figure 11.23(a), however, the two solutions have different osmotic pressures, π_{NaCl} and $\pi_{seawater}$. Let's derive an equation for the difference between the osmotic pressure $\Delta\pi$ in terms of M, R, and T for the situation depicted in Figure 11.23.

Earlier in this section we said that $\Delta\pi = \pi_{seawater} - \pi_{NaCl}$ and that the osmotic pressures of both solutions are described by Equation 11.14:

$$\pi_{seawater} = iM_{seawater}RT \quad \text{and} \quad \pi_{NaCl} = iM_{NaCl}RT$$

Taking the difference in π expressions:

$$\Delta\pi = \pi_{\text{seawater}} - \pi_{\text{NaCl}} = iM_{\text{seawater}}RT - iM_{\text{NaCl}}RT$$

$$\Delta\pi = (iM_{\text{seawater}} - iM_{\text{NaCl}})RT \tag{11.15}$$

The total ion concentration in seawater is about 1.15 M; therefore, iM_{seawater} = 1.15 M. If a solution of that concentration is put on one side of a semipermeable membrane and a solution of 0.10 M NaCl on the other side, then iM_{NaCl} = $(2 \times 0.10\ M)$, and the osmotic pressure difference between the two solutions at 25°C is

$$\Delta\pi = (iM_{\text{seawater}} - iM_{\text{NaCl}})RT$$

$$= \left(1.15\ \frac{\text{mol}}{\text{L}} - 2 \times 0.10\ \frac{\text{mol}}{\text{L}}\right)\left(0.0821\ \frac{\text{L} \cdot \text{atm}}{\text{mol} \cdot \text{K}}\right)(298\ \text{K})$$

$$= 23\ \text{atm}$$

SAMPLE EXERCISE 11.14 Calculating Osmotic Pressure II

Red blood cells placed in seawater shrivel as shown in Figure 11.22(b). Calculate the osmotic pressure across the semipermeable cell membrane, separating the solution inside a red blood cell from seawater at 25°C if the total concentration of all the particles inside the cell is 0.30 M and the total concentration of ions in seawater is 1.15 M. Compare the result with the answer from Sample Exercise 11.13.

Collect and Organize We are to calculate an osmotic pressure across a membrane that separates two solutions with different total concentrations of particles at a constant temperature of 25°C. We can use Equation 11.15 to calculate the difference in osmotic pressure between the two solutions. As in Sample Exercise 11.13, we must convert the temperature to units of kelvins.

Analyze We know that the concentration of particles in seawater is greater than inside a red blood cell, so solvent will flow from the cell to the seawater. Thus, the osmotic pressure in the cell (π_{cell}) will be less than the osmotic pressure of the seawater (π_{seawater}) and the osmotic pressure across the membrane will be the difference between π_{seawater} and π_{cell}, or $\Delta\pi$. The value of $\Delta\pi$ should be greater than the π calculated in Sample Exercise 11.13, where red blood cells were bathed in pure water because the difference in solute concentrations is greater.

Solve First we need to convert the temperature in degrees Celsius to kelvins: $T(°C) + 273 = T(\text{K})$. Inserting the values of iM, R, and T in Equation 11.15 for seawater and cells we have

$$\Delta\pi = (iM_{\text{seawater}} - iM_{\text{cell}})RT$$

$$= \left(1.15\ \frac{\text{mol}}{\text{L}} - 0.30\ \frac{\text{mol}}{\text{L}}\right)\left(0.0821\ \frac{\text{L} \cdot \text{atm}}{\text{mol} \cdot \text{K}}\right)(298\ \text{K})$$

$$= 21\ \text{atm}$$

Think about It As predicted, the calculated pressure across a semipermeable membrane separating seawater from red blood cells, 21 atm, is greater than the osmotic pressure across a semipermeable membrane separating red blood cells from pure water (7.3 atm).

Practice Exercise Calculate the osmotic pressure at 25°C across a semipermeable membrane separating seawater (1.15 M total particles) from a 0.50 M solution of aqueous NaCl.

The osmotic pressures across cell membranes in Sample Exercises 11.13 and 11.14 are very large. The pressure of over 7 atm calculated in Sample Exercise 11.13 for red blood cells immersed in pure water is more than three times the air pressure in a typical automobile tire and about the same as the water pressure experienced by a diver at a depth of 82 m. A pressure of 21 atm across the walls of a steel-reinforced concrete building is sufficient to cause major structural damage or even collapse the building. The possibility of serious structural damage to cells as a result of such huge pressure differentials across membranes is one reason why solutions dispensed intravenously (IV) or intramuscularly (IM) must be carefully constituted.

During a medical emergency, medication may need to be administered to a patient intravenously (Figure 11.24), and it is crucial that the osmotic pressure exerted by the intravenous solution on the body's cells be identical to the osmotic pressure exerted by the solution inside the cells. The solute concentrations in such solutions are said to be *isotonic* because they exert the same osmotic pressure as the blood exerts. As we saw in Figure 11.22, solutions with higher (*hypertonic*) or lower (*hypotonic*) solute concentrations cause the body's cells to either shrink as water leaves or swell as water enters the cell.

Two solutions are widely used to administer intravenous medications depending on the clinical situation. One is physiological saline, which contains 0.92% NaCl by mass: 0.92 g NaCl for every 100 g of solution. The density of dilute aqueous solutions is close to 1.00 g/mL, so 100 g of this solution has a volume of 100 mL, which means a concentration of 0.92 g/100 g is nearly the same as 0.92 g/100 mL = 9.2 g/L.

Another common IV solution is called D5W. The acronym stands for a 5.5% solution by mass of dextrose (another name for glucose; molar mass 180.16 g/mol) in water. Because this solution must be isotonic with blood and therefore isotonic with physiological saline solution, it must contain about the same concentration of solute particles as saline solution. A 5.5% by mass concentration means 5.5 g dextrose/100 g water, or 5.5 g dextrose/100 mL water, or 55 g dextrose/L.

To compare the solute levels of these two solutions, we compare molarities:

Physiological saline: $\dfrac{9.2\ \cancel{g}}{1\ L} \times \dfrac{1\ mol\ NaCl}{58.44\ \cancel{g}\ NaCl} = 0.16\ M$

D5W: $\dfrac{55\ \cancel{g}}{1\ L} \times \dfrac{1\ mol\ D}{180.16\ \cancel{g}\ D} = 0.31\ M$

With the molar concentration of dextrose about twice that of the saline solution, how can both solutions match the solute level in blood? The answer comes again from the definition of a colligative property and the role of the i factor. We must look at the number of particles in solution in both cases. Sodium chloride is a salt with $i = 2$. Dextrose is a sugar, a molecular material, and its molar concentration

FIGURE 11.24 A solution of physiological saline has a concentration of 0.92 g NaCl per 100 g of solution. The concentration of ions in this solution is equal to the concentration of ions in blood plasma. This solution is *isotonic* with blood plasma.

reverse osmosis a water purification process in which water is forced through semipermeable membranes, leaving dissolved impurities behind.

directly reflects the number of sugar molecules in solution; $i = 1$. The total particle concentration in the NaCl solution is two times the molarity, or about $0.32\ M$, which is very close to the molar concentration of particles (molecules) in the dextrose solution. Therefore, the solute concentration in 0.92% NaCl is the same as the solute concentration in 5.5% D5W, which means that the two solutions exert the same osmotic pressure.

Reverse Osmosis

The sea is the source of drinking water in several desert countries bordering the Persian Gulf. To make the seawater drinkable, it must be *desalinated*, which means the salts must be removed. One way to desalinate saltwater is to distill it, but distillation requires a great deal of energy to heat seawater to its boiling point and convert it to steam.

The process of osmosis can also be applied to the desalination of seawater. As we have seen, osmosis is the movement of water from a region of low solute concentration to a region of high solute concentration. However, if a sufficiently high pressure is applied to the region of high solute concentration, the water can be forced to move from the region of high solute concentration to the region of low solute concentration. This technique, called **reverse osmosis**, is another way of desalinating seawater to make it potable. The apparatus shown in Figure 11.25 consists of an outer metal tube containing a large number of inner tubes made of a semipermeable membrane. Seawater is forced through the outer tube so that it washes over the exterior of all the inner tubes, which are initially filled with flowing pure water. Ordinarily, the direction of osmosis would be from the inner tubes (zero solute concentration) into the seawater (very high solute concentration), and indeed, the pure water does exert an osmotic pressure on the interior of the tube membranes. However, when an external pressure—a *reverse osmotic pressure*—greater than the osmotic pressure is exerted on the seawater side of the membranes, water molecules in the seawater move across the membranes into the inner tubes. The desalinated water entering the inner tubes flows into a collector and is ready for use.

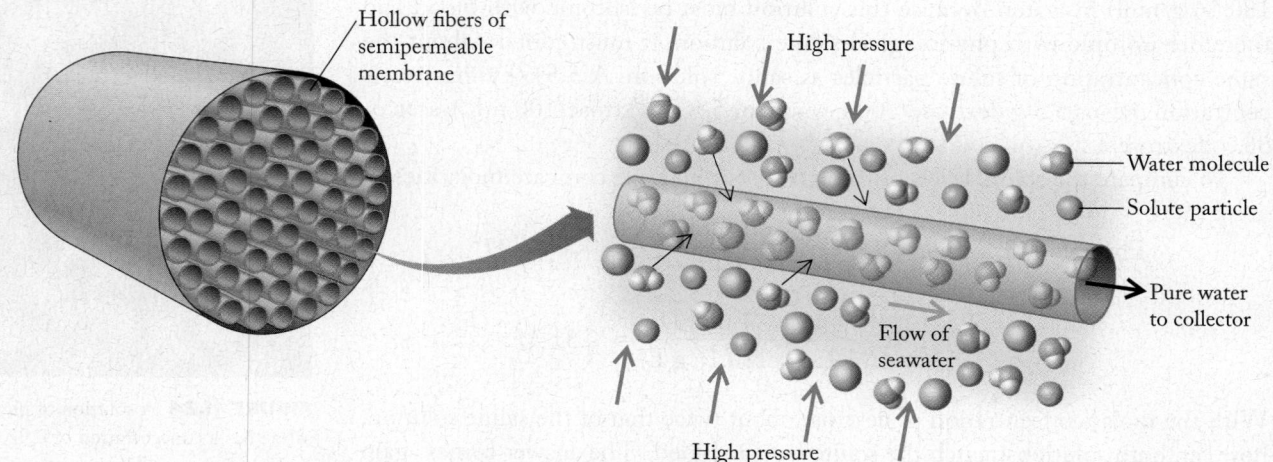

FIGURE 11.25 Seawater being desalinated by reverse osmosis flows at a pressure greater than its osmotic pressure around bundles of tubes with semipermeable walls. Water molecules pass from the seawater into the tubes and flow through the tubes to a collection vessel.

Some municipal water-supply systems use reverse osmosis to make saline (brackish) water fit to drink. Some industries use this method to purify conventional tap water, and it is a common way for ships at sea to desalinate ocean water. However, very tough semipermeable membranes are necessary because reverse osmosis systems operate, as we have discussed, at very high pressures.

SAMPLE EXERCISE 11.15   Calculating Pressure for Reverse Osmosis

What is the reverse osmotic pressure required at 20°C to purify brackish well water containing 0.355 M dissolved particles if the product water is to contain no more than 87 mg of dissolved solids (as NaCl) per liter?

Collect and Organize We are asked to calculate the reverse osmotic pressure needed to purify water to a stated solute concentration. We are given the temperature, the solute concentration in the water to be purified, and the amount of solute (expressed as NaCl) tolerable in the product water. We can adapt Equation 11.15 to calculate the difference in osmotic pressure between the two solutions.

Analyze We need the molarity of the less concentrated solution, the drinkable water, to calculate the osmotic pressure exerted by the drinkable water and we can calculate that molarity from the information given. We can also calculate the osmotic pressure once we know the difference in the molarities of the solutions on the two sides of the semipermeable membrane.

Solve First we convert 87 mg NaCl/L to molarity:

$$\frac{87 \text{ mg NaCl}}{1 \text{ L}} \times \frac{1.000 \text{ g}}{1000 \text{ mg}} \times \frac{1 \text{ mol NaCl}}{58.44 \text{ g NaCl}} = \frac{1.5 \times 10^{-3} \text{ mol NaCl}}{1 \text{ L}}$$

$$= 1.5 \times 10^{-3} \, M \text{ NaCl}$$

The total ion concentration for a $1.5 \times 10^{-3} \, M$ NaCl solution is $2(1.5 \times 10^{-3} \, M)$ = 0.0030 M. Using Equation 11.15:

$$\Delta \pi = \pi_{\text{brackish water}} - \pi_{\text{drinkable water}}$$

$$= (iM_{\text{brackish water}} - iM_{\text{drinkable water}})RT$$

$$= (0.355 \, M - 0.0030 \, M)\left(0.0821 \, \frac{\text{L} \cdot \text{atm}}{\text{mol} \cdot \text{K}}\right)(293 \text{ K})$$

$$= 8.47 \text{ atm}$$

If we maintain an osmotic pressure of exactly 8.47 atm on the side of the membrane with the brackish well water, no net flow of solvent will occur between the two solutions. Pressures greater than 8.47 atm will force water molecules from the well water through the membrane to give water containing less than 87 mg NaCl per liter. (The reason we can never obtain absolutely pure water in this process is that, in practice, some Na^+ and Cl^- ions pass through the membrane from the well water to the drinkable water.)

Think about It In the absence of any external pressure, solvent will flow from the product water to the well water, driven by the concentration difference between the two solutions as shown in Figure 11.23. However, if we apply sufficient external pressure we can reverse the flow of water. The value of the external pressure, about 8.5 atm, represents the minimum external pressure that must be applied to make this device function.

Practice Exercise Calculate the minimum external pressure that must be applied in a reverse osmosis system to seawater with a total ion concentration of 1.15 M at 20°C if the maximum concentration allowed in the product water is 174 mg NaCl per liter.

■ ┄┄┄┄┄┄┄┄┄┄┄┄┄┄┄┄┄┄┄┄┄┄┄┄┄┄┄┄┄┄

CONCEPT TEST

If you were trying to purify water by reverse osmosis using the apparatus in Figure 11.25, what advantage might there be in running the system at 50°C rather than 20°C?

11.5 Measuring the Molar Mass of a Solute Using Colligative Properties

CONNECTION In Chapter 3 we used molar masses determined by mass spectrometry to convert empirical formulas derived from elemental analyses to molecular formulas. In Chapter 6, we used measurements of density and the ideal gas law to calculate molar masses of gases.

In principle, the molar mass of any solute can be determined by dissolving a known quantity of the solute in a known quantity of solvent and then measuring the effect the dissolved solute has on any colligative property of the solvent. In practice, this method works only for nonelectrolytes, which have a van't Hoff factor of 1. Freezing-point-depression measurements, for example, can be used to find the molar mass of a molecular compound if it is sufficiently soluble in a solvent whose K_f value is known, as illustrated in Sample Exercise 11.16.

SAMPLE EXERCISE 11.16 **Using Freezing Point Depression to Determine Molar Mass**

Eicosene is a molecular compound and nonelectrolyte with the empirical formula CH_2. The freezing point of a solution prepared by dissolving 100 mg of eicosene in 1.00 g of benzene was 1.75°C lower than the freezing point of pure benzene. What is the molar mass of eicosene? (K_f for benzene is 4.90°C/m.)

Collect and Organize We are asked to determine the molar mass of a compound. We are given the mass of the compound that lowers the freezing point of a solvent by a known amount, and we have the K_f of the solvent. Because the solute is a nonelectrolyte, $i = 1$. Equation 11.13 relates concentration to the change in freezing point.

Analyze The molar mass of a compound is expressed in units of grams per mole. Our sample consists of 100 mg, or 0.100 g. To find the molar mass we need to determine how many moles of eicosene are contained in the sample. The concentration term in Equation 11.13 is molality, or moles of solute per kilogram of solvent, so we can use the experimental data to first calculate molality. When we know the molality of the eicosene, we know the number of moles of eicosene in 1 kg of benzene; we are given the mass of benzene used, so we can calculate the number of moles of eicosene in the sample, from which we may calculate the molecular mass of eicosene. The empirical formula of eicosene enables us to check the validity of our answer as the molar mass of eicosene must be a whole-number multiple of the mass of CH_2 (14 g/mol).

Solve The molality of the eicosene solution is

$$\Delta T_f = iK_f m = K_f m$$

$$m = \frac{\Delta T_f}{K_f} = \frac{1.75°\text{C}}{4.90°\text{C}/m} = 0.357 \ m \ \text{eicosene}$$

which means 0.357 mol of eicosene is dissolved per kilogram of solvent. Only 1.00 g (1.00×10^{-3} kg) was used in this sample. Calculating the moles of eicosene in the sample:

$$m = \frac{\text{moles of solute}}{\text{kilograms of solvent}}$$

$$0.357 \ m = \frac{\text{moles of eicosene}}{1.00 \times 10^{-3} \ \text{kg benzene}}$$

$$\text{Moles of eicosene} = \frac{0.357 \ \text{mol eicosene}}{1 \ \text{kg benzene}} \times 1.00 \times 10^{-3} \ \text{kg benzene}$$

$$= 3.57 \times 10^{-4} \ \text{mol eicosene}$$

Because the molar mass is the mass of one mole of eicosene, and 100 mg of eicosene was used to prepare the solution,

$$\text{Molar mass} = \frac{\text{mass of eicosene}}{\text{moles of eicosene}} = \frac{0.100 \ \text{g eicosene}}{3.57 \times 10^{-4} \ \text{mol}} = 280 \ \text{g/mol}$$

Think about It The molar mass of eicosene, 280 g/mol, is reasonable since it corresponds to $(CH_2)_n$ where $n = 20$. The molecular formula of eicosene is therefore $C_{20}H_{40}$.

Practice Exercise A solution prepared by dissolving 360 mg of a sugar (a molecular compound and a nonelectrolyte) in 1.00 g of water froze at $-3.72°C$. What is the molar mass of this sugar? The value of K_f of water is $1.86°C/m$.

For determining the molar mass of water-soluble substances, measuring osmotic pressure is a better choice than measuring either boiling point elevation or freezing point depression for a number of reasons. First, the K_f and K_b values for water are much smaller than those of other solvents (Table 11.4). Thus, in order to have a solution that gives a measurable boiling point or freezing point change for an aqueous solution, the concentration of the solution has to be much higher than readily achievable. Second, biomaterials such as proteins and carbohydrates are nearly always available only in small quantities, and they often are not very soluble in nonaqueous solvents that have larger K_f or K_b values. Furthermore, these biomaterials often have high molar masses, which means that large quantities are needed to give high

TABLE 11.4 Molal Freezing-Point-Depression and Boiling-Point-Elevation Constants for Selected Solvents

Solvent	Freezing Point (°C)	K_f (°C/m)	Boiling Point (°C)	K_b (°C/m)
Water (H_2O)	0.0	1.86	100.0	0.52
Benzene (C_6H_6)	5.5	4.90	80.1	2.53
Ethanol (CH_3CH_2OH)	−114.6	1.99	78.4	1.22
Carbon tetrachloride (CCl_4)	−22.3	29.8	76.8	5.02

enough molal concentrations for reliable ΔT_f or ΔT_b measurements. Third, a solute might need to be recovered unchanged for other uses, and so boiling-point-elevation measurements are ruled out for heat-sensitive solutes.

On the other hand, very small osmotic pressures can be measured precisely, the measurement equipment can be miniaturized so that only minute quantities of solute are needed, and the measurements can be made at room temperature.

SAMPLE EXERCISE 11.17 Using Osmotic Pressure to Determine Molar Mass

A molecular compound that is a nonelectrolyte was isolated from a South African tree. A 47 mg sample was dissolved in water to make 2.50 mL of solution at 25°C, and the osmotic pressure of the solution was 0.489 atm. Calculate the molar mass of the compound.

Collect and Organize We are given the mass of a substance, the volume of its aqueous solution, the temperature, and its osmotic pressure. We can relate these parameters using Equation 11.14, using the value $i = 1$ because this is a nonelectrolyte.

Analyze We can calculate the molar concentration of the solution from the information given and Equation 11.14. Since we know the solution volume, we can calculate the number of moles of the solute from the molarity. Because we know the mass of this number of moles, we can calculate the molar mass of the solute.

Solve Rearranging Equation 11.14 to isolate M and substituting the given values:

$$M = \pi/iRT = \frac{0.489 \text{ atm}}{(1)(0.0821 \text{ L} \cdot \text{atm}/\text{mol} \cdot \text{K})(298 \text{ K})}$$

$$= \frac{2.00 \times 10^{-2} \text{ mol}}{\text{L}} = 2.00 \times 10^{-2} \text{ M}$$

Next we solve the defining equation for molarity, $M = n/V$, for n, the number of moles of solute:

$$n = MV = \frac{2.00 \times 10^{-2} \text{ mol}}{1 \text{ L}} \times 2.50 \times 10^{-3} \text{ L} = 5.00 \times 10^{-5} \text{ mol}$$

We know that this number of moles of solute has a mass of 47 mg. The molar mass of the solute is therefore

$$\text{Molar mass} = \frac{\text{g solute}}{\text{moles of solute}}$$

$$= \frac{47 \times 10^{-3} \text{ g}}{5.00 \times 10^{-5} \text{ mol}} = 9.4 \times 10^2 \text{ g/mol}$$

Think about It One of the advantages of determining molar mass by osmotic pressure is that only a small amount of material is required. With only a 47 mg sample of a rather large molecule (molar mass 940 g/mol), the osmotic pressure is sufficiently large (0.489 atm) to enable us to calculate the molar mass accurately.

Practice Exercise A solution was made by dissolving 5.00 mg of a polysaccharide in water to give a final volume of 1.00 mL. The osmotic pressure of this solution was 1.91×10^{-3} atm at 25°C. Calculate the molar mass of the polysaccharide, which is a nonelectrolyte.

We very rarely deal with pure liquids in the real world. Milk, gasoline, tap water, shampoo, olive oil, cough syrup, and countless other fluids that we use in our daily lives are all solutions whose physical properties depend on the amounts of solutes dissolved in them. By understanding the role solutes play in determining the properties of solutions, we can make many extraordinarily useful materials that display exactly the behaviors we desire. The boiling and freezing points of water can be extended above 100°C and below 0°C to make antifreeze that takes heat away from an operating engine and also protects the fluid from freezing in the winter. Adding the right amount of dextrose to water for an IV drip keeps trauma patients hydrated. Having too much or too little solute in a solution can have catastrophic effects: engine blocks can crack if insufficient antifreeze is dissolved in the water in a radiator to protect the fluid from freezing; a patient can die if the concentration of dextrose in an IV drip is too low or too high. It is important to remember that changes in boiling point elevation, freezing point depression, and osmotic pressure—as well as other properties we have discussed in this chapter—depend solely on the concentration of solute particles present in the solutions and not on their identity.

SUMMARY

Section 11.1 The **enthalpy of solution** for an ionic compound is the sum of the **lattice energy** and the **enthalpy of hydration**. Lattice energies can be calculated with a **Born–Haber cycle**, an application of Hess's law.

Section 11.2 Molecules at the surface of a liquid evaporate by breaking intermolecular interactions with neighboring molecules and entering the gas phase. The **vapor pressure** of a liquid is proportional to the fraction of its molecules that enter the gas phase. A greater proportion of liquid molecules enter the gas phase as the temperature increases, leading to higher vapor pressures at higher temperatures. The vapor pressure of a solution is a **colligative property**, which is any property of a solution that depends only on the concentration of solutes, not on their identity. **Raoult's law** relates the vapor pressure of a solution to its composition and to the vapor pressure of the solvent.

Section 11.3 **Crude oil** is composed primarily of hydrocarbons that can be separated by **fractional distillation** into gasoline and other useful products. The vapor pressure of a pure substance as a function of temperature is determined by the **Clausius–Clapeyron equation**. The vapor pressure of an ideal solution of volatile compounds follows Raoult's law.

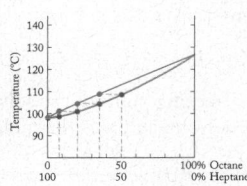

Section 11.4 The concentration units used for colligative property measurements include molarity and **molality**, which is the number of moles of solute per kilogram of solvent. Solutes in solution elevate the solvent's boiling point and depress its freezing point. The **van't Hoff factor** accounts for the colligative properties of electrolytes and the formation of solute **ion pairs** in concentrated solutions. In **osmosis**, solvent flows through a semipermeable membrane from a solution of lower solute concentration into a solution of higher solute concentration. **Osmotic pressure**, a colligative property, is defined as the pressure required to halt the flow of solvent from the more dilute solution across the membrane. **Reverse osmosis** is used to purify water. The more concentrated a solution, the higher its boiling point, the lower its freezing point, and the higher its osmotic pressure.

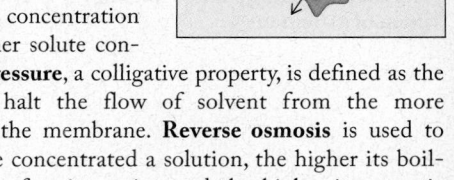

Section 11.5 The molar mass of a compound can be determined by measuring the freezing point depression, boiling point elevation, or osmotic pressure of a solution of the compound.

PROBLEM-SOLVING SUMMARY

TYPE OF PROBLEM	CONCEPTS AND EQUATIONS	SAMPLE EXERCISES
Ranking lattice energies and melting points	Predict lattice energies and relative melting points using	11.1

$$U = \frac{kQ_1Q_2}{d} \qquad (11.3)$$

TYPE OF PROBLEM	CONCEPTS AND EQUATIONS	SAMPLE EXERCISES
Calculating lattice energy with a Born–Haber cycle	The enthalpy of reaction ΔH_{rxn} when an ionic solid forms from its constituent elements is equal to the sum of the enthalpies of reaction for every step in the process. Solve for lattice energy U or any one unknown enthalpy change term.	11.2
Calculating lattice energy using enthalpies of hydration	Use enthalpies of hydration ($\Delta H_{hydration}$) and enthalpies of solution ($\Delta H_{solution}$) to calculate lattice energy: $$\Delta H_{solution} = \Delta H_{hydration} - U$$ where $$\Delta H_{hydration} = \Delta H_{hydration,cation} + \Delta H_{hydration,anion}$$	11.3
Calculating vapor pressure of a solution	Use Raoult's law, $$P_{solution} = X_{solvent}P^\circ_{solvent} \qquad (11.6)$$ where $P_{solution}$ is the vapor pressure of the solution at a given temperature, $X_{solvent}$ is the mole fraction of the solvent in the solution, and $P^\circ_{solvent}$ is the vapor pressure of the pure solvent at the same temperature.	11.4
Calculating vapor pressure, temperature, or enthalpy of vaporization using the Clausius–Clapeyron equation	Substitute the appropriate given values into the Clausius–Clapeyron equation: $$\ln\left(\frac{P_{vap,T_1}}{P_{vap,T_2}}\right) = \frac{\Delta H_{vap}}{R}\left(\frac{1}{T_2} - \frac{1}{T_1}\right) \qquad (11.7)$$ and solve for any one unknown value.	11.5
Interpreting fractional distillation data	Draw an idealized graph showing distillation temperature versus volume of distillate.	11.6
Calculating the vapor pressure of a solution of volatile substances	Determine the mole fractions and vapor pressures of each component of the solution. Use Raoult's law to calculate the vapor pressure of the solution: $$P_{total} = X_1 P^\circ_1 + X_2 P^\circ_2 + X_3 P^\circ_3 + \cdots \qquad (11.8)$$	11.7
Calculating molal concentrations	Molality is defined as $$m = \frac{n_{solute}}{kg\ solvent} \qquad (11.9)$$ where n is the number of moles of solute.	11.8
Calculating boiling point or freezing point of a solution	For nonelectrolyte solutes, use $$\Delta T_b = K_b m \qquad (11.10)$$ where ΔT_b is the elevation in the boiling point of the solvent, K_b is a constant that depends only on the solvent, and m is the molality of the solution. For the freezing point, use $$\Delta T_f = K_f m \qquad (11.11)$$ where ΔT_f is the depression in the freezing point of the solvent, K_f is a constant that depends only on the solvent, and m is the molality of the solution.	11.9, 11.10
Assessing interactions among particles in solution by comparing the theoretical value of the van't Hoff factor i with the experimentally measured value	For electrolytes, use $$\Delta T_b = iK_b m \qquad (11.12)$$ and $$\Delta T_f = iK_f m \qquad (11.13)$$ where the theoretical value of i is the number of particles created when an electrolyte dissociates completely: $i = 2$ for NaCl, 3 for $CaCl_2$, 4 for Na_3PO_4. For real solutions of electrolytes where $\Delta T_{b/f}$, $K_{b/f}$, and m are known, rearrange $\Delta T_b = iK_b m$ and $\Delta T_f = iK_f m$ to calculate the value of i and compare the result with the theoretical value.	11.11, 11.12

TYPE OF PROBLEM	CONCEPTS AND EQUATIONS	SAMPLE EXERCISES
Calculating osmotic pressure and reverse osmotic pressure	Use $$\pi = iMRT \qquad (11.14)$$ where π is the osmotic pressure, i is the van't Hoff factor, M is the molar concentration of the solution, R is the ideal gas constant, and T is the absolute temperature of the solution.	11.13–11.15
Determining molar mass from boiling point elevation, freezing point depression, or osmotic pressure	Use $\Delta T_b = K_b m$, $\Delta T_f = K_f m$, or $\pi = MRT$ to calculate the molality m or molarity M of the solution. Then use m or M to determine the moles of solute and molar mass of the solute.	11.16, 11.17

VISUAL PROBLEMS ··

(Answers to boldface end-of-chapter questions and problems are in the back of the book.)

11.1. Use the graph in Figure P11.1 to estimate the normal boiling points of substances X and Y. Which substance has the stronger intermolecular forces?

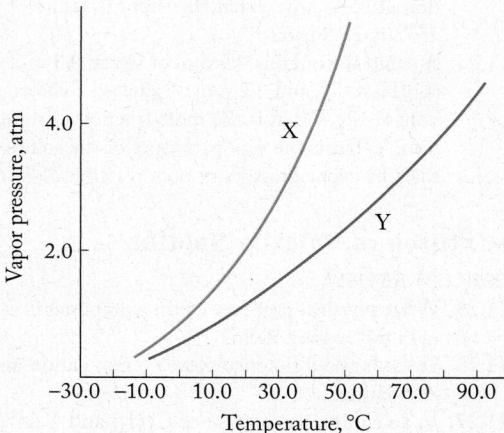

FIGURE P11.1

11.2. The graph in Figure P11.2 shows the decrease in the freezing point of water ΔT_f for solutions of two different substances A (triangles) and B (circles) in water. Explain how you can reasonably conclude that (a) A and B are nonelectrolytes and (b) the freezing point depression constant K_f of water is independent of the solute's identity.

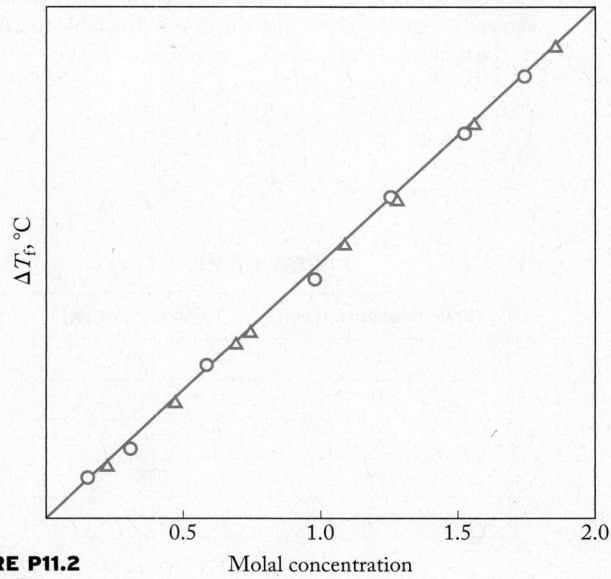

FIGURE P11.2

11.3. The arrow in Figure P11.3 indicates the direction of solvent flow through a semipermeable membrane in equipment designed to measure osmotic pressure. Which solution, A or B, is the more concentrated? Explain your answer.

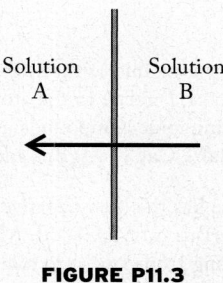

FIGURE P11.3

11.4. Kidney Dialysis Semipermeable membranes of the sort used in kidney dialysis do not allow large molecules and cells to pass but do allow small ions and water to pass. Figure P11.4 shows such a membrane separating fluids of various compositions.

a. In which direction does the water flow in each apparatus?
b. In which direction do sodium ions flow in each apparatus?
c. In which direction do the potassium ions flow in each apparatus?

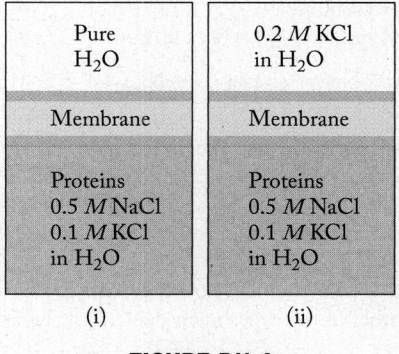

FIGURE P11.4

● ■

Energy Changes during Dissolution; Vapor Pressure

CONCEPT REVIEW

11.5. Explain why the term for enthalpy of hydration ($\Delta H_{hydration}$) contains two terms: one for ion–dipole interactions and one for dipole–dipole interactions.

11.6. Explain why trends in lattice energies for ionic compounds parallel trends in melting points.

11.7. Explain the term *nonvolatile solute*.

11.8. Which has the higher vapor pressure at constant temperature, pure water or seawater? Explain your answer.

11.9. Why does the vapor pressure of a liquid increase with increasing temperature?

11.10. What happens when the vapor pressure of a liquid is equal to atmospheric pressure?

11.11. Generally speaking, how is the vapor pressure of a liquid affected by the strength of intermolecular forces?

11.12. Is vapor pressure an intensive or extensive property of a substance?

PROBLEMS

11.13. How do the melting points of the series of sodium halides NaX (X = F, Cl, Br, I) relate to the atomic number of X?

11.14. Rank the following ionic compounds in order of increasing melting point: BaF_2, $CaCl_2$, $MgBr_2$, and SrI_2.

11.15. Which substance has the least negative lattice energy? (a) MgI_2; (b) $MgBr_2$; (c) $MgCl_2$; (d) MgF_2

11.16. Rank the following from lowest to highest lattice energy: $NaBr$, $MgBr_2$, $CaBr_2$, KBr.

11.17. Use a Born–Haber cycle to calculate the lattice energy of potassium chloride (KCl) from the following data:
Ionization energy of K(g) = 425 kJ/mol
Electron affinity of Cl(g) = −349 kJ/mol
Energy to sublime K(s) = 89 kJ/mol
Bond energy of $Cl_2(g)$ = 240 kJ/mol

$$\Delta H_{rxn} \text{ for } K(s) + \tfrac{1}{2}Cl_2(g) \rightarrow KCl(s) = -438 \text{ kJ/mol}$$

11.18. Calculate the lattice energy of sodium oxide (Na_2O) from the following data:
Ionization energy of Na(g) = 495 kJ/mol
Electron affinity of O(g) for 2 electrons = 603 kJ/mol
Energy to sublime Na(s) = 109 kJ/mol
Bond energy of $O_2(g)$ = 499 kJ/mol

$$\Delta H_{rxn} \text{ for } 2 Na(s) + \tfrac{1}{2}O_2(g) \rightarrow Na_2O(s) = -416 \text{ kJ/mol}$$

11.19. An experiment like that shown in Figure 11.6 is set up with the beaker containing pure water full to the brim and the beaker containing seawater half-full. Explain why the beaker that contained seawater will eventually overflow.

11.20. Explain to a nonscientist how the water gets from one beaker to the other in the experiment depicted in Figure 11.6.

11.21. Rank the following compounds in order of increasing vapor pressure at 298 K: (a) CH_3CH_2OH, (b) CH_3OCH_3, and (c) $CH_3CH_2CH_3$.

11.22. Rank the compounds in Figure P11.22 in order of increasing vapor pressure at 298 K.

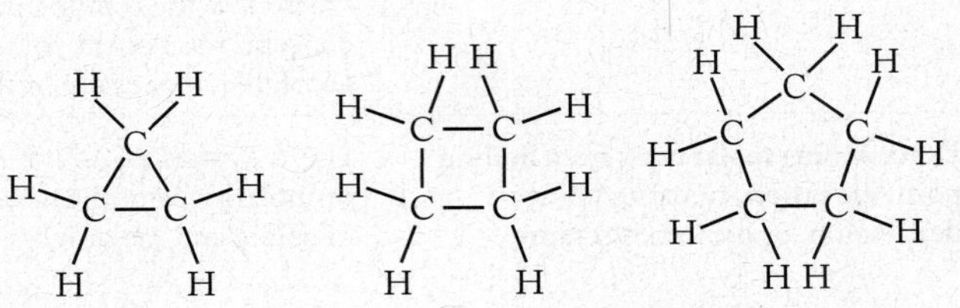

(a)   (b)   (c)

FIGURE P11.22

11.23. A solution contains 3.5 mol of water and 1.5 mol of nonvolatile glucose ($C_6H_{12}O_6$). What is the mole fraction of water in this solution? What is the vapor pressure of the solution at 25°C, given that the vapor pressure of pure water at 25°C is 23.8 torr?

11.24. A solution contains 4.5 mol of water, 0.3 mol of sucrose ($C_{12}H_{22}O_{11}$), and 0.2 mol of glucose. Sucrose and glucose are nonvolatile. What is the mole fraction of water in this solution? What is the vapor pressure of the solution at 35°C, given that the vapor pressure of pure water at 35°C is 42.2 torr?

Mixtures of Volatile Solutes

CONCEPT REVIEW

11.25. What physical property of the components of crude oil is used to separate them?

11.26. What is the difference between distillation and fractional distillation?

11.27. In an equimolar mixture of C_5H_{12} and C_7H_{16}, which compound is present in higher concentration in the vapor above the solution?

11.28. Why does the boiling point of a mixture of volatile hydrocarbons increase over time during a distillation?

PROBLEMS

11.29. Pine Oil The smell of fresh cut pine is due in part to the cyclic alkene pinene, whose carbon-skeleton structure is shown in Figure P11.29. Use the data in the table to calculate the heat of vaporization, ΔH_{vap}, of pinene.

Pinene

FIGURE P11.29

Vapor Pressure (torr)	Temperature (K)
760	429
515	415
340	401
218	387
135	373

11.30. **Almonds and Cherries** Almonds and almond extracts are common ingredients in baked goods. Almonds contain the compound benzaldehyde (shown in Figure P11.30), which accounts for the odor of the nut. Benzaldehyde is also responsible for the aroma of cherries. Use the data in the table to calculate the heat of vaporization, ΔH_{vap}, of benzaldehyde.

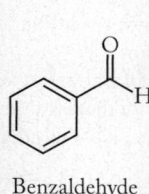

Benzaldehyde

FIGURE P11.30

Vapor Pressure (torr)	Temperature (K)
50	373
111	393
230	413
442	433
805	453

11.31. At 20°C, the vapor pressure of ethanol is 45 torr and the vapor pressure of methanol is 92 torr. What is the vapor pressure at 20°C of a solution prepared by mixing 25 g of methanol and 75 g of ethanol?

11.32. A bottle is half-filled with a 50:50 (mole-to-mole) mixture of *n*-heptane (C_7H_{16}) and *n*-octane (C_8H_{18}) at 25°C. What is the mole ratio of heptane vapor to octane vapor in the air space above the liquid in the bottle? The vapor pressures of heptane and octane at 25°C are 31 torr and 11 torr, respectively.

11.33. **High-Octane Gasoline** Gasoline is a complex mixture of hydrocarbons. Gasoline is sold with a variety of octane ratings that are based on the comparison of the gasoline with the combustion properties of isooctane, a compound with the molecular formula C_8H_{18}. The Lewis structures of isooctane and another compound with molecular formula C_8H_{18} are shown in Figure P11.33, along with their normal boiling points and heats of vaporization. Determine the vapor pressure of each isomer on a day when the temperature is 38°C.

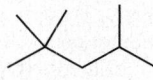

Isooctane
bp = 98.2°C
ΔH_{vap} = 35.8 kJ/mol

Tetramethylbutane
bp = 106.5°C
ΔH_{vap} = 43.3 kJ/mol

FIGURE P11.33

11.34. **Stove Fuel** Portable lanterns and stoves used for camping and backpacking often use a mixture of C_5 and C_6 hydrocarbons known as white gas. White gas is easy to transport but stoves that burn white gas are harder to light in the cold. Figure P11.34 shows the carbon-skeleton structure of *n*-pentane, C_5H_{12}, along with its normal boiling point and heat of vaporization. Determine the vapor pressure of *n*-pentane on a morning when the temperature is 5°C.

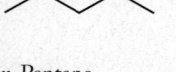

n-Pentane
bp = 36.0°C
ΔH_{vap} = 27.6 kJ/mol

FIGURE P11.34

Colligative Properties of Solutions

CONCEPT REVIEW

11.35. What is the difference between molarity and molality?

11.36. As a solution of NaCl becomes more concentrated, does the difference between its molarity and its molality increase or decrease?

11.37. Explain why seawater has a lower freezing point than freshwater.

11.38. The thermostat in a refrigerator filled with cans of soft drinks malfunctions and the temperature of the refrigerator drops below 0°C. The contents of the cans of diet soft drinks freeze, rupturing many of the cans and causing an awful mess. However, none of the cans containing regular, nondiet soft drinks rupture. Why?

11.39. Why is it important to know if a substance is a strong electrolyte before predicting its effect on the boiling and freezing points of a solvent?

11.40. What role does the van't Hoff factor play in describing colligative properties of solutions?

11.41. Explain how the theoretical value of the van't Hoff factor *i* for substances such as CH_3OH, NaBr, and K_2SO_4 can be predicted from their formulas.

11.42. Explain why it is possible for an experimentally measured value of a van't Hoff factor to be less than the theoretical value.

11.43. What is a semipermeable membrane?

11.44. A pure solvent is separated from a solution containing the same solvent by a semipermeable membrane. In which direction does the solvent flow across the membrane, and why?

11.45. A dilute solution is separated from a more concentrated solution containing the same solvent by a semipermeable membrane. In which direction does the solvent tend to flow across the membrane, and why?

11.46. How is the osmotic pressure of a solution related to its molar concentration and its temperature?

11.47. What is reverse osmosis? List the basic components of equipment used to purify seawater by reverse osmosis.

11.48. Explain how the minimum pressure for purification of seawater by reverse osmosis can be estimated from its composition.

PROBLEMS

11.49. Calculate the molality of each of the following solutions:
  a. 0.875 mol of glucose ($C_6H_{12}O_6$) in 1.5 kg of water
  b. 11.5 mmol of acetic acid (CH_3COOH) in 65 g of water
  c. 0.325 mol of baking soda ($NaHCO_3$) in 290.0 g of water

*11.50. Table 4.1 lists molar concentrations of major ions in seawater. Using a density of 1.022 g/mL for seawater, convert the concentrations into molalities.

11.51. What mass of the following solutions contains 0.100 mol of solute? (a) 0.334 m NH_4NO_3; (b) 1.24 m ethylene glycol, $HOCH_2CH_2OH$; (c) 5.65 m $CaCl_2$

11.52. How many moles of solute are there in the following solutions?
  a. 0.150 m glucose solution made by dissolving the glucose in 100.0 kg of water
  b. 0.028 m Na_2CrO_4 solution made by dissolving the Na_2CrO_4 in 1000.0 g of water
  c. 0.100 m urea solution made by dissolving the urea in 500.0 g of water

11.53. Fish Kills High concentrations of ammonia (NH_3), nitrite ion, and nitrate ion in water can kill fish. Lethal concentrations of these species for rainbow trout are 1.1 mg/L, 0.40 mg/L, and 1361 mg/L, respectively. Express these concentrations in molality units, assuming a solution density of 1.00 g/mL.

11.54. The concentrations of six important elements in a sample of river water are 0.050 mg/kg of Al^{3+}, 0.040 mg/kg of Fe^{3+}, 13.4 mg/kg of Ca^{2+}, 5.2 mg/kg of Na^+, 1.3 mg/kg of K^+, and 3.4 mg/kg of Mg^{2+}. Express each of these concentrations in molality units.

11.55. Cinnamon Cinnamon owes its flavor and odor to cinnamaldehyde (C_9H_8O). Determine the boiling point elevation of a solution of 100 mg of cinnamaldehyde dissolved in 1.00 g of carbon tetrachloride ($K_b = 5.02°C/m$).

11.56. Spearmint Determine the boiling point elevation of a solution of 125 mg of carvone ($C_{10}H_{14}O$, oil of spearmint) dissolved in 1.50 g of carbon disulfide ($K_b = 2.34°C/m$).

11.57. What molality of a nonvolatile, nonelectrolyte solute is needed to lower the melting point of camphor by 1.000°C ($K_f = 39.7°C/m$)?

11.58. What molality of a nonvolatile, nonelectrolyte solute is needed to raise the boiling point of water by 7.60°C ($K_b = 0.52°C/m$)?

11.59. Saccharin Determine the melting point of an aqueous solution made by adding 186 mg of saccharin ($C_7H_5O_3NS$) to 1.00 mL of water (density = 1.00 g/mL, $K_f = 1.86°C/m$).

11.60. Determine the boiling point of an aqueous solution that is 2.50 m ethylene glycol ($HOCH_2CH_2OH$); K_b for water is 0.52°C/m. Assume that the boiling point of pure water is 100.00°C.

11.61. Which aqueous solution has the lowest freezing point: 0.5 m glucose, 0.5 m NaCl, or 0.5 m $CaCl_2$?

11.62. Which aqueous solution has the highest boiling point: 0.5 m glucose, 0.5 m NaCl, or 0.5 m $CaCl_2$?

11.63. Which one of the following aqueous solutions should have the highest boiling point: 0.0200 m CH_3OH, 0.0125 m KCl, or 0.0100 m $Ca(NO_3)_2$?

11.64. Which one of the following aqueous solutions should have the lowest freezing point: 0.0500 m $C_6H_{12}O_6$, 0.0300 m KBr, or 0.0150 m Na_2SO_4?

11.65. Arrange the following aqueous solutions in order of increasing boiling point:
  a. 0.06 m $FeCl_3$ ($i = 3.4$)
  b. 0.10 m $MgCl_2$ ($i = 2.7$)
  c. 0.20 m KCl ($i = 1.9$)

11.66. Arrange the following solutions in order of increasing freezing point depression:
  a. 0.10 m $MgCl_2$ in water, $i = 2.7$, $K_f = 1.86°C/m$
  b. 0.20 m toluene in diethyl ether, $i = 1.00$, $K_f = 1.79°C/m$
  c. 0.20 m ethylene glycol in ethanol, $i = 1.00$, $K_f = 1.99°C/m$

11.67. The following pairs of aqueous solutions are separated by a semipermeable membrane. In which direction will the solvent flow?
  a. A = 1.25 M NaCl; B = 1.50 M KCl
  b. A = 3.45 M $CaCl_2$; B = 3.45 M NaBr
  c. A = 4.68 M glucose; B = 3.00 M NaCl

11.68. The following pairs of aqueous solutions are separated by a semipermeable membrane. In which direction will the solvent flow?
  a. A = 0.48 M NaCl; B = 55.85 g of NaCl dissolved in 1.00 L of solution
  b. A = 100 mL of 0.982 M $CaCl_2$; B = 16 g of NaCl in 100 mL of solution
  c. A = 100 mL of 6.56 mM $MgSO_4$; B = 5.24 g of $MgCl_2$ in 250 mL of solution

11.69. Calculate the osmotic pressure of each of the following aqueous solutions at 20°C:
  a. 2.39 M methanol (CH_3OH)
  b. 9.45 mM $MgCl_2$
  c. 40.0 mL of glycerol ($C_3H_8O_3$) in 250.0 mL of aqueous solution (density of glycerol = 1.265 g/mL)
  d. 25 g of $CaCl_2$ in 350 mL of solution

11.70. Calculate the osmotic pressure of each of the following aqueous solutions at 27°C:
  a. 10.0 g of NaCl in 1.50 L of solution
  b. 10.0 mg/L of $LiNO_3$
  c. 0.222 M glucose
  d. 0.00764 M K_2SO_4

11.71. Determine the molarity of each of the following solutions from its osmotic pressure at 25°C. Include the van't Hoff factor for the solution when the factor is given.
  a. π = 0.674 atm for a solution of ethanol (CH_3CH_2OH)
  b. π = 0.0271 atm for a solution of aspirin ($C_9H_8O_4$)
  c. π = 0.605 atm for a solution of $CaCl_2$, $i = 2.47$

11.72. Determine the molarity of each of the following solutions from its osmotic pressure at 25°C. Include the van't Hoff factor for the solution when the factor is given.
  a. π = 0.0259 atm for a solution of urea (CH_4N_2O)
  b. π = 1.56 atm for a solution of sucrose ($C_{12}H_{22}O_{11}$)
  c. π = 0.697 atm for a solution of KI, $i = 1.90$

11.73. Is the following statement true or false? For solutions of the same reverse osmotic pressure at the same temperature, the molarity of a solution of NaCl will always be less than the molarity of a solution of $CaCl_2$. Explain your answer.

11.74. Suppose you have 1.00 M aqueous solutions of each of the following solutes: glucose ($C_6H_{12}O_6$), NaCl, and acetic acid (CH_3COOH). Which solution has the highest pressure requirement for reverse osmosis?

Measuring the Molar Mass of a Solute Using Colligative Properties

CONCEPT REVIEW

11.75. What effect does dissolving a solute have on the following properties of a solvent? (a) its osmotic pressure; (b) its freezing point; (c) its boiling point

11.76. How can measurements of osmotic pressure, freezing point depression, and boiling point elevation be used to find the molar mass of a solute? Why is it important to know whether the solute is an electrolyte or a nonelectrolyte?

PROBLEMS

11.77. Throat Lozenges A 188 mg sample of a nonelectrolyte isolated from throat lozenges was dissolved in enough water to make 10.0 mL of solution at 25°C. The osmotic pressure of the resulting solution was 4.89 atm. Calculate the molar mass of the compound.

*__11.78.__ An unknown compound (152 mg) was dissolved in water to make 75.0 mL of solution. The solution did not conduct electricity and had an osmotic pressure of 0.328 atm at 27°C. Elemental analysis revealed the substance to be 78.90% C, 10.59% H, and 10.51% O. Determine the molecular formula of this compound.

*__11.79. Cloves__ Eugenol is one of the compounds responsible for the flavor of cloves. A 111 mg sample of eugenol was dissolved in 1.00 g of chloroform (K_b =3.63°C/m), increasing the boiling point of chloroform by 2.45°C. Calculate eugenol's molar mass. Eugenol is 73.17% C, 7.32% H, and 19.51% O by mass. What is the molecular formula of eugenol?

*__11.80. Caffeine__ The freezing point of a solution prepared by dissolving 150 mg of caffeine in 10.0 g of camphor is lower by 3.07°C than that of pure camphor (K_f = 39.7°C/m). What is the molar mass of caffeine? Elemental analysis of caffeine yields the following results: 49.49% C, 5.15% H, 28.87% N, and the remainder O. What is the molecular formula of caffeine?

Additional Problems

11.81. Which substance has the least negative lattice energy? (a) SrI_2; (b) $CaBr_2$; (c) $CaCl_2$; (d) MgF_2

11.82. Explain why the boiling point of pure sodium chloride is much higher than the boiling point of an aqueous solution of sodium chloride.

*__11.83. Melting Ice__ $CaCl_2$ is often used to melt ice on sidewalks. Could $CaCl_2$ melt ice at −20°C? Assume that the solubility of $CaCl_2$ at this temperature is 70.1 g $CaCl_2$/100.0 g of H_2O and that the van't Hoff factor for a saturated solution of $CaCl_2$ is 2.5.

*__11.84. Making Ice Cream__ A mixture of table salt and ice is used to chill the contents of hand-operated ice-cream makers. What is the melting point of a mixture of 2.00 lb of NaCl and 12.00 lb of ice if exactly half of the ice melts? Assume that all the NaCl dissolves in the melted ice and that the van't Hoff factor for the resulting solution is 1.44.

11.85. The freezing points of 0.0935 m ammonium chloride and 0.0378 m ammonium sulfate in water were found to be −0.322°C and −0.173°C, respectively. What are the values of the van't Hoff factors for these salts?

11.86. The following data were collected for three compounds in aqueous solution. Determine the value of the van't Hoff factor for each salt (K_f for water = 1.86°C/m).

Compound	Concentration	Experimentally Measured ΔT_f
LiCl	5.0 g/kg	0.410°C
HCl	5.0 g/kg	0.486°C
NaCl	5.0 g/kg	0.299°C

11.87. Physiological Saline 100.0 mL of a solution of physiological saline (0.92% NaCl by mass) is diluted by the addition of 250.0 mL of water. What is the osmotic pressure of the final solution at 37°C? Assume that NaCl dissociates completely into $Na^+(aq)$ and $Cl^-(aq)$.

11.88. 100.0 mL of 2.50 mM NaCl is mixed with 80.0 mL of 3.60 mM $MgCl_2$ at 20°C. Calculate the osmotic pressure of each starting solution and that of the mixture, assuming that the volumes are additive and that both salts dissociate completely into their component ions.

11.89. A solution of 7.50 mg of a small protein in 5.00 mL aqueous solution has an osmotic pressure of 6.50 torr at 23.1°C. What is the molar mass of the protein?

11.90. Kidney Dialysis Hemodialysis, a method of removing waste products from the blood if the kidneys have failed, uses a tube made of a cellulose membrane that is immersed in a large volume of aqueous solution. Blood is pumped through the tube and is then returned to the patient's vein. The membrane does not allow passage of large protein molecules and cells but does allow small ions, urea, and water to pass through it. Assume that a physician wants to decrease the concentration of sodium ion and urea in a patient's blood while maintaining the concentration of potassium ion and chloride ion in the blood. What materials must be dissolved in the aqueous solution in which the dialysis tube is immersed? How must the concentrations of ions in the immersion fluid compare to those in blood?

Additional study materials including ChemTours and Diagnostic Quizzes are available at StudySpace at www.wwnorton.com/studyspace.

12

The Chemistry of Solids

Learning Outcomes

- Describe the two bonding models for metals in the solid state
- Recognize the differences in packing schemes for atoms, molecules, and ions in the solid state
- Identify face-centered, body-centered, and simple cubic unit cells
- Describe the differences between substitutional and interstitial alloys
- Relate the physical properties of solids to their crystalline structures
- Calculate the interlayer spacing in crystalline solids using the Bragg equation

A LOOK AHEAD

Stronger, Tougher, Harder

Did you know that most gold rings and other gold jewelry are not made of pure gold? Because gold is one of the softer metals, it is very easy to bend jewelry made of pure gold. In contrast, jewelry made of gold blended with other metals is more resistant to physical damage and can last a lifetime. The amount of gold in jewelry is measured in *karats*, with 24 karat (24K) being pure gold. The 22-karat gold—yellow gold—used in Europe and Asia for most wedding rings is a homogeneous mixture that is 92% by mass gold (22/24 = 0.92) and 8% other metals. Different colors of gold result from different blends: White gold typically includes from 25% to 60% palladium and silver, and rose gold contains from 20% to 55% copper. An especially beautiful alloy of 96% copper and 4% gold called shakudo is prized in Japan for its dark purplish-blue sheen.

Any mixture of two or more metals in the solid state is called an *alloy*, and over 500,000 alloys have been made and characterized. Some are solid solutions, which means they are homogeneous at the atomic level just as liquid solutions are, and others are heterogeneous mixtures. The purpose of making alloys is that their properties can be controlled by varying the proportions of their constituent metals.

The many types of steel are alloys of iron, the nonmetal carbon, and other elements. Steel that is 0.4% by mass carbon is much harder and more durable than pure iron. Other elements may be added to give steel certain properties. For example, stainless steel that is 18% chromium by mass resists rusting and corrosion. Stainless steel is used in cookware, in knives, forks, and spoons of standard tableware, and in other products where appearance matters and chemical reactivity is undesirable.

Alloys for Strength The metal cables in this bridge over the Charles River in ▶ Boston are made of steel: an alloy of iron, carbon, and other metals. Each cable can support over 5 million kilograms.

The construction industry uses alloys in many ways, from steel I-beams in building skeletons to other metals for architectural features and interior fixtures. Brass is an alloy of copper and zinc that is harder and stronger than either element alone. It is commonly used in doorknobs and hinges, but over time they tarnish, looking dull, even dirty. Because stainless steel doorknobs look clean and sleek, they are frequently used in public places such as hospitals and schools, where the appearance of cleanliness is important. Recent studies have found, however, that brass may be the better choice because it kills bacteria that come in contact with it. This bactericidal effect is due to the presence of copper, which has long been known to exhibit antimicrobial properties. This property of pure copper metal is retained in the alloy.

Alloys are of great importance in the transportation industry. Aluminum alloys are ideal for making aircraft because they are strong and lightweight. The wing of a 747-400 airplane is about 1.8 m longer than the wing of a regular 747 but about 2 metric tons lighter because it is made of an alloy of 95% aluminum blended with copper, manganese, and iron.

Even sports equipment benefits from the development of new alloys. A patented five-metal alloy of nickel, zirconium, titanium, copper, and beryllium is advertised as possibly changing the game of golf because the alloy produces a stronger, lighter, more resilient club that enables a golfer to transfer more energy from the swing into the ball.

In this chapter we explore the links between the physical properties of solids at the macroscopic level and the structures of these solids at the atomic level. We start with metals and their alloys and conclude with compounds called ceramics, seeking answers to such questions as these: Why do metals bend? Why are they such good conductors of electricity compared to ceramics or salts? Why are metal alloys so much stronger, tougher, and harder than the pure metals from which they are made? ∎

◉◉ **CONNECTION** In Chapter 9 we discussed the valence bond theory of chemical bond formation as well as combining atomic orbitals into molecular orbitals.

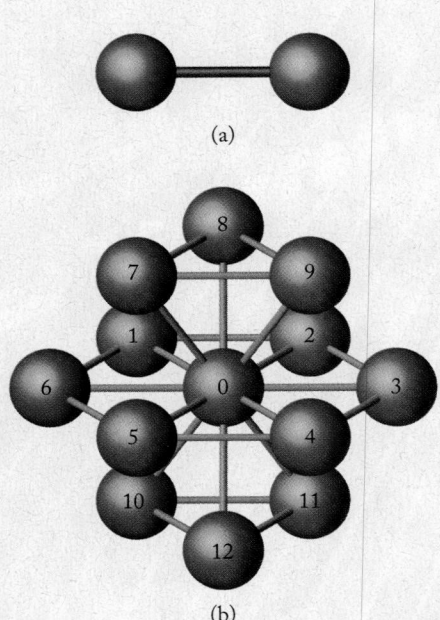

FIGURE 12.1 Covalent bonds differ from metallic bonds. (a) A ball-and-stick model of the molecule Cu_2 is based on the assumption that the two atoms share their $4s$ electrons to form a covalent bond. (b) In solid copper, the atom labeled 0 shares its $4s$ electron with 12 other atoms. As a result, the bonds in copper and other metals are much more diffuse than the covalent bonds in small molecules.

12.1 Metallic Bonds and Conduction Bands

Most elements are metals, which means they are typically hard, shiny, malleable (easily shaped), ductile (easily drawn out), and able to conduct electricity. In this section we explore *why* they have these properties by examining metals at the atomic level and by exploring models describing the bonds that hold metal atoms together.

According to valence bond theory, a covalent bond forms between two atoms when partially filled atomic orbitals—one from each atom—overlap. Our focus in Chapter 9 was on covalent bonding in gas-phase molecules. In this section we explore the bonds that form between the densely packed atoms in metallic solids. Dense packing means that the valence orbitals of atoms overlap with orbitals of many nearby atoms. This large number of interactions makes metals strong. At the same time, sharing a limited number of valence electrons with many bonding partners makes the bond linking any two metal atoms relatively weak.

To understand this point, consider the bond that would result if two Cu atoms came together and formed a molecule of Cu_2. Copper atoms have the electron configuration $[Ar]3d^{10}4s^1$. When the partially filled $4s$ orbitals of the two atoms overlap, they form a diatomic molecule held together by a single covalent bond (Figure 12.1a). However, the Cu atoms in solid copper are each surrounded by *a total of 12* other Cu atoms and each is bonded to all 12 of its neighbors (Figure 12.1b).

This means that each Cu atom must share its 4s electron with 12 other atoms, not just one other atom. Inevitably, this dispersion of bonding electrons weakens the Cu—Cu bond between each pair of Cu atoms.

Adding to the diffuse nature of this bonding is the fact that metallic elements have lower electronegativities than nonmetals. Recall from Chapter 8 that differences in electronegativity are important for defining how bonding electrons are distributed between pairs of atoms. In a pure metal there are no such differences in electronegativity. However, the low electronegativities of metals mean that the bonding electrons are not held tightly to the nuclei of individual atoms; they have higher energy and hence are more readily removed.

In Chapter 8 we described metal atoms "floating" in seas of mobile bonding electrons where the electrons are shared by all of the nuclei in the sample. The diffuse nature of metallic bonding described in the preceding paragraphs certainly fits the sea-of-electrons model, but a more sophisticated approach, called **band theory**, better explains the bonding in metals and other solids. Let's apply band theory, which is an extension of molecular orbital theory, to explain the bonding between the atoms in solid copper. When the 4s atomic orbitals on two Cu atoms overlap to form Cu_2, the atomic orbitals combine to form two molecular orbitals with different energies equally spaced above and below the initial value (Figure 12.2a). This is analogous to the formation of low-energy bonding and high-energy antibonding molecular orbitals (Section 9.7). If another two Cu atoms join the first two to form a molecule of Cu_4, the 4s atomic orbitals of four Cu atoms combine to form four molecular orbitals. If we add another four atoms to make Cu_8, a total of eight copper 4s atomic orbitals combine to form eight molecular orbitals. In all these molecules the lower-energy orbitals are filled with the available 4s electrons and the upper orbitals are empty (Figure 12.2b). If we apply this model to the enormous number of atoms in a piece of copper wire, an equally enormous number of molecular orbitals is created. The lower-energy half of them are occupied by electrons; the higher-energy half are empty. There are so many of these orbitals that they form a continuous *band* of energies with no gap between the occupied lower half and the empty upper half. Because this band of MOs was formed by combining valence-shell orbitals, it is called a **valence band**.

Band theory explains the conductivity of copper and many other metals by assuming that there is essentially no gap between the energy of the occupied lower portion of the valence band and the empty upper portion. Therefore, valence electrons can move easily from the filled lower portion to the empty upper portion, where they are free to move from one empty orbital to the next and flow throughout the solid.

The model of a partially filled valence shell explains the conductivity of many metals, but not all. Consider, for example, zinc, copper's neighbor in the periodic table. Its electron configuration, $[Ar]3d^{10}4s^2$, tells us that all its valence-shell electrons reside in filled orbitals, which means the valence band in solid zinc is filled (Figure 12.3). With no empty space in the valence band to accommodate additional electrons, it might seem that the valence-shell electrons in Zn would be immobile, making Zn a poor electrical conductor. However, electrons in the valence band of Zn *do* migrate through the solid; zinc is a good conductor of electricity, and band theory explains why. The theory assumes that *all* atomic orbitals of comparable shape and energy, including the empty

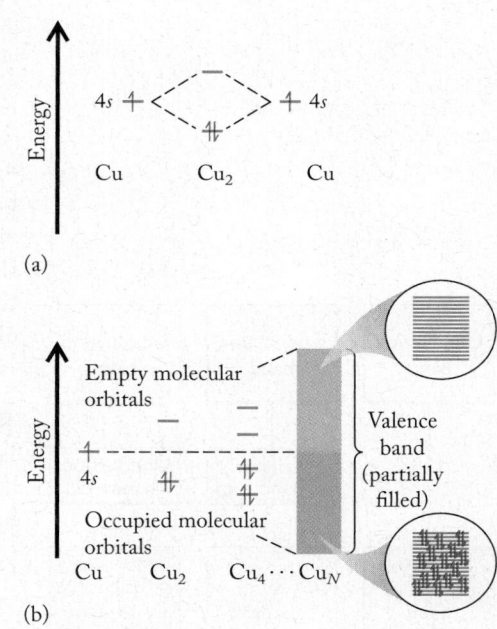

FIGURE 12.2 (a) The 4s atomic orbitals combine to form two molecular orbitals in the molecule Cu_2. (b) As the half-filled 4s atomic orbitals of an increasing number of Cu atoms overlap, more and more molecular orbitals are formed; half of them are occupied, the other half are empty. As more and more MOs form, their energies get closer and closer together until a continuous energy band is formed—a valence band that is only half-filled with electrons. The electrons can move from the filled half (purple) to the slightly higher-energy upper half (orange), where they are free to migrate through delocalized empty orbitals throughout the entire solid.

conduction band an unoccupied band higher in energy than a valence band in which electrons are free to migrate.

band gap (E_g) the energy gap between the valence and conduction bands.

semiconductor a semimetal (metalloid) with electrical conductivity between that of metals and insulators that can be chemically altered to increase its electrical conductivity.

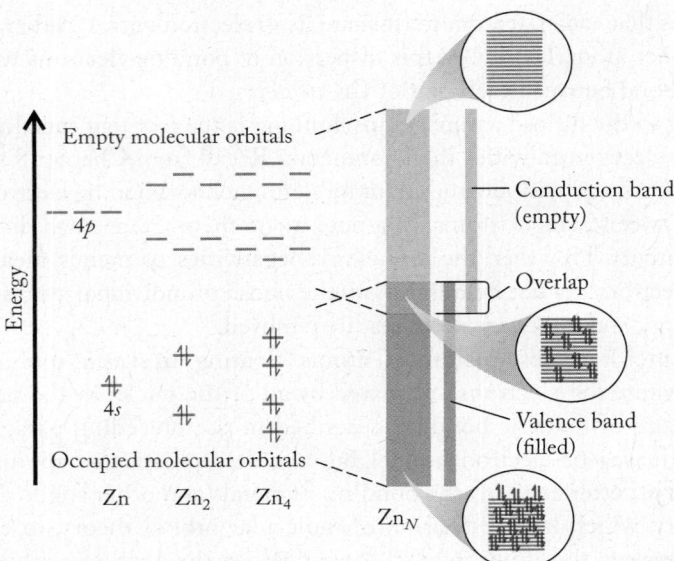

FIGURE 12.3 As the filled 4s atomic orbitals of an increasing number of Zn atoms overlap, they form a filled valence band (purple). An empty conduction band (gray) is produced by combining the empty 4p orbitals. The valence and conduction bands overlap each other and electrons move easily from the filled valence band to the empty conduction band.

4p orbitals on zinc, can combine to form additional energy bands. The energy band produced by combining empty 4p orbitals, called a **conduction band**, is also empty and is broad enough to overlap the valence band. This overlap means that electrons from the valence band can move to the conduction band, where they are free to migrate from atom to atom in solid zinc, thereby conducting electricity.

CONCEPT TEST

Is the electrical conductivity of magnesium metal best explained in terms of overlapping conduction and valence bands or in terms of a partially filled valence band? Explain your answer.

(Answers to Concept Tests are in the back of the book.)

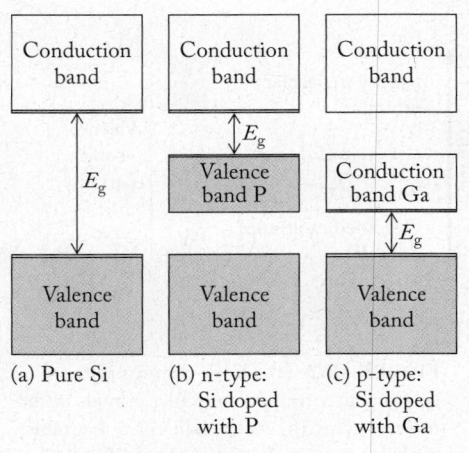

FIGURE 12.4 The electrical conductivity of semiconductors can be greatly enhanced by doping. (a) Pure Si has a band gap E_g of about 100 kJ/mol. (b) Adding a few atoms of P, which have more valence electrons than Si, creates a narrow, filled valence band in the band gap of an n-type semiconductor. (c) Adding a few atoms of Ga, which have fewer valence electrons than Si, creates an empty conduction band in the band gap of a p-type semiconductor.

12.2 Semiconductors

To the right of the metals in the periodic table there is a "staircase" of elements that tend to have the physical properties of metals and the chemical properties of nonmetals. These semimetals (or metalloids) are not as good at conducting electricity as metals, but they are much better at it than nonmetals. We can use band theory to explain this intermediate behavior. In semimetals, conduction and valence bands do not overlap but instead are separated by an energy gap. In silicon, the most abundant semimetal, band theory predicts an energy gap, or **band gap (E_g)**, of about 100 kJ/mol (Figure 12.4a).

Generally, only a few valence-band electrons in Si have sufficient energy to move to the conduction band, which limits silicon's ability to conduct electricity and makes it a **semiconductor**. However, we can enhance the conductivity of solid Si, or of any other elemental semimetal, by replacing some of the Si atoms with atoms of an element of similar atomic radius but with a different number of valence electrons. The replacement process is called *doping*, and the added element is called a *dopant*. Suppose the dopant is a group 15 element such as phosphorus. Each P atom has one more electron than the atom of Si that it replaced. The energy of these additional electrons is different from the energy of the silicon electrons. They populate a narrow band located in the silicon band gap (Figure 12.4b). This arrangement effectively reduces the size of the band gap and increases electrical conductivity because electrons can move more easily across the remaining gaps between the valence and conduction bands. Phosphorus-doped silicon is an example of an **n-type semiconductor** because the dopant contributes extra negative charges (electrons) to the structure of the host element.

The conductivity of solid silicon can also be enhanced by replacing some Si atoms with atoms of a group 13 element, such as gallium (Figure 12.4c). Because Ga atoms have one fewer valence electron than Si atoms, substituting them in the Si structure means fewer valence electrons in the solid. The result is the creation of a narrow Ga conduction band (*acceptor band*, Figure 12.5a) in the Si band gap. Because the gap between the Si valence band and the acceptor band is smaller than the band gap in pure Si, electrons from the valence band more easily move to the acceptor band, increasing electrical conductivity. This array of bands makes Si doped with Ga a **p-type semiconductor** because a reduction in the number of negatively charged electrons in the valence band is equivalent to the presence of positively charged "holes" (Figure 12.5b). The semiconductors used in solid-state electronics are combinations of n-type and p-type.

Doping is not the only way to change the conductivity of semimetals. Compounds prepared from combinations of group 13 and group 15 elements may also behave as semiconductors. For example, gallium arsenide (GaAs) is a semiconductor that emits infrared radiation ($\lambda = 874$ nm) when connected to an electrical circuit. This emission is used in devices such as bar-code readers and CD players. Like silicon, solid gallium arsenide has both a valence band and a conduction band separated by a characteristic band gap. The energy of each photon of light corresponds to the energy gap between the valence band and the conduction band. When electrical energy is applied to the material, electrons are raised to the conduction band. When they fall back to the valence band, they emit radiation. If aluminum is substituted for gallium in GaAs, the band gap increases, and predictably the wavelength of emitted light decreases. For example, a material with the empirical formula $AlGaAs_2$ emits orange-red light ($\lambda = 620$ nm). Many of the multicolored indicator lights in electronic devices use $AlGaAs_2$ semiconductors.

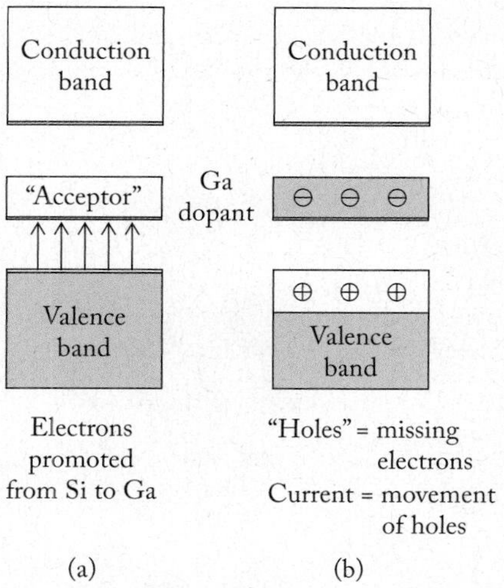

FIGURE 12.5 (a) Band structure of a p-type semiconductor with no electrons in the conduction band of the dopant. The valence band is filled. (b) In a p-type semiconductor, some electrons have enough energy to move from the valence band to the empty conduction band of the dopant, leaving behind positively charged vacancies or "holes." The presence of holes makes the valence band partially filled, increasing electrical conductivity.

CONNECTION We introduced the semimetals (metalloids) in Section 2.4 when we described the structure of the periodic table.

CONNECTION Emission spectra obtained from gas discharge tubes were discussed in Chapter 7.

n-type semiconductor semiconductor containing electron-rich dopant atoms that contribute excess electrons.

p-type semiconductor semiconductor containing electron-poor dopant atoms that cause a reduction in the number of electrons, which is equivalent to the presence of positively charged holes.

SAMPLE EXERCISE 12.1 **Distinguishing p- and n-Type Semiconductors**

Which kind of semiconductor—n-type or p-type—does doping germanium with arsenic create?

Collect and Organize We are asked to determine the type of semiconductor formed when germanium (Ge, group 14) is doped with arsenic (As, group 15). Figure 12.4 helps us distinguish between n- and p-type semiconductors.

Analyze When the dopant has more valence electrons than the host semimetal, the two form an n-type semiconductor, as in Figure 12.4(b). If the dopant has fewer valence electrons than the host, the result is a p-type semiconductor (Figure 12.4c).

Solve Arsenic is in group 15, which means that its atoms have one more valence electron than the atoms of Ge, a group 14 element. This makes As-doped Ge an n-type semiconductor.

Think about It Arsenic is a good candidate for a dopant to make an n-type semiconductor with germanium because As atoms are nearly the same size as Ge atoms and fit easily into the structure of solid Ge.

Practice Exercise Gallium arsenide (GaAs) is a semiconductor used in optical scanners in retail stores. GaAs can be made an n-type or a p-type semiconductor by replacing some of the As atoms with another element. Which element—Se or Sn—would form an n-type semiconductor with GaAs?

(Answers to Practice Exercises are in the back of the book.)

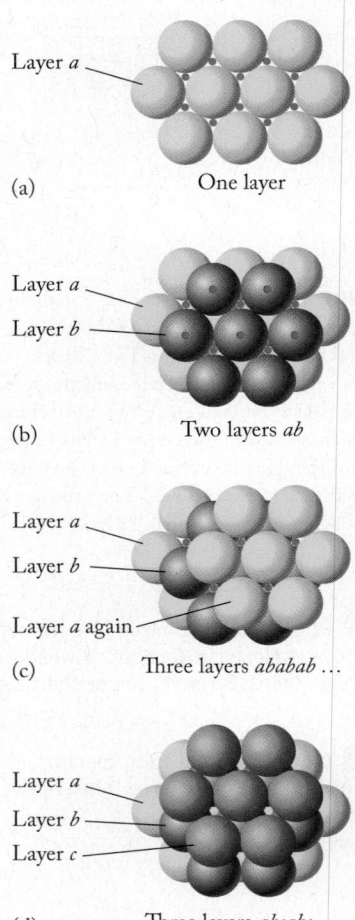

Layer *a*

(a)   One layer

Layer *a*

Layer *b*

(b)   Two layers *ab*

Layer *a*

Layer *b*

Layer *a* again

(c)   Three layers *ababab* ...

Layer *a*

Layer *b*

Layer *c*

(d)   Three layers *abcabc* ...

FIGURE 12.6 Two equally efficient ways to stack layers of atoms (or any particles of equal size). In both stacking patterns the atoms in all layers (shown in different colors to distinguish one layer from another) are packed as closely together as possible. In both stacking patterns the first *a* layer (a) is represented by yellow spheres and (b) the second layer, *b*, of purple spheres is nestled into some of the spaces (marked by green dots) between the spheres in the *a* layer below. (c) In an *ababab* ... stacking pattern, atoms in the third layer are nestled directly above the atoms in the first layer, creating a second *a* layer shown in yellow. (d) In an *abcabc* ... stacking pattern, atoms in the third layer *are not* directly above the atoms in the first layer, creating a *c* layer of atoms shown in red.

12.3 Structures of Metals

In Section 12.1 we learned that the atoms in a piece of copper metal are arranged in such a way that each Cu atom touches 12 other Cu atoms (Figure 12.1b). In this section we take a more detailed look at how the atoms in copper and other metals are stacked together.

Stacking Patterns

When a metal is heated above its melting point and then slowly allowed to cool, it solidifies into a **crystalline solid**, that is, a solid in which the atoms are arranged in an ordered three-dimensional array called a **crystal lattice**. Think of a crystal lattice as stacked layers (designated *a, b, c,* . . .) of metal atoms packed together as tightly as possible. Each atom in layer *a* touches 6 others in that layer, as shown in Figure 12.6(a). The atoms in layer *b* nestle into some of the spaces created by the atoms of layer *a* (Figure 12.6b), just like oranges in a fruit-stand display or cannonballs at an 18th-century fort (Figure 12.7). Similarly, the atoms in a third

(a)

(b)

FIGURE 12.7 (a) Stacks of oranges in a grocery store and (b) cannonballs at an 18th-century fort illustrate closest-packed arrays of spherical objects.

layer, *c*, nestle among those in layer *b*. However, two different alignments are possible for the atoms in layer *c*. They can align directly above the atoms in layer *a* (Figure 12.6c), or they can nestle into the atoms of layer *b* in such a way that they are not aligned directly above the layer *a* atoms (Figure 12.6d). When a fourth layer is nestled into the spaces of layer *c*, the fourth layer atoms lie directly above the layer *a* atoms. In Figure 12.6(c), we have a stacking pattern *ababab* . . . throughout the crystal, and in Figure 12.6(d) we show the stacking pattern *abcabc* . . . .

Let's revisit the bonding pattern between the atoms in a piece of copper metal shown in Figure 12.1(b). Note how the central Cu atom in the pattern, labeled "0" is bonded to 12 other Cu atoms. Are either of the stacking patterns we have just discussed consistent with the bonding pattern in Figure 12.1(b)? To answer this question, think of the three Cu atoms numbered 10–12 at the bottom of Figure 12.1(b) as a part of an *a* stacking layer. This makes Cu atoms 0–6 part of a *b* layer. Now, are atoms 7–9 in the third layer directly above atoms 10–12? Close inspection discloses that they are not: the triangle formed by atoms 10–12 is pointed toward us, but the triangle formed by atoms 7–9 is pointed away from us. This means that the third layer is not the same as the first, making it a *c* layer. It turns out that a fourth layer of Cu atoms would fit directly above the first, giving us another *a* layer and an *abcabc* . . . stacking pattern.

Stacking Spheres and Unit Cells

Whether the atoms in a metal like copper are stacked in an *ababab* or an *abcabc* pattern determines the shape of the crystals the metal forms when it slowly cools and solidifies from the molten state. To see how crystal structures are linked to stacking patterns, let's take a closer look at a cluster of atoms in the *ababab* . . . stacking pattern (Figure 12.8). This cluster forms a *hexagonal* (six-sided) prism of closely packed atoms. In fact, they are as tightly packed as they can be, so the crystal structure is called **hexagonal closest-packed (hcp)**. As Figure 12.9 shows, the atoms in 16 metallic elements have an hcp crystal structure. In these metals, the 17-atom cluster in Figure 12.8(c) serves as an atomic-scale building

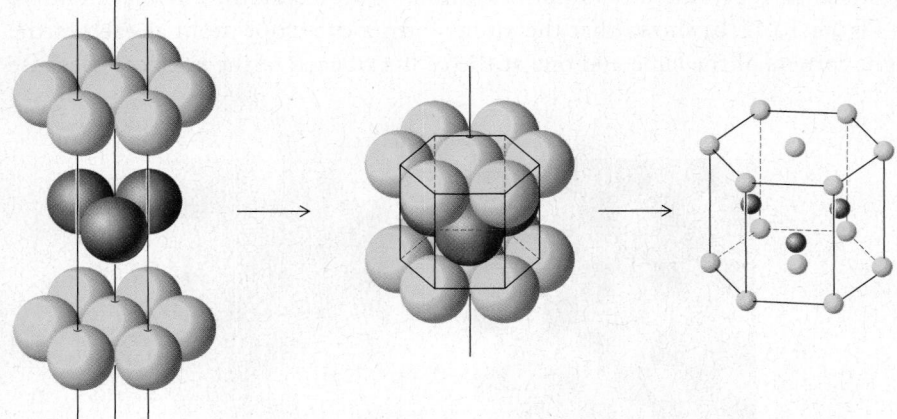

(a) *ababab* . . . layering   (b) Hexagonal closest-packed structure   (c) Hexagonal unit cell

FIGURE 12.8 (a, b) A hexagonal closest-packed (hcp) crystal structure and (c) its hexagonal unit cell represent a highly efficient way to pack atoms in a crystalline solid. Sixteen metals, including all those in groups 3 and 4, have the hcp crystal structure.

crystalline solid a solid made of an ordered array of atoms, ions, or molecules.

crystal lattice a three-dimensional array of particles (atoms, ions, or molecules) in a crystalline solid.

hexagonal closest-packed (hcp) a crystal lattice in which the layers of atoms or ions in hexagonal unit cells have an *ababab* . . . stacking pattern.

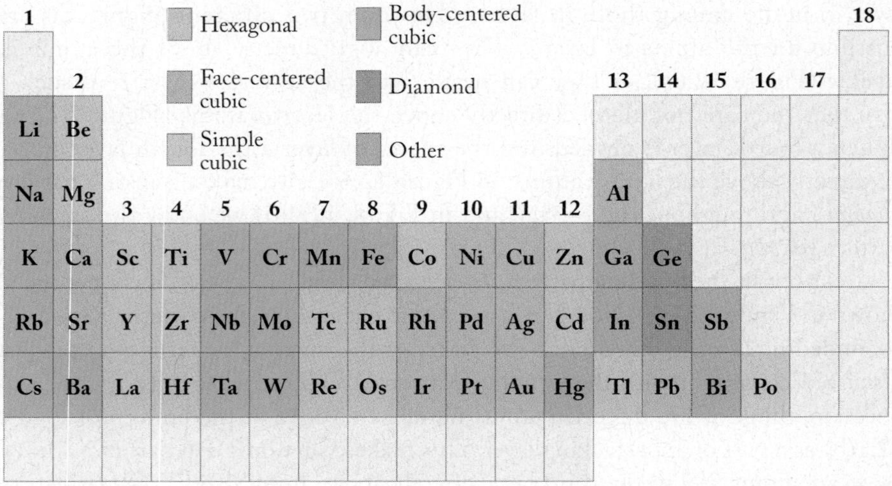

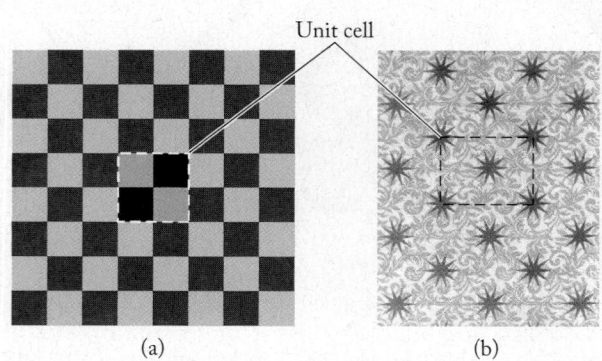

FIGURE 12.9 Unit cells of metals in the periodic table. The five metals designated "Other" have unit cells more complicated than can easily be described in this book.

CHEMTOUR Crystal Packing

CHEMTOUR Unit Cell

unit cell the basic repeating unit of the arrangement of atoms, ions, or molecules in a crystalline solid.

hexagonal unit cell an array of closest-packed particles that has seven of the particles on the top and bottom faces of a hexagonal prism and three of them in a middle layer.

crystal structure an ordered arrangement in three-dimensional space of the particles (atoms, ions, or molecules) that make up a crystalline solid.

cubic closest-packed (ccp) a crystal structure composed of face-centered cubic unit cells and layers of particles having an *abcabc* . . . stacking pattern.

face-centered cubic (fcc) unit cell an array of closest-packed particles that has eight of the particles at the corners of a cube and six of them at the centers of each face of the cube.

packing efficiency percentage of the total volume of a unit cell occupied by the spheres.

block—a pattern of atoms repeated over and over again in all three dimensions in the metal.

We call each of these building blocks a **unit cell**, and in the example in Figure 12.8 we have a **hexagonal unit cell**. A unit cell represents the minimum repeating pattern that describes the three-dimensional array of atoms forming the crystal lattice of any crystalline solid, including metals. Think of unit cells as three-dimensional microscopic analogs of the two-dimensional repeating pattern in fabrics, wrapping paper, or even a checkerboard. Look carefully at Figure 12.10 to confirm that the outlined portion represents the minimum repeating pattern in the checkerboard and in the paper. A unit cell plays the same role in the **crystal structure** of a solid. The crystal structure of an element or compound describes the location of the atoms in three-dimensional space.

Atoms in solid copper and 11 other metals adopt the *abcabc* . . . stacking pattern when they solidify. What is the unit cell in this stacking pattern? Consider what happens when we take the 14-atom cluster in Figure 12.11(a), rotate it and tip it 45° to get to the orientation shown in Figure 12.11(b). The black outline in Figure 12.11(b) shows that the atoms form a cube: one atom at each of the eight corners of the cube and one at the center of each of the six faces. Because

FIGURE 12.10 (a) The highlighted "unit cell" of a checkerboard is the smallest set of squares that defines the pattern repeated over the entire board. (b) This wrapping paper has a more complex pattern. One unit cell is highlighted. Can you locate others?

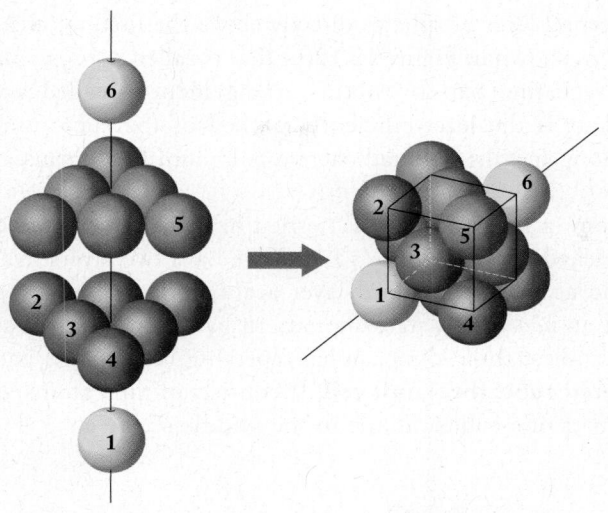

(a) *abcabc*... layering   (b) Face-centered cubic unit cell

FIGURE 12.11 (a) The stacking pattern *abcabc* . . . has a face-centered cubic (fcc) unit cell. (b) The shape of the unit cell can be seen when the layers are tipped 45° and rotated. Note that atoms at adjacent corners do not touch each other, but the three atoms along the diagonal of any face of the cube—atoms 2, 3, 4 here—do touch each other. (c) The unit cell is defined by 14 atoms: one at each of the eight corners of the cube and one in the center of each of the six faces.

the atoms are stacked together as closely as possible, this crystal lattice is called **cubic closest-packed (ccp)**, and the corresponding unit cell is called a **face-centered cubic (fcc) unit cell**. In Figure 12.11(c) the atoms in Figure 12.11(b) have been reduced in size to better show the cubic array. Because this unit cell is a cube, the edges are all of equal length, and the angle between any two edges is 90°. Note that in Figure 12.11(b), the corner atoms do not touch one another but adjacent atoms along the face diagonal do touch each other.

So far we have introduced two *closest-packed* crystal lattices—hexagonal and cubic—along with their associated unit cells (hexagonal and face-centered cubic). The hcp and ccp crystal lattices represent the most efficient ways of arranging solid spheres of equal radius. We can express the **packing efficiency** as the percentage of the total volume of the unit cell occupied by the spheres:

$$\text{Packing efficiency (\%)} = \frac{\text{volume occupied by spheres}}{\text{volume of unit cell}} \times 100\% \qquad (12.1)$$

For both hcp and ccp crystal lattices, the packing efficiency is approximately 74%. We will come back to this calculation at the end of the next subsection, but first, let's look at some other packing arrangements.

Stacking patterns also exist, two of which are shown in Figures 12.12 and 12.13, in which the atoms are arranged close together but not as efficiently as in hcp and ccp lattices. We can arrange the atoms in an *a* layer so that each atom touches four adjacent atoms, an arrangement called *square packing* (Figure 12.12a).

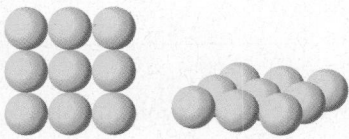      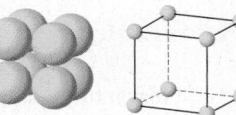

(a) Square-packed *a* layer   (b) Cubic packing   (c) Simple cubic unit cell

FIGURE 12.12 (a) In a square-packed *a* layer, each atom (like the one in the middle) touches four others. (b) If the atoms in all other layers are directly above those in the *a* layer, the stacking pattern is called *cubic packing*. (c) The repeating unit of this pattern is called a *simple cubic* (sc) unit cell with eight atoms at the eight corners of a cube.

simple cubic (sc) unit cell a cell with atoms only at the eight corners of a cube.

body-centered cubic (bcc) unit cell a cell with atoms at the eight corners of a cube and at the center of the cell.

If we add a second layer of spheres directly above the first, we create the *aaa...* stacking pattern shown in Figure 12.12(b) that is called *cubic packing*. The three-dimensional repeating pattern of this arrangement is called a **simple cubic (sc) unit cell**. It is the least-efficiently packed of the cubic unit cells and is quite rare among metals: only radioactive polonium (Po) forms a simple cubic unit cell.

If each atom in a second layer is nestled in the space created by four atoms in a square-packed *a* layer (Figure 12.13a), we have two layers in an *ab* stacking pattern. If the atoms in the third layer are directly above those in the first, then we have an *ababab...* stacking pattern based on layers of square-packed atoms. The simplest three-dimensional repeating unit of this pattern is called a **body-centered cubic (bcc) unit cell**. It consists of nine atoms, one at each of the eight corners of a cube and one in the middle of the cube (Figure 12.13b).

  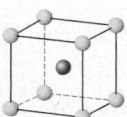

(a) Two square-packed layers in an *ab* pattern   (b) Body-centered cubic unit cell

FIGURE 12.13 (a) The atoms represented by purple spheres in the second *b* layer nestle into the spaces between the square-packed atoms (yellow spheres) in the *a* layer. (b) Atoms in the third layer are directly above those in the first, producing an *ababab...* stacking pattern and a *body-centered cubic* (bcc) unit cell.

All the group 1 metals and many transition metals have bcc unit cells (Figure 12.9). Table 12.1 summarizes the different stacking patterns, packing efficiencies, and unit cells described in this section.

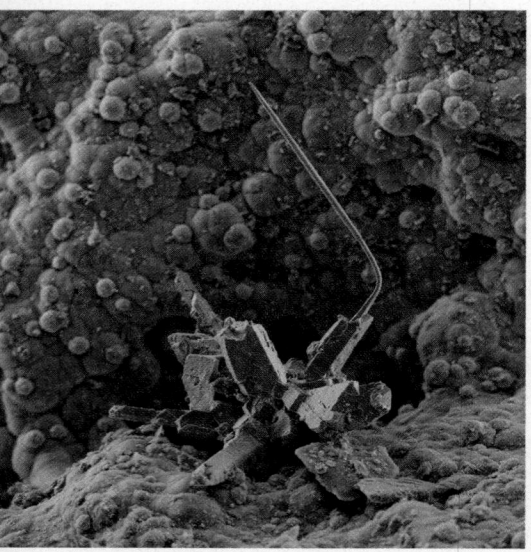

FIGURE 12.14 Copper metal forms cubic crystals in which its atoms are arranged in a cubic closest-packed pattern.

TABLE 12.1	**Summary of Unit Cells, Stacking Patterns, and Packing Efficiency for Solid Spheres**				
Lattice Name	Unit Cell	Type of Packing	Stacking Pattern	Number of Nearest Neighbors	Packing Efficiency
Hexagonal closest-packed (hcp)	Hexagonal	Close packing	*ababab...*	12	74%
Cubic closest-packed (ccp)	Face-centered cubic (fcc)	Close packing	*abcabc...*	12	74%
Body-centered cubic packing	Body-centered cubic (bcc)	Square packing	*abab...*	8	68%
Cubic packing	Simple cubic (sc)	Square packing	*aaaa...*	6	52%

FIGURE 12.15 Some metals form hexagonal shaped crystals. Tungsten is one such metal: it has an hcp crystal lattice and a hexagonal unit cell.

Crystalline solids with cubic unit cells form crystals that are cubic in appearance on a macroscopic scale (Figure 12.14), whereas crystalline solids with hexagonal unit cells tend to form hexagonal crystals (Figure 12.15).

CONCEPT TEST ....

What is the difference between a crystal lattice and a unit cell?

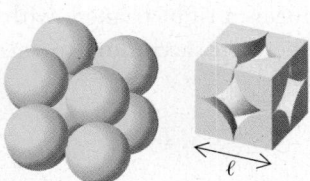 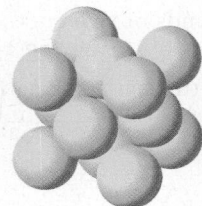 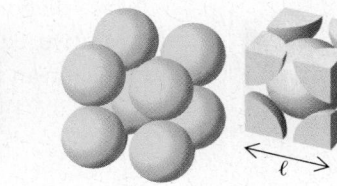

(a) Simple cubic:
Atoms touch along edge

(b) Face-centered cubic:
Atoms touch along face diagonal

(c) Body-centered cubic:
Atoms touch along body diagonal

FIGURE 12.16 Whole-atom and cutaway views of cubic unit cells. (a) In a simple cubic unit cell each corner atom of the unit cell is part of eight unit cells. Atoms along each edge touch. (b) In a face-centered cubic unit cell the face atoms are part of two unit cells. Atoms along the face diagonal touch. (c) In a body-centered cubic unit cell, one atom in the center lies entirely in one unit cell. The atoms along the body diagonal touch.

Unit-Cell Dimensions

Figure 12.16 shows whole-atom and cutaway views of sc, fcc, and bcc unit cells. These views also provide us with a way to determine how many equivalent atoms are in each type of cubic unit cell.

Let's start with the simple cubic unit cell (Figure 12.16a). Note how only a fraction of each corner atom is inside the unit-cell boundary. In a crystal lattice with this unit cell, each atom is a corner atom in eight unit cells (Figure 12.17a). Thus each atom contributes the equivalent of one-eighth of an atom to the unit cell. There are eight corners in a cube, so there is a total of

$\frac{1}{8}$ corner atom/~~corner~~ $\times$ 8 ~~corners~~/unit cell = 1 atom/unit cell

This calculation applies to the corner atoms in any type of cubic unit cell. Note that the two corner atoms along each edge in Figure 12.16(a) touch each other. Therefore the edge length ℓ in the simple cubic unit cell is equal to twice the atomic radius:

$$\ell = 2r$$

In an fcc unit cell (Figure 12.16b) there are eight corner atoms and one atom in the center of each of the six faces (Figure 12.17b). Each face atom is shared by the two unit cells that abut each other at that face. Therefore each unit cell "owns" half of each face atom, making a total of

$\frac{1}{2}$ face atom/~~face~~ $\times$ 6 ~~faces~~/fcc unit cell
= 3 face atoms/fcc unit cell

As just noted, every cubic unit cell owns the equivalent of one corner atom. Therefore, an fcc unit cell consists of

1 corner atom + 3 face atoms = 4 atoms per fcc unit cell

To relate the size of these atoms to the dimensions of the fcc unit cell, note in the cutaway view of Figure 12.16(b) that the corner atoms do not touch one another but adjacent atoms along the face diagonal do touch each other. Therefore, a face diagonal spans the radius r of two corner atoms and the diameter (2 radii = 2r) of a face atom. Therefore the length of a face diagonal is 1 + 2 + 1 = 4 atomic radii = 4r.

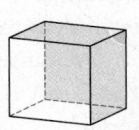 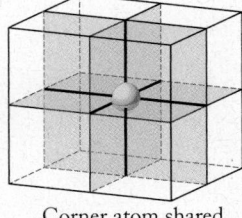

Corner atom in 1 unit cell | Corner atom shared by 8 unit cells
(a)

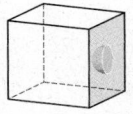 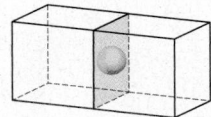

Face-centered atom in 1 unit cell | Face-centered atom shared by 2 unit cells
(b)

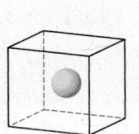 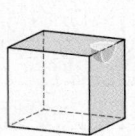 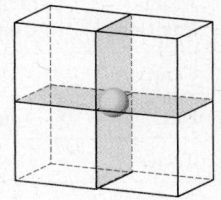

Body-centered atom in 1 unit cell | Edge atom in 1 unit cell | Edge atom shared by 4 unit cells
(c) | (d)

FIGURE 12.17 Crystal lattices illustrating (a) corner atoms shared by eight unit cells, (b) face atoms shared by two unit cells, (c) center atoms entirely in one unit cell, and (d) edge atoms shared by four unit cells.

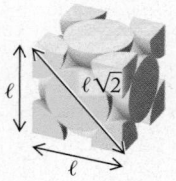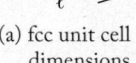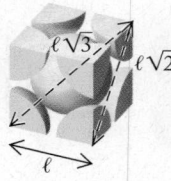

(a) fcc unit cell dimensions   (b) bcc unit cell dimensions

FIGURE 12.18 (a) In a face-centered cubic unit cell, the face diagonal forms a right triangle with two adjoining edges each of length ℓ. Applying the Pythagorean theorem to this triangle yields $r = 0.3536\ell$. (b) In a body-centered cubic unit cell, the atoms touch along a body diagonal. Applying the Pythagorean equation to the right triangle formed by an edge, a face diagonal, and a body diagonal yields $r = 0.4330\ell$.

A face diagonal connects the ends of two edges and forms a right triangle with those two edges, each of length ℓ (Figure 12.18a). Therefore, according to the Pythagorean theorem, the length of a face diagonal is

$$\text{Face diagonal} = 4r = \sqrt{\ell^2 + \ell^2} = \sqrt{2\ell^2} = \ell\sqrt{2}$$

$$r = \frac{\ell\sqrt{2}}{4} = 0.3536\ell \qquad (12.2)$$

Now let's focus on the bcc unit cell in Figure 12.16(c). In addition to the one atom from one-eighth of an atom at each of the eight corners, there is also one atom in the center of the cell that is entirely within the cell (Figure 12.17c). This means a bcc unit cell consists of

1 corner atom + 1 center atom = 2 atoms per bcc unit cell

Relating unit-cell edge length to atomic radius in a bcc cell is complicated by the fact that, in addition to not touching along the edges, adjacent atoms along any face diagonal do not touch each other. However, each corner atom does touch the atom in the center of the cell, which means that the atoms touch along a *body diagonal*, which runs between opposite corners through the center of the cube. In the cutaway view in Figure 12.18(b), the body diagonal runs from the bottom left corner of the front face to the top right corner of the rear face. It spans (1) the radius of the front-face bottom left corner atom, (2) the diameter (2 radii) of the central atom, and (3) the radius of the rear-face top right atom, making the length of the body diagonal equivalent to $4r$.

We can again use the Pythagorean theorem to determine the relationship between ℓ and r. A right triangle is formed by an edge, a face diagonal, and a body diagonal serving as the hypotenuse of the triangle (Figure 12.18b). Using the face-diagonal value $\ell\sqrt{2}$, we get

$$\text{Body diagonal} = 4r = \sqrt{(\text{edge length})^2 + (\text{face diagonal})^2}$$

$$= \sqrt{\ell^2 + (\ell\sqrt{2})^2} = \sqrt{\ell^2 + \ell^2(2)} = \sqrt{3\ell^2} = \ell\sqrt{3}$$

so

$$r = \frac{\ell\sqrt{3}}{4} = 0.4330\ell \qquad (12.3)$$

Table 12.2 summarizes how atoms in different locations in sc, bcc, and fcc unit cells contribute to the total number of atoms in each unit cell, while Table 12.3 summarizes the number of equivalent atoms and the relationship between r and ℓ for the three cubic unit cells.

TABLE 12.2	Contributions of Atoms to Cubic Unit Cells	
Atom Position	**Contribution to Unit Cell**	**Unit-Cell Type**
Center	1 atom	bcc
Face	$\frac{1}{2}$ atom	fcc
Corner	$\frac{1}{8}$ atom	bcc, fcc, sc

TABLE 12.3	Summary of Unit Cells, Equivalent Atoms, and the Relationship between Radius and Edge Length of Atoms to Cubic Unit Cells	
Unit Cell	**Number of Equivalent Atoms per Unit Cell**	**Relationship between r and ℓ**
Simple cubic	1	$r = \dfrac{\ell}{2} = 0.50\ell$
Body-centered cubic	2	$r = \dfrac{\ell\sqrt{3}}{4} = 0.4330\ell$
Face-centered cubic	4	$r = \dfrac{\ell\sqrt{2}}{4} = 0.3536\ell$

SAMPLE EXERCISE 12.2 **Calculating Atomic Radius from Unit Cell Dimensions**

The most stable form of iron at room temperature, called *ferrite*, is shown in Figure 12.19. Its unit cell has an edge length of 287 pm. Calculate the radius in picometers of the iron atoms in it. Check your answer against the data in Appendix 3.

Collect and Organize The unit cell of ferrite consists of eight corner atoms and one atom in the center, a bcc unit cell. We know the edge length of the cell: 287 pm.

Analyze The iron atoms do not touch along the unit cell edges or along any face diagonal but do touch along the body diagonals. Table 12.3 gives the relationship between r and ℓ for a bcc unit cell. We know that the radius is a little less than half the edge length ($\ell/2$), so we expect a value less than 144 pm for r.

Solve
$$r = 0.4330 \times 287 \text{ pm} = 124 \text{ pm}$$

Think about It The value of r, 124 pm, is indeed less than half the value of the edge length ($\ell/2 = 144$ pm). From Appendix 3, the average atomic radius of iron atoms is 126 pm, so the result of this calculation is reasonable.

Practice Exercise At 1070°C the most stable form of iron is *austenite* (shown in Figure 12.20). The edge length of its unit cell is 361 pm. What is the atomic radius of iron in austenite?

(a)

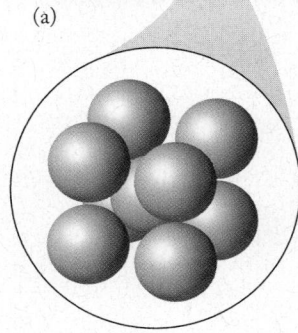

(b) bcc unit cell

FIGURE 12.19 (a) The pattern in this newly cut surface of the meteorite that caused the Odessa crater in Texas about 50,000 years ago is due to crystals of ferrite, which grew when the meteor formed from molten iron 4.5 billion years ago. (b) The unit cell of ferrite.

SAMPLE EXERCISE 12.3 **Using Unit Cell Dimensions to Calculate Density**

Calculate the density of iron (ferrite) in grams per cubic centimeter at 25°C, given that its bcc unit cell has an edge length of 287 pm.

Collect and Organize We are to calculate the density of iron at 25°C knowing its unit-cell geometry and edge length. The fact that ferrite has a bcc unit cell means there are two Fe atoms per cell. The molar mass of Fe is 55.84 g/mol, and the formula for the volume of a cube of edge length ℓ is $V = \ell^3$.

Analyze We assume that the density of the unit cell is the same as the density of solid Fe. The density of the Fe bcc unit cell is the mass of two Fe atoms divided by the volume of the cell. We need to calculate the mass of two Fe atoms starting with the molar mass. The conversion includes dividing by Avogadro's number to calculate the mass of each Fe atom in grams. A piece of iron sinks when immersed in water so we predict that the density of iron should be greater than 1.0 g/cm^3.

Solve We calculate first the mass m of two Fe atoms:

$$m = \frac{55.84 \text{ g Fe}}{1 \text{ mol Fe}} \times \frac{1 \text{ mol Fe}}{6.022 \times 10^{23} \text{ atoms Fe}} \times 2 \text{ atoms Fe}$$
$$= 1.855 \times 10^{-22} \text{ g Fe}$$

and then the volume of the cell in cubic centimeters:

$$V = \ell^3 = (287 \text{ pm})^3 \times \frac{(10^{-10} \text{ cm})^3}{1 \text{ pm}^3} = 2.364 \times 10^{-23} \text{ cm}^3$$

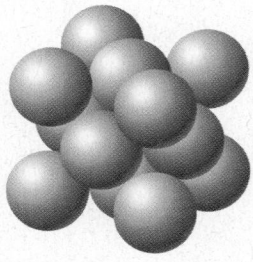

FIGURE 12.20 The fcc unit cell of austenite.

The density is

$$d = \frac{m}{V} = \frac{1.855 \times 10^{-22}\ \text{g}}{2.364 \times 10^{-23}\ \text{cm}^3} = 7.85\ \text{g/cm}^3$$

Think about It According to the data in Appendix 3, the density of iron is 7.874 g/mL, so the result of the calculation is reasonable and reflects the prediction we made. The density of the unit cell should equal the density of a bulk sample because the latter is composed of a crystal lattice of bcc unit cells.

Practice Exercise Silver and gold both crystallize in face-centered cubic unit cells with edge lengths of 407.7 and 407.0 pm, respectively. Calculate the density of each metal and compare your answers with the densities listed in Appendix 3.

Earlier in this section we mentioned that ccp and hcp are the most efficient packing schemes for spheres. Now that we know more about the fcc unit cell, let's take another look at the calculation of packing efficiency using Equation 12.1. The unit cell for a ccp arrangement of solid spheres of radius r is an fcc unit cell that contains four equivalent spheres. The volume occupied by these spheres is

$$V_{\text{spheres}} = 4\left(\tfrac{4}{3}\pi r^3\right) = \tfrac{16}{3}\pi r^3$$

From Table 12.3, $r = \left(\ell\sqrt{2}\right)/4$, and the unit-cell edge (ℓ) is

$$\ell = \frac{4r}{\sqrt{2}} = 2.828r$$

This equation enables us to express the volume of the unit cell in terms of r:

$$V_{\text{unit cell}} = \ell^3 = (2.828r)^3 = 22.62r^3$$

Substituting V_{spheres} and $V_{\text{unit cell}}$ into Equation 12.1:

$$\text{Packing efficiency (\%)} = \frac{V_{\text{spheres}}}{V_{\text{unit cell}}} \times 100\% = \frac{\tfrac{16}{3}\pi r^3}{22.62r^3} \times 100\% = 74.1\%$$

This means that the most efficient packing of spheres results in about 26% of the total volume being empty.

12.4 Alloys

People have been using metals for tools, weapons, currency, and jewelry for tens of thousands of years. For most of that time, only three elements—copper, silver, and gold—were used because these are the only ones found as free metals in Earth's crust. Even so, most silver and copper occur not as pure metals but in ores such as argentite (Ag_2S), chalcopyrite ($CuFeS_2$), and chalcocite (Cu_2S). **Ores** are naturally occurring compounds or mixtures of compounds from which elements can be extracted.

Around 6000 years ago, metal technology took a giant leap forward when artisans in areas that are now Iraq and Pakistan discovered how to convert copper ore, principally $CuFeS_2$, to copper metal. The process involved pulverizing the ore and then baking it in ovens. Baking initiated a chemical reaction with O_2 (from air), which converted the Cu in $CuFeS_2$ to CuO. In the second step in the process, CuO was reacted with carbon monoxide, which was produced by burning wood or charcoal (mostly carbon) in furnaces with insufficient supplies of air:

$$CuO(s) + CO(g) \rightarrow Cu(s) + CO_2(g) \qquad (12.4)$$

ore a mineral that contains one or more metals valuable enough to be mined.

alloy a blend of a host metal and one or more other elements, which may or may not be metals, that are added to change the properties of the host metal.

substitutional alloy an alloy in which atoms of the nonhost metal replace host atoms in the crystal lattice.

One disadvantage of copper tools and weapons is that the metal is very malleable, which means that copper objects are easily bent and damaged. We can explain the malleability of Cu (and other metals) in terms of the metallic bonds between its atoms and their cubic closest-packed crystal structure. We have noted that copper atoms are weakly bonded to their nearest neighbors. This arrangement gives the atoms in one layer the ability, under stress, to slip past atoms in an adjacent layer (Figure 12.21). When the stress is relieved and the atoms stop slipping, many have different atoms as nearest neighbors but the overall crystal structure is still cubic closest-packed. The ease with which copper atoms slip past each other made it easy for prehistoric metalworkers to hammer copper metal into spear points and shields, but it also meant that those objects could easily be damaged in battle.

Substitutional Alloys

About 5500 years ago, people living around the Aegean Sea discovered that mixing molten tin and copper produced bronze, a material that was much stronger than either tin or copper alone. Its discovery ushered in the Bronze Age. Bronze is an **alloy**, a metallic material made when a host metal is blended with one or more other elements, which may or may not be metals, thereby changing the properties of the host metal. Like the mixtures discussed in Chapter 1, alloys can be classified according to their composition as homogeneous or heterogeneous mixtures. Bronze is a *homogeneous alloy*, a solid solution in which the atoms of the added element(s) (in this case tin) are randomly but uniformly distributed among the atoms of the host (copper in this example). *Heterogeneous alloys* consist of matrices of atoms of host metals interspersed with small "islands" made up of individual atoms of other elements. In both cases, the compositions may vary over a limited range. In contrast, *intermetallic compounds* have a reproducible stoichiometry and constant composition (just like chemical compounds) but are still commonly referred to as alloys and are considered to be a subgroup within homogeneous alloys. An example of an intermetallic compound is Ag_3Sn, an alloy of silver and tin used in dental fillings. It is a homogeneous mixture of silver and tin atoms in exactly a 3:1 ratio.

If alloys are mixtures of metals, and pure metals have crystal lattices as described in Section 12.3, how are the metal atoms arranged in the crystal lattices and unit cells of alloys? The answer to this question gives rise to another classification system: a **substitutional alloy** is one in which atoms of the nonhost metal replace host atoms in the crystal lattice. Bronze is a *homogeneous, substitutional alloy*, in which the tin concentration can be as high as 30% by mass. Substitutional alloys may form between metals that have the same crystal lattice and atomic radii that are within about 15% of each other.

CONCEPT TEST •

Is it accurate to call bronze a *solution* of tin *dissolved* in copper? Explain why or why not.

• •

Figure 12.22 illustrates one layer of the crystal lattice of bronze. The radii of copper and tin atoms are similar—128 pm and 140 pm, respectively. Inserting the slightly larger Sn atoms in the cubic closest-packed Cu crystal lattice disturbs the structure a little, making the planes of copper atoms "bumpy" instead of uniform

External force

Metal is deformed

FIGURE 12.21 Copper and other metals are malleable because their atoms are stacked in layers that can slip past each other under stress. Slippage is possible because of the diffuse nature of metallic bonds and the relatively weak interactions between pairs of atoms in adjoining layers.

∞ **CONNECTION** The classification of homogeneous and heterogeneous mixtures and the difference between compounds and mixtures were described in Chapter 1.

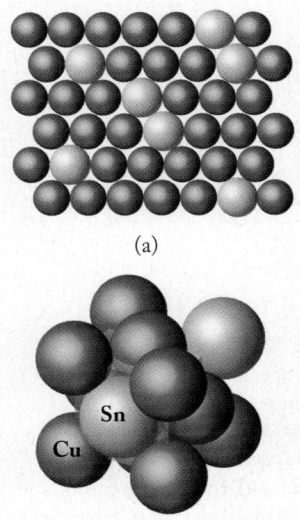

(a)

Sn

Cu

(b)

FIGURE 12.22 Two atomic-scale views of one type of bronze, a substitutional alloy. (a) A layer of close-packed copper (Cu) atoms interspersed with a few atoms of tin (Sn). (b) One possible unit cell for bronze. In this case tin atoms have replaced one corner Cu atom and one face Cu atom.

FIGURE 12.23 The larger Sn atoms in bronze disturb the Cu crystal lattice, producing atomic-scale bumps in the slip plane (wavy line) between layers of Cu atoms. These bumps make it more difficult for Cu atoms to slide by each other when an external force is applied.

(Figure 12.23). This atomic-scale roughness makes it more difficult for the copper atoms to slip past one another. Less slippage makes bronze less malleable than copper, but being less malleable also means that bronze is harder and stronger.

There are many other substitutional alloys. Some are also copper based, including brass (zinc alloyed with copper) and pewter (tin alloyed with copper and antimony). However, most substitutional alloys in the modern world are *ferrous alloys*, so called because the host metal is iron. An important class of ferrous alloys is that of the rust-resistant *stainless steels*, which contain about 10% nickel and up to 20% chromium. When atoms of Cr on the surface of a piece of stainless steel combine with oxygen, they form a layer of Cr_2O_3 that bonds tightly to the surface and protects the metallic material beneath from further oxidation. This resistance to surface discoloration due to corrosion means that these alloys "stain less" than pure iron.

Interstitial Alloys

The Bronze Age began to wane about 3000 years ago with the discovery that iron oxides could be reduced to iron metal by limiting the air supplied to a wood or charcoal fire. As in the conversion of CuO to Cu (Equation 12.4), the reducing agent for iron oxide is carbon monoxide:

$$Fe_2O_3(s) + 3\,CO(g) \rightarrow 2\,Fe(s) + 3\,CO_2(g) \qquad (12.5)$$

Iron quickly replaced bronze as the metallic material of choice for fabricating tools and weapons both because iron ore is much more abundant in Earth's crust than the ores of copper and tin and because tools and weapons made of iron or ferrous alloys are much stronger than those made of bronze.

Today, the reduction of iron ore is done in blast furnaces, enormous reaction vessels that operate at about 1600°C (Figure 12.24). Iron ore, hot carbon (*coke*), and limestone are added to the top of the vessel. The solid by-products (*slag*) float on top of the molten iron, which is harvested from the bottom. Blast furnaces get their name from blasts of hot air (up to 1000°C) that are injected through nozzles near the bottom of the furnace and that suspend the reactants until iron reduction is complete. It may take as long as 8 hours for a batch of reactants to fall to the bottom of a blast furnace. On their way down, O_2 in the hot-air blasts partially

FIGURE 12.24 (a) Blast furnaces operate continuously at temperatures near 1600°C to convert iron ore into iron. Blasts of hot air inject O_2 into the furnace, which converts C to CO. Limestone is added to react with Si and P impurities. The products of these reactions become part of the slag layer. (b) Molten iron from a blast furnace is further purified in a second furnace, where $O_2(g)$ is injected instead of air.

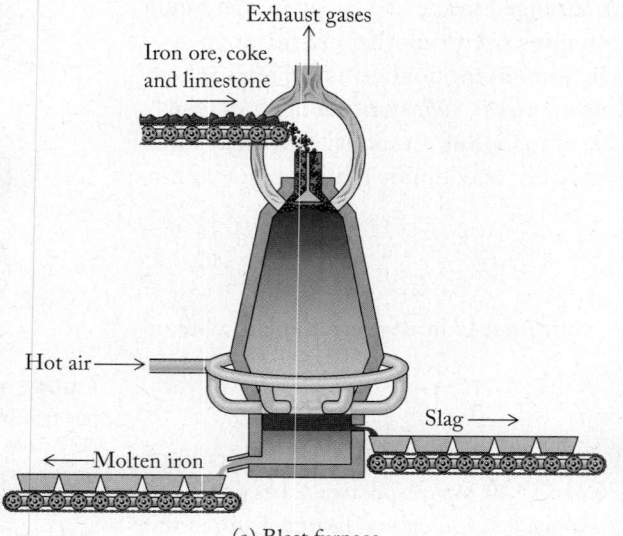

(a) Blast furnace

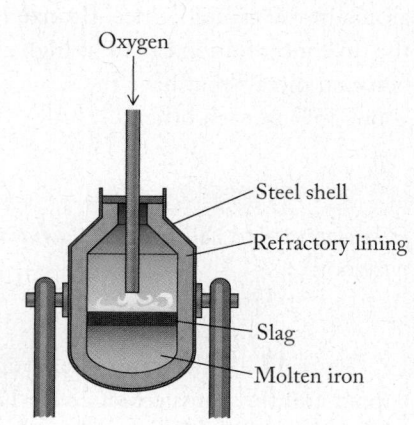

(b) The basic oxygen process

oxidizes the coke to carbon monoxide, and the CO reduces the iron in iron ore as described in Equation 12.5. The limestone ($CaCO_3$) added along with the iron ore decomposes to calcium oxide (CaO, also called lime):

$$CaCO_3(s) \rightarrow CaO(s) + CO_2(g) \qquad (12.6)$$

The lime then reacts with silica impurities in the ore to form calcium silicate:

$$CaO(s) + SiO_2(s) \rightarrow CaSiO_3(\ell) \qquad (12.7)$$

Calcium silicate becomes part of the slag that floats on the denser molten iron at the bottom of the furnace.

Molten iron produced in a blast furnace may contain up to 5% carbon. To reduce the carbon content, the molten iron is transferred to a second furnace. When hot O_2 and additional CaO are injected, some of the carbon is oxidized to CO_2 and any remaining silicon impurities form more $CaSiO_3$ slag.

When the molten iron coming out of this second furnace cools to its melting point of 1538°C, it crystallizes in a body-centered cubic structure before undergoing a phase transition at around 1390°C to austenite, a form of solid iron made up of face-centered cubic unit cells. The spaces, or *holes*, between iron atoms in austenite can accommodate carbon atoms, forming an **interstitial alloy**, so named because the carbon atoms occupy spaces, or *interstices*, between the iron atoms (Figure 12.25).

CONCEPT TEST ....

Is a substitutional alloy a homogeneous or a heterogeneous alloy, or could it be both? Is an interstitial alloy a homogeneous or a heterogeneous alloy, or could it be both?

All interstices in a crystal lattice are not equivalent. Indeed, holes of two different sizes occur between the atoms in any closest-packed crystal lattice (Figure 12.26). The larger holes are surrounded by clusters of six host atoms in the shape of an octahedron and are called *octahedral holes*. The smaller holes are located between clusters of four host atoms and are called *tetrahedral holes*. The data in Table 12.4 show which holes are more likely to be occupied based on the relative sizes of nonhost and host atoms. According to Appendix 3, the atomic radii of C and Fe are 77 and 126 pm, respectively. According to Table 12.4, the ratio 77/126 = 0.61 means that C atoms should fit in the octahedral holes of austenite, as shown in Figure 12.25, but not in the tetrahedral holes.

interstitial alloy atoms of one element occupy the spaces between atoms of the host.

FIGURE 12.25 Carbon steel is an interstitial alloy of carbon in iron. The fcc form of iron (austenite) that forms at high temperatures can accommodate carbon atoms in its octahedral holes.

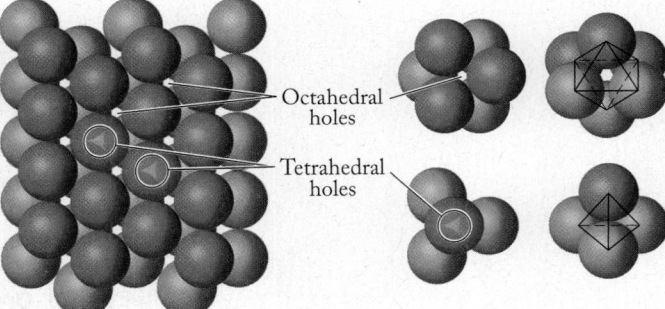

FIGURE 12.26 Close-packed atoms in adjacent layers of a crystal lattice produce octahedral holes surrounded by six host atoms and tetrahedral holes surrounded by four host atoms. Octahedral holes are larger than tetrahedral holes and so can accommodate larger nonhost atoms in interstitial alloys. Note that all the atoms are identical here; the colors are only to distinguish one atom from another.

TABLE 12.4	Atomic Radius Ratios and Location of Nonhost Atoms in Unit Cells of Interstitial Alloys	
Lattice Name	**Hole Type**	**Atomic Radius Ratio** $r_{nonhost}/r_{host}$[a]
hcp or ccp	Tetrahedral	0.22–0.41
hcp or ccp	Octahedral	0.41–0.73
Cubic packing	Cubic	0.73–1.00

[a] Radius ratios as predictors of the crystal lattice in crystalline solids are of limited value because atoms are not truly solid spheres with a constant radius; because the radius of an atom may differ in different compounds, these ranges are approximate.

As austenite with its fcc unit cell cools to room temperature, it converts into the crystalline solid form of iron called ferrite, which is body-centered cubic. The octahedral holes in ferrite are smaller than those in austenite, and they are too small to accommodate carbon atoms. As a result, carbon precipitates as clusters of carbon atoms, or it may react with iron to form iron carbide, Fe_3C. The clusters of carbon and Fe_3C disrupt ferrite's body-centered cubic lattice and inhibit the host iron atoms from slipping past each other when a stress is applied. This resistance to slippage, which is much like that experienced by the copper atoms in bronze (Figure 12.23), makes iron–carbon alloys, known as *carbon steel*, much harder and stronger than pure iron. In general, the higher the carbon concentration, the stronger the steel. However, there is a trade-off in this relationship. As Table 12.5 notes, increased strength and hardness comes at the cost of increased brittleness.

TABLE 12.5	Effect of Carbon Content on the Properties of Steel		
Carbon Content (%)	Designation	Properties	Used to Make
0.05–0.19	Low carbon	Malleable, ductile	Nails, cables
0.20–0.49	Medium carbon	High strength	Construction girders
0.50–3.0	High carbon	Hard but brittle	Cutting tools

CONCEPT TEST

In Chapter 7 we learned that atomic radius increases down a group and decreases across a period. Which would you predict to have the larger radius ratio, a lithium–aluminum or a lithium–magnesium alloy?

SAMPLE EXERCISE 12.4   Predicting the Crystal Structure of a Two-Element Alloy

Sterling silver, which is 93% Ag and 7% Cu by mass, is widely used in jewelry. The presence of Cu inhibits tarnishing and strengthens the alloy. Is this copper–silver alloy a substitutional or an interstitial alloy? Silver has a cubic closest-packed crystal lattice with face-centered cubic unit cells.

Collect and Organize We are asked whether in sterling silver the Cu–Ag alloy is substitutional or interstitial. Atoms of two or more elements form a substitutional alloy when all the atoms are of similar size. We are told that the atoms in solid Ag form fcc unit cells and have a cubic closest-packed crystal lattice.

Analyze The highest $r_{nonhost}/r_{host}$ ratio in Table 12.4 for a ccp lattice, 0.73, means that an interstitial alloy can form only when the radius of the nonhost atoms is less than 73% of the radius of the host atoms. The atomic radii found in Appendix 3 are 144 pm for Ag and 128 pm for Cu. The ratio of the atomic radii of Cu to Ag is 128 pm/144 pm = 0.89. Elements with similar radii are more likely to form substitutional alloys, as we saw with bronze.

Solve The $r_{nonhost}/r_{host}$ ratio of 0.89 indicates that the copper atoms are too big to fit into interstices in the lattice of silver atoms, regardless of whether the holes are tetrahedral or octahedral. Thus an interstitial alloy is impossible. Atoms of Cu can substitute for atoms of Ag in the Ag ccp lattice, however, with some room to spare. Figure 12.9 tells us that both elements form fcc unit cells, so little disruption to the Ag lattice should result from incorporating Cu atoms. Thus, the Cu–Ag alloy, sterling silver, is a substitutional alloy.

Think about It The guidelines for two metals forming a substitutional alloy are that they have the same type of crystal lattice and that their atomic radii are within 15% of each other. The radii of Ag and Cu differ by only 11%, so we would expect copper and silver to form a substitutional alloy.

Practice Exercise Would you expect gold (atomic radius 144 pm) to form a substitutional alloy with silver (atomic radius 144 pm)? With copper (atomic radius 128 pm)?

Aluminum forms alloys with many other elements. Alloys with Mg, Si, Cu, and Zn are the most common and are widely used in aircraft construction. The tabs on beverage cans are manufactured from an aluminum alloy containing Mg and Mn. Because Li has the smallest molar mass (6.941 g/mol) and lowest density (0.534 g/mL) of all the metallic elements, aluminum alloys containing Li are attractive for applications where low mass is essential.

Aluminum and aluminum alloys are widely used as building materials and in fuselages of airplanes because they are corrosion resistant. This resistance is noteworthy because aluminum is actually a very reactive metal. However, when the surface of a piece of aluminum starts to oxidize in air, a thin layer of Al_2O_3 is produced that adheres strongly to the aluminum metal below it and acts as a protective shield against further oxidation.

12.5 Structures of Some Crystalline Nonmetals

In Figure 12.9, the color key identifies the unit cell of germanium and tin as "diamond." We introduced the structure of diamond in the descriptive chemistry feature at the end of Chapter 5 where we described the allotropes of carbon. Carbon, of course, is a nonmetal, but the specific arrangement of carbon atoms in diamond (Figure 12.27a) is also found in some metallic materials. In addition, some diamonds may include metal impurities in the crystal lattice that give rise to distinctive colors.

FIGURE 12.27 Two of carbon's three allotropes. (a) Diamond is a three-dimensional covalent network solid made of carbon atoms, each connected by σ bonds to four adjacent carbon atoms. (b) Graphite is a collection of layers of carbon atoms connected by σ bonds and delocalized π bonds.

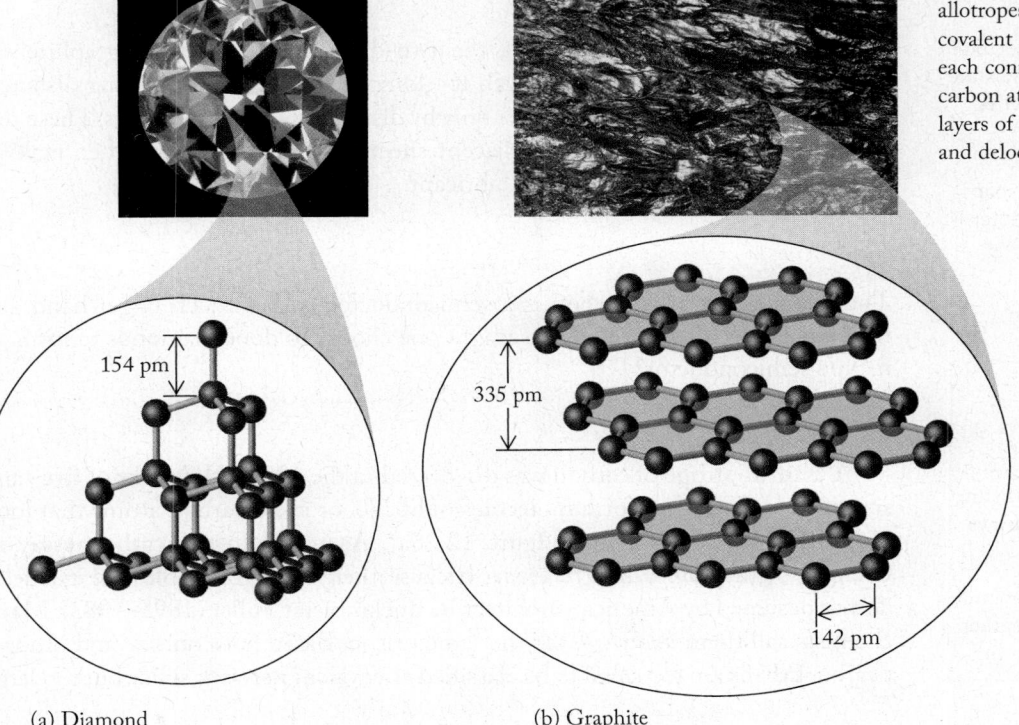

154 pm

335 pm

142 pm

(a) Diamond

(b) Graphite

▶❚❚ **CHEMTOUR** Allotropes of Carbon

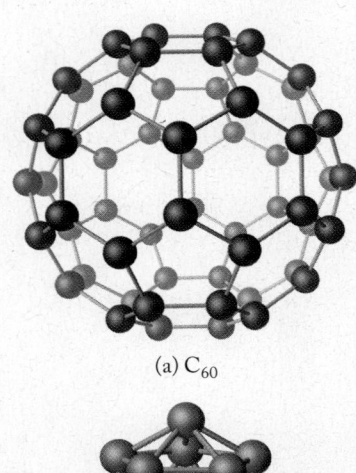

(a) C_{60}

(b) B_{12}

FIGURE 12.28 Some solids are described as clusters, a category between covalent network solids and molecular solids. (a) Fullerenes, the third allotrope of carbon, are clusters. One of them, buckminsterfullerene (C_{60}), is made up of 60 sp^2-hybridized carbon atoms. Both five- and six-membered rings are required to construct the nearly spherical molecule. (b) One crystalline form of boron is a cluster of 12 boron atoms. The clusters are arranged in a close-packed array.

♾ **CONNECTION** Allotropes are structurally different forms of the same physical state of an element, as explained in Chapter 8.

♾ **CONNECTION** We introduced dispersion forces between molecules in Chapter 10.

covalent network solid a solid consisting of atoms held together by extended arrays of covalent bonds.

molecular solid a solid formed by neutral, covalently bonded molecules held together by intermolecular attractive forces.

Diamond is one of three allotropes of carbon, the other two being graphite and fullerenes (Figures 12.27b and 12.28a). Diamond is classified as a crystalline **covalent network solid** because it consists of atoms held together in an extended three-dimensional network of covalent bonds. Each carbon atom in diamond bonds by overlapping one of its sp^3 orbitals with an sp^3 orbital in each of four neighboring carbon atoms, creating a network of carbon tetrahedra. The atoms in these tetrahedra are connected by localized σ bonds, making diamond a poor electrical conductor. The sigma-bond network is extremely rigid, making diamond the hardest natural material known. The atoms of other group 14 elements, particularly silicon, germanium, and tin, also form covalent network solids based on the diamond crystal lattice.

Natural diamond forms from graphite under intense heat (>1700 K) and pressure (>50,000 atm) deep in Earth. Industrial diamonds are synthesized at high temperatures and pressures from graphite or any other source rich in carbon. Synthetic diamonds are used as abrasives and for coating the tips and edges of cutting tools. Diamond has the highest thermal conductivity of any natural substance (five times higher than copper and silver, the most thermally conductive metals), so tools made from diamond do not become overheated. The conditions used to manufacture industrial diamonds mostly yield stones that lack the size and optical clarity of gemstones. The inclusion of impurities in natural diamonds leads to rare and valuable colored diamonds, a process that is difficult to duplicate under laboratory conditions.

By far the most abundant allotrope of carbon is graphite, another covalent network solid, frequently the principal ingredient in soot and smoke and used to make pencils, lubricants, and gunpowder. Graphite contains sheets of carbon atoms in which each atom is connected by sp^2 orbitals to a like orbital in each of three neighboring carbon atoms in a two-dimensional covalent network of six-membered rings (Figure 12.27b). Each carbon–carbon σ bond is 142 pm, which is shorter than the C—C σ bond in diamond (154 pm). Overlapping unhybridized p orbitals on the carbon atoms form a network of π bonds that are delocalized across the plane defined by the rings. The mobility of these delocalized electrons makes graphite a conductor of electricity.

As shown in Figure 12.27(b), the two-dimensional sheets in graphite are 335 pm apart. This distance is much too long to be a covalent bonding distance. Instead, the sheets are held together only by dispersion (London) forces. These relatively weak interactions allow adjacent sheets to slide past each other, making graphite soft, flexible, and a good lubricant.

▸ **CONCEPT TEST** ···

The diamond form of carbon is a semiconductor with a much larger band gap than silicon. With what elements might one choose to dope diamonds to form an n-type semiconductor?

··

A third allotrope of carbon was discovered in the 1980s. Networks of five- and six-atom carbon rings form molecules of 60, 70, or more carbon atoms that look like miniature soccer balls (Figure 12.28a). As noted in the carbon essay in Chapter 5, they are called *fullerenes* because their shape resembles the geodesic domes designed by American architect R. Buckminster Fuller (1895–1983). Many chemists call them *buckyballs* for the same reason. Based both on size and properties, buckyballs are too small to be classified as covalent network solids but too large

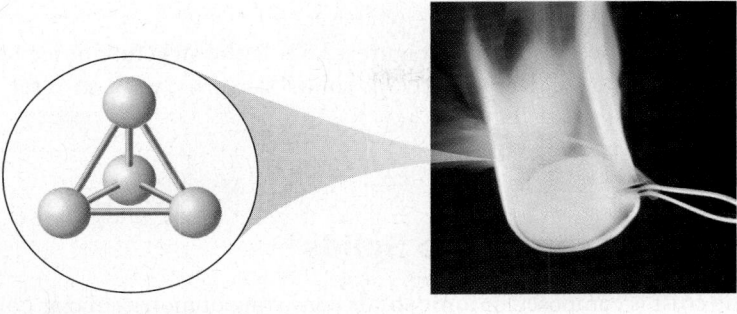

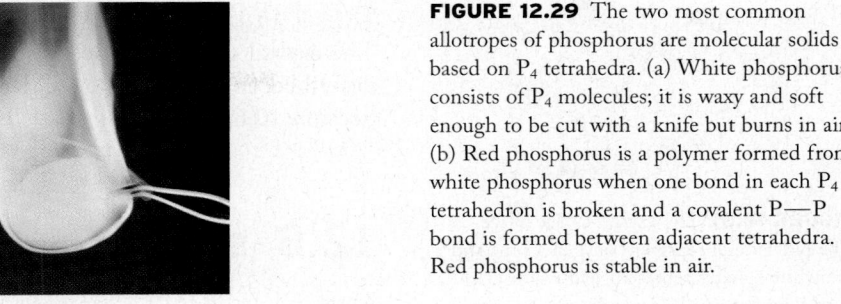

(a) White phosphorus

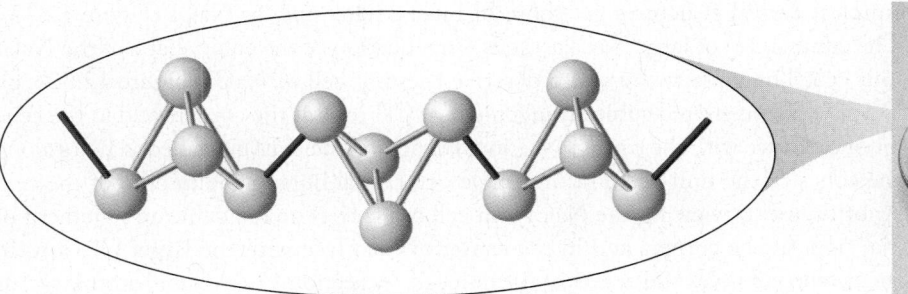

(b) Red phosphorus

FIGURE 12.29 The two most common allotropes of phosphorus are molecular solids based on P_4 tetrahedra. (a) White phosphorus consists of P_4 molecules; it is waxy and soft enough to be cut with a knife but burns in air. (b) Red phosphorus is a polymer formed from white phosphorus when one bond in each P_4 tetrahedron is broken and a covalent P—P bond is formed between adjacent tetrahedra. Red phosphorus is stable in air.

to be molecular solids (discussed below). They fall in an ambiguous zone between small molecules and large networks and are classified as *clusters*.

When fullerenes were discovered, they were believed to be a form of carbon rarely found in nature. In recent years, however, analyses of soot and emission spectra from giant stars have disclosed that fullerenes are present in trace amounts throughout the universe.

Some nonmetals, like boron, have structures like C_{60} and may be considered clusters. For example, one form of boron contains closest-packed arrays of 12-vertex, 20-sided *icosahedra* (singular *icosahedron*; Figure 12.28b) composed of 12 boron atoms.

Other nonmetals form crystalline **molecular solids**, which consist of molecules held together by intermolecular forces. Ice, CO_2, glucose, and all organic molecules crystallize as molecular solids. One of the two most common allotropes of phosphorus, white phosphorus, is a molecular solid consisting of P_4 tetrahedra arranged in a cubic array (Figure 12.29a). White phosphorus is a waxy material that oxidizes rapidly under standard conditions and gives off a yellow-green light in a phenomenon called *phosphorescence*.

The other common phosphorus allotrope, red phosphorus, is not a molecular solid but rather a covalent network solid made of chains of P_4 tetrahedra connected by covalent phosphorus–phosphorus bonds (Figure 12.29b). Both red and white phosphorus melt to give the same liquid consisting of symmetrical P_4 tetrahedral molecules.

Sulfur has more allotropic forms than any other element. Most are molecular solids. This variety of forms arises because sulfur atoms form cyclic (ring) compounds of different sizes, which means that different crystalline arrangements of the molecules are possible. The most common allotropes of sulfur consist of puckered rings (Figure 12.30) containing eight covalently bonded sulfur atoms. Dispersion forces hold one ring to another in solid sulfur. The weakness of these interactions is the reason elemental sulfur is soft and melts at only 115.21°C. The packing of S_8 rings together in the molecular solid results in a complex crystal structure.

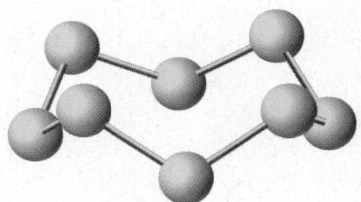

FIGURE 12.30 One form of sulfur is a molecular solid based on puckered S_8 rings. Different allotropes of sulfur have different stacking patterns of these rings.

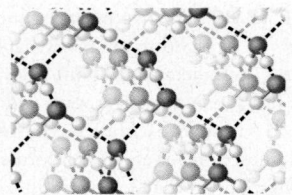

FIGURE 12.31 The crystal lattice of ice. The red spheres represent oxygen atoms and the white spheres represent hydrogen. The solid lines are covalent bonds and the dotted lines are hydrogen bonds.

CONCEPT TEST

The crystal structure of ice is shown in Figure 12.31. Is the structure of ice best described as a molecular solid or a network solid? You may wish to refer to Section 10.6.

12.6 Salt Crystals: Ionic Solids

Most of Earth's crust is composed of **ionic solids** consisting of monatomic or polyatomic ions held together by ionic bonds. Most of these solids are crystalline. The simplest crystal structures are those of binary salts, such as NaCl (Figure 12.32). The cubic shape of large NaCl crystals is a reflection of the cubic shape of the NaCl unit cell. There are two ways to describe the unit cell of NaCl (Figure 12.32b, c). One is a face-centered cubic arrangement of Cl⁻ ions at the corners and in the center of each face with the smaller Na⁺ ions occupying the twelve octahedral holes along the edges of the unit cell and the single octahedral hole in the middle of the cell. Another way of viewing the NaCl unit cell is a face-centered cubic arrangement of Na⁺ ions at the corners and in the center of each face with the larger Cl⁻ ions in the octahedral holes and center of the unit cell. In Section 12.4 we used atomic radius ratios to determine the location of atoms in the interstices of a crystal lattice. Since all of the radius ratios in Table 12.4 are ≤1, it makes more sense to focus on the unit cell of NaCl on the left in Figure 12.32(b), where the smaller Na⁺ ions occupy the interstices in the cubic closest-packed lattice of Cl⁻ ions. The Na⁺ ions fit better into the octahedral holes because the radius ratio of Na⁺ to Cl⁻ is

$$\frac{r_+}{r_-} = \frac{102 \text{ pm}}{181 \text{ pm}} = 0.564$$

The radius ratio 0.564 is too large for Na⁺ to occupy a tetrahedral hole but well within the range for occupying an octahedral hole.

Let's take an inventory of the ions in a unit cell of NaCl. Like the metal atoms in the fcc unit cell in Figure 12.16(b), an isolated unit cell contains portions of fourteen Cl⁻ ions: one at each corner and one in each of the six faces of the cube. As in the analysis of Figure 12.16, accounting for partial ions gives us a total of four Cl⁻ ions

ionic solid a solid consisting of monatomic or polyatomic ions held together by ionic bonds.

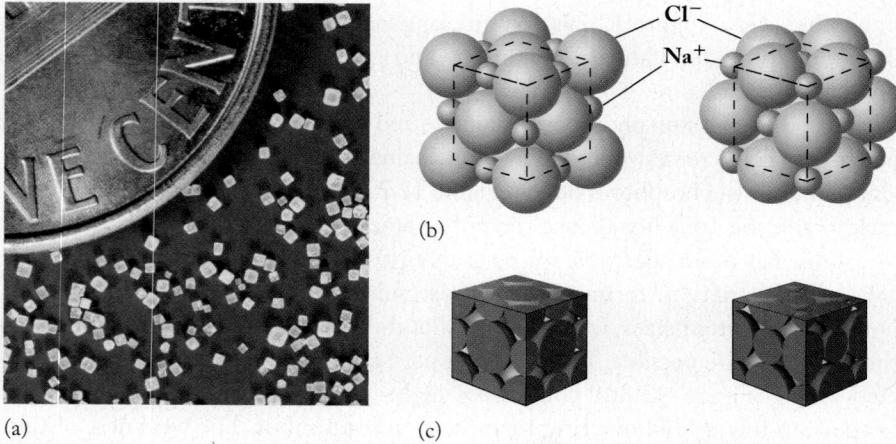

FIGURE 12.32 (a) Sodium chloride forms cubic crystals that reach various sizes. (b) Two views of a NaCl crystal lattice based on cubic closest-packing: (left) an fcc unit cell made of Cl⁻ ions, showing the Na⁺ ions in the octahedral holes; (right) an fcc unit cell made of Na⁺ ions with the larger Cl⁻ ions in octahedral holes. (c) Cutaway views of the two unit cells in part b.

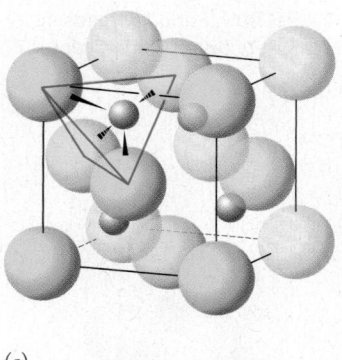

(a)   (b)   (c)

FIGURE 12.33 (a) Many crystals of the mineral sphalerite (ZnS), like the largest one in this photograph, have a tetrahedral shape. (b) The crystal lattice of sphalerite is based on an fcc unit cell of S^{2-} ions with Zn^{2+} ions in four of the eight tetrahedral holes. (c) An expanded view of the sphalerite unit cell. Each Zn^{2+} ion is in a tetrahedral hole shown in red formed by one corner S^{2-} ion and three face-centered S^{2-} ions.

in the unit cell on the left in Figure 12.32(c). To count the Na^+ ions we note that one Na^+ ion occupies the central octahedral hole, and one Na^+ ion fits into each of the twelve octahedral holes along the edges of the cell. Because each Na^+ ion along an edge is shared by four unit cells, only one-fourth of each Na^+ ion on an edge is in each cell (see Figure 12.17d). Only the Na^+ in the center belongs completely to the unit cell. Therefore, the total number of Na^+ ions in the unit cell is

$$\left(12 \times \tfrac{1}{4}\right) + 1 = 4 \, Na^+ \text{ ions}$$

The ratio of Na^+ to Cl^- ions in the unit cell is therefore 4:4, consistent with the chemical formula NaCl. Because the four Na^+ ions occupy all the octahedral holes in the unit cell, the result of this calculation also means that each fcc unit cell contains the equivalent of four octahedral holes.

Note in Figure 12.32 that adjacent Cl^- ions along any face diagonal do not touch each other the way they do in Figure 12.16(b), because the Cl^- ions have to spread out a little to accommodate the Na^+ ions in the octahedral holes. Sodium ions and chloride ions touch along each edge of the unit cell, however, which means that each Na^+ ion touches six Cl^- ions and each Cl^- ion touches six Na^+ ions. This arrangement of positive and negative ions is common enough among binary ionic compounds to be assigned its own name: the *rock salt structure.*

In other binary ionic solids, the smaller ion is small enough to fit into the tetrahedral holes formed by the larger ions. For example, in the unit cell of the mineral sphalerite (zinc sulfide) the S^{2-} anions (ionic radius 184 pm) are arranged in an fcc unit cell (Figure 12.33), and half of the eight tetrahedral holes inside the cell are occupied by Zn^{2+} cations (74 pm). Therefore the unit cell contains four Zn^{2+} ions that balance the charges on the four S^{2-} ions. This pattern of half-filled tetrahedral holes in an fcc unit cell is sometimes called the *sphalerite structure.* Many other compounds, particularly those formed between transition metal cations with a 2+ charge and anions of the group 6 elements with a 2− charge, have a sphalerite structure.

The crystal structure of the mineral fluorite (CaF_2) is based on an fcc unit cell of smaller Ca^{2+} ions at the eight cube corners and six face centers with all eight tetrahedral holes formed by neighboring Ca^{2+} ions filled by larger F^- ions. Because there are a total of four Ca^{2+} ions in the unit cell and the eight F^- ions are all completely inside the cell, this arrangement satisfies the 1:2 mole ratio of Ca^{2+} ions to F^- ions. This structure is so common that it too has its own name: the *fluorite structure* (Figure 12.34). Other compounds having this structure are SrF_2, $BaCl_2$, and PbF_2.

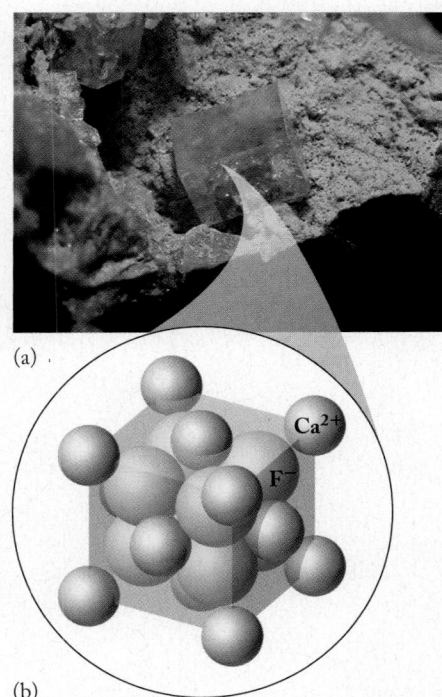

(a)

(b)

FIGURE 12.34 (a) The mineral fluorite (CaF_2) forms cubic crystals. (b) The crystal lattice of CaF_2 is based on an fcc array of Ca^{2+} ions, with F^- ions occupying all eight tetrahedral holes. Because they are bigger than Ca^{2+} ions, the F^- ions do not fit in the tetrahedral holes of a cubic closest-packed array of Ca^{2+} ions. Instead, the Ca^{2+} ions, while maintaining an fcc unit cell arrangement, spread out to accommodate the larger F^- ions. Note how adjacent Ca^{2+} ions along any face diagonal do not touch each other the way they do in the ideal fcc unit cell in Figure 12.16(b).

CONNECTION Periodic trends in atomic radii were discussed in Chapter 7.

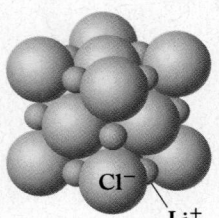

Cl⁻

Li⁺

FIGURE 12.35

Some compounds in which the cation-to-anion mole ratio is 2:1 have an *antifluorite structure*. In the crystal lattices of these compounds, which include Li_2O and K_2S, the smaller cations occupy the tetrahedral holes in an fcc unit cell formed by cubic closest-packing of the larger anions.

SAMPLE EXERCISE 12.5   **Calculating an Ionic Radius from a Unit-Cell Dimension**

The unit cell of lithium chloride (LiCl) contains an fcc arrangement of Cl^- ions (Figure 12.35). In LiCl the Li^+ cations (radius 76 pm) are small enough to allow adjacent Cl^- ions to touch along any face diagonal.

    a. If the edge length of the LiCl fcc cell is 513 pm, what is the radius of the Cl^- ion?

    b. Use that value to predict the type of hole the Li^+ ion occupies.

Collect and Organize We are given the edge length of a LiCl unit cell, the radius of the Li^+ cation, and a picture of the unit cell. The unit cell is an fcc array of Cl^- ions that touch along the face diagonal.

Analyze Equation 12.2 relates the edge length (ℓ) of an fcc cell to the radius (r) of the atoms or ions that touch along its diagonals: $r = 0.3536\,\ell$. The radius ratio of Li^+ to Cl^- ions and Table 12.4 allow us to predict which type of hole Li^+ occupies.

Solve

    a. Substituting the edge length into Equation 12.2, we calculate the radius of Cl^-:

$$r = 0.3536 \times 513 \text{ pm} = 181 \text{ pm}$$

    b. The ratio of the radius of a Li^+ ion to the radius of a Cl^- ion is

$$76 \text{ pm}/181 \text{ pm} = 0.42$$

According to Table 12.4, the Li^+ cations occupy octahedral holes in the lattice formed by the larger Cl^- anions.

Think about It The average ionic radius value in Figure 10.2 for Cl^- ions is 181 pm, so our calculation is correct. Figure 12.35 shows the structure of LiCl to be similar to the rock salt structure of NaCl in Figure 12.32(b).

Practice Exercise Assuming the Cl^- radius in NaCl is also 181 pm, what is the radius of the Na^+ ion in NaCl if the edge length of the NaCl unit cell is 564 pm but the anions do not touch along any face diagonal?

■ •••

SAMPLE EXERCISE 12.6   **Calculating the Density of a Salt from Its Unit-Cell Dimensions**

What is the density of LiCl at 25°C if the edge length of its fcc unit cell is 513 pm?

Collect and Organize We are given the edge length of an fcc unit cell of a crystalline salt and are asked to calculate its density. As in Sample Exercise 12.3, we have to assume that the density of the unit cell is the same as the density of the overall solid. The fact that LiCl has an fcc unit cell means that there are four Cl^- ions and four Li^+ ions in the cell. The molar masses of these ions are 35.45 and 6.941 g/mol, respectively. Density is the ratio of mass to volume, and the volume of a cubic cell is the cube of its edge length: $V = \ell^3$.

Analyze The density of an fcc unit cell of LiCl is the sum of the masses of four Cl^- ions and four Li^+ ions divided by the volume of the cell, $(513 \text{ pm})^3$. As in Sample Exercise 12.3, we need to convert from molar masses to the masses of individual particles by dividing by Avogadro's number.

Solve Calculating the mass of four Cl^- ions:

$$m = \frac{35.45 \text{ g } Cl^-}{1 \text{ mol } Cl^-} \times \frac{1 \text{ mol } Cl^-}{6.022 \times 10^{23} \text{ ions } Cl^-} \times 4 \text{ ions } Cl^-$$
$$= 2.355 \times 10^{-22} \text{ g } Cl^-$$

and the mass of four Li^+ ions:

$$m = \frac{6.941 \text{ g } Li^+}{1 \text{ mol } Li^+} \times \frac{1 \text{ mol } Li^+}{6.022 \times 10^{23} \text{ ions } Li^+} \times 4 \text{ ions } Li^+$$
$$= 0.4610 \times 10^{-22} \text{ g } Li^+$$

Combining the masses of the two kinds of ions in the unit cell:

$$2.355 \times 10^{-22} \text{ g} + 0.4610 \times 10^{-22} \text{ g} = 2.816 \times 10^{-22} \text{ g}$$

The volume of the cell in cubic centimeters is

$$V = \ell^3 = (513 \text{ pm})^3 \times \frac{(10^{-10} \text{ cm})^3}{(1 \text{ pm})^3} = 1.35 \times 10^{-22} \text{ cm}^3$$

Taking the ratio of mass to volume, we have

$$d = \frac{m}{V} = \frac{2.816 \times 10^{-22} \text{ g}}{1.35 \times 10^{-22} \text{ cm}^3} = 2.09 \text{ g/cm}^3$$

Think about It The result is reasonable in that most minerals are more dense than water but less dense than common metals.

Practice Exercise What is the density of NaCl at 25°C if the edge length of its fcc unit cell is 564 pm?

12.7 Ceramics: Insulators to Superconductors

A **ceramic** is a solid inorganic compound or mixture of compounds that has been heated to transform it into a harder and more heat-resistant material. The use of ceramic materials preceded metal technology by many thousands of years. The first ceramics were probably made of *clay*, the fine-grained soil produced by the physical and chemical weathering of igneous rocks (rocks of volcanic origin). Moist clay is easily molded into a desired shape and then hardened over fires or in wood-burning kilns, an ancient process still in use today.

In this section we examine some of the physical properties of both primitive earthenware (ceramics fired at low temperatures) and modern ceramic materials (typically fired at high temperatures) and relate those properties to the chemical composition of the materials and to the chemical changes that occur when they are heated.

ceramic a solid inorganic compound or mixture that has been transformed into a harder, more heat-resistant material by heating.

(a)

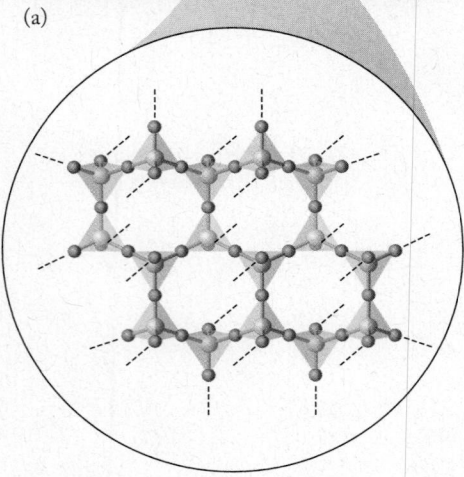

(b) SiO_2

FIGURE 12.36 (a) Quartz crystals are hexagonal. (b) Their crystal structures feature hexagonal arrays of silicon–oxygen tetrahedra that form an extended three-dimensional network. The dashed lines indicate how the lattice extends in three dimensions.

Polymorphs of Silica

One of the most abundant families of minerals found in igneous rocks has the chemical composition SiO_2. The correct chemical name is silicon dioxide, but the more common name is *silica*. Silica is a covalent network solid in which each silicon atom is covalently bonded to four oxygen atoms, forming a tetrahedron with an oxygen atom at each corner and the silicon atom at the center (Figure 12.36). Each oxygen atom is covalently bonded to two silicon atoms, thereby linking the tetrahedra into an extended three-dimensional network. Because each corner oxygen atom is bonded to two silicon atoms, each silicon atom gets only half "ownership" of the four oxygen atoms to which it is covalently bonded—hence the formula SiO_2.

At least eight different minerals have the empirical formula SiO_2. The members of a family of substances with the same empirical formula but different crystal structures and properties are called *polymorphs*. The most abundant silica polymorph is quartz, a type of SiO_2 that can form impressively large, nearly transparent crystals (Figure 12.36a). Note how the hexagonal ordering of the SiO_2 tetrahedra (Figure 12.36b) translates into hexagonal crystals.

Most, but not all, silica polymorphs are crystalline. When lava containing molten SiO_2 flows from a volcano into the sea or a lake, it cools so quickly that the Si and O atoms may not have enough time to achieve an ordered crystal lattice as the lava solidifies. The solid formed in this way is an *amorphous* (disordered, noncrystalline) polymorph of silica known as either volcanic glass or obsidian (Figure 12.37). *Glass* is a term scientists and engineers use to describe any solid that has either no crystalline structure or only very tiny crystals surrounded by disordered arrays of atoms. This definition applies to laboratory glassware and the drinking glasses we use at home, as well as to more exotic solids such as obsidian.

Ionic Silicates

In addition to covalent silica, igneous rocks also contain ionic minerals made of silicon and oxygen. These minerals have some of the tetrahedral crystal structure of silica, but not all the oxygen corner atoms are bonded to two Si atoms. Instead, some of the O atoms have an extra electron. The result is a *silicate* anion. One of the common ionic silicates is chrysotile, a type of asbestos that consists of sheets of linked silicon–oxygen tetrahedra that form hexagonal clusters of six tetrahedra each (Figure 12.38). Each tetrahedron has three O atoms that it shares with other tetrahedra and one O atom—the one with the extra electron—that it does not share. Thus the basic tetrahedral unit consists of one Si atom and $[1 + 3(\frac{1}{2})] = 2.5$ oxygen atoms as well as a negative charge. This gives the sheet the empirical formula $SiO_{2.5}^-$. We generally use whole-number subscripts in chemical formulas when possible. In this case, multiplying $SiO_{2.5}^-$ by 2 gives us the empirical formula $Si_2O_5^{2-}$ for this silicate anion.

FIGURE 12.37 (a) Obsidian (volcanic glass) is an unusual form of silica in that it is not crystalline. (b) Obsidian contains mostly amorphous silica with random arrangements of silicon and oxygen atoms. (All atoms are drawn undersized relative to the volume they actually occupy to make the structure easier to see.)

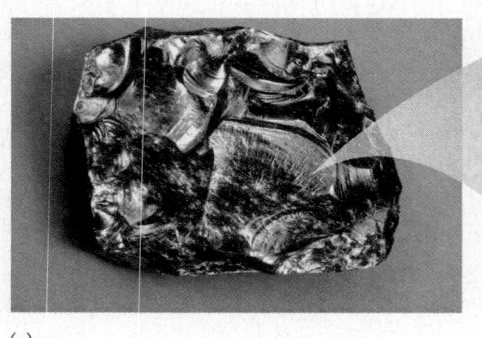

(a)

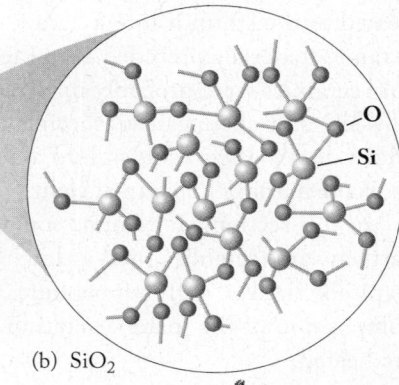

O

Si

(b) SiO_2

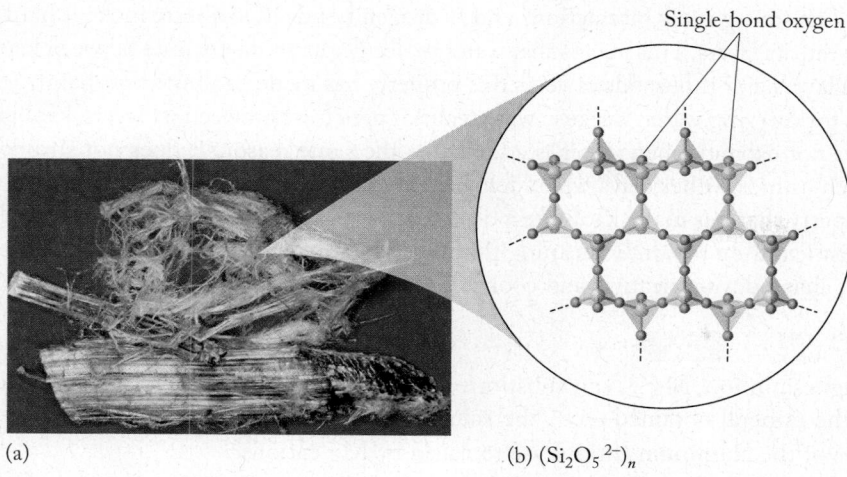

Single-bond oxygen

(a)

(b) $(Si_2O_5{}^{2-})_n$

FIGURE 12.38 (a) Chrysotile, one of the two principal forms of asbestos formerly used in building construction as thermal insulation, is an ionic silicate compound. (b) The ease with which thin fibers of chrysotile can flake off is related to its layered crystal lattice and the relatively weak intermolecular interactions between layers. The O atoms that are bonded to only one Si atom have an extra electron and a negative charge.

The subscript n in the formula in Figure 12.38(b) indicates that there are many empirical formula units in a single crystal.

Silicate minerals are neutral materials, so they must contain cations to balance the negative charges on the silicate layers. When this cation is Al^{3+}, the minerals are called *aluminosilicates*. One of the most common aluminosilicates is the clay mineral kaolinite (Figure 12.39). At least a little kaolinite is found in practically every soil, but rich deposits of nearly pure, brilliantly white kaolinite are found in highly weathered soils. For centuries these deposits have been mined to make fine china and white porcelain. Today the greatest demand for kaolinite is in the production of glossy white paper, which is used in most magazines and books (including this one).

Common metal ions found in igneous rocks—including Na^+, K^+, Ca^{2+}, Mg^{2+}, and Fe^{3+}—are largely absent in kaolinite. Their absence indicates that kaolinite deposits form under acidic weathering conditions. Under these conditions H^+ ions displace other cations from ion-exchange sites. For example, $-O^-Na^+$ sites exchange H^+ for Na^+, leaving behind $-OH$ groups such as those shown in Figure 12.39.

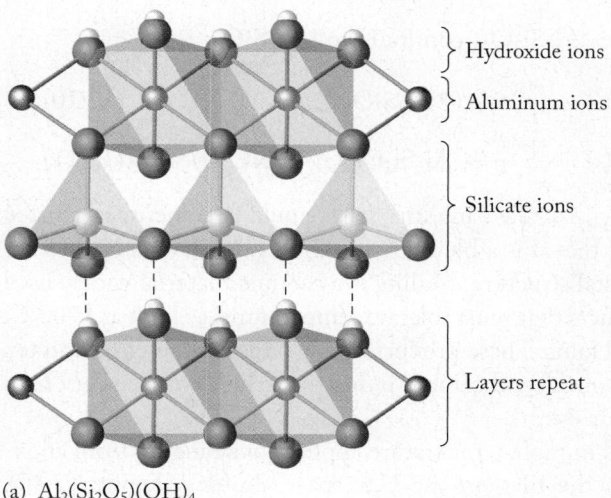

Hydroxide ions

Aluminum ions

Silicate ions

Layers repeat

(a) $Al_2(Si_2O_5)(OH)_4$

(b)

FIGURE 12.39 (a) Edge-on view of the structure of kaolinite shows a top layer of OH^- ions bonded to a middle layer of Al^{3+} ions followed by a bottom layer of silicate ions ($Si_2O_5{}^{2-}$). The structure is repeated in subsequent layers. The empirical formula of this crystal lattice is $Al_2(Si_2O_5)(OH)_4$. (b) This enormous kaolinite mine in Bulgaria contributes to a worldwide production of about 44 trillion metric tons of the mineral per year. Kaolinite is used in many products, from ceramics to glossy white paper.

∞○ **CONNECTION** Ion-exchange reactions were discussed in Chapter 4.

The strong ionic interactions and hydrogen bonds in kaolinite make it hard to separate its layers. This means that water molecules cannot penetrate between kaolinite layers. For thousands of years this property has made kaolinite pots handy vessels for carrying water. Because water cannot penetrate between its layers, kaolinite does not expand when water is added. For the same reason, it does not shrink as much as most other clays when dehydrated at high temperatures. This is another property that has made kaolinite a desirable starting material for ceramics. Finally, moist kaolinite is *plastic,* meaning that it can be molded into a shape, and it keeps that shape during heating and cooling.

CONCEPT TEST ●

Magnesium ion, Mg^{2+}, can substitute for Al^{3+} in kaolinite. What is the formula of the mineral obtained—i.e., the values of x and y in $Mg_xAl_y(Si_2O_5)(OH)_4$—if 50% of the aluminum cations are replaced by Mg cations?

● ●

From Clay to Ceramic

Creating ceramic objects from kaolinite and other clays takes several steps. First, moist clay is formed into pots, bricks, and other objects on a potter's wheel or in molds or presses. Drying at just above 100°C removes much of the water that made the clay plastic. Further heating to about 450°C removes water that is adsorbed onto the surfaces of the clay particles or between the layers of non-kaolinite clays. Between 450°C and 650°C the hydroxide ions react with each other to form water and a solid in which the Al and Si content is higher than in kaolinite:

$$Al_2Si_2O_5(OH)_4(s) \xrightarrow{450-650°C} 2\,H_2O(g) + Al_2Si_2O_7(s) \quad (12.8)$$

The next structural change occurs just below 1000°C when $Al_2Si_2O_7$ is converted into another aluminosilicate, $Al_4Si_3O_{12}$, and another solid, $SiO_2(s)$, is created:

$$2\,Al_2Si_2O_7(s) \xrightarrow{\sim950°C} Al_4Si_3O_{12}(s) + SiO_2(s) \quad (12.9)$$

At even higher temperatures $Al_4Si_3O_{12}$ continues to lose SiO_2:

$$Al_4Si_3O_{12}(s) \xrightarrow{>950°C} 2\,Al_2SiO_5(s) + SiO_2(s) \quad (12.10)$$

$$3\,Al_2SiO_5(s) \xrightarrow{1350°C} Al_6Si_2O_{13}(s) + SiO_2(s) \quad (12.11)$$

The last product, $Al_6Si_2O_{13}$, is called mullite. Its formula is sometimes written $3\,Al_2O_3 \cdot 2\,SiO_2$ indicating that it is a blend of an Al_2O_3 (alumina) crystal structure and a SiO_2 (silica) crystal structure. Mullite is a ceramic material widely used in the manufacture of products that must tolerate temperatures as high as 1700°C: furnaces, boilers, ladles, and kilns. These products are used as containers of molten metals and in the glass, chemical, and cement industries. Mullite is also very hard and is widely used as an abrasive.

Other ceramics are used in high-temperature applications ranging from cookware to fireplace bricks to the tiles on the U.S. space shuttles (Figure 12.40). Ceramics are well suited to these uses because of their high melting points and because they are good thermal and electrical insulators. For example, the thermal conductivity of aluminum metal at 100°C is over eight times that of alumina (Al_2O_3).

FIGURE 12.40 About 25,000 heat-resistant tiles cover much of the surface of each U.S. space shuttle, protecting it from re-entry temperatures above 1100°C. Most tiles are made of high-purity, amorphous SiO_2 fibers. Ninety percent of each tile is open space between the fibers, which means that the tiles are both lightweight and highly resistant to high temperatures.

Superconductors

Band theory can be used to describe the properties of, and bonding in, all solids, including the compounds in ceramic materials. Earlier in the chapter we noted that metals are good conductors because of the ease with which their valence electrons can move to conduction bands, whereas semimetals are semiconductors because they have significant energy gaps between their valence and conduction bands (though doping can shrink the gaps). Similarly, the insulating properties of ceramics can be explained by the large energy gaps between their valence and conduction bands. As a result, their valence electrons have very limited mobility.

In conventional metallic conductors the vibration of the metal atoms in the crystal lattice can interfere with the flow of free electrons: the higher the temperature, the greater the atomic vibration and the greater the resistance to electron flow. However, at temperatures approaching absolute zero these vibrations become very small. In the early 20th century, scientists discovered that the lattice vibrations in mercury and some metal alloys are so small at temperatures below about 20 K, called the **critical temperature (T_c)** for each substance, that electrons can pass freely through these materials and they become **superconductors**. Today superconducting alloys, such as Nb_3Sn, are widely used in devices that require very high electrical currents, such as the electromagnets of magnetic-resonance-imaging (MRI) instruments.

Unfortunately it is difficult and expensive to chill materials to 20 K. In 1986 it was discovered that $YBa_2Cu_3O_7$ and several other ceramic materials become superconductors when cooled to temperatures just above the boiling point of liquid nitrogen (77 K). This represented an important economical advance because it is much easier and cheaper to cool a material with liquid nitrogen than with liquid helium (boiling point = 4 K), which must be used to achieve superconductivity in Hg and metal alloys. Since the 1986 discovery, scientists have produced superconducting ceramic materials with critical temperatures as high as 133 K.

Let's take a closer look at the crystal structure of $YBa_2Cu_3O_7$, which has the unit cell shown in Figure 12.41. The unit cell is a stack of three cubes with a Y^{3+} ion in the center of the middle cube, Ba^{2+} ions in the center of the top and bottom cubes, sixteen copper ions at the corners of each cube, and twenty O^{2-} ions along the edges of the three stacked cubes.

Applying our usual practice of assigning fractions of atoms and ions to unit cells, we find that the stack in Figure 12.41 has 7 oxide ions ($\frac{1}{4}$ of the 12 that occupy edges of the top and bottom cubes in the stacked array $+ \frac{1}{2}$ of the remaining 8 ions that occupy faces—7 in all) and 3 copper ions ($\frac{1}{8}$ of the 8 ions on the top and bottom surfaces $+ \frac{1}{4}$ of the 8 ions around the middle—3 in all). The two Ba^{2+} ions and the Y^{3+} ion lie entirely within the unit cell. Thus, the material with the crystal structure shown in Figure 12.41 has the chemical formula $YBa_2Cu_3O_7$.

Yttrium–barium–copper (YBC) oxides and related materials are superconductors because of the formation of electron pairs called *Cooper pairs* after American physicist Leon Cooper (1930–). In an undisturbed crystal lattice, the ions are spaced

▶‖ **CHEMTOUR** Superconductors

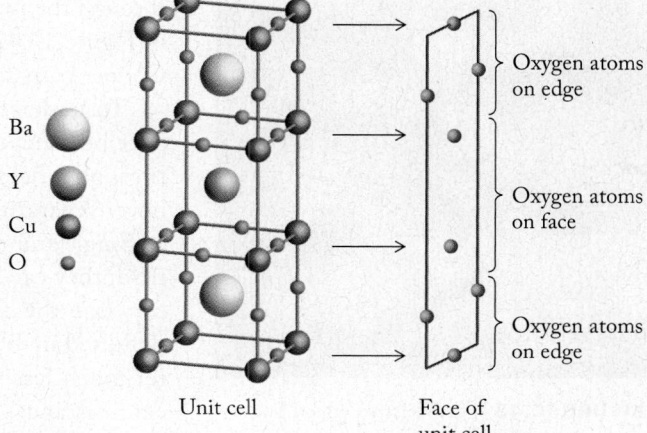

Ba
Y
Cu
O

Oxygen atoms on edge

Oxygen atoms on face

Oxygen atoms on edge

Unit cell

Face of unit cell

FIGURE 12.41 The unit cell of $YBa_2Cu_3O_7$, a high-temperature superconductor, contains a central barium ion in the top and bottom cubes, a central yttrium ion in the middle cube, and oxide ions on the edges of the unit cell.

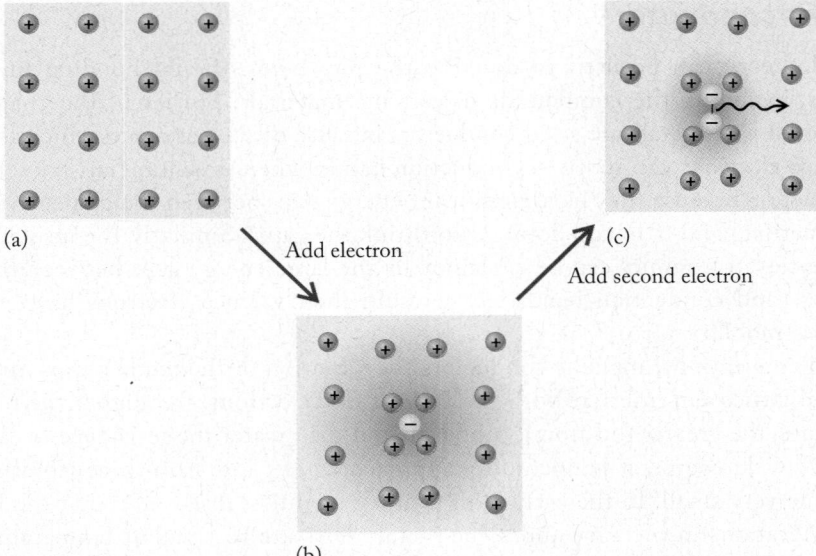

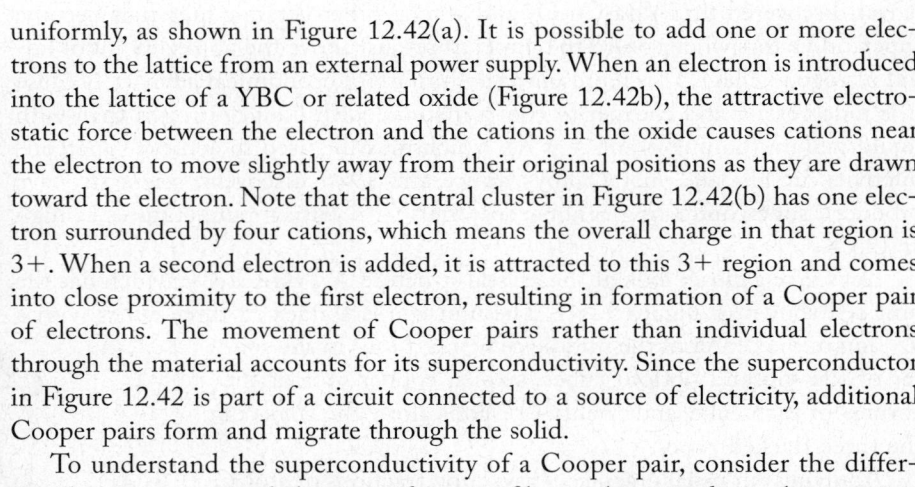

FIGURE 12.42 (a) In an undisturbed crystal lattice of an ionic solid, the cations are spaced uniformly. (b) An electron inserted into the lattice distorts the uniform spacing of the cations because the cations are attracted to the negative electrical charge of the electron. (c) A second electron in the lattice is attracted to the area of localized positive charge. It and the first electron create a Cooper pair.

FIGURE 12.43 The Meissner effect. The magnetic field produced by the magnet (background) cannot penetrate the small cylinder of superconductive material that has been chilled below its critical temperature. As a result, the magnet repels the superconductor and the superconductor floats above the magnet.

uniformly, as shown in Figure 12.42(a). It is possible to add one or more electrons to the lattice from an external power supply. When an electron is introduced into the lattice of a YBC or related oxide (Figure 12.42b), the attractive electrostatic force between the electron and the cations in the oxide causes cations near the electron to move slightly away from their original positions as they are drawn toward the electron. Note that the central cluster in Figure 12.42(b) has one electron surrounded by four cations, which means the overall charge in that region is 3+. When a second electron is added, it is attracted to this 3+ region and comes into close proximity to the first electron, resulting in formation of a Cooper pair of electrons. The movement of Cooper pairs rather than individual electrons through the material accounts for its superconductivity. Since the superconductor in Figure 12.42 is part of a circuit connected to a source of electricity, additional Cooper pairs form and migrate through the solid.

To understand the superconductivity of a Cooper pair, consider the difference between two single horses and a pair of horses harnessed together running through a field of boulders. The harnessed horses must travel forward together, never separating from each other, and they reach the far end of the field together. Single animals, however, can choose completely different paths across the same field; they can be far apart from each other and scatter over a wide area, and they can exit the field at different locations. Similarly, single electrons are easily deflected in different directions by the atoms or ions in a solid, but electrons scatter much less when harnessed in a Cooper pair. Less scattering lowers electrical resistance.

Superconductors are of technological interest because, in principle, large amounts of electric charge can flow through them with no resistance. This property is attractive for the design of extremely fast computers. Superconductors also have the ability to exclude magnetic fields—a property known as the *Meissner effect* after German physicist Walther Meissner (1882–1974). In Figure 12.43 a small cylinder of superconducting material cooled below its critical temperature floats in a magnetic field because magnetic lines of force are excluded from it. Because of this exclusion, the magnet below the superconductor repels it, suspending it in air. This magnetic levitation could in principle be used to both float a train above electromagnetic tracks and propel it along the tracks at relatively low cost and at much higher speeds than a conventional train. An experimental train using superconductor technology has been tested in Japan; however, widespread use of levitating trains remains, for now, a goal for the future.

> **X-ray diffraction (XRD)** a technique for determining the arrangement of atoms or ions in a crystal by analyzing the pattern that results when X-rays are scattered after bombarding the crystal.

12.8 X-ray Diffraction: How We Know Crystal Structures

Unit-cell dimensions are determined using **X-ray diffraction (XRD)**. X-rays are well suited for the task of crystal structure determination because the wavelengths of X-rays (10^2 to 10^3 pm) are similar to the radii of atoms and ions and, therefore, the distances between them in crystals. To see how XRD works we will look at atoms in a crystalline metal, but keep in mind that our description also applies to the particles in any crystalline solid.

A narrow beam of X-rays is directed at some angle of incidence θ at the surface of the metal, as shown in Figure 12.44(a). Some of the X-rays hit atoms in the surface layer, and any X-ray that does so is reflected off the atom at an angle

▶❚❚ **CHEMTOUR** X-ray Diffraction

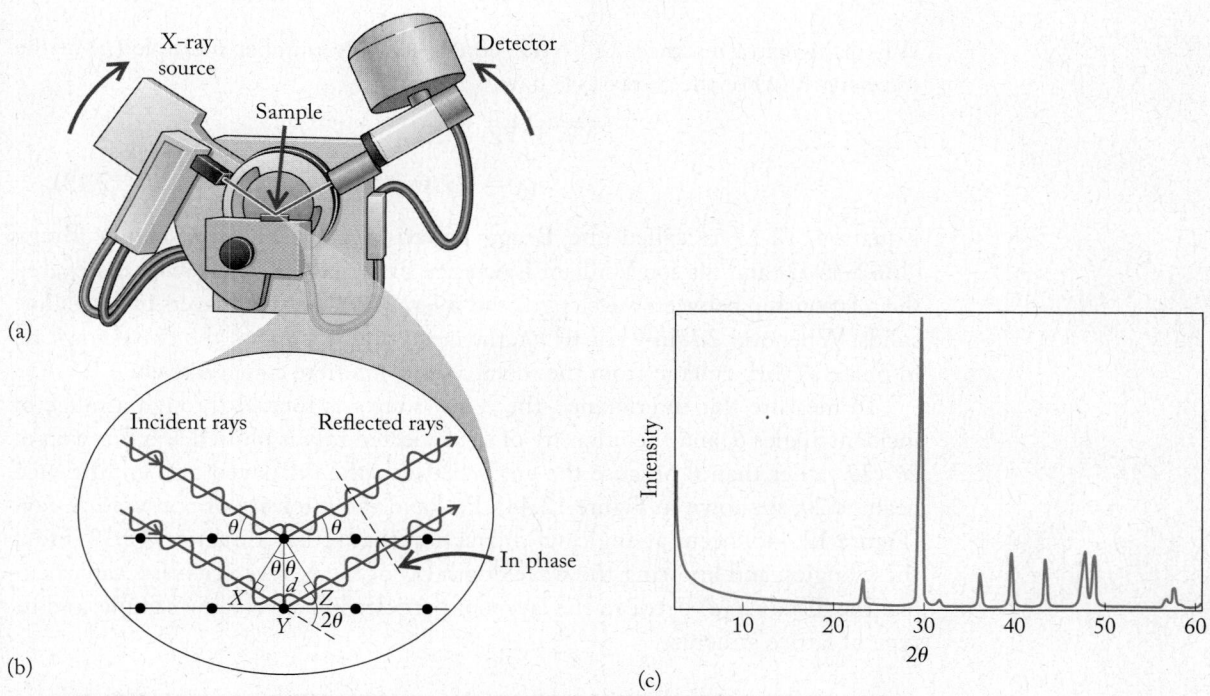

(a)

(b)

(c)

FIGURE 12.44 X-ray diffraction. (a) An X-ray diffractometer is used to determine the crystal lattices of solids. A source of X-rays and a detector are mounted so that they can rotate around the sample. (b) X-rays reflected from the surface layer of atoms and rays reflected from the next layer interfere constructively when they are in phase, as they are at the distance d shown here. The angle between the incident and reflected X-rays is 2θ. (c) Moving the source and detector around the sample produces a scan such as this one for quartz. The peaks at different values of 2θ represent reflections from different planes of atoms in the solid.

of reflection equal to the angle of incidence θ. Therefore the total change in direction of the reflected X-ray is the sum of the two angles, or 2θ. The value 2θ is the *angle of diffraction*.

Some of the incident X-rays pass through the surface layer and hit atoms in a second layer a distance d below the surface. For any X-ray reaching this layer, the angle of incidence and angle of reflection are again both θ, and this ray is also diffracted through an angle 2θ. The two reflected rays can interfere with each other either constructively or destructively. They interfere constructively when they are in phase, as in Figure 12.44(b). To be in phase, the extra distance traveled by the X-ray reflecting off the second layer must be some whole-number multiple of the wavelength of the X-rays. This extra distance traveled is the sum of the lengths of line segments $\overline{XY}$ and $\overline{YZ}$ in Figure 12.44(b). The two right triangles incorporating these line segments share a hypotenuse d. Geometry tells us that the angles opposite $\overline{XY}$ and $\overline{YZ}$ are both equal to θ. According to trigonometry, the ratio of either $\overline{XY}$ or $\overline{YZ}$ to d is

$$\frac{\overline{XY}}{d} = \sin\theta \qquad \frac{\overline{YZ}}{d} = \sin\theta$$

which means

$$d\sin\theta = \overline{XY} \qquad d\sin\theta = \overline{YZ} \qquad (12.12)$$

As noted above, we are interested in $\overline{XY} + \overline{YZ}$, the extra distance traveled by the second ray. We therefore add Equations 12.12 to get

$$\overline{XY} + \overline{YZ} = d\sin\theta + d\sin\theta = 2d\sin\theta$$

When this extra distance $\overline{XY} + \overline{YZ}$ equals a whole-number multiple (n) of the wavelength (λ) of the X-rays, we have

$$\overline{XY} + \overline{YZ} = n\lambda \qquad \text{or}$$

$$n\lambda = 2d\sin\theta \qquad (12.13)$$

Equation 12.13 is called the **Bragg equation** after William Henry Bragg (1862–1942) and his son William Lawrence Bragg (1890–1971), who discovered the relationship between wavelength and the spacings between layers in crystalline solids. Whenever $2d\sin\theta$ equals $n\lambda$, the crests and troughs of the two X-rays are in phase as they emerge from the metal and so interfere constructively.

To measure this interference, the X-ray source is rotated through a range of incident angles θ, and the intensity of the reflected rays is plotted as a function of 2θ (2θ rather than θ because the angle between the diffracted and undiffracted beam is 2θ, as shown in Figure 12.44). Peaks in the intensity of scattered X-rays (Figure 12.44c) occur at angles of diffraction that satisfy Equation 12.13. From these angles, and knowing the wavelength (λ) of the X-rays, scientists can calculate the distance (d) between the layers of particles in a crystalline sample, and its type of lattice structure.

◐◐ **CONNECTION** We discussed the constructive and destructive interference of waves in Chapter 7.

Bragg equation relates the angle of diffraction (2θ) of X-rays to the spacing (d) between the layers of ions or atoms in a crystal: $n\lambda = 2d\sin\theta$.

> **SAMPLE EXERCISE 12.7** **Determining Interlayer Distances by X-ray Diffraction**

An XRD analysis of a sample of copper has peaks at $2\theta = 24.64°, 50.54°$, and $79.62°$. What distance d between layers of Cu atoms can produce this diffraction pattern if the wavelength of the X-rays is 154 pm?

Collect and Organize We are given a series of diffraction angles 2θ and asked to find the distance between layers of atoms in a sample of copper metal. Equation 12.13 enables us to calculate this distance when we know the angles of diffraction associated with constructive interference in a beam of reflected X-rays.

Analyze We must rearrange Equation 12.13 to solve for the parameter we are after: d.

$$d = \frac{n\lambda}{2 \sin \theta}$$

We are given the value of λ and can get values of θ from the given values of 2θ. The additional unknown in Equation 12.13 is the wavelength multiplier n, which we must know before we can determine the value of d. Since we know that the radius of a copper atom is on the order of 10^2 pm, we expect the distance between the layers to also be on the order of 10^2 pm.

Solve The key to determining n is to look for a pattern in the θ values. Notice that the higher values of θ are approximately two and three times the lowest value:

$$24.64°/2 = 12.32° \qquad 50.54°/2 = 25.27° \qquad 79.62°/2 = 39.81°$$

$$\frac{25.27°}{12.32°} = 2.051 \qquad \frac{39.81°}{12.32°} = 3.231$$

This pattern suggests that the values of n for this set of data are 1, 2, and 3, so let's use these combinations of n and θ to see whether they all give the same value of d:

$$d = \frac{(1)(154 \text{ pm})}{2 \sin 12.32°} = \frac{154 \text{ pm}}{(2)(0.2134)} = 361 \text{ pm}$$

$$d = \frac{(2)(154 \text{ pm})}{2 \sin 25.27°} = \frac{308 \text{ pm}}{(2)(0.4269)} = 361 \text{ pm}$$

$$d = \frac{(3)(154 \text{ pm})}{2 \sin 39.81°} = \frac{462 \text{ pm}}{(2)(0.6402)} = 361 \text{ pm}$$

We do indeed get the same value of d, so $d = 361$ pm is the distance between Cu atoms that produced the three peaks.

Think about It These consistent results mean that our assumption about the values of n for the three values of θ was correct. Also, the value of d is in the range of the edge lengths of unit cells we used in several Sample Exercises earlier in the chapter and so is reasonable.

Practice Exercise An X-ray diffraction analysis of crystalline CsCl using X-rays of wavelength 71.2 pm has a prominent peak at $2\theta = 19.9°$. If this peak corresponds to $n = 2$, what is the spacing between the ion layers? What are the values of 2θ for $n = 3$ and $n = 4$?

Silicon, Silica, Silicates, Silicone: What's in a Name?

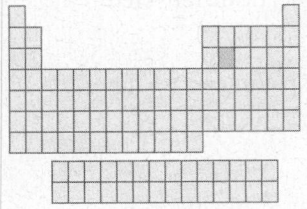

Group 14 of the periodic table is headed by carbon, an element so important to life, energy, and commerce that it merited its own descriptive chemistry section in Chapter 5. The element below carbon, silicon, is central both to ancient and modern technology and to the entire solid-state revolution that brought us computers and many of our electronic devices; it too deserves to be highlighted.

Silicon is the second most abundant element in Earth's crust; oxygen is the first. Elemental silicon never occurs free in nature. Mostly it is present as *silica* (SiO_2, a covalent network solid) and *silicate minerals*, also called just *silicates* (solids containing silicon–oxygen groups and metals). Silicates almost invariably consist of silicon–oxygen tetrahedral units bonded together in chains, rings, sheets, and three-dimensional arrays similar to those found in carbon (Section 12.5). Silica constitutes about 60% by mass of Earth's crust and is of great geological and commercial importance. It is most familiar to us as quartz crystals and beach sand.

Humans have used silica and silicates since prehistoric times in simple tools, arrowheads, and flint knives. The glass most familiar to us as window panes and bottles, called soda-lime glass, is principally a mixture of SiO_2 (73%), CaO (11%), and Na_2O (13%) with the balance composed of other metal oxides. Leaded crystal stemware is made by adding PbO (24%) and K_2O (15%) to silica (60%). Laboratory glassware is composed primarily of silica (81%), B_2O_3 (11%), and Na_2O (5%).

Natural silicates are also the raw materials of the ceramic and concrete industries. Silica is also used in detergents; in the pulp and paper industry; as anticaking agents in powdered foods such as cocoa, sugar, and spices; and as an ingredient in paints to provide a matte finish. Small packs of material labeled "desiccant" in packaged electronics, pharmaceuticals, and some foods may contain porous silica because it can absorb about 40% of its mass in water and thereby keep moisture-sensitive materials dry. Silica is a major component in toothpaste, functioning as both a mild abrasive and a thickening agent.

Most silica is now used in the production of highly purified elemental silicon, the most basic material of the microelectronics industry. The name "Silicon Valley" to describe the semiconductor industry around Palo Alto, CA, is an acknowledgment of the importance of the element to the industry. Computers, calculators, liquid-crystal displays, cell phones, video games, solar cells, and space vehicles are all products of the solid-state revolution that began in 1947 with the discovery of the transistor effect at Bell Laboratories in Murray Hill, NJ. The theoretical work and practical development of transistors and semiconductors by Walter H. Brattain, John Bardeen, and William B. Shockley resulted in their sharing the Nobel Prize in Physics in 1956.

Silicon is the element most widely used as a starting material in the production of semiconductors. Germanium, the element below silicon in the periodic table, is second. To function correctly in electronic devices, silicon must contain less than 1 ppb of impurities. Such ultrapure silicon is produced in a series of reactions starting with

$$SiO_2(s) + C(s) \rightarrow Si(s) + CO_2(g)$$

Silicon prepared in this way is about 98% pure. Before it can be used to make microchips, it must be refined further, through the reaction

$$Si(s) + 2\,Cl_2(g) \rightarrow SiCl_4(g)$$

Impurities that do not form volatile chlorides are removed, and the $SiCl_4$ is heated above 700°C, where it decomposes to silicon and chlorine gas:

$$SiCl_4(g) \rightarrow Si(s) + 2\,Cl_2(g)$$

The final step in processing silicon results in both an ultrapure material (one that contains only silicon atoms) and a solid that is free of crystalline imperfections (every position in the lattice is occupied by a silicon atom, and no gaps or vacancies exist). A semiconductor is usually made from a single crystal of silicon grown by a process called "pulling" (Figure 12.45). A small seed crystal is allowed to touch the surface of a pool of melted silicon. The seed crystal is withdrawn from the melt at a rate that allows material from the melt to solidify on the seed crystal. Single crystals of silicon 25 cm long or longer are made in this fashion, and semiconductor devices are made from pieces of these single crystals. Crystalline silicon has the same crystal structure as diamond.

Silicones are another commercially important family of silicon compounds; they are made of carbon, hydrogen, oxygen, and silicon and are hence called organosilicon compounds.

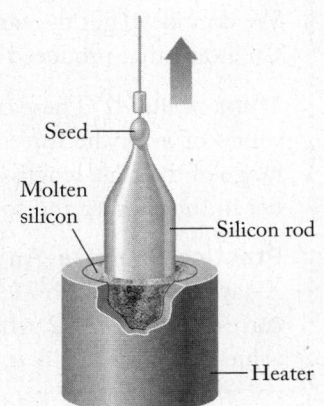

FIGURE 12.45 "Pulling" a silicon seed crystal from a melt yields ultrapure and perfectly crystalline silicon.

Seed — Molten silicon — Silicon rod — Heater

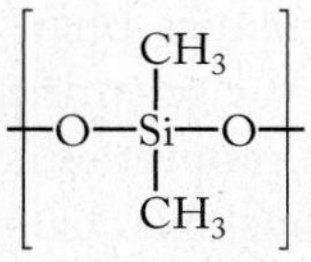

$$\left[-O-\underset{\underset{\displaystyle CH_3}{|}}{\overset{\overset{\displaystyle CH_3}{|}}{Si}}-O- \right]$$

FIGURE 12.46

No natural organosilicon compounds exist, but they are the basis of a major industry. A silicone polymer is applied to the windshields of automobiles and aircraft to prevent bugs and grime from clouding vision, to reduce glare, and to disperse rain, sleet, and snow. The most common silicone used in this application, poly(dimethylsiloxane) or PDMS, has the monomer repeating unit shown in Figure 12.46.

Glass contains silicates, which interact with water, as we discussed in Chapter 10 in terms of meniscus formation. PDMS molecules adhere readily to glass because their structure is similar to that of the silicates, but the $-CH_3$ groups make PDMS much more hydrophobic. When water contacts the PDMS coating, it does not adhere to the surface; raindrops are repelled (Figure 12.47).

Silicones are available as liquids, gels, and flexible solids. They are stable over a range of temperatures from below $-40°C$ to greater than $150°C$. Thousands of applications for silicones beyond coatings for glass surfaces include their use as moisture-proof sealants for ignition cables, spark plugs, medical appliances, and catheters, and as protective surfaces for everything from integrated circuits to textiles and skyscrapers. PDMS is also a key ingredient in Silly Putty.

The chemical properties of the other elements in group 14—germanium, tin, and lead—vary with increasing atomic number down the family (Figure 12.48). Germanium (like silicon) is a metalloid, whereas tin and lead are two of the

FIGURE 12.47 Rain beads on metal treated with a silicone polymer.

oldest known metals. All three form oxides with the formula MO_2 (where M is the group 14 metal or metalloid) and react with halogens to form tetrahalides (MX_4, where X = F, Cl, Br, or I). The dichlorides of tin and lead, $SnCl_2$ and $PbCl_2$, also are stable. Both germanium and one of the allotropes of tin, α-tin, have the diamond structure, but β-tin has a more complicated structure. Elemental lead crystallizes in an fcc unit cell. Germanium and α-tin are semiconductors, but β-tin and lead are conductors. Tin is used extensively in metallurgy in combination with other metals. Tin oxides are used to make ceramics. Lead and lead(IV) oxide are the electrode materials in most automobile batteries.

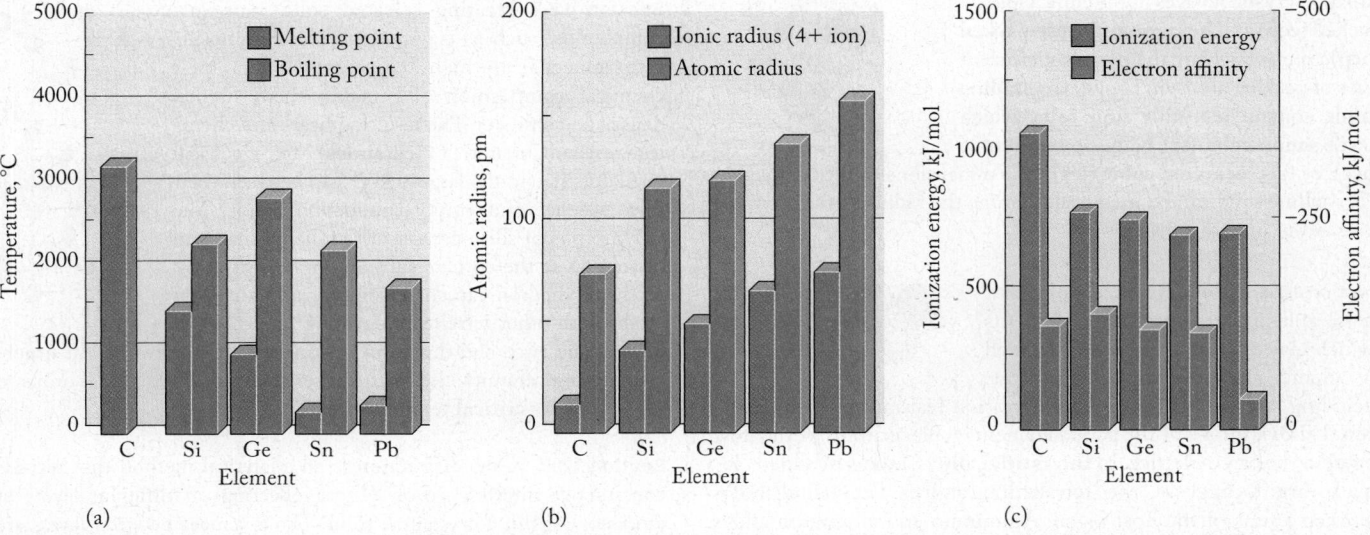

FIGURE 12.48 (a) Melting points and boiling points of the group 14 elements; the boiling point of carbon has not yet been determined. (b) Atomic and ionic radii of the group 14 elements. (c) The ionization energies and electron affinities of the group 14 elements.

As we have seen throughout this text, the arrangement of atoms, ions, and molecules in a sample of matter is responsible for many of the properties of materials. XRD is a powerful tool for analyzing structure at the atomic level of a wide range of materials, including metals, minerals, polymers, plastics, ceramics, pharmaceuticals, and semiconductors. The technique is indispensable for scientific research and industrial production. The link between structure and properties is robust, and understanding the ordered arrangements of atoms and ions that characterize pure metals, alloys, semiconductors, ionic compounds, and ceramics provides tremendous insight into the behavior of bulk materials. Knowing the structure of solids at the atomic level enables us to explain properties of materials and to design new materials with improved properties. The increasingly rapid development of stronger, tougher, harder, and more heat-resistant solid materials is made possible by an awareness of the role of size, shape, and arrangement of atoms in determining these properties.

SUMMARY

Section 12.1 Most metals are malleable and ductile. The electrical conductivity of metals can be explained by **band theory** as the easy movement of electrons from the **valence band** to the **conduction band**.

Section 12.2 Semimetals are **semiconductors**, intermediate in electrical conducting ability between metals and nonmetals. In semiconductors, the conduction band and the valence band are separated by a **band gap (E_g)**. Substituting electron-rich atoms into a semiconductor results in **n-type semiconductors**. Substituting electron-poor atoms results in **p-type semiconductors**. Both types of substitution increase the conductivity of the semiconductor by decreasing its band gap.

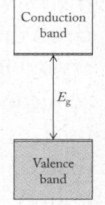

Section 12.3 Many metallic crystals are based on **crystal lattices** of the **cubic closest-packed (ccp)** and **hexagonal closest-packed (hcp)** types, which are the two most efficient ways of packing atoms in a solid. **Crystalline solids** contain repeating **unit cells**, which can be **simple cubic (sc)**, **body-centered cubic (bcc)**, or **face-centered cubic (fcc)**. The dimensions of the unit cell in a crystalline solid can be used to determine the radius of the atoms or ions and to predict density.

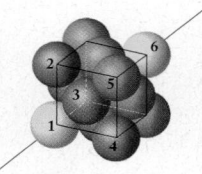

Section 12.4 **Alloys** are blends of a host metal and one or more other elements (which may or may not be metals) added to enhance the properties of the host, including strength, hardness, and corrosion resistance. In **substitutional alloys**, atoms of the added elements replace atoms of the host metal in the crystal lattice. In **interstitial alloys**, atoms of added elements are located in the tetrahedral and/or octahedral holes between atoms of the host metal. Aluminum and aluminum alloys are highly desirable for applications requiring corrosion resistance and low mass.

Section 12.5 Two allotropes of carbon are the **covalent network solids** graphite and diamond. Many nonmetals form **molecular solids**, including sulfur, which forms puckered rings of eight sulfur atoms, and phosphorus, which forms P_4 tetrahedra.

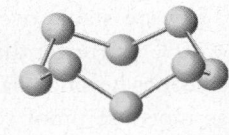

Section 12.6 Many **ionic solids** consist of crystals with some number of either cations or anions forming the unit cell and the opposite ion occupying octahedral and tetrahedral holes in the unit cell. The unit-cell edge lengths of ionic solids can be used to calculate ionic radii and to predict densities.

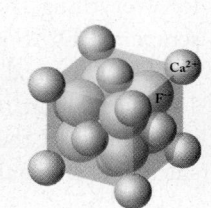

Section 12.7 Heating selected solid inorganic compounds (such as clays, which are aluminosilicate minerals) to high temperature alters their chemical composition and makes them harder, denser, and stronger. The resulting heat- and chemical-resistant materials (**ceramics**) are electrical insulators due to the large energy gap between their filled valence and empty conduction bands. The polymorphs of silica consist of tetrahedra made up of four O at the corners surrounding a central Si. Each tetrahedron can share some or all its oxygen atoms with other tetrahedra, forming Si—O—Si bridges and two- and three-dimensional covalent networks. In **superconducting** ceramics, the electrical resistance of the material drops to zero below its **critical temperature (T_c)**.

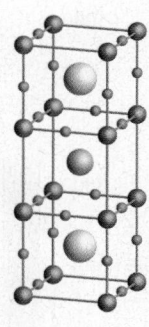

Section 12.8 **X-ray diffraction** is an analytical method that records constructive interference of X-rays reflecting off different layers of atoms or ions in a crystalline solid. The distances between layers are calculated using the **Bragg equation**. X-ray diffraction makes it possible to determine the crystal structure of crystalline solids.

PROBLEM-SOLVING SUMMARY

TYPE OF PROBLEM	CONCEPTS AND EQUATIONS	SAMPLE EXERCISES
Distinguishing p- and n-type semiconductors	Determine whether the dopant has more (n-type) or fewer (p-type) valence electrons than the host semiconductor.	12.1
Calculating atomic radii from a unit cell	Determine the length ℓ of a unit-cell edge, face diagonal, or body diagonal along which adjacent atoms touch. Use the relationship between ℓ and the atomic radius r to calculate the value of r: $r = \ell/2$ for sc unit cells, $r = 0.3536\ell$ for fcc unit cells, and $r = 0.4330\ell$ for bcc unit cells.	12.2
Using unit cells to calculate density of a solid	Determine the mass of the atoms in the unit cell from their molar mass and the volume of the unit cell from the unit-cell edge length, then calculate the density: $$d = \frac{m}{V}$$	12.3
Predicting the crystal structure of two-element alloys	Compare the radii of the alloying elements. Similarly sized radii (within 15%) predict a substitutional alloy. When the radius of the smaller atom in an hcp or ccp lattice is <73% of the radius of the larger atom, an interstitial alloy forms.	12.4
Calculating ionic radii from unit-cell dimensions	Determine the length of a unit-cell edge, face diagonal, or body diagonal along which adjacent ions touch. Use the relationship between edge length ℓ and the ionic radius r to calculate the value of r: $r = \ell/2$ for sc unit cells, $r = 0.3536\ell$ for fcc unit cells, and $r = 0.4330\ell$ for bcc unit cells.	12.5
Calculating density of a salt from its unit-cell dimensions	Determine the mass of the ions in the unit cell from their molar masses and the volume of the unit cell from the unit-cell edge length. Then apply $$d = \frac{m}{V}$$	12.6
Determining interlayer distances by X-ray diffraction	Apply the Bragg equation: $$n\lambda = 2d \sin\theta \qquad (12.13)$$	12.7

VISUAL PROBLEMS

(Answers to boldface end-of-chapter questions and problems are in the back of the book.)

12.1. In Figure P12.1, which drawings are analogous to crystalline solids, and which are analogous to amorphous solids?

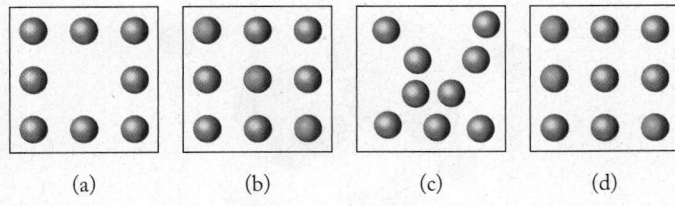

(a)   (b)   (c)   (d)

FIGURE P12.1

12.2. The unit cells in Figure P12.2 continue infinitely in two dimensions. Draw a box around the unit cell in each pattern. How many light squares and how many dark squares are in each unit cell?

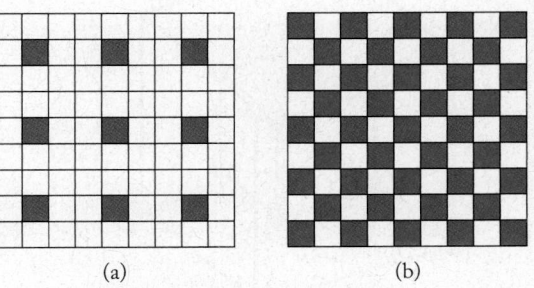

(a)   (b)

FIGURE P12.2

12.3. The pattern in Figure P12.3 continues indefinitely in three dimensions. Draw a box around the unit cell. If the red circles represent element A and the blue circles element B, what is the chemical formula of the compound?

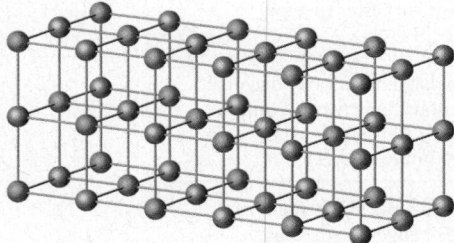

FIGURE P12.3

12.4. How many total cations (A and B) and anions (X) are there in the unit cell in Figure P12.4?

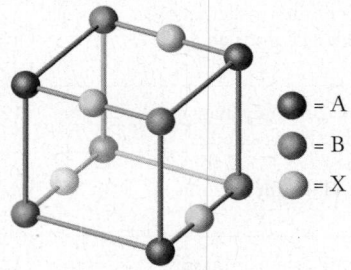

= A
= B
= X

FIGURE P12.4

12.5. How many equivalent atoms of elements A and B are there in the portion of a unit cell in Figure P12.5?

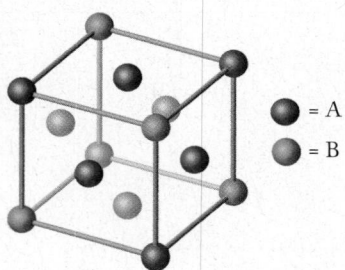

= A
= B

FIGURE P12.5

12.6. What is the chemical formula of the compound a portion of whose unit cell is shown in Figure P12.6?

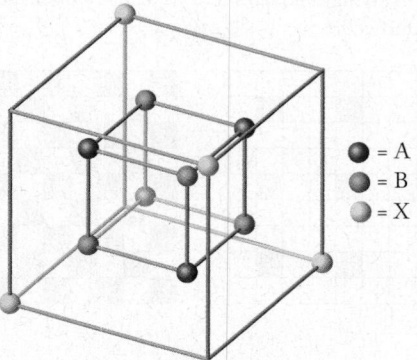

= A
= B
= X

FIGURE P12.6

12.7. What is the chemical formula of the ionic compound a portion of whose unit cell is shown in Figure P12.7? (A and B are cations, X is an anion.)

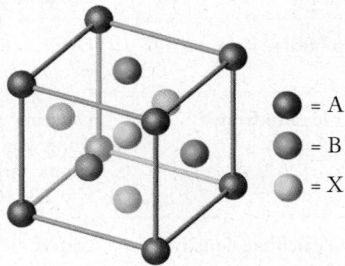

= A
= B
= X

FIGURE P12.7

12.8. When amorphous red phosphorus is heated at high pressure, it is transformed into the allotrope black phosphorus, which can exist in one of several forms. One form consists of six-membered rings of phosphorus atoms (Figure P12.8a). Why are the six-atom rings in black phosphorus puckered, whereas the six-atom rings in graphite (Figure P12.8b) are planar?

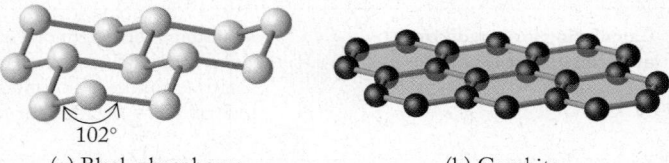

102°

(a) Black phosphorus   (b) Graphite

FIGURE P12.8

12.9. The distance between atoms in a cubic form of phosphorus is 238 pm (Figure P12.9). Calculate the density of this form of phosphorus.

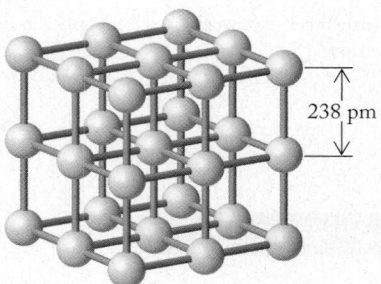

238 pm

FIGURE P12.9

12.10. How many of the large spheres and how many of the small ones are assignable to the unit cell in Figure P12.10?

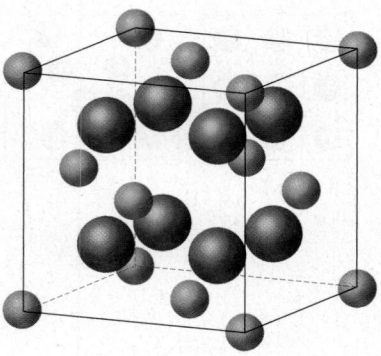

FIGURE P12.10

12.11. What is the formula of the compound that crystallizes with aluminum ions occupying half of the octahedral holes and magnesium ions occupying one-eighth of the tetrahedral holes in a cubic closest-packed arrangement of oxide ions? See Figure P12.11.

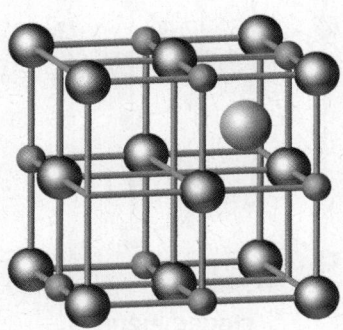

FIGURE P12.11

12.12. What is the formula of the compound that crystallizes with zinc ions occupying half of the tetrahedral holes in a cubic closest-packed arrangement of sulfur ions? See Figure P12.12.

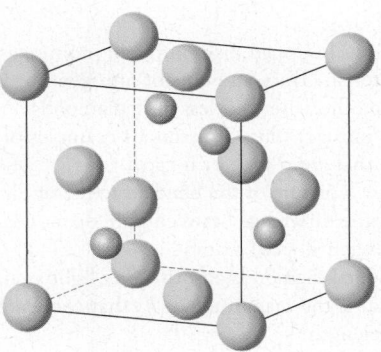

FIGURE P12.12

12.13. What is the formula of the compound that crystallizes with lithium ions occupying all of the tetrahedral holes in a cubic closest-packed arrangement of sulfide ions? See Figure P12.13.

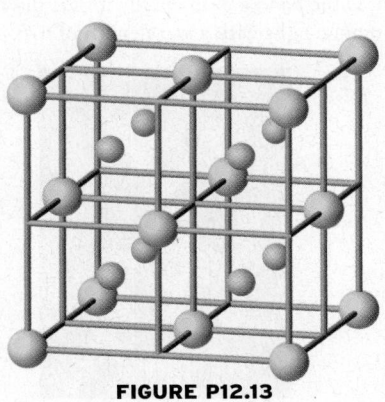

FIGURE P12.13

12.14. Figure P12.14 shows the unit cell of CsCl. From the information given and the radius of the chloride (corner) ions of 181 pm, calculate the radius of Cs^+ ions.

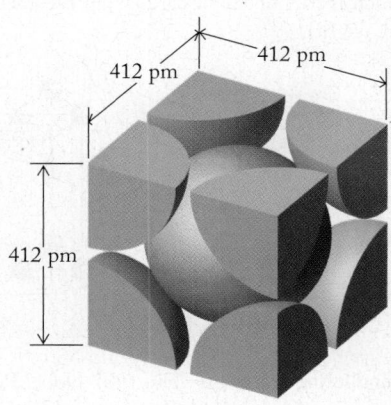

FIGURE P12.14

12.15. A number of metal chlorides adopt the rock salt crystal structure, in which the metal ions occupy all the octahedral holes in a face-centered cubic array of chloride ions. For at least one (maybe more) of the metallic elements highlighted in Figure P12.15, this crystal structure is not possible. Which one(s) cannot form a rock salt structure with Cl^- ions?

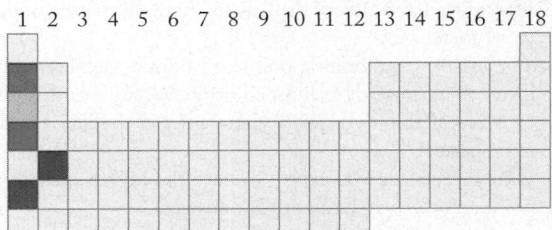

FIGURE P12.15

12.16. A number of metal fluorides adopt the fluorite crystal structure, in which the fluoride ions occupy all the tetrahedral holes in a face-centered cubic array of metal ions. For at least one (maybe more) of the highlighted metallic elements in Figure P12.16, this crystal structure is not possible. Which one(s)?

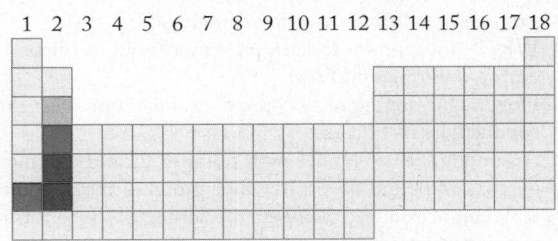

FIGURE P12.16

*12.17. Superconducting Materials In 2000, magnesium boride was observed to behave as a superconductor. Its unit cell is shown in Figure P12.17. What is the formula of magnesium boride? A boron atom is in the center of the unit cell (on the left), which is part of the hexagonal closest-packed crystal structure (right).

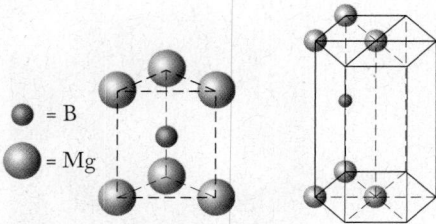

FIGURE P12.17

*12.18. Superconducting Materials The 1987 Nobel Prize in Physics was awarded to J. G. Bednorz and K. A. Müller for their discovery of superconducting ceramic materials such as $YBa_2Cu_3O_7$. Figure P12.18 shows the unit cell of another yttrium–barium–copper oxide.

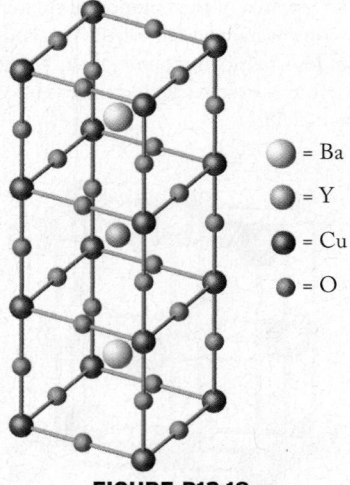

= Ba
= Y
= Cu
= O

FIGURE P12.18

a. What is the chemical formula of this compound?
b. Eight oxygen atoms must be removed from the unit cell shown here to produce the unit cell of $YBa_2Cu_3O_7$. Does it make a difference which oxygen atoms are removed?

QUESTIONS AND PROBLEMS ■

Metallic Bonds and Conduction Bands

CONCEPT REVIEW

12.19. How does the sea-of-electrons model (Chapter 8) explain the high electrical conductivity of gold?

*12.20. How does band theory explain the high electrical conductivity of mercury?

12.21. The melting and boiling points of sodium metal are much lower than those of sodium chloride. What does this difference reveal about the relative strengths of metallic bonds and ionic bonds?

*12.22. Which metal do you expect to have the higher melting point—Al or Na? Explain your answer.

12.23. Some scientists believe that the solid hydrogen that forms at very low temperatures and high pressures may conduct electricity. Is this hypothesis supported by band theory?

12.24. Would you expect solid helium to conduct electricity?

Semiconductors

CONCEPT REVIEW

12.25. Which groups in the periodic table contain metals with filled valence bands?

12.26. Insulators are materials that do not conduct electricity; conductors are substances that allow electricity to flow through them easily. Rank the following in order of increasing band gap: semiconductor, insulator, conductor.

12.27. Why is it important to keep phosphorus out of silicon chips during their manufacture?

12.28. How might doping of silicon with germanium affect the conductivity of silicon?

*12.29. Antimony (Sb) combines with sulfur to form the semiconductor compound Sb_2S_3. In which group of the periodic table might you find elements for doping Sb_2S_3 to form a p-type semiconductor?

*12.30. In which group of the periodic table might you find elements for doping Sb_2S_3 to form an n-type semiconductor?

PROBLEMS

12.31. Thin films of doped diamond hold promise as semiconductor materials. Trace amounts of nitrogen impart a yellow color to otherwise colorless pure diamonds.

a. Are nitrogen-doped diamonds examples of semiconductors that are p-type or n-type?
b. Draw a picture of the band structure of diamond to indicate the difference between pure diamond and N-doped (nitrogen-doped) diamond.
*c. N-doped diamonds absorb violet light at about 425 nm. What is the magnitude of E_g that corresponds to this wavelength?

12.32. Hope Diamond Trace amounts of boron give diamonds (including the Smithsonian's Hope Diamond) a blue color (Figure P12.32).

a. Are boron-doped diamonds examples of semiconductors that are p-type or n-type?
b. Draw a picture of the band structure of diamond to indicate the difference between pure diamond and B-doped diamond.
*c. What is the band gap in energy if blue diamonds absorb red-orange light with a wavelength of 675 nm?

FIGURE P12.32

*12.33. The nitride ceramics AlN, GaN, and InN are all semiconductors used in the microelectronics industry. Their band gaps are 580.6, 322.1, and 192.9 kJ/mol, respectively. Which, if any, of these energies correspond to radiation in the visible region of the spectrum?

*12.34. Calculate the wavelengths of light emitted by the semiconducting phosphides AlP, GaP, and InP, which have band gaps of 241.1, 216.0, and 122.5 kJ/mol, respectively, and are used in the type of light source shown in Figure P12.34.

FIGURE P12.34

Structures of Metals

CONCEPT REVIEW

12.35. Explain the difference between cubic closest-packed and hexagonal closest-packed arrangements of identical spheres.

*12.36. Is it possible to have a closest-packed crystal lattice with four different repeating layers, *abcdabcd* . . . ?

12.37. Which unit cell has the greater packing efficiency, simple cubic or body-centered cubic?

12.38. Consult Figure 12.16 to predict which unit cell has the greater packing efficiency: body-centered cubic or face-centered cubic.

12.39. The unit cell in iron metal is either fcc or bcc, depending on temperature (see Sample Exercise 12.2). Are the fcc form of iron and the bcc form allotropes? Explain your answer.

*12.40. At low temperatures, the unit cell of calcium metal is found to be fcc, a closest-packed crystal lattice. At higher temperatures, the unit cell of calcium metal is found to be bcc, a crystal lattice that is not a closest-packed structure. What might be a reason for this difference?

PROBLEMS

12.41. Derive the edge length in bcc and fcc unit cells in terms of the radius (r) of the atoms in the cells in Figure 12.16.

*12.42. Derive the length of the body diagonal in simple cubic and fcc unit cells in terms of the radius (r) of the atoms in the unit cells in Figure 12.16.

12.43. Europium, one of the lanthanide elements used in television screens, crystallizes in a crystal lattice built on bcc unit cells, with a unit-cell edge of 240.6 pm. Calculate the radius of a europium atom.

12.44. Nickel has an fcc unit cell with an edge length of 350.7 pm. Calculate the radius of a nickel atom.

12.45. What is the length of an edge of the unit cell when barium (atomic radius 222 pm) crystallizes in a crystal lattice of bcc unit cells?

12.46. What is the length of an edge of the unit cell when aluminum (atomic radius 143 pm) crystallizes in a crystal lattice of fcc unit cells?

12.47. A crystalline form of copper has a density of 8.95 g/cm³. If the radius of copper atoms is 127.8 pm, is the copper unit cell (a) simple cubic; (b) body-centered cubic; or (c) face-centered cubic?

12.48. A crystalline form of molybdenum has a density 10.28 g/cm³ at a temperature at which the radius of a molybdenum atom is 139 pm. Which unit cell is consistent with these data? (a) simple cubic; (b) body-centered cubic; (c) face-centered cubic

Alloys

CONCEPT REVIEW

12.49. Is there a difference between a solid solution and a homogeneous alloy?

12.50. White gold was originally developed to give the appearance of platinum. One formulation of white gold contains 25% nickel and 75% gold. Which is more malleable, white gold or pure gold?

*12.51. Explain why an alloy that is 28% Cu and 72% Ag melts at a lower temperature than the melting points of either Cu or Ag.

12.52. Is it possible for an alloy to be both substitutional and interstitial?

12.53. The interstitial alloy tungsten carbide (WC) is one of the hardest materials known. It is used on the tips of cutting tools. Without consulting a table of radii, decide which element you think is the host and which occupies the holes.

12.54. Why are the alloys that second-row nonmetals—such as B, C, and N—form with transition metals more likely to be interstitial than substitutional?

PROBLEMS

12.55. The unit cell of a substitutional alloy consists of a face-centered cube that has an atom of element X at each corner and an atom of element Y at the center of each face.
a. What is the formula of the alloy?
b. What would the formula of the alloy be if the positions of the two elements were reversed in the unit cell?

12.56. The bcc unit cell of a substitutional alloy has atoms of element A at the corners of the unit cell and an atom of element B at the center of the unit cell.
a. What is the formula of the alloy?
b. What would the formula of the alloy be if the positions of the two elements were reversed in the unit cell?

12.57. Vanadium reacts with carbon to form vanadium carbide, an interstitial alloy. Given the atomic radii of V (135 pm) and C (77 pm), which holes in a cubic closest-packed array of vanadium atoms do you think the carbon atoms are more likely to occupy—octahedral or tetrahedral?

12.58. What is the minimum atomic radius required for a cubic closest-packed metal to accommodate boron atoms (radius 88 pm) in its octahedral holes?

12.59. **Dental Fillings** Dental fillings are mixtures of several alloys including one with the formula Ag₃Sn. Silver (radius 144 pm) and tin (140 pm) both crystallize in an fcc unit cell. Is this alloy likely to be a substitutional alloy or an interstitial alloy?

12.60. An alloy used in dental fillings has the formula Sn₃Hg. The radii of tin and mercury atoms are 140 pm and 151 pm, respectively. Which alloy has a smaller mismatch (percent difference in atomic radii), Sn₃Hg or bronze (Cu/Sn alloys)?

*12.61. **Hardening Metal Surfaces** Plasma nitriding is a process for embedding nitrogen atoms in the surfaces of metals that hardens the surfaces and makes them more corrosion resistant. Do the nitrogen atoms in the nitrided surface of a sample of cubic closest-packed iron fit in the octahedral holes of the crystal lattice? (Assume that the atomic radii of N and Fe are 75 and 126 pm, respectively.)

*12.62. **Hydrogen Storage** A number of crystalline transition metals (including titanium, zirconium, and hafnium) can store hydrogen as metal hydrides for use as fuel in a hydrogen-powered vehicle. Which metal or metals are most likely to accommodate H atoms (radius 37 pm) with a radius ratio closest to the ideal value in Table 12.4, given that their atomic radii are 147, 160, and 159 pm, respectively?

12.63. An interstitial alloy is prepared from metals A and B where B has the smaller atomic radius. The unit cell of metal A is fcc. What is the formula of the alloy if B occupies (a) all of the octahedral holes; (b) half of the octahedral holes; (c) half of the tetrahedral holes?

12.64. An interstitial alloy was prepared from two metals. Metal A with the larger atomic radius has a hexagonal closest-packed crystal lattice. What is the formula of the alloy if atoms of metal B occupy (a) all of the tetrahedral holes; (b) half of the tetrahedral holes; (c) half of the octahedral holes?

12.65. An interstitial alloy with an fcc unit cell contains one atom of B for every five atoms of host element A. What fraction of the octahedral holes is occupied in this alloy?

12.66. If the B atoms in the alloy described in Problem 12.65 occupy tetrahedral holes in A, what percentage of the holes would they occupy?

Structures of Some Crystalline Nonmetals

CONCEPT REVIEW

12.67. S_8 is not a flat octagon—why?

*12.68. Selenium exists either as Se_8 rings or in a structure with helical chains of Se atoms. Are these two structures of selenium allotropes? Explain your answer.

*12.69. If the carbon atoms in graphite are replaced by alternating B and N atoms, would the resulting structure contain puckered rings like black phosphorus or flat ones like graphite (see Figure P12.8)?

*12.70. Cyclic allotropes of sulfur containing up to 20 sulfur atoms have been isolated and characterized. Propose a reason why the bond angles in S_n (where $n = 10, 12, 18$, and 20) are all close to $106°$.

PROBLEMS

12.71. Ice is a network solid. However, theory predicts that, under high pressure, ice (solid H_2O) becomes an ionic compound composed of H^+ and O^{2-} ions. The proposed unit cell for ice under these conditions is a bcc unit cell of oxygen ions with hydrogen ions in holes.
   a. How many H^+ and O^{2-} ions are in each unit cell?
   b. Draw a Lewis structure for "ionic" ice.

12.72. **Ice under Pressure** Kurt Vonnegut's novel *Cat's Cradle* describes an imaginary, high-pressure form of ice called "ice nine." With the assumption that ice nine has a cubic closest-packed arrangement of oxygen atoms with hydrogen atoms in the appropriate holes, what type of hole will accommodate the H atoms?

12.73. A chemical reaction between H_2S_4 and S_2Cl_2 produces cyclic S_6. What are the bond angles in S_6?

12.74. Reaction between S_8 and six equivalents of AsF_5 yields $[S_4^{2+}][AsF_6^-]_2$ by the reaction

$$S_8 + 6\,AsF_5 \rightarrow 2[S_4^{2+}][AsF_6^-]_2 + 2\,AsF_3$$

(The brackets identify complex ions in the formula, in this case indicating that the S_4 unit has a 2+ charge and the AsF_6 unit has a 1− charge.) The S_4^{2+} ion has a cyclic structure. Are all four sulfur atoms in one plane?

Salt Crystals: Ionic Solids

CONCEPT REVIEW

12.75. Crystals of both LiCl and KCl have the rock salt structure. In the unit cell of LiCl adjacent Cl^- ions touch each other. In KCl they don't. Why?

12.76. Can $CaCl_2$ have the rock salt structure?

*12.77. In some books the unit cell of CsCl is described as being body-centered cubic (Figure P12.77); in others, as simple cubic (see Figure 12.16a). Explain how CsCl crystals might be described by either unit cell type.

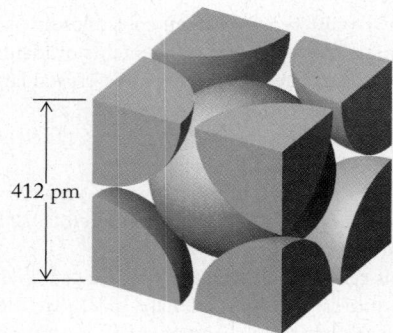

412 pm

FIGURE P12.77

12.78. In the crystals of ionic compounds, how do the relative sizes of the ions influence the location of the smaller ions?

*12.79. Instead of describing the unit cell of NaCl as an fcc array of Cl^- ions with Na^+ ions in octahedral holes, might we describe it as an fcc array of Na^+ ions with Cl^- ions in octahedral holes? Explain why or why not.

12.80. Why isn't crystalline sodium chloride considered a network solid?

12.81. If the unit cell of a substitutional alloy of copper and tin has the same unit-cell edge as the unit cell of copper, will the alloy have a greater density than copper?

12.82. If the unit cell of an interstitial alloy of vanadium and carbon has the same unit-cell edge as the unit cell of vanadium, will the alloy have a greater density than vanadium?

*12.83. As the cation–anion radius ratio increases for an ionic compound with the rock salt crystal structure, is the calculated density more likely to be greater than, or less than, the measured value?

*12.84. As the cation–anion radius ratio increases for an ionic compound with the rock salt crystal structure, is the length of the unit-cell edge calculated from ionic radii likely to be greater than, or less than, the observed unit-cell edge length?

PROBLEMS

12.85. What is the formula of the oxide that crystallizes with Fe^{3+} ions in one-fourth of the octahedral holes, Fe^{3+} ions in one-eighth of the tetrahedral holes, and Mg^{2+} in one-fourth of the octahedral holes of a cubic closest-packed arrangement of oxide ions (O^{2-})?

12.86. What is the chemical formula of the compound that crystallizes a simple cubic arrangement of fluoride ions with Ba^{2+} ions occupying half of the cubic holes?

12.87. **The Vinland Map** At Yale University there is a map, believed to date from the 1400s, of a landmass labeled "Vinland" (Figure P12.87). The map is thought to be evidence of early Viking exploration of North America. Debate over the map's authenticity centers on yellow stains on the map paralleling the black ink lines. One analysis suggests the yellow color is from the mineral anatase, a form of TiO_2 that was not used in 15th-century inks.
   a. The crystal structure of anatase is approximated by a ccp arrangement of oxide ions with titanium(IV) ions in holes. Which type of hole are Ti^{4+} ions likely to occupy?
   b. What fraction of these holes are likely to be occupied? (The radius of Ti^{4+} is 60.5 pm.)

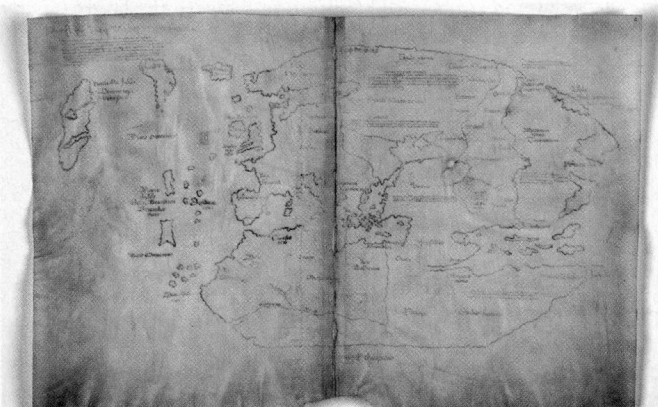

FIGURE P12.87

*12.88. The crystal structure of olivine—M_2SiO_4 (M = Mg, Fe)—can be viewed as a ccp arrangement of oxide ions with Si(IV) in tetrahedral holes and the metal ions in octahedral holes.
   a. What fraction of each type of hole is occupied?
   b. The unit-cell volumes of Mg_2SiO_4 and Fe_2SiO_4 are 2.91×10^{-26} cm^3 and 3.08×10^{-26} cm^3. Why is the unit-cell volume of Fe_2SiO_4 larger?

*12.89. In nature, cadmium(II) sulfide (CdS) exists as two minerals. One of them, called *hawleyite*, has a sphalerite structure, and its density at 25°C is 4.83 g/cm^3. A hypothetical form of CdS with the rock salt structure would have a density of 5.72 g/cm^3. Why should the rock salt structure of CdS be denser? The ionic radii of Cd^{2+} and S^{2-} are 95 pm and 184 pm, respectively.

12.90. There are two crystalline forms of manganese(II) sulfide (MnS): the α form has a NaCl structure; the β form has a sphalerite structure.
   a. Describe the differences between the two structures of MnS.
   b. The ionic radii of Mn^{2+} and S^{2-} are 67 and 184 pm. Which type of hole in a ccp lattice of sulfide ions could theoretically accommodate a Mn^{2+} ion?

12.91. The unit cell of rhenium trioxide (ReO_3) consists of a cube with rhenium atoms at the corners and an oxygen atom on each of the 12 edges. The atoms touch along the edge of the unit cell. The radii of Re and O atoms in ReO_3 are 137 and 73 pm. Calculate the density of ReO_3.

12.92. With reference to Figure P12.77, calculate the density of simple cubic CsCl.

12.93. Magnesium oxide crystallizes in the rock salt structure. Its density is 3.60 g/cm^3. What is the edge length of the fcc unit cell of MgO?

12.94. Crystalline potassium bromide (KBr) has a rock salt structure and a density of 2.75 g/cm^3. Calculate its unit-cell edge length.

Ceramics: Insulators to Superconductors

CONCEPT REVIEW

12.95. Which of the following properties are associated with ceramics and which are associated with metals? Ductile; thermal insulator; electrically conductive; malleable

12.96. Many ceramics such as TiO_2 are electrical insulators. What differences are there in the band structure of TiO_2 compared with Ti metal that account for the different electrical properties?

12.97. Replacement of Al^{3+} ions in kaolinite [$Al_2(Si_2O_5)(OH)_4$] with Mg^{2+} ions yields the mineral antigorite. What is its formula?

12.98. What is the formula of the silicate mineral talc, obtained by the replacement of Al^{3+} ions in pyrophyllite [$Al_4Si_8O_{20}(OH)_4$] with Mg^{2+} ions?

PROBLEMS

12.99. Kaolinite [$Al_2(Si_2O_5)(OH)_4$] is formed by weathering of the mineral $KAlSi_3O_8$ in the presence of carbon dioxide and water, as described by the following unbalanced equation:

$$KAlSi_3O_8(s) + H_2O(\ell) + CO_2(aq) \rightarrow$$
$$Al_2(Si_2O_5)(OH)_4(s) + SiO_2(s) + K_2CO_3(aq)$$

Balance the equation and determine whether or not this is a redox reaction.

12.100. Albite, a feldspar mineral with an ideal composition of $NaAlSi_3O_8$, can be converted to jadeite ($NaAlSi_2O_6$) and quartz. Write a balanced chemical equation describing this transformation.

12.101. Under the high pressures in Earth's crust, the mineral anorthite ($CaAl_2Si_2O_8$) is converted to a mixture of three minerals: grossular [$Ca_3Al_2(SiO_4)_3$], kyanite (Al_2SiO_5), and quartz (SiO_2).
   a. Write a balanced chemical equation describing this transformation.
   b. Determine the charges and formulas of the silicate anions in anorthite, grossular, and kyanite.

*12.102. The calcium silicate mineral grossular is also formed under pressure in a reaction between anorthite ($CaAl_2Si_2O_8$), gehlenite ($Ca_2Al_2SiO_7$), and wollastonite ($CaSiO_3$):

$$\underset{\text{Anorthite}}{CaAl_2Si_2O_8} + \underset{\text{Gehlenite}}{Ca_2Al_2SiO_7} + \underset{\text{Wollastonite}}{CaSiO_3} \rightarrow \underset{\text{Grossular}}{Ca_3Al_2(SiO_4)_3}$$

   a. Balance this chemical equation.
   b. Express the composition of gehlenite the way mineralogists often do: as the percentage of the metal and semi-metal oxides in it, that is, %CaO, %Al_2O_3, and %SiO_2.

12.103. The ceramic material barium titanate ($BaTiO_3$) is used in devices that measure pressure. The radii of Ba^{2+}, Ti^{4+}, and O^{2-} are 135, 60.5, and 140 pm, respectively. If the O^{2-} ions are in a closest-packed structure, which hole(s) can accommodate the metal cations?

12.104. The mixed metal oxide $LiMnTiO_4$ has a structure with cubic closest-packed oxide ions and metal ions in both octahedral and tetrahedral holes. Which metal ion is most likely to be found in the tetrahedral holes? The ionic radii of Li^+, Mn^{3+}, Ti^{4+}, and O^{2-} are 76, 67, 60.5, and 140 pm, respectively.

X-ray Diffraction: How We Know Crystal Structures

CONCEPT REVIEW

12.105. Why does an amorphous solid not produce an XRD scan with sharp peaks?

12.106. X-ray diffraction cannot be used to determine the structures of compounds in solution—why?

12.107. Why are X-rays rather than microwaves chosen for diffraction studies of crystalline solids?

12.108. The radiation sources used in X-ray diffraction can be changed. Figure P12.108 shows a diffraction pattern made by a short-wavelength source. How would changing to a longer-wavelength source affect the pattern?

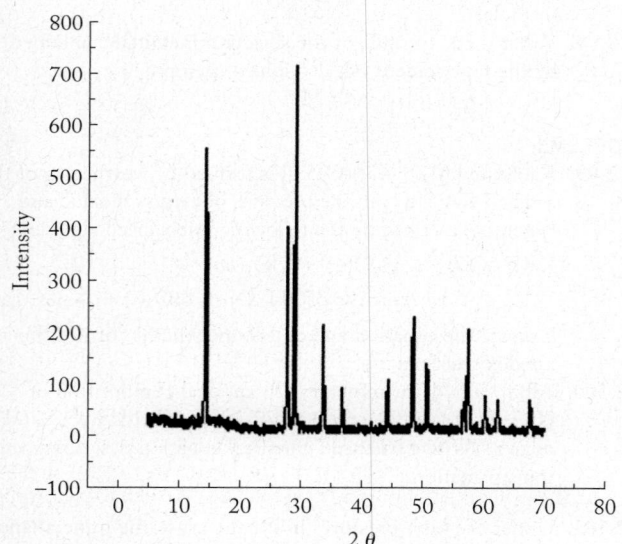

FIGURE P12.108

*12.109. Why might a crystallographer (a scientist who studies crystal structures) use different X-ray wavelengths to determine a crystal structure? (*Hint*: Consider what mechanical limits are inherent in the design of the instrument depicted in Figure 12.44, and how those limits impact the 2θ scanning range.)

*12.110. Where in earlier chapters have we seen diffraction used to acquire structural information?

PROBLEMS

12.111. The spacing between the layers of ions in sylvite (the mineral form of KCl) is larger than in halite (NaCl). Which crystal will diffract X-rays of a given wavelength through larger 2θ values?

12.112. Silver halides are used in black-and-white photography. In which compound would you expect to see a larger distance between ion layers, AgCl or AgBr? Which compound would you expect to diffract X-rays through larger values of 2θ if the same wavelength of X-ray were used?

12.113. Galena, Illinois, is named for the rich deposits of lead(II) sulfide (PbS) found nearby. When PbS is exposed to X-rays with $\lambda = 71.2$ pm, strong reflections from a single crystal of PbS are observed at 13.98° and 21.25°. Determine the values of n to which these reflections correspond, and calculate the spacing between the crystal layers.

12.114. **Pigments in Ceramics** Cobalt(II) oxide is used as a pigment in ceramics. It has the same type of crystal structure as NaCl. When cobalt(II) oxide is exposed to X-rays with $\lambda = 154$ pm, reflections are observed at 42.38°, 65.68°, and 92.60°. Determine the values of n to which these reflections correspond, and calculate the spacing between the crystal layers.

12.115. Pyrophyllite $[Al_2Si_4O_{10}(OH)_2]$ is a silicate mineral with a layered structure. The distances between the layers is 1855 pm. What is the smallest angle of diffraction of X-rays with $\lambda = 154$ pm from this solid?

12.116. Minnesotaite $[Fe_3Si_4O_{10}(OH)_2]$ is a silicate mineral with a layered structure similar to that of kaolinite. The distance between the layers in minnesotaite is 1940 ± 10 pm. What is the smallest angle of diffraction of X-rays with $\lambda = 154$ pm from this solid?

Additional Problems

12.117. A unit cell consists of a cube that has an ion of element X at each corner, an ion of element Y at the center of the cube, and an ion of element Z at the center of each face. What is the formula of the compound?

12.118. The unit cell of an oxide of uranium consists of cubic closest-packed uranium ions with oxide ions in all the tetrahedral holes. What is the formula of the oxide?

*12.119. The packing efficiency for a unit cell can be calculated using Equation 12.1 (repeated here):

$$\text{Packing efficiency (\%)} = \frac{\text{volume occupied by spheres}}{\text{volume of unit cell}} \times 100\%$$

What is the packing efficiency of the Si atoms in pure Si if the radius of one Si atom is 117 pm? The density of pure silicon is 2.33 g/mL.

*12.120. **The Composition of Light-Emitting Diodes** The colored lights on many electronic devices are light-emitting diodes (LEDs). One of the compounds used to make them is aluminum phosphide (AlP), which crystallizes in a sphalerite crystal structure.

a. If AlP were an ionic compound, would the ionic radii of Al^{3+} and P^{3-} be consistent with the size requirements of the ions in a sphalerite crystal structure?

b. If AlP were a covalent compound, would the atomic radii of Al and P be consistent with the size requirements of atoms in a sphalerite crystal structure?

12.121. Under the appropriate reaction conditions, small cubes of molybdenum, 4.8 nm on a side, can be deposited on carbon surfaces. These "nanocubes" are made of bcc arrays of Mo atoms.

    a. If the edge of each nanocube corresponds to 15 unit-cell lengths, what is the effective radius of a molybdenum atom in these structures?

    b. What is the density of each molybdenum nanocube?

    c. How many Mo atoms are in each nanocube?

12.122. In the fullerene known as buckminsterfullerene, C_{60}, molecules of C_{60} form a cubic closest-packed array of spheres with a unit-cell edge length of 1410 pm.

    a. What is the density of crystalline C_{60}?

    b. If we treat each C_{60} molecule as a sphere of 60 carbon atoms, what is the radius of the C_{60} molecule?

    c. C_{60} reacts with alkali metals to form M_3C_{60} (where M = Na or K). The crystal structure of M_3C_{60} contains cubic closest-packed spheres of C_{60} with metal ions in holes. If the radius of a K^+ ion is 138 pm, which type of hole is a K^+ ion likely to occupy? What fraction of the holes will be occupied?

    d. Under certain conditions, a different substance, M_6C_{60}, can be formed in which the C_{60} molecules have a bcc unit cell. Calculate the density of a crystal of M_6C_{60}.

12.123. The center of Earth is composed of a solid iron core within a molten iron outer core. When molten iron cools, it crystallizes in different ways depending on pressure—in a bcc unit cell at low pressure and in a hexagonal unit cell at high pressures like those at Earth's center.

    a. Calculate the density of bcc iron given that the radius of an iron atom is 126 pm.

    b. Calculate the density of hexagonal iron given a unit-cell volume of 5.414×10^{-23} cm^3.

    *c. Seismic studies suggest that the density of Earth's solid core is only about 90% of that of hexagonal Fe. Laboratory studies have shown that up to 4% by mass of Si can be substituted for Fe without changing the hcp crystal structure built on hexagonal unit cells. Calculate the density of such a crystal.

12.124. The unit cell of an alloy with a 1:1 ratio of magnesium and strontium is identical to the unit cell of CsCl. The unit-cell edge of MgSr is 390 pm. What is the density of MgSr?

12.125. Gold and silver can be separately alloyed with zinc to form AuZn (unit-cell edge 319 pm) and AgZn (unit-cell edge 316 pm). The two alloys have the same unit cell. Which alloy is more dense?

*12.126.** Removing two electrons from cyclo-S_8 yields the dication cyclo-S_8^{2+}. Will all of the sulfur atoms be in one plane in the S_8^{2+} cation?

12.127. Use Equation 12.1 to calculate the packing efficiency in a simple cubic unit cell.

12.128. Use Equation 12.1 to calculate the packing efficiency in a body-centered cubic unit cell.

12.129. Manganese steels are a mixture of iron, manganese, and carbon. Is the manganese likely to occupy holes in the austenite lattice, or are manganese steels substitutional alloys?

12.130. Aluminum forms alloys with lithium (LiAl), gold ($AuAl_2$), and titanium (Al_3Ti). Based on their crystal lattices, each of these alloys is considered to be a substitutional alloy.

    a. Do these alloys fit the general size requirements for substitutional alloys? The atomic radii for Li, Al, Au, and Ti are 152, 143, 144, and 147 pm, respectively.

    b. If the unit cell of LiAl is bcc, what is the density of LiAl?

***12.131.** The aluminum alloy Cu_3Al crystallizes in a bcc unit cell. Propose a way that the Cu and Al atoms could be allocated between bcc unit cells that is consistent with the formula of the alloy.

Silicon, Silica, Silicates, Silicone: What's in a Name?

12.132. Write a balanced chemical equation for each of the following reactions: (a) silicon dioxide reacts with carbon; (b) silicon reacts with chlorine; (c) germanium reacts with bromine.

12.133. Write balanced chemical equations for each of the following reactions: (a) tin reacts with chlorine to make a tetrahalide; (b) lead reacts with chlorine to make a dihalide; (c) silicon(IV) chloride is heated above 700°C.

12.134. Calcium silicide ($CaSi_2$) has a structure consisting of graphite-like layers of Si atoms with Ca atoms between the layers. In an X-ray diffraction analysis of a sample of $CaSi_2$, X-rays with a wavelength of 154 pm produced signals at $2\theta = 29.86°$, $45.46°$, and $62.00°$ for $n = 2, 3$, and 4.

    a. What is the distance between the Si layers?

    b. If the Ca^{2+} ion lies exactly halfway between the layers, what is the Ca–Si distance?

    c. $CaSi_2$ is sometimes used as a "deoxidizer" in converting iron ore to steel. Explain how $CaSi_2$ might fill this role. (*Hint*: What is the most common oxidation state of Si?)

12.135. The ionization energies and electron affinities of the group 14 elements have values close to the corresponding average values for all the elements in their rows in the periodic table. How does this "average" behavior explain the tendency for the group 14 elements to form covalent bonds rather than ionic bonds?

Additional study materials including ChemTours and Diagnostic Quizzes are available at StudySpace at www.wwnorton.com/studyspace.

13

Organic Chemistry: Fuels, Pharmaceuticals, and Materials

Learning Outcomes

- Describe the differences among alkanes, alkenes, alkynes, and aromatic compounds

- Name alkanes, alkenes, and alkynes

- Identify structural isomers and stereoisomers

- Discuss the differences between addition and condensation polymers

- Explain the differences among alcohols, ethers, amines, ketones, and aldehydes

- Explain the differences among carboxylic acids, esters, and amides

- Identify chiral molecules

A LOOK AHEAD

The Stuff of Daily Life

Think for a moment about the stuff of everyday life. We drive cars powered by gasoline or diesel fuel. We cook chicken on a grill that burns propane. We live in homes heated either by a furnace that burns natural gas or fuel oil or by electricity, most of which is produced by burning coal. All of these fuels are organic compounds—which means they are carbon-based compounds—and they are used as sources of heat because their combustion reactions are highly exothermic.

The great majority of medicines used to treat everything from headaches to strokes, from diabetes to cancer are organic compounds. Whether they are natural products derived from living plants or animals or are synthetic materials prepared in the laboratory, the products of the pharmaceutical industry are compounds of carbon.

Your favorite soft T-shirt is made of cotton, a natural organic fiber. Because of the composition, shape, and orientation of the large molecules that form cotton fibers, the material is soft and absorbent. In contrast, the soles of your running shoes are made of a synthetic rubber, a polymer manufactured from compounds of carbon and hydrogen derived from crude oil. Because of the size, shape, and orientation of the polymer molecules in synthetic rubber, the material is springy, non-absorbent, and resistant to wear.

Fans at a hockey game sit on molded plastic seats watching goalies wearing helmets made of impact-resistant Kevlar and munch on hot dogs wrapped in plastic wrap. Plastics, Kevlar, and cling wrap are made from compounds derived from crude oil—in other words, they are all organic compounds. The plastic seat is strong, lightweight, and capable

Olympic Polymers The elastic outerwear worn by Olympic speed skaters is made of synthetic organic polymers. ▶

of bearing considerable mass without breaking. The goalie's helmet must be light-weight but still able to withstand the impact of a high-speed hockey puck, and the sandwich wrap must be flexible and prevent oxygen from reaching the hot dog and bun before your purchase. All these physical properties can be designed into the materials because they are directly linked to their structure and composition at the molecular level.

The human body is 18% by mass carbon—making this element the second most abundant in us (after oxygen, most of which is present in molecules of H_2O). Most of the carbon in our cells is in the form of organic compounds. The same is true of the food we eat.

In short, compounds of carbon are everywhere, and they are so varied in size, shape, and properties that an entire field of chemistry is devoted to their study. This chapter is a brief introduction to that field, which is known as organic chemistry. A knowledge of organic chemistry is fundamental and essential for a scientific understanding of fuels, foods, pharmaceutical agents, plastics, fibers, living creatures, plants—indeed almost everything we are, almost everything we need to survive, and almost everything we produce to make our lives easier in the modern world. As you read, think about questions such as these: Why do we burn some organic compounds as fuel? What makes Kevlar strong but allows nylon to stretch? What kinds of molecules do we find in common headache and fever remedies? ■

13.1 Carbon: The Scope of Organic Chemistry

The designation *organic* for carbon-containing compounds was once limited to substances produced by living organisms, but that definition has been broadened for two reasons. First, since 1828, when Friedrich Wöhler (1800–1882) discovered how to prepare urea (Figure 13.1) in the laboratory "without the intervention of a kidney," scientists have learned to synthesize many materials previously thought to be the products only of living systems. Second, chemists have learned how to synthesize many carbon-based materials that have never been produced by living systems. Thus today the study of **organic chemistry** encompasses the chemistry of all compounds containing carbon–carbon and/or carbon–hydrogen bonds, regardless of their origin.

Families Based on Functional Groups

Much of the variety in the chemistry of carbon arises from a carbon atom's ability to form covalent bonds to other carbon atoms, which in turn bond to additional carbon atoms or atoms of other elements to make compounds containing a few to many thousands of atoms. Additional variety is introduced because, in addition to long chains, these molecules may contain a multitude of branched structures as well as rings. Organic compounds can also contain a wide range of nonmetallic elements in addition to carbon and hydrogen.

Managing the wealth of information about the millions of organic compounds that exist requires some organizing concepts. Chemists group organic compounds into families on the basis of **functional groups**, which are subunits in a molecule that confer particular chemical and physical properties.

We have already discussed two functional groups: the –OH group in alcohols (Section 4.4) and the –COOH group in carboxylic acids (Section 2.6). Although we have yet to identify them as functional groups, we have seen carbon atoms make double bonds to oxygen in formaldehyde, CH_2O, and to carbon in ethylene, C_2H_4, and form delocalized π bonds in benzene, C_6H_6. In this chapter, we examine

---

CONNECTION In Chapter 10, we learned that compounds with the same molecular formula but different arrangements of atoms differ in physical properties such as melting point and boiling point.

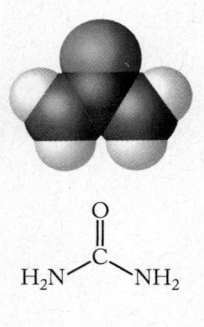

Urea

FIGURE 13.1 Urea was the first naturally occurring organic compound to be synthesized in the laboratory. It was prepared in 1828 by Friedrich Wöhler.

organic chemistry the study of compounds containing C—C and/or C—H bonds.

functional group a structural subunit in organic molecules that imparts characteristic chemical and physical properties.

TABLE 13.1	Functional Groups of Organic Compounds	
Name	**Structural Formula of Group**	**Example and Name**
Alkane	R—H	$CH_3CH_2CH_3$ Propane
Alkene	C=C	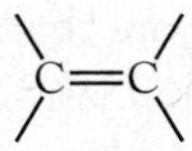 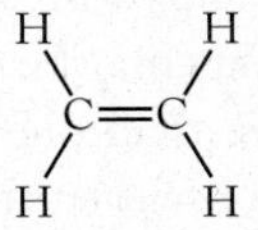 H₂C=CH₂ Ethylene (ethene)
Alkyne	—C≡C—	H—C≡C—H Acetylene (ethyne)
Aromatic	 e.g,  (benzene ring)	(benzene ring) Benzene
Amine	R—NH₂ R—NHR R—NR₂	$H_3C—NH_2$ Methylamine
Alcohol	R—OH	CH_3CH_2OH Ethanol
Ether	R—O—R	$CH_3CH_2OCH_2CH_3$ Diethyl ether
Aldehyde	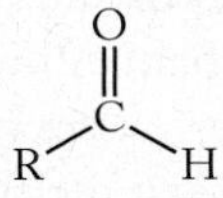 O‖C R—C—H	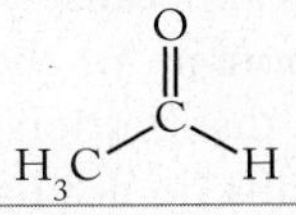 O‖C H₃C—C—H Acetaldehyde
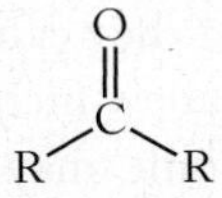 Ketone	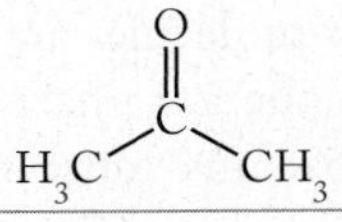 O‖C R—C—R	O‖C H₃C—C—CH₃ Acetone
Carboxylic acid	O‖C R—C—OH	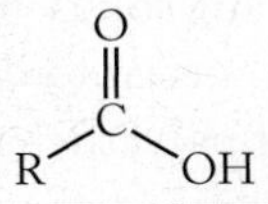 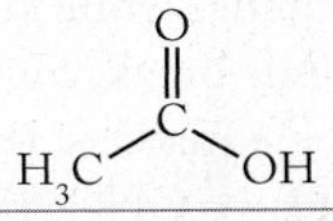 O‖C H₃C—C—OH Acetic acid
Ester	O‖C R—C—OR	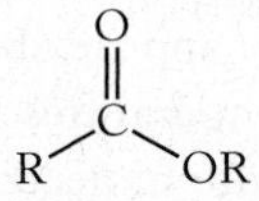 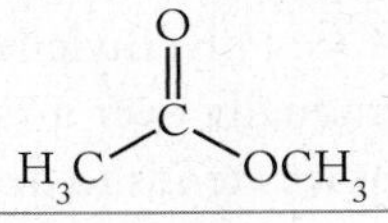 O‖C H₃C—C—OCH₃ Methyl acetate
Amide	O‖C R—C—NH₂	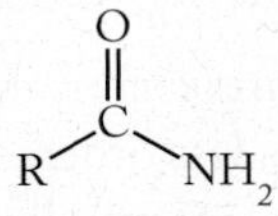 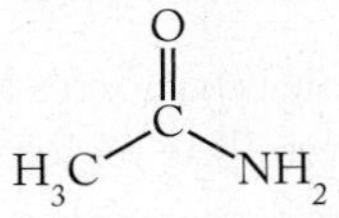 O‖C H₃C—C—NH₂ Acetamide

these and the other functional groups summarized in Table 13.1. When discussing functional groups, the convention is to use **R** to represent all of the molecule except the functional group. Think of R as standing for "the rest of the molecule" (and don't confuse it with the italic *R* standing for the universal gas constant in the ideal gas equation).

R symbol in a general formula standing for an organic group that has one available bond; it is used to indicate the variable part of a molecule so that the focus is placed on the functional group.

polymer a very large molecule with high molar mass; the root word *meros* is Greek for "part" or "unit," so *polymer* literally means "many units"; also known as *macromolecule*.

Monomers and Polymers

A second organizing principle for organic molecules is based on size. We have two categories: all organic molecules having a molar mass typically less than 1000 g/mol are one category, and all larger ones, called either **polymers** or *macromolecules*, are

monomer a small molecule that bonds with others like it to form polymers.

CONNECTION We have already introduced many organic molecules in Chapters 4 through 11: acetic acid; methane, ethane, and other hydrocarbons as fuels; acetylene used in welding; ethanol and other alcohols.

FIGURE 13.2 Items made of synthetic polymers are ubiquitous in modern society.

the second category with molar masses up to and exceeding 1,000,000 g/mol. These size boundaries are somewhat arbitrary, and in between small molecules and polymers is the realm of *oligomers*.

We have already seen examples of polymers in previous chapters, including poly-vinyl chloride used to make drain pipes and other building materials and Teflon used in nonstick cookware. They are composed of small structural units; called **monomers**. Polymers may be formed from more than one type of monomer unit; they may have more than one functional group, and they may have shapes other than long chains. However, the single feature that distinguishes them as a class is their large size.

Small Molecules versus Polymers: Physical Properties

Before looking at how the composition and structure of polymers influence their behavior, let's first discuss the differences between the physical properties of small organic molecules and the physical properties of polymers. For now, we restrict our discussion to synthetic polymers made in the laboratory from small organic molecules (Figure 13.2).

All small organic molecules have well-defined properties. They have constant composition, for example, and their phase transitions take place at well-defined temperatures. In contrast, many synthetic polymers do not have constant composition and well-defined properties because they are mixtures of large molecules that are similar but not identical because they differ in the number of monomers and the arrangement of monomers in the polymer chain. Their physical properties depend on the range and distribution of molecular masses in the sample.

Thus, polymers typically do not have well-defined melting points and boiling points. Indeed, polymer molecules are so large that they cannot acquire sufficient kinetic energy to enter the gas phase. As temperature increases, some polymers gradually soften and become more malleable. Some may eventually melt and become very viscous liquids, while others remain solids up to temperatures at which their covalent bonds break, causing the material to decompose. For example, plastic grocery bags are made of the polymer polyethylene, the monomer unit of which is ethylene ($CH_2 = CH_2$). The melting point of ethylene is $-169°C$; its boiling point is $-104°C$. Polyethylene is a tough and flexible solid at room temperature. It softens gradually over a range of temperatures from $85°C$ to $110°C$, and at higher temperatures tends to decompose into molecules of low-molar-mass organic gases.

In addition, intermolecular forces between long chains in polymers can lead to both highly ordered, crystalline regions as well as less ordered, amorphous regions. When a polymer is heated, rearrangements of the intermolecular forces contribute to a broad range of melting temperatures. For now, we conclude that matter composed of polymers behaves very differently from matter composed of small organic molecules.

CONCEPT TEST

Which types of intermolecular forces are most likely to dominate between large molecules (polymers) containing long chains of carbon atoms bound to hydrogen atoms?

(Answers to Concept Tests are in the back of the book.)

13.2 Alkanes

The first three functional groups in Table 13.1 define the families of organic compounds called alkanes, alkenes, and alkynes. We have discussed some members of these families as compounds found in fuels.

Alkanes, alkenes, and alkynes are collectively called *hydrocarbons* because they are made of only carbon and hydrogen. The three classes are distinguished by the types of carbon–carbon bonds in their molecules: **alkanes** have only carbon–carbon single bonds; **alkenes** have one or more carbon–carbon double bonds; **alkynes** have one or more carbon–carbon triple bonds. Alkanes are classified as **saturated hydrocarbons** because they contain the maximum ratio of hydrogen atoms to carbon atoms. The general molecular formula of alkanes is C_nH_{2n+2}, where n is the number of carbon atoms per molecule.

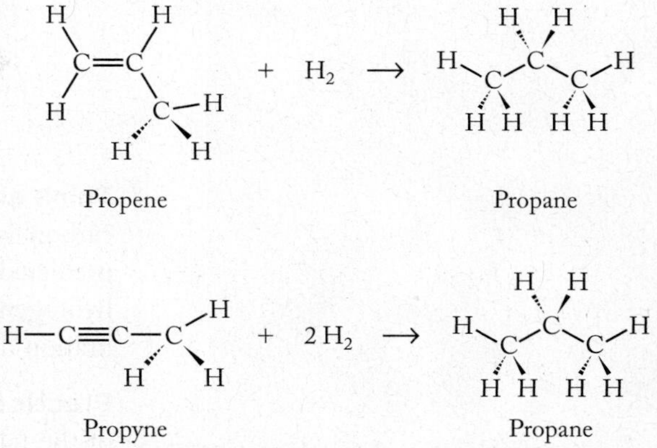

Alkenes and alkynes are **unsaturated hydrocarbons** that can combine with H_2 to form alkanes in a process called **hydrogenation**. A molecule with one C=C bond is described as having one *degree of unsaturation*. It combines with one molecule of H_2 to form one molecule of alkane (Figure 13.3). A molecule with one C≡C bond has two degrees of unsaturation and combines with two molecules of H_2, and a molecule with one double bond and one triple bond has three degrees of unsaturation. Determining the degree of unsaturation of an unknown compound using a hydrogenation reaction can help identify its structure.

FIGURE 13.3 Hydrogenation of propene and propyne requires 1 mole H_2 per mole propene and 2 moles H_2 per mole propyne.

∞ **CONNECTION** In Chapter 3, we defined hydrocarbons as compounds composed of only carbon and hydrogen atoms.

SAMPLE EXERCISE 13.1 **Distinguishing among Alkanes, Alkenes, and Alkynes**

Three hydrocarbons, each containing four carbon atoms, are stored in separate, unlabeled flasks. If one is an alkane, one an alkene with one double bond, and one an alkyne with one triple bond, design an experiment based on their reactivity with H_2 gas to determine which is the alkane, which is the alkene, and which is the alkyne given that each flask contains one mole of compound.

Collect and Organize We are reacting three compounds with hydrogen. The compounds contain the same number of carbon atoms, but one compound contains a carbon–carbon double bond and another contains a carbon–carbon triple bond.

Analyze The alkane is a saturated hydrocarbon, which means it already contains the maximum number of hydrogen atoms possible and so does not react with hydrogen. The alkene and alkyne are unsaturated, so they can both react with H_2 gas. One mole of the alkene with one double bond (or one degree of unsaturation) reacts with 1 mole of H_2. One mole of the alkyne with one triple bond (or two degrees of unsaturation) reacts with 2 moles of H_2. The flask containing the alkyne will require the most H_2 to react completely, while the flask containing the alkane will not react at all with H_2.

Solve If we allow 1 mole of each hydrocarbon to react with 2 moles of hydrogen and measure how much hydrogen is consumed, we can identify the compounds.

alkane hydrocarbon in which all the bonds are single bonds with the general formula C_nH_{2n+2}.

alkene hydrocarbon containing one or more carbon–carbon double bonds.

alkyne hydrocarbon containing one or more carbon–carbon triple bonds.

saturated hydrocarbon an alkane.

unsaturated hydrocarbon an alkene or alkyne.

hydrogenation the reaction of an unsaturated hydrocarbon with hydrogen.

We pick 2 moles of H_2 because that is the maximum amount that any of the three samples requires for complete reaction:

$$1\,\text{mol alkane} + 2\,\text{mol H}_2 \xrightarrow{\text{no reaction}} 2\,\text{mol H}_2 \text{ left}$$

$$1\,\text{mol alkene} + 2\,\text{mol H}_2 \rightarrow 1\,\text{mol alkane} + 1\,\text{mol H}_2 \text{ left}$$

$$1\,\text{mol alkyne} + 2\,\text{mol H}_2 \rightarrow 1\,\text{mol alkane} + 0\,\text{mol H}_2 \text{ left}$$

Think about It Measuring the amount of hydrogen that reacts with a hydrocarbon is a way to distinguish saturated from unsaturated hydrocarbons. As predicted, the alkyne, with two degrees of unsaturation, reacted with the most hydrogen. Unsaturated hydrocarbons may be distinguished if their degree of saturation differs.

Practice Exercise The labels have fallen off two containers in the lab. One of the fallen labels has structure A printed on it, and the other has structure B printed on it as shown below:

$$CH_3{-}CH_2{-}CH{=}CH{-}CH{=}CH{-}CH_3$$
Compound A

$$CH_3{-}CH{=}CH{-}CH{=}CH{-}CH{=}CH_2$$
Compound B

Can you use hydrogenation reactions to decide which label belongs on which container? What is the structure of the product of hydrogenation in each case?

(Answers to Practice Exercises are in the back of the book.)

CONCEPT TEST

Could we use information from hydrogenation reactions to distinguish a hydrocarbon with one triple bond from a hydrocarbon with two double bonds?

In the remainder of this section we look at the properties of the alkanes, deferring until Section 13.3 our coverage of the alkenes and alkynes.

Physical Properties and Structures of Alkanes

Alkanes are also known as *paraffins*, a name derived from Latin meaning "little affinity." This is a perfect description of the alkanes, which tend to be much less reactive than the other hydrocarbon families. Despite their lack of reactivity, alkanes are compounds of great importance because of their use as fuels, oils, and lubricants. *Unreactive* may not seem like the correct term to apply to compounds that are fuels, but alkanes do not react readily, even with oxygen. As evidence of this, consider that most fuels need some source of energy, such as a spark from a spark plug in an engine, to initiate combustion.

In the language of organic chemistry, a **homologous series** is defined as a series of compounds in which members can be described by a general formula and have similar chemical properties. Table 13.2 shows a homologous series of alkanes, with the general formula C_nH_{2n+2} and in which each member differs from the next by one $-CH_2-$ unit, called a **methylene group**. The terminal $-CH_3$ groups are **methyl groups**. All of the alkanes in the table are **straight-chain alkanes**, which means a continuous sequence of carbon atoms with no branching.

homologous series a set of related organic compounds that differ from one another by the number of common subgroups, such as $-CH_2-$, in their molecular structures.

methylene group $(-CH_2-)$ a structural unit that can make two bonds.

methyl group $(-CH_3)$ a structural unit that can make only one bond.

straight-chain alkane a hydrocarbon in which the carbon atoms are bonded together in one continuous line. Linear alkane chains have a methyl group at each end with methylene groups connecting them.

Kekulé structure a structure showing all of the bonds in a covalently bonded molecule using lines but not showing lone pairs on the atoms.

Condensed Structure[a]	Use	Melting Point (°C)	Normal Boiling Point (°C)
$CH_3CH_2CH_3$		−190	−42
$CH_3(CH_2)_2CH_3$	Gaseous fuels	−138	−0.5
$CH_3(CH_2)_3CH_3$		−130	36
$CH_3(CH_2)_4CH_3$		−95	69
$CH_3(CH_2)_5CH_3$	Gasoline	−91	98
$CH_3(CH_2)_6CH_3$		−57	126
$CH_3(CH_2)_7CH_3$		−54	151
$CH_3(CH_2)_{10}CH_3$ through $CH_3(CH_2)_{16}CH_3$	Diesel fuel and heating oil	−10 28	216 316
$CH_3(CH_2)_{18}CH_3$ through $CH_3(CH_2)_{32}CH_3$	Paraffin candle wax	37 72–75	343 na[b]
$CH_3(CH_2)_{34}CH_3$ and higher homologs	Asphalt	72–76	na

TABLE 13.2 Melting Points and Boiling Points for Selected Straight-Chain Alkanes

[a] See Figure 13.4.
[b] na = not available; compound decomposes before boiling at 1 atm pressure.

Straight-chain alkanes are identified by the letter *n*- (for *normal*) preceding the chemical name.

The data in Table 13.2 show similar trends in melting and boiling points: as straight-chain alkanes increase in molar mass, their melting and boiling points increase. This trend in physical properties is typical of all homologous series.

CONCEPT TEST

Alkanes have the general formula C_nH_{2n+2}. If an alkane has a molar mass of 114 g/mol, what is the value of *n*?

Drawing Organic Molecules

Figure 13.4(a) shows the Lewis structure for the five-carbon alkane *n*-pentane. In a Lewis structure, all of the bonds in the molecule are shown as well as any lone pairs on the atoms. Alkanes do not have lone pairs on any of the atoms, but lone pairs are common in other organic molecules. When we draw structures for organic molecules showing all the bonds with lines but leaving off any lone pairs, the structural formulas are called **Kekulé structures** after August Kekulé (1829–1896), who first used this method for illustrating molecules. As you might imagine, it soon gets tedious to write Lewis or Kekulé structures for alkanes or any other organic molecule. For this reason, chemists use various shorter notations to convey structures in organic chemistry, such as those shown in Figure 13.4. The notation in Figure 13.4(b) is called a *condensed structure*, a name that makes perfect sense when you compare the structure with the Lewis structure in Figure 13.4(a): a condensed structure is condensed in that it does not show the individual bonds between atoms the way a Lewis structure does.

Structures that use subscripts to indicate the number of times a particular subgroup is repeated are also considered condensed structures. For example, the condensed structure of *n*-pentane can also be written as $CH_3(CH_2)_3CH_3$ (Figure 13.4b).

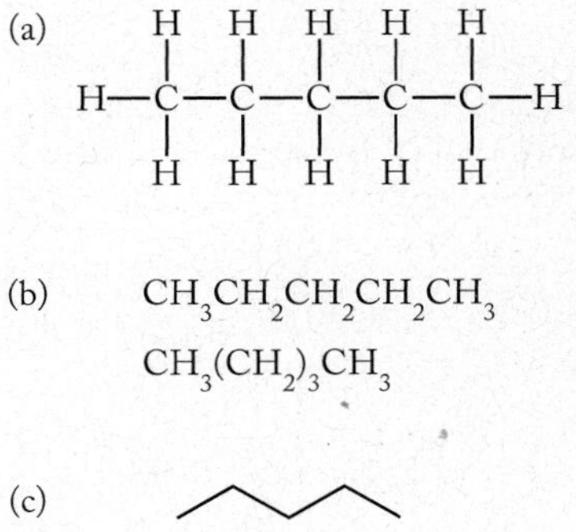

FIGURE 13.4 Several representations of the molecular structure of *n*-pentane: (a) Lewis structure; (b) two condensed structures; (c) carbon-skeleton structure.

CONNECTION Kekulé structures are the same as the structural formulas we have been drawing throughout the text since Chapter 1.

(a) Condensed structural formula:

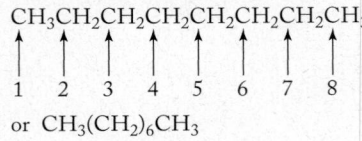

or $CH_3(CH_2)_6CH_3$

(b) Carbon-skeleton structure:

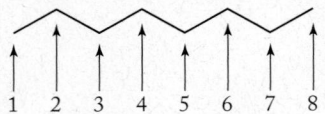

FIGURE 13.5 When converting from (a) a condensed structure to (b) a carbon-skeleton structure, the carbon atoms are represented by junctions between lines and each junction is assumed to have a sufficient number of H atoms to give that carbon atom four bonds.

The numerical subscript after the parenthetical methylene group means that three of these groups connect the two terminal methyl groups in this compound. (Note that the subscript indicating number of methylene groups comes *after* the closing parenthesis. The subscript 2 inside the parentheses is for the two H atoms on the C atom of each methylene group.)

The most minimal notation, shown in Figure 13.4(c), is the *carbon-skeleton structure*, where no alphabetic symbols are used for carbon and hydrogen atoms. (Atoms other than C and H are shown in carbon-skeleton structures, as we see in Section 13.5.) Figure 13.5 shows how a carbon-skeleton structure is created. Short line segments are drawn at angles to one another, and each line segment symbolizes one carbon–carbon bond in the molecule. The angles represent the bond angle between the two carbon atoms (109.5° in the case of sp^3 hybridized carbon atoms in alkanes). Each end of the zigzag line is a –CH_3 group, and every intersection of two line segments is a –CH_2– group. In other words, it is understood that each carbon atom has the appropriate number of hydrogen atoms to give it a steric number of four. Hydrogen atoms are not shown in a carbon-skeleton structure because all carbon atoms are known to make four bonds, and any bonds not shown are understood to be C—H bonds.

SAMPLE EXERCISE 13.2   Drawing Alkane Structures

(a) Write the condensed structure and the carbon-skeleton structure for *n*-propane. (b) Write the condensed structure and the carbon-skeleton structure for the 12-carbon *n*-dodecane.

Collect and Organize Figure 13.4 summarizes the differences between condensed and carbon-skeleton structures, which we apply to (a) a 3-carbon and (b) a 12-carbon hydrocarbon.

Analyze Both compounds are normal alkanes, so all the carbon atoms in each molecule are in one chain. Condensed structures show the symbol for each element in a molecule and subscripts indicating the numbers of atoms of each element, but they do not show C—H or single C—C bonds. We can group all methylene groups into one item by using parentheses followed by a subscript showing the number of –CH_2– groups. Carbon-skeleton structures use short line segments to represent the carbon–carbon bonds in molecules. In the carbon-skeleton structures, we must use zigzag line segments because the intersections of these segments represent the carbon atoms in the skeleton.

Solve

a. The condensed structure for propane comes from its Lewis structure (on the left), and then we write each carbon followed by H plus a subscript showing the number of H atoms bonded to that carbon (structure in the center):

$$
\begin{array}{c}
\quad\; H \;\; H \;\; H \\
\quad\; | \;\;\;\; | \;\;\;\; | \\
H-C-C-C-H \\
\quad\; | \;\;\;\; | \;\;\;\; | \\
\quad\; H \;\; H \;\; H
\end{array}
\qquad CH_3CH_2CH_3
$$

Lewis structure   Condensed structural formula   Carbon-skeleton structure

To draw the propane carbon-skeleton structure (structure on the right), we first draw a short line slanted upward to the right; this represents the bond between the first and second carbons. Then without removing our pencil from the paper, we draw a short line slanted downward to the right to represent the bond between the second and third carbons. This simple inverted V represents propane because any chemist who looks at it knows that there is a CH_3- group on each end and a $-CH_2-$ group at the peak.

b. *n*-Dodecane is a continuous chain of 12 carbon atoms. There must be two methyl groups at the ends with ten methylene groups between them. The condensed structure is therefore

$$CH_3CH_2CH_2CH_2CH_2CH_2CH_2CH_2CH_2CH_2CH_2CH_3 \text{ or } CH_3(CH_2)_{10}CH_3$$

or, gathering the ten $-CH_2-$ groups: $CH_3(CH_2)_{10}CH_3$. The carbon-skeleton structure must contain as many ends and intersections as there are carbon atoms in the molecule:

Here we have 2 ends and 10 intersections for a total of 12 carbon atoms.

Think about It Condensed and carbon-skeleton structures must both contain the same number of atoms and the same number of bonds, even though the bonds are not shown in the former and not all of the atoms are shown in the latter.

Practice Exercise Draw the carbon-skeleton structure of *n*-hexane, $CH_3(CH_2)_4CH_3$, and a condensed structure of *n*-heptane:

n-Heptane

Structural Isomers

The alkane family would be huge if it consisted of only straight-chain molecules. However, another structural possibility arises that makes the family even larger. We first saw this possibility in Chapter 10 when we compared the boiling points of compounds having the same molecular formula, and hence the same molar mass, but different shapes. For example, the four-carbon alkane has the molecular formula C_4H_{10}. However, the straight-chain molecule shown on the left in Table 13.3 is not the only shape possible, because with four or more carbons, a *branched* structure also exists, as shown on the right in the table. A *branch* is a side chain attached to the main carbon chain.

The molecular formula of both structures in Table 13.3 is C_4H_{10}, but they represent different compounds with different properties. In recognition of these differences, the two compounds have different formal names (*n*-butane and 2-methylpropane), but at this point, it is not important to be able to name them but to recognize them as **structural isomers**: compounds with the same molecular formula but with their atoms connected in different ways. This difference

structural isomers molecules having the same molecular formula but different arrangements of atoms; they are different compounds and have different chemical and physical properties.

TABLE 13.3	**Comparing Structural Isomers**	
	***n*-Butane**	**2-Methylpropane**
Condensed structure:	$CH_3CH_2CH_2CH_3$	$CH_3CH(CH_3)CH_3$ or CH_3CHCH_3 | CH_3
Carbon-skeleton structure:		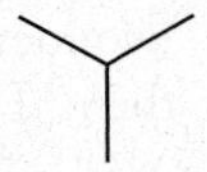
Melting point (°C):	−138	−160
Normal boiling point (°C):	0	−12
Density of gas (g/L):	0.5788	0.5934

makes structural isomers chemically distinct from one another. 2-Methylpropane is a **branched-chain hydrocarbon,** which means that its molecular structure contains a main chain (the longest carbon chain in the molecule) and at least one side chain. We can illustrate structural isomerism with either condensed or carbon-skeleton structures as shown in Table 13.3. There are two ways to show the location of the methyl group. One way is to use condensed structures and to put the CH_3 group in parentheses next to the carbon to which it is bonded: $CH_3CH(CH_3)CH_3$. We can also show the carbon–carbon bond between the CH_3 and the center carbon of the three-carbon main chain. In this instance, it may actually be easier to see the structural isomers of C_4H_{10} using the carbon-skeleton structures.

SAMPLE EXERCISE 13.3   Identifying the Longest Chain in Organic Molecules

Determine the number of carbon atoms in the longest chain in the following two branched, saturated hydrocarbons:

(a) 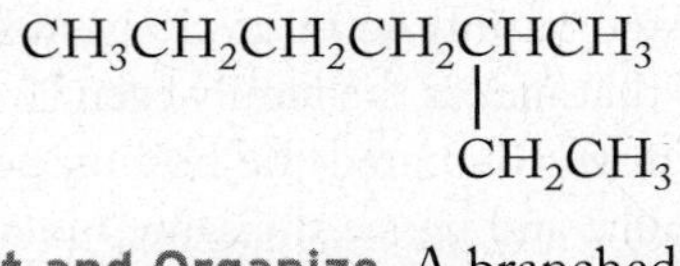 $CH_3CH_2CH_2CH_2CHCH_3$
$\qquad\qquad\qquad\qquad\quad |$
$\qquad\qquad\qquad\qquad\ CH_2CH_3$

(b) 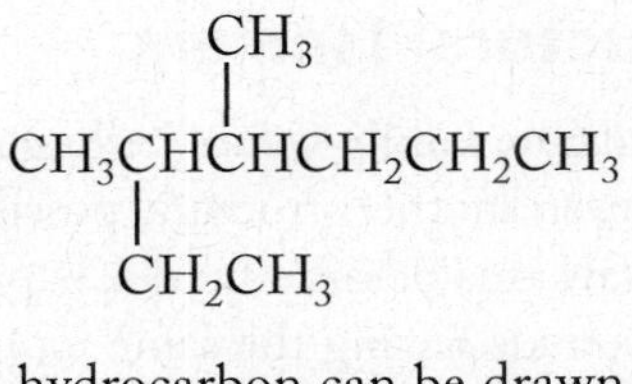 $\overset{\displaystyle CH_3}{\underset{\displaystyle CH_2CH_3}{CH_3CHCHCH_2CH_2CH_3}}$

Collect and Organize A branched, saturated hydrocarbon can be drawn in a variety of ways. We are asked to identify the longest carbon chain or *main chain* in two compounds. Both compounds are represented by condensed structures.

Analyze We assume that there is a bond between each carbon atom in the condensed structures. Carbon bonds to additional chains are indicated by a single line. We need to determine which of these branches belong to the *main chain* and which belong to the *side chains*.

Solve We start at one end of any branch and assign numbers to each carbon atom. If the chain branches, we must choose one branch to follow. Later, we may wish to return to the compound and number the carbons starting from a different end of the molecule or taking a different branch.

branched-chain hydrocarbon an organic molecule in which the chain of carbon atoms is not linear.

a. Starting with the carbon atom furthest to the left and numbering consecutively from left to right, we reach a branch at carbon atom 5. If we continue to the right, we find that the chain contains six carbons. If we take the branch at carbon atom 5, however, we find that the chain contains seven carbon atoms. This makes the longest chain in the compound seven carbon atoms long.

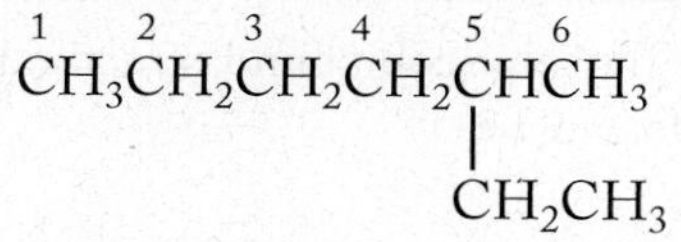

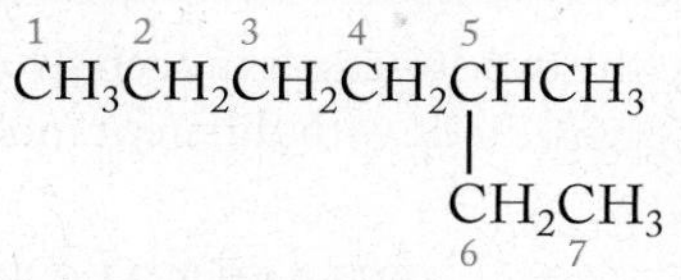

$$\overset{1}{C}H_3\overset{2}{C}H_2\overset{3}{C}H_2\overset{4}{C}H_2\overset{5}{C}H\overset{6}{C}H_3 \qquad \overset{1}{C}H_3\overset{2}{C}H_2\overset{3}{C}H_2\overset{4}{C}H_2\overset{5}{C}H\overset{}{C}H_3$$
$$\underset{}{C}H_2CH_3 \qquad\qquad\qquad \underset{6\quad7}{C}H_2CH_3$$

If we start at the right end of the molecule, we do not find a longer chain; the only other chain in the molecule is four carbon atoms long.

b. There are four places to start counting carbon atoms in structure b. Some possible ways to count them are as follows:

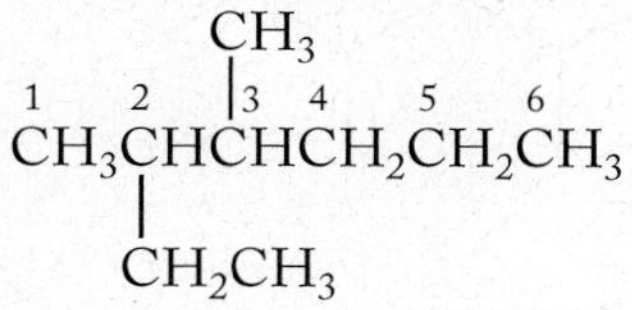

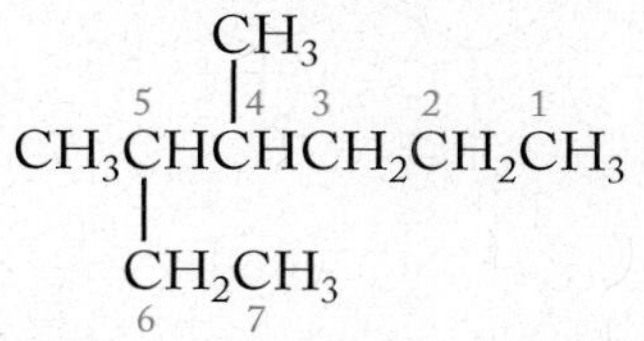

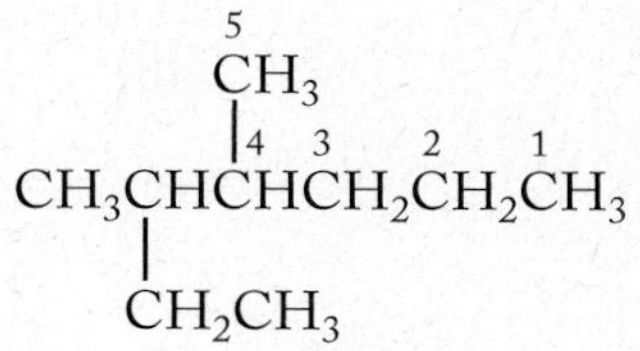

$$\begin{array}{ccc}
\overset{}{C}H_3 & \overset{}{C}H_3 & \overset{5}{C}H_3\\
\overset{1}{C}H_3\overset{2}{C}H\overset{3}{C}H\overset{4}{C}H_2\overset{5}{C}H_2\overset{6}{C}H_3 & \overset{5}{C}H_3\overset{}{C}H\overset{4}{C}H\overset{3}{C}H_2\overset{2}{C}H_2\overset{1}{C}H_3 & \overset{}{C}H_3\overset{}{C}H\overset{4}{C}H\overset{3}{C}H_2\overset{2}{C}H_2\overset{1}{C}H_3\\
\underset{}{C}H_2CH_3 & \underset{6\quad7}{C}H_2CH_3 & \underset{}{C}H_2CH_3
\end{array}$$

The longest chain in the molecule (the main chain) also has seven carbon atoms.

Think about It The main chain in an organic compound may not always be the one running horizontally across the page.

Practice Exercise Determine the number of carbon atoms in the longest chain in the following two branched, saturated hydrocarbons:

(a) 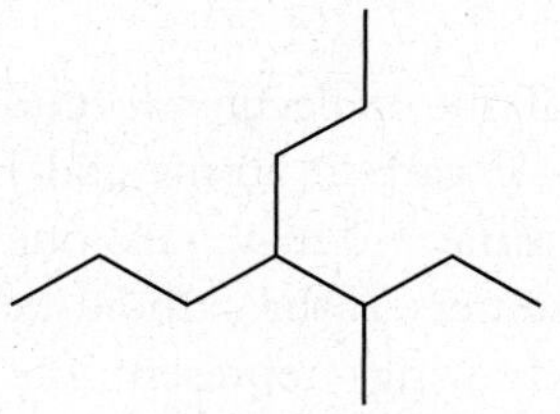   (b)

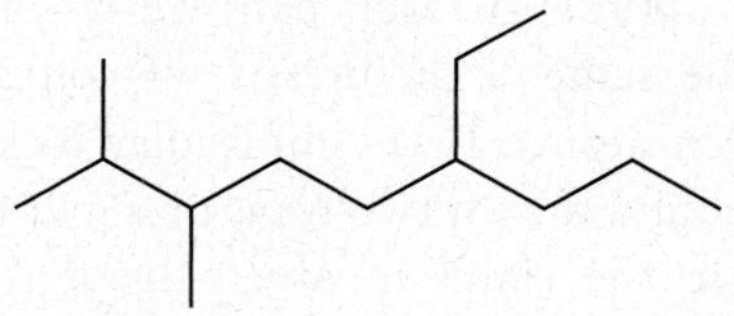

To determine whether two condensed structures or carbon-skeleton structures represent two different compounds, a single compound, or two structural isomers, our first step is to translate each structure into a molecular formula. If the molecular formulas are different, the structures represent two different compounds. If the molecular formulas are the same, we must compare the way the atoms are connected in the two structures. In doing so, we may have to reverse or rotate one of the structures. For example, structures 1 and 2 below may seem to be different from each other, but rotating structure 2 by 180° shows that it is the same as structure 1. The two represent the same compound: a four-carbon main chain with a one-carbon side chain connected to the carbon next to a terminal carbon atom.

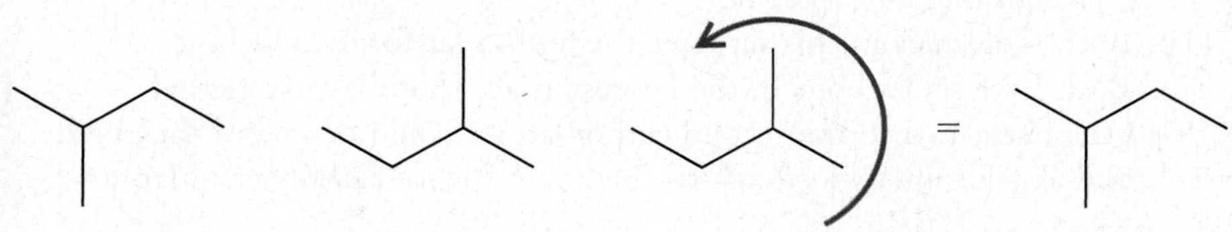

(1)   (2)   180° rotation   After rotation

To determine whether two structures are identical or not, draw the structures with the longest chain horizontal, and then check to see whether the same side chains are attached at the same positions along the longest chain.

SAMPLE EXERCISE 13.4   Recognizing Structural Isomers

Do the two structures in each set describe the same compound, structural isomers, or compounds with different molecular formulas?

(a) $(CH_3)_2CHCH_2CH(CH_3)_2$

(b)

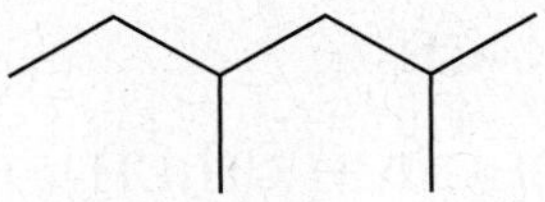

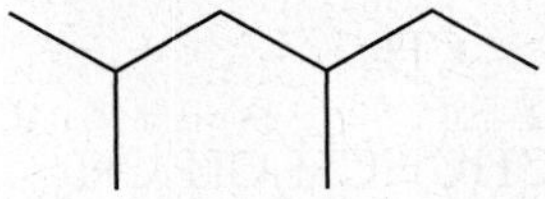

(c)

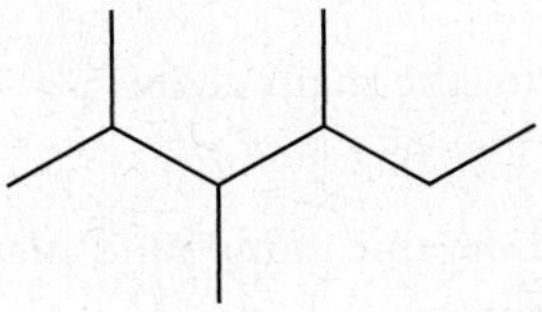

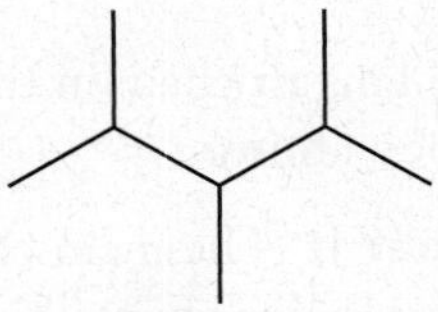

Collect and Organize Identical compounds have the same molecular formula and the same connectivity of the atoms. Structural isomers have the same molecular formula but different connectivity.

Analyze In each pair, we check first to see if the molecular formulas are the same. That means we count the number of carbon atoms and hydrogen atoms. If the molecular formulas are the same, we may have one compound drawn two ways or a pair of structural isomers. If the carbon skeletons are the same in any pair of drawings, the drawings represent the same hydrocarbon.

Solve

a. The molecular formulas are the same: C_7H_{16}. Converting the condensed structure to a carbon-skeleton structure gives us

$(CH_3)_2CHCH_2CH(CH_3)_2 \longrightarrow$

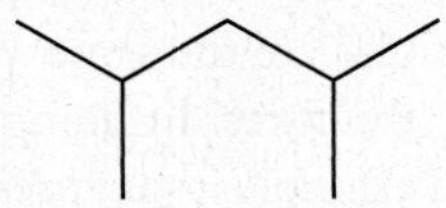

This structure is identical to the carbon-skeleton structure in this set. Therefore the condensed structure and carbon-skeleton structure represent the same molcule.

b. Both molecules in this set have the molecular formula C_8H_{18}. Both have six carbons in the longest chain. Both have a methyl group attached to the second carbon atom from one end of the chain and another methyl group attached to the third carbon atom from the other end. Therefore the two structures represent the same compound:

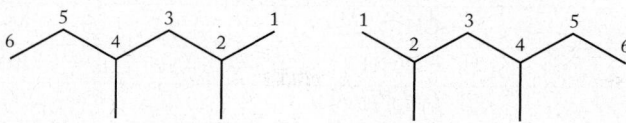

c. The left structure contains nine carbon atoms, but the right structure contains only eight. Therefore the two structures represent different compounds.

Think about It Just because two compounds have the same molecular formula, they are not necessarily identical. We must determine whether they are structural isomers.

Practice Exercise Do the two structures in each set describe the same compound, two different compounds, or a pair of structural isomers?

(a)

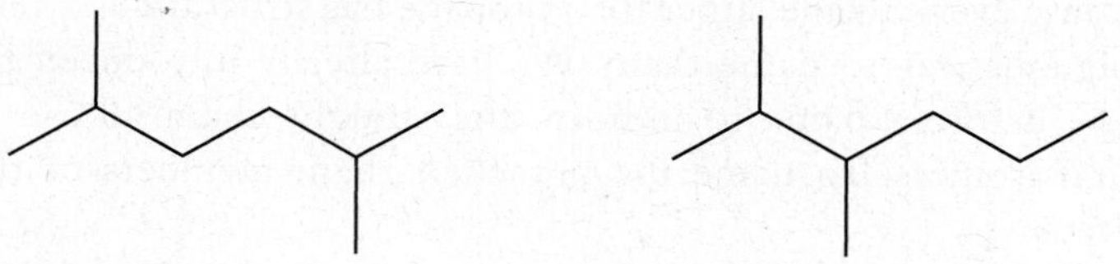

(b)

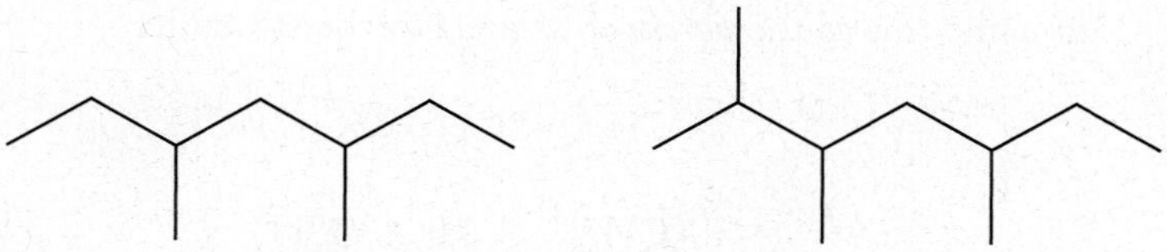

(c)

Naming Alkanes

Now that we know how to draw alkanes, let's look at a few simple rules for naming them. The nomenclature system follows the same pattern for all families of organic compounds, so we begin with rules for naming alkanes and will add a few more for alkenes and alkynes. These rules are called IUPAC rules after the International Union of Pure and Applied Chemistry, the organization that defined them. Appendix 7 contains a more extensive treatment of nomenclature for organic compounds.

The first four members of the alkane family are methane (1 carbon atom: a C_1 alkane), ethane (2 carbon atoms: a C_2 alkane), propane (a C_3 alkane), and butane (a C_4 alkane). The names of alkanes with more than 4 carbon atoms are derived from the Latin or Greek prefix for the number of carbon

TABLE 13.4	Prefixes for Naming *n*-Alkanes	
Prefix	**Condensed Structure**	**Name**
Meth-	CH_4	Methane
Eth-	CH_3CH_3	Ethane
Prop-	$CH_3CH_2CH_3$	Propane
But-	$CH_3(CH_2)_2CH_3$	Butane
Pent-	$CH_3(CH_2)_3CH_3$	Pentane
Hex-	$CH_3(CH_2)_4CH_3$	Hexane
Hept-	$CH_3(CH_2)_5CH_3$	Heptane
Oct-	$CH_3(CH_2)_6CH_3$	Octane
Non-	$CH_3(CH_2)_7CH_3$	Nonane
Dec-	$CH_3(CH_2)_8CH_3$	Decane

atoms per molecule followed by *-ane*. Table 13.4 lists the prefixes for C_1 through C_{10} alkanes.

Because every alkane larger than propane has structural isomers, we need an unambiguous way to name them. We have already introduced the use of the prefix *n-* before a name to indicate the straight-chain alkane in any set of structural isomers. To name the branched-chain members of the set, follow these steps:

1. Select the longest chain of carbon atoms and use the prefixes in Table 13.4 to name this as the *parent structure*. For the structure

$$CH_3CH(CH_3)CH_2CH_2CH_3$$

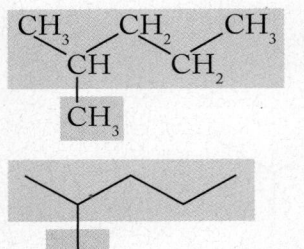

<table>
<tr><td>■</td><td>Parent structure</td></tr>
<tr><td>■</td><td>Branch</td></tr>
</table>

the parent name is pentane because the longest chain is 5 carbon atoms long.

2. Identify each branch and name it with the prefix from Table 13.4 that defines the number of carbons in the branch; append the suffix *-yl* to the prefix. A $-CH_3$ group is methyl, a $-CH_2CH_3$ group is ethyl, and so forth. The structure in step 1 has a methyl group branch. The name of the branch comes before the name of the parent structure, and the two are written together as one word: methylpentane.

3. To indicate the point where the branch is attached, we number the carbon atoms in the parent chain so that the branch (or branches, if there is more

than one) has the lowest possible number. In the molecule in step 1, we start numbering from the left so that the methyl group is on carbon atom 2 in the parent chain.

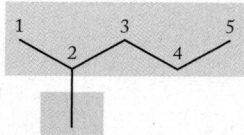

We indicate the position of the branch by a 2 followed by a hyphen in front of the name: 2-methylpentane.

4. If the same group is attached more than once to the parent structure, we use the prefixes di-, tri-, tetra-, and so forth to indicate the number of groups present. The position of each group is indicated by the appropriate number before the group name, with the numbers separated by commas. Thus the name is 2,4-dimethylpentane for the structure

$$\underset{1}{CH_3}\underset{2}{CH}(\underset{3}{CH_3})\underset{3}{CH_2}\underset{4}{CH}(\underset{3}{CH_3})\underset{5}{CH_3}$$

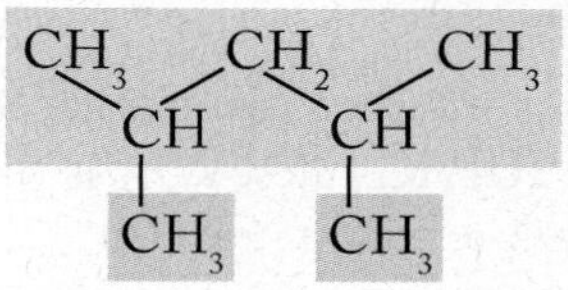

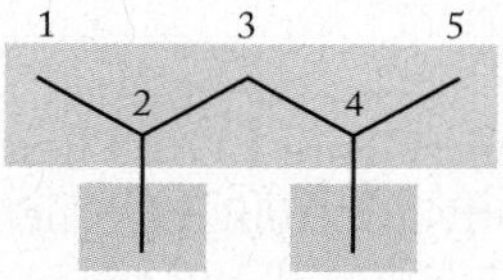

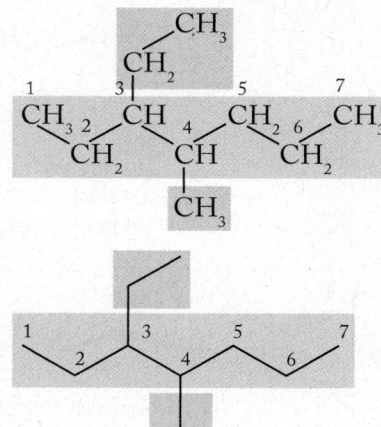

5. If different groups are attached to a parent chain, they are named in alphabetical order, as in 3-ethyl-4-methylheptane for the following structure (achieving the lowest numbering, however, takes precedence over alphabetization of the substituents):

$$\underset{1}{CH_3}\underset{2}{CH_2}\underset{3}{CH}(\underset{2}{CH_2}\underset{3}{CH_3})\underset{4}{CH}(\underset{3}{CH_3})\underset{5}{CH_2}\underset{6}{CH_2}\underset{7}{CH_3}$$

Note that we begin numbering the parent chain in this structure at the carbon on the far left, in accord with the instruction in step 3, so that the branch locations have the lowest possible numbers. For example, the compound shown here is 5-ethyl-2-methyloctane, not 4-ethyl-7-methyloctane:

Correct numbering

Incorrect numbering

CONCEPT TEST ••

A useful way of determining whether two similar structures are the same is to name each according to the IUPAC rules. Why is this the case?

Cycloalkanes

Alkanes can form ring structures (Figure 13.6). These **cycloalkanes** have the general formula C_nH_{2n}, which is different from the general formula for straight-chain alkanes (C_nH_{2n+2}) because cycloalkanes have one more carbon–carbon bond and two fewer hydrogen atoms per molecule than the *n*-alkanes with the same number of carbon atoms. Although cycloalkanes have no $C{=}C$ or $C{\equiv}C$ bonds, 1 mole of a cycloalkane can, at least in theory, react with 1 mole of hydrogen to

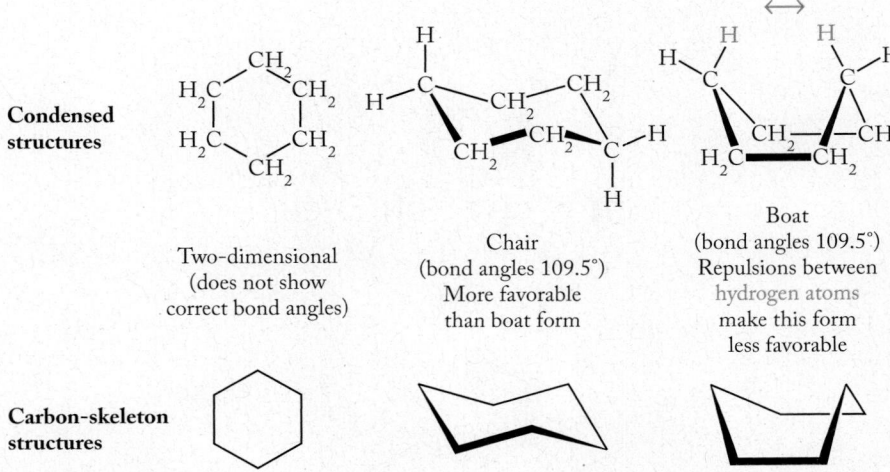

Condensed structures

Two-dimensional (does not show correct bond angles)

Chair (bond angles 109.5°) More favorable than boat form

Boat (bond angles 109.5°) Repulsions between hydrogen atoms make this form less favorable

Carbon-skeleton structures

cycloalkane ring-containing alkane with the general formula C_nH_{2n}.

FIGURE 13.6 The six-membered ring of cyclohexane is drawn either flat or in styles that show its three-dimensional puckered forms.

make 1 mole of a straight-chain alkane. However, in practice only cycloalkanes with three-carbon and four-carbon rings react with hydrogen.

The left column of Figure 13.6 shows the condensed structure and carbon-skeleton structure of cyclohexane drawn in two dimensions, where the C—C—C bond angles appear to be 120°. Because the carbon atoms have sp^3 hybrid orbitals, we expect the angles to be 109.5°, however, and indeed they are in this molecule. A more accurate representation of the ring is shown in the adjacent structures, in which the ring is puckered instead of flat and the bond angles are 109.5°.

There are two possible pucker patterns: a *chair* form and a *boat* form. In the boat form, repulsion between the two hydrogens across the ring from each other causes this form to be less stable than the chair form, and so the chair form is the preferred, lower-energy form.

Ring structures are possible for other alkanes, but smaller rings are less stable than the six-membered one. One of the important features that determines the relative stability of rings is the size of the C—C—C bond angle. For example, cyclopropane (a C_3 ring) exists but is a more reactive species because the interior ring angles are 60°, far from the ideal bond angle of 109.5° for an sp^3 hybridized carbon atom (Figure 13.7). No puckering is possible in a three-membered ring to relieve the strain in this system, and cyclopropane tends to react in a fashion that opens up the ring and relieves the strain. The C—C—C bond angles in cyclobutane and cyclopentane are larger, so there is less ring strain. Rings of six sp^3 hybridized carbons and beyond are essentially the same as straight-chain alkanes in terms of bond angle and have no ring strain. Six-membered rings are the most favored, because other thermodynamic factors have an impact on the formation of rings with seven or more carbons.

▶❚❚ **CHEMTOUR** Structure of Cyclohexane

▶❚❚ **CHEMTOUR** Cyclohexane in 3-D

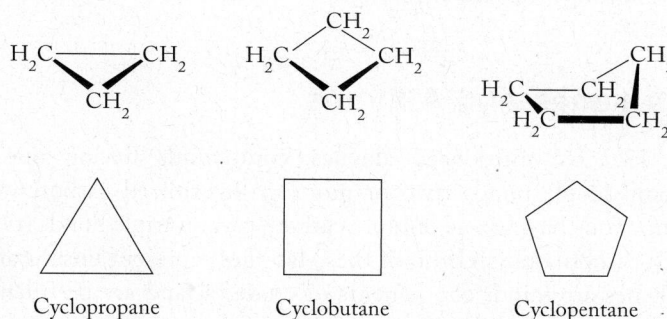

Cyclopropane   Cyclobutane   Cyclopentane

FIGURE 13.7 Cyclic alkanes have the general formula C_nH_{2n} and are considered to be unsaturated hydrocarbons. Shown here are cyclopropane (n = 3), cyclobutane (n = 4), and cyclopentane (n = 5).

Sources and Uses of Alkanes

The principal source of liquid alkanes on Earth is crude oil. Natural gas is the major source for the simplest alkane, methane, as well as smaller quantities of low-molar-mass alkanes including ethane, propane, and perhaps some butanes. Methane is often associated with oil deposits but is also produced during bacterial decomposition of vegetable matter in the absence of air, a condition that frequently arises in swamps. Hence, methane's common name is swamp gas or marsh gas (Figure 13.8). Frequently in the reducing environment of a marsh, a molecule known as phosphine (PH_3) is also formed. Phosphine and methane

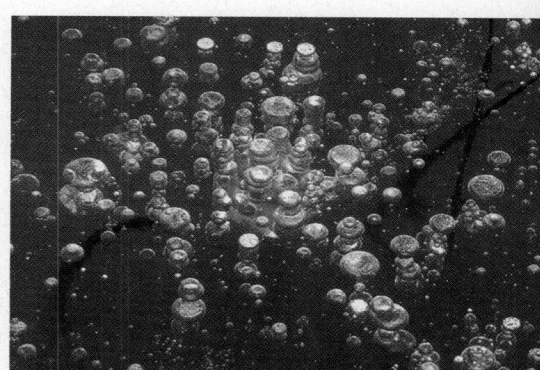

FIGURE 13.8 Methane bubbles trapped in a frozen pond. Methane is produced by rotting organic matter at the bottom of the pond.

together spontaneously ignite; this produces a ghostly flame known as will-o'-the-wisp that features prominently in some legends and gothic mysteries. Methane can also be produced in coal mines, where it can be especially dangerous because it forms explosive mixtures with humid air, giving rise to another common name for the gas: firedamp.

By far the most common use of alkanes in our lives is as fuels. Combustion reactions between alkanes and oxygen provide energy to power vehicles, generate electricity, warm our homes, and prepare our meals. Gasoline, kerosene, and diesel fuels are mostly mixtures of alkanes containing up to 20 carbon atoms per molecule as summarized in Table 13.2. Alkanes with higher boiling points are viscous liquids used as lubricating oils. Low-melting solid alkanes (C_{20}–C_{40}) are used in candles and in manufacturing matches. Very heavy hydrocarbon gums and solid residues (C_{36} and up) are used for paving roads.

The combustion of fossil fuels is an important topic in the 21st century for several reasons. The production of CO_2, a greenhouse gas, leads to the potential for climate change. The supplies of crude oil are also limited and with increasing world demand for fossil fuels, society faces choices in how best to allocate finite supplies of oil. The alkanes distilled from crude oil serve not only as fuels but also as the starting materials for building more complex molecules. They are the major source of carbon for the chemical manufacturing industry.

Some alkanes have therapeutic value. Mineral oil is a mixture of C_{15}–C_{24} alkanes and is used as a skin ointment ("baby oil") to treat diaper rash and to alleviate some forms of eczema. It is used in many cosmetics, creams, and ointments. Taken orally, mineral oil acts as a laxative. As we continue our study of organic compounds, we will return to these dual themes of fuels and pharmaceuticals with the functional groups that we encounter.

13.3 Alkenes and Alkynes

In Section 13.2 we introduced alkenes, compounds having one or more carbon–carbon double bonds (two or more sp^2 hybridized carbon atoms), and alkynes, compounds having one or more carbon–carbon triple bonds (two or more sp hybridized carbon atoms). Both of these families represent unsaturated hydrocarbons. Alkenes are minor components of crude oil and are prevalent in many plants including pine needles, ginger, and celery (Figure 13.9).

CONNECTION In Chapter 11 we discussed the distillation of crude oil to produce gasoline, kerosene, and other hydrocarbon mixtures useful as fuels and as feedstocks for the chemical industry.

CONNECTION In Chapter 8 we discussed methane's role as one of the greenhouse gases.

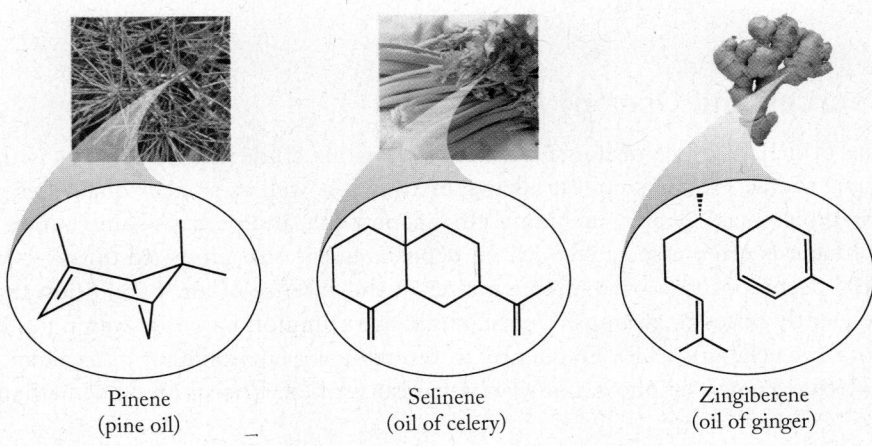

FIGURE 13.9 Some naturally occurring alkenes include (a) pinene in pine resin, (b) selinene in celery, and (c) zingiberene, in oil of ginger.

Pinene
(pine oil)

Selinene
(oil of celery)

Zingiberene
(oil of ginger)

Alkynes are found in crude oil as well, but they are generally more difficult to find in nature because the carbon–carbon triple bond is quite reactive. Nevertheless, carbon–carbon triple bonds are found in some drugs along with other functional groups. Some examples are shown in Figure 13.10: capillin, an antifungal drug; pargyline, used to treat hypertension; and panaxytriol, a potent antitumor compound isolated from ginseng. The simplest alkyne, C_2H_2, also known as acetylene, is used in welding. Acetylene and some other alkynes are manufactured by the controlled oxidation of alkanes. When the simplest alkane, methane, is oxidized completely, the products are CO_2 and H_2O. But if this oxidation is carried out in a highly controlled process, acetylene can be produced:

$$6\,CH_4(g) + O_2(g) \rightarrow 2\,HC\equiv CH(g) + 2\,CO(g) + 10\,H_2(g) \qquad (13.1)$$

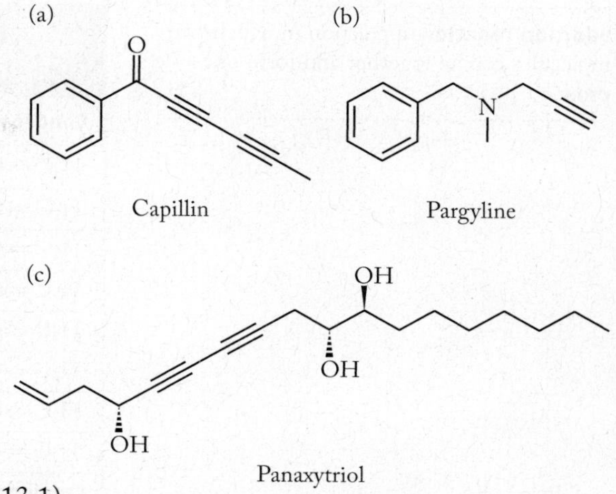

This reaction illustrates an important difference between alkanes and the unsaturated hydrocarbons: alkanes are the most reduced form of carbon. The sp^2 hybridized carbons in alkenes are in a higher oxidation state than the sp^3 hybridized carbon atoms of alkanes, and the sp hybridized carbons in alkynes are in an even higher oxidation state. Their capacity to be reduced makes alkenes and alkynes more reactive than alkanes, and the carbon–carbon double bond in alkenes is one of the most versatile functional groups in organic chemistry. Structures of a C_5 alkene and a C_5 alkyne are shown in Figure 13.11.

Alkenes and alkynes share many features of alkanes. The melting and boiling points of homologous series of these compounds (Table 13.5) vary with molar mass and size, just as with the alkanes.

Molecules may contain more than one alkene or alkyne group, as shown in Figure 13.10. In particular, many molecules found in crude oil or produced by living systems have several double bonds. We discuss molecules with multiple double bonds in Chapter 20, but for now we concentrate on the properties associated with small molecules containing only one or a small number of double or triple bonds.

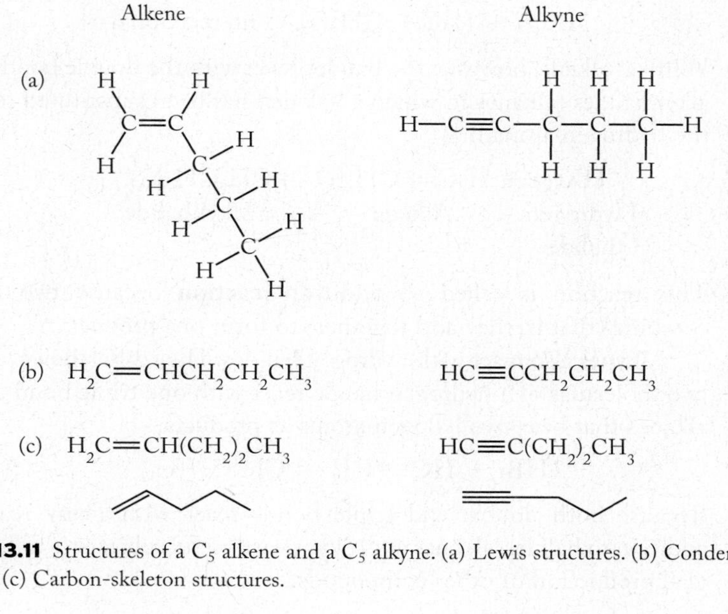

FIGURE 13.10 (a) Capillin, an antifungal drug, (b) pargyline, used to treat hypertension, and (c) panaxytriol, a potential antitumor drug, all contain the alkyne functional group.

FIGURE 13.11 Structures of a C_5 alkene and a C_5 alkyne. (a) Lewis structures. (b) Condensed structures. (c) Carbon-skeleton structures.

addition reaction a reaction in which two molecules couple together and form one product.

TABLE 13.5	Melting Points and Normal Boiling Points of Homologous Series of Alkenes and Alkynes	
Condensed Structure: Alkene	**Melting Point (°C)**[a]	**Normal Boiling Point (°C)**
$H_2C=CHCH_3$	−185	−47
$H_2C=CHCH_2CH_3$	−185	−6
$H_2C=CH(CH_2)_2CH_3$	−138	30
$H_2C=CH(CH_2)_3CH_3$	−140	63
$H_2C=CH(CH_2)_4CH_3$	−119	94
$H_2C=CH(CH_2)_5CH_3$	−104	123
$H_2C=CH(CH_2)_6CH_3$	−81	146
$H_2C=CH(CH_2)_7CH_3$	−87	171
Condensed Structure: Alkyne		
$HC\equiv CCH_3$	−102	−23
$HC\equiv CCH_2CH_3$	−126	8
$HC\equiv C(CH_2)_2CH_3$	−90	40
$HC\equiv C(CH_2)_3CH_3$	−132	71
$HC\equiv C(CH_2)_4CH_3$	−81	100

[a] Melting points increase with molar mass but also depend on how molecules fit into crystal lattices. Melting points of alkenes with even numbers of carbon atoms form one series that follows this trend; alkenes with odd numbers of carbon atoms (gray shading) form another series.

Chemical Reactivities of Alkenes and Alkynes

Figure 13.12 shows the electron distributions in the π bonding orbitals in alkenes and alkynes. The electrons in these orbitals, where electron density is greatest above and below the plane of the carbon skeleton, are more accessible to reactants than the electrons in the σ bonds of alkanes, where electron density is greatest between the atoms. For this reason, unsaturated hydrocarbons are more reactive than saturated ones.

As an illustration of this difference in reactivity, consider how alkanes and alkenes react with the hydrogen halides HCl, HBr, and HI. With alkanes, there is no reaction:

$$HX(g) + H_3C-CH_3(g) \rightarrow \text{no reaction} \qquad (13.2)$$

With an alkene, however, the halides react with the double bond to make alkyl halides (alkanes in which a halogen has been substituted for one of the hydrogen atoms):

$$HX(g) + H_2C=CH_2(g) \rightarrow CH_3CH_2X(\ell) \qquad (13.3)$$

Hydrogen   Alkene   Alkyl halide
halide

This reaction is called an **addition reaction** because two reactants combine (that is, they add together) to form one product.

Alkynes also react with hydrogen halides. They differ, however, in that two molecules of a hydrogen halide react with one triple bond and yield alkanes that bear two halogen atoms as products:

$$2\,HBr + HC\equiv CH \rightarrow CH_3CHBr_2 \qquad (13.4)$$

Because both double and triple bonds react with many reagents in addition to halides, alkenes and alkynes are useful substances in the industrial production of other compounds.

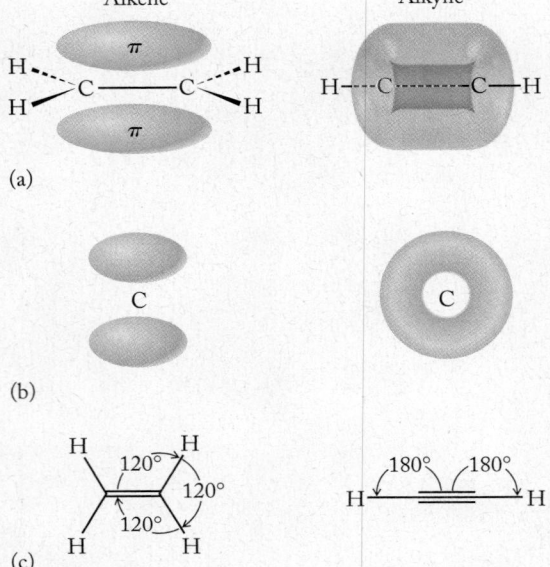

FIGURE 13.12 The characteristic structural feature of alkenes and alkynes is the presence of electrons in π orbitals. (a) The electron distribution in a C=C double bond and a C≡C triple bond. (b) View of electron distribution looking down the C—C bond axis. (c) The idealized H—C—C bond angles in double and triple bonds.

Isomers of Alkenes and Alkynes

Molecules that contain alkene and alkyne functional groups can have straight or branched chains, with the same types of structural isomers we saw with alkanes. One additional facet of structural isomerization involves the location of the double or triple bond in a molecule. For example, consider the straight-chain isomers of the alkene that contains five carbon atoms and one double bond (Figure 13.13).

$$\overset{1}{H_2}C\!=\!\overset{2}{C}H\overset{3}{C}H_2\overset{4}{C}H_2\overset{5}{C}H_3$$

(a)

$$\overset{1}{C}H_3\overset{2}{C}H\!=\!\overset{3}{C}H\overset{4}{C}H_2\overset{5}{C}H_3$$

(b)

$$\overset{5}{C}H_3\overset{4}{C}H_2\overset{3}{C}H\!=\!\overset{2}{C}H\overset{1}{C}H_3$$

(c)

$$\overset{5}{C}H_3\overset{4}{C}H_2\overset{3}{C}H_2\overset{2}{C}H\!=\!\overset{1}{C}H_2$$

(d)

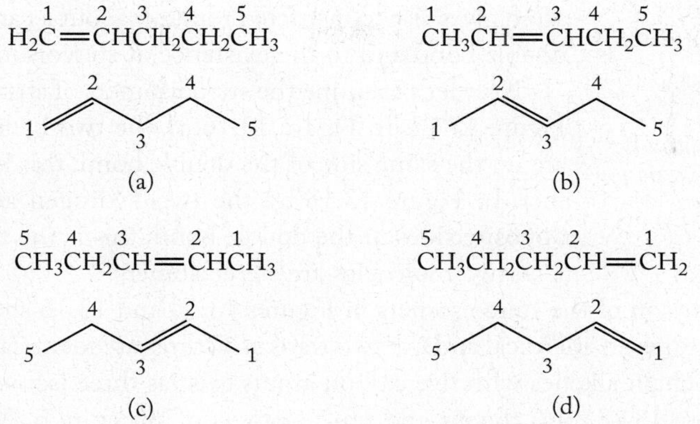

FIGURE 13.13 Four possible structural isomers of pentene, C_5H_{10}. Notice that structures a and d are identical, as are b and c.

Applying the test used in Section 13.2 for alkanes, we see that structures a and d in Figure 13.13 are the same. This is easier to see if we look at the carbon-skeleton structures in the figure. In both cases the double bond is between carbon atoms 1 and 2, that is, between C1 and C2. Remember, with branched-chain alkanes, we learned to number the carbons from whichever end gives the carbon attached to the branch the lowest number. The same holds for functional groups, as shown here, where the carbons in structure d must be numbered from right to left.

Structures b and c are equivalent and structural isomers of a and d because they have the same chemical formula but their double bond is in a different location. Drawing the carbon-skeleton structures of b and c, however, presents us with a new situation. After we draw the first three atoms of structure b, as in Figure 13.14(a), and draw a straight dashed line through the double bond, we see that we have two options for how to orient the rest of the molecule relative to the double bond. We can place the bond between C3 and C4 on the same side as the methyl group at C (Figure 13.14b) or on the opposite side (Figure 13.14c).

These two molecules are isomers of each other because they have the same molecular formula but different structures and therefore different properties. The isomer in Figure 13.14(b) is called either the **Z isomer** (Z stands for the German word *zusammen* or "together") or the **cis isomer** (cis is Latin for "on this side"), which in this case translates to "the methyl group and the chain after the double bond are both *together* or *on this side* of the structure." The isomer in Figure 13.14(c) is called either the **E isomer** (E for *entgegen* or "opposite") or **trans isomer** (trans is Latin for "across"). These are called **stereoisomers**, and they exist because there is no free rotation about the double bond. The system of naming using cis and trans is in wide use and is sufficient for the simple molecules discussed in this text. The *E/Z* system is routinely used for more complex molecules in which more than two different substituents are attached to a double bond.

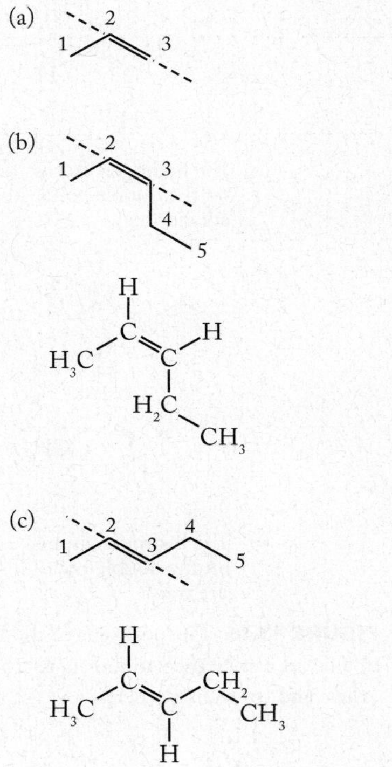

cis isomer (also called *Z* isomer) molecule with two like groups (such as two R groups or two hydrogen atoms) on the same side of the molecule.

trans isomer (also called *E* isomer) molecule with two like groups (such as two R groups or two hydrogen atoms) on opposite sides of the molecule.

stereoisomers molecules with the same formulas and the same connectivities between their atoms, but with different spatial arrangements of their atoms.

FIGURE 13.14 (a) The first three atoms of a carbon chain with a double bond between C2 and C3. (b) The chain continues on the same side of the double bond as C1 in the cis isomer. (c) The chain continues on the opposite side of the double bond as C1 in the trans isomer.

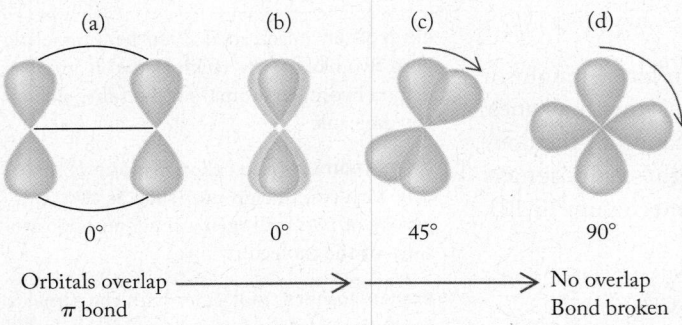

(a)   (b)   (c)   (d)

0°   0°   45°   90°

Orbitals overlap ⟶ ⟶ No overlap
π bond   Bond broken

FIGURE 13.15 (a) To form a π bond, p_z orbitals overlap to establish a region of shared electron density above and below the plane of the C—C bond. (b) If you look down the C—C bond axis, the orbitals line up. (c) If you rotate one carbon atom while keeping the other fixed, the orbitals are no longer parallel and do not overlap. (d) If you rotate one of the two bonded C atoms far enough, the π bond breaks.

CONNECTION In Chapter 5 we introduced rotations about bonds as one of the types of motion molecules experience as part of their overall kinetic energy.

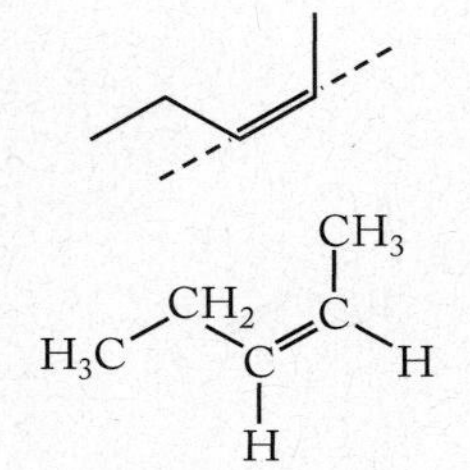

(a) The hydrogen atoms on the double bond are cis

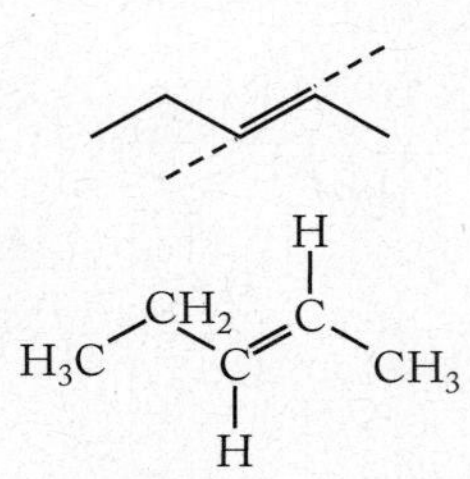

(b) The hydrogen atoms on the double bond are trans

FIGURE 13.16 The positions of the H atoms in alkenes can be used to distinguish between (a) cis and (b) trans isomers.

Recall from Chapter 9 that a double bond is formed from the overlap of two unhybridized p orbitals on adjacent carbon atoms. As Figure 13.15 shows, for the carbon atoms joined in a double bond to rotate freely, they have to twist about the bond axis, eliminating the orbital overlap and breaking the bond. Breaking a π bond in 1 mole of an alkene costs about 290 kJ of energy, and that much energy is not available to the molecules at room temperature. This situation gives rise to restricted rotation about a carbon–carbon double bond and to the existence of stereoisomers.

Now let's examine the stereoisomers of structure c from Figure 13.13. In Figure 13.16(a), the two hydrogen atoms are on the same side of the double bond; this is the cis isomer. In Figure 13.16(b), the two hydrogen atoms are on opposite sides of the double bond; this is the trans isomer. The two molecules are stereoisomers.

A comparison of the stereoisomers in Figures 13.14 and 13.16 shows that the two cis structures are identical and the two trans structures are identical. Therefore, the straight-chain alkenes with five carbon atoms exist as three isomers: structure a from Figure 13.13, plus the cis and trans isomers of structure b. All three isomers are chemically distinct compounds.

CONCEPT TEST

Which of the following alkenes has cis and trans isomers?

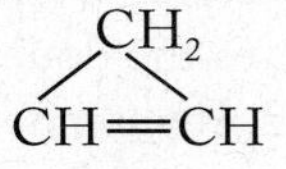

$$CH_2\!\!=\!\!CHCH_2CH_3$$
(a)

$$CH=CH \quad \overset{CH_2}{\overbrace{}}$$
(b)

$$(CH_3)_2C\!\!=\!\!C(CH_3)_2$$
(c)

$$(CH_3)_2C\!\!=\!\!CH_2$$
(d)

$$CH_3CH\!\!=\!\!CHCH_3$$
(e)

Naming Alkenes and Alkynes

To name straight-chain alkenes and alkynes, the prefixes in Table 13.4 are used to identify the length of the chain. The suffix *-ene* is appended if the compound is an alkene and *-yne* if the substance is an alkyne. The carbon atoms in the chain are numbered so that the first carbon atom in the double or triple bond has the lowest number possible, and that number precedes the name, followed by a dash. Stereoisomers are identified by writing *cis-* or *trans-* before the number. Thus the compounds in Figure 13.16(a) and (b) are *cis*-2-pentene and *trans*-2-pentene, respectively.

SAMPLE EXERCISE 13.5   **Identifying and Naming Stereoisomers and Structural Isomers**

Draw both the condensed structure and the carbon-skeleton structure of the five isomers of the six-carbon straight-chain alkene containing one double bond, and name each isomer.

Collect and Organize We are to draw and name five isomers made of molecules that each have six carbon atoms in a single chain and one C=C double bond. Figures 13.11, 13.13, 13.14, and 13.16 show examples of condensed and carbon-skeleton structures of alkenes.

Analyze We must consider two kinds of isomers: structural isomers, which depend on where the double bond is located in the chain, and stereoisomers, which depend on the orientation of groups about the double bond (cis/trans isomers). A straight-chain six-carbon alkene has a maximum of five places where a C=C can be placed; however, some locations may result in identical molecules. Not all alkenes have cis and trans isomers.

Solve Let's start with the structural isomers, which have the double bond at different locations, and then draw the stereoisomers (cis/trans) where possible. If all six carbons are in one straight chain, then a double bond can be between C1 and C2, C2 and C3, or C3 and C4. These isomers are named 1-hexene, 2-hexene, and 3-hexene, respectively, where the number indicates the location of the double bond along the chain. Chains with a double bond between C4 and C5 or C5 and C6 are identical to the isomers with a double bond between C2 and C3 or C1 and C2, respectively.

1-Hexene does not have stereoisomers because the carbon atoms that form the double bond have three H atoms and only one nonhydrogen atom attached. Only 2-hexene and 3-hexene can have stereoisomers. *cis*-2-Hexene and *cis*-3-hexene have the two R groups on the same side of the double bond. *trans*-2-Hexene and *trans*-3-hexene have R groups on opposite sides of the double bond.

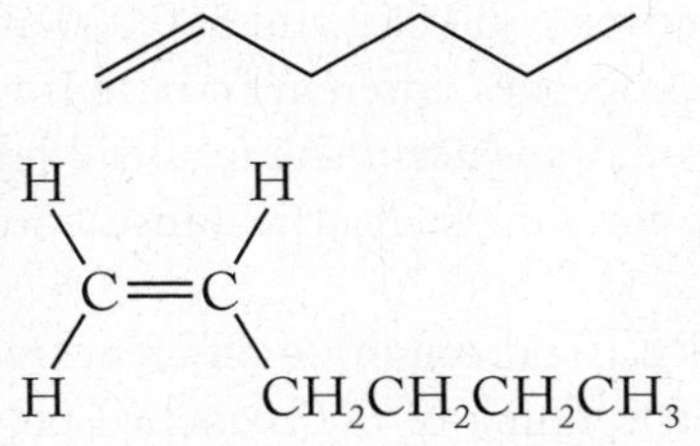

1-Hexene 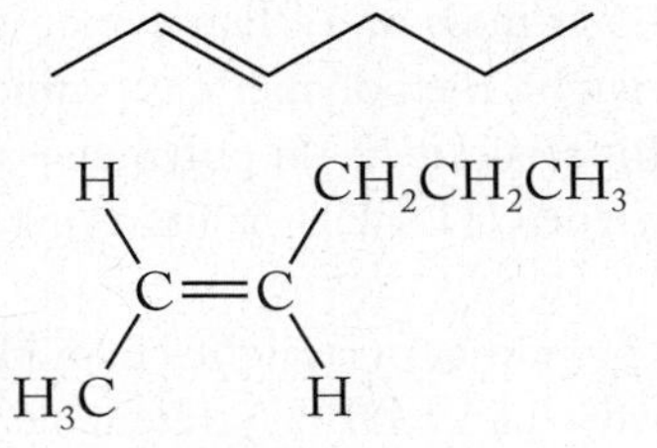   *trans*-2-Hexene 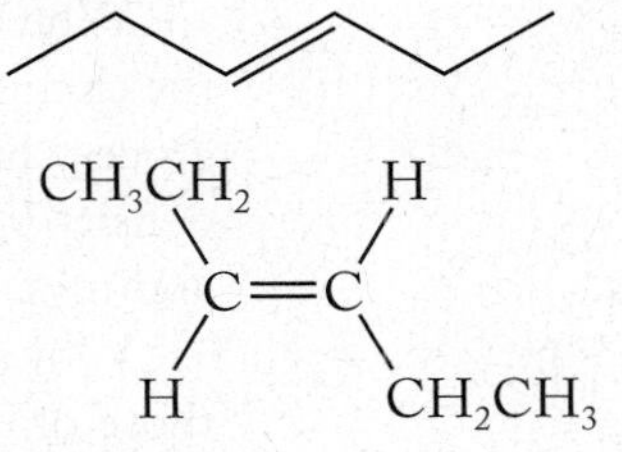   *trans*-3-Hexene

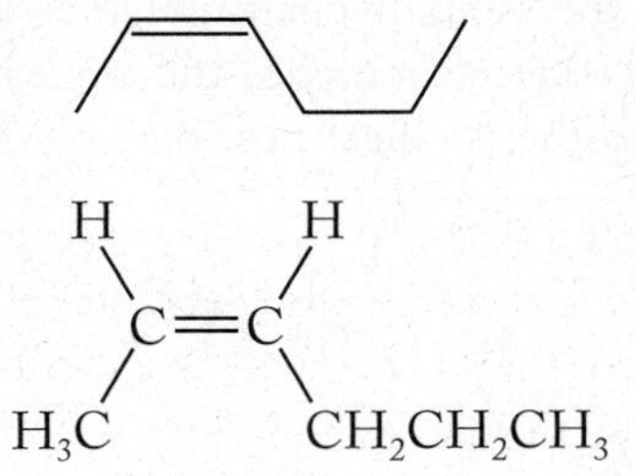

cis-2-Hexene 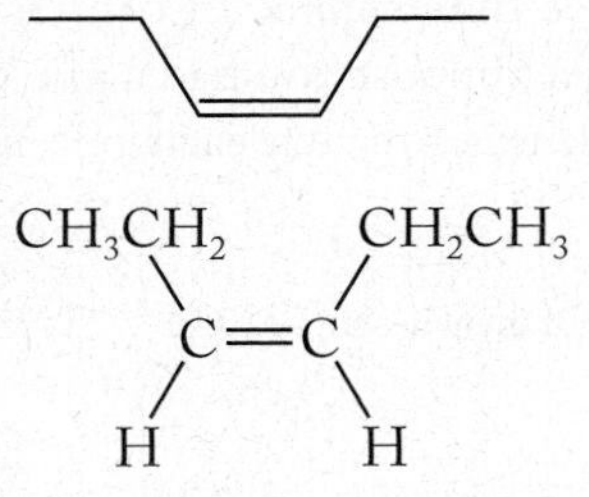   *cis*-3-Hexene

Think about It The number of isomers depends on the number of independent locations for a double bond in an alkene. Not all alkenes have stereoisomers.

Practice Exercise Draw and name the five isomers of the molecule with this carbon skeleton and one carbon–carbon double bond:

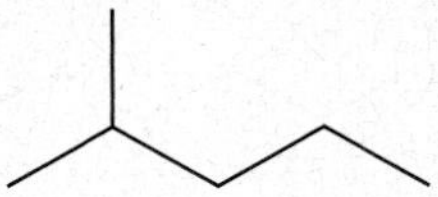

homopolymer a polymer composed of only one kind of monomer unit.

addition polymer macromolecule prepared by adding monomers to a growing polymer chain.

CONCEPT TEST

(a) Does a straight-chain hydrocarbon with a terminal double bond, such as 1-pentene, $CH_2\!=\!CHCH_2CH_2CH_3$, have stereoisomers? (b) Do alkynes have stereoisomers?

Polymers of Alkenes

Some widely used polymeric alkanes are produced industrially from small alkenes. The alkane polymer with the simplest structure is linear polyethylene (PE), produced from ethylene ($CH_2\!=\!CH_2$) at high temperature and pressure:

$$n\,CH_2\!=\!CH_2 \rightarrow \!-\!\!\left[CH_2\!-\!CH_2\right]_n \qquad (13.5)$$

PE is a **homopolymer**, which means it is composed of only one type of monomer. Its condensed structure is $CH_3(CH_2)_nCH_3$, but there are so many more methylene groups than methyl groups that the structure is frequently written $-\!\!\left[CH_2CH_2\right]_n$ to highlight the structure and composition of the monomer. Polyethylene is also an example of an **addition polymer**, a polymer constructed by adding many molecules together to form the polymer chain.

In most products made of polyethylene, n is a very large number ranging from 1000 to almost 1 million. Polyethylene has a wide range of properties that depend on the value of n and on whether the polymer chains are straight or branched. In low-density PE (LDPE), a stretchable, soft plastic used in films and wrappers, n ranges from 350 to 3500 and the chains are branched. When the bagger at the grocery store asks, "Paper or plastic?" the plastic in question is LDPE.

When the molar mass of a PE polymer is between 100,000 and 500,000 and the chains are straight, the polymer has physical properties different from those of grocery bags. This straight-chain polymer—a rigid, translucent solid called high-density polyethylene (HDPE)—is used in milk containers, electrical insulation, and toys.

Why are the properties of straight-chain HDPE (rigid, tough) so different from those of branched-chain LDPE (stretchable, soft)? Think of the branched polymer (Figure 13.17) as a tree branch with lots of smaller branches attached to it. The polymer can have branches that come out of the plane of the paper, so it has three dimensions. In contrast, the straight-chain polymer is like a long, straight pole. Suppose you had a pile of 100 tree branches and a pile of 100 poles, and your task was to stack each pile into the smallest possible volume to fit into a truck.

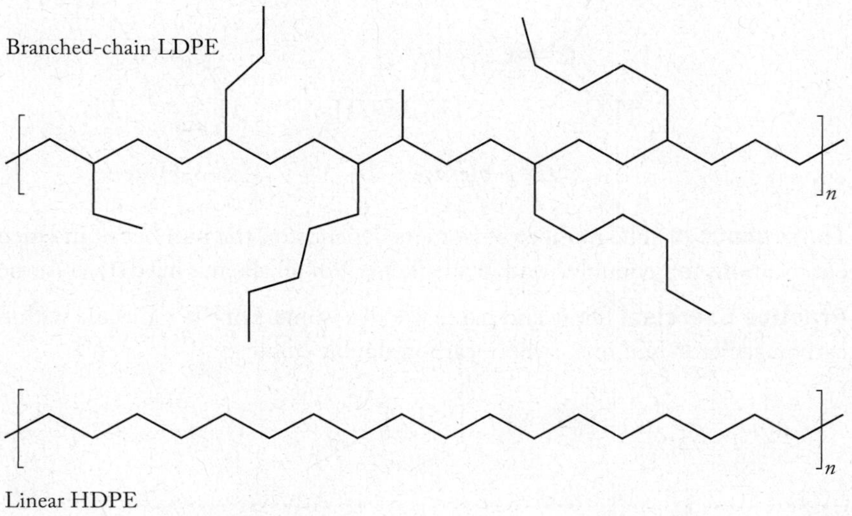

Branched-chain LDPE

Linear HDPE

FIGURE 13.17 Low-density polyethylene consists of branched chains, but high-density polyethylene consists of straight chains. Efficient stacking—and thus high density—is possible in large polymer molecules of HDPE but not in large polymer molecules of LDPE.

You can certainly pile the branches on top of one another, but they will not fit together neatly, and probably the best you can do is to make the pile a bit more compact. In contrast, you can stack the poles into a very compact pile.

The same situation arises with the branched and linear molecules of polyethylene. Branched PE is low density because the molecules do not line up neatly. Their density is low compared to HDPE, primarily because the branched molecules stack less efficiently. This makes LDPE more deformable and softer; HDPE is more rigid and even has some regions that are crystalline because the packing is so uniform. Ultrahigh molecular weight PE (UHMWPE; $n > 100,000$) is an even tougher material, because not only are the molecules straight, they are significantly larger than the molecules in HDPE. UHMWPE is used as a coating on some artificial ball-and-socket joints (Figure 13.18) because it is extremely resistant to abrasion and makes the joints last longer. The different forms of polyethylene illustrate how the size and shape of its molecules affect the physical properties of a material.

> **CONCEPT TEST**
>
> During recycling, articles made from LDPE are separated from those made from HDPE (Figure 13.19). Why?

FIGURE 13.18 This prosthetic hip consists of a metal shaft (black) that fits into the thigh bone and ends in a silver ball embedded in a (white) plastic socket that is cemented into the pelvis. To make the joint last longer, the inner surface of the socket may be coated with a relatively new polymer called ultrahigh molecular weight PE (UHMWPE), which is a very tough material and highly resistant to wear. UHMWPE consists of linear molecules with an average molar mass over 3 million.

$$-\!\!\left[CH_2\!-\!CH_2\right]_n$$
Polyethylene

FIGURE 13.19 Products made of polyethylene bear a recycle symbol that identifies them as straight-chain molecules (high-density) or branched-chain molecules (low-density).

$$-\!\!\left[CF_2\!-\!CF_2\right]_n$$
Teflon

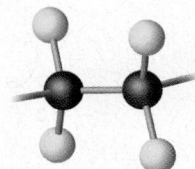

FIGURE 13.20 Repeating units of polytetrafluoroethylene (Teflon).

Of course, chemical composition also plays a role in determining physical properties. If the hydrogen atoms in PE are all replaced with fluorine atoms, the resultant polymer is chemically very unreactive, capable of withstanding high temperatures, and has a very low coefficient of friction, which means other things do not stick to it. This polymer is Teflon, $-\!\!\left[CF_2CF_2\right]_n$, (Figure 13.20), most familiar for its use as a nonstick surface in cookware. However, Teflon tubing is

$+CH_2 - CCl_2 +_n$

FIGURE 13.21 Repeating unit of poly(1,1-dichloroethylene).

also used in the grafts inserted into small-diameter blood vessels during vascular surgery on limbs. The analogous material cannot be formed with chlorine, but a polymer does exist in which every other $-CH_2-$ in the polymer chain is $-CCl_2-$. Its repeating monomer unit is $+CH_2CCl_2+_n$, (Figure 13.21), and the polymer is the familiar thin, flexible plastic used as Saran wrap.

Hydrocarbons containing two or more double bonds are frequently used to manufacture polymers. For example, polymerization of butadiene yields a stretchy, synthetic rubber useful in rubber bands (Figure 13.22). Polyisoprene, prepared from 2-methyl-1,3-butadiene, is used in surgical gloves.

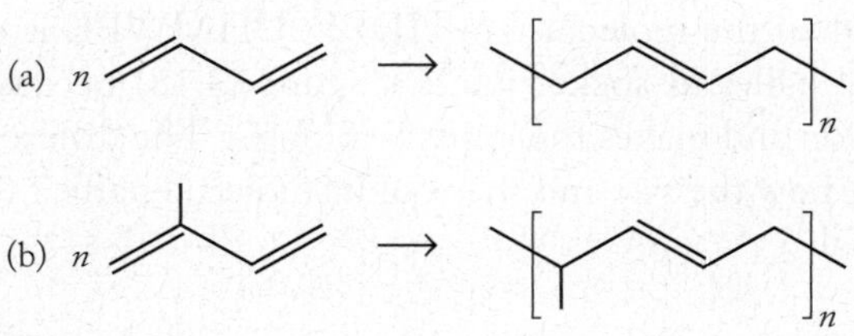

FIGURE 13.22 Monomer units and polymer structures for (a) butadiene and (b) a methylated butadiene known as isoprene.

SAMPLE EXERCISE 13.6   Identifying Monomers

Polypropylene, $+CH_2CH(CH_3)+_n$, is an addition polymer used in the manufacture of fabrics, ropes, and other materials (Figure 13.23). Draw condensed and carbon-skeleton structures of the monomer used to prepare polypropylene and name it.

FIGURE 13.23 Polypropylene is a common material for furniture, containers, clothing, lighting fixtures, and even objects of art. In addition to being moldable into many shapes, polypropylene is a good thermal insulator and does not absorb water easily.

The HON Company

Collect and Organize We are given a condensed structure of a polymer and asked to identify the monomer used in its preparation. We know that polypropylene is an addition polymer, so the monomer must be an alkene.

Analyze To understand the relationship between the polymer and the monomer from which it is made, let's look at Equation 13.5 in the reverse direction (Equation 13.6):

$$-\!\!-\!\!CH_2\!-\!\!CH_2\!-\!\!\big]_n \rightarrow n\,CH_2\!\!=\!\!CH_2 \qquad (13.6)$$

Breaking the blue bonds in Equation 13.6 and making the red bond a double bond illustrates the relationship between polyethylene and ethylene, the alkene monomer. We need to apply a similar analysis to polypropylene.

Solve The relationship between polypropylene and its monomer is illustrated by Equation 13.7:

$$-\!\!-\!\!CH_2\!-\!\!CH(CH_3)\!-\!\!\big]_n \rightarrow n\,CH_2\!\!=\!\!CH(CH_3) \qquad (13.7)$$

Breaking the two blue bonds and making the red C—C bond a C=C double bond yields the alkene shown in Figure 13.24. This three-carbon alkene has the name propene. There is no need to precede the name of the monomer with a number to indicate the position of the C=C bond, because it has to be between the C1 and C2 carbon atoms.

Think about It The structural difference between propene and ethylene is the presence of a CH_3- group bonded to one of the two sp^2 carbon atoms in propene instead of an H atom in ethylene.

Practice Exercise Draw the condensed and carbon-skeleton structures of the monomer used to make poly(methyl methacrylate), PMMA, a polymer used in shatterproof transparent plastic that can be used in place of glass:

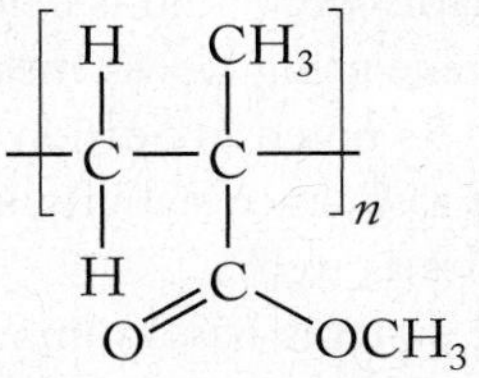

$$\left[\begin{array}{cc} H & CH_3 \\ | & | \\ C - C \\ | & | \\ H & C \\ & O^{\diagdown}\;\;OCH_3 \end{array}\right]_n$$

$CH_3CH\!\!=\!\!CH_2$

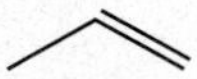

FIGURE 13.24 The monomer propene is polymerized to make polypropylene.

All addition polymers based on addition reactions of monosubstituted ethylene are called **vinyl polymers** because the $CH_2\!\!=\!\!CH-$ subunit is called the **vinyl group**, a name derived from *vinum* (Latin: wine). The name was given to the group by 18th-century chemists who prepared ethylene ($CH_2\!\!=\!\!CH_2$) from ethanol (CH_3CH_2OH), the alcohol in wine and other liquors.

Polyvinyl chloride, PVC, is widely used in commercial articles ranging from plastic pipes for plumbing to computer cases. Classic vinyl phonograph records are made from PVC. The condensed structures for the monomer and the polymer are shown in Equation 13.8:

$$n\,CH_2\!\!=\!\!CHCl \rightarrow -\!\!-\!\!CH_2CHCl\!-\!\!\big]_n \qquad (13.8)$$

CONCEPT TEST

Suggest a structural reason why Saran wrap (Figure 13.21), with two chlorine atoms on every other carbon atom, is a soft, flexible polymer, while PVC, with one chlorine on every other carbon atom, is more rigid.

Alkenes derived from crude oil are the primary source of monomers used to make most of the polymers used in construction, in fabrics, as wrapping and packaging material, and in medical devices. In an age of limited supply, the availability of crude oil impacts more than our ability to drive automobiles. Vinyl polymers are

vinyl polymer one of the family of polymers formed from monomers containing the subgroup $CH_2\!\!=\!\!CH-$.

vinyl group the subgroup $CH_2\!\!=\!\!CH-$.

aromatic compound a cyclic, planar compound with delocalized π (pi) electrons above and below the plane of the molecule.

the world's second largest selling plastics materials, and polymers in this category are extraordinarily versatile. You probably encounter five to ten vinyl polymers before you leave your room in the morning: vinyl shower curtains, vinyl drain pipes, vinyl flooring, vinyl insulation around electrical conduits. As we explore more organic functional groups, the basic concepts developed for the vinyl polymers will apply to polymers in other categories: the features of functional group, size, and shape determine the chemical and physical properties of these extraordinarily useful materials.

13.4 Aromatic Compounds

Among the components of gasoline that play an important role in increasing octane ratings (a measure of the ignition temperature of the fuel and its ability to resist engine "knock") is the class of compounds called *aromatic* hydrocarbons. Benzene is an example of an **aromatic compound**, a cyclic, planar molecule with delocalized π electrons above and below the plane of its six carbon and six hydrogen atoms.

As their class name implies, aromatic hydrocarbons have distinctive odors. However, *aromaticity* from a chemist's perspective is associated with the molecular and electronic structure of cyclic, planar molecules with sp^2 hybridized carbon atoms joined by a combination of alternating σ and π bonds (Figure 13.25). Aromatic compounds are relatives of alkenes because they contain carbon–carbon double bonds. However, because their chemical and physical properties are unique and distinct from those of alkenes, they merit designation as a separate family.

The most common aromatic compound is benzene, C_6H_6. The different ways we view the bonding in benzene using Lewis theory and valence bond theory are summarized in Figure 13.25. As noted in Section 8.5, the delocalized electrons in benzene lead to considerable resonance stability in this molecule, and this is true of all other aromatic molecules as well.

The stability of aromatic systems has an impact on their chemical reactivity. In Section 13.3 we saw that alkenes react rapidly with hydrogen halides to make halogenated alkanes. In contrast, benzene does not react at all if HBr gas is bubbled through it. Alkenes and aromatic compounds differ with respect to many other reactions, so the classification of aromatic systems as a unique family is justified.

▶❙❙ **CHEMTOUR** Structure of Benzene

◐◑ **CONNECTION** In Chapters 8 and 9 we introduced benzene as an aromatic compound and described bonding in the benzene molecule using Lewis theory and valence bond theory.

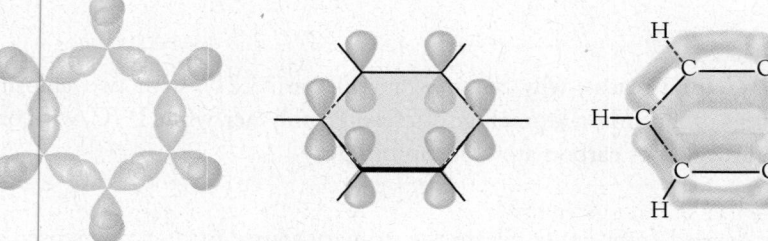

(a) Carbon-skeleton structures showing resonance forms of benzene

Skeletal symbol of benzene ring

(b) Sigma bonds in benzene

(c) Unhybridized p orbitals of carbon atoms

(d) Delocalized π cloud of electrons above and below plane of ring

FIGURE 13.25 Different views of the bonding in benzene. (a) Carbon-skeleton structures showing resonance and double-bond delocalization. (b) Hexagonal array of σ bonds. (c) The unhybridized p_z orbitals on the sp^2 hybridized carbons. (d) Delocalized π electrons above and below the ring.

Structural Isomers of Aromatic Compounds

Many compounds can be formed by replacing the hydrogen atoms in an aromatic ring with other substituents. For example, when one methyl group replaces a hydrogen atom in benzene, we get methylbenzene, also known by its common name, toluene:

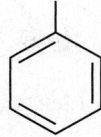

Toluene

All the positions around the benzene ring are equivalent, so it does not matter which of its six carbon atoms is bonded to the methyl group. This is why we do not have to write a number in the name methylbenzene.

There are three options for attaching two methyl groups to a benzene ring, so there are three structural isomers of dimethylbenzene, also known as xylene. We distinguish between the three structural isomers by numbering the carbon atoms to give the substituents the lowest possible numbers. From left to right below, the three dimethylbenzenes are 1,2-dimethylbenzene; 1,3-dimethylbenzene; and 1,4-dimethylbenzene. Toluene and xylenes are used in inks, glues, and disinfectants.

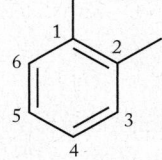   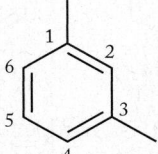   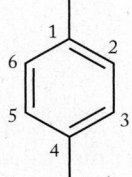

1,2-Dimethylbenzene   1,3-Dimethylbenzene   1,4-Dimethylbenzene

The existence of isomers and the variety of substituents that can be attached to an aromatic system result in the occurrence of a large number of aromatic compounds.

Benzene rings can share one or more of their hexagonal sides and thereby form polycyclic aromatic molecules. Three such compounds are naphthalene, anthracene, and phenanthrene:

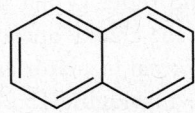   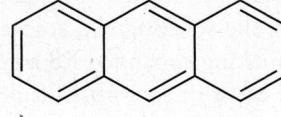   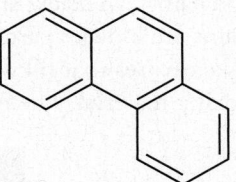

Naphthalene   Anthracene   Phenanthrene

Extensive delocalization of the π electrons over all the rings makes these structures particularly stable. In addition to being found in fossil fuels, they may be formed during the incomplete combustion of hydrocarbons and are present in particularly high concentrations in the soot from incinerators and diesel engines.

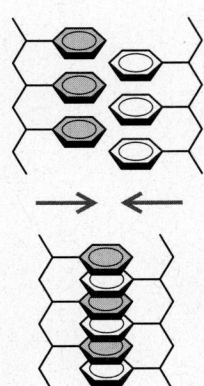

FIGURE 13.26 The aromatic rings on neighboring chains in polystyrene stack together and provide strength to the material.

They have also been identified in interstellar dust clouds and in blackened portions of grilled meat. When introduced into the environment, these compounds persist and are among the most long-lived of hydrocarbons.

CONCEPT TEST ....

Why isn't hexatriene considered an aromatic compound?

$$\underset{1}{\overset{2}{\diagup}}\underset{3}{\overset{4}{\diagup}}\underset{5}{\overset{6}{\diagup}}$$

1,3,5-Hexatriene

....

Polymers Containing Aromatic Rings

Individual aromatic rings as well as fused rings are flat molecules. They tend to stack neatly (Figure 13.26), and this tendency gives rise to useful properties in materials that incorporate aromatic systems. Replacing one hydrogen in ethylene with a benzene ring gives the monomer styrene, and the polymer made from this monomer is polystyrene (PS):

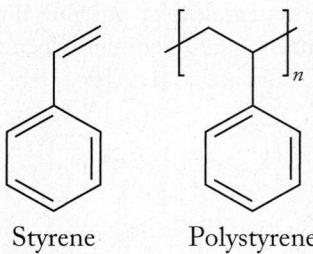

Styrene   Polystyrene

FIGURE 13.27 Polystyrene can be made into either rigid or foamed products. In its nonexpanded form, it is rigid and strong, suitable for making such products as plastic knives, forks, and spoons. In its expanded form, it is Styrofoam, used in carry-out food containers, packing materials, and thermal insulation in buildings.

Solid PS is a transparent, colorless, hard, inflexible plastic. In this form it is used for compact disc cases and plastic cutlery. A more common form of PS, however, is the *expanded solid* made by blowing CO_2 or pentane gas into molten polystyrene, which then expands and retains voids in its structure when it solidifies. One form of this expanded PS is Styrofoam, the familiar material of coffee cups and take-out food containers (Figure 13.27).

The difference in properties between transparent, colorless, inflexible nonexpanded PS and opaque, white, pliable Styrofoam can be explained by considering the role of the aromatic ring in aligning the polymer chains. Branches in the chains have the same effect that we saw with polyethylene, but the aromatic rings and their tendency to stack (Figure 13.26) provide additional interactions that make chain alignment more favorable energetically. The aromatic rings along two neighboring chains can stack, and this stacking makes nonexpanded PS rigid. When the chains are blown apart by a gas, the stacking is disrupted and the chains open to make cavities that fill with air, making expanded PS a good thermal insulator and packing material. The presence of air in Styrofoam is illustrated in Figure 13.28.

FIGURE 13.28 When the Styrofoam coffee cup on the left is placed under pressure, some of the air between the polystyrene chains is forced out. The cup shrinks to the size on the right but retains its overall shape.

13.5 Amines

Nitrogen atoms are the defining components of functional groups in another important family of organic molecules called **amines**. In organic compounds nitrogen atoms—and any atoms other than carbon, hydrogen, or metals—are called **heteroatoms** and are shown in Kekulé structures (without showing the lone pairs), in condensed structures, and in carbon-skeleton structures of organic compounds. Notice how there is an N at the junction of three lines in Benadryl and pargyline in Figure 13.29.

Amines containing an alkyl or aromatic group are thought of as being derived from ammonia, NH_3. If one hydrogen atom in ammonia is replaced by an R group, the compound is called a *primary amine*. If two hydrogen atoms are replaced by R groups, the compound is a *secondary amine;* if all three hydrogen atoms are replaced by R groups, a *tertiary amine*:

$$RNH_2 \qquad R_2NH \qquad R_3N$$
Primary amine   Secondary amine   Tertiary amine

The R groups in a secondary or tertiary amine may be the same or different organic subunits. You may be familiar with the odor of trimethylamine, $(CH_3)_3N$, the compound responsible for the smell of decaying fish. The amine functional group is found in many natural products and drugs. Some relatively simple examples include amphetamine (Figure 13.29), a stimulant that also contains an aromatic group, and Benadryl, an antihistamine used to treat the symptoms associated with allergies. Another amine, adrenaline, is produced by our bodies in glands near the kidneys and plays an important role in our nervous system.

Amines are organic bases, and their basicity is their defining chemical characteristic. They all react with water to some extent to produce hydroxide ions and protonated cations, just like ammonia:

$$NH_3(aq) + H_2O(\ell) \rightleftharpoons NH_4^+(aq) + OH^-(aq) \qquad (13.9)$$

$$RNH_2(aq) + H_2O(\ell) \rightleftharpoons RNH_3^+(aq) + OH^-(aq) \qquad (13.10)$$

Amines also react readily with acids like hydrochloric acid to form salts:

$$RNH_2(\ell) + HCl(aq) \rightarrow RNH_3^+(aq) + Cl^-(aq) \qquad (13.11)$$

In fact, many pharmaceuticals (like Benadryl) containing the amine functional group are sold as hydrochloride salts to improve their solubility in water.

Amines are also polar and form hydrogen bonds with water. Hydrogen bonding is possible with primary and secondary amines but not with tertiary amines, because the nitrogen atom in a tertiary amine bears no hydrogen atom.

A particularly important type of amine is one in which the nitrogen atom is part of a ring. We will explore the properties of these cyclic amines in Chapter 20.

Bacteria of the genus *Methanosarcina* convert primary, secondary, and tertiary methylamines to methane, carbon dioxide, and ammonia:

$$4\,CH_3NH_2(aq) + 2\,H_2O(\ell) \rightarrow 3\,CH_4(g) + CO_2(g) + 4\,NH_3(aq) \qquad (13.12)$$

$$2\,(CH_3)_2NH(aq) + 2\,H_2O(\ell) \rightarrow 3\,CH_4(g) + CO_2(g) + 2\,NH_3(aq) \qquad (13.13)$$

$$4\,(CH_3)_3N(aq) + 6\,H_2O(\ell) \rightarrow 9\,CH_4(g) + 3\,CO_2(g) + 4\,NH_3(aq) \qquad (13.14)$$

These reactions describe a pathway by which methane can be produced from the decay of biomass, serving as potential sources of methane for heating and as fuel

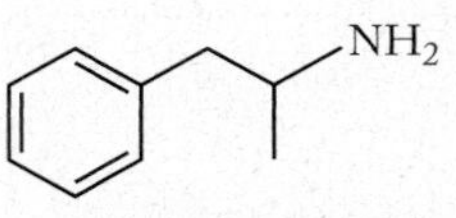

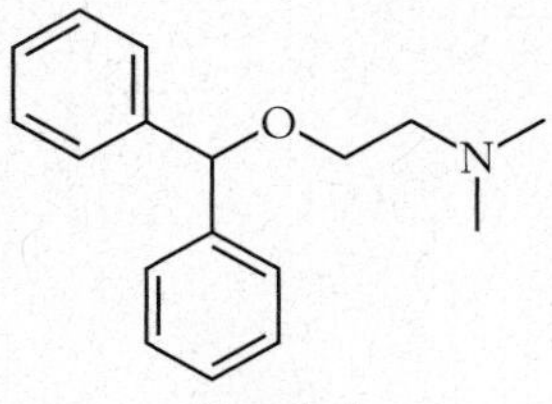

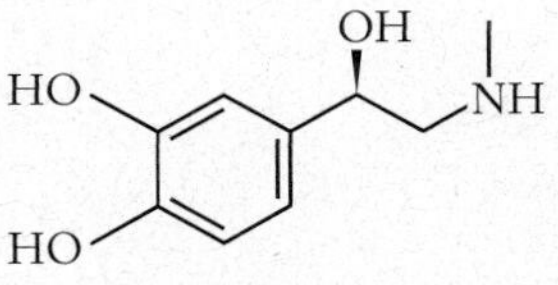

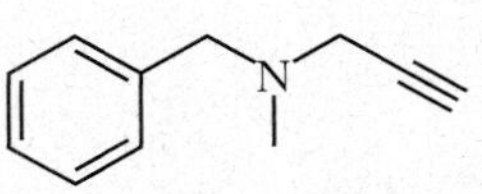

FIGURE 13.29 Amphetamine, a drug known to produce increased wakefulness and focus; Benadryl, an antihistamine; adrenaline, a hormone produced in our bodies and involved in the fight-or-flight response; and pargyline, a drug used to treat hypertension, are examples of physiologically active compounds that contain the amine functional group.

amine organic compound that contains a group with the general formula RNH_2, R_2NH, or R_3N, where R is any organic subgroup.

heteroatom any atom other than carbon or hydrogen in an organic compound.

alcohol organic compound containing the –OH functional group.

for transportation. Amines represent only a small fraction of the biomass of plants, however, and the industrial development of fuel production via amine-based reactions has been slow, mainly because fossil fuels are still plentiful enough and cheap enough to make amine pathways not cost-effective. This economic imbalance may change as fossil fuels are depleted and become more expensive.

CONCEPT TEST

Are pargyline, amphetamine, adrenaline, and Benadryl (Figure 13.29) primary, secondary, or tertiary amines?

13.6 Alcohols, Ethers, and Reformulated Gasoline

In January 1995, air-quality regulations went into effect in many U.S. cities mandating reductions in atmospheric pollutants from gasoline-fueled engines. The regulations led to the widespread use of "reformulated" gasoline containing additives to promote complete combustion and boost octane ratings. These additives are often organic compounds that contain oxygen in addition to hydrogen and carbon. Oxygen atoms in an organic compound are also heteroatoms and are components of functional groups in two important families of organic molecules: alcohols and ethers.

CONCEPT TEST

Explain why the solubility of low-molar-mass alcohols and ethers in water is greater than the solubility of hydrocarbons in water.

Alcohols: Methanol and Ethanol

Alcohols have the general formula R—OH, where R is any alkyl group. The R group can be a straight chain, branched chain, or ring. Like the N atoms in amines, the O atoms in alcohols are shown in carbon-skeleton structures. The chemical and physical properties of alcohols can be understood if we recognize that an alcohol looks like a combination of an alkane and water:

$$\text{R—H} \qquad \text{H—OH} \qquad \text{R—OH}$$
Alkane   Water   Alcohol

As we saw in Section 10.4, if the R group in the molecule is small, the alcohol behaves like water; as the R group gets larger, the alcohol behaves more like a hydrocarbon. As an illustration of this behavior, Table 13.6 shows the water solubilities of a homologous series of alcohols. The polar –OH group makes one end of these molecules "water-like." However, as the number of carbon atoms increases, a greater proportion of these alcohols are "oil-like." Therefore, their solubilities in water decrease until about C_8, beyond which their solubilities are comparable to those of the corresponding hydrocarbons.

As the names of the two simplest alcohols—methanol and ethanol—indicate, the chemical names of alcohols end in -*ol*; this suffix identifies the compound as an alco<u>hol</u>. Methanol (CH_3OH) is also known as methyl alcohol or wood alcohol.

TABLE 13.6	Solubilities of a Homologous Series of Alcohols in Water at 20°C
Condensed Structure	**Water Solubility (g/100 mL)**
CH_3OH	Miscible
CH_3CH_2OH	Miscible
$CH_3(CH_2)_2OH$	Miscible
$CH_3(CH_2)_3OH$	7.9
$CH_3(CH_2)_4OH$	2.3
$CH_3(CH_2)_5OH$	0.6
$CH_3(CH_2)_6OH$	0.2
$CH_3(CH_2)_7OH$	0.05

The latter name comes from one former source of this alcohol: it was made by collecting the vapors given off when wood is heated to the point of decomposition in the absence of oxygen. Methanol is a widely used industrial organic chemical. It is the starting material in the preparation of several organic compounds used to make polymers. Its industrial synthesis is based on reducing carbon monoxide with hydrogen:

$$CO(g) + 2H_2(g) \rightarrow CH_3OH(\ell)$$

The CO and H_2 used to make methanol come from the reaction of methane and water in the steam reforming process we discussed in Chapter 5:

$$CH_4(g) + H_2O(g) \rightarrow CO(g) + 3H_2(g)$$

Methanol burns according to the thermochemical equation:

$$2CH_3OH(\ell) + 3O_2(g) \rightarrow 2CO_2(g) + 4H_2O(\ell) \qquad \Delta H^\circ_{comb} = -1454 \text{ kJ}$$

If we divide the absolute value of ΔH°_{comb} by twice the molar mass of methanol (because the reaction consumes 2 moles of methanol), we get a fuel value for methanol of

$$\frac{1454 \text{ kJ}}{\left(\dfrac{32.04 \text{ g}}{\text{mol}}\right)(2 \text{ mol})} = 22.7 \text{ kJ/g}$$

Let's compare the fuel value of methanol with that of octane:

$$2C_8H_{18}(\ell) + 25O_2(g) \rightarrow 16CO_2(g) + 18H_2O(\ell) \quad \Delta H^\circ_{comb} = -1.091 \times 10^4 \text{ kJ}$$

$$\frac{1.091 \times 10^4 \text{ kJ}}{\left(\dfrac{114.22 \text{ g}}{\text{mol}}\right)(2 \text{ mol})} = 47.8 \text{ kJ/g}$$

The fuel value of methanol is less than half that of octane (and most of the other hydrocarbons in gasoline). Why is this the case? The answer involves the composition of methanol. The amount of energy released during combustion depends on the number of carbon atoms available for forming $C{=}O$ bonds in CO_2 and the number of hydrogen atoms available for forming $O{-}H$ bonds in H_2O. The presence of oxygen in CH_3OH adds significantly to its mass (methanol is 50% oxygen by mass) but adds nothing to its fuel value. The oxygen content of a combustible substance essentially dilutes its energy value. The higher the oxygen content of a fuel, the lower is its fuel value.

As noted earlier, ethanol (CH_3CH_2OH), also known as ethyl alcohol, is the alcohol in alcoholic beverages. It is formed by the fermentation of sugar from an amazing variety of vegetable sources. Indeed, any plant matter containing sufficient sugar may be used to produce ethanol. Grains are commonly used, from which ethanol derives its trivial name *grain alcohol*. Ethanol may be the oldest organic chemical used by humans, and it is still one of the most important. For industrial purposes, ethanol is prepared by the reaction of water and ethylene.

Most gasoline in the United States currently contains 10% ethanol, with efforts underway to increase this value to 15%. Most of the ethanol used in gasoline is produced by fermentation of sugar derived from corn. Ethanol burns readily in air:

$$CH_3CH_2OH(\ell) + 3O_2(g) \rightarrow 2CO_2(g) + 3H_2O(\ell) \qquad \Delta H^\circ_{comb} = -1367 \text{ kJ}$$

∞ CONNECTION We introduced fuel values in Chapter 5 as the amount of heat given off when one gram of fuel is burned.

SAMPLE EXERCISE 13.7   **Comparing the Energy Content of Fuel Mixtures**

Suppose the hydrocarbons in gasoline can be represented by *n*-nonane, $CH_3(CH_2)_7CH_3$ (ΔH°_{comb} −6160 kJ/mol). The ΔH°_{comb} of ethanol is −1367 kJ/mol. How much energy in the form of heat is available from a mixture of 10.0% ethanol / 90.0% gasoline by mass compared with the amount available from pure gasoline? Carry out the calculation based on 2.70×10^3 g of mixture, which is about the mass of 1.00 gallon of gasoline.

Collect and Organize In this exercise, we are given the enthalpies of combustion of nonane and ethanol, the total mass of the fuels, and the composition of the fuel as a percentage by mass. Calculations of the heat produced by a reaction (*q*) given the enthalpy change for the process and the amount of reactant were illustrated in Section 5.6 for the recycling of aluminum.

Analyze The heats of combustion are given in terms of moles of material, and the mass of the mixture is in grams, so to use the thermochemical values we need to calculate the number of moles of each component. To convert grams to moles, we need molar masses, which we can determine from the molecular formulas: CH_3CH_2OH, 46.07 g/mol; $CH_3(CH_2)_7CH_3$, 128.25 g/mol. The heat produced by each component is the product of the number of moles of the component and ΔH°_{comb}. The molar mass of ethanol is about one-third the molar mass of nonane, but the mass of nonane present in the mixture is 9 times the mass of the ethanol. This means that we have many more moles of nonane than ethanol in the mixture. The ΔH°_{comb} of nonane is about 4–5 times greater than the ΔH°_{comb} for ethanol. This allows us to predict that the heat produced from burning the mixture will be lower but not much lower than the heat produced by burning pure nonane.

Solve Let's calculate how many grams of each component are in the mixture and then convert that to moles:

10.0% ethanol: $0.100(2.70 \times 10^3 \text{ g}) = 2.70 \times 10^2$ g ethanol

90.0% *n*-nonane: $0.900(2.70 \times 10^3 \text{ g}) = 2.43 \times 10^3$ g *n*-nonane

Moles ethanol: $2.70 \times 10^2 \text{ g} \times \dfrac{1 \text{ mol}}{46.07 \text{ g}} = 5.861$ mol

Moles *n*-nonane: $2.43 \times 10^3 \text{ g} \times \dfrac{1 \text{ mol}}{128.25 \text{ g}} = 18.95$ mol

The amount of heat given off by the mixture is the sum of the heat given off by the two components:

$$5.861 \text{ mol} \left(\frac{-1367 \text{ kJ}}{\text{mol}} \right) + 18.95 \text{ mol} \left(\frac{-6160 \text{ kJ}}{\text{mol}} \right) = -124{,}750 \text{ kJ} =$$
$$-1.25 \times 10^5 \text{ kJ}$$

When 2.70×10^3 g of *n*-nonane is burned, the heat given off is

$$2.70 \times 10^3 \text{ g} \times \frac{1 \text{ mol}}{128.25 \text{ g}} \times \frac{-6160 \text{ kJ}}{\text{mol}} = -129{,}700 \text{ kJ} = -1.30 \times 10^5 \text{ kJ}$$

Expressing the difference as a percentage gives us

$$\frac{-(129{,}700 - 124{,}750) \text{ kJ}}{-129{,}700 \text{ kJ}} \times 100\% = 3.82\%$$

The ethanol/gasoline mixture produces about 4% less energy than gasoline by itself.

Think about It The presence of oxygen in ethanol adds to its mass but does not add to its fuel value. It is logical that a blend of a hydrocarbon and an alcohol has a slightly lower energy content than the hydrocarbon itself. Among the reasons we use ethanol is that it produces less air pollution and, in principle, is a renewable resource.

Practice Exercise A fuel called E-85 is a mixture of 85% ethanol and 15% gasoline by volume. How much energy in the form of heat is available from this mixture compared with the amount from pure gasoline? The densities of ethanol and gasoline are 0.789 and 0.737 g/mL, respectively, and *n*-nonane may be used as a model hydrocarbon for gasoline.

The use of ethanol as a gasoline additive resulted in a sharp increase in ethanol production at the beginning of the 21st century, with the annual production in the United States estimated as 9 billion gallons in 2008. However, several challenges limit the wide use of ethanol as an automobile fuel. Like methanol, ethanol has a fuel value that is less than that of a comparable mass or volume of gasoline. Furthermore, considerable energy, irrigation water, and valuable farm land are required in its production: growing and harvesting corn, converting corn starch into sugar, converting the sugar into alcohol, and finally distilling the alcohol from the fermentation mixture. It is estimated that more than two-thirds of the energy released in the combustion of ethanol derived from corn is consumed in its production. Ethanol produced in this way is more expensive than gasoline, even at today's prices for fossil fuels. Fuels are extraordinarily complex in terms of their composition, their combustion, the emissions they produce, and the issue of renewability, and this brief discussion addresses a very small part of a challenging problem.

Alcohols are also prevalent in natural products and in pharmaceuticals. The distinctive aroma of mint leaves comes from menthol, which is an alcohol, as is terpineol (oil of turpentine), an oil distilled from the resin of pine trees (Figure 13.30). Notice the similarity in the carbon-skeleton structures of these two compounds—both contain a six-carbon ring and an –OH group. Only the location of the –OH group and the presence of a C=C bond distinguish these two compounds.

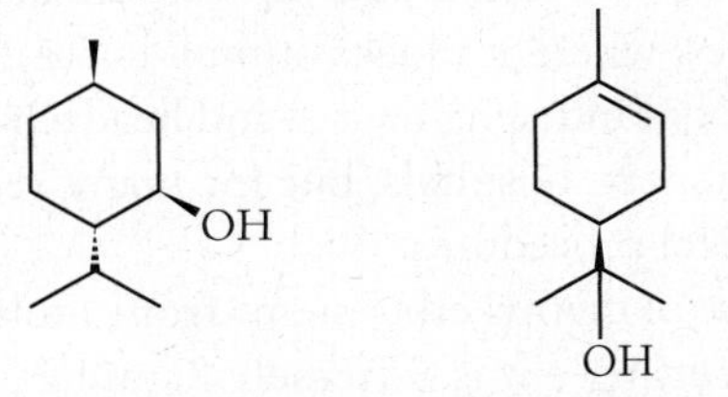

(a) Menthol
(oil of mint)

(b) Terpineol
(oil of turpentine)

FIGURE 13.30 (a) Menthol and (b) terpineol are two naturally occurring alcohols present in mint leaves and pine needles, respectively.

Ethers: Diethyl Ether

Ethers have the general formula R—O—R, where R is any alkyl group or an aromatic ring. Just as with alcohols, we can think of an ether structurally as a water molecule in which the two H atoms have been replaced by two organic groups (R and R′, which may be the same or different):

R—H	H—O—H	H—R′	R—O—R′
Alkane	Water	Alkane	Ether

Because the C—O—C bond angle is close to the tetrahedral bond angle of 109.5°, the dipole moments of the two C—O bonds in an ether do not cancel, which means that ethers are polar molecules. This structural feature gives rise to the properties of typical ethers: their water solubility is comparable to that of

ether organic compound with the general formula R—O—R, where R is any alkyl group or aromatic ring; the two R groups may be different.

TABLE 13.7	Functional Groups Affect Physical Properties		
	Molar Mass (g/mol)	Normal Boiling Point (°C)	Solubility in Water (g/100 mL at 20°C)
CH_3CH_2—O—CH_2CH_3 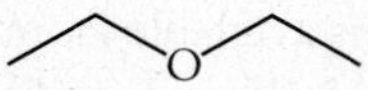 Diethyl ether	74	35	6.9
$CH_3CH_2CH_2CH_2CH_3$  n-Pentane	72	36	0.0038
$CH_3CH_2CH_2CH_2OH$  n-Butanol	74	117	7.9

alcohols of similar molar mass, but their boiling points are about the same as alkanes of comparable molar mass (Table 13.7). Note that Benadryl in Figure 13.29 also contains an ether functional group.

The most important ether industrially is diethyl ether, $CH_3CH_2OCH_2CH_3$. You may have first heard of this as the material simply called "ether" that has had wide use in medicine as an anesthetic since 1842. Although exactly how an anesthetic dulls nerves and puts patients to sleep is still unknown, certain properties of diethyl ether play a role in determining its behavior as a medicinal agent. Because diethyl ether has a low boiling point, 35°C, it vaporizes easily, and a patient can inhale it. Because diethyl ether has a significant solubility in water, it is soluble in blood, which means that once inhaled, it can be easily transported throughout the body. Its low polarity and short saturated hydrocarbon chains combine to make it soluble in cell membranes, where it blocks stimuli coming into nerves. Ether has the unfortunate side effect of inducing nausea and headaches and has been replaced by new anesthetics in modern hospitals, but for many years ether was the anesthetic of choice for surgical procedures.

A second common use of diethyl ether stems from another property that caused difficulty in the clinical setting—it is extremely flammable. Flammability is a liability in an operating room, but this property is used to our advantage when we spray ether in diesel engines to start them in cold weather when it is too cold for diesel fuel to ignite.

One ether widely used as a gasoline additive in the 1990s was methyl tert-butyl ether (MTBE, Figure 13.31), added to promote complete combustion (tert- is an abbreviation for tertiary, referring to a carbon atom bonded to three other carbon atoms). Unlike the nonpolar hydrocarbons in gasoline, MTBE is soluble in water. Consequently, gasoline spills, leakage from storage tanks, and release from watercraft produces extensive MTBE contamination of groundwater and drinking water. After toxicity tests showed MTBE to be a possible carcinogen (cancer-causing agent), several states—including California, where more than 25% of the world's production of MTBE was used in gasoline—banned the use of MTBE as a gasoline additive. Most oil companies stopped adding MTBE to their gasolines in 2006. These changes raised the question of which additives would replace MTBE. The leading candidate to date has been ethanol.

FIGURE 13.31 MTBE, methyl tert-butyl ether, was used in the early 1990s as a fuel additive but its use was curtailed when it was found to be an environmental pollutant.

CONCEPT TEST

Rank the following compounds in order of decreasing fuel value: diethyl ether, MTBE, methanol, and ethanol.

copolymer a macromolecule formed from the chemical combination of two different monomers.

heteropolymer a polymer made of three or more different monomer units.

Polymers of Alcohols and Ethers

Over 400,000 tons of the addition polymer poly(vinyl alcohol) (PVAL) are produced annually in the United States. It is used in fibers, in adhesives, and in materials known as sizing, which change the surface properties of textiles and paper to make them less porous, less able to absorb liquids, and smooth. PVAL is the material of choice for laboratory gloves that are resistant to organic solvents. Because its chains are studded with –OH groups (Figure 13.32a), its surface is very polar and very water-like, and hydrocarbon solvents that are not soluble in water do not penetrate PVAL barriers.

PVAL is also impenetrable to carbon dioxide, and this property has led to its use in soda bottles, in which it is blended with the polymer poly(ethylene terephthalate) (PETE, Figure 13.32b). The two polymers do not mix but separate into layers (Figure 13.32c). The PETE makes the bottle strong enough to bear pressure changes due to temperature changes and survive the impact of falling off tables. However, CO_2, the dissolved gas that makes soda fizz, passes readily through PETE but not through the PVAL layers, with the result that the soda does not go flat. Polymers with different properties are frequently combined to create new materials with desired properties.

The monomer from which poly(vinyl alcohol) is made is not the one you might expect based on our discussion of how addition polymers are synthesized. There is no such compound as "vinyl alcohol":

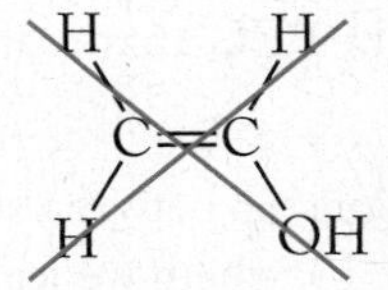

Instead, the monomer vinyl acetate is polymerized to make poly(vinyl acetate) (PVAC), which is then reacted with water to replace the acetate group with an –OH group. This replacement reaction turns PVAC into PVAL as shown in Figure 13.32(a).

The blend of PVAL and PETE in early soda bottles was a physical mixture of the two polymers. New materials can also be made by combining different monomer units in one polymer molecule. This type of molecule is called a **copolymer** when two different monomers are combined and a **heteropolymer** when three or more different monomers are combined. One example of an addition copolymer is a material called EVAL, made from ethylene and vinyl acetate (Figure 13.33). EVAL is used in food wrappings when preservation of aroma and flavor are required. Food usually deteriorates in the presence of oxygen, and packages made of EVAL provide an excellent barrier to the entry of oxygen while retaining the flavor and fragrance of the packaged food.

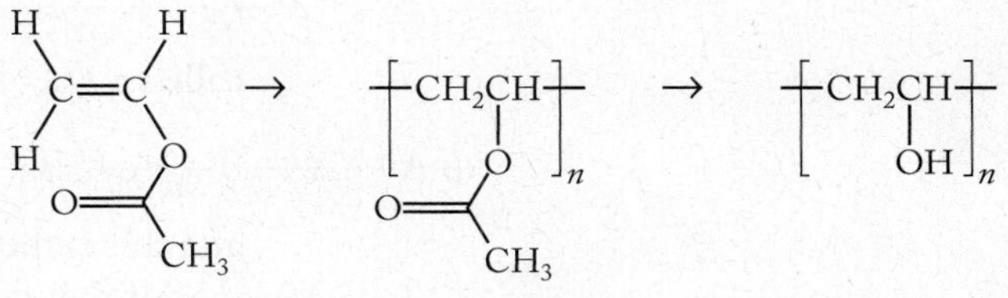

Vinyl acetate | PVAC | PVAL

(a)

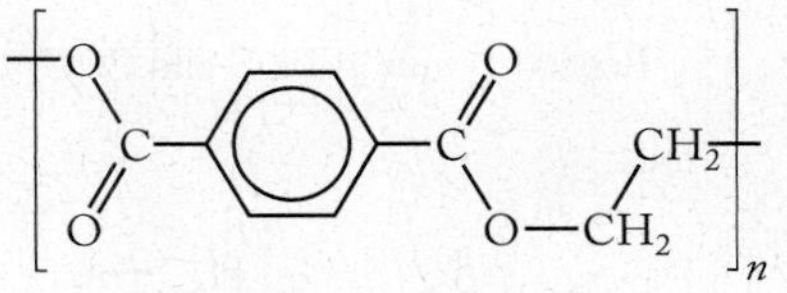

Repeating unit in PETE

(b)

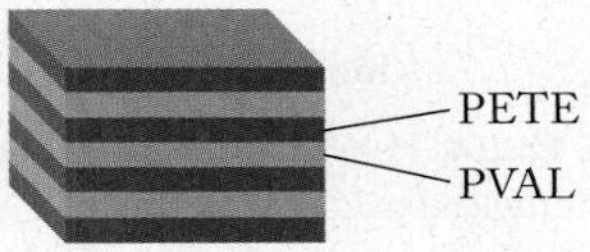

(c)

FIGURE 13.32 (a) Poly(vinyl alcohol), or PVAL, in synthesized from vinyl acetate. The intermediate polymer, poly(vinyl acetate), or PVAC, is reacted with water to produce PVAL. (b) The repeating unit in PETE. (c) Layers of the polymers PVAL and PETE are used to make soda bottles. PETE makes the bottle strong, and PVAL keeps the carbon dioxide in.

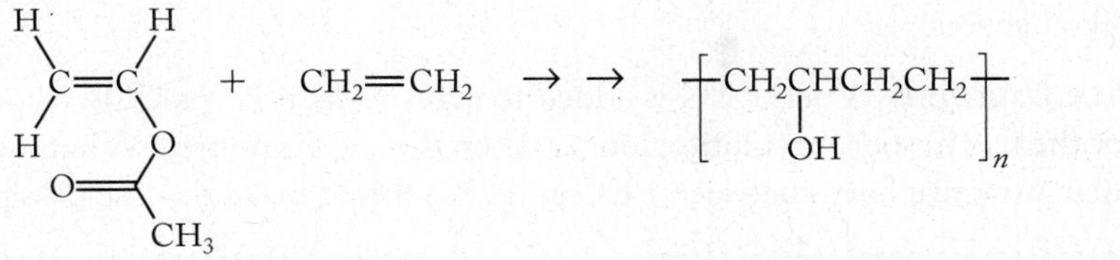

Vinyl acetate | Ethylene | Poly(ethylene-co-vinyl alcohol) = EVAL

FIGURE 13.33 EVAL is a copolymer of ethylene and vinyl acetate.

Monomers forming hetero- or copolymers can combine in different ways. If we represent the monomer units making up a copolymer with the letters A and B, one possible way they can combine is

$$\text{---}\!\!\text{[}A\text{---}B\text{---}A\text{---}B\text{---}A\text{---}B\text{---}A\text{---}B\text{---}A\text{---}B\text{]}\!\!\text{---}$$

an arrangement called an *alternating copolymer*. Another possibility is

$$\text{---}\!\!\text{[}A\text{---}A\text{---}A\text{---}A\text{---}B\text{---}B\text{---}B\text{---}B\text{---}A\text{---}A\text{---}A\text{---}A\text{---}B\text{---}B\text{---}B\text{---}B\text{]}\!\!\text{---}$$

called a *block copolymer*. Finally, what is called a *random copolymer* is possible:

$$\text{---}\!\!\text{[}A\text{---}A\text{---}B\text{---}A\text{---}B\text{---}B\text{---}A\text{---}B\text{---}A\text{---}A\text{---}A\text{---}A\text{---}B\text{---}B\text{---}A\text{---}B\text{---}B\text{---}B\text{]}\!\!\text{---}$$

and this is the arrangement we see in the copolymer EVAL: it is a random copolymer of the monomers A = ethylene, B = vinyl acetate.

Commercially important polymers made from ethers include poly(ethylene glycol) (PEG) and poly(oxyethylene) (POE), which contain the same subunit (Figure 13.34). PEG is a low-molar-mass liquid made from ethylene glycol, and POE is a higher-molar-mass solid made from ethylene oxide. As a polyether, PEG has properties closely related to those of diethyl ether, in that it is soluble in both polar and nonpolar liquids. It is a common component in toothpaste because it interacts both with water and with the water-insoluble materials in the paste and keeps the toothpaste uniform both in the tube and during use. PEGs of many lengths are finding increasing use as attachments to pharmaceutical agents to improve their solubility and biodistribution.

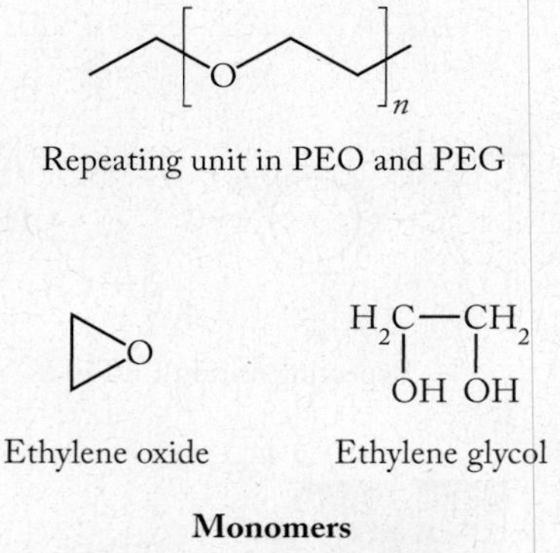

Repeating unit in PEO and PEG

Ethylene oxide   Ethylene glycol

Monomers

FIGURE 13.34 Poly(ethylene glycol) (PEG) and poly(ethylene oxide) (PEO) have the same repeating unit. The two polymers differ only in their molar masses. PEG is typically made from ethylene glycol, and ethylene oxide is the monomer of choice for making PEO.

SAMPLE EXERCISE 13.8   Assessing Properties of Polymers

The polymer PEG (Figure 13.34) is used to blend materials that are not soluble in each other. It is soluble both in water and in benzene, a nonpolar solvent. Describe the structural features of PEG that make it soluble in these two liquids of very different polarities.

Collect and Organize The relationship between structure and solubility of compounds was discussed in Chapter 10, where we learned that "like dissolves like."

Analyze The statement "like dissolves like" refers to the polarity of the molecules and to the attractive and repulsive forces between molecules. We need to describe the intermolecular forces in PEG and see if one part is water-like and another benzene-like. Water is a polar molecule and benzene is a nonpolar molecule, so we predict that PEG contains both polar and nonpolar regions.

Solve The structure of PEG consists of $-CH_2CH_2-$ groups connected by oxygen atoms. The oxygen atoms are capable of hydrogen bonding with water molecules, so the attractive force between PEG and water is due to hydrogen bonding. Benzene is nonpolar and is attracted to the $-CH_2CH_2-$ groups in the polymer. Nonpolar groups interact via dispersion forces, so those forces must be responsible for the solubility of PEG in nonpolar benzene.

Think about It As predicted, PEG contains both polar and nonpolar regions, allowing it to be solvated by both polar solvents like water and nonpolar solvents like benzene.

Practice Exercise When PEG is added to soft drinks it keeps CO_2, responsible for the fizz in soda, in solution longer when the soda is poured. What intermolecular attractive forces between PEG and CO_2 might make this use possible?

CONCEPT TEST

A polymer chemist decides to make a series of derivatives of PEG using the following alcohols in place of ethylene glycol:

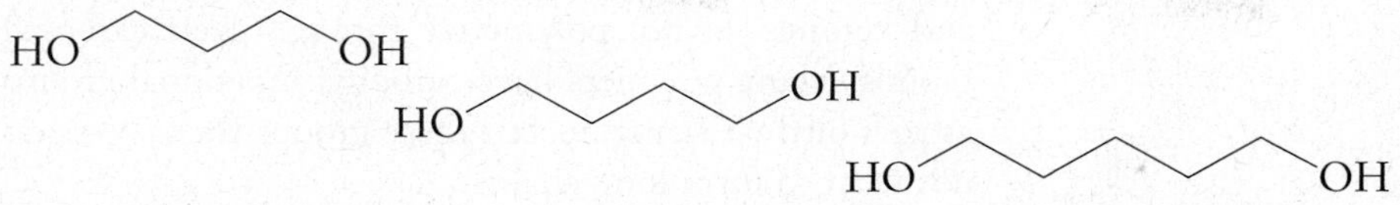

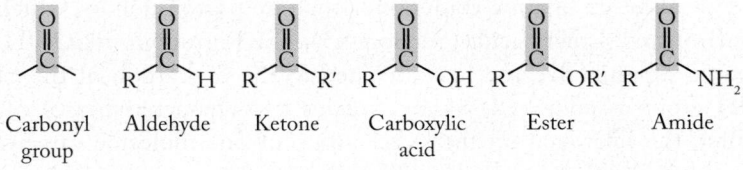

What effect will this have on the solubility of the resulting polymer?

carbonyl group a carbon atom with a double bond to an oxygen atom.

aldehyde organic compound containing a carbonyl group bonded to one R group and one hydrogen; its general formula is RCHO.

ketone organic molecule containing a carbonyl group bonded to two R groups; its general formula is R(C═O)R.

13.7 Aldehydes, Ketones, Carboxylic Acids, Esters, and Amides

Five functional groups—aldehydes, ketones, carboxylic acids, esters, and amides—all contain a subunit called the **carbonyl group**: a carbon atom double-bonded to an oxygen atom (Figure 13.35). The R groups in the figure may be any organic group. Aldehydes and ketones are collectively referred to as *carbonyl compounds* because the carbonyl group determines their chemistry. Carboxylic acids, as their name implies, are acidic, and their chemistry is determined by the –COOH subunit, referred to as a *carboxylic acid group*. Esters and amides can be made from carboxylic acids by reacting them with alcohols and amines.

⊙⊙ **CONNECTION** We first introduced carboxylic acids in our discussion of acids in Chapter 2.

Carbonyl group   Aldehyde   Ketone   Carboxylic acid   Ester   Amide

FIGURE 13.35 The carbonyl group is found in five important functional groups: aldehydes, ketones, carboxylic acids, esters, and amides.

Aldehydes and Ketones

An **aldehyde** contains a carbonyl group bound to one R group and one hydrogen atom; its general formula is RCHO or RC(O)H. A **ketone** contains a carbonyl group bound to two R groups; its general formula is RCOR or R(C═O)R. The R groups may be the same as in acetone, $CH_3C(O)CH_3$, or different as in 2-heptanone (Figure 13.36), which is found in cloves, blue cheese, and many fruits and dairy products.

The double bond in the carbonyl group accounts for the reactivity of aldehydes and ketones. It is different from the double bond in an alkene, however, because it is polar (Figure 13.37). The electronegative oxygen pulls electron density toward itself, and the chemistry of aldehydes and ketones is linked to the polarity of the C═O bond. Other polar species tend to react with carbonyls when electron-rich regions of their molecules approach the δ^+ carbon atoms of the carbonyl groups.

Aldehydes and ketones are polar, and they tend to parallel the ethers with respect to water solubility. They cannot hydrogen bond with other aldehyde and ketone molecules because they contain only carbon-bonded hydrogen atoms, so they have lower boiling points than alcohols of comparable molar mass.

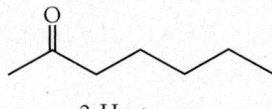

2-Heptanone

FIGURE 13.36 The ketone 2-heptanone is found in many plants and dairy products.

FIGURE 13.37 The electron distribution in a carbonyl group is skewed toward the oxygen end of the bond because oxygen is more electronegative than carbon.

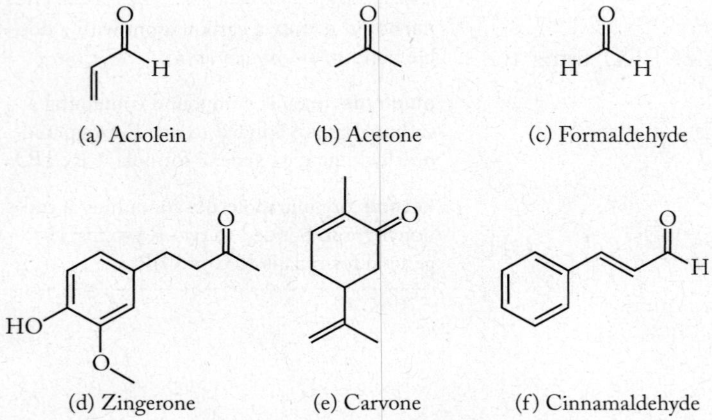

FIGURE 13.38 The ketone and aldehyde functional groups are common among organic compounds: (a) acrolein is found in barbeque smoke, (b) acetone is used in nail polish remover, (c) aqueous solutions of formaldehyde are used to preserve biological specimens, (d) zingerone is found in the spice ginger, (e) carvone is found in the leaves of spearmint, and (f) cinnamon owes its flavor and odor to cinnamaldehyde.

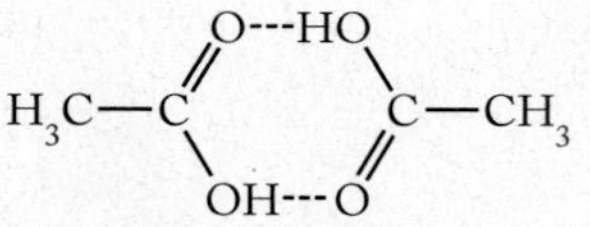

FIGURE 13.39 The high boiling points of carboxylic acids are the result of strong hydrogen bonds between neighboring molecules.

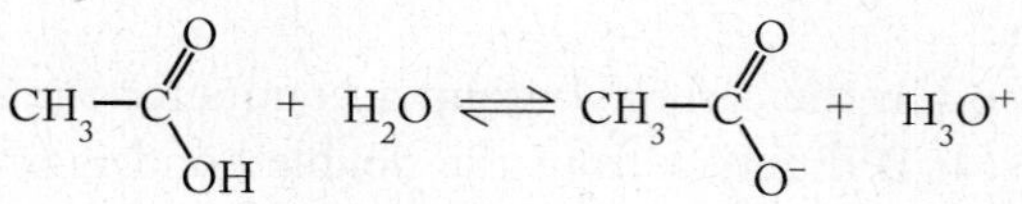

FIGURE 13.40 Carboxylic acids such as acetic acid (vinegar) are weak acids in water.

Because of the C=O bond, aldehydes and ketones are in a higher oxidation state than alcohols, and indeed many of the smaller aldehydes and ketones are made by oxidizing alcohols of the same carbon number. Aldehydes and ketones do not polymerize through their carbonyl groups. Many polymers have carbonyl functional groups as part of their structure, but these groups themselves do not react to form long chains.

We have already seen several examples of aldehydes and ketones in earlier chapters. In Chapter 9, we were introduced to formaldehyde and acrolein, two aldehydes shown in Figure 13.38. Acetone, $CH_3C(O)CH_3$, is used in nail polish remover and is a widely used solvent. The flavors of ginger (zingerone), spearmint (carvone), and cinnamon (cinnamaldehyde) all come from compounds that contain carbonyl groups.

> **CONCEPT TEST**
>
> What other functional groups are present in zingerone, carvone, and cinnamaldehyde besides the carbonyl group?

Carboxylic Acids

Carboxylic acids are organic compounds that are proton donors, which means they are Brønsted–Lowry acids (Section 4.5). The R group in RCOOH may be any organic subunit. Because it is attached to the δ^+ carbon of the carbonyl, the –OH group is polarized, which explains two characteristics of carboxylic acids. First, the hydrogen on the –OH group of one molecule can hydrogen-bond to a neighboring carboxylic acid, either at the oxygen atom in the –OH group or at the carbonyl oxygen atom (Figure 13.39). This interaction results in high boiling points relative to those of other organic compounds of comparable molar mass.

Second, donating a proton leaves a negatively charged oxygen on the carboxylic acid that is delocalized over the whole carbonyl group. This delocalization contributes to the stability of the carboxylate anion. The common carboxylic acids are weak acids, which means that they are present in aqueous solutions as mostly neutral molecules, a small fraction of which are ionized, donating H^+ ions to molecules of water, as shown in Figure 13.40 for acetic acid.

Vinegar is a dilute aqueous solution of the carboxylic acid acetic acid. Large quantities of vinegar are produced commercially by the air oxidation of ethanol in the presence of enzymes from *Acetobacter* bacteria. Bacteria can also convert acetic acid and other constituents in biomass to methane. For example, the digestive systems of cows introduce significant amounts of methane to the atmosphere, about 100–200 liters per day per animal. Translating this process to an industrial scale is an attractive future source of hydrocarbons, provided the complexities of bacterial action can be adapted for large-scale production. Recall that we described the timescale for the formation of fossil fuels in Chapter 3. Converting organic matter into hydrocarbon fuel may be possible without waiting millennia for the anaerobic processes deep within Earth to do so.

The production of methane from plant residues that are mostly cellulose (a carbohydrate) requires the sequential action of several types of bacteria. In the first stages, selected bacteria break up cellulose into mixtures of small molecules. Depending on the bacterial strain, these small-molecule products include H_2 and CO_2, acetic acid, formic acid, or methanol or other small alcohol. All these products then undergo reactions promoted by the metabolism of **methanogenic** (methane-producing) **bacteria**, which consume hydrogen and simple organic compounds for energy and produce methane gas in the process:

$$4\,H_2(g) \,+\, CO_2(g) \rightarrow CH_4(g) \,+\, 2\,H_2O(\ell)$$

$$CH_3COOH(aq) \rightarrow CH_4(g) \,+\, CO_2(g)$$
Acetic acid

$$4\,HCOOH(aq) \rightarrow CH_4(g) \,+\, 3\,CO_2(g) \,+\, 2\,H_2O(\ell)$$
Formic acid

$$4\,CH_3OH(aq) \rightarrow 3\,CH_4(g) \,+\, CO_2(g) \,+\, 2\,H_2O(\ell)$$
Methanol

The actions of methanogenic bateria have a measurable effect on Earth's atmosphere and climate. Methane, like CO_2, is a greenhouse gas but is much more potent, trapping about 20 times more heat per molecule than carbon dioxide.

Esters and Amides

A number of chemical families are closely related to the carboxylic acids. We consider only two of them here: esters and amides. As Figure 13.41(a) shows, in **esters**, the –OH of the carboxylic acid is replaced by –OR, where R can be any organic group. The presence of the carbonyl group makes esters polar, and their boiling points are comparable to those of aldehydes and ketones of similar size.

In **amides**, the –OH is replaced by an amine group—either $-NH_2$, –NHR, or $-NR_2$. Amides are also polar and capable of intermolecular hydrogen bonding. The hydrogen atoms on the $-NH_2$ group can hydrogen-bond with the oxygen in the carbonyl of an adjacent molecule. This causes their boiling points to be considerably higher than those of esters of comparable size.

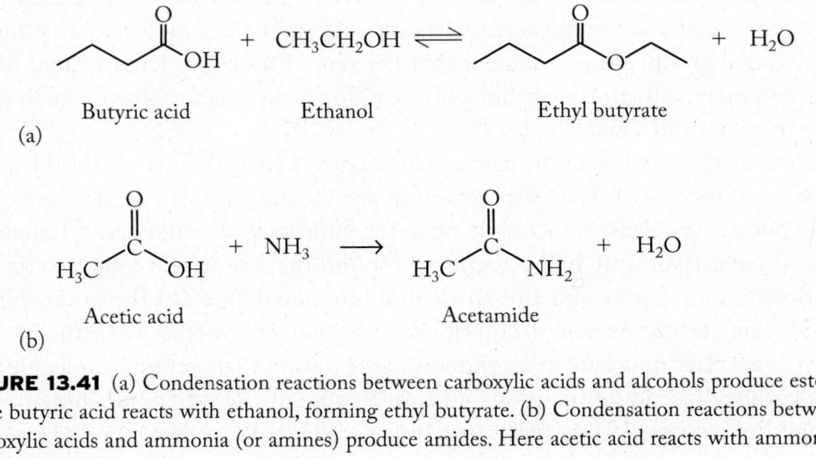

(a) Butyric acid + CH_3CH_2OH ⇌ Ethyl butyrate + H_2O

(b) Acetic acid + NH_3 → Acetamide + H_2O

FIGURE 13.41 (a) Condensation reactions between carboxylic acids and alcohols produce esters. Here butyric acid reacts with ethanol, forming ethyl butyrate. (b) Condensation reactions between carboxylic acids and ammonia (or amines) produce amides. Here acetic acid reacts with ammonia, forming acetamide.

carboxylic acid an organic compound containing the –COOH functional group.

methanogenic bacteria bacteria using simple organic compounds and hydrogen for energy; their respiration produces methane, carbon dioxide, and water, depending on the compounds they consume.

ester organic compound in which the –OH of a carboxylic acid group is replaced by –OR, where R can be any organic group.

amide organic compound in which the same carbon atoms are single bonded to nitrogen atoms and double bonded to oxygen atoms.

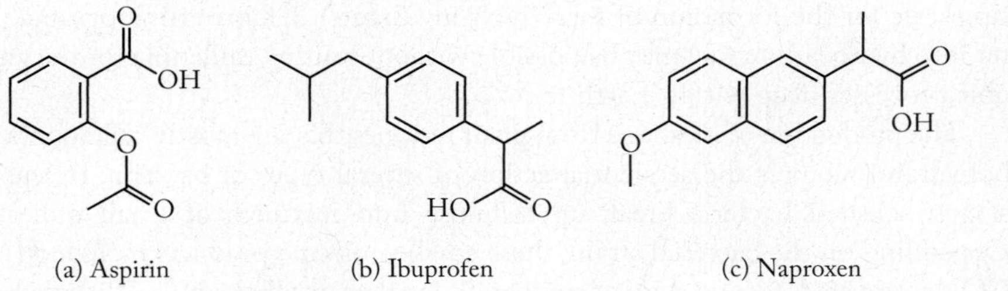

FIGURE 13.42 Three pain medications, all of which contain carboxylic acid functional groups. (a) Aspirin was the first medication to be available in tablet form. (b) Ibuprofen has fewer side effects than aspirin. (c) Naproxen belongs to a class of compounds called non-steroidal anti-inflammatory drugs (NSAIDs) and is used for the management of mild to moderate pain.

(a) Aspirin   (b) Ibuprofen   (c) Naproxen

Esters frequently have very pleasant fragrances, much different from the acids from which they are derived. For example, the carboxylic acid butyric acid, with a straight chain of 4 carbon atoms, is responsible for the odor of rancid butter. The ethyl ester of this carboxylic acid (ethyl butyrate) is the ester responsible for the aroma of ripe pineapples.

An ester is prepared by the *esterification* of an acid with an alcohol (Figure 13.41a). Esterification is a **condensation reaction**: two molecules combine ("condense") to create a larger molecule while a small molecule (typically water) is also formed. Esters are widely used in the personal products industry to provide pleasant scents for products like shampoos and soaps. Esters and carboxylic acids are also common in over-the-counter medications. Figure 13.42 illustrates the carbon-skeleton structures of three common pain relievers, aspirin, ibuprofen, and naproxen, each of which contains a carboxylic acid. Aspirin also contains an ester functional group.

Amides are made from carboxylic acids by several methods, but the net result is a condensation reaction in which the −OH of the carboxylic acid is replaced by the −NH$_2$ group of ammonia or by −NHR or −NR$_2$ groups of amines. Water is also formed in the process as shown in Figure 13.41(b).

> **CONCEPT TEST** ....
>
> What is the principal structural difference between an amine and an amide?
>
> ....

Polyesters and Polyamides

Prior to this point, many of the compounds we have examined have been monofunctional, which means they have only one functional group that identifies their family. With the polymers of carboxylic acids and their derivatives, we enter the world of *difunctional* molecules, which are molecules with two functional groups. The key point to remember is that the functional groups for the most part still retain their individual chemical reactivity even if they are in a molecule with another functional group. Also remember that the same features we enumerated for all other polymers still apply here: for polymers, function is determined by composition, structure, and size.

Look at the esterification reaction in Figure 13.41(a). The −COOH group of the acid reacts with the −OH group of the alcohol to form a carbon–oxygen single bond and release a molecule of water. Think about what could happen at the molecular level if we had a single compound that contained a carboxylic acid functional group at one end and an alcohol functional group at the other (Figure 13.43). The carboxylic acid group of one molecule could react with the alcohol group of another molecule in a condensation reaction to generate a molecule that has a carboxylic acid group at one end, an alcohol at the other end, and an ester linkage in between. If this reaction happens repeatedly, a monomer containing one carboxylic acid and one hydroxyl group (a hydroxy acid) polymerizes, as

▶❙❙ **CHEMTOUR** Polymers

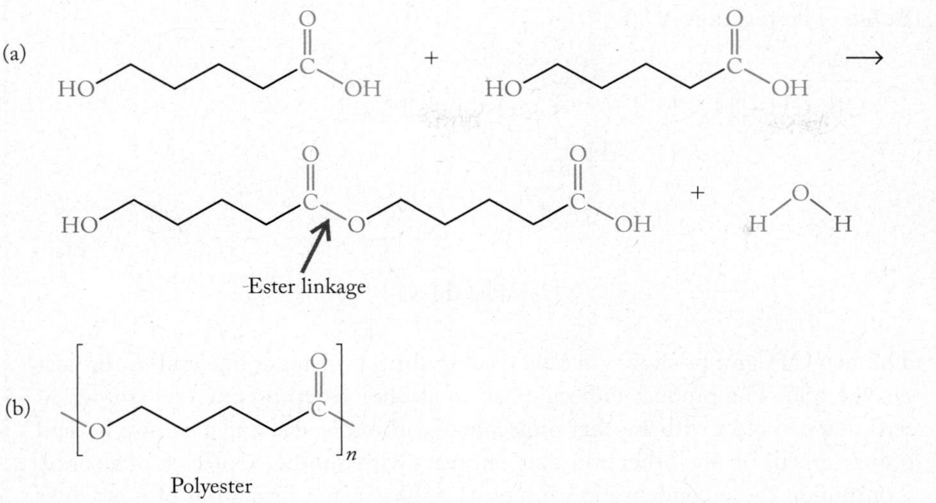

(a)

(b)

FIGURE 13.43 (a) Synthesis of an ester from a condensation reaction between two identical difunctional molecules, each one containing an alcohol group and a carboxylic acid group. The diester can then react with additional difunctional molecules at its –OH and –COOH ends, forming a triester, and the reaction repeats over and over, forming (b) the polyester made up of the repeating monomer unit shown.

Ester linkage

Polyester

shown in Figure 13.43, to form a polyester, a **condensation polymer**. We have already encountered one condensation polymer, PETE, in Section 13.6. In general, condensation polymers are formed by the reaction of monomers yielding a polymer and water as a by-product of the reaction. In addition to its use in plastic soda bottles, PETE is used extensively in medicine. Artificial heart valves and grafts for arteries are made from PETE.

A copolymer formed by reaction of glycolic acid and lactic acid (Figure 13.44) is used to support the growth of skin cells for burn victims. The polymer in Figure 13.44 is also used in making dissolving sutures. Esterification reactions used to make polyesters can be reversed by the addition of water, breaking their ester linkages and forming alcohol and acid functional groups. We examine this process in greater detail in Chapter 20.

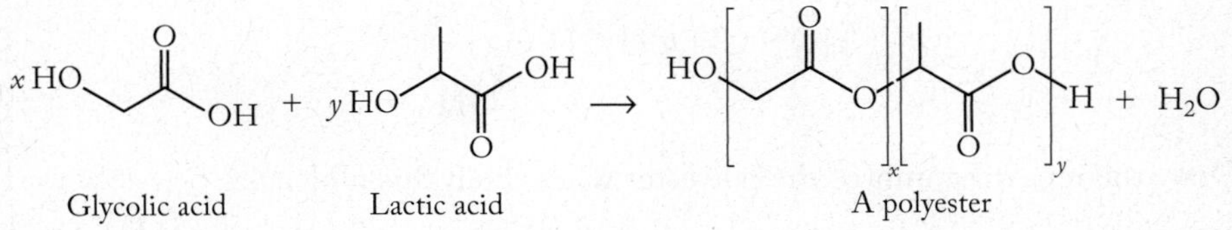

Glycolic acid   Lactic acid   A polyester

(a)

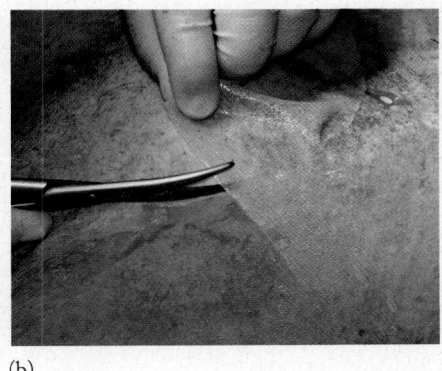

(b)

FIGURE 13.44 (a) The condensation polymer prepared from glycolic acid and lactic acid is used to make sutures that dissolve and as artificial skin that protects against infection while promoting the regrowth of skin cells. (b) Synthetic skin being applied to a burn patient.

SAMPLE EXERCISE 13.9   Making a Polyester

Show how a polyester can be synthesized from the difunctional alcohol $HO(CH_2)_3OH$ and the difunctional carboxylic acid $HOOC(CH_2)_3COOH$.

Collect and Organize An ester is the product of a reaction between a carboxylic acid and an alcohol. A polyester is a polymer with a repeating unit containing an ester functional group. We are given an alcohol and a carboxylic acid to react to make the ester repeating monomer unit.

Analyze Because the starting materials are difunctional, the alcohol can react with two molecules of carboxylic acid, and the acid can react with two molecules of alcohol.

condensation reaction two molecules combining to form a larger molecule and a small molecule (typically water).

condensation polymer macromolecule formed by the reaction of monomers yielding a polymer and water or other small molecule as a by-product of the reaction.

Solve The reaction is

$$HOCH_2CH_2CH_2OH \; + \; \underset{HO}{\overset{O}{\underset{\|}{C}}}CH_2CH_2CH_2\underset{OH}{\overset{O}{\underset{\|}{C}}} \; \rightarrow$$

$$\underset{HOCH_2CH_2CH_2O}{\overset{O}{\underset{\|}{C}}}CH_2CH_2CH_2\underset{OH}{\overset{O}{\underset{\|}{C}}} \; + \; H_2O$$

The two OH groups shown in blue react to form an ester at one end of the carboxylic acid. The product molecule has an alcohol group on one end (shown in red) that can react with another molecule of carboxylic acid and a carboxylic acid group (green) on the other end that can react with another molecule of alcohol. Continuing these condensation reactions results in the formation of a polymer whose repeating unit is

$$\left[-CH_2CH_2CH_2O \underset{}{\overset{O}{\underset{\|}{C}}}CH_2CH_2CH_2\overset{O}{\underset{\|}{C}}O- \right]_n$$

Think about It The repeating unit in the polyester contains one section that came from the alcohol and a second section that came from the carboxylic acid because esters are formed from alcohols and acids.

Practice Exercise A difunctional molecule may contain two different functional groups, such as this one with an alcohol group and a carboxylic acid group:

$$HO{-}CH_2CH_2CH_2\underset{OH}{\overset{O}{\underset{\|}{C}}}$$

Draw the repeating unit of the polyester made from this molecule.

(a)

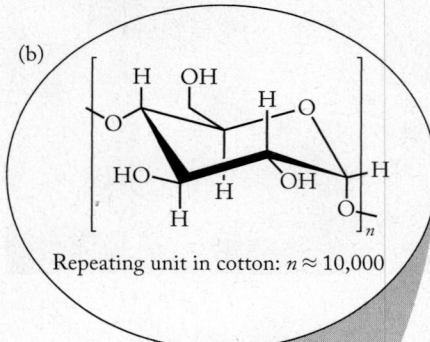

Repeating unit in Dacron

(b)

Repeating unit in cotton: $n \approx 10{,}000$

FIGURE 13.45 Based on the molecular structures of (a) Dacron and (b) cotton, why is cotton the better material for making tee shirts worn during strenuous exercise?

SAMPLE EXERCISE 13.10 **Comparing Properties of Polymers**

Clothes made from the polyester fabric known as Dacron can be less comfortable in hot weather than clothes made of cotton (also a polymer) because Dacron does not absorb perspiration as effectively as cotton. Based on the repeating units of these two polymers (Figure 13.45), suggest a structural reason why cotton absorbs perspiration (water) better than Dacron.

Collect and Organize Cotton and Dacron are polymers that differ in the functional groups in their repeating units. We need to identify the different functional groups and the interactions between these functional groups and water, a polar molecule.

Analyze The absorption of water by a polymer depends on the intermolecular forces present. Water is a polar molecule and is attracted to polar groups. We need to compare the groups in each monomer to see which has the greatest number of polar functional groups or atoms that can form hydrogen bonds. This will be the polymer more likely to absorb water.

Solve Each monomer unit in cotton has three −OH groups attached to it, all polar and capable of hydrogen-bond formation, which means they are likely to interact with the water in perspiration and thereby draw it away from the body. The monomer unit in Dacron has oxygen atoms in it, but no −OH groups. Although regions in the Dacron monomer are polar, they are not nearly as polar as the −OH groups in the cotton monomer unit.

Think about It The principle of like interacting with like works for polymers just as it does for small molecules.

Practice Exercise Gloves made of a woven blend of cotton and polyester fibers protect the hands from exposure to oil and grease and are comfortable to wear because they "breathe"—they allow perspiration to evaporate and pass through them, thereby cooling the skin. Suggest how these gloves work at the molecular level. ∎

Combining difunctional molecules in a condensation reaction can be used to make *polyamides*, another class of very useful synthetic polymers. The functional groups are a carboxylic acid and an amine. The difunctional monomer units can be identical (Figure 13.46a), each containing one carboxylic acid group and one amine group, or they can be different (Figure 13.46b), with one monomer containing two carboxylic acid groups (a dicarboxylic acid) and the other containing two amine groups (a diamine).

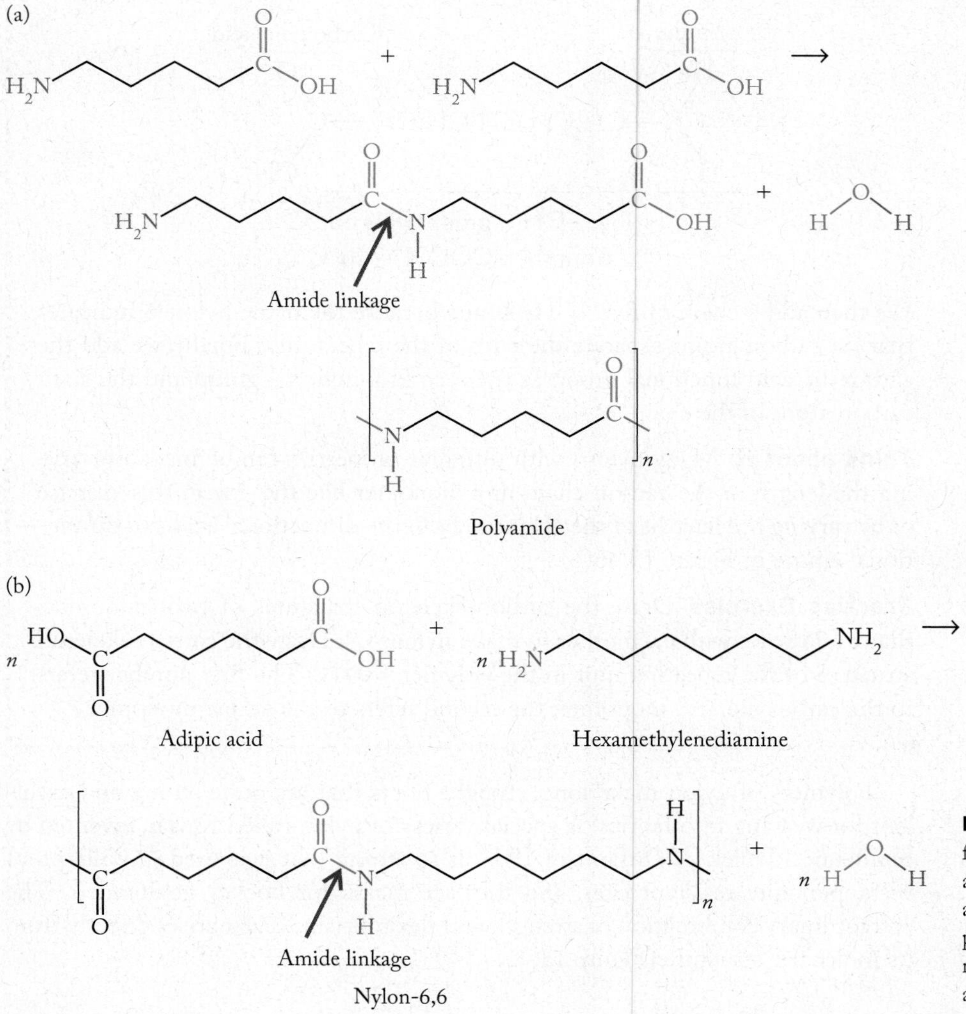

(a)

Amide linkage

Polyamide

(b)

Adipic acid

Hexamethylenediamine

Amide linkage

Nylon-6,6

FIGURE 13.46 (a) Synthesis of a polyamide from two identical monomer, each containing a carboxylic acid functional group and an amine functional group. (b) Synthesis of the polyamide nylon-6,6 from nonidentical monomers: adipic acid (a dicarboxylic acid) and hexamethylenediamine (a diamine).

Probably the most familiar polyamide is nylon-6,6, made from the monomers shown in Figure 13.46(b). Each monomer contains six carbon atoms, which is what the digits in the name represent.

SAMPLE EXERCISE 13.11   **Identifying Monomers**

Another form of nylon is nylon-6, with the single 6 indicating that the polymer is made from the reaction of a series of identical six-carbon monomers. By analogy with the polyester in Sample Exercise 13.9 and the accompanying Practice Exercise, draw the condensed structure of a six-carbon molecule that could polymerize to make nylon-6.

Collect and Organize Amides form from carboxylic acids and amines. We are asked to draw the condensed structure for a monomer that contains both functional groups and that could react with identical monomers to form a polyamide with a repeating unit six carbon atoms long. The exercise refers us to an example with polyesters to use as an analogy.

Analyze By analogy to Sample Exercise 13.9 and its accompanying Practice Exercise, we should suggest a difunctional molecule that has a carboxylic acid on one end and an amine on the other because these are the two functional groups that react to form the amide linkage in a polyamide.

Solve We can build the required monomer by starting with one of the functional groups. It doesn't matter which one, so let's begin with the amine:

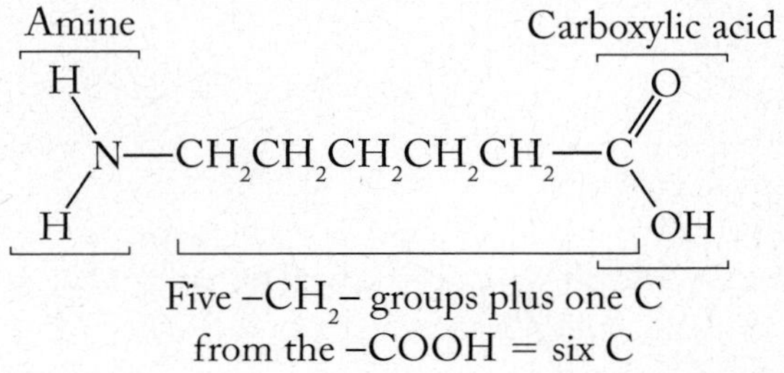

Five $-CH_2-$ groups plus one C
from the $-COOH$ = six C

We then add a chain of five $-CH_2-$ units because the name nylon-6 indicates that six carbon atoms separate the ends of the repeat unit. Finally, we add the carboxylic acid functional group as the second functional group and the sixth carbon atom in the chain.

Think about It Many nylons with different properties can be made by varying the length of the carbon chain in a monomer like the one in this exercise or by varying the lengths of the chains in both the difunctional acid and difunctional amine in Figure 13.46(b).

Practice Exercise Draw the carbon-skeleton structures of two monomers that could react with each other to make nylon-5,4. Draw the carbon-skeleton structure of the repeating unit in the polymer. (NOTE: The first number refers to the carboxylic acid monomer; the second refers to the amine monomer.)

Benzene-1,4-dicarboxylic acid
(terephthalic acid)

1,4-Diaminobenzene

Repeating unit in Kevlar

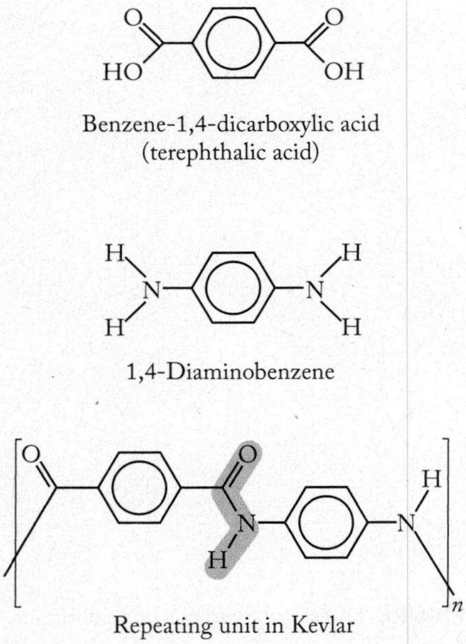

FIGURE 13.47 The monomers used to make Kevlar are a dicarboxylic acid and a diamine. The amide bond in the repeating unit is highlighted.

Polymers of nylon make long, straight fibers that are quite strong and excellent for weaving into fabrics. A special variety of nylon called Kevlar, invented by Stephanie Kwolek of DuPont in 1965, is so strong that it is used in bulletproof vests, puncture-resistant tires, and the face masks for hockey goaltenders. The extraordinary combination of strength and flexibility in Kevlar arises directly from its molecular structure (Figure 13.47).

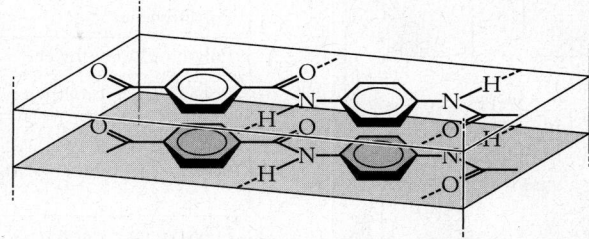

FIGURE 13.48 Interactions between groups in Kevlar.

Hydrogen bonding between chains in the same plane

Stacking of benzene rings between chains in layered planes

Nylon is flexible and stretchable because the hydrocarbon chains can bend and curl, much like a telephone cord or a Slinky spring toy. To produce a stronger nylon, researchers recognized they had to find some way to reduce the ability of the chains to form coils. They discovered this could be done using monomer units containing functional groups that made it difficult for the chains to bend. The result of this work was Kevlar, a polyamide formed from a dicarboxylic acid of benzene and a diamine of benzene. When these two monomers polymerize, the flat, rigid aromatic rings keep the chains straight.

Two additional intermolecular interactions orient the chains and hold them together very tightly (Figure 13.48). First, the –NH hydrogen atoms form hydrogen bonds with the oxygen atoms of carbonyl groups on adjacent chains. Second, the rings stack on top of one another (just as in polystyrene) and provide additional interactions that hold the chains together in parallel arrays. The result is a fiber that is very strong but still flexible. Fabrics and helmets made of Kevlar resist puncture, even by bullets fired at them, and they are also resistant to flames and reactive chemicals (Figure 13.49).

FIGURE 13.49 A bullet fired point-blank at a sheet of Kevlar does not puncture the fabric.

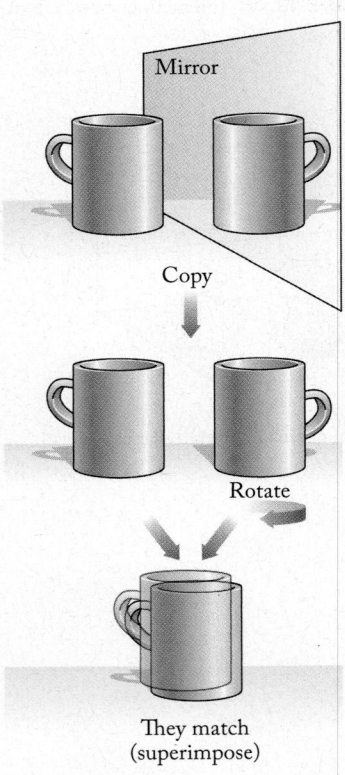

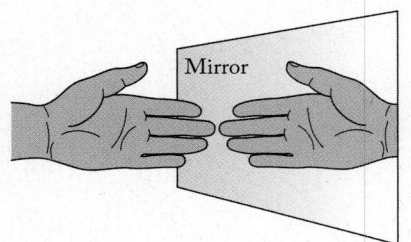

They do not match (they do not superimpose)

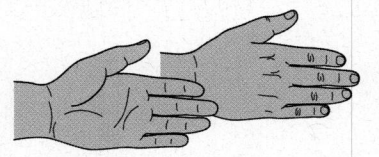

FIGURE 13.50 A plain coffee mug is superimposable on its mirror image and is achiral. The mirror image of your left hand is your right hand, and the two are not superimposable.

---

optical isomers molecules that are not superimposable on their mirror images.

chiral compounds having nonsuperimposable mirror images.

achiral not chiral; describes compounds that can be superimposed on their mirror images.

TABLE 13.8	Summary of Common Polymers and Their Uses		
Name	**Abbreviation**	**Functional Group**	**Use**
Addition Polymers			
Polyethylene	PE	Alkane	Plastic bags and films
Polytetrafluoroethylene	Teflon	Fluoroalkane	Nonstick coatings
Poly(1,1-dichloroethylene)	Saran	Chloroalkane	Plastic wrap
Poly(vinyl chloride)	PVC	Chloroalkane	Drain pipes
Poly(methyl methacrylate)	PMMA	Alkane and ester	Shatter-resistant glass, e.g., Plexiglas, Lucite
Polystyrene	PS	Aromatic hydrocarbon	Cups, dishes, insulation
Poly(vinyl alcohol)	PVAL	Alcohol	Gloves, bottles
Condensation Polymers			
Poly(ethylene glycol)	PEG	Ether	Pharmaceuticals, consumer products
Poly(ethylene oxide)	PEO	Ether	As for PEG
Poly(ethylene terephthalate)	PETE	Ester	Plastic bottles
Nylon		Amide	Clothing
Kevlar		Amide	Protective equipment
Dacron		Ester	Clothing

In this chapter we have seen many polymers and mentioned their uses in common products. The classification of these polymeric materials as addition or condensation polymers is summarized in Table 13.8. Table 13.8 also identifies the polymers by their distinctive functional group and mentions some common uses in our lives.

13.8 Chirality

In this chapter we have encountered two types of isomerism: structural isomers and stereoisomers. Cis and trans isomers of alkenes are stereoisomers: molecules having the same formula and the same bonds but differing in the orientation in space of groups within the molecules. Another type of stereoisomerism called *optical isomerism* is especially important in organic molecules involved in biological processes.

Optical Isomerism

Optical isomers are molecules that are not superimposable on their mirror images, just as your left hand is not superimposable on your right hand. As we explored in Section 9.6, such molecules are **chiral**, and their optical isomerism comes from their molecular structures containing *chiral centers*—most commonly a carbon atom that has four different groups attached to it. Figure 13.50 illustrates what is meant by mirror images and superimposition. The reflection of an object like a plain coffee mug that is **achiral** (not chiral) in a mirror is an image that can be superimposed on the original object. Your two hands, however, are mirror images that cannot be superimposed. Molecules that have this same property—molecules that cannot be superimposed on their mirror images—exist as optical isomers and so are chiral. Please revisit Section 9.6 and the discussion surrounding Figure 9.39 for a simple molecular example.

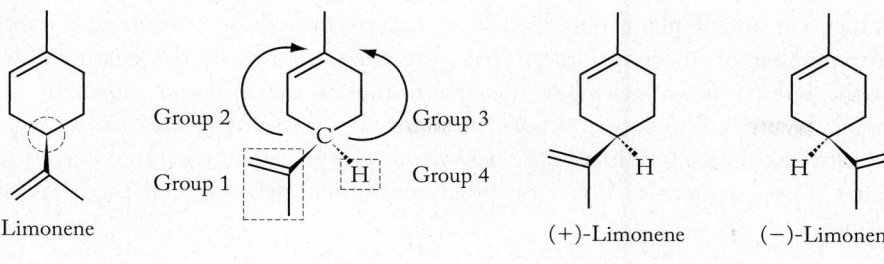

FIGURE 13.51 Limonene is a chiral molecule with two enantiomeric forms. To show why it is chiral, we have highlighted its chiral carbon with a dashed circle and numbered the four groups bonded to it. Groups 1 and 4 are clearly different. Groups 2 and 3 are two halves of the same ring. They are different because group 2 has a C=C bond and group 3 does not. Therefore, the circled carbon atom is bonded to four different groups and is chiral.

∞ CONNECTION Chirality and optical isomerism were introduced in Section 9.6.

Let's now look at a slightly more complicated molecule, limonene (Figure 13.51), an alkene found in oranges and turpentine. Its chiral center is identified by the dotted circle. This chiral carbon atom is part of a six-membered ring, and if we examine the structure of limonene closely, we see that this carbon atom has four different groups bound to it. The three-carbon alkene group and the hydrogen atom are easy to see, but if we follow the carbon atoms around the ring, we find a C=C bond three bonds from the circled carbon on the left side, but only C—C single bonds on the right side. The presence of the C=C bond makes the groups attached to the circled carbon different, making limonene chiral. Therefore optical isomers exist. The two optical isomers are called **enantiomers**, nonsuperimposable mirror images. In no orientation do all of the groups on the chiral center of superimposed enantiomers coincide, because the molecules have different shapes in three dimensions.

Chirality is especially important in living systems, where molecules interact with other molecules in a process called **recognition**. Recognition in biochemistry means identifying a molecule based on its interaction with another molecule because of its shape. Just as your right hand fits into the correct glove, a molecule may fit into a three-dimensional cavity in a protein and cause an event based on that recognition. The protein that forms such a cavity is called a **receptor**.

▶‖ CHEMTOUR Chirality

Structural isomers and stereoisomers differ in their physical and chemical properties, but optical isomers have the same physical and chemical properties except for those that relate to a few specialized types of behavior. Recognition is one of those special behaviors. Limonene has two enantiomers, distinguished by the (+) and (−) signs in front of their names. One enantiomer smells like oranges; the other smells like turpentine. Part of the process of sensing different aromas involves recognition by receptors in your nasal passages. One receptor is shaped like (+)-limonene, the other like (−)-limonene, and the enantiomers bind to their own receptor just like your right and left hands fit into right and left gloves.

Enantiomers are also called **optically active molecules**. The (+) and (−) signs that distinguish the two isomers of limonene refer to the specific effect each isomer has on polarized light. In plane-polarized light, the electric fields that compose the beam oscillate in only one plane (Figure 13.52).

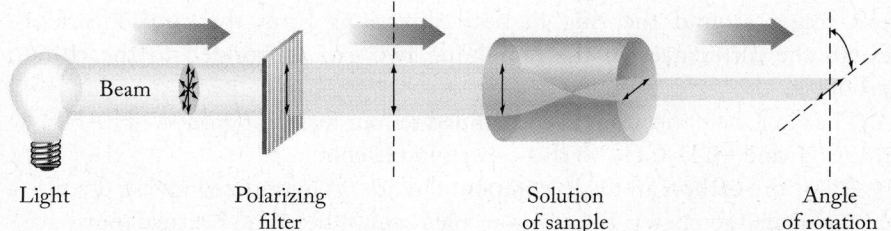

FIGURE 13.52 A beam of plane-polarized light consists of electric field vectors that oscillate in only one direction. The plane of oscillation rotates if the beam passes through a solution of one enantiomer of an optically active compound. The (+) enantiomer causes the beam to rotate clockwise; the (−) enantiomer causes the beam to rotate counterclockwise.

enantiomer one of a pair of optical isomers of a compound.

recognition process by which molecules in living systems interact with one another.

receptor a cavity in a protein molecule that fits a particular molecule.

optically active molecule a chiral compound that causes rotation of a beam of plane-polarized light when it passes through a solution.

When a beam of plane-polarized light passes through a solution containing one member of an enantiomeric pair, the beam rotates. If the beam rotates to the left (counterclockwise), the enantiomer is the *levorotary* form of the molecule and a (−) sign precedes its name. If the beam rotates to the right (clockwise), the enantiomer is the *dextrorotary* form and a (+) sign precedes its name. These molecules' effect on polarized light is why we call them *optical isomers*.

▶II **CHEMTOUR** Chiral Centers

SAMPLE EXERCISE 13.12   Recognizing Chiral Molecules

Identify which of the molecules shown are chiral, and circle the chiral centers. Structures may have more than one chiral center.

(a)

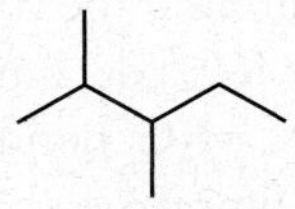

(d)

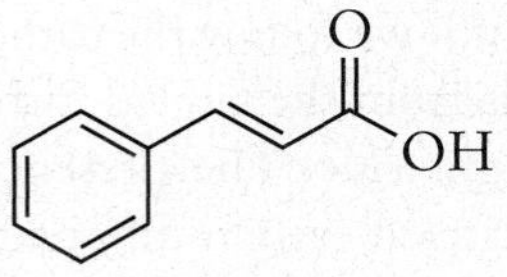

(b)

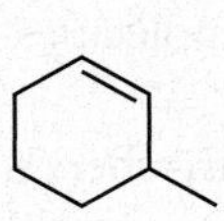

(e)

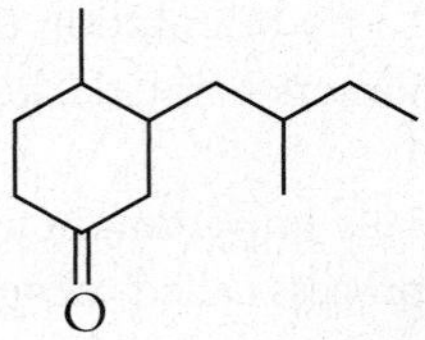

(c)

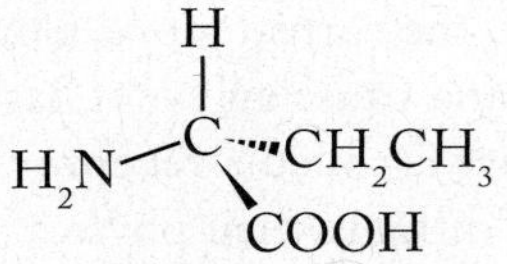

H
|
H_2N—C—CH_2CH_3
|
COOH

Collect and Organize Chiral molecules are nonsuperimposable on their mirror images. A chiral molecule has one or more chiral centers.

Analyze If molecules have carbon atoms with four different groups attached, they are chiral.

Solve The chiral carbons in each structure are circled on the next page.

The C3 carbon of the pentane chain in compound a is bonded to a –CH_3 group, a –C_2H_5 group, a –C_3H_7 group, and an H atom, so the compound, 2,3-dimethylpentane, is chiral.

The chiral center in compound b is similar to the chiral center in limonene. The chiral carbon is bonded to a methyl group and an H. Tracing your finger around the ring in both directions from the circled carbon reveals the differences in the remaining two groups bonded to the chiral carbon.

The circled carbon atom in c is bonded to four different groups: –H, –NH_2, –COOH, and –CH_2CH_3, so this compound is chiral.

All of the carbon atoms in compound d are sp^2 hybridized giving d a planar molecular geometry. Planar molecules cannot be chiral because they have a superimposable mirror image. Think of a mirror plane that contains all nine carbons in compound d: the mirror image is exactly the same.

Compound e has three chiral centers. Working from right to left, the first chiral center is similar to the chiral center in compound a. The other two chiral centers are in the cyclohexane ring. In each case, there are four different groups around the carbon atom, so compound e is chiral.

(a)

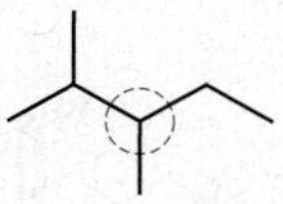

(d)

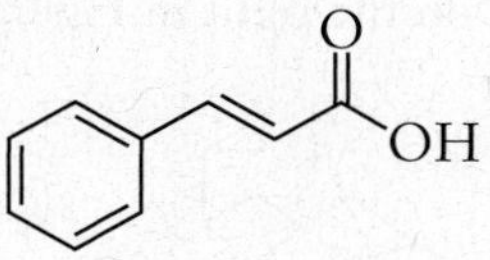

(b)

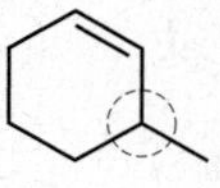

(e)

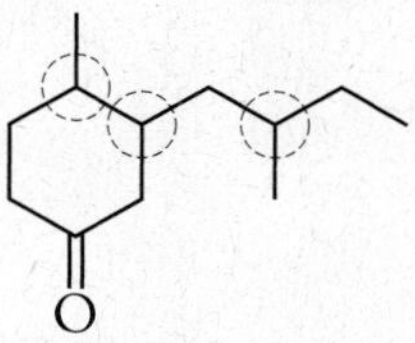

(c)

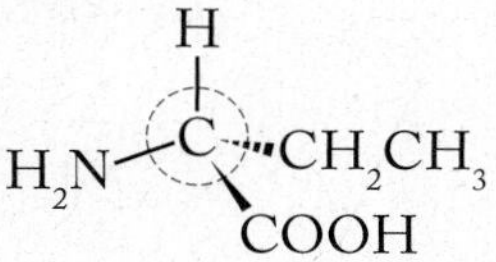

Think about It The presence of a chiral center in a molecule means that the molecule has two enantiomeric forms. The molecule is not superimposable on its mirror image. We should also note that although structure d has no chiral center, it would have cis–trans isomers about the double bond in the chain. The isomer shown has the trans configuration.

Practice Exercise Identify which of the molecules below are chiral. Circle the chiral centers in each structure.

(a)

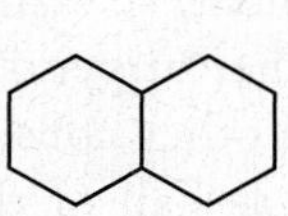

(d)

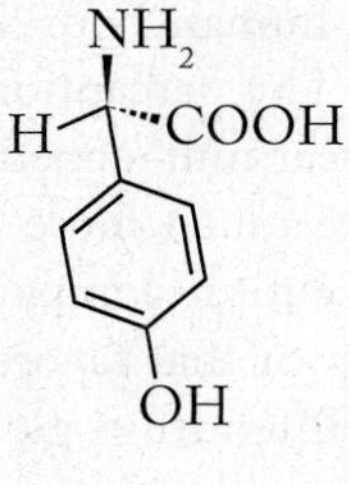

(b)

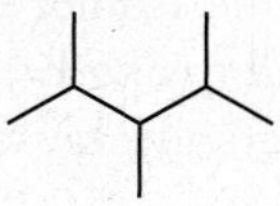

(e)

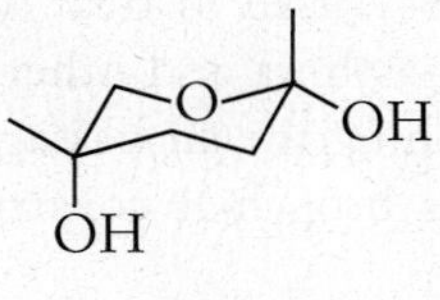

(c)

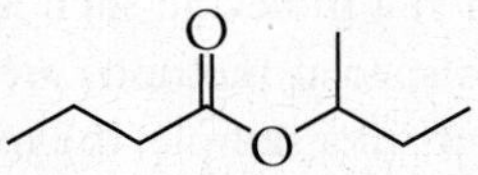

Quinine is a bitter-tasting substance extracted from the bark of the cinchona tree. It was the first effective substance used to treat malaria. Its enantiomer quinidine is a synthetic compound that suppresses cardiac arrhythmias (fast-beating heart). Describe the recognition between these molecules and their receptors in terms of the illustration on the right in Figure 13.53.

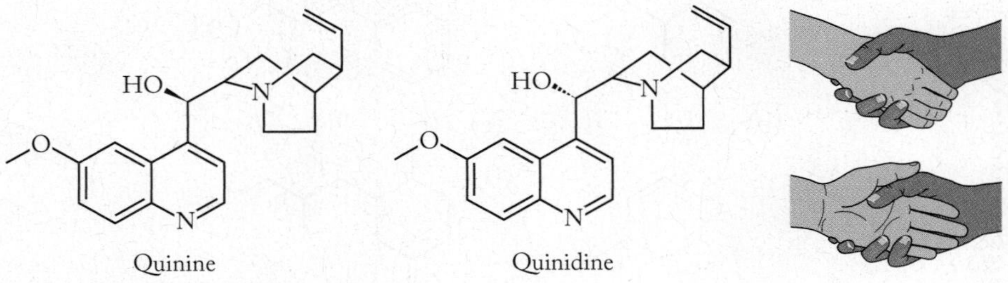

Quinine   Quinidine

FIGURE 13.53

Chirality in Nature

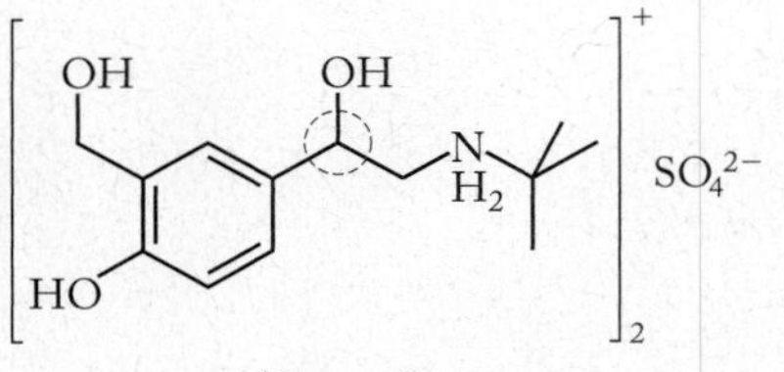

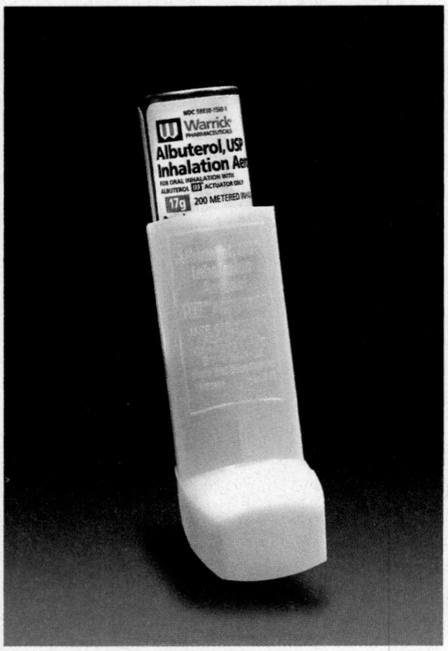

Chirality is ubiquitous in the organic compounds formed by living systems, and in most cases only one optical isomer occurs naturally in a particular organism. A molecule might even have more than one chiral center.

The origin of the fundamental preference of life for one enantiomer over another is unknown. Ongoing studies on the origin of chiral preference in living systems have produced no definitive answers but are providing increasingly interesting suggestions. Perhaps the origin of the preference involves the effect of slightly polarized sunlight.

Whatever the origin of these preferences, processes requiring molecular recognition in living systems often depend on the selectivity conveyed by chirality. The human body is a chiral environment, so handedness of molecules matters. The perception of an aroma as either turpentine or orange or as caraway or spearmint depends on chirality, but other consequences of chiral recognition are much more dramatic. As many as half of the drugs made by large pharmaceutical companies are chiral and owe their function to recognition by a receptor that favors one enantiomer over the other. In 2008, eight of the ten top-selling drugs globally were chiral. It is typical that only one of the enantiomers of a chiral drug is active. A classic example is the drug albuterol (Figure 13.54), used to treat wheezing and shortness of breath in people suffering from asthma and other lung disorders. One isomer causes bronchodilation (widening the air passages of the lungs and easing breathing), while the other causes bronchial constriction and may actually be detrimental to the patient.

When chiral compounds are produced by living systems, typically only one stereoisomer is produced. For example, the aroma of spearmint leaves is due to (+)-carvone (Figure 13.38). However, when a molecule such as albuterol is made in a laboratory, both isomers result unless special methods are used. When both optical isomers are present in equal amounts in a sample, the material is known as a **racemic mixture**. Because one isomer rotates the plane of polarized light in one

Albuterol

FIGURE 13.54 The antiasthma drug albuterol; the chiral center is circled.

direction, and the other to the same extent in the opposite direction, a racemic mixture does not rotate the plane of polarized light at all. Because the two isomers interact differently with receptors, the pharmaceutical industry routinely faces two choices: devise a special synthetic procedure that yields only the isomer of interest, or separate the two isomers at the end of the manufacturing process. Both approaches are widely used.

racemic mixture a sample containing equal amounts of both optical isomers of a compound.

SAMPLE EXERCISE 13.13 **Recognizing Optical Properties of Chiral Molecules**

The structure of the antidepressant drug bupropion is

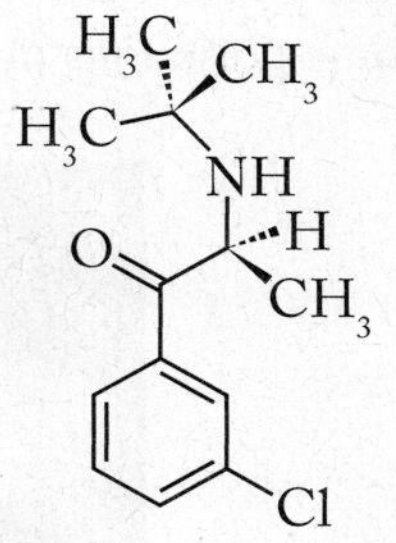

Bupropion

When tested, a solution of a sample of the drug that comes directly from the laboratory does not rotate the plane of polarized light, so the analyst rejects the sample, saying it is not the pure drug, which is known to cause a beam of polarized light to rotate when it passes through a solution of the compound. The chemist who made the sample tells you that other analytical data (percent composition and mass spectrometry) prove the material is 100% bupropion. Both people are correct in what they say. Explain why.

Collect and Organize We are asked to explain how a sample that is 100% chemically pure is not 100% pure active drug. We are given the structure of the compound, and the information that a solution of the material does not rotate the plane of polarized light.

Analyze The sample was rejected by one analyst because a solution did not rotate the plane of polarized light. We must examine the structure to see if the molecule has a chiral center, that is, a carbon atom bonded to four different groups.

Solve Bupropion is a chiral molecule. The chiral center is circled in this structure:

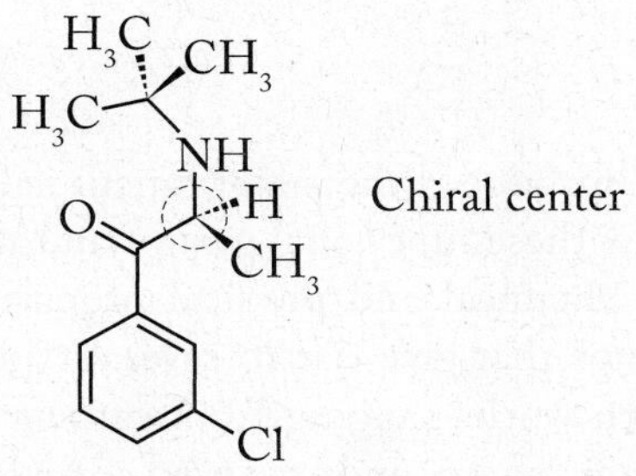

Chiral center

Bupropion

The sample must contain both enantiomers of bupropion in equal amounts. They have exactly the same chemical formula, so the sample is chemically pure, but one enantiomer rotates polarized light to the left and the other to the right. No net rotation is observed when polarized light passes through the sample. Only one enantiomer is the active drug. Therefore, the sample is chemically pure but not optically pure.

Think about It Enantiomers have the same chemical composition and the same molar mass. Their physical properties are identical except for the direction in which their solutions cause plane-polarized light to rotate when it passes through them.

Practice Exercise Identify the chiral carbon atoms in the cationic natural product muscarine, found in some poisonous mushrooms (Figure 13.55).

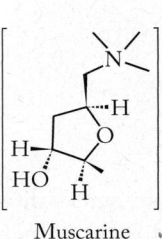

Muscarine

FIGURE 13.55 Several varieties of mushrooms, including *Amanita muscaria*, contain the toxic substance muscarine.

CONCEPT TEST

The Documents in the Case, a mystery written by Dorothy L. Sayers in 1930, involves the suspicious death of an authority on wild, edible mushrooms. Allegedly, the victim ate a stew made of poisonous mushrooms that contained muscarine, a toxic natural product. A forensic specialist evaluated the contents of the victim's stomach and observed that the fluid contained muscarine but did not rotate the plane of polarized light. Because of this fact, the coroner concluded that the man was murdered. Why did the coroner in the story reach this conclusion?

In this chapter, we introduced the major functional groups in organic chemistry and discussed how these functional groups and the structure of the molecules determine their chemical and physical properties. We examined a few reactions of those groups that give rise to several types of polymeric materials common in the modern world, ranging from carpets and insulation to bullet-proof vests. Other organic compounds are used as fuels, as pharmaceuticals, and as building materials. Despite the range of materials we have discussed, this treatment can give only a brief taste of the importance of organic compounds.

Chemical Abstracts Service (CAS), an organization that tracks and collects all chemical information published worldwide, announced in September 2009 that it had recorded the 50 millionth unique chemical substance (a new drug with pain-relieving properties) in its registry. CAS registered the 40 millionth substance just nine months earlier. In contrast, it took 33 years for CAS to register the 10 millionth compound in 1990. Typically, more than 95% of the new compounds registered in a given year contain carbon, which gives you some idea of why the study of organic chemistry occupies a special place within the discipline.[1]

SUMMARY

Section 13.1 Organic chemistry encompasses the study of all carbon compounds, classified on the basis of **functional groups**—subunits of structure that confer on molecules specific and characteristic chemical and physical properties. Organic compounds are also differentiated based on size. **Polymers**, or macromolecules, have molar masses from several thousand to over 1,000,000 g/mol and are composed of repeating **monomer** units.

Section 13.2 The carbon atoms in **alkanes** (or **saturated hydrocarbons**) bear the maximum number of hydrogen atoms possible and contain carbon–carbon single bonds. A **homologous series** of alkanes is generated by sequential addition of $-CH_2-$ units (**methylene groups**) into the chain, which has a $-CH_3$ (**methyl group**) at each end. **Straight-chain alkanes** may have **structural isomers**, hydrocarbons with the same molecular formula but different arrangements of C—C bonds (**branched-chain hydrocarbons**) and physical properties. **Cycloalkanes** are alkanes containing rings of carbon atoms.

Section 13.3 Alkenes and alkynes are unsaturated hydrocarbons because the double bonds in alkenes and triple bonds in alkynes can be hydrogenated to incorporate more hydrogen into their molecular structures. In addition to having structural isomers, alkenes also have **stereoisomers**: *E*, or **trans, isomers** and *Z*, or **cis, isomers**, depending on the arrangement of the groups around the double bond. Alkenes undergo **addition reactions** with hydrogen and hydrogen halides. Alkenes can also be polymerized to **homopolymers** used in construction, in fabrics, as wrapping and packaging material, and in medical devices. Most of these polymers are **addition polymers**.

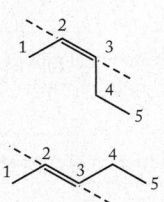

Section 13.4 Aromatic hydrocarbons are characterized by planar rings in which sp^2 hybridized carbon atoms are joined by a combination of σ and π bonds. The π bond electrons are delocalized over all the carbon atoms in the ring. Compounds with two or more aromatic rings are polycyclic aromatic hydrocarbons.

Section 13.5 Amines are organic compounds with the general formula RNH_2, R_2NH, or R_3N, where R is any organic subgroup.

Section 13.6 The **alcohol** (R—OH) and **ether** (R—O—R) functional groups (where R is an alkyl group or an aromatic ring) represent two ways of incorporating the heteroatom oxygen into organic compounds. Different monomer units derived from alcohols or ethers can be chemically combined to make **heteropolymers** or **copolymers** whose properties depend on the arrangement of the monomer units in the polymer.

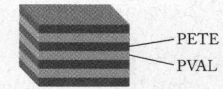

Section 13.7 The **carbonyl group**, C=O, is found in **aldehydes** (RCHO), **ketones** (RCOR), **carboxylic acids** (RCOOH), **esters** (RCOOR), and **amides** (RCONH$_2$). The chemical reactivity of aldehydes and ketones centers on the C=O bond. In carboxylic acids, the COOH group imparts acidic properties to the molecules. Via **condensation reactions**, carboxylic acids react with alcohols to form esters and with ammonia or amines to form amides. Condensation reactions are used to prepare **condensation polymers** such as polyesters and polyamides from difunctional compounds for use in fabrics under the familiar names Dacron and nylon.

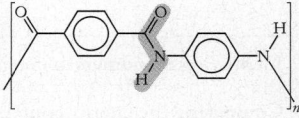

Section 13.8 Many biologically important molecules exhibit chirality, or optical isomerism, a type of stereoisomerism. **Optical isomers** (**enantiomers** or **optically active molecules**) are nonsuperimposable on their mirror images and result from the presence of a **chiral** atom in their structures. Molecules lacking a chiral atom are **achiral**. Chirality is important in biological systems because **recognition** of molecules by **receptors** is often based on shape. Chiral molecules can rotate a beam of plane-polarized light. If two optical isomers of a compound are present in equal ratios in a sample, the material is known as a **racemic mixture** and there is no rotation of polarized light.

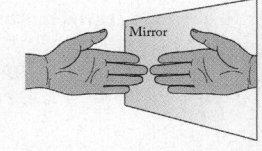

[1] William G. Schulz, "CAS Registers 50 Millionth Compound," *Chemical and Engineering News*, http://pubs.acs.org/cen/email/html/8737news2.html (accessed February 3, 2010).

PROBLEM-SOLVING SUMMARY

TYPE OF PROBLEM	CONCEPTS AND EQUATIONS	SAMPLE EXERCISES
Distinguishing among alkanes, alkenes, and alkynes	Alkanes contain only C—C single bonds; they are saturated hydrocarbons and do not react with H_2. Alkenes contain at least one C=C double bond that can combine with a molecule of H_2. Alkynes contain at least one C≡C triple bond that can combine with two molecules of H_2.	13.1
Drawing alkane structures	Use the Lewis structure to create a condensed structure, then gather all methylene groups inside parentheses to create the final condensed structure. To create a carbon-skeleton structure, use lines to depict single covalent bonds between carbon atoms. A sufficient number of H atoms to complete the valency of the carbon atoms is assumed.	13.2
Identifying the longest chain in organic molecules	Start at one end of any branch and assign numbers to each carbon atom. If the chain branches, choose one branch to follow. Repeat the process to explore other side chains and identify the longest chain.	13.3
Recognizing structural isomers	Establish that the compounds have the same molecular formula, and if they do, look for different arrangements of C—C bonds.	13.4
Identifying and naming stereoisomers and structural isomers	After establishing that the compounds have the same molecular formula, look for different arrangements of C—C bonds. Molecules with two like groups on the same side of a C=C bond are cis; those with two like groups on opposite sides are trans isomers.	13.5
Identifying monomers in polymers	Find the smallest portion of the polymer that is repeated.	13.6, 13.11
Comparing the energy content of fuel mixtures	Determine the number of moles of each component, then multiply that number by the corresponding value of $\Delta H°_{comb}$. The total amount of energy available equals the sum of the heats available from each component.	13.7
Assessing properties of polymers	Evaluate the polarity of the functional groups in the polymer, and assess the relative importance of all types of intermolecular forces possible in the molecules.	13.8, 13.10
Making a polyester	Combine monomers with alcohol functional groups (ROH) and carboxylic acid functional groups (RCOOH) to form water (H_2O) and ester groups (RCOOR).	13.9
Recognizing chiral molecules	Identify carbon atoms with four different groups attached.	13.12
Recognizing optical properties of chiral molecules	Racemic mixtures contain equal amounts of two enantiomers and do not rotate the plane of polarized light.	13.13

VISUAL PROBLEMS

(Answers to boldface end-of-chapter questions and problems are in the back of the book.)

13.1. How many degrees of unsaturation are in each of the hydrocarbons shown in Figure P13.1?

13.2. Which of the hydrocarbons in Figure P13.2 are structural isomers of each other?

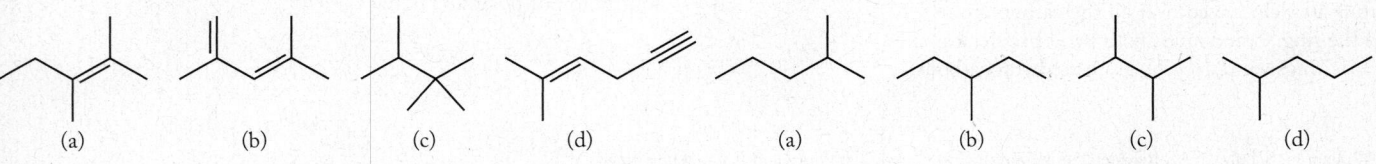

FIGURE P13.1   **FIGURE P13.2**

13.3. In Figure P13.3 are the carbon-skeleton structures of four organic compounds found in nature as fragrant oils. Which are alkenes?

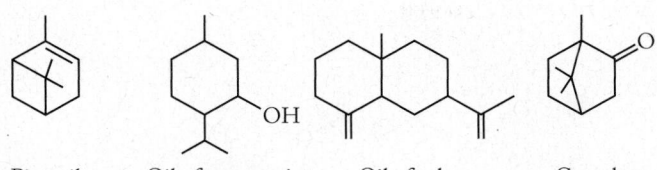

Pine oil Oil of peppermint Oil of celery Camphor

FIGURE P13.3

13.4. Figure P13.4 shows three molecules: acrylonitrile (found in barbeque smoke), capillin (an antifungal drug), and pargyline (an antihypertensive drug). Which of these molecules does not contain the alkyne functional group?

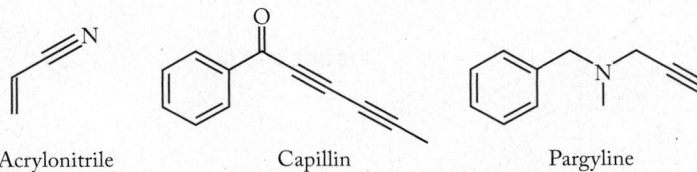

Acrylonitrile Capillin Pargyline

FIGURE P13.4

13.5. Which molecules in Figure P13.5 are considered to be aromatic compounds?

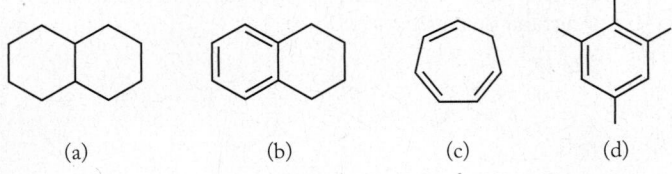

(a) (b) (c) (d)

FIGURE P13.5

13.6. Benzyl acetate, carvone, and cinnamaldehyde are all naturally occurring oils. Their carbon-skeleton structures are shown in Figure P13.6. Which ones contain an aromatic ring?

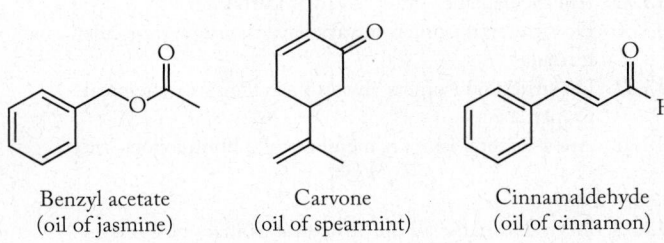

Benzyl acetate Carvone Cinnamaldehyde
(oil of jasmine) (oil of spearmint) (oil of cinnamon)

FIGURE P13.6

13.7. In addition to the aromatic ring, what other functional groups can you identify in the molecules in Problem 13.6?

*13.8. The three polymers shown in Figure P13.8 are widely used in the plastics industry. In which of them are the intermolecular forces per mole of monomer the strongest?

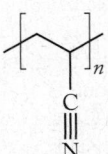

Polyethylene Poly(vinyl chloride) Poly(1,1-dichloroethylene)
(a) (b) (c)

FIGURE P13.8

*13.9. **Silly Putty** Silly Putty is a condensation polymer of dihydroxydimethylsilane (Figure P13.9). Draw the condensed structure of the repeating monomer unit in Silly Putty.

$$HO-\underset{\underset{CH_3}{|}}{\overset{\overset{CH_3}{|}}{Si}}-OH$$

Dihydroxydimethylsilane

FIGURE P13.9

13.10. **Orlon and Acrilon Fibers** Figure P13.10 shows the carbon-skeleton structure of polyacrylonitrile, which is marketed as Orlon and Acrilon. Draw the Lewis structure of the monomeric reactant that produces this polymer.

Polyacrylonitrile

FIGURE P13.10

13.11. Rubber is a polymer of isoprene. It is sometimes called polyisoprene. There are two forms of polyisoprene (Figure P13.11): *cis*-polyisoprene is the soft, flexible material we associate with the term "rubber"; gutta-percha, or *trans*-polyisoprene, is a much harder material. Draw the monomeric units of *cis*- and *trans*-polyisoprene.

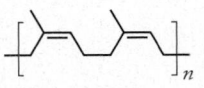 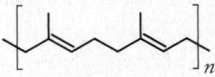

cis-Polyisoprene *trans*-Polyisoprene

FIGURE P13.11

13.12. Drugs and Enantiomeric Purity Thalidomide was marketed in the late 1950s as a drug to relieve morning sickness. Unfortunately, one isomer caused birth defects. Circle the chiral carbon atom(s) in the thalidomide molecule (Figure P13.12).

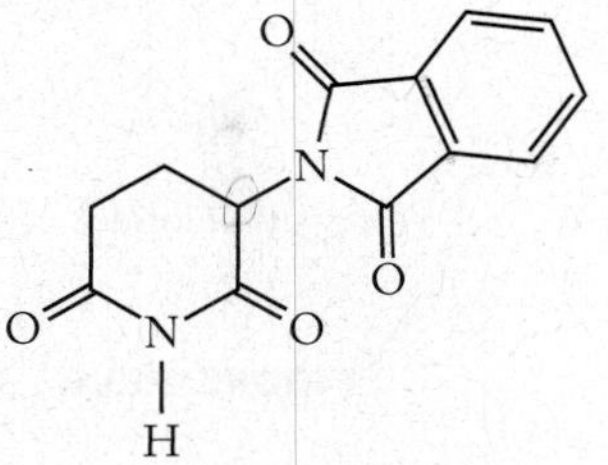

Thalidomide

FIGURE P13.12

13.13. Cholesterol Lowering Drugs High serum cholesterol levels often correlate with increased risk of heart attacks. The drug sold under the trade name Mevacor has proven to be effective in lowering serum cholesterol.

How many chiral carbon atoms are there in Mevacor (Figure P13.13)?

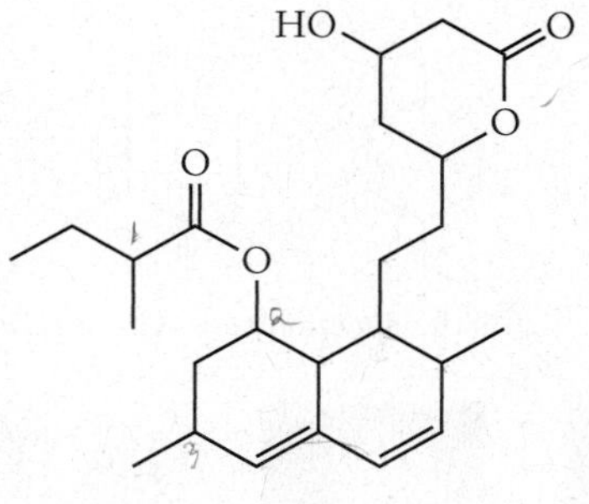

Mevacor

FIGURE P13.13

*13.14. The two compounds shown in Figure P13.14 are both considered to be amino acids. Identify the structural difference between them and explain why they are both amino acids.

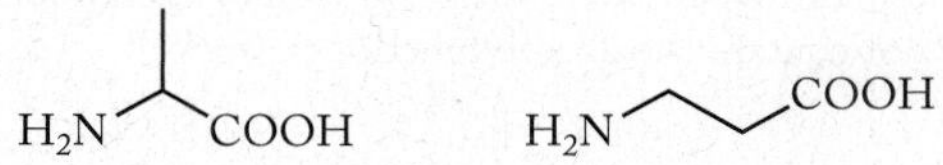

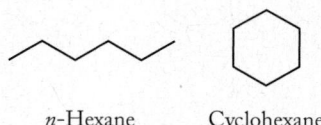

FIGURE P13.14

QUESTIONS AND PROBLEMS

Carbon: The Scope of Organic Chemistry

CONCEPT REVIEW

13.15. Describe three ways in which carbon atoms can form bonds to other carbon atoms using hybrid orbitals (valence bond theory).

13.16. Can a macromolecule be composed of more than one type of monomer?

*13.17. Is the interstitial alloy tungsten carbide (WC) considered to be an organic compound?

*13.18. Calcium carbide, CaC_2 was used in miner's lamps. Reaction of CaC_2 with water yields acetylene which, when ignited, gives light. Is calcium carbide considered an organic compound?

PROBLEMS

13.19. What functional groups were introduced in Chapter 8?

13.20. Find an example of a small molecule with more than one functional group in Chapter 9.

13.21. Polyethylene is prepared from the monomer ethylene, C_2H_4. About how many monomers are needed to make a polymer with a molar mass of 100,000 g/mol?

13.22. Synthetic rubber is prepared from butadiene, C_4H_6. About how many monomers are needed to make a polymer with a molar mass of 100,000 g/mol?

Alkanes

CONCEPT REVIEW

13.23. Do linear and branched alkanes with the same number of carbon atoms all have the same empirical formula?

13.24. If an alkane and a cycloalkane have equal numbers of carbon atoms per molecule, do they have the same number of hydrogen atoms?

13.25. What is the hybridization of carbon in alkanes?

13.26. Figure P13.26 shows the carbon-skeleton structures of n-hexane and cyclohexane. Are n-hexane and cyclohexane structural isomers?

n-Hexane   Cyclohexane

FIGURE P13.26

13.27. Why isn't cyclohexane a planar molecule?

13.28. Which of the simple cycloalkanes (C_nH_{2n}, $n = 3–8$) has a nearly planar geometry?

13.29. Are cycloalkanes saturated hydrocarbons?

13.30. Do structural isomers always have the same molecular formula?

13.31. Do structural isomers always have the same chemical properties?

13.32. Are structural isomers members of a homologous series?

PROBLEMS

13.33. Draw and name all the structural isomers of C_5H_{12}.

13.34. Draw and name all the structural isomers of C_6H_{14}.

13.3[...] of the molecules in Figure P13.35 are structural [...]s of *n*-octane (C_8H_{18})? Name these molecules.

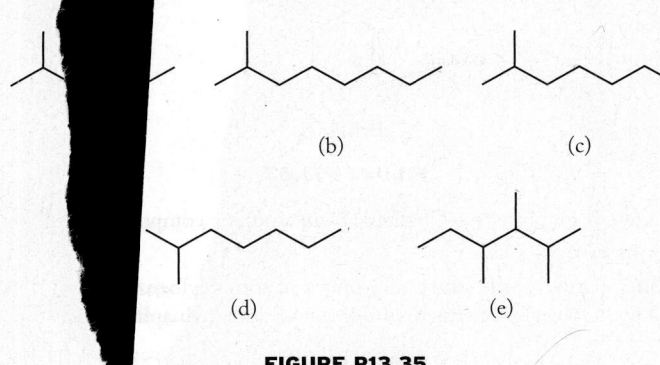

(b) (c)

(d) (e)

FIGURE P13.35

13.36. Which of the molecules in Figure P13.36 are structural isomers of *n*-heptane (C_7H_{16})? Name these molecules.

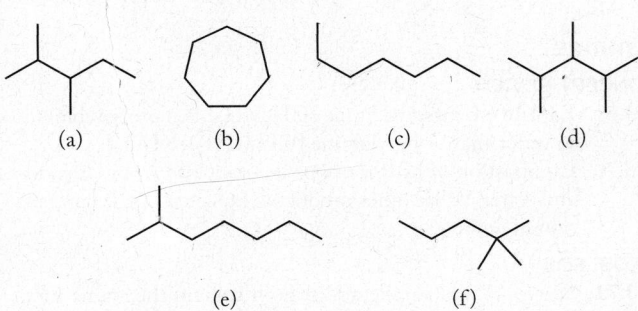

(a) (b) (c) (d)

(e) (f)

FIGURE P13.36

13.37. Convert the carbon-skeleton structures in Problem 13.35 to molecular formulas.

13.38. Convert the carbon-skeleton structures in Problem 13.36 to molecular formulas.

13.39. Using the average bond strengths given in Appendix 4, estimate the molar heat of hydrogenation, $\Delta H_{hydrogenation}$, for the conversion of C_2H_4 to C_2H_6.

$$CH_2{=}CH_2(g) + H_2(g) \rightarrow CH_3CH_3(g)$$

13.40. Using the average bond strengths given in Appendix 4, estimate the molar heat of hydrogenation, $\Delta H_{hydrogenation}$, for the conversion of C_2H_2 to C_2H_6.

$$CH{\equiv}CH(g) + 2H_2(g) \rightarrow CH_3CH_3(g)$$

13.41. Place the following molecules in order of increasing boiling point: C_3H_8, $C_{14}H_{30}$, cyclooctane (C_8H_{16}).

*13.42. Rank the molecules in Figure P13.42 in order of decreasing van der Waals forces.

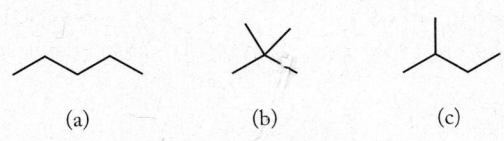

(a) (b) (c)

FIGURE P13.42

Alkenes and Alkynes

CONCEPT REVIEW

13.43. How do structural isomers differ from stereoisomers?

13.44. Explain why alkanes don't have stereoisomers.

13.45. Can combustion analysis distinguish between an alkene and a cycloalkane containing the same number of carbon atoms?

13.46. Can combustion analysis data be used to distinguish between an alkyne and a cycloalkene containing the same number of carbon atoms?

13.47. Why don't the alkenes in Figure P13.47 have cis and trans isomers?

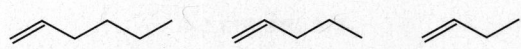

FIGURE P13.47

13.48. Why don't alkynes have cis and trans isomers?

*13.49. Figure P13.49 shows the carbon-skeleton structure of carvone, which is found in oil of spearmint. Why doesn't the molecule carvone have cis and trans isomers?

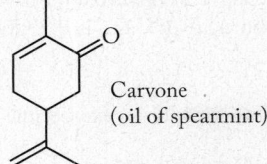

Carvone
(oil of spearmint)

FIGURE P13.49

*13.50. Figure P13.50 shows the carbon-skeleton structure of the antifungal compound capillin. Are the π electrons in capillin delocalized?

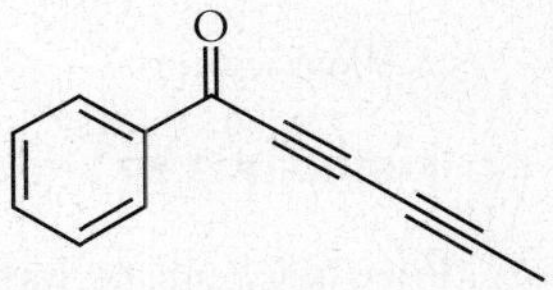

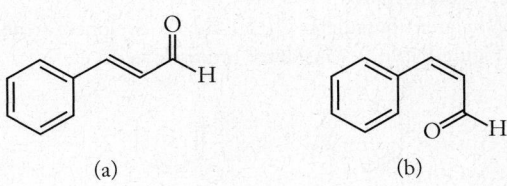

Capillin

FIGURE P13.50

13.51. Ethylene reacts quickly with HBr at room temperature, but polyethylene is chemically unreactive toward HBr. Explain why these related substances have such different properties.

*13.52. Polymerization of butadiene ($CH_2{=}CHCH{=}CH_2$) does not yield the same polymer as polymerization of ethylene ($CH_2{=}CH_2$). How could we convert polybutadiene into polyethylene?

PROBLEMS

13.53. **Cinnamon** Label the isomers of cinnamaldehyde (oil of cinnamon) in Figure P13.53 as cis or trans and *E* or *Z*.

(a) (b)

FIGURE P13.53

13.54. Prostaglandins, naturally occurring compounds in our bodies that cause inflammation and other physiological responses, are formed from arachidonic acid, an unsaturated hydrocarbon containing four C=C double bonds and a carboxylic acid functional group. The stereoisomer containing all cis double bonds is shown in Figure P13.54. How many stereoisomers other than this one are possible? Draw the isomer containing all trans double bonds.

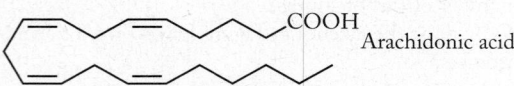

Arachidonic acid

FIGURE P13.54

13.55. Using data in Appendix 4, calculate ΔH_{rxn} for the production of acetylene from the controlled combustion of methane:

$$6\,CH_4(g) + O_2(g) \rightarrow 2\,C_2H_2(g) + 2\,CO(g) + 10\,H_2(g)$$

Is this an endothermic or an exothermic reaction?

13.56. Using data in Appendix 4, calculate ΔH_{rxn} for the production of acetylene from the reaction between calcium carbide and water given ΔH_f° of CaC_2 is -59.8 kJ/mol:

$$CaC_2(s) + 2\,H_2O(\ell) \rightarrow C_2H_2(g) + Ca(OH)_2(s)$$

Is this an endothermic or an exothermic reaction?

13.57. **Making Glue** Wood glue or "carpenter's glue" is made of poly(vinyl acetate). Draw the carbon-skeleton structure of this polymer. The monomer is shown in Figure P13.57.

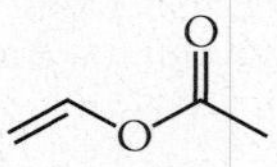

Vinyl acetate

FIGURE P13.57

13.58. The 2000 Nobel Prize in Chemistry was awarded for research on the electrically conductive polymer poly-acetylene.
   a. Draw the carbon-skeleton structure of three monomeric units of the addition polymer that results from polymerization of acetylene, HC≡CH.
   *b. There are two possible stereoisomers of polyacetylene. Describe the two isomeric forms.

Aromatic Compounds

CONCEPT REVIEW

13.59. Why is benzene a planar molecule?

13.60. Why are aromatic molecules stable?

13.61. Do tetramethylbenzene and pentamethylbenzene have structural isomers?

13.62. Why aren't butadiene, C_4H_6, and 1,3-cyclohexadiene, C_6H_8 (Figure P13.62), considered aromatic molecules?

Butadiene   1,3-Cyclohexadiene

FIGURE P13.62

*13.63. Pyridine (Figure P13.63) has the mo... Which ......H_5N. Is pyridine an aromatic molecule?   isomer...

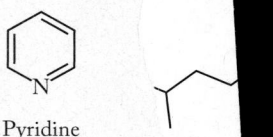

Pyridine   (a)

FIGURE P13.63

*13.64. Is graphite (see Chapter 12) an aromati...

PROBLEMS

13.65. Draw all the structural isomers of trimeth...

13.66. Draw all the structural isomers of dimethy..........ne.

13.67. Calculate the fuel values of gaseous benzene (......) and ethylene gas (C_2H_4). Does 1 mole of benzene ... a higher or lower fuel value than 3 moles of ethylene?

13.68. Does 1 mole of gaseous benzene (C_6H_6) have a higher or lower fuel value than 3 moles of acetylene gas (C_2H_2)?

Amines

CONCEPT REVIEW

13.69. Explain why methylamine (CH_3NH_2) is more soluble in water than n-butylamine [$CH_3(CH_2)_3NH_2$].

13.70. Combustion of hydrocarbons in air yields carbon dioxide and water. What other product is expected in the combustion of amines?

PROBLEMS

13.71. Serotonin and amphetamine both contain the amine functional group (Figure P13.71). Serotonin is responsible, in part, for signaling that we have had enough to eat. Amphetamine, an addictive drug, can be used as an appetite suppressant. Identify the primary and secondary amine functional groups in these molecules.

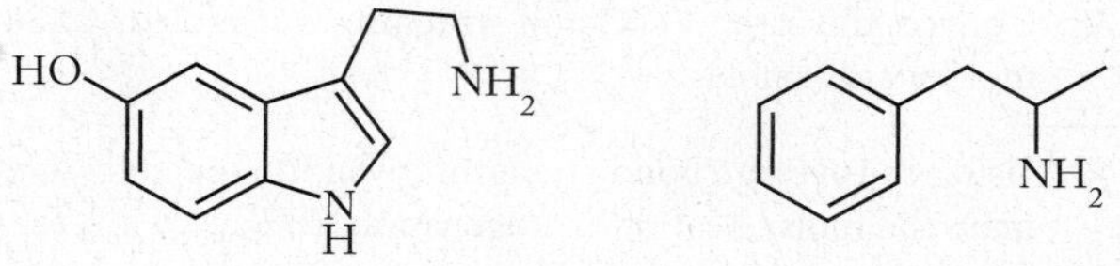

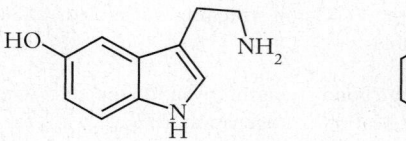

Serotonin   Amphetamine

FIGURE P13.71

13.72. **Coffee** Caffeine, the active ingredient in coffee, contains four nitrogen atoms per molecule. Aspartame is an artificial sweetener containing two nitrogen atoms. Structures of caffeine and aspartame are shown in Figure P13.72. Which nitrogen atoms represent secondary amines and which ones represent tertiary amines?

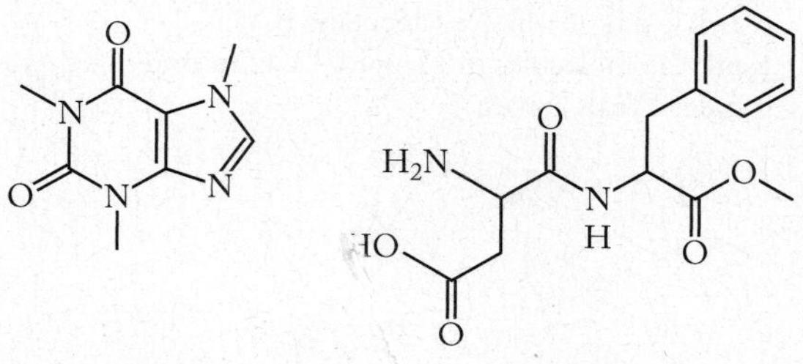

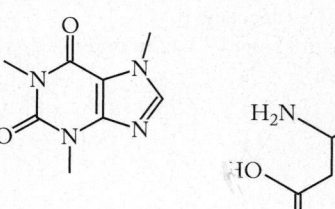

Caffeine   Aspartame

FIGURE P13.72

13.73. Bacteria of the genus *Methanosarcina* convert amines to methane. Their action helps make methane a renewable energy source. Determine the standard enthalpy of the following reaction from the appropriate standard enthalpies of formation ($\Delta H^{\circ}_{f,CH_3NH_2} = -23.0$ kJ/mol):

$$4\,CH_3NH_2(g) + 2\,H_2O(\ell) \rightarrow 3\,CH_4(g) + CO_2(g) + 4\,NH_3(g)$$

13.74. Determine the ΔH°_{rxn} values of these combustion reactions of methylamine ($\Delta H^{\circ}_{f,CH_3NH_2} = -23.0$ kJ/mol):

$$4\,CH_3NH_2(g) + 13\,O_2(g) \rightarrow 4\,CO_2(g) + 4\,NO_2(g) + 10\,H_2O(\ell)$$

$$4\,CH_3NH_2(g) + 6\,O_2(g) \rightarrow 4\,CO_2(g) + 4\,NH_3(g) + 4\,H_2O(\ell)$$

Alcohols, Ethers, and Reformulated Gasoline

CONCEPT REVIEW

13.75. Why are the fuel values of dimethyl ether and ethanol (Figure P13.75) lower than that of ethane?

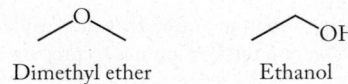

Dimethyl ether   Ethanol

FIGURE P13.75

13.76. Would you expect the fuel value of alcohols to increase or decrease as the number of carbon atoms in the alcohol increases?

13.77. Why do ethers typically boil at lower temperatures than alcohols with the same molecular formula?

13.78. Which do you expect to be more soluble in water, MTBE or 2,2-dimethylbutane (Figure P13.78)? Explain your answer.

MTBE   2, 2-Dimethylbutane

FIGURE P13.78

*13.79. Disposable wipes used to clean the skin prior to a getting an immunization shot contain ethanol. After wiping your arm, your skin feels cold. Why?

*13.80. During the winter months in cold climates, water condensing in a vehicle's gas tank reduces engine performance. An auto mechanic recommends adding "dry gas" to the tank during your next fill-up. Dry gas is typically an alcohol that dissolves in gasoline and absorbs water. Based on the structures shown in Figure P13.80, which product would you predict would do a better job—methanol or 2-propanol?

CH₃OH

Methanol   2-Propanol

FIGURE P13.80

13.81. Which of the compounds in Figure P13.81 are alcohols and which ones are ethers? Place them in order of increasing boiling point.

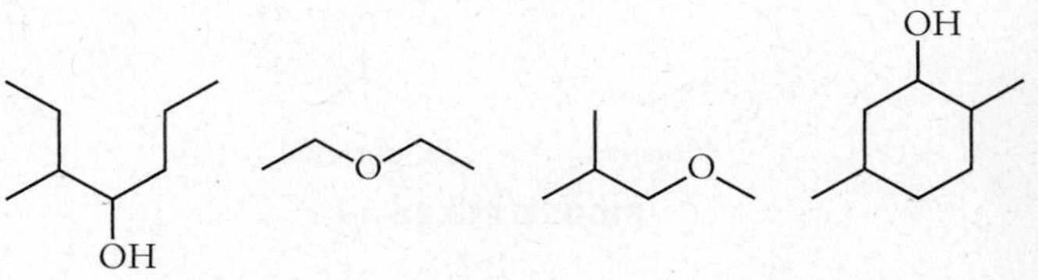

(a)   (b)   (c)   (d)

FIGURE P13.81

13.82. Which of the compounds in Figure P13.82 are alcohols and which ones are ethers? Place them in order of increasing vapor pressure at 25°C.

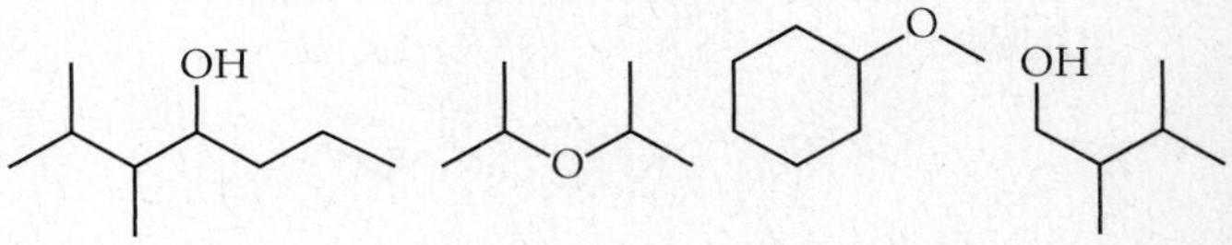

(a)   (b)   (c)   (d)

FIGURE P13.82

Consult tables of thermochemical data in Appendix 4 for any values you may need to solve Problems 13.83 through 13.86.

13.83. Calculate the fuel value of diethyl ether and butanol (Figure P13.83). Which has the higher fuel value?

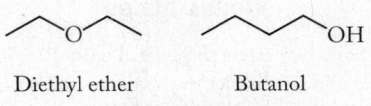

Diethyl ether   Butanol

FIGURE P13.83

13.84. Calculate the fuel value of liquid diethyl ether and methyl propyl ether (Figure P13.84). Which has the higher fuel value? ($\Delta H^{\circ}_{f,methyl\,propyl\,ether} = -266.0$ kJ/mol.)

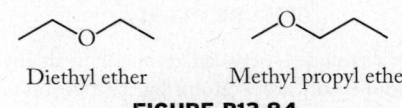

Diethyl ether   Methyl propyl ether

FIGURE P13.84

13.85. Problem 13.76 asked you to predict whether the fuel value of alcohols increased or decreased with the number of carbon atoms in the alcohol. Calculate the fuel values of liquid methanol and ethanol (Figure P13.85). Does your answer support the prediction you made in Problem 13.76?

CH₃OH   OH

Methanol   Ethanol

FIGURE P13.85

13.86. Calculate the fuel values of liquid propanol and isopropanol (Figure P13.86). Which has the higher fuel value? ($\Delta H^{\circ}_{f,propanol,\ell} = -302.5$ kJ/mol and $\Delta H^{\circ}_{f,isopropanol,\ell} = -317.0$ kJ/mol.)

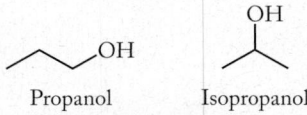

Propanol   Isopropanol

FIGURE P13.86

Aldehydes, Ketones, Carboxylic Acids, Esters, and Amides

CONCEPT REVIEW

13.87. Explain why carboxylic acids tend to be more soluble in water than aldehydes with the same number of carbon atoms.

13.88. In reference books, diethyl ether (Figure P13.88) is usually listed as "slightly soluble" in water, but 2-butanone is listed as "very soluble." Why do you suppose 2-butanone is more soluble?

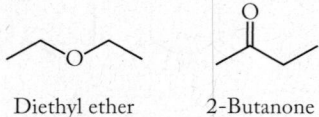

Diethyl ether   2-Butanone

FIGURE P13.88

13.89. Are butanal and 2-butanone (Figure P13.89) structural isomers?

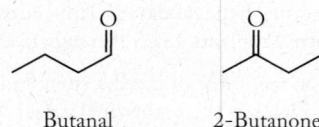

Butanal   2-Butanone

FIGURE P13.89

13.90. **Apples** The two esters shown in Figure P13.90 are both found in apples and contribute to the flavor and aroma of the fruit. Are the two compounds identical, structural isomers, or stereoisomers, or do they have different molecular formulas?

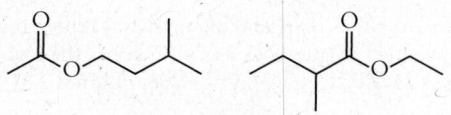

FIGURE P13.90

13.91. Can we distinguish between ketones and aldehydes with the same number of carbon atoms by combustion analysis?

13.92. Can we distinguish between ethers and ketones with the same number of carbon atoms by combustion analysis?

13.93. Resonance forms for acetic acid are shown in Figure P13.93. Which one contributes more to the bonding picture? Explain your choice.

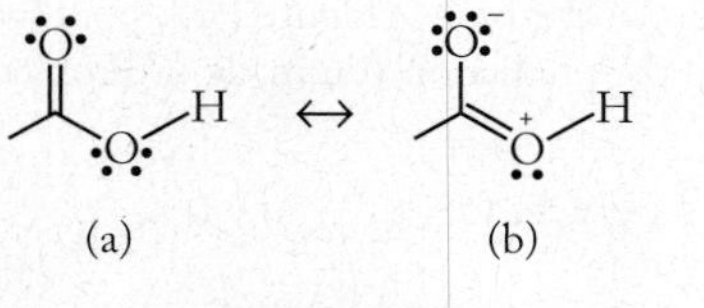

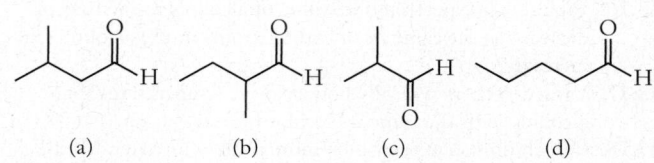

(a)   (b)

FIGURE P13.93

13.94. Figure P13.94 shows resonance forms for acetamide and acetic acid. Does resonance form a containing the C=N double bond contribute more to the bonding picture of acetamide than does resonance form b to the bonding picture in acetic acid? Explain your answer.

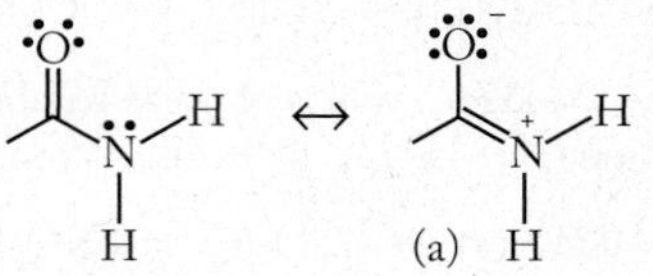

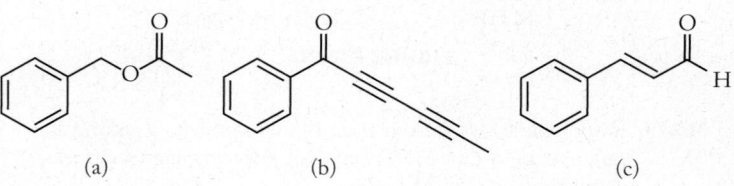

Acetamide

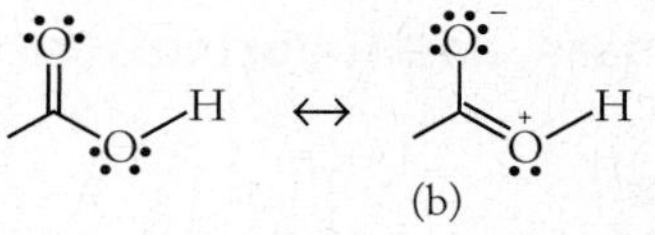

Acetic acid

FIGURE P13.94

13.95. What distinguishes an amine from an amide?

*13.96. Why can't we use tertiary amines to prepare amides?

PROBLEMS

13.97. Which of the compounds in Figure P13.97 are structural isomers of the aldehyde $C_5H_{10}O$?

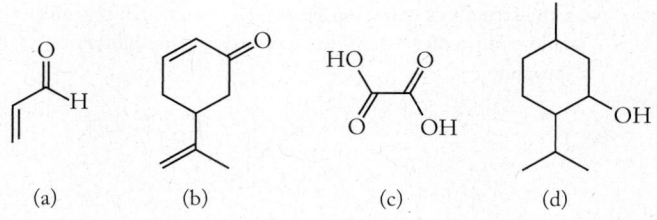

(a)   (b)   (c)   (d)

FIGURE P13.97

13.98. Each of the natural products in Figure P13.98 contains more than one functional group. Which of the compounds is an aldehyde?

(a)   (b)   (c)

FIGURE P13.98

13.99. Which of the compounds in Figure P13.99 is a ketone?

(a)   (b)   (c)   (d)

FIGURE P13.99

13.100. Propanal and acetone (2-propanone) have the same molecular formula, C_3H_6O, but different structures (Figure P13.100). Which compound is a ketone?

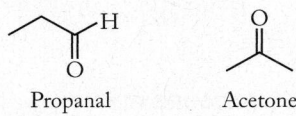

Propanal   Acetone

FIGURE P13.100

13.101. Plot the carbon-to-hydrogen ratio in aldehydes with one to six carbons as a function of the number of carbon atoms. Does this graph correlate better with the plot of C:H ratios for alkanes or for alkenes?

13.102. Plot the carbon-to-hydrogen ratio in carboxylic acids with one to six carbons as a function of the number of carbon atoms. Does this graph correlate better with the plot of C:H ratios for alkanes or for alkenes?

13.103. Esters are responsible for the odors of fruits, including apples, bananas, and pineapples. Figure P13.103 shows three esters from these fruits. Identify the alcohol and carboxylic acid that react to form these compounds.

(a) Pineapples   (b) Bananas   (c) Apples

FIGURE P13.103

13.104. Beeswax (Figure P13.104) is an ester composed of an alcohol and a carboxylic acid with a long hydrocarbon chain. Identify the alcohol and acid in beeswax.

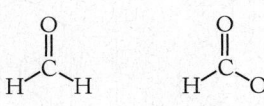

FIGURE P13.104

Consult tables of thermochemical data in Appendix 4 for any values you may need to solve Problems 13.105 through 13.108.

13.105. Calculate the fuel values of formaldehyde and formic acid (Figure P13.105). The ΔH_f° of formaldehyde is -108.6 kJ/mol and the ΔH_f° of formic acid is -378.7 kJ/mol. Which has the higher fuel value?

Formaldehyde   Formic acid

FIGURE P13.105

13.106. Calculate the fuel values of formamide and methyl formate (Figure P13.106), which have ΔH_f° values of -251 and -391 kJ/mol, respectively. Assume $NO_2(g)$ is a product of formamide combustion.

Formamide   Methyl formate

FIGURE P13.106

13.107. Calculate ΔH_{rxn}° for the following reactions of methanogenic bacteria given $\Delta H_{f,HCOOH,g}^\circ = -378.7$ kJ/mol:
(1) $CH_3COOH(\ell) \rightarrow CH_4(g) + CO_2(g)$
(2) $4HCOOH(g) \rightarrow CH_4(g) + 3CO_2(g) + 2H_2O(\ell)$

13.108. Calculate ΔH_{rxn}° for the following reactions of methanogenic bacteria:
(1) $4H_2(g) + CO_2(g) \rightarrow CH_4(g) + 2H_2O(\ell)$
(2) $4CH_3OH(\ell) \rightarrow 3CH_4(g) + CO_2(g) + 2H_2O(\ell)$

13.109. Reactions between 1,6-diaminohexane, $H_2N(CH_2)_6NH_2$, and different dicarboxylic acids, $HOOC(CH_2)_2COOH$, are used to prepare polymers that have a structure similar to that of nylon. How many carbon atoms (n) were in the dicarboxylic acids used to prepare the polymers with the repeating units shown in Figure P13.109?

(a)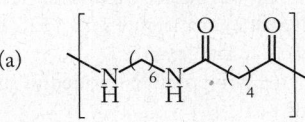

(b)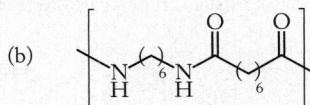

(c)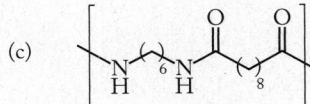

FIGURE P13.109

13.110. The two polymers in Figure P13.110 have the same empirical formula.
a. What pairs of monomers could be used to make each of them?
b. How might the physical properties of these two polymers differ?

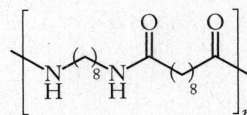

Polymer I

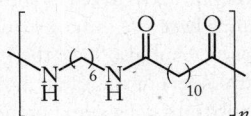

Polymer II

FIGURE P13.110

13.111. The polyester called Kodel is made with polymeric strands prepared by the reaction of dimethyl terephthalate with 1,4-di(hydroxymethyl) cyclohexane (Figure P13.111).

a. Is Kodel a condensation polymer or an addition polymer? What is the other product of the reaction?

*b. Dacron (see Sample Exercise 13.10) is made from dimethyl terephthalate and ethylene glycol. What properties of Kodel fibers might make them better than Dacron as a clothing material?

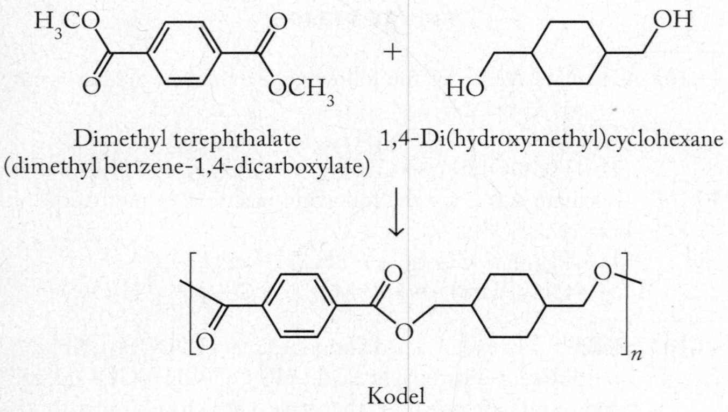

Dimethyl terephthalate
(dimethyl benzene-1,4-dicarboxylate)   1,4-Di(hydroxymethyl)cyclohexane

Kodel

FIGURE P13.111

13.112. Lexan is a polymer belonging to the class of materials called polycarbonates. Figure P13.112 shows the polymerization reaction for Lexan.

a. What other compound is formed in the polymerization reaction?

*b. Why is Lexan called a "polycarbonate"?

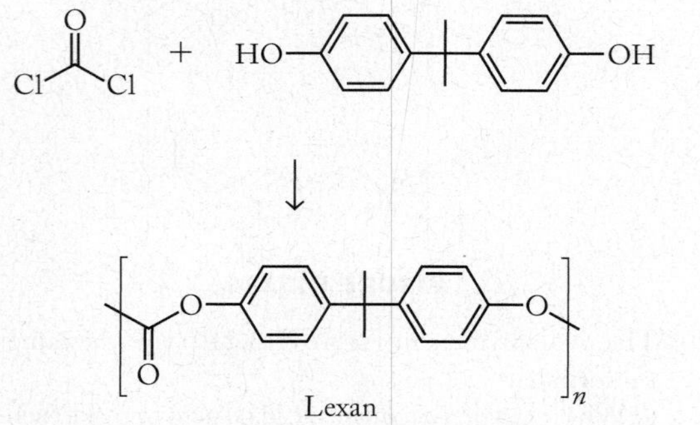

Lexan

FIGURE P13.112

Chirality

CONCEPT REVIEW

13.113. Can all of the terms *enantiomer*, *achiral*, and *optically active* be used to describe a single compound? Explain.

13.114. Two compounds have the same structure and the same physical properties but also have the same optical activity. Are they enantiomers or the same molecule?

13.115. Are racemic mixtures considered homogeneous or heterogeneous mixtures?

*13.116. Could a racemic mixture be distinguished from an achiral compound based on optical activity? Explain your answer.

13.117. Why is the amino acid glycine (Figure P13.117) achiral?

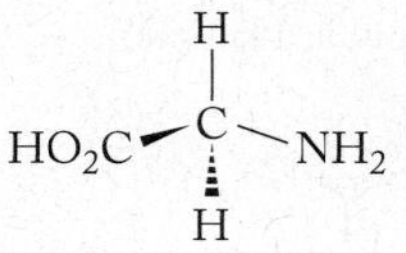

FIGURE P13.117

13.118. Can stereoisomers of molecules such as cis and trans RCH=CHR also have optical isomers? (R may be any of the functional groups we have encountered in this textbook.) Explain your answer.

13.119. Which type of hybrid orbitals on a carbon atom, *sp*, *sp²*, or *sp³*, can give rise to enantiomers?

*13.120. Could an oxygen atom in an alcohol, ketone, or ether ever be a chiral center in the molecule?

13.121. Which of the following objects are chiral? (a) a golf club; (b) a tennis racket; (c) a glove; (d) a shoe

13.122. Which of the following objects are chiral? (a) a key; (b) a screwdriver; (c) a lightbulb; (d) a baseball

PROBLEMS

13.123. Which of the molecules in Figure P13.123 are chiral?

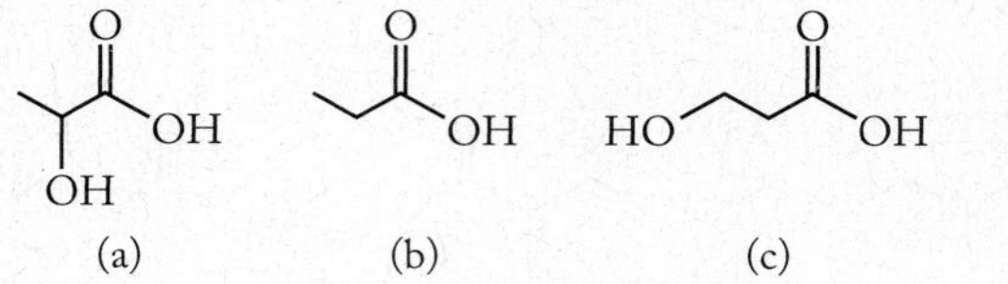

(a)   (b)   (c)

FIGURE P13.123

13.124. Which, if any, of the molecules shown in Figure P13.124 contains a chiral center?

(a)   (b)   (c)

FIGURE P13.124

13.125. **Artificial Sweeteners** Artificial sweeteners are fundamental to the diet food industry. Figure P13.125 shows three artificial sweeteners that have been used in food. Saccharin is the oldest, dating to 1879. Cyclamates were banned in 1969 following research suggesting they led to tumors. Aspartame may be more familiar to you under the name NutraSweet. Each of these sweeteners contain between zero and two chiral carbon atoms. Circle the chiral center in each compound.

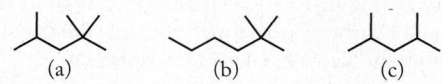

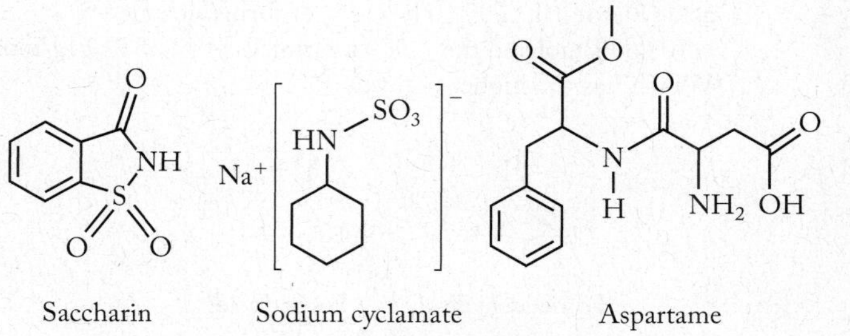

Saccharin   Sodium cyclamate   Aspartame

FIGURE P13.125

13.126. Identify the chiral centers in each of the molecules in Figure P13.126.

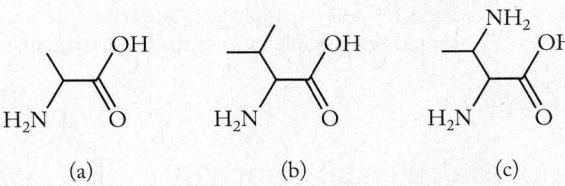

(a)   (b)   (c)

FIGURE P13.126

13.127. The Smell of Raspberries The compound 3-(*p*-hydroxyphenyl)-2-butanone is a major contributor to the smell of raspberries. One enantiomer is shown in Figure P13.127. Identify the single chiral center in the molecule and draw the mirror image of this enantiomer.

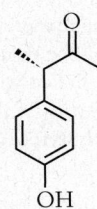

3-(*p*-Hydroxyphenyl)-2-butanone

FIGURE P13.127

13.128. The scent associated with pine trees is derived from the molecule terpineol. One enantiomer of terpineol is shown in Figure P13.128. Draw the other optical isomer.

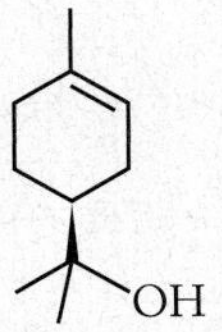

Terpineol

FIGURE P13.128

13.129. Nicotine is a stimulant found in tobacco. Valium is a tranquilizer. Both molecules contain two nitrogen atoms in addition to other functional groups. In Figure P13.129 identify the nitrogen atoms shown in blue as belonging to an amine or an amide.

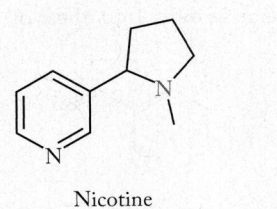

Nicotine 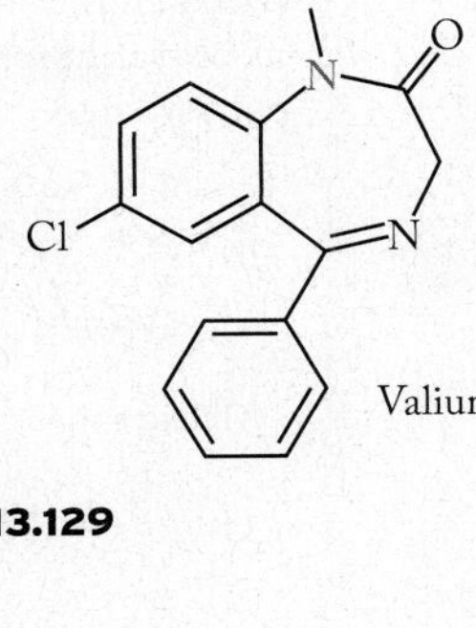   Valium

FIGURE P13.129

13.130. Piperine and capsaicin are ingredients of peppers that give "hotness" to foods. In Figure P13.130 identify the nitrogen atoms shown in blue as belonging to an amine or an amide.

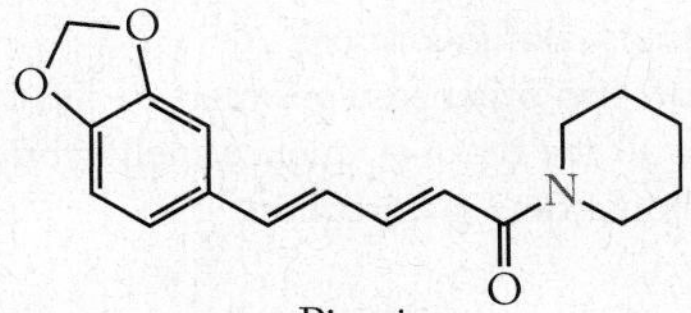

Piperine

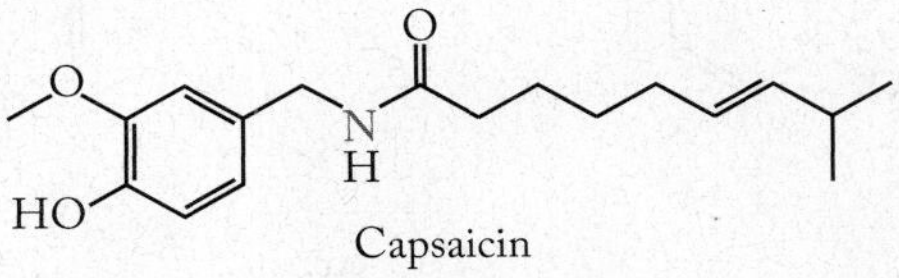

Capsaicin

FIGURE P13.130

Additional Problems

13.131. How many grams of liquid methanol must be combusted to raise the temperature of 454 g of water from 20.0°C to 50.0°C? Assume that the transfer of heat to the water is 100% efficient. How many grams of carbon dioxide are produced in this combustion reaction?

13.132. How many grams of methylamine ($\Delta H_f^\circ = -23.0\,\text{kJ/mol}$) must be combusted to raise the temperature of 454 g of water from 20°C to 50°C? Assume that the transfer of heat to the water is 100% efficient. Also assume $NO_2(g)$ is a product of the combustion of methylamine. How many grams of carbon dioxide are produced in this combustion reaction?

*13.133. Two compounds, both with molar masses of 74.12 g/mol, were combusted in a bomb calorimeter with $C_{\text{calorimeter}} = 3.640$ kJ/°C. Combustion of 0.9842 g of compound A led to an increase in temperature of 10.33°C, while combustion of 1.110 g of compound B caused the temperature to rise 11.03°C. Which compound is butanol and which is diethyl ether?

13.134. Why should methane be more soluble in decane ($C_{10}H_{22}$) than in water?

13.135. Salsa Salsa has antibacterial properties because it contains dodecenal (Figure P13.135), a compound found in the cilantro used to make salsa.
   a. How many carbon atoms are in dodecenal?
   b. What functional groups are present in dodecenal?
   c. What types of isomerism are possible in dodecenal?

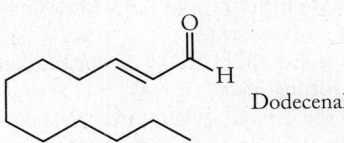

Dodecenal

FIGURE P13.135

13.136. **Turmeric, a Spice** Turmeric is commonly used as a spice in Indian and Southeast Asian dishes. Turmeric contains a high concentration of curcumin (Figure P13.136), a potential anticancer drug and a possible treatment for cystic fibrosis.
   a. Are the substituents on the C=C double bonds in cis or trans configurations?
   b. Draw two other stereoisomers of this compound.
   c. List all the types of valence shell hybridization of the carbon atoms in curcumin.

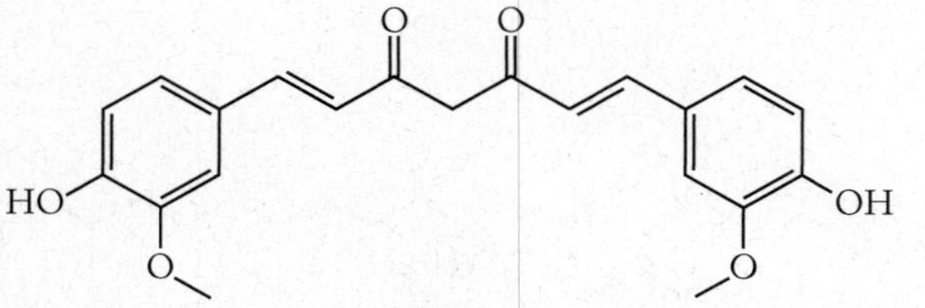

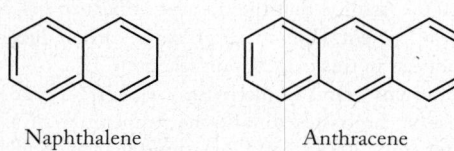

Curcumin

FIGURE P13.136

13.137. Polycyclic aromatic hydrocarbons are potent carcinogens. They are produced during combustion of fossil fuels and have also been found in meteorites. Can we use combustion analysis to distinguish between naphthalene and anthracene (Figure P13.137)?

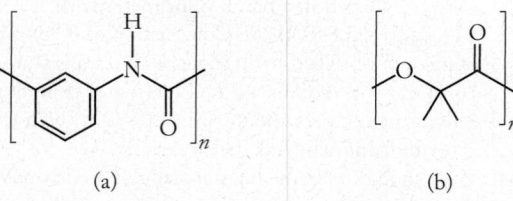

Naphthalene   Anthracene

FIGURE P13.137

13.138. Identify the reactants in the polymerization reactions that produce the polymers shown in Figure P13.138.

(a)   (b)

FIGURE P13.138

*13.139. **Raincoats** "Waterproof" nylon garments have a coating to prevent water from penetrating the hydrophilic fibers. Which functional groups in the nylon molecule make it hydrophilic?

13.140. Draw the carbon-skeleton structure of the condensation polymer of $H_2N(CH_2)_6COOH$. How does this polymer compare with nylon-6?

*13.141. Putrescine, $H_2N(CH_2)_4NH_2$, is one of the compounds that form in rotting meat.
   a. Draw the carbon-skeleton structures of all the trimers (a molecule formed from three monomers) that can be formed from putrescine, adipic acid, and terephthalic

acid (Figure P13.141). The three monomers forming the trimer do not have to be different from one another.
   b. A chemist wishes to make a putrescine polymer containing a 1:1 ratio of adipic acid to terephthalic acid. What should be the mole ratio of the three reactants?

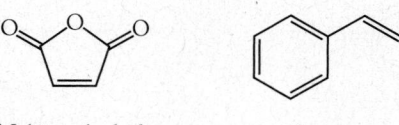

$HOOCCH_2CH_2CH_2CH_2COOH$

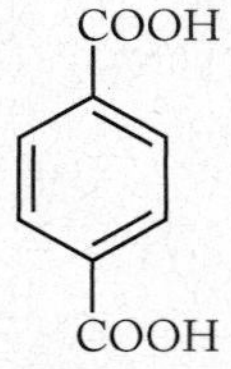

Adipic acid   Terephthalic acid
                     (benzene-1,4-dicarboxylic acid)

FIGURE P13.141

*13.142. Polymer chemists can modify the physical properties of polystyrene by copolymerizing divinylbenzene with styrene (Figure P13.142). The resulting polymer has strands of polystyrene cross-linked with divinylbenzene. Predict how the physical properties of the copolymer might differ from those of 100% polystyrene.

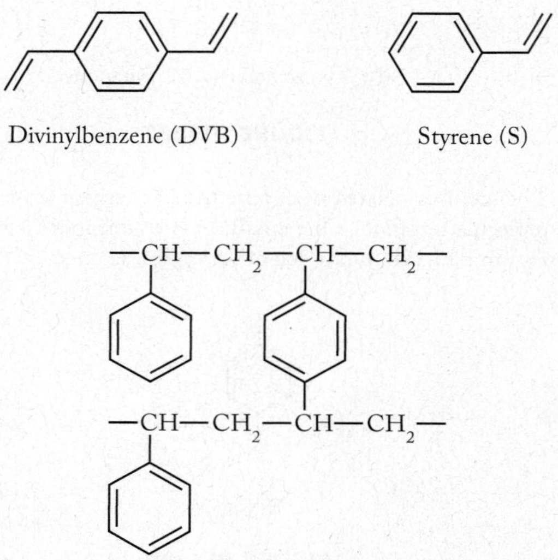

Divinylbenzene (DVB)   Styrene (S)

S cross-linked with DVB

FIGURE P13.142

13.143. Maleic anhydride and styrene (Figure P13.143) form a polymer with alternating units of each monomer.
   a. Draw two repeating monomer units of the polymer.
   b. Based on the structure of the copolymer, predict how its physical properties might differ from those of polystyrene.

Maleic anhydride   Styrene

FIGURE P13.143

13.144. Superglue The active ingredient in superglue is methyl 2-cyanoacrylate (Figure P13.144). The liquid glue hardens rapidly when methyl 2-cyanoacrylate polymerizes. This happens when it contacts a surface containing traces of water or other compounds containing –OH or –NH– groups. Draw the carbon-skeleton structure of two repeating monomer units of poly(methyl 2-cyanoacrylate).

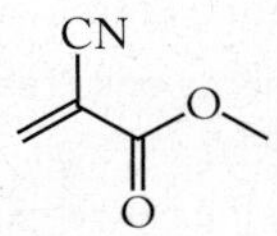

Methyl 2-cyanoacrylate

FIGURE P13.144

*13.145. Silicones are polymeric materials with the formula $[R_2SiO]_n$ (Figure P13.145). They are prepared by reaction of R_2SiCl_2 with water yielding the polymer and aqueous HCl. Consider this reaction as taking place in two steps: (1) water reacts with 1 mole of R_2SiCl_2 to produce a new monomer and 1 mole of $HCl(aq)$; (2) one new monomer molecule reacts with another new monomer molecule to eliminate one molecule of H_2O and make a dimer with a Si—O—Si bond.

   a. Suggest two balanced equations describing these reactions that occur over and over again to produce a silicone polymer.

   b. Why are silicones water repellent?

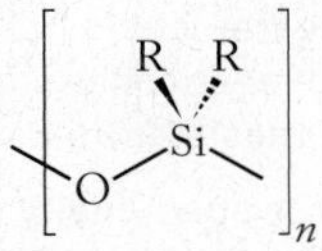

Silicone

FIGURE P13.145

*13.146. Piperine and capsaicin are the spicy ingredients of black and red pepper, respectively (Figure P13.146). Both compounds contain an amide functional group.

   a. Draw the amine and the carboxylic acid that could react to form these two compounds.

   b. Are the double bonds in these molecules cis or trans?

   c. Name the functional groups that contain the oxygen atoms in these compounds.

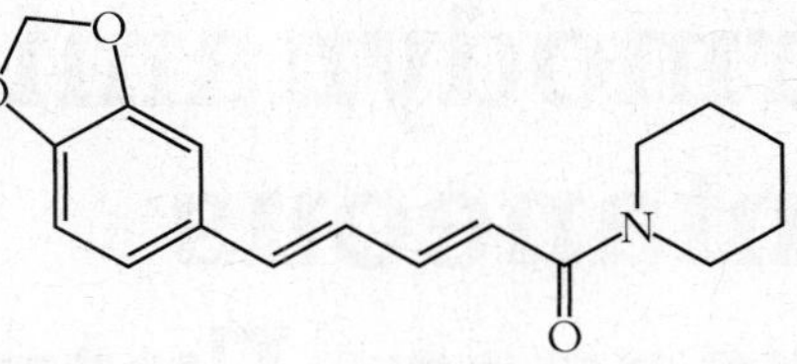

Piperine

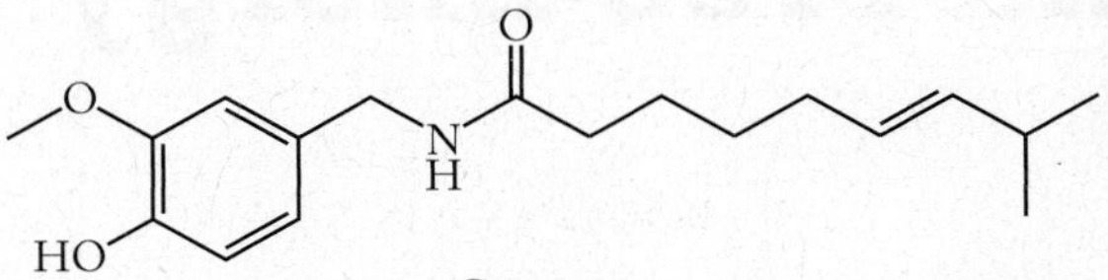

Capsaicin

FIGURE P13.146

13.147. The heats of combustion of the two structural isomers in Table 13.3 calculated from bond energies have the same value, but the two experimentally determined values for $\Delta H°_{comb}$ are not the same. Why?

13.148. We studied the energy changes associated with the combustion of hydrocarbon fuels in Chapter 5. In principle, we could generate tremendous quantities of fuel by combining CO_2 and water vapor and chemically converting them back into CH_4 or other hydrocarbon fuels. Why is this not possible in practice?

Additional study materials including ChemTours and Diagnostic Quizzes are available at StudySpace at www.wwnorton.com/studyspace.

14

Thermodynamics: Spontaneous Processes, Entropy, and Free Energy

Learning Outcomes

- Distinguish between spontaneous and nonspontaneous processes using the second law of thermodynamics

- Explain entropy changes in terms of changes in the number of accessible microstates

- Predict the sign of entropy changes for chemical reactions and physical processes

- Calculate entropy changes in chemical reactions using standard molar entropies

- Calculate free-energy changes in chemical reactions using standard free energies of formation

- Predict the spontaneity of a chemical reaction as a function of temperature

- Explain how free-energy changes are used to drive coupled reactions, including biological pathways

A LOOK AHEAD

The Game of Energy

Some events are so familiar to us that they hardly seem worth our attention. If a car tire is punctured, the air inside rushes out of the hole and the tire goes flat. No one has ever seen air rush into the hole and reinflate the tire. If you leave an iron nail on the ground, it rusts; left alone, it will never turn back into a shiny nail. An ice cube melts if left on a countertop at room temperature; an ice cube in this situation never becomes colder while the countertop gets warmer.

These events are all very different from one another: a gas moves from one place to another, a chemical reaction converts iron into iron oxide, a solid becomes a liquid. However, they have an important thing in common: they are all *spontaneous*, which means they all happen without any ongoing intervention from us, and in each case the reverse process is *nonspontaneous*. We cannot simply say that it is "obvious" or "natural" that such events occur spontaneously. To understand how the world works, science demands a satisfactory explanation for these events. We find that explanation in the second law of thermodynamics and the concept called entropy.

Remember that the first law of thermodynamics tells us that energy cannot be created or destroyed. In the game of energy, when we use energy to do work, the best we can do is to break even—we cannot create new energy, nor can we do more work than the amount of energy available. Furthermore, the second law of thermodynamics says that the amount of energy available to do useful work is constantly decreasing; in other words, in the game of energy, not only can we not win, we can't break even.

Chateau de Villandry, France Plants do not spontaneously arrange themselves into the highly ordered patterns of a formal garden. ▶

spontaneous process a process that occurs without outside intervention.

nonspontaneous process a process that occurs only as long as energy is continually added to the system.

entropy (S) a measure of the distribution of energy in a system at a specific temperature.

second law of thermodynamics the total entropy of the universe increases in any spontaneous process.

If energy cannot be destroyed, where does the energy go that is unavailable to do useful work? The answer to this question lies in the fact that energy naturally spreads out, becoming less concentrated over time. Entropy is a measure of the energy that has spread out in such a way that gathering it up again—concentrating it so we could use it to do useful work—requires more energy than we can recover in the process. In this chapter we explore entropy, the second law of thermodynamics, how energy spreads out, and spontaneity. We begin with some questions about everyday occurrences that allow us to develop the core ideas behind the flow of energy in all spontaneous processes: why doesn't ice freeze above 0°C at atmospheric pressure? Why are some endothermic reactions spontaneous? What drives the many nonspontaneous reactions in living organisms that sustain life? ■

14.1 Spontaneous Processes and Entropy

The rusting of iron is a **spontaneous**, exothermic process, a process that proceeds without outside intervention. The reverse process—converting rust, $Fe_2O_3(s)$, into iron, $Fe(s)$—is not *impossible*, and indeed it is carried out routinely on a very large scale to produce iron from iron ore. It is, however, **nonspontaneous**, which means it requires the continuous input of energy. When we stop adding energy, the conversion of rust to iron stops, and the spontaneous rusting reaction begins again. This relationship is a general one; any time a process is spontaneous, the reverse process is nonspontaneous.

A spontaneous reaction may need a little boost to proceed rapidly. For example, a fire may need to be ignited by some energy source like a spark or an open flame, but then it continues without the input of additional energy as long as both fuel and oxidizer are available. Once a spontaneous reaction starts, it keeps going on its own. A spontaneous reaction is not necessarily a rapid reaction: rust forms relatively slowly on a piece of iron.

Spontaneous is also *not* a synonym for exothermic. Some endothermic reactions are spontaneous. A good example is the reaction between acetic acid (vinegar) and sodium bicarbonate (baking soda) at room temperature (Figure 14.1), an endothermic process with a positive enthalpy change (ΔH). Whether a reaction is endothermic ($\Delta H > 0$) or exothermic ($\Delta H < 0$) does not tell us whether or not the reaction is spontaneous.

∞ **CONNECTION** We introduced enthalpy change (ΔH) in Chapter 5 as a thermodynamic quantity that describes heat flow into or out of a system.

CONCEPT TEST

Each of the following processes is spontaneous at room temperature. Which ones are endothermic?

   a. A glass of water evaporates.

   b. Fog forms over a pond on a cool autumn evening.

   c. Dissolving a solid in water causes the temperature of the water to decrease.

   d. Dry leaves ignite when touched by a burning ember.

(Answers to Concept Tests are in the back of the book.)

▶‖ **CHEMTOUR** Entropy

Why are some chemical and physical processes spontaneous? The answer to this question lies in the dispersion (spreading out) of energy that occurs in any spontaneous process. **Entropy (S)**, a thermodynamic property, is a measure of the distribution of energy at a specific temperature. The **second law of thermodynamics**

$$NaHCO_3(s) \quad + \quad CH_3COOH(aq) \quad \longrightarrow \quad Na^+(aq) + CH_3COO^-(aq) + H_2O(aq) + CO_2(g)$$

FIGURE 14.1 Adding baking soda to vinegar produces sodium acetate, water, and carbon dioxide in an endothermic ($\Delta H^\circ_{rxn} = 48.5$ kJ/mol) yet spontaneous reaction.

states that the total entropy *of the universe* increases in any spontaneous process. Thus, we have the criterion for a spontaneous process—an increase in the entropy of the universe—as well as a criterion for a nonspontaneous process: a decrease in the entropy of the universe. To develop a clearer picture of entropy and the second law and of what is meant by energy dispersion, let's examine some spontaneous processes. In doing so, we'll make a few assumptions that are not entirely true but simplify the initial discussion. We will correct these assumptions later in the chapter once the language of thermodynamics is clearer.

CONNECTION In Chapter 5 we introduced the idea of the universe in thermodynamics. The *universe* means the system we are studying and its surroundings.

Statistical Entropy and Microstates: Flat Tires

Let's start by examining the event in which air leaves a punctured tire by applying the following assumptions: (1) The air behaves as an ideal gas. (2) The tire, the air inside it, and the outside air are all at the same temperature. (3) The temperature of the air in the tire does not change as it escapes.[1] Because the internal energy contained in the gas depends only on its temperature, that energy does not change as the gas escapes and ΔH is zero. When the air escapes from the tire, however, it expands to fill a larger volume at a lower pressure. Because the kinetic energy of the molecules is spread throughout this larger volume, the entropy of the air that came from the tire increases. The deflation of the tire was spontaneous, and a spontaneous process, by definition, must lead to an increase in the entropy of the universe. The dispersion of the kinetic energy of gas molecules through a larger volume accompanies the dispersion of those molecules themselves. But the question remains: what does an increase in entropy mean on a molecular level?

To answer this question, we need to consider the internal and translational energies of the molecules in air as they leave the tire. Of course, air is a mixture of gases, but to keep our story simple, let's focus on the energy states available to one molecule of oxygen escaping from the tire.

As we discussed in Chapter 5, and as shown in Figure 14.2, a molecule of O_2 undergoes three types of motion: (1) *translational motion* as the molecule travels from one location to another; (2) *rotational motion* as the molecule spins about an imaginary axis perpendicular to the bond; (3) *vibrational motion* as its two atoms move toward and away from each other, like balls on the ends of a spring.

CONNECTION In Chapter 5 we defined internal energy as the sum of the kinetic and potential energies of all the components of a system.

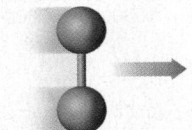

(a) Translational motion

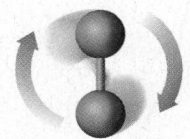

(b) Rotational motion

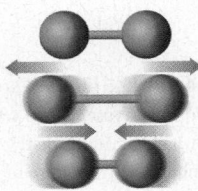

(c) Vibrational motion

FIGURE 14.2 A diatomic molecule has three fundamental types of motion: (a) translational motion, (b) rotational motion, and (c) vibrational motion.

[1] Assumptions 1 and 3 are linked: the temperature of an ideal gas does not change as its volume expands. However, the temperature of a real gas does decrease as it expands; this is the principle by which air conditioners function.

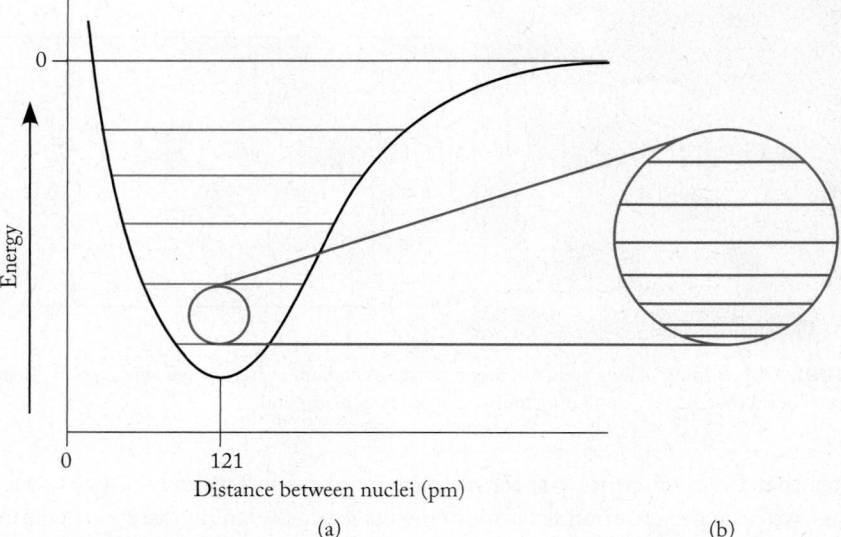

(a) (b)

FIGURE 14.3 Vibrational and rotational energy levels in a molecule of O_2 are quantized. (a) The electrostatic potential energy between two oxygen atoms reaches a minimum when their nuclei are 121 pm apart—the length of an O==O bond. (b) Quantized rotational energy levels (represented by the blue horizontal lines) are superimposed on each vibrational energy level, represented by a red horizontal line in a. (Not all the available rotational and vibrational states are shown, and the gaps between them are not necessarily to scale.)

All three modes of motion have one thing in common: the greater the thermal energy of the molecule, the greater each type of motion. Thus, a molecule of O_2 (or any gas), moves, vibrates, and rotates more rapidly and with more energy at higher temperatures.

In Chapter 7 quantum mechanics taught us that energy is not continuous on the atomic scale. Instead, the energies and motion of atoms and molecules are quantized. At temperatures near room temperature, the differences between the translational energy levels of atoms and molecules in the gas phase are so small they may be considered to be a continuum of energy. However, the gaps in vibrational and rotational energy levels are large enough that we must take quantization into account.

To investigate these gaps, let's revisit the change in electrostatic potential energy that occurs when two atoms approach each other and a covalent bond forms between them (see Figure 8.1 on page 369). Energy reaches a minimum (Figure 14.3a) when the nuclei of two oxygen atoms are a distance apart that corresponds to the length of the O==O bond, or 121 pm. Figure 14.3(a) also contains additional energy levels represented by horizontal red lines. These are vibrational energy levels. Note how these red lines lengthen with increasing vibrational energy. Increases in length correspond to greater variations in intermolecular distances, that is, more energetic oscillations of the O==O bond, with increasing vibrational energy.

Superimposed on each vibrational energy level is a set of rotation energy levels, shown for the first vibrational energy level in Figure 14.3(b). Similar sets of rotational energy levels are associated with all the other vibrational energy levels, creating a multitude of energy states that are accessible to an O_2 molecule.

In Section 6.8 we noted that the molecules in any gas sample have a range of speeds. Figure 14.4(a) shows this range, which is characterized by a *Boltzmann distribution,* so named because it was developed by Ludwig Boltzmann (1844–1906). An individual oxygen molecule in a sample changes speeds depending on its

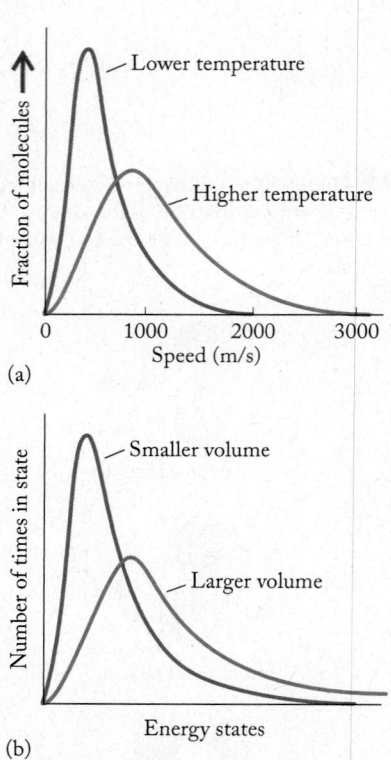

FIGURE 14.4 (a) The range of speeds of O_2 molecules in a gas sample at two temperatures is described by a Boltzmann distribution. As temperature increases, the range of speeds of the gas molecules increases. (b) The number of times a molecule occupies a particular accessible energy level is also described by a Boltzmann distribution. As the volume of the system increases, molecules can access a greater number of energy levels.

collisions with other molecules, but the overall distribution of speeds for the entire population of molecules stays the same at a given temperature.

With this condition in mind, let's use Figure 14.3 to consider a hypothetical situation for one oxygen molecule in a sample of many oxygen molecules at room temperature. The figure shows only a few of the energy levels available to the individual molecule, but these are nevertheless sufficient to illustrate the situation. To follow the fate of one specific oxygen molecule, imagine that we can take photographs of it over time. As it collides and exchanges energy with other molecules in the system, it gains and loses kinetic energy. Each gain or loss corresponds to a transition to a new state, and so our photographs are images of all the different states that one molecule occupies. We take many photos when the molecule is in the most accessible states and fewer photos of it in less accessible states. Very occasionally the molecule picks up a lot of energy, and we might have just one or two photos of the molecule in high-energy states. Correspondingly, the molecule may stop moving completely and enter the lowest of the accessible states, but that would also be a very rare event that would yield only one or two photos. If we count the number of photographs we have of the O_2 molecule in each of the states it occupies over a period of time and plot the number of photos as a function of the energy of the states, the resulting curve fits the Boltzmann distribution of Figure 14.4(a), except that instead of showing the number of molecules moving at different speeds, our new curve describes the number of times a molecule occupies a particular state among a range of accessible states (Figure 14.4b).

Now let's think about what happens with a lot of molecules—say, an entire mole of them—at room temperature. The energy corresponding to the vibrational, rotational, and translational motions of all the molecules is distributed among the accessible levels in much the same way it is distributed for one molecule: a few molecules are in the lowest-energy states, more are in the intermediate levels, and fewer (but not zero) molecules are in the highest-energy states. As with the single molecule, we can imagine taking photographs to capture how all the molecules move among the accessible levels at a series of instants in time. Each photograph shows the same overall distribution of molecules among the energy levels, because each photograph represents the *same total energy in the system*, but the individual molecules in each photograph differ in their position among the accessible states. Each of the photographs shows the same Boltzmann distribution for the population of molecules at the specified temperature; each individual photograph shows different locations of individual particles and represents a **microstate**. A microstate is a unique distribution of particles among energy levels. Because a mole contains so many particles, each experiencing millions of collisions per second, a mole of particles has a huge number of microstates, each microstate showing the individual particles in different locations throughout the same overall Boltzmann distribution. Microstates are the key to understanding entropy at the molecular level.

CONCEPT TEST

Imagine you have four identical chairs to arrange on four steps leading up to a stage, one chair on each step. The chairs have numbers on their backs: 1, 2, 3, and 4. How many different microstates for the chairs are possible? (Notice that when viewed from the front, all the microstates look the same. Viewed from the back, you can identify the different microstates because you can distinguish the chairs by their numbers.)

microstate a unique distribution of particles among energy levels.

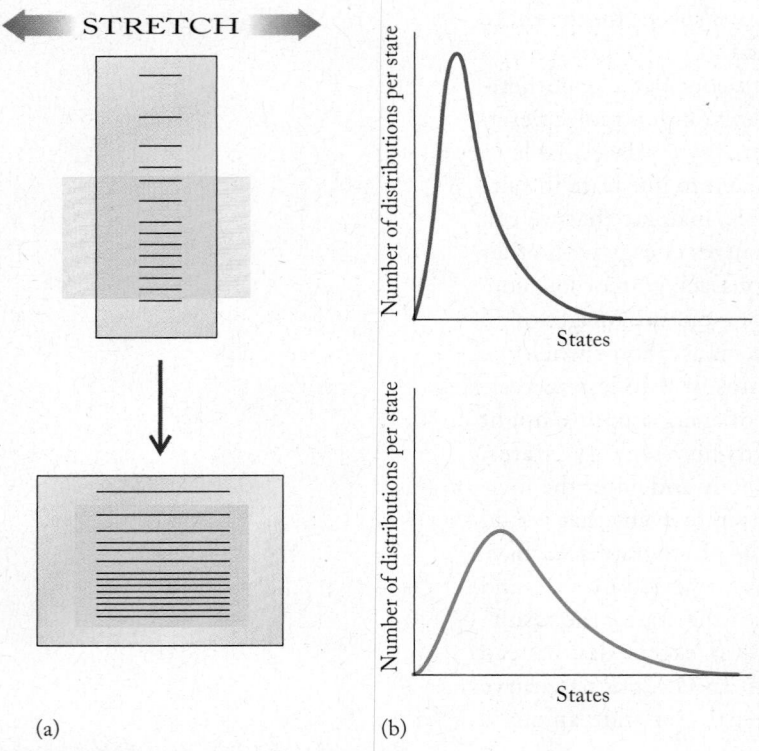

STRETCH

States

Number of distributions per state

States

Number of distributions per state

(a)   (b)

FIGURE 14.5 (a) Accessible translational energy levels in this system are those covered by the yellow shading. When the size of the system is increased by stretching it, the energy levels move closer together vertically. The size of the window does not change, but the distance between adjacent levels becomes smaller, and more levels fall within the yellow area and become accessible. (b) The increased number of states is shown by the change in the Boltzmann distribution. The area under the curve is the same in both cases, but the curve corresponding to the larger system is shorter (fewer molecules per state) but extends farther to the right (more accessible states).

We now go back to the example of air escaping from a punctured tire and consider the process in light of the concept of microstates. Whenever the volume of a system increases, as when air initially confined inside a tire leaks out, the translational energy levels of the system move closer together. To get a mental image of how this works, imagine printing horizontal lines representing translational energy levels on a rubber sheet, as shown in Figure 14.5. If you pull on the sides of the sheet, the width increases, the height decreases, and the lines (which represent the energy levels) get closer together.

When the separation between energy levels decreases, the number of energy levels to which the molecules have access, represented by the number of lines inside the yellow area, increases. If the temperature does not change, then the population of molecules has the same total energy as it did before the volume increased, but the molecules carrying that energy have access to a greater number of levels. That is what "energy spreading out" means. The energy is less localized because the molecules carrying it can be in a greater number of accessible energy levels in the larger space. The larger the number of accessible energy levels, the larger is the number of *different distributions of molecules*, and the larger is the number of microstates. In terms of the tire, the gas that was previously confined within a small volume expands to occupy a larger volume, thereby increasing the number of energy levels over which the same amount of total energy can be distributed. The resulting increase in the number of distributions of particles means an increase in the number of microstates, which is what an increase in entropy means at a molecular level.

CONCEPT TEST

As in the previous Concept Test, you are asked to distribute four identical chairs on steps leading up to a stage, one chair per step. Because someone felt the steps were too large, a new set of five steps is available on which the chairs can be placed. The chairs are numbered 1, 2, 3, and 4. How many different microstates for the chairs are possible?

A Mathematical View of Entropy Microstates

So far in this section we have described both the macroscopic view of entropy (energy spreading out) and the microscopic view (increase in number of microstates)—but only in words. To relate entropy and microstates mathematically, Boltzmann derived the equation

$$S = k \ln W \qquad (14.1)$$

where S is entropy, W is the number of microstates, and $k = 1.38 \times 10^{-23}$ J/K, the *Boltzmann constant*, which is equal to R/N_A (where R is the universal gas constant and N_A is Avogadro's number). As we discuss later in the chapter, the entropy of some substances has been determined experimentally, which means that for these substances we can use Equation 14.1 to calculate the number of microstates.

In the next section, we will see how Boltzmann's equation provides a connection between the macroscopic view of entropy, which depends on temperature, and the microscopic view, which depends on the position and motion of every particle in a system. Microstates provide a theoretical link between these two views. Equation 14.1 tells us that entropy increases as the number of microstates increases.

14.2 Thermodynamic Entropy

When anything cold comes into contact with warmer surroundings, the cold object spontaneously absorbs energy from the surroundings. This is another example of an endothermic process that is spontaneous. Let's look at what happens at the microscopic level when a simple spontaneous process—the melting of ice—takes place at room temperature.

Isothermal Processes

When an ice cube spontaneously melts at room temperature, energy flows from the surroundings into the ice cube. The molecules of H_2O in ice occupy fixed positions and experience only vibrational motion (Figure 14.6a), which means that molecules in ice have fewer accessible microstates than molecules in liquid water.

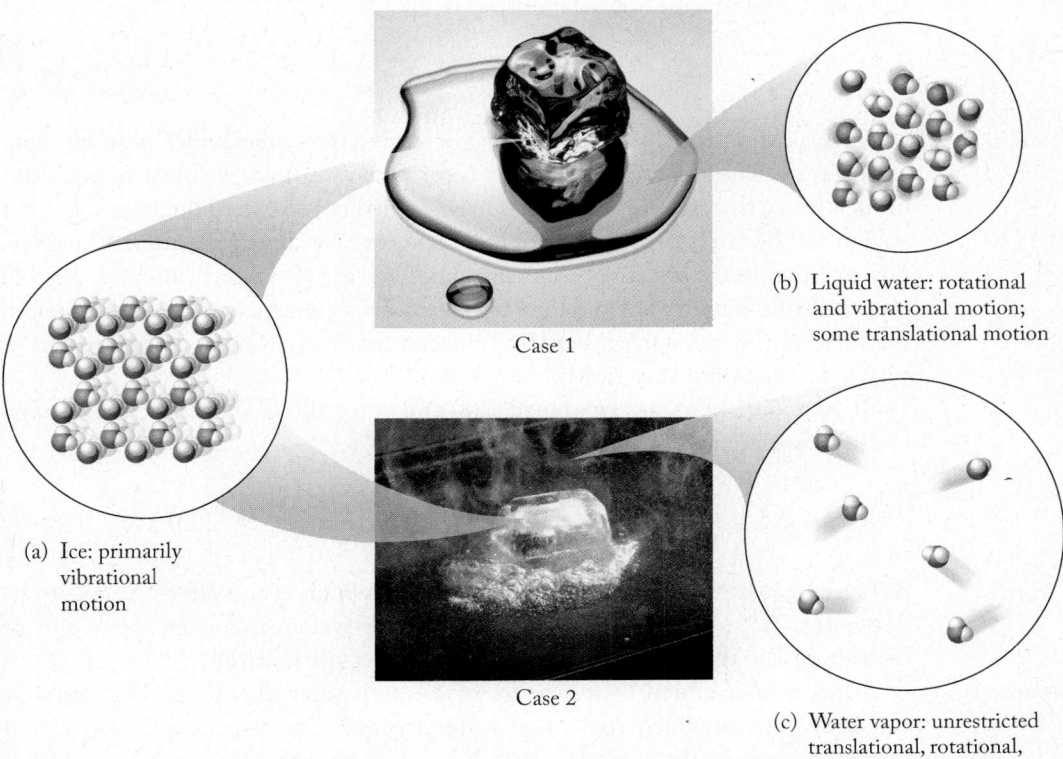

Case 1

Case 2

(a) Ice: primarily vibrational motion

(b) Liquid water: rotational and vibrational motion; some translational motion

(c) Water vapor: unrestricted translational, rotational, and vibrational motion

FIGURE 14.6 (a) Molecules in ice exist in a rigid crystal lattice; they have vibrational motion only. (b) Molecules in liquid water have translational, rotational, and vibrational motion, but their translational motion is restricted. (c) When an ice cube is placed on a hot surface, it rapidly melts, then boils. Molecules in water vapor have translational, rotational, and vibrational motion, and their translational motion is unrestricted.

isothermal process a process that takes place at constant temperature.

reversible process a process that can be run in the reverse direction in such a way that, once the system has been restored to its original state, no net heat has flowed either to the system or to its surroundings.

As the cube melts, the attractive forces that hold the molecules together in the crystal lattice are overcome, and the molecules acquire rotational and limited translational motion (Figure 14.6b), which means the number of microstates increases and therefore the entropy of the system increases. When the liquid evaporates and the molecules enter the gas phase (Figure 14.6c), their translational motion is unrestricted, so many more microstates become available and the entropy increases even further. Thus we have both our macroscopic view of what happens as H_2O changes phase—entropy increases because energy spreads out—and our microscopic view: entropy increases because the number of microstates increases.

As we saw in Chapter 5, ice melts at 0°C and remains at 0°C while changing from solid to liquid. When heated to 100°C, liquid water boils and enters the gas state at that temperature. Any process that takes place at a constant temperature is an **isothermal process**, which means that melting and boiling a pure substance are isothermal processes. Entropy is a state function (just as enthalpy H is), which means that the change in entropy in a system undergoing any change depends only on the initial and final states of the system, not on the path taken for the change. Thus we can say that the *entropy change* ΔS for a process carried out by (or on) a system is defined as

$$\Delta S_{sys} = S_{final} - S_{initial}$$

From Chapter 5 we recall that the enthalpy change for a process is related to the heat q for the process. The entropy change for a process is also related to q. For any isothermal process, the relationship is

$$\Delta S_{sys} = \frac{q_{rev}}{T} \tag{14.2}$$

CONNECTION State functions were defined in Chapter 5 along with q as a measure of heat.

where q_{rev} is the quantity of heat entering or leaving the system and T is the absolute temperature. The subscript on q_{rev} stands for *reversible*. A **reversible process** is run in such a way that, once a change has been carried out (forward direction) and the system has been restored to original state (reverse direction), no net heat has flowed either to the system or to its surroundings. In other words, after a reversible process is run first forward and then backward, everything is *exactly* as it was before the forward process started. For this to occur, the changes must be carried out in very small steps and very slowly.

It takes 6.01 kJ of energy to melt 1.00 mol of ice at 0°C. Therefore the entropy change associated with this phase change is

$$\Delta S_{sys} = \frac{q_{sys}}{T} = \frac{(1.00 \text{ mol})(6.01 \times 10^3 \text{ J/mol})}{273 \text{ K}} = 22.0 \text{ J/K}$$

Because heat flows into the ice, q_{sys} is positive, which also makes ΔS_{sys} positive. A positive ΔS_{sys} means that the entropy of the system increases as a result of melting. (Note that the units of entropy are joules per kelvin.)

CONNECTION In Chapter 10 we defined the melting point as the temperature at which a solid transforms into a liquid at the same rate at which the liquid transforms into the solid.

If no heat is added to a mixture of ice and water at 0°C, the ice melts at the same rate at which the liquid water freezes. The two states coexist, and no net change in the quantity of either ice or water occurs over time. What does this mean in terms of the entropy of the system (ice and water)? We just calculated that melting 1.00 mol of ice at 0°C results in an entropy increase of +22.0 J/K. The reverse process, freezing 1.00 mol of water at 0°C, results in an entropy decrease of −22.0 J/K. Therefore the net entropy change in the system is 0 J/K. Because the two processes occur isothermally, the surroundings experience no change in temperature and therefore no change in entropy. If we define our *universe* here in the narrow sense introduced in Section 5.2—the system and its immediate surroundings—the entropy change of the universe is

$$\Delta S_{univ} = \Delta S_{sys} + \Delta S_{surr} \tag{14.3}$$

Let's explore how the temperature of a system's surroundings can influence the value of ΔS_{univ} by revisiting the flows of energy and resulting entropy changes that occur when ice melts in the two cases illustrated in Figure 14.6: (1) melting 1.00 mole of ice at 0°C on a countertop at 20°C (293 K) to form a pool of water at 0°C that then warms to room temperature and (2) melting 1.00 mole of ice at 0°C on a stove at 100°C (373 K) to form a pool of water at 0°C that then warms to 100°C. We can consider the changes in each case as taking place in two steps: first, the isothermal melting of the solid, then the warming of the liquid (which is of course not an isothermal process).

In the isothermal step of case 1, the flow of heat required to melt 1.00 mole of ice is $q_{sys} = +6.01 \times 10^3$ J, and the entropy change of the system (the ice) is $\Delta S_{sys} = +22.0$ J/K. The quantity of heat q_{surr} lost by the countertop in contact with the ice as the ice melts at 0°C must be the same magnitude as q_{sys} but opposite in sign: $q_{surr} = -6.01 \times 10^3$ J/mol. Therefore the entropy change of the surroundings is, from Equation 14.2,

$$\Delta S_{surr} = \frac{q_{surr}}{T} = \frac{(1.00 \text{ mol})(-6.01 \times 10^3 \text{ J/mol})}{293 \text{ K}} = -20.5 \text{ J/K}$$

The total entropy change as a result of this process of 1.00 mole of ice melting to liquid water at 0°C is positive:

$$\Delta S_{univ} = \Delta S_{sys} + \Delta S_{surr} = (+22.0 \text{ J/K}) + (-20.5 \text{ J/K})$$
$$= 1.5 \text{ J/K}$$

Notice the relationship between ΔS_{sys} and ΔS_{surr}: The magnitude of the entropy increase in the cooler object (the system in this case) is larger than the magnitude of the entropy drop in the warmer object (the surroundings). Our experience tells us that the melting of ice at room temperature is spontaneous. We now see that the overall change in entropy of the universe ΔS_{univ} is positive for this spontaneous process.

What about warming the water from 0°C to 20°C? This step is not an isothermal process (because the temperature changes). As the water warms, it experiences another increase in entropy because a rise in temperature increases both the average kinetic energy of the molecules in the sample and the number of accessible microstates.

You may wonder why we do not bother to consider the change in temperature of the surroundings in any of these examples. The answer lies within your experience: how much does the temperature of the surroundings change as one ice cube (1 mol = 18.0 g) melts or freezes? It does not change, at least not by a measurable amount. The surroundings may be treated in these three cases, and indeed in most cases, as a huge constant-temperature *heat source* or *heat sink*, depending on the direction in which heat flows in the process being examined. The surroundings are so vast relative to the size of the systems studied in chemical thermodynamics that we can always take small quantities of heat away from the surroundings, or dump small quantities of heat into them, and not change their temperature.

In the case we just considered, the difference between the temperature of the system and the temperature of the surroundings was only 20°C. What is the effect on the entropy change during an isothermal process when the temperature difference during the isothermal step is larger, as in our case 2, where the temperature of the system is again 0°C but the temperature of the surroundings is 100°C? In this case,

$$\Delta S_{surr} = \frac{(1.00 \text{ mol})(-6.01 \times 10^3 \text{ J/mol})}{373 \text{ K}} = -16.1 \text{ J/K}$$

and the total entropy change is larger than in case 1:

$$\Delta S_{univ} = \Delta S_{sys} + \Delta S_{surr} = (+22.0 \text{ J/K}) + (-16.1 \text{ J/K})$$
$$= 5.9 \text{ J/K}$$

Comparing these isothermal steps with similar situations we explored earlier, we can conclude that:

➤ Whenever a system is cooler than its surroundings, heat transfer into the system is a spontaneous process.
➤ The greater the temperature difference between system and surroundings, the greater is the entropy increase in any isothermal process that occurs in the system.

The water formed at 0°C warms up to stove temperature and thereby experiences another increase in entropy, but via a nonisothermal process, as the increasing average kinetic energy of the molecules is accompanied by an increase in the number of accessible energy levels. The greater the temperature change, the greater is the number of accessible energy levels, and the greater is the number of microstates, so we conclude that the water warmed to 100°C experiences a greater increase in entropy than the water warmed to 20°C.

In the examples thus far, we have considered spontaneous processes involving heat flowing from warmer surroundings into a cooler system. How does entropy change when the temperature of the surroundings is *lower* than the temperature of the system? Suppose a tray containing 1.00 mole of liquid water at 0°C is placed in a freezer at −10°C. We know the water will freeze. Because the temperature of the surroundings (the freezer) is lower than the temperature of the liquid water, heat spontaneously flows from the water into the surroundings. The net entropy change for this process is

$$\Delta S_{univ} = \Delta S_{sys} + \Delta S_{surr}$$
$$= \frac{(1.00 \text{ mol})(-6.01 \times 10^3 \text{ J/mol})}{273 \text{ K}} + \frac{(1.00 \text{ mol})(+6.01 \times 10^3 \text{ J/mol})}{263 \text{ K}}$$
$$= (-22.0 \text{ J/K}) + (+22.9 \text{ J/K})$$
$$= 0.9 \text{ J/K}$$

Once again there is an increase in the entropy of the universe as heat flows spontaneously from the warmer object (liquid water at 0°C) into the colder surroundings (the freezer at −10°C). The ice that forms at 0°C then cools further to the temperature of the freezer.

Is ΔS_{univ} greater than, less than, or equal to zero when water vapor exhaled by the Inuit hunter in Figure 14.7 condenses to droplets of water that then turn to crystals of ice?

A Closer Look at Reversible Processes

In Equation 14.2, we specified that q is the heat from a reversible process. *Reversibility* comes up frequently in thermodynamics, so it is helpful to understand what it means and why it is used.

Equation 14.1 relates entropy to the number of microstates W, but another way to calculate entropy is to consider entropy changes in terms of the heat q_{rev} exchanged in an isothermal process carried out at some absolute temperature T:

$$\Delta S = \frac{q_{rev}}{T} \tag{14.2}$$

It is important to understand that reversible processes are idealizations; real processes never work this way. But the *idea* of a reversible process provides us with a starting point for calculations involving real processes.

Real processes approach reversibility if they are carried out in a series of infinitesimally small steps. For example, if we transfer heat from surroundings at 273.0001 K to an object at 273.0000 K, that small step approximates a reversible process; the change in temperature is so small that we can report that no change has occurred. We could in principle heat an object reversibly by going through a series of small steps like this one. Of course, it would take a very long time, but in thermodynamics time is not important.

Why do we bother to think about a method that is essentially not possible? The answer is that considering reversible processes makes analysis simpler. Just remember, a reversible process is ideal, not real, and evaluating the ideal can be a very useful and insightful exercise. (Recall how in our discussion of air escaping from a tire we assumed no temperature change. In effect, we carried out a thought experiment in which the process occurred reversibly.) Entropy changes calculated for reversible processes are always idealized and always give the *minimum* value of ΔS for a given process. When actual processes take place in the real world, the changes in entropy are larger. In the ideal world of reversible processes in thermodynamics, entropy increases and we always lose; in the real world, we lose even more.

Entropy Changes for Some Common Processes

Let's consider the meaning of Equation 14.3 in the context of another spontaneous process: the dissolution of ammonium nitrate in a cold pack (Figure 14.8). Among the first-aid supplies carried by athletic trainers are cold packs: plastic bags containing water in one compartment and solid ammonium nitrate in a second compartment. When an athlete is injured, a trainer activates the cold pack by breaking the partition separating the compartments, mixing the $NH_4NO_3(s)$ and water. The dissolution of the salt is spontaneous, but the process is endothermic, so the water cools considerably, providing a cold compress to ease pain and reduce swelling.

Why is this endothermic process spontaneous? From Section 14.1, we know that spontaneous processes are accompanied by an increase in the entropy of the universe, $\Delta S_{univ} > 0$. How does this happen in our example of a cold pack? The

FIGURE 14.7 On a cold day, the water vapor in your breath forms a cloud of condensed water droplets in the cold air. Spending an extended period of time outdoors under these conditions may even lead to frost forming on facial hair, as shown on this Inuit hunter in the Northwest Territories of Canada.

▶‖ **CHEMTOUR** Dissolution of Ammonium Nitrate

◉◉ **CONNECTION** The ion-dipole interactions that promote the solubility of ionic compounds in water were described in Chapter 10.

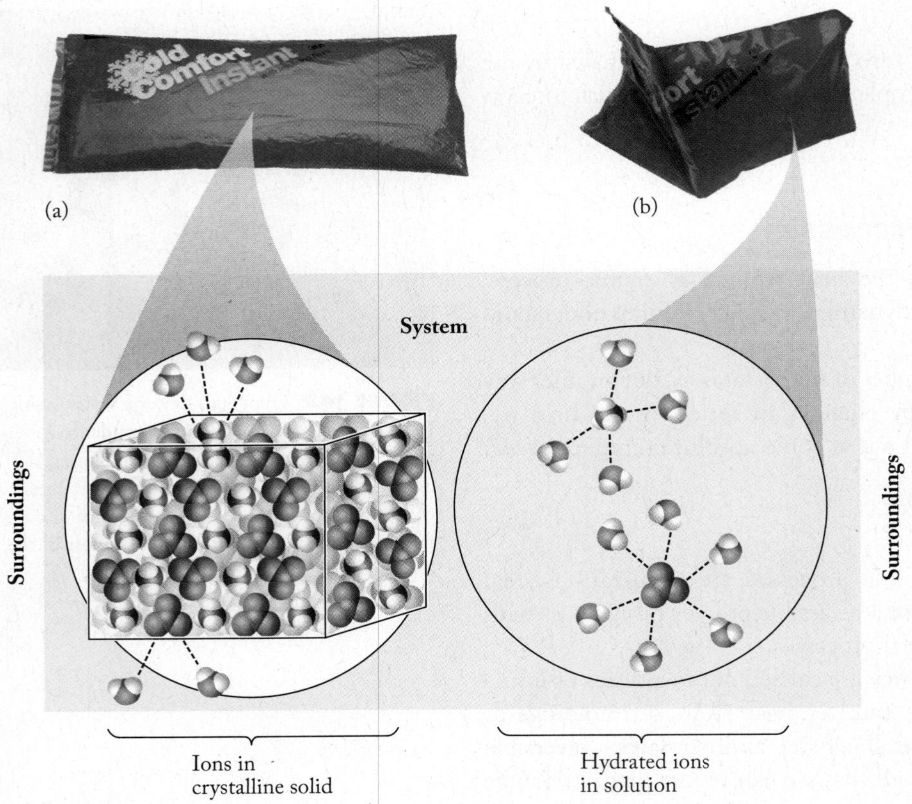

(a)

(b)

System

Surroundings

Surroundings

Ions in
crystalline solid

Hydrated ions
in solution

FIGURE 14.8 When a supply of ice is unavailable, a chemical cold pack may be used. (a) Many of these packs consist of a bag of water plus a sealed pouch containing solid ammonium nitrate. (b) The pack is typically activated by breaking the pouch, causing the NH_4NO_3 to dissolve in the water. Cold packs are cold because of the endothermic dissolution of ammonium nitrate in water. Dissolving the ionic solute increases the freedom of motion of the ions, which are hydrated by water molecules. Even though the dissolution is endothermic ($\Delta H > 0$), it is spontaneous because the result is a positive change in the entropy of the universe.

dissolution of any solid solute increases the *freedom of motion* of the atoms, ions, or molecules that make up the solid. We can think of entropy in terms of an increased number of microstates for a group of particles as well as an increase in the randomness of a system. Even though the originally ordered arrangement of dipoles in the water in the cold pack before the compartments are broken is disrupted by the dissolved ions, the water molecules reorder as they orient themselves around the ions. Because the ions have much greater freedom of motion in solution than in the solid state, the net effect is an overall positive change in the entropy of the system ($\Delta S_{sys} > 0$) because the molecules and ions in a solution of ammonium nitrate are in a more random arrangement than when the solute and solvent are separated.

Similarly, dilution of a concentrated solution when solvent is added is a spontaneous process. If you add water to antifreeze, the water spontaneously dilutes the antifreeze. No significant change in temperature occurs as the two liquids dissolve in each other, so spontaneity is not related to a flow of heat either into or out of the solution. No significant heat flow means no significant energy dispersion into the surroundings, which means no significant change in the entropy of the surroundings. Therefore, spontaneity ($\Delta S_{univ} > 0$) must be linked to the increase in entropy that happens when solute and solvent molecules mix together and each molecule disperses into a larger overall volume ($\Delta S_{sys} > 0$).

Spontaneous processes *must* result in an increase in the entropy of the universe. No violation of this concept has ever been observed, and it is so widely accepted as true that most people and virtually all scientists have stopped looking for violations. This is why so-called perpetual motion machines (also called free-energy devices), which supposedly either store or put out more energy than is put into them, are all hoaxes; everything ultimately runs down. The combinations of ΔS_{sys} and ΔS_{surr} that produce a positive value for ΔS_{univ} are shown in Table 14.1 and Figure 14.9.

TABLE 14.1 Spontaneity of Reaction as a Function of ΔS_{sys} and ΔS_{surr}

ΔS_{sys}	ΔS_{surr}	Spontaneity of Process	Reference
>0	>0	Always spontaneous	Figure 14.9(a)
<0	>0	Spontaneous if $|\Delta S_{sys}| < |\Delta S_{surr}|$	Figure 14.9(b)
		Nonspontaneous if $|\Delta S_{sys}| > |\Delta S_{surr}|$	Figure 14.9(e)
>0	<0	Spontaneous if $|\Delta S_{sys}| > |\Delta S_{surr}|$	Figure 14.9(c)
		Nonspontaneous if $|\Delta S_{sys}| < |\Delta S_{surr}|$	Figure 14.9(f)
<0	<0	Always nonspontaneous	Figure 14.9(d)

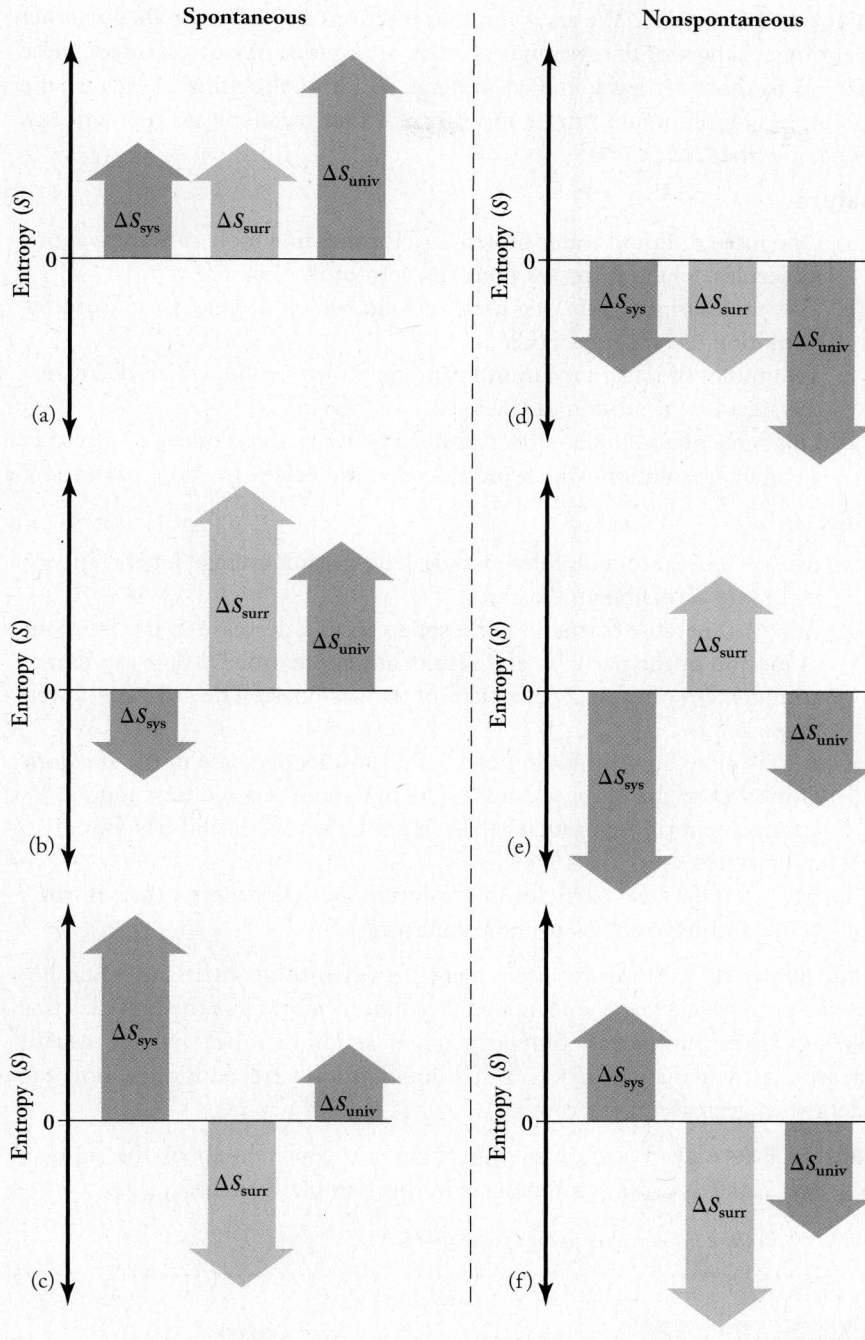

FIGURE 14.9 The relationships between ΔS_{sys}, ΔS_{surr}, and ΔS_{univ} in spontaneous and nonspontaneous processes. A process is spontaneous whenever the entropy of the universe increases: $\Delta S_{univ} > 0$. A process is nonspontaneous whenever the entropy of the universe decreases: $\Delta S_{univ} < 0$.

SAMPLE EXERCISE 14.1   Predicting the Sign of Entropy Changes

Predict whether ΔS_{sys} is positive or negative for each of these spontaneous isothermal processes:

   a. $H_2O(\ell) \rightarrow H_2O(g)$

   b. $NH_3(g) + HCl(g) \rightarrow NH_4Cl(s)$

   c. $Ag^+(aq) + Cl^-(aq) \rightarrow AgCl(s)$

   d. $C_{12}H_{22}O_{11}(s) \xrightarrow{H_2O} C_{12}H_{22}O_{11}(aq)$

Collect and Organize We are given four reactions and are to predict whether the entropy change for the system is positive or negative. We can compare these processes to those we have studied in detail so far: the melting of ice and the dissolution of ammonium nitrate in water. All four reactions are spontaneous, so we know that $\Delta S_{univ} > 0$.

Analyze

a. One mole of liquid water molecules becomes one mole of water vapor molecules, which increases their freedom of motion.

b. Two moles of gas form one mole of solid, which decreases the freedom of motion of NH_3 and HCl.

c. Two moles of ions in solution form one mole of solid, which decreases the freedom of motion of the ions.

d. One mole of a solid dissolves, forming one mole of molecules dispersed in an aqueous solution, which increases the molecules' freedom of motion.

Solve

a. $\Delta S_{sys} > 0$ because molecules of a gas have greater average kinetic energy and more accessible microstates.

b. $\Delta S_{sys} < 0$ because formation of a solid causes a decrease in the freedom of motion of the particles and results in a more ordered (less random) arrangement of the particles than in the gas phase. The solid has fewer microstates.

c. $\Delta S_{sys} < 0$ because formation of a solid causes a decrease in the freedom of motion of the particles and results in a more ordered (less random) arrangement of the particles than in a solution. The solid has fewer microstates.

d. $\Delta S_{sys} > 0$ because particles in a solution are less ordered than in the solid, and have access to more microstates.

Think about It Entropy generally increases when solids melt and when liquids vaporize because of the increased freedom of motion of the particles that make up these substances. Similarly, when solids dissolve, entropy usually increases, but when a gas dissolves in a liquid, it loses freedom of motion and undergoes a decrease in entropy.

Practice Exercise What, if anything, can you conclude about the signs of ΔS_{sys}, ΔS_{surr}, and ΔS_{univ} in each process in this Sample Exercise?

(Answers to Practice Exercises are in the back of the book.)

■

CONCEPT TEST

In which one of the following processes are ΔS_{sys} and ΔS_{surr} both positive, and in which one are ΔS_{sys} and ΔS_{surr} both negative? (a) combustion of methane; (b) photosynthesis of glucose from CO_2 and water

14.3 Absolute Entropy, the Third Law of Thermodynamics, and Structure

We know that the entropy of a system depends on temperature. Higher temperature means higher kinetic energy for the particles making up any substance, which means more vibrational, rotational, and translational motion, which means more microstates, which means more entropy. Conversely, lower temperatures mean less of all these quantities. If we lower the temperature of a substance to absolute zero, in principle all motion ceases. If the particles of a crystalline solid are perfectly aligned and motionless at 0 K, only one distribution of particles is possible and hence only one microstate exists, and from Equation 14.1 we calculate that the entropy is zero:

$$S = k(\ln W) = k(\ln 1) = k(0) = 0$$

This situation is described by the **third law of thermodynamics**: the entropy of a perfect crystal is zero at absolute zero (0 K). By *perfect* we mean that all particles are exactly aligned with one another in the crystal lattice (Figure 14.10) and there are no imperfections in the crystal. Only one arrangement that is perfect can exist.

Setting a zero point on the entropy scale—$S = 0$ at 0 K—allows scientists to compute **absolute entropy** values for pure substances. For example, the absolute entropy of solid NaCl is 72.1 J/(mol · K) at 298 K. Absolute entropies are determined from careful measurements of either the molar heat capacity or the specific heat of the material as a function of temperature. Our ability to measure absolute entropies contrasts with the observation in Chapter 5 that the measurement of *absolute enthalpy* (*H*) is difficult. While we can only measure changes in enthalpy, Δ*H*, we can measure absolute entropies where the term *absolute* refers to the true value of *S*, not a change in entropy from some arbitrary reference.

The absolute entropy of a substance is usually reported as the **standard molar entropy (*S*°)**, the value of *S* for one mole of a pure substance at 298 K and 1 bar (~1 atm) of pressure (Table 14.2) in its standard state (Table 14.3). Unless otherwise indicated, when we speak of entropies in this book, we are speaking about standard molar entropies.

The values in Table 14.2 for liquid water and water vapor illustrate an important difference in the entropies of liquids and gases: the molecules in a gas under standard conditions are much more widely dispersed than the molecules in a

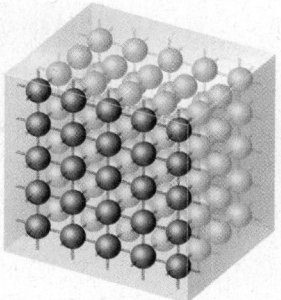 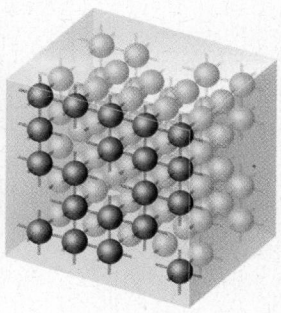

(a) Perfect crystal at 0 K: $S = 0$

All particles in exact positions; all positions occupied

(b) Imperfect crystal at 0 K: $S > 0$

Each empty site is an imperfection in crystal

FIGURE 14.10 A perfect crystal at 0 K is the basis for defining absolute entropy. (a) In a perfect crystal at absolute zero, all the atoms, ions, or molecules are arranged perfectly in the crystal lattice, and $S = 0$. (b) If a crystal has any defects, even at absolute zero, $S > 0$. Here the defects are empty sites.

CONNECTION Molar heat capacity and specific heat were defined in Chapter 5.

CONNECTION In Chapter 5 we defined the *standard state* of a pure substance as its most stable form at 1 atm and some specified temperature, often 25.0°C.

third law of thermodynamics the entropy of a perfect crystal is zero at absolute zero.

absolute entropy the entropy change of a substance taken from $S = 0$ (at $T = 0$ K) to some other temperature. Absolute entropies are determined from the temperature dependence of the molar heat capacity.

standard molar entropy (*S*°) the absolute entropy of 1 mole of a substance in its standard state.

TABLE 14.2	Selected Standard Molar Entropy Values[a]			
Formula	*S*°, J/(mol · K)	Formula	Name	*S*°, J/(mol · K)
$Br_2(g)$	245.5	$CH_4(g)$	Methane	186.2
$Br_2(\ell)$	152.2	$CH_3CH_3(g)$	Ethane	229.5
$C_{diamond}(s)$	2.4	$CH_3OH(g)$	Methanol	239.9
$C_{graphite}(s)$	5.7	$CH_3OH(\ell)$		126.8
$CO(g)$	197.7	$CH_3CH_2OH(g)$	Ethanol	282.6
$CO_2(g)$	213.8	$CH_3CH_2OH(\ell)$		160.7
$H_2(g)$	130.6	$CH_3CH_2CH_3(g)$	Propane	269.9
$N_2(g)$	191.5	$CH_3(CH_2)_2CH_3(g)$	*n*-Butane	310.0
$O_2(g)$	205.0	$CH_3(CH_2)_2CH_3(\ell)$		231.0
$H_2O(g)$	188.8	$C_6H_6(g)$	Benzene	269.2
$H_2O(\ell)$	69.9	$C_6H_6(\ell)$		172.9
$NH_3(g)$	192.5	$C_{12}H_{22}O_{11}(s)$	Sucrose	360.2

[a] Values for additional substances are given in Appendix 4.

TABLE 14.3	Standard States of Substances and Solutions		
Physical State	**Standard State**	**Pressure**[a]	**Temperature**[b]
Solid	Pure solid, most stable allotrope of an element	1 bar	25°C
Liquid	Pure liquid	1 bar	25°C
Gas	Pure gas	1 bar	25°C
Solution	1 M	1 bar	25°C

[a] Since 1982 1 bar has been the standard pressure for tabulating all thermodynamic data. Prior to 1982 standard pressure was 1 atmosphere (atm) = 1.01325 bar.
[b] The thermodynamic data in Appendices 4–6 and used elsewhere in this book are based on a temperature of 25°C (298 K). NOTE: this temperature is not the STP temperature we use for gases (see Chapter 6), which is 273 K.

liquid, and the entropies of the different phases of a given substance at a given temperature follow the order $S_{solid} < S_{liquid} < S_{gas}$.

CONCEPT TEST ....

Table 14.2 lists standard molar entropies for liquid water and water vapor at 298 K. Why is $S°$ for ice not listed?

....

CONCEPT TEST ....

Does an *amorphous* (noncrystalline) solid have an absolute entropy equal to zero at 0 K?

....

The entropy changes that occur as ice is heated are shown in Figure 14.11. Note the jump in entropy as the ice melts and the even bigger jump as the liquid water

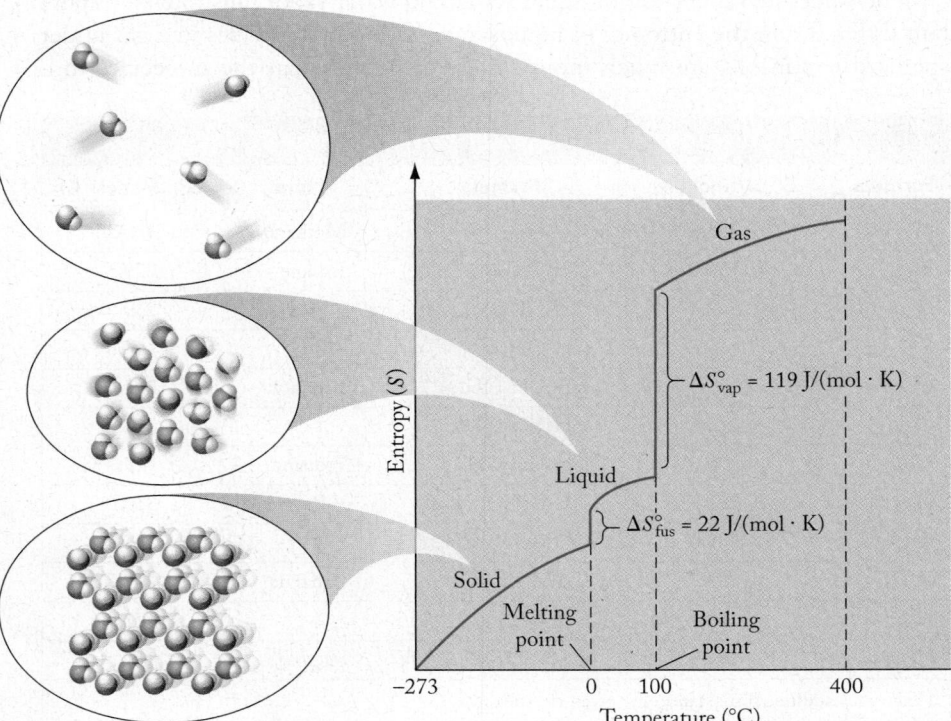

FIGURE 14.11 Abrupt increases in entropy accompany changes of state, with the greater increase occurring during the transition from liquid to gas.

vaporizes. Also note that the lines between the phase changes are curved. The change in entropy with temperature is not linear because heating a substance at a higher temperature produces a smaller entropy increase than adding the same quantity of heat to the same substance at a lower temperature, as predicted by Equation 14.2.

The data in Table 14.2 suggest another important factor that influences standard molar entropy: molecular structure. In the liquids listed, note that $S°_{water} < S°_{methanol} < S°_{benzene}$. To explore this observation, consider these standard molar entropies for the C_1 to C_4 alkanes in natural gas:

Compound:	CH_4	CH_3CH_3	$CH_3CH_2CH_3$	$CH_3(CH_2)_2CH_3$
$S°$, J/(mol · K):	186	230	270	310

Note that the standard molar entropy increases with increasing molecular size (Figure 14.12). We can explain this trend by considering the internal motion of the atoms in these molecules. The more bonds there are in a molecule, the more opportunities for internal motion, and the greater the standard molar entropy of the molecule.

Another structural feature that influences standard molar entropy is rigidity. Diamond and graphite are both polymeric network solids with a large number of carbon–carbon bonds; however, they have different standard molar entropies. The more-rigid diamond molecule has less entropy [$S° = +2.4$ J/(mol · K)] than the less-rigid graphite form of carbon [$S° = +5.7$ J/(mol · K)].

CH_4

CH_3CH_3

$CH_3CH_2CH_3$

$CH_3CH_2CH_2CH_3$

FIGURE 14.12 Methane, ethane, propane, and butane have different standard molar entropies related to their size and structure.

> **CONCEPT TEST**
>
> Does Figure 14.11 illustrate any isothermal processes? If so, which ones?

Let's summarize the factors that affect entropy. Entropy increases when (1) temperature increases, (2) volume increases, or (3) the number of independently moving particles increases. Entropy is increased in these cases because each change increases the number of microstates. We can often make qualitative predictions about entropy changes based on these three factors, even if we have no thermodynamic data about the system under examination.

When evaluating a chemical reaction, remember that the reaction is the system. For example, when a hydrocarbon such as propane burns in air,

$$C_3H_8(g) + 5\,O_2(g) \rightarrow 3\,CO_2(g) + 4\,H_2O(g)$$

6 moles of gaseous reactants combine to form 7 moles of gaseous products. The number of moles of gas increases. When the reactants and products are at the same temperature and pressure (which we assume they are), 7 moles of product gases occupy a greater volume than 6 moles of reactant gases. In other words, volume increases. Thus, the reaction produces an increase in the number of particles and an increase in the volume they occupy. Both of these changes cause an increase in entropy.

⊗⊗ **CONNECTION** In Chapter 6 Avogadro's law told us that the number of moles of gas is directly proportional to the volume occupied by the gas at constant temperature and pressure.

> **SAMPLE EXERCISE 14.2** **Predicting the Sign of ΔS_{sys}**
>
> Predict whether each reaction results in an increase or decrease in the entropy of the system. Assume the reactants and products are at the same temperature and pressure.

a. $CaCO_3(s) + 2\,HCl(aq) \rightarrow CaCl_2(aq) + CO_2(g) + H_2O(\ell)$

b. $NH_3(g) + BF_3(g) \rightarrow NH_3BF_3(s)$

Collect and Organize We are given two reactions and are to predict whether they result in an increase or a decrease in the entropy of the system. The reaction in each case is the system.

Analyze Entropy increases if temperature, volume, or the number of independently moving particles increases. The reactants and products are at the same temperature, so temperature is not a factor. We must evaluate each reaction in terms of volume and number of particles.

Solve

a. A total of 3 moles of reactants in the solid or liquid phase (condensed phases) yields 2 moles of products in the liquid phase (a condensed phase) and, more significantly from the point of view of entropy, 1 mole of gaseous product. Recall that molecules in the gas phase (Figure 14.6c) have many more microstates available and have large entropies, ensuring that this reaction has a positive ΔS_{sys}.

b. There are 2 moles of gas in the reactants but no gaseous products. Fewer microstates are available to the product (a solid) than to the reactants (gases). Therefore this reaction results in a loss in entropy by the system ($\Delta S_{sys} < 0$).

Think about It In Section 14.1 we remarked that entropy changes are reflected in changes in the randomness of the system. When a chemical reaction produces a gas—as in part a—the gas represents a more random arrangement (more microstates). The opposite situation exists in part b, where the degree of randomness decreases when gases react to form a solid.

Practice Exercise Which processes result in an increase in entropy of the system? (a) amorphous sulfur crystallizes; (b) solid carbon dioxide sublimes at room temperature; (c) iron rusts

14.4 Calculating Entropy Changes

We can calculate the change in entropy under standard conditions for any chemical reaction from the difference in the standard molar entropies of $n_{reactants}$ moles of reactants and $n_{products}$ moles of products:

$$\Delta S_{rxn}^{\circ} = \sum n_{products} S_{products}^{\circ} - \sum n_{reactants} S_{reactants}^{\circ} \qquad (14.4)$$

Each individual S° value for a product or reactant is multiplied by the appropriate number of moles from the balanced chemical equation. In other words, just as we discovered about ΔH° in Chapter 5, entropy is an extensive thermodynamic property that depends on the amount of a substance consumed or produced in a reaction. Standard molar entropies of selected substances are listed in Appendix 4.

SAMPLE EXERCISE 14.3 Calculating Entropy Changes

Calculate ΔS_{rxn}° for the dissolution of ammonium nitrate (Figure 14.13), given the following standard molar entropy values:

$$NH_4NO_3(s) \rightarrow NH_4^+(aq) + NO_3^-(aq)$$

| S° [J/(mol·K)] | 151.1 | 113.4 | 146.4 |

Collect and Organize Entropy changes associated with a chemical reaction depend on the entropies of the reactants and products. We are also given the standard molar entropies for the species involved in the reaction. The reaction takes place under standard conditions.

Analyze We can calculate ΔS_{rxn}° using the given S° values and Equation 14.4. Dissolving an ionic solid in water increases the randomness of the system, so we predict that ΔS_{rxn}° will be positive.

Solve

$$\Delta S_{rxn}^{\circ} = \sum n_{products} S_{products}^{\circ} - \sum n_{reactants} S_{reactants}^{\circ}$$

$$= \left[1.000 \ \text{mol} \times \left(\frac{113.4 \text{ J}}{\text{mol} \cdot \text{K}} \right) + 1.000 \ \text{mol} \times \left(\frac{146.4 \text{ J}}{\text{mol} \cdot \text{K}} \right) \right]$$

$$- 1.000 \ \text{mol} \times \left(\frac{151.1 \text{ J}}{\text{mol} \cdot \text{K}} \right)$$

$$= 108.7 \text{ J/K}$$

Think about It As predicted, the value of ΔS_{rxn}° is positive, so the entropy of the system increases. Because the number of independently moving particles increases, we expect the entropy of the system to increase, so our answer is logical.

Practice Exercise Calculate the standard molar entropy change for the combustion of methane gas using S° values from Appendix 4. Before carrying out the calculation, predict whether the entropy of the system increases or decreases. Assume that liquid water is one of the products.

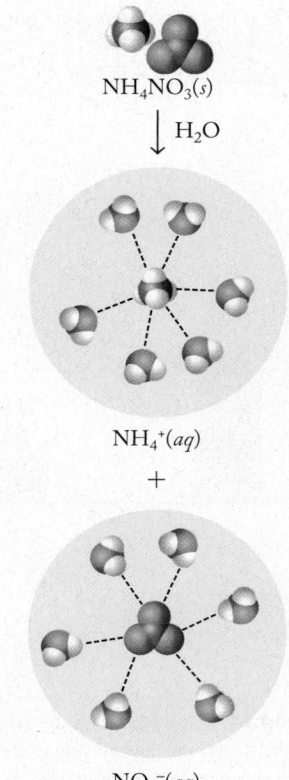

FIGURE 14.13 Dissolution of solid ammonium nitrate into NH_4^+ and NO_3^- ions is endothermic but spontaneous because $\Delta S_{univ} > 0$.

All values calculated using Equation 14.4 are standard molar entropy changes *of the system*. As we have repeatedly seen, however, the surroundings matter as well, even though we tend to treat them as a large heat source or sink. The entropy change experienced by the surroundings of a chemical reaction depends on whether the reaction is exothermic or endothermic.

If we assume the reactions we are studying are isothermal processes occurring at constant pressure, the heat flow is reversible (q_{rev}), allowing us to use Equation 14.2 to determine the change in entropy of the surroundings. Under these conditions, q_{rev} is equal to the enthalpy change of the reaction:

$$\Delta H_{rxn} = q_{rev} \tag{14.5}$$

CONNECTION In Chapter 5 we defined a change in enthalpy (ΔH) as the heat gained or lost in a reaction carried out at constant pressure.

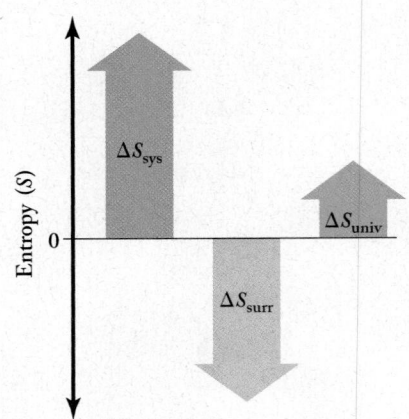

FIGURE 14.14 If the surroundings experience a decrease in entropy, the system must experience an increase in entropy that more than offsets the decrease if the process is spontaneous.

If the reaction takes place under standard conditions, then we can calculate ΔH°_{rxn} for the reaction in Sample Exercise 14.3 by using values from Appendix 4 in Equation 14.6:

$$\Delta H^\circ_{rxn} = \sum n_{products}\Delta H^\circ_{f,products} - \sum n_{reactants}\Delta H^\circ_{f,reactants} \quad (14.6)$$

For the dissolution of 1.00 mole of $NH_4NO_3(s)$—recall that ammonium nitrate is the solid in a cold pack—Equation 14.6 gives $\Delta H^\circ_{rxn} = 28.1$ kJ. The positive value confirms that the process is endothermic. Dissolving 1.00 mole of $NH_4NO_3(s)$ absorbs 28.1 kJ of heat from the surroundings. To calculate the entropy change of the surroundings for the dissolution taking place at constant pressure and 25°C, we use Equation 14.2:

$$\Delta S_{surr} = \frac{q_{surr}}{T} = -\frac{q_{sys}}{T} = -\frac{(1.00 \ \text{mol}) \times \left(\dfrac{28.1 \times 10^3 \ \text{J}}{\text{mol}}\right)}{298 \ \text{K}}$$
$$= -94.3 \ \text{J/K}$$

Note that ΔS_{surr} is negative but ΔS_{sys} (calculated as ΔS°_{rxn} in Sample Exercise 14.3) is positive. Both terms must be considered to determine how the entropy of the universe changes during this process:

$$\Delta S_{univ} = \Delta S_{sys} + \Delta S_{surr} = (108.7 \ \text{J/K}) + (-94.3 \ \text{J/K})$$
$$= 14.4 \ \text{J/K}$$

The entropy change of the universe is positive, which means that the process is spontaneous even though it is endothermic. The value of ΔS_{univ} is positive because the decrease in entropy of the surroundings (negative ΔS_{surr}) is more than offset by an increase in entropy of the system (positive ΔS_{sys}, Figure 14.14).

CONCEPT TEST ··

In which of the following spontaneous processes is there a negative ΔS_{surr} and a positive ΔS_{sys}?

a. A supersaturated solution precipitates a solid to form a saturated solution.

b. Sodium hydroxide dissolves in water, causing the temperature of the solution to increase.

c. Dew forms on a lawn overnight.

d. The valve on a tank of nitrogen is left open, causing frost to form on the outside of the cylinder.

··

Notice one more detail about the dissolution of ammonium nitrate. Because the reaction is endothermic, it takes energy from the surroundings. As discussed in Chapter 5, the amount of heat that the reaction absorbs equals the amount of heat taken from the surroundings:

$$\Delta H_{sys} = -\Delta H_{surr}$$

As we learned earlier in this chapter, however, the entropy change in the system is not the opposite of the entropy change in the surroundings:

$$\Delta S_{sys} \neq -\Delta S_{surr}$$

SAMPLE EXERCISE 14.4 **Predicting the Spontaneity of a Reaction**

Consider the reaction of nitrogen gas and hydrogen gas (Figure 14.15) at 298 K to make ammonia at the same temperature:

$$N_2(g) + 3\,H_2(g) \rightarrow 2\,NH_3(g)$$

a. Before doing any calculations, predict the sign of ΔS°_{sys}.

b. Use data from Table 14.2 to calculate ΔS°_{sys}.

c. Use data from Appendix 4 to calculate ΔS_{surr}.

d. Is the reaction, as written, spontaneous at 298 K and 1 bar pressure?

Collect and Organize We are to predict the sign and calculate the value of the standard molar entropy change for a reaction. We are also to calculate the entropy change for the surroundings and to determine whether the reaction is spontaneous under the stated conditions. We are to use the tabulated data for standard molar entropies found in Table 14.2 and Appendix 4.

Analyze

a. From the balanced equation, we can determine the sign of the change in the number of moles of gaseous reactants and products, and from that determine the sign of ΔS°_{sys}.

b. We need to look up S° values for the reactants and products and use Equation 14.4 to calculate ΔS°_{rxn}, which equals ΔS°_{sys}.

c. To calculate ΔS_{surr}, we need to know how much heat the reaction gives off to the surroundings. We can calculate that from the heat of formation (ΔH°_f) values in Appendix 4.

d. We can use Equation 14.3 to calculate ΔS_{univ} and predict spontaneity based on its sign: the reaction is spontaneous if $\Delta S_{univ} > 0$. It is difficult to predict whether this reaction is spontaneous when reactants and products are in their standard states without knowing the magnitudes of ΔH_{surr} and ΔS_{surr}.

Solve

a. According to the balanced chemical equation, four moles of gaseous reactants are converted into two moles of gaseous products. We predict that this decrease in the number of moles of gases means $\Delta S^{\circ}_{rxn} < 0$.

b. Using data from Table 14.2 in Equation 14.4 gives

$$\Delta S^{\circ}_{rxn} = \sum n_{products} S^{\circ}_{products} - \sum n_{reactants} S^{\circ}_{reactants} = \Delta S^{\circ}_{sys}$$

$$= \left\{ \left[2.000\ \text{mol} \times \left(\frac{192.5\ \text{J}}{\text{mol} \cdot \text{K}} \right) \right] \right.$$

$$\left. - \left[1.000\ \text{mol} \times \left(\frac{191.5\ \text{J}}{\text{mol} \cdot \text{K}} \right) + 3.000\ \text{mol} \times \left(\frac{130.6\ \text{J}}{\text{mol} \cdot \text{K}} \right) \right] \right\}$$

$$= -198.3\ \text{J/K}$$

The entropy change for the reaction is negative, as predicted in part a.

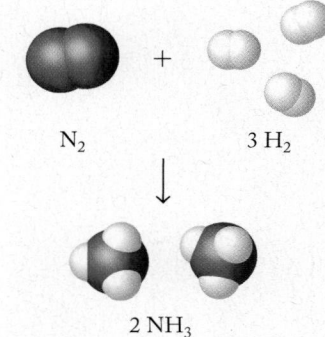

N_2 $3\,H_2$

$2\,NH_3$

FIGURE 14.15 In the synthesis of ammonia, one molecule of nitrogen reacts with three molecules of hydrogen to yield two molecules of ammonia. All substances are gases.

c. If we consider the reaction to be reversible, we can calculate the entropy change of the surroundings from the enthalpy change of the reaction:

$$\Delta H^\circ_{surr} = -\Delta H^\circ_{rxn} = -\left(\sum n_{products} \Delta H^\circ_{f,products} - \sum n_{reactants} \Delta H^\circ_{f,reactants}\right)$$

$$= -\left\{\left[2.000 \text{ mol} \times \left(\frac{-46.1 \text{ kJ}}{\text{mol}}\right)\right]\right.$$

$$\left. - \left[1.000 \text{ mol} \times \left(\frac{0.0 \text{ kJ}}{\text{mol}}\right) + 3.000 \text{ mol} \times \left(\frac{0.0 \text{ kJ}}{\text{mol}}\right)\right]\right\}$$

$$= 92.2 \text{ kJ of heat absorbed by the surroundings}$$

Because we were given the temperature (298 K), we now have all we need to calculate ΔS_{surr}. Combining Equations 14.2 and 14.5 ($\Delta H_{rxn} = q_{rev}$), and changing 92.2 kJ to joules so that the units match, from the perspective of the surroundings, we have

$$\Delta S_{surr} = \frac{\Delta H_{surr}}{T} = \frac{92,200 \text{ J}}{298 \text{ K}} = 309 \text{ J/K}$$

The surroundings experience an increase in entropy.

d. $\Delta S_{univ} = \Delta S_{sys} + \Delta S_{surr} = \Delta S^\circ_{rxn} + \Delta S_{surr}$
$$= -198.3 \text{ J/K} + (+309 \text{ J/K}) = 111 \text{ J/K}$$

The entropy change of the universe is positive, so the reaction is spontaneous as written.

Think about It At the start of this exercise, we were unable to make a prediction about the spontaneity of the reaction because we didn't have an intuitive feel for the magnitudes of ΔH_{surr} and ΔS_{rxn}. As long as the formation of ammonia is exothermic ($\Delta H_{rxn} < 0$), ΔH_{surr} and ΔS_{surr} will be positive. Since $\Delta S_{univ} > 0$, $\Delta S_{surr} > 0$, and $\Delta S_{sys} < 0$, the reaction is an example of the situation in Figure 14.9(b).

Practice Exercise For the reaction $2 H_2(g) + O_2(g) \rightarrow 2 H_2O(\ell)$,

a. Predict the sign of the entropy change for the reaction.
b. Calculate ΔS°_{sys}.
c. Calculate ΔS_{surr} at 298 K.
d. Is the reaction as written spontaneous at 298 K and 1 bar?

CONCEPT TEST

The preparation of ammonia from nitrogen and hydrogen in Sample Exercise 14.4 is predicted to be spontaneous under standard conditions, yet if we mix the two gases at 298 K, no reaction is observed. Explain this observation.

14.5 Free Energy and Free-Energy Change

So far in this text, we have identified two driving forces that make chemical reactions happen:

1. The formation of lower-energy products from higher-energy reactants in exothermic reactions ($\Delta H_{sys} < 0$).

CONNECTION Determining the enthalpy change of reactions was discussed in Chapter 5.

2. The formation of products that have more entropy than the reactants ($\Delta S_{\text{sys}} > 0$).

An exothermic reaction ($\Delta H_{\text{sys}} < 0$) that has a positive entropy change ($\Delta S_{\text{sys}} > 0$) must be spontaneous because heat flowing from the system to the surroundings means $\Delta S_{\text{surr}} > 0$. As we saw in Figure 14.9(a) and Table 14.1, ΔS_{univ} is positive when ΔS_{sys} and ΔS_{surr} are both positive, and a positive ΔS_{univ} is what defines a spontaneous reaction (Figure 14.16a). A reaction that cools its surroundings ($\Delta S_{\text{surr}} < 0$) because it is endothermic ($\Delta H_{\text{sys}} > 0$) and also results in a loss of entropy in the system ($\Delta S_{\text{sys}} < 0$) is never spontaneous because ΔS_{univ} is negative in this case (Figure 14.16b).

Not addressed in this analysis is whether either an endothermic reaction with a positive entropy change ($\Delta S_{\text{sys}} > 0$) or an exothermic reaction with a negative entropy change ($\Delta S_{\text{sys}} < 0$) is spontaneous. Either combination may be spontaneous or not, depending on the relative magnitudes of ΔS_{surr} and ΔS_{sys} as described in Figure 14.9 and Table 14.1.

Because it is not practical to measure changes in the entropy of the universe, we need to measure some other quantity that enables us to predict when a process is spontaneous. The American mathematical physicist J. Willard Gibbs (1839–1903) is credited with developing the mathematical foundation of modern thermodynamics. Among his accomplishments was the development of a way to use ΔH_{sys} and ΔS_{sys} to predict whether a reaction will occur spontaneously at constant pressure and temperature. In doing so, he defined a state function that he named **free energy**, now assigned the symbol G in his honor. The *free* in *free energy* does not mean "at no cost"; it means energy available to do useful work.

We can predict whether or not a reaction is spontaneous by relating G to the entropy and enthalpy changes of the system. Let's derive that relation. We know from Equation 14.3 that $\Delta S_{\text{univ}} = \Delta S_{\text{sys}} + \Delta S_{\text{surr}}$. Furthermore, we know that $-q_{\text{sys}} = +q_{\text{surr}}$ (see, for example, Equation 5.15: $q_{\text{rxn}} = -q_{\text{calorimeter}}$, where the calorimeter is the surroundings), and we have learned in this chapter that $\Delta S_{\text{sys}} = q_{\text{rev}}/T$. Since $-q_{\text{sys}} = +q_{\text{surr}}$, we can say that

$$\Delta S_{\text{surr}} = \frac{-q_{\text{rev}}}{T}$$

Because Equation 14.5 tells us that $\Delta H_{\text{sys}} = q_{\text{rev}}$ at constant pressure,

$$\Delta S_{\text{surr}} = \frac{-q_{\text{rev}}}{T} = \frac{-\Delta H_{\text{sys}}}{T}$$

Therefore, substituting for ΔS_{surr} in Equation 14.3, we see that

$$\Delta S_{\text{univ}} = \Delta S_{\text{sys}} + \frac{(-\Delta H_{\text{sys}})}{T} \qquad (14.7)$$

Equation 14.7 is important because it identifies the change in entropy of the universe exclusively in terms of the system. Rearranging Equation 14.7, we find that

$$-T\,\Delta S_{\text{univ}} = -T\,\Delta S_{\text{sys}} + \Delta H_{\text{sys}}$$
$$= \Delta H_{\text{sys}} - T\,\Delta S_{\text{sys}}$$

The **free-energy change (ΔG)** for a process occurring at constant pressure and temperature is defined as $\Delta G = -T\,\Delta S_{\text{univ}}$, and so we can say

$$\Delta G = \Delta H - T\,\Delta S \qquad (14.8)$$

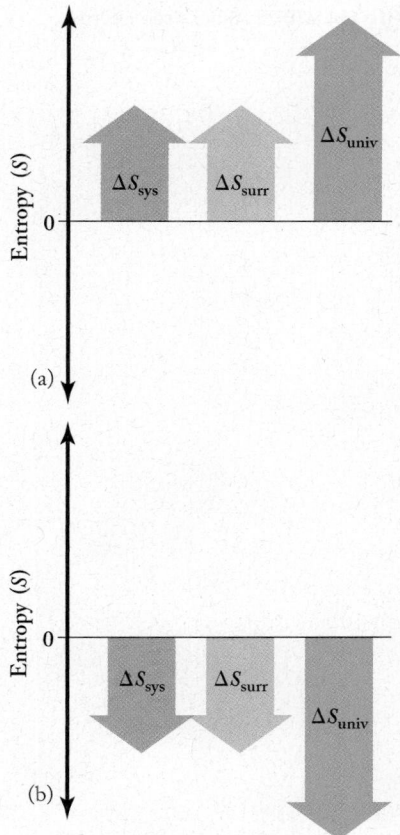

FIGURE 14.16 The magnitudes and signs of ΔS_{sys} and ΔS_{surr} determine the sign and magnitude of ΔS_{univ}. (a) When ΔS_{sys} and ΔS_{surr} are both positive, ΔS_{univ} is always positive and the process is spontaneous. (b) When ΔS_{sys} and ΔS_{surr} are both negative, ΔS_{univ} is always negative and the process is nonspontaneous.

free energy (G) a thermodynamic state function that provides a criterion for spontaneous change; an indication of the energy available to do useful work.

free-energy change (ΔG) the change in free energy of a process; $\Delta G < 0$ for spontaneous processes at constant temperature and pressure.

▶❚❚ **CHEMTOUR** Gibbs Free Energy

This expression defines the change in Gibbs free energy. The lack of subscripts means that all the terms refer to the system, and from now on any variable without a subscript is understood to belong to the system we are considering. If ΔG is negative, ΔS_{univ} must be positive (remember the definition: $\Delta G = -T\,\Delta S_{univ}$). Therefore the following conditions hold:

1. If ΔG for a process is negative, ΔS_{univ} is positive and the process is spontaneous.
2. If ΔG for a process is positive, ΔS_{univ} is negative and the process is non-spontaneous. Instead, the reverse of the process is spontaneous.
3. If ΔG is zero, no net change in the system occurs.

Equation 14.8 expresses our understanding of the two thermodynamic forces that drive chemical reactions: enthalpy and entropy. Table 14.4 summarizes the effects of the signs of ΔH and ΔS on ΔG and on reaction spontaneity.

TABLE 14.4		Effects of ΔH, ΔS, and T on ΔG and Spontaneity	
	SIGN OF		
ΔH	**ΔS**	**ΔG**	**Spontaneity**
−	+	Always <0	Always spontaneous
−	−	<0 at lower temperature	Spontaneous at lower temperature
+	+	<0 at higher temperature	Spontaneous at higher temperature
+	−	Always >0	Never spontaneous

CONCEPT TEST

(a) Given the thermodynamic data in Table A4.3, is the conversion of diamond to graphite spontaneous? Explain your answer. (b) If yes, can knowing that the conversion is spontaneous tell you anything about how fast or slow the conversion is?

The Meaning of Free Energy

We defined free energy as the energy available to do useful work. Let's explore what this means using the combustion of gasoline in an automobile engine. The combustion releases energy in the form of heat q and work w as defined by Equation 5.5:

$$\Delta E = q + w$$

Much of this heat is wasted energy because it flows from the engine to the surroundings. Automobiles have cooling systems to manage this heat, but the key issue is that this wasted energy does nothing to move the car. The useful energy that propels the car is derived from the rapid expansion of the gaseous products of combustion in the cylinders of the engine. As Figure 14.17 shows, this expansion pushes down on the piston of a cylinder to increase the cylinder volume. The product of the pressure exerted by these gases on the piston and the volume change is $P\,\Delta V$ work (discussed in Section 5.2) that moves the car:

$$w = -P\,\Delta V$$

The concept of free energy relates to our attempts to do useful work with the energy from chemical reactions. The free energy released by a spontaneous reaction run at constant T and P is a measure of the maximum amount of energy that is available to do useful work. To see how this amount compares with the total energy released, let's rearrange Equation 14.8 to

$$\Delta H = \Delta G + T\,\Delta S \qquad (14.9)$$

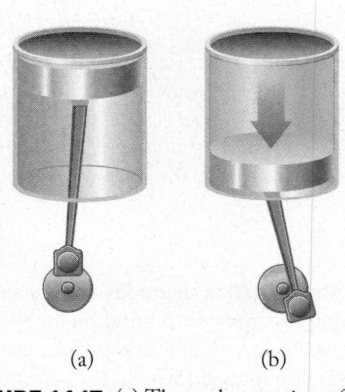

FIGURE 14.17 (a) Thermal expansion of the gases in a cylinder in a car engine (b) pushes down on a piston with a pressure P represented by the blue arrow. The product of P and the change in volume of the gases ΔV is the work done by the expanding gases that propels the car.

a form that tells us that the enthalpy change that comes from the making and breaking of bonds during a chemical process may be divided into two parts. One part, ΔG, is the energy that devices, such as internal combustion engines and steam generators, batteries, and fuel cells, convert into motion, light, or some other manifestation of the interaction of energy with matter. Such conversions are what is described by the statement, "ΔG is energy available to do useful work." The other part of the enthalpy change, $T\,\Delta S$, is not usable: it is the portion of energy that disperses and thereby increases the entropy of the universe.

In any engine, the conversion of the chemical energy stored in bonds (ΔH) into useful mechanical energy (ΔG) that can move a car or perform any other mechanical work is never 100% efficient. Some portion of the ΔG energy released by spontaneous processes such as burning fuels is always wasted. Remember the axiom given at the outset of the chapter: in the game of energy, we cannot win (energy cannot be created), and we cannot even break even (every time we use energy, some is wasted).

We can think of the *efficiency* of an engine as the ratio of work done to energy consumed. Ideally, energy efficiency would be 100%, but in reality it is always less. Efficiency is usually inversely related to the rate at which a spontaneous reaction takes place. The slower the reaction, the greater the amount of free energy likely to be harvested and available to do work; the faster the reaction, the smaller the amount of free energy likely to be harvested. Recall our earlier discussion of reversible processes: the more slowly a process is carried out, the more closely it approaches complete reversibility, at which point it would be at maximum efficiency. That is why calculated efficiencies of mechanical devices are considered maximum efficiencies. If an engine does its work reversibly (ideally), we get maximum useful work. Because no engine works ideally, devices are always less efficient than we calculate them to be.

Reporting the efficiencies of energy-consuming devices is becoming increasingly common (Figure 14.18). We can now understand what these numbers mean in terms of this discussion. Mechanical efficiency is frequently expressed as a percentage obtained from the efficiency ratio:

$$\text{Efficiency} = \frac{\text{work done}}{\text{energy consumed}} \times 100\%$$

Water wheels, for example, are 90% efficient; fuel cells 80%, jet engines 60%, and car engines 30%. Remember: Not only is all the $T\,\Delta S$ energy from a reaction wasted, but the amount of ΔG actually available to do work in an irreversible process is always less than the ideal amount calculated because some portion of ΔG always flows into the surroundings without doing anything useful.

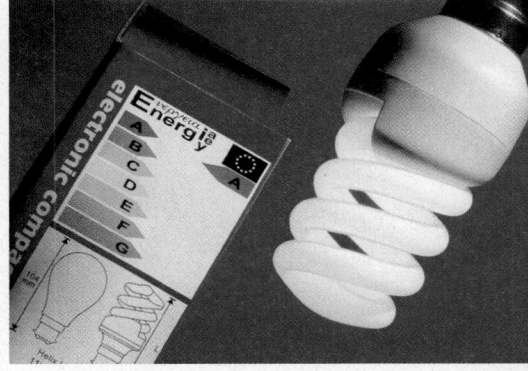

FIGURE 14.18 The packaging of this fluorescent lightbulb displays the European Union Energy Efficiency label. The "A" rating indicates that this 15-watt lightbulb produces as much light as a 75-watt incandescent bulb.

> **CONCEPT TEST** ···
>
> Why might fluorescent lightbulbs be more efficient than incandescent lightbulbs? (*Hint*: What happens if you place your hand close to each type of lightbulb?)
>
> ···

Calculating Free-Energy Changes

The free-energy change in a reaction can be calculated using Equation 14.8. It can also be calculated from the standard free energies of formation of the products and reactants. The **standard free energy of formation (ΔG_f°)** is the change in free energy associated with the formation of 1 mole of a compound in its standard

standard free energy of formation (ΔG_f°) the change in free energy associated with the formation of 1 mole of a compound in its standard state from its elements.

state from its elements in their standard states (Section 5.6 and Table 14.3). See Appendix 4 for a list of ΔG_f° values.

We can use standard free energies of formation to calculate the free energy change ΔG_{rxn}° for a chemical reaction at 1 bar and 298 K using the equation

$$\Delta G_{rxn}^\circ = \sum n_{products} \, \Delta G_{f,products}^\circ - \sum n_{reactants} \, \Delta G_{f,reactants}^\circ \qquad (14.10)$$

which tells us that the standard free-energy change for a chemical reaction forming $n_{products}$ moles of products from $n_{reactants}$ moles of reactants is the difference in the sums of the standard free energies of formation of the products and the reactants. By definition, the standard free energy of formation of the most stable form of an element in its standard state is zero. We use the subscript "rxn" here to remind ourselves that we are calculating ΔG_{sys}°, where the system refers to a specific chemical reaction. Note the similarities between Equation 14.10 and the expressions for ΔH_{rxn}° (Equation 14.6) and ΔS_{rxn}° (Equation 14.4).

SAMPLE EXERCISE 14.5   **Effect of Structure on ΔG° of a Reaction**

The standard free energies of formation for three structural isomers of the alkane containing eight carbons are shown in Figure 14.19. All three isomers burn in air by the combustion reaction

$$2\,C_8H_{18}(\ell) + 25\,O_2(g) \rightarrow 16\,CO_2(g) + 18\,H_2O(g)$$

Predict whether the ΔG_{rxn}° values for the combustion reactions of these isomers are all the same or different.

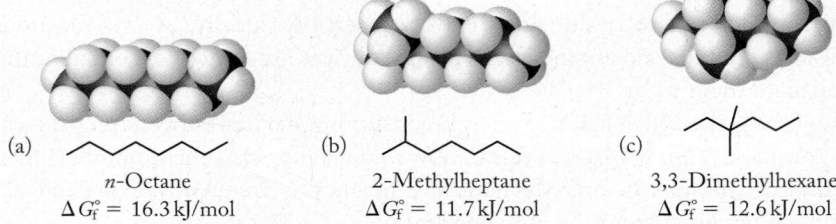

(a)   *n*-Octane   (b)   2-Methylheptane   (c)   3,3-Dimethylhexane

n-Octane	2-Methylheptane	3,3-Dimethylhexane
$\Delta G_f^\circ = 16.3 \,\text{kJ/mol}$	$\Delta G_f^\circ = 11.7 \,\text{kJ/mol}$	$\Delta G_f^\circ = 12.6 \,\text{kJ/mol}$

FIGURE 14.19

Collect and Organize Structural isomers have the same molecular formula but different arrangements of bonds. A combustion reaction converts all of the carbon in a hydrocarbon to CO_2 and all of the hydrogen to water. We are to predict whether the ΔG° values for the combustion of a set of structural isomers are the same or different.

Analyze The standard free energy of combustion can be calculated using Equation 14.10; however, there is actually no need to do the calculation. Because the products of the reaction are the same for all three isomers, $\Delta G_{f,products}^\circ$ is the same in each combustion reaction. We predict that the values are related to the values for ΔG_f° for the three isomers.

Solve Let's use Equation 14.10 to express ΔG° for the combustion of all three isomers:

$$\Delta G_{rxn}^\circ = \sum n_{products} \, \Delta G_{f,products}^\circ - \sum n_{reactants} \, \Delta G_{f,reactants}^\circ$$

(a) $\Delta G_{rxn}^\circ = [(16\,\Delta G_{f,CO_2}^\circ + 18\,\Delta G_{f,H_2O}^\circ) - (2\,\Delta G_{f,octane}^\circ + 25\,\Delta G_{f,O_2}^\circ)]$

(b) $\Delta G_{rxn}^\circ = [(16\,\Delta G_{f,CO_2}^\circ + 18\,\Delta G_{f,H_2O}^\circ) - (2\,\Delta G_{f,methylheptane}^\circ + 25\,\Delta G_{f,O_2}^\circ)]$

(c) $\Delta G_{rxn}^\circ = [(16\,\Delta G_{f,CO_2}^\circ + 18\,\Delta G_{f,H_2O}^\circ) - (2\,\Delta G_{f,dimethylhexane}^\circ + 25\,\Delta G_{f,O_2}^\circ)]$

Notice that the term for the products is the same in all three equations. The difference in ΔG_{rxn}° values are determined by the differences in ΔG_f° between the isomers in the reactant term. Since ΔG_f° is different for each isomer, ΔG_{rxn}° will also be different.

Think about It As predicted, the value of ΔG_{rxn}° depends on the values for ΔG_f° when we compare structural isomers in a combustion reaction.

Practice Exercise Using the ΔG_f° value for *n*-octane given in Sample Exercise 14.5 and obtaining other values as needed from Appendix 4, calculate ΔG_{rxn}° for the combustion of 1 mole of *n*-octane.

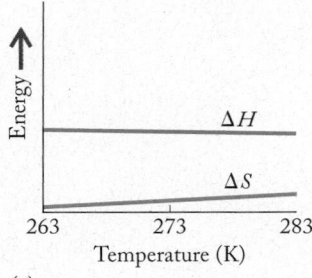
(a)

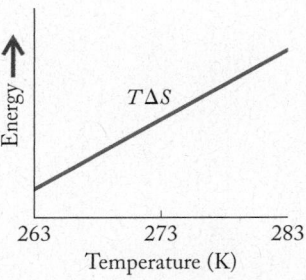
(b)

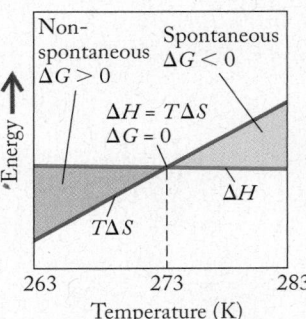
(c)

FIGURE 14.20 Changes in (a) entropy (ΔS), enthalpy (ΔH), (b) the quantity $T \Delta S$, and (c) free energy (ΔG) for ice melting at temperatures from $-10°C$ (263 K) to $+10°C$ (283 K).

Temperature and Spontaneity

Let's reconsider the isothermal process of ice melting, this time in light of the full complement of thermodynamic quantities discussed in this chapter. The influence of temperature on ΔH, ΔS, $T \Delta S$, and ΔG as ice is warmed from $-10°C$ to $+10°C$ is shown in Figure 14.20. As the temperature rises, the enthalpy of fusion ΔH changes very slightly and the entropy ΔS of the system gradually increases, but $T \Delta S$ increases rapidly (Figures 14.20a and 14.20b). Figure 14.20(c), which shows the combined influence of ΔH and $T \Delta S$ on ΔG (remember: $\Delta G = \Delta H - T \Delta S$), defines the conditions under which melting is nonspontaneous or spontaneous.

Above 0°C (273 K), ice melts spontaneously ($\Delta G < 0$). Because melting is an endothermic process, ΔH is positive, but for melting ΔS is also positive and the $T \Delta S$ term in Equation 14.8 is large enough to more than offset the positive ΔH term. As a result, ΔG is negative, which is the condition necessary for spontaneity.

Below 0°C, ice does not melt spontaneously ($\Delta G > 0$). The entropy change ΔS is still positive, but the $T \Delta S$ term in Equation 14.8 is smaller than the positive ΔH term, and the free-energy change for melting is positive. The opposite process—liquid water freezing—*is* spontaneous ($\Delta G < 0$) below 0°C. The important point here is the same as with the thermodynamic quantities ΔH and ΔS: the free-energy change ΔG of a process or reaction is equal in magnitude but opposite in sign to the free-energy change of the reverse process or reaction.

CONCEPT TEST

A given process is spontaneous at lower temperature and nonspontaneous at higher temperature. What does the graph in Figure 14.20(c) look like for such a process?

The temperature at the point in Figure 14.20(c) where $\Delta G = 0$ defines the melting (or freezing) point, which by definition is the temperature at which the solid melts at the same rate as the liquid freezes. The two phases are in *equilibrium*. At 0°C, no net change takes place, and both phases coexist.

A similar situation exists for water at 100°C and 1 atm pressure, the temperature and pressure at which $\Delta G_{vaporization}$ and $\Delta G_{condensation}$ both equal zero. At 100°C, liquid water vaporizes and water vapor condenses at the same rate; the two phases exist in equilibrium with each other.

CONNECTION In Chapter 10 we discussed the equilibrium between phases of a substance and used phase diagrams to illustrate which physical states are stable at various combinations of temperature and pressure.

The temperature at which $\Delta G = 0$ for a process can be calculated from Equation 14.8 if we know the values of ΔH and ΔS. For example, for the fusion of water (melting of ice; forward reaction as written), we have

$$H_2O(s) \rightleftharpoons H_2O(\ell) \qquad \Delta H^\circ = 6.01 \times 10^3 \, \text{J/mol}; \, \Delta S^\circ = 22.0 \, \text{J/(mol} \cdot \text{K)}$$

In doing this calculation, we assume that ΔH° and ΔS° do not change significantly with small changes in temperature. For this process, as well as for most other physical and chemical processes, this assumption is acceptable. Therefore we can assume that $\Delta G = \Delta H - T \Delta S \approx \Delta H^\circ - T \Delta S^\circ$. Inserting the values of ΔH° and ΔS° and using the fact that $\Delta G = 0$ for a process at equilibrium, we get

$$\Delta G = (6.01 \times 10^3 \, \text{J/mol}) - T[22.0 \, \text{J/(mol} \cdot \text{K)}] = 0$$

$$T = \frac{6.01 \times 10^3 \, \cancel{\text{J/mol}}}{22.0 \, \cancel{\text{J/(mol}} \cdot \text{K)}} \approx 273 \, \text{K} = 0°\text{C}$$

which is the familiar value for the melting point of ice.

SAMPLE EXERCISE 14.6   Calculating Standard Free-Energy Changes

Calculate ΔG° for the dissolution of 1 mole of ammonium nitrate in water (total volume = 1 liter) at 298 K, given $\Delta H^\circ = 28.1$ kJ/mol and $\Delta S^\circ = 108.7$ J/(mol $\cdot$ K).

Collect and Organize We are to find the standard free-energy change for $NH_4NO_3(s)$ dissolving in water. We are given the change in enthalpy and the change in entropy for the process under standard conditions (1 M solution, 1 bar, and 298 K).

Analyze We can use Equation 14.8 for this calculation. Since the dissolution of ammonium nitrate (a chemical cold pack) is spontaneous at 298 K (room temperature), we predict that ΔG° will be negative.

Solve Substituting the values of ΔH° and ΔS° for ΔH and ΔS in Equation 14.8 and converting 108.7 J/(mol $\cdot$ K) to kJ allows us to calculate ΔG°:

$$\Delta G^\circ = \Delta H^\circ - T\Delta S^\circ = 28.1 \, \text{kJ} - (298 \, \cancel{\text{K}})\left(\frac{0.1087 \, \text{kJ}}{\cancel{\text{K}}}\right)$$

$$= -4.3 \, \text{kJ}$$

Think about It As predicted, ΔG° is negative, which is consistent with a spontaneous process. A chemical cold pack is only useful if the dissolution of ammonium nitrate is spontaneous at ambient temperature. This sample exercise illustrates a situation where a positive value for ΔH (endothermic process) is more than offset by the increase in entropy, a positive value for ΔS.

Practice Exercise Predict the signs of ΔH°, ΔS°, and ΔG° for the combustion of methane:

$$CH_4(g) + 2O_2(g) \rightarrow CO_2(g) + 2H_2O(\ell)$$

Like physical processes, chemical reactions can have zero change in free energy. As a spontaneous reaction proceeds ($\Delta G < 0$), reactants become products and ΔG becomes less negative (for reasons that will become clear in Chapter 16). In fact,

ΔG may reach zero before all the reactants are consumed. In this case, no more products form and no more reactants are consumed. Rather, reactants and products coexist in equilibrium with each other. Reactants still react, and products are still formed, but the reverse reaction, in which products become reactants, proceeds at the same rate as the forward reaction. This state, called *chemical equilibrium*, is the subject of Chapters 16 and 17. No net change in the amounts of reactants and products occurs once equilibrium is reached.

SAMPLE EXERCISE 14.7   **Relating Reaction Spontaneity to ΔH and ΔS**

A certain chemical reaction is spontaneous at low temperatures but not at high temperatures. Use Equation 14.8 to determine the sign of the enthalpy and entropy changes for this reaction.

Collect and Organize We are to determine the signs of ΔH (enthalpy change) and ΔS (entropy change) based on the change in spontaneity of a reaction as temperature changes. This means we need to think about how the signs of ΔH and ΔS determine how ΔG varies with temperature.

Analyze There are three possible combinations for ΔH and ΔS that lead to spontaneous reactions: $\Delta H < 0$ and $\Delta S > 0$; $\Delta H > 0$ and $\Delta S > 0$; $\Delta H < 0$ and $\Delta S < 0$. Only one of these combinations of ΔH and ΔS will satisfy the condition that $\Delta G > 0$ at higher temperature and $\Delta G < 0$ at lower temperature.

Solve The importance of ΔS increases with increasing temperature because the product $T\Delta S$ appears in Equation 14.8. The reaction is nonspontaneous at higher temperatures, where the magnitude of $T\Delta S$ is more likely to be larger than the magnitude of ΔH. The reaction is spontaneous at low temperatures, however, where the impact of a negative ΔS value is more than offset by a decrease in enthalpy, a change that favors the reaction. The reaction must have negative ΔS and negative ΔH values.

Think about It Table 14.4 confirms our prediction that a process that is spontaneous only at low temperatures is one in which there is a decrease in both entropy and enthalpy.

Practice Exercise At high temperatures, ammonia decomposes to nitrogen and hydrogen gases:

$$2\,NH_3(g) \rightarrow N_2(g) + 3\,H_2(g)$$

ΔH for the reaction is positive and ΔS is positive. Predict whether the reaction is spontaneous at all temperatures, at only high temperatures, or at only low temperatures.

Free-Energy Changes Under Nonstandard Conditions

So far, most of our calculations involving free-energy changes have taken place under standard conditions, but Equation 14.8 is not limited to the conditions specified in Table 14.3. Do Equations 14.4 and 14.10 also hold for cases where the reaction is not under standard conditions? The answer is yes, but in order to use these equations, we must either assume that the values of ΔG_f° and S° do not change much with temperature, or calculate how much they do change with temperature.

In the previous section we made the assumption that $\Delta H°$ and $\Delta S°$ do not change significantly with small changes in temperature when calculating the melting point of water.

How do ΔG and ΔS vary with temperature? From Equation 14.8, we know that ΔG depends on temperature. Equation 14.2 illustrates how ΔS depends on temperature:

$$\Delta S = \frac{q_{rev}}{T} = \frac{\Delta H}{T}$$

To use this equation, however, we need to know how ΔH varies with temperature. As we noted in Chapter 5, we typically assume that ΔH does not change much with temperature. In fact, ΔH does depend on changes in the molar heat capacities of the reactants and products; however, the equations describing the temperature dependence of ΔH are beyond the scope of this course. Thus we will not explore the temperature dependence of free energy, enthalpy, and entropy in a quantitative fashion.

14.6 Driving the Human Engine: Coupled Reactions

The laws of thermodynamics that govern chemical reactions in the laboratory also govern all the chemical reactions that take place in living systems (Figure 14.21). Organisms carry out reactions that release the energy contained in the chemical bonds of food molecules and then use that energy to do work and sustain an array of other essential biological functions. Just like mechanical engines, however, humans and other life-forms are far from 100% efficient, which means that life requires a continuous input of energy in terms of the caloric content of the food we eat. Thus we humans must constantly absorb energy in the form of food and then release heat and waste products into our surroundings. Young women have an average daily nutritional need of 2100 Cal; for young men the figure is 2900 Cal. This level of caloric intake provides the energy we need in order to function at all levels, from thinking to getting out of bed in the morning.

∞ CONNECTION In Chapter 5 we defined 1 Calorie, the "calorie" used in discussing food, as equivalent to 1 kcal.

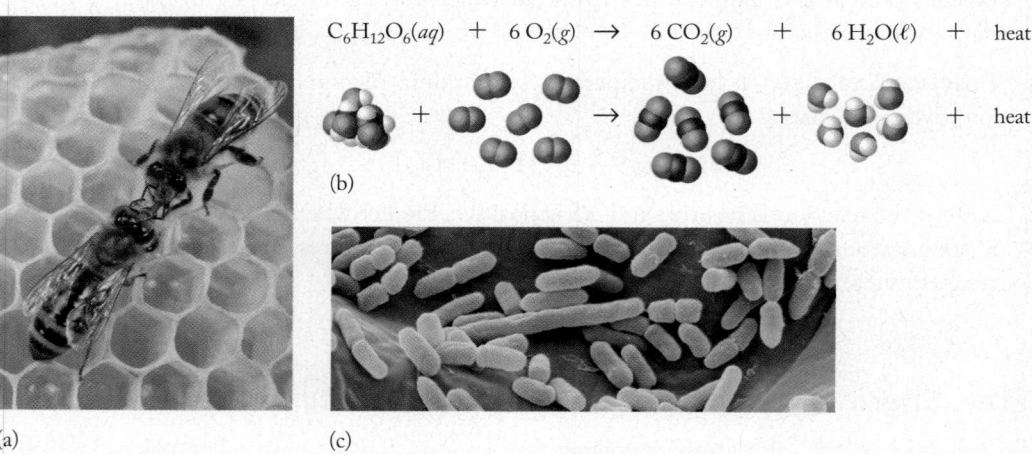

$$C_6H_{12}O_6(aq) \;+\; 6\,O_2(g) \;\rightarrow\; 6\,CO_2(g) \;+\; 6\,H_2O(\ell) \;+\; heat$$

(b)

(a)   (c)

FIGURE 14.21 The rules of thermodynamics apply to all living systems. (a) Honeybees extract energy from nutrients to support life. The bees store this energy as honey, which is then consumed by the bees, by humans, or by other animals to generate energy. (b) When honey is consumed, heat, water, and carbon dioxide are released, increasing the entropy of the universe. (c) Microscopic organisms—such as the *E. coli* that live in our gastrointestinal tracts—consume nutrients to live and generate heat in the process. Their expenditure of energy also increases the entropy of the universe.

In living systems, spontaneous reactions ($\Delta G < 0$) typically involve breaking food down, while nonspontaneous reactions ($\Delta G > 0$) involve building molecules needed by the body. Living systems use the energy from spontaneous reactions to run nonspontaneous reactions; we say that the spontaneous reactions are *coupled* to the nonspontaneous reactions. Part of the study of biochemistry involves deciphering the molecular mechanisms that enable reaction coupling. This topic is covered in Chapter 20, but for now it is sufficient to know that elegant molecular processes have evolved to enable living systems to couple chemical reactions so that the energy obtained from spontaneous reactions can be used to drive the nonspontaneous reactions that maintain life. The metabolic chemical reactions we look at here are *not* presented for you to memorize. Rather, the intent is to aid your understanding of how changes in free energy allow spontaneous reactions to drive nonspontaneous reactions and to illustrate operationally what the phrase *coupled reactions* actually means.

As noted in the essay at the beginning of Chapter 5, all the energy contained in the food we eat has sunlight as its ultimate source. Green plants store energy from sunlight in their tissues as molecules such as glucose ($C_6H_{12}O_6$), which they produce from CO_2 and H_2O during photosynthesis.

Production of glucose by green plants is a nonspontaneous process, which is why the plants require the energy of sunlight to carry out this reaction. Animals that consume plants use the energy stored in the chemical bonds of glucose and other molecules, and release CO_2 and H_2O back into the environment. That reaction,

$$C_6H_{12}O_6(s) + 6\,O_2(g) \rightarrow 6\,CO_2(g) + 6\,H_2O(\ell)$$

also releases energy, an event that increases the entropy of the universe. Thus the processes of life increase the entropy of the universe by converting chemical energy into heat that flows into the surroundings.

Therefore the free-energy change for the breakdown of glucose must be less than zero: $\Delta G° = -880$ kJ/mol. This spontaneous process is highly controlled in living systems, however, so that the energy it produces can be directed into the nonspontaneous processes essential to organisms that don't carry out photosynthesis. Here we look at one portion of glucose metabolism—**glycolysis**—in order to investigate coupled reactions.

In glycolysis, each mole of glucose is converted into 2 moles of pyruvate ion (CH_3COCOO^-, Figure 14.22). An early step is conversion of glucose into glucose 6-phosphate (Figure 14.23), an example of a **phosphorylation** reaction.

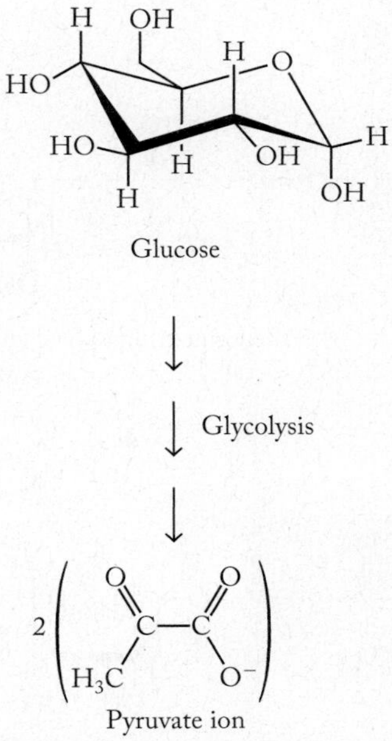

Glucose

Glycolysis

Pyruvate ion

FIGURE 14.22 In glycolysis, one molecule of glucose is converted into two pyruvate ions.

∞ **CONNECTION** Photosynthesis and the carbon cycle were introduced in Chapter 3.

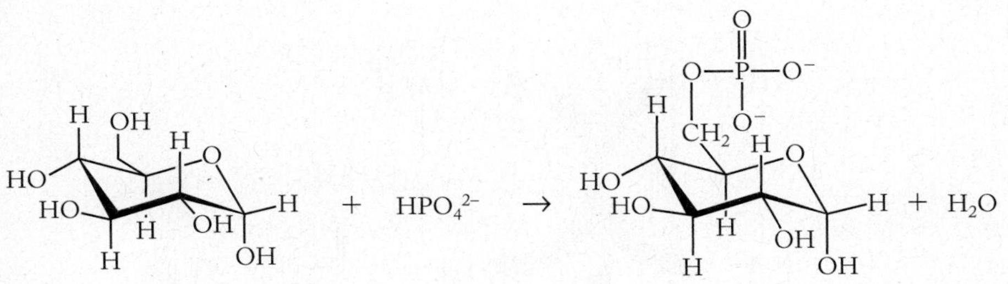

Glucose(*aq*) + HPO_4^{2-}(*aq*) → (glucose 6-phosphate)$^{2-}$(*aq*) + $H_2O(\ell)$

FIGURE 14.23 The conversion of glucose into glucose 6-phosphate is an early step in glycolysis. This reaction is nonspontaneous, which means energy must be added to make the reaction go: $\Delta G°_{rxn} = 13.8$ kJ/mol.

glycolysis a series of reactions that converts glucose into pyruvate; a major anaerobic (no oxygen required) pathway for the metabolism of glucose in the cells of almost all living organisms.

phosphorylation a reaction resulting in the addition of a phosphate group to an organic molecule.

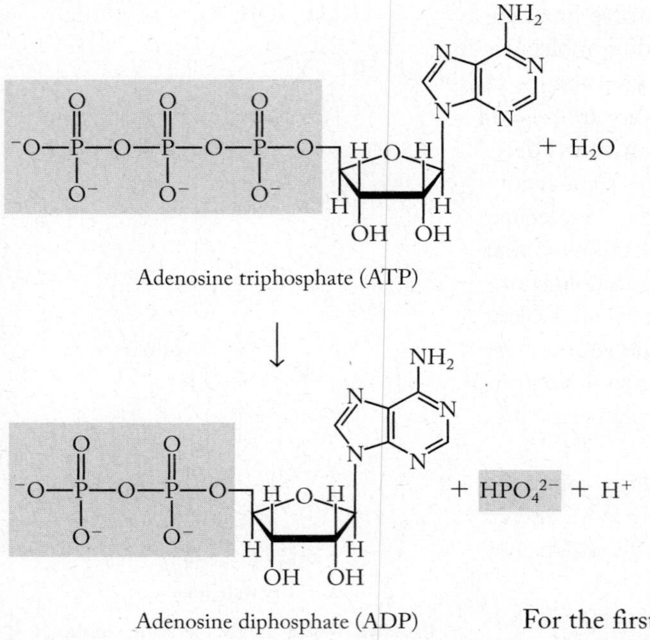

Adenosine triphosphate (ATP)

↓

Adenosine diphosphate (ADP)

FIGURE 14.24 The hydrolysis of ATP to ADP is a spontaneous reaction: $\Delta G° = -30.5$ kJ/mol. The body couples this reaction to nonspontaneous reactions so that the energy released can drive them (Figure 14.25).

∞ CONNECTION In Chapter 5 we used Hess's law (the enthalpy change of a reaction that is the sum of two or more reactions equals the sum of the enthalpy changes of the constituent reactions) to calculate ΔH. We apply a similar principle here when adding $\Delta G°$ values of coupled reactions.

Glucose reacts with the hydrogen phosphate ion (HPO_4^{2-}), producing glucose 6-phosphate and water. This reaction is not spontaneous ($\Delta G° = +13.8$ kJ/mol), and the energy needed to make it happen comes from a compound called adenosine triphosphate (ATP). ATP functions in our cells both as a storehouse of energy and as an energy-transfer agent: it hydrolyzes to adenosine diphosphate (ADP) in a reaction (Figure 14.24) that produces a hydrogen phosphate ion and energy: $\Delta G°_{rxn} = -30.5$ kJ/mol.

In a living system, spontaneous hydrolysis of ATP consumes water and produces HPO_4^{2-} and H^+, whereas the nonspontaneous phosphorylation of glucose consumes HPO_4^{2-} and produces water. The two reactions are coupled: the spontaneous ATP → ADP reaction supplies the energy that drives the nonspontaneous formation of glucose 6-phosphate (Figure 14.25).

This example illustrates another important general point about reactions and $\Delta G°$ values. The $\Delta G°$ values for coupled reactions (and also for sequential reactions) are additive. This is true for any set of reactions, not just those occurring in living systems. For the first two steps in glycolysis:

$$(1)\ ATP^{4-}(aq) + H_2O(\ell) \rightarrow ADP^{3-}(aq) + HPO_4^{2-}(aq) + H^+(aq)$$
$$\Delta G° = -30.5\text{kJ}$$

$$(2)\ C_6H_{12}O_6(aq) + HPO_4^{2-}(aq) \rightarrow C_6H_{11}O_6PO_3^{2-}(aq) + H_2O(\ell)$$

Glucose　　　Hydrogen　　　　　Glucose
　　　　　　phosphate　　　　6–phosphate

$$\Delta G° = 13.8\text{ kJ}$$

If we add these reactions and their free energies, we get

$$ATP^{4-}(aq) + \cancel{H_2O(\ell)} + C_6H_{12}O_6(aq) + \cancel{HPO_4^{2-}(aq)} \rightarrow$$
$$ADP^{3-}(aq) + \cancel{HPO_4^{2-}(aq)} + C_6H_{11}O_6PO_3^{2-}(aq) + \cancel{H_2O(\ell)} + H^+(aq)$$
$$\Delta G° = (-30.5 + 13.8)\text{ kJ} = -16.7\text{ kJ}$$

Because equal quantities of H_2O and HPO_4^{2-} appear on both sides of the combined equation, they cancel out, leaving the net reaction

$$C_6H_{12}O_6(aq) + ATP^{4-}(aq) \rightarrow ADP^{3-}(aq) + C_6H_{11}O_6PO_3^{2-}(aq) + H^+(aq)$$
$$\Delta G° = -16.7\text{ kJ}$$

Since $\Delta G°$ for the net reaction is negative, the reaction is spontaneous.

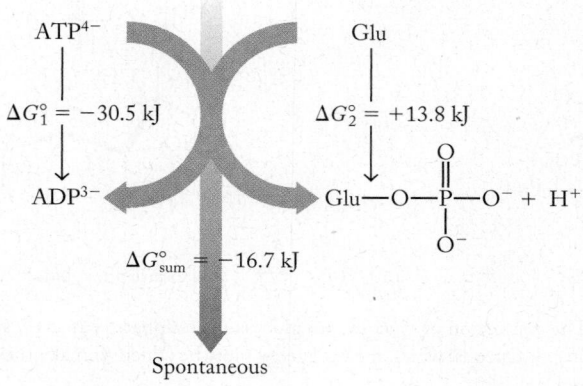

FIGURE 14.25 The spontaneous hydrolysis of ATP is coupled to the nonspontaneous phosphorylation of glucose. The overall reaction—the sum of the two individual reactions—is spontaneous.

1,3-Diphosphoglycerate^{4-}
(1,3-DPG^{4-})

SAMPLE EXERCISE 14.8 Calculating $\Delta G°$ of Coupled Reactions

The body would rapidly run out of ATP if there were not some process for regenerating it from ADP, and that process is the hydrolysis of 1,3-diphosphoglycerate^{4-} (1,3-DPG^{4-}) to 3-phosphoglycerate^{3-} (3-PG^{3-}) (Figure 14.26):

$$ADP^{3-} + \text{1,3-diphosphoglycerate}^{4-} \rightarrow \text{3-phosphoglycerate}^{3-} + ATP^{4-}$$

This hydrolysis is spontaneous. Calculate its $\Delta G°$ value from these values:

(1) $\text{1,3-DPG}^{4-}(aq) + H_2O(\ell) \rightarrow \text{3-PG}^{3-}(aq) + HPO_4^{2-}(aq) + H^+(aq)$
$$\Delta G° = -49.0 \text{ kJ}$$

(2) $ADP^{3-}(aq) + HPO_4^{2-}(aq) + H^+(aq) \rightarrow ATP^{4-}(aq) + H_2O(\ell)$
$$\Delta G° = 30.5 \text{ kJ}$$

Collect and Organize We can calculate $\Delta G°$ for a reaction that is the sum of two reactions. If the reactions in equations (1) and (2) add up to the overall reaction, then overall $\Delta G°$ is the sum of the $\Delta G°$ values for the individual reactions.

Analyze First, we add the reactions described by equations (1) and (2). Assuming that the overall reaction between ADP and 1,3-diphosphoglycerate^{4-} is the sum of the reactions describing the hydrolysis of 1,3-diphosphoglycerate^{4-} and the phosphorylation of ADP, we know that the sum of $\Delta G_1°$ and $\Delta G_2°$ will be less than zero because the overall reaction is spontaneous.

Solve Summing the reactions in equations (1) and (2) confirms that they equal the overall reaction:

(1) $\text{1,3-DPG}^{4-}(aq) + H_2O(\ell) \rightarrow \text{3-PG}^{3-}(aq) + HPO_4^{2-}(aq) + H^+(aq)$

(2) $ADP^{3-}(aq) + HPO_4^{2-}(aq) + H^+(aq) \rightarrow ATP^{4-}(aq) + H_2O(\ell)$

$\text{1,3-DPG}^{4-}(aq) + \cancel{H_2O(\ell)} + ADP^{3-}(aq) + \cancel{HPO_4^{2-}(aq)} + \cancel{H^+(aq)} \rightarrow$
$\quad \text{3-PG}^{3-}(aq) + \cancel{HPO_4^{2-}(aq)} + \cancel{H^+(aq)} + ATP^{4-}(aq) + \cancel{H_2O(\ell)}$

We sum the $\Delta G°$ values for steps 1 and 2 to determine $\Delta G°$ for the overall reaction:

$$\Delta G°_{overall} = \Delta G_1° + \Delta G_2° = [(-49.0) + (30.5)] \text{ kJ} = -18.5 \text{ kJ}$$

Think about It The hydrolysis of 1,3-diphosphoglycerate^{4-} provides more than sufficient energy for the conversion of ADP into ATP.

Practice Exercise The conversion of glucose into lactic acid drives the phosphorylation of 2 moles of ADP to ATP:

$C_6H_{12}O_6(aq) + 2\,HPO_4^{2-}(aq) + 2\,ADP^{3-}(aq) + 2\,H^+(aq) \rightarrow$
(Glucose)

$\quad\quad 2\,CH_3CH(OH)COOH(aq) + 2\,ATP^{4-}(aq) + 2\,H_2O(\ell)$
(Lactic acid)
$$\Delta G° = -135 \text{ kJ/mol}$$

What is $\Delta G°$ for the conversion of glucose into lactic acid?

$$C_6H_{12}O_6(aq) \rightarrow 2\,CH_3CH(OH)COOH(aq)$$

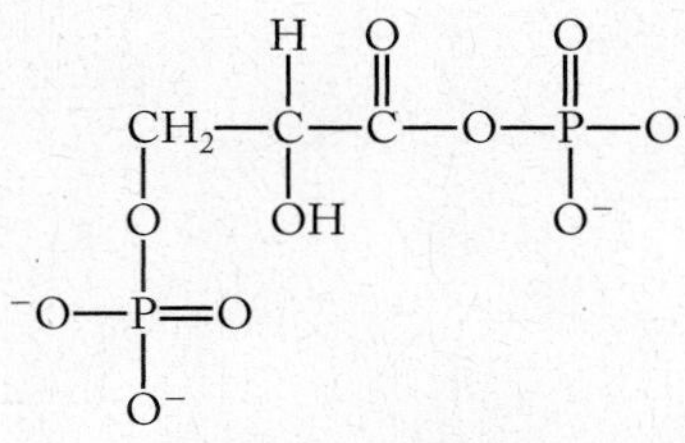

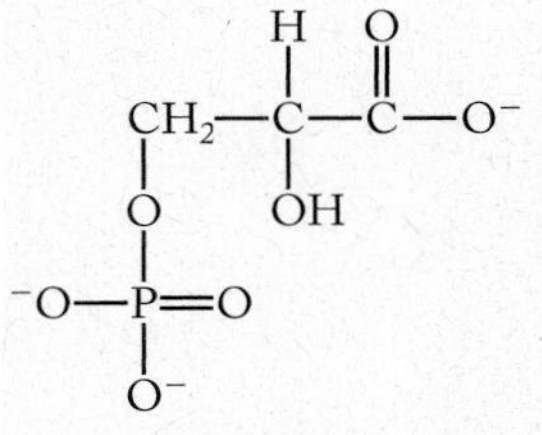

3-Phosphoglycerate^{3-}
(3-PG^{3-})

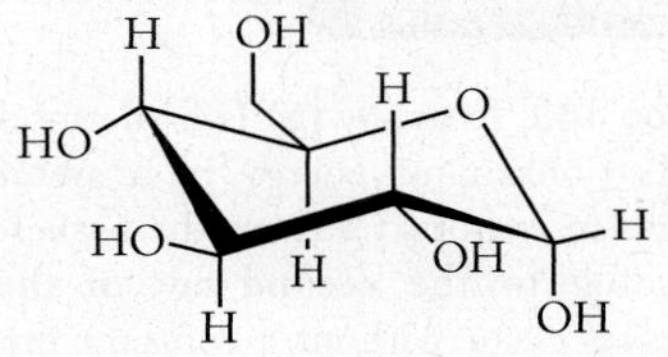

Glucose

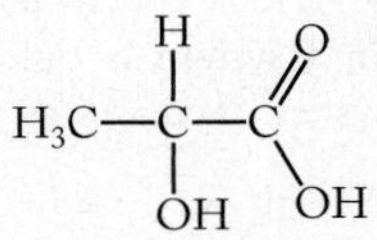

Lactic acid

FIGURE 14.26 Structures of 1,3-diphosphoglycerate, phosphoglycerate, glucose, and lactic acid.

The ATP produced from the breakdown of glucose (as, for example, via the first reaction in the previous Practice Exercise) is used to drive nonspontaneous reactions in cells. The metabolism of fats and proteins relies on a series of cycles, all of which involve coupled reactions. The ATP–ADP system is a carrier of chemical energy because ADP requires energy to accept a phosphate group and thus is coupled to reactions that yield energy, while ATP donates a phosphate group, releases energy, and is coupled to reactions that require energy. All energy changes in living systems are governed by the first and second laws of thermodynamics, as are all the energy changes in the inanimate world.

In this chapter we addressed the reasons why some reactions and processes are spontaneous while others are not. We have learned that the free-energy change (ΔG) determines the spontaneity of a chemical reaction or process. Together, enthalpy and entropy changes for chemical reactions allow us to determine the ΔG for a chemical reaction or process. The free-energy change represents the maximum amount of work that can be done by the energy associated with a change. For changes carried out in the real world, the maximum amount of work is always actually less than ΔG. Paraphrasing the second law of thermodynamics, in the game of energy, not only can you not win, you can't even break even.

SUMMARY

Section 14.1 **Entropy (S)** is a thermodynamic property that reflects the distribution of energy in a system at a specific temperature; changes in entropy (ΔS) are the basis for understanding spontaneity. According to the **second law of thermodynamics**, **spontaneous processes** occur without a constant input of energy and result in an increase in entropy of the universe, whereas **nonspontaneous processes** require continuous input of energy to occur. Most exothermic reactions and some endothermic ones are spontaneous. All particles in a system occupy various energy levels. Each different arrangement of the particles in a system is called a **microstate** of the system. The entropy of a system increases as the number of available microstates increases.

Section 14.2 Phase changes of pure substances take place at a constant temperature and are **isothermal processes**. For a spontaneous process, the entropy change of the universe, $\Delta S_{univ} = \Delta S_{sys} + \Delta S_{surr}$, must be positive: $\Delta S_{univ} > 0$. A **reversible process** is an ideal process that takes place in very small steps and very slowly. No change in entropy takes place in a reversible process. If $\Delta S_{sys} < 0$ in a spontaneous process, then $\Delta S_{surr} > 0$ and must be large enough to ensure that $\Delta S_{sys} + \Delta S_{surr} > 0$. If $\Delta S_{sys} > 0$, the magnitude of ΔS_{surr} must be such that $\Delta S_{sys} + \Delta S_{surr} > 0$.

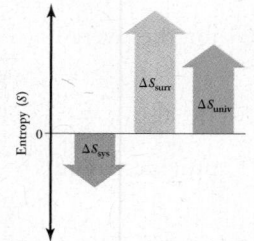

Section 14.3 According to the **third law of thermodynamics**, perfect crystals of a pure substance have zero entropy at absolute zero. All substances have positive entropies at temperatures above absolute zero. **Standard molar entropies ($S°$)** are entropy values for substances in their standard states. The entropy of a system increases with increasing molecular complexity and with increasing temperature.

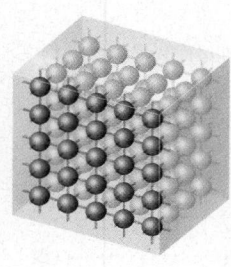

Section 14.4 The entropy change in a reaction under standard conditions can be calculated from the standard entropies of the products and reactants and their coefficients in the balanced chemical equation. For an isothermal process carried out reversibly, we can calculate ΔS_{surr} by dividing the heat exchanged (q_{rev}) by the absolute temperature T at which the process occurs. We can then calculate ΔS_{univ} and determine whether the process is spontaneous by noting the sign of ΔS_{univ}.

Section 14.5 **Free energy (G)** is defined as the energy available to do useful work. The **free-energy change** ($\Delta G = \Delta H - T\Delta S$) of a process is a state function defining the maximum useful work the system can do on its surroundings. The sign of ΔG determines whether a process is spontaneous (Table 14.4): $\Delta G < 0$ means a spontaneous process, $\Delta G > 0$ means a nonspontaneous process.

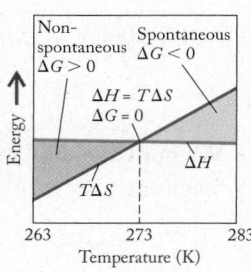

Reversing a process changes the sign of ΔG, and for a system at equilibrium, $\Delta G = 0$. For any process, the standard free-energy change $\Delta G°$ can be calculated either from the **standard free energies of formation ($\Delta G_f°$)** of the products and reactants or from the enthalpy and entropy changes. The temperature range over which a process is spontaneous depends on the relative magnitudes of ΔH and ΔS. Maximum efficiency is the most work available from a given quantity of energy.

Section 14.6 Many important biochemical processes, including **glycolysis** and **phosphorylation**, are made possible by coupled spontaneous and nonspontaneous reactions. The free energy released in the spontaneous processes going on in the body is used to drive nonspontaneous processes.

PROBLEM-SOLVING SUMMARY

TYPE OF PROBLEM	CONCEPTS AND EQUATIONS	SAMPLE EXERCISES
Predicting the sign of the entropy change in a physical or chemical process (system)	Look for the net removal of gas molecules or precipitation of a solute from solution ($\Delta S < 0$ for both). For the reverse processes, $\Delta S > 0$.	14.1, 14.2
Calculating the entropy change of a chemical reaction	Use $$\Delta S^\circ_{rxn} = \sum n_{products} S^\circ_{products} - \sum n_{reactants} S^\circ_{reactants} \qquad (14.4)$$ where $S^\circ_{products}$ and $S^\circ_{reactants}$ are the standard molar entropies and $n_{products}$ and $n_{reactants}$ are the stoichiometric coefficients for the process.	14.3
Predicting the spontaneity of a reaction	A reaction is spontaneous if $$\Delta S_{univ} = (\Delta S_{sys} + \Delta S_{surr}) > 0$$	14.4
Calculating the free-energy change of a physical or chemical process	Use $$\Delta G^\circ_{rxn} = \sum n_{products} \, \Delta G^\circ_{f,products} - \sum n_{reactants} \, \Delta G^\circ_{f,reactants} \qquad (14.10)$$ where $\Delta G^\circ_{f,products}$ and $\Delta G^\circ_{f,reactants}$ are the standard molar free energies of formation, and $n_{products}$ and $n_{reactants}$ are the stoichiometric coefficients for the process. Alternatively, use $$\Delta G^\circ = \Delta H^\circ - T\,\Delta S^\circ$$ where ΔH° and ΔS° are the standard enthalpy and entropy changes for the reaction, respectively.	14.5, 14.6
Relating reaction spontaneity to ΔH and ΔS	Use $$\Delta G = \Delta H - T\,\Delta S \qquad (14.8)$$	14.7
Calculating ΔG° of coupled reactions	Free-energy changes are additive. If adding two reactions gives the overall reaction you need, add the free-energy changes of the two reactions to obtain the free-energy change of the overall reaction.	14.8

VISUAL PROBLEMS

(Answers to boldface end-of-chapter questions and problems are in the back of the book.)

14.1. More air is added to the partially filled party balloon in Figure P14.1. Does the entropy of the balloon and its contents increase or decrease?

FIGURE P14.1

14.2. The two cubic containers in Figure P14.2 contain two gas samples at the same temperature and pressure. Which has the higher entropy—the sample in a or the sample in b?

If the sample in b is left unchanged but the sample in a is cooled so that it condenses, which sample has the higher entropy?

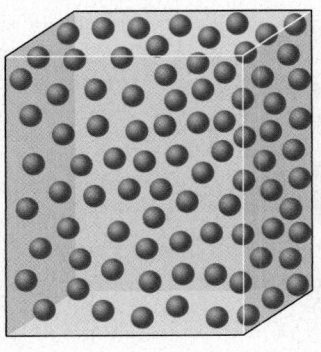
(a)

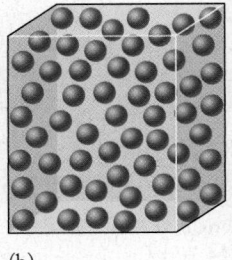
(b)

FIGURE P14.2

14.3. Figure P14.3 shows a glass tube connecting two bulbs containing a mixture of ideal gases A (red spheres) and B (blue spheres). Is the probability high or low that gas A will collect in the left-hand bulb and gas B will collect in the right-hand bulb? Explain your answer in terms of the entropy changes involved.

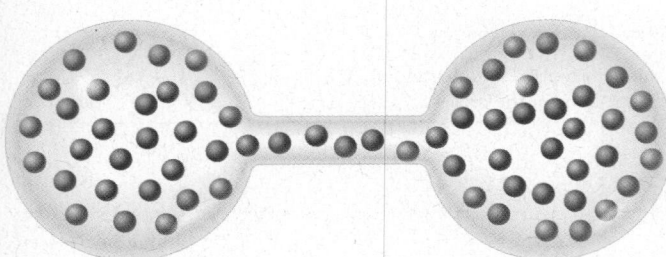

FIGURE P14.3

14.5. Figure P14.5 shows the plots of ΔH and $T\Delta S$ for a reaction as a function of temperature. What is the significance of their point of intersection? Over what temperature range is the reaction nonspontaneous?

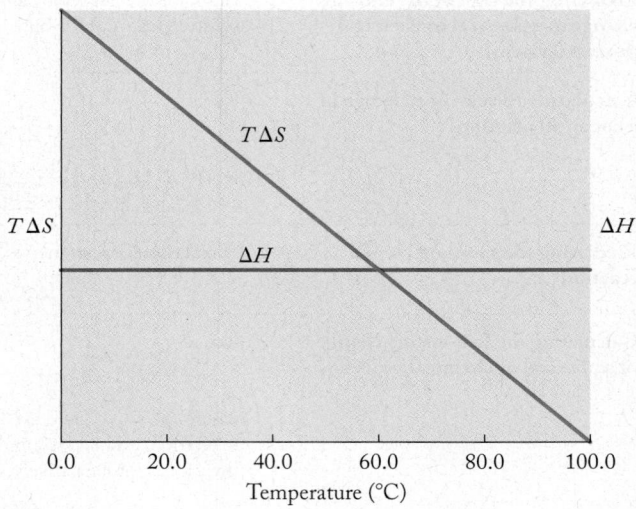

FIGURE P14.5

14.4. The box in Figure P14.4(a) contains a gas. Complete the box in Figure P14.4(b) to illustrate deposition of the gas as a solid. Suggest whether ΔS, ΔH, and/or ΔG are positive or negative for this spontaneous process. Does the entropy of the surroundings increase or decrease?

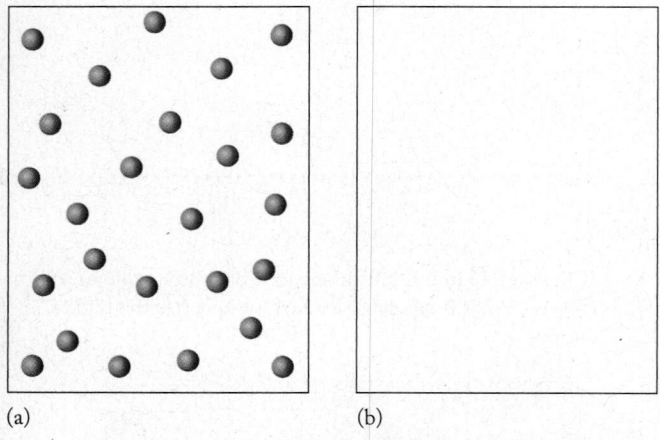

(a)   (b)

FIGURE P14.4

14.6. Figure P14.6 is based on the ΔG_f° values of elements and compounds selected from Appendix 4. Which of the following conversions are spontaneous? (a) $C_6H_6(\ell)$ to $CO_2(g)$ and $H_2O(\ell)$; (b) $CO_2(g)$ to $C_2H_2(g)$; (c) $H_2(g)$ and $O_2(g)$ to $H_2O(\ell)$. Explain your reasoning.

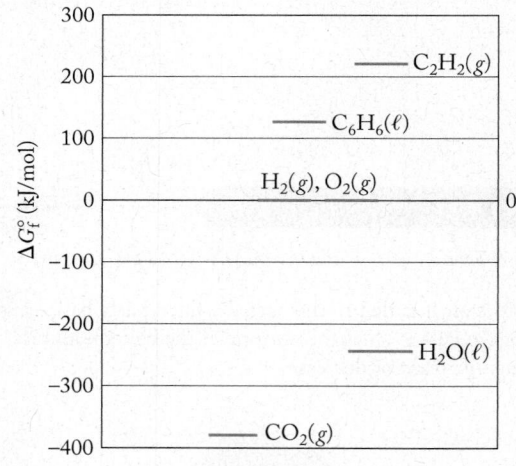

FIGURE P14.6

QUESTIONS AND PROBLEMS

Spontaneous Processes and Entropy; Thermodynamic Entropy

CONCEPT REVIEW

14.7. What happens to the sign of the entropy change when a process is reversed?

14.8. Identify the following processes as spontaneous or nonspontaneous, and explain your choice: (a) the battery in a cell phone discharging; (b) a helium balloon rising above Earth's surface; (c) radioactive decay.

14.9. You flip three coins, assigning the values +1 for heads and −1 for tails. How many different microstates are possible from flipping the three coins? Which value or values for the sums in the microstates are most likely? *Hint*: The sequence HHT $(+1 +1 -1)$ is one possible outcome, or microstate. Note, however, that this outcome differs from THH $(-1 +1 +1)$, even though the two sequences sum to the same value.

14.10. Adding Drano to water causes the temperature of the water to increase. If Drano is the system, what are the signs of ΔS_{sys} and ΔS_{surr}?

14.11. Ice cubes melt in a glass of lemonade, cooling the lemonade from 10.0°C to 0.0°C. If the ice cubes are the system, what are the signs of ΔS_{sys} and ΔS_{surr}?

PROBLEMS

14.12. Which of the following combinations of entropy changes for a process are possible?
 a. $\Delta S_{sys} < 0, \Delta S_{surr} > 0, \Delta S_{univ} > 0$
 b. $\Delta S_{sys} < 0, \Delta S_{surr} < 0, \Delta S_{univ} > 0$
 c. $\Delta S_{sys} < 0, \Delta S_{surr} > 0, \Delta S_{univ} < 0$

14.13. Which of the following combinations of entropy changes for a process are possible?
 a. $\Delta S_{sys} > 0, \Delta S_{surr} > 0, \Delta S_{univ} > 0$
 b. $\Delta S_{sys} > 0, \Delta S_{surr} < 0, \Delta S_{univ} > 0$
 c. $\Delta S_{sys} > 0, \Delta S_{surr} > 0, \Delta S_{univ} < 0$

14.14. The spontaneous reaction A → B + C increases the system entropy by 132.0 J/K. What is the minimum value of the entropy change of the surroundings?

14.15. The nonspontaneous reaction D + E → F decreases the system entropy by 48.0 J/K. What is the maximum value of the entropy change of the surroundings?

14.16. In each of the following pairs, which alternative has the greater entropy?
 a. A pound of ice cubes or a pound of liquid water.
 b. A spoonful of sugar or a spoonful of sugar dissolved in a cup of coffee.
 c. A cup of hot water or a cup of cold water.
 d. A mole of cyclohexane, C_6H_{12}, or a mole of 1-hexene, $CH_2{=}CH(CH_2)_3CH_3$.

14.17. In each of the following pairs, which alternative has the greater entropy?
 a. Cyclopropane or propene.
 b. Wet paint or dry paint.
 c. One mole of $SO_2(g)$ or 1 mole of $SO_3(g)$.
 d. An aquarium with fish or the same aquarium without fish.

Absolute Entropy, the Third Law of Thermodynamics, and Structure

CONCEPT REVIEW

14.18. In living cells, small molecules react together to make much larger ones. Are these processes accompanied by increases or decreases in entropy of the molecules?

14.19. Which physical state of a substance—solid, liquid, or gas—has the highest standard molar entropy?

14.20. Does the dissolution of a gas in a liquid result in an increase or a decrease in entropy?

*14.21. Diamond and the fullerenes are two allotropes of carbon. On the basis of their different structures and properties, predict which has the higher standard molar entropy.

14.22. **Superfluids** The 1996 Nobel Prize in Physics was awarded to Douglas Osheroff, Robert Richardson, and David Lee for discovering *superfluidity* (apparently frictionless flow) in ^{3}He. When ^{3}He is cooled to 2.7 mK, the liquid settles into an *ordered* superfluid state. What is the predicted sign of the entropy change for the conversion of liquid ^{3}He into its superfluid state?

PROBLEMS

14.23. Rank the compounds in each of the following groups in order of increasing standard molar entropy ($S°$):
 a. $CH_4(g)$, $CF_4(g)$, and $CCl_4(g)$
 b. $CH_3OH(\ell)$, $CH_3CH_2OH(\ell)$, and $CH_3CH_2CH_2OH(\ell)$
 c. $HF(g)$, $H_2O(g)$, and $NH_3(g)$

14.24. Without referring to either Table 14.2 or Appendix 4, rank the compounds in each of the following groups in order of increasing standard molar entropy ($S°$):
 a. $CH_4(g)$, $CH_3CH_3(g)$, and $CH_3CH_2CH_3(g)$
 b. $CCl_4(\ell)$, $CHCl_3(\ell)$, and $CH_2Cl_2(\ell)$
 c. $CO_2(\ell)$, $CO_2(g)$, and $CS_2(g)$

14.25. Predict the sign of ΔS_{sys} for each of the following processes:
 a. A bricklayer builds a wall out of a random pile of bricks.
 b. You rake a yard full of leaves into a single pile.
 c. $Ag^+(aq) + Cl^-(aq) \rightarrow AgCl(s)$
 d. $Zn(s) + 2HCl(aq) \rightarrow H_2(g) + ZnCl_2(aq)$

14.26. Predict the sign of ΔS_{sys} for each of the following processes:
 a. Sweat evaporates.
 b. Solid silver chloride dissolves in aqueous ammonia.
 c. $CH_3CH_2CH_3(g) + 5O_2(g) \rightarrow 3CO_2(g) + 4H_2O(\ell)$
 d. $N_2O_5(g) \rightarrow NO_2(g) + NO_3(g)$

Calculating Entropy Changes

CONCEPT REVIEW

14.27. The products of a process have lower entropy than the reactants. Is the entropy change of the process positive or negative?

*14.28. Why might the entropy change of a process increase with increasing temperature?

*14.29. A reaction of A(s) to make B(ℓ) is studied at different temperatures. Would the entropy change of the reaction be markedly different at temperatures higher than the melting point of reactant A?

*14.30. What stoichiometric and molecular factors affect the standard entropy change for the conversion of ozone to oxygen?

$$2O_3(g) \rightarrow 3O_2(g)$$

PROBLEMS

14.31. **Smog** Use the standard molar entropies in Appendix 4 to calculate $\Delta S°$ values for each of the following atmospheric reactions that contribute to the formation of photochemical smog, a haze caused by reactions of NO released from car and truck engines with molecules in the atmosphere:
 a. $N_2(g) + O_2(g) \rightarrow 2NO(g)$
 b. $2NO(g) + O_2(g) \rightarrow 2NO_2(g)$
 c. $NO(g) + \frac{1}{2}O_2(g) \rightarrow NO_2(g)$
 d. $2NO_2(g) \rightarrow N_2O_4(g)$

14.32. Use the standard molar entropies in Appendix 4 to calculate the $\Delta S°$ value for each of the following reactions of sulfur compounds.

a. $H_2S(g) + \frac{3}{2}O_2(g) \rightarrow H_2O(g) + SO_2(g)$

b. $2SO_2(g) + O_2(g) \rightarrow 2SO_3(g)$

c. $SO_3(g) + H_2O(\ell) \rightarrow H_2SO_4(aq)$

d. $S(g) + O_2(g) \rightarrow SO_2(g)$

14.33. Ozone Layer The following reaction plays a key role in the destruction of ozone in the atmosphere:

$$Cl(g) + O_3(g) \rightarrow ClO(g) + O_2(g)$$

The standard entropy change ($\Delta S°_{rxn}$) is 19.9 J/(mol · K). Use the standard molar entropies ($S°$) in Appendix 4 to calculate the $S°$ value of $ClO(g)$.

14.34. Calculate the $\Delta S°$ value for the conversion of ozone to oxygen

$$2O_3(g) \rightarrow 3O_2(g)$$

in the absence of Cl atoms, and compare it with the $\Delta S°$ value in Problem 14.33.

Free Energy and Free-Energy Change

CONCEPT REVIEW

14.35. The 19th-century scientist Marcellin Berthelot stated that all exothermic reactions are spontaneous. Is this statement correct?

14.36. Under what conditions does an increase in temperature turn a nonspontaneous process into a spontaneous one?

14.37. If a reaction has a negative free-energy change, is it spontaneous?

14.38. If a reaction has a positive free-energy change, does it proceed slowly?

14.39. What can we say if the calculated free-energy change of a process is positive?

14.40. What can we say if the calculated free-energy change of a process is zero?

14.41. Are exothermic reactions spontaneous only at low temperature? Explain your answer.

14.42. Are endothermic reactions never spontaneous at low temperature? Explain your answer.

PROBLEMS

14.43. What are the signs of ΔS, ΔH, and ΔG for the sublimation of dry ice (solid CO_2) at 25°C?

14.44. What are the signs of ΔS, ΔH, and ΔG for the formation of dew on a cool night?

14.45. Indicate whether each of the following processes is spontaneous:

a. The fragrance of a perfume spreads through a room.

b. A broken clock is mended.

c. An iron fence rusts.

d. An ice cube melts in a glass of water.

14.46. Indicate whether each of the following processes is spontaneous:

a. Charcoal is converted to carbon dioxide when ignited in air.

b. Steam condenses on a cold window.

c. Sugar dissolves in hot water.

d. $CH_4(g)$ and $O_2(g)$ are formed from $CO_2(g)$ and $H_2O(\ell)$.

14.47. Calculate the free-energy change for the dissolution in water of 1 mole of NaBr and 1 mole of NaI at 298 K, given that the values of $\Delta H°_{soln}$ are -1 and -7 kJ/mol for NaBr and NaI. The corresponding values of $\Delta S°_{soln}$ are 57 and 74 J/(mol · K).

14.48. The values of $\Delta H°_{rxn}$ and $\Delta S°_{rxn}$ for the reaction

$$2NO(g) + O_2(g) \rightarrow 2NO_2(g)$$

are -12 kJ and -146 J/K. Calculate $\Delta G°$ at 298 K for this reaction. Why do you think the value of $\Delta S°$ is negative?

*14.49. A mixture of $CO(g)$ and $H_2(g)$ is produced by passing steam over charcoal:

$$H_2O(g) + C(s) \rightarrow H_2(g) + CO(g)$$

Assuming that the values of $\Delta H°$ and $\Delta S°$ do not change appreciably with temperature, calculate the $\Delta G°$ value for the reaction from the data in Appendix 4, and predict the lowest temperature at which the reaction is spontaneous.

14.50. Consider the following combustion reactions:

(1) $2CH_3OH(g) + 3O_2(g) \rightarrow 2CO_2(g) + 4H_2O(g)$

(2) $CH_4(g) + 2O_2(g) \rightarrow CO_2(g) + 2H_2O(\ell)$

(3) $2H_2(g) + O_2(g) \rightarrow 2H_2O(g)$

a. For each reaction, predict the sign of $\Delta S°$ before calculating its value.

b. For each reaction, calculate $\Delta S°$ and $\Delta G°$ from the data in Appendix 4.

14.51. Use the data in Appendix 4 to calculate $\Delta H°$ and $\Delta S°$ for the following process:

$$H_2O(\ell) \rightarrow H_2O(g)$$

Assume that the calculated values are independent of temperature, and calculate the boiling point of water at $P = 1.00$ atm.

14.52. Chlorofluorocarbons (CFCs) are no longer used as refrigerants because they help destroy the ozone layer. Trichlorofluoromethane (CCl_3F) boils at 23.8°C and its molar heat of vaporization is 24.8 kJ/mol. Calculate the molar entropy of evaporation of $CCl_3F(\ell)$.

*14.53. Deposits of elemental sulfur are often seen near active volcanoes. Their presence there may be due to the following reaction of SO_2 with H_2S:

$$SO_2(g) + 2H_2S(g) \rightarrow \frac{3}{8}S_8(s) + 2H_2O(g)$$

Calculate $\Delta H°$ and $\Delta S°$ for this reaction. Assuming that the values of $\Delta H°$ and $\Delta S°$ do not change appreciably with temperature, predict the temperature range over which the reaction is spontaneous.

14.54. Methanogenic bacteria convert acetic acid (CH_3COOH) into $CO_2(g)$ and $CH_4(g)$.

a. Is this process endothermic or exothermic under standard conditions?

b. Is the reaction spontaneous under standard conditions?

14.55. Use the data in Appendix 4 to calculate $\Delta G°$ for each of the following reactions. Assuming that the values of $\Delta H°$ and $\Delta S°$ do not change appreciably with temperature, is each reaction spontaneous at 355 K?
a. $N_2(g) + O_2(g) \rightarrow 2NO(g)$
b. $2NO(g) + O_2(g) \rightarrow 2NO_2(g)$
c. $NO(g) + \frac{1}{2}O_2(g) \rightarrow NO_2(g)$
d. $2NO_2(g) \rightarrow N_2O_4(g)$

14.56. Use the data in Appendix 4 to calculate $\Delta G°$ for each of the following reactions. Assuming that the values of $\Delta H°$ and $\Delta S°$ do not change appreciably with temperature, is each reaction spontaneous at 211 K?
a. $2H_2S(g) + 3O_2(g) \rightarrow 2H_2O(g) + 2SO_2(g)$
b. $2SO_2(g) + O_2(g) \rightarrow 2SO_3(g)$
c. $SO_3(g) + H_2O(\ell) \rightarrow H_2SO_4(\ell)$
d. $S(g) + O_2(g) \rightarrow SO_2(g)$

14.57. Assuming that the values of $\Delta H°$ and $\Delta S°$ do not change appreciably with temperature, which of the reactions in Problem 14.55 are spontaneous at (a) high temperature? (b) low temperature? (c) all temperatures?

14.58. Assuming that the values of $\Delta H°$ and $\Delta S°$ do not change appreciably with temperature, which of the reactions in Problem 14.56 are spontaneous at (a) high temperature? (b) low temperature? (c) all temperatures?

14.59. Use the free energies of formation from Appendix 4 to calculate the standard free-energy change for the decomposition of ammonia in the following reaction:

$$2NH_3(g) \rightarrow N_2(g) + 3H_2(g)$$

Is the reaction spontaneous under standard conditions?

14.60. Use heats of formation and entropies of reactants and products from Appendix 4 to calculate the standard free-energy change for the decomposition reaction in Problem 14.59 at 298 K. Compare these results with those you obtained in Problem 14.59. Assume that ΔH_{rxn} and ΔS_{rxn} are independent of temperature, and calculate the temperature at which ΔG_{rxn} is zero. What is the significance of this temperature for this reaction?

14.61. A reaction

$$A(g) \rightarrow B(g) + C(g)$$

has the following standard thermodynamic parameters: $\Delta H°_{rxn} = 40.0$ kJ/mol and $\Delta S°_{rxn} = 80.0$ J/(mol · K).
a. Is the reaction exothermic or endothermic?
b. Does the positive entropy change make sense?
c. Is the reaction spontaneous at all temperatures? If not, explain.
d. Calculate the temperature at which the reaction becomes spontaneous.

14.62. A reaction $C(g) + D(g) \rightarrow E(g)$ has the following standard thermodynamic parameters: $\Delta H°_{rxn} = 35.0$ kJ/mol and $\Delta S°_{rxn} = -35.0$ J/(mol · K).
a. Is the reaction exothermic or endothermic?
b. Does the negative entropy change make sense?
c. Is the reaction spontaneous at all temperatures? If not, explain.
d. Calculate the temperature at which the reaction becomes nonspontaneous.

Driving the Human Engine: Coupled Reactions

CONCEPT REVIEW

14.63. Glycolysis The second step in glycolysis converts glucose 6-phosphate into fructose 6-phosphate (Figure P14.63). Suggest a reason why $\Delta G°$ for this reaction is close to zero.

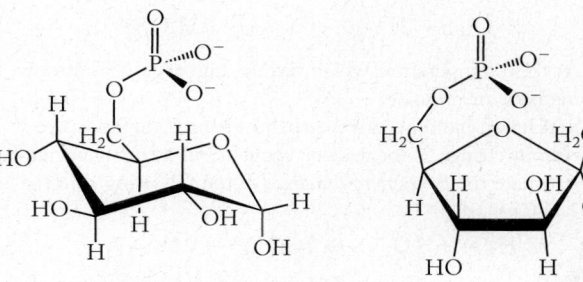

Glucose 6-phosphate

Fructose 6-phosphate

FIGURE P14.63

14.64. Why is it important that at least some of the spontaneous steps in glycolysis convert ADP to ATP?

14.65. How do we calculate the overall free-energy change of a process consisting of two steps?

14.66. Make a statement in the style of Hess's law that describes how to calculate $\Delta G°$ for two coupled reactions.

PROBLEMS

14.67. The hydrolysis of 1 mole of the sugar maltose ($\Delta G°_f = -910.1$ kJ/mol) produces 2 moles of glucose ($\Delta G°_f = -695.65$ kJ/mol):

$$\text{Maltose} + H_2O \rightarrow 2\text{ glucose}$$

If the value of $\Delta G°_f$ of water is -237.2 kJ/mol, what is the standard free-energy change of the hydrolysis reaction?

14.68. The standard free-energy change in the reaction of water with ethyl acetate ($CH_3COOCH_2CH_3$) to give ethanol (CH_3CH_2OH) and acetic acid (CH_3COOH) is -19.7 kJ. From this value and the values of the standard free energy of formation of liquid ethanol, acetic acid, and water (see Appendix 4), calculate $\Delta G°_f$ for ethyl acetate.

Additional Problems

14.69. Explain what part or parts of the following statement are wrong and why: "Almost all substances contract when you cool them down. At some point of cooling they freeze, and continuing to cool the sample increases its entropy." Now revise the statement so that it is correct.

14.70. Rewrite the following statement so that it is factually correct: "A dollar bill and a penny in a cash box have higher entropy than 101 penny coins in the same box at the same temperature."

14.71. At what temperature is the free-energy change for the following reaction equal to zero?

$$NH_4Cl(s) \rightarrow NH_3(g) + HCl(g)$$

14.72. Which of these processes result in an entropy decrease of the system?
  a. Diluting hydrochloric acid with water
  b. Boiling water
  c. $2 NO(g) + O_2(g) \rightarrow 2 NO_2(g)$
  d. Making ice cubes in the freezer

*14.73. Calculate the standard free-energy change of the following reaction. Is it spontaneous?

$$2 NO(g) + 2 H_2(g) \rightarrow N_2(g) + 2 H_2O(g)$$

14.74. At fixed temperature, which has the higher molar entropy, methane or propane?

14.75. A reaction has a negative enthalpy change and a positive entropy change. Is the reaction spontaneous at any temperature?

*14.76. Estimate the free-energy change of the following reaction at 225°C:

$$C_2H_4(g) + 3 O_2(g) \rightarrow 2 CO_2(g) + 2 H_2O(g)$$

14.77. Explain why the standard molar entropies of elements are positive and not zero.

14.78. Explain why the standard free energies of formation of elements are zero.

*14.79. Show that hydrogen cyanide (HCN) is a gas at 25°C by estimating its normal boiling point from the following data:

	ΔH_f°, kJ/mol	S°, J/(mol · K)
HCN(ℓ):	108.9	113
HCN(g):	135.1	202

*14.80. Show that hydrogen peroxide (H_2O_2) is a liquid at 25°C by estimating its normal boiling point from the following data:

	ΔH_f°, kJ/mol	S°, J/(mol · K)
$H_2O_2(\ell)$:	−187.8	110
$H_2O_2(g)$:	−136.3	233

14.81. Keeping Cool Carbon tetrachloride (CCl_4) was a favored refrigerant until it was found to be carcinogenic. The heat of vaporization of $CCl_4(\ell)$ is 32.5 kJ/mol and its normal boiling point is 76.7°C. Estimate the entropy of vaporization of $CCl_4(\ell)$. Should the answer be positive or negative?

14.82. Making Methanol The element hydrogen is not abundant in nature, but it is a useful reagent in, for example, the potential synthesis of the liquid fuel methanol from gaseous carbon monoxide:

$$2 H_2(g) + CO(g) \rightarrow CH_3OH(\ell)$$

Under what temperature conditions is this reaction spontaneous?

14.83. Lightbulb Filaments Tungsten (W) is the favored metal for lightbulb filaments, in part because of its high melting point of 3422°C. The enthalpy of fusion of tungsten is 35.4 kJ/mol. What is its entropy of fusion?

*14.84. Over what temperature range is the reduction of tungsten(VI) oxide by hydrogen to give metallic tungsten and water spontaneous? The standard heat of formation of $WO_3(s)$ is −843 kJ/mol, and its standard molar entropy is 76 J/(mol · K).

14.85. Two allotropes (A and B) of sulfur interconvert at 369 K and 1 atm pressure:

$$S_8(s, A) \rightleftharpoons S_8(s, B)$$

The enthalpy change in this transition is 297 J/mol. What is the entropy change?

*14.86. Copper forms two oxides, Cu_2O and CuO.
  a. Name these oxides.
  b. Predict over what temperature range this reaction is spontaneous using the following thermodynamic data:

$$Cu_2O(s) \rightarrow CuO(s) + Cu(s)$$

	ΔH_f°, kJ/mol	S°, J/(mol · K)
$Cu_2O(s)$:	−170.7	92.4
$CuO(s)$:	−156.1	42.6

  c. Why is the standard molar entropy of $Cu_2O(s)$ larger than that of $CuO(s)$?

*14.87. Lime Enormous amounts of lime (CaO) are used in steel industry blast furnaces to remove impurities from iron. Lime is made by heating limestone and other solid forms of $CaCO_3(s)$. Why is the standard molar entropy of $CaCO_3(s)$ higher than that of CaO(s)? At what temperature is the pressure of $CO_2(g)$ over $CaCO_3(s)$ equal to 1.0 atm?

	ΔH_f°, kJ/mol	S°, J/(mol · K)
$CaCO_3(s)$:	−1207	93
$CaO(s)$:	−636	40
$CO_2(g)$:	−394	214

*14.88. *Trouton's rule* says that the ratio $\Delta H_{vap}^\circ/T_b$ for a liquid is approximately 80 J/K. Here, ΔH_{vap}° is the molar enthalpy of vaporization of a liquid and T_b is its normal boiling point.
  a. What idea suggests that $\Delta H_{vap}^\circ/T_b$ for a range of liquids should be approximately constant?
  b. Check Trouton's rule against the data in Figure P14.88. Which liquids deviate from Trouton's rule, and why?

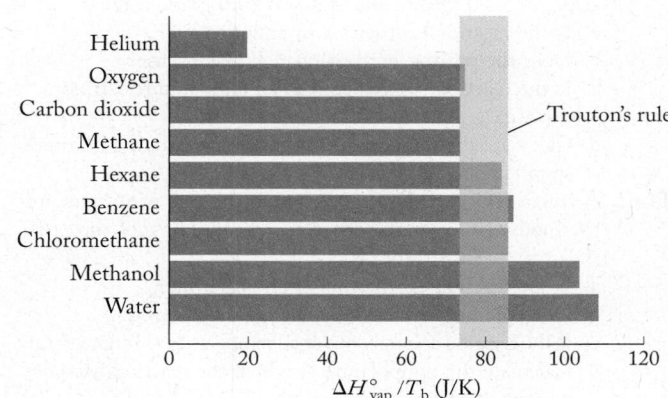

FIGURE P14.88

*14.89. Melting DNA When a solution of DNA in water is heated, the DNA double helix separates into two single strands:

$$1 \text{ DNA double helix} \rightleftharpoons 2 \text{ single strands}$$

a. What is the sign of ΔS for the forward process as written?
b. The DNA double helix re-forms as the system cools. What is the sign of ΔS for the process by which two single strands re-form the double helix?
c. The melting point of DNA is defined as the temperature at which $\Delta G = 0$. At that temperature, the forward reaction produces two single strands as fast as two single strands recombine to form the double helix. Write an equation that defines the melting temperature (T) of DNA in terms of ΔH and ΔS.

*14.90. Melting Organic Compounds When dicarboxylic acids (compounds with two –COOH groups in their structures) melt, they frequently decompose to produce 2 moles of CO_2 gas for every 1 mole of dicarboxylic acid melted (shown in Figure P14.90).

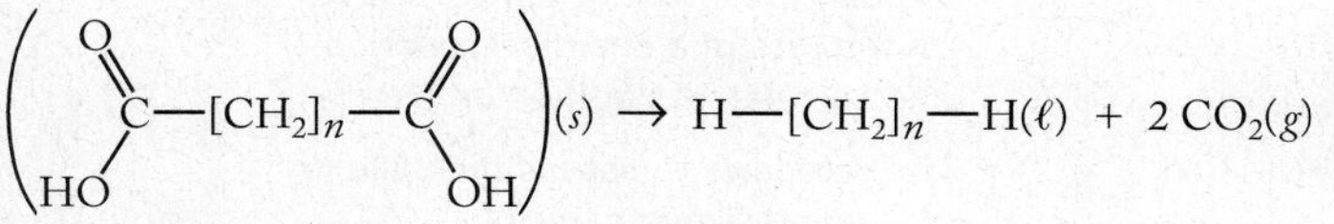

 $(s) \rightarrow H-[CH_2]_n-H(\ell) + 2\, CO_2(g)$

FIGURE P14.90

a. What are the signs of ΔH and ΔS for the process as written?
b. Problem 14.89 describes the DNA double helix re-forming when the system cools after melting. Do you think the dicarboxylic acid will re-form when the melted material cools? Why or why not?

*14.91. Draw all the vibrational motions possible in a H_2 molecule and a H_2O molecule. How does the number of vibrational motions possible in an NH_3 molecule compare with the number for H_2 and H_2O? (*Hint*: You may wish to review Section 8.4.)

14.92. Lead(II) chloride and lithium hydroxide are two common ionic compounds.
a. Predict the sign of $\Delta S°$ for these reactions:

$$Pb^{2+}(aq) + 2\,Cl^-(aq) \rightarrow PbCl_2(s)$$
$$Li^+(aq) + OH^-(aq) \rightarrow LiOH(s)$$

b. Using the values for $S°$ from Appendix 4 calculate $\Delta S°$ for these reactions. Note that $S°_{LiOH} = 3.8 \text{ J/(mol · K)}$.
c. Do your calculations support your prediction?

Chemical Kinetics

Learning Outcomes

- Determine the rate law and overall order for a chemical reaction using initial rate data

- Use integrated rate laws to identify first- and second-order reactions

- Calculate rate constants from kinetic data

- Explain the effect of temperature on the rate of a chemical reaction and calculate its activation energy

- Use rate laws to assess the validity of a reaction mechanism

- Identify the rate-determining step in a mechanism using reaction rate data

- Explain the effect of catalysts on reaction rates and mechanisms

A LOOK AHEAD

Clearing the Air

Take a look at the photographs to the right. What is that yellow-brown haze that colors the left side of the picture and obscures the mountains in the distance? That murky color is the result of *smog*, so called because it combines the effects of smoke and fog, reducing visibility and causing respiratory difficulties. *Photochemical smog* forms when sunny weather and light winds allow vehicle emissions to collect over urban centers. The interaction of sunlight with nitrogen and oxygen compounds in the air initiates a host of chemical reactions that produce the brown haze. Some approaches to cleaning up this smog are futuristic, such as replacing our current vehicles with electric cars powered by high-performance batteries and/or fuel cells. However, one device has removed many billions of tons of smog-forming pollutants from the air above U.S. cities since it was developed in the 1970s: the catalytic converter. While many other nations also use this technology, developing countries have been slower to make catalytic converters mandatory on vehicles.

The development of catalytic converters required extensive study of the chemical reactions taking place in engines burning fossil fuels. The gaseous products of these reactions include unburned hydrocarbons, carbon monoxide, and various oxides of nitrogen. Pollution due to these emissions can be reduced if the hydrocarbons and CO are oxidized to CO_2 and H_2O and if the NO_x compounds are converted to N_2 and O_2. Unfortunately, these redox reactions all take place at different rates, so researchers had to find ways to make all the required reactions take place in the short time interval between generation in the engine and departure from the vehicle through the exhaust system. Understanding how factors such as temperature and catalysts influence *reaction rates* and what happens to these pollutants at the molecular level led to the development of successful catalytic converters.

Clearing the Air A light snowfall and change in wind direction dramatically ▶ affect visibility and air quality over Salt Lake City, Utah, clearing the photochemical smog from the air. These photos were taken on January 28, 2007, at 4:15 p.m. (left) and January 31, 2007, at 3:53 p.m. (right).

photochemical smog a mixture of gases formed in the lower atmosphere when sunlight interacts with compounds produced in internal combustion engines and other pollutants.

Air pollution caused by vehicles is ongoing, but so too is our commitment to cleaning the air. By studying chemical reactions both in internal combustion engines and in the atmosphere, we have learned much about the air-quality problems generated from automobile exhaust. In addition, we now understand the molecular basis of other atmospheric problems, such as the ozone (O_3) hole in the stratosphere (upper atmosphere).

A crucial part of these studies has been the examination of rates of reactions—how rapidly they proceed under a variety of conditions. Rates of reactions are the focus of this chapter. As we begin the study of reaction rates, consider the following questions: How do we measure the rates of smog formation and other chemical reactions? How does the study of reaction rates allow us to understand chemical reactions on a molecular basis? How can we change the rate of a reaction to suit our needs? ■

15.1 Cars, Trucks, and Air Quality

Photochemical smog consists of various nitrogen oxides (NO_x), ozone (O_3), and organic molecules such as peroxyacetyl nitrate, a strong respiratory and eye irritant (Figure 15.1). This smog is created by the interaction of sunlight with nitrogen oxides from vehicle exhaust and volatile organic compounds (VOCs), from natural and man-made sources.

One of the first reactions leading to photochemical smog takes place inside an automobile engine—the formation of nitrogen monoxide from N_2 and O_2:

$$N_2(g) + O_2(g) \rightarrow 2\,NO(g) \qquad \Delta H° = 180.6 \text{ kJ} \qquad (15.1)$$

CONCEPT TEST ....

The reaction shown in Equation 15.1 is endothermic ($\Delta H°_{rxn} > 0$) but it is spontaneous ($\Delta G_{rxn} < 0$) at very high temperatures.

a. What is the sign of $\Delta S°_{rxn}$?
b. Why is the sign of $\Delta S°_{rxn}$ difficult to predict by looking at the reaction in Equation 15.1 alone?

(Answers to Concept Tests are in the back of the book.)

FIGURE 15.1 Photochemical smog, like the orange-brown layer over Denver in this photograph, contains a mixture of compounds including NO, NO_2, O_3, and peroxyacetyl nitrate ($CH_3CO_2NO_3$).

Once NO enters the atmosphere, it reacts with more oxygen, producing brown nitrogen dioxide gas:

$$2\,NO(g) + O_2(g) \rightarrow 2\,NO_2(g) \qquad \Delta H° = -114.2 \text{ kJ} \qquad (15.2)$$

Sunlight provides sufficient energy to break the bonds in NO_2, forming NO and very reactive oxygen atoms:

$$NO_2(g) \xrightarrow{\text{sunlight}} NO(g) + O(g) \qquad (15.3)$$

This photochemically generated atomic oxygen combines with molecular oxygen, producing ozone,

$$O_2(g) + O(g) \rightarrow O_3(g) \qquad (15.4)$$

and with water vapor to produce hydroxyl radicals:

$$O(g) + H_2O(g) \rightarrow 2\,OH(g) \qquad (15.5)$$

If we examine just these few reactions, we see immediately how challenging is the task of understanding smog formation, hinging on the fact that most of the substances in photochemical smog are both reactants and products. The relationships among the concentrations, even in this simplified view, are more complex than others we have seen in previous chapters.

In the atmosphere, VOCs react with ozone, hydroxyl radicals, and atomic oxygen to form a series of oxygen compounds that subsequently react with NO_2 to form peroxyacetyl nitrate. One typical reaction involves the VOC acetaldehyde:

$$N_2(g) + 3\,O_2(g) + OH(g) + CH_3CHO(g) \rightarrow$$
$$\text{Acetaldehyde}$$
$$\qquad (15.6)$$
$$CH_3C(O)O_2NO_2(g) + H_2O(g) + NO_2(g)$$
$$\text{Peroxyacetyl nitrate}$$

To understand the production of NO_x in engines and the environment, we need to understand how the reactions producing NO_x are linked. In particular, we need to know their **reaction rates**, which are the rates at which their reactants are consumed and their products are formed.

In the reactions involved in smog formation, the products of some reactions are the reactants in others. Therefore, the relative rates of these reactions influence when during the day pollutants appear, how long they persist, and what their concentrations are. Note in Figure 15.2, for instance, that the maximum NO concentration occurs during the morning rush hour. Later in the morning, the concentration of NO_2 reaches a maximum. This sequence makes sense because NO is a precursor of NO_2. The highest ozone concentrations are reached in the middle of the afternoon, when the reactions shown in Equations 15.2 and 15.3 are in full swing, producing a supply of free O atoms for the formation of ozone (Equation 15.4), hydroxyl radicals (Equation 15.5), and, indirectly, strong respiratory and eye irritants that include peroxyacetyl nitrate (Equation 15.6).

The ozone formed in the reaction in Equation 15.4 also reacts with NO to form NO_2 and O_2:

$$O_3(g) + NO(g) \rightarrow O_2(g) + NO_2(g) \qquad (15.7)$$

The NO_2 concentration drops in the afternoon, because NO_2, ozone, and hydrocarbons react to form an array of other compounds.

◯◯ CONNECTION We introduced species like OH and O in Chapter 8 in the discusson of odd-electron molecules and free radicals.

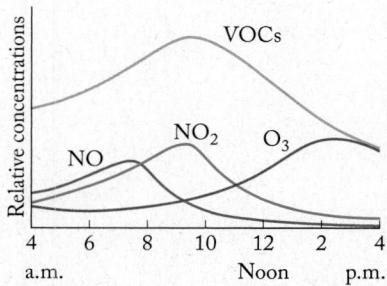

FIGURE 15.2 In photochemical smog, NO from engine exhaust builds up in the early morning, then decreases as the NO reacts with atmospheric O_2, forming NO_2, the concentration of which is highest in late morning. Photodecomposition of NO_2 leads to the formation of high levels of O_3 in the afternoon.

reaction rate how rapidly a reaction occurs; it is related to rates of change in the concentrations of reactants and products over time.

chemical kinetics the study of the rates of change of concentrations of substances involved in chemical reactions.

The catalytic converter, now a standard feature on automobiles in many countries, helps combat the production of photochemical smog by removing NO and unburned or partially oxidized hydrocarbons from exhaust gases. Knowledge of **chemical kinetics**, the study of the rates at which reactant and product concentrations change during a chemical reaction, has provided the understanding required to develop catalytic converters and other devices aimed at improving our air quality. At least 75% of the NO produced in the engines of vehicles equipped with catalytic converters is converted back to N_2 and O_2 before being emitted into the air.

15.2 Reaction Rates

The amount of NO that forms inside an automobile engine in a given time interval depends on how rapidly the reaction in Equation 15.1

$$N_2(g) + O_2(g) \rightarrow 2\,NO(g)$$

▶❚❚ **CHEMTOUR** Reaction Rate

proceeds. We can express the reaction rate as the change in product (or reactant) concentration that occurs over some interval of time. For example, the change in concentration of NO is $\Delta[NO] = [NO]_{final} - [NO]_{initial}$ over the time interval $\Delta t = t_{final} - t_{initial}$. The reaction rate, expressed as the rate of formation of NO, is

$$\text{Rate of NO formation} = \frac{\Delta[NO]}{\Delta t} = \frac{[NO]_{final} - [NO]_{initial}}{t_{final} - t_{initial}} \quad (15.8)$$

We use square brackets in Equation 15.8 and throughout this chapter and beyond to indicate concentration, typically in units of moles/liter.

The rate of the reaction between N_2 and O_2 to form NO can also be expressed as the rate of consumption of either reactant. However, the rate of change in the concentration of a reactant ($\Delta[\text{reactant}]/\Delta t$) has a negative value because reactant concentrations decrease as a reaction proceeds. The measured rate of any reaction is defined as a positive quantity because it describes the rate at which reactants form products, so a minus sign is used with $\Delta[\text{reactant}]/\Delta t$ values to obtain an overall positive value for the reaction rate. Thus the rate at which the N_2 is consumed is

$$\text{Rate of } N_2 \text{ consumption} = -\frac{\Delta[N_2]}{\Delta t} = -\frac{[N_2]_{final} - [N_2]_{initial}}{t_{final} - t_{initial}} \quad (15.9)$$

Similarly for O_2:

$$\text{Rate of } O_2 \text{ consumption} = -\frac{\Delta[O_2]}{\Delta t} = -\frac{[O_2]_{final} - [O_2]_{initial}}{t_{final} - t_{initial}} \quad (15.10)$$

The rate of formation or consumption in all three cases is a change in concentration divided by a change in time, so the units are concentration per unit time, such as molarity per second (M/s).

CONCEPT TEST

A reaction rate is analogous to the speed at which you drive your car. If reaction rates reflect the "speed" of a reaction, why can't a reaction have a negative rate?

Are the rates of N_2 consumption, O_2 consumption, and NO formation the same? The answer to this question lies in the balanced equation for the reaction. The fact that 2 moles of NO are formed from 1 mole of N_2 and 1 mole of O_2 means that the rate of consumption of N_2 ($-\Delta[N_2]/\Delta t$) is the same as the rate of consumption of O_2 ($-\Delta[O_2]/\Delta t$), and the rate of formation of NO ($\Delta[NO]/\Delta t$), is twice the rate of consumption of either N_2 or O_2. These relations are expressed in equation form as

$$-2\frac{\Delta[N_2]}{\Delta t} = -2\frac{\Delta[O_2]}{\Delta t} = \frac{\Delta[NO]}{\Delta t} \qquad (15.11)$$

or, if we divide through by 2,

$$-\frac{\Delta[N_2]}{\Delta t} = -\frac{\Delta[O_2]}{\Delta t} = \frac{1}{2}\frac{\Delta[NO]}{\Delta t} \qquad (15.12)$$

Note that in Equation 15.12 the number 2 in the denominator of the $\Delta[NO]/\Delta t$ term matches the coefficient of NO in the balanced chemical equation. This pattern applies to all chemical reactions: the coefficient of each species in the balanced chemical equation appears in the **denominator** of its term in this rate expression. Notice that the **numerators** in Equation 15.13 all have a value of 1:

$$1N_2(g) + 1O_2(g) \rightarrow 2NO(g)$$

$$\text{Rate} = -\frac{1}{1}\frac{\Delta[N_2]}{\Delta t} = -\frac{1}{1}\frac{\Delta[O_2]}{\Delta t} = \frac{1}{2}\frac{\Delta[NO]}{\Delta t} \qquad (15.13)$$

(We have added "1" coefficients to the N_2 and O_2 terms in the chemical equation only to clarify the origin of the denominators in the rate equation.)

A balanced chemical equation enables us to predict the *relative* rates at which reactants are consumed and products are formed: the rate of formation of NO is twice the rate of consumption of N_2. However, Equations 15.11 to 15.13 provide no information on the numerical values of the rates, which are typically obtained experimentally. Although computational tools at our disposal today allow us to predict the rates of very simple reactions, we limit our discussion of reaction rates in this chapter to rates based on experimental data.

SAMPLE EXERCISE 15.1   Predicting a Relative Reaction Rate

The synthesis of ammonia via the reaction

$$N_2(g) + 3H_2(g) \rightarrow 2NH_3(g)$$

is an important reaction in the production of agricultural fertilizers. How is the rate of formation of NH_3 related to the rates of consumption of N_2 and H_2?

Collect and Organize The rate of formation of NH_3 is related to the rates of consumption of N_2 and H_2 by the balanced chemical equation. Rates are expressed as $-\Delta[\text{reactant}]/\Delta t$ or $\Delta[\text{product}]/\Delta t$.

Analyze Our task is to use the coefficients of the balanced chemical equation to determine the relative rates at which the concentrations of N_2, H_2, and NH_3 change during the reaction. Because nitrogen and hydrogen are consumed in the reaction, we include minus signs in their rate terms. Because ammonia is generated, its rate term is positive. Since the coefficients for the reactants and products in the balanced chemical equation are all different, the rate of formation of product and the rates of consumption of reactants are all different.

Solve We write the expressions for reactants and product with the correct signs and insert the coefficients from the balanced equation in the denominators:

$$-\frac{\Delta[N_2]}{\Delta t} = -\frac{1}{3}\frac{\Delta[H_2]}{\Delta t} = \frac{1}{2}\frac{\Delta[NH_3]}{\Delta t}$$

Think about It The balanced equation indicates that 2 moles of NH_3 are formed for every 1 mole of N_2 consumed. That means the rate of consumption of N_2 is half the rate of formation of ammonia. Similarly, the balanced equation tells us that 3 moles of H_2 are consumed for every 1 mole of N_2 consumed. Therefore, the rate at which N_2 is consumed is one-third the rate at which H_2 is consumed.

Practice Exercise In the oxidation of carbon monoxide to carbon dioxide,

$$2\,CO(g) + O_2(g) \rightarrow 2\,CO_2(g)$$

which reactant is consumed at the higher rate? How is the rate of change in the concentration of CO_2 related to the rate of change in the concentration of O_2?

(Answers to Practice Exercises are in the back of the book.)

Experimentally Determined Rates: Actual Values

Up to this point we have worked only with expressions for *relative* rates of reactions. Values for reaction rates are determined experimentally, and once we know this value for one reactant or product, we can use that value and information from the balanced chemical equation to express the reaction rate in terms of any other reactant or product.

The reaction in Sample Exercise 15.1 can serve as an example. If the rate at which ammonia forms $(\Delta[NH_3]/\Delta t)$ under a given set of conditions is determined experimentally to be 0.472 *M*/s, we can calculate a numerical value for the rate of consumption of N_2 by using the relationships we derived in the sample exercise:

$$-\frac{\Delta[N_2]}{\Delta t} = \frac{1}{2}\frac{\Delta[NH_3]}{\Delta t} = \frac{1}{2}(0.472\ M/s) = 0.236\ M/s$$

Because the coefficient of N_2 in the balanced equation is 1, the rate of N_2 consumption is equal to one-half the rate of the formation of ammonia, and thus we can state that the rate of consumption of N_2 is 0.236 *M*/s.

Which value for the rate of the reaction is correct? Both values are correct as long as we have referenced the compound in the chemical equation whose concentration was monitored as a function of time. However, it is conventional to use the rate of change of the reactant or product with a coefficient of "1" in the reaction equation as the basis for expressing the rate of the overall reaction.

SAMPLE EXERCISE 15.2   Converting Reaction Rates

Suppose that during the reaction between NO and O_2 to form NO_2,

$$2\,NO(g) + O_2(g) \rightarrow 2\,NO_2(g)$$

the rate of consumption of O_2 ($-\Delta[O_2]/\Delta t$) is measured as 0.033 M/s. What is the rate of formation of NO_2?

Collect and Organize We are asked to determine the rate of formation of a product in a reaction. We are given two items that we can use to determine this: the balanced chemical equation and the rate of consumption of one of the reactants. The rate of formation of NO_2 is related to the rates of consumption of NO and O_2 by the balanced chemical equation. The coefficients of NO_2 and O_2 are 2 and 1, respectively. Rates are expressed as $-\Delta[\text{reactant}]/\Delta t$ or $\Delta[\text{product}]/\Delta t$.

Analyze We can write an equation that expresses the relative rates of change in $[NO_2]$ and $[O_2]$ from the coefficients in the balanced chemical equation:

$$\frac{1}{2}\frac{\Delta[NO_2]}{\Delta t} = -\frac{\Delta[O_2]}{\Delta t}$$

The negative sign is needed because the concentration of O_2 decreases as the concentration of NO_2 increases. This stoichiometric relation means that the rate of consumption of O_2 is half the rate of formation of NO_2 because 2 moles of NO_2 are formed from every 1 mole of O_2 consumed.

Solve Solving for the rate of change of $[NO_2]$, we get

$$\frac{\Delta[NO_2]}{\Delta t} = -2\frac{\Delta[O_2]}{\Delta t} = 2(0.033\ M/\text{s}) = 0.066\ M/\text{s}$$

Think about It This result is twice the magnitude of $\Delta[O_2]/\Delta t$, which is consistent with the stoichiometry of the reaction: 2 moles of NO_2 produced for every 1 mole of O_2 consumed.

Practice Exercise The gas NO reacts with H_2, forming N_2 and H_2O:

$$2\,NO(g) + 2\,H_2(g) \rightarrow 2\,H_2O(g) + N_2(g)$$

If $-\Delta[NO]/\Delta t = 21.5\ M/\text{s}$ under a given set of conditions, what are the rates of change of $[N_2]$ and $[H_2O]$?

TABLE 15.1 Changing Concentrations of Reactants and Products for the Reaction $N_2(g) + O_2(g) \rightarrow 2\,NO(g)$

Time (μs)	$[N_2]$, $[O_2]$ (μM)	$[NO]$ (μM)
0	17.0	0.0
5.0	13.1	7.8
10.0	9.6	14.8
15.0	7.6	18.6
20.0	5.8	22.2
25.0	4.5	24.8
30.0	3.6	26.7

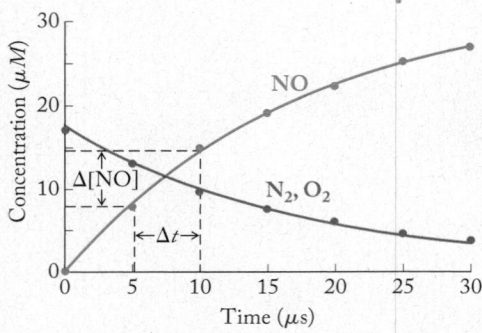

FIGURE 15.3 Concentrations of N_2, O_2, and NO over 30.0 μs for the reaction $N_2(g) + O_2(g) \rightarrow 2\,NO(g)$, plotted from data in Table 15.1.

instantaneous rate the rate of a reaction at a specific instant during the course of the reaction.

Average and Instantaneous Rates of Formation of NO

Suppose we run an experiment to determine the rate of formation of NO in an automobile engine. In the laboratory, we use a reaction vessel as hot as the combustion chambers in the engine and obtain the data in Table 15.1, which are plotted in Figure 15.3. The data can be used to calculate the reaction rate based on the change in the concentration of any participant in the reaction over a particular time interval. Calculations of reaction rates based on $-\Delta[\text{reactant}]/\Delta t$ or $\Delta[\text{product}]/\Delta t$ are *average* reaction rates. An average reaction rate could be the difference in NO concentration, $\Delta[\text{NO}] = [\text{NO}]_{\text{final}} - [\text{NO}]_{\text{initial}}$, at the beginning and end of the selected time interval $\Delta t = t_{\text{final}} - t_{\text{initial}}$. For example, the average rate of change in [NO] between 5.0 and 10.0 μs is

$$\frac{\Delta[\text{NO}]}{\Delta t} = \frac{[\text{NO}]_{10.0\,\mu s} - [\text{NO}]_{5.0\,\mu s}}{(10.0 - 5.0)\,\mu s} = \frac{(14.8 - 7.8)\,\mu M}{5.0\,\mu s}$$
$$= 1.4\ M/s$$

During the same interval the average rate of consumption of N_2 is

$$-\frac{\Delta[\text{N}_2]}{\Delta t} = -\frac{[\text{N}_2]_{10.0\,\mu s} - [\text{N}_2]_{5.0\,\mu s}}{(10.0 - 5.0)\,\mu s} = -\frac{(9.6 - 13.1)\,\mu M}{5.0\,\mu s}$$
$$= 0.70\ M/s$$

These results give us two values for expressing the rate of the reaction. Recall that we use the rate of change of the reactant or product with a coefficient of 1 in the reaction equation as the basis for expressing the rate of the reaction—in this example, we can choose the rate of consumption of either N_2 or O_2, but not the rate of formation of NO. The average rate of this reaction is

$$\text{Rate} = -\frac{\Delta[\text{N}_2]}{\Delta t} = 0.70\ M/s$$

The curvature of the lines in Figure 15.3 tells us that this value applies only to the interval from $t = 5.0\ \mu$s to $t = 10.0\ \mu$s and not over the entire 30.0 μs interval. Any other 5.0 μs interval has a different average rate. For instance, from $t = 25.0\ \mu$s to $t = 30.0\ \mu$s, the rate of the reaction is

$$-\frac{\Delta[\text{N}_2]}{\Delta t} = -\frac{[\text{N}_2]_{30.0\,\mu s} - [\text{N}_2]_{25.0\,\mu s}}{(30.0 - 25.0)\,\mu s}$$
$$= -\frac{(3.6 - 4.5)\,\mu M}{5.0\,\mu s} = 0.18\ M/s$$

Clearly, different reaction rates for the same reaction can be confusing, so why would we ever calculate average rates for a reaction? The usefulness of average rates is limited, but if we are comparing the rates of two different reactions over the same time period, an average rate can be sufficient to establish a relationship between the two rates. For example, if the average rate of reaction of nitrogen is 0.70 M/s between 5.0 and 10.0 μs, and the average rate of reaction for a different reaction at the same temperature is 2.10 M/s between 5.0 and 10.0 μs, then we know that the second reaction is three times faster than the first.

We can also determine an **instantaneous rate** of a reaction. An instantaneous rate is the reaction rate at a particular instant, which means the rate at a particular point on a curve of concentration as a function of time. The difference between average and instantaneous reaction rates is analogous to the difference between the average and instantaneous speeds of a runner. If a competitor in a marathon runs from mile 10 to mile 20 in 1 hr, her average speed over that distance is 10 mi/hr. At a given instant during the run, however, her instantaneous speed could be 12 mi/hr while going downhill and 8 mi/hr while going uphill.

Let's again consider the conversion of NO in engine exhaust into NO_2 (Equation 15.2), the gas responsible for much of the brown color in the photographs of photochemical smog at the beginning of this chapter. The conversion takes place when NO reacts with oxygen in the air:

$$2NO(g) + O_2(g) \rightarrow 2NO_2(g)$$

The rate of this reaction is described by the data in Table 15.2, specifically in terms of the rate of consumption of O_2, the reactant with a coefficient of 1. The data are plotted in Figure 15.4, where we calculate the instantaneous rate by drawing the tangent to a particular point on the curve.

TABLE 15.2	Changing Concentrations of Reactants and Products for the Reaction $2NO(g) + O_2(g) \rightarrow 2NO_2(g)$ at 25°C		
Time (s)	[NO] (*M*)	[O₂] (*M*)	[NO₂] (*M*)
0	0.0100	0.0100	0.0000
285	0.0090	0.0095	0.0010
660	0.0080	0.0090	0.0020
1175	0.0070	0.0085	0.0030
1895	0.0060	0.0080	0.0040
2975	0.0050	0.0075	0.0050
4700	0.0040	0.0070	0.0060
7800	0.0030	0.0065	0.0070

CONCEPT TEST

Which of the following statements is/are true about the instantaneous rate for the chemical reaction $A \rightarrow B$ as the reaction progresses?

a. $-\Delta[A]/\Delta t$ increases, $\Delta[B]/\Delta t$ decreases
b. $-\Delta[A]/\Delta t$ decreases, $\Delta[B]/\Delta t$ increases
c. $-\Delta[A]/\Delta t$ and $\Delta[B]/\Delta t$ both increase
d. $-\Delta[A]/\Delta t$ and $\Delta[B]/\Delta t$ both decrease

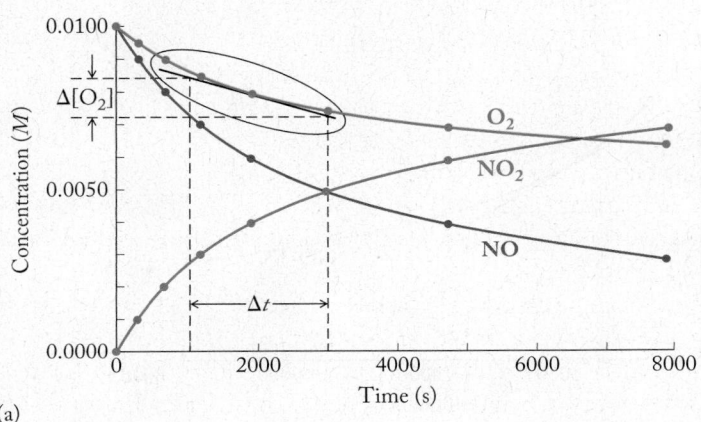

(a)

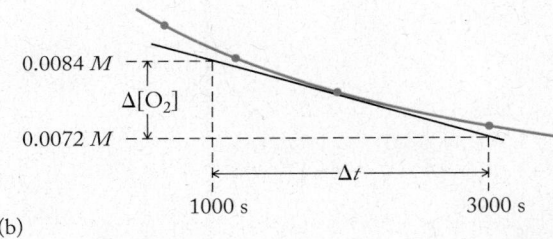

(b)

FIGURE 15.4 (a) The instantaneous rate of change in $[O_2]$ in the reaction $2NO(g) + O_2(g) \rightarrow 2NO_2(g)$ is equal to the slope of a tangent to the curve of $[O_2]$ versus time. (b) An expanded view of the instantaneous rate of change in $[O_2]$ at $t = 2000$ s.

Any instantaneous reaction rate of a reaction can be determined from a graph of the concentration-versus-time data for the participants. If for the NO_2 reaction we wish to determine the instantaneous rate of change of $[O_2]$ at $t = 2000$ s, we draw a tangent to the point on the green curve in Figure 15.4 corresponding to $t = 2000$ s and then select two convenient points, for example, $t = 1000$ s and $t = 3000$ s, along that tangent. From the differences in the coordinates on the y-axis for $[O_2]_{3000\,s}$ and $[O_2]_{1000\,s}$ at these two points and the difference in the x-axis coordinates: 3000 s and 1000 s, we calculate the slope of the line, which is a measure of the instantaneous rate of change in $[O_2]$, the negative value of which is the instantaneous rate of the reaction at $t = 2000$ s:

$$\text{Slope} = \frac{\Delta[O_2]}{\Delta t} = \frac{(0.0072 - 0.0084)\,M}{(3000 - 1000)\,s} = -6.0 \times 10^{-7}\,M\,s^{-1}$$

$$\text{Rate} = -\frac{\Delta[O_2]}{\Delta t} = -(-6.0 \times 10^{-7}\,M\,s^{-1}) = 6.0 \times 10^{-7}\,M\,s^{-1}$$

Note that we substitute the concentration values corresponding to time points along the tangent line, not along the curve drawn from the data points.

SAMPLE EXERCISE 15.3 Determining an Instantaneous Rate

(a) What is the instantaneous rate of change of [NO] at $t = 2000$ s in the experiment that produced the data in Table 15.2? (b) What is the rate of the reaction based on your result in part a?

Collect and Organize We are asked to determine the instantaneous rate of change of [NO] and the corresponding reaction rate at $t = 2000$ s. The coefficient of NO is 2 in the balanced chemical equation:

$$2\,NO(g) + O_2(g) \rightarrow 2\,NO_2(g)$$

Analyze The instantaneous rate of change in [NO] at the stated time can be determined from a graph of the data. The corresponding reaction rate will be $\left(-\frac{1}{2}\right)$ this value.

Solve

a. First we plot [NO] versus time and draw a tangent to the curve at the point $t = 2000$ s (Figure 15.5). We then choose two points along the tangent, $t = 1000$ and 3000 s, and determine the concentrations corresponding to those times along the vertical axis. By using those values, we calculate the slope of the line:

$$\frac{\Delta[NO]}{\Delta t} = \frac{(0.0046 - 0.0070)\,M}{(3000 - 1000)\,s} = -1.2 \times 10^{-6}\,M\,s^{-1}$$

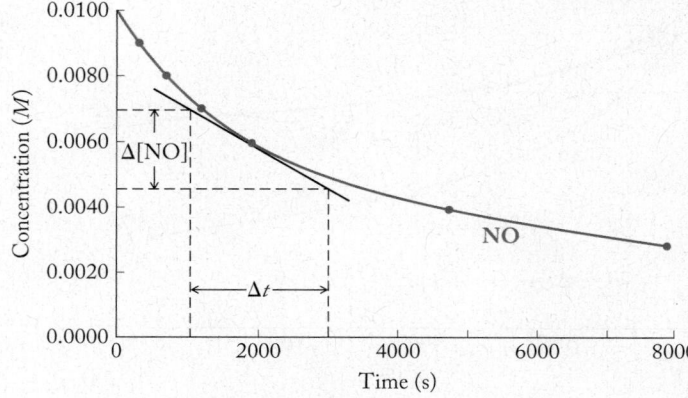

FIGURE 15.5

b. The corresponding reaction rate is

$$\text{Rate} = -\frac{1}{2}\frac{\Delta[NO]}{\Delta t} = -\frac{1}{2}(-1.2 \times 10^{-6}\,M\,s^{-1}) = 6.0 \times 10^{-7}\,M\,s^{-1}$$

Think about It The sign of $\Delta[NO]/\Delta t$ is negative because NO is a reactant whose concentration decreases with time. However, the rates of chemical reactions have positive values. Therefore, we needed a minus sign in front of the $\Delta[NO]/\Delta t$ term in the solution to part b. Note that the instantaneous reaction rate calculated in this Sample Exercise is the same as the one we calculated on page 710 based on the rate of change of $[O_2]$.

Practice Exercise What is the instantaneous rate of change in $[NO_2]$ at $t = 2000$ s in the experiment that produced the data in Table 15.2? ∎

> **initial rate** the rate of a reaction at $t = 0$, immediately after the reactants are mixed.
>
> **reaction order** an experimentally determined number defining the dependence of the reaction rate on the concentration of a reactant.

15.3 Effect of Concentration on Reaction Rate

Figure 15.6 shows a typical result for a reactant concentration plotted as a function of time. Tangents have been drawn to the line at three points: (a) at the instant the reaction begins, (b) when the reaction is about halfway to completion, and (c) when the reaction is nearly over. Point (a) defines the **initial rate** of the reaction, which is the rate that occurs at the instant the reactants are mixed at $t = 0$.

The key observation from Figure 15.6 is that the slopes of the tangents approach zero as the reaction proceeds. When the reaction is over, no more change occurs in the concentration of any remaining reactant or in the concentration of any product, and in this region of the curve the slope of any tangent is zero. This behavior is typical: the most rapid changes in reactant and product concentrations take place early in most reactions.

Kinetic molecular theory and our picture of molecules in the gas phase provide us with a way to explain this trend. If we assume that most reactions take place as a result of collisions between reactant molecules, then the more reactant molecules there are in a given space, such as a flask, the more collisions per unit time and the more opportunities for reactants to turn into products. As reactant concentrations decrease, fewer reactant molecules occupy the space in the flask, so the frequency of collisions and the rate of conversion of reactants to products slow down.

What happens to the rate of some of the reactions we have discussed when we change the initial concentrations of the reactants? Doubling the concentration of O_2 in Equation 15.2 doubles the rate of the reaction. However, doubling the concentration of NO *quadruples* the rate. Why does the rate of the reaction have different dependencies on the concentrations of the two reactants? We will explore the reason why in this section.

Reaction Order and Rate Constants

Experimental observations and theoretical considerations tell us that reaction rates depend on reactant concentrations. However, they do not tell us *to what extent* rates depend on reactant concentrations. For example, if the concentration of a reactant doubles, does the reaction rate double? The answer is expressed in the **reaction order**, a parameter derived from experiments that tells us how reaction rate depends on reactant concentrations. Knowing the order of a reaction provides insights into *how* the reaction takes place—which molecules collide with which other molecules as bonds break, new bonds form, and reactants are converted into products.

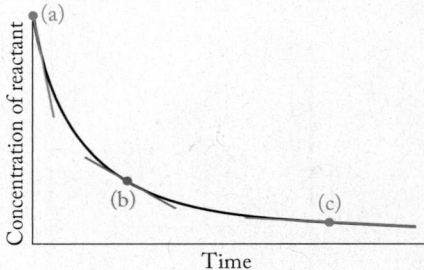

FIGURE 15.6 Typical plot of reactant concentration as a function of time: (a) tangent at $t = 0$; (b) tangent at the midpoint of the reaction; (c) tangent close to the end of the reaction.

⦾⦾ **CONNECTION** We discussed kinetic molecular theory, a model that describes the behavior of gases, in Chapter 6.

▶❚❚ **CHEMTOUR** Reaction Order

Let's look at one way in which reaction order is determined by revisiting the reaction between oxygen and nitrogen monoxide:

$$2\,NO(g) + O_2(g) \rightarrow 2\,NO_2(g)$$

In experiments to evaluate the kinetics of this reaction, different initial concentrations of NO and O_2 are introduced into a reaction vessel at 25°C, and the initial reaction rate is determined in each case. Table 15.3 shows four of these initial reaction rates. Remember that the initial rate is determined from the slope of a line tangent to the curve of concentration versus time. Why do we choose to calculate the initial rate (instantaneous rate at $t = 0$)? In the previous section we observed that reaction rates depend on collisions between reactants. Choosing $t = 0$ as our point on the concentration-versus-time curve means that the concentration of products is essentially zero. This choice is important because reactions can also run in reverse. If the product concentration is essentially zero, then no collisions occur between product molecules and the rate of the reverse reaction can be largely ignored. If we choose to work with rates at a point where $t \neq 0$, then we must account for the different rates of the forward and reverse reactions.

To interpret the data in Table 15.3, we select pairs of experiments in which the concentrations of one reactant differ but the concentrations of the other are the same. For example, $[NO]_0$ is the same in experiments 1 and 2, but $[O_2]_0$ in experiment 1 is twice $[O_2]_0$ in experiment 2. (The zero subscripts indicate that the concentrations of NO and O_2 are the values at the start of the experiments, when $t = 0$.) The initial reaction rate in experiment 1 is twice that in experiment 2, allowing us to state that doubling the concentration of O_2 while holding the NO concentration constant doubles the initial reaction rate. We conclude that the initial reaction rate is proportional to O_2 concentration:

$$\text{Rate} \propto [O_2]$$

In experiments 1 and 3, $[O_2]_0$ is the same but $[NO]_0$ in experiment 1 is twice $[NO]_0$ in experiment 3. Comparing the initial reaction rates for these two experiments, we find that the rate in experiment 1 is four times the rate in experiment 3. Thus, the reaction rate is proportional to [NO] squared:

$$\text{Rate} \propto [NO]^2$$

We combine these two rate expressions to get an overall rate expression for the reaction by multiplying the right sides of the rate expressions for the two reactants:

$$\text{Rate} \propto [NO]^2[O_2]$$

To understand why we multiply the concentration terms together, let's consider another reaction between oxygen and nitrogen monoxide. This one involves ozone (O_3) and NO and produces NO_2 and O_2:

$$NO(g) + O_3(g) \rightarrow NO_2(g) + O_2(g)$$

TABLE 15.3 Effect of Reactant Concentrations on Initial Reaction Rates at 25°C for the Reaction $2\,NO(g) + O_2(g) \rightarrow 2\,NO_2(g)$

Experiment	$[NO]_0$ (*M*)	$[O_2]_0$ (*M*)	Initial Reaction Rate, $-\dfrac{1}{2}\dfrac{\Delta[NO]}{\Delta t}$ (*M* s^{-1})
1	0.0100	0.0100	1.0×10^{-6}
2	0.0100	0.0050	0.5×10^{-6}
3	0.0050	0.0100	2.5×10^{-7}

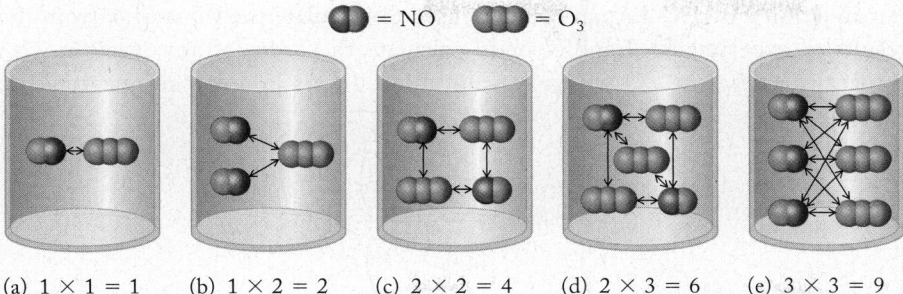

(a) $1 \times 1 = 1$ (b) $1 \times 2 = 2$ (c) $2 \times 2 = 4$ (d) $2 \times 3 = 6$ (e) $3 \times 3 = 9$

FIGURE 15.7 Increasing the concentration increases the number of possible collisions (double-headed arrows) and therefore the number of potential reaction events. Reaction rate depends on the number of collisions, which are shown for the reaction between NO and ozone (O_3) that produces NO_2 and O_2. With only one NO molecule and one O_3 molecule, as in a, each molecule can collide only with the other, giving a relative reaction rate of $1 \times 1 = 1$. In e, three molecules of NO can collide with three molecules of O_3 for a relative reaction rate of $3 \times 3 = 9$ times the rate in a.

Figure 15.7 shows how different numbers of NO and O_3 molecules in a reaction vessel might collide together and react with each other. Note how increasing the numbers of molecules in the containers in Figure 15.7(a–e) produces increasing numbers of collisions that are proportional to the *product* of the number of molecules of each reactant. Increasing the numbers of molecules of each type in the vessels is equivalent to increasing the concentrations of the two gases. Therefore, the rate of the reaction should (and does) depend on the product of the concentrations of NO and O_3:

$$\text{Rate} \propto [\text{NO}][\text{O}_3] \qquad (15.14)$$

Similar patterns occur with all chemical reactions whose rate depends on the concentration of more than one reactant. We can modify Equation 15.14 to obtain a **rate law**, an equation that defines the relation between reactant concentrations and reaction rate. We convert the proportionality to an equation by inserting a proportionality constant k, called the reaction **rate constant**:

$$\text{Rate} = k[\text{NO}][\text{O}_3] \qquad (15.15)$$

Now let's consider a generic chemical reaction with two reactants, A and B:

$$\text{A} + \text{B} \rightarrow \text{C}$$

The rate law expression for this reaction may be written

$$\text{Rate} = k[\text{A}]^m[\text{B}]^n \qquad (15.16)$$

where m is the reaction order with respect to A and n is the reaction order with respect to B. We determine reaction order values by comparing differences in reaction rates to differences in reactant concentrations. Sometimes these comparisons are easy to make, as they are using the data for the reaction between NO and O_2 on page 712. Other times the comparisons are not so easy to make. On those occasions we can use a more mathematical approach. For example, suppose we run three experiments to solve for the values of m and n in Equation 15.16. In experiments 1 and 2 the value of [A] is the same, but the values of [B] are different. In experiments 2 and 3 we keep [B] constant and vary the concentrations of [A]. The ratio of the reaction rates in experiments 1 (Rate_1) and 2 (Rate_2), are related to the ratio of the concentrations of B: $[\text{B}]_1/[\text{B}]_2$, and to the dependence of reaction rate on [B]; that is, on the value of n. Expressing this relationship in equation form and then solving for n by taking the logarithm of both sides:

$$\frac{\text{Rate}_1}{\text{Rate}_2} = \left(\frac{[\text{B}]_1}{[\text{B}]_2}\right)^n \qquad \log\left(\frac{\text{Rate}_1}{\text{Rate}_2}\right) = n\log\left(\frac{[\text{B}]_1}{[\text{B}]_2}\right)$$

and rearrange the terms:

$$n = \frac{\log\left(\dfrac{\text{Rate}_1}{\text{Rate}_2}\right)}{\log\left(\dfrac{[\text{B}]_1}{[\text{B}]_2}\right)}$$

rate law an equation that defines the experimentally determined relation between the concentrations of reactants in a chemical reaction and the rate of that reaction.

rate constant the proportionality constant that relates the rate of a reaction to the concentrations of reactants.

overall reaction order the sum of the exponents of the concentration terms in the rate law.

An equation with this format could be used to calculate the value of m from the results of experiments 2 and 3, or to calculate the order of any reaction with respect to a reactant (X) whose concentration differs in a pair of reaction rate experiments:

$$n = \frac{\log\left(\dfrac{\text{Rate}_1}{\text{Rate}_2}\right)}{\log\left(\dfrac{[X]_1}{[X]_2}\right)} \tag{15.17}$$

CONCEPT TEST

In the reaction A → B, the rate of the reaction triples when [A] is tripled.

a. What is the correct value of m in the rate law: Rate = $k[A]^m$?

b. What is the value of m if the rate is unchanged when [A] is tripled?

The power to which a concentration term is raised is the order of the reaction in terms of that reactant. Thus in our example, the reaction of NO and O_2 is *second order* in NO and *first order* in O_2. The **overall reaction order** for a reaction is the sum of the powers in the rate equation, so the reaction of NO and O_2 is *third order* overall.

An exponent in a rate law may be a fraction, zero, or, in rare cases, negative. It is important to remember that rate laws and reaction orders are different from relative rates and must be determined experimentally. They cannot be predicted from the coefficients in a balanced chemical equation. The significance of reaction order in describing how a reaction takes place is addressed in Section 15.5.

The value of the rate constant k is unique to each particular reaction at a given temperature, and right now we may consider k simply as a proportionality constant. It does not change with concentration; in other words, *reaction rate depends on the concentration of the reactants but the rate constant does not.* The rate constant changes only with changing temperature or in the presence of a catalyst.

We can calculate k for a reaction run at some specified temperature from initial reaction rate data. Let's do so for the reaction of O_2 and NO by selecting the results of one experiment in Table 15.3. Which experiment we use doesn't matter. As long as the temperature is the same, the value of k is the same. To determine k, let's use Equation 15.15 and insert the data from experiment 1:

$$1.0 \times 10^{-6} \; M/s = k(0.0100 \; M)^2(0.0100 \; M)$$

$$k = \frac{1.0 \times 10^{-6} \; M/s}{(0.0100 \; M)^2(0.0100 \; M)} = 1.0 \; M^{-2} \, s^{-1} \tag{15.18}$$

Keep in mind that the value calculated from experimental data for a rate constant is valid only at the temperature at which the experiments were carried out.

The units of k differ from one reaction to another. Remember that the units of reaction rates always change in concentration per unit time. Therefore, in a first-order reaction in which concentration is expressed in molarity and time in seconds, the units of k must be per second (s^{-1}):

$$\text{Rate} = k[X]^1$$

$$k = \frac{\text{rate}}{[X]} = \frac{M/s}{M} = \frac{1}{s} = s^{-1}$$

The units of the rate constant for a reaction that is second order overall can be derived in a similar way. If the reaction is first order in reactants X and Y, we have

$$\text{Rate} = k[\text{X}][\text{Y}]$$

$$k = \frac{\text{rate}}{[\text{X}][\text{Y}]} = \frac{\cancel{M}/s}{M\cancel{M}} = M^{-1}\,s^{-1}$$

The units of k for a reaction that is third order overall are $M^{-2}\,s^{-1}$, as we determined in Equation 15.18.

CONCEPT TEST ....

The reaction $A + 2B \rightarrow C$ is found to be third order overall. How many possible rate laws, Rate $= k[\text{A}]^m[\text{B}]^n$, could we write assuming that m and n are positive integers?

....

SAMPLE EXERCISE 15.4   **Deriving a Rate Law from Initial Reaction Rate Data**

Write the rate law for the reaction of N_2 with O_2

$$N_2(g) + O_2(g) \rightarrow 2\,NO(g)$$

using the data in Table 15.4. Determine the overall reaction order and the value of the rate constant.

TABLE 15.4	**Initial Reaction Rates for the Formation of NO from the Reaction of N_2 with O_2 at Constant Temperature**		
Experiment	$[N_2]_0$ (*M*)	$[O_2]_0$ (*M*)	Initial Reaction Rate (*M*/s)
1	0.040	0.020	707
2	0.040	0.010	500
3	0.010	0.010	125

Collect and Organize We are to determine the rate law and rate constant for a reaction. We are given the initial reaction rate (note the subscript zero on the concentration terms in Table 15.4) for each of three sets of initial concentrations.

Analyze The general form of the rate law for any reaction between reactants A and B is

$$\text{Rate} = k[\text{A}]^m[\text{B}]^n$$

We can use the experimental data given to find the values of k, m, and n for the reaction in which $A = N_2$ and $B = O_2$. The overall order of the reaction is the sum of the orders of the individual reactants. Once we have established the rate law, the rate constant is calculated from the rate law using concentrations of reactants from any row in Table 15.4. The rate constant must have units that express the reaction rate in $M\,s^{-1}$.

Solve To determine the order of the reaction (m) with respect to N_2, we use the data from experiments 2 and 3 because in these two experiments the values of $[N_2]$ are different but the $[O_2]$ values are the same. When the concentration N_2 is increased by a factor of 4, the rate increases by a factor of 4. Thus, the reaction rate is proportional to N_2 concentration, which means that $m = 1$. There are different values of $[O_2]$ in Experiments 1 and 2, but $[N_2]$ is the same. Therefore we can use these data to calculate the value of n.

The ratio of the reaction rates (707/500) is not a whole number, so let's apply Equation 15.17:

$$n = \frac{\log\left(\dfrac{\text{Rate}_1}{\text{Rate}_2}\right)}{\log\left(\dfrac{[O_2]_1}{[O_2]_2}\right)} = \frac{\log\left(\dfrac{1000}{500}\right)}{\log\left(\dfrac{0.040\ M}{0.010\ M}\right)} = \frac{\log(2)}{\log(4)} = 0.50$$

Thus, the reaction is $\frac{1}{2}$ order with respect to O_2, first order with respect to N_2, and $(\frac{1}{2} + 1) = \frac{3}{2}$ order overall:

$$\text{Rate} = k[N_2][O_2]^{1/2}$$

We can use the data from any experiment to obtain the value of k. Let's use experiment 1:

$$707\ M/\text{s} = k(0.040\ M)(0.020\ M)^{1/2} = k(0.0051)$$

$$k = 1.4 \times 10^5\ M^{-1/2}\ \text{s}^{-1}$$

Think about It An order of $\frac{1}{2}$ for oxygen in this reaction is determined from experimental data, not from the balanced chemical equation. The units of the rate constant, $M^{-1/2}\ \text{s}^{-1}$, are appropriate because when we substitute the units for each term into the rate law, we end up with the correct units for reaction rate: $(M^{-1/2}\ \text{s}^{-1})(M)(M^{1/2}) = M\ \text{s}^{-1}$.

Practice Exercise Nitric oxide reacts rapidly with unstable nitrogen trioxide (NO_3) to form NO_2:

$$NO(g) + NO_3(g) \rightarrow 2\,NO_2(g)$$

Determine the rate law for the reaction and calculate the rate constant from the data in Table 15.5.

	TABLE 15.5	Initial Reaction Rates for the Formation of NO_2 from the Reaction of NO with NO_3 at 25°C		
Experiment	**$[NO]_0$ (M)**	**$[NO_3]_0$ (M)**	**Initial Reaction Rate (M/s)**
1	1.25×10^{-3}	1.25×10^{-3}	2.45×10^4
2	2.50×10^{-3}	1.25×10^{-3}	4.90×10^4
3	2.50×10^{-3}	2.50×10^{-3}	9.80×10^4

Integrated Rate Laws: First-Order Reactions

Determining a rate law using initial reaction rate data has two distinct disadvantages. The method of initial rates requires several experiments with different concentrations of reactants, varied in a systematic fashion. We also must accurately determine the reaction rate at the instant the reaction begins. It would be much easier if we could determine the rate law and calculate the rate constant for a reaction from the plot of concentration versus time alone. In fact, this can be accomplished for reactions in which the reaction rate depends on the concentration of only one substance. One such reaction is the photochemical decomposition of ozone:

$$O_3(g) \xrightarrow{\text{sunlight}} O_2(g) + O(g)$$

This decomposition reaction can be studied in the laboratory using high-intensity ultraviolet light to simulate sunlight; one such study yielded the results listed in Table 15.6 and plotted in Figure 15.8(a). Because ozone is the only reactant, the

TABLE 15.6 Rate of Photochemical Decomposition of Ozone

Time (s)	$[O_3]$ (M)	$\ln[O_3]$
0	1.000×10^{-4}	-9.2103
100	0.896×10^{-4}	-9.320
200	0.803×10^{-4}	-9.430
300	0.719×10^{-4}	-9.540
400	0.644×10^{-4}	-9.650
500	0.577×10^{-4}	-9.760
600	0.517×10^{-4}	-9.870

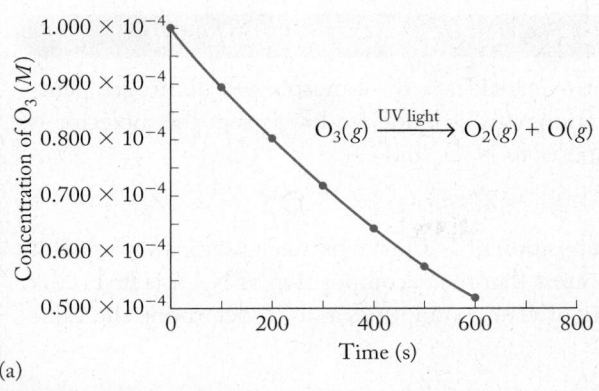

FIGURE 15.8 Data for the decomposition of O_3: (a) plot of $[O_3]$ versus time; (b) plot of $\ln[O_3]$ versus time. The line in b is straight, indicating that the decomposition reaction is first order in O_3.

rate law for the reaction should depend only on the ozone concentration. In our interpretation of the data in Table 15.6, we can start with the assumption that the reaction is first order in O_3, which means that the rate law can be written

$$\text{Rate} = k[O_3]$$

Because the coefficient of O_3 in the balanced equation is 1, the O_3 consumption rate $-\Delta[O_3]/\Delta t$ is equal to the reaction rate, and we can write

$$\text{Rate} = -\frac{\Delta[O_3]}{\Delta t} = k[O_3]$$

This rate law can be transformed using calculus into an expression that relates the concentration of ozone $[O_3]$ at any instant during the reaction to the initial concentration $[O_3]_0$:

$$\ln\frac{[O_3]}{[O_3]_0} = -kt \qquad (15.19)$$

This version of the rate law is called an **integrated rate law** because integral calculus is used to derive it, and it describes the change in reactant concentration with time. The general integrated rate law for any reaction that is first order in reactant X is

$$\ln\frac{[X]}{[X]_0} = -kt \qquad (15.20)$$

Using the identity $\ln(a/b) = \ln a - \ln b$, we can rearrange this equation to

$$\ln[X] = -kt + \ln[X]_0 \qquad (15.21)$$

which is the equation of a straight line of the form

$$y = mx + b$$

where $\ln[X]$ is the y variable and t is the x variable. The slope of the line (m) is $-k$, and the y intercept (b) is $\ln[X]_0$. Rearranging Equation 15.20 to fit the format of Equation 15.21 gives

$$\ln[O_3] = -kt + \ln[O_3]_0 \qquad (15.22)$$

Note in Figure 15.8(b) that a graph of the natural logarithm of $[O_3]$ versus time is indeed a straight line. This linearity means that our assumption was correct and the reaction is first order in O_3. Calculating the slope of the line, $\Delta(\ln[O_3])/\Delta t$, from the two points labeled in Figure 15.8(b) yields the value of the rate constant:

$$\text{Slope} = \frac{-9.650 - (-9.320)}{400 \text{ s} - 100 \text{ s}} = -1.10 \times 10^{-3}\,\text{s}^{-1}$$

$$k = 1.10 \times 10^{-3}\,\text{s}^{-1}$$

integrated rate law a mathematical expression that describes the change in concentration of a reactant in a chemical reaction with time.

TABLE 15.7(a) Concentration of N_2O_5 as a Function of Time	
Time (s)	[N_2O_5] (M)
0	0.1000
50	0.0707
100	0.0500
200	0.0250
300	0.0125
400	0.00625

TABLE 15.7(b) Concentration and Natural Logarithm of Concentration of N_2O_5 as a Function of Time		
Time (s)	[N_2O_5] (M)	ln[N_2O_5]
0	0.1000	−2.303
50	0.0707	−2.649
100	0.0500	−2.996
200	0.0250	−3.689
300	0.0125	−4.382
400	0.00625	−5.075

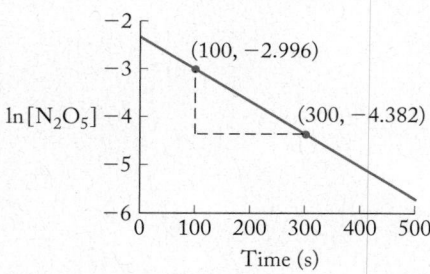

FIGURE 15.9

TABLE 15.8 Concentration of Hydrogen Peroxide as a Function of Time	
Time (s)	[H_2O_2] (M)
0	0.500
100	0.460
200	0.424
500	0.330
1000	0.218
1500	0.144

SAMPLE EXERCISE 15.5   Using an Integrated Rate Law

One of the least abundant nitrogen oxides in the atmosphere is dinitrogen pent-oxide. One reason concentrations of this oxide are low is that the molecule is unstable and rapidly decomposes to N_2O_4 and O_2:

$$2\,N_2O_5(g) \rightarrow 2\,N_2O_4(g) \;+\; O_2(g)$$

A kinetic study of the decomposition of N_2O_5 at a particular temperature yields the data in Table 15.7(a). Assume that the decomposition of N_2O_5 is first order in N_2O_5. (a) Test the validity of your assumption, and (b) determine the value of the rate constant.

Collect and Organize We are given experimental data showing the concentration of a single reactant as a function of time and are told to assume a first-order reaction. We can verify the assumption using the integrated rate law for a first-order reaction (Equation 15.21) and then calculate the rate constant.

Analyze Equation 15.21 has the form $y = mx + b$, which means that a plot of ln[N_2O_5] (y) versus time (x) should be linear if the decomposition of N_2O_5 is first order. The slope (m) of the graph corresponds to $−k$, the negative of the rate constant.

Solve Our first step is to determine ln[N_2O_5] values (Table 15.7b).
  a. The plot of ln[N_2O_5] versus t is shown in Figure 15.9. The fact that the curve in Figure 15.9 is a straight line indicates that the reaction is first order in N_2O_5.
  b. Arbitrarily choosing $t = 100$ s and $t = 300$ s as two points for calculating the slope:

$$\text{Slope} = \frac{\Delta y}{\Delta x} = \frac{-4.382 - (-2.996)}{300 \text{ s} - 100 \text{ s}} = \frac{-1.386}{200 \text{ s}} = -0.00693 \text{ s}^{-1}$$

The slope of the line equals $−k$. Therefore, the rate constant $k = 0.00693$ s^{-1}. Note that we could also use a graphing calculator to plot ln[N_2O_5] versus t and have the calculator determine the best fit of the data to a straight line. The equation of the line gives us the slope, which we then convert into the rate constant as described above.

Think about It We can test whether any reaction with a single reactant (x) is first order in x by determining whether a plot of ln[X] versus t is linear.

Practice Exercise Hydrogen peroxide (H_2O_2) decomposes into water and oxygen:

$$H_2O_2(\ell) \rightarrow H_2O(\ell) + \tfrac{1}{2}O_2(g)$$

Use the data in Table 15.8 to determine whether the decomposition of H_2O_2 is first order in H_2O_2, and calculate the value of the rate constant at the temperature of the experiment that produced the data.

Reaction Half-Lives

A parameter frequently cited in kinetic studies is the **half-life ($t_{1/2}$)** of a reaction, which is the interval during which the concentration of a reactant decreases by half. Half-life is inversely related to the rate constant of a reaction: the higher the reaction rate, the shorter the half-life.

Let's consider reaction half-life in the context of another nitrogen oxide found in the atmosphere: dinitrogen monoxide, also called nitrous oxide or laughing gas, an anesthetic sometimes used by dentists. Atmospheric concentrations of this potent greenhouse gas have been increasing in recent years, although the principal source is not automotive emissions but rather bacterial degradation of nitrogen compounds in soil. Dinitrogen monoxide is not a product of combustion reactions in internal combustion engines because at typical engine temperatures any N_2O formed rapidly decomposes into nitrogen and oxygen:

$$N_2O(g) \rightarrow N_2(g) + \tfrac{1}{2}O_2(g)$$

The reaction rate for this first-order reaction is very high, which means that the reaction has a short half-life (Figure 15.10).

We can derive a mathematical relation between half-life $t_{1/2}$ and rate constant k for this or any other first-order reaction by starting with the Equation 15.20:

$$\ln \frac{[X]}{[X]_0} = -kt$$

After one half-life has passed, $t = t_{1/2}$, the concentration of X is half its original value: $[X] = \tfrac{1}{2}[X]_0$. Inserting these values for $[X]$ and t into the equation yields

$$\ln \frac{\tfrac{1}{2}[\cancel{X}]_0}{[\cancel{X}]_0} = -kt_{1/2}$$

$$\ln \left(\tfrac{1}{2}\right) = -kt_{1/2}$$

The natural log of $\tfrac{1}{2}$ is -0.693, so

$$-0.693 = -kt_{1/2}$$

$$t_{1/2} = \frac{0.693}{k} \qquad (15.23)$$

half-life ($t_{1/2}$) the time in the course of a chemical reaction during which the concentration of a reactant decreases by half.

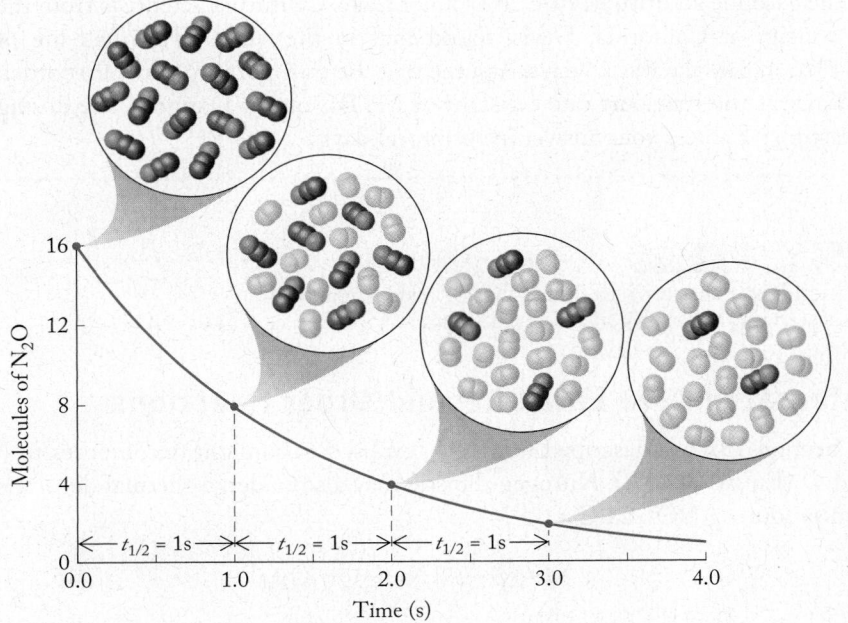

16

12

8

4

0

Molecules of N_2O

$t_{1/2}$ = 1s $t_{1/2}$ = 1s $t_{1/2}$ = 1s

0.0   1.0   2.0   3.0   4.0

Time (s)

FIGURE 15.10 The decomposition of $N_2O(g)$ is first order in N_2O. At a particular temperature the half-life of the reaction is 1.0 s, which means that, on average, half of a population of 16 N_2O molecules decomposes in 1.0 s, half of the remaining 8 molecules decompose in the next 1.0 s, and so on.

Thus, the half-life of a first-order reaction is inversely proportional to the rate constant, as noted at the beginning of this discussion. The absence of any concentration term in Equation 15.23 means that no matter the initial concentration of reactant, half of it is consumed in one half-life.

> ### SAMPLE EXERCISE 15.6 Calculating the Half-Life of a First-Order Reaction
>
> The rate constant for the decomposition of N_2O_5 at a particular temperature is 7.8×10^{-3} s^{-1}. What is the half-life of N_2O_5 at that temperature?
>
> **Collect and Organize** The half-life is the time required for the concentration of a reactant to decrease by half. We are asked to determine the half-life of N_2O_5 for its decomposition reaction. We know from Sample Exercise 15.5 that the decomposition of N_2O_5 is a first-order process.
>
> **Analyze** Equation 15.23 relates k for a first-order reaction to the reaction half-life. We are given no information about concentrations of reactants or products, but we know from Equation 15.23 that the half-life of a first-order reaction is independent of concentration. Because half-life is inversely proportional to k, and the value of k is on the order of 10^{-2} s^{-1}, we predict that the half-life will be on the order of 10^2 s.
>
> **Solve**
>
> $$t_{1/2} = \frac{0.693}{k} = \frac{0.693}{7.8 \times 10^{-3} \text{ s}^{-1}} = 89 \text{ s}$$
>
> **Think about It** The calculated value of $t_{1/2}$ is a little less than 100, the predicted value. Remember that Equation 15.23 is valid only for first-order reactions.
>
> **Practice Exercise** Environmental scientists calculating half-lives of pollutants often define a *transport rate constant* that is analogous to a reaction rate constant and describes how a pollutant moves out of an ecosystem. In a study of the gasoline additive MTBE in Donner Lake, California, scientists from the University of California, Davis, found that in the summer the half-life of MTBE in the lake was 28 days. Assume that the transport process is first order. What was the transport rate constant of MTBE out of Donner Lake during the study? Express your answer in reciprocal days.

> **CONCEPT TEST**
>
> Which has a shorter half-life, a fast or a slow reaction?

Integrated Rate Laws: Second-Order Reactions

In Section 15.1 we described how NO_2 exposed to sunlight decomposes to NO and O (Equation 15.3). Nitrogen dioxide may also undergo thermal decomposition producing NO and O_2:

$$2\,NO_2(g) \rightarrow 2\,NO(g) + O_2(g) \tag{15.24}$$

Because it has only one reactant, like the reactions of N_2O_5 and H_2O_2 in Sample Exercise 15.5 and its Practice Exercise, we might expect this reaction to be first order. However, when we use the data in Table 15.9 to evaluate the reaction order and rate constant, we find that the plot of $\ln[NO_2]$ versus time (Figure 15.11a) is clearly not linear, which tells us that the thermal decomposition of NO_2 is *not* first order.

TABLE 15.9	Rate of Decomposition of NO_2 to NO and O_2		
Time (s)	$[NO_2]$ (*M*)	$\ln[NO_2]$	$1/[NO_2]$ (1/*M*)
0	1.00×10^{-2}	-4.605	100
100	6.48×10^{-3}	-5.039	154
200	4.79×10^{-3}	-5.341	209
300	3.80×10^{-3}	-5.573	263
400	3.15×10^{-3}	-5.760	317
500	2.69×10^{-3}	-5.918	372
600	2.35×10^{-3}	-6.057	426

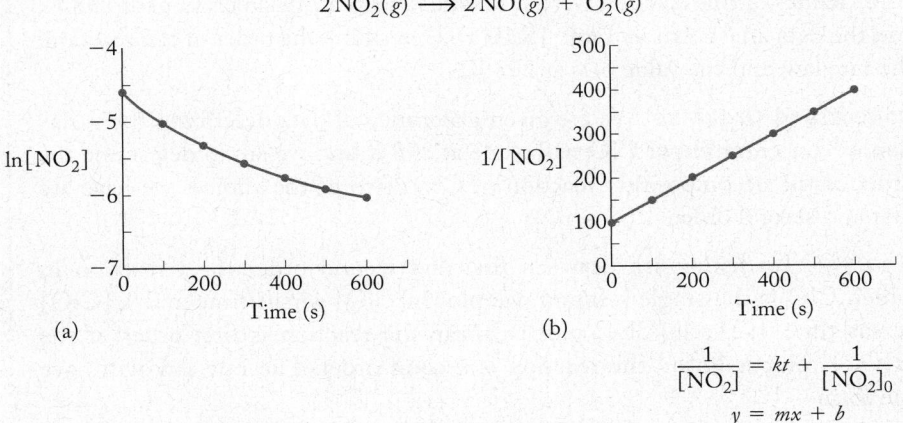

$$2\,NO_2(g) \longrightarrow 2\,NO(g) + O_2(g)$$

(a)

(b)

$$\frac{1}{[NO_2]} = kt + \frac{1}{[NO_2]_0}$$
$$y = mx + b$$

FIGURE 15.11 At high temperatures, NO_2 slowly decomposes into NO and O_2. (a) The plot of $\ln[NO_2]$ versus time is not linear, indicating the reaction is not first order. (b) The plot of $1/[NO_2]$ versus time is linear, indicating that the reaction is second order in NO_2. The slope of the line in this graph equals the rate constant.

What is the reaction order? The answer to this question is hidden in how the reaction takes place. If each NO_2 molecule simply fell apart, the reaction would be first order, much like the decomposition of N_2O_5. However, if the reaction happened as a result of collisions between pairs of NO_2 molecules, the reaction would be first order in each and second order overall. In other words, if the reaction were second order, the decomposition described by Equation 15.24 would depend on collisions between pairs of molecules that just happen to be molecules of the same substance, NO_2. The rate law expression is

$$\text{Rate} = k[NO_2]^2 \tag{15.25}$$

How can we determine whether this decomposition is really second order? One way is to assume that it is and then test that assumption. The test entails transforming the rate law in Equation 15.25 into the integrated rate law for a second-order reaction, again using calculus. The result of the transformation is

$$\frac{1}{[NO_2]} = kt + \frac{1}{[NO_2]_0} \tag{15.26}$$

which, like Equation 15.22, has the form $y = mx + b$ and is the equation of a straight line, this time with $1/[NO_2]$ as the y variable and t as the x variable.

The graph obtained using data from columns 1 and 4 of Table 15.9 is shown in Figure 15.11(b). That the curve is linear tells us that decomposition of NO_2 is second order. The slope of the line provides a direct measure of k, which is $0.544\ M^{-1}\text{s}^{-1}$.

A general form of Equation 15.26 that applies to any reaction that is second order in a single reactant (X) is

$$\frac{1}{[X]} = kt + \frac{1}{[X]_0} \tag{15.27}$$

SAMPLE EXERCISE 15.7 Distinguishing between First- and Second-Order Reactions

Chlorine monoxide accumulates in the stratosphere above Antarctica each winter and plays a key role in the formation of the ozone hole above the South Pole each spring. Eventually, ClO decomposes according to the equation

$$2\,ClO(g) \rightarrow Cl_2(g) + O_2(g)$$

The kinetics of this reaction were studied in a laboratory experiment at 298 K, and the data are shown in Table 15.10(a). Determine the order of the reaction, the rate law, and the value of k at 298 K.

Collect and Organize We are given experimental data describing the variation of concentration of ClO with time at 298 K and we are to determine the order of the decomposition reaction of ClO. Two of the choices we have are first and second order.

Analyze To distinguish between first and second order for a reaction in which ClO is the single reactant, we plot $\ln[ClO]$ versus time and $1/[ClO]$ versus time. If the $\ln[ClO]$ plot is linear, the reaction is first order; if the $1/[ClO]$ plot is linear, the reaction is second order. The rate law will have the form

$$\text{Rate} = k[ClO]^m$$

where $m = 1$ or 2. The overall order of the reaction will equal the exponent for [ClO] in the rate law. We determine the rate constant from the slope of whichever slot is linear.

Solve To evaluate the two possibilities, we need to calculate $\ln[ClO]$ natural logarithms and $1/[ClO]$ values; sets of both these values are given in Table 15.10(b).

TABLE 15.10(a) Concentration of Chlorine Monoxide as Function of Time

Time (ms)	[ClO] (M)
0	1.50×10^{-8}
10	7.19×10^{-9}
20	4.74×10^{-9}
30	3.52×10^{-9}
40	2.81×10^{-9}
100	1.27×10^{-9}
200	0.66×10^{-9}

TABLE 15.10(b) Concentration of Chlorine Monoxide, ln[ClO], and 1/[ClO] as Function of Time

Time (ms)	[ClO] (M)	ln[ClO]	1/[ClO] (1/M)
0	1.50×10^{-8}	-18.015	6.67×10^7
10	7.19×10^{-9}	-18.751	1.39×10^8
20	4.74×10^{-9}	-19.167	2.11×10^8
30	3.52×10^{-9}	-19.465	2.84×10^8
40	2.81×10^{-9}	-19.690	3.56×10^8
100	1.27×10^{-9}	-20.484	7.89×10^8
200	0.66×10^{-9}	-21.139	1.51×10^9

The graphs for ln[ClO] and 1/[ClO] versus time are shown in Figure 15.12. The ln[ClO] plot is not linear, but the 1/[ClO] plot is, meaning that the reaction is second order in ClO. Thus, $m = 2$ and the rate law is

$$\text{Rate} = k[\text{ClO}]^2$$

Substituting [ClO] into the generic integrated rate law (Equation 15.27) gives

$$\frac{1}{[\text{ClO}]} = kt + \frac{1}{[\text{ClO}]_0}$$

From the general equation $y = mx + b$, we know that k is the slope of the graph of 1/[ClO] versus time. Arbitrarily choosing two convenient data points (at 0 and 100 ms), we can calculate k:

$$k = \text{slope} = \frac{\Delta y}{\Delta x} = \frac{\Delta\left(\dfrac{1}{[\text{ClO}]}\right)}{\Delta t}$$

$$= \frac{(7.89 - 0.67) \times 10^8 \ M^{-1}}{(100 - 0) \times 10^{-3} \ s}$$

$$= 7.22 \times 10^9 \ M^{-1} \ s^{-1}$$

As in Sample Exercise 15.5, we could also plot the data on a graphing calculator. By fitting the data to a straight line we would discover that the graph of 1/[ClO] versus t is a better fit and then find the slope of the line. This result allows us to conclude that the reaction is second order and calculate the rate constant k.

Think about It Sample Exercise 15.5 and this one show we can use integrated rate laws to distinguish between first- and second-order reactions in a single reactant.

Practice Exercise Experimental evidence shows that in the reaction

$$\text{NO}_2(g) + \text{CO}(g) \rightarrow \text{NO}(g) + \text{CO}_2(g)$$

the reaction rate depends only on the concentration of NO_2. Determine whether the reaction is first or second order in NO_2, and calculate the rate constant from the data in Table 15.11, which were obtained at 488 K. ■

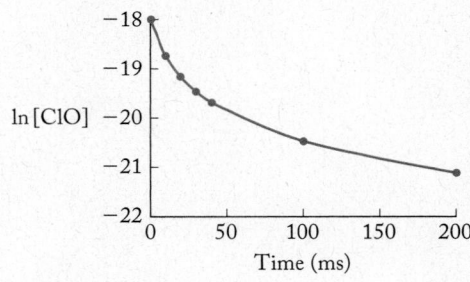

(a)

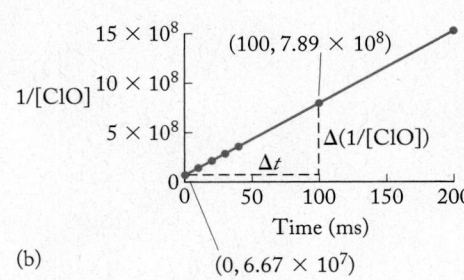

(b)

FIGURE 15.12

TABLE 15.11   Concentration of NO_2 as a Function of Time

Time (hr)	[NO₂] (M)
0.00	0.250
1.39	0.198
3.06	0.159
4.72	0.132
6.39	0.114
8.06	0.099
9.72	0.088
11.39	0.080

The concept of half-life can also be applied to second-order reactions. The relation between rate constant and half-life for the decomposition of NO_2 can be derived from Equation 15.27 if we first rearrange the terms to solve for kt:

$$kt = \frac{1}{[\text{X}]} - \frac{1}{[\text{X}]_0}$$

After one half-life has elapsed ($t = t_{1/2}$), [X] has decreased to half its initial concentration. Substituting this information into the preceding equation, we have

$$kt_{1/2} = \frac{1}{\frac{1}{2}[\text{X}]_0} - \frac{1}{[\text{X}]_0}$$

$$= \frac{2}{[\text{X}]_0} - \frac{1}{[\text{X}]_0} = \frac{1}{[\text{X}]_0}$$

or

$$t_{1/2} = \frac{1}{k[\text{X}]_0} \qquad (15.28)$$

Note that this value of $t_{1/2}$ is inversely proportional to the initial concentration of X. This dependence on concentration is unlike the $t_{1/2}$ values of first-order reactions, which are independent of concentration.

SAMPLE EXERCISE 15.8   Calculating the Half-Life of a
                        Second-Order Reaction

Calculate the half-life of the second-order decomposition of NO_2 (Equation 15.24) if the rate constant is $0.544\ M^{-1}\ s^{-1}$ at a particular temperature and the initial concentration of NO_2 is $0.0100\ M$.

Collect and Organize The half-life represents the time required for the concentration of NO_2 to decrease by a half. We are given the rate constant and initial concentration of NO_2 and told the reaction is second order. The half-life of a second-order reaction in which there is only one reactant is given by Equation 15.28.

Analyze The half-life is inversely proportional to the rate constant k and to the initial concentration of NO_2. The value of k is about 0.5 at an initial concentration of $10^{-2}\ M$, so $t_{1/2}$ is inversely proportional to a quantity $\approx 5 \times 10^{-3}$. Thus we predict a half-life of about 200 s.

Solve

$$t_{1/2} = \frac{1}{k[NO_2]_0} = \frac{1}{(0.544\ M^{-1}\ s^{-1})(1.00 \times 10^{-2}\ M)} = 184\ s$$

Think about It The calculated value of $t_{1/2}$ is very close to the predicted value of 200 s. We needed to know $[NO_2]$ to calculate the value of $t_{1/2}$ for this second-order reaction in NO_2.

Practice Exercise The rate constant for the decomposition of ClO (Sample Exercise 15.7) is $7.2 \times 10^9\ M^{-1}s^{-1}$ at 298 K. Determine the half-life of ClO when its initial concentration is $1.50 \times 10^{-8}\ M$.

Pseudo-First-Order Reactions

The integrated rate law in Equation 15.27:

$$\frac{1}{[X]} = kt + \frac{1}{[X]}$$

applies only to reactions that are second order in a single reactant. It does not apply to reactions that are second order overall but are first order in two reactants, such as

$$NO(g) + O_3(g) \rightarrow NO_2(g) + O_2(g)$$

or

$$NO_2(g) + O_3(g) \rightarrow NO_3(g) + O_2(g)$$

Because the integrated rate law for a reaction that is first order in two reactants is complicated, kineticists frequently adjust reaction conditions so that a simpler rate law can be used. One approach is to have one of the reactants present at a much higher concentration than the other. This condition is common for many components in the urban atmosphere where, for example, ozone concentrations are often hundreds to thousands of times greater than NO concentrations. With such a large excess, the ozone concentration remains virtually constant over the course of the reaction

$$NO(g) + O_3(g) \rightarrow NO_2(g) + O_2(g)$$

Thus, the rate law for the reaction

$$\text{Rate} = k[\text{NO}][\text{O}_3]$$

may be simplified to

$$\text{Rate} = k'[\text{NO}] \qquad (15.29)$$

where

$$k' = k[\text{O}_3]_0 \qquad (15.30)$$

and where $[\text{O}_3]_0$ is the initial concentration of ozone, which remains virtually constant throughout the reaction.

Equation 15.29 looks like the rate law for a first-order reaction (Rate = $k[\text{X}]$). It is considered a **pseudo-first-order** rate law because the reaction *appears* to obey first-order kinetics. A pseudo-first-order reaction has the same integrated rate law as a first-order reaction, but the rate of the pseudo-first-order reaction depends on the concentration of more than one reactant.

pseudo-first-order a reaction in which all the reactants but one are present at such high concentrations that they do not decrease significantly during the course of the reaction, so that reaction rate is controlled by the concentration of the limiting reactant.

SAMPLE EXERCISE 15.9 **Deriving a Pseudo-First-Order Rate Law**

The data in Table 15.12(a) were obtained in a study of the oxidation of trace levels of NO in the presence of a large excess of ozone at 298 K:

$$\text{NO}(g) + \text{O}_3(g) \rightarrow \text{NO}_2(g) + \text{O}_2(g)$$

(Note the NO concentration units in Table 15.12a. When the molar concentrations of reactants are very low, as in the case of many atmospheric pollutants, it is easier to express them in units of molecules/cm^3.)

  a. Verify that the reaction is pseudo-first-order and determine the pseudo-first-order rate constant k'.
  b. If the ozone concentration is 100 times the NO concentration at $t = 0$, what is the second-order rate constant k? Express this k in $M^{-1}\,s^{-1}$.

Collect and Organize Under pseudo-first-order conditions, a large excess of one reactant is maintained so that we can use the integrated rate law for a first-order reation. We are to verify that the reaction is pseudo-first-order and to determine the pseudo-first-order rate constant. We are given experimental data showing the change of concentration of NO with time. We are also given the initial concentration of O_3 and asked to calculate the second-order rate constant for the reaction.

Analyze We verify our assumption by treating the data as we would for a first-order reaction and plotting ln[NO] versus time. The plot should be a straight line with a slope equal to $-k'$. We can solve for the second-order rate constant k using Equation 15.30. We expect the second-order rate constant to be rather large since the data (concentration changes) were collected on a microsecond scale.

Solve

  a. The plot of ln[NO] values (Table 15.12b) versus time is linear as shown in Figure 15.13, which indicates that the reaction is indeed

TABLE 15.12(a) Concentration of NO as a Function of Time	
Time (μs)	Concentration of NO (molecules/cm^3)
0	1.00×10^9
100	8.36×10^8
200	6.98×10^8
300	5.83×10^8
400	4.87×10^8
500	4.07×10^8
1000	1.65×10^8

pseudo-first-order. We get the pseudo-first-order rate constant k' by calculating the slope from the two points labeled in Figure 15.13:

$$\text{Slope} = -k' = \frac{\Delta y}{\Delta x} = \frac{(18.915 - 20.184)}{(1000 - 300)\,\mu s} = -1.81 \times 10^3$$

$$k' = 1.81 \times 10^3 \text{ s}^{-1}$$

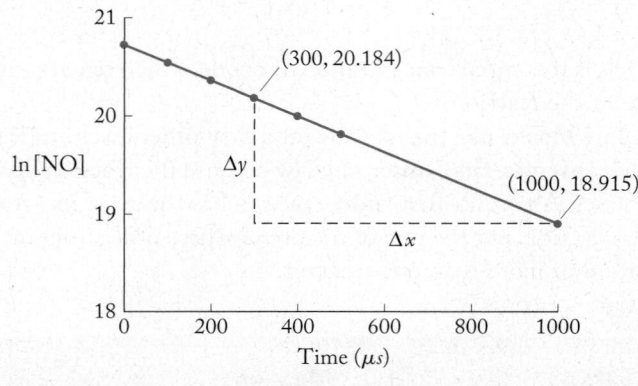

FIGURE 15.13

TABLE 15.12(b) Concentration of NO and ln[NO] as a Function of Time

Time (μs)	[NO] (molecules/cm^3)	ln[NO]
0	1.00×10^9	20.723
100	8.36×10^8	20.544
200	6.98×10^8	20.364
300	5.83×10^8	20.184
400	4.87×10^8	20.004
500	4.07×10^8	19.824
1000	1.65×10^8	18.915

b. We calculate the second-order rate constant k using Equation 15.30, $k' = k[O_3]_0$, and

$$[O_3]_0 = 100[NO]_0 = 100(1.00 \times 10^9 \text{ molecules/cm}^3)$$

solving for k:

$$k = \frac{k'}{[O_3]_0} = \frac{1.81 \times 10^3 \text{ s}^{-1}}{1.00 \times 10^{11} \text{ molecules/cm}^3} = \frac{1.81 \times 10^{-8} \text{ cm}^3}{\text{molecules} \cdot \text{s}}$$

To convert the units to $M^{-1}\,\text{s}^{-1}$, we need to convert the reciprocal concentration units of cm^3/molecule into M^{-1}:

$$k = \frac{1.81 \times 10^{-8} \cancel{\text{cm}^3}\,\text{s}^{-1}}{\cancel{\text{molecules}}} \times \frac{6.022 \times 10^{23} \cancel{\text{molecules}}}{1 \text{ mol}} \times \frac{1 \text{ L}}{1000 \cancel{\text{cm}^3}}$$

$$= 1.09 \times 10^{13} \frac{\text{L} \cdot \text{s}^{-1}}{\text{mol}} = 1.09 \times 10^{13} \, M^{-1}\,\text{s}^{-1}$$

Think about It As we predicted, the second-order rate constant for the reaction is large. The pseudo-first-order (k') and second-order (k) rate constants are very different because k' contains a term for the initial concentration of ozone, which is very small when expressed in moles per liter. The value of k' will be different for every initial concentration of O_3, but the value of k will be independent of $[O_3]_0$.

Practice Exercise The reaction

$$Cl(g) + O_3(g) \rightarrow ClO(g) + O_2(g)$$

is first-order in both reactants. Determine the pseudo-first-order and second-order rate constants for the reaction from the data in Table 15.13 if the initial ozone concentration is $8.5 \times 10^{-11} M$.

TABLE 15.13 Concentration of Chlorine Atoms as a Function of Time

Time (μs)	[Cl] (M)
0	5.60×10^{-14}
100	5.27×10^{-14}
600	3.89×10^{-14}
1200	2.69×10^{-14}
1850	1.81×10^{-14}

A student measured the pseudo-first-order rate constant for the reaction of NO with O_3 in Sample Exercise 15.9 at four initial concentrations of ozone. Assuming that all four $[O_3]_0$ values were much greater than [NO], how could the student determine the second-order rate constant graphically?

Zero-Order Reactions

In the Practice Exercise accompanying Sample Exercise 15.7, we were introduced to the reaction

$$NO_2(g) + CO(g) \rightarrow NO(g) + CO_2(g)$$

and the solution revealed that the rate law for the reaction is

$$\text{Rate} = k[NO_2]^2 \qquad (15.31)$$

The rate of the reaction does not change with changing [CO], even when the concentrations of CO and NO_2 are comparable. This situation is not the same as in pseudo-first-order reactions, where the rate does not depend on the concentration of a reactant because that reactant is present in large excess.

One interpretation of Equation 15.31 is that it contains a [CO] term to the zeroth power, making the reaction *zero order* in that reactant. Because any value raised to the zeroth power is 1, we have

$$\text{Rate} = k[NO_2]^2[CO]^0 = k[NO_2]^2(1) = k[NO_2]^2$$

Reactions with a true zero-order rate law are rare, but let's consider a generic reaction involving a single reactant A, with the reaction zero order in reactant A and zero order overall. For the reaction

$$A \rightarrow B$$

the rate law is

$$\text{Rate} = -\Delta[A]/\Delta t = k[A]^0 = k$$

and the integrated rate law is

$$[A] = -kt + [A]_0$$

The slope of a plot of reactant concentration versus time (Figure 15.14) equals the negative of the zero-order rate constant k.

We can calculate the half-life of a zero-order reaction by substituting $t = t_{1/2}$ and $[A] = [A]_0/2$ into the integrated rate law:

$$[A]_0/2 = -kt_{1/2} + [A]_0$$
$$kt_{1/2} = [A]_0 - [A]_0/2 = [A]_0/2$$
$$t_{1/2} = [A]_0/2k$$

For now we leave the discussion of zero-order reactions with this purely mathematical treatment. We return to these reactions and examine their meaning at the molecular level in Section 15.5.

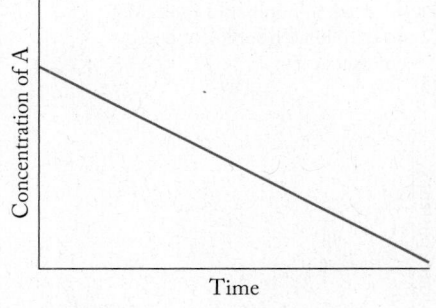

FIGURE 15.14 The change in concentration of the reactant A in the zero-order reaction $A \rightarrow B$ is constant over time.

Which of the following situations represent a zero-order reaction in one or more reactants?

a. In the reaction $A + 2B \rightarrow C$, the rate law is Rate $= k[B]^2$.
b. In the reaction $X + Y \rightarrow Z$, the plot of ln[X] versus t is linear when a large excess of Y is used.
c. The rate of the reaction $E \rightarrow F$ is independent of [E].

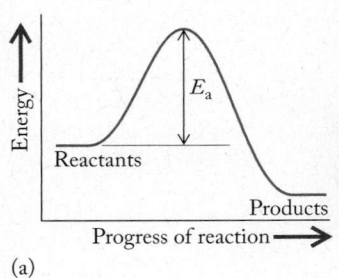

(a)

(b)

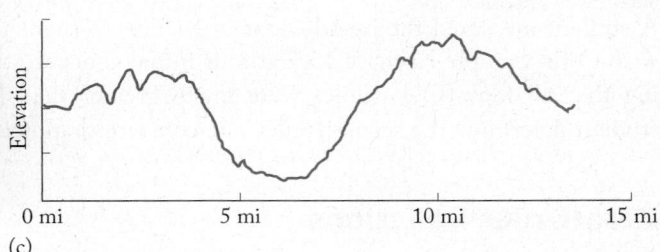

(c)

FIGURE 15.15 (a) The energy profile of a reaction includes an activation energy barrier E_a that must be overcome before the reaction can proceed. (b) A real-world analogy confronts a hiker climbing a series of hills. The Beartree Gap Trail forms part of the Iron Mountain Loop in the Mount Rogers National Recreation Area in southwestern Virginia. (c) From the starting point at mile zero, hikers must climb over two "energy barriers": Straight Mountain (at 3 mi) and Iron Mountain (at 10 mi) before returning to their starting point after a 13.5-mile (22-km) hike.

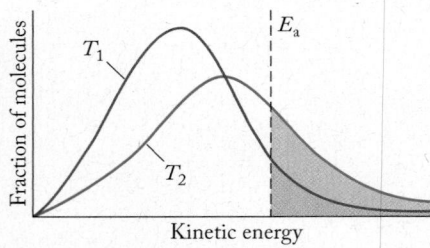

Fraction of molecules in sample with sufficient energy to react at T_1

Increase in number of molecules in sample with sufficient energy to react at T_2; $T_2 > T_1$

(a)

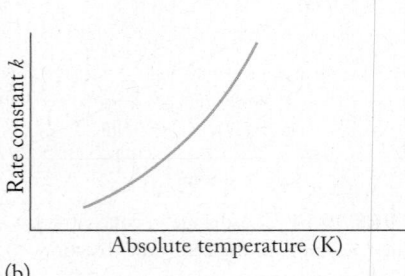

(b)

FIGURE 15.16 (a) According to kinetic molecular theory, a fraction of reactant molecules have kinetic energies equal to or greater than the activation energy (E_a) of the reaction. As temperature increases from T_1 to T_2, the number of molecules with energies exceeding E_a increases, leading to an increase in reaction rate. (b) The rate constant for any reaction increases with increasing temperature.

activation energy (E_a) the minimum energy molecules need to react when they collide.

Arrhenius equation relates the rate constant of a reaction to absolute temperature (T), the activation energy of the reaction (E_a), and the frequency factor (A).

15.4 Reaction Rates, Temperature, and the Arrhenius Equation

Why do rate constants for different reactions have different values? In Section 15.3, we introduced the idea that chemical reactions take place in the gas phase when molecules collide with sufficient energy to break bonds in reactants and allow bonds in products to form. The minimum amount of energy that enables this to happen is called the **activation energy (E_a)**, and every chemical reaction has a characteristic activation energy, usually expressed in kilojoules per mole. Activation energy is an energy barrier that must be overcome if a reaction is to proceed—like the mountains that must be climbed to hike the trail shown in Figure 15.15. Generally, the greater the activation energy, the slower the reaction.

According to kinetic molecular theory, the fraction of molecules with kinetic energies greater than a given activation energy increases with increasing temperature, as shown in Figure 15.16(a). Therefore, the rates of chemical reactions should increase with increasing temperature, and indeed they do (Figure 15.16b).

CONCEPT TEST •••••••••••••••••••••••••••••••

Why does increasing temperature increase the frequency of collisions between molecules in the gas phase?

••

In the late 19th century, experiments carried out in the laboratories of Jacobus van't Hoff and the Swedish chemist Svante Arrhenius (1859–1927) led to a fundamental advance in understanding how temperature affects the rates of chemical reactions. The mathematical connection between temperature, the rate constant k for a reaction, and its activation energy is given by the **Arrhenius equation**:

$$k = Ae^{-E_a/RT} \qquad (15.32)$$

where R is the gas constant in J/(mol · K) and T is the reaction temperature in kelvins. The factor A, called the **frequency factor**, is the product of collision frequency and a term that accounts for the fact that not every collision results in a chemical reaction.

Some collisions do not lead to products because the colliding molecules are not oriented relative to each other in the right way. To examine the importance of

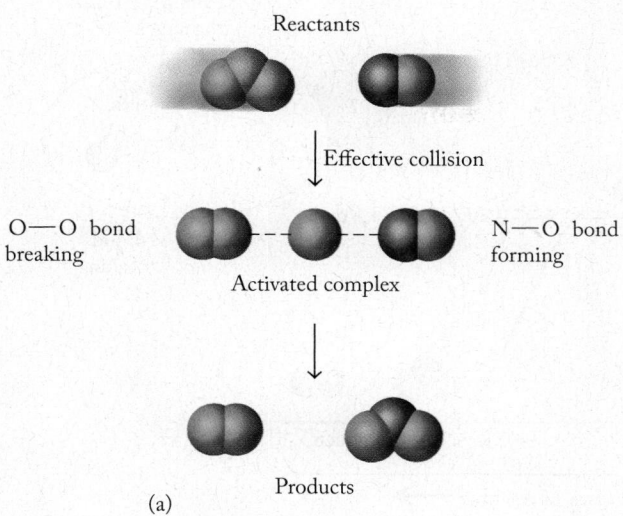

Reactants

Effective collision

O—O bond breaking

Activated complex

N—O bond forming

Products

(a)

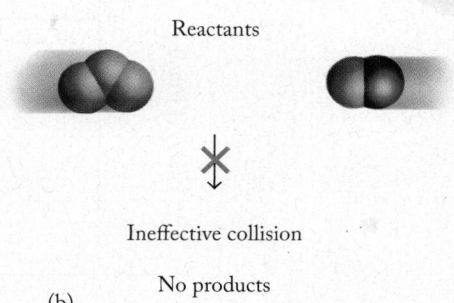

FIGURE 15.17 The effect of molecular orientation on reaction rate. (a) When an O_3 and a NO molecule are oriented such that the collision is between an O_3 oxygen and the NO nitrogen, the collision is effective and an activated complex forms and then yields the two product molecules NO_2 and O_2. (b) When the reactant molecules are oriented such that the collision is between an O_3 oxygen and the NO oxygen, there is no activated complex and no reaction.

Reactants

Ineffective collision

No products

(b)

molecular orientation during collisions, let's revisit the reaction (Equation 15.7) between O_3 and NO:

$$O_3(g) + NO(g) \rightarrow O_2(g) + NO_2(g)$$

Two ways that ozone and nitric oxide molecules might approach each other are shown in Figure 15.17. Only one of these orientations, the one in which an O_3 molecule collides with the nitrogen atom of NO, leads to a chemical reaction between the two molecules.

A collision between O_3 and NO molecules with the correct orientation and enough kinetic energy may result in the formation of the **activated complex** shown in Figure 15.17(a). In this species, one of the O—O bonds in the O_3 molecule has started to break and the new N—O bond is beginning to form. Activated complexes represent midway points in chemical reactions. They have extremely brief lifetimes and fall apart rapidly, either forming products or re-forming reactants. Activated complexes are formed by reacting species that have acquired enough energy to react with each other. The internal energy of an activated complex represents a high-energy **transition state** of the reaction. In fact, the energy of an activated complex for a reaction defines the height of the activation energy barrier for the reaction. The magnitudes of activation energies can vary from a few kilojoules to hundreds of kilojoules per mole.

We can draw an **energy profile** for a chemical reaction that shows the changes in energy for the reaction as a function of the *progress of the reaction* from reactants to products. We have already seen one energy profile in Figure 15.15(a). Now consider the energy profile for the reaction between nitric oxide and ozone, shown in Figure 15.18(a). The *x*-axis represents the progress of the reaction and the *y*-axis represents chemical energy. The activation energy is equivalent to the difference in energy between the transition state and either the reactants or the products. The size of the activation energy barrier depends on the direction from which it is approached. In the forward direction (NO + $O_3 \rightarrow NO_2 + O_2$, Figure 15.18a), E_a is smaller than in the reverse direction ($NO_2 + O_2 \rightarrow NO + O_3$, Figure 15.18b). A smaller activation energy barrier would mean that the forward reaction would proceed at a higher rate than the reverse reaction if we had equal concentrations of reactants and products.

What is the significance of the difference between the energies of the reactants and products in Figure 15.18? At temperatures below 500 K, the value of ΔE in

CONNECTION The van't Hoff factor in Chapter 11 is named after the same Jacobus van't Hoff who worked on the temperature dependence of reaction rates.

CHEMTOUR Arrhenius Equation

CHEMTOUR Collision Theory

frequency factor (A) the product of the frequency of molecular collisions and a factor that expresses the probability that the orientation of the molecules is appropriate for a reaction to occur.

activated complex a species formed in a chemical reaction when molecules have enough energy to react with each other.

transition state a high-energy state between reactants and products in a chemical reaction.

energy profile graph showing the changes in potential energy for a reaction as a function of the progress of the reaction from reactants to products.

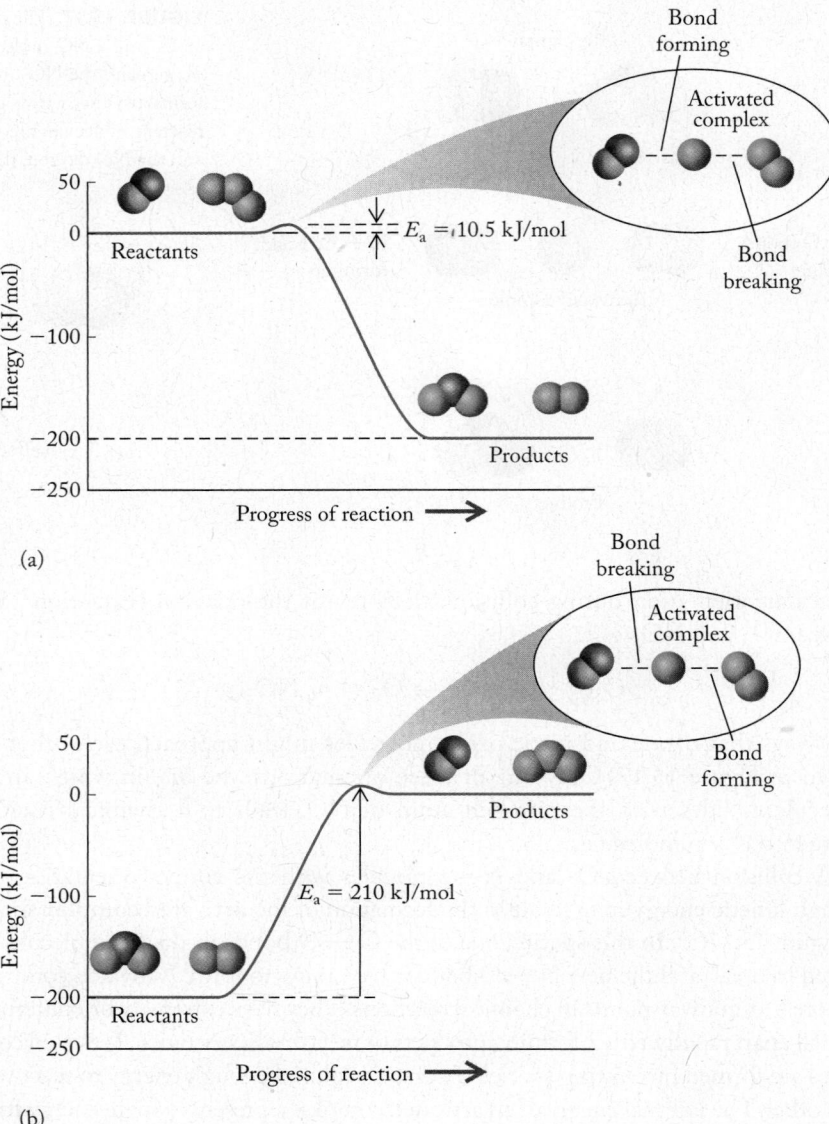

FIGURE 15.18 (a) The energy profile for the spontaneous reaction $NO(g) + O_3(g) \rightarrow NO_2(g) + O_2(g)$ includes an activation energy barrier of 10.5 kJ/mol. (b) The reverse reaction has a much larger activation energy of 210 kJ/mol.

⦿⦿ **CONNECTION** The equation that relates free energy (ΔG), enthalpy (ΔH), and entropy (ΔS) was introduced in Chapter 14.

Figure 15.18 is approximately equal to ΔH, the enthalpy change for the reaction.[1] In Chapter 14 we learned that the sign of ΔG, the free-energy change of a reaction, indicates whether it is spontaneous. A negative free-energy change ($\Delta G < 0$) represents a spontaneous reaction, while a positive free-energy change ($\Delta G > 0$) indicates a nonspontaneous process. The difference between ΔG and ΔH is the $T \Delta S$ term in the equation:

$$\Delta G = \Delta H - T \Delta S$$

In the reaction above and many others, the $T \Delta S$ term is small compared to ΔH, so the energy difference between reactants and products is approximately equal to ΔG in Figures 15.15 and 15.18.[2] This means that the reaction between NO and O_3 in Figure 15.18(a) is spontaneous while the reverse reaction is nonspontaneous.

[1] $\Delta H = \Delta E + RT \Delta n$ for a gas-phase reaction, where Δn is the difference between the number of moles of gaseous product and moles of gaseous reactant. At 298 K and $\Delta n = 1$, $RT \Delta n = 2.5$ kJ/mol, which is small compared to typical ΔH_{rxn} values.

[2] The same is not true for the activation energies shown in Figures 15.15 and 15.18. The free energy of activation can differ significantly from E_a. The details of how we find the free energy of activation are beyond the scope of this book.

Which of the following statements is/are true?

a. Activation energies are always greater than ΔG.
b. Activation energies are always greater than ΔG for nonspontaneous reactions.
c. The magnitudes of activation energies are independent of ΔG.

One of the many uses of the Arrhenius equation is to calculate the value of E_a for a chemical reaction. When we take the natural logarithm of both sides of Equation 15.32:

$$\ln k = -\frac{E_a}{R}\left(\frac{1}{T}\right) + \ln A \qquad (15.33)$$

the result fits the general equation of a straight line ($y = mx + b$) if we make ($\ln k$) the y variable and ($1/T$) the x variable. We can calculate E_a by determining the rate constant k for the reaction at several temperatures. Plotting $\ln k$ versus $1/T$ should give a straight line, the slope of which is $-E_a/R$. Table 15.14 and Figure 15.19 show data for the reaction between NO and O_3 at six temperatures. Arbitrarily picking two points on the line in Figure 15.19, we calculate its slope:

$$\text{Slope} = \frac{\Delta y}{\Delta x} = \frac{(23.814 - 24.264)}{(2.86 \times 10^{-3} - 2.50 \times 10^{-3})\ \text{K}^{-1}} = -1.25 \times 10^3\ \text{K}$$

TABLE 15.14	Temperature Dependence of the Rate of Reaction for the Reaction $NO(g) + O_3(g) \rightarrow NO_2(g) + O_2(g)$		
T (K)	**k ($M^{-1}\ s^{-1}$)**	**$\ln k$**	**$1/T$ (K^{-1})**
300	1.21×10^{10}	23.216	3.33×10^{-3}
325	1.67×10^{10}	23.539	3.08×10^{-3}
350	2.20×10^{10}	23.814	2.86×10^{-3}
375	2.79×10^{10}	24.052	2.67×10^{-3}
400	3.45×10^{10}	24.264	2.50×10^{-3}
425	4.15×10^{10}	24.449	2.35×10^{-3}

The slope equals $-E_a/R$, so

$$E_a = -\text{slope} \times R$$

$$= -(-1.25 \times 10^3\ \text{K}) \times \left(\frac{8.314\ \text{J}}{\text{mol} \cdot \text{K}}\right) = 1.04 \times 10^4\ \text{J/mol}$$

$$= 10.45\ \text{kJ/mol}$$

The y intercept ($1/T = 0$) in Figure 15.19 is 27.41. From Equation 15.33, we know that this value represents $\ln A$, which means

$$A = e^{27.41} = 8.0 \times 10^{11}$$

Now we can use the values of E_a and A to calculate k at any temperature. For example, at $T = 250$ K:

$$k = Ae^{-E_a/RT}$$

$$= (8.0 \times 10^{11})\ e^{-\left(\dfrac{1.04 \times 10^4\ \cancel{\text{J/mol}}}{8.314\ \dfrac{\cancel{\text{J}}}{\cancel{\text{mol} \cdot \text{K}}} \cdot 250\ \cancel{\text{K}}}\right)}$$

$$= 5.4 \times 10^9\ M^{-1}\ s^{-1}$$

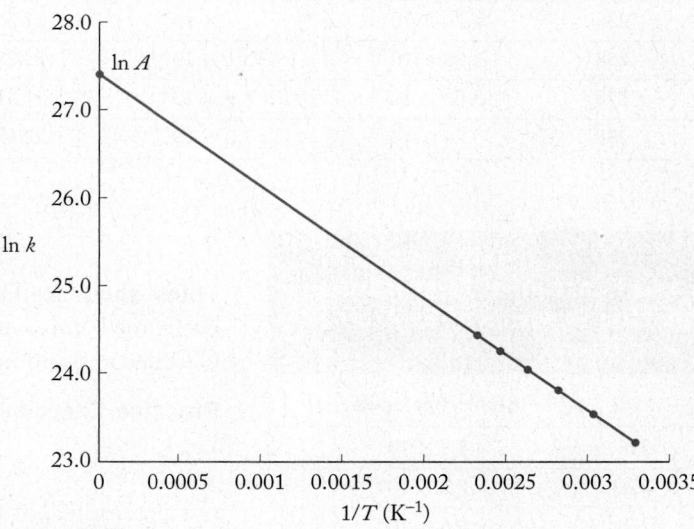

FIGURE 15.19 A graph of $\ln k$ versus $1/T$ yields a straight line with a slope equal to $-E_a/R$ and a y intercept equal to $\ln A$, the natural logarithm of the frequency factor.

TABLE 15.15(a) Rate Constant as a Function of Temperature for the Decomposition of ClO

$k\,(M^{-1}\,s^{-1})$	$T\,(K)$
1.9×10^9	238
3.1×10^9	258
4.9×10^9	278
7.2×10^9	298

SAMPLE EXERCISE 15.10 Calculating an Activation Energy from Rate Constants

The data in Table 15.15(a) were collected in a study of the effect of temperature on the rate of the decomposition reaction

$$2\,ClO(g) \rightarrow Cl_2(g) + O_2(g)$$

Determine the activation energy for the reaction.

Collect and Organize Activation energy is calculated using the Arrhenius equation. We are given values of the rate constant k as a function of absolute temperature. According to the Arrhenius equation, the slope of a plot of $\ln k$ against $1/T$ is equal to $-E_a/R$, where E_a is the activation energy and R is the ideal gas constant with units of $J/(mol \cdot K)$.

Analyze First, we need to convert the T and k values from Table 15.15(a) to $1/T$ and $\ln k$, respectively. We predict a positive value for E_a because activation energy represents a barrier that costs energy to overcome. The rate constant for the reaction is fairly large, about $10^9\,M^{-1}\,s^{-1}$, so we predict that the activation energy barrier will be relatively low.

Solve Expanding the data table to include columns for $1/T$ and $\ln k$ yields Table 15.15(b). A plot of $\ln k$ versus $1/T$ gives us a straight line, the slope of which is -1590 K (Figure 15.20). Using the values of the slope and R [8.314 $J/(mol \cdot K)$] to calculate the value of E_a:

$$E_a = -\text{slope} \times R$$

$$= -(-1590\ \cancel{K}) \times \left(\frac{8.314\ J}{mol \cdot \cancel{K}}\right) = 1.32 \times 10^4\ J/mol$$

$$= 13.2\ kJ/mol$$

TABLE 15.15(b) Summary of T, $1/T$, k, and $\ln k$ for the Decomposition of ClO

$T\,(K)$	$1/T\,(K^{-1})$	$k\,(M^{-1}\,s^{-1})$	$\ln k$
238	4.20×10^{-3}	1.9×10^9	21.365
258	3.88×10^{-3}	3.1×10^9	21.855
278	3.60×10^{-3}	4.9×10^9	22.313
298	3.36×10^{-3}	7.2×10^9	22.697

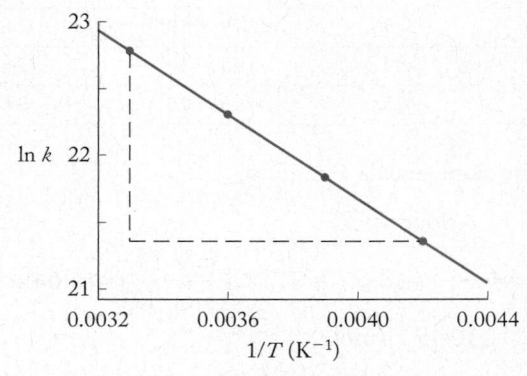

FIGURE 15.20

TABLE 15.16 Rate Constants as a Function of Temperature for the Reaction of Br Atoms with Ozone

$T\,(K)$	$k\,[cm^3/(molecule \cdot s)]$
238	5.9×10^{-13}
258	7.7×10^{-13}
278	9.6×10^{-13}
298	1.2×10^{-12}

Think about It The activation energy is the height of a barrier that has to be overcome before a reaction can proceed. As predicted, the activation energy for ClO decomposition has a small positive value.

Practice Exercise The rate constant for the reaction

$$Br(g) + O_3(g) \rightarrow BrO(g) + O_2(g)$$

was determined at the four temperatures shown in Table 15.16. Calculate the activation energy for this reaction.

Calculation of activation energies using the graphical method generally requires measurement of the rate constant at a minimum of three different temperatures to verify that the plot of $\ln k$ vs $1/T$ is a straight line. Once we know the value of E_a we can use it and the value of the rate constant (k_1) of a reaction at one temperature (T_1) to calculate the value of the rate constant (k_2) at another temperature (T_2). We start by substituting k_1, k_2, T_1, and T_2 into Equation 15.33:

$$\ln k_1 = -\frac{E_a}{R}\left(\frac{1}{T_1}\right) + \ln A \qquad \ln k_2 = -\frac{E_a}{R}\left(\frac{1}{T_2}\right) + \ln A$$

Subtracting these two expressions, $\ln k_1 - \ln k_2$, gives

$$\ln k_1 - \ln k_2 = \left[-\frac{E_a}{R}\left(\frac{1}{T_1}\right) + \ln A\right] - \left[-\frac{E_a}{R}\left(\frac{1}{T_2}\right) + \ln A\right]$$

Using the mathematical properties of logarithms, we can rearrange this equation:

$$\ln \frac{k_1}{k_2} = \frac{E_a}{R}\left(\frac{1}{T_2}\right) - \frac{E_a}{R}\left(\frac{1}{T_1}\right)$$

$$= \frac{E_a}{R}\left(\frac{1}{T_2} - \frac{1}{T_1}\right) \qquad\qquad (15.34)$$

CONCEPT TEST ....

Which of the following statements is/are true about activation energies?

a. Exothermic reactions have negative activation energies.
b. Fast reactions have large rate constants *and* large activation energies.
c. The forward reaction sometimes has a lower activation energy than the reverse reaction.
d. Endothermic reactions always have large activation energies.

15.5 Reaction Mechanisms

In this section we explore how reactions proceed at the molecular level. Let's start by revisiting the thermal decomposition of NO_2:

$$2\,NO_2(g) \rightarrow 2\,NO(g) + O_2(g)$$

We noted in Section 15.3 that this reaction is second order in NO_2 because it takes place as a result of the collisions of *pairs* of NO_2 molecules. How do the atoms in two colliding NO_2 molecules rearrange themselves to form two NO molecules and one O_2 molecule? The answer to this question is contained in the mechanism of the reaction. A **reaction mechanism** describes the stepwise manner in which the bonds in reactant molecules break and the bonds in product molecules form. Chemists use the results of reaction rate measurements to develop explanations of how reactions actually happen. Sometimes they can test for the presence of the products formed in preliminary steps, but, until recently, this was not possible for most reactions. Today, technological advances allow chemists to follow the transformation of reactants to products in the time that it takes for individual covalent bonds to break and new ones to form. These processes occur as rapidly as the bonds vibrate, that is, in femtoseconds (10^{-15} s), and the area of research based on monitoring these ultrafast processes is called *femtochemistry*.[3]

▶❚❚ **CHEMTOUR** Reaction Mechanisms

reaction mechanism a set of steps that describe how a reaction occurs at the molecular level; the mechanism must be consistent with the rate law for the reaction.

[3] In 1999 Ahmed H. Zewail received the Nobel Prize in Chemistry for his pioneering research in the development of femtochemistry.

Elementary Steps

A reaction mechanism proposed for the decomposition of NO_2 is shown in Figure 15.21. In the first step of the mechanism, a collision between two NO_2 molecules produces a molecule of NO and a molecule of NO_3. In a second step, the NO_3 decomposes to NO and O_2. Both steps in the mechanism involve very short-lived activated complexes. In the activated complex of the first step, two molecules share an oxygen atom. The bonds in the activated complex of the second step rearrange so that two oxygen atoms become bonded together, forming a molecule of O_2 and leaving behind a molecule of NO. The molecule NO_3 is an **intermediate** in this mechanism because it is produced in one step and consumed in the next. Intermediates are not considered reactants or products and do not appear in the equation describing a reaction. In some cases, intermediates in chemical reactions are sufficiently long-lived to be isolated. In contrast, activated complexes have never been isolated although they have been detected.

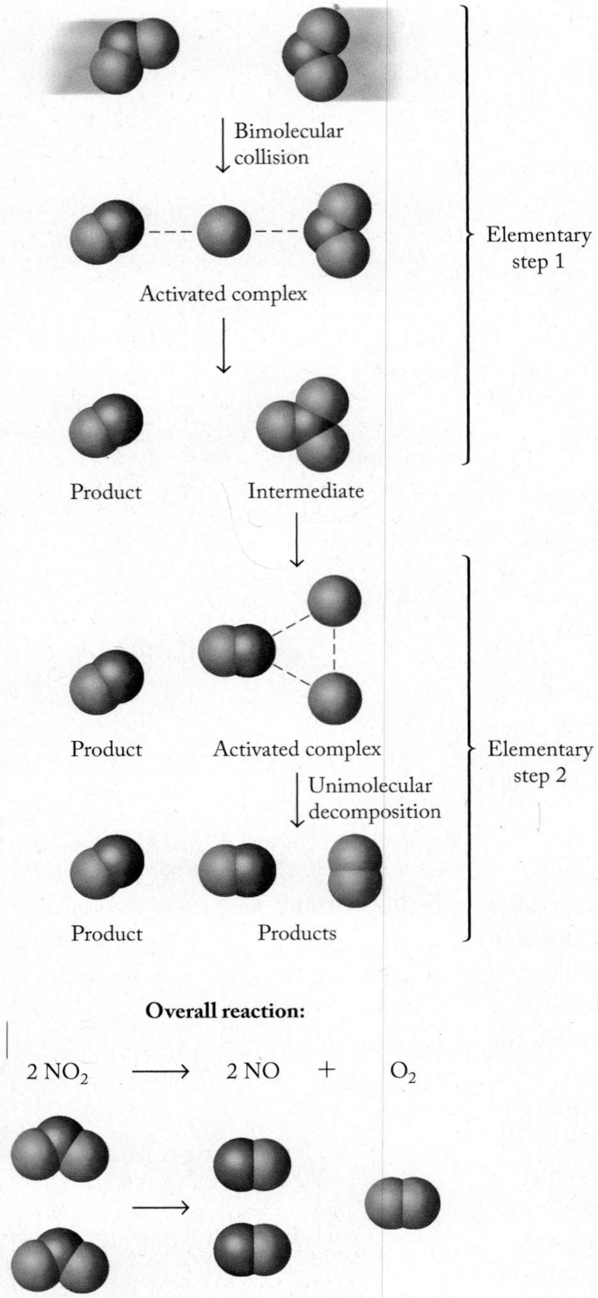

This reaction mechanism is a combination of two **elementary steps**. An elementary step that involves a single molecule is called **unimolecular**, and one that involves a collision between two molecules is **bimolecular**. Bimolecular elementary steps are much more common than **termolecular** (three-molecule) elementary steps because the chance of three molecules colliding at exactly the same time is much smaller. The terms *uni-*, *bi-*, and *termolecular* are used by chemists to describe the **molecularity** of an elementary step, which refers to the number of atoms, ions, or molecules involved in that step.

A valid reaction mechanism must be consistent with the stoichiometry of the reaction. In other words, the sum of the elementary steps in Figure 15.21 must be consistent with the observed proportions of reactants and products as defined in the balanced chemical equation. In this case, the sum matches the overall stoichiometry:

Elementary step 1 $\qquad 2\,NO_2(g) \rightarrow NO(g) + NO_3(g)$

Elementary step 2 $\qquad NO_3(g) \rightarrow NO(g) + O_2(g)$

Summing the two elementary steps and simplifying by cancelling out the intermediate (NO_3) terms:

$$2\,NO_2(g) + \cancel{NO_3(g)} \rightarrow 2\,NO(g) + \cancel{NO_3(g)} + O_2(g)$$

we get the overall reaction

$$2\,NO_2(g) \rightarrow 2\,NO(g) + O_2(g)$$

Before ending this discussion, let's consider how activation energy applies to a two-step reaction such as this one. The two elementary steps produce an energy profile with two maxima. In elementary step 1, collisions between pairs of NO_2 molecules result in the formation of an activated complex associated with the first transition state in Figure 15.22. As this activated complex transforms into NO and NO_3, the energy of the system drops to the bottom of the trough between the two maxima. In elementary step 2, NO_3 forms the activated complex associated with the second transition state. As this complex transforms into the final products NO and O_2, the energy of the system drops to its final level.

Bimolecular collision

Activated complex

Product   Intermediate

Elementary step 1

Product   Activated complex

Elementary step 2

Unimolecular decomposition

Product   Products

Overall reaction:

$2\,NO_2 \longrightarrow 2\,NO + O_2$

FIGURE 15.21 The decomposition of NO_2 begins when two NO_2 molecules collide, producing NO and NO_3 (elementary step 1). The NO_3 then rapidly decomposes into NO and O_2 (elementary step 2).

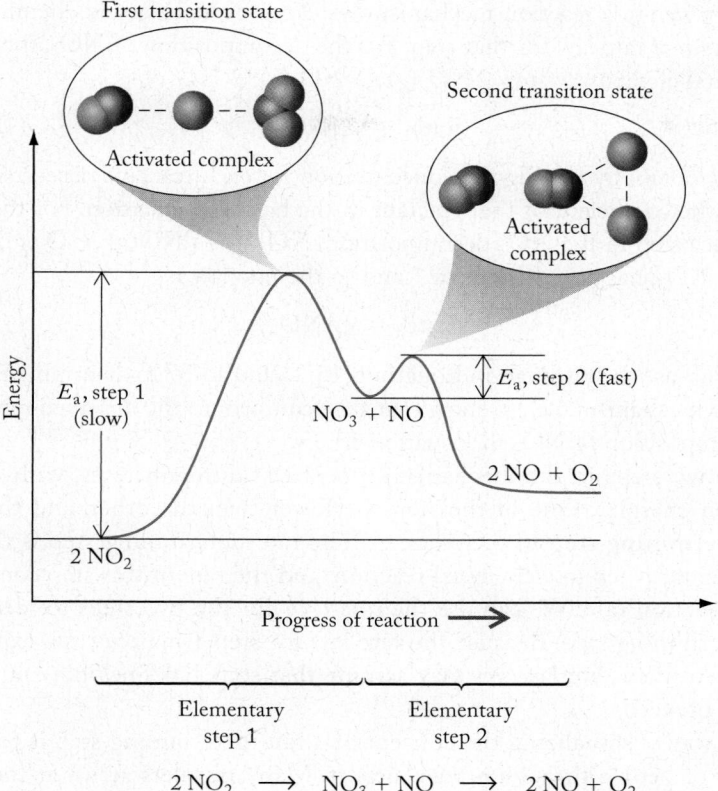

FIGURE 15.22 The energy profile for the decomposition of NO_2 to NO and O_2 shows activation energy barriers for both elementary steps. The activation energy of the first step is larger than that of the second step, so the first step is the slower of the two.

Figure 15.22 shows that the energy barrier for elementary step 1 is much greater than that for elementary step 2. This difference is consistent with the relative rates of the two steps: step 1 is slower than step 2. If the reaction were to proceed in the reverse direction (as NO and O_2 react, forming NO_2), the first energy barrier would be the smaller of the two, and the first elementary step would be the more rapid one. Experimental evidence supports these expectations.

For a reaction mechanism with n elementary steps, how many activation energies does the overall reaction have?

Rate Laws and Reaction Mechanisms

Any mechanism proposed for a reaction must be consistent with the rate law derived from experimental data. Is the mechanism proposed in Figure 15.21 consistent with the reaction's second-order rate law? For a reaction mechanism to be consistent with a rate law, the molecularity of one of the elementary steps in the mechanism must be the same as the reaction order expressed in the rate law.

As noted in Section 15.3, the experimentally determined rate law for NO_2 decomposition is second order in NO_2:

$$\text{Rate} = k[NO_2]^2 \qquad (15.35)$$

Remember that the balanced chemical equation for the overall reaction does not give us enough information to write the rate law for the reaction. However, for any

intermediate a species produced in one step of a reaction and consumed in a subsequent step.

elementary step a molecular-level view of a single process taking place in a chemical reaction.

unimolecular step a step in a reaction mechanism involving only one molecule on the reactant side.

bimolecular step a step in a reaction mechanism involving a collision between two molecules.

termolecular step a step in a reaction mechanism involving a collision among three molecules.

molecularity the number of ions, atoms, or molecules involved in an elementary step in a reaction.

rate-determining step the slowest step in a multistep chemical reaction.

(1) Formation of intermediate: elementary step 1

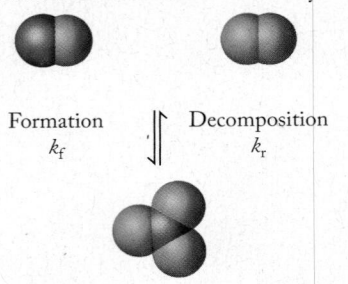

Formation
k_f

Decomposition
k_r

(2) Reaction of intermediate: elementary step 2

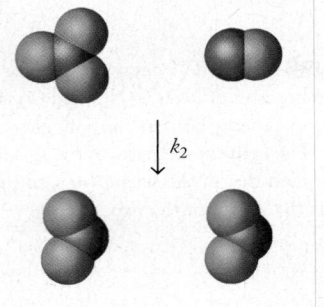

k_2

Overall reaction:

$2\,NO \quad + \quad O_2 \quad \longrightarrow \quad 2\,NO_2$

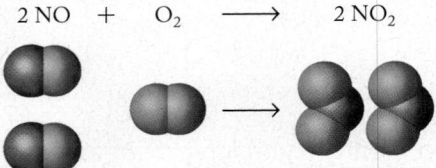

FIGURE 15.23 A mechanism for the formation of NO_2 from NO and O_2 has two elementary steps: (1) a fast, reversible bimolecular reaction in which NO and O_2 form NO_3 and (2) a slower, rate-determining bimolecular reaction in which NO_3 reacts with a molecule of NO to form two molecules of NO_2.

elementary step in a reaction mechanism we can use the balanced chemical equation to write a rate law for that step. For the decomposition of NO_2, the rate law for the first elementary step, $2\,NO_2(g) \rightarrow NO_3(g) + NO(g)$, is

$$\text{Rate}_1 = k_1[NO_2]^2 \qquad (15.36)$$

which we obtain by writing the concentration for each reactant raised to a power equal to the coefficient of that reactant in the balanced equation. For the second elementary step in the NO_2 decomposition, $NO_3(g) \rightarrow NO(g) + O_2(g)$, the sole reactant, NO_3, has a coefficient of 1 and so the rate law is

$$\text{Rate}_2 = k_2[NO_3] \qquad (15.37)$$

How do we use the rate laws in Equations 15.36 and 15.37 to determine the validity of the mechanism, i.e., to show that they conform to the observed rate law for the decomposition of NO_2 in Equation 15.35?

The two steps in the mechanism proceed at different rates, with different activation energies. One of the steps is slower than the other, and this is the **rate-determining step** in the reaction. The rate-determining step is the slowest elementary step in a chemical reaction, and the rate of this step controls the overall reaction rate. We can use the rate laws for the two steps to identify the rate-determining step. Because the rate law for step 1 matches the experimentally determined rate law, we may assume that step 1 defines how rapidly the reaction proceeds.

One way of visualizing the concept of a rate-determining step is to analyze the flow of people through a busy airport. Many travelers arrive at the airport with their boarding passes in hand or print them from convenient kiosks that allow them to avoid lines at ticket counters and move quickly. The next step in the process is passing through security. Typically, the number of people in the line outside the security point greatly exceeds the number of available security gates so that the time required to make it to a flight depends mostly on the time needed to pass through security. Security screening is the rate-determining step on the way through the airport.

If the first step in the mechanism in Figure 15.21 is the rate-determining step, then the value of k in Equation 15.35 is equal to k_1 from Equation 15.36. In addition, the value of k_1 must be smaller than the value of k_2. NO_3 is sufficiently stable that we can make small amounts of it and test whether $k_1 < k_2$. Experiments run at 300 K starting with NO_2 or NO_3 yield these values: $k_1 \approx 1 \times 10^{-10}\ M^{-1}\ s^{-1}$ and $k_2 \approx 6.3 \times 10^4\ s^{-1}$. Therefore, as soon as any NO_3 forms in step 1, it rapidly falls apart to NO and O_2 in step 2.

Now let's consider the reaction that is the reverse of NO_2 decomposition, namely the formation of NO_2 from NO and O_2:

Overall reaction $\qquad 2\,NO(g) + O_2(g) \rightarrow 2\,NO_2(g)$

We determined in Section 15.3 that this reaction is second order in NO and first order in O_2:

$$\text{Rate} = k[NO]^2[O_2] \qquad (15.38)$$

One proposed mechanism is shown in Figure 15.23. It has two elementary steps:

Step 1 $\quad NO(g) + O_2(g) \rightarrow NO_3(g) \qquad$ Rate $= k_1[NO][O_2] \qquad (15.39)$

Step 2 $\quad NO_3(g) + NO(g) \rightarrow 2\,NO_2(g) \quad$ Rate $= k_2[NO_3][NO] \qquad (15.40)$

The rate laws in Equations 15.39 and 15.40 are obtained by writing the concentration for each reactant raised to a power equal to the reactant's coefficient in the balanced equation. If step 1 were the rate-determining step, the reaction would be first order in NO and O_2, but that is not what the experimentally determined rate law indicates. If step 2 were the rate-determining step, the reaction would be first order in NO and NO_3, but that is not consistent with the rate law either. So, how can we account for the experimental rate law?

Consider what happens if step 2 is slow while step 1 is fast *and reversible*, which means NO_3 forms rapidly from NO and O_2 but decomposes just as rapidly back into NO and O_2. Expressing this equality in equation form:

$$\text{Rate of forward reaction} = k_f[NO][O_2] = \text{fast}$$

$$\text{Rate of reverse reaction} = k_r[NO_3] = \text{equally fast}$$

The subscripts "f" and "r" refer to the forward and reverse reactions, respectively.

Combining these two expressions, we have

$$k_f[NO][O_2] = k_r[NO_3]$$

$$[NO_3] = \frac{k_f}{k_r}[NO][O_2] \tag{15.41}$$

However, if we replace the $[NO_3]$ term in the rate law of Equation 15.40 (which we select because step 2 is the rate-determining step) with the right side of Equation 15.41, we get

$$\text{Rate} = k_2\frac{k_f}{k_r}[NO]^2[O_2]$$

The three rate constants can be combined,

$$k_{\text{overall}} = k_2\frac{k_f}{k_r}$$

and the rate law for the overall reaction becomes

$$\text{Rate} = k_{\text{overall}}[NO]^2[O_2]$$

This expression matches the overall rate law in Equation 15.38, so the proposed mechanism, a fast and reversible step 1 followed by a slow step 2, may be valid. As a final point regarding reaction mechanisms, even though the proposed reaction mechanism is consistent with the overall stoichiometry of the reaction and with the experimentally derived rate law, these consistencies do not *prove* that the proposed mechanism is correct. On the other hand, *not* finding a reactive (and transient) intermediate would not necessarily disprove a reaction mechanism.

CONCEPT TEST ....

Could the products of an elementary step in a chemical reaction include activated complexes?

....

SAMPLE EXERCISE 15.11 **Linking Reaction Mechanisms to Experimental Rate Laws**

The experimentally determined rate law for the reduction reaction

$$2\,NO(g) + 2\,H_2(g) \rightarrow N_2(g) + 2\,H_2O(g)$$

which occurs at high temperatures, is

$$\text{Rate} = k[NO]^2[H_2]$$

A proposed mechanism for the reaction is

Elementary step 1 $2\,NO(g) + H_2(g) \rightarrow N_2O(g) + H_2O(g)$

Elementary step 2 $N_2O(g) + H_2(g) \rightarrow N_2(g) + H_2O(g)$

Is this reaction mechanism consistent with the stoichiometry of the reaction and with the rate law? If so, which is the rate-determining step?

Collect and Organize We are first asked if the proposed mechanism is consistent with the reaction stoichiometry. We can use the two elementary steps to answer that question. We are then asked if the mechanism is consistent with the experimentally determined rate law. A rate-determining step has the smallest rate constant of all steps in the mechanism.

Analyze To answer the questions posed in this exercise we need to do the following:

1. Determine whether the chemical equations of the elementary steps add up to the overall reaction equation.
2. Write rate laws for each elementary step.
3. Compare the rate laws of the elementary steps to the experimental rate law of the overall reaction to assess the validity of the mechanism.
4. Determine which step is rate determining by matching its rate law to the observed rate law.

Solve Let's test whether the elementary steps add up to the overall reaction:

$$(1) \qquad 2\,NO(g) + H_2(g) \rightarrow \cancel{N_2O(g)} + H_2O(g)$$

$$(2) \qquad \underline{\cancel{N_2O(g)} + H_2(g) \rightarrow N_2(g) + H_2O(g)}$$

$$2\,NO(g) + 2\,H_2(g) \rightarrow N_2(g) + 2\,H_2O(g)$$

This is indeed the equation of the overall reaction. The elementary steps are consistent with the stoichiometry of the overall reaction.

Next we need to focus on the reaction mechanism. Elementary step 1 involves two molecules of NO colliding with one molecule of H_2 in a termolecular reaction. Elementary step 2 is a bimolecular reaction between the N_2O produced in step 1 and another molecule of H_2. We apply the fact that the rate law for any elementary step can be written directly from the balanced equations, using the equation coefficients as exponents in the rate law:

Step 1 $2\,NO(g) + H_2(g) \rightarrow N_2O(g) + H_2O(g)$ Rate $= k_1[H_2][NO]^2$

Step 2 $N_2O(g) + H_2(g) \rightarrow N_2(g) + H_2O(g)$ Rate $= k_2[H_2][N_2O]$

The rate law of step 1 matches the observed rate law of the overall reaction. Therefore, the proposed two-step mechanism is consistent with the experimental rate law, and step 1 is the rate-determining step.

Think about It We do not have direct proof that this is the correct mechanism, though it is consistent with the available data. A plausible mechanism for a reaction must yield a balanced equation whose stoichiometry matches the overall reaction and must be consistent with the rate law for the overall process. Both of these conditions are met in the solution to this Sample Exercise.

Practice Exercise The following mechanism is proposed for a reaction between compounds A and B:

Step 1	$2\,A(g) + B(g) \rightleftharpoons C(g)$	fast and reversible
Step 2	$B(g) + C(g) \rightarrow D(g)$	slow
Overall	$2\,A(g) + 2\,B(g) \rightarrow D(g)$	

What is the rate law for the overall reaction based on the proposed mechanism? ∎

SAMPLE EXERCISE 15.12 **Testing a Proposed Reaction Mechanism**

One proposed mechanism for the decomposition of N_2O_5 to NO_2, shown below, involves three elementary steps:

Step 1	$2\,N_2O_5(g) \rightleftharpoons N_4O_{10}(g)$	fast and reversible
Step 2	$N_4O_{10}(g) \rightarrow N_2O_3(g) + 2\,NO_2(g) + O_3(g)$	slow
Step 3	$N_2O_3(g) + O_3(g) \rightarrow 2\,NO_2(g) + O_2(g)$	fast
Overall	$2\,N_2O_5(g) \rightarrow 4\,NO_2(g) + O_2(g)$	

What is the rate law for the overall reaction based on the proposed mechanism?

Collect and Organize The rate law for the overall reaction reflects the rate laws for the elementary reactions preceding and including the rate-determining (slowest) step in the mechanism. We are given three elementary steps and their relative rates.

Analyze The rate law for an elementary step can be written directly from the balanced chemical equation for that step. We are told that step 1 is fast *and reversible*, which means that as the reaction proceeds and $[N_4O_{10}]$ increases, eventually the rate of step 1 in the forward direction is matched by the rate of step 1 in the reverse direction. Only elementary steps 1 and 2 will contribute to the rate law for the overall reaction since step 2 is the rate-determining step.

Solve The rate laws for step 1 in the forward and reverse directions are:

$$\text{Rate of forward step 1} = k_f[N_2O_5]^2$$
$$\text{Rate of reverse step 1} = k_r[N_4O_{10}]$$

The rate law for step 2 is

$$\text{Rate of step 2} = k_2[N_4O_{10}]$$

To find the overall rate law, we need to express $[N_4O_{10}]$ in terms of $[N_2O_5]$ by making the step 1 rates equal to each other and rearranging the terms:

$$[N_4O_{10}] = \frac{k_f}{k_r}[N_2O_5]^2$$

Substituting this expression for $[N_4O_{10}]$ into the rate law for step 2:

$$\text{Rate} = \frac{k_f k_2}{k_r}[N_2O_5]^2$$

Think about It The rate law for the overall reaction must depend on the concentration of N_2O_5. The order of the reaction in N_2O_5 depends on the proposed mechanism for the reaction. Notice that the rate of step 3 has no impact on the rate law for this mechanism because it follows the slowest step. In addition, note that the rate law for the mechanism in this Sample Exercise does not match the experimentally determined rate law in Sample Exercise 15.5. Therefore, the mechanism that starts with the dimerization of N_2O_5 cannot be correct.

Practice Exercise Here is another proposed mechanism for the reaction of NO with H_2 (Sample Exercise 15.11):

Elementary step 1	$H_2(g) + NO(g) \rightarrow N(g) + H_2O(g)$
Elementary step 2	$N(g) + NO(g) \rightarrow N_2(g) + O(g)$
Elementary step 3	$H_2(g) + O(g) \rightarrow H_2O(g)$

Is this a valid mechanism?

Mechanisms and Zero-Order Reactions

Before we leave reaction mechanisms, let's revisit the reaction between NO_2 and CO, which has an experimentally determined rate law that is zero order in CO, second order in NO_2, and second order overall:

$$NO_2(g) + CO(g) \rightarrow NO(g) + CO_2(g) \qquad Rate = k[NO_2]^2$$

What does this overall rate law tell us about the reaction? Remember that the overall rate depends on the concentrations of the reactants in the rate-determining step, which means that CO is not a reactant in the rate-determining step. Carbon monoxide is clearly involved in the reaction—it is converted into CO_2—but whatever step involves CO must not be the rate-determining step. This leads us to conclude that the reaction must have at least two elementary steps, one rate-determining and one not.

It has been proposed that the reaction mechanism is

(1)	$2\,NO_2(g) \rightarrow NO_3(g) + NO(g)$	$Rate = k_1[NO_2]^2$
(2)	$NO_3(g) + CO(g) \rightarrow NO_2(g) + CO_2(g)$	$Rate = k_2[NO_3][CO]$

The experimentally determined overall rate law matches the rate law for the first step, which must be the slower, rate-determining step. The overall reaction is zero order in CO because CO is not a reactant in that step.

CONCEPT TEST

The rate law for the reaction $XO_2(g) + M(g) \rightarrow XO(g) + MO(g)$, where X and M represent metallic elements, is second order in $[XO_2]$ and independent of $[M]$ for a wide variety of compounds XO_2 and M. Why can we conclude that these reactions likely proceed by the same mechanism?

15.6 Catalysis

We noted in Section 15.4 that spontaneous reactions may be slow if they have a high activation energy. Suppose we wanted to increase the rate of such a reaction. How could we do it? One way is to increase the temperature of the reaction mixture. However, in some chemical reactions, elevated temperatures can lead to undesired products or to lower yields. Another way is to add a **catalyst**, a substance that increases the rate of a reaction but is not consumed in the process.

catalyst a substance added to a reaction that increases the rate of the reaction but is not consumed in the process.

Catalysts and the Ozone Layer

As we discussed in Chapter 8, the way we think about ozone depends on where the ozone is located. Ozone in the stratosphere between 10 and 40 km above Earth's surface is necessary to protect us from UV radiation, but ozone at ground level is hazardous to our health. In this section, we discuss the role of catalysis in the loss of stratospheric ozone that has led to the annual formation of ozone holes over Antarctica.

The natural photodecomposition of ozone in the stratosphere

$$2\,O_3(g) \rightarrow 3\,O_2(g) \tag{15.42}$$

begins with the absorption of UV radiation from the sun and the generation of atomic oxygen:

$$(1) \qquad O_3(g) \rightarrow O_2(g) + O(g)$$

The oxygen atom may react with another ozone molecule to form two more molecules of oxygen:

$$(2) \qquad O_3(g) + O(g) \rightarrow 2\,O_2(g)$$

The rate of the second elementary step is low because its activation energy is relatively high: 17.7 kJ/mol.

In 1974, two American scientists, Sherwood Rowland and Mario Molina, predicted significant depletion of stratospheric ozone because of the release of a class of volatile compounds called chlorofluorocarbons (CFCs) into the atmosphere at ground level that ultimately enter the stratosphere. This prediction was later supported by experimental evidence of a thinning of the ozone layer—and formation of ozone holes—over Antarctica. By 2000, stratospheric ozone concentrations over Antarctica were less than half of what they were in 1980 (Figure 15.24), and the ozone hole covered nearly all of Antarctica and the tip of South America. Less severe thinning of stratospheric ozone was observed in the Northern Hemisphere.

FIGURE 15.24 A 30-year trend in stratospheric ozone depletion over the South Pole. The data points represent the minimum ozone concentration observed each year, expressed in Dobson units (DU). The colors of the six satellite images show the size of the holes in the ozone layer on a scale in which purple represents the most ozone depletion (see color chart). Values below 220 are considered dangerously low.

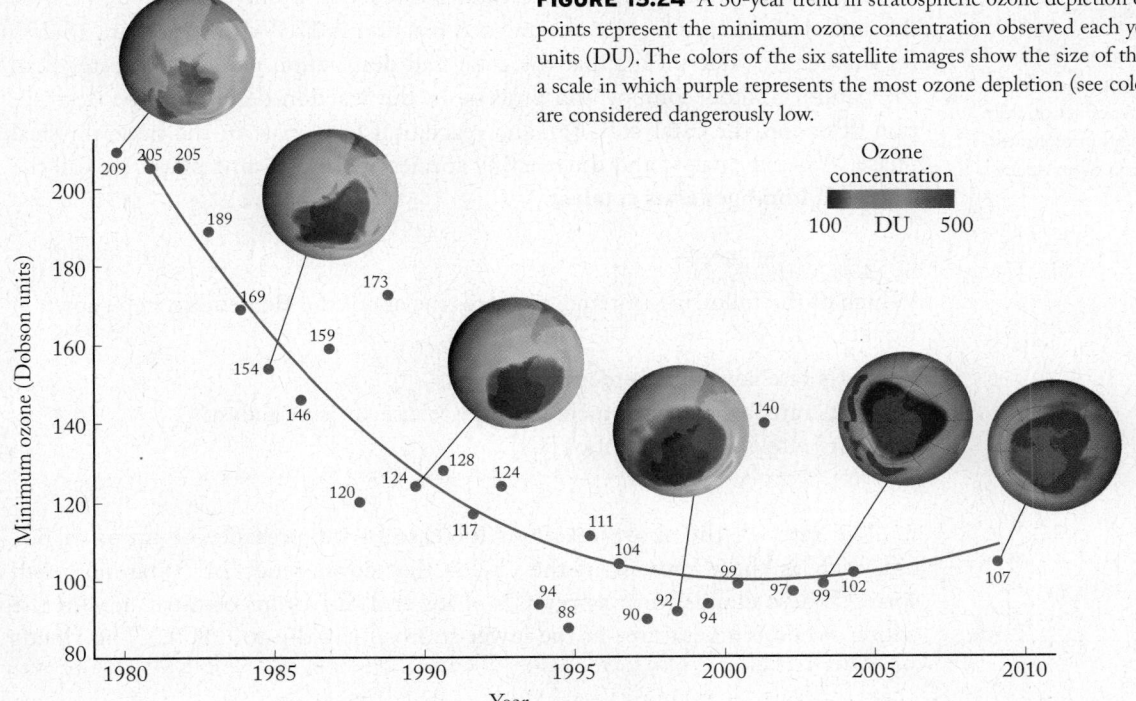

An international agreement known as the Montreal Protocol called for an end to the production of ozone-depleting CFCs. The Montreal Protocol has had a dramatic effect on CFC production and emission into the atmosphere. However, these compounds last for many years in the atmosphere, and recovery of the ozone layer may take most of the 21st century.

How do CFCs contribute to the destruction of ozone? Three of the more widely used CFCs were CCl_2F_2, CCl_3F, and $CClF_3$. In the stratosphere, they encounter UV radiation with enough energy to break C—Cl bonds, releasing chlorine atoms.

$$CCl_3F(g) \xrightarrow{h\nu} CCl_2F(g) + Cl(g) \qquad (15.43)$$

Free chlorine atoms react with ozone, forming chlorine monoxide:

$$Cl(g) + O_3(g) \rightarrow ClO(g) + O_2(g) \qquad (15.44)$$

Chlorine monoxide then reacts with ozone producing oxygen and regenerating atomic chlorine:

$$ClO(g) + O_3(g) \rightarrow Cl(g) + 2 O_2(g) \qquad (15.45)$$

If we add Equations 15.44 and 15.45 and cancel species as needed, we get

$$2 O_3(g) \rightarrow 3 O_2(g)$$

The overall reaction is exactly the same as the natural photodecomposition of ozone. The difference is the presence of chlorine atoms in Equations 15.44 and 15.45, which dramatically increase the rate of the overall reaction. Chlorine atoms act as a catalyst for the destruction of ozone because they speed up the reaction but are not consumed by it. Rather, they are consumed in one elementary step but then regenerated in a later elementary step. Chlorine is a catalyst, not an intermediate, because in a reaction mechanism, a catalyst is consumed in an early step and then regenerated in a later one, whereas an intermediate is produced before it is consumed. A single chlorine atom can catalyze the destruction of hundreds to thousands of stratospheric O_3 molecules before it combines with other atoms and forms a less reactive molecule.

The activation energy for the Cl-catalyzed reaction is only 2.2 kJ/mol, whereas the activation energy for the uncatalyzed reaction is 17.7 kJ/mol (Figure 15.25). Its smaller E_a value means that the catalyzed destruction of ozone is faster than the natural photodecomposition process. In the reaction describing the destruction of ozone, the catalyst, $Cl(g)$, and reactant, $O_3(g)$, exist in the same physical phase. When a catalyst and the reacting species are in the same phase, we call the catalyst a **homogeneous catalyst**.

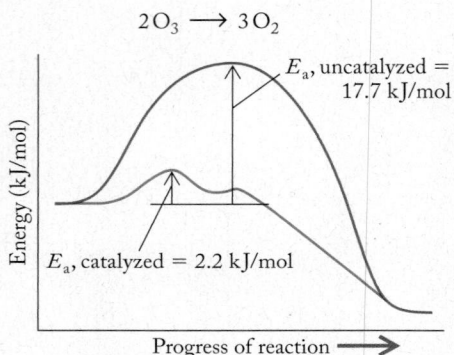

FIGURE 15.25 The decomposition of O_3 in the presence of chlorine atoms has a smaller activation energy (2.2 kJ/mol) than the naturally occurring photodecomposition of O_3 to O_2 (17.7 kJ/mol). The catalytic effect of chlorine is a key factor in the depletion of stratospheric ozone and the formation of an ozone hole over the South Pole.

CONCEPT TEST

Which of the following statements is/are true about the elementary step shown in Equation 15.44?

a. Its rate law is first order in [Cl].
b. Its rate law does not include [Cl] because Cl is a catalyst.
c. Its rate is independent of [Cl].

The rates of the above reactions increase in the presence of drops of liquid, such as those present in the clouds that cover much of Antarctica each spring. These clouds form as crystals of ice and tiny drops of nitric acid in the winter, when temperatures in the lower stratosphere dip to −80°C. The clouds

homogeneous catalyst a catalyst in the same phase as the reactants.

become collection sites for HCl, ClO, and other compounds containing chlorine. In August (the end of the Antarctic winter), the ice in the clouds melts and ClO is free to react on the surface of the drops of liquid water. These drops catalyze reactions by *adsorbing* (binding to the surface) the reactants. The drops are not reactants, but they provide a surface on which the reactants collect. The resulting proximity of the reactants increases the likelihood of reaction and, coupled with a decrease in activation energy, increases the reaction rate. When the catalyst is a liquid drop on which gas molecules adsorb, the drop is in a different phase than the reacting species and is called a **heterogeneous catalyst**. Both homogeneous and heterogeneous catalysts play a role in the reactions that diminish the amount of ozone in the stratosphere.

heterogeneous catalyst a catalyst in a different phase than the reactants.

CONCEPT TEST

Is the ClO produced in Equation 15.44 a catalyst or an intermediate?

Evidence supporting the occurrence of these reactions in the stratosphere is presented in Figure 15.26. The concentration of ClO decreases sharply over the South Pole at the end of the Antarctic winter as shown in satellite images taken one month apart, consistent with the reaction between ClO and O_3 shown in Equation 15.45. Although it is difficult to track the concentration of Cl atoms directly, some of them end up in molecules of HCl present in the icy polar clouds that form during the winter and melt in August. The images in Figure 15.26 show that a decrease in ClO concentration between late August and the Antarctic spring (late September) is accompanied by an increase in HCl concentration.

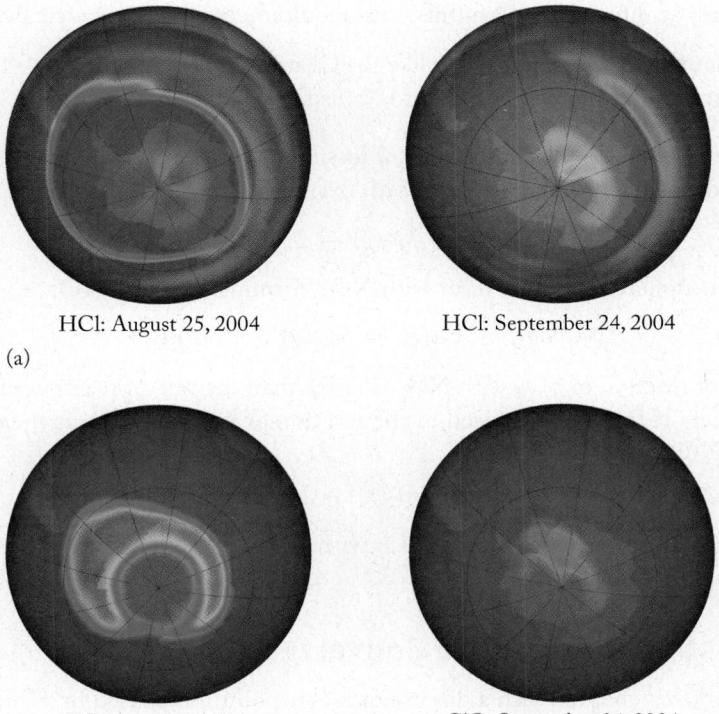

HCl: August 25, 2004   HCl: September 24, 2004

(a)

ClO: August 25, 2004   ClO: September 24, 2004

(b)

FIGURE 15.26 (a) NASA satellite images of stratospheric concentrations of hydrogen chloride in August and September 2004 over the South Pole. Blue color at the pole indicates low levels in August; red color at the pole indicates high levels in September. (b) Stratospheric concentrations of chlorine monoxide in August and September 2004. Red indicates high levels in August and blue indicates low levels in September. The shifts in the concentrations of the two compounds are due in part to the reaction $OH + ClO \rightarrow HCl + O_2$.

> ### SAMPLE EXERCISE 15.13 — Identifying Catalysts in Reaction Mechanisms
>
> A reaction mechanism proposed for the decomposition of ozone in the presence of NO at high temperatures consists of three elementary steps:
>
> $$\text{(1)} \qquad O_3(g) + NO(g) \rightarrow O_2(g) + NO_2(g)$$
> $$\text{(2)} \qquad NO_2(g) \rightarrow NO(g) + O(g)$$
> $$\text{(3)} \qquad O(g) + O_3(g) \rightarrow 2\,O_2(g)$$
>
> If the rate of the overall reaction is higher than the rate of the uncatalyzed decomposition of ozone to oxygen (Equation 15.42), is NO a catalyst in the reaction?
>
> **Collect and Organize** We are asked to determine whether NO is a catalyst in a reaction. A catalyst increases the rate of a reaction and is not consumed by the overall reaction. We are given the elementary steps of the reaction and are told that the reaction is more rapid in the presence of NO.
>
> **Analyze** We can sum the reactions to determine the overall reaction. If NO is consumed in an early step before it is regenerated in a later one and not consumed in the overall process, it is a catalyst.
>
> **Solve** Summing the three elementary steps:
>
> $$O_3(g) + \cancel{NO(g)} + \cancel{NO_2(g)} + \cancel{O(g)} + O_3(g) \rightarrow$$
> $$O_2(g) + \cancel{NO_2(g)} + \cancel{NO(g)} + \cancel{O(g)} + 2\,O_2(g)$$
>
> gives the overall reaction:
>
> $$2\,O_3(g) \rightarrow 3\,O_2(g)$$
>
> This equation does not include NO, and the rate of the reaction is higher when NO is present. Thus the NO fulfills both requirements for being a catalyst.
>
> **Think about It** NO behaves much like the Cl atoms in Equations 15.44–15.45. NO is not an intermediate because it is consumed before it is produced.
>
> **Practice Exercise** The combustion of fossil fuels results in the release of SO_2 into the atmosphere, where it reacts with oxygen to form SO_3:
>
> $$2\,SO_2(g) + O_2(g) \rightarrow 2\,SO_3(g)$$
>
> In the atmosphere, SO_2 may react with NO_2, forming SO_3 and NO:
>
> $$NO_2(g) + SO_2(g) \rightarrow NO(g) + SO_3(g)$$
>
> The rate of reaction of SO_2 with NO_2 is faster than the rate of reaction of SO_2 with oxygen. If the NO produced in the reaction of NO_2 and SO_2 is then oxidized to NO_2:
>
> $$2\,NO(g) + O_2(g) \rightarrow 2\,NO_2(g)$$
>
> is NO_2 a catalyst in the reaction of SO_2 with O_2?

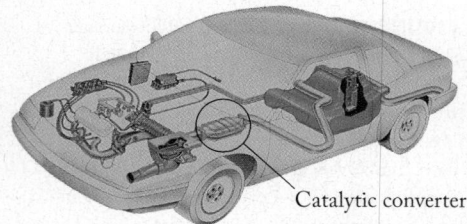

Catalytic converter

FIGURE 15.27 In an automobile the converter is located close to the engine because it works best at high temperatures before the exhaust gases have a chance to cool.

Catalysts and Catalytic Converters

We started this chapter with a discussion of air pollution caused by vehicles and of the technology that has been developed to clean the air. Figures 15.27 and 15.28 show a catalytic converter in a car's exhaust system and the reactions that take place in the converter to remove one representative pollutant, NO, from the engine exhaust. Hot exhaust gases flowing through the converter pass through a fine honeycomb mesh coated with one or more of the transition metals palladium, platinum, and rhodium. These metals are the catalysts, and they have two roles: (1) to speed up oxidation of carbon monoxide to CO_2 and of unburned hydrocarbons to CO_2 and water vapor, and (2) to convert NO and NO_2 into N_2 and O_2.

The metals used in making catalytic converters are dissolved as metal salts and dispersed on the mesh, and are then reduced to clusters 2 to 10 nm in diameter. The large surface area of the metal clusters provides sites where the oxidation and reduction of the gases take place.

Catalysts not only speed up reactions, but also allow them to take place at a lower temperatures. For example, CO reacts rapidly with O_2 above 700°C, but the presence of a Pd or Pt/Rh catalyst enables this reaction to take place rapidly at the much lower temperature of automobile exhaust, about 250°C. The catalysts have similar effects on the reduction reactions taking place in the converters.

The catalysts are selective in terms of the molecules they interact with and specific in the reactions they promote. What makes the catalysts selective is that several reactions are possible for each pollutant but one reaction proceeds more rapidly than the others. For example, the preferred reduction of the nitrogen in NO is to N_2 rather than to N_2O or to NH_3, and carbon monoxide and hydrocarbons are potential reducing agents for this reaction. Recall, however, that oxidizing CO and hydrocarbons is one of the two primary goals of a catalytic converter. If these compounds are oxidized, no NO reduction can take place, and the converter only does half its job. Fortunately the reduction of NO by CO and hydrocarbons is much faster than the reactions of CO and hydrocarbons with O_2. As a result, NO reduction is essentially complete before any of the necessary reducing agents are consumed by reaction with oxygen.

In this chapter we have explored the connection between the spontaneity of chemical reactions introduced in Chapter 14 and the rates of chemical reactions. As we have seen with those reactions responsible for the production of photochemical smog, determinations of the rates of chemical reactions are crucial for us to understand processes on a molecular level. The mechanisms of the reactions of volatile oxides produced during combustion and by other natural events must be thoroughly understood so that problems arising because of the presence of these substances in the environment can be effectively managed. The continued development of catalytic converters for vehicles to diminish the problems caused by burning fossil fuels and to clean our air requires a thorough knowledge of the kinetics and mechanisms of many of the reactions discussed in this chapter.

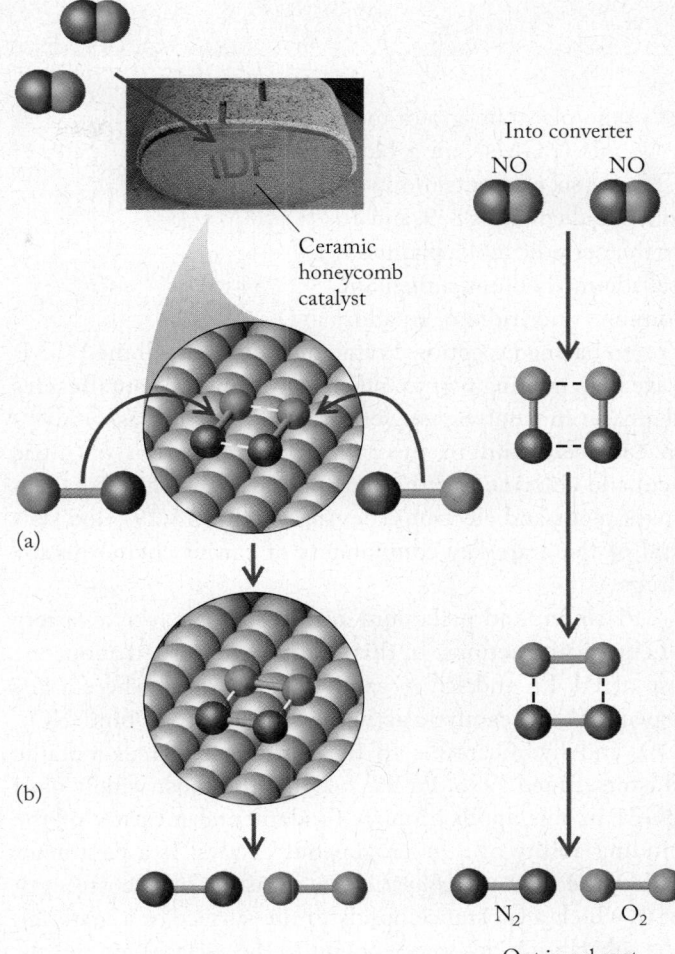

FIGURE 15.28 Catalytic converters in automobiles reduce emissions of NO by lowering the activation energy of its decomposition into N_2 and O_2. Metal catalysts are supported on a porous ceramic honeycomb. (a) NO molecules are adsorbed onto the surface of metal clusters where their NO bonds are broken, and (b) pairs of O atoms and N atoms form O_2 and N_2. The O_2 and N_2 desorb from the surface and are released to the atmosphere.

The Platinum Group: Catalysts, Jewelry, and Investment

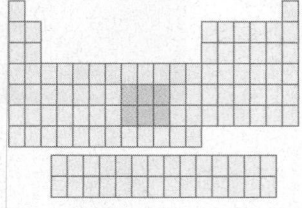

The platinum group metals (PGMs) are a block of six elements in rows 5 and 6 and columns 8, 9, and 10 of the periodic table: platinum, palladium, rhodium, ruthenium, osmium, and iridium. In addition to having exceptional catalytic properties, all the PGMs have high melting points and are corrosion resistant. Besides their use in catalytic converters, they function as catalysts in fuel cells and in the industrial production of nitric acid and chlorine gas; they are alloyed with other metals in spark plugs and electronic devices (Figure 15.29); and several of them are key components of cancer chemotherapy drugs.

Platinum and palladium are used to catalyze a variety of chemical reactions. In this chapter, we learned about the use of Pd, Pt, and Rh in catalytic converters, where a key aspect of their catalytic activity is the ability to bind NO_x, CO, and hydrocarbons to the surfaces of small metallic clusters. Since 1996, Pd has become the most widely used PGM in this application. A fundamental measure of the binding ability of a heterogeneous catalyst is a parameter called the *heat of adsorption,* a measure of the strength with which a substance bonds to the surface of a material. If a substance is adsorbed (bound to the surface) too tightly, it cannot react with other adsorbed species. If it is bonded too weakly, it will desorb (move away from the surface) before it has a chance to react. Platinum and palladium operate between these two extremes, binding a wide range of substances with moderate strength. Both metals bind many gases with a moderate heat of adsorption, and this is the key to their catalytic activity.

The technological importance of PGMs extends far beyond their roles as heterogeneous catalysts of gas-phase reactions. They are very resistant to corrosion by acids, alkalis, and salts; they are stable at high temperatures; and they do not oxidize readily. As a result, they are frequently used as linings or coatings to protect other metals. The rigidity and hardness of individual PGMs can be improved by alloying them with other metals in the PGM family. Platinum is frequently alloyed with Rh, Ir, and Ru to produce hard, highly corrosion-resistant solids.

The PGMs are rare, so their heavy use in technologically important applications like automotive catalytic converters has led to recycling efforts to recover the metals. In the United States the most important source of Pt, Pd, and Rh is their collection from scrapped catalytic converters. It can be challenging to recover PGMs from ceramic support structures. One technique involves the selective dissolution of the metals by treatment with *aqua regia* ("water of kings"), a mixture of hydrochloric and nitric acids so named because it is capable of dissolving gold.

All PGMs are used in medical devices and implants. Because of their lack of reactivity and tolerance for solutions that corrode most metals, they are highly biocompatible. Platinum alloyed with iridium is used in the electrodes of pacemakers and defibrillators and also in the tips of the guide wires surgeons use to direct the placement of catheters. One mode of cancer therapy uses radioactive ^{192}Ir wire wrapped in platinum as an implant to deliver radiation doses in the body.

Platinum is the PGM most commonly used for jewelry. In Japan it is actually preferred over gold in decorative items (Figure 15.30). As we saw with gold, pure platinum is too soft for jewelry, so it is usually alloyed with iridium or ruthenium to increase its wear resistance. Palladium is used in small quantities as a whitener in gold jewelry. Pure platinum is also made into coins and ingots and has considerable value as an investment.

FIGURE 15.29 Discs for computer drives are coated with platinum.

FIGURE 15.30 This platinum model of the Japanese robot Gundam is 12.5 cm tall and is valued at $250,000.

SUMMARY

Section 15.1 The reactions that follow the formation of NO in vehicle engines can produce **photochemical smog**, a major air pollutant in urban areas. Familiarity with **chemical kinetics**, the study of **reaction rates**, is important in understanding how natural processes and those caused by human activity occur.

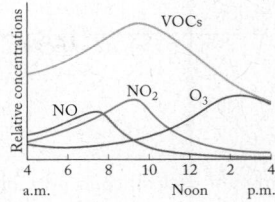

Section 15.2 The overall rate of a reaction, which is proportional to the rate of change in the concentrations of reactants or products, can be expressed as either an average rate or an **instantaneous rate**. The relative rates of disappearance of reactants and appearance of products are related by the stoichiometry of the reaction and are typically expressed in molarity per unit time. Reaction rates are typically determined from experimental measurements. In most reactions, the reaction rate decreases as the reaction proceeds.

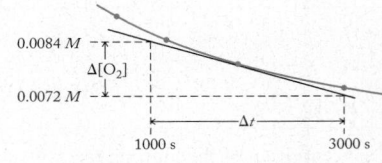

Section 15.3 The dependence of the rate of a reaction A + B → C on reactant concentrations is expressed in the **rate law** for the reaction: Rate = $k[A]^m[B]^n$, where m and n are the **reaction order** with respect to reactants A and B, respectively, and k is the **rate constant**. The units of a rate constant depend on the **overall reaction order**, which is the sum of the reaction orders with respect to individual reactants. The order of a reaction and the rate law for the reaction can be determined from differences in the **initial rates** of reaction observed with different concentrations of reactants or derived from the results of single kinetics experiments using **integrated rate laws**. The **half-life** of a reaction is the time required for the concentration of a reactant to decrease to one-half its starting concentration.

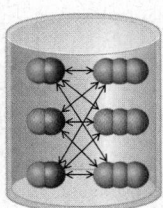

Section 15.4 Increasing the temperature of a chemical reaction increases its rate. The **activation energy** of a reaction is a barrier that separates the sum of the internal energies of the reactants from the energies of the products. The top of the energy barrier is the **transition state** related to the internal energy of a short-lived **activated complex**. Reactions with large activation energies are usually slow. Measuring the rate constant of a reaction at different temperatures allows the calculation of its activation energy using the **Arrhenius equation**.

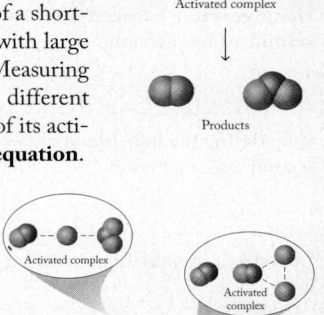

Section 15.5 Rate studies give insight into **reaction mechanisms**, which describe what is happening at a molecular level. A reaction mechanism consists of one or more **elementary steps** that describe how the reaction takes place on a molecular level. The overall reaction is the sum of these elementary steps. Elementary steps that involve one, two, or three molecules are said to be **unimolecular**, **bimolecular**, and **termolecular**, respectively. The rate law for a reaction applies to the slowest elementary step, which is called the **rate-determining step**. The proposed mechanism for any reaction must be consistent with the observed rate law and with the stoichiometry of the overall reaction.

Section 15.6 **Catalysts** increase reaction rates by changing the mechanism of a reaction and decreasing activation energies. A **homogeneous catalyst** is one that is in the same phase as the reactants in the reaction being catalyzed. A **heterogeneous catalyst** is one that is in a phase different from the phase of the reactants.

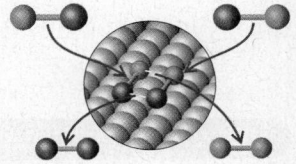

PROBLEM-SOLVING SUMMARY

TYPE OF PROBLEM	CONCEPTS AND EQUATIONS	SAMPLE EXERCISES
Predicting a relative reaction rate	A relative rate is determined from the stoichiometry of the balanced chemical equation.	15.1
Converting reaction rates	Use the balanced chemical equation and the known consumption/formation rate to determine the consumption/formation rate of another reaction participant.	15.2
Determining an instantaneous reaction rate	Determine the slope of a line tangent to a point on the plot of concentration versus time.	15.3
Deriving a rate law from initial rate data	Compare the change in rate when the concentration of one reactant is changed (while the concentrations of other reactants are kept constant) to determine the reaction order (usually whole numbers) with respect to that reactant: $$n = \frac{\log\left(\dfrac{\text{Rate}_1}{\text{Rate}_2}\right)}{\log\left(\dfrac{[X]_1}{[X]_2}\right)} \quad (15.17)$$	15.4

TYPE OF PROBLEM	CONCEPTS AND EQUATIONS	SAMPLE EXERCISES
Using an integrated rate law	A linear plot of the natural logarithm of concentration versus time indicates a first-order reaction with a slope of $-k$.	15.5
Calculating the half-life of a first-order reaction	In a first-order reaction, $$t_{1/2} = \frac{0.693}{k} \qquad (15.23)$$	15.6
Distinguishing between first- and second-order reactions	A linear plot of the natural logarithm of reactant concentration versus time indicates a first-order reaction, whereas a linear plot of the reciprocal of reactant concentration $1/[X]$ versus time indicates a second-order reaction.	15.7
Calculating the half-life of a second-order reaction	In a second-order reaction, $$t_{1/2} = \frac{1}{k[X]_0} \qquad (15.28)$$	15.8
Deriving a pseudo-first-order rate law	A plot of the natural logarithm of concentration of the limiting reactant versus time is linear. The slope of the plot gives $k[\text{excess reactant}] = k'$.	15.9
Calculating an activation energy from rate constants	Using the logarithmic form of the Arrhenius equation, $$\ln k = -\frac{E_a}{R}\left(\frac{1}{T}\right) + \ln A \qquad (15.33)$$ plot $\ln k$ versus $1/T$. The slope is $-E_a/R$.	15.10
Linking reaction mechanisms to experimental rate laws	The order of each reactant in an elementary step equals its coefficient in that step. The rate law for the mechanism must be the same as the observed rate law and does not include intermediates.	15.11, 15.12
Identifying catalysts in reaction mechanisms	Determine whether or not a potential catalyst is present by summing the elementary-step reactions to get the overall reaction. If that procedure reveals a potential catalyst, determine whether it increases the rate of reaction and whether it is initially consumed and then regenerated in the process.	15.13

VISUAL PROBLEMS ■

(Answers to boldface end-of-chapter questions and problems are in the back of the book.)

15.1. Nitrous oxide decomposes to nitrogen and oxygen in the following reaction:

$$2\,N_2O(g) \rightarrow 2\,N_2(g) + O_2(g)$$

In Figure P15.1, which curve represents $[N_2O]$ and which curve represents $[O_2]$?

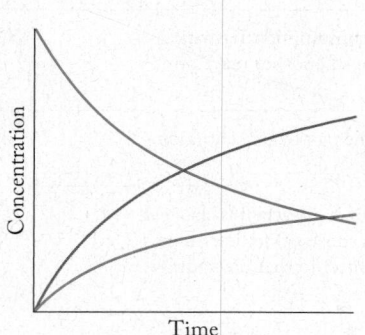

FIGURE P15.1

15.2. Sulfur trioxide is formed in the reaction

$$SO_2(g) + \tfrac{1}{2}O_2(g) \rightarrow SO_3(g)$$

In Figure P15.2, which curve represents $[SO_2]$ and which curve represents $[O_2]$? All three gases are present initially.

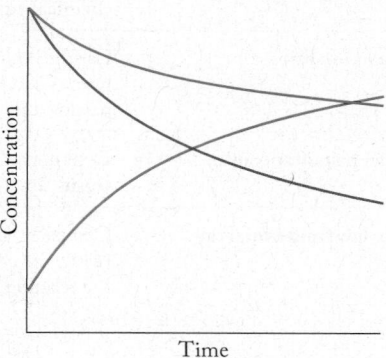

FIGURE P15.2

15.3. The rate law for the reaction $2A \rightarrow B$ is second order in A. Figure P15.3 represents samples with different concentrations of A; the red spheres represent molecules of A. In which sample will the reaction $A \rightarrow B$ proceed most rapidly?

(a) 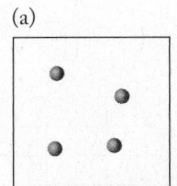 (b) 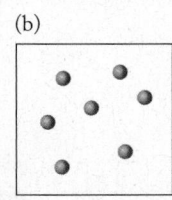 (c)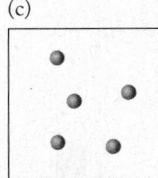

FIGURE P15.3

15.4. The rate law for the reaction $A + B \rightarrow C$ is first order in both A and B. Figure P15.4 represents samples with different concentrations of A (red spheres) and B (blue spheres). In which sample will the reaction $A + B \rightarrow C$ proceed most rapidly?

(a) 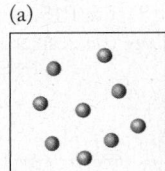 (b) 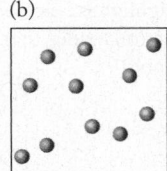 (c)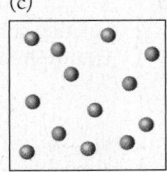

FIGURE P15.4

15.5. Which of the reaction profiles in Figure P15.5 represents the slowest reaction?

(a) 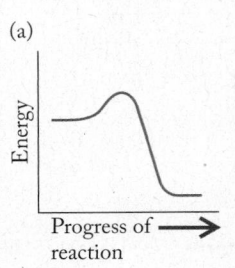 (b) 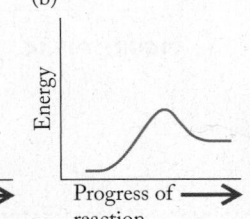 (c)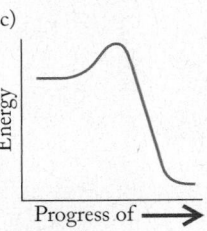

FIGURE P15.5

15.6. Which of the reaction profiles in Figure P15.6 represents the fastest reaction?

(a) 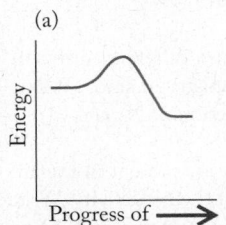 (b) 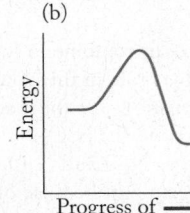 (c)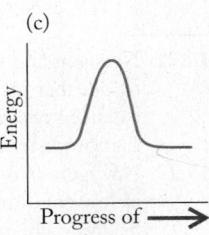

FIGURE P15.6

15.7. Which of the following mechanisms is consistent with the reaction profile shown in Figure P15.7?

a. $2A \xrightarrow{\text{slow}} B$
 $B \xrightarrow{\text{fast}} C$
b. $A + B \rightarrow C$
c. $2A \underset{}{\overset{\text{fast}}{\rightleftharpoons}} B$
 $B \xrightarrow{\text{slow}} C$

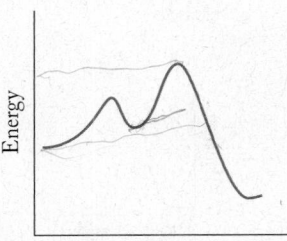

FIGURE P15.7

15.8. Which of the following mechanisms is consistent with the reaction profile shown in Figure P15.8?

a. $A + B \xrightarrow{\text{slow}} C$
 $C \xrightarrow{\text{fast}} D$
b. $A + B \rightarrow C$
c. $2A \xrightarrow{\text{fast}} B$
 $B + C \xrightarrow{\text{slow}} D$

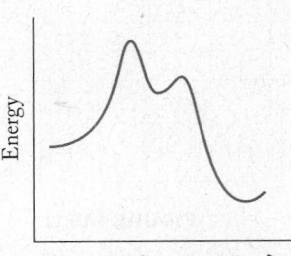

FIGURE P15.8

15.9. Which of the reaction profiles in Figure P15.9 represents the effect of a catalyst on the rate of a reaction?

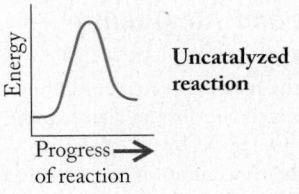

(a) 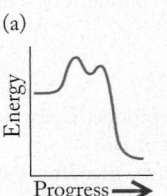 (b) 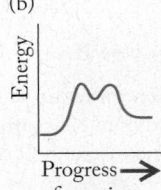 (c)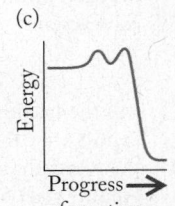

FIGURE P15.9

*15.10. Which of the reaction profiles in Figure P15.10 represents the effect of a catalyst on the rate of a reaction?

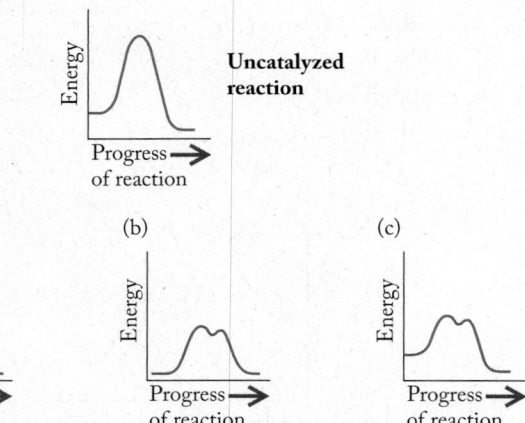

FIGURE P15.10

15.11. Which of the highlighted elements in Figure P15.11 forms volatile oxides associated with photochemical smog formation?

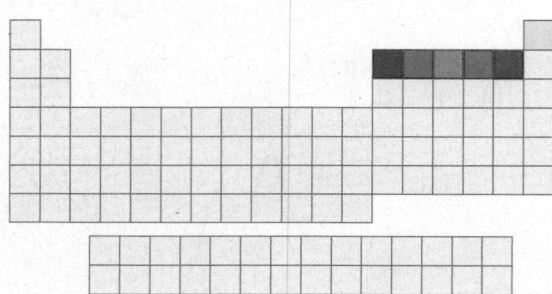

FIGURE P15.11

15.12. Which of the highlighted elements in Figure P15.11 forms noxious oxides that are removed from automobile exhaust as it passes through a catalytic converter?

15.13. Which of the highlighted elements in Figure P15.13 are widely used as heterogeneous catalysts?

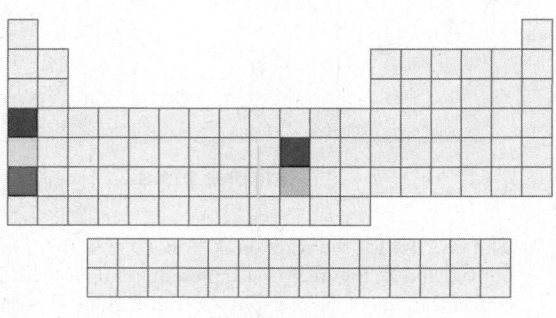

FIGURE P15.13

15.14. Which of the highlighted elements in Figure P15.14 forms volatile, odd-electron oxides that catalyze the destruction of stratospheric ozone?

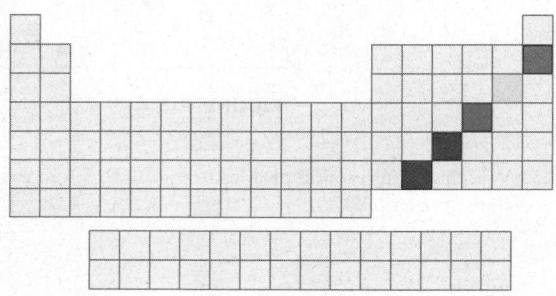

FIGURE P15.14

QUESTIONS AND PROBLEMS

Cars, Trucks, and Air Quality

CONCEPT REVIEW

15.15. Why does the maximum concentration of ozone in Figure 15.2 occur much later in the day than the maximum concentration of NO and NO_2?

15.16. If we plot the concentration of reactants and products as a function of time for any sequence of two spontaneous chemical reactions, such as

$$A \rightarrow B \rightarrow C$$

will the maximum concentration of final product C always appear after the maximum concentration of B?

15.17. Why isn't there an increase in NO concentration after the evening rush hour?

15.18. If ozone can react with NO to form NO_2, why does the ozone concentration reach a maximum in the early afternoon?

PROBLEMS

15.19. Using data in Appendix 4, calculate ΔH° for the reaction

$$2\,NO(g) + O_2(g) \rightarrow 2\,NO_2(g)$$

15.20. Using data in Appendix 4, calculate ΔH° for the reaction

$$O_3(g) + NO(g) \rightarrow O_2(g) + NO_2(g)$$

15.21. Nitrogen and oxygen can combine to form different nitrogen oxides that play a minor role in the chemistry of smog. Write balanced chemical equations for the reaction of N_2 and O_2 that produce (a) N_2O and (b) N_2O_5.

15.22. Nitrogen oxides such as N_2O and N_2O_5 are present in the air in low concentrations, in part because of their reactivity. Write balanced chemical equations for (a) the conversion of N_2O to NO_2 in the presence of oxygen and (b) the decomposition of N_2O_5 to NO_2 and O_2.

Reaction Rates

CONCEPT REVIEW

15.23. Explain the difference between the average rate and the instantaneous rate of a chemical reaction.

15.24. Can the average rate and instantaneous rate of a chemical reaction ever be the same?

15.25. Why do the average rates of most reactions change with time?

15.26. Does the instantaneous rate of a chemical reaction change with time?

PROBLEMS

15.27. Bacterial Degradation of Ammonia *Nitrosomonas* bacteria convert ammonia into nitrite in the presence of oxygen by the following reaction:

$$2\,NH_3(aq) + 3\,O_2(g) \rightarrow 2\,H^+(aq) + 2\,NO_2^-(aq) + 2\,H_2O(\ell)$$

a. How are the rates of formation of H^+ and NO_2^- related to the rate of consumption of NH_3?

b. How is the rate of formation of NO_2^- related to the rate of consumption of O_2?

c. How is the rate of consumption of NH_3 related to the rate of consumption of O_2?

15.28. Catalytic Converters and Combustion Catalytic converters in automobiles combat air pollution by converting NO and CO into N_2 and CO_2:

$$2\,CO(g) + 2\,NO(g) \rightarrow N_2(g) + 2\,CO_2(g)$$

a. How is the rate of formation of N_2 related to the rate of consumption of CO?

b. How is the rate of formation of CO_2 related to the rate of consumption of NO?

c. How is the rate of consumption of CO related to the rate of consumption of NO?

15.29. Write expressions for the rate of formation of products and the rate of consumption of reactants in each of the following reactions:

a. $H_2O_2(g) \rightarrow 2\,OH(g)$
b. $ClO(g) + O_2(g) \rightarrow ClO_3(g)$
c. $N_2O_5(g) + H_2O(g) \rightarrow 2\,HNO_3(g)$

15.30. Write expressions for the rate of formation of products and the rate of consumption of reactants in each of the following reactions:

a. $Cl_2O_2(g) \rightarrow 2\,ClO(g)$
b. $N_2O_5(g) \rightarrow NO_2(g) + NO_3(g)$
c. $2\,INO(g) \rightarrow I_2(g) + 2\,NO(g)$

15.31. Power-Plant Emissions Sulfur dioxide emissions in power-plant stack gases may react with carbon monoxide as follows:

$$SO_2(g) + 3\,CO(g) \rightarrow 2\,CO_2(g) + COS(g)$$

Write an equation relating the rates for each of the following:

a. The rate of formation of CO_2 to the rate of consumption of CO

b. The rate of formation of COS to the rate of consumption of SO_2

c. The rate of consumption of CO to the rate of consumption of SO_2

15.32. Reducing Nitric Oxide Emissions from Power Plants Nitric oxide (NO) can be removed from gas-fired power-plant emissions by reaction with methane as follows:

$$CH_4(g) + 4\,NO(g) \rightarrow 2\,N_2(g) + CO_2(g) + 2\,H_2O(g)$$

Write an equation relating the rates for each of the following:

a. The rate of formation of N_2 to the rate of formation of CO_2

b. The rate of formation of CO_2 to the rate of consumption of NO

c. The rate of consumption of CH_4 to the rate of formation of H_2O

15.33. Stratospheric Ozone Depletion Chlorine monoxide (ClO) plays a major role in the creation of the ozone holes in the stratosphere over Earth's polar regions.

a. If $\Delta[ClO]/\Delta t$ at 298 K is -2.3×10^7 M/s, what is the rate of change in $[Cl_2]$ and $[O_2]$ in the following reaction?

$$2\,ClO(g) \rightarrow Cl_2(g) + O_2(g)$$

b. If $\Delta[ClO]/\Delta t$ is -2.9×10^4 M/s, what is the rate of formation of oxygen and ClO_2 in the following reaction?

$$ClO(g) + O_3(g) \rightarrow O_2(g) + ClO_2(g)$$

15.34. The chemistry of smog formation includes NO_3 as an intermediate in several reactions.

a. If $\Delta[NO_3]/\Delta t$ is -2.2×10^5 mM/min in the following reaction, what is the rate of formation of NO_2?

$$NO_3(g) + NO(g) \rightarrow 2\,NO_2(g)$$

b. What is the rate of change of $[NO_2]$ in the following reaction if $\Delta[NO_3]/\Delta t$ is -2.3 mM/min?

$$2\,NO_3(g) \rightarrow 2\,NO_2(g) + O_2(g)$$

15.35. Nitrite ion reacts with ozone in aqueous solution, producing nitrate ion and oxygen:

$$NO_2^-(aq) + O_3(g) \rightarrow NO_3^-(aq) + O_2(g)$$

The following data were collected for this reaction at 298 K. Calculate the average reaction rate between 0 and 100 μs (microseconds) and between 200 and 300 μs.

Time (μs)	$[O_3]$ (M)
0	1.13×10^{-2}
100	9.93×10^{-3}
200	8.70×10^{-3}
300	8.15×10^{-3}

15.36. Dinitrogen pentoxide (N_2O_5) decomposes as follows to nitrogen dioxide and nitrogen trioxide:

$$N_2O_5(g) \rightarrow NO_2(g) + NO_3(g)$$

Calculate the average rate of this reaction between consecutive measurement times in the following table.

Time (s)	$[N_2O_5]$ (molecules/cm^3)
0	1.500×10^{12}
1.45	1.357×10^{12}
2.90	1.228×10^{12}
4.35	1.111×10^{12}
5.80	1.005×10^{12}

15.37. The following data were collected for the dimerization of ClO to Cl_2O_2 at 298 K.

Time (s)	[ClO] (molecules/cm³)
0	2.60×10^{11}
1	1.08×10^{11}
2	6.83×10^{10}
3	4.99×10^{10}
4	3.93×10^{10}
5	3.24×10^{10}
6	2.76×10^{10}

Plot [ClO] and $[Cl_2O_2]$ as a function of time and determine the instantaneous rates of change in both at 1 s.

15.38. **Tropospheric Ozone** Tropospheric (lower atmosphere) ozone is rapidly consumed in many reactions, including

$$O_3(g) + NO(g) \rightarrow NO_2(g) + O_2(g)$$

Use the following data to calculate the instantaneous rate of the preceding reaction at $t = 0.000$ s and $t = 0.052$ s.

Time (s)	[NO] (*M*)
0.000	2.0×10^{-8}
0.011	1.8×10^{-8}
0.027	1.6×10^{-8}
0.052	1.4×10^{-8}
0.102	1.2×10^{-8}

Effect of Concentration on Reaction Rate

CONCEPT REVIEW

15.39. Can two different chemical reactions have the same rate law?

15.40. Why are the units of the rate constants different for reactions of different order?

15.41. Does the half-life of a second-order reaction have the same units as the half-life for a first-order reaction?

15.42. Does the half-life of a first-order reaction depend on the concentration of the reactants?

15.43. What effect does doubling the initial concentration of a reactant have on the half-life of a reaction that is second order in the reactant?

15.44. Two first-order decomposition reactions of the form $A \rightarrow B + C$ have the same rate constant at a given temperature. Do the reactants in the two reactions have the same half-lives at this temperature?

PROBLEMS

15.45. For each of the following rate laws, determine the order with respect to each reactant and the overall reaction order.
  a. Rate = $k[A][B]$
  b. Rate = $k[A]^2[B]$
  c. Rate = $k[A][B]^3$

15.46. Determine the overall order of the following rate laws and the order with respect to each reactant.
  a. Rate = $k[A]^2[B]^{1/2}$
  b. Rate = $k[A]^2[B][C]$
  c. Rate = $k[A][B]^3[C]^{1/2}$

15.47. Write rate laws and determine the units of the rate constant (by using the units M for concentration and s for time) for the following reactions:
  a. The reaction of oxygen atoms with NO_2 is first order in both reactants.
  b. The reaction between NO and Cl_2 is second order in NO and first order in Cl_2.
  c. The reaction between Cl_2 and chloroform ($CHCl_3$) is first order in $CHCl_3$ and one-half order in Cl_2.
  *d. The decomposition of ozone (O_3) to O_2 is second order in O_3 and an order of -1 in O atoms.

15.48. Compounds A and B react to give a single product, C. Write the rate law for each of the following cases and determine the units of the rate constant by using the units M for concentration and s for time:
  a. The reaction is first order in A and second order in B.
  b. The reaction is first order in A and second order overall.
  c. The reaction is independent of the concentration of A and second order overall.
  d. The reaction is second order in both A and B.

15.49. Predict the rate law for the reaction $2\,BrO(g) \rightarrow Br_2(g) + O_2(g)$ if the following conditions hold true:
  a. The rate doubles when [BrO] doubles.
  b. The rate quadruples when [BrO] doubles.
  c. The rate is halved when [BrO] is halved.
  d. The rate is unchanged when [BrO] is doubled.

15.50. Predict the rate law for the reaction $NO(g) + Br_2(g) \rightarrow NOBr_2(g)$ if the following conditions apply:
  a. The rate doubles when [NO] is doubled and $[Br_2]$ remains constant.
  b. The rate doubles when $[Br_2]$ is doubled and [NO] remains constant.
  c. The rate increases by 1.56 times when [NO] is increased 1.25 times and $[Br_2]$ remains constant.
  d. The rate is halved when [NO] is doubled and $[Br_2]$ remains constant.

15.51. In the reaction of NO with ClO,

$$NO(g) + ClO(g) \rightarrow NO_2(g) + Cl(g)$$

the initial rate of reaction quadruples when the concentrations of both reactants are doubled. What additional information do we need to determine whether the reaction is first order in each reactant?

15.52. The reaction between chlorine monoxide and nitrogen dioxide

$$ClO(g) + NO_2(g) + M(g) \rightarrow ClONO_2(g) + M(g)$$

produces chlorine nitrate ($ClONO_2$). A third molecule (M) takes part in the reaction but is unchanged by it. The reaction is first order in NO_2 and in ClO.
  a. Write the rate law for this reaction.
  b. What is the reaction order with respect to M?

15.53. Rate Laws for Destruction of Tropospheric Ozone The reaction of NO_2 with ozone produces NO_3 in a second-order reaction overall:

$$NO_2(g) + O_3(g) \rightarrow NO_3(g) + O_2(g)$$

a. Write the rate law for the reaction if the reaction is first order in each reactant.
b. The rate constant for the reaction is $1.93 \times 10^4 \, M^{-1} s^{-1}$ at 298 K. What is the rate of the reaction when $[NO_2] = 1.8 \times 10^{-8} \, M$ and $[O_3] = 1.4 \times 10^{-7} \, M$?
c. What is the rate of formation of NO_3 under these conditions?
d. What happens to the rate of the reaction if the concentration of $O_3(g)$ is doubled?

15.54. Sources of Nitric Acid in the Atmosphere The reaction between N_2O_5 and water

$$N_2O_5(g) + H_2O(g) \rightarrow 2\,HNO_3(g)$$

is a source of nitric acid in the atmosphere.
a. The reaction is first order in each reactant. Write the rate law for the reaction.
b. When $[N_2O_5]$ is 0.132 mM and $[H_2O]$ is 230 mM, the rate of the reaction is 4.55×10^{-4} mM^{-1} min^{-1}. What is the rate constant for the reaction?

15.55. Each of the following reactions is first order in the reactants and second order overall. Which reaction is fastest if the initial concentrations of the reactants are the same? All reactions are at 298 K.
a. $ClO_2(g) + O_3(g) \rightarrow ClO_3(g) + O_2(g)$
$$k = 3.0 \times 10^{-19} \text{ cm}^3/(\text{molecule} \cdot \text{s})$$
b. $ClO_2(g) + NO(g) \rightarrow NO_2(g) + ClO(g)$
$$k = 3.4 \times 10^{-13} \text{ cm}^3/(\text{molecule} \cdot \text{s})$$
c. $ClO(g) + NO(g) \rightarrow Cl(g) + NO_2(g)$
$$k = 1.7 \times 10^{-11} \text{ cm}^3/(\text{molecule} \cdot \text{s})$$
d. $ClO(g) + O_3(g) \rightarrow ClO_2(g) + O_2(g)$
$$k = 1.5 \times 10^{-17} \text{ cm}^3/(\text{molecule} \cdot \text{s})$$

15.56. Two reactions in which there is a single reactant have nearly the same magnitude rate constant. One is first order; the other is second order.
a. If the initial concentrations of the reactants are both 1.0 mM, which reaction will proceed at the higher rate?
b. If the initial concentrations of the reactants are both 2.0 M, which reaction will proceed at the higher rate?

15.57. In the presence of water, the species NO and NO_2 react to form nitrous acid (HNO_2) by the following reaction:

$$NO(g) + NO_2(g) + H_2O(\ell) \rightarrow 2\,HNO_2(aq)$$

When the concentration of NO or NO_2 is doubled, the initial rate of reaction doubles. If the rate of the reaction does not depend on $[H_2O]$, what is the rate law for this reaction?

15.58. Hydroperoxyl Radicals in the Atmosphere During a smog event, trace amounts of many highly reactive substances are present in the atmosphere. One of these is the hydroperoxyl radical, HO_2, which reacts with sulfur trioxide, SO_3. The rate constant for the reaction

$$2\,HO_2(g) + SO_3(g) \rightarrow H_2SO_3(g) + 2\,O_2(g)$$

at 298 K is $2.6 \times 10^{11} \, M^{-1} \, s^{-1}$. The initial rate of the reaction doubles when the concentration of SO_3 or HO_2 is doubled. What is the rate law for the reaction?

15.59. Disinfecting Municipal Water Supplies Chlorine dioxide (ClO_2) is a disinfectant used in municipal water-treatment plants (Figure P15.59). It dissolves in basic solution, producing ClO_3^- and ClO_2^-:

$$2\,ClO_2(g) + 2\,OH^-(aq) \rightarrow ClO_3^-(aq) + ClO_2^-(aq) + H_2O(\ell)$$

FIGURE P15.59

The following kinetic data were obtained at 298 K for the reaction:

Experiment	$[ClO_2]_0$ (M)	$[OH^-]_0$ (M)	Initial Rate (M/s)
1	0.060	0.030	0.0248
2	0.020	0.030	0.00827
3	0.020	0.090	0.0247

Determine the rate law and the rate constant for this reaction at 298 K.

15.60. The following kinetic data were collected at 298 K for the reaction of ozone with nitrite ion, producing nitrate and oxygen:

$$NO_2^-(aq) + O_3(g) \rightarrow NO_3^-(aq) + O_2(g)$$

Experiment	$[NO_2^-]_0$ (M)	$[O_3]_0$ (M)	Initial Rate (M/s)
1	0.0100	0.0050	25
2	0.0150	0.0050	37.5
3	0.0200	0.0050	50.0
4	0.0200	0.0200	200.0

Determine the rate law for the reaction and the value of the rate constant.

15.61. Hydrogen gas reduces NO to N_2 in the following reaction:

$$2\,H_2(g) + 2\,NO(g) \rightarrow 2\,H_2O(g) + N_2(g)$$

The initial reaction rates of four mixtures of H_2 and NO were measured at 900°C with the following results:

Experiment	$[H_2]_0$ (M)	$[NO]_0$ (M)	Initial Rate (M/s)
1	0.212	0.136	0.0248
2	0.212	0.272	0.0991
3	0.424	0.544	0.793
4	0.848	0.544	1.59

Determine the rate law and the rate constant for the reaction at 900°C.

15.62. The rate of the reaction

$$NO_2(g) + CO(g) \rightarrow NO(g) + CO_2(g)$$

was determined in three experiments at 225°C. The results are given in the following table:

Experiment	$[NO_2]_0$ (M)	$[CO]_0$ (M)	Initial Rate, $-\Delta[NO_2]/\Delta t$ (M/s)
1	0.263	0.826	1.44×10^{-5}
2	0.263	0.413	1.44×10^{-5}
3	0.526	0.413	5.76×10^{-5}

a. Determine the rate law for the reaction.
b. Calculate the value of the rate constant at 225°C.
c. Calculate the rate of formation of CO_2 when $[NO_2] = [CO] = 0.500\ M$.

15.63. Nitrogen trioxide decomposes to NO_2 and O_2 in the following reaction:

$$2\,NO_3(g) \rightarrow 2\,NO_2(g) + O_2(g)$$

The following data were collected at 298 K:

Time (min)	$[NO_3]$ (μM)
0	1.470×10^{-3}
10	1.463×10^{-3}
100	1.404×10^{-3}
200	1.344×10^{-3}
300	1.288×10^{-3}
400	1.237×10^{-3}
500	1.190×10^{-3}

Calculate the value of the second-order rate constant at 298 K.

15.64. Two structural isomers of ClO_2 are shown in Figure P15.64. The isomer with the Cl–O–O skeletal arrangement is unstable and rapidly decomposes according to the reaction $2\,ClOO(g) \rightarrow Cl_2(g) + 2\,O_2(g)$. The following data were collected for the decomposition of ClOO at 298 K:

Time (μs)	[ClOO] (M)
0.0	1.76×10^{-6}
0.7	2.36×10^{-7}
1.3	3.56×10^{-8}
2.1	3.23×10^{-9}
2.8	3.96×10^{-10}

FIGURE P15.64

Determine the rate law for the reaction and the value of the rate constant at 298 K.

15.65. At high temperatures, ammonia spontaneously decomposes into N_2 and H_2. The following data were collected at one such temperature:

Time (s)	$[NH_3]$ (M)
0	2.56×10^{-2}
12	2.47×10^{-2}
56	2.16×10^{-2}
224	1.31×10^{-2}
532	5.19×10^{-3}
746	2.73×10^{-3}

Determine the rate law for the decomposition of ammonia and the value of the rate constant at the temperature of the experiment.

15.66. Atmospheric Chemistry of Hydroperoxyl Radicals Atmospheric chemistry involves highly reactive, odd-electron molecules such as the hydroperoxyl radical HO_2, which decomposes into H_2O_2 and O_2. Determine the rate law for the reaction and the value of the rate constant at 298 K by using the following data obtained at 298 K.

Time (μs)	$[HO_2]$ (μM)
0.0	8.5
0.6	5.1
1.0	3.6
1.4	2.6
1.8	1.8
2.4	1.1

15.67. Laughing Gas Nitrous oxide (N_2O) is used as an anesthetic (laughing gas) and in aerosol cans to produce whipped cream. It is a potent greenhouse gas and decomposes slowly to N_2 and O_2:

$$2\,N_2O(g) \rightarrow 2\,N_2(g) + O_2(g)$$

a. If the plot of $\ln[N_2O]$ as a function of time is linear, what is the rate law for the reaction?

b. How many half-lives will it take for the concentration of the N_2O to reach 6.25% of its original concentration? [*Hint*: The amount of reactant remaining after time t (A_t) is related to the amount initially present (A_0) by the equation $A_t/A_0 = (0.5)^n$, where n is the number of half-lives in time t.]

15.68. The unsaturated hydrocarbon butadiene (C_4H_6) dimerizes to 4-vinylcyclohexene (C_8H_{12}). When data collected in studies of the kinetics of this reaction were plotted against reaction time, plots of $[C_4H_6]$ or $\ln[C_4H_6]$ produced curved lines, but the plot of $1/[C_4H_6]$ was linear.

a. What is the rate law for the reaction?

b. How many half-lives will it take for the $[C_4H_6]$ to decrease to 3.1% of its original concentration?

15.69. Tracing Phosphorus in Organisms Radioactive isotopes such as ^{32}P are used to follow biological processes. The following radioactivity data (in relative radioactivity values) were collected for a sample containing ^{32}P:

Time (days)	Radioactivity (relative radioactivity values)
0	10.0
1	9.53
2	9.08
5	7.85
10	6.16
20	3.79

a. Write the rate law for the decay of ^{32}P.

b. Determine the value of the first-order rate constant.

c. Determine the half-life of ^{32}P.

15.70. Nitrous acid slowly decomposes to NO, NO_2, and water in the following second-order reaction:

$$2\,HNO_2(aq) \rightarrow NO(g) + NO_2(g) + H_2O(\ell)$$

a. Use the data below to determine the rate constant for this reaction at 298 K:

Time (min)	$[HNO_2]$ (μM)
0	0.1560
1000	0.1466
1500	0.1424
2000	0.1383
2500	0.1345
3000	0.1309

b. Determine the half-life for the decomposition of HNO_2.

15.71. The dimerization of ClO,

$$2\,ClO(g) \rightarrow Cl_2O_2(g)$$

is second order in ClO. Use the following data to determine the value of k at 298 K:

Time (s)	[ClO] (molecules/cm^3)
0	2.60×10^{11}
1	1.08×10^{11}
2	6.83×10^{10}
3	4.99×10^{10}
4	3.93×10^{10}

Determine the half-life for the dimerization of ClO.

15.72. Kinetic data for the reaction $Cl_2O_2(g) \rightarrow 2\,ClO(g)$ are summarized in the following table. Determine the value of the first-order rate constant.

Time (μs)	$[Cl_2O_2]$ (M)
0	6.60×10^{-8}
172	5.68×10^{-8}
345	4.89×10^{-8}
517	4.21×10^{-8}
690	3.62×10^{-8}
862	3.12×10^{-8}

Determine the half-life for the decomposition of Cl_2O_2.

15.73. Kinetics of Sucrose Hydrolysis The metabolism of table sugar (sucrose, $C_{12}H_{22}O_{11}$) begins with the hydrolysis of the disaccharide to glucose and fructose (both $C_6H_{12}O_6$):

$$C_{12}H_{22}O_{11}(aq) + H_2O(\ell) \rightarrow 2\,C_6H_{12}O_6(aq)$$

The kinetics of the reaction were studied at 24°C in a reaction system with a large excess of water, so the reaction was pseudo-first-order in sucrose. Determine the rate law and the pseudo-first-order rate constant for the reaction from the following data:

Time (s)	$[C_{12}H_{22}O_{11}]$ (M)
0	0.562
612	0.541
1600	0.509
2420	0.484
3160	0.462
4800	0.4417

15.74. Hydroperoxyl radicals react rapidly with ozone to produce oxygen and OH radicals:

$$HO_2(g) + O_3(g) \rightarrow OH(g) + 2O_2(g)$$

The rate of this reaction was studied in the presence of a large excess of ozone. Determine the pseudo-first-order rate constant and the second-order rate constant for the reaction from the following data:

Time (ms)	$[HO_2]$ (M)	$[O_3]$ (M)
0	3.2×10^{-6}	1.0×10^{-3}
10	2.9×10^{-6}	1.0×10^{-3}
20	2.6×10^{-6}	1.0×10^{-3}
30	2.4×10^{-6}	1.0×10^{-3}
80	1.4×10^{-6}	1.0×10^{-3}

Reaction Rates, Temperature, and the Arrhenius Equation

CONCEPT REVIEW

15.75. Why are some spontaneous chemical reactions slow?

15.76. Is the activation energy of a spontaneous reaction in the forward direction greater than, less than, or equal to the activation energy for the reverse reaction?

15.77. Which, if any, of the following statements is true?
  a. Exothermic reactions are always fast.
  b. Reactions with $\Delta G > 0$ are slow.
  c. Endothermic reactions are always slow.
  d. Reactions accompanied by an increase in entropy are fast.

*15.78. Which, if any, of the following statements is true?
  a. Reactions with $\Delta G < 0$ are always fast.
  b. Reactions with $\Delta H > 0$ are always fast.
  c. Reactions with $\Delta S < 0$ are always slow.
  d. Reactions with $\Delta H < 0$ are fast only at low temperature.

*15.79. The order of a reaction is independent of temperature, but the value of the rate constant varies with temperature. Why?

15.80. Why is the value of E_a for a spontaneous reaction less than the E_a value for the same reaction running in reverse?

*15.81. Two first-order reactions have activation energies of 15 and 150 kJ/mol. Which reaction will show the larger increase in rate as temperature is increased?

15.82. According to the Arrhenius equation, does the activation energy of a chemical reaction depend on temperature? Explain your answer.

PROBLEMS

15.83. The rate constant for the reaction of ozone with oxygen atoms was determined at four temperatures. Calculate the activation energy and frequency factor A for the reaction

$$O(g) + O_3(g) \rightarrow 2O_2(g)$$

given the following data:

T (K)	k [cm³/(molecule · s)]
250	2.64×10^{-4}
275	5.58×10^{-4}
300	1.04×10^{-3}
325	1.77×10^{-3}

15.84. The rate constant for the reaction

$$NO_2(g) + O_3(g) \rightarrow NO_3(g) + O_2(g)$$

was determined over a temperature range of 40 K, with the following results:

T (K)	k (M^{-1} s⁻¹)
203	4.14×10^5
213	7.30×10^5
223	1.22×10^6
233	1.96×10^6
243	3.02×10^6

  a. Determine the activation energy for the reaction.
  b. Calculate the rate constant of the reaction at 300 K.

15.85. Activation Energy for Smog-Forming Reactions The initial step in the formation of smog is the reaction between nitrogen and oxygen. The activation energy of the reaction can be determined from the temperature dependence of the rate constants. At the temperatures indicated, values of the rate constant of the reaction

$$N_2(g) + O_2(g) \rightarrow 2NO(g)$$

are as follows:

T (K)	k ($M^{-1/2}$ s⁻¹)
2000	318
2100	782
2200	1770
2300	3733
2400	7396

  a. Calculate the activation energy of the reaction.
  b. Calculate the frequency factor for the reaction.
  c. Calculate the value of the rate constant at ambient temperature, $T = 300$ K.

15.86. Values of the rate constant for the decomposition of N_2O_5 gas at four different temperatures are as follows:

T (K)	k (s⁻¹)
658	2.14×10^5
673	3.23×10^5
688	4.81×10^5
703	7.03×10^5

  a. Determine the activation energy of the decomposition reaction.
  b. Calculate the value of the rate constant at 300 K.

15.87. Activation Energy of Stratospheric Ozone Destruction Reactions The kinetics of the reaction between chlorine dioxide and ozone is relevant to the study of atmospheric ozone destruction. The activation energy of the reaction can be determined from the temperature dependence of the rate constant. The value of the rate constant for the reaction between chlorine dioxide and ozone was measured at four temperatures between 193 and 208 K. The results are as follows:

T (K)	k (M^{-1} s^{-1})
193	34.0
198	62.8
203	112.8
208	196.7

Calculate the values of the activation energy and the frequency factor for the reaction.

15.88. Chlorine atoms react with methane, forming HCl and CH_3. The rate constant for the reaction is 6.0×10^7 $M^{-1}s^{-1}$ at 298 K. When the experiment was repeated at three other temperatures, the following data were collected

T (K)	k (M^{-1} s^{-1})
303	6.5×10^7
308	7.0×10^7
313	7.5×10^7

Calculate the values of the activation energy and the frequency factor for the reaction.

Reaction Mechanisms

CONCEPT REVIEW

15.89. The rate law for the reaction between NO and H_2 is second order in NO and third order overall, whereas the reaction of NO with Cl_2 is first order in each reactant and second order overall. Do these reactions proceed by similar mechanisms?

15.90. The rate law for the reaction of NO with Cl_2 (rate = k[NO][Cl_2]) is the same as that for the reaction of NO_2 with F_2 (rate = k[NO_2][F_2]). Is it possible that these reactions have similar mechanisms?

*15.91. Under what reaction conditions does a bimolecular reaction obey pseudo-first-order reaction kinetics?

*15.92. If a reaction is zero order in a reactant, does that mean the reactant is never involved in collisions with other reactants? Explain your answer.

PROBLEMS

15.93. The hypothetical reaction A → B has an activation energy of 50.0 kJ/mol. Draw a reaction profile for each of the following mechanisms:
a. A single elementary step.
b. A two-step reaction in which the activation energy of the second step is 15 kJ/mol.
c. A two-step reaction in which the activation energy of the second step is the rate-determining barrier.

15.94. For the spontaneous reaction A + B → C → D + E, draw three reaction profiles, one for each of the following mechanisms:
a. C is an activated complex.
b. The reaction has two elementary steps; the first step is rate determining and C is an intermediate.
c. The reaction has two elementary steps; the second step is rate determining and C is an intermediate.

15.95. Write the rate laws for the following elementary steps and identify them as uni-, bi-, or termolecular steps:
a. $SO_2Cl_2(g) \rightarrow SO_2(g) + Cl_2(g)$
b. $NO_2(g) + CO(g) \rightarrow NO(g) + CO_2(g)$
c. $2\,NO_2(g) \rightarrow NO_3(g) + NO(g)$

15.96. Write the rate laws for the following elementary steps and identify them as uni-, bi-, or termolecular steps:
a. $Cl(g) + O_3(g) \rightarrow ClO(g) + O_2(g)$
b. $2\,NO_2(g) \rightarrow N_2O_4(g)$
c. $^{14}_{6}C \rightarrow {}^{14}_{7}N + {}^{0}_{-1}\beta$

15.97. Write the overall reaction that consists of the following elementary steps:

$$N_2O_5(g) \rightarrow NO_3(g) + NO_2(g)$$
$$NO_3(g) \rightarrow NO_2(g) + O(g)$$
$$2\,O(g) \rightarrow O_2(g)$$

15.98. What overall reaction consists of the following elementary steps?

$$ClO^-(aq) + H_2O(\ell) \rightarrow HClO(aq) + OH^-(aq)$$
$$I^-(aq) + HClO(aq) \rightarrow HIO(aq) + Cl^-(aq)$$
$$OH^-(aq) + HIO(aq) \rightarrow H_2O(\ell) + IO^-(aq)$$

*15.99. In the following mechanism for NO formation, oxygen atoms are produced by breaking O=O bonds at high temperature in a fast reversible reaction. If $\Delta[NO]/\Delta t = k[N_2][O_2]^{1/2}$, which step in the mechanism is the rate-determining step?

(1)	$O_2(g) \rightleftharpoons 2\,O(g)$
(2)	$O(g) + N_2(g) \rightarrow NO(g) + N(g)$
(3)	$N(g) + O(g) \rightarrow NO(g)$
Overall	$N_2(g) + O_2(g) \rightarrow 2\,NO(g)$

15.100. A proposed mechanism for the decomposition of hydrogen peroxide consists of three elementary steps:

$$H_2O_2(g) \rightarrow 2\,OH(g)$$
$$H_2O_2(g) + OH(g) \rightarrow H_2O(g) + HO_2(g)$$
$$HO_2(g) + OH(g) \rightarrow H_2O(g) + O_2(g)$$

If the rate law for the reaction is first order in H_2O_2, which step in the mechanism is the rate-determining step?

15.101. At a given temperature, the rate of the reaction between NO and Cl_2 is proportional to the product of the concentrations of the two gases: $[NO][Cl_2]$. The following two-step mechanism was proposed for the reaction:

(1) $\qquad NO(g) + Cl_2(g) \rightarrow NOCl_2(g)$

(2) $\qquad NOCl_2(g) + NO(g) \rightarrow 2\,NOCl(g)$

Overall $\quad 2\,NO(g) + Cl_2(g) \rightarrow 2\,NOCl(g)$

Which step must be the rate-determining step if this mechanism is correct?

15.102. Mechanism of Ozone Destruction Ozone decomposes thermally to oxygen in the following reaction:

$$2\,O_3(g) \rightarrow 3\,O_2(g)$$

The following mechanism has been proposed:

$$O_3(g) \rightarrow O(g) + O_2(g)$$

$$O(g) + O_3(g) \rightarrow 2\,O_2(g)$$

The reaction is second order in ozone. What properties of the two elementary steps (specifically, relative rate and reversibility) are consistent with this mechanism?

15.103. Mechanism of NO_2 Destruction The rate laws for the thermal and photochemical decomposition of NO_2 are different. Which of the following mechanisms are possible for the thermal decomposition of NO_2, and which are possible for the photochemical decomposition of NO_2? For thermal decomposition, rate $= k[NO_2]^2$ and for photochemical decomposition, rate $= k[NO_2]$.

a. $\qquad NO_2(g) \xrightarrow{\text{slow}} NO(g) + O(g)$

$\qquad O(g) + NO_2(g) \xrightarrow{\text{fast}} NO(g) + O_2(g)$

b. $NO_2(g) + NO_2(g) \xrightarrow{\text{fast}} N_2O_4(g)$

$\qquad N_2O_4(g) \xrightarrow{\text{slow}} NO(g) + NO_3(g)$

$\qquad NO_3(g) \xrightarrow{\text{fast}} NO(g) + O_2(g)$

c. $NO_2(g) + NO_2(g) \xrightarrow{\text{slow}} NO(g) + NO_3(g)$

$\qquad NO_3(g) \xrightarrow{\text{fast}} NO(g) + O_2(g)$

15.104. The rate laws for the thermal and photochemical decomposition of NO_2 are different. Which of the following mechanisms are possible for the thermal decomposition of NO_2, and which are possible for the photochemical decomposition of NO_2? For thermal decomposition, rate $= k[NO_2]^2$ and for photochemical decomposition, rate $= k[NO_2]$.

a. $NO_2(g) + NO_2(g) \xrightarrow{\text{slow}} N_2O_4(g)$

$\qquad N_2O_4(g) \xrightarrow{\text{fast}} N_2O_3(g) + O(g)$

$\qquad N_2O_3(g) + O(g) \xrightarrow{\text{fast}} N_2O_2(g) + O_2(g)$

$\qquad N_2O_2(g) \xrightarrow{\text{fast}} 2\,NO(g)$

b. $NO_2(g) + NO_2(g) \xrightarrow{\text{slow}} NO(g) + NO_3(g)$

$\qquad NO_3(g) \xrightarrow{\text{fast}} NO(g) + O_2(g)$

c. $\qquad NO_2(g) \xrightarrow{\text{slow}} N(g) + O_2(g)$

$\qquad N(g) + NO_2(g) \xrightarrow{\text{fast}} N_2O_2(g)$

$\qquad N_2O_2(g) \xrightarrow{\text{slow}} 2\,NO(g)$

Catalysis

CONCEPT REVIEW

15.105. Does a catalyst affect both the rate and the rate constant of a reaction?

15.106. Is the rate law for a catalyzed reaction the same as that for the uncatalyzed reaction?

15.107. Does a substance that increases the rate of a reaction also increase the rate of the reverse reaction?

15.108. The rate of the reaction between NO_2 and CO is independent of [CO]. Does this mean that CO is a catalyst for the reaction?

15.109. Why doesn't the concentration of a homogeneous catalyst appear in the rate law for the reaction it catalyzes?

*__15.110.__ The rate of a chemical reaction is too slow to measure at room temperature. We could either raise the temperature or add a catalyst. Which would be a better solution for making an accurate determination of the rate constant?

PROBLEMS

15.111. Is NO a catalyst for the decomposition of N_2O in the following two-step reaction mechanism, or is N_2O a catalyst for the conversion of NO to NO_2?

(1) $\qquad NO(g) + N_2O(g) \rightarrow N_2(g) + NO_2(g)$

(2) $\qquad 2\,NO_2(g) \rightarrow 2\,NO(g) + O_2(g)$

15.112. NO as a Catalyst for Ozone Destruction Explain why NO is a catalyst in the following two-step process that results in the depletion of ozone in the stratosphere:

(1) $\qquad NO(g) + O_3(g) \rightarrow NO_2(g) + O_2(g)$

(2) $\qquad O(g) + NO_2(g) \rightarrow NO(g) + O_2(g)$

Overall $\quad O(g) + O_3(g) \rightarrow 2\,O_2(g)$

15.113. On the basis of the frequency factors and activation energy values of the following two reactions, determine which one will have the larger rate constant at room temperature (298 K).

$$O_3(g) + O(g) \rightarrow O_2(g) + O_2(g)$$

$A = 8.0 \times 10^{-12}$ cm^3/(molecules $\cdot$ s) $\qquad E_a = 17.1$ kJ/mol

$$O_3(g) + Cl(g) \rightarrow ClO(g) + O_2(g)$$

$A = 2.9 \times 10^{-11}$ cm^3/(molecules $\cdot$ s) $\qquad E_a = 2.16$ kJ/mol

15.114. On the basis of the frequency factors and activation energy values of the following two reactions, determine which one will have the larger rate constant at room temperature (298 K).

$$O_3(g) + Cl(g) \rightarrow ClO(g) + O_2(g)$$

$A = 2.9 \times 10^{-11}$ cm^3/(molecules $\cdot$ s) $\qquad E_a = 2.16$ kJ/mol

$$O_3(g) + NO(g) \rightarrow NO_2(g) + O_2(g)$$

$A = 2.0 \times 10^{-12}$ cm^3/(molecules $\cdot$ s) $\qquad E_a = 11.6$ kJ/mol

Additional Problems

15.115. A student inserts a glowing wood splint into a test tube filled with O_2. The splint quickly catches on fire (Figure P15.115). Why does the splint burn so much faster in pure O_2 than in air?

FIGURE P15.115

*15.116. A backyard chef turns on the propane gas to a barbecue grill. Even though the reaction between propane and oxygen is spontaneous, the gas does not begin to burn until the chef pushes an igniter button to produce a spark. Why is the spark needed?

15.117. On average, someone who falls through the ice covering a frozen lake is less likely to experience anoxia (lack of oxygen) than someone who falls into a warm pool and is underwater for the same length of time. Why?

*15.118. Why doesn't a quadrupling of the rate correspond to a reaction order of 4, e.g., Rate $\propto [NO]^4$?

15.119. If the rate of the reverse reaction is much slower than the rate of the forward reaction, does the method used to determine a rate law from initial concentrations and initial rates also work at some other time t? What concentrations would we use in the case where we use the rate when $t \neq 0$?

15.120 What is wrong with the following statement: The reaction rate and the rate constant for a reaction both depend on the number of collisions and on the concentrations of the reactants.

15.121. How do we find k if we plot $1/[X] - 1/[X]_0$ as a function of t?

15.122. Many reactions are first order, fewer are second order in a single reactant, and third-order reactions in a single reactant are practically nonexistent. Can you suggest why?

15.123. Why can't an elementary step in a mechanism have a rate law that is zero order in a reactant?

15.124. During the decomposition of dinitrogen pentoxide,

$$2\,N_2O_5(g) \rightarrow 4\,NO_2(g) + O_2(g)$$

how is the rate of consumption of N_2O_5 related to the rate of formation of NO_2 and O_2?

15.125. In the reaction between nitrogen dioxide and ozone,

$$2\,NO_2(g) + O_3(g) \rightarrow N_2O_5(g) + O_2(g)$$

how are the rates of change in the concentrations of the reactants and products related?

15.126. Determine the order of the decomposition reaction of N_2O_5, by using the initial rate data from the following table:

Experiment	$[N_2O_5]_0$ (*M*)	Initial Rate (*M*/s)
1	0.050	1.8×10^{-5}
2	0.100	3.6×10^{-5}

15.127. At the temperature at which the experiments were carried out in the previous problem, what is the rate constant for the decomposition of N_2O_5? Write the complete rate law for the decomposition reaction.

15.128. The table below contains reaction rate data for the reaction

$$2\,NO(g) + Cl_2(g) \rightarrow 2\,NOCl(g)$$

Experiment	$[NO]_0$ (*M*)	$[Cl_2]_0$ (*M*)	Initial Rate (*M*/s)
1	0.20	0.10	0.63
2	0.20	0.30	5.70
3	0.80	0.10	2.58
4	0.40	0.20	?

Predict the initial rate of reaction in experiment 4.

15.129. An important reaction in the formation of photochemical smog is the reaction between ozone and NO:

$$NO(g) + O_3(g) \rightarrow NO_2(g) + O_2(g)$$

The reaction is first order in NO and O_3. The rate constant of the reaction is 80 M^{-1} s^{-1} at 25°C and 3000 M^{-1} s^{-1} at 75°C.

a. If this reaction were to occur in a single step, would the rate law be consistent with the observed order of the reaction for NO and O_3?

b. What is the value of the activation energy of the reaction?

c. What is the rate of the reaction at 25°C when $[NO] = 3 \times 10^{-6}$ M and $[O_3] = 5 \times 10^{-9}$ M?

d. Predict the values of the rate constant at 10°C and 35°C.

15.130. Ammonia reacts with nitrous acid to form an intermediate, ammonium nitrite (NH_4NO_2), which decomposes to N_2 and H_2O:

$$NH_3(g) + HNO_2(aq) \rightarrow NH_4NO_2(aq) \rightarrow N_2(g) + 2H_2O(\ell)$$

a. The reaction is first order in ammonia and second order in nitrous acid. What is the rate law for the reaction? What are the units of the rate constant if concentrations are expressed in molarity and time in seconds?

b. The rate law for the reaction has also been written as Rate = $k[NH_4^+][NO_2^-][HNO_2]$ Is this expression equivalent to the one you wrote in part a?

c. With the data in Appendix 4, calculate the value of ΔH_{rxn}° of the overall reaction (ΔH_f°, $HNO_2 = -43.1$ kJ/mol).

d. Draw a reaction-energy profile for the process with the assumption that E_a of the first step is lower than E_a of the second step.

*15.131. When ionic compounds such as NaCl dissolve in water, the sodium ions are surrounded by six water molecules. The bound water molecules exchange with those in bulk solution as described by the reaction involving ^{18}O-enriched water:

$$Na(H_2O)_6^+(aq) + H_2^{18}O(\ell) \rightarrow Na(H_2O)_5(H_2^{18}O)^+(aq) + H_2O(\ell)$$

a. The following reaction mechanism has been proposed:

(1) $Na(H_2O)_6^+(aq) \rightarrow Na(H_2O)_5^+(aq) + H_2O(\ell)$

(2) $Na(H_2O)_5^+(aq) + H_2^{18}O(\ell) \rightarrow Na(H_2O)_5(H_2^{18}O)^+(aq)$

What is the rate law if the first step is the rate-determining step?

b. If you were to sketch a reaction-energy profile, which would you draw with the higher energy, the reactants or the products?

15.132. Lachrymators in Smog The combination of ozone, volatile hydrocarbons, nitrogen oxide, and sunlight in urban environments produces peroxyacetyl nitrate (PAN), a potent lachrymator. PAN decomposes to acetyl radicals and nitrogen dioxide in a process that is second order in PAN, as shown in Figure P15.132:

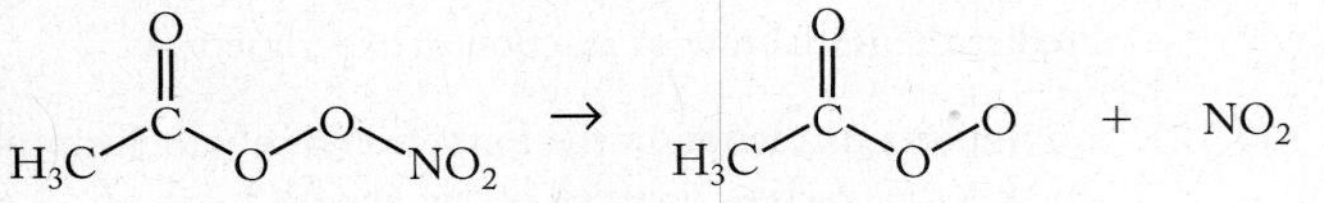

FIGURE P15.132

a. The half-life of the reaction, at 23°C and $P_{CH_3CO_3NO_2} = 10.5$ torr, is 100 hr. Calculate the rate constant for the reaction.

b. Determine the rate of the reaction at 23°C and $P_{CH_3CO_3NO_2} = 10.5$ torr.

c. Draw a graph showing P_{PAN} as a function of time from 0 to 200 hr starting with $P_{CH_3CO_3NO_2} = 10.5$ torr.

15.133. Nitric Oxide in the Human Body Nitric oxide (NO) is a gaseous free radical that plays many biological roles including regulating neurotransmission and the human immune system. One of its many reactions involves the peroxynitrite ion ($ONOO^-$):

$$NO(g) + ONOO^-(aq) \rightarrow NO_2(g) + NO_2^-(aq)$$

a. Use the following data to determine the rate law and rate constant of the reaction at the experimental temperature at which these data were generated.

Experiment	$[NO]_0$ (M)	$[ONOO^-]_0$ (M)	Rate (M/s)
1	1.25×10^{-4}	1.25×10^{-4}	2.03×10^{-11}
2	1.25×10^{-4}	0.625×10^{-4}	1.02×10^{-11}
3	0.625×10^{-4}	2.50×10^{-4}	2.03×10^{-11}
4	0.625×10^{-4}	3.75×10^{-4}	3.05×10^{-11}

b. Draw the Lewis structure of peroxynitrite ion (including all resonance forms) and assign formal charges. Note which form is preferred.

c. Use the average bond energies in Table 8.2 to estimate the value of ΔH_{rxn} using the preferred structure from part b.

15.134. Kinetics of Protein Chemistry In the presence of O_2, NO reacts with sulfur-containing proteins to form S-nitrosothiols, such as $C_6H_{13}SNO$. This compound decomposes to form a disulfide and NO:

$$2C_6H_{13}SNO(aq) \rightarrow 2NO(g) + C_{12}H_{26}S_2(aq)$$

a. The following data were collected for the decomposition reaction at 69°C.

Time (min)	$[C_6H_{13}SNO]$ (M)
0	1.05×10^{-3}
10	9.84×10^{-4}
20	9.22×10^{-4}
30	8.64×10^{-4}
60	7.11×10^{-4}

Calculate the value of the first-order rate constant for the reaction.

b. Which amino acids might act as sources of S-nitrosothiols?

15.135. Solutions of nitrous acid, HNO_2, in ^{18}O-labeled water undergo isotope exchange:

$$HNO_2(aq) + H_2^{18}O(\ell) \rightarrow HN^{18}O_2(aq) + H_2O(\ell)$$

a. Use the following data at 24°C to determine the dependence of the reaction rate on the concentration of HNO_2.

Time (min)	$[HN^{18}O_2]$ (M)
0	5.4×10^{-2}
20	1.5×10^{-3}
40	7.7×10^{-4}
60	5.2×10^{-4}

b. Does the reaction rate depend on the concentration of $H_2^{18}O$?

15.136. Ethylene (C_2H_4) reacts with ozone to form 2 mol of formaldehyde (a probable human carcinogen) per mole of ethylene as shown in Figure P15.136. The following kinetic data were collected at 298 K.

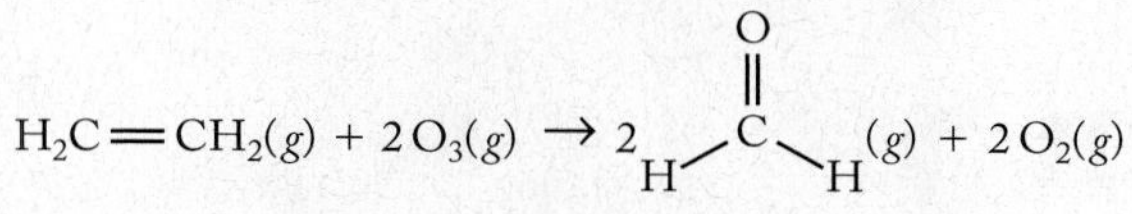

$$H_2C{=}CH_2(g) + 2\,O_3(g) \rightarrow 2\,\underset{H}{\overset{O}{\underset{\displaystyle}{\overset{\|}{C}}}}\diagdown_H (g) + 2\,O_2(g)$$

FIGURE P15.136

Experiment	$[O_3]_0$ (M)	$[C_2H_4]_0$ (M)	Rate (M/s)
1	0.86×10^{-2}	1.00×10^{-2}	0.0877
2	0.43×10^{-2}	1.00×10^{-2}	0.0439
3	0.22×10^{-2}	0.50×10^{-2}	0.0110

a. Determine the rate law and the value of the rate constant of the reaction at 298 K.

b. The rate constant was determined at several additional temperatures. Calculate the activation energy of the reaction from the following data.

T (K)	k ($M^{-1}\,s^{-1}$)
263	3.28×10^2
273	4.73×10^2
283	6.65×10^2
293	9.13×10^2

15.137. **Reducing NO Emissions** Adding NH_3 to the stack gases at an electric power-generating plant can reduce NO_x emissions. This selective noncatalytic reduction (SNR) process depends on the reaction between NH_2 (an odd-electron molecule) and NO:

$$NH_2(g) + NO(g) + N_2(g) + H_2O(g)$$

The following kinetic data were collected at 1200 K.

Experiment	$[NH_2]_0$ (M)	$[NO]_0$ (M)	Rate (M/s)
1	1.00×10^{-5}	1.00×10^{-5}	0.12
2	2.00×10^{-5}	1.00×10^{-5}	0.24
3	2.00×10^{-5}	1.50×10^{-5}	0.36
4	2.50×10^{-5}	1.50×10^{-5}	0.45

a. What is the rate law for the reaction?

b. What is the value of the rate constant at 1200 K?

16

Chemical Equilibrium

Learning Outcomes

- Describe the dynamic nature of chemical equilibria

- Write mass action/equilibrium constant expressions for reversible reactions including those involving heterogeneous equilibria

- Interconvert the K_c and K_p values of gas-phase reactions

- Calculate the value of a reaction quotient and use it to predict the direction of a reversible chemical reaction

- Calculate the concentrations or partial pressures of reactants and products in a reaction mixture at equilibrium from their starting values and the value of K

- Describe how a reaction at equilibrium responds to the stresses induced by changing the quantities of reactants and products in the reaction mixture

- Relate the standard free-energy change of a reversible reaction to its equilibrium constant

- Predict the value of K at any temperature from thermodynamic data

A LOOK AHEAD

Feeling the Pressure

Consider an unopened can of soda. The contents of the can include carbon dioxide gas—some of it dissolved in the soda, and some of it in the small space in the can above the soda's surface. A dynamic equilibrium exists between dissolved and gaseous CO_2 inside the can. Molecules of dissolved CO_2 constantly leave the liquid and enter the gas phase while CO_2 molecules leave the gas phase and dissolve in the liquid. At a given temperature and pressure, these two opposing processes are in balance and the concentration of dissolved CO_2 doesn't change.

When we open the can, this balance is disrupted. Some of the CO_2 gas above the soda's surface escapes into the air as the pressure inside the can is suddenly reduced. Immediately the soda fizzes as bubbles of CO_2 form. These bubbles form because the solubility of a gas in a liquid is proportional to its partial pressure as we discussed in Chapter 10. Given time, most of the CO_2 that was dissolved escapes from the liquid and the soda goes "flat."

The solubility of CO_2 or any gas in a liquid involves a physical equilibrium. So too does another process involving pressure: the regulation of pressure inside the human eye. The eye is filled with a fluid called the *vitreous humor*. Pressure exerted by this liquid maintains the shape of the eyeball and holds the retina in its proper place. That pressure is maintained by a dynamic process in which liquid is constantly produced by cells behind the iris and then drained out of the eyeball into the bloodstream. In a healthy eye, the amount of liquid being produced equals the amount draining out, so the system maintains a constant pressure and the eye functions properly. If too much liquid is produced or not enough drains away, then the equilibrium is disturbed, the pressure increases, and a new equilibrium may be established at a pressure that represents a danger to good vision.

These physical equilibria in our eyes and in soft drinks are like chemical equilibria in that they are all dynamic processes. A chemical reaction achieves dynamic equilibrium when the rate at which the

Disrupting an Equilibrium Opening a can of soda relieves the pressure that ▶ kept CO_2 in solution.

chemical equilibrium a dynamic process in which the concentrations of reactants and products remain constant over time and the rate of a reaction in the forward direction matches its rate in the reverse direction.

reactants in a reversible reaction become products is equal to the rate at which products turn back into reactants. Thus, the reaction never stops, but rather proceeds in the forward and reverse directions at the same rate. Therefore, the concentrations of reactants and products remain constant over time.

To understand how a chemical reaction comes to equilibrium, we need to know about the free energy that drives it, how rapidly the reaction proceeds, and under what conditions equilibrium may be achieved. In Chapter 14 we saw that chemical reactions happen for a reason: high-energy reactants form lower-energy products. In a spontaneous reaction there is always a decrease in free energy ($\Delta G < 0$). In Chapter 15 we learned that *spontaneous* does not necessarily mean *rapid*: kinetics is as important as thermodynamics in determining whether or not a reaction happens in a reasonable length of time. In this chapter we discuss how the free energy that drives spontaneous reactions is dissipated when reactants turn into products. As ΔG approaches zero, the reaction approaches the state of chemical equilibrium. This chapter, together with Chapters 14 and 15, gives us the tools to answer three major questions about chemical reactions: Do they occur spontaneously? How rapidly do they occur? Do they go to completion? ■

16.1 The Dynamics of Chemical Equilibrium

Chemical equilibrium is a dynamic process in which the reactants in a reversible chemical reaction are constantly converted into products, while at the same time products are constantly converted into reactants. This reversibility is represented by a pair of arrows pointed in opposite directions:

$$\text{Reactants} \rightleftharpoons \text{products}$$

At equilibrium, the rate at which reactants become products equals the rate at which products turn back into reactants; no net change in the concentrations of the reactants and products occurs over time. Said another way, the rate of the forward reaction is the same as the rate of the reverse reaction:

$$\text{Rate}_{forward} = \text{rate}_{reverse}$$

Some reactions reach equilibrium only after nearly all the reactants have formed products. We say that these equilibria *lie far to the right* (the direction of the forward reaction arrow). In other reactions little product is formed. We say that these equilibria favor reactants and that they *lie far to the left* (the direction of the reverse reaction arrow). Keep in mind that nothing can be inferred from the position of the equilibrium regarding how much time the reaction takes to reach equilibrium. Studies of equilibrium reveal only the extent to which a reaction proceeds, not how rapidly it proceeds.

To explore the dynamics of chemical equilibrium, let's look at a two-step industrial process for making H_2 gas. The initial reactants are methane and steam. In the first step, methane reacts with steam in the presence of a nickel or iron oxide catalyst at temperatures near 1000°C:

$$CH_4(g) + H_2O(g) \rightleftharpoons CO(g) + 3\,H_2(g) \tag{16.1}$$

This reaction, called the steam–methane reforming reaction, is followed by another, called the water–gas shift reaction, in which the carbon monoxide formed in the

▶❚❚ **CHEMTOUR** Equilibrium

∞ **CONNECTION** We introduced the steam-methane reforming reaction and the water-gas shift reaction in the descriptive chemistry box in Chapter 3.

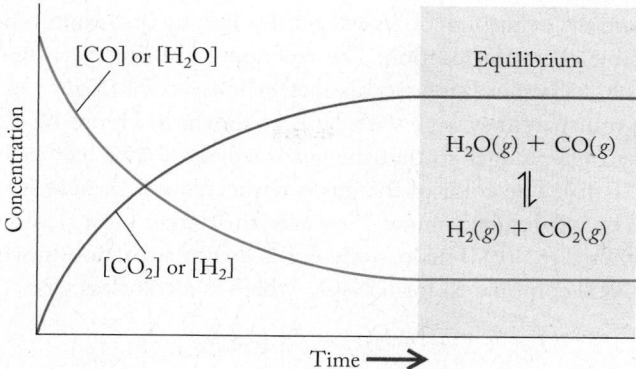

$$H_2O(g) + CO(g)$$

$$H_2(g) + CO_2(g)$$

FIGURE 16.1 Concentrations of reactants and products in the water–gas shift reaction change over time until equilibrium is reached. At equilibrium in the water–gas shift reaction, H_2 and CO_2 are being formed at the same rate at which they are reacting to reform H_2O and CO.

first step is reacted with more steam in the presence of a Cu/ZnO catalyst at around 200°C:

$$H_2O(g) + CO(g) \rightleftharpoons H_2(g) + CO_2(g) \qquad (16.2)$$

Let's explore the dynamics of the second reaction. If we put an equal number of moles of water vapor and carbon monoxide in a closed chamber and allow them to react, the concentrations of CO and H_2O initially fall as the concentrations of H_2 and CO_2 increase, as shown in Figure 16.1. Because H_2O and CO react in a 1:1 stoichiometric ratio, their concentrations decrease at the same rate and are always the same in the reaction chamber. Similarly, H_2 and CO_2 are formed in a 1:1 ratio, and so their concentrations increase at the same rate and are always equal.

The water–gas shift reaction is reversible. Reversibility, as we discussed in Chapter 14, means that if we intervene in the course of the reaction by, for example, changing reaction temperature or removing or adding a reactant or product, we might be able to force the reaction to run in reverse, reforming reactants from products. One requirement for reversibility is that the products must remain in contact with each other. If one or more of the products is a gas, then we need to run the reaction in a sealed chamber.

The graph in Figure 16.1 shows that eventually the concentrations of reactants and products in the water–gas shift reaction no longer change with time, at which point the reaction has reached chemical equilibrium. Note, however, that the concentrations of the reactants (CO and H_2O) do not go to zero. The presence of reactants in the mixture when the reaction has reached equilibrium means that the yield of the reaction never reaches 100%.

Now let's think about the changes in the *rates* of the forward and reverse reactions during the course of the water–gas shift reaction. When the CO and H_2O are initially mixed, their concentrations are at their maximum and the rate of the forward reaction, as indicated by the slope of the [CO] and [H_2O] curve at $t = 0$ in Figure 16.1, is also at its maximum. As the reaction proceeds, reactant concentrations decrease, and the rate of the forward reaction also decreases because the likelihood of collisions between reactant molecules decreases.

The concentrations of products are initially zero and so is the rate of the reverse reaction. As product molecules form, the likelihood of their colliding with one another to reform reactant molecules increases and so does the rate of the reverse reaction. Ultimately the rate of the reverse reaction equals the rate of the forward reaction as shown in Figure 16.2. At this point equilibrium has been achieved.

CONNECTION The concept of reaction reversibility was described in Section 14.2.

CONNECTION The method for calculating the percent yield of a reaction was described in Section 3.9.

CONNECTION In Chapter 15 we discussed how reactions occur when molecules collide; the higher the concentrations of reactants, the more frequently molecules collide, and the faster a reaction proceeds.

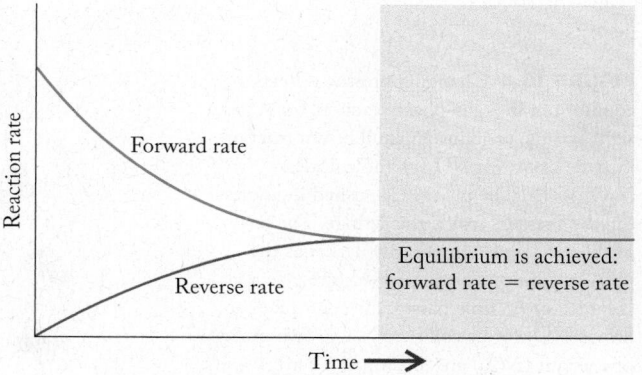

FIGURE 16.2 The rates of forward and reverse reactions are the same when equilibrium is achieved.

FIGURE 16.3 Air quality in Los Angeles has improved since this photograph of "brown LA haze" was taken. The color was caused by high concentrations of NO_2.

∞ **CONNECTION** We discussed the rates of reactions of nitrogen oxides in the atmosphere in Chapter 15 in our discussion of the chemistry of photochemical smog.

Let's explore the concept of chemical equilibrium by first examining the kinetics of a reversible chemical reaction. The reaction has a single reactant: NO_2, the gas that gives photochemical smog its distinctive brown color (Figure 16.3). Suppose we fill a large transparent syringe with NO_2 as shown in Figure 16.4(a). Then we push on the syringe plunger so that the gas is squeezed into half its original volume (Figure 16.4b). The color of the gas is darker brown because we doubled its concentration by halving its volume. However, the darker brown color soon fades, as shown in Figure 16.4(c). The loss of color is due to a reaction in which pairs of molecules of NO_2 combine to form N_2O_4, which is a colorless gas:

$$2\,NO_2(g) \rightleftharpoons N_2O_4(g)$$
$$\text{(brown)} \qquad \text{(colorless)} \qquad \qquad (16.3)$$

The color never fades completely, however, because the reaction reaches an equilibrium in which some NO_2 is still present. Chemists know from experimental data that the rate laws of the forward and reverse reactions are

$$\text{Rate}_f = k_f[NO_2]^2 \qquad \qquad (16.4)$$

$$\text{Rate}_r = k_r[N_2O_4] \qquad \qquad (16.5)$$

where "f" represents the forward reaction and "r" represents the reverse reaction. When the reaction achieves chemical equilibrium,

$$\text{Rate}_f = \text{rate}_r \qquad \qquad (16.6)$$

Replacing the terms in Equation 16.6 with the right sides of Equations 16.4 and 16.5:

$$k_f[NO_2]^2 = k_r[N_2O_4]$$

which we can rearrange to

$$\frac{k_f}{k_r} = \frac{[N_2O_4]}{[NO_2]^2} \qquad \qquad (16.7)$$

FIGURE 16.4 Changing pressure affects equilibrium in a gas-phase reaction. (a) A gas-tight syringe contains an equilibrium reaction mixture of brown $NO_2(g)$ and colorless $N_2O_4(g)$. (b) The plunger is pushed in, increasing the pressure inside the syringe. The color of the mixture is temporarily darker as the NO_2 molecules are compressed into a smaller volume. (c) As time passes, the color fades as brown NO_2 forms colorless N_2O_4. Two moles of reactant (NO_2) are consumed for every one mole of N_2O_4 formed. The total number of moles of gas in the syringe is reduced, partly relieving the increase in pressure.

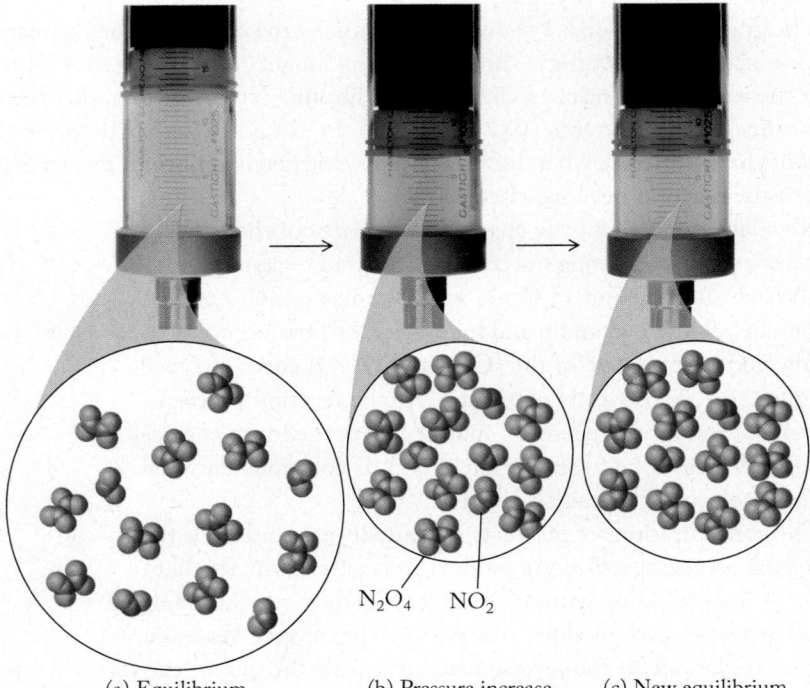

N_2O_4 NO_2

(a) Equilibrium   (b) Pressure increase   (c) New equilibrium

The ratio k_f/k_r is the ratio of two constants, which is simply another constant. This constant has a special name: it is called an **equilibrium constant (K)** and equating it to the ratio of product to reactant concentrations on the right side of Equation 16.7 gives us the **equilibrium constant expression** for this reaction:

$$K = \frac{[N_2O_4]}{[NO_2]^2} \qquad (16.8)$$

We will shortly explore other ways to formulate equilibrium constant expressions.

> **equilibrium constant (K)** the value of the ratio of concentration (or partial pressure) terms in the equilibrium constant expression at a specific temperature.
>
> **equilibrium constant expression** the ratio of the equilibrium concentrations or partial pressures of *products to reactants, each term raised to a power equal to the coefficient of that substance* in the balanced chemical equation for the reaction.

CONCEPT TEST

In the reversible reaction $A \rightleftharpoons B$, the rate constant of the forward reaction at a particular temperature is 3.0 times the value of the rate constant of the reverse reaction. What is the value of the equilibrium constant at that temperature?

(Answers to Concept Tests are in the back of the book.)

The equilibrium constants can have values that range from those that are almost too large to determine to those that are almost too small to determine. If K is very large ($K \gg 1$), the concentrations of products in a reaction mixture at equilibrium are much larger than the concentrations of reactants. If K is very small ($K \ll 1$), the concentrations of products at equilibrium are much smaller than the concentrations of reactants. Always keep in mind that the equilibrium constant says nothing about how fast a reaction reaches equilibrium; it only tells us the extent of the reaction once equilibrium is reached.

16.2 Writing Equilibrium Constant Expressions

Let's revisit the water–gas shift reaction:

$$H_2O(g) + CO(g) \rightleftharpoons H_2(g) + CO_2(g) \qquad (16.2)$$

The data in Table 16.1 describe the results of four experiments in which different quantities of H_2O gas, CO, H_2, and CO_2 are injected into a sealed reaction vessel heated to 500 K. The four gases are allowed to react and the final concentrations of all four are determined when the reaction has reached equilibrium.

TABLE 16.1	**Initial and Equilibrium Concentrations of the Reactants and Products in the Water-Gas Shift Reaction [$H_2O(g) + CO(g) \rightleftharpoons H_2(g) + CO_2(g)$] at 500 K**							
Experiment	**INITIAL CONCENTRATION (M)**				**EQUILIBRIUM CONCENTRATION (M)**			
	[H_2O]	**[CO]**	**[H_2]**	**[CO_2]**	**[H_2O]**	**[CO]**	**[H_2]**	**[CO_2]**
1	0.0200	0.0200	0	0	0.0034	0.0034	0.0166	0.0166
2	0	0	0.0200	0.0200	0.0034	0.0034	0.0166	0.0166
3	0.0100	0.0200	0.0300	0.0400	0.0046	0.0146	0.0354	0.0454
4	0.0200	0.0100	0.0200	0.0100	0.0118	0.0018	0.0282	0.0182

In Experiment 1 the reaction vessel initially contains equimolar concentrations of H_2O and CO but no H_2 or CO_2. In Experiment 2 the vessel initially contains equimolar concentrations of H_2 and CO_2 but no H_2O or CO. The data in Table 16.1 indicate that when the reaction mixtures in Experiments 1 and 2 achieve chemical equilibrium, the concentrations of H_2O and CO are the same (0.0034 M) in both experiments and so too are the concentrations of H_2 and CO_2 (0.0166 M). (Note that these statements about equilibrium concentrations are true in this case because the stoichiometric coefficients are all "1.") These results indicate that the composition of a reaction mixture at equilibrium is independent of the direction in which a particular reaction ran to achieve equilibrium.

Additional data in Table 16.1 show that the forward reaction takes place when the concentrations of the products are initially the same as the concentrations of the reactions (Experiment 4), or even higher (Experiment 3). The significance of these observations, taken with those from Experiments 1 and 2, can be appreciated if we do the following math: multiply the equilibrium concentrations of the products (H_2 and CO_2) together and divide that product by the product of the equilibrium concentrations of the reactants (H_2O and CO):

$$\text{Experiments 1 and 2} \qquad \frac{[H_2][CO_2]}{[H_2O][CO]} = \frac{(0.0166)(0.0166)}{(0.0034)(0.0034)} = 24$$

$$\text{Experiment 3} \qquad \frac{[H_2][CO_2]}{[H_2O][CO]} = \frac{(0.0354)(0.0454)}{(0.0046)(0.0146)} = 24$$

$$\text{Experiment 4} \qquad \frac{[H_2][CO_2]}{[H_2O][CO]} = \frac{(0.0282)(0.0182)}{(0.0118)(0.0018)} = 24$$

In every experiment this calculation yields the same result.

As you might guess, we would get the same ratio of product to reactant concentrations at equilibrium from *any* combination of initial concentrations of these four gases at 500 K. This constancy applies to other reaction mixtures, and has been known since the mid-19th century, when Norwegian chemists Cato Guldberg (1836–1902) and Peter Waage (1833–1900) discovered that any reversible reaction eventually reaches a state in which the ratio of the concentrations of products to reactants, with each value raised to a power corresponding to the coefficient for that substance in the balanced chemical equation for the reaction, has a characteristic value at a given temperature. They called this phenomenon the **law of mass action**. This ratio of concentration terms is the equilibrium constant expression for the reaction. It is also called the **mass action expression**, but we will limit the use of that term to those situations in which the ratio of products to reactants does not match the value of the equilibrium constant. For the water–gas shift reaction at 500 K, the value of K and the equilibrium constant expression are

$$K = \frac{[CO_2][H_2]}{[CO][H_2O]} = 24$$

Remember that the exponents in the equilibrium constant expression must be the same as the coefficients in the balanced equation. This is in contrast to rate law expressions, where the exponents are frequently not the same as the coefficients in the balanced equation because they reflect the stoichiometry of the rate-determining step, not necessarily the overall reaction.

law of mass action the ratio of the concentrations or partial pressures of products to reactants at equilibrium has a characteristic value at a given temperature when each term is raised to a power equal to the coefficient of that substance in the balanced chemical equation for the reaction.

mass action expression equivalent to the equilibrium constant expression, but applied to reaction mixtures that may, or may not, be at equilibrium.

Refer to the water–gas shift reaction on p. 767. Equal numbers of moles of water vapor, carbon dioxide, carbon monoxide, and hydrogen gas are injected into a rigid, sealed reaction vessel and heated to 500 K. Which expression about the composition of the reaction mixture at equilibrium is true?

    a. $[CO] = [H_2] = [CO_2] = [H_2O]$
    b. $[CO_2] = [H_2] > [CO] = [H_2O]$
    c. $[CO_2] = [H_2] < [CO] = [H_2O]$
    d. $[H_2O] = [H_2] > [CO_2] = [CO]$
    e. $[H_2O] = [H_2] < [CO_2] = [CO]$

In the preceding paragraphs we used equilibrium constant expressions based on the molar concentrations of products and reactants. For the generic reaction in which a moles of reactant A react with b moles of B to form c moles of substance C and d moles of D:

$$aA + bB \rightleftharpoons cC + dD$$

the equilibrium constant expression is

$$K_c = \frac{[C]^c[D]^d}{[A]^a[B]^b} \tag{16.9}$$

where the subscript "c" of the equilibrium constant represents *concentration*. If substances A, B, C, and D are gases, then the equilibrium constant may also be expressed in terms of their partial pressures:

$$K_p = \frac{(P_C)^c(P_D)^d}{(P_A)^a(P_B)^b} \tag{16.10}$$

As we shall see later in this chapter, the values of K_c and K_p for a given reaction and temperature may or may not be the same. It depends on whether the number of moles of gaseous reactants is the same as the number of moles of gaseous products. Also, *equilibrium constants never have units*. This is true even when there are different numbers of moles of reactants and products. Values of K have no units because the concentrations and partial pressures used to calculate K_c and K_p values are actually *ratios* of concentrations and partial pressures to an *ideal standard* concentration (1.000 M) or partial pressure (1.000 atm). These ratios take into account the nonideal behavior of substances. Because the terms are ratios of quantities with the same units, their units cancel out, leaving unitless equilibrium constants. When we use concentration and partial pressure values directly in equilibrium calculations in this text, we are making the assumption that all the reactants and products are behaving ideally.

▶❙❙ **CHEMTOUR** Equilibrium in the Gas Phase

♾ **CONNECTION** We introduced the concept of nonideal behavior in Section 6.9 when we described the reasons why *real* gases may not behave exactly like *ideal* gases.

SAMPLE EXERCISE 16.1   **Writing Equilibrium Constant Expressions**

A key reaction in the formation of acid rain involves the reversible combination of SO_2 and O_2 in the atmosphere, producing SO_3:

$$2\,SO_2(g) + O_2(g) \rightleftharpoons 2\,SO_3(g)$$

Write the K_c and K_p expressions for this reaction.

Collect and Organize We are given the balanced chemical equation for a reaction and asked to write K_c and K_p expressions for the reaction. Equilibrium constant expressions are ratios of the concentrations (K_c) or partial pressures (K_p) of products to reactants, with each term raised to the power equal to its coefficient in the balanced chemical equation of the reaction.

Analyze In the reaction of interest, the coefficients of SO_2 and SO_3 are both 2, so the SO_2 and SO_3 terms in the K_c and K_p expressions will be squared.

Solve

$$K_c = \frac{[SO_3]^2}{[SO_2]^2[O_2]}$$

$$K_p = \frac{(P_{SO_3})^2}{(P_{SO_2})^2(P_{O_2})}$$

Think about It The K_c and K_p expressions have the same format: their numerators and denominators contain terms for the same products and reactants, each raised to the same power. The difference between them is the nature of the terms: molar concentrations in the K_c expression and partial pressures in the K_p expression.

Practice Exercise Write the equilibrium constant expressions K_c and K_p for this reaction:

$$CH_4(g) + H_2O(g) \rightleftharpoons CO(g) + 3\,H_2(g)$$

(Answers to Practice Exercises are in the back of the book.)

SAMPLE EXERCISE 16.2   Calculating the Value of K_c

Table 16.2 contains data from four experiments on the dimerization of NO_2:

$$2\,NO_2(g) \rightleftharpoons N_2O_4(g)$$

The experiments were run at 100°C in a rigid, closed container. Use the data from each experiment to calculate a value of the equilibrium constant K_c for the dimerization reaction.

TABLE 16.2   Data for the Reaction $2\,NO_2(g) \rightleftharpoons N_2O_4(g)$ at 100°C

Experiment	INITIAL CONCENTRATION (M)		EQUILIBRIUM CONCENTRATION (M)	
	[NO$_2$]	[N$_2$O$_4$]	[NO$_2$]	[N$_2$O$_4$]
1	0.0200	0.0000	0.0172	0.00139
2	0.0300	0.0000	0.0244	0.00280
3	0.0400	0.0000	0.0310	0.00452
4	0.0000	0.0200	0.0310	0.00452

Collect and Organize We are given four sets of data that contain initial and equilibrium concentrations of a reactant and product. We are asked to determine the value of the equilibrium constant K_c in each experiment. The equilibrium constant expression (Equation 16.8) for this reaction is

$$K_c = \frac{[N_2O_4]}{[NO_2]^2}$$

Analyze In each of the four experiments the concentration of NO_2 is nearly 10 times the concentration of N_2O_4 at equilibrium. However, both values in each experiment are much less than unity, and the $[NO_2]$ term is squared. These two factors taken together mean that the values of the numerators in the equilibrium constant expressions will probably be greater than the denominators, so the calculated values of K_c will be greater than 1.

Solve

$$\text{Experiment 1} \qquad K_c = \frac{[N_2O_4]}{[NO_2]^2} = \frac{0.00139}{(0.0172)^2} = 4.70$$

$$\text{Experiment 2} \qquad K_c = \frac{0.00280}{(0.0244)^2} = 4.70$$

$$\text{Experiment 3} \qquad K_c = \frac{0.00452}{(0.0310)^2} = 4.70$$

$$\text{Experiment 4} \qquad K_c = \frac{0.00452}{(0.0310)^2} = 4.70$$

Think about It The values calculated for K_c are the same, as they should be for the same reaction at the same temperature, and, as predicted, are greater than 1.

Practice Exercise A mixture of gaseous CO and H_2, called *synthesis gas,* is used commercially to prepare methanol (CH_3OH), a compound considered an alternative fuel to gasoline. Under equilibrium conditions at 700 K, $[H_2] = 0.074$ mol/L, $[CO] = 0.025$ mol/L, and $[CH_3OH] = 0.040$ mol/L. What is the value of K_c for this reaction at 700 K?

SAMPLE EXERCISE 16.3   Calculating the Value of K_p

A sealed chamber contains an equilibrium mixture of NO_2 and N_2O_4 at 300°C and partial pressures $P_{NO_2} = 0.101$ atm and $P_{N_2O_4} = 0.074$ atm. What is the value of K_p for the following reaction under these conditions?

$$2\,NO_2(g) \rightleftharpoons N_2O_4(g)$$

Collect and Organize We are asked to determine the value of the K_p equilibrium constant for the dimerization reaction of NO_2 to form N_2O_4. We are given the partial pressure values of both gases at equilibrium. The concentration-based equilibrium constant expression for the reaction (Equation 16.8) is

$$K_c = \frac{[N_2O_4]}{[NO_2]^2}$$

Analyze The format of the K_p expression for the reaction is the same as the K_c expression. The only difference is that the concentration terms in the K_c expression are replaced with partial pressure terms:

$$K_p = \frac{P_{N_2O_4}}{(P_{NO_2})^2}$$

The partial pressure of NO_2 is slightly greater than the partial pressure of N_2O_4 at equilibrium, but both values are less than one. Moreover, the P_{NO_2} term is squared. These two factors taken together mean that the value of the numerator in the K_p expression will probably be greater than that of the denominator, so the calculated value of K_p will be greater than 1.

Solve

$$K_p = \frac{0.074}{(0.101)^2} = 7.3$$

Think about It As we predicted, the value of K_p is greater than 1, even though the equilibrium partial pressure of the product ($P_{N_2O_4} = 0.074$ atm) is actually less than that of the reactant ($P_{NO_2} = 0.101$ atm). Relatively little N_2O_4 forms in this case because P_{NO_2} is so low, and two moles of NO_2 are required to make one mole of N_2O_4. As we will see in Section 16.7, low pressures favor the side of a gas-phase reaction that has more moles of gas. To illustrate this point, let's calculate the value of $P_{N_2O_4}$ that would be in equilibrium with 1 atm of NO_2 at 300°C:

$$K_p = \frac{(P_{N_2O_4})}{(P_{NO_2})^2} = 7.3 = \frac{(P_{N_2O_4})}{1^2}$$

$$P_{N_2O_4} = 7.3 \text{ atm}$$

The value of $P_{N_2O_4}$ is 7.3 times the P_{NO_2} value at these higher overall pressures, whereas $P_{N_2O_4}$ was lower than P_{NO_2} at the lower pressures of the Sample Exercise.

Practice Exercise A reaction vessel contains an equilibrium mixture of SO_2, O_2, and SO_3. Given the partial pressures $P_{SO_2} = 0.0018$ atm, $P_{O_2} = 0.0032$ atm, and $P_{SO_3} = 0.0166$ atm, calculate the value of K_p for the reaction:

$$2\,SO_2(g) + O_2(g) \rightleftharpoons 2\,SO_3(g)$$

The value of K indicates how far a reaction proceeds at a given temperature. The values we have seen thus far—$K_c = 24$ for the water–gas shift reaction at 500 K, and $K_p = 7.3$, $K_c = 4.7$ for the dimerization of NO_2 at 300°C and 100°C, respectively—are considered intermediate values. Because the range of K values is so large ($0 < K < \infty$), all three of these values are considered *close* to 1, which means that comparable concentrations of reactants and products are likely to be present at equilibrium.

In contrast, the reaction between H_2 and O_2 to form water proceeds until one of the reactants is completely consumed. This observation is consistent with a large value of K at 25°C:

$$2\,H_2(g) + O_2(g) \rightleftharpoons 2\,H_2O(g) \qquad K_c = 3 \times 10^{81}$$

On the other hand, the decomposition of CO_2 to CO and O_2 at 25°C proceeds hardly at all and is consistent with a very small value of K and virtually no product being formed:

$$2\,CO_2(g) \rightleftharpoons 2\,CO(g) + O_2(g) \qquad K_c = 3 \times 10^{-92}$$

16.3 Relationships between K_c and K_p Values

As we noted in Section 16.2, the values of K_c and K_p for a given reaction and temperature may or may not be the same, depending on the numbers of moles of gaseous reactants and products. To better understand this relationship, we begin with the ideal gas law:

$$PV = nRT$$

If we solve for P and express volume in liters, then n/V has units of moles per liter, which is the same as molarity (M):

$$P = \frac{n}{V} RT$$

$$P = MRT \qquad (16.11)$$

Let's apply Equation 16.11 to the gases in the NO_2/N_2O_4 equilibrium from Sample Exercise 16.3:

$$P_{NO_2} = \frac{n_{NO_2}}{V} RT = [NO_2]RT$$

$$P_{N_2O_4} = \frac{n_{N_2O_4}}{V} RT = [N_2O_4]RT$$

Substituting these values into the expression for K_p from Sample Exercise 16.3, we get

$$K_p = \frac{(P_{N_2O_4})}{(P_{NO_2})^2} = \frac{[N_2O_4]RT}{([NO_2]RT)^2} = \frac{[N_2O_4]RT}{[NO_2]^2 R^2 T^2}$$

The ratio of concentration terms in the expression on the right, $[N_2O_4]/[NO_2]^2$, is the same as the K_c expression for this reaction. Substituting K_c for those terms and simplifying the RT terms:

$$K_p = K_c \frac{1}{RT}$$

This last equation defines the specific relationship between K_c and K_p for this reaction. A more general expression can be derived for the generic reaction of gases A and B forming gases C and D:

$$a\mathrm{A} + b\mathrm{B} \rightleftharpoons c\mathrm{C} + d\mathrm{D}$$

As noted on p. 769, the K_p expression for this reaction is

$$K_p = \frac{(P_C)^c (P_D)^d}{(P_A)^a (P_B)^b} \qquad (16.10)$$

Replacing each partial pressure term in Equation 16.10 with the corresponding molar concentration term $\times RT$ (from Equation 16.11) gives us the general expression

$$K_p = \frac{([C]RT)^c ([D]RT)^d}{([A]RT)^a ([B]RT)^b} \qquad (16.12)$$

Combining the RT terms, we get

$$K_p = \frac{[C]^c [D]^d}{[A]^a [B]^b} \times (RT)^{[(c+d)-(a+b)]} \qquad (16.13)$$

The concentration ratio on the right side of this equation matches the K_c expression for this generic reaction (see Equation 16.9). Substituting this equality into Equation 16.13 gives us

$$K_p = K_c(RT)^{[(c+d)-(a+b)]} \qquad (16.14)$$

To simplify Equation 16.14, consider this: $(c + d)$ represents the sum of the coefficients of the gaseous products in the reaction—the sum of the number of moles of gases produced. Similarly, $(a + b)$ represents the sum of the number of moles of gaseous reactants consumed. The difference between the two sums, $(c + d) - (a + b)$, represents the *change in the number of moles of gases* between the product and reactant sides of the balanced chemical equation. We use the symbol Δn to represent this change. Substituting Δn for $(c + d) - (a + b)$ in Equation 16.14 gives us

$$K_p = K_c(RT)^{\Delta n} \qquad (16.15)$$

Equation 16.15 provides a quantitative interpretation of the opening statement of this section: the relationship between the K_p and K_c values of a chemical reaction involving gases depends on the number of moles of gaseous reactants and products. In reactions such as the water–gas shift reaction:

$$H_2O(g) + CO(g) \rightleftharpoons H_2(g) + CO_2(g) \qquad (16.2)$$

in which the number of moles of gas on both sides of the reaction arrow is the same, $\Delta n = 0$ and $K_p = K_c$. However, in the steam–methane reforming reaction:

$$CH_4(g) + H_2O(g) \rightleftharpoons CO(g) + 3\,H_2(g) \qquad (16.1)$$

2 moles of gaseous reactants form 4 moles of gaseous products. Therefore,

$$\Delta n = 4\text{ mol} - 2\text{ mol} = 2\text{ mol}$$

Inserting this value for Δn in Equation 16.15 gives us this reaction's relationship between K_p and K_c:

$$K_p = K_c(RT)^2$$

A final point about the relative sizes of K_p and K_c values: One mole of an ideal gas at STP occupies a volume of 22.4 liters. Therefore, its molar concentration is 1 mol/22.4 L = 0.0446 M. Thus the value of the pressure of a pure gas at STP (1.00 atm) is 22.4 times its molar concentration. The RT term in Equation 16.15 essentially corrects for this difference in how we express how much of a gas is present in a reaction mixture. It is a *conversion factor* for changing molar concentrations into partial pressures. The value of RT at STP is

$$0.08206 \, \frac{\text{L} \cdot \text{atm}}{\text{mol} \cdot \text{K}} \times 273 \, \text{K} = 22.4 \, \frac{\text{L} \cdot \text{atm}}{\text{mol}}$$

SAMPLE EXERCISE 16.4   Calculating K_c from K_p

In Sample Exercise 16.3 we calculated the value of K_p (7.3) for the dimerization of NO_2 to N_2O_4 at 300°C. What is the value of K_c for this reaction at 300°C?

Collect and Organize We are given a K_p value and asked to calculate the corresponding K_c value for the same reaction at the same temperature. Equation 16.15 relates K_p and K_c values:

$$K_p = K_c(RT)^{\Delta n}$$

where Δn represents the change in the number of moles of gas when going from reactant to product.

Analyze We need a balanced chemical equation to determine the value of Δn. From Sample Exercise 16.2 we know that the equation is

$$2\,NO_2(g) \rightleftharpoons N_2O_4(g)$$

Because the number of moles of gas is not the same on both sides of the reaction arrow, we predict that the values of K_c and K_p will not be the same.

Solve According to the balanced equation, 2 moles of gaseous reactants yield 1 mole of gaseous product. Therefore,

$$\Delta n = 1\,mol - 2\,mol = -1\,mol$$

Inserting this value, the given value of K_p, and temperature into Equation 16.15,

$$K_p = K_c(RT)^{\Delta n}$$
$$7.3 = K_c[0.08206 \times (273 + 300)]^{-1}$$
$$K_c = (7.3)(0.08206)(573) = 3.4 \times 10^2$$

Think about It The value of K_c differs from the value of K_p because two moles of gaseous reactants produce only one mole of gaseous product. Actually, the value of K_c is nearly 50 times larger than that of K_p. The K_c/K_p ratio is more than twice the STP molar volume factor (22.4 L · atm/mol) because the absolute temperature of the reaction is more than twice the temperature at STP.

Practice Exercise An important industrial process for synthesizing the ammonia used in agricultural fertilizers involves the combination of N_2 and H_2:

$$N_2(g) + 3\,H_2(g) \rightleftharpoons 2\,NH_3(g) \qquad K_c = 2.8 \times 10^{-9} \text{ at } 30°C$$

What is the value of K_p of this reaction at 30°C?

16.4 Manipulating Equilibrium Constant Expressions

We can write equilibrium constant expressions for reactions running in reverse, for chemical equations that have been multiplied or divided by a value that gives a key component a coefficient of 1, and for overall reactions that are combinations of other reactions.

K for Reverse Reactions

The K values for the forward and reverse directions of a reversible reaction are related. For example, in Sample Exercise 16.2, we wrote the K_c expression for the dimerization reaction

$$2\,NO_2(g) \rightleftharpoons N_2O_4(g)$$

this way:

$$K_c = \frac{[N_2O_4]}{[NO_2]^2}$$

In the reverse of this process, the decomposition of N_2O_4:

$$N_2O_4(g) \rightleftharpoons 2\,NO_2(g)$$

the reactant becomes the product and the product becomes the reactant, and this reversal is reflected in the equilibrium constant expression for the decomposition reaction:

$$K_c = \frac{[NO_2]^2}{[N_2O_4]}$$

where numerator has become denominator, and the denominator, the numerator. Expressing the relation between the forward (K_f) and reverse (K_r) equilibrium constants mathematically, we have

$$K_f = \frac{1}{K_r} \qquad (16.16)$$

To explore the meaning of Equation 16.16, consider a generic reversible reaction in which the equilibrium favors the formation of product, which means the equilibrium lies far to the right and is reflected in a large K value:

$$A \rightleftharpoons B \qquad K \gg 1 \qquad (16.17)$$

Because the reciprocal of a large value is a small one, Equation 16.16 tells us that the equilibrium constant for the reverse reaction must be small:

$$B \rightleftharpoons A \qquad K \ll 1 \qquad (16.18)$$

If we apply Equation 16.16 to the dimerization of NO_2 at 300°C (Sample Exercise 16.3):

$$2\,NO_2(g) \rightleftharpoons N_2O_4(g) \qquad K_p = 7.3$$

we can calculate the K_p value for the decomposition of N_2O_4:

$$N_2O_4(g) \rightleftharpoons 2\,NO_2(g) \qquad K_p = 1/7.3 = 0.14$$

SAMPLE EXERCISE 16.5 **Calculating the Value of K for a Reverse Reaction**

Atmospheric NO combines with O_2 to form NO_2. The reverse reaction is the decomposition of NO_2 to NO and O_2. At 184°C, the value of K_c for the forward reaction is 1.48×10^4. Write the equilibrium constant expressions for both reactions and calculate the value of K_c for the decomposition of NO_2 at 184°C.

Collect and Organize We are to write equilibrium constant expressions for the forward and reverse reactions for a process at equilibrium and to calculate the equilibrium constant for the reverse reaction. We know the identities of the reactants and products and the K_c value of the forward reaction. We need to write balanced chemical equations for the forward and reverse reactions and then use the coefficients from those equations to write equilibrium constant expressions.

Analyze We start by writing balanced chemical equations for the forward and reverse reactions:

Forward reaction $2\,NO(g) + O_2(g) \rightleftharpoons 2\,NO_2(g)$

Reverse reaction $2\,NO_2(g) \rightleftharpoons 2\,NO(g) + O_2(g)$

The coefficient of 2 in front of NO and NO_2 means that the concentration terms for these compounds are squared in the K_c expressions. The large ($>10^4$) value of K_c of the forward reaction means that the K_c value of the reverse reaction should be less than 10^{-4}.

Solve

$$K_f = \frac{[NO_2]^2}{[NO]^2[O_2]}$$

$$K_r = \frac{[NO]^2[O_2]}{[NO_2]^2}$$

$$K_r = \frac{1}{K_f} = \frac{1}{1.48 \times 10^4} = 6.76 \times 10^{-5}$$

Think about It The value of K_f is large, so it makes sense that its reciprocal, K_r, is small.

Practice Exercise At 300°C, the value of K_p for the combination reaction of N_2 and H_2 that produces NH_3 gas is 4.3×10^{-3}. What is the equilibrium constant at the same temperature for the decomposition reaction of NH_3 that produces N_2 and H_2?

K for an Equation Multiplied by a Number

As we saw in Chapter 5, sometimes we must represent a reaction by an equation containing fractional coefficients to have exactly 1 mole of a particular reactant or product. For example, we can rewrite the forward reaction from Sample Exercise 16.5:

$$2\,NO(g) + O_2(g) \rightleftharpoons 2\,NO_2(g)$$

to reflect the preparation of only 1 mole of NO_2 by multiplying each coefficient by $\frac{1}{2}$:

$$NO(g) + \tfrac{1}{2}O_2(g) \rightleftharpoons NO_2(g) \qquad (16.19)$$

Because the equilibrium constant expression for any reaction must be written to match the balanced chemical equation, the appropriate K_c expression for Equation 16.19 is

$$K_c = \frac{[NO_2]}{[NO][O_2]^{1/2}} \qquad (16.20)$$

Comparing this K_c expression with the K_c expression for the forward reaction from Sample Exercise 16.5:

$$K_c = \frac{[NO_2]^2}{[NO]^2[O_2]} \qquad (16.21)$$

we see that the right side of Equation 16.20 is equal to the square root of the right side of Equation 16.21: the NO and NO_2 terms are squared in Equation 16.21 but are raised to only the first power in Equation 16.20, and the O_2 term is raised to the first power in Equation 16.21 but to the $\frac{1}{2}$ power in Equation 16.20. Therefore, we may conclude that the value of the equilibrium constant K_c in Equation 16.20 is the square root of the value of K_c in Equation 16.21. We can extend this pattern to all chemical equilibria with the following rule: If the balanced chemical equation of a reaction is multiplied by some factor n, then the value of K is raised to the nth power.

SAMPLE EXERCISE 16.6   Calculating *K* for Different Coefficients

An important reaction in the formation of atmospheric aerosols of sulfuric acid (acid rain) is the oxidation of SO_2 to SO_3. One way to write a chemical equation for the oxidation reaction is

$$\text{Equation A} \qquad SO_2(g) + \tfrac{1}{2}O_2(g) \rightleftharpoons SO_3(g)$$

If the value of K_c for this reaction at 298 K is 2.8×10^{12}, what is the value of K_c at 298 K for the following reaction?

$$\text{Equation B} \qquad 2\,SO_2(g) + O_2(g) \rightleftharpoons 2\,SO_3(g)$$

Collect and Organize We are given the value of K_c for a reaction that describes the production of 1 mole of SO_3 and are asked to recalculate the value of K_c for the same reaction but based on producing 2 moles of SO_3.

Analyze Equation A is multiplied by 2 to generate Equation B. This means that the K_c for Equation A must be raised to the second power to generate K_c for Equation B. The value of K_c for Equation A is large; the value of K_c for Equation B should be even larger.

Solve

$$K_B = (K_A)^2$$
$$= (2.8 \times 10^{12})^2 = 7.8 \times 10^{24}$$

Think about It Doubling the coefficients means squaring the value of *K*. Our prediction that the second *K* would have a very large value is correct.

Practice Exercise If $K_c = 2.4 \times 10^{-3}$ for the reaction

$$N_2(g) + 3\,H_2(g) \rightleftharpoons 2\,NH_3(g)$$

at 1000 K, what is K_c at 1000 K for this reaction?

$$\tfrac{1}{3}N_2(g) + H_2(g) \rightleftharpoons \tfrac{2}{3}NH_3(g)$$

Given two equally legitimate equilibrium constant expressions for the same reaction, you may wonder how the same reaction having the same reactants and products can have two or more equilibrium constant values. Surely the same ingredients should be present in the same proportions at equilibrium no matter how we choose to write a balanced equation describing their reaction. In fact, they are. The difference in *K* values is not chemical; it is mathematical. It is related to how we use the equilibrium concentrations to calculate *K* values. It is, for example, affected by our choice to use the value of [NO] in Equation 16.20 and not the value of $[NO]^2$ as in Equation 16.21. However, that choice does not affect the concentration of NO at equilibrium. Always remember that whenever you write an expression for either K_c or K_p, you must identify the specific balanced equation you used to obtain the expression.

Combining *K* Values

In Chapter 5, we applied Hess's law to calculate the enthalpies of combined reactions. We carry out a similar process here to determine the overall *K* for a reaction that is the sum of two or more other reactions.

Consider two reactions from Chapter 15 involved in the formation of photochemical smog, wherein the NO produced in a car's engine at high temperatures is oxidized to NO_2 in the atmosphere:

(1) $\quad\quad N_2(g) + O_2(g) \rightleftharpoons \cancel{2NO(g)}$

(2) $\quad\quad \cancel{2NO(g)} + O_2(g) \rightleftharpoons 2NO_2(g)$

Overall $\quad N_2(g) + 2O_2(g) \rightleftharpoons 2NO_2(g)$

The equilibrium constant expression for the overall reaction is

$$K_c = \frac{[NO_2]^2}{[N_2][O_2]^2}$$

We can derive this expression from the equilibrium constant expressions for reactions 1 and 2,

$$K_1 = \frac{[NO]^2}{[N_2][O_2]} \quad \text{and} \quad K_2 = \frac{[NO_2]^2}{[NO]^2[O_2]}$$

if we multiply K_1 by K_2:

$$K_1 \times K_2 = \frac{\cancel{[NO]^2}}{[N_2][O_2]} \times \frac{[NO_2]^2}{\cancel{[NO]^2}[O_2]} = \frac{[NO_2]^2}{[N_2][O_2]^2} = K_{overall}$$

This approach works for all series of reactions, and as a general rule

$$K_{overall} = K_1 \times K_2 \times K_3 \times K_4 \times \ldots \times K_n \quad\quad (16.22)$$

The overall equilibrium constant for a sum of two or more reactions is the product of the equilibrium constants of the individual reactions. Thus, the value of K_c for the overall reaction for the formation of NO_2 from N_2 and O_2 at 1000 K is the product of the equilibrium constants for reaction 1:

$$K_1 = \frac{[NO]^2}{[N_2][O_2]} = 7.2 \times 10^{-9}$$

and reaction 2:

$$K_2 = \frac{[NO_2]^2}{[NO]^2[O_2]} = 0.020$$

Using Equation 16.22:

$$K_{overall} = K_1 \times K_2 = 7.2 \times 10^{-9} \times 0.020 = 1.4 \times 10^{-10}$$

Remember that the equilibrium constant expression for the overall reaction must contain the appropriate terms for the products and reactants of that reaction. Just as with Hess's law in thermochemical calculations, we may need to reverse an equation or multiply an equation by a factor when we combine it with another to create the equation of interest. If we reverse a reaction, we must take the reciprocal of its K. If we multiply a reaction by a constant, we must raise its K to that power.

SAMPLE EXERCISE 16.7   **Calculating Overall *K* of Combined Reactions**

At 1000 K, the K_c value of the following reaction is 1.5×10^6:

(1) $\quad N_2O_4(g) \rightleftharpoons 2NO_2(g)$

At 1000 K, the K_c value of the following reaction is 1.4×10^{-10}:

(2) $\quad N_2(g) + 2O_2(g) \rightleftharpoons 2NO_2(g)$

Calculate the K_c value at 1000 K of the reaction

$$N_2(g) + 2O_2(g) \rightleftharpoons N_2O_4(g)$$

Collect and Organize We are given two reactions and their K_c values. We need to combine the two reactions in such a way that N_2 and O_2 are on the reactant side of the overall equation and N_2O_4 is on the product side. We can then calculate the K_c value of the overall reaction. When two reactions are added, the value of K_c of the overall reaction is the product of the K_c values of the two reactions.

Analyze The overall reaction is the sum of the reserve of reaction 1 and reaction 2 as written:

$$\begin{array}{ll} \text{Reaction 1 reversed} & \cancel{2NO_2(g)} \rightleftharpoons N_2O_4(g) \\ +\ \text{Reaction 2} & N_2(g) + 2O_2(g) \rightleftharpoons \cancel{2NO_2(g)} \\ \hline \text{Overall} & N_2(g) + 2O_2(g) \rightleftharpoons N_2O_4(g) \end{array}$$

Reversing a chemical reaction requires taking the reciprocal of its K_c value. The K_c value of reaction 1 is large ($>10^6$), which means its reciprocal is small ($<10^{-6}$). The product of the latter value times the even smaller K_c value of reaction 2 ($\sim 10^{-10}$) should be a very small value—about 10^{-16}.

Solve

$$K_{overall} = \frac{1}{K_1} \times K_2 = \frac{1}{1.5 \times 10^6} \times 1.4 \times 10^{-10} = 9.3 \times 10^{-17}$$

Think about It The result of the calculation is close to the ballpark value we predicted. The overall equilibrium of the combined reactions lies far to the left; little N_2O_4 forms from N_2 and O_2 at 1000 K.

Practice Exercise Calculate the value of K_c for the hypothetical reaction

$$Q(g) + X(g) \rightleftharpoons M(g)$$

from the following information:

$$\begin{array}{ll} 2M(g) \rightleftharpoons Z(g) & K_c = 6.2 \times 10^{-4} \\ Z(g) \rightleftharpoons 2Q(g) + 2X(g) & K_c = 5.6 \times 10^{-2} \end{array}$$

To summarize the key points for manipulating equilibrium constants:

➤ The K value of a reaction running in reverse is the reciprocal of the K of the forward reaction.
➤ If the original chemical equation describing an equilibrium is multiplied by a factor n, the value of K of the new equilibrium constant expression is the value of the original K raised to the nth power.
➤ If an overall chemical reaction is the sum of two or more other reactions, the overall value of K is the product of the K values of the other reactions.

16.5 Equilibrium Constants and Reaction Quotients

In Sections 16.1 and 16.2 we introduced two key terms: *equilibrium constant expression* and *mass action expression*. Until now we have used the first term almost exclusively because we have been dealing with chemical reactions that have achieved equilibrium. Now it is time to reintroduce the concept of a mass action expression because we can apply it not only to concentrations (or partial pressures) of products and reactants in reaction mixtures that have reached equilibrium, but also to reaction mixtures that are on their way to equilibrium but are not there yet.

Even if a reversible chemical reaction has not reached equilibrium, we can still insert reactant and product concentrations (or partial pressures) into its mass action expression. The mathematical result is not a K value because the reaction is not yet at equilibrium. Instead it is a Q value, where Q stands for **reaction quotient**.

The value of Q provides us with a kind of status report on how a reaction is proceeding. To see how, let's revisit the water–gas shift reaction

$$H_2O(g) + CO(g) \rightleftharpoons H_2(g) + CO_2(g) \qquad K_c = 24 \text{ at } 500 \text{ K}$$

and the data from Experiment 3 in Table 16.1. In Experiment 3, the initial concentrations of reactants and products are

[H₂O] (*M*)	[CO] (*M*)	[H₂] (*M*)	[CO₂] (*M*)
0.0100	0.0200	0.0300	0.0400

Inserting these values into the mass action expression for the reaction yields a value for Q based on concentration, that is, Q_c:

$$Q_c = \frac{[H_2][CO_2]}{[H_2O][CO]} = \frac{(0.0300)(0.0400)}{(0.0100)(0.0200)} = 6.00$$

Now we compare this value of Q_c to the reaction's K_c value, which is 24. Clearly Q_c is less than K_c, which means there are proportionately smaller concentrations of products and larger concentrations of reactants in the initial reaction mixture than there will be at equilibrium. To achieve equilibrium, some of the reactants must react and form products, increasing the value of Q_c until it matches the value of K_c and the following equilibrium concentrations of reactants and products are present in the reaction vessel:

[H₂O] (*M*)	[CO] (*M*)	[H₂] (*M*)	[CO₂] (*M*)
0.0046	0.0146	0.0354	0.0454

To put the results from the data in Table 16.1 in context, let's consider the curves in Figure 16.5. Starting at the left side (zone a) of the graph we have the

reaction quotient (Q) the numerical value of the mass action expression for *any values* of the concentrations (or partial pressures) of reactants and products; at equilibrium, $Q = K$.

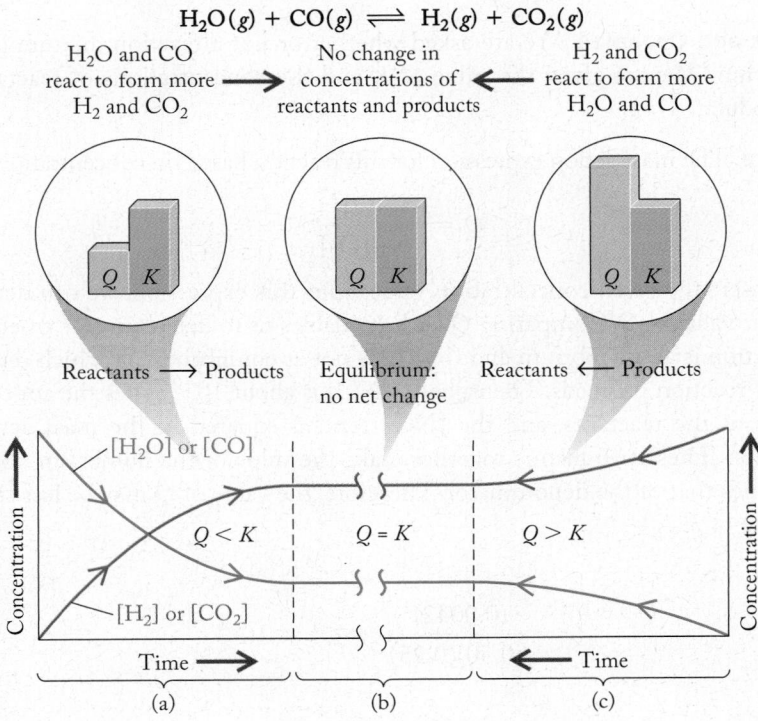

FIGURE 16.5 The value of the reaction quotient Q relative to the equilibrium constant K for the water–gas shift reaction. (a) Reactant concentrations (red) are higher than they are once equilibrium is reached, and product concentrations (blue) are lower than at equilibrium; $Q < K$, and the reactants form more products. (b) Equilibrium concentrations are achieved; $Q = K$, and no net change in concentrations takes place. (c) Product concentrations are higher than what they are at equilibrium, and reactant concentrations are lower than at equilibrium; $Q > K$, and products form more reactants as the reaction runs in reverse.

initial conditions of Experiment 1 from Table 16.1: equal concentrations of reactants are present, but no products. Over time, reactant concentrations (the red curve) decrease as product concentrations (the blue curve) increase. At the right side of the graph in zone (c), we have the initial conditions of Experiment 2: products are present, but no reactants. Over time, reactant concentrations increase as product concentrations decrease. In zone (b) in the middle of the graph, no net change occurs in the composition because the reaction is at equilibrium.

We can also characterize the three zones on the basis of the value of Q compared to K. In zone (a), Q values are less than K and there is a net conversion of reactants into products as the forward reaction dominates. In zone (c), Q values are greater than K and a net conversion of products into reactants takes place as the reverse reaction dominates. In the middle zone (b), $Q = K$ and no change in the composition of the reaction mixture occurs over time. The relative values of Q and K and their consequences are summarized in Table 16.3.

TABLE 16.3	Comparison of Q and K Values	
Value of Q	**What It Means**
$Q < K$	Reaction as written proceeds in forward direction ($\rightarrow$)
$Q = K$	Reaction is at equilibrium ($\rightleftharpoons$)
$Q > K$	Reaction as written proceeds in reverse direction ($\leftarrow$)

CONCEPT TEST

In which of the three zones do the initial reaction conditions of Experiments 3 and 4 from Table 16.1 (p. 767) fall?

SAMPLE EXERCISE 16.8 Using Q and K Values to Predict the Direction of a Reaction

At 2300 K the value of K of the following reaction is 1.5×10^{-3}:

$$N_2(g) + O_2(g) \rightleftharpoons 2\,NO(g)$$

At the instant when a reaction vessel at 2300 K contains 0.50 M N_2, 0.25 M O_2, and 0.0042 M NO, is the reaction mixture at equilibrium? If not, in which direction will the reaction proceed to reach equilibrium?

Collect and Organize We are asked whether or not a reaction mixture is at equilibrium. We are given the value of K and the concentrations of reactants and product.

Analyze The mass action expression for this reaction based on concentrations is

$$Q = \frac{[NO]^2}{[N_2][O_2]}$$

If we insert the given concentration values into this expression, we can determine the value of Q. Comparing Q with K enables us to determine (a) whether the reaction is at equilibrium and (b) if it is not at equilibrium, in which direction the reaction proceeds. The value of [NO] is about 10^{-2} times the concentrations of the reactants, and the [NO] term is squared in the mass action expression. These two factors together make the value of the numerator about 10^{-4} times that of the denominator. Therefore, the value of Q may be less than the value of K.

Solve

$$Q = \frac{(0.0042)^2}{(0.50)(0.25)} = 1.4 \times 10^{-4}$$

homogeneous equilibria involve reactants and products in the same phase.

heterogeneous equilibria involve reactants and products in more than one phase.

The value of Q is indeed less than that of K; so, the reaction mixture is not at equilibrium. To achieve equilibrium, more reactant must form product so that the numerator of the mass action expression increases and the denominator decreases. This happens if the reaction proceeds in the forward direction.

Think about It Our guess that Q would be less than K was correct. The value of K is small, but the value of Q is even smaller.

Practice Exercise The value of K_c for the reaction

$$2\,NO_2(g) \rightleftharpoons N_2O_4(g)$$

is 4.7 at 373 K. Is a mixture of the two gases in which $[NO_2] = 0.025\ M$ and $[N_2O_4] = 0.0014\ M$ in chemical equilibrium? If not, in which direction does the reaction proceed to achieve equilibrium?

CONCEPT TEST

In Sample Exercise 16.8 we did not use a subscript "c" with the symbols Q and K. Suggest a reason why.

(a)

(b)

FIGURE 16.6 (a) For centuries primitive kilns were used to decompose limestone ($CaCO_3$) into lime (CaO) and CO_2. These kilns were charged with layers of fuel (originally wood, later coal) and crushed limestone, and the fuel was ignited. Farmers use lime to "sweeten" (neutralize the acidity of) soil. It is also used to make mortar and cement. (b) A modern limestone kiln.

16.6 Heterogeneous Equilibria

Thus far in this chapter we have focused on reactions in the gas phase. However, the principles of chemical equilibrium also apply to reactions in the liquid phase, particularly reactions in solution. Equilibria in which products and reactants are all in the same phase are called **homogeneous equilibria**. Equilibria in which reactants and products are in different phases are **heterogeneous equilibria**.

In Sample Exercise 16.6 we considered the equilibrium associated with the oxidation of SO_2 to SO_3, a key step in forming aerosols of H_2SO_4 in the atmosphere. One way to prevent this reaction from happening is to "scrub" SO_2 from the exhaust gases emitted at factories where sulfur-containing fuels are burned. Solid lime, CaO, is a widely used scrubbing agent. Sprayed into the exhaust gases, it combines with SO_2 to form calcium sulfite:

$$CaO(s) + SO_2(g) \rightleftharpoons CaSO_3(s)$$

The large quantities of lime needed for this reaction and for many other industrial and agricultural uses come from heating pulverized limestone, which is mostly $CaCO_3$, in kilns (Figure 16.6) operated at temperatures near 900°C. At these temperatures, $CaCO_3$ decomposes into lime (CaO) and CO_2 gas:

$$CaCO_3(s) \rightleftharpoons CaO(s) + CO_2(g) \qquad \Delta H° = 178.1\ kJ/mol$$

We might write the concentration-based equilibrium constant expression for this reaction as follows:

$$K_c = \frac{[CaO][CO_2]}{[CaCO_3]}$$

This expression contains concentration terms for two solids: CaO and $CaCO_3$. But what do we mean by the concentration of a solid? Any pure solid has a constant

CONNECTION Scrubbing exhaust gases with CaO was discussed in the descriptive chemistry box in Chapter 4.

(a)

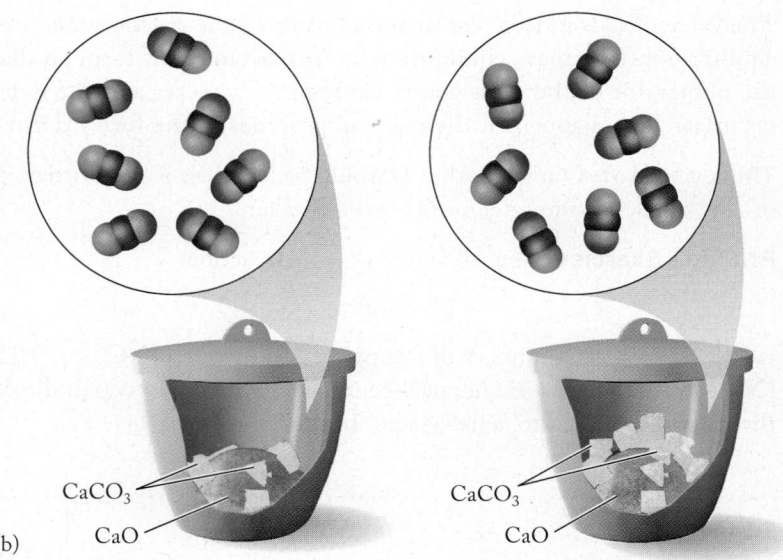

(b)

CaCO$_3$
CaO

CaCO$_3$
CaO

FIGURE 16.7 The position of a heterogeneous equilibrium between CaCO$_3$, CaO, and CO$_2$ at constant temperature depends only on the concentration of CO$_2$ gas present. As long as some of each solid is in the system, the equilibrium concentration of CO$_2$ remains the same. (a) Muffle furnace for heating crucibles containing CaCO$_3$ and CaO. (b) Two crucibles containing different amounts of CaCO$_3$ and CaO have the same concentration of CO$_2$ gas.

concentration because its mass (and number of moles) per unit volume is always the same. As long as there is *any* CaO or CaCO$_3$ present, there is no change in the "concentration" of either substance. Consequently, we remove them from the equilibrium constant expression, leaving us with

$$K_c = [CO_2]$$

This expression means that, as long as some CaO and CaCO$_3$ are present, the concentration of CO$_2$ gas does not vary at a given temperature, as shown in Figure 16.7. Instead, the concentration of CO$_2$ is the same as the value of K_c at that temperature.

The same concept of constant concentration applies to pure liquids that are involved in reversible chemical reactions. As long as the liquid is present, its "concentration" is considered constant during the course of the reaction and does not appear in the equilibrium constant expression. Similarly, K_c expressions for most reactions in aqueous solutions do not include a term for [H$_2$O] even when water is a reactant or product because its concentration does not change significantly. In writing equilibrium constant expressions for heterogeneous equilibria, we follow the rules we learned earlier, with the additional rule that pure liquids and solids do not appear in the expression.

SAMPLE EXERCISE 16.9 **Writing Equilibrium Constant Expressions for Heterogeneous Equilibria**

Write K_c expressions for

    a. CaO(s) + SO$_2$(g) $\rightleftharpoons$ CaSO$_3$(s)
    b. CO$_2$(g) + H$_2$O(ℓ) $\rightleftharpoons$ H$_2$CO$_3$(aq)

Collect and Organize We are to write equilibrium constant expressions for equilibria involving reactants and products in more than one phase. We need to identify the pure liquids and solids involved in the equilibrium and to exclude terms for them in the equilibrium constant expressions.

Analyze The first equilibrium involves two solids: CaO and CaSO₃. The second involves liquid H₂O.

Solve

a. An expression with terms for all reactants and products is

$$K_c = \frac{[CaSO_3]}{[CaO][SO_2]}$$

but once we remove the terms representing pure solids, we have

$$K_c = \frac{1}{[SO_2]}$$

b. In the equilibrium constant expression for reaction b, [H₂O] is a constant and is not included, leaving

$$K_c = \frac{[H_2CO_3]}{[CO_2]}$$

Think about It The equilibrium constant expressions we have written in this Sample Exercise and in previous exercises included terms for reactants and products whose concentrations or partial pressures were likely to change significantly during the course of the reaction. We exclude terms for pure solids and liquids because their concentrations do not change.

Practice Exercise Write K_p expressions for the reactions

a. $C(s) + CO_2(g) \rightleftharpoons 2\,CO(g)$

b. $CO_2(g) + H_2(g) \rightleftharpoons CO(g) + H_2O(\ell)$

CONCEPT TEST

Explain why $K_c = [H_2O(g)]$ is the equilibrium constant expression for the equilibrium $H_2O(\ell) \rightleftharpoons H_2O(g)$.

16.7 Le Châtelier's Principle

We can perturb chemical reactions at equilibrium in several ways—for example, by changing the concentration or partial pressure of a reactant or product, or by changing the temperature of the system. Adding or removing an ingredient in the system alters the value of the reaction quotient Q so that it is no longer equal to the value of K. On the other hand, changing the temperature of a system changes the value of K. Either way, the perturbed system is not at equilibrium and the composition of the system must change to restore equilibrium.

One of the first scientists to study and then successfully predict how chemical equilibria respond to such perturbations was French chemist Henri Louis Le Châtelier (1850–1936). He articulated **Le Châtelier's principle**, which states that, if a system at equilibrium is perturbed (or *stressed*), the position of the equilibrium shifts in the direction that relieves that stress. Through the years, chemists have used Le Châtelier's principle to increase the yields of chemical reactions that would otherwise have produced very little of a desired compound.

▶‖ **CHEMTOUR** Le Châtelier's Principle

Le Châtelier's principle a system at equilibrium responds to a stress in such a way that it relieves that stress.

Effects of Adding or Removing Reactants or Products

When a reactant or product is added or removed, a system at chemical equilibrium is perturbed. Following Le Châtelier's principle, the system responds in such a way as to restore equilibrium (Figure 16.8).

To explore how industrial chemists exploit Le Châtelier's principle, let's revisit the water–gas shift reaction for making hydrogen:

$$H_2O(g) + CO(g) \rightleftharpoons H_2(g) + CO_2(g) \qquad (16.2)$$

To shift the equilibrium toward the production of more H_2, chemists pass the reaction mixture through a scrubber containing a concentrated aqueous solution of K_2CO_3. Doing this removes CO_2 from the gaseous mixture because of the following reaction:

$$CO_2(g) + H_2O(\ell) + K_2CO_3(aq) \rightleftharpoons 2\,KHCO_3(s)$$

Removing CO_2 means fewer molecules of it are available to collide with molecules of H_2 to drive the reverse reaction in Equation 16.2. As a result, the rate of the reverse reaction becomes slower than the rate of the forward reaction. This means the system is no longer in equilibrium. To return to equilibrium, the reaction proceeds in the forward direction (chemists say the reaction *shifts to the right*), making more product to restore some of what was removed until a new equilibrium is achieved. The new equilibrium is like the old one in that the value of the mass action expression:

$$\frac{[H_2][CO_2]}{[H_2O][CO]}$$

is equal to the value of K. This is true even though the concentrations of the individual reactants and products have changed; the *overall* ratio in the K expression is restored.

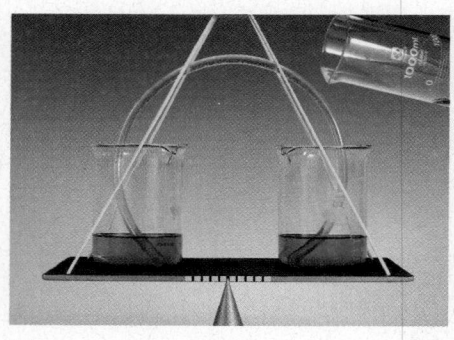**CONCEPT TEST** ••

The reaction mixture in the water–gas shift reaction is at equilibrium and some CO_2 is rapidly removed.

a. How does this removal affect the value of the reaction quotient Q?

b. When equilibrium is restored, which of the four compounds in the system is present at a higher concentration, which is present at a lower concentration, and which, if any, has the same concentration as in the original equilibrium mixture?

•••

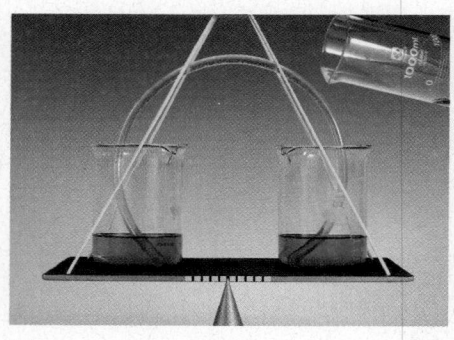

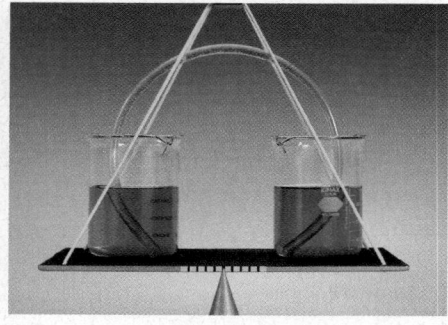

(a)   (b)   (c)

FIGURE 16.8 Chemical equilibrium is a bit like two beakers of water connected by a siphon. (a) The system is at equilibrium when the levels of water in both beakers are the same. (b) If we add water to one beaker, the added water drives a flow of water through the siphon into the opposite beaker until (c) equilibrium is restored. The oppositte response would occur if we removed some water from one of the beakers; the flow through the siphon would be toward that beaker.

SAMPLE EXERCISE 16.10 **Adding or Removing Reactants or Products to Stress an Equilibrium**

Suggest three ways the production of ammonia via the reaction

$$N_2(g) + 3H_2(g) \rightleftharpoons 2NH_3(g)$$

could be increased without changing the reaction temperature.

Collect and Organize We are given the balanced chemical equation of a reversible reaction and are to suggest three ways to increase its yield, that is, to shift its equilibrium to the right.

Analyze Equilibria can be shifted to the right by either (1) removing NH_3, which reduces the value of the numerator of the mass action expression:

$$Q_p = \frac{(P_{NH_3})^2}{(P_{H_2})^3(P_{N_2})}$$

or (2) increasing the partial pressure of N_2 or H_2, which increases the value of the denominator. Either approach will have the effect of producing a smaller reaction quotient Q_p. If $Q_p < K_p$, the reaction system will respond by consuming N_2 and H_2 and forming more NH_3.

Solve If we (1) increase the partial pressure of N_2, (2) increase the partial pressure of H_2, or (3) remove NH_3 from the system, the equilibrium will shift to the right.

Think about It The industrial synthesis of ammonia relies on shifting the equilibrium to the right by (1) running the reaction at high partial pressures of the reactants, and (2) removing the product NH_3 by passing the reaction mixture through chilled condensers. Ammonia can be removed because it condenses at a higher temperature than N_2 or H_2.

Practice Exercise Describe the changes that occur in a gas-phase equilibrium based on the reaction

$$2H_2S(g) + 3O_2(g) \rightleftharpoons 2SO_2(g) + 2H_2O(g)$$

if (a) the mixture is cooled and water vapor condenses; (b) SO_2 gas dissolves in liquid water as it condenses; (c) more O_2 is added.

We can express the effects of stressing equilibria in general terms. First, increasing the partial pressure (or concentration) of a reactant or product shifts the equilibrium so that more of that substance is consumed in the reaction. Second, decreasing the partial pressure (or concentration) of a reactant or product shifts the equilibrium toward the production of more of that substance.

Effects of Pressure and Volume Changes

A reaction involving gaseous reactants or products may be perturbed by altering the volume of the system, thereby changing the partial pressures of the reactants and products. To see how, let's revisit the equilibrium between NO_2 and its dimer, N_2O_4,

$$2NO_2(g) \rightleftharpoons N_2O_4(g) \tag{16.3}$$

and the equilibrium shown in Figure 16.4. Suppose a syringe is filled with brown NO_2 gas. A fraction of the NO_2 combines to form colorless N_2O_4 according to the reaction in Equation 16.3 and as shown in the molecular view of Figure 16.4(a). When the mixture is compressed at constant T (Figure 16.4b), the pressure of the system increases as its volume decreases. More importantly, the partial pressures of both gases increase. These increases in partial pressure stress the equilibrium and change the value of the reaction quotient.

To understand this behavior, let's suppose that we have an equilibrium mixture of these gases in which $P_{NO_2} = X$ and $P_{N_2O_4} = Y$. Then the value of K_p is

$$K_p = \frac{(P_{N_2O_4})}{(P_{NO_2})^2} = \frac{Y}{X^2}$$

When this equilibrium mixture is squeezed into half its initial volume, the partial pressure of each gas is doubled. Inserting these new values into the mass action expression gives us a reaction quotient that is half the value of K_p:

$$Q_p = \frac{2Y}{(2X)^2} = \frac{2}{4}\frac{Y}{X^2} = \frac{1}{2}K_p$$

When the value of Q_p for any reaction is less than the value of K_p, the reaction proceeds in the forward direction, consuming reactants and forming products. In this case, brown NO_2 is consumed, colorless N_2O_4 forms, and the color of the compressed mixture of gases fades as shown in Figure 16.4(c).

Another way to explain this response to changing the pressure of a mixture of reacting gases involves the ideal gas law, $PV = nRT$. If we rearrange the terms to solve for P:

$$P = \left(\frac{RT}{V}\right)n \tag{16.23}$$

we obtain an equation that shows the direct proportionality between the pressure of a gas mixture at constant temperature and volume and the number of moles of gas in the mixture. Equation 16.23 tells us that one way to relieve the stress of an increase in the pressure of a mixture of reacting gases is for the reaction to proceed in a direction that reduces the number of moles of gas. Because 2 moles of NO_2 form 1 mole of N_2O_4, producing more N_2O_4 reduces the overall pressure of the system.

In any reaction between gases in which the number of moles of gas changes as the reaction proceeds, changing the overall pressure exerted by the reactant and product gases shifts the equilibrium of the mixture. *Increasing* the pressure shifts the equilibrium toward the side of the reaction with *fewer* moles of gas. *Decreasing* the pressure shifts the equilibrium toward the side of the reaction with *more* moles of gas.

CONCEPT TEST $\cdots\cdots\cdots\cdots\cdots\cdots\cdots\cdots\cdots\cdots\cdots\cdots\cdots\cdots\cdots\cdots\cdots$

Changing the overall pressure on the reaction of the water–gas shift reaction at equilibrium:

$$H_2O(g) + CO(g) \rightleftharpoons H_2(g) + CO_2(g)$$

does not shift the equilibrium. Why?

$\cdots\cdots\cdots\cdots\cdots\cdots\cdots\cdots\cdots\cdots\cdots\cdots\cdots\cdots\cdots\cdots\cdots\cdots\cdots\cdots\cdots$

SAMPLE EXERCISE 16.11 **Assessing Pressure Effects on Gas-Phase Equilibria**

In which of the following equilibria would an increase in pressure promote the formation of more product(s)?

   a. $N_2(g) + O_2(g) \rightleftharpoons 2\,NO(g)$
   b. $2\,NO(g) + O_2(g) \rightleftharpoons 2\,NO_2(g)$
   c. $N_2O_4(g) \rightleftharpoons 2\,NO_2(g)$
   d. $H_2O(\ell) + CO_2(g) \rightleftharpoons H_2CO_3(aq)$
   e. $CaCO_3(s) \rightleftharpoons CaO(s) + CO_2(g)$

Collect and Organize We are asked to identify the reactions for which an increase in pressure causes an increase in product formation. We know that increasing pressure shifts a chemical equilibrium involving gases toward the side of the reaction with fewer moles of gas.

Analyze We need to identify those reactions in which there are fewer moles of gaseous products than there are gaseous reactants.

Solve Summing the number of moles of gas on the reactant side and product side in each reaction, we have

Reaction	Moles of Gaseous Reactants	Moles of Gaseous Products
a	2	2
b	3	2
c	1	2
d	1	0
e	0	1

The only two reactions with fewer moles of gaseous products than reactants are reactions b and d. Therefore, they are the only two in which an increase in the total pressure of the reacting gases increases product formation.

Think about It In the case of reaction a, the number of moles of gas on the reactant side is the same as the number of moles on the product side, so changing pressure does not cause the equilibrium to shift. In reactions c and e, the number of moles on the product side is greater than on the reactant side, so increasing pressure favors the reverse of the reaction as written and decreases product formation.

Practice Exercise How does compressing a reaction mixture of N_2, H_2, and NH_3 affect the following equilibrium?

$$N_2(g) + 3\,H_2(g) \rightleftharpoons 2\,NH_3(g)$$

Reaction d in Sample Exercise 16.11 is the subject of the essay that opened this chapter. Because 1 mole of gas appears on the reactant side and no moles of gas appear on the product side, increasing pressure makes CO_2 more soluble in the water in a soda can. When the top on a can is popped, the rapid decrease in pressure favors the reactants and causes gaseous CO_2 to fizz out of the soda.

Effect of Temperature Changes

In Chapter 5 we explored the flow of heat that accompanies many chemical reactions. Now we explore the stresses produced on chemical equilibria by adding or removing heat (increasing or decreasing temperature). Let's start with an exothermic reaction, the synthesis of ammonia:

$$N_2(g) + 3H_2(g) \rightleftharpoons 2NH_3(g) + \text{heat}$$

If we think of heat as a product in the forward reaction, then raising the temperature of a reaction mixture favors the reverse reaction; lowering the temperature favors the forward reaction.

One major difference arises, however, between applying Le Châtelier's principle to concentration or pressure changes and applying it to temperature changes: *temperature changes change the value of K.* Increasing temperature reduces the yield of ammonia synthesis because increasing temperature reduces the value of K.

In general, the value of K decreases as temperature increases for exothermic reactions and increases as temperature increases for endothermic reactions. We will look more closely at the influence of temperature on K values in Section 16.10, but for now this general analysis enables us to predict the direction of a shift in equilibrium with changing temperature.

SAMPLE EXERCISE 16.12 **Predicting Changes in Equilibrium with Temperature**

The color of an aqueous acidic solution of cobalt(II) chloride depends on the temperature (Figure 16.9). In aqueous HCl, the solution is pink at 0°C, magenta at 25°C, and dark blue at 75°C. Is the reaction producing the pink-to-blue color change exothermic or endothermic?

Collect and Organize We are asked to determine whether a reversible reaction is exothermic or endothermic. Asked another way, is heat a reactant or product?

Analyze If the reaction is exothermic, then increasing the temperature is the equivalent of adding a product, the impact of which will be to shift the reaction toward the pink reactant side. If the reaction is endothermic, then increasing the temperature is the equivalent of adding a reactant, the impact of which will be to shift the reaction toward the side of the blue product.

Solve When heat is added to the system, its color changes from pink to blue. This means heat is a reactant and the reaction as written must be endothermic:

$$\text{Heat} + \text{pink} \rightleftharpoons \text{blue}$$

Think about It This Sample Exercise illustrates how determining the effect of changing the temperature of an equilibrium reaction mixture can tell us whether the reaction is exothermic or endothermic. The fact that the reaction mixture in this exercise is magenta at room temperature tells us that both the pink and blue forms are present. This must mean that the value of K at room temperature is close to 1.

Temperature = 5°C   Temperature = 75°C

$$\underbrace{Co(H_2O)_6{}^{2+}(aq)}_{\text{Pink}} + 4\,Cl^-(aq) \rightleftharpoons \underbrace{CoCl_4{}^{2-}(aq)}_{\text{Royal blue}} + 6\,H_2O(\ell)$$

FIGURE 16.9 Two forms of cobalt, one pink and one blue, are in equilibrium in aqueous hydrochloric acid solution. The position of equilibrium shifts to the right as the temperature changes, causing the color of the solution to change as more and more of the blue $CoCl_4{}^{2-}$ ion forms.

Practice Exercise Predict how the value of the equilibrium constant of the reaction

$$N_2(g) + O_2(g) \rightleftharpoons 2NO(g) \qquad \Delta H^\circ_{rxn} = 181 \text{ kJ/mol}$$

changes with increasing temperature.

Table 16.4 summarizes how an exothermic system at equilibrium responds to various stresses.

TABLE 16.4	Responses of an Exothermic Reaction [2 A(g) $\rightleftharpoons$ B(g)] at Equilibrium to Different Kinds of Stress	
Kind of Stress	How Stress Is Relieved	Direction of Shift
Add A	Remove A	To the right
Remove A	Add A	To the left
Remove B	Add B	To the right
Add B	Remove B	To the left
Increase temperature by adding heat	Consume some of the heat	To the left
Decrease temperature by removing heat	Generate heat	To the right
Increase pressure by compressing the reaction mixture	Reduce moles of gas to relieve pressure increase	To the right
Decrease pressure by expanding volume	Increase moles of gas to maintain equilibrium pressure	To the left

Catalysts and Equilibrium

The industrial production of ammonia (see Sample Exercise 16.10 and the descriptive chemistry box in Chapter 6) was developed by the German chemists Fritz Haber (1868–1934) and Carl Bosch (1874–1940) in the early 20th century and is still widely referred to as the Haber–Bosch process. What makes the process commercially feasible is the use of catalysts. As discussed in Chapter 15, a catalyst increases the rate of a chemical reaction by lowering its activation energy. The question is this: If a catalyst increases the rate of a reaction, does that catalyst affect the equilibrium constant of the reaction?

To answer this question, consider the energy profiles of the catalyzed and uncatalyzed reaction in Figure 16.10. The catalyst increases the rate of the reaction by decreasing the height of the energy barrier. However, the barrier height is reduced by the same amount whether the reaction proceeds in the forward direction or in reverse. As a result, the increase in reaction rate produced by the catalyst is the same in both directions. Therefore a catalyst has no effect on the equilibrium constant of a reaction or on the composition of an equilibrium reaction mixture. A catalyst does, however, decrease the amount of time needed for a reaction to reach equilibrium.

CONNECTION In Chapter 15 we discovered that a catalyst simultaneously increases the rate of a reaction in both the forward and the reverse directions, when we discussed the effect of temperature on the rates of reactions occurring in a catalytic converter.

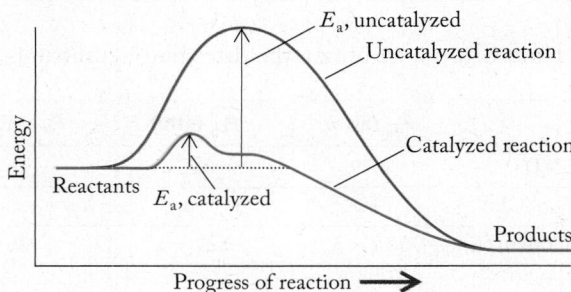

FIGURE 16.10 The effect of a catalyst on a reaction. A catalyst lowers the activation energy barrier, and as a result the rate of the reaction increases. However, because both the forward reaction and the reverse reaction occur more rapidly, the position of equilibrium (that is, the value of K) does not change. The system comes to equilibrium more rapidly, but the relative amounts of product and reactant present at equilibrium do not change.

16.8 Calculations Based on *K*

Reference books and the tables in Appendix 5 of this book contain lists of equilibrium constants for chemical reactions. These values are used in several kinds of calculations, including those in which:

1. We want to determine whether a reaction mixture has reached equilibrium (Sample Exercise 16.8).
2. We know the value of *K* and the starting concentrations or partial pressures of reactants and/or products, and we want to calculate their equilibrium concentrations or pressures.

▶❙❙ **CHEMTOUR** Solving Equilibrium Problems

In this section we focus on the second type of calculation and introduce a useful way of handling such problems: a table of reactant and product concentrations (or partial pressures) called an *ICE table*. The acronym ICE means that there is a row of *I*nitial values, a row of *C*hanges in those initial values as the reaction proceeds toward equilibrium, and a third row of *E*quilibrium values.

In our first example, we calculate how much nitrogen monoxide forms in a sample of air heated to a temperature at which the K_p of the following reaction is 1.00×10^{-5}:

$$N_2(g) + O_2(g) \rightleftharpoons 2\,NO(g)$$

The initial partial pressures are $P_{N_2} = 0.79$ atm and $P_{O_2} = 0.21$ atm, and we assume no NO is present.

We start by writing the given (initial) information in our ICE table:

	P_{N_2} (atm)	P_{O_2} (atm)	P_{NO} (atm)
Initial (I):	0.79	0.21	0
Change (C):			
Equilibrium (E):			

We know the reaction will proceed in the forward direction because there is no product initially present, so $Q_p = 0 < K_p$.

We need to use algebra to fill in rows C and E. We don't know how much N_2 or O_2 will be consumed or how much NO will be made. We can define the change in partial pressure of N_2 as $-x$ because N_2 is consumed during the reaction. Because the mole ratio of N_2 to O_2 in the reaction is 1:1, the change in P_{O_2} is also $-x$. Two moles of NO are produced from each mole of N_2 and O_2, so the change in P_{NO} is $+2x$. Inserting these values in the C row, we have

	P_{N_2} (atm)	P_{O_2} (atm)	P_{NO} (atm)
Initial (I):	0.79	0.21	0
Change (C):	$-x$	$-x$	$+2x$
Equilibrium (E):			

Combining the I and C rows, we obtain the three partial pressures at equilibrium:

	P_{N_2} (atm)	P_{O_2} (atm)	P_{NO} (atm)
Initial (I):	0.79	0.21	0
Change (C):	$-x$	$-x$	$+2x$
Equilibrium (E):	$0.79 - x$	$0.21 - x$	$2x$

The next step is to substitute the terms from the E row into the K_p expression for the reaction:

$$K_p = \frac{(P_{NO})^2}{(P_{N_2})(P_{O_2})}$$

$$= \frac{(2x)^2}{(0.79 - x)(0.21 - x)} \qquad (16.24)$$

Expanding the terms in the numerator and denominator of Equation 16.24 gives

$$K_p = \frac{4x^2}{0.1659 - 1.00x + x^2} = 1.00 \times 10^{-5}$$

Cross-multiplying, we get

$$1.659 \times 10^{-6} - (1.00 \times 10^{-5})x + (1.00 \times 10^{-5})x^2 = 4x^2$$

Combining the x^2 terms and rearranging, we have

$$3.99999x^2 + (1.00 \times 10^{-5})x - 1.659 \times 10^{-6} = 0$$

You may recognize this equation as one that fits the general form of a quadratic equation:

$$ax^2 + bx + c = 0$$

which can be solved for x either with a scientific calculator or using the quadratic formula

$$x = \frac{-b \pm \sqrt{b^2 - 4ac}}{2a}$$

Two values are possible for x, but only one is positive: 6.428×10^{-4} atm. We focus on this one because a gas (NO in this example) cannot have a negative partial pressure. Calculating equilibrium partial pressures to two significant figures:

$$P_{O_2} = 0.21 - x$$
$$= 0.21 - (6.428 \times 10^{-4}) = 0.21 \text{ atm}$$
$$P_{N_2} = 0.79 - x$$
$$= 0.79 - (6.428 \times 10^{-4}) = 0.79 \text{ atm}$$
$$P_{NO} = 2x$$
$$= 2(6.428 \times 10^{-4}) = 1.286 \times 10^{-3} = 0.0013 \text{ atm}$$

The small quantity of NO produced by the reaction means that there is no significant change in the partial pressures of N_2 or O_2.

Because the x terms in the denominator of Equation 16.24 are very much smaller than the initial partial pressures, a simpler approach to calculating P_{NO} is possible: let's ignore the x terms in the denominator and use the initial values for P_{N_2} and P_{O_2} instead:

$$K_p = 1.0 \times 10^{-5} = \frac{(P_{NO})^2}{(P_{N_2})(P_{O_2})}$$

$$= \frac{4x^2}{(0.79 - x)(0.21 - x)} \approx \frac{4x^2}{(0.79)(0.21)}$$

$$4x^2 = (0.79)(0.21)(1.0 \times 10^{-5})$$
$$= 1.659 \times 10^{-6}$$
$$x^2 = 4.148 \times 10^{-7}$$
$$x = 6.44 \times 10^{-4} \text{ atm}$$

This value does not differ significantly from that obtained by solving the quadratic equation (6.43×10^{-4}), given that we know the initial partial pressures to only two significant figures. Generally speaking, we can ignore the x component of an equilibrium concentration or partial-pressure if the value of x is less than 5% of the initial value.

The calculations associated with the terms in an ICE table can also be simplified when the initial concentrations of the reactants are the same. For example, suppose we have a vessel containing 0.100 M N_2 and 0.100 M O_2 at a temperature where K_c for the NO formation reaction is 0.100. What is the equilibrium concentration of NO? The ICE table in this case is

	$[N_2]$ (M)	$[O_2]$ (M)	$[NO]$ (M)
Initial (I):	0.100	0.100	0
Change (C):	$-x$	$-x$	$+2x$
Equilibrium (E):	$0.100 - x$	$0.100 - x$	$2x$

Inserting the values from the E row into the expression for K_c gives

$$\frac{(2x)^2}{(0.100 - x)(0.100 - x)} = 0.100$$

Taking the square root of each side:

$$\frac{2x}{0.100 - x} = 0.316$$

Solving for x, we get $x = 0.0136$ M, and the equilibrium concentrations are $[N_2] = [O_2] = 0.100$ $M - 0.0136$ $M = 0.086$ M and $[NO] = 2x = 0.0272$ M.

It's a good idea to check that these concentrations are consistent with the known value of K_c. Substituting into the K_c expression, we get

$$K_c = \frac{(0.0272)^2}{(0.086)^2} = 0.100$$

which is the value of K given for the reaction in question.

SAMPLE EXERCISE 16.13 Calculating an Equilibrium Partial Pressure I

Some of the H_2 used in the Haber–Bosch process is produced by the water–gas shift reaction:

$$CO(g) + H_2O(g) \rightleftharpoons CO_2(g) + H_2(g) \qquad (16.2)$$

If a reaction vessel at 400°C is filled with an equimolar mixture of CO and steam such that $P_{CO} = P_{H_2O} = 2.00$ atm, what is the partial pressure of H_2 at equilibrium? The equilibrium constant $K_p = 10$ at 400°C.

Collect and Organize We are asked to find the partial pressure of a gaseous product in an equilibrium mixture given the initial partial pressures of reactants and the value of K_p. One way to approach this problem involves (1) setting up an ICE table, (2) using the partial pressures from the E row in the equilibrium constant expression, and (3) solving for P_{H_2}.

Analyze The system initially contains no product. This means that the reaction quotient Q_p is equal to zero and thus less than K. Therefore, the reaction proceeds in the forward direction, decreasing the partial pressures of the reactants while increasing those of the products. A K_p value of 10 means that most of the reactants should be converted into products, but there should be significant partial pressures of both reactants and products at equilibrium.

Solve Let x be the increase in partial pressure of H_2 as a result of the reaction. The stoichiometry of the reaction tells us that the change in P_{CO_2} is also x and that the changes in both P_{CO} and P_{H_2O} are $-x$:

	P_{CO} (atm)	P_{H_2O} (atm)	P_{CO_2} (atm)	P_{H_2} (atm)
Initial (I):	2.00	2.00	0.00	0.00
Change (C):	$-x$	$-x$	$+x$	$+x$
Equilibrium (E):	$2.00 - x$	$2.00 - x$	x	x

Inserting these equilibrium terms into the equilibrium constant expression for the reaction gives

$$K_p = \frac{(P_{CO_2})(P_{H_2})}{(P_{CO})(P_{H_2O})}$$

$$= \frac{(x)(x)}{(2.00 - x)(2.00 - x)} = 10$$

The latter equation can be simplified by taking the square root of both sides:

$$\frac{x}{2.00 - x} = \sqrt{10} = 3.16$$

Solving for x gives $x = 1.52$ atm, which is the equilibrium partial pressure of H_2 (and CO_2).

Think about It Our prediction that most, but far from all, of the reactants would form products was correct. It is a good idea to substitute the results into the equilibrium constant expression as a check on the validity of the solution. The calculated partial pressures are: $P_{H_2} = P_{CO_2} = 1.52$ atm and $P_{H_2O} = P_{CO} = 2.00 - 1.52 = 0.48$ atm. Inserting these values in the equilibrium constant expression gives $K_p = (1.52)^2/(0.48)^2 = 10.03$, which is not significantly different from the given value of 10 and which confirms that our calculation is correct.

Practice Exercise The chemical equation for the formation of hydrogen iodide from H_2 and I_2 is

$$H_2(g) + I_2(g) \rightleftharpoons 2\,HI(g)$$

The value of K_p for the reaction is 50 at 450°C. What is the partial pressure of HI in a sealed reaction vessel at 450°C if the initial partial pressures of H_2 and I_2 are both 0.100 atm and initially there is no HI present?

SAMPLE EXERCISE 16.14 Calculating an Equilibrium Partial Pressure II

Suppose that in a reaction vessel running the water–gas shift reaction,

$$CO(g) + H_2O(g) \rightleftharpoons CO_2(g) + H_2(g)$$

at 400°C the initial partial pressures are $P_{CO} = 2.00$ atm, $P_{H_2O} = 2.00$ atm, $P_{H_2} = 0.15$ atm, and $P_{CO_2} = 0.00$ atm. What is the partial pressure of H_2 at equilibrium, given $K_p = 10$ at 400°C?

Collect and Organize We are asked to calculate the partial pressure of a product at equilibrium. We know the initial partial pressures of all reactants and of one product as well as the value of K_p. The difference between this problem and the preceding Sample Exercise is that here we have product present before the reaction starts. Comparing the reaction quotient Q with the value of K lets us know in which direction the reaction proceeds to attain equilibrium.

Analyze There is no CO_2 initially present, so $Q = 0$. Therefore, the reaction as written proceeds in the forward direction. Our strategy is to solve the problem by setting up an ICE table.

Solve Let x be the increase in P_{H_2}. The change in P_{CO_2} is also x, and the changes in P_{CO} and P_{H_2O} are $-x$:

	P_{CO} (atm)	P_{H_2O} (atm)	P_{CO_2} (atm)	P_{H_2} (atm)
Initial (I):	2.00	2.00	0.00	0.15
Change (C):	$-x$	$-x$	$+x$	$+x$
Equilibrium (E):	$2.00 - x$	$2.00 - x$	x	$0.15 + x$

$$K_p = \frac{(P_{CO_2})(P_{H_2})}{(P_{CO})(P_{H_2O})}$$

$$= \frac{(x)(0.15 + x)}{(2.00 - x)(2.00 - x)} = 10$$

Solving for x gives two values, $x = 1.50$ and 2.96. The value $x = 2.96$ is not possible because using it in the equilibrium terms results in negative partial pressures of CO and H_2O, a physical impossibility. Therefore, $x = 1.50$ and at equilibrium $P_{H_2} = 1.50 + 0.15 = 1.65$ atm.

Think about It Using the value of x in the terms in the E row of the ICE table, we find that the equilibrium partial pressures of CO, H_2O, and CO_2 are 0.50, 0.50, and 1.50, respectively. These values result in a K_p value of $(1.65)(1.50)/(0.5)(0.5) = 9.9$, which is acceptably close to the given value of 10. Comparing the results of this Sample Exercise to the previous one, we find that having an initial P_{H_2} of 0.15 atm results in a higher final P_{H_2} value (1.65 vs 1.52 atm); however, the presence of H_2 at the start of the reaction resulted in slightly less conversion of reactants to products than when no H_2 was present initially. This result makes sense because the presence of some product before the reaction starts means that less product has to form before $Q_p = K_p$.

Practice Exercise The value of K_c for the reaction

$$N_2O_4(g) \rightleftharpoons 2NO_2(g)$$

is 0.21 at 373 K. If a reaction vessel at that temperature initially contains 0.030 M NO_2 and 0.030 M N_2O_4, what are the concentrations of the two gases at equilibrium?

16.9 Equilibrium and Thermodynamics

In Section 14.5 we explored how the change in free energy, ΔG, of a chemical reaction provides us with an indication of whether or not it will happen. If its ΔG is negative, a reaction is spontaneous as written and proceeds in the forward direction. If ΔG is positive, the reaction as written is nonspontaneous; the reverse reaction *is* spontaneous, and the reaction as written proceeds in the reverse direction. As a spontaneous reaction under constant temperature and pressure proceeds, the concentrations of reactants and products change, and the free energy of the system changes as well. Eventually ΔG reaches zero. When it does, no free energy is left to do useful work. The reaction has achieved chemical equilibrium.

CHEMTOUR Equilibrium and Thermodynamics

The magnitude of ΔG—how far it is from zero in either a negative or positive direction—indicates how far a system is from its equilibrium position. Similarly, when Q is much larger or smaller than K, we know that a chemical system is far from equilibrium. It is reasonable, then, to think that the separation between the values of Q and K and the sign and magnitude of ΔG are somehow related. Indeed they are, and their mathematical relationship is one of the most important connections in chemistry because it enables us to relate the thermodynamics of a chemical reaction to the composition of a reaction mixture at equilibrium.

CONNECTION In Chapter 14 we defined the change in free energy of a reaction, ΔG_{rxn}, as the energy available to do useful work.

The thermodynamic view of equilibrium and the relationship between ΔG and Q are described by the equation

$$\Delta G = \Delta G^\circ + RT \ln Q \qquad (16.25)$$

where ΔG° is the change in free energy under standard conditions. Let's explore what this means by looking at a reaction we have examined several times in this chapter, the decomposition of N_2O_4:

$$N_2O_4(g) \rightleftharpoons 2NO_2(g)$$

We can calculate the change in standard free energy for the reaction (ΔG°_{rxn}) as we did in Chapter 14, using standard free energy of formation (ΔG°_f) values from Table A4.3 in Appendix 4 and the formula

$$\Delta G^\circ_{rxn} = \sum n_{products}\,\Delta G^\circ_{f,products} - \sum n_{reactants}\,\Delta G^\circ_{f,reactants} \qquad (14.10)$$

$$= 2\,\cancel{mol}(51.3\text{ kJ}/\cancel{mol}) - 1\,\cancel{mol}(97.8\text{ kJ}/\cancel{mol}) = +4.8\text{ kJ}$$

The positive value of ΔG°_{rxn} indicates that the reaction as written is not spontaneous at 298 K. However, this value of ΔG°_{rxn} applies only under standard conditions when $P_{NO_2} = P_{N_2O_4} = 1$ atm.

Let's explore what might happen in a reaction vessel at 298 K that initially contained 1 mole of N_2O_4 at a partial pressure of 1 atm and no (or hardly any) NO_2. Under these conditions the value of the reaction quotient (Q_p) is zero:

$$Q_p = \frac{(P_{NO_2})^2}{P_{N_2O_4}} = \frac{0}{1} = 0$$

Although we do not know the value of K, it has to be greater than zero. Therefore, $Q_p < K_p$ and the system should respond by spontaneously forming NO_2 from N_2O_4.

If the reaction in the forward direction is spontaneous, then the change in free energy of the reaction (ΔG) must be less than zero, even though the value of $\Delta G°$ is positive (4.8 kJ/mol). The value of ΔG *is* negative because the $(RT \ln Q)$ term in Equation 16.25 approaches $-\infty$ as Q approaches zero. Actually, any reversible reaction regardless of its $\Delta G°$ value has a negative ΔG value and is spontaneous when there is only reactant and no product (or practically none) in the system.

The opposite situation would exist if we had 2 moles of NO_2 in the reaction vessel and essentially no N_2O_4. Under these conditions the denominator of the reaction quotient is nearly zero, making the value of Q_p enormous. Likewise the $RT \ln Q_p$ term in Equation 16.25 has a large positive value, guaranteeing that $\Delta G > 0$. This means that the forward reaction is not spontaneous, but that the reverse reaction is. We also predict that NO_2 in the reaction vessel should combine to form N_2O_4 based on the fact that the enormous Q_p must be greater than K_p. Therefore, the reaction should run in reverse until $Q_p = K_p$.

Figure 16.11(a) shows how free energy changes as the quantities of N_2O_4 and NO_2 in the reaction mixture change. The minimum of the curve (point 3)

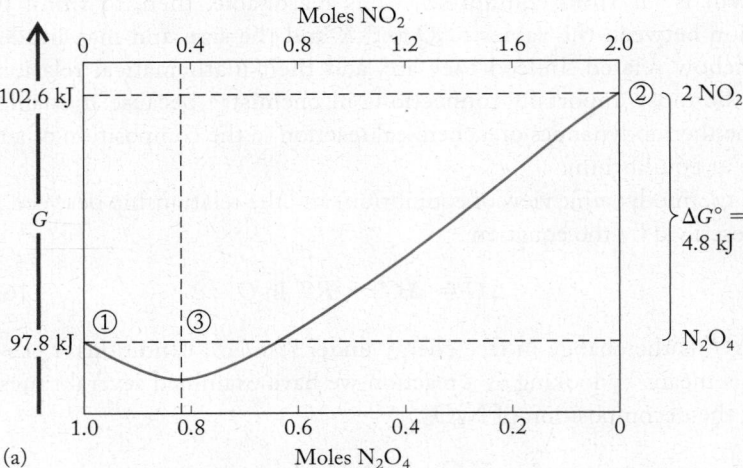

(a)

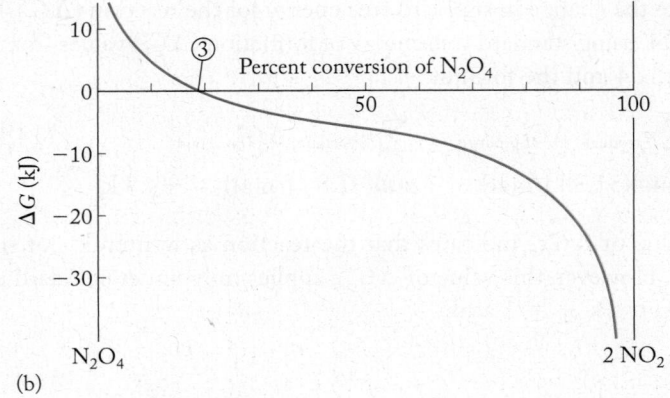

(b)

FIGURE 16.11 The change in standard free energy $\Delta G°_{rxn}$ is a constant for a given reaction. (a) Point 3, the point of minimum free energy, defines the composition of a system at equilibrium. At point 1, the system consists of reactants only; at point 2, the system consists of products only. (b) ΔG is the "distance" from equilibrium in terms of free energy. As a spontaneous reaction proceeds, the composition of the system changes and ΔG approaches 0 (point 3), at which point no further change in composition occurs.

corresponds to a free energy that is lower than that of either pure N_2O_4 (point 1) or pure NO_2 (point 2). The intersection of the vertical dashed line marking the minimum in free energy crosses the top and bottom axes at values that tell us the reaction mixture at equilibrium contains about 0.83 mol N_2O_4 and about 0.34 mol NO_2.

The curve in Figure 16.11(b) delivers the same message as in part a. In b, the y-axis represents how far from equilibrium are N_2O_4/NO_2 reaction mixtures of different composition. Clearly mixtures that are either pure reactant or product are very far away as we discussed above. The reaction curve crosses the x-axis ($\Delta G = 0$) at point 3, which defines the composition of the reaction mixture at equilibrium. This value, expressed as the percentage of N_2O_4 that has dissociated into NO_2, corresponds to the same N_2O_4/NO_2 ratio as the minimum in the curve in Figure 16.11(a).

CONCEPT TEST ··

Suppose the ΔG°_{rxn} value of the hypothetical chemical reaction $A \rightleftharpoons B$ is -3.0 kJ/mol. Which of the following statements about an equilibrium mixture of A and B at 298 K is true?

a. There is only A present.
b. There is only B present.
c. There is an equimolar mixture of A and B present.
d. There is more A than B present.
e. There is more B than A present.

··

Once a reaction has reached equilibrium, $Q = K$, $\Delta G = 0$, and Equation 16.25 becomes

$$\Delta G = \Delta G^\circ + RT \ln K = 0$$

or

$$\Delta G^\circ = -RT \ln K \qquad (16.26)$$

Rearranging Equation 16.26 allows us to calculate the K value for a reaction from its change in standard free energy and absolute temperature. First, we rearrange the terms:

$$\ln K = \frac{-\Delta G^\circ}{RT} \qquad (16.27)$$

and then take the antilogarithm of both sides:

$$K = e^{-\Delta G^\circ/RT} \qquad (16.28)$$

Equation 16.28 provides the following interpretation of reaction spontaneity under standard conditions. Whenever ΔG° is negative, the exponent $-\Delta G^\circ/RT$ in Equation 16.28 is positive, and $e^{-\Delta G^\circ/RT} > 1$, making $K > 1$. Therefore, any reversible reaction with an equilibrium constant greater than 1 is spontaneous under standard conditions as shown in Figure 16.12(a). This spontaneity has its limits. As reactants are consumed and products are formed, the value of the reaction quotient increases, making the value of ΔG less negative. When it reaches zero, there is no further change in the composition of the reaction mixture because chemical equilibrium has been achieved.

It follows that a reversible reaction with a less negative value of ΔG° (Figure 16.12b) is still spontaneous, but has a smaller equilibrium constant, so less reactant is consumed and less product is formed before the value of ΔG reaches zero. Finally, a reaction that has a positive value of ΔG° (Figure 16.12c), has an equilibrium constant that is less than zero, and is not spontaneous under standard conditions. Instead, the reverse of the reaction is spontaneous.

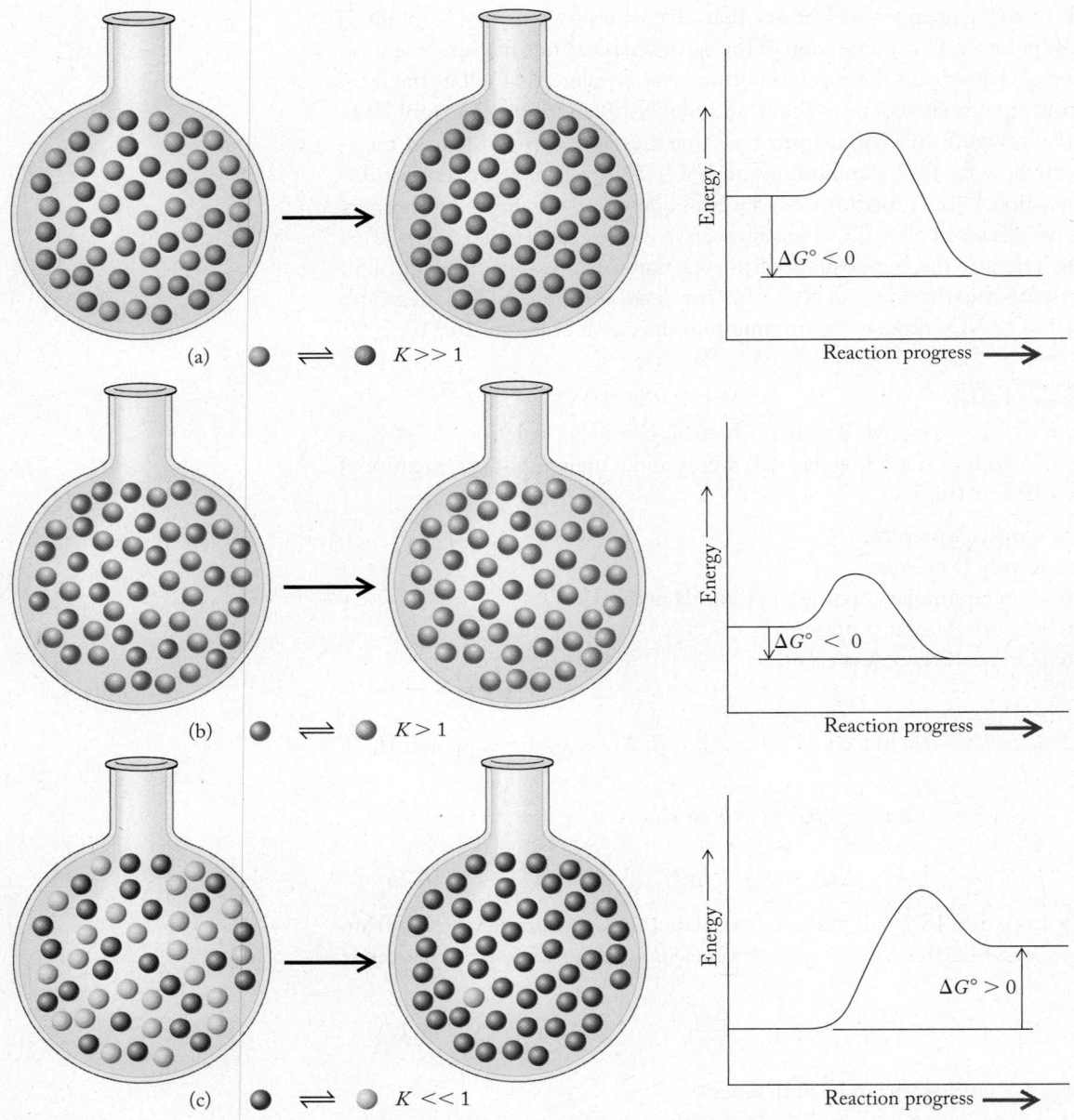

FIGURE 16.12 The equilibrium constant of a chemical reaction is linked to its $\Delta G°$ value. The three reaction vessels on the left each contain equimolar mixtures of different pairs of gases. The initial partial pressure of each gas is 1 atm. (a) The value of $\Delta G°$ for the formation of "red" gas from "green" gas has a large negative value, which makes K much greater than 1. The reaction proceeds in the forward direction leaving little green gas left over. (b) If the value of $\Delta G°$ for the formation of "orange" gas from "blue" gas has a less negative value than in part a, the value of K will be smaller, though still greater than 1, and there is more orange gas present at equilibrium than blue gas. (c) If the value of $\Delta G°$ for the formation of "yellow" gas from "purple" gas is positive, K is less than 1 and the reaction runs in the reverse forming purple gas from yellow gas.

Let's apply this concept to the equilibrium between N_2O_4 and NO_2:

$$N_2O_4(g) \rightleftharpoons 2\,NO_2(g) \qquad \Delta G°_{rxn} = 4.8 \text{ kJ/mol}$$

and focus on the composition of the reaction mixture of these two gases at equilibrium in Figure 16.11. We start by calculating the value of the exponent in Equation 16.28:

$$-\frac{\Delta G°}{RT} = -\frac{\left(\dfrac{4.8 \text{ kJ}}{\text{mol}}\right)\left(\dfrac{1000 \text{ J}}{1 \text{ kJ}}\right)}{\left(\dfrac{8.314 \text{ J}}{\text{mol} \cdot \text{K}}\right)(298 \text{ K})} = -1.94$$

Inserting this value into Equation 16.28 gives

$$K = e^{-\Delta G^\circ / RT} = e^{-1.94} = 0.144$$

This result is consistent with the composition of the reaction mixture at equilibrium in Figure 16.11 where $P_{N_2O_4} \approx 0.83$ atm and $P_{NO_2} \approx 0.34$ atm. Inserting these values into the equilibrium constant expression for the reaction gives us an approximate value of K_p that is close to the one we calculated from ΔG°_{rxn}:

$$K_p = \frac{(P_{NO_2})^2}{P_{N_2O_4}} \approx \frac{(0.34)^2}{0.83} = 0.14$$

In this example, a positive ΔG°_{rxn} value of only few kilojoules per mole corresponds to an equilibrium constant value that is less than one but still greater than 0.1. As a result, the equilibrium reaction mixture contains more reactant than product, but less than an order of magnitude more. Similarly, an equilibrium reaction mixture produced by a reaction with a negative ΔG°_{rxn} value of only few kilojoules per mole is likely to contain more products than reactants, but with significant quantities of both.

SAMPLE EXERCISE 16.15   Calculating K from ΔG°_f

Use ΔG°_f values from Table A4.3 to calculate ΔG°_{rxn} and the value of K_p for the formation of NO_2 from NO and O_2 at 298 K:

$$NO(g) + \tfrac{1}{2}O_2(g) \rightleftharpoons NO_2(g)$$

Collect and Organize We are to calculate the values of ΔG°_{rxn} and K for a reaction starting with ΔG°_f values from Table A4.3: 51.3 kJ/mol for NO_2 and 86.6 kJ/mol for NO. Because O_2 gas is the most stable form of the element, its ΔG°_f value is 0.0 kJ/mol.

Analyze We can use Equation 14.10 to calculate ΔG°_{rxn} and then Equation 16.28 to calculate the value of K.

Solve

$$\begin{aligned}
\Delta G^\circ_{rxn} &= [\Delta G^\circ_f(NO_2)] - [\Delta G^\circ_f(NO) + \tfrac{1}{2}\Delta G^\circ_f(O_2)] \\
&= [1 \text{ mol}(51.3 \text{ kJ/mol})] - [1 \text{ mol}(86.6 \text{ kJ/mol}) + \tfrac{1}{2}\text{ mol}(0.0 \text{ kJ/mol})] \\
&= (51.3 - 86.6) \text{ kJ} \\
&= -35.3 \text{ kJ or } -35,300 \text{ J per mole of } NO_2 \text{ produced}
\end{aligned}$$

The exponent in Equation 16.28 is

$$-\frac{\Delta G^\circ}{RT} = -\frac{\left(\dfrac{-35,300 \text{ J}}{\text{mol}}\right)}{\left(\dfrac{8.314 \text{ J}}{\text{mol} \cdot \text{K}}\right)(298 \text{ K})} = 14.2$$

The corresponding value of K_p is

$$K_p = e^{-\Delta G^\circ / RT} = e^{14.2} = 1.5 \times 10^6$$

Think about It The exponential relationship between ΔG°_{rxn} and K_p means that a moderately negative free-energy change of -35.3 kJ/mol corresponds to a very large value of K_p: in this case, greater than 10^6.

Practice Exercise The standard free energy of formation of ammonia at 298 K is −16.5 kJ/mol. What is the value of K for the reaction

$$N_2(g) + 3\,H_2(g) \rightleftharpoons 2\,NH_3(g)$$

at 298 K?

We have yet to explain which kind of equilibrium constant, K_c or K_p, is related to $\Delta G°$ by Equation 16.28. The symbol $\Delta G°$ represents a change in free energy under standard conditions. The standard state of a gaseous reactant or product is one in which its *partial pressure* is 1 atm. Thus, the $\Delta G°$ of a reaction *in the gas phase* is linked by Equation 16.28 to its K_p value. However, standard conditions for reactions in solution (the focus of Chapter 17) mean that all dissolved reactants and products are present at a concentration of 1.00 *M*. Thus, the $\Delta G°$ of a reaction *in solution* is related by Equation 16.28 to its K_c value.

16.10 Changing *K* with Changing Temperature

We have noted often in this chapter that the value of K changes with changing temperature. In this section we look more closely at that relationship, developing several important equations that link K and T.

Temperature, *K*, and Δ*G*°

Let's begin by combining a key equation from Chapter 14:

$$\Delta G° = \Delta H° - T\Delta S° \tag{14.8}$$

with one from this chapter:

$$\ln K = \frac{-\Delta G°}{RT} \tag{16.27}$$

to derive an equation that relates K to $\Delta H°$ and $\Delta S°$:

$$\ln K = \frac{-\Delta G°}{RT} = -\frac{\Delta H°}{RT} + \frac{T\Delta S°}{RT}$$

$$= -\frac{\Delta H°}{RT} + \frac{\Delta S°}{R} \tag{16.29}$$

Note how a negative value of $\Delta H°$ or a positive value of $\Delta S°$ contributes to a large value of K. These dependencies make sense because negative values of $\Delta H°$ and positive values of $\Delta S°$ are the two factors that contribute to making reactions spontaneous.

Because we are discussing the influence of temperature on K, let's identify the factors affected by changes in T in Equation 16.29. The $\Delta S°$ term is not affected, but the influence of $\Delta H°$ does depend on temperature: the higher the temperature, the larger the denominator of the $\Delta H°$ term and the smaller the influence of a favorable $\Delta H°$, that is, a $\Delta H°$ value that is negative. This temperature dependence makes sense based on Le Châtelier's principle and the notion that heat is a product of exothermic reactions and a reactant in endothermic reactions. Increasing temperature shifts a reaction toward the side opposite the heat term: it promotes endothermic reactions and inhibits exothermic reactions.

CONNECTION In Chapter 5 we defined exothermic reactions as those giving off heat. They have a negative enthalpy change ($\Delta H < 0$), while endothermic reactions absorb heat ($\Delta H > 0$).

If ΔH° and ΔS° do not vary much with temperature, then Equation 16.29 predicts that $\ln K$ will be a linear function of $1/T$. Furthermore, we expect a graph of $\ln K$ versus $1/T$ to have a positive slope for an exothermic process and a negative slope for an endothermic process. We can determine ΔH° from the slope of the line and ΔS° from its y-intercept. Thus, we can calculate fundamental thermodynamic values of a reaction at equilibrium by determining its equilibrium constant at different temperatures. For example, let's determine the values of ΔH° and ΔS° for the reaction

$$2\,CO_2(g) \rightleftharpoons 2\,CO(g) + O_2(g)$$

starting with its K_c values of 5.5×10^{-9} at 1500 K, 4.0×10^{-1} at 2500 K, and 40.3 at 3000 K. The fact that K_c increases as temperature increases tells us that heat is a reactant and the reaction is endothermic. We can use these data to determine the values of ΔH° and ΔS° by plotting $\ln K_c$ values versus $1/T$. To see how this graphical method works, let's rewrite Equation 16.29 so that it fits the form of the equation for a straight line ($y = mx + b$):

$$\ln K = -\frac{\Delta H^\circ}{R}\left(\frac{1}{T}\right) + \frac{\Delta S^\circ}{R} \tag{16.30}$$

The graph of $\ln K$ versus $1/T$ (Figure 16.13) is indeed a straight line. The slope of this line is $-68{,}070$ K, which is equal to $-\Delta H^\circ_{rxn}/R$. The corresponding value of ΔH°_{rxn} is

$$\Delta H^\circ_{rxn} = -\text{slope} \times R$$

$$= -(-68{,}070\ \cancel{K})\left(\frac{8.314\ J}{mol \cdot \cancel{K}}\right)$$

$$= 566{,}000\ J/mol = 566\ kJ/mol$$

The y-intercept ($\Delta S^\circ_{rxn}/R$) of the graph is 26.4. The corresponding value of ΔS°_{rxn} is:

$$\Delta S^\circ_{rxn} = (26.4)\left(\frac{8.314\ J}{mol \cdot K}\right) = 219\ J/(mol \cdot K)$$

This positive entropy change is logical because the forward reaction converts 2 moles of gaseous CO_2 into 3 moles of gaseous products.

We can use a modified version of Equation 16.30 to relate the values of K at two different temperatures:

$$\ln\left(\frac{K_2}{K_1}\right) = -\frac{\Delta H^\circ}{R}\left(\frac{1}{T_2} - \frac{1}{T_1}\right) \tag{16.31}$$

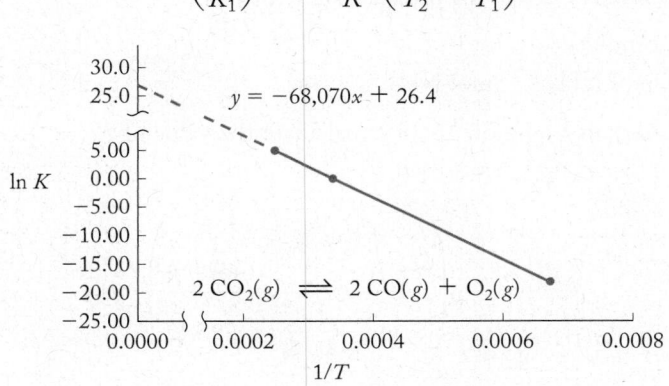

FIGURE 16.13 The graph of $\ln K$ versus $1/T$ is a straight line. We calculate ΔH°_{rxn} from the slope and extrapolate the line to the y-intercept, from which we calculate ΔS°_{rxn}.

This equation looks very much like the *Clausius–Clapeyron equation* we developed in Chapter 11 (Equation 11.7), which described the relationship between the vapor pressures of a liquid at two different temperatures:

$$\ln\left(\frac{P_{vap,T_1}}{P_{vap,T_2}}\right) = \frac{\Delta H_{vap}}{R}\left(\frac{1}{T_2} - \frac{1}{T_1}\right) \tag{11.7}$$

Indeed the Clausius–Clapeyron equation is just a special case of Equation 16.31, which relates two temperatures (T_1 and T_2) to the values of the equilibrium constants at those temperatures. The pressures in Equation 11.7 are actually equilibrium vapor pressures, and the enthalpy of reaction is the enthalpy of vaporization (ΔH_{vap}). Equation 16.31 is called the *van't Hoff equation*, because it was first derived by Jacobus van't Hoff. It is particularly useful for calculating the value of K at a very high or low temperature if we know what it is at a standard reference temperature, e.g., 298 K. We use such a calculation in the following Sample Exercise.

SAMPLE EXERCISE 16.16   **Calculating a Value of *K* at a Specific Temperature**

Use data from Appendix 4 to calculate the equilibrium constant K_p for the exothermic reaction

$$N_2(g) + 3\,H_2(g) \rightleftharpoons 2\,NH_3(g)$$

at 298 K and at 773 K, a typical temperature used in the Haber–Bosch process for synthesizing ammonia.

Collect and Organize We are to calculate the K_p value for a reaction at two temperatures. One of the temperatures is 298 K, which is the reference temperature for the standard thermodynamic data in Appendix 4. The ΔG_f° value of NH_3 is -16.5 kJ/mol.

Analyze We can use Equation 16.28 to calculate the value of K_p at 298 K from the value of ΔG_{rxn}°. Then we can use Equation 16.31 to calculate the value of K_p at 773 K from the value of K_p at 298 K and the value of ΔH_{rxn}°. The value of ΔH_{rxn}° can be calculated from the enthalpy of formation of NH_3 (-46.1 kJ/mol). The negative value of ΔH_f° means that ΔH_{rxn}° is also negative, which means that the reaction is exothermic. Since heat is a product of the reaction, we can predict that the value of K_p will be smaller at 773 K than at 298 K.

Solve The value of ΔG_{rxn}° for the reaction is

$$\frac{2 \;\text{mol NH}_3}{\text{mol NH}_2} \times \frac{-16.5 \;\text{kJ}}{\text{mol NH}_3} \times \frac{1000 \;\text{J}}{\text{kJ}} = -33,000 \;\frac{\text{J}}{\text{mol N}_2}$$

Using this value in Equation 16.28 to calculate the value of K_p:

$$K_p = e^{-\Delta G^\circ/RT}$$

$$= \exp\left(\frac{-\left(-33,000\;\dfrac{\text{J}}{\text{mol}}\right)}{8.314\dfrac{\text{J}}{\text{mol}\cdot\text{K}} \times 298\;\text{K}}\right)$$

$$= 6.1 \times 10^5$$

Once we know the value of K_p at 298 K, we can calculate K_p at 773 K by using Equation 16.31. To use Equation 16.31 we must first calculate the value of ΔH°_{rxn}, which is twice the ΔH°_f of NH_3:

$$\Delta H^{\circ}_{rxn} = \frac{2 \text{ mol NH}_3}{\text{mol N}_2} \times \frac{-46.1 \text{ kJ}}{\text{mol NH}_3} \times \frac{1000 \text{ J}}{\text{kJ}}$$

$$= -92.2 \text{ kJ, or } -92,200 \frac{\text{J}}{\text{mol N}_2}$$

After substituting $K_1 = 6.1 \times 10^5$, $T_1 = 298$ K, and $T_2 = 773$ K into Equation 16.31, we solve for K_2:

$$\ln\left(\frac{K_2}{6.1 \times 10^5}\right) = -\frac{-\left(-92,200 \dfrac{\text{J}}{\text{mol}}\right)}{8.314 \dfrac{\text{J}}{\text{mol} \cdot \text{K}}}\left(\frac{1}{773 \text{ K}} - \frac{1}{298 \text{ K}}\right)$$

$$\ln\left(\frac{K_2}{6.1 \times 10^5}\right) = -22.87$$

$$\frac{K_2}{6.1 \times 10^5} = e^{-22.87} = 1.2 \times 10^{-10}$$

$$K_2 = 7.3 \times 10^{-5} \text{ at } 773 \text{ K}$$

Think about It The equilibrium constant decreases markedly when the temperature of the reaction is raised from 298 K to 773 K. This decrease fits our prediction for this exothermic reaction. Note that the standard thermodynamic data for the reaction (ΔH°_{rxn} and ΔG°_{rxn}) are expressed per mole of the reactant with a coefficient of 1 in the balanced chemical equation: N_2.

Practice Exercise Use data from Appendix 4 to calculate the value of K_p for the reaction

$$2 N_2(g) + O_2(g) \rightleftharpoons 2 N_2O(g)$$

at 298 K and 2000 K.

The equilibrium constant determined in Sample Exercise 16.16 for the ammonia synthesis reaction is greater than 10^5 at room temperature but about 10^{-5} at 773 K. That's ten orders of magnitude smaller. We now have a more complete picture of this reaction, which was the basis for Sample Exercise 16.10. Ammonia is usually synthesized from N_2 and H_2 at temperatures near 400°C. These high temperatures *increase the rate* of this exothermic reaction even though they significantly *decrease* K_p and decrease the amount of ammonia produced at equilibrium. In this case, the practical benefit of a more favorable reaction rate outweighs the less favorable thermodynamics of running the reaction at high temperature. In practice, the ammonia produced in the reaction is continuously removed from the reaction vessel through condensation, shifting the equilibrium toward the formation of more product as predicted by Le Châtelier's principle.

We have seen throughout this chapter that every reversible chemical reaction has a unique equilibrium constant at a specific temperature. If we know the value of that equilibrium constant, we can calculate the concentrations of reactants and products at equilibrium from any combination of initial concentrations. We can carry out other calculations that enable us to optimize the yields of a chemical process by adjusting factors such as temperature, and to predict the response of the process to changing reaction conditions.

SUMMARY

Section 16.1 Chemical equilibrium can be approached from either reaction direction and is achieved when the forward and reverse reaction rates are the same. At equilibrium, a reaction vessel may contain comparable amounts of reactants and prod-

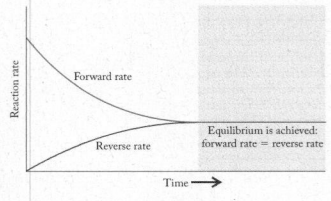

ucts, may contain mostly reactants (the equilibrium *lies to the left*), or may contain mostly products (the equilibrium *lies to the right*).

Section 16.2 According to the **law of mass action**, the equilibrium constant expression for K_c is the ratio of the equilibrium molar concentrations of the products divided by the equilibrium molar concentrations of the reactants, each raised to the respective stoichiometric coefficient in the balanced equation. If the substances involved in an equilibrium are gases, then an equilibrium constant (K_p) may also be written based on the partial pressures of the gases. Whether expressed as K_c or K_p, equilibrium constants have no units.

Section 16.3 The relationship between K_c and K_p depends on the number of moles of gaseous reactants and products in the balanced chemical equation.

Section 16.4 The reverse of a reaction has an equilibrium constant that is the reciprocal of K for the forward reaction. If the balanced equation for a reaction is multiplied by some factor n, the value of K for that reaction is raised to the nth power. If reactions are summed to give an overall reaction, their equilibrium constants are multiplied together to obtain an overall equilibrium constant.

Section 16.5 The **reaction quotient (Q)** is the value of the mass action expression at any instant during a reaction.

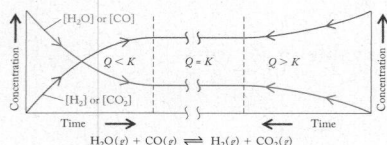

Section 16.6 Heterogeneous equilibria involve more than one phase. The concentrations of pure liquids and solids do not change during a reaction and so are omitted from equilibrium constant expressions.

Section 16.7 According to **Le Châtelier's principle**, chemical reactions at equilibrium respond to stress by shifting position to relieve the stress. Adding or removing a reactant or product, chang-

ing the partial pressures of reacting gases, and changing temperature all create stresses that shift the equilibrium position. A catalyst decreases the time it takes a system to achieve equilibrium but does not change the value of the equilibrium constant.

Section 16.8 Equilibrium concentrations or partial pressures of reactants and products can be calculated from initial concentrations or pressures, the reaction stoichiometry, and the value of the equilibrium constant.

Section 16.9 A negative value of $\Delta G°$ corresponds to $K > 1$ (products favored), and a positive value of $\Delta G°$ corresponds to $K < 1$ (reactants favored). Concentrations of reactants and products at equilibrium are similar if the value of K is near 1.

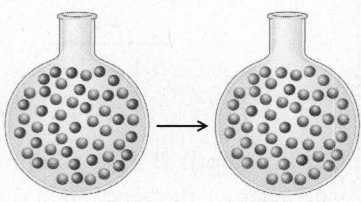

Section 16.10 Higher reaction temperatures increase the equilibrium constant of an endothermic reaction but decrease the equilibrium constant of an exothermic reaction. The slope of a plot of $\ln K$ versus $1/T$ for an equilibrium system is used to determine the standard enthalpy for the equilibrium, and the y-intercept of the plot is used to determine the standard entropy change.

PROBLEM-SOLVING SUMMARY

TYPE OF PROBLEM	CONCEPTS AND EQUATIONS		SAMPLE EXERCISES
Writing equilibrium constant expressions	For the reaction $$a\,A + b\,B \rightleftharpoons c\,C + d\,D$$ $$K_c = \frac{[C]^c[D]^d}{[A]^a[B]^b}$$ and $$K_p = \frac{(P_C)^c(P_D)^d}{(P_A)^a(P_B)^b}$$	(16.9) (16.10)	16.1
Calculating K_c or K_p	Insert equilibrium molar concentrations or partial pressures into the equilibrium constant expression.		16.2, 16.3
Interconverting K_c and K_p	Use the relation $$K_p = K_c(RT)^{\Delta n}$$ where $R = 0.08206$ L · atm/(mol · K), T is the absolute temperature, and Δn is the number of moles of product gas minus the number of moles of reactant gas in the balanced chemical equation.	(16.15)	16.4

TYPE OF PROBLEM	CONCEPTS AND EQUATIONS	SAMPLE EXERCISES
Recalculating K	To calculate K for the reverse of a reaction, take the reciprocal of K of the forward reaction. If all the coefficients in a chemical equation are multiplied by n, the value of K increases by the power of n. If reactions are summed to give an overall reaction, their equilibrium constants are multiplied together to obtain an overall K.	16.5–16.7
Using Q and K values to predict the direction of a reaction	If $Q < K$, the reaction proceeds in the forward direction to make more products; if $Q = K$, the reaction is at equilibrium; if $Q > K$, the reaction proceeds in the reverse direction to make more reactants.	16.8
Writing equilibrium constant expressions for heterogeneous equilibria	Molar concentrations of pure liquids and pure solids are omitted from equilibrium constant expressions because such concentrations are constant.	16.9
Adding or removing reactants or products to stress an equilibrium	Decreasing the concentration of a substance involved in an equilibrium shifts the equilibrium toward the production of more of that substance. Increasing the concentration of a substance shifts the equilibrium so that some of that substance is consumed in the reaction.	16.10
Predicting the effect of changing volume on gas-phase equilibria	Equilibria involving different numbers of moles of gaseous reactants and products shift in response to an increase (or decrease) in pressure caused by a decrease (or increase) in volume toward the side with fewer (or more) moles of gases.	16.11
Predicting changes in equilibrium with temperature	The value of K for an endothermic reaction increases with increasing temperature; the value of K for an exothermic reaction decreases with increasing temperature.	16.12
Calculating quantities of reactants and products at equilibrium	Use an ICE table to develop algebraic terms for each reactant's and product's partial pressure or concentration at equilibrium. Let x be the change in concentration or partial pressure of one component of the reaction. Express the changes in the other components in terms of x. Substitute these terms into the expression for K and solve for x.	16.13, 16.14
Calculating K from ΔG_f°	Calculate ΔG_{rxn}° from $$\Delta G_{rxn}^\circ = \sum n_{products}\, \Delta G_{f,products}^\circ - \sum n_{reactants}\, \Delta G_{f,reactants}^\circ \quad (14.10)$$ Then use $$\Delta G^\circ = -RT \ln K \quad (16.26)$$ Be sure to convert ΔG° to joules per mole and use $R = 8.314$ J/(mol · K).	16.15
Calculating a value of K at a specific temperature	Use $$\ln\left(\frac{K_2}{K_1}\right) = -\frac{\Delta H^\circ}{R}\left(\frac{1}{T_2} - \frac{1}{T_1}\right) \quad (16.31)$$ Convert ΔH° from kilojoules per mole to joules per mole to match the units of R.	16.16

VISUAL PROBLEMS

(Answers to boldface end-of-chapter questions and problems are in the back of the book.)

16.1. Figure P16.1 shows the energy profiles of reactions A $\rightleftharpoons$ B and C $\rightleftharpoons$ D, respectively. Which reaction has the larger forward rate constant? Which reaction has the smaller reverse rate constant? Which reaction has the larger equilibrium constant K_c?

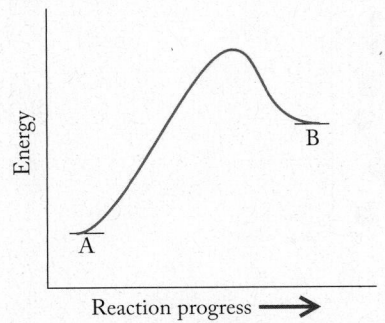
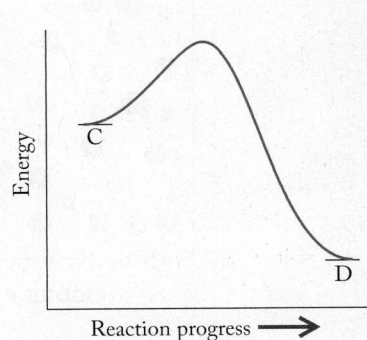

FIGURE P16.1

16.2. The progress with time of a reaction system is depicted in Figure P16.2. Red spheres represent the molar concentration of substance A and blue spheres represent the molar concentration of substance B.
a. Does the system reach equilibrium?
b. In which direction (A → B or B → A) is equilibrium attained?
c. What is the value of the equilibrium constant K_c?

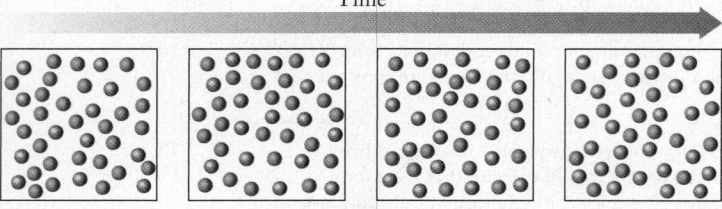

FIGURE P16.2

16.3. In Figure P16.3 the red spheres represent reactant A and the blue spheres represent product B in equilibrium with A.
a. Write a chemical equation that describes the equilibrium.
b. What is the value of the equilibrium constant K_c?

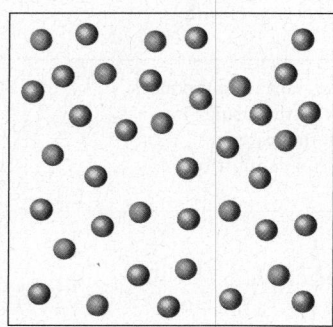

FIGURE P16.3

16.4. The equilibrium constant K_c for the reaction

A (red spheres) + B (blue spheres) ⇌ AB

is 3.0 at 300.0 K. Does the situation depicted in Figure P16.4 correspond to equilibrium? If not, in what direction (to the left or to the right) will the system shift to attain equilibrium?

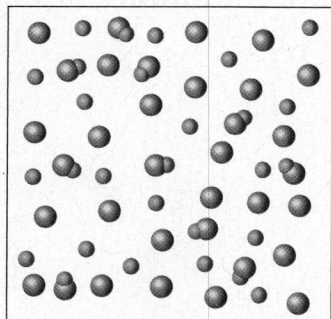

FIGURE P16.4

16.5. The top and bottom diagrams in Figure P16.5 represent equilibrium states of the reaction

A (red spheres) + B (blue spheres) ⇌ AB

at 300 K and 400 K, respectively. Is this reaction endothermic or exothermic? Explain.

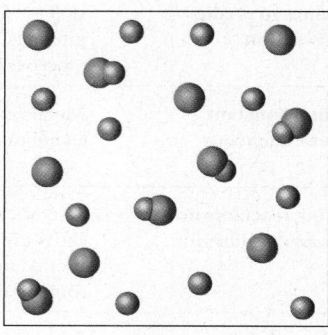

300 K

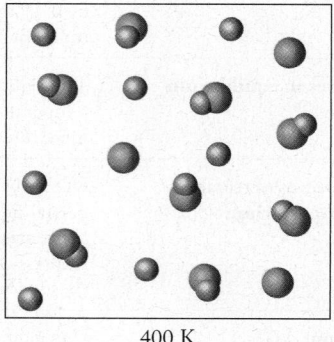

400 K

FIGURE P16.5

16.6. Figure P16.6 shows a plot of ln K_c versus $1/T$ for the reaction A + 2 B ⇌ AB$_2$. Is the reaction endothermic or exothermic?

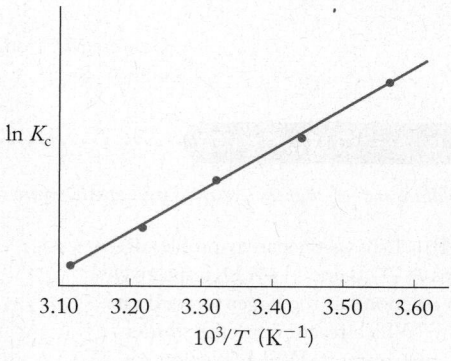

FIGURE P16.6

QUESTIONS AND PROBLEMS

The Dynamics of Chemical Equilibrium

CONCEPT REVIEW

16.7. How are forward and reverse reaction rates related in a system at chemical equilibrium?

16.8. Describe an example of a dynamic equilibrium that you experienced today.

16.9. Does the reaction $A \rightarrow 2B$ represented in Figure P16.9 reach equilibrium in 20 μs? Explain your answer.

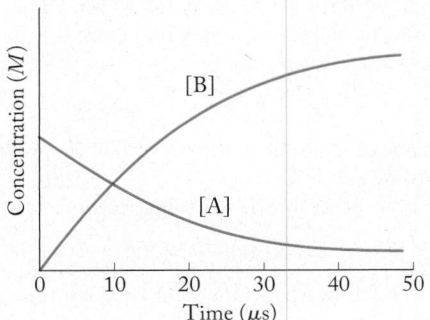

FIGURE P16.9

16.10. At equilibrium, is the sum of the concentrations of all the reactants always equal to the sum of the concentrations of the products? Explain.

16.11. Suppose the forward rate constant of the reaction $A \rightleftharpoons B$ is greater than the rate constant of the reverse reaction at a given temperature. Is the value of the equilibrium constant less than, greater than, or equal to 1?

16.12. Explain how it is possible for a reaction to have a large equilibrium constant but small forward and reverse rate constants.

PROBLEMS

16.13. In a study of the reaction

$$2N_2O(g) \rightleftharpoons 2N_2(g) + O_2(g)$$

quantities of all three gases were injected into a reaction vessel. The N_2O consisted entirely of isotopically labeled $^{15}N_2O$. Analysis of the reaction mixture after 1 day revealed the presence of compounds with molar masses 28, 29, 30, 32, 44, 45, and 46 g/mol. Identify the compounds and account for their appearance.

16.14. A mixture of ^{13}CO, $^{12}CO_2$, and O_2 in a sealed reaction vessel was used to follow the reaction

$$2CO(g) + O_2(g) \rightleftharpoons 2CO_2(g)$$

Analysis of the reaction mixture after 1 day revealed the presence of compounds with molar masses 28, 29, 32, 44, and 45 g/mol. Identify the compounds and account for their appearance.

16.15. Suppose the reaction $A \rightleftharpoons B$ in the forward direction is first order in A and the rate constant is $1.50 \times 10^{-2} \text{ s}^{-1}$. The reverse reaction is first order in B and the rate constant is $4.50 \times 10^{-2} \text{ s}^{-1}$ at the same temperature. What is the value of the equilibrium constant for the reaction $A \rightleftharpoons B$ at this temperature?

16.16. At 700 K the equilibrium constant K_c for the gas-phase reaction between NO and O_2 forming NO_2 is 8.7×10^6. The rate constant for the reverse reaction at this temperature is $0.54 \, M^{-1} \text{ s}^{-1}$. What is the value of the rate constant for the forward reaction at 700 K?

Writing Equilibrium Constant Expressions; Relationships between K_c and K_p Values

CONCEPT REVIEW

16.17. Under what conditions are the numerical values of K_c and K_p equal?

16.18. At 298 K, is K_p greater than or less than K_c if there is a net increase in the number of moles of gas in the reaction and if $K_c > 1$? Explain your answer.

16.19. Nitrogen oxides play important roles in air pollution. Write expressions for K_c and K_p for the following reactions involving nitrogen oxides.
   a. $N_2(g) + 2O_2(g) \rightleftharpoons N_2O_4(g)$
   b. $3NO(g) \rightleftharpoons NO_2(g) + N_2O(g)$
   c. $2N_2O(g) \rightleftharpoons 2N_2(g) + O_2(g)$

16.20. Write expressions for K_c and K_p for the following reactions, which contribute to the destruction of stratospheric ozone.
   a. $Cl(g) + O_3(g) \rightleftharpoons ClO(g) + O_2(g)$
   b. $2ClO(g) \rightleftharpoons 2Cl(g) + O_2(g)$
   c. $2O_3(g) \rightleftharpoons 3O_2(g)$

PROBLEMS

16.21. Use the graph in Figure P16.21 to estimate the value of the equilibrium constant K_c for the reaction

$$N_2O(g) \rightleftharpoons N_2(g) + \tfrac{1}{2}O_2(g)$$

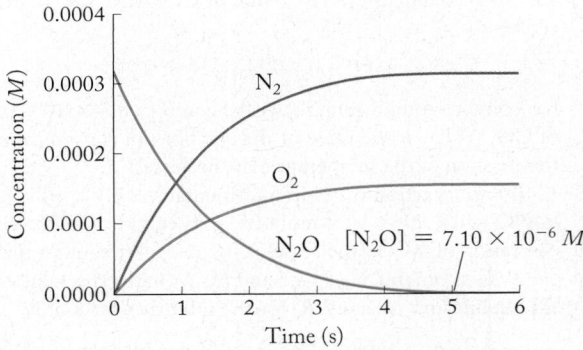

FIGURE P16.21

16.22. Estimate the value of the equilibrium constant K_c for the reaction

$$2NO(g) + O_2(g) \rightleftharpoons 2NO_2(g)$$

from the data in Figure P16.22.

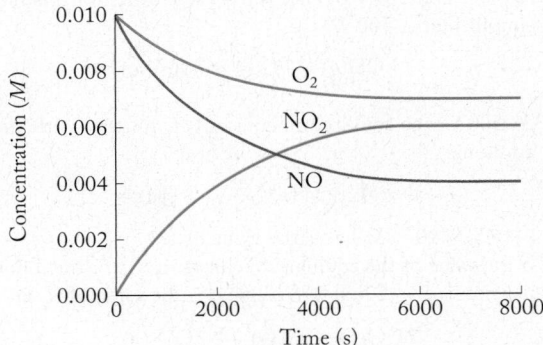

FIGURE P16.22

16.23. At 1000 K the partial pressures of an equilibrium mixture of $BrCl$, Br_2, and Cl_2 are 0.15, 0.33, and 0.33 atm, respectively. What is the value of K_p for the reaction

$$2\,BrCl(g) \rightleftharpoons Br_2(g) + Cl_2(g)$$

at 1000 K?

16.24. At 1045 K the partial pressures of an equilibrium mixture of H_2O, H_2, and O_2 are 0.040, 0.0045, and 0.0030 atm, respectively. Calculate the value of the equilibrium constant K_p at 1045 K.

$$2\,H_2O(g) \rightleftharpoons 2\,H_2(g) + O_2(g)$$

16.25. At equilibrium, the concentrations of gaseous N_2, O_2, and NO in a sealed reaction vessel are $[N_2] = 3.3 \times 10^{-3}\ M$, $[O_2] = 5.8 \times 10^{-3}\ M$, and $[NO] = 3.1 \times 10^{-3}\ M$. What is the value of K_c for the reaction

$$N_2(g) + O_2(g) \rightleftharpoons 2\,NO(g)$$

at the temperature of the reaction mixture?

16.26. Analyses of an equilibrium mixture of gaseous N_2O_4 and NO_2 gave the following results: $[NO_2] = 4.2 \times 10^{-3}\ M$ and $[N_2O_4] = 2.9 \times 10^{-3}\ M$. What is the value of the equilibrium constant K_c for the following reaction at the temperature of the mixture?

$$2\,NO_2(g) \rightleftharpoons N_2O_4(g)$$

16.27. A sealed reaction vessel initially contains 1.50×10^{-2} mol of water vapor and 1.50×10^{-2} mol of CO. After the following reaction

$$H_2O(g) + CO(g) \rightleftharpoons H_2(g) + CO_2(g)$$

has come to equilibrium, the vessel contains 8.3×10^{-3} mol of CO_2. What is the value of the equilibrium constant K_c of the reaction at the temperature of the vessel?

*16.28. A 100 mL reaction vessel initially contains 2.60×10^{-2} mol of NO and 1.30×10^{-2} mol of H_2. At equilibrium, the concentration of NO in the vessel is 0.161 M. At equilibrium the vessel also contains N_2, H_2O, and H_2. What is the value of the equilibrium constant K_c for the following reaction?

$$2\,H_2(g) + 2\,NO(g) \rightleftharpoons 2\,H_2O(g) + N_2(g)$$

16.29. The equilibrium constant K_p for the following equilibrium is 32 at 298 K. What is the value of K_c for this same equilibrium at 298 K?

$$A(g) + B(g) \rightleftharpoons AB(g)$$

16.30. The equilibrium constant K_c for the following equilibrium is 6.0×10^4 at 500 K. What is the value of K_p for this same equilibrium at 500 K?

$$CD(g) \rightleftharpoons C(g) + D(g)$$

16.31. At 500°C, the equilibrium constant K_p for the synthesis of ammonia

$$N_2(g) + 3\,H_2(g) \rightleftharpoons 2\,NH_3(g)$$

is 1.45×10^{-5}. What is the value of K_c?

16.32. If the value of the equilibrium constant, K_c, for the following reaction is 5×10^5 at 298 K, what is the value of K_p at 298 K?

$$2\,CO(g) + O_2(g) \rightleftharpoons 2\,CO_2(g)$$

16.33. For which of the following reactions are the values of K_c and K_p equal?
a. $2\,SO_2(g) + O_2(g) \rightleftharpoons 2\,SO_3(g)$
b. $Fe(s) + CO_2(g) \rightleftharpoons FeO(s) + CO(g)$
c. $H_2O(g) + CO(g) \rightleftharpoons H_2(g) + CO_2(g)$

16.34. For which of the following reactions are the values of K_c and K_p different?
a. $SO_2Cl_2(g) \rightleftharpoons SO_2(g) + Cl_2(g)$
b. $2\,NO(g) + O_2(g) \rightleftharpoons 2\,NO_2(g)$
c. $2\,O_3(g) \rightleftharpoons 3\,O_2(g)$

16.35. Bulletproof Glass Phosgene ($COCl_2$) is used in the manufacture of foam rubber and bulletproof glass. It is formed from carbon monoxide and chlorine in the following reaction:

$$Cl_2(g) + CO(g) \rightleftharpoons COCl_2(g)$$

The value of K_c for the reaction is 5.0 at 327°C. What is the value of K_p at 327°C?

16.36. If the value of K_p for the following reaction

$$SO_2(g) + NO_2(g) \rightleftharpoons NO(g) + SO_3(g)$$

is 3.45 at 298 K, what is the value of K_c for the reverse reaction?

Manipulating Equilibrium Constant Expressions

CONCEPT REVIEW

16.37. How is the value of the equilibrium constant affected by scaling up or down the coefficients of the reactants and products in the chemical equation describing the reaction?

16.38. Why must a written form of a reaction and the temperature be given when reporting the value of an equilibrium constant?

PROBLEMS

16.39. The equilibrium constant for the reaction

$$I_2(g) + Br_2(g) \rightleftharpoons 2\,IBr(g)$$

is 120 at 425 K. What is the value of K_c for the equilibrium

$$\tfrac{1}{2}I_2(g) + \tfrac{1}{2}Br_2(g) \rightleftharpoons IBr(g)$$

at 425 K?

16.40. The equilibrium constant K_p for the synthesis of ammonia:

$$N_2(g) + 3\,H_2(g) \rightleftharpoons 2\,NH_3(g)$$

is 4.3×10^{-3} at 300°C. What is the value of K_p for the equilibrium

$$\tfrac{1}{2}N_2(g) + \tfrac{3}{2}H_2(g) \rightleftharpoons NH_3(g)$$

at 300°C?

16.41. The following reaction is one of the elementary steps in the oxidation of NO:

$$NO(g) + NO_3(g) \rightleftharpoons 2\,NO_2(g)$$

Write an expression for the equilibrium constant K_c for this reaction and for the reverse reaction:

$$2\,NO_2(g) \rightleftharpoons NO(g) + NO_3(g)$$

How are the two K_c expressions related?

16.42. Making Ammonia The value of the equilibrium constant K_p for the formation of ammonia,

$$N_2(g) + 3\,H_2(g) \rightleftharpoons 2\,NH_3(g)$$

is 4.5×10^{-5} at 450°C. What is the value of K_p for the following reaction?

$$2\,NH_3(g) \rightleftharpoons N_2(g) + 3\,H_2(g)$$

16.43. Air Pollutants Sulfur oxides are major air pollutants. The reaction between sulfur dioxide and oxygen can be written in two ways:

$$SO_2(g) + \tfrac{1}{2}O_2(g) \rightleftharpoons SO_3(g)$$

and

$$2\,SO_2(g) + O_2(g) \rightleftharpoons 2\,SO_3(g)$$

Write expressions for the equilibrium constants for both reactions. How are they related?

16.44. At a given temperature, the equilibrium constant K_c for the reaction

$$2\,NO(g) + 2\,H_2(g) \rightleftharpoons N_2(g) + 2\,H_2O(g)$$

is 0.11. What is the equilibrium constant for the following reaction?

$$NO(g) + H_2(g) \rightleftharpoons \tfrac{1}{2}N_2(g) + H_2O(g)$$

16.45. At a given temperature, the equilibrium constant K_c for the reaction

$$2\,SO_2(g) + O_2(g) \rightleftharpoons 2\,SO_3(g)$$

is 2.4×10^{-3}. What is the value of the equilibrium constant for each of the following reactions at that temperature?
a. $SO_2(g) + \tfrac{1}{2}O_2(g) \rightleftharpoons SO_3(g)$
b. $2\,SO_3(g) \rightleftharpoons 2\,SO_2(g) + O_2(g)$
c. $SO_3(g) \rightleftharpoons SO_2(g) + \tfrac{1}{2}O_2(g)$

16.46. If the equilibrium constant K_c for the reaction

$$2\,NO(g) + O_2(g) \rightleftharpoons 2\,NO_2(g)$$

is 5×10^{12}, what is the value of the equilibrium constant of each of the following reactions at the same temperature?
a. $NO(g) + \tfrac{1}{2}O_2(g) \rightleftharpoons NO_2(g)$
b. $2\,NO_2(g) \rightleftharpoons 2\,NO(g) + O_2(g)$
c. $NO_2(g) \rightleftharpoons NO(g) + \tfrac{1}{2}O_2(g)$

16.47. Calculate the value of the equilibrium constant K for the hypothetical reaction

$$2\,D \rightleftharpoons A + 2\,B$$

from the following information:

$$A + 2\,B \rightleftharpoons C \qquad K_c = 3.3$$
$$C \rightleftharpoons 2\,D \qquad K_c = 0.041$$

16.48. Calculate the value of the equilibrium constant K for the hypothetical reaction

$$E + F \rightleftharpoons G$$

from the following information:

$$2\,G \rightleftharpoons H \qquad K_c = 3.1 \times 10^{-4}$$
$$H \rightleftharpoons 2\,E + 2\,F \qquad K_c = 2.8 \times 10^{22}$$

Equilibrium Constants and Reaction Quotients

CONCEPT REVIEW

16.49. What is a reaction quotient?

16.50. How is an equilibrium constant different from a reaction quotient?

16.51. What does it mean when the reaction quotient Q is numerically equal to the equilibrium constant K?

16.52. Explain how knowing Q and K for an equilibrium system enables you to say whether it is at equilibrium or whether it will shift in one direction or another.

PROBLEMS

16.53. If the equilibrium constant K_c for the hypothetical reaction $A(g) \rightleftharpoons B(g)$ is 22 at a given temperature, and if $[A] = 0.10\ M$ and $[B] = 2.0\ M$ in a reaction mixture at that temperature, is the reaction at chemical equilibrium? If not, in which direction will the reaction proceed to reach equilibrium?

16.54. The equilibrium constant K_c for the hypothetical reaction

$$2\,C \rightleftharpoons D + E$$

is 3×10^{-3}. At a particular time, the composition of the reaction mixture is $[C] = [D] = [E] = 5 \times 10^{-4}\ M$. In which direction will the reaction proceed to reach equilibrium?

16.55. Suppose the value of the equilibrium constant K_p of the following hypothetical reaction

$$A(g) + B(g) \rightleftharpoons C(g)$$

is 1.00 at 300 K. Are either of the following reaction mixtures at chemical equilibrium at 300 K?
a. $P_A = P_B = P_C = 1.0$ atm
b. $[A] = [B] = [C] = 1.0\ M$

16.56. In which direction will the following hypothetical reaction proceed to reach equilibrium under the conditions given?

$$A(g) + B(g) \rightleftharpoons C(g) \qquad K_p = 1.00 \text{ at } 300\text{ K}$$

a. $P_A = P_C = 1.0$ atm, $P_B = 0.50$ atm
b. $[A] = [B] = [C] = 1.0\ M$

16.57. If the equilibrium constant K_c for the reaction

$$N_2(g) + O_2(g) \rightleftharpoons 2\,NO(g)$$

is 1.5×10^{-3}, in which direction will the reaction proceed if the partial pressures of the three gases are all 1.00×10^{-3} atm?

16.58. At 650 K, the value of the equilibrium constant K_p for the ammonia synthesis reaction

$$N_2(g) + 3\,H_2(g) \rightleftharpoons 2\,NH_3(g)$$

is 4.3×10^{-4}. If a vessel contains a reaction mixture in which $[N_2] = 0.010\ M$, $[H_2] = 0.030\ M$, and $[NH_3] = 0.00020\ M$, will more ammonia form?

16.59. The hypothetical equilibrium $X + Y \rightleftharpoons Z$ has $K_c = 1.00$ at 350 K. If the initial molar concentrations of X, Y, and Z in a solution are all $0.2\ M$, in which direction will the reaction shift to reach equilibrium?
a. To the left, making more X and Y
b. To the right, making more Z
c. The system is at equilibrium and the concentrations will not change.

16.60. In Problem 16.59, when the equilibrium shifts, does the concentration of X increase or decrease?

Heterogeneous Equilibria

CONCEPT REVIEW

16.61. Write the K_c expression for the following reaction:

$$CuS(s) \rightleftharpoons Cu^{2+}(aq) + S^{2-}(aq)$$

16.62. Write the K_c expression for the following reaction:

$$Al_2O_3(s) + 3H_2O(\ell) \rightleftharpoons 2Al^{3+}(aq) + 6OH^-(aq)$$

16.63. Why does the K_c expression for the reaction

$$CaCO_3(s) \rightleftharpoons CaO(s) + CO_2(g)$$

not contain terms for the concentrations of $CaCO_3$ and CaO?

*16.64. Why does the value of K_p for the reaction

$$CaCO_3(s) \rightleftharpoons CaO(s) + CO_2(g)$$

increase with increasing temperature?

Le Châtelier's Principle

CONCEPT REVIEW

16.65. Does adding reactants to a system at equilibrium increase the value of the equilibrium constant?

16.66. Increasing the concentration of a reactant shifts the position of chemical equilibrium toward formation of more products. What effect does adding a reactant have on the rates of the forward and reverse reactions?

16.67. **Carbon Monoxide Poisoning** Patients suffering from carbon monoxide poisoning are treated with pure oxygen to remove CO from the hemoglobin (Hb) in their blood. The two relevant equilibria are

$$Hb + 4CO(g) \rightleftharpoons Hb(CO)_4$$

$$Hb + 4O_2(g) \rightleftharpoons Hb(O_2)_4$$

The value of the equilibrium constant for CO binding to Hb is greater than that for O_2. How, then, does this treatment work?

16.68. Is the equilibrium constant K_p for the reaction

$$2NO_2(g) \rightleftharpoons N_2O_4(g)$$

in air the same in Los Angeles as in Denver if the atmospheric pressure in Denver is lower but the temperature is the same?

16.69. Henry's law (Chapter 10) predicts that the solubility of a gas in a liquid increases with its partial pressure. Explain Henry's law in relation to Le Châtelier's principle.

*16.70. For the reaction

$$2CO(g) + O_2(g) \rightleftharpoons 2CO_2(g)$$

why does adding an inert gas such as argon to an equilibrium mixture of CO, O_2, and CO_2 in a sealed vessel increase the total pressure of the system but not affect the position of the equilibrium?

PROBLEMS

16.71. Which of the following equilibria will shift toward formation of more products if an equilibrium mixture is compressed into half its volume?
a. $2N_2O(g) \rightleftharpoons 2N_2(g) + O_2(g)$
b. $2CO(g) + O_2(g) \rightleftharpoons 2CO_2(g)$
c. $N_2(g) + O_2(g) \rightleftharpoons 2NO(g)$
d. $2NO(g) + O_2(g) \rightleftharpoons 2NO_2(g)$

16.72. Which of the following equilibria will shift toward formation of more products if the volume of a reaction mixture at equilibrium increases by a factor of 2?
a. $2SO_2(g) + O_2(g) \rightleftharpoons 2SO_3(g)$
b. $NO(g) + O_3(g) \rightleftharpoons NO_2(g) + O_2(g)$
c. $2N_2O_5(g) \rightleftharpoons 4NO_2(g) + O_2(g)$
d. $N_2O_4(g) \rightleftharpoons 2NO_2(g)$

16.73. What would be the effect of the changes listed on the equilibrium concentrations of reactants and products in the following reaction?

$$2O_3(g) \rightleftharpoons 3O_2(g)$$

a. O_3 is added to the system.
b. O_2 is added to the system.
c. The mixture is compressed to one-tenth its initial volume.

16.74. How will the changes listed affect the position of the following equilibrium?

$$2NO_2(g) \rightleftharpoons NO(g) + NO_3(g)$$

a. The concentration of NO is increased.
b. The concentration of NO_2 is increased.
c. The volume of the system is allowed to expand to 5 times its initial value.

16.75. How would reducing the partial pressure of $O_2(g)$ affect the position of the equilibrium in the following reaction?

$$2SO_2(g) + O_2(g) \rightleftharpoons 2SO_3(g)$$

*16.76. Ammonia is added to a gaseous reaction mixture containing H_2, Cl_2, and HCl that is at chemical equilibrium. How will the addition of ammonia affect the relative concentrations of H_2, Cl_2, and HCl if the equilibrium constant of reaction 2 is much greater than the equilibrium constant of reaction 1?

(1) $\qquad H_2(g) + Cl_2(g) \rightleftharpoons 2HCl(g)$
(2) $\qquad HCl(g) + NH_3(g) \rightleftharpoons NH_4Cl(s)$

16.77. In which of the following hypothetical equilibria does the product yield increase with increasing temperature?
a. $A + 2B \rightleftharpoons C \qquad \Delta H > 0$
b. $A + 2B \rightleftharpoons C \qquad \Delta H = 0$
c. $A + 2B \rightleftharpoons C \qquad \Delta H < 0$

16.78. In which of the following hypothetical equilibria does the product yield decrease with increasing temperature?
a. $2X + Y \rightleftharpoons Z \qquad \Delta H > 0$
b. $2X + Y \rightleftharpoons Z \qquad \Delta H = 0$
c. $2X + Y \rightleftharpoons Z \qquad \Delta H < 0$

Calculations Based on K

CONCEPT REVIEW

16.79. Why are calculations based on K often simpler when the value of K is very small?

16.80. Could the quadratic equation be used to solve for the equilibrium concentration of NO_2 in the following reaction?

$$2\,NO + O_2 \rightarrow 2\,NO_2$$

PROBLEMS

16.81. For the reaction

$$PCl_5(g) \rightleftharpoons PCl_3(g) + Cl_2(g) \qquad K_p = 23.6 \text{ at } 500 \text{ K}$$

a. Calculate the equilibrium partial pressures of the reactants and products if the initial pressures are $P_{PCl_5} = 0.560$ atm and $P_{PCl_3} = 0.500$ atm.

b. If more chlorine is added after equilibrium is reached, how will the concentrations of PCl_5 and PCl_3 change?

16.82. Enough NO_2 gas is injected into a cylindrical vessel to produce a partial pressure, P_{NO_2}, of 0.900 atm at 298 K. Calculate the equilibrium partial pressures of NO_2 and N_2O_4, given

$$2\,NO_2(g) \rightleftharpoons N_2O_4(g) \qquad K_p = 4 \text{ at } 298 \text{ K}$$

16.83. The value of K_c for the reaction between water vapor and dichlorine monoxide

$$H_2O(g) + Cl_2O(g) \rightleftharpoons 2\,HOCl(g)$$

is 0.0900 at 25°C. Determine the equilibrium concentrations of all three compounds if the starting concentrations of both reactants are 0.00432 M and no HOCl is present.

16.84. The value of K_p for the reaction

$$3\,H_2(g) + N_2(g) \rightleftharpoons 2\,NH_3(g)$$

is 4.3×10^{-4} at 648 K. Determine the equilibrium partial pressure of NH_3 in a reaction vessel that initially contained 0.900 atm N_2 and 0.500 atm H_2 at 648 K.

16.85. The value of K_p for the reaction

$$NO(g) + \tfrac{1}{2}O_2(g) \rightleftharpoons NO_2(g)$$

is 2×10^6 at 25°C. At equilibrium, what is the ratio of P_{NO_2} to P_{NO} in air at 25°C? Assume that $P_{O_2} = 0.21$ atm and does not change.

*16.86. Water Gas The water–gas reaction is a source of hydrogen. Passing steam over hot carbon produces a mixture of carbon monoxide and hydrogen:

$$H_2O(g) + C(s) \rightleftharpoons CO(g) + H_2(g)$$

The value of K_c for the reaction at 1000°C is 3.0×10^{-2}.

a. Calculate the equilibrium partial pressures of the products and reactants if $P_{H_2O} = 0.442$ atm and $P_{CO} = 5.0$ atm at the start of the reaction. Assume that the carbon is in excess.

b. Determine the equilibrium partial pressures of the reactants and products after sufficient CO and H_2 are added to the equilibrium mixture in part a to initially increase the partial pressures of both gases by 0.075 atm.

16.87. The value of K_p for the reaction

$$CO_2(g) + C(s) \rightleftharpoons 2\,CO(g)$$

is 1.5 at 700°C. Calculate the equilibrium partial pressures of CO and CO_2 if initially $P_{CO_2} = 5.0$ atm and $P_{CO} = 0.0$. Pure graphite is present initially and when equilibrium is achieved.

16.88. Jupiter's Atmosphere Ammonium hydrogen sulfide (NH_4SH) was detected in the atmosphere of Jupiter subsequent to its collision with the comet Shoemaker–Levy. The equilibrium between ammonia, hydrogen sulfide, and NH_4SH is described by the following equation:

$$NH_4SH(s) \rightleftharpoons NH_3(g) + H_2S(g)$$

The value of K_p for the reaction at 24°C is 0.126. Suppose a sealed flask contains an equilibrium mixture of NH_4SH, NH_3, and H_2S. At equilibrium, the partial pressure of H_2S is 0.355 atm. What is the partial pressure of NH_3?

*16.89. A flask containing pure NO_2 was heated to 1000 K, a temperature at which the value of K_p for the decomposition of NO_2 is 158.

$$2\,NO_2(g) \rightleftharpoons 2\,NO(g) + O_2(g)$$

The partial pressure of O_2 at equilibrium is 0.136 atm.

a. Calculate the partial pressures of NO and NO_2.

b. Calculate the total pressure in the flask at equilibrium.

16.90. The equilibrium constant K_p of the reaction

$$2\,SO_3(g) \rightleftharpoons 2\,SO_2(g) + O_2(g)$$

is 7.69 at 830°C. If a vessel at this temperature initially contains pure SO_3 and if the partial pressure of SO_3 at equilibrium is 0.100 atm, what is the partial pressure of O_2 in the flask at equilibrium?

*16.91. NO$_x$ Pollution In a study of the formation of NO_x air pollution, a chamber heated to 2200°C was filled with air (0.79 atm N_2, 0.21 atm O_2). What are the equilibrium partial pressures of N_2, O_2, and NO if $K_p = 0.050$ for the following reaction at 2200°C?

$$N_2(g) + O_2(g) \rightleftharpoons 2\,NO(g)$$

*16.92. The equilibrium constant K_p for the thermal decomposition of NO_2

$$2\,NO_2(g) \rightleftharpoons 2\,NO(g) + O_2(g)$$

is 6.5×10^{-6} at 450°C. If a reaction vessel at this temperature initially contains 0.500 atm NO_2, what will be the partial pressures of NO_2, NO, and O_2 in the vessel when equilibrium has been attained?

16.93. The value of K_c for the thermal decomposition of hydrogen sulfide

$$2\,H_2S(g) \rightleftharpoons 2\,H_2(g) + S_2(g)$$

is 2.2×10^{-4} at 1400 K. A sample of gas in which $[H_2S] = 6.00$ M is heated to 1400 K in a sealed high-pressure vessel. After chemical equilibrium has been achieved, what is the value of $[H_2S]$? Assume that no H_2 or S_2 was present in the original sample.

16.94. Urban Air On a very smoggy day, the equilibrium concentration of NO_2 in the air over an urban area reaches 2.2×10^{-7} M. If the temperature of the air is 25°C, what is the concentration of the dimer N_2O_4 in the air? Given:

$$N_2O_4(g) \rightleftharpoons 2\,NO_2(g) \qquad K_c = 6.1 \times 10^{-3}$$

*16.95. **Chemical Weapon** Phosgene, $COCl_2$, gained notoriety as a chemical weapon in World War I. Phosgene is produced by the reaction of carbon monoxide with chlorine

$$CO(g) + Cl_2(g) \rightleftharpoons COCl_2(g)$$

The value of K_c for this reaction is 5.0 at 600 K. What are the equilibrium partial pressures of the three gases if a reaction vessel initially contains a mixture of the reactants in which $P_{CO} = P_{Cl_2} = 0.265$ atm and $P_{COCl_2} = 0.000$ atm?

*16.96. At 2000°C, the value of K_c for the reaction

$$2CO(g) + O_2(g) \rightleftharpoons 2CO_2(g)$$

is 1.0. What is the ratio of [CO] to $[CO_2]$ in an atmosphere in which $[O_2] = 0.0045\ M$?

*16.97. The water–gas shift reaction is an important source of hydrogen. The value of K_c for the reaction

$$CO(g) + H_2O(g) \rightleftharpoons CO_2(g) + H_2(g)$$

at 700 K is 5.1. Calculate the equilibrium concentrations of the four gases if the initial concentration of each of them is 0.050 M.

*16.98. Sulfur dioxide reacts with NO_2, forming SO_3 and NO:

$$SO_2(g) + NO_2(g) \rightleftharpoons SO_3(g) + NO(g)$$

If the value of K_c for the reaction is 2.50, what are the equilibrium concentrations of the products if the reaction mixture was initially 0.50 M SO_2, 0.50 M NO_2, 0.0050 M SO_3, and 0.0050 M NO?

Equilibrium and Thermodynamics

CONCEPT REVIEW

16.99. Do all reactions with equilibrium constants < 1 have values of $\Delta G° > 0$?

*16.100. The equation $\Delta G° = -RT \ln K$ relates the value of K_p, not K_c, to the change in standard free energy for a reaction in the gas phase. Explain why.

16.101. Starting with pure reactants, in which direction will an equilibrium shift if $\Delta G° < 0$?

16.102. Starting with pure products, in which direction will an equilibrium shift if $\Delta G° < 0$?

PROBLEMS

16.103. Which of the following reactions has the largest equilibrium constant at 25°C?
a. $Cl_2(g) + F_2(g) \rightleftharpoons 2ClF(g)$ $\qquad \Delta G° = 115.4$ kJ
b. $Cl_2(g) + Br_2(g) \rightleftharpoons 2ClBr(g)$ $\qquad \Delta G° = -2.0$ kJ
c. $Cl_2(g) + I_2(g) \rightleftharpoons 2ICl(g)$ $\qquad \Delta G° = -27.9$ kJ

16.104. **Glycolysis** Problems 16.104–16.107 focus on glycolysis, the multistep biochemical process by which sugar is metabolized to create energy in the body. Which of the following steps in glycolysis has the largest equilibrium constant?
a. Fructose 1,6-diphosphate $\rightleftharpoons$ 2 glyceraldehyde-3-phosphate
$\qquad\qquad\qquad\qquad\qquad\qquad\qquad \Delta G° = 24$ kJ
b. 3-Phosphoglycerate $\rightleftharpoons$ 2-phosphoglycerate
$\qquad\qquad\qquad\qquad\qquad\qquad\qquad \Delta G° = 4.4$ kJ
c. 2-Phosphoglycerate $\rightleftharpoons$ phosphoenolpyruvate
$\qquad\qquad\qquad\qquad\qquad\qquad\qquad \Delta G° = 1.8$ kJ

16.105. The value of $\Delta G°$ for the phosphorylation of glucose in glycolysis is 13.8 kJ/mol. What is the value of the equilibrium constant for the reaction at 298 K?

16.106. In glycolysis, the hydrolysis of ATP to ADP drives the phosphorylation of glucose:

$$Glucose + ATP \rightleftharpoons ADP + glucose\ 6\text{-phosphate}$$
$$\Delta G° = -17.7\ kJ$$

What is the value of K_c for this reaction at 298 K?

16.107. Sucrose enters the series of reactions in glycolysis after its hydrolysis into glucose and fructose:

$$Sucrose + H_2O \rightleftharpoons glucose + fructose$$
$$K_c = 5.3 \times 10^{12}\ at\ 298\ K$$

What is the value of $\Delta G°$ for this process?

16.108. The value of the equilibrium constant K_p for the reaction

$$H_2(g) + CO_2(g) \rightleftharpoons H_2O(g) + CO(g)$$

is 0.534 at 700°C.
a. Calculate the value of ΔG of the reaction.
b. Using values from Appendix 4, calculate the value of $\Delta G°$ of the reaction and compare the result with that obtained in part a.

16.109. Use the following data to calculate the value of K_p at 298 K for the reaction

$$N_2(g) + 2O_2(g) \rightleftharpoons 2NO_2(g)$$
$$N_2(g) + O_2(g) \rightleftharpoons 2NO(g) \qquad \Delta G° = 173.2\ kJ$$
$$2NO(g) + O_2(g) \rightleftharpoons 2NO_2(g) \qquad \Delta G° = -69.7\ kJ$$

16.110. Under the appropriate conditions, NO forms N_2O and NO_2:

$$3NO(g) \rightleftharpoons N_2O(g) + NO_2(g)$$

Use the values for $\Delta G°$ for the following reactions to calculate the value of K_p for the above reaction at 500°C.

$$2NO(g) + O_2(g) \rightleftharpoons 2NO_2(g) \qquad \Delta G° = -69.7\ kJ$$
$$2N_2O(g) \rightleftharpoons 2NO(g) + N_2(g) \qquad \Delta G° = -33.8\ kJ$$
$$N_2(g) + O_2(g) \rightleftharpoons 2NO(g) \qquad \Delta G° = 173.2\ kJ$$

Changing *K* with Changing Temperature

CONCEPT REVIEW

16.111. The value of the equilibrium constant of a reaction decreases with increasing temperature. Is this reaction endothermic or exothermic?

16.112. The reaction

$$2CO(g) + O_2(g) \rightleftharpoons 2CO_2(g)$$

is exothermic. Does the value of K_p increase or decrease with increasing temperature?

16.113. The value of K_p for the water–gas shift reaction

$$CO(g) + H_2O(g) \rightleftharpoons H_2(g) + CO_2(g)$$

increases as the temperature decreases. Is the reaction exothermic or endothermic?

16.114. Does the value of K_p for the reaction

$$CH_4(g) + H_2O(g) \rightleftharpoons 3H_2(g) + CO(g) \qquad \Delta H° = 206 \text{ kJ}$$

increase, decrease, or remain unchanged as temperature increases?

PROBLEMS

16.115. **Air Pollution** Automobiles and trucks pollute the air with NO. At 2000°C, K_c for the reaction

$$N_2(g) + O_2(g) \rightleftharpoons 2NO(g)$$

is 4.10×10^{-4}, and $\Delta H° = 180.6$ kJ. What is the value of K_c at 25°C?

16.116. At 400 K the value of K_p for the reaction

$$N_2(g) + 3H_2(g) \rightleftharpoons 2NH_3(g)$$

is 41, and $\Delta H° = -92.2$ kJ. What is the value of K_p at 700 K?

16.117. The equilibrium constant for the reaction

$$NO(g) + O_2(g) \rightleftharpoons 2NO_2(g)$$

decreases from 1.5×10^5 at 430°C to 23 at 1000°C. From these data, calculate the value of $\Delta H°$ for the reaction.

16.118. The value of K_c for the reaction $A \rightleftharpoons B$ is 0.455 at 50°C and 0.655 at 100°C. Calculate $\Delta H°$ for the reaction.

Additional Problems

*16.119. **CO as a Fuel** Is carbon dioxide a viable source of the fuel CO? Pure carbon dioxide ($P_{CO_2} = 1$ atm) decomposes at high temperatures. For the system

$$2CO_2(g) \rightleftharpoons 2CO(g) + O_2(g)$$

the percentage of decomposition of $CO_2(g)$ changes with temperature as follows:

Temperature (K)	Decomposition (%)
1500	0.048
2500	17.6
3000	54.8

Is the reaction endothermic? Calculate the value of K_p at each temperature and discuss the results. Is the decomposition of CO_2 an antidote for global warming?

16.120. Ammonia decomposes at high temperatures. In an experiment to explore this behavior, 2.00 moles of gaseous NH_3 is sealed in a rigid 1-liter vessel. The vessel is heated at 800 K and some of the NH_3 decomposes in the following reaction:

$$2NH_3(g) \rightleftharpoons N_2(g) + 3H_2(g)$$

The system eventually reaches equilibrium and is found to contain 1.74 moles of NH_3. What are the values of K_p and K_c for the decomposition reaction at 800 K?

*16.121. Elements of group 16 form hydrides with the generic formula H_2X. When gaseous H_2X is bubbled through a solution containing 0.3 M hydrochloric acid, the solution

becomes saturated and $[H_2X] = 0.1$ M. The following equilibria exist in this solution:

$$H_2X(aq) + H_2O(\ell) \rightleftharpoons HX^-(aq) + H_3O^+(aq)$$
$$K_1 = 8.3 \times 10^{-8}$$

$$HX^-(aq) + H_2O(\ell) \rightleftharpoons X^{2-}(aq) + H_3O^+(aq)$$
$$K_2 = 1 \times 10^{-14}$$

Calculate the concentration of X^{2-} in the solution.

*16.122. Nitrogen dioxide reacts with SO_2 to form SO_3 and NO:

$$NO_2(g) + SO_2(g) \rightleftharpoons NO(g) + SO_3(g)$$

An equilibrium mixture is analyzed at a certain temperature and found to contain $[NO_2] = 0.100$ M, $[SO_2] = 0.300$ M, $[NO] = 2.00$ M, and $[SO_3] = 0.600$ M. At the same temperature, extra $SO_2(g)$ is added to make $[SO_2] = 0.800$ M. Calculate the composition of the mixture when equilibrium has been reestablished.

*16.123. Carbon disulfide is a foul-smelling solvent that dissolves sulfur and other nonpolar substances. It can be made by heating sulfur in an atmosphere of methane:

$$4CH_4(g) + S_8(s) \rightleftharpoons 4CS_2(g) + 8H_2(g)$$

Starting with the appropriate data in Appendix 4, calculate the values of K_p for the reaction at 25°C and 500°C.

*16.124. **Making Hydrogen** Debate continues on the practicality of H_2 gas as a fuel for vehicles. The equilibrium constant K_c for the reaction

$$CO(g) + H_2O(g) \rightleftharpoons CO_2(g) + H_2(g)$$

is 1.0×10^5 at 25°C. Starting with this value, calculate the value of $\Delta G°_{rxn}$ at 25°C, and, without doing any calculations, guess the sign of $\Delta H°_{rxn}$.

*16.125. **Air Pollution Control** Calcium oxide is used to remove the pollutant SO_2 from smokestack gases. The $\Delta G°$ of the overall reaction

$$CaO(s) + SO_2(g) + \tfrac{1}{2}O_2(g) \rightleftharpoons CaSO_4(s)$$

is -418.6 kJ. What is P_{SO_2} in equilibrium with air ($P_{O_2} = 0.21$ atm) and solid CaO?

*16.126. **Volcanic Eruptions** During volcanic eruptions, gases as hot as 700°C and rich in SO_2 are released into the atmosphere. As air mixes with these gases, the following reaction converts some of this SO_2 into SO_3:

$$2SO_2(g) + O_2(g) \rightleftharpoons 2SO_3(g)$$

Calculate the value of K_p for this reaction at 700°C. What is the ratio of P_{SO_2} to P_{SO_3} in equilibrium with $P_{O_2} = 0.21$ atm?

Additional study materials including ChemTours and Diagnostic Quizzes are available at StudySpace at www.wwnorton.com/studyspace.

17

Equilibrium in the Aqueous Phase

Learning Outcomes

- Identify the Brønsted-Lowry acids and bases and their conjugate bases and acids in chemical reactions

- Relate the strengths of acids and bases to their K_a, K_b, pK_a, and pK_b values and to their percent ionization or dissociation in water

- Predict whether a salt is acidic, basic, or neutral

- Calculate the pH values of solutions of weak acids and bases and the salts of their conjugate bases and acids

- Explain how pH buffers control pH and calculate the pH of a conjugate acid-base pair

- Relate the solubility of an ionic compound to its solubility product

- Interpret the results of an acid-base titration

A LOOK AHEAD

A Balancing Act

For centuries, farmers have added lime (CaO) to their fields to "sweeten" the soil, which means making the soil less acidic. Adding a basic substance like lime neutralizes the organic acids that are produced when biological matter, such as the plowed-under remnants of last year's crop, decays. In more recent times, farmers and gardeners have also added lime to soil to neutralize acidic precipitation seeping into it.

Normal rainwater is slightly acidic because atmospheric CO_2 dissolves in it, forming a dilute solution of weak carbonic acid. *Acid rain*, however, contains dissolved oxides of nitrogen and sulfur that enter the air mostly as by-products of fossil fuel combustion. These oxides form both weak and strong acids when they dissolve in water. As a result, acid rain may be from 10 to 1000 times more acidic than uncontaminated rainwater. Though some plants, such as blueberries, actually thrive in acidic soil, most crops do not. At the very least their growth is stunted. Sometimes they don't grow at all.

Farmers seek to achieve a balanced soil chemistry that is *neutral*—neither acidic nor basic—or nearly so. Most plants grow best in neutral soil because the nutrients they need to grow are more available to them under such conditions. Sometimes this availability depends on microorganisms that thrive under near-neutral conditions and transform nutrients into species that are more readily absorbed by plant roots.

The bioavailability of nutrients and nonessential elements may also be a matter of their simply being more soluble and more easily absorbed by plants. Consider, for example, the color of hydrangea blossoms, which is controlled by the availability of Al^{3+} ions in the soil. In acidic soil, hydrangeas can absorb Al^{3+} ions because the ions are soluble in acidic solutions. As a result the plants form blue blossoms.

Shades of Purple, Pink, and Blue Flowers The color of hydrangea flowers depends on the acid content of the soil in which they are grown. ▶

▶‖ **CHEMTOUR** Acid Rain

∞ CONNECTION We introduced acids, bases, and neutralization reactions in Chapter 4, where we defined acids as proton donors and bases as proton acceptors. Neutralization reactions between acids and bases produce water and a salt.

▶‖ **CHEMTOUR** Acid-Base Ionization

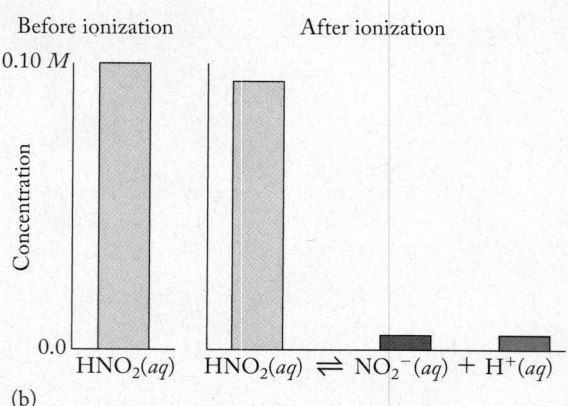

FIGURE 17.1 Difference in degree of ionization for strong and weak acids. (a) The strong acid HNO_3 ionizes completely to H^+ and NO_3^- ions. (b) The weak acid HNO_2 ionizes very little.

In neutral to slightly basic soil, aluminum ions precipitate as aluminum hydroxide and are not available to plants. The blossoms of hydrangeas grown in these soils lack the blue aluminum-containing pigment and are pink.

Other biological systems—including the human body—are highly sensitive to even small changes in acid–base balance. For example, we produce CO_2 during metabolism, and we eliminate it every time we exhale. Our breathing maintains an acid–base balance by controlling the concentration of dissolved CO_2 and the carbonic acid that it forms in our blood. When lung function is impaired by a chronic disease, such as emphysema, CO_2 concentrations in the blood and tissue fluids increase, making them more acidic, which can lead to coma and even death.

In this chapter we explore processes that influence the acid–base balance in the world around us and within our bodies. We address questions such as why these changes happen, how extensive they are, and what they mean to our own health and that of our planet. ■

17.1 Acids and Bases: The Brønsted-Lowry Model

In Chapter 4 we defined acids as substances that donate H^+ ions and bases as substances that accept H^+ ions. These descriptions of acids and bases were developed independently by Danish chemist Johannes Brønsted (1879–1947) and English chemist Thomas Lowry (1874–1936) and published in the same year, 1923. Today these descriptions are known as the **Brønsted–Lowry model** of acids and bases.

Strong and Weak Acids

As we discussed in Section 4.5, acids in aqueous solution are classified as strong or weak depending on the extent to which they ionize, donating H^+ ions to molecules of water and forming hydronium (H_3O^+) ions. Strong acids are completely ionized in water; weak acids are only partially ionized. Figure 17.1 shows data on the ionization of two acids with similar chemical structures: nitric acid (HNO_3), which is strong, and nitrous acid (HNO_2), which is weak. As shown by the bar graphs in Figure 17.1(a), HNO_3 in a 0.10 M solution is completely ionized; there are no intact molecules of HNO_3 in the solution, only nitrate (NO_3^-) and H^+ ions that combine with H_2O to form H_3O^+ ions as a result of the following reaction:

$$HNO_3(aq) + H_2O(\ell) \rightarrow NO_3^-(aq) + H_3O^+(aq) \quad (17.1)$$
$$\text{(H}^+\text{ donor)} \quad \text{(H}^+\text{ acceptor)}$$

In this equation we have highlighted (in red) nitric acid as the H^+ ion donor (Brønsted–Lowry acid) and water (in blue) as the H^+ ion acceptor (Brønsted–Lowry base). All the molecules of HNO_3 in an aqueous solution of the acid are ionized; chemists say that the reaction in Equation 17.1 *goes to completion*. The single reaction arrow in Equation 17.1 conveys this message, indicating that, even though the reaction is reversible, the equilibrium lies very far to the right.

In contrast, only a small percentage of the HNO_2 molecules in 0.10 M nitrous acid donate H^+ ions to molecules of water to form NO_2^- and H_3O^+ ions (Figure 17.1b):

$$HNO_2(aq) + H_2O(\ell) \rightleftharpoons NO_2^-(aq) + H_3O^+(aq) \quad (17.2)$$

TABLE 17.1	Strong Acids and Their Ionization Reactions in Water
Strong Acid	**Reaction in Water**
Hydrobromic	$HBr(aq) + H_2O(\ell) \rightarrow Br^-(aq) + H_3O^+(aq)$
Hydrochloric	$HCl(aq) + H_2O(\ell) \rightarrow Cl^-(aq) + H_3O^+(aq)$
Hydroiodic	$HI(aq) + H_2O(\ell) \rightarrow I^-(aq) + H_3O^+(aq)$
Nitric	$HNO_3(aq) + H_2O(\ell) \rightarrow NO_3^-(aq) + H_3O^+(aq)$
Perchloric	$HClO_4(aq) + H_2O(\ell) \rightarrow ClO_4^-(aq) + H_3O^+(aq)$
Sulfuric	$H_2SO_4(aq) + H_2O(\ell) \rightarrow HSO_4^-(aq) + H_3O^+(aq)$
	$HSO_4^-(aq) + H_2O(\ell) \rightleftharpoons SO_4^{2-}(aq) + H_3O^+(aq)$

At equilibrium, molecules of HNO_2 donate H^+ ions to molecules of H_2O at the same rate that H_3O^+ ions donate H^+ ions to NO_2^- ions, re-forming HNO_2 and H_2O. Both the molecular form of the acid and the ions arising from its dissociation coexist in solution.

Two types of strong acids are listed in Table 17.1. First there are three binary acids with the generic formula HX, where X is Cl, Br, or I. Next there are three *oxoacids* with the generic formula H_mXO_n, where X is a nonmetal. Several factors promote the transfer of H^+ ions from molecules of these acids (or any acid) to molecules of water. We will focus on two of them: (1) the polarity of the bonds to the ionizable H atoms and (2) the strength of the dipole–dipole interactions between these H atoms and O atoms in molecules of water. In all strong acids, ionizable H atoms are bonded to atoms of one of these electronegative elements: O, Cl, Br, or I. Large differences in electronegativity mean that these bonds are all polar covalent, which sets up strong dipole–dipole interactions between their H atoms and the O atoms of H_2O molecules. These interactions are so strong in aqueous solutions of HCl, for example, that all H—Cl bonds break, and O—H bonds form, producing Cl^- and H_3O^+ ions as shown in Figure 17.2.

CONNECTION In Chapter 4 we introduced three ways to refer to the hydrogen ions produced by acids in aqueous solution: as the proton [$H^+(aq)$], the hydrogen ion [also $H^+(aq)$], or the hydronium ion [$H_3O^+(aq)$]. The hydronium ion most closely represents the species present in solution.

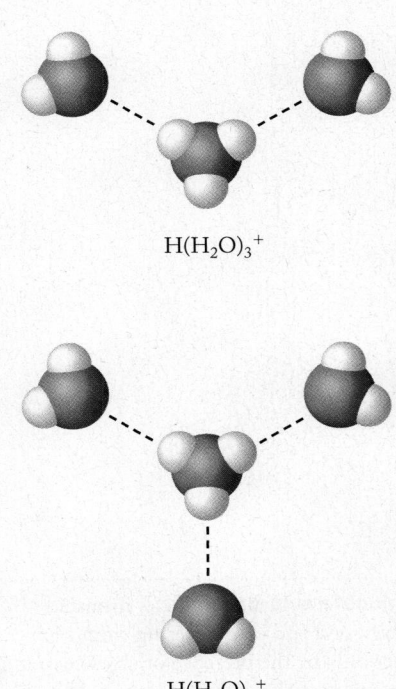

$H(H_2O)_3^+$

$H(H_2O)_4^+$

FIGURE 17.3 Water molecules cluster around hydronium ions, forming species with the general formula $H(H_2O)_n^+$. The clusters shown here have the formulas $H(H_2O)_3^+$ and $H(H_2O)_4^+$.

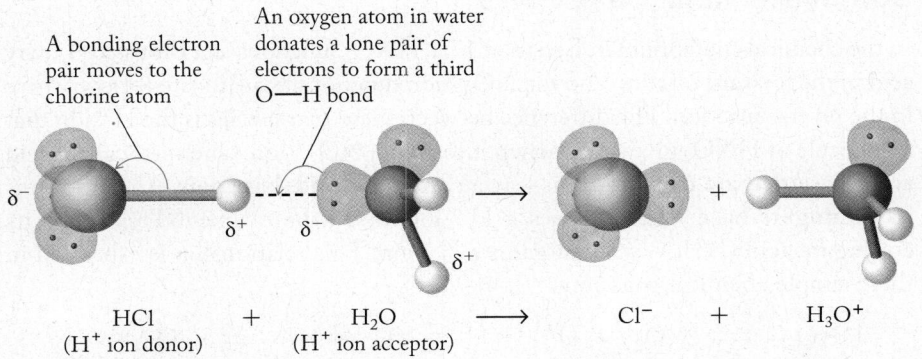

A bonding electron pair moves to the chlorine atom

An oxygen atom in water donates a lone pair of electrons to form a third O—H bond

δ^-	δ^+	δ^-	δ^+				
HCl (H^+ ion donor)	+	H_2O (H^+ ion acceptor)	$\longrightarrow$	Cl^-	+	H_3O^+

FIGURE 17.2 When hydrogen chloride dissolves in water, HCl ionizes and forms Cl^- and H_3O^+ ions.

We often describe acid–base reactions in terms of the donation and acceptance of H^+ ions, but we need to keep in mind that free protons (H^+ ions) do not exist in aqueous solutions. Rather they exist as H_3O^+ ions. In addition, ion–dipole interactions between water molecules and H_3O^+ ions cause H_2O molecules to cluster around central H_3O^+ ions, as shown in Figure 17.3. As we describe concentrations of H^+ ions in acid–base reactions, remember that we really mean H_3O^+ ions with water molecules clustered around them.

As with the gas-phase equilibria we discussed in Chapter 16, the extent to which a reaction takes place before it reaches equilibrium in the aqueous phase is characterized by the value of its equilibrium constant. The equilibrium constants that describe the ionization of acids in water are given the symbol K_a. They are concentration-based equilibrium constants, but the subscript "c" is replaced with "a" to indicate that the equilibrium involves the ionization of an *acid*. Table 17.2 lists some common weak acids, their ionization reactions in water, and their K_a values at 25°C. Note that the K_a values of these weak acids are all much less than one. These small values contrast with the large K_a values of the strong acids in Table 17.1, which are often expressed "$K_a \gg 1$" to indicate that ionization of the acid is essentially complete in aqueous solutions.

TABLE 17.2	**Some Common Weak Acids and Their Ionization Reactions in Water**	
Weak Acid	**Reaction in Water**	$K_a = \dfrac{[A^-][H_3O^+]}{[HA]}$
Acetic	$CH_3COOH(aq) + H_2O(\ell) \rightleftharpoons CH_3COO^-(aq) + H_3O^+(aq)$	1.76×10^{-5}
Formic	$HCOOH(aq) + H_2O(\ell) \rightleftharpoons HCOO^-(aq) + H_3O^+(aq)$	1.77×10^{-4}
Hydrofluoric	$HF(aq) + H_2O(\ell) \rightleftharpoons F^-(aq) + H_3O^+(aq)$	6.8×10^{-4}
Hypochlorous	$HClO(aq) + H_2O(\ell) \rightleftharpoons ClO^-(aq) + H_3O^+(aq)$	2.9×10^{-8}
Nitrous	$HNO_2(aq) + H_2O(\ell) \rightleftharpoons NO_2^-(aq) + H_3O^+(aq)$	4.0×10^{-4}

CONCEPT TEST ···

Rank the acids in Table 17.2 in order of decreasing acid strength.

(Answers to Concept Tests are in the back of the book.)

···

Conjugate Acid–Base Pairs

In the chemical equilibrium in Equation 17.2, HNO_2 functions as a Brønsted–Lowry acid in the forward reaction and the NO_2^- ion functions as a Brønsted–Lowry base in the reverse reaction. The difference between these two species is the H^+ ion that a molecule of HNO_2 gives away when it forms an NO_2^- ion. Chemists call an acid and a base that are related in this way a **conjugate acid–base pair**. An acid forms its **conjugate base** when it loses a H^+ ion, and a base (like NH_3) forms its **conjugate acid** (NH_4^+) when it gains a H^+ ion. This relationship is expressed in these simple chemical equations:

$$\underset{\text{Acid}}{HNO_2} \rightleftharpoons \underset{\substack{\text{Conjugate} \\ \text{base}}}{NO_2^-} + H^+ \qquad \underset{\text{Base}}{NH_3} + H^+ \rightleftharpoons \underset{\substack{\text{Conjugate} \\ \text{acid}}}{NH_4^+}$$

conjugate acid–base pair a Brønsted–Lowry acid and base differing from each other only by the presence or absence of a H^+ ion: acid $\rightleftharpoons$ conjugate base + H^+.

conjugate base formed when a Brønsted–Lowry acid donates a H^+ ion.

conjugate acid formed when a Brønsted–Lowry base accepts a proton.

The corresponding general equations for equilibria in aqueous solutions are

$$\text{Acid}(aq) + H_2O(\ell) \rightleftharpoons \text{conjugate base}(aq) + H_3O^+(aq) \quad (17.3)$$

$$\text{Base}(aq) + H_2O(\ell) \rightleftharpoons \text{conjugate acid}(aq) + OH^-(aq) \quad (17.4)$$

Water and the hydronium ion form another conjugate acid–base pair: H_2O is the base and H_3O^+ is its conjugate acid in the reaction with nitrous acid; H_2O is the acid and OH^- its conjugate base in the reaction with ammonia.

SAMPLE EXERCISE 17.1 **Identifying Conjugate Acid-Base Pairs**

Identify the conjugate acid–base pairs in the reactions that result when perchloric acid, $HClO_4$, and formic acid, $HCOOH$, dissolve in water.

Collect and Organize We are to identify the conjugate acid–base pairs that form in aqueous solutions of $HClO_4$ and $HCOOH$. We know from Table 17.1 that perchloric acid is a strong acid and from Table 17.2 that formic acid is a weak acid.

Analyze Acids in aqueous solutions form their conjugate bases by donating H^+ ions to molecules of H_2O as described by Equation 17.3. Therefore, the formulas of their conjugate bases are the formulas of the original acids minus a H^+ ion. The formula of perchloric acid has only one H atom in it, which must be the one that it loses as a H^+ ion. The formula of formic acid has two H atoms. The ionizable one is bonded to an O atom in the carboxylic acid group.

Solve Rewriting Equation 17.3 for aqueous solutions of these two acids, we have

Perchloric acid: $HClO_4(aq) + H_2O(\ell) \rightarrow ClO_4^-(aq) + H_3O^+(aq)$
                     Acid   Conjugate base

Formic acid: $HCOOH(aq) + H_2O(\ell) \rightleftharpoons HCOO^-(aq) + H_3O^+(aq)$
                Acid   Conjugate base

In both reactions H_3O^+ and H_2O are also a conjugate acid–base pair; H_2O is the base, and H_3O^+ is its conjugate acid.

Think about It A single arrow is used in the $HClO_4$ equation because $HClO_4$ is a strong acid that ionizes completely in water. Equilibrium arrows are used for $HCOOH$ because it is a weak acid and both $HCOOH$ and $HCOO^-$ are likely to be present in solution at equilibrium.

Practice Exercise Identify the conjugate acid–base pairs in the reaction that takes place when the weak organic acid acetic acid (CH_3COOH) dissolves in water.

(Answers to Practice Exercises are in the back of the book.)

Strong and Weak Bases

The most common strong bases are hydroxides of group 1 and 2 metals. Table 17.3 shows how these ionic compounds dissociate when they dissolve in water. Their corresponding equilibrium constants all have values much greater than one ($K_b \gg 1$, where the "b" subscript indicates that the reactant functions as a *base*). The hydroxide ions produced when these bases dissolve in water are very effective H^+ acceptors and so these compounds are strong Brønsted–Lowry bases.

TABLE 17.3 Strong Bases and Their Ionization Reactions in Water

Strong Base	Reaction in Water
Lithium hydroxide	$LiOH(aq) \rightarrow Li^+(aq) + OH^-(aq)$
Sodium hydroxide	$NaOH(aq) \rightarrow Na^+(aq) + OH^-(aq)$
Potassium hydroxide	$KOH(aq) \rightarrow K^+(aq) + OH^-(aq)$
Calcium hydroxide	$Ca(OH)_2(aq) \rightarrow Ca^{2+}(aq) + 2\,OH^-(aq)$
Barium hydroxide	$Ba(OH)_2(aq) \rightarrow Ba^{2+}(aq) + 2\,OH^-(aq)$
Strontium hydroxide	$Sr(OH)_2(aq) \rightarrow Sr^{2+}(aq) + 2\,OH^-(aq)$

TABLE 17.4	Some Common Weak Bases and Their Ionization Reactions in Water	
Weak Base	**Reaction in Water**	K_b
Ammonia	$NH_3(aq) + H_2O(\ell) \rightleftharpoons NH_4^+(aq) + OH^-(aq)$	1.76×10^{-5}
Aniline	$C_6H_5NH_2(aq) + H_2O(\ell) \rightleftharpoons C_6H_5NH_3^+(aq) + OH^-(aq)$	4.0×10^{-10}
Dimethylamine	$(CH_3)_2NH(aq) + H_2O(\ell) \rightleftharpoons (CH_3)_2NH_2^+(aq) + OH^-(aq)$	5.9×10^{-4}
Methylamine	$CH_3NH_2(aq) + H_2O(\ell) \rightleftharpoons CH_3NH_3^+(aq) + OH^-(aq)$	4.4×10^{-4}
Pyridine	$C_5H_5N(aq) + H_2O(\ell) \rightleftharpoons C_5H_5NH^+(aq) + OH^-(aq)$	1.7×10^{-9}

The Brønsted–Lowry model can also be used to explain what happens when a weak base (Table 17.4) dissolves in water. Let's use ammonia as an example. In aqueous solution, NH_3 molecules accept H^+ ions from molecules of water to form NH_4^+ and OH^- ions:

$$\overset{\text{(Acid)} \qquad\qquad \text{(Conjugate base)}}{\underset{\text{(Base)} \qquad\qquad \text{(Conjugate acid)}}{NH_3(aq) + H_2O(\ell) \rightleftharpoons NH_4^+(aq) + OH^-(aq)}} \qquad (17.5)$$

Note how the NH_4^+ ion is the conjugate acid of NH_3, a relationship that may be clearer if we consider the reaction that occurs when an ammonium salt dissolves in water:

$$NH_4^+(aq) + H_2O(\ell) \rightleftharpoons NH_3(aq) + H_3O^+(aq) \qquad (17.6)$$

Here NH_4^+, acting as an acid, donates a H^+ ion and thereby forms NH_3, its conjugate base. Also note that water molecules are proton donors in Equation 17.5, making water a Brønsted–Lowry acid in this reaction. However, water molecules are proton acceptors in Equation 17.6, making water a Brønsted–Lowry base. We revisit this acid–base duality of water throughout this chapter.

Relative Strengths of Acids and Bases

We can use the concepts we have developed for describing equilibria to evaluate the relative strengths of acids and bases. To do so, let's revisit the ionization of HCl:

$$HCl(aq) + H_2O(\ell) \rightarrow Cl^-(aq) + H_3O^+(aq)$$

Because HCl is a strong acid, this reaction goes to completion, which means the reverse reaction essentially does not happen at all. This in turn means that a Cl^- ion (the conjugate base of HCl) must be a very weak base because it has no tendency to accept a proton from H_3O^+. This contrast in relative strengths applies to all conjugate pairs: strong acids have very weak conjugate bases and strong bases have very weak conjugate acids (Figure 17.4).

In between these extremes are many substances that are weak acids with weak conjugate bases. For example, HNO_2 is a weak acid ($K_a = 4.0 \times 10^{-4}$), which means that its conjugate base, NO_2^-, is a weak base. This pairing of weakly acidic and weakly basic strengths applies to all conjugate acid–base pairs.

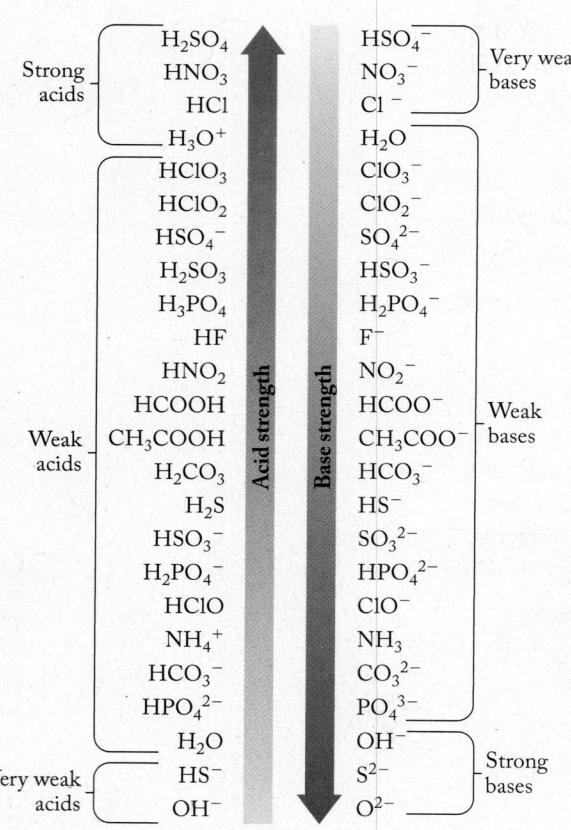

FIGURE 17.4 Opposing trends characterize the relative strengths of acids and their conjugate bases: the stronger the acid, the weaker its conjugate base. The same is true for bases: the stronger the base, the weaker its conjugate acid.

leveling effect the observation that strong acids all have the same strength in water and are completely converted into solutions of H_3O^+ ions; strong bases are likewise leveled in water and are completely converted into solutions of OH^- ions.

autoionization the process that produces equal and very small concentrations of H_3O^+ and OH^- ions in pure water.

All the strong acids in Figure 17.4 ionize completely in water. The H_2O molecules in their solutions readily accept H^+ ions from 100% of the acid molecules. In this context, water is said to *level* the strengths of these acids; they all are equally strong because they cannot be more than 100% ionized. This **leveling effect** means that the conjugate acid of H_2O, H_3O^+, is the strongest H^+ donor that can exist in water. An even stronger acid, no matter how much stronger it is, simply donates all its ionizable H atoms to water molecules, forming H_3O^+ ions.

On the other hand, weak acids are differentiated by their ability to donate their ionizable H atoms to water molecules. The weak acids higher on the list in Figure 17.4 form more acidic aqueous solutions than acids lower on the list.

A similar pattern is evident in the strengths of bases. The strongest base that can exist in water is the conjugate base of H_2O, which is the OH^- ion. Any base that is stronger than OH^- hydrolyzes in water to produce OH^- ions. The oxide ion (O^{2-}), for example, is a very strong base, and reacts with water to produce two OH^- ions:

$$O^{2-}(aq) + H_2O(\ell) \rightarrow 2\,OH^-(aq)$$

The strengths of bases weaker than OH^- ions can be differentiated by the fraction of their molecules that accept H^+ ions from water molecules in aqueous solutions. Weaker bases are higher on the list in Figure 17.4; stronger bases are lower on the list.

> 🔗 **CONNECTION** Hydrolysis reactions, which are the reactions of substances with water, were introduced in Chapters 3 and 4.

CONCEPT TEST ···

List the following anions in order of decreasing strength as Brønsted–Lowry bases: F^-, Cl^-, OH^-, $HCOO^-$, NO_2^-.

17.2 pH and the Autoionization of Water

We have seen that the acidity of a solution is directly related to the concentration of H_3O^+ ions in it. In this section we examine another way to express acidity. To understand this alternative we need to first understand the **autoionization** of water, which is the process that produces equal and very small concentrations of H_3O^+ and OH^- ions in pure water:

$$H_2O(\ell) + H_2O(\ell) \rightleftharpoons H_3O^+(aq) + OH^-(aq) \qquad (17.7)$$

One water molecule, acting as an acid, donates a hydrogen ion to another, which acts as a base (Figure 17.5). The donor H_2O forms its conjugate base (OH^-), and the acceptor H_2O forms its conjugate acid (H_3O^+). We have already encountered this dual nature of water: molecules of H_2O act as H^+ ion acceptors in solutions of acidic solutes, and as H^+ ion donors in solutions of basic solutes. As we discussed in Chapter 4, any substance that can act as either an acid or a base is said to be amphiprotic. The autoionization of water is an example of amphiprotic behavior.

> 🔗 **CONNECTION** We defined *amphiprotic* compounds in Chapter 4 as having both acidic and basic properties.

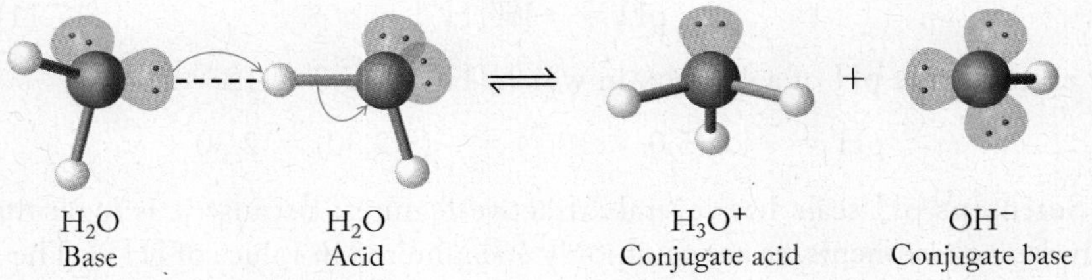

| H_2O | H_2O | | H_3O^+ | | OH^- |
| Base | Acid | | Conjugate acid | | Conjugate base |

FIGURE 17.5 The autoionization of water takes place when a proton is transferred from one water molecule to another. Both molecules are converted to ions.

The equilibrium constant for the autoionization of water shown in Equation 17.7 is written as

$$K_c = \frac{[H_3O^+][OH^-]}{[H_2O][H_2O]} \qquad (17.8)$$

Because water is a pure liquid, its concentration does not appear in the equilibrium constant expression. To further simplify the expression, we substitute $[H^+]$ for $[H_3O^+]$ because these terms represent the same species. This reduces Equation 17.8 to an equilibrium constant expression that is given the symbol K_w:

$$K_w = [H^+][OH^-] \qquad (17.9)$$

In pure water at 25°C $[H^+] = [OH^-] = 1.00 \times 10^{-7}$ M. Inserting these values into Equation 17.9 gives

$$K_w = [H^+][OH^-] = (1.00 \times 10^{-7})(1.00 \times 10^{-7}) = 1.00 \times 10^{-14} \quad (17.10)$$

Such a tiny value of K_w confirms that a very tiny fraction of water molecules undergoes autoionization. The reverse of autoionization—the reaction between $[H^+]$ and $[OH^-]$ to produce H_2O—has an equilibrium constant of $1/K_w = 1.00 \times 10^{14}$ and essentially goes to completion:

$$H^+(aq) + OH^-(aq) \rightleftharpoons H_2O(\ell) \qquad K = 1/K_w = 1.00 \times 10^{14}$$

The value $K_w = 1.00 \times 10^{-14}$ applies to all aqueous solutions at 25°C, not just pure water. (Because this value depends on temperature, throughout this and subsequent discussions, we assume a temperature of 25°C.) Equation 17.10 tells us that there is an inverse relation between $[H^+]$ and $[OH^-]$ in any aqueous sample: as the value of one increases, the value of the other must decrease so that the product of the two is always 1.0×10^{-14}. A solution in which $[H^+] > [OH^-]$ is acidic, a solution in which $[H^+] < [OH^-]$ is basic, and a solution in which $[H^+] = [OH^-] = 1.00 \times 10^{-7}$ M is neutral (neither acidic nor basic).

The tiny value of K_w means that autoionization of water does not contribute significantly to $[H^+]$ in solutions of most acids or to $[OH^-]$ in solutions of most bases, and so we can ignore the contribution of autoionization in most calculations of acid or base strength. However, if acids or bases are extremely weak or if their concentrations are extremely low, H_2O autoionization may need to be taken into account.

The pH Scale

In the early 1900s, scientists developed a device called the *hydrogen electrode* to determine the $[H^+]$ of solutions. The electrical voltage, or *potential*, produced by the hydrogen electrode is a linear function of the logarithm of $[H^+]$. This relation led Danish biochemist Søren Sørensen (1868–1939) to propose a scale for expressing acidity and basicity based on what he termed "the *p*otential of the hydrogen ion," abbreviated **pH**. Mathematically, we define pH as the negative logarithm of $[H^+]$:

$$pH = -\log[H^+] \qquad (17.11)$$

For example, the pH of a solution in which $[H^+] = 5.0 \times 10^{-3}$ M is

$$pH = -\log(5.0 \times 10^{-3}) = -(-2.30) = 2.30$$

Sørensen's pH scale has several attractive features. Because it is logarithmic, there are no exponents, as are commonly encountered in values of $[H^+]$. The logarithmic scale also means that a change of one pH unit corresponds to a 10-fold

---

▶II **CHEMTOUR** Autoionization of Water

◉◉ **CONNECTION** In Chapter 16, we learned that any pure solid or liquid has a constant concentration because its mass per unit volume is always the same. As long as there is *any* solid or liquid substance present, there is no change in its concentration and it does not appear in the equilibrium constant expression.

▶II **CHEMTOUR** pH Scale

pH the negative logarithm of the hydrogen ion concentration in an aqueous solution.

change in $[H^+]$, so that a solution with a pH of 5.0 has 10 times the $[H^+]$ of a solution with a pH of 6.0 and is 10 times as acidic. Similarly, a solution with a pH of 12.0 has 1/10 the $[H^+]$, or 10 times the $[OH^-]$, as a solution with a pH of 11.0.

The negative sign in front of the logarithmic term means that most pH values, except for concentrated solutions of strong acids or bases, are positive numbers between 0 and 14. It also means that *large pH values* correspond to *small values of $[H^+]$*. Acidic solutions have pH values less than 7.00 ($[H^+] > 1.00 \times 10^{-7}$ *M*), and basic solutions have pH values greater than 7.00 ($[H^+] < 1.00 \times 10^{-7}$ *M*). A solution with a pH of exactly 7 is neutral. The pH values for some common aqueous solutions are shown in Figure 17.6.

CONCEPT TEST •••

Match the pH values on the left with the descriptors on the right:

13.77	strongly acidic
10.03	weakly acidic
7.00	weakly basic
4.37	strongly basic
0.22	neutral

CONCEPT TEST •••

Suppose Solution A has a pH of 6.0 and Solution B has a pH of 7.0. Which of the following statements about the two solutions is/are true?

a. Solution A is 10 times more acidic than Solution B.
b. Solution B is neither acidic nor basic.
c. The concentration of OH^- ions in Solution B is 10 times their concentration in Solution A.
d. $[OH^-] = [H^+]$ in Solution B.
e. The value of $[H^+]$ in Solution A is 10 times that in Solution B.

A note about expressing pH values to the appropriate number of significant figures is necessary. Because any pH value is the negative logarithm of the hydrogen ion concentration, the first number in the value defines the location of the decimal point in the concentration term. As such, it is not considered when determining the number of significant figures. For example, a hydrogen ion concentration of 2.7×10^{-4} has two significant figures in the coefficient of the power of 10. The corresponding pH value with two significant figures is 3.57. The 3 in pH 3.57 is not considered a significant figure because it just means that the $[H^+]$ is between 10^{-3} and 10^{-4} *M*.

SAMPLE EXERCISE 17.2   Interconverting pH and $[H^+]$

Is Solution A with a pH of 9.58 more or less acidic than Solution B in which $[H^+] = 4.3 \times 10^{-10}$ *M*?

Collect and Organize We are asked to compare the acidities of two solutions. We know the pH of one and the $[H^+]$ of the other. These two parameters are related by Equation 17.11:

$$pH = -\log[H^+]$$

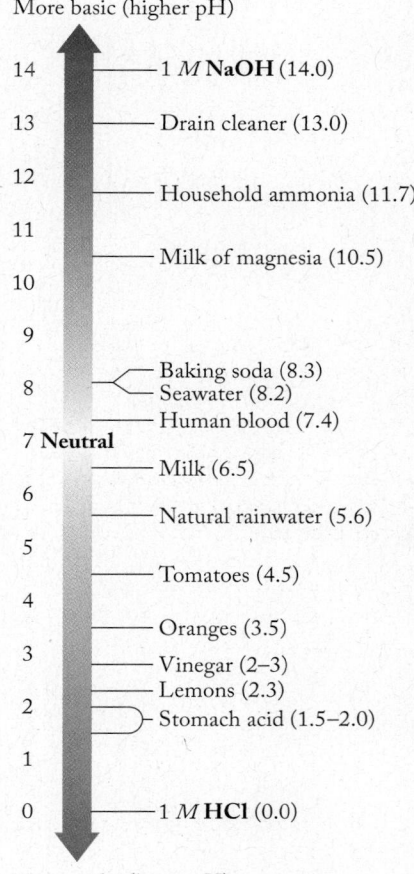

More basic (higher pH)

14 —— 1 *M* **NaOH** (14.0)

13 —— Drain cleaner (13.0)

12 —— Household ammonia (11.7)

11 —— Milk of magnesia (10.5)

10

9

8 —— Baking soda (8.3)
—— Seawater (8.2)
—— Human blood (7.4)

7 **Neutral**

—— Milk (6.5)

6

—— Natural rainwater (5.6)

5

—— Tomatoes (4.5)

4

—— Oranges (3.5)

3 —— Vinegar (2–3)
—— Lemons (2.3)

2 —— Stomach acid (1.5–2.0)

1

0 —— 1 *M* **HCl** (0.0)

More acidic (lower pH)

FIGURE 17.6 The pH scale is a convenient way to express the range of acidic or basic properties of some common materials.

Analyze We can either convert the given pH value into the corresponding $[H^+]$ value, or the given $[H^+]$ value into pH. Doing both will provide us with a check on our calculations. The given $[H^+]$ value is between 10^{-9} and 10^{-10} M, which means that the corresponding pH value will have a 9 in front of the decimal place.

Solve

$$(1) \text{ pH of Solution B:} \qquad pH = -\log(4.3 \times 10^{-10})$$
$$= -(-9.37) = 9.37$$

$$(2) \text{ } [H^+] \text{ of Solution A:} \qquad 9.58 = -\log[H^+]$$
$$[H^+] = 10^{-9.58} = 2.6 \times 10^{-10} \text{ } M$$

1. The pH of Solution B (9.37) is lower than the pH of Solution A (9.58). Therefore, Solution A is less acidic.
2. The $[H^+]$ of Solution B (4.3×10^{-10} M) is greater than that of Solution A (2.6×10^{-10} M). Therefore, Solution A is less acidic.

Think about It Our two results agree. Solution A is slightly less acidic than Solution B in both calculations. Actually, both solutions have pH values greater than 7 and are weakly basic. Therefore, a better way of expressing their relative acid–base balance would be to say that Solution A is slightly more basic than Solution B.

Practice Exercise What is the pH of 6.92×10^{-3} M HCl?

CONCEPT TEST

Is the pH of a 1.00 M solution of a weak acid higher or lower than the pH of a 1.00 M solution of a strong acid?

---

pOH

The letter p as used in pH is also used with other symbols to mean *the negative logarithm* of the variable that follows it. For example, just as every aqueous solution has a pH value, it also has a **pOH** value, defined as

$$pOH = -\log[OH^-] \qquad (17.12)$$

We can use Equation 17.10 to relate pOH to pH. We start by taking the negative logarithm of both sides of the equation:

$$K_w = [H^+][OH^-] = 1.00 \times 10^{-14}$$
$$-\log K_w = -\log([H^+][OH^-]) = -\log(1.00 \times 10^{-14})$$
$$pK_w = -(\log[H^+] + \log[OH^-]) = -(-14.00)$$
$$pK_w = pH + pOH = 14.00 \qquad (17.13)$$

Many tables of equilibrium constants list pK values rather than K values because doing so does not require the use of exponential notation and is more convenient. The tables in Appendix 5 of this book contain both sets of values. Use them whenever you need a K or pK value that is not provided in a problem.

pOH the negative logarithm of the hydroxide ion concentration in an aqueous solution.

SAMPLE EXERCISE 17.3   **Relating [H⁺], [OH⁻], pH, and pOH**

The carbonic acid that forms when atmospheric CO_2 dissolves in rainwater gives rain a normal pH of about 5.6. The pH of acid rain (Figure 17.7), however, can be one or more pH units lower than 5.6. Calculate the values of $[H^+]$, pOH, and $[OH^-]$ in pH 5.6 rainwater and in a sample of acid rain with a pH of 4.3.

Collect and Organize We are given pH values of two samples of rainwater and asked to determine the corresponding $[H^+]$, pOH, and $[OH^-]$ values. These variables are related by the following equations:

$$K_w = [H^+][OH^-] = 1.00 \times 10^{-14} \qquad (17.10)$$
$$pH = -\log[H^+] \qquad (17.11)$$
$$pOH = -\log[OH^-] \qquad (17.12)$$
$$pK_w = pH + pOH = 14.00 \qquad (17.13)$$

Analyze We can start with Equation 17.11 to convert pH into $[H^+]$, use Equation 17.13 to convert pH into pOH, and then use Equation 17.12 to convert pOH into $[OH^-]$. Both samples are weakly acidic with pH values below 7. This means, according to Equation 17.13, that the corresponding pOH values must be greater than 7. The acid rain sample has a lower pH than natural rainwater and should have a higher $[H^+]$ value, which means a smaller $[OH^-]$ value and a higher pOH.

Solve For $[H^+]$,

Normal rain: $pH = 5.6 = -\log[H^+]$
$$[H^+] = 10^{-5.6} = 3 \times 10^{-6}\, M$$
Acid rain: $pH = 4.3 = -\log[H^+]$
$$[H^+] = 10^{-4.3} = 5 \times 10^{-5}\, M$$

For pOH,

$$pK_w = pH + pOH = 14.00$$
$$pOH = 14.00 - pH$$

Normal rain: $pOH = 14.00 - 5.6 = 8.4$
Acid rain: $pOH = 14.00 - 4.3 = 9.7$

For $[OH^-]$,

$$pOH = -\log[OH^-]$$
$$[OH^-] = 10^{-pOH}$$

Normal rain: $[OH^-] = 10^{-8.4} = 4 \times 10^{-9}\, M$
Acid rain: $[OH^-] = 10^{-9.7} = 2 \times 10^{-10}\, M$

Think about It The $[H^+]$ in the acid rain sample is nearly 20 times higher than its concentration in normal rain, and its pH value differs by 1.3 units. These differences make sense because (1) the pH scale is logarithmic, so one unit difference in pH means a 10-fold difference in $[H^+]$ or $[OH^-]$, and (2) pH values decrease as $[H^+]$ increases. The differences in pOH values make sense for the same reasons. All calculated concentration values were rounded off to only one significant figure because each starting pH value had only one significant figure—the single digit after the decimal point.

Practice Exercise What are the values of $[H^+]$ and $[OH^-]$ in household ammonia, an aqueous solution of NH_3 that has a pH of 11.7?

◯◯ **CONNECTION** In Chapter 3 we identified nonmetal oxides like $CO_2(g)$, $NO_2(g)$, and $SO_3(g)$ as acid anhydrides, so called because they hydrolyze in water to produce acidic solutions.

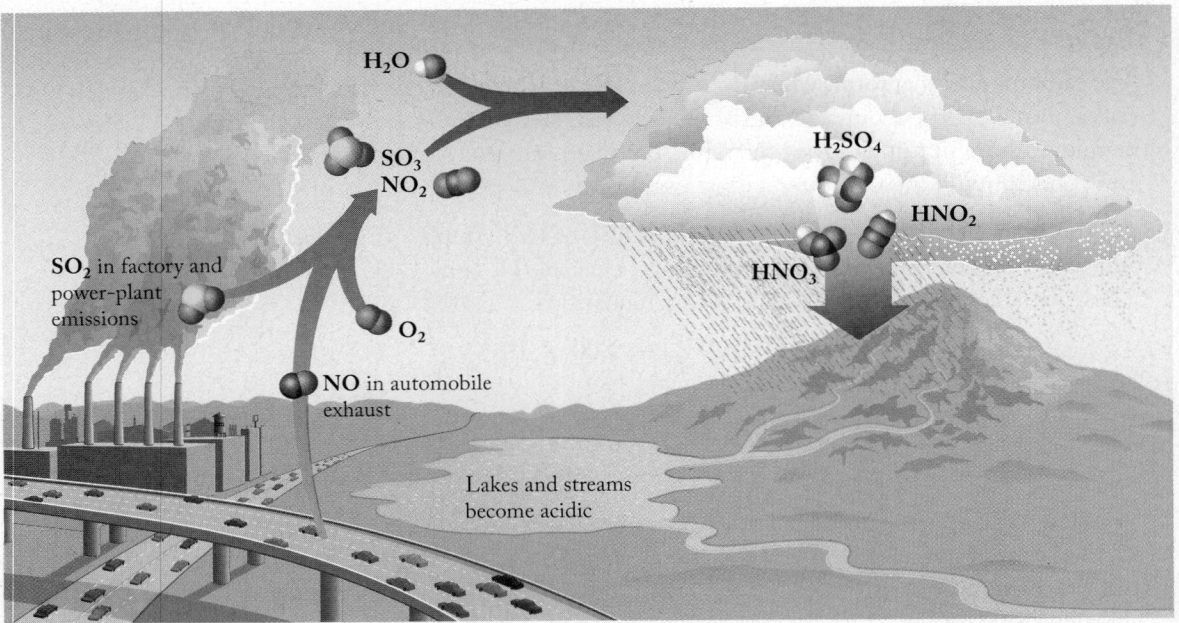

FIGURE 17.7 Acid rain forms when volatile nonmetal oxides such as NO and SO_2 are further oxidized in the atmosphere and dissolve in rain to form nitric (HNO_3), nitrous (HNO_2), and sulfuric (H_2SO_4) acids. These acids ionize, forming NO_3^-, NO_2^-, and SO_4^{2-} ions, respectively, and the hydronium (H_3O^+) ions that make acid rain acidic.

17.3 Calculations Involving pH, K_a, and K_b

If we know the pH and concentration of a solution of a weakly acidic or basic substance, we can calculate the acid or base equilibrium constant K_a or K_b for that substance. Of more practical importance, if we know the value of K_a or K_b we can calculate the pH of an aqueous solution of a weak acid or weak base.

Weak Acids

The vast majority of the acids on our planet are weak acids. Among them, as we saw in Section 17.1, is nitrous acid:

$$HNO_2(aq) + H_2O(\ell) \rightleftharpoons NO_2^-(aq) + H_3O^+(aq) \qquad (17.2)$$

Let's begin our quantitative analysis of this equilibrium by writing the equilibrium constant expression for the reaction:

$$K_a = \frac{[NO_2^-][H_3O^+]}{[HNO_2][H_2O]}$$

Water is the solvent in this case and is present in great abundance. Consequently, the concentration of water does not significantly change during the course of reactions in aqueous solutions, so we do not include the $[H_2O]$ term in the equilibrium expression. We also replace $[H_3O^+]$ with $[H^+]$ to simplify the expression. With these changes we have

$$K_a = \frac{[NO_2^-][H^+]}{[HNO_2]}$$

A generic form of this K_a expression that applies to the ionization reaction of any acid (HA):

$$HA(aq) \rightleftharpoons A^-(aq) + H^+(aq) \qquad (17.14)$$

is written

$$K_a = \frac{[A^-][H^+]}{[HA]} \qquad (17.15)$$

We can use Equation 17.15 to calculate the value of the K_a of an unknown weak acid if we know the pH of a solution of the acid and the value of [HA]. Suppose we measure the pH of a 0.100 M solution and find that it is 2.20. To calculate K_a we first convert the pH value to [H$^+$]:

$$pH = -\log[H^+] = 2.20$$
$$[H^+] = 10^{-2.20} = 6.3 \times 10^{-3} \ M$$

Assuming the only source of H$^+$ ions is ionization of HA, then [A$^-$] must also be $6.3 \times 10^{-3} \ M$ because the stoichiometry of the reaction is that 1 mole of HA ionizes to form 1 mole of H$^+$ and 1 mole of A$^-$. If the ionization reaction yields a [H$^+$] of $6.3 \times 10^{-3} \ M$, then [HA] must have decreased by the same amount. Therefore, at equilibrium:

$$[HA] = (0.100 - 6.3 \times 10^{-3}) \ M = 0.094 \ M$$

Using the three calculated equilibrium concentrations in the K_a expression:

$$K_a = \frac{[A^-][H^+]}{[HA]} = \frac{(6.3 \times 10^{-3})(6.3 \times 10^{-3})}{(0.094)} = 4.2 \times 10^{-4}$$

The small value of K_a confirms that HA is a weak acid.

The ratio of the concentration of H$^+$ ions at equilibrium to the initial concentration of HA represents the **degree of ionization** of HA, which is usually expressed as a percentage of the initial acid concentration. It is also called **percent ionization**. In equation form this relationship is

$$Percent\ ionization = \frac{[H^+]_{equilibrium}}{[HA]_{initial}} \times 100\% \qquad (17.16)$$

Inserting the data from the previous calculation into Equation 17.16 gives

$$Percent\ ionization = \frac{6.3 \times 10^{-3}\ \cancel{M}}{0.100\ \cancel{M}} \times 100\% = 6.3\%$$

CONCEPT TEST ••

Describe how the percent ionization of a weak acid is related to the value of its K_a.

•••

A plot of percent ionization as a function of initial concentration of nitrous acid is shown in Figure 17.8. This same pattern is observed for all weak acids: the degree to which they ionize increases as their concentration decreases. The following Sample Exercise provides a mathematical perspective on this trend.

degree of ionization the ratio of the quantity of a substance that is ionized to the concentration of the substance before ionization; when expressed as a percentage, called **percent ionization**.

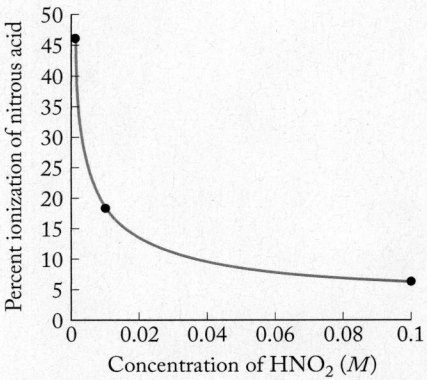

FIGURE 17.8 The degree of ionization of a weak acid increases with decreasing acid concentration. Here the degree of ionization of nitrous acid increases from about 6% in a 0.100 M solution to 18% in a 0.010 M solution to 46% in a 0.001 M solution.

SAMPLE EXERCISE 17.4 **Relating pH, K_a, and Percent Ionization of a Weak Acid**

The pH of a 1.00 M solution of formic acid (HCOOH), a weak organic acid found in red ants and responsible for the sting of their bite, is 1.88.

a. What is the percent ionization of 1.00 M HCOOH?
b. What is the K_a value of the acid?
c. What is the percent ionization of 0.0100 M HCOOH?

Collect and Organize We are asked to determine the K_a value of formic acid and its percent ionization in two solutions of known concentration. We know the pH of one of the solutions. Equation 17.11,

$$pH = -\log[H^+]$$

relates pH to $[H^+]$. Equation 17.15,

$$K_a = \frac{[A^-][H^+]}{[HA]}$$

is the generic equilibrium constant expression for a weak acid, and Equation 17.16,

$$\text{Percent ionization} = \frac{[H^+]_{\text{equilibrium}}}{[HA]_{\text{initial}}} \times 100\%$$

is the formula for calculating percent ionization.

Analyze We can use Equation 17.11 to convert pH to $[H^+]$ and then use Equation 17.16 to calculate percent ionization of the 1.00 M solution. The 1:1:1 stoichiometry of the ionization reaction:

$$HCOOH(aq) \rightleftharpoons HCOO^-(aq) + H^+(aq)$$

tells us that in a solution of HCOOH at equilibrium, $[HCOO^-] = [H^+]$, and [HCOOH] is equal to the initial concentration of the acid minus the portion of it that ionized, or

$$[HCOOH]_{\text{equilibrium}} = [HCOOH]_{\text{initial}} - [H^+]_{\text{equilibrium}}$$

Inserting these concentration values into the equilibrium constant expression for formic acid based on Equation 17.15:

$$K_a = \frac{[HCOO^-][H^+]}{[HCOOH]}$$

enables us to calculate the value of K_a. Once we know K_a, we can use the equilibrium constant expression to calculate $[H^+]$ in any solution of formic acid, and, from $[H^+]$, the percent ionization of the acid in that solution.

The pH value of the 1.00 M solution is close to 2, which corresponds to $[H^+] = 10^{-2}$ M. Therefore the percent ionization of formic acid in this solution should be about 1%. The more dilute solution should be more extensively ionized.

Solve

a. The pH of the 1.00 M solution is 1.88. The corresponding $[H^+]$ is

$$[H^+] = 10^{-1.88} = 1.32 \times 10^{-2}\ M$$

Inserting this value and the initial concentration of HCOOH in Equation 17.16:

$$\text{Percent ionization} = \frac{[H^+]_{\text{equilibrium}}}{[HCOOH]_{\text{initial}}} \times 100\%$$

$$= \frac{1.32 \times 10^{-2}\ M}{1.00\ M} \times 100\% = 1.32\%$$

b. At equilibrium $[HCOO^-] = [H^+] = 1.32 \times 10^{-2}\ M$, and the equilibrium concentration of HCOOH is

$$(1.00 - 1.32 \times 10^{-2})\ M = 0.99\ M$$

Inserting these values in the expression for K_a:

$$K_a = \frac{[H^+][HCOO^-]}{[HCOOH]} = \frac{(1.32 \times 10^{-2})(1.32 \times 10^{-2})}{(0.99)} = 1.76 \times 10^{-4}$$

c. To calculate $[H^+]$ in 0.0100 M HCOOH, we use an ICE table, as in the equilibrium calculations in Chapter 16, to solve for $[H^+]$ at equilibrium. Since it is the unknown in the calculation, we give it the symbol x. Filling in the other cells in the ICE table:

	[HCOOH]	**[HCOO$^-$] (*M*)**	**[H$^+$] (*M*)**
Initial (I):	0.0100	0.0000	0.0000
Change (C):	$-x$	$+x$	$+x$
Equilibrium (E):	$(0.0100 - x)$	x	x

Inserting the equilibrium terms into the K_a expression and using K_a from part b,

$$K_a = \frac{(x)(x)}{(0.0100 - x)} = 1.76 \times 10^{-4}$$

and solving for x using the quadratic equation or an equation-solver program:

$$x = 1.24 \times 10^{-3}\ M$$

The x value is equal to $[H^+]$ at equilibrium. Therefore, the percent ionization of formic acid in a 0.0100 M solution is

$$\text{Percent ionization} = \frac{[H^+]_{\text{equilibrium}}}{[HCOOH]_{\text{initial}}} \times 100\%$$

$$= \frac{1.24 \times 10^{-3}}{0.0100} \times 100\% = 12.4\%$$

Think about It The solutions to parts a and c agree with our estimates: the percent ionization of HCOOH in the 1.00 M solution was indeed near 1%, and the percent ionization in the 0.0100 M solution was significantly higher: 12.4%. The calculated K_a value also agrees with the tabulated value for formic acid (Appendix 5). Finally, we did not attempt to simplify the calculation of $[H^+]$ in part c by eliminating "$-x$" from the denominator of the equilibrium constant expression. Had we done so, the value of x would have been

$$\frac{(x)(x)}{0.0100} \approx 1.76 \times 10^{-4}$$

$$x \approx 1.33 \times 10^{-3}\ M$$

The relative difference between this value and the correct one:

$$\frac{(1.33 \times 10^{-3}) - (1.24 \times 10^{-3})}{1.24 \times 10^{-3}} \times 100\% = 7\%$$

represents a $+7\%$ error. In addition, the value of x is 12.4% of the initial [HCOOH], which is greater than the 5% value we used to judge the appropriateness of equilibrium calculation approximations in Chapter 16, so we were correct in not simplifying this calculation.

Practice Exercise The value of $[H^+]$ in a 0.050 M solution of an organic acid is 5.9×10^{-3} M. What is the pH of the solution, the percent ionization of the acid, and its K_a value?

■ ....

CONCEPT TEST

Three weak acids have these K_a values:

Acid	K_a
A	3.6×10^{-5}
B	4.9×10^{-4}
C	9.2×10^{-4}

Which of the three acids is the most extensively ionized in a 0.100 M solution of the acid? Which of the three acids has the lowest percent ionization in a 1.00 M solution of the acid?

Weak Bases

Now let's consider what happens when a weakly basic compound dissolves in water. As noted in Section 17.1, the weak base ammonia accepts hydrogen ions from water as described by the chemical equation

$$NH_3(aq) + H_2O(\ell) \rightleftharpoons NH_4^+(aq) + OH^-(aq) \qquad (17.17)$$

This reaction is the result of strong intermolecular forces that lead to covalent bonds breaking and new bonds forming, as shown in Figure 17.9. Not all the NH_3 molecules in an ammonia solution accept hydrogen ions. Instead, the reaction reaches an equilibrium in which most ammonia molecules are present as NH_3 rather than NH_4^+. The limited strength of ammonia as a base is reflected in its small K_b value at 25°C:

$$K_b = \frac{[NH_4^+][OH^-]}{[NH_3]} = 1.76 \times 10^{-5} \qquad (17.18)$$

Note that no $[H_2O]$ term appears in the expression for K_b. As with the ionization of weak acids, the concentration of water does not change significantly in the course of the reaction in Equation 17.17. We can calculate K_b if we know the initial base concentration and the solution pH, or we can work in the other direction and calculate pH from a known K_b value. One additional step is necessary: the unknown in this equilibrium calculation is $[OH^-]$ instead of $[H^+]$, but once we know $[OH^-]$ we can take the negative logarithm of it to calculate pOH and then use that value to calculate pH using Equation 17.13.

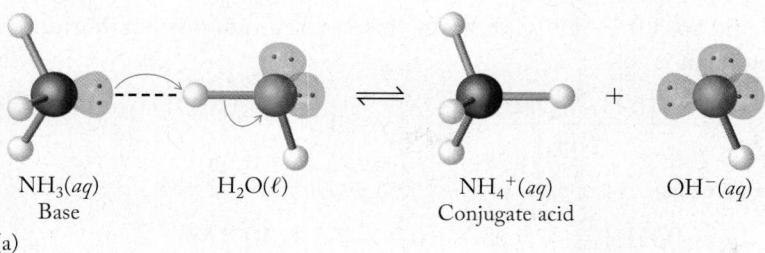

NH₃(aq)   H₂O(ℓ)   NH₄⁺(aq)   OH⁻(aq)

$NH_3(aq)$ $H_2O(\ell)$ $NH_4^+(aq)$ $OH^-(aq)$
Base   Conjugate acid

(a)

FIGURE 17.9 (a) Ammonia reacts with water to produce ammonium ions and hydroxide ions in solution. The lone pair of electrons on the nitrogen of NH_3 is shared with a transferred H^+ ion to make the fourth N—H bond in NH_4^+. (b) Ammonia is a weak base, as illustrated by the very small change in the height of the purple NH_3 bar at equilibrium and the small amount of hydroxide ion produced, representing the small extent to which the reaction proceeds.

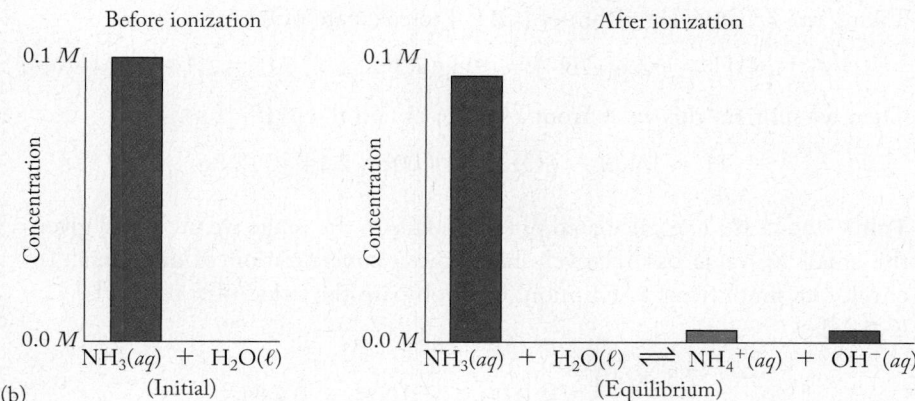

(b)

Before ionization

After ionization

$NH_3(aq) + H_2O(\ell)$ (Initial)

$NH_3(aq) + H_2O(\ell) \rightleftharpoons NH_4^+(aq) + OH^-(aq)$ (Equilibrium)

SAMPLE EXERCISE 17.5   **Calculating the pH of a Solution of a Weak Base**

The concentration of NH_3 in household ammonia ranges between 50 and 100 g/L, or from about 3 M to almost 6 M. What is the pH of a 3.0 M solution of NH_3?

Collect and Organize We are asked to determine the pH of a 3.0 M solution of ammonia. Equation 17.17 describes the basic behavior of ammonia in aqueous solutions, and Equation 17.18 provides the equilibrium constant expression and K_b value.

Analyze The hydrolysis of ammonia produces OH^- ions. We can calculate the equilibrium concentration of OH^- ions using the K_b value and expression and then convert $[OH^-]$ to pOH and finally to pH. Given the 1:1:1 stoichiometry of NH_3, NH_4^+, and OH^- in Equation 17.18, $[NH_4^+] = [OH^-]$ at equilibrium, and if that value is x, then the change in $[NH_3]$ during the course of the reaction is $-x$. The pH value of a fairly concentrated solution of a base with a K_b value near 10^{-5} should be well above 7 but below 14.

Solve We begin by setting up an ICE table based on Equation 17.18 and letting $[NH_4^+] = [OH^-] = x$ at equilibrium:

	[NH₃] (*M*)	[NH₄⁺] (*M*)	[OH⁻] (*M*)
Initial:	3.0	0.0	0.0
Change:	$-x$	$+x$	$+x$
Equilibrium:	$3.0 - x$	x	x

Because K_b is small (1.76×10^{-5}) relative to the initial concentration of base (3.0 M), we can make the simplifying assumption that x will be small compared with 3.0 M, and so $3.0 - x \approx 3.0$. With this assumption, our equilibrium constant expression is

$$K_b = \frac{[NH_4^+][OH^-]}{[NH_3]} = \frac{(x)(x)}{(3.0)} = 1.76 \times 10^{-5}$$

Solving for x gives us

$$x = [OH^-] = \sqrt{5.28 \times 10^{-5}} = 7.3 \times 10^{-3}\ M$$

Taking the negative logarithm of $[OH^-]$ to calculate pOH:

$$pOH = -\log[OH^-] = -\log(7.3 \times 10^{-3}\ M) = 2.14$$

Then we subtract this value from 14.00 to obtain the pH:

$$pH = 14.00 - pOH = 14.00 - 2.14 = 11.86$$

Think about It The calculated pH value falls in the range we predicted given the small K_b value but relatively high initial concentration of ammonia. To check our simplifying assumption, let's compare the value of x to $[NH_3]_{initial}$ (3.0 M):

$$\frac{7.3 \times 10^{-3}}{3.0} = 0.0024 \times 100\% = 0.24\%$$

This small percentage is acceptable, which means our simplifying assumption was justified.

Practice Exercise What is the pH of a 0.200 M solution of methylamine (CH_3NH_2, $K_b = 4.4 \times 10^{-4}$)?

pH of Very Dilute Solutions

In the pH calculations thus far, we have not had to consider how much the autoionization of water contributes to the concentration of H^+ or OH^-. Let's now look at one case when we do have to take this into account. Suppose we want to calculate the pH of $1.00 \times 10^{-8}\ M$ HCl. The acid is completely ionized so that $[H^+] = 1.00 \times 10^{-8}$ and pH (from Equation 17.11) is

$$pH = -\log[H^+] = -\log(1.00 \times 10^{-8}) = 8.000$$

This answer is not reasonable: how could a solution of a strong acid, no matter how dilute, have a weakly basic pH? We would expect the solution to be at least slightly acidic (pH < 7).

To calculate the pH of a solution this dilute, we must consider two sources of H^+ ions: ionization of the acid ($1.00 \times 10^{-8}\ M$) and the autoionization of water. Let's use x to represent $[H^+]$ and $[OH^-]$ resulting from autoionization. The $[H^+]$ term in the K_w expression is the sum of x and $1.00 \times 10^{-8}\ M$:

$$K_w = [H^+][OH^-]$$

$$1.00 \times 10^{-14} = (x + 1.00 \times 10^{-8})(x)$$

Rearranging this equation to solve for x gives:

$$x^2 + (1.00 \times 10^{-8}x) - (1.00 \times 10^{-14}) = 0$$

$$x = 9.5 \times 10^{-8}\ M$$

The concentration of hydrogen ion in the solution is therefore

$$[H^+] = (9.5 \times 10^{-8} M) + (1.00 \times 10^{-8} M) = 10.5 \times 10^{-8} M = 1.05 \times 10^{-7} M$$

and pH is

$$pH = -\log(1.05 \times 10^{-7} M) = 6.98$$

This value agrees with our prediction that the solution should be slightly acidic.

17.4 Polyprotic Acids

Up to this point we have dealt with **monoprotic acids**, which have only one ionizable hydrogen atom per molecule. Acids that contain more than one ionizable hydrogen—such as sulfuric acid (H_2SO_4) and phosphoric acid (H_3PO_4)—are called **polyprotic acids**. For molecules with two and three ionizable hydrogen atoms, we use the more specific terms *diprotic acids* and *triprotic acids*, respectively. Let's consider the acidic properties of a strong diprotic acid, sulfuric acid. Sulfuric acid is a strong acid (Table 17.1) because ionization of the first proton:

$$H_2SO_4(aq) \rightarrow HSO_4^-(aq) + H^+(aq) \qquad (17.19)$$

is complete ($K_{a_1} \gg 1$). However, the second ionization step is not; it has $K_{a_2} < 1$:

$$HSO_4^-(aq) \rightleftharpoons SO_4^{2-}(aq) + H^+(aq) \qquad K_{a_2} = 1.2 \times 10^{-2} \qquad (17.20)$$

Note that these equilibrium constant symbols have an additional subscript, 1 or 2, corresponding to the loss of first one and then a second H^+ ion per molecule.

The combination of one complete and one incomplete ionization means that many solutions of $H_2SO_4(aq)$ contain more than 1 mole but less than 2 moles of $H^+(aq)$ for every mole of H_2SO_4 dissolved. Let's be more quantitative about this and determine the pH of a 0.100 M solution of H_2SO_4. The starting point in this calculation is a solution in which all the H_2SO_4 has ionized as described in Equation 17.19. Therefore, as the second step begins, $[HSO_4^-] = [H^+] = 0.100$ M. Ionization of HSO_4^- (Equation 17.20) produces additional H^+ ions. To analyze the effect of the second ionization, we set up an ICE table in which $+x$ is the change in $[H^+]$ produced by the second ionization. Given the 1:1:1 stoichiometry of the balanced equation for this step, the change in $[SO_4^{2-}]$ is also $+x$, and the change in $[HSO_4^-]$ is $-x$. Inserting these values in the ICE table and completing the third row:

	$[HSO_4^-]$ (M)	$[SO_4^{2-}]$ (M)	$[H^+]$ (M)
Initial:	0.100	0.000	0.100
Change:	$-x$	$+x$	$+x$
Equilibrium:	$0.100 - x$	x	$0.100 + x$

Inserting the equilibrium concentrations in the equilibrium constant expression for K_{a_2}:

$$K_{a_2} = \frac{[H^+][SO_4^{2-}]}{[HSO_4^-]} = \frac{(0.100 + x)(x)}{(0.100 - x)} = 1.2 \times 10^{-2}$$

Solving for x using the quadratic equation or a solver program, we get

$$x = +0.010 \quad \text{or} \quad -0.122$$

monoprotic acid has one ionizable hydrogen atom per molecule.

polyprotic acid has two or more ionizable hydrogen atoms per molecule.

The negative value for x has no physical meaning because it gives us a negative $[SO_4^{2-}]$ value. Therefore, we use only the positive x value. At equilibrium,

$$[H^+] = (0.100 + x)\,M = (0.100 + 0.010)\,M = 0.110\,M$$

The corresponding pH is

$$pH = -\log[H^+] = -\log(0.110\,M) = 0.96$$

As predicted, the value of $[H^+]$ is between one and two times the initial concentration of H_2SO_4. The degree of ionization of HSO_4^- is

$$\frac{[SO_4^{2-}]_{equilibrium}}{[HSO_4^-]_{initial}} = \frac{0.010\,M}{0.100\,M} = 0.10 \times 100\% = 10\%$$

so we were correct in our decision not to use the simplifying assumption to avoid solving a quadratic equation.

Calculating the pH of a solution of a weak diprotic acid, such as carbonic or sulfurous acid, is actually easier than the above calculation for sulfuric acid. To see why, look closely at the K_a values in Table 17.5. In each of the two pairs of values, K_{a_2} is much smaller than K_{a_1}. We can rationalize the difference on the basis of electrostatic attractions between oppositely charged ions. The first ionization produces a negatively charged oxoanion—HCO_3^- or HSO_3^-. The second ionization requires that a positive ion (H^+) dissociate from a negative ion to produce an even more negative oxoanion. Separating oppositely charged ions that are naturally attracted to each other is not a process that we would expect to be favored, and the smaller values for K_{a_2} confirm our expectations. In general, the K_{a_2} of any diprotic acid is less, and often much less, than K_{a_1}. The consequence of these large differences is that essentially all of the limited strength of weak polyprotic acids is due to the first ionization reaction.

TABLE 17.5   Ionization Equilibria for Two Diprotic Acids

Acid	Ionization Equilibria		K_a
Carbonic acid	Step 1:	$H_2CO_3(aq) \rightleftharpoons HCO_3^-(aq) + H^+(aq)$	$K_{a_1} = 4.3 \times 10^{-7}$
	Step 2:	$HCO_3^-(aq) \rightleftharpoons CO_3^{2-}(aq) + H^+(aq)$	$K_{a_2} = 4.7 \times 10^{-11}$
Sulfurous acid	Step 1:	$H_2SO_3(aq) \rightleftharpoons HSO_3^-(aq) + H^+(aq)$	$K_{a_1} = 1.7 \times 10^{-2}$
	Step 2:	$HSO_3^-(aq) \rightleftharpoons SO_3^{2-}(aq) + H^+(aq)$	$K_{a_2} = 6.2 \times 10^{-8}$

To see how this separation of ionization steps plays out, let's focus on carbonic acid (H_2CO_3), which is present in every drop of rain that falls from the sky. It gets there because the atmosphere is about 0.039% (by volume) CO_2. Carbon dioxide is slightly soluble in water, and when it dissolves it forms carbonic acid:

$$CO_2(g) + H_2O(\ell) \rightleftharpoons H_2CO_3(aq)$$

Actually, the H_2CO_3 molecule is not stable in aqueous solutions, but we write it as a convenience to show how dissolving CO_2 in water produces an acidic solution. The net result of CO_2 dissolving in water is

$$CO_2(g) + H_2O(\ell) \rightleftharpoons HCO_3^-(aq) + H^+(aq)$$

In the following Sample Exercise we will use this equilibrium to calculate the natural pH of rainwater.

SAMPLE EXERCISE 17.6 **Calculating the pH of a Solution of a Weak Diprotic Acid**

What is the pH of rainwater at 25°C in which atmospheric CO_2 has dissolved, producing a constant $[H_2CO_3]$ of 1.2×10^{-5} M?

Collect and Organize We are asked to determine the pH of a dilute solution of H_2CO_3. There are two ionizable H atoms in H_2CO_3. The K_{a_1} and K_{a_2} values are given in Table 17.5. Any H_2CO_3 consumed by the reaction is replaced by the dissolution of more CO_2 so that $[H_2CO_3]$ remains a constant 1.2×10^{-5} M.

Analyze The large difference between the K_{a_1} and K_{a_2} values indicates that the pH of the solution is controlled by the first ionization equilibrium:

$$H_2CO_3(aq) \rightleftharpoons HCO_3^-(aq) + H^+(aq) \qquad K_{a_1} = 4.3 \times 10^{-7}$$

Because of the small value of K_{a_1} and the small concentration of H_2CO_3, we should obtain a pH value that is less than 7 but closer to 7 than to 0.

Solve First we set up an ICE table in which $x = [H^+] = [HCO_3^-]$ at equilibrium and the value of $[H_2CO_3]$ at equilibrium is 1.2×10^{-5} M.

	$[H_2CO_3]$	$[HCO_3^-]$ (M)	$[H^+]$ (M)
Initial:	1.2×10^{-5}	0	0
Change:	0	$+x$	$+x$
Equilibrium:	1.2×10^{-5}	x	x

$$K_{a_1} = \frac{[HCO_3^-][H^+]}{[H_2CO_3]} = \frac{(x)(x)}{1.2 \times 10^{-5}} = 4.3 \times 10^{-7}$$
$$x = [H^+] = 2.3 \times 10^{-6} \, M$$

Taking the negative logarithm of $[H^+]$ to calculate pH:

$$pH = -\log[H^+] = -\log(2.3 \times 10^{-6} \, M) = 5.64$$

Think about It Carbonic acid is a weak acid, and its concentration here is small, so obtaining a pH value that is only about 1.4 units below neutral pH (7.00) is reasonable.

Practice Exercise The pH value in Sample Exercise 17.6 is not far from 7.00, and it raises the question of whether the autoionization of water that produces a $[H^+]$ of 1.00×10^{-7} M contributes significantly to $[H^+]$ in the rainwater sample. Recalculate the pH of the rainwater sample taking into account the autoionization of water.

Some acids have three ionizable H atoms per molecule. Two important ones are phosphoric acid, H_3PO_4, and citric acid, the acid responsible for the tart flavor of citrus fruits. Note in Table 17.6 how $K_{a_1} > K_{a_2} > K_{a_3}$ for both acids. This pattern is much like that for the $K_{a_1} > K_{a_2}$ values of diprotic acids and for the same reason: it is more difficult to remove a second H^+ ion from the negatively charged ion formed after the first H^+ ion is removed, and it is even more difficult to remove a third H^+ ion from an ion with a 2− charge.

TABLE 17.6   Ionization Equilibria for Two Triprotic Acids

Phosphoric Acid

(1) $\text{HO}-\underset{\underset{\text{OH}}{|}}{\overset{\overset{\text{O}}{\|}}{\text{P}}}-\text{OH} \rightleftharpoons \text{HO}-\underset{\underset{\text{OH}}{|}}{\overset{\overset{\text{O}}{\|}}{\text{P}}}-\text{O}^- + \text{H}^+$ $K_{a_1} = 7.11 \times 10^{-3}$

(2) $\text{HO}-\underset{\underset{\text{OH}}{|}}{\overset{\overset{\text{O}}{\|}}{\text{P}}}-\text{O}^- \rightleftharpoons {}^-\text{O}-\underset{\underset{\text{OH}}{|}}{\overset{\overset{\text{O}}{\|}}{\text{P}}}-\text{O}^- + \text{H}^+$ $K_{a_2} = 6.32 \times 10^{-8}$

(3) ${}^-\text{O}-\underset{\underset{\text{OH}}{|}}{\overset{\overset{\text{O}}{\|}}{\text{P}}}-\text{O}^- \rightleftharpoons {}^-\text{O}-\underset{\underset{\text{O}^-}{|}}{\overset{\overset{\text{O}}{\|}}{\text{P}}}-\text{O}^- + \text{H}^+$ $K_{a_3} = 4.5 \times 10^{-13}$

Citric Acid

(1) $\text{HO}-\underset{\underset{\text{CH}_2\text{COOH}}{|}}{\overset{\overset{\text{CH}_2\text{COOH}}{|}}{\text{C}}}-\text{COOH} \rightleftharpoons \text{HO}-\underset{\underset{\text{CH}_2\text{COOH}}{|}}{\overset{\overset{\text{CH}_2\text{COO}^-}{|}}{\text{C}}}-\text{COOH} + \text{H}^+$ $K_{a_1} = 7.44 \times 10^{-4}$

(2) $\text{HO}-\underset{\underset{\text{CH}_2\text{COOH}}{|}}{\overset{\overset{\text{CH}_2\text{COO}^-}{|}}{\text{C}}}-\text{COOH} \rightleftharpoons \text{HO}-\underset{\underset{\text{CH}_2\text{COOH}}{|}}{\overset{\overset{\text{CH}_2\text{COO}^-}{|}}{\text{C}}}-\text{COO}^- + \text{H}^+$ $K_{a_2} = 1.73 \times 10^{-5}$

(3) $\text{HO}-\underset{\underset{\text{CH}_2\text{COOH}}{|}}{\overset{\overset{\text{CH}_2\text{COO}^-}{|}}{\text{C}}}-\text{COO}^- \rightleftharpoons \text{HO}-\underset{\underset{\text{CH}_2\text{COO}^-}{|}}{\overset{\overset{\text{CH}_2\text{COO}^-}{|}}{\text{C}}}-\text{COO}^- + \text{H}^+$ $K_{a_3} = 4.02 \times 10^{-7}$

▶❚❚ **CHEMTOUR** Acid Strength and Molecular Structure

CONCEPT TEST

Do you expect the second or third acid ionization steps in phosphoric acid and citric acid to influence the pH of 0.100 M solutions of either acid?

17.5 Acid Strength and Molecular Structure

In Section 17.1, we noted that nitric acid (HNO_3) is a strong acid but nitrous acid (HNO_2) is weak. Similarly, sulfuric acid (H_2SO_4) is a strong acid, but sulfurous acid (H_2SO_3) is weak. The reason for these differences in strength lies in subtle differences in molecular structure (Figure 17.10). The ionizable hydrogen atoms in both molecules are bonded to oxygen atoms that are also bonded to the central sulfur atoms. The difference between the two is that the central sulfur atom is also bonded to either one (in H_2SO_3) or two (in H_2SO_4) other oxygen atoms.

Recall from Section 8.3 that oxygen is the second most electronegative element (after fluorine). This means that oxygen atoms bonded to the central atom of an oxoacid attract electron density toward themselves. The more electron density that is drawn away from the O—H groups, the more spread out (or *delocalized*) is the negative charge on the anion that is formed when a H^+ ion is lost. Spreading out charge over more atoms has a stabilizing effect on the anion.

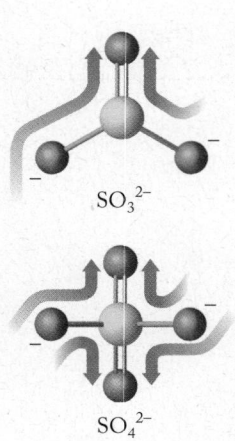

FIGURE 17.10 Sulfuric acid (H_2SO_4) is a stronger acid than sulfurous acid (H_2SO_3) because of the greater stability that comes with delocalizing the negative charge of a SO_4^{2-} ion over more atoms (shown by the curved blue arrows).

Thus, SO_4^{2-} ions are more stable than SO_3^{2-} ions, making H_2SO_4 a stronger acid than H_2SO_3.

This trend of increasing acid strength with increasing numbers of oxygen atoms bonded to the central atom (that is, with increasing oxidation number of the central atom) is true for all oxoacids. The trend is illustrated by the strong acidity of HNO_3 and the weak acidity of HNO_2, and by the strengths of the oxoacids of chlorine, shown in Figure 17.11.

The strength of an oxoacid is also related to the electron-withdrawing power of the central atom. Consider, for example, the relative strengths of the three hypohalous acids in Figure 17.12. The most electronegative of the three halogen atoms (Cl) has the greatest attraction for the pair of electrons it shares with oxygen. This attraction draws electron density away from hydrogen toward chlorine and toward the oxygen end of the already polar O—H bond. These shifts in electron density make the hypochlorite (ClO^-) ions better able to bear a negative charge because the charge is more delocalized. Thus, $HClO(aq)$ is the strongest of the three acids, followed by hypobromous acid [$HBrO(aq)$] and hypoiodous acid [$HIO(aq)$].

CONCEPT TEST •••••••••••••••••••••••••••••••••••

Rank the following compounds in order of decreasing acid strength: H_3PO_4, H_3AsO_4, H_3SbO_4, and H_3BiO_4.

Acid	Structure	Oxidation Number of Cl	K_a
Hypochlorous HClO	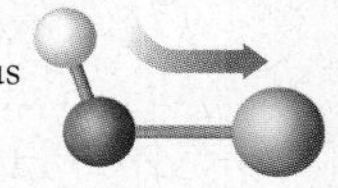	+1	2.9×10^{-8}
Chlorous $HClO_2$	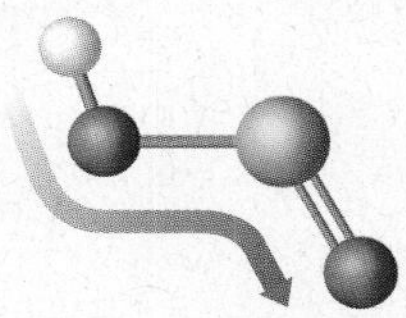	+3	1.1×10^{-2}
Chloric $HClO_3$	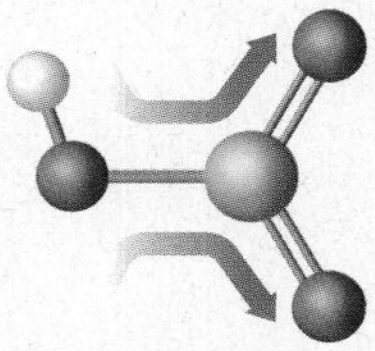	+5	~1
Perchloric $HClO_4$	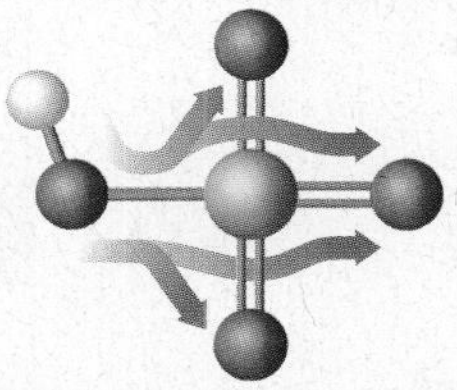	+7	Strong acid

FIGURE 17.11 In the oxoacids of chlorine, acid strength increases with increasing Cl oxidation number. The higher the oxidation number, the greater the number of O atoms bonded to the Cl. The greater the number of O bonded to Cl, the greater the ability to delocalize the negative charge on the anion created when each acid loses its H.

17.6 pH of Salt Solutions

Seawater and the freshwater in many rivers and lakes have pH values that range from weakly basic to weakly acidic. How can these waters be more basic than the acidic rainwater (pH ≤ 5.6) that serves, directly or indirectly, as their water supply? The answer is that, when rain soaks into the ground, its pH changes as it flows through soils that contain basic components. To understand the chemical processes that produce neutral or slightly basic groundwater, we first need to examine the acid–base properties of some common ionic compounds present in these waters.

As we discussed in Chapter 4, soluble ionic compounds separate into their component ions when they dissolve in water. For example, a 0.01 M solution of NaCl contains 0.01 M Na^+ ions and 0.01 M Cl^- ions. It is also a neutral solution. Neither Na^+ ions nor Cl^- ions hydrolyze to form either H_3O^+ or OH^- ions. Recall that the Cl^- ion is the conjugate base of a strong acid (HCl). Therefore, the Cl^- ion must be a very weak Brønsted–Lowry base (Figure 17.4).

Now let's consider another sodium salt, NaF. When it dissolves in water it produces F^- ions, which are the conjugate base of the *weak* acid, HF. Therefore, F^- ions should be weakly basic, producing at least some OH^- ions when they dissolve in water:

$$F^-(aq) + H_2O(\ell) \rightleftharpoons HF(aq) + OH^-(aq)$$

Thus, solutions of NaF are weakly basic.

CONNECTION The electronegativities of the elements were given in Chapter 8 (Figure 8.5)

Acid	Structure	Electronegativity of Halogen Atom	K_a
Hypochlorous HClO	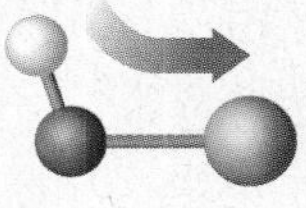	3.0	2.9×10^{-8}
Hypobromous HBrO	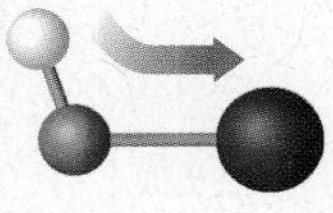	2.8	2.3×10^{-9}
Hypoiodous HIO	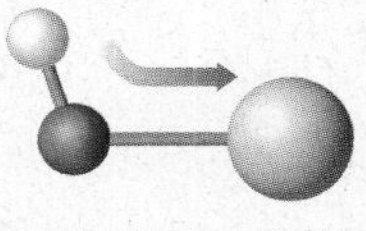	2.5	2.3×10^{-11}

FIGURE 17.12 The strengths of these three hypohalous acids are related to the electronegativities of their halogen atoms. The more electronegative the halogen atom, the more it pulls electron density (blue arrows) away from the hydrogen end of the molecule. The less electron density at the H atom, the more easily the H ionizes and so the stronger the acid.

If salts that contain the conjugate bases of weak acids can be basic, then it is logical that salts that contain the conjugate acids of weak bases can be acidic. An example of such a salt is NH_4Cl. We have seen that the Cl^- ions that are produced when NH_4Cl dissolves have negligible strengths as Brønsted–Lowry bases. However, NH_4^+ ions are the conjugate acid of a weak base, NH_3. Therefore, NH_4^+ ions should be weakly acidic, producing at least some H_3O^+ ions:

$$NH_4^+(aq) + H_2O(\ell) \rightleftharpoons NH_3(aq) + H_3O^+(aq)$$

Consequently, solutions of NH_4Cl are weakly acidic.

Table 17.7 summarizes how salts can be acidic, basic, or neutral depending on whether they include cations that are the conjugate acids of weak bases, or anions that are the conjugate bases of weak acids, or both. Note that salts that contain both the conjugate base of a weak acid *and* the conjugate acid of a weak base may be acidic, basic, or neutral, depending on the relative strengths of the acid and base. Ammonium acetate represents the rare example of a salt in which the strengths of the acid (acetic acid) and the base (ammonia) happen to be *exactly the same* [K_a (acetic acid) = K_b (ammonia) = 1.76×10^{-5}]. As a result, ammonium acetate is a neutral salt.

TABLE 17.7   Acid-Base Properties of Some Common Salts

Anion Is Derived from a	Cation Is Derived from a	pH of Aqueous Solutions	Example
Strong acid	Strong base	7	NaCl
Strong acid	Weak base	<7	NH_4Cl
Weak acid	Strong base	>7	NaF
Weak acid	Weak base	Depends on relative values of pK_a and pK_b	$pK_a < pK_b$, acidic; NH_4F $pK_b > pK_a$, basic; NH_4HCO_3 $pK_a = pK_b$, neutral; CH_3COONH_4

CONCEPT TEST ··

When ammonium fluoride is heated to about 100°C it decomposes, forming ammonia and ammonium hydrogen fluoride:

$$2\,NH_4F(s) \rightarrow NH_3(g) + (NH_4)HF_2(s)$$

Do you think $(NH_4)HF_2$ is more acidic than NH_4F? Why?

···

SAMPLE EXERCISE 17.7   Distinguishing Acidic, Basic, and Neutral Salts

Is an aqueous solution of NaClO acidic, basic, or neutral?

Collect and Organize We are asked whether a solution of NaClO is acidic, basic, or neutral. When this salt dissolves in water, it dissociates into Na^+ ions and ClO^- ions.

Analyze Sodium ions do not hydrolyze and hence do not produce acidic solutions in water. However, ClO^- ions are the conjugate base of HClO, which is a weak acid (Table 17.2). Therefore, ClO^- ions should partially hydrolyze in water, forming OH^- ions:

$$ClO^-(aq) + H_2O(\ell) \rightleftharpoons HClO(aq) + OH^-(aq)$$

Solve Because hydrolysis of ClO^- ions produces OH^- ions, solutions of $NaClO$ are weakly basic.

Think about It Any sodium salt that contains an anion that is the conjugate base of a weak acid produces weakly basic aqueous solutions.

Practice Exercise Write a chemical equation for the hydrolysis reaction that explains why an aqueous solution of K_2SO_4 is basic.

Having established that salts can be acidic, basic, or neutral, let's now explore a strategy for calculating the pH values of their aqueous solutions. Our strategy is much like the approach we have taken to calculate the pH values of solutions of weak acids and bases. Let's start with a 0.100 M solution of sodium carbonate, Na_2CO_3. We start with a carbonate salt because the pH of many natural waters (and biological systems) is controlled by the presence of CO_3^{2-} ions and their conjugate acid, HCO_3^- ions. The $[CO_3^{2-}]$ and $[HCO_3^-]$ values in a solution are linked by the second acid ionization reaction of H_2CO_3 (see Table 17.5):

$$HCO_3^-(aq) + H_2O(\ell) \rightleftharpoons CO_3^{2-}(aq) + H_3O^+(aq)$$

$$K_{a_2} = 4.7 \times 10^{-11} \tag{17.21}$$

Carbonate and bicarbonate are also linked by the chemical reaction in which the carbonate ion acts like a Brønsted–Lowry base:

$$CO_3^{2-}(aq) + H_2O(\ell) \rightleftharpoons HCO_3^-(aq) + OH^-(aq) \tag{17.22}$$

None of the tables in this chapter or Appendix 5 contains the K_b value for the reaction in Equation 17.22, so we have to calculate it. We start with the equilibrium constant expression for the reaction:

$$K_{b_1} = \frac{[HCO_3^-][OH^-]}{[CO_3^{2-}]}$$

We have labeled this constant K_{b_1} because it describes the first of two possible hydrolysis reactions that release OH^- ions. In the second, the bicarbonate ion produced in the first reaction also acts like a Brønsted–Lowry base:

$$HCO_3^-(aq) + H_2O(\ell) \rightleftharpoons H_2CO_3(aq) + OH^-(aq)$$

We give the equilibrium constant for this reaction the symbol K_{b_2}:

$$K_{b_2} = \frac{[H_2CO_3][OH^-]}{[HCO_3^-]}$$

To calculate the values of K_{b_1} and K_{b_2}, we begin with the K_{a_1} and K_{a_2} equilibrium constant expressions for H_2CO_3:

$$K_{a_1} = \frac{[HCO_3^-][H^+]}{[H_2CO_3]} = 4.3 \times 10^{-7} \qquad K_{a_2} = \frac{[CO_3^{2-}][H^+]}{[HCO_3^-]} = 4.7 \times 10^{-11}$$

Now let's compare the K_{b_1} expression for the carbonate ion and the K_{a_2} expression for carbonic acid:

$$K_{b_1} = \frac{[HCO_3^-][OH^-]}{[CO_3^{2-}]} \qquad K_{a_2} = \frac{[CO_3^{2-}][H^+]}{[HCO_3^-]} = 4.7 \times 10^{-11}$$

CONNECTION We noted in Chapter 2 that *bicarbonate* is a more common name for the HCO_3^- ion than *hydrogen carbonate*.

Their similarity becomes more apparent when we write the reciprocal of the K_{a_2} expression:

$$K_{b_1} = \frac{[HCO_3^-][OH^-]}{[CO_3^{2-}]} \qquad \frac{1}{K_{a_2}} = \frac{[HCO_3^-]}{[CO_3^{2-}][H^+]} = \frac{1}{4.7 \times 10^{-11}}$$

The only difference between the two is the $[OH^-]$ term in the K_{b_1} expression and the $[H^+]$ term in the $1/K_{a_2}$ expression. As we have seen, $[H^+]$ and $[OH^-]$ are linked by K_w (Equation 17.10). Consider what happens when we multiply $1/K_{a_2}$ by K_w:

$$\left(\frac{1}{K_{a_2}}\right) \times K_w = \left(\frac{[HCO_3^-]}{[\cancel{H^+}][CO_3^{2-}]}\right)([\cancel{H^+}][OH^-]) = \frac{[HCO_3^-][OH^-]}{[CO_3^{2-}]}$$

The resulting expression is the K_{b_1} expression for the carbonate ion. We know the values of K_w and K_{a_2}, so we can calculate K_{b_1}:

$$K_{b_1} = \frac{K_w}{K_{a_2}} = \frac{1.0 \times 10^{-14}}{4.7 \times 10^{-11}} = 2.1 \times 10^{-4}$$

The inverse relation between the K_{a_2} of H_2CO_3 and the K_{b_1} of CO_3^{2-} also holds for the K_{a_1} of H_2CO_3 and the K_{b_2} of CO_3^{2-}. The connection between the acidic strength of H_2CO_3 and the basic strength of its conjugate base, HCO_3^-, is given by

$$K_{b_2} = \frac{K_w}{K_{a_1}} = \frac{1.0 \times 10^{-14}}{4.3 \times 10^{-7}} = 2.3 \times 10^{-8}$$

The K_w connection between K_a and K_b values for carbonic acid equilibria applies to any conjugate acid–base pair:

$$K_b = \frac{K_w}{K_a}$$

$$\text{or} \quad K_a \times K_b = K_w \tag{17.23}$$

Equation 17.23 reinforces the complementary nature of an acid and its conjugate base: as the strength (K_a) of the acid increases, the strength (K_b) of its conjugate base decreases, and vice versa (Figure 17.4). In the carbonate and carbonic acid examples, the CO_3^{2-} ion is a stronger base ($K_{b_1} = 2.1 \times 10^{-4}$) than its conjugate acid, the HCO_3^- ion, is an acid ($K_{a_2} = 4.7 \times 10^{-11}$), but the HCO_3^- ion is a weaker base ($K_{b_2} = 2.3 \times 10^{-8}$) than H_2CO_3 is an acid ($K_{a_1} = 4.3 \times 10^{-7}$).

Now that we have a value for $K_{b_1} = 2.1 \times 10^{-4}$ for the carbonate ion, we can calculate the pH of the 0.100 M solution of Na_2CO_3. In setting up an ICE table, we assume that the only important source of OH^- is hydrolysis of the carbonate ion and that the autoionization of water does not contribute significantly to $[OH^-]$ at equilibrium. Let x be the equilibrium value of $[OH^-]$. Then $[HCO_3^-]$ also is x. The changes in the two must both be $+x$, and the change in $[CO_3^{2-}]$ must be $-x$. Completing the ICE table, we have

	$[CO_3^{2-}]$ (M)	$[HCO_3^-]$ (M)	$[OH^-]$
Initial:	0.100	0	0
Change:	$-x$	$+x$	$+x$
Equilibrium:	$0.100 - x$	x	x

Solving for x:

$$K_{b_1} = 2.1 \times 10^{-4} = \frac{[HCO_3^-][OH^-]}{[CO_3^{2-}]} = \frac{(x)(x)}{0.100 - x}$$

$$x = 4.5 \times 10^{-3} \, M = [OH^-]$$

The calculated $[OH^-]$ is much greater than $[OH^-]$ in pure water (1.0×10^{-7} M), so the assumption that water autoionization is unimportant in this calculation is valid. To calculate pH from $[OH^-]$, we first calculate pOH:

$$pOH = -\log[OH^-] = -\log(4.5 \times 10^{-3}) = 2.35$$

and then use Equation 17.13 to calculate pH:

$$pH = pK_w - pOH = 14.00 - 2.35 = 11.65$$

A pH of 11.65 is quite basic. If you swam in a pool of water at that pH, you would experience skin irritation and painful burning in your eyes. Sodium carbonate *is* often used to adjust the pH of pools, but to make sure just the right amount is used, the pH of the water is tested to ensure that the pool is not too basic. We will see how this is done later in this chapter.

CONCEPT TEST

In the pH calculations in this chapter we routinely ignore the concentrations of H_3O^+ and OH^- ions produced by the autoionization of water. Suppose calculations of the pH of six different salt solutions produced the results shown in the table. Which, if any, of the calculations should have taken into account the autoionization of water to obtain an accurate result?

Solution	pH
A	2.66
B	4.12
C	6.39
D	7.27
E	9.10
F	12.88

SAMPLE EXERCISE 17.8 **Calculating the pH of a Solution of an Acidic Salt**

What is the pH of 0.25 M NH_4Cl?

Collect and Organize We are to calculate the pH of a solution of NH_4Cl. When NH_4Cl dissolves in water, NH_4^+ and Cl^- ions are released into solution. The NH_4^+ ion is the conjugate acid of NH_3; the Cl^- ion is the conjugate base of HCl.

Analyze We have seen that the Cl^- ion has negligible strength as a Brønsted–Lowry base, so it does not contribute to the acid–base properties of NH_4Cl. Ammonia is a weak base, which means that its conjugate acid, NH_4^+, is a weak acid. Therefore, some of the ammonium ions in solution donate H^+ ions to molecules of water:

$$NH_4^+(aq) + H_2O(\ell) \rightleftharpoons NH_3(aq) + H_3O^+(aq)$$

The K_a value of NH_4^+ is not given in the problem and is not listed in Appendix 5. However, the K_b value of its conjugate base, NH_3, is in Appendix 5 (Table A5.3): 1.76×10^{-5}. The value of K_a can be calculated from K_b using Equation 17.23.

The K_b value of ammonia is close to 10^{-5}. The product of $K_a \times K_b$ of a conjugate acid–base pair is 10^{-14}; therefore, the K_a value of the ammonium ion will be close to 10^{-9}. Because the ammonium ion is a very weak acid, we can anticipate a pH value that is less than 7 but probably closer to 7 than to 0.

Solve The simplified K_a expression for the NH_4^+ ion is

$$K_a = \frac{[NH_3][H^+]}{[NH_4^+]}$$

Rearranging Equation 17.23 to solve for K_a:

$$K_a = \frac{K_w}{K_b} = \frac{1.00 \times 10^{-14}}{1.76 \times 10^{-5}} = 5.68 \times 10^{-10} = \frac{[NH_3][H^+]}{[NH_4^+]}$$

We set up an ICE table in which we make the usual assumptions that the reaction is the only significant source of H^+ and that $x = [H^+] = [NH_3]$ at equilibrium:

	$[NH_4^+]$ (M)	$[NH_3]$ (M)	$[H^+]$ (M)
Initial:	0.25	0	0
Change:	$-x$	$+x$	$+x$
Equilibrium:	$0.25 - x$	x	x

$$K_a = 5.68 \times 10^{-10} = \frac{[NH_3][H^+]}{[NH_4^+]} = \frac{(x)(x)}{0.25 - x}$$

Given the very small value of K_a, we can make the simplifying assumption that $(0.25\ M - x) \approx 0.25\ M$, which gives us

$$\frac{x^2}{0.25} = 5.68 \times 10^{-10}$$
$$x^2 = 1.42 \times 10^{-10}$$
$$x = 1.19 \times 10^{-5} = [H^+]$$
$$pH = -\log[H^+] = -\log(1.19 \times 10^{-5}) = 4.92$$

Think about It This result matches our prediction: the pH of the solution is less than 7, but closer to 7 than to 0. The calculated $[H^+]$ is less than 5% of the initial concentration of NH_4^+, so our simplifying assumption was valid.

Practice Exercise What is the pH of a 0.25 M solution of sodium acetate? (*Hint*: The acetate ion is the conjugate base of acetic acid.)

The reactions of carbonate minerals in soils and rocks with acidic groundwater are critically important in mitigating the effects of acidic precipitation. When, for example, rain containing dilute sulfuric acid soaks into soil containing $CaCO_3$, in the form of limestone, marble, or shellfish shells, the acid is converted into either environmentally more benign carbonic acid

$$CaCO_3(s) + H_2SO_4(aq) \rightleftharpoons CaSO_4(s) + H_2CO_3(aq)$$

or, if enough $CaCO_3$ is available, into calcium sulfate and soluble calcium hydrogen carbonate:

$$2\,CaCO_3(s) + H_2SO_4(aq) \rightleftharpoons CaSO_4(s) + Ca(HCO_3)_2(aq)$$

As long as carbonates and other basic substances are present in soils and in the sediments of rivers and lakes, nature has the capacity to neutralize the acid in acid rain and maintain pH in a range that supports aquatic life.

CONNECTION The reaction of acidic groundwater with calcium carbonate and its connection to the formation of limestone caves were described in Chapter 4.

CONCEPT TEST

A mineral called dolomite contains a mixture of calcium carbonate and magnesium carbonate. Which can neutralize more acid: $CaCO_3$ or the same mass of $MgCO_3$?

17.7 The Common-Ion Effect

Thus far in this chapter we have worked with reaction systems consisting of a single acidic or basic reactant. Natural systems are often more complicated than that: they typically have multiple reactants that can influence pH or resist pH change.

Suppose, for example, that a sample of river water contains 1.2×10^{-5} M H_2CO_3 as a result of atmospheric carbon dioxide dissolving in the water. Suppose also that river sediment suspended in the water contains tiny particles of solid calcium carbonate ($CaCO_3$). We have seen how $CaCO_3$ can neutralize strong acids; it can also neutralize weak acids, as shown in the following reaction that produces soluble calcium bicarbonate:

$$CaCO_3(s) + H_2CO_3(aq) \rightleftharpoons Ca(HCO_3)_2(aq)$$

Other acidic substances in the river water could also be neutralized by calcium carbonate, producing additional soluble bicarbonate salts. Suppose that the total concentration of bicarbonate ions in the river water from these reactions is 1.0×10^{-4} M. How would the presence of this much HCO_3^- affect the pH of the water, assuming that the water also contains 1.2×10^{-5} M H_2CO_3? To answer this question, we need to keep in mind that H_2CO_3 and HCO_3^- represent a conjugate acid–base pair. They are related by the first acid ionization of carbonic acid:

$$H_2CO_3(aq) \rightleftharpoons HCO_3^-(aq) + H^+(aq) \qquad (17.24)$$

and its equilibrium constant expression:

$$K_{a_1} = \frac{[HCO_3^-][H^+]}{[H_2CO_3]} = 4.3 \times 10^{-7}$$

Let's insert the given values of $[H_2CO_3]$ and $[HCO_3^-]$ into the K_{a_1} expression, letting $x = [H^+]$:

$$K_{a_1} = \frac{[HCO_3^-][H^+]}{[H_2CO_3]} = \frac{(1.0 \times 10^{-4})(x)}{1.2 \times 10^{-5}} = 4.3 \times 10^{-7}$$

Solving for x:

$$x = 4.3 \times 10^{-7} \times \frac{1.2 \times 10^{-5}}{1.0 \times 10^{-4}} = 5.16 \times 10^{-8}\ M = [H^+]$$

Taking the negative logarithm to obtain pH:

$$pH = -\log[H^+] = -\log(5.16 \times 10^{-8}\ M) = 7.29$$

In Sample Exercise 17.6 we calculated that the pH of 1.2×10^{-5} M H_2CO_3 is 5.64. When H_2CO_3 is the only solute, ionization of H_2CO_3 is the only source of HCO_3^-. Dissolution of carbonate minerals provides a second source. The additional HCO_3^- causes the pH of the carbonic acid solution to increase by nearly 2 units. This increase in pH corresponds to a *decrease in [H+]* of nearly 2 orders of magnitude.

Is that the sort of change we should have expected? It does make sense because $[HCO_3^-]$ in equilibrium with 1.2×10^{-5} M H_2CO_3 alone is only $10^{-5.64} = 2.3 \times 10^{-6}$ M. By increasing $[HCO_3^-]$ to 1.0×10^{-4} M, we drive the equilibrium in Equation 17.24 to the left, as predicted by Le Châtelier's principle. The shift to the left lowers $[H^+]$ and raises pH.

This phenomenon illustrates a principle known as the **common-ion effect**: in any ionic equilibrium, a reaction that produces an ion is suppressed when another source of the same ion is added to the system. In the river water sample, ionization of H_2CO_3 (Equation 17.24) is suppressed when HCO_3^- from carbonate minerals is added.

We can apply the common-ion effect to any equilibrium involving a weak acid and its conjugate base:

$$Acid(aq) \rightleftharpoons H^+(aq) + base(aq)$$

common-ion effect the shift in the position of an equilibrium caused by the addition of an ion taking part in the reaction.

Henderson–Hasselbalch equation used to calculate the pH of a solution in which the concentrations of acid and conjugate base are known.

which has the equilibrium constant expression:

$$K_a = \frac{[H^+][base]}{[acid]}$$

If we take the negative logarithm of both sides of this expression, we transform $[H^+]$ into pH and K_a into pK_a:

$$pK_a = pH - \log \frac{[base]}{[acid]}$$

$$pH = pK_a + \log \frac{[base]}{[acid]} \qquad (17.25)$$

Equation 17.25 is particularly useful in calculating the pH of a solution in which there are independent sources of both an acid and its conjugate base (or a base and its conjugate acid). It is called the **Henderson–Hasselbalch equation**.

Consider what happens to the logarithmic term in the Henderson–Hasselbalch equation when the concentrations of the acid and base are the same. Then the numerator and denominator in the log term are equal, and the value of the fraction is 1. The log of 1 is 0, and $pH = pK_a$. This equality serves as a handy reference point in an acid–conjugate base system. If the concentration of the basic component is greater than that of the acid, the logarithmic term is greater than zero and $pH > pK_a$. If the concentration of the basic component is less than that of the acid, the logarithmic term is less than zero, and $pH < pK_a$.

Consider the case in which the concentration of the base is 10 times the concentration of the acid, that is, $[base] = 10\,[acid]$. Substituting this equality into Equation 17.25, we have

$$pH = pK_a + \log \frac{10[acid]}{[acid]}$$

$$= pK_a + \log 10 = pK_a + 1$$

A 10-fold higher concentration of base produces a pH one unit above the pK_a value. Similarly, if the concentration of the acid component is 10 times that of the base, then $pH = pK_a - 1$.

To demonstrate how the Henderson–Hasselbalch equation simplifies pH calculations when we know the concentrations of a weak acid and its conjugate base, let's use that equation to recalculate the pH of our river water sample. Our starting information includes the concentrations of a weak acid, $[H_2CO_3] = 1.2 \times 10^{-5}\ M$, and its conjugate base, $[HCO_3^-] = 1.0 \times 10^{-4}\ M$. They are linked by the equilibrium

$$H_2CO_3(aq) \rightleftharpoons HCO_3^-(aq) + H^+(aq) \qquad K_{a_1} = 4.3 \times 10^{-7}$$

First we take the negative log of K_{a_1}:

$$pK_{a_1} = -\log K_{a_1} = -\log(4.3 \times 10^{-7}) = 6.37$$

and then insert that value and the concentrations into the Henderson–Hasselbalch equation:

$$pH = pK_a + \log \frac{[base]}{[acid]}$$

$$= 6.37 + \log \frac{1.0 \times 10^{-4}}{1.2 \times 10^{-5}}$$

$$= 7.29$$

This result matches the one we calculated earlier.

The Henderson–Hasselbalch equation can also be used to calculate the pH of a solution of a weak base and its conjugate acid. An extra step may be involved in that calculation because we may know the value of the K_b of the base but not of the K_a of its conjugate acid, and only K_a can be used in the Henderson–Hasselbalch equation. Equation 17.23 allows us to interconvert K_a and K_b values. Interconverting pK_a and pK_b values is even easier. All we have to do is a logarithmic transformation of Equation 17.23:

$$-\log K_b = -\log(1.00 \times 10^{-14}) - (-\log K_a)$$
$$pK_b = 14.00 - pK_a$$
$$pK_b + pK_a = 14.00 \tag{17.26}$$

Thus, converting a pK_b value into the pK_a of its conjugate acid is simply a matter of subtracting the pK_b value from 14.00.

SAMPLE EXERCISE 17.9   **Calculating the pH of a Solution of a Weak Base and Its Conjugate Acid**

Calculate the pH of a solution that is 0.200 M in NH_3 and 0.300 M in NH_4Cl.

Collect and Organize We are to calculate the pH of a solution containing known concentrations of a weak base (NH_3) and a salt of its conjugate acid (NH_4^+). The Henderson–Hasselbalch equation may be used to calculate the pH of such a solution from the concentrations of the two components and the pK_a of the acid. Table A5.3 in Appendix 5 contains K_b values of common bases. The pK_a and pK_b values of a conjugate acid–base pair are related by Equation 17.26:

$$pK_b + pK_a = 14.00$$

Analyze Our approach involves converting the pK_b value from Appendix 5 (4.75) into pK_a. Addition of ammonium ion to a solution of ammonia should produce a solution that is still basic, but not as basic as a solution of only ammonia.

Solve Inserting the value of pK_b in Equation 17.26 and solving for pK_a:

$$pK_a = 14.00 - pK_b = 14.00 - 4.75 = 9.25$$

Using this value and the given concentrations of NH_3 and NH_4^+ in Equation 17.25:

$$pH = pK_a + \log \frac{[\text{base}]}{[\text{acid}]}$$

$$= 9.25 + \log \frac{0.200}{0.300} = 9.07$$

Think about It We predicted a result that would be a basic pH, but not as basic as a solution of ammonia alone. To confirm this prediction, we can calculate the pH of 0.200 M NH_3 using the approach in Sample Exercise 17.5. The result is a pH of 11.27—over 2 pH units higher and more than 100 times more basic than the solution of ammonia and ammonium chloride in this exercise.

Practice Exercise Calculate the pH of a solution that is 0.150 M in benzoic acid and 0.100 M in sodium benzoate.

pH buffer a solution that resists changes in pH when acids or bases are added to it; typically a solution of a weak acid and its conjugate base.

17.8 pH Buffers

A **pH buffer** is a solution that has the capacity to resist pH change by neutralizing small additions of acid or base. Typically it is a solution of a weak acid and its conjugate base. Buffers are an important component of natural water systems and of living organisms. In both settings, the role of a buffer is to maintain pH within a desired range. The internal pH of most living cells is regulated by buffers, including the carbonic acid–bicarbonate system we have been exploring. When the pH stabilization provided by these buffers is disturbed in a living system, protein function is impaired, and the health of individual cells or the entire organism may be in jeopardy.

An Environmental Buffer

▶II **CHEMTOUR** Buffers

The presence of bicarbonate ion in river water in the preceding section gives the water a capacity to resist pH change when, for example, acid rain falls into it. We can use the Henderson–Hasselbalch equation to determine the effect of adding a small amount of a strong acid or base to a buffered solution. Consider, for example, what happens when a quantity of strong acid is added to our model river water in which $[H_2CO_3] = 1.2 \times 10^{-5}\ M$ and $[HCO_3^-] = 1.0 \times 10^{-4}\ M$. For reference purposes we will also evaluate the impact of adding the same quantity of acid to pure (unbuffered) pH 7.00 water. We start with 1.00 L each of river water and pure water and add 10.0 mL of $1.0 \times 10^{-3}\ M$ HNO_3 to both.

Adding acid to pure water is an exercise in dilution, a concept introduced in Section 4.3. In this case, 10.0 mL of $1.0 \times 10^{-3}\ M$ HNO_3 is diluted to a final volume of 1.01 L. To calculate the final concentration of HNO_3 (and $[H^+]$), we use Equation 4.3:

$$V_{\text{initial}} \times M_{\text{initial}} = V_{\text{diluted}} \times M_{\text{diluted}}$$

Solving for M_{diluted}:

$$M_{\text{diluted}} = [H^+] = \frac{V_{\text{initial}} \times M_{\text{initial}}}{V_{\text{diluted}}}$$

$$= \frac{0.0100\ \text{L} \times (1.0 \times 10^{-3}\ M)}{1.01\ \text{L}} = 9.9 \times 10^{-6}\ M$$

The corresponding pH is $-\log(9.9 \times 10^{-6}\ M) = 5.00$. Thus, the addition of acid dropped the pH of the water by 2.00 pH units.

To calculate the change in pH when the same quantity of acid is added to 1.00 L of pH 7.29 river water, we need to focus on the carbonic acid–bicarbonate equilibrium and the pK_{a_1} of H_2CO_3:

$$H_2CO_3(aq) \rightleftharpoons HCO_3^-(aq) + H^+(aq) \qquad pK_{a_1} = 6.37 \quad (17.27)$$

Any acid added to an equilibrium mixture of H_2CO_3 and HCO_3^- reacts with HCO_3^-, producing H_2CO_3 as the reaction runs in reverse. The number of moles of H^+ added is

$$(0.0100\ \cancel{\text{L}}) \times \left(1.0 \times 10^{-3}\ \frac{\text{mol H}^+}{\cancel{\text{L}}}\right) = 1.0 \times 10^{-5}\ \text{mol H}^+$$

This quantity of HCO_3^- in the river water sample is consumed as the reaction in Equation 17.27 runs in the reverse direction. The number of moles of bicarbonate present initially is

$$\left(1.0 \times 10^{-4}\ \frac{\text{mol}}{\cancel{\text{L}}}\right) \times (1.00\ \cancel{\text{L}}) = 1.0 \times 10^{-4}\ \text{mol}$$

After subtracting the number of moles of HCO_3^- consumed, we have 9×10^{-5} mol left in a final volume of 1.01 L. As we have done in previous calculations, we assume that the value of $[H_2CO_3]$ is controlled by the solubility of CO_2 in water and does not change from the initial value of 1.2×10^{-5}. With this information we can calculate the pH of the river water using the Henderson–Hasselbalch equation:

$$pH = pK_a + \log\left(\frac{[\text{base}]}{[\text{acid}]}\right) = 6.37 + \log\left[\frac{\left(\dfrac{9 \times 10^{-5}\ \cancel{\text{mol}}}{1.01\ \cancel{\text{L}}}\right)}{1.2 \times 10^{-5}\ \dfrac{\cancel{\text{mol}}}{\cancel{\text{L}}}}\right]$$

$$= 7.24$$

The original pH was 7.29. Therefore, the addition of 10.0 mL of strong acid lowered the pH by only 0.05 pH units because of the action of the carbonic acid–hydrogen carbonate buffer. This stands in contrast to a decrease of 2.00 pH units in unbuffered water.

SAMPLE EXERCISE 17.10 | **Calculating Buffer Response to Addition of Acid or Base**

Calculate the change in pH when 1.0 mL of 1.00 M HCl is added to 100 mL of a solution that is 0.100 M sodium acetate and 0.100 M acetic acid.

Collect and Organize We are to determine by how much the pH of a solution of acetic acid and sodium acetate changes when a quantity of strong acid is added to it. This solution functions as a buffer. We are given the volume and composition of the buffer and of the strong acid added to it. The pH of a solution with known concentrations of a conjugate acid–base pair can be calculated using the Henderson–Hasselbalch equation:

$$pH = pK_a + \log\left(\frac{[\text{base}]}{[\text{acid}]}\right)$$

The pK_a value of acetic acid is 4.75 (Appendix 5).

Analyze The pH of this buffer is controlled by the ionization equilibrium of acetic acid:

$$CH_3COOH(aq) \rightleftharpoons CH_3COO^-(aq) + H^+(aq)$$

When a strong acid is added to a solution containing acetic acid and an acetate salt, some of the CH_3COO^- ions react with the added H^+ ions, forming more CH_3COOH as the ionization reaction runs in reverse. Initially $[CH_3COOH] = [CH_3COO^-]$, which means the log term in the Henderson–Hasselbalch equation is zero and so $pH = pK_a = 4.75$. Addition of a quantity of strong acid that does not consume all the acetate ion should result in a pH that is slightly lower than 4.75.

Solve The initial quantities of CH_3COO^- and CH_3COOH in the buffer are both

$$100\ \cancel{\text{mL}} \times \frac{0.100\ \cancel{\text{mol}}}{\cancel{\text{L}}} \times \frac{1\ \cancel{\text{L}}}{1000\ \cancel{\text{mL}}} \times \frac{1000\ \text{mmol}}{1\ \cancel{\text{mol}}} = 10.0\ \text{mmol}$$

The quantity of H^+ added:

$$1.0 \; \text{mL} \times \frac{1.00 \; \text{mol}}{\text{L}} \times \frac{1 \; \text{L}}{1000 \; \text{mL}} \times \frac{1000 \; \text{mmol}}{1 \; \text{mol}} = 1.0 \; \text{mmol}$$

represents the quantity of CH_3COO^- converted into CH_3COOH. The impact of this conversion is summarized in the following table:

	CH₃COOH (mmol)	CH₃COO⁻ (mmol)
Initial:	10.0	10.0
Change:	+1.0	−1.0
Final:	11.0	9.0

Both the final quantities are in the same total volume, 101 mL. Therefore we can use the ratio of quantities in lieu of concentration values in Equation 17.25 to calculate pH. Inserting these quantities and the pK_a value of acetic acid in Equation 17.25 gives

$$pH = pK_a + \log \frac{[CH_3COO^-]}{[CH_3COOH]} = 4.75 + \log \frac{9.0}{11.0} = 4.66$$

The change in the pH of the solution after adding the acid is $4.75 - 4.66 = 0.09$ pH units.

Think about It The result is reasonable because adding strong acid produces a buffer solution with more acid and less conjugate base, so it has a slightly lower final pH (4.66) than it had initially (4.75).

Practice Exercise Calculate the change in pH when 10.0 mL of a 0.100 M solution of NaOH is added to 1.00 L of a solution that is 1.00 M in sodium acetate and 1.00 M in acetic acid.

A Physiological Buffer

Buffers are vital components of living systems because most biochemical reactions involved in life-sustaining processes, like metabolism, respiration, and transmission of nerve impulses, take place only within a narrow pH range. The pH of blood, for instance, needs to be buffered against perturbations caused by the ingestion or internal production of acidic or basic substances. One of the buffer systems the human body relies on to control pH is the same one we have discussed in the context of environmental waters: the carbonic acid/bicarbonate system. This system is intimately tied to respiration, and a key feature of pH control in this system is the role of breathing to maintain pH balance by regulating the concentration of dissolved CO_2.

Carbon dioxide is a product of cell metabolism, and one of the jobs of the blood is to carry CO_2 from cells to the lungs, where it is eliminated as we exhale. If lung function is impaired through disease or injury, the concentration of CO_2 dissolved in the blood may increase, which increases the concentration of carbonic acid, which in turn increases $[H^+]$ and lowers pH. This sequence of events is summarized by shifts from left to right in the following equilibria:

$$CO_2(aq) + H_2O(\ell) \rightleftharpoons H_2CO_3(aq) \rightleftharpoons HCO_3^-(aq) + H^+(aq)$$

The resulting drop in pH produces a condition called *respiratory acidosis*. On the other hand, hyperventilation—fast, overly deep breathing—may cause too much CO_2 to be exhaled, shifting the above equilibria to the left and causing a decrease in $[H^+]$ and a rise in pH. The result is a condition called *respiratory alkalosis*.

CONCEPT TEST

The mice used in biomedical experiments are sometimes euthanized by exposing them to high concentrations of CO_2 in the air they breathe. Why would this exposure be deadly?

Buffer Range and Capacity

We have seen how buffers resist pH change when either acid or base is added. When scientists select buffers for particular applications, they need to answer two questions:

1. What is the desired pH range to be maintained? The appropriate buffer is one whose weak acid has a pK_a that is within one pH unit of the desired pH. The Henderson–Hasselbalch equation tells us that over this pH range the ratio [base]/[acid] varies from 10:1 to 1:10. Expressed another way, if this condition is met:

$$0.1 < \frac{[\text{base}]}{[\text{acid}]} < 10$$

both components are available to neutralize additions of either acid or base and maintain the desired pH.

2. How much acid or base can the system consume without a large change in pH? The answer to this question defines the **buffer capacity**. A buffer is best able to resist changes in pH when the initial concentrations of acid and conjugate base are comparable to each other and greater than the concentration of acid or base that might be added. The greater the concentration of the buffer components, the greater the buffer capacity (Figure 17.13).

CONCEPT TEST

Select from the list in Appendix 5 a weak acid that, when mixed with the sodium salt of its conjugate base in approximately equimolar proportions, produces a buffer with a pH of exactly 2.80. Indicate whether the buffer will contain *exactly* the same concentrations of acid and conjugate base, or slightly more acid or base.

17.9 Solubility Equilibria

In Section 17.1 we noted that the common strong bases are all hydroxides of alkali or alkaline earth elements. One exception is $Mg(OH)_2$, which is a weak base because it has limited solubility in water. Magnesium hydroxide is the active ingredient in a product found in many medicine cabinets: the antacid called *milk of magnesia*. This liquid appears "milky" because it is an aqueous *suspension* (not solution) of solid, white $Mg(OH)_2$. We can express the limited solubility of solid $Mg(OH)_2$ by the following equation:

$$Mg(OH)_2(s) \rightleftharpoons Mg^{2+}(aq) + 2\,OH^-(aq)$$

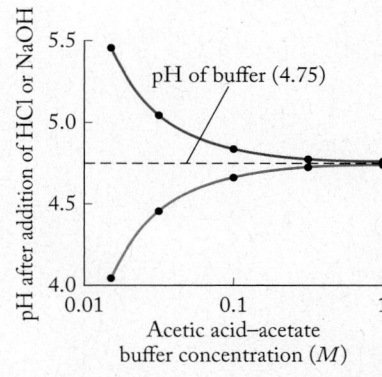

FIGURE 17.13 When strong acid (red line) or strong base (blue line) is added to a buffer solution, the extent to which the pH changes is inversely proportional to buffer concentration: the higher the concentrations of the buffer components, the smaller the change in pH. In this illustration, five 100 mL solutions that are 0.015, 0.030, 0.100, 0.300, and 1.000 M acetic acid and sodium acetate all have an initial pH of 4.75. The graph shows the pH values of these solutions after 1.00 mL of 1.00 M HCl or 1.00 M NaOH has been added.

buffer capacity the quantity of acid or base that a pH buffer can neutralize while maintaining its pH within a desired range.

solubility-product constant (also called **solubility product, K_{sp}**) an equilibrium constant that describes the formation of a saturated solution of a slightly soluble salt.

Because $Mg(OH)_2$ is a solid, its effective concentration does not change as long as some of it is present in the system. As was the case for pure liquids (Section 17.2), pure solids do not appear in the equilibrium constant expression. Therefore, the equilibrium constant for the dissolution of $Mg(OH)_2$ is

$$K_{sp} = [Mg^{2+}][OH^-]^2$$

where K_{sp} represents an equilibrium constant called either the **solubility-product constant** or simply the **solubility product**.

The K_{sp} values of $Mg(OH)_2$ and other slightly soluble compounds are listed in Table A5.4 of Appendix 5. We can use these values to calculate concentrations of these compounds in aqueous solutions. Two terms are used to describe how much of a solid dissolves in a solvent: *solubility*, which is expressed in grams of solute per liter of solution, and *molar solubility*, which is expressed in moles of solute per liter of solution. Let's use the K_{sp} of $Mg(OH)_2$ from Appendix 5 (5.6×10^{-12}) to calculate the molar solubility of $Mg(OH)_2$ at 25°C and how many grams of it dissolve in 50.0 mL at 25°C. We start with the K_{sp} expression:

$$K_{sp} = [Mg^{2+}][OH^-]^2 = 5.6 \times 10^{-12}$$

If we let x be the number of moles of $Mg(OH)_2$ that dissolves in 1 L of solution, then x moles of Mg^{2+} ions and $2x$ moles of OH^- ions are produced:

$$K_{sp} = (x)(2x)^2 = (x)(4x^2) = 4x^3 = 5.6 \times 10^{-12}$$

$$x = 1.1 \times 10^{-4}\ M$$

Note that the entire algebraic expression for $[OH^-]$, $2x$, is squared in this calculation. Forgetting to square the coefficient is a common mistake. Also note that the molar solubility of $Mg(OH)_2$ is much greater than its solubility product. This difference is true for all sparingly soluble ionic compounds because each K_{sp} is the product of small concentration values multiplied together, producing an even smaller overall K_{sp} value.

Using the molar solubility of $Mg(OH)_2$ that we just calculated (1.1×10^{-4} mol/L), we can determine how many grams of $Mg(OH)_2$ dissolve in 50.0 mL of a saturated solution. We convert molar solubility into an equivalent number of grams of $Mg(OH)_2$ per liter using the molar mass of $Mg(OH)_2$, 58.32 g/mol, and then into the mass that dissolves in 50.0 mL of solution:

$$\frac{1.1 \times 10^{-4}\ \cancel{\text{mol}}}{1\ \cancel{\text{L}}} \times \frac{58.32\ \text{g}}{1\ \cancel{\text{mol}}} \times \frac{1\ \cancel{\text{L}}}{1000\ \cancel{\text{mL}}} \times 50.0\ \cancel{\text{mL}} = 3.2 \times 10^{-4}\ \text{g}$$

We can derive a general equation relating the molar solubility of any ionic compound M_mZ_z to its K_{sp} value. We start with the dissolution equilibrium

$$M_mZ_z(s) \rightleftharpoons m\,M^{n+}(aq) + z\,Z^{y-}(aq)$$

If S represents the molar solubility of M_mZ_z, then there are $(m \times S)$ mol/L M^{n+} ions and $(z \times S)$ mol/L Z^{y-} ions in solution. Inserting these molar concentrations into the equilibrium constant expression for M_mZ_z:

$$K_{sp} = [M^{n+}]^m[Z^{y-}]^z = (m \times S)^m(z \times S)^z$$

$$K_{sp} = (m^mz^z)S^{(m+z)} \qquad (17.28)$$

Before we get lost in all of these letters, let's apply Equation 17.28 to calculate the molar solubility of calcium phosphate, $Ca_3(PO_4)_2$:

$$K_{sp} = (m^mz^z)S^{(m+z)} = (3^32^2)S^{(3+2)} = 108\ S^5$$

Note that there are really only two terms in Equation 17.28. The first contains each subscript raised to the same power as its value: every "2" is squared, every "3" is cubed, and so on. The second term is simply molar solubility (S) raised to a power equal to the sum of the subscripts. One warning about subscripts and polyatomic ions: use only the subscript outside the parentheses, not the subscripts that are part of the formula of the ion.

Now let's insert the K_{sp} value listed for $Ca_3(PO_4)_2$ in Appendix 5 and solve for S.

$$K_{sp} = 2.1 \times 10^{-33} = 108S^5$$

$$S = \sqrt[5]{\frac{2.1 \times 10^{-33}}{108}} = 1.1 \times 10^{-7}\ M$$

CONCEPT TEST ....

Use Equation 17.28 to write an equation that relates the K_{sp} value for Na_2CO_3 to its molar solubility S.

....

SAMPLE EXERCISE 17.11   **Calculating Molar Solubility from K_{sp}**

The mineral barite is mostly barium sulfate ($BaSO_4$) and is widely used in industry and in medical imaging of the digestive system. Calculate the molar solubility at 25°C of $BaSO_4$ in (a) pure water and (b) seawater in which the concentration of sulfate ions is 2.8 g/L.

Collect and Organize We are to calculate the molar solubility of $BaSO_4$ in both pure water and in seawater that already contains sulfate ions. The K_{sp} value of $BaSO_4$ given in Appendix 5 is 9.1×10^{-11}.

Analyze The dissolution reaction:

$$BaSO_4(s) \rightleftharpoons Ba^{2+}(aq) + SO_4^{2-}(aq)$$

indicates that 1 mole of Ba^{2+} ions and 1 mole of SO_4^{2-} ions form from each mole of $BaSO_4$ that dissolves. If S mol/L of $BaSO_4$ dissolves in pure water, then $[Ba^{2+}] = [SO_4^{2-}] = S$. However, seawater has a background concentration of SO_4^{2-}. According to Le Châtelier's principle and the common-ion effect, the sulfate ion already in seawater should shift the dissolution equilibrium to the left, which means that less $BaSO_4$ should dissolve in seawater than in pure water.

Solve

a. In pure water,

$$K_{sp} = [Ba^{2+}][SO_4^{2-}] = (S)(S) = 9.1 \times 10^{-11}$$

$$S^2 = 9.1 \times 10^{-11}$$

$$S = 9.5 \times 10^{-6}\ M$$

b. In seawater, we first need to calculate the value of $[SO_4^{2-}]$ before any $BaSO_4$ dissolves:

$$[SO_4^{2-}]_{initial} = \frac{2.8\ g}{L} \times \frac{1\ mol}{96.06\ g} = \frac{0.029\ mol}{L}$$

The value of $[SO_4^{2-}]$ at equilibrium is the sum of the background concentration (0.029 mol/L) and the additional SO_4^{2-} ions from the dissolution of $BaSO_4$ (S). Incorporating this value into the $[SO_4^{2-}]$ term in the K_{sp} expression gives

$$K_{sp} = [Ba^{2+}][SO_4^{2-}] = (S)(0.029 + S) = 9.1 \times 10^{-11}$$

Solving for S is simplified if we assume that the K_{sp} of $BaSO_4$ is so small that we can ignore its contribution to the total SO_4^{2-} concentration. This assumption is reasonable because S is likely to be much less than 0.029 M given the limited solubility of $BaSO_4$ in pure water. Therefore,

$$(S)(0.029 + S) \approx (S)(0.029) = 9.1 \times 10^{-11}$$
$$S = 3.1 \times 10^{-9} \, M$$

Think about It The calculated molar solubility of $BaSO_4$ in seawater is much less than the initial $[SO_4^{2-}]$ value, so our simplifying assumption was justified. The lower solubility of $BaSO_4$ in seawater is another illustration of the common-ion effect: the dissolution of $BaSO_4$ is suppressed by the SO_4^{2-} ions already present in seawater.

Practice Exercise What is the molar solubility of $MgCO_3$, a component of the mineral dolomite, at 25°C?

In the preceding Sample Exercises we saw how the common-ion effect can suppress the solubility of an ionic compound. Other perturbations to solubility equilibria can actually promote solubility, as happens when the anion of the compound is the conjugate base of a weak acid. The molar solubilities of such compounds increase when strong acid is added and pH is lowered. We will see why as we solve the following Sample Exercise.

SAMPLE EXERCISE 17.12 **Calculating the Effect of pH on Solubility**

What is the molar solubility of CaF_2 at 25°C in (a) pure water and (b) an acidic buffer in which $[H^+]$ is a constant 0.050 M?

Collect and Organize We are asked to calculate the solubility of CaF_2 in both pure water and in an acidic buffer. The dissolution process is described by the equilibrium:

$$(1) \qquad CaF_2(s) \rightleftharpoons Ca^{2+}(aq) + 2\,F^-(aq)$$

The equilibrium constant expression for the process and the K_{sp} value from Appendix 5 are

$$K_{sp} = [Ca^{2+}][F^-]^2 = 3.9 \times 10^{-11}$$

Calcium fluoride is a basic salt because the F^- ion is the conjugate base of the weak acid HF ($K_a = 6.8 \times 10^{-4}$).

Analyze To account for the effect of acid on the solubility of a fluoride salt, we need to consider the chemical equilibrium in which the fluoride ion acts as a Brønsted–Lowry base (H^+ ion acceptor):

$$(2) \qquad F^-(aq) + H^+(aq) \rightleftharpoons HF(aq)$$

This reaction is the reverse of the acid ionization reaction:

$$HF(aq) \rightleftharpoons F^-(aq) + H^+(aq)$$

Therefore, the equilibrium constant for reaction 2 is the reciprocal of the K_a of HF:

$$K_2 = \frac{[HF]}{[H^+][F^-]} = \frac{1}{6.8 \times 10^{-4}} = 1.47 \times 10^3$$

As reaction 2 proceeds, F^- ions are consumed, which shifts the equilibrium in reaction 1 to the right and so increases CaF_2 solubility. Therefore we can anticipate an increase in the solubility of CaF_2 when acid is present.

Solve

a. Let S be the molar solubility of CaF_2 in pure water. According to the stoichiometry of reaction 1, $[Ca^{2+}] = S$ mol/L and $[F^-] = 2S$ mol/L. Inserting these symbols in the K_{sp} expression:

$$K_{sp} = [Ca^{2+}][F^-]^2 = (S)(2S)^2 = 4S^3$$

$$4S^3 = 3.9 \times 10^{-11}$$

$$S = 2.1 \times 10^{-4}\, M$$

b. In the acidic buffer, $[H^+] = 0.050\, M$. The F^- and H^+ ions combine as in reaction 2. Assuming the buffer pH is constant, $[H^+] = 0.050\, M$ and the equilibrium constant expression for reaction 2 is

$$K_2 = \frac{[HF]}{[0.050][F^-]} = 1.47 \times 10^3$$

$$\frac{[HF]}{[F^-]} = 73.5$$

$$(3) \qquad [HF] = 73.5[F^-]$$

This calculation tells us that most of the F^- produced when calcium fluoride dissolves is converted into HF. The tiny fraction of all the F^- ions produced by CaF_2 that remain as free F^- ions is defined by this ratio:

$$(4) \qquad \frac{[F^-]}{[F^-] + [HF]}$$

where the numerator is the concentration of free F^- ions at equilibrium and the denominator is the concentration of all the F^- ions produced when CaF_2 dissolved. Combining expression 3 and 4:

$$\frac{[F^-]}{[F^-] + [HF]} = \frac{[F^-]}{[F^-] + 73.5[F^-]} = \frac{[F^-]}{74.5[F^-]} = 0.0134$$

If S is the molar solubility of CaF_2 in the acid, S mol/L Ca^{2+} and $2S$ mol/L F^- are produced. However, most of the fluoride ions are converted into HF, and, as we just calculated, the free F^- ion concentration is only 0.0134 of that produced when CaF_2 dissolved. Therefore, the $[F^-]$ term in the K_{sp} expression is not $2S$, but only $(0.0134 \times 2S) = 0.0268\ S$. Inserting this value in the K_{sp} expression,

$$K_{sp} = [\text{Ca}^{2+}][\text{F}^-]^2 = (S)(0.0268\ S)^2 = 7.18 \times 10^{-4}\ S^3$$
$$7.18 \times 10^{-4}\ S^3 = 3.9 \times 10^{-11}$$
$$S = 3.8 \times 10^{-3}\ M$$

Think about It A comparison of the results from parts a and b reveals that the molar solubility of CaF_2 is about 10 times higher in the acidic buffer, as we predicted, because most of the F^- ions produced when CaF_2 dissolves in the buffer form HF. This conversion of F^- ions to HF means a product (free F^- ion) is being removed from the dissolution equilibrium. Le Châtelier's principle tells us that the result will be a shift in the position of the equilibrium to the right, in favor of forming product. In this case greater product formation means greater solubility.

Practice Exercise What is the solubility of ZnCO_3 at 25°C in a buffer solution with a pH value of 10.33?

CONCEPT TEST

In part b of Sample Exercise 17.12, we assumed that $[\text{H}^+]$ did not change significantly as a result of the reaction $\text{F}^-(aq) + \text{H}^+(aq) \rightleftharpoons \text{HF}(aq)$. If $[\text{H}^+]$ had dropped significantly, how would the solubility of CaF_2 have been affected?

K_{sp} and Q

In the preceding Sample Exercises we used calculations involving the K_{sp} of slightly soluble salts in cases where the solubility of the salt was either diminished or enhanced. We can also use K_{sp} values to determine if a precipitate will form when two solutions are mixed. In answering such questions, we can use the concept of the reaction quotient Q that we developed in Chapter 16. When applied to the equilibrium governing a slightly soluble salt, Q is sometimes called the *ion product*, a name that describes exactly what Q is in this case: the product of the concentrations of the ions in the precipitate raised to powers equal to their coefficients in the balanced equation. If the calculated Q value is greater than the K_{sp} of a salt ($Q > K_{sp}$), then that salt will precipitate when the solutions are mixed. If $Q < K_{sp}$, no precipitate will form.

SAMPLE EXERCISE 17.13 **Determining if a Precipitate Forms when Solutions Are Mixed**

Lead(II) chloride ($K_{sp} = 1.60 \times 10^{-5}$) is a white pigment used in 15th-century European sculpture. Will PbCl_2 precipitate when 275 mL of a 0.134 M solution of $\text{Pb(NO}_3)_2$ is added to 125 mL of a 0.0339 M solution of NaCl?

Collect and Organize We are asked if PbCl_2 will precipitate when two solutions containing Pb^{2+} ions and Cl^- ions are mixed. The process is described by the equilibrium

$$\text{Pb}^{2+}(aq) + 2\,\text{Cl}^-(aq) \rightleftharpoons \text{PbCl}_2(s) \qquad K_{sp} = 1.60 \times 10^{-5}$$

Analyze To determine if a precipitate forms, we need to calculate Q and compare its value to K_{sp}. If $Q > K_{sp}$, $PbCl_2$ will precipitate; if $Q < K_{sp}$, it will not precipitate. Q has the same form as the equilibrium constant:

$$K_{sp} = [Pb^{2+}][Cl^-]^2$$

but the concentrations used in calculating it are the values given for the system, which are probably not equilibrium values.

Solve First we must calculate the concentrations of the lead ions and chloride ions in the two solutions immediately upon mixing. Remember that mixing two solutions dilutes both, and that the volumes of the solutions (275 mL and 125 mL) may be added together to get the final solution volume (400 mL = 0.400 L).

$$Pb^{2+}(aq): \quad 0.134\ \frac{mol}{L} \times 0.275\ L = 0.0369\ mol;$$

$$[Pb^{2+}] = \frac{0.0369\ mol}{0.400\ L} = 0.0921\ M$$

$$Cl^-(aq): \quad 0.0339\ \frac{mol}{L} \times 0.125\ L = 0.00424\ mol;$$

$$[Cl^-] = \frac{0.00424\ mol}{0.400\ L} = 0.0106\ M$$

The value of Q is

$$Q = [Pb^{2+}][Cl^-]^2 = (0.0921)(0.0106)^2 = 1.03 \times 10^{-5}$$

which is smaller than K_{sp}, so no precipitate forms.

Think about It To predict if a precipitate forms when solutions of known ion concentration are combined, we compare the Q value based on the concentrations of the ions present immediately after the solutions are mixed to the K_{sp} value of the salt. If Q equals K_{sp}, then the solution is saturated and contains the maximum quantity of solute.

Practice Exercise Will calcium fluoride ($K_{sp} = 3.9 \times 10^{-11}$) precipitate when 175 mL of a $4.78 \times 10^{-3}\ M$ solution of $Ca(NO_3)_2$ is added to 135 mL of a $7.35 \times 10^{-3}\ M$ solution of KF?

We can use differences in the solubilities of ionic compounds to selectively remove ions from solution. For example, suppose a solution contains 0.10 M Ca^{2+} ion and 0.020 M Mg^{2+} ion. The hydroxide salts of both ions are slightly soluble. Is it possible to remove the Mg^{2+} ions from solution by precipitating them as $Mg(OH)_2$ while leaving the Ca^{2+} ions in solution? Consider the following solubility equilibria and K_{sp} values:

$$Ca(OH)_2 \rightleftharpoons Ca^{2+}(aq) + 2OH^-(aq) \quad K_{sp} = [Ca^{2+}][OH^-]^2 = 4.7 \times 10^{-6}$$

$$Mg(OH)_2 \rightleftharpoons Mg^{2+}(aq) + 2OH^+(aq) \quad K_{sp} = [Mg^{2+}][OH^-]^2 = 5.6 \times 10^{-12}$$

Note that $Ca(OH)_2$ has a larger K_{sp} than $Mg(OH)_2$. We can therefore ask: What is the maximum concentration (x) of OH^- ions that will *not* cause the

0.10 M Ca^{2+} ion to precipitate? We can calculate that directly from the K_{sp} of calcium hydroxide:

$$K_{sp} = 4.7 \times 10^{-6} = [Ca^{2+}][OH^-]^2 = (0.10)(x)^2$$

$$x = \sqrt{\frac{4.7 \times 10^{-6}}{0.10}} = 6.9 \times 10^{-3} \ M$$

How much of the Mg^{2+} in the solution would precipitate if $[OH^-] = 6.9 \times 10^{-3}$ M? The concentration x of magnesium ion in the presence of 6.9×10^{-3} M hydroxide ion may be calculated from its K_{sp}:

$$K_{sp} = 5.6 \times 10^{-12} = [Mg^{2+}][OH^-]^2 = (x)(6.9 \times 10^{-3})^2$$

$$x = \frac{5.6 \times 10^{-12}}{(6.9 \times 10^{-3})^2} = 1.2 \times 10^{-7} \ M$$

Because the original solution was 0.020 M in Mg^{2+} ions, a concentration of 1.2×10^{-7} M Mg^{2+} means that only $[(1.2 \times 10^{-7})/0.020] \times 100\%$, or only 0.00060%, of the original amount of Mg^{2+} remains in solution. This corresponds to virtually complete precipitation of Mg^{2+} and successful separation of the solution's calcium ions from its magnesium ions. Indeed, magnesium can be separated from the less abundant calcium ions in seawater by this technique.

CONCEPT TEST •

In the analysis of using a precipitation reaction to separate magnesium ion from calcium ion in solution, we did not specify the concentration of the hydroxide ion solution used to form the precipitates. Why did we not need this value?

• •

SAMPLE EXERCISE 17.14 **Separating Ions in Solution**

Both lead(II) chloride and lead(II) fluoride are slightly soluble salts. A solution of lead(II) nitrate is added to a solution that is 0.275 M in both $Cl^-(aq)$ and $F^-(aq)$. Can we use this method to separate the two halide ions? If "complete precipitation" is defined as there being less than 0.10% of a particular ion left in solution, is the precipitation of the first salt complete before the second salt begins to precipitate? (K_{sp} $PbCl_2 = 1.6 \times 10^{-5}$; K_{sp} $PbF_2 = 3.2 \times 10^{-8}$.)

Collect and Organize We are given a solution that contains two ions that form slightly soluble lead(II) salts and asked if one ion can be completely removed before the second one starts to precipitate when lead(II) ion is added to the solution. We have the K_{sp} values for both salts and the initial concentrations of both ions.

Analyze The equilibrium constant expressions for both ions are

$$PbCl_2(s) \rightleftharpoons Pb^{2+}(aq) + 2\,Cl^-(aq) \qquad K_{sp} = [Pb^{2+}][Cl^-]^2 = 1.6 \times 10^{-5}$$

$$PbF_2(s) \rightleftharpoons Pb^{2+}(aq) + 2\,F^-(aq) \qquad K_{sp} = [Pb^{2+}][F^-]^2 = 3.2 \times 10^{-8}$$

In both equilibrium expressions, the concentration of lead ion is raised to the first power and the concentration of the halide ion is raised to the second power, so we can compare the influence of the ion concentrations on the K_{sp} values directly. The K_{sp} of $PbCl_2(s)$ is almost 10^3 times larger than the K_{sp} of $PbF_2(s)$, so we can assume the $PbF_2(s)$ will precipitate first. We can therefore ask: What is the maximum Pb^{2+} concentration in the solution that will not cause the $PbCl_2(s)$ to precipitate? When we determine that value, we can calculate how much $F^-(aq)$ is in solution under those conditions and determine if the precipitation of F^- is complete.

Solve The maximum amount of Pb^{2+} in the solution that will not cause the chloride ion to precipitate is

$$K_{sp} = 1.6 \times 10^{-5} = [Pb^{2+}][Cl^-]^2 = (x)(0.275)^2$$

$$x = \frac{1.6 \times 10^{-5}}{(0.275)^2} = 2.12 \times 10^{-4}\ M$$

The concentration of $F^-(aq)$ in the solution at this concentration of lead(II) ion is

$$K_{sp} = 3.2 \times 10^{-8} = (2.12 \times 10^{-4})(x)^2$$

$$x = \sqrt{\frac{3.2 \times 10^{-8}}{2.12 \times 10^{-4}}} = 0.0123\ M$$

The original solution was 0.275 M in $F^-(aq)$, and a concentration of 0.0123 M means that (0.0123/0.275) × 100% = 4.5% of the original amount of $F^-(aq)$ remains in solution, so the precipitation of $PbF_2(s)$ is not complete and we could not use this method to separate the two ions.

Think about It Determining which salt precipitates first was easy in this example because the concentrations of both ions were the same and both were raised to the same power in the K_{sp} expressions. If the concentrations differ, or if the concentrations of the ions in the K_{sp} expressions are raised to different powers, determining which species precipitates first may have to be determined by calculation rather than simple inspection.

Practice Exercise An aqueous solution of sodium fluoride is slowly added to a water sample that contains barium ion (0.0375 M) and calcium ion (0.0667 M). Consult Appendix 5 to help you answer these questions:
  a. Are both barium fluoride and calcium fluoride slightly soluble salts?
  b. Applying the definition of complete precipitation given in Sample Exercise 17.14, determine if Ba^{2+} and Ca^{2+} ions in solution can be completely separated by selective precipitation with F^- ions.

17.10 Indicators and Acid-Base Titrations

In Section 17.6 we noted that swimming pool operators routinely check pool pH. They often use a test kit that includes a **pH indicator** (Figure 17.14), a substance that changes color as pH changes. One such substance is phenol red. It is a weak acid ($pK_a = 7.6$) that is yellow in its un-ionized form (which, for convenience, we assign the generic formula HIn) and violet in its ionized (In^-) form. At a pH one unit above the pK_a—at pH 8.6—the ratio $[In^-]/[HIn]$ is 10:1 according to the Henderson–Hasselbalch equation, and a phenol red solution is violet. At a pH less than 6.6, phenol red is largely un-ionized, and so a solution of the indicator is

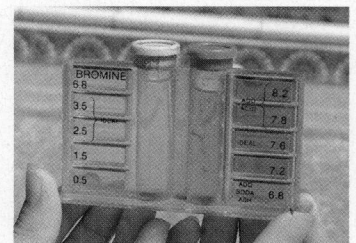
(a)

(b)

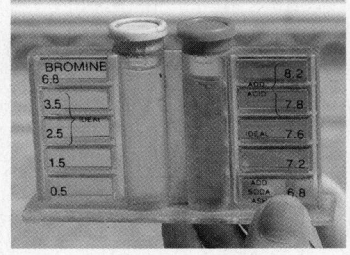
(c)

FIGURE 17.14 Many pool test kits include the pH indicator phenol red. A few drops are added to a sample of pool water collected in the tube with the red cap. (a) After a rainstorm, the pH of the pool water is 6.8 (or less), as indicated by the yellow color of the sample. (b) Sodium carbonate is added to the pool to raise the pH. (c) A follow-up test produces a red-orange color, indicating the pH of the pool has been properly adjusted.

pH indicator a water-soluble weak organic acid that changes color as pH changes.

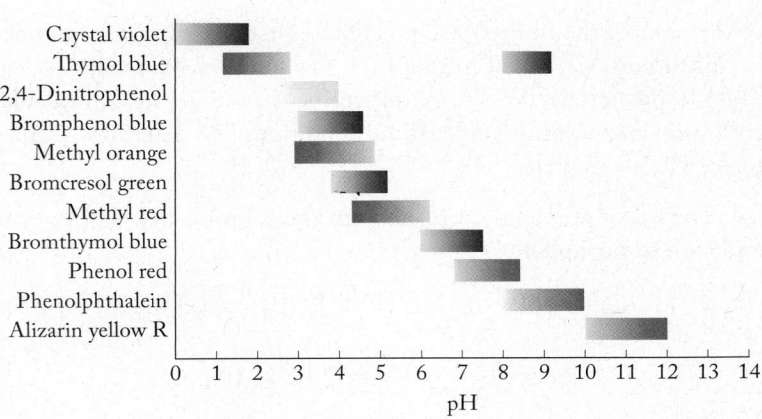

FIGURE 17.15 A pH indicator is useful within a range of 1 pH unit above and below the pK_a value of the indicator. This array of indicators could be used to determine pH values from 0 to 12.

yellow. In the pH range from about 6.8 to 8.6, the color of a phenol red solution changes from yellow to orange to red to violet with increasing pH (Figure 17.15).

As with buffers, different pH indicators are useful in different pH ranges. For example, phenol red is the best choice if the pH values being monitored fall between pH 6.6 and 8.6; one of the other indicators shown in Figure 17.15 would be a better choice in another pH range. Every indicator has a useful pH range defined by its $pK_a \pm 1.0$ pH unit. In addition to their role in determining pH values, indicators are also used to detect the large changes in pH that occur in acid–base titrations.

Acid-Base Titrations

We discussed titration methods in Section 4.6 and summarize here how they are carried out. Figure 17.16 shows a typical titration apparatus. There are four steps in its use:

1. Accurately transfer a known volume of sample to a flask or beaker.
2. Either add a few drops of an indicator solution to the sample or insert the probe of a pH meter.
3. Fill a buret with a solution, the *titrant*, of known concentration of a substance that will react with a solute, the *analyte*, in the sample.
4. Slowly add titrant to the sample, and monitor the change in pH. The volume of titrant needed to completely consume the analyte is indicated by either a change in indicator color or a rapid change in the pH meter reading.

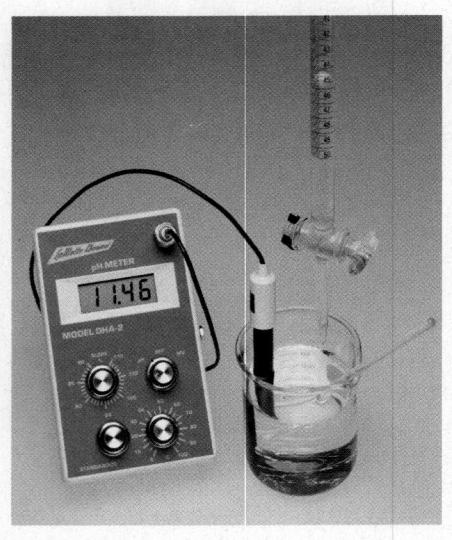

FIGURE 17.16 A digital pH meter is used to measure pH during a titration.

We use the recorded titrant volume to calculate the concentration of analyte in the sample.

The neutralization titrations in Chapter 4 involved titrating strong acids with strong bases and vice versa. Here we look at another type, called an *alkalinity titration*. The term *alkalinity* is used to indicate the buffer capacity of a solution or sample of natural water against additions of acid. Although the buffers found in natural waters have more species involved than we included in our model of river water, the same carbonic acid–hydrogen carbonate buffer system that we have discussed several times is usually the most important one.

An alkalinity titration is more complex than a strong acid/strong base titration in that an alkalinity titration may have two equivalence points as we titrate first carbonate ions and then bicarbonate ions. Let's work our way up to dealing with the complexity of an alkalinity titration by starting with titrations of artificial samples containing known concentrations of weak and strong monoprotic acids and *monobasic* bases. (A monobasic base accepts one hydrogen ion per molecule.) In these examples we monitor the change in pH with addition of titrant using a pH electrode.

CONNECTION When we discussed titrations in Chapter 4, we introduced the equivalence point as that point in a titration when just enough standard solution has been added to completely react with all the solute.

In the first example, let's compare the titration curves of two 20.0 mL samples. One contains 0.100 M NaOH; the other 0.100 M NH$_3$. Both are titrated with 0.100 M HCl. The two titration curves are shown in Figure 17.17. The initial pH of the NaOH solution is higher than that of the NH$_3$ solution because NaOH is a strong base.

In the titration of the strong base NaOH with the strong acid HCl, the pH of the sample as acid is added does not change much until it is close to the equivalence point. As the equivalence point is approached, the principal ions present in the reaction mixture are Na$^+$(aq), Cl$^-$(aq), and OH$^-$(aq). Sample pH is determined only by the [OH$^-$] still present. When enough acid has been added to completely consume all the OH$^-$ ions in the sample, the equivalence point is reached. The solution now consists of water and NaCl. The ions in solution (Na$^+$ and Cl$^-$) do not hydrolyze and do not influence pH; therefore, at the equivalence point, pH = 7.00.

The pH of the weak base NH$_3$ changes abruptly with the first few drops of added acid, but then the changes become smaller and the titration curve levels out. In this nearly flat region additions of acidic titrant are consumed as NH$_3$ combines with H$^+$ ions to form NH$_4$$^+$ ions. In this region of the titration curve the sample acts like a pH buffer (a solution of a weak base and its conjugate acid), and as long as there are significant concentrations of NH$_3$ and NH$_4$$^+$ in the sample, the changes in pH with added titrant are small.

When enough titrant has been added to consume all the base in either of the two samples—in other words, at the equivalence point of each titration—pH drops sharply. At the equivalence point, the same volume of acid has been added to both samples, because the volumes of the two samples and their concentrations of base are the same. Whether the base is strong or weak does not affect the volume of titrant needed to reach the equivalence point: only the number of moles of base present determines the number of moles of acid required to neutralize it. Beyond the equivalence point, the curves are identical because the pH is determined only by the amount of HCl added and the total volume.

The pH values at the equivalence points of the two curves, however, are different: 7.00 (neutral) in the strong acid–strong base titration, but 5.27 (slightly acidic) in the strong acid–weak base titration. The neutralization of NH$_3$ with HCl does not produce a neutral solution because the product of the titration reaction is NH$_4$Cl:

$$\text{HCl}(aq) + \text{NH}_3(aq) \rightarrow \text{NH}_4^+(aq) + \text{Cl}^-(aq)$$

and, as we saw in Sample Exercise 17.8, ammonium ions are weakly acidic:

$$\text{NH}_4^+(aq) + \text{H}_2\text{O}(\ell) \rightleftharpoons \text{NH}_3(aq) + \text{H}_3\text{O}^+(aq)$$

As a general rule, the pH at the equivalence point in a titration of a weak base with a strong acid is less than 7 because the cation of the salt produced in the neutralization reaction hydrolyzes to produce an acidic solution.

One other important point in the NH$_3$ titration curve lies halfway to the equivalence point. At this *midpoint*, half of the NH$_3$ initially in the sample has been converted into NH$_4$$^+$ ions. Therefore, [NH$_3$] = [NH$_4$$^+$]. Because we know the relative concentrations of an acid and its conjugate base, we can use the Henderson–Hasselbalch equation to determine its pH. Because [NH$_3$] = [NH$_4$$^+$], the logarithmic term is zero and pH = pK_a. The K_a of NH$_4$$^+$ is 5.68×10^{-10} (see Sample Exercise 17.8), so

$$\text{pH} = -\log(5.68 \times 10^{-10}) = 9.25$$

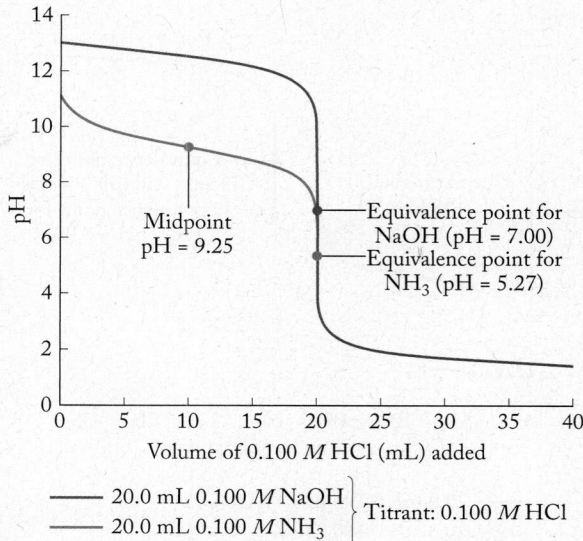

— 20.0 mL 0.100 M NaOH
— 20.0 mL 0.100 M NH$_3$ } Titrant: 0.100 M HCl

FIGURE 17.17 The red curve describes a titration of 20.0 mL of 0.100 M NH$_3$ with 0.100 M HCl titrant. The blue curve describes a titration of 20.0 mL of 0.100 M NaOH with 0.100 M HCl titrant. Before the equivalence point, the pH of the weak base (NH$_3$) solution is always lower than the pH of the strong base (NaOH) solution. Once the equivalence point is reached, the two solutions have the same pH as additional titrant is added.

▶❚❚ **CHEMTOUR** Acid/Base Titrations

▶❚❚ **CHEMTOUR** Titrations of Weak Acids

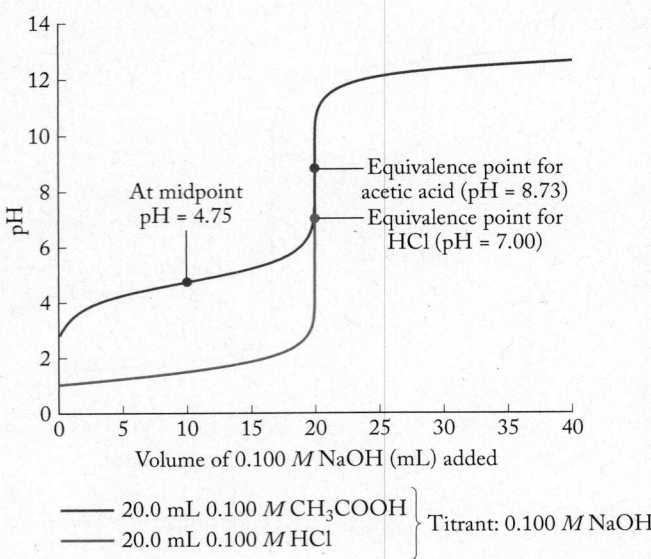

20.0 mL 0.100 M CH$_3$COOH
20.0 mL 0.100 M HCl
} Titrant: 0.100 M NaOH

FIGURE 17.18 The blue curve describes a titration of 20.0 mL of 0.100 M acetic acid (CH$_3$COOH) with 0.100 M NaOH titrant. The red curve describes a titration of 20.0 mL of 0.100 M HCl with 0.100 M NaOH titrant. Before the equivalence point, the pH of the weak acid (acetic acid) solution is always higher than the pH of the strong acid (HCl) solution. After the equivalence point is reached, the two curves overlap.

Just as the concentration of basic compounds in aqueous solution can be determined by titration with known quantities of a strong acid, the concentration of acidic compounds in aqueous solution can be determined by titration with known quantities of a strong base. Figure 17.18 illustrates two such titrations: 20.0 mL of 0.100 M HCl and 20.0 mL of 0.100 M acetic acid (CH$_3$COOH), each titrated with 0.100 M NaOH. The two titration curves differ until the equivalence points are reached, and the pH values are consistently higher for the weak acid. The curves overlap beyond the equivalence points, where pH is controlled only by the increasing concentration of NaOH titrant. The titration curve for acetic acid prior to the equivalence point is nearly flat for the same reason the ammonia curve is nearly flat in Figure 17.17: the partially titrated sample acts like a pH buffer. In this case the buffer consists of a weak acid—the acetic acid in the original sample—and its conjugate base, the acetate ions produced as a result of the titration reaction.

The pH at the equivalence point in the acetic acid titration is 8.73. This value is well above 7 even though it is the point in the titration at which just enough NaOH has been added to exactly neutralize all the acetic acid in the sample. The product of the titration reaction

$$CH_3COOH(aq) + NaOH(aq) \rightarrow CH_3COONa(aq) + H_2O(\ell)$$

is a solution of sodium acetate (CH$_3$COONa). Aqueous solutions of sodium acetate are basic because the acetate ion hydrolyzes:

$$CH_3COO^-(aq) + H_2O(\ell) \rightleftharpoons CH_3COOH(aq) + OH^-(aq)$$

As a general rule, the pH at the equivalence point in the titration of a weak acid with a strong base is greater than 7 because the hydrolysis of the anion of the salt produced in the neutralization produces a basic solution.

CONCEPT TEST

What is the pH at the midpoint in the titration with a strong base of an aqueous sample that contains an unknown concentration of benzoic acid?

CONCEPT TEST

The titration of a weak acid with a weak base is usually not done. Why?

To review the acid–base titration calculation described in Chapter 4, suppose we titrate 10.00 mL of vinegar and find that it requires 16.24 mL of 0.1050 M NaOH to reach the equivalence point. To determine the concentration of acetic acid in the vinegar, we start with the stoichiometry of the balanced equation for the titration reaction:

$$CH_3COOH(aq) + NaOH(aq) \rightarrow CH_3COONa(aq) + H_2O(\ell)$$

One mole of CH$_3$COOH is consumed per mole of NaOH added. Because we are working with volumes in milliliters, a more useful quantity is the millimole (mmol), which is 10^{-3} mole. The definition of molarity (M) is moles of solute per liter of solution, and we can convert moles to millimoles and liters to milliliters by dividing both by 1000:

$$M = \frac{\text{mol}}{\text{L}} = \frac{\text{mol}/1000}{\text{L}/1000} = \frac{\text{mmol}}{\text{mL}}$$

so moles per liter (mol/L) is equivalent to millimoles per milliliter (mmol/mL). The number of millimoles of a solute in solution is equal to the volume of the solution in milliliters times the molarity of the solute. Thus we can write

$$(V_{CH_3COOH})(M_{CH_3COOH}) = (V_{NaOH})(M_{NaOH}) \qquad (17.29)$$

$$M_{CH_3COOH} = \frac{(V_{NaOH})(M_{NaOH})}{(V_{CH_3COOH})}$$

and solve for the concentration of acetic acid by inserting the experimental data:

$$M_{CH_3COOH} = \frac{(16.24~\text{mL})(0.1050~M)}{(10.00~\text{mL})} = 0.1705~M$$

We can write an equation for calculating the concentration of any acidic or basic solute from the results of a titration by using a general version of Equation 17.29:

$$V_A M_A = \frac{n_A}{n_B} V_B M_B \qquad (17.30)$$

in which V_A and M_A represent the volume and molarity of the acid, V_B and M_B represent the volume and molarity of the base, and n_A/n_B is the ratio of the moles of acid to moles of base in the balanced chemical equation describing the reaction. For example, in a titration of a solution of sulfuric acid with NaOH,

$$H_2SO_4(aq) + 2\,NaOH(aq) \rightarrow Na_2SO_4(aq) + 2\,H_2O(\ell)$$

the number of moles of NaOH consumed is twice the number of moles of H_2SO_4 consumed. Therefore,

$$\frac{n_A}{n_B} = \frac{1}{2}$$

and Equation 17.30 becomes

$$(V_{H_2SO_4})(M_{H_2SO_4}) = \frac{1}{2}(V_{NaOH})(M_{NaOH})$$

CONCEPT TEST ···

What is the n_A/n_B ratio when a sample containing an unknown concentration of $Ca(OH)_2$ is titrated with 0.0100 M H_2SO_4?

···

Alkalinity Titrations

Let's return to the application with which we began this section: determination of the alkalinity of a sample of natural water, by which we mean the capacity of the water to neutralize additions of acid. We start with a known volume of the sample and slowly add a strongly acidic titrant from a buret, monitoring the change in pH produced by each drop. If carbonate is present in the sample, the first additions of titrant convert carbonate into bicarbonate:

$$CO_3{}^{2-}(aq) + H^+(aq) \rightarrow HCO_3{}^-(aq)$$

This reaction is the first stage of the alkalinity titration. In the second stage, the bicarbonate formed in the first stage plus any bicarbonate present in the original sample reacts with additional titrant,

$$HCO_3^-(aq) + H^+(aq) \rightarrow H_2CO_3(aq)$$

to form carbonic acid. If more carbonic acid is produced than is soluble (remember, carbonic acid is a solution of CO_2 in water), the carbonic acid leaves the solution in the form of bubbles of carbon dioxide:

$$H_2CO_3(aq) \rightarrow H_2O(\ell) + CO_2(g)$$

In the first stage of the alkalinity titration, the titration curve has a region in which added acid has little effect on pH. This is a buffering region where HCO_3^- and CO_3^{2-} function as a weak acid–conjugate base pair. When the CO_3^{2-} in the sample is completely consumed, the pH drops sharply, producing the first equivalence point. During the second stage of the titration the conversion of HCO_3^- ions to H_2CO_3 produces a second pH buffer and a plateau of nearly constant pH before the HCO_3^- ions are completely consumed and pH drops sharply for a second time, creating a second equivalence point.

Note that the initial pH of the sample in Figure 17.19 is slightly above 10, which is quite basic and above the pH range tolerated by many species of aquatic life. Such highly basic water may be found in arid regions such as the U.S. Southwest, where rocks containing $CaCO_3$ and other basic compounds are in contact with water. The pH of seawater is about 8.2, which is near the first equivalence point on the alkalinity titration curve in Figure 17.19. This means that the dominant carbonate species in seawater is actually bicarbonate. For this reason, alkalinity titration curves for seawater (and most freshwater samples) have only one equivalence point, coinciding with the pH of the second equivalence point in Figure 17.19.

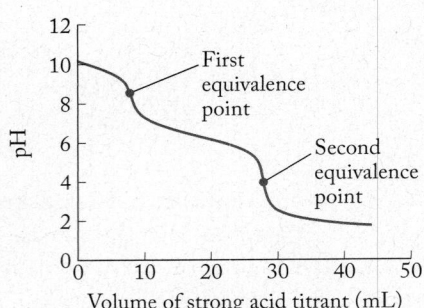

FIGURE 17.19 The titration curve for an alkalinity titration has two equivalence points. The first marks the complete conversion of carbonate into bicarbonate, and the second marks the conversion of bicarbonate into carbonic acid.

CONCEPT TEST

In the alkalinity titration of a sample that initially contains both CO_3^{2-} and HCO_3^-, the volume of titrant required to reach the first equivalence point is less than that required to titrate from the first equivalence point to the second. Why?

We can detect both equivalence points in an alkalinity titration with the appropriate color indicators. Phenol red would not be a good choice because it changes color between pH 6.8 and 8.4. This range is just below the pH of the first equivalence point and well above the pH of the second equivalence point. To detect the first equivalence point, we need an indicator with a pK_a near the pH of the solution at the first equivalence point (8.5). Looking at Figure 17.15, we see that one candidate is phenolphthalein ($pK_a = 8.8$), which is pink in its basic form and colorless at low pH.

To detect the second equivalence point, we could add bromcresol green ($pK_a = 4.6$) after the first equivalence point has been reached. We would not add it before then because its blue-green color in basic solutions would obscure the pink-to-colorless transition of phenolphthalein. We do not need to be concerned about the phenolphthalein obscuring the bromcresol green color change because phenolphthalein is colorless in acidic solutions.

The Chemistry of Two Strong Acids: Sulfuric and Nitric Acids

Sulfuric acid (H_2SO_4) and nitric acid (HNO_3) are among the most widely used industrial chemicals in the world (Figure 17.20). Sulfuric acid ranks first, with a worldwide production of more than 100 million metric tons each year, about one-third of that coming from North America (U.S. and Canada). Nitric acid production ranks 10th on the North American industrial chemicals list. About 70% of the H_2SO_4 and 75% of the HNO_3 produced in the United States is used to make fertilizer. The rest is used in a variety of chemical manufacturing processes, including the preparation of synthetic fibers described in Chapter 13.

Pure sulfuric acid is a dense, colorless, oily liquid. When heated, it fumes as it partially decomposes into H_2O and SO_3. The residual solution is 98.3% H_2SO_4 and 1.7% H_2O. This solution, which is 18 M H_2SO_4, is the liquid sold as concentrated sulfuric acid. It is very hygroscopic (that is, it absorbs water) and is used as a drying agent and to remove water from many compounds. It can dehydrate sugar, turning the carbohydrate into carbon. Sulfuric acid dissolves in water in a process so exothermic that the solution may boil, which is why concentrated sulfuric acid must be diluted by slowly adding it to cold water. *Never* add water to concentrated sulfuric acid.

The synthesis of sulfuric acid starts with the combustion of sulfur to sulfur dioxide followed by the oxidation of SO_2 to SO_3:

$$S_8(s) + 8\,O_2(g) \rightarrow 8\,SO_2(g)$$
$$+ \quad 8\,SO_2(g) + 4\,O_2(g) \rightleftharpoons 8\,SO_3(g)$$
$$\overline{\text{Overall} \qquad S_8(s) + 12\,O_2(g) \rightleftharpoons 8\,SO_3(g)}$$

Both reactions are equilibrium processes, but the equilibrium constant for the formation of SO_2 is large and the reaction essentially goes to completion. The reaction is exothermic but slow at room temperature. Higher temperatures speed the rate of the reaction but decrease the equilibrium concentration of the product. The yield is improved by (1) increasing the pressure of the reactants, (2) using an excess of O_2, and (3) harvesting SO_3 during the reaction. Vanadium(V) oxide, V_2O_5, is used as a catalyst, allowing the reaction to proceed at an acceptable rate at moderate temperatures.

Sulfuric acid is produced by reacting sulfur trioxide with water:

$$SO_3(g) + H_2O(\ell) \rightarrow H_2SO_4(\ell)$$

Note that the steps in the production of sulfuric acid are the same as those that lead to the formation of acid rain in the environment.

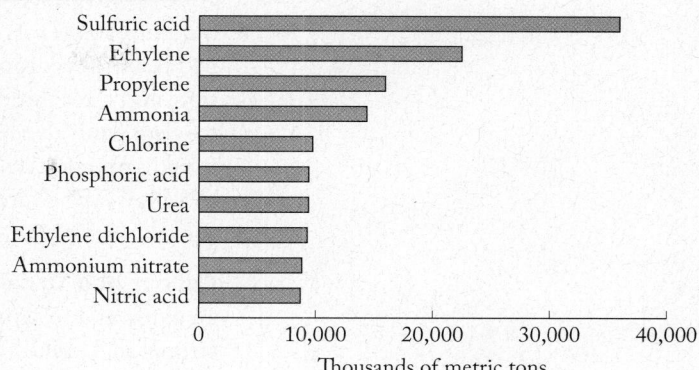

FIGURE 17.20 The top 10 industrial chemicals produced in the United States and Canada in 2008 included sulfuric acid (1st) and nitric acid (10th). (*Source: Chemical and Engineering News*, July 6, 2009, pp. 53, 56.)

The production of nitric acid is linked to the production of ammonia because NH_3 is a reactant in the synthesis of HNO_3. The controlled, selective oxidation of ammonia to NO and subsequent conversion into nitric acid is known as the Ostwald process. Developing it earned Wilhelm Ostwald (1853–1932) the Nobel Prize in Chemistry in 1909 (Figure 17.21). The three steps in the process are

$$(1) \qquad 4\,NH_3(g) + 5\,O_2(g) \rightarrow 4\,NO(g) + 6\,H_2O(g)$$

$$(2) \qquad 2\,NO(g) + O_2(g) \rightarrow 2\,NO_2(g)$$

$$(3) \qquad 3\,NO_2(g) + H_2O(\ell) \rightarrow 2\,HNO_3(\ell) + NO(g)$$

Note that the oxidation number of nitrogen increases from -3 (in NH_3) to $+2$ (in NO) to $+4$ (in NO_2) to $+5$ (in HNO_3) during the process. A catalyst composed of platinum or platinum and rhodium and a reaction temperature of 850°C are needed to achieve a rapid-rate conversion of ammonia into nitric acid.

FIGURE 17.21 Wilhelm Ostwald was a professor of physical chemistry at Leipzig University in Germany from 1887 until 1906. During that time he mentored several brilliant students. Among them were Jacobus Henricus van't Hoff, who won the Nobel Prize in Chemistry in 1901, and Svante August Arrhenius, who won the Nobel Prize in Chemistry in 1903.

SAMPLE EXERCISE 17.15 Interpreting Results of an Alkalinity Titration

Suppose 50.00 mL of pH 10.0 water from a hot spring is titrated with 0.02075 M HCl. A few drops of phenolphthalein are added at the beginning of the titration, and the solution turns pink. It takes 11.21 mL of titrant to reach the pink-to-clear equivalence point. Then a few drops of bromcresol green are added, and it takes an additional 32.28 mL of titrant before the blue-green color changes to yellow. What are the initial concentrations of carbonate and bicarbonate in the sample?

Collect and Organize We are asked to determine the concentrations of two analytes in one sample from the results of a single titration with a monoprotic strong acid, HCl. These determinations are based on the volumes of titrant needed to reach two equivalence points: an initial one at which any CO_3^{2-} in the sample has been converted to HCO_3^-:

$$(1) \quad H^+(aq) + CO_3^{2-}(aq) \rightarrow HCO_3^-(aq)$$

and a second one at which HCO_3^- has been converted to H_2CO_3:

$$(2) \quad H^+(aq) + HCO_3^-(aq) \rightarrow H_2CO_3(aq)$$

Analyze The HCO_3^- titrated in reaction 2 includes any HCO_3^- in the original sample plus all the HCO_3^- produced in reaction 1. If there were no HCO_3^- present initially, the volume of titrant needed to react with the HCO_3^- in reaction 2 would be exactly the same as the volume needed to react with CO_3^{2-}, 11.21 mL, in reaction 1. However, the volume of titrant required to reach the second equivalence point is much greater: 32.28 mL. The difference between these two volumes, (32.28 − 11.21) = 21.07 mL, is the volume of acid required to react with any HCO_3^- that was present in the original sample. Because this difference is nearly twice the volume of titrant required to react with the CO_3^{2-} in the sample, the value of $[HCO_3^-]$ that we calculate for the original sample should be nearly twice its $[CO_3^{2-}]$ value.

The stoichiometry of the reaction tells us that the titrant and carbonate react in a 1:1 mole ratio, so, at the first equivalence point,

$$\text{mol HCl added} = \text{mol } CO_3^{2-} \text{ consumed}$$

This means that the ratio n_A/n_B for this reaction is 1:1. The same is true for the conversion of HCO_3^- into H_2CO_3:

$$\text{mol HCl added} = \text{mol } HCO_3^- \text{ consumed}$$

Solve We use Equation 17.30 for both calculations:

$$M_{CO_3^{2-}} = \frac{(11.21 \text{ mL})(0.02075 \text{ M})}{50.00 \text{ mL}} = 4.652 \times 10^{-3} \text{ M}$$

$$M_{HCO_3^-} = \frac{(21.07 \text{ mL})(0.02075 \text{ M})}{50.00 \text{ mL}} = 8.744 \times 10^{-3} \text{ M}$$

Think about It The titration results confirm that the bicarbonate concentration in the original sample was almost twice the carbonate concentration. To check our two values from the calculations, we can insert the results into the

Henderson–Hasselbalch equation and calculate what the initial pH of the sample should have been. To do that, we need the pK_{a_2} value for carbonic acid from Appendix 5: 10.33.

$$pH = pK_a + \log \frac{[base]}{[acid]}$$

$$= 10.33 + \log \frac{4.652 \times 10^{-3} \, M}{8.744 \times 10^{-3} \, M}$$

$$= 10.06$$

This calculated pH value agrees with the pH given in the problem statement.

Practice Exercise Your job is to determine the concentration of ammonia in a commercial window cleaner. In the titration of a 25.00 mL sample, the equivalence point is reached after 10.49 mL of 0.155 M HCl has been added.

    a. What is the concentration of ammonia in the solution?
    b. Which pH indicator in Figure 17.15 would be suitable for this
       titration?

SUMMARY

Section 17.1 The **Brønsted–Lowry model** of acids and bases defines acids as H^+ ion donors and bases as H^+ ion acceptors. Strong acids include binary acids HX, where X is 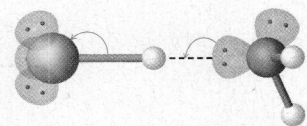 any group 17 element other than fluorine, and some oxoacids $H_m XO_n$, where X is a nonmetal. All strong acids are completely ionized in water. The H^+ ions they release combine with water molecules to form hydronium (H_3O^+) ions. Most acids are weak acids, which means they ionize only partially in water. When weak acid HA ionizes, it forms its **conjugate base**, A^-. When base B acquires a H^+ ion, it forms its **conjugate acid**, HB^+. Strong bases are those group 1 and 2 hydroxides that are soluble in water and dissociate completely when they dissolve. The strongest acid in water is the H_3O^+ ion; the strongest base is the OH^- ion.

Section 17.2 Water is an amphiprotic substance in that it is capable of behaving both as an acid and as a base. This behavior is evident in the **autoionization** of water in which strong hydrogen bonding between water molecules results in formation of a hydronium ion and a hydroxide ion:

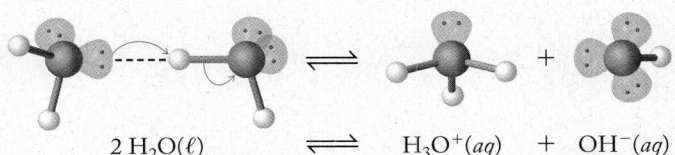

$$2 \, H_2O(\ell) \quad \rightleftharpoons \quad H_3O^+(aq) \; + \; OH^-(aq)$$

In a neutral solution, $[H_3O^+] = [OH^-] = 1.00 \times 10^{-7}$. The **pH** scale is a logarithmic scale for expressing the acidic or basic strength of solutions. Acidic solutions have pH values less than 7; basic solutions have pH values greater than 7. Because pH is the negative logarithm of H^+ concentration, the higher the pH, the lower the H^+ concentration. An increase in one pH unit represents a decrease in $[H^+]$ to 1/10 of its initial value.

Section 17.3 The **degree of ionization** of a weak acid HA is the ratio of $[H^+]$ to the initial (total) acid concentration and is usually expressed as a percentage.

Section 17.4 Weak **polyprotic acids** can undergo more than one acid-ionization reaction, but the first one is the one that usually controls pH.

Section 17.5 The strength of an oxoacid is related to the stability of the anion formed when the acid ionizes. This stability increases as the electronegativity of the central atom increases and as the number of oxygen atoms bonded to the central atom increases.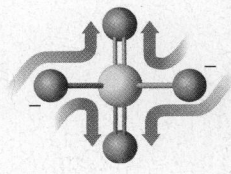

Section 17.6 A salt solution is acidic if the cation in the salt is the conjugate acid of a weak base and the anion is the conjugate base of a strong acid. A salt solution is basic if the anion in the salt is the conjugate base of a weak acid and the cation is the conjugate acid of a strong base.

Section 17.7 Adding the salt of a weak acid HA to a solution of the acid provides a second source of the conjugate base A^-. As predicted by Le Châtelier's principle, the added A^- inhibits the acid-ionization reaction, causing pH to rise. Adding a salt of a weak base B to a solution of the base provides a second source of its conjugate acid HB^+, which lowers pH. These shifts are examples of the **common-ion effect**.

Section 17.8 A **pH buffer** is a solution that contains either a weak acid and a salt of its conjugate base or a weak base and a salt of its conjugate acid. Buffer solutions resist pH change when acids or bases are added. Additions of acid are neutralized by the basic component of a buffer; additions of base are neutralized by the acidic component. The result is a change in the concentration ratio in the Henderson–Hasselbalch equation, but the changes in the logarithm of the ratio, and the impact on pH, tend to be small as long as neither component of the buffer is completely consumed.

Section 17.9 The solubility of slightly soluble ionic compounds is described by their K_{sp} or **solubility product**, which is the value of the equilibrium constant for their dissolution.

Section 17.10 A **pH indicator** is a weak acid or base that has a color that is different from that of its conjugate base or acid. Indicators can be used to determine the pH of a solution that is within 1.0 pH unit of the pK_a or pK_b of the indicator. These indicators are also used to detect the equivalence points in pH titrations, which are used to determine the concentration of an acid or a base in an aqueous sample.

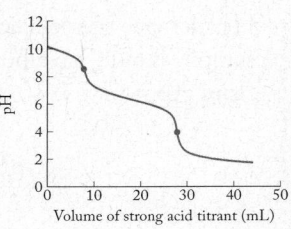

PROBLEM-SOLVING SUMMARY

TYPE OF PROBLEM	CONCEPTS AND EQUATIONS	SAMPLE EXERCISES
Identifying conjugate acid–base pairs	The formula of the base in a conjugate pair is the formula of the acid less one H^+ ion.	17.1
Interconverting $[H^+]$, $[OH^-]$, pH, and pOH	Use the following: $$pH = -\log[H^+] \qquad (17.11)$$ $$pOH = -\log[OH^-] \qquad (17.12)$$ $$pK_w = pH + pOH = 14.00 \qquad (17.13)$$	17.2, 17.3
Converting pH into K_a and calculating percent ionization of a weak acid HA	Use the following: $$[H^+] = \text{antilog}(-pH)$$ $$K_a = \frac{[A^-][H^+]}{[HA]} \qquad (17.15)$$ $$\text{Percent ionization} = \frac{[H^+]_{equilibrium}}{[HA]_{initial}} \times 100\% \qquad (17.16)$$	17.4
Calculating the pH of a solution of a weak base B	Set up an ICE table based on the equilibrium $$B(aq) + H_2O(\ell) \rightleftharpoons HB^+(aq) + OH^-(aq)$$ Let $x = [OH^-] = [HB^+]$ at equilibrium. Calculate x using $$K_b = \frac{x^2}{[B] - x}$$ Then use $$pOH = -\log[OH^-]$$ $$pH = 14.00 - pOH$$	17.5
Calculating the pH of a solution of a weak diprotic acid	Set up an ICE table based on the K_{a_1} equilibrium $$H_2A(aq) \rightleftharpoons H^+(aq) + HA^-(aq)$$ Let $x = [H^+] = [HA^-]$ at equilibrium. Calculate x using $$K_{a_1} = \frac{x^2}{[H_2A] - x}$$ Then calculate $pH = -\log[H^+]$.	17.6
Distinguishing acidic, basic, and neutral salts	The cations in acidic salts are the conjugate acids of weak bases. The anions in basic salts are the conjugate bases of weak acids.	17.7

TYPE OF PROBLEM	CONCEPTS AND EQUATIONS	SAMPLE EXERCISES
Calculating the pH of a solution of an acidic salt	Assume the salt completely dissociates into BH^+ and X^-. Set up an ICE table for the equilibrium $$BH^+(aq) \rightleftharpoons B(aq) + H^+(aq)$$ Let $x = [H^+] = [B]$ at equilibrium. Calculate x using $$K_a = \frac{K_w}{K_b} = \frac{x^2}{[BH^+] - x}$$ Then calculate pH.	17.8
Calculating pH of a solution of a weak base and its conjugate acid (or a weak acid and its conjugate base)	Use the relation $$pH = pK_a + \log \frac{[base]}{[acid]} \qquad (17.25)$$	17.9, 17.10
Calculating molar solubility (S) of M_mZ_z from K_{sp}	Use the relation $$K_{sp} = (m^m z^z)S^{(m+z)} \qquad (17.28)$$	17.11
Calculating the effect of pH on the molar solubility (S) of MZ	If Z^- is the conjugate base of a weak acid, calculate the fraction of Z^- that remains as the free ion. Use this fraction as the coefficient of S in the K_{sp} expression.	17.12
Determining if a precipitate forms when solutions are mixed and which precipitate forms if more than one is possible	Compare the ion product Q to K_{sp} to determine if a precipitate will form; use K_{sp} expressions to calculate maximum concentrations of one ion in solution that will not cause another ion to precipitate.	17.13, 17.14
Interpreting results of an alkalinity titration	Use the relation $$V_A M_A = \frac{n_A}{n_B} V_B M_B \qquad (17.30)$$ where V_A and M_A are the volume and molarity of the acid, V_B and M_B are the volume and molarity of the base, and n_A and n_B are the coefficients of the acid and base, respectively, in the balanced equation.	17.15

VISUAL PROBLEMS

(Answers to boldface end-of-chapter questions and problems are in the back of the book.)

17.1. Which of the lines in Figure P17.1 best represents the dependence of the degree of ionization of acetic acid on its concentration in aqueous solution?

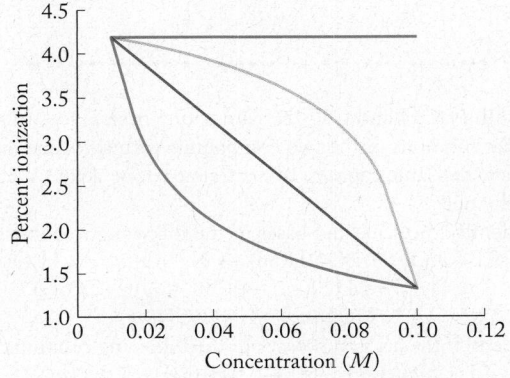

FIGURE P17.1

17.2. The bar graph in Figure P17.2 shows the degree of ionization of $1 \times 10^{-3}\ M$ solutions of three hypohalous acids: HOCl, HOBr, and HOI. Which bar is the one for HOI?

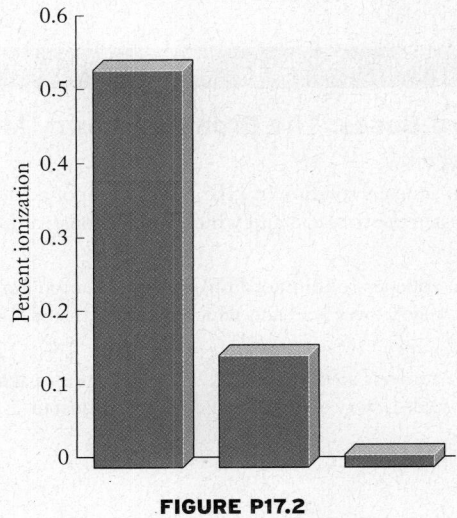

FIGURE P17.2

17.3. The graph in Figure P17.3 shows the titration curves of a 1 *M* solution of a weak acid with a strong base and a 1 *M* solution of a strong acid with the same base. Which curve is which?

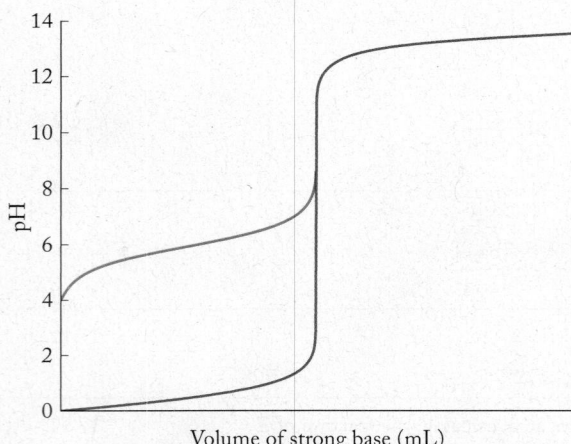

FIGURE P17.3

17.4. Estimate to within one pH unit the pH of a 0.5 *M* solution of the sodium salt of the weak acid in Problem 17.3.

17.5. Suppose you have four color indicators to choose from for detecting the equivalence point of the titration reaction represented by the red curve (upper curve on the left side of the plot) in Figure P17.3. The pK_a values of the four indicators are 3.3, 5.0, 7.0, and 9.0. Which indicator would be the best one to choose?

17.6. What is the pK_a value of the weak acid in Figure P17.3?

17.7. One of the titration curves in Figure P17.7 represents the titration of an aqueous sample of Na_2CO_3 with strong acid; the other represents the titration of an aqueous sample of $NaHCO_3$ with the same acid. Which curve is which?

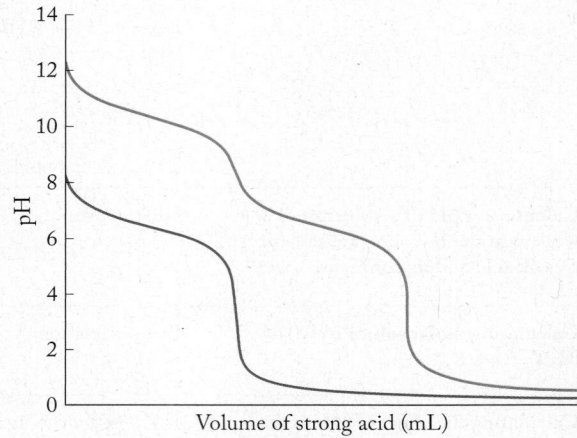

FIGURE P17.7

17.8. Consider the three beakers in Figure P17.8. Each contains a few drops of the color indicator bromthymol blue, which is yellow in acidic solutions and blue in basic solutions. One beaker contains a solution of ammonium chloride, one contains ammonium acetate, and the third contains sodium acetate. Which beaker contains which salt?

FIGURE P17.8

Acids and Bases: The Brønsted-Lowry Model

CONCEPT REVIEW

17.9. In an aqueous solution of HBr, which compound acts as a Brønsted–Lowry acid and which is the Brønsted–Lowry base?

17.10. In an aqueous solution of HNO_3, which compound acts as a Brønsted–Lowry acid and which is the Brønsted–Lowry base?

17.11. In an aqueous solution of NaOH, which species acts as a Brønsted–Lowry acid and which is the Brønsted–Lowry base?

17.12. Both NaOH and $Ca(OH)_2$ are strong bases. Does this mean that solutions of the two compounds with the same molarity have the same capacity to neutralize strong acids? Why or why not?

17.13. Identify the acids and bases in the following reactions:
a. $HNO_3(aq) + NaOH(aq) \rightarrow NaNO_3(aq) + H_2O(\ell)$
b. $CaCO_3(s) + 2\,HCl(aq) \rightarrow CaCl_2(aq) + CO_2(g) + H_2O(\ell)$
c. $NH_3(aq) + HCN(aq) \rightarrow NH_4CN(aq)$

17.14. Identify the acids and bases in the following reactions:
a. $NH_2^-(aq) + H_2O(\ell) \rightleftharpoons NH_3(aq) + OH^-(aq)$
b. $HClO_4(aq) + H_2O(\ell) \rightleftharpoons ClO_4^-(aq) + H_3O^+(aq)$
c. $HSO_4^-(aq) + CO_3^{2-}(aq) \rightleftharpoons SO_4^{2-}(aq) + HCO_3^-(aq)$

17.15. Identify the conjugate base of each of the following compounds: HNO_2, $HClO$, H_3PO_4, and NH_3.

17.16. Identify the conjugate acid of each of the following species: NH_3, ClO_2^-, SO_4^{2-}, and OH^-.

PROBLEMS

17.17. What is the concentration of H^+ ions in a 1.50 M solution of HNO_3?

17.18. What is the concentration of H^+ ions in a solution of hydrochloric acid that was prepared by diluting 20.0 mL of concentrated (11.6 M) HCl to a final volume of 500 mL?

17.19. What is the value of $[OH^-]$ in a 0.0800 M solution of $Sr(OH)_2$?

17.20. A particular drain cleaner contains NaOH. What is the value of $[OH^-]$ in a solution produced when 5.0 g of NaOH dissolves in enough water to make 250 mL of solution?

17.21. Describe how you would prepare 2.50 L of a NaOH solution in which $[OH^-] = 0.70$ M, starting with solid NaOH.

17.22. How many milliliters of a 1.00 M solution of NaOH do you need to prepare 250 mL of a solution in which $[OH^-] = 0.0200$ M?

pH and the Autoionization of Water

CONCEPT REVIEW

17.23. Explain why pH values decrease as acidity increases.

17.24. Solution A is 100 times more acidic than solution B. What is the difference in the pH values of solution A and solution B?

17.25. Under what conditions is the pH of a solution negative?

*__17.26.__ In principle, ethanol (CH_3CH_2OH) can undergo autoionization. Propose an explanation for why the value of the equilibrium constant for the autoionization of ethanol is much less than that of water.

PROBLEMS

17.27. Calculate the pH and pOH of the solutions with the following hydrogen ion or hydroxide ion concentrations. Indicate which solutions are acidic, basic, or neutral.
a. $[H^+] = 3.45 \times 10^{-8}$ M
b. $[H^+] = 2.0 \times 10^{-5}$ M
c. $[H^+] = 7.0 \times 10^{-8}$ M
d. $[OH^-] = 8.56 \times 10^{-4}$ M

17.28. Calculate the pH and pOH of the solutions with the following hydrogen ion or hydroxide ion concentrations. Indicate which solutions are acidic, basic, or neutral.
a. $[OH^-] = 7.69 \times 10^{-3}$ M
b. $[OH^-] = 2.18 \times 10^{-9}$ M
c. $[H^+] = 4.0 \times 10^{-8}$ M
d. $[H^+] = 3.56 \times 10^{-4}$ M

17.29. Calculate the pH of stomach acid in which $[HCl] = 0.155$ M.

17.30. Calculate the pH of a 0.00500 M solution of HNO_3.

17.31. Calculate the pH and pOH of a 0.0450 M solution of NaOH.

17.32. Calculate the pH and pOH of a 0.160 M solution of KOH.

17.33. Calculate the pH of a 1.33 M solution of HNO_3.

*__17.34.__ Calculate the pH of a 6.9×10^{-8} M solution of HBr.

Calculations Involving pH, K_a, and K_b

CONCEPT REVIEW

17.35. One-molar solutions of the following acids are prepared: CH_3COOH, HNO_2, $HClO$, and HCl. Rank them in order of decreasing $[H^+]$.

17.36. On the basis of the following degree-of-ionization data for 0.100 M solutions, select which acid has the smallest K_a.

Acid	Degree of Ionization (%)
C_6H_5COOH	2.5
HF	8.5
HN_3	1.4
CH_3COOH	1.3

17.37. A 1.0 M aqueous solution of $NaNO_2$ is a much better conductor of electricity than is a 1.0 M solution of HNO_2. Explain why.

17.38. Hydrogen chloride and water are molecular compounds, yet a solution of HCl dissolved in H_2O is an excellent conductor of electricity. Explain why.

17.39. Hydrofluoric acid is a weak acid. Write the mass action expression for its acid-ionization reaction.

17.40. In the formula of formic acid, HCOOH, one H atom is ionizable. Write the mass action expression for the acid-ionization equilibrium of formic acid.

*__17.41.__ The K_a values of weak acids depend on the solvent in which they dissolve. For example, the K_a of alanine in aqueous ethanol is less than its K_a in water.
a. In which solvent does alanine ionize to the largest degree?
b. Which is the stronger Brønsted–Lowry base: water or ethanol?

*__17.42.__ The K_a of proline is 2.5×10^{-11} in water, 2.8×10^{-11} in an aqueous solution that is 28% ethanol, and 1.66×10^{-8} in aqueous formaldehyde at 25°C.
a. In which solvent is proline the strongest acid?
b. Rank these compounds on the basis of their strengths as Brønsted–Lowry bases: water, ethanol, and formaldehyde.

17.43. When methylamine, CH_3NH_2, dissolves in water, the resulting solution is slightly basic. Which compound is the Brønsted–Lowry acid and which is the base?

*__17.44.__ When 1,2-diaminoethane, $H_2NCH_2CH_2NH_2$, dissolves in water, the resulting solution is basic. Write the formula of the ionic compound that is formed when hydrochloric acid is added to a solution of 1,2-diaminoethane.

PROBLEMS

17.45. Sore Muscles The muscle fatigue felt during strenuous exercise is caused by the buildup of lactic acid in muscle tissues. In a 1.00 M aqueous solution, 2.94% of lactic acid is ionized. What is the value of its K_a?

17.46. Rancid Butter The odor of spoiled butter is due in part to butanoic acid, which results from the chemical breakdown of butter fat. A 0.100 M solution of butanoic acid is 1.23% ionized. Calculate the value of K_a for butanoic acid.

17.47. At equilibrium, the value of $[H^+]$ in a 0.250 M solution of an unknown acid is 4.07×10^{-3} M. Determine the degree of ionization and the K_a of this acid.

17.48. Nitric acid (HNO_3) is a strong acid that is completely ionized in aqueous solutions of concentrations ranging from 1% to 10% (1.5 M). However, in more concentrated solutions, part of the nitric acid is present as un-ionized molecules of HNO_3. For example, in a 50% solution (7.5 M) at 25°C, only 33% of the molecules of HNO_3 dissociate into H^+ and NO_3^-. What is the K_a value of HNO_3?

17.49. Ant Bites The venom of biting ants contains formic acid, HCOOH, $K_a = 1.8 \times 10^{-4}$ at 25°C. Calculate the pH of a 0.060 M solution of formic acid.

17.50. Gout Uric acid can collect in joints, giving rise to a medical condition known as gout. If the pK_a of uric acid is 3.89, what is the pH of a 0.0150 M solution of uric acid?

17.51. Acid Rain A weather system moving through the American Midwest produced rain with an average pH of 5.02. By the time the system reached New England, the rain it produced had an average pH of 4.66. How much more acidic was the rain falling in New England?

17.52. Acid Rain II A newspaper reported that the "level of acidity" in a sample taken from an extensively studied watershed in New Hampshire in February 1998 was "an astounding 200 times lower than the worst measurement" taken in the preceding 23 years. What is this difference expressed in units of pH?

17.53. The K_b of dimethylamine $[(CH_3)_2NH]$ is 5.9×10^{-4} at 25°C. Calculate the pH of a 1.20×10^{-3} M solution of dimethylamine.

17.54. Painkiller Morphine is an effective painkiller but is also highly addictive. Calculate the pH of a 0.115 M solution of morphine if its $pK_b = 5.79$.

17.55. Painkiller II Codeine is a popular prescription painkiller because it is much less addictive than morphine. Codeine contains a basic nitrogen atom that can be protonated to give the conjugate acid of codeine. Calculate the pH of a 3.42×10^{-4} M solution of codeine if the pK_a of the conjugate acid is 8.21.

17.56. Pyridine (C_5H_5N) is a particularly foul-smelling substance used in manufacturing pesticides and plastic resins. Calculate the pH of a 0.125 M solution of pyridine.

Polyprotic Acids

CONCEPT REVIEW

17.57. Why is the K_{a_2} value of phosphoric acid less than its K_{a_1} value but greater than its K_{a_3} value?

17.58. In calculating the pH of a 1.0 M solution of sulfurous acid, we can ignore the H^+ ions produced by the ionization of the bisulfite ion; however, in calculating the pH of a 1.0 M solution of sulfuric acid, we cannot ignore the H^+ ions produced by the ionization of the bisulfate ion. Why?

PROBLEMS

17.59. What is the pH of a 0.300 M solution of H_2SO_4?

17.60. What is the pH of a 0.150 M solution of sulfurous acid?

17.61. Ascorbic acid (vitamin C) is a diprotic acid. What is the pH of a 0.250 M solution of ascorbic acid?

17.62. The leaves of the rhubarb plant contain high concentrations of diprotic oxalic acid (HOOCCOOH) and must be removed before the stems are used to make rhubarb pie. What is the pH of a 0.0288 M solution of oxalic acid?

17.63. Addiction to Tobacco Nicotine is responsible for the addictive properties of tobacco. What is the pH of a 1.00×10^{-3} M solution of nicotine?

17.64. 1,2-Diaminoethane, $H_2NCH_2CH_2NH_2$, is used extensively in the synthesis of compounds containing transition metals in water. If $pK_{b_1} = 3.29$ and $pK_{b_2} = 6.44$, what is the pH of a 2.50×10^{-4} M solution of 1,2-diaminoethane?

17.65. Malaria Treatment Quinine occurs naturally in the bark of the cinchona tree. For centuries it was the only treatment for malaria. Calculate the pH of a 0.01050 M solution of quinine in water.

17.66. Dozens of pharmaceuticals ranging from Cyclizine for motion sickness to Viagra for impotence are derived from the organic compound piperazine, whose structure is shown in Figure P17.66:

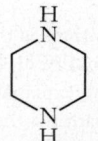

FIGURE P17.66

a. Solutions of piperazine are basic ($K_{b_1} = 5.38 \times 10^{-5}$; $K_{b_2} = 2.15 \times 10^{-9}$). What is the pH of a 0.0133 M solution of piperazine?

*b. Draw the structure of the ionic form of piperazine that would be present in stomach acid (about 0.15 M HCl).

Acid Strength and Molecular Structure

CONCEPT REVIEW

17.67. Explain why the K_{a_1} of H_2SO_4 is much greater than the K_{a_1} of H_2SeO_4.

17.68. Explain why the K_{a_1} of H_2SO_4 is much greater than the K_{a_1} of H_2SO_3.

17.69. Predict which acid in the following pairs of acids is the stronger acid: (a) H_2SO_3 or H_2SeO_3; (b) H_2SeO_4 or H_2SeO_3.

17.70. Predict which acid in the following pairs of acids is the stronger acid: (a) HBrO or HOBrO; (b) HClO or HBrO.

pH of Salt Solutions

CONCEPT REVIEW

*17.71. The pK_a values of the conjugate acids of pyridine derivatives shown in Figure P17.71 increase as more methyl groups are added. Do more methyl groups increase or decrease the strength of the parent pyridine bases?

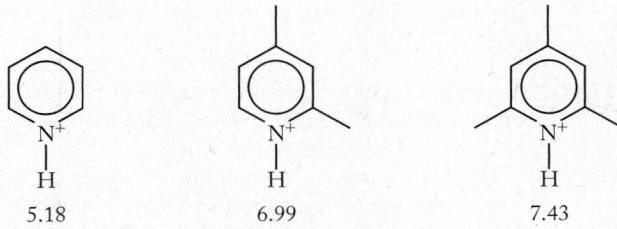

5.18   6.99   7.43

FIGURE P17.71

17.72. Why is it unnecessary to publish tables of K_b values of the conjugate bases of weak acids whose K_a values are known?

17.73. Which of the following salts produces an acidic solution in water: ammonium acetate, ammonium nitrate, or sodium formate?

17.74. Which of the following salts produces a basic solution in water: NaF, KCl, NH_4Cl?

17.75. Neutralizing the Smell of Fish Trimethylamine, $(CH_3)_3N$, $K_b = 6.5 \times 10^{-5}$ at 25°C, is a contributor to the "fishy" odor of not-so-fresh seafood. Some people squeeze fresh lemon juice (which contains a high concentration of citric acid) on cooked fish to reduce the fishy odor. Why is this practice effective?

17.76. Nutritional Value of Beets Beets contain high concentrations of the calcium salt of a dicarboxylic acid with the common name malonic acid and the formula $HOOCCH_2COOH$. Could the presence of the calcium salt of malonic acid affect the pH balance of beets? If so, in which direction? Explain.

PROBLEMS

17.77. If the K_a of the conjugate acid of the artificial sweetener saccharin is 2.1×10^{-11}, what is the pK_b for saccharin?

17.78. If the K_{a_1} value of chromic acid (H_2CrO_4) is 0.16 and its K_{a_2} value is 3.2×10^{-7}, what are the values of K_{b_1} and K_{b_2} of the CrO_4^{2-} anion?

17.79. Dental Health Sodium fluoride is added to many municipal water supplies to reduce tooth decay. Calculate the pH of a 0.00339 M solution of NaF at 25°C.

17.80. Calculate the pH of a 1.25×10^{-2} M solution of the decongestant ephedrine hydrochloride if the pK_b of ephedrine (its conjugate base) is 3.86.

The Common-Ion Effect and pH Buffers

CONCEPT REVIEW

17.81. Why is a solution of sodium acetate and acetic acid a much better pH buffer than is a solution of sodium chloride and hydrochloric acid?

17.82. Why does a solution of a weak base and its conjugate acid act as a better buffer than does a solution of the weak base alone?

PROBLEMS

17.83. Calculate the pH of a buffer that is 0.244 M acetic acid and 0.122 M sodium acetate at 25°C. What is the pH of this mixture at 0°C ($K_a = 1.64 \times 10^{-5}$)?

17.84. Calculate the pH of a buffer that is 0.100 M pyridine and 0.275 M pyridinium chloride at 25°C.

17.85. Calculate the pH and pOH of 500.0 mL of a phosphate buffer that is 0.225 M HPO_4^{2-} and 0.225 M PO_4^{3-} at 25°C.

17.86. Determine the pH and pOH of 0.250 L of a buffer that is 0.0200 M boric acid and 0.0250 M sodium borate at 25°C.

17.87. What is the ratio of acetate ion to acetic acid in a buffer containing these compounds at pH = 3.56?

17.88. What is the ratio of lactic acid to lactate in a solution with pH = 4.00?

17.89. What is the pH at 25°C of a solution that results from mixing together equal volumes of a 0.05 M solution of ammonia and a 0.025 M solution of hydrochloric acid?

17.90. What is the pH at 25°C of a solution that results from mixing together equal volumes of a 0.05 M solution of acetic acid and a 0.025 M solution of sodium hydroxide?

*17.91. How much 10 M HNO_3 must be added to 1.00 L of a buffer that is 0.010 M acetic acid and 0.10 M sodium acetate to reduce the pH to 5.00 at 25°C?

*17.92. How much 6.0 M NaOH must be added to 0.500 L of a buffer that is 0.0200 M acetic acid and 0.0250 M sodium acetate to raise the pH to 5.75 at 25°C?

*17.93. Calculate the pH at 25°C of 1.00 L of a buffer that is 0.120 M HNO_2 and 0.150 M $NaNO_2$ before and after the addition of 1.00 mL of 12.0 M HCl.

*17.94. Calculate the pH at 25°C of 100.0 mL of a buffer that is 0.100 M NH_4Cl and 0.100 M NH_3 before and after the addition of 1.0 mL of 6 M HNO_3.

Solubility Equilibria

CONCEPT REVIEW

17.95. What is the difference between *molar solubility* and *solubility product*?

17.96. Give an example of how the common-ion effect limits the dissolution of a sparingly soluble ionic compound.

17.97. Which of the following cations will precipitate first as a carbonate mineral from an equimolar solution of Mg^{2+}, Ca^{2+}, and Sr^{2+}?

17.98. If the solubility of a compound increases with increasing temperature, does K_{sp} increase or decrease?

17.99. The K_{sp} of strontium sulfate increases from 2.8×10^{-7} at 37°C to 3.8×10^{-7} at 77°C. Is the dissolution of strontium sulfate endothermic or exothermic?

17.100. How will adding concentrated NaOH(*aq*) affect the solubility of an Al(III) salt?

17.101. Chemistry of Tooth Decay Tooth enamel is composed of a mineral known as hydroxyapatite with the formula $Ca_5(PO_4)_3(OH)$. Explain why tooth enamel can be eroded by acidic substances released by bacteria growing in the mouth.

17.102. Fluoride and Dental Hygiene Fluoride ions in drinking water and toothpaste convert hydroxyapatite in tooth enamel into fluorapatite:

$$Ca_5(PO_4)_3(OH)(s) + F^-(aq) \rightleftharpoons Ca_5(PO_4)_3F(s) + OH^-(aq)$$

Why is fluorapatite less susceptible than hydroxyapatite to erosion by acids?

PROBLEMS

17.103. At a particular temperature the value of $[Ba^{2+}]$ in a saturated solution of barium sulfate is 1.04×10^{-5} *M*. Starting with this information, calculate the K_{sp} value of barium sulfate at this temperature.

17.104. Suppose a saturated solution of barium fluoride contains 1.5×10^{-2} *M* F^-. What is the K_{sp} value of BaF_2?

17.105. What are the equilibrium concentrations of Cu^+ and Cl^- in a saturated solution of copper(I) chloride if $K_{sp} = 1.02 \times 10^{-6}$?

17.106. What are the equilibrium concentrations of Pb^{2+} and F^- in a saturated solution of lead fluoride if the K_{sp} value of PbF_2 is 3.2×10^{-8}?

17.107. What is the solubility of calcite ($CaCO_3$) in grams per milliliter at a temperature at which its $K_{sp} = 9.9 \times 10^{-9}$?

17.108. What is the solubility of silver iodide in grams per milliliter at a temperature at which its $K_{sp} = 1.50 \times 10^{-16}$?

17.109. What is the pH at 25°C of a saturated solution of silver hydroxide?

17.110. pH of Milk of Magnesia What is the pH at 25°C of a saturated solution of magnesium hydroxide (the active ingredient in the antacid milk of magnesia)?

17.111. Suppose you have 100 mL of each of the following solutions. In which will the most $CaCO_3$ dissolve? (a) 0.1 *M* NaCl; (b) 0.1 *M* Na_2CO_3; (c) 0.1 *M* NaOH; (d) 0.1 *M* HCl

17.112. In which of the following solutions will CaF_2 be most soluble? (a) 0.010 *M* $Ca(NO_3)_2$; (b) 0.01 *M* NaF; (c) 0.001 *M* NaF; (d) 0.10 *M* $Ca(NO_3)_2$

17.113. Composition of Seawater The average concentration of sulfate in surface seawater is about 0.028 *M*. The average concentration of Sr^{2+} is 9×10^{-5} *M*. If the K_{sp} value of strontium sulfate is 3.4×10^{-7}, is the concentration of strontium in the sea probably controlled by the insolubility of its sulfate salt?

17.114. Fertilizing the Sea to Combat Global Warming Some scientists have proposed adding Fe(III) compounds to large expanses of the open ocean to promote the growth of phytoplankton that would in turn remove CO_2 from the atmosphere through photosynthesis. The average pH of open ocean water is 8.1. What is the maximum value of $[Fe^{3+}]$ in seawater if the K_{sp} value of $Fe(OH)_3$ is 1.1×10^{-36}?

17.115. Will calcium fluoride precipitate when 125 mL of 0.375 *M* $Ca(NO_3)_2$ is added to 245 mL of 0.255 *M* NaF at 25°C?

17.116. Will strontium sulfate precipitate at 25°C when 345 mL of a solution that is 0.0100 *M* in $Sr^{2+}(aq)$ is added to 75 mL of 0.175 *M* K_2SO_4?

17.117. A solution is 0.010 *M* in both Br^- and SO_4^{2-}. A 0.250 *M* solution of lead(II) nitrate is slowly added to it.
   a. Which anion will precipitate first?
   b. What is the concentration in the solution of the first ion when the second one starts to precipitate at 25°C?

*17.118. Solution A is 0.0250 *M* in Ag^+ ion and Pb^{2+} ion. You have access to two other solutions: (B) 0.500 *M* NaCl and (C) 0.500 *M* NaBr.
   a. Which would be the better solution to add to separate lead from silver by precipitation? (The better solution is the one that has less lead remaining in solution when the silver begins to precipitate.)
   b. Using the solution you selected in part a, is the separation of the two ions complete? ("Complete" is defined as the point when less than 0.10% of the silver ion is left in the solution when the lead ion begins to precipitate.)

Indicators and Acid-Base Titrations
CONCEPT REVIEW

17.119. What are the differences between the titration curve of a strong acid titrated with a strong base and that of a weak acid titrated with a strong base?

17.120. Do all titrations of a strong base with a strong acid have the same pH at the equivalence point?

17.121. Do all titrations of a weak acid with a strong base have the same pH at the equivalence point?

17.122. What properties must a compound have to serve as an acid–base indicator?

PROBLEMS

17.123. A 25.0 mL sample of 0.100 *M* acetic acid is titrated with 0.125 *M* NaOH. Calculate the pH at 25°C of the titration mixture after 10.0, 20.0, and 30.0 mL of the base have been added.

17.124. A 25.0 mL sample of a 0.100 *M* solution of aqueous trimethylamine is titrated with a 0.125 *M* solution of HCl. Calculate the pH of the solution after 10.0, 20.0, and 30.0 mL of acid have been added; pK_b of $(CH_3)_3N = 4.19$ at 25°C.

17.125. What is the concentration of ammonia in a solution if 22.35 mL of 0.1145 *M* HCl is needed to titrate a 100.0 mL sample of the solution?

17.126. In an alkalinity titration of a 100.0 mL sample of water from a hot spring, 2.56 mL of a 0.0355 *M* solution of HCl is needed to reach the first equivalence point (pH = 8.3) and another 10.42 mL is needed to reach the second equivalence point (pH = 4.0). If the alkalinity of the spring water is due only to the presence of carbonate and bicarbonate, what are the concentrations of each?

17.127. What volumes of 0.0100 M HCl are required to titrate 250 mL of 0.0100 M Na_2CO_3 to the first equivalence point?

17.128. How much 0.0100 M HCl is required to titrate 250 mL of 0.0100 M Na_2CO_3 and 250 mL of 0.0100 M HCO_3^-?

17.129. In the titration of a solution of a weak monoprotic acid with a 0.1025 M solution of NaOH, the pH halfway to the equivalence point was 4.44. In the titration of a second solution of the same acid, exactly twice as much of a 0.1025 M solution of NaOH was needed to reach the equivalence point. What was the pH halfway to the equivalence point in this titration?

17.130. A 125.0 mg sample of an unknown, monoprotic acid was dissolved in 100.0 mL of distilled water and titrated with a 0.050 M solution of NaOH. The pH of the solution was monitored throughout the titration, and the following data were collected. Determine the K_a of the acid.

Volume of OH⁻ Added (mL)	pH
0	3.09
5	3.65
10	4.10
15	4.50
17	4.55
18	4.71
19	4.94
20	5.11
21	5.37
22	5.93
22.2	6.24
22.6	9.91
22.8	10.2
23	10.4
24	10.8
25	11.0
30	11.5
40	11.8

17.131. Sketch a titration curve for the titration of 50.0 mL of 0.250 M HNO_2 with 1.00 M NaOH. What is the pH at the equivalence point?

17.132. Red cabbage juice is a sensitive acid–base indicator; its colors range from red at acidic pH to yellow in alkaline solutions. What color would red cabbage juice have when 25 mL of a 0.10 M solution of acetic acid is titrated with 0.10 M NaOH to its equivalence point?

17.133. Sketch a titration curve for the titration of the malaria drug quinine if 40.0 mL of a 0.100 M solution of quinine is titrated with a 0.100 M solution of HCl.

17.134. Sketch a titration curve for the titration of 100 mL of 1.25×10^{-2} M ascorbic acid with 1.00×10^{-2} M NaOH. How many equivalence points should the curve have, and what color indicator(s) could be used?

Additional Problems

17.135. Describe the intermolecular forces and changes in bonding that lead to the formation of a basic solution when methylamine (CH_3NH_2) dissolves in water.

17.136. Describe the chemical reactions of sulfur that begin with the burning of high-sulfur fossil fuel and that end with the reaction between acid rain and building exteriors made of marble ($CaCO_3$).

17.137. The value of K_{a_1} of phosphorous acid, H_3PO_3, is nearly the same as the K_{a_1} of phosphoric acid, H_3PO_4.
a. Draw the Lewis structure of phosphorous acid.
b. Identify the ionizable hydrogen atoms in the structure.
c. Explain why the K_{a_1} values of phosphoric and phosphorous acid are similar.

17.138. The value of K_{a_1} of hypophosphorous acid, H_3PO_2, is nearly the same as the K_{a_1} of phosphoric acid, H_3PO_4.
a. Draw a Lewis structure for phosphoric acid that is consistent with this behavior.
b. Identify the ionizable hydrogen atoms in the structure.

*17.139. **pH of Baking Soda** A cook dissolves a teaspoon of baking soda ($NaHCO_3$) in a cup of water, then discovers that the recipe calls for a tablespoon, not a teaspoon. So, the cook adds two more teaspoons of baking soda to make up the difference. Does the additional baking soda change the pH of the solution? Explain why or why not.

*17.140. **Antacid Tablets** Antacids contain a variety of bases such as $NaHCO_3$, $MgCO_3$, $CaCO_3$, and $Mg(OH)_2$. Only $NaHCO_3$ has appreciable solubility in water.
a. Write a net ionic equation for the reaction of each base with aqueous HCl.
b. Explain how insoluble substances can act as effective antacids.

*17.141. **pH of Natural Waters** In a 1985 study of Little Rock Lake in Wisconsin, 400 gallons of 18 M sulfuric acid were added to the lake over six years. The initial pH of the lake was 6.1 and the final pH was 4.7. If none of the acid was consumed in chemical reactions, estimate the volume of the lake.

17.142. **pH of Natural Waters II** Between 1993 and 1995, sodium phosphate was added to Seathwaite Tarn in the English Lake District to increase its pH. Explain why addition of this compound increased pH.

17.143. Acid-Base Properties of Pharmaceuticals Zoloft is a common prescription drug for the treatment of depression. It is sold as a salt of HCl.

a. In the reaction shown in Figure P17.143, which structure is that of the acid salt?

b. When Zoloft dissolves in water, will the resulting solution be acidic or basic?

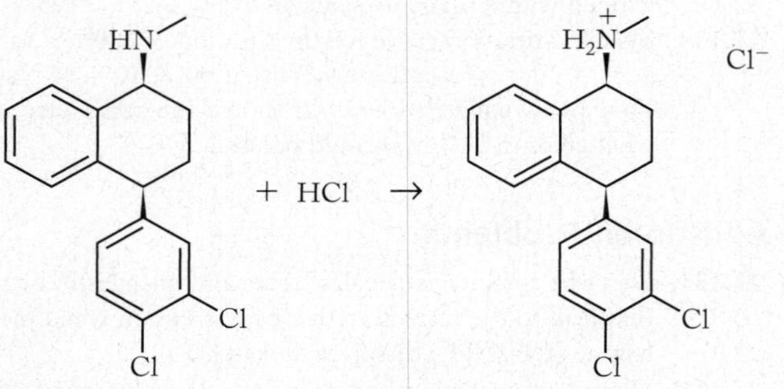

FIGURE P17.143

17.144. Acid-Base Properties of Pharmaceuticals II Prozac is a popular antidepressant drug. Its structure is given in Figure P17.144.

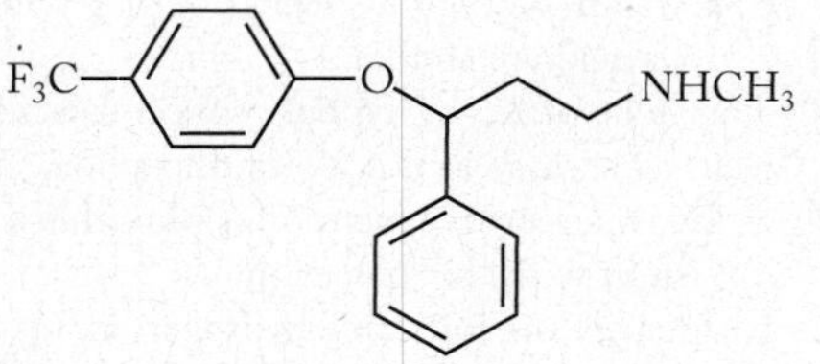

Prozac

FIGURE P17.144

a. Is a solution of Prozac in water likely to be slightly basic or slightly acidic? Explain your answer.

b. Prozac is also sold as a salt of HCl. Which atom, N or O, is most likely to react with HCl?

c. Prozac is sold as a salt of HCl because the solubility of the salt in water is higher than Prozac itself. Why is the salt more soluble?

17.145. Hydrogen fluoride (HF) behaves as a weak acid in aqueous solution. Two equilibria influence which fluorine-containing species are present in solution.

$$HF(aq) + H_2O(\ell) \rightleftharpoons H_3O^+(aq) + F^-(aq) \qquad K_a = 1.1 \times 10^{-3}$$

$$F^-(aq) + HF(g) \rightleftharpoons HF_2^-(aq) \qquad K = 2.6 \times 10^{-1}$$

a. Is fluoride in pH 7.00 drinking water more likely to be present as F^- or HF_2^-?

b. What is the equilibrium constant for this equilibrium?

$$2\,HF(aq) + H_2O(\ell) \rightleftharpoons H_3O^+(aq) + HF_2^-(aq)$$

c. What is the pH and equilibrium concentration of HF_2^- in a 0.150 M solution of HF?

*17.146. Pentafluorocyclopentadiene, which has the structure shown in Figure P17.146, is a strong acid.

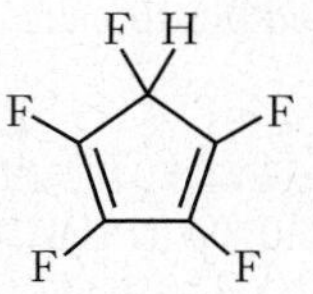

FIGURE P17.146

a. Draw the conjugate base of C_5F_5H.

b. Why is the compound so acidic when most organic acids are weak?

17.147. Naproxen (a.k.a. Aleve) is an anti-inflammatory drug used to reduce pain, fever, inflammation, and stiffness caused by conditions such as osteoarthritis and rheumatoid arthritis. Naproxen is an organic acid; its structure is shown in Figure P17.147. Naproxen has limited solubility in water, so it is sold as its sodium salt.

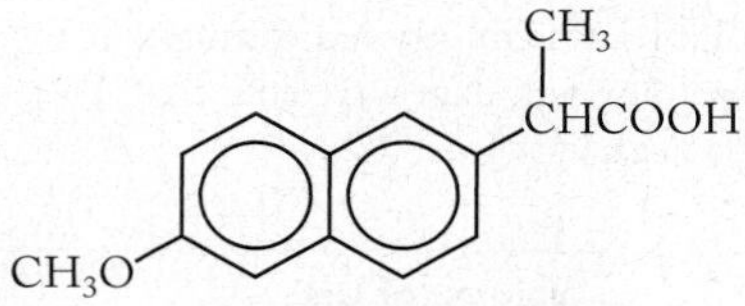

FIGURE P17.147

a. Draw the molecular structure of the sodium salt.

b. Should a solution of the salt be acidic or basic? Explain why.

c. Explain why the salt is more soluble than Naproxen itself.

*17.148. **Greenhouse Gases and Ocean pH** Some climate models predict a decrease in the pH of the oceans of 0.77 pH units because of increases in atmospheric carbon dioxide.

a. Explain, by using the appropriate chemical reactions and equilibria, how an increase in atmospheric CO_2 could produce a decrease in oceanic pH.

b. How much more acidic (in terms of $[H^+]$) would the oceans be if their pH dropped this much?

c. Oceanographers are concerned about the impact of a drop in oceanic pH on the survival of coral reefs. Why?

The Chemistry of Two Strong Acids: Sulfuric and Nitric Acids

17.149. Complete the following chemical equations with the appropriate product(s).

a. $SO_3(g) + H_2O(\ell) \rightarrow$

b. $3\,NO_2(g) + H_2O(\ell) \rightarrow$

c. $4\,NH_3(g) + ? \rightarrow ? + 6\,H_2O(g)$

17.150. In the Ostwald process, which steps should have a higher yield at higher total pressure?

17.151. In the Ostwald process, which steps should have higher yields at higher temperature?

17.152. Given ΔH_f° for $SO_2(g)$ (-296.8 kJ/mol) and $SO_3(g)$ (-395.7 kJ/mol), calculate ΔH_{rxn}° for the conversion of SO_2 to SO_3.

17.153. Write balanced chemical equations that correspond to $\Delta H_{f,SO_2}^{\circ}$ and $\Delta H_{f,SO_3}^{\circ}$. Show how Hess's law can be used to determine ΔH° for the reaction

$$2\,SO_2(g) + O_2(g) \rightleftharpoons 2\,SO_3(g)$$

*17.154. Reaction of sodium nitrate with sodium oxide at high temperature produces Na_3NO_4, which contains the NO_4^{3-} (orthonitrate) anion. However, the corresponding acid, H_3NO_4, is unknown. Would you expect H_3NO_4 to be a stronger or weaker acid than HNO_3?

*17.155. The Henry's law constant for CO_2 dissolved in water is 3.5×10^{-2} M/atm. Do you expect the corresponding constants for SO_3 and NO_2 to be greater than or less than this value? Explain your answer.

17.156. Write a balanced chemical equation to describe the following reactions of sulfuric acid and nitric acid:
a. Nitric acid reacts with ammonia.
b. Sulfuric acid reacts with ammonia.
c. Sulfuric acid dissolves in water.

*17.157. Thiosulfuric acid, $H_2S_2O_3$, can be prepared by the reaction of H_2S with HSO_3Cl:

$$HSO_3Cl(\ell) + H_2S(g) \rightarrow HCl(g) + H_2S_2O_3(\ell)$$

a. Draw a Lewis structure for $H_2S_2O_3$, given that it is isostructural with H_2SO_4.
b. Do you expect $H_2S_2O_3$ to be a stronger or weaker acid than H_2SO_4? Explain your answer.

*17.158. Sulfuric acid reacts with nitric acid as shown below:

$$HNO_3(aq) + 2\,H_2SO_4(aq) \rightarrow NO_2^+(aq) + H_3O^+(aq) + 2\,HSO_4^-(aq)$$

a. Is the reaction a redox process?
b. Identify the acid, base, conjugate acid, and conjugate base in the reaction. (*Hint*: Draw the Lewis structures for each.)

Additional study materials including ChemTours and Diagnostic Quizzes are available at StudySpace at www.wwnorton.com/studyspace.

18

The Colorful Chemistry of Metals

Learning Outcomes

- Describe the similarities and differences between Brønsted-Lowry and Lewis acids and bases

- Describe coordinate bond formation and the general structure of complex ions and coordination compounds

- Use formation constants to calculate the concentrations of free and complexed metal ions in solution

- Explain the acidic pH of solutions of soluble transition metal salts

- Explain the chelate effect and its importance

- Use crystal field theory to explain why many transition metal compounds and solutions are colored

- Describe the factors that lead to high-spin or low-spin magnetic states of complex ions

A LOOK AHEAD

The Company They Keep

Many of the metallic elements in the periodic table are essential to good health. For example, copper, zinc, and cobalt play key roles in protein function; iron is needed to transport oxygen from our lungs to all the cells of our body; and nearly everyone knows the importance of calcium in building strong teeth and bones.

These and other essential metallic elements should be present either in our diets or in the supplements many of us rely on for balanced nutrition. However, the mere presence of these elements is not sufficient—they must be in a form that we can digest. Swallowing an 18 mg steel pellet as if it were an aspirin tablet would not be a good way for a young woman to get her recommended daily allowance of iron. Chewing a gram of calcium metal would be even less pleasant. If we are to benefit from consuming essential metals in food and nutritional supplements, the metals need to be in compounds, not free elements, and these compounds must be absorbable by the body.

All the metallic elements essential to human health occur in nature in ionic compounds, but not all ionic forms are absorbed equally well. For example, most of the iron in fish, poultry, and red meat is readily absorbed because it is present in a form called *heme iron*. However, the iron in plants is mostly nonheme and not as readily absorbed. Eating a meal that includes both meat and vegetables improves the absorption of the nonheme iron in the vegetables, as does consuming foods high in vitamin C. All these dietary factors work together at the molecular level to provide us with the nutrients we need to survive.

Metal ions are essential to good health, but the nonmetal ions and molecules that accompany them are equally important because they make the metals chemically reactive and biologically available. Interactions between metals and these ions and molecules influence the solubility of the metals, which is a key factor that determines their activity in plants and animals. These interactions also influence other properties

The Green Leaves of Summer The pigments in green plants include ▶ chlorophyll *a*, whose molecules each contain a ring of nitrogen atoms surrounding a Mg^{2+} ion.

Lewis base a substance that *donates* a lone pair of electrons in a chemical reaction.

Lewis acid a substance that *accepts* a lone pair of electrons in a chemical reaction.

including the wavelengths of visible light the metals absorb and therefore their color in solution. In this chapter we explore how the chemical environment of metal ions in solids and solutions affects their physical, chemical, and biological properties. We answer such questions as why many, but not all, metal compounds have distinctive colors, and how, through the formation of complex ions with biomolecules, metals play key roles in many biological processes. ■

18.1 Lewis Acids and Bases

In this chapter we focus on the interactions between metal ions and the other ions and molecules that surround them in solids and solutions. To understand these interactions, we need to reconsider the definitions of acids and bases we used in Chapter 17. Let's begin by revisiting what happens when ammonia gas dissolves in water:

$$NH_3(g) + H_2O(\ell) \rightleftharpoons NH_4^+(aq) + OH^-(aq)$$

Figure 18.1(a) shows a Brønsted–Lowry interpretation of this reaction: in donating H^+ ions to ammonia, H_2O acts as a Brønsted–Lowry acid, and in accepting the protons NH_3 acts as a Brønsted–Lowry base.

Another way to view this reaction is illustrated in Figure 18.1(b). Instead of focusing on the transfer of hydrogen ions, consider the two reactants as a donor and an acceptor of *electron pairs*. In this view, the N atom in NH_3 donates its lone pair of electrons to one of the H atoms in H_2O. In the process, one of the H—O bonds in H_2O is broken in such a way that the bonding pair of electrons remains with the O atom. The donated lone pair from the N atom forms a fourth N—H covalent bond. The result is the same as in the Brønsted–Lowry model: a molecule of NH_3 bonds to a H^+ ion, forming an NH_4^+ ion. The molecule of H_2O that lost the H^+ ion becomes a OH^- ion.

Viewing this process as the donation and acceptance of an electron pair provides the following basis for defining acids and bases:

➤ A **Lewis base** is a substance that *donates* a lone pair of electrons in a chemical reaction.

➤ A **Lewis acid** is a substance that *accepts* a lone pair of electrons in a chemical reaction.

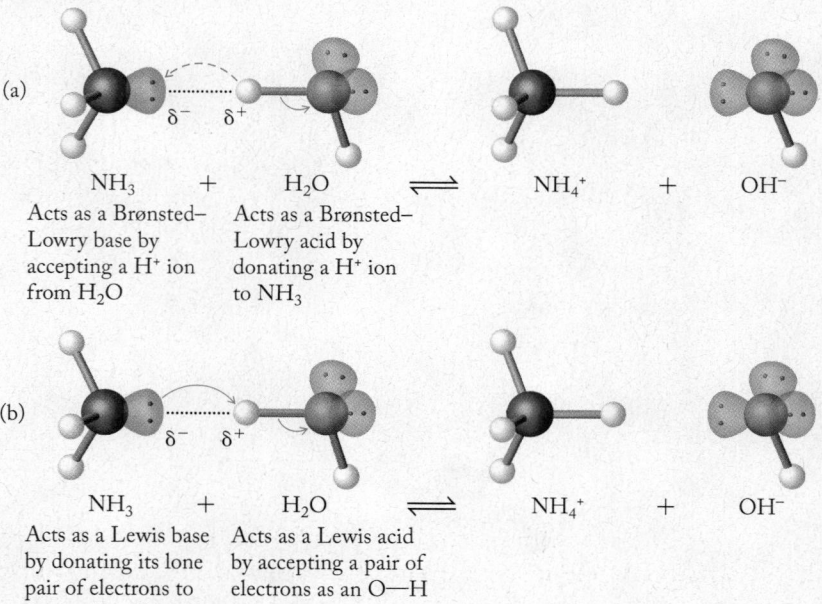

FIGURE 18.1 (a) Brønsted–Lowry view of the reaction between H_2O (proton donor) and NH_3 (proton acceptor). (b) Lewis view of the reaction: H_2O acts as a Lewis acid (electron-pair acceptor) and NH_3 acts as a Lewis base (electron-pair donor).

(a)

NH_3 + H_2O $\rightleftharpoons$ NH_4^+ + OH^-

Acts as a Brønsted–Lowry base by accepting a H^+ ion from H_2O

Acts as a Brønsted–Lowry acid by donating a H^+ ion to NH_3

(b)

NH_3 + H_2O $\rightleftharpoons$ NH_4^+ + OH^-

Acts as a Lewis base by donating its lone pair of electrons to form a N—H bond

Acts as a Lewis acid by accepting a pair of electrons as an O—H bond breaks

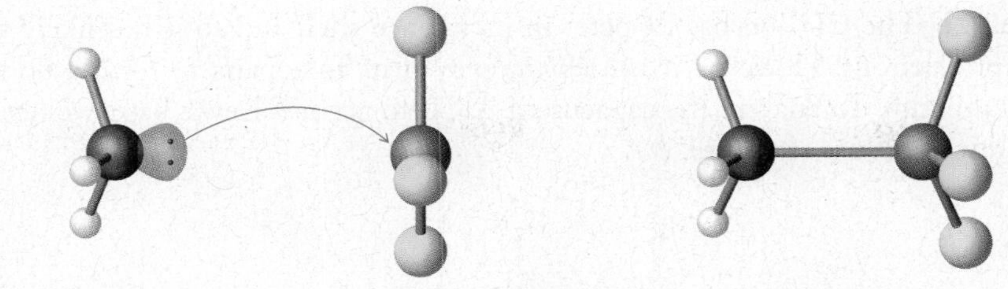

FIGURE 18.2 In the reaction between NH_3 and BF_3, NH_3 acts as a Lewis base and BF_3 acts as a Lewis acid.

NH_3 + BF_3 $\longrightarrow$ $H_3N—BF_3$

Acts as a Lewis base by donating a pair of electrons to BF_3 Acts as a Lewis acid by accepting a pair of electrons from NH_3

These definitions are named after their developer, Gilbert N. Lewis, who pioneered research into the nature of chemical bonds (Section 8.2). The Lewis definition of a base is consistent with the Brønsted–Lowry model we explored in Chapter 17 because a substance must be able to donate a pair of electrons if it is to bond with a H^+ ion. However, the same parallelism is not true for acids. The Brønsted–Lowry model defines an acid as a hydrogen-ion donor, but the Lewis definition encompasses species that have no hydrogen ions to donate but that can still accept electrons. One such compound is boron trifluoride, BF_3.

With only six valence electrons, the boron atom in BF_3 can accept another pair to complete its octet. NH_3 is a suitable electron-pair donor, as shown in Figure 18.2. There is no transfer of H^+ ions in this reaction, and so it is not an acid–base reaction according to the Brønsted–Lowry model. However, NH_3 donates a lone pair of electrons and BF_3 accepts them, so it is an acid–base reaction according to the broader Lewis model.

Many important Lewis bases are anions, including the halide ions, OH^-, and O^{2-}. To see how O^{2-} functions as a Lewis base, let's revisit the reaction described in Section 16.6 between SO_2 and CaO that is used to reduce SO_2 emissions from power stations and smelters:

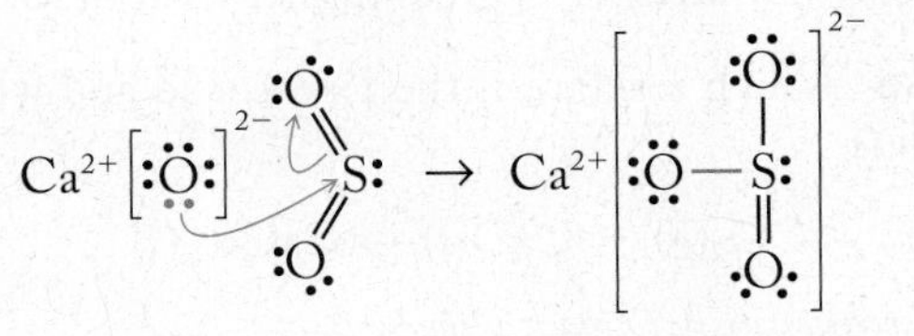

The oxide ion in CaO is the electron-pair donor, and so O^{2-} is a Lewis base. Sulfur dioxide is the electron-pair acceptor and therefore a Lewis acid. To accommodate the third O atom in SO_3^{2-}, one of the S=O double bonds becomes a single bond, and a bonding pair from that bond becomes a third lone pair on an O atom in SO_3^{2-}. These bonding changes are necessary to produce a structure in which the formal charges on the S atom and the double-bonded O atom are both zero and those on the two single-bonded O atoms are -1, giving an overall charge of $2-$.

CONNECTION Lewis's pioneering theories of the nature of covalent bonding were described in Section 8.2.

CONNECTION The concept of formal charge and its calculation were described in Section 8.6.

SAMPLE EXERCISE 18.1   **Identifying Lewis Acids and Bases**

Which species is a Lewis acid and which is a Lewis base in this reaction?

$$AlCl_3 + Cl^- \rightarrow AlCl_4^-$$

Collect and Organize We are asked to identify which of two reactants is acting as a Lewis acid, meaning an electron-pair acceptor, and which is acting as a Lewis base, meaning an electron-pair donor.

Analyze The Cl⁻ ion has an octet in its valence shell and so is not likely to accept electrons. However, it can *donate* one of its four pairs to form a bond to aluminum. To analyze the capacity of $AlCl_3$ to act as a Lewis base, we need to draw its Lewis structure:

This structure accounts for all the valence electrons (3 × 7 from 3 Cl atoms + 3 from Al = 24) with three Al—Cl single bonds and *no lone pairs*. This leaves Al with only 6 valence electrons and thus the capacity to accept one more pair, i.e., to act as a Lewis acid.

Solve In this reaction, $AlCl_3$ is a Lewis acid and the Cl⁻ ion is a Lewis base. We can represent the reaction using Lewis structures:

Think about It Drawing the Lewis structure of $AlCl_3$ is the key to solving the problem. With its incomplete octet, $AlCl_3$ has the capacity to accept an additional pair of electrons, much like BF_3, so it may act as a Lewis acid.

These Lewis structures assume that $AlCl_3$ is a molecular compound and not, like most other metal chlorides, an ionic compound. This assumption is supported by the physical properties of $AlCl_3$ and particularly by the fact that it sublimes at 178°C and 1 atm. Most metal halides do not even melt until heated to temperatures many hundreds of degrees higher than that.

Practice Exercise Which reactant is the Lewis acid and which is the Lewis base in this reaction?

$$CO_2(g) + CaO(s) \rightarrow CaCO_3(s)$$

(Answers to Practice Exercises are in the back of the book.)

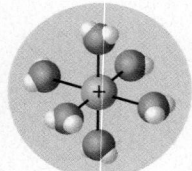

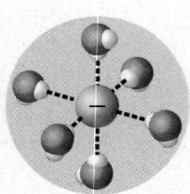

FIGURE 18.3 In Chapter 10 we explored how ions in aqueous solutions are surrounded by water molecules oriented with their positive dipoles (H atoms) directed toward anions and their negative dipoles (O atoms) directed toward cations. When these O atoms donate a lone pair of electrons to an empty orbital of a cation, a coordinate bond is formed.

18.2 Complex Ions

In Chapter 10, we described how ions dissolved in water are *hydrated*, that is, surrounded by water molecules oriented with their positive dipoles directed toward anions and their negative dipoles directed toward cations (Figure 18.3). In some hydrated cations, ion–dipole interactions lead to the sharing of lone-pair electrons on the oxygen atoms of H_2O with empty valence-shell orbitals on the cations. These shared electron pairs meet our definition of covalent bonds, but these particular bonds are called *coordinate* covalent bonds, or simply **coordinate bonds**. Such bonds form when either a molecule or an anion donates a lone pair of electrons to an empty valence-shell orbital of an atom, molecule, or cation. Once formed, a coordinate bond is indistinguishable from any other kind of covalent bond.

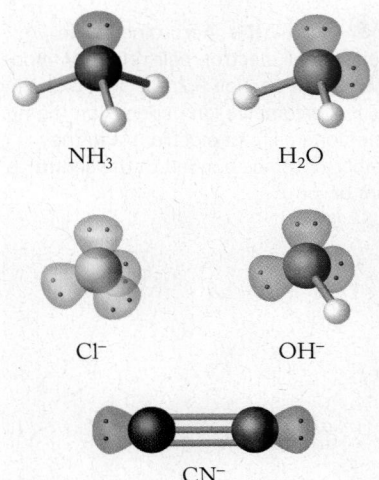

CONCEPT TEST •

Is the N—B bond that forms between NH_3 and BF_3 (as shown in Figure 18.2) a coordinate bond? Explain your answer.

(Answers to Concept Tests are in the back of the book.)

• •

Molecules or anions (Figure 18.4) that function as Lewis bases and form coordinate bonds with metal cations are called **ligands**. The resulting species, which are composed of central metal ions and the surrounding ligands, are called **complex ions** or simply *complexes*. Direct bonding to a central cation means that the ligands in a complex occupy the **inner coordination sphere** of the cation.

To better understand the meaning of these terms and others related to complex formation, let's consider what happens when $Zn(NO_3)_2$ dissolves in a solution containing ammonia. As the salt dissolves, each Zn^{2+} ion is surrounded by four NH_3 molecules. Each molecule donates the lone pair of electrons on its N atom to an empty valence orbital of the Zn^{2+} ion, forming a coordinate bond and resulting in a complex with the formula $Zn(NH_3)_4{}^{2+}$ (Figure 18.5).

FIGURE 18.4 Anions or molecules with lone pairs of electrons have the capacity to donate those electrons to metal cations. When these compounds do this, they are called ligands. All these ligands except NH_3 have more than one lone pair, but only one pair at a time can be directed toward a single metal cation.

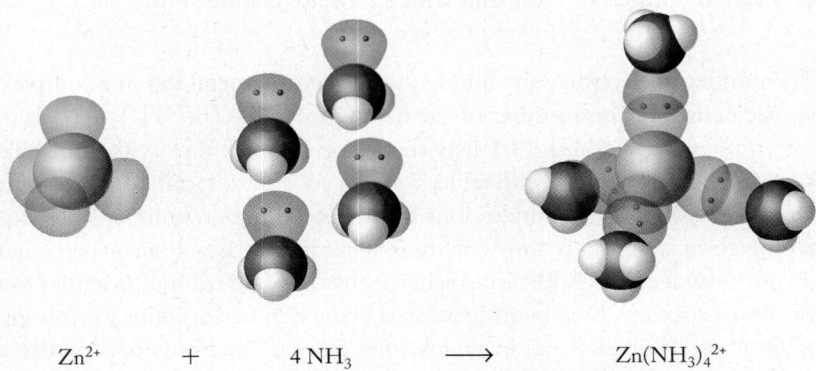

Zn^{2+} $\quad$ + $\quad$ $4\,NH_3$ $\quad \longrightarrow \quad$ $Zn(NH_3)_4{}^{2+}$

FIGURE 18.5 The complex ion $Zn(NH_3)_4{}^{2+}$ is produced when four molecules of ammonia form coordinate bonds with empty sp^3 orbitals on a Zn^{2+} ion.

Complex ions have characteristic shapes depending on the number of ligands surrounding the central cation. For example, the four NH_3 molecules in $Zn(NH_3)_4{}^{2+}$ occupy the four corners of a tetrahedron. We can explain its tetrahedral shape this way: each Zn^{2+} ion has the electron configuration $[Ar]3d^{10}$ as shown in the left-hand orbital diagram below. Its $3d$ orbitals are full, but its $4s$ and $4p$ orbitals are empty. According to valence bond theory (see Section 9.4), these orbitals can hybridize, forming four empty sp^3 hybrid orbitals:

coordinate bond formed when one anion or molecule donates a pair of electrons to another ion or molecule to form a covalent bond.

ligand a Lewis base bonded to the central metal ion of a complex ion.

complex ion an ionic species consisting of a metal ion bonded to one or more Lewis bases.

inner coordination sphere the ligands that are bound directly to a metal via coordinate bonds.

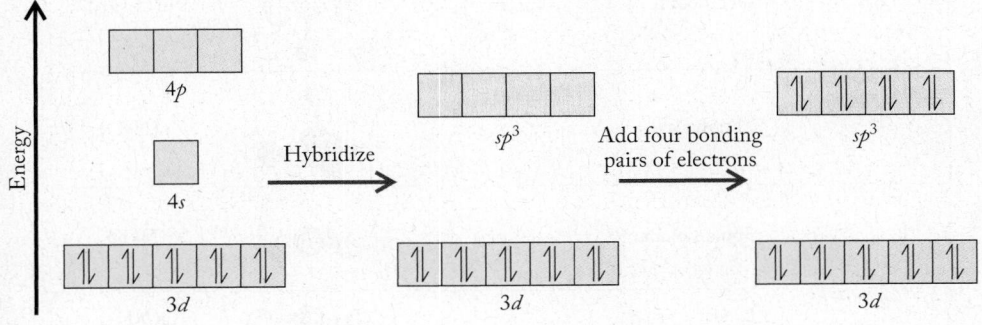

As with all sets of sp^3 orbitals, these are directed toward the four corners of a tetrahedron as shown in the image of the Zn^{2+} ion in Figure 18.5. The four lone pairs of the N atoms from four molecules of NH_3 are shared by these sp^3 orbitals when the complex forms, yielding a tetrahedral $Zn(NH_3)_4^{2+}$ ion.

The positive charges on the $Zn(NH_3)_4^{2+}$ ions in the solution are balanced by the NO_3^- ions that were released when $Zn(NO_3)_2$ dissolved. The nitrate ions in this solution serve as **counter ions**, a term describing ions of opposite charge that are not part of the inner coordination sphere. These NO_3^- ions are not bonded directly to the Zn^{2+} ions but are electrostatically attracted to them. For this reason, the formula of the solute in this solution is written

$$[Zn(NH_3)_4](NO_3)_2$$

where the brackets separate the formula of the complex ion from the formula of its counter ions. This formula is that of a **coordination compound**—a label that applies to any compound that contains a complex ion.

CONCEPT TEST •

In the coordination compound $Na_3[Fe(CN)_6]$, which ions occupy the inner coordination sphere of the Fe^{3+} ion, and which ions are counter ions?

• •

The number of electron pairs bonded to the central metal ion in a complex ion defines the **coordination number** of the metal ion. Thus $Zn(NH_3)_4^{2+}$ has a coordination number of 4. Table 18.1 lists shapes of complex ions with coordination numbers of 6, 4, and 2. From the table, you can see that a tetrahedral geometry is not the only possibility for complex ions having coordination number 4. To explain the geometry of a complex ion with four ligands that is square planar, such as $Pt(NH_3)_4^{2+}$, we need a hybridization scheme that has four orbitals oriented toward the corners of a square. One hybridization scheme that can account for this geometry is sp^3d^2 (see Figure 9.32) in which lone pairs of electrons occupy the axial positions of an octahedral orientation of the 6 hybridized orbitals, and four bonding pairs are oriented to the four equatorial corners.

A complex ion that has a coordination number of 6 is octahedral. This is also a shape we associate with sp^3d^2 hybridization. Consider $Fe(H_2O)_6^{3+}$, the first octahedral complex listed in Table 18.1. It consists of one Fe^{3+} ion with coordinate

counter ions provide electrical balance of charges of complex ions in coordination compounds.

coordination compound made up of at least one complex ion.

coordination number identifies the number of electron pairs surrounding a metal ion in a complex.

TABLE 18.1	Common Coordination Numbers and Shapes for Complex Ions			
Coordination Number	Shape	Hybridization	Structure	Examples
6	Octahedral	sp^3d^2 or d^2sp^3	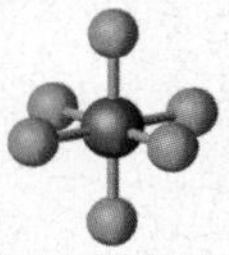	$Fe(H_2O)_6^{3+}$ $Ni(H_2O)_6^{2+}$ $Co(H_2O)_6^{3+}$
4	Tetrahedral	sp^3	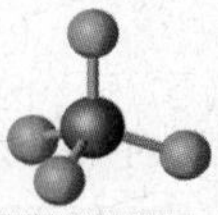	$Zn(H_2O)_4^{2+}$
4	Square planar	sp^3d^2 or dsp^2	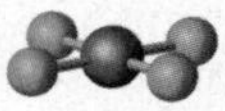	$Pt(NH_3)_4^{2+}$
2	Linear	sp	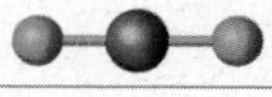	$Ag(NH_3)_2^+$

bonds to the O atoms in six H_2O molecules. The electron configuration of Fe^{3+} is $[Ar]3d^5$. According to Hund's rule, each $3d$ orbital is half-filled:

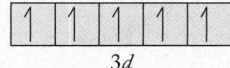

which means that none of them can accept *two* electrons to form a coordinate bond. To explain the octahedral shape of $Fe(H_2O)_6^{3+}$, we need to include two empty $4d$ orbitals in our hybridization scheme. Mixing one $4s$, three $4p$, and two $4d$ orbitals creates six sp^3d^2 orbitals:

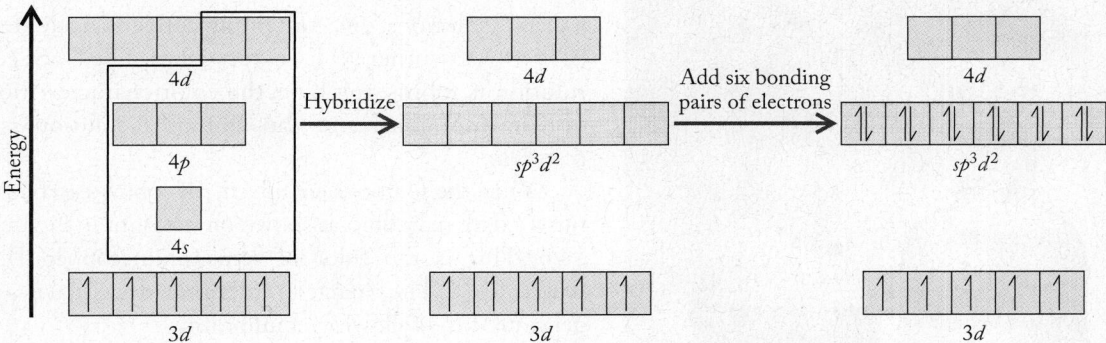

Not all complex ions having a coordination number of 6 and an octahedral shape have a central metal ion with sp^3d^2 hybrid orbitals in its valence shell. Consider, for instance, the Fe(III) complex ion $Fe(CN)_6^{3-}$. The $1-$ charge on each CN^- gives the complex ion an overall charge of $3-$. Given the electron configuration of Fe^{3+}, $[Ar]3d^5$, and the octahedral shape of the complex ion, you might assume sp^3d^2 hybridization for the reasons discussed in the previous paragraph. However, measurements of the magnetic properties of $Fe(CN)_6^{3-}$ indicate that there is only one unpaired electron per ion. We can explain this result by assuming that Hund's rule is not obeyed in $Fe(CN)_6^{3-}$. Instead, four of the five $3d$ electrons are paired:

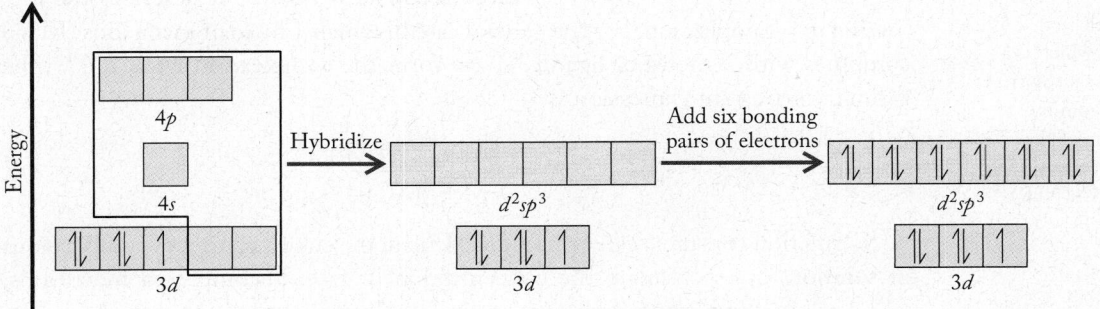

This pairing means that two empty $3d$ orbitals are available for hybridization. Hybridization involving these two $3d$ orbitals with the $4s$ and three $4p$ orbitals produces a set of six d^2sp^3 hybrid orbitals around Fe^{3+} in $Fe(CN)_6^{3-}$. In this case we write "d^2" before "sp^3" in the hybridization symbol because the d orbitals come from an electron shell with a smaller principal quantum number n (3 instead of 4) than the s and p orbitals.

You may be wondering why Hund's rule is obeyed in $Fe(H_2O)_6^{3+}$ but not in $Fe(CN)_6^{3-}$. The answer lies in the different strengths of the interactions between lone pairs of electrons on different ligands and valence electrons of the central metal ions. We explore these differences in Section 18.8, and we will see in Section 18.9 why these differences sometimes lead to electron distributions that appear to disobey Hund's rule.

Is the hybridization of Ni in $Ni(H_2O)_6^{2+}$ d^2sp^3 or sp^3d^2? Explain your answer.

18.3 Complex-Ion Equilibria

Let's investigate the formation of complex ions using mathematical tools we developed in Chapters 16 and 17. These tools are appropriate because complex formation processes are reversible, and many of them reach chemical equilibrium rapidly. We start this investigation with two aqueous solutions, one containing copper(II) sulfate ($CuSO_4$), the other NH_3 (Figure 18.6). The $CuSO_4$ solution is robin's-egg blue, the color characteristic of $Cu^{2+}(aq)$ ions, and the ammonia solution is colorless.

When the solutions are mixed, the robin's-egg blue turns a dark navy blue, as shown on the right in Figure 18.6. This is the color of tetraamminecopper(II), $Cu(NH_3)_4^{2+}$. The change in color provides visual evidence that the following equilibrium

$$Cu^{2+}(aq) + 4NH_3(aq) \rightleftharpoons Cu(NH_3)_4^{2+}(aq)$$

lies far to the right, favoring complex formation. This conclusion is supported by the large equilibrium constant for the reaction:

$$K_f = \frac{[Cu(NH_3)_4^{2+}]}{[Cu^{2+}][NH_3]^4} = 5.0 \times 10^{13}$$

This equilibrium constant K_f is called a **formation constant** because it describes the formation of a complex ion. For the general case in which 1 mole of metal ions (M^{m+}) combines with n moles of ligand X^{x-} to form the complex ion $MX_n^{(m-nx)+}$, the formation constant expression is

$$K_f = \frac{[MX_n^{(m-nx)+}]}{[M^{m+}][X^{x-}]^n}$$

Formation constants can be used to calculate the concentration of complex ions in solution, or to calculate the concentration of free, uncomplexed metal ions, $M^{m+}(aq)$, in equilibrium with a given (usually larger) concentration of ligand, as in the following Sample Exercise.

FIGURE 18.6 The beaker on the left contains a solution of $Cu^{2+}(aq)$, which is a characteristic robin's-egg blue. As a colorless solution of ammonia is added (from the bottle in the middle), the mixture of the two solutions turns dark blue (beaker on the right), which is the color of the $Cu(NH_3)_4^{2+}$ complex ion.

formation constant (K_f) equilibrium constant describing the formation of a metal complex from a free metal ion and its ligands.

Calculating the Concentration of a Free Metal Ion in Equilibrium with a Complex

Ammonia gas is dissolved in a 1.00×10^{-4} M solution of $CuSO_4$, so that initially $[NH_3] = 2.00 \times 10^{-3}$ M. Calculate the concentration of $Cu^{2+}(aq)$ ions in the solution after the reaction mixture has come to chemical equilibrium.

Collect and Organize The concentration of $CuSO_4$ means that $[Cu^{2+}]$ before complex formation is 1.00×10^{-4} M. The initial concentration of the ligand (NH_3) is 2.00×10^{-3} M. We know from the text preceding this Sample Exercise that the reaction involves the formation of $Cu(NH_3)_4^{2+}$:

$$Cu^{2+}(aq) + 4\,NH_3(aq) \rightleftharpoons Cu(NH_3)_4^{2+}(aq)$$

and that the values of $[NH_3]$, $[Cu^{2+}]$, and $[Cu(NH_3)_4^{2+}]$ are related by the formation constant expression

$$K_f = \frac{[Cu(NH_3)_4^{2+}]}{[Cu^{2+}][NH_3]^4} = 5.0 \times 10^{13}$$

Analyze The stoichiometric ratio of NH_3 to Cu^{2+} is 4:1, but initially $[NH_3]/[Cu^{2+}] = 20$, so there is more than enough NH_3 to convert all the Cu^{2+} into $Cu(NH_3)_4^{2+}$. Because K_f is large, we can assume that nearly all the Cu^{2+} ions are converted and that only a tiny concentration of free Cu^{2+} ions (x) remains uncomplexed at equilibrium. Therefore, at equilibrium

$$[Cu^{2+}] = x$$

and

$$[Cu(NH_3)_4^{2+}] = (1.00 \times 10^{-4}) - x$$

The balanced equation tells us that 4 moles of NH_3 are consumed for every mole of $Cu(NH_3)_4^{2+}$ produced. This means that if the change in $[Cu(NH_3)_4^{2+}]$ is $(1.00 \times 10^{-4} - x)$, the change in $[NH_3]$ is a *decrease* equal to four times that value: $4(1.00 \times 10^{-4} - x)$.

Given the large value of K_f, we should obtain a $[Cu^{2+}]$ value at equilibrium that is much less than 1.00×10^{-4} M.

Solve Completing the ICE table for the equilibrium:

	$[Cu^{2+}]$ (M)	$[NH_3]$ (M)	$[Cu(NH_3)_4^{2+}]$ (M)
Initial (I):	1.00×10^{-4}	2.00×10^{-3}	0
Change (C):	$-(1.00 \times 10^{-4} - x)$	$-4(1.00 \times 10^{-4} - x)$	$+(1.00 \times 10^{-4} - x)$
Equilibrium (E):	x	$(1.60 \times 10^{-3}) + 4x$	$1.00 \times 10^{-4} - x$

Now we make the simplifying assumption that x is much smaller than 1.00×10^{-4} M. If it is, then $4x$ must be much smaller than 1.60×10^{-3} M. Therefore, we can ignore the x terms in the equilibrium values of $[NH_3]$ and $[Cu(NH_3)_4^{2+}]$ and write

$$K_f = \frac{[Cu(NH_3)_4^{2+}]}{[Cu^{2+}][NH_3]^4}$$

$$= \frac{1.00 \times 10^{-4}}{(x)(1.60 \times 10^{-3})^4} = 5.0 \times 10^{13}$$

$$x = \frac{1.00 \times 10^{-4}}{(1.60 \times 10^{-3})^4(5.0 \times 10^{13})}$$

$$= 3.1 \times 10^{-7} \, M = [Cu^{2+}]$$

Think about It This result confirms our simplifying assumption and also our prediction that $[Cu^{2+}]$ at equilibrium is much less than $[Cu^{2+}]$ initially. In fact, more than 99% of the Cu(II) in the solution is present as $Cu(NH_3)_4^{2+}$.

Practice Exercise Calculate the equilibrium concentration of $Ag^+(aq)$ in a solution that is initially 0.100 M $AgNO_3$ and 0.800 M NH_3 after this reaction takes place:

$$Ag^+(aq) + 2\,NH_3(aq) \rightleftharpoons Ag(NH_3)_2^+(aq) \qquad K_f = 1.7 \times 10^7$$

18.4 Naming Complex Ions and Coordination Compounds

The names of complex ions and coordination compounds tell us the identity and oxidation state of the central ion, the names and numbers of ligands, the charge in the case of complex ions, and the identity of counter ions. To convey all this information, we need to follow some simple naming rules.

Complex Ions with a Positive Charge

1. Start with the identities of the ligand(s). Names of common ligands appear in Table 18.2. If there is more than one kind of ligand, list the names alphabetically.

TABLE 18.2   Names and Structures of Common Ligands

Ligand	Name within Complex Ion	Structure	Charge	Number of Donor Groups
Iodide	Iodo	I$^-$	1−	1
Bromide	Bromo	Br$^-$	1−	1
Chloride	Chloro	Cl$^-$	1−	1
Fluoride	Fluoro	F$^-$	1−	1
Nitrate	Nitrato	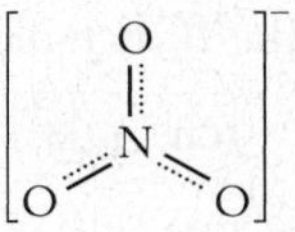	1−	1
Hydroxide	Hydroxo	[O—H]$^-$	1−	1
Water	Aqua	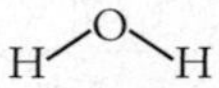	0	1
Pyridine	Pyridyl	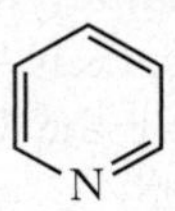	0	1
Ammonia	Ammine	NH$_3$	0	1
Ethylenediamine (en)	(same)^a	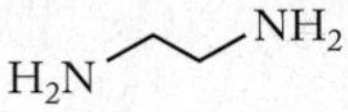	0	2
2,2′-Bipyridine (bipy)	Bipyridyl	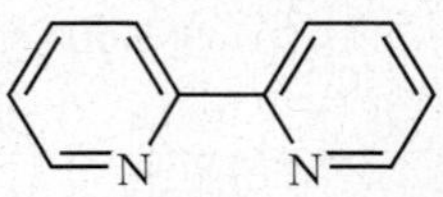	0	2
1,10-Phenanthroline (phen)	(same)^a	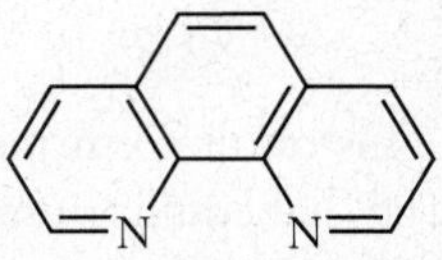	0	2
Cyanide^b	Cyano	[C≡N]$^-$	1−	1
Carbon monoxide^b	Carbonyl	C≡O	0	1

^{a} The names of some electrically neutral ligands in complexes are the same as the names of the molecules.
^{b} Carbon atoms are the lone pair donors in these ligands.

2. In front of the name(s) written in step 1, use the usual prefix(es) to indicate the number of each type of ligand:

Number of Ligands	Prefix
2	Di–
3	Tri–
4	Tetra–
5	Penta–
6	Hexa–

3. Write the name of the transition metal ion with a Roman numeral indicating its oxidation state.

Examples:

$Ni(H_2O)_6^{2+}$	Hexaaquanickel(II)
$Co(NH_3)_6^{3+}$	Hexaamminecobalt(III)
$[Cu(NH_3)_4(H_2O)_2]^{2+}$	Tetraamminediaquacopper(II)

It may seem strange having two *a*'s together in these names, but it is permitted under current naming rules. In the third name, prefixes are ignored in determining alphabetical order: the order is *ammine* before *aqua* rather than *di* before *tetra*.

In these three examples the ligands are all electrically neutral. This makes determining the oxidation state of the central metal ion a simple task because the charge on the complex ion is the same as the charge on the metal ion, which is the oxidation state of the metal. When the ligands include anions, determining the oxidation state of the central metal ion requires us to account for these charges.

Complex Ions with a Negative Charge

1. Follow the steps for naming positively charged complexes.
2. Add *-ate* to the name of the central metal ion to indicate that the complex ion carries a negative charge (just as we use *-ate* to end the names of oxoanions). For some metals, the base name changes, too. The two most common examples are iron, which becomes *ferrate*, and copper, which becomes *cuprate*.

Examples:

$Fe(CN)_6^{3-}$	Hexacyanoferrate(III)
$[Fe(H_2O)(CN)_5]^{3-}$	Aquapentacyanoferrate(II)
$[Al(H_2O)_2(OH)_4]^-$	Diaquatetrahydroxoaluminate

In the first two examples we must determine the oxidation state of Fe. We start with the charge on the complex ion and then take into account the charges on the ligand anions to calculate the charge on the metal ion. For example, the overall charge of the aquapentacyanoferrate(II) ion is $3-$. It contains five CN^- ions. To reduce the combined charge of $5-$ from these cyanide ions to an overall charge of $3-$, the charge on Fe must be $2+$.

CONCEPT TEST ∙∙

What is the name of the complex anion with the formula $PtCl_4^{2-}$?

Coordination Compounds

1. If the counter ion of the complex ion is a cation, the cation name goes first, followed by the name of the anionic complex ion.
2. If the counter ion of the complex ion is an anion, the name of the cationic complex ion goes first, followed by the name of the anion.

Examples:

$[Ni(NH_3)_6]Cl_2$	Hexaamminenickel(II) chloride
$K_3[Fe(CN)_6]$	Potassium hexacyanoferrate(III)
$[Co(NH_3)_5(H_2O)]Br_2$	Pentaammineaquacobalt(II) bromide

A key to naming coordination compounds is to recognize from their formulas that they *are* coordination compounds. For help with this, look for formulas that have the atomic symbol of a metallic element and one or more ligands all in brackets, either followed by the symbol of an anion, as in $[Co(NH_3)_5(H_2O)]Br_2$ or preceded by the symbol of a cation, as in $K_3[Fe(CN)_6]$.

SAMPLE EXERCISE 18.3   Naming Coordination Compounds

Name the coordination compounds (a) $Na_4[Co(CN)_6]$ and (b) $[Co(NH_3)_5Cl](NO_3)_2$.

Collect and Organize We are asked to write a name for each compound that unambiguously identifies its composition. The formulas of the complex ions appear in brackets in both compounds. Because cobalt, the central metal ion in both, is a transition metal, we express its oxidation state using Roman numerals. The names of common ligands are given in Table 18.2.

Analyze It is useful to take an inventory of the ligands and counter ions:

Compound	Counter Ion	LIGAND			
		Formula	Name	Number	Prefix
$Na_4[Co(CN)_6]$	Na^+	CN^-	Cyano	6	Hexa-
$[Co(NH_3)_5Cl](NO_3)_2$	NO_3^-	NH_3	Ammine	5	Penta-
		Cl^-	Chloro	1	—

The oxidation state of each cobalt ion can be calculated by setting the sum of the charges on all the ions in both compounds equal to zero:

a. Ions: (4 Na^+ ions) + (1 Co ion) + (6 CN^- ions)
   Charges: 4+   + x +   6−   = 0
                                                  $x = 2+$

b. Ions: (1 Co ion) + (1 Cl^- ion) + (2 NO_3^- ions)
   Charges: x +   1−   +   2−   = 0
                                                  $x = 3+$

Solve

a. Because the counter ion, *sodium*, is a cation, its name comes first. The complex ion is an anion in this compound. To name it, we begin with the ligand *cyano*, to which we add the prefix *hexa-* and write *hexacyano*. This is followed by the name of the transition metal ion: hexacyano*cobalt*.

We then change the ending of the name of the complex ion to *-ate* because it is an anion: hexacyanocobalt*ate*. We then add a Roman numeral to indicate the oxidation state of the cobalt: hexacyanocobaltate(*II*). Putting it all together, we obtain the name: sodium hexacyanocobaltate(II).

b. The complex ion is the cation in this compound, and we begin by naming the ligands directly attached to the metal ion in alphabetical order: *ammine* and *chloro*. We indicate the number (5) of NH_3 ligands with the appropriate prefix: *penta*amminechloro. We name the metal next and indicate its oxidation state with a Roman numeral: pentaamminechloro*cobalt(III)*. Finally we name the anionic counter ion: *nitrate*. Putting it all together, we obtain the name: pentaamminechlorocobalt(III) nitrate.

Think about It Naming coordination compounds requires you to (1) distinguish between ligands and counter ions and (2) recall which ligands are electrically neutral and which are anions.

Practice Exercise Identify the ligands and counter ions in (a) $[Zn(NH_3)_4]Cl_2$ and (b) $[Co(NH_3)_4(H_2O)_2](NO_2)_2$, and name each compound.

18.5 Hydrated Metal Ions as Acids

In Chapter 17 we explored how electronegative atoms in a molecule containing a –OH group stabilize the anion formed when the H atom ionizes (Figures 17.10–17.12). A similar stabilization occurs for a hydrated metal ion having the generic formula $M(H_2O)_6{}^{n+}$ when $n \geq 2$. The electrons in the H—O bonds in the water molecules of the inner coordination sphere are attracted to the positively charged central metal ion and repelled by the hydrogen atoms. This delocalization stabilizes the charge of the hydroxide ion formed when one of the water molecules in the inner coordination sphere acts as a Brønsted–Lowry acid and donates a H^+ ion to a water molecule outside that sphere:

$$M(H_2O)_6{}^{n+}(aq) + H_2O(\ell) \rightleftharpoons M(H_2O)_5(OH)^{(n-1)+}(aq) + H_3O^+(aq)$$

Many hydrated metal ions, particularly those with charges of 2+ and higher, are Brønsted–Lowry acids. The acidic properties of these ions are reflected in the K_a values in Table 18.3.

To illustrate this acidic behavior, let's consider what happens when $FeCl_3$ dissolves in water. The compound dissociates as it dissolves, forming $Fe(H_2O)_6{}^{3+}$ and Cl^- ions. As we saw in Chapter 17, Cl^- ions are very weak bases and do not influence the pH of the solution. On the other hand, the $Fe(H_2O)_6{}^{3+}$ ion behaves as an acid as one of the water molecules of its inner coordination sphere donates a H^+ ion to a water molecule in the bulk solution as shown in Figure 18.7.

CONNECTION In Chapter 17 we learned that the solutions of some salts are acidic or basic because of the reaction of their ions with water.

TABLE 18.3 K_a Values of Hydrated Metal Ions

Ion	K_a	
$Fe^{3+}(aq)$	3×10^{-3}	
$Cr^{3+}(aq)$	1×10^{-4}	
$Al^{3+}(aq)$	1×10^{-5}	
$Cu^{2+}(aq)$	3×10^{-8}	Acid strength
$Pb^{2+}(aq)$	3×10^{-8}	
$Zn^{2+}(aq)$	1×10^{-9}	
$Co^{2+}(aq)$	2×10^{-10}	
$Ni^{2+}(aq)$	1×10^{-10}	

FIGURE 18.7 A hydrated Fe^{3+} cation draws electron density away from the water molecules of its inner coordination sphere, which makes it possible for one or more of these molecules to donate a H^+ ion to a water molecule outside the sphere. The hydrated Fe^{3+} ion is left with one fewer H_2O molecule and one OH^- ion.

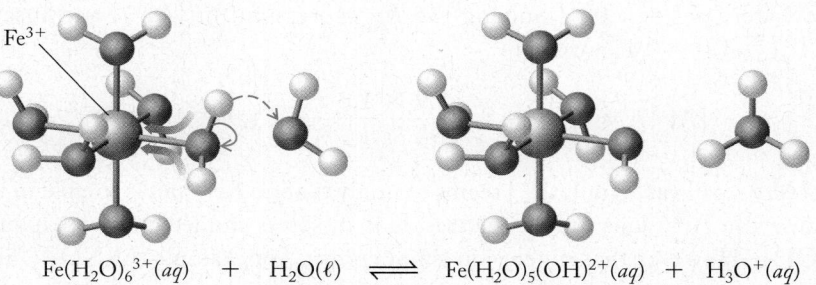

$$Fe(H_2O)_6{}^{3+}(aq) + H_2O(\ell) \rightleftharpoons Fe(H_2O)_5(OH)^{2+}(aq) + H_3O^+(aq)$$

Note that in the product complex ion, one of the original six water molecules has been converted into a hydroxide ion, reducing the charge on the complex from 3+ to 2+. The $Fe(H_2O)_5(OH)^{2+}$ ion can then act as a Brønsted–Lowry acid (H^+ donor) in a second acid ionization:

$$Fe(H_2O)_5(OH)^{2+}(aq) + H_2O(\ell) \rightleftharpoons Fe(H_2O)_4(OH)_2^+(aq) + H_3O^+(aq)$$

A third acid ionization is possible, producing solid iron(III) hydroxide:

$$Fe(H_2O)_4(OH)_2^+(aq) + H_2O(\ell) \rightleftharpoons Fe(H_2O)_3(OH)_3(s) + H_3O^+(aq)$$

For simplicity we usually write the formula of iron(III) hydroxide $Fe(OH)_3(s)$ even though the neutral material contains water of hydration.

Other 3+ cations, including Cr^{3+} and Al^{3+}, display similar behavior. However, these two trivalent ions form electrically neutral hydroxide compounds that have an unusual solubility pattern. Whereas $Fe(OH)_3(s)$ and most other transition metal hydroxides have very limited solubility in basic solutions, $Cr(OH)_3$ and $Al(OH)_3$ are more soluble in strongly basic solutions (pH > 11) than in weakly basic solutions (pH $\approx$ 8) because solid $Cr(OH)_3$ and $Al(OH)_3$ may accept additional OH^- ions at high pH, forming soluble anionic complex ions:

$$Cr(OH)_3(s) + OH^-(aq) \rightleftharpoons Cr(OH)_4^-(aq) = Cr(H_2O)_2(OH)_4^-(aq)$$

$$Al(OH)_3(s) + OH^-(aq) \rightleftharpoons Al(OH)_4^-(aq) = Al(H_2O)_2(OH)_4^-(aq)$$

Formation of complex ions explains the solubility of Cr^{3+} and Al^{3+} ions at low and high pH. These ions are soluble in strongly acidic solutions (pH $\approx$ 2), where they exist as $Cr(H_2O)_6^{3+}(aq)$ and $Al(H_2O)_6^{3+}(aq)$. In less acidic solutions they exist as positively charged complex ions with generic formulas such as $M(H_2O)_5(OH)^{2+}(aq)$ and $M(H_2O)_4(OH)_2^+(aq)$, and in strongly basic solutions they exist as $M(H_2O)_2(OH)_4^-(aq)$. Zinc hydroxide, $Zn(OH)_2$, is the only other transition metal hydroxide that is soluble at high pH.

Nearly all transition metals share one common characteristic: They exist as $M^{n+}(aq)$ ions only in strongly acidic solutions. They occur as complex ions in which at least one of their ligands is a OH^- ion in aqueous solutions that range from slightly acidic to slightly basic (3 < pH < 9). This pH range includes most environmental waters and biological fluids.

We just discussed how little $Fe(OH)_3$ or $Al(OH)_3$ dissolves in pure water. The minuscule solubilities of these compounds are reflected in their K_{sp} values: 1.1×10^{-36} and 1.9×10^{-33}, respectively. However, we must keep in mind that these tiny K_{sp} values apply to equilibrium concentrations of *free metal ions*, that is, the concentrations of $Fe^{3+}(aq)$ and $Al^{3+}(aq)$:

$$K_{sp} = [Fe^{3+}][OH^-]^3 = 1.1 \times 10^{-36}$$

$$K_{sp} = [Al^{3+}][OH^-]^3 = 1.9 \times 10^{-33}$$

Let's consider the implication of the extremely small K_{sp} of $Al(OH)_3$ by answering the question: What is the maximum $[Al^{3+}]$ that can exist in pure water (pH = 7.00)? Solving the K_{sp} expression for $[Al^{3+}]$ and inserting $[OH^-] = 1.0 \times 10^{-7}$, we get

$$[Al^{3+}] = \frac{K_{sp}}{[OH^-]^3} = \frac{1.9 \times 10^{-33}}{(1.0 \times 10^{-7})^3} = 1.9 \times 10^{-12}\ M$$

This very small value for $[Al^{3+}]$ seems to imply that no Al^{3+} salt is soluble in water because the Al^{3+} ions that it releases as it dissolves immediately precipitate as $Al(OH)_3$. However, that conclusion is not correct. For example, $Al(NO_3)_3$, like all nitrates, is quite soluble in water. This solubility can be explained by the acidic properties of $Al^{3+}(aq)$, i.e., $Al(H_2O)_6^{3+}$. This acidity means that the concentration of the acid $Al(H_2O)_6^{3+}$ at pH = 7.00 is very small for the same reason that the concentration of any weak acid is small compared to the concentration of its conjugate base at a pH above its pK_a. Most of the Al^{3+} ions will be present in solution as the conjugate base of $Al(H_2O)_6^{3+}$—that is, $Al(H_2O)_5(OH)^{2+}$—or even as the conjugate base of $Al(H_2O)_5(OH)^{2+}$: $Al(H_2O)_4(OH)_2^{+}$. The formation of these complexes means that the overall concentration of all soluble Al(III) species is much greater than $[Al(H_2O)_6^{3+}]$ alone. Therefore, the molar solubility of $Al(NO_3)_3$ is actually much greater than 1.9×10^{-12} M. Moreover, its solubility increases if enough strong base is added, because soluble $Al(H_2O)_2(OH)_4^{-}(aq)$ is the dominant Al(III) species in alkaline solutions:

$$Al(H_2O)_3(OH)_3(s) + OH^-(aq) \rightleftharpoons Al(H_2O)_2(OH)_4^{-}(aq) + H_2O(\ell)$$

Among the representative elements, Sn(II) and Sn(IV) form soluble complexes in strongly basic solutions, as does the transition metal ion Cr(III).

CONCEPT TEST

Is it correct to say that $Al(OH)_3$ has the capacity to act as both a Brønsted–Lowry acid and a Brønsted–Lowry base? Use balanced chemical equations to support your answer.

18.6 Polydentate Ligands

The ligands in Figure 18.4 and many of those in Table 18.2 can donate only one pair of electrons to a metal ion. Even atoms with more than one lone pair can donate only one pair at a time to a given metal ion because the other lone pair or pairs are oriented away from the metal ion. Because these ligands have effectively only one donor group, they are called **monodentate ligands**, which literally means "single-toothed."

Molecules larger than those in Figure 18.4 may be able to donate more than one lone pair of electrons and therefore form more than one coordinate bond to a central metal ion. Ligands in this category are called **polydentate ligands**, or more specifically *bidentate*, *tridentate*, and so on. One group of polydentate ligands is the polyamines, which include the compound ethylenediamine, a bidentate ligand that has the structure

$$\overset{\displaystyle ..}{H_2N} \qquad \overset{\displaystyle ..}{N}H_2$$
$$H_2C-CH_2$$

Note the lone pairs on the two –NH_2 groups separated from each other by two –CH_2– groups. This combination means that a molecule of ethylenediamine can partially encircle a metal ion so that both lone pairs can bond to the metal ion.

Formation of the ethylenediamine complex of $Ni^{2+}(aq)$ is shown in Figure 18.8(a). The two bonding orbitals that accept lone pairs of electrons from a molecule of ethylenediamine must be on the same side of the Ni^{2+} ion. However, two more ethylenediamine molecules can bond to other pairs of bonding sites, displacing additional pairs of water molecules and forming a complex in which the Ni^{2+} ion is surrounded by three bidentate ethylenediamine molecules, as shown in Figure 18.8(b).

monodentate ligand a species that forms only a single coordinate bond to a metal ion in a complex.

polydentate ligand a species that can form more than one coordinate bond per molecule.

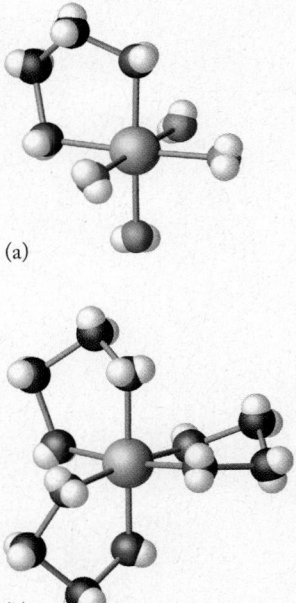

(a)

(b)

FIGURE 18.8 (a) The bidentate ligand ethylenediamine has two N atoms that can each donate a pair of electrons to the empty orbitals of adjacent octahedral bonding sites on the same $Ni^{2+}(aq)$ ion (gold sphere), displacing two molecules of H_2O. (b) Three ethylenediamine molecules occupy all six octahedral coordination sites of a Ni^{2+} ion.

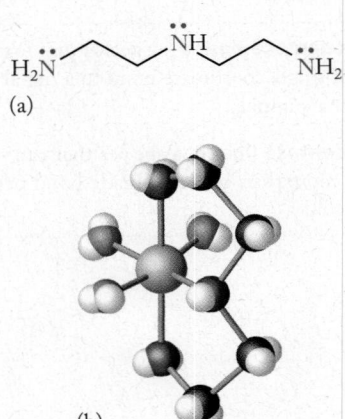

(a)

(b)

FIGURE 18.9 Tridentate chelation. (a) The three amine groups in the tridentate ligand diethylenetriamine are all potential electron-pair donor groups. (b) When these groups donate their lone pairs of electrons to a $Ni^{2+}(aq)$ ion (gold sphere) in solution, they occupy three of the six coordination sites on the ion.

Note that each ethylenediamine forms a five-atom ring with the metal ion. If the ring were a perfect pentagon (meaning all bond lengths and bond angles were exactly the same), each of its bond angles would be 108°. These pentagons are not perfect, but each ring's preferred octahedral bond angles of 90° for the N–Ni–N bond and of 107° to 109° for all the other bonds are accommodated with only a little strain on the ideal bond angles.

An even larger ligand, diethylenetriamine ($H_2NCH_2CH_2NHCH_2CH_2NH_2$), is shown in Figure 18.9(a). The lone pairs of electrons on its three nitrogen atoms give diethylenetriamine the capacity to form three coordinate bonds to a metal ion, meaning this is a *tridentate* ligand (Figure 18.9b).

As you may imagine, larger molecules may have even more atoms per molecule that can bond to a single metal ion. The interaction of a metal ion with a ligand having multiple donor atoms is called **chelation** (pronounced *key-LAY-shun*). The word comes from the Greek *chele*, meaning "claw." The polydentate ligands that take part in these interactions are called *chelating agents*.

Many chelating agents have more than one kind of electron-pair-donating group. One family of such compounds is called *aminocarboxylic acids*. The most important of them is ethylenediaminetetraacetic acid, EDTA, the molecular structure of which is shown in Figure 18.10(a). Note that one molecule of EDTA contains two amine groups and four carboxylic acid groups. When the acid groups release their H^+ ions, they form four carboxylate anions, $-COO^-$, in which either of the O atoms can donate a pair of electrons to a central metal ion. When O atoms on all four groups do so and the two amine groups do as well, six octahedral bonding sites around the metal ion can be occupied, as shown in Figure 18.10(b).

EDTA forms very stable complex ions and is used as a metal ion *sequestering agent*, that is, a chelating agent that binds metal ions so tightly that they are "sequestered" and prevented from reacting with other substances. For example, EDTA is used as a preservative in many beverages and prepared foods because it sequesters iron, copper, zinc, manganese, and other transition metal ions often present in these foods that can catalyze the degradation of ingredients in the foods. Many foods are fortified with ascorbic acid (vitamin C), which is particularly vulnerable to metal-catalyzed degradation because it is also a polydentate ligand and is more likely to be oxidized when chelated to one of the above metal ions. EDTA effectively shields vitamin C from these ions. We explore the preferential binding of metal ions to different ligands in the next section.

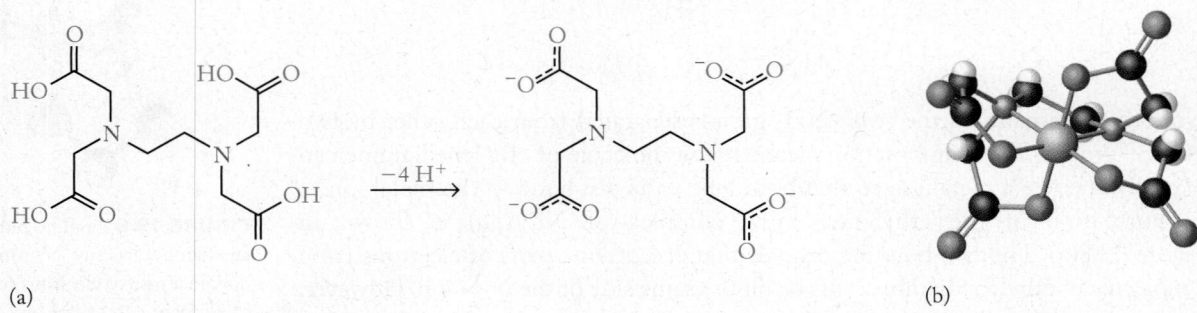

(a)

$\xrightarrow{-4\,H^+}$

(b)

FIGURE 18.10 (a) In the hexadentate ligand EDTA, the six donor groups are the two amine groups and the four carboxylic acid groups. The acid groups ionize to form coordinate bonding sites. (b) All six Lewis base groups in ionized EDTA can form a coordinate bond with the same metal ion, such as Co^{3+} (the gold sphere) shown here. In the process they form four 5-membered rings.

SAMPLE EXERCISE 18.4 **Identifying the Potential Electron-Pair Donor Groups in a Molecule**

How many donor groups does this polydentate ligand, nitrilotriacetic acid (NTA), have?

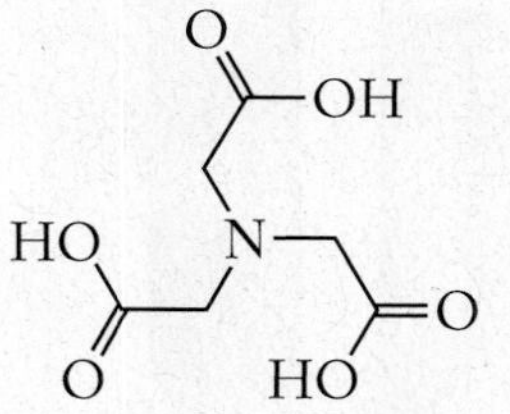

Collect and Organize We are asked to examine this molecular structure to find electron pairs that can be donated. The molecule has a nitrogen atom with three single bonds and three carboxylic acid groups.

Analyze Because there are three single bonds around the N atom, the atom's fourth sp^3 orbital must contain a lone pair of electrons. When all three carboxylic acid groups are ionized, there are three carboxylate groups in the molecule. Each carboxylate group can donate one nonbonding pair of electrons from one of its oxygen atoms to a metal atom. When one of these O atoms or the center N atom simultaneously forms coordinate bonds with a metal ion, a five-atom ring is produced, as in EDTA complexes.

Solve The central N atom and an O atom from each of the three carboxylate groups can form a total of four coordinate bonds. Therefore, NTA is potentially a tetradentate ligand with four donor groups.

Think about It The tetradentate capacity of NTA is reasonable because, like EDTA, it is an aminocarboxylic acid. It has one fewer amino group and one fewer carboxylic acid group than the hexadentate EDTA.

Practice Exercise How many potential donor groups are there in citric acid, a component of citrus fruits and a widely used preservative in the food industry?

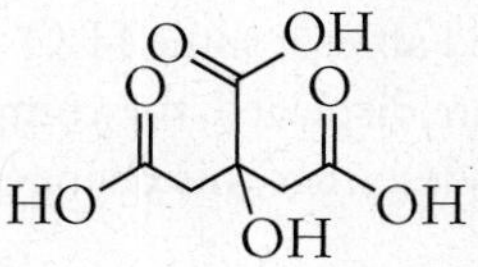

18.7 Ligand Strength and the Chelate Effect

Let's begin our discussion of the relative strengths of ligands as Lewis bases by exploring the relative affinity of $Ni^{2+}(aq)$ ions for two common monodentate ligands, H_2O and NH_3. We start by dissolving crystals of nickel(II) chloride hexahydrate in water. The formula of this compound is $NiCl_2 \cdot 6\ H_2O$. The dot connecting the two halves of the formula and the prefix *hexa* tell chemists that crystals of nickel(II) chloride contain Ni^{2+} ions that are each surrounded by six water molecules. Brilliant green crystals of $NiCl_2 \cdot 6\ H_2O$ form green solutions (Figure 18.11). The fact that the solid and its aqueous solution have the same color tells us that the same clustering of water molecules around Ni^{2+} ions occurs in both solid $NiCl_2 \cdot 6\ H_2O$ and aqueous solutions of Ni^{2+} ions. Thus, the Ni^{2+} ion in a solution of nickel(II)chloride is $Ni(H_2O)_6^{2+}$.

chelation the interaction of a metal with a polydentate ligand (chelating agent); pairs of electrons on one molecule of the ligand occupy two or more coordination sites on the central metal.

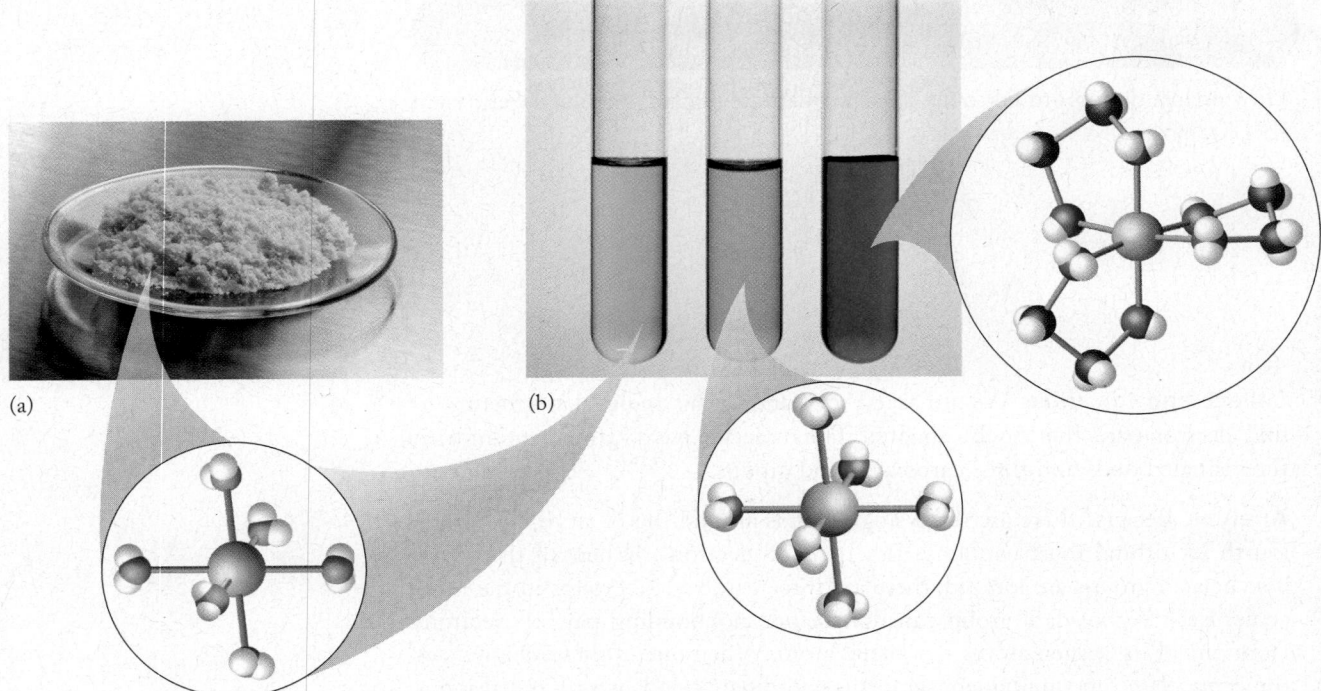

FIGURE 18.11 (a) Solid nickel(II) chloride hexahydrate is green. (b) When it dissolves in water, the resulting solution has the same green color (left), telling us that each Ni^{2+} ion (gold sphere) must be surrounded by H_2O molecules both in the solid and in the solution. When ammonia gas is bubbled through a solution of $Ni(H_2O)_6^{2+}$, the color changes to blue as NH_3 replaces H_2O in the Ni^{2+} ion's inner coordination sphere. When ethylenediamine is added to a solution of $Ni(NH_3)_6^{2+}$, the color turns from blue to purple as the ethylenediamine displaces the ammonia ligands and the $Ni(en)_3^{2+}$ complex forms.

Now let's bubble colorless ammonia gas through a green solution of $Ni(H_2O)_6^{2+}$ ions. As shown in Figure 18.11(b), the green solution turns blue. The color change means that different ligands are bonded to the Ni^{2+} ions. We may conclude that NH_3 molecules have displaced at least some H_2O molecules around the Ni^{2+} ions. If all the molecules of H_2O are displaced, the complex $Ni(NH_3)_6^{2+}$ is formed. The following chemical equation describes this change:

$$Ni(H_2O)_6^{2+}(aq) + 6\,NH_3(g) \rightleftharpoons Ni(NH_3)_6^{2+}(aq) + 6\,H_2O(\ell) \quad K_f = 5 \times 10^8$$

Keep in mind that the hydrated ion $Ni(H_2O)_6^{2+}$ is often expressed as $Ni^{2+}(aq)$.

This *ligand displacement* reaction illustrates that Ni^{2+} ions have a greater affinity for molecules of NH_3 than for molecules of H_2O. This affinity is reflected in the large formation constant for the reaction. Many other transition metal ions also have a greater affinity for ammonia than for water. We may conclude that ammonia is inherently a better electron-pair donor and so a stronger Lewis base than water. This conclusion is reasonable because we saw in Chapter 17 that NH_3 is also a stronger Brønsted–Lowry base than H_2O.

Next we add ethylenediamine to the blue solution of $Ni(NH_3)_6^{2+}$ ions. The solution changes color again: from blue to purple (Figure 18.11b), indicating yet another change in the ligands surrounding the Ni^{2+} ions. Molecules of ethylenediamine displace ammonia molecules from the inner coordination sphere of Ni^{2+} ions. This affinity of Ni^{2+} ions for ethylenediamine molecules is reflected in the formation constant for $Ni(en)_3^{2+}$ (where "en" represents ethylenediamine):

$$Ni(H_2O)_6^{2+}(aq) + 3\,en(aq) \rightleftharpoons Ni(en)_3^{2+}(aq) + 6\,H_2O(\ell) \qquad K_f = 1.1 \times 10^{18}$$

which is more than 10^9 times the K_f value for $Ni(NH_3)_6^{2+}$.

Why should the affinity of Ni^{2+} ions for ethylenediamine be so much greater than their affinity for ammonia? After all, in both ligands the coordinate bonds are formed by lone pairs of electrons on N atoms. Let's begin our explanation with the chemical equation describing the displacement of ammonia ligands in $Ni(NH_3)_6^{2+}$ ions by ethylenediamine:

$$Ni(NH_3)_6^{2+}(aq) + 3\,en(aq) \rightleftharpoons Ni(en)_3^{2+}(aq) + 6\,NH_3(aq) \qquad (18.1)$$

The color change from blue to purple tells us that this reaction as written is spontaneous. As we discussed in Chapter 14, spontaneous reactions are those in which free energy decreases ($\Delta G < 0$). Furthermore, under standard conditions the change in free energy is related to the changes in enthalpy and entropy that accompany the reaction:

$$\Delta G° = \Delta H° - T\,\Delta S°$$

A negative value for $\Delta G°$ means that $\Delta H°$ must be less than $T\,\Delta S°$. The displacement of NH_3 by ethylenediamine is exothermic, but only slightly ($\Delta H° = -12$ kJ/mol). More important, $\Delta S° = +185$ J/(mol · K). This means that at 25°C:

$$T\,\Delta S° = 298\ \cancel{K} \times \frac{185\ \cancel{J}}{mol \cdot \cancel{K}} \times \frac{1\ kJ}{1000\ \cancel{J}} = 55.1\ kJ/mol$$

To understand why there is such a large increase in entropy, consider that there are 4 moles of reactants but 7 moles of products in Equation 18.1. Nearly doubling the number of moles of aqueous products over reactants translates into a large gain in entropy. It is this positive $\Delta S°$ more than the negative $\Delta H°$ value that drives the reaction and makes it spontaneous. Entropy gains drive many complexation reactions that involve polydentate ligands. The entropy-driven affinity of metal ions for polydentate ligands is called the **chelate effect**.

> **CONCEPT TEST**
>
> The reaction
>
> $$Ni(H_2O)_6^{2+}(aq) + 6\,NH_3(aq) \rightleftharpoons Ni(NH_3)_6^{2+}(aq) + 6\,H_2O(\ell)$$
>
> is spontaneous. Is the spontaneity due principally to a negative ΔH or to a positive ΔS? Explain your answer.

18.8 Crystal Field Theory

We have seen that formation of complex ions can change the color of solutions of transition metals. Why is this? The colors of transition metal compounds and ions in solution are due to transitions of d-orbital electrons. Let's explore these transitions using Cr^{3+} as our model transition metal ion (Figure 18.12).

The Cr^{3+} ion has the electron configuration $[Ar]3d^3$. When a Cr^{3+} ion (or any atom or ion) is in the gas phase, all the orbitals in a given subshell have the same energy. However, when a Cr^{3+} ion is in an aqueous solution and surrounded by an octahedral array of water molecules in $Cr(H_2O)_6^{3+}$, the energies of its $3d$ orbitals are

FIGURE 18.12 When chromium(III) nitrate dissolves in water, the resulting solution has a distinctive violet color due to the presence of $Cr(H_2O)_6^{3+}$ ions.

▶❚❚ **CHEMTOUR** Crystal Field Splitting

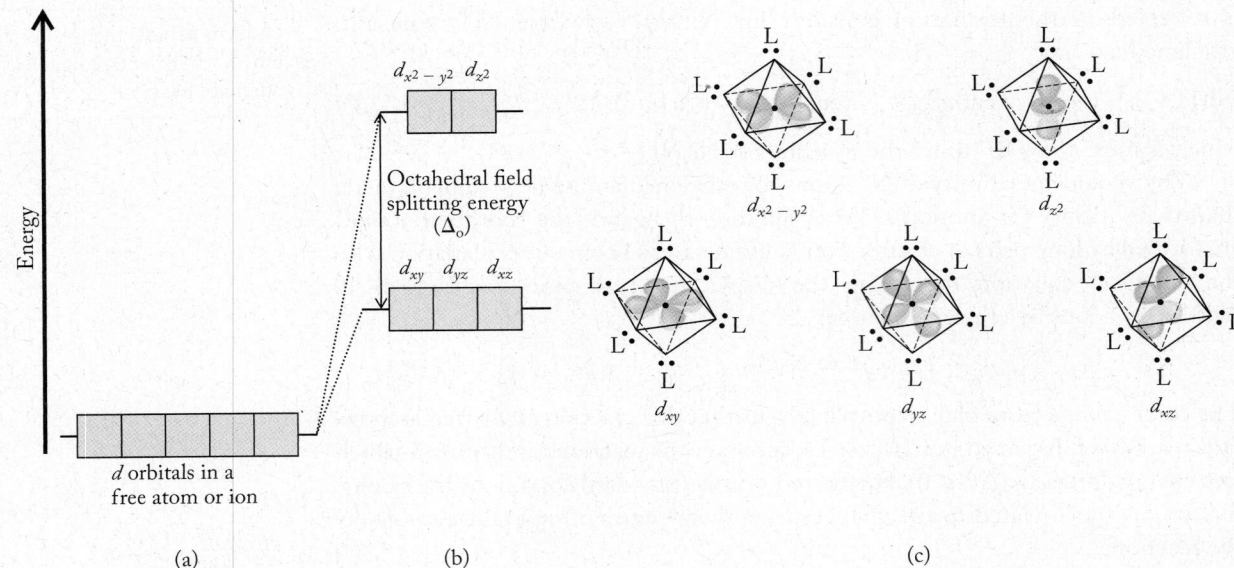

(a) (b) (c)

FIGURE 18.13 Octahedral crystal field splitting. (a) In an atom or ion in the gas phase, all orbitals in a subshell are degenerate, as shown here for the five $3d$ orbitals. (b) When an ion is part of a complex ion in a compound or solution, repulsion between electrons in the ion's d orbitals and ligand electrons raise the energies of the orbitals as shown here for an octahedral field. (c) The greatest repulsion is experienced by electrons in the d_{z^2} and $d_{x^2-y^2}$ orbitals because the lobes are directed toward the corners of the octahedron and so are closest to the lone pairs on the ligands (L). The lobes of the lower-energy d_{xy}, d_{xz}, and d_{yz} orbitals are directed toward points that lie between the corners of the octahedron, so electrons in them experience less repulsion.

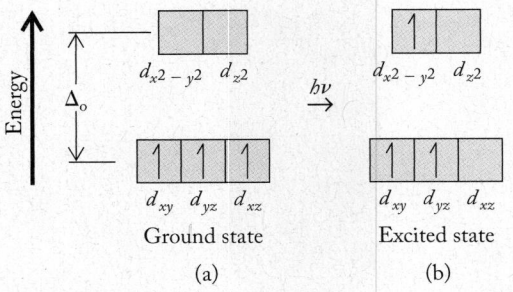

(a) (b)

FIGURE 18.14 A Cr^{3+} ion, [Ar]$3d^3$, in an octahedral field can absorb a photon of light that has energy ($h\nu$) equal to Δ_o. This energy raises a $3d$ electron from one of the lower-energy d orbitals (a) to one of the higher-energy d orbitals (b).

no longer all the same. As the six oxygen atoms of the water molecules approach the Cr^{3+} ion during coordinate bond formation, the $3d$ electrons of Cr^{3+} and the lone pairs of electrons on the ligands repel one another. These repulsions raise the energies of all the d orbitals, but to different extents. The $3d_{xy}$, $3d_{xz}$, and $3d_{yz}$ orbitals experience some increase in energy, but the energies of the $3d_{z^2}$ and $3d_{x^2-y^2}$ orbitals increase even more (Figure 18.13a) because the lobes of the $3d_{z^2}$ and $3d_{x^2-y^2}$ orbitals point directly toward the H_2O oxygen atoms at the corners of the octahedron formed by the ligands, and are repelled by the electrons on those O atoms (Figure 18.13b). The energies of the $3d_{xy}$, $3d_{xz}$, and $3d_{yz}$ orbitals are not raised as much because the lobes of these three orbitals do not point directly toward the corners of the octahedron, so their electron repulsions are weaker.

This process of changing *degenerate* (equal energy) orbitals to orbitals of different energies is known as **crystal field splitting**, and the difference in energy created by crystal field splitting is called **crystal field splitting energy (Δ)**. The name was originally used to describe splitting of d-orbital energies in ionic crystals, but the theory also applies to species in aqueous solutions.

In a $Cr(H_2O)_6^{3+}$ ion, three electrons are distributed among five $3d$ orbitals: three with lower energy than the other two. According to Hund's rule, each of the three electrons should occupy one of the three lower-energy orbitals, leaving the two higher-energy orbitals unoccupied, as shown in Figure 18.14(a). The energy difference between the two subsets of orbitals is symbolized by Δ_o, where the subscript indicates that the energy split was caused by an *o*ctahedral array of electron repulsions.

Now let's consider what happens when a Cr^{3+} ion absorbs a photon whose energy is exactly equal to Δ_o. As the photon is absorbed a $3d$ electron moves from a lower-energy orbital to a higher-energy orbital (Figure 18.14b). The wavelength λ of the absorbed photon is related to the energy difference between the two orbitals—in other words, to the crystal field splitting energy—by Equation 18.2:

$$E = \frac{hc}{\lambda} = \Delta_o \qquad (18.2)$$

As we discussed in Chapter 7, the energy and wavelength of a photon are inversely proportional to each other. Therefore, the larger the crystal field splitting in a complex ion, the shorter the wavelength of the photons the ion absorbs.

The size of the energy gap between split *d* orbitals often corresponds to radiation in the visible region of the electromagnetic spectrum. This means that the colors of solutions of metal complexes depend on the strengths of metal–ligand interactions. When white light (which contains all colors of visible light) passes through a solution containing complex ions, the ions may absorb energy corresponding to a particular color. The light leaving the solution and reaching our eyes is missing that color.

The color we perceive for any transparent object is not the color it absorbs but rather the color(s) that it transmits. To relate the color of a solution to the wavelengths of light it absorbs, we need to consider complementary colors as defined by a simple color wheel (Figure 18.15). For example, red and green are complementary colors; therefore, a solution that absorbs green light appears red to us because our eyes and brains process the transmitted colors—red, orange, yellow, blue, and violet—as the average of those colors, which is red.

Now we have enough information to explain why the solution of $Cr^{3+}(aq)$ in Figure 18.12 is violet. The radiation absorbed by the electron transition shown in Figure 18.14 happens to be yellow-orange light. Because yellow-orange is the complement of violet, $Cr^{3+}(aq)$ solutions appear to be violet.

Color-averaging also occurs when a substance absorbs more than one color, as many transition metal solutions do. For example, a solution of $Cu(NH_3)_4^{2+}$ ions has a distinctive deep blue color, as we saw in Figure 18.6. The light transmitted by such a solution features an absorption band that spans yellow, orange, and red wavelengths with a minimum transmission of light at 620 nm (Figure 18.16). Our eyes sense the range that is transmitted, which spans violet to blue-green. Then our brain processes this band of transmitted colors and signals to us the average of these colors, a deep navy blue.

In two of the three colored solutions we have examined up to this point—Ni^{2+} complexes and Cr^{3+} complexes—the complex ions are octahedral and involve six ligands. However, in the solution of $Cu(NH_3)_4^{2+}$ the complex ions have only four ligands. This means that the deep blue color of this complex ion is caused by a different type of ligand–metal interaction.

Four ligands around a central metal have either a tetrahedral arrangement or a square planar arrangement (Table 18.1). Square planar geometries tend to be limited to the transition metal ions with nearly filled valence-shell *d* orbitals,

CONNECTION In Chapter 7, we first used the equation $E = h\nu = hc/\lambda$ in discussing the energy of light in the electromagnetic spectrum.

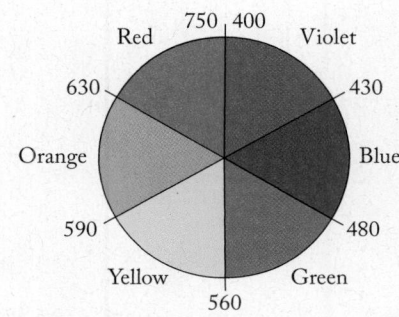

FIGURE 18.15 A color wheel. Colors on opposite sides of the wheel are complementary to each other. When we look at a solution that absorbs light corresponding to a given color, we see the complementary color. Wavelengths are in nanometers.

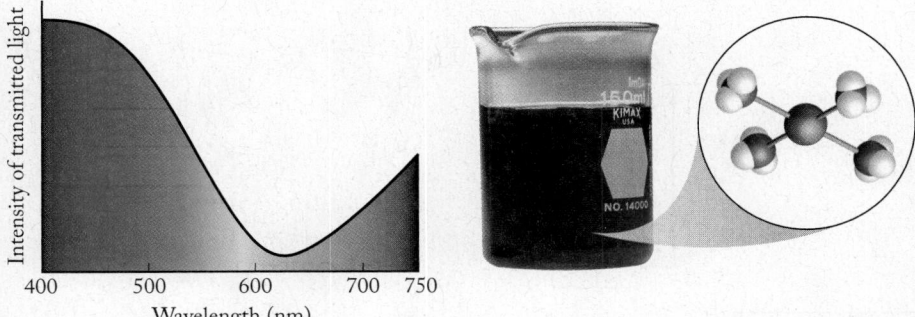

FIGURE 18.16 The visible light transmitted by a solution of $Cu(NH_3)_4^{2+}$ ions is missing much of the yellow, orange, and red portions of the visible spectrum because of a broad absorption band centered at 620 nm. Our eyes and brain perceive the transmitted colors as navy blue.

crystal field splitting the separation of a set of *d* orbitals into subsets with different energies as a result of interactions between electrons in those orbitals and lone pairs of electrons in ligands.

crystal field splitting energy (Δ) the difference in energy between subsets of *d* orbitals split by interactions in a crystal field.

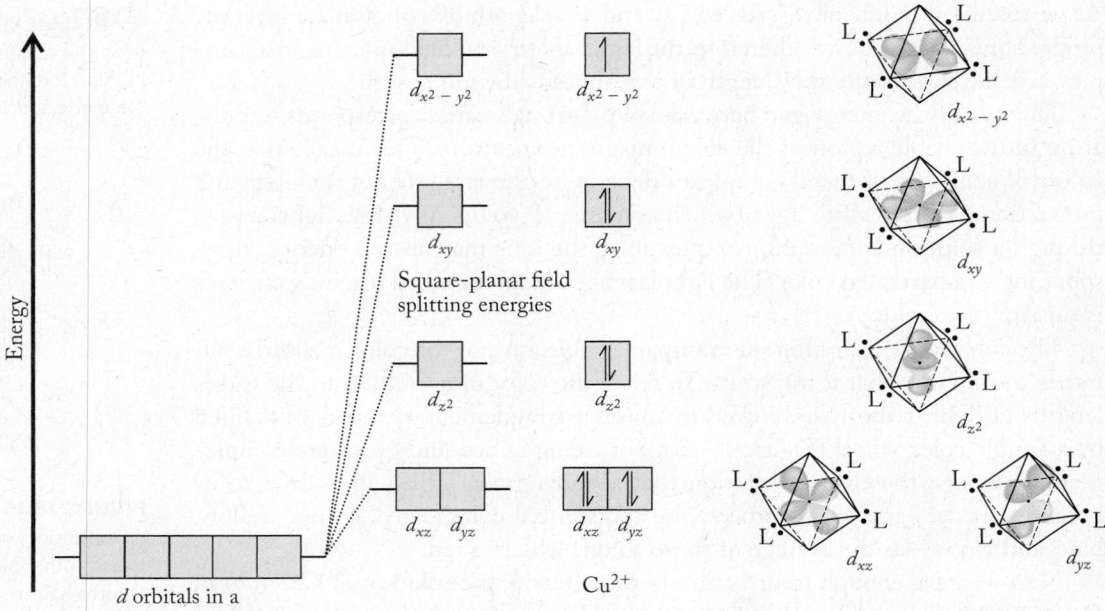

FIGURE 18.17 Square planar crystal field splitting. The d orbitals of a transition metal ion in a square planar field are split into several energy levels depending on how close the orbital lobes are to the ligand electrons that are located at the four corners of the square. The $d_{x^2-y^2}$ orbital has the highest energy because its lobes are directed right at the four corners of the square.

particularly those with d^8 or d^9 electron configurations. One such ion is Cu^{2+}, which has the electron configuration $[Ar]3d^9$. The $Cu(NH_3)_4^{2+}$ complex (Figure 18.6) is square planar, which means that the strongest interactions occur between the $3d$ orbitals on Cu^{2+} and the nitrogen atom lone pairs at the four corners of the equatorial plane of the octahedron, as shown in Figure 18.17. In the Cu^{2+} ion, the $3d$ orbital with the strongest interactions and so the highest energy is the $d_{x^2-y^2}$ orbital because its lobes are oriented directly at the four corners of the plane. The d_{xy} orbital has slightly less energy because its lobes, though in the xy plane, are directed 45° away from the corners. Electrons in the three d orbitals with lobes out of the xy plane interact even less with the lone pairs of the ligand and thus have even lower energies.

Finally, let's consider the d-orbital crystal field splitting that occurs in a tetrahedral complex (Figure 18.18). In this geometry, the greatest electron–electron repulsions are experienced in the d_{xy}, d_{xz}, and d_{yz} orbitals because the lobes of these orbitals are oriented most directly to the corners of the tetrahedron that are occupied by ligand electron pairs. The two other d orbitals are less affected because their lobes do not point toward the corners. The difference in energy

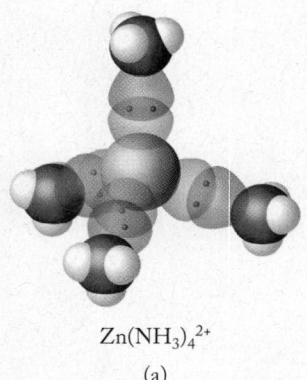

$Zn(NH_3)_4^{2+}$

(a)

FIGURE 18.18 (a) In a tetrahedral complex ion, such as $Zn(NH_3)_4^{2+}$, the d orbitals of the metal ion undergo tetrahedral crystal field splitting. (b) The lobes of the higher-energy orbitals—d_{xy}, d_{xz}, and d_{yz}—are closer to the ligands at the four corners of the tetrahedron than the lobes of the lower-energy orbitals are. (One of the four corners of the tetrahedron is hidden in these drawings.)

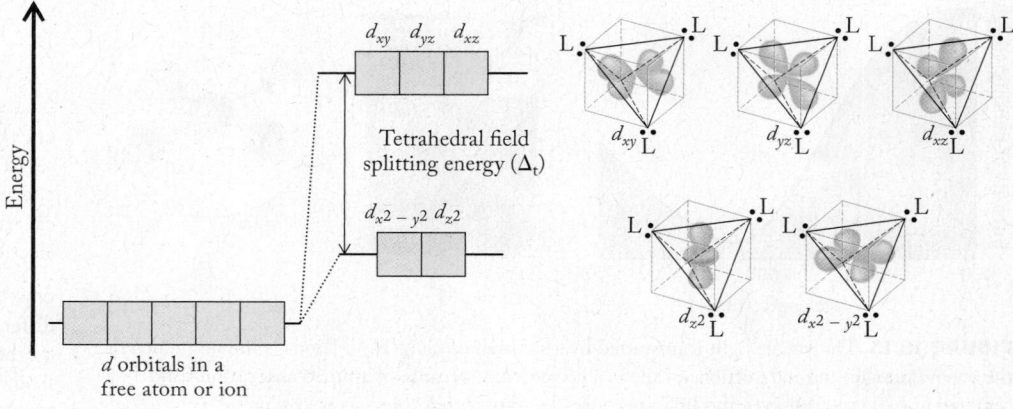

(b)

between the two subsets of *d* orbitals resulting from the tetrahedral interactions is labeled Δ_t.

Before ending this discussion on the colors of transition metal ions, let's revisit the color changes we saw in Figure 18.11, when first ammonia and then ethylenediamine were added to a solution of Ni^{2+} ions. Let's think in terms of the colors these solutions *absorb*. A solution of $Ni^{2+}(aq)$ ions is green because it absorbs colors at the red end of the spectrum. Similarly, a solution of $Ni(NH_3)_6^{2+}$ is blue because it absorbs colors opposite blue, which are centered on orange, and a solution of $Ni(en)_3^{2+}$ is violet because it absorbs colors centered on the complement of violet, which is yellow. Note how the colors these three solutions absorb are in the sequence red, orange, and yellow. This sequence runs from longest wavelength to shortest, from about 730 nm for red to about 570 nm for yellow. Radiant energy is inversely proportional to wavelength (Equation 18.2); therefore, the ability of these ligands to split the energies of the *d* orbitals of Ni^{2+} ions is en > NH_3 > H_2O. Table 18.4 summarizes the observed colors of these Ni^{2+} ion complexes and the meaning we can infer from their colors.

spectrochemical series a list of ligands rank-ordered by their ability to split the energies of the *d* orbitals of transition metal ions.

TABLE 18.4	Light Transmitted and Absorbed by Three Ni^{2+} Complexes				
Complex	$Ni(H_2O)_6^{2+}$	$\xrightarrow{NH_3}$	$Ni(NH_3)_6^{2+}$	$\xrightarrow{en}$	$Ni(en)_3^{2+}$
Appearance:	Green		Blue		Violet
Absorbs:	Red		Orange		Yellow
Absorbed λ (nm):	730	>	600	>	570
$E = hc/\lambda$ (J × 10^{19}):	3.0	<	3.3	<	3.5

Chemists use the parameter *field strength* to describe the relative abilities of ligands to split the energies of *d* orbitals in metal ions, ranking ligands in what is called a **spectrochemical series**. Table 18.5 contains one such series. The ligands at the top of the chart create the strongest repulsions with central-ion electrons, and those at the bottom create the weakest. As the field strength of the ligand increases, the crystal field splitting energy (Δ) increases. Consequently, high-field-strength ligands form complexes that absorb short-wavelength, high-energy light, whereas complexes of low-field-strength ligands absorb long-wavelength, low-energy light.

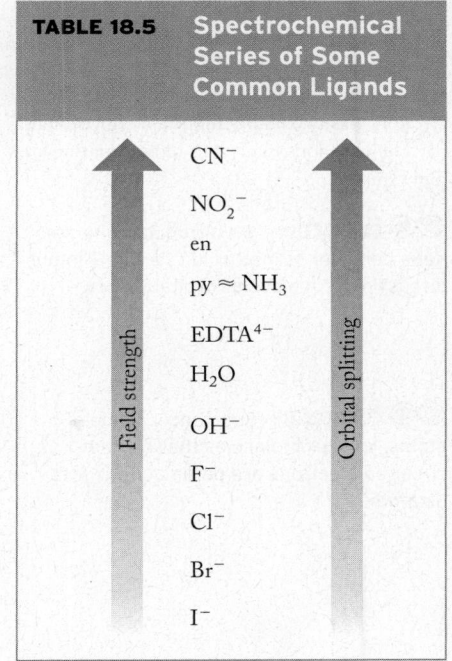

TABLE 18.5	Spectrochemical Series of Some Common Ligands

Field strength / Orbital splitting:

CN^-
NO_2^-
en
py ≈ NH_3
$EDTA^{4-}$
H_2O
OH^-
F^-
Cl^-
Br^-
I^-

> **CONCEPT TEST**
>
> Figure 18.19 is the absorption spectrum of a complex ion found in nature. What color is it?

18.9 Magnetism and Spin States

In addition to determining the color of transition metal ions, crystal field splitting can influence their magnetic properties because these properties depend on the number of unpaired electrons in the valence-shell *d* orbitals. The larger this number,

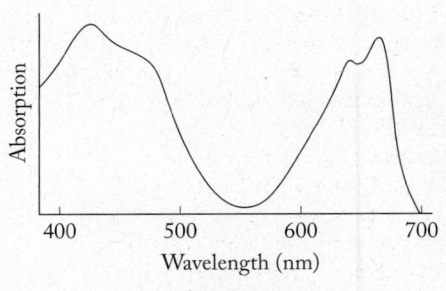

FIGURE 18.19

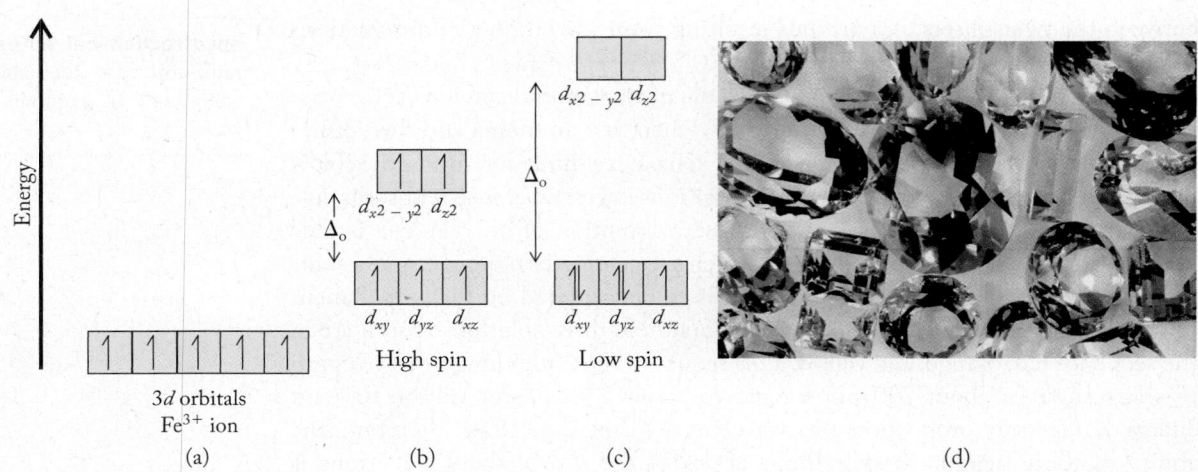

FIGURE 18.20 Low-spin and high-spin complexes. (a) The ground state of a free Fe^{3+} ion has a degenerate, half-filled set of $3d$ orbitals. (b) A weak octahedral field (Δ_o < electron pairing energy) produces the high-spin state: five unpaired electrons, each in its own orbital. (c) In a strong octahedral field (Δ_o > electron pairing energy), the energies of the $3d$ orbitals are split enough to produce the low-spin state: two sets of paired electrons, one unpaired electron, and two empty higher-energy orbitals. (d) The Fe^{3+} ions in crystals of aquamarine are high-spin.

∞ **CONNECTION** We introduced the magnetic behavior of matter in Chapter 9 in our discussion of molecular orbital theory.

∞ **CONNECTION** Substances made of atoms, ions, or molecules that contain unpaired electrons are paramagnetic (see Section 9.7).

the more paramagnetic the ion. For example, an Fe^{3+} ion has five $3d$ electrons (Figure 18.20a). In an octahedral field there are two ways to distribute the five $3d$ electrons: one way conforms to Hund's rule and has a single electron in each orbital, leaving them all unpaired (Figure 18.20b). However, when Δ_o is large, as shown in Figure 18.20(c), all five electrons may go into the three lower-energy orbitals. This pattern of electron distribution occurs when the energy of repulsion between two electrons in the same orbital is less than the energy needed to promote an electron to a higher-energy orbital. In this configuration, only one electron is unpaired.

The configuration with all five electrons unpaired is called the *high-spin state* because the spin on all five electrons is in the same direction, resulting in the maximum magnetic field produced by the spins. The configuration with only one electron unpaired is called the *low-spin state*. Both configurations are paramagnetic because both have at least one unpaired electron. However, the high-spin state is much more paramagnetic.

Not all transition metal ions can have both high-spin and low-spin states. Consider, for example, Cr^{3+} ions in an octahedral field. Because each ion has only three $3d$ electrons (Figure 18.14), each electron is unpaired whether the orbital energies are split a lot or only a little. Therefore Cr^{3+} ions have only one spin state. Other metal ions with full or nearly full sets of d orbitals have only one spin state because there cannot be more than two unpaired electrons in these orbitals no matter how they are distributed.

SAMPLE EXERCISE 18.5 **Predicting Spin States**

Determine which of these ions can have high-spin and low-spin configurations when part of an octahedral complex: (a) Mn^{4+}; (b) Mn^{2+}; (c) Cu^{2+}.

Collect and Organize To determine whether high-spin and low-spin states are possible, we need to determine the number of d electrons in each ion. Then we need to distribute them among sets of d orbitals split by an octahedral field to see if it is possible for the ions to have different spin states. Mn and Cu are in groups 7 and 11 of the periodic table and so their atoms have 7 and 11 valence electrons, respectively. Also, in an octahedral field, a set of five d orbitals splits into a low-energy subset of three orbitals and a high-energy subset of two orbitals.

Analyze The electron configurations of the atoms of the three elements are $[Ar]3d^54s^2$ for Mn and $[Ar]3d^{10}4s^1$ for Cu. When they form cations, the atoms of these transition metals lose their $4s$ electrons first and then their $3d$ electrons. Therefore, the numbers of d electrons in the ions are 3 in Mn^{4+}, 5 in Mn^{2+}, and 9 in Cu^{2+}. It is likely that the ion with the fewest d electrons (Mn^{4+}) and the one with nearly the most d electrons possible (Cu^{2+}) will each have only one spin state.

Solve

   a. Mn^{4+}: Putting three electrons into the lowest-energy $3d$ orbitals available and keeping them as unpaired as possible gives this orbital distribution of electrons:

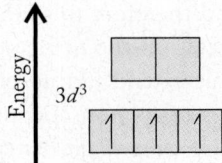

      There is no low-spin option for Mn^{4+} assuming Hund's rule is obeyed.

   b. Mn^{2+}: There are two options for distributing five electrons among the five $3d$ orbitals:

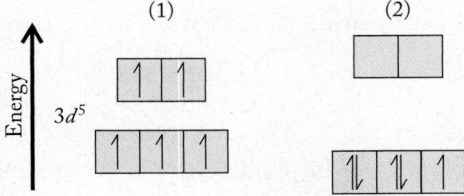

      Thus Mn^{2+} can have a high-spin (on the left) or a low-spin (on the right) configuration when part of an octahedral complex.

   c. Cu^{2+}: The nine $3d$ electrons completely fill the lower-energy orbitals and nearly fill the higher-energy ones. No arrangement is possible other than the one shown, and so Cu^{2+} has only one spin state:

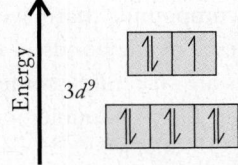

Think about It In an octahedral field, metal ions with 4, 5, 6, or 7 d electrons can exist in high-spin and low-spin states. Those ions with 3 or fewer d electrons have only one spin state, in which all the electrons are unpaired and in the lower-energy set of orbitals. Ions with 8 or more d electrons have only one spin state because their lower-energy set of orbitals is completely filled. The magnitude of the crystal field splitting energy, Δ_o, determines which spin state an ion with 4, 5, 6, or 7 d electrons occupies.

Practice Exercise Which of these ions can have high-spin and low-spin configurations when part of an octahedral complex: (a) V^{4+}; (b) Cr^{3+}; (c) Ni^{3+}? Are any of the possible spin states diamagnetic?

As noted earlier, whether a transition metal ion is in a high-spin state or a low-spin state depends on whether less energy is needed to promote an electron to a higher-energy orbital than to overcome the repulsion experienced by two electrons sharing the same lower-energy orbital. Several factors affect the size of Δ_o. We have already discussed a major one in the context of the spectrochemical series shown in Table 18.5: the different field strengths of different ligands. Because N-containing molecules are stronger field-splitting ligands than H_2O molecules, hydrated metal ions (small Δ_o) are more likely to be in high-spin states, and metals surrounded by nitrogen-containing ligands (large Δ_o) are more likely to be in low-spin states. Another factor affecting spin state is the oxidation state of the metal ion. The higher the oxidation number (and ionic charge), the stronger the attraction of the electron pairs on the ligands for the ion. Greater attraction leads to more ligand–d orbital interaction and therefore to a larger Δ_o.

Complexes of transition metals in the fifth and sixth rows tend to be low spin because their $4d$ and $5d$ orbitals extend farther out from the nucleus than do $3d$ orbitals. These larger d orbitals overlap more and interact more strongly with the lone pairs of electrons on the ligands, leading to greater crystal field splitting.

Our discussion of high-spin and low-spin states has focused entirely on d orbitals split by octahedral fields. What about spin states in tetrahedral fields? Most tetrahedral complexes are high spin because tetrahedral fields, which are created by only four ligands, are weaker than octahedral fields created by six. Weaker field strength means less d-orbital splitting—not enough to offset the energies associated with pairing two electrons in the same orbitals. Therefore, Hund's rule is obeyed.

CONCEPT TEST •

Explain the following: (a) $Mn(pyridine)_6^{2+}$ is a high-spin complex ion, but $Mn(CN)_6^{4-}$ is low spin; (b) $Fe(NH_3)_6^{2+}$ is high spin, but $Ru(NH_3)_6^{2+}$ is low spin.

• •

18.10 Isomerism in Coordination Compounds

We introduced the concepts of structural and stereoisomerism in Chapter 13. To review:

➤ Structural isomers are compounds that have the same chemical formula but different arrangements of the bonds in their molecules. For example, these two hydrocarbons are structural isomers because they have the same formula (C_4H_8) but their $C{=}C$ double bonds are in different locations:

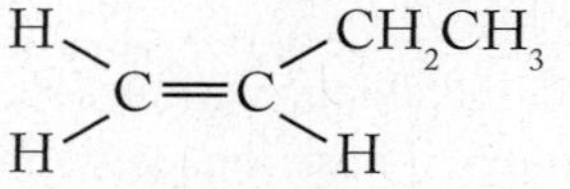

1-Butene 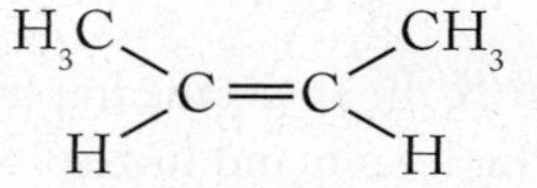   2-Butene

➤ Stereoisomers are compounds with the same formula *and* the same bonding pattern, but the spatial orientation of the groups connected by those bonds is different. For example, these two hydrocarbons are stereoisomers:

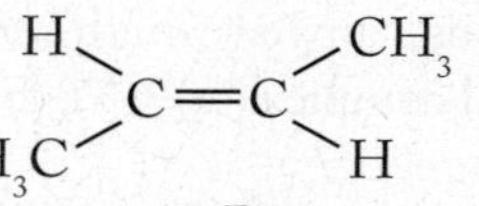

trans-2-Butene 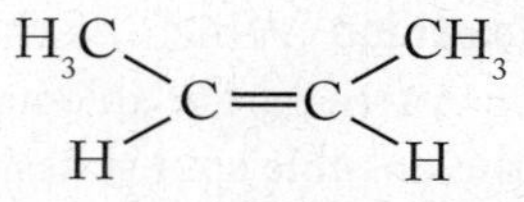   *cis*-2-Butene

In *trans*-2-butene the –CH₃ groups are on opposite sides of the C=C double bond, but in *cis*-2-butene they are on the same side. Remember that C=C bonds are rigid, so the ends of molecules cannot rotate around the C=C axis as they can around C—C single bonds.

CONNECTION The rigidity of double bonds was described in Section 13.3.

Stereoisomers of Complex Ions

Both structural isomerism and stereoisomerism occur in complex ions, but stereoisomerism is our focus here. Let's begin with the square planar Pt²⁺ coordination compound Pt(NH₃)₂Cl₂. Both molecules of ammonia and the two chloride ions are coordinately bonded to the Pt²⁺ ion, so the name of the compound is: diamminedichloroplatinum(II). No counter ions are present because the sum of the charges on the other ions is zero and the complex is neutral.

Two stereoisomers of diamminedichloroplatinum(II) are possible because there are two ways to orient the two pairs of ligands in the square plane of the complex (Figure 18.21): the two members of each pair can be at adjacent corners of the square or at opposite corners. The isomer with the two members of each pair at adjacent corners is called *cis*-diamminedichloroplatinum(II), and the isomer with the pairs at opposite corners is *trans*-diamminedichloroplatinum(II). Note that cis and trans have the same meaning as in the nomenclature of organic compounds.

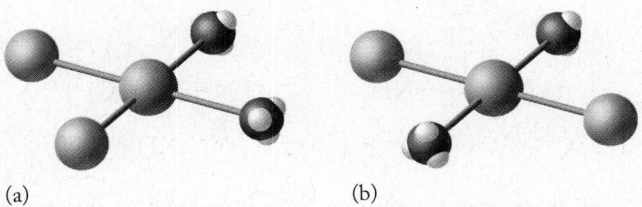

(a)   (b)

FIGURE 18.21 Two ways to orient the Cl⁻ ions and NH₃ molecules around a Pt²⁺ ion (gold sphere) in the square planar coordination compound Pt(NH₃)₂Cl₂. (a) The two members of each pair of ligands are on the same side of the square in *cis*-diamminedichloroplatinum(II). (b) The two members of each pair are at opposite corners in *trans*-diamminedichloroplatinum(II).

To illustrate the importance of stereoisomerism, consider this: *cis*-diamminedichloroplatinum(II) is a widely used anticancer drug with the common name *cisplatin*, but the trans isomer is ineffective in fighting cancer. The therapeutic power of cisplatin comes from its structurally specific reactions with DNA. During these reactions the two Cl atoms of Pt(NH₃)₂Cl₂ are replaced by nitrogen-containing bases along a strand of DNA in the nucleus of a cancerous cell. This ability of cisplatin to cross-link the bases distorts the molecular shape of the DNA and prevents it from replicating. Eventually the cell dies. The trans isomer cannot form links to adjacent bases on a DNA strand and is ineffective as an anticancer agent.

CONCEPT TEST ••••••••••••••••••••••••••••••••••••••

How many isomers does the square planar coordination compound Pt(NH₃)₃Cl have?

(a) *cis*-Tetraamminedichlorocobalt(III) chloride

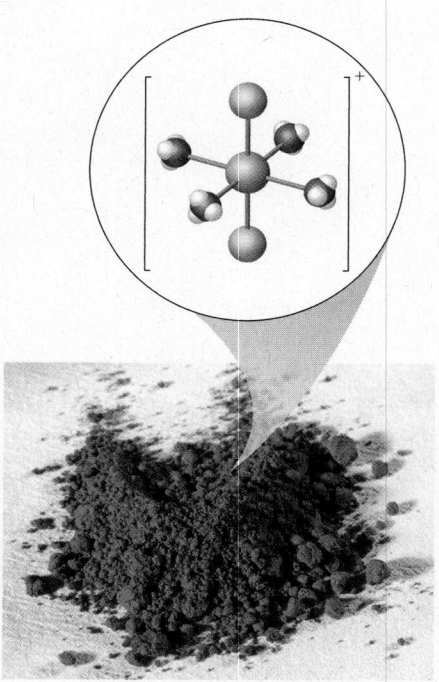

(b) *trans*-Tetraamminedichlorocobalt(III) chloride

FIGURE 18.22 The two stereoisomers of the coordination compound with the formula $[Co(NH_3)_4Cl_2]Cl$: (a) *cis*-tetraamminedichlorocobalt(III) chloride and (b) *trans*-tetraamminedichlorocobalt(III) chloride.

Stereoisomerism is also possible in octahedral complexes containing more than one type of ligand. For example, there are two possible stereoisomers of $[Co(NH_3)_4Cl_2]Cl$ as shown in Figure 18.22. The two chloro ligands in $[Co(NH_3)_4Cl_2]^+$ are either on the same side of the complex with a 90° Cl–Co–Cl bond angle (the cis isomer), or across from each other so that the Cl–Co–Cl bond angle is 180° (the trans isomer). *cis*-Tetraamminedichlorocobalt(III) chloride is violet, and *trans*-tetraamminedichlorocobalt(III) chloride is green.

SAMPLE EXERCISE 18.6 **Identifying Stereoisomers of Coordination Compounds**

Sketch the structures and name the stereoisomers of $Ni(NH_3)_4Cl_2$.

Collect and Organize We are given the formula and are to name and draw the structures of the stereoisomers of a coordination compound. The Ni^{2+} complexes we have seen so far in the chapter have all been octahedral. Ammonia molecules and Cl^- ions are both monodentate ligands.

Analyze The formula contains no brackets, so the Cl^- ions are not counter ions; they must be covalently bonded to the Ni ion. There are two of them, so the charge on Ni must be 2+. There are a total of six ligands, which confirms that the compound is octahedral.

Solve There are two ways to orient the chloride ions: opposite each other with a Cl–Ni–Cl bond angle of 180° or on the same side of the octahedron with a Cl–Ni–Cl bond angle of 90°:

$$\begin{array}{ccc} & Cl & NH_3 \\ & | & \diagup \\ H_3N — & Ni & —NH_3 \\ & | & \\ H_3N & Cl & \end{array} \qquad \begin{array}{ccc} H_3N & & Cl \\ & \diagdown & \diagup \\ H_3N — & Ni & —Cl \\ & \diagup & \\ H_3N & & NH_3 \end{array}$$

The first isomer is *trans*-tetraamminedichloronickel(II); the second is *cis*-tetraamminedichloronickel(II).

Think about It We can draw other tetraamminedichloronickel(II) structures that do not look exactly like these two structures. If we flip or rotate them, however, we will see that they match one of the two structures shown above.

Practice Exercise Sketch the stereoisomers of $[CoBr_2(en)(NH_3)_2]^+$ and name them.

Enantiomers

Another kind of stereoisomerism is possible in complex ions and coordination compounds. Consider the octahedral Co(III) complex ion containing two ethylenediamine molecules and two chloride ions. There are two ways to arrange the chloride ions: on adjacent bonding sites in a cis isomer or on opposite sides of the octahedron in a trans isomer. The cis isomer is shown in Figure 18.23.

This complex ion is called *cis*-dichlorobis(ethylenediamine)cobalt(III). Note the prefix *bis–* just before "(ethylenediamine)." In naming complex ions containing a polydentate ligand, we use *bis–* instead of *di–* to indicate that two molecules of the ligand are present and avoid the use of two *di–* prefixes in the same ligand name. The corresponding prefix for three polydentate ligands is *tris–*.

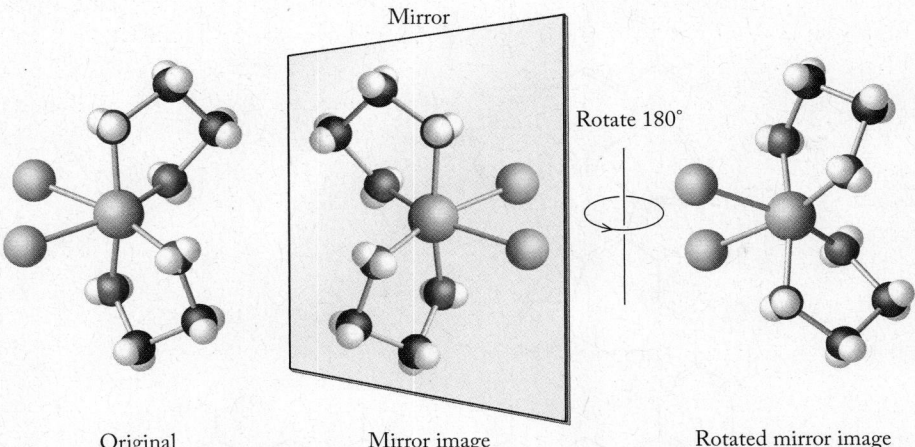

Mirror

Rotate 180°

Original   Mirror image   Rotated mirror image

FIGURE 18.23 The complex ion *cis*-dichlorobis(ethylenediamine)cobalt(III) is chiral, which means that its mirror image is not superimposable on the original complex. To illustrate this point, we rotate the mirror image 180° about its vertical axis so that it looks as much like the original as possible. However, note that the top ethylenediamine ligand is located behind the plane of the page in the original but in front of the plane of the page in the rotated mirror image. Thus, the mirror images are not superimposable.

Figure 18.23 shows that *cis*-Co(en)$_2$Cl$_2^+$ is chiral: it has a mirror image that is not identical to the original. The difference is demonstrated by the fact that there is no way to rotate the mirror image so that its atoms align exactly with those in the original. In other words, the two structures are not *superimposable*. We encountered this phenomenon in Chapter 13 and noted that chemists call such nonsuperimposable stereoisomers *enantiomers*.

CONCEPT TEST

Would four different ligands arranged in a square planar geometry produce a chiral complex ion? What about four different ligands in a tetrahedral geometry?

18.11 Complex Ions in Biomolecules

At the beginning of the chapter, we noted that metals essential to human health must be present in foods in forms the body can absorb. In this section we explore some biological polydentate ligands that help metal ions participate in processes that are essential to nutrition and good health.

Let's begin with photosynthesis, a chemical process at the base of our food chain. Green plants can harness solar energy because they contain large biomolecules we collectively call *chlorophyll*. All molecules of chlorophyll contain ring-shaped tetradentate ligands called *chlorins* (Figure 18.24a). The structures of chlorins are similar to those of **porphyrins**, another class of tetradentate ligands found in biological systems (Figure 18.24b). Chlorins and porphyrins are members of a larger category of compounds known as **macrocyclic ligands**. (*Macrocycle* means, literally, "big ring.")

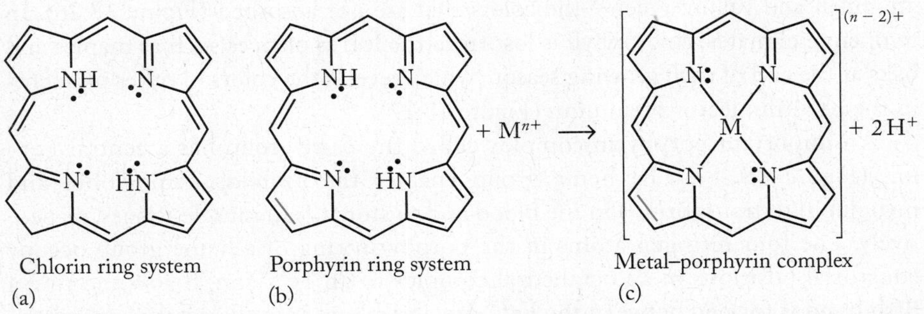

$+ M^{n+} \longrightarrow$ $+ 2 H^+$

Chlorin ring system   Porphyrin ring system   Metal–porphyrin complex
(a)   (b)   (c)

FIGURE 18.24 (a) Chlorin and (b) porphyrin rings are biologically important tetradentate ligands that have similar core structures. The principal difference is a C=C double bond in the porphyrin structure (shown in red) that is a single bond in chlorin rings. The innermost atoms in each ring are four nitrogen atoms with lone pairs of electrons. All four N atoms form coordinate bonds with a metal ion, as shown with the porphyrin ring in (c).

porphyrin a type of tetradentate macrocyclic ligand.

macrocyclic ligand a ring containing multiple electron-pair donors that bind to a metal ion.

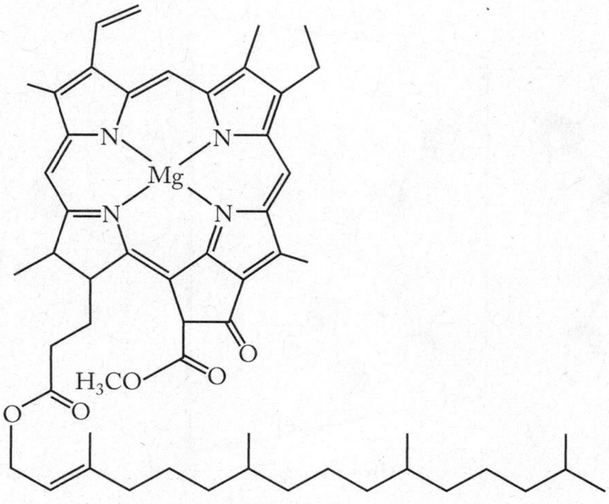

FIGURE 18.25 Chlorophyll absorbs sunlight and initiates a series of reactions that converts sunlight, carbon dioxide, and water into chemical energy stored in the bonds of carbohydrates. All forms of chlorophyll, including chlorophyll *a* shown here, have a Mg^{2+} ion in a chlorin ring as part of their structure.

Chlorophyll *a*

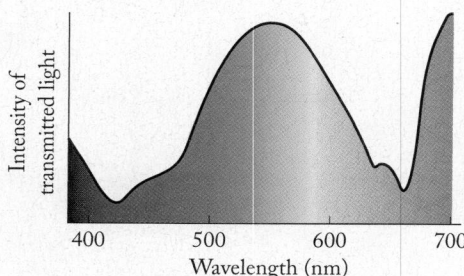

FIGURE 18.26 Combined transmission spectrum of chlorophyll and other pigments in a typical green leaf. The pigments absorb most of the visible radiation emitted by the sun except yellow-green.

FIGURE 18.27 The colors of fall in southern Vermont are characterized by the reds, yellows, and golds of leaves that have lost their green pigments and are about to fall.

Two of the four nitrogen atoms in porphyrins and chlorins are sp^3 hybridized and bound to hydrogen atoms, whereas the other two are sp^2 hybridized with no hydrogen atoms. When either ring forms a coordinate bond with a metal ion M^{n+} (Figure 18.24c), the two hydrogen atoms ionize, giving the ring a charge of 2− and the complex ion an overall charge of $(n − 2)+$. The lone pairs of electrons on the N atoms in the ionized structure are oriented toward the ring center. These lone pairs can occupy either the four equatorial coordination sites in an octahedral complex ion or all four coordination sites in a square planar complex ion. In octahedral complex ions, each central metal ion still has its two axial sites available for bonding to other ligands. Depending on the charge of the central ion, the complex ion may be either ionic or electrically neutral.

Porphyrin and chlorin rings are widespread in nature and play many biochemical roles. Their chemical and physical properties depend on

1. The identity of the central metal ion.
2. The species that occupy the axial coordination sites of octahedral complexes.
3. The number and identity of organic groups attached to the outside of the ring.

The chlorin ring in chlorophyll *a* has a Mg^{2+} ion at its center (Figure 18.25). Delocalized *p* electrons in the conjugated double bonds in and around the ring stabilize it and give it and other plant pigments the ability to absorb wavelengths of red and blue-violet light. Because plant leaves absorb these colors, most of them are green and yellow-green—the colors that are not absorbed (Figure 18.26). In temperate climates, chlorophyll is lost from the leaves of trees such as maples and oaks at the end of each growing season, which reveals the colors of other pigments in the beautiful leaves of autumn (Figure 18.27).

An important porphyrin complex called the *heme* group has a central Fe^{2+} ion (Figure 18.28). The heme group enables the proteins hemoglobin and myoglobin to transport O_2 in the blood and to store O_2 in muscle tissues, respectively. The four nitrogen atoms in the porphyrin ring of a heme group occupy equatorial positions in an octahedral complex of an Fe^{2+} ion. Below the ring a fifth bond is formed between the Fe^{2+} ion and a lone pair of electrons on another nitrogen atom in the protein. The sixth ligand, located above the porphyrin ring, is typically a molecule of O_2, as in the oxygenated forms of hemoglobin in blood leaving the lungs.

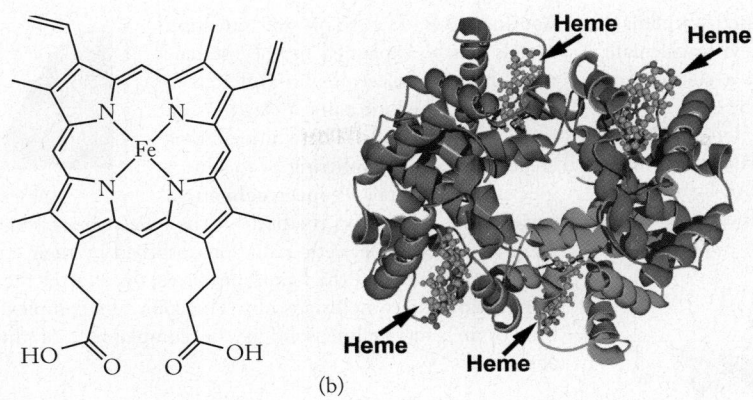

(a) (b)

FIGURE 18.28 (a) In the porphyrin known as heme, the four nitrogen atoms occupy equatorial positions in an octahedral complex in which the central metal ion is Fe^{2+}. (b) Hemoglobin contains four heme groups, each of which is bound to a protein chain. The four protein chains are held together by intermolecular forces.

Each O_2 molecule can act as a Lewis base and donate one of its lone pairs of electrons to the iron in heme. This coordinate covalent bond is strong enough to carry oxygen from the lungs to the cells in our bodies but weak enough to break easily when the oxygen reaches a cell. Other ligands of similar size can bind to the Fe^{2+} ion in heme, and problems arise when some of them do. Carbon monoxide is such a ligand. Similar in size to O_2, a CO molecule easily fits into the sixth binding site. Unfortunately, it binds about 200 times more strongly than O_2. If a person breathes air containing carbon monoxide, CO prevents O_2 from being taken up by blood flowing through the lungs, causing the symptoms of CO poisoning and, ultimately, death.

Proteins called *cytochromes* also contain heme groups (Figure 18.29). Cytochromes mediate oxidation and reduction processes connected with energy production in cells. The heme group conveys electrons as the half-reaction

$$Fe^{3+} + e^- \rightleftharpoons Fe^{2+}$$

rapidly and reversibly consumes or releases electrons needed in the biochemical reactions that sustain life. There are many kinds of cytochrome proteins with different substituents on the porphyrin rings and different axial ligands, each of which influences the function of the complex. This last point has been repeated several times in this chapter: the chemical properties and biological functions of transition metals that are essential to living organisms are linked to their molecular environments and to the formation of stable complex ions with ligands that are strong electron donors, that is, strong Lewis bases.

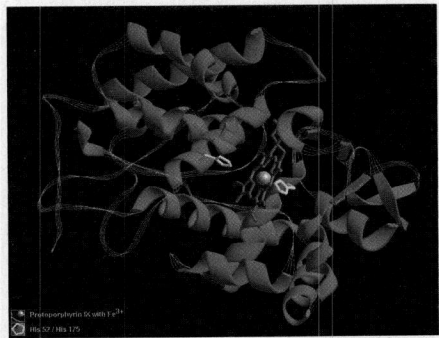

FIGURE 18.29 The structure of cytochrome proteins, such as cytochrome *c* shown here, includes heme complexes (shown in red) that mediate energy production and redox reactions in living cells. Different cytochromes have different ligands (shown in yellow) occupying the fifth and sixth octahedral coordination sites, and different groups in the protein may be attached to the porphyrin ring.

SUMMARY

Section 18.1 A **Lewis base** is a substance that donates pairs of electrons to a **Lewis acid**, defined as an electron-pair acceptor. The donated electron pair forms a covalent bond. In some Lewis acid–Lewis base reactions, other bonds must break to accommodate the new one.

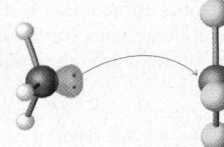

Section 18.2 Transition metal ions form **complex ions** when **ligands** donate pairs of electrons to empty valence-shell orbitals on the metal ion, forming **coordinate bonds**. The number of coordinate bonds in a complex defines the **coordination number** of the metal ion. The metal ion acts as a Lewis acid, and the ligands act as Lewis bases. Compounds that contain complex ions are called **coordination compounds**; the net charges on complex ions are balanced by charges from **counter ions**. Ligands occupy binding sites in the **inner coordination sphere** of a metal ion.

Section 18.3 The stability of any complex ion is expressed mathematically by its **formation constant (K_f)**, which can be used to calculate the equilibrium concentration of free metal ions in a solution of complex ions.

Section 18.4 The names of complex ions and coordinate compounds provide information about the identities and numbers of ligands, the identity and oxidation state of the central metal ion, and the identity and number of counter ions.

Section 18.5 Hydrated metal ions with charges of 2+ or greater can act as Brønsted–Lowry acids, which is why solutions of their soluble salts are acidic. Most transition metal ions have limited solubility in concentrated basic solutions. Metals that are exceptions to this rule form anionic complexes, such as $Cr(OH)_4^-$. The solubility of an ionic compound may be enhanced if the cation forms stable complex ions in solution.

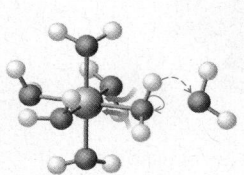

Section 18.6 A **monodentate ligand** donates one pair of electrons in a complex ion; a **polydentate ligand** donates more than one in a process called **chelation**. EDTA is a particularly effective sequestering agent, which is a chelating agent that prevents metal ions in solution from reacting with other substances.

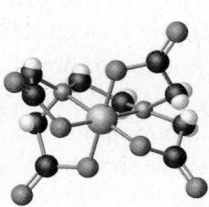

Section 18.7 Polydentate ligands are particularly effective at forming complex ions. This phenomenon is called the **chelate effect** and can be explained by the increase in entropy that accompanies the chelation process.

Section 18.8 The colors of transition metals can be explained by the interactions between electrons in different d orbitals and the lone pairs of electrons on surrounding ligands. These interactions create **crystal field splitting** of the energies of the d orbitals. A **spectrochemical series** ranks ligands on the basis of their *field strength* and the wavelengths of electromagnetic radiation absorbed by their complex ions; the stronger the field the ligand produces, the shorter the wavelength of radiation the complex absorbs. The color of a complex ion in solution or in a crystalline solid is the complement of the color(s) it absorbs.

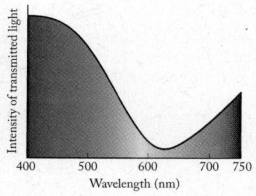

Section 18.9 Strong repulsions and large values of crystal field splitting energy can lead to electron pairing in lower-energy orbitals and an electron configuration called a low-spin state. Metals and their ions are less paramagnetic in low-spin states than when their d electrons are evenly distributed across all the d orbitals in the valence shell—a configuration called a high-spin state.

Section 18.10 Complex metal ions containing more than one type of ligand may form stereoisomers. When one type occupies two adjacent corners of a square planar complex, the complex is a cis isomer; when the same ligand occupies opposite corners, it is a trans isomer.

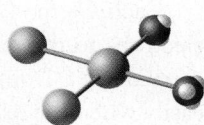

Section 18.11 Complex ions play key roles in many biochemical processes. Photosynthesis is mediated by chlorophyll, a molecule that contains tetradentate chlorin rings coordinately bonded to central Mg^{2+} ions. Oxygen transport in the body is based on the reversible bonding of O_2 molecules to heme groups of Fe^{2+} ions in **porphyrin** rings. Energy production in cells is mediated by cytochromes, proteins that contain heme groups and metals in different oxidation states.

PROBLEM-SOLVING SUMMARY

TYPE OF PROBLEM	CONCEPTS AND EQUATIONS	SAMPLE EXERCISES
Identifying Lewis acids and bases	Determine which reactant donates a pair of electrons (the Lewis base) and which one accepts them (the Lewis acid).	18.1
Calculating the concentration of a free metal ion in equilibrium with a complex	Set up an ICE table based on formation of the complex. Let x = the concentration of metal ions that *does not* form the complex. If the value of K_f is large (it usually is), assume x is much less than the other concentrations.	18.2
Naming coordination compounds	Follow the naming rules in Section 18.4.	18.3
Identifying the potential electron-pair donor groups in a molecule	Examine the molecular structure of a compound and find lone pairs that can be donated to a metal.	18.4
Predicting spin states	Sketch a d-orbital diagram based on crystal field splitting. Fill the lowest-energy orbitals with the valence electrons. If the number of electron pairs in the diagram is greater than the number of pairs you would have if you distributed the electrons evenly over all five d orbitals, multiple spin states are possible.	18.5
Identifying stereoisomers of coordination compounds	Isomers are possible only if there are at least two types of ligands. If ligands of one type are all on the same side of the complex ion, it is a cis isomer. If ligands of one type are on opposite sides, it is a trans isomer.	18.6

VISUAL PROBLEMS ••• ■

(Answers to boldface end-of-chapter questions and problems are in the back of the book.)

18.1. The chlorides of two of the four highlighted elements in Figure P18.1 are colored. Which ones?

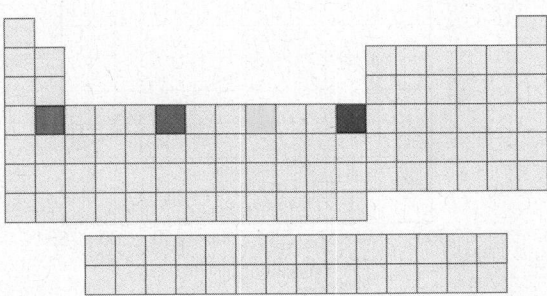

FIGURE P18.1

18.2. Which of the highlighted transition metals in Figure P18.2 form M^{2+} cations that cannot have high-spin and low-spin states?

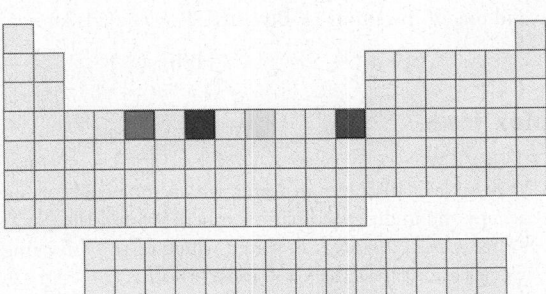

FIGURE P18.2

18.3. Which of the highlighted transition metals in Figure P18.3 have M^{2+} cations that form colorless tetrahedral complex ions?

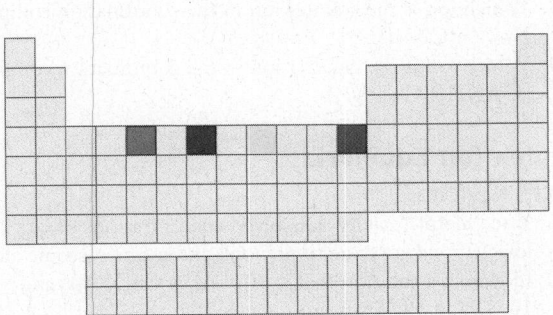

FIGURE P18.3

18.4. Smoky quartz has distinctive lavender and purple colors due to the presence of manganese impurities in crystals of silicon dioxide. Which of the orbital diagrams in Figure P18.4 best describes the Mn^{2+} ion in a tetrahedral field?

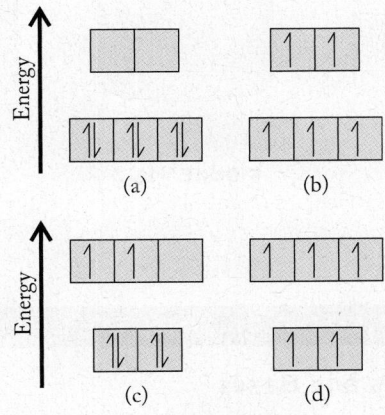

FIGURE P18.4

18.5. Chelation Therapy The compound with the structure shown in Figure P18.5 is widely used in chelation therapy to remove excessive lead or mercury in patients exposed to these metals. How many electron-pair donor groups ("teeth") does the sequestering agent have when the carboxylic acid groups are ionized?

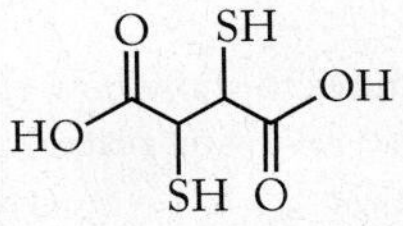

FIGURE P18.5

18.6. Chelation Therapy II The compound with the structure shown in Figure P18.6 has been used to treat people exposed to plutonium, americium, and other actinide metal ions. How many donor groups does the sequestering agent have when the carboxylic acid groups are ionized?

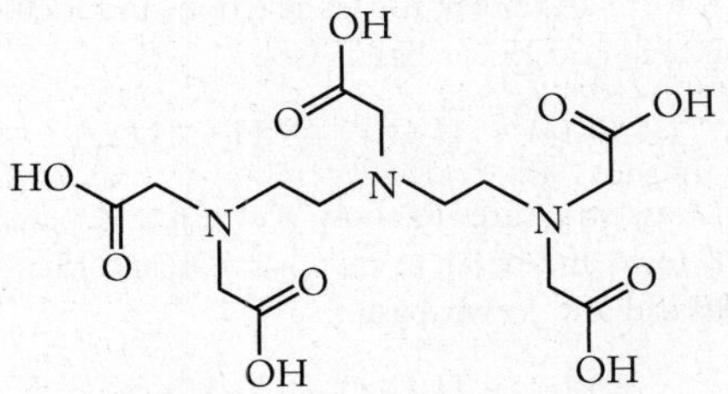

FIGURE P18.6

18.7. The three beakers in Figure P18.7 contain solutions of $[CoF_6]^{3-}$, $[Co(NH_3)_6]^{3+}$, and $[Co(CN)_6]^{3-}$. Based on the colors of the three solutions, which of the complex ions is present in each of the beakers?

(a)   (b)   (c)

FIGURE P18.7

18.8. Figure P18.8 shows the absorption spectrum of a solution of $Ti(H_2O)_6^{3+}$. What color is the solution?

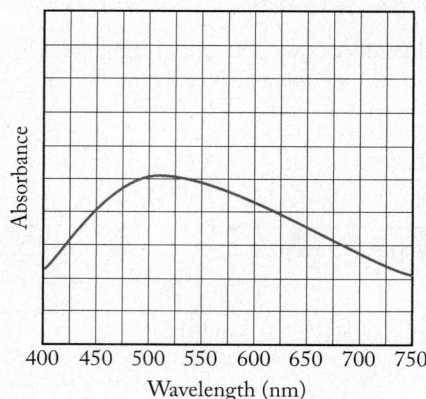

FIGURE P18.8

QUESTIONS AND PROBLEMS

Lewis Acids and Bases

CONCEPT REVIEW

18.9. Can a substance be a Lewis base and not a Brønsted–Lowry base? Explain why or why not.

18.10. Can a substance be a Brønsted–Lowry acid and not be a Lewis acid? Explain why or why not.

18.11. Why is BF_3 a Lewis acid but not a Brønsted–Lowry acid?

18.12. Would you expect NH_3 or H_2O to be a stronger Lewis base? Explain your selection.

PROBLEMS

18.13. Use Lewis structures to show how electron pairs move and bonds form in the following reaction, and identify the Lewis acid and Lewis base.

$$BF_3(g) + F^-(aq) \rightarrow BF_4^-(aq)$$

18.14. Use Lewis structures to show how electron pairs move and bonds form and break in this reaction, and identify the Lewis acid and Lewis base.

$$MgO(s) + CO_2(g) \rightarrow MgCO_3(s)$$

18.15. Use Lewis structures to show how electron pairs move and bonds form and break in this reaction, and identify the Lewis acid and Lewis base.

$$CO_2(g) + H_2O(\ell) \rightleftharpoons H_2CO_3(aq)$$

18.16. Use Lewis structures to show how electron pairs move and bonds form and break in this reaction, and identify the Lewis acid and Lewis base.

$$SO_3(g) + H_2O(\ell) \rightarrow H_2SO_4(aq)$$

18.17. Use Lewis structures to show how electron pairs move and bonds form and break in this reaction, and identify the Lewis acid and Lewis base.

$$B(OH)_3(aq) + H_2O(\ell) \rightleftharpoons B(OH)_4^-(aq) + H^+(aq)$$

*18.18.** Use Lewis structures to show how electron pairs move and bonds form and break in this reaction, and identify the Lewis acid and Lewis base. NOTE: $HSbF_6$ is an ionic compound and one of the strongest Brønsted–Lowry acids known.

$$SbF_5(s) + HF(g) \rightleftharpoons HSbF_6(s)$$

Complex Ions

CONCEPT REVIEW

18.19. When NaCl dissolves in water, which molecules or ions occupy the inner coordination sphere around the Na^+ ions?

18.20. When $CrCl_3$ dissolves in water, which of the following species are nearest the Cr^{3+} ions: (a) other Cr^{3+} ions, (b) Cl^- ions, (c) molecules of H_2O with their O atoms closest to the Cr^{3+} ions, (d) molecules of H_2O with their H atoms closest to the Cr^{3+} ions?

18.21. When $Ni(NO_3)_2$ dissolves in water, what molecules or ions occupy the inner coordination sphere around the Ni^{2+} ions?

18.22. When $[Ni(NH_3)_6]Cl_2$ dissolves in water, which molecules or ions occupy the inner coordination sphere around the Ni^{2+} ions?

18.23. Which ion is the counter ion in the coordination compound $Na_2[Zn(CN)_4]$?

18.24. Which ion is the counter ion in the coordination compound $[Co(NH_3)_4Cl_2]NO_3$?

Complex-Ion Equilibria

CONCEPT REVIEW

18.25. **Trace Metal Toxicity** Dissolved concentrations of Cu^{2+} as low 10^{-6} M are toxic to phytoplankton (microscopic algae). However, a solution that is 10^{-3} M in $Cu(NO_3)_2$ and 10^{-3} M in EDTA is not toxic. Why?

18.26. When a strong base is added to a solution of $CuSO_4$, which is pale blue, a precipitate forms and the solution above the precipitate is colorless. When ammonia is added, the precipitate dissolves and the solution turns a deep navy blue. Use appropriate chemical equations to explain why the observed changes occur.

18.27. A lab technician cleaning glassware that contains residues of AgCl washes the glassware with an aqueous solution of ammonia. The AgCl, which is insoluble in water, rapidly dissolves in the ammonia solution. Why?

18.28. The procedure used in the previous question dissolves AgCl but not AgI. Why?

PROBLEMS

NOTE: Appendix 5 contains formation constant (K_f) values that may be useful in solving the following problems.

18.29. One millimole of $Ni(NO_3)_2$ dissolves in 250 mL of a solution that is 0.500 M in ammonia.

a. What is the concentration of $Ni(NO_3)_2$ in the solution?

b. What is the concentration of $Ni^{2+}(aq)$ in the solution?

18.30. A 1.00 L solution contains 3.00×10^{-4} M $Cu(NO_3)_2$ and 1.00×10^{-3} M ethylenediamine. What is the concentration of $Cu^{2+}(aq)$ in the solution?

18.31. Suppose a 500 mL solution contains 1.00 mmol of $Co(NO_3)_2$, 100 mmol of NH_3, and 100 mmol of ethylenediamine. What is the concentration of $Co^{2+}(aq)$ in the solution?

***18.32.** To a 250 mL volumetric flask are added 1.00 mL volumes of three solutions: 0.0100 M $AgNO_3$, 0.100 M NaBr, and 0.100 M NaCN. The mixture is diluted with deionized water to the mark and shaken vigorously. Are the contents of the flask cloudy or clear? Support your answer with the appropriate calculations. (*Hint*: The K_{sp} of AgBr is 5.4×10^{-13}.)

Naming Complex Ions and Coordination Compounds

PROBLEMS

18.33. What are the names of the following complex ions?

a. $Cr(NH_3)_6^{3+}$

b. $Co(H_2O)_6^{3+}$

c. $[Fe(NH_3)_5Cl]^{2+}$

18.34. What are the names of the following complex ions?

a. $Cu(NH_3)_2^+$

b. $Ti(H_2O)_4(OH)_2^{2+}$

c. $Ni(NH_3)_4(H_2O)_2^{2+}$

18.35. What are the names of the following complex ions?

a. $CoBr_4^{2-}$

b. $Zn(H_2O)(OH)_3^-$

c. $Ni(CN)_5^{3-}$

18.36. What are the names of the following complex ions?

a. CoI_4^{2-}

b. $CuCl_4^{2-}$

c. $[Cr(en)(OH)_4]^-$

18.37. What are the names of the following coordination compounds?

a. $[Zn(en)_2]SO_4$

b. $[Ni(NH_3)_5(H_2O)]Cl_2$

c. $K_4Fe(CN)_6$

18.38. What are the names of the following coordination compounds?

a. $(NH_4)_3[Co(CN)_6]$

b. $[Co(en)_2Cl](NO_3)_2$

c. $[Fe(H_2O)_4(OH)_2]Cl$

Hydrated Metal Ions as Acids

CONCEPT REVIEW

18.39. Which, if any, aqueous solutions of the following compounds are acidic: (a) $CaCl_2$, (b) $CrCl_3$, (c) NaCl, (d) $FeCl_2$?

18.40. If 0.1 M aqueous solutions of each of these compounds were prepared, which one has the lowest pH? (a) $BaCl_2$; (b) $NiCl_2$; (c) KCl; (d) $TiCl_4$

18.41. When ozone is bubbled through a solution of iron(II) nitrate, dissolved Fe(II) ions are oxidized to Fe(III) ions. How does the oxidation process affect the pH of the solution?

18.42. As an aqueous solution of KOH is slowly added to a stirred solution of $AlCl_3$, the mixture becomes cloudy, but then clears when more KOH is added.

a. Explain the chemical changes responsible for the changes in the appearance of the mixture.

b. Would you expect to observe the same changes if KOH were added to a solution of $FeCl_3$? Explain why or why not.

18.43. Chromium(III) hydroxide is amphiprotic. Write chemical equations showing how an aqueous suspension of this compound reacts to the addition of a strong acid and a strong base.

18.44. Zinc hydroxide is amphiprotic. Write chemical equations showing how an aqueous suspension of this compound reacts to the addition of a strong acid and a strong base.

18.45. Refining Aluminum To remove impurities such as calcium and magnesium carbonates and Fe(III) oxides from aluminum ore (which is mostly Al_2O_3), the ore is treated with a strongly basic solution. In this treatment Al(III) dissolves but the other metal ions do not. Why?

***18.46.** The same quantity of $FeCl_3$ dissolves in 1 M aqueous solutions of each of these acids: HNO_3, HNO_2, H_2SO_3, and CH_3COOH. Is the concentration of $Fe^{3+}(aq)$ the same in all four solutions? Explain why or why not.

PROBLEMS

18.47. What is the pH of 0.50 M $Al(NO_3)_3$?

18.48. What is the pH of 0.25 M $CrCl_3$?

18.49. What is the pH of 0.100 M $Fe(NO_3)_3$?

18.50. What is the pH of 1.00 M $Cu(NO_3)_2$?

18.51. Sketch the titration curve (pH versus volume of 0.50 M NaOH) for a 25 mL sample of 0.5 M $FeCl_3$.

18.52. Sketch the titration curve that results from the addition of 0.50 M NaOH to a sample containing 0.5 M $KFe(SO_4)_2$.

Polydentate Ligands; Ligand Strength and the Chelate Effect

CONCEPT REVIEW

18.53. What is meant by the term *sequestering agent*? What properties make a substance an effective sequestering agent?

18.54. The condensed molecular structures of two compounds that each contain two $-NH_2$ groups are shown in Figure P18.48. The one on the left is ethylenediamine, a bidentate ligand. Does the molecule on the right have the same ability to donate two pairs of electrons to a metal ion? Explain why you think it does or does not.

FIGURE P18.48

18.55. How does the chelating ability of an *aminocarboxylate* vary with changing pH?

*18.56. The EDTA that is widely used as a food preservative is added to food, not as the undissociated acid, but rather as the calcium disodium salt: $Na_2[CaEDTA]$. This salt is actually a coordination compound with a Ca^{2+} ion at the center of a complex ion. Draw a line structure of this compound.

Crystal Field Theory

CONCEPT REVIEW

18.57. Explain why the compounds of most of the first-row transition metals are colored.

18.58. Unlike the compounds of most transition metal ions, those of Ti^{4+} are colorless. Why?

18.59. Why is the d_{xy} orbital higher in energy than the d_{xz} and d_{yz} orbitals in a square planar crystal field?

18.60. On average, the d orbitals of a transition metal ion in an octahedral field are higher in energy than they are when the ion is in the gas phase. Why?

PROBLEMS

18.61. Aqueous solutions of one the following complex ions of Cr(III) are violet; solutions of the other are yellow. Which is which? (a) $Cr(H_2O)_6^{3+}$; (b) $Cr(NH_3)_6^{3+}$

18.62. Which of the following complex ions should absorb the shortest wavelengths of electromagnetic radiation? (a) $Cu(Cl)_4^{2-}$; (b) $Cu(F)_4^{2-}$; (c) $Cu(I)_4^{2-}$; (d) $Cu(Br)_4^{2-}$

18.63. The octahedral crystal field splitting energy Δ_o of $Co(phen)_3^{3+}$ is 5.21×10^{-19} J/ion. What is the color of a solution of this complex ion?

18.64. The octahedral crystal field splitting energy Δ_o of $Co(CN)_6^{3-}$ is 6.74×10^{-19} J/ion. What is the color of a solution of this complex ion?

18.65. Solutions of $NiCl_4^{2-}$ and $NiBr_4^{2-}$ absorb light at 702 and 756 nm, respectively. In which ion is the split of d-orbital energies greater?

18.66. Chromium(III) chloride forms six-coordinate complexes with bipyridine, including *cis*-$[Cr(bipy)_2Cl_2^+]$, which reacts slowly with water to produce two products, *cis*-$[Cr(bipy)_2(H_2O)Cl]^{2+}$ and *cis*-$[Cr(bipy)_2(H_2O)_2]^{3+}$. In which of these complexes should Δ_o be the largest?

Magnetism and Spin States

CONCEPT REVIEW

18.67. What determines whether a transition metal ion is in a *high-spin* configuration or a *low-spin* configuration?

*18.68. In Section 18.2 we noted that the bonding in $Fe(H_2O)_6^{3+}$ involves sp^3d^2 hybrid orbitals, but that the bonding in $Fe(CN)_6^{3-}$ comes from coordinate bonds with d^2sp^3 hybrid orbitals on Fe^{3+}. What is the reason for this difference in hybridization?

PROBLEMS

18.69. How many unpaired electrons are there in the following transition metal ions in an octahedral field? High-spin Fe^{2+}, Cu^{2+}, Co^{2+}, and Mn^{3+}

18.70. Which of the following cations can have either a high-spin or a low-spin electron configuration in an octahedral field: Fe^{2+}, Co^{3+}, Mn^{2+}, and Cr^{3+}?

18.71. Which of the following cations can, in principle, have either a high-spin or a low-spin electron configuration in a tetrahedral field: Co^{2+}, Cr^{3+}, Ni^{2+}, and Zn^{2+}?

18.72. How many unpaired electrons are in the following transition metal ions in an octahedral crystal field? High-spin Fe^{3+}, Rh^+, and V^{3+}, and low-spin Mn^{3+}

18.73. The manganese minerals pyrolusite, MnO_2, and hausmannite, Mn_3O_4, contain Mn ions in octahedral holes formed by oxide ions.
   a. What are the charges of the Mn ions in each mineral?
   b. In which of these compounds could there be high-spin and low-spin Mn ions?

18.74. **Dietary Supplement** Chromium picolinate is an over-the-counter diet aid sold in many pharmacies. The Cr^{3+} ions in this coordination compound are in an octahedral field. Is the compound paramagnetic or diamagnetic?

*18.75. One method for refining cobalt involves the formation of the complex ion $CoCl_4^{2-}$. This anion is tetrahedral. Is this complex paramagnetic or diamagnetic?

*18.76. Why is it that $Ni(CN)_4^{2-}$ is diamagnetic, but $NiCl_4^{2-}$ is paramagnetic?

Isomerism in Coordination Compounds; Complex Ions in Biomolecules

CONCEPT REVIEW

18.77. What do the prefixes *cis-* and *trans-* mean in the context of an octahedral complex ion?

18.78. What do the prefixes *cis-* and *trans-* mean in the context of a square planar complex?

18.79. How many different types of donor groups are required to have stereoisomers of a square planar complex?

18.80. With respect to your answer to the previous question, do all square planar complexes with this many different types of donor groups have stereoisomers?

PROBLEMS

18.81. Does $Co(en)(H_2O)_2Cl_2$ have stereoisomers?

18.82. Does the complex ion $Fe(en)_3^{3+}$ have stereoisomers?

*18.83. Sketch the stereoisomers of the square planar complex ion $CuCl_2Br_2^{2-}$. Are any of these isomers chiral?

*18.84. Sketch the stereoisomers of the complex ion $Ni(en)Cl_2(CN)_2^{2-}$. Are any of these isomers chiral?

Additional Problems

18.85. **Photographic Film Processing** During the processing of black-and-white photographic film, excess silver(I) halides are removed by washing the film in a bath containing sodium thiosulfate. This treatment is based on the following complexation reaction:

$$Ag^+(aq) + 2\,S_2O_3^{2-}(aq) \rightleftharpoons Ag(S_2O_3)_2^{3-}(aq) \qquad K_f = 5 \times 10^{13}$$

What is the ratio of $[Ag^+]$ to $[Ag(S_2O_3)_2^{3-}]$ in a bath in which $[S_2O_3^{2-}] = 0.233$ *M*?

18.86. **Lead Poisoning** Children used to be treated for lead poisoning with intravenous injections of EDTA. If the concentration of EDTA in the blood of a patient is 2.5×10^{-8} *M* and the formation constant for the complex $[Pb(EDTA)]^{2-}$ is 2.0×10^{18}, what is the concentration ratio of the free (and potentially toxic) $Pb^{2+}(aq)$ in the blood to the much less toxic Pb^{2+}–EDTA complex?

18.87. Dissolving cobalt(II) nitrate in water gives a beautiful purple solution. There are three unpaired electrons in this cobalt(II) complex. When cobalt(II) nitrate is dissolved in aqueous ammonia and oxidized with air, the resulting yellow complex has no unpaired electrons. Which cobalt complex has the larger crystal field splitting energy Δ_o?

18.88. A solid compound containing Fe(II) in an octahedral crystal field has four unpaired electrons at 298 K. When the compound is cooled to 80 K, the same sample appears to have no unpaired electrons. How do you explain this change in the compound's properties?

18.89. When Ag_2O reacts with peroxodisulfate ($S_2O_8^{2-}$) ion (a powerful oxidizing agent), AgO is produced. Crystallographic and magnetic analyses of AgO suggest that it is not simply Ag(II) oxide, but rather a blend of Ag(I) and Ag(III) in a square planar environment. The Ag^{2+} ion is paramagnetic but, like AgO, Ag^+ and Ag^{3+} are diamagnetic. Explain why.

18.90. The iron(II) compound $Fe(bipy)_2(SCN)_2$ is paramagnetic, but the corresponding cyanide compound $Fe(bipy)_2(CN)_2$ is diamagnetic. Why do these two compounds have different magnetic properties?

18.91. Aqueous solutions of copper(II)–ammonia complexes are dark blue. Will the color of the series of complexes $Cu(H_2O)_{(6-x)}(NH_3)_x^+$ shift toward shorter or longer wavelengths as the value of x increases from 0 to 6?

Additional study materials including ChemTours and Diagnostic Quizzes are available at StudySpace at www.wwnorton.com/studyspace.

19

Electrochemistry and the Quest for Clean Energy

Learning Outcomes

- Combine the appropriate half-reactions to write net ionic equations of spontaneous redox reactions
- Draw cell diagrams and describe the components of electrochemical cells and their roles in interconverting chemical and electrical energy
- Calculate standard cell potentials from standard reduction potentials
- Interconvert a cell's potential and the change in free energy of the cell reaction
- Use the Nernst equation to calculate cell potentials
- Compare and contrast voltaic and electrolytic cells
- Describe the structure and cell reactions of nickel-metal hydride batteries, lithium-ion batteries, and fuel cells

A LOOK AHEAD

Running on Electricity

The first decade of the 21st century saw remarkable swings in the price of gasoline. These swings, coupled with growing concerns over climate change, invigorated development of innovative propulsion systems for vehicles that consume less fossil fuel. Some of these vehicles, called "hybrids," are propelled by combinations of electric motors powered by rechargeable batteries and small gasoline engines. Others are completely electric, powered by banks of high-performance batteries or fuel cells.

To give all-electric cars the driving range and performance many motorists want, scientists and engineers continue to develop lighter, higher-capacity, and more reliable batteries to power these vehicles. Batteries convert chemical energy into electrical energy; motors then convert that electrical energy into mechanical energy. These two processes together are more efficient than the conversion of chemical energy directly to mechanical energy in gasoline engines. Moreover, electric motors consume no energy when the vehicles they power are stopped in traffic. A vehicle's electric motor can even recharge its batteries when the brakes are applied. As the car slows, the motor becomes an electric *generator*, turning the vehicle's kinetic energy into electrical energy.

Ultimately, the driving range of a battery-powered vehicle is limited by the capacity of its battery to store and deliver electrical energy. The sports car in the photo can make trips as long as 400 km on a single charge, though this range has a cost: its battery pack, which contains over 6,800 individual cells each about the size of a AA battery, accounts for half the weight of the car. The smaller battery packs in vehicles known as plug-in hybrids (their batteries can be recharged using electricity from the power grid) give them driving ranges of about 50 km on battery power alone.

An Electrifying Sports Car A bank of lithium–ion batteries and a 185-kilowatt motor power this high-performance electric sports car. ▶

electrochemistry the branch of chemistry that examines the transformations between chemical and electrical energy.

In this chapter we examine the chemistry of modern batteries and of another source of mobile electrical power, fuel cells. Batteries and fuel cells harness the chemical energy released by spontaneous redox reactions, so we begin the chapter by reviewing what we learned about redox chemistry in Chapter 4. Then we explore how the electrochemical energy released inside batteries can be harnessed to pump electrons through devices like electric motors, and how a battery's chemical energy can be restored by forcing its spontaneous cell reaction to run in reverse. We also investigate how the energy released in fuel cells when H_2 and O_2 combine to form H_2O is used to do electrical work and propel automobiles more efficiently and with less environmental impact than the cars we drive today. ■

19.1 Redox Chemistry Revisited

In Chapters 15 through 17 we examined some of the environmental problems associated with the combustion of fossil fuels and the operation of internal combustion engines. In this chapter, we explore alternative propulsion technologies that are more efficient than internal combustion engines at converting chemical energy into mechanical energy and that have the potential to dramatically reduce air pollution. These technologies are based on **electrochemistry**, the branch of chemistry that links chemical reactions to the production or consumption of electrical energy. At the heart of electrochemistry are chemical reactions in which electrons are gained and lost at electrode surfaces. In other words, electrochemistry is based on *red*uction and *ox*idation, or *redox,* chemistry.

The principles of redox reactions were introduced in Section 4.9. Let's briefly review them here:

➤ A redox reaction is the sum of two half-reactions: a reduction half-reaction in which a reactant gains electrons, and an oxidation half-reaction in which a reactant loses electrons.

➤ Reduction and oxidation half-reactions happen simultaneously so that the number of electrons gained during reduction exactly matches the number lost during oxidation.

➤ A substance that is easily oxidized is one that readily gives up electrons. This electron-donating power makes the substance an effective reducing agent. In any redox reaction the *reducing agent is always oxidized.*

➤ A substance that readily accepts electrons and is thereby reduced is an effective oxidizing agent. In any redox reaction the *oxidizing agent is always reduced.*

Let's begin our exploration of electrochemistry with an examination of the redox properties of two neighbors in the periodic table: copper ($Z = 29$) and zinc ($Z = 30$). Elemental zinc has considerable electron-donating power, and Cu^{2+} ions tend to be willing acceptors of donated electrons. These complementary properties are captured in Figure 19.1. When a strip of Zn metal is placed in a solution of $CuSO_4$, that is, a solution of Cu^{2+} ions and SO_4^{2-} ions, Zn atoms spontaneously donate electrons to Cu^{2+} ions, forming Zn^{2+} ions and Cu atoms. The shiny zinc surface turns dark brown as a textured layer of copper metal accumulates on it, and the distinctive blue color of $Cu^{2+}(aq)$ ions fades as these ions acquire electrons and become atoms of copper metal.

In this spontaneous reaction, each mole of Zn atoms that is oxidized *loses* 2 moles of electrons in this oxidation half-reaction:

$$Zn(s) \rightarrow Zn^{2+}(aq) + 2\,e^-$$

(a)   (b)   (c)

FIGURE 19.1 (a) A strip of zinc is immersed in an aqueous solution of blue copper(II) sulfate. (b) Over time the zinc strip becomes encrusted with a dark layer of copper. (c) Eventually, the blue color of the solution fades as Cu^{2+} ions are reduced to Cu atoms. In the process, Zn atoms are oxidized to colorless Zn^{2+} ions.

and every mole of Cu^{2+} ions that is reduced *accepts* 2 moles of electrons in this reduction half-reaction:

$$Cu^{2+}(aq) + 2\,e^- \rightarrow Cu(s)$$

Because the number of moles of electrons lost and gained is the same in the two half-reactions, writing a net ionic equation to describe the overall redox reaction is simply a matter of adding the two half-reactions together:

$$Zn(s) \rightarrow Zn^{2+}(aq) + 2\,e^-$$

$$\underline{Cu^{2+}(aq) + 2\,e^- \rightarrow Cu(s)}$$

$$Zn(s) + Cu^{2+}(aq) + \cancel{2\,e^-} \rightarrow Cu(s) + Zn^{2+}(aq) + \cancel{2\,e^-}$$

Canceling out the equal numbers of electrons gained and lost, we get

$$Zn(s) + Cu^{2+}(aq) \rightarrow Cu(s) + Zn^{2+}(aq) \qquad (19.1)$$

Combining half-reactions is a convenient way to write net ionic equations for redox reactions, as noted in Section 4.9. In this chapter we use a valuable resource in this equation writing process, a table of common half-reactions (Table A6.1) in Appendix 6. Note that all the half-reactions are written as reduction half-reactions. There is no need for a separate table of *oxidation* half-reactions because any reduction half-reaction can always be reversed to obtain the corresponding oxidation half-reaction. Also, nearly all the half-reactions in Table A6.1 are written as if they occur in acidic solutions. We know this because H^+ ions are used to balance the number of H atoms in most of the half-reactions in which water is a reactant or product. For example, the reduction of O_2 to H_2O is written

$$O_2(g) + 4\,H^+(aq) + 4\,e^- \rightarrow 2\,H_2O(\ell)$$

However, a few half-reactions in Table A6.1 contain OH^- ions, which tells us that they apply to reactions in basic solutions. One of them also involves the reduction of O_2:

$$O_2(g) + 2\,H_2O(\ell) + 4\,e^- \rightarrow 4\,OH^-(aq)$$

In the following Sample and Practice Exercises you will need to select the appropriate half-reaction that matches the pH conditions of the overall redox reaction.

CONNECTION Keep in mind that hydrogen ions in aqueous solutions are really hydronium ions, $H_3O^+(aq)$, as described in Chapters 4 and 17.

SAMPLE EXERCISE 19.1   **Writing Net Ionic Equations of Redox Reactions by Combining Half-Reactions**

Write a net ionic equation describing the oxidation of $Fe^{2+}(aq)$ by O_2 gas dissolved in an acidic solution. *Hint*: Water is a component of the O_2 reduction half-reaction.

Collect and Organize We find these half-reactions for O_2 reduction in Table A6.1:

$$O_2(g) + 4H^+(aq) + 4e^- \rightarrow 2H_2O(\ell)$$

$$O_2(g) + 2H_2O(\ell) + 4e^- \rightarrow 4OH^-(aq)$$

All of the half-reactions in Table A6.1 are reductions, so there is no half-reaction for the oxidation of Fe^{2+} ions. However, Fe^{2+} ions appear as a product in the reduction of Fe^{3+}:

$$Fe^{3+}(aq) + e^- \rightarrow Fe^{2+}(aq)$$

We can reverse this reduction half-reaction to make Fe^{2+} the reactant in an oxidation half-reaction:

$$Fe^{2+}(aq) \rightarrow Fe^{3+}(aq) + e^-$$

Analyze The problem specifies acidic conditions, so we should use the O_2 half-reaction with H^+ rather than OH^- ions in it. In any redox reaction the number of electrons gained by the substance that is reduced must equal the number lost by the substance that is oxidized. There is a gain of 4 moles of electrons in the O_2 half-reaction and a loss of 1 mole of electrons in the Fe^{2+} half-reaction. Therefore, we will need to multiply the Fe^{2+} half-reaction by 4 to obtain a balanced overall redox reaction equation.

Solve Multiplying the Fe^{2+} half-reaction by 4 and adding it to the O_2 half-reaction containing H^+ ions, we get

$$4Fe^{2+}(aq) \rightarrow 4Fe^{3+}(aq) + 4e^-$$

$$O_2(g) + 4H^+(aq) + 4e^- \rightarrow 2H_2O(\ell)$$

$$\overline{4Fe^{2+}(aq) + O_2(g) + 4H^+(aq) + \cancel{4e^-} \rightarrow 4Fe^{3+}(aq) + 2H_2O(\ell) + \cancel{4e^-}}$$

Simplifying gives

$$4Fe^{2+}(aq) + O_2(g) + 4H^+(aq) \rightarrow 4Fe^{3+}(aq) + 2H_2O(\ell)$$

Think about It We can verify that this is a balanced net ionic equation by confirming that the numbers of Fe, O, and H atoms are the same on the two sides of the reaction arrow, which they are, and that the total electrical charge is the same on the two sides (both are 12+). Note that the result of combining these two half-reactions is not a molecular equation but rather a net ionic equation.

Practice Exercise Write a net ionic equation describing the oxidation of NO_2^- to NO_3^- by O_2 in a basic solution.

(Answers to Practice Exercises are in the back of the book.)

19.2 Electrochemical Cells

Having shown how the redox reaction between Zn metal and Cu^{2+} ions is the net result of two separate half-reactions, one involving the oxidation of Zn and the other the reduction of Cu^{2+} ions, let's now physically separate the two half-reactions using a device called an **electrochemical cell**. Figure 19.2 provides a view of an electrochemical cell based on the Zn/Cu^{2+} reaction. Like nearly all electrochemical cells, it is made of two compartments. One of the compartments in the Zn/Cu^{2+} cell contains a strip of zinc metal immersed in 1.00 M $ZnSO_4$, the other a strip of copper metal immersed in 1.00 M $CuSO_4$. Sulfate ions are spectator ions in this reaction and so are not included in the net ionic equation nor in Figure 19.2, though we shall see that they do play a role in cell function. The two metal strips serve as the cell's *electrodes*, providing pathways along which the electrons produced and consumed in the two half-reactions flow to and from an external circuit.

As the cell reaction proceeds, oxidation of Zn atoms produces electrons, which travel from the Zn electrode through the external circuit to the surface of the Cu electrode where they combine with Cu^{2+} ions, forming atoms of Cu metal. In an electrochemical cell the electrode at which the oxidation half-reaction takes place (the zinc electrode in this case) is called the **anode**, and the electrode at which the reduction half-reaction takes place is called the **cathode**.

You might think that production of Zn^{2+} ions in the left compartment in Figure 19.2 would result in a buildup of positive charge on that side of the cell, and that conversion of Cu^{2+} ions to Cu metal would create an excess of SO_4^{2-} ions and thus a negative charge in the Cu compartment. However, no such charge buildup occurs because the two compartments are connected by a permeable bridge made of porous glass or plastic, allowing ions to migrate between compartments. For example, when SO_4^{2-} ions are in excess in the Cu compartment, they can flow through the bridge to the Zn compartment where they are needed to balance the charges of the newly formed Zn^{2+} ions.

In most electrochemical cells ionic compounds are added to supply ions that migrate back and forth between the two compartments to compensate for the flow of electrons in the external circuit. It is important that these electrolytes not interfere with the redox reactions at the electrodes. Their function is simply to produce a flow of charges that completes the electrical circuit.

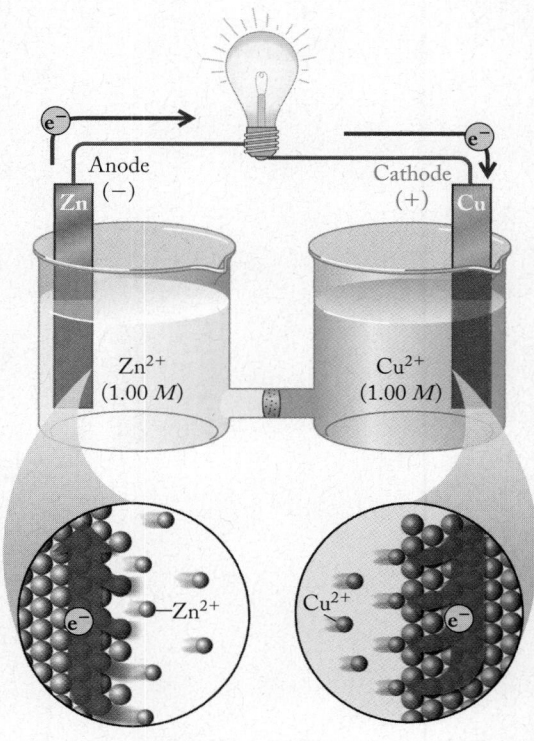

FIGURE 19.2 This electrochemical cell consists of two compartments: the one on the left contains a zinc metal anode immersed in a 1.00 M solution of $ZnSO_4$; the one on the right contains a copper metal cathode immersed in a 1.00 M solution of $CuSO_4$. A porous bridge made of either glass or plastic provides an electrical connection through which ions (and their charges) can migrate from one compartment to the other.

▶❚❚ **CHEMTOUR** Zinc-Copper Cell

CONCEPT TEST ∙∙∙∙∙∙∙∙∙∙∙∙∙∙∙∙∙∙∙∙∙∙∙∙∙∙∙∙∙∙∙∙∙∙∙∙∙

As the cell reaction in Figure 19.2 proceeds, is the increase in mass of the copper strip the same as the decrease in mass of the zinc strip?

(Answers to Concept Tests are in the back of the book.)
∙∙∙

Figure 19.2 provides a view of the physical reality of a Zn/Cu^{2+} electrochemical cell, but we could use a more compact way to represent the components of the cell. A **cell diagram** uses a string of chemical formulas and symbols to show how the components of the cell are connected. A cell diagram does not convey stoichiometry, and so any coefficients in the balanced equation for the cell reaction do not appear in the cell diagram.

electrochemical cell an apparatus that converts chemical energy into electrical work or electrical work into chemical energy.

anode an electrode at which an oxidation half-reaction (loss of electrons) takes place.

cathode an electrode at which a reduction half-reaction (gain of electrons) takes place.

cell diagram symbols that show how the components of an electrochemical cell are connected.

When writing a cell diagram,

1. Write the chemical symbol of the anode at the far left of the diagram, the symbol of the cathode at the far right, and double vertical lines for the connecting bridge halfway between them.
2. Work inward from the electrodes toward the connecting bridge using vertical lines to indicate phase changes (like that between a solid metal electrode and an aqueous solution). Represent the electrolytes surrounding the electrode using the symbols of the ions or compounds that are changed by the cell reaction. Use commas to separate species in the same phase.
3. If known, use the concentrations of the dissolved species in place of (*aq*) phase symbols, and add the partial pressures of any gases within their (*g*) phase symbols.

Following these steps for the Zn/Cu^{2+} electrochemical cell in Figure 19.2:

(1) $\text{Zn}(s) \ldots \ldots \ldots \| \ldots \ldots \ldots \text{Cu}(s)$

(2) $\text{Zn}(s) \,|\, \text{Zn}^{2+}(aq) \,\|\, \text{Cu}^{2+}(aq) \,|\, \text{Cu}(s)$

(3) $\text{Zn}(s) \,|\, \text{Zn}^{2+} \,(1.00\ M) \,\|\, \text{Cu}^{2+} \,(1.00\ M) \,|\, \text{Cu}(s)$

CONCEPT TEST

Describe in your own words the meaning of the above cell diagram. Start your description with: "The cell consists of a zinc anode, which is oxidized to . . .".

SAMPLE EXERCISE 19.2 **Diagramming an Electrochemical Cell**

Figure 19.3 depicts an electrochemical cell in which a copper electrode immersed in a 1.00 *M* solution of Cu^{2+} ions is connected to a silver electrode immersed in a 1.00 *M* solution of Ag^{+} ions. Write a balanced chemical equation for this cell reaction and write a cell diagram for this cell.

Collect and Organize We have a cell reaction in which electrons spontaneously flow from a copper electrode in contact with a solution of Cu^{2+} ions through an external circuit to a silver electrode in contact with a solution of Ag^{+} ions. The half-reactions in Table A6.1 involving these metals and ions are

$$\text{Cu}^{2+}(aq) + 2\,\text{e}^{-} \rightarrow \text{Cu}(s)$$

$$\text{Ag}^{+}(aq) + \text{e}^{-} \rightarrow \text{Ag}(s)$$

In a cell diagram, the anode and the species involved in the oxidation half-reaction are on the left and the cathode and the species involved in the reduction half-reaction are on the right. We use single lines to separate phases and a double line to represent the porous bridge separating the two compartments of the cell.

Analyze In an electrochemical cell, electrons flow from the anode through an external circuit to the cathode. Therefore, in the cell in Figure 19.3, copper is the anode and silver is the cathode. This means that the Cu reduction half-reaction must run in reverse as an oxidation half reaction:

$$\text{Cu}(s) \rightarrow \text{Cu}^{2+}(aq) + 2\,\text{e}^{-}$$

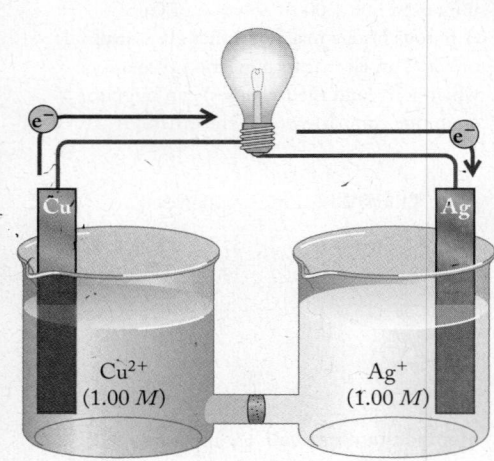

FIGURE 19.3 A Cu/Ag voltaic cell.

Two moles of electrons are produced in the anode half-reaction, but only 1 mole of electrons is consumed in the cathode half-reaction at the Ag electrode. We therefore need to multiply the silver half-reaction by 2 before combining the two equations.

Solve Multiplying the Ag^+ half-reaction by 2 and adding it to the Cu half-reaction, we get

$$2\,Ag^+(aq) + 2\,e^- \rightarrow 2\,Ag(s)$$
$$Cu(s) \rightarrow Cu^{2+}(aq) + 2\,e^-$$
$$\overline{2\,Ag^+(aq) + \cancel{2\,e^-} + Cu(s) \rightarrow 2\,Ag(s) + Cu^{2+}(aq) + \cancel{2\,e^-}}$$
$$2\,Ag^+(aq) + Cu(s) \rightarrow 2\,Ag(s) + Cu^{2+}(aq)$$

The equation is balanced, and we are finished with this portion of our task.
  Applying the rules for writing a cell diagram:
1. Anode on the left, cathode on the right, bridge in the middle:

$$Cu(s) \quad \| \quad Ag(s)$$

2. Adding electrode/solution boundaries and the formulas of the ions produced and consumed in the cell reaction:

$$Cu(s)\,|\,Cu^{2+}(aq)\,\|\,Ag^+(aq)\,|\,Ag(s)$$

3. Adding concentration terms:

$$Cu(s)\,|\,Cu^{2+}\,(1.00\ M)\,\|\,Ag^+\,(1.00\ M)\,|\,Ag(s)$$

Think about It To test the validity of the cell diagram, let's translate it into a sentence: *a copper anode is oxidized to aqueous Cu^{2+} ions, which are separated by a permeable bridge from an aqueous solution of Ag^+ ions that are reduced to Ag atoms at a silver cathode.* This description matches the net ionic equation for the reaction and is consistent with the layout of, and flow of electrons in, Figure 19.3.

Practice Exercise Write a balanced chemical equation and the cell diagram for an electrochemical cell that has a copper cathode immersed in a solution of Cu^{2+} ions and an aluminum anode immersed in a solution of Al^{3+} ions.

19.3 Standard Potentials

Table A6.1 lists half-reactions in order of a parameter called the **standard reduction potential (E°_{red})**. These potentials are expressed in volts (V). As we noted in Section 19.1, reduction half-reactions can be reversed, making them oxidation half-reactions. Reversing a reduction half-reaction does not affect the magnitude of its standard potential but does change its sign. In this respect, E°_{red} is like other thermodynamic parameters, such as ΔG°, ΔH°, and ΔS°. Recall that reversing a chemical reaction changes the sign of these parameters but does not change their magnitude.

  The E°_{red} value of a reduction half-reaction is an indication of how likely the half-reaction is to occur. The more positive the value of E°_{red}, the greater the probability that the reduction half-reaction will couple with an oxidation half-reaction to produce a spontaneous redox reaction. The most positive E°_{red} value in Table A6.1 is that for the reduction of fluorine:

$$F_2(g) + 2\,e^- \rightarrow 2\,F^-(aq) \qquad E^\circ_{red} = +2.866\ V$$

standard reduction potential (E°_{red}) the potential of a reduction half-reaction in which all reactants and products are in their standard states at 25°C.

FIGURE 19.4 Italian physicist Alessandro Volta (1745–1827) is credited with building the first battery in 1798. It consisted of a stack of alternating layers of zinc, blotter paper soaked in salt water, and silver.

▶❙❙ **CHEMTOUR** Cell Potential

standard cell potential (E_{cell}°) a measure of how forcefully an electrochemical cell, in which all reactants and products are in their standard states, can pump electrons through an external circuit.

electromotive force (emf) also called voltage, the force pushing electrons through an electrical circuit.

voltaic cell an electrochemical cell in which chemical energy is transformed into electrical work by a spontaneous redox reaction.

cell potential (E_{cell}) the electromotive force expressed in volts (V) with which an electrochemical cell can push electrons through an external circuit connected to its terminals.

Fluorine's top position means that it is the most easily reduced reactant in Table A6.1. It also means that F_2 is the strongest oxidizing agent in the table. It is capable of oxidizing any of the substances on the product side of the half-reactions lower in the table.

The reactants with very negative E_{red}° values at the bottom of Table A6.1 include the major cations in biological systems and environmental waters: Na^+, K^+, Mg^{2+}, and Ca^{2+}. Their negative E_{red}° values tell us that these ions in aqueous solutions are not easily reduced to their free metals. They are chemically stable (which explains the presence of these elements as cations in nature).

Now let's consider what happens when the half-reactions in Table A6.1 run in reverse. This means that the products of reduction half-reactions become the reactants of oxidation half-reactions. It also means that the half-reaction at the very bottom of the table with the most negative E_{red}° value:

$$Li^+(aq) + e^- \rightarrow Li(s) \qquad E^{\circ} = -3.05 \text{ V}$$

is likely to run in reverse as an oxidation half-reaction:

$$Li(s) \rightarrow Li^+(aq) + e^-$$

Thus, Li^+ ions in aqueous solution have little ability to oxidize anything, but Li metal is a very powerful reducing agent that can reduce any of the substances on the reactant side of the half-reactions in Table A6.1.

⬤ **CONCEPT TEST** ••

Given the positions of the reactants and products in Table A6.1, predict which of the following reactions is/are spontaneous under standard conditions:

a. $Cu(s) + 2\,Fe^{3+}(aq) \rightarrow Cu^{2+}(aq) + 2\,Fe^{2+}(aq)$

b. $2\,Ag(s) + 2\,Zn^{2+}(aq) \rightarrow 2\,Ag^+(aq) + 2\,Zn(s)$

c. $Hg(\ell) + 2\,H^+(aq) \rightarrow Hg^{2+}(aq) + H_2(g)$

••

We can use the standard reduction potentials in Table A6.1 to calculate the **standard cell potentials (E_{cell}°)** of electrochemical cells. Standard cell potentials are measures of the **electromotive force (emf)** generated by cell reactions, which reflects how forcefully cells pump electrons out from their anodes through external circuits and into their cathodes. In the process of pumping electrons these electrochemical cells convert the chemical energy of a spontaneous cell reaction into electrical work. Cells that can do this are called **voltaic cells**. We can combine the standard reduction potentials of a voltaic cell's cathode and anode to calculate the force of its electron pump under standard conditions, that is, the value of E_{cell}°. The value of E_{cell}° is the difference in the standard reduction potentials of its cathode and anode:

$$E_{cell}^{\circ} = E_{cathode}^{\circ} - E_{anode}^{\circ} \qquad (19.2)$$

To understand why we subtract E_{anode}° from $E_{cathode}^{\circ}$, keep in mind that E_{anode}°, like *all electrode potentials*, is a standard *reduction* potential. However, the anode is the electrode in a cell where *oxidation* takes place. The half-reaction happening at the anode is essentially a reduction half-reaction running in reverse. As we saw in Chapters 5 and 14, when we reverse a reaction we change the signs of its thermodynamic properties, such as ΔH°, ΔS°, and ΔG°. Standard cell potential is another thermodynamic property, which is, as we shall see in Section 19.4, directly linked to the change in standard free energy, ΔG_{cell}°. Therefore, it is only reasonable that the sign of E_{anode}° change when the direction of its half-reaction is reversed. That change in sign is the reason for the minus sign in Equation 19.2.

The superscript (°) in Equation 19.2 has its usual thermodynamic meaning—all reactants and products are in their standard states, that is, the concentrations of all dissolved substances are 1 M and the partial pressures of all gases are 1 atm. If we want to denote a generalized cell potential in which reactants and products are not necessarily in their standard states, we simply refer to the **cell potential (E_{cell})**.

Let's use Equation 19.2 to calculate E_{cell}° for the Zn/Cu^{2+} voltaic cell in Figure 19.2. The standard reduction potential of the cathode half-reaction is

$$Cu^{2+} + 2e^- \rightarrow Cu \qquad E° = 0.342 \text{ V}$$

To obtain the standard potential for the oxidation half-reaction at the zinc anode we start with the standard reduction potential of Zn^{2+} ions in Table A6.1:

$$Zn^{2+} + 2e^- \rightarrow Zn \qquad E° = -0.762 \text{ V}$$

Now we use Equation 19.2 to calculate E_{cell}°:

$$E_{cell}^\circ = E_{cathode}^\circ - E_{anode}^\circ$$
$$= 0.342 - (-0.762) = 1.104 \text{ V}$$

This is the cell potential we measure if we connect a device called a voltmeter across the two electrodes in Figure 19.2, as shown in Figure 19.5.

To use Equation 19.2 we need to know which half-reaction occurs at the cathode (reduction) and which occurs at the anode (oxidation). In other words, we need to know which component of the spontaneous cell reaction is more likely to be oxidized and which is more likely to be reduced. As we have seen, this decision can be made based on the data in Table A6.1. In our Zn/Cu^{2+} cell, for instance, the value of $E°$ for the reduction of Cu^{2+} ions to Cu metal is 0.342 V, which is greater than the value of $E°$ for reducing Zn^{2+} ions to Zn metal (−0.762 V). Therefore, in the Zn/Cu^{2+} voltaic cell, Cu^{2+} ions are reduced and Zn metal is oxidized. We can generalize this observation to the cell reaction of any voltaic cell: the half-reaction with the more positive value of $E°$ runs as a reduction and the other one runs in reverse as an oxidation. This way we are assured that E_{cell}° is, overall, a positive value.

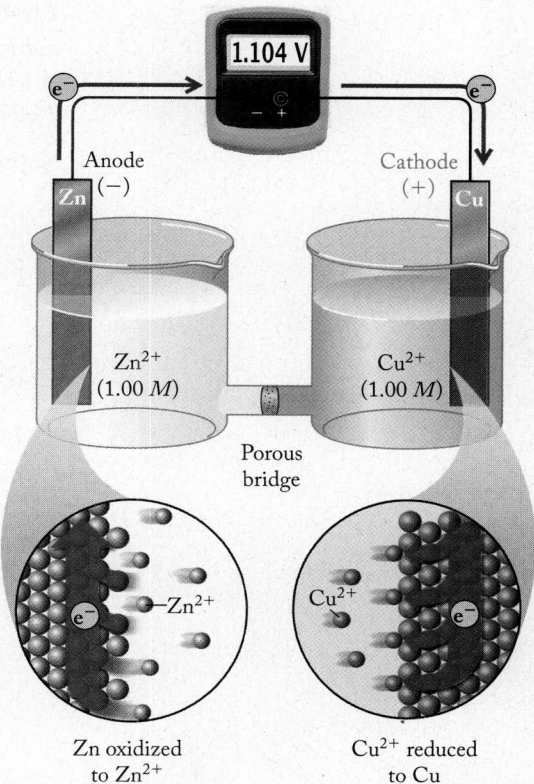

FIGURE 19.5 A voltmeter displays a cell potential of 1.104 V between a Zn electrode immersed in a 1.00 M solution of Zn^{2+} ions and a Cu electrode immersed in a 1.00 M solution of Cu^{2+} ions.

SAMPLE EXERCISE 19.3   **Identifying Anode and Cathode Half-Reactions and Calculating the Value of E_{cell}°**

The standard reduction potentials of the half-reactions in single-use alkaline batteries are

$$ZnO(s) + H_2O(\ell) + 2e^- \rightarrow Zn(s) + 2OH^-(aq) \qquad E° = -1.25 \text{ V}$$
$$2MnO_2(s) + H_2O(\ell) + 2e^- \rightarrow Mn_2O_3(s) + 2OH^-(aq) \qquad E° = 0.15 \text{ V}$$

▶❚❚   **CHEMTOUR**   Alkaline Battery

What is the net ionic equation for the cell reaction and the value of E_{cell}°?

Collect and Organize We can calculate E_{cell}° using Equation 19.2:

$$E_{cell}^\circ = E_{cathode}^\circ - E_{anode}^\circ$$

but first we need to decide which half-reaction occurs at the cathode and which at the anode. The equation for a cell reaction is written by combining half-reactions once the loss or gain of electrons in the two half-reactions is balanced.

Analyze The MnO_2 half-reaction has the more positive $E°$, making it our reduction half-reaction. We must reverse the ZnO half-reaction, turning it into an oxidation half-reaction. The two half-reactions both involve the transfer of 2 electrons and so they may be combined by simply adding them together.

Solve The oxidation half-reaction at the anode is

$$Zn(s) + 2OH^-(aq) \rightarrow ZnO(s) + H_2O(\ell) + 2e^-$$

and the reduction half-reaction at the cathode is

$$2MnO_2(s) + H_2O(\ell) + 2e^- \rightarrow Mn_2O_3(s) + 2OH^-(aq)$$

Combining these half-reactions to obtain the overall cell reaction, we get

$$2MnO_2(s) + \cancel{H_2O(\ell)} + Zn(s) + \cancel{2OH^-(aq)} + \cancel{2e^-} \rightarrow$$
$$Mn_2O_3(s) + \cancel{2OH^-(aq)} + ZnO(s) + \cancel{H_2O(\ell)} + \cancel{2e^-}$$

Simplifying gives us the net ionic equation for the cell reaction:

$$2MnO_2(s) + Zn(s) \rightarrow Mn_2O_3(s) + ZnO(s)$$

The overall $E°_{cell}$ for this reaction is obtained by using Equation 19.2:

$$E°_{cell} = E°_{cathode} - E°_{anode}$$
$$= 0.15\ V - (-1.25\ V)$$
$$= 1.40\ V$$

Think about It The $E°_{cell}$ value is reasonable because the potential of most alkaline batteries is nominally 1.5 V. In this particular cell reaction the net ionic equation is also the complete molecular equation.

Practice Exercise The half-reactions in nicad (nickel–cadmium) batteries are

$$Cd(OH)_2(s) + 2e^- \rightarrow Cd(s) + 2OH^-(aq) \qquad E° = -0.403\ V$$
$$2NiO(OH)(s) + 2H_2O(\ell) + 2e^- \rightarrow 2Ni(OH)_2(s) + 2OH^-(aq)$$
$$E° = 1.32\ V$$

Write the net ionic equation for the cell reaction and calculate the value of $E°_{cell}$.

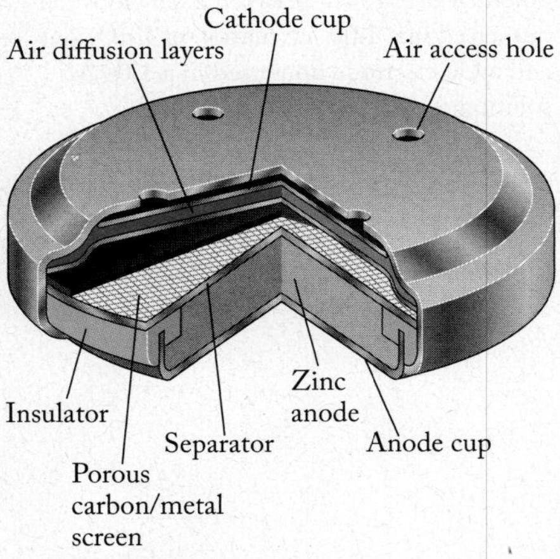

■ ⋯⋯⋯⋯⋯⋯⋯⋯⋯⋯⋯⋯⋯⋯⋯

FIGURE 19.6 Most of the internal volume of a zinc–air battery is occupied by the anode: a slurry of Zn particles in an aqueous solution of KOH surrounded by a metal cup that serves as the negative terminal of the battery. Oxygen from the air is the reactant at the cathode. Air enters through holes in an inverted metal cup that serves as the positive terminal of the battery. Once inside the battery, air diffuses through layers of gas-permeable plastic film that let air in but keep electrolyte from leaking out. Oxygen in the air is reduced at the porous carbon/metal cathode to OH^- ions that migrate toward the anode, where they are consumed in the Zn oxidation half-reaction.

Before closing this discussion of standard cell potentials, let's combine two half-reactions in which different numbers of electrons are gained and lost. This combination occurs in a type of battery that has a limitless supply of one of its reactants. It is called the zinc–air battery (Figure 19.6), and it powers devices in which small battery size and low mass are high priorities, such as hearing aids. Most of the internal volume of one of these batteries is occupied by an anode consisting of a paste of zinc particles packed in an aqueous solution of KOH. As in alkaline batteries (Sample Exercise 19.3), the anode half-reaction is

$$Zn(s) + 2OH^-(aq) \rightarrow ZnO(s) + H_2O(\ell) + 2e^- \qquad E°_{anode} = -1.25\ V$$

The cathode consists of porous carbon supported by a metal screen. Air diffuses through small holes in the battery and across a layer of Teflon that lets gases pass through but keeps electrolyte from leaking out. As air passes through the cathode, oxygen is reduced to hydroxide ions:

$$O_2(g) + 2H_2O(\ell) + 4e^- \rightarrow 4OH^-(aq) \qquad E^\circ_{cathode} = 0.401 \text{ V}$$

To write the overall cell reaction, we need to multiply the oxidation half-reaction by 2 before combining it with the reduction half-reaction:

$$2\,[Zn(s) + 2OH^-(aq) \rightarrow ZnO(s) + H_2O(\ell) + 2e^-]$$

$$+ O_2(g) + 2H_2O(\ell) + 4e^- \rightarrow 4OH^-(aq)$$

$$2Zn(s) + 4OH^-(aq) + O_2(g) + 2H_2O(\ell) + 4e^- \rightarrow$$
$$2ZnO(s) + 2H_2O(\ell) + 4OH^-(aq) + 4e^-$$

or

$$2ZnO(s) + O_2(g) \rightarrow 2ZnO(s)$$

$$E^\circ_{cell} = E^\circ_{cathode} - E^\circ_{anode} = 0.401 \text{ V} - (-1.25 \text{ V}) = 1.65 \text{ V}$$

Note that when we multiplied the anode half-reaction by 2 and added it to the cathode half-reaction *we did not multiply the E° of the anode half-reaction by 2*. The reason we did not is that E° is an *intensive* property of a half-reaction or a complete cell reaction. It does not change when the quantities of reactants and products change. Thus, a zinc–air battery the size of a pea has the same E°_{cell} as one the size of a book (like those being developed for electric vehicles). On the other hand, the amount of electrical work a zinc–air battery can do *does* depend on how much zinc is inside it because, as we are about to see, the electrical work that a voltaic cell can do depends on both cell potential *and* the quantity of charge it can deliver at that potential.

19.4 Chemical Energy and Electrical Work

When we connect the Zn and Cu electrodes from Figure 19.2 to a digital voltmeter as in Figure 19.5—the Zn electrode to the negative terminal of the meter and the Cu electrode to the positive terminal—the meter reads 1.104 V. These connections tell us that the battery is pumping electrons from the Zn electrode through the external circuit to the Cu electrode. The reading tells us how much electromotive force (emf) is pushing the electrons through the circuit. Under standard conditions this emf is the same as the cell's standard cell potential (E°_{cell}); under any other conditions it is simply E_{cell}.

When a voltaic cell pumps electrons through an external circuit, those moving electrons can do electrical work, like lighting a lightbulb or turning an electric motor. The decrease in chemical free energy (ΔG_{cell}) that accompanies a spontaneous cell reaction is a measure of the electrical work (w_{elec}) that may result:

$$\Delta G_{cell} = w_{elec} \qquad (19.3)$$

▶❙❙ **CHEMTOUR** Free Energy

The sign of ΔG_{cell} is negative because free energy always decreases as a spontaneous reaction proceeds. The sign of w_{elec} is also negative because it represents work the cell (the "system") does *on* its surroundings, which has a negative value according to the sign convention summarized in Figure 5.18.

The work done by a voltaic cell on its surroundings is the product of the quantity of electrical charge (C) the cell pushes through an external circuit times the force (emf) pushing that charge. That force is the same as the cell potential, so the connection between E_{cell} and w_{elec} is

$$w_{elec} = -CE_{cell} \qquad (19.4)$$

where the negative sign reflects the fact that work done *by* a voltaic cell on its surroundings (the external circuit) corresponds to energy lost by the cell.

◑◐ **CONNECTION** The sign conventions used for work done *on* a thermodynamic system (+) and the work done *by* the system (−) were explained in Section 5.2.

Faraday's constant (F) the magnitude of electrical charge in 1 mole of electrons. Its value to three significant figures is 9.65×10^4 C/mol.

The quantity of charge is proportional to the number of electrons flowing through the circuit. As noted in Chapter 2, the magnitude of the charge on a single electron is 1.602×10^{-19} coulombs (C). Note the distinction between italic C, a symbol for the variable "charge," and nonitalic C, the abbreviation for the unit "coulomb." The magnitude of electrical charge on 1 mole of electrons is

$$\frac{1.602 \times 10^{-19}\ \text{C}}{e^-} \times \frac{6.022 \times 10^{23}\ e^-}{\text{mol}\,e^-} = \frac{9.65 \times 10^4\ \text{C}}{\text{mol}\,e^-}$$

This quantity of charge, 9.65×10^4 C/mol, is called **Faraday's constant (F)** after Michael Faraday (1791–1867), the English chemist and physicist who discovered that redox reactions take place when electrons are transferred from one species to another. The quantity of charge C flowing through an electrical circuit is the product of the number of moles of electrons n times the Faraday constant:

$$C = nF \qquad (19.5)$$

Combining Equations 19.4 and 19.5 gives us an equation relating w_{elec} and E_{cell}:

$$w_{\text{elec}} = -nFE_{\text{cell}} \qquad (19.6)$$

If we combine Equations 19.3 and 19.6, we connect the quantity of electrical work a voltaic cell does on its surroundings with the change in free energy in the cell:

$$\Delta G_{\text{cell}} = -nFE_{\text{cell}} \qquad (19.7)$$

Perhaps you are wondering how the product on the right side of Equation 19.7 is the equivalent of energy. It works out that way because the units on the right side are

$$\cancel{\text{moles}} \times \frac{\text{coulombs}}{\cancel{\text{mole}}} \times \text{volts} = \text{coulombs-volts}$$

A coulomb-volt is the same quantity of energy as a joule:

$$1\ \text{coulomb-volt} = 1\ \text{joule}$$

$$1\ \text{C} \cdot \text{V} = 1\ \text{J}$$

The negative sign on the right side of Equation 19.7 tells us that the E_{cell} of any voltaic cell must have a positive value because the sign of ΔG_{cell} for the spontaneous chemical reaction inside the cell must be negative.

Let's use Equation 19.7 to calculate the change in standard free energy of the Zn/Cu^{2+} cell reaction. Let's start with the standard cell potential we calculated in Section 19.3:

$$E^{\circ}_{\text{cell}(Zn/Cu^{2+})} = 1.104\ \text{V}$$

We can convert this standard cell potential into a change in standard free energy $(\Delta G^{\circ}_{\text{cell}})$ using Equation 19.7 under standard conditions so that $\Delta G = \Delta G^{\circ}$ and $\Delta E = \Delta E^{\circ}$:

$$\Delta G^{\circ}_{\text{cell}} = -nFE^{\circ}_{\text{cell}}$$

$$= -\left(2\ \cancel{\text{mol}} \times \frac{9.65 \times 10^4\ \text{C}}{\cancel{\text{mol}}} \times 1.10\ \text{V} \right) = -2.12 \times 10^5\ \text{C} \cdot \text{V}$$

$$= -2.12 \times 10^5\ \text{J} = -212\ \text{kJ}$$

To put this value in perspective, on a mole-for-mole basis, the Zn/Cu^{2+} reaction produces nearly as much useful energy as the combustion of hydrogen gas:

$$H_2(g) + \tfrac{1}{2}O_2(g) \rightarrow H_2O(g) \qquad \Delta G^{\circ} = -228.6\ \text{kJ}$$

When a rechargeable battery, such as the one used to start a car's engine, is recharged, an external source of electrical power forces the voltaic cell reaction to run in reverse. What are the signs of ΔG_{cell} and E_{cell} during the recharging process?

SAMPLE EXERCISE 19.4   Relating ΔG°_{cell} and E°_{cell}

Many of the "button" batteries used in electric watches consist of a Zn anode and Ag_2O cathode separated by a membrane soaked in a concentrated solution of KOH (Figure 19.7). At the cathode, Ag_2O is reduced to Ag metal, and at the anode Zn is oxidized to solid $Zn(OH)_2$. Write the net ionic equation for the reaction, and, using the appropriate standard reduction potentials in Table A6.1, calculate the values of E°_{cell} and ΔG°_{cell}.

Collect and Organize We know the reactants and products of the anode and cathode reactions and that the reaction occurs in a basic solution. The following equations should be useful in calculating E°_{cell} and ΔG°_{cell} from the appropriate standard potentials:

$$E^{\circ}_{cell} = E^{\circ}_{cathode} - E^{\circ}_{anode}$$

$$\Delta G^{\circ}_{cell} = -nFE^{\circ}_{cell}$$

Analyze The half-reaction at the cathode is based on the reduction of Ag_2O to Ag. The appropriate half-reaction in Table A6.1 is

$$Ag_2O(s) + H_2O(\ell) + 2e^- \rightarrow 2Ag(s) + 2OH^-(aq) \qquad E^{\circ}_{cathode} = 0.342\ V$$

We must reverse the anode oxidation half-reaction to find an entry in Table A6.1 in which $Zn(OH)_2$ is the reactant and Zn is the product:

$$Zn(OH)_2(s) + 2e^- \rightarrow Zn(s) + 2OH^-(aq) \qquad E^{\circ}_{anode} = -1.249\ V$$

We need to reverse this half-reaction before combining it with the cathode half-reaction. The two half-reactions involve the transfer of the same number of electrons, and so combining them simply means adding them together. The value of E°_{cell} will be about $[0.35 - (-1.25)]$ or about 1.60 V. This value is half again as large as the E°_{cell} of the Zn/Cu^{2+} cell. Therefore, the magnitude of its ΔG°_{cell} value should be half again as large as -212 kJ/mol, or about -300 kJ/mol.

Solve Reversing the $Zn(OH)_2$ half-reaction and adding it to the Ag_2O half-reaction, we get

$$Zn(s) + 2OH^-(aq) \rightarrow Zn(OH)_2(s) + 2e^-$$
$$Ag_2O(s) + H_2O(\ell) + 2e^- \rightarrow 2Ag(s) + 2OH^-(aq)$$

$$Ag_2O(s) + H_2O(\ell) + Zn(s) + \cancel{2OH^-(aq)} + \cancel{2e^-} \rightarrow$$
$$2Ag(s) + \cancel{2OH^-(aq)} + Zn(OH)_2(s) + \cancel{2e^-}$$

or

$$Ag_2O(s) + H_2O(\ell) + Zn(s) \rightarrow 2Ag(s) + Zn(OH)_2(s)$$

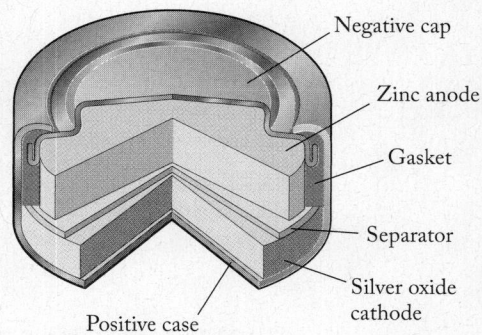

FIGURE 19.7 Many of the button batteries that power small electronic devices incorporate a Zn anode and a Ag_2O cathode separated by a membrane containing KOH electrolyte.

Negative cap
Zinc anode
Gasket
Separator
Silver oxide cathode
Positive case

Calculating E°_{cell}:

$$E^\circ_{cell} = E^\circ_{cathode} - E^\circ_{anode}$$
$$= 0.342 \text{ V} - (-1.249 \text{ V})$$
$$= 1.591 \text{ V}$$

Calculating ΔG°_{cell}:

$$\Delta G^\circ_{cell} = -nFE^\circ_{cell}$$
$$= -(2 \text{ mol} \times 9.65 \times 10^4 \text{ C/mol} \times 1.591 \text{ V})$$
$$= -3.07 \times 10^5 \text{ C} \cdot \text{V} = -3.07 \times 10^5 \text{ J} = -307 \text{ kJ}$$

Think about It The positive value of E°_{cell} and negative value of ΔG°_{cell} are expected because voltaic cell reactions are spontaneous, and the calculated values are close to those we estimated.

Practice Exercise If an alkaline battery produces a cell potential of 1.50 V, what is the value of ΔG_{cell}?

Some final thoughts about the ΔG°_{cell} value calculated in Sample Exercise 19.4: First, it is based on the reaction of 1 mol Ag_2O and 1 mol Zn, which correspond to 232 g Ag_2O and 65 g Zn. The energy stored in a button battery (Figure 19.7), which has a mass of only 1 or 2 g, would be a tiny fraction of the calculated value. Also, note that no ions appear in the net ionic equation because all the reactants and products in the silver oxide battery reaction are solids, and so the net ionic equation and molecular equation are identical.

19.5 A Reference Point: The Standard Hydrogen Electrode

We can measure the value of E_{cell} using a voltmeter, but how do we measure the individual electrode potentials of the cathode and anode? The answer to this question is that we arbitrarily assign a value of zero volts to the standard potential for the reduction of hydrogen ions to hydrogen gas:

$$2H^+(aq) + 2e^- \rightarrow H_2(g) \qquad E^\circ = 0.000 \text{ V} \qquad (19.8)$$

An electrode that generates this reference potential, called the **standard hydrogen electrode (SHE)**, consists of a platinum electrode in contact with a solution of a strong acid ($[H^+] = 1.00 M$) and hydrogen gas at a pressure of 1.00 atm (Figure 19.8). The platinum is not changed by the electrode reaction. Rather, it serves as a chemically inert conveyor of electrons. Electrons are conveyed to the electrode surface if H^+ ions are being reduced to hydrogen gas and away from the electrode surface if hydrogen gas is being oxidized to H^+ ions. The potential of the SHE is the same for both half-reactions, 0.000 V.

To write the cell diagram for a cell in which the SHE serves as the anode, we represent the SHE half of the cell as follows:

$$Pt \mid H_2(g, 1.00 \text{ atm}) \mid H^+(1.00 M) \parallel$$

to indicate that the anode half-reaction involves the oxidation of H_2 gas to H^+ ions. If the SHE is the cathode, then we diagram its half of the cell this way:

$$\parallel H^+ (1.00 M) \mid H_2(g, 1.00 \text{ atm}) \mid Pt$$

to indicate that the cathode half-reaction involves the reduction of H^+ ions to H_2 gas.

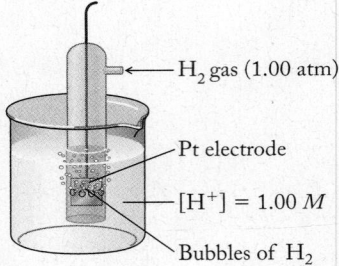

FIGURE 19.8 The standard hydrogen electrode consists of a platinum electrode immersed in a 1.00 M solution of $H^+(aq)$ and bathed in a stream of pure H_2 gas at a pressure of 1.00 atm. Its potential is the same (0.000 V) whether $H^+(aq)$ ions are reduced or H_2 gas is oxidized.

H$_2$ gas (1.00 atm)

Pt electrode

$[H^+] = 1.00 M$

Bubbles of H$_2$

standard hydrogen electrode (SHE) a reference electrode based on the half-reaction $2H^+(aq) + 2e^- \rightarrow H_2(g)$ that produces a standard electrode potential of 0.000 V.

Because the standard reduction (or oxidation) potential of the SHE is 0.000 V, the measured E_{cell} of any voltaic cell in which a SHE is one of the two electrodes—either cathode or anode—is the potential produced by the other electrode. This means that if we attach a voltmeter to the cell, the meter reading is the electrode potential of the other electrode. Suppose, for example, that a voltaic cell consists of a strip of zinc metal immersed in a 1.00 M solution of Zn^{2+} ions in one compartment and a SHE in the other (Figure 19.9a). Also suppose that a voltmeter is connected to the cell so that it measures the potential at which the cell pumps electrons from the zinc electrode to the SHE. This direction of electron flow means that the zinc electrode is the cell's anode and the SHE is the cathode of the cell. At 25°C the meter reads 0.762 V. We know that the value of E_{red}(cathode) is that of the SHE (0.000 V), and that E_{ox}(anode) is the standard oxidation potential for

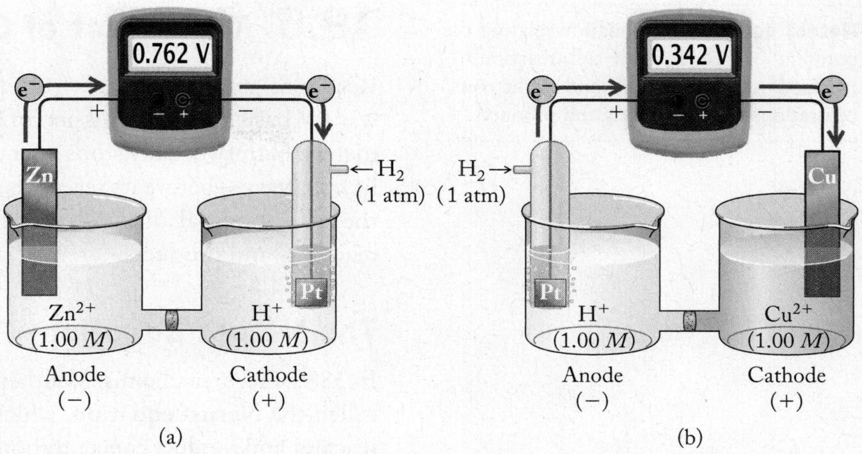

$$Zn(s) \rightarrow Zn^{2+}(aq) + 2\,e^-$$

Inserting these values and symbols into Equation 19.2,

$$E^\circ_{cell} = E^\circ_{cathode} - E^\circ_{anode}$$
$$E^\circ_{cell} = E^\circ_{SHE} - E^\circ_{Zn}$$
$$0.762\ V = 0.000\ V - E^\circ_{Zn}$$
$$E^\circ_{Zn} = -0.762\ V$$

This value is equal to the standard reduction potential of Zn^{2+} in Table A6.1:

$$Zn^{2+}(aq) + 2\,e^- \rightarrow Zn(s) \qquad E^\circ = -0.762\ V$$

In Figure 19.9(b), the SHE is coupled to a copper electrode immersed in a 1.00 M solution of Cu^{2+} ions. In this cell, electrons flow from the SHE through an external circuit to the copper electrode at a cell potential of 0.342 V at 25°C. The value of E°_{red} for the copper half-reaction is calculated as follows:

$$E^\circ_{cell} = E^\circ_{cathode} - E^\circ_{anode}$$
$$= E^\circ_{Cu} - E^\circ_{SHE}$$
$$0.342\ V = E^\circ_{Cu} - 0.000\ V$$
$$E^\circ_{Cu} = 0.342\ V$$

This half-reaction potential matches the value of E° for the reduction of Cu^{2+} to Cu metal in Table A6.1.

CONCEPT TEST ....

A cell consists of a SHE in one compartment and a Ni electrode immersed in a 1.00 M solution of Ni^{2+} ions in the other. If a voltmeter is connected to the electrode as shown in Figure 19.10, what will be the value in the voltmeter's display?

FIGURE 19.9 The standard hydrogen electrode allows us to determine the standard potential of any half-reaction. (a) When coupled to a Zn electrode under standard conditions, the SHE is the cathode and the Zn electrode is the anode. When the SHE is connected to the "COM" (negative) terminal of a voltmeter and the Zn electrode to the positive terminal, the meter measures a cell potential of 0.762 V. (b) When coupled to a Cu electrode under standard conditions, the SHE is the anode, the Cu electrode is the cathode, and the meter measures a cell potential of 0.342 V.

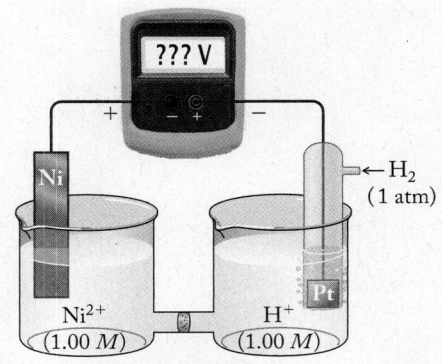

FIGURE 19.10

Nernst equation an equation relating the potential of a cell (or half-cell) reaction to its standard potential ($E°$) and to the concentrations of its reactants and products.

19.6 The Effect of Concentration on E_{cell}

Reactions stop when one of the reactants is completely consumed. This concept was the basis for our discussion of limiting reactants in Chapter 3. However, a commercial battery usually stops operating at its rated cell potential—1.5 V for a flashlight battery—before its reactants are completely consumed. This happens because the cell potential of a voltaic cell is determined by the concentrations of the reactants and products.

The Nernst Equation

In 1889, German chemist Walther Nernst (1864–1941) derived an expression, now called the **Nernst equation**, which describes the dependence of cell potentials on reactant and product concentrations. We can reconstruct his derivation starting with Equation 16.21, which relates the change in free energy ΔG of any reaction to its change in free energy under standard conditions $\Delta G°$:

$$\Delta G = \Delta G° + RT \ln Q$$

As a spontaneous reaction proceeds, concentrations of products increase and concentrations of reactants decrease until the positive value of $RT \ln Q$ offsets the negative value of $\Delta G°$. At that point $\Delta G = 0$ and the reaction has reached chemical equilibrium.

Now let's write an expression analogous to Equation 16.21 that relates E_{cell} to $E°_{cell}$. We start by substituting $-nFE_{cell}$ for ΔG_{cell} and $-nFE°_{cell}$ for $\Delta G°_{cell}$:

$$-nFE_{cell} = -nFE°_{cell} + RT \ln Q$$

Dividing all terms by $-nF$ gives

$$E_{cell} = E°_{cell} - \frac{RT \ln Q}{nF} \qquad (19.9)$$

This is the equation Walther Nernst developed in 1889. We can obtain a very useful form of it if we insert values for R [8.314 J/(mol · K)] and F (9.65×10^4 C/mol), make $T = 298$ K, and convert the natural logarithm to a base-10 logarithm: $\ln Q = 2.303 \log Q$. With these changes the Nernst equation becomes

$$E_{cell} = E°_{cell} - \frac{0.0592}{n} \log Q \qquad (19.10)$$

Equation 19.10 allows us to predict how the potential of a voltaic cell changes as the concentrations of products inside the cell increase and the concentrations of reactants decrease. As they do, Q increases and so does the value of $0.0592/n \times \log Q$. The negative sign in front of this term in Equation 19.10 means that the value of E_{cell} decreases as reactants are converted into products. Eventually E_{cell} approaches zero. When it reaches zero, the cell reaction has achieved chemical equilibrium. The cell can no longer pump electrons through an external circuit. In other words, it's dead.

CONNECTION In Chapter 16 we discussed the relationship between change in free energy and the reaction quotient Q.

CONCEPT TEST

We can also use Equation 19.10 to calculate the potential of a single electrode. Consider the half-reaction at the Ag/Ag$^+$ electrode:

$$Ag^+ + e^- \rightarrow Ag \qquad E° = 0.799 \text{ V}$$

What is the value of E_{red} at 25°C when [Ag$^+$] = 0.100 M?

Batteries are voltaic cells, so their cell potential should drop with usage. Let's consider how much they drop by focusing on the *lead–acid* battery used to start most car engines. These batteries each contain six electrochemical cells. Their anodes are made of Pb and their cathodes are made of PbO_2. Both electrodes are immersed in 4.5 *M* H_2SO_4 (Figure 19.11). The value of a fully charged cell is about 2.0 V. The six cells are connected in series so that the operating potential of the battery is the sum of the six cell potentials, or 12.0 V.

As the battery discharges, $PbO_2(s)$ is reduced to $PbSO_4(s)$ at the cathodes:

$$PbO_2(s) + 3H^+(aq) + HSO_4^-(aq) + 2e^- \rightarrow PbSO_4(s) + 2H_2O(\ell)$$
$$E° = 1.685 \text{ V}$$

and Pb(s) is oxidized to $PbSO_4(s)$ at the anodes:

$$Pb(s) + HSO_4^-(aq) \rightarrow PbSO_4(s) + H^+(aq) + 2e^- \quad E° = -0.356 \text{ V}$$

The reduction half-reaction consumes 2 moles of electrons, and the oxidation half-reaction involves the loss of 2 moles of electrons for each mole of lead.

The net ionic equation for the overall cell reaction is the sum of the two half-reactions:

$$PbO_2(s) + Pb(s) + 2H^+(aq) + 2HSO_4^-(aq) \rightarrow 2PbSO_4(s) + 2H_2O(\ell).$$

The molecular form of this chemical equation is

$$Pb(s) + PbO_2(s) + 2H_2SO_4(aq) \rightarrow 2PbSO_4(s) + 2H_2O(\ell)$$

and the value of $E°_{cell}$ is

$$E°_{cell} = E°_{cathode} - E°_{anode} = 1.685 - (-0.356) = 2.041 \text{ V}$$

As the battery discharges, $[H_2SO_4]$ decreases, and so does the value of E_{cell} calculated from the Nernst equation:

$$E_{cell} = 2.041 \text{ V} - \frac{0.0592}{2} \log \frac{1}{[H_2SO_4]^2}$$

However, the decrease in E_{cell} is very gradual, not falling below 2.0 V until the battery is about 97% discharged, as shown in Figure 19.12. The gradual decrease makes sense because of the logarithmic relationship between Q and E_{cell}. If, for example, $[H_2SO_4]$ decreased by an order of magnitude, say, from 1.00 *M* to 0.100 *M*, the value of E_{cell} would decrease by less than 3%—from 2.041 V to

$$E_{cell} = 2.041 \text{ V} - \frac{0.0592}{2} \log \frac{1}{0.100^2} = 1.982 \text{ V}$$

The logarithmic relationship between Q and E_{cell} means that most batteries can deliver current at a cell potential close to their "design" potential until they are nearly completely discharged.

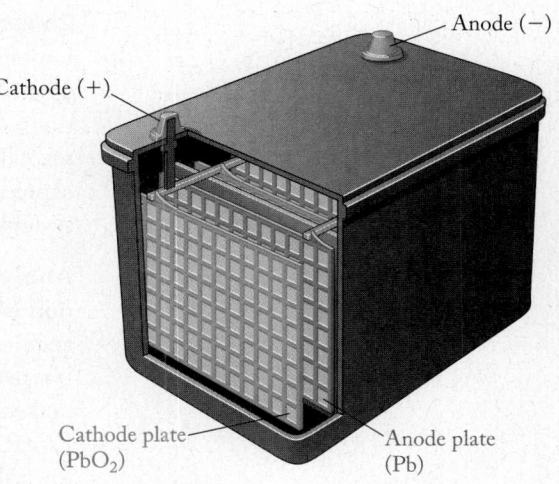

FIGURE 19.11 The lead–acid battery that provides power to start most motor vehicles contains six cells. Each has an anode made of lead and a cathode made of PbO_2 immersed in a background electrolyte of 4.5 *M* H_2SO_4. The electrodes are formed into plates and held in place by grids made of a lead alloy. The grids connect the cells together in series so that the operating potential of the battery (12.0 V) is the sum of six E_{cell} values (each 2.0 V).

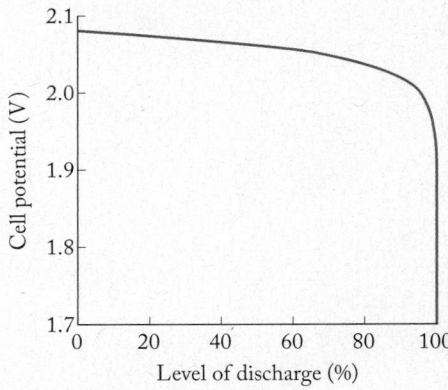

FIGURE 19.12 The potential of a cell in a lead–acid battery decreases as reactants are converted into products, but the change in potential is small until the battery is nearly completely discharged.

> **SAMPLE EXERCISE 19.5** **Calculating E_{cell} from $E°_{cell}$ and the Concentrations of Reactants and Products**

The standard potential ($E°_{cell}$) of a voltaic cell based on the Zn/Cu^{2+} ion reaction:

$$Zn(s) + Cu^{2+}(aq) \rightarrow Zn^{2+}(aq) + Cu(s)$$

is 1.104 V. What is the value of E_{cell} at 25°C when $[Cu^{2+}] = 0.100$ *M* and $[Zn^{2+}] = 1.90$ *M*?

Collect and Organize We are given the standard cell potential and are asked to determine the value of E_{cell} when $[Cu^{2+}] = 0.100$ M and $[Zn^{2+}] = 1.90$ M. The Nernst equation enables us to calculate E_{cell} values for different concentrations of reactants and products. This equation requires us to work with the reaction quotient Q, which we know from Section 16.5 to be the mass action expression for the reaction. Solid copper and zinc are also part of the reaction system, but no terms for pure solids appear in reaction quotients.

Analyze The only term in the numerator of the Q expression for this cell reaction is $[Zn^{2+}]$, and the only one in the denominator is $[Cu^{2+}]$. Each Cu^{2+} ion acquires 2 electrons, and each Zn atom donates 2, so the value of n in the Nernst equation is 2. The value of $[Zn^{2+}]$ is greater than $[Cu^{2+}]$, which makes $Q > 1$. The negative sign in front of the $0.0592/n \times \log Q$ term in Equation 19.10 means that the calculated value of E_{cell} should be less than the value of E_{cell}°.

Solve Substituting the values of $[Zn^{2+}]$ and $[Cu^{2+}]$ in the Nernst equation gives

$$E_{cell} = E_{cell}^{\circ} - \frac{0.0592 \log Q}{n} = 1.104 \text{ V} - \frac{0.0592}{2} \log \frac{1.90}{0.100}$$

$$E_{cell} = 1.104 \text{ V} - \frac{0.0592}{2}(1.279) = 1.066 \text{ V}$$

Think about It The calculated E_{cell} value is only 0.038 V less than E_{cell}° because the logarithmic dependence of cell potential on reactant and product concentrations minimizes the impact of changing concentrations.

Practice Exercise The standard cell potential of the zinc–air battery (Figure 19.6) is 1.65 V. If at 25.0°C the partial pressure of oxygen in the air diffusing through its cathode is 0.21 atm, what is the cell potential? Assume the cell reaction is

$$2\,Zn(s) + O_2(g) \rightarrow 2\,ZnO(s)$$

$E°$ and K

When the cell reaction of a voltaic cell reaches chemical equilibrium, $\Delta G_{cell} = E_{cell} = 0$ and $Q = K$. Therefore, Equation 19.10 becomes

$$0 = E_{cell}^{\circ} - \frac{0.0592}{n} \log K$$

which we can rearrange to

$$\log K = \frac{nE_{cell}^{\circ}}{0.0592} \tag{19.11}$$

We can use Equation 19.11 to calculate the equilibrium constant for any redox reaction at 25°C, not just those in electrochemical cells. For the more general case, we substitute E_{rxn}° for E_{cell}°:

$$\log K = \frac{nE_{rxn}^{\circ}}{0.0592} \tag{19.12}$$

SAMPLE EXERCISE 19.6 **Calculating *K* for a Redox Reaction from the Standard Potentials of Its Half-Reactions**

Many procedures for determining mercury levels in environmental samples begin by oxidizing the mercury to Hg^{2+} and then reducing the Hg^{2+} to elemental Hg with Sn^{2+}. Use the appropriate standard reduction potentials from Table A6.1 to calculate the equilibrium constant at 25°C for the reaction

$$Sn^{2+}(aq) + Hg^{2+}(aq) \rightarrow Sn^{4+}(aq) + Hg(\ell)$$

Collect and Organize We can calculate the equilibrium constant for a redox reaction from the standard potentials of its half-reactions. Equation 19.12 relates the equilibrium constant for any redox reaction to the standard potential E_{rxn}°. To calculate E_{rxn}°, we need to combine the appropriate standard potentials. Table A6.1 lists two reduction half-reactions involving our reactants and products:

$$Hg^{2+}(aq) + 2\,e^- \rightarrow Hg(\ell) \qquad E^{\circ} = 0.851 \text{ V}$$
$$Sn^{4+}(aq) + 2\,e^- \rightarrow Sn^{2+}(aq) \qquad E^{\circ} = 0.154 \text{ V}$$

Analyze The problem states that Hg^{2+} is reduced by Sn^{2+}, so Sn^{2+} is the reducing agent in the reaction, which means that it must be oxidized. Therefore, the second reaction runs in reverse, as an oxidation, and we must subtract its standard potential from that of the mercury half-reaction. The difference in the two half-reaction potentials is about +0.7 V. Therefore the right side of Equation 19.12 will be about $(2 \times 0.7)/0.06 \approx 23$, and the value of *K* should be about 10^{23}.

Solve We obtain the standard potential for the reaction from a modified version of Equation 19.2:

$$E_{rxn}^{\circ} = E_{Hg}^{\circ} - E_{Sn}^{\circ} = 0.851 \text{ V} - 0.154 \text{ V} = 0.697 \text{ V}$$

Using this value for E_{rxn}° in Equation 19.12 and a value of 2 for *n*, we have

$$\log K = \frac{nE_{rxn}^{\circ}}{0.0592} = \frac{2(0.697)}{0.0592} = 23.5$$
$$K = 10^{23.5} = 3 \times 10^{23}$$

Think about It The calculated value is quite close to what we estimated. Note how a relatively small (<1 V) positive value of E_{rxn}° corresponds to a huge equilibrium constant, indicating that the reaction essentially goes to completion. That the reaction goes to completion is one reason it can be reliably used to determine the concentrations of mercury in samples containing Hg^{2+} ions.

Practice Exercise Use the appropriate standard reduction potentials in Table A6.1 to calculate the value of *K* at 25°C for the reaction

$$5\,Fe^{2+}(aq) + MnO_4^-(aq) + 8\,H^+(aq) \rightarrow 5\,Fe^{3+}(aq) + Mn^{2+}(aq) + 4\,H_2O(\ell)$$

Before ending our discussion of how to derive equilibrium constant values at 25°C from E_{cell}° values, we should note that measuring the potential of an electrochemical reaction allows us to calculate equilibrium constant values that may be too large or too small to determine from the equilibrium concentrations of reactants and products. A value of *K* as large as that calculated in Sample

Exercise 19.6 could not be obtained by analyzing the composition of an equilibrium reaction mixture because the concentrations of the reactants would be too small to be determined accurately. Similarly, a cell potential of about −1 V would correspond to a tiny K value and concentrations of products that are too small to be determined quantitatively.

Table 19.1 summarizes how the values of K and $E°_{cell}$ are related to each other and to the change in free energy ($\Delta G°_{cell}$) of a cell reaction under standard conditions. Note how spontaneous electrochemical reactions are those with $E°_{cell}$ values greater than zero and K values greater than one. The connection between positive cell potential (E_{cell}) and reaction spontaneity applies even under nonstandard conditions. Also keep in mind that small positive values of $E°_{cell}$ (only a fraction of a volt, for example) correspond to very large K values and to cell reactions that go nearly to completion.

TABLE 19.1	Relations between K, $E°_{cell}$, and $\Delta G°_{cell}$ Values of Electrochemical Reactions		
K	$E°_{cell}$	$\Delta G°_{cell}$	Favors Formation of:
<1	<0	>0	Reactants
>1	>0	<0	Products
1	0	0	Neither

19.7 Relating Battery Capacity to Quantities of Reactants

An important performance characteristic of a battery is its *capacity* to do electrical work, that is, to deliver electrical charge at the designed cell potential. This capacity—the amount of electrical work done—is defined by Equation 19.4,

$$w_{elec} = -CE_{cell}$$

where C is the quantity of electrical charge delivered in coulombs (C).

Another important unit in electricity is the *ampere* (A), which is the SI base unit of electrical current. An ampere is defined as a current of 1 coulomb per second:

$$1 \text{ ampere} = 1 \text{ coulomb/second}$$

which we can rearrange to

$$1 \text{ coulomb} = 1 \text{ ampere-second}$$

Multiplying both sides of this latter equation by volts, and recalling that 1 joule of electrical energy is equivalent to 1 coulomb-volt of electrical work, we get

$$1 \text{ (coulomb)(volt)} = 1 \text{ joule} = 1 \text{ (ampere-second)(volt)} \qquad (19.13)$$

However, joules are small energy units, so battery capacities are usually expressed in energy units with time intervals longer than seconds. For example, the energy ratings of rechargeable AA batteries (Figure 19.13) are often expressed in ampere-hours at the rated cell potential.

We need even bigger units to express the power and energy capacities of the large battery packs used in hybrid vehicles. They are the *watt* (W), the SI unit of

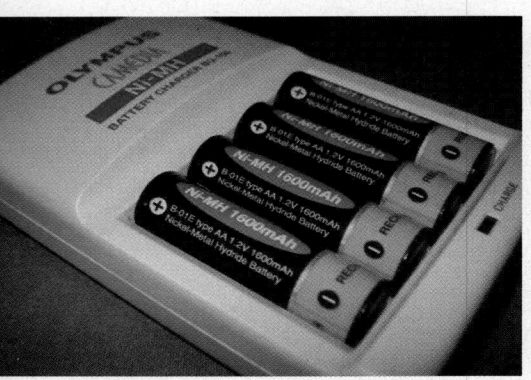

FIGURE 19.13 The electrical energy rating of these rechargeable nickel–metal hydride AA batteries is 1.6 A · hr at 1.2 V.

power, and the *kilowatt-hour*, a unit of energy equal to over 3 million joules as shown in the following unit conversions:

$$1 \text{ watt} = 1 \text{ joule/second}$$
$$1 \text{ kilowatt} = 1000 \text{ W} = 1000 \text{ J/s}$$
$$1 \text{ kilowatt-hour} = (1000 \text{ W})(1 \text{ hr})$$
$$= (1000 \text{ J/s} \times 60 \text{ s/min} \times 60 \text{ min/hr})(1 \text{ hr})$$
$$= 3.6 \times 10^6 \text{ J}$$

CONCEPT TEST

Which of the energy units discussed in this section would be appropriate in expressing (a) the capacity of a cell phone battery, (b) the annual electrical energy consumed by an Energy Star dishwasher, and (c) the monthly electricity bill for a student apartment?

Nickel-Metal Hydride Batteries

As we noted in A Look Ahead, hybrid vehicles such as the Toyota Prius (Figure 19.14) are powered by combinations of small gasoline engines and electric motors. The motors are powered by battery packs made of dozens of nickel–metal hydride (NiMH) cells (Figure 19.15). At the cathodes in these cells NiO(OH) is reduced to $Ni(OH)_2$, and at the anodes, made of one or more transition metals, hydrogen atoms are oxidized to H^+ ions. The electrodes are separated by aqueous KOH.

The cathode half-reaction is

$$NiO(OH)(s) + H_2O(\ell) + e^- \rightarrow Ni(OH)_2(s) + OH^-(aq) \qquad E° = 1.32 \text{ V}$$

At the anode, hydrogen is present as a *metal hydride*. To write the anode half-reaction, we use the generic formula MH, where M stands for a transition metal

FIGURE 19.14 The 2010 Toyota Prius is powered by a combination of a 73 kW (98 horsepower) gasoline engine and a 60 kW electric motor. Electricity for the motor comes from a 1.3 kW · hr nickel–metal hydride battery pack located behind the back seat. The battery pack is recharged when the engine is running or when the brakes are applied as the car's electric motor acts like a generator converting the car's kinetic energy into electric energy.

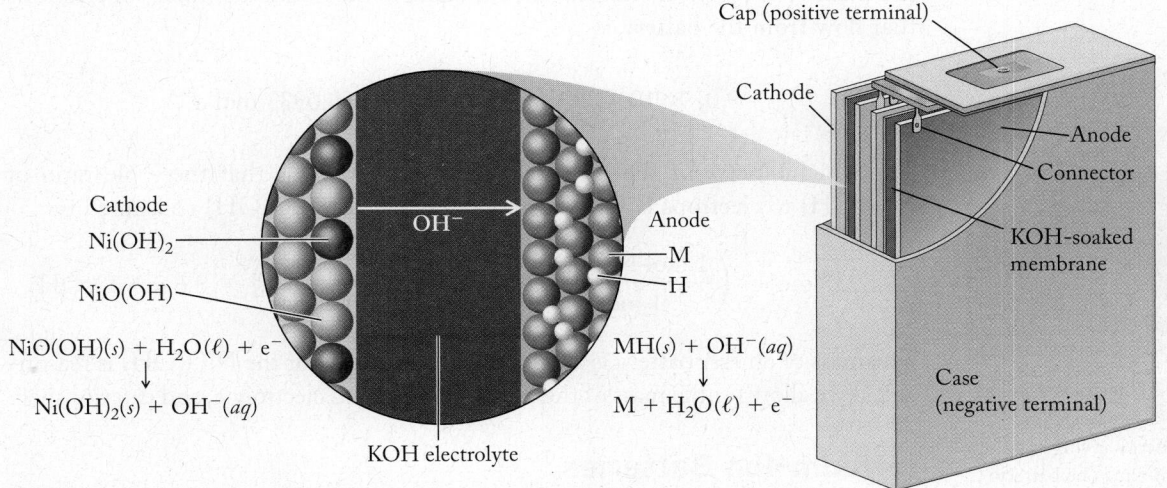

FIGURE 19.15 In the cells of a nickel–metal hydride battery pack, H atoms are oxidized to H^+ ions at the anodes (blue plates) and NiO(OH) is reduced to $Ni(OH)_2$ at the cathodes (green plates). The OH^- ions produced by the cathode half-reaction migrate across a KOH-soaked porous membrane and are consumed in the anode half-reaction. The anodes are connected to the case of the battery pack, which serves as the (−) terminal, and the cathodes are connected to the cap, which is the (+) terminal.

or metal alloy that forms a hydride. In a basic background electrolyte, the anode oxidation half-reaction is

$$MH(s) + OH^-(aq) \rightarrow M(s) + H_2O(\ell) + e^-$$

The standard potential of this half-reaction depends on the chemical properties of MH, but generally the value is near that of the SHE, or about 0.0 V.

The overall cell reaction from these two half-reactions is

$$MH(s) + NiO(OH)(s) \rightarrow M(s) + Ni(OH)_2(s)$$

The value of E°_{cell} for the NiMH battery cannot be calculated precisely because we have only an approximate value of $E^\circ_{ox}(anode)$. Most NiMH cells are rated at about 1.2 V.

CONCEPT TEST

In a NiMH battery what are the oxidation states of (a) Ni in NiO(OH), (b) H in MH, (c) M in MH, and (d) H in H_2O?

Now let's relate the electrical energy stored in a battery (in other words, its capacity) to the quantities of reactants needed to produce that energy. Consider a rechargeable AA NiMH battery rated to deliver 2.5 ampere-hours of electrical charge at 1.2 V. How much NiO(OH) has to be converted to $Ni(OH)_2$ to deliver this much charge? To answer this question, we need to relate the quantity of charge to a number of moles of electrons and then convert that to an equivalent number of moles of reactant and finally to a mass of reactant. Let's begin by recalling that an ampere is defined as a coulomb per second, which means the quantity of electrical charge delivered is

$$2.5 \ \cancel{A \cdot hr} \times \frac{1\,C}{\cancel{A \cdot s}} \times \frac{60 \ \cancel{min}}{1 \ \cancel{hr}} \times \frac{60\,\cancel{s}}{1\,\cancel{min}} = 9.0 \times 10^3 \ C$$

Faraday's constant tells us that 1 mole of charge is equivalent to 9.65×10^4 C, so the number of moles of charge, which is equal to the number of moles of electrons that flow from the battery, is

$$9.0 \times 10^3 \ \cancel{C} \left(\frac{1 \, mol \ e^-}{9.65 \times 10^4 \ \cancel{C}} \right) = 0.0933 \ mol \ e^-$$

The stoichiometry of the cathode half-reaction tells us that the mole ratio of NiO(OH) to electrons is 1:1. Therefore, the mass of NiO(OH) consumed is

$$0.0933 \ \cancel{mol \ e^-} \left(\frac{1 \ \cancel{mol \ NiO(OH)}}{1 \ \cancel{mol \ e^-}} \right) \left(\frac{91.70 \ g \ NiO(OH)}{1 \ \cancel{mol \ NiO(OH)}} \right) = 8.6 \ g \ NiO(OH)$$

The mass of an AA battery is about 30 g, so this mass for the NiO(OH) is reasonable if we allow for the mass of the anode, background electrolyte, and exterior shell.

Lithium-Ion Batteries

The NiMH batteries used in hybrid vehicles do not have the capacity to power them at highway speeds nor for extended distances. Nor do these batteries have the energy capacity to power all-electric vehicles such as the Tesla Roadster on the opening page of this chapter or plug-in hybrids such as the Chevrolet Volt (Figure 19.16).

FIGURE 19.16 The 2011 Chevrolet Volt has a lithium–ion battery pack that can store 16 kW · hr of electrical energy, giving the Volt a driving range of about 60 km. The battery pack is recharged by either plugging it into an electrical outlet or by running an onboard gasoline-powered generator. The generator gives the Volt an overall driving range of about 1000 km.

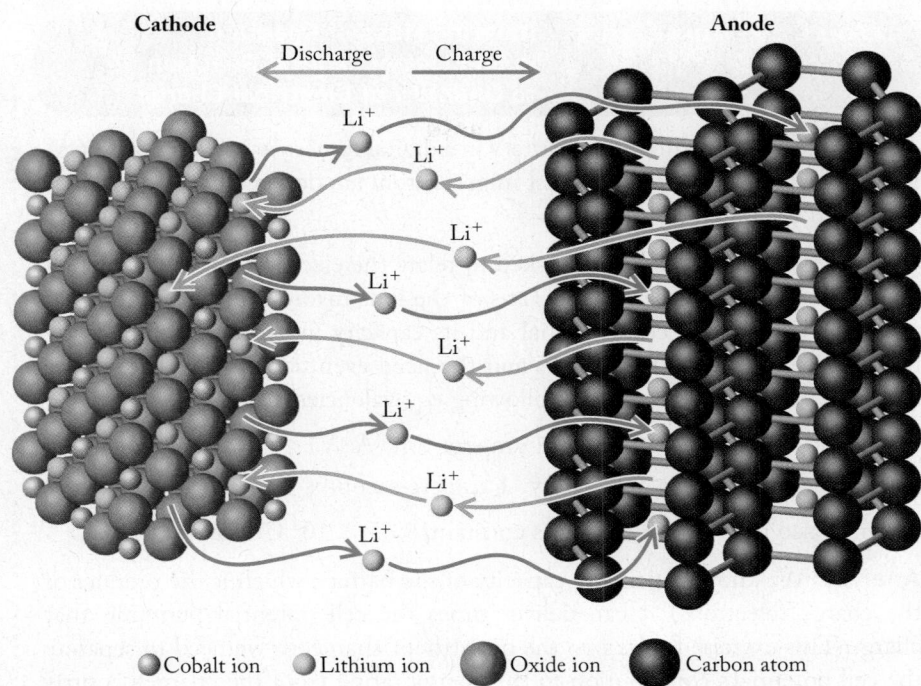

Cathode   Anode

Discharge   Charge

○ Cobalt ion   ○ Lithium ion   ● Oxide ion   ● Carbon atom

FIGURE 19.17 In a discharging lithium–ion battery, Li^+ ions stored in graphite layers of the anode travel to the cathode, which is made of CoO_2 in this example. During recharging the direction of ion migration reverses. The crystal structure of the cathode is a cubic closest-packed array of oxide ions in which Co^{4+} ions occupy half the octahedral holes. Li^+ ions move in and out of the remaining holes.

The electrical power demands of these vehicles require batteries with much greater ratios of energy capacity to battery size. The technology of choice in these applications is the lithium–ion battery (Figure 19.17), the same kind of battery that powers laptop computers, cell phones, and digital cameras.

In a lithium–ion battery, Li^+ ions are stored in a graphite anode. During discharge, these ions migrate through a nonaqueous electrolyte to a porous cathode. These cathodes are made of transition metal oxides or phosphates that can form stable complexes with Li^+ ions. One popular cathode material is cobalt(IV) oxide. Lithium–ion batteries with these cathodes have cell potentials of about 3.6 V (three times that of a NiMH battery). The cell reaction for a lithium–ion battery with this cathode is

$$Li_{1-x}CoO_2(s) + Li_xC_6(s) \rightarrow 6\,C(s) + LiCoO_2(s) \qquad (19.14)$$

In a fully charged cell, $x = 1$, which makes the cathode lithium-free CoO_2. As the cell discharges and Li^+ ions migrate from the carbon anode to the cobalt oxide cathode, the value of x falls toward zero. To balance this flow of positive charges inside the cell, electrons flow from the anode to the cathode through an external circuit. When fully discharged, the cathode is $LiCoO_2$, and the oxidation number of Co is reduced to +3. The electrodes in a lithium–ion battery may react with oxygen and water, so the background electrolytes (for example, $LiPF_6$) are dissolved in polar organic solvents, such as tetrahydrofuran, ethylene carbonate, or propylene carbonate (Figure 19.18).

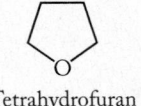

Tetrahydrofuran

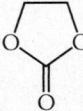

Ethylene carbonate

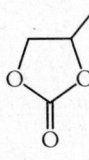

Propylene carbonate

FIGURE 19.18 Polar organic compounds such as these are the solvents for the background electrolytes in a lithium–ion battery.

CONCEPT TEST

Which element is oxidized and which is reduced in the Li^+ ion cell reaction (Equation 19.14)?

SAMPLE EXERCISE 19.7 **Relating Mass of Reactant in an Electrochemical Reaction to Quantity of Electrical Charge**

The capacity of the lithium–ion battery in a digital camera is $3.4 \, W \cdot hr$ at $3.6 \, V$. How many grams of Li^+ ions must migrate from anode to cathode to produce this much electrical energy?

Collect and Organize We are asked to relate the electrical energy generated by an electrochemical cell to the mass of the ions involved in generating that energy. We know the cell potential and its capacity in the energy unit watt-hours. Given these starting points and the need eventually to calculate moles and then grams of Li^+ ions, the following equivalencies may be useful:

$$1 \text{ watt} = 1 \text{ ampere-volt} (A \cdot V)$$

$$1 \text{ coulomb} (C) = 1 \text{ ampere-second} (A \cdot s)$$

We may also need to use Faraday's constant, $9.65 \times 10^4 \, C/mol$.

Analyze We know the energy capacity of the battery, which is the product of the charge (electrons) it can deliver times the cell potential pumping that charge. This exercise focuses on the quantity of charge, so we need to separate the cell potential's contribution to the energy rating from the charge's contribution. To do that we need to divide the energy rating in watt-hours by the battery's cell potential in volts, and then follow that division with these unit conversions to get to moles of charge:

$$\frac{\text{Watt} \cdot \text{hour}}{\text{volt}} \times \frac{1 \, \text{ampere} \cdot \text{volt}}{1 \, \text{watt}} \times \frac{1 \, \text{coulomb}}{1 \, \text{ampere} \cdot \text{second}} \times \frac{3600 \, \text{seconds}}{1 \, \text{hour}}$$

$$\times \frac{1 \, \text{mol e}^-}{9.65 \times 10^4 \, \text{coulombs}}$$

Solve Using the given energy and cell-potential values in the above unit conversion series, we get

$$\frac{3.4 \, W \cdot hr}{3.6 \, V} \times \frac{1 \, A \cdot V}{1 \, W} \times \frac{1 \, C}{1 \, A \cdot s} \times \frac{3600 \, s}{1 \, hr} \times \frac{1 \, \text{mol e}^-}{9.65 \times 10^4 \, C} = 0.0352 \, \text{mol e}^-$$

Converting 0.0352 moles of electrons into an equivalent mass of Li^+ ions:

$$0.0352 \, \text{mol e}^- \left(\frac{1 \, \text{mol Li}}{1 \, \text{mol e}^-}\right)\left(\frac{6.941 \, g \, Li}{1 \, \text{mol Li}^+}\right) = 0.24 \, g \, Li$$

Think about It The calculated mass of Li^+ ions is close to our estimate and so is reasonable. The battery that is the subject of this exercise has a mass of about 22 g, so Li^+ ions make up only about 1% of the mass of the battery. This small percentage is reasonable given the masses of the other required components of the cell, including an anode where Li^+ ions are surrounded by hexagons of six carbon atoms and a cathode made of CoO_2, which has 13 times the molar mass of Li.

Practice Exercise Magnesium metal is produced by passing an electric current through molten $MgCl_2$. The reaction at the cathode is

$$Mg^{2+}(\ell) + 2 \, e^- \rightarrow Mg(\ell)$$

How many grams of magnesium metal are produced if an average current of 63.7 A flows for 4.50 hr? Assume all of the current is consumed by the half-reaction shown.

19.8 Electrolytic Cells and Rechargeable Batteries

Lead–acid, NiMH, and lithium–ion batteries are rechargeable, which means that their spontaneous ($\Delta G < 0$) cell reactions that convert chemical energy into electrical energy can be forced to run in reverse. Recharging happens when external sources of electrical energy are applied to the batteries. This electrical energy is converted into chemical energy as it drives nonspontaneous ($\Delta G > 0$) reverse cell reactions, re-forming reactants from products. To make this possible, the products of the original cell reactions must be substances that either adhere to or are embedded in the electrodes, and so are available to react with the electrons supplied to the cathodes and drawn away from the anodes by the external power supply.

Recharging a battery is an example of **electrolysis**, which is defined as any chemical reaction driven by electricity. During recharging, a battery is transformed from a voltaic cell into an **electrolytic cell** (Figure 19.19). To explore this transformation, let's revisit the lead–acid battery used to start car engines. Its discharge and recharge cycles are shown in Figure 19.20. Note that electrons flow in one direction when the battery discharges—out of the negative terminal and into the positive terminal—but in the opposite direction when the battery is recharging. Thus, the Pb electrodes, which are connected to the negative battery terminal in Figure 19.20, serve as anodes during discharge but as cathodes during recharge. Any $PbSO_4$ that forms on these Pb electrodes during discharge is reduced back to Pb metal during recharge:

$$PbSO_4(s) + H^+(aq) + 2\,e^- \rightarrow Pb(s) + HSO_4{}^-(aq)$$

Similarly, the PbO_2 electrodes at the positive terminal serve as cathodes during discharge but as anodes during recharge. Any $PbSO_4$ that forms on these PbO_2 electrodes during discharge is oxidized back to PbO_2 during recharge:

$$PbSO_4(s) + 2\,H_2O(\ell) \rightarrow PbO_2(s) + 3\,H^+(aq) + HSO_4{}^-(aq) + 2\,e^-$$

electrolysis a process in which electrical energy is used to drive a nonspontaneous chemical reaction.

electrolytic cell a device in which an external source of electrical energy does work on a chemical system, turning reactant(s) into higher-energy product(s).

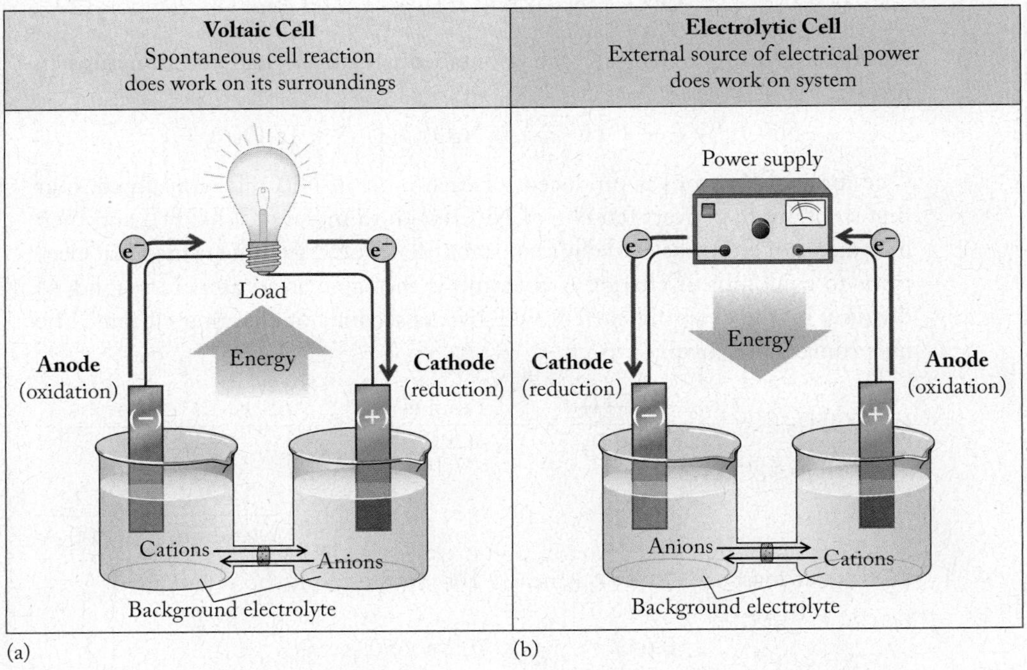

Voltaic Cell Spontaneous cell reaction does work on its surroundings	**Electrolytic Cell** External source of electrical power does work on system

(a) (b)

FIGURE 19.19 Voltaic versus electrolytic cells. (a) In a voltaic cell, a spontaneous reaction produces electrical energy and does electrical work on its surroundings, such as lighting a lightbulb. (b) In an electrolytic cell, an external supply of electrical energy does work on the chemical system in the cell, driving a nonspontaneous reaction.

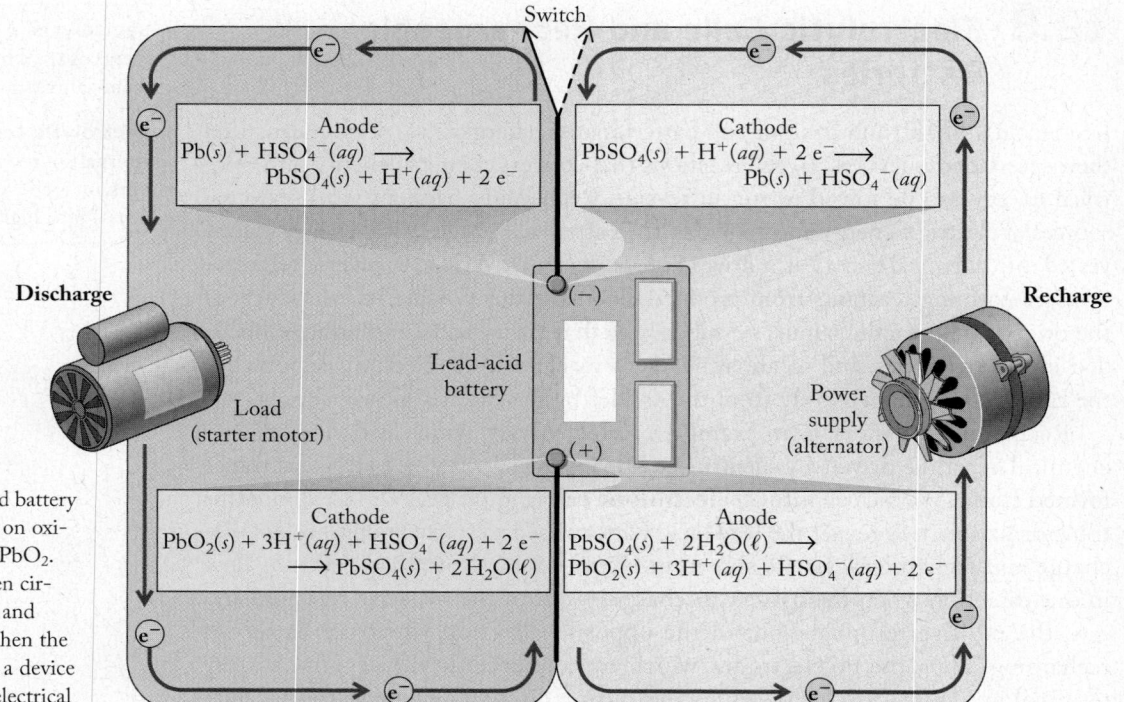

FIGURE 19.20 The lead–acid battery used in many vehicles is based on oxidation of Pb and reduction of PbO_2. As the battery discharges (green circuit), Pb is oxidized to $PbSO_4$ and PbO_2 is reduced to $PbSO_4$. When the engine is running (tan circuit), a device called an alternator generates electrical energy that flows into the battery, recharging it as both electrode reactions are reversed: $PbSO_4$ is oxidized to PbO_2 and $PbSO_4$ is reduced to Pb.

SAMPLE EXERCISE 19.8   **Calculating the Time Required to Oxidize a Quantity of Reactant**

If a battery charger for AA NiMH batteries supplies a charging current of 1.00 A, how long does it take to oxidize 0.649 g of $Ni(OH)_2$ to NiO(OH)?

Collect and Organize We are asked to calculate the time required for a charging current to oxidize a given mass of $Ni(OH)_2$ to NiO(OH). During the discharge of a NiMH battery the spontaneous cathode half-reaction is

$$NiO(OH)(s) + H_2O(\ell) + e^- \rightarrow Ni(OH)_2(s) + OH^-(aq) \qquad E° = 1.32 \text{ V}$$

Analyze During recharging the spontaneous cathode half-reaction runs in reverse:

$$Ni(OH)_2(s) + OH^-(aq) \rightarrow NiO(OH)(s) + H_2O(\ell) + e^-$$

One mole of electrons is produced for each mole of $Ni(OH)_2$ consumed. Our first steps are to convert 0.649 g of $Ni(OH)_2$ into moles of $Ni(OH)_2$ and then into moles of electrons. Faraday's constant can be used to convert moles of electrons to coulombs of charge. A coulomb is the same as an ampere-second, so dividing by the charging current will give us seconds of charging current. The unit conversions involved are

$$\text{g Ni(OH)}_2 \times \frac{1 \text{ mol Ni(OH)}_2}{\mathcal{M} \text{ g Ni(OH)}_2} \times \frac{1 \text{ mol e}^-}{1 \text{ mol Ni(OH)}_2} \times \frac{9.65 \times 10^4 \text{ coulombs}}{1 \text{ mol e}^-}$$

$$\times \frac{1 \text{ A} \cdot \text{s}}{1 \text{ coulomb}} \times \frac{1}{1.00 \text{ A}}$$

The molar mass of $Ni(OH)_2$ is nearly 100, so the result should be about

$$0.649 \times \frac{1}{10^2} \times 10^5 \approx 650 \text{ seconds}$$

Solve

$$0.649 \text{ g Ni(OH)}_2 \times \frac{1 \text{ mol Ni(OH)}_2}{92.71 \text{ g Ni(OH)}_2} \times \frac{1 \text{ mol e}^-}{1 \text{ mol Ni(OH)}_2} \times \frac{9.65 \times 10^4 \text{ C}}{1 \text{ mol e}^-}$$

$$\times \frac{1 \text{ A} \cdot \text{s}}{1 \text{ C}} \times \frac{1}{1.00 \text{ A}} = 676 \text{ s}$$

Thus, the charger must deliver 1.00 A of current for 676 s or

$$(676 \text{ s}) \times \frac{1 \text{ min}}{60 \text{ s}} = 11.3 \text{ min}$$

Think about It The calculated result agrees well with our estimate. A charging time of 11.3 min may seem short, but the quantity of the $Ni(OH)_2$ to be oxidized (0.649 g) is much less than the quantity of NiO(OH) in a fully charged AA NiMH battery (8.6 g, as calculated on page 938), so this battery was only slightly discharged.

Practice Exercise Suppose that a car's starter motor draws 230 A of current for 6.0 s to start the car. What mass of Pb is oxidized in the battery to supply this much electricity?

Electrolysis is used in many processes other than recharging batteries. Electrolytic cells are used to electroplate thin layers of silver, gold, and other metals onto objects, giving these objects the appearance, resistance to corrosion, and other properties of the electroplated metal but at a fraction of the cost of fabricating the entire object out of the metal (Figure 19.21).

In the chemical industry, electrolysis of molten salts is used to produce highly reactive substances such as sodium, chlorine, and fluorine; alkali and alkaline earth metals, and aluminum. When NaCl, for instance, is heated to just above its melting point (above 800°C), it becomes an ionic liquid that can conduct electricity. If a sufficiently large potential is applied to carbon electrodes immersed in the molten NaCl, sodium ions are attracted to the negative electrode and are reduced to sodium metal and chloride ions are attracted to the positive electrode and oxidized to Cl_2 gas:

$$2\,Na^+(\ell) + 2\,Cl^-(\ell) \rightarrow 2\,Na(\ell) + Cl_2(g)$$

A final note about anode and cathode polarity is in order. The reactions in voltaic cells are spontaneous. These cells pump electric current through external circuits and electrical devices with a force equal to their cell potentials. The anode in a voltaic cell is negative because an oxidation half-reaction supplies negatively charged electrons to the device powered by the cell. Electrons flow from the device into the positive battery terminal which is connected to the cathode, where these electrons are consumed in a reduction half-reaction.

The reactions in electrolytic cells are nonspontaneous. They require electrical energy from an external power supply. When the negative terminal of such a power supply is connected to the cathode of the battery, the power supply pumps electrons into the cathode, where they are consumed in reduction half-reactions. Electrons are pumped away from the anode, where they must have been generated in oxidation half-reactions, toward the positive terminal of the power supply. Thus, the cathode of an electrolytic cell is the negative electrode, but the cathode of a voltaic cell is the positive electrode. Similarly, the anode of an electrolytic cell is the positive electrode, but the anode of a voltaic cell is the negative electrode. These "pole reversals" make sense if we keep in mind the fundamental definitions:

Anodes are electrodes where oxidation takes place.

Cathodes are electrodes where reduction takes place.

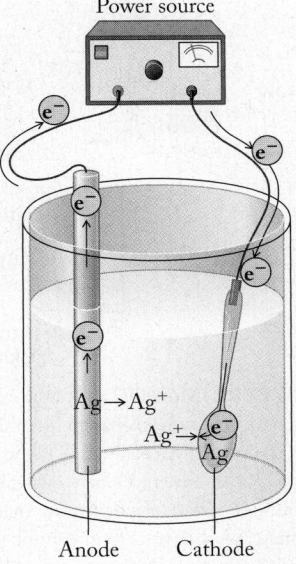

FIGURE 19.21 The cutlery known as silverplate has only a thin coating of silver, applied over a base metal core by a type of electrolysis known as electroplating. The positive terminal of an electric power supply is attached to a piece of pure silver, and the negative terminal is connected to the piece to be electroplated, such as the spoon shown here. The oxidation half-reaction at the silver electrode (anode), $Ag \rightarrow Ag^+ + e^-$, runs in reverse at the cathode (spoon) as Ag^+ ion is reduced to the free metal on the surface of the spoon.

fuel cell a voltaic cell based on the oxidation of a continuously supplied fuel. The reaction is the equivalent of combustion, but chemical energy is converted directly into electrical energy.

CONCEPT TEST

The electrolysis of molten NaCl produces liquid Na metal at the cathode and Cl_2 gas at the anode. However, the electrolysis of an aqueous solution of NaCl produces gases at both cathode and anode. On the basis of the standard potentials in Table A6.1, predict the gas formed at each electrode.

19.9 Fuel Cells

Fuel cells are promising energy sources for many applications, from powering office buildings to cruise ships to electric vehicles. Fuel cells are voltaic cells, but they are different from batteries in that their supplies of reactants are constantly renewed. Therefore, they do not "discharge"; they never run down, and they don't die unless their fuel supply is cut off. From a thermodynamic perspective, batteries are closed systems and fuel cells are open systems.

In a typical fuel cell, electrons are supplied to an external circuit by the oxidation of H_2 at the anode. Electrons are consumed by the reduction of O_2 at the cathode, so the fuel cell reaction is

$$2\,H_2(g) + O_2(g) \rightarrow 2\,H_2O(\ell)$$

The fuel cells used to power electric vehicles consist of metallic or graphite electrodes separated by a hydrated polymeric material called a *proton-exchange membrane* (PEM) that serves as both an electrolyte and a barrier to the flow of electrons (Figure 19.22). The surfaces of the electrodes are coated with transition-metal catalysts to speed up the electrode half-reactions. Platinum catalysts promote H—H bond breaking during the oxidation of H_2 gas to H^+ ions at the anode:

$$H_2(g) \rightarrow 2\,H^+(aq) + 2\,e^- \qquad E° = 0.000\text{ V}$$

and a platinum–nickel alloy with the formula Pt_3Ni is particularly effective in catalyzing the formation of free O atoms from O_2 molecules that is part of the reduction half-reaction at the cathode:

$$O_2(g) + 4\,H^+(aq) + 4\,e^- \rightarrow 2\,H_2O(\ell) \qquad E° = +1.229\text{ V}$$

Hydrogen ions that form at the anode migrate through the PEM to the cathode, where they combine with O_2. This migration of positive charges inside the fuel cell drives the flow of electrons in the electrical device attached to it.

A single PEM fuel cell typically has a cell potential of about 1.0 V. When hundreds of these cells are assembled into fuel cell *stacks*, they are capable of producing 100 kW of electrical power. That is enough to give a mid-size car such as the one in Figure 19.23 a top speed of 160 km/hr (99 mi/hr).

PEM fuel cells are well suited for use in vehicles because they are compact, lightweight, and operate at fairly low temperatures of 60 to 80°C. It turns out that the performance of nearly all fuel cells is better at above-ambient temperatures because the higher rates of their half-reactions at these temperatures means that they generate more electrical power.

Some fuel cells, such as those that supply electrical power (and drinking water) to the U.S. space shuttles, use basic electrolytes such as concentrated KOH. Pure O_2 is supplied to a cathode made of porous graphite containing a nickel catalyst,

▶❚❚ **CHEMTOUR** Fuel Cell

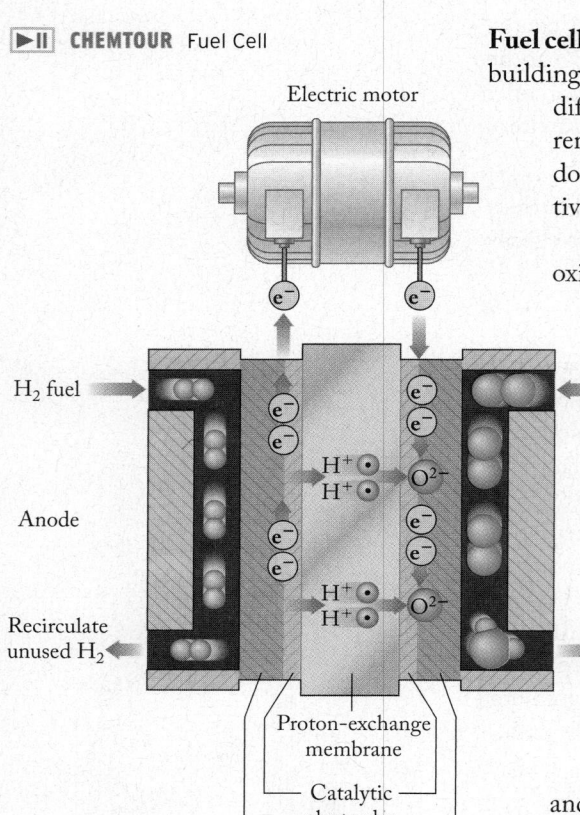

FIGURE 19.22 Most fuel cells used in vehicles have a proton-exchange membrane between the two halves of the cell. Hydrogen gas diffuses to the anode, and oxygen gas diffuses to the cathode. These electrodes are made of porous material, such as carbon nanofibers, that have a relatively high surface area for a given mass of material. Catalysts on the electrode surfaces also increase the rate of the half-reactions at the anode ($H_2 \rightarrow 2\,H^+ + 2\,e^-$) and the cathode ($O_2 + 4\,H^+ + 4\,e^- \rightarrow 2\,H_2O$).

and H_2 gas is supplied to a graphite anode containing nickel(II) oxide. Hydroxide ions formed during O_2 reduction at the cathode,

$$O_2(g) + 2H_2O(\ell) + 4e^- \rightarrow 4OH^-(aq) \qquad E° = 0.401 \text{ V}$$

migrate through the cell to the anode, where they combine with H_2 as it is oxidized to water:

$$H_2(g) + 2OH^-(aq) \rightarrow 2H_2O(\ell) + 2e^-$$

This oxidation half-reaction is the reverse of the following reduction half-reaction in Table A6.1:

$$2H_2O(\ell) + 2e^- \rightarrow H_2(g) + 2OH^-(aq) \qquad E° = -0.828 \text{ V}$$

Note that this basic pair of standard electrode potentials yields the same $E°_{cell}$ value:

$$E°_{cell} = 0.401 \text{ V} - (-0.828 \text{ V}) = 1.229 \text{ V}$$

as the acidic pair described earlier:

$$E°_{cell} = 1.229 \text{ V} - 0.000 \text{ V} = 1.229 \text{ V}$$

This equality is logical because the energy $\Delta G°$ released under standard conditions by the oxidation of hydrogen gas to form liquid water should have only one value, which means that $E°_{cell}$ should have only one value independent of electrolyte pH.

CONCEPT TEST

During the operation of molten alkali metal carbonate fuel cells, carbonate ions are generated at one electrode, migrate across the cell, and are consumed at the other electrode. Do the carbonate ions migrate toward the cathode or the anode?

The same chemical energy that is released in fuel cells could also be obtained by burning hydrogen gas in an internal combustion engine. However, typically only about 20–25% of the chemical energy in the fuel burned in such an engine is converted into mechanical energy; most is lost to the surroundings as heat. In contrast, fuel-cell technologies can convert up to about 80% of the energy released in a fuel-cell redox reaction into electrical energy. The electric motors they power are also about 80% efficient at converting electrical energy into mechanical energy. Thus, the overall conversion efficiency of a fuel cell–powered car is theoretically as high as (80% × 80%) or 64%. Actually, the measured efficiency of the propulsion system of the car in Figure 19.23 is about 60%, which is still more than twice that of an internal combustion engine. In addition, H_2-fueled vehicles emit only water vapor; they produce no oxides of nitrogen, no carbon monoxide, and no CO_2.

As fuel cells become even more efficient and less expensive, the principal limit on their use in passenger cars will be the availability and cost of hydrogen fuel. Some people worry about the safety of storing hydrogen in a high-pressure tank in a car. Actually H_2 has a higher ignition temperature than gasoline and spreads through the air more quickly, reducing the risk of fire. Still, hydrogen in air burns over a much wider range of concentrations than gasoline, and its flame is almost invisible.

Because of the lack of an extensive hydrogen distribution network, it is likely that most fuel cell–powered vehicles will be buses and fleet vehicles operating from a central location where hydrogen gas is available. With their very large fuel tanks, buses powered by fuel cells have an operating range of 400 km. Longer ranges would be possible if better methods for storing hydrogen gas were available. Currently under development are mobile chemical-processing plants that can extract hydrogen from gasoline, methane, or methanol. On-board production of hydrogen from fossil fuels also generates carbon dioxide as a by-product, so these fuels still produce greenhouse gases. However, the efficiencies of fuel cells and the electric motors they power mean that much less of these fuels will be needed and less CO_2 will be emitted.

FIGURE 19.23 The Honda FCX Clarity is powered by a 100 kW fuel-cell stack and 100 kW electric motor that give the car a top speed of 160 km/hr (99 mi/hr).

CONNECTION Processes for producing hydrogen gas by reacting methane and other fuels with steam were discussed in Chapter 16.

Power, Pigments, and Phosphors: The Group 12 Elements

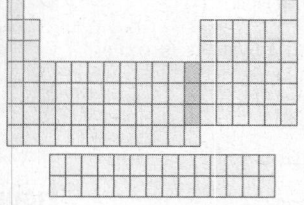

Among the group 12 transition metals (Zn, Cd, Hg), zinc is the most abundant in Earth's crust (78 mg/kg) and mercury is the least (67 μg/kg). All three are found in sulfide ores, including sphalerite (ZnS), greenockite (CdS), and cinnabar (HgS). Zinc is also found in carbonates and silicates. Most zinc ores contain cadmium impurities.

The cathode-ray tubes used by J. J. Thomson in his pioneering work with electrons (see Chapter 2) were coated with ZnS because this compound emits light when bombarded by electrons. If another metal called an *activator* is added to ZnS, the mixture phosphoresces, which means that the energy absorbed by the material is emitted slowly as visible light. The color of the phosphorescence depends on the activator: traces of silver produce blue light, manganese yields red-orange, and copper produces a long-lasting dark green that is used in electroluminescent panels.

Zinc oxide is obtained from ore containing ZnS by roasting the ore in air:

$$2\,ZnS(s) + 3\,O_2(g) \rightarrow 2\,ZnO(s) + 2\,SO_2(g)$$

Zinc oxide is a common ingredient in antibiotic creams and opaque sunscreens (Figure 19.24) and is used to coat the insides of fluorescent lightbulbs. A special grade of zinc oxide called Chinese white is used in artists' pigments. Zinc metal is recovered from ZnO by reacting the oxide with hot carbon:

$$2\,ZnO(s) + C(s) \rightarrow 2\,Zn(\ell) + CO_2(g)$$

FIGURE 19.24 Zinc oxide is used to make topical creams that are also effective sunscreens.

Cadmium is recovered from CdS impurities in ZnS ore and is separated by distilling the molten metals. Cadmium is used in making nickel–cadmium (nicad) batteries, but large quantities are also used in producing pigments ranging in color from lemon yellow (CdS) to deep maroon (CdSe). Cadmium pigments are popular with artists because of their bright colors and their stability when exposed to light.

Cinnabar (HgS) has been mined in Spain since Roman times. Also called vermilion or Chinese red, cinnabar was widely used in Chinese decorative art and lacquerware (Figure 19.25). Liquid mercury is recovered from HgS by reacting the sulfide with oxygen:

$$HgS(s) + O_2(g) \rightarrow Hg(\ell) + SO_2(g)$$

Mercury is the only metal, and one of only two elements (bromine is the other), that is a liquid at room temperature. Zinc and cadmium are low-melting, silvery solids that react rapidly with moist air to form the corresponding oxides. The oxides and sulfides of zinc and cadmium are semiconductors (see Chapter 12).

The group 12 metals have an $(n-1)d^{10}ns^2$ electron configuration in their valence shells and, like the metals in group 2, lose two valence electrons and form M^{2+} cations. Mercury also forms a Hg_2^{2+} cation, which contains a Hg—Hg covalent bond. The tendency of Hg to form covalent bonds is consistent with its electronegativity (Figure 19.26), which is higher than that of most metallic elements. A high electronegativity value means that the compounds Hg forms with nonmetals have more molecular and less ionic character (see Section 8.3).

Zinc is an essential element in living organisms, but cadmium and mercury have no known nutritional value. Indeed, they and many of their compounds are quite toxic. Their toxicity is related to the ability of their ions to form strong bonds to sulfur groups in proteins. If these metals reach the brain, interactions with proteins there can lead to mental retardation and other neurological disorders. The toxicity of mercury is made worse when it forms covalent alkyl mercury compounds such as methyl mercuric chloride, CH_3HgCl, and dimethyl mercury, $(CH_3)_2Hg$. These compounds can penetrate hydrophobic cell membranes more easily than inorganic mercury compounds and can more severely disrupt internal cell functions.

Microbial methylation of inorganic mercury wastes dumped in Minamata Bay, Japan, in the 1950s led to 52 sudden deaths and severe neurological damage to more than 3000 people. This incident sparked recognition of the environmental hazards of improperly disposing of mercury and other toxic metals.

FIGURE 19.25 The deep red color of Chinese lacquerware comes from the mineral cinnabar, which is principally HgS.

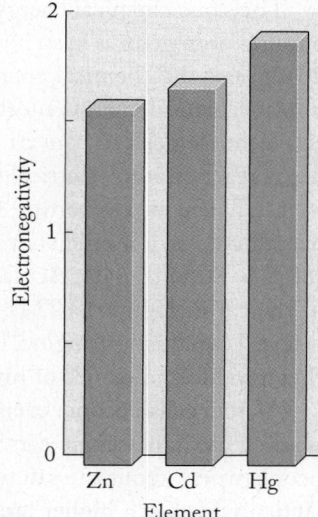

FIGURE 19.26 The electronegativities of the group 12 elements increase as their atomic numbers increase.

SUMMARY

Section 19.1 Electrochemistry is the branch of chemistry that links redox reactions to the production or consumption of electrical energy. Any redox reaction can be broken down into oxidation and reduction half-reactions.

Section 19.2 In an **electrochemical cell** the oxidation half-reaction occurs at the **anode** and the reduction half-reaction occurs at the **cathode**. Migration of the ions in the cell's electrolyte allows electrical charges to flow between the cathode and anode compartments as electrons flow through an external electrical circuit. A **cell diagram** shows how the components of the cathodic and anodic compartments of the cell are connected.

Section 19.3 The difference in the **standard reduction potentials** of a cell's cathode and anode half-reactions is equal to the **standard cell potential ($E°_{cell}$)**. The value of $E°_{cell}$ is a measure of the **electromotive force (emf)** with which a **voltaic cell** can pump electrons through an external circuit under standard conditions.

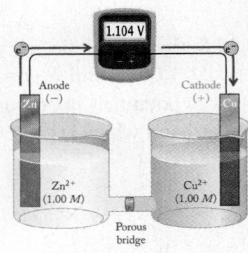

Section 19.4 A voltaic cell has a positive cell potential ($E_{cell} > 0$) and its cell reaction has a negative change in free energy ($\Delta G_{cell} < 0$). This decrease in free energy in a voltaic cell is available to do work in an external electrical circuit. Faraday's constant relates the quantity of electrical charge to the number of moles of electrons and indirectly to the number of moles of reactants.

Section 19.5 All standard cell potentials are referenced to the cell potential of the **standard hydrogen electrode** ($E_{SHE} = 0.000$ V).

Section 19.6 The potential of a voltaic cell decreases as reactants turn into products. The **Nernst equation** describes how cell potential changes with concentration changes. The potential of a voltaic cell approaches zero as the cell reaction approaches chemical equilibrium, at which point $E_{cell} = 0$ and $Q = K$.

Section 19.7 The quantities of reactants consumed in a voltaic cell reaction are directly proportional to the coulombs of electrical charge delivered by the cell. Nickel–metal hydride batteries supply electricity when H atoms are oxidized to H^+ ions at the anodes and NiO(OH) is reduced to $Ni(OH)_2$ at the cathodes. In Li^+ ion batteries electricity is produced when Li^+ ions stored in graphite anodes migrate toward and are incorporated into transition metal oxide or phosphate cathodes.

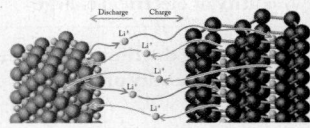

Section 19.8 A spontaneous cell reaction can be reversed by applying an opposing potential greater than E_{cell} using an external power supply. Using electrical power to force a nonspontaneous cell reaction is called **electrolysis**, and this process turns the cell into an **electrolytic cell**. The polarities of cathodes and anodes in electrolytic cells are opposite the polarities of these electrodes in voltaic cells.

Section 19.9 Fuel cells directly convert the chemical free energy released during the reaction $2H_2 + O_2 \rightarrow 2H_2O$ into electrical energy. The electrodes in fuel cells often incorporate catalysts to speed up the half-reactions involving H_2 and O_2 gas. Fuel cells with proton-exchange membranes have been developed as power supplies for electric vehicles.

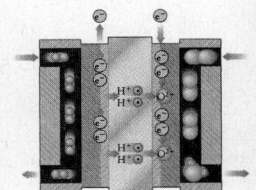

PROBLEM-SOLVING SUMMARY

TYPE OF PROBLEM	CONCEPTS AND EQUATIONS	SAMPLE EXERCISES
Writing net ionic equations for redox reactions	Combine the reduction and oxidation half-reactions after balancing the gain and loss of electrons.	19.1
Diagramming an electrochemical cell	Start with the anode half-reaction on the left; use single lines to separate phases and a double line to separate the two half-reactions. Separate species in the same phase with commas. Insert concentrations if known.	19.2
Identifying anode and cathode half-reactions and calculating the value of $E°_{cell}$	The half-reaction with the more positive standard reduction potential is the cathode half-reaction. $$E°_{cell} = E°_{cathode} - E°_{anode} \qquad (19.2)$$	19.3
Relating $\Delta G°_{cell}$ and $E°_{cell}$	$$\Delta G°_{cell} = -nFE°_{cell} \qquad (19.7)$$ where n is the number of moles of electrons transferred in the cell reaction and F is Faraday's constant, 9.65×10^4 C/mol.	19.4
Calculating E_{cell} from $E°_{cell}$ and the concentrations of reactants and products	E_{cell} at 25°C is related to $E°_{cell}$ and the cell reaction quotient by the Nernst equation: $$E_{cell} = E°_{cell} - \frac{0.0592}{n} \log Q \qquad (19.10)$$	19.5

TYPE OF PROBLEM	CONCEPTS AND EQUATIONS	SAMPLE EXERCISES
Calculating K for a redox reaction from the standard potentials of its half-reactions	$E°_{rxn}$ is related to K at 298 K by $$\log K = \frac{nE°_{rxn}}{0.0592} \qquad (19.12)$$	19.6
Relating the mass of a reactant in an electrochemical reaction to a quantity of electrical charge	Determine the ratio of moles of reactants to moles of electrons transferred; use Faraday's constant to relate coulombs of charge to moles of electrons.	19.7
Calculating the time required to oxidize a quantity of reactant	Use Faraday's constant and the relation between moles of electrons and moles of reactants to describe an electrolytic process.	19.8

VISUAL PROBLEMS

(Answers to boldface end-of-chapter questions and problems are in the back of the book.)

19.1. In the voltaic cell shown in Figure P19.1, the greater density of a concentrated solution of $CuSO_4$ allows a less concentrated solution of $ZnSO_4$ solution to be (carefully) layered on top of it. Why is a porous separator not needed in this cell?

FIGURE P19.1

19.2. In the voltaic cell shown in Figure P19.2, the concentrations of Cu^{2+} and Cd^{2+} are 1.00 M. On the basis of the standard potentials in Appendix 6, identify which electrode is the anode and which is the cathode. Indicate the direction of electron flow.

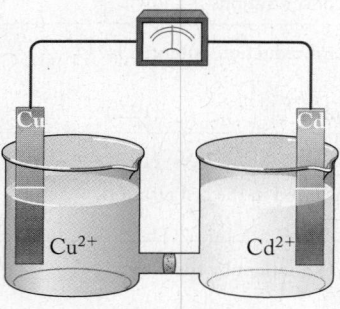

FIGURE P19.2

19.3. In the voltaic cell shown in Figure P19.3, $[Ag^+] = [H^+] = 1.00$ M and $P_{H_2} = 1.00$ atm. On the basis of the standard potentials in Appendix 6, identify which electrode is the anode and which is the cathode. Indicate the direction of electron flow.

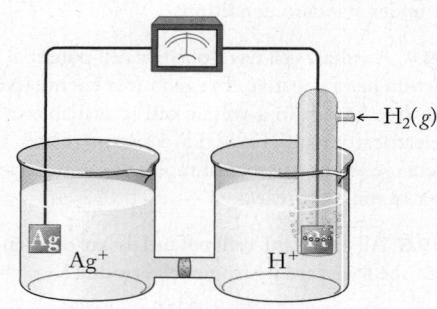

FIGURE P19.3

19.4. In many electrochemical cells the electrodes are metals that carry electrons to and from the cell but are not chemically changed by the cell reaction. Each of the highlighted clusters in the periodic table in Figure P19.4 consists of three metals. Which of the highlighted clusters is best suited to form inert electrodes?

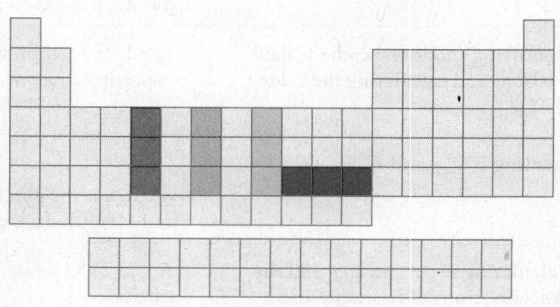

FIGURE P19.4

19.5. Which of the four curves in Figure P19.5 best represents the dependence of the potential of a lead–acid battery on the concentration of sulfuric acid? Note that the scale of the *x*-axis is logarithmic.

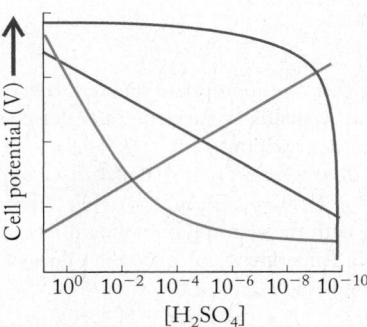

FIGURE P19.5

19.6. Consider the four types of batteries in Figure P19.6. From top to bottom the sizes are AAA, AA, C, and D. The performance of batteries like these is often expressed in units such as (a) volts, (b) watt-hours, or (c) milliampere-hours. Which of the values differ significantly between the four batteries?

FIGURE P19.6

19.7. The apparatus in Figure P19.7 is used for the electrolysis of water. Hydrogen and oxygen gas are collected in the two inverted burettes. An inert electrode at the bottom of the left burette is connected to the negative terminal of a 6-volt battery; the electrode in the burette on the right is connected to the positive terminal. A small quantity of sulfuric acid is added to speed up the electrolytic reaction.
  a. What are the half-reactions at the left and right electrodes and their standard potentials?
  b. Why does sulfuric acid make the electrolysis reaction go more rapidly?

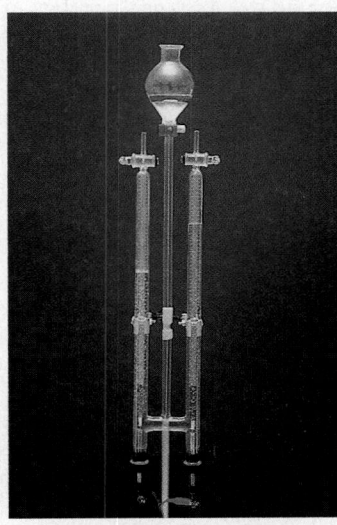

Overall cell reaction
$H_2O(\ell) \rightarrow H_2(g) + \frac{1}{2}O_2(g)$

FIGURE P19.7

19.8. An electrolytic apparatus identical to the one shown in Problem 19.7 is used to electrolyze water, but the reaction is speeded up by the addition of sodium carbonate instead of sulfuric acid.
  a. What are the half-reactions and the standard potentials for the electrodes on the left and right?
  b. Why does sodium carbonate make the electrolysis reaction go more rapidly?

QUESTIONS AND PROBLEMS • ∎

Redox Chemistry and Electrochemical Cells

CONCEPT REVIEW

19.9. What is the role of the porous separator in an electrochemical cell?

19.10. The Zn/Cu^{2+} reactions in Figures 19.1 and 19.2 are the same. However, the reaction in the cell in Figure 19.2 generates electricity; the reaction in the beaker in Figure 19.1 does not. Why?

19.11. Why can't a wire perform the same function as a porous separator in an electrochemical cell?

19.12. In a voltaic cell, why is the cathode labeled the positive terminal and the anode the negative terminal?

PROBLEMS

19.13. A voltaic cell with an aqueous electrolyte is based on the reaction between $Pb^{2+}(aq)$ and $Zn(s)$, producing $Pb(s)$ and $Zn^{2+}(aq)$.
  a. Write half-reactions for the anode and cathode.
  b. Write a balanced cell reaction.
  c. Diagram the cell.

19.14. A voltaic cell is based on the reaction between $Ag^+(aq)$ and $Ni(s)$, producing $Ag(s)$ and $Ni^{2+}(aq)$.
  a. Write the anode and cathode half-reactions.
  b. Write a balanced cell reaction.
  c. Diagram the cell.

19.15. A voltaic cell with a basic aqueous background electrolyte is based on the oxidation of $Cd(s)$ to $Cd(OH)_2(s)$ and the reduction of $MnO_4^-(aq)$ to $MnO_2(s)$.
   a. Write half-reactions for the cell's anode and cathode.
   b. Write a balanced cell reaction.
   c. Diagram the cell.

19.16. A voltaic cell is based on the reduction of $Ag^+(aq)$ to $Ag(s)$ and the oxidation of $Sn(s)$ to $Sn^{2+}(aq)$.
   a. Write half-reactions for the cell's anode and cathode.
   b. Write a balanced cell reaction.
   c. Diagram the cell.

19.17. Super Iron Batteries In 1999, scientists in Israel developed a battery based on the following cell reaction with iron(VI), nicknamed "super iron":

$$2\,K_2FeO_4(aq) + 3\,Zn(s) \rightarrow$$
$$Fe_2O_3(s) + ZnO(s) + 2\,K_2ZnO_2(aq)$$

   a. Determine the number of electrons transferred in the cell reaction.
   b. What are the oxidation states of the transition metals in the reaction?
   c. Diagram the cell.

19.18. Aluminum–Air Batteries In recent years engineers have been working on an aluminum–air battery as an alternative energy source for electric vehicles. The battery consists of an aluminum anode, which is oxidized to solid aluminum hydroxide, immersed in an electrolyte of aqueous KOH. At the cathode oxygen from the air is reduced to hydroxide ions on an inert metal surface. Write the two half-reactions for the battery and diagram the cell. Use the generic $M(s)$ symbol for the metallic cathode material.

Standard Potentials

CONCEPT REVIEW

19.19. What is the function of platinum in the standard hydrogen electrode?

19.20. Is it possible to build a battery in which the anode chemistry is based on a half-reaction in which none of the species is a solid conductor, for example:

$$Fe^{2+}(aq) \rightarrow Fe^{3+}(aq) + e^-$$

19.21. In some textbooks the formula used to calculate standard cell is written this way:

$$E°_{cell} = E°_{red}(\text{cathode}) + E°_{ox}(\text{anode})$$

Show how this equation is equivalent to Equation 19.2.

19.22. Suppose there were a scale for expressing electrode potentials in which the standard potential for the reduction of water in base

$$2\,H_2O(\ell) + 2\,e^- \rightarrow H_2(g) + 2\,OH^-(aq)$$

is assigned an $E°$ value of 0.000 V. How would the standard potential values on this new scale differ from those in Appendix 6?

19.23. Is O_2 a stronger oxidizing agent in acid or in base? Use standard reduction potentials in Appendix 6 to support your answer.

*19.24. To inhibit corrosion of steel structures in contact with seawater, pieces of other metals (often zinc) are attached to the structures to serve as "sacrificial anodes." Explain how these attached pieces of metal might protect the structures and describe which properties of zinc make it a good selection.

PROBLEMS

19.25. Starting with the appropriate standard free energies of formation in Appendix 4, calculate the value of $\Delta G°$ and $E°_{cell}$ of the following reactions:
   a. $2\,Cu^+(aq) \rightarrow Cu^{2+}(aq) + Cu(s)$
   b. $Ag(s) + Fe^{3+}(aq) \rightarrow Ag^+(aq) + Fe^{2+}(aq)$

19.26. Starting with the appropriate standard free energies of formation in Appendix 4, calculate the values of $\Delta G°$ and $E°_{cell}$ of the following reactions:
   a. $FeO(s) + H_2(g) \rightarrow Fe(s) + H_2O(\ell)$
   b. $2\,Pb(s) + O_2(g) + 2\,H_2SO_4(aq) \rightarrow$
$$2\,PbSO_4(s) + 2\,H_2O(\ell)$$

19.27. If a piece of silver is placed in a solution in which $[Ag^+] = [Cu^{2+}] = 1.00\ M$, will the following reaction proceed spontaneously?

$$2\,Ag(s) + Cu^{2+}(aq) \rightarrow 2\,Ag^+(aq) + Cu(s)$$

19.28. A piece of cadmium is placed in a solution in which $[Cd^{2+}] = [Sn^{2+}] = 1.00\ M$. Will the following reaction proceed spontaneously?

$$Cd(s) + Sn^{2+}(aq) \rightarrow Cd^{2+}(aq) + Sn(s)$$

19.29. Sometimes the anode half-reaction in the zinc–air battery (Figure 19.16) is written with the zincate ion, $Zn(OH)_4^{2-}$, as the product. Write a balanced equation for the cell reaction based on this product.

*19.30. Sometimes the cell reaction of nickel–cadmium batteries is written with Cd metal as the anode and solid NiO_2 as the cathode. Assuming that the products of the reactions are a solid hydroxide of Cd(II) at the anode and a solid hydroxide of Ni(II) at the cathode, write balanced equations for the cathode and anode half-reactions and the overall cell reaction.

19.31. In a voltaic cell similar to the Cu–Zn cell in Figure 19.2, the Cu electrode is replaced with one made of Ni immersed in a solution of $NiSO_4$. Will the standard potential of this Ni–Zn cell be greater than, the same as, or less than 1.10 V?

19.32. Suppose the copper half of the Cu–Zn cell in Figure 19.2 were replaced with a silver wire in contact with $1\ M\ Ag^+(aq)$.
   a. What would be the value of $E°_{cell}$?
   b. Which electrode would be the anode?

19.33. Starting with standard potentials listed in Appendix 6, calculate the values of $E°_{cell}$ and $\Delta G°$ of the following reactions.
   a. $Cu(s) + Sn^{2+}(aq) \rightarrow Cu^{2+}(aq) + Sn(s)$
   b. $Zn(s) + Ni^{2+}(aq) \rightarrow Zn^{2+}(aq) + Ni(s)$

19.34. Starting with the standard potentials listed in Appendix 6, calculate the values of $E°_{cell}$ and $\Delta G°$ of the following reactions.
   a. $Fe(s) + Cu^{2+}(aq) \rightarrow Fe^{2+}(aq) + Cu(s)$
   b. $Ag(s) + Fe^{3+}(aq) \rightarrow Ag^+(aq) + Fe^{2+}(aq)$

19.35. Voltaic cells based on the following pairs of half-reactions are prepared so that all reactants and products are in their standard states. For each pair, write a balanced equation for the cell reaction, and identify which half-reaction takes place at the anode and which at the cathode.

a. $Hg^{2+}(aq) + 2e^- \rightarrow Hg(\ell)$
$Zn^{2+}(aq) + 2e^- \rightarrow Zn(s)$

b. $ZnO(s) + H_2O(\ell) + 2e^- \rightarrow Zn(s) + 2OH^-(aq)$
$Ag_2O(s) + H_2O(\ell) + 2e^- \rightarrow 2Ag(s) + 2OH^-(aq)$

c. $Ni(OH)_2(s) + 2e^- \rightarrow Ni(s) + 2OH^-(aq)$
$O_2(g) + 2H_2O(\ell) + 4e^- \rightarrow 4OH^-(aq)$

19.36. Voltaic cells based on the following pairs of half-reactions are constructed. For each pair, write a balanced equation for the cell reaction, and identify which half-reaction takes place at each anode and cathode.

a. $Cd^{2+}(aq) + 2e^- \rightarrow Cd(s)$
$Ag^+(aq) + e^- \rightarrow Ag(s)$

b. $AgBr(s) + e^- \rightarrow Ag(s) + Br^-(aq)$
$MnO_2(s) + 4H^+(aq) + 2e^- \rightarrow Mn^{2+}(aq) + 2H_2O(\ell)$

c. $PtCl_4^{2-}(aq) + 2e^- \rightarrow Pt(s) + 4Cl^-(aq)$
$AgCl(s) + e^- \rightarrow Ag(s) + Cl^-(aq)$

19.37. The half-reactions and standard potentials for a nickel–metal hydride battery with a titanium–zirconium anode are as follows:

Cathode: $NiO(OH)(s) + H_2O(\ell) + e^- \rightarrow$
$Ni(OH)_2(s) + OH^-(aq)$ $E° = 1.32$ V

Anode: $TiZr_2H(s) + OH^-(aq) \rightarrow$
$TiZr_2(s) + H_2O(\ell) + e^-$ $E° = 0.00$ V

a. Write the overall cell reaction for this battery.
b. Calculate the standard cell potential.

*__19.38.__ **Lithium-Ion Batteries** Scientists at the University of Texas, Austin, and at MIT developed a cathode material for lithium–ion batteries based on $LiFePO_4$, which is the composition of the cathode when the battery is fully discharged. Batteries with this cathode are more powerful than those of the same mass with $LiCoO_2$ cathodes. They are also more stable at high temperatures.

a. What is the formula of the $LiFePO_4$ cathode when the battery is fully charged?
b. Is Fe oxidized or reduced as the battery discharges?
c. Is the cell potential of a lithium–ion battery with an iron phosphate cathode likely to differ from one with a cobalt oxide cathode? Explain your answer.

Chemical Energy and Electrical Work

CONCEPT REVIEW

19.39. How is it that a voltaic cell with a *positive* cell potential does *negative* work?

*__19.40.__ Mechanical work (w) is done by exerting a force (F) to move an object through a distance (d) according to the equation $w = F \times d$. Explain how this definition of work relates to electrical work ($w_{elec} = C \times E$).

PROBLEMS

19.41. For many years the 1.50 V batteries used to power flashlights were based on the following cell reaction:

$$Zn(s) + 2NH_4Cl(s) + 2MnO_2(s) \rightarrow$$
$$Zn(NH_3)_2Cl_2(s) + Mn_2O_3(s) + H_2O(\ell)$$

What is the value of ΔG_{cell}?

19.42. **Laptop Battery** The first generation of laptop computers was powered by nickel–cadmium (nicad) batteries, which generated 1.20 V based on the following cell reaction:

$$Cd(s) + 2NiO(OH)(s) + 2H_2O(\ell) \rightarrow$$
$$Cd(OH)_2(s) + 2Ni(OH)_2(s)$$

What is the value of ΔG_{cell}?

19.43. The cells in the nickel–metal hydride battery packs used in many hybrid vehicles produce 1.20 V based on the following cell reaction:

$$MH(s) + NiO(OH)(s) \rightarrow M(s) + Ni(OH)_2(s)$$

What is the value of ΔG_{cell}?

19.44. A cell in a lead–acid battery delivers exactly 2.00 V of cell potential based on the following cell reaction:

$$Pb(s) + PbO_2(s) + 2H_2SO_4(aq) \rightarrow 2PbSO_4(s) + 2H_2O(\ell)$$

What is the value of ΔG_{cell}?

A Reference Point: The Standard Hydrogen Electrode; The Effect of Concentration on E_{cell}

CONCEPT REVIEW

19.45. Why does the operating cell potential of most batteries change little until the battery is nearly discharged?

19.46. The standard potential of the Cu–Zn cell reaction

$$Zn(s) + Cu^{2+}(aq) \rightarrow Zn^{2+}(aq) + Cu(s)$$

is 1.10 V. Would the potential of the Cu–Zn cell differ from 1.10 V if the concentrations of both Cu^{2+} and Zn^{2+} were 0.25 M?

PROBLEMS

19.47. Calculate the E_{cell} value at 298 K for the cell based on the reaction

$$Fe^{3+}(aq) + Cr^{2+}(aq) \rightarrow Fe^{2+}(aq) + Cr^{3+}(aq)$$

when $[Fe^{3+}] = [Cr^{2+}] = 1.50 \times 10^{-3} M$ and $[Fe^{2+}] = [Cr^{3+}] = 2.5 \times 10^{-4} M$.

19.48. Calculate the E_{cell} value at 298 K for the cell based on the reaction

$$Cu(s) + 2Ag^+(aq) \rightarrow Cu^{2+}(aq) + 2Ag(s)$$

when $[Ag^+] = 2.56 \times 10^{-3} M$ and $[Cu^{2+}] = 8.25 \times 10^{-4} M$.

19.49. Using the appropriate standard potentials in Appendix 6, determine the equilibrium constant for the following reaction at 298 K:

$$Fe^{3+}(aq) + Cr^{2+}(aq) \rightarrow Fe^{2+}(aq) + Cr^{3+}(aq)$$

19.50. Using the appropriate standard potentials in Appendix 6, determine the equilibrium constant at 298 K for the following reaction between MnO_2 and Fe^{2+} in acid solution:

$$4H^+(aq) + MnO_2(s) + 2Fe^{2+}(aq) \rightarrow$$
$$Mn^{2+}(aq) + 2Fe^{3+}(aq) + 2H_2O(\ell)$$

19.51. If the potential of a hydrogen electrode based on the half-reaction

$$2H^+(aq) + 2e^- \rightarrow H_2(g)$$

is 0.000 V at pH = 0.00, what is the potential of the same electrode at pH = 7.00?

19.52. Glucose Metabolism The standard potentials for the reduction of nicotinamide adenine dinucleotide (NAD^+) and oxaloacetate (reactants in the multistep metabolism of glucose) are as follows:

$$NAD^+(aq) + H^+(aq) + 2e^- \rightarrow NADH(aq)$$
$$E° = -0.320 \text{ V}$$

$$Oxaloacetate^-(aq) + 2H^+(aq) + 2e^- \rightarrow malate^-(aq)$$
$$E° = -0.166 \text{ V}$$

a. Calculate the standard potential for the following reaction:

$$Oxaloacetate^-(aq) + NADH(aq) + H^+(aq) \rightarrow$$
$$malate^-(aq) + NAD^+(aq)$$

b. Calculate the equilibrium constant for the reaction at 298 K.

19.53. Permanganate ion can oxidize sulfite to sulfate in basic solution as follows:

$$2MnO_4^-(aq) + 3SO_3^{2-}(aq) + H_2O(\ell) \rightarrow$$
$$2MnO_2(s) + 3SO_4^{2-}(aq) + 2OH^-(aq)$$

Determine the potential for the reaction (E_{rxn}) at 298 K when the concentrations of the reactants and products are as follows: $[MnO_4^-] = 0.150 \, M$, $[SO_3^{2-}] = 0.256 \, M$, $[SO_4^{2-}] = 0.178 \, M$, and $[OH^-] = 0.0100 \, M$. Will the value of E_{rxn} increase or decrease as the reaction proceeds?

*19.54. Manganese dioxide is reduced by iodide ion in acid solution as follows:

$$MnO_2(s) + 2I^-(aq) + 4H^+(aq) \rightarrow$$
$$Mn^{2+}(aq) + I_2(aq) + 2H_2O(\ell)$$

Determine the electrical potential of the reaction at 298 K when the initial concentrations of the components are as follows: $[I^-] = 0.225 \, M$, $[H^+] = 0.900 \, M$, $[Mn^{2+}] = 0.100 \, M$, and $[I_2] = 0.00114 \, M$. If the solubility of iodine in water is approximately 0.114 M, will the value of E_{rxn} increase or decrease as the reaction proceeds?

19.55. A copper penny dropped into a solution of nitric acid produces a mixture of nitrogen oxides. The following reaction describes the formation of NO, one of the products:

$$3Cu(s) + 8H^+(aq) + 2NO_3^-(aq) \rightarrow$$
$$2NO(g) + 3Cu^{2+}(aq) + 4H_2O(\ell)$$

a. Starting with the appropriate standard potentials in Appendix 6, calculate $E_{rxn}°$ for this reaction.
b. Calculate $E_{rxn}°$ at 298 K when $[H^+] = 0.100 \, M$, $[NO_3^-] = 0.0250 \, M$, $[Cu^{2+}] = 0.0375 \, M$, and the partial pressure of NO = 0.00150 atm.

19.56. Chlorine dioxide (ClO_2) is produced by the following reaction of chlorate (ClO_3^-) with Cl^- in acid solution:

$$2ClO_3^-(aq) + 2Cl^-(aq) + 4H^+(aq) \rightarrow$$
$$2ClO_2(g) + Cl_2(g) + 2H_2O(\ell)$$

a. Determine $E°$ for the reaction.
b. The reaction produces an atmosphere in the reaction vessel in which $P_{ClO_2} = 2.0$ atm and $P_{Cl_2} = 1.00$ atm. Calculate $[ClO_3^-]$ if, at equilibrium ($T = 298$ K), $[H^+] = [Cl^-] = 10.0 \, M$.

*19.57. The oxidation of NH_4^+ to NO_3^- in acid solution is described by the following equation:

$$NH_4^+(aq) + 2O_2(g) \rightarrow NO_3^-(aq) + 2H^+(aq) + H_2O(\ell)$$

a. Calculate $E°$ for the overall reaction.
b. If the reaction is in equilibrium with air ($P_{O_2} = 0.21$ atm) at pH 5.60, what is the ratio of $[NO_3^-]$ to $[NH_4^+]$ at 298 K?

19.58. What is the value of $E°$ for the following reaction?

$$2AgCl(s) + H_2(g) \rightarrow 2Ag(s) + 2HCl(aq)$$

Relating Battery Capacity to Quantities of Reactants

CONCEPT REVIEW

19.59. One 12-volt lead–acid battery has a higher ampere-hour rating than another. Which of the following parameters are likely to be different for the two batteries?
a. individual cell potentials
b. anode half-reactions
c. total masses of electrode materials
d. number of cells
e. electrolyte composition
f. combined surface areas of their electrodes

19.60. In a voltaic cell based on the Cu–Zn cell reaction

$$Zn(s) + Cu^{2+}(aq) \rightarrow Cu(s) + Zn^{2+}(aq)$$

there is exactly 1 mole of each reactant and product. A second cell based on the Cd–Cu cell reaction

$$Cd(s) + Cu^{2+}(aq) \rightarrow Cu(s) + Cd^{2+}(aq)$$

also has exactly 1 mole of each reactant and product. Which of the following statements about these two cells is true?
a. Their cell potentials are the same.
b. The masses of their electrodes are the same.
c. The quantities of electrical charge that they can produce are the same.
d. The quantities of electrical energy that they can produce are the same.

PROBLEMS

19.61. Which of the following voltaic cells will produce the greater quantity of electrical charge per gram of anode material?

$$Cd(s) + 2NiO(OH)(s) + 2H_2O(\ell) \rightarrow$$
$$2Ni(OH)_2(s) + Cd(OH)_2(s)$$

or

$$4Al(s) + 3O_2(g) + 6H_2O(\ell) + 4OH^-(aq) \rightarrow$$
$$4Al(OH)_4^-(aq)$$

19.62. Which of the following voltaic cells will produce the greater quantity of electrical charge per gram of anode material?

$$Zn(s) + MnO_2(s) + H_2O(\ell) \rightarrow ZnO(s) + Mn(OH)_2(s)$$

or

$$Li(s) + MnO_2(s) \rightarrow LiMnO_2(s)$$

---

*19.63. Which of the following voltaic cell reactions delivers more electrical energy per gram of anode material at 298 K?

$$Zn(s) + 2NiO(OH)(s) + 2H_2O(\ell) \rightarrow$$
$$2Ni(OH)_2(s) + Zn(OH)_2(s) \quad E°_{cell} = 1.20 \text{ V}$$

or

$$Li(s) + MnO_2(s) \rightarrow LiMnO_2(s) \quad E°_{cell} = 3.15 \text{ V}$$

*19.64. Which of the following voltaic cell reactions delivers more electrical energy per gram of anode material at 298 K?

$$Zn(s) + Ni(OH)_2(s) \rightarrow Zn(OH)_2(s) + Ni(s) \quad E°_{cell} = 1.50 \text{ V}$$

or

$$2Zn(s) + O_2(g) \rightarrow 2ZnO(s) \quad E°_{cell} = 2.08 \text{ V}$$

Electrolytic Cells and Rechargeable Batteries

CONCEPT REVIEW

19.65. The positive terminal of a voltaic cell is the cathode. However, the cathode of an electrolytic cell is connected to the negative terminal of a power supply. Explain this difference in polarity.

19.66. The anode in an electrochemical cell is defined as the electrode where oxidation takes place. Why is the anode in an electrolytic cell connected to the positive (+) terminal of an external supply, whereas the anode in a voltaic cell battery is connected to the negative (−) terminal?

19.67. The salts obtained from the evaporation of seawater can act as a source of halogens, principally Cl_2 and Br_2, through the electrolysis of the molten alkali metal halides. As the potential of the anode in an electrolytic cell is increased, which of these two halogens forms first?

19.68. In the electrolysis described in Problem 19.67, why is it necessary to use molten salts rather than seawater itself?

19.69. Quantitative Analysis Electrolysis can be used to determine the concentration of Cu^{2+} in a given volume of solution by electrolyzing the solution in a cell equipped with a platinum cathode. If all of the Cu^{2+} is reduced to Cu metal at the cathode, the increase in mass of the electrode provides a measure of the concentration of Cu^{2+} in the original solution. To ensure the complete (99.99%) removal of the Cu^{2+} from a solution in which $[Cu^{2+}]$ is initially about 1.0 M, will the potential of the cathode (versus SHE) have to be more negative or less negative than 0.34 V (the standard potential for $Cu^{2+} + 2e^- \rightarrow Cu$)?

19.70. A high school chemistry student wishes to demonstrate how water can be separated into hydrogen and oxygen by electrolysis. She knows that the reaction will proceed more rapidly if an electrolyte is added to the water. She has access to 2.00 M solutions of these compounds: H_2SO_4, HBr, NaI, Na_2SO_4, and Na_2CO_3. Which one(s) should she use? Explain your selection(s).

PROBLEMS

19.71. Suppose the current from a battery is used to electroplate an object with silver. Calculate the mass of silver that would be deposited by a battery that delivers 1.7 A · hr of charge.

19.72. A battery charger used to recharge the NiMH batteries used in a digital camera can deliver as much as 0.50 A of current to each battery. If it takes 100 min to recharge one battery, how much $Ni(OH)_2$ (in grams) is oxidized to NiO(OH)?

19.73. A NiMH battery containing 4.10 g of NiO(OH) was 50% discharged when it was connected to a charger with an output of 2.00 A at 1.3 V. How long does it take to recharge the battery?

*19.74. How long does it take to deposit a coating of gold 1.00 μm thick on a disk-shaped medallion 4.0 cm in diameter and 2.0 mm thick at a constant current of 85 A? The density of gold is 19.3 g/cm^3. The gold solution contains Au(III).

---

*19.75. Oxygen Supply in Submarines Nuclear submarines can stay under water nearly indefinitely because they can produce their own oxygen by the electrolysis of water.
   a. How many liters of O_2 at 298 K and 1.00 bar are produced in 1 hr in an electrolytic cell operating at a current of 0.025 A?
   b. Could seawater be used as the source of oxygen in this electrolysis? Explain why or why not.

19.76. In the electrolysis of water, how long will it take to produce 1.00×10^2 L of H_2 at STP (273 K and 1.00 atm) using an electrolytic cell through which the current is 52 mA?

19.77. Calculate the minimum (least negative) cathode potential (versus SHE) needed to begin electroplating nickel from 0.35 M Ni^{2+} onto a piece of iron.

*19.78. What is the minimum (least negative) cathode potential (versus SHE) needed to electroplate silver onto cutlery in a solution of Ag^+ and NH_3 in which most of the silver ions are present as the complex, $Ag(NH_3)_2^+$, and the concentration of $Ag^+(aq)$ is only 3.50×10^{-5} M?

Fuel Cells

CONCEPT REVIEW

19.79. Describe two advantages of hybrid (gasoline engine–electric motor) power systems over all-electric systems based on fuel cells. Describe two disadvantages.

19.80. Describe three factors limiting widespread use of cars powered by fuel cells.

19.81. Methane can serve as the fuel for electric cars powered by fuel cells. Carbon dioxide is a product of the fuel cell reaction. All cars powered by internal combustion engines burning natural gas (mostly methane) produce CO_2. Why are electric vehicles powered by fuel cells likely to produce less CO_2 per mile?

19.82. To make the refueling of fuel cells easier, several manufacturers offer converters that turn readily available fuels—such as natural gas, propane, and methanol—into H_2 for the fuel cells and CO_2. Although vehicles with such power systems are not truly "zero emission," they still offer significant environmental benefits over vehicles powered by internal combustion engines. Describe a few of them.

PROBLEMS

19.83. Fuel cells with molten alkali metal carbonates as electrolytes can use methane as a fuel. The methane is first converted into hydrogen in a two-step process:

$$CH_4(g) + H_2O(g) \rightarrow CO(g) + 3H_2(g)$$
$$CO(g) + H_2O(g) \rightarrow H_2(g) + CO_2(g)$$

a. Assign oxidation numbers to carbon and hydrogen in the reactants and products.
b. Using the standard free energy of formation values in Appendix 4, calculate the standard free-energy changes in the two reactions and the overall $\Delta G°$ for the formation of $H_2 + CO_2$ from methane and steam.

*19.84. Molten carbonate fuel cells fueled with H_2 convert as much as 60% of the free energy released by the formation of water from H_2 and O_2 into electrical energy. Determine the quantity of electrical energy obtained from converting 1 mole of H_2 into $H_2O(\ell)$ in such a fuel cell.

Additional Problems

*19.85. **Electrolysis of Seawater** Magnesium metal is obtained by the electrolysis of molten Mg^{2+} salts from evaporated seawater.
a. Would elemental Mg form at the cathode or anode?
b. Do you think the principal ingredient in sea salt (NaCl) would need to be separated from the Mg^{2+} salts before electrolysis? Explain your answer.
c. Would electrolysis of an aqueous solution of $MgCl_2$ also produce elemental Mg?
d. If your answer to part c was no, what would be products of electrolysis?

*19.86. **Silverware Tarnish** Low concentrations of hydrogen sulfide in air react with silver to form Ag_2S, more familiar to us as tarnish. Silver polish contains aluminum metal powder in a basic suspension.

a. Write a balanced net ionic equation for the redox reaction between Ag_2S and Al metal that produces Ag metal and $Al(OH)_3$.
b. Calculate $E°$ for the reaction.

19.87. A magnesium battery can be constructed from an anode of magnesium metal and a cathode of molybdenum sulfide, Mo_3S_4. The half-reactions are

Anode: $Mg(s) \rightarrow Mg^{2+}(aq) + 2e^- \qquad E° = 2.37\ V$

Cathode: $Mg^{2+}(aq) + Mo_3S_4(s) + 2e^- \rightarrow MgMo_3S_4(s)$
$$E°_{red} = ?$$

a. If the standard cell potential for the battery is 1.50 V, what is the value of $E°_{red}$ for the reduction of Mo_3S_4?
b. What are the apparent oxidation states of Mo in Mo_3S_4 and in $MgMo_3S_4$?
*c. The electrolyte in the battery contains a complex magnesium salt, $Mg(AlCl_3CH_3)_2$. Why is it necessary to include Mg^{2+} ions in the electrolyte?

19.88. **Clinical Chemistry** The concentration of Na^+ ions in red blood cells (11 mM) and in the surrounding plasma (140 mM) are quite different. Calculate the electrochemical potential (emf) across the cell membrane as a result of this concentration gradient.

*19.89. The element fluorine, F_2, was first produced in 1886 by electrolysis of HF. Chemical syntheses of F_2 did not happen until 1986 when Karl O. Christe successfully prepared F_2 by the reaction

$$K_2MnF_6(s) + 2SbF_5(\ell) \rightarrow 2KSbF_6(s) + MnF_3(s) + \tfrac{1}{2}F_2(g)$$

a. Assign oxidation numbers to each compound and determine the number of electrons involved in the process.
b. Using the following $\Delta H_f°$ values calculate $\Delta H°$ for the reaction.

$\Delta H°_{f,SbF_5(\ell)} = -1324\ kJ/mol \qquad \Delta H°_{f,K_2MnF_6(s)} = -2435\ kJ/mol$

$\Delta H°_{f,MnF_3(s)} = -1579\ kJ/mol \qquad \Delta H°_{f,KSbF_6(s)} = -2080\ kJ/mol$

c. If we assume that ΔS is relatively small, such that $\Delta G \approx \Delta H$, estimate $E°$ for this reaction.
d. If ΔS for the reaction is greater than zero, is our value for $E°$ in part c too high or too low?
e. The electrochemical synthesis of F_2 is described by the electrolytic cell reaction

$$2KHF_2(\ell) \rightarrow 2KF(\ell) + H_2(g) + F_2(g)$$

Assign oxidation numbers and determine the number of electrons involved in this process.

19.90. **Corrosion of Copper Pipes** The copper pipes frequently used in household plumbing may corrode and eventually leak. The corrosion reaction is believed to involve the formation of copper(I) chloride:

$$2Cu(s) + Cl_2(aq) \rightarrow 2CuCl(s)$$

a. Write balanced equations for the half-reactions in this redox reaction.
b. Calculate $E°_{rxn}$ and $\Delta G°_{rxn}$ for the reaction.

Power, Pigments, and Phosphors: The Group 12 Elements

19.91. Write balanced chemical equations for the following processes.
  a. Zinc(II) sulfide reacts with oxygen.
  b. Zinc(II) oxide reacts with carbon.
  c. Mercury(II) sulfide reacts with oxygen.

19.92. Write balanced chemical equations for the following processes:
  a. Zinc reacts with oxygen.
  b. Cadmium reacts with sulfur.
  c. Zinc metal reacts with aqueous copper(II) nitrate.

19.93. Zinc(II) sulfide and cadmium(II) sulfide crystallize in two different structures, an hcp (hexagonal closest-packed) and a ccp (cubic closest-packed) arrangement of sulfide ions with the cations in tetrahedral holes.
  a. Do these structures have the same packing efficiency?
  b. Are the tetrahedral holes the same size in both structures? Explain your answer.

19.94. Zinc(II) oxide and zinc(II) sulfide (wurtzite) have the same structure, an fcc (face-centered cubic) arrangement of anions with zinc cations in tetrahedral holes.
  a. Are the tetrahedral holes in ZnO the same size as in ZnS?
  b. How many tetrahedral holes are in the unit cell?

19.95. Calculate the $E°_{cell}$ value of the following reactions.
  a. $Zn(s) + Cd^{2+}(aq) \rightarrow Cd(s) + Zn^{2+}(aq)$
  b. $Hg(\ell) + Zn^{2+}(aq) \rightarrow Zn(s) + Hg^{2+}(aq)$
  c. $Cd(s) + Hg_2^{2+}(aq) \rightarrow Cd^{2+}(aq) + 2\,Hg(\ell)$

19.96. Which of the reactions in Problem 19.97 are spontaneous?

19.97. Which of these elements is the best reducing agent: Zn, Cd, or Hg?

19.98. Which cation, Hg^{2+} or Hg_2^{2+}, is the better oxidizing agent?

19.99. Why is the chemistry of zinc similar to the chemistry of magnesium?

19.100. Predict the products of the following reactions:
  a. $Zn(s) + Cl_2(g) \rightarrow$
  b. $ZnCO_3(s) + heat \rightarrow$
  c. $Zn(OH)_2(s) + heat \rightarrow$

19.101. The aqueous solutions of many transition metal compounds are colored, but solutions of zinc compounds are colorless. Why are solutions of $Zn^{2+}(aq)$ colorless?

Additional study materials including ChemTours and Diagnostic Quizzes are available at StudySpace at www.wwnorton.com/studyspace.

Biochemistry: The Compounds of Life

Learning Outcomes

- Describe the molecular structures and chemical properties of amino acids and the nature of the bonds that link them together in peptides and proteins

- Name and draw the structures of small peptides

- Describe the four levels of protein structure and how intermolecular forces and covalent bonds stabilize these structures

- Describe the molecular structures of simple sugars and polysaccharides and how these compounds are used as energy sources and for energy storage

- Describe the types of bonds that link saccharides

- Describe the molecular structure and the physical and chemical properties of saturated and unsaturated glycerides

- Describe the structures of DNA and RNA and how they function together to translate genetic information

A LOOK AHEAD

Function Follows Form

For all the stunning diversity of the biosphere, from single-cell organisms to elephants, whales, and giant redwoods, all life-forms consist of substances made from only about 40 or 50 different small molecules. Huge variations among life-forms are possible when unique sequences of these few starting materials link together to form larger molecules and biopolymers. Unfortunately, subtle defects in these sequences can have devastating effects on health and even survival.

It is estimated, for example, that 70% of inherited diseases in humans are caused by the absence of certain proteins or the production of proteins that have the wrong composition, which leads to the wrong molecular shape. For example, the malformation or the total absence of the protein dystrophin causes a number of debilitating diseases referred to as muscular dystrophy (MD). In mild forms of MD, misshapen molecules of dystrophin are unable to build muscle fibers of sufficient strength to function normally. As a result, muscle tissue wastes away and the patient becomes disabled. In a severe form of the disease, Duchenne MD (DMD), functional dystrophin is absent, and the disability often results in early death. Knowledge of the mechanisms that produce dystrophin has led to promising therapies, some involving grafting of healthy muscle cells that can synthesize normal dystrophin into the tissues of MD patients. Some treatments involve injection of stem cells that fuse with and genetically complement dystrophic muscle in DMD patients. Similar approaches have resulted in the development of new drugs and cell therapies for other inherited and contagious diseases.

In this chapter we explore the composition, structure, and function of proteins and three other major classes of biomolecular compounds: carbohydrates, lipids, and nucleic acids. Each class performs similar

The Origins of Life Research into the origins of life on Earth has recently ▶ focused on the chemical and thermal energy that is available at geothermal vents: energy that could have been used to synthesize amino acids and RNA, and support primordial life.

protein biological polymer made of amino acids.

biomolecule an organic molecule present naturally in a living system.

amino acid molecule that contains at least one amine group and one carboxylic acid group; in an *α-amino acid*, the two groups are attached to the same (*α*) carbon atom.

functions in all the life-forms in which they occur. Many of these molecules are large and complex, but our knowledge about the chemical behavior of small molecules will serve as a useful framework on which to build an understanding of the behavior of the larger assemblies of molecules that form cells. As we explore the relationships between the structures and functions of these classes of compounds, we can begin to address such questions as these: How do nucleic acids convey genetic information from one generation to the next? How do some proteins provide structure to living organisms while others mediate biochemical processes? How do carbohydrates and lipids provide the energy we need and allow us to store the energy we don't need? ■

20.1 The Composition of Proteins

We begin this chapter with a discussion of **proteins**, the most abundant class of **biomolecules** in all animals, including us. Proteins account for about half of the mass of the human body that is not water. They are the major component in skin, muscles, cartilage, hair, and nails. Most of the enzymes that catalyze biochemical reactions are proteins, as are the molecules that transport oxygen to our cells and many of the hormones that regulate cell function and growth. Most proteins are large, with molar masses of 10^5 g/mol or more. Thus, they are examples of *biopolymers*.

Amino Acids

The molecular structure and biological function of big biomolecules depend on the identities of their small-molecule building blocks and the sequence in which those small molecules occur. The molecular building blocks of proteins are **amino acids**, so named because each of them contains at least one amine (–NH₂) group and at least one carboxylic acid (–COOH) group. The amino acids in proteins are *α*-amino acids because in their structures one carbon atom, called the *α*-carbon, is bonded to both an –NH₂ and a –COOH group (Figure 20.1).

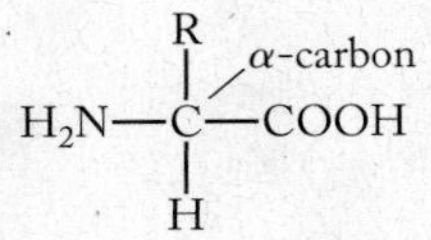

FIGURE 20.1 General structure of an *α*-amino acid.

CONCEPT TEST

According to valence bond theory, what is the hybridization of the *α*-carbon atom in the amino acid in Figure 20.1, and what are the approximate angles between the four bonds surrounding it?

(Answers to Concept Tests are in the back of the book.)

👓 **CONNECTION** As in Chapter 13, we use the generic symbol R to represent any group of covalently bonded atoms that are connected to the main structure of a molecule via a C—C bond.

👓 **CONNECTION** The weakly basic properties of amines were described in Section 17.3.

In addition to its single bonds to the –NH₂ and –COOH groups, the *α*-carbon atom in each of the amino acids that make up human proteins is bonded to a hydrogen atom and to one of 20 different *R groups*. The structures of these R groups, often called *side-chain groups*, are highlighted in red in the amino acid structures in Table 20.1. The amino acids are arranged in four categories based on their R groups. In the first category, containing 9 amino acids, the R groups contain mostly carbon and hydrogen atoms, and are nonpolar. The R groups of the remaining 11 amino acids contain at least one *heteroatom* (O, N, or S) bonded to a H atom and are polar. Of these amino acids, 2 (aspartic acid and glutamic acid) have R groups that contain carboxylic acid functional groups, and 3 (histidine, lysine, and arginine) contain amine groups, which are weakly basic.

TABLE 20.1	**Structures and Abbreviations of the 20 Common Amino Acids**

Nonpolar R Groups

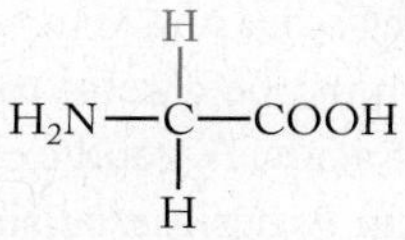

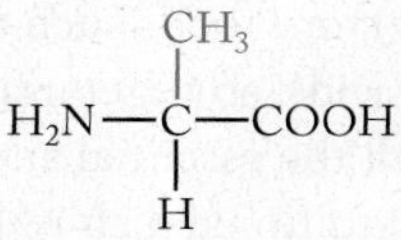

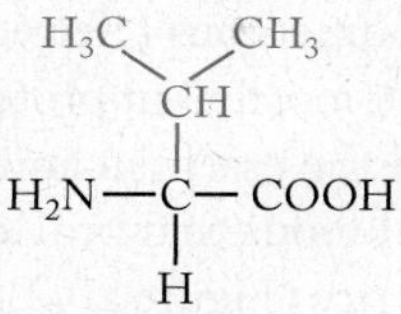

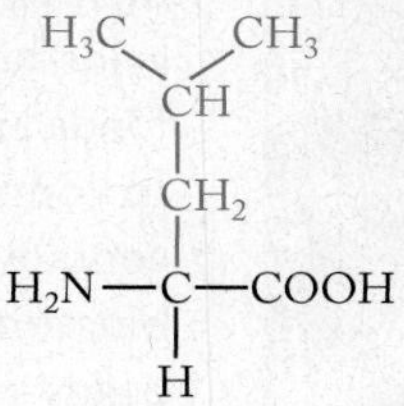

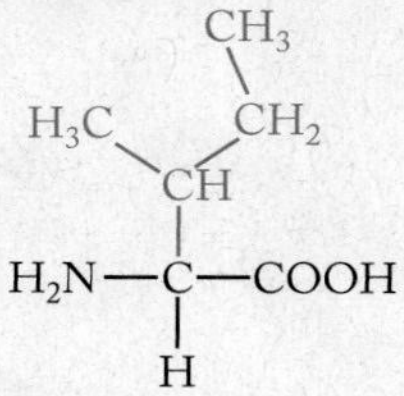

Glycine
(Gly)

Alanine
(Ala)

Valine[a]
(Val)

Leucine[a]
(Leu)

Isoleucine[a]
(Ile)

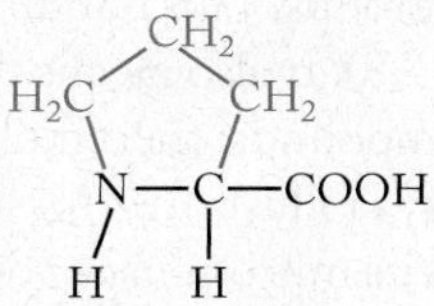

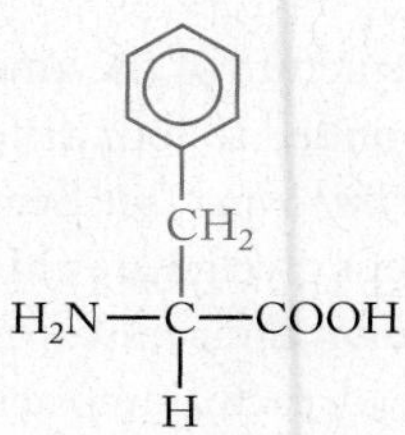

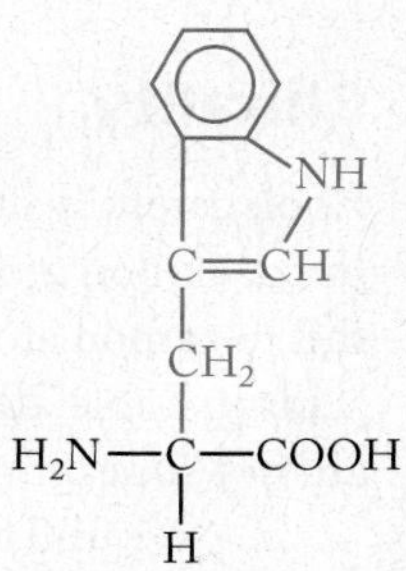

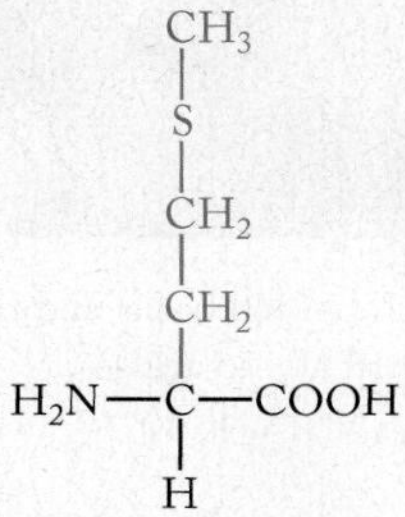

Proline
(Pro)

Phenylalanine[a]
(Phe)

Tryptophan
(Trp)

Methionine[a]
(Met)

Polar R Groups

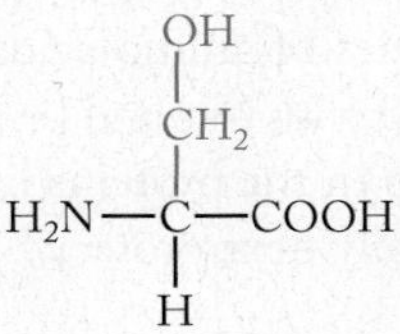

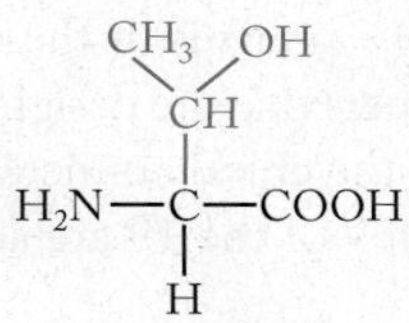

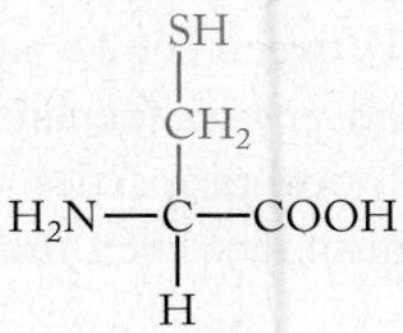

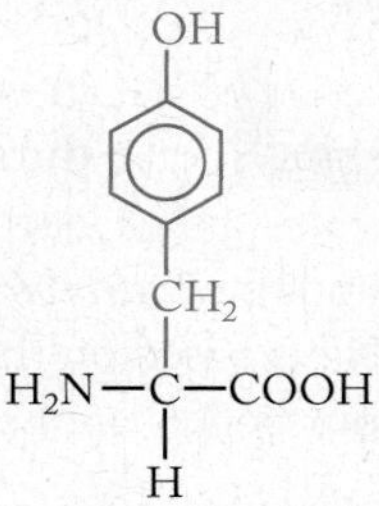

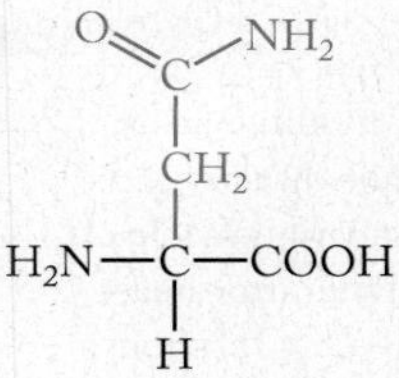

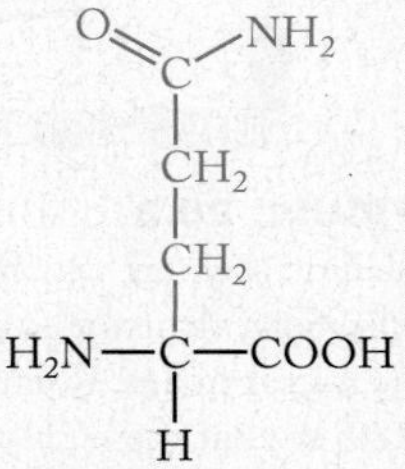

Serine
(Ser)

Threonine[a]
(Thr)

Cysteine
(Cys)

Tyrosine[a]
(Tyr)

Asparagine
(Asn)

Glutamine
(Gln)

Acid R Groups

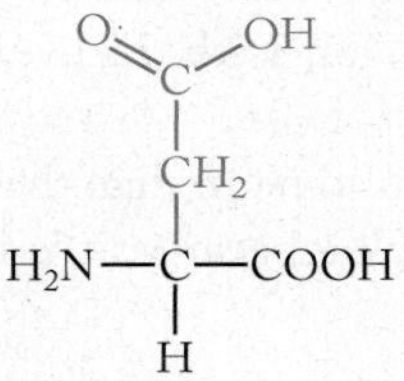

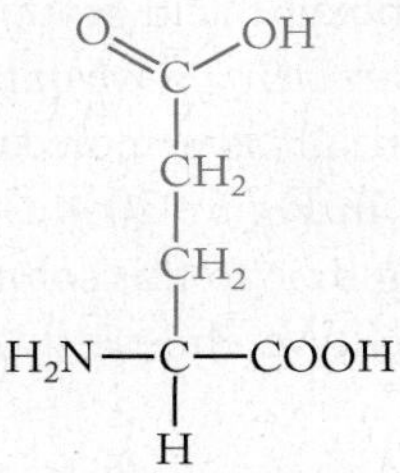

Aspartic acid
(Asp)

Glutamic acid
(Glu)

Basic R Groups

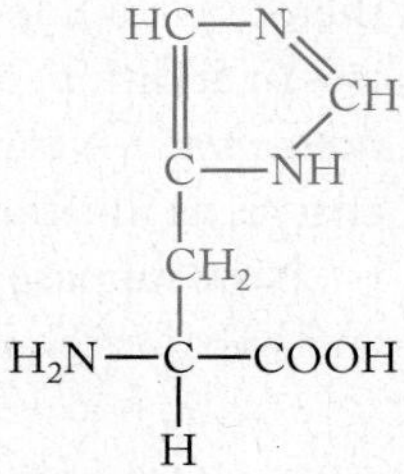

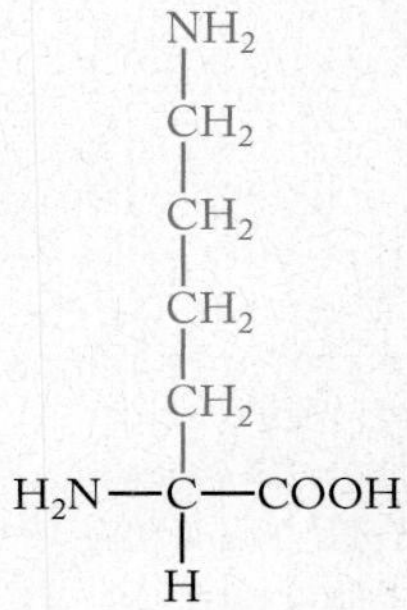

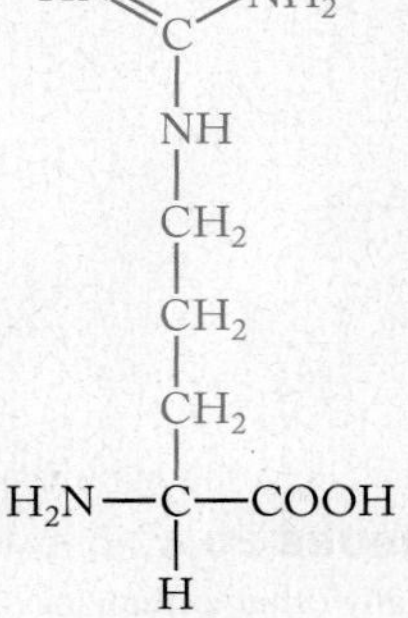

Histidine
(His)

Lysine[a]
(Lys)

Arginine
(Arg)

[a] The eight essential amino acids for adults (histidine is essential for children).

FIGURE 20.2 A meal of red beans and rice provides all the essential amino acids.

Our bodies can synthesize 12 of the 20 amino acids in Table 20.1, but the other 8 must be present in the food we eat. These 8 are marked with a superscript *a* and are referred to as the **essential amino acids**. Most proteins from animal sources, including those in meats, eggs, and dairy products, contain all the essential amino acids needed by the human body, and in close to the correct proportions. These foods are sometimes referred to as *perfect foods* or, more precisely, *complete proteins*. In contrast, most plant proteins from foods such as legumes and vegetables do not contain all the essential amino acids, so vegetarians must be careful to eat a combination of foods that provide all the essential amino acids. A good example is red beans and rice (Figure 20.2), a traditional dish in Latin American cuisine that provides a balance of essential amino acids: rice has all of them but lysine, and beans lack only methionine.

Chirality

Look carefully at the structure of the amino acid alanine in Figure 20.3. Note how the α-carbon atom is bonded to four different groups. This bonding pattern makes this α-carbon atom a *chiral center* (see Section 13.8). It also means that all the amino acids in Table 20.1 except glycine are chiral compounds: their molecular structures are not superimposable on their mirror images, as illustrated for alanine in Figure 20.3. Instead, each amino acid and its mirror image constitute an enantiomeric pair.

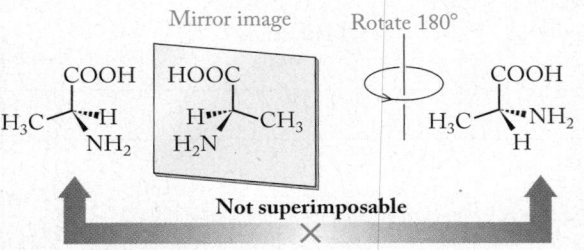

FIGURE 20.3 Like most α-amino acids, alanine is chiral, which means that its molecular structure is not superimposable on its mirror image. Rotating the mirror image 180° so that the –CH$_3$ and –COOH groups overlap does not produce a structure in which –NH$_2$ and –H overlap.

CONNECTION Optical isomerism and the concept of chirality were described in Section 13.8.

For historical reasons, amino acid enantiomers are designated by the prefixes D- (for *dextro-*, right) and L- (*levo-*, left). These labels refer to how the four groups bonded to each chiral carbon atom are oriented in three-dimensional space. They *do not* refer to the direction in which plane-polarized light rotates (see Figure 13.52) as it passes through a solution of the amino acid. In other words, there is no connection between the D- and L- prefixes in the names of amino acids and the *dextrorotary* and *levorotary* enantiomers that are designated with (+) and (−) signs based on their optical properties. All the chiral amino acids in the proteins in our bodies are L-enantiomers, even though 9 of the 19 are actually dextrorotary.

CONCEPT TEST

One of the 20 amino acids in Table 20.1, glycine, is not chiral. Why?

Zwitterions

If we dissolve an amino acid in a solution that already contains a strong acid (and therefore has a low pH), the carboxylic acid group on each molecule does not ionize. In addition, each amine group, being a weak base, accepts a H$^+$ ion forming a *protonated* –NH$_3^+$ group. The result is a molecular ion with an overall positive charge, as illustrated for alanine in Figure 20.4(a).

Now suppose we add a strong base to this solution to neutralize the strong acid and to raise the pH to about 7.4. We choose this value because it is close to the

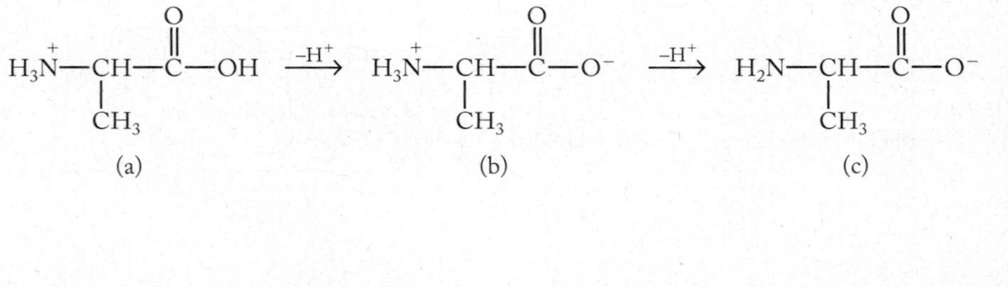

FIGURE 20.4 (a) At low pH, alanine (like many other amino acids) exists as a 1+ ion. (b) At physiological pH (7.4), the –COOH group is ionized and the molecule becomes a zwitterion with an overall charge of zero. (c) In basic solutions the charge decreases to 1− as the –NH$_3^+$ group deprotonates.

pH of human blood and of the fluids in most of our tissues. At this *physiological pH* the –COOH groups in amino acids are mostly ionized, but nearly all the amine groups are still in the protonated $-NH_3^+$ form because $-NH_3^+$ is the conjugate acid of a weak base and a *very* weak acid. As a result, most of the protonated amine groups still have positive charges and most of the ionized carboxylic acid groups have negative charges. A molecule with this distribution of charges is called a **zwitterion** (literally, a *hybrid ion*) because it has both a positive and a negative functional group (Figure 20.4b). Its net charge is zero.

If we add more base, alanine loses a H^+ from its NH_3^+ group (we say that it *deprotonates*), and we have a molecular ion with an overall charge of $1-$ (Figure 20.4c). The titration curve for alanine in Figure 20.5 shows this two-step neutralization process: the carboxylic acid is neutralized first, the protonated amine deprotonates second.

CONCEPT TEST

Each amino acid has a characteristic *p*I *value* representing the pH at which molecules of the amino acid have, on average, zero charge. Rank the following amino acids in order of increasing *p*I value: (a) aspartic acid; (b) lysine; (c) glycine.

SAMPLE EXERCISE 20.1   **Interpreting Acid–Base Titration Curves of Amino Acids**

The $-NH_2$ groups in α-amino acids are weak bases ($pK_b \approx 4-5$), so their conjugate acids ($-NH_3^+$) are even weaker acids ($pK_a \approx 9-10$). Figure 20.6 shows that the titration curve for aspartic acid has three equivalence points. Draw the molecular structures of the principal form of aspartic acid that is in solution at the start of the titration and at each equivalence point.

Collect and Organize We are to draw the molecular structures of aspartic acid at three different characteristic pH values. The molecular structure of the un-ionized form of aspartic acid is in Table 20.1.

Analyze There are two carboxylic acid groups and one amine group in aspartic acid. At low pH the carboxylic acid groups are not ionized and the amine group is protonated. As pH is raised the –COOH groups ionize and then, under basic conditions, the protonated amine group releases its proton. The stronger of the two acid groups will ionize first. The –COOH group bonded to the α-carbon atom is only one carbon atom away from an electropositive NH_3^+ group. The side-chain –COOH group is two carbon atoms away, and so the carboxylate ion that it forms when it ionizes will be less stabilized by delocalization of its negative charge toward the cationic group. Therefore, the –COOH group bonded to the α-carbon atom should be the stronger acid and ionize first.

Solve At the beginning of the titration, the fully protonated $1+$ ion is present:

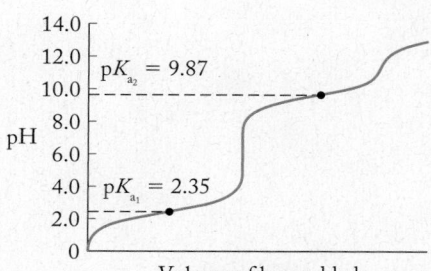

FIGURE 20.5 The titration curve of alanine resembles that of a weak diprotic acid. The carboxylic acid group ($pK_a = 2.35$) is neutralized first. The protonated amine group ($pK_a = 9.87$) is neutralized in the second step of the titration.

CONNECTION The inverse relationship between the strengths of acid–base conjugate pairs was discussed in Chapter 17 and illustrated in Figure 17.4.

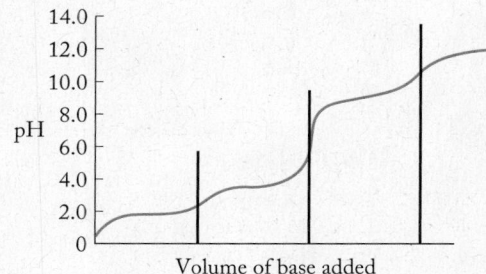

FIGURE 20.6 Titration curve of aspartic acid.

essential amino acid any of the 8 amino acids that make up peptides and proteins but are not synthesized in the human body and must be obtained through the food we eat.

zwitterion a molecule that has both positively and negatively charged groups in its structure.

At the first equivalence point, the –COOH group bonded to the α-carbon atom is ionized:

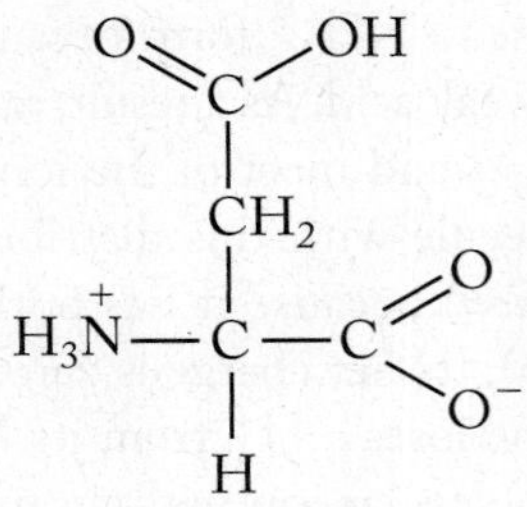

At the second equivalence point, the side-chain –COOH group is ionized:

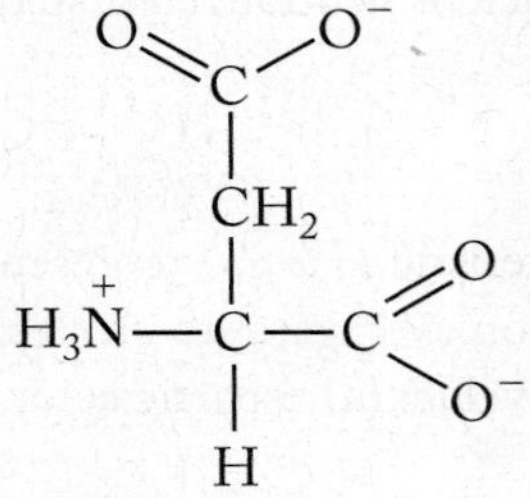

And at the third equivalence point in the titration, the amine group is deprotonated:

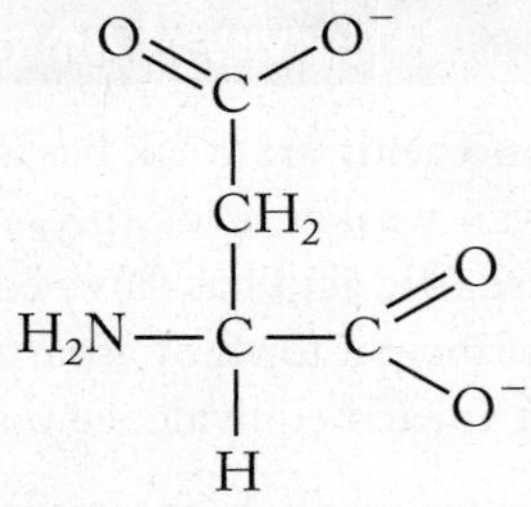

Think about It We can estimate the pK_a values of the two carboxylic acid groups from the midpoint pH values in the first two stages of the titration: about 1.9 and 3.7. They differ by nearly 2 pH units, which means that the α-COOH group is almost 10^2 times stronger an acid than the side-chain –COOH group, due to the presence of the α-NH$_3{}^+$ group.

Practice Exercise Sketch the titration curve for lysine starting at pH = 1.0, and draw the molecular structures of the principal form of lysine that is in solution at each equivalence point.

(Answers to Practice Exercises are in the back of the book.)

Peptides

Amino acids bond together to form chainlike molecules with a wide range of sizes. Amino acid *residues* (the name we give to amino acids that are part of a larger molecule) make up each link in the chain. The longest chains, some with molar masses as high as 3 million g/mol, are called proteins, but the shortest chains, called **peptides**, are only a few links long. The smallest peptides contain only two or three amino acid residues. These molecules are called *dipeptides* and *tripeptides*. Peptides up to 20 residues long are called *oligopeptides*. Those made of more than 20 are called *polypeptides*. The size at which a polypeptide becomes a protein is arbitrary but is typically set around 50–75 amino acid residues.

peptide a compound of two or more amino acids joined by peptide bonds. Small peptides containing up to 20 amino acids are *oligopeptides*; and the term *polypeptide* is used for chains longer than 20 amino acids but shorter than proteins.

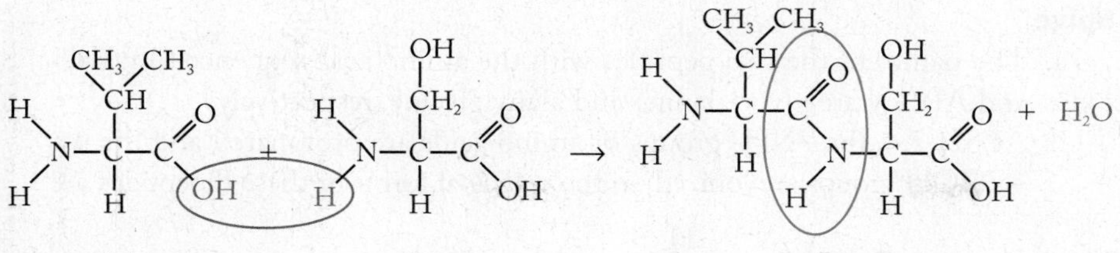

FIGURE 20.7 When the carboxylic acid group of valine reacts to form a peptide bond (red oval) with the amine group of serine, the products are the dipeptide valylserine and water.

The type of bond linking the amino acids in peptides and proteins is called a **peptide bond** (in the red oval in Figure 20.7) or *peptide linkage*. They form when the α-carboxylic acid group of one amino acid condenses with the α-amine group of another, with the loss of a molecule of water. Note that peptide bonds have the same structure as the amide bonds that hold together the monomeric units of synthetic polyamides such as nylon.

The convention for drawing the structures of peptides begins by placing the amino acid that has a free α-amine group at the left end of the peptide chain and the amino acid with a free α-carboxylic acid at the right end. The left end is called the *amine* (or *N-*) *terminus* of the peptide and the right end is called the *carboxylic acid* (or *C-*) *terminus*. The name of a peptide is based on the names of its amino acids, starting with the one at the N-terminus. The names of all the amino acids except the one at the C-terminus are changed to end in *-yl*. For example, if a peptide bond forms between the α-COOH group of valine (Val) and the α-NH$_2$ group of serine (Ser) as in Figure 20.7, the dipeptide that is produced is called valylserine (ValSer).

The artificial sweetener aspartame is the methyl ester of the dipeptide aspartylphenylalanine (Figure 20.8). At pH 7.4, aspartame exists as a zwitterion because the aspartic acid amine group is protonated and the carboxylic acid in its R group is ionized. In contrast, its parent dipeptide has a net charge of $1-$ because both –COOH groups are ionized at that pH. In an amino acid that has an amine group in its R group, such as lysine or arginine, that amine group is probably protonated at physiological pH, giving the amino acid a net charge of $1+$. To sum up, the overall charge on a peptide at physiological pH is the sum of the positive charges on protonated amine groups and the negative charges of ionized carboxylic acid groups.

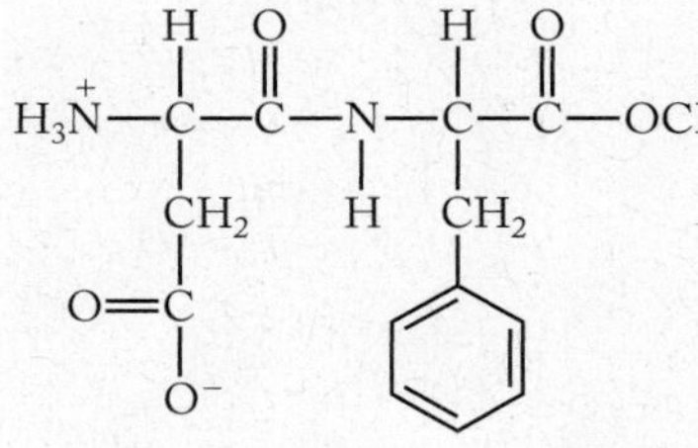

▶‖ **CHEMTOUR** Condensation of Biological Polymers

CONNECTION The formation of amide bonds in reactions between carboxylic acids and amines was described in Chapter 13 and illustrated in Figure 13.46.

Aspartame

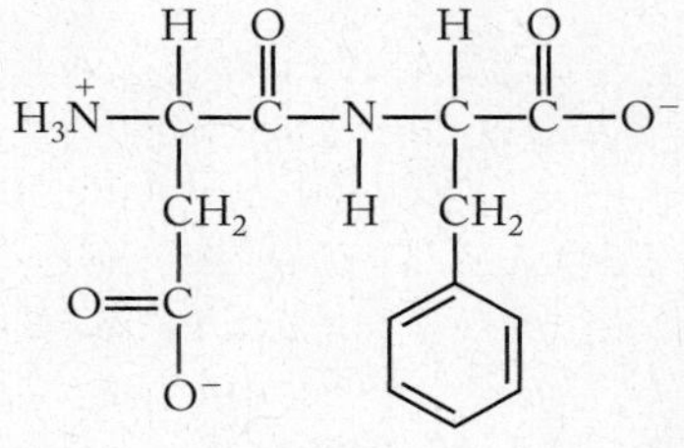

Aspartylphenylalanine

FIGURE 20.8 Aspartame is the methyl ester of the dipeptide aspartylphenylalanine and so has one fewer –COOH group than its parent dipeptide. Therefore, the overall charge of a molecule of aspartame at pH 7.4 is zero, whereas the charge on aspartylphenylalanine is 1–.

SAMPLE EXERCISE 20.2 **Drawing and Naming Peptides**

(a) Name all the dipeptides that can be made by reacting alanine with glycine, and (b) draw their molecular structures in solution at pH 7.4.

Collect and Organize We are to draw and name the dipeptides that are produced when alanine and glycine are linked by a peptide bond. The structures of these amino acids are in Table 20.1.

Analyze Two different peptides can be made from two amino acids by changing their sequence from the N-terminus to the C-terminus. For the two dipeptides, the sequences are AlaGly and GlyAla. The R groups in alanine (–CH$_3$) and glycine (–H) have no acidic or basic properties. At pH 7.4, the carboxylic acid terminus –COOH groups should be ionized and the amine terminus –NH$_2$ groups should be protonated.

peptide bond the result of a condensation reaction between the carboxylic acid group of one amino acid and the amine group of another.

Solve

a. The names of the two peptides with the amino acid sequence GlyAla and AlaGly are glycylalanine and alanylglycine, respectively.

b. At pH 7.4, the –NH$_2$ groups of amino acids are protonated and the –COOH groups are ionized, so the principal forms of these dipeptides are

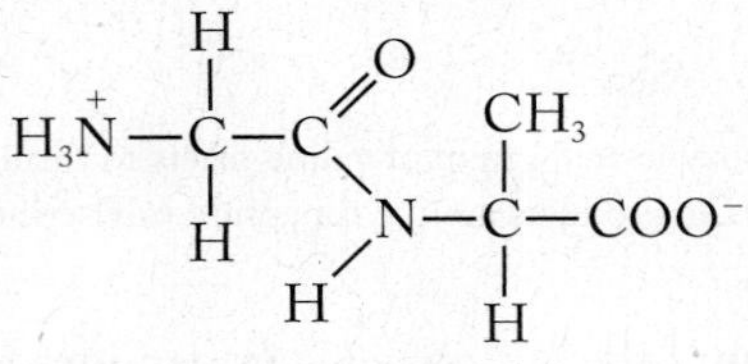

Glycylalanine
(GlyAla)

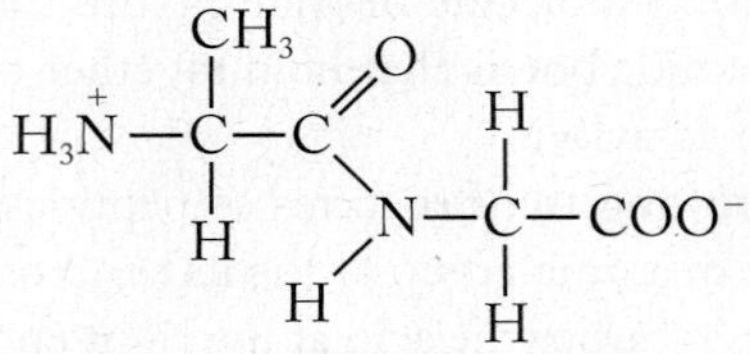

Alanylglycine
(AlaGly)

Think about It Only the N-terminal –NH$_2$ and C-terminal –COOH groups can be protonated or ionized in these dipeptides. Thus, AlaGly and GlyAla are both zwitterionic with net charges of $(1+) + (1-) = 0$ at pH 7.4.

Practice Exercise How many different tripeptides can be synthesized from one molecule of each of three different amino acids A, B, and C?

⸻⸻⸻⸻⸻⸻⸻

CONCEPT TEST ⋯⋯⋯⋯⋯⋯⋯⋯⋯⋯⋯⋯⋯⋯⋯⋯⋯⋯⋯⋯⋯⋯⋯⋯⋯⋯⋯

Rank the following amino acids in order of their net charges at pH 7.4, starting with the most positive net charge: (a) alanine; (b) arginine; (c) aspartic acid.

20.2 Protein Structure and Function

The structure of a protein is crucial to its function. In A Look Ahead at the beginning of this chapter, we mentioned the catastrophic impact of a malformed muscle protein. The functions of nonstructural proteins, such as the ability of enzymes to catalyze biochemical reactions, are also closely linked to their structures. These large biomolecules must assume particular three-dimensional conformations to interact with other molecules and to function properly.

Primary Structure

The **primary (1°) structure** of a protein is the sequence of the amino acids in it, starting with the N-terminus (Figure 20.9a). If two proteins are made up of the same number and type of amino acids but have different amino acid sequences, they are different proteins.

Changing only one amino acid can dramatically alter a protein's function. For example, in hemoglobin the sixth amino acid from the N-terminus of a protein strand that is 146 amino acids long is normally glutamic acid. In some people, valine substitutes

primary (1°) structure the sequence in which amino acid monomers occur in a protein chain.

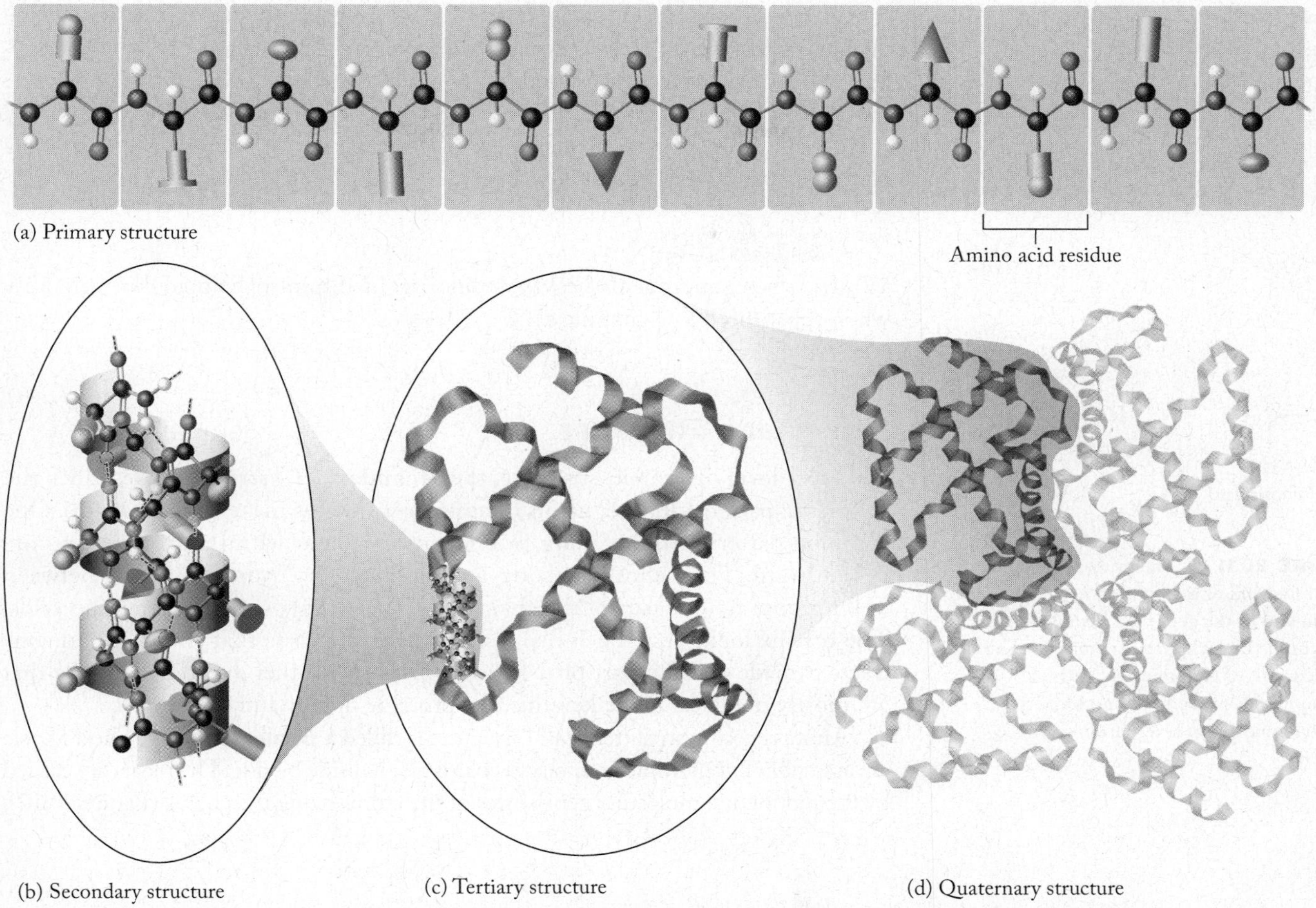

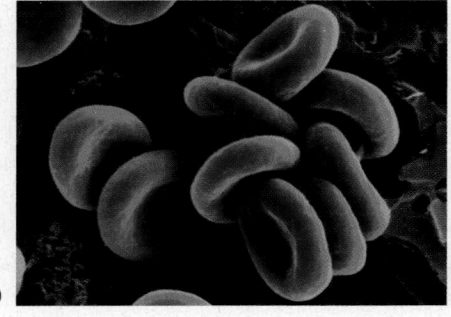

(a) Primary structure

Amino acid residue

(b) Secondary structure

(c) Tertiary structure

(d) Quaternary structure

for glutamic acid at this position. This one substitution alters the solubility of the protein and causes the red blood cell to take on a sickle shape instead of the normal, plump disc shape (Figure 20.10). These sickled cells do not pass through capillaries easily and may impede blood circulation. They also break readily and so do not last as long as normal blood cells. These factors lead to a diminished capacity of the blood to carry oxygen, which is one of the characteristics of the disease called sickle-cell anemia.

Why does switching valine for glutamic acid affect the solubility of hemoglobin? The answer lies in the R groups of these two amino acids (Figure 20.11). The side-chain –COOH group of glutamic acid is ionized at physiological pH, and strong ion–dipole interactions with water molecules enhance the protein's solubility. However, if valine with its nonpolar isopropyl R group replaces glutamic acid, that strong intermolecular interaction with water is lost. The valine creates a hydrophobic patch on the surface of the protein that results in the polymer assuming a different shape.

FIGURE 20.9 The four levels of protein structure. (a) A protein's primary structure is its amino acid sequence. The green shapes represent different R groups. (b) Secondary structure (here, an α helix) describes the three-dimensional pattern adopted by segments of the protein strand. (c) Tertiary structure is the overall shape of the molecule as segments of it bend and fold. (d) Quaternary structure refers to the overall shape adopted by multiple protein strands that assemble into a single unit.

∞ CONNECTION We discussed in Chapter 10 how ion-dipole and dipole-dipole interactions are the key to the solubility of solutes in water.

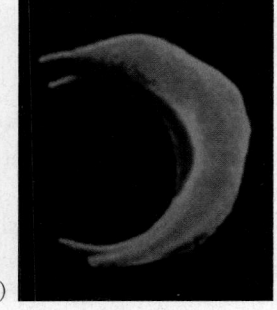

(a) (b)

FIGURE 20.10 (a) Normal red blood cells are plump discs, whereas (b) those in patients suffering from sickle-cell anemia are distorted and incomplete.

Primary structure

Normal protein:

Val - His - Leu - Thr - Pro - [Glu] - Lys - . . .

Abnormal protein:

Val - His - Leu - Thr - Pro - [Val] - Lys - . . .

(a)

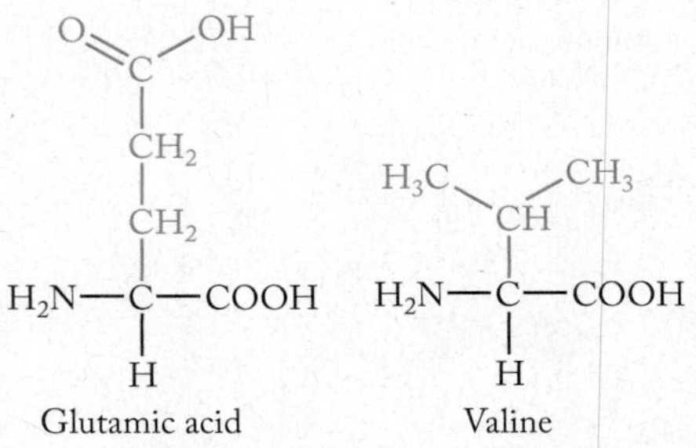

Glutamic acid   Valine

(b)

FIGURE 20.11 (a) The primary structure of the amine end of a protein in normal hemoglobin and in the abnormal hemoglobin responsible for sickle-cell anemia. (b) The replacement of glutamic acid with its hydrophilic R group by valine with its hydrophobic group is responsible for the disease.

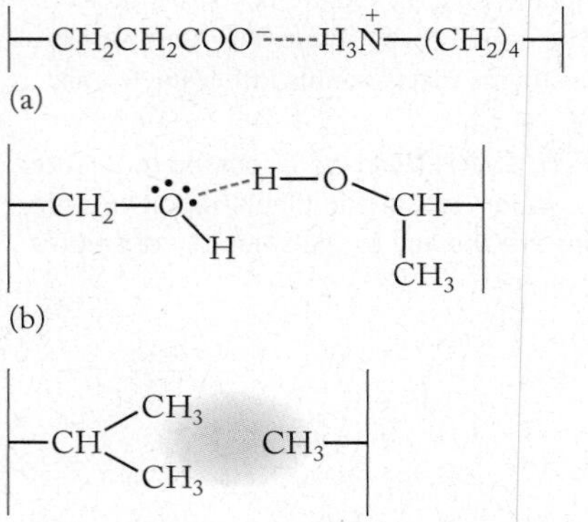

$$\left[-CH_2CH_2COO^- \cdots \cdot H_3\overset{+}{N}-(CH_2)_4- \right]$$

(a)

$$\left[-CH_2-\overset{\cdots}{\underset{H}{O}}\cdots\cdot H-O-CH- \atop \qquad\qquad\qquad CH_3 \right]$$

(b)

$$\left[-CH\overset{CH_3}{\underset{CH_3}{\big\langle}} \quad CH_3- \right]$$

(c)

FIGURE 20.13 Intramolecular interactions that influence the secondary and tertiary structures of proteins include (a) ion–ion interactions between acidic and basic R groups, (b) hydrogen bonding, and (c) van der Waals interactions between nonpolar side chains.

Sickle-cell anemia is a debilitating disease, but it provides a survival advantage in regions where malaria is endemic. Having sickle-cell anemia does not protect people from contracting malaria or make them invulnerable to the parasite that causes it. However, children infected with malaria are more likely to survive the illness if they have sickle-cell anemia. Exactly why sickle-cell anemia has this effect is not understood.

CONCEPT TEST •

Which other amino acids besides valine might diminish hemoglobin solubility when substituted for glutamic acid?

• •

Secondary Structure

The next level of protein structure, the **secondary (2°) structure**, describes the geometric patterns that segments of amino acid chains make (Figure 20.9b). One common pattern is the **α helix**, a coiled arrangement with the R groups pointing outward. The helical structure is maintained by hydrogen bonds between –NH groups on one part of the chain and C=O groups on nearby amino acids. The α helix looks very much like a spring. Muscle tissue that stretches and contracts is made in large part of α-helical proteins. Another group of proteins that are mostly α-helical is the keratins, the proteins in hair and fingernails.

Another common pattern of 2° structure is called a **β-pleated sheet**. These sheets are assemblies of multiple amino acid chains aligned side by side. The pleats are caused by the tetrahedral molecular geometries of the atoms along the chains (Figure 20.12).

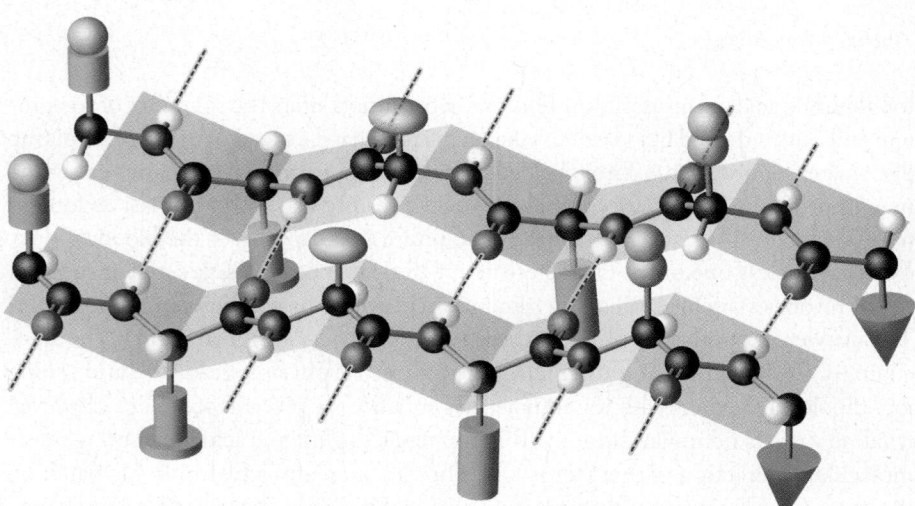

FIGURE 20.12 Each amino acid chain in a β-pleated sheet is folded in a zigzag pattern. Adjacent chains in a sheet are held together by hydrogen bonds (blue dashed lines). The R groups (green structures) extend above and below the sheet, linking it to adjacent sheets via noncovalent intermolecular interactions.

Adjacent chains are linked together by hydrogen bonds, and the collection of side-by-side zigzag chains forms a continuous β-pleated sheet. R groups extend above and below the pleats. Sheets may stack on top of one another like two pieces of corrugated roofing. Stacked sheets are held together by the same interactions that hold all proteins together, including ion–ion, hydrogen-bonding, and van der Waals interactions, depending on the pairs of R groups involved (Figure 20.13).

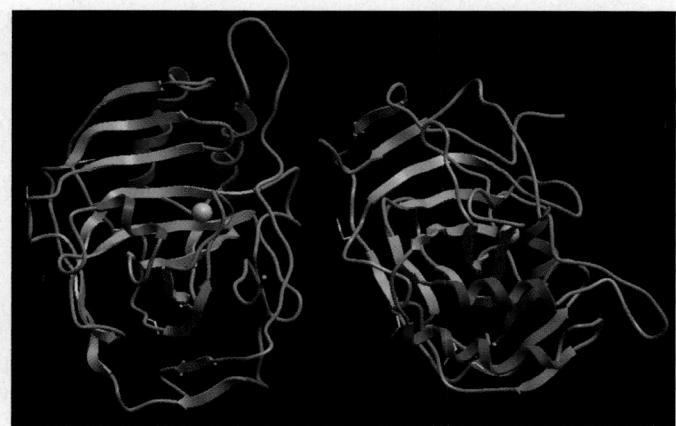

FIGURE 20.14 The structure of the protein carbonic anhydrase has α-helical regions (red), β-pleated sheet regions (green), and random coil (blue). The light blue sphere in the center is a zinc ion.

▶II **CHEMTOUR** Fiber Strength and Elasticity

The proteins that make up strands of silk form thin, planar crystals of β-pleated sheets that are only a few nanometers on a side. Enormous numbers of these crystals form long arrays of sheets stacked together like nanoscale pancakes. Hydrogen bonds hold the stacks together and reinforce adjacent sheets. As a result of these interactions strands of silk are stronger than strands of steel with the same mass; indeed, silk is one of the toughest known materials—natural or synthetic.

Some single-stranded proteins exist as α helices on their own but form β-pleated sheets when they clump together in multistrand aggregates. One consequence of this clumping is the formation of insoluble protein deposits called plaque. Abnormal accumulation of plaque can be a serious health risk: plaque formed by a protein called amyloid β has been linked to the onset of Alzheimer's disease.

If part of a protein is characterized by an irregular or rapidly changing structure, it is said to have a **random coil** 2° structure. The amino acid chain may fold back on itself and around itself, but it has no regular features the way an α helix or a β-pleated sheet does. When proteins *denature*, losing their secondary structure because of heat or change in pH, they may become random coils.

Large protein molecules may contain all three types of 2° structure. In describing a protein, scientists may indicate the percentage of amino acids involved in each type—for example, 50% α-helical, 30% β-pleated sheet, and 20% random coil. Figure 20.14 shows a model of a protein called carbonic anhydrase, which illustrates all three types.

CONCEPT TEST

The aqueous solution of proteins in egg whites turns into a solid mass when eggs are cooked or are dropped into an organic solvent such as acetone.

a. What is the likely secondary structure of the proteins in cooked eggs?
b. Do you think the solidification process occurs as a result of a change in the primary structure of the proteins?

Tertiary and Quaternary Structure

Large proteins have structure beyond the 1° and 2° levels. Larger molecules can fold back on themselves as a result of interactions between R groups on amino acids that are considerable distances apart along the protein chain. These interactions may be ion–ion, ion–dipole, or van der Waals forces. They may even involve the formation of covalent bonds. For example, the –SH groups on two cysteine residues may combine to form a disulfide linkage that holds two parts of the protein strand together via an intrastrand –S—S– covalent bond (Figure 20.15).

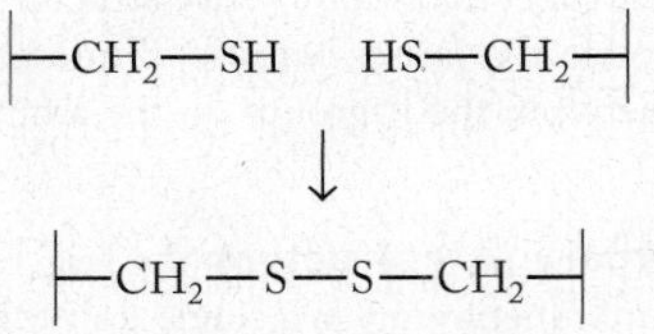

—CH₂—SH   HS—CH₂—
↓
—CH₂—S—S—CH₂—

FIGURE 20.15 The tertiary structure of some proteins is stabilized by intrastrand –S—S– bonds that form between the –SH groups on the side chains of cysteine residues.

CONNECTION All the types of intermolecular forces described in Chapter 10 are involved in the *intra*molecular interactions that give proteins their unique 2° and 3° structures.

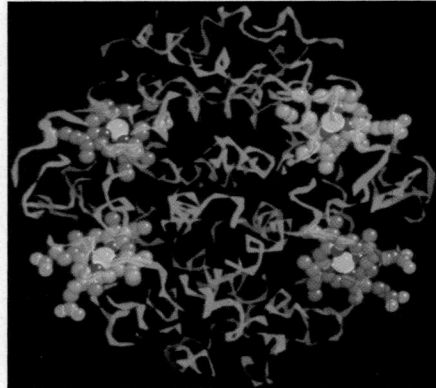

(a) Hemoglobin

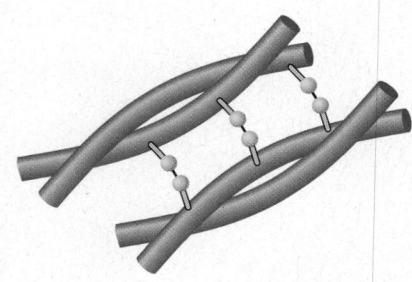

(b) Keratin

FIGURE 20.16 Quaternary structure of proteins. (a) Four protein chains form a single unit in the quaternary structure of hemoglobin. The iron-containing porphyrins are bright green. (b) Pairs of α-helical chains (blue) wound together and linked by interstrand –S—S– bonds (yellow) stabilize the quaternary structure of the keratin in hair and fingernails.

tertiary (3°) structure the three-dimensional, biologically active structure of the protein that arises because of interactions between the R groups on the amino acids.

quaternary (4°) structure the larger structure functioning as a single unit that results when two or more proteins associate.

enzyme a protein that catalyzes a reaction.

Interactions and reactions such as these determine a protein's **tertiary (3°) structure**, the overall three-dimensional shape of the protein that is key to its biological activity (Figure 20.9c). Because the proteins in living systems exist in an aqueous environment, hydrophobic R groups tend to reside in the interiors of their 3° structures, whereas hydrophilic groups (as we saw in normal hemoglobin) are oriented toward the outside, where they interact with nearby molecules of water. Hydrophobic interactions are the primary force that causes protein folding and compaction, but all the other modes of interaction help a large protein find its unique 3° structure.

Hemoglobin and some other proteins exhibit an even higher order of structure. One hemoglobin unit (Figure 20.16a) contains four protein strands, each of which enfolds a porphyrin ring containing one Fe^{2+} ion. The combination of four protein strands to make one hemoglobin assembly is an example of **quaternary (4°) structure** (Figure 20.9d). In the 4° structures of keratins (Figure 20.16b), protein strands with α-helical 2° structures coil around each other to make even larger coils. When the protein strands in keratin structures are held together mostly by van der Waals forces, as they are in the keratin in skin tissue, the structures are flexible and elastic. If they are also restrained by many covalent bonds as in Figure 20.16(b), they produce tissues that are hard and less flexible, like fingernails and the beaks of birds.

Enzymes: Proteins as Catalysts

The chemical reactions involved with metabolism—both *catabolism* (breaking down of molecules) and *anabolism* (synthesis of complex materials from simple feedstocks)—are mediated in large part by proteins called **enzymes**. Enzymes are biological catalysts. Both catabolism and anabolism are organized in sequences of reactions called *metabolic pathways*, and each step in a metabolic pathway is catalyzed by a specific enzyme. For example, carbonic anhydrase (Figure 20.14) is an enzyme that speeds up the hydrolysis of CO_2:

$$CO_2(aq) + H_2O(\ell) \rightleftharpoons HCO_3^-(aq) + H^+(aq) \qquad (20.1)$$

In the presence of carbonic anhydrase, this reaction proceeds about 10 million times faster than in its absence. Without carbonic anhydrase, we would not be able to expel carbon dioxide fast enough to survive (as the reaction in Equation 20.1 runs in reverse). One molecule of carbonic anhydrase can hydrolyze approximately 10,000 molecules of CO_2 in 1 s. This value is called the *turnover number* for the enzyme; in general, the higher the turnover number, the faster the enzyme-catalyzed reaction proceeds. Turnover numbers for enzymes typically range from 10^3 to 10^7. The higher the turnover number, the lower the activation energy of the catalyzed reaction. As we learned in Chapter 15, for reactions to proceed, molecules must collide with the proper orientation. Carbonic anhydrase is essentially a perfect enzyme because it catalyzes the hydrolysis reaction every time it collides with CO_2.

Unlike the inorganic catalysts discussed in Chapter 15, enzymes are highly selective: each catalyzes a particular reaction involving a particular reactant. For example, an enzyme called lactase catalyzes only the reaction by which lactose, the sugar in milk, is broken down during digestion. People who lack this enzyme cannot metabolize this sugar; they are said to be *lactose intolerant*. If they consume dairy products, unmetabolized lactose passes into their large intestines where bacteria ferment it, and unpleasant and painful abdominal disturbances result.

Even in the simplest organism, hundreds of enzyme-catalyzed chemical reactions are constantly underway. Most of these enzymes are effective only under limited reaction conditions: in aqueous media, at temperatures between 4 and 37°C, and over a narrow pH range. These hundreds of catalyzed reactions require hundreds of different enzymes, many operating with exquisite efficiency—sometimes on only one member of a pair of enantiomers, for example, producing only one enantiomerically pure product.

Enzyme selectivity is the main reason natural products tend to be optically pure materials while the same products synthesized in the laboratory tend to be racemic mixtures. In the pharmaceutical industry, research is currently focused on using enzymes outside living systems to produce enantiomerically pure materials needed as drugs. This research area, called **biocatalysis**, uses enzymes to catalyze chemical reactions run in industrial-sized reactors.

Synthetic reaction pathways catalyzed by enzymes are usually more rapid and involve fewer steps than uncatalyzed pathways to make the same product. These advantages can significantly reduce the cost of production of, for example, biological pharmaceuticals. However, the biocatalytic reactions often run best in very dilute solutions, which limits production. A key issue driving interest in such processes is that an enantiomerically pure pharmaceutical is likely to be more potent and produce fewer side effects than a racemic mixture of the same product. As an example, the drug thalidomide is a racemic mixture of two enantiomeric forms: one is very effective in treating morning sickness, but the second is a teratogen (an agent that disturbs the development of a fetus). In perhaps the worst medical tragedy in modern times, more than 10,000 children were born in the late 1950s and early 1960s with serious, frequently fatal, deformities as a result of their mothers taking the racemic drug.

The molecular structure of enzymes contains a region called an **active site** that binds the reactant molecule, called the **substrate**. The action of enzymes was originally explained by a lock-and-key analogy in which the substrate is the key and the active site is the lock (Figure 20.17). The substrate is held in the active site by the same kinds of intermolecular interactions that hold any biomolecules together. Some enzymes become covalently bonded to intermediates in the catalytic process. Once in the active site, the substrate is converted into product via a reaction having a lower-energy transition state than it would without the enzyme. The reaction of a substrate S with an enzyme E produces an *enzyme–substrate* (E–S) *complex* that decomposes, forming a product P and regenerating the enzyme:

$$E + S \rightarrow E\text{–}S \rightarrow E + P$$

The lock-and-key analogy does not fully account for enzyme behavior. A more accurate view is provided by the *induced-fit model*, which assumes that the substrate does more than just fit into the existing shape of an active site.

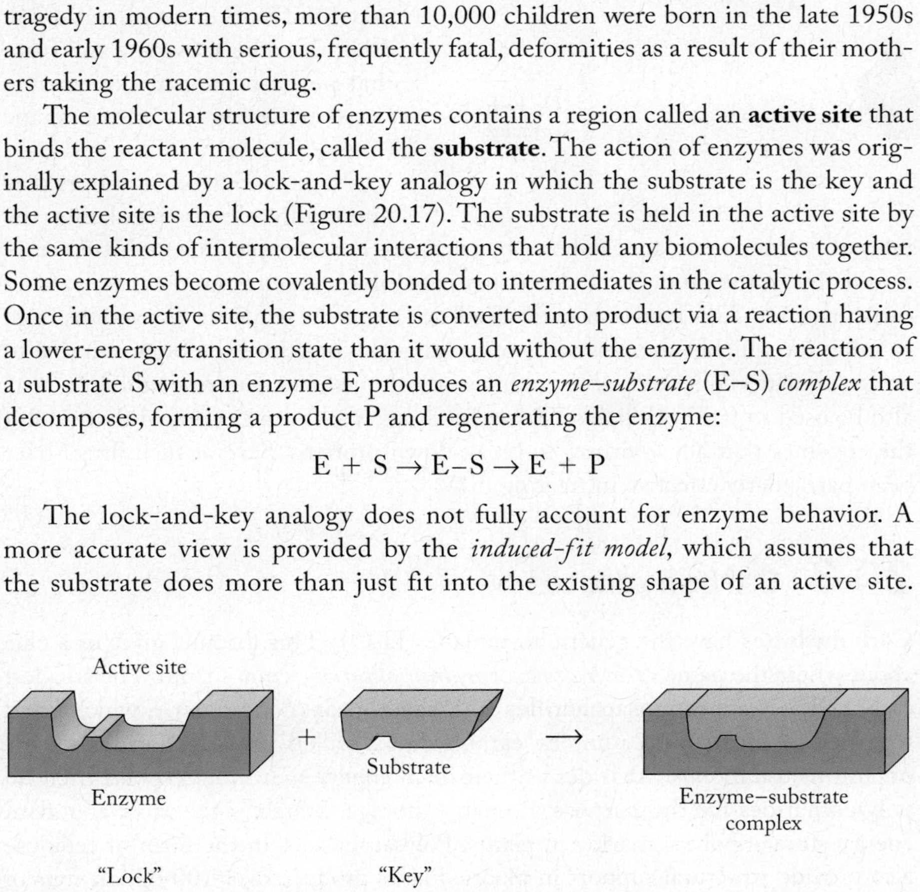

CONNECTION In Section 15.6 we described heterogeneous catalysts used in automobile exhaust systems. Enzymes are homogeneous catalysts that selectively speed up biochemical reactions.

CONNECTION A racemic mixture contains equal proportions of the two enantiomers of a chiral compound, as we discussed in Section 13.8.

FIGURE 20.17 In the lock-and-key model of enzyme activity, the substrate fits exactly into the active site of the enzyme that catalyzes a chemical reaction involving the substrate.

Active site

Enzyme   +   Substrate   →   Enzyme–substrate complex

"Lock"   "Key"

biocatalysis the strategy of using enzymes to catalyze reactions on a large scale; it is becoming especially important in processes that involve chiral materials.

active site the location on an enzyme where a reactive substance binds.

substrate the reactant that binds to the active site in an enzyme-catalyzed reaction.

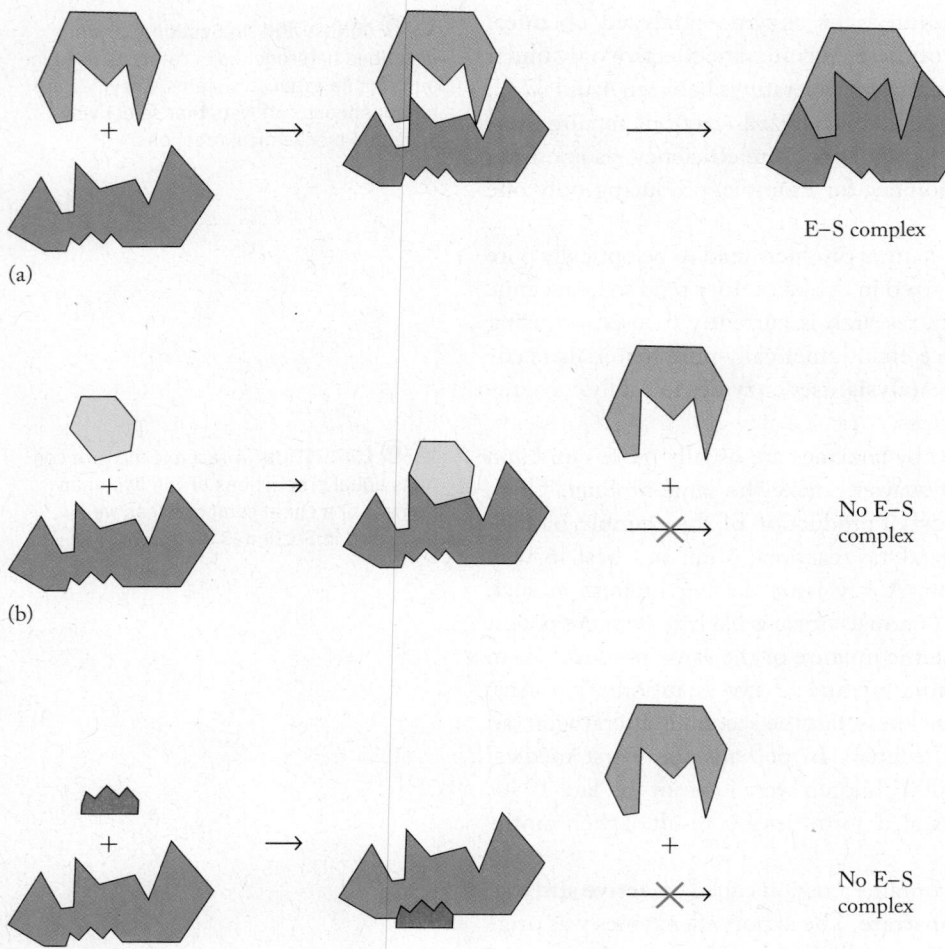

(a)

(b)

(c)

E–S complex

No E–S complex

No E–S complex

FIGURE 20.18 (a) The induced-fit model assumes that the shape of the enzyme changes to accommodate the substrate and form the enzyme–substrate (E–S) complex. Inhibitors (yellow molecules) may (b) block the enzyme's binding site or (c) cause a change elsewhere in the enzyme's structure that prevents the active site from attaining the shape it needs to form the E–S complex.

This model assumes that as the E–S complex forms, the binding site undergoes subtle changes in its shape to more precisely fit the three-dimensional structure of the transition state. The binding energy between the enzyme and the transition state lowers the activation energy barrier, thereby speeding up the reaction. A simplified illustration of such an interaction is shown in Figure 20.18(a).

The induced-fit model helps explain the behavior of compounds called **inhibitors** that can diminish or destroy the effectiveness of enzymes. An inhibitor may bind to an active site and block it from interacting with the substrate (Figure 20.18b), or it may disable the enzyme by preventing it from assuming its active shape. In the latter process the inhibitor may bind to the enzyme at a site other than the active site and from that position prevent the enzyme from achieving its active shape (Figure 20.18c).

Natural enzyme inhibitors play important roles in regulating the rates of reactions that are catalyzed by enzymes. For example, in a multistep reaction pathway, the product of a later step may inhibit an enzyme that catalyzes an earlier reaction. This kind of negative feedback keeps the sequence of reactions from running too quickly and perhaps jeopardizing the health of the organism due to an accumulation of undesirable products or intermediates. Enzyme inhibitors may also be used to fight off disease. Powerful drugs have been developed that inhibit the enzymes that allow viruses to break down proteins. Several such drugs have been particularly effective in treating HIV.

20.3 Carbohydrates

Carbohydrates have the generic formula $C_x(H_2O)_y$. This formula gives us a clue about where the name *carbohydrate*, or *hydrate of carbon*, comes from. The smallest carbohydrates are **monosaccharides** (the name means "one sugar"), which bond together to form more complex carbohydrates called **polysaccharides**. Many organisms use monosaccharides as their main energy source but convert them to polysaccharides for the purpose of energy storage. Starch is the most abundant energy-storage polysaccharide in plants. Polysaccharides in the form of cellulose also provide structural support in plants. Plants produce over 100 billion tons of cellulose each year—the woody parts of trees are over 50% cellulose, and cotton is 99% cellulose.

inhibitor a compound that diminishes or destroys the ability of an enzyme to catalyze a reaction.

carbohydrate an organic molecule with the generic formula $C_x(H_2O)_y$.

monosaccharide a single-sugar unit and the simplest carbohydrate.

polysaccharide a polymer of monosaccharides.

The principal building block of both starch and cellulose is the monosaccharide glucose. The different properties and functions of these polysaccharides come from the subtle differences in the molecular geometries of the glucose monomers in their structures and how the monomers are bonded together. Molecules of most monosaccharides, including glucose, contain several chiral centers, so multiple optical isomers exist. This complexity gives rise to another major function of carbohydrates: molecular recognition. For example, combinations of carbohydrates and proteins, called *glycoproteins*, on the surfaces of blood cells determine the blood type of an individual.

Molecular Structures of Glucose and Fructose

Glucose is the most abundant monosaccharide in nature and in the human body. It is also called *dextrose*—a kind of abbreviation for *dextro*-glucose. Fructose is the principal monosaccharide in many fruits and root vegetables. Given the importance and abundance of these sugars, let's examine their structures and properties more closely.

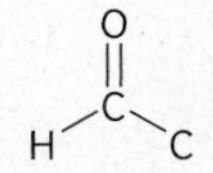

Three molecular views of glucose are provided by the structures in Figure 20.19. All three have the same chemical formula ($C_6H_{12}O_6$), so they are all structural isomers of one another. Note that the middle structure contains an aldehyde (–CH=O) group and all three structures contain multiple alcohol (–C—OH) groups. The polarities of these groups and their capacities to form hydrogen bonds give glucose its high molar solubility in water (5.0 M at 25°C).

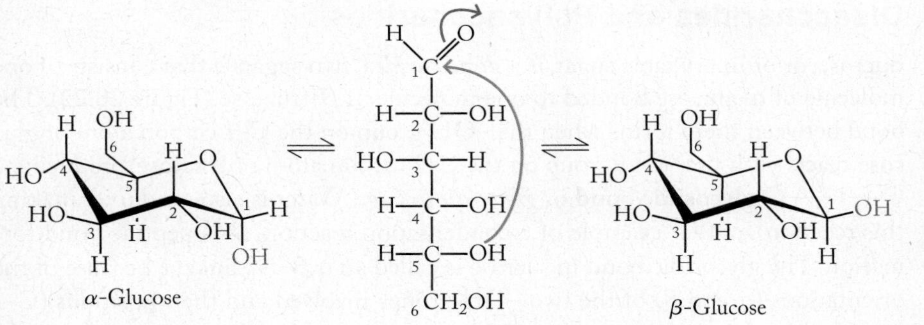

CONNECTION An aldehyde group (Chapter 13) includes these atoms and bonds:

FIGURE 20.19 An equilibrium exists between the linear structure of glucose and the two cyclic forms α-glucose and β-glucose. The difference between the two cyclic structures is the orientation of the –OH group on C-1 (highlighted in blue).

The cyclic structures of glucose form when the carbon backbone of the linear form curls around so that the –OH group bonded to the carbon atom labeled number 5 (C-5) comes close to the aldehyde group on C-1 (as shown by the red arrow in Figure 20.19). The aldehyde and alcohol groups react to form a six-membered ring made up of five carbon atoms (C-1 to C-5) and the oxygen atom of the C-5 alcohol. Note a small but significant difference in the two cyclic structures. In the molecule on the left, called α-glucose, the –OH group on C-1 points down. In the molecule on the right, β-glucose, the C-1 –OH group points up. These two orientations are possible because there are two ways that the carbon chain in the middle structure can form a cyclic structure: the –OH group on C-5 can approach C-1 from either of the two sides of the plane defined by the C-1 carbon atom and the C and H atoms bonded to it. When the C-5 –OH group approaches C-1 as shown in Figure 20.20, the α isomer is produced. Approaching C-1 from the other side yields the β isomer. Notice that in the linear form, C-1 is an aldehyde and is not a chiral center. The cyclization creates a new chiral center in the cyclic molecule.

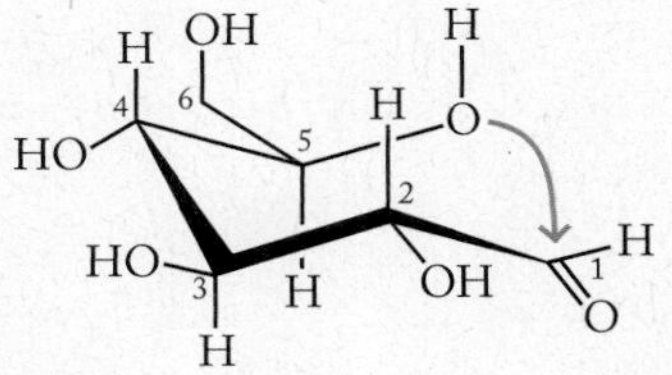

The β form is slightly more stable than the α form and accounts for 64% of glucose molecules in aqueous solution; the α form accounts for the remaining 36%.

FIGURE 20.20 The –OH group on C-5 may approach the aldehyde on C-1 from either side of the plane defined by the C, H, and O atoms in the aldehyde group. In the approach shown here the product will be the α isomer.

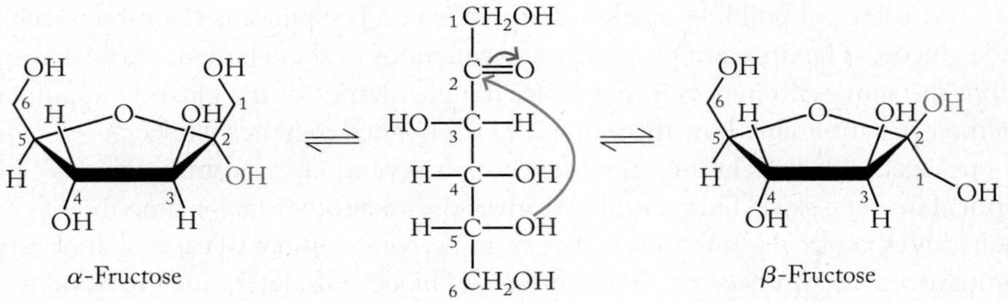

FIGURE 20.21 The cyclization of fructose proceeds via the –OH group on C-5 and the ketone on C-2.

α-Fructose

β-Fructose

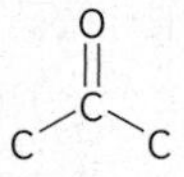

Both cyclic forms are more stable than the straight-chain form, which exists only as an intermediate between the two cyclic forms. The energy differences are small, however, so glucose molecules in solution are constantly opening and closing in a dynamic structural equilibrium.

Figure 20.21 shows the structures of the linear and cyclic forms of fructose. Note that the C=O group in the linear form is not on the terminal carbon atom. In other words, fructose is a ketone rather than an aldehyde. It forms a five-membered ring, not a six-membered ring, when the –OH group at C-5 reacts with the carbonyl carbon atom at the C-2 position. The product is a ring with two –CH$_2$OH groups that are either on the same side of the ring (α-fructose) or on opposite sides (β-fructose).

CONNECTION A ketone group (Chapter 13) contains these atoms and bonds:

CONCEPT TEST

Explain how the linear form of fructose can wrap around to produce two different cyclic forms.

CHEMTOUR Formation of Sucrose

Disaccharides and Polysaccharides

Sucrose, or ordinary table sugar, is a *disaccharide* ("two sugars") that consists of one molecule of α-glucose bonded to one molecule of β-fructose (Figure 20.22). The bond between them forms when the –OH group on the C-1 carbon atom of glucose reacts with the –OH group on the C-2 carbon atom of fructose, producing a C—O—C **glycosidic bond** or *glycosidic linkage*. Water is also produced, making this reaction another example of a condensation reaction, as is peptide bond formation. The glycosidic bond in sucrose is called an α,β-1,2 linkage because of the orientations (α and β) of the two –OH groups involved and their positions (C-1 and C-2) in the cyclic structures of the two monosaccharides.

Another important glycosidic bond involves the –OH groups on the C-1 and C-4 carbon atoms of glucose molecules. When glucose molecules link at these positions they can form long-chain polysaccharides. If the starting monomer is α-glucose, the

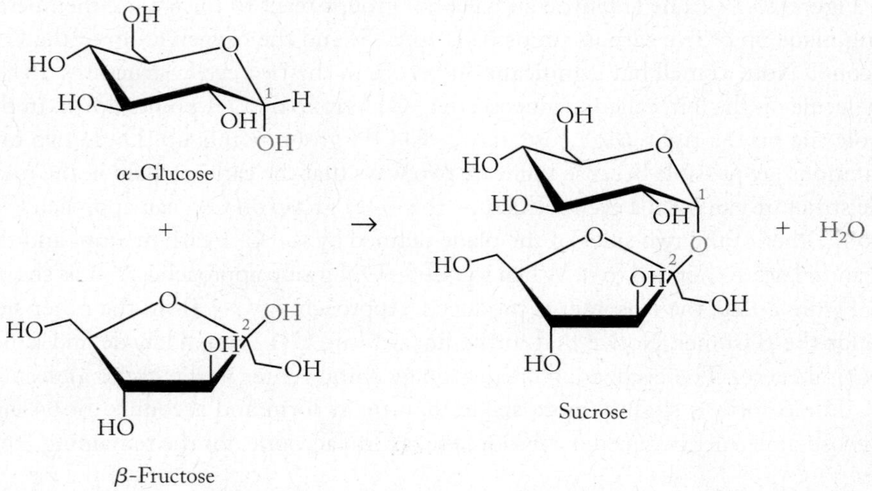

α-Glucose

+

β-Fructose

→

Sucrose

+ H$_2$O

FIGURE 20.22 α-Glucose and β-fructose combine to form a molecule of sucrose, ordinary table sugar.

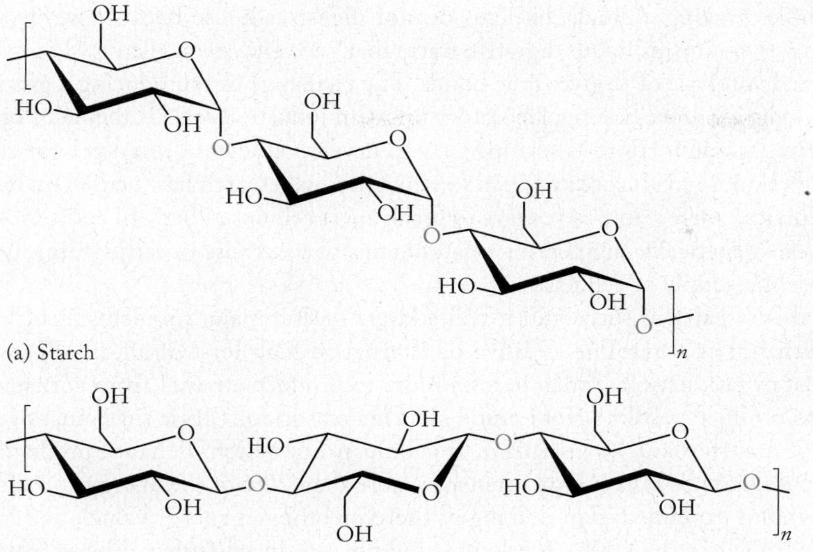

(a) Starch

(b) Cellulose

FIGURE 20.23 (a) Starch is a polysaccharide of α-glucose molecules joined by α-1,4 glycosidic bonds. (b) Cellulose is a polysaccharide of β-glucose molecules joined by β-1,4 glycosidic bonds.

bonds between them are α-1,4 glycosidic linkages, and the product of bond formation is starch (Figure 20.23a). The conversion of α-glucose into starch is an effective way for plants to store energy because formation of the α-1,4 linkage is reversible. With the aid of digestive enzymes, α-1,4 bonds can be hydrolyzed and starch converted back into glucose by plants that make the starch, or by animals that eat the plants.

Carbohydrates in plants are a major part of the total organic matter in any given ecological system, that is, they make up much of the system's **biomass**. Humans have used various forms of biomass, including wood and animal dung, as fuel for thousands of years. More recently, we have begun to convert biomass into a liquid *biofuel*, ethanol, for use as a gasoline additive and substitute. Because ethanol contains oxygen, it improves the combustion of a hydrocarbon fuel like gasoline and reduces carbon monoxide emissions. An important industrial application of starch hydrolysis is the conversion of cornstarch into glucose and then, through fermentation, into ethanol:

$$C_6H_{12}O_6(aq) \rightarrow 2\,CH_3CH_2OH(aq) + 2\,CO_2(g) \qquad (20.2)$$

This exothermic reaction provides energy to the yeast cells whose biological processes drive fermentation.

The cellulose that plants synthesize to build stems and other organs has a structure (Figure 20.23b) slightly different from that of starch because the building blocks of cellulose are β-glucose instead of α-glucose, so the monomers are linked by β-1,4 glycosidic bonds. This structural difference is important because it enables starch to coil and make granules for efficient energy storage while cellulose forms structural fibers.

CONCEPT TEST

Cellobiose is a disaccharide made from the degradation of cellulose. We cannot digest cellobiose. Which of the two structures in Figure 20.24 represents a molecule of cellobiose?

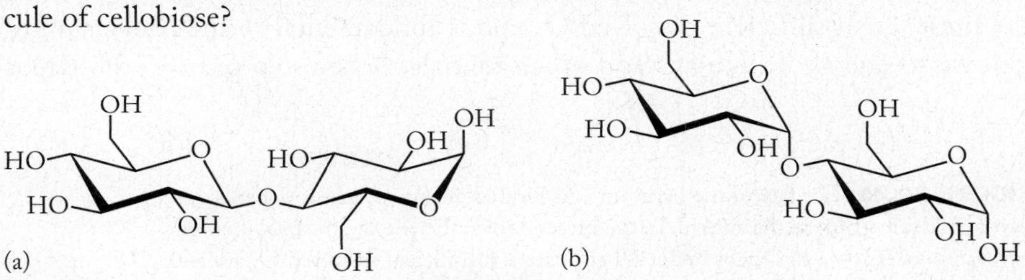

(a)

(b)

FIGURE 20.24

glycosidic bond a C—O—C bond between sugar molecules.

biomass the sum total of the mass of organic matter in any given ecological system.

Krebs cycle a series of reactions that continue the oxidation of pyruvate formed in glycolysis.

Unlike grazing animals, humans cannot digest cellulose because we do not have microorganisms in our digestive tracts that have enzymes called *cellulases* that catalyze hydrolysis of β-glycosidic bonds. The challenge of reproducing a grazing cow's digestive processes in a laboratory or on an industrial scale is the focus of an enormous research effort as scientists try to develop efficient procedures for converting cellulose to glucose and then to ethanol. This research has focused on more efficient, less energy-intensive ways to break apart cellulose fibers. In addition scientists are genetically engineering microorganisms like those in cattle stomachs to increase the supply of cellulases.

If this research is successful it will address several major problems associated with ethanol as a gasoline additive or alternative fuel. First of all, it will lower the cost of production. Today it costs more to produce ethanol from cornstarch than to produce gasoline from crude oil. One reason for this is that most of the mass of a corn plant, or any plant, is cellulose, not starch. Ethanol production from cornstarch is also energy intensive: more than 70% of the energy contained in ethanol is expended in producing it; therefore, the net energy value of ethanol from corn is less than 30%. If ethanol could be produced from cellulose instead of starch, its net energy value could be as high as 80%. Finally, the use of edible cornstarch in fuel production has driven up the price of foods derived from corn. The impacts of this inflation have been felt worldwide and have been particularly painful in developing countries. The use of agricultural land and consumption of increasingly scarce water resources for ethanol production raise further concerns.

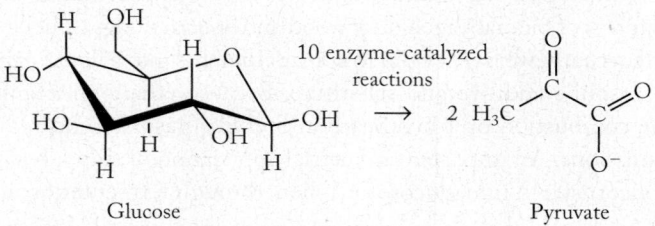

FIGURE 20.25 In glycolysis, molecules of glucose are broken down to pyruvate ions.

Glycolysis Revisited

Most cells including those in our bodies use glucose as a fuel in a series of reactions, collectively called glycolysis, that oxidize glucose to pyruvate ions (Figure 20.25), the conjugate base of pyruvic acid. Pyruvate sits at a metabolic crossroads and can be converted into different products depending on the type of cell in which it is generated, the enzymes present, and the availability of oxygen (Figure 20.26). In yeast cells growing under low-oxygen conditions, pyruvate is converted into ethanol and CO_2. Another series of reactions, called the **Krebs cycle**, occurs in the presence of sufficient dissolved O_2 and is fundamental to the conversion of glucose to energy in humans and other animals. A key step prior to the Krebs

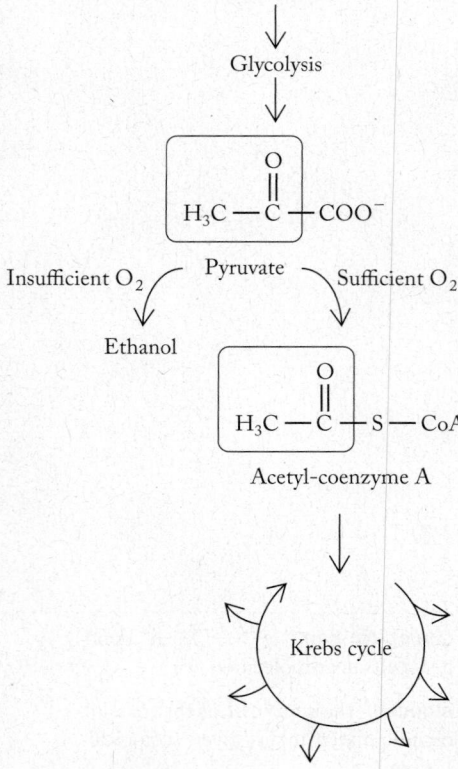

FIGURE 20.26 The fate of the pyruvate ions formed during glycolysis depends on the partial pressure of O_2 in the system. In the presence of sufficient O_2, the oxidation of pyruvate proceeds via the Krebs cycle. When there is insufficient dissolved O_2 available, as in fermentation, the pyruvate may be converted to ethanol.

cycle occurs when pyruvate loses CO_2 and forms an acetyl group, which then combines with coenzyme A. The resulting product, acetyl-coenzyme A, is a reactant in many biosynthetic pathways, including the production and breakdown of cholesterol.

Cholesterol is a key component in the structure of cell membranes. It is also a precursor of the bile acids that aid in digestion and of steroid hormones, which regulate the development of the sex organs and secondary sexual traits, stimulate the biosynthesis of proteins, and regulate the balance of electrolytes in the kidneys. In a healthy person, synthesis and use of cholesterol are tightly regulated to prevent overaccumulation and consequent deposition of cholesterol in coronary arteries. We clearly need cholesterol, but deposition in the arteries can lead to serious coronary disease (Figure 20.27).

CONCEPT TEST ···

A newspaper article contains the wording, "Primarily made by the liver, cholesterol begins with tiny pieces of sugar. . . ." What does this statement mean at the molecular level?

···

20.4 Lipids

Lipids differ from carbohydrates and proteins in that they are not biopolymers. Lipids are best described by their physical properties rather than any common structural subunit: lipids do not dissolve in water but are soluble in nonpolar solvents, and they are oily to the touch. Because they are insoluble in water, they are ideal components of cell membranes, which separate the aqueous solutions within cells from the aqueous environments outside them. An important class of lipids called **glycerides** are esters formed between glycerol and long-chain fatty acids (Figure 20.28). The three –OH groups on glycerol allow for mono-, di-, and triglycerides, with the latter being the most abundant. Glycerides account for over 98% of the lipids in the fatty tissues of mammals.

CONNECTION In Chapter 14 we introduced glycolysis in a discussion of the thermodynamics of coupled reactions.

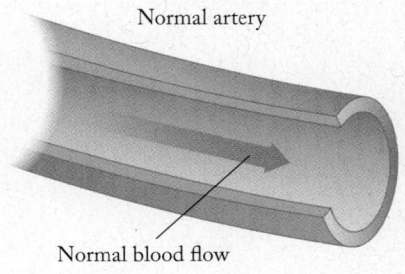

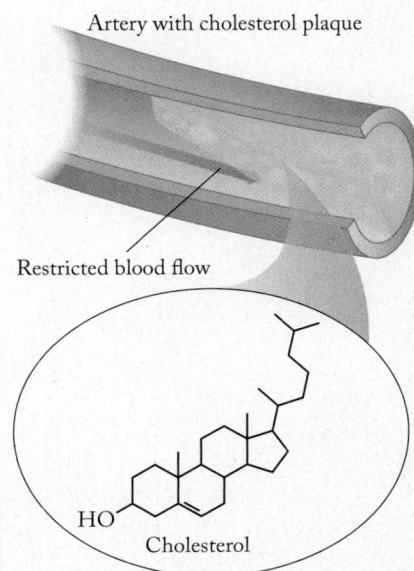

FIGURE 20.27 Cholesterol deposits called plaques are responsible for restricted blood flow, which results in a variety of sometimes catastrophic cardiovascular problems.

···

lipid a class of water-insoluble, oily organic compounds that are common structural materials in cells.

glyceride lipid consisting of esters formed between fatty acids and the alcohol glycerol.

···

FIGURE 20.28 Glycerides are esters that form when glycerol combines with fatty acids. When all three –OH groups on glycerol react to form ester bonds, the product is a *tri*glyceride.

Glycerol

Fatty acid

Triglyceride

Table 20.2 lists some common fatty acids. The most abundant ones have an even number of carbon atoms because the fatty acids are built from two-carbon subunits. Their biosynthesis begins, as does the biosynthesis of cholesterol, with the conversion of pyruvate to acetyl-coenzyme A. Most fatty acids contain between 14 and 22 carbon atoms.

TABLE 20.2	Names and Structural Formulas of Common Fatty Acids
Common Name **(chemical name)** **(source)**	**Formula**
SATURATED FATTY ACIDS	
Lauric acid (dodecanoic acid) (coconut oil)	$CH_3(CH_2)_{10}COOH$
Myristic acid (tetradecanoic acid) (nutmeg butter)	$CH_3(CH_2)_{12}COOH$
Palmitic acid (hexadecanoic acid) (animal and vegetable fats)	$CH_3(CH_2)_{14}COOH$
Stearic acid (octadecanoic acid) (animal and vegetable fats)	$CH_3(CH_2)_{16}COOH$
UNSATURATED FATTY ACIDS	
Oleic acid (*cis*-9-octadecenoic acid) (animal and vegetable fats)	$CH_3(CH_2)_7CH{=}CH(CH_2)_7COOH$
Linoleic acid (*cis,cis*-9,12-octadecadienoic acid) (linseed oil, cottonseed oil)	$CH_3(CH_2)_4CH{=}CHCH_2CH{=}CH(CH_2)_7COOH$
Linolenic acid (*cis,cis,cis*-9,12,15-octadecatrienoic acid) (linseed oil)	$CH_3CH_2CH{=}CHCH_2CH{=}CHCH_2CH{=}CH(CH_2)_7COOH$

SAMPLE EXERCISE 20.3   **Identifying Triglycerides**

How many different triglycerides (including structural isomers and stereoisomers) can be made from glycerol combining with two different fatty acids (X and Y) if each molecule of triglyceride contains at least one molecule of each fatty acid?

Collect and Organize We are asked to determine how many different triglycerides can be made from glycerol and two different fatty acids. A triglyceride contains three fatty acid units.

Analyze Each fatty acid may bond to one of three —OH groups in glycerol. Each triglyceride has at least one X and one Y residue, which makes two formulas possible: X_2Y and Y_2X. Each of these formulas has two structural isomers, depending on whether the single fatty acid in the formula is bonded to the middle carbon or to one of the end carbon atoms. Finally, if a structure has an X on one end carbon atom and a Y on the other, then the middle carbon atom is a chiral center, which means that there are two enantiomeric forms of that compound.

Solve We can generate four different molecular structures by attaching X and Y in four different sequences to the glycerol –OH groups:

$$
\begin{array}{cccc}
H_2C-O-X & H_2C-O-X & H_2C-O-Y & H_2C-O-Y \\
| & | & | & | \\
HC-O-X & HC-O-Y & HC-O-Y & HC-O-X \\
| & | & | & | \\
H_2C-O-Y & H_2C-O-X & H_2C-O-X & H_2C-O-Y \\
(1) & (2) & (3) & (4)
\end{array}
$$

Structures (1) and (3) have chiral centers, so each of these isomers has two enantiomeric forms. Therefore, there are a total of six different triglycerides possible.

Think about It The central carbon atoms in structures (1) and (3) are chiral because they are each bonded to a H atom, an O atom, and to two C atoms that are themselves bonded to different fatty acids: X and Y.

Practice Exercise How many different triglycerides can be made from glycerol and one molecule each of three different fatty acids A, B, and C?

■

Function and Metabolism of Lipids

Lipids are an important energy source in our diets, providing more energy per gram than carbohydrates or proteins. As indicated in Table 20.2, fatty acids are either saturated, containing no carbon–carbon double bonds, or unsaturated, containing one or more carbon–carbon double bonds. **Fats** are glycerides composed primarily of saturated fatty acids. They are solids at room temperature. **Oils** are glycerides composed predominantly of unsaturated fatty acids and are liquids at room temperature. Oils can be converted into solid, saturated glycerides by hydrogenation. For example, in the hydrogenation of corn oil, which is a mixture of mostly two unsaturated fatty acids (oleic and linoleic acids), hydrogen is added to convert some or all the $-CH=CH-$ subunits into $-CH_2-CH_2-$ subunits. Hydrogenation converts the oil into a solid at room temperature that is whipped with skim milk, coloring agents, and vitamins to produce the food spread we know as margarine.

The consumption of too much saturated fat is associated with coronary heart disease. One of the purported advantages of the so-called Mediterranean diet is that olive oil, a liquid composed of glycerides containing over 80% oleic acid, is used in cooking rather than animal fats like butter and lard. In addition, this diet tends to be richer in fish and vegetables, both of which contain unsaturated fats. Most animal fats, like the marbling in beef that enhances its flavor, are saturated.

Another problem with hydrogenating vegetable oil arises when the oils are only partially hydrogenated. Partial hydrogenation alters the molecular structure around their remaining C=C double bonds, changing them from their natural cis isomers into trans isomers (Figure 20.29). Unsaturated trans fats like elaidic acid tend to be solids at room temperature because their molecules pack together more uniformly than do molecules of cis unsaturated fatty acids. Consumption of trans fatty acids is associated with increased levels of cholesterol in the blood and other health risks.

fat solid triglyceride containing primarily saturated fatty acids.

oil liquid triglyceride containing primarily unsaturated fatty acids.

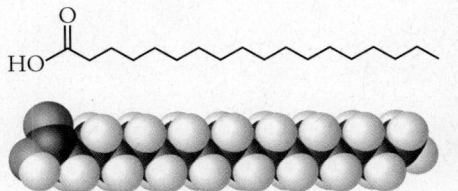

(a) Stearic acid: a saturated fatty acid

(b) Elaidic acid: a trans unsaturated fatty acid

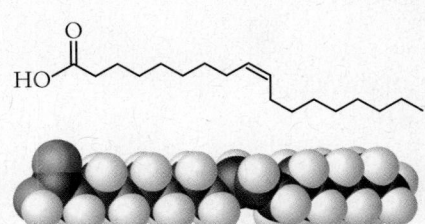

(c) Oleic acid: a cis unsaturated fatty acid

FIGURE 20.29 Types of fatty acids. (a) Stearic acid, a saturated C_{18} fatty acid. (b) Elaidic acid, an unsaturated C_{18} fatty acid (trans isomer). (c) Oleic acid, an unsaturated C_{18} fatty acid (cis isomer).

The mixture of lipids in the legs of reindeer living close to the Arctic Circle changes as a function of distance from the hoof: The closer to the hoof, the higher the percentage of unsaturated fatty acids. Why would this be an advantage for these reindeer?

The lipids in some prepared foods have been modified to reduce their caloric content but still provide the taste, aroma, and "mouth feel" we associate with lipid-rich foods. The active sites of enzymes that break down natural lipids accommodate triglycerides formed from glycerol and fatty acids. However, chemically modified esters made from the same fatty acids but attached to an alcohol other than glycerol cannot be metabolized by these enzymes. Such molecules, if they have the appropriate physical properties and are nontoxic, can be incorporated into foods without adding any calories because they are not metabolized.

Olestra is one such product (Figure 20.30). It is an ester made from long-chain fatty acids and the carbohydrate sucrose. (Remember, sugars have –OH groups and technically are alcohols.) Each of the eight –OH groups in a sucrose molecule reacts with a molecule of fatty acid to make the ester in olestra. The resultant material is used to deep-fry potato chips. Any olestra that remains on the chip does not add calories because it cannot be processed by enzymes that recognize only fatty acid esters on a glycerol scaffold.

Olestra may be "calorie-free" as a food subject to metabolism in the living system, but how would it compare with a common triglyceride in terms of kilojoules of heat released per mole in a calorimeter experiment?

The enzymes that metabolize triglycerides hydrolyze the esters and release glycerol and the fatty acids that were bonded to it. Glycerol enters the metabolic pathway for glucose. The fatty acids are oxidized in a series of reactions known as β-oxidation: a process that removes two carbon atoms at a time. For example, stearic acid (the saturated C_{18} fatty acid) is transformed into a C_2 fragment and the C_{16} acid, palmitic acid. Palmitic acid yields another C_2 fragment and myristic acid, and

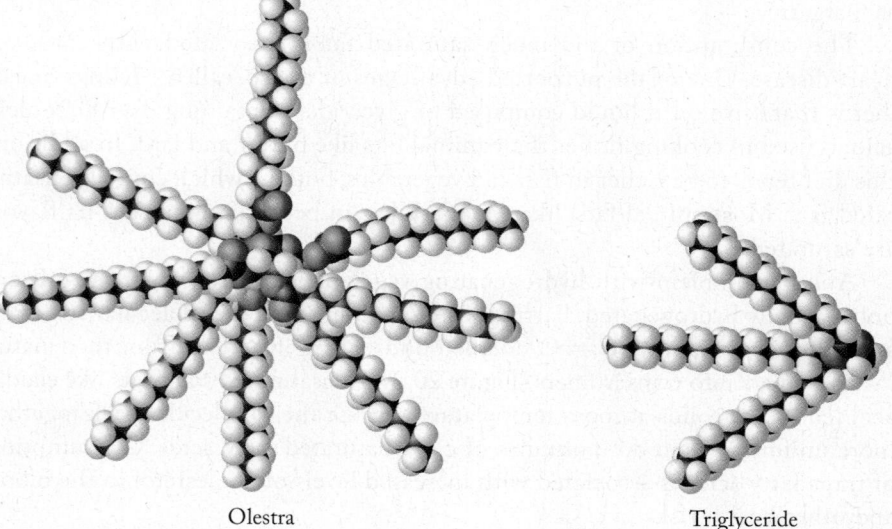

FIGURE 20.30 Olestra has a shape very different from that of the triglycerides typically metabolized by our bodies. Consequently, it cannot be processed by the enzymes that digest triglycerides.

Olestra   Triglyceride

Choline

$$CH_3 - ^+N - CH_2 - CH_2 - O - P - O - CH_2$$

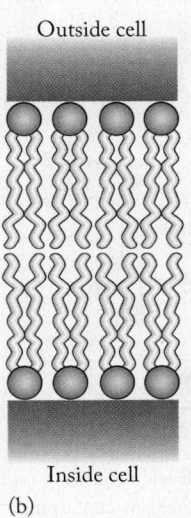

Polar			Nonpolar
Ionic or polar group	Phosphate	Glycerol	Fatty acids

(a)

Outside cell

Inside cell

(b)

FIGURE 20.31 Phospholipids are major constituents of cell membranes. (a) The presence of a polar group (here, choline) attached to a phosphate unit on glycerol changes the properties of the resulting diglyceride (here, phosphatidylcholine). (b) In the lipid bilayer that forms cell membranes, phospholipids orient themselves so that the polar groups in one half of the bilayer face the aqueous environment outside the cell while the polar groups in the other half of the bilayer face the aqueous environment of the cell interior. This arrangement leaves the nonpolar fatty acid part of the phospholipids in the interior of the bilayer.

so forth until the fatty acid is completely degraded. Electrons released from this oxidative process eventually are donated to O_2. The energy released by this process powers metabolism.

Other Types of Lipids

Cells contain other types of lipids in addition to triglycerides. One type, **phospholipids** (Figure 20.31a), plays a key role in cell structure. A phospholipid molecule consists of a glycerol molecule bonded to two fatty acid chains and to one phosphate group that is also bonded to polar substituents. The presence of nonpolar fatty acid chains and a polar region in the same molecule makes phospholipids ideal for forming cell membranes. In an aqueous medium, phospholipids form a **lipid bilayer**, a double layer enclosing each cell and isolating its interior from the outside environment. The phospholipid molecules of the bilayer align so that the nonpolar groups interact with each other inside the membrane while the polar groups interact with water molecules outside it (Figure 20.31b). Membranes exist both to isolate the contents of cells and to serve as the locus of communication between processes that occur within and outside the cells.

20.5 Nucleotides and Nucleic Acids

Nucleic acids are our fourth class of biomolecules and third class of biopolymers. We focus on two types: deoxyribonucleic acid (DNA) and ribonucleic acid (RNA). Even though nucleic acids make up only about 1% of an organism's mass, they control the metabolic activity of all its cells. DNA carries the genetic blueprint of an organism, and a variety of RNAs use that DNA blueprint to guide the production of proteins.

phospholipid a molecule of glycerol with two fatty acid chains and one polar group containing a phosphate; phospholipids are major constituents of cell membranes.

lipid bilayer a double layer of molecules whose polar head groups interact with water molecules and whose nonpolar tails interact with each other.

nucleic acid one of a family of large molecules, which includes deoxyribonucleic acid (DNA) and ribonucleic acid (RNA), that stores the genetic blueprint of an organism and controls the production of proteins.

nucleotide a monomer unit from which nucleic acids are made.

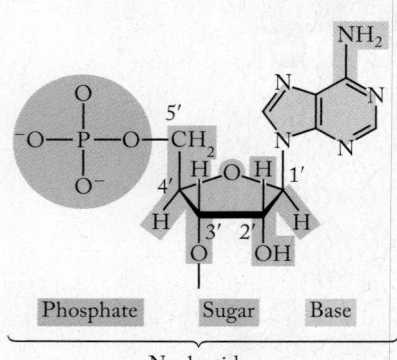

FIGURE 20.32 A nucleotide consists of a phosphate group and a nitrogen-containing base that are both bonded to a 5-carbon sugar.

A nucleic acid is a polymer composed of monomeric units called **nucleotides**. Each nucleotide unit is in turn composed of three subunits: a five-carbon sugar, a phosphate group, and a nitrogen-containing base (Figure 20.32). The phosphate group in each nucleotide is attached to a carbon in the side chain of the sugar called the 5′ carbon atom (the prime number refers to the position of the carbon atom in the sugar molecule). The nitrogen-containing base is attached to the 1′ carbon atom in each sugar molecule. The sugar in Figure 20.32 is called *ribose*, which makes this a nucleotide in a strand of *ribo*nucleic acid, or RNA. If the sugar were *deoxyribose* instead, there would be a H atom in place of the −OH group on the 2′ carbon atom, and the nucleotide would be a building block of *deoxyribo*nucleic acid, or DNA.

The nitrogen-containing base in Figure 20.32 is called adenine (A). Structural formulas of adenine and the other four bases in nucleic acids—cytosine (C), guanine (G), thymine (T), and uracil (U)—are shown in Figure 20.33. The point of attachment of the sugar–phosphate groups on each base is indicated by −R. In addition to the difference in their sugars, RNA and DNA also differ in one of the bases in their nucleotides: RNA contains A, C, G, and U, but DNA contains A, C, G, and T.

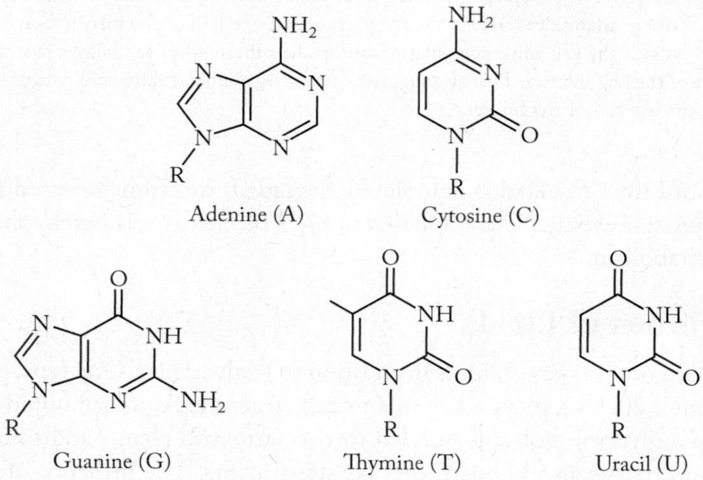

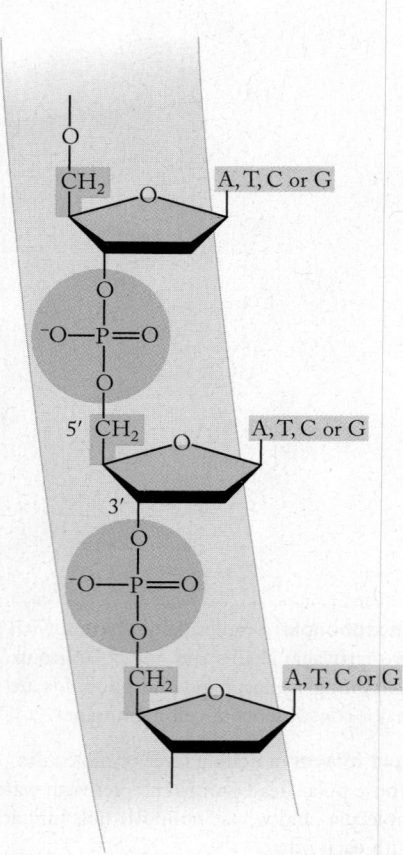

FIGURE 20.34 The backbone of the polymer chain in DNA consists of alternating sugar units (yellow) and phosphate units (pink). The bases (blue) are attached to the backbone through the C-1′ atom of the sugar unit.

Adenine (A)

Cytosine (C)

Guanine (G)

Thymine (T)

Uracil (U)

FIGURE 20.33 Structural formulas of the five nitrogen-containing bases in nucleotides. The nucleotides in DNA contain A, C, G, and T; those in RNA contain A, C, G, and U. The R group identifies the point of attachment of the sugar residue.

In a polymeric strand of nucleic acid, each phosphate is also linked to the 3′ carbon atom in the sugar of the monomer that precedes it in the chain, as shown for a strand of DNA in Figure 20.34. Both DNA and RNA strands are synthesized in the cell from the 5′ to the 3′ direction (downward in Figure 20.34). The structures of DNA and RNA are frequently written using only the single-letter labels of their bases, beginning with the free phosphate group on the 5′ end of the chain and reading toward the free 3′ hydroxyl group at the other terminus.

When scientists first isolated DNA and began to analyze its composition, they made a pivotal observation about the abundance of the nitrogen-containing bases. A typical molecule of DNA consists of thousands of nucleotides, and the percentages of the four bases in different samples of DNA can vary over a wide range. However, the percentage of A in a sample always matches the percentage of T.

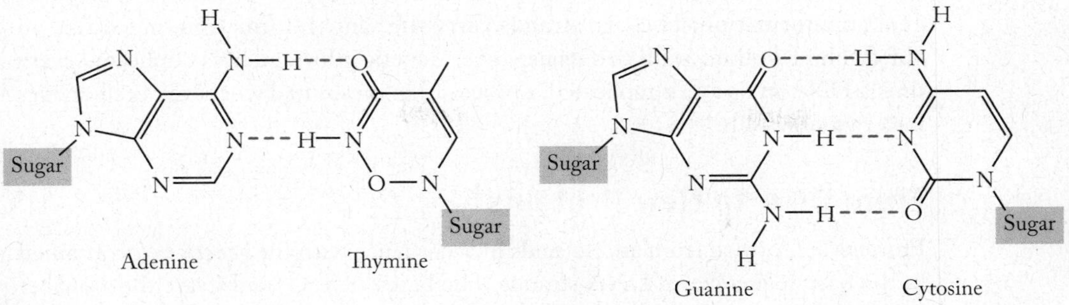

(a)

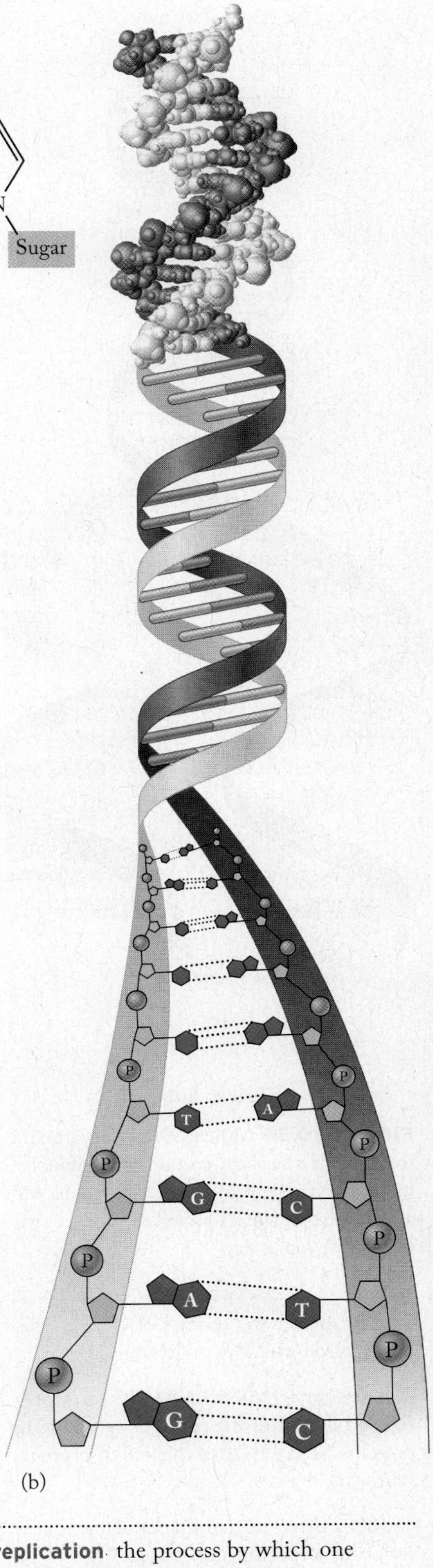

FIGURE 20.35 The nitrogen-containing bases on one strand of DNA pair with the bases on a second strand by hydrogen bonding. (a) Adenine and thymine pair via two hydrogen bonds; guanine and cytosine pair via three hydrogen bonds. (b) DNA as a double helix with the sugar–phosphate backbone on the outside and the base pairs on the inside.

Likewise, the percentage of C always matches that of G. This result makes sense if the bases are paired because a molecule of A can form two hydrogen bonds to a molecule of T, whereas a molecule of C can form three hydrogen bonds with a molecule of G. Therefore, A–T and G–C pairings maximize the number of hydrogen bonds possible (Figure 20.35a).

The normal structure of DNA has two strands of nucleotides wrapped around each other in a form that is called a *double helix* (Figure 20.35b). The nucleotide backbone is on the outside of the spiraling strands, with hydrogen bonds between the complementary bases keeping the two strands together. Notice also that the base pairs are parallel to each other and perpendicular to the helical axis. The fidelity of this base-pairing—A always with T, and C always with G—gives DNA the ability to copy itself. If a pair of complementary strands is unzipped into two single strands, each strand provides a template on which a new complementary strand can be synthesized via the process called **replication** (Figure 20.36).

SAMPLE EXERCISE 20.4 **Using Base Complementarity in DNA**

If 31.6% of the nucleotides in a sample of DNA are adenine, what are the percentages of cytosine, guanine, and thymine?

Collect and Organize We know how much adenine is in a DNA sample and need to calculate the remainder of the nucleotide composition. Nucleotides are paired: A always pairs with T, and C always pairs with G.

Analyze Because of base-pairing, the percentage of T must equal the percentage of A, and the percentage of C must equal the percentage of G.

Solve If A = 31.6%, then T = 31.6%. This leaves (100 − 2 × 31.6) = 36.8% left to be equally distributed between G and C. Therefore G = C = 18.4%.

Think about It The percentages should total 100%, and they do.

Practice Exercise Indicate the sequence of the complementary strand on the double helix formed by each of these sequences of nucleotides:
    a. CGGTATCCGAT
    b. TTAAGCCGCTAG

(b)

replication the process by which one double-stranded DNA forms two new DNA molecules, each one containing one strand from the original molecule and one new strand.

Original DNA molecule

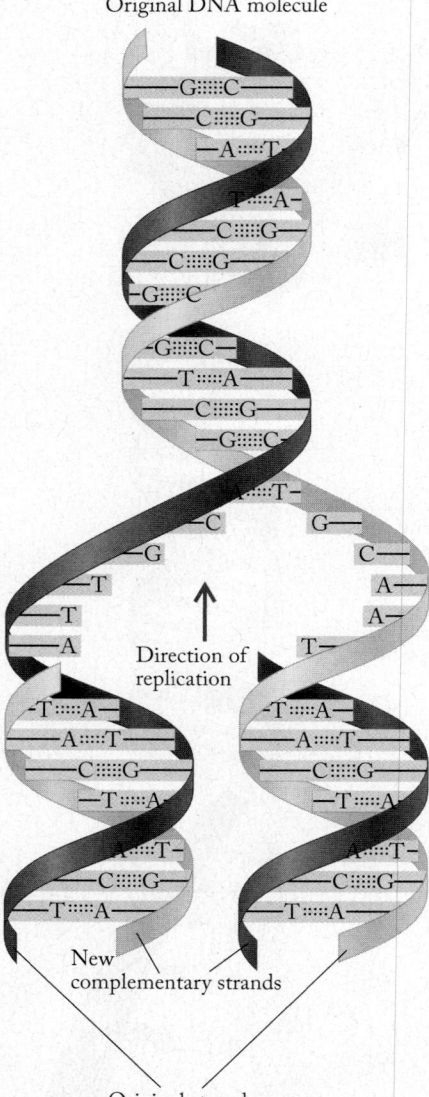

FIGURE 20.36 When DNA replicates, the two strands of a short portion of the double helix are unzipped, and each strand joins with a new complementary strand to produce two new DNA molecules.

....

transcription the process of copying the information in DNA to RNA.

messenger RNA (mRNA) the form of RNA that carries the code for synthesizing proteins from DNA to the site of protein synthesis in a cell.

codon a three-nucleotide sequence that codes for a specific amino acid.

translation the process of assembling proteins from the information encoded in RNA.

....

DNA's double-stranded structure is also the key to its ability to preserve genetic information. The two strands carry the same information, much like an old-fashioned photograph and its negative. Genetic information is duplicated every time a DNA molecule is replicated, a process that is essential whenever a cell divides into two new cells.

From DNA to New Proteins

Proteins are formed from amino acids in accordance with the *genetic code* contained in the base sequences of DNA strands. The bases A, T, G, and C are the alphabet in this code, and the "words" in the code are three-letter combinations of these four letters, with each word representing a particular amino acid or a signal to begin or end protein synthesis. Using four letters to write three-letter words means there are $4^3 = 64$ combinations possible, more than enough to encode for the 20 amino acids found in cells. The function of the genetic code is to specify the protein's primary structure—the sequence of amino acids in proteins.

Protein synthesis begins with a process called **transcription** (Figure 20.37a), in which double-stranded DNA unwinds and its genetic information guides the synthesis of a single strand of a molecule called **messenger RNA (mRNA)**. This strand of mRNA has the complementary base sequence of the original DNA. It carries the 3-letter words of that DNA, in the form of three-base sequences called **codons** (Table 20.3), from the nucleus of the cell into the cytoplasm, where the mRNA binds with a cellular structure called a ribosome. Keep in mind that a sequence of A, C, G, and T in the original DNA is transcribed into the following sequence in mRNA:

$$\text{DNA:} \quad \dots \text{ACGT} \dots$$
$$\text{mRNA:} \quad \dots \text{UGCA} \dots$$

At the ribosome the genetic information in the messenger RNA directs the synthesis of particular proteins in a process called **translation**. Another type of RNA, called **transfer RNA (tRNA)**, plays a key role in translation. There are 20 different forms of tRNA in the cell, one for each amino acid. To see how tRNA works, let's look at Figure 20.37(b). The first codon in this piece of an mRNA strand is AUG, which codes for the amino acid methionine (see Table 20.3). In the cytoplasm surrounding the ribosome, molecules of tRNA are reversibly bonded to molecules of every amino acid. The particular tRNA molecules that are

TABLE 20.3	**mRNA Codons**		
Ala	GCU, GCC, GCA, GCG	**Leu**	UUA, UUG, CUU, CUC, CUA, CUG
Arg	CGU, CGC, CGA, CGG, AGA, AGG	**Lys**	AAA, AAG
Asn	AAU, AAC	**Met**	AUG
Asp	GAU, GAC	**Phe**	UUU, UUC
Cys	UGU, UGC	**Pro**	CCU, CCC, CCA, CCG
Gln	CAA, CAG	**Ser**	UCU, UCC, UCA, UCG, AGU, AGC
Glu	GAA, GAG	**Thr**	ACU, ACC, ACA, ACG
Gly	GGU, GGC, GGA, GGG	**Trp**	UGG
His	CAU, CAC	**Tyr**	UAU, UAC
Ile	AUU, AUC, AUA	**Val**	GUU, GUC, GUA, GUG
Start	AUG	**Stop**	UAG, UGA, UAA

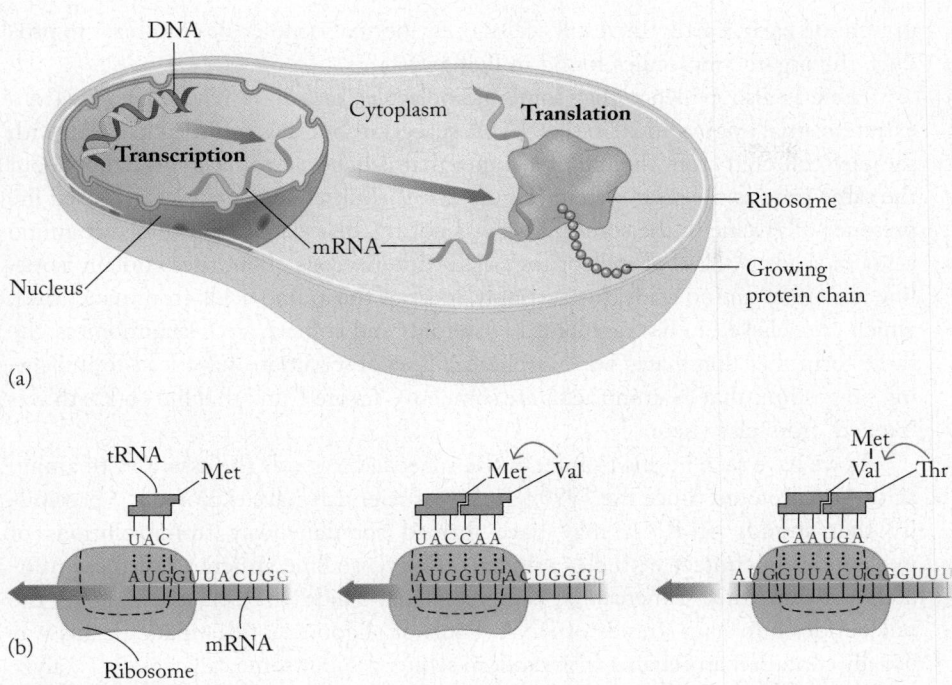

(a)

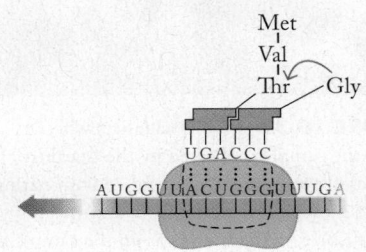

transfer RNA (tRNA) the form of the nucleic acid RNA that delivers amino acids, one at a time, to polypeptide chains being assembled by the ribosome–mRNA complex.

(b)

FIGURE 20.37 In protein synthesis, DNA is *transcribed* into mRNA, and the mRNA is then *translated* into proteins on the cell's ribosomes.

bonded to methionine also contain the sequence UAC, the complement of AUG, at a site that allows the tRNA to interact with mRNA. As Figure 20.37(b) shows, the segment of mRNA with the AUG codon links with the complementary strand on the tRNA molecule bonded to methionine. In doing so, the methionine is put into a position to unlink from the tRNA and to be the first amino acid residue in the protein being synthesized.

The sequence of events in the preceding paragraph is repeated many times. In this illustration the next codon, GUU, links up with a molecule of tRNA that has a CAA binding site and a molecule of valine in tow. In this way valine moves into position to become the next amino acid residue in the protein and to form a peptide bond with the N-terminal methionine. Valine is followed by threonine, which is followed by glycine, and so on until a Stop codon finally signals the end of the translation process.

CONCEPT TEST

If a GUU codon attracts valine to the translation site, does a UUG codon do the same thing? Explain your answer.

20.6 From Biomolecules to Living Cells

We end this chapter by addressing two fundamental questions about the major classes of biomolecules and their roles in sustaining life: (1) how were they first formed on prebiotic Earth, and (2) how did they assemble into living cells? Experiments conducted in the 1950s at the University of Chicago by Professor Harold Urey (1893–1981) and his student Stanley Miller (1930–2007) showed that amino acids could form from H_2O, CH_4, NH_3, and H_2 (Figure 20.38). Although the reactants the two scientists chose are now thought to be different from those

FIGURE 20.38 The apparatus used by Miller and Urey to simulate the synthesis of amino acids in the atmosphere of early (prebiotic) Earth. Discharges between the tungsten electrodes were meant to provide the sort of energy that might have come from lightning.

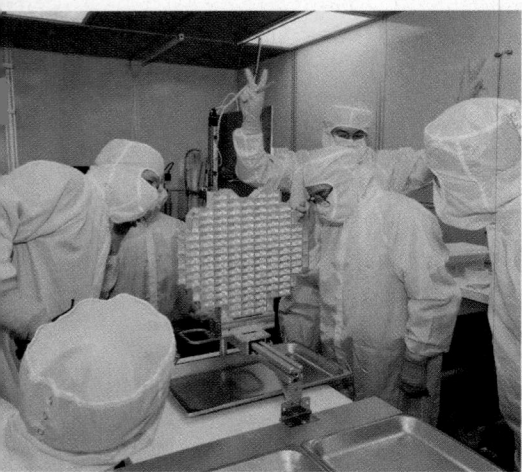

FIGURE 20.39 NASA scientists who analyzed the samples collected by the Stardust spacecraft were careful to avoid contaminating it with biological material from Earth. This meant isolating themselves from the sample as they prepared it for analysis.

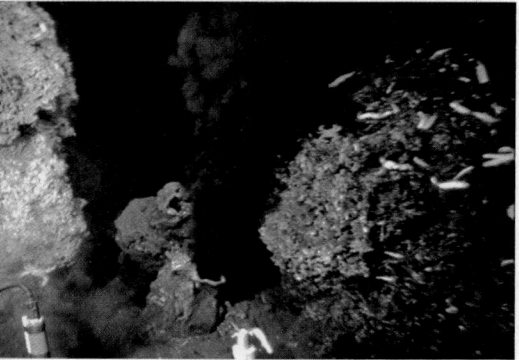

FIGURE 20.40 Black clouds of transition metal oxides and sulfides flow into the sea through chimneys like this one at a deep-ocean hydrothermal vent. Some scientists believe that these particles may have guided and catalyzed the formation of the first self-replicating molecules on Earth.

present on early Earth, the result still stands: inorganic molecules can react to produce the organic molecules found in living systems.

There is also evidence that some biomolecules may have reached Earth from extraterrestrial origins. In 2006 the NASA spacecraft Stardust returned to Earth with samples collected from the tail of a comet that is believed to have formed at about the same time as the solar system (Figure 20.39). Subsequent analyses disclosed the presence of glycine in the comet. This was not the first experiment to detect amino acids in space. A class of meteorites called carbonaceous chondrites contain isovaline and other amino acids. Interestingly, most of the amino acids from meteorites, which are believed to be fragments of asteroids and comets, are L-enantiomers, the same form that dominates our biosphere. These observations have lead to intriguing suggestions that L-amino acids are somehow "favored" and that life on Earth was "seeded" from elsewhere.

As we have seen in this chapter, RNA is needed to guide the assembly of amino acids into proteins. Since the 1990s, groups of scientists have explored the possibility that strands of RNA may have formed spontaneously from solutions of nucleotides in contact with clay minerals. The crystalline structures of these minerals provide three-dimensional templates that guide the self-assembly of the nucleotides into long strands of RNA. Pools of oligonucleotides made in this way usually contain many chains with random sequences, but some can actually catalyze their own replication. This ability to self-replicate is crucial to life, and the observation that molecules are capable of speeding up their own replication on a clay surface suggests that processes essential to the formation of living cells could have happened spontaneously. Much controversy still exists about these ideas, but the fact that RNA can act as both a source of information and a catalyst is part of the *RNA world hypothesis*. This hypothesis proposes that a world filled with life based on RNA predates the current world of life based on DNA and proteins. The capacity of RNA to both store information like DNA *and* act as a catalyst like an enzyme suggests that RNA alone could have supported cellular or precellular life-forms.

Current research is also testing the theory that life on Earth may have evolved near deep-ocean hydrothermal vents. Entire ecosystems have been discovered at these locations since they were first explored in the 1970s. They are sustained by geothermal and chemical energy rather than energy from the sun. It may be that life actually began in such environments with hydrothermal energy driving reactions in which inorganic compounds like carbon dioxide and hydrogen sulfide formed small organic compounds. As with the reactions on the surfaces of clay minerals, the synthesis reactions at hydrothermal vents may have been catalyzed and guided by reactants adsorbed on solid compounds such as FeS and MnO_2, which pour into the sea in dense black clouds near some vents (Figure 20.40). Among the known products of these reactions are acetate ions (CH_3COO^-). Acetate is a key intermediate in many biosynthetic pathways in living organisms. In modern bacteria, the systems that make acetate depend on a catalyst made of iron, nickel, and sulfur that has a structure much like that of particles produced by "black smokers" on the ocean floor.

To take the next step toward forming living cells, large biomolecules must have assembled themselves into even larger structures, such as membranes, that allow cells and structures within them to collect materials and retain them at concentrations different from those in the surrounding medium. Molecules in these assemblies are not necessarily connected by covalent bonds, but rather are held together by the intermolecular interactions we have discussed in this chapter.

SUMMARY

Section 20.1 The **proteins** and **peptides** in the human body are composed of 20 **α-amino acids** reversibly linked together by **peptide bonds**. At physiological pH (7.4), the amino acids exist as **zwitterions**.

Section 20.2 Protein molecules have four levels of structure. **Primary (1°) structure** is the amino acid sequence. **Secondary (2°) structure** is the shape the amino acid chain takes (**α helices, β-pleated sheets**, or **random coils**). **Tertiary (3°) structure** is the three-dimensional shape that results from attractive forces between amino acids located in various parts of the chain. **Quaternary (4°) structure** results when two or more proteins associate with each other to make larger functional units. Proteins called **enzymes** mediate the chemical reactions involved in metabolism. The structure of an enzyme contains a distinct region called an **active site** that binds to a **substrate**. The *induced-fit model* suggests that the binding of a substrate to an enzyme changes the shape of the enzyme so that the reaction can take place. **Biocatalysis** seeks to replicate the selectivity of enzymes in industrial settings.

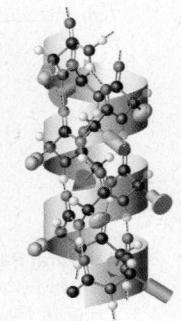

Section 20.3 Carbohydrates are produced from CO_2 and H_2O. **Monosaccharides** are joined into **polysaccharides** through **glycosidic bonds**. The isomers α-glucose and β-glucose form the polysaccharides starch and cellulose, respectively. Organisms derive energy from glycolysis and the **Krebs cycle**, the reaction pathways by which glucose is oxidized to CO_2 and H_2O.

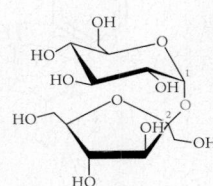

Section 20.4 Lipids are a major source of energy in our diet. The lipids known as **glycerides** are esters of the alcohol glycerol and up to three long-chain fatty acids. **Fats** are solid glycerides composed primarily of saturated fatty acids; **oils** are mostly unsaturated fatty acids and are liquids at room temperature. **Phospholipids** form **lipid bilayer** cell membranes.

Section 20.5 The **nucleic acids** DNA and RNA contain an organism's genetic information and control protein synthesis through **transcription** and **translation**. Genetic information is transmitted to new cells during cell division in a process called **replication**. DNA and RNA are made of chains of **nucleotides**, molecules composed of a five-carbon sugar, a phosphate group, and one of five nitrogen-containing organic bases. DNA consists of two nucleotide chains coiled into a double helix and connected via hydrogen bonds between complementary bases—A with T, and C with G. RNA, in which the base pairings are A with U and C with G, is a single-stranded nucleic acid involved in protein synthesis through **messenger RNA (mRNA)** and **transfer RNA (tRNA)**.

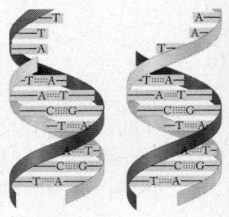

Section 20.6 Living cells may have formed as a result of chemical reactions in which inorganic molecules combined to form small organic molecules such as amino acids and nucleotides. The latter may have self-assembled, perhaps in polymerization reactions catalyzed by adsorption onto the surface of clay minerals or transition metal sulfides or oxides, into self-replicating biopolymers.

PROBLEM-SOLVING SUMMARY

TYPE OF PROBLEM	CONCEPTS AND EQUATIONS	SAMPLE EXERCISES
Interpreting acid–base titration curves of amino acids	At low pH all amino acids have at least two ionizable H atoms, one each from –COOH and $-NH_3^+$. Side-chain carboxylic acid and amine groups may also impart acidic and basic strength to amino acids.	20.1
Drawing and naming peptides	Connect the α-amine of one amino acid to the α-carboxylic acid of another with a peptide bond. Starting with the free amine (N-) terminus on the left, name each amino acid residue by changing the ending of the name of the parent amino acid to -*yl* in all but the last (C-terminal) amino acid.	20.2
Identifying triglycerides	The –COOH groups of fatty acids react with the –OH groups in glycerol to form triglycerides and water.	20.3
Using base complementarity in DNA	Identify the base pairs: A pairs with T; G pairs with C. The percentage of T should equal the percentage of A, and the percentage of C should equal the percentage of G.	20.4

VISUAL PROBLEMS ••• ■

(Answers to boldface end-of-chapter questions and problems are in the back of the book.)

20.1. The photochemical reaction of sodium hydrogen phosphite with formaldehyde is shown in Figure P20.1. It may have played a role in the formation of nucleic acids before life existed on Earth. Draw the Lewis structure for the hydrogen phosphite (HPO_3^{2-}) ion.

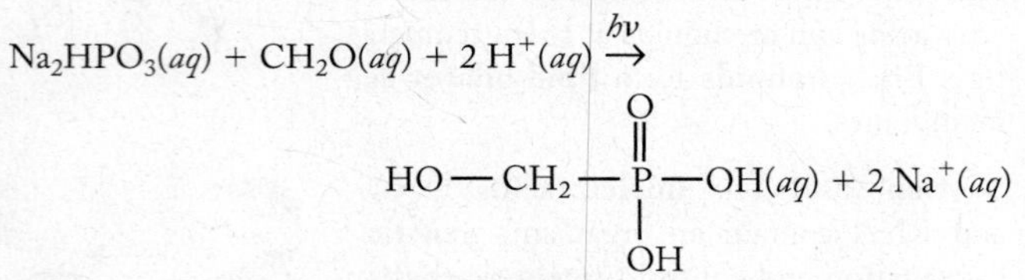

$$Na_2HPO_3(aq) + CH_2O(aq) + 2\,H^+(aq) \xrightarrow{h\nu}$$

FIGURE P20.1

20.2. The nucleotides in DNA contain the bases with the structures shown in Figure P20.2. Identify the basic functional groups in the structures.

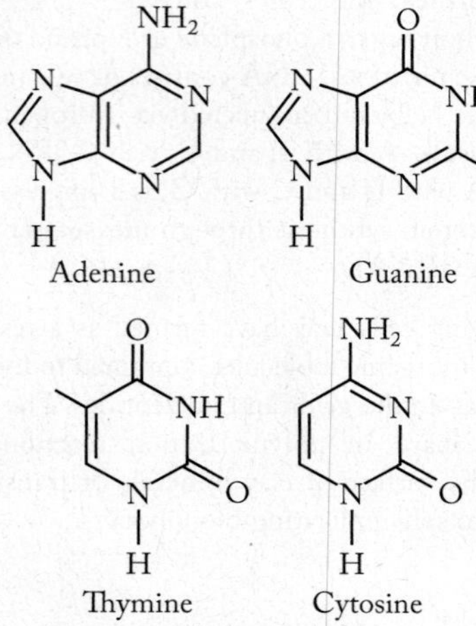

Adenine   Guanine

Thymine   Cytosine

FIGURE P20.2

20.3. Olive Oil Olive oil contains triglycerides such as those shown in Figure P20.3. Which of the fatty acids in these triglycerides is/are saturated?

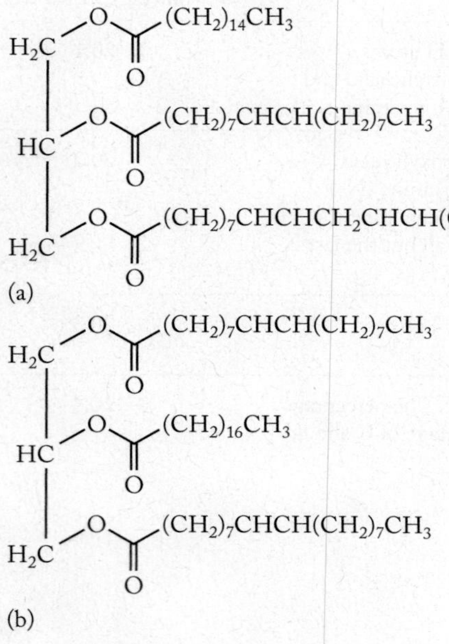

(a)

(b)

20.4. The two compounds shown in Figure P20.4 are both considered to be amino acids. Identify the structural difference between them and explain why they are both amino acids.

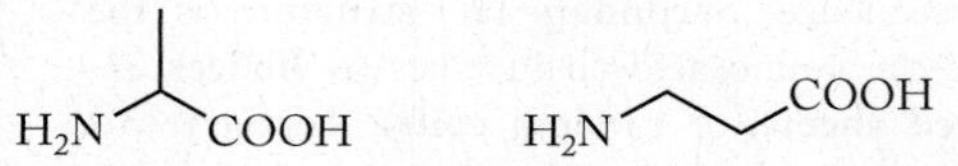

FIGURE P20.4

20.5. Natural Painkillers The human brain produces polypeptides called *endorphins* that help in controlling pain. The pentapeptide in Figure P20.5 is called enkephalin. Identify the five amino acids that make up enkephalin.

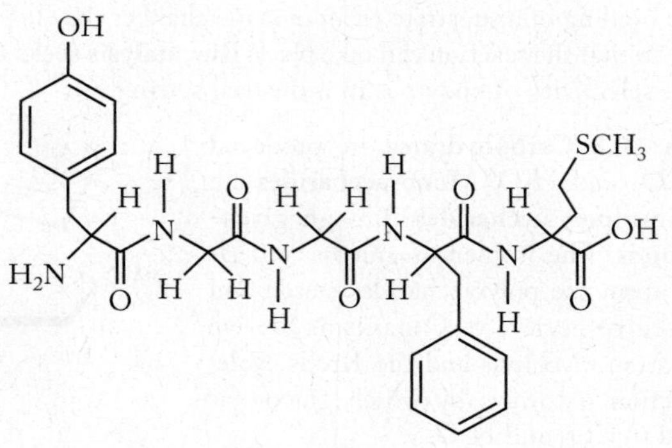

Enkephalin

FIGURE P20.5

20.6. Regulating Blood Pressure Angiotensin II is a polypeptide that regulates blood pressure. The structure of angiotensin II is shown in Figure P20.6. Which amino acids are in the structure?

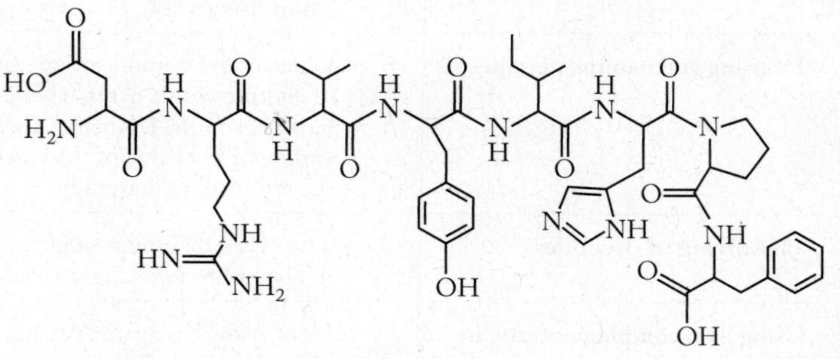

Angiotensin II

FIGURE P20.6

20.7. Trans Fats The role of "trans fats" in human health has been extensively debated both in the scientific community and in the popular press. What type of isomerism does the word "trans fat" refer to? Which of the molecules in Figure P20.7 are considered trans fats?

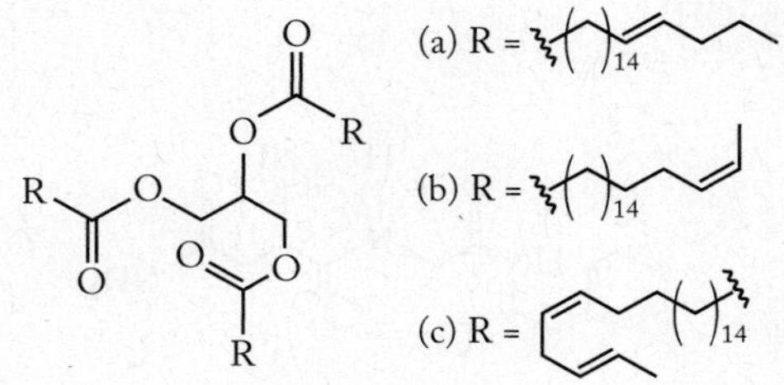

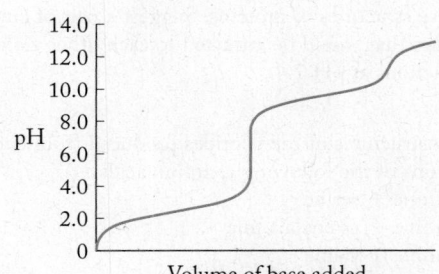

(a) R = ⌇⌇⟋⟍₁₄⟋⟍⟋

(b) R = ⌇⌇⟋⟍₁₄⟋⟍⟋

(c) R = ⟍⟋⟍⟋⟍₁₄⟋⟍⟋

FIGURE P20.7

20.8. Cocoa Butter Cocoa butter (Figure P20.8) is a key ingredient in chocolate. Cocoa butter is a triglyceride that results from esterification of glycerol with three different fatty acids. Identify the fatty acids produced by hydrolysis of cocoa butter.

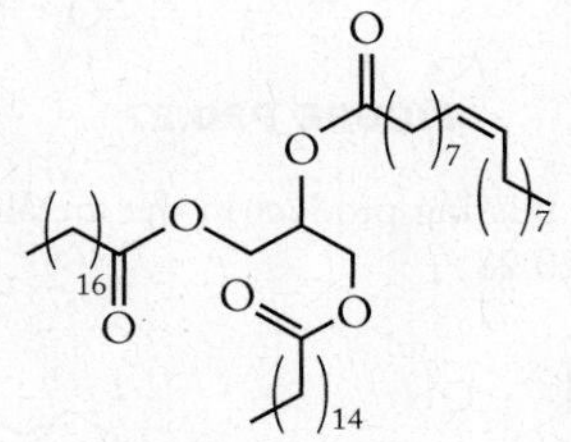

Cocoa butter

FIGURE P20.8

20.9. Sucralose The molecular structure of the artificial sweetner sucralose (trade name Splenda) is shown in Figure P20.9. Advertising for this product claims that it is made from sugar, implying that it is a natural product. What sugar might it be made from? Comment on the implication that it is a "natural" product.

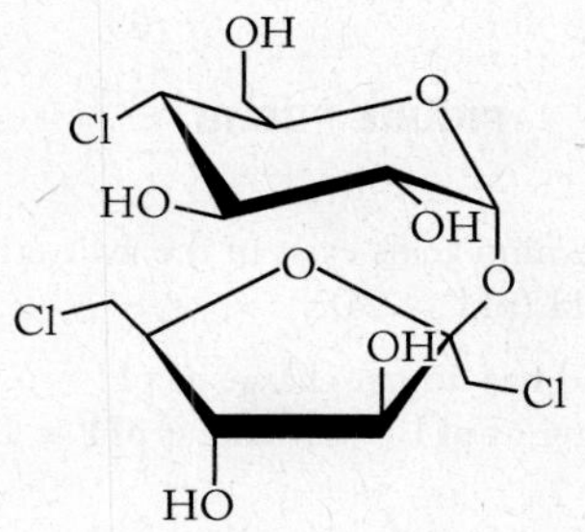

Sucralose

FIGURE P20.9

20.10. Figure P20.10 contains the titration curve of which of these amino acids: leucine, histidine, or lysine?

pH vs. Volume of base added

FIGURE P20.10

QUESTIONS AND PROBLEMS

The Composition of Proteins

CONCEPT REVIEW

20.11. In living cells, amino acids combine to make peptides and proteins. Are these processes accompanied by increases or decreases in entropy of the reaction system?

20.12. What is the difference between a peptide bond and an amide bond?

20.13. What does the alpha mean in α-amino acid?

20.14. In 1806 French scientists were the first to isolate an amino acid. The source was asparagus shoots. The compound had a negative charge in neutral aqueous solutions. Can you identify it?

20.15. Meteorites contain more L-amino acids, which are the forms that make up the proteins in our bodies, than D-amino acids. What do the prefixes L- and D- mean?

20.16. Do any of the amino acid in Table 20.1 have more than one chiral carbon atom per molecule?

20.17. Which of the compounds in Figure P20.17 is not an α-amino acid?

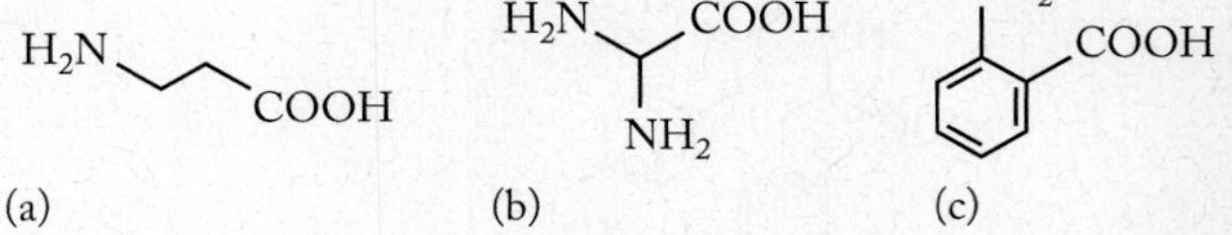

(a)   (b)   (c)

FIGURE P20.17

20.18. Which of the compounds in Figure P20.18 are α-amino acids?

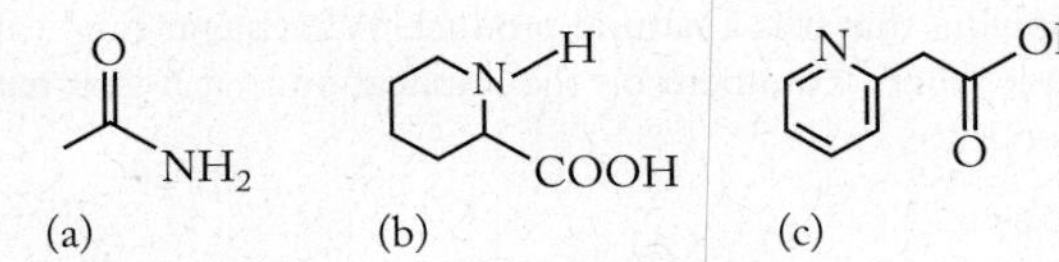

FIGURE P20.18

20.19. Why do most amino acids exist in the zwitterionic form at physiological pH (pH ≈ 7.4)?

20.20. If an amino acid has no net charge at pH = 7.4, what charge is it likely to have at pH = 2.3 and at pH = 10.3? Explain your answers.

Protein Structure and Function

CONCEPT REVIEW

20.21. When protein strands fold back on themselves in forming stable tertiary structures, lysine residues are often paired up with glutamic acid residues. Why?

20.22. Ion–ion interactions are particularly effective at stabilizing tertiary structures of proteins. Suggest a pair of amino acid residues that would be attracted to each other via ion–ion interactions at pH 7.4.

PROBLEMS

20.23. Draw structures of the peptides produced from condensation reactions of the following L-amino acids:
a. alanine + serine
b. alanine + phenylalanine
c. alanine + valine

20.24. Draw structures of the peptides produced from condensation reactions of the following L-amino acids:
a. methionine + alanine + glycine
b. methionine + valine + alanine
c. serine + glycine + tyrosine

20.25. Identify the amino acids in the dipeptides shown in Figure P20.25.

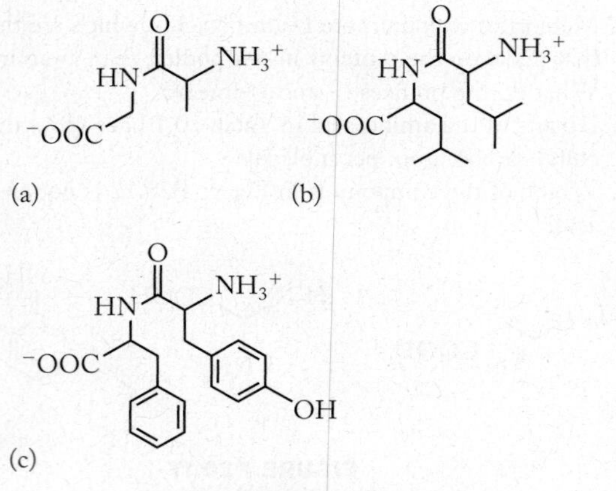

FIGURE P20.25

20.26. Identify the amino acids in the tripeptides in Figure P20.26.

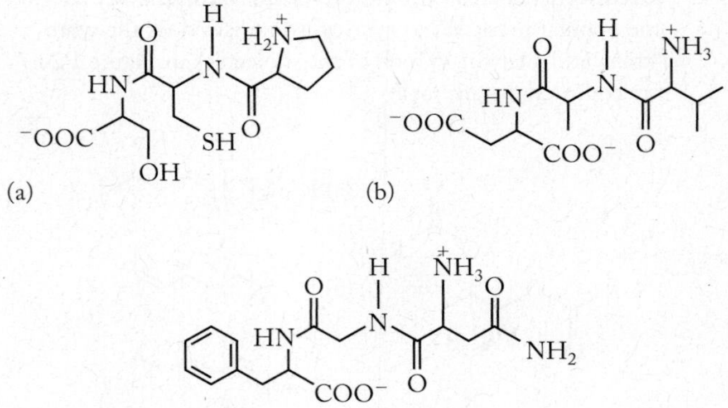

FIGURE P20.26

20.27. Identify the missing product in the metabolic reaction shown in Figure P20.27.

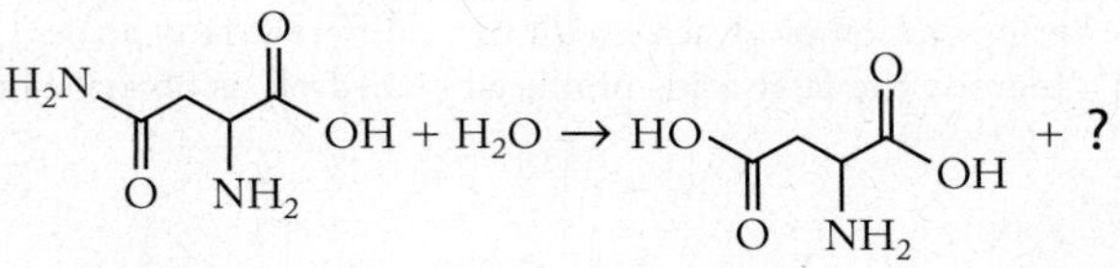

FIGURE P20.27

20.28. Identify the missing product in the metabolic reaction shown in Figure P20.28.

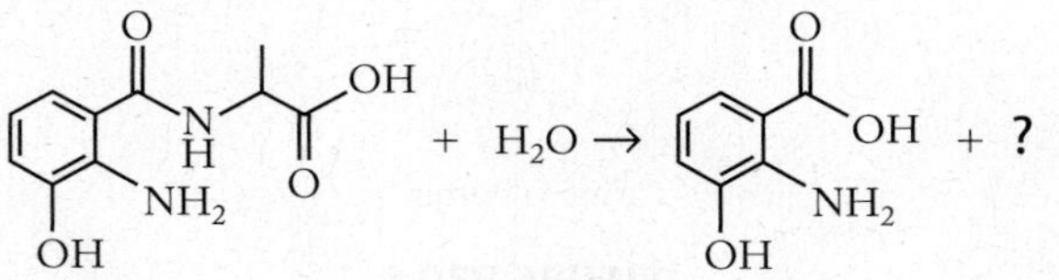

FIGURE P20.28

Carbohydrates

CONCEPT REVIEW

20.29. What are the structural differences between starch and cellulose?

20.30. Why is the discovery of enzymes that catalyze cellulose hydrolysis a worthwhile objective?

20.31. Is the fuel value of glucose in the linear form the same as that in the cyclic form?

*__20.32.__ Without doing the actual calculation, estimate the fuel values of glucose and starch by considering average bond energies. Do you predict the fuel values of the two substances to be the same or different?

20.33. The second step in glycolysis converts glucose 6-phosphate into fructose 6-phosphate. Can you think of a reason why $\Delta G°$ for this reaction is close to zero?

*__20.34.__ Why is it important that at least some of the steps in glycolysis convert ADP to ATP?

20.35. How do we calculate the overall free-energy change of a process consisting of two steps?

20.36. During glycolysis a monosaccharide is converted to pyruvate. Do you think this process produces an increase or decrease in the entropy of the system? Explain your answer.

PROBLEMS

20.37. Draw a diagram that shows how isomers of the sugar shown as a flat Lewis structure in Figure P20.37 result when the linear molecule forms a 6-atom ring.

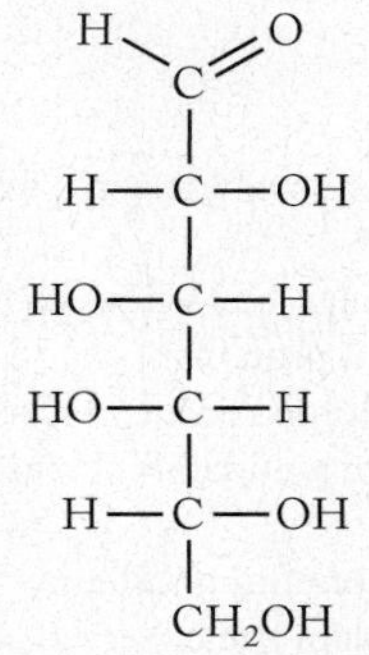

Galactose

FIGURE P20.37

20.38. Draw a diagram that shows how isomers of the sugar shown as a flat Lewis structure in Figure P20.38 result when the linear molecule forms a 5-atom ring.

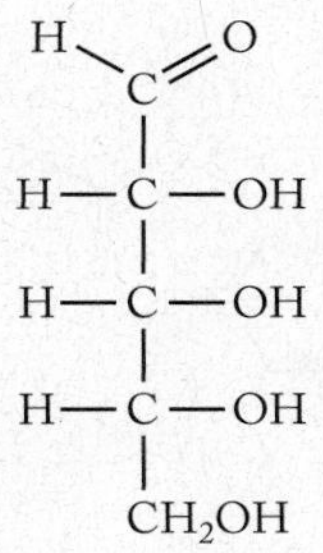

Ribose

FIGURE P20.38

20.39. Which, if any, of the structures in Figure P20.39 are β isomers of a monosaccharide?

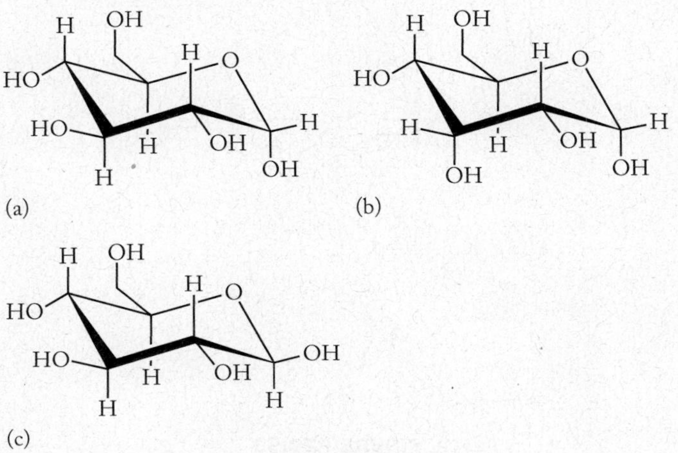

(a)   (b)

(c)

FIGURE P20.39

20.40. Identify which, if any, of the structures in Figure P20.40 are β isomers.

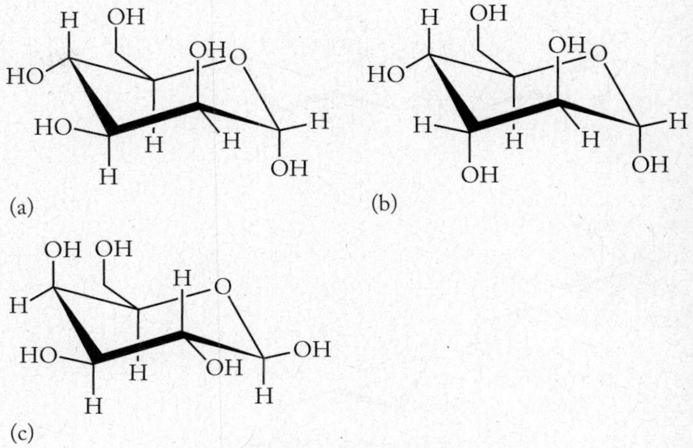

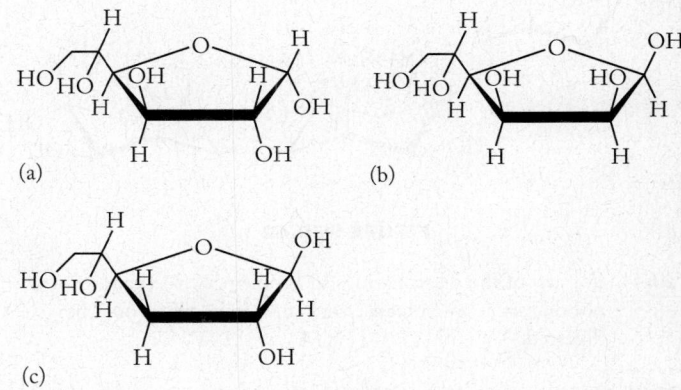

(a)   (b)

(c)

FIGURE P20.40

20.41. Which, if any, of the structures in Figure P20.41 are α isomers?

(a)   (b)

(c)

FIGURE P20.41

20.42. Which, if any, of the structures in Figure P20.42 are α isomers?

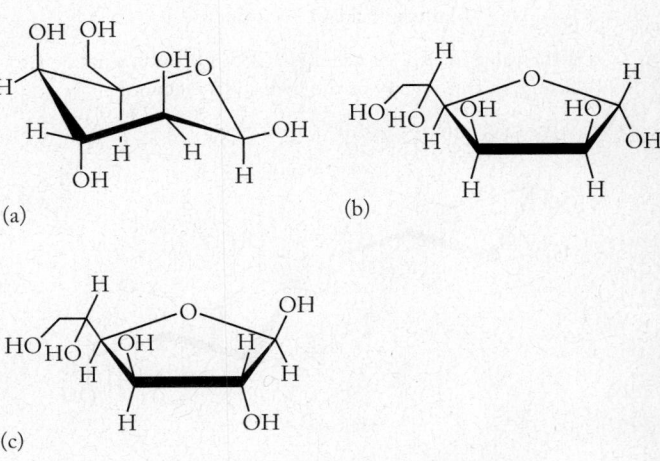

(a)   (b)

(c)

FIGURE P20.42

20.43. Which of the saccharides in Figure P20.43 is digestible by humans?

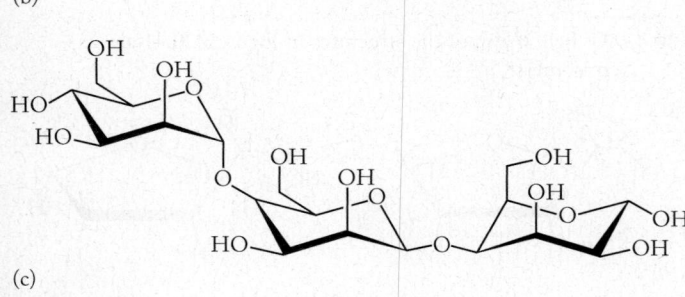

(a)

(b)

(c)

FIGURE P20.43

*20.44. For any of the disaccharides in Problem 20.43 that are not digestible by humans, draw an isomer that would be digested.

20.45. The structure of the disaccharide maltose appears in Figure P20.45. Hydrolysis of 1 mol of maltose ($\Delta G_f^\circ = -2246.6$ kJ/mol) produces 2 mol of glucose ($\Delta G_f^\circ = -1274.4$ kJ/mol):

$$\text{Maltose} + \text{H}_2\text{O} \rightarrow 2\,\text{glucose}$$

If the value of ΔG_f° of water is -285.8 kJ/mol, what is the change in free energy of the hydrolysis reaction?

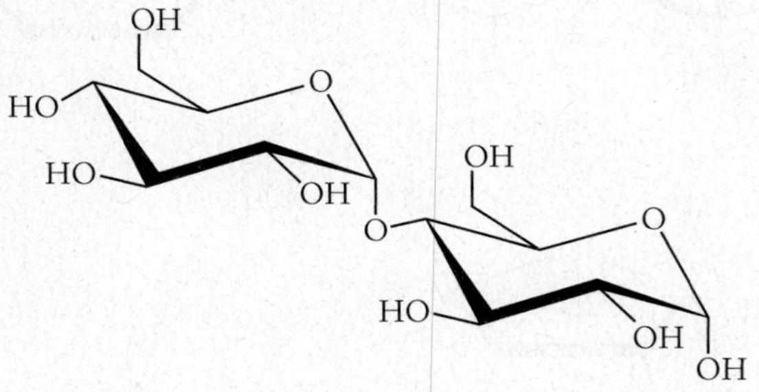

Maltose

FIGURE P20.45

20.46. If the maltose in Problem 20.45 were replaced by another disaccharide, would you expect the free-energy change for the hydrolysis to be exactly the same or just similar in value? Explain your answer.

Lipids

CONCEPT REVIEW

20.47. What is the difference between a saturated and an unsaturated fatty acid?

20.48. Why are the average fuel values of fats higher than those of carbohydrates and proteins?

20.49. Some Arctic explorers have eaten sticks of butter on their explorations. Give a nutritional reason for this unusual cuisine.

20.50. Salad dressings containing oil and vinegar quickly separate on standing. Explain the observed separation of layers based on the structure and properties of aqueous vinegar and oil.

20.51. Do triglycerides have a chiral center? Explain your answer.

*20.52. Using your knowledge of molecular geometry and intermolecular forces, why might polyunsaturated triglycerides be more likely to be liquid than saturated triglycerides?

PROBLEMS

20.53. Which of the triglycerides in Figure P20.53 are unsaturated fats?

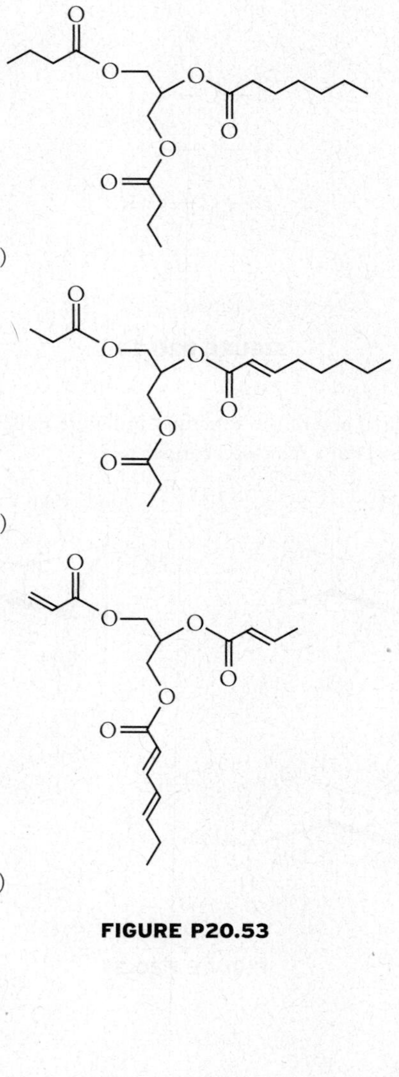

(a)

(b)

(c)

FIGURE P20.53

20.54. For each of the pairs of fatty acids in Figure P20.54, indicate whether they are structural isomers, geometric isomers, or unrelated compounds.

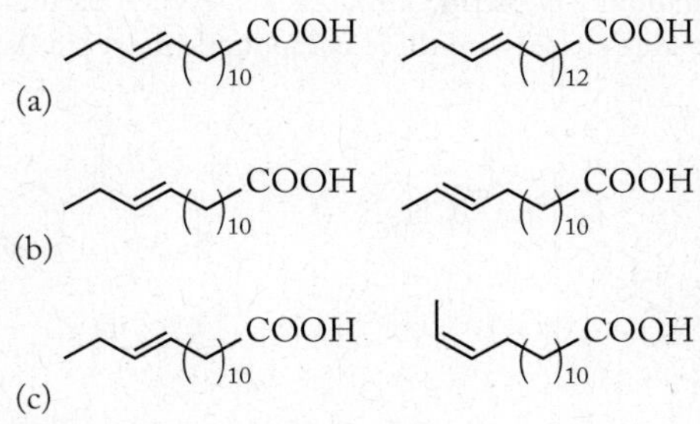

FIGURE P20.54

20.55. Draw the structures of the three fats formed by reaction of glycerol with (a) octanoic acid ($C_7H_{15}COOH$), (b) decanoic acid ($C_9H_{19}COOH$), and (c) dodecanoic acid ($C_{11}H_{23}COOH$).

20.56. **Oil-Based Paints** Oil-based paints contain linseed oil, a triglyceride formed by esterification of glycerol with linolenic acid (Figure P20.56).
  a. Draw the line structure of linolenic acid.
  *b. Are the double bonds in linolenic acid conjugated?

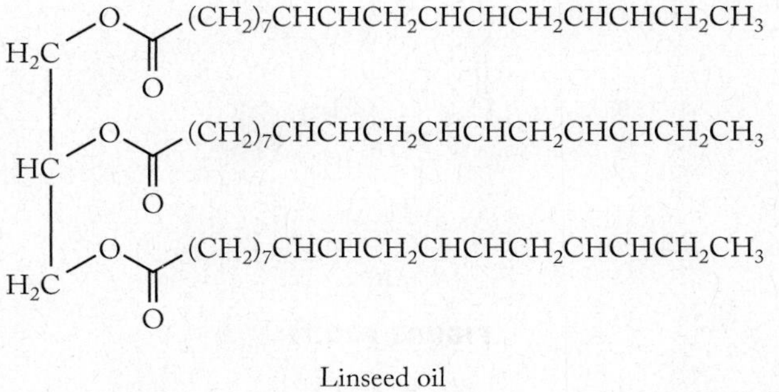

Linseed oil

FIGURE P20.56

Nucleotides and Nucleic Acids

CONCEPT REVIEW
20.57. What are the three kinds of molecular subunits in DNA? Which two form the "backbone" of DNA strands?

20.58. Why does a codon consist of a sequence of three, and not two, ribonucleotides?

20.59. What kind of intermolecular force holds together the strands of DNA in the double-helix configuration?

20.60. What is meant by "base-pairing" in DNA? Which bases are paired?

PROBLEMS
20.61. Draw the structure of adenosine 5′-monophosphate, one of the four ribonucleotides in a strand of RNA.

20.62. Draw the structure of deoxythymidine 5′-monophosphate, one of the four nucleotides in a strand of DNA.

20.63. In the replication of DNA, a segment of an original strand has the sequence T-C-G-G-T-A. What is the sequence of the double-stranded helix formed in replication?

20.64. In transcription, a segment of the strand of DNA that is transcribed has the sequence T-C-G-G-T-A. What is the corresponding sequence of nucleotides on the messenger RNA that is produced in transcription?

Additional Problems

20.65. Olestra is a calorie-free fat substitute. The core of the olestra molecule (Figure P20.65) is a disaccharide that has reacted with a carboxylic acid; this results in the conversion of hydroxyl groups on the disaccharide into the depicted structure.
  a. What is the name of the disaccharide core of the olestra molecule?
  b. What functional group has replaced the hydroxyl groups on the disaccharide?
  c. What is the formula of the carboxylic acid used to make olestra?

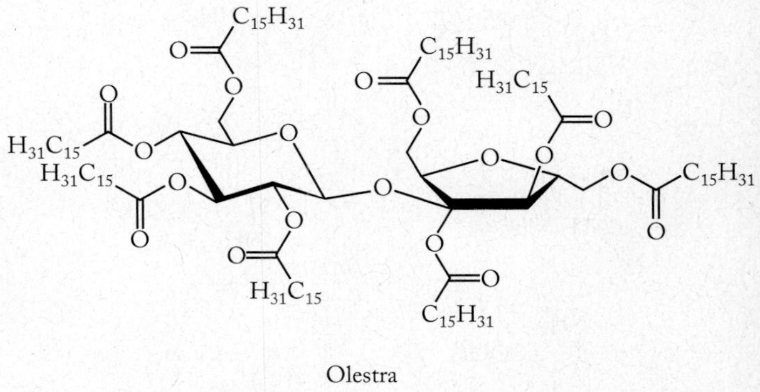

Olestra

FIGURE P20.65

20.66. When scientists at UC Santa Cruz directed UV radiation at an ice crystal containing methanol, ammonia, and hydrogen cyanide, three amino acids (glycine, alanine, and serine) were detected among the products of photochemical reactions. The formation of these amino acids suggests that they may also be synthesized in comets approaching the sun (and Earth). Determine the standard free-energy change of the hypothetical formation of glycine in comets, using standard free energies of formation of the reactants and products in this reaction [ΔG_f° for HCN(g) is +125 kJ/mol and for solid glycine is −368.4 kJ/mol; other ΔG_f° values are in Appendix 4].

$$CH_3OH(\ell) + HCN(g) + H_2O(\ell) \rightarrow$$
$$H_2NCH_2COOH(s) + H_2(g)$$

20.67. Homocysteine (Figure P20.67) is formed during the metabolism of amino acids. A mutation in some people's genes leads to high concentrations of homocysteine in the blood and a consequent increase in their risk of heart disease and incidence of bone fractures in old age.

a. What is the structural difference between homocysteine and cysteine?

b. Cysteine is a chiral compound. Is homocysteine chiral?

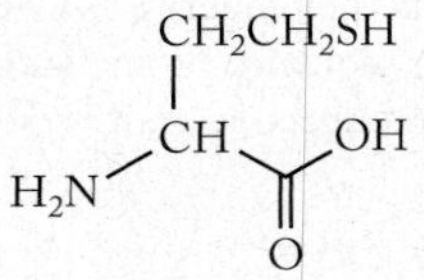

Homocysteine

FIGURE P20.67

20.68. Some scientists believe life on Earth can be traced to amino acids and other molecules brought to Earth by comets and meteorites. In 2004, a new class of amino acids called diamino acids (Figure P20.68) were found in the Murchison meteorite.

a. Which of these diamino acids is not an α-amino acid?

b. Which of these amino acids is chiral?

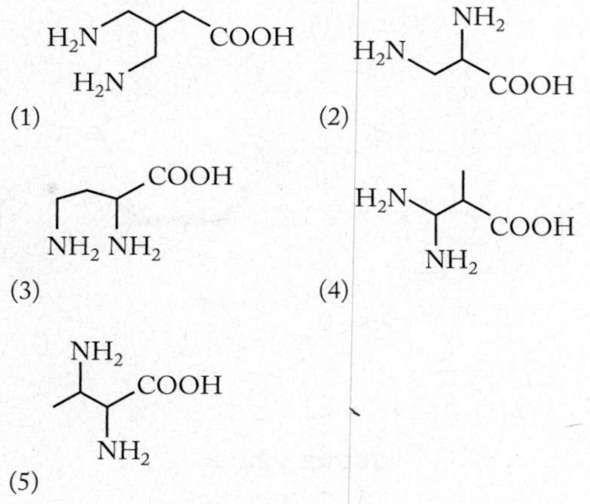

FIGURE P20.68

20.69. Ackee, the national fruit of Jamaica, is a staple in many Jamaican diets. Unfortunately, a potentially fatal sickness known as Jamaican vomiting disease is caused by the consumption of unripe ackee fruit, which contains the amino acid hypoglycin (Figure P20.69). Is hypoglycin an α-amino acid?

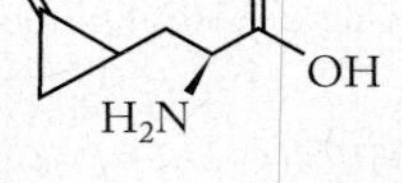

Hypoglycin

FIGURE P20.69

20.70. In late 2003, researchers at the Scripps Research Institute reported the development of genetically modified *E. coli* that could incorporate five new amino acids into proteins. These five amino acids, shown in Figure P20.70, are not among the 20 naturally occurring amino acids. Which naturally occurring amino acids are these compounds most similar to?

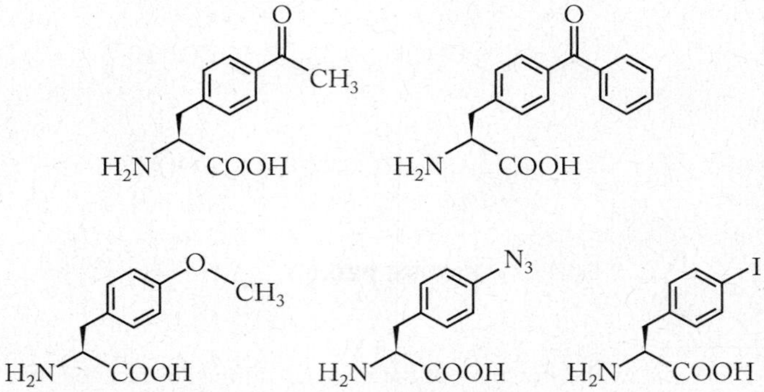

FIGURE P20.70

20.71. Creatine (Figure P20.71) is an amino acid produced by the human body. Body builders sometimes take creatine supplements to help gain muscle strength. A 2003 study reported that creatine may boost memory and cognitive thinking.

a. Is creatine an α-amino acid?

b. Draw the two dipeptides that can be formed from glycine and creatine.

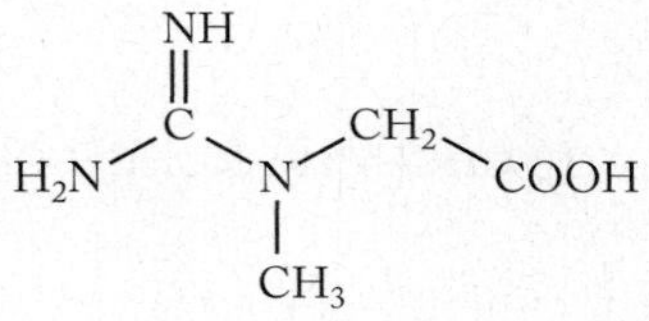

Creatine

FIGURE P20.71

20.72. In response to specific neural messages, the human hypothalamus may secrete a number of polypeptides including the tripeptide shown in Figure P20.72.

a. Sketch the structures of the three amino acids that combine to make this tripeptide.

b. Which, if any, of the constituent amino acids are among the 20 α-amino acids in proteins?

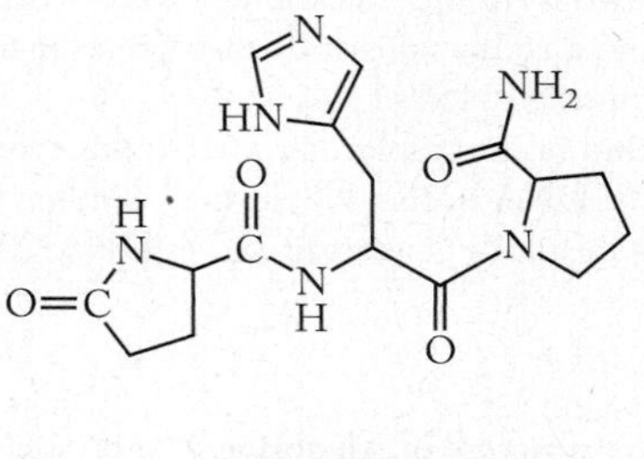

Thyrotropin-releasing factor

FIGURE P20.72

20.73. Glutathione (Figure P20.73) is an essential molecule in the human body. It acts as an activator for enzymes and protects lipids from oxidation. Which three amino acids combine to make glutathione?

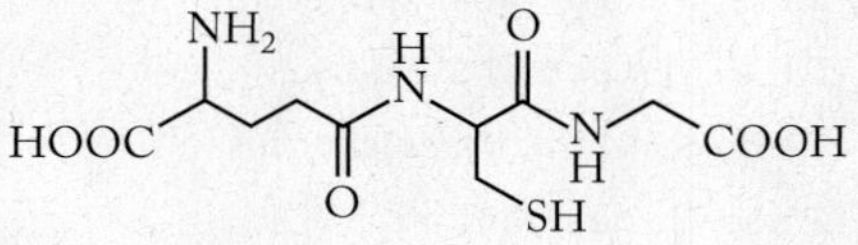

Glutathione

FIGURE P20.73

*20.74. Polyunsaturated fats have a tendency to polymerize. What kind of polymerization reaction might account for this observation?

*20.75. Without doing the actual calculation, estimate the fuel values of leucine and isoleucine by considering average bond energies. Should the fuel values of the two amino acids be the same? Actual calorimetric measurements show that isoleucine has a lower fuel value than leucine. Explain why.

20.76. Sucralose (see Figure P20.9) is about 600 times sweeter than sucrose (see Figure 20.22). All substances that taste sweet have functional groups that form hydrogen bonds with "sweetness" receptor sites on the tongue. What does the difference in sweetness between sucralose and sucrose tell you about additional intermolecular interactions between sweet compounds and receptor sites that contribute to their sweet taste?

Additional study materials including ChemTours and Diagnostic Quizzes are available at StudySpace at www.wwnorton.com/studyspace.

Nuclear Chemistry

A LOOK AHEAD

Back to the Beginning

I n Section 2.7 we described how the matter in the universe may have formed following an event we call the Big Bang. Scientists believe that the enormous energy released at the beginning of space and time rapidly transformed into particles of matter called quarks and that these quarks rapidly combined to form neutrons and protons. Presumably no free quarks have existed in nature since that time, and none have ever been produced or observed in a laboratory—until now.

On March 30, 2010, an international team of scientists working at the world's newest and largest particle accelerator, the Large Hadron Collider (LHC) near Geneva, Switzerland (see photo on p. 43), achieved a major scientific milestone: they crossed two streams of protons, each with over 3.5 trillion electron volts (TeV) of energy. And that was just for starters. The team hopes to accelerate an even higher-energy stream of 7-TeV protons around the LHC's 27 km underground track and collide it with another stream going just as fast in the opposite direction. In other experiments nuclei of heavy elements such as lead will be accelerated to speeds very near the speed of light and smashed together. By the time you read this, the results of these endeavors may well be known.

There is so much energy in the collisions of these heavy nuclei that they and the neutrons and protons inside them may be blown apart, producing small regions of 21st-century space that resemble the universe as it existed 13.7 billion years ago, a few microseconds after the Big Bang. The quarks inside protons and neutrons are held together by a fundamental force of nature called the *strong force*, which operates over only very short distances, such as those inside atoms. It is mediated by force "carriers" called *gluons*. By colliding high-speed nuclei, LHC scientists hope to produce plasmas of free quarks and gluons and that these particles will recombine in ways that lead to the

Learning Outcomes

- Write balanced equations describing radiochemical reactions

- Calculate the quantity of a radionuclide remaining after a defined decay time and the radiometric age of an object

- Calculate the energy released in a nuclear reaction from the masses of the products and reactants

- Calculate the binding energies of nuclei

- Predict the decay modes of radionuclides

- Describe nuclear fission and how it can be controlled

- Describe the dangers of exposure to nuclear radiation and its uses in medical imaging and treatment

Positron-Emission Tomography (PET) In nuclear medicine PET is a ▶ powerful imaging technique based on administering trace concentrations of radioactive nuclides that emit positrons—the antimatter version of electrons. The distribution of these compounds enables physicians to detect abnormal cell activity in, for example, this patient's lymph nodes.

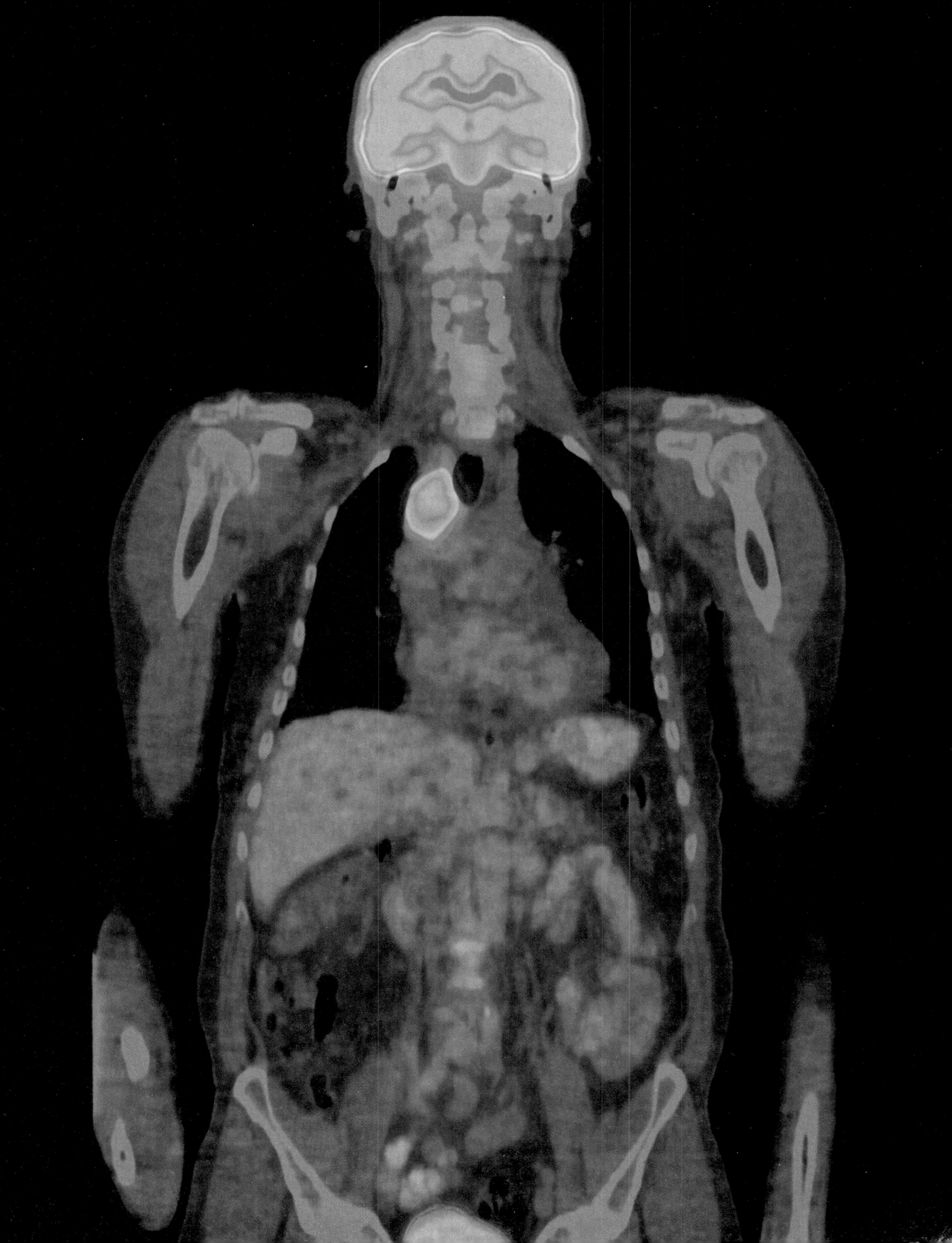

nuclear chemistry the study of reactions that involve changes in the nuclei of atoms.

radioactive decay the spontaneous disintegration of unstable particles accompanied by the release of radiation.

formation of other particles—ones that scientists believe may exist but that have never been found.

Among them is a particle called the *Higgs boson*. Theoretically this particle is a component of all other particles of matter, and it gives them their mass. It may play a key role in confining the quarks inside neutrons and protons, and it may account for the fact that the masses of quarks account for only about 1% of the mass of each neutron and proton. In their research LHC scientists are exploring the fundamental laws that govern the interaction of matter and energy at the atomic and subatomic level.

In this chapter we examine what is already known about the interactions inside nuclei, quantifying the changes in mass that accompany the release of enormous quantities of energy in nuclear reactions. We will shift our focus from the transformations of matter and energy that occurred at the moment of creation, and that may soon happen again under the Swiss–French border, and answer questions about nuclear processes occurring elsewhere on our own planet. How can the rates of decay of selected radionuclides be used to develop historical records of events over geological time and since the dawn of civilization? How are radionuclides used to diagnose and treat disease? How can the energy released by spontaneous nuclear processes be better used to meet our needs for clean energy? ■

21.1 Radioactive Decay

Throughout this book we have stressed that, in every chemical reaction, the identities of the atoms involved remain unchanged. Now we turn to *nuclear* reactions, in which atoms' identities *do* change—because their nuclei change. The field of chemistry that studies such reactions is called **nuclear chemistry**.

We introduced some of the basic principles of nuclear chemistry in Chapter 2 as we discussed the rapid transformation of energy into matter following the Big Bang. The proportions involved in this transformation are described by Einstein's famous equation

$$E = mc^2$$

where E is the amount of energy transformed into matter, m is the mass of that matter, and c is the speed of light. The subatomic particles that first formed were not all stable. Free neutrons, for example, spontaneously decompose to protons and electrons as described by the following nuclear equation:

$$_{0}^{1}\text{n} \rightarrow _{1}^{1}\text{p} + _{-1}^{0}\beta \qquad (21.1)$$

Recall from Section 2.7 that in any nuclear equation each subscript represents the electrical charge of a particle and the superscript represents its mass number. When the particle is the nucleus of an atom, then its electrical charge is equal to its number of protons, that is, its atomic number.

▶❙❙ **CHEMTOUR** Balancing Nuclear Equations

CONCEPT TEST

Is Equation 21.1 balanced? Explain why you think it is, or is not.

(Answers to Concept Tests are in the back of the book.)

The conversion of neutrons to protons is an example of **radioactive decay**, the spontaneous disintegration of unstable particles accompanied by the release of nuclear radiation. All radioactive decay processes follow first-order kinetics

(Section 15.3), and so each has a characteristic *half-life* ($t_{1/2}$), the time interval during which the quantity of radioactive particles decreases by one-half (Figure 21.1). The faster the decay process, the shorter the half-life.

The half-life of a free neutron is 12 min. In other words, 12 min after the Big Bang, half of the neutrons created at the instant of the Big Bang had decayed into protons and electrons via the process shown in Equation 21.1. To calculate what fraction of those neutrons remained at a time t after the Big Bang, we can convert this time into an equivalent number of half-lives n:

$$n = \frac{t}{t_{1/2}} \qquad (21.2)$$

Once we know the value of n, we can calculate the fraction of neutrons remaining at an instant t by writing the ratio of the number of neutrons at instant t (N_t) to the number present at $t = 0$ (N_0):

$$\frac{N_t}{N_0} = 0.5^n \qquad (21.3)$$

Actually, it is more convenient to relate quantities of radioactive particles directly to time by combining Equations 21.2 and 21.3:

$$\frac{N_t}{N_0} = 0.5^{t/t_{1/2}} \qquad (21.4)$$

In the following exercises and elsewhere in this chapter we will apply Equation 21.4 to various radioactive decay processes. We can do this because all of these processes follow first-order reaction kinetics.

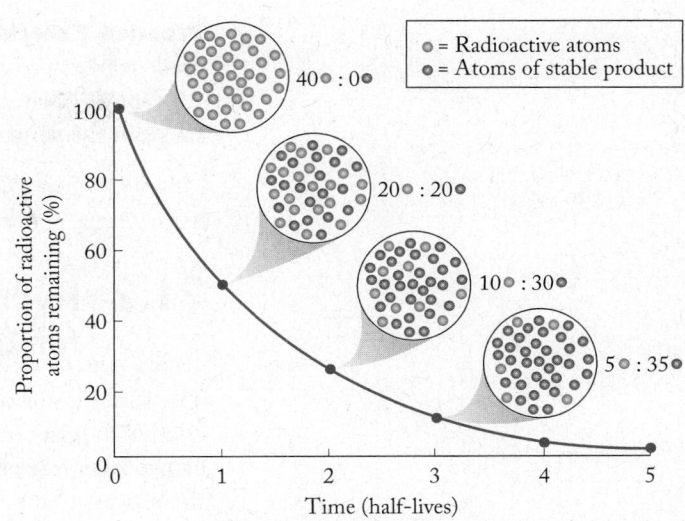

FIGURE 21.1 Radioactive decay follows first-order kinetics, which means, for example, that if a sample initially contains 40 radioactive atoms, it will contain only one-half that number after a time interval equal to one half-life. Half of the remaining half, or 10 radioactive atoms, remain after two half-lives, and so on.

▶❚❚ **CHEMTOUR** Half-Life

SAMPLE EXERCISE 21.1 **Calculations Involving Half-Lives**

Starting with a population of 6.6×10^5 free neutrons, how many remain after 2.0 min?

Collect and Organize The problem gives an initial quantity of free neutrons and asks how many are left after 2.0 minutes. Free neutrons are radioactive with a half-life of 12 minutes. Equation 21.4 relates quantities of radioactive particles to decay times.

Analyze In this problem the initial number of neutrons (N_0) is 6.6×10^5, $t = 2.0$ min, $t_{1/2} = 12$ min, and we need to solve for N_t. The value of t is only a fraction of $t_{1/2}$; far fewer than half of the initial number of neutrons will have decayed after 2.0 min.

Solve Substituting the values from Analyze into Equation 21.4 gives

$$\frac{N_t}{N_0} = 0.5^{t/t_{1/2}}$$

$$\frac{N_t}{6.6 \times 10^5} = 0.5^{2.0 \text{ min}/12 \text{ min}}$$

$$N_t = 5.9 \times 10^5$$

Think about It The value of N_t is reasonable because, as we predicted, only a small fraction of the initial quantity of free neutrons decayed in 2.0 minutes.

Practice Exercise Cesium-131 is a short-lived radionuclide ($t_{1/2}$ = 9.7 d) used to treat prostate cancer. How much therapeutic strength does a cesium-131 source lose over 60 days? Express your answer as a percentage of the strength the source had at the beginning of the first day.

(Answers to Practice Exercises are in the back of the book.)

21.2 Hydrogen Fusion and the Quest for Limitless Energy

▶❙❙ **CHEMTOUR** Fusion of Hydrogen

The sun is a star composed primarily of hydrogen, with a small amount of helium and much lesser quantities of other elements. Its energy is derived from the fusion of hydrogen nuclei to form helium. This fusion process involves more steps than the process that probably produced helium nuclei within minutes of the Big Bang. During primordial nucleosynthesis, protons (hydrogen nuclei) and neutrons fused together, forming deuterons:

$$_1^1p + {}_0^1n \rightarrow {}_1^2d$$

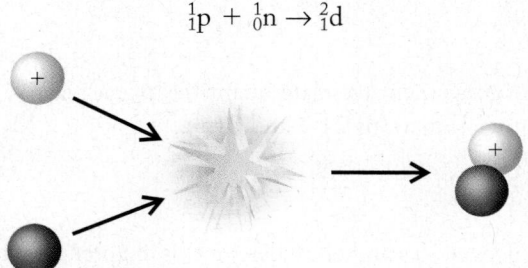

Then deuterons fused together to form alpha particles, which are the nuclei of helium-4 atoms:

$$_1^2d + {}_1^2d \rightarrow {}_2^4\alpha$$

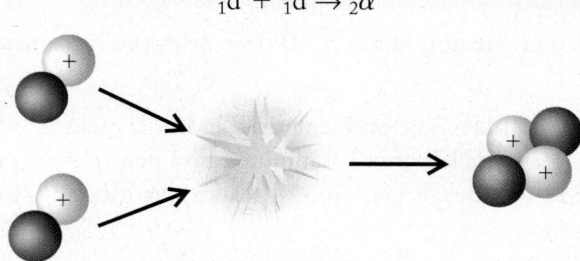

Hydrogen fusion in our sun follows a different pathway because free neutron concentrations are far lower there than they were in the primordial universe. In the sun two colliding protons may fuse together, forming a deuteron and a particle with hardly any mass and a charge of 1+:

$$_1^1p + {}_1^1p \rightarrow {}_1^2d + {}_1^0\beta \qquad (21.5)$$

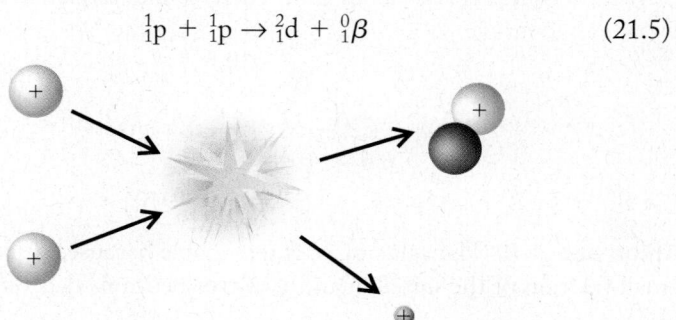

We use the electron (β particle) symbol "β" for this tiny positive particle, but the positive charge on it means that it is actually a "positive electron" or **positron**. Positrons belong to a group of subatomic particles that are the charge opposites of particles typically found in atoms. In addition to these electrons with positive charges, there are protons with negative charges, called *antiprotons*, $_{-1}^{1}p$. These charge opposites are particles of **antimatter**.

Particles of matter and their antimatter opposites are like mortal enemies. If they collide, they instantly annihilate each other. In their mutual destruction, they cease to exist as matter, and all of their mass is converted into energy in an amount predicted by Einstein's equation. The positrons produced from hydrogen fusion are rapidly annihilated in reactions with electrons. The sole product of the reaction is energy in the form of two or more gamma (γ) rays:

$$_{1}^{0}\beta + _{-1}^{0}\beta \rightarrow 2\,\gamma \tag{21.6}$$

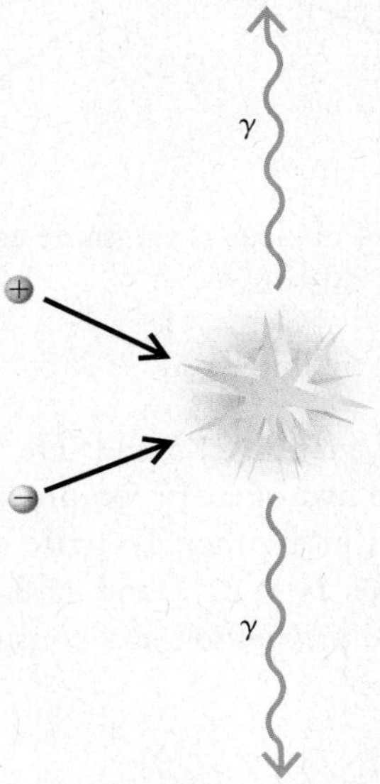

> **positron** a particle with the mass of an electron but with a positive charge.
>
> **antimatter** particles that are the charge opposites of normal subatomic particles.

Gamma ray emission accompanies all nuclear reactions. Gamma rays are generated by the nuclear furnaces of stars and so they permeate outer space. Those rays that reach Earth are absorbed by the gases in our atmosphere. In the process, molecular gases are broken up into their component atoms and atomic nuclei may be broken up into their subatomic particles. Radioactive sources that emit gamma rays are also used by oncologists to break up the molecules in, and kill, cancer cells.

CONCEPT TEST

What particle is formed when a proton fuses with an electron?

In the second stage of solar fusion, protons fuse with deuterons to form helium-3 nuclei:

$$_{1}^{2}d + _{1}^{1}p \rightarrow _{2}^{3}He \tag{21.7}$$

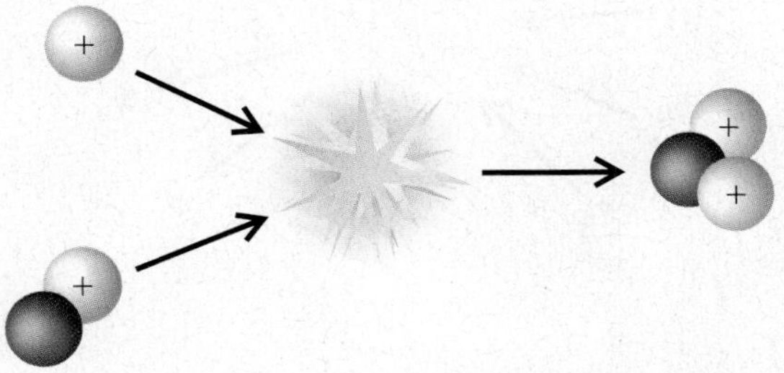

The superscript 3 indicates that the particle has 3 nucleons (2 protons and 1 neutron). Recall from Chapter 2 that any atom with 2 protons in its nucleus is by definition a helium atom and that atoms of the same element with different numbers of nucleons are called isotopes. Finally, fusion of two helium-3 nuclei produces a helium-4 nucleus and 2 protons:

$$2{}_{2}^{3}\text{He} \rightarrow {}_{2}^{4}\text{He} + 2{}_{1}^{1}\text{p} \qquad (21.8)$$

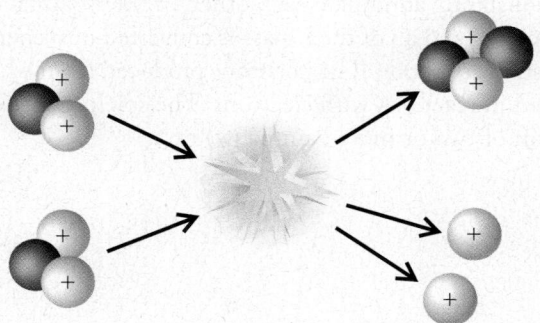

Keep in mind that a helium-4 nucleus is the same as an α particle. Therefore, we can also write Equation 21.8 this way:

$$2{}_{2}^{3}\text{He} \rightarrow {}_{2}^{4}\alpha + 2{}_{1}^{1}\text{p}$$

Deuterons produced in Equation 21.5 and ${}^{3}\text{He}$ nuclei produced in Equation 21.7 are intermediates in the hydrogen-fusion process because they are made in one step but then consumed in another. To write an overall equation for solar fusion, we combine Equations 21.5, 21.7, and 21.8, multiplying Equations 21.5 and 21.7 by 2 to balance the production and consumption of the intermediate particles:

$$2[{}_{1}^{1}\text{p} + {}_{1}^{1}\text{p} \rightarrow {}_{1}^{2}\text{d} + {}_{1}^{0}\beta]$$
$$+\ 2[{}_{1}^{1}\text{p} + {}_{1}^{2}\text{d} \rightarrow {}_{2}^{3}\text{He}]$$
$$+\ \ \ \ \ 2{}_{2}^{3}\text{He} \rightarrow {}_{2}^{4}\text{He} + 2{}_{1}^{1}\text{p}$$

$$\overline{4{}_{1}^{1}\text{p} + 2{}_{1}^{2}\cancel{\text{d}} + 2{}_{2}^{3}\cancel{\text{He}} \rightarrow 2{}_{1}^{2}\cancel{\text{d}} + 2{}_{2}^{3}\cancel{\text{He}} + {}_{2}^{4}\text{He} + 2{}_{1}^{1}\cancel{\text{p}} + 2{}_{1}^{0}\beta}$$

which reduces to

$$4{}_{1}^{1}\text{p} \rightarrow {}_{2}^{4}\text{He} + 2{}_{1}^{0}\beta \qquad (21.9)$$

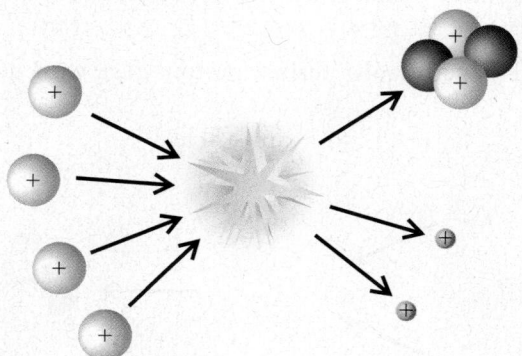

A proton is also a ^{1}H nucleus, so we can write the equation for the overall solar fusion reaction this way:

$$4\,{}^1_1\text{H} \rightarrow {}^4_2\text{He} + 2\,{}^0_1\beta$$

Annihilation reactions between the positrons produced in Equation 21.9 and electrons in the matter surrounding the reactants release considerable energy, but most of the energy from hydrogen fusion comes from the loss in mass as four protons are transformed into an alpha particle and two positrons (Table 21.1). In Sample Exercise 21.2 we will use Einstein's equation, $E = mc^2$, to calculate how much energy this is. For now, be assured that it is a lot.

TABLE 21.1 Symbols and Masses of Subatomic Particles and Small Nuclei

Particle	Symbol	Mass (kg)
Neutron	^1_0n	1.67493×10^{-27}
Proton	^1_1p or ^1_1H	1.67262×10^{-27}
Electron (β particle)	$^0_{-1}\beta$ or $^0_{-1}\text{e}$	9.10939×10^{-31}
Deuteron	^2_1d or ^2_1H	3.34370×10^{-27}
α particle	$^4_2\alpha$ or ^4_2He	6.64465×10^{-27}
Positron	$^0_1\beta$	9.10939×10^{-31}

Since the 1950s scientists and engineers have sought to harness the enormous energy released during hydrogen fusion for peaceful purposes. In 2009 construction began on the world's largest hydrogen fusion facility in southern France as part of an international collaboration that includes the European Union, Japan, India, China, Russia, and the United States. When it is operational the facility will use a device called a *tokamak* (Figure 21.2) to heat a mixture of deuterium ($^2_1$H) and tritium ($^3_1$H) to temperatures near 1.5×10^8 K. At such temperatures all these atoms are ionized, forming an incandescent plasma that is confined by the tokamak's powerful magnets to the center of a circular tunnel. High-speed collisions between deuterium and tritium nuclei in the plasma produce nuclei of helium-4:

$$^2_1\text{H} + {}^3_1\text{H} \rightarrow {}^4_2\text{He} + {}^1_0\text{n} \tag{21.10}$$

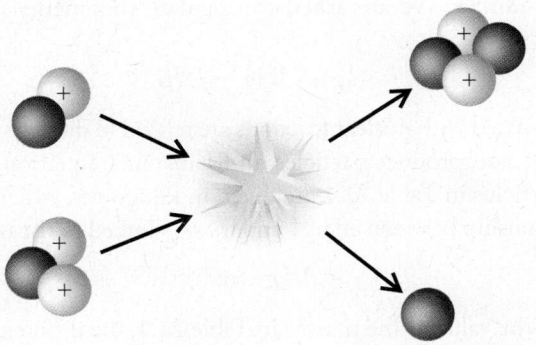

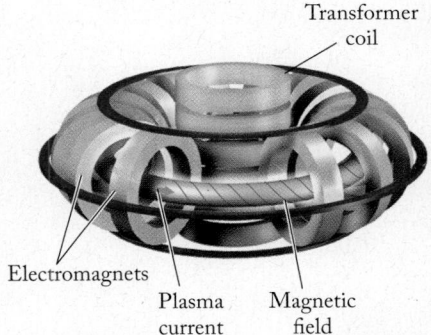

FIGURE 21.2 A tokamak transmits electrical energy into a toroidal (donut-shaped) chamber containing deuterium and tritium, causing these isotopes of hydrogen to ionize and form a plasma of nuclei and free electrons with temperatures above 10^8 K. Combinations of electromagnets confine the plasma to the interior of the torus, where collisions between ^{2}H and ^{3}H nuclei result in their fusing together, forming ^{4}He nuclei and free neutrons. The neutrons then collide with Li atoms in the walls of the chamber, initiating additional nuclear reactions that produce more tritium fuel.

The neutrons produced in the reaction collide with the nuclei of Li atoms in "breeder" blankets surrounding the hydrogen plasma. Two nuclear reactions are initiated by these collisions depending on which isotope of Li is involved:

$$\ce{^1_0n + ^6_3Li -> ^4_2He + ^3_1H} \qquad (21.11)$$

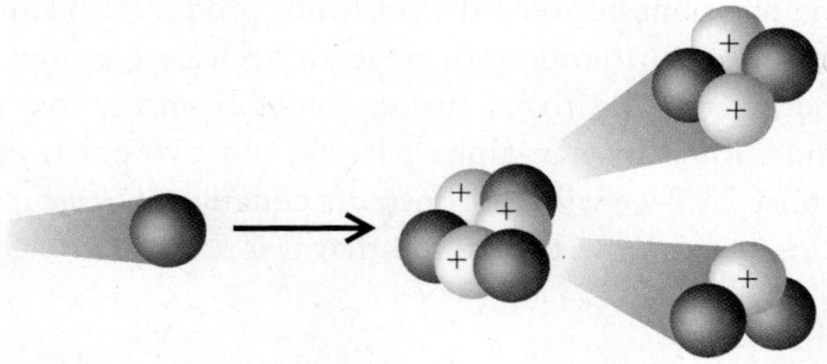

$$\ce{^1_0n + ^7_3Li -> ^4_2He + ^3_1H + ^1_0n} \qquad (21.12)$$

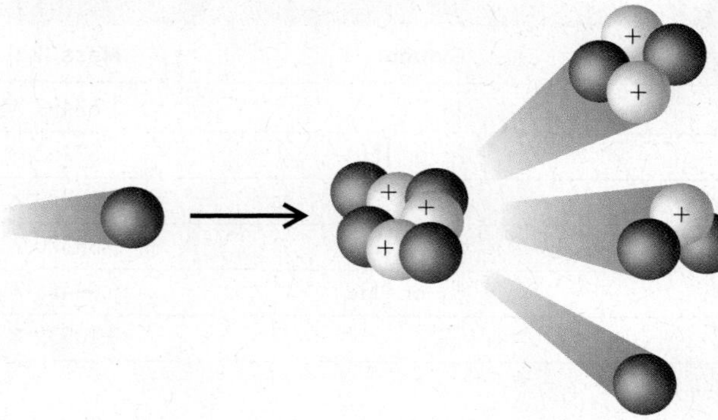

Note that the reactions in Equations 21.11 and 21.12 produce tritium nuclei. Thus, these reactions supply more fuel for the primary fusion reaction. The world's supply of the other fuel, deuterium, is enormous (seawater contains about 15 mg of deuterium per kilogram). On the other hand, tritium is not abundant in nature because it is radioactive with a half-life of only 12.3 years. One disadvantage of these reactions is their reliance on lithium during a time when expanding production of lithium–ion batteries (see Section 19.7) is increasing our demand for the element.

SAMPLE EXERCISE 21.2   Calculating the Energy Released in a Nuclear Reaction

How much energy in joules is released by the overall fusion process in which four protons undergo nuclear fusion, producing an α particle and two positrons (Equation 21.9)?

Collect and Organize We are asked to calculate the energy released in the nuclear reaction:

$$\ce{4^1_1p -> ^4_2He + 2^0_1\beta} \qquad (21.9)$$

The energies associated with nuclear reactions are related to differences in the masses of the reactant and product particles and Einstein's equation, $E = mc^2$. The masses of the particles in Table 21.1 are given in kilograms, which is convenient because the relationship between energy and mass is linked to the unit conversion:

$$1\,\text{J} = 1\,\text{kg} \cdot (\text{m/s})^2$$

Analyze Given the value of the masses in Table 21.1, the difference in mass will probably be less than 10^{-27} kg. When multiplied by the square of the speed of light $(2.998 \times 10^8 \text{ m/s})^2 \approx 10^{17}$, the calculated value of E should be less than 10^{-10} J.

Solve First we calculate the change in mass:

$$\Delta m = (m_{\alpha\ particle} + 2m_{positron}) - 4m_{proton}$$

$$= (6.64465 \times 10^{-27} + 2 \times 9.10939 \times 10^{-31})\ kg - 4 \times 1.67262 \times 10^{-27}\ kg$$

$$= -4.40081 \times 10^{-29}\ kg$$

The energy corresponding to this loss in mass is calculated using Einstein's equation where $m = -4.4008 \times 10^{-29}$ kg:

$$E = mc^2$$

$$= -4.4008 \times 10^{-29}\ kg \times (2.998 \times 10^8\ m/s)^2$$

$$= -3.955 \times 10^{-12}\ kg \cdot (m/s)^2 = -3.955 \times 10^{-12}\ J$$

Think about It The decrease in the mass translates into energy lost by the reaction system to its surroundings. As we predicted, the absolute value of this energy is less (actually much less) than 10^{-10} J, which seems like an awfully small value compared to the world's energy needs. However, this value applies to the formation of a single α particle. If we multiply it by Avogadro's number and convert to a value in kilojoules per mole, a unit we typically use in thermochemistry, we get

$$-3.955 \times 10^{-12}\ J \times \frac{6.022 \times 10^{23}}{1\ mol} \times \frac{1\ kJ}{1000\ J} = -2.382 \times 10^9\ kJ/mol$$

To put this value in perspective, it is about 10^7 times the change in free energy from the combustion of one mole of hydrogen gas.

Practice Exercise How much energy is released in the nuclear reaction described by Equation 21.10? Express your answer in kJ/mol. NOTE: The mass of a tritium nucleus is 5.00827×10^{-27} kg.

21.3 Nuclear Binding Energies

In the 1930s, scientists discovered that the mass of a stable nucleus is always less than the sum of the free masses (see Table 21.1) of its nucleons. For example, a helium-4 nucleus consists of 2 neutrons and 2 protons and has a mass of 6.64465×10^{-27} kg. The mass of a neutron is 1.67493×10^{-27} kg, and the mass of a proton is 1.67262×10^{-27} kg; therefore, the total mass of the nucleons is

$$\text{Mass of 2 neutrons} = 2(1.67493 \times 10^{-27}\ kg)$$
$$\underline{\text{Mass of 2 protons} = 2(1.67262 \times 10^{-27}\ kg)}$$
$$\text{Total mass} = 6.69510 \times 10^{-27}\ kg$$

The difference between this value and the mass of the ^{4}He nucleus is 5.045×10^{-29} kg. This difference is called the **mass defect (Δm)** of this nucleus. The energy equivalent to the mass defect represents the **binding energy (BE)** of the helium nucleus. This is the energy that would be released if free nucleons combined to form the nucleus. It is also the energy needed to split the nucleus into free nucleons.

Nucleons are held together in nuclei by the **strong force**. This is the force that also holds quarks together in neutrons and protons (Section 2.7). It operates over very small distances, but it is 10^2 times stronger than the coulombic repulsions the protons experience.

mass defect (Δm) the difference between the mass of a stable nucleus and the masses of the individual nucleons that comprise it.

binding energy (BE) the energy released when nucleons combine to form a nucleus.

strong force the fundamental force of nature that keeps quarks together in subatomic particles and nucleons together in atomic nuclei.

As in Sample Exercise 21.2, Einstein's equation gives us the energy/mass relationship we need to convert mass defect quantities into binding energies. Let's calculate the binding energy of an α particle ($^4_2$He nucleus) based on its mass defect of 5.045×10^{-29} kg:

$$
\begin{aligned}
\text{BE} &= (\Delta m)c^2 \\
&= 5.045 \times 10^{-29} \text{ kg} \times (2.998 \times 10^8 \text{ m/s})^2 \\
&= 4.534 \times 10^{-12} \text{ kg} \cdot (\text{m/s})^2 = 4.534 \times 10^{-12} \text{ J}
\end{aligned}
$$

As with the energies involved in hydrogen fusion, this value may seem very small, but it is for the formation of a single particle. It is equivalent to billions of kilojoules per mole.

Stable nuclei with larger numbers of nucleons than helium-4 have larger binding energies. This makes sense because a nucleus with many protons in close proximity to each other must be held together by an enormous energy to overcome the coulombic repulsion that all of these positively charged particles exert on each other. To make comparisons of nuclear binding energies meaningful for nuclei of varying sizes, the energies are usually divided by the number of nucleons in each nucleus.

Expressing binding energy on a per nucleon basis allows us to compare the relative stabilities of different nuclides. When we plot binding energy per nucleon values against atomic number we get the curve in Figure 21.3. Note that these values reach a maximum with ^{56}Fe. This pattern means that ^{56}Fe is the most stable nuclide in the universe, which fits what we know about the nuclear processes that synthesize the elements in the cores of giant stars. Recall from our discussion of stellar nucleosynthesis in Chapter 2 that the nuclear furnaces of giant stars are fueled by the energy released when lighter nuclei fuse together to form heavier nuclei, but only if the heavier ones are more stable than the lighter ones. Energy is not released if one of the reacting particles is ^{56}Fe or a heavier nuclide, because, according to Figure 21.3, an even heavier fusion product would be *less stable*, not more stable. Therefore, the fusion process would not release energy. This is why an aged giant star whose core has turned into iron as a result of multiple nuclear fusion processes has essentially run out of fuel. Without the heat of its nuclear furnace, the enormous gravity of the cooling star forces it to collapse into itself. The result is rapid heating as the star's mass is compressed, and a gigantic explosion called a supernova.

CONNECTION The lighter elements from helium to iron are synthesized by fusion reactions in the cores of giant stars. Heavier elements are formed and dispersed throughout galaxies by a combination of neutron capture and β decay during the supernova events as described in Section 2.7.

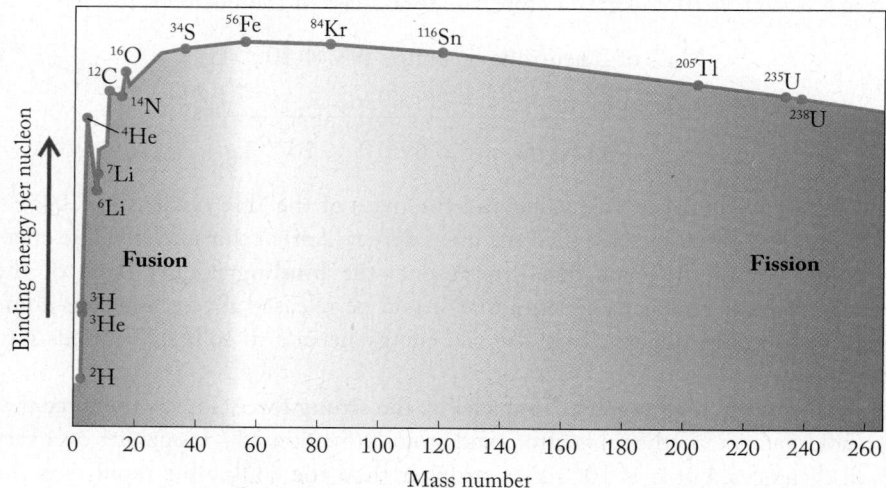

FIGURE 21.3 The stability of a nucleus is directly proportional to its binding energy per nucleon. For all nuclides up to ^{56}Fe, fusion reactions lead to products that have greater binding energy per nucleon than the reactants. This means the fusion reactions release energy. However, fusion of nuclei with atomic numbers greater than 26 produces nuclei that have less binding energy per nucleon than the reactants, and so consumes energy.

Why does a negative change in mass when nucleons combine to form a nucleus produces a positive binding energy?

belt of stability the region on a graph of number-of-neutrons versus number-of-protons that includes all stable nuclei.

21.4 The Belt of Stability

The values of the atomic masses and mass numbers of the elements in the periodic table tell us about the ratios of neutrons to protons in the nuclei of their stable isotopes. The lighter elements have atomic masses that are about twice their atomic numbers and have neutron-to-proton ratios close to unity. For example, ^{12}C has 6 neutrons and 6 protons, and most oxygen atoms have 8 neutrons and 8 protons.

However, with increasing values of Z, the ratios of neutrons to protons increase. This trend is illustrated in Figure 21.4, where the green dots represent combinations of neutrons and protons that form stable nuclides. The band of green dots runs diagonally through the graph, defining the **belt of stability**.

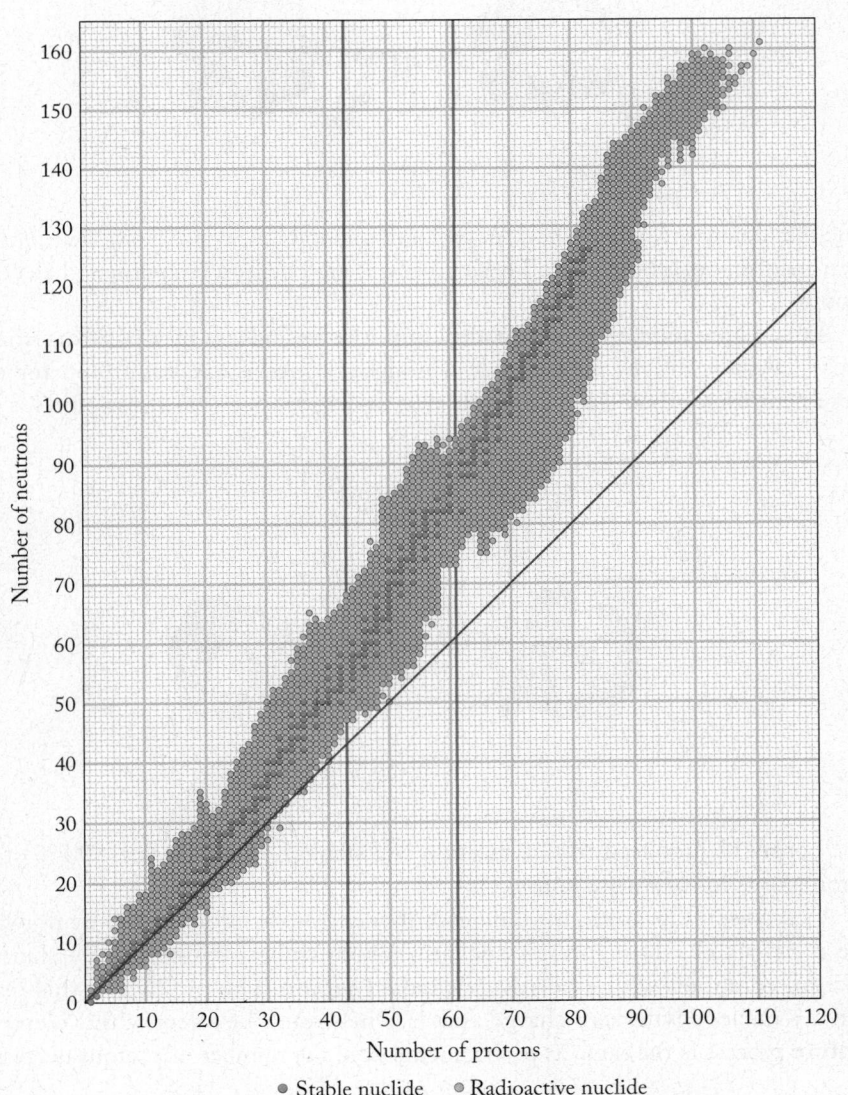

● Stable nuclide ● Radioactive nuclide

FIGURE 21.4 The belt of stability. Green dots represent stable combinations of protons and neutrons. Orange dots represent known radioactive (unstable) nuclides. Nuclides that fall along the purple line have equal numbers of neutrons and protons. Note that there are no stable nuclides (no green dots) for $Z = 43$ (technetium) and $Z = 61$ (promethium). These elements are the only two among the first 83 that are not found in nature.

Note how the belt curves upward away from the purple (n = p) straight line. This curvature shows how the neutron-to-proton ratios increase from about 1:1 for the lightest stable nuclides to about 1.5:1 for the most massive ones.

Nuclides, represented by the orange dots in Figure 21.4, are outside the belt of stability and are radioactive; that is, they are **radionuclides**. Their mode of radioactive decay depends on whether they are above or below the belt of stability. Those above the belt of stability, such as carbon-14, are *neutron rich* and tend to undergo decay reactions that reduce their neutron-proton ratio, i.e., β decay. For example, when ^{14}C undergoes β decay, its atomic number increases by one while its mass number stays the same. This is the result of a neutron in its nucleus becoming a proton according to this nuclear reaction:

$$^{14}_{6}\text{C} \rightarrow \, ^{14}_{7}\text{N} + \, ^{0}_{-1}\beta$$

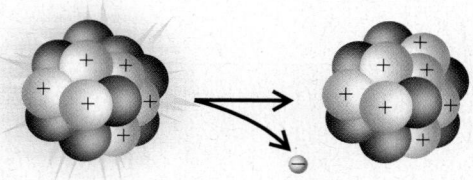

Nuclides below the belt are *neutron poor* and undergo decay processes that *increase* their neutron-to-proton ratio. These processes are positron emission and electron capture.

As its name implies, **positron emission** involves release of a positron from a nucleus. As a result, the nucleus that is produced contains one less proton and one more neutron. Carbon-11 is a nuclide that undergoes positron emission:

$$^{11}_{6}\text{C} \rightarrow \, ^{11}_{5}\text{B} + \, ^{0}_{1}\beta$$

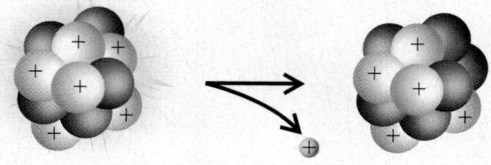

radionuclide an unstable nuclide that undergoes radioactive decay.

positron emission the spontaneous emission of a positron from a proton-rich nucleus.

electron capture a neutron-poor nucleus draws in one of its surrounding electrons, which transforms a proton in the nucleus into a neutron.

The boron-11 produced in this reaction is a stable isotope. In fact, 80.2% of all boron atoms in nature are boron-11.

A carbon-11 atom can also increase its neutron-to-proton ratio by drawing one of its six electrons into its nucleus. When it does, the negatively charged electron combines with a positively charged proton. The product of this reaction is a nucleon with zero charge, that is, a neutron. The effect of this **electron capture** process is the same as positron emission: the number of protons decreases by one and the number of neutrons increases by one, so that again the product is boron-11.

$$^{11}_{6}\text{C} + ^{0}_{-1}\text{e} \rightarrow ^{11}_{5}\text{B}$$

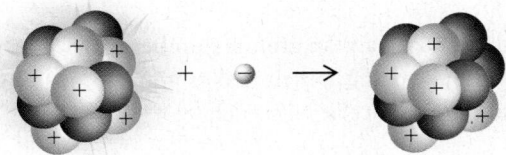

Table 21.2 illustrates the impact of being neutron rich or neutron poor on some isotopes of carbon. Note that carbon has two stable isotopes, ^{12}C and ^{13}C. The isotopes with mass numbers greater than 13 are neutron rich and undergo beta decay. Those with mass numbers less than 12 are neutron poor and undergo either positron emission or electron capture.

TABLE 21.2		**Isotopes of Carbon and Their Radioactive Decay Products**			
Name	**Symbol**	**Mass (amu)**	**Mode(s) of Decay**	**Half-Life**	**Natural Abundance (%)**
Carbon-10	$^{10}_{6}\text{C}$		Positron emission	19.45 s	
Carbon-11	$^{11}_{6}\text{C}$		Positron emission, electron capture	20.3 min	
Carbon-12	$^{12}_{6}\text{C}$	12.00000		(Stable)	98.89
Carbon-13	$^{13}_{6}\text{C}$	13.00335		(Stable)	1.11
Carbon-14	$^{14}_{6}\text{C}$		β decay	5730 yr	
Carbon-15	$^{15}_{6}\text{C}$		β decay	2.4 s	
Carbon-16	$^{16}_{6}\text{C}$		β decay	0.74 s	

CONCEPT TEST •••

Manganese has only one stable isotope. Predict which of the following isotopes are (a) radioactive and undergo electron capture decay, (b) radioactive and undergo β decay, or (c) stable: ^{51}Mn, ^{53}Mn, ^{55}Mn, ^{57}Mn.
••

SAMPLE EXERCISE 21.3   Predicting the Modes and Products of Radioactive Decay

Predict the mode of radioactive decay of ^{32}P, which is one of the most widely used radionuclides in biomedical research and treatment. Identify the nuclide that is produced in the decay process.

Collect and Organize We are to predict the mode of decay of a radionuclide, which depends on whether it is neutron rich or neutron poor. Phosphorus-32 has 17 neutrons and 15 protons per nucleus. According to the belt of stability (Figure 21.5), phosphorus ($Z = 15$) has only one stable nuclide, which has 16 neutrons and a mass number of 31.

Analyze Phosphorus-32 is represented by the orange dot directly above the ^{31}P green dot, which means that it is (1) radioactive and (2) neutron rich.

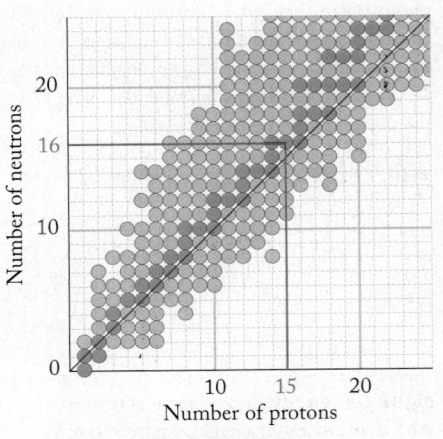

FIGURE 21.5

Solve Neutron-rich radioisotopes of the lighter elements undergo beta decay. Making a beta particle one of the products of the decay reaction gives the incomplete nuclear equation

$$^{32}_{15}P \rightarrow ? + {}^{0}_{-1}\beta$$

The missing product must have an atomic number of 16 (so that the subscripts on the right side add up to 15), which makes it an isotope of S. Its mass number must be 32, so the product is sulfur-32:

$$^{32}_{15}P \rightarrow {}^{32}_{16}S + {}^{0}_{-1}\beta$$

Think about It By emitting a beta particle, the nucleus increased its number of protons by one and decreased its number of neutrons by one, thereby reducing its neutron "richness" and forming a stable isotope of sulfur.

Practice Exercise What is the mode of radioactive decay of ^{28}Al? Identify the nuclide produced by the decay process.

All known nuclides with more than 83 protons are radioactive. Because there is no stable reference point in the pattern of green dots in Figure 21.4, it is hard to say whether any given $Z > 83$ nuclide is neutron rich or neutron poor. We can make one general statement, though: these most massive nuclides tend to undergo either β decay or **α decay**. In the latter process they produce nuclides with two fewer protons and two fewer neutrons, as in the case of uranium-238:

$$^{238}_{92}U \rightarrow {}^{234}_{90}Th + {}^{4}_{2}\alpha \qquad (21.14)$$

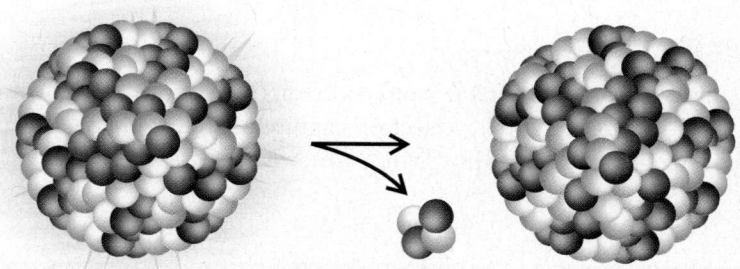

Figure 21.6 summarizes the changes in atomic number and mass number caused by the various modes of decay.

For the most massive radioactive isotopes, one radioactive decay process often leads to another in what is referred to as a *radioactive decay series*. Consider, for example, the decay series that begins with the α decay of ^{238}U to ^{234}Th (Figure 21.7). Thorium has no stable isotopes and undergoes two β decay steps to produce ^{234}U. In a series of subsequent alpha decay steps, the ^{234}U turns into thorium-230, radium-226, radon-222, polonium-218, and then lead-214. Although some isotopes of lead ($Z = 82$) are stable, ^{214}Pb is not one of them. Therefore, the radioactive decay series continues as shown at the bottom left of Figure 21.7 and does not end until the stable nuclide ^{206}Pb is produced.

FIGURE 21.6 Radioactive decay results in predictable changes in the number of protons and neutrons in a nucleus. In alpha decay the nucleus loses 2 neutrons and 2 protons, so there is a decrease of 2 in atomic number and 4 in mass number. Beta decay leads to an increase of 1 proton at the expense of 1 neutron. Therefore, the atomic number increases by 1, but the mass number is unchanged. In positron emission and electron capture, the number of protons decreases by 1 and the number of neutrons increases by 1. As a result the atomic number decreases, but the mass number remains the same.

▶❙❙ **CHEMTOUR** Radioactive Decay Modes

alpha (α) decay a nuclear reaction in which an unstable nuclide spontaneously emits an alpha particle.

[Figure: graph with Number of neutrons on vertical axis and Number of protons on horizontal axis, showing β decay, α decay, Electron capture, and Positron emission]

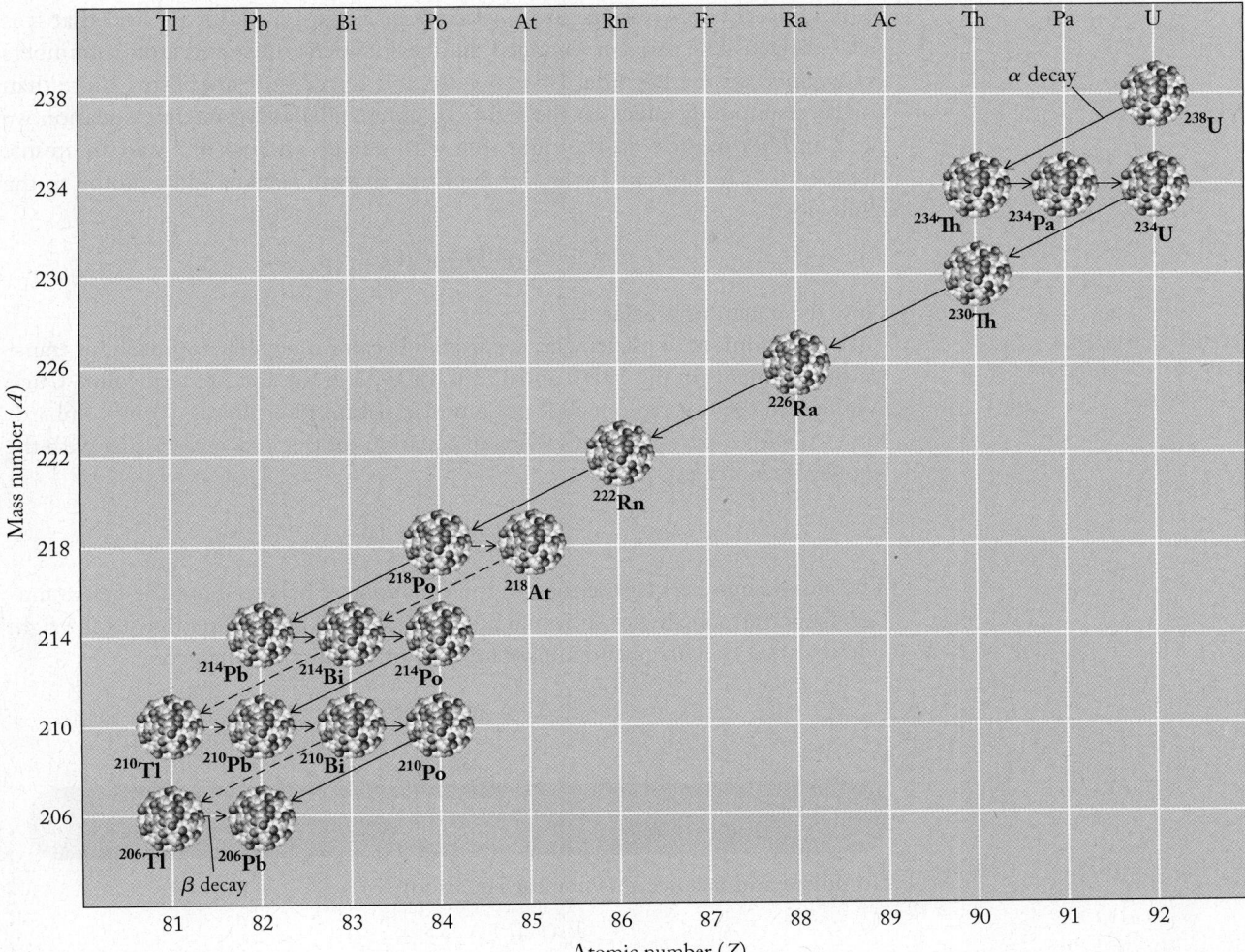

FIGURE 21.7 Uranium-238 radioactive decay series. The long diagonal arrows represent α decay events; the short horizontal ones represent β decay events. The dashed arrows are alternative pathways representing less than 1% of the decay events in this series. Note that whether decay proceeds by the solid-line pathways or the dashed-line pathways, the end product is always stable lead-206.

CONCEPT TEST

In the ^{238}U radioactive decay series, five α decay steps in a row transform ^{234}U into ^{214}Pb. Given the shape of the belt of stability, why does it make sense that the product of these α decay steps would be a neutron-rich nuclide that undergoes β decay?

21.5 Synthesizing Nuclides

Scientists have been synthesizing nuclides since 1919. In that year, Ernest Rutherford reported the synthesis of oxygen-17 by bombardment of nitrogen-14 with α particles:

$$^{14}_{7}N + ^{4}_{2}\alpha \rightarrow ^{17}_{8}O \qquad (21.15)$$

Adding up the superscripts and subscripts in Equation 21.15, we find that it is not balanced. The sums of the mass numbers (superscripts) and atomic numbers (subscripts) on the left side: $14 + 4 = 18$ and $7 + 2 = 9$, are both 1 more than the corresponding values on the right. To obtain a balanced nuclear equation we need another particle on the right side with a mass number of 1 and an atomic number of 1. A check of Table 21.1 confirms that we need to add a proton to the right side:

$$^{14}_{7}\text{N} + \,^{4}_{2}\alpha \rightarrow \,^{17}_{8}\text{O} + \,^{1}_{1}\text{p}$$

Now the equation is balanced.

Bombardment of nuclei with α particles became a popular approach for transmuting elements in the 1920s and 1930s. In 1933, Irène and Frédéric Joliot-Curie synthesized the first radionuclide not found in nature, phosphorus-30, by bombarding ^{27}Al with α particles. Let's write an equation for this process, starting with the known reactants and product:

$$^{27}_{13}\text{Al} + \,^{4}_{2}\alpha \rightarrow \,^{30}_{15}\text{P}$$

The atomic numbers (subscripts) in the equation are balanced but the mass numbers (superscripts) are not. It would be balanced if we added a particle to the right side that had no charge and a mass number of 1, which is a neutron:

$$^{27}_{13}\text{Al} + \,^{4}_{2}\alpha \rightarrow \,^{30}_{15}\text{P} + \,^{1}_{0}\text{n}$$

> **SAMPLE EXERCISE 21.4**   **Completing and Balancing Nuclear Equations**

Complete and balance this nuclear equation:

$$^{27}_{13}\text{Al} + \,^{1}_{0}\text{n} \rightarrow \,? + \,^{0}_{-1}\beta$$

Collect and Organize We are to complete and balance a nuclear equation, which means the sum of subscripts of the particles on the left of the reaction arrow must equal the sum of the subscripts on the right. The same is true for the mass numbers denoted by superscripts.

Analyze The sums of the subscripts and superscripts on the left side of the equation are 13 and 28, respectively. The only particle on the right has zero mass and a charge of $1-$. This means that the missing particle must have a mass number of 28 and an atomic number of 14 (which sums with -1 to equal the 13 on the left side).

Solve The element with an atomic number of 14 is silicon. Therefore, the unknown particle is silicon-28:

$$^{27}_{13}\text{Al} + \,^{1}_{0}\text{n} \rightarrow \,^{28}_{14}\text{Si} + \,^{0}_{-1}\beta$$

Think about It As we discussed in Section 2.7, a stable nuclide that captures one or more neutrons may become so neutron rich that it will undergo β decay. ^{28}Al is such a nuclide.

Practice Exercise Naturally occurring chlorine is a mixture of chlorine-35 and chlorine-37. Write a balanced equation describing the nuclear reaction that occurs when chlorine-37 captures a neutron and undergoes beta decay.

∞ **CONNECTION** Rutherford's early characterization of nuclear radiation is described in Section 2.1.

To successfully bombard a nucleus with an α particle, the electrostatic repulsion between the positively charged particle and the positively charged nucleus must be overcome. Overcoming this repulsion requires that the particle be shot toward the nucleus at a very high speed. The *linear accelerator* drawn in Figure 21.8(a) uses alternating electrical fields to accelerate charged particles in a straight line. The *cyclotron* of Figure 21.8(b) uses a combination of magnetic and electrical fields to accelerate particles in a spiral pathway before smashing into target nuclei. Since 1933, more than a thousand radionuclides have been synthesized using cyclotrons and linear accelerators.

In 1940 elements 93, neptunium, and 94, plutonium, were produced at the University of California, Berkeley, by bombarding uranium-238 with neutrons. Transmutation by neutron bombardment is easier than by α bombardment because there is no electrostatic repulsion between neutrons and nuclei. Neutron capture by uranium-238 initiates two β decay events leading to formation of plutonium-239:

$$^{238}_{92}\text{U} + ^{1}_{0}\text{n} \rightarrow ^{239}_{94}\text{Pu} + 2\,^{0}_{-1}\beta$$

Between 1944 and 1961, the Berkeley research team synthesized elements from $Z = 95$ through $Z = 103$ by using combinations of neutron and α-particle bombardment of actinide nuclei. These methods are of limited use in the synthesis

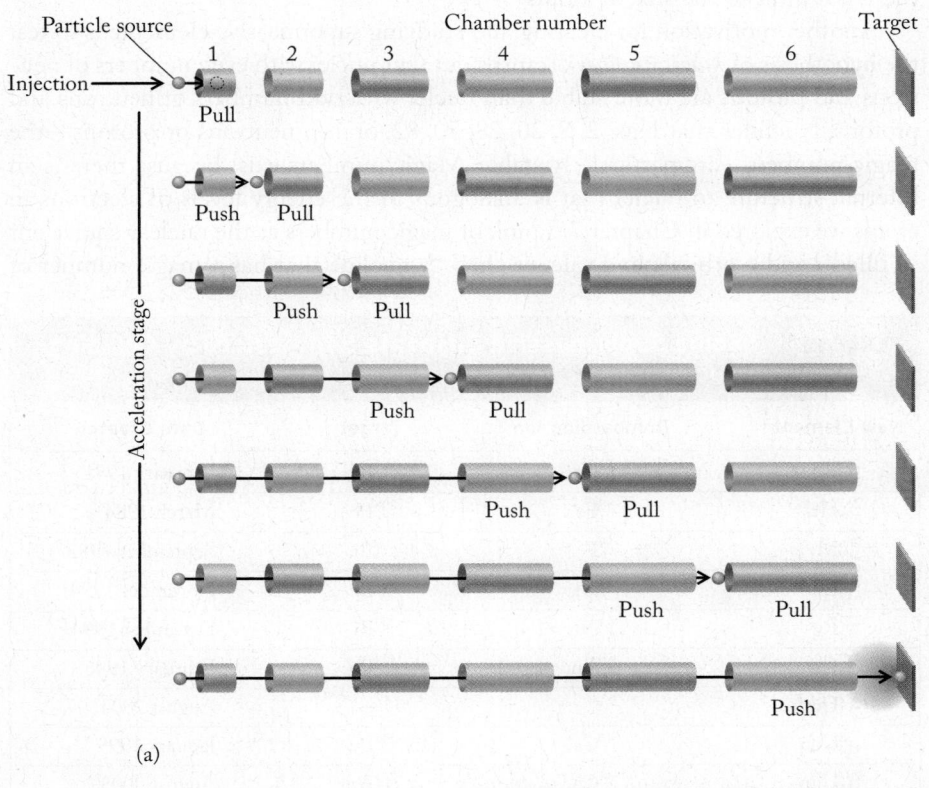

(a)

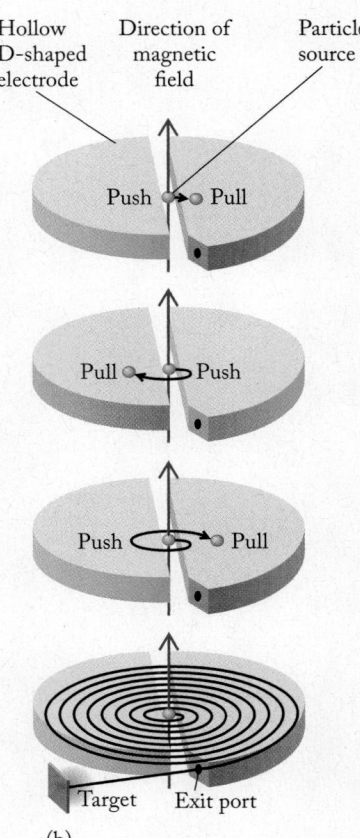

(b)

■ Negative charge   ■ Positive charge

FIGURE 21.8 Two devices for accelerating nuclear particles by subjecting them to alternating positively and negatively charged environments. (a) In a linear accelerator, electrical fields of alternating charge are used to accelerate positively charged particles to very high speeds. (b) In a cyclotron, particles are accelerated between two D-shaped hollow half-cylinders of alternating charge, and their path is bent by magnets above and below the half-cylinders as the particles spiral outward at increasing speed.

of elements with $Z > 103$ because nuclides more massive than californium-249 have extremely short half-lives and so are not useful target materials. However, using carbon, nitrogen, and oxygen nuclei in place of α particles, scientists were able to synthesize rutherfordium ($Z = 104$), dubnium ($Z = 105$), and seaborgium ($Z = 106$), as described by these nuclear equations:

$$\ce{^{249}_{98}Cf} + \ce{^{12}_{6}C} \rightarrow \ce{^{257}_{104}Rf} + 4\ce{^{1}_{0}n}$$

$$\ce{^{249}_{98}Cf} + \ce{^{15}_{7}N} \rightarrow \ce{^{260}_{105}Db} + 4\ce{^{1}_{0}n}$$

$$\ce{^{249}_{98}Cf} + \ce{^{18}_{8}O} \rightarrow \ce{^{263}_{106}Sg} + 4\ce{^{1}_{0}n}$$

In recent years, scientists have reported synthesizing nuclides that have as many as 118 protons by bombarding ^{209}Bi or ^{208}Pb targets with medium-mass nuclei. For example, in January 1999 an atom with 114 protons and 175 neutrons was synthesized and lasted for 30 s before undergoing a series of α decays that yielded isotopes of elements 112, 110, and 108. Some of these "supermassive" nuclides are listed in Table 21.3.

Why bother to make such short-lived nuclei? The answer is that their mere existence, no matter for how brief a time, can be a source of insight into the nature of nuclear structure and into the competition between the strong force that holds nucleons together and the electrostatic repulsion that drives them apart. Supermassive elements are pieces of a puzzle that someday may tell us whether there is a limit to the size of atoms.

Another motivation for creating and studying supermassive elements is to test the hypothesis of *magic numbers*. It turns out that nuclei with even numbers of neutrons and protons are more stable than nuclei with odd numbers of neutrons and protons. Nuclides that have 2, 8, 20, 28, 50, 82, or 126 neutrons or protons—the magic numbers—are particularly stable. Magic numbers exist because there is an internal structure to nuclei that is analogous to the energy levels of electrons in atoms we explored in Chapter 7. Think of magic numbers as the nuclear equivalent of filled s and p orbitals in a valence shell. A nuclide that has a magic number of

TABLE 21.3	Some Supermassive Elements Synthesized by Colliding Nuclei		
New Element[a]	**Bombarding Ion**	**Target**	**Date Created**
$^{262}_{107}$Bh	^{54}Cr	^{209}Bi	February 1981
$^{265}_{108}$Hs	^{58}Fe	^{208}Pb	March 1984
$^{266}_{109}$Mt	^{58}Fe	^{209}Bi	September 1982
$^{269}_{110}$Ds	^{62}Ni	^{208}Pb	November 1994
$^{272}_{111}$Rg	^{64}Ni	^{209}Bi	December 1994
$^{277}_{112}$Cn	^{69}Zn	^{208}Pb	February 1996
$^{283}_{113}$Uut[b]			August 2003
$^{288}_{114}$Uuq	^{48}Ca	^{244}Pu	January 1999
$^{287}_{115}$Uup	^{48}Ca	^{243}Am	August 2003
$^{292}_{116}$Uuh	^{48}Ca	^{248}Cm	December 2000
$^{294}_{118}$Uuo	^{48}Ca	^{249}Cf	April 2002

[a] No names or symbols have been adopted for elements 113 through 118. The names of elements 107 through 112—bohrium, hassium, meitnerium, darmstadtium, roentgenium, and copernicium, respectively—are based on the recommendations of the scientists who created them.
[b] Element Uut was formed by the α decay of element Uup.

neutrons *and* a magic number of protons is said to be *doubly magic*, and is exceptionally stable. Lead-208, for example, has a magic number of both protons (82) and neutrons (126), and is the heaviest stable nuclide known. The nuclides listed in Table 21.3 show the progress international teams of scientists are making toward synthesizing a nuclide with the magic number of 126 protons.

21.6 Nuclear Fission

When an atom of uranium-235 captures a neutron, the nucleus of the unstable product, uranium-236, splits into two lighter nuclei in a process called **nuclear fission**. There are several uranium-235 fission reactions, including these three:

$$^{235}_{92}U + ^{1}_{0}n \rightarrow ^{141}_{56}Ba + ^{92}_{36}Kr + 3^{1}_{0}n$$

$$^{235}_{92}U + ^{1}_{0}n \rightarrow ^{137}_{52}Te + ^{97}_{40}Zr + 2^{1}_{0}n$$

$$^{235}_{92}U + ^{1}_{0}n \rightarrow ^{138}_{55}Cs + ^{96}_{37}Rb + 2^{1}_{0}n$$

In all these reactions, the sums of the masses of the products are slightly less than the sums of the masses of the reactants. As we observed for hydrogen fusion, this loss of mass produces a release of energy in accordance with Einstein's equation ($E = mc^2$).

These reactions also produce additional neutrons, which can smash into other uranium-235 nuclei and initiate more fission events in a **chain reaction** (Figure 21.9).

nuclear fission a nuclear reaction in which the nucleus of an element splits into two lighter nuclei. The process is usually accompanied by the release of one or more neutrons and energy.

chain reaction a self-sustaining series of fission reactions in which the neutrons released when nuclei split apart initiate additional fission events and sustain the reaction.

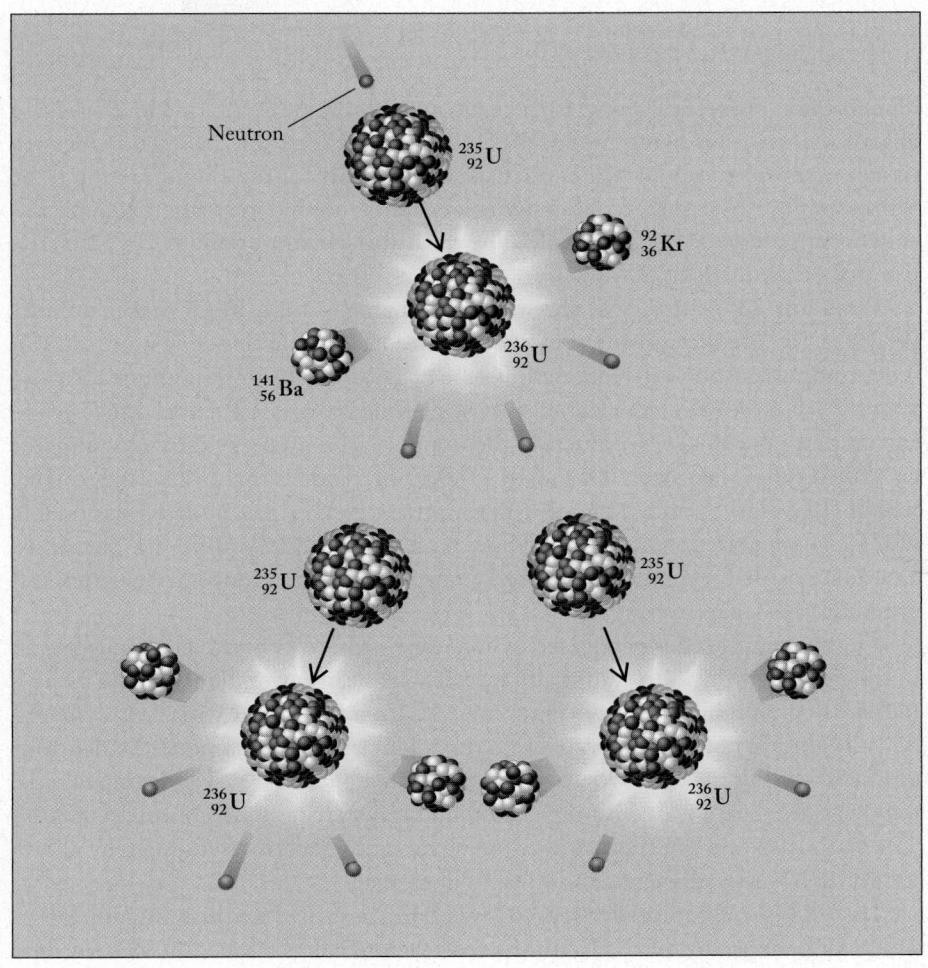

FIGURE 21.9 Each fission event in the chain reaction of a uranium-235 nucleus begins when the nucleus captures a neutron, forming an unstable uranium-236 nucleus that then splits apart (fissions) in one of several ways. In the first process shown here, the uranium-236 nucleus splits into krypton-92, barium-141, and three neutrons. If, on average, at least one of the three neutrons from each fission event causes the fission of another uranium-235 nucleus, then the process is sustained in a chain reaction.

(a)

(b)

(c)

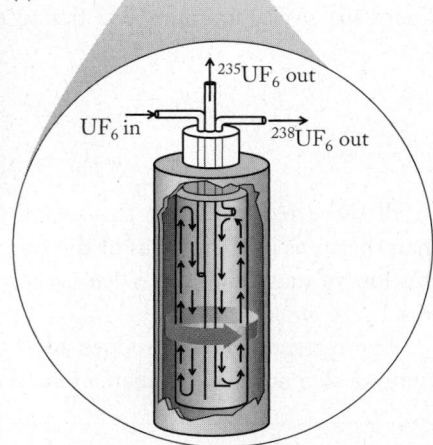

FIGURE 21.10 Preparing uranium fuel. (a) A piece of pitchblende, source of the uranium fuel for nuclear reactors. (b) Pitchblende ore is ground up and extracted with strong acid. The uranium compounds (mostly U_3O_8) obtained from the extract are called yellowcake. (c) Uranium oxides are converted to volatile UF_6, which is centrifuged at very high speed to separate $^{235}UF_6$ from $^{238}UF_6$. The less dense and less abundant $^{235}UF_6$ is enriched near the center of the centrifuge cylinder and separated from the heavier $^{238}UF_6$.

The reaction proceeds as long as there are enough uranium-235 nuclei present to absorb the neutrons being produced. On average, at least one neutron from each fission event must cause another nucleus to split apart for the chain reaction to be self-sustaining. The quantity of fissionable material needed to assure that every fission event produces another is called the **critical mass**. For uranium-235, the critical mass is about 1 kg of the pure isotope.

Uranium-235 is the most abundant fissionable isotope, but it makes up only 0.72% of the uranium in the principal uranium ore, pitchblende (Figure 21.10). The uranium in nuclear reactors must be at least 3% to 4% uranium-235, and above 7% is needed for nuclear weapons. The most common method for *enriching* uranium ore involves extracting the uranium in a process that yields a material called yellowcake, which is mostly U_3O_8. This oxide is then converted to UF_6, which is a volatile solid. The solid is vaporized to UF_6 gas, and the fissionable $^{235}UF_6$ is separated from $^{238}UF_6$ based on their slightly different densities. Elaborate centrifuge systems are used to exploit this difference in densities and speed up the separation.

Harnessing the energy released by nuclear fission to generate electricity began in the middle of the 20th century. In a typical nuclear power plant (Figure 21.11), fuel rods containing 3% to 4% uranium-235 are interspersed with rods of boron or cadmium that control the rate of the fission by absorbing some of the neutrons produced during fission. Pressurized water flows around the fuel and control rods, removing the heat created during fission and transferring it to a steam generator. The water also acts as a moderator, slowing down the neutrons and thereby allowing for their more efficient capture by ^{235}U atoms.

In 1952, the first **breeder reactor** was built, so called because in addition to producing energy to make electricity, the reactor makes ("breeds") its own fuel.

critical mass the minimum quantity of fissionable material needed to sustain a chain reaction.

breeder reactor a nuclear reactor in which fissionable material is produced during normal reactor operation.

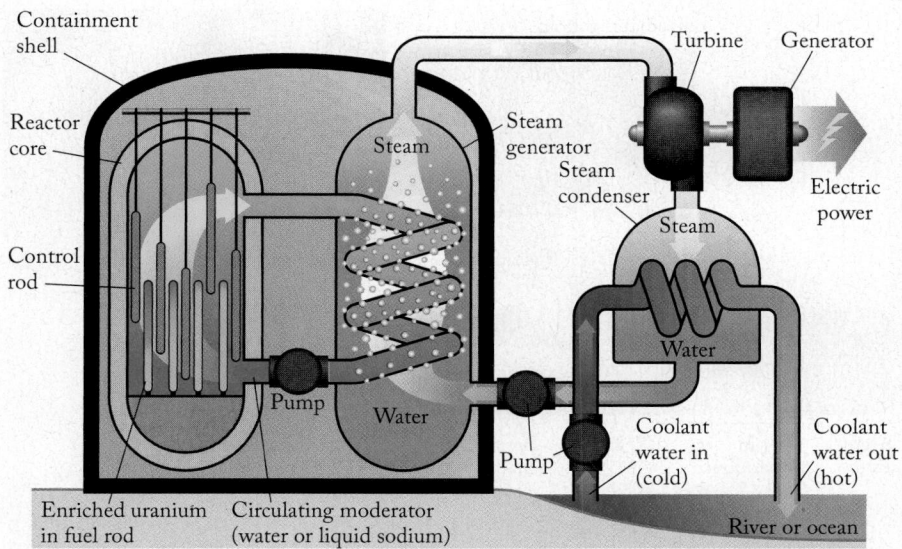

FIGURE 21.11 A pressurized, water-cooled nuclear power plant uses fuel rods containing uranium enriched to uranium-235. The fission chain reaction is regulated with control rods and a moderator such as water. The moderator slows down the neutrons released by fission so that they can be more efficiently captured by other uranium-235 nuclei. It also transfers the heat produced by the fission reaction to a steam generator. The steam generated by this heat drives a turbine that generates electricity.

The reactor starts out with a mixture of plutonium-239 and uranium-238. As plutonium fissions and the energy from those reactions is collected to produce electricity, some of the neutrons that are produced sustain the fission chain reaction just as in the reactor of Figure 21.11, but others convert the uranium into more plutonium fuel:

$$^{238}_{92}U + {}^{1}_{0}n \rightarrow {}^{239}_{92}U + \gamma \rightarrow {}^{239}_{94}Pu + 2\,{}^{0}_{-1}\beta$$

In less than 10 years of operation, a breeder reactor can make enough plutonium-239 to refuel itself *and* another reactor. Unfortunately, plutonium-239 is a carcinogen and one of the most toxic substances known. Only about half a kilogram is needed to make an atomic bomb, and it has a long half-life: 2.4×10^4 years. Understandably, extreme caution and tight security surround the handling of plutonium fuel and the transportation and storage of nuclear wastes containing even small amounts of plutonium. Health and safety matters related to reactor operation and spent-fuel disposal are the principal reasons there are no breeder power stations in the United States, although they have been built in at least seven other countries.

CONCEPT TEST ··

Nuclear reactors powered by the energy released by the fission of uranium-235 have been operating since the 1950s, but a reactor powered by the energy released by the fusion of hydrogen has yet to be built. Why is it taking so long to build a fusion reactor?

··

21.7 Measuring Radioactivity

French scientist Henri Becquerel (1852–1908) discovered radioactivity in 1896 when he observed that uranium and other substances produce radiation that fogs photographic film. Photographic film is still used to detect radioactivity, as in the film dosimeter badges worn by people working with radioactive materials to record their exposure to radiation. Detectors called **scintillation counters** use *phosphors* to absorb energy released during radioactive decay. The phosphors then

scintillation counter an instrument that determines the level of radioactivity in samples by measuring the intensity of light emitted by phosphors in contact with the samples.

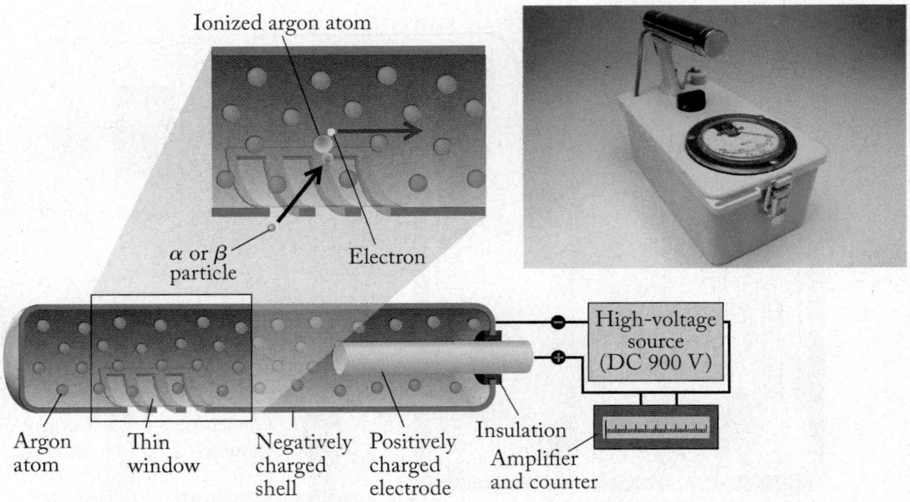

FIGURE 21.12 In a Geiger counter, a particle produced by radioactive decay passes through a thin window usually made of beryllium or a plastic film. Inside the tube, the particle collides with atoms of argon gas and ionizes them. The resulting argon cations migrate toward the negatively charged tube housing, and the electrons migrate toward a positive electrode, creating a pulse of current through the tube. The current pulses are amplified and recorded via a meter and a speaker that produces an audible "click" for each pulse.

release some of the absorbed energy as visible light, the intensity of which is a measure of the amount of radiation initially emitted.

Radioactivity also can be measured with a **Geiger counter**, which detects the common products of radioactivity—α particles, β particles, and γ rays—on the basis of their abilities to ionize atoms (Figure 21.12). A sealed metal cylinder filled with gas (usually argon) has a positively charged electrode at its center and a window that allows α particles, β particles, and γ rays to enter. Once inside the cylinder, these particles ionize argon atoms into Ar^+ ions and free electrons. If an electrical potential difference is applied between the cylinder shell and the central electrode, free electrons migrate toward the positive electrode and argon ions migrate toward the negatively charged shell. This ion migration produces a pulse of electrical current whenever radiation enters the cylinder. The current is amplified and read out to a meter and a microphone that makes a clicking sound.

One measure of radioactivity in a sample is the number of decay events per unit time. This parameter is called the radioactivity (A) of the sample. The SI unit of radioactivity is the **becquerel (Bq)**, named in honor of Henri Becquerel and equal to one decay event per second. An older radioactivity unit is the **curie (Ci)**, named in honor of Marie and Pierre Curie, where

$$1 \text{ Ci} = 3.70 \times 10^{10} \text{ Bq} = 3.70 \times 10^{10} \text{ decay events/s}$$

Both the becquerel and the curie depend on the *rate* at which a radioactive substance decays and on how much of it is in the sample. As noted in Section 21.1, all decay processes follow first-order kinetics. Recall from Section 15.3 that for a first-order reaction the rate constant k is related to the concentration of a reactant R according to the equation

$$\text{Rate} = k[\text{R}]$$

The same mathematical relationship applies to radioactive decay processes except that we refer to radioactivity instead of rate of decay, and to the number of atoms (N) of a radionuclide in a sample instead of its concentration:

$$A = kN \tag{21.16}$$

Because radioactivity is the number of decay events per second, the units of the *decay rate constant k* are decay events per atom per second.

Geiger counter a portable device for determining nuclear radiation levels by measuring how much the radiation ionizes the gas in a sealed detector.

becquerel (Bq) the SI unit of radioactivity. One becquerel equals one decay event per second.

curie (Ci) non-SI unit of radioactivity; $1 \text{ Ci} = 3.70 \times 10^{10}$ decay events per second.

The decay rate constant is related to the half-life of the reaction by Equation 15.23:

$$t_{1/2} = \frac{0.693}{k}$$

Scientists usually express the quantity of a radionuclide in a sample not in terms of its mass but rather in terms of its radioactivity, because the latter value is much more important in determining how the substance is used and how we handle it safely. This substitution is valid because the quantity of a radionuclide is directly proportional to its activity.

**SAMPLE EXERCISE 21.5   Calculating the Activity
of a Radioactive Sample**

Radium-223 undergoes β decay with a half-life of 11.4 d. What is the radioactivity of a sample that contains 1.00 μg of radioactive ^{223}Ra? Express your answer in becquerels and in curies.

Collect and Organize We are given the half-life and quantity of a radioactive substance and are to determine its radioactivity, that is, its rate of decay. We know the decay must be first-order because all radioactive decay processes are. Therefore, the decay rate constant k is related to half-life by Equation 15.23:

$$t_{1/2} = \frac{0.693}{k}$$

and radioactivity is the product of the rate constant and the number of atoms of radionuclide in the sample:

$$A = kN \qquad (21.16)$$

Analyze Before using Equation 15.23, we must convert the half-life into seconds because both radioactivity units we need to calculate are based on decay events per second. To calculate radioactivity we determine the number of atoms in 1.00 μg of radium. This will likely be a very large number, which should translate into a large number of decay events per second given the relatively short half-life of ^{223}Ra.

Solve The half-life is

$$11.4 \text{ d} \times \frac{24 \text{ hr}}{1 \text{ d}} \times \frac{60 \text{ min}}{1 \text{ hr}} \times \frac{60 \text{ s}}{1 \text{ min}} = 98.5 \times 10^5 \text{ s}$$

Using this value in Equation 15.23 and solving for k:

$$k = \frac{0.693}{9.85 \times 10^5 \text{ s}}$$

$$= 7.04 \times 10^{-7} \text{ s}^{-1} = 7.04 \times 10^{-7} \text{ decay events/(atom} \cdot \text{s)}$$

The number of atoms (N) of ^{223}Ra is

$$N = 1.00 \text{ } \mu g \times \frac{10^{-6} \text{ g}}{1 \text{ } \mu g} \times \frac{1 \text{ mol}}{223 \text{ g}} \times \frac{6.022 \times 10^{23} \text{ atoms}}{1 \text{ mol}}$$

$$= 2.70 \times 10^{15} \text{ atoms}$$

Inserting these values of k and N into Equation 21.16:

$$A = kN$$

$$= 7.04 \times 10^{-7} \, \frac{\text{decay events}}{\cancel{\text{atoms}} \cdot \text{s}} \times 2.70 \times 10^{15} \, \cancel{\text{atoms}}$$

$$= 1.90 \times 10^9 \, \text{decay events/s}$$

Because 1 Bq = 1 decay event/s, the radioactivity of the sample is 1.90×10^9 Bq. Expressing activity in curies:

$$1.90 \times 10^9 \, \cancel{\text{decay events/s}} \times \frac{1 \, \text{Ci}}{3.70 \times 10^{10} \, \cancel{\text{decay events/s}}} = 0.051 \, \text{Ci}$$

Think about It The large number of decay events per second that we calculated meets our expectation of a high level of activity given the large number of radioactive atoms in the sample.

Practice Exercise Determine the activity in 1.00 mg of radium-226 ($t_{1/2} = 1.6 \times 10^3$ yr) in becquerels and in millicuries.

21.8 Biological Effects of Radioactivity

The γ rays and many of the α and β particles produced by nuclear reactions have more than enough energy to tear chemical bonds apart, producing odd-electron radicals or free electrons and cations. Consequently, these rays and particles are classified as **ionizing radiation**. Other examples are X-rays and short-wavelength ultraviolet rays. The ionization of atoms (and molecules) in living tissue can lead to radiation sickness, cancer, birth defects, and death. The scientists who first worked with radioactive materials were not aware of these hazards, and some of them suffered for it. Marie Curie died of leukemia caused by her many years of radiation exposure, and the same disease claimed her daughter Irène Joliot-Curie, who continued the research program started by her parents.

In medicine, the term *ionizing radiation* is limited to photons and particles that have sufficient energy to remove an electron from water:

$$H_2O(\ell) \xrightarrow{\text{1216 kJ/mol}} H_2O^+(aq) + e^-$$

The logic behind this definition is that the human body is composed largely of water. Therefore water molecules are the most abundant ionizable targets in an organism exposed to nuclear radiation. The cation H_2O^+ reacts with another water molecule to form a hydronium ion and a hydroxyl free radical:

$$H_2O^+(aq) + H_2O(\ell) \rightarrow H_3O^+(aq) + \cdot OH(aq)$$

The rapid reactions of free radicals such as $\cdot OH$ with biomolecules can threaten the life of the cell and the host organism.

Radiation-induced alterations to the biochemical machinery that controls cell growth are most likely to occur in tissues in which cells grow and divide rapidly. One such tissue is bone marrow, where billions of white blood cells are produced

ionizing radiation high-energy products of radioactive decay that can ionize molecules.

each day to fortify the body's immune system. Molecular damage to bone marrow can lead to leukemia, an uncontrolled production of nonfunctioning white blood cells that spread throughout the body, crowding out healthy cells. Ionizing radiation can also cause molecular alterations in the genes and chromosomes of sperm and egg cells, increasing the chances of birth defects in offspring.

> **CONCEPT TEST**
>
> Rank the following types of electromagnetic radiation in decreasing order of their ability to ionize molecular bonds: radio waves, gamma rays, microwaves, X-rays, UV radiation.

Radiation Dosage

The biological impact of ionizing radiation depends on how much of it is absorbed by an organism. If the radiation is coming from one radioactive source, then the amount absorbed depends on the activity of the source and the energy of the radiation that is produced per decay event. Tables of radioactive isotopes often include information about their modes of decay and the energies of the particles and gamma rays they emit.

Absorbed dose is the quantity of ionizing radiation absorbed by a unit mass of living tissue. The SI unit of absorbed dose is the **gray (Gy)**. One gray is equal to the absorption of 1 joule of radiation energy per kilogram of body mass:

$$1 \text{ Gy} = 1 \text{ J/kg}$$

Grays express dosage, but they do not indicate the amount of *tissue damage* caused by that dosage. Different products of nuclear reactions affect living tissue differently. Exposure to 1 Gy of γ rays produces about the same amount of tissue damage as exposure to 1 Gy of β particles. However, 1 Gy of α particles, which move about 10 times slower than β particles but have nearly 10^4 times the mass, causes as much as 20 times as much damage as 1 Gy of γ rays. Neutrons cause 3 to 5 times as much damage. To account for these differences, values of **relative biological effectiveness (RBE)** have been established for the various forms of ionizing radiation (Table 21.4). When absorbed dose in grays is multiplied by an RBE factor, the product is called *effective dose*, a measure of tissue damage. The SI unit of effective dose is the **sievert (Sv)**.

Table 21.5 summarizes the various units used to express quantities of radiation and their biological impact. Two non-SI units are listed that predate their SI counterparts but are still often used. They are radiation absorbed dose, or rad, which is equivalent to 0.01 Gy, and the rem for tissue damage, which is an acronym for *roentgen equivalent man*. One rem is the product of one rad of ionization times the appropriate RBE factor. There are 100 rems in 1 Sv.

gray (Gy) the SI unit of absorbed radiation; 1 Gy = 1 J/kg of tissue.

relative biological effectiveness (RBE) a factor that accounts for the differences in physical damage caused by different types of radiation.

sievert (Sv) SI unit used to express the amount of biological damage caused by ionizing radiation.

CONNECTION Gamma rays are the highest-energy form of electromagnetic radiation (see Figure 7.1).

TABLE 21.4	RBE Values of Nuclear Radiation
Radiation	**RBE**
γ rays	1.0
β particles	1.0–1.5
Neutrons	3–5
Protons	10
α particles	20

TABLE 21.5   Units for Expressing Quantities of Ionizing Radiation

Parameter	SI Unit	Description	Alternative Common Units	Description
Radioactivity	Becquerel (Bq)	1 decay event/s	Curie (Ci)	3.70×10^{10} decay events/s
Ionizing energy absorbed	Gray (Gy)	1 J/kg of tissue	Rad	0.01 J/kg of tissue
Amount of tissue damage	Sievert (Sv)	1 Gy × RBE[a]	Rem	1 rad × RBE[a]
[a] RBE, relative biological effectiveness.				

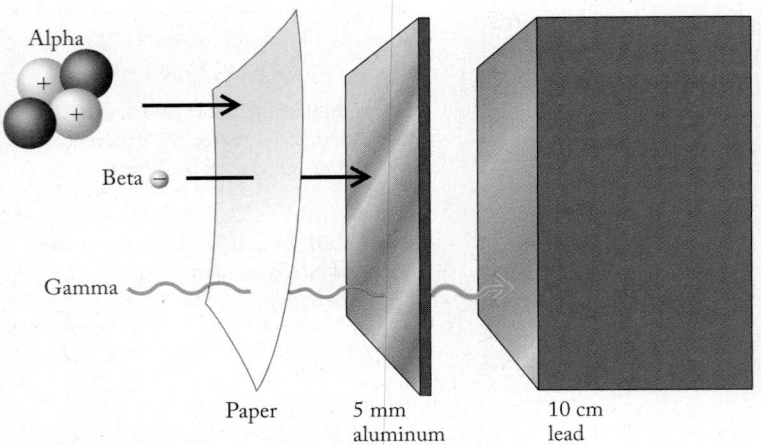

Alpha

Beta

Gamma

Paper

5 mm aluminum

10 cm lead

FIGURE 21.13 The tissue damage caused by α particles, β particles, and γ rays depends on their ability to penetrate materials shielding us from their source. Alpha particles are stopped by paper or clothing but are extremely dangerous if formed inside the body because their low penetrating power traps them inside. Stopping gamma rays requires a thick layer of lead or several meters of concrete or soil.

The RBE of 20 for α particles may lead you to believe that these particles pose the greatest health threat from radioactivity. Not exactly. Alpha particles are so big that they have little penetrating power; they are stopped by a sheet of paper, clothing, or even a layer of dead skin (Figure 21.13). However, if you ingest or inhale an α emitter, tissue damage can be severe because the relatively massive α particles do not have to travel far to cause cell damage. Gamma rays are considered the most dangerous form of radiation emanating from a source outside the body because they have the greatest penetrating power.

The effects of exposure to different single effective doses of radiation are summarized in Table 21.6. To put these data in perspective, the effective dose from a typical dental X-ray is about 25 μSv, or about 2,000 times smaller than the lowest exposure level cited in the table.

TABLE 21.6	Acute Effects of Single Whole-Body Effective Doses of Ionizing Radiation
Effective Dose (Sv)	**Toxic Effect**
0.05–0.25	No acute effect, possible carcinogenic or mutagenic damage to DNA
0.25–1.0	Temporary reduction in white blood cell count
1.0–2.0	Radiation sickness: fatigue, vomiting, diarrhea, impaired immune system
2.0–4.0	Severe radiation sickness: intestinal bleeding, bone marrow destruction
4.0–10.0	Death, usually through infection, within weeks
>10.0	Death within hours

Widespread exposure to very high levels of radiation occurred after the 1986 explosion at the Chernobyl nuclear reactor in what is now Ukraine (Figure 21.14). Many plant workers and first responders were exposed to more than 1.0 Sv of radiation. At least 30 of them died in the weeks after the accident. Many of the more

FIGURE 21.14 The ruins of the nuclear reactor at Chernobyl, Ukraine, that exploded in 1986.

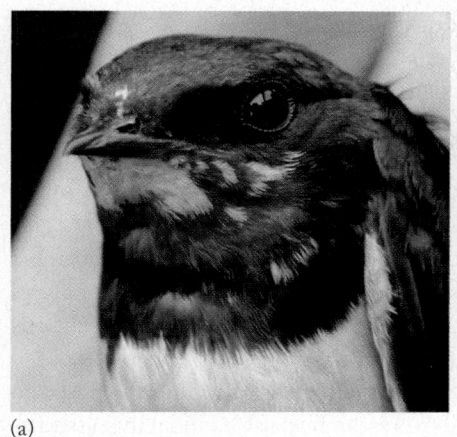

(a) (b)

FIGURE 21.15 Wildlife surrounding the destroyed nuclear reactor at Chernobyl, Ukraine, were exposed to intense ionizing radiation, which led to deaths and sublethal biological effects such as genetic mutations. One example of the latter is the partially albino barn swallow (a). A normal swallow (b) has no white feathers directly beneath its beak.

than 300,000 workers who cleaned up the area around the reactor exhibited symptoms of radiation sickness, and at least 5 million people in Ukraine, Belarus, and Russia were exposed to fallout in the days following the accident. Studies conducted in the early 1990s uncovered high incidences of thyroid cancer in children in southern Belarus due to ^{131}I released in the Chernobyl accident, and children born in the region nearly a decade after the accident had unusually high rates of mutations in their DNA because of their parents' exposure to ionizing radiation. Genetic damage was also widespread among plants and animals living in the region (Figure 21.15).

Radiation exposure was not confined to Ukraine and Belarus. After the accident a cloud of radioactive material spread rapidly across northern Europe, and within 2 weeks increased levels of radioactivity were detected throughout the Northern Hemisphere (Figure 21.16). The accident produced a global increase in human exposure to ionizing radiation estimated to be equivalent to 0.05 mSv per year.

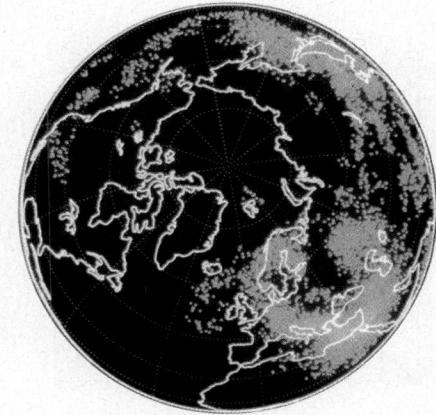

FIGURE 21.16 Radioactive fallout (shown in pink) from the Chernobyl accident in 1986 was detected throughout the Northern Hemisphere.

Evaluating the Risks of Radiation

To put global radiation exposure from the Chernobyl disaster in perspective, we need to consider typical annual exposure levels. For many people the principal source of radiation is radon gas in indoor air and well water (Figure 21.17). Like all the noble gases, radon is chemically inert. Unlike the others, all of its isotopes are radioactive. The most common isotope, radon-222, is produced when uranium-238 in rocks and soil decays to lead-206 (Figure 21.7). The radon gas formed in this decay series percolates upward and can enter a building through cracks and pores in its foundation.

If you breathe radon-contaminated air and then exhale before it decays, no harm is done. However, if radon-222 decays inside the lungs, it emits an α particle that can attack lung tissue. The nuclide produced by the α decay of ^{222}Rn is radioactive polonium-218 that may become attached to tissue in the respiratory system and undergo a second α decay, forming lead-214:

$$^{222}_{86}\text{Rn} \rightarrow {}^{218}_{84}\text{Po} + {}^{4}_{2}\alpha \qquad t_{1/2} = 3.8 \text{ d}$$

$$^{218}_{84}\text{Po} \rightarrow {}^{214}_{82}\text{Pb} + {}^{4}_{2}\alpha \qquad t_{1/2} = 3.1 \text{ min}$$

As we have seen, α particles are the most damaging product of nuclear decay when formed *inside the body*. How big a threat does radon pose to human health?

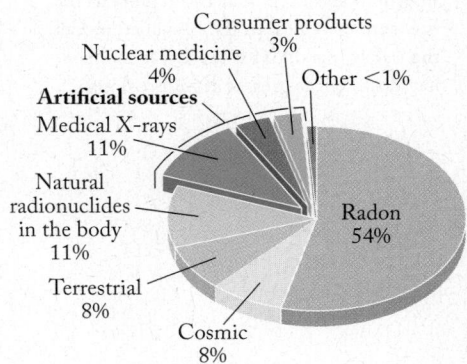

FIGURE 21.17 Sources of radiation exposure of the U.S. population. On average, a person living in the United States is exposed to 0.0036 Sv of radiation each year. More than 80% of this exposure comes from natural sources, mainly radon in the air and water. Artificial sources account for about 18% of the total exposure.

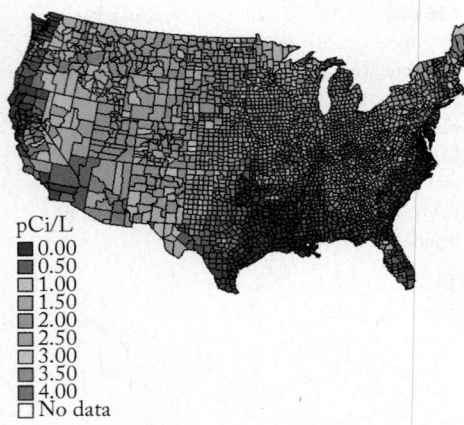

pCi/L
- 0.00
- 0.50
- 1.00
- 1.50
- 2.00
- 2.50
- 3.00
- 3.50
- 4.00
- No data

FIGURE 21.18 Levels of radon gas in soils and rocks across the United States.

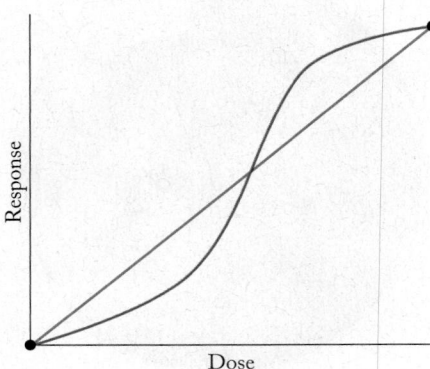

Response

Dose

FIGURE 21.19 The risk of death from radiation-induced cancer may follow one of two models. In the linear response model (red line), risk is directly proportional to the radiation exposure. In the S-shaped model (blue line), risk remains low below a critical threshold and then increases rapidly as the exposure increases. In the S-shaped model, the risk is less than for the linear model at low doses but is greater at higher doses.

Concentrations of indoor radon depend on local geology (Figure 21.18) and on how gastight building foundations are. The air in many buildings contains concentrations of radon in the range of 1 pCi per liter of air. How hazardous are such tiny concentrations? There appears to be no simple answer. The U.S. Environmental Protection Agency has established 4 pCi/L as an "action level," meaning that people occupying houses with higher concentrations should take measures to minimize their exposure.

This action level is based on the results of studies of the incidence of lung cancer in workers in uranium mines. These workers are exposed to radon concentrations (and concentrations of other radionuclides) that are much higher than the concentrations in homes and other buildings. However, many scientists believe that people exposed to very low levels of radon for many years are as much at risk as miners exposed to high levels of radiation for shorter periods. Some researchers use a model that assumes a linear relation between radon exposure and incidence of lung cancer. This model is represented by the red line in Figure 21.19, which graphs cancer deaths as a function of radiation absorbed. On the basis of this dose–response model, an estimated 15,000 Americans die of lung cancer each year because of exposure to indoor radon. This number comprises 10% of all lung-cancer fatalities and 30% of those among nonsmokers.

Is this linear model valid? Perhaps—but some scientists believe that there may be a threshold of exposure below which radon poses no significant threat to public health. They advocate an S-shaped dose–response curve, shown by the blue line in Figure 21.19. Notice that the risk of death from cancer in the S-shaped curve is much lower than in the linear response model at low radiation exposure but rises rapidly above a critical value.

CONCEPT TEST

In the early 20th century, wristwatches were sold with dials that glowed in the dark because they were painted with compounds that contained radioactive radium. Like radon, radium isotopes emit α particles. Compare the health risks of wearing a radium-containing watch versus breathing in air containing radon.

SAMPLE EXERCISE 21.6   Calculating Effective Dose

It has been estimated that a person living in a home where the air radon concentration is 4.0 pCi/L receives an annual absorbed dose of ionizing radiation equivalent to 0.40 mGy. What is the person's annual effective dose from this radon in millisieverts? Use information from Figure 21.17 to compare this annual effective dose of radon with the average annual effective dose of radon estimated for persons living in the United States.

Collect and Organize We are given an absorbed radon dose of 0.40 mGy. Radon isotopes emit alpha particles, which we know from Table 21.4 have a relative biological effectiveness of 20.

Analyze The effective dose caused by an absorbed dose of ionizing radiation is the absorbed dose multiplied by the RBE of the radiation.

Solve

$$0.40 \text{ mGy} \times 20 = 8.0 \text{ mSv}$$

Figure 21.17 tells us that the average American is exposed to 3.6 mSv of radiation per year, with 54% of that amount, or 1.9 mSv, from radon. The person living in the home described in the problem has an effective dose slightly more than four times the average value.

Think about It The calculated value is more than twice the average annual effective dose of 3.6 mSv from all sources of radiation. The U.S. National Research Council has estimated that a nonsmoker living in air contaminated with 4.0 pCi/L of radon has a 1% chance of dying from lung cancer due to this exposure. The cancer risk for a smoker is close to 5%.

Practice Exercise A dental X-ray for imaging impacted wisdom teeth produces an effective dose of 10 μSv. If a dental X-ray machine emits X-rays with an energy of 6.0×10^{-17} J each, how many of these X-rays must be absorbed per kilogram of tissue to produce an effective dose of 10 μSv? Assume the RBE of these X-rays is 1.2.

21.9 Medical Applications of Radionuclides

Radionuclides are used in both the detection and the treatment of diseases, and they are key agents in the medical fields of *diagnostic radiology* and *therapeutic radiology*. In diagnostic radiology, radionuclides are used alongside magnetic resonance imaging (MRI) and other imaging systems that involve only nonionizing radiation. Therapeutic radiology, however, is based almost entirely on the ionizing radiation that comes from radionuclides.

Therapeutic Radiology

Because ionizing radiation causes the most damage to cells that grow and divide rapidly, it is a powerful tool in the fight *against* cancer. Radiation therapy consists of exposing cancerous tissue to γ radiation. Often the radiation source is external to the patient, but sometimes it is encased in a platinum capsule and surgically implanted in a cancerous tumor. The platinum provides a chemically inert outer layer and acts as a filter, absorbing α and β particles emitted by the radionuclide but allowing γ rays to pass into the tumor.

A nuclide's chemical properties can be used to direct it to a tumor site. For example, most iodine in the body is concentrated in the thyroid gland, so an effective therapy against thyroid cancer starts with ingestion of potassium iodide containing radioactive iodine-123. Other radionuclides used in cancer therapy are listed in Table 21.7.

TABLE 21.7 Some Radionuclides Used in Radiation Therapy

Nuclide	Radiation	Half-Life	Treatment
^{32}P	β	14.3 d	Leukemia therapy
^{60}Co	β, γ	5.3 yr	Cancer therapy
^{131}I	β	8.1 d	Thyroid therapy
^{131}Cs	γ	9.7 d	Prostate cancer therapy
^{192}Ir	β, γ	74 d	Coronary disease

Surgically inaccessible tumors can be treated with beams of γ rays from a radiation source outside the body. Unfortunately, γ radiation destroys both cancer cells and healthy ones. Thus, patients receiving radiation therapy frequently suffer symptoms of radiation sickness, including nausea and vomiting (the tissues that make up intestinal walls are especially susceptible to radiation-induced damage), fatigue, weakened immune response, and hair loss. To reduce the severity of these side effects, radiologists must carefully control the dose a patient receives.

Diagnostic Radiology

The movement of radionuclides in the body and their accumulation in certain organs provide ways to assess organ function. A tiny quantity of a radioactive isotope is used, together with a much larger amount of a stable isotope of the same element. The radioactive isotope is called a *tracer*, and the stable isotope is the *carrier*. For example, the circulatory system can be imaged by injecting into the blood a solution of sodium chloride containing a trace amount of ^{24}NaCl. Circulation is monitored by measuring the γ rays emitted by ^{24}Na as it decays.

The ideal tracer for medical imaging is one that has a half-life about equal to the length of time required to perform the imaging measurements. It should emit moderate-energy γ rays but no α particles or β particles that might cause tissue damage. Table 21.8 lists a few of the radionuclides used in medical imaging.

TABLE 21.8	**Selected Radionuclides Used for Medical Imaging**		
Nuclide	**Radiation**	**Half-Life (hr)**	**Use**
^{99}Tc	γ	6.0	Bones, circulatory system, various organs
^{67}Ga	γ	78	Tumors in the brain and other organs
^{201}Tl	γ	73	Coronary arteries, heart muscle
^{123}I	γ	13.3	Thyroxine production in thyroid gland

Positron-emission tomography (PET) is a powerful tool for diagnosing organ and cell function. PET uses short-lived, neutron-poor radionuclides that emit positrons such as carbon-11, oxygen-15, and fluorine-18. A patient might be administered a solution of glucose in which some of the sugar molecules contain atoms of ^{11}C, ^{15}O, or ^{18}F. The rate at which glucose is metabolized in various regions of the brain is monitored by detecting the γ rays produced by positron–electron annihilations (Equation 21.6). Unusual patterns in PET images of brains (Figure 21.20) can indicate damage from strokes, schizophrenia, manic depression, Alzheimer's disease, and even nicotine addiction in tobacco smokers.

21.10 Radiometric Dating

Radiometric dating is a term used to describe methods for determining the age of objects based on their naturally occurring radionuclides and/or the nuclides' decay products. The concept was invented in the early 1900s by Ernest Rutherford, who had already recognized in his pioneering studies on radioactivity that radioactive decay processes have characteristic half-lives. Rutherford proposed to use this concept to determine the age of rocks and even the age of Earth itself. The basis for his initial attempt was the emission of α particles from uranium ore. He suspected (correctly) that alpha particles were part of helium atoms, and so he proposed to determine the age of uranium ore samples by determining the concentration of helium gas trapped inside them.

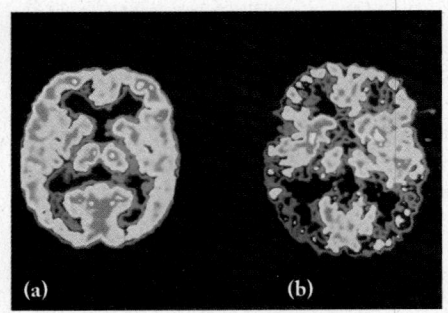

(a)   (b)

FIGURE 21.20 Positron-emission tomography (PET) is used to monitor cell activity in organs such as the brain. (a) Brain function in a healthy person. The red and yellow regions indicate high brain activity; blue and black indicate low activity. (b) Brain function in a patient suffering from Alzheimer's disease.

radiometric dating a method for determining the age of an object based on the quantity of a radioactive nuclide and/or the products of its decay that the object contains.

Rutherford's helium method did not yield accurate results, but it did inspire a young American chemist, Bertram Boltwood (1870–1927), who had determined that the decay of radioactive uranium involves a series of decay events ending with the formation of stable lead (illustrated for ^{238}U in Figure 21.7). In 1907 Boltwood published the results of dating 43 samples of uranium-containing minerals based on the ratio of lead to uranium in them. The ages he reported spanned hundreds of millions to over a billion years and probably represent the first successful attempt at radiometric dating.

CONCEPT TEST

Suggest a reason why Rutherford's method based on the He content of uranium minerals might not have produced accurate ages of these minerals.

In recent years the development of the mass spectrometer for accurately determining the abundances of individual isotopes of elements coupled with more accurate half-life values for decay events such as those in Figure 21.7 have allowed scientists to use the ratio of ^{206}Pb to ^{238}U in geological samples to determine their ages with a precision of about $\pm 1\%$. Other methods, including one based on the decay of ^{235}U to ^{207}Pb ($t_{1/2} = 7.0 \times 10^6$ yr), may be used to analyze the same samples, providing multiple independent determinations that mutually assure more accurate results. These analyses have shown that the oldest rocks on Earth are over 4.0 billion years old and that meteorites that formed as the solar system formed are 4.5 billion years old.

The radiometric methods described above yield reliable results only when the sample is a *closed system*, which means that the only loss of the radionuclide is via radioactive decay, and that all of the nuclides produced by the decay processes remain in the sample. In addition, those decay processes must be the *only* source of the product nuclides. For these reasons, scientists must exercise care in selecting the types of samples they subject to radiometric dating analysis. For example, the presence of the mineral zircon ($ZrSiO_4$) in a geological sample is good news for scientists interested in dating it because U^{4+} ions readily substitute for Zr^{4+} ions as crystals of $ZrSiO_4$ solidify from the molten state, but Pb^{2+} ions do not. Therefore, the only source of ^{206}Pb and ^{207}Pb in a zircon sample should be the decay of ^{238}U and ^{235}U, respectively.

In 1947, American chemist Willard Libby (1908–1980) developed a radiometric dating technique, called **radiocarbon dating**, for determining the age of artifacts from prehistory and early civilizations. The method is based on determining the carbon-14 content of samples derived from plants (or the animals that consumed them). Carbon-14 originates in the upper atmosphere where cosmic rays break apart the nuclei of atoms, forming free protons and neutrons. When one of these neutrons collides with a nitrogen-14 atom, they form an atom of radioactive carbon-14 and a proton:

$$^{14}_{7}N + {}^{1}_{0}n \rightarrow {}^{14}_{6}C + {}^{1}_{1}p$$

Atmospheric carbon-14 combines with oxygen, forming $^{14}CO_2$. The atmospheric concentration of $^{14}CO_2$ amounts to only about 10^{-12} of all the molecules of CO_2 in the air. These traces of radioactive CO_2 (along with the stable forms $^{12}CO_2$ and $^{13}CO_2$) are incorporated into the structures of green plants during photosynthesis. The tiny fraction of the plant's mass that is ^{14}C gets even tinier after a plant dies, or after a part of it stops growing (and photosynthesizing) because ^{14}C undergoes β decay as we described in Section 21.4:

$$^{14}_{6}C \rightarrow {}^{14}_{7}N + {}^{0}_{-1}\beta$$

The half-life of the decay process is 5730 years.

CONNECTION Francis Aston's "positive ray analyzer" (Section 2.2) was the forerunner of the modern-day mass spectrometers used to determine isotopic abundances and radiometric ages of samples.

radiocarbon dating a method for establishing the age of a carbon-containing object by measuring the activity of carbon-14 remaining in the object.

If we could determine the ^{14}C content (N_t) of an object of historical interest, such as a piece of wood from an ancient building, charcoal from a prehistoric campfire, or papyrus from an early Egyptian scroll, and if we knew (or could predict) its ^{14}C content when the material in it was alive (N_0), then we could apply Equation 21.4:

$$\frac{N_t}{N_0} = 0.5^{t/t_{1/2}} \tag{21.4}$$

to determine its age. Predicting the value of N_0 is usually done by analyzing samples from growing plants—that is, samples for which the ^{14}C decay time is zero.

To facilitate radiocarbon dating calculations, let's solve Equation 21.4 for t by first taking the natural log of both sides (keeping in mind that $\ln 0.5 = -0.693$):

$$\ln \frac{N_t}{N_0} = -0.693 \frac{t}{t_{1/2}} \tag{21.17}$$

Rearranging the terms to solve for t:

$$t = -\frac{t_{1/2}}{0.693} \ln \frac{N_t}{N_0} \tag{21.18}$$

gives us a useful equation for determining the radiocarbon age t.

SAMPLE EXERCISE 21.7   Radiocarbon Dating

The ^{14}C content of a wooden harpoon handle found recently in the remains of an Inuit encampment in western Alaska was 61.9% of the ^{14}C content of the same type of wood from a recently cut tree. How old is the harpoon?

Collect and Organize We are asked to calculate the age of a sample that contains 61.9% of the ^{14}C in a modern sample of the same material. The half-life of carbon-14 is 5730 years. Equation 21.18 relates the age t of the artifact to the ratio of the ^{14}C in it today to its initial ^{14}C content.

Analyze The ^{14}C content of the modern sample can be used as a surrogate for the initial ^{14}C content of the artifact. Therefore 61.9% (or 0.619) represents the ratio N_t/N_0. This value is greater than 0.5, which means that the age of the sample is less than one half-life (5730 yr).

Solve

$$t = -\frac{t_{1/2}}{0.693} \ln \frac{N_t}{N_0}$$

$$= -\frac{5730 \text{ yr}}{0.693} \ln(0.619)$$

$$= 3966 \text{ yr} = 3.97 \times 10^3 \text{ yr}$$

Think about It The resulting age is less than one half-life, which is reasonable because it contained more than half the original carbon-14 content. The result is expressed with three significant figures to match that of the starting composition (61.9%).

Practice Exercise The carbon-14 activity in papyrus growing along the Nile River today is 231 Bq per kilogram of carbon. If a papyrus scroll found near the Great Pyramid at Cairo has a carbon-14 activity of 127 Bq per kilogram of carbon, how old is the scroll?

The accuracy of radiocarbon dating can be checked by determining the ^{14}C content of the annual growth rings of very old trees, such as the bristlecone pines that grow in the American Southwest (Figure 21.21). When scientists plot the radio-

FIGURE 21.21 Radiocarbon dating relies on knowing the atmospheric concentration of carbon-14 over time. Ancient living trees, such as the bristlecone pines in the American Southwest, act as a check of the atmospheric carbon-14 levels over thousands of years. The ages of the rings can be determined by counting them, and their carbon-14 content can be determined by mass spectrometry.

carbon ages of these rings against their actual ages obtained by counting rings starting from the outer growth layer of the tree (representing $t = 0$), they find that the two sets of ages do not agree exactly, as shown in Figure 21.22. There are several reasons for this lack of agreement, including variability in the rates of ^{14}C production due to changing intensity of the cosmic rays striking Earth's upper atmosphere. To assure accurate ^{14}C results, scientists must correct for these and other variations, and they must be careful to avoid contaminating ancient samples with modern carbonaceous material. With proper analytical technique, radiocarbon dating results are generally accurate to within ± 40 yr for samples that are 500–50,000 years old.

CONCEPT TEST ....

How might the increased consumption of fossil fuels over the last century affect the ^{14}C content of growing plant tissues?

....

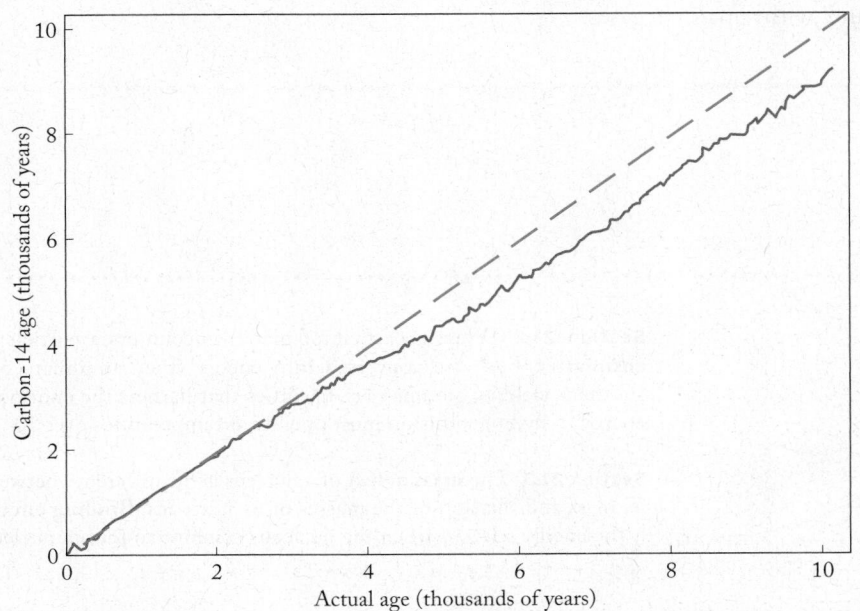

FIGURE 21.22 Calibration curves for radiocarbon dating allow scientists to calculate accurately the ages of archaeological objects. If the rate of ^{14}C production in the upper atmosphere were constant, then the age of objects based on their ^{14}C content would match their actual age—a condition represented by the red dashed line. However, analyses of tree rings, corals, and lake sediments tell us that the rate of ^{14}C production in the upper atmosphere is variable, so that a real plot of ^{14}C age versus actual age produces the jagged blue line. This plot allows scientists to convert ^{14}C ages into actual ages.

Radon, Radium, and Uranium: Too Hot to Handle

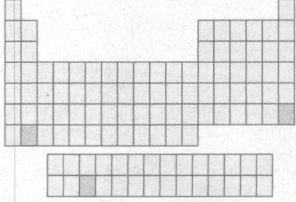

Uranium is the most massive naturally occurring element on Earth. All isotopes of uranium are radioactive including the two most abundant ones: ^{238}U and ^{235}U. Uranium is found primarily in combination with oxygen as the black mineral pitchblende (U_3O_8) and as a complex mineral called uraninite [$K_2(UO_2)_2(VO_4)_2 \cdot 3H_2O$].

Isolating uranium metal involves converting uranium ore first to UO_3 and then to UO_2. Conversion of UO_2 to UF_4 followed by reaction with magnesium metal produces silvery uranium metal. Uranium is one of the densest metals known (19.04 g/mL) and reacts with most elements in the periodic table. The most significant use of uranium is in nuclear reactors, where the fission of uranium-235 is used to produce energy. Both uranium metal and UO_2 are used as reactor fuel.

The spontaneous decay of ^{238}U and ^{235}U produces other massive elements, including thorium, protactinium, and radium, as shown in Figure 21.7.

Because it is a product of uranium-238 decay, radium-226 is typically found in uranium-containing ores. Radium belongs to the same family as magnesium and calcium and has similar chemical properties. For example, radium forms Ra^{2+} ions, whose common salts dissolve in water. In the 19th century, before the hazards of radioactivity were understood, hot springs rich in radium were thought to be useful in curing diseases (Figure 21.23). Radium was also used in watches in the early 20th century to provide glowing dials. The radium-containing paint was applied by hand with small

FIGURE 21.23

brushes. The painters would often pass the tip of the brush between their lips to sharpen its point. As a result, they ingested this emitter of alpha particles. Radium tended to concentrate in tissues like bone that are normally high in Ca^{2+} and Mg^{2+} and dramatically increased the incidence of cancer tumors in those tissues.

Radium-226 decays by emitting alpha particles, yielding the radioactive gas radon-222. The hazards of radon are discussed in Section 21.8. A simple radon detector containing a form of carbon that absorbs radon is available for homeowners who wish to test the air in their homes. The radioactive decay of radon-222 leaves solid lead-206 and bismuth-214 on the carbon. The radiation from these isotopes provides an indirect measure of the amount of radon originally present in the sample. Radon is a group 18 element, one of the noble gases.

SUMMARY

Section 21.1 Nuclear chemistry is the study and application of reactions that involve changes in atomic nuclei. Unstable nuclei undergo **radioactive decay**, which follows first-order kinetics, so the half-life ($t_{1/2}$) of a radionuclide is a characteristic value of the decay process.

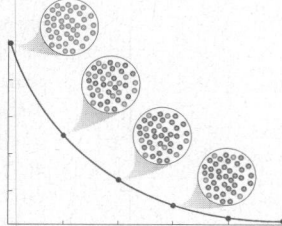

Section 21.2 When a particle of matter encounters a particle of **antimatter**, both are converted into energy (they annihilate one another), yielding gamma rays. Facilities that harness the enormous energy of hydrogen fusion must operate at temperatures over 10^8 K.

Section 21.3 The **mass defect** of a nucleus is the difference between its mass and the sum of the masses of its nucleons. **Binding energy** is the energy released when the nucleons combine to form a nucleus.

It is also the energy needed to split the nucleus into its nucleons. Binding energy per nucleon is a measure of the relative stability of a nucleus.

Section 21.4 Stable nuclei have neutron-to-proton ratios that fall within a range of values called the **belt of stability**. Unstable nuclides undergo radioactive decay. Neutron-rich nuclides (mass number greater than the average atomic mass) undergo β decay; neutron-poor nuclides undergo **positron emission** or **electron capture**. Very large nuclides ($Z > 83$) may undergo α **decay**.

Section 21.5 Nuclides not found in nature can be synthesized in particle accelerators and nuclear reactors in which atoms and subatomic particles collide at high speeds.

Section 21.6 Neutron absorption by uranium-235 and a few other massive isotopes may lead to **nuclear fission** into lighter nuclei accompanied by the release of energy that can be harnessed to generate electricity. A **chain reaction** happens when the neutrons released during fission collide with other fissionable nuclei. They require a **critical mass** of a fissionable isotope. A **breeder reactor** is used to make plutonium-239 from uranium-238 while also producing energy to make electricity.

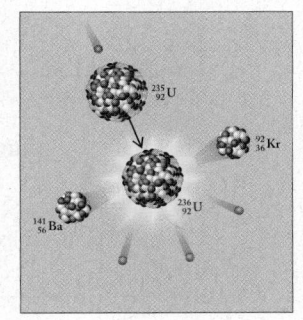

Section 21.7 **Scintillation counters** and **Geiger counters** are used to measure levels of nuclear radiation. The activity of a radioactive sample is the number of decay events per unit time. Common units of radioactivity are the **becquerel (Bq)** (1 decay event/s) and the **curie (Ci)** (1 Ci = 3.70×10^{10} Bq).

Section 21.8 Alpha particles, β particles, and γ rays have enough energy to break up molecules into electrons and cations and so are examples of **ionizing radiation** that can damage body tissue and DNA. The quantity of ionizing radiation energy absorbed per kilogram of body mass is called the *absorbed dose* and is expressed in **grays (Gy)**: 1 Gy = 1.00 J/kg. The effective dose of any type of ionizing radiation is the product of the absorbed dose in grays and the **relative biological effectiveness (RBE)** of the radiation; the unit of effective dose is the **sievert (Sv)**. Alpha particles have a larger RBE than β particles and γ rays but have the least penetrating power of these three types of ionizing radiation.

Section 21.9 Selected radioactive isotopes are useful as tracers in the human body to map biological activity and diagnose diseases. Other radioactive isotopes are used to treat cancers.

Section 21.10 Radiometric dating is used to determine the age of an object based on its content of a radionuclide and/or its decay product. Radiocarbon dating involves measuring the amount of radioactive carbon-14 that remains in an object derived from plant or animal tissue to calculate the age of the object. To improve the accuracy of the technique, scientists calibrate the results of radiometric analysis of samples of known age, such as growth rings in ancient trees.

PROBLEM-SOLVING SUMMARY

TYPE OF PROBLEM	CONCEPTS AND EQUATIONS	SAMPLE EXERCISES
Calculations involving half-lives	$N_t/N_0 = 0.5^n$ where $n = t/t_{1/2}$ and N_t/N_0 is the ratio of the quantity of radionuclide present in a sample at time t (N_t) to the quantity at $t = 0$ (N_0).	21.1
Calculating the energy released in a nuclear reaction	$E = mc^2$, where m is loss in mass as reactants form products.	21.2
Predicting the modes and products of radioactive decay	Neutron-rich nuclides tend to undergo β decay; neutron-poor nuclides undergo positron emission or electron capture.	21.3
Completing and balancing nuclear equations	Add products to balance the mass numbers and atomic numbers (charges) of reactants and products.	21.4
Calculating the activity of a radioactive sample	Activity = kN where $k = 0.693/t_{1/2}$ (Equation 21.16).	21.5
Calculating effective dose	Effective dose = absorbed dose × RBE	21.6
Radiocarbon dating	$t = -\dfrac{t_{1/2}}{0.693} \ln \dfrac{N_t}{N_0}$ (21.18)	21.7

VISUAL PROBLEMS

(Answers to boldface end-of-chapter questions and problems are in the back of the book.)

21.1. Which of the highlighted elements in Figure P21.1 has a stable isotope with no neutrons in its nucleus?

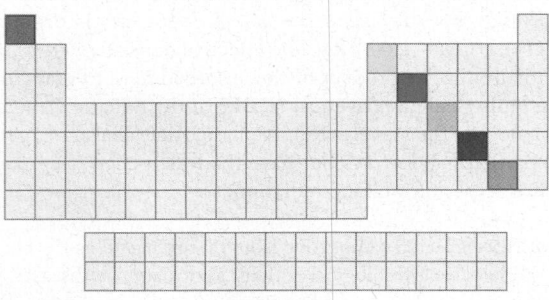

FIGURE P21.1

21.2. Which highlighted element in Figure P21.1 has stable isotopes with the largest ratios of neutrons to protons in their nuclei?

21.3. Which of the highlighted elements in Figure P21.1 has no stable isotopes?

21.4. Which of the graphs in Figure P21.4 illustrates α decay? Which decay pathway does the other graph illustrate?

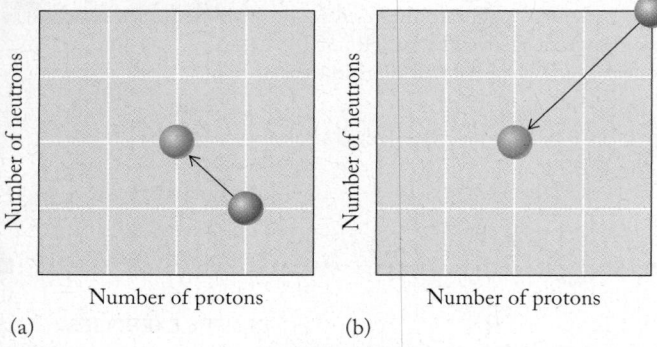

FIGURE P21.4

21.5. Which of the graphs in Figure P21.5 illustrates β decay?

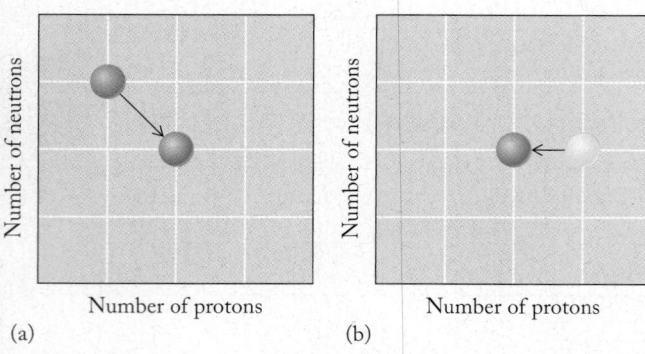

FIGURE P21.5

21.6. Which of the graphs in Figure P21.6 illustrates the overall effect of neutron capture followed by β decay?

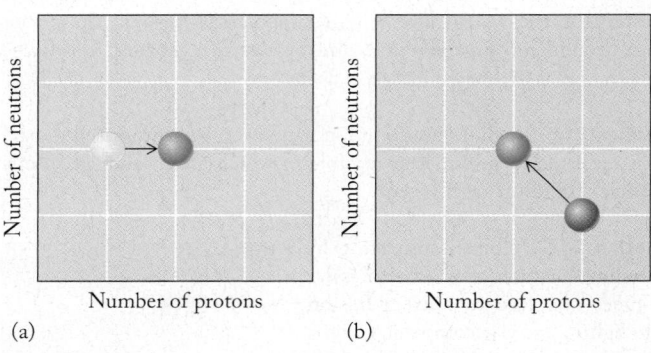

FIGURE P21.6

21.7. Which of the curves in Figure P21.7 represents the decay of an isotope that has a half-life of 2.0 d?

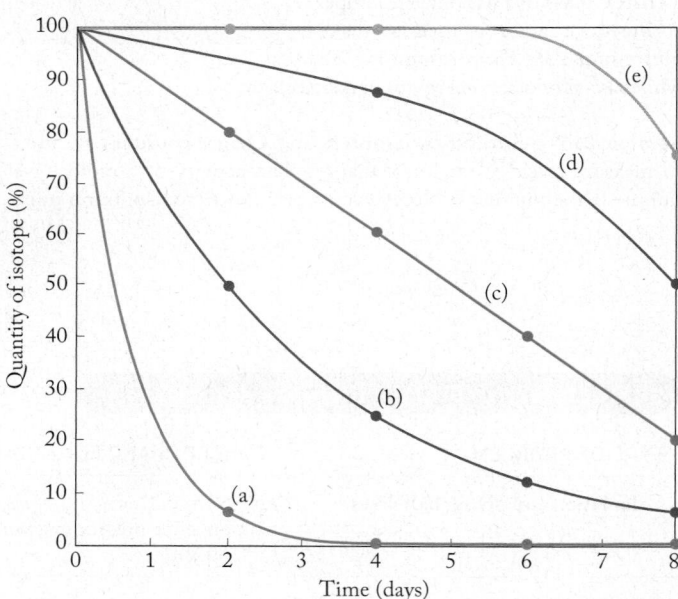

FIGURE P21.7

21.8. Which of the curves in Figure P21.7 do not represent a radioactive decay curve?

21.9. Which of the models in Figure P21.9 represents fission and which represents fusion?

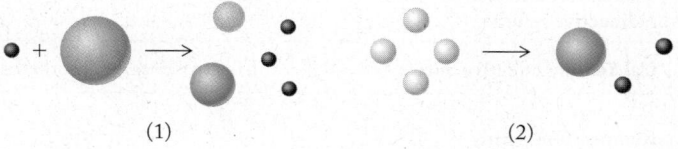

FIGURE P21.9

21.10. Isotopes in a nuclear decay series emit particles with positive and negative charges. The two kinds of particles penetrate a column of water as shown in Figure P21.10. Is the "X" particle the positive or the negative one?

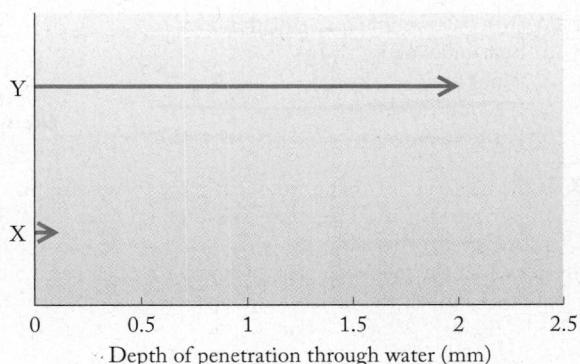

FIGURE P21.10

QUESTIONS AND PROBLEMS ···································■

Radioactive Decay

CONCEPT REVIEW

21.11. Arrange the following particles in order of increasing mass: electron, β particle, positron, proton, neutron, α particle, deuteron.

21.12. Electromagnetic radiation is emitted when a neutron and proton fuse to make a deuteron. In which region of the electromagnetic spectrum is the radiation?

21.13. Scientists at the Fermi National Accelerator Laboratory in Illinois announced in the fall of 1996 that they had created "antihydrogen." How does antihydrogen differ from hydrogen?

21.14. Describe an antiproton.

PROBLEMS

21.15. Calculate the energy and wavelength of the two gamma rays released by the annihilation of a proton and an antiproton.

21.16. Calculate the energy released and the wavelength of the two photons emitted in the annihilation of an electron and a positron.

21.17. What percentage of a sample's original radioactivity remains after two half-lives?

21.18. What percentage of a sample's original radioactivity remains after five half-lives?

Nuclear Binding Energies

CONCEPT REVIEW

21.19. What do the terms *mass defect* and *binding energy* mean?

21.20. Why is energy released in a nuclear fusion process when the product is an element preceding iron in the periodic table?

PROBLEMS

21.21. What is the binding energy of ^{51}V? The mass of ^{51}V is 50.9440 amu.

21.22. What is the binding energy of ^{50}Ti? The mass of ^{50}Ti is 49.9448 amu.

21.23. All of the following fusion reactions produce ^{28}Si. Calculate the energy released in each reaction from the masses of the isotopes: ^2H (2.0146 amu), ^4He (4.00260 amu), ^{10}B (10.0129 amu), ^{12}C (12.000 amu), ^{14}N (14.00307 amu), ^{16}O (15.99491 amu), ^{24}Mg (23.98504 amu), and ^{28}Si (27.97693 amu).
   a. $^{14}N + ^{14}N \rightarrow ^{28}Si$
   b. $^{10}B + ^{16}O + ^2H \rightarrow ^{28}Si$
   c. $^{16}O + ^{12}C \rightarrow ^{28}Si$
   d. $^{24}Mg + ^4He \rightarrow ^{28}Si$

21.24. All of the following fusion reactions produce ^{32}S. Calculate the energy released in each reaction from the masses of the isotopes: ^4He (4.00260 amu), ^6Li (6.01512 amu), ^{12}C (12.000 amu), ^{14}N (14.00307 amu), ^{16}O (15.99491 amu), ^{24}Mg (23.98504 amu), ^{28}Si (27.97693 amu), and ^{32}S (31.97207 amu).
   a. $^{16}O + ^{16}O \rightarrow ^{32}S$
   b. $^{28}Si + ^4He \rightarrow ^{32}S$
   c. $^{14}N + ^{12}C + ^6Li \rightarrow ^{32}S$
   d. $^{24}Mg + 2\,^4He \rightarrow ^{32}S$

21.25. Our sun is a fairly small star that has barely enough mass to fuse hydrogen. Calculate the binding energy per nucleon for ^4He, given the exact masses of ^4He (4.00260 amu), ^1_1p (1.00728 amu), and ^1_0n (1.00866 amu).

21.26. What is the binding energy per nucleon of ^{12}C, the mass of which is 12.00000 amu?

The Belt of Stability

CONCEPT REVIEW

21.27. How can the belt of stability be used to predict the likely decay mode of an unstable nuclide?

21.28. Compare and contrast positron-emission and electron-capture processes.

21.29. The ratio of neutrons to protons in stable nuclei increases with increasing atomic number. Use this trend to explain why multiple α decay steps in the ^{238}U decay series are often followed by β decay.

21.30. If a ^{10}B nucleus absorbs a proton, the radionuclide that is produced may undergo α decay. What nuclide is produced in the decay process? Is it stable or radioactive?

PROBLEMS

21.31. Iodine-137 decays to give xenon-137, which decays to give cesium-137. What are the modes of decay in these two reactions?

21.32. Write a balanced nuclear equation for:
   a. Beta emission by ^{28}Mg
   b. Alpha emission by ^{255}Lr
   c. Electron capture by ^{129}Cs
   d. Positron emission by ^{25}Al

21.33. If the mass number of an isotope is more than twice the atomic number, is the neutron-to-proton ratio less than, greater than, or equal to 1?

21.34. In each of the following pairs of isotopes, which has more protons and which one has more neutrons? (a) ^{127}I or ^{131}I; (b) ^{188}Re or ^{188}W; (c) ^{14}N or ^{14}C

21.35. Aluminum is found on Earth exclusively as ^{27}Al. However, ^{26}Al is formed in stars. It decays to ^{26}Mg with a half-life of 7.4×10^5 years. Write an equation describing the decay of ^{26}Al to ^{26}Mg.

21.36. Which nuclide is produced by the β decay of ^{131}I?

21.37. Predict the modes of decay for the following radioactive isotopes: (a) ^{10}C; (b) ^{19}Ne; (c) ^{50}Ti.

21.38. Predict the decay pathways of the following radioactive isotopes: (a) ^{56}Ni; (b) ^{90}Sr; (c) ^{116}Sb.

21.39. **Elements in a Supernova** The isotopes ^{56}Co and ^{44}Ti were detected in supernova SN 1987A. Predict the decay pathway for these radioactive isotopes.

21.40. Nine isotopes of sulfur have mass numbers ranging from 30 to 38. Five of the nine are radioactive: ^{30}S, ^{31}S, ^{35}S, ^{37}S, and ^{38}S. Which of these isotopes do you expect to decay by β decay?

Synthesizing Nuclides

CONCEPT REVIEW

21.41. What factor makes it easier to bombard nuclei with neutrons than protons?

21.42. Explain how the product of β decay has a higher atomic number than the radionuclide from which the product forms.

21.43. Elements with atomic numbers between 27 and 83 are not produced by fusion reactions in giant stars. How are they formed?

21.44. Our sun contains carbon even though it is too small to synthesize carbon by nuclear fusion. Where may the carbon have come from?

21.45. How are linear accelerators and cyclotrons used to make artificial elements?

21.46. Why must the velocity of the nuclide that is fired at a target nuclide to form a supermassive element be not too fast and not too slow, but "just right"?

PROBLEMS

21.47. During a supernova, a ^{56}Fe nucleus absorbs 3 neutrons.
   a. Write a balanced nuclear equation describing this process and the nuclide that forms.
   b. The nuclide that is formed is radioactive. Write a balanced nuclear equation describing its decay.

21.48. A ^{206}Pb nucleus absorbs 4 neutrons.
   a. Write a balanced nuclear equation describing this process and the nuclide that forms.
   b. The nuclide that is formed is radioactive. Write a balanced nuclear equation describing its decay.

21.49. If a nucleus of ^{96}Mo absorbs 3 neutrons and then undergoes β decay, will the nuclide that is produced be stable? Explain your answer.

21.50. During a supernova, a ^{212}Po nucleus absorbs 4 neutrons and then undergoes α decay. What nuclide is produced by these events?

21.51. Complete the following nuclear equations for reactions used in the preparation of isotopes for nuclear medicine:
   a. ^{32}S + ^{1}n $\rightarrow$? + ^{1}H
   b. ^{55}Mn + ^{1}H $\rightarrow$ ^{52}Fe + ?
   c. ^{75}As + ? $\rightarrow$ ^{77}Br
   d. ^{124}Xe + ^{1}n $\rightarrow$? $\rightarrow$ ^{125}I + ?

21.52. Complete the following nuclear equations for reactions used in the preparation of isotopes for nuclear medicine:
   a. ^{6}Li + ^{1}n $\rightarrow$ ^{3}H + ?
   b. ^{16}O + ^{3}H $\rightarrow$ ^{18}F + ?
   c. ^{56}Fe + ? $\rightarrow$ ^{57}Co + ^{1}n
   d. ^{121}Sb + ^{4}He $\rightarrow$? + $2\,^1$n

21.53. Complete the following nuclear equations:
   a. ? $\rightarrow$ ^{122}Xe + $_{-1}^{0}\beta$
   b. ? + ^{4}He $\rightarrow$ ^{13}N + ^{1}n
   c. ? + ^{1}n $\rightarrow$ ^{59}Fe
   d. ? + ^{1}H $\rightarrow$ ^{67}Ga + $2\,^1$n

21.54. Complete the following nuclear equations:
   a. ^{210}Po $\rightarrow$ ^{206}Pb + ?
   b. ^{3}H $\rightarrow$ ^{3}He + ?
   c. ^{11}C $\rightarrow$ ^{11}B + ?
   d. ^{111}In $\rightarrow$ ^{111}Cd + ?

*21.55. **Uranium Mining** The presence of uranium-containing ores has made part of the Northern Territory of Australia a battleground between those seeking to mine the uranium and the indigenous aborigine population. An article in *Outside* magazine in March 1999 described the dangers of a proposed mine as follows:

> Thorium 230 becomes radium 226. . . . Radium 226 goes to radon 222. Radon 222, a heavy gas that will flow downhill, goes to polonium 218 when one alpha pops out of the nucleus. . . . Polonium 218 goes to lead 214, lead 214 to bismuth 214, bismuth 214 to polonium 214, and then that goes to lead 210, all within minutes, amid a crackle of alphas and betas and gammas.

   a. Write balanced nuclear reactions for the decay of thorium-230 and determine how many alpha and beta particles are produced.
   b. Using an appropriate reference such as the *CRC Handbook of Chemistry and Physics* (CRC Press, Boca Raton, FL), find the half-lives for each isotope and comment on the statement that all these processes take place "within minutes."

21.56. It was incorrectly reported in 1999 that bombardment of a ^{208}Pb target with ^{86}Kr nuclei may produce a nucleus with 118 protons and 175 neutrons. Write a balanced equation describing this process.

21.57. Describe how a ^{209}Bi target might be bombarded with subatomic particles to form ^{211}At. Use balanced equations for the required nuclear reactions.

*21.58. Bombardment of ^{239}Pu with α particles produces ^{242}Cm and another particle. Use a balanced nuclear reaction to determine the identity of the missing particle. The synthesis of which other nuclei in this chapter involves the same subatomic particles?

Nuclear Fission

CONCEPT REVIEW

21.59. How is the rate of energy release controlled in a nuclear reactor?

21.60. How does a breeder reactor create fuel and energy at the same time?

*21.61. Why are neutrons always by-products of the fission of most massive nuclides? (*Hint*: Look closely at the neutron-to-proton ratios shown in Figure 21.4.)

21.62. Seaborgium (Sg, element 106) is prepared by the bombardment of curium-248 with neon-22, which produces two isotopes, ^{265}Sg and ^{266}Sg. Write balanced nuclear reactions for the formation of both isotopes. Are these reactions better described as fusion or fission processes?

PROBLEMS

21.63. The fission of uranium produces dozens of isotopes. For each of the following fission reactions, determine the identity of the unknown nuclide:

a. $^{235}U + {}^1n \longrightarrow {}^{96}Zr + ? + 2\,{}^1n$

b. $^{235}U + {}^1n \longrightarrow {}^{99}Nb + ? + 4\,{}^1n$

c. $^{235}U + {}^1n \longrightarrow {}^{90}Rb + ? + 3\,{}^1n$

21.64. For each of the following fission reactions, determine the identity of the unknown nuclide:

a. $^{235}U + {}^1n \longrightarrow {}^{137}I + ? + 2\,{}^1n$

b. $^{235}U + {}^1n \longrightarrow {}^{94}Kr + ? + 2\,{}^1n$

c. $^{235}U + {}^1n \longrightarrow {}^{95}Sr + ? + 2\,{}^1n$

Measuring Radioactivity; Biological Effects of Radioactivity

CONCEPT REVIEW

21.65. What is the difference between a *level* of radioactivity and a *dose* of radioactivity?

21.66. What are some of the molecular effects of exposure to radioactivity?

21.67. Describe the dangers of exposure to radon-222.

21.68. **Food Safety** Periodic outbreaks of food poisoning from *E. coli* contaminated meat have renewed the debate about irradiation as an effective treatment of food. In one newspaper article on the subject, the following statement appeared: "Irradiating food destroys bacteria by breaking apart their molecular structure." How would you improve or expand on this explanation?

PROBLEMS

21.69. **Radiation Exposure from Dental X-Rays** Dental X-rays expose patients to about 5 μSv of radiation. Given an RBE of 1 for X-rays, how many grays of radiation does 5 μSv represent? For a 50 kg person, how much energy does 5 μSv correspond to?

*21.70. **Radiation Exposure at Chernobyl** Some workers responding to the explosion at the Chernobyl nuclear power plant were exposed to 5 Sv of radiation, resulting in death for many of them. If the exposure was primarily in the form of γ rays with an energy of 3.3×10^{-14} J and an RBE of 1, how many γ rays did an 80 kg person absorb?

*21.71. **Strontium-90 in Milk** In the years immediately following the explosion at the Chernobyl nuclear power plant, the concentration of ^{90}Sr in cows' milk in southern Europe was slightly elevated. Some samples contained as much as 1.25 Bq/L of ^{90}Sr radioactivity. The half-life of strontium-90 is 28.8 years.

a. Write a balanced nuclear equation describing the decay of ^{90}Sr.

b. How many atoms of ^{90}Sr are in a 200 mL glass of milk with 1.25 Bq/L of ^{90}Sr radioactivity?

c. Why would strontium-90 be more concentrated in milk than other foods, such as grains, fruits, or vegetables?

*21.72. **Radium and Cancer** Early in the last century paint containing ^{226}Ra was used to make the dials on watches glow in the dark. The watch painters often passed the tips of their paint brushes between their lips to maintain fine points to paint the tiny numbers on the watches. Many of these painters died from bone cancer. Why?

21.73. In 1999, the U.S. Environmental Protection Agency set a maximum radon level for drinking water at 4.0 pCi per milliliter.

a. How many decay events occur per second in a milliliter of water for this level of radon radioactivity?

b. If the above radioactivity were due to decay of ^{222}Rn ($t_{1/2} = 3.8$ d), how many ^{222}Rn atoms would there be in 1.0 mL of water?

21.74. A former Russian spy died from radiation sickness in 2006 after dining at a London restaurant where he apparently ingested polonium-210. The other people at his table did not suffer from radiation sickness, even though they were very near the radioactive food the victim ate. Why were they not affected?

Medical Applications of Radionuclides

CONCEPT REVIEW

21.75. How does the selection of an isotope for radiotherapy relate to (a) its half-life, (b) its mode of decay, and (c) the properties of the products of decay?

21.76. Are the same radioactive isotopes likely to be used for both imaging and cancer treatment? Why or why not?

PROBLEMS

21.77. Predict the most likely mode of decay for the following isotopes used as imaging agents in nuclear medicine: (a) ^{197}Hg (kidney); (b) ^{75}Se (parathyroid gland); (c) ^{18}F (bone).

21.78. Predict the most likely mode of decay for the following isotopes used as imaging agents in nuclear medicine: (a) ^{133}Xe (cerebral blood flow); (b) ^{57}Co (tumor detection); (c) ^{51}Cr (red blood cell mass); (d) ^{67}Ga (tumor detection).

21.79. A 1.00 mg sample of ^{192}Ir was inserted into the artery of a heart patient. After 30 days, 0.756 mg remained. What is the half-life of ^{192}Ir?

21.80. In a treatment that decreases pain and reduces inflammation of the lining of the knee joint, a sample of dysprosium-165 with an activity of 1100 counts per second was injected into the knee of a patient suffering from rheumatoid arthritis. After 24 hr, the activity had dropped to 1.14 counts per second. Calculate the half-life of ^{165}Dy.

21.81. **Treatment of Tourette's Syndrome** Tourette's syndrome is a condition whose symptoms include sudden movements and vocalizations. Iodine isotopes are used in brain imaging of people suffering from Tourette's syndrome. To study the uptake and distribution of iodine in cells, mammalian brain cells in culture were treated with a solution containing ^{131}I with an initial activity of 108 counts per minute. The cells were removed after 30 days, and the remaining solution was found to have an activity of 4.1 counts per minute. Did the brain cells absorb any ^{131}I ($t_{1/2} = 8.1$ d)?

21.82. A patient is administered mercury-197 to evaluate kidney function. Mercury-197 has a half-life of 65 hr. What fraction of an initial dose of mercury-197 remains after 6 days?

21.83. Carbon-11 is an isotope used in positron-emission tomography and has a half-life of 20.4 min. How long will it take for 99% of the ^{11}C injected into a patient to decay?

21.84. Sodium-24 is used to treat leukemia and has a half-life of 15 hr. A patient was injected with a salt solution containing sodium-24. What percentage of the ^{24}Na remained after 48 hr?

*21.85. **Boron Neutron-Capture Therapy** In boron neutron-capture therapy (BNCT), a patient is given a compound containing ^{10}B that accumulates inside cancer tumors. Then the tumors are irradiated with neutrons, which are absorbed by ^{10}B nuclei. The product of neutron capture is an unstable form of ^{11}B that undergoes α decay to ^{7}Li.
  a. Write a balanced nuclear equation for the neutron absorption and α decay process.
  b. Calculate the energy released by each nucleus of boron-10 that captures a neutron and undergoes α decay, given the following masses of the particles in the process: ^{10}B (10.0129 amu), ^{7}Li (7.01600 amu), ^{4}He (4.00260 amu), and ^{1}n (1.00866 amu).
  c. Why is the formation of a nuclide that undergoes α decay a particularly effective cancer therapy?

21.86. **Balloon Angioplasty and Arteriosclerosis** Balloon angioplasty is a common procedure for unclogging arteries in patients suffering from arteriosclerosis. Iridium-192 therapy is being tested as a treatment to prevent reclogging of the arteries. In the procedure, a thin ribbon containing pellets of ^{192}Ir is threaded into the artery. The half-life of ^{192}Ir is 74 days. How long will it take for 99% of the radioactivity from 1.00 mg of ^{192}Ir to disappear?

Radiometric Dating

CONCEPT REVIEW

21.87. Explain why radiocarbon dating is reliable only for artifacts and fossils younger than about 50,000 years.

21.88. Which of the following statements about ^{14}C dating are true?
  a. The amount of ^{14}C in all objects is the same.
  b. Carbon-14 is unstable and is readily lost from the atmosphere.
  c. The ratio of ^{14}C to ^{12}C in the atmosphere is a constant.
  d. Living tissue will absorb ^{12}C but not ^{14}C.

21.89. Why is ^{40}K dating ($t_{1/2} = 1.28 \times 10^9$ yr) useful only for rocks older than 300,000 years?

21.90. Where does the ^{14}C found in plants come from?

PROBLEMS

21.91. **First Humans in South America** Archeologists continue to debate the origins and dates of arrival of the first humans in the Western Hemisphere. Radiocarbon dating of charcoal from a cave in Chile was used to establish the earliest date of human habitation in South America as 8700 years ago. What fraction of the ^{14}C present initially remained in the charcoal after 8700 years?

21.92. For thousands of years native Americans living along the north coast of Peru used knotted cotton strands called *quipu* (Figure P21.92) to record financial transactions and governmental actions. A particular quipu sample is 4800 years old. Compared with the fibers of cotton plants growing today, what is the ratio of carbon-14 to carbon-12 in the sample?

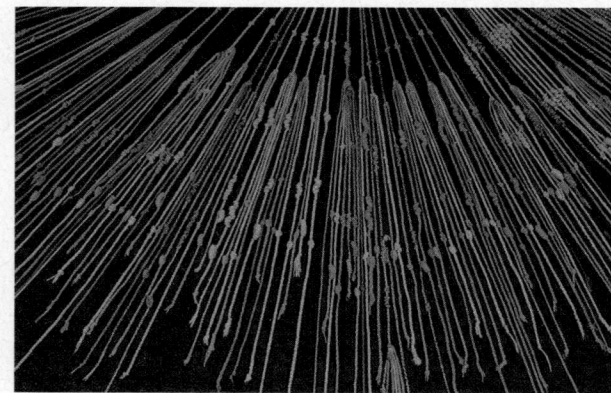

FIGURE P21.92

*21.93. Figure P21.93 is a close-up of the center of a giant sequoia tree cut down in 1891 in what is now Kings Canyon National Park. It contained 1342 annual growth rings. If samples of the tree were removed for radiocarbon dating today, what would be the difference in the ^{14}C/^{12}C ratio in the innermost (oldest) ring compared with that ratio in the youngest ring?

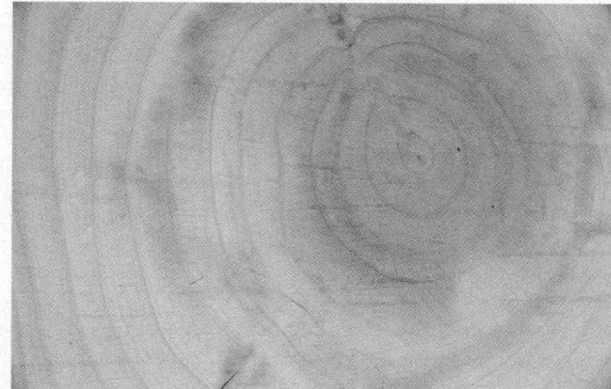

FIGURE P21.93

*21.94. Geologists who study volcanoes can develop historical profiles of previous eruptions by determining the ^{14}C/^{12}C ratios of charred plant remains entrapped in old magma and ash flows. If the uncertainty in determining these ratios is 0.1%, could radiocarbon dating distinguish between debris from the eruptions of Mt. Vesuvius that occurred in AD 472 and AD 512? (*Hint:* Calculate the ^{14}C/^{12}C ratios for samples from the two dates.)

21.95. Figure P21.95 shows a carved mammoth tusk that was uncovered at an ancient camp site in the Ural Mountains in 2001. The $^{14}C/^{12}C$ ratio in the tusk was only 1.19% of that in modern elephant tusks. How old is the mammoth tusk?

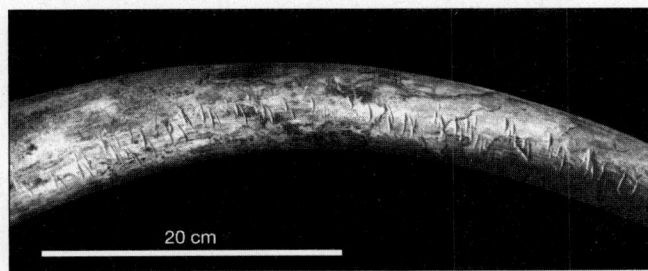

20 cm

FIGURE P21.95

21.96. The Destruction of Jericho The Bible describes the Exodus as a period of 40 years that began with plagues in Egypt and ended with the destruction of Jericho. Archeologists seeking to establish the exact dates of these events have proposed that the plagues coincided with a huge eruption of the volcano Thera in the Aegean Sea.

    a. Radiocarbon dating suggests that the eruption occurred around 1360 BCE, though other records place the eruption of Thera in the year 1628 BCE. (BCE is the abbreviation for "before common era." BCE dates have the same values as BC dates). What is the percent difference in the ^{14}C decay rate in biological samples from these two dates?

    b. Radiocarbon dating of blackened grains from the site of ancient Jericho provides a date of 1315 BCE ± 13 years for the fall of the city. What is the $^{14}C/^{12}C$ ratio in the blackened grains compared with that of grain harvested last year?

Additional Problems

21.97. Thirty years before the creation of antihydrogen, television producer Gene Roddenberry (1921–1991) proposed to use this form of antimatter to fuel the powerful "warp" engines of the fictional starship *Enterprise*. Why would antihydrogen have been a particularly suitable fuel?

21.98. Why are much higher energies required in the cores of giant stars for the fusion of carbon and helium nuclei to form oxygen than for the fusion of hydrogen nuclei to form helium?

21.99. Tiny concentrations of radioactive tritium (^3_1H) occur naturally in rain and groundwater. The half-life of ^3_1H is 12 years. Assuming that tiny concentrations of tritium can be determined accurately, could the isotope be used to determine whether a bottle of wine with the year 1969 on its label actually contained wine made from grapes that were grown in 1969? Explain your answer.

21.100. In the summer of 2003 a team of American and Russian scientists synthesized isotopes of element 115 by bombarding a target of ^{243}Am with ^{48}Ca. One of the isotopes had a mass number of 288.

    a. Write a balanced equation describing the nuclear reaction that produced this isotope of element 115. (You may use the symbol "X" for nuclides of $Z > 112$.)

    b. About 100 ms after it was produced this isotope underwent α decay. Write a balanced equation describing the decay process.

    c. Within about 20 s the isotope formed by the α decay process emitted four more α particles. Write the symbols of the nuclides that were produced after each of these four additional decay events.

21.101. The energy released during the fission of ^{235}U is about 3.2×10^{-11} J per atom of the isotope. Compare this quantity of energy with that released by the fusion of four hydrogen atoms to make an atom of helium-4:

$$4^1_1H \rightarrow {}^4_2He + 2^0_1\beta$$

Assume that the positrons are annihilated in collisions with electrons so that the masses of the positrons are converted into energy. In your comparison, express the energies released by the fission and fusion processes in joules per nucleon for ^{235}U and ^4He, respectively.

21.102. How much energy is required to remove a neutron from the nucleus of an atom of carbon-13 (mass = 13.00335 amu)? (*Hint*: The mass of an atom of carbon-12 is exactly 12.00000 amu.)

21.103. The absorption of a neutron by ^{11}B produces ^{12}B, which decays by two pathways: α decay and β decay.

    a. Write balanced nuclear reactions for these processes.

    b. Which, if either, of the nuclides produced by these decay processes is stable?

*21.104. Colorectal Cancer Treatment Cancer therapy with radioactive rhenium-188 shows promise in patients suffering from colorectal cancer.

    a. Write the symbol for rhenium-188 and determine the number of neutrons, protons, and electrons.

    b. Are most rhenium isotopes likely to have fewer neutrons than rhenium-188?

    c. The half-life of rhenium-188 is 17 hours. If it takes 30 minutes to bind the isotope to an antibody that delivers the rhenium to the tumor, what percentage of the rhenium remains after binding to the antibody?

    d. The effectiveness of rhenium-188 is thought to result from penetration of β particles as deep as 8 mm into the tumor. Why wouldn't an α emitter be more effective?

    e. Using an appropriate reference text, such as the *CRC Handbook of Chemistry and Physics* (CRC Press, Boca Raton, FL), pick out the two most abundant isotopes of rhenium. List their natural abundances and explain why the one that is radioactive decays by the pathway that it does.

21.105. The following nuclear equations are based on successful attempts to synthesize supermassive elements. Complete each equation by filling in the symbol of the supermassive nuclide that was synthesized.

    a. $^{58}_{26}Fe + {}^{209}_{83}Bi \rightarrow ? + {}^1_0n$

    b. $^{64}_{28}Ni + {}^{209}_{83}Bi \rightarrow ? + {}^1_0n$

    c. $^{62}_{28}Ni + {}^{208}_{82}Pb \rightarrow ? + {}^1_0n$

    d. $^{22}_{10}Ne + {}^{249}_{97}Bk \rightarrow ? + 4^1_0n$

    e. $^{58}_{26}Fe + {}^{208}_{82}Pb \rightarrow ? + {}^1_0n$

21.106. **Smoke Detectors** Americium-241 ($t_{1/2} = 433$ yr) is used in smoke detectors. The α particles from this isotope ionize nitrogen and oxygen in the air, creating an electric current. When smoke is present, the current decreases, setting off the alarm.

a. Does a smoke detector bear a closer resemblance to a Geiger counter or to a scintillation counter?

b. How long will it take for the activity of a sample of ^{241}Am to drop to 1% of its original activity?

c. Why are smoke detectors containing ^{241}Am safe to handle without protective equipment?

21.107. In 2006 an international team of scientists confirmed the synthesis of a total of three atoms of $^{294}_{118}$Uuo in experiments run in 2002 and 2005. They bombarded a ^{249}Cf target with ^{48}Ca nuclei.

a. Write a balanced nuclear equation describing the synthesis of $^{294}_{118}$Uuo.

b. The synthesized isotope of Uuo undergoes α decay ($t_{1/2} = 0.9$ ms). What nuclide is produced by the decay process?

c. The nuclide produced in part b also undergoes α decay ($t_{1/2} = 10$ ms). What nuclide is produced by this decay process?

d. The nuclide produced in part c also undergoes α decay ($t_{1/2} = 0.16$ s). What nuclide is produced by this decay process?

e. If you had to select an element that occurs in nature and that has physical and chemical properties similar to Uuo, which element would it be?

*21.108. Consider the following decay series:

$$A\ (t_{1/2} = 4.5\ s) \rightarrow B\ (t_{1/2} = 15.0\ d) \rightarrow C$$

If we start with 10^6 atoms of A, how many atoms of A, B, and C are there after 30 days?

21.109. Which element in the following series will be present in the greatest amount after 1 year?

$$^{214}_{83}\text{Bi} \xrightarrow{\alpha} {}^{210}_{81}\text{Tl} \xrightarrow{\beta} {}^{210}_{82}\text{Pb} \xrightarrow{\beta} {}^{210}_{83}\text{Bi} \rightarrow$$

$$t_{1/2} = \quad 20\ \text{min} \quad 1.3\ \text{min} \quad 20\ \text{yr} \quad 5\ \text{d}$$

*21.110. **Dating Cave Paintings** Cave paintings in Gua Saleh Cave in Borneo have been dated by measuring the amount of ^{14}C in calcium carbonate that formed over the pigments used in the paint. The source of the carbonate ion was atmospheric CO_2.

a. What is the ratio of the ^{14}C activity in calcium carbonate that formed 9900 years ago to that in calcium carbonate formed today?

b. The archeologists also used a second method, uranium–thorium dating, to confirm the age of the paintings by measuring trace quantities of these elements present as contaminants in the calcium carbonate. Shown below are two candidates for the U–Th dating method. Which isotope of uranium do you suppose was chosen? Explain your answer.

$$^{235}_{92}\text{U} \quad \rightarrow \quad ^{231}_{90}\text{Th} \quad \rightarrow \quad ^{231}_{91}\text{Pa} \quad \rightarrow$$

$$t_{1/2} = \quad 7.04 \times 10^8\ \text{yr} \quad 25.6\ \text{hr} \quad 3.25 \times 10^4\ \text{yr}$$

$$^{234}_{92}\text{U} \quad \rightarrow \quad ^{230}_{90}\text{Th} \quad \rightarrow \quad ^{226}_{88}\text{Pa} \quad \rightarrow$$

$$t_{1/2} = \quad 2.44 \times 10^5\ \text{yr} \quad 7.7 \times 10^4\ \text{hr} \quad 1600\ \text{yr}$$

21.111. The synthesis of new elements and specific isotopes of known elements in linear accelerators involves the fusion of smaller nuclei.

a. An isotope of platinum can be prepared from nickel-64 and tin-124. Write a balanced equation for this nuclear reaction. (You may assume that no neutrons are ejected in the fusion reaction.)

b. Substitution of tin-132 for tin-124 increases the rate of the fusion reaction 10 times. Which isotope of Pt is formed in this reaction?

21.112. A sample of drinking water collected from a suburban Boston municipal water system in 2002 contained 0.5 pCi/L of radon. Assume that this level of radioactivity was due to the decay of ^{222}Rn ($t_{1/2} = 3.8$ d).

a. What was the level of radioactivity (Bq/L) of this nuclide in the sample?

b. How many decay events per hour would occur in 2.5 L of the water?

21.113. An atom of darmstadtium-269 was synthesized in 2003 by bombardment of a ^{208}Pb target with ^{62}Ni nuclei. Write a balanced nuclear reaction describing the synthesis of ^{269}Ds.

21.114. There was once a plan to store radioactive waste that contained plutonium-239 in the reefs of the Marshall Islands. The planners claimed that the plutonium would be "reasonably safe" after 240,000 years. If the half-life is 24,400 years, what percentage of the ^{239}Pu would remain after 240,000 years?

21.115. **Dating Prehistoric Bones** In 1997 anthropologists uncovered three partial skulls of prehistoric humans in the Ethiopian village of Herto. Based on the amount of ^{40}Ar in the volcanic ash in which the remains were buried, their age was estimated at between 154,000 and 160,000 years old.

a. ^{40}Ar is produced by the decay of ^{40}K ($t_{1/2} = 1.28 \times 10^9$ yr). Propose a decay mechanism for ^{40}K to ^{40}Ar.

b. Why did the researchers choose ^{40}Ar rather than ^{14}C as the isotope for dating these remains?

*21.116. **Biblical Archeology** The Old Testament describes the construction of the Siloam Tunnel, used to carry water into Jerusalem under the reign of King Hezekiah (727–698 BCE). An inscription on the tunnel has been interpreted as evidence that the tunnel was not built until 200–100 BCE. ^{14}C dating (in 2003) indicated a date close to 700 BCE. What is the ratio of ^{14}C in a wooden object made in 100 BCE to one made from the same kind of wood in 700 BCE?

21.117. **Stone Age Skeletons** The discovery of six skeletons in an Italian cave at the beginning of the 20th century was considered a significant find in Stone Age archaeology. The age of these bones has been debated. The first attempt at radiocarbon dating indicated an age of 15,000 years. Redetermination of the age in 2004 indicated an older age for two bones, between 23,300 and 26,400 years. What is the ratio of ^{14}C in a sample 15,000 years old to one 25,000 years old?

*21.118. The origin of the two naturally occurring isotopes of boron, ^{11}B and ^{10}B, are unknown. Both isotopes may be formed from collisions between protons and carbon, oxygen, or nitrogen in the aftermath of supernova explosions. Propose nuclear reactions for the formation of ^{10}B from such collisions with ^{12}C and ^{14}N.

21.119. Thorium-232 slowly decays to lead-208 (Figure P21.119). The half-lives of the intermediate decay products range from seconds to years. Note that bismuth-212 decays to lead-208 by two pathways: first β and then α decay, or α and then β decay. Thallium-208 is an intermediate nuclide in the second pathway. This radionuclide can be separated from a sample of thorium nitrate by passing a solution of the sample through a filter pad containing ammonium phosphomolybdate. The radioactivity of ^{208}Tl trapped on the filter is measured as a function of time. In one such experiment, the following data were collected:

$$^{232}_{90}\text{Th} \xrightarrow{\alpha} {}^{228}_{88}\text{Ra} \xrightarrow{\beta} {}^{228}_{89}\text{Ac} \xrightarrow{\beta} {}^{228}_{90}\text{Th} \xrightarrow{\alpha}$$

$$t_{1/2} = 1.41 \times 10^{10} \text{ yr} \quad 5.8 \text{ yr} \quad 6.13 \text{ hr} \quad 1.91 \text{ yr}$$

$$^{224}_{88}\text{Ra} \xrightarrow{\alpha} {}^{220}_{86}\text{Rn} \xrightarrow{\alpha} {}^{216}_{84}\text{Po} \xrightarrow{\alpha} {}^{212}_{82}\text{Pb} \xrightarrow{\beta}$$

$$t_{1/2} = \quad 3.64 \text{ d} \quad\quad 55 \text{ s} \quad\quad 0.15 \text{ s} \quad 10.6 \text{ hr}$$

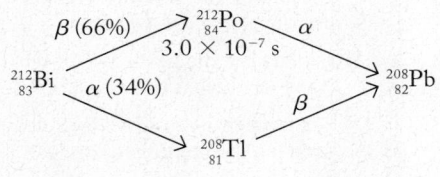

$$t_{1/2} = 60.6 \text{ min} \quad\quad\quad ?$$

FIGURE P21.119

Time (s)	Counts/min
60	62
120	40
180	35
240	22
300	16
360	10

Use the data in the table to determine the half-life of ^{208}Tl.

Radon, Radium, and Uranium: Too Hot to Handle

21.120. Some radon detectors contain a thin plastic film that is sensitive to α particles. An α particle striking the film leaves a track. Why is the number of tracks more important than the length of the track in determining how much radon is present?

21.121. One type of radon detector uses charcoal to collect radioactive ^{214}Pb and ^{214}Bi. The concentration of Rn is calculated by measuring γ rays produced by those two isotopes.
a. How are ^{214}Pb and ^{214}Bi related to ^{222}Rn?
b. Why are lead and bismuth collected rather than collecting radon directly?

21.122. Why is it a bad idea to bathe repeatedly in hot springs containing dissolved radium?

21.123. Both ^{226}Rn and ^{222}Rn are produced from radioactive decay of uranium in soils and rocks. Why are these radon isotopes a greater risk to health than the parent uranium?

21.124. Which of the following elements is most likely to have chemical properties similar to radium: uranium, potassium, barium, or krypton?

 Additional study materials including ChemTours and Diagnostic Quizzes are available at StudySpace at www.wwnorton.com/studyspace.

22

Life and the Periodic Table

22.1 The Periodic Table of Life
22.2 Major Essential Elements
22.3 Trace and Ultratrace Essential Elements
22.4 Nonessential Elements
22.5 Elements for Diagnosis and Therapy

Learning Outcomes

- Distinguish between essential and nonessential elements
- Identify the major essential elements
- Identify atomic properties associated with the toxicity of certain elements
- Summarize the pathways for ion transport across cell membranes
- Explain how radioactive isotopes are used in the diagnosis of disease
- Describe how metals and alloys are used in medical devices

A LOOK AHEAD

Which Elements Do We Find in Our Bodies?

Have you ever wondered how many of the elements in the periodic table are found in the human body? Or wondered which of them are important to the health of humans or other organisms?

Roughly one-third of the 90 naturally occurring elements have an identifiable role in human health and in organisms in general. Some—like carbon, hydrogen, oxygen, nitrogen, sulfur, and phosphorus—are the principal constituents of all plants and animals. The alkali metal cations Na^+ and K^+ act as charge carriers, maintain osmotic pressure, and transmit nerve impulses. The alkaline earth cation Mg^{2+} is important in photosynthesis, and its family member Ca^{2+} forms structural materials such as bones and teeth. Chloride ions balance the charge of Na^+ and K^+ ions to maintain electrical neutrality in living cells. Other main group elements, such as iodine and selenium, are required in tiny amounts in our bodies, and a number of transition metal ions, such as Fe^{2+}, Zn^{2+}, and Ni^{2+}, are key components of the enzymes that catalyze biochemical reactions.

Many elements with no known biological function are useful in medicine as either diagnostic tools or therapeutic agents. Some radioactive isotopes act as effective imaging agents for organs and tumors, for example, and others are used to kill cancer cells. Compounds containing platinum, silver, or gold are effective drugs for treating cancer, burns, and arthritis, respectively. Drugs containing lithium ions are used to treat depression, and corrosion-resistant metals such as tantalum are used in artificial joints.

In this chapter we survey the roles of various elements in the human body and their importance to good health. At the same time we call on the knowledge and skills you have acquired in your study of general chemistry to solve problems that link concepts from prior chapters to the central question of this chapter: What are the roles of the elements in the chemistry of life? ∎

MRI Magnetic resonance imaging is a valuable diagnostic tool in medicine. ▶ Some MRI studies rely on compounds containing gadolinium to enhance the image.

1038

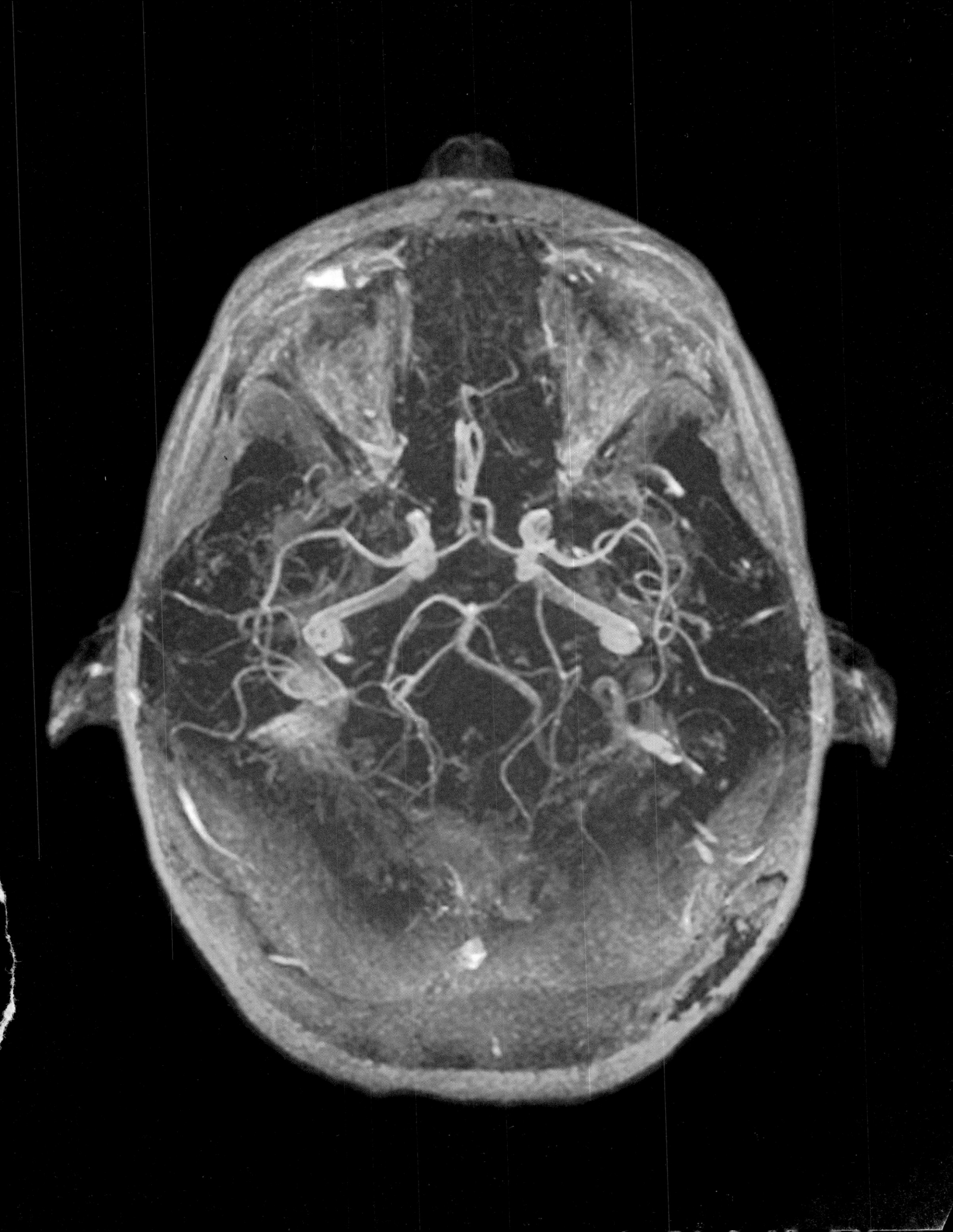

essential element element present in tissue, blood, or other body fluids that has a physiological function.

nonessential element element present in humans that has no known function.

stimulatory effect increased growth or other biological response to the presence of a nonessential element.

major essential element essential element present in the body in average concentrations greater than 1 mg of element per gram of body mass.

trace essential element essential element present in the body in average concentrations between 1 and 1000 μg of element per gram of body mass.

ultratrace essential element essential element present in the body in average concentrations less than 1 μg of element per gram of body mass.

TABLE 22.1	Essential Elements Found in the Human Body	
Major	**Trace**	**Ultratrace**
Calcium	Fluorine	Chromium
Carbon	Iodine	Cobalt
Chlorine	Iron	Copper
Hydrogen	Silicon	Manganese
Magnesium	Zinc	Molybdenum
Nitrogen		Nickel
Oxygen		Selenium
Phosphorus		Vanadium
Potassium		
Sodium		
Sulfur		

TABLE 22.2	Nonessential Elements Found in the Human Body	
Stimulatory	**Unknown Role**	**No Role**
Boron	Antimony	Barium
Titanium	Arsenic	Cesium
		Germanium
		Rubidium
		Strontium

22.1 The Periodic Table of Life

The elements found in the human body can be classified as either essential or nonessential to life (Tables 22.1 and 22.2). **Essential elements** are defined as those that have a beneficial physiological function, including those whose absence impairs functioning of the organism. **Nonessential elements** are present in the body but have no known function. In some cases, the presence of a nonessential element has a **stimulatory effect**, which means that the consumption of small amounts of the element leads to enhanced function via a mechanism that is not yet understood. For example, small amounts of the nonessential element antimony promote growth in some mammals when added to their diets. Nonessential elements and even toxic elements are often incorporated into our bodies because their chemical properties are similar to those of an essential element. For example, Pb^{2+} ions are incorporated into teeth and bones because they are similar to Ca^{2+} ions in size and charge. The alkali metal rubidium is the most abundant nonessential element in humans; Rb^+ is retained by the body because of the similarity of its size and chemistry to K^+.

Oxygen in the form of O_2 gas is the only element that occurs in the body in elemental form though oxygen is also incorporated into many compounds (e.g., H_2O) or ions, such as HCO_3^-. When we speak of any other element in the body, we are always referring to an ion or compound containing that element rather than the pure element. For example, when we describe zinc as an essential element, we are referring to zinc ions, Zn^{2+}, rather than zinc metal.

The essential elements are further classified as **major, trace,** or **ultratrace essential elements**. Major essential elements are present in gram quantities in the human body and are required in large amounts in our diet. Almost all foods are rich in compounds containing carbon, hydrogen, oxygen, nitrogen, sulfur, and phosphorus. Salt is perhaps the most familiar dietary source of sodium and chloride ions, although both are ubiquitous in food. Vegetables like broccoli and Brussels sprouts or fruits like bananas are rich in potassium. Calcium is found in dairy products and is often added to orange juice as a dietary supplement.

Table 22.3 compares the elemental compositions of the human body, the universe, Earth's crust, and seawater. Note that the composition of our bodies most closely resembles the composition of seawater. The match would be even closer if it were not for the biological processes in the sea that remove essential elements such as nitrogen and phosphorus, and that store others in solid structures like the $CaCO_3$ that makes up corals and mollusk shells.

Our diet should supply us with sufficient quantities of all essential elements. In the United States and Canada, these quantities are called *dietary reference intake* (DRI) values. They are based on the recommendations of the Food and Nutrition Board of the National Academy of Sciences and are frequently updated in response to research. For many essential elements, DRI values have replaced the *recommended dietary allowance* (RDA) values you may be familiar with from labels on food and vitamin packages (Figure 22.1). Table 22.4 compares the DRI and RDA values for a number of major, trace, and ultratrace essential elements. Among the major essential elements, DRI/RDA values range from 0.32–0.42 g of magnesium per day to 4.7 g/d of potassium. DRI/RDA values for trace essential elements, including iron and zinc, are in the 10–20 mg/d range. There are DRI/RDA values for some, but not all, of the ultratrace essential elements, ranging from 55 μg/d for selenium and 45 μg/d for molybdenum up to 5 mg/day for manganese.

Some of the elements in the periodic table are toxic. These include radon, beryllium, cadmium, mercury, and lead. As described in Chapter 21, inhaled radon gas poses serious health hazards from α decay taking place inside the body.

TABLE 22.3	Comparative Composition^a of the Universe, Earth's Crust, Seawater, and the Human Body			

Element	Universe (%)	Crust (%)	Seawater (%)	Human Body (%)
Hydrogen	91	0.22	66	63
Oxygen	0.57	47	33	25.5
Carbon	0.021	0.019	0.0014	9.5
Nitrogen	0.042			1.4
Calcium		3.5	0.006	0.31
Phosphorus				0.22
Chlorine			0.33	0.03
Potassium		2.5	0.006	0.06
Sulfur	0.001	0.034	0.017	0.05
Sodium		2.5	0.28	0.01
Magnesium	0.002	2.2	0.033	0.01
Helium	9.1			
Silicon	0.003	28		
Aluminum		7.9		
Neon	0.003			
Iron	0.002	6.2		
Bromine			0.0005	
Titanium		0.46		
All other elements	< 0.1	< 0.1	< 0.1	< 0.1

^a Compositions are expressed as the percentage of the total number of atoms. Because of rounding, the totals do not equal exactly 100%.

TABLE 22.4	Recommended Dietary Allowances (RDA) and Dietary Reference Intakes (DRI) for Selected Essential Elements^a	

Element	DRI	RDA
Calcium	1000 mg	1200 mg
Chlorine	2300 mg	2300 mg
Chromium	25–35 μg	35 μg
Copper	900 μg	900 μg
Fluorine	3–4 mg	4 mg
Iodine	150 μg	150 μg
Iron	8–18 mg	18 mg
Magnesium	420 mg	320–400 mg
Manganese	1.8–2.3 mg	2–5 mg
Molybdenum	45 μg	45 μg
Phosphorus	700 mg	700 mg
Potassium	4700 mg	4700 mg
Selenium	55 μg	55 μg
Sodium	1500 mg	1500 mg
Zinc	8–11 mg	11 mg

^a RDA and DRI values in mg or μg per day from the U.S. Department of Agriculture (2009) and from the Council on Responsible Nutrition (CRN) for 19- to 30-year-olds.

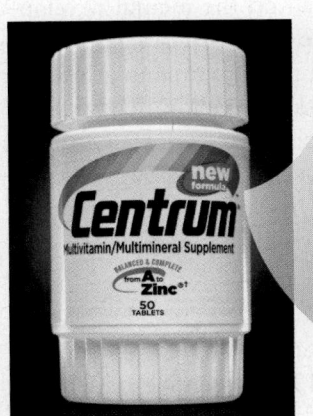

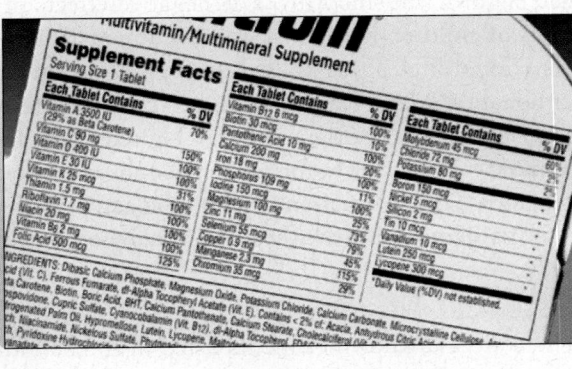

FIGURE 22.1 The labels on multivitamin supplements may not list DRI or RDA values but rather *% daily values* (DVs). Daily values are based on RDA or DRI values, but there can be inconsistencies, particularly among the ultratrace essential elements.

Beryllium toxicity is most often encountered in industrial settings where beryllium-contaminated dust is inhaled. The Be^{2+} ion replaces Mg^{2+} in the body, where it inhibits Mg^{2+}-catalyzed RNA and DNA synthesis in cells.

Prior to the development of nickel–cadmium batteries in the latter part of the 20th century, the toxicity of cadmium compounds was of minimal concern to humans. Nicad batteries thrown into landfills may introduce Cd^{2+} ions into groundwater. Cadmium(II) ions have an ionic radius (95 pm) close to that of calcium ions (100 pm). Their similarity in size leads to cadmium accumulation in bones, which can weaken them.

ion channel group of helical proteins that penetrate cell membranes and allow selective transport of ions.

♾️ **CONNECTION** The vapor pressures of liquids were discussed in Chapter 11. Pressure units such as the pascal (Pa) were defined in Chapter 6.

♾️ **CONNECTION** The applications of the group 12 elements Zn, Cd, and Hg in electrochemistry and battery technology were discussed in Chapter 19.

Mercury poisoning has a longer history. The Mad Hatter in *Alice in Wonderland* exhibits symptoms that were common among hat makers who used mercury(II) nitrate to make felt easier to work. Mercury(II) ions bind to sulfur-containing amino acids and are readily transported throughout the body. Some of the effects of mercury poisoning include memory loss, tremors, and impaired coordination. Although no longer used in hat manufacture, mercury was used in other ways throughout the 20th century, dental amalgams and thermometers being two familiar applications. Mercury spills from broken thermometers are difficult to clean up because the dense liquid finds its way into cracks and corners. Mercury has a small but significant vapor pressure, so anyone in a room with an open container of Hg metal will inhale its vapor. In our bodies, mercury can be transformed into methylmercury(II) ion, CH_3Hg^+, a potent neurotoxin. The methyl group increases the solubility of mercury in the nonpolar portion of cell membranes and allows the CH_3Hg^+ to diffuse through the lipid bilayer. In particular, the membranes separating blood from the brain are vulnerable to penetration by CH_3Hg^+, which leads to the symptoms of mercury poisoning cited above.

The conversion of mercury into CH_3Hg^+ in the environment (*biomethylation*) was responsible for the health problems experienced in the 1950s by people living around Minamata Bay in Japan. Mercury-containing waste from industrial plants was routinely dumped into the bay, and the mercury was converted to CH_3Hg^+. Fish accumulated the ion in their tissues. Local residents ate the fish and suffered the effects of mercury poisoning.

The history of lead toxicity dates back thousands of years. Lead compounds including PbS, $PbCO_3$, and PbCl(OH) were used by the ancient Egyptians in cosmetics. The Romans used lead in pipes for plumbing and in wine carafes and suffered from lead poisoning as a result. In the Middle Ages, lead acetate was used to sweeten wine, a practice with dire consequences. In the 20th century, tetraethyl lead $[(CH_3CH_2)_4Pb]$ was used as a gasoline additive to help automobile engines run smoothly. The negative effects of lead on the mental development of children have been well documented, and the use of lead in gasoline in the United States was phased out by 1986. Similarly, paints containing lead(II) carbonate ("white lead") and lead(II) chromate ("chrome yellow") have been banned from residential use in the United States since 1978. The toxicity of lead is traced to its ability to form strong bonds with oxygen and sulfur groups in many enzymes. This interaction inhibits the activity of enzymes, including those that catalyze hemoglobin synthesis, and many other physiological processes.

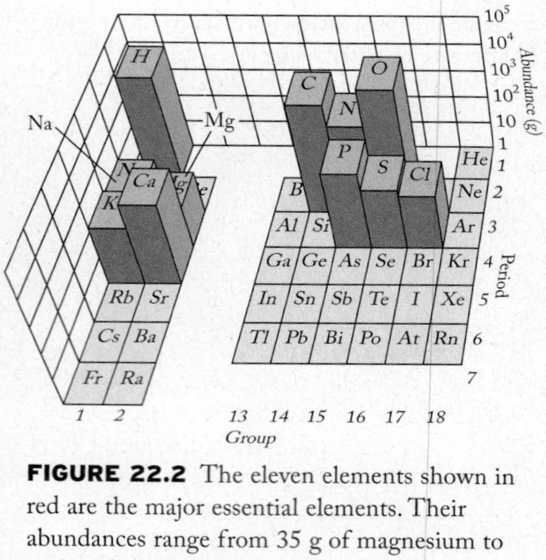

CONCEPT TEST

There are 112 known elements listed in the periodic table. Why do you suppose that none of the elements with atomic numbers 93–112 are essential to life?

(Answers to Concept Tests are in the back of the book.)

22.2 Major Essential Elements

The 11 elements shown in red in Figure 22.2 and listed in the first column of Table 22.1 are the major essential elements. Together they account for more than 99% of the mass of the human body. Oxygen is the most abundant element by mass, followed by carbon and hydrogen. Although life depends on the presence of elemental oxygen in the form of O_2 gas, much of the oxygen in our bodies is combined with hydrogen in water molecules.

FIGURE 22.2 The eleven elements shown in red are the major essential elements. Their abundances range from 35 g of magnesium to 46 kg of oxygen in a 70 kg adult human.

The most abundant elements in the human body are all nonmetals: C, H, O, S, P, and N. Together they make up most of the mass of the human body and are the building blocks of most of its molecular compounds and the principal polyatomic ions: HCO_3^-, SO_4^{2-}, and $H_2PO_4^-$ that are dissolved in body fluids. The average concentrations of the next five most abundant elements in the human body are listed in Table 22.5. Let's explore some of the roles these five elements and nitrogen play in the biochemistry of the human body. As we do, we will revisit several of the chemical principles discussed in prior chapters.

Sodium and Potassium

Regulated concentrations of sodium and potassium ions are crucial to cell function. For example, too much Na^+ has been linked to hypertension (high blood pressure). To maintain a constant concentration of these two alkali metal (group 1) ions in body fluids, the ions must be able to move in and out of cells. As noted in Section 20.4, the membrane surrounding a typical cell is a lipid bilayer with polar groups containing phosphate on the two surfaces of the cell membrane and nonpolar fatty acids oriented toward the middle of the membrane. Direct diffusion of Na^+ and K^+ through the lipid bilayer is difficult because these polar cations do not dissolve in the nonpolar interior of the membrane.

As Figure 22.3 shows, the cell membrane is pierced by **ion channels**, groups of protein complexes that allow selective transport of ions. The cell membrane's ion channels control which ions pass through the membrane based on the size and charge of

TABLE 22.5	Average Concentration of Ca, K, Na, Mg, and Cl in the Human Body
Element	**mg/g of Body Mass**
Calcium	15.0
Potassium	2.0
Sodium	1.5
Magnesium	0.5
Chlorine	1.5

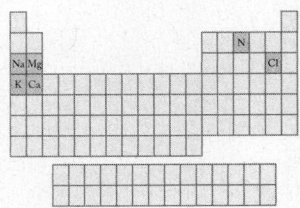

CONNECTION We introduced phospholipids in Chapter 20 as consisting of glycerol with two fatty acid chains and a polar region containing a phosphate group.

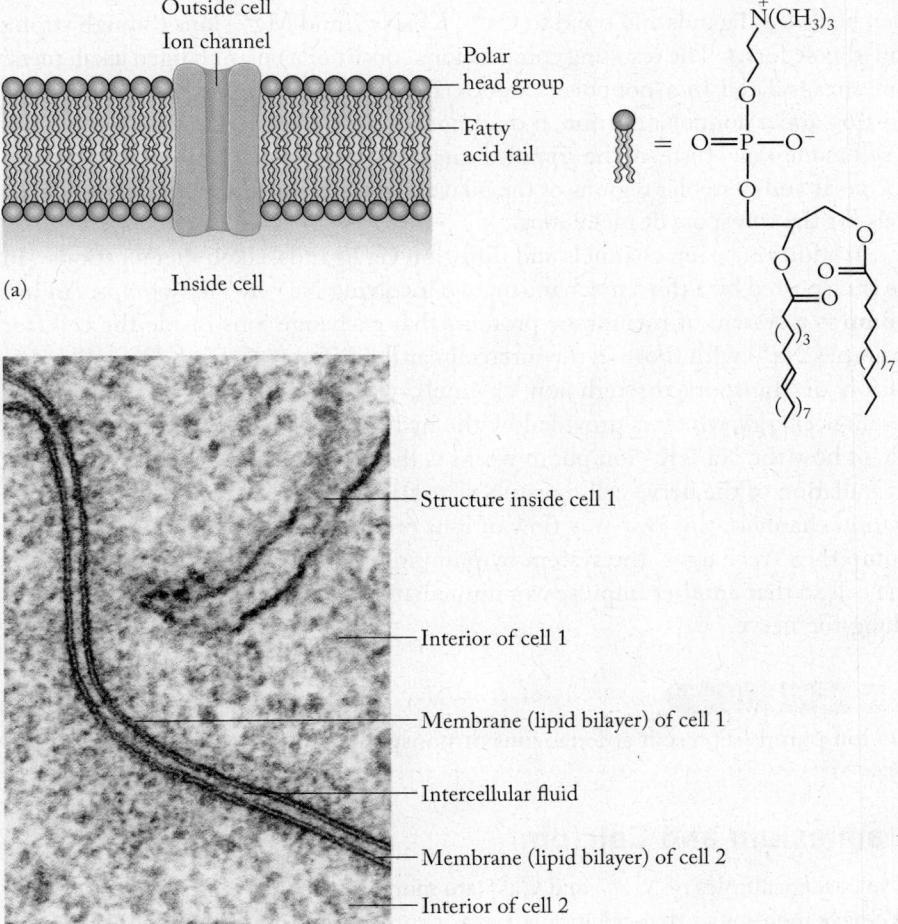

(a)

(b)

FIGURE 22.3 (a) Cell membranes consist of a bilayer of phospholipids pierced by ion channels. The polar groups of the phospholipids face the aqueous solutions inside and outside the cell, whereas the fatty acids form a nonpolar region within the membrane. (b) An electron micrograph of the membranes separating two adjacent cells.

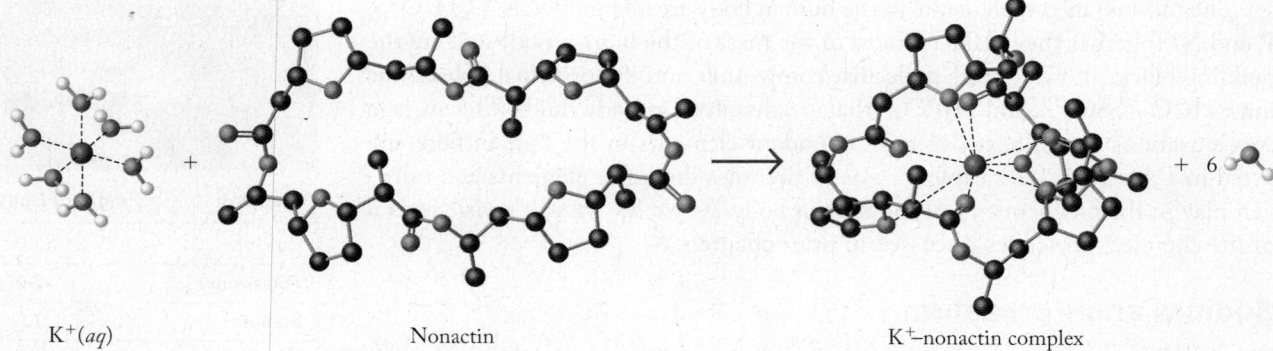

K⁺(aq)   Nonactin   K⁺–nonactin complex   + 6

FIGURE 22.4 In living organisms, ligands such as nonactin can form a complex with any one of the four alkali metal and alkaline earth major essential ions and carry the ion directly through a cell membrane. No ion channel is required in this transport pathway.

the ion as well as the shape of the protein. For example, the protein of the ion channel for potassium ions has its amino acids oriented in such a fashion that favorable ion–dipole interactions occur only for ions with the radius of a K^+ ion (138 pm) and not Na^+ (102 pm) or any other cation. The sodium ion channel is also selective, excluding K^+ and Ca^{2+} even though the radii of Na^+ and Ca^{2+} (100 pm) differ by only 2 pm. Another difference between the Na^+ and K^+ channels is the ability of H_3O^+ (hydronium ion, radius 113 pm) to pass through sodium ion channels but not potassium ion channels.

Living organisms contain oxygen-rich molecules like nonactin (Figure 22.4) that behave as ligands and bond to Ca^{2+}, K^+, Na^+, and Mg^{2+} ions through strong ion–dipole forces. The resulting complex ions consist of a polar, charged alkali metal ion encapsulated in a nonpolar exterior. Because the complex has both a polar portion and a nonpolar portion, it does not require a channel for passage through a cell membrane. Instead, the complex carries its alkali metal cation through both the polar and nonpolar regions of the bilayer, providing an alternative to ion channels for the transport of metal ions.

In addition to ion channels and diffusion via ligands, alkali metal cations can be transported by a third mechanism, one involving Na^+–K^+ ion pumps. An **ion pump** is a system of membrane proteins that exchange ions inside the cell (for example, Na^+) with those in the intercellular fluid (for example, K^+). Unlike diffusion or transport through ion channels, transport via the Na^+–K^+ pump requires energy, which is provided by the hydrolysis of ATP to ADP. An example of how the Na^+–K^+ ion pump works is the response of a nerve cell to touch. Stimulation of the nerve cell causes Na^+ to flow into the cell and K^+ to flow out via ion channels; this two-way flow of ions produces the nerve impulse. The ion pump then "recharges" the system by pumping Na^+ out of the cell and K^+ into the cell so that another impulse can immediately be transmitted from cell to cell along the nerve.

CONNECTION In Chapter 10 we learned that alkali metal cations dissolved in water are surrounded by six water molecules. Each water molecule is oriented so that the oxygen atoms point toward the cation. In Chapter 18 we described this interaction as an example of a Lewis acid (cation, electron-pair acceptor) interacting with a Lewis base or ligand (water, electron-pair donor).

CONNECTION The role of ATP and ADP in metabolism was described in Chapter 20.

> **CONCEPT TEST**
>
> Do ion pumps represent spontaneous or nonspontaneous processes?

Magnesium and Calcium

ion pump system of membrane proteins that exchange ions inside the cell with those in the intercellular fluid.

The biological roles of Mg^{2+} and Ca^{2+} are more varied than those of Na^+ and K^+. We have mentioned that calcium is a major component of teeth and bones. A prolonged deficiency of calcium can lead to osteoporosis, whereas high concentrations of calcium in muscle cells contribute to cramps. Most kidney stones are made of calcium oxalate or calcium phosphate. Magnesium deficiencies can reduce physical and mental capacity because of the role of Mg^{2+} in the transfer of phosphate groups to and from ATP; slowing this transfer diminishes the amount of energy available to cells. The cellular concentrations of Mg^{2+} and Ca^{2+} are maintained by ion pumps.

Magnesium is a component of chlorophyll, which is one of a number of molecules used by plants to collect and capture light energy across the visible portion (400 to 700 nm) of the electromagnetic spectrum (Figure 22.5). Chlorophylls from different plants vary slightly in composition, but all of them contain magnesium coordinated to four nitrogen atoms. The presence of magnesium in chlorophyll does not account for the green color of the molecule, nor does it play a direct role in

CONNECTION The inorganic chemistry of calcium was described in Chapter 4.

CONNECTION The stability of complexes formed between metal ions and polydentate ligands, such as the chlorin ring in chlorophyll, was described in Chapter 18.

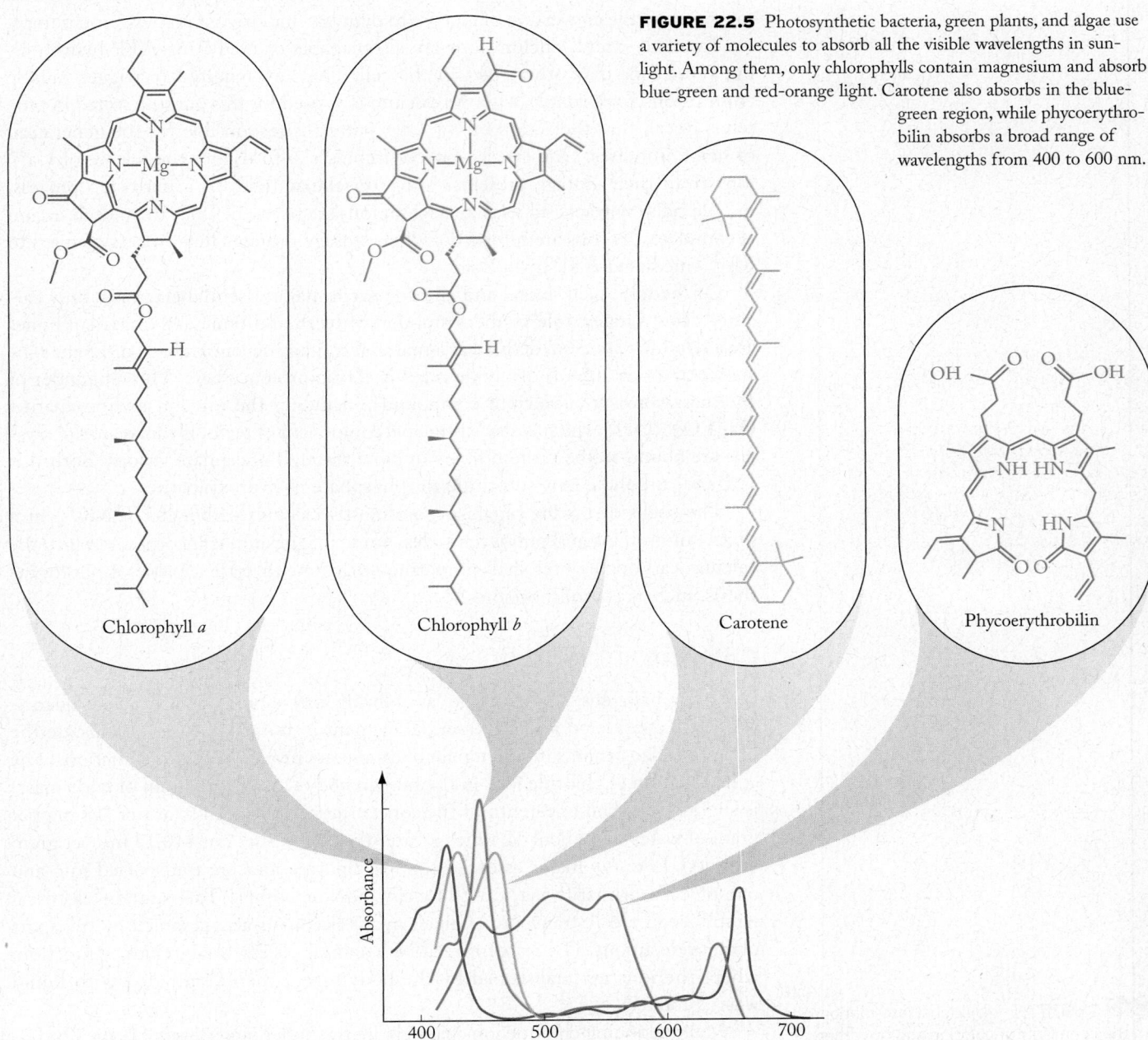

FIGURE 22.5 Photosynthetic bacteria, green plants, and algae use a variety of molecules to absorb all the visible wavelengths in sunlight. Among them, only chlorophylls contain magnesium and absorb blue-green and red-orange light. Carotene also absorbs in the blue-green region, while phycoerythrobilin absorbs a broad range of wavelengths from 400 to 600 nm.

Chlorophyll *a*

Chlorophyll *b*

Carotene

Phycoerythrobilin

absorption of sunlight. The function of the Mg^{2+} ion is to orient the molecules in positions that allow energy to be transferred to the reaction centers where H_2O is consumed and O_2 is produced during photosynthesis. Carotene (and related compounds) in Figure 22.5 are responsible for the orange colors of autumn leaves on deciduous trees when chlorophyll production ceases.

> **CONCEPT TEST** ....
>
> The colors of metal compounds were discussed in Chapter 18. Most Mg^{2+} compounds are white, not green like chlorophyll. Why?
>
> ....

Mg^{2+} ions play important roles in ATP hydrolysis and ADP phosphorylation. The many Mg^{2+}-mediated ATP $\rightarrow$ ADP processes include transferring phosphate to glucose in the conversion of glucose to pyruvate and driving Na^+–K^+ ion pumps.

To some extent, calcium ions are also capable of mediating ATP hydrolysis, but these ions play other roles in the cell. They are required to trigger muscle contractions, for example, with the calcium ions used for this purpose stored in proteins. Recall that the action of Na^+–K^+ pumps is responsible for the generation of nerve impulses. One effect of nerve impulses is to trigger the release of Ca^{2+} ions from their storage proteins into intracellular fluid. In a multistep process, muscle cells contract and relax as calcium ions are released. Once the muscle action is complete, the ions are returned to their storage proteins in a process coupled to Mg^{2+}-mediated ATP hydrolysis.

Of the four alkali metal and alkaline earth major essential elements, only calcium plays a major role in the formation of teeth and bones. Mammalian bones are a *composite material*, defined as a material containing a mixture of different substances. About 30% of dry bone mass is elastic protein fibers. The remainder of the mass consists of calcium compounds, including the mineral hydroxyapatite, $Ca_5(PO_4)_3(OH)$, which is also a principal component of teeth. Hydroxyapatite crystals are bound to the protein fibers in bone through phosphate groups. Both sulfate and carbonate may substitute for phosphate in hydroxyapatite.

The shells of marine organisms are mostly calcium carbonate ($CaCO_3$) in a matrix of proteins and polysaccharides. Some magnesium is incorporated into the calcium carbonate outer shell of marine organisms that are capable of photosynthesis, such as coccolithophores.

Chlorine

Of all the halogens, only chlorine (as chloride ion) is present in sufficient quantities to be considered a major essential element in humans. Chloride ions are the most abundant anions in the human body and are involved in many functions. The concentration of chloride ions in the human body (1.5 mg per gram of body mass) is slightly less than one-tenth of the concentration of Cl^- in seawater (19 mg per gram of water) but about 12 times greater than in Earth's crust (0.13 mg per gram of crust). Like the major essential cations, chloride ions are transported into and out of cells primarily via ion channels and ion pumps. To maintain electrical neutrality in a cell, transport of alkali metal cations is accompanied by transport of chloride anions. The *cotransport* of Na^+ and Cl^- is essential in kidney function, where the ions are reabsorbed by the body rather than eliminated with liquid waste products.

Malfunctioning chloride ion channels are the underlying cause of cystic fibrosis, a lethal genetic disease that causes patients to accumulate mucus in their airways

CONNECTION Additional information on the chemistry of chlorine and the other halogens can be found in Chapters 8 and 10.

so that breathing is difficult. The discovery of high concentrations of Na^+ and Cl^- in the sweat of cystic fibrosis patients led to an understanding of the role of chloride ion transport in patients with this disease.

Chloride ions also play a major role in the elimination of CO_2 from the body. Because it is nonpolar, carbon dioxide produced during catabolism of glucose can pass from muscle cells (for example) into red blood cells, moving easily through the largely nonpolar cell membranes of these cells. Inside the red blood cells, CO_2 is converted to bicarbonate ion, HCO_3^-. When HCO_3^- is pumped out of the cell, Cl^- enters the cell through an ion channel to maintain charge balance in the cell.

Chloride ion concentrations are high in gastric juices because of the presence of hydrochloric acid, which catalyzes digestive processes in the stomach. In response to food in the digestive system, cells tap ATP for the needed energy to pump hydrochloric acid into the stomach.

CONNECTION The catabolism of glucose was described in Chapter 20.

SAMPLE EXERCISE 22.1 **Calculating the Concentration of HCl in Stomach Acid**

Acid reflux (sometimes called heartburn, though the heart is not involved) affects many people. Acid reflux results from acid in the stomach leaking into the esophagus and causing discomfort. Stomach acid is primarily an aqueous solution of HCl.

a. Calculate the molarity of hydrochloric acid in gastric juice that has a pH of 0.80.

b. One treatment for the symptoms of acid reflux is to take an antacid tablet. What volume of gastric juice can be neutralized by a 750 mg tablet of calcium carbonate (a typical size for an over-the-counter antacid)?

CONNECTION An introduction to the strengths of acids and the calculation of pH can be found in Chapters 4 and 17.

Collect and Organize We are given the pH of a solution and are to calculate the concentration of HCl that corresponds to that pH. According to Equation 17.11, $pH = -\log[H^+]$. Hydrochloric acid is a strong acid and ionizes completely to H^+ and Cl^- in water:

$$HCl(aq) \rightarrow H^+(aq) + Cl^-(aq)$$

We are also asked to calculate the volume of HCl solution that can be neutralized by a 750 mg tablet of calcium carbonate, $CaCO_3$. We need to write a balanced chemical equation for the neutralization reaction. Remember that in aqueous solution $H^+(aq)$ is actually present as $H_3O^+(aq)$.

Analyze The equation describing the ionization of hydrochloric acid indicates that 1 mole of H^+ ions is formed for every mole of HCl present. The pH of gastric juice falls between 1 and 0, so $[H^+]$ will be between 10^{-1} $(= 0.1)$ M and 10^0 $(= 1)$ M.

The neutralization reaction is

$$CaCO_3(s) + 2\,H^+(aq) \rightarrow Ca^{2+}(aq) + CO_2(g) + H_2O(\ell)$$

The equation describing the neutralization reaction indicates that 2 moles of H^+ are consumed for every one mole of $CaCO_3$. We are told that the tablet size is typical of an antacid tablet, so common sense leads us to predict that the volume of acid this tablet can neutralize will not be excessively large (greater than 1 L) or small (less than 10 mL): too large a tablet would be a waste of antacid, and too small a tablet wouldn't relieve the symptoms.

Solve

a. Substitution into Equation 17.11 gives

$$pH = -\log[H^+] = 0.80$$

We take the antilog of both sides to solve for $[H^+]$:

$$[H^+] = 10^{-0.80} = 0.16 \; M \, H^+$$

Therefore, the concentration of HCl is

$$0.16 \; M \, H^+ \times \frac{1 \; mol \; HCl}{1 \; mol \; H^+} = 0.16 \; M \, HCl$$

b. First we calculate the number of moles of $CaCO_3$ present in 750 mg:

$$0.750 \; \cancel{g \; CaCO_3} \times \frac{1 \; mol \; CaCO_3}{100.09 \; \cancel{g \; CaCO_3}} = 7.49 \times 10^{-3} \; mol \; CaCO_3$$

Next we use the stoichiometry of the neutralization reaction to calculate the volume of 0.16 M HCl this quantity of $CaCO_3$ can neutralize:

$$7.49 \times 10^{-3} \; \cancel{mol \; CaCO_3} \times \frac{2 \; \cancel{mol \; H^+}}{1 \; \cancel{mol \; CaCO_3}} \times \frac{1 \; L}{0.16 \; \cancel{mol \; H^+}}$$

$$= 9.36 \times 10^{-2} \; L = 94 \; mL \; 0.16 \; M \, HCl$$

Think about It A concentration of 0.16 M seems reasonable because it is indeed within the range of values predicted for a solution with pH < 1. The volume of 0.16 M acid that a 750 mg tablet of $CaCO_3$ can neutralize is also reasonable; 94 mL represents about 3 ounces of gastric juice.

Practice Exercise Calculate the pH of a solution prepared by mixing 10.0 mL of 0.160 M HCl with 15.0 mL of water. How much antacid containing 4.00×10^3 mg of $Mg(OH)_2$ in 5.00 mL of water is needed to neutralize this volume of acid?

(Answers to Practice Exercises are in the back of the book.)

■ •••

CONCEPT TEST •••••••••••••••••••••••••••••••••••

Taking an antacid tablet is often sufficient to treat an occasional case of mild acid reflux. Another remedy is a drug like Prilosec that inhibits a cell's proton pumps by binding to the proton site of the pump and disabling it for more than 24 hours. Is the equilibrium constant for the binding of a proton pump inhibitor likely to be less than or greater than 1?

•••

Nitrogen

Nitrogen is a major essential element found primarily in proteins but also in DNA and RNA. Nitrogen is available in the atmosphere as N_2, and soil and water contain nitrate ions, but neither of these forms of nitrogen can be directly incorporated into amino acids, the building blocks of proteins. The biosynthesis of amino acids requires ammonia or ammonium ions. For example, glycine (NH_2CH_2COOH), the simplest amino acid, is formed by reaction of CO_2 and ammonia in the presence of the appropriate enzyme. Certain bacteria use enzymes called *nitrogenases* to convert N_2 to ammonia. Plants convert NO_3^- ions to NO_2^- and then to NH_3 by using enzymes called *reductases*. Ultimately, the chemical reactions in these organisms begin a food chain that supplies the essential amino acids for human diets.

SAMPLE EXERCISE 22.2   **Balancing the Reaction of Nitrate Reductases**

Nitrate ion can be reduced to ammonia by enzymes called nitrate reductases. The first step is conversion of nitrate ion to nitrite ion. Assign oxidation numbers to the elements in these ions and write a balanced equation for the half-reaction in a basic solution:

$$NO_3^-(aq) \rightarrow NO_2^-(aq)$$

Collect and Organize We need to assign oxidation numbers based on the guidelines in Section 4.9.

Analyze The oxidation numbers of nitrogen and oxygen in a polyatomic ion must add up to the charge on the ion. Oxygen in compounds usually has an oxidation number of -2, whereas the oxidation number of nitrogen is unknown, so we begin by calling it x.

Solve The oxidation number of nitrogen in NO_3^- is

$$x + 3(-2) = -1$$
$$x = +5$$

The oxidation number of nitrogen in NO_2^- is

$$x + 2(-2) = -1$$
$$x = +3$$

Nitrogen is reduced going from the $+5$ oxidation state to the $+3$ oxidation state, so the half-reaction given in the problem is a reduction half-reaction, and 2 electrons are gained by each nitrate ion.

To balance the half-reaction, we start with what we know:

$$NO_3^-(aq) + 2\,e^- \rightarrow NO_2^-(aq)$$

Because the half-reaction takes place in base, we can balance the charges by adding $2\,OH^-$ ions to the product side:

$$NO_3^-(aq) + 2\,e^- \rightarrow NO_2^-(aq) + 2\,OH^-(aq)$$

We balance hydrogen and oxygen atoms by adding H_2O to the reactant side:

$$NO_3^-(aq) + H_2O(\ell) + 2\,e^- \rightarrow NO_2^-(aq) + 2\,OH^-(aq)$$

Think about It Assigning oxidation numbers is a convenient way of identifying which element is reduced or oxidized in a half-reaction and of determining how many electrons are gained or lost. The balanced half-reaction confirms that electrons are added to nitrate to reduce it to nitrite ion.

Practice Exercise The reduction half-reaction catalyzed by one type of nitrogenase produces 1 mole of $H_2(g)$ for every 2 moles of $NH_4^+(aq)$ under acidic conditions. Write a balanced equation for this half-reaction.

CONNECTION Another approach, based on balancing the number of O and H atoms with water molecules and H^+ ions and then balancing charges with OH^- ions, is described in Chapter 4.

CONNECTION The inorganic chemistry of nitrogen was discussed in Chapter 6.

In humans and other mammals, excess nitrogen is converted to urea in the liver and excreted via the kidneys. Plants use urea as a source of ammonia by the action of *ureases* via the reaction:

$$\underset{\substack{\\ H_2N \qquad NH_2}}{\overset{\overset{\displaystyle O}{\parallel}}{C}} + H_2O \rightarrow 2\,NH_3 + CO_2$$

Unlike reactions catalyzed by nitrogenases and nitrate reductases, the conversion of urea to ammonia and carbon dioxide is not a redox reaction. It is a hydrolysis reaction, similar to the reaction of nonmetal oxides with water described in Chapter 3.

22.3 Trace and Ultratrace Essential Elements

Figure 22.6 shows the DRI values of trace (red) and ultratrace (blue) essential elements including four main group elements (silicon, selenium, fluorine, and iodine) and nine transition elements (chromium, cobalt, copper, iron, manganese, molybdenum, nickel, vanadium, and zinc). Note that eight of the ten transition metals are from the fourth row of the periodic table. Some transition metals in the fifth and sixth periods are also found in enzymes, for example, Mo and W.

Trace Essential Main Group Elements

Silicon

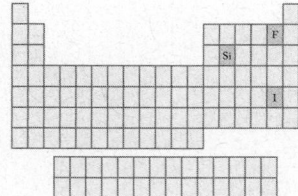

In mammals, a lack of the trace essential element silicon stunts growth. The presence of silicon as silicic acid [$Si(OH)_4$] is believed to reduce the toxicity of Al^{3+} ions in organisms by precipitating the aluminum as aluminosilicate minerals. Amorphous silica (SiO_2) is found in the exoskeletons of diatoms and in the cell membranes of some plants, such as the tips of stinging nettles.

Fluorine and Iodine

Fluoride ions have significant benefits for dental health. Tooth enamel is composed of the mineral hydroxyapatite, $Ca_5(PO_4)_3(OH)$, which is essentially insoluble in water:

$$Ca_5(PO_4)_3(OH)(s) \rightleftharpoons Ca_5(PO_4)_3^+(aq) + OH^-(aq) \qquad K_{sp} \approx 1 \times 10^{-59}$$

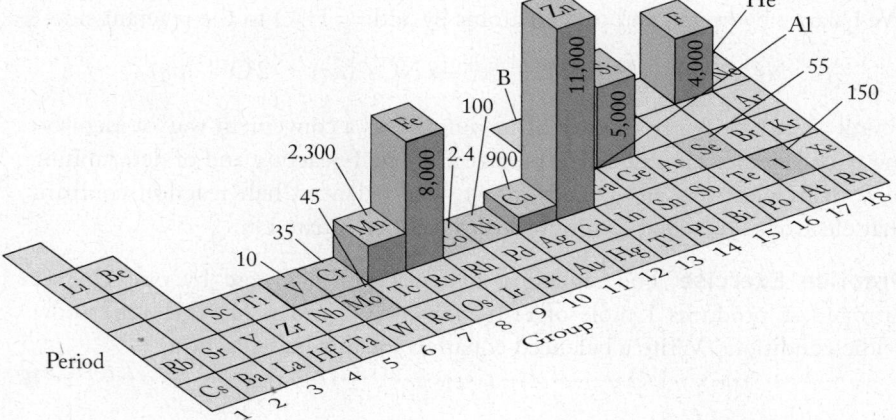

FIGURE 22.6 The elements shown in red are trace essential elements, and those shown in blue are ultratrace essential elements. The remaining labeled elements are nonessential. The vertical bars show DRI values for these elements in micrograms per day. The value for cobalt reflects the DRI for vitamin B_{12}, the principal source of cobalt in our diets.

FIGURE 22.7 Thyroxine and 3,5,3'-triiodothyronine, two iodine-containing hormones found in the thyroid gland, regulate metabolism.

When hydroxyapatite comes into contact with weak acids in your mouth, this equilibrium shifts to the right as the acid reacts with the hydroxide ions. This shift effectively increases the solubility of hydroxyapatite, so that your tooth enamel becomes pitted and dental caries form. This is an example of Le Châtelier's principle. Fluoride ions reduce the likelihood of caries by displacing the OH^- ions in hydroxyapatite to form fluorapatite:

$$Ca_5(PO_4)_3(OH)(s) + F^-(aq) \rightleftharpoons Ca_5(PO_4)_3F(s) + OH^-(aq) \qquad K = 8.48$$

The solubility of fluorapatite is less dependent on pH than the solubility of hydroxyapatite is, so changing tooth enamel to fluorapatite makes your teeth more resistant to decay. This is why toothpaste contains fluoride compounds and why fluoride is added to drinking water in many communities in North America and Europe.

Of all the trace essential elements, iodine may have the best-defined role in human health. The body concentrates iodide ions in the thyroid gland, where they are incorporated into two hormones—thyroxine and 3,5,3'-triiodothyronine (Figure 22.7)—whose role is to regulate energy production and use. The conversion of thyroxine to 3,5,3'-triiodothyronine is catalyzed by selenocysteine-containing proteins. A deficiency of iodine or of either hormone can cause fatigue or feeling cold and can ultimately lead to an enlarged thyroid gland, a condition known as goiter. To help prevent iodine deficiency, table salt sold in the United States and many other countries is "iodized" with a small amount of sodium iodide. An excess of either hormone can cause one to feel hot and is linked to Graves' disease, an autoimmune disease. The immune system in a patient with Graves' disease attacks the thyroid gland and causes it to overproduce the two hormones.

CONNECTION Solubility products, or K_{sp}, were introduced in Chapter 17. Le Châtelier's principle was discussed in Chapter 16.

CONNECTION In Chapter 10 we discussed some of the roles of the halogens in biology and medicine.

Trace Essential Transition Elements

Zinc and iron are transition metals that are also trace essential elements. Transition metal cations can form complexes (coordination compounds) by bonding to the nitrogen atoms of the amino acids in proteins and other Lewis bases present in biological systems. Many of the enzymes in our bodies contain zinc or iron ions; these enzymes, along with others containing transition metal ions, are called *metalloenzymes*. Table 22.6 lists some of the more important ones and the reactions they catalyze.

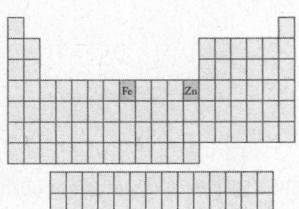

Zinc

The enzyme carbonic anhydrase catalyzes the reaction between water and carbon dioxide to form bicarbonate ions:

$$H_2O(\ell) + CO_2(aq) \rightleftharpoons HCO_3^-(aq) + H^+(aq)$$

TABLE 22.6   Selected Metalloenzymes and Some Reactions and Half-Reactions They Catalyze

Metal[a]	Enzyme	Reaction or Half-Reaction Catalyzed
V(Fe)	Nitrogenase	$N_2 + 10H^+ + 8e^- \rightarrow H_2 + 2NH_4^+$
	Haloperoxidase	$CH_4 + H_2O_2 + Cl^- + H^+ \rightarrow CH_3Cl + 2H_2O$
Mo	Nitrate reductase	$NO_3^- + 3H^+ + 2e^- \rightarrow H_2O + HNO_2$
	Sulfite oxidase	$SO_3^{2-} + H_2O \rightarrow SO_4^{2-} + 2e^- + 2H^+$
W(Fe)	Formate dehydrogenase	$HCOO^- \rightarrow CO_2 + 2e^- + H^+$
Fe	Cytochrome P450	$R{-}H + O_2 + 2e^- + 2H^+ \rightarrow R{-}OH + H_2O$
	Peroxidase	$RCH_2COOH + 2H_2O_2 \rightarrow 3H_2O + RCHO + CO_2$
Co	Coenzyme B_{12}	Required as coenzyme for many reactions
Ni	Urease	(urea) $+ H_2O \rightarrow 2NH_3 + CO_2$
Ni(Fe)	Hydrogenase	$H_2 \rightarrow 2H^+ + 2e^-$
Cu	N_2O reductase	$N_2O + 2e^- + 2H^+ \rightarrow N_2 + H_2O$
	Amine oxidase	$CH_3NH_2 + O_2 + H_2O \rightarrow CH_2O + H_2O_2 + NH_3$
Zn	Carbonic anhydrase	$H_2O + CO_2 \rightleftharpoons HCO_3^- + H^+$
	Carboxypeptidase	(peptide) $+ H_2O \longrightarrow$ (carboxylic acid) $+$ (amine)

[a]The metal in parentheses is also present in the enzyme and is essential to its function.

The α-form of carbonic anhydrase contains 260 amino acid residues and a zinc ion at the active site. Note in Figure 22.8 that the zinc ion is coordinately bonded to three nitrogen atoms on histidine side chains and to one molecule of water. The presence of these ligands and a fourth histidine nearby facilitate ionization of the water molecule. Ionization leaves an OH^- ion attached to the Zn^{2+} ion and an H^+ ion bonded to the side-chain nitrogen atom of the fourth histidine. In addition, a pocket just the right size and shape to accept a CO_2 molecule is next to the active site. A CO_2 molecule in the pocket combines with the hydroxide ion, forming a HCO_3^- ion. As the bicarbonate ion pulls away, another water molecule occupies the fourth coordination site on the Zn^{2+} ion, another CO_2 molecule enters the pocket, the histidine is protonated, and the catalytic cycle can be repeated. This reaction is important because it helps eliminate CO_2 from cells during respiration and mediates the uptake of CO_2 during photosynthesis in some plants.

Iron

In Chapter 20 we were introduced to the role of iron in hemoglobin's ability to transport oxygen throughout the body. In hemoglobin, Fe^{2+} is coordinated to four nitrogen atoms in the porphyrin ring of a *heme group*. Two additional Lewis bases, water and the amino acid histidine complete the coordination sphere of Fe^{2+}. The heme group is found in a number of iron-containing enzymes.

Iron-containing peroxidases and catalases are integral to the transfer of oxygen to biomolecules. For example, plants use a fatty acid peroxidase to catalyze the step-

∞ **CONNECTION** Examples of coordination complexes of transition metals with biologically important ligands can also be found in Chapter 18.

∞ **CONNECTION** The hemoglobin molecule was described in Section 20.3 and shown in Figure 20.14(a).

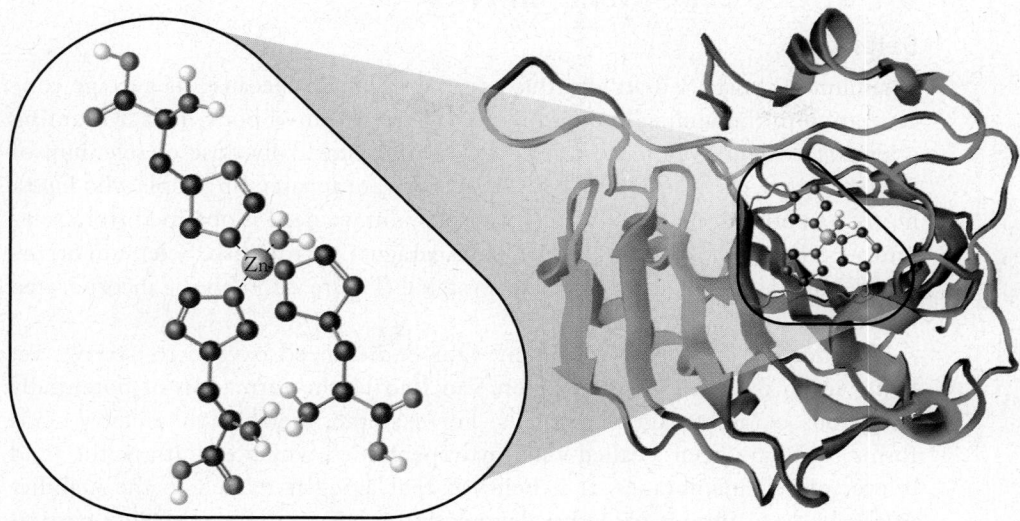

FIGURE 22.8 The active site of one form of carbonic anhydrase consists of a zinc atom bonded to three histidine molecules and one H_2O molecule. The OH^- ion produced when this H_2O ionizes combines with a molecule of CO_2, forming a HCO_3^- ion.

wise degradation of fatty acids. One CH_2 group at a time is removed from the fatty acid using hydrogen peroxide:

$$R\!-\!CH_2\!-\!COOH(aq) + 2\,H_2O_2(aq) \xrightarrow{\substack{\text{fatty acid}\\ \text{peroxidase}}}$$

$\quad\quad$ Fatty acid $\quad\quad\quad\quad\quad\quad$ Hydrogen
$\quad\quad\quad\quad\quad\quad\quad\quad\quad\quad\quad$ peroxide

$$3\,H_2O(\ell) + R\!-\!CHO(aq) + CO_2(aq)$$
$$\text{Aldehyde}$$

Subsequent oxidation of the aldehyde back to a carboxylic acid yields a new fatty acid with one fewer CH_2 group:

$$2\,R\!-\!CHO(aq) + O_2(aq) \rightarrow 2\,RCOOH(aq)$$
$$\quad\quad\text{Aldehyde} \quad\quad\quad\quad\quad\quad\quad\quad \text{Fatty acid}$$

Iron-containing enzymes called cytochromes catalyze electron transport in photosynthesis and the metabolism of glucose to CO_2 and water. The conversion of glucose, $C_6H_{12}O_6$, to CO_2 involves the oxidation of carbon. The oxidation number of carbon in CO_2 is $+4$, but the average oxidation number of the carbon atoms in $C_6H_{12}O_6$ is zero.

Iron-containing enzymes also catalyze the reduction of nitrite (NO_2^-) and sulfite (SO_3^{2-}) ions:

$$NO_2^-(aq) + 6\,e^- + 8\,H^+(aq) \xrightarrow{\text{nitrite reductase}} NH_4^+(aq) + 2\,H_2O(\ell)$$

$$SO_3^{2-}(aq) + 6\,e^- + 7\,H^+(aq) \xrightarrow{\text{sulfite reductase}} HS^-(aq) + 3\,H_2O(\ell)$$

In each case, the iron in the enzyme is oxidized, providing the electrons needed for reduction.

◯◯ **CONNECTION** The assignment of oxidation numbers to organic compounds was discussed in Chapter 4.

CONCEPT TEST •

Which of the following small molecules could not behave as a Lewis base and hence would not be expected to bond to iron in a heme protein: CO, H_2, NO_2, or H_2S?

• •

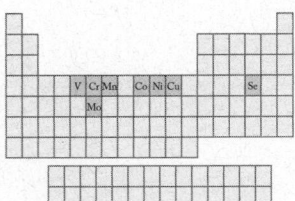

FIGURE 22.9 Selenocysteine is the selenium-containing analog of the amino acid cysteine. Much of the selenium in the human body is found in proteins containing selenocysteine.

⚮ **CONNECTION** The inorganic chemistry of oxygen and sulfur, which like selenium are in group 16 of the periodic table, was discussed in Chapter 9.

Ultratrace Essential Elements

Selenium

Selenium is considered an ultratrace essential element because its average concentration in the human body is only 0.3 μg per gram of body mass. Mounting scientific evidence points to a need for a minimum daily dose of selenium of 55 μg. The effects of selenium toxicity, however, are apparent in people who ingest more than 500 μg per day. Most of the selenium we need is obtained from selenium-rich produce (garlic, mushrooms, asparagus) or from fish. Selenium occurs in the body as the amino acid selenocysteine (Figure 22.9) and is incorporated into enzymes.

Selenocysteine is an antioxidant. Our bodies need oxygen to survive, yet living in an oxygen-rich atmosphere can lead to the formation of potentially dangerous oxidizing agents in cells. For example, metabolism of fatty acids forms oxidizing agents called alkyl hydroperoxides, which can attack the lipid bilayer of cell membranes. It is believed that aging is related to the inability of the body to inhibit oxidative degradation of tissue. Selenocysteine participates in a series of reactions that result in the decomposition of these alkyl hydroperoxides.

Molybdenum and Vanadium

Many of the metalloenzymes in Table 22.6 are involved in transformations of nitrogen. Molybdenum-containing reductases are responsible for converting NO_3^- ions to NO_2^- and then to NH_3 (Section 22.2). The active site of sulfite oxidase, which converts SO_3^{2-} to SO_4^{2-}, also contains molybdenum. Sometimes more than one type of transition metal is found in a metalloenzyme. Both molybdenum and iron are required by xanthine oxidase, an important enzyme along the pathway for degradation of excess nucleic acids (adenine and guanine) to xanthine and then to uric acid for elimination through the kidneys.

Xanthine + H_2O + O_2 $\xrightarrow{\text{xanthine oxidase}}$ Uric acid + H_2O_2

The combination of iron and vanadium is essential to the function of haloperoxidases, a class of enzymes found in some algae, lichens, and fungi that replace C—H bonds with carbon–halogen bonds. The reaction products may function in the defense systems of these organisms.

Copper

Copper-containing proteins perform a number of functions in both plants and animals, including oxygen transport in mollusks such as clams and oysters. Copper is also an essential element in the enzymes azurin and plastocyanin, which mediate electron transfer during photosynthesis.

Reactions catalyzed by xanthine oxidase may produce other reactive oxygen species besides H_2O_2. One of them is the superoxide ion, O_2^-. Superoxide ion is a strong oxidizing agent and must be eliminated to prevent cell damage. The removal of superoxide begins with the action of superoxide dismutases that convert superoxide to hydrogen peroxide:

$$2\,O_2^-(aq)\ +\ 2\,H^+(aq)\ \xrightarrow[\text{dismutase}]{\text{superoxide}}\ H_2O_2(aq)\ +\ O_2(aq)$$

Researchers have isolated superoxide dismutases containing a variety of transition metals, including a copper–zinc enzyme. The hydrogen peroxide produced in this reaction is decomposed to water and oxygen by iron-containing catalase.

> **coenzyme** organic molecule that, like an enzyme, accelerates the rate of biochemical reactions.

CONCEPT TEST ···

Why is superoxide ion a good oxidizing agent?

···

Nickel

Ureases, the enzymes responsible for the conversion of urea to ammonia in plants, contain nickel(II). Nickel is also found with iron in enzymes called *hydrogenases*. Hydrogenases oxidize hydrogen gas to protons:

$$H_2(g)\ \xrightarrow{\text{hydrogenase}}\ 2\,H^+(aq)\ +\ 2\,e^-$$

Nickel and iron also combine in CO dehydrogenase, an enzyme that catalyzes the formation of acetyl-CoA, which is a key component in the Krebs cycle discussed in Section 20.3. Finally, nickel-containing enzymes are found among the enzymes responsible for methane generation by bacteria as discussed in Section 13.5.

Cobalt and Coenzymes

Many enzymes require the presence of a **coenzyme**, which is an organic molecule that cocatalyzes a biochemical reaction. The coenzyme B_{12} (Figure 22.10) contains the ultratrace essential element cobalt(III) and is a derivative of vitamin B_{12}.

The cobalt(III) in coenzyme B_{12} is easily reduced to cobalt(II) and even cobalt(I) in the course of enzyme-catalyzed redox reactions. The change in oxidation state of Co allows for facile transfer of methyl groups, as in the conversion of methionine to homocysteine:

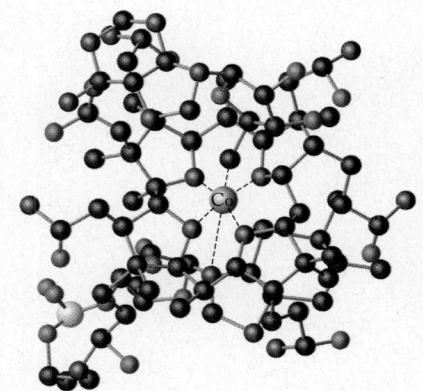

FIGURE 22.10 Coenzyme B_{12} contains cobalt, an ultratrace essential metal.

Methionine → Homocysteine

Coenzyme B_{12} is also critical to the function of *mutases*, which are enzymes that catalyze the rearrangement of the skeleton of a molecule, as in the interconversion of glutamate and methylaspartate:

Glutamate $\xrightleftharpoons[\text{coenzyme }B_{12}]{\text{glutamate mutase}}$ Methylaspartate

Manganese and Photosynthesis

We noted in discussing the chlorophyll structures in Figure 22.5 that the role of the major essential element Mg^{2+} is to orient the chlorophyll molecules in the proper way. Another metal, the ultratrace essential element manganese, also plays a role in the production of oxygen during photosynthesis. To examine that role, let's write an equation for photosynthesis that is slightly different from the equation we are used to seeing. Instead of writing the formula $C_6H_{12}O_6$ for glucose, we use the generic carbohydrate formula $(CH_2O)_n$ so that the coefficient is 1 for all other species in the reaction (rather than the 6):

$$H_2O(\ell) + CO_2(g) \rightarrow \tfrac{1}{n}(CH_2O)_n(aq) + O_2(g)$$

Writing the equation in this form makes it easier to see how the overall reaction between water and carbon dioxide involves electron transfer. Photosynthesis is a redox reaction. As usual, the oxidation number for O is -2 in H_2O and CO_2, and 0 in O_2. This increase in oxidation number means the O is oxidized. The CO_2 oxygen atom ends up in the $(CH_2O)_n$ molecule, and the H_2O oxygen atom ends up in the O_2. We can write two half-reactions to describe this overall redox reaction:

$$4e^- + 4H^+ + CO_2 \rightarrow CH_2O + H_2O$$
$$2H_2O \rightarrow O_2 + 4H^+ + 4e^-$$

While the O atom in H_2O is oxidized to O_2, what is reduced? If the oxidation number of O is -2 in CO_2, then the oxidation number of C must be $+4$. All of the carbon in CO_2 ends up as CH_2O, in which the oxidation number of carbon is zero. Thus photosynthesis is accompanied by the reduction of CO_2, since the oxidation number of carbon decreases from $+4$ to 0. The redox process requires metalloenzymes. Manganese-containing biomolecules in which Mn is in the $+3$ and $+4$ oxidation states mediate the transfer of electrons from water in photosynthesis. Although the exact structure of these manganese compounds remains undetermined, it is believed that two Mn(III) ions and two Mn(IV) ions are present at the site of O_2 production. Recall that copper-containing enzymes are also involved in the series of reactions that make up photosynthesis.

> **CONCEPT TEST** ••
>
> Manganese ions involved in photosynthesis are often surrounded by six Lewis base ligands in an octahedral geometry. Is reduction of a Mn(IV) ion to Mn(III) in an octahedral complex (see Section 18.2) accompanied by a change in spin state of the Mn ion?
>
> ••

Chromium

Chromium in the $+3$ oxidation state is an ultratrace essential element in our diets. It is involved in regulating glucose levels in the blood through a molecule called chromodulin. Chromodulin is a polypeptide incorporating only 4 of the 20 naturally occurring amino acids: glycine, cysteine, glutamic acid, and aspartic acid. Four Cr^{3+} ions are bound to the peptide chain. Cereals and grains contain enough chromium for our daily needs, but certain plants (such as shepherd's purse) concentrate chromium and have been used as herbal remedies in diabetes treatment. However, chromium in the $+6$ oxidation state, as found in chromate ions (CrO_4^{2-}), is acutely toxic and also carcinogenic. Chromate ion enters cells through ion channels that transport SO_4^{2-} ions. Once inside, CrO_4^{2-} is reduced to Cr^{3+}, which binds to the phosphate backbone of DNA.

22.4 Nonessential Elements

Table 22.2 lists a number of elements that are found in the human body but are classified as nonessential. In this section, we discuss how some of these elements may end up in our bodies, working our way from left to right across the periodic table.

Rubidium and Cesium

Rubidium is generally regarded as nonessential, yet it is the 15th most abundant element in the body. It is believed that Rb^+ is retained by the body because of the similarity of its size and chemistry to K^+. Like the other cations of group 1, cesium ions (Cs^+) are also readily absorbed by the body. Cesium cations have no known function, although they can substitute for K^+ and interfere with potassium-dependent functions. In most cases, the concentration of cesium in the environment is low, and so exposure to Cs^+ is not a health concern. The nuclear accident at Chernobyl in 1986, however, released significant quantities of radioactive ^{137}Cs into the environment. The ability of Cs^+ to substitute for K^+ led to the incorporation of $^{137}Cs^+$ into plants, which rendered crops grown in the immediate area unfit for human consumption because of the radiation hazard posed by this long-lived ($t_{1/2} \approx 30$ yr) β emitter.

CONNECTION The biological effects of different types of radiation were described in Chapter 21.

Strontium and Barium

Some single-celled organisms build exoskeletons made with $SrSO_4$ and $BaSO_4$, but the human body appears to have no use for Sr^{2+} and Ba^{2+} ions. These ions do find their way into human bones, where they replace Ca^{2+} ions. At the low concentrations of Sr^{2+} and Ba^{2+} that are typically present in the human body, these elements appear to be benign. However, as in the case of radioactive ^{137}Cs, incorporation of ^{90}Sr ($t_{1/2} = 29$ yr) in bones can lead to leukemia. Atmospheric testing of nuclear weapons over the Pacific Ocean and in sparsely populated regions of the American West in the 1950s released ^{90}Sr into the environment. The full extent of the toxic effects of the fallout from these tests did not become apparent for several decades.

Germanium

It is generally agreed that germanium is a nonessential element and is barely detectable in the human body. The use of bis(carboxyethyl)germanium sesquioxide (Figure 22.11) has been touted as a nutritional supplement, but its efficacy remains controversial.

Antimony

The role of antimony is also poorly understood. Most antimony compounds are toxic; they cause liver damage. However, ultratrace amounts of antimony may have a stimulatory effect, and selected antimony compounds have been used medically as antiparasitic agents, as discussed in the next section.

$O_3(GeCH_2CH_2COOH)_2$

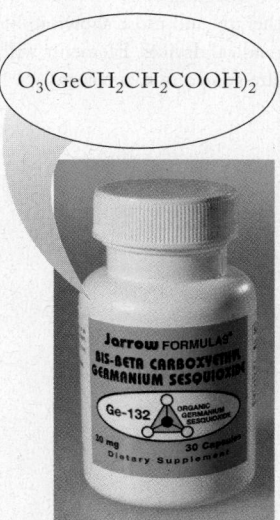

FIGURE 22.11 Bis(carboxyethyl)germanium sesquioxide has been sold as a nutrition supplement, but its benefits are not well established.

CONCEPT TEST

Looking at groups 1, 2, 14, and 15, what periodic trend do you see in the location of the nonessential elements relative to the essential elements in the same group?

22.5 Elements for Diagnosis and Therapy

So far we have talked about the biological roles of approximately 30 essential and nonessential elements found in our bodies. A number of these 30 elements are also useful in diagnosing and/or treating diseases, as are some of the 60 other elements in the periodic table (Figure 22.12). In this section, we describe some of the applications of radioactive isotopes in the diagnosis of diseases. We also explore how compounds of essential and nonessential elements have found application in the treatment of a wide range of illnesses.

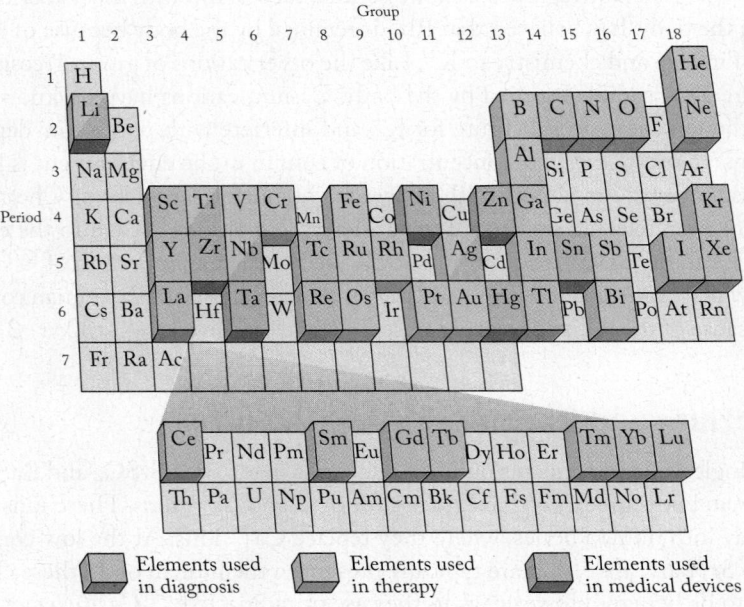

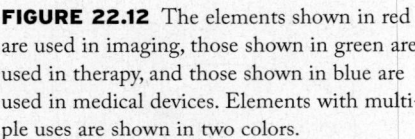

FIGURE 22.12 The elements shown in red are used in imaging, those shown in green are used in therapy, and those shown in blue are used in medical devices. Elements with multiple uses are shown in two colors.

Any diagnostic or therapeutic compound that is injected intravenously must be sufficiently soluble in blood to be delivered to the target. While in transit, the compound must be stable enough not to undergo chemical reactions that result in its precipitation or rapid elimination from the body. Occasionally, the compound can be in the form of a simple salt, but more often than not a metal ion is introduced as a coordination complex or coordination compound. Ligands used in forming biologically active coordination complexes include amino acids and simple anions like the citrate ion. Chelating ligands like diethylenetriaminepentaacetate ($DTPA^{5-}$, Figure 22.13)

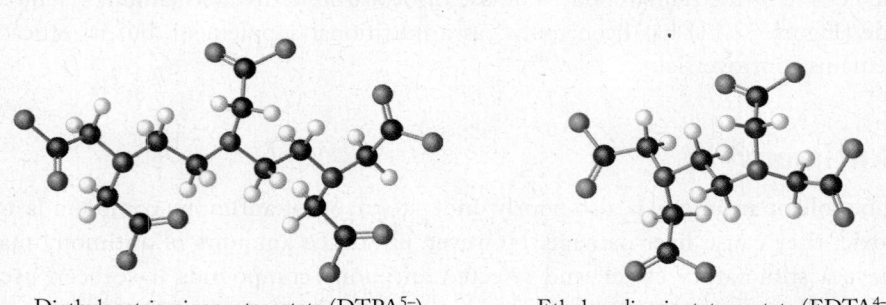

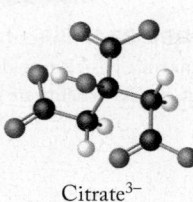

Citrate^{3-} Diethylenetriaminepentaacetate ($DTPA^{5-}$)   Ethylenediaminetetraacetate ($EDTA^{4-}$)

FIGURE 22.13 Citrate^{3-}, diethylenetriaminepentaacetate ($DTPA^{5-}$), and ethylenediaminetetraacetate ($EDTA^{4-}$) are often used as chelating ligands in diagnostic and therapeutic agents based on transition metals. These ions form stable complex ions with 2+ and 3+ metal cations. The solubilities of the complex ions are typically much greater than those of the hydrated ions at physiological pH (7.4).

are often used in biological applications. A medicinal chemist can also take advantage of substances that occur naturally in the body, such as antibodies, to carry a diagnostic or therapeutic metal ion to its target.

Diagnostic Applications

Physicians in the 21st century have an array of imaging agents to help in diagnosing disease. Some methods use radionuclides with short half-lives that emit easily detectable gamma rays. Examples cited in Chapter 21 include the use of iodine-131 to image the thyroid gland (Figure 22.14) and of neutron-poor isotopes like carbon-11 and fluorine-18 for positron-emission tomography (PET). Not all imaging depends on radionuclides, however. In magnetic resonance imaging (MRI), for instance, which can diagnose soft-tissue injuries, stable isotopes of gadolinium are used to enhance images.

Imaging with Radionuclides

The radionuclides used in medicine have short half-lives to limit the patient's exposure to ionizing radiation. If the half-life is too short, however, the nuclide may either decay before it can be administered or not reach the target organ rapidly enough to provide an image. Emission of relatively low-energy γ rays is essential to preventing collateral tissue damage.

Nuclide selection is also governed by the toxicity of both the parent element and the daughter nuclide. The speed at which the imaging agent is eliminated from the body can help mitigate toxic effects. Naturally, the cost and availability of a particular nuclide also factor into its utility in the clinical setting.

> **CONCEPT TEST**
>
> Why is it important to consider the nature of the decay products—α, β, or γ particles, or positrons—when choosing a radionuclide for medical imaging?

Gallium, Indium, and Thallium

Gallium-66, gallium-67, gallium-68, and indium-111 are used as imaging agents for tumors and leukemia. All four nuclides decay by electron capture, and it is the γ radiation emitted in this nuclear reaction that produces the images. In addition, the three gallium isotopes also decay by positron emission, which makes compounds containing these isotopes attractive for positron-emission tomography. The half-lives range from just over 1 hr for gallium-68 to 78 hr for gallium-67. The discovery that indium-111-containing compounds can image a range of cancers has led to the development of the drug Zevalin™, currently used to treat some forms of non-Hodgkin's lymphoma.

The use of the gamma emitter thallium-201 ($t_{1/2} = 73$ hr) in diagnosing heart disease presents an interesting case for considering the risks and benefits of using a particular isotope in medicine. Although thallium compounds are among the most toxic metal-containing compounds known, the nanogram quantities required for diagnosis pose few if any health hazards, meaning that the benefits outweigh the risks.

Technetium and Rhenium

Technetium and rhenium are transition metals in group 7 of the periodic table, just below manganese. Technetium is unusual in that it does not exist naturally

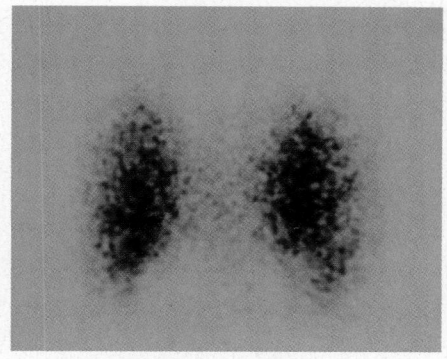

FIGURE 22.14 Gamma radiation that accompanies the decay of iodine-131 can be used to image the two butterfly-shaped lobes of the thyroid gland.

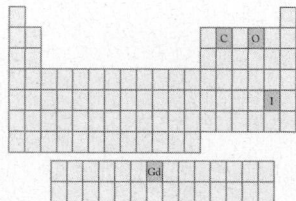

CONNECTION Magnetic resonance imaging (MRI) was mentioned in Chapter 3 in our discussion of the use of cryogenic helium for superconducting magnets.

CONNECTION Chapter 21 gave a more detailed discussion of nuclear chemistry and nuclear medicine, including an assessment of the effects of different types of radiation on living tissue.

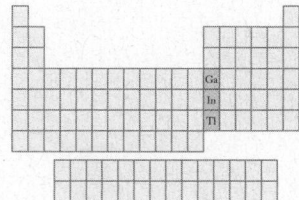

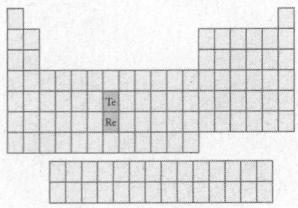

on Earth because it has no stable isotopes; in other words, all technetium isotopes are radioactive. These isotopes can be produced in nuclear reactors for use in medicine. The ^{99}Tc used in hospitals for imaging is prepared in technetium generators in which a stable isotope of molybdenum, ^{98}Mo (23.78% natural abundance), is bombarded with neutrons. Technetium-99 has a half-life greater than 20,000 years, making ^{99}Tc useless in nuclear medicine; however, when it is produced in a nuclear reactor, its nucleus is in an excited state that quickly decays to a more stable nucleus of technetium-99 ($t_{1/2}$ = 6.0 hr). The excited state is designated by adding the letter "m" to the mass number to indicate a *metastable* nucleus: ^{99m}Tc, technetium-99m.

> **SAMPLE EXERCISE 22.3** **Identifying Particles in Nuclear Reactions**

Technetium-99m produced from molybdenum-98 decays by γ emission followed by β decay. Complete the nuclear equations representing this sequence by identifying the missing particles P1, P2, and P3:

$$^{98}_{42}\text{Mo} + \text{P1} \rightarrow ^{99}_{42}\text{Mo} \rightarrow ^{99m}_{43}\text{Tc} + \text{P2} \tag{A}$$

$$^{99m}_{43}\text{Tc} \rightarrow ^{99}_{43}\text{Tc} + \gamma \tag{B}$$

$$^{99}_{43}\text{Tc} \rightarrow \text{P3} + ^{\ 0}_{-1}\beta \tag{C}$$

Collect and Organize We are provided with three nuclear equations, two of them are missing one or more particles and are not balanced.

Analyze When an unstable nucleus decays, it can eject a β particle ($^{\ 0}_{-1}\beta$), a positron, ($^{0}_{1}\beta$), or an alpha particle ($^{4}_{2}\text{He}$) from its nucleus, along with gamma rays. The sum of the electrical charges denoted by the subscripts and the sum of the masses denoted by the superscripts to the left of the reaction arrow must equal the comparable sums to the right of the arrow. To find each missing particle, we must determine its electrical charge and mass. Because Tc and Mo account for most of the mass in Equation A, we predict that the subscripts and superscripts of P1 and P2 will be small. Particle P3, however, will have an atomic number and mass number close to that of Tc.

Solve In the first step of Equation A,

$$^{98}_{42}\text{Mo} + \text{P1} \rightarrow ^{99}_{42}\text{Mo}$$

the atomic number is unchanged, meaning the molybdenum nucleus stays a molybdenum nucleus. The mass changes by one mass unit, from ^{98}Mo to ^{99}Mo. The particle with a mass number equal to 1 and a charge of 0 is a neutron ($^{1}_{0}\text{n}$). Thus, P1 is a neutron, and we conclude that molybdenum-98 is bombarded by neutrons to form molybdenum-99.

In the second step of Equation A,

$$^{99}_{42}\text{Mo} \rightarrow ^{99m}_{43}\text{Tc} + \text{P2}$$

the mass number does not change, telling us that P2 must have negligible mass. The atomic number (number of protons) increases from 42 to 43, and so the charge on P2 must be -1:

$$42 = 43 + x$$
$$x = -1$$

The particle that has negligible mass and a charge of -1 is a β particle ($_{-1}^{0}\beta$). P2 is an electron.

$$_{-1}^{0}e$$

To balance Equation C,

$$_{43}^{99}Tc \rightarrow P3 + _{-1}^{0}\beta$$

P3 must have a mass number of 99. Its charge (that is, the number of protons it contains), determined from the subscripts, must be 44:

$$43 = x + (-1)$$
$$x = 44$$

The element with an atomic number of 44 is ruthenium, so P3 is ruthenium-99, and our sequence is

$$_{42}^{98}Mo + _{0}^{1}n \rightarrow _{42}^{99}Mo \rightarrow _{43}^{99m}Tc + _{-1}^{0}\beta \qquad (A)$$

$$_{43}^{99m}Tc \rightarrow _{43}^{99}Tc + \gamma \qquad (B)$$

$$_{43}^{99}Tc \rightarrow _{44}^{99}Ru + _{-1}^{0}\beta \qquad (C)$$

Think about It As predicted, the subscripts (charges) and superscripts (mass numbers) of particles 1 and 2 in Equation A are small. Because γ rays have no mass or charge, there is no change in the atomic or mass number of technetium-99m when it emits a γ ray in Equation B. We also confirm our prediction that the beta decay of ^{99}Tc in Equation C produces an isotope with atomic and mass numbers comparable to those of ^{99}Tc. Remember that β decay leads to an increase of 1 in atomic number (number of protons).

Practice Exercise Gamma rays emitted by thallium-201 can be used for cardiac imaging. Thallium-201 decays by electron capture (a proton and an electron form a neutron). Write a balanced nuclear equation for this decay.

Technetium-99m has been widely used as an imaging agent because it has a short half-life and emits low-energy γ rays. Patients can be injected with a variety of technetium compounds, depending on the target organ. For imaging the heart, the coordination compounds shown in Figure 22.15 are used.

The ability to deliver radionuclides to a particular organ opens the possibility for selective irradiation of a tumor located in that organ. Therefore, some radionuclides can be used to both image and treat cancers. Rhenium, for example, is being studied as both an imaging and a therapeutic agent for many tumors, including breast, liver, and skin cancer. Rhenium-186 and rhenium-188 undergo β decay with half-lives of 3.72 d and 17.0 hr, respectively. Certain tumors have highly selective receptor sites for particular molecules on their surfaces. By including a radioactive rhenium ion in a molecule that binds strongly and specifically to these receptors, physicians can deliver both an imaging agent and a therapeutic agent to the tumor. In principle, the β particles destroy the tumor. Several patents have been issued for the use of compounds containing rhenium isotopes, but therapies based on these compounds remain in the experimental stage.

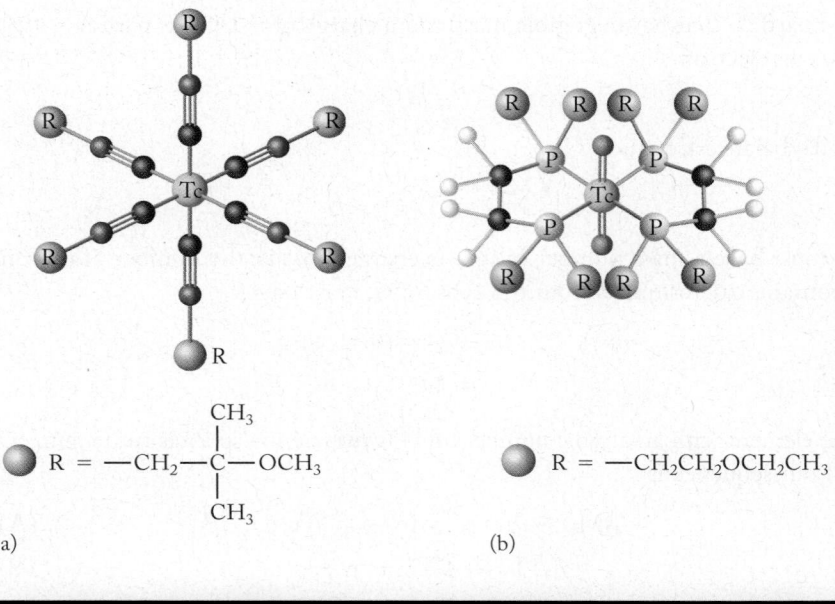

$$R = -CH_2-\overset{\displaystyle CH_3}{\underset{\displaystyle CH_3}{\overset{|}{\underset{|}{C}}}}-OCH_3$$

(a)

$$R = -CH_2CH_2OCH_2CH_3$$

(b)

(c)

FIGURE 22.15 (a, b) Cardiolite™ [Tc(CNR)₆] and Myoview™ [TcO₂(RPCH₂CH₂PR)₂] are used for imaging the heart. (c) Images of a patient's heart after intravenous injection with a technetium-containing drug. The images along the bottom show four measurements of heart function. Radioactive technetium compounds emit gamma rays that allow the blood to be tracked as it is pumped through the heart.

One example of a rhenium compound used in these applications is shown in Figure 22.16.

Scandium, Yttrium, and Lanthanide Elements

Scandium, yttrium, and lanthanum are in group 3 in the periodic table, the first group in the transition metal series. Scandium-46 has been used to image the spleen, and scandium-47 shows promise in diagnosing breast cancer. Among the available radioactive isotopes of yttrium, yttrium-90 has been applied to imaging a variety of tumors including intestinal, breast, and thyroid cancers. Chelating ligands are used to complex the $^{90}Y^{3+}$ cation and deliver it to the tumor where it binds strongly to receptors on the surface of the tumor. Yttrium-90 agents for treatment of non-Hodgkin's lymphoma are currently in clinical trials.

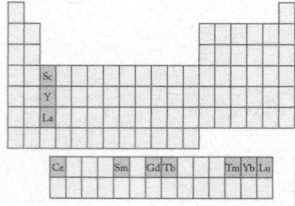

Radioactive isotopes of the lanthanides have seen extensive use in scintigraphic imaging, a procedure that uses a scintillation counter (Section 21.7) to measure the light emitted when radiation strikes a phosphor-coated screen in the instrument. The intensity of the emitted light is translated into a three-dimensional image of the organ of interest, a method similar to the PET technique described in Section 21.9. Among the nuclides used for this purpose are cerium-141, samarium-153, gadolinium-153, terbium-160, thulium-170, ytterbium-169, and lutetium-177. They have an advantage over gallium isotopes in having lower retention times in blood, thereby reducing the possibility of potentially harmful biological side effects. In some cases, lutetium-177 derivatives have replaced yttrium-90 compounds under investigation as radioactive antitumor agents.

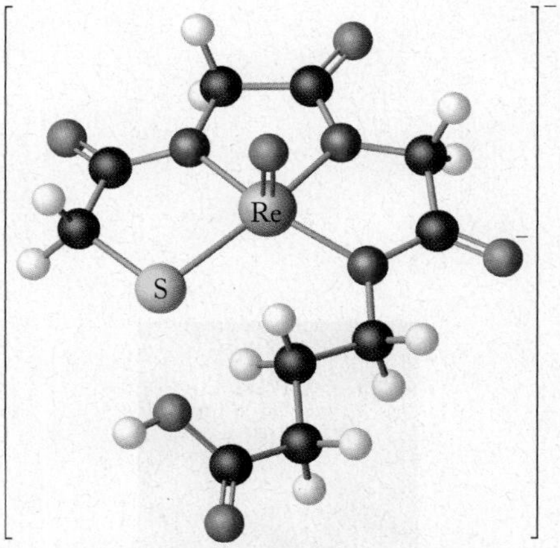

FIGURE 22.16 This rhenium–mercaptoacetylglycylglycyl-γ-amino acid complex is used to attach ^{186}Re or ^{188}Re to monoclonal antibodies or peptides.

CONCEPT TEST

What is wrong with the statement, "A radioactive isotope with twice the half-life of another will allow an image to be collected in half the time"?

SAMPLE EXERCISE 22.4   **Calculating Quantities of Radioactive Isotopes**

Two isotopes of yttrium, yttrium-86 ($t_{1/2}$ = 14.6 hr) and yttrium-90 ($t_{1/2}$ = 64 hr), have been used in radioimaging. Which decays faster? If we start with 10.0 mg of each isotope, how much of each remains after 24 hr?

Collect and Organize We are given the half-lives of two radionuclides and asked to predict which one will decay faster. We are also asked to calculate how much of the 10.0 mg of each isotope remains after 24 hr.

Analyze An isotope with a shorter half-life decays faster. Radioactive decay follows first-order kinetics. Quantitatively, the relationship between half-life and amount of material remaining is described by Equation 21.17:

$$\ln \frac{N_t}{N_0} = \frac{-0.693t}{t_{1/2}}$$

where N_0 and N_t refer to the amount of material present initially and the amount at time t, respectively. If our prediction for the relative decay rates of the two yttrium isotopes is correct, then more of the isotope with the longer half-life should remain after 24 hr. We need to complete two calculations to determine the amount of each sample present after 24 hr.

CONNECTION We discussed first-order reactions in Chapter 15 and applied the same equations to the first-order kinetics of radioactive decay in Chapter 21.

Solve For the amount of yttrium-86 remaining after 24 hr, we have

$$\ln \frac{N_t}{10.0 \text{ mg}} = \frac{(-0.693)(24 \text{ hr})}{14.6 \text{ hr}} = -1.139$$

Taking the antilog of both sides, we get

$$\frac{N_t}{10.0 \text{ mg}} = 0.320$$

$$N_t = (0.320)(10.0 \text{ mg}) = 3.20 \text{ mg}$$

For yttrium-90,

$$\ln \frac{N_t}{10.0 \text{ mg}} = \frac{(-0.693)(24 \text{ hr})}{64 \text{ hr}} = -0.2599$$

$$\frac{N_t}{10.0 \text{ mg}} = 0.771$$

$$N_t = 7.7 \text{ mg}$$

Think about It We predicted that yttrium-86 would decay faster, which means that after 24 hr the quantity of this isotope should be less than the quantity of yttrium-90, and it is.

Practice Exercise Two radioactive isotopes of rhenium are used to treat breast, liver, and skin cancer. The half-lives are 89 hr for rhenium-186 and 17 hr for rhenium-188. If we start with 25.0 mg of each isotope, how much of each sample remains after 24 hr?

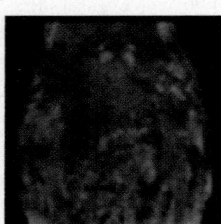

(unenhanced)

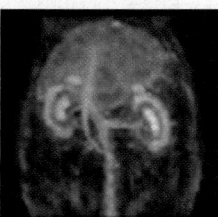

(enhanced)

FIGURE 22.17 MRI scans made without a contrast agent and enhanced by a gadolinium contrast agent.

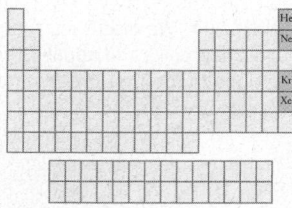

Imaging with Stable Isotopes: Lanthanides, Noble Gases, and MRI

As noted earlier, the quality of an MRI scan (Figure 22.17) can be improved when gadolinium is used as a contrast agent. Prior to the procedure, the patient is injected with a gadolinium compound. Many ligands have been investigated in Gd^{3+} MRI contrast agents. The gadolinium used is a mixture of naturally occurring isotopes of gadolinium. Coordination compounds of other lanthanide ions (Dy^{3+} and Ho^{3+}) have been evaluated but do not work as well as those of gadolinium.

So far in this chapter we have had little opportunity to mention the noble gas elements. None of these elements are essential to the human body. Their lack of chemical reactivity and ease of introduction into the body by inhalation, however, make selected isotopes of the noble gases, including helium-3, krypton-83, and xenon-129, attractive as agents for enhancing MRI images, particularly of the lungs. Krypton-83 provides greater sensitivity than xenon-129, allowing for better resolution in the images and, in principle, requiring the use of less gas.

Helium-3 has no known side effects and is preferable to xenon-129 for MRI, but it is present in only trace natural abundance. This isotope is obtained from β decay of tritium (^{3}H):

$$^3_1H \rightarrow \, ^3_2He + \, ^{\ 0}_{-1}\beta$$

Neon-19 has been used in PET despite its short half-life (17.5 s). A patient positioned in a PET scanner breathes air containing a small amount of this isotope. Positron emission from the neon is recorded, and an image is created. Of the group 18 elements, only argon and radon have not found direct medical applications.

Therapeutic Applications

In this section we examine therapeutic agents that contain metallic elements, including alkali metals, transition elements, and heavier main group elements in addition to carbon, hydrogen, nitrogen, oxygen, and sulfur.

∞ **CONNECTION** Tritium is produced in nuclear reactors by bombarding ^{6}Li and ^{7}Li with high-energy neutrons. The reactions involved are discussed in Chapter 21.

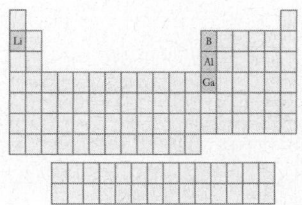

Lithium, Boron, Aluminum, and Gallium

The similar size of Li^+ (76 pm) and Mg^{2+} (72 pm) means that lithium ions can compete with magnesium ions in biological systems. The substitution of lithium for magnesium may account for its toxicity at high concentrations. Nevertheless, lithium carbonate is used to treat bipolar disorder, and other lithium compounds have been used to treat hyperactivity. In all cases, however, the use of lithium-containing drugs must be carefully monitored.

Of the elements of group 13, only boron and aluminum have been detected in humans. The role of boron in our bodies is not fully understood, but this element appears to play a role in nucleic acid synthesis and carbohydrate metabolism. Selected boron compounds appear to concentrate in human brain tumors. This property has opened the door to a treatment known as boron neutron-capture therapy (BNCT). Once a suitable boron compound has been injected and has made its way to a tumor, irradiation of the tumor with low-energy neutrons leads to the nuclear reaction

$$^{10}_{5}B \rightarrow {}^{1}_{0}n \rightarrow {}^{7}_{3}Li + {}^{4}_{2}He$$

The α particles generated in the reaction have a short penetration depth but high relative biological effectiveness (RBE), so they can kill the tumor cells without harming surrounding tissue. The identification of compounds suitable for BNCT remains an area of active research.

Aluminum is found in some antacids as aluminum hydroxide, $Al(OH)_3$, or aluminum carbonate, $Al_2(CO_3)_3$; aluminum sodium sulfate, $AlNa(SO_4)_2 \cdot 12 H_2O$, is an ingredient in baking powder. Most of the aluminum in the human body can be traced to these sources. Aluminum is not considered essential to humans, but low-aluminum diets have been observed to harm goats and chickens. High concentrations of aluminum are clearly toxic; the effects are most noticeable in patients with impaired kidney function. The role of aluminum in Alzheimer's disease has been extensively debated but remains unresolved.

Simple gallium compounds like gallium(III) nitrate and gallium(III) chloride, either alone or in combination with other drugs, have shown activity toward bladder and ovarian cancers. The similar ionic radii of Ga^{3+} (62 pm) and Fe^{3+} (64.5 pm) allow gallium to block DNA synthesis by replacing iron in a protein called transferrin and in other enzymes. Because gallium compounds accumulate in tumors at a higher rate than in healthy tissue, the disruption of DNA synthesis in the tumor cells prevents the tumor from growing.

Antimony and Bismuth

Antimony compounds are generally considered to be toxic. It has been reported, for instance, that exposure of infants to antimony compounds used as fire retardants in mattresses may contribute to sudden infant death syndrome (SIDS). However, this element does appear to have a medical use. Leishmaniasis, an insect-borne disease characterized by the formation of boils or skin lesions, is resistant to most treatments, but patients have been successfully treated with sodium stibogluconate, one of the few applications of antimony compounds in human health.

Popular over-the-counter remedies for indigestion, diarrhea, and other gastrointestinal disorders contain bismuth subsalicylate (Figure 22.18). The bismuth in these compounds acts as a mild antibacterial agent that reduces the number of diarrhea-causing bacteria.

CONNECTION The relative biological effectiveness (RBE) of radioactive particles was introduced in Chapter 21.

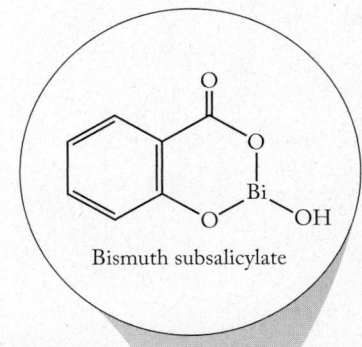

Bismuth subsalicylate

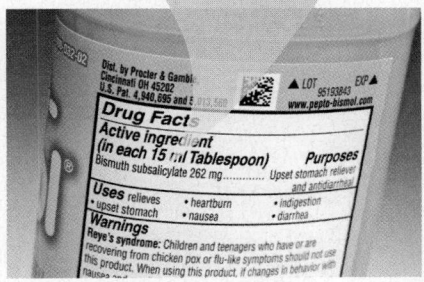

FIGURE 22.18 Bismuth subsalicylate is found in some antacids.

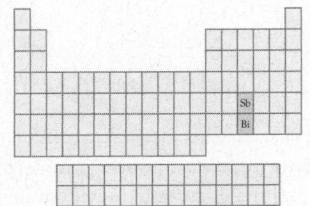

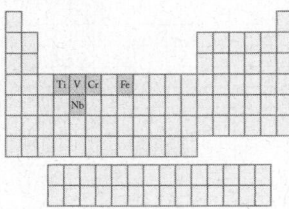

Titanium, Vanadium, and Niobium Cancer Drugs

Compounds of some transition metals demonstrate antitumor activity. Titanium(IV) complexes such as budotitane (Figure 22.19) show promise against colon cancer. Encouraging results against breast, lung, and colon cancers were observed with a series of compounds with formulas $(C_5H_5)_2TiCl_2$, $(C_5H_5)_2VCl$, and $(C_5H_5)_2NbCl_2$. Unfortunately, at therapeutically useful doses the potential for liver damage outweighs the benefits of these compounds. These compounds contain carbon–metal bonds and belong to a class of substances called **organometallic compounds**.

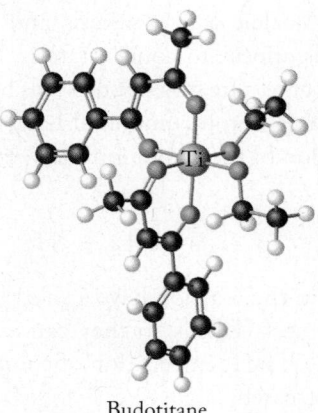

Budotitane

FIGURE 22.19 Budotitane is a titanium(IV) complex that shows activity against colon cancer.

Vanadium and Chromium

Insulin is a hormone needed for proper glucose metabolism and protein synthesis. People suffering from diabetes either are unable to produce insulin or produce the hormone but are unable to use it effectively. The suggestion that vanadium plays a role in insulin production has prompted investigation of vanadium compounds as diabetes drugs. Encouraging results from animal studies using bis(allixinato)oxovanadium(IV) (Figure 22.20a) were reported in 2009, but the

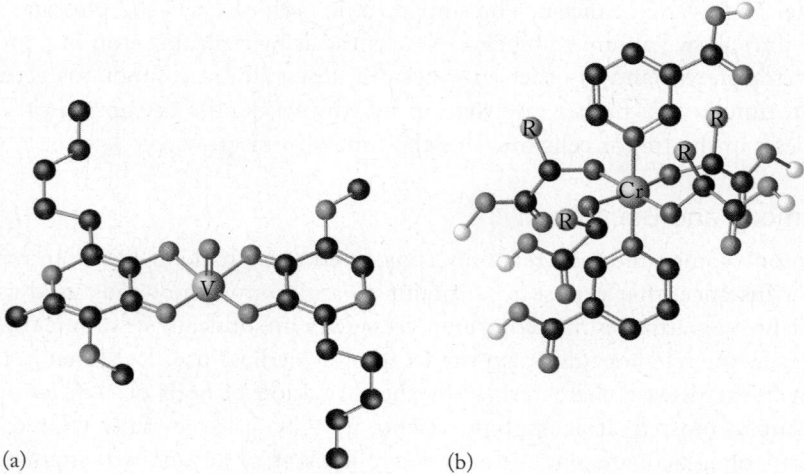

(a)   (b)

FIGURE 22.20 (a) Vanadium(IV) compounds like bis(allixinato)oxovanadium(IV) can act as insulin mimics. (b) The chromium(III) complex in a glucose tolerance factor contributes to our bodies' ability to regulate insulin levels. The green spheres are the side chain R groups of the amino acids in the structure. (To simplify the structures, the hydrogen atoms bonded to carbon atoms are not shown.)

organometallic compound a molecule containing direct carbon–metal covalent bonds.

compound is not currently approved for use in humans. Mixed results have been reported in clinical trials using chromium(III) compounds to reduce the amount of insulin required by some diabetics (Figure 22.20b).

Iron

The body must regulate the amount of iron in cells in order to produce enough hemoglobin to maintain good health. Deficiencies in iron lead to a number of diseases broadly classified as anemia. Mild forms of anemia are common among women of child-bearing age and are treated with oral iron supplements. A more serious genetic form of anemia known as thalassemia must be treated with blood transfusions that leave patients with too much iron in their blood. To remove the excess iron, these patients are treated with chelating ligands that complex some of the iron and transport it out of the red blood cells and eventually out of the body in what is known as chelation therapy. Chelation therapy relies on the equilibrium constant for the reaction:

$$FeL(aq) + L'(aq) \rightleftharpoons FeL'(aq) + L(aq)$$

Iron coordinated to ligand L (such as hemoglobin) in the blood is treated with ligand L', which also binds iron. If the formation constant for the complex between iron and L' is greater than for iron and L, then the equilibrium constant for the equation will be greater than one, and formation of the iron–L' complex will be favored. The ligand L' is chosen so that the complex between iron and L' is eliminated from the body, reducing the concentration of iron and relieving the symptoms caused by thalassemia treatments. Chelation therapy is also the treatment of choice for heavy metal poisoning from toxic metals including lead or mercury.

The use of chelating ligands to form coordination complexes of iron and other transition metals introduces the possibility of forming optical isomers. Consider the two iron complexes in Figure 22.21; they are nonsuperimposable mirror images of each other and hence are enantiomers. You may wish to build models of the complex and its mirror image to see for yourself that they are not superimposable. One enantiomer of an iron complex often works better than the other in the treatment of diseases like thalassemia.

∞ CONNECTION We encountered enantiomers of metal complexes in Chapter 18, where we saw that cis-Co(en)$_2$Cl$_2^+$ has two enantiomers.

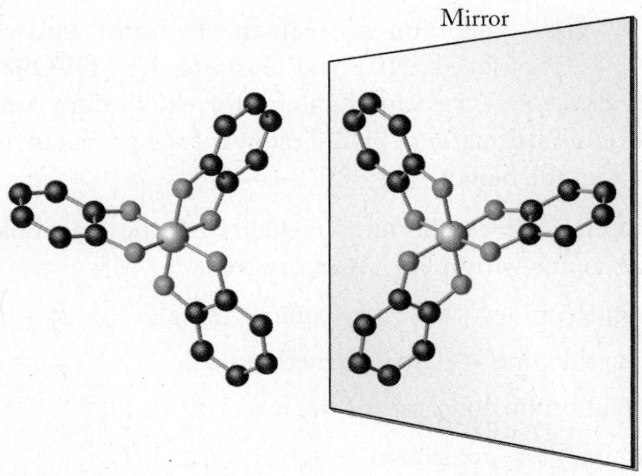

Mirror

FIGURE 22.21 Some octahedral transition metal coordination complexes containing polydentate ligands, such as this iron(II) compound, exist as optical isomers.

SAMPLE EXERCISE 22.5   **Calculating the Equilibrium Constant of a Ligand Exchange Reaction**

The protein transferrin is involved in the transport of iron into cells. Iron accumulation in the human body has been implicated in diseases such as Parkinson's, Alzheimer's, and thalassemia. The chelating ligand deferoxamine (DFO) is used to treat thalassemia; the complex between iron and DFO is eliminated from the body. Given the formation constants for the complexation of Fe(III) by transferrin (Equation A) and the reaction of deferoxamine with iron (Equation B), calculate the equilibrium constant for the ligand exchange reaction between deferoxamine and transferrin (Equation C).

$$Fe^{3+}(aq) + transferrin(aq) \rightleftharpoons$$
$$Fe(transferrin)^{3+}(aq) \qquad K_{f,A} = 4.7 \times 10^{20} \quad (A)$$
$$Fe^{3+}(aq) + DFO(aq) \rightleftharpoons$$
$$Fe(DFO)^{3+}(aq) \qquad K_{f,B} = 4.0 \times 10^{30} \quad (B)$$
$$Fe(transferrin)^{3+}(aq) + DFO(aq) \rightleftharpoons$$
$$Fe(DFO)^{3+}(aq) + transferrin(aq) \qquad K_C = ? \quad (C)$$

Collect and Organize We are given the formation constants of two reactions involving iron and different ligands. These data allow us to predict the position of the equilbrium and to calculate the equilibrium constant for the ligand exchange reaction. You may wish to refer to Chapter 18 for calculations involving formation constants of complexes, as well as to Chapters 16 and 17, where we manipulated equilibrium constants.

Analyze Transferrin is a reactant in Equation A, and a product in the exchange reaction (C). Therefore, we would need to reverse Equation A before adding it to Equation B to obtain Equation C. Reversing a reaction means taking the reciprocal of its equilibrium constant. Combining the reverse of Equation A with Equation B to obtain Equation C means multiplying $1/K_{f,A}$ and $K_{f,B}$ together to obtain K_C. The reciprocal of $K_{f,A}$ has a value of about 10^{-20}. Therefore, the value of K_C should be about $10^{-20} \times 10^{30}$, or about 10^{10}.

Solve Multiplying $1/K_{f,A}$ by $K_{f,B}$ to obtain K_C:

$$K_C = \frac{1}{4.7 \times 10^{20}} \times (4.0 \times 10^{30}) = 8.5 \times 10^9$$

Think about It The equilibrium between the Fe(transferrin) complex and deferoxamine (DFO) is indeed $\sim 10^{10}$ and illustrates why DFO is an effective treatment for thalassemia—the large value for the equilibrium constant means that the equilibrium in Equation C will lie toward the products (to the right), removing iron from the blood.

Practice Exercise The equilibrium constants for the reactions of penicillamine and methionine with methyl mercury are

$$CH_3Hg^+ + penicillamine \rightleftharpoons CH_3Hg(penicillamine)^+ \qquad K_f = 6.3 \times 10^{13}$$
$$CH_3Hg^+ + methionine \rightleftharpoons CH_3Hg(methionine)^+ \qquad K_f = 2.5 \times 10^7$$

Calculate the equilibrium constant for the reaction:

$$CH_3Hg(methionine)^+ + penicillamine \rightleftharpoons$$
$$CH_3Hg(penicillamine)^+ + methionine$$

⨳⨳ CONNECTION In Chapter 18 we discussed formation constants for metal complexes. Formation constants are equilibrium constants that describe the formation of a metal-ligand complex.

Platinum Group and Coinage Metals

The period 5 and period 6 elements in groups 8, 9, and 10 (Ru, Rh, Pd, Os, Ir, and Pt) are often referred to as the *platinum group metals*. The group 11 elements of these two periods (Ag and Au) along with Cu are called the *coinage metals*. In this section we explore examples of soluble compounds of these metals used in medications for arthritis, cancer, and other diseases.

The serendipitous discovery in the 1960s that *cis*-diamminedichloro-platinum(II) (cisplatin in Figure 22.22) is effective in treating testicular, ovarian, and other cancers has spurred development of a host of cancer drugs based on both platinum group metals and coinage metals. The results of this research include a compound known as carboplatin that shows the same activity as cisplatin but has fewer side effects. In addition to platinum compounds, the antitumor activities of complexes of gold, rhodium, ruthenium, and silver have been explored. The effectiveness of all of these drugs lies in their ability to bind to the nitrogen-containing bases in DNA and inhibit cell replication. If their cells cannot divide, tumors cannot grow. The greater toxicity of rhodium compounds relative to those of platinum and ruthenium has limited their clinical application.

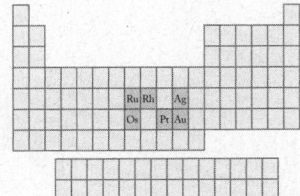

CONNECTION The use of the platinum group metals (groups 8–10) as catalysts for inorganic chemical reactions was discussed in Chapter 15.

FIGURE 22.22 Cisplatin and carboplatin are effective antitumor agents. The ruthenium and rhodium compounds show activity against leukemia but have not yet seen widespread use.

Selected osmium compounds reduce inflammation in joints resulting from arthritis, although the use of such compounds has diminished with the development of other anti-inflammatory agents. The therapeutic effects of aqueous solutions of osmium tetroxide, OsO_4, were first investigated in the 1950s. The use of osmium tetroxide was superseded by the use of glucose polymers containing osmium, known as osmarins. The use of osmarins reduces the toxic effects of osmium and illustrates how even toxic metals can be adapted to therapy.

Although the historic use of gold for medicinal purposes dates back millennia, the effective use of gold-containing pharmaceuticals originated with the discovery in the 1920s and 1930s that a gold thiosulfate compound, sanochrysin, alleviates the symptoms of rheumatoid arthritis (Figure 22.23). The most commonly used gold drugs for the treatment of arthritis today are sold under the names myochrysine and auranofin (Figure 22.23). Myochrysine is injected; auranofin can be taken orally.

Without knowing it, most of us experienced the bactericidal effects of silver nitrate within minutes of birth when our eyes were washed with a dilute aqueous solution of $AgNO_3$. This practice prevents infantile blindness by killing any bacteria in the eyes that accompany birth. Eyedrops containing silver nitrate have also been used for other infectious eye diseases. Silver sulfadiazine is a

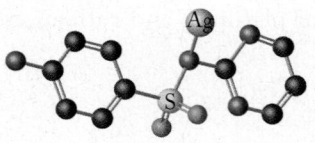

(a) Sanochrysin

[R = —CH₂(CH₂COONa)COONa]
(b) Myochrysine

[L = =P(CH₂CH₃)₃]
(c) Auranofin

FIGURE 22.23 (a) Sanochrysin, (b) myochrysine, and (c) auranofin are three gold-containing drugs used to treat arthritis.

broad-spectrum "sulfa" drug used to prevent and treat bacterial and fungal infections (Figure 22.24). It is an active ingredient in creams used to treat thermal and chemical burns.

Silver sulfadiazine

FIGURE 22.24 Silver sulfadiazine is an effective antibiotic when applied to burns. (H atoms are not shown.)

Medical Devices and Materials

Dental Alloys

When the fluoride treatment described in Section 22.3 fails to prevent dental caries, a dentist will clean the cavity and fill it, often with a metal alloy that could contain mercury, silver, tin, copper, and traces of zinc. An alloy containing mercury is called an *amalgam*. The most common dental amalgam contains Ag_2Hg_3, Ag_3Sn, and Sn_8Hg. Although mercury and mercury compounds are highly toxic, when combined with silver or tin the mercury is chemically unreactive.

Pacemakers

Pacemakers are lifesaving devices implanted in the chests of patients with certain heart conditions. A pacemaker delivers electrical pulses to stimulate the heart to beat at the proper rate. The batteries used in pacemakers must deliver reliable power for long periods because, without additional surgery, the batteries cannot be replaced once the pacemaker is implanted. A solid-state zinc/mercury oxide battery (Figure 22.25)

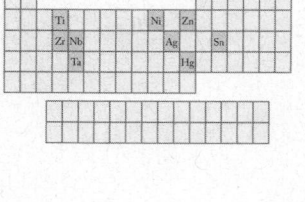

CONNECTION Alloys were described in Chapter 12. The cell potentials of simple voltaic cells were discussed in Chapter 19.

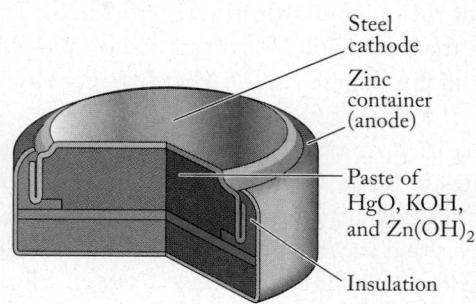

Steel cathode

Zinc container (anode)

Paste of HgO, KOH, and Zn(OH)₂

Insulation

FIGURE 22.25 The Zn/HgO battery used in pacemakers.

meets these stringent requirements and was the battery of choice in the first pace-makers. The zinc container acts as the anode, where oxidation takes place. A steel cathode extends into a paste of mercury(II) oxide, potassium hydroxide, and zinc(II) hydroxide. The half-reactions and overall cell reaction are

Anode: $\qquad$ $Zn(s) + 2\,OH^-(aq) \rightarrow ZnO(s) + H_2O(\ell) + 2\,e^-$

Cathode: $\quad HgO(s) + H_2O(\ell) + 2\,e^- \rightarrow Hg(\ell) + 2\,OH^-(aq)$

Overall: $\qquad$ $Zn(s) + HgO(s) \rightarrow ZnO(s) + Hg(s) \qquad E^\circ_{cell} = 1.347 \text{ V}$

The hazards associated with mercury and other performance issues have led to the use of lithium/iodine batteries in pacemakers in the last 30 years. The cathode is an electrically conductive material prepared by heating iodine with poly(vinylpyridine) (PVP). Lithium metal acts as the anode, and the reactions are

Anode: $\qquad$ $2\,Li(s) \rightarrow 2\,Li^+(s) + 2\,e^-$

Cathode: $\qquad$ $I_2(PVP)(s) + 2\,e^- \rightarrow 2\,I^-(s) + (PVP)(s)$

Overall: $\quad 2\,Li(s) + I_2(PVP)(s) \rightarrow 2\,LiI(s) + (PVP)(s) \qquad E^\circ_{cell} = 3.59 \text{ V}$

∞ CONNECTION In Chapter 19 we learned how to use reduction potentials for half-reactions to calculate overall potentials of reactions.

∞ CONNECTION Poly(vinylpyridine) is a polymer. The preparation and properties of synthetic polymers like poly(vinylpyridine) were described in Chapter 13.

SAMPLE EXERCISE 22.6 **Calculating the Electrochemical Potential of a Battery**

Hearing aid batteries are based on the zinc/air electrochemical cell. Given the following standard reduction potentials, (a) determine the overall equation for the cell reaction and (b) calculate the standard cell potential.

$$Zn(OH)_2(s) + 2\,e^- \rightarrow Zn(s) + 2\,OH^-(aq) \qquad E^\circ = -1.249 \text{ V}$$
$$O_2(g) + 2\,H_2O(\ell) + 4\,e^- \rightarrow 4\,OH^-(aq) \qquad E^\circ = 0.401 \text{ V}$$

Collect and Organize We are given two half-reactions and their potentials. The cell reaction is the sum of the oxidation and reduction half-reactions, each multiplied by a coefficient to make the number of electrons gained equal the number of electrons lost. The overall electrochemical cell potential is the difference in the potentials of the cathode and anode half-reactions (Equation 19.2). You may wish to review Sections 19.2 and 19.3.

Analyze Both half-reactions in the zinc/air battery are written as reductions: that is, the reactant gains electrons. We need to determine which half-reaction actually runs in reverse, i.e., as an oxidation at the anode. The electrochemical potential for the overall reaction must be positive ($E^\circ > 0$) in order to have a functioning battery. We also need to determine each half-reaction's coefficient to ensure that the number of moles of electrons gained equals the number of moles of electrons lost.

Solve

a. The reduction of $Zn(OH)_2$ has the smaller E°. Therefore, the anode half-reaction will be the reverse of $Zn(OH)_2$ reduction, that is, oxidation of Zn metal in alkaline solution:

$$Zn(s) + 2\,OH^-(aq) \rightarrow Zn(OH)_2(s) + 2\,e^-$$

The reduction of O_2 to OH^- involves four moles of electrons, so we must multiply the oxidation half-reaction by 2:

$$2\,Zn(s) + 4\,OH^-(aq) \rightarrow 2\,Zn(OH)_2(s) + 4\,e^-$$

Adding the two half-reactions:

$$2\,Zn(s) + 4\,OH^-(aq) \rightarrow 2\,Zn(OH)_2(s) + 4\,e^-$$
$$O_2(g) + 2\,H_2O(\ell) + 4\,e^- \rightarrow 4\,OH^-(aq)$$

$$\overline{2\,Zn(s) + 4\,\cancel{OH^-(aq)} + O_2(g) + 2\,H_2O(\ell) + 4\,\cancel{e^-} \rightarrow}$$
$$4\,\cancel{OH^-(aq)} + 2\,Zn(OH)_2(s) + 4\,\cancel{e^-}$$

and the overall reaction is

$$2\,Zn(s) + O_2(g) + 2\,H_2O(\ell) \rightarrow 2\,Zn(OH)_2(s)$$

b. The standard cell potential is the difference in the standard potentials for the half-reactions:

$$E^\circ_{cell} = E^\circ_{cathode} - E^\circ_{anode} = 0.401\ V - (-1.249\ V) = 1.650\ V$$

Think about It The standard cell potential, 1.650 V, assumes that $P_{O_2} = 1$ atm. In air P_{O_2} is about 0.21 atm; so the potential of the Zn/air battery will likely be less than 1.650 V.

Practice Exercise Implantable cardioverter-defibrillators (ICDs) are devices that detect certain types of cardiac arrhythmias and provide a shock directly to the patient's heart, stopping ventricular fibrillation. Lithium silver vanadium oxide batteries are used to power these devices. The equation for the overall reaction is

$$7\,Li(s) + Ag_2V_4O_{11}(s) \rightarrow Li_7Ag_2V_4O_{11}(s) \qquad E^\circ_{cell} = 2.00\ V$$

Using the data in Appendix 6 for the oxidation potential for Li metal, calculate the reduction potential for silver vanadium oxide.

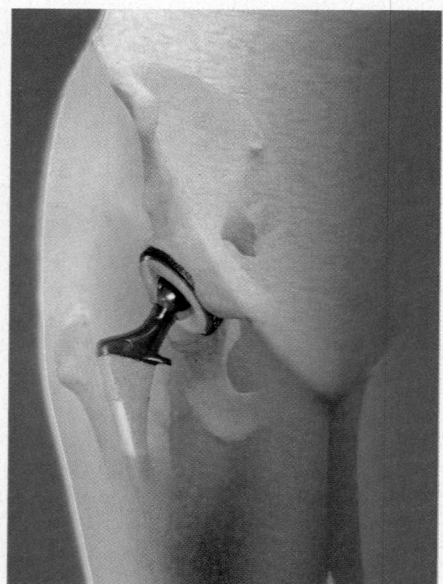

∞ **CONNECTION** In Chapter 13 we learned about the development of high-density polyethylene polymers for use in artificial joints.

FIGURE 22.26 Tantalum is used to coat artificial joints such as this hip joint.

Artificial Joints

The degeneration of knees and hips that often accompanies aging necessitates their replacement with artificial joints. The biomaterials for these joints must be chemically nonreactive, and they must provide a smooth surface that allows the joint to move. Thin coatings of pure tantalum or niobium, or of alloys of these two metals, are used to make surfaces on artificial joints smooth and abrasion-resistant (Figure 22.26). Titanium and zirconium have also been used in alloys to coat implanted joints.

Stents

Shape-memory alloys are an unusual class of materials. Like wire made from most metals, a wire made from a shape-memory alloy can be deformed or bent into a variety of shapes. When it is heated, however, the shape-memory wire reverts to its original shape—it "remembers" that original shape.

(a)   (b)   (c)

FIGURE 22.27 Shape-memory alloys are used in stents for heart patients. (a) The S shape on the right is the shape a stent has both when it is manufactured and after it has been inserted in an artery. The left and middle stents have been coiled into a compressed shape, ready to be inserted in an artery. The white circle in the background is a hair dryer heating the coiled stents. (b) The heated middle stent has returned to its original, "remembered" S shape. (c) The heated left stent has returned to its original shape. With a stent inserted in an artery, the same return to a remembered shape happens at body temperature, and the S-shaped stent forces the vessel open, allowing better blood flow.

The first shape-memory alloy was a 1:1 nickel–titanium alloy. The "memory" property of shape-memory alloys depends on changes in crystal structure as a function of temperature. The crystal structure of NiTi at room temperature is slightly different from its crystal structure at temperatures above 500°C. When a wire made of NiTi is bent into a shape at room temperature and then heated, the crystal structure changes to the high-temperature form, removing the bend.

The shape-memory property of selected alloys leads to their use in *stents*, medical devices used to prop open weakened and clogged arteries (Figure 22.27). The stent is formed into a coil or tube that can be surgically inserted into a partially blocked artery. Once the stent is in place, it changes back to its original shape as it is warmed by body heat, propping open the artery.

At least 15 shape-memory alloys have been identified, ranging from the most commonly used nickel–titanium alloy to alloys containing hafnium, titanium, and nickel. Perhaps hafnium-containing shape-memory alloys will join the other group 4 elements as useful materials for medical applications. If so, such alloys would represent one of the very few medical applications of hafnium.

CONCEPT TEST

The table of standard reduction potentials in Appendix 6 does not include values for many of the metals and alloys used in artificial joints and in stents. Do you expect their standard reduction potentials to be greater than or less than that of $Fe^{2+} + 2\,e^- \rightarrow Fe$?

The human body requires about 30 elements to function properly. These elements span the periodic table from group 1 elements like sodium, to transition metals like iron, to halogens like chlorine. Our bodies need different amounts of these essential elements, ranging from gram to microgram quantities. Through chemical, biochemical, and clinical research, many of the nonessential elements have found application in the diagnosis and treatment of disease.

··· ■

Section 22.1 **Essential elements** have a physiological function in the body. **Nonessential elements** are present in the body but have no known functions. Some may have **stimulatory effects**. Essential elements are categorized as **major, trace,** or **ultratrace essential elements** depending on their concentrations in the body. Some elements, such as mercury, cadmium, and lead, have chemical properties that make them highly toxic.

Section 22.2 Transport of Na^+ and K^+ across cell membranes involves **ion pumps** or selective transport through **ion channels** formed by groups of helical proteins. Magnesium ions facilitate

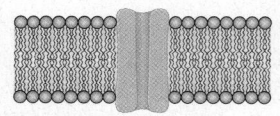

the transfer of phosphate groups between ATP and ADP and are found in chlorophyll, the green pigment in plants that mediates photosynthesis. Calcium is a major component of teeth and bones. A deficiency of calcium can lead to osteoporosis, whereas high concentrations in intracellular fluids contribute to muscle cramps. Bones are primarily elastic protein fibers and calcium compounds including the mineral hydroxyapatite, $Ca_5(PO_4)_3(OH)$, which is also the principal component of teeth. Chloride ion is the most abundant anion in the human body, facilitating transport of alkali metal cations and elimination of CO_2. Malfunctioning chloride channels are an underlying cause of cystic fibrosis. Nitrogen is found in proteins, DNA, and RNA.

Section 22.3 Amorphous silica, SiO_2, makes up the exoskeletons of diatoms. Silicon scavenges aluminum in mammals, and silicon deficiency stunts growth. Fluoride ions may replace hydroxide ions in hydroxyapatite in teeth and inhibit the formation of cavities. The body concentrates iodide ions in the thyroid gland where they are incorporated into the

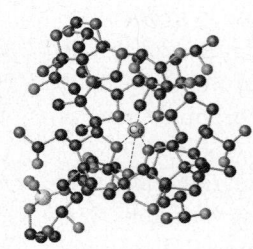

hormones that regulate energy production and use. Many enzymes and **coenzymes** contain transition metal ions. The amino acid selenocysteine is a constituent of proteins that act as catalysts for biochemical reactions. Manganese is important in photosynthesis. Chromium helps regulate glucose levels in blood.

Section 22.4 Rubidium and cesium ions are probably nonessential, yet are retained by the body because of their similarity in size and chemistry to K^+. Strontium and barium ions are incorporated into human bones by replacing Ca^{2+} ions but have no known function in the human body. However, Sr^{2+} and Ba^{2+} are found in the exoskeletons of certain single-celled organisms. Germanium is a nonessential element and barely detectable in the human body.

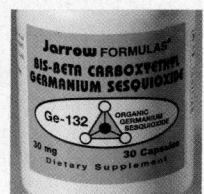

Section 22.5 Radionuclides with short half-lives that emit low-energy γ rays are used in diagnosing disease. The selection of a radionuclide is also governed by the toxicities of the element and its daughter nuclides, and the speed at which it is eliminated from the body. A useful diagnostic or therapeutic agent must be sufficiently

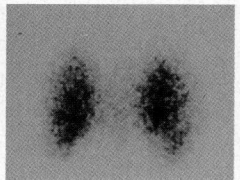

soluble in blood to be delivered to the target. Corrosion-resistant metals and alloys are used in artificial joints and stents.

···························· ■

TYPE OF PROBLEM	CONCEPTS AND EQUATIONS	SAMPLE EXERCISES
Calculating an acid concentration from its pH	Relate the pH of a solution to the $[H^+]$ by the equation $$pH = -\log[H^+]$$	22.1
Assigning oxidation numbers and writing half-reactions	Assign oxidation numbers using the guidelines in Chapter 4: oxidation numbers for pure elements are zero; for monatomic ions O.N. equals the ionic charge; H and O typically have O.N. equal to +1 and, −2, respectively. Then use the numbers to determine the number of electrons gained or lost; balance charges with H^+ ions in acids or OH^- ions in base; add water molecules to balance the numbers of H and O atoms.	22.2

TYPE OF PROBLEM	CONCEPTS AND EQUATIONS	SAMPLE EXERCISES
Identifying particles in nuclear reactions	Sum the subscripts and superscripts on both sides of the nuclear equation. The difference in the superscripts equals the mass of the missing particle, and the difference in the subscripts equals the charge on the missing particle.	22.3
Calculating quantities of radioactive isotopes	Use the equation $$\ln\frac{N_t}{N_0} = \frac{-0.693t}{t_{1/2}} \qquad (21.17)$$ where N_0 and N_t are the amounts of material present initially and at time t, respectively.	22.4
Calculating the equilibrium constant of a ligand exchange reaction	Combine the equilibrium constant expressions for the formation of the two species, after reversing the one that dissociates in the overall reaction, to generate an equivalent equilibrium constant expression for the exchange reaction.	22.5
Calculating the electrochemical potential of a battery	Identify the cathode and anode half-reactions. Then use the equation $$E^\circ_{cell} = E^\circ_{cathode} - E^\circ_{anode} \qquad (19.2)$$	22.6

VISUAL PROBLEMS ∙∙∙ ■

(Answers to boldface end-of-chapter questions and problems are in the back of the book.)

22.1. Which part of Figure P22.1 best describes the periodic trend in monatomic cation radii moving up or down a group or across a period in the periodic table? (Arrows point in the direction of increasing radii.)

22.2. Which part of Figure P22.1 best describes the periodic trend in monatomic anion radii moving up or down a group or across a period in the periodic table? (Arrows point in the direction of increasing radii.)

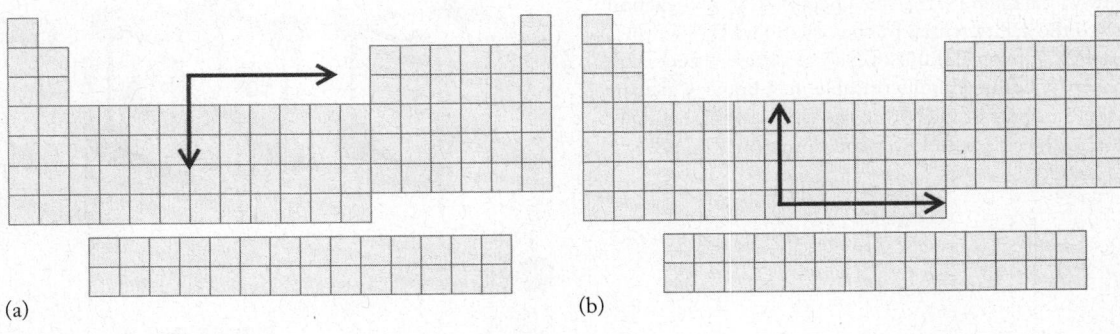

(a)   (b)

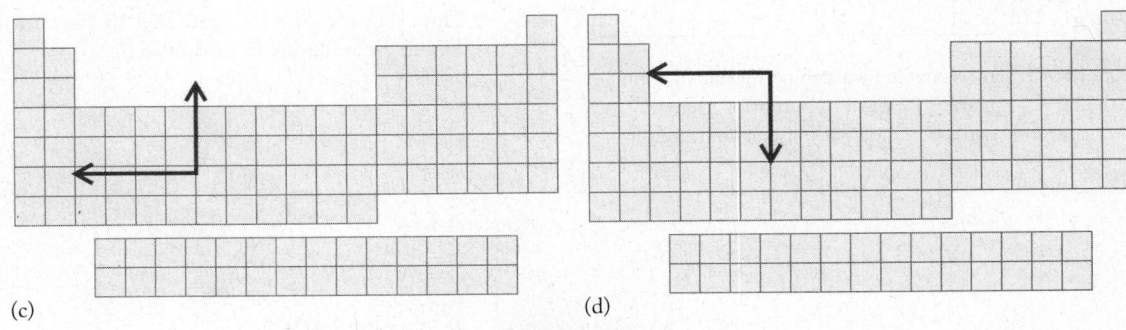

(c)   (d)

FIGURE P22.1

22.3. Which group highlighted in the periodic table in Figure P22.3 typically forms ions that have larger radii than the corresponding neutral atoms?

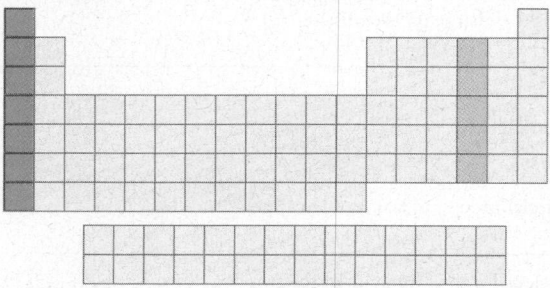

FIGURE P22.3

22.4. Which group highlighted in the periodic table in Figure P22.4 typically forms ions that have smaller radii than the corresponding neutral atoms?

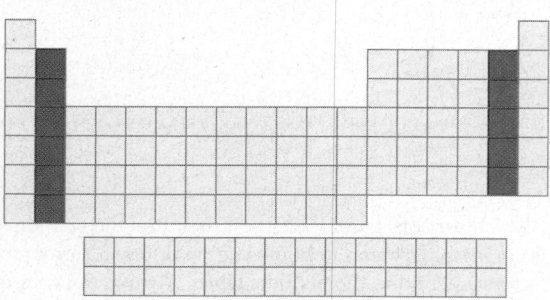

FIGURE P22.4

22.5. As we saw in Chapter 19, the free energy (ΔG) of a reaction is related to the cell potential by the equation $\Delta G = -nFE$. In Figure P22.5, two solutions of Na^+ of different concentrations are separated by a semipermeable membrane. Calculate ΔG for the transport of Na^+ from the side with higher concentration to the side with lower concentration. *Hint:* See Problem 22.35.

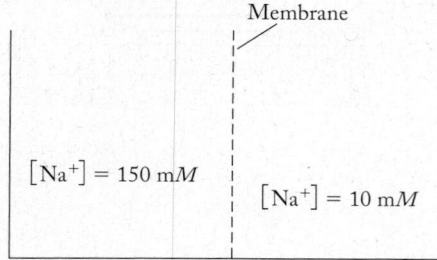

Membrane

$[Na^+] = 150$ mM $[Na^+] = 10$ mM

FIGURE P22.5

22.6. Two solutions of K^+ are separated by a semipermeable membrane in Figure P22.6. Calculate ΔG for the transport of K^+ from the side with lower concentration to the side with higher concentration.

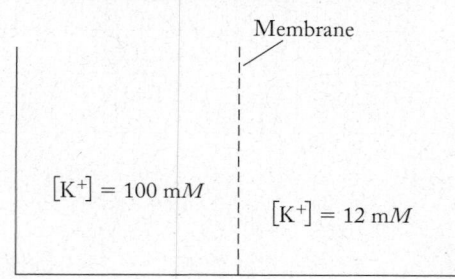

Membrane

$[K^+] = 100$ mM $[K^+] = 12$ mM

FIGURE P22.6

22.7. Describe the molecular geometry around each germanium atom in the structure of the germanium compound shown in Figure P22.7.

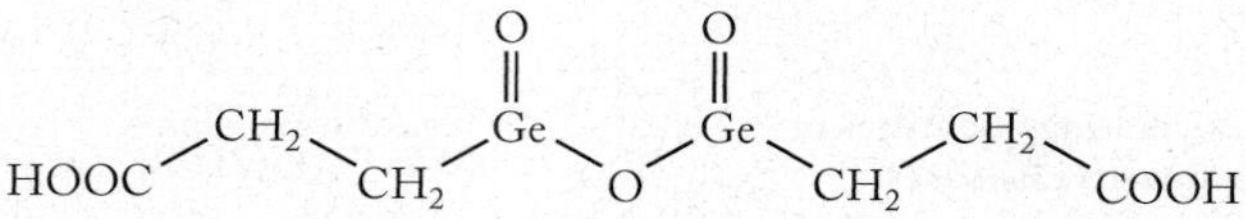

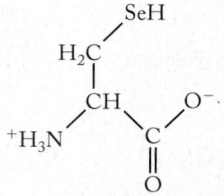

FIGURE P22.7

22.8. Selenocysteine can exist as two enantiomers (stereoisomers). Identify the atom in Figure P22.8 responsible for the two enantiomers.

FIGURE P22.8

22.9. The austenite structure for NiTi is similar to the CsCl structure. Which crystal structure shown in Figure P22.9 is correct for the austenite form of the shape-memory alloy NiTi?

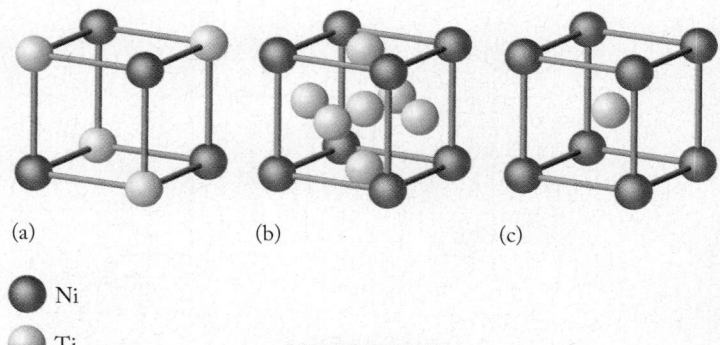

(a)   (b)   (c)

● Ni
○ Ti

FIGURE P22.9

22.10. The unit cell for the martensite form of the shape-memory alloy NiTi is shown in Figure P22.10. How many equivalent Ti and Ni atoms are in the unit cell?

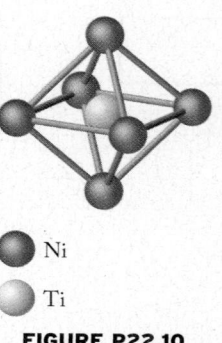

● Ni
○ Ti

FIGURE P22.10

QUESTIONS AND PROBLEMS

The Periodic Table of Life

CONCEPT REVIEW

22.11. What is the difference between an essential element and a nonessential element?

22.12. Are all essential elements major essential elements?

22.13. What is the main criterion that distinguishes (a) major, (b) trace, and (c) ultratrace essential elements from one another?

22.14. Should trace essential elements also be considered to be stimulatory?

22.15. Why is methylmercury ion, CH_3Hg^+, more toxic than mercury metal?

*__22.16.__ Why is Cd^{2+} more likely than Cr^{2+} to replace Zn^{2+} in an enzyme like carbonic anhydrase?

22.17. Why is Be^{2+} more likely than Ca^{2+} to displace Mg^{2+} in biomolecules?

22.18. PbS, $PbCO_3$, and $PbCl(OH)$ have limited solubility in water. Which of them is/are more likely to dissolve in acidic solutions?

PROBLEMS

22.19. Which ion is larger: K^+ or Na^+?

22.20. Which ion is larger: Cl^- or I^-?

22.21. The concentrations of very dilute solutions are sometimes expressed as parts per million. Express the concentration of each of the following trace and ultratrace essential elements in parts per million:
  a. copper, 110 mg in 70 kg
  b. zinc, 3.3×10^{-2} g/kg
  c. iodine, 0.043 g in 100 kg

22.22. In the human body, the concentrations of ultratrace essential elements are even lower than those of trace essential elements and therefore are sometimes expressed in parts per billion. Express the concentrations of each of the following elements in parts per billion:
  a. cobalt, 4.3×10^{-5} g/kg
  b. boron, 0.014 g/100 kg
  c. chromium, 5.0 mg/70 kg

22.23. In the following pairs, which element is more abundant in the human body: (a) silicon or oxygen; (b) iron or oxygen; (c) carbon or aluminum?

22.24. In the following pairs, which element is more abundant in the human body: (a) H or Si; (b) Ca or Fe; (c) N or Cr?

Major Essential Elements

CONCEPT REVIEW

22.25. Ion Transport in Cells Describe three ways in which ions of major essential elements (such as Na^+ and K^+) enter and exit cells.

22.26. Which transport mechanism for ions requires ATP: diffusion, ion channels, or ion pumps?

22.27. Why is it difficult for ions to diffuse across cell membranes?

22.28. Why does Sr^{2+} substitute for Ca^{2+} in bones?

22.29. Which alkali metal ion is Rb^+ most likely to substitute for?

22.30. Why don't alkaline earth metal cations substitute for alkali metal cations in cases where the ionic radii are similar?

*__22.31.__ Why might nature have chosen calcium carbonate over calcium sulfate as the major exoskeleton material in shells?

22.32. Bromide ion and fluoride ion are nonessential elements in the body. Do you expect their concentrations to be more similar to the concentrations of major essential elements or to the concentrations of ultratrace essential elements?

PROBLEMS

22.33. Osmotic Pressure of Red Blood Cells One of the functions of the alkali metal cations Na^+ and K^+ in cells is to maintain the cells' osmotic pressure. The concentration of NaCl in red blood cells is approximately 11 mM. Calculate the osmotic pressure of this solution at body temperature (37°C). (*Hint*: See Equation 11.14.)

22.34. Calculate the osmotic pressure exerted by a 92 mM solution of KCl in a red blood cell at body temperature (37°C). (*Hint*: See Equation 11.14.)

*__22.35.__ **Electrochemical Potentials across Cell Membranes** Very different concentrations of Na^+ ions exist in red blood cells (11 mM) and the blood plasma (160 mM) surrounding those cells. Solutions with two different concentrations separated by a membrane constitute a concentration cell of the type described in Chapter 19.

$$E = E° - (0.0592/n)\, \log([Na^+]_{cell}/[Na^+]_{plasma})$$

Calculate the electrochemical potential created by the unequal concentrations of Na^+.

22.36. The concentration of K^+ in red blood cells is 92 mM, and the concentration of K^+ in plasma is 10 mM. Calculate the electrochemical potential created by the two concentrations of K^+.

$$E = E° - (0.0592/n)\, \log([K^+]_{plasma}/[K^+]_{cell})$$

22.37. If the transport of K^+ across a cell membrane requires 5 kJ/mol, how many moles of ATP must be hydrolyzed to provide the necessary energy? The hydrolysis of ATP is described by the equation

$$ATP^{4-} + H_2O \rightarrow ADP^{3-} + HPO_4^{2-} + H^+ \quad \Delta G° = -34.5 \text{ kJ}$$

*__22.38.__ Removing excess Na^+ from a cell by an ion pump requires energy. How many moles of ATP must be hydrolyzed to overcome a cell potential of -0.07 V? The hydrolysis of 1 mol of ATP provides 34.5 kJ of energy.

22.39. Plankton Exoskeletons Exoskeletons of planktonic acantharia contain strontium sulfate. Calculate the solubility in moles per liter of $SrSO_4$ in water at 25°C given that $K_{sp} = 3.4 \times 10^{-7}$.

22.40. Algae in the genus *Closterium* contain structures built from barium sulfate (barite). Calculate the solubility in moles per liter of $BaSO_4$ in water at 25°C given that $K_{sp} = 1.1 \times 10^{-10}$.

Trace and Ultratrace Essential Elements

CONCEPT REVIEW

22.41. What danger to human health is posed by ^{137}Cs ($t_{1/2} \approx 30$ yr)?

22.42. Why is ^{137}Cs ($t_{1/2} \approx 30$ yr) considered to be dangerous to human health when naturally occurring ^{40}K ($t_{1/2} = 1.28 \times 10^6$ yr) is benign?

22.43. What is the origin of most of the ^{137}Cs and ^{90}Sr in our environment?

22.44. Why does fluorapatite resist acid better than hydroxyapatite if both are insoluble in water?

22.45. What is the function of enzymes?

22.46. Enzymes are large proteins. Are all proteins enzymes?

*22.47. What effect does an enzyme have on the activation energy of a biochemical reaction?

*22.48. Why might reductases also be described as reducing agents?

*22.49. When a transition metal ion like Cu^{2+} is incorporated into a metalloenzyme, is the formation constant likely to be much greater than one ($K \gg 1$) or much less than one ($K < 1$)?

$$Cu^{2+} + \text{protein} \rightleftharpoons \text{metalloenzyme} \qquad K = \frac{[\text{metalloenzyme}]}{[Cu^{2+}][\text{protein}]}$$

*22.50. When transition metals bind to proteins to form enzymes, is the process likely to be nonspontaneous ($\Delta G > 0$) or spontaneous ($\Delta G < 0$)?

PROBLEMS

22.51. What are the products of radioactive decay of ^{137}Cs? Write a balanced equation for the nuclear decay reaction.

22.52. Potassium-40 decays by three pathways: β decay, positron emission, and electron capture. Write balanced equations for each of these processes.

22.53. Calculate the pH of a 1.00×10^{-3} M solution of selenocysteine ($pK_{a_1} = 2.21$, $pK_{a_2} = 5.43$).

22.54. Calculate the pH of a 1.00×10^{-3} M solution of cysteine ($pK_{a_1} = 1.7$, $pK_{a_2} = 8.3$). Is selenocysteine a stronger acid than cysteine?

22.55. **Composition of Tooth Enamel** Tooth enamel contains the mineral hydroxyapatite. Hydroxyapatite reacts with fluoride ion in toothpaste to form fluorapatite. The equilibrium constant for the reaction between hydroxyapatite and fluoride ion is $K = 8.48$. Write the equilibrium constant expression for the following reaction. In which direction does the equilibrium lie?

$$Ca_5(PO_4)_3(OH)(s) + F^-(aq) \rightleftharpoons Ca_5(PO_4)_3(F)(s) + OH^-(aq)$$

*22.56. **Effects of Excess Fluoridation on Teeth** Too much fluoride might lead to the formation of calcium fluoride according to the reaction

$$Ca_5(PO_4)_3(OH)(s) + 10\,F^-(aq) \rightleftharpoons$$
$$5\,CaF_2(s) + 3\,PO_4^{3-}(aq) + OH^-(aq)$$

Write the equilibrium constant expression for the reaction. Given the K_{sp} values for the following two reactions, calculate K for the reaction between $Ca_5(PO_4)_3(OH)$ and fluoride ion that forms CaF_2.

$$Ca_5(PO_4)_3(OH)(s) \rightleftharpoons 5\,Ca^{2+}(aq) + 3\,PO_4^{3-}(aq) + OH^-(aq)$$
$$K_{sp} = 2.3 \times 10^{-59}$$

$$CaF_2(s) \rightleftharpoons Ca^{2+}(aq) + 2\,F^-(aq) \qquad K_{sp} = 3.9 \times 10^{-11}$$

22.57. Tooth enamel is actually a composite material containing both hydroxyapatite and a calcium phosphate, $Ca_8(HPO_4)_2(PO_4)_4 \cdot 6\,H_2O$ ($K_{sp} = 1.1 \times 10^{-47}$). Is this calcium mineral more or less soluble than hydroxyapatite ($K_{sp} = 2.3 \times 10^{-59}$)?

22.58. The K_{sp} of actual tooth enamel is reported to be 1×10^{-58}. Does this mean that tooth enamel is more soluble than pure hydroxyapatite ($K_{sp} = 2.3 \times 10^{-59}$)? Does the measured value of K_{sp} for tooth enamel support the idea that tooth enamel is a mixture of hydroxyapatite, $Ca_5(PO_4)_3(OH)$, and a calcium phosphate $Ca_8(HPO_4)_2(PO_4)_4 \cdot 6\,H_2O$ ($K_{sp} = 1.1 \times 10^{-47}$)?

22.59. Calculate the solubility in moles per liter of hydroxyapatite, $Ca_5(PO_4)_3(OH)$, $K_{sp} = 2.3 \times 10^{-59}$, and fluorapatite, $Ca_5(PO_4)_3F$, $K_{sp} = 3.2 \times 10^{-60}$, in water at 25°C and pH = 7.0.

22.60. Calculate the solubility in moles per liter of hydroxyapatite, $Ca_5(PO_4)_3(OH)$, $K_{sp} = 2.3 \times 10^{-59}$, in water at 25°C and pH = 5.0.

*22.61. Some sources give the formula of hydroxyapatite as $Ca_{10}(PO_4)_6(OH)_2$. If the K_{sp} of $Ca_5(PO_4)_3(OH)$ is 2.3×10^{-59}, what is the K_{sp} of $Ca_{10}(PO_4)_6(OH)_2$?

*22.62. The same sources mentioned in the previous problem cite the formula of fluorapatite as $Ca_{10}(PO_4)_6F_2$. If the K_{sp} of $Ca_5(PO_4)_3F$ is 3.2×10^{-60}, what is the K_{sp} of $Ca_{10}(PO_4)_6F_2$?

*22.63. The activation energy for the uncatalyzed decomposition of hydrogen peroxide at 20°C is 75.3 kJ/mol. In the presence of the enzyme catalase, the activation energy is reduced to 29.3 kJ/mol. By using the following form of the Arrhenius equation, $RT \ln(k_1/k_2) = E_{a_2} - E_{a_1}$, how much faster is the catalyzed reaction?

*22.64. **Enzymatic Activity of Urease** Urease catalyzes the decomposition of urea to ammonia and carbon dioxide (Figure P22.64). The rate constant for the uncatalyzed reaction at 20°C and pH 8 is $k = 3 \times 10^{-10}$ s^{-1}. A urease isolated from the jack bean increases the rate constant to $k = 3 \times 10^4$ s^{-1}. By using the $RT \ln(k_1/k_2) = E_{a_2} - E_{a_1}$ form of the Arrhenius equation, calculate the difference between the activation energies.

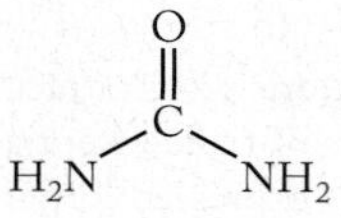

$$\underset{H_2N}{\overset{O}{\underset{\,}{\overset{\|}{C}}}}\!\!\!_{NH_2} \;(aq) + H_2O(\ell) \;\longrightarrow\; 2\,NH_3(aq) + CO_2(g)$$

FIGURE P22.64

Elements for Diagnosis and Therapy

CONCEPT REVIEW

22.65. List some of the considerations in choosing a radioisotope for imaging.

22.66. What advantages does an isotope like ^{99m}Tc have over ^{201}Tl for imaging the circulatory system?

22.67. Why might an α emitter be a good choice for chemotherapy?

*22.68. What advantage might a β emitter have over an α emitter for imaging?

22.69. Gadolinium-153 decays by electron capture. What type of radiation does ^{153}Gd produce that makes it useful for imaging?

22.70. Gadolinium-153 and samarium-153 both have the same mass number. Why might ^{153}Gd decay by electron capture whereas ^{153}Sm decays by emitting β particles?

22.71. How do platinum- and ruthenium-containing drugs fight cancer?

*22.72. Many transition metal complexes are brightly colored. Why might the titanium(IV) compound budotitane be colorless? (*Hint*: See Chapter 18.)

*22.73. Is the glucose tolerance factor that contains chromium(III) paramagnetic or diamagnetic? (*Hint*: See Chapter 18.)

22.74. Why might lithium be an attractive anode material in batteries used in pacemakers?

22.75. Mercury compounds are generally toxic to humans. Why can we use mercury in dental amalgams?

22.76. Tantalum is used in artificial joints. What property of Ta makes it attractive for this purpose?

PROBLEMS

22.77. The lanthanide isotopes cerium-141, terbium-160, thulium-170, and lutetium-177 all undergo β decay. Write a balanced nuclear equation for each decay reaction. Are all of the product isotopes in the same group of the periodic table as the reactants?

22.78. The lanthanide isotopes gadolinium-153 and ytterbium-169 decay by electron capture. Write a balanced nuclear equation for each decay reaction. Are both product isotopes in the same group of the periodic table?

22.79. **PET Imaging with Gallium** A patient is injected with a 5 μM solution of gallium citrate containing ^{68}Ga ($t_{1/2}$ = 9.4 hr) for a PET study. How long is it before the activity of the ^{68}Ga drops to 5% of its initial value?

22.80. Indium-111 ($t_{1/2}$ = 2.81 d) has been used in imaging. How long is it before the activity of ^{111}In drops to 15% of its initial value?

22.81. The bismuth in over-the-counter antacids is found as BiO^+. Draw the Lewis structure for the BiO^+ cation.

22.82. Some medicines used in treating depression contain lithium carbonate. Draw the Lewis structure for Li_2CO_3.

22.83. The complexation of mercury(II) ion with methionine

$$Hg^{2+} + \text{methionine} \rightleftharpoons \text{Hg(methionine)}^{2+}$$

has a formation constant of log K = 14.2, whereas the formation constant for the Hg^{2+} complex with penicillamine

$$Hg^{2+} + \text{penicillamine} \rightleftharpoons \text{Hg(penicillamine)}^{2+}$$

is log K = 16.3. Calculate the equilibrium constant for the reaction

$$\text{Hg(methionine)}^{2+} + \text{peinicillamine} \rightleftharpoons \text{Hg(penicillamine)}^{2+} + \text{methionine}$$

22.84. The complexation of mercury(II) ion with cysteine in aqueous solution

$$Hg^{2+} + \text{cysteine} \rightleftharpoons \text{Hg(cysteine)}^{2+}$$

has a formation constant of log K = 14.2, whereas the formation constant for the Hg^{2+} complex with glycine

$$Hg^{2+} + \text{glycine} \rightleftharpoons \text{Hg(glycine)}^{2+}$$

is log K = 10.3. Calculate the equilibrium constant for the reaction

$$\text{Hg(cysteine)}^{2+} + \text{glycine} \rightleftharpoons \text{Hg(glycine)}^{2+} + \text{cysteine}$$

22.85. The equilibrium constant of the reaction

$$CH_3\text{Hg(penicillamine)}^+(aq) + \text{cysteine}(aq) \rightleftharpoons CH_3\text{Hg(cysteine)}^+(aq) + \text{penicillamine}(aq)$$

is K = 0.633. Calculate the equilibrium concentrations of cysteine and penicillamine if we start with a 1.00 M solution of cysteine and a 1.00 mM solution of CH_3Hg(penicillamine)$^+$.

22.86. The equilibrium constant of the reaction

$$CH_3\text{Hg(glutathione)}^+(aq) + \text{cysteine}(aq) \rightleftharpoons CH_3\text{Hg(cysteine)}^+(aq) + \text{glutathione}(aq)$$

is K = 5.0. Calculate the equilibrium concentrations of cysteine and glutathione if we start with a 1.20 mM solution of cysteine and a 1.20 mM solution of CH_3Hg(glutathione)$^+$.

22.87. Draw the Lewis structure for the citrate ion by using the skeletal drawing in Figure P22.87 as a guide.

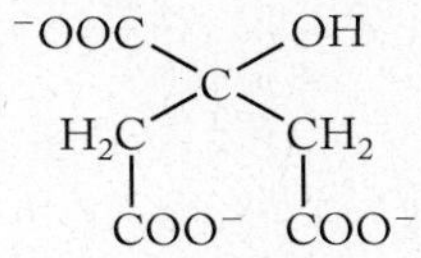

$^-$OOC
OH
C
H_2C
CH_2
COO$^-$
COO$^-$

FIGURE P22.87

22.88. Draw the Lewis structure for the thiosulfate ion, $S_2O_3^{2-}$, which is found in some arthritis drugs containing gold atoms.

22.89. Aluminum hydroxide is used in some antacids. Write a balanced net ionic equation for the reaction of aluminum hydroxide with HCl.

22.90. Aluminum carbonate is used in some antacids. Write a balanced net ionic equation for the reaction of aluminum carbonate with hydrochloric acid.

22.91. A silver/zinc (Ag/Zn) battery has the advantage of being mercury-free. The half-reactions are as follows:

$$ZnO(s) + H_2O(\ell) + 2e^- \rightarrow Zn(s) + 2OH^-(aq) \qquad E° = -1.258 \text{ V}$$

$$Ag_2O(s) + H_2O(\ell) + 2e^- \rightarrow 2Ag(s) + 2OH^-(aq) \qquad E° = 0.342 \text{ V}$$

What is the overall reaction of an electrochemical cell based on these materials? Using the $E°$ values provided, determine the standard cell potential of the Ag/ZnO battery.

22.92. **Power Sources for Pacemakers** Another power source considered for pacemakers is the lithium/copper sulfide battery. Use the half-reactions shown here to determine the overall reaction for an electrochemical cell based on these materials. With the $E°$ values provided, calculate the cell potential of the Li/CuS battery.

$$Li^+(s) + e^- \rightarrow Li(s) \qquad E° = -3.05 \text{ V}$$

$$CuS(s) + 2e^- \rightarrow Cu(s) + S^{2-}(aq) \qquad E° = -0.851 \text{ V}$$

Additional study materials including ChemTours and Diagnostic Quizzes are available at StudySpace at www.wwnorton.com/studyspace.

Mathematical Procedures

Working with Scientific Notation

Quantities that scientists work with often are very large, such as Earth's mass, or very small, such as the mass of an electron. It is easier to work with these numbers if they are expressed in scientific notation.

The general form of standard scientific notation is a value between 1 and 10 multipled by 10 raised to an integral power. According to this definition, 598×10^{22} kg (Earth's mass) is not in standard scientific notation, but 5.98×10^{24} kg is. It is good practice to use and report data or parameters in standard scientific notation.

1. **To convert an "ordinary" number to standard scientific notation,** move the decimal point to the left for a large number, or to right for a small one, so that the decimal point is located after the first nonzero digit.

   A. For example, to express Earth's average density ($5,517$ kg/m^3) in scientific notation requires moving the decimal three places to the left. Doing so is the same as dividing the number by 1000, or 10^3. To keep the value the same we add an exponent to multiply it by 10^3. So, Earth's density in standard scientific notation is 5.517×10^3 kg/m^3.

   B. If you move the decimal point of a value less than one to the right to express it in scientific notation, then the exponent is a negative integer equal to the number of places you moved the decimal point to the right. For example, the value of R used in solving ideal gas law problems is 0.08206 L · atm/(mol · K). Moving the decimal point two places to the right converts the value of R to scientific notation: 8.206×10^{-22} L · atm/(mol · K).

   C. Another value of R, 8.314 J/(mol · K), does not need an exponent, though it could be written 8.314×10^0 J/(mol · K).

2. **To add or subtract numbers in scientific notation,** their exponents must be the same. This may require you to change the exponents (and thus the decimal point position) of some values. (NOTE: This step is not necessary when using a scientific calculator because the calculator will make all necessary conversions.)

Sample Exercise 1 Calculate the sum of the masses of the subatomic particles in an atom of lithium-7, given the following masses in grams:

Particle	Mass (g)
proton	1.67262×10^{-24}
neutron	1.67493×10^{-24}
electron	9.10939×10^{-28}

Solution An atom of ^{7}Li has three protons, three electrons, and four neutrons. Therefore, the total mass of the subatomic particles in an atom is the sum of three times the masses of a proton and an electron and four times the mass of a neutron. Doing the multiplication steps first yields the following:

$$\text{mass of three protons} = 3(1.67262 \times 10^{-24} \text{ g}) = 5.01786 \times 10^{-24} \text{ g}$$

$$\text{mass of four protons} = 4(1.67493 \times 10^{-24} \text{ g}) = 6.69972 \times 10^{-24} \text{ g}$$

$$\text{mass of three electrons} = 3(9.10939 \times 10^{-28} \text{ g}) = 2.732817 \times 10^{-27} \text{ g}$$

Before adding these masses together, we must express them all with the same exponent. The most convenient is 10^{-24}. To express the combined mass of three electrons using this exponent requires shifting the decimal point in 2.732817×10^{-27} g three places to the left, which makes the number before the exponent 1000 times smaller and compensates for the exponent becoming 1000 times larger.

Adding the resulting mass of the three electrons, $0.002732817 \times 10^{-24}$ g, to the other two masses, we have

$$
\begin{array}{r}
5.01786 \times 10^{-24} \text{ g} \\
+\ 6.69972 \times 10^{-24} \text{ g} \\
+\ 0.002732817 \times 10^{-24} \text{ g} \\
\hline
11.720312817 \times 10^{-24} \text{ g}
\end{array}
$$

We need to round off this sum so that we have only five digits to the right of the decimal place, because that is the smallest number of digits to the right in two of the three values being summed. Therefore, the final answer is

$$11.72031 \times 10^{-24} \text{ g}$$

In scientific notation, the masses of the subatomic particles in an atom of ^{7}Li sum to 1.172031×10^{-23} g.

3. **To multiply values with exponents,** the values in front of the exponents are multiplied together, but the exponents are added (these steps happen automatically with scientific calculators).

Sample Exercise 2 American Steve Fossett (1944–2007) circumnavigated the globe in early summer 2002 in the *Spirit of Freedom* balloon, which was partially filled with 5.5×10^5 ft^3 of helium. What is this volume in liters? Given: 1 ft^3 = 28.3 L.

Solution We convert the starting value by multiplying 5.5 by 2.83 and adding the exponents (1 + 5):

$$(5.5 \times 10^5 \text{ ft}^3)\left(\frac{2.83 \times 10^1 \text{ L}}{\text{ft}^3}\right) = 15.6 \times 10^6 \text{ L or } 1.6 \times 10^7 \text{ L}$$

4. **To divide values with exponents,** the values in front of the exponents are divided, but the exponents are subtracted (again, these steps happen automatically with scientific calculators).

Sample Exercise 3 The speed of light is 2.998×10^8 m/s. What is the equivalent speed in miles per second? Given: 1 mile = 1.609×10^3 m.

Solution Expressing the speed of light in miles per second requires dividing the speed of light in meters per second by the conversion factor given in the exercise. We divide the values in front of the exponents (2.998/1.609) and subtract to get their exponents (8 − 3):

$$\frac{2.998 \times 10^8 \text{ m/s}}{1.609 \times 10^3 \text{ m/mi}} = 1.863 \times 10^5 \text{ mi/s}$$

Working with Logarithms

A logarithm to the base 10 has the following form:

$$\log_{10} x = \log x = p, \text{ where } x = 10^p$$

We usually abbreviate the logarithm function "log" if the logarithm is to the base 10, which means the scale in which the log 10 = 1.

A logarithm to the base e, called a *natural logarithm*, has the following form:

$$\log_e x = \ln x = q, \text{ where } x = e^q$$

Scientific calculators have "log" and "ln" buttons, so it is easy to convert a number into its log or ln form. The directions below apply to most nongraphing calculators.

Sample Exercise 4 Find the logarithm to the base 10 of 4.5 (log 4.5).

Solution Enter 4.5 into your calculator and press the "LOG" button.[a] The answer should be 0.6532 (to four significant figures).

Sample Exercise 5 Find the logarithm to the base 10 of 100 (log 100).

Solution Enter 100 into your calculator and press the "LOG" button. The answer should be 2. This answer is as expected, because 10 (the base) raised to the power of the log (2) is $10^2 = 100$.

Sample Exercise 6 Find the natural logarithm of 4.5.

Solution Enter 4.5 into your calculator and press the "LN" button. The answer should be 1.504.

Sample Exercise 7 Find the natural logarithm of 100.

Solution Enter 100 into your calculator and press the "LN" button. The answer should be 4.61.

Let's compare the results of the four previous exercises. In both pairs of ln and log values, the ln value is 2.303 times the log value. These examples fit the general equation:

$$\ln x = 2.303 \log x$$

It is reasonable that the ln of a value is greater than the log of the same value because ln is based on e (2.718), whereas log is to the base 10. The smaller base of ln units means that there are more of them than log units in a given value.

Sample Exercise 8 Calculate $\ln 1.2 \times 10^{-3}$.

Solution Enter 1.2 into your calculator, and then press the "EXP" button, and then enter 3 and "+/−". Finally press "LN." The corresponding keystrokes with a graphing calculator are "LN," 1.2, "x," "^," "(−)," 3, "ENTER." The result should be −6.725, which is negative because the original number is less than 1. Keep in mind that values greater than zero have positive logarithm values; those less than zero have negative logarithm values.

5. **Combining logs:** The following equations summarize how logarithms of the products or quotients of two or more values are related to the individual logs of those values:

$$\text{logarithm } ab = \text{logarithm } a + \text{logarithm } b$$

and

$$\text{logarithm } a/b = \text{logarithm } a - \text{logarithm } b$$

Converting Logarithms into Numbers

If we know the value of log x, what is the value of x? This question frequently arises when working with pH (see Chapter 17), which is the negative log of the concentration of hydrogen ions, $[H^+]$, in solution:

$$pH = -\log[H^+]$$

Suppose the pH of a solution of a weak acid is 2.50. The concentration of H^+ is related to this pH value as follows:

$$2.50 = -\log[H^+]$$

or

$$-2.50 = \log[H^+]$$

[a] If you have a graphing calculator such as a TI84+, press the "LOG" button, enter 4.5, and then press the "ENTER" button.

To find the value of $[H^+]$, enter 2.5 into your calculator and press the "+/−" button to change the value's sign to −2.5. The next step depends on the type of calculator you have. If yours has a "10^x" button, push it to find the value of $10^{-2.5}$, which is the number we are looking for. The corresponding keystrokes with a graphing calculator are "10^x," "(−)," 2.5, "ENTER." On some calculators there is no "10^x" key, but there is an inverse function, or "INV" key, that is used to invert other function keys. Hitting the "INV" key followed by the "log" key takes the inverse of a log, called an *antilog*, which is the same as raising 10 to the power (−2.5 in this case) that was entered. Some calculators, including the virtual one in many Windows operating systems, have an "x^y" key. To use it you enter 10, push the "x^y" key, enter 2.5, and then push the "+/−" key followed by the equals sign. All of these approaches do the same calculation, taking 10 to the −2.50 power, and give the same answer, $[H^+] = 3.2 \times 10^{-3}$.

Sample Exercise 9 Calculate the hydrogen ion concentration in rainwater in which pH = 5.62.

Solution If you use one of the methods described above, you should find that the value of $10^{-5.62}$ is 2.4×10^{-6}.

Solving Quadratic Equations

If the terms in an equation can be rearranged so that they take the form

$$ax^2 + bx + c = 0$$

they have the form of a quadratic equation. The value(s) of x can be determined from the values of the coefficients a, b, and c by using the equation

$$x = \frac{-b \pm \sqrt{b^2 - 4ac}}{2a}$$

For example, if the solution to a problem yields the following expression where x is the concentration of a solute:

$$x^2 + 0.112x - 1.2 \times 10^{-3} = 0$$

Then the value of x can be determined as follows:

$$
\begin{aligned}
x &= \frac{-b \pm \sqrt{b^2 - 4ac}}{2a} \\[2mm]
&= \frac{-0.112 \pm \sqrt{(0.112)^2 - 4(1)(-1.2 \times 10^{-3})}}{2(1)} \\[2mm]
&= \frac{-0.112 \pm \sqrt{0.01254 + 0.0048}}{2} \\[2mm]
&= \frac{-0.112 \pm 0.132}{2} = +0.010 \text{ or } -0.122
\end{aligned}
$$

In this example, the negative value for x satisfies the equation, but it has no meaning because we cannot have negative concentration values; therefore we use only the +0.010 value.

Expressing Data in Graphical Form

Fitting curves to plots of experimental data is a powerful tool in determining the relationships between variables. Many natural phenomena obey exponential functions. For example, the rate constant (k) of a chemical reaction increases exponentially with increasing absolute temperature (T). This relationship is described by the Arrhenius equation (see Chapter 15):

$$k = A\,e^{-E_a/RT}$$

where A is a constant for a particular reaction (called the frequency factor), E_a is the activation energy of the reaction, and R is the ideal gas constant. Taking the natural logarithms of both sides of the Arrhenius equation gives

$$\ln k = \ln A - \left(\frac{E_a}{RT}\right)$$

This equation fits the general equation of a straight line ($y = mx + b$) if ($\ln k$) is the y-variable and ($1/T$) is the x-variable. Plotting ($\ln k$) versus ($1/T$) should give a straight line with a slope equal to $-E_a/R$. The slopes of these plots are negative because the activation energies, E_a, of chemical reactions are positive. The data for a reaction given in columns 2 and 4 of Table A1.1 are plotted in Figure A1.1. The slope of the straight line (−1281 K) is used to calculate the value of E_a:

$$-1281\ \text{K} = -\frac{E_a}{R}$$

$$E_a = -(-1281\ \text{K})[8.31\ \text{J/(mol·K)}]$$

$$= 10{,}645\ \text{J/mol} = 10.6\ \text{kJ/mol}$$

TABLE A1.1	Rate Constant k as a Function of Temperature T		
Temperature T (K)	**$1/T$ (K^{-1})**	**Rate Constant k**	**ln k**
500	0.0020	0.030	−3.5
550	0.0018	0.38	−0.97
600	0.0017	2.9	1.1
650	0.0015	17	2.8
700	0.0014	75	4.3

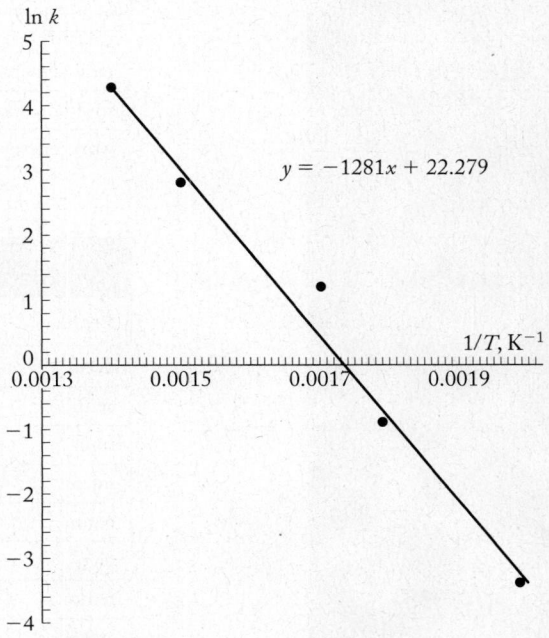

FIGURE A1.1

SI Units and Conversion Factors

TABLE A2.1   SI Base Units

SI Base Quantity	Unit	Symbol
length	meter	m
mass	kilogram	kg
time	second	s
amount of substance	mole	mol
temperature	kelvin	K
electric current	ampere	A

TABLE A2.2   Some SI-Derived Units

SI-Derived Quantity	Unit	Symbol	Dimensions
electric charge	coulomb	C	$A \cdot s$
electric potential	volt	V	J/C
force	newton	N	$kg \cdot m/s^2$
frequency	hertz	Hz	s^{-1}
momentum	newton-second	—	$kg \cdot m/s$
power	watt	W	J/s
pressure	pascal	Pa	N/m^2
radioactivity	becquerel	Bq	s^{-1}
speed or velocity	meter per second	—	m/s
work, energy, or heat	joule (newton-meter)	J	$kg \cdot m^2/s^2$

TABLE A2.3   SI Prefixes

Prefix	Symbol	Multiplier	Prefix	Symbol	Multiplier
deci	d	10^{-1}	deka	da	10^1
centi	c	10^{-2}	hecto	h	10^2
milli	m	10^{-3}	kilo	k	10^3
micro	μ	10^{-6}	mega	M	10^6
nano	n	10^{-9}	giga	G	10^9
pico	p	10^{-12}	tera	T	10^{12}
femto	f	10^{-15}	peta	P	10^{15}
atto	a	10^{-18}	exa	E	10^{18}

TABLE A2.4	Special Units and Conversion Factors		
Quantity	**Unit**	**Symbol**	**Conversion**
energy	electron-volt	eV	$1\ eV = 1.60217733 \times 10^{-19}\ J$
mass	pound	lb	$1\ lb = 453.592\ g$
mass	atomic mass unit	amu	$1\ amu = 1.6605402 \times 10^{-27}\ kg$
heat	calorie	cal	$1\ cal = 4.184\ J$
length	angstrom	Å	$1\ Å = 10^{-8}\ cm = 10^{-10}\ m$
length	inch	in	$1\ in = 2.54\ cm$
length	mile	mi	$1\ mi = 5280\ ft = 1.6093\ km$
pressure	atmosphere	atm	$1\ atm = 1.01325 \times 10^{5}\ Pa$
pressure	torr	torr	$1\ torr = 1/760\ atm$
temperature	Celsius scale	°C	$°C = K - 273.15$
temperature	Fahrenheit scale	°F	$°F = \frac{9}{5}(°C) + 32$
time	minute	min	$1\ min = 60\ s$
time	hour	hr	$1\ hr = 60\ min = 3600\ s$
time	day	d	$1\ d = 24\ hr = 86,400\ s$
time	year	yr	$1\ yr = 365.25\ d = 8766\ hr$
volume	liter	L	$1\ L = 1\ dm^3 = 10^{-3}\ m^3$
volume	cubic centimeter	cm³, cc	$1\ cm^3 = 1\ mL = 10^{-3}\ L$
volume	cubic foot	ft³	$1\ ft^3 = 7.4805\ gal$
volume	gallon (U.S.)	gal	$1\ gal = 3.7854\ L$

TABLE A2.5	Physical Constants	
Quantity	**Symbol**	**Value**
acceleration due to gravity (Earth)	g	$9.80665\ m/s^2$ (exact)
Avogadro's number	N_A	$6.0221367 \times 10^{23}\ mol^{-1}$
Bohr radius	a_0	$5.29177249 \times 10^{-11}\ m$
Boltzmann's constant	k_B	$1.380658 \times 10^{-23}\ J/K$
electron charge-to-mass ratio	$-e/m_e$	$1.75881962 \times 10^{11}\ C/kg$
elementary charge	e	$1.60217733 \times 10^{-19}\ C$
Faraday's constant	F	$9.6485309 \times 10^{4}\ C/mol$
mass of an electron	m_e	$9.10939 \times 10^{-28}\ g$
mass of a neutron	m_n	$1.67493 \times 10^{-24}\ g$
mass of a proton	m_p	$1.67262 \times 10^{-24}\ g$
molar volume of ideal gas at STP	V_m	$22.41410\ L/mol$
Planck's constant	h	$6.6260755 \times 10^{-34}\ J \cdot s$
speed of light in vacuum	c	$2.99792458 \times 10^{8}\ m/s$ (exact)
universal gas constant	R	$8.3145\ J/(mol \cdot K)$ $0.082058\ L \cdot atm/(mol \cdot K)$

The Elements and Their Properties

TABLE A3.1	Ground-State Electron Configurations, Atomic Radii, and First Ionization Energies of the Elements				
Element	Symbol	Atomic Number Z	Ground-State Configuration	Atomic Radius (pm)	Ionization Energy (kJ/mol)
hydrogen	H	1	$1s^1$	37	1312.0
helium	He	2	$1s^2$	32	2372.3
lithium	Li	3	$[He]2s^1$	152	520.2
beryllium	Be	4	$[He]2s^2$	112	899.5
boron	B	5	$[He]2s^22p^1$	88	800.6
carbon	C	6	$[He]2s^22p^2$	77	1086.5
nitrogen	N	7	$[He]2s^22p^3$	75	1402.3
oxygen	O	8	$[He]2s^22p^4$	73	1313.9
fluorine	F	9	$[He]2s^22p^5$	71	1681.0
neon	Ne	10	$[He]2s^22p^6$	69	2080.7
sodium	Na	11	$[Ne]3s^1$	186	495.3
magnesium	Mg	12	$[Ne]3s^2$	160	737.7
aluminum	Al	13	$[Ne]3s^23p^1$	143	577.5
silicon	Si	14	$[Ne]3s^23p^2$	117	786.5
phosphorus	P	15	$[Ne]3s^23p^3$	110	1011.8
sulfur	S	16	$[Ne]3s^23p^4$	103	999.6
chlorine	Cl	17	$[Ne]3s^23p^5$	99	1251.2
argon	Ar	18	$[Ne]3s^23p^6$	97	1520.6
potassium	K	19	$[Ar]4s^1$	227	418.8
calcium	Ca	20	$[Ar]4s^2$	197	589.8
scandium	Sc	21	$[Ar]4s^23d^1$	162	633.1
titanium	Ti	22	$[Ar]4s^23d^2$	147	658.8
vanadium	V	23	$[Ar]4s^23d^3$	135	650.9
chromium	Cr	24	$[Ar]4s^13d^5$	128	652.9
manganese	Mn	25	$[Ar]4s^23d^5$	127	717.3
iron	Fe	26	$[Ar]4s^23d^6$	126	762.5
cobalt	Co	27	$[Ar]4s^23d^7$	125	760.4
nickel	Ni	28	$[Ar]4s^23d^8$	124	737.1
copper	Cu	29	$[Ar]4s^13d^{10}$	128	745.5
zinc	Zn	30	$[Ar]4s^23d^{10}$	134	906.4
gallium	Ga	31	$[Ar]4s^23d^{10}4p^1$	135	578.8

TABLE A3.1 Ground-State Electron Configurations, Atomic Radii, and First Ionization Energies of the Elements (*Continued*)

Element	Symbol	Atomic Number Z	Ground-State Configuration	Atomic Radius (pm)	Ionization Energy (kJ/mol)
germanium	Ge	32	$[Ar]4s^23d^{10}4p^2$	122	762.2
arsenic	As	33	$[Ar]4s^23d^{10}4p^3$	121	947.0
selenium	Se	34	$[Ar]4s^23d^{10}4p^4$	119	941.0
bromine	Br	35	$[Ar]4s^23d^{10}4p^5$	114	1139.9
krypton	Kr	36	$[Ar]4s^23d^{10}4p^6$	110	1350.8
rubidium	Rb	37	$[Kr]5s^1$	247	403.0
strontium	Sr	38	$[Kr]5s^2$	215	549.5
yttrium	Y	39	$[Kr]5s^24d^1$	180	599.8
zirconium	Zr	40	$[Kr]5s^24d^2$	160	640.1
niobium	Nb	41	$[Kr]5s^14d^4$	146	652.1
molybdenum	Mo	42	$[Kr]5s^14d^5$	139	684.3
technetium	Tc	43	$[Kr]5s^24d^5$	136	702.4
ruthenium	Ru	44	$[Kr]5s^14d^7$	134	710.2
rhodium	Rh	45	$[Kr]5s^14d^8$	134	719.7
palladium	Pd	46	$[Kr]4d^{10}$	137	804.4
silver	Ag	47	$[Kr]5s^14d^{10}$	144	731.0
cadmium	Cd	48	$[Kr]5s^24d^{10}$	151	867.8
indium	In	49	$[Kr]5s^24d^{10}5p^1$	167	558.3
tin	Sn	50	$[Kr]5s^24d^{10}5p^2$	140	708.6
antimony	Sb	51	$[Kr]5s^24d^{10}5p^3$	141	833.6
tellurium	Te	52	$[Kr]5s^24d^{10}5p^4$	143	869.3
iodine	I	53	$[Kr]5s^24d^{10}5p^5$	133	1008.4
xenon	Xe	54	$[Kr]5s^24d^{10}5p^6$	130	1170.4
cesium	Cs	55	$[Xe]6s^1$	265	375.7
barium	Ba	56	$[Xe]6s^2$	222	502.9
lanthanum	La	57	$[Xe]6s^25d^1$	187	538.1
cerium	Ce	58	$[Xe]6s^24f^15d^1$	182	534.4
praseodymium	Pr	59	$[Xe]6s^24f^3$	182	527.2
neodymium	Nd	60	$[Xe]6s^24f^4$	181	533.1
promethium	Pm	61	$[Xe]6s^24f^5$	183	535.5
samarium	Sm	62	$[Xe]6s^24f^6$	180	544.5
europium	Eu	63	$[Xe]6s^24f^7$	208	547.1
gadolinium	Gd	64	$[Xe]6s^24f^75d^1$	180	593.4
terbium	Tb	65	$[Xe]6s^24f^9$	177	565.8
dysprosium	Dy	66	$[Xe]6s^24f^{10}$	178	573.0
holmium	Ho	67	$[Xe]6s^24f^{11}$	176	581.0
erbium	Er	68	$[Xe]6s^24f^{12}$	176	589.3
thulium	Tm	69	$[Xe]6s^24f^{13}$	176	596.7
ytterbium	Yb	70	$[Xe]6s^24f^{14}$	193	603.4
lutetium	Lu	71	$[Xe]6s^24f^{14}5d^1$	174	523.5

Continued on next page

TABLE A3.1 Ground-State Electron Configurations, Atomic Radii, and First Ionization Energies of the Elements (*Continued*)

Element	Symbol	Atomic Number Z	Ground-State Configuration	Atomic Radius (pm)	Ionization Energy (kJ/mol)
hafnium	Hf	72	$[Xe]6s^2 4f^{14} 5d^2$	159	658.5
tantalum	Ta	73	$[Xe]6s^2 4f^{14} 5d^3$	146	761.3
tungsten	W	74	$[Xe]6s^2 4f^{14} 5d^4$	139	770.0
rhenium	Re	75	$[Xe]6s^2 4f^{14} 5d^5$	137	760.3
osmium	Os	76	$[Xe]6s^2 4f^{14} 5d^6$	135	839.4
iridium	Ir	77	$[Xe]6s^2 4f^{14} 5d^7$	136	878.0
platinum	Pt	78	$[Xe]6s^1 4f^{14} 5d^9$	139	868.4
gold	Au	79	$[Xe]6s^1 4f^{14} 5d^{10}$	144	890.1
mercury	Hg	80	$[Xe]6s^2 4f^{14} 5d^{10}$	151	1007.1
thallium	Tl	81	$[Xe]6s^2 4f^{14} 5d^{10} 6p^1$	170	589.4
lead	Pb	82	$[Xe]6s^2 4f^{14} 5d^{10} 6p^2$	154	715.6
bismuth	Bi	83	$[Xe]6s^2 4f^{14} 5d^{10} 6p^3$	150	703.3
polonium	Po	84	$[Xe]6s^2 4f^{14} 5d^{10} 6p^4$	167	812.1
astatine	At	85	$[Xe]6s^2 4f^{14} 5d^{10} 6p^5$	140	924.6
radon	Rn	86	$[Xe]6s^2 4f^{14} 5d^{10} 6p^6$	145	1037.1
francium	Fr	87	$[Rn]7s^1$	270^a	380
radium	Ra	88	$[Rn]7s^2$	223^a	509.3
actinium	Ac	89	$[Rn]7s^2 6d^1$	188	499
thorium	Th	90	$[Rn]7s^2 6d^2$	179	587
protactinium	Pa	91	$[Rn]7s^2 5f^2 6d^1$	163	568
uranium	U	92	$[Rn]7s^2 5f^3 6d^1$	156	587
neptunium	Np	93	$[Rn]7s^2 5f^4 6d^1$	155	597
plutonium	Pu	94	$[Rn]7s^2 5f^6$	159	585
americium	Am	95	$[Rn]7s^2 5f^7$	173	578
curium	Cm	96	$[Rn]7s^2 5f^7 6d^1$	174	581
berkelium	Bk	97	$[Rn]7s^2 5f^9$	170	601
californium	Cf	98	$[Rn]7s^2 5f^{10}$	186	608
einsteinium	Es	99	$[Rn]7s^2 5f^{11}$	186	619
fermium	Fm	100	$[Rn]7s^2 5f^{12}$	—	627
mendelevium	Md	101	$[Rn]7s^2 5f^{13}$	—	635
nobelium	No	102	$[Rn]7s^2 5f^{14}$	—	642
lawrencium	Lr	103	$[Rn]7s^2 5f^{14} 6d^1$	—	—
rutherfordium	Rf	104	$[Rn]7s^2 5f^{14} 6d^2$	—	—
dubnium	Db	105	$[Rn]7s^2 5f^{14} 6d^3$	—	—
seaborgium	Sg	106	$[Rn]7s^2 5f^{14} 6d^4$	—	—
bohrium	Bh	107	$[Rn]7s^2 5f^{14} 6d^5$	—	—
hassium	Hs	108	$[Rn]7s^2 5f^{14} 6d^6$	—	—
meitnerium	Mt	109	$[Rn]7s^2 5f^{14} 6d^7$	—	—
darmstadtium	Ds	110	$[Rn]7s^2 5f^{14} 6d^8$	—	—
roentgenium	Rg	111	$[Rn]7s^2 5f^{14} 6d^9$	—	—
copernicium	Cn	112	$[Rn]7s^2 5f^{14} 6d^{10}$	—	—

^{a}These values are estimates.

TABLE A3.2	Miscellaneous Physical Properties of the Elements[a]					

Element	Symbol	Atomic Number	Physical State[b,c]	Density[d] (g/mL)	Melting Point (°C)	Boiling Point (°C)
hydrogen	H	1	gas	0.000090	−259.14	−252.87
helium	He	2	gas	0.000179	<−272.2	−268.93
lithium	Li	3	solid	0.534	180.5	1347
beryllium	Be	4	solid	1.848	1283	2484
boron	B	5	solid	2.34	2300	3650
carbon	C	6	solid (gr)	1.9–2.3	~3350	sublimes
nitrogen	N	7	gas	0.00125	−210.00	−195.8
oxygen	O	8	gas	0.00143	−218.8	−182.95
fluorine	F	9	gas	0.00170	−219.62	−188.12
neon	Ne	10	gas	0.00090	−248.59	−246.08
sodium	Na	11	solid	0.971	97.72	883
magnesium	Mg	12	solid	1.738	650	1090
aluminum	Al	13	solid	2.6989	660.32	2467
silicon	Si	14	solid	2.33	1414	2355
phosphorus	P	15	solid (wh)	1.82	44.15	280
sulfur	S	16	solid	2.07	115.21	444.60
chlorine	Cl	17	gas	0.00321	−101.5	−34.04
argon	Ar	18	gas	0.00178	−189.3	−185.9
potassium	K	19	solid	0.862	63.28	759
calcium	Ca	20	solid	1.55	842	1484
scandium	Sc	21	solid	2.989	1541	2380
titanium	Ti	22	solid	4.54	1668	3287
vanadium	V	23	solid	6.11	1910	3407
chromium	Cr	24	solid	7.19	1857	2671
manganese	Mn	25	solid	7.3	1246	1962
iron	Fe	26	solid	7.874	1538	2750
cobalt	Co	27	solid	8.9	1495	2870
nickel	Ni	28	solid	8.902	1455	2730
copper	Cu	29	solid	8.96	1084.6	2562
zinc	Zn	30	solid	7.133	419.53	907
gallium	Ga	31	solid	5.904	29.76	2403
germanium	Ge	32	solid	5.323	938.25	2833
arsenic	As	33	solid (gy)	5.727	614	sublimes
selenium	Se	34	solid (gy)	4.79	221	685
bromine	Br	35	liquid	3.12	−7.2	58.78
krypton	Kr	36	gas	0.00373	−157.36	−153.22
rubidium	Rb	37	solid	1.532	39.31	688
strontium	Sr	38	solid	2.54	777	1382
yttrium	Y	39	solid	4.469	1526	3336
zirconium	Zr	40	solid	6.506	1855	4409
niobium	Nb	41	solid	8.57	2477	4744

Continued on next page

TABLE A3.2 **Miscellaneous Physical Properties of the Elements^a (Continued)**

Element	Symbol	Atomic Number	Physical State^{b,c}	Density^d (g/mL)	Melting Point (°C)	Boiling Point (°C)
molybdenum	Mo	42	solid	10.22	2623	4639
technetium	Tc	43	solid	11.50	2157	4538
ruthenium	Ru	44	solid	12.41	2334	3900
rhodium	Rh	45	solid	12.41	1964	3695
palladium	Pd	46	solid	12.02	1555	2963
silver	Ag	47	solid	10.50	961.78	2212
cadmium	Cd	48	solid	8.65	321.07	767
indium	In	49	solid	7.31	156.60	2072
tin	Sn	50	solid (wh)	7.31	231.9	2270
antimony	Sb	51	solid	6.691	630.63	1750
tellurium	Te	52	solid	6.24	449.5	998
iodine	I	53	solid	4.93	113.7	184.4
xenon	Xe	54	gas	0.00589	−111.75	−108.0
cesium	Cs	55	solid	1.873	28.44	671
barium	Ba	56	solid	3.5	727	1640
lanthanum	La	57	solid	6.145	920	3455
cerium	Ce	58	solid	6.770	799	3424
praseodymium	Pr	59	solid	6.773	931	3510
neodymium	Nd	60	solid	7.008	1016	3066
promethium	Pm	61	solid	7.264	1042	~3000
samarium	Sm	62	solid	7.520	1072	1790
europium	Eu	63	solid	5.244	822	1596
gadolinium	Gd	64	solid	7.901	1314	3264
terbium	Tb	65	solid	8.230	1359	3221
dysprosium	Dy	66	solid	8.551	1411	2561
holmium	Ho	67	solid	8.795	1472	2694
erbium	Er	68	solid	9.066	1529	2862
thulium	Tm	69	solid	9.321	1545	1946
ytterbium	Yb	70	solid	6.966	824	1194
lutetium	Lu	71	solid	9.841	1663	3393
hafnium	Hf	72	solid	13.31	2233	4603
tantalum	Ta	73	solid	16.654	3017	5458
tungsten	W	74	solid	19.3	3422	5660
rhenium	Re	75	solid	21.02	3186	5596
osmium	Os	76	solid	22.57	3033	5012
iridium	Ir	77	solid	22.42	2446	4130
platinum	Pt	78	solid	21.45	1768.4	3825
gold	Au	79	solid	19.3	1064.18	2856
mercury	Hg	80	liquid	13.546	−38.83	356.73
thallium	Tl	81	solid	11.85	304	1473
lead	Pb	82	solid	11.35	327.46	1749
bismuth	Bi	83	solid	9.747	271.4	1564

TABLE A3.2 Miscellaneous Physical Properties of the Elements[a] (Continued)

Element	Symbol	Atomic Number	Physical State[b,c]	Density[d] (g/mL)	Melting Point (°C)	Boiling Point (°C)
polonium	Po	84	solid	9.32	254	962
astatine	At	85	solid	unknown	302	337
radon	Rn	86	gas	0.00973	−71	−61.7
francium	Fr	87	solid	unknown	27	677
radium	Ra	88	solid	5	700	1737
actinium	Ac	89	solid	10.07	1051	~3200
thorium	Th	90	solid	11.72	1750	4788
protactinium	Pa	91	solid	15.37	1572	unknown
uranium	U	92	solid	18.95	1132	3818

[a] For relative atomic masses and alphabetical listing of the elements, see the flyleaf at the front of this volume.
[b] Normal state at 25°C and 1 atm.
[c] Allotropes: gr = graphite, gy = gray, wh = white.
[d] Liquids and solids at 25°C and 1 atm; gases at 0°C and 1 atm (STP).

TABLE A3.3 A Selection of Stable Isotopes[a]

Isotope $^A X$	Natural Abundance (%)	Atomic Number Z	Neutron Number N	Mass Number A	Atomic Mass (amu)	Binding Energy per Nucleon (MeV)[b]
^{1}H	99.985	1	0	1	1.007825	—
^{2}H	0.015	1	1	2	2.014000	1.160
^{3}He	0.000137	2	1	3	3.016030	2.572
^{4}He	99.999863	2	2	4	4.002603	7.075
^{6}Li	7.5	3	3	6	6.015121	5.333
^{7}Li	92.5	3	4	7	7.016003	5.606
^{9}Be	100.0	4	5	9	9.012182	6.463
^{10}B	19.9	5	5	10	10.012937	6.475
^{11}B	80.1	5	6	11	11.009305	6.928
^{12}C	98.90	6	6	12	12.000000	7.680
^{13}C	1.10	6	7	13	13.003355	7.470
^{14}N	99.634	7	7	14	14.003074	7.476
^{15}N	0.366	7	8	15	15.000108	7.699
^{16}O	99.762	8	8	16	15.994915	7.976
^{17}O	0.038	8	9	17	16.999131	7.751
^{18}O	0.200	8	10	18	17.999160	7.767
^{19}F	100.0	9	10	19	18.998403	7.779
^{20}Ne	90.48	10	10	20	19.992435	8.032
^{21}Ne	0.27	10	11	21	20.993843	7.972
^{22}Ne	9.25	10	12	22	21.991383	8.081
^{23}Na	100.0	11	12	23	22.989770	8.112
^{24}Mg	78.99	12	12	24	23.985042	8.261
^{25}Mg	10.00	12	13	25	24.985837	8.223
^{26}Mg	11.01	12	14	26	25.982593	8.334
^{27}Al	100.0	13	14	27	26.981538	8.331

Continued on next page

TABLE A3.3   A Selection of Stable Isotopes[a] (*Continued*)

Isotope ^{A}X	Natural Abundance (%)	Atomic Number Z	Neutron Number N	Mass Number A	Atomic Mass (amu)	Binding Energy per Nucleon (MeV)[b]
^{28}Si	92.23	14	14	28	27.976927	8.448
^{29}Si	4.67	14	15	29	28.976495	8.449
^{30}Si	3.10	14	16	30	29.973770	8.521
^{31}P	100.0	15	16	31	30.973761	8.481
^{32}S	95.02	16	16	32	31.972070	8.493
^{33}S	0.75	16	17	33	32.971456	8.498
^{34}S	4.21	16	18	34	33.967866	8.584
^{36}S	0.02	16	20	36	35.967080	8.575
^{35}Cl	75.77	17	18	35	34.968852	8.520
^{37}Cl	24.23	17	20	37	36.965903	8.570
^{36}Ar	0.337	18	18	36	35.967545	8.520
^{38}Ar	0.063	18	20	38	37.962732	8.614
^{40}Ar	99.600	18	22	40	39.962384	8.595
^{39}K	93.258	19	20	39	38.963707	8.557
^{41}K	6.730	19	22	41	40.961825	8.576
^{40}Ca	96.941	20	20	40	39.962591	8.551
^{42}Ca	0.647	20	22	42	41.958618	8.617
^{43}Ca	0.135	20	23	43	42.958766	8.601
^{44}Ca	2.086	20	24	44	43.955480	8.658
^{46}Ca	0.004	20	26	46	45.953689	8.669
^{48}Ca	0.187	20	28	48	47.952533	8.666
^{45}Sc	100.0	21	24	45	44.955910	8.619
^{46}Ti	8.0	22	24	46	45.952629	8.656
^{47}Ti	7.3	22	25	47	46.951764	8.661
^{48}Ti	73.8	22	26	48	47.947947	8.723
^{49}Ti	5.5	22	27	49	48.947871	8.711
^{50}Ti	5.4	22	28	50	49.944792	8.756
^{51}V	99.750	23	28	51	50.943962	8.742
^{50}Cr	4.345	24	26	50	49.946046	8.701
^{52}Cr	83.789	24	28	52	51.940509	8.776
^{53}Cr	9.501	24	29	53	52.940651	8.760
^{54}Cr	2.365	24	30	54	53.938882	8.778
^{55}Mn	100.0	25	30	55	54.938049	8.765
^{54}Fe	5.9	26	28	54	53.939612	8.736
^{56}Fe	91.72	26	30	56	55.934939	8.790
^{57}Fe	2.1	26	31	57	56.935396	8.770
^{58}Fe	0.28	26	32	58	57.933277	8.792
^{59}Co	100.0	27	32	59	58.933200	8.768
^{204}Pb	1.4	82	122	204	203.973020	7.880
^{206}Pb	24.1	82	124	206	205.974440	7.875
^{207}Pb	22.1	82	125	207	206.975872	7.870
^{208}Pb	52.4	82	126	208	207.976627	7.868
^{209}Bi	100.0	83	126	209	208.980380	7.848

[a] Selection is complete through cobalt-59. Where natural abundances do not add to 100%, the differences are made up by radioactive isotopes with exceedingly long half-lives: potassium-40 (0.0117%, $t_{1/2} = 1.3 \times 10^9$ yr); vanadium-50 (0.250%, $t_{1/2} > 1.4 \times 10^{17}$ yr).
[b] 1 MeV (mega electron-volts) = $1.60217733 \times 10^{-13}$ J.

TABLE A3.4   A Selection of Radioactive Isotopes

Isotope $^A X$	Decay Mode[a]	Half-Life $t_{1/2}$	Atomic Number Z	Neutron Number N	Mass Number A	Atomic Mass (amu)	Binding Energy per Nucleon (MeV)[b]
^{3}H	β^-	12.3 yr	1	2	3	3.01605	2.827
^{8}Be	α	$\sim 7 \times 10^{-17}$ s	4	4	8	8.005305	7.062
^{14}C	β^-	5.7×10^3 yr	6	8	14	14.003241	7.520
^{22}Na	β^+	2.6 yr	11	11	22	21.994434	7.916
^{24}Na	β^-	15.0 hr	11	13	24	23.990961	8.064
^{32}P	β^-	14.3 d	15	17	32	31.973907	8.464
^{35}S	β^-	87.2 d	16	19	35	34.969031	8.538
^{59}Fe	β^-	44.5 d	26	33	59	58.934877	8.755
^{60}Co	β^-	5.3 yr	27	33	60	59.933819	8.747
^{90}Sr	β^-	29.1 yr	38	52	90	89.907738	8.696
^{99}Tc	β^-	2.1×10^5 yr	43	56	99	98.906524	8.611
^{109}Cd	EC	462 d	48	61	109	108.904953	8.539
^{125}I	EC	59.4 d	53	72	125	124.904620	8.450
^{131}I	β^-	8.04 d	53	78	131	130.906114	8.422
^{137}Cs	β^-	30.3 yr	55	82	137	136.907073	8.389
^{222}Rn	α	3.82 d	86	136	222	222.017570	7.695
^{226}Ra	α	1600 yr	88	138	226	226.025402	7.662
^{232}Th	α	1.4×10^{10} yr	90	142	232	232.038054	7.615
^{235}U	α	7.0×10^8 yr	92	143	235	235.043924	7.591
^{238}U	α	4.5×10^9 yr	92	146	238	238.050784	7.570
^{239}Pu	α	2.4×10^4 yr	94	145	239	239.052157	7.560

[a] Modes of decay include alpha emission (α), beta emission (β^-), positron emission (β^+), electron capture (EC).
[b] 1 MeV (mega electron-volts) = $1.60217733 \times 10^{-13}$ J.

Chemical Bonds and Thermodynamic Data

TABLE A4.1	Average Lengths and Strengths of Covalent Bonds		
Atom	**Bond**	**Bond Length (pm)**	**Bond Strength (kJ/mol)**
H	H—H	75	436
	H—F	92	567
	H—Cl	127	431
	H—Br	141	366
	H—I	161	299
C	C—C	154	348
	C=C	134	614
	C≡C	120	839
	C—H	110	413
	C—N	143	293
	C=N	138	615
	C≡N	116	891
	C—O	143	358
	C=O[a]	123	743
	C≡O	113	1072
	C—F	133	485
	C—Cl	177	328
	C—Br	179	276
	C—I	215	238
N	N—N	147	163
	N=N	124	418
	N≡N	110	941
	N—H	104	388
	N—O	136	201
	N=O	122	607
	N≡O	106	463
O	O—O	148	146
	O=O	121	495
	O—H	96	463
S	S—O	151	265
	S=O	143	523
	S—S	204	266
	S—H	134	347
F	F—F	143	155
Cl	Cl—Cl	200	243
Br	Br—Br	228	193
I	I—I	266	151

[a] The bond strength of C=O in CO_2 is 799 kJ/mol.

TABLE A4.2 Critical Temperatures (T_c) and van der Waals Parameters (a, b) of Real Gases

Gas[a]	Molar Mass (g/mol)	T_c (K)	a ($L^2 \cdot atm/mol^2$)	b (L/mol)
H_2O	18.015	647.14	5.46	0.0305
Br_2	159.808	588	9.75	0.0591
CCl_3F	137.367	471.2	14.68	0.1111
Cl_2	70.906	416.9	6.343	0.0542
CO_2	44.010	304.14	3.59	0.0427
Kr	83.798	209.41	2.325	0.0396
CH_4	16.043	190.53	2.25	0.0428
O_2	31.999	154.59	1.36	0.0318
Ar	39.948	150.87	1.34	0.0322
F_2	37.997	144.13	1.171	0.0290
CO	28.010	132.91	1.45	0.0395
N_2	28.013	126.21	1.39	0.0391
H_2	2.016	32.97	0.244	0.0266
He	4.003	5.19	0.0341	0.0237

[a] Listed in descending order of critical temperature.

TABLE A4.3 Thermodynamic Properties at 25°C

Substance[a,b]	Molar Mass (g/mol)	ΔH_f° (kJ/mol)	S° [J/(mol·K)]	ΔG_f° (kJ/mol)
Elements and Monatomic Ions				
$Ag^+(aq)$	107.868	105.6	72.7	77.1
$Ag(g)$	107.868	284.9	173.0	246.0
$Ag(s)$	107.868	0.0	42.6	0.0
$Al^{3+}(aq)$	26.982	−531	−321.7	−485
$Al(g)$	26.982	330.0	164.6	289.4
$Al(s)$	26.982	0.0	28.3	0.0
$Ar(g)$	39.948	0.0	154.8	0.0
$Au(g)$	196.967	366.1	180.5	326.3
$Au(s)$	196.967	0.0	47.4	0.0
$B(g)$	10.811	565.0	153.4	521.0
$B(s)$	10.811	0.0	5.9	0.0
$Ba^{2+}(aq)$	137.327	−537.6	9.6	−560.8
$Ba(g)$	137.327	180.0	170.2	146.0
$Ba(s)$	137.327	0.0	62.8	0.0
$Be(g)$	9.012	324.0	136.3	286.6
$Be(s)$	9.012	0.0	9.5	0.0
$Br^-(aq)$	79.904	−121.6	82.4	−104.0
$Br(g)$	79.904	111.9	175.0	82.4
$Br_2(g)$	159.808	30.9	245.5	3.1
$Br_2(\ell)$	159.808	0.0	152.2	0.0
$C(g)$	12.011	716.7	158.1	671.3
$C(s, diamond)$	12.011	1.9	2.4	2.9
$C(s, graphite)$	12.011	0.0	5.7	0.0

Continued on next page

TABLE A4.3 Thermodynamic Properties at 25°C (*Continued*)

Substance[a,b]	Molar Mass (g/mol)	ΔH_f° (kJ/mol)	S° [J/(mol · K)]	ΔG_f° (kJ/mol)
$Ca^{2+}(aq)$	40.078	−542.8	−55.3	−553.6
$Ca(g)$	40.078	177.8	154.9	144.0
$Ca(s)$	40.078	0.0	41.6	0.0
$Cl^-(aq)$	35.453	−167.2	56.5	−131.2
$Cl(g)$	35.453	121.3	165.2	105.3
$Cl_2(g)$	70.906	0.0	223.0	0.0
$Co^{2+}(aq)$	58.933	−58.2	−113	−54.4
$Co^{3+}(aq)$	58.933	92	−305	134
$Co(g)$	58.933	424.7	179.5	380.3
$Co(s)$	58.933	0.0	30.0	0.0
$Cr(g)$	51.996	396.6	174.5	351.8
$Cr(s)$	51.996	0.0	23.8	0.0
$Cs^+(aq)$	132.905	−258.3	133.1	−292.0
$Cs(g)$	132.905	76.5	175.6	49.6
$Cs(s)$	132.905	0.0	85.2	0.0
$Cu^+(aq)$	63.546	71.7	40.6	50.0
$Cu^{2+}(aq)$	63.546	64.8	−99.6	65.5
$Cu(g)$	63.546	337.4	166.4	297.7
$Cu(s)$	63.546	0.0	33.2	0.0
$F^-(aq)$	18.998	−332.6	−13.8	−278.8
$F(g)$	18.998	79.4	158.8	62.3
$F_2(g)$	37.997	0.0	202.8	0.0
$Fe^{2+}(aq)$	55.845	−89.1	−137.7	−78.9
$Fe^{3+}(aq)$	55.845	−48.5	−315.9	−4.7
$Fe(g)$	55.845	416.3	180.5	370.7
$Fe(s)$	55.845	0.0	27.3	0.0
$H^+(aq)$	1.0079	0.0	0.0	0.0
$H(g)$	1.0079	218.0	114.7	203.3
$H_2(g)$	2.0158	0.0	130.6	0.0
$He(g)$	4.0026	0.0	126.2	0.0
$Hg_2^{2+}(aq)$	401.18	172.4	84.5	153.5
$Hg^{2+}(aq)$	200.59	171.1	−32.2	164.4
$Hg(g)$	200.59	61.4	175.0	31.8
$Hg(\ell)$	200.59	0.0	75.9	0.0
$I^-(aq)$	126.904	−55.2	111.3	−51.6
$I(g)$	126.904	106.8	180.8	70.2
$I_2(g)$	253.808	62.4	260.7	19.3
$I_2(s)$	253.808	0.0	116.1	0.0
$K^+(aq)$	39.098	−252.4	102.5	−283.3
$K(g)$	39.098	89.0	160.3	60.5
$K(s)$	39.098	0.0	64.7	0.0
$Li^+(aq)$	6.941	−278.5	13.4	−293.3
$Li(g)$	6.941	159.3	138.8	126.6
$Li^+(g)$	6.941	685.7	133.0	648.5
$Li(s)$	6.941	0.0	29.1	0.0
$Mg^{2+}(aq)$	24.305	−466.9	−138.1	−454.8

TABLE A4.3   Thermodynamic Properties at 25°C (*Continued*)

Substance[a,b]	Molar Mass (g/mol)	ΔH_f° (kJ/mol)	S° [J/(mol · K)]	ΔG_f° (kJ/mol)
Mg(g)	24.305	147.1	148.6	112.5
Mg(s)	24.305	0.0	32.7	0.0
Mn^{2+}(aq)	54.938	−220.8	−73.6	−228.1
Mn(g)	54.938	280.7	173.7	238.5
Mn(s)	54.938	0.0	32.0	0.0
N(g)	14.0067	472.7	153.3	455.5
N_2(g)	28.0134	0.0	191.5	0.0
Na^+(aq)	22.990	−240.1	59.0	−261.9
Na(g)	22.990	107.5	153.7	77.0
Na^+(g)	22.990	609.3	148.0	574.3
Na(s)	22.990	0.0	51.3	0.0
Ne(g)	20.180	0.0	146.3	0.0
Ni^{2+}(aq)	58.693	−54.0	−128.9	−45.6
Ni(g)	58.693	429.7	182.2	384.5
Ni(s)	58.693	0.0	29.9	0.0
O(g)	15.999	249.2	161.1	231.7
O_2(g)	31.999	0.0	205.0	0.0
P(g)	30.974	314.6	163.1	278.3
P_4(s, red)	123.895	−17.6	22.8	−12.1
P_4(s, white)	123.895	0.0	41.1	0.0
Pb^{2+}(aq)	207.2	−1.7	10.5	−24.4
Pb(g)	207.2	195.2	162.2	175.4
Pb(s)	207.2	0.0	64.8	0.0
Rb^+(aq)	85.468	−251.2	121.5	−284.0
Rb(g)	85.468	80.9	170.1	53.1
Rb(s)	85.468	0.0	76.8	0.0
S(g)	32.065	277.2	167.8	236.7
S_8(g)	256.52	102.3	430.2	49.1
S_8(s)	256.52	0.0	32.1	0.0
Sc(g)	44.956	377.8	174.8	336.0
Si(g)	28.086	450.0	168.0	405.5
Si(s)	28.086	0.0	18.8	0.0
Sn(g)	118.710	301.2	168.5	266.2
Sn(s, gray)	118.710	−2.1	44.1	0.1
Sn(s, white)	118.710	0.0	51.2	0.0
Sr^{2+}(aq)	87.62	−545.8	−32.6	−559.5
Sr(g)	87.62	164.4	164.6	130.9
Sr(s)	87.62	0.0	52.3	0.0
Ti(g)	47.867	473.0	180.3	428.4
Ti(s)	47.867	0.0	30.7	0.0
V(g)	50.942	514.2	182.2	468.5
V(s)	50.942	0.0	28.9	0.0
W(s)	183.84	0.0	32.6	0.0
Zn^{2+}(aq)	65.38	−153.9	−112.1	−147.1
Zn(g)	65.38	130.4	161.0	94.8
Zn(s)	65.38	0.0	41.6	0.0

Continued on next page

TABLE A4.3 Thermodynamic Properties at 25°C (Continued)

Substance[a,b]	Molar Mass (g/mol)	ΔH_f° (kJ/mol)	S° [J/(mol · K)]	ΔG_f° (kJ/mol)
Polyatomic Ions				
$CH_3COO^-(aq)$	59.045	−486.0	86.6	−369.3
$CO_3^{2-}(aq)$	60.009	−677.1	−56.9	−527.8
$C_2O_4^{2-}(aq)$	88.020	−825.1	45.6	−673.9
$CrO_4^{2-}(aq)$	115.994	−881.2	50.2	−727.8
$Cr_2O_7^{2-}(aq)$	215.988	−1490.3	261.9	−1301.1
$HCOO^-(aq)$	45.018	−425.6	92	−351.0
$HCO_3^-(aq)$	61.017	−692.0	91.2	−586.8
$HSO_4^-(aq)$	97.072	−887.3	131.8	−755.9
$MnO_4^-(aq)$	118.936	−541.4	191.2	−447.2
$NH_4^+(aq)$	18.038	−132.5	113.4	−79.3
$NO_3^-(aq)$	62.005	−205.0	146.4	−108.7
$OH^-(aq)$	17.007	−230.0	−10.8	−157.2
$PO_4^{3-}(aq)$	94.971	−1277.4	−222	−1018.7
$SO_4^{2-}(aq)$	96.064	−909.3	20.1	−744.5
Inorganic Compounds				
$AgCl(s)$	143.321	−127.1	96.2	−109.8
$AgI(s)$	234.773	−61.8	115.5	−66.2
$AgNO_3(s)$	169.873	−124.4	140.9	−33.4
$Al_2O_3(s)$	101.961	−1675.7	50.9	−1582.3
$B_2H_6(g)$	27.669	35.0	232.0	86.6
$B_2O_3(s)$	69.622	−1263.6	54.0	−1184.1
$BaCO_3(s)$	197.34	−1216.3	112.1	−1137.6
$BaSO_4(s)$	233.39	−1473.2	132.2	−1362.2
$CaCO_3(s)$	100.087	−1206.9	92.9	−1128.8
$CaCl_2(s)$	110.984	−795.4	108.4	−748.8
$CaF_2(s)$	78.075	−1228.0	68.5	−1175.6
$CaO(s)$	56.077	−634.9	38.1	−603.3
$Ca(OH)_2(s)$	74.093	−985.2	83.4	−897.5
$CaSO_4(s)$	136.142	−1434.5	106.5	−1322.0
$CO(g)$	28.010	−110.5	197.7	−137.2
$CO_2(g)$	44.010	−393.5	213.8	−394.4
$CO_2(aq)$	44.010	−412.9	121.3	−386.2
$CS_2(g)$	76.143	115.3	237.8	65.1
$CS_2(\ell)$	76.143	87.9	151.0	63.6
$CsCl(s)$	168.358	−443.0	101.2	−414.6
$CuSO_4(s)$	159.610	−771.4	109.2	−662.2
$FeCl_2(s)$	126.750	−341.8	118.0	−302.3
$FeCl_3(s)$	162.203	−399.5	142.3	−334.0
$FeO(s)$	71.844	−271.9	60.8	−255.2
$Fe_2O_3(s)$	159.688	−824.2	87.4	−742.2
$HBr(g)$	80.912	−36.3	198.7	−53.4
$HCl(g)$	36.461	−92.3	186.9	−95.3
$HF(g)$	20.006	−273.3	173.8	−275.4
$HI(g)$	127.912	26.5	206.6	1.7
$HNO_3(g)$	63.013	−135.1	266.4	−74.7
$HNO_3(\ell)$	63.013	−174.1	155.6	−80.7
$HNO_3(aq)$	63.013	−206.6	146.0	−110.5

TABLE A4.3   Thermodynamic Properties at 25°C (Continued)

Substance[a,b]	Molar Mass (g/mol)	ΔH_f° (kJ/mol)	S° [J/(mol · K)]	ΔG_f° (kJ/mol)
$HgCl_2(s)$	271.50	−224.3	146.0	−178.6
$Hg_2Cl_2(s)$	472.09	−265.4	191.6	−210.7
$H_2O(g)$	18.015	−241.8	188.8	−228.6
$H_2O(\ell)$	18.015	−285.8	69.9	−237.2
$H_2S(g)$	34.082	−20.17	205.6	−33.01
$H_2O_2(g)$	34.015	−136.3	232.7	−105.6
$H_2O_2(\ell)$	34.015	−187.8	109.6	−120.4
$H_2SO_4(\ell)$	98.079	−814.0	156.9	−690.0
$H_2SO_4(aq)$	98.079	−909.2	20.1	−744.5
$KBr(s)$	119.002	−393.8	95.9	−380.7
$KCl(s)$	74.551	−436.5	82.6	−408.5
$LiBr(s)$	86.845	−351.2	74.3	−342.0
$LiCl(s)$	42.394	−408.6	59.3	−384.4
$Li_2CO_3(s)$	73.891	−1215.9	90.4	−1132.1
$MgCl_2(s)$	95.211	−641.3	89.6	591.8
$Mg(OH)_2(s)$	58.320	−924.5	63.2	−833.5
$MgSO_4(s)$	120.369	−1284.9	91.6	−1170.6
$MnO_2(s)$	86.937	−520.0	53.1	−465.1
$NaCH_3OO(s)$	82.034	−708.8	123.0	−607.2
$NaBr(s)$	102.894	−361.1	86.82	−349.0
$NaCl(s)$	58.443	−411.2	72.1	−384.2
$NaCl(g)$	58.443	−181.4	229.8	−201.3
$Na_2CO_3(s)$	105.989	−1130.7	135.0	−1044.4
$NaHCO_3(s)$	84.007	−950.8	101.7	−851.0
$NaNO_3(s)$	84.995	−467.9	116.5	−367.0
$NaOH(s)$	39.997	−425.6	64.5	−379.5
$Na_2SO_4(s)$	142.043	−1387.1	149.6	−1270.2
$NF_3(g)$	71.002	−132.1	260.8	−90.6
$NH_3(aq)$	17.031	−80.3	111.3	−26.50
$NH_3(g)$	17.031	−46.1	192.5	−16.5
$NH_4Cl(s)$	53.491	−314.4	94.6	−203.0
$NH_4NO_3(s)$	80.043	−365.6	151.1	−183.9
$N_2H_4(g)$	32.045	95.40	238.5	159.4
$NiCl_2(s)$	129.60	−305.3	97.7	−259.0
$NiO(s)$	74.60	−239.7	38.0	−211.7
$NO(g)$	30.006	90.3	210.7	86.6
$NO_2(g)$	46.006	33.2	240.0	51.3
$N_2O(g)$	44.013	82.1	219.9	104.2
$N_2O_4(g)$	92.011	9.2	304.2	97.8
$NOCl(g)$	65.459	51.7	261.7	66.1
$O_3(g)$	47.998	142.7	238.8	163.2
$PCl_3(g)$	137.33	−288.07	311.7	−269.6
$PCl_3(\ell)$	137.33	−319.6	217	−272.4
$PF_5(g)$	125.96	−1594.4	300.8	−1520.7
$PH_3(g)$	33.998	5.4	210.2	13.4

Continued on next page

TABLE A4.3 Thermodynamic Properties at 25°C (*Continued*)

Substance[a,b]	Molar Mass (g/mol)	ΔH_f° (kJ/mol)	S° [J/(mol·K)]	ΔG_f° (kJ/mol)
$PbCl_2(s)$	278.1	−359.4	136.0	−314.1
$PbSO_4(s)$	303.3	−920.0	148.5	−813.0
$SO_2(g)$	64.065	−296.8	248.2	−300.1
$SO_3(g)$	80.064	−395.7	256.8	−371.1
$ZnCl_2(s)$	136.30	−415.1	111.5	−369.4
$ZnO(s)$	81.37	−348.0	43.9	−318.2
$ZnSO_4(s)$	161.45	−982.8	110.5	−871.5
Organic Molecules				
$CCl_4(g)$	153.823	−102.9	309.7	−60.6
$CCl_4(\ell)$	153.823	−135.4	216.4	−65.3
$CH_4(g)$	16.043	−74.8	186.2	−50.8
$CH_3COOH(g)$	60.053	−432.8	282.5	−374.5
$CH_3COOH(\ell)$	60.053	−485.8	159.8	−389.9
$CH_3OH(g)$	32.042	−200.7	239.9	−162.0
$CH_3OH(\ell)$	32.042	−238.7	126.8	−166.4
$C_2H_2(g)$	26.038	226.7	200.8	209.2
$C_2H_4(g)$	28.054	52.3	219.5	68.1
$C_2H_6(g)$	30.070	−84.7	229.5	−32.9
$CH_3CH_2OH(g)$	46.069	−235.1	282.6	−168.6
$CH_3CH_2OH(\ell)$	46.069	−277.7	160.7	−174.9
$CH_3CHO(g)$	44.05	−166	266	−133.7
$C_3H_8(g)$	44.097	−103.9	269.9	−23.5
$n\text{-}CH_3(CH_2)_2CH_3(g)^c$	58.123	−125.6	310.0	−15.7
$n\text{-}CH_3(CH_2)_2CH_3(\ell)^c$	58.123	−147.6	231.0	−15.0
$CH_3COCH_3(\ell)$	46.07	−248.4	199.8	
$CH_3COCH_3(g)$	46.07	−217.1	295.3	−152.7
$CH_3(CH_2)_2CH_2OH(\ell)$	74.12	−327.3	225.8	
$(CH_3CH_2)_2O(\ell)$	74.12	−279.6	172.4	
$(CH_3CH_2)_2O(g)$	74.12	−252.1	342.7	
$(CH_3)_2C{=}C(CH_3)_2(\ell)$	84.16	66.6	362.6	−69.2
$(CH_3)_2NH(\ell)$	45.09	−43.9	182.3	
$(CH_3)_2NH(g)$	45.09	−18.5	273.1	
$(C_2H_5)_2NH(\ell)$	73.14	−103.3		
$(C_2H_5)_2NH(g)$	73.14	−71.4		
$(CH_3)_3N(\ell)$	59.11	−46.0	208.5	
$(CH_3)_3N(g)$	59.11	−23.6	287.1	
$(CH_3CH_2)_3N(\ell)$	101.19	−134.3		
$(CH_3CH_2)_3N(g)$	101.19	−95.8		
$C_6H_6(g)$	78.114	82.9	269.2	129.7
$C_6H_6(\ell)$	78.114	49.0	172.9	124.5
$C_6H_{12}O_6(s)$	180.158	−1274.4	212.1	−910.1
$n\text{-}C_8H_{18}(\ell)^c$	114.231	−249.9	361.1	6.4
$n\text{-}C_8H_{18}(g)$	114.231	−208.6	466.7	16.4
$C_{12}H_{22}O_{11}(s)$	342.300	−2221.7	360.2	−1543.8
$HCOOH(\ell)$	46.026	−424.7	129.0	−361.4

[a] Substances are arranged alphabetically by chemical formula within each class: (1) elements and monatomic ions; (2) polyatomic ions; (3) inorganic compounds (including CO and CO_2); (4) organic molecules (hydrocarbon-based).
[b] Symbols denote standard enthalpy of formation (ΔH_f°), standard third-law entropy (S°), and standard Gibbs free energy of formation (ΔG_f°). Entropies in aqueous solution are referred to $S^\circ[H^+(aq)] = 0$, not to absolute zero.
[c] The symbol n denotes the "normal" unbranched alkane.

TABLE A4.4 Vapor Pressure of Water as a Function of Temperature

T (°C)	P (torr)
0.0	4.579
10.0	9.209
20.0	17.535
25.0	23.756
30.0	31.824
40.0	55.324
60.0	149.4
70.0	233.7
90.0	525.8
100	760.0
105	906.0

Equilibrium Constants

TABLE A5.1   Ionization Constants of Selected Acids at 25°C

Acid	Step	Aqueous Equilibrium[a]	K_a	pK_a
acetic	1	$CH_3COOH(aq) \rightleftharpoons H^+(aq) + CH_3COO^-(aq)$	1.76×10^{-5}	4.75
arsenic	1	$H_3AsO_4(aq) \rightleftharpoons H^+(aq) + H_2AsO_4^-(aq)$	5.5×10^{-3}	2.26
	2	$H_2AsO_4^-(aq) \rightleftharpoons H^+(aq) + AsO_4^{2-}(aq)$	1.7×10^{-7}	6.77
	3	$HAsO_4^{2-}(aq) \rightleftharpoons H^+(aq) + AsO_4^{3-}(aq)$	5.1×10^{-12}	11.29
ascorbic	1	$H_2C_6H_6O_6(aq) \rightleftharpoons H^+(aq) + HC_6H_6O_6^-(aq)$	1.0×10^{-5}	5.00
	2	$HC_6H_6O_6^-(aq) \rightleftharpoons H^+(aq) + C_6H_6O_6^{2-}(aq)$	5×10^{-12}	11.3
benzoic	1	$C_6H_5COOH(aq) \rightleftharpoons H^+(aq) + C_6H_5COO^-(aq)$	6.46×10^{-5}	4.19
boric	1	$H_3BO_3(aq) \rightleftharpoons H^+(aq) + H_2BO_3^-(aq)$	5.4×10^{-10}	9.27
	2	$H_2BO_3^-(aq) \rightleftharpoons H^+(aq) + HBO_3^{2-}(aq)$	$<10^{-14}$	$>.14$
bromoacetic	1	$CH_2BrCOOH(aq) \rightleftharpoons H^+(aq) + CH_2BrCOO^-(aq)$	2.0×10^{-3}	2.70
butanoic	1	$CH_3CH_2CH_2COOH(aq) \rightleftharpoons$ $H^+(aq) + CH_3CH_2CH_2COO^-(aq)$	1.5×10^{-5}	4.82
carbonic	1	$H_2CO_3(aq) \rightleftharpoons H^+(aq) + HCO_3^-(aq)$	4.3×10^{-7}	6.37
	2	$HCO_3^-(aq) \rightleftharpoons H^+(aq) + CO_3^{2-}(aq)$	4.7×10^{-11}	10.33
chloric	1	$HClO_3(aq) \rightleftharpoons H^+(aq) + ClO_3^-(aq)$	~ 1	~ 0
chloroacetic	1	$CH_2ClCOOH(aq) \rightleftharpoons H^+(aq) + CH_2ClCOO^-(aq)$	1.4×10^{-3}	2.85
chlorous	1	$HClO_2(aq) \rightleftharpoons H^+(aq) + ClO_2^-(aq)$	1.1×10^{-2}	1.96
citric	1	$HOC(CH_2)_2(COOH)_3(aq) \rightleftharpoons$ $H^+(aq) + HOC(CH_2)_2(COOH)_2COO^-(aq)$	7.4×10^{-4}	3.13
	2	$HOC(CH_2)_2(COOH)_2COO^-(aq) \rightleftharpoons$ $H^+(aq) + HOC(CH_2)_2(COOH)(COO^-)_2(aq)$	1.7×10^{-5}	4.77
	3	$HOC(CH_2)_2(COOH)(COO^-)_2(aq) \rightleftharpoons$ $H^+(aq) + HOC(CH_2)_2(COO^-)_3(aq)$	4.0×10^{-7}	6.49
dichloroacetic	1	$CHCl_2COOH(aq) \rightleftharpoons H^+(aq) + CHCl_2COO^-(aq)$	5.5×10^{-2}	1.26
ethanol	1	$CH_3CH_2OH(aq) \rightleftharpoons H^+(aq) + CH_3CH_2O^-(aq)$	1.3×10^{-16}	15.9
fluoroacetic	1	$CH_2FCOOH(aq) \rightleftharpoons H^+(aq) + CH_2FCOO^-(aq)$	2.6×10^{-3}	2.59
formic	1	$HCOOH(aq) \rightleftharpoons H^+(aq) + HCOO^-(aq)$	1.77×10^{-4}	3.75
germanic	1	$H_2GeO_3(aq) \rightleftharpoons H^+(aq) + HGeO_3^-(aq)$	9.8×10^{-10}	9.01
	2	$HGeO_3^-(aq) \rightleftharpoons H^+(aq) + GeO_3^{2-}(aq)$	5×10^{-13}	12.3
hydr(o)azoic	1	$HN_3(aq) \rightleftharpoons H^+(aq) + N_3^-(aq)$	1.9×10^{-5}	4.72
hydrobromic	1	$HBr(aq) \rightleftharpoons H^+(aq) + Br^-(aq)$	$\gg 1$ (strong)	<0
hydrochloric	1	$HCl(aq) \rightleftharpoons H^+(aq) + Cl^-(aq)$	$\gg 1$ (strong)	<0
hydrocyanic	1	$HCN(aq) \rightleftharpoons H^+(aq) + CN^-(aq)$	6.2×10^{-10}	9.21
hydrofluoric	1	$HF(aq) \rightleftharpoons H^+(aq) + F^-(aq)$	6.8×10^{-4}	3.17

Continued on next page

TABLE A5.1 Ionization Constants of Selected Acids at 25°C (*Continued*)

Acid	Step	Aqueous Equilibrium[a]	K_a	pK_a
hydr(o)iodic	1	$HI(aq) \rightleftharpoons H^+(aq) + I^-(aq)$	$\gg 1$ (strong)	< 0
hydrosulfuric	1	$H_2S(aq) \rightleftharpoons H^+(aq) + HS^-(aq)$	8.9×10^{-8}	7.05
	2	$HS^-(aq) \rightleftharpoons H^+(aq) + S^{2-}(aq)$	$\sim 10^{-19}$	~ 19
hypobromous	1	$HBrO(aq) \rightleftharpoons H^+(aq) + BrO^-(aq)$	2.3×10^{-9}	8.64
hypochlorous	1	$HClO(aq) \rightleftharpoons H^+(aq) + ClO^-(aq)$	2.9×10^{-8}	7.54
hypoiodous	1	$HIO(aq) \rightleftharpoons H^+(aq) + IO^-(aq)$	2.3×10^{-11}	10.64
iodic	1	$HIO_3(aq) \rightleftharpoons H^+(aq) + IO_3^-(aq)$	1.7×10^{-1}	0.77
iodoacetic	1	$CH_2ICOOH(aq) \rightleftharpoons H^+(aq) + CH_2ICOO^-(aq)$	7.6×10^{-4}	3.12
lactic	1	$CH_3CHOHCOOH(aq) \rightleftharpoons$ $H^+(aq) + CH_3CHOHCOO^-(aq)$	1.4×10^{-4}	3.85
maleic	1	$HOOCCH{=}CHCOOH(aq) \rightleftharpoons$ $H^+(aq) + HOOCCH{=}CHCOO^-(aq)$	1.2×10^{-2}	1.92
	2	$HOOCCH{=}CHCOO^-(aq) \rightleftharpoons$ $H^+(aq) + {}^-OOCCH{=}CHCOO^-(aq)$	4.7×10^{-7}	6.33
malonic	1	$HOOCCH_2COOH(aq) \rightleftharpoons$ $H^+(aq) + HOOCCH_2COO^-(aq)$	1.5×10^{-3}	2.82
	2	$HOOCCH_2COO^-(aq) \rightleftharpoons$ $H^+(aq) + {}^-OOCCH_2COO^-(aq)$	2.0×10^{-6}	5.70
nitric	1	$HNO_3(aq) \rightleftharpoons H^+(aq) + NO_3(aq)$	$\gg 1$ (strong)	< 0
nitrous	1	$HNO_2(aq) \rightleftharpoons H^+(aq) + NO_2^-(aq)$	4.0×10^{-4}	3.40
oxalic	1	$HOOCCOOH(aq) \rightleftharpoons H^+(aq) + HOOCCOO^-(aq)$	5.9×10^{-2}	1.23
	2	$HOOCCOO^-(aq) \rightleftharpoons H^+(aq) + {}^-OOCCOO^-(aq)$	6.4×10^{-5}	4.19
perchloric	1	$HClO_4(aq) \rightleftharpoons H^+(aq) + ClO_4^-(aq)$	$\gg 1$ (strong)	< 0
periodic	1	$HIO_4(aq) \rightleftharpoons H^+(aq) + IO_4^-(aq)$	2.3×10^{-2}	1.64
phenol	1	$C_6H_5OH(aq) \rightleftharpoons H^+(aq) + C_6H_5O^-(aq)$	1.3×10^{-10}	9.89
phosphoric	1	$H_3PO_4(aq) \rightleftharpoons H^+(aq) + H_2PO_4^-(aq)$	7.11×10^{-3}	2.12
	2	$H_2PO_4^-(aq) \rightleftharpoons H^+(aq) + HPO_4^{2-}(aq)$	6.32×10^{-8}	7.21
	3	$HPO_4^{2-}(aq) \rightleftharpoons H^+(aq) + PO_4^{3-}(aq)$	4.5×10^{-13}	12.66
propanoic	1	$CH_3CH_2COOH(aq) \rightleftharpoons$ $H^+(aq) + CH_3CH_2COO^-(aq)$	1.4×10^{-5}	4.86
pyruvic	1	$CH_3C(O)COOH(aq) \rightleftharpoons$ $H^+(aq) + CH_3C(O)COO^-(aq)$	2.8×10^{-3}	2.55
sulfuric	1	$H_2SO_4(aq) \rightleftharpoons H^+(aq) + HSO_4^-(aq)$	$\gg 1$ (strong)	< 0
	2	$HSO_4^-(aq) \rightleftharpoons H^+(aq) + SO_4^{2-}(aq)$	1.2×10^{-2}	1.92
sulfurous	1	$H_2SO_3(aq) \rightleftharpoons H^+(aq) + HSO_3^-(aq)$	1.7×10^{-2}	1.9
	2	$HSO_3^-(aq) \rightleftharpoons H^+(aq) + SO_3^{2-}(aq)$	6.2×10^{-8}	7.1
thiocyanic	1	$HSCN(aq) \rightleftharpoons H^+(aq) + SCN^-(aq)$	$\gg 1$ (strong)	< 0
trichloroacetic	1	$CCl_3COOH(aq) \rightleftharpoons H^+(aq) + CCl_3COO^-(aq)$	2.3×10^{-1}	0.64
trifluoroacetic	1	$CF_3COOH(aq) \rightleftharpoons H^+(aq) + CF_3COO^-(aq)$	5.9×10^{-1}	0.23
water	1	$H_2O(aq) \rightleftharpoons H^+(aq) + OH^-(aq)$	1.0×10^{-14}	14.00

[a] The formulas of the carboxylic acids are written in an RCOOH format to highlight their molecular structures.

TABLE A5.2	Acid Ionization Constants of Hydrated Metal Ions at 25°C	
Free Ion	**Hydrated Ion**	K_a
Fe^{3+}	$Fe(H_2O)_6^{3+}$	3×10^{-3}
Sn^{2+}	$Sn(H_2O)_6^{2+}$	4×10^{-4}
Cr^{3+}	$Cr(H_2O)_6^{3+}$	1×10^{-4}
Al^{3+}	$Al(H_2O)_6^{3+}$	1×10^{-5}
Cu^{2+}	$Cu(H_2O)_6^{2+}$	3×10^{-8}
Pb^{2+}	$Pb(H_2O)_6^{2+}$	3×10^{-8}
Zn^{2+}	$Zn(H_2O)_6^{2+}$	1×10^{-9}
Co^{2+}	$Co(H_2O)_6^{2+}$	2×10^{-10}
Ni^{2+}	$Ni(H_2O)_6^{2+}$	1×10^{-10}

TABLE A5.3 Ionization Constants of Selected Bases at 25°C

Base	Aqueous Equilibrium	K_b	pK_b
ammonia	$NH_3(aq) + H_2O(aq) \rightleftharpoons NH_4^+(aq) + OH^-(aq)$	1.76×10^{-5}	4.75
aniline	$C_6H_5NH_2(aq) + H_2O(aq) \rightleftharpoons C_6H_5NH_3^+(aq) + OH^-(aq)$	4.0×10^{-10}	9.4
diethylamine	$(CH_3CH_2)_2NH(aq) + H_2O(aq) \rightleftharpoons (CH_3CH_2)_2NH_2^+(aq) + OH^-(aq)$	8.6×10^{-4}	3.1
dimethylamine	$(CH_3)_2NH(aq) + H_2O(aq) \rightleftharpoons (CH_3)_2NH_2^+(aq) + OH^-(aq)$	5.9×10^{-4}	3.2
methylamine	$CH_3NH_2(aq) + H_2O(aq) \rightleftharpoons CH_3NH_3^+(aq) + OH^-(aq)$	4.4×10^{-4}	3.4

nicotine

(1)

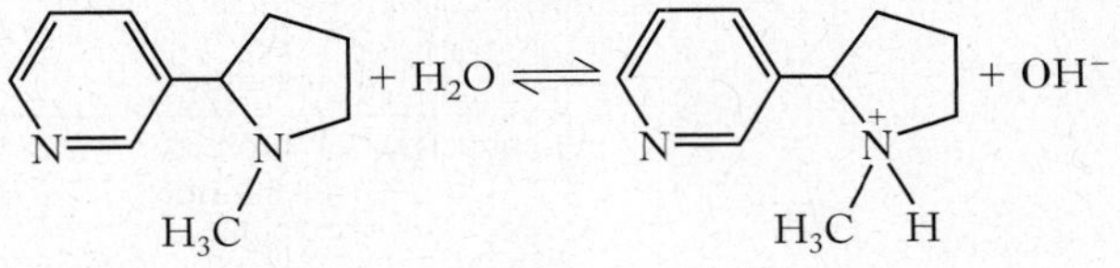

1.0×10^{-6} 6.0

(2)

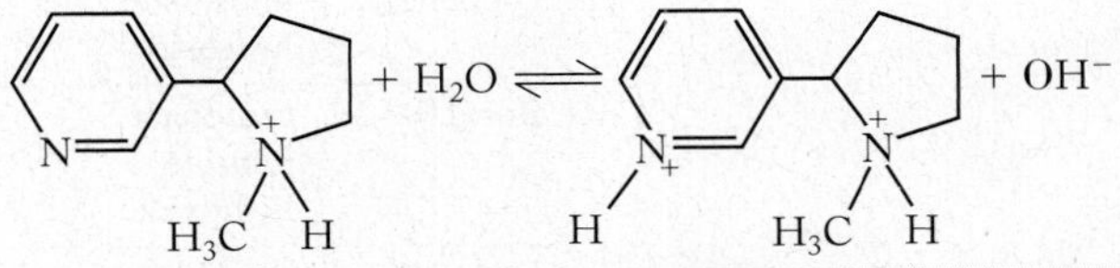

1.3×10^{-11} 10.9

| pyridine | $C_5H_5N(aq) + H_2O(aq) \rightleftharpoons C_5H_5NH^+(aq) + OH^-(aq)$ | 1.7×10^{-9} | 8.8 |

quinine

(1)

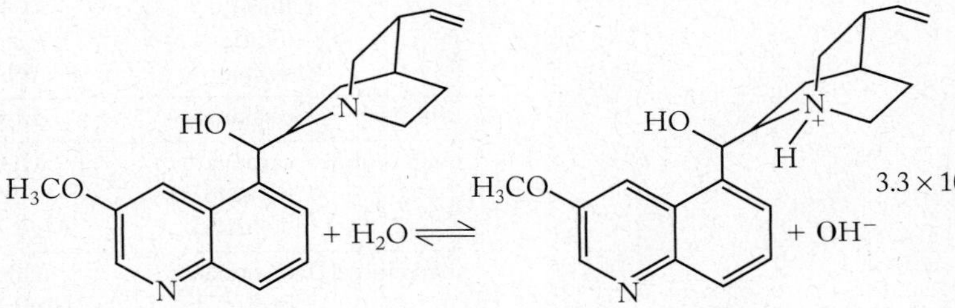

3.3×10^{-6} 5.5

(2)

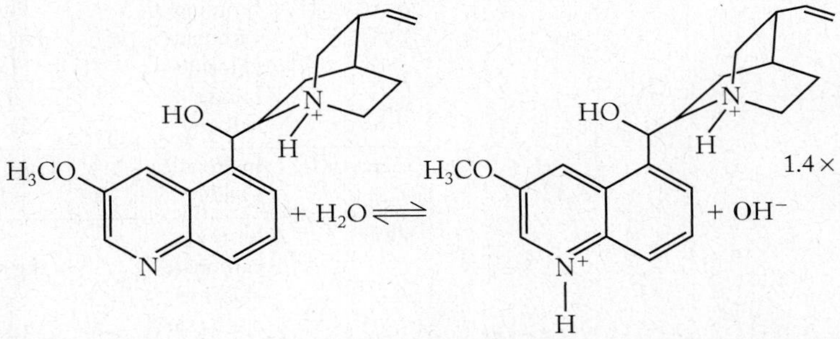

1.4×10^{-10} 9.9

| urea | $H_2NCONH_2(aq) + H_2O(aq) \rightleftharpoons H_2NCONH_3^+(aq) + OH^-(aq)$ | 1.3×10^{-14} | 13.9 |

TABLE A5.4 Solubility-Product Constants at 25°C

Cation	Anion	Heterogeneous Equilibrium[a]	K_{sp}
aluminum	hydroxide	$Al(OH)_3(s) \rightleftharpoons Al^{3+}(aq) + 3\,OH^-(aq)$	1.9×10^{-33}
	phosphate	$AlPO_4(s) \rightleftharpoons Al^{3+}(aq) + PO_4^{3-}(aq)$	9.8×10^{-21}
barium	carbonate	$BaCO_3(s) \rightleftharpoons Ba^{2+}(aq) + CO_3^{2-}(aq)$	2.6×10^{-9}
	fluoride	$BaF_2(s) \rightleftharpoons Ba^{2+}(aq) + 2\,F^-(aq)$	1.0×10^{-6}
	sulfate	$BaSO_4(s) \rightleftharpoons Ba^{2+}(aq) + SO_4^{2-}(aq)$	9.1×10^{-11}
calcium	carbonate	$CaCO_3(s) \rightleftharpoons Ca^{2+}(aq) + CO_3^{2-}(aq)$	5.0×10^{-9}
	fluoride	$CaF_2(s) \rightleftharpoons Ca^{2+}(aq) + 2\,F^-(aq)$	3.9×10^{-11}
	hydroxide	$Ca(OH)_2(s) \rightleftharpoons Ca^{2+}(aq) + 2\,OH^-(aq)$	4.7×10^{-6}
	phosphate	$Ca_3(PO_4)_2(s) \rightleftharpoons 3\,Ca^{2+}(aq) + 2\,PO_4^{3-}(aq)$	2.1×10^{-33}
	sulfate	$CaSO_4(s) \rightleftharpoons Ca^{2+}(aq) + SO_4^{2-}(aq)$	7.1×10^{-5}
copper(I)	bromide	$CuBr(s) \rightleftharpoons Cu^+(aq) + Br^-(aq)$	6.3×10^{-9}
	chloride	$CuCl(s) \rightleftharpoons Cu^+(aq) + Cl^-(aq)$	1.0×10^{-6}
	iodide	$CuI(s) \rightleftharpoons Cu^+(aq) + I^-(aq)$	1.3×10^{-12}
copper(II)	phosphate	$Cu_3(PO_4)_2(s) \rightleftharpoons 3\,Cu^{2+}(aq) + 2\,PO_4^{3-}(aq)$	1.4×10^{-37}
	hydroxide	$Cu(OH)_2(s) \rightleftharpoons Cu^{2+}(aq) + 2\,OH^-(aq)$	4.8×10^{-20}
iron(II)	carbonate	$FeCO_3(s) \rightleftharpoons Fe^{2+}(aq) + CO_3^{2-}(aq)$	3.1×10^{-11}
	fluoride	$FeF_2(s) \rightleftharpoons Fe^{2+}(aq) + 2\,F^-(aq)$	2.4×10^{-6}
	hydroxide	$Fe(OH)_2(s) \rightleftharpoons Fe^{2+}(aq) + 2\,OH^-(aq)$	4.9×10^{-17}
lead	bromide	$PbBr_2(s) \rightleftharpoons Pb^{2+}(aq) + 2\,Br^-(aq)$	6.6×10^{-6}
	carbonate	$PbCO_3(s) \rightleftharpoons Pb^{2+}(aq) + CO_3^{2-}(aq)$	1.5×10^{-13}
	chloride	$PbCl_2(s) \rightleftharpoons Pb^{2+}(aq) + 2\,Cl^-(aq)$	1.6×10^{-5}
	fluoride	$PbF_2(s) \rightleftharpoons Pb^{2+}(aq) + 2\,F^-(aq)$	3.2×10^{-8}
	iodide	$PbI_2(s) \rightleftharpoons Pb^{2+}(aq) + 2\,I^-(aq)$	8.5×10^{-9}
	sulfate	$PbSO_4(s) \rightleftharpoons Pb^{2+}(aq) + SO_4^{2-}(aq)$	1.8×10^{-8}
lithium	carbonate	$Li_2CO_3(s) \rightleftharpoons 2\,Li^+(aq) + CO_3^{2-}(aq)$	8.2×10^{-4}
magnesium	carbonate	$MgCO_3(s) \rightleftharpoons Mg^{2+}(aq) + CO_3^{2-}(aq)$	6.8×10^{-6}
	fluoride	$MgF_2(s) \rightleftharpoons Mg^{2+}(aq) + 2\,F^-(aq)$	6.5×10^{-9}
	hydroxide	$Mg(OH)_2(s) \rightleftharpoons Mg^{2+}(aq) + 2\,OH^-(aq)$	5.6×10^{-12}
manganese(II)	carbonate	$MnCO_3(s) \rightleftharpoons Mn^{2+}(aq) + CO_3^{2-}(aq)$	2.2×10^{-11}
	hydroxide	$Mn(OH)_2(s) \rightleftharpoons Mn^{2+}(aq) + 2\,OH^-(aq)$	5.6×10^{-12}
mercury(I)	bromide	$Hg_2Br_2(s) \rightleftharpoons Hg_2^{2+}(aq) + 2\,Br^-(aq)$	6.4×10^{-23}
	carbonate	$Hg_2CO_3(s) \rightleftharpoons Hg_2^{2+}(aq) + CO_3^{2-}(aq)$	3.7×10^{-17}
	chloride	$Hg_2Cl_2(s) \rightleftharpoons Hg_2^{2+}(aq) + 2\,Cl^-(aq)$	1.5×10^{-18}
	iodide	$Hg_2I_2(s) \rightleftharpoons Hg_2^{2+}(aq) + 2\,I^-(aq)$	5.3×10^{-29}
	sulfate	$Hg_2SO_4(s) \rightleftharpoons Hg_2^{2+}(aq) + SO_4^{2-}(aq)$	8.0×10^{-7}
mercury(II)	hydroxide	$Hg(OH)_2(s) \rightleftharpoons Hg^{2+}(aq) + 2\,OH^-(aq)$	3.1×10^{-26}
	iodide	$HgI_2(s) \rightleftharpoons Hg^{2+}(aq) + 2\,I^-(aq)$	2.8×10^{-29}
silver	bromide	$AgBr(s) \rightleftharpoons Ag^+(aq) + Br^-(aq)$	5.4×10^{-13}
	carbonate	$Ag_2CO_3(s) \rightleftharpoons 2\,Ag^+(aq) + CO_3^{2-}(aq)$	8.5×10^{-12}
	chloride	$AgCl(s) \rightleftharpoons Ag^+(aq) + Cl^-(aq)$	1.8×10^{-10}
	chromate	$Ag_2CrO_4(s) \rightleftharpoons 2\,Ag^+(aq) + CrO_4^{2-}(aq)$	1.1×10^{-12}
	hydroxide	$AgOH(s) \rightleftharpoons Ag^+(aq) + OH^-(aq)$	1.52×10^{-8}
	iodide	$AgI(s) \rightleftharpoons Ag^+(aq) + I^-(aq)$	8.3×10^{-17}
	phosphate	$Ag_3PO_4(s) \rightleftharpoons 3\,Ag^+(aq) + PO_4^{3-}(aq)$	8.9×10^{-17}
	sulfate	$Ag_2SO_4(s) \rightleftharpoons 2\,Ag^+(aq) + SO_4^{2-}(aq)$	1.2×10^{-5}
strontium	carbonate	$SrCO_3(s) \rightleftharpoons Sr^{2+}(aq) + CO_3^{2-}(aq)$	5.6×10^{-10}
	fluoride	$SrF_2(s) \rightleftharpoons Sr^{2+}(aq) + 2\,F^-(aq)$	4.3×10^{-9}
	sulfate	$SrSO_4(s) \rightleftharpoons Sr^{2+}(aq) + SO_4^{2-}(aq)$	3.4×10^{-7}
zinc	carbonate	$ZnCO_3(s) \rightleftharpoons Zn^{2+}(aq) + CO_3^{2-}(aq)$	1.2×10^{-10}
	hydroxide	$Zn(OH)_2(s) \rightleftharpoons Zn^{2+}(aq) + 2\,OH^-(aq)$	3.0×10^{-16}

[a] Equilibrium is between solid phase and aqueous solution.

TABLE A5.5 Formation Constants of Complex Ions at 25°C

Complex Ion	Aqueous Equilibrium	K_f
$[Ag(NH_3)_2]^+$	$Ag^+(aq) + 2NH_3(aq) \rightleftharpoons Ag(NH_3)_2^+(aq)$	1.7×10^7
$[AgCl_2]^-$	$Ag^+(aq) + 2Cl^-(aq) \rightleftharpoons AgCl_2^-(aq)$	2.5×10^5
$[Ag(CN)_2]^-$	$Ag^+(aq) + 2CN^-(aq) \rightleftharpoons Ag(CN)_2^-(aq)$	1.0×10^{21}
$[Ag(S_2O_3)_2]^{3-}$	$Ag^+(aq) + 2S_2O_3^{2-}(aq) \rightleftharpoons Ag(S_2O_3)_2^{3-}(aq)$	4.7×10^{13}
$[AlF_6]^{3-}$	$Al^{3+}(aq) + 6F^-(aq) \rightleftharpoons AlF_6^{3-}(aq)$	4.0×10^{19}
$[Al(OH)_4]^-$	$Al^{3+}(aq) + 4OH^-(aq) \rightleftharpoons Al(OH)_4^-(aq)$	7.7×10^{33}
$[Au(CN)_2]^-$	$Au^+(aq) + 2CN^-(aq) \rightleftharpoons Au(CN)_2^-(aq)$	2.0×10^{38}
$[Co(NH_3)_6]^{2+}$	$Co^{2+}(aq) + 6NH_3(aq) \rightleftharpoons Co(NH_3)_6^{2+}(aq)$	7.7×10^4
$[Co(NH_3)_6]^{3+}$	$Co^{3+}(aq) + 6NH_3(aq) \rightleftharpoons Co(NH_3)_6^{3+}(aq)$	5.0×10^{31}
$[Co(en)_3^{2+}]$	$Co^{2+}(aq) + 3en(aq) \rightleftharpoons Co(en)_3^{2+}(aq)$	8.7×10^{13}
$[Cu(NH_3)_4]^{2+}$	$Cu^{2+}(aq) + 4NH_3(aq) \rightleftharpoons Cu(NH_4)_4^{2+}(aq)$	5.0×10^{13}
$[Cu(en)_2^{2+}]$	$Cu^{2+}(aq) + 2en(aq) \rightleftharpoons Cu(en)_2^{2+}(aq)$	3.2×10^{19}
$[Cu(CN)_4]^{2-}$	$Cu^{2+}(aq) + 4CN^-(aq) \rightleftharpoons Cu(CN)_4^{2-}(aq)$	1.0×10^{25}
$[HgCl_4]^{2-}$	$Hg^{2+}(aq) + 4Cl^-(aq) \rightleftharpoons HgCl_4^{2-}(aq)$	1.2×10^{15}
$[Ni(NH_3)_6]^{2+}$	$Ni^{2+}(aq) + 6NH_3(aq) \rightleftharpoons Ni(NH_3)_6^{2+}(aq)$	5.5×10^8
$[PbCl_4]^{2-}$	$Pb^{2+}(aq) + 4Cl^-(aq) \rightleftharpoons PbCl_4^{2-}(aq)$	2.5×10^1
$[Zn(NH_3)_4]^{2+}$	$Zn^{2+}(aq) + 4NH_3(aq) \rightleftharpoons Zn(NH_3)_4^{2+}(aq)$	2.9×10^9
$[Zn(OH)_4]^{2-}$	$Zn^{2+}(aq) + 4OH^-(aq) \rightleftharpoons Zn(OH)_4^{2-}(aq)$	2.8×10^{15}

Standard Reduction Potentials

TABLE A6.1	Standard Reduction Potentials at 25°C		
Half-Reaction		n	E° (V)
$F_2(g) + 2\,e^- \rightarrow 2\,F^-(aq)$		2	2.866
$H_2N_2O_2(s) + 2\,H^+(aq) + 2\,e^- \rightarrow N_2(g) + 2\,H_2O(\ell)$		2	2.65
$O(g) + 2\,H^+(aq) + 2\,e^- \rightarrow H_2O(\ell)$		2	2.421
$Cu^{3+}(aq) + e^- \rightarrow Cu^{2+}(aq)$		1	2.4
$XeO_3(s) + 6\,H^+(aq) + 6\,e^- \rightarrow Xe(g) + 3\,H_2O(\ell)$		6	2.10
$O_3(g) + 2\,H^+(aq) + 2\,e^- \rightarrow O_2(g) + H_2O(\ell)$		2	2.076
$OH(g) + e^- \rightarrow OH^-(aq)$		1	2.02
$Co^{3+}(aq) + e^- \rightarrow Co^{2+}(aq)$		1	1.92
$H_2O_2(\ell) + 2\,H^+(aq) + 2\,e^- \rightarrow 2\,H_2O(\ell)$		2	1.776
$N_2O(g) + 2\,H^+(aq) + 2\,e^- \rightarrow N_2(g) + H_2O(\ell)$		2	1.766
$Ce(OH)^{3+}(aq) + H^+(aq) + e^- \rightarrow Ce^{3+}(aq) + H_2O(\ell)$		1	1.70
$Au^+(aq) + e^- \rightarrow Au(s)$		1	1.692
$PbO_2(s) + SO_4^{2-}(aq) + 4\,H^+(aq) + 2\,e^- \rightarrow$ $\quad\quad\quad\quad\quad PbSO_4(s) + 2\,H_2O(\ell)$		2	1.6913
$PbO_2(s) + HSO_4^-(aq) + 3\,H^+(aq) + 2\,e^- \rightarrow$ $\quad\quad\quad\quad\quad PbSO_4(s) + 2\,H_2O(\ell)$		2	1.685
$MnO_4^-(aq) + 4\,H^+(aq) + 3\,e^- \rightarrow$ $\quad\quad\quad\quad\quad MnO_2(s) + 2\,H_2O(\ell)$		3	1.673
$NiO_2(s) + 4\,H^+(aq) + 2\,e^- \rightarrow Ni^{2+}(aq) + 2\,H_2O(\ell)$		2	1.678
$HClO(\ell) + H^+(aq) + e^- \rightarrow \frac{1}{2}Cl_2(g) + H_2O(aq)$		1	1.63
$Ce^{4+}(aq) + e^- \rightarrow Ce^{3+}(aq)$		1	1.61
$Mn^{3+}(aq) + e^- \rightarrow Mn^{2+}(aq)$		1	1.542
$MnO_4^-(aq) + 8\,H^+(aq) + 5\,e^- \rightarrow Mn^{2+}(aq) + 4\,H_2O(\ell)$		5	1.507
$BrO_3^-(aq) + 6\,H^+(aq) + 5\,e^- \rightarrow \frac{1}{2}Br_2(\ell) + 3\,H_2O(\ell)$		5	1.52
$ClO_3^-(aq) + 6\,H^+(aq) + 5\,e^- \rightarrow \frac{1}{2}Cl_2(g) + 3\,H_2O(\ell)$		5	1.47
$PbO_2(s) + 4\,H^+(aq) + 2\,e^- \rightarrow Pb^{2+}(aq) + 2\,H_2O(\ell)$		2	1.455
$Au^{3+}(aq) + 3\,e^- \rightarrow Au(s)$		3	1.40
$Cl_2(g) + 2\,e^- \rightarrow 2\,Cl^-(aq)$		2	1.3583
$Cr_2O_7^{2-}(aq) + 14\,H^+(aq) + 6\,e^- \rightarrow$ $\quad\quad\quad\quad\quad 2\,Cr^{3+}(aq) + 7\,H_2O(\ell)$		6	1.33
$2\,NiO(OH)(s) + 2\,H_2O(\ell) + 2\,e^- \rightarrow$ $\quad\quad\quad\quad\quad 2\,Ni(OH)_2(s) + 2\,OH^-(aq)$		2	1.32
$MnO_2(s) + 4\,H^+(aq) + 2\,e^- \rightarrow Mn^{2+}(aq) + 2\,H_2O(\ell)$		2	1.23
$O_2(g) + 4\,H^+(aq) + 4\,e^- \rightarrow 2\,H_2O(\ell)$		4	1.229
$IO_3^-(aq) + 6\,H^+(aq) + 5\,e^- \rightarrow \frac{1}{2}I_2(s) + 3\,H_2O(\ell)$		5	1.195

TABLE A6.1 Standard Reduction Potentials at 25°C (*Continued*)

Half-Reaction	n	$E°$ (V)	Half-Reaction	n	$E°$ (V)
$IO_3^-(aq) + 6H^+(aq) + 6e^- \rightarrow I^-(aq) + 3H_2O(\ell)$	6	1.085	$Ag(S_2O_3)_2^{3-}(aq) + e^- \rightarrow Ag(s) + 2S_2O_3^{2-}(aq)$	1	0.01
$Br_2(\ell) + 3e^- \rightarrow 3Br^-(aq)$	2	1.066	$NO_3^-(aq) + H_2O(\ell) + 2e^- \rightarrow NO_2^-(aq) + 2OH^-(aq)$	2	0.01
$HNO_2(\ell) + H^+(aq) + e^- \rightarrow NO(g) + H_2O(\ell)$	1	1.00	$2H^+(aq) + 2e^- \rightarrow H_2(g)$	2	0.000
$VO_2^+(aq) + 2H^+(aq) + e^- \rightarrow VO^{2+}(aq) + H_2O(\ell)$	1	1.00	$Pb^{2+}(aq) + 2e^- \rightarrow Pb(s)$	2	−0.126
$NO_3^-(aq) + 4H^+(aq) + 3e^- \rightarrow NO(g) + 2H_2O(\ell)$	3	0.96	$CrO_4^{2-}(aq) + 4H_2O(\ell) + 3e^- \rightarrow Cr(OH)_3(s) + 5OH^-(aq)$	3	−0.13
$2Hg^{2+}(aq) + 2e^- \rightarrow Hg_2^{2+}(aq)$	2	0.92	$Sn^{2+}(aq) + 2e^- \rightarrow Sn(s)$	2	−0.136
$ClO^-(aq) + H_2O(\ell) + 2e^- \rightarrow Cl^-(aq) + 2OH^-(aq)$	2	0.89	$AgI(s) + e^- \rightarrow Ag(s) + I^-(aq)$	1	−0.1522
$HO_2^-(aq) + H_2O(\ell) + 2e^- \rightarrow 3OH^-(aq)$	2	0.88	$CuI(s) + e^- \rightarrow Cu(s) + I^-(aq)$	1	−0.185
$Hg^{2+}(aq) + 2e^- \rightarrow Hg(\ell)$	2	0.851	$N_2(g) + 5H^+(aq) + 4e^- \rightarrow N_2H_5^+(aq)$	4	−0.23
$Ag^+(aq) + e^- \rightarrow Ag(s)$	1	0.7996	$Ni^{2+}(aq) + 2e^- \rightarrow Ni(s)$	2	−0.257
$Hg_2^{2+}(aq) + 2e^- \rightarrow 2Hg(\ell)$	2	0.7973	$PbSO_4(s) + H^+(aq) + 2e^- \rightarrow Pb(s) + HSO_4^-(aq)$	2	−0.356
$Fe^{3+}(aq) + e^- \rightarrow Fe^{2+}(aq)$	1	0.770	$Co^{2+}(aq) + 2e^- \rightarrow Co(s)$	2	−0.277
$PtCl_4^{2-}(aq) + 2e^- \rightarrow Pt(s) + 4Cl^-(aq)$	2	0.73	$Ag(CN)_2^-(aq) + e^- \rightarrow Ag(s) + 2CN^-(aq)$	1	−0.31
$O_2(g) + 2H^+(aq) + 2e^- \rightarrow H_2O_2(\ell)$	2	0.68	$Cd^{2+}(aq) + 2e^- \rightarrow Cd(s)$	2	−0.403
$MnO_4^-(aq) + 2H_2O(\ell) + 3e^- \rightarrow MnO_2(s) + 4OH^-(aq)$	3	0.59	$Cd(OH)_2(s) + 2e^- \rightarrow Cd(s) + 2OH^-(aq)$	2	−0.403
$H_3AsO_4(s) + 2H^+(aq) + 2e^- \rightarrow H_3AsO_3(aq) + H_2O(\ell)$	2	0.559	$Cr^{3+}(aq) + e^- \rightarrow Cr^{2+}(aq)$	1	−0.41
$I_2(s) + 2e^- \rightarrow 2I^-(aq)$	2	0.5355	$Fe^{2+}(aq) + 2e^- \rightarrow Fe(s)$	2	−0.447
$Cu^+(aq) + e^- \rightarrow Cu(s)$	1	0.521	$2CO_2(g) + 2H^+(aq) + 2e^- \rightarrow H_2C_2O_4(s)$	2	−0.49
$H_2SO_3(\ell) + 4H^+(aq) + 4e^- \rightarrow S(s) + 3H_2O(\ell)$	4	0.449	$Ni(OH)_2(s) + 2e^- \rightarrow Ni(s) + 2OH^-(aq)$	2	−0.72
$Ag_2CrO_4(s) + 2e^- \rightarrow 2Ag(s) + CrO_4^{2-}(aq)$	2	0.4470	$Cr^{3+}(aq) + 3e^- \rightarrow Cr(s)$	3	−0.74
$O_2(g) + 2H_2O(\ell) + 4e^- \rightarrow 4OH^-(aq)$	4	0.401	$Zn^{2+}(aq) + 2e^- \rightarrow Zn(s)$	2	−0.7618
$Fe(CN)_6^{3-}(aq) + e^- \rightarrow Fe(CN)_6^{4-}(aq)$	1	0.36	$2H_2O(\ell) + 2e^- \rightarrow H_2(g) + 2OH^-(aq)$	2	−0.8277
$PbSO_4(s) + H^+(aq) + 2e^- \rightarrow Pb(s) + HSO_4^-(aq)$	2	0.356	$SO_4^{2-}(aq) + H_2O(\ell) + 2e^- \rightarrow SO_3^{2-}(aq) + 2OH^-(aq)$	2	−0.92
$Ag_2O(s) + H_2O(\ell) + 2e^- \rightarrow 2Ag(s) + 2OH^-(aq)$	2	0.342	$N_2(g) + 4H_2O(\ell) + 4e^- \rightarrow 4OH^-(aq) + N_2H_4(\ell)$	4	−1.16
$Cu^{2+}(aq) + 2e^- \rightarrow Cu(s)$	2	0.3419	$Mn^{2+}(aq) + 2e^- \rightarrow Mn(s)$	2	−1.185
$BiO^+(aq) + 2H^+(aq) + 3e^- \rightarrow Bi(s) + H_2O(\ell)$	3	0.32	$Zn(OH)_2(s) + 2e^- \rightarrow Zn(s) + 2OH^-(aq)$	2	−1.249
$AgCl(s) + e^- \rightarrow Ag(s) + Cl^-(aq)$	1	0.2223	$ZnO(s) + H_2O(\ell) + 2e^- \rightarrow Zn(s) + 2OH^-(aq)$	2	−1.25
$HSO_4^-(aq) + 3H^+(aq) + 2e^- \rightarrow H_2SO_3(\ell) + H_2O(\ell)$	2	0.17	$Al^{3+}(aq) + 3e^- \rightarrow Al(s)$	3	−1.662
$Sn^{4+}(aq) + 2e^- \rightarrow Sn^{2+}(aq)$	2	0.154	$Mg^{2+}(aq) + 2e^- \rightarrow Mg(s)$	2	−2.37
$Cu^{2+}(aq) + e^- \rightarrow Cu^+(aq)$	1	0.153	$Na^+(aq) + e^- \rightarrow Na(s)$	1	−2.71
$2MnO_2(s) + H_2O(\ell) + 2e^- \rightarrow Mn_2O_3(s) + 2OH^-(aq)$	2	0.15	$Ca^{2+}(aq) + 2e^- \rightarrow Ca(s)$	2	−2.868
$S(s) + 2H^+(aq) + 2e^- \rightarrow H_2S(g)$	2	0.141	$Ba^{2+}(aq) + 2e^- \rightarrow Ba(s)$	2	−2.912
$HgO(s) + H_2O(\ell) + 2e^- \rightarrow Hg(\ell) + 2OH^-(aq)$	2	0.0977	$K^+(aq) + e^- \rightarrow K(s)$	1	−2.95
$AgBr(s) + e^- \rightarrow Ag(s) + Br^-(aq)$	1	0.095	$Li^+(aq) + e^- \rightarrow Li(s)$	1	−3.05

Naming Organic Compounds

While organic chemistry was becoming established as a discipline within chemistry, many compounds were given trivial names that are still commonly used and recognized. We refer to many of these compounds by their nonsystematic names throughout this book, and their names and structures are listed in Table A7.1.

TABLE A7.1	Organic Compounds and Their Commonly Used Nonsystematic Names	
Name	**Formula**	**Structure**
ethylene	C_2H_4	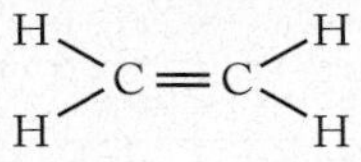
acetylene	C_2H_2	$HC\equiv CH$
benzene	C_6H_6	
toluene	$C_6H_5CH_3$	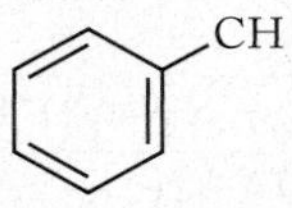
ethyl alcohol	CH_3CH_2OH	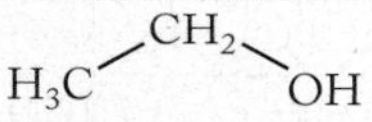
acetone	CH_3COCH_3	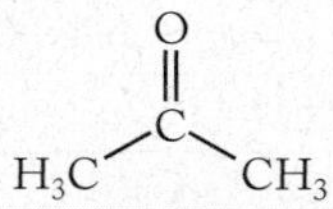
acetic acid	CH_3COOH	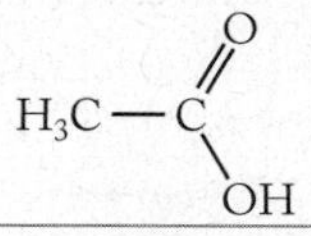
formaldehyde	CH_2O	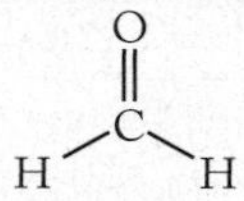

The International Union of Pure and Applied Chemistry (IUPAC) has proposed a set of rules for the systematic naming of organic compounds. The basic principles for naming alkanes, alkenes, and alkynes are presented in Chapter 13. These rules are summarized here and extended to include compounds containing other functional groups. When naming compounds or drawing structures based on names, we need to keep in mind that the IUPAC system of nomenclature is based on two fundamental ideas: (1) the name of a compound must indicate how the carbon atoms in the skeleton are bonded together, and (2) the name must identify the location of any functional groups in the molecule.

Alkanes

Table A7.2 contains the prefixes used for carbon chains ranging in size from C_1 to C_{20} and gives the names for compounds consisting of unbranched chains. The name of a compound consists of a prefix identifying the number of carbons in the chain and a suffix defining the type of hydrocarbon. The suffix *-ane* indicates that the compounds are alkanes and that all carbon–carbon bonds are single bonds.

TABLE A7.2		Prefixes for Naming Carbon Chains			
Prefix	**Example**	**Name**	**Prefix**	**Example**	**Name**
meth	CH_4	methane	undec	$C_{11}H_{24}$	undecane
eth	C_2H_6	ethane	dodec	$C_{12}H_{26}$	dodecane
pro	C_3H_8	propane	tridec	$C_{13}H_{28}$	tridecane
but	C_4H_{10}	butane	tetradec	$C_{14}H_{30}$	tetradecane
pent	C_5H_{12}	pentane	pentadec	$C_{15}H_{32}$	pentadecane
hex	C_6H_{14}	hexane	hexadec	$C_{16}H_{34}$	hexadecane
hept	C_7H_{16}	heptane	heptadec	$C_{17}H_{36}$	heptadecane
oct	C_8H_{18}	octane	octadec	$C_{18}H_{38}$	octadecane
non	C_9H_{20}	nonane	nonadec	$C_{19}H_{40}$	nonadecane
dec	$C_{10}H_{22}$	decane	eicos	$C_{20}H_{42}$	eicosane

Branched-Chain Alkanes

The alkane drawn here is used to illustrate each step in the naming rules:

$$CH_3CH_2CHCH_2CHCHCH_2CH_2CH_3$$
with CH_3 on top, CH_3 and CH_2CH_3 below

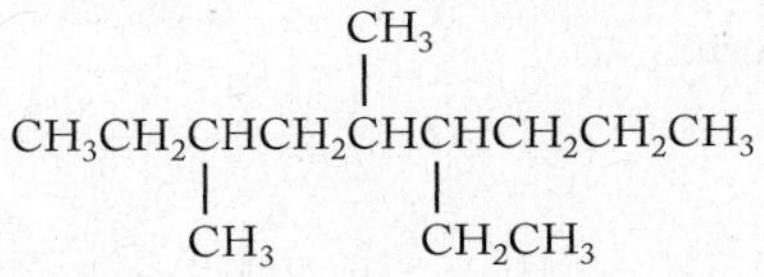

1. **Identify and name the longest continuous carbon chain.**

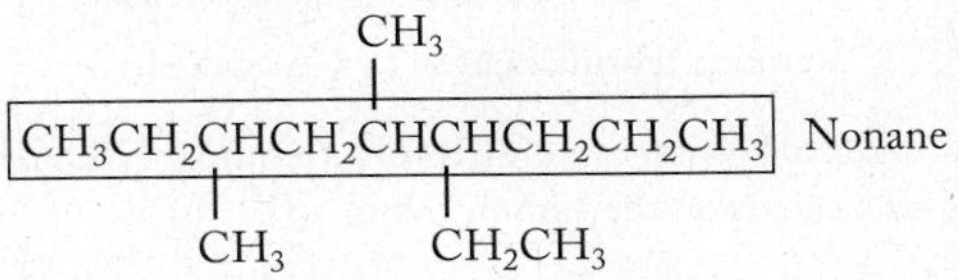

$$\boxed{CH_3CH_2CHCH_2CHCHCH_2CH_2CH_3} \quad \text{Nonane}$$
with CH_3 on top, CH_3 and CH_2CH_3 below

2. **Identify the groups attached to this chain and name them.** Names of substituent groups consist of the prefix from Table A7.2 that identifies the length of the group and the suffix *-yl* that identifies it as an alkyl group.

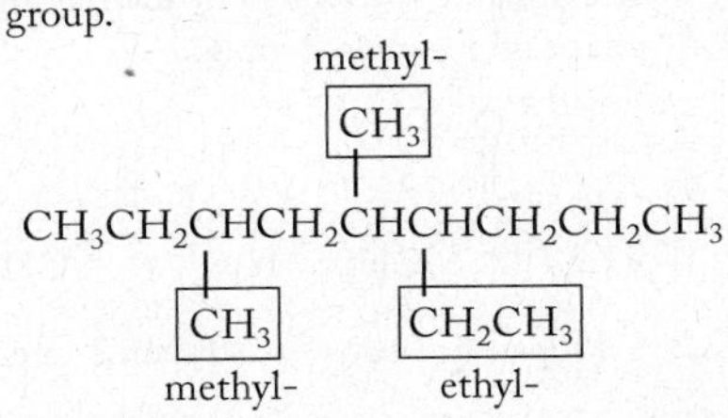

methyl-

$$CH_3CH_2CHCH_2CHCHCH_2CH_2CH_3$$
with CH_3 on top labeled methyl-, and CH_3 (methyl-) and CH_2CH_3 (ethyl-) below

3. **Number the carbon atoms in the longest chain,** starting at the end nearest a substituent group. Doing this identifies the points of attachment of the alkyl groups with the lowest possible numbers.

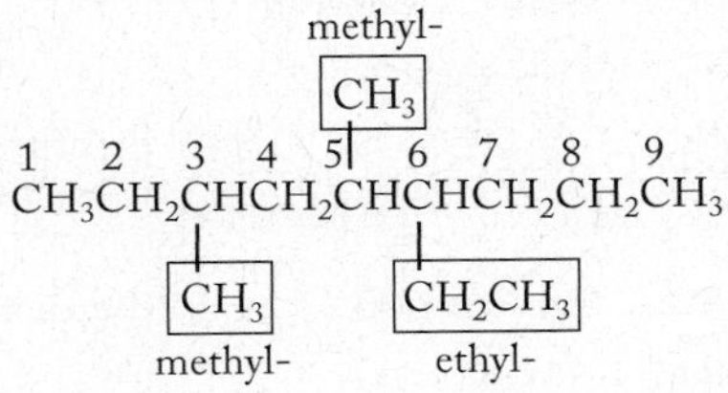

methyl-

$$\overset{1\ \ 2\ \ 3\ \ 4\ \ 5\ \ 6\ \ 7\ \ 8\ \ 9}{CH_3CH_2CHCH_2CHCHCH_2CH_2CH_3}$$
with CH_3 on top labeled methyl-, and CH_3 (methyl-) and CH_2CH_3 (ethyl-) below

4. **Designate the location and identity of each substituent group with a number,** followed by a hyphen, and its name.

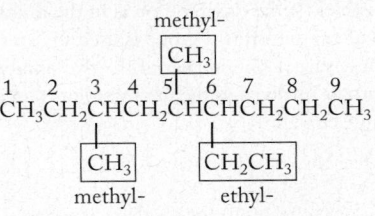

3-methyl-, 5-methyl-, 6-ethyl-

5. **Put together the complete name by listing the substituent groups in alphabetical order.** If more than one of a given type of substituent group is present, prefixes *di-*, *tri-*, *tetra-*, and so forth are appended to the names, but these numerical prefixes are not considered when determining the alphabetical order. The name of the last substituent group is written together with the name identifying the longest carbon chain.

$$CH_3CH_2CHCH_2CHCHCH_2CH_2CH_3$$
with CH_3 on top, CH_3 and CH_2CH_3 below

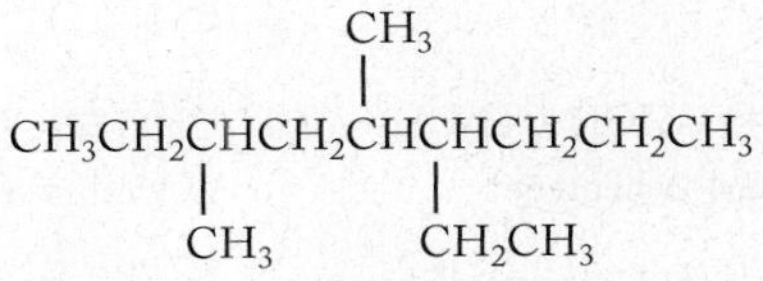

6-Ethyl-3,5-dimethylnonane

Cycloalkanes

The simplest examples of this class of compounds consist of one unsubstituted ring of carbon atoms. The IUPAC names of these compounds consist of the prefix *cyclo-* followed by the parent name from Table A7.2 to indicate the number of carbon atoms in the ring. As an illustration, the names, formulas, and line structures of the first three cycloalkanes in the homologous series are

△ C_3H_6 ☐ C_4H_8 ⬠ C_5H_{10}

Cyclopropane   Cyclobutane   Cyclopentane

Alkenes and Alkynes

Alkenes have carbon–carbon double bonds and alkynes have carbon–carbon triple bonds as functional groups. The names of these types of compounds consist of (1) a parent name that identifies the longest carbon chain that includes the double or triple bond, (2) a suffix that identifies the class of compound, and (3) names of any substituent groups attached to the longest carbon chain. The suffix *-ene* identifies an alkene; *-yne* identifies an alkyne.

The alkene and alkyne drawn here are used to illustrate each step in the naming rules:

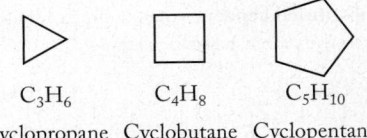

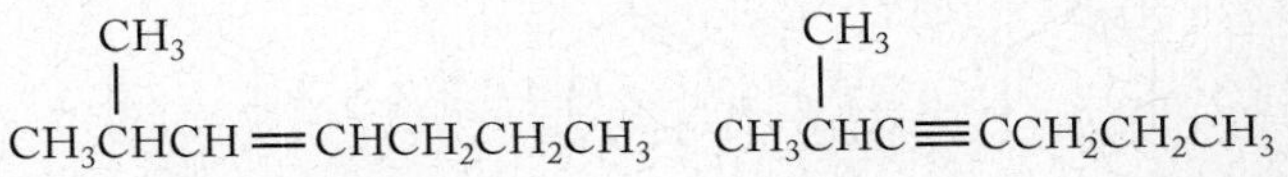

1. **To determine the parent name,** identify the longest chain that contains the unsaturation. Name the parent compound with the prefix that defines the number of carbons in that chain and the suffix that identifies the class of compound.

CH_3 CH_3

$\boxed{CH_3CHCH=CHCH_2CH_2CH_3}$ $\boxed{CH_3CHC\equiv CCH_2CH_2CH_3}$

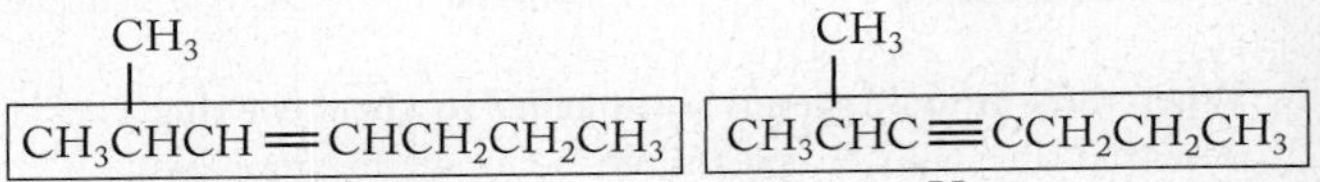

Heptene   Heptyne

2. **Number the parent chain from the end nearest the unsaturation so that the first carbon in the double or triple bond has the lowest number possible.** (If the unsaturation is in the middle of a chain, the location of any substituent group is used to determine where the numbering starts.) The smaller of the two numbers identifying the carbon atoms involved in the unsaturation is used as the locator of the multiple bond.

$$
\begin{array}{c}
\underset{2}{\overset{CH_3}{|}} \\
\underset{CH_3CHCH}{1\ 2\ 3} = \underset{CHCH_2CH_2CH_3}{4\ 5\ 6\ 7}
\end{array}
\qquad
\begin{array}{c}
\underset{2}{\overset{CH_3}{|}} \\
\underset{CH_3CHC}{1\ 2\ 3} \equiv \underset{CCH_2CH_2CH_3}{4\ 5\ 6\ 7}
\end{array}
$$
3-Heptene   3-Heptyne

3. **Stereoisomers of alkenes are named by writing *cis-* or *trans-* before the number identifying the location of the double bond.** Section 13.3 in the text addresses naming stereoisomers.

4. **The rules for naming substituted alkanes are followed to name and locate any other groups on the chain.**

$$
\begin{array}{c}
\overset{CH_3}{|} \\
CH_3CHCH = CHCH_2CH_2CH_3
\end{array}
\qquad
\begin{array}{c}
\overset{CH_3}{|} \\
CH_3CHC \equiv CCH_2CH_2CH_3
\end{array}
$$
2-Methyl-3-heptene   2-Methyl-3-heptyne

Halogens attached to an alkane, alkene, or alkyne are named as fluoro- (F-), chloro- (Cl-), bromo- (Br-), or iodo- (I-) and are located by using the same numbering system described for alkyl groups.

Benzene Derivatives

Naming compounds containing substituted benzene rings is less systematic than naming hydrocarbons. Many compounds have common names that are incorporated into accepted names, but for simple substituted benzene rings, the following rules may be applied.

1. **For monosubstituted benzene rings,** a prefix identifying the group is appended to the parent name benzene:

Chlorobenzene   Nitrobenzene   Ethylbenzene

2. **For disubstituted benzene rings,** three isomers are possible. The relative position of the substituent groups is indicated by numbers in IUPAC nomenclature, but the set of prefixes shown are very commonly used as well:

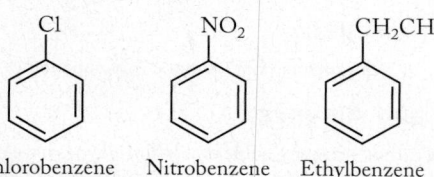

IUPAC:
1,2-Dichlorobenzene   1,3-Dichlorobenzene   1,4-Dichlorobenzene
Common:
ortho-Dichlorobenzene   *meta*-Dichlorobenzene   *para*-Dichlorobenzene
o-Dichlorobenzene   *m*-Dichlorobenzene   *p*-Dichlorobenzene

3. **When three or more groups are attached to a benzene ring,** the lowest possible numbers are assigned to locate the groups with respect to each other.

1,2,3-Trichlorobenzene   1,2,4-Trichlorobenzene   1,2,3,5-Tetrachlorobenzene
(NOTE: Not 1,3,4-trichlorobenzene; and not 1,3,4,5-tetrachlorobenzene.)

Hydrocarbons Containing Other Functional Groups

The same basic principles developed for naming alkanes apply to naming hydrocarbons with functional groups other than alkyl groups (Table 13.1 in the text). The name must identify the carbon skeleton, locate the functional group, and contain a suffix that defines the class of compound. The following examples give the suffixes for some common functional groups; when suffixes are used, they replace the final −*e* in the name of the parent alkane. Other functional groups may be identified by including the name of the class of compounds in the name of the molecule.

Alcohols: Suffix -*OL*

$$
CH_3CH_2CH_2OH \qquad
\begin{array}{c}
CH_3CHCH_3 \\
| \\
OH
\end{array}
\qquad
\begin{array}{c}
CH_3CH_2CH_2CHCH_3 \\
| \\
OH
\end{array}
$$
1-Propanol   2-Propanol   2-Pentanol

Aldehydes: Suffix -*AL*

IUPAC: Methanal   Ethanal
Common: Formaldehyde   Acetaldehyde

Because the aldehyde group can only be on a terminal carbon, no number is necessary to locate it on the carbon chain.

Ketones: Suffix -*ONE*

The location of the carbonyl is given by a number, and the chain is numbered so that the carbonyl carbon has the lowest possible value. Many ketones also have common names generated by identifying the hydrocarbon groups on both sides of the carbonyl group.

IUPAC: Propan-2-one   Butan-2-one
Common: Acetone   Methyl ethyl ketone

Carboxylic Acids: Suffix -*OIC ACID*

The carboxylic acid group is by definition carbon 1, so no number identifying its location is included in the name.

IUPAC: Ethanoic acid   *trans*-2-Butenoic acid
Common: Acetic acid

Salts of Carboxylic Acids

Salts are named with the cation first, followed by the anion name of the acid from which –*ic acid* is dropped and the suffix –*ate* is added. The sodium salt of acetic acid is sodium acetate.

Acetic acid   Acetate ion   Sodium acetate

Esters

Esters are viewed as derivatives of carboxylic acids. They are named in a manner analogous to that of salts. The alkyl group comes first followed by the name of the carboxylate anion.

Alkyl Carboxylate   Ethyl acetate

Amides

Amides are also derivatives of carboxylic acids. They are named by replacing –*ic acid* (of the common names) or –*oic acid* of the IUPAC names with –*amide*.

Parent acid -amide   Acetamide

Ethers

Ethers are frequently named by naming the two groups attached to the oxygen and following those names by the word *ether*.

$$CH_3OCH_3 \qquad CH_3CH_2OCH_2CH_3$$
Dimethyl ether   Diethyl ether

Amines

Aliphatic amines are usually named by listing the group or groups attached to the nitrogen and then appending –*amine* as a suffix. They may also be named by prefixing *amino-* to the name of the parent chain.

Methylamine   Ethylmethylamine   2-Aminoethanol

This brief summary will enable you to understand the names of organic compounds used in this book. IUPAC rules are much more extensive than this and can be applied to all varieties of carbon compounds including those with multiple functional groups. It is important to recognize that the rules of systematic nomenclature do not necessarily lead to a unique name for each compound, but they do always lead to an unambiguous one. Furthermore, common names are still used frequently in organic chemistry because the systematic alternatives do not improve communication. Remember that the main purpose of chemical nomenclature is to identify a chemical species by means of written or spoken words. Anyone who reads or hears the name should be able to deduce the structure and thereby the identity of the compound.

A

absolute entropy   The entropy change of a substance taken from $S = 0$ (at $T = 0$ K) to some other temperature. Absolute entropies are determined from the temperature dependence of the molar heat capacity.

absolute temperature   Temperature expressed in kelvins on the absolute temperature scale, on which 0 K is the lowest possible temperature.

absolute zero (0 K)   The zero point on the Kelvin temperature scale; theoretically the lowest temperature possible.

accuracy   The agreement between an experimental value and the true value.

achiral   Not chiral; describes compounds that can be superimposed on their mirror images.

acid (Brønsted–Lowry acid)   A proton donor.

activated complex   A species formed in a chemical reaction when molecules have enough energy to react with each other.

activation energy (E_a)   The minimum energy molecules need to react when they collide.

active site   The location on an enzyme where a reactive substance binds.

actual yield   The amount of product obtained from a chemical reaction, which is often less than the theoretical yield.

addition polymer   Macromolecule prepared by adding monomers to a growing polymer chain.

addition reaction   A reaction in which two molecules couple together and form one product.

alcohol   Organic compound containing the −OH functional group.

aldehyde   Organic compound containing a carbonyl group bonded to one R group and one hydrogen; its general formula is RCHO.

alkali metal   An element in group 1 of the periodic table.

alkaline earth metal   An element in group 2 of the periodic table.

alkane   Hydrocarbon in which all the bonds are single bonds with the general formula C_nH_{2n+2}.

alkene   Hydrocarbon containing one or more carbon–carbon double bonds.

alkyne   Hydrocarbon containing one or more carbon–carbon triple bonds.

allotropes   Different molecular forms of the same element, such as oxygen (O_2) and ozone (O_3).

alloy   A blend of a host metal and one or more other elements, which may or may not be metals, that are added to change the properties of the host metal.

alpha (α) decay   A nuclear reaction in which an unstable nuclide spontaneously emits an alpha particle.

α helix   A coil in a protein chain's secondary structure.

amide   Organic compound in which the same carbon atoms are single bonded to nitrogen atoms and double bonded to oxygen atoms.

amine   Organic compound that contains a group with the general formula RNH_2, R_2NH, or R_3N, where R is any organic subgroup.

amino acid   Molecule that contains at least one amine group and one carboxylic acid group; in an *α-amino acid*, the two groups are attached to the same (α) carbon atom.

Amontons's law   As the absolute temperature of a fixed amount of gas increases, the pressure increases as long as the volume and quantity of gas remain constant.

amphiprotic   A substance that can behave as either a proton acceptor or a proton donor.

amplitude   The height of the crest or the depth of the trough of a wave with respect to the center line of the wave.

angular momentum quantum number (ℓ)   An integer having any value from 0 to $n - 1$ that defines the shape of an orbital.

anion   Negatively charged particle created when an atom or molecule gains one or more electrons.

anode   An electrode at which an oxidation half-reaction (loss of electrons) takes place.

antibonding orbital   Term in MO theory describing regions of electron density in a molecule that destabilize the molecule because they do not increase the electron density between nuclear centers.

antimatter   Particles that are the charge opposites of normal subatomic particles.

aromatic compound   A cyclic, planar compound with delocalized π (pi) electrons above and below the plane of the molecule.

Arrhenius equation   Relates the rate constant of a reaction to absolute temperature (T), the activation energy of the reaction (E_a), and the frequency factor (A).

atmospheric pressure (P_{atm})   The force exerted by the gases surrounding Earth on Earth's surface and on all surfaces of all objects.

atom The smallest particle of an element that retains the chemical characteristics of the element.

atomic absorption spectrum   (also called *dark-line spectrum*) A characteristic series of dark lines produced when free, gaseous atoms are illuminated by an external source of radiation.

atomic emission spectrum   (also called *bright-line spectrum*) A characteristic series of bright lines produced by excited-state atoms.

atomic mass unit (amu)   Unit used to express the relative masses of atoms and subatomic particles that is exactly 1/12 the mass of one atom of carbon with 6 protons and 6 neutrons in its nucleus.

atomic number (Z)   The number of protons in the nucleus of an atom.

atomic radius   (also called *covalent radius*) Half the distance between identical nuclear centers in a molecule.

aufbau principle   The method of building electron configurations of atoms by adding one electron at a time as atomic number increases across the rows of the periodic table.

autoionization   The process that produces equal and very small concentrations of H_3O^+ and OH^- ions in pure water.

average atomic mass   A weighted average of masses of all isotopes of an element, calculated by multiplying the natural abundance of each isotope by its mass in atomic mass units and then summing these products.

Avogadro's law   The volume of a gas at a given temperature and pressure is proportional to the quantity of the gas.

Avogadro's number (N_A)   The number of carbon atoms in exactly 12 grams of the carbon-12 isotope; $N_A = 6.022 \times 10^{23}$. It is the number of particles in one mole.

B

band gap (E_g)   The energy gap between the valence and conduction bands.

band theory   An extension of molecular orbital theory that describes bonding in solids.

barometer   An instrument that measures atmospheric pressure.

base (Brønsted–Lowry base)   A proton acceptor.

becquerel (Bq)   The SI unit of radioactivity. One becquerel equals one decay event per second.

belt of stability   The region on the graph of number-of-neutrons versus number-of-protons that includes all stable nuclei.

beta (β) decay   The process by which a neutron decays into a proton and a β particle.

beta (β) particle   A radioactive emission that is a high-energy electron.

β-pleated sheet   A puckered two-dimensional array of protein strands held together by hydrogen bonds.

bimolecular step   A step in a reaction mechanism involving a collision between two molecules.

binding energy (BE)   The energy released when nucleons combine to form a nucleus.

biocatalysis   The strategy of using enzymes to catalyze reactions on a large scale; it is becoming especially important in processes that involve chiral materials.

biomass   The sum total of the mass of organic matter in any given ecological system.

biomolecule   An organic molecule present naturally in a living system.

body-centered cubic (bcc) unit cell   A cell with atoms at the eight corners of a cube and at the center of the cell.

bomb calorimeter   A constant-volume device used to measure the energy released during a combustion reaction.

bond angle   The angle (in degrees) defined by lines joining the centers of two atoms to a third atom to which they are chemically bonded.

bond dipole   Separation of electrical charge created when atoms with different electronegativities form a covalent bond.

bond energy   The energy needed to break 1 mole of a particular covalent bond in a molecule or polyatomic ion in the gas phase.

bond length   The distance between the nuclear centers of two atoms joined together in a bond.

bond order   The number of bonds between atoms: 1 for a single bond, 2 for a double bond, and 3 for a triple bond.

bond polarity   A measure of the extent to which bonding electrons are unequally shared due to differences in electronegativity of the bonded atoms.

bonding capacity   The number of covalent bonds an atom forms to have an octet of electrons in its valence shell.

bonding orbital   Term in MO theory describing regions of increased electron density between nuclear centers that serve to hold atoms together in molecules.

bonding pair   A pair of electrons shared between two atoms.

Born–Haber cycle   A series of steps with corresponding enthalpy changes that describes the formation of an ionic solid from its constituent elements.

Boyle's law   The volume of a given amount of gas at constant temperature is inversely proportional to its pressure.

Bragg equation   Relates the angle of diffraction (2θ) of X-rays to the spacing (d) between the layers of ions or atoms in a crystal: $n\lambda = 2d \sin \theta$.

branched-chain hydrocarbon   An organic molecule in which the chain of carbon atoms is not linear.

breeder reactor   A nuclear reactor in which fissionable material is produced during normal reactor operation.

bright-line spectrum   (also called *atomic emission spectrum*) A characteristic series of bright lines produced by excited-state atoms.

Brønsted–Lowry acid   A proton donor.

Brønsted–Lowry base   A proton acceptor.

Brønsted–Lowry model   Defines acids as H^+ ion donors and bases as H^+ ion acceptors.

buffer capacity   The quantity of acid or base that a pH buffer can neutralize while maintaining its pH within a desired range.

C

calorie (cal)   The amount of energy necessary to raise the temperature of 1 g of water by 1°C.

calorimeter   A device used to measure the absorption or release of energy by a physical change or chemical process.

calorimeter's constant ($C_{calorimeter}$)   The heat capacity of a calorimeter.

calorimetry   The measurement of the quantity of heat transferred during a physical change or chemical process.

capillary action   The rise of a liquid in a narrow tube as a result of adhesive forces between the liquid and the tube and cohesive forces within the liquid.

carbohydrate   An organic molecule with the generic formula $C_x(H_2O)_y$.

carbonyl group   A carbon atom with a double bond to an oxygen atom.

carboxylic acid   An organic compound containing the –COOH functional group.

catalyst   A substance added to a reaction that increases the rate of the reaction but is not consumed in the process.

cathode   An electrode at which a reduction half-reaction (gain of electrons) takes place.

cathode rays   Streams of electrons emitted by the cathode in a partially evacuated tube.

cation   Positively charged particle created when an atom or molecule loses one or more electrons.

cell diagram   Symbols that show how the components of an electro-chemical cell are connected.

cell potential (E_{cell})   The electromotive force expressed in volts (V) with which an electrochemical cell can push electrons through an external circuit connected to its terminals.

ceramic   A solid inorganic compound or mixture that has been transformed into a harder, more heat-resistant material by heating.

chain reaction   A self-sustaining series of fission reactions in which the neutrons released when nuclei split apart initiate additional fission events and sustain the reaction.

Charles's law   The volume of a fixed quantity of gas at constant pressure is directly proportional to its absolute temperature.

chelate effect   The greater affinity of metal ions for polydentate ligands than for monodentate ligands.

chelation   The interaction of a metal with a polydentate ligand (chelating agent); pairs of electrons on one molecule of the ligand occupy two or more coordination sites on the central metal.

chemical bond   The energy that holds two atoms in a molecule together.

chemical equation   Notation in which chemical formulas express the identities and their coefficients express the quantities of substances involved in a chemical reaction; a description of the identities and quantities of reactants (substances consumed during a chemical reaction) and products (substances formed).

chemical equilibrium   A dynamic process in which the concentrations of reactants and products remain constant over time and the rate of a reaction in the forward direction matches its rate in the reverse direction.

chemical formula   A notation for representing elements and compounds; consists of the symbols of the constituent elements and subscripts identifying the number of atoms of each element in one molecule.

chemical kinetics   The study of the rates of change of concentrations of substances involved in chemical reactions.

chemical property   A property of a substance that can be observed only by reacting it to form another substance.

chemical reaction   The transformation of one or more substances into different substances.

chemistry   The study of the composition, structure, and properties of matter and of the energy consumed or given off when matter undergoes a change.

chiral   Compounds having nonsuperimposable mirror images.

chirality   Property of a molecule that is not superimposable on its mirror image.

cis isomer   (also called *Z isomer*) Molecule with two like groups (such as two R groups or two hydrogen atoms) on the same side of the molecule.

Clausius–Clapeyron equation   Relates the vapor pressures of a substance at different temperatures to its heat of vaporization.

closed system   A system that exchanges energy but not matter with the surroundings.

codon   A three-nucleotide sequence that codes for a specific amino acid.

coenzyme   Organic molecule that, like an enzyme, accelerates the rate of biochemical reactions.

colligative properties   Characteristics of solutions that depend on the concentration and not the identity of particles dissolved in the solvent.

combination reaction   A reaction in which two (or more) substances combine to form one product.

combined gas law   (also called *general gas equation*) Based on the ideal gas law and used when one or more of the four gas variables are held constant while the remaining variables change.

combustion analysis   A laboratory procedure for determining the composition of a substance by burning it completely in oxygen to produce known compounds whose masses are used to determine the composition of the original material.

combustion reaction   A reaction between oxygen and another element or compound that produces heat.

common-ion effect   The shift in the position of an equilibrium caused by the addition of an ion taking part in the reaction.

complex ion   An ionic species consisting of a metal ion bonded to one or more Lewis bases.

compound   A pure substance that is composed of two or more elements linked together in fixed proportions and that can be broken down to those elements by some chemical process.

condensation polymer   Macromolecule formed by the reaction of monomers yielding a polymer and water or other small molecule as a by-product of the reaction.

condensation reaction   Two molecules combining to form a larger molecule and a small molecule (typically water).

conduction band   An unoccupied band higher in energy than a valence band in which electrons are free to migrate.

conjugate acid   Formed when a Brønsted–Lowry base accepts a proton.

conjugate acid–base pair   A Brønsted–Lowry acid and base differing from each other only by the presence or absence of a H^+ion: acid $\rightleftharpoons$ conjugate base + H^+.

conjugate base   Formed when a Brønsted–Lowry acid donates a H^+ ion.

conversion factor   A fraction in which the numerator is equivalent to the denominator but is expressed in different units, making the value of the fraction one.

coordinate bond   Formed when one anion or molecule donates a pair of electrons to another ion or molecule to form a covalent bond.

coordination compound   Made up of at least one complex ion.

coordination number   Identifies the number of electron pairs surrounding a metal ion in a complex.

copolymer   A macromolecule formed from the chemical combination of two different monomers.

core electrons   Electrons in the filled, inner shells in an atom or ion that are not involved in chemical reactions.

counter ions   Provide electrical balance of charges of complex ions in coordination compounds.

covalent bond   A bond between two atoms created by sharing one or more pairs of electrons.

covalent network solid   A solid consisting of atoms held together by extended arrays of covalent bonds.

covalent radius   (also called *atomic radius*) Half the distance between identical nuclear centers in a molecule.

critical mass   The minimum quantity of fissionable material needed to sustain a chain reaction.

critical point   A specific temperature and pressure at which the liquid and gas phases of a substance have the same density and are indistinguishable from each other.

critical temperature (T_c) The temperature below which a material becomes a superconductor.

crude oil A combustible liquid mixture of hydrocarbons and other organic molecules formed under Earth's surface.

crystal field splitting The separation of a set of *d* orbitals into subsets with different energies as result of interactions between electrons in those orbitals and lone pairs of electrons in ligands.

crystal field splitting energy (Δ) The difference in energy between subsets of *d* orbitals split by interactions in a crystal field.

crystal lattice A three-dimensional array of particles (atoms, ions, or molecules) in a crystalline solid.

crystal structure An ordered arrangement in three-dimensional space of the particles (atoms, ions, or molecules) that make up a crystalline solid.

crystalline solid A solid made of an ordered array of atoms, ions, or molecules.

cubic closest-packed (ccp) A crystal structure composed of face-centered cubic unit cells and layers of particles having an *abcabc* . . . stacking pattern.

curie (Ci) Non-SI unit of radioactivity; 1 Ci = 3.70×10^{10} decay events per second.

cycloalkane Ring-containing alkane with the general formula C_nH_{2n}.

D

dalton (Da) A unit of mass identical to 1 atomic mass unit.

Dalton's law of partial pressures The total pressure of any mixture of gases equals the sum of the partial pressures of all the gases in the mixture.

dark-line spectrum (also called *atomic absorption spectrum*) A characteristic series of dark lines produced when free, gaseous atoms are illuminated by an external source of radiation.

degenerate Describes orbitals of the same energy.

degree of ionization The ratio of the quantity of a substance that is ionized to the concentration of the substance before ionization; when expressed as a percentage, it is called *percent ionization*.

delocalization (*adjective*: **delocalized**) When electrons in alternating single and double bonds are spread over the three or more atoms in a molecule.

density (*d*) The ratio of the mass (*m*) of an object to its volume (*V*).

deposition Transformation of a vapor (gas) directly into a solid.

diamagnetic Describes a substance with no unpaired electrons that is weakly repelled by a magnetic field.

diffraction Bending of electromagnetic radiation as it passes around an edge of an object or through a narrow opening.

diffusion The spread of one substance (usually a gas or liquid) through another.

dilution The process of lowering the concentration of a solution by adding more solvent.

dipole moment (μ) A measure of the degree to which a molecule aligns itself in an applied electric field; a quantitative expression of the polarity of a molecule.

dipole–dipole interaction An attractive force between polar molecules.

dispersion force (also called *London force*) An intermolecular force between nonpolar molecules caused by the presence of temporary dipoles in the molecules.

distillation A separation technique in which the more *volatile* (more easily vaporized) components of a mixture are vaporized and then condensed, thereby separating them from the less volatile components.

double bond Results when two atoms share two pairs of electrons.

E

***E* isomer** (also called *trans isomer*) Molecule with two like groups (such as two R groups or two hydrogen atoms) on opposite sides of the molecule.

effective nuclear charge (Z_{eff}) The attractive force toward the nucleus experienced by an electron in an atom; the positive charge on the nucleus reduced by the extent to which other electrons in the atom shield the electron from the nucleus.

effusion The process by which a gas escapes from its container through a tiny hole into a region of lower pressure.

electrochemical cell An apparatus that converts chemical energy into electrical work or electrical work into chemical energy.

electrochemistry The branch of chemistry that examines the transformations between chemical and electrical energy.

electrolysis A process in which electrical energy is used to drive a nonspontaneous chemical reaction.

electrolyte A substance that dissociates into ions when it dissolves, enhancing the conductivity of the solvent.

electrolytic cell A device in which an external source of electrical energy does work on a chemical system, turning reactant(s) into higher-energy product(s).

electromagnetic radiation Any form of radiant energy in the electromagnetic spectrum.

electromagnetic spectrum A continuous range of radiant energy that includes radio waves, infrared radiation, visible light, ultraviolet radiation, X-rays, and gamma rays.

electromotive force (emf) Also called voltage, the force pushing electrons through an electrical circuit.

electron A subatomic particle that has a negative charge and essentially zero mass.

electron affinity (EA) The energy change that occurs when 1 mole of electrons combines with 1 mole of atoms or ions in the gas phase.

electron capture A neutron-poor nucleus draws in one of its surrounding electrons, which transforms a proton in the nucleus into a neutron.

electron configuration The distribution of electrons among the orbitals of an atom or ion.

electron transition Movement of an electron between energy levels.

electronegativity A relative measure of the ability of an atom in a bond to attract electrons to itself.

electron-pair geometry The three-dimensional arrangement of bonding pairs and lone pairs of electrons about a central atom.

electrostatic potential energy (E_{el}) The energy a particle has because of its position relative to another particle; it is directly proportional to the product of the charges of the particles and inversely proportional to the distance between them.

element A pure substance that cannot be separated into simpler substances by any chemical process.

elementary step A molecular-level view of a single process taking place in a chemical reaction.

empirical formula A formula showing the smallest whole-number ratio of elements in a compound.

enantiomer One of a pair of optical isomers of a compound.

end point The point in a titration that is reached when just enough standard solution has been added to cause the indicator to change color.

endothermic process One in which energy flows from the surroundings into the system.

energy The capacity to transfer heat or do work.

energy profile   Graph showing the changes in potential energy for a reaction as a function of the progress of the reaction from reactants to products.

enthalpy (H)   The sum of the internal energy and the pressure–volume product of a system; $H = E + PV$.

enthalpy change (ΔH)   The energy absorbed by the reactants (endothermic reaction) or the energy given off by the products (exothermic reaction) for a reaction carried out at constant pressure.

enthalpy of hydration ($\Delta H_{hydration}$)   The heat change when gas-phase ions dissolve in a solvent.

enthalpy of reaction (ΔH_{rxn})   (also called *heat of reaction*) The energy absorbed or given off by a chemical reaction.

enthalpy of solution ($\Delta H_{solution}$)   The overall heat change when a solute is dissolved in a solvent.

entropy (S)   A measure of the distribution of energy in a system at a specific temperature.

enzyme   A protein that catalyzes a reaction.

equilibrium constant (K)   The value of the ratio of concentration (or partial pressure) terms in the equilibrium constant expression at a specific temperature.

equilibrium constant expression   the ratio of the equilibrium concentrations or partial pressures of *products to reactants, each term raised to a power equal to the coefficient of that substance* in the balanced chemical equation for the reaction.

equivalence point   The point in a titration where the number of moles of titrant added is stoichiometrically equal to the number of moles of the substance being analyzed.

essential amino acid   Any of the 8 amino acids that make up peptides and proteins but are not synthesized in the human body and must be obtained through the food we eat.

essential element   Element present in tissue, blood, or other body fluids that has physiological functions.

ester   Organic compound in which the –OH of a carboxylic acid group is replaced by –OR, where R can be any organic group.

ether   Organic compound with the general formula R—O—R, where R is any alkyl group or aromatic ring; the two R groups may be different.

excited state   Any energy state in an atom or ion above the ground state.

exothermic process   One in which energy flows from a system into its surroundings.

extensive property   A property that varies with the quantity of the substance present.

F

face-centered cubic (fcc) unit cell   An array of closest-packed particles that has eight of the particles at the corners of a cube and six of them at the centers of each face of the cube.

family   (also called *group*) All elements in the same column of the periodic table.

Faraday's constant (F)   The magnitude of electrical charge in 1 mole of electrons. Its value to three significant figures is 9.65×10^4 C/mol.

fat   Solid triglyceride containing primarily saturated fatty acids.

filtration   A process for separating particles suspended in a liquid or a gas from that liquid or gas by passing the mixture through a medium that retains the particles.

first law of thermodynamics   The energy gained or lost by a system must equal the energy lost or gained by the surroundings.

food value   The quantity of energy produced when a material consumed by an organism for sustenance is burned completely; it is typically reported in Calories (kilocalories) per gram of food.

formal charge (FC)   Value calculated for an atom in a molecule or polyatomic ion by determining the difference between the number of valence electrons in the free atom and the sum of lone-pair electrons plus half of the electrons in the atom's bonding pairs.

formation constant (K_f)   Equilibrium constant describing the formation of a metal complex from a free metal ion and its ligands.

formation reaction   A reaction in which 1 mole of a substance is formed from its component elements in their standard states.

formula mass   The mass in atomic mass units of one formula unit of an ionic compound.

formula unit   The smallest electrically neutral unit of an ionic compound.

fractional distillation   A method of separating a mixture of compounds on the basis of their different boiling points.

Fraunhofer lines   A set of dark lines in the otherwise continuous solar spectrum.

free energy (G)   A thermodynamic state function that provides a criterion for spontaneous change; an indication of the energy available to do useful work.

free radical   An odd-electron molecule with an unpaired electron in its Lewis structure.

free-energy change (ΔG)   The change in free energy of a process; $\Delta G < 0$ for spontaneous processes at constant temperature and pressure.

frequency (ν)   The number of crests of a wave that pass a stationary point of reference per second.

frequency factor (A)   The product of the frequency of molecular collisions and a factor that expresses the probability that the orientation of the molecules is appropriate for a reaction to occur.

fuel cell   A voltaic cell based on the oxidation of a continuously supplied fuel. The reaction is the equivalent of combustion, but chemical energy is converted into electrical energy.

fuel density   The amount of energy released during the complete combustion of 1 liter of a liquid fuel.

fuel value   The energy released during complete combustion of 1 g of a substance.

functional group   A structural subunit in organic molecules that imparts characteristic chemical and physical properties.

G

gas   A form of matter that has neither definite volume nor shape, and that expands to fill its containers; also known as *vapor*.

Geiger counter   A portable device for determining nuclear radiation levels by measuring how much the radiation ionizes the gas in a sealed detector.

general gas equation   (also called *combined gas law*) Based on the ideal gas law and used when one or more of the four gas variables are held constant while the remaining variables change.

glyceride   Lipid consisting of esters formed between fatty acids and the alcohol glycerol.

glycolysis   A series of reactions that converts glucose into pyruvate; a major anaerobic (no oxygen required) pathway for the metabolism of glucose in the cells of almost all living organisms.

glycosidic bond   A C—O—C bond between sugar molecules.

Graham's law of effusion   The rate of effusion of a gas is inversely proportional to the square root of its molar mass.

gray (Gy)   The SI unit of absorbed radiation; 1 Gy = 1 J/kg of tissue.

ground state   The most stable, lowest energy state available to atom or ion.

group   (also called *family*) All elements in the same column of the periodic table.

H

half-life ($t_{1/2}$)   The time in the course of a chemical reaction during which the concentration of a reactant decreases by half.

half-reactions   One of the two halves of an oxidation–reduction reaction; one half-reaction is the oxidation component, and the other is the reduction component.

halogen   An element in group 17 of the periodic table.

heat   The energy transferred between objects because of a difference in their temperatures.

heat capacity (C_P)   The quantity of energy needed to raise the temperature of an object 1°C at constant pressure.

heat of reaction   (also called *enthalpy of reaction [ΔH_{rxn}]*) The energy absorbed or given off by a chemical reaction.

heat transfer   The process of heat energy flowing from one object into another.

Heisenberg uncertainty principle   One cannot determine both the position and the momentum of an electron in an atom at the same time.

Henderson–Hasselbalch equation   Used to calculate the pH of a solution in which the concentrations of an acid and conjugate base are known.

Henry's law   The concentration of a sparingly soluble, chemically unreactive gas in a liquid is proportional to the partial pressure of the gas.

hertz (Hz)   The SI unit of frequency with units of reciprocal seconds: $1 \text{ Hz} = 1 \text{ s}^{-1} = 1$ cycle per second (cps).

Hess's law   The standard enthalpy of reaction ΔH°_{rxn} for a reaction that is the sum of two or more reactions is equal to the sum of the ΔH°_{rxn} values of the constituent reactions; also known as *Hess's law of constant heat of summation*.

heteroatom   Any atom other than carbon or hydrogen in an organic compound.

heterogeneous catalyst   A catalyst in a phase different than the reactants.

heterogeneous equilibria   Involve reactants and products in more than one phase.

heterogeneous mixture   A mixture in which the components are not distributed uniformly, so that the mixture contains distinct regions of different compositions.

heteropolymer   A polymer made of three or more different monomer units.

hexagonal closest-packed (hcp)   A crystal lattice in which the layers of atoms or ions in hexagonal unit cells have an *ababab* . . . stacking pattern.

hexagonal unit cell   An array of closest-packed particles that has seven of the particles on the top and bottom faces of a hexagonal prism and three of them in a middle layer.

homogeneous catalyst   A catalyst in the same phase as the reactants.

homogeneous equilibria   Involve reactants and products in the same phase.

homogeneous mixture   A mixture in which the components are distributed uniformly throughout and have no visible boundaries or regions.

homologous series   A set of related organic compounds that differ from one another by the number of common subgroups, such as $-CH_2-$, in their molecular structures.

homopolymer   A polymer composed of only one kind of monomer unit.

Hund's rule   The lowest-energy electron configuration of an atom has the maximum number of unpaired electrons, all of which have the same spin, in degenerate orbitals.

hybrid atomic orbital   In valence bond theory one of a set of equivalent orbitals about an atom created when specific atomic orbitals are mixed.

hybridization   In valence bond theory the mixing of atomic orbitals to generate new sets of orbitals that then are available to form covalent bonds with other atoms.

hydrocarbons   A class of organic compounds containing molecular compounds composed of only hydrogen and carbon.

hydrogen bond   The strongest dipole–dipole interaction. It occurs between a hydrogen atom bonded to a small, highly electronegative element (O, N, F) and an atom of oxygen or nitrogen in another molecule. Molecules of HF also form hydrogen bonds.

hydrogenation   The reaction of an unsaturated hydrocarbon with hydrogen.

hydrolysis   The reaction of water with another material. The hydrolysis of nonmetal oxides produces acids

hydronium ion (H_3O^+)   An H^+ ion plus a water molecule, H_2O; the form in which the hydrogen ion is found in an aqueous solution.

hydrophilic   A "water-loving" or attractive interaction between a solute and water that promotes water solubility.

hydrophobic   A "water-fearing" or repulsive interaction between a solute and water that diminishes water solubility.

hypothesis   A tentative and testable explanation for an observation or a series of observations.

I

i factor   (also called *van't Hoff factor*) The ratio of the experimentally measured value of a colligative property to the theoretical value expected for that property if the solute were a nonelectrolyte.

ideal gas   A gas whose behavior is predicted by the linear relations defined by Boyle's, Charles's, Avogadro's, and Amontons's laws.

ideal gas equation   (also called *ideal gas law*) Relates the pressure, volume, number of moles, and temperature of an ideal gas; expressed as $PV = nRT$, where R is the universal gas constant.

ideal gas law   (also called *ideal gas equation*) Relates the pressure, volume, number of moles, and temperature of an ideal gas; expressed as $PV = nRT$, where R is the universal gas constant.

ideal solution   One that obeys Raoult's law.

induced dipole   (also called *temporary dipole*) The separation of charge produced in an atom or molecule by a momentary uneven distribution of electrons.

inhibitor   A compound that diminishes or destroys the ability of an enzyme to catalyze a reaction.

initial rate   The rate of a reaction at $t = 0$, immediately after the reactants are mixed.

inner coordination sphere   The ligands that are bound directly to a metal via coordinate bonds.

instantaneous rate   The rate of a reaction at a specific instant during the course of the reaction.

integrated rate law   A mathematical expression that describes the change in concentration of a reactant in a chemical reaction with time.

intensive property   A property that is independent of the amount of substance present.

interference   The interaction of waves that results in either reinforcing their amplitudes (constructive interference) or canceling them out (destructive interference).

intermediate   A species produced in one step of a reaction and consumed in a subsequent step.

internal energy (E)   The sum of all the kinetic and potential energies of all of the components of a system.

interstitial alloy   Atoms of one element occupy the spaces between atoms of the host.

ion   An atom or group of atoms that has a positive or negative charge.

ion channel   Group of helical proteins that penetrate cell membranes and allow selective transport of ions.

ion exchange   A process by which one ion is displaced by another.

ion pair   A cluster formed when a cation and an anion associate with each other in solution.

ion pump   System of membrane proteins that exchange ions inside the cell with those in the intercellular fluid.

ion–dipole interaction   An attractive force between an ion and a molecule that has a permanent dipole moment.

ionic bond   Results from the electrostatic attraction of a cation for an anion.

ionic compound   A compound composed of positively and negatively charged ions held together by electrostatic attraction.

ionic radius   Radius derived from the distance between nuclear centers in ionic crystals.

ionic solid   A solid consisting of monatomic or polyatomic ions held together by ionic bonds.

ion–ion attraction   An attractive force between ions of opposite charge that results in an ionic bond.

ionization energy (IE)   The amount of energy needed to remove 1 mole of electrons from 1 mole of ground-state atoms or ions in the gas phase.

ionizing radiation   High-energy products of radioactive decay that can ionize molecules.

isoelectronic   Describes atoms of ions that have identical electron configurations.

isolated system   A system that exchanges neither energy nor matter with the surroundings.

isothermal process   A process that takes place at constant temperature.

isotopes   Atoms of an element containing the same number of protons but different numbers of neutrons.

J

joule (J)   The SI unit of energy; 4.184 J = 1 cal.

K

Kekulé structure   A structure showing all of the bonds in a covalently bonded molecule using lines but not showing lone pairs on the atoms.

kelvin (K)   The SI unit of temperature.

ketone   Organic molecule containing a carbonyl group bonded to two R groups; its general formula is $R(C{=}O)R$.

kinetic energy (KE)   The energy of an object in motion due to its mass (m) and its speed (u): $KE = \frac{1}{2}mu^2$.

kinetic molecular theory   A model that describes the behavior of gases; all equations defining relationships between pressure, volume, temperature, and number of moles of gases can be derived from the theory.

Krebs cycle   A series of reactions that continue the oxidation of pyruvate formed in glycolysis.

L

lattice energy (U)   The energy released when 1 mole of an ionic compound forms from its free ions in the gas phase.

law of conservation of energy   Energy cannot be created or destroyed.

law of conservation of mass   The sum of the masses of the reactants in a chemical reaction is equal to the sum of the masses of the products.

law of constant composition   All samples of a particular compound contain the same elements combined in the same proportions.

law of mass action   The ratio of the concentrations or partial pressures of products to reactants at equilibrium has a characteristic value at a given temperature when each term is raised to a power equal to the coefficient of that substance in the balanced chemical equation for the reaction.

law of multiple proportions   The ratio of the two masses of one element that react with a given mass of another element to form two different compounds is the ratio of two small whole numbers.

Le Châtelier's principle   A system at equilibrium responds to a stress in such a way that it relieves that stress.

leveling effect   The observation that strong acids all have the same strength in water and are completely converted into solutions of H_3O^+ ions; strong bases are likewise leveled in water and are completely converted into solutions of OH^- ions.

Lewis acid   A substance that *accepts* a lone pair of electrons in a chemical reaction.

Lewis base   A substance that *donates* a lone pair of electrons in a chemical reaction.

Lewis dot symbol   (also called *Lewis symbol*) The chemical symbol for an atom surrounded by one or more dots representing the valence electrons.

Lewis structure   A two-dimensional representation of the bonds and lone pairs of valence electrons in a molecule or polyatomic ion.

Lewis symbol   (also called *Lewis dot symbol*) The chemical symbol for an atom surrounded by one or more dots representing the valence electrons.

ligand   A Lewis base bonded to the central metal ion of a complex ion.

limiting reactant   A reactant that is consumed completely in a chemical reaction. The amount of product formed depends on the amount of the limiting reactant available.

lipid   A class of water-insoluble, oily organic compounds that are common structural materials in cells.

lipid bilayer   A double layer of molecules whose polar head groups interact with water molecules and whose nonpolar tails interact with each other.

liquid   A form of matter that occupies a definite volume but flows to assume the shape of its containers.

London force   (also called *dispersion force*) An intermolecular force between nonpolar molecules caused by the presence of temporary dipoles in the molecules.

lone pair   A pair of electrons that is not shared.

M

macrocyclic ligand   A ring containing multiple electron-pair donors that bind to a metal ion.

magnetic quantum number (m_ℓ)   Defines the orientation of an orbital in space; an integer that may have any value from $-\ell$ to $+\ell$, where ℓ is the angular momentum quantum number.

main group elements   (also called *representative elements*) The elements in groups 1, 2, and 13 through 18 of the periodic table.

major essential element   Essential elements present in the body in average concentrations greater than 1 mg of element per gram of body mass.

manometer   An instrument for measuring the pressure exerted by a gas.

mass   The property that defines the quantity of matter in an object.

mass action expression   Equivalent to the equilibrium constant expression, but applied to reaction mixtures that may, or may not, be at equilibrium.

mass defect (Δ*m*) The difference between the mass of a stable nucleus and the masses of the individual nucleons that comprise it.

mass number (*A*) The number of nucleons in an atom.

mass spectrum A graph of the data from a mass spectrometer, where m/z ratios of the deflected particles are plotted against the number of particles with a particular mass. Because the charge on the ions typically is 1+, $m/z = m/1 = m$, and the mass of the particle may be read directly from the m/z axis.

matter Anything that has mass and occupies space.

matter wave The wave associated with any particle.

meniscus The concave or convex surface of a liquid.

messenger RNA (mRNA) The form of RNA that carries the code for synthesizing proteins from DNA to the site of protein synthesis in a cell.

metallic bond Consists of the nuclei of metal atoms surrounded by a "sea" of shared electrons.

metallic radius Half the distance between nuclear centers in the crystal of a metal.

metalloids (also called *semimetals*) Elements along the border of the periodic table between metals and nonmetals; they have some metallic and some nonmetallic properties.

metals The elements on the left side of the periodic table that are typically shiny solids that conduct heat and electricity well and are malleable and ductile.

meter The standard unit of length, named after the Greek *metron*, which means "measure," and equivalent to 39.37 inches.

methanogenic bacteria Bacteria using simple organic compounds and hydrogen for energy; their respiration produces methane, carbon dioxide, and water, depending on the compounds they consume.

methyl group (–CH₃) A structural unit that can make only one bond.

methylene group (–CH₂–) A structural unit that can make two bonds.

microstate A unique distribution of particles among energy levels.

millimeters of mercury (mmHg) (also called *torr*) A unit of pressure where 1 atm = 760 mmHg = 760 torr.

miscible Capable of being mixed in any proportion (without reacting chemically).

mixture A combination of pure substances in variable proportions in which the individual substances retain their chemical identities and can be separated from one another by a physical process.

molality (*m*) Concentration expressed as the number of moles of solute per kilogram of solvent.

molar heat capacity (*c*ₚ) The energy required at constant pressure to raise the temperature of 1 mole of a substance by 1°C.

molar heat of fusion (Δ*H*fus) The energy required to convert 1 mole of a solid substance at its melting point into the liquid state.

molar heat of vaporization (Δ*H*vap) The energy required to convert 1 mole of a liquid substance at its boiling point to the vapor state.

molar mass (ℳ) The mass of 1 mole of a substance. The molar mass of an element in grams per mole is numerically equal to that element's average atomic mass in atomic mass units.

molar volume Volume occupied by 1 mole of an ideal gas at STP; 22.4 L.

molarity (*M*) The number of moles of solute divided by solution volume in liters: $M = n/V$. A 1.0 *M* solution contains 1.0 mol of solute per liter of solution; also known as *molar concentration*.

mole (mol) An amount of material (atoms, ions, or molecules) that contains Avogadro's number ($N_A = 6.022 \times 10^{23}$) of particles.

mole fraction (𝒳ₓ) The ratio of the number of moles of a component in a mixture to the total number of moles in the mixture.

molecular compound A compound composed of atoms held together in molecules by covalent bonds.

molecular equation A balanced equation that describes a reaction in solution in which the reactants are written as undissociated molecules.

molecular formula A notation showing the number and type of atoms present in one molecule of a molecular compound.

molecular geometry The three-dimensional arrangement of the atoms in a molecule.

molecular ion (M⁺) An ion formed in a mass spectrometer when a molecule loses an electron after being bombarded with high-energy electrons. The molecular ion has a charge of 1+ and has essentially the same molecular mass as the molecule from which it came.

molecular mass The mass of one molecule of a molecular compound.

molecular orbital A region of characteristic shape and energy where electrons in a molecule are located.

molecular orbital diagram In MO theory, an energy-level diagram showing the relative energies and electron occupancy of the molecular orbitals for a molecule.

molecular orbital (MO) theory A bonding theory based on the mixing of atomic orbitals of similar shapes and energies to form molecular orbitals that belong to the molecule as a whole.

molecular recognition The process by which molecules interact with other molecules to produce a biological effect.

molecular solid A solid formed by neutral, covalently bonded molecules held together by intermolecular attractive forces.

molecularity The number of ions, atoms, or molecules involved in an elementary step in a reaction.

molecule A collection of atoms chemically bonded together in characteristic proportions.

monodentate ligand A species that forms only a single coordinate bond to a metal ion in a complex.

monomer A small molecule that bonds with others like it to form polymers.

monoprotic acid Has one ionizable hydrogen atom per molecule.

monosaccharide A single-sugar unit and the simplest carbohydrate.

N

natural abundance The proportion of a particular isotope, usually expressed as a percentage, relative to all the isotopes of that element in a natural sample.

Nernst equation An equation relating the potential of a cell (or half-cell) reaction to its standard potential ($E°$) and to the concentrations of its reactants and products.

net ionic equation A balanced equation that describes the actual reaction taking place in aqueous solution; it is obtained by eliminating the spectator ions from the overall ionic equation.

neutralization reaction A reaction that takes place when an acid reacts with a base and produces a solution of a salt in water.

neutron An electrically neutral (uncharged) subatomic particle found in the nucleus of an atom.

neutron capture The absorption of a neutron by a nucleus.

noble gases The elements in group 18 of the periodic table.

node A location in a standing wave that experiences no displacement.

nonelectrolyte A substance that does not dissociate into ions and therefore does not enhance the conductivity of water when it dissolves.

nonessential element Element present in humans that has no known function.

nonmetals Elements with properties opposite those of metals including poor conductors of heat and electricity.

nonpolar covalent bond A bond characterized by an even distribution of charge; electrons in the bonds are shared equally by the two atoms; pure covalent bonds give rise to nonpolar diatomic molecules.

nonspontaneous process A process that occurs only as long as energy is continually added to the system.

normal boiling point The temperature at which the vapor pressure of a liquid equals 1 atm (760 torr).

n-type semiconductor Semiconductor containing electron-rich dopant atoms that contribute excess electrons.

nuclear chemistry The study of reactions that involve changes in the nuclei of atoms.

nuclear fission A nuclear reaction in which the nucleus of an element splits into two lighter nuclei. The process is usually accompanied by the release of one or more neutrons and energy.

nucleic acid One of a family of large molecules, which includes deoxyribonucleic acid (DNA) and ribonucleic acid (RNA), that stores the genetic blueprint of an organism and controls the production of proteins.

nucleon Either a proton or a neutron in a nucleus.

nucleosynthesis The natural formation of nuclei as a result of fusion and other nuclear processes.

nucleotide A monomer unit from which nucleic acids are made.

nucleus (of an atom) The positively charged center of an atom that contains nearly all the atom's mass.

nuclide A specific isotope of an element.

O

octahedral Molecular geometry about a central atom with a steric number of 6 and no lone pairs of electrons in which all six sites are equivalent.

octet rule Atoms of main group elements make bonds by gaining, losing, or sharing electrons to achieve a valence shell containing 8 electrons, or four electron pairs.

oil Liquid triglyceride containing primarily unsaturated fatty acids.

open system A system that exchanges both energy and matter with the surroundings.

optical isomers Molecules that are not superimposable on their mirror images.

optically active molecule A chiral compound that causes rotation of a beam of plane-polarized light when it passes through a solution.

orbital diagram Depiction of the arrangement of electrons in an atom or ion using boxes to represent orbitals.

orbital penetration The probability that an electron in an outer orbital will be as close to the nucleus as an electron in an inner shell.

orbitals Defined by the square of the wave function (ψ^2); regions around the nucleus of an atom where the probability of finding an electron is high and identified by a unique combination of three quantum numbers.

ore A mineral that contains one or more metals valuable enough to be mined.

organic chemistry The study of compounds containing C—C and/or C—H bonds.

organic compounds Compounds containing carbon, and commonly including certain other elements such as hydrogen, oxygen, and nitrogen.

organometallic compound A molecule containing direct carbon–metal covalent bonds.

osmosis The flow of a fluid through a semipermeable membrane to balance the concentration of solutes in solutions on the two sides of the membrane. The solvent molecules' flow proceeds from the more dilute solution into the more concentrated one.

osmotic pressure (π) The pressure applied across a semipermeable membrane to stop the flow of solvent from the compartment containing pure solvent or a less concentrated solution to the compartment containing a more concentrated solution. The osmotic pressure of a solution increases with solute concentration M and with solution temperature T.

overall ionic equation A balanced equation that shows all the species, both ionic and molecular, present in a reaction occurring in aqueous solution.

overall reaction order The sum of the exponents of the concentration terms in the rate law.

overlap A term in valence bond theory describing bonds arising from two orbitals on different atoms that occupy the same region of space.

oxidation A chemical change in which a species loses electrons; the oxidation number of the species increases.

oxidation number (O.N.) (also called *oxidation state*) A positive or negative number based on the number of electrons the atom gains or loses when it forms an ion, or that it shares when it forms a covalent bond with another element; pure elements have an oxidation number of zero.

oxidation state (also called *oxidation number [O.N.]*) A positive or negative number based on the number of electrons the atom gains or loses when it forms an ion, or that it shares when it forms a covalent bond with another element; pure elements have an oxidation number of zero.

oxidizing agent A substance in a redox reaction that accepts electrons from another species, thereby oxidizing that species; the oxidizing agent is reduced in the reaction.

oxoanions Polyatomic ions that contain oxygen in combination with one or more other elements.

P

packing efficiency Percentage of the total volume of a unit cell occupied by the spheres.

paramagnetic Describes a substance with unpaired electrons that is attracted to a magnetic field.

partial pressure The contribution to the total pressure made by a component in a mixture of gases.

Pauli exclusion principle No two electrons in an atom can have the same set of four quantum numbers.

peptide A compound of two or more amino acids joined by peptide bonds. Small peptides containing up to 20 amino acids are *oligopeptides;* and the term *polypeptide* is used for chains longer than 20 amino acids but shorter than proteins.

peptide bond The results of a condensation reaction between the carboxylic acid group of one amino acid and the amine group of another.

percent composition The composition of a compound expressed in terms of the percentage by mass of each element in the compound.

percent ionization The ratio of the quantity of a substance that is ionized to the concentration of the substance before ionization, expressed as a percentage.

percent yield The ratio, expressed as a percentage, of the actual yield of a chemical reaction to the theoretical yield.

periodic table of the elements A chart of the elements in order of their atomic numbers and in a pattern based on their physical and chemical properties.

periods The horizontal rows in the periodic table.

pH The negative logarithm of the hydrogen ion concentration in an aqueous solution.

pH buffer   A solution that resists changes in pH when acids or bases are added to it; typically a solution of a weak acid and its conjugate base.

pH indicator   A water-soluble weak organic acid that changes color as pH changes.

phase diagram   A graphical representation of the dependence of the stabilities of the physical states of a substance on temperature and pressure.

phospholipid   A molecule of glycerol with two fatty acid chains and one polar group containing a phosphate; phospholipids are major constituents of cell membranes.

phosphorylation   A reaction resulting in the addition of a phosphate group to an organic molecule.

photochemical smog   A mixture of gases formed in the lower atmosphere when sunlight interacts with compounds produced in internal combustion engines and other pollutants.

photoelectric effect   When light striking a metal surface produces an electric current (a flow of electrons).

photon   A quantum of electromagnetic radiation.

physical process   A transformation of a sample of matter, such as a change in its physical state, that does not alter the chemical identity of any substance in the sample.

physical property   A property of a substance that can be observed without changing it into another substance.

pi (π) bond   A covalent bond in which electron density is greatest on opposite sides of the bonding axis.

pi (π) molecular orbitals   In MO theory, formed by the mixing of atomic orbitals oriented above and below, or in front of and behind the bonding axis.

Planck's constant (h)   The proportionality constant between the energy and frequency of electromagnetic radiation expressed in $E = h\nu$; $h = 6.626 \times 10^{-34}$ J · s.

pOH   The negative logarithm of the hydroxide ion concentration in an aqueous solution.

polar covalent bond   Results from unequal sharing of bonding pairs of electrons between atoms.

polarizability   The relative ease with which the electron cloud in a molecule, ion, or atom can be distorted, inducing a temporary dipole.

polyatomic ions   Charged groups of two or more atoms joined together by covalent bonds.

polydentate ligand   A species that can form more than one coordinate bond per molecule.

polymer   A very large molecule with high molar mass; the root word *meros* is Greek for "part" or "unit," so *polymer* literally means "many units"; also known as *macromolecule*.

polyprotic acid   Has two or more ionizable hydrogen atoms per molecule.

polysaccharide   A polymer of monosaccharides.

porphyrin   A type of tetradentate macrocyclic ligand.

positron   A particle with the mass of an electron but with a positive charge.

positron emission   The spontaneous emission of a positron from a proton-rich nucleus.

potential energy (PE)   The energy stored in an object because of its position.

precipitate   A solid product formed from a reaction in solution.

precision   The extent to which repeated measurements of the same variable agree.

pressure (P)   The ratio of force to surface area over which the force is applied.

pressure–volume ($P–V$) work   The work associated with the expansion or compression of a gas.

primary (1°) structure   The sequence in which the amino acid monomers occur in a polymer chain.

principal quantum number (n)   A positive integer describing the relative size and energy of an atomic orbital or group of orbitals in an atom.

product   Substance formed as a result of a chemical reaction.

protein   Biological polymer made of amino acids.

proton   A positively charged subatomic particle present in the nucleus of an atom.

pseudo-first-order   A reaction in which all the reactants but one are present at such high concentrations that they do not decrease significantly during the course of the reaction, so that reaction rate is controlled by the concentration of the limiting reactant.

p-type semiconductor   Semiconductor containing electron-poor dopant atoms that cause a reduction in the number of electrons, which is equivalent to the presence of positively charged holes.

Q

quantized   Having values restricted to whole-number multiples of a specific base value.

quantum   (plural *quanta*) The smallest discrete quantity of a particular form of energy.

quantum mechanics   (also called *wave mechanics*) Mathematical description of the wavelike behavior of particles on the atomic level.

quantum number   One of four related numbers that specify the shape and energy of orbitals in an atom; the "address" of an electron in an atom or ion.

quantum theory   A model based on the idea that energy is absorbed and emitted in discrete quantities of energy called quanta.

quarks   Elementary particles that combine to form neutrons and protons.

quaternary (4°) structure   The larger structure functioning as a single unit that results when two or more proteins associate.

R

R   Symbol in a general formula standing for an organic group that has one available bond; it is used to indicate the variable part of a molecule so that the focus is placed on the functional group.

racemic mixture   A sample containing equal amounts of both optical isomers of a compound.

radioactive decay   The spontaneous disintegration of unstable particles accompanied by the release of radiation.

radioactivity   The spontaneous emission of high-energy radiation and particles by materials.

radiocarbon dating   A method for establishing the age of a carbon-containing object by measuring the activity of carbon-14 remaining in the object.

radiometric dating   A method for determining the age of an object based on the quantity of a radioactive nuclide and/or the products of its decay that the object contains.

radionuclide   An unstable nuclide that undergoes radioactive decay.

random coil   An irregular or rapidly changing part of the secondary structure of a protein.

Raoult's law   The vapor pressure of a solution containing nonvolatile solutes is proportional to the mole fraction of the solvent.

rate constant   The proportionality constant that relates the rate of a reaction to the concentrations of reactants.

rate law   An equation that defines the experimentally determined relation between the concentrations of reactants in a chemical reaction and the rate of that reaction.

rate-determining step   The slowest step in a multistep chemical reaction.

reactant   Substance consumed during a chemical reaction.

reaction mechanism   A set of steps that describe how a reaction occurs at the molecular level; the mechanism must be consistent with the rate law for the reaction.

reaction order   An experimentally determined number defining the dependence of the reaction rate on the concentration of a reactant.

reaction quotient (Q)   The numerical value of the mass action expression for *any values* of the concentrations (or partial pressures) of reactants and products; at equilibrium, $Q = K$.

reaction rate   How rapidly a reaction occurs; it is related to rates of change in the concentrations of reactants and products over time.

receptor   A cavity in a protein molecule that fits a particular molecule.

recognition   Process by which molecules in living systems interact with one another.

reducing agent   A substance in a redox reaction that gives up electrons to another species, thereby reducing that species; the reducing agent is oxidized in the reaction.

reduction   A chemical change in which a species gains electrons; the oxidation number of the species decreases.

refraction   The bending of light as it passes from one medium to another of different density.

relative biological effectiveness (RBE)   A factor that accounts for the differences in physical damage caused by different types of radiation.

replication   The process by which one double-stranded DNA forms two new DNA molecules, each one containing one strand from the original molecule and one new strand.

representative elements   (also called *main group elements*) The elements in groups 1, 2, and 13 through 18 of the periodic table.

resonance   Characteristic of electron distributions when two or more equivalent Lewis structures can be drawn for one compound.

resonance structure   One of two or more Lewis structures with the same arrangement of atoms but different arrangements of bonding pairs of electrons.

reverse osmosis   A water purification process in which water is forced through semi-permeable membranes, leaving dissolved impurities behind.

reversible process   A process that can be run in the reverse direction in such a way that, once the system has been restored to its original state, no net heat has flowed either to the system or to its surroundings.

root-mean-square speed (u_{rms})   The square root of the average of the squared speeds of all the molecules in a population of gas molecules; a molecule possessing the average kinetic energy moves at this speed.

S

salt   The product of a neutralization reaction; it is made up of the cation of the base in the reaction plus the anion of the acid.

saturated hydrocarbon   An alkane.

saturated solution   A solution that contains the maximum concentration of a solute possible at a given temperature.

Schrödinger wave equation   A description of how the electron matter wave varies with location and time around the nucleus of a hydrogen atom.

scientific method   An approach to acquiring knowledge based on observation of phenomena, development of a testable hypothesis, and additional experiments that test the validity of the hypothesis.

scientific theory (model)   A general explanation of a widely observed phenomenon that has been extensively tested and validated.

scintillation counter   An instrument that determines the level of radioactivity in samples by measuring the intensity of light emitted by phosphors in contact with the samples.

screening   (also called *shielding*) The effect when inner-shell electrons protect outer-shell electrons from experiencing the total nuclear charge.

second law of thermodynamics   The total entropy of the universe increases in any spontaneous process.

secondary (2°) structure   The pattern of arrangement of segments of a protein chain.

seesaw   Molecular geometry about a central atom with a steric number of 5 and one lone pair of electrons in an equatorial position.

semiconductor   A semimetal (metalloid) with electrical conductivity between that of metals and insulators that can be chemically altered to increase its electrical conductivity.

semimetals   (also called *metalloids*) Elements along the border of the periodic table between metals and nonmetals; they have some metallic and some nonmetallic properties.

shielding   (also called *screening*) The effect when inner-shell electrons protect outer-shell electrons from experiencing the total nuclear charge.

sievert (Sv)   SI unit used to express the amount of biological damage caused by ionizing radiation.

sigma (σ) bond   A covalent bond in which the highest electron density lies between the two atoms along the bond axis.

sigma (σ) molecular orbital   In MO theory, the lowest-energy orbital that forms when atomic orbitals mix; electrons in σ molecular orbitals form sigma (σ) bonds.

significant figures   All the certain digits in a measured value plus one estimated digit. The greater the number of significant figures, the greater the certainty with which the value is known.

simple cubic (sc) unit cell   A cell with atoms only at the eight corners of a cube.

single bond   Results when two atoms share one pair of electrons.

solid   A form of matter that has a definite shape and volume.

solubility   The maximum amount of a substance that dissolves in a given quantity of solvent at a given temperature.

solubility product, K_{sp}   (also called *solubility-product constant*) An equilibrium constant that describes the formation of a saturated solution of a slightly soluble salt.

solubility-product constant   (also called *solubility product*, K_{sp}) An equilibrium constant that describes the formation of a saturated solution of a slightly soluble salt.

solute   Any component in a solution other than the solvent. A solution may contain one or more solutes.

solution   Another name for homogeneous mixture. Solutions are often liquids, but they may also be solids or gases.

solvent   The component of a solution that is present in the largest amount.

***sp* hybrid orbitals**   Two hybrid orbitals on opposite sides of the hybridized atom formed by mixing one *s* and one *p* orbital.

***sp²* hybrid orbitals**   Three hybrid orbitals in a trigonal planar orientation formed by mixing of one *s* and two *p* orbitals.

sp^3 hybrid orbitals A set of four hybrid orbitals with a tetrahedral orientation produced by mixing one *s* and three *p* atomic orbitals.

sp^3d hybrid orbitals Five equivalent hybrid orbitals with lobes pointing toward the vertices of a trigonal bipyramid that form by mixing one *s* orbital, three *p* orbitals, and one *d* orbital from the same shell.

sp^3d^2 hybrid orbitals Six equivalent hybrid orbitals that point toward the vertices of an octahedron form from mixing one *s* orbital, three *p* orbitals, and two *d* orbitals from the same shell.

specific heat (c_s) The energy required to raise the temperature of 1 g of a substance 1°C at constant pressure.

spectator ion An ion that is present in a reaction vessel when a chemical reaction takes place but is unchanged by the reaction; spectator ions appear in an overall ionic equation but not in a net ionic equation.

spectrochemical series A list of ligands rank-ordered by their ability to split the energies of the *d* orbitals of transition metal ions.

sphere of hydration The cluster of water molecules surrounding an ion in aqueous medium; the general term applied to such a cluster forming in any solvent is *sphere of solvation*.

spin magnetic quantum number (m_s) Either $+\frac{1}{2}$ or $-\frac{1}{2}$, indicating that the spin orientation of an electron is either up or down.

spontaneous process A process that occurs without outside intervention.

square planar Molecular geometry about a central atom with a steric number of 6 and two lone pairs of electrons that occupy axial sites; the atoms occupy four equatorial positions.

square pyramidal Molecular geometry about a central atom with a steric number of 6 and one lone pair of electrons; as typically drawn, the atoms occupy four equatorial and one axial site.

standard atmosphere (1 atm) The pressure capable of supporting a column of mercury 760 mm high in a barometer.

standard cell potential (E°_{cell}) A measure of how forcefully an electrochemical cell, in which all reactants and products are in their standard states, can pump electrons through an external circuit.

standard conditions In thermodynamics: a pressure of 1 bar (~1 atm) and some specified temperature, assumed to be 25°C unless otherwise stated; for solutions, a concentration of 1 *M* is specified.

standard enthalpy of formation (ΔH°_f) The enthalpy change of a formation reaction; also known as *standard heat of formation* or *heat of formation*.

standard enthalpy of reaction (ΔH°_{rxn}) The energy associated with a reaction that takes place under standard conditions; also known as *standard heat of reaction*.

standard free energy of formation (ΔG°_f) The change in free energy associated with the formation of 1 mole of a compound in its standard state from its elements.

standard hydrogen electrode (SHE) A reference electrode based on the half-reaction $2\,H^+(aq) + 2\,e^- \rightarrow H_2(g)$ that produces a standard electrode potential of 0.000 V.

standard molar entropy (S°) The absolute entropy of 1 mole of a substance in its standard state.

standard reduction potential (E°_{red}) The potential of a reduction half-reaction in which all reactants and products are in their standard states at 25°C.

standard solution A solution of known concentration used in titrations.

standard state The most stable form of a substance under 1 bar pressure and some specified temperature (25°C unless otherwise stated).

standard temperature and pressure (STP) 0°C and 1 bar as defined by IUPAC; in the United States, 0°C and 1 atm.

standing wave A wave confined to a given space with a wavelength λ related to the length L of the space by $L = n(\lambda/2)$, where n is a whole number.

state function A property of an entity based solely on its chemical or physical state or both, but not on how it achieved that state.

stereoisomers Molecules with the same formulas and the same connectivities between their atoms, but with different spatial arrangements of their atoms.

steric number (SN) The sum of the number of atoms bonded to a central atom plus the number of lone pairs of electrons on the central atom.

stimulatory effect Increased growth or other biological response to the presence of a nonessential element.

stock solution A concentrated solution of a substance used to prepare solutions of lower concentration.

stoichiometry The quantitative relation between the reactants and products in a chemical reaction.

straight-chain alkane A hydrocarbon in which the carbon atoms are bonded together in one continuous line. Linear alkane chains have a methyl group at each end with methylene groups connecting them.

strong acid An acid that completely dissociates into ions in aqueous solution.

strong base A base that completely dissociates into ions in aqueous solution.

strong electrolyte A substance that dissociates completely into ions when it dissolves in water.

strong force The fundamental force of nature that keeps quarks together in subatomic particles and nucleons together in atomic nuclei.

structural isomers Molecules having the same molecular formula but different arrangements of atoms; they are different compounds and have different chemical and physical properties.

subatomic particles The neutrons, protons, and electrons in an atom.

sublimation Transformation of a solid directly into a vapor (gas).

substance Matter that cannot be broken down to simpler matter by any physical process; also known as *pure substance*.

substitutional alloy An alloy in which atoms of the nonhost metal replace host atoms in the crystal lattice.

substrate The reactant that binds to the active site in an enzyme-catalyzed reaction.

superconductor A material that has zero resistance to the flow of electric current.

supercritical fluid A substance at conditions above its critical temperature and pressure, where the liquid and vapor phases are indistinguishable and have some characteristics of both a liquid and a gas.

supersaturated solution A solution that contains more than the maximum quantity of solute predicted to be soluble in a given volume of solution at a given temperature.

surface tension The energy needed to separate the molecules at the surface of the liquid.

surroundings Everything that is not part of the system.

synthetic polymer Macromolecule made in the laboratory and often produced industrially for commercial use.

system The part of the universe that is the focus of a thermochemical study.

T

temporary dipole (also called *induced dipole*) The separation of charge produced in an atom or molecule by a momentary uneven distribution of electrons.

termolecular step   A step in a reaction mechanism involving a collision among three molecules.

tertiary (3°) structure   The three-dimensional, biologically active structure of the protein that arises because of interactions between the R groups on the amino acids.

tetrahedral   Molecular geometry about a central atom with a steric number of 4 and no lone pairs of electrons.

theoretical yield   The maximum amount of product possible in a chemical reaction for given quantities of reactants; also known as *stoichiometric yield*.

thermal energy   The kinetic energy of atoms, ions, and molecules.

thermal equilibrium   A condition in which temperature is constant throughout a material and no heat flows from one point to another.

thermochemical equation   The chemical equation of a reaction that includes heat as a reactant or a product.

thermochemistry   The study of the relation between chemical reactions and changes in energy.

thermodynamics   The study of energy and its transformations.

third law of thermodynamics   The entropy of a perfect crystal is zero at absolute zero.

threshold frequency (ν_0)   The minimum frequency of light required to produce the photoelectric effect.

titrant   The standard solution added to the sample in a titration.

titration   An analytical method for determining the concentration of a solute in a sample by reacting the solute with a standard solution of known concentration.

torr   (also called *millimeters of mercury [mmHg]*) A unit of pressure, where 1 atm = 760 mmHg = 760 torr.

trace essential element   Essential element present in the body in average concentrations between 1 and 1,000 μg of element per gram of body mass.

trans isomer   (also called *E isomer*) Molecule with two like groups (such as two R groups or two hydrogen atoms) on opposite sides of the molecule.

transcription   The process of copying the information in DNA to RNA.

transfer RNA (tRNA)   The form of the nucleic acid RNA that delivers amino acids, one at a time, to polypeptide chains being assembled by the ribosome–mRNA complex.

transition metals   The elements in groups 3 through 12 of the periodic table.

transition state   A high-energy state between reactants and products in a chemical reaction.

translation   The process of assembling proteins from the information encoded in RNA.

trigonal bipyramidal   Molecular geometry about a central atom with a steric number of 5 and no lone pairs of electrons in which three atoms occupy equatorial sites and two other atoms occupy axial sites above and below the equatorial plane.

trigonal planar   Molecular geometry about a central atom with a steric number of 3 and no lone pairs of electrons.

triple bond   Results when two atoms share three pairs of electrons.

triple point   The temperature and pressure where all three phases of a substance coexist. Freezing and melting, boiling and liquefaction, and sublimation and deposition all proceed at the same rate, so no net change takes place in the system.

T-shaped   Molecular geometry about a central atom with a steric number of 5 and two lone pairs of electrons that occupy equatorial positions; the atoms occupy two axial sites and one equatorial site.

U

ultratrace essential element   Essential element present in the body in average concentrations less than 1 μg of element per gram of body mass.

unimolecular step   A step in a reaction mechanism involving only one molecule on the reactant side.

unit cell   The basic repeating unit of the arrangement of atoms, ions, or molecules in a crystalline solid.

universal gas constant   The constant R in the ideal gas equation; its value and units depend on the units used for the variables in the equation.

unsaturated hydrocarbon   An alkene or alkyne.

V

valence band   A band of orbitals that are filled or partially filled by valence electrons.

valence bond theory   A quantum mechanics–based theory of bonding that assumes covalent bonds form when half-filled orbitals on different atoms overlap or occupy the same region in space.

valence electrons   Electrons in the outermost occupied shell of an atom having the most influence on the atom's chemical behavior.

valence-shell electron-pair repulsion theory (VSEPR)   A model predicting the arrangement of valence electron pairs around a central atom that minimizes their mutual repulsion to produce the lowest-energy orientations.

van der Waals equation   An equation that includes experimentally determined factors a and b that quantify the contributions of nonnegligible molecular volume and nonnegligible intermolecular interactions to the behavior of real gases with respect to changes in P, V, and T.

van der Waals forces   All types of attractive forces possible between molecules: hydrogen bonds, other dipole–dipole interactions, and dispersion forces. The term applies only to interactions between molecules; ion–ion and ion–dipole interactions are *not* van der Waals forces.

van't Hoff factor   (also called *i factor*) The ratio of the experimentally measured value of a colligative property to the theoretical value expected for that property if the solute were a nonelectrolyte.

vapor pressure   The pressure exerted by a gas at a given temperature in equilibrium with its liquid phase.

vinyl group   The subgroup $CH_2{=}CH-$.

vinyl polymer   One of the family of polymers formed from monomers containing the subgroup $CH_2{=}CH-$.

viscosity   The measure of the resistance to flow of a liquid.

voltaic cell   An electrochemical cell in which chemical energy is transformed into electrical work by a spontaneous redox reaction.

W

wave function (ψ)   A solution to the Schrödinger wave equation.

wave mechanics   (also called *quantum mechanics*) Mathematical description of the wavelike behavior of particles on the atomic level.

wavelength (λ)   The distance from crest to crest or trough to trough on a wave.

weak acid   An acid that only partially dissociates in aqueous solution and so has a limited capacity to donate protons to the medium.

weak base   A base that only partially dissociates in aqueous solution and so has a limited capacity to accept protons in the medium.

weak electrolyte   A substance that only partly dissociates into ions when it dissolves in water.

work A form of energy: the energy required to move an object through a given distance.

work function (Φ) The amount of energy needed to dislodge an electron from the surface of a metal.

X

X-ray diffraction (XRD) A technique for determining the arrangement of atoms or ions in a crystal by analyzing the pattern that results when X-rays are scattered after bombarding the crystal.

Z

Z isomer (also called *cis isomer*) Molecule with two like groups (such as two R groups or two hydrogen atoms) on the same side of the molecule.

zeolites Natural crystalline minerals or synthetic materials consisting of three-dimensional networks of channels that contain sodium or other 1+ cations.

zwitterion A molecule that has both positively and negatively charged groups in its structure.

Answers to Concept Tests and Practice Exercises

CHAPTER 1

Concept Tests

p. 6　1:1

p. 8　CH_2O

p. 11　(top) Even though it appears to be a uniform mixture, milk is not a true solution because it contains tiny particles of fat that are not dissolved. These tiny particles make milk "milky" instead of transparent.

p. 11　(bottom) b and d

p. 12　(top) Solubility is a physical property because the sugar is not changed by dissolving it. It can be recovered by evaporating the water.

p. 12　(bottom) Sodium metal combines with chlorine gas producing sodium chloride.

p. 16　Less than (b), because fewer intermolecular interactions are broken in melting ice than in boiling water

p. 18　Decreasing (b)

p. 22　1, 3 significant figures; 2, 3 significant figures; 4, 4 significant figures

p. 23　Fails to disprove: The density value matches that of gold, but the density alone does not prove conclusively that it *is* gold.

p. 25　Distance. At a track like the one in the figure, distance could vary depending on whether a car stays closer to the inside or the outside of the oval.

p. 35　Hypothesis, because his explanation had not been thoroughly tested yet

Practice Exercises

1.1.　Properties a, b, and c are physical properties; property d is a chemical property.

1.2.　(a) The particles in the box on the left represent a gas because the particles are widely spaced and fill the box. The particles in the box on the right represent a solid because its particles are ordered and do not fill the box. The change of state represented is deposition.
(b) Sublimation

1.3.　1.14

1.4.　Statistics a and e are exact numbers; statistics b, c, and d have inherent uncertainty.

1.5.　0.324 km; 3.24×10^4 cm

1.6.　1.5×10^2 cm

1.7.　9.46×10^{12} km/y

1.8.　$K_{low} = 40$ K and $K_{high} = 396$ K; $°F_{low} = -387°F$ and $°F_{high} = 253°F$

CHAPTER 2

Concept Tests

p. 53　The concept of atomic number was unknown in the mid-19th century.

p. 58　(top) CsN

p. 58　(bottom) H_2O_2 and HO

Practice Exercises

2.1. (a) $^{56}_{26}Fe$
(b) $^{15}_{7}N$
(c) $^{37}_{17}Cl$
(d) $^{39}_{19}K$

2.2. $^{107}Ag = 51.5\%$; $^{109}Ag = 48.5\%$

2.3. (a) As, arsenic
(b) Ca, calcium
(c) Hg, mercury
(d) S, sulfur

2.4. 40.0 g

2.5. a, b, c, and d are molecular; e is ionic.

2.6. (a) Tetraphosphorus decoxide
(b) Carbon monoxide
(c) Nitrogen trichloride

2.7. (a) $SrCl_2$
(b) MgO
(c) NaF
(d) $CaBr_2$

2.8. $MnCl_2$ and MnO_2

2.9. (a) $Sr(NO_3)_2$
(b) K_2SO_3

2.10. (a) Calcium phosphate
(b) Magnesium perchlorate
(c) Lithium nitrite
(d) Sodium hypochlorite
(e) Potassium permanganate

2.11. (a) Hypochlorous acid
(b) Chlorous acid
(c) Carbonic acid

2.12. Se

CHAPTER 3

Concept Tests

p. 83 Both the mole and a gross represent a specific number of particles independent of their size, mass, or identity.

p. 85 One gram of Ag has more atoms than 1 g Au because Ag has a smaller molar mass.

p. 89 The flashbulb contains a fixed mass of reactants in a sealed system so the products and remaining reactants can't escape.

p. 93 When balancing equations we can't change the subscripts because doing so changes the identity of the substance.

p. 103 C_2H_4 and $C_{20}H_{40}$ have the same empirical formula; C_2H_2 and C_6H_6 have the same empirical formula.

p. 106 All three are molecular formulas but only C_3H_8O is also an empirical formula.

p. 109 Molecular formula

p. 114 We need an equal number of nuts and bolts, but they do not weigh the same.

Practice Exercises

3.1. 1.5×10^{10} atoms

3.2. 0.0541 mol

3.3. 49.2 g

3.4. CO_2 = 44.01 g/mol; O_2 = 32.00 g/mol; $C_6H_{12}O_6$ = 180.2 g/mol

3.5. 9.18×10^{21} Na^+ ions

3.6. $P_4(s) + 5\,O_2(g) \rightarrow P_4O_{10}(s)$; $P_4O_{10}(s) + 6\,H_2O(\ell) \rightarrow 4\,H_3PO_4(\ell)$

3.7. $C_3H_8(g) + 5\,O_2(g) \rightarrow 3\,CO_2(g) + 4\,H_2O(\ell)$

3.8. $2\,C_4H_{10}(g) + 13\,O_2(g) \rightarrow 8\,CO_2(g) + 10\,H_2O(\ell)$; 3.30 g CO_2 produced

3.9. Mg = 24.20%; Si = 27.98%; O = 47.81%

3.10. Cu_2S

3.11. $FeCr_2O_4$

3.12. CH is the empirical formula for both acetylene and benzene.

3.13. Empirical formula: P_2O_5; molecular formula: P_4O_{10}

3.14. $C_8H_8O_3$

3.15. This fuel–oxygen mixture is rich.

3.16. 89.9%

CHAPTER 4

Concept Tests

p. 135 (c) clear cough syrup; (d) filtered dry air

p. 139 (d) 56,000 nM NaCl is the least concentrated solution

p. 145 (top) Solute is not removed on dilution; rather, solvent is added, which decreases the concentration.

p. 145 (bottom) a < d < c < e < b (a is the most dilute solution)

p. 147 Equivalent molar concentrations mean that the same number of solute moles are dissolved in a given solution volume. Differences in conductivity are due to the different numbers of ions the solutes make when they dissolve.

p. 148 Drawing a describes a weak electrolyte and c describes a strong electrolyte.

p. 149 NaCl is a neutral, strong electrolyte; hydrochloric acid is a strong electrolyte and an acid; NaOH is a strong electrolyte and a base; acetic acid is a weak electrolyte and an acid.

p. 152 (top) (a) H_2E^{2-}; (b) H_4E

p. 152 (bottom) (b) HSO_4^- is amphiprotic.

p. 158 React aqueous lead(II) nitrate with the stoichiometric amount of aqueous potassium dichromate. Stir the mixture for a minute or so, let it stand for 10 min, filter off the yellow $PbCr_2O_7$ precipitate, wash it with water in the filter, and allow the washed solid to air-dry overnight.

p. 165 The oxidation states of iron in Fe_3O_4 are two Fe(III) and one Fe(II) per formula unit, whereas the oxidation state of iron in Fe_2O_3 is Fe(III). The Fe(II) in magnetite has been oxidized. The oxygen atoms in the element O_2 are in oxidation state zero, but in Fe_2O_3 the oxidation state is −2. The oxygen atoms in O_2 have been reduced.

p. 168 Reactions a and b are redox reactions. In reaction a, Br_2 is reduced to Br^-, so Br_2 is the oxidizing agent and Sn^{2+} is the reducing agent. In reaction b, F_2 (O.N. zero) is reduced (O.N. in HF is −1), so F_2 is the oxidizing agent. The oxygen atom in H_2O has O.N. = −2, but in O_2 O.N. is zero, so H_2O is the reducing agent.

Practice Exercises

4.1. Well water is 120 times more concentrated in arsenic.

4.2. 1.88 M $MgCl_2$

4.3. 0.109 M KCl

4.4. 4.48 g $NaC_3H_3O_5$

4.5. $V_{initial} = 2.88 \times 10^{-2}$ mL

4.6. (a) $H_3PO_4(aq) + 3\,NaOH(aq) \rightarrow 3\,H_2O(\ell) + Na_3PO_4(aq)$
(b) $3\,H^+(aq) + PO_4^{3-}(aq) + 3\,Na^+(aq) + 3\,OH^-(aq) \rightarrow$
$3\,H_2O(\ell) + 3\,Na^+(aq) + PO_4^{3-}(aq)$
(c) $3\,H^+(aq) + 3\,OH^-(aq) \rightarrow 3\,H_2O(\ell)$, or $H^+(aq) + OH^-(aq) \rightarrow$
$H_2O(\ell)$

4.7 The lemon juice is 0.416 M $C_6H_8O_7$; 100 mL of juice contains 8.00 g $C_6H_8O_7$.

4.8. (a) No precipitate forms.
(b) Hg_2Cl_2 precipitates from the mixture.
(c) $Hg_2^{2+}(aq) + 2 Cl^-(aq) \rightarrow Hg_2Cl_2(s)$

4.9. 0.174 g HgS

4.10. 1.31×10^{-4} M SO_4^{2-}

4.11. (a) +4
(b) +1
(c) +5

4.12. Oxygen is reduced and is the oxidizing agent; SO_2 is oxidized and is the reducing agent.

4.13. (a) Because the oxidation number for iron and palladium change from reactants to products, this reaction is a redox reaction.
(b) $2 Fe(s) + 3 Pd^{2+}(aq) \rightarrow 2 Fe^{3+}(aq) + 3 Pd(s)$

4.14. $3 HO_2^-(aq) + H_2O(\ell) + 2 MnO_4^-(aq) \rightarrow$
$2 MnO_2(s) + 3 O_2(g) + 5 OH^-(aq)$

CHAPTER 5

Concept Tests

p. 193 (top) No; skier 1 has more PE: $m_1gh = (PE)_{skier\ 1} > (PE)_{skier\ 2} = m_2gh$

p. 193 (bottom) The PE of skier 2 is less than the PE of skier 1; skier 1 has greater KE:

$$\tfrac{1}{2} m_1 u^2 = (KE)_{skier\ 1} > (KE)_{skier\ 2} = \tfrac{1}{2} m_2 u^2$$

p. 195 There is far more water in the pool than in the cup. Therefore, the correct answer is "less than," even if the pool temperature is, say, 20°C.

p. 198 (a) open; (b, c) closed because the bottle and the sandwich wrap can conduct heat; (d) open

p. 213 The mass of the aluminum is much less than the mass of water and the molar heat capacity of Al is much less than that of water ($C_{P,Al} < C_{P,H_2O}$).

p. 217 Measuring the temperature to ±0.001°C allows more significant figures in the value of specific heat.

p. 224 +2219.9 kJ

p. 227 (a) 1 mol CH_4; (b) 1 g H_2

Practice Exercises

5.1. (a) The match is the system, $q < 0$, and the process is exothermic.
(b) The wax is the system, $q < 0$, and the process is exothermic.
(c) The liquid is the system, $q > 0$, and the process is endothermic.

5.2. $\Delta E = -68$ J

5.3. $w = 1.56 \times 10^7$ L · atm

5.4. 1.13×10^3 g or 1.13 kg

5.5. −321 kJ

5.6. 0.0°C

5.7. When 0.500 g of the hydrocarbon mixture is burned, the energy released is 24.6 kJ. When 1.000 g of the hydrocarbon mixture is burned, the energy released is 49.2 kJ.

5.8. (a) $Ca(s) + C(s) + \tfrac{3}{2} O_2(g) \rightarrow CaCO_3(s)$
(b) $2 C(s) + 2 H_2(g) + O_2(g) \rightarrow CH_3COOH(\ell)$
(c) $K(s) + Mn(s) + 2 O_2(g) \rightarrow KMnO_4(s)$

5.9. $\Delta H^\circ_{rxn} = -41.2$ kJ

5.10. Fuel value of kerosene = 41.40 kJ/g; fuel density of kerosene = 3.10×10^4 kJ/L

5.11. $C_{calorimeter} = 11.2$ kJ/°C

5.12. $2 CH_4(g) + 3 O_2(g) \rightarrow$
$2 CO(g) + 4 H_2O(g)$ $\quad \Delta H_{comb} = -1038$ kJ
$2 CO(g) + O_2(g) \rightarrow 2 CO_2(g)$ $\quad \Delta H_{comb} = -566$ kJ
$\overline{2 CH_4(g) + 4 O_2(g) \rightarrow 2 CO_2(g) + 4 H_2O(g)}$
$2 \times \Delta H_{comb} = -1604$ kJ
$\Delta H_{comb} = -802$ kJ

CHAPTER 6

Concept Tests

p. 249 (c) NO_2; (a) Br_2 in CCl_4

p. 252 Mars

p. 257 c

p. 259 The two graphs will have different slopes because the slope depends on P. Slope of graph b = $\tfrac{1}{2}$ slope of graph a.

p. 262 c

p. 266 b and d

p. 272 Kr

p. 274 (top) (a) i; (b) ii

p. 274 (bottom) (c) Low T and high P give greatest d.

p. 277 Because $P = F/A$, the force results from all components combined so one cannot measure the partial pressure since one cannot measure the partial force.

p. 279 Yes

p. 283 (top) Lower

p. 283 (bottom) c

p. 286 (top) $UF_6 < SF_6 < Kr < CO_2 < Ar < H_2$

p. 286 (bottom) $T_1 > T_2$

p. 288 T cancels out

p. 290 c

p. 292 CO_2, CH_4

Practice Exercises

6.1. $P = 7.71 \times 10^2$ Pa

6.2. At the end of the experiment the mercury levels will be

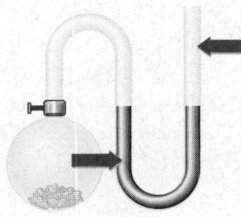

$\Delta h = 144$ mmHg

6.3. $V_2 = 10.5$ L

6.4. $V_1/V_2 = 0.622$

6.5. $P_2 = 32$ psi

6.6. $V_2 = 2.10 \times 10^3$ L

6.7. $V = 1.8 \times 10^3$ L

6.8. 111 g

6.9. The balloon will sink to the floor.

6.10. $\mathcal{M} = 44.0$ g/mol, CO_2

6.11. $X_{O_2} = 0.120$, $X_{He} = 0.882$

6.12. $X_{O_2} = 0.042$

6.13. 2.2×10^{-3} g H_2

6.14. $u_{rms,\ He} = 1.37 \times 10^3$ m/s, or 2.65 times faster than N_2

6.15. Ar

6.16. N_2 behaves more ideally.

CHAPTER 7

Concept Tests

p. 312 Higher frequency

p. 314 No change

p. 317 c

p. 319 (a) continuous; (b) discrete; (c) continuous; (d) discrete

p. 321 Yes

p. 324 $c > a > b > d$

p. 325 Yes because He^+ has only one electron

p. 328 The vibrations of the strings have wavelengths that are equal to $2L/n$, where n is a whole number.

p. 336 5

p. 347 A half-filled set of f orbitals is more stable.

p. 349 $1s > 2s > 2p > 3s > 3p > 4s > 4p$

p. 353 Because the valence electrons are farther from the nucleus and shielded from it by more inner shell electrons

p. 354 The magnitude of IE and EA both increase with increasing Z across a row (except for group 18). EA values do not display clear trends within groups whereas IE values decrease with increasing Z.

Practice Exercises

7.1. $\lambda = 3.30$ m

7.2. $\lambda = 1.99$ nm or 1.99×10^{-9} m

7.3. $\lambda = 2.62 \times 10^{-7}$ m or 262 nm

7.4. 486.27 nm

7.5. Prediction: Less energy is required to remove an electron from the hydrogen atom in the $n = 3$ state versus that for a hydrogen atom in the $n = 1$ state. Calculated value: 2.420×10^{-19} J

7.6. $\lambda = 3.3 \times 10^{-10}$ m

7.7. $\Delta x \geq 7 \times 10^{-11}$ m

7.8. Four

7.9.

n	ℓ	m_ℓ	m_s
3	1	−1	$\frac{1}{2}$
3	1	0	$\frac{1}{2}$
3	1	1	$\frac{1}{2}$

7.10. $Co = [Ar]3d^7 4s^2$

7.11. $K^+ = [Ar]$; $Rb^+ = [Kr]$; $Ba^{2+} = [Xe]$; $Al^{3+} = [Ne]$; $I^- = [Kr]4d^{10}5s^2 5p^6 = [Xe]$; $Cl^- = [Ne]3s^2 3p^6 = [Ar]$; $O^{2-} = [He]2s^2 2p^6 = [Ne]$. K^+ and Cl^- are isoelectronic with Ar.

7.12. $Mn = [Ar]3d^5 4s^2$; $Mn^{3+} = [Ar]3d^4$; $Mn^{4+} = [Ar]3d^3$

7.13. (a) $Li^+ < F^- < Cl^-$
        (b) $Al^{3+} < Mg^{2+} < P^{3-}$

7.14. $Ne > Ca > Cs$

CHAPTER 8

Concept Tests

p. 371 $BC = 18 - GN$

p. 379 (top) Most noble gases don't form covalent bonds.

p. 379 (bottom) :O=C=O:

p. 381 No, stretching in N≡N or O=O does not result in IR absorptions because the bonds are nonpolar and no change in polarity occurs when they stretch.

p. 389 −1

p. 395 Xe and Kr can expand their octets to form covalent bonds. He and Ne have no empty valence shell orbitals.

p. 396 $N_2O > NO_2 > NO$

p. 401 The O=O is not as strong as the N≡N bond and is more easily broken.

Practice Exercises

8.1.

8.2.

8.3. :O=C=O:

8.4. $[:\!F\!:]^- \ Mg^{2+} \ [:\!F\!:]^-$

8.5. Be—Cl; the bond is considered to be a polar covalent bond.

8.6. :O=S—O: ↔ :O—S=O:

8.7. Resonance forms for N_3^-:

Resonance forms for NO_2^+:

8.8. $[:\!N\!=\!N\!=\!N\!:]^-$ and $[:\!O\!=\!N\!=\!O\!:]^+$

8.9.

8.10.

8.11. $\Delta H_{rxn} = -79$ kJ

CHAPTER 9

Concept Tests

p. 415 (top) The central O atom in O_3 has a lone pair of electrons that repels the bonding pairs.

p. 415 (bottom) Molecular geometry is derived from electron-pair geometry. They are the same if there are no lone pairs but different if there are.

p. 416 Less: 2, 3 value; expanded: 5, 6; octet: 4

p. 419 $b > a > c$

p. 422 One lone pair on PH_3 and two lone pairs on H_2S repel bonding pairs and reduce bond angles.

p. 423 Slightly smaller

p. 425 All the bond dipoles offset each other, as in CH_4.

p. 428 Because the difference in electronegativity between H and S is less than between H and O

p. 434 It needs unhybridized p orbitals to form π bonds.

p. 438 a and c, because the double bonds are conjugated

p. 441 b and c are chiral

p. 442 Similar in that they are the products of mixing atomic orbitals; different in that MOs are delocalized

p. 448 Yes, because O_2 is paramagnetic and N_2 is not.

p. 451 No, bond order is lower because one more electron is in an antibonding orbital and one fewer is in a bonding orbital.

Practice Exercises

9.1. Tetrahedral:

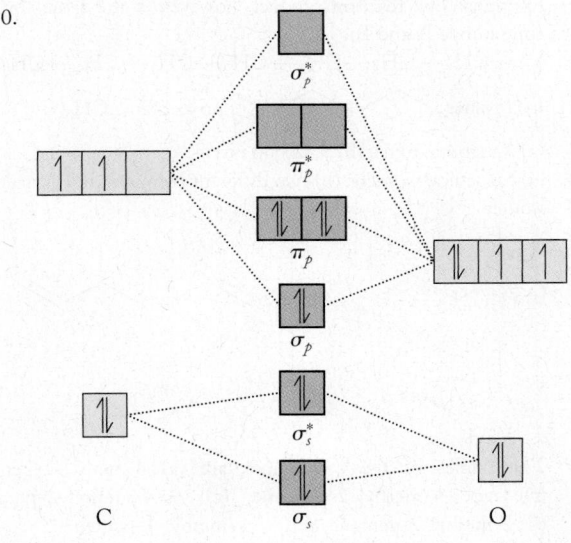

9.2. The O–S–O angle in SO_3 is greater than the O–S–O angle in SO_2.

9.3. Tetrahedral, bond angles ~109.5°

9.4. No

9.5. CCl_4 and PH_3

9.6. sp^3d^2

9.7. Each N in diazene is trigonal planar. With one lone pair on each N atom, the molecular geometry around the N atoms is bent with H–N–N bond angles of less than 120°. The N atoms are sp^2 hybridized and the molecule is flat. Each N in hydrazine is tetrahedral. With one lone pair on each N atom, the molecular geometry around the N atoms is trigonal pyramidal with H–N–N bond angles of less than 109.5°. The N atoms are sp^3 hybridized and the molecule is three-dimensional.

9.8. H_2^+ may exist.

9.9. The bond order increases on the addition of an electron for Be_2 to Be_2^-, B_2 to B_2^-, C_2 to C_2^-, and Ne_2 to Ne_2^-.

9.10.

Bond order = 3

CHAPTER 10

Concept Tests

p. 465 (a) LiF < NaF < KF; (b) CaF_2 < $CaCl_2$ < $CaBr_2$

p. 469 Acetone

p. 470 (a) CH_3Br; (b) CH_3CH_2OH; (c) CH_3NH_2

p. 474 H_2 < Ne < Ar < Kr

p. 475 Yes, ion–induced dipole, slightly stronger than dipole–induced dipole and weaker than dipole–dipole

p. 476 CCl_4 is larger and experiences larger dispersion forces.

p. 484 Yes, it's possible that increased pressure causes the ice to melt based on the slope of the blue line in Figure 10.22 and the footprint of the tires.

p. 485 No

p. 488 (top) No, the adhesive forces are weak.

p. 488 (bottom) Yes, because of the strength of ion–dipole interactions

Practice Exercises

10.1. BaO < $CaCl_2$ < NaCl

10.2. To enter the vapor phase from the liquid phase (to boil), ethylene glycol would break two hydrogen bonds compared to isopropanol's one hydrogen bond. Therefore, ethylene glycol has a higher boiling point.

10.3. Largest dipole–dipole forces: H_2NNH_2; largest dispersion forces: $CH_3CH_2CH_2CH_2CH_3$; lowest boiling point: Ne

10.4. O_2 < CO < H_2O

10.5. Helium, being smaller and with fewer electrons than nitrogen, is less soluble in blood because it is less polarizable in its interaction with the polar water (blood) molecules.

10.6. $C_{gas} = 9.5 \times 10^{-5}$ mol/L

10.7. At 25 atm pressure and –100°C, the sample of CO_2 is a solid. As the temperature is increased to about –50°C the solid melts into a liquid, and as the temperature is raised further (to about –20°C) the liquid CO_2 boils to form a gas.

CHAPTER 11

Concept Tests

p. 506 One needs to take into account the formation of the perchlorate ion as well as its EA, which may not be known.

p. 508 Both ethylene glycol and triethylene glycol

p. 509 The vapor pressure of a pure solvent is an intensive property. The vapor pressure of solution is an extensive property.

p. 513 Gasoline

p. 515 Dimethyl ether

p. 518 c

p. 519 The solution is mostly water which has a density of 1 kg/L.

p. 524 b

p. 526 (a) possible; (b) not possible; (c) possible; (d) not possible

p. 527 Water flows from the cells of the cucumbers into the brine.

p. 529 (top) Toward the KCl

p. 529 (bottom) Doubling the concentration

p. 534 Although more opposing pressure is needed at 50°C than at 20°C, the greater molecular motion would cause osmosis to occur faster.

Practice Exercises

11.1. TiO_2

11.2. $U = -3792$ kJ

11.3. $U_{NaClO_4} = -658$ kJ/mol

11.4. $P_{solution} = 0.848$ atm or 644 torr

11.5. $\Delta H_{vap} = 28.4$ kJ/mol

11.6.

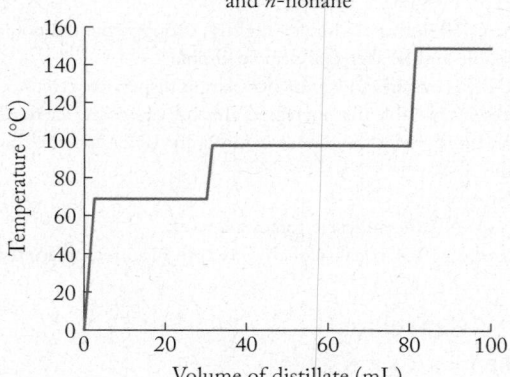

Distillation of mixture of *n*-hexane, *n*-heptane, and *n*-nonane

11.7. 1.4
11.8. 0.840 *m*
11.9. 4.4 *m*
11.10. 109.4°C
11.11. $i = 3$; 102.7°C
11.12. −0.12°C
11.13. 28.1 atm
11.14. 3.7 atm
11.15. 27.5 atm
11.16. 180 g/mol
11.17. 6.40×10^4 g/mol

CHAPTER 12

Concept Tests

p. 548 Overlapping conduction and valence bands
p. 554 Unit cell is the smallest repeating pattern in a crystal lattice.
p. 559 Yes, because it is homogeneous.
p. 561 Both alloys could be either.
p. 562 LiAl
p. 564 A group 15 element such as N or P
p. 566 Ice is a molecular solid in which the molecules are linked by a network of hydrogen bonds.
p. 572 $x = 1.5$; $y = 1$

Practice Exercises

12.1. Selenium
12.2. 128 pm
12.3. For silver, $d = 10.57$ g/mL, close to the 10.50 g/mL value for the density of silver from Appendix 3; for gold, $d = 19.41$ g/mL, close to the 19.3 g/mL value for the density of gold from Appendix 3
12.4. Gold forms substitutional alloys with both silver and copper.
12.5. 101 pm
12.6. 2.16 g/cm^3
12.7. $d = 412$ pm; for $n = 3$, $2\theta = 30.0°$; for $n = 4$, $2\theta = 40.4°$

CHAPTER 13

Concept Tests

p. 594 Dispersion forces
p. 596 No, they both consume 2 moles of H$_2$ per mole
p. 597 $n = 8$
p. 606 The same name means the same compound.
p. 612 e

p. 614 (a) No; (b) no
p. 615 Different molecular structures and properties mean LDPE and HDPE must be recycled separately.
p. 617 The 2 Cl atoms per monomer unit in Saran do not allow the chains to pack tightly; less interaction makes them more flexible than PVC.
p. 620 The π bonds are localized in the linear molecule.
p. 622 (top) Paraglyine: tertiary; amphetamine: primary; Benadryl: tertiary; adrenaline: secondary
p. 622 (bottom) Alcohols and ethers are both polar and interact with polar water molecules better than hydrocarbons do.
p. 627 MTBE, diethyl ether, ethanol, methanol
p. 629 The longer the hydrocarbon portion of the chain, the lower the water solubility and the higher the solubility in nonpolar organic solvents
p. 630 Zingerone: ether, aromatic rings, –OH; carvone: C=C; cinnamaldehyde: aromatic ring, C=C

p. 632 Amine: 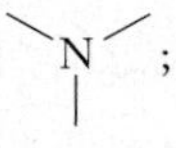 ; Amide:

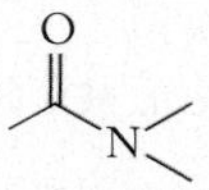

The nitrogen in the amide is attached to a carbonyl group.
p. 642 The carbon atom bonded to the –OH group is a chiral center. The different orientations of the parts of the molecule on either side of this center result in the enantiomers interacting with differently shaped receptor sites like the different shapes of left and right hands.
p. 644 If the muscarine came from mushrooms, it would be one enantiomer and its solution would be optically active. Because the solution did not rotate the plane of polarized light, the muscarine present was a racemic mixture, which had to be the product of a laboratory synthesis. The coroner concluded that the victim was poisoned by someone who had access to synthetic muscarine.

Practice Exercises

13.1. Yes, we can differentiate between them using hydrogenation reactions. The reaction product, however, is the same for both compounds A and B:
CH$_3$—CH$_2$—CH$_2$—CH$_2$—CH$_2$—CH$_2$—CH$_3$
13.2. *n*-Hexane: ; *n*-heptane: CH$_3$(CH$_2$)$_5$CH$_3$
13.3. (a) 7 carbon atoms; (b) 9 carbon atoms
13.4. (a) Structural isomers; (b) two different compounds; (c) structural isomers
13.5.

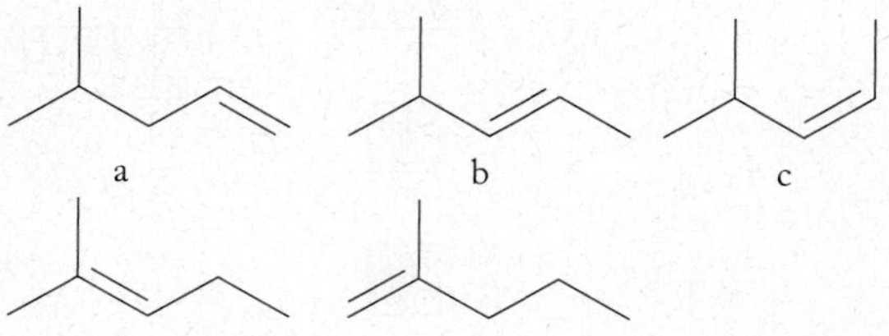

a   b   c

d   e

The names of the compounds are (a) 4-methyl-1-pentene, (b) *trans*-4-methyl-2-pentene, (c) *cis*-4-methyl-2-pentene, (d) 2-methyl-2-pentene, and (e) 2-methyl-1-pentene

13.6. The carbon skeleton of the monomer is

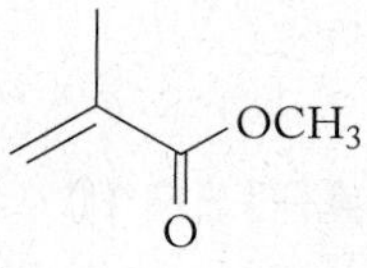

The condensed structure of the monomer is H$_2$C=C(CH$_3$)C(O)OCH$_3$
13.7. The heat released from the E-85 fuel is less than that released from the same volume of pure *n*-nonane.

13.8. London dispersion and dipole–induced dipole interactions.

13.9.

13.10. The polar fibers of cotton and polyester repel very nonpolar greases and oils but attract water molecules so perspiration wicks out of the gloves to cool the skin.

13.11. The carbon skeleton structures of the monomers are

The repeating unit in the polymer is

13.12. (a)

achiral

(b)

achiral

(c)

chiral

(d)

chiral

(e)

chiral

13.13.

CHAPTER 14

Concept Tests

p. 660   a and c
p. 663 $4! = 4 \times 3 \times 2 \times 1 = 24$
p. 664 $5! = 120$
p. 669 $\Delta S_{univ} > 0$
p. 672   Both are (+) in (a) and (−) in (b).
p. 674   (top) At 298 K water exists as liquid with a significant vapor pressure but is too warm to be ice.
p. 674   (bottom) No, a substance must be a perfect crystal to have 0 entropy.

p. 675   Yes, melting and boiling
p. 678   d
p. 680   The reaction is very slow.
p. 682   (a) Yes, it is spontaneous ($\Delta G < 0$); (b) thermodynamics says nothing about the rate of reaction.
p. 683   Incandescent bulbs operate at much higher temperatures, wasting energy as heat.

p. 685

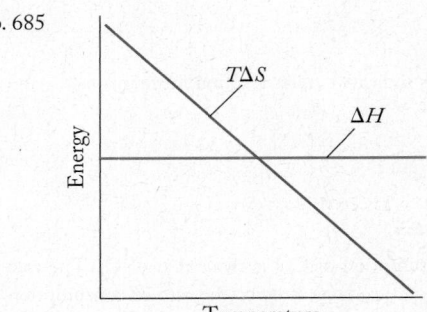

The line for $T\Delta S$ has a negative slope.

Practice Exercises

14.1. (a) For the evaporation of water, $\Delta S_{sys} > 0$ and $\Delta S_{univ} > 0$. Therefore, ΔS_{surr} may be either positive or negative. If $\Delta S_{surr} < 0$ it must be the case that $|\Delta S_{surr}| < \Delta S_{sys}$ so that the process remains spontaneous with $\Delta S_{univ} > 0$.
(b) For the reaction of gaseous ammonia with hydrogen chloride to produce solid ammonium chloride, $\Delta S_{sys} < 0$ and $\Delta S_{univ} > 0$. Therefore, ΔS_{surr} must be positive and greater in magnitude than ΔS_{sys} ($\Delta S_{surr} > |\Delta S_{sys}|$).
(c) For the precipitation of solid silver chloride from an aqueous solution of Ag^+ and Cl^-, $\Delta S_{sys} < 0$ and $\Delta S_{univ} > 0$. Therefore, ΔS_{surr} must be positive and greater in magnitude than ΔS_{sys} ($\Delta S_{surr} > |\Delta S_{sys}|$).
(d) For the dissolution of $C_{12}H_{22}O_{11}$ in water, $\Delta S_{sys} > 0$ and $\Delta S_{univ} > 0$. Therefore, ΔS_{surr} may be either positive or negative. If $\Delta S_{surr} < 0$ it must be the case that $|\Delta S_{surr}| < \Delta S_{sys}$ so that the proces remains spontaneous with $\Delta S_{univ} > 0$.

14.2. b
14.3. Prediction: S_{sys} decreases; $\Delta S^{\circ}_{rxn} = -243$ J/K
14.4. (a) ΔS_{rxn} is expected to be negative.
(b) $\Delta S^{\circ}_{sys} = -326$ J/K
(c) $\Delta S^{\circ}_{surr} = 1.92$ kJ/K or 1920 J/K
(d) $\Delta S_{univ} = 1592$ J/K. The ΔS_{univ} is positive so this reaction is spontaneous.
14.5. $\Delta G^{\circ}_{rxn} = -5230$ kJ
14.6. $\Delta H^{\circ} < 0, \Delta S^{\circ} < 0, \Delta G^{\circ} < 0$
14.7. The reaction is spontaneous at high temperatures.
14.8. $\Delta G^{\circ} = -196$ kJ

CHAPTER 15

Concept Tests

p. 702   (a) $\Delta S^{\circ}_{rxn} > 0$; (b) same number of moles of gaseous reactants and products
p. 704   Because a negative value would mean the reaction is running in reverse, not as written
p. 709   d
p. 714   (a) one; (b) zero

p. 715   Four: Rate = $k[A][B]^2$; Rate = $k[A]^2[B]$; Rate = $k[A]^3$; Rate = $k[B]^3$

p. 720   Fast

p. 727   (top) Plot k as $[O_3]_0$; slope = k

p. 727   (bottom) a and c

p. 728   Molecules are moving faster.

p. 731   b

p. 733   c

p. 735 n

p. 737   No

p. 740   Similar rate laws indicate similar reaction mechanisms.

p. 742   a

p. 743   Intermediate

Practice Exercises

15.1.   The rate of consumption of CO is twice that of O_2. The rate of formation of O_2 (a product) is twice the rate of consumption of CO_2.

15.2. $\dfrac{\Delta[N_2]}{\Delta t} = 10.8\ M/s$; $\dfrac{\Delta[H_2O]}{\Delta t} = 21.5\ M/s$

15.3. $1.2 \times 10^{-6}\ M/s$

15.4.   Rate = $k[NO][NO_3]$; $k = 1.57 \times 10^{10}\ /(M \cdot s)$

15.5.   The decomposition of H_2O_2 is first order; $k = 8.30 \times 10^{-4}\ M/s$

15.6. $k = 2.5 \times 10^{-2}\ /\text{day}$

15.7.   This reaction is second order in $[NO_2]$; $k = 0.757\ /(M \cdot s)$

15.8. $t_{1/2} = 9.23 \times 10^{-3}\ s$

15.9.   The pseudo-first-order rate constant is $k' = 6.11 \times 10^{-4}\ /\mu s$. The second-order rate constant is $k = 7.2 \times 10^6\ /(M \cdot \mu s)$ or $7.2 \times 10^{12}\ /(M \cdot s)$.

15.10.   6.9 kJ

15.11.   Rate = $k_{\text{overall}}[A]^2[B]^2$

15.12.   Because none of the rate laws match the experimental rate law, this proposed mechanism cannot be valid.

15.13.   Yes, NO_2 acts as a catalyst in this reaction.

CHAPTER 16

Concept Tests

p. 767   3.0

p. 769   (b) $[CO_2] = [H_2] > [CO] = [H_2O]$

p. 782   Zone a in Figure 16.5 where $Q < K$

p. 783   The values of K_c and K_p are the same.

p. 785   There is no term for liquid water because its concentration is considered constant.

p. 786   (a) The value of Q will decrease. (b) After equilibrium is reached the concentrations of CO_2, H_2O, and CO will all have decreased and $[H_2]$ will have increased.

p. 788   The number of moles of gaseous reactants and products are the same.

p. 799   (e) There is more B than A present.

Practice Exercises

16.1. $K_c = \dfrac{[CO][H_2]^3}{[CH_4][H_2O]}$ $K_p = \dfrac{(P_{CO})(P_{H_2})^3}{(P_{CH_4})(P_{H_2O})}$

16.2. $K_c = [CH_3OH]/[CO][H_2]^2 = 290$

16.3. $K_p = 2.7 \times 10^4$

16.4. $K_p = 4.5 \times 10^{-12}$

16.5. $K_{p,\text{reverse}} = 2.3 \times 10^2$

16.6. $K_c = 0.13$

16.7. $K_{c,\text{overall}} = 1.7 \times 10^2$

16.8.   This reaction is not at equilibrium and proceeds to the right.

16.9.   (a) $K_p = \dfrac{(P_{CO})^2}{(P_{CO_2})}$

(b) $K_p = \dfrac{(P_{CO})}{(P_{CO_2})(P_{H_2})}$

16.10.   (a) When the reaction is cooled and water vapor condenses, one product is removed from the reaction mixture and the equilibrium shifts to the right, forming more SO_2.

(b) When SO_2 gas dissolves in liquid water as it condenses, products are removed and the equilibrium shifts to the right, forming more products.

(c) When O_2 is added, the concentration of one reactant increases and the equilibrium shifts to the right, forming more products.

16.11.   Increasing the pressure shifts the equilibrium in the reaction to the products, the side of the reaction that has the fewest moles of gas.

16.12.   The value of K for the endothermic reaction increases with increasing reaction temperature.

16.13. $P_{HI} = 0.16$ atm

16.14. $[NO_2] = 0.058\ M$ and $[N_2O_4] = 0.016\ M$ at equilibrium

16.15. $K = 6.09 \times 10^5$

16.16. $K_{p,298\ K} = 2.95 \times 10^{-37}$ and $K_{p,2000\ K} = 9.20 \times 10^{-13}$

CHAPTER 17

Concept Tests

p. 820 $HF > HNO_2 > HCOOH > CH_3COOH > HClO$

p. 823 $OH^- > HCOO^- > NO_2^- > F^- > Cl^-$

p. 825   (top) pH 0.22 = strongly acidic; 4.37 = weakly acidic; 10.03 = weakly basic; 13.77 = strongly basic; 7.00 = neutral

p. 825   (bottom) (a) T; (b) T; (c) T; (d) T; (e) T

p. 826   Higher

p. 829   The larger its K_a, the greater is the percent ionization for a given concentration of HA.

p. 832   Most: C; least: A

p. 838   No for phosphoric acid, but the second ionization step of citric acid will influence pH

p. 839 $H_3PO_4 > H_3AsO_4 > H_3SbO_4 \approx H_3BiO_4$

p. 840   More acidic because the decomposition reaction released NH_3, which is a base. The solid left behind should be more acidic than the reactant.

p. 843   C and D

p. 844 $MgCO_3$, because it has the smaller molar mass

p. 851   (top) The equilibrium would shift to produce more acid in their blood; they would die of extreme acidosis.

p. 851   (bottom) Chloroacetic acid has a pK_a of 2.85. A buffer with a pH of 2.80 has slightly more of the acid than a salt of its conjugate base.

p. 853 $K_{sp} = 4S^3$

p. 856   If $[H^+]$ decreases, less F^- combines with H^+ to form HF. Therefore, the solubility of CaF_2 decreases.

p. 858   The value $[OH^-]$ is defined by the initial $[Ca^{2+}]$ and the K_{sp} of $Ca(OH)_2$.

p. 862   (top) pH = pK_a = 4.19

p. 862   (bottom) There would be too small a change in pH at the equivalence point to detect it precisely.

p. 863   1:1

p. 864 Because neutralization of the CO_3^{2-} produces more HCO_3^- which adds to the HCO_3^- present initially in the sample to make the second plateau wider than the first

Practice Exercises

17.1.

$$CH_3COOH(aq) + H_2O(\ell) \longrightarrow CH_3COO^-(aq) + H_3O^+(aq)$$

acid   base   conjugate base   conjugate acid

17.2. pH = 2.160
17.3. $[H^+] = 2 \times 10^{-12}\ M$ and $[OH^-] = 5 \times 10^{-3}\ M$
17.4. pH = 2.23; percent ionization = 12%; $K_a = 7.9 \times 10^{-4}$
17.5. pH = 11.97
17.6. pH = 5.69
17.7. $SO_4^{2-}(aq) + H_2O(\ell) \rightleftharpoons HSO_4^-(aq) + OH^-(aq)$
17.8. pH = 9.08
17.9. pH = 4.01
17.10. pH = 4.75. There is essentially no change in the pH.
17.11. $S = 2.6 \times 10^{-3}\ M$
17.12. $S = 1.5 \times 10^{-5}\ M$
17.13. Yes
17.14. (a) Yes, both BaF_2 and CaF_2 are slightly soluble.
(b) Yes, Ba^{2+} and Ca^{2+} ions in solution can be completely separated by selective precipitation with F^-.
17.15. (a) Ammonia concentration = 0.0650 M
(b) Methyl red (pH at equivalence point = 5.29)

CHAPTER 18

Concept Tests

p. 883 Yes, N donates a pair of electrons to form the N—B bond.
p. 884 CN^- ions are in the inner coordination sphere, and the Na^+ are counter ions.
p. 886 sp^3d^2, because there are no empty $3d$ orbitals
p. 889 Tetracyanoplatinate(II)
p. 893 Yes; as a base: $Al(OH)_3(s) \rightleftharpoons Al(OH)_2^+(aq) + OH^-(aq)$; as an acid: $Al(OH)_3(s) + H_2O(\ell) \rightleftharpoons Al(OH)_4^-(aq) + H^+(aq)$
p. 897 Due to a negative ΔH because ΔS is probably close to zero
p. 901 Green
p. 904 (a) CN^- is a stronger field ligand than pyridine. (b) Ru^{2+} ions are larger than Fe^{2+} and their $4d$ electrons interact more with ligand lone pairs than do the $3d$ electrons of Fe^{2+}.
p. 905 Just one form, no isomers
p. 907 No for square planar, yes for tetrahedral

Practice Exercises

18.1. CaO acts as a Lewis base and CO_2 acts as a Lewis acid.
18.2. $[Ag^+(aq)] = 1.6 \times 10^{-8}\ M$
18.3.

(a) $[Zn(NH_3)_4]Cl_2$ = tetraamminezinc(II) chloride
(b) $[Co(NH_3)_4(H_2O)_2](NO_2)_2$ = tetraamminediaquacobalt(II) nitrite
18.4. Four
18.5. Ni^{3+} can have either a high-spin or a low-spin configuration; none of the ions are diamagnetic.
18.6.

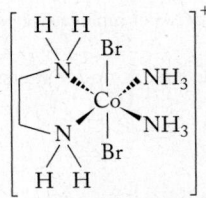

cis-Diammine-*cis*-dibromo-
ethylenediaminecobalt(III) ion

cis-Diammine-*trans*-dibromo-
ethylenediaminecobalt(III) ion

trans-Diammine-*cis*-dibromo-
ethylenediaminecobalt(III) ion

CHAPTER 19

Concept Tests

p. 921 No because the atomic masses of Cu and Zn are different
p. 922 ... Zn^{2+} ions as Cu^{2+} ions are reduced to Cu metal at the cathode.
p. 924 a
p. 929 $\Delta G_{cell} > 0$, so positive; $E_{cell} < 0$, so negative
p. 931 0.257 V
p. 932 $E_{red} = E_{red}^\circ + 0.0592\ V = 0.740\ V$
p. 937 (a) A-hr; (b) kW-hr; (c) kW-hr
p. 938 (a) +3; (b) −1; (c) +1; (d) +1
p. 939 C is oxidized; Co is reduced.
p. 944 Cathode: H_2; Anode: Cl_2
p. 945 Anode

Practice Exercises

19.1. $O_2(g) + 2\,NO_2^-(aq) \rightarrow 2\,NO_3^-(aq)$
19.2. The balanced redox reaction is $3\,Cu^{2+}(aq) + 2\,Al(s) \rightarrow 3\,Cu(s) + 2\,Al^{3+}(aq)$. The cell diagram is $Al(s)\,|\,Al^{3+}(aq)\,||\,Cu^{2+}(aq)\,|\,Cu(s)$.
19.3. The net ionic equation is $Cd(s) + 2\,NiO(OH)(s) + 2\,H_2O(\ell) \rightarrow Cd(OH)_2(s) + 2\,Ni(OH)_2(s)$. $E_{cell}^\circ = 1.72\ V$
19.4. $\Delta G_{cell} = -290\ kJ$
19.5. $E_{cell} = 1.64\ V$
19.6. $K = 1.8 \times 10^{62}$
19.7. 130 g
19.8. 1.5 g

| Compound | Counter Ion | LIGAND | | | | | | | | | M^{n+} |
		Formula	Name	Number	Prefix	Formula	Name	Number	Prefix	
$[Zn(NH_3)_4]Cl_2$	Cl^- (chloride)	NH_3	ammine	4	tetra-					2+
$[Co(NH_3)_4$ $(H_2O)_2](NO_2)_2$	NO_2^- (nitrite)	NH_3	ammine	4	tetra-	H_2O	aqua	2	di-	2+

CHAPTER 20

Concept Tests

p. 958 sp^3; 109.5°

p. 960 Two of the groups on the sp^3 C atom are identical.

p. 961 Aspartic acid (a) < glycine (c) < lysine (b)

p. 964 Arginine (b) > alanine (a) > aspartic acid (c)

p. 966 Those with nonpolar R groups

p. 967 (a) β-pleated sheet; (b) no, a change in the 2° structure is involved.

p. 972 As with glucose, the C-5 –OH group can approach C-2 from either side.

p. 973 Figure 20.24a

p. 975 Glycolysis of sugar produces the smaller molecule acetyl-CoA, which is a precursor in the synthesis of cholesterol.

p. 978 (top) At low temperatures the unsaturated fatty acids remain liquid.

p. 978 (bottom) Olestra is a much larger molecule with many more C—C and C—H bonds, so on a per mole basis it would give off much more energy.

p. 983 No, the direction in which the code is read matters; UUG codes for leucine.

Practice Exercises

20.1.

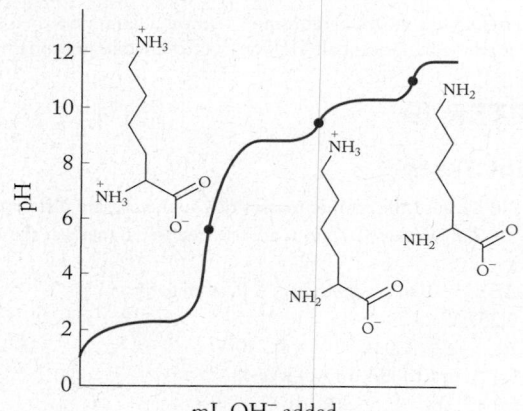

20.2. 6

20.3. 6

20.4. (a) GCCATAGGCTA; (b) AATTCGGCGATC

CHAPTER 21

Concept Tests

p. 996 Yes, because the sums of the superscripts and subscripts are the same on both sides of the reaction arrow

p. 999 Neutron

p. 1005 The loss in mass becomes an energy that must be added to separate the nucleons that are bound together.

p. 1007 (a) ^{51}Mn and ^{53}Mn undergo electron capture. (b) ^{57}Mn undergoes β decay. (c) ^{55}Mn is stable.

p. 1009 The n–p ratio for stable isotopes increases with atomic number. For heavy isotopes it is about 1.5 to 1. Losing α particles increases this ratio, creating neutron-rich nuclides that undergo β decay.

p. 1015 To fuse nuclei their coulombic repulsion must be overcome which requires they collide at very high velocities, which require very high temperatures. Man-made fusion has not been carried out except in a hydrogen bomb. One cannot make a fusion reactor until the process itself can be safely carried out.

p. 1019 Gamma rays > X-rays > UV radiation > microwaves > radio waves

p. 1022 The watch dial protected the wearer from most α particles emitted by Ra. There is no protection from inhaled Rn, so it is more hazardous.

p. 1025 He gas might escape over time.

p. 1027 Increasing fossil (C-14 depleted) CO_2 in the air will reduce the $^{14}C/^{12}C$ ratio in the air and in plants.

Practice Exercises

21.1. 98.6%

21.2. 1.753×10^9 kJ/mol

21.3. Beta decay, $^{28}_{14}$Si

21.4. $^{37}_{17}$Cl $+ ^1_0 n \rightarrow ^0_{-1}\beta + ^{38}_{18}$Ar

21.5. $A = 3.7 \times 10^7$ Bq; $A = 1.0$ mCi

21.6. 1.4×10^{11} X-rays

21.7. 4950 years old

CHAPTER 22

Concept Tests

p. 1042 They are not found in nature and so have no natural biological function.

p. 1044 Nonspontaneous, they require energy to pump ions

p. 1046 Because it has no d electrons

p. 1048 $K > 1$ for the drug to be effective

p. 1053 H_2 because it has no lone pairs of electrons

p. 1055 In O_2^- the oxidation number of O is $-1/2$; it is readily reduced to the more stable 2− ion.

p. 1056 Yes, the spin state changes.

p. 1057 They are generally toward the bottom of the periodic table. The nonessential elements have larger atomic numbers and are less abundant than the essential elements in the same group.

p. 1059 To avoid tissue damage, the nuclide should not emit α particles or high-energy β particles.

p. 1063 Longer half-life means slower decay and longer time to get the image.

p. 1073 The reduction potentials should be less negative than Fe so that the metals are less likely to oxidize than iron.

Practice Exercises

22.1. pH = 1.194; the volume of $Mg(OH)_2$ solution required to neutralize the acid solution is 5.83×10^{-2} mL.

22.2. $N_2(g) + 10 H^+(aq) + 8 e^- \rightarrow 2 NH_4^+(aq) + H_2(g)$

22.3. $^{201}_{81}$Tl $+ ^0_{-1}\beta \rightarrow ^{201}_{80}$Hg

22.4. The amount of ^{186}Re is 21 mg after 24 hr. The amount of ^{188}Re is 9.4 mg after 24 hr.

22.5. $K_{overall} = 2.5 \times 10^6$

22.6. $E^\circ_{cat} = 1.05$ V

Answers to Selected End-of-Chapter Questions and Problems

CHAPTER 1

1.1. a. A pure compound in the gas phase.
      b. A mixture of blue element atoms and red element atoms: blue atoms are in the gas phase, red spheres are in the liquid phase.

1.3. b

1.5. H_3COH or CH_4O

1.7. The sun is an example where matter is being changed into energy through nuclear fusion reactions. Therefore, both students are correct.

1.9. One chemical property of gold is its resistance to corrosion (oxidation). Gold's physical properties include its density, color, melting temperature, and electrical and thermal conductivity.

1.11. Add water to the salt–sand mixture to dissolve the salt. Passing the sand–solution mixture through a filter will leave the sand on the filter. The salt can be recovered by evaporating the water from the solution that passed through the filter.

1.13. b. Combustion

1.15. A Snickers bar (b) and an uncooked hamburger (d)

1.17. Orange juice (with pulp)

1.19. We can distinguish between table sugar, water, and oxygen by examining their physical states (sugar is a solid, water is a liquid, and oxygen is a gas) and by their densities, melting points, and boiling points.

1.21. Density, melting point, thermal and electrical conductivity, and softness (a–d) are all physical properties while tarnishing and reaction with water (e and f) are both chemical properties.

1.23. Distillation will separate the water and the dissolved proteins that have formed a homogeneous solution with the water. This would have to be accomplished, however, at low temperature, because heating the solution of the enzyme might cause a chemical change.

1.25. At ordinary temperatures Fe is a solid, O_2 is a gas, and Hg is a liquid.

1.27. Extensive properties will change with the size of the sample and therefore cannot be used to identify a substance.

1.29. To form a hypothesis we need at least one observation, experiment, or idea (from examining nature).

1.31. Yes

1.33. *Theory* in normal conversation is someone's idea or opinion or speculation that can be changed.

1.35. SI units can be easily converted into a larger or smaller unit by multiplying or dividing by multiples of 10. English units are based on other number multiples and thus are more complicated to manipulate.

1.37. 1.0792×10^9 km/hr

1.39. 93.2%

1.41. 4.1×10^{13} km

1.43. 1330 Cal

1.45. 2.5 mi

1.47. 4.0 m/s

1.49. 23 g

1.51. 19.0 mL

1.53. 26.5 g; 0.0265 kg

1.55. 58.0 cm^3

1.57. 73.8 mL

1.59. 5.1 g/cm^3

1.61. Yes

1.63. 0.28 cm^3

1.65. a. Accuracy means that the experimental value agrees with the true value of the measurement. Precision in measurements means that several repeated measurements agree with each other with little variability.
b. No, the lawyer confuses the two. He is saying that if his weight is not known exactly to the ounce compared to the true value, then even the pound value is in error.
c. Yes, to be precisely accurate a series of measurements would be each very close to each other and also in agreement with the true value.
d. The sign "Precise Weight" means that the weight as measured by the scale is within the smallest unit of scale (presumably an ounce) of the true value.

1.67. a. Manufacturer #1 has a range of $0.516 - 0.504 = 0.012$ μm; Manufacturer #2 has a range of $0.514 - 0.512 = 0.002$ μm; and Manufacturer #3 has a range of $0.502 - 0.500 = 0.002$ μm.
b. Yes, Manufacturers #2 and #3 can justify the claim.
c. Yes in the case of Manufacturer #2 the lines are printed at wider widths than the widths specified.

1.69. (b) 0.08206, (c) 8.314, (f) 3.752×10^{-5}, and (d) 5420 if the 0 is significant

1.71. a. 17.4
b. 1×10^{-13}
c. 5.70×10^{-23}
d. 3.58×10^{-3}

1.73. Yes, $-40°C$ is equal to $-40°F$.

1.75. $-269.0°C$

1.77. 285.4 K; 54.1°F

1.79. 39.2°C

1.81. $-89.2°C$; 183.9 K

1.83. $-38°F$; 230°F

1.85. The T_c for YBa$_2$Cu$_3$O$_7$ is already expressed in kelvins, $T_c = 93.0$ K. The T_c of Nb$_3$Ge converted to K is 23.2 K. The T_c of HgBa$_2$CaCu$_2$O$_6$ converted to K is 127.0 K. The superconductor with the highest T_c is HgBa$_2$CaCu$_2$O$_6$.

1.87. 0.031 mg/L

1.89. Both mixtures a and b react so that there is neither sodium nor chlorine left over.

1.91. (a) No; (b) 83 plates

1.93. 17 bicycles

1.95. Day 11

CHAPTER 2

2.1. (c) A mixture of NO$_2$ and NO

2.3. The element shaded dark blue—helium

2.5. a. Chlorine (Cl$_2$) (yellow)
b. Neon (Ne) (red)
c. Sodium (Na) (dark blue)

2.7. a. Mg (green) will form MgO.
b. K (red) will form K$_2$O.
c. Ti (yellow) will form TiO$_2$.
d. Al (dark blue) will form Al$_2$O$_3$.

2.9. Rutherford concluded that the positive charge in the atom could not be spread out (the pudding) in the atom, but must result from a concentration of charge in the center of the atom (the nucleus). Most of the particles were deflected only slightly or passed directly through the gold foil; so he reasoned that the nucleus must be small compared to the size of the entire atom. The negatively charged electrons do not deflect the particles, and Rutherford reasoned that the electrons took up the remainder of the space of the atom outside the nucleus.

2.11. The fact that cathode rays were deflected by a magnetic field indicated that the rays were streams of charged particles.

2.13. A *weighted average* takes into account the proportion of each value in the group of values to be averaged.

2.15. Greater than 1

2.17. a and b

2.19. 35.45 amu

2.21. Yes

2.23. 47.95 amu

2.25. Mendeleev knew only the masses of the elements at the time he arranged the elements into his periodic table.

2.27.

Atom	Mass Number	Atomic Number = Number of Protons	Number of Neutrons = Mass Number − Atomic Number	Number of Electrons = Number of Protons
(a) ^{14}C	14	6	8	6
(b) ^{59}Fe	59	26	33	26
(c) ^{90}Sr	90	38	52	38
(d) ^{210}Pb	210	82	128	82

2.29.

Symbol	^{23}Na	^{89}Y	^{118}Sn	^{197}Au
Number of protons	11	39	50	79
Number of neutrons	12	50	68	118
Number of electrons	11	39	50	79
Mass number	23	89	118	197

2.31.

Symbol	$^{37}Cl^-$	$^{23}Na^+$	$^{81}Br^-$	$^{226}Ra^{2+}$
Number of protons	17	11	35	88
Number of neutrons	20	12	46	138
Number of electrons	18	10	36	86
Mass number	37	23	81	226

2.33. (c) Be

2.35. (c) S^{2-}

2.37. (a) Ar, (b) P^{3-}, and (d) Ca^{2+}

2.39. (b) Br

2.41. NaCl and $NaSO_4$; KCl and K_2SO_4; $CaCl_2$ and $CaSO_4$; $MgCl_2$ and $MgSO_4$

2.43. Dalton's atomic theory states that, because atoms are indivisible, the ratio of the atoms (elements) in a compound is a ratio of whole numbers. Thus, in water, the ratio volumes of hydrogen to oxygen is 2:1, a whole-number ratio, because the atoms in water are in the ratio of 2:1.

2.45. 1.5

2.47. 7.5 g

2.49. NaCl, $MgCl_2$, $CaCl_2$, KCl, $SrCl_2$; Na_2SO_4, $MgSO_4$, $CaSO_4$, K_2SO_4, $SrSO_4$

2.51. Compounds a and d are molecules; compounds b and c consist of ions.

2.53. $XO_2{}^{2-}$

2.55. Roman numerals indicate the charge on the transition metal cation.

2.57. a. NO_3, nitrogen trioxide
b. N_2O_5, dinitrogen pentoxide
c. N_2O_4, dinitrogen tetroxide
d. NO_2, nitrogen dioxide
e. N_2O_3, dinitrogen trioxide
f. NO, nitrogen monoxide
g. N_2O, dinitrogen monoxide
h. N_4O, tetranitrogen monoxide

2.59. a. Na_2S, sodium sulfide
b. $SrCl_2$, strontium chloride
c. Al_2O_3, aluminum oxide
d. LiH, lithium hydride

2.61. a. cobalt(II) oxide
b. cobalt(III) oxide
c. cobalt(IV) oxide

2.63. a. BrO^-
b. $SO_4{}^{2-}$
c. $IO_3{}^-$
d. $NO_2{}^-$

2.65. a. nickel(II) carbonate
b. sodium cyanide
c. lithium hydrogen carbonate
d. calcium hypochlorite

2.67. a. hydrofluoric acid
b. bromic acid
c. H_3PO_4
d. HNO_2

2.69. a. sodium oxide
b. sodium sulfide
c. sodium sulfate
d. sodium nitrate
e. sodium nitrite

2.71. a. K_2S
b. K_2Se
c. Rb_2SO_4
d. $RbNO_2$
e. $MgSO_4$

2.73. a. manganese(II) sulfide
b. vanadium(II) nitride
c. chromium(III) sulfate
d. cobalt(II) nitrate
e. iron(III) oxide

2.75. (b) Na_2SO_3

2.77. (b) Cl_2

2.79. (a) Na

2.81. Chemistry is the study of the composition, structure, properties, and reactivity of matter. Cosmology is the study of the history, structure, and dynamics of the universe. A few of the ways that these two sciences are related might be (1) because the universe is composed of matter and the study of matter is chemistry, the study of the universe is really chemistry; (2) the changing universe is driven by chemical and atomic or nuclear reactions, which are also studied in chemistry; and (3) cosmology often asks what the universe (including stars, black holes, etc.) are made of at the atomic level.

2.83. Because quarks combine to make up the three particles that are important to the properties and reactivity of atoms: protons, neutrons, and electrons.

2.85. The density of the universe is decreasing.

2.87. The higher the charge (the number of protons), the more repulsion the nuclei feel for each other and the higher the temperature needed to overcome that repulsion.

2.89. The electron is twice as hard to remove from a helium atom compared to a hydrogen atom because it is being held by a nucleus of 2+ charge rather than one of 1+ charge.

2.91. The expanding universe was cooling and therefore could not support the high temperatures needed for fusion. Also, the expanding universe was not dense enough for nuclei to fuse.

2.93. a. $^{16}_{8}O$
b. $^{24}_{12}Mg$
c. $^{36}_{18}Ar$

2.95. a. $^{59}_{27}Co$
b. $^{121}_{51}Sb$
c. $^{110}_{80}Hg$

2.97. The transformation of ^{137}I to ^{137}Xe can be balanced as
$^{137}_{53}I \rightarrow {}^{137}_{54}Xe + {}^{0}_{-1}\beta$
The transformation of ^{137}Xe to ^{137}Cs can be balanced as
$^{137}_{54}Xe \rightarrow {}^{137}_{53}Cs + {}^{0}_{-1}\beta$
Both of these nuclear reactions involve β emission.

2.99. a. Electrons
   b. The negatively charged electrons were attracted to the positively charged plate as the electrons passed through the electric field.
   c. If the polarities of the plates were switched, the electron would still be deflected toward the positively charged plate, which would now be at the bottom of the tube.
   d. The position of the light spot on the fluorescent screen would be halfway between the position where it was before the voltage was reduced and the "zero" spot position when there is no voltage between the plates.

2.101. a. Two
   b. Both spots would be seen below the center of the screen. The spot lowest on the screen (farthest from the middle) would be the α particle, and the one closest to the middle would be the proton.

2.103. 1:2:4

2.105. 17 Cu : 1 Sn

2.107. a. Sc, Ga, Ge
   b. Ekaaluminum is gallium, ekaboron is scandium, and ekasilicon is germanium.
   c. Scandium was discovered in 1879, gallium was discovered in 1875, and germanium was discovered in 1886.

2.109. a. 52.93%
   b. 0.599 cm^3

2.111. 60.11%

2.113. a. ^{79}Br–^{79}Br = 157.8366 amu
   ^{79}Br–^{81}Br = 159.8346 amu
   ^{81}Br–^{81}Br = 161.8326 amu
   b. ^{79}Br–^{79}Br = 25.41%
   ^{79}Br–^{81}Br = 50.00%
   ^{81}Br–^{81}Br = 24.59%

CHAPTER 3

3.1. a. $4X(g) + 4Y(g) \rightarrow 4XY(g)$
   b. $4X(g) + 4Y(g) \rightarrow 4XY(s)$
   c. $4X(g) + 4Y(g) \rightarrow 2XY_2(g) + 2X(g)$
   d. $4X_2(g) + 4Y_2(g) \rightarrow 8XY(g)$

3.3. $Fe(\ell)$

3.5. Less than

3.7. It is too small a unit to express the very large number of atoms, ions, or molecules present in laboratory quantities such as a mole.

3.9. No, the molar mass of a substance does not directly correlate to the number of atoms in a molecular compound. The statement would be true only if the two compounds were composed of the same element.

3.11. (a) 7.3×10^{-10} mol Ne; (b) 7.0×10^{-11} mol CH_4;
   (c) 4.2×10^{-12} mol O_3; (d) 8.1×10^{-15} mol NO_2

3.13. (a) 2.5×10^{-12} mol bytes; (b) 3.3×10^{-15} mol bytes

3.15. (a) 7.53×10^{22} Ti atoms; (b) 7.53×10^{22} Ti atoms;
   (c) 1.51×10^{23} Ti atoms; (d) 2.26×10^{23} Ti atoms

3.17. (a) Both contain the same; (b) N_2O_4; (c) CO_2

3.19. (a) 3.00 mol; (b) 4.50 mol; (c) 1.50 mol

3.21. 41.63 mol

3.23. 0.25 mol; 10 g

3.25. (a) 1 mol; (b) 2 mol; (c) 1 mol; (d) 3 mol

3.27. (a) 64.06 g/mol; (b) 48.00 g/mol; (c) 44.01 g/mol;
   (d) 108.02 g/mol

3.29. (a) 152.16 g/mol; (b) 164.22 g/mol; (c) 148.22 g/mol;
   (d) 132.17 g/mol

3.31. (a) NO; (b) CO_2; (c) O_2

3.33. 0.752 mol SiO_2

3.35. 10.3 g

3.37. Diamond

3.39. No

3.41. No

3.43. a. $CH_4(g) + H_2O(g) \rightarrow CO(g) + 3H_2(g)$
   b. $2NH_3(g) \rightarrow N_2(g) + 3H_2(g)$
   c. $CO(g) + H_2O(g) \rightarrow CO_2(g) + H_2(g)$

3.45. a. $3FeSiO_3(s) + 4H_2O(\ell) \rightarrow Fe_3Si_2O_5(OH)_4(s) + H_4SiO_4(aq)$
   b. $Fe_2SiO_4(s) + 2CO_2(g) + 2H_2O(\ell) \rightarrow 2FeCO_3(s) + H_4SiO_4(aq)$
   c. $Fe_3Si_2O_5(OH)_4(s) + 3CO_2(g) + 2H_2O(\ell) \rightarrow$
   $3FeCO_3(s) + 2H_4SiO_4(aq)$

3.47. a. $N_2(g) + O_2(g) \rightarrow 2NO(g)$
   b. $2NO(g) + O_2(g) \rightarrow 2NO_2(g)$
   c. $NO(g) + NO_3(g) \rightarrow 2NO_2(g)$
   d. $2N_2(g) + O_2(g) \rightarrow 2N_2O(g)$

3.49. a. $N_2O_5(g) + Na(s) \rightarrow NaNO_3(s) + NO_2(g)$
   b. $N_2O_4(g) + H_2O(\ell) \rightarrow HNO_3(aq) + HNO_2(aq)$
   c. $3NO(g) \rightarrow N_2O(g) + NO_2(g)$

3.51. $2C_2H_2(g) + 5O_2(g) \rightarrow 4CO_2(g) + 2H_2O(\ell)$

3.53. Yes

3.55. (a) 4.5×10^{11} mol C; (b) 2.0×10^{10} kg CO_2

3.57. a. $2NaHCO_3(s) \rightarrow CO_2(g) + H_2O(g) + Na_2CO_3(s)$;
   b. 6.55 g CO_2

3.59. 1.17 kg

3.61. 1.5 t

3.63. (a) 1.48 kg; (b) 1.11 kg

3.65. 346 g

3.67. An empirical formula shows the lowest whole-number ratio of atoms in a substance. A molecular formula shows the actual numbers of each kind of atom that compose one molecule of the substance.

3.69. No

3.71. (a) 74.19% Na, 25.81% O; (b) 57.48% Na, 40.00% O, 2.52% H;
   (c) 27.37% Na, 1.20% H, 14.30% C, 57.13% O; (d) 43.38% Na, 11.33% C, 45.28% O

3.73. Pyrene, $C_{16}H_{10}$

3.75. NO, N_2O_3, and NO_2

3.77. No

3.79. Ti_6Al_4V

3.81. (a) MgO; (b) $2\,Mg(s) + O_2(g) \rightarrow 2\,MgO(s)$

3.83. $Mg_3Si_2H_4O_9$

3.85. $CuCl_2O_8$

3.87. The excess of oxygen is required in combustion analysis to ensure the complete reaction of the hydrogen and carbon to form water and carbon dioxide.

3.89. Yes

3.91. The empirical formula is C_2H_3. The molecular formula is $C_{20}H_{30}$.

3.93. $C_{10}H_{18}O$

3.95. (c) Less than the sum of the masses of Fe and S to start

3.97. Theoretical yield is the greatest amount of a product possible from a reaction and assumes that the reaction goes to 100% completion. The percent yield is the observed experimental yield divided by the theoretical yield and multiplied by 100.

3.99. Reactions do not always go to completion because the reaction may be slow or may have, for a portion of the reaction, yielded different products than expected.

3.101. 3 cups

3.103. 0.844 g O_2

3.105. $NH_3(g) + HCl(g) \rightarrow NH_4Cl(s)$; 0.7 g NH_3

3.107. 59%

3.109. (a) $C_6H_{12}O_6(aq) \rightarrow 2\,C_2H_5OH(\ell) + 2\,CO_2(g)$; (b) 77.1%

3.111. (a) calcium triphosphate hydroxide; (b) 39.89%; (c) decreases slightly

3.113. (a) 529.3 kg; (b) $Al_2O_3(s) + 3\,C(s) \rightarrow 2\,Al(s) + 3\,CO(g)$; (c) 824.1 kg

3.115. (a) 45 g; (b) 15 g; (c) 1.67 cm

3.117. (a) $a = 1$, $b = 3$, charge on U is 6+; (b) $c = 3$, $d = 8$, charge on U is 5.33+; (c) $x = 2$, $y = 2$, $z = 6$

3.119. (a) 5.838×10^{20} molecules of $C_{13}H_{18}O_2$; (b) 3.008×10^{21} molecules of $CaCO_3$; (c) 9×10^{18} molecules of $C_{16}H_{19}N_2Cl$

3.121. (a) No (b) $C_5H_{10}O_5(s) + 5\,O_2(g) \rightarrow 5\,CO_2(g) + 5\,H_2O(\ell)$; $2\,C_7H_{12}O_7(s) + 13\,O_2(g) \rightarrow 14\,CO_2(g) + 12\,H_2O(\ell)$

3.123. (a) FeS is iron(II) sulfide with Fe^{2+} and S^{2-}; FeS_2 is iron(IV) sulfide with Fe^{4+} and S^{2-} based on your knowledge so far in this course with Fe^{4+} and S^{2-}. Actually, this compound is Fe^{2+} with S_2^{2-} and is named iron(II) persulfide. (b) 0.26 g HCO_2H

3.125. (a) $3\,FeO(s) + H_2O(\ell) \rightarrow Fe_3O_4(s) + H_2(g)$; (b) $12\,FeO(s) + 2\,H_2O(\ell) + CO_2(g) \rightarrow 4\,Fe_3O_4(s) + CH_4(g)$

3.127. 1×10^{-8} mol

3.129. 55 mol ethanol

3.131. A is $CaCO_3$, B is CO_2, and C is CaO.

3.133. Re

3.135. 82.4%

3.137. a. 6.0 metric tons
b. $2\,SO_2(g) + 2\,H_2O(g) + O_2(g) \rightarrow 2\,H_2SO_4(\ell)$
c. 9.2 metric tons

3.139. 3.06 g H_2SO_4

3.141. Mg_2SiO_4

CHAPTER 4

4.1. Yellow

4.3. a. Cl (purple)
b. S (orange)
c. N (green)
d. P (blue)

4.5. The solvent is usually the liquid component of the solution. If both the solvent and solute are liquids or solids, the solvent is that component present in the greatest amount.

4.7. 1.00 M

4.9. a. 5.6 M $BaCl_2$
b. 1.00 M Na_2CO_3
c. 1.30 M $C_6H_{12}O_6$
d. 5.92 M KNO_3

4.11. a. 0.14 M Na^+
b. 0.11 M Cl^-
c. 0.096 M SO_4^{2-}
d. 0.20 M Ca^{2+}

4.13. a. 11.7 g NaCl
b. 4.99 g $CuSO_4$
c. 6.41 g CH_3OH

4.15. 2.72 g

4.17. a. 9.6×10^{-3} mol
b. 7.80×10^{-4} mol
c. 8.8×10^{-2} mol
d. 4.22 mol

4.19. Orchard sample: 3.4×10^{-4} mmol/L
Residential area sample: 5.6×10^{-5} mmol/L
After storm sample: 3.2×10^{-2} mmol/L

4.21. 1.5×10^{-4} M

4.23. $AgNO_3$, $Fe(NO_3)_2$; $6\,H_2O$, and $Ca(OH)_2$

4.25. 4.57×10^{-2} M Mg^{2+}

4.27. a. The final concentration after diluting will be 1.81×10^{-2} M Na^+.
b. The final concentration after diluting will be 2.7×10^{-1} mM LiCl.
c. The final concentration after diluting will be 1.28×10^{-2} mM Zn^{2+}.

4.29. 1.95 M

4.31. The concentration of the adult-strength medication is 1.8 mg/mL; 23 mL is needed to prepare the child-strength cough syrup.

4.33. Table salt produces Na^+ and Cl^- ions in solution when it dissolves. Sugar does not dissociate into ions because it is not a salt. Ions are required to conduct electricity.

4.35. The lack of ions in methanol means that the liquid is nonconductive. Molten NaOH, however, has freely moving Na^+ and OH^- ions, which can conduct electricity.

4.37. In order of decreasing conductivity, 1.0 M Na_2SO_4 (c) > 1.2 M KCl (b) > 1.0 M NaCl (a) > 0.75 M LiCl (d).

4.39. a. 0.025 M
   b. 0.050 M
   c. 0.075 M

4.41. (b) 1 M $CaCl_2$

4.43. Acid

4.45. Strong acids include HCl, HNO_3, $HClO_4$, H_2SO_4, HI, HBr; weak acids include CH_3COOH, HCOOH, HF, H_3PO_4.

4.47. Base

4.49. Strong bases include NaOH, KOH, CsOH, LiOH, RbOH, $Ba(OH)_2$, $Sr(OH)_2$, $Ca(OH)_2$; weak bases include NH_3, CH_3NH_2, C_5H_5N.

4.51. a. Ionic and net ionic equation: $2H^+(aq) + SO_4^{2-}(aq) + Ca^{2+}(aq) + 2OH^-(aq) \rightarrow CaSO_4(s) + 2H_2O(\ell)$
   The acid is H_2SO_4; the base is $Ca(OH)_2$.
   b. Ionic and net ionic equation: $PbCO_3(s) + 2H^+(aq) + SO_4^{2-}(aq) \rightarrow PbSO_4(s) + CO_2(g) + H_2O(\ell)$
   $PbCO_3$ is the base; sulfuric acid is the acid.
   c. Ionic equation: $Ca^{2+}(aq) + 2OH^-(aq) + 2CH_3COOH(aq) \rightarrow Ca^{2+}(aq) + 2CH_3COO^-(aq) + 2H_2O(\ell)$
   Calcium is a spectator ion. $Ca(OH)_2$ is the base; CH_3COOH is the acid.
   Net ionic equation:
   $OH^-(aq) + CH_3COOH(aq) \rightarrow CH_3COO^-(aq) + H_2O(\ell)$

4.53. a. Molecular equation:
   $Mg(OH)_2(s) + H_2SO_4(aq) \rightarrow MgSO_4(aq) + 2H_2O(\ell)$
   Net ionic equation:
   $Mg(OH)_2(s) + 2H^+(aq) \rightarrow Mg^{2+}(aq) + 2H_2O(\ell)$
   b. Molecular equation:
   $MgCO_3(s) + 2HCl(aq) \rightarrow MgCl_2(aq) + H_2CO_3(aq)$
   Net ionic equation:
   $MgCO_3(s) + 2H^+(aq) \rightarrow Mg^{2+}(aq) + H_2O(\ell) + CO_2(g)$
   c. Molecular equation: $NH_3(g) + HCl(g) \rightarrow NH_4Cl(s)$
   This is also the net ionic equation.

4.55. $PbCO_3(s) + 2H^+(aq) \rightarrow Pb^{2+}(aq) + CO_2(g) + H_2O(\ell)$;
   $Pb(OH)_2(s) + 2H^+(aq) \rightarrow Pb^{2+}(aq) + 2H_2O(\ell)$

4.57. a. 5.00 mL
   b. 31.5 mL
   c. 21.5 mL

4.59. 500 mL

4.61. 290 mL

4.63. A saturated solution contains the maximum concentration of a solute. A supersaturated solution *temporarily* contains *more* than the maximum concentration of a solute at a given temperature.

4.65. A precipitation reaction occurs when two solutions are mixed to form an insoluble compound.

4.67. A saturated solution may not be a concentrated solution if the solute is only sparingly or slightly soluble in the solution. In that case, the solution is a saturated dilute solution.

4.69. (a) Barium sulfate is insoluble; (e) lead hydroxide is insoluble; (f) calcium phosphate is insoluble.

4.71. a. Balanced reaction:
   $Pb(NO_3)_2(aq) + Na_2SO_4(aq) \rightarrow PbSO_4(s) + 2NaNO_3(aq)$.
   Net ionic equation: $Pb^{2+}(aq) + SO_4^{2-}(aq) \rightarrow PbSO_4(s)$
   b. No precipitation reaction occurs.
   c. Balanced reaction:
   $FeCl_2(aq) + Na_2S(aq) \rightarrow FeS(s) + 2NaCl(aq)$.
   Net ionic equation: $Fe^{2+}(aq) + S^{2-}(aq) \rightarrow FeS(s)$
   d. Balanced reaction:
   $MgSO_4(aq) + BaCl_2(aq) \rightarrow MgCl_2(aq) + BaSO_4(s)$.
   Net ionic equation: $Ba^{2+}(aq) + SO_4^{2-}(aq) \rightarrow BaSO_4(s)$

4.73. $CaCO_3$

4.75. 2.11×10^{-2} g

4.77. 5.4×10^{-2} g

4.79. 130 kg

4.81. To deionize water, cations such as Na^+ and Ca^{2+} are exchanged for H^+ at cation-exchange sites. Anions such as Cl^- and SO_4^{2-} are exchanged for OH^- at the anion-exchange sites. The released ions (H^+ and OH^-) at these sites combine to form H_2O.

4.83. To deionize water, the cation at the cation-exchange site must be H^+ and the anion at the anion-exchange site must be OH^-. When these combine, they form H_2O.

4.85. The number of electrons gained or lost is directly related to the change in oxidation number of a species.

4.87. a. −1
   b. +1
   c. −2
   d. −3

4.89. Silver

4.91. $Na^+ + e^- \rightarrow Na(s)$
   $2Cl^- \rightarrow Cl_2(g) + 2e^-$

4.93. a. +1
   b. +5
   c. +7

4.95. a. $2e^- + Br_2(\ell) \rightarrow 2Br^-(aq)$; reduction
   b. $Pb(s) + 2Cl^-(aq) \rightarrow PbCl_2(s) + 2e^-$; oxidation
   c. $2e^- + O_3(g) + 2H^+(aq) \rightarrow O_2(g) + H_2O(\ell)$; reduction
   d. $H_2S(g) \rightarrow S(s) + 2H^+(aq) + 2e^-$; oxidation

4.97. $H_2O(\ell) + 2Fe_3O_4(s) \rightarrow 3Fe_2O_3(s) + 2H^+(aq) + 2e^-$

4.99. a. Reactants   Products
   SiO_2: Si = +4, O = −2 Fe_2SiO_4: Fe = +2, Si = +4, O = −2
   Fe_3O_4: Fe = +8/3, O = −2 O_2: O = 0
   Oxygen is oxidized (O^{2-} to O_2) and iron is reduced (Fe^{3+} to Fe^{2+}).
   b. Reactants   Products
   SiO_2: Si = +4, O = −2 Fe_2SiO_4: Fe = +2, Si = +4, O = −2
   Fe: Fe = 0
   O_2: O = 0
   Iron is oxidized (Fe^0 to Fe^{2+}) and oxygen is reduced (O_2 to O^{2-}).

c. Reactants   Products
FeO: Fe = +2, O = −2   Fe(OH)$_3$: Fe = +3, O = −2, H = +1
O$_2$: O = 0
H$_2$O: H = +1, O = −2
Iron is oxidized (Fe^{2+} to Fe^{3+}) and oxygen is reduced (O$_2$ to O^{2-}).

4.101. a. $O_2(aq) + 4\,FeCO_3(s) \rightarrow 2\,Fe_2O_3(s) + 4\,CO_2(g)$
    b. $O_2(aq) + 6\,FeCO_3(s) \rightarrow 2\,Fe_3O_4(s) + 6\,CO_2(g)$
    c. $O_2(aq) + 4\,Fe_3O_4(s) \rightarrow 2\,Fe_2O_3(s)$

4.103. $NH_4^+(aq) + 2\,O_2(g) \rightarrow NO_3^-(aq) + 2\,H^+(aq) + H_2O(\ell)$

4.105. $2\,Fe(OH)_2^+(aq) + Mn^{2+}(aq) \rightarrow$
             $2\,Fe^{2+}(aq) + 2\,H_2O(\ell) + MnO_2(s)$

4.107. $2\,H_2O(\ell) + 4\,Ag(s) + 8\,CN^-(aq) + O_2(g) \rightarrow$
             $4\,Ag(CN)_2^-(aq) + 4\,OH^-(aq)$

4.109. a. $2\,ClO_3^-(aq) + SO_2(g) \rightarrow 2\,ClO_2(g) + SO_4^{2-}(aq)$
    b. $4\,H^+(aq) + 2\,ClO_3^-(aq) + 2\,Cl^-(aq) \rightarrow$
             $2\,ClO_2(g) + 2\,H_2O(\ell) + Cl_2(g)$
    c. $2\,ClO_3^-(aq) + Cl_2(g) \rightarrow 2\,ClO_2(g) + 2\,Cl^-(aq) + O_2(g)$

4.111. $7.98 \times 10^{-4}\ M\ SO_4^{2-}$

4.113. a. 11.7 M
    b. 42.7 mL
    c. 1.72 kg

4.115. a. $2\,OH^-(aq) + 2\,H_2O(\ell) + 3\,S_2O_4^{2-}(aq) + 2\,CrO_4^{2-}(aq) \rightarrow$
             $6\,SO_3^{2-}(aq) + 2\,Cr(OH)_3(s)$
    b. Sulfur is oxidized; chromium is reduced.
    c. Oxidizing agent = CrO_4^{2-}; reducing agent = $S_2O_4^{2-}$.
    d. 38.7 g

4.117. a. Balanced equation: $2\,Ag(s) + H_2S(g) \rightarrow Ag_2S(s) + H_2(g)$
    Oxidation numbers:
    Ag = 0 Ag_2S : Ag = +1, S = −2
    H_2S: H = +1, S = −2 H_2: H = 0
    One mole of electrons is transferred per mole of Ag.
    b. $3\,Ag_2S(s) + 12\,H_2O(\ell) + 4\,Al(s) \rightarrow$
             $6\,Ag(s) + 3\,H_2S(g) + 3\,H_2(g) + 4\,Al(OH)_3(s)$

4.119. a. H_3PO_4; phosphoric acid
    b. H_2SeO_3; selenous acid
    c. H_3BO_3; boric acid

4.121. $2\,H^+(aq) + ClO^-(aq) + 2\,I^-(aq) \rightarrow Cl^-(aq) + H_2O(\ell) + I_2(aq)$
    $I_2(aq) + 2\,S_2O_3^{2-}(aq) \rightarrow 2\,I^-(aq) + S_4O_6^{2-}(aq)$

4.123. a. $NaClO_4$, NH_4ClO_4
    b. 427 kg
    c. 2.80×10^{10} gal
    d. The MA lab

4.125. a. $3\,CH_2O \rightarrow CO_2 + C_2H_5OH$
    b. $C_2H_5OH + O_2 \rightarrow HC_2H_3O_2 + H_2O$
    c. CH_2O: C = 0
    CO_2: C = +4
    C_2H_5OH: C = −4 over two carbon atoms, so oxidation number on each carbon = −2
    $HC_2H_3O_2$: C = 0
    d. 66.7 g acetic acid

4.127. a. $2\,H^+(aq) + SO_4^{2-}(aq) + Ba^{2+}(aq) + 2\,OH^-(aq) \rightarrow$
             $BaSO_4(s) + 2\,H_2O(\ell)$
    b. c

4.129. a. The first reaction is a redox reaction; eight electrons are transferred.
    b. $2\,H^+(aq) + SO_4^{2-}(aq) + CaCO_3(s) \rightarrow$
             $CaSO_4(s) + H_2O(\ell) + CO_2(g)$
    c. $SO_4^{2-}(aq) + CaCO_3(s) \rightarrow CaSO_4(s) + CO_3^{2-}(aq)$

4.131. c and d

CHAPTER 5

5.1. At 35 ft above street level KE = 150 J; just before hitting the street KE = 500 J

5.3. a.

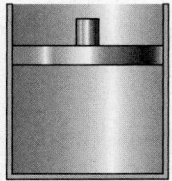

    b. The piston is higher in the cylinder.
    c. Yes
    d. The system did work on the surroundings.

5.5. a. A closed system
    b. The internal energy of the system will increase.
    c. No

5.7. a. Because the heat of formation of an element in its standard state is defined as zero
    b. Because the enthalpy of its formation is positive
    c. By subtracting the sum of the enthalpies of formation of the reactants (multiplied by the number of moles in the balanced equation for each) from the sum of enthalpies of formation of the products (again multiplied by their molar amounts from the balanced equation).

5.9. Energy makes work possible.

5.11. The value of a state function is independent of the path; only the initial and final values are important.

5.13. a. The potential energy in a battery consists of the chemicals that can react via a redox reaction.
    b. The potential energy in a gallon of gasoline consists of the chemical bonds in the fuel that release heat as the fuel is combusted.
    c. The potential energy of the crest of a wave is due to its position above the ground.

5.15. The system is that part of the universe that we are interested in. The surroundings are everything else, extending to the entire universe.

5.17. The internal energy of the gas sample can be increased by raising the temperature or by increasing the pressure through compression.

5.19. a. Exothermic
    b. Exothermic
    c. Endothermic

5.21. Energy is absorbed from the surroundings. Thus, q increases and therefore ΔE increases.

5.23. $w = -0.500 \text{ L} \cdot \text{atm} = -50.7 \text{ J}$

5.25. a. 50 J
b. 6.2 kJ
c. −940 J

5.27. −276 kJ

5.29. b

5.31. A change in enthalpy is the sum of the change of internal energy and the product of the system's pressure and change in volume.

5.33. If the system transfers energy to the surroundings its energy will be less after the process than at the start of the process.

5.35. Negative

5.37. Positive

5.39. Negative

5.41. Specific heat is specified for a gram of the substance. Heat capacity does not take into account how much of a substance there is; it is defined for a given object.

5.43. No

5.45. Water's high heat capacity compared to air means that water carries away more energy from the engine for every Celsius degree rise in temperature, so water is a good choice to cool automobile engines.

5.47. 29.3 kJ

5.49.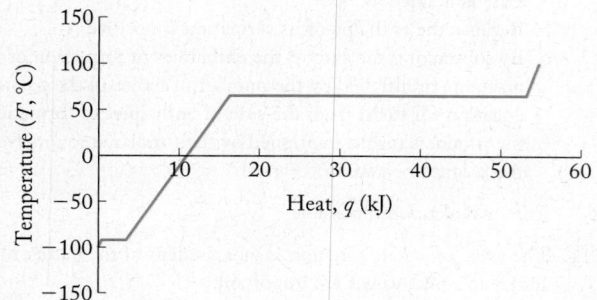

5.51. 886 g

5.53. −47.5°C

5.55. To know how much energy (generated or absorbed by the system) is required to change the temperature of the surroundings (the calorimeter) in order to calculate the heat capacity or final temperature of the system in an experiment

5.57. Yes

5.59. 8.044 kJ/°C

5.61. −5129 kJ/mol

5.63. 23.29°C

5.65. No

5.67. a and d

5.69. −252.9 kJ

5.71. −35.9 kJ

5.73. −7198 kJ

5.75. The energy per gram a fuel releases on burning

5.77. The fuel value (kJ/g) is obtained by dividing the molar heat of combustion (kJ/mol) by the molar mass (mol/g).

5.79. 201 kg

5.81. a. 48.99 kJ/g
b. 4.90×10^4 kJ
c. 5.97 g

5.83. When we apply Hess's law all the heat is accounted for in the reaction; energy is neither created nor destroyed when using Hess's law.

5.85. If we write out the chemical equations for the ΔH_f° for the reactants and products, these formation reactions will add up to the overall reaction.

5.87.
$$CO_2(g) \rightarrow CO(g) + \tfrac{1}{2}O_2(g)$$
$$C(s) + O_2(g) \rightarrow CO_2(g)$$
$$\overline{C(s) + \tfrac{1}{2}O_2(g) \rightarrow CO(g)}$$

5.89. −297 kJ/mol

5.91. 28.0 kJ/mol

5.93. a. The first reaction, which produces NO from N_2 and O_2
b. −103 kJ

5.95. 2.24 J/(g · K)

5.97. 342 kJ

5.99. 3.60 kJ/°C

5.101. We are given that ΔE for this process is less than q absorbed ($\Delta E < q$). If this is the case, for the equality $\Delta E = q + w$ to be maintained, then w must be negative. A negative value of w must mean that work is done by the system on the surroundings.

5.103. a. $2\,NaOH(aq) + H_2SO_4(aq) \rightarrow 2\,H_2O(\ell) + Na_2SO_4(aq)$
b. No
c. −114 kJ/mol H_2SO_4

5.105. 26.0°C

5.107. d

5.109. a. Inorganic compounds: B_2H_6, CS_2, HI, N_2O, O_3; organic compounds; C_2H_4, C_6H_6, $(CH_3)_2C\!=\!C(CH_3)_2$
b. −1255.5 kJ/mol

5.111. $Zn(s) + \tfrac{1}{4}S_8(s) + 3\,O_2(g) \rightarrow ZnSO_4(s) + SO_2(g)$; −1279.6 kJ

5.113. −272 J; 4.53×10^{-2} mol

5.115. $\Delta H_3 = -\Delta H_1 + -\Delta H_2$

5.117. a. $CH_3OH(g) + N_2(g) \rightarrow HCN(g) + NH_3(g) + \tfrac{1}{2}O_2(g)$
b. As a reactant
c. 307 kJ

5.119. 175 kJ

5.121. −841 kJ/mol

5.123. 31.5 kJ

5.125. Hydrogen

5.127. a. Carbon is both oxidized and reduced.
b. 636.0 kJ

CHAPTER 6

6.1. Barometer a

6.3. a

6.5. c

6.7. Line 1

6.9. Line 2

6.11. The molar mass of helium is 4 g/mol. The line on the graph in Figure P6.10 for this gas goes below line 1. The molar mass of NO is 30 g/mol so the line on the graph for this gas goes above line 2. All lines will converge at $P = 0$ and $d = 0$.

6.13. The total pressures for flasks a, b, and d are the same. Flask c has a lower pressure. The flask with the highest partial pressure of N_2 is b.

6.15. The maximum of the distribution curve will be at lower molecular speeds and the distribution will not be as wide.

6.17. Br_2 (orange)

6.19. a

6.21. Force is the product of the mass of an object and the acceleration due to gravity. Pressure uses force in its definition: It is the force an object exerts over a given area.

6.23. 760 torr = 1 atmosphere

6.25. The ethanol barometer

6.27. A sharpened blade has a smaller area over which the force is distributed compared to a dull blade.

6.29. As we go up in altitude, the overlying mass of the atmosphere above us decreases, so the pressure also decreases.

6.31. 3.9×10^3 Pa

6.33. a. 0.020 atm
b. 0.739 atm

6.35. a. 814.6 mmHg
b. 1.072 atm
c. 1086 mbar

6.37. The higher the temperature, the faster the gas molecules move. The faster they move, the more often they collide with the walls of the container and the greater the force with which gas molecules hit the walls. Both of these result in increased pressure as temperature is raised.

6.39. The balloonist should decrease the temperature.

6.41. The gas pressure increases.

6.43. 2.00 atm

6.45. 2.30 atm; 13.0 m

6.47.

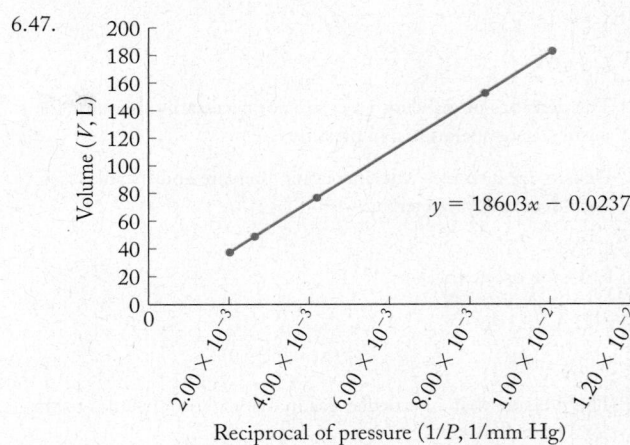

$y = 18603x − 0.0237$

Reciprocal of pressure ($1/P$, 1/mm Hg)

Yes, the graph is exactly the same for the same number of moles of argon gas.

6.49.

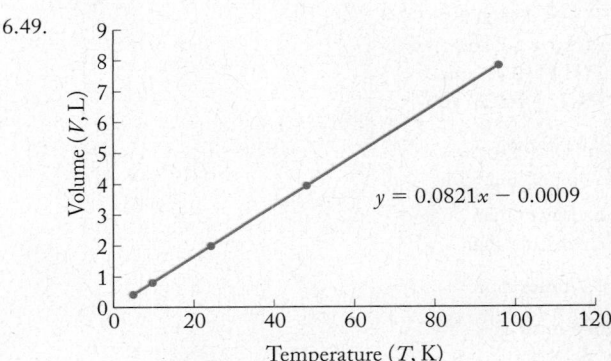

$y = 0.0821x − 0.0009$

Temperature (T, K)

If the amount of gas were halved, the graph would still be linear, but the slope of the line would be halved.

6.51. 596 K or 323°C

6.53. 4.27 L

6.55. 1.7 L

6.57. b

6.59. a. No change
b. Decrease to 1/4 the original volume
c. Increase of 17%

6.61. 144 L

6.63. 6.6 atm

6.65. STP is defined as 1 atm and 0°C (273 K); V = 22.4 L

6.67. The product of the number of moles of gas in the sample, the temperature, and the gas constant

6.69. 0.67 mol

6.71. 1.50 atm

6.73. 1730 L

6.75. 814 g

6.77. (a) 0.0419 mol/hr; (b) 10.7 g

6.79. 6.2×10^2 g

6.81. 715 g

6.83. The densities of different gases are not necessarily the same for a particular temperature and pressure.

6.85. Density (a) increases with increasing pressure and (b) increases with decreasing temperature.

6.87. a. 9.08 g/L
b. In the basement

6.89. SO_2

6.91. CO

6.93. The pressure that a particular gas individually contributes to the total pressure

6.95. Sample c

6.97. 0.20

6.99. $P_{total} = 2.46$ atm
$P(N_2) = 1.7$ atm
$P(H_2) = 0.49$ atm
$P(CH_4) = 0.25$ atm

6.101. 0.0190 mol

6.103. a. Greater than
b. Lower than
c. Greater than

6.105. 1.7 times more

6.107. 680 mmHg

6.109. 25%

6.111. The speed of a molecule in a gas that has the average kinetic energy of all the molecules of the sample

6.113. a. As the molar mass increases, u_{rms} decreases
b. As temperature increases, the u_{rms} increases

6.115. To determine the molar mass of an unknown gas, measure the rate of its effusion (r_x) relative to the rate of effusion of a known gas (r_y). Since we know the molar mass of the known gas, $\mathcal{M}_y$, we can use the equation to solve for the unknown $\mathcal{M}_x$.

6.117. Diffusion is the spread of one substance into another. Effusion is the escape of a gas from its container through a tiny hole. The gas is escaping from a region of higher pressure to one of lower pressure.

6.119. The rank order in terms of increasing root-mean-square speed is $SO_2 < NO_2 < CO_2$.

6.121. Gas C

6.123. 717 m/s

6.125. 0.711

6.127. 32.3 g/mol

6.129. 18.2 g/mol

6.131. a. $r(^{12}CO_2)/r(^{13}CO_2) = 1.01$
b. $^{12}CO_2$ diffuses faster than $^{13}CO_2$.

6.133. The smaller balloon contains hydrogen.

6.135. At low temperatures the gas particles move more slowly, and their collisions become inelastic; they stick together due to the weak attractive forces between them. The particles, therefore, do not act separately to contribute to the pressure in the container, and the pressure is lower than would be expected by the ideal gas law. Also, the gas particles take up real volume in the container and as the pressure increases the volume of the particles takes up a greater volume of the free space in the container. This has the effect of raising the pressure–volume product above what we would expect from the ideal gas law (in a plot of PV/RT versus P).

6.137. Since b is a measure of the volume that the gas particles occupy, b increases as the sizes of the particles increase.

6.139. H_2

6.141. a. $P = 910$ atm
b. $P = 476$ atm

6.143. 126 L

6.145. 27.3 L

6.147. 21.1 atm

6.149. 3.31 L

6.151. 0.25 m/s

6.153. 18 kg

6.155. Xe

6.157. a. $NH_3(g) + HCl(g) \rightarrow NH_4Cl(s)$
b. The ring of NH_4Cl should appear closer to the end with HCl because its molar mass is greater and it diffuses more slowly along the tube compared to NH_3.
c. 0.594 m

6.159. 0.0078 g

6.161. 1.70

6.163. 26.7 g

6.165. 5.56 atm

6.167. 137 g/mol

6.169. A bubble at high pressure must be smaller than one at lower pressure in order to maintain the equality in the equation $P_1/V_1 = P_2/V_2$. Therefore, as the bubble rises to lower pressure, it expands its volume.

6.171. 2.7 atm

6.173. 38.4 L N_2, 192 L CO_2, 1.74 g/L

6.175. 6.5×10^4 Pa

6.177. 54.9 L

6.179. a. $4 NH_3(g) + 5 O_2(g) \rightarrow 4 NO(g) + 6 H_2O(g)$
b. $NH_3(g) + HNO_3(\ell) \rightarrow NH_4NO_3(s)$

6.181. a. $NH_4NO_2(s) \rightarrow N_2(g) + 2 H_2O(g)$
b. Yes

6.183. Sodium is a highly reactive element. It reacts quickly with moist air to form NaOH and H_2. NaOH is caustic since it is a strong base and the H_2 produced could form an explosive mixture with oxygen in the air.

6.185. -642.2 kJ

CHAPTER 7

7.1. a. True purple (Na), red (Cr), and orange (Au)
    b. Dark purple (Ne)
    c. Orange (Au)
    d. Red (Cr)
    e. Dark purple (Ne) and green (Cl)

7.3. Green (Cl)

7.5. Blue (Rb), green (Sr), and orange (Y)

7.7. Orange (Y^{3+}) < green (Sr^{2+}) < blue (Rb^+) < gray (I^-) < red (Te^{2-})

7.9. Blue (Rb)

7.11. All these forms of light have perpendicular, oscillating electric and magnetic fields that travel together through space.

7.13. The lead shield must protect the parts of our bodies that might be exposed to X-rays but are not being imaged. Lead is a very high density metal with many electrons, which interact with X-rays and absorb nearly all the X-rays before they can reach our bodies.

7.15. X-rays and ultraviolet radiation

7.17. No, this wavelength is in the infrared portion of the electromagnetic spectrum

7.19. 4.87×10^{14} s^{-1}

7.21. a. 2.88 m
    b. 2.95 m
    c. 2.98 m

7.23. The radio station has the lower frequency.

7.25. 8.3 min

7.27. The hydrogen absorption spectrum consists of dark lines at wavelengths specific to hydrogen. The emission spectrum has bright lines on a dark background with the lines appearing at the exact same wavelengths as the dark lines in the absorption spectrum.

7.29. Because each element shows distinctive and unique absorption and emission lines, the bright emission lines observed for the pure elements could be matched to the many dark absorption lines in the spectrum of sunlight. This approach can be used to deduce the sun's elemental composition.

7.31. The quantum is the smallest indivisible amount of radiant energy that an atom can absorb or emit.

7.33. At a low power setting, only the low wavelength–emitting excited states are reached. Once, however, the power supplied allows the atoms to reach higher excited states, these become populated and can emit the higher energy red light. At higher power settings, white light is emitted. Because white light consists of several colors, this means the emission is coming from many excited states that emit in the visible range. The emission of light has therefore shifted away from the red to include emissions of shorter wavelength.

7.35. b

7.37. 6.93×10^{-19} J

7.39. Potassium; 8.04×10^5 m/s

7.41. No

7.43. 3.17×10^{18} photons/s

7.45. Because the single electron interacts only with the proton in the nucleus; there are no other electrons to repel it.

7.47. It is the difference between n levels that determines emission energy.

7.49. a

7.51. No

7.53. At $n = 7$, the wavelength of the electron's transition ($n = 7$ to $n = 2$) has moved out of the visible region.

7.55. 1875 nm; infrared

7.57. (a) Decreases; (b) no

7.59. 72.9 nm

7.61. In the de Broglie equation, λ is the wavelength the particle of mass m exhibits as it travels at speed u, where h is Planck's constant. This equation states that (1) any moving particle has wavelike properties because a wavelength can be calculated through the equation, and (2) the wavelength of the particle is inversely related to its momentum (mass multiplied by velocity).

7.63. No

7.65. a. 10.8 nm
    b. 0.180 nm
    c. 1.24×10^{-27} nm
    d. 3.68×10^{-54} nm

7.67. c

7.69. $\Delta x \geq 1.3 \times 10^{-13}$ m

7.71. The Bohr model orbit showed the quantized nature of the electron in the atom as a particle moving around the nucleus in concentric orbits. In quantum theory, an orbital is a region of space where the probability of finding the electron is high. The electron is not viewed as a particle, but as a wave, and it is not confined to a clearly defined orbit; rather, we refer to the probability of the electron being at various locations around the nucleus.

7.73. Three: n, ℓ, and m_ℓ.

7.75. a. 1
    b. 4
    c. 9
    d. 16
    e. 25

7.77. 3, 2, 1, 0

7.79. a. 2s
    b. 3p
    c. 4d
    d. 1s

7.81. a. 2
    b. 2
    c. 10
    d. 2

7.83. b

7.85. Degenerate orbitals have the same energy and are indistinguishable from each other.

7.87. As we start from an argon core of electrons, we move to potassium and calcium, which are located in the s block on the periodic table. It is not until Sc, Ti, V, etc., that we begin to fill electrons into the 3d shell.

7.89. (c) $3s$ < (a) $3d$ < (d) $4p$ < (b) $5g$

7.91. Li: [He]$2s^1$ Na$^+$: [He]$2s^2 2p^6$ or [Ne]
     Li$^+$: $1s^2$ or [He]   Mg^{2+}: [He]$2s^2 2p^6$ or [Ne]
     Ca: [Ar]$4s^2$ Al^{3+}: [He]$2s^2 2p^6$ or [Ne]
     F$^-$: [He]$2s^2 2p^6$ or [Ne]

7.93. K: [Ar]$4s^1$ Ba: [Xe]$6s^2$
     K$^+$: [Ar]   Ti^{4+}: [Ne]$3s^2 3p^6$ or [Ar]
     S^{2-}: [Ar] or [Ne]$3s^2 3p^6$ Al: [Ne]$3s^2 3p^1$
     N: [He]$2s^2 2p^3$

7.95. Na: [Ne]$3s^1$ Mn: [Ar]$3d^5 4s^2$
     Cl: [Ne]$3s^2 3p^5$ Mn^{2+}: [Ar]$3d^5$

7.97. a. 3
     b. 2
     c. 0
     d. 0

7.99. Ti, two unpaired electrons.

7.101. Cl$^-$, no unpaired electrons

7.103. Al^{3+}, N^{3-}, Mg^{2+}, and Cs$^+$

7.105. a and d

7.107. $5p$, yes

7.109. If electrons do not repel each other as much in Na$^+$ as they do in Na, they will have lower energy and be, on average, closer to the nucleus, resulting in a smaller size. When electrons are added to an atom (Cl), the e$^-$–e$^-$ repulsion increases, so the electrons have higher energy and they will be, on average, farther from the nucleus, thereby creating a larger size species (Cl$^-$).

7.111. Rb. The size of atoms increases down a group because electrons have been added to higher n levels.

7.113. a. As the atomic number increases down a group, electrons are added to higher n levels, leading to a decrease in ionization energy.
     b. As the atomic number increases across a period, the effective nuclear charge increases. This means that the ionization energy increases across a period of elements.

7.115. Fluorine, with a higher nuclear charge, exerts a higher Z_{eff} on the $2p$ electrons than boron, resulting in higher ionization energy.

7.117. Sr

7.119. a. -1.11×10^{-26} J
     b. 17.9 m
     c. A radio telescope

7.121. a, c, and d

7.123. a. Yes. It is generally observed that as Z increases so does the IE$_2$. However, Ge's second IE$_2$ is lower than Ga's because to ionize the second electron in Ga, we need to remove an electron from a lower energy $4s$ orbital. Also, Br's IE$_2$ is lower than Se's because the electron pairing ($4p^4$) in one of the p orbitals for the Br$^+$ ion lowers its IE$_2$ slightly.
     b. Rubidium's second ionization would occur from the electron configuration [Ar]$3d^{10}4s^2 4p^6$ while krypton's would occur from [Ar]$3d^{10}4s^2 4p^5$. Both would have an electron lost from the $4p$ orbital. Since Rb has a higher Z, it exerts a higher Z_{eff} on the $4p$ electron being lost, so it has the higher IE$_2$ compared to krypton.

7.125. a. Sn^{2+}: [Kr]$4d^{10}5s^2$
        Sn^{4+}: [Kr]$4d^{10}$
        Mg^{2+}: [He]$2s^2 2p^6$ or [Ne]
     b. Cadmium has the same electron configuration as Sn^{2+} and neon has the same electron configuration as Mg^{2+}.
     c. Cd^{2+}

7.127. a. Ne, 5.76; Ar, 6.76
     b. The outermost electron in argon is a $3p$ electron which is mostly shielded by the electrons in the $n = 2$ level (10 electrons) and the $n = 1$ level (2 electrons), whereas the outermost electron in neon is a $2p$ electron which is shielded only by the electrons in the $n = 1$ level (2 electrons).

7.129. When we think of the electron as a wave, we can envision the node between the two lobes as a wave of zero amplitude and the p orbital as a standing wave.

7.131. The heavier noble gases are easier to ionize (IE decreases down a group in the periodic table) and therefore can combine with oxygen and fluorine.

7.133. The high velocity of helium atoms means that once helium atoms are released into the atmosphere, they can escape Earth's gravitational pull. Therefore, Earth's atmosphere contains very little helium.

CHAPTER 8

8.1. a. Group 1 (red)
    b. Group 14 (blue)
    c. Group 16 (purple)

8.3. Mg^{2+}

8.5. Group 14 (blue, carbon)

8.7. Lithium (red) and fluorine (lilac)

8.9. b

8.11. The arrangement of the atoms in two of the structures is S—O—S and in the other two structures it is S—S—O. Because the arrangement of atoms differs, they are not resonance structures. Also, for each arrangement, the structures do not show a different arrangement of electrons on the atoms; only the bonds are drawn bent, not straight. The "bent form" and "linear form" are not resonance forms of each other if the numbers of lone pairs and bonding pairs of electrons on each atom are the same.

8.13. a

8.15. Fluorine (purple) and oxygen (light blue)

8.17. Group 17 (blue)

8.19. Yes, for hydrogen and helium

8.21. Yes

8.23. In the diatomic molecule XY shown here

$$:\overset{..}{\underset{..}{X}}:\overset{..}{\underset{..}{Y}}:$$

Lewis counts 6 e⁻ in 3 lone pairs on both X and Y. He also counts the 2 e⁻ shared between X and Y separately (2 e⁻ for X and 2 e⁻ for Y). However, there are not 4 e⁻ being shared, only 2 e⁻. It seems that the Lewis counting scheme counts the shared electrons twice.

8.25. For the H—O—H bonding pattern, the oxygen of the central atom forms bonds to the two hydrogen atoms. This uses 4 of the 8 e⁻ leaving 4 e⁻ left over for the 2 lone pairs. Each hydrogen atom has a duet of electrons, so the lone pairs reside on oxygen and form an octet on oxygen.

$$H-\overset{..}{\underset{..}{O}}-H$$

For H—H—O bonding, the two covalent bonds again use 4 of the 8 e⁻, leaving 4 e⁻ for 2 lone pairs. If these are placed on the oxygen atom as shown here,

$$H-H-\overset{..}{\underset{..}{O}}$$

oxygen does not complete its octet and the central hydrogen atom has 4 e⁻, not a duet. This structure would violate the Lewis structure formalism.

8.27. Li· ·Mg· ·Ȧl·

8.29. Na⁺ [·In·]⁺ Ca²⁺ [:S:]²⁻

8.31. I⁻ and Ca²⁺

8.33. a. [X·]⁺

b. [X]³⁺

8.35. a. 8
b. 8
c. 8
d. 10

8.37. a. :C≡O:
b. Ö=Ö:
c. [:Cl—Ö:]⁻
d. [:C≡N:]⁻

8.39. a. 3
b. 2
c. 1
d. 3

8.41. a.

$$\begin{matrix} & :\overset{..}{Cl}: \\ :\overset{..}{F}-&C&-\overset{..}{Cl}: \\ & :\overset{..}{F}: \end{matrix}$$

b.

$$\begin{matrix}:\overset{..}{Cl}:&:\overset{..}{F}:\\:F-C-C-F:\\:\overset{..}{Cl}:&:\overset{..}{Cl}:\end{matrix}$$

c.

$$\begin{matrix}:\overset{..}{Cl}:&&:\overset{..}{F}:\\&C=C\\:\overset{..}{F}:&&:\overset{..}{F}:\end{matrix}$$

8.43.

$$H-C-C-C-C-\overset{..}{S}-H$$ (with H's)

$$H-\overset{..}{S}-H$$

8.45. :Cl—Cl—Ö:

$$\left[:\overset{..}{O}-Cl-\overset{..}{O}: \right]^{-}$$ with O on top

8.47. If there is an electronegativity difference of 2.0 or greater, the bond between the atoms is ionic; below 2.0, the bond is covalent.

8.49. The size of an atom is the result of the nucleus pulling on the electrons. The higher the nuclear charge, the stronger the pull on the electrons within a given valence shell. This is why the size of atoms generally decreases across a period. A small atom will form a shorter bond with another atom and the electrons in the bond will feel a strong pull from the nucleus of a smaller atom since the bonding electrons will be "closer" to the nucleus. This stronger pull results in a higher electronegativity for smaller atoms.

8.51. A polar covalent bond is one in which the electrons are shared, but not equally, by the atoms.

8.53. The polar bonds and the atoms with the greater electronegativity (underlined) are C—Se, C—O, N—H, and C—H.

8.55. Binary compounds of (b) C and O and (c) Al and Cl have polar covalent bonds. The binary compound of (d) Ca and O has ionic bonds.

8.57. Like the panes of glass in a greenhouse the greenhouse gases in the atmosphere are transparent to visible light. Once the visible light warms the surface of the earth and is reemitted as infrared (lower energy) light the greenhouse gases absorb the infrared light, in the same way that the panes of glass do not allow the heat from inside the greenhouse to escape.

8.59. The N–O bond would be expected to absorb IR radiation on stretching.

8.61. CO does absorb IR radiation because stretching its linear bond gives rise to a fluctuating electric field.

8.63. Infrared radiation with its longer wavelengths and lower energy than UV radiation cause chemical bonds only to stretch and bend, but not to break.

8.65. More

8.67. Resonance occurs when two or more valid Lewis structures may be drawn for a molecular species. The true structure of the species is a hybrid of the structures drawn.

8.69. A molecule or ion shows resonance when there is more than one correct Lewis structure. That is, when the electrons in the correct Lewis structure may be distributed in more than one way. Often, when the central atom has both a single and a double bond resonance is possible.

8.71. Either N–O bond in the NO_2 structure could be double-bonded and the formal charges for each structure are identical so there is more than one correct Lewis structure and NO_2 will exhibit resonance.

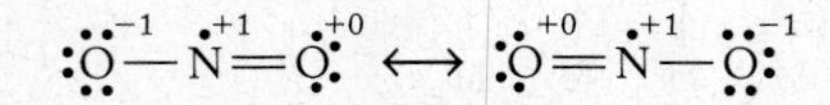

The resonance forms of CO_2 show that one is dominant (the one in which all formal charges are zero) and so the other forms contribute little to the true structure of CO_2.

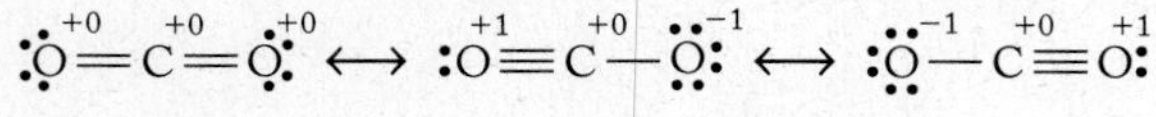

8.73.

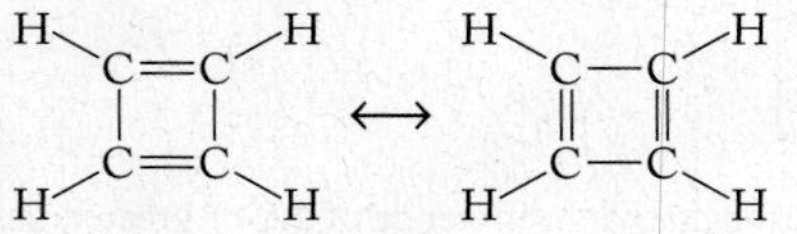

8.75. N_2O_2:

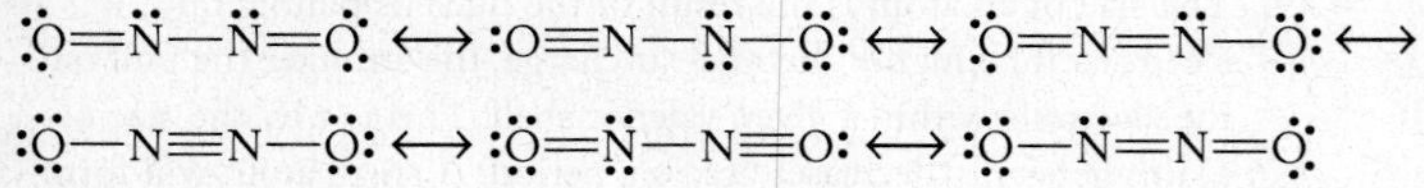

N_2O_3:

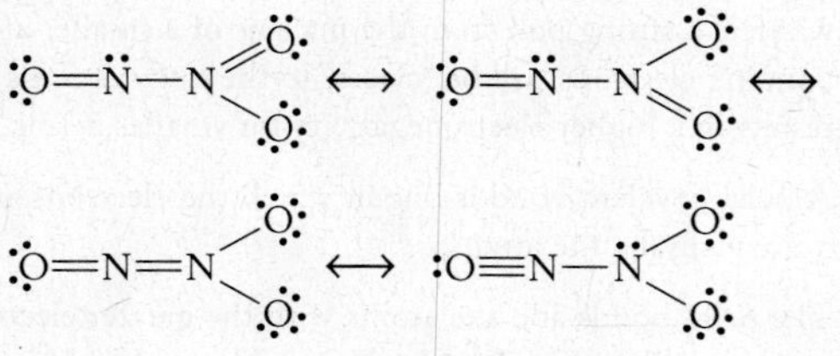

8.77.

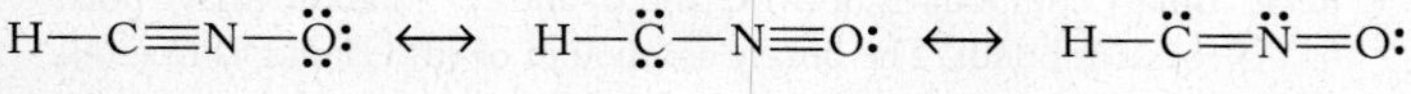

8.79.

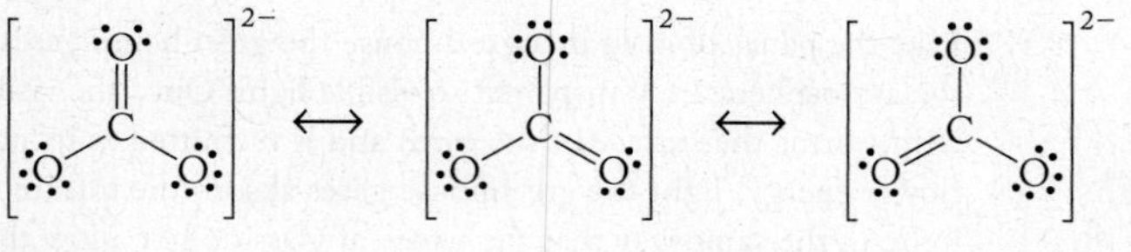

8.81. The best possible structure for a molecule judging by formal charges is the structure in which the formal charges are minimized and the negative formal charges are on the most electronegative atoms in the structure.

8.83. No

8.85.

The formal charges are zero for all the atoms in HCN, whereas in HNC the carbon atom, with a lower electronegativity than N, has a −1 formal charge.

8.87.

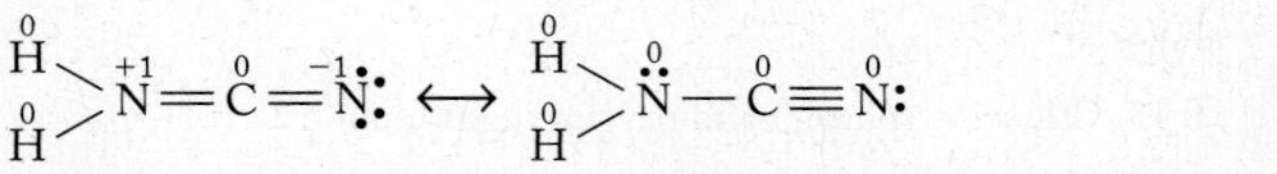

The preferred structure is the one with the C triple bonded to N.

8.89.

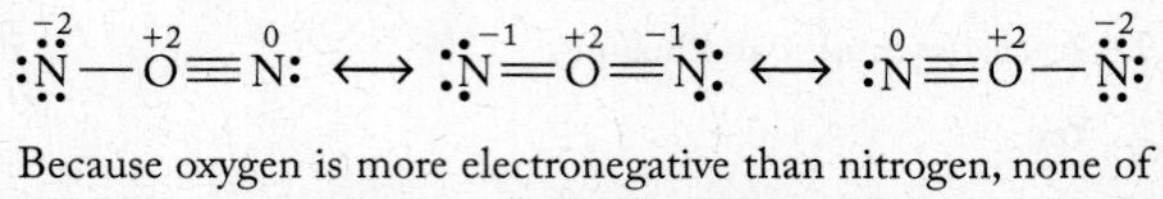

Because oxygen is more electronegative than nitrogen, none of these structures is likely to be stable because the formal charge on O is positive.

8.91. a.

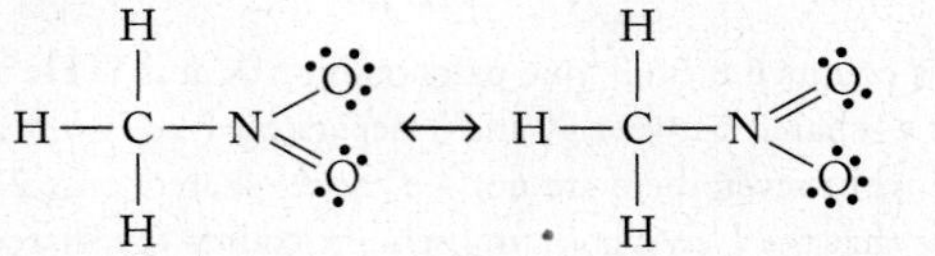

b.

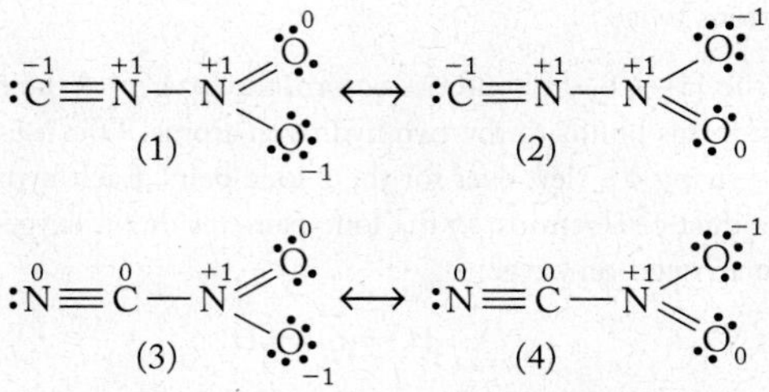

(1)   (2)

(3)   (4)

Formal charges are minimized in structures 3 and 4, so they are preferred.

c. No

8.93. Yes

8.95. In order for the atom to accommodate more than 8 e⁻ in covalently bonded molecules, it would require the use of orbitals beyond s and p. The d orbitals are not available to the small elements in the second period.

8.97. (a) SF_6, (b) SF_5, and (c) SF_4

8.99. a. 12
b. 8
c. 12
d. 10

8.101.

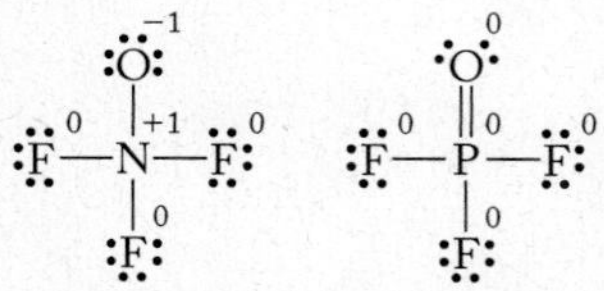

In POF_3 there is a double bond and no formal charges; in NOF_3 there are only single bonds and formal charges are present on N and O.

8.103.

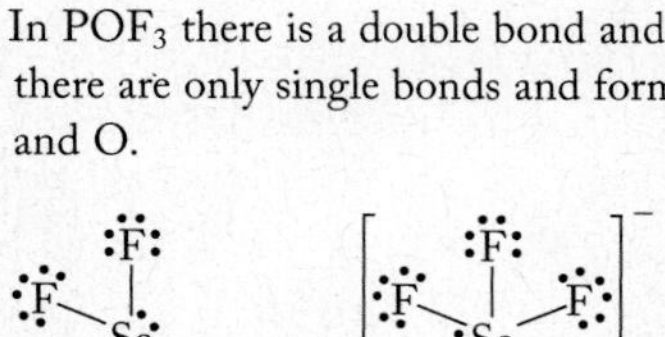

In both structures Se has more than 8 valence electrons.

8.105.

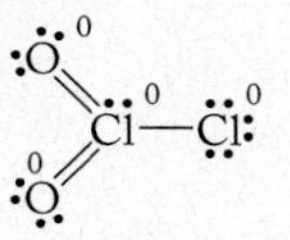

The central chlorine atom has an expanded octet.

8.107. (c) ClO_4, (d) ClO_3, and (e) ClO_2

8.109. a. S
b. N
c. C
d. O

8.111. d

8.113. No

8.115. The nitrogen–oxygen bond in N_2O_4 has a bond order of 1.5 due to four equivalent resonance forms:

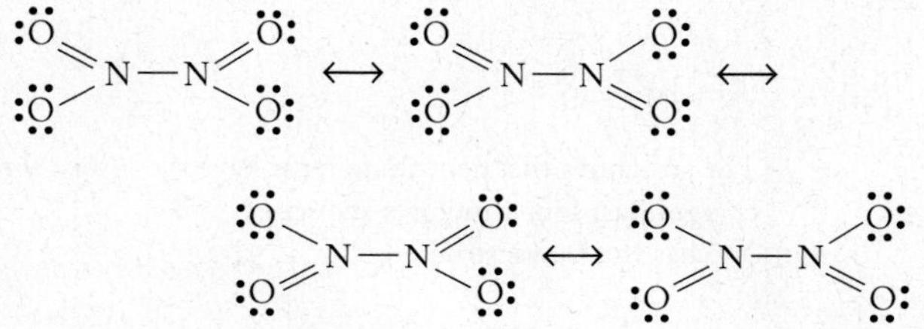

The nitrogen–oxygen bond in N_2O has a bond order of 1.5 due to resonance between three resonance forms (where the last resonance structure shown does not significantly contribute to the structure of the molecule because of the buildup of too much formal charge):

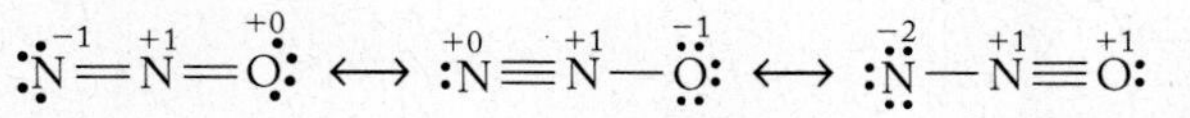

Therefore, owing to resonance, N_2O_4 and N_2O are expected to have nearly equal bond lengths.

8.117. $NO^+ < NO_2^- < NO_3^-$

8.119. $NO_3^- < NO_2^- < NO^+$

8.121. We must account for all the bonds that break and all the bonds that form in the reaction. In order to do so we must have a balanced chemical reaction.

8.123. If the compounds are in the solid or liquid phase, interactions between molecules may slightly change the bond energy for a given bond.

8.125. a. 862 kJ
b. 98 kJ
c. 93 kJ

8.127. 1068 kJ/mol

8.129. The incomplete combustion reaction releases 278 kJ less than the complete combustion reaction.

8.131. −667 kJ

8.133. 552 kJ/mol

8.135.

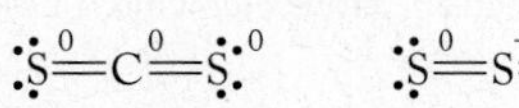

Both carbon–oxygen bonds are equal.

8.137. (a) ·Be· (b) ·Ȧl· (c) ·Ċ· (d) He:

8.139.

The preferred structure for carbon disulfide is when C is the central atom.

8.141. a.

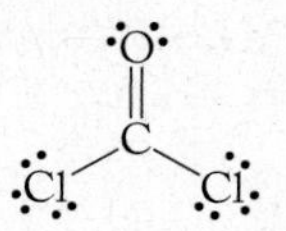

b.

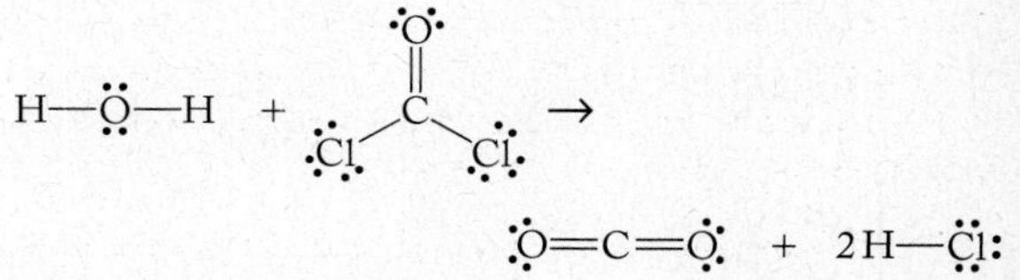

8.143. a.

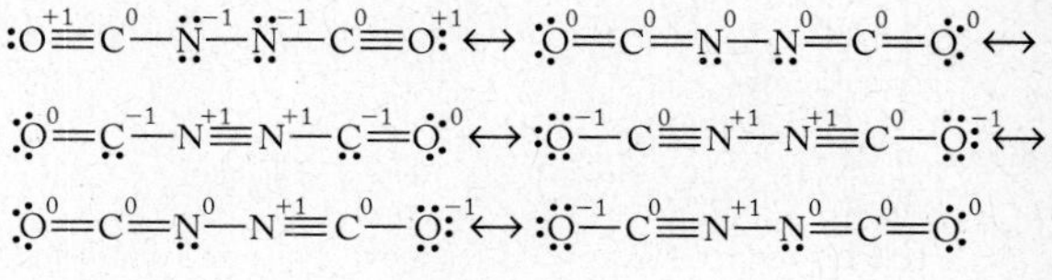

b.

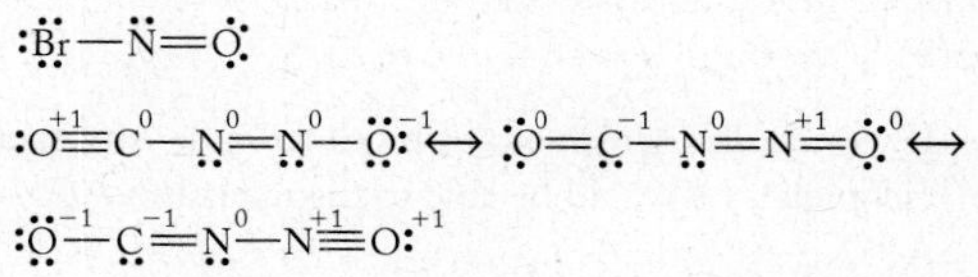

c.

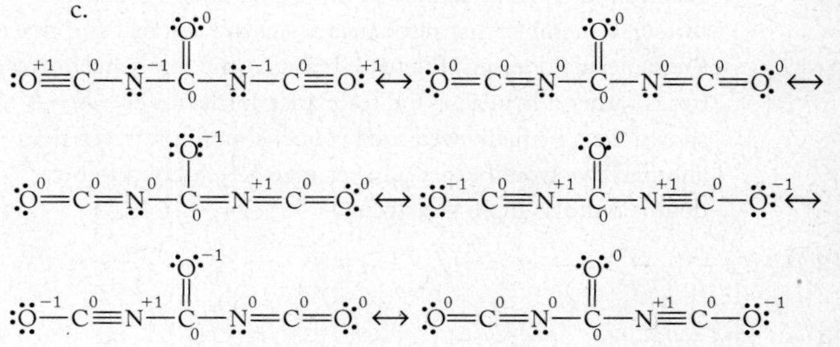

8.145. For Cl_2O_6 with a Cl—Cl bond

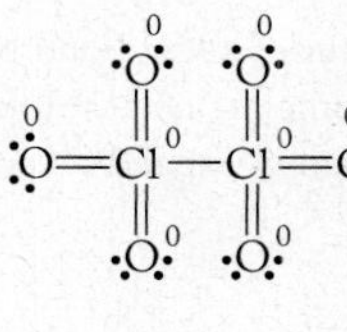

For Cl_2O_6 with a Cl—O—Cl bond

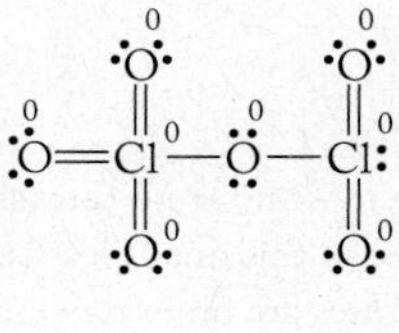

For ClO_2

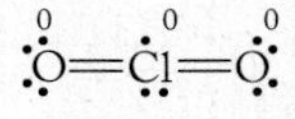

8.147. a. ·C≡N:; the more likely structure for cyanogen is the one that contains the C—C bond
b. It would be expected that oxalic acid would retain the C—C bond from the cyanogen from which it is formed in the reaction of cyanogen with water. This is consistent with the structure for cyanogen predicted by formal charge analysis.

8.149.

$$\begin{array}{c} \overset{0}{\underset{\|}{\text{N}}} \\ \overset{\|}{\underset{}{\text{C}}} \\ \overset{0}{:}\text{F}\!-\!\overset{0}{\underset{|}{\text{S}}}\!-\!\text{F}\overset{0}{:} \\ \overset{|}{\underset{}{:}}\!\text{F}\!: \\ \overset{}{0} \end{array}$$

8.151.

$$\left[\begin{array}{c} :\overset{-1}{\text{O}}: \\ \text{F} \quad | \quad \text{F} \\ \overset{-1}{\text{F}}\!-\!\text{Te}\!-\!\text{F} \\ \text{F} \quad \quad \text{F} \end{array}\right]^{2-}$$

8.153. a. 2
  b. 4
  c. 1
  d. 5
  e. 8

8.155. a. If we can distinguish atoms by electron density using this technique, we would be able to distinguish X—A—A from A—X—A.
  b. Electron diffraction cannot distinguish among resonance forms. Remember that resonance forms are not real and that the molecule does not fluctuate between the resonance forms, but is rather a hybrid. If the resonance forms for A—X—A shown are all equally weighted (none is more preferred than another), we would expect the average X—A bond to be a double bond as in A=X=A.

8.157.

$$\left[\begin{array}{c} \text{H} \\ | \\ \text{H}\!-\!\text{N}\!-\!\text{H} \\ | \\ \text{H} \end{array}\right]^{+} \qquad \left[:\!\ddot{\text{S}}\!-\!\text{H}\right]^{-}$$

There cannot be a nitrogen–sulfur covalent bond because the nitrogen atom in NH_4^+ has a complete octet through its bonding with hydrogen and because it cannot expand its octet since it is a second period element.

8.159. (a, b)

$$:\overset{0}{\text{N}}\!\equiv\!\overset{+1}{\text{N}}\!-\!\overset{0}{\text{N}}\!=\!\overset{-1}{\ddot{\text{N}}}: \longleftrightarrow \overset{-1}{:\ddot{\text{N}}}\!=\!\overset{+1}{\text{N}}\!=\!\overset{+1}{\text{N}}\!=\!\overset{-1}{\ddot{\text{N}}}: \longleftrightarrow$$

$$\overset{-1}{\ddot{\text{N}}}\!=\!\overset{0}{\text{N}}\!-\!\overset{+1}{\text{N}}\!\equiv\!\overset{0}{\text{N}}:$$

The middle structure has the most nonzero formal charges separated over three bond lengths, so this one is least preferred. The first and last resonance structures are preferred and are indistinguishable from each other.

(c)

$$\begin{array}{cc} \overset{0}{\text{N}}\!-\!\overset{0\cdot}{\text{N}} & \overset{0}{\text{N}}\!=\!\overset{0}{\text{N}} \\ \|\quad\quad\| & |\quad\quad| \\ \overset{0}{\text{N}}\!-\!\overset{0}{\text{N}} & \overset{0}{\text{N}}\!=\!\overset{0}{\text{N}} \end{array} \longleftrightarrow$$

8.161.

$$\begin{array}{c} \overset{0}{:}\ddot{\text{Cl}}: \\ | \\ \overset{0}{:}\text{F}\!-\!\overset{}{\underset{0}{\text{Al}}}\!-\!\overset{0}{\ddot{\text{Cl}}}: \end{array}$$

8.163. (b) and (c)

8.165. (a, b)

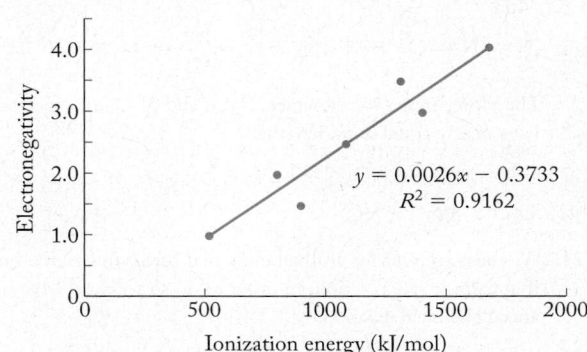

The structures that contribute most have the lowest formal charges (last four structures shown).

(c) N_3^- has the Lewis structures

$$\left[\overset{-2}{:\ddot{\text{N}}}\!-\!\overset{+1}{\text{N}}\!\equiv\!\overset{0}{\text{N}}:\right]^{+} \longleftrightarrow \left[\overset{-1}{:\ddot{\text{N}}}\!=\!\overset{+1}{\text{N}}\!=\!\overset{-1}{\ddot{\text{N}}}:\right]^{-} \longleftrightarrow \left[\overset{0}{:}\text{N}\!\equiv\!\overset{+1}{\text{N}}\!-\!\overset{-2}{\ddot{\text{N}}}:\right]^{-}$$

From these resonance structures we see that each bond is predicted to be of double bond character in N_3^-. Therefore, in N_5^- there are two longer N–N bonds than in N_3^-. N_3^- has the higher average bond order.

8.167.

[Graph: Electronegativity (y-axis, 0 to 4.0) versus Ionization energy (kJ/mol) (x-axis, 0 to 2000) with scattered points and a best-fit line]

$$y = 0.0026x - 0.3733$$
$$R^2 = 0.9162$$

Using the equation for the best-fit line where x = the ionization energy of neon gives a value of y (electronegativity) of neon: $y = 5.0$

8.169. (a) Isoelectronic means that the two species have the same number of electrons.
  (b–d)

$$\left[:\overset{0}{\text{N}}\!\equiv\!\overset{+1}{\text{N}}\!=\!\overset{0}{\ddot{\text{F}}}:\right]^{+} \longleftrightarrow \left[:\overset{-1}{\ddot{\text{N}}}\!=\!\overset{+1}{\text{N}}\!=\!\overset{+1}{\ddot{\text{F}}}:\right]^{+} \longleftrightarrow \left[:\overset{-2}{\ddot{\text{N}}}\!-\!\overset{+1}{\text{N}}\!\equiv\!\overset{+2}{\text{F}}:\right]^{+}$$

The central nitrogen atom in all the resonance structures always carries a +1 formal charge. The second and third resonance forms shown are unacceptable because they have greater than the minimal formal charges on the atoms.
  (e) Yes, the fluorine could be the central atom in the molecule, but this would place significant positive formal charge on the fluorine atom (the most electronegative element). These structures are unlikely:

$$\left[:\overset{0}{\text{N}}\!\equiv\!\overset{+3}{\text{F}}\!-\!\overset{-2}{\ddot{\text{N}}}:\right]^{+} \longleftrightarrow \left[:\overset{-1}{\ddot{\text{N}}}\!=\!\overset{+3}{\text{F}}\!=\!\overset{-1}{\ddot{\text{N}}}:\right]^{+} \longleftrightarrow \left[:\overset{-2}{\ddot{\text{N}}}\!-\!\overset{+3}{\text{F}}\!\equiv\!\overset{0}{\text{N}}:\right]^{+}$$

8.171. $5 F_2(g) + 5 H_2O(\ell) \rightarrow 8 HF(g) + O_2(g) + H_2O_2(\ell) + OF_2(g)$

(a) F_2 is the oxidizing agent. H_2O is the reducing agent.

(b)

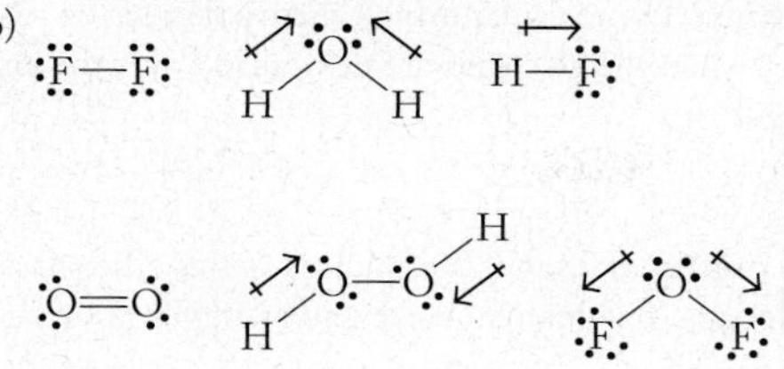

8.173.

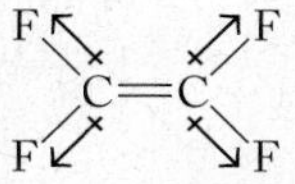

This molecule is overall nonpolar because the individual bond dipoles are equal in magnitude and, as vectors, they cancel each other out.

CHAPTER 9

9.1. Yes

9.3. N_2F_2 and NCCN are planar; there are no delocalized π electrons in any of these molecules.

9.5. More

9.7. The axial F–Re–axial F bond angle is 180°. The axial F–Re–equatorial F angle is 90°. The equatorial F–Re–equatorial F bonds are all 72°.

9.9. Because the electrons take up most of the space in the atom and because the nucleus is located in the center of the electron cloud, the electron clouds repel each other before the nuclei get close enough to each other.

9.11. Both have three atoms bonded with no lone pairs on the central atom.

9.13. Because the lone pair feels attraction from only one nucleus, it is less confined than bonding pairs and therefore occupies more space around the central N atom in ammonia.

9.15. The seesaw geometry has only two lone pair–bond pair interactions at 90° (compared to trigonal pyramidal's three), so it has lower energy.

9.17. (b) Octahedral < (c) tetrahedral < (a) trigonal planar.

9.19. Trigonal bipyramidal, seesaw, T-shaped, octahedral, square pyramidal, and square planar

9.21. Tetrahedral

9.23. Pentagonal pyramidal and distorted octahedral

9.25. a. Tetrahedral
b. Trigonal pyramidal
c. Bent
d. Tetrahedral

9.27. a. Tetrahedral
b. Trigonal planar
c. Bent
d. Square pyramidal

9.29. a. Tetrahedral
b. Tetrahedral
c. Trigonal planar
d. Linear

9.31. O_3 and SO_2

9.33. SCN^- and CNO^-

9.35.

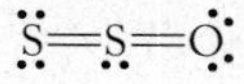

Bent

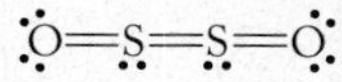

Bent at each S atom

or

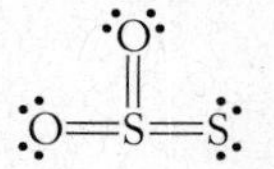

Trigonal planar

9.37.

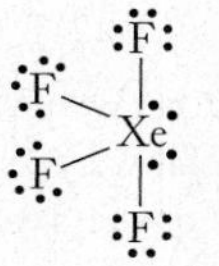

Square planar

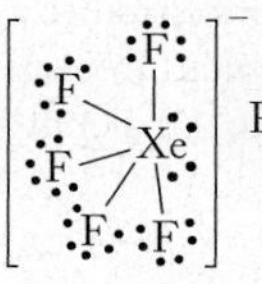

Pentagonal bipyramidal

9.39.

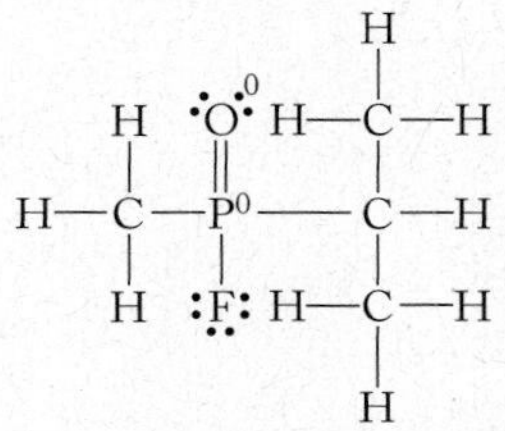

The geometry around the P atom in Sarin is tetrahedral.

9.41. A polar bond is only between two atoms in a molecule. Molecular polarity takes into account all the individual bond polarities and the geometry of the molecule. A polar molecule has a permanent, measurable dipole moment.

9.43. Yes

9.45. Polar molecules are (b) $CHCl_3$, (d) H_2S, and (e) SO_2. Nonpolar molecules are (a) CCl_4 and (c) CO_2.

9.47. All of the molecules (a–c) are polar.

9.49. a. $CBrF_3$
b. CHF_2Cl
c. Cl_2FCCF_2Cl

9.51. $COCl_2 < COBr_2 < COI_2$; in the molecule the CX bonds are pulling opposite the CO bond. As the electronegativity of the halogen (X) decreases therefore the overall molecular polarity increases.

9.53. No

9.55. Yes

9.57. All are sp^2 hybridized.

9.59. Both Lewis structures of N_2F_2 have sp^2 hybridized orbitals on N. Each F atom is sp^3 hybridized. In acetylene, C_2H_2, the carbon atoms are sp hybridized.

9.61.

CO_2	NO_2	O_3	ClO_2
sp	sp^2	sp^2	sp^3

9.63.

$$\left[\begin{array}{c} \ddot{O} \\ \| \\ :\ddot{O}=Cl-\ddot{O}: \\ \| \\ :\ddot{O}: \end{array} \right]^- \longleftrightarrow \left[\begin{array}{c} :\ddot{O}:^{-1} \\ \| \\ :\ddot{O}=\overset{+1}{Cl}-\ddot{O}:^{-1} \\ \| \\ :\ddot{O}: \end{array} \right]^- \longleftrightarrow$$

$$\left[\begin{array}{c} :\ddot{O}:^{-1} \\ | \\ :\ddot{O}=\overset{+2}{Cl}-\ddot{O}:^{-1} \\ \| \\ :\ddot{O}:^{-1} \end{array} \right]^- \longleftrightarrow \left[\begin{array}{c} :\ddot{O}:^{-1} \\ | \\ :\ddot{O}-\overset{+3}{Cl}-\ddot{O}:^{-1} \\ | \\ :\ddot{O}:^{-1} \end{array} \right]^-$$

Tetrahedral molecular geometry. At first glance this would mean that the hybridization would be assigned as sp^3. However, notice that the Cl forms three π bonds to three of the oxygen atoms. This requires that three of the p orbitals on Cl not be involved in the hybridization so that it can form parallel π bonds. Therefore, Cl must use low-lying d orbitals in place of the p orbitals for sd^3 hybridization to form the 4 σ bonds to oxygen.

9.65. $H-\ddot{A}r-\ddot{F}:$ sp^3d hybridized

9.67.

$$\left[\begin{array}{c} :\ddot{F}: \\ | \\ :\ddot{F}-S-\ddot{O}: \\ | \\ :\ddot{F}: \end{array} \right]^- \quad sp^3d \text{ hybridized}$$

Molecular geometry = seesaw

9.69. Yes

9.71. Yes, in resonance structures the electron distribution is blurred across all the resonance forms, which, in essence, defines the delocalization of electrons

9.73. b and c

9.75. One N atom has trigonal pyramidal geometry. The other N atom has trigonal planar geometry. No, the hybridization of both N atoms is not the same.

9.77. Both the S and N atoms have SN = 4 for an electron-pair geometry of tetrahedral. The presence of a lone pair on N gives this atom trigonal pyramidal geometry and the nitrogen atom is sp^3 hybridized. The steric number for S is also 4, which, at first glance, would also mean that the hybridization would be assigned as sp^3. However, notice that the S forms two π bonds to two of the oxygen atoms. This requires that two of the p orbitals on S not be involved in the hybridization so that it can form parallel π bonds. Therefore, S must use two low-lying d orbitals in place of two of the p orbitals for spd^2 hybridization to form the four bonds σ to oxygen and nitrogen.

9.79. a and c

9.81. Valence bond theory better explains molecular geometry because the hybridization on an atom has associated angles for that hybridization. In molecular orbital theory there is no associated geometry when the electrons are delocalized over the entire molecule.

9.83. No

9.85. No. The overlap of $1s$ and $2s$ orbitals is not as efficient as $1s-1s$ or $2s-2s$ overlaps. The match in size and energy is poor.

9.87.

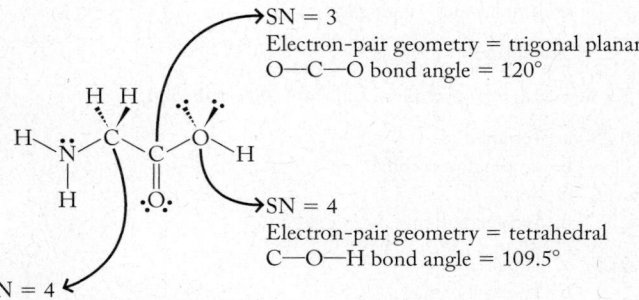

9.89. N_2^+ BO = 2.5
O_2^+ BO = 2.5
C_2^+ BO = 1.5
Br_2^{2-} BO = 0

All species with nonzero bond order (N_2^+, O_2^+, and C_2^+) are expected to exist.

9.91. The species with one or more unpaired electrons are (a) N_2^+, (b) O_2^+, and (c) C_2^+.

9.93. The species with electrons in π^* orbitals are (b) N_2^{2-}, (c) O_2^{2-}, and (d) Br_2^{2-}.

9.95. (a) B_2 and (b) C_2

9.97. No

9.99.

$$\left[\begin{array}{c} H \\ | \\ H-N-H \\ | \\ H \end{array} \right]^+ \quad \text{Tetrahedral}$$

$$\left[\begin{array}{c} :\ddot{O}: \\ \| \\ :\ddot{O}=Cl-\ddot{O}:^{-1} \\ | \\ :\ddot{O}: \end{array} \right]^- \quad \text{Tetrahedral}$$

9.101.

SN = 3
Electron-pair geometry = trigonal planar
O—C—O bond angle = 120°

SN = 4
Electron-pair geometry = tetrahedral
C—O—H bond angle = 109.5°

SN = 4
Electron-pair geometry = tetrahedral
N—C—C bond angle = 109.5°

9.103. No, neither of the two molecules is linear

9.105. (a) $\left[:\ddot{Cl}=\ddot{O}: \right]^+$

(b) BO = 2

9.107.

Molecular geometry around P = tetrahedral

9.109. $:B-B=C=\ddot{O}:$ Both B atoms have incomplete octets.
Molecular geometry = linear

$:\ddot{O}=C=B-B=C=\ddot{O}:$ Both B atoms have incomplete octets.
Molecular geometry = linear

9.111.

(1)   (2)

(3)

Structure 1 is likely to contribute the most to bonding. The methyl (CH₃) carbon is tetrahedral. The isothiocyanate (NCS) carbon is linear.

9.113. Yes

9.115. BO = 0

9.117. N_2O_5 and N_2O_3; N_2O_2 depending on its actual structure.

9.119. This species is diamagnetic.

9.121. sp^3

9.123. This molecule is polar because, although the oxygen–oxygen bonds themselves are nonpolar, the lone pair has its own "pull" on the electrons in the molecule. Also, the π bonds between the oxygen atoms places slightly more electron density on the terminal O atoms and makes the "nonpolar O—O bond" actually polar.

9.125. a. Alliin

→ SN = 4, trigonal pyramidal molecular geometry

Allicin

→ SN = 4, bent molecular geometry

→ SN = 4, trigonal pyramidal molecular geometry

b. The C–S–S bond angle is predicted to be the same as in H_2S, CH_3SH, and $(CH_3)_2S$.

CHAPTER 10

10.1. KF

10.3. XH_3

10.5. b

10.7. Increases

10.9. (c) KI

10.11. The ion–ion bond in $CaSO_4$ is stronger than in NaCl because of the higher charges on the cation and anion. For $CaSO_4$, this is greater than the ion–dipole interactions that would occur when Ca^{2+} and SO_4^{2-} dissolve, so $CaSO_4$ is not very soluble in water. NaCl has a lower ion–ion bond strength and its ion–dipole interactions with water are strong, so it dissolves in water.

10.13. $CsBr < KBr < SrBr_2$

10.15. The water molecule is oriented around Cl^- so as to point the partially positive hydrogen atoms toward the Cl^- ion.

10.17. Because of the full positive or negative charge on the ion, the ion–dipole interaction is stronger than the dipole–dipole interaction.

10.19. The charge buildup on H (partially positive) and the electronegative element (partially negative) means that the X—H bond is polar. It is still a dipole–dipole interaction except that its strength is noticeably higher than other dipole–dipole interactions.

10.21. Cl^-

10.23. CH_3F is a polar molecule and therefore has stronger intermolecular forces than the nonpolar molecules of CH_4, which have only the weak dispersion forces. As it takes more energy to overcome strong intermolecular forces, CH_3F has a higher melting point than CH_4.

10.25. The H in methane has just a single bond to the relatively low electronegativity C atom and, therefore, the carbon–hydrogen bond is not polar enough to exhibit hydrogen bonding. In methanol, however, one of the H atoms is bonded to oxygen, which is second to fluorine in electronegativity. It is this H that shows hydrogen bonding in methanol.

10.27. (b) CF_2Cl_2

10.29. Dispersion forces

10.31. Because real gases have nonzero volume and when cooled have less kinetic energy to overcome their intermolecular forces

10.33. The greater dispersion forces of CH_2Cl_2 add to the dipole–dipole interactions to give stronger intermolecular forces between the CH_2Cl_2 molecules compared to those of CH_2F_2 molecules. Also, the molar mass of CH_2Cl_2 is higher than that of CH_2F_2, so it takes more energy to vaporize.

10.35. (a) CCl_4; (b) C_3H_8

10.37. Ar has more electrons that are easily polarizable compared to He. Therefore, Ar has a greater attraction between its molecules and a larger value for a than He.

10.39. Miscible solutes and solvents dissolve completely in each other; an insoluble solute does not dissolve at all.

10.41. When we increase the pressure of a gas above a solvent, we increase the number of gas molecules above the solvent. More gas molecules striking the surface of the solvent means more dissolve into the solution.

10.43. k_H

10.45. The reaction of CO_2 with water changes dissolved CO_2 into carbonic acid. Once this occurs, more CO_2 can dissolve in water.

10.47. Hydrophilic substances dissolve in water. Hydrophobic substances do not dissolve, or are immiscible, in water.

10.49. a. $CHCl_3$
b. CH_3OH
c. NaF
d. BaF_2

10.51. 3.7×10^{-2} mol/(L · atm)

10.53. a. 2.74×10^{-3} M
b. 2.34×10^{-2} M

10.55. At 10°C, $k_H = 1.6 \times 10^{-3}$ mol/(L · atm)
At 20°C, $k_H = 1.3 \times 10^{-3}$ mol/(L · atm)
At 30°C, $k_H = 1.1 \times 10^{-3}$ mol/(L · atm)

10.57. (b) KI

10.59. (d) $CH_3(CH_2)_8CH_2OH$

10.61. In sublimation the solid does not first liquefy before evaporating. In evaporation a liquid becomes a gas.

10.63. If you are along the equilibrium line in a phase diagram, the two phases that border that line are stable and coexist at that pressure–temperature combination.

10.65. a. Solid phase
b. Gas phase

10.67. Yes

10.69. Reduce the temperature from 25°C to 0.01°C, then reduce the pressure from 1 atm to 0.006 atm.

10.71. The water vaporizes from liquid to gas.

10.73. −57°C

10.75. a. Liquid
b. Gas
c. Gas

10.77. A needle floats on water but not on methanol because of the high surface tension of water. This is because water can hydrogen bond through two O—H bonds with other water molecules whereas methanol has only one O—H bond through which to form strong hydrogen bonds.

10.79. The expansion of water in the pipes on freezing may create sufficient pressure on the wall of the pipes to cause them to burst.

10.81. The cohesive forces in mercury are stronger than the adhesive forces of the mercury to the glass.

10.83. Molecules in the bulk liquid are "pulled" by all the other liquid molecules surrounding them and they are, therefore, "suspended" in the bulk liquid. Molecules on the surface of a liquid, however, are only pulled by molecules under and beside them, creating a tight film of molecules on the surface.

10.85. As temperature increases, the surface tension decreases because the molecular "film" on the surface of the liquid has fewer molecules held together by tight intermolecular forces. Likewise, the viscosity decreases as the temperature increases because molecules have more energy to readily break the intermolecular forces to enable them to slide past each other more freely.

10.87. Water

10.89. Liquid B is expected to have the higher surface tension and higher viscosity because its higher melting melting point is indicative of stronger intermolecular forces.

10.91. (d) NaBr

10.93. Although the dispersion forces between methanol molecules are greater than those between water molecules because methanol has more electrons and greater molar mass, water can form two hydrogen bonds compared to methanol's one hydrogen bond. This greater number of stronger interactions between water molecules raises the boiling point of water above that of methanol.

10.95. Increases

10.97. Contracts

10.99. Water in the solid (ice) form has a lower density than in the liquid form due to the open lattice formed by extensive hydrogen bonding between the water molecules.

10.101. b

10.103. Seawater, natural brine sources, and rock salt deposits

10.105. F_2 (−188°C) < Cl_2 (−34°C) < Br_2 (59°C) < I_2 (185°C)

10.107. a. Chlorine is used in water purification, disinfectants, bleaches, the manufacture of organic chlorine compounds, and the synthesis of the other halogens.
b. Bromine is used to manufacture bromine compounds (as in ethylene dibromide), to purify water, to produce dyes and disinfectants, and to brominate vegetable oil.

10.109. a. Iodide
b. Bromide
c. Chloride
d. Iodide

10.111. a. HOCl
b. $HClO_3$
c. $HClO_3$
d. $HClO_4$

CHAPTER 11

11.1. $bp_X \approx 5°C$ and $bp_Y \approx 20°C$; Y has stronger intermolecular forces

11.3. Solution A must be the more concentrated solution because solvent flows through the membrane from the least to the most concentrated side.

11.5. When a gaseous ion, M^+ or X^-, dissolves in a water, new ion–dipole interactions are formed between the water molecules and the ions. In order for these interactions to form, some water–water (dipole–dipole) interaction must be broken. Both of these terms combine to give the enthalpy of hydration.

11.7. A nonvolatile solute is a compound that dissolves into a solvent and does not, under conditions to maintain the solution, enter appreciably into the gas phase.

11.9. When the average kinetic energy of the liquid molecules increases, more of the molecules can escape the liquid phase and enter the gas phase. More molecules in the gas phase increase the vapor pressure.

11.11. As intermolecular forces increase in strength, the vapor pressure decreases.

11.13. Melting point decreases as the atomic number of X increases.

11.15. (a) MgI_2

11.17. −723 kJ/mol

11.19. The vapor pressure of pure water is greater than the vapor pressure of seawater, which contains dissolved solutes. The rate of evaporation will be higher for the pure water. On the other hand, the rate of condensation in the sea water beaker is higher than the rate of evaporation, so eventually the seawater beaker will overflow.

11.21. (a) $CH_3CH_2OH <$ (b) $CH_3OCH_3 <$ (c) $CH_3CH_2CH_3$

11.23. Mole fraction of water in solution: $X = 0.70$; vapor pressure of solution at 25°C: $P_{solution} = 17$ torr

11.25. The components of crude oil can be separated by fractional distillation, which uses differences in boiling points of the compounds.

11.27. C_5H_{12}

11.29. 41.0 kJ/mol

11.31. 60 torr

11.33. For isooctane, 0.105 atm or 80.0 torr
For tetramethylbutane, 0.0483 atm or 36.7 torr

11.35. Molarity is the moles of the solute in one liter of solution. Molality is the moles of solute in one kilogram of solvent.

11.37. The greater the concentration of dissolved solutes in water (as present in seawater), the lower the freezing point of the water. The presence of nonvolatile solutes shifts the solid–liquid line on the phase diagram to lower temperature.

11.39. A strong electrolyte completely dissociates in the solvent. This dissociation yields two or more particles in solution from one dissolved solute particle. This results in greater changes in the melting and boiling points compared to that of a solute that does not dissociate.

11.41. The theoretical value of i for CH_3OH is 1 because methanol is molecular and does not dissociate in a solvent such as water. NaBr has a theoretical value of $i = 2$ because it dissociates into two particles on dissolution (Na^+ and Br^-). K_2SO_4 has a theoretical value of $i = 3$ because it dissociates into three particles on dissolution ($2 K^+$ and SO_4^{2-}).

11.43. A semipermeable membrane is a boundary between two solutions through which some molecules may pass through but others cannot. Usually, small molecules may pass through but large molecules are excluded.

11.45. Solvent flows across a semipermeable membrane from the more dilute solution side to the more concentrated solution side to balance the concentration of solutes on both sides of the membrane.

11.47. Reverse osmosis transfers solvent across a semipermeable membrane from a region of higher solute concentration to a region of lower solute concentration. Because reverse osmosis goes against the natural flow of solvent across the membrane, the key component needed is a pump to apply pressure to the more concentrated side of the membrane. Other components needed include a containment system, piping to introduce and remove the solutions, and a tough semipermeable membrane that can withstand the high pressures needed.

11.49. a. 0.58 m
b. 0.18 m
c. 1.12 m

11.51. a. 307 g
b. 86.8 g
c. 28.8 g

11.53. 6.5×10^{-5} m NH_3, 8.7×10^{-6} m NO_2^-, 2.195×10^{-2} m NO_3^-

11.55. 3.81°C

11.57. 2.52×10^{-2} m

11.59. −1.89°C

11.61. 0.5 m $CaCl_2$

11.63. 0.0100 m $Ca(NO_3)_2$

11.65. (a) 0.06 m $FeCl_3 <$ (b) 0.10 m $MgCl_2 <$ (c) 0.20 m KCl

11.67. a. From side A to side B
b. From side B to side A
c. From side A to side B

11.69. a. 57.5 atm
b. 0.682 atm
c. 52.9 atm
d. 46.5 atm

11.71. a. 2.75×10^{-2} M
b. 1.11×10^{-3} M
c. 1.00×10^{-2} M

11.73. False, the molarity of the NaCl solution would be greater by 1.5 times than the molarity of $CaCl_2$

11.75. a. Osmotic pressure increases
b. Freezing point decreases
c. Boiling point increases

11.77. 94.1 g/mol

11.79. Molar mass = 164 g/mol. The molecular formula of eugenol is $C_{10}H_{12}O_2$.

11.81. (a) SrI_2

11.83. Yes

11.85. For 0.0935 m NH_4Cl, $i = 1.85$
For 0.0378 m $(NH_4)_2SO_4$, $i = 2.46$

11.87. 2.3 atm

11.89. 4270 g/mol

CHAPTER 12

12.1. b and d are crystalline; a and c are amorphous.

12.3. The chemical formula is A_4B_4 or AB.

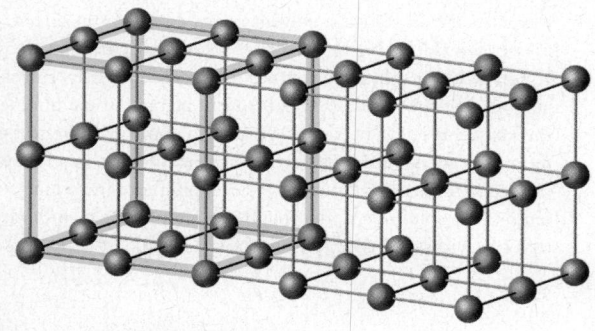

12.5. 3.5 A atoms, 0.5 B atom

12.7. The chemical formula is AB_3X.

12.9. 3.81 g/cm^3

12.11. $MgAl_2O_4$

12.13. Li_2S

12.15. Cs (blue) and Sr (purple)

12.17. MgB_2

12.19. Application of an electrical potential across a metal causes its mobile valence electrons to move toward the positive potential.

12.21. Ionic bonds are stronger than metallic bonds.

12.23. Yes

12.25. Groups 2 and 12

12.27. Phosphorus gives silicon a higher conductivity because it has one more valence electron making an n-type semiconductor.

12.29. Group 14

12.31. a. n-type
    b.

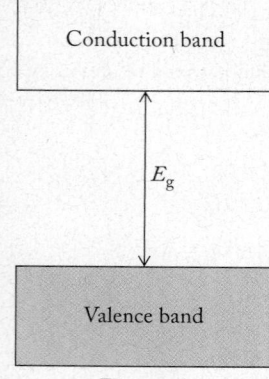

Diamond
insulator

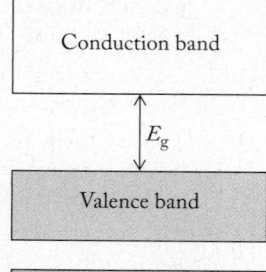

N-doped diamond
semiconductor

c. 4.68×10^{-19} J

12.33. InN

12.35. Cubic closest-packed structures have an *abcabc* . . . pattern and hexagonal closest-packed structures have an *abab* . . . pattern.

12.37. Body-centered cubic

12.39. These structural forms are not allotropes because iron is not molecular.

12.41. For bcc $\ell = \dfrac{4r}{\sqrt{3}}$; for fcc $\ell = \dfrac{4r}{\sqrt{2}}$

12.43. 104.2 pm

12.45. 513 pm

12.47. c

12.49. No

12.51. Because the metallic bonding between Cu and Ag in the alloy is weaker due to a mismatch of their atomic sizes

12.53. Tungsten is the host and carbon occupies the holes.

12.55. (a) XY_3; (b) YX_3

12.57. Octahedral holes

12.59. Substitutional alloy

12.61. Yes

12.63. (a) AB; (b) A_2B; (c) AB

12.65. One-fifth

12.67. Each S atom has a bent geometry due to sp^3 hybridization, and therefore the ring is not flat.

12.69. The ring of BN atoms is flat due to sp^2 hybridization of the B and N atoms.

12.71. a. $2 O^{2-}$ ions and $4 H^+$ ions

    b. $\left[:\ddot{O}: \right]^{2-} 2 H^+$

12.73. 109.5°

12.75. K^+ is large and so does not fit well into the octahedral holes of the fcc lattice.

12.77. The radius of Cl^- is 181 pm and the radius of Cs^+ is 170 pm and so their radii are very similar. The Cs^+ ion at the center of Figure P12.77 occupies the center of the cubic cell, so CsCl could be viewed as a body-centered cubic structure when taking into account the ions' slight difference in size. However, if we look at the ions as roughly equal in size, the unit cell becomes two interpenetrating simple cubic unit cells.

12.79. No, the Cl^- radius is so much larger than that of Na^+ that the Na^+ would not be closest-packed.

12.81. Yes, the alloy is more dense.

12.83. Less than

12.85. $MgFe_2O_4$

12.87. a. Octahedral
    b. Half

12.89. This rock salt arrangement is more dense than the sphalerite arrangement because in sphalerite the lattice of S^{2-} ions must expand to accommodate the Cd^{2+} ions.

12.91. 5.25 g/cm^3

12.93. 421 pm

12.95. Ductility, electrical and thermal conductivity, and malleability describe metals. Ceramics are electrical and thermal insulators and brittle.

12.97. $Mg_3(Si_2O_5)(OH)_4$

12.99. $2\,KAlSi_3O_8(s) + 2\,H_2O(\ell) + CO_2(g) \rightarrow$ $Al_2(Si_2O_5)(OH)_4(s) + 4\,SiO_2(s) + K_2CO_3(aq)$; this is not a redox reaction

12.101. a. $3\,CaAl_2Si_2O_8(s) \rightarrow Ca_3Al_2(SiO_4)_3(s) + 2\,Al_2SiO_5(s) + SiO_2(s)$
    b. In anorthite, the silicate anion is $Si_2O_8^{8-}$.
    In grossular, the silicate anion is SiO_4^{4-}.
    In kyanite, the silicate anion is SiO_5^{6-}.

12.103. Cubic holes can accommodate Ba^{2+}. Octahedral holes can accommodate Ti^{4+}.

12.105. An amorphous solid has no regular, repeating lattice to diffract X-rays.

12.107. X-rays have wavelengths of the order of the separation of atoms in crystals. Microwaves have wavelengths too long to be diffracted by crystal lattices.

12.109. If a crystallographer uses a shorter λ wavelength, the data set can be collected over a smaller scanning range.

12.111. Halite

12.113. The values of n are 2 ($\theta = 6.99°$) and 3 ($\theta = 10.62°$). The average lattice spacing is $d = 582$ pm.

12.115. 4.76°

12.117. XYZ_3

12.119. 33.5%

12.121. a. 139 pm
    b. 9.96 g/cm^3
    c. 3375 unit cells, 6750 Mo atoms

12.123. a. 7.53 g/cm^3
    b. 3.42 g/cm^3
    c. 3.36 g/cm^3

12.125. AuZn

12.127. 52.4%

12.129. Substitutional alloy

12.131. A cluster of three simple cubic unit cells with an Al atom in the center of one of them consistent with the formula (Cu_3Al)

12.133. a. $Sn(s) + 2\,Cl_2(g) \rightarrow SnCl_4(\ell)$
    b. $Pb(s) + Cl_2(g) \rightarrow PbCl_2(s)$
    c. $SiCl_4(\ell) \xrightarrow{>700°C} Si(s) + 2\,Cl_2(g)$

12.135. Group 14 elements do not have low ionization energies or electron affinities so they tend to neither easily lose nor easily gain an electron. As a result, they form bonds by sharing electrons, not by forming ionic bonds.

CHAPTER 13

13.1. a. One degree of unsaturation
    b. Two degrees of unsaturation
    c. No degrees of unsaturation
    d. Three degrees of unsaturation

13.3. Pine oil and oil of celery

13.5. b and d

13.7. Benzyl acetate contains an ester group; carvone contains two alkene groups and a ketone group; cinnamaldehyde contains an alkene and an aldehyde group.

13.9.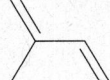

13.11. For *cis*-polyisoprene:

For *trans*-polyisoprene:

13.13. 8

13.15. An *sp* hybridized carbon atom can form two double bonds or one triple bond; an *sp^2* hybridized carbon atom can form one double bond; an *sp^3* hybridized carbon atom can form only single bonds.

13.17. No

13.19. Carboxylic acids, aldehydes, alkenes, alcohols, alkynes, and aromatic rings.

13.21. 3565 monomers

13.23. Yes

13.25. *sp^3*

13.27. The structure of cyclohexane shows that C atoms are *sp^3* hybridized with bond angles of 109.5°. It cannot be a planar molecule.

13.29. No

13.31. No

13.33.

 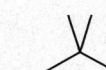

n-Pentane   2-Methylbutane   2,2-Dimethylpropane

13.35. (a) 2,3-dimethylhexane, (c and d) 2-methylheptane

13.37. a. C_8H_{18}
    b. C_9H_{20}
    c. C_8H_{18}
    d. C_8H_{18}
    e. C_9H_{20}

13.39. −124 kJ

13.41. $C_3H_8 < C_8H_{16} < C_{14}H_{30}$

13.43. Structural isomers have different connectivity of the atoms; geometric isomers have the same connectivity of the atoms but a different spatial arrangement.

13.45. No

13.47. When the double bond is "terminal" (occurs at the end or beginning of the carbon chain) there are three like groups (H) so no cis and trans isomers are possible.

13.49. The C=C double bond outside of the ring does not show cis–trans isomerism because there are not two dissimilar groups on the terminal carbon atom. The C=C double bond in the ring of carbon atoms is cis in the structure of carvone. This bond cannot be trans or the ring of 6 carbon atoms would not be possible.

13.51. Ethylene has a C=C bond with which HBr is reactive but polyethylene has only saturated C—C bonds that do not react with HBr.

13.53. a is trans, E; b is cis, Z

13.55. 681.2 kJ, endothermic

13.57.

13.59. In benzene, each C atom is sp^2 hybridized with bond angles of 120°. This geometry at each of the carbon atoms in the ring makes benzene a planar molecule.

13.61. Tetramethylbenzene has three structural isomers; pentamethylbenzene has no structural isomers.

13.63. Yes

13.65.

13.67. Fuel value for 1 mol benzene = 41.83 kJ/g
Fuel value for 3 mol ethylene = 50.30 kJ/g
1 mol benzene has a lower energy content than 3 mol ethylene.

13.69. Methylamine has a smaller nonpolar hydrocarbon chain compared to n-butylamine and so it is more soluble in water.

13.71.

13.73. −138.7 kJ

13.75. The more oxygenated the fuel, the lower the fuel value.

13.77. Ethers have lower boiling points compared to alcohols because they have weaker dipole–dipole forces compared to the alcohols, which have hydrogen bonding between the molecules.

13.79. Evaporation of ethanol from the skin is an endothermic process (phase change from liquid to vapor). The heat transfers from the skin to the ethanol so the skin feels cold.

13.81. a and d are alcohols, b and c are ethers; b < c < d < a

13.83. Fuel value for diethyl ether = 36.74 kJ/g
Fuel value for n-butanol = 36.10 kJ/g
Diethyl ether has a slightly higher fuel value.

13.85. Fuel value for methanol = 22.67 kJ/g
Fuel value for ethanol = 29.67 kJ/g
Yes, the answer supports the prediction made in Problem 13.76

13.87. Both carboxylic acids and aldehydes have polar functional groups. Carboxylic acids, however, are more soluble in water because they form strong hydrogen bonds with water.

13.89. Yes

13.91. No

13.93. Structure a because all of the formal charges are zero

13.95. An amide includes a carbonyl (C=O) as part of its functional group in addition to the −NH₂ group.

13.97. a, b, and d

13.99. b

13.101.

The plot of C:H ratio versus number of C atoms for aldehydes correlates exactly to that of alkenes and poorly to that of alkanes.

13.103. a. Pineapples

b. Bananas

c. Apples

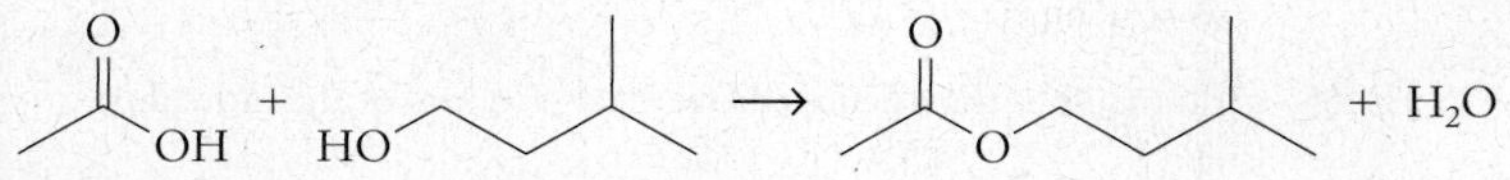

Acetic acid   3-Methylbutanol

13.105. Fuel value for formaldehyde = 19.00 kJ/g
Fuel value for formic acid = 6.531 kJ/g
Formaldehyde has a significantly higher fuel value than formic acid.

13.107. For reaction 1, $\Delta H^\circ_{rxn} = 17.5$ kJ
For reaction 2, $\Delta H^\circ_{rxn} = -312.1$ kJ

13.109. (a) 4; (b) 6; (c) 8

13.111. a. Condensation; methanol
b. Because of the presence of the six-membered ring, Kodel might be better able to accept nonpolar organic dyes.

13.113. No; enantiomer and optically active can describe the same chiral molecule, but achiral cannot.

13.115. Homogeneous

13.117. Glycine has no chiral carbon centers.

13.119. sp^3

13.121. a, c, and d

13.123. a

13.125.

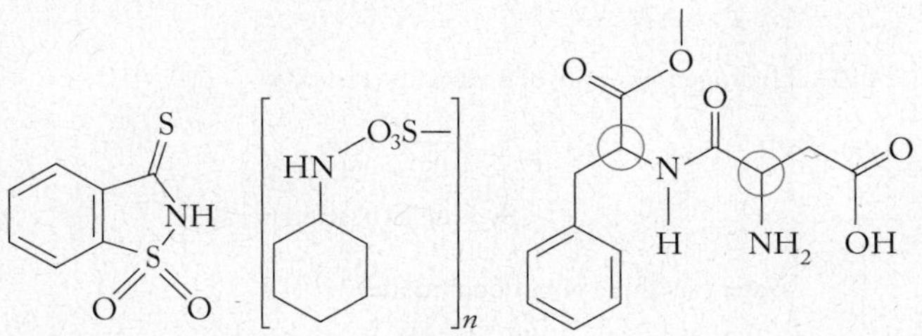

Saccharin   Cyclamate   Aspartame

13.127.

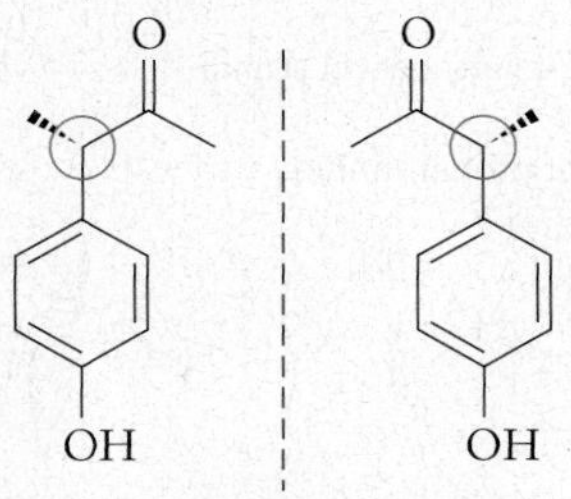

13.129. Nicotine's highlighted N atom is in a tertiary amine group; valium's highlighted N atom is in an amide group

13.131. 2.52 g methanol; 3.45 g carbon dioxide

13.133. Compound A is diethyl ether; compound B is butanol

13.135. a. 12
b. Alkene and aldehyde
c. Stereoisomers (cis and trans) and structural isomers

13.137. Yes

13.139. Amide groups

13.141. a.

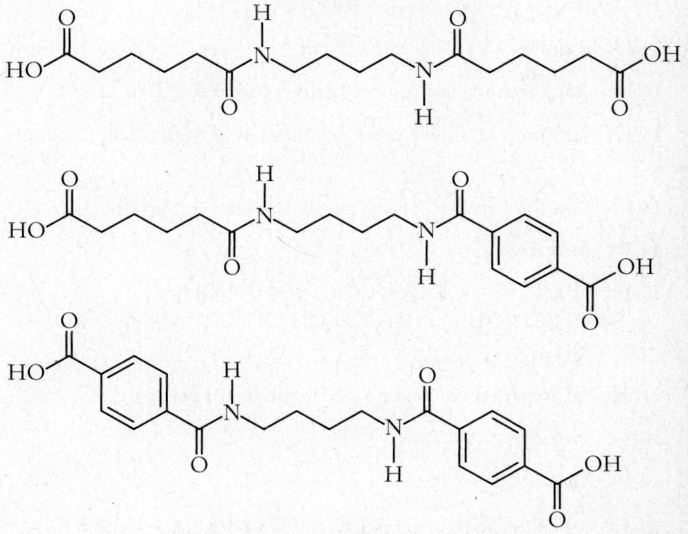

b. 1 mol adipic acid : 1 mol terephthalic acid : 2 mol putrescine

13.143. a.

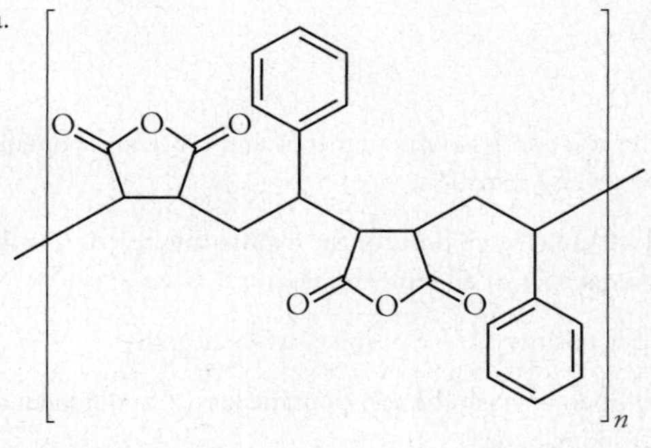

b. It would be more hydrophilic and less rigid.

13.145. a. $R_2SiCl_2(aq) + H_2O(\ell) \rightarrow R_2SiCl(OH)(aq) + HCl(aq)$
$R_2SiCl(OH)(aq) + (HO)SiClR_2(aq) \rightarrow$
$R_2ClSi\!-\!O\!-\!SiClR_2(aq) + H_2O(\ell)$
b. The side chains (R groups) are nonpolar.

13.147. The heat of combustion determined using experimental means is different from that calculated from average bond energies because the bond energy of a particular bond depends on the structure of the rest of the molecule.

CHAPTER 14

14.1. Increases

14.3. Low probability because each gas would then be confined to a smaller volume and would have more order, so this change would involve a decrease in entropy.

14.5. At the intersection $\Delta G = 0$; the reaction is nonspontaneous at $T > 60°C$.

14.7. The sign is reversed.

14.9. Eight microstates; the most likely microstates have sums of +1 and −1.

14.11. ΔS_{sys} is positive; ΔS_{surr} is negative.

14.13. a and b

14.15. ΔS_{surr} must be less (more negative) than +48.0 J/K.

14.17. (a) Propene; (b) wet paint; (c) 1 mol of $SO_3(g)$; (d) an aquarium with fish

14.19. Gas

14.21. Fullerenes

14.23. (a) $CH_4(g) < CF_4(g) < CCl_4(g)$; (b) $CH_3OH(\ell) < CH_3CH_2OH(\ell) < CH_3CH_2CH_2OH(\ell)$; (c) $HF(g) < H_2O(g) < NH_3(g)$

14.25. (a) Negative; (b) negative; (c) negative; (d) positive

14.27. Negative

14.29. Yes

14.31. (a) 24.9 J/K; (b) −146.4 J/K; (c) −73.2 J/K; (d) −175.8 J/K

14.33. 218.9 J/(mol · K)

14.35. No

14.37. Yes

14.39. The reaction is nonspontaneous and proceeds spontaneously in the reverse direction.

14.41. No, if ΔS_{rxn} were positive the exothermic reaction would be spontaneous at all temperatures

14.43. ΔS is positive, ΔH is positive, ΔG is negative

14.45. (a) Spontaneous; (b) nonspontaneous; (c) spontaneous; (d) spontaneous

14.47. For NaBr, −18 kJ/mol; for NaI, −29 kJ/mol

14.49. $\Delta G^{\circ}_{rxn} = 91.4$ kJ; above 981.3 K

14.51. $\Delta H_{rxn} = 44.0$ kJ; $\Delta S_{rxn} = 118.9$ J/K or 0.1189 kJ/K; 370.1 K or 96.9°C

14.53. $\Delta H^{\circ}_{rxn} = -146.5$ kJ; $\Delta S^{\circ}_{rxn} = -185.5$ J/K; spontaneous at temperatures below 789.8 K

14.55. a. $\Delta G^{\circ}_{rxn} = 173.2$ kJ, the reaction is not spontaneous at 355 K
b. $\Delta G^{\circ}_{rxn} = -70.6$ kJ, the reaction is spontaneous at 355 K
c. $\Delta G^{\circ}_{rxn} = -35.3$ kJ, the reaction is spontaneous at 355 K
d. $\Delta G^{\circ}_{rxn} = -4.8$ kJ, the reaction is not spontaneous at 355 K

14.57. Reaction a is spontaneous at high temperature (a).
Reaction b is spontaneous at low temperature (b).
Reaction c is spontaneous at low temperature (b).
(d) Reaction d is spontaneous at low temperature (b).

14.59. $\Delta G^{\circ}_{rxn} = 33.0$ kJ; no

14.61. a. Endothermic
b. Yes
c. No, because the reaction is not favored by enthalpy but is favored by entropy.
d. At temperatures higher than 500 K

14.63. The bond arrangements are only slightly different between the two structures.

14.65. Add together the calculated free-energy values of the individual steps.

14.67. −887.4 kJ

14.69. The entropy of a system does not increase when a substance is cooled. The second sentence should read, "At some point of cooling they freeze, and continuing to cool the sample *decreases* its entropy."

14.71. $T = 618$ K

14.73. −630.4 kJ; spontaneous

14.75. Spontaneous at all temperatures

14.77. Because standard conditions are not defined as 0 K, all substances, even the elements, above this temperature have positive entropy values.

14.79. $T_b \approx 294$ K

14.81. We expect ΔS_{vap} to be positive; $\Delta S_{vap} = 92.9$ J/(mol · K)

14.83. 9.58 J/(mol · K)

14.85. 0.805 J/(mol · K)

14.87. There are more atoms in $CaCO_3$ than in CaO, so the S° is more positive; 1099 K

14.89. (a) Positive
(b) Negative
(c) $T = \Delta H/\Delta S$

14.91. Hydrogen has only one vibrational mode,

Symmetrical stretch

Water has three vibrational modes:

Symmetrical stretch   Asymmetrical stretch   Bend

Ammonia has these vibrational modes:

Symmetrical   Asymmetrical   Asymmetrical
stretch   stretch   stretch

Bend   Bend   Deformation

Both water and hydrogen, with fewer atoms in their structures, have fewer possible vibrational modes than ammonia, with more atoms in its structure.

CHAPTER 15

15.1. The $[N_2O]$ is represented by the green line and $[O_2]$ is represented by the red line.

15.3. b

15.5. b

15.7. c

15.9. b

15.11. Nitrogen (light blue)

15.13. Palladium (blue) and platinum (orange)

15.15. The presence of NO_2 in the atmosphere and ample sunlight allows the O atoms to react with O_2 to generate O_3. The reactant NO_2 is present in the atmosphere due to automobile exhausts, which build up during the day. The buildup of O_3 lags behind until later in the day, until $[NO_2]$ increases and the sunlight becomes stronger as midday approaches.

15.17. In the evening the sunlight (and UV radiation) is less intense, so the photochemical breakdown of NO_2 does not occur to as great an extent as after the morning rush hour.

15.19. −114.2 kJ

15.21. a. $2 N_2(g) + O_2(g) \rightarrow 2 N_2O(g)$
   b. $2 N_2(g) + 5 O_2(g) \rightarrow 2 N_2O_5(g)$

15.23. The average rate is the rate averaged over a defined time interval, whereas the instantaneous rate is the rate at a specific moment.

15.25. As the reaction proceeds, the concentrations of the reactants decrease. Because most reactions depend on the availability (i.e., concentration) of reactants to proceed, the decrease in reactant concentrations lowers the reaction rate.

15.27. a. The rates are the same.
   b. The rate of formation of NO_2^- and of H^+ is two-thirds the rate of consumption of O_2.
   c. The rate of consumption of NH_3 is two-thirds the rate of consumption of O_2.

15.29. a. Rate $= -\dfrac{\Delta[H_2O_2]}{\Delta t} = \dfrac{1}{2}\dfrac{\Delta[OH]}{\Delta t}$

   b. Rate $= -\dfrac{\Delta[ClO]}{\Delta t} = -\dfrac{\Delta[O_2]}{\Delta t} = \dfrac{\Delta[ClO_3]}{\Delta t}$

   c. Rate $= -\dfrac{\Delta[N_2O_5]}{\Delta t} = -\dfrac{\Delta[H_2O]}{\Delta t} = \dfrac{1}{2}\dfrac{\Delta[HNO_3]}{\Delta t}$

15.31. a. Rate $= \dfrac{\Delta[CO_2]}{\Delta t} = -\dfrac{2}{3}\dfrac{\Delta[CO]}{\Delta t}$

   b. Rate $= \dfrac{\Delta[COS]}{\Delta t} = -\dfrac{\Delta[SO_2]}{\Delta t}$

   c. Rate $= \dfrac{\Delta[CO]}{\Delta t} = 3\dfrac{\Delta[SO_2]}{\Delta t}$

15.33. a. 1.2×10^7 M/s
   b. 2.9×10^4 M/s

15.35. Between 0 and 100 μs: 1.4×10^{-5} M/μs
   Between 200 and 300 μs: 5.5×10^{-6} M/μs

15.37. For the change in concentration of ClO versus time we obtain the following plot:

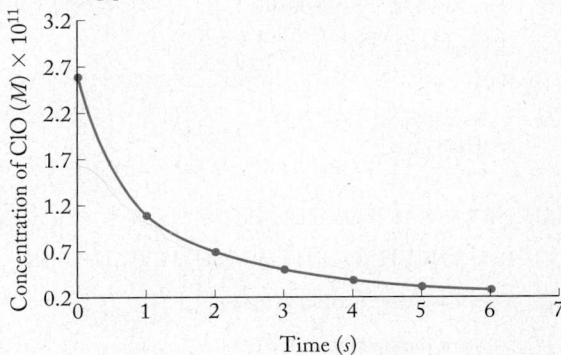

The instantaneous rate at 1 s is 8.28×10^{10} molecules $\cdot$ cm$^{-3} \cdot$ s^{-1}. For the change in concentration of Cl_2O_2 versus time we obtain the following plot:

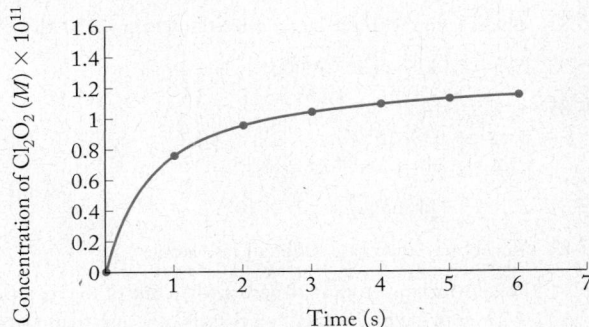

The instantaneous rate at 1s is 4.13×10^{10} molecules $\cdot$ cm$^{-3} \cdot$ s^{-1}.

15.39. Yes

15.41. Yes

15.43. The half-life will be halved.

15.45. a. First order in both A and B, and second order overall
   b. Second order in A, first order in B, and third order overall
   c. First order in A, third order in B, and fourth order overall

15.47. a. Rate $= k[O][NO_2]$; k units $= M^{-1}s^{-1}$
   b. Rate $= k[NO]^2[Cl_2]$; k units $= M^{-2}s^{-1}$
   c. Rate $= k[CHCl_3][Cl_2]^{\frac{1}{2}}$; k units $= M^{-\frac{1}{2}}s^{-1}$
   d. Rate $= k[O_3]^2[O]^{-1}$; k units $= s^{-1}$

15.49. a. Rate $= k[BrO]$
   b. Rate $= k[BrO]^2$
   c. Rate $= k[BrO]$
   d. Rate $= k[BrO]^0 = k$

15.51. We need to determine the change in the rate when only [NO] or [ClO] is changed.

15.53. a. Rate $= k[NO_2][O_3]$
   b. 4.9×10^{-11} M/s
   c. 4.9×10^{-11} M/s
   d. The rate doubles.

15.55. c

15.57. Rate $= k[NO][NO_2]$

15.59. Rate $= k[ClO_2][OH^-]$; $k = 14\ M^{-1}s^{-1}$

15.61. Rate $= k[NO]^2[H_2]$; $k = 6.32\ M^{-2}s^{-1}$

15.63. $0.32 \ \mu M^{-1} \cdot min^{-1}$

15.65. Rate = $k[NH_3]$; $k = 0.0030 \ s^{-1}$

15.67. a. Rate = $k[N_2O]$
b. 4

15.69. a. Rate = $k[^{32}P]$
b. $0.0485 \ day^{-1}$
c. 14.3 days

15.71. $k = 5.40 \times 10^{-12} \ cm^3 \ molecules^{-1} \ s^{-1}$; $t_{1/2} = 0.712 \ s$

15.73. Rate = $k[C_{12}H_{22}O_{11}][H_2O] = k'[C_{12}H_{22}O_{11}]$; $k' = 6.21 \times 10^{-5} \ s^{-1}$

15.75. They have large activation energies.

15.77. None of the statements are true.

15.79. An increase in temperature increases the frequency and the kinetic energy at which the reactants collide. This speeds up the reaction. The order of the reaction is unaffected.

15.81. The reaction with the larger activation energy (150 kJ/mol).

15.83. $E_a = 17.1 \ kJ/mol$; $A = 1.002$

15.85. a. $E_a = 314 \ kJ/mol$
b. $A = 5.03 \times 10^{10}$
c. $k = 1.06 \times 10^{-44} \ M^{-1/2}s^{-1}$

15.87. $E_a = 39.1 \ kJ/mol$; $A = 1.27 \times 10^{12}$

15.89. No, because they have different rate laws

15.91. Pseudo-first-order kinetics occurs when one of the reactants is in sufficiently high concentration that its concentration does not change appreciably over the course of the reaction.

15.93.

15.95. a. Rate = $k[SO_2Cl_2]$; unimolecular

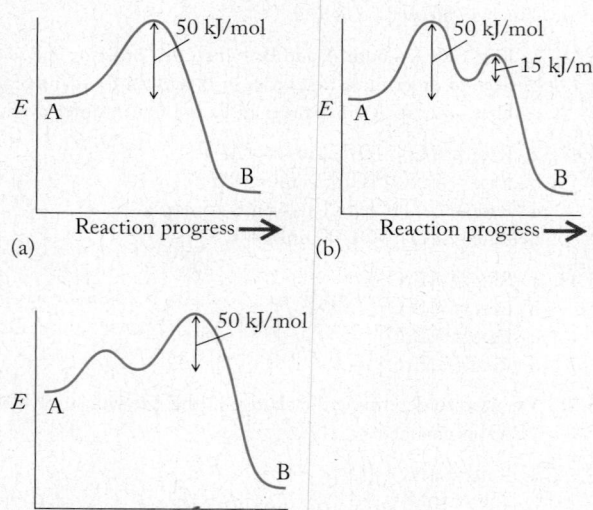

(a)

(b)

(c)

b. Rate = $k[NO_2][CO]$; bimolecular
c. Rate = $k[NO_2]^2$; bimolecular

15.97. $N_2O_5(g) + O(g) \rightarrow 2\,NO_2(g) + O_2(g)$

15.99. The second step

15.101. The first step

15.103. Photochemical decomposition: a
Thermal decomposition: b or c

15.105. Yes

15.107. Yes

15.109. Because the catalyst itself is not involved in the rate-limiting step

15.111. NO is the catalyst.

15.113. The reaction of O_3 with Cl has the larger rate constant.

15.115. When the concentration of a reactant (O_2 for the combustion reaction) increases, the rate of combustion increases.

15.117. The bodily reactions that use O_2 are slower at colder temperatures.

15.119. Yes, we could use other times, not just $t = 0$, as long as the rate of the reverse reaction is still much slower than the forward reaction.

15.121. In this plot $1/[X] - 1/[X]_0$ divided by $t - t_0$ is the slope of the line which corresponds to k, the reaction rate constant.

15.123. Zero order means that a reactant is not involved in a particular step in a reaction. However, the reactants in an elementary step ARE involved in that step.

15.125. The rate of consumption of O_3 is the same as the rate of formation of N_2O_5 and O_2 and one-half the rate of consumption of NO_2.

15.127. $k = 3.6 \times 10^{-4} \ s^{-1}$; Rate = $(3.6 \times 10^{-4} \ s^{-1})[N_2O_5]$

15.129. a. Yes
b. $E_a = 62.5 \ kJ/mol$
c. Rate = $1.2 \times 10^{-12} \ M/s$
d. At 10°C (283 K), $k = 21 \ M^{-1}s^{-1}$; at 35°C (308 K), $k = 1.8 \times 10^2 \ M^{-1}s^{-1}$

15.131. a. Rate = $k[Na(H_2O)_6^+]$
b. Neither

15.133. a. Rate = $k[NO][ONOO^-]$; $k = 1.30 \times 10^{-3} \ M^{-1}s^{-1}$
b.

$$\left[\ddot{O}\!\!=\!\!\overset{0}{N}\!-\!\overset{0}{\ddot{O}}\!-\!\overset{-1}{\ddot{N}\!\!:}\right]^- \longleftrightarrow \left[\overset{-1}{:\!\ddot{O}}\!-\!\overset{0}{N}\!\!=\!\!\overset{+1}{\ddot{O}}\!-\!\overset{-1}{\ddot{O}\!\!:}\right]^- \longleftrightarrow$$

$$\left[\overset{-1}{:\!\ddot{O}}\!-\!\overset{-1}{\ddot{N}}\!-\!\overset{+1}{\ddot{O}}\!\!=\!\!\overset{0}{\ddot{O}}\right]^-$$

c. −55 kJ

15.135. a. Second order
b. No

15.137. a. Rate = $k[NH_2][NO]$
b. $1.2 \times 10^9 \ M^{-1} \ s^{-1}$

CHAPTER 16

16.1. Reaction $C \rightleftharpoons D$ has the larger k_f, the smaller k_r, and the larger K_c.

16.3. (a) $A \rightleftharpoons B$; (b) 2.0

16.5. The reaction is endothermic. As temperature increases, K increases, indicating that more products form at higher temperatures.

16.7. A system is at equilibrium when the rate of the forward reaction equals the rate of the reverse reaction.

16.9. No, because at 20 μs the concentrations of A and B are still changing.

16.11. Greater than 1

16.13.

Molar Mass	Compound	How Present
28	$^{14}N_2$	Originally present
29	$^{15}N^{14}N$	From decomposition of $^{15}N^{14}NO$
30	$^{15}N_2$	From decomposition of $^{15}N_2O$
32	O_2	Originally present
44	$^{14}N_2O$	From combination of $^{14}N_2$ and O_2
45	$^{15}N^{14}NO$	From combination of $^{15}N^{14}N$ and O_2
46	$^{15}N_2O$	Originally present

16.15. 0.333

16.17. When $\Delta n = 0$; when the number of moles of gaseous products equals the number of moles of gaseous reactants

16.19. a. $K_c = \dfrac{[N_2O_4]}{[N_2][O_2]^2}$ and $K_p = \dfrac{(P_{N_2O_4})}{(P_{N_2})(P_{O_2})^2}$

b. $K_c = \dfrac{[NO_2][N_2O]}{[NO]^3}$ and $K_p = \dfrac{(P_{NO_2})(P_{N_2O})}{(P_{NO})^3}$

c. $K_c = \dfrac{[N_2]^2[O_2]}{[N_2O]^2}$ and $K_p = \dfrac{(P_{N_2})^2(P_{O_2})}{(P_{N_2O})^2}$

16.21. 0.50

16.23. 4.8

16.25. 0.50

16.27. 1.5

16.29. 780

16.31. 0.0583

16.33. b and c

16.35. 0.10

16.37. When scaling the coefficients of a reaction up or down the new value of the equilibrium constant is the first K raised to the power of the scaling constant.

16.39. 11.0

16.41. $K_{c,forward} = \dfrac{[NO_2]^2}{[NO][NO_3]}$; $K_{c,reverse} = \dfrac{[NO][NO_3]}{[NO_2]^2}$;

$K_{c,reverse} = \dfrac{1}{K_{c,forward}}$

16.43. $K_c = \dfrac{[SO_3]}{[SO_2][O_2]^{\frac{1}{2}}}$; $K'_c = \dfrac{[SO_3]^2}{[SO_2]^2[O_2]}$; $K'_c = (K_c)^2$

16.45. a. 0.049
b. 420
c. 20

16.47. 7.4

16.49. The reaction quotient Q is the ratio of the concentrations or partial pressures of the products of a reaction raised to their stoichiometric coefficients to the concentrations of reactants raised to their stoichiometric coefficients. The reaction quotient has the same form as the equilibrium constant K expression, but the reaction is not necessarily at equilibrium.

16.51. The system is at equilibrium.

16.53. No, $Q < K$ so the reaction proceeds to the right to reach equilibrium

16.55. Mixture a is at equilibrium.

16.57. $Q > K$, so the reaction will proceed to the left.

16.59. a

16.61. $K_c = [Cu^{2+}][S^{2-}]$

16.63. The concentrations of pure solids ($CaCO_3$ and CaO) do not change during the reaction and so they do not appear in the equilibrium constant expression.

16.65. No

16.67. As the concentration of O_2 increases the reaction shifts to the right and the CO on the hemoglobin is displaced.

16.69. According to Le Châtelier's principle an increase in the partial pressure (or concentration) of O_2 above the water shifts the equilibrium to the right so that more oxygen becomes dissolved in the water. This is consistent with Henry's law.

16.71. b and d

16.73. a. Increasing the concentration of the reactant O_3 shifts the equilibrium to the right, increasing the concentration of the product O_2.
b. Increasing the concentration of the product O_2 shifts the equilibrium to the left, increasing the concentration of the reactant O_3.
c. Decreasing the volume of the reaction to 1/10 its original volume shifts the equilibrium to the left, increasing the concentration of the reactant O_3.

16.75. The equilibrium shifts to the left.

16.77. a

16.79. When K is small the amount of reactants that are transformed into products may be so small that at equilibrium the concentrations of the reactants is approximately equal to the initial concentrations. This means that we can make an approximation in the K expression to make our calculations easier.

16.81. a. $P_{PCl_5} = 0.024$ atm, $P_{PCl_3} = 1.036$ atm, $P_{Cl_2} = 0.536$ atm
b. The partial pressure of PCl_3 decreases and the partial pressure of PCl_5 increases.

16.83. $[H_2O] = [Cl_2O] = 3.76 \times 10^{-3}\ M$; $[HOCl] = 1.13 \times 10^{-3}\ M$

16.85. 9×10^5

16.87. $P_{CO} = 2.4$ atm, $P_{CO_2} = 3.8$ atm

16.89. a. $P_{NO} = 0.272$ atm; $P_{NO_2} = 7.98 \times 10^{-3}$ atm
b. $P_T = 0.416$ atm

16.91. $P_{O_2} = 0.17$ atm, $P_{N_2} = 0.75$ atm, $P_{NO} = 0.080$ atm

16.93. 5.75 M

16.95. $P_{CO} = P_{Cl_2} = 0.258$ atm, $P_{COCl_2} = 0.00680$ atm

16.97. $[CO] = [H_2O] = 0.031 \ M$, $[CO_2] = [H_2] = 0.069 \ M$

16.99. Yes

16.101. To the right

16.103. c

16.105. 3.81×10^{-3}

16.107. -73 kJ/mol

16.109. 3.0×10^{-20}

16.111. Exothermic

16.113. Exothermic

16.115. 1.3×10^{-31}

16.117. -115 kJ/mol

16.119. The reaction is endothermic:
At 1500 K, $K_p = 5.5 \times 10^{-11}$
At 2500 K, $K_p = 4.0 \times 10^{-3}$
At 3000 K, $K_p = 0.40$
This reaction does not favor products even at very high temperature, so this is not a viable source of CO and is not a remedy to decrease CO_2 as a contributor to global warming. Also, the process produces poisonous CO gas.

16.121. $9 \times 10^{-22} M$

16.123. $K_{p,25°C} = 3.15 \times 10^{-59}$; $K_{p,500°C} = 5.14 \times 10^{26}$

16.125. $P_{SO_2} = 9.2 \times 10^{-74}$ atm

CHAPTER 17

17.1. Red line

17.3. The blue titration curve represents the titration of a 1 M solution of strong acid. The red titration curve represents the titration of a 1 M solution of weak acid.

17.5. The indicator with a pK_a of 9.0

17.7. The red titration curve represents the titration of Na_2CO_3; the blue titration curve represents the titration of $NaHCO_3$.

17.9. HBr is the acid; H_2O is the base.

17.11. OH^- is the base; H_2O is the acid.

17.13. a. HNO_3 is the acid; NaOH is the base.
b. HCl is the acid; $CaCO_3$ is the base.
c. HCN is the acid; NH_3 is the base.

17.15. NO_2^-; OCl^-; $H_2PO_4^-$; NH_2^-

17.17. 1.50 M

17.19. 0.160 M

17.21. Dissolve 70.0 g of NaOH(s) in water and dilute to a total volume of 2.50 L.

17.23. Because the pH function is a $-\log$ function, as $[H^+]$ increases, the value of $-\log[H^+]$ decreases.

17.25. When $[H^+]$ is greater than 1 M

17.27. a. pH = 7.462; pOH = 6.538; basic
b. pH = 4.70; pOH = 9.30; acidic
c. pH = 7.15; pOH = 6.85; basic
d. pH = 10.932; pOH = 3.068; basic

17.29. 0.810

17.31. pOH = 1.347; pH = 12.653

17.33. -0.124

17.35. In order of largest K_a (strongest acid) to smallest K_a (weakest acid): $HCl > HNO_2 > CH_3COOH > HClO$

17.37. $NaNO_2$ is soluble in water, separating into Na^+ and NO_2^- ions, each in 1.0 M concentration for a total ion concentration of 2.0 M. HNO_2, however, only weakly dissociates in water and so produces just slightly greater than 1.0 M ions in solution. $NaNO_2$, therefore, with more dissolved ions in solution, is a better conductor of electricity.

17.39. $K_a = \dfrac{[H^+][F^-]}{[H]}$

17.41. (a) Water
(b) Water

17.43. H_2O is the acid, CH_3NH_2 is the base.

17.45. 8.91×10^{-4}

17.47. 1.63%; $K_a = 6.74 \times 10^{-5}$

17.49. 2.49

17.51. 2.3 times

17.53. 10.77

17.55. 9.36

17.57. With each successive ionization, it becomes more difficult to remove H^+ from a species that is more negatively charged.

17.59. 0.51

17.61. 2.80

17.63. 9.50

17.65. 10.27

17.67. Sulfur is more electronegative than selenium. The higher electronegativity on the sulfur atom stabilizes the anion HSO_4^- more than the anion $HSeO_4^-$.

17.69. a. H_2SO_3
b. H_2SeO_4

17.71. Increase

17.73. Ammonium nitrate

17.75. The citric acid in the lemon juice neutralizes the volatile trimethylamine to make a nonvolatile dissolved salt.

17.77. 3.32

17.79. 7.35

17.81. A solution of acetic acid and acetate ions can neutralize additions of acid or base. However, a solution of HCl and NaCl has no acid-neutralizing power because the Cl^- ion is too weak a base.

17.83. At 25°C, pH = 4.453; At 0°C, pH = 4.484

17.85. pH = 12.34; pOH = 1.65

17.87. 0.064

17.89. 9.25

17.91. 3.0 mL

17.93. pH = 3.50 before adding HCl; pH = 3.42 after adding HCl

17.95. Molar solubility is the quantity (moles) of substance that dissolves in a liter of solution. The solubility product is the equilibrium constant for the dissolution of a substance.

17.97. Sr^{2+}

17.99. Endothermic

17.101. Acidic substances react with the OH^- released on dissolution of hydroxyapatite. The equilibrium is shifted to the right, dissolving more hydroxyapatite.

17.103. 1.08×10^{-10}

17.105. $[Cu^+] = [Cl^-] = 1.01 \times 10^{-3}\ M$

17.107. 9.96×10^{-6} g/mL

17.109. 10.091

17.111. d

17.113. No

17.115. Yes

17.117. (a) SO_4^{2-}
(b) $1.34 \times 10^{-4}\ M$

17.119. The weak acid titration curve has an initial pH that is higher (less acidic) than that of an equimolar solution of a strong acid (lower pH, more acidic). The pH at the equivalence point in the titration of a strong acid is 7.00 whereas the pH at the equivalence point for a weak acid is basic.

17.121. No

17.123. After 10.0 mL of OH^- has been added, pH = 4.754; after 20.0 mL of OH^- has been added, pH = 8.750; after 30.0 mL of OH^- has been added, pH = 12.356

17.125. 0.02559 M

17.127. 250 mL

17.129. 4.44

17.131.

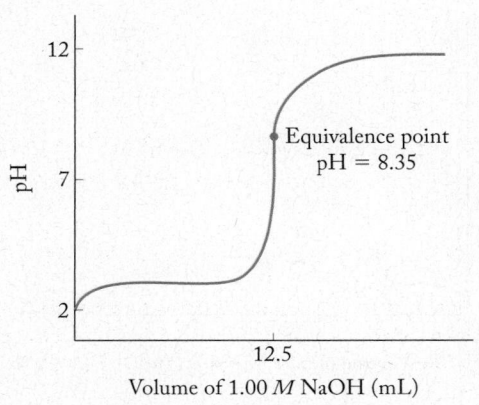

Volume of 1.00 M NaOH (mL)

17.133.

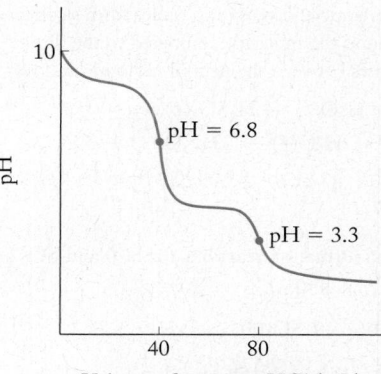

Volume of 0.100 M HCl (mL)

17.135. The hydrogen bonds between some of the water molecules must break and re-form around the species CH_3NH_2. Also, the amine hydrolyzes and forms $CH_3NH_3^+$ and OH^-; resulting in ion–dipole forces between these ions and the surrounding water molecules.

17.137. (a, b)

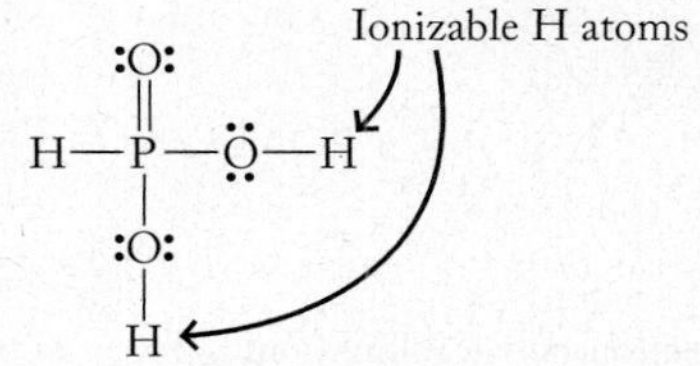

:O:
‖
H—P—Ö—H
|
:O:
|
H

Ionizable H atoms

(c) Phosphoric acid has a similar structure with a single P=O bond and ionizable H atoms also bonded to oxygen atoms.

17.139. Subsequent additions of HCO_3^- react with water to form bicarbonate's conjugate acid (H_2CO_3) and its conjugate base (CO_3^{2-}) in the same proportions as the first addition, so pH does not change.

17.141. 1.4×10^9 L

17.143. a. The acid salt form has an H^+ ion on the amine (R_2NH) group with Cl^- as a counter ion. This structure is shown on the right of Figure P17.143.
b. Acidic

17.145. a. Because HF is weak, the $[F^-]$ is low compared to that of water and so HF reacts with H_2O to form F^- as the major anionic species.
b. $K_{overall} = 2.9 \times 10^{-4}$
c. pH = 1.91; $[HF_2^-]_{eq} = 4.40 \times 10^{-4}\ M$

17.147. a.

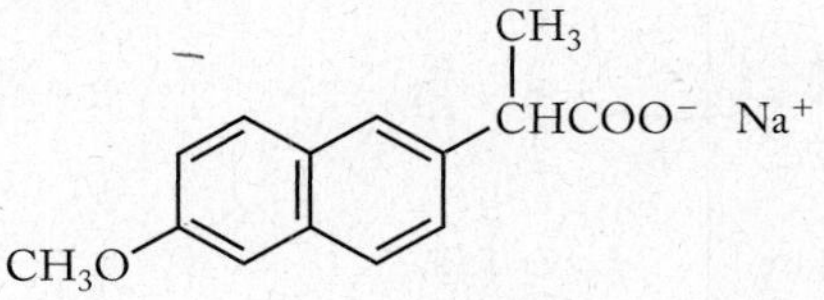

b. Basic

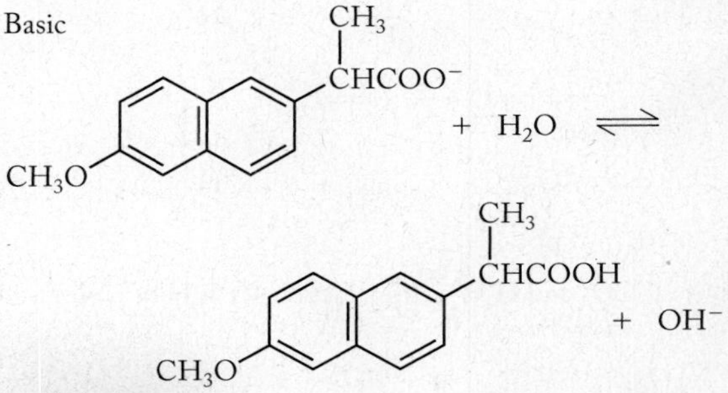

c. The salt is ionic and water molecules form stronger ion–dipole forces around the molecule compared to the dipole–induced dipole forces between the neutral molecule and water.

17.149. a. $SO_3(g) + H_2O(\ell) \rightarrow H_2SO_4(\ell)$

b. $3\,NO_2(g) + H_2O(\ell) \rightarrow 2\,HNO_3(\ell) + NO(g)$

c. $4\,NH_3(g) + 5\,O_2(g) \rightarrow 4\,NO(g) + 6\,H_2O(g)$

17.151. None

17.153. The balanced formation reactions for SO_2 and SO_3 are

$\frac{1}{8}S_8(s) + O_2(g) \rightarrow SO_2(g) \qquad \Delta H^\circ_{f,SO_2}$

$\frac{1}{8}S_8(s) + \frac{3}{2}O_2(g) \rightarrow SO_3(g) \qquad \Delta H^\circ_{f,SO_3}$

$\Delta H^\circ_{rxn} = 2\,\Delta H^\circ_{f,SO_3} - 2\,\Delta H^\circ_{f,SO_2}$

17.155. The corresponding constant for NO_2 should be greater than the Henry's law constant for CO_2 because NO_2 is polar while CO_2 is nonpolar. When SO_3 gas comes in contact with water it rapidly reacts with it instead of simply dissolving in it, as SO_3 and H_2O combine to form sulfuric acid.

17.157. a.

b. When a less electronegative sulfur atom replaces an oxygen atom in the acid, the acidity decreases. Therefore, $H_2S_2O_3$ is less acidic than H_2SO_4.

CHAPTER 18

18.1. Chromium (green) and cobalt (yellow)

18.3. Zinc (blue)

18.5. 4

18.7. a. $[Co(CN)_6]^{3-}$
b. $[CoF_6]^{3-}$
c. $[Co(NH_3)_6]^{3+}$

18.9. Yes because it may donate an electron pair but not accept a proton.

18.11. BF_3 can accept electron pairs but has no H atoms to donate to be a Brønsted–Lowry acid.

18.13.

Lewis base   Lewis acid

18.15.

CO_2 and H_2O in this reaction act as both Lewis acids and Lewis bases.

18.17.

$B(OH)_3$ is the Lewis acid and H_2O is the Lewis base.

18.19. Water

18.21. Water

18.23. Na^+

18.25. Because the Cu^{2+} ion is fully complexed by the EDTA

18.27. Ag^+ forms a soluble complex with NH_3, removing Ag^+ from solution and shifting the equilibrium for the dissolution of AgCl to the right.

18.29. (a) 4.0×10^{-3} M; (b) 6.3×10^{-10} M

18.31. 2.6×10^{-15} M

18.33. a. Hexaamminechromium(III)
b. Hexaaquacobalt(III)
c. Pentaamminechloroiron(III)

18.35. a. Tetrabromocolbaltate(II)
b. Aquatrihydroxozincate(II)
c. Pentacyanonickelate(II)

18.37. a. Ethylenediaminezinc(II) sulfate
b. Pentaammineaquanickel(II) chloride
c. Potassium hexacyanoferrate(II)

18.39. b and d

18.41. The solution will become more acidic.

18.43. In basic solution: $Cr(OH)_3(s) + OH^-(aq) \rightleftharpoons Cr(OH)_4^-(aq)$

In acidic solution: $Cr(OH)_3(s) + 3\,H^+(aq) \rightleftharpoons$
$$Cr^{3+}(aq) + 3\,H_2O(\ell)$$

18.45. $Al(OH)_3$ reacts with OH^- in solution to form soluble $Al(OH)_4^-$. The other ions do not form this type of soluble complex ion.

18.47. 2.65

18.49. 1.80

18.51.

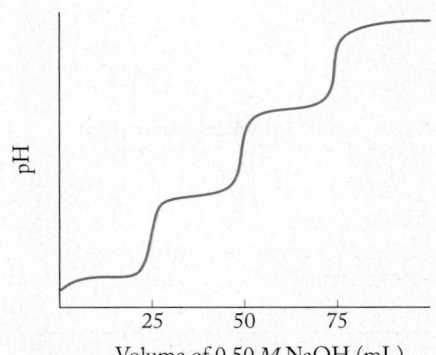

18.53. A sequestering agent is a multidentate ligand that separates metal ions from other substances so that they can no longer react. Properties that make a sequestering agent effective include strong bonds formed between the metal and the ligand and large formation constants.

18.55. As pH increases the chelating ability increases because OH^- removes the H on the carboxylic acid groups, providing an additional site for binding to the metal cation.

18.57. When the transition metals bond to ligands the d orbitals split in energy. If there is a d to d transition possible for the ion, the compound is likely to be colored.

18.59. The repulsions due to the ligands in a square-planar crystal field are highest for the d_{xy} orbital and so it is raised in energy because this orbital lies in the plane of the ligands.

18.61. The yellow solution contains (b) $Cr(NH_3)_6{}^{3+}$. The violet solution contains (a) $Cr(H_2O)_6{}^{3+}$.

18.63. Colorless

18.65. $NiCl_4{}^{2-}$

18.67. The magnitude of the crystal field splitting energy compared to the pairing energy of the electrons in a lower energy d orbital.

18.69. Fe^{2+} has 4 unpaired electrons.
Cu^{2+} has 1 unpaired electron.
Co^{2+} has 3 unpaired electrons.
Mn^{3+} has 4 unpaired electrons.

18.71. Cr^{3+}

18.73. a. Mn^{4+} in MnO_2; $2\,Mn^{3+}$ and $1\,Mn^{2+}$ in Mn_3O_4
b. Both low-spin and high-spin configurations are possible in Mn_3O_4 (d^4 and d^5) but not in MnO_2 (d^3).

18.75. Paramagnetic

18.77. For an octahedral geometry cis- means that two ligands are side by side and have a 90° bond angle between them. Ligands that are $trans$- to each other have a 180° bond angle between them.

18.79. At least two different ligands

18.81. Yes

18.83.

Cis   Trans

No, neither isomer is chiral

18.85. 4×10^{-13}

18.87. The yellow complex containing Co^{3+} in aqueous ammonia has the larger Δ_o.

18.89. Ag^{2+} has 9 d electrons, leaving an unpaired electron in the $d_{x^2-y^2}$ orbital to make it paramagnetic. Ag^{3+} has 8 d electrons and Ag^+ has 10 d electrons. Both have all electrons paired, so those silver ions are diamagnetic.

18.91. To longer wavelengths

CHAPTER 19

19.1. Because of the careful layering, each half-cell has its metal in contact with its cation solution. The solutions are not mixing, but nevertheless the layers allow the ions needed to balance the charge in each half-cell to pass.

19.3. Ag is the cathode; Pt in the SHE is the anode; electrons flow from the SHE to Ag

19.5. Blue line

19.7. a. $2\,H_2O(\ell) + 2\,e^- \rightarrow H_2(g) + 2\,OH^-(aq)$
$$E^\circ_{cathode} = -0.8277\text{ V}$$
$2\,H_2O(\ell) \rightarrow O_2(g) + 4\,H^+(aq) + 4\,e^- \quad E^\circ_{anode} = 1.229\text{ V}$
b. To increase the conductivity of the solution

19.9. To allow nonreactive ions to pass through the separator to maintain electrical neutrality

19.11. A wire can only pass electrons through it, not ions.

19.13. a. $Pb^{2+}(aq) + 2\,e^- \rightarrow Pb(s)$ cathode
$Zn(s) \rightarrow Zn^{2+}(aq) + 2\,e^-$ anode
b. $Pb^{2+}(aq) + Zn(s) \rightarrow Zn^{2+}(aq) + Pb(s)$
c. $Zn(s)\,|\,Zn^{2+}(aq)\,||\,Pb^{2+}(aq)\,|\,Pb(s)$

19.15. a. $MnO_4{}^-(aq) + 2\,H_2O(\ell) + 3\,e^- \rightarrow$
$MnO_2(s) + 4\,OH^-(aq)$ cathode
$Cd(s) + 2\,OH^-(aq) \rightarrow Cd(OH)_2(s) + 2\,e^-$ anode
b. $2\,MnO_4{}^-(aq) + 4\,H_2O\,(\ell) + 3\,Cd(s) \rightarrow$
$2\,MnO_2(s) + 3\,Cd(OH)_2(s) + 2\,OH^-(aq)$
c. $Cd(s)\,|\,Cd(OH)_2(s)\,||\,MnO_4{}^-(aq)\,|\,MnO_2(s)\,|\,Pt(s)$

19.17. a. 6
b. $FeO_4{}^{2-}$ has Fe^{6+}, Fe_2O_3 has Fe^{3+}, Zn has Zn^0, ZnO and $ZnO_2{}^{2-}$ have Zn^{2+}
c. $Zn(s)\,|\,ZnO(s)\,|\,ZnO_2{}^{2-}(aq)\,||\,FeO_4{}^{2-}(aq)\,|\,Fe_2O_3(s)\,|\,Pt(s)$

19.19. The platinum electrode transfers electrons to the half-cell; it is inert and not involved in the reaction.

19.21. Because $E^\circ_{ox}(\text{anode}) = -E^\circ_{red}$ we can substitute $-E^\circ_{red}(\text{anode})$ for $E^\circ_{ox}(\text{anode})$ in the expression
$$E^\circ_{cell} = E^\circ_{red}(\text{cathode}) + E^\circ_{ox}(\text{anode})$$
to obtain
$$E^\circ_{cell} = E^\circ_{red}(\text{cathode}) + [E^\circ_{red}(\text{anode})]$$
$$= E^\circ_{red}(\text{cathode}) - E^\circ_{red}(\text{anode})$$
This is equal to the expression for E°_{cell} in Equation 19.2.

19.23. E°_{red} (acid, 1.229 V) is greater than E°_{red} (base, 0.401 V).

19.25. a. $\Delta G^\circ = -34.5$ kJ; $\Delta E^\circ_{cell} = 0.358$ V
b. $\Delta G^\circ = 2.9$ kJ; $E^\circ_{cell} = -0.030$ V

19.27. No

19.29. $O_2(g) + 2\,H_2O(\ell) + 2\,Zn(s) + 4\,OH^-(aq) \rightarrow 2\,Zn(OH)_4{}^{2-}(aq)$

19.31. Less than 1.10 V

19.33. a. $\Delta E^\circ_{cell} = -0.478$ V; $\Delta G^\circ = 92.2$ kJ
b. $\Delta E^\circ_{cell} = 0.548$ V; $\Delta G^\circ = -97.4$ kJ

19.35. a.

Anode: $\quad\quad\quad\quad Zn(s) \rightarrow Zn^{2+}(aq) + 2e^{-} \quad\quad E^{\circ}_{anode} = -0.7618\ V$

Cathode: $Hg^{2+}(aq) + 2e^{-} \rightarrow Hg(\ell) \quad\quad\quad E^{\circ}_{cathode} = 0.851\ V$

$\quad\quad\overline{Zn(s) + Hg^{2+}(aq) \rightarrow Zn^{2+}(aq) + Hg(\ell) \quad E^{\circ}_{cell} = E^{\circ}_{cathode} - E^{\circ}_{anode} = 1.613\ V}$

b. Anode: $\quad\quad Zn(s) + 2\,OH^{-}(aq) \rightarrow ZnO(s) + H_2O(\ell) + 2e^{-} \quad E^{\circ}_{anode} = -1.25\ V$

Cathode: $\quad Ag_2O(s) + H_2O(\ell) + 2e^{-} \rightarrow 2\,Ag(s) + 2\,OH^{-}(aq) \quad\quad E^{\circ}_{cathode} = 0.342\ V$

$\quad\quad\overline{Zn(s) + Ag_2O(s) \rightarrow ZnO(s) + 2\,Ag(s) \quad\quad\quad\quad E^{\circ}_{cell} = E^{\circ}_{cathode} - E^{\circ}_{anode} = 1.59\ V}$

c. Anode: $\quad\quad 2 \times [Ni(s) + 2\,OH^{-}(aq) \rightarrow Ni(OH)_2(s) + 2e^{-}] \quad E^{\circ}_{anode} = -0.72\ V$

Cathode: $\quad O_2(g) + 2\,H_2O(\ell) + 4e^{-} \rightarrow 4\,OH^{-}(aq) \quad\quad\quad\quad E^{\circ}_{cathode} = 0.401\ V$

$\quad\quad\overline{2\,Ni(s) + O_2(g) + 2\,H_2O(\ell) \rightarrow 2\,Ni(OH)_2(s) \quad\quad E^{\circ}_{cell} = E^{\circ}_{cathode} - E^{\circ}_{anode} = 1.12\ V}$

19.37. a. $NiO(OH)(s) + TiZr_2H(s) \rightarrow TiZr_2(s) + Ni(OH)_2(s)$
b. 1.32 V

19.39. A positive E_{cell} indicates a spontaneous reaction that does electrical work on the surroundings, so the sign of w is negative.

19.41. −290 kJ

19.43. −116 kJ

19.45. Voltage of a battery (a voltaic cell) is governed by the Nernst equation: $E_{cell} = E^{\circ}_{cell} - \dfrac{RT}{nF} \ln Q$. As a battery discharges, the value of Q, the reaction quotient, changes: $Q = \dfrac{[\text{products}]^x}{[\text{reactants}]^y}$. At the start of the reaction, Q is very small because [reactants] $\gg$ [products]. As the reaction proceeds [products] grows and Q increases but does not increase significantly until significant amounts of products form, that is, when the battery is nearly discharged.

19.47. 1.27 V

19.49. 8.56×10^{19}

19.51. −0.414 V

19.53. $E_{cell} = 1.54\ V$; E_{cell} will decrease

19.55. a. 0.62 V
b. 0.61 V

19.57. a. 0.349 V
b. 1.02×10^{57}

19.59. c and f

19.61. Al–O_2

19.63. Li–MnO_2

19.65. In a voltaic cell, the electrons are produced at the anode so a negative (−) charge builds up there; in an electrolytic cell, electrons are being forced onto the cathode so that it builds up negative (−) charge. The flow of electrons in the outside circuit is reversed in an electrolytic cell compared to the flow in a voltaic cell.

19.67. Br_2

19.69. More negative

19.71. 6.8 g

19.73. 18.0 minutes

19.75. a. 5.78×10^{-3} L
b. No, some Cl_2 and Br_2 would be produced

19.77. −0.270 V

19.79. A hybrid vehicle uses a relatively inexpensive fuel (gasoline) in the internal combustion engine and has good fuel economy, but still gives off emissions. A fuel-cell vehicle does not give off emissions (the reaction produces H_2O) but requires a more expensive and explosive fuel (hydrogen); moreover, current battery technologies incorporate materials that are still very expensive and bulky.

19.81. Electric engines are more efficient by converting more of the energy into motion instead of losing it as heat.

19.83. a. $\overset{-4+1}{CH_4}(g) + \overset{+1}{H_2}O(g) \rightarrow \overset{+2}{C}O(g) + 3\,\overset{0}{H_2}(g)$
$\overset{+2}{C}O(g) + \overset{+1}{H_2}O(g) \rightarrow \overset{+1}{H_2}(g) + \overset{+4}{C}O_2(g)$
b. For the reaction of CH_4 with H_2O, $\Delta G^{\circ}_{rxn} = 142.2\ kJ$. For the reaction of CO with H_2O, $\Delta G^{\circ}_{rxn} = -28.6\ kJ$. For the overall reaction, $\Delta G^{\circ}_{overall} = \Delta G^{\circ}_{rxn_1} + \Delta G^{\circ}_{rxn_2} = 113.6\ kJ$.

19.85. a. Cathode
b. No. Mg^{2+}, with a higher positive charge, has a lower (less negative) reduction potential than Na^+.
c. No
d. H_2 and O_2

19.87. a. −0.87 V
b. Mo_3S_4: Mo = +2.67; $MgMoS_4$: Mo = +2
c. Mg^{2+} is added to the electrolyte to better carry the charge in the cell. This cation is produced at the anode and consumed at the cathode.

19.89. a. In K_2MnF_6: K = +1, Mn = +4, F = −1
In SbF_5: Sb = +5, F = −1
In $KSbF_6$: K = +1, Sb = +5, F = −1
In MnF_3: Mn = +3, F = −1
In F_2: F = 0

This is a one-electron process.
b. −656 kJ
c. 6.80 V
d. Too low
e. In H_2: H = 0
In F_2: F = 0
In KF: K = +1, F = −1
In KHF_2: K = +1, H = +1, F = −1

This is a two-electron process.

19.91. a. $2\,ZnS(s) + 3\,O_2(g) \rightarrow 2\,ZnO(s) + 2\,SO_2(g)$
  b. $2\,ZnO(s) + C(s) \rightarrow 2\,Zn(s) + CO_2(g)$
  c. $HgS(s) + O_2(g) \rightarrow Hg(\ell) + SO_2(g)$

19.93. a. Yes
  b. If the S^{2-} are closest packed (as given in the statement of the problem) in both structures, yes, the size of the tetrahedral holes is the same in both structures.

19.95. a. 0.359 V
  b. −1.613 V
  c. 1.200 V

19.97. Zn

19.99. Because the filled d orbitals in Zn are considered to be core electrons, both Zn and Mg have the same number of valence electrons (two $s\,e^-$) and therefore have the same oxidation state in compounds (+2).

19.101. Because Zn^{2+} has a d^{10} configuration there are no d–d transitions possible.

CHAPTER 20

20.1.

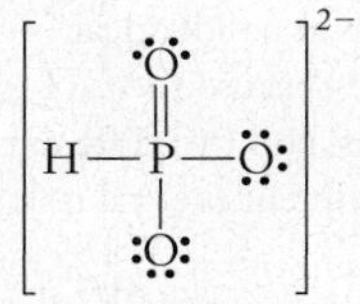

20.3. a. Palmitic acid
  b. Stearic acid

20.5. Tyrosine, glycine, glycine, phenylalanine, and methionine

20.7. Trans fats exhibit geometric isomerism around the C=C bond where similar groups on the two carbon atoms are situated on opposite sides of the double bond. Structures a and c contain trans fats.

20.9. Sucrose. The difference in the structures is that in sucralose, three –OH groups on sucrose have been replaced by Cl atoms. Being derived from sucrose implies that the sugar is natural, but the presence of Cl atoms on sugars is not natural.

20.11. Decreases

20.13. The "α" refers to the single carbon atom in amino acids to which both –NH$_2$ and –COOH groups are bonded.

20.15. D- and L- refer to how the four groups on a chiral carbon are oriented.

20.17. a and c

20.19. Most amino acids are zwitterions at pH ≈ 7.4 because the amino group will be protonated and the carboxylic acid group will be deprotonated, giving $RC\overset{+}{N}H_3COO^-$.

20.21. Lysine contains two amino groups, one of which is on a long carbon tail. This can react with the carboxylic acid on the carbon tail of glutamic acid to form a salt bridge.

20.23. a.

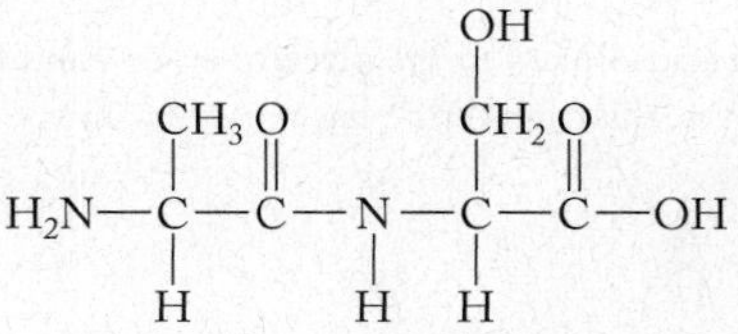

b.

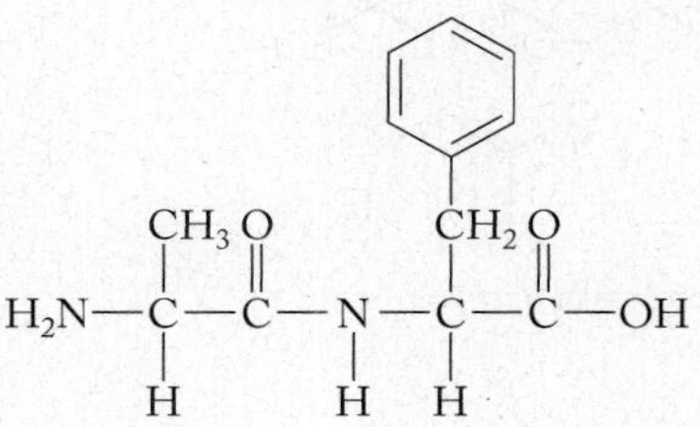

c.

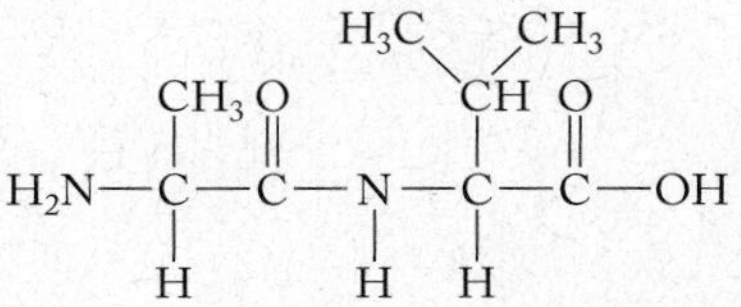

20.25. a. Alanine + glycine
  b. Leucine + leucine
  c. Tyrosine + phenylalanine

20.27. NH$_3$

20.29. Starch has α-glycosidic bonds and cellulose has β-glycosidic bonds. Starch coils into granules and cellulose forms linear molecules.

20.31. No

20.33. The bonding in fructose and glucose is nearly the same.

20.35. To calculate the free-energy change for a two-step process we need only to sum the individual ΔG values for each reaction.

20.37.

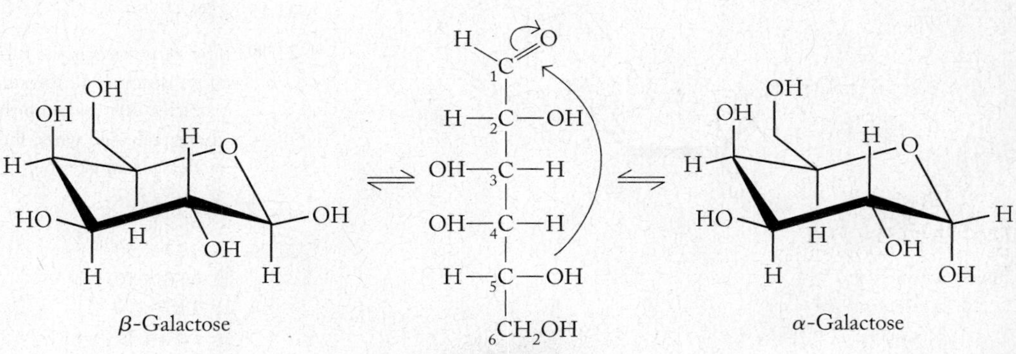

20.39. c

20.41. a

20.43. b

20.45. −16.4 kJ

20.47. Saturated fatty acids have all C—C single bonds in their structure; unsaturated fatty acids have C=C double bonds.

20.49. Fatty acids have a high fuel value (see Problem 20.48) and eating sticks of butter affords Arctic explorers with more energy per gram of food compared to carbohydrates or proteins.

20.51. If the two fatty acids linked to the glycerol at C-1 and C-3 are different then, yes, the triglyceride has a chiral center.

20.53. b and c

20.55.

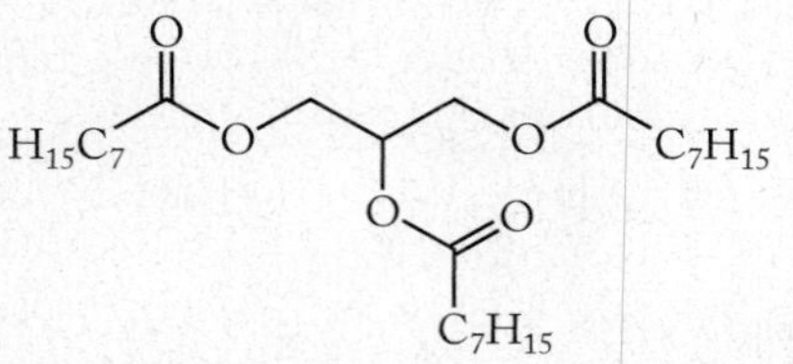

(a) Gycerol with octanoic acid

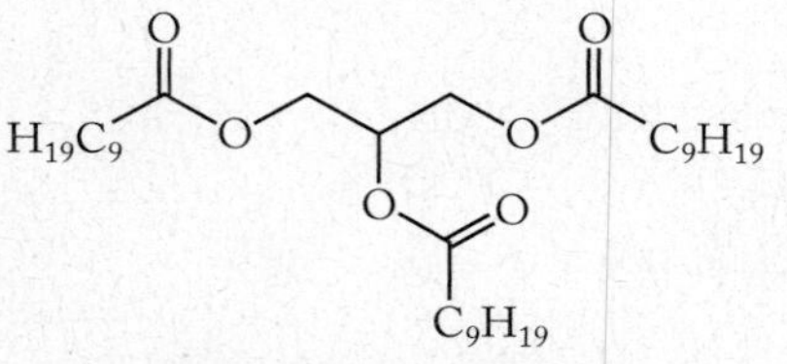

(b) Gycerol with decanoic acid

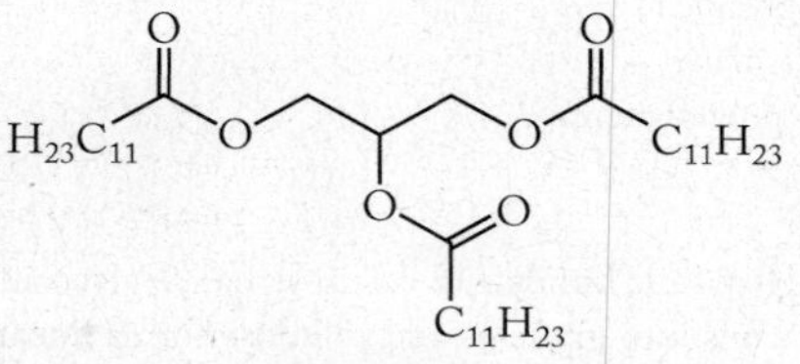

(c) Gycerol with dodecanoic acid

20.57. Phosphate group, a five-carbon sugar, and a nitrogen base; the backbone of DNA is composed of alternating sugar residues and phosphate groups.

20.59. Hydrogen bonds

20.61.

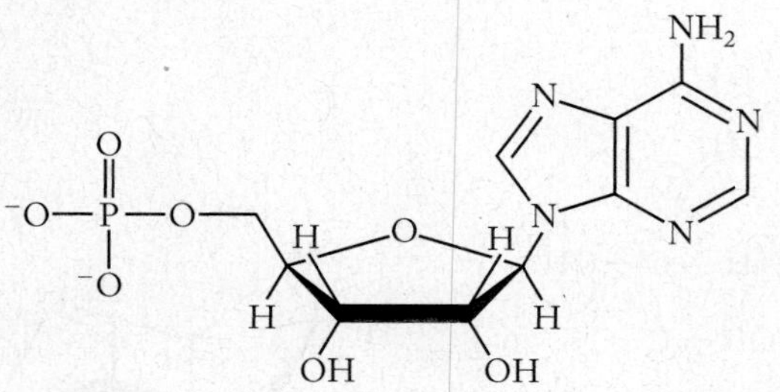

20.63. A-G-C-C-A-T.

20.65. a. Sucrose
b. Esters
c. $C_{15}H_{31}COOH$

20.67. a. There is an extra –CH_2– group in homocysteine's sulfur-containing side chain.
b. Yes

20.69. Yes

20.71. a. No
b. When the –NH_2 group of glycine reacts with the –COOH group of creatine:

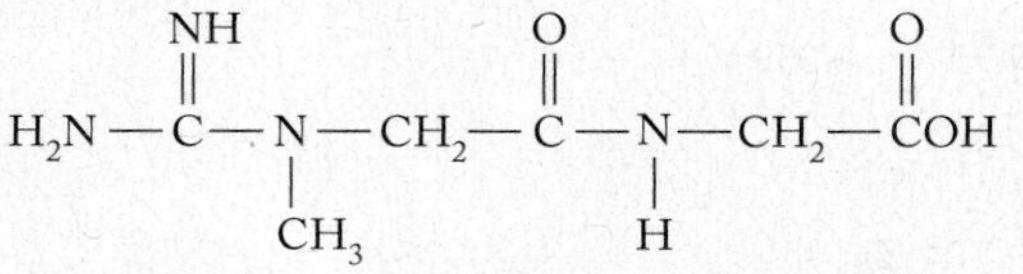

When the –NH_2 group of creatine reacts with the –COOH group of glycine:

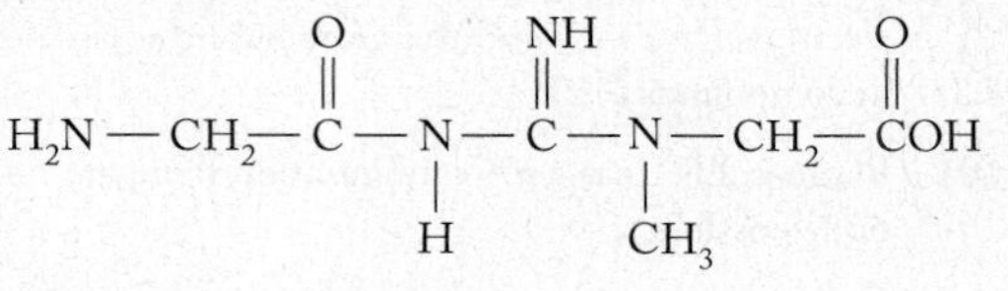

20.73. Glutamic acid, cysteine, and glycine

20.75. Yes. Because there is no difference in the number of C—C, C—H, C=O, C—O, or N—H bonds between the two compounds, we expect on the basis of average bond energies that the fuel values of leucine and isoleucine should be identical. Isoleucine might have a lower fuel value because the CH_3 group is closer to the COOH and NH_2 groups, and this difference in shape must contribute to the slightly different fuel values.

CHAPTER 21

21.1. Red (hydrogen)

21.3. Orange (astatine)

21.5. a

21.7. Blue line (b)

21.9. Process 1 represents fission; process 2 represents fusion

21.11. Electron = β particle = positron < proton < neutron < deuteron < α particle

21.13. Antihydrogen has the same mass as hydrogen but its nucleus has a negative charge with a positively charged electron. It contains the antiproton in the nucleus and a positron in place of the electron

21.15. $\Delta E = 3.01 \times 10^{-10}$ J; $\lambda = 1.32 \times 10^{-15}$ m or 1.32×10^{-6} nm

21.17. 25%

21.19. The *mass defect* is the difference between the mass of the nucleus of an isotope and the sum of the masses of the individual nuclear particles that make up that isotope. The *binding energy* is the energy released when individual nucleons combine to form the nucleus of an isotope.

21.21. 6.97×10^{-11} J

21.23. a. 4.37×10^{-12} J
b. 6.80×10^{-12} J
c. 2.69×10^{-12} J
d. 1.60×10^{-12} J

21.25. 1.09×10^{-12} J/nucleon

21.27. If the nuclide lies in the belt of stability (green dots on the plot in Figure 21.4), it is not radioactive and is stable. If it lies above the belt of stability, then it is neutron-rich and tends to undergo β decay to increase the number of protons and reduce the number of neutrons in its nucleus. If it lies below the belt of stability, it is neutron-poor and tends to undergo positron emission or electron capture to increase the number of neutrons and reduce the number of protons in its nucleus.

21.29. Alpha decay increases the neutron-to-proton ratio to produce less stable isotopes, which can then be made more stable through β emission to decrease the neutron-to-proton ratio.

21.31. Both of these processes are β decays.

21.33. Greater than 1

21.35. $^{26}_{13}\text{Al} \rightarrow {}^{0}_{1}\beta + {}^{26}_{12}\text{Mg}$

21.37. a. Electron capture or positron emission
   b. Electron capture or positron emission
   c. This isotope is stable.

21.39. ^{56}Co has 27 protons and 29 neutrons and is neutron-poor; it may undergo electron capture or positron emission. ^{44}Ti has 22 protons and 22 neutrons and is neutron-poor; it may undergo electron capture or positron emission.

21.41. Neutrons have no charge and are not repelled by the positively charged nucleus like the proton is.

21.43. The elements heavier than iron are formed by fusion reactions in supernovae.

21.45. Small particles are accelerated to high velocities (high energies) so that the particles overcome the repulsive forces to fuse atoms or nuclear particles together.

21.47. a. $^{56}_{26}\text{Fe} + 3\,{}^{1}_{0}\text{n} \rightarrow {}^{59}_{26}\text{Fe}$
   b. $^{59}_{26}\text{Fe} \rightarrow {}^{0}_{1-}\beta + {}^{59}_{27}\text{Co}$

21.49. The neutron-to-proton ratio in ^{99}Tc is $56/43 = 1.3$, which lies above the belt of stability. This nuclide is not stable.

21.51. a. $^{32}_{15}\text{P}$
   b. $4\,{}^{1}_{0}\text{n}$
   c. $2\,{}^{1}_{1}\text{H}$
   d. $^{125}_{54}\text{Xe}$

21.53. a. $^{122}_{53}\text{I}$
   b. $^{10}_{5}\text{B}$
   c. $^{58}_{26}\text{Fe}$
   d. $^{68}_{30}\text{Zn}$

21.55. a. There are 5 alpha and 2 beta decays:

   $^{230}_{90}\text{Th} \rightarrow {}^{4}_{2}\alpha + {}^{226}_{88}\text{Ra}$ $t_{1/2} = 7.54 \times 10^4$ years

   $^{226}_{88}\text{Ra} \rightarrow {}^{4}_{2}\alpha + {}^{226}_{88}\text{Rn}$ $t_{1/2} = 1600$ years

   $^{222}_{86}\text{Rn} \rightarrow {}^{4}_{2}\alpha + {}^{218}_{84}\text{Po}$ $t_{1/2} = 3.82$ days

   $^{218}_{84}\text{Po} \rightarrow {}^{4}_{2}\alpha + {}^{214}_{82}\text{Pb}$ $t_{1/2} = 3.10$ min

   $^{214}_{82}\text{Pb} \rightarrow {}^{0}_{1-}\beta + {}^{214}_{83}\text{Bi}$ $t_{1/2} = 26.8$ min

   $^{214}_{83}\text{Bi} \rightarrow {}^{0}_{1-}\beta + {}^{214}_{84}\text{Po}$ $t_{1/2} = 19.9$ min

   $^{214}_{84}\text{Po} \rightarrow {}^{4}_{2}\alpha + {}^{210}_{88}\text{Pb}$ $t_{1/2} = 164.3\ \mu\text{s}$

b. Once ^{218}Po is produced this statement is true, but the rate of the entire process is determined by the rate-limiting step, the α decay of ^{230}Th.

21.57. $^{209}_{83}\text{Bi} + {}^{4}_{2}\alpha \rightarrow {}^{211}_{85}\text{At} + 2\,{}^{1}_{0}\text{n}$

21.59. Control rods made of boron or cadmium are used to absorb the excess neutrons to control the rate of energy release.

21.61. The neutron-to-proton ratio for heavy nuclei is high and when the nuclide undergoes fission to form smaller nuclides, it must emit neutrons because the fission products require a lower neutron-to-proton ratio for stability.

21.63. a. $^{138}_{52}\text{Te}$
   b. $^{133}_{51}\text{Sb}$
   c. $^{143}_{55}\text{Cs}$

21.65. The level of radioactivity is the amount of radioactive particles present in a given instant of time. The dose is the accumulation of exposure over a length of time.

21.67. When radon-222 decays to polonium-218 while in the lungs, the ^{218}Po, a reactive solid that is chemically similar to oxygen, lodges in the lung tissue where it continues to emit α radiation. Alpha radiation is one of the most damaging kinds of radiation when in contact with biological tissues. The result of exposure to high levels of radon is an increased risk for lung cancer.

21.69. $5\ \mu\text{Sv} = 5\ \mu\text{Gy}$; $250\ \mu\text{J}$

21.71. a. $^{90}_{38}\text{Sr} \rightarrow {}^{0}_{1-}\beta + {}^{90}_{39}\text{Y}$
   b. $3.28 \times 10^8\ {}^{90}Sr$ atoms
   c. Strontium-90 is found in milk and not other foods because it is chemically similar to calcium and milk is rich in calcium.

21.73. a. 0.15 decays/s
   b. 7.0×10^4

21.75. a. The half-life should be long enough to effect treatment of the cancerous cells but not so long as to cause damage to healthy tissues.
   b. Because α radiation does not penetrate far beyond a tumor, the α decay mode is best.
   c. Products should be nonradioactive, if possible, or have short half-lives and be able to be flushed from the body by normal cellular and biological processes.

21.77. a. Positron emission or electron capture
   b. Positron emission or electron capture
   c. Positron emission or electron capture

21.79. 74.3 days

21.81. Yes

21.83. 136 min

21.85. a. $^{10}_{5}\text{B} + {}^{1}_{0}\text{n} \rightarrow {}^{7}_{3}\text{Li} + {}^{4}_{2}\alpha$
   b. 4.43×10^{-13} J
   c. Alpha particles have a high RBE and they do not penetrate into healthy tissue if the radionuclide is placed inside a tumor.

21.87. After 8.726 half-lives the ratio of ^{14}C present to that originally in an artifact is $N_t/N_0 = 0.50^{8.726} = 0.00236$ or 0.236%. This is too little to detect.

21.89. After 0.00023 half-lives the ratio of ^{40}K present to that originally in a sample is $N_t/N_0 = 0.50^{0.00023} = 0.9998$ or 99.98%. This level is just when we can detect the difference in amounts of ^{40}K.

21.91. 35%

21.93. 85%

21.95. 36,640 y

21.97. Besides releasing a large amount of energy to power the starship *Enterprise*, hydrogen is an abundant fuel in the universe and therefore could easily react with any antihydrogen produced.

21.99. No, we cannot discriminate using the tritium activity between grapes grown in 1969 versus those in 1968 or 1970.

21.101. The energy released in the fusion reaction is $\Delta E = 9.91 \times 10^{-13}$ J/nucleon. The energy released in the fission reaction is $\Delta E = 1.4 \times 10^{-13}$ J/nucleon. On a per nucleon basis, the fusion reaction generates more energy.

21.103. a. $^{11}_{5}B + ^{1}_{0}n \rightarrow ^{12}_{5}B$

   $^{12}_{5}B \rightarrow ^{12}_{6}C + ^{0}_{-1}\beta$

   $^{12}_{5}B \rightarrow ^{8}_{3}Li + ^{4}_{2}\alpha$

   b. ^{12}C is stable.

21.105. a. $^{266}_{109}Mt$

   b. $^{272}_{111}Rg$

   c. $^{269}_{110}Ds$

   d. $^{267}_{107}Bh$

   e. $^{265}_{108}Hs$

21.107. a. $^{249}_{98}Cf + ^{48}_{20}Ca \rightarrow ^{294}_{118}Uuo + 3^{1}_{0}n$

   b. $^{290}_{116}Uuh$

   c. $^{286}_{114}Uuq$

   d. $^{282}_{112}Cn$

   e. Because ^{294}Uuo is a member of the noble gas family, it has chemical and physical properties similar to naturally occurring radon.

21.109. ^{210}Pb

21.111. a. $^{64}_{28}Ni + ^{124}_{50}Sn \rightarrow ^{188}_{78}Pt$

   b. $^{196}_{78}Pt$

21.113. $^{208}_{82}Pb + ^{62}_{28}Ni \rightarrow ^{269}_{110}Ds + ^{1}_{0}n$

21.115. a. $^{40}_{19}K \rightarrow ^{40}_{18}Ar + ^{0}_{1}\beta$

   b. Because the half-life of ^{40}K is so much longer than that of ^{14}C

21.117. 3.35

21.119. 118 s

21.121. a. Both ^{214}Pb and ^{214}Bi are formed in the nuclear decay of ^{222}Rn. ^{214}Pb is produced after two α decays and ^{214}Bi is produced when ^{214}Pb undergoes β decay.

   b. Radon is an inert gas and does not stick to the charcoal like solid Pb and Bi do.

21.123. Because radon is a gas it seeps up from the ground to fill enclosed spaces such as basements. Uranium stays locked in the uranium ore.

CHAPTER 22

22.1. d

22.3. Group 16 (lavender)

22.5. −6.72 kJ

22.7. Trigonal planar

22.9. c

22.11. Without an essential element biological processes that rely on that element would shut down or deteriorate. If a nonessential element is missing, there would not be severe deleterious effects.

22.13. a. Present in milligram per gram quantities
   b. Present in milligram per gram quantities
   c. Present in nanogram to microgram per gram quantities

22.15. The methyl group on CH_3Hg^+ is relatively nonpolar and this helps the cation be soluble in nonpolar environments. Its charge, on the other hand allows it to be soluble in polar environments. Mercury metal, however, is neutral and has no nonpolar substituents and so is less soluble in biological systems.

22.17. The difference in behavior must be due to size. Ca^{2+} must simply be too large to fit into the biomolecules where Mg^{2+} is important.

22.19. K^+

22.21. a. 1.6 ppm
   b. 33 ppm
   c. 0.43 ppm

22.23. a. Oxygen
   b. Oxygen
   c. Carbon

22.25. Osmosis, ion channels, and ion pumps

22.27. The hydrophobic interior of the cell membrane makes it difficult to transport charged ions through the cell membrane.

22.29. Potassium

22.31. The greater insolubility of $CaCO_3$ compared to $CaSO_4$ makes calcium carbonate a better structural material. Also, the partial pressure of CO_2 in the atmosphere is higher than SO_3, so the carbonate solubility equilibrium is shifted more to the left by Le Châtelier's principle than the sulfate equilibrium.

22.33. 0.56 atm

22.35. 0.0688 V

22.37. 0.15 mol

22.39. $5.8 \times 10^{-4} M$

22.41. $^{137}Cs^+$ may substitute for K^+ in cells; as a β emitter with a relatively long half-life, it may cause cancer.

22.43. Nuclear test explosions and nuclear reactor accidents (e.g., Chernobyl)

22.45. To catalyze (lower the E_a of) biological processes

22.47. Lowers the activation energy

22.49. K has a value much greater than 1.

22.51. $^{137}_{55}\text{Cs} \rightarrow \, ^{0}_{-1}\beta + \, ^{137}_{56}\text{Ba}$

22.53. 3.06

22.55. $K = [\text{OH}^-]/[\text{F}^-] = 8.48$; the equilibrium lies to the right.

22.57. More soluble

22.59. For $\text{Ca}_5(\text{PO}_4)_3\text{OH}$, the solubility ($s$) is $8.5 \times 10^{-8}\ M$. For $\text{Ca}_5(\text{PO}_4)_3\text{F}$, the solubility ($s$) is $7.0 \times 10^{-8}\ M$.

22.61. 5.3×10^{-118}

22.63. 1.59×10^8

22.65. Type of radioactive decay; the half-life, chemical reactivity with tissues, and mechanism by which the radioactive compound can be eliminated by the body

22.67. Low tissue penetration and high RBE

22.69. γ radiation

22.71. By binding to the nitrogen atoms in DNA to stop the division of cells

22.73. Paramagnetic

22.75. The amalgam renders the mercury insoluble and chemically unreactive.

22.77. $^{141}_{58}\text{Ce} \rightarrow \, ^{141}_{59}\text{Pr} + \, ^{0}_{-1}\beta$

$^{160}_{65}\text{Tb} \rightarrow \, ^{160}_{66}\text{Dy} + \, ^{0}_{-1}\beta$

$^{170}_{69}\text{Tm} \rightarrow \, ^{170}_{70}\text{Yb} + \, ^{0}_{-1}\beta$

$^{177}_{71}\text{Lu} \rightarrow \, ^{177}_{72}\text{Hf} + \, ^{0}_{-1}\beta$

No, Pr, Dy, and Yb are lanthanides and Hf is a transition metal

22.79. 41 hr

22.81. $\left[\text{:Bi} \equiv \text{O:} \right]^+$

22.83. 130

22.85. $[\text{penicillamine}]_{\text{eq}} = 4.43 \times 10^{-4}\ M$ and $[\text{cysteine}]_{\text{eq}} = 5.6 \times 10^{-4}\ M$

22.87.

22.89. $\text{Al(OH)}_3(s) + 3\,\text{H}^+(aq) \rightarrow 3\,\text{H}_2\text{O}(\ell) + \text{Al}^{3+}(aq)$

22.91. $\text{Zn}(s) + \text{Ag}_2\text{O}(s) \rightarrow \text{ZnO}(s) + 2\,\text{Ag}(s); E° = 1.600\ \text{V}$

Chapter 1

Pages 2–3: NASA/ESA/Hubble Heritage; p. 4 (all): Courtesy NASA/JPL; p. 5 (from left to right): Daniel Smith/Corbis; Photographer's Choice/Punchstock; (2): NRH Photography; p. 6: Lester V. Bergman/Corbis; p. 7 (left): Mason Morfit/Taxi/Getty Images; (right): Phototake/Alamy; p. 9; (top left): Courtesy GEOEYE; (bottom from left to right): © 2009 Richard Megna/Fundamental Photographs; (inset): Courtesy Markus Geisen/NHMPL; © 2009 Richard Megna/Fundamental Photographs; p. 10: Aquacone courtesy Solar Solutions, Inc.; p. 15: (from left to right): Owen Franken/Corbis; David Wrobel/Visuals Unlimited; Larry Stepanowicz/Visuals Unlimited; p. 17: © Todd Jason Baker; p. 23: © 2009 Richard Megna/Fundamental Photographs; p. 25: Aeropix/Alamy; p. 28: Crown ©/The Royal Collection © 2007, Her Majesty Queen Elizabeth II; p. 29: Courtesy Chris Joosen/White Mountain National Forest; p. 33 (top right): Courtesy Special Collections, Princeton University Library; (bottom from left to right): Jim Pickerell/Stock Collection Blue/Alamy; Peter Arnold/Alamy; Andrew Holt/Photographer's Choice/Getty Images; Brian Whitney/Photonica/Getty Images; Digital Vision/Getty Images; p. 34 (top): AP Images; (bottom): Courtesy NASA/WMAP Science Team; p. 35 (left): Aquacone courtesy Solar Solutions, Inc.; (top right): David Wrobel/Visuals Unlimited; p. 38: (from left to right): Car Culture/Corbis; AP Images; p. 39: Non Sequitur © 1996 Wiley Miller. Dist. Universal Press Syndicate. Reprinted with permission. All rights reserved.

Chapter 2

Pages 42–43: Michael Hoch/© CERN; p. 44: The Royal Institution/Bridgeman Art Library; p. 46: Courtesy Jacob Lewis Bourjaily; p. 57: Dirk Wiersma/Science Photo Library; p. 68: NASA/HST/J. Morse/K. Davidson; p. 69: Courtesy NASA/CXC/SAO; p. 70: NASA/HST/J. Morse/K. Davidson; p. 77 (left): Taxi/Getty Images; (right): Harry Taylor/Getty Images.

Chapter 3

Pages 78–79: Courtesy Nate Smith and Mel Halbach; p. 81: Austin Post, USGS/CVO/Glaciology Project; p. 82: Richard Megna/Fundamental Photographs; p. 83: © 2009 Richard Megna/Fundamental Photographs; p. 84: Moodboard/Corbis; p. 89 (bottom, both): Michael Dalton/Fundamental Photographs; p. 104 (top): Stephen Earle, PhD, Geology Department, Malaspina University College, Naimo, Canada; (bottom): © M. Dini; p. 116: Courtesy NASA/ESA; p. 118: AP Images; p. 119 (top): Peter Saloutos/Corbis; (bottom): AP Photo; p. 121 (left): Richard Megna/Fundamental Photographs; (right): Courtesy NASA/ESA; p. 130: Smithsonian Institution/Corbis.

Chapter 4

Pages 132–133: Copyright © MC CORMICK, VICTORIA/Animals Animals - Earth Scenes — All rights reserved; p. 134 (from top to bottom): NASA; Time Life Pictures/Getty Images; Steve Schmeissner/Photo Researchers, Inc.; p. 135 (top left): Owen Franklin/Corbis; (bottom, all): Courtesy NASA/JPL/Malin Space Systems; (top right): NASA; p. 136 (left): Richard Megna/Fundamental Photographs; (right): Dr. E. R. Degginger/www.color-pic.com; p. 139: Leigh Smith Images/Alamy; p. 142 (all): Richard Megna/Fundamental Photographs, NYC; p. 145: © 2009 Richard Megna, Fundamental Photographs, NYC; p. 146 (all): © 2009 Richard Megna, Fundamental Photographs, NYC; p. 150: Richard Thom/Visuals Unlimited; p. 151 (left): ©1994 NYC Parks Photo Archive/Fundamental Photographs NYC; (right): © 1994 Kristen Brochmann/Fundamental Photographs NYC; p. 153 (both): Richard Megna/Fundamental Photographs; p. 156 (both): Richard Megna/Fundamental Photographs; p. 161 (both): © 2009 Richard Megna, Fundamental Photographs, NYC; p. 162 (all): Fundamental Photographs; p. 163: Joel Arem/Photo Researchers, Inc.; p. 164: Frans Lanting/Corbis; p. 169 (both): Peticolas/Megna/Fundamental Photographs; p. 171 (all): © 2009 Richard Megna, Fundamental Photographs, NYC; p. 172 (top): Bill Ross/Corbis; (center, bottom): From Wetlands Field Manual/Courtesy USDA; p. 173 (all): © 2009 Richard Megna, Fundamental Photographs, NYC; p. 176 (top): Robert Holmes/Corbis; (bottom): Gianni Dagli Orti/Corbis; p. 178 (top left): © 1996 Richard Megna, Fundamental Photographs, NYC; (bottom left): © 2009 Richard Megna, Fundamental Photographs, NYC; (right): Peticolas/Megna/Fundamental Photographs;

p. 182: David R. Frazier/Photo Library/Alamy; p. 183: P. Rona/NOAA; p. 186: Courtesy Richard Sugarek/Environmental Protection Agency.

Chapter 5

Pages 188–189: Dennis Flaherty/The Image Bank/Getty Images; p. 193 (left): David Bergman/Corbis; (right): Matthias Schrader/epa/Corbis; p. 194: Kristen Brochmann/ Fundamental Photographs, NYC; p. 196 (left): Pasacal Le Segretain/Corbis Sygma; (right): AP Photo; p. 204: Reuters/Corbis; p. 207: © Jon Gnass; p. 222: AP Photo; p. 225 (all): Courtesy of Alcoa, Inc; p. 232: (top left): K-Photos/Alamy; (bottom right): Joseph Sohm/Visions of America/Corbis; (center): NRH Photography; (bottom left): Courtesy Jochen Gechnaller, Universat Innsbruck; p. 233: David Hall/Photo Researchers, Inc.

Chapter 6

Pages 246–247: Peter Scoones/Photo Researchers, Inc.; p. 250 (top): Galen Rowell/Corbis; (center): Hubert Stadler/Corbis; (bottom): Bill Ross/Corbis; p. 251: Courtesy NOAA; p. 253: Sam Ogden/Photo Researchers Inc.; p. 255: David R. Frazier Photography/Alamy; p. 266: © 2001 Richard Megna, Fundamental Photographs, NYC; p. 272 (top): Photography, Inc./Alamy; (bottom): David Samuel Robbins/Corbis; p. 274: © Joel Gordon; p. 275 (all): © 2009 Richard Megna, Fundamental Photographs, NYC; p. 280: (top): Galen Rowell/Corbis; (bottom): © 2008 Richard Megna, Fundamental Photographs, NYC; p. 294: Visuals Unlimited/Corbis; p. 296: (top left): Courtesy NOAA; (bottom left): Trip/Alamy; (top right): © 2008 Richard Megna, Fundamental Photographs, NYC; p. 300: Imageshop/Corbis; p. 301: Jonathan Blair/Corbis; p. 303: F. Jack Jackson/Alamy; p. 304: Didrik Johnck/Corbis; p. 307: Corbis.

Chapter 7

Pages 308–309: Will & Deni McIntyre/Photo Researchers, Inc.; p. 312: Bettmann/Corbis: p. 313: (top right): Richard Megna/Fundamental Photographs, NYC; p. 316 (all): Richard Megna/Fundamental Photographs, NYC; p. 317: (all): Richard Megna/Fundamental Photographs, NYC; p. 318: Science Source/Photo Researchers, Inc.; p. 319: James Leynse/Corbis; p. 328: Lebrecht Music and Arts Photo Library/Alamy; p. 334: Margrethe Bohr Collection/American Institute of Physics/Science Photo Library; p. 356 (left): Bettmann/Corbis; (right): Science Faction/Getty Images; p. 357: (all): Richard Megna/Fundamental Photographs, NYC; p. 358: Richard Megna/Fundamental Photographs, NYC.

Chapter 8

Page 366–367: Albert Lleal/Minden Pictures; p. 369: (all): Charles D. Winters/Photo Researchers, Inc.; p. 382: W. Perry Conway/Corbis; p. 390: Steve Nichols/Alamy; p. 391: Bob Rowan/Progressive Image/Corbis; p. 398: Alexandra Grablewski/Getty Images; p. 405 (all): NRH Photography; p. 406: Rick Scott and Joe Orman, Courtesy Night of the Comet; p. 410: (left): NRH Photography; (right): NASA/JPL/University of Arizona.

Chapter 9

Pages 412–413: Tracey Thompson/Corbis; p. 417: Richard Megna/Fundamental Photographs, NYC; p. 441: E.R. Degginger/www.color-pic.com; p. 448: © Yoav Levy/Phototake; p. 453 (right): Tom Brakefield/Stockbyte/Getty Images; (left): Dreamstime; p. 454: Richard Megna/Fundamental Photographs, NYC; p. 457: The Magellan mission was managed for NASA by Jet Propulsion Laboratory (JPL), Pasadena, CA. Data processed by JPL, the Massachusetts Institute of Technology, Cambridge, MA, and the U.S. Geological Survey, Flagstaff, AZ; p. 460: Charles E. Rotkin/Corbis.

Chapter 10

Pages 462–463: Micro Discovery/Corbis; p. 480: Robert Yin/Corbis; p. 483: Courtesy of Presto; p. 486 (left): Jeff Daly/Visuals Unlimited; (right): Martin Shields/Photo Researchers Inc.; p. 487: (top): Sinclair Stammers/Science Photo Library/Photo Researchers, Inc.; (bottom left): © Tom Pantages; (bottom right): Phil Degginger/www.color-pic.com; p. 488: Larry Stepanowicz/Visuals Unlimited; p. 489: © Phil Souders; p. 490: (both): Charles D. Winters/Photo Researchers, Inc.; p. 491: Richard T. Nowitz/Corbis; p. 492: Jeff Daly/Visuals Unlimited.

Chapter 11

Pages 498–499: Comstock Select/Corbis; p. 502: (left): Charles D. Winters/Science Photo Library/Photo Researchers, Inc; (right): Paul Whitehill/Science Photo Library/Photo Researchers, Inc.; p. 527: David M. Phillips/Visuals Unlimited; p. 531: Richard T. Nowitz/Corbis; p. 537: David M. Phillips/Visuals Unlimited; p. 541 (left): G. Flayols/PhotoCuisine/Corbis; (right): OSH/Alamy; (center): Mark Bolton/Corbis.

Chapter 12

Pages 544–545 © Andy Ryan Photography; p. 550 (left): Dorling Kindersley/Getty Images; (right): Karlene Lowell Schwartz; p. 554 (top): Eye of Science/Photo Researchers, Inc.; (bottom): Andrew Syred/Photo Researchers Inc.; p. 557: David Parker/Photo Researchers Inc.; p. 559: Kris Mercer/Alamy; p. 563 (left): Jon Stokes/Photo Researchers Inc.; (right): Ken Lucas/Visuals Unlimited; p. 565 (top): 1994 Richard Megna, Fundamental Photographs, NYC; (bottom): Charles D. Winters/Photo Researchers Inc.; p. 566: Richard Megna/Fundamental Photographs, NYC; p. 567 (left): Andrew Silver/U.S. Geological Survey; (bottom): Courtesy www.VassichkoMinerals.com © 2007 JWV; p. 570 (top): Jose Manuel Sachis Calvete/Corbis; (bottom): E. R. Degginger/Photo Researchers Inc.; p. 571: (top): Arthur Hill/Visuals Unlimited; (bottom): Ashley Cooper/Alamy; p. 572: NASA/epa/Corbis; p. 574: © Tom Pantages; p. 579: Richard Megna/Fundamental Photographs, NYC; p. 584: Smithsonian Institution/Bridgeman Art Library; p. 585: Courtesy Ledtronics, Inc.; p. 587: Beinecke Rare Book and Manuscript Library, Yale University.

Chapter 13

Pages 590–591: Chang W. Lee/The New York Times/Redux; p. 594: Bernd Vogel/Solus-Veer/Corbis; p. 607: Dr. Keith Wheeler/Photo Researchers Inc.; p. 608 (from left to right): Eric and David Hosking/Corbis; Digital Vision/Jupiterimages; Cabannes/PhotoCuisine/Corbis; p. 615 (top): Tek Image/Photo Researchers Inc.; (bottom): Jeffrey Hamilton/Getty Images; p. 616 (clockwise from top left): Emily Spence/Lexington Herald-Leader/MCT/Newscom; Fotosearch; The HON Company; Studiomode/Alamy; p. 620: © 2010 Richard Megna/ Fundamental Photographs NYC; p. 633: Barry Slaven/The Medical File/Peter Arnold; p. 634: Creatas/Jupiterimages; p. 637: Courtesy DuPont; p. 642: Phil Degginger/www.color-pic.com; p. 644: Phil Dotson/Science Photo Researchers, Inc.; p. 645: Bernd Vogel/Solus-Veer/Corbis; p. 647 (from left to right): Eric and David Hosking/Corbis; John Madere/Corbis; Cabannes/Photo Cuisine/Corbis; Douglas Peebles/Corbis.

Chapter 14

Pages 658–659: Jose Fuste Raga/Corbis; p. 661: Phil Degginger/Alamy; p. 665 (top): TongRo/Beateworks/Corbis; (bottom): Richard Megna/Fundamental Photographs, NYC; p. 669: B. & C. Alexander/Photo Researchers, Inc.; p. 670: © 1987 Richard Megna, Fundamental Photographs, NYC; p. 683: Sheila Terry/Photo Researchers, Inc.; p. 688 (left): Hans Reinhard/zefa/Corbis; (right): SCIMAT/Science Photo Library/Photo Researchers, Inc.

Chapter 15

Pages 700–701: Chris Detrick/Salt Lake Tribune; p. 702: Courtesy Bob Burkhart; p. 728: Wikimedia Commons; p. 741: NASA.GSFC/Scientific Visualization Studio; p. 743: NASA.GSFC/Scientific Visualization Studio; p. 744: Document General Motors/Reuters/Corbis; p. 745: Courtesy HowStuff Works.com; p. 746 (left): Courtesy Platinum Today (www.platinum.matthey.com); (right): Yoshitazu Tsuno/AFP/Getty Images; p. 753: Lawrence Migdale/Photo Researchers, Inc; p. 759: Andrew Lambert Photography/Science Photo Library/Photo Researchers, Inc.

Chapter 16

Pages 762–763: Phil Degginger/www.color-pic.com; p. 766 (top): Robert Landau/Corbis; (bottom, all): © 1993 Richard Megna, Fundamental Photographs, NYC; p. 783 (top): Stan Pritchard/Alamy; (bottom): Robert Jones/Alamy; p. 784: Thermo Scientific Model 48000/Courtesy of Thermofisher; p. 786 (all): © 2010 Richard Megna/ Fundamental Photographs NYC; p. 790 (both): Richard Megna/Fundamental Photographs, NYC; p. 806: Richard Megna/Fundamental Photographs, NYC.

Chapter 17

Pages 816–817: Purestock/Alamy; p. 859 (all): Larry Stepanowicz/Visuals Unlimited; p. 860: © 1994 Richard Megna, Fundamental Photographs, NYC; p. 865: Hulton-Deutsch Collection/Corbis; p. 870: Richard Megna, Fundamental Photographs, NYC.

Chapter 18

Pages 878–879 Kevin Schafer/Corbis; p. 886: Richard Megna, Fundamental Photographs, NYC; p. 896 (both) © 1994 Richard Megna, Fundamental Photographs, NYC; p. 897: © 1997 Richard Megna, Fundamental Photographs, NYC; p. 899: © 1990 Richard Megna, Fundamental Photographs, NYC; p. 902: Vaughan Fleming/Photo Researchers, Inc.; p. 906: (both): Phil Degginger/www.color-pic.com; p. 908: Terry W. Eggers/Corbis; p. 909: European Bioinformatics Institute, Protein Data Bank.

Chapter 19

Pages 916–917: Tesla Motors; p. 919 (all): Phil Degginger/www.color-pic.com; p. 924: Mary Evans Picture Library/Alamy; p. 936: Tom Gilbert; 937: SHNS photo courtesy Toyota/Newscom; p. 938: SHNS photo courtesy General Motors/Newscom; p. 945: AP Photos; p. 946 (left): Richard G. Bingham II/Alamy; (right): Heini Schneebeli/The Bridgeman Art Library; p. 947: Phil Degginger/www.color-pic.com; p. 949 (left): Alix/Photo Researchers, Inc; (right): no credit.

Chapter 20

Page 957: © Frans Lanting/Corbis; p. 960: Valueline/Punchstock; p. 965: (left): Dennis Kunkel Microscopy, Inc./Visuals Unlimited; (right): Omikron/Photo Researchers, Inc.; p. 967: Courtesy of N.I.S.T.; p. 968: Chemical Design/Science Photo Library/Photo Researchers Inc.; p. 983: Roger Ressmeyer/Corbis; p. 984 (top): NASA; (bottom): © W.R. Normak, courtesy USGS.

Chapter 21

Page 995: © Centre Jean Perrin/ISM/PhototakeUSA.com; p. 1014: (from left to right): Astrid & Hanns-Frieder Michler/Science Photo Library/Photo Researchers, Inc.; Tom Tracey Photography/Alamy; © Urenco; p. 1016: Photodisc/Alamy; p. 1020: Igor Kostin/Corbis Sygma; p. 1021 (top, both): © 2006 T. A. Mousseau and A. P. Moller; (center): DOE Photo; p. 1024: Dr. Robert Friedland/Photo Researchers, Inc.; p. 1027: Tom Bean/Corbis; p. 1028: Photographers Blais/Turnbull; Parks Canada; p. 1029: Tom Bean/Corbis; p. 1034 (top): Mireille Vautier/Alamy; (bottom): Blickwinkel/Alamy; p. 1035: Reprinted by permission from Macmillan Publishers Ltd.: "Human Presence in the European Arctic Nearly 40,000 Years Ago," *Nature* 413 (September 6, 2001): 64–67, Figure 4—mammoth tusk showing human markings, © 2001.

Chapter 22

Page 1039: ISM/Phototake, Inc.; p. 1041 (both): Phil Degginger www.color-pic.com; p. 1043: Don W. Fawcett/Photo Researchers, Inc.; p. 1057: Phil Degginger www.color-pic.com; p. 1059: SIU/Visuals Unlimited, Inc.; p. 1062: Zephry/Science Photo Library/Photo Researchers, Inc.; p. 1064: From "Small molecular gadolinium (III) complexes as MRI contrast agents for diagnostic imaging," Chan Kannie Wai-Yan and Wong wing-Tak, Coordination Chemistry Reviews, Sept., 2007, Elsevier B.V., Copyright Clearance Center; p. 1065: Phil Degginger www.color-pic.com; p. 1072: no credit; p. 1073: Philippe Plailly/Photo Researchers, Inc.

NOTE: Material in figures or tables is indicated by *italic* page numbers. Footnotes are indicated by *n* after the page number.